Geometry

(A = area, B = area of base, C = circumference, S = lateral area or surface area, V = volume)

1. Triangle

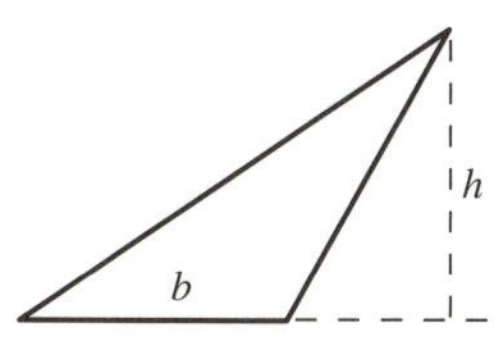

$A = \frac{1}{2}bh$

2. Similar Triangles

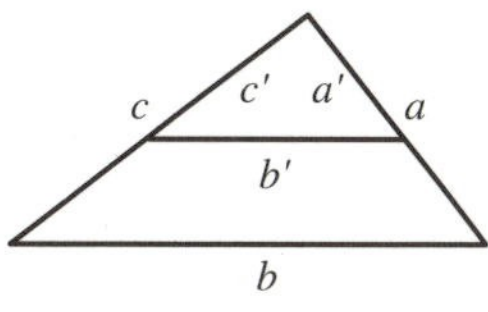

$\frac{a'}{a} = \frac{b'}{b} = \frac{c'}{c}$

3. Pythagorean Theorem

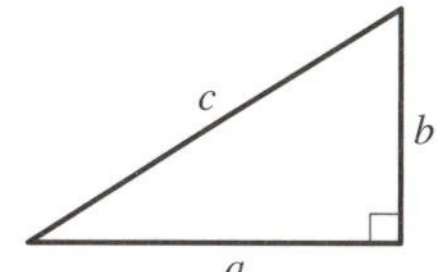

$a^2 + b^2 = c^2$

4. Parallelogram

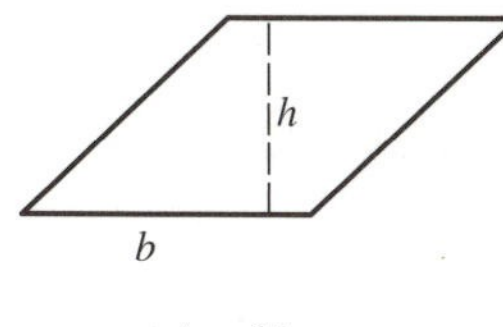

$A = bh$

5. Trapezoid

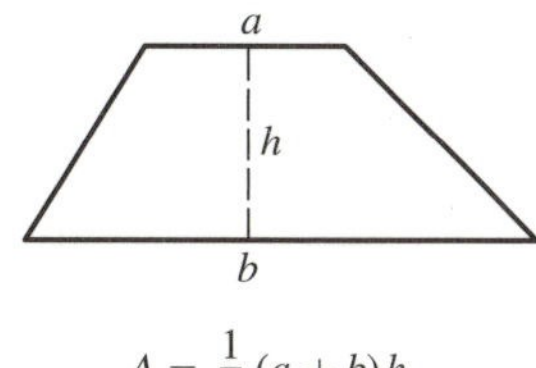

$A = \frac{1}{2}(a + b)h$

6. Circle

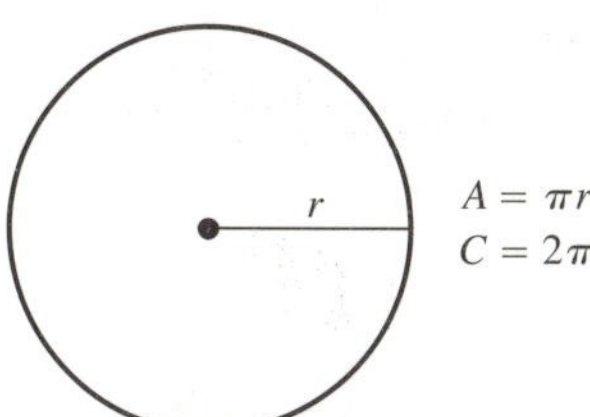

$A = \pi r^2,$
$C = 2\pi r$

7. Any Cylinder or Prism with Parallel Bases

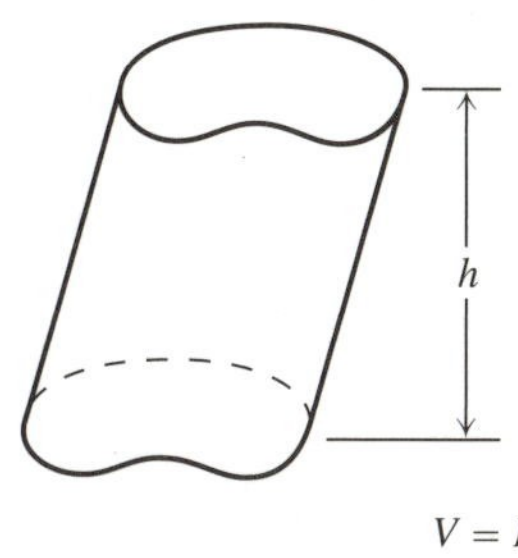

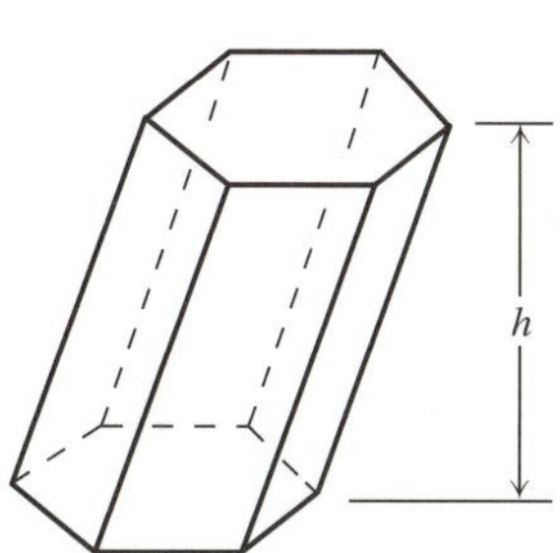

$V = Bh$

8. Right Circular Cylinder

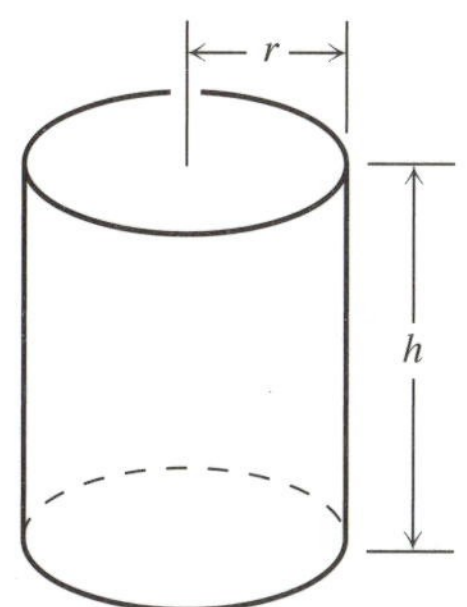

$V = \pi r^2 h,\ S = 2\pi rh$

9. Any Cone or Pyramid

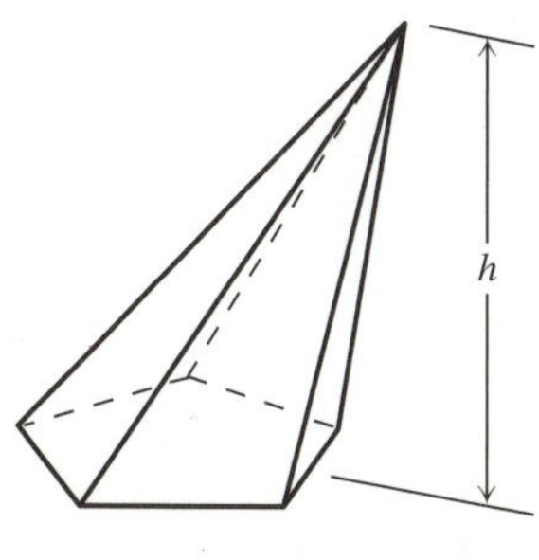

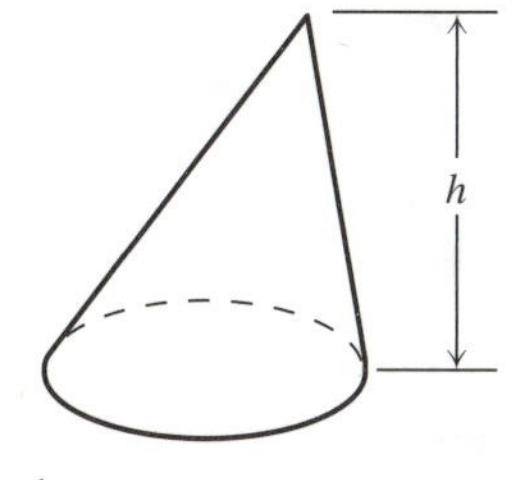

$V = \frac{1}{3}Bh$

10. Right Circular Cone

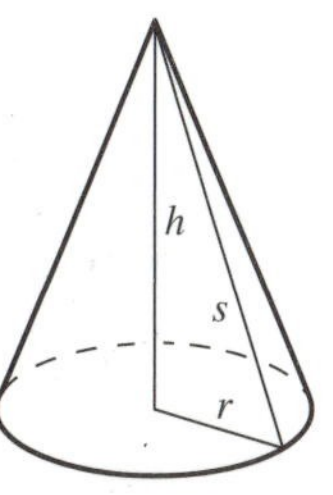

$V = \frac{1}{3}\pi r^2 h,\ S = \pi rs$

11. Sphere

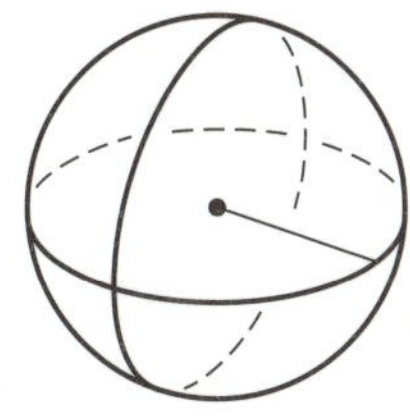

$V = \frac{4}{3}\pi r^3,\ S = 4\pi r^2$

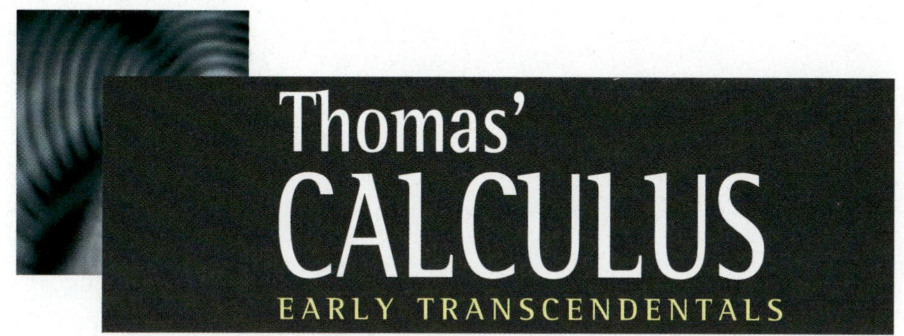

UPDATED TENTH EDITION

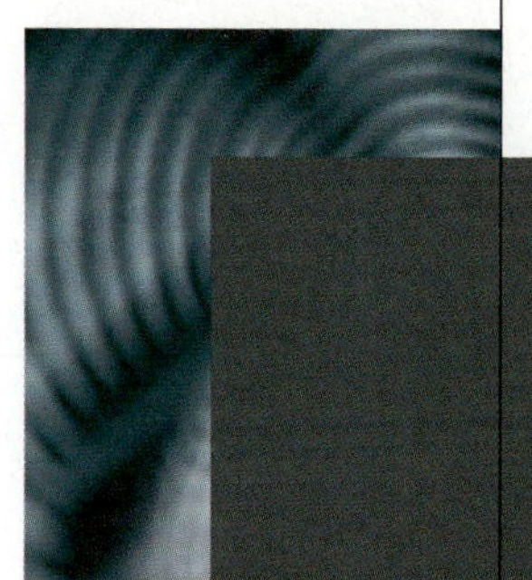

Thomas' CALCULUS

EARLY TRANSCENDENTALS

UPDATED TENTH EDITION

Based on the original work by

George B. Thomas, Jr.

Massachusetts Institute of Technology

As revised by

Ross L. Finney

Maurice D. Weir

Naval Postgraduate School

and

Frank R. Giordano

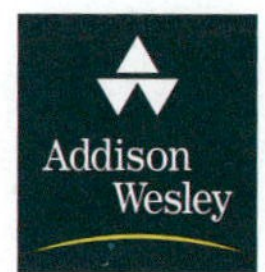

Boston San Francisco New York
London Toronto Sydney Tokyo Singapore Madrid
Mexico City Munich Paris Cape Town Hong Kong Montreal

Publisher	Greg Tobin
Project Editor	Ellen Keohane
Editorial Assistant	Stefanie Borge
Managing Editor	Karen Guardino
Production Supervisor	Julie LaChance
Editorial and Production Services	UG / GGS Information Services, Inc.
Art Editor	Geri Davis/The Davis Group, Inc.
Marketing Manager	Weslie Lewis
Illustrators	Tech Graphics\cs
Technical Art Consultants	Scott Silva and Joe Vetere
Senior Prepress Supervisor	Caroline Fell
Compositor	UG / GGS Information Services, Inc.
Text Designer	Geri Davis/The Davis Group, Inc.
Cover Designer	Barbara T. Atkinson
Senior Manufacturing Buyer	Evelyn Beaton
Media Buyer	Ginny Michaud

Photo Credits: 47, Courtesy of Agilent Technologies, Inc. **124,** John Elk III/Bruce Coleman, Inc. **139,** NASA/JPL **171, 245, 472, 631, 759, 864, 1196,** PSSC Physics 2/e, 1965; D.C. Heath & Co. with Educational Development Center, Inc., Newton, MA **236,** "Differentiation" by W.U. Walton et. Al., Project CALC, Educational Development Center, Inc., Newton, MA **239,** AP/Wide World Photos **323,** Scott Burns, Urbana, IL., www.designbyalgorithm.com **324,** COMAP, Inc. **466, 754,** Corbis **854,** Richard F. Voss/IBM Research **877,** Appalachian Mountain Club **911,** Department of History, US Military Academy, West Point, New York **929,** ND Roger-Viollet **1060, 1061,** Adapted from "NCFMF Book of Film Notes", 1974, MIT Press with Educational Development Center, Inc., Newton, MA **1061,** InterNetwork Media, Inc. and NASA/JPL

Library of Congress Cataloging-in-Publication Data

Finney, Ross L.

Thomas' calculus : early transcendentals.—10th ed. / based on the original work by George B. Thomas, Jr., as revised by Ross L. Finney, Maurice D. Weir, and Frank R. Giordano.

p. cm.

Rev. ed. of Calculus and analytic geometry / George B. Thomas, Jr., Ross L. Finney, 9th ed. c1996.

Includes bibliographical references and index.

ISBN 0-321-16957-3

1. Calculus. 2. Geometry, Analytic. I. Title: Calculus. II. Weir, Maurice D. III. Giordano, Frank R. IV. Thomas, George Brinton, 1914– Calculus and analytic geometry. V. Title.

QA303.F48 2000

515′.15—dc21 00-032790

1 2 3 4 5 6 7 8 9 VH 05 04 03 02

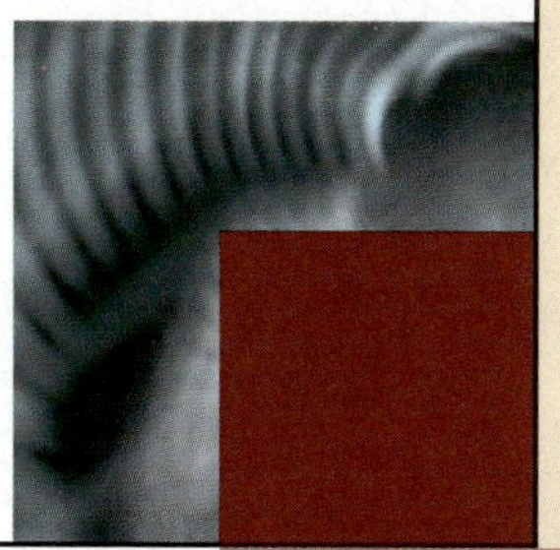

Contents

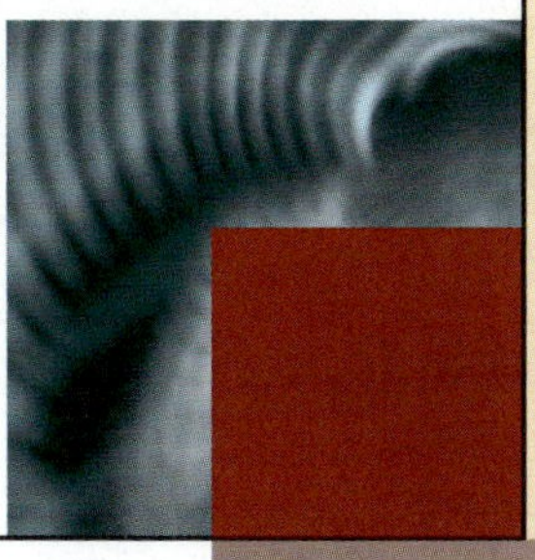

Computer Algebra System (CAS) Exercises

P Preliminaries

P.7 Fitting curves to observed data, analyzing the error, making predictions, and improving the model if appropriate.

1 Limits and Continuity

1.1 Comparing graphical estimates of limits with CAS symbolic limit calculations. Exploring the formal definition of limit by finding deltas for specific epsilons graphically.

1.3 Exploring asymptotes and graphical behavior as $x \to \pm\infty$.

1.5 Exploring graphically and numerically average rates of change and tangents.

2 Derivatives

2.1 Exploring the convergence of the secant lines graphically. Using the definition to find a function's derivative. Exploring the relationship between the graphs of f and f'. Plotting selected tangent lines.

2.2 Exploring animated visualizations of the derivatives for velocity and acceleration.

2.4 Exploring harmonic and decaying oscillations.

2.5 Exploring trigonometric "polynomial" approximations to sawtooth and square wave functions. Plotting curves defined parametrically together with specified tangent lines.

2.6 Finding a derivative implicitly and plotting implicit curves together with specified tangent lines.

2.8 Exploring inverse functions and their derivatives graphically.

3 Applications of Derivatives

3.1 Finding absolute extrema by analyzing f and f' numerically and graphically.

3.2 Graphing solutions of differential equations.

3.3 Exploring families of cubic and quartic polynomials and logistic functions.

3.5 Exploring the volumes of cones constructed from a circular disk. Exploring a triangle circumscribing an ellipse.

3.6 Finding linearizations and exploring the absolute error in a linearization by comparing its graph with that of the function.

3.7 Finding zeros of functions using Newton's method. Approximating the numbers $\sqrt{2}$, π, and e.

4 Integration

4.1 Solving initial value problems.

4.3 Finding the average value of $f(x)$ and the point or points where it is assumed by f. Estimating volumes with finite sums.

4.4 Exploring Riemann sums and their limits.

4.5 Investigating the relationship of $F(x) = \int_a^x f(t)\,dt$ to $f(x)$ and $f'(x)$. Analyzing $F(x) = \int_a^{u(x)} f(t)\,dt$.

4.6 Finding the area between curves in the plane when their intersection points cannot be found using simple algebra.

4.7 Evaluating definite integrals numerically.

5 Applications of Integrals

5.1 Finding the volumes of solids of revolution about the x-axis for circular and washer cross sections.

5.2 Estimating the lengths of curves defined explicitly or parametrically.

5.4 Investigating the differential equation modeling the change over time in glucose fed intravenously. Plotting slope fields and solution curves for separable differential equations.

5.5 Exploring the relationship between work and kinetic energy.

6 Transcendental Functions and Differential Equations

6.1 Exploring the linearization of $\ln(1 + x)$ at $x = 0$.
6.2 Exploring the linearizations of e^x, 2^x, and $\log_3 x$. Exploring inverse functions and their derivatives.
6.6 Plotting the slope field and investigating solutions of a modified logistic equation. Finding numerical solutions using Euler's and improved Euler's methods. Exploring solutions to initial value problems graphically, analytically, and numerically, and comparing the results.

7 Integration Techniques, L'Hôpital's Rule, and Improper Integrals

7.5 Using a CAS to integrate. An example of a CAS-resistant integral. Monte Carlo integration.
7.7 Exploring the convergence of improper integrals involving $x^P \ln x$.

8 Infinite Series

8.1 Plotting sequences to explore their convergence or divergence. For convergent sequences, finding a tail that lies within a specified interval centered at the limit.
8.2 Exploring the convergence of sequences defined recursively. Compound interest with deposits and withdrawals. The logistic difference equation and chaotic behavior.
8.4 Exploring $\sum_{n=1}^{\infty} (1/(n^3 \sin^2 n))$, a series whose convergence or divergence has not yet been determined.
8.7 Comparing functions' linear, quadratic, and cubic approximations.
8.9 Finding Fourier series expansions. Using a Fourier series to show that $\sum_{n=1}^{\infty} 1/n^2 = \pi^2/6$.
8.10 Finding Fourier sine and cosine series for $f(x) = |2x - \pi|, 0 < x < \pi$.

9 Vectors in the Plane and Polar Functions

9.6 Exploring a figure skater tracing a polar plot.

10 Vectors and Motion in Space

10.3 Putting a three-dimensional scene on a two-dimensional canvas.
10.4 Plotting three-dimensional lines, planes, cylinders, and quadric surfaces.
10.5 Plotting tangents to space curves. Exploring a general helix curve.
10.6 Analyzing the motion of a particle moving along a space curve.
10.7 Finding κ, τ, **T**, **N**, and **B** for curves in space. Finding and plotting circles of curvature for curves in the plane.

11 Multivariable Functions and Their Derivatives

11.1 Plotting surfaces $z = f(x, y)$ and associated level curves. Plotting implicit and parametrized surfaces.
11.5 Exploring directional derivatives.
11.7 Finding and classifying critical points for functions of two independent variables using information gathered from surface plots, level curves, and discriminant values. Looking for patterns in data and applying the method of least squares.
11.8 Implementing the method of Lagrange multipliers for functions of three and four independent variables.

12 Multiple Integrals

12.1 Using a double-integral evaluator to find values of integrals.
12.3 Changing Cartesian integrals into equivalent polar integrals for evaluation.
12.4 Evaluating triple integrals over solid regions. Using volumes to measure rainfall and ensure adequate drainage from a satellite dish.
12.5 Exploring moments and means to determine if a buoy will overturn. Exploring new plotting techniques.

13 Integration in Vector Fields

13.1 Evaluating line integrals along different paths.
13.2 Estimating the work done by a vector field along a given path in space.
13.3 Visualizing force fields. Verifying that a force field is conservative.
13.4 Applying Green's Theorem to find counterclockwise circulation. Finding the path through a force field that maximizes the work done. Comparing conservative and nonconservative force fields.
13.8 Visualizing and interpreting flux and divergence in three dimensions. Evaluating integrals on surfaces defined parametrically. Evaluating the divergence integral.

To the Instructor

Throughout its illustrious history, *Thomas' Calculus* has been used to support a variety of courses and teaching methods, from traditional to experimental. This tenth edition is a substantial revision, yet it retains the traditional strengths of the text: sound mathematics, relevant and important applications to the sciences and engineering, and excellent exercises. This flexible and modern text contains all the elements needed to teach the many different kinds of courses that exist today.

A book does not make a course; the instructor and the students do. This text is a resource to support your course. With this in mind, we have added a number of features to the tenth edition making it even more flexible and useful, both for teaching and learning calculus.

Features of the Tenth Edition

- For the first time, this classic text is available in *both* standard and Early Transcendentals versions.
- The new *Annotated Instructor's Edition* contains suggestions for the incorporation of technology, highlighting how the Web site and CD-ROM can be used to enhance the presentation of chapter topics.
- As always, this text continues to be easy to read, conversational, and mathematically rich. Each new topic is motivated by clear, easy-to-understand examples and is then reinforced by its application to real-world problems of immediate interest to students.
- Each section now begins with a list of subsection headings, making key concepts readily apparent.
- Within the tenth edition is an increased emphasis on modeling and applications using real data. As a result, there is an improved balance of graphical, numerical, and analytic methods and techniques, accomplished without compromising the mathematical integrity of the book.
- Vectors and projectile motion in the plane are now covered separately from vectors in space, concluding the treatment of single-variable calculus. Three-dimensional vectors are then treated in conjunction with multivariable calculus.
- Exercise sets continue to be grouped within appropriate headings. Titles that indicate the content or application have been added for most word problems, and those requiring the use of a graphing utility are identified throughout the

text by the icon T. Computer Algebra System (CAS) exercises also appear in every chapter and are grouped in special subsections labeled "Computer Explorations."

- Together, the CD-ROM and Web site provide students and instructors with even more support:
  - A collection of Maple® and *Mathematica*® modules, videos, and Java applets are available to help students visualize key calculus concepts.
  - Interactive online tutorials help students review precalculus and textbook-specific material, take practice tests, and receive diagnostic feedback on their performance.
  - Chapter-by-chapter quizzes are also provided. These quizzes can be administered and graded online for skills-based mastery assessment.
  - Downloadable technology resources are provided for specific computer algebra systems and graphing calculators.
  - Expanded historical biographies are now on the Web site and CD-ROM, leaving more room in the margin of the text for notes, observations, and annotations and giving the book a more open look.

With all these changes, we have not compromised our belief that the fundamental goal of calculus is helping prepare students to enter the worlds of mathematics, science, and engineering.

Mastering Skills and Concepts

As always, this text continues to maintain a strong skill-building emphasis. Throughout this edition, we have included examples and discussions encouraging students to think visually, analytically, and numerically. Almost every exercise set contains problems requiring students to generate and interpret graphs as a tool for understanding mathematical or real-world relationships. Many sections also contain problems to extend the range of applications, mathematical ideas, and rigor.

Students are asked to explore and explain a variety of calculus concepts and applications in writing exercises placed throughout the text. In addition, each chapter ends with a list of questions to help students review and summarize what they have learned. Many of these review questions make great writing assignments.

Problem-Solving Strategies

We believe students learn best when procedural techniques are laid out as clearly and simply as possible. To this end, stepwise problem-solving summaries are included as appropriate, especially for the more difficult or complicated procedures. As always, we are especially careful that examples in the text illustrate the steps outlined by the summaries.

Exercises

Exercise sets have been carefully reviewed and revised in this new edition. They are grouped by topic, with special sections for computer explorations. These sections contain CAS explorations and projects.

Within exercise sets are practice and applied problems, critical thinking and challenging exercises (in subsections marked "Applications and Theory"), and exercises requiring students to write about important calculus concepts. Writing exercises appear throughout exercise sets. Exercises generally follow the order of presentation in the text, and those requiring a graphing utility (such as a graphing calculator) are identified throughout the text by the icon T.

Chapter End Support Material

At the end of each chapter are three features summarizing the chapter contents:

"Questions to Guide Your Review" ask students to think about key chapter concepts and then verbalize their understanding of them and include illustrative examples. These questions are also suitable for writing exercises.

"Practice Exercises" provide a review of the techniques, computational and numerical skills, and key applications.

"Additional Exercises: Theory, Examples, and Applications" provide students with more theoretical or challenging applications and problems to further deepen their understanding of the mathematical ideas.

Applications and Examples

A hallmark of this book has been the application of calculus to science and engineering. These applied problems have been updated, improved, and extended continually over the last several editions. With this edition, we include more problems based on real data requiring graphical and numerical techniques for their solution. Throughout the text, we cite sources for the data or articles from which these applications are drawn, helping students understand that calculus is a current, dynamic field requiring a multiplicity of different techniques and approaches. Most of these applications are directed toward the physical sciences and engineering, but there are many from biology and the social sciences as well.

Technology: Graphing Utility and Computer Explorations

Virtually every section of the text contains exercises to explore numerical patterns or graphing utility exercises that ask students to generate and interpret graphs as a tool to understanding mathematical and real-world relationships. Many of the graphing utility exercises are suitable for classroom demonstration or for group work by students in or out of class. These exercises are identified throughout the text by the icon T or the heading "Computer Explorations."

Computer Explorations

Numbering more than 200, the computer explorations exercises have been solved using both *Mathematica* and Maple. In addition, *Mathematica* and Maple modules are available on the Web site and CD-ROM. These modules have been carefully designed to help students develop a geometric intuition and a deeper understanding and appreciation of calculus concepts, methodologies, and applications. CD/Web site icons mark the locations in the text where material related to these modules is covered.

Notes also appear throughout the text that encourage students to explore with graphing utilities and help them assess when the use of technology is helpful and when it may be misleading.

Expanded History and Biographies

Any student is enriched by seeing the human side of mathematics through its historical development. In previous editions, we featured history boxes describing the origins of ideas, conflicts concerning ownership of these ideas, and interesting sidelights into modern topics such as fractals and chaos. For the tenth edition, we have expanded and written more biographies and historical essays. These essays are now available on the CD-ROM and Web site; they are referenced by icons throughout the text, leaving more room in the margins for student notes, observations, and annotations.

The Many Faces of This Book

Mathematics Is a Formal and Beautiful Language

Calculus is one of the most powerful of human intellectual achievements. One goal of this book is to give students an appreciation of the beauty of calculus. As in previous editions, we have been careful to say only what is true and mathematically sound. Every definition, theorem, corollary, and proof has been reviewed for clarity and mathematical correctness.

Whether calculus is taught in a traditional lecture format or entirely in labs with individual and group learning focusing on numerical and graphical experimentation, its ideas and techniques need to be articulated clearly and accurately.

Students Will Learn from This Book for Many Years to Come

We intentionally provide far more material than any one instructor would want to teach. Students can continue to learn calculus from this book long after the class has ended. It provides an accessible review of the calculus a student has already studied and is a resource for the working engineer or scientist.

Highlights of New Content Features, by Chapter

Preliminaries

- All the familiar precalculus functions are covered completely.
- Parametric equations are introduced.
- Inverses of familiar functions, including inverse trigonometric functions, are also covered.
- Mathematical modeling, with modeling exercises, is introduced.
- New examples and exercises employ real data and regression analysis using a calculator.

Chapter 1 Limits and Continuity

- Limits are introduced by way of rates of change, with a concluding section on tangent lines to connect and complete the initial discussion.
- All the fundamental ideas on limits are now together in a single chapter, including finite limits, infinite limits, asymptotes, limit rules, and $\lim_{\theta\to 0} ((\sin\theta)/\theta)$.
- Both informal and precise definitions of the limit concept are given, but there is less emphasis on using the precise definition to prove theorems.

Chapter 2 Derivatives

- The derivative as a rate of change is presented earlier to stress its importance in studying motion along a line in modeling real-world phenomena.
- Differentiation rules are presented in two sections to enhance the clarity and flow of the presentation.
- First and second derivatives for parametric equations are included as an application of the Chain Rule.
- The Early Transcendentals version of this Chapter differs from the standard edition in that it has two additional sections, the first devoted to the derivatives of inverse trigonometric functions, and the second to derivatives of exponential and logarithmic functions. The number e is defined as the value of a that makes $\lim_{h \to 0} ((a^h - 1)/h)$ equal 1. The natural logarithm is defined as the inverse of e^x.
- With exponential and logarithmic functions now available, we study exponential change and verify the derivative power rule for arbitrary real powers.

Chapter 3 Applications of Derivatives

- The availability of the derivatives of inverse trigonometric, exponential, and logarithmic functions enriches the examples, exercises, and applications in this chapter.
- The treatment of using the first and second derivatives to determine the shape of a graph is more focused and streamlined.
- A new section on using the first and second derivative to produce graphical solutions to autonomous first-order differential equations acts as a graphical prelude to Chapters 4 and 6.
- The new section includes an introduction to population modeling.

Chapter 4 Integration

- As before, indefinite integrals are presented first, stressing their importance for solving elementary differential equations. The rules for antiderivatives and the substitution method follow next.
- The list of available integrable funtions now includes exponential functions and functions leading to logarithmic and inverse trigonometric functions. We no longer have to wait until Chapter 6, as in the standard edition, to integrate the tangent and cotangent functions.
- As in the previous edition, estimating with finite sums in a variety of application settings motivates the ideas of Riemann sums and definite integrals. Students see the definite integral early as more than just a tool for finding area.
- The section defining the definite integral as a limit of Riemann sums has been streamlined and now focuses on continuous functions. Piecewise-continuous functions are treated in the Additional Exercises at the end of the chapter.
- All the material on single integral area calculations (including areas between curves) is now treated in this chapter.

Chapter 5 Applications of Integrals

- The treatment of volumes has been combined from three into two sections.
- Arc length formulas are developed for both explicit function and parametric curves in the plane.
- Surface area has been moved to Chapter 13, where it is needed for surface integrals. There, it is treated in a unified fashion rather than as a special case of surfaces of revolution.
- The availability of exponential and logarithmic functions makes it possible to treat applications of separable first-order differential equations here instead of waiting until the next chapter as in the standard edition. Among the applications are modeling growth and decay, heat transfer, falling with resistance proportional to velocity, coasting to a stop, and torricelli's law.
- The important applications to springs, pumping and lifting, fluid forces, and moments have all been retained from the previous edition.

Chapter 6 Transcendental Functions and Differential Equations

- With so much of the standard material on transcendental functions now in earlier chapters, this chapter is considerably shorter than its standard-edition counterpart. It still begins with defining the natural logarithm as an integral, however, then defining the natural exponential function as its inverse and reconfirming the laws and differentiation rules derived at the end of Chapter 2.

- Linear first-order differential equations follow, modeling mixture problems and *RL* circuits.
- Euler's method and the improved Euler's method are combined with additional material on population models, illustrating graphical, numerical, and analytic solution methods.
- This chapter concludes with a short treatment of hyperbolic and inverse hyperbolic functions.

Chapter 7 Integration Techniques, L'Hôpital's Rule, and Improper Integrals

- Monte Carlo integration is now included with the use of integral tables or computer algebra systems (CAS) to find integrals.
- L'Hôpital's Rule is covered in this chapter just prior to its use for calculating some improper integrals and limits of sequences (in Chapter 8).

Chapter 8 Infinite Series

- The basic ideas concerning sequences of numbers and their limits are covered in the first section. The next section, which is optional, treats the more theoretical ideas involving subsequences and bounded monotonic sequences.
- Most of the important series convergence tests are presented together in a single, streamlined section.
- Two new optional sections at the end of the chapter cover the basics of Fourier series. This inclusion allows for an earlier introduction to these important concepts for students requiring their use right away in their applied science and engineering courses. Completing the elementary introduction to series, these sections illustrate important representations of functions by series other than power series.

Chapter 9 Vectors in the Plane and Polar Functions

- This is a new chapter on vectors and projectile motion in the plane, with two sections at the end covering polar coordinates and graphs and the calculus of polar curves to prepare students for their use in multivariable calculus. It permits an earlier self-contained treatment of planar vectors, if desired. The chapter can be covered any time after the coverage of the integral and the calculus of exponential and logarithmic functions.
- Chapters P through 9 now form a complete package treating the ideas of single variable calculus. Three-dimensional vectors are presented independently along with multivariable calculus, beginning in Chapter 10.
- Vector ideas are motivated by their application to studying paths, velocities, accelerations, and forces associated with bodies moving along planar paths.
- The detailed analytic geometry of conic sections and quadratic equations has been eliminated. These ideas are thoroughly covered in high school and precalculus courses, but we nevertheless review many of the basics throughout the text as needed.
- Parametrizations of plane curves has been moved to earlier chapters.

Chapter 10 Vectors and Motion in Space

- Three-dimensional vectors, the geometry of space, and vector-valued functions defining space curves are now organized together in this single chapter with fresh introductions and examples. This chapter now constitutes, and clearly delineates, the entry point for the multivariable calculus.
- Letters representing vectors have been changed from uppercase letters to the now more standard lowercase letters.
- Vectors in the plane are reviewed along with the development of the algebra and geometry of three-dimensional vectors to help students bridge any possible gap between Calculus II and Calculus III courses.
- The logical treatment and organization of motion along space curves and the **TNB** frame has been retained from the previous edition.

Chapter 11 Multivariable Functions and Their Derivatives

- The chapter has been reorganized to improve efficiency and flow. The treatment of partial derivatives with constrained variables has been moved toward the end of the chapter to follow the introduction to Lagrange multipliers. The treatment of linearization and differentials now follows the treatment of directional derivatives, gradient vectors, and tangent planes.

- The treatment of gradients and tangent planes is shorter and more direct.
- A new introduction to extreme values and saddle points compares and contrasts the multivariable case with the single-variable case.
- The exercise sets have been streamlined and all applications exercises labeled for quick identification.

Chapter 12 Multiple Integrals

- The treatment of the calculation of masses, moments, and centers of mass with multiple integrals is now self-contained. It no longer assumes previous exposure to the single-integral calculations in Chapter 5, which may now be bypassed entirely.
- Again, the practice of titling exercises makes them noticeably easier to select than before.

Chapter 13 Integration in Vector Fields

- In the treatment of Green's Theorem in the plane, circulation density at a point is introduced as the **k**-component of a more general circulation vector called the curl, which is treated in detail in the later section on Stokes' Theorem. This arrangement resolves the apparent inconsistency of having circulation in the plane represented by a scalar while circulation in space is represented by a vector.

Supplements for the Instructor

TestGen-EQ with QuizMaster-EQ

Windows and Macintosh CD (dual platform)
ISBN 0-201-70288-6
TestGen-EQ's friendly graphical interface enables instructors to view, edit, and add questions, transfer questions to tests, and print tests in a variety of fonts and forms easily. Search and sort features let the instructor quickly locate questions and arrange them in a preferred order. Six question formats are available, including short answer, true–false, multiple choice, essay, matching, and bimodal formats. A built-in question editor gives the user power to create graphs, import graphics, insert mathematical symbols and templates, and insert variable numbers or text. Computerized test banks include algorithmically defined problems organized according to each version of the textbook (standard and Early Transcendentals). An "Export to HTML" feature allows instructors to create practice tests for the Web.

QuizMaster-EQ enables instructors to create and save tests using *TestGen-EQ* so that students can take them for either practice or a grade on a computer network. Instructors can set preferences for how and when tests are administered. *QuizMaster-EQ* automatically grades exams, stores results on disk, and allows the instructor to view or print a variety of reports for individual students, classes, or courses.

This software is free to adopters of the text. Consult your Addison-Wesley representative for details.

Instructor's Solutions Manual

Volume I (Chapters P–9), ISBN 0-201-71010-2
Volume II (Chapters 8–13), ISBN 0-201-50404-9
The *Instructor's Solutions Manual* by Maurice D. Weir and John L. Scharf contains complete worked-out solutions to all the exercises in the text.

Answer Book

ISBN 0-201-71009-9
The *Answer Book* by Maurice D. Weir and John L. Scharf contains short answers to most of the exercises in the text.

Technology Resource Manuals for Computer Algebra Systems and Graphing Calculators

TI-Graphing Calculator Manual ISBN 0-201-72198-8
Maple Manual ISBN 0-201-72197-X
Mathematica Manual ISBN 0-201-72196-1
Each manual provides detailed guidance for integrating a specific software package or graphing calculator throughout the course, including syntax and commands.

Transparency Masters

Instructors may download from the CD-ROM a full set of color PowerPoint art transparencies featuring a number of the more complex figures from the text for use in the classroom.

Supplements for the Student

Student's Study Guide

Volume I (Chapters P–9), ISBN 0-201-71008-0
Volume II (Chapters 8–13), ISBN 0-201-50406-5
Organized to correspond with the text, the *Student's Study Guide* reinforces important concepts and provides study tips and additional practice problems.

Student's Solutions Manual

Volume I (Chapters P–9), ISBN 0-201-66211-6
Volume II (Chapters 8–13), ISBN 0-201-50402-2
The Student's Solutions Manual by Maurice D. Weir and John L. Scharf is designed for the student and contains carefully worked-out solutions to all the odd-numbered exercises in the text.

Just-in-Time Algebra and Trigonometry for Students of Calculus, Second Edition

ISBN 0-201-66974-9
Sharp algebra and trigonometry skills are critical to mastering calculus, and *Just-in-Time Algebra and Trigonometry for Students of Calculus, Second Edition* by Guntram Mueller and Ronald I. Brent is designed to bolster these skills while students study calculus. As students make their way through calculus, this text is with them every step of the way, showing them the necessary algebra or trigonometry topics and pointing out potential problem spots. The easy-to-use contents has algebra and trigonometry topics arranged in the order in which students will need them as they study calculus.

AWL Math Tutor Center

The AWL Math Tutor Center (www.awl.com/tutorcenter) provides assistance to students who take calculus and purchase a mathematics textbook published by Addison Wesley Longman. Help is provided via phone, fax, and e-mail. Students who use the service will be helped by tutors who are qualified mathematics instructors.

CD-ROM and Web Site

Maple and Mathematica Modules

Over 35 modules have been written by John L. Scharf and Marie M. Vanisko of Carroll College in Montana and Colonel D. Chris Arney of the U.S. Military Academy. These modules have been carefully designed to help students develop their geometric intuition and deepen their understanding of calculus concepts and methods. Based on real-world applications, they encourage students to visualize calculus and to discover its importance in everyday life. Users will need *Mathematica* or Maple to access these modules. Icons reference these modules throughout the text.

Interactive Calculus (Java Applets)

These unique interactive calculus Java applets are easy to use, with no syntax or special languages to learn. Students can manipulate equations and graphs in "real time." Topics span limits, projectile motion, slopes, tangents, derivatives, integrals, **TNB** frames, and the concept of curl. By bringing these applets into classroom demonstration and discussion, laboratory and homework assignments, or independent study, teachers and students can explore the mathematics of time and motion. These applets are designed to build a clear understanding of concepts when they are first encountered and to help students over the hurdles of abstraction that have often confused other students in the past.

Video Clips

Video clips of real-world situations provide motivation for learning and applying calculus. These videos have been developed specifically to accompany several of the calculus modules described above.

Expanded History and Biographies

CD-ROM
WEBsite
Historical Biography

Icons throughout the book refer to expanded historical biographies and notes on the Web site and CD-ROM. These materials have been written by Colonel D. Chris Arney of the U.S. Military Academy in collaboration with Joe B. Albree of Auburn University.

Just-in-Time Online Algebra and Trigonometry

Compiled by Ronald I. Brent and Guntram Mueller of the University of Massachusetts, Lowell, this interactive Web-based testing and tutorial system allows students to practice the algebra and trigonometry skills critical to mastering calculus. *Just-in-Time Online* tracks student progress and provides personalized study plans to help students succeed. The registration coupon at the back of this text provides access to this feature of the Web site.

Interactive Calculus Tutorial

Written by G. Donald Allen, Michael Stecher, and Philip B. Yasskin of Texas A&M University, this interactive online calculus tutorial lets students review textbook-specific material by chapter via practice quizzes and receive diagnostic feedback on their performance.

Skill Mastery Quizzes

A collection of chapter-by-chapter quizzes is also provided on the Web site. These quizzes can be administered and graded online for skills-based mastery assessment.

Transparency Masters

Instructors may download from the CD-ROM a full set of color PowerPoint art transparencies featuring a number of the more complex figures from the text for use in the classroom.

Downloadable Technology Resources for Specific Computer Algebra Systems and Graphing Calculators

Each manual provides detailed guidance for integrating a specific software package or graphing calculator throughout the course, including syntax and commands. These manuals are available on the Web site for downloading in PDF form.

Collaborative Network

The Collaborative Network is a suite of online communication tools, which includes message boards and i-chat. These tools can be used to deliver courses in a distance learning environment. Message boards allow users to post messages and check back periodically for responses. Students can also use message boards to obtain peer support for study guide activities, freeing up instructor time. I-chat is a perfect arena for instructor-led live discussions with groups of students. The i-chat auditorium allows instructors to post a series of slides in the upper part of the screen while fielding questions (text only) from students in the bottom portion. This feature is particularly useful either as a review of lectures or for class meetings of geographically dispersed students.

Syllabus Manager™

Syllabus Manager™ is a free online syllabus creation and management tool for instructors and students who use this text. It can be used by a nontechnical person to build and maintain one or more syllabi on the Web. Students may "turn on" an instructor's syllabus from the Web site.

Acknowledgments

We would like to express our thanks for the many valuable contributions of the people who reviewed this edition as it developed through its various stages:

Manuscript Reviewers

Tuncay Aktosun, North Dakota State University
Andrew G. Bennett, Kansas State University
Terri A. Bourdon, Virginia Polytechnic Institute and State University
Mark Brittenham, University of Nebraska, Lincoln
Bob Brown, Essex Community College
David A. Edwards, University of Delaware
Mark Farris, Midwestern State University
Kim Jongerius, Northwestern College
Jeff Knisley, East Tennessee State University
Slawomir Kwasik, Tulane University
Jeuel LaTorre, Clemson University
Daniel G. Martinez, California State University, Long Beach
Sandra E. McLaurin, University of North Carolina, Wilmington
Stephen J. Merrill, Marquette University
Shai Neumann, Brevard Community College
Linda Powers, Virginia Polytechnic Institute and State University
William L. Siegmann, Rensselaer Polytechnic Institute
Rick L. Smith, University of Florida
James W. Thomas, Colorado State University
Abraham Ungar, North Dakota State University
Harvey E. Wolff, University of Toledo

Technology Reviewers

Mark Brittenham, University of Nebraska, Lincoln
Warren J. Burch, Brevard Community College, Cocoa
Lyle Cochran, Whitworth College
Philip S. Crooke III, Vanderbilt University
Linda Powers, Virginia Polytechnic Institute and State University
David Ruch, Metropolitan State College of Denver
Paul Talaga, Weber State University
James W. Thomas, Colorado State University
Robert L. Wheeler, Virginia Polytechnic Institute and State University

Other Contributors

We especially thank Colonel D. Chris Arney, John L. Scharf, and Marie M. Vanisko for sharing their insights in using technology to help make calculus come alive for the student, and Colonel D. Chris Arney and Joe B. Albree for their contributions on the history of calculus. Their dedication, encouragement, and team efforts in working with us to conceive and create the technology modules and historical biographies and essays is deeply appreciated. Further thanks to John L. Scharf for his contributions to the solutions manuals.

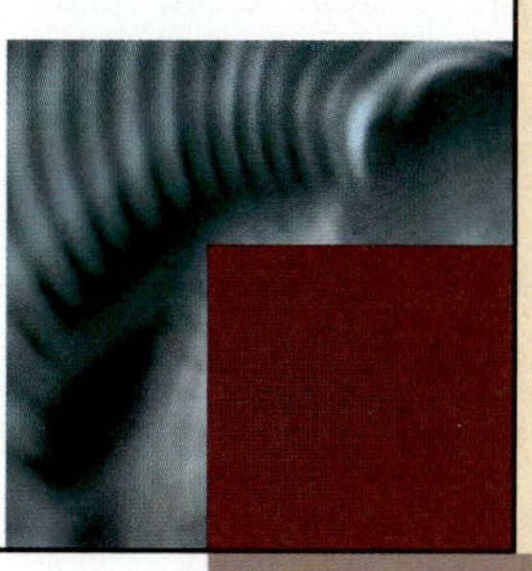

To the Student

What Is Calculus?

Calculus is the mathematics of motion and change. Where there is motion or growth, where variable forces are at work producing acceleration, calculus is the mathematics to apply. This was true in the beginnings of the subject, and it is true today.

Calculus was first invented to meet the mathematical needs of the scientists of the sixteenth and seventeenth centuries, needs that were mainly mechanical in nature. Differential calculus dealt with the problem of calculating rates of change. It enabled people to define slopes of curves, to calculate velocities and accelerations of moving bodies, to find firing angles that would give cannons their greatest range, and to predict the times when planets would be closest together or farthest apart. Integral calculus dealt with the problem of determining a function from information about its rate of change. It enabled people to calculate the future location of a body from its present position and a knowledge of the forces acting on it, to find the areas of irregular regions in the plane, to measure the lengths of curves, and to find the volumes and masses of arbitrary solids.

Today, calculus and its extensions in mathematical analysis are far-reaching indeed, and the physicists, mathematicians, and astronomers who first invented the subject would surely be amazed and delighted, as we hope you will be, to see what a profusion of problems it solves and what a range of fields now use it in the mathematical models that bring understanding about the universe and the world around us. The goal of this edition is to present a modern view of calculus enhanced by the use of technology.

How to Learn Calculus

Learning calculus is not the same as learning arithmetic, algebra, and geometry. In those subjects, you learn primarily how to calculate with numbers; how to simplify algebraic expressions and calculate with variables; and how to reason about points, lines, and figures in the plane. Calculus involves those techniques and skills but develops others as well, with greater precision and at a deeper level. Calculus introduces so many new concepts and computational operations, in fact, that you will no longer be able to learn everything you need in class. You will have to learn a fair amount on your own or by working with other students. What should you do to learn?

1. *Read the text.* You will not be able to learn all the meanings and connections you need just by attempting the exercises. You will need to read relevant passages in

the book and work through examples step by step. Speed reading will not work here. You are reading and searching for detail in a step-by-step, logical fashion. This kind of reading, required by any deep and technical content, takes attention, patience, and practice.

2. *Do the homework*, keeping the following principles in mind.
   (a) *Sketch diagrams* whenever possible.
   (b) *Write your solutions in a connected step-by-step, logical fashion*, as if you were explaining it to someone else.
   (c) *Think about why* each exercise is there. Why was it assigned? How is it related to the other assigned exercises?

3. *Use your graphing calculator and computer* whenever possible. Complete as many grapher and Computer Exploration exercises as you can, even if they are not assigned. Graphs provide insight and visual representation of important concepts and relationships. Numbers can reveal important patterns. A graphing calculator or computer gives you the freedom to explore realistic problems and examples that involve calculations too difficult or lengthy to do by hand.

4. Try on your own to *write short descriptions of the key points* each time you complete a section of the text. If you succeed, you probably understand the material. If you do not, you will know where there is a gap in your understanding.

Learning calculus is a process; it does not come all at once. Be patient, persevere, ask questions, discuss ideas and work with classmates, and seek help when you need it, right away. The rewards of learning calculus will be very satisfying, both intellectually and professionally.

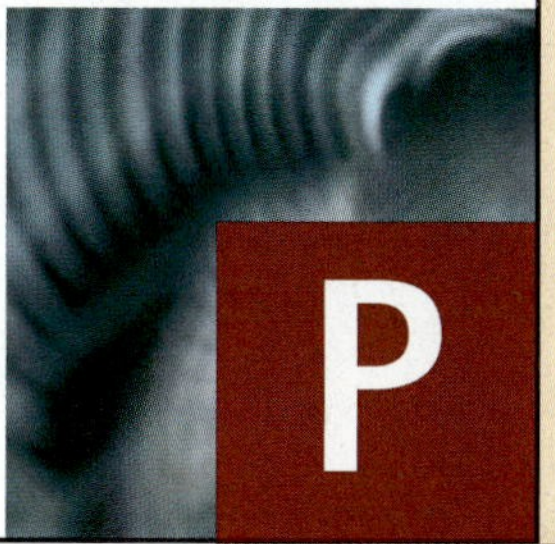

P Preliminaries

OVERVIEW This chapter reviews the most important things you need to know to start learning calculus. It also introduces the use of a graphing utility as a tool to investigate mathematical ideas, to support analytical work, and to solve problems with numerical and graphical methods. The emphasis is on functions and graphs, the main building blocks of calculus.

Functions and parametric equations are the major tools for describing the real world in mathematical terms, from temperature variations to planetary motions, from brain waves to business cycles, and from heartbeat patterns to population growth. Many functions have particular importance because of the behavior they describe. Trigonometric functions describe cyclic, repetitive activity; exponential, logarithmic, and logistic functions describe growth and decay; and polynomial functions can approximate these and most other functions.

1 Lines

Increments • Slope of a Line • Parallel and Perpendicular Lines • Equations of Lines • Applications • Regression Analysis with a Calculator

One reason calculus has proved to be so useful is that it is the right mathematics for relating the rate of change of a quantity to the graph of the quantity. Explaining that relationship is one goal of this book. It all begins with the slopes of lines.

Increments

When a particle in the plane moves from one point to another, the net changes or *increments* in its coordinates are found by subtracting the coordinates of its starting point from the coordinates of its stopping point.

The symbols Δx and Δy are read "delta x" and "delta y." The letter Δ is a Greek capital d for "difference." Neither Δx nor Δy denotes multiplication; Δx is not "delta times x" nor is Δy "delta times y."

Definition Increments

If a particle moves from the point, (x_1, y_1) to the point (x_2, y_2), the **increments** in its coordinates are

$$\Delta x = x_2 - x_1 \qquad \text{and} \qquad \Delta y = y_2 - y_1.$$

Increments can be positive, negative, or zero, as in Example 1.

Example 1 Finding Increments

The coordinate increments from $(4, -3)$ to $(2, 5)$ are

$$\Delta x = 2 - 4 = -2, \qquad \Delta y = 5 - (-3) = 8.$$

From $(5, 6)$ to $(5, 1)$, the increments are

$$\Delta x = 5 - 5 = 0, \qquad \Delta y = 1 - 6 = -5.$$

Slope of a Line

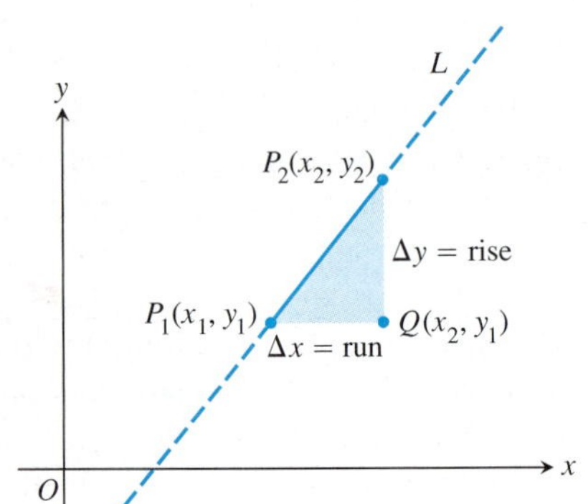

FIGURE 1 The slope of line L is $m = \dfrac{\text{rise}}{\text{run}} = \dfrac{\Delta y}{\Delta x}$.

It is conventional to denote the slope by the letter m.

Each nonvertical line L has a *slope,* which we calculate as elevation change per unit of run in the following way. Let $P_1(x_1, y_1)$ and $P_2(x_2, y_2)$ be any two points on L (Figure 1). We call $\Delta y = y_2 - y_1$ the **rise** from P_1 to P_2, $\Delta x = x_2 - x_1$ the **run** from P_1 to P_2, and we define the slope of L to be $\Delta y/\Delta x$.

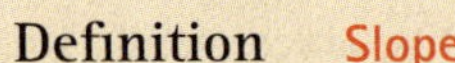

Definition Slope

Let $P_1(x_1, y_1)$ and $P_2(x_2, y_2)$ be points on a nonvertical line, L. The **slope** of L is

$$m = \frac{\text{rise}}{\text{run}} = \frac{\Delta y}{\Delta x} = \frac{y_2 - y_1}{x_2 - x_1}.$$

A line that goes uphill as x increases has a positive slope. A line that goes downhill as x increases has a negative slope. A horizontal line has slope zero since all its points have the same y-coordinate, making $\Delta y = 0$. For vertical lines, $\Delta x = 0$ and the ratio $\Delta y/\Delta x$ is undefined. We express this fact by saying *vertical lines have no slope.*

Parallel and Perpendicular Lines

Parallel lines form equal angles with the x-axis (Figure 2). Hence, nonvertical parallel lines have the same slope. Conversely, lines with equal slopes form equal angles with the x-axis and are therefore parallel.

If two nonvertical lines L_1 and L_2 are perpendicular, their slopes m_1 and m_2 satisfy $m_1 m_2 = -1$, so each slope is the *negative reciprocal* of the other:

$$m_1 = -\frac{1}{m_2}, \qquad m_2 = -\frac{1}{m_1}.$$

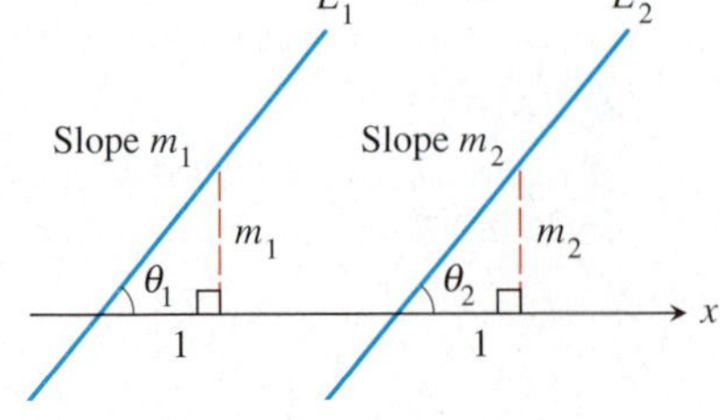

FIGURE 2 If $L_1 \parallel L_2$, then $\theta_1 = \theta_2$ and $m_1 = m_2$. Conversely, if $m_1 = m_2$, then $\theta_1 = \theta_2$ and $L_1 \parallel L_2$.

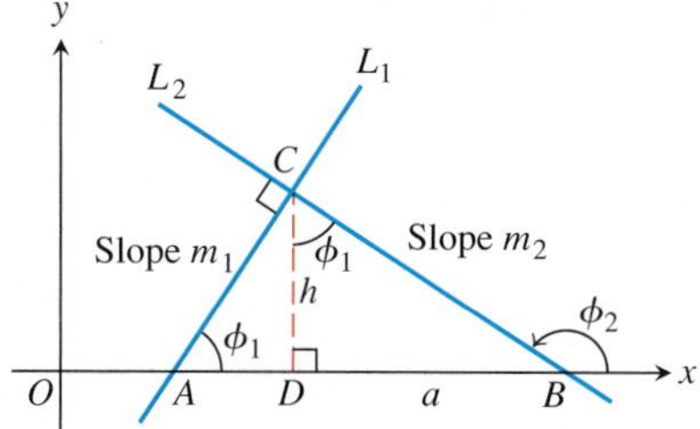

FIGURE 3 ΔADC is similar to ΔCDB. Hence, ϕ_1 is also the upper angle in ΔCDB, where $\tan \phi_1 = a/h$.

The argument goes like this: In the notation of Figure 3, $m_1 = \tan \phi_1 = a/h$, whereas $m_2 = \tan \phi_2 = -h/a$. Hence, $m_1 m_2 = (a/h)(-h/a) = -1$.

Example 2 Determining Perpendicularity from Slope

If L is a line with slope $3/4$, any line with slope $-4/3$ is perpendicular to L.

Equations of Lines

The vertical line through the point (a, b) has equation $x = a$ since every x-coordinate on the line has the value a. Similarly, the horizontal line through (a, b) has equation $y = b$.

Example 3 Finding Equations for Vertical and Horizontal Lines

The vertical and horizontal lines through the point (2, 3) have equations $x = 2$ and $y = 3$, respectively (Figure 4).

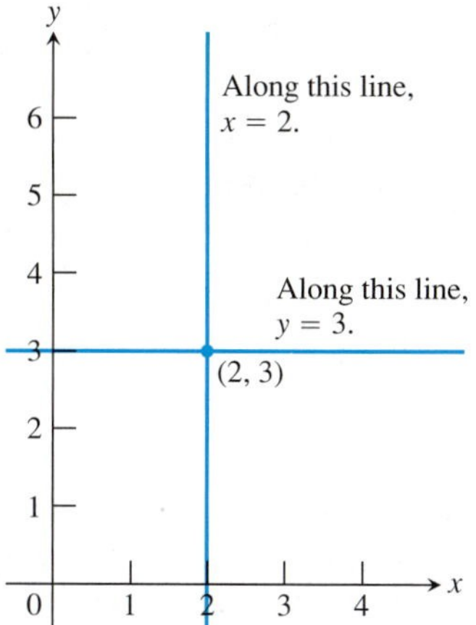

FIGURE 4 The standard equations for the vertical and horizontal lines through the point (2, 3) are $x = 2$ and $y = 3$. (Example 3)

We can write an equation for any nonvertical line if we know its slope m and the coordinates of one point $P_1(x_1, y_1)$ on it. For if $P(x, y)$ is *any* other point on the line then

$$\frac{y - y_1}{x - x_1} = m,$$

so that

$$y - y_1 = m(x - x_1) \qquad \text{or} \qquad y = m(x - x_1) + y_1.$$

Definition Point-Slope Equation

The equation

$$y = m(x - x_1) + y_1$$

is the **point-slope equation** of the line through the point (x_1, y_1) with slope m.

Example 4 Using the Point-Slope Equation

Write an equation for the line through the point (2, 3) with slope $-3/2$.

Solution We substitute $x_1 = 2$, $y_1 = 3$, and $m = -3/2$ into the point-slope equation and obtain

$$y = -\frac{3}{2}(x - 2) + 3 \qquad \text{or} \qquad y = -\frac{3}{2}x + 6.$$

Example 5 Using the Point-Slope Equation

Write an equation for the line through $(-2, -1)$ and $(3, 4)$.

Solution The line's slope is

$$m = \frac{4 - (-1)}{3 - (-2)} = \frac{5}{5} = 1.$$

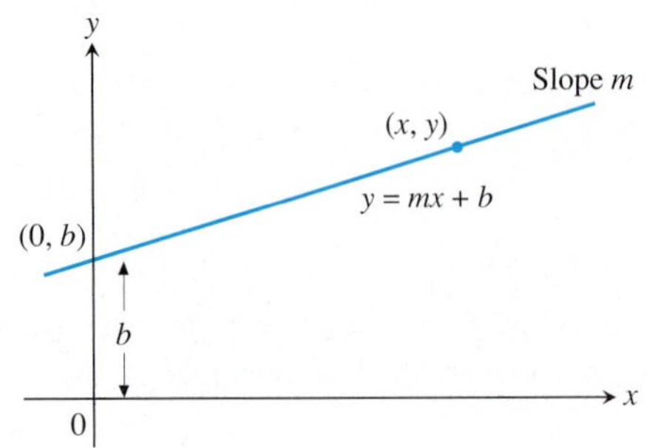

FIGURE 5 A line with slope m and y-intercept b.

We can use this slope with either of the two given points in the point-slope equation. Using $(x_1, y_1) = (-2, -1)$, we obtain

$$y = 1 \cdot (x - (-2)) + (-1)$$
$$y = x + 2 + (-1)$$
$$y = x + 1.$$

The y-coordinate of the point where a nonvertical line intersects the y-axis is the **y-intercept** of the line. Similarly, the x-coordinate of the point where a nonhorizontal line intersects the x-axis is the **x-intercept** of the line. A line with slope m and y-intercept b passes through $(0, b)$ (Figure 5), so

$$y = m(x - 0) + b, \qquad \text{or, more simply,} \qquad y = mx + b.$$

Definition Slope-Intercept Equation

The equation

$$y = mx + b$$

is the **slope-intercept equation** of the line with slope m and y-intercept b.

Example 6 Writing Equations for Lines

Write an equation for the line through the point $(-1, 2)$ that is **(a)** parallel and **(b)** perpendicular to the line $L: y = 3x - 4$.

Solution The line L, $y = 3x - 4$, has slope 3.

(a) The line $y = 3(x + 1) + 2$, or $y = 3x + 5$, passes through the point $(-1, 2)$ and is parallel to L because it has slope 3.

(b) The line $y = (-1/3)(x + 1) + 2$, or $y = (-1/3)x + 5/3$, passes through the point $(-1, 2)$ and is perpendicular to L because it has slope $-1/3$.

If A and B are not both zero, the graph of the equation $Ax + By = C$ is a line. Every line has an equation in this form, even lines with undefined slopes.

Definition General Linear Equation

The equation

$$Ax + By = C \qquad (A \text{ and } B \text{ not both } 0)$$

is a **general linear equation** in x and y.

Although the general linear form helps in the quick identification of lines, the slope-intercept form is the one to enter into a calculator for graphing.

Example 7 Analyzing and Graphing a General Linear Equation

Find the slope and y-intercept of the line $8x + 5y = 20$. Graph the line.

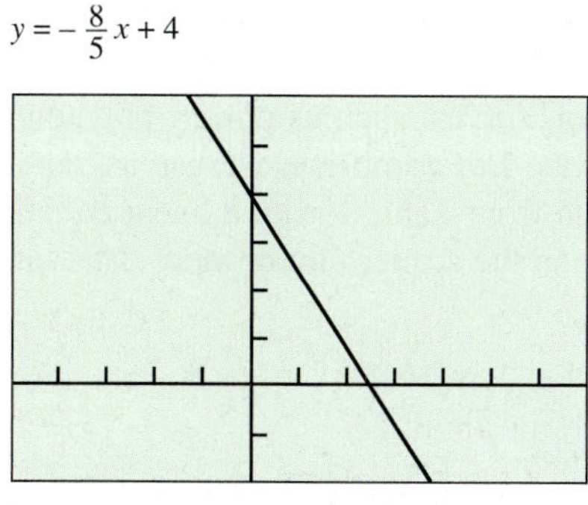

[–5, 7] by [–2, 6]

FIGURE 6 The line $8x + 5y = 20$.

Solution Solve the equation for y to put the equation in slope-intercept form.

$$8x + 5y = 20$$
$$5y = -8x + 20$$
$$y = -\frac{8}{5}x + 4.$$

This form reveals the slope ($m = -8/5$) and y-intercept ($b = 4$), and it puts the equation in a form suitable for graphing on a graphing calculator (Figure 6).

Applications

Many important variables are related by linear equations. For example, the relationship between Fahrenheit temperature and Celsius temperature is linear, a fact we use to advantage in the next example.

Example 8 Temperature Conversion

Find a formula relating Fahrenheit and Celsius temperature. Then find the Celsius equivalent of 90°F and the Fahrenheit equivalent of −5°C.

Solution Because the relationship between the two temperature scales is linear, it has the form $F = mC + b$. The freezing point of water is $F = 32°$ or $C = 0°$, whereas the boiling point is $F = 212°$ or $C = 100°$. Thus,

$$32 = m \cdot 0 + b \qquad \text{and} \qquad 212 = m \cdot 100 + b,$$

so $b = 32$ and $m = (212 - 32)/100 = 9/5$. Therefore,

$$F = \frac{9}{5}C + 32 \qquad \text{or} \qquad C = \frac{5}{9}(F - 32).$$

These relationships let us find equivalent temperatures. The Celsius equivalent of 90°F is

$$C = \frac{5}{9}(90 - 32) \approx 32.2°.$$

The Fahrenheit equivalent of −5°C is

$$F = \frac{9}{5}(-5) + 32 = 23°.$$

Regression Analysis with a Calculator

It can be difficult to see patterns or trends in lists of paired numbers. For this reason, we sometimes begin by plotting the pairs (such a plot is called a **scatter plot**) to see whether the corresponding points have a pattern or trend of some kind. If they do, and if we can find an equation $y = f(x)$ for a curve that approximates the trend, then we have a formula that

1. summarizes the data with a simple expression, and
2. lets us predict values of y for other values of x.

Table 1 Price of a U.S. postage stamp

Year x	Cost y
1885	0.02
1917	0.03
1919	0.02
1932	0.03
1958	0.04
1963	0.05
1968	0.06
1971	0.08
1974	0.10
1975	0.13
1977	0.15
1981	0.18
1981	0.20
1985	0.22
1987	0.25
1991	0.29
1995	0.32
1998	0.33

The process of finding a specific curve type to fit data is **regression analysis,** and the curve is a **regression curve.**

There are many useful types of regression curves, such as power, polynomial, exponential, logarithmic, and sinusoidal curves. In Example 9, we use a graphing calculator's linear regression feature to fit data from Table 1 with a linear equation. This process is equivalent to fitting the points in the scatter plot of these data with a line.

Example 9 Regression Analysis with a Calculator

Starting with the data in Table 1, build a model for the price of a postage stamp as a function of time. After verifying that the model is "reasonable," use it to predict the price in 2010.

Solution

Interpreting the Data

There is little change in the price of a U.S. stamp prior to 1968. Because we are really interested in the trend of the more recent data, we begin with that year. There were two increases in 1981, one of three cents followed by another of two cents. To make 1981 comparable with the other listed years, we lump them together as a single five-cent increase, giving the data in Table 2. Figure 7a gives the scatter plot for Table 2.

Table 2 Price of a U.S. postage stamp since 1968

x	0	3	6	7	9	13	17	19	23	27	30
y	6	8	10	13	15	20	22	25	29	32	33

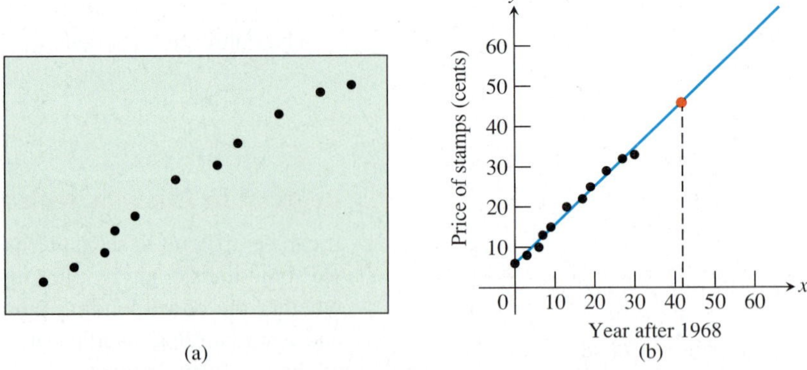

FIGURE 7 (a) Scatter plot of (x, y) data in Table 2. (b) Using the regression line to estimate the price of a stamp in 2010.

Model

Since the plot is fairly linear, we investigate a linear model. Upon entering the data into a graphing calculator (or spreadsheet) and selecting the linear regression option, we find the regression line to be

$$y = 0.96185x + 5.8978 . \tag{1}$$

Figure 7b shows the line and scatter plot together. The fit is remarkably good, so the model seems reasonable.

Solve Graphically

Our goal is to predict the price of a stamp in the year 2010. Reading from the graph in Figure 7b, we conclude that in 2010 ($x = 42$), the value of y is about 46.

Interpret

In the year 2010, a postage stamp will cost about 46 cents.

Confirm Algebraically

Evaluating Equation (1) for $x = 42$ gives

$$y = 0.96185(42) + 5.8978 \approx 46.3 .$$

Regression Analysis

Regression analysis has four steps:

Step 1. Plot the data (scatter plot).

Step 2. Find a regression equation. For a line, it has the form $y = mx + b$.

Step 3. Superimpose the graph of the regression equation on the scatter plot to see the fit.

Step 4. If the fit is satisfactory, use the regression equation to predict y-values for values of x not in the table.

EXERCISES 1

In Exercises 1 and 2, find the coordinate increments from A to B.

1. **(a)** $A(1, 2)$, $B(-1, -1)$ **(b)** $A(-3, 2)$, $B(-1, -2)$

2. **(a)** $A(-3, 1)$, $B(-8, 1)$ **(b)** $A(0, 4)$, $B(0, -2)$

In Exercises 3 and 4, let L be the line determined by points A and B.

(i) Plot A and B. **(ii)** Find the slope of L.

(iii) Draw the graph of L.

3. **(a)** $A(1, -2)$, $B(2, 1)$ **(b)** $A(-2, -1)$, $B(1, -2)$

4. **(a)** $A(2, 3)$, $B(-1, 3)$ **(b)** $A(1, 2)$, $B(1, -3)$

In Exercises 5 and 6, write an equation for **(i)** the vertical line and **(ii)** the horizontal line through the point P.

5. **(a)** $P(2, 3)$ **(b)** $P(-1, 4/3)$

6. **(a)** $P(0, -\sqrt{2})$ **(b)** $P(-\pi, 0)$

In Exercises 7 and 8, write the point-slope equation for the line through the point P with slope m.

7. **(a)** $P(1, 1)$, $m = 1$ **(b)** $P(-1, 1)$, $m = -1$

8. **(a)** $P(0, 3)$, $m = 2$ **(b)** $P(-4, 0)$, $m = -2$

In Exercises 9 and 10, write a general linear equation for the line through the two points.

9. **(a)** $(0, 0)$, $(2, 3)$ **(b)** $(1, 1)$, $(2, 1)$

10. **(a)** $(-2, 0)$, $(-2, -2)$ **(b)** $(-2, 1)$, $(2, -2)$

In Exercises 11 and 12, write the slope-intercept equation for the line with slope m and y-intercept b.

11. **(a)** $m = 3$, $b = -2$ **(b)** $m = -1$, $b = 2$

12. **(a)** $m = -1/2$, $b = -3$ **(b)** $m = 1/3$, $b = -1$

In Exercises 13 and 14, the line contains the origin and the point in the upper right corner of the grapher screen. Write an equation for the line. In Exercise 13, a tick mark on the x-axis represents 1 unit, and on the y-axis, it represents 5 units. In Exercise 14, a tick mark on either axis represents 1 unit.

13.

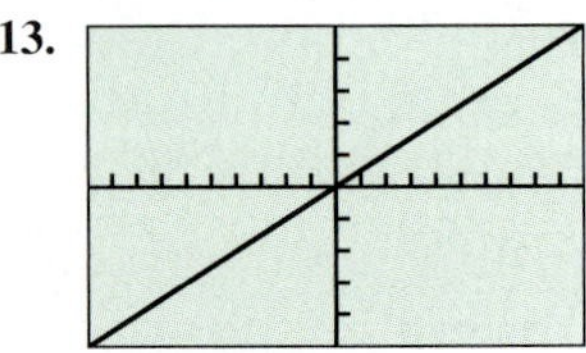

[–10, 10] by [–25, 25]

14.

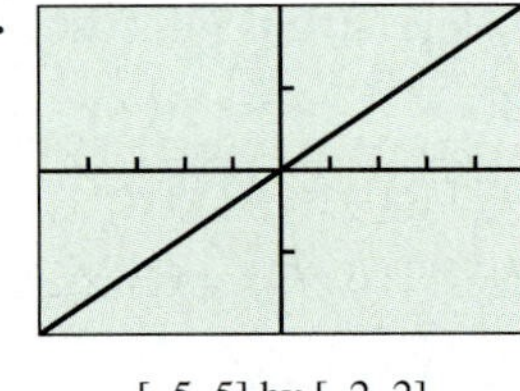

[–5, 5] by [–2, 2]

In Exercises 15 and 16, find the **(i)** slope and **(ii)** y-intercept and **(iii)** graph the line.

15. **(a)** $3x + 4y = 12$ **(b)** $x + y = 2$

16. **(a)** $\frac{x}{3} + \frac{y}{4} = 1$ **(b)** $y = 2x + 4$

In Exercises 17 and 18, write an equation for the line through P that is **(i)** parallel to L and **(ii)** perpendicular to L.

17. **(a)** $P(0, 0)$, $L: y = -x + 2$

(b) $P(-2, 2)$, $L: 2x + y = 4$

18. **(a)** $P(-2, 4)$, $L: x = 5$

(b) $P(-1, 1/2)$, $L: y = 3$

In Exercises 19 and 20, a table of values is given for the linear function $f(x) = mx + b$. Determine m and b.

19.

x	$f(x)$
1	2
3	9
5	16

20.

x	$f(x)$
2	−1
4	−4
6	−7

In Exercises 21 and 22, find the value of x or y for which the line through A and B has the given slope m.

21. $A(-2, 3)$, $B(4, y)$, $m = -2/3$

22. $A(-8, -2)$, $B(x, 2)$, $m = 2$

23. *Revisiting Example 5* Show that you get the same equation in Example 5 if you use the point $(3, 4)$ to write the equation.

24. *Writing to Learn: x-and y-intercepts*

(a) Explain why c and d are the x-intercept and y-intercept, respectively, of the line

$$\frac{x}{c} + \frac{y}{d} = 1.$$

(b) How are the x-intercept and y-intercept related to c and d in the line

$$\frac{x}{c} + \frac{y}{d} = 2?$$

25. *Parallel and perpendicular lines* For what value of k are the two lines $2x + ky = 3$ and $x + y = 1$ **(a)** parallel? **(b)** perpendicular?

In Exercises 26–28, *work in groups of two or three* to solve the problem.

26. *Insulation* By measuring slopes in the figure, find the temperature change in degrees per inch for the following materials.

(a) gypsum wallboard

(b) fiberglass insulation

(c) wood sheathing

(d) *Writing to Learn* Which of the materials in parts **(a)** through **(c)** is the best insulator? The poorest? Explain.

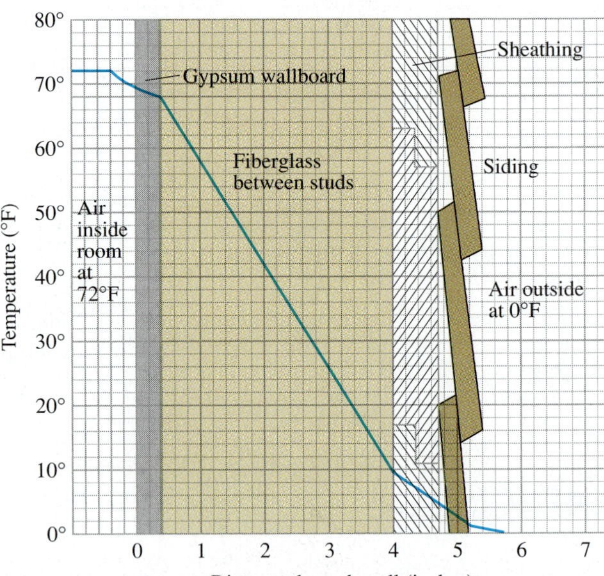

27. *Pressure under water* The pressure p experienced by a diver under water is related to the diver's depth d by an equation of the form $p = kd + 1$ (k a constant). When $d = 0$ meters, the pressure is 1 atmosphere. The pressure at 100 meters is 10.94 atmospheres. Find the pressure at 50 meters.

28. *Modeling distance traveled* A car starts from point P at time $t = 0$ and travels at 45 mph.

(a) Write an expression $d(t)$ for the distance the car travels from P in t hours.

(b) Graph $y = d(t)$.

(c) What is the slope of the graph in part (b)? What does it have to do with the car?

(d) *Writing to Learn* Create a scenario in which t could have negative values.

(e) *Writing to Learn* Create a scenario in which the y-intercept of $y = d(t)$ could be 30.

Extending the Ideas

29. *Fahrenheit versus Celsius* We found a relationship between Fahrenheit temperature and Celsius temperature in Example 8.

(a) Is there a temperature at which a Fahrenheit thermometer and a Celsius thermometer give the same reading? If so, what is it?

T **(b)** *Writing to Learn* In the same viewing window graph $y_1 = (9/5)x + 32$, $y_2 = (5/9)(x - 32)$, and $y_3 = x$. Explain how this figure is related to the question in part (a).

30. *Parallelogram* Three different parallelograms have vertices at $(-1, 1)$, $(2, 0)$, and $(2, 3)$. Draw the three and give the coordinates of the missing vertices.

31. *Parallelogram* Show that if the midpoints of consecutive sides of any quadrilateral are connected, the result is a parallelogram.

32. *Tangent line* Consider the circle of radius 5 centered at $(0, 0)$. Find an equation of the line tangent to the circle at the point $(3, 4)$.

33. *Distance from a point to a line* This activity investigates how to find the distance from a point $P(a, b)$ to a line $L: Ax + By = C$. We suggest that students *work in groups of two or three*.

(a) Write an equation for the line M through P perpendicular to L.

(b) Find the coordinates of the point Q in which M and L intersect.

(c) Find the distance from P to Q.

34. *Reflected light* A ray of light comes along the line $x + y = 1$ from the second quadrant and reflects off the x-axis. The angle of incidence is equal to the angle of reflection as measured from the perpendicular. Write an equation for the line along which the departing light travels.

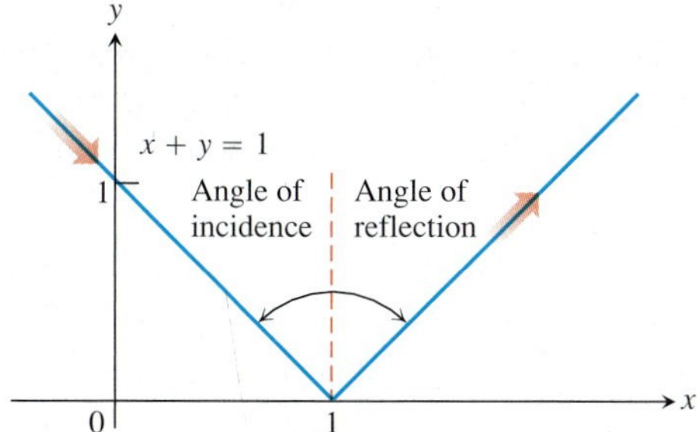

35. *The Mt. Washington cog railway* Civil engineers calculate the slope of roadbed as the ratio of the distance it rises or falls to the distance it runs horizontally. They call this ratio the **grade** of the roadbed, usually written as a percentage. Along the coast, commercial railroad grades are usually less than 2%. In the mountains, they may go as high as 4%. Highway grades are usually less than 5%.

The steepest part of the Mt. Washington cog railway in New Hampshire has an exceptional 37.1% grade. Along this part of the track, the seats in the front of the car are 14 ft above those in the rear. About how far apart are the front and rear rows of seats?

36. A 90° rotation counterclockwise about the origin takes $(2, 0)$ to $(0, 2)$, and $(0, 3)$ to $(-3, 0)$, as shown. Where does it take each of the following points?

(a) $(4, 1)$ **(b)** $(-2, -3)$ **(c)** $(2, -5)$

(d) $(x, 0)$ **(e)** $(0, y)$ **(f)** (x, y)

(g) What point is taken to $(10, 3)$?

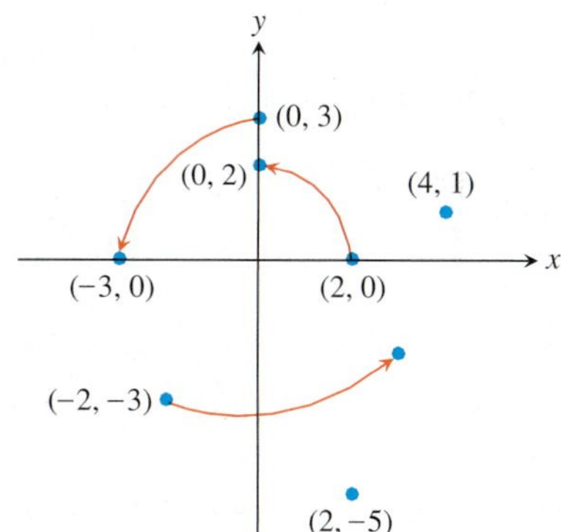

In Exercises 37 and 38, use linear regression analysis.

T **37.** Table 3 lists the ages and weights of nine girls.

Table 3 Girls' ages and weights

Age (months)	Weight (pounds)
19	22
21	23
24	25
27	28
29	31
31	28
34	32
38	34
43	39

(a) Find a linear regression equation for the data.

(b) Find the slope of the regression line. What does the slope represent?

(c) Superimpose the graph of the linear regression equation on a scatter plot of the data.

(d) Use the regression equation to predict the approximate weight of a 30-month old girl.

T **38.** Table 4 shows the mean annual compensation of construction workers.

Table 4 Construction workers' average annual compensation

Year	Annual compensation (dollars)
1980	22,033
1985	27,581
1988	30,466
1989	31,465
1990	32,836

Source: U.S. Bureau of Economic Analysis.

(a) Find a linear regression equation for the data.

(b) Find the slope of the regression line. What does the slope represent?

(c) Superimpose the graph of the linear regression equation on a scatter plot of the data.

(d) Use the regression equation to predict the construction workers' average annual compensation in 2000.

T **39.** The median price of existing single-family homes has increased consistently since 1970. The data in Table 5, however, show that there have been differences in various parts of the country.

(a) Find a linear regression equation for home cost in the Northeast.

(b) What does the slope of the regression line represent?

(c) Find a linear regression equation for home cost in the Midwest.

(d) Where is the median price increasing more rapidly, in the Northeast or the Midwest?

Table 5 Median price of single-family homes

Year	Northeast (dollars)	Midwest (dollars)
1970	25,200	20,100
1975	39,300	30,100
1980	60,800	51,900
1985	88,900	58,900
1990	141,200	74,000

Source: National Association of Realtors®, *Home Sales Yearbook* (Washington DC, 1990).

2 Functions and Graphs

Functions • Domains and Ranges • Viewing and Interpreting Graphs • Increasing versus Decreasing Functions • Even Functions and Odd Functions: Symmetry • Functions Defined in Pieces • The Absolute Value Function • How to Shift a Graph • Composite Functions

Functions are the major tools for describing the real world in mathematical terms. This section discusses the basic ideas of functions, their graphs, and ways of shifting or combining them. Several important types of functions that arise in calculus are presented.

Functions

The values of one variable often depend on the values for another:

- The temperature at which water boils depends on elevation (the boiling point drops as you go up).
- The amount by which your savings will grow in a year depends on the interest rate offered by the bank.

In each case, the value of one variable quantity depends on the value of another. The boiling point of water, b, depends on the elevation, e; the amount of

FIGURE 8 A "machine" diagram for a function.

interest, I, depends on the interest rate, r. We call b and I **dependent variables** because they are determined by the values of the variables e and r on which they depend. The variables e and r are **independant variables.**

A rule that assigns to each element in one set a unique element in another set is called a *function*. The sets may be sets of any kind and do not have to be the same. A function is like a machine that assigns a unique output to every allowable input. The inputs make up the **domain** of the function; the outputs make up the **range** (Figure 8).

Definition Function

A **function** from a set D to a set R is a rule that assigns a unique element in R to each element in D.

In this definition, D is the domain of the function and R is a set *containing* the range (Figure 9).

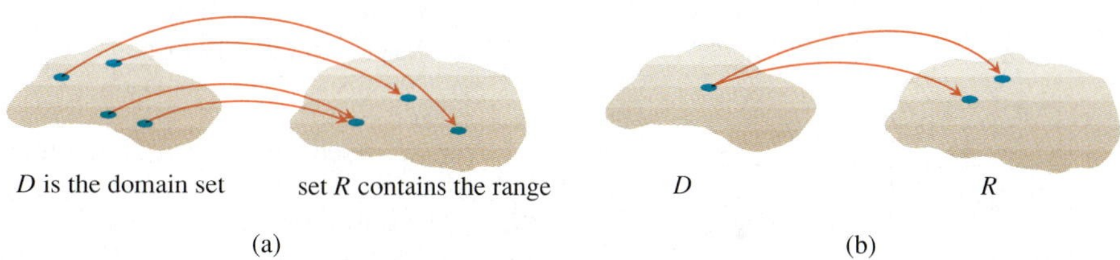

FIGURE 9 (a) A function from set D to set R. (b) *Not* a function. The assignment is not unique.

Years ago, Swiss mathematician Leonhard Euler invented a way to say "y is a function of x" symbolically:

$$y = f(x),$$

which we read as "y equals f of x." This notation enables us to give different functions different names by changing the letters we use. To say the boiling point of water is a function of elevation, we can write $b = f(e)$. To say the area of a circle is a function of the radius, we can write $A = A(r)$, giving the function the same name as the dependent variable.

The notation $y = f(x)$ also gives a way to denote specific values of a function. The value of f at a can be written as $f(a)$, read "f of a."

Example 1 The Circle-Area Function

The domain of the circle-area function $A(r) = \pi r^2$ is the set of all possible radii, the set of all positive real numbers. The range is also the set of all positive real numbers.

The value of A at $r = 2$ is

$$A(2) = \pi(2)^2 = 4\pi.$$

The area of a circle of radius 2 is 4π.

Domains and Ranges

In Example 1, the domain of the function is restricted by context: The independent variable is a radius and must be positive. When we define a function $y = f(x)$ with a formula and the domain is not stated explicitly or restricted by context, the domain is assumed to be the largest set of x-values for which the formula gives real y-values, the so-called **natural domain.** If we want to restrict the domain in some way, we must say so. The domain of $y = x^2$ is the entire set of real numbers. To restrict the function to, say, positive values of x, we would write "$y = x^2, x > 0$."

The domains and ranges of many real-valued functions of a real variable are intervals or combinations of intervals. The intervals may be open, closed, or half open (Figures 10 and 11) and finite or infinite (Figure 12).

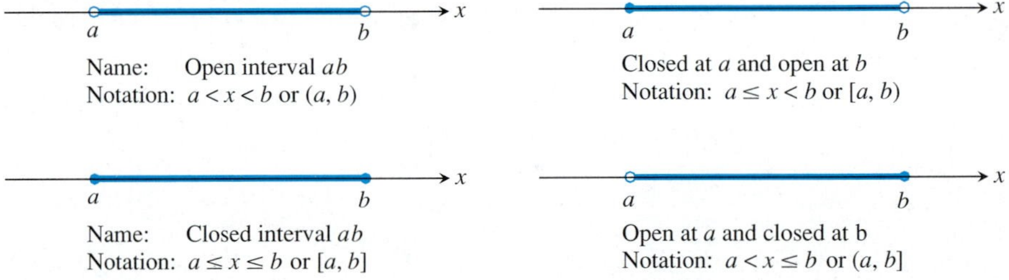

FIGURE 10 Open and closed finite intervals. FIGURE 11 Half-open finite intervals.

The endpoints of an interval are called **boundary points.** They make up the interval's **boundary.** The remaining points are **interior points,** and they make up the interval's **interior.** Intervals containing all their boundary points are **closed.** Intervals containing no boundary points are **open.** Every point of an open interval is an interior point of the interval.

0

Name: The set of all real numbers
Notation: $-\infty < x < \infty$ or $(-\infty, \infty)$

a

Name: The set of numbers greater than a
Notation: $a < x$ or (a, ∞)

a

Name: The set of numbers greater than or equal to a
Notation: $a \leq x$ or $[a, \infty)$

b

Name: The set of numbers less than b
Notation: $x < b$ or $(-\infty, b)$

b

Name: The set of numbers less than or equal to b
Notation: $x \leq b$ or $(-\infty, b]$

FIGURE 12 Infinite intervals: rays on the number line and the number line itself. The symbol ∞ (infinity) is used merely for convenience; it does not mean that there is a number ∞.

Example 2 Identifying Domain and Range

Verify the domains of these functions.

Function	Domain (x)	Range (y)
$y = x^2$	$(-\infty, \infty)$	$[0, \infty)$
$y = 1/x$	$(-\infty, 0) \cup (0, \infty)$	$(-\infty, 0) \cup (0, \infty)$
$y = \sqrt{x}$	$[0, \infty)$	$[0, \infty)$
$y = \sqrt{4 - x}$	$(-\infty, 4]$	$[0, \infty)$
$y = \sqrt{1 - x^2}$	$[-1, 1]$	$[0, 1]$

Solution The formula $y = x^2$ gives a real y-value for any real number x, so the domain is $(-\infty, \infty)$.

The formula $y = 1/x$ gives a real y-value for every real x-value except $x = 0$. *We cannot divide any number by* 0.

The formula $y = \sqrt{x}$ gives a real y-value only when x is positive or zero.

The formula $y = \sqrt{4 - x}$ gives a real y-value only when $4 - x$ is greater than or equal to zero. So, $0 \leq 4 - x$, or $x \leq 4$.

The formula $y = \sqrt{1 - x^2}$ gives a real y-value for every value of x in the closed interval from -1 to 1. Outside this interval, $1 - x^2$ is negative and its square root is not a real number. The domain is $[-1, 1]$.

Viewing and Interpreting Graphs

The points (x, y) in the plane whose coordinates are the input-output pairs of a function $y = f(x)$ make up the function's **graph.** The graph of the function $y = x + 2$, for example, is the set of points with coordinates (x, y) for which $y = x + 2$.

Graphing with pencil and paper requires that you develop graph *drawing* skills. Graphing with a grapher requires that you develop graph *viewing* skills.

Graph Viewing Skills

Step 1. Recognize that the graph is reasonable.

Step 2. See all the important characteristics of the graph.

Step 3. Interpret those characteristics.

Step 4. Recognize grapher failure.

Being able to recognize that a graph is reasonable comes with experience. You need to know the basic functions, their graphs, and how changes in their equations affect the graphs.

Grapher failure occurs when the graph produced by a grapher is less than precise—or even incorrect—which is usually due to the limitations of the screen resolution of the grapher.

$y = \frac{1}{\sqrt{4-x^2}}$

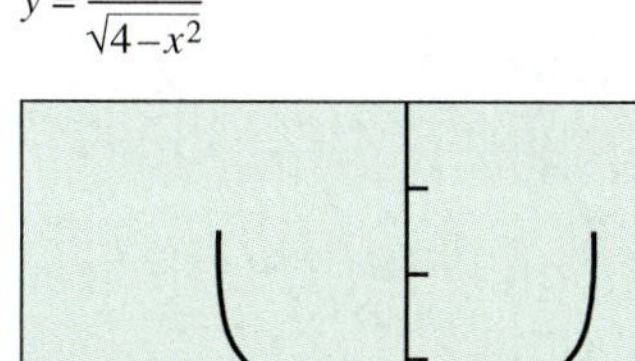

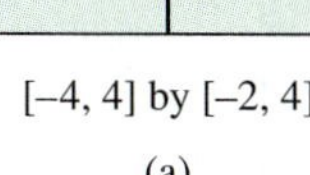

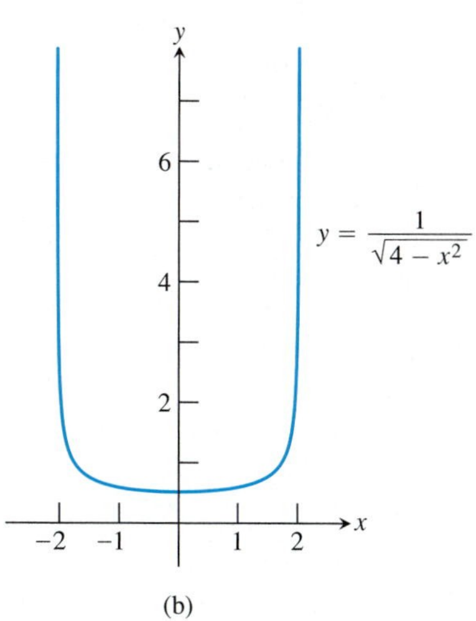

FIGURE 13 (a) Grapher failure. (b) A more accurate graph of $y = 1/\sqrt{4 - x^2}$. (Example 3)

Example 3 Recognizing Grapher Failure

Find the domain and range of $y = f(x) = 1/\sqrt{4 - x^2}$.

Solution The graph of f in Figure 13a seems to suggest that the domain of f is an interval between -2 and 2 and that the range is also a finite interval. The latter observation is the result of grapher failure; in this case, we can recognize the failure using algebra.

Solve Algebraically

The expression $4 - x^2$ must be greater than zero.

$$4 - x^2 > 0$$
$$x^2 < 4$$

Thus, $-2 < x < 2$, and the domain is $(-2, 2)$.

The smallest value of f is $1/2$ and occurs when $x = 0$. The values of f get very large as x approaches 2 from the left or -2 from the right, as suggested by the following table. (The values of f are rounded to three decimal places.)

x	± 1.99	± 1.999	± 1.9999	± 1.99999
$f(x)$	5.006	15.813	50.001	158.114

The range of f is $[0.5, \infty)$.

Figure 14 shows graphs of *power functions* that arise frequently in calculus. Knowing the general shapes of these graphs will help you recognize grapher failure. We will review other functions as the chapter continues.

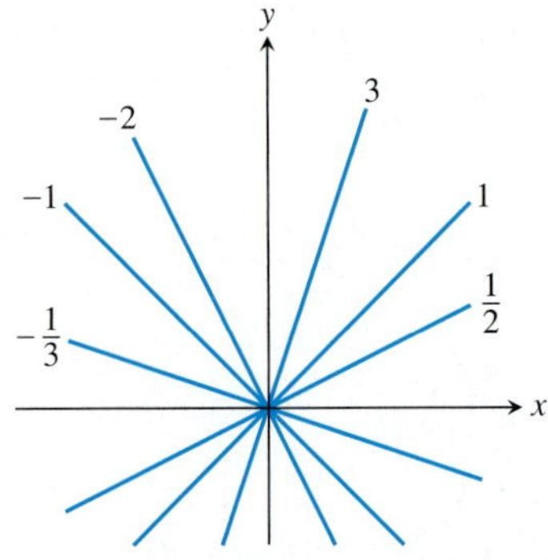

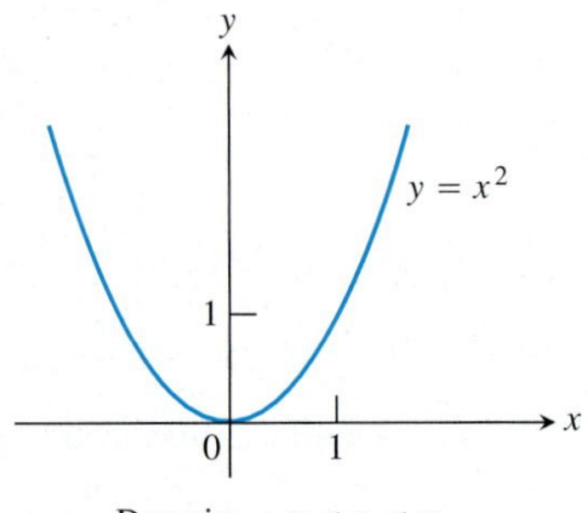

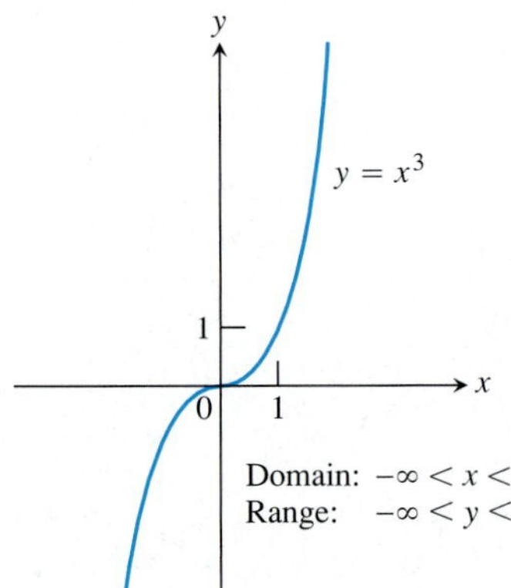

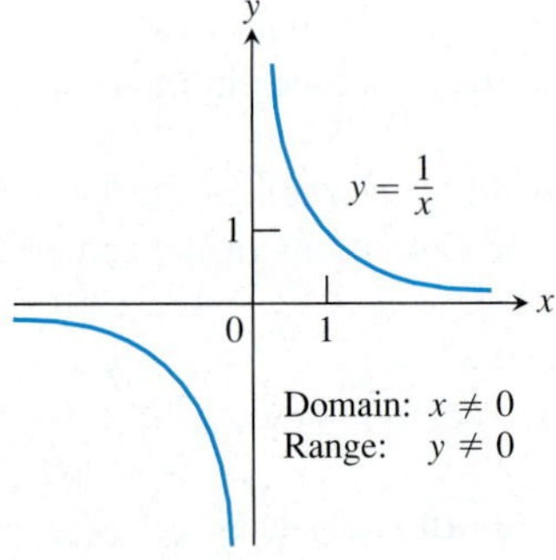

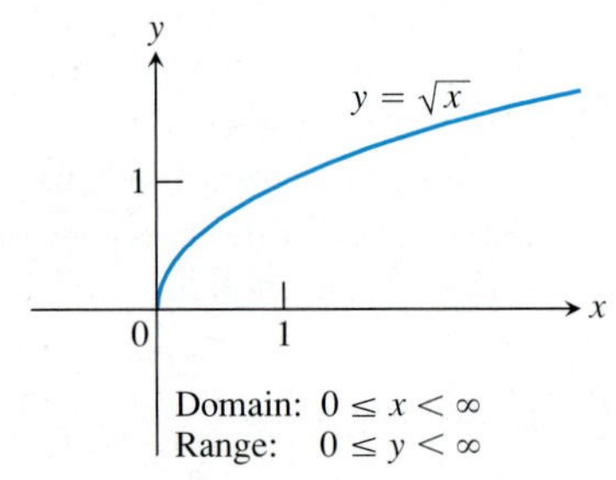

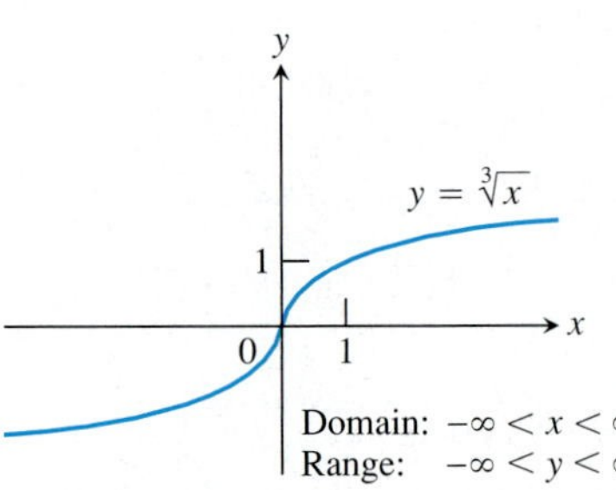

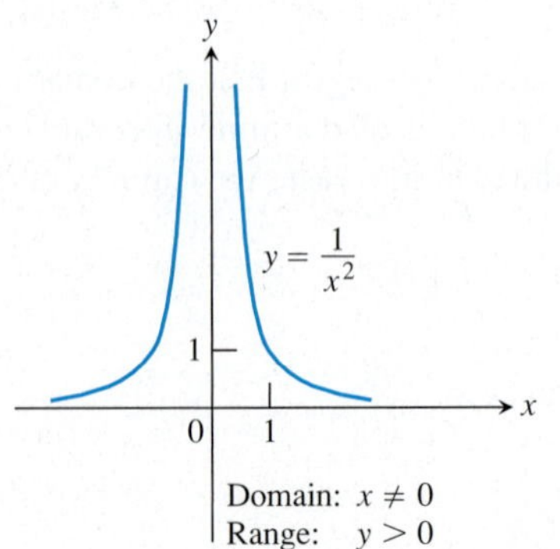

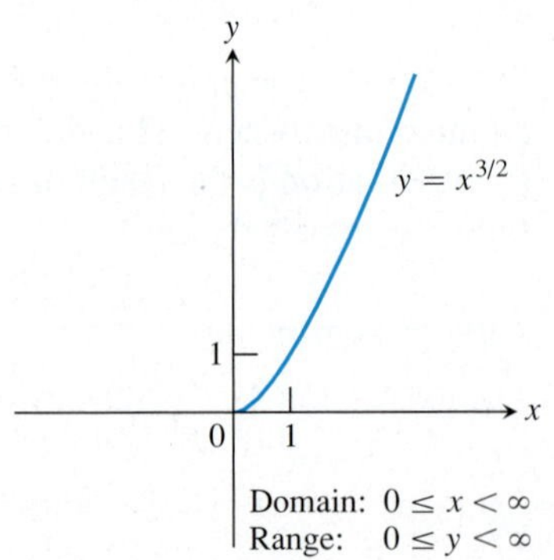

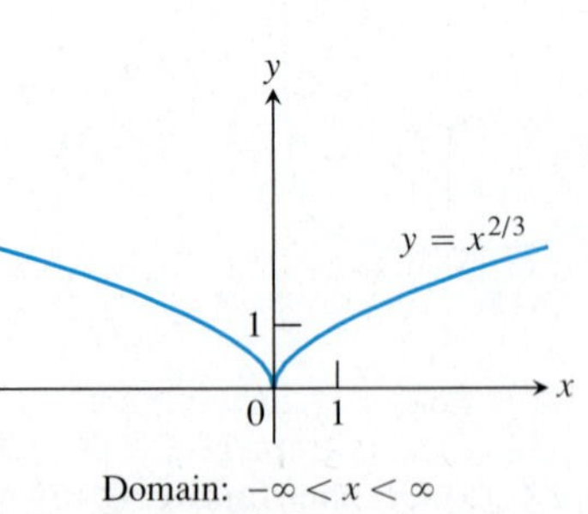

FIGURE 14 Useful power functions.

Increasing versus Decreasing Functions

If the graph of a function *climbs* or *rises* as you move from left to right, we say that the function is *increasing*. If the graph *descends* or *falls* as you move from left to right, the function is *decreasing*. We give formal definitions of increasing functions and decreasing functions in Section 3.3. In that section, you will learn how to find the intervals over which a function is increasing and the intervals where it is decreasing. Here are examples from Figure 14.

Function	Where Increasing	Where Decreasing
$y = x^2$	$0 \le x < \infty$	$-\infty < x \le 0$
$y = x^3$	$-\infty < x < \infty$	Nowhere
$y = 1/x$	Nowhere	$-\infty < x < 0, 0 < x < \infty$
$y = 1/x^2$	$-\infty < x < 0$	$0 < x < \infty$
$y = \sqrt{x}$	$0 \le x < \infty$	Nowhere
$y = x^{2/3}$	$0 \le x < \infty$	$-\infty < x \le 0$

Even Functions and Odd Functions: Symmetry

The graphs of *even* and *odd* functions have characteristic symmetry properties.

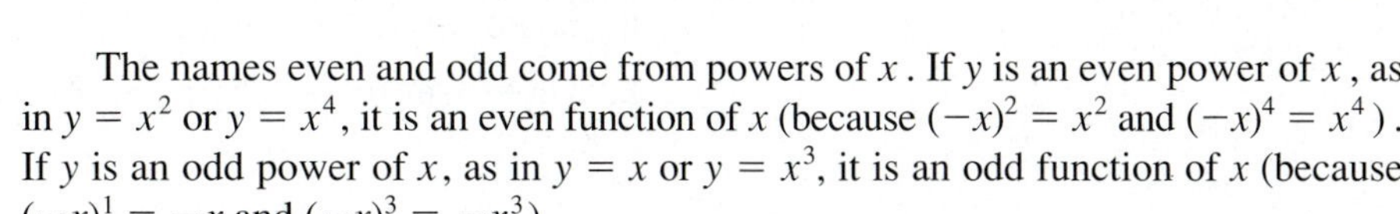

Definitions Even Function, Odd Function

A function $y = f(x)$ is an

even function of x if $f(-x) = f(x)$,
odd function of x if $f(-x) = -f(x)$,

for every x in the function's domain.

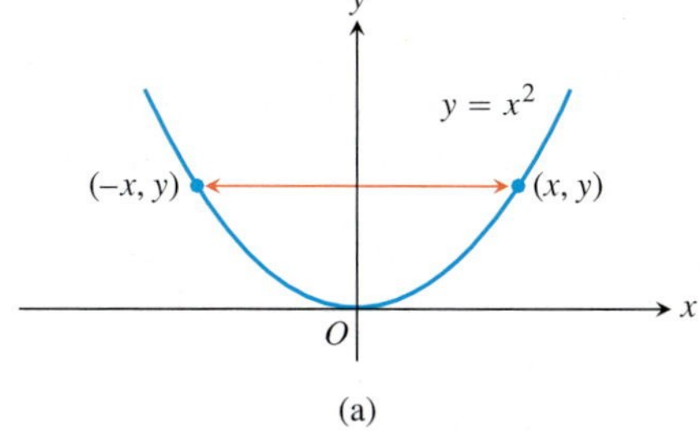

(a)

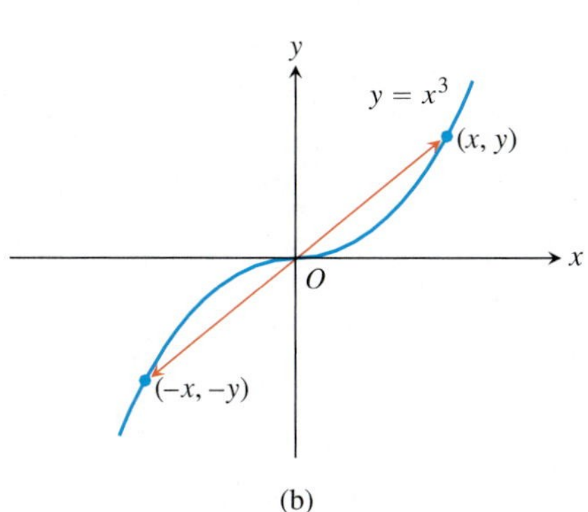

(b)

FIGURE 15 (a) The graph of $y = x^2$ (an even function) is symmetric about the y-axis. (b) The graph of $y = x^3$ (an odd function) is symmetric about the origin.

The names even and odd come from powers of x. If y is an even power of x, as in $y = x^2$ or $y = x^4$, it is an even function of x (because $(-x)^2 = x^2$ and $(-x)^4 = x^4$). If y is an odd power of x, as in $y = x$ or $y = x^3$, it is an odd function of x (because $(-x)^1 = -x$ and $(-x)^3 = -x^3$).

The graph of an even function is **symmetric about the y-axis.** Since $f(-x) = f(x)$, a point (x, y) lies on the graph if and only if the point $(-x, y)$ lies on the graph (Figure 15a).

The graph of an odd function is **symmetric about the origin.** Since $f(-x) = -f(x)$, a point (x, y) lies on the graph if and only if the point $(-x, -y)$ lies on the graph (Figure 15b). Equivalently, a graph is symmetric about the origin if a rotation of 180° about the origin leaves the graph unchanged.

Example 4 Recognizing Even and Odd Functions

$f(x) = x^2$	Even function: $(-x)^2 = x^2$ for all x; symmetry about y-axis.
$f(x) = x^2 + 1$	Even function: $(-x)^2 + 1 = x^2 + 1$ for all x; symmetry about y-axis (Figure 16a).
$f(x) = x$	Odd function: $(-x) = -x$ for all x; symmetry about the origin.

$f(x) = x + 1$ Not odd: $f(-x) = -x + 1$, but $-f(x) = -x - 1$. The two are not equal.

Not even: $(-x) + 1 \neq x + 1$ for all $x \neq 0$ (Figure 16b).

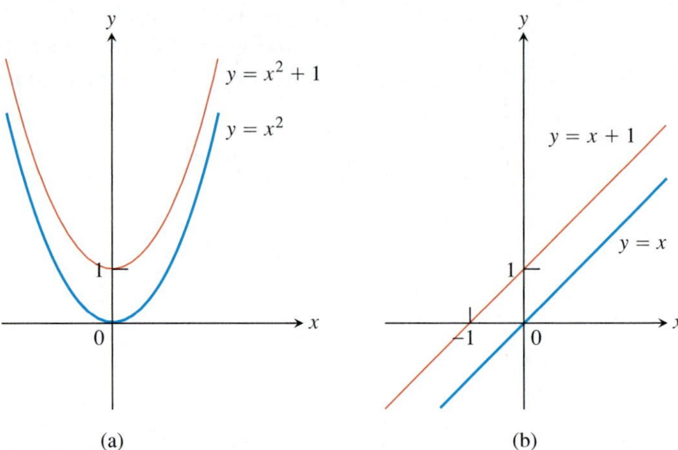

FIGURE 16 (a) When we add the constant term 1 to the function $y = x^2$, the resulting function $y = x^2 + 1$ is still even and its graph is still symmetric about the y-axis. (b) When we add the constant term 1 to the function $y = x$, the resulting function $y = x + 1$ is no longer odd. The symmetry about the origin is lost. (Example 4)

It is useful in graphing to recognize even and odd functions. Once we know the graph of either type of function on one side of the y-axis, we automatically know its graph on the other side.

Functions Defined in Pieces

Functions can be defined by applying different formulas to different parts of the domain.

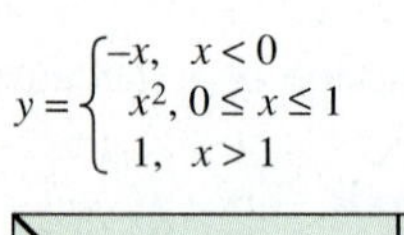

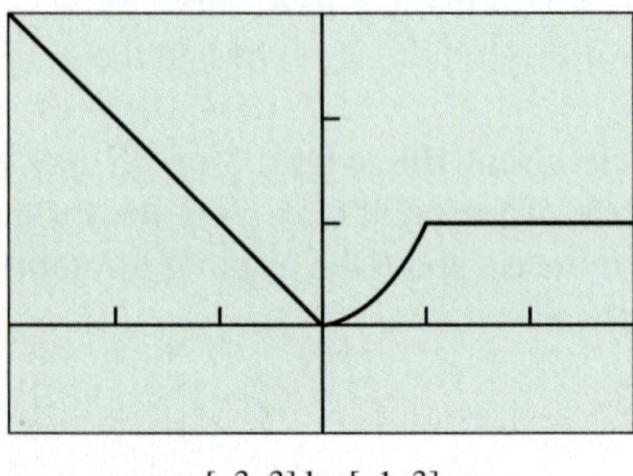

[–3, 3] by [–1, 3]

FIGURE 17 The graph of a piecewise defined function. (Example 5)

Example 5 Graphing Piecewise-Defined Functions

Graph

$$y = f(x) = \begin{cases} -x, & x < 0 \\ x^2, & 0 \leq x \leq 1 \\ 1, & x > 1. \end{cases}$$

Solution The values of f are given by three separate formulas: $y = -x$ when $x < 0$, $y = x^2$ when $0 \leq x \leq 1$, and $y = 1$ when $x > 1$. The function, however, is *just one function* whose domain is the entire set of real numbers (Figure 17).

Example 6 Writing Formulas for Piecewise-Defined Functions

Write a formula for the function $y = f(x)$ whose graph consists of the two line segments in Figure 18.

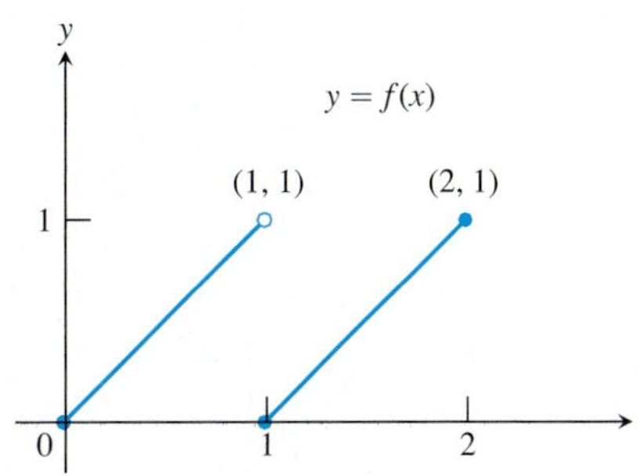

FIGURE 18 The segment on the left contains (0, 0) but not (1, 1). The segment on the right contains both of its endpoints. (Example 6)

Solution We find formulas for the segments from (0, 0) to (1, 1) and from (1, 0) to (2, 1) and piece them together in the manner of Example 5.

Segment from (0, 0) to (1, 1) The line through (0, 0) and (1, 1) has slope $m = (1 - 0)/(1 - 0) = 1$ and y-intercept $b = 0$. Its slope-intercept equation is $y = x$. The segment from (0, 0) to (1, 1) that includes the point (0, 0) but not the point (1, 1) is the graph of the function $y = x$ restricted to the half-open interval $0 \le x < 1$, namely,

$$y = x, \qquad 0 \le x < 1.$$

Segment from (1, 0) to (2, 1) The line through (1, 0) and (2, 1) has slope $m = (1 - 0)/(2 - 1) = 1$ and passes through the point (1, 0). The corresponding point-slope equation for the line is

$$y = 1(x - 1) + 0, \qquad \text{or} \qquad y = x - 1.$$

The segment from (1, 0) to (2, 1) that includes both endpoints is the graph of $y = x - 1$ restricted to the closed interval $1 \le x \le 2$, namely,

$$y = x - 1, \qquad 1 \le x \le 2.$$

Piecewise formula Combining the formulas for the two pieces of the graph, we obtain

$$f(x) = \begin{cases} x, & 0 \le x < 1 \\ x - 1, & 1 \le x \le 2. \end{cases}$$

Remember that $\sqrt{a^2} = |a|$. Do not write $\sqrt{a^2} = a$ unless you already know that $a \ge 0$.

The Absolute Value Function

The absolute value function $y = |x|$ is defined piecewise by the formula

$$|x| = \begin{cases} -x, & x < 0 \\ x, & x \ge 0. \end{cases}$$

Absolute Value Properties

1. $|-a| = |a|$
2. $|ab| = |a||b|$
3. $\left|\dfrac{a}{b}\right| = \dfrac{|a|}{|b|}$
4. $|a+b| \le |a| + |b|$

The function is even, and its graph (Figure 19) is symmetric about the y-axis. Since the symbol $\sqrt{a}$ denotes the *nonnegative* square root of a, an alternate definition of $|x|$ is

$$|x| = \sqrt{x^2}.$$

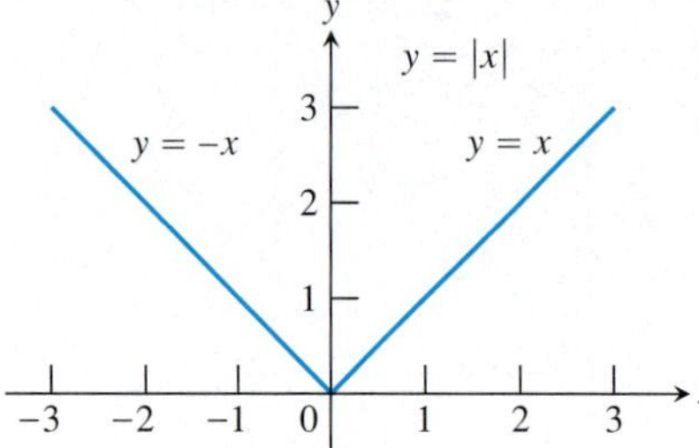

FIGURE 19 The absolute value function has domain $(-\infty, \infty)$ and range $[0, \infty)$.

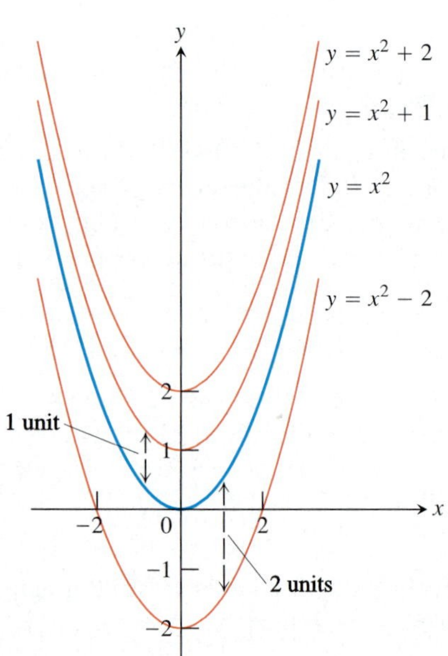

FIGURE 20 To shift the graph of $f(x) = x^2$ up (or down), we add positive (or negative) constants to the formula for f.

How to Shift a Graph

To shift the graph of a function $y = f(x)$ straight up, add a positive constant to the right-hand side of the formula $y = f(x)$.

To shift the graph of a function $y = f(x)$ straight down, add a negative constant to the right-hand side of the formula $y = f(x)$.

Example 7 Shifting a Graph Vertically

Adding 1 to the right-hand side of the formula $y = x^2$ to get $y = x^2 + 1$ shifts the graph up 1 unit (Figure 20). Adding -2 to the right-hand side of the formula $y = x^2$ to get $y = x^2 - 2$ shifts the graph down 2 units (Figure 20).

To shift the graph of $y = f(x)$ to the left, add a positive constant to x. To shift the graph of $y = f(x)$ to the right, add a negative constant to x.

Example 8 Shifting a Graph Horizontally

Adding 3 to x in $y = x^2$ to get $y = (x+3)^2$ shifts the graph 3 units to the left (Figure 21). Adding -2 to x in $y = x^2$ to get $y = (x-2)^2$ shifts the graph 2 units to the right (Figure 21).

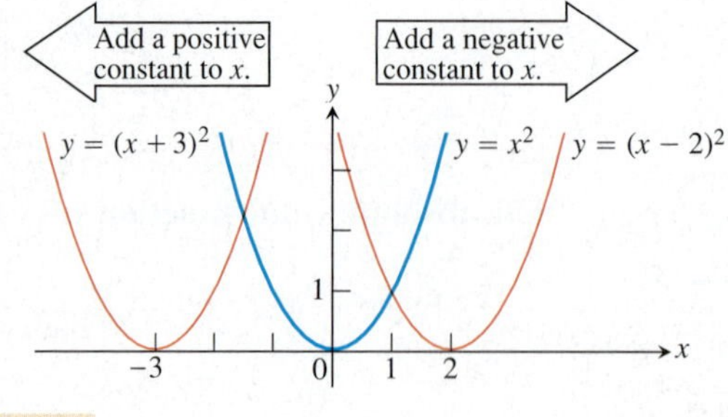

FIGURE 21 To shift the graph of $y = x^2$ to the left, we add a positive constant to x. To shift the graph to the right, we add a negative constant to x.

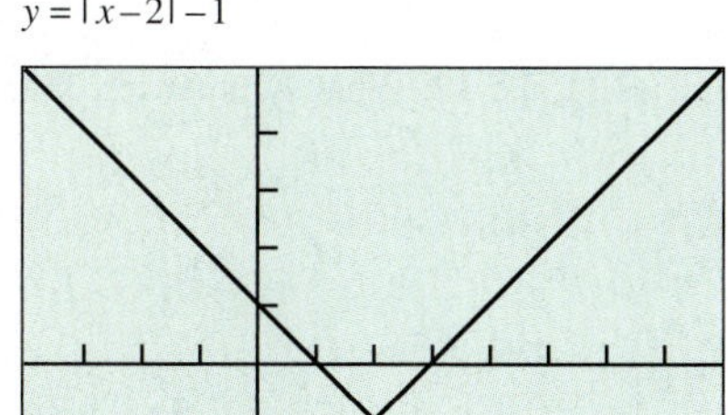

FIGURE 22 The lowest point of the graph of $f(x) = |x - 2| - 1$ is $(2, -1)$. (Example 9)

Shift Formulas

VERTICAL SHIFTS

$y = f(x) + k$	Shifts the graph *up* k units if $k > 0$
	Shifts it *down* $\vert k \vert$ units if $k < 0$

HORIZONTAL SHIFTS

$y = f(x + h)$	Shifts the graph *left* h units if $h > 0$
	Shifts it *right* $\vert h \vert$ units if $h < 0$

Example 9 Combining Shifts

Find the domain and range, and draw the graph of $f(x) = |x - 2| - 1$.

Solution The graph of f is the graph of the absolute value function shifted 2 units horizontally to the right and 1 unit vertically downward (Figure 22). The domain of f is $(-\infty, \infty)$, and the range is $[-1, \infty)$.

Composite Functions

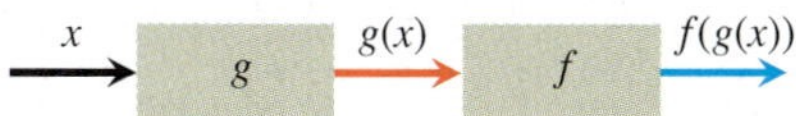

FIGURE 23 Two functions can be composed at x whenever the value of one at x lies in the domain of the other. The composite is denoted by $f \circ g$.

Suppose that some of the outputs of a function g can be used as inputs of a function f. We can then link g and f to form a new function whose inputs x are inputs of g and whose outputs are the numbers $f(g(x))$, as in Figure 23. We say the function $f(g(x))$ (read "f of g of x") is the **composite of g and f.** It is made by *composing* g and f in the order of first g, then f. The usual "stand-alone" notation for this composite is $f \circ g$, which is read as "f of g." The value of $f \circ g$ at x is $(f \circ g)(x) = f(g(x))$. Notice that in the notation $f \circ g$, we first apply g, and then the function f, to the input variable x.

Example 10 Viewing a Function as a Composite

The function $y = \sqrt{1 - x^2}$ in Example 2 can be thought of as first calculating $1 - x^2$ followed by taking the square root of the result. The function y is the composite of the function $g(x) = 1 - x^2$ and the function $f(x) = \sqrt{x}$. Notice that $1 - x^2$ cannot be negative. The domain of the composite is $[-1, 1]$.

Example 11 Finding a Formula for a Composite and Evaluating It

Find a formula for $f(g(x))$ if $g(x) = x^2$ and $f(x) = x - 7$. Then find $f(g(2))$.

Solution To find $f(g(x))$, we replace x in the formula $f(x) = x - 7$ by the expression given for $g(x)$.

$$f(x) = x - 7$$
$$f(g(x)) = g(x) - 7 = x^2 - 7$$

We then find the value of $f(g(2))$ by substituting 2 for x.

$$f(g(2)) = (2)^2 - 7 = -3$$

EXERCISES 2

Finding Formulas for Functions

1. Express the area and perimeter of an equilateral triangle as a function of the triangle's side length x.

2. Express the side length of a square as a function of the length d of the square's diagonal. Then express the area as a function of the diagonal length.

3. Express the edge length of a cube as a function of the cube's diagonal length d. Then express the surface area and volume of the cube as a function of the diagonal length.

4. A point P in the first quadrant lies on the graph of the function $f(x) = \sqrt{x}$. Express the coordinates of P as functions of the slope of the line joining P to the origin.

In Exercises 5 and 6, which of the graphs are graphs of functions of x, and which are not? Give reasons for your answers.

5. (a)

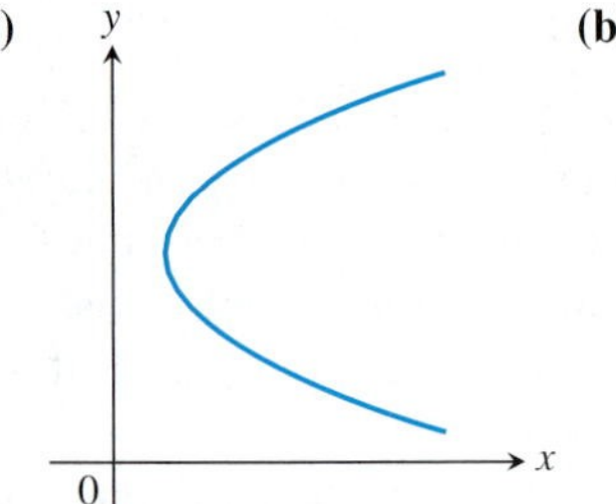

(b)

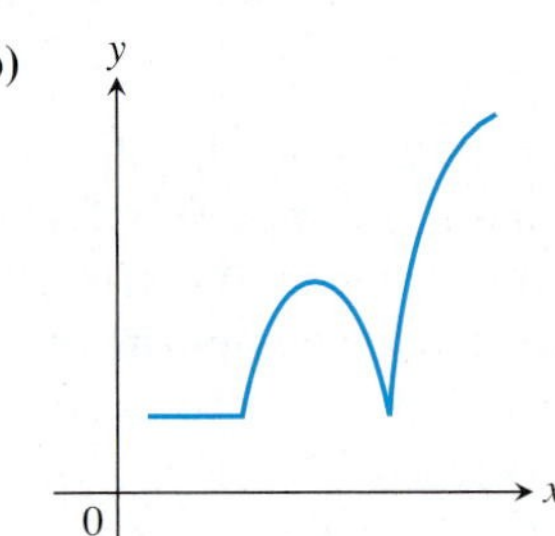

6. (a)

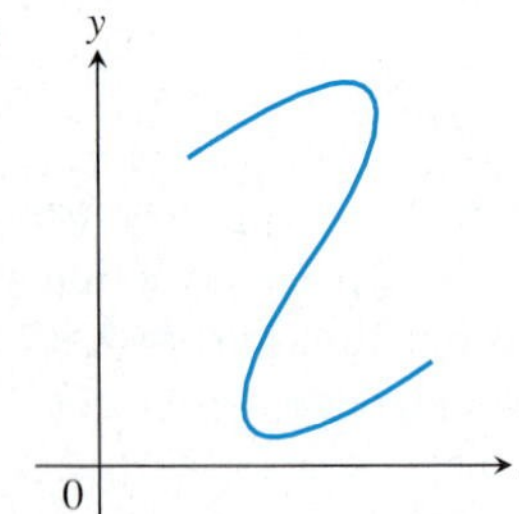

(b)

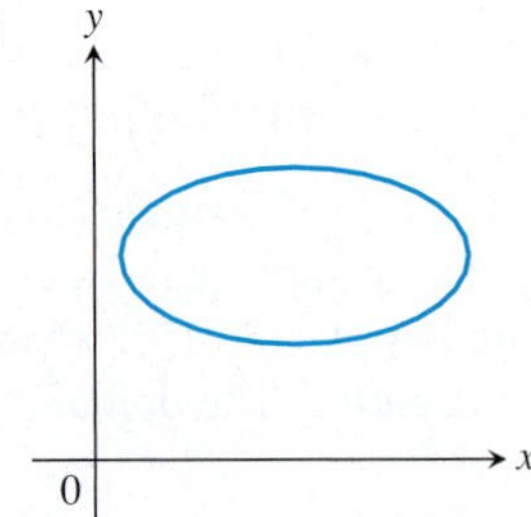

Domain and Range

In Exercises 7–10, find the domain and range of each function.

7. (a) $f(x) = 1 + x^2$ (b) $f(x) = 1 - \sqrt{x}$

8. (a) $F(t) = \dfrac{1}{\sqrt{t}}$ (b) $F(t) = \dfrac{1}{1 + \sqrt{t}}$

9. $g(z) = \sqrt{4 - z^2}$

10. $g(z) = \sqrt[3]{z - 3}$

Functions and Graphs

Graph the functions in Exercises 11 and 12. What symmetries, if any, do the graphs have?

11. (a) $y = -x^3$ (b) $y = -\dfrac{1}{x^2}$

12. (a) $y = \sqrt{|x|}$ (b) $y = -\dfrac{1}{x}$

13. Graph the following equations and explain why they are not graphs of functions of x.

(a) $|y| = x$ (b) $y^2 = x^2$

14. Graph the following equations and explain why they are not graphs of functions of x.

(a) $|x| + |y| = 1$ (b) $|x + y| = 1$

Even and Odd Functions

In Exercises 15–20, say whether the function is even, odd, or neither.

15. (a) $f(x) = 3$ (b) $f(x) = x^{-5}$

16. (a) $f(x) = x^2 + 1$ (b) $f(x) = x^2 + x$

17. (a) $g(x) = x^3 + x$ (b) $g(x) = x^4 + 3x^2 - 1$

18. (a) $g(x) = \dfrac{1}{x^2 - 1}$ (b) $g(x) = \dfrac{x}{x^2 - 1}$

19. (a) $h(t) = \dfrac{1}{t - 1}$ (b) $h(t) = |t^3|$

20. (a) $h(t) = \sqrt{t^2 + 3}$ (b) $h(t) = 2|t| + 1$

Piecewise-Defined Functions

In Exercises 21–24, **(a)** draw the graph of the function. Then find its **(b)** domain and **(c)** range.

21. (a) $f(x) = -|3 - x| + 2$ (b) $f(x) = 2|x + 4| - 3$

22. (a) $f(x) = \begin{cases} 3 - x, & x \le 1 \\ 2x, & 1 < x \end{cases}$ (b) $f(x) = \begin{cases} 1, & x < 0 \\ \sqrt{x}, & x \ge 0 \end{cases}$

23. $f(x) = \begin{cases} 4 - x^2, & x < 1 \\ (3/2)x + 3/2, & 1 \le x \le 3 \\ x + 3, & x > 3 \end{cases}$

24. $f(x) = \begin{cases} x^2, & x < 0 \\ x^3, & 0 \le x \le 1 \\ 2x - 1, & x > 1 \end{cases}$

25. *Writing to Learn* The *vertical line test* to determine whether a curve is the graph of a function states: If every vertical line in the

xy-plane intersects a given curve in at most one point, then the curve is the graph of a function. Explain why this statement is true.

26. *Writing to Learn* For a curve to be *symmetric about the x-axis,* the point (x, y) must lie on the curve if and only if the point $(x, -y)$ lies on the curve. Explain why a curve that is symmetric about the x-axis is not the graph of a function, unless the function is $y = 0$.

In Exercises 27 and 28, write a piecewise formula for the function.

27. **(a)**

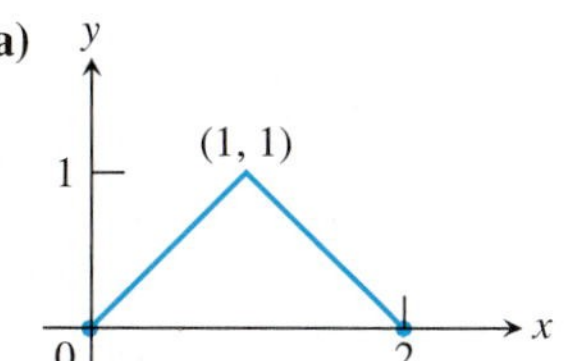

(b)

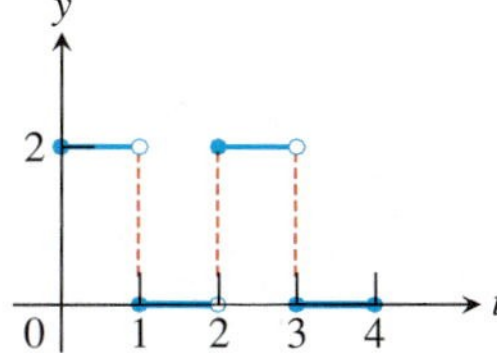

(c)

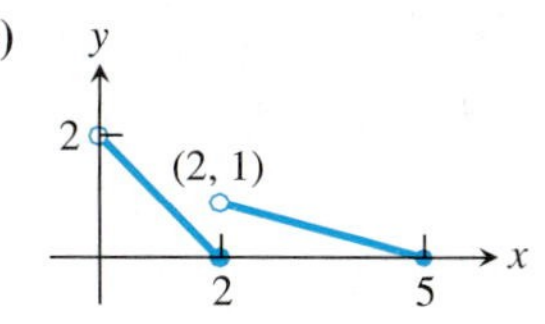

(d)

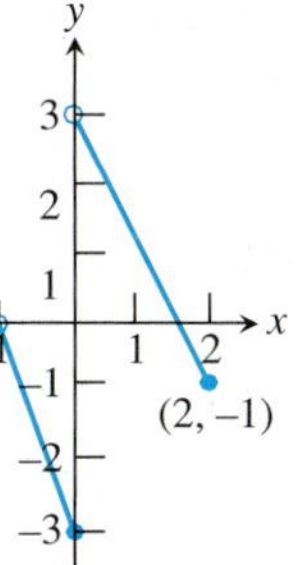

28. **(a)**

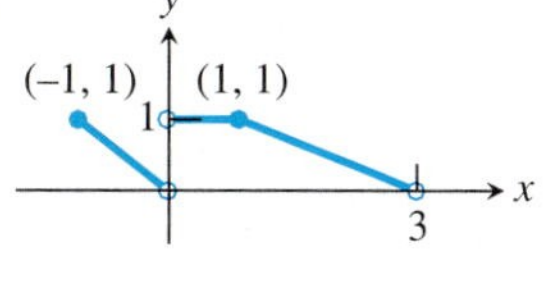

(b)

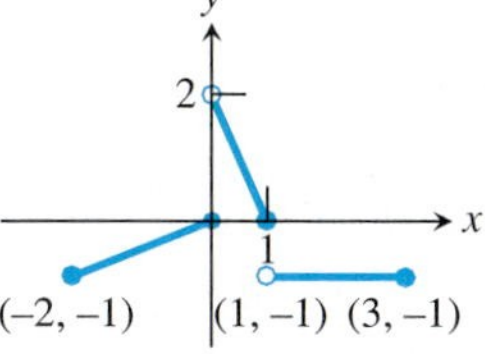

(c)

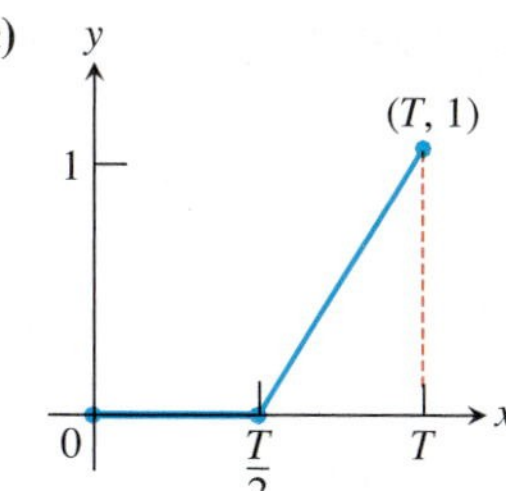

(d)

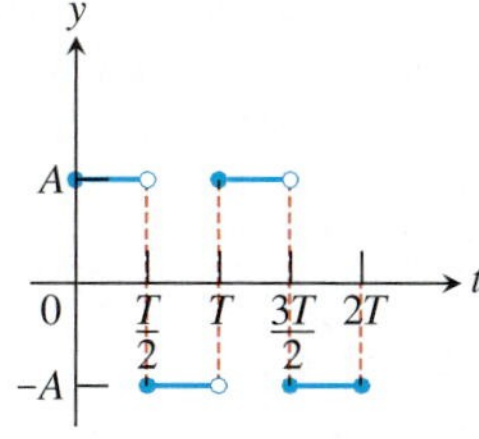

Shifting Graphs

29. Match the equations listed in **(a)** through **(d)** to the positions on the graphs in the accompanying figure.

(a) $y = (x - 1)^2 - 4$ **(b)** $y = (x - 2)^2 + 2$

(c) $y = (x + 2)^2 + 2$ **(d)** $y = (x + 3)^2 - 2$

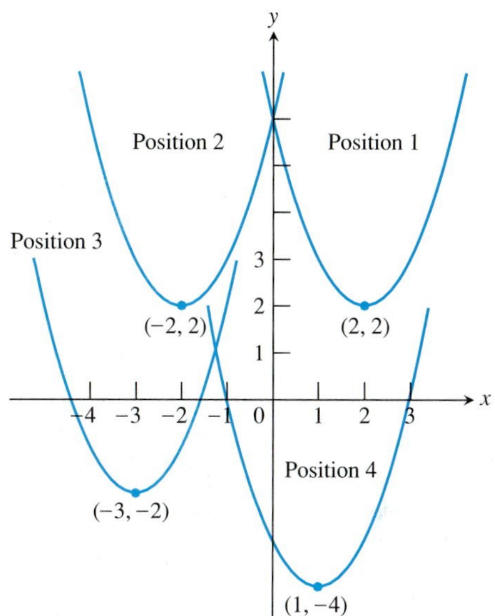

30. The figure shows the graph of $y = -x^2$ shifted to four new positions. Write an equation for each new graph.

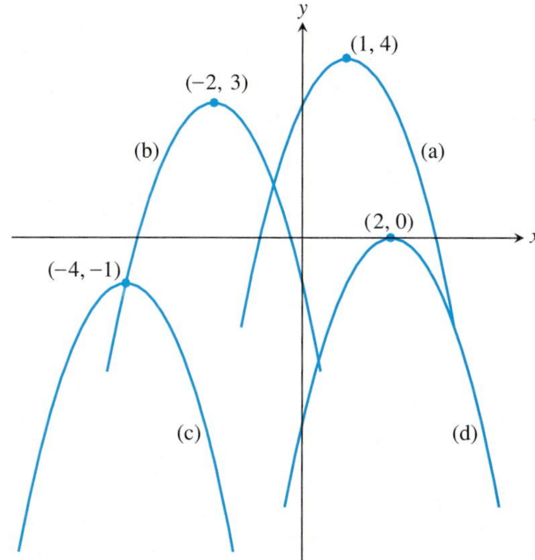

Exercises 31–36 tell how many units and in what directions the graphs of the given equations are to be shifted. Give an equation for the shifted graph. Then sketch the original and shifted graphs together, labeling each graph with its equation.

31. $x^2 + y^2 = 49$ Down 3, left 2

32. $y = x^3$ Left 1, down 1

33. $y = x^{2/3}$ Right 1, down 1

34. $y = -\sqrt{x}$ Right 3

35. $y = (1/2)(x + 1) + 5$ Down 5, right 1

36. $x = y^2$ Left 1

Composites of Functions

37. If $f(x) = x + 5$ and $g(x) = x^2 - 3$, find the following.

(a) $f(g(0))$ **(b)** $g(f(0))$

(c) $f(g(x))$ **(d)** $g(f(x))$

(e) $f(f(-5))$ **(f)** $g(g(2))$

(g) $f(f(x))$ **(h)** $g(g(x))$

38. If $f(x) = x - 1$ and $g(x) = 1/(x + 1)$, find the following.

(a) $f(g(1/2))$ **(b)** $g(f(1/2))$

(c) $f(g(x))$ **(d)** $g(f(x))$

(e) $f(f(2))$ **(f)** $g(g(2))$

(g) $f(f(x))$ **(h)** $g(g(x))$

39. If $u(x) = 4x - 5$, $v(x) = x^2$, and $f(x) = 1/x$, find formulas for the following.

(a) $u(v(f(x)))$ **(b)** $u(f(v(x)))$

(c) $v(u(f(x)))$ **(d)** $v(f(u(x)))$

(e) $f(u(v(x)))$ **(f)** $f(v(u(x)))$

40. If $f(x) = \sqrt{x}$, $g(x) = x/4$, and $h(x) = 4x - 8$, find formulas for the following.

(a) $h(g(f(x)))$ **(b)** $h(f(g(x)))$

(c) $g(h(f(x)))$ **(d)** $g(f(h(x)))$

(e) $f(g(h(x)))$ **(f)** $f(h(g(x)))$

Let $f(x) = x - 3$, $g(x) = \sqrt{x}$, $h(x) = x^3$, and $j(x) = 2x$. Express each of the functions in Exercises 41 and 42 as a composite involving one or more of f, g, h, and j.

41. **(a)** $y = \sqrt{x - 3}$ **(b)** $y = 2\sqrt{x}$

(c) $y = x^{1/4}$ **(d)** $y = 4x$

(e) $y = \sqrt{(x - 3)^3}$ **(f)** $y = (2x - 6)^3$

42. **(a)** $y = 2x - 3$ **(b)** $y = x^{3/2}$

(c) $y = x^9$ **(d)** $y = x - 6$

(e) $y = 2\sqrt{x - 3}$ **(f)** $y = \sqrt{x^3 - 3}$

43. Copy and complete the following table.

	$g(x)$	$f(x)$	$(f \circ g)(x)$
(a)	?	$\sqrt{x - 5}$	$\sqrt{x^2 - 5}$
(b)	?	$1 + 1/x$	x
(c)	$1/x$	?	x
(d)	$\sqrt{x}$	?	$\|x\|$

44. Copy and complete the following table.

	$g(x)$	$f(x)$	$(f \circ g)(x)$
(a)	$x - 7$	$\sqrt{x}$	?
(b)	$x + 2$	$3x$	?
(c)	?	$\sqrt{x - 5}$	$\sqrt{x^2 - 5}$
(d)	$\dfrac{x}{x - 1}$	$\dfrac{x}{x - 1}$	?
(e)	?	$1 + \dfrac{1}{x}$	x
(f)	$\dfrac{1}{x}$	?	x

45. The accompanying figure shows the graph of a function $f(x)$ with domain $[0, 2]$ and range $[0, 1]$. Find the domains and ranges of the following functions and sketch their graphs.

(a) $f(x) + 2$ **(b)** $f(x) - 1$

(c) $2f(x)$ **(d)** $-f(x)$

(e) $f(x + 2)$ **(f)** $f(x - 1)$

(g) $f(-x)$ **(h)** $-f(x + 1) + 1$

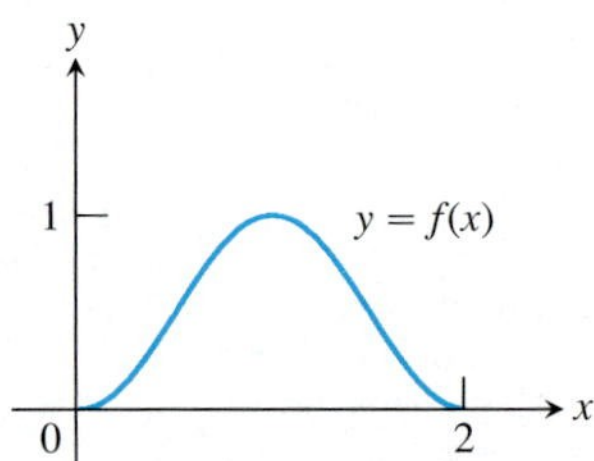

46. The accompanying figure shows the graph of a function $g(t)$ with domain $[-4, 0]$ and range $[-3, 0]$. Find the domains and ranges of the following functions and sketch their graphs.

(a) $g(-t)$ **(b)** $-g(t)$

(c) $g(t) + 3$ **(d)** $1 - g(t)$

(e) $g(-t + 2)$ **(f)** $g(t - 2)$

(g) $g(1 - t)$ **(h)** $-g(t - 4)$

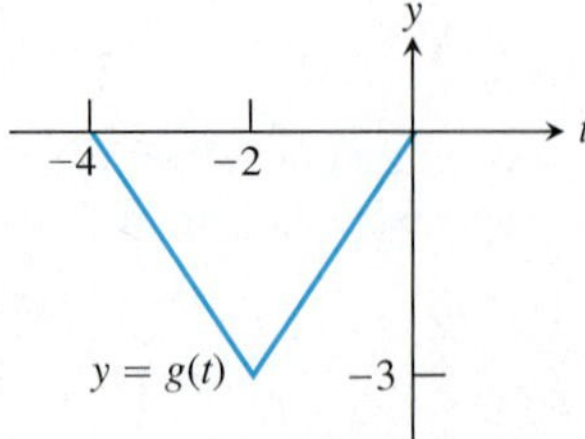

Theory and Examples

47. *The cone problem* Begin with a circular piece of paper with a 4 in. radius as shown in part (a). Cut out a sector with an arc length of x. Join the two edges of the remaining portion to form a cone with radius r and height h, as shown in part (b).

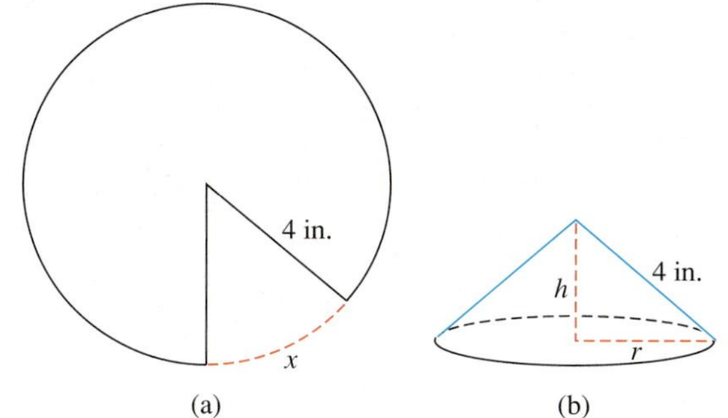

(a) Explain why the circumference of the base of the cone is $8\pi - x$.

(b) Express the radius r as a function of x.

(c) Express the height h as a function of x.

(d) Express the volume V of the cone as a function of x.

48. *Industrial costs* Dayton Power and Light, Inc., has a power plant on the Miami River where the river is 800 ft wide. To lay a new cable from the plant to a location in the city 2 mi downstream on the opposite side costs \$180 per foot across the river and \$100 per foot along the land.

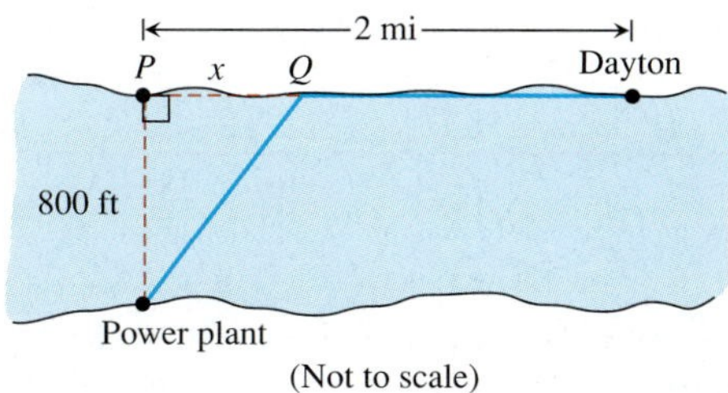

(a) Suppose that the cable goes from the plant to a point Q on the opposite side that is x ft from the point P directly opposite the plant. Write a function $C(x)$ that gives the cost of laying the cable in terms of the distance x.

(b) Generate a table of values to determine if the least expensive location for point Q is less than 2000 ft or greater than 2000 ft from point P.

49. *Even and odd functions*

(a) Must the product of two even functions always be even? Give reasons for your answer.

(b) Can anything be said about the product of two odd functions? Give reasons for your answer.

(c) Can a function be both even and odd? Give reasons for your answer.

50. *A magic trick* You may have heard of a magic trick that goes like this: Take any number. Add 5. Double the result. Subtract 6. Divide by 2. Subtract 2. Now tell me your answer, and I'll tell you what you started with.

(a) Pick a number and try it.

(b) Why does it work with any number?

T **51.** Graph the functions $f(x) = \sqrt{x}$ and $g(x) = \sqrt{1 - x}$ together with their **(a)** sum, **(b)** product, **(c)** two differences, and **(d)** two quotients.

T **52.** Let $f(x) = x - 7$ and $g(x) = x^2$. Graph f and g together with $f \circ g$ and $g \circ f$.

Some graphers allow a function such as y_1 to be used as the independent variable of another function. With such a grapher, we can compose functions.

T **53.** **(a)** Enter the functions $y_1 = f(x) = 4 - x^2$, $y_2 = g(x) = \sqrt{x}$, $y_3 = y_2(y_1(x))$, and $y_4 = y_1(y_2(x))$. Which of y_3 and y_4 corresponds to $f \circ g$? To $g \circ f$?

(b) Graph y_1, y_2, and y_3 to make conjectures about the domain and range of y_3.

(c) Graph y_1, y_2, and y_4 to make conjectures about the domain and range of y_4.

(d) Confirm your conjectures algebraically by finding formulas for y_3 and y_4.

T **54.** Enter $y_1 = \sqrt{x}$, $y_2 = \sqrt{1 - x}$ and $y_3 = y_1 + y_2$ on your grapher.

(a) Graph y_3 in $[-3, 3]$ by $[-1, 3]$.

(b) Compare the domain of the graph of y_3 with the domains of the graphs of y_1 and y_2.

(c) Replace y_3 by

$$y_1 - y_2, \quad y_2 - y_1, \quad y_1 \cdot y_2, \quad y_1/y_2, \quad \text{and} \quad y_2/y_1,$$

in turn, and repeat the comparison of part (b).

(d) Based on your observations in parts (b) and (c), what would you conjecture about the domains of sums, differences, products, and quotients of functions?

Regression Analysis: Stern Waves and Stopping Distance

See page 5 for an introduction to regression analysis with a calculator.

T **55.** *Stern waves* Observations of the stern waves that follow a boat at right angles to its course have disclosed that the distance between the crests of these waves (their *wave length*) increases with the speed of the boat. Table 6 shows the relationship between wave length and the speed of the boat.

(a) Find a power regression equation $y = ax^b$ for the data in Table 6, where x is the wave length, and y the speed of the boat.

(b) Superimpose the graph of the power regression equation on a scatter plot of the data.

(c) Use the graph of the power regression equation to predict the speed of the boat when the wave length is 11 m. Confirm algebraically.

(d) Now use *linear* regression to predict the speed when the wave length is 11 m. Superimpose the regression line on a scatter plot of the data. Which gives the better fit, the line here or the curve in part (b)?

Table 6 Wave lengths

Wave length (m)	Speed (km/h)
0.20	1.8
0.65	3.6
1.13	5.4
2.55	7.2
4.00	9.0
5.75	10.8
7.80	12.6
10.20	14.4
12.90	16.2
16.00	18.0
18.40	19.8

T **56.** *Vehicular stopping distance* Table 7 shows the total stopping distance of a car as a function of its speed.

(a) Find the quadratic regression equation for the data in Table 7.

(b) Superimpose the graph of the quadratic regression equation on a scatter plot of the data.

(c) Use the graph of the quadratic regression equation to predict the average total stopping distance for speeds of 72 and 85 mph. Confirm algebraically.

(d) Now use *linear* regression to predict the average total stopping distance for speeds of 72 and 85 mph. Superimpose the regression line on a scatter plot of the data. Which gives the better fit, the line here or the graph in part (b)?

Table 7 Vehicular stopping distance

Speed (mph)	Average total stopping distance (ft)
20	42
25	56
30	73.5
35	91.5
40	116
45	142.5
50	173
55	209.5
60	248
65	292.5
70	343
75	401
80	464

Source: U.S. Bureau of Public Roads.

3 Exponential Functions

Exponential Growth • Population Growth • The Exponential Function e^x • What Happened to a^x?

Exponential functions are of particular importance in science and engineering applications. We review your experience with this class of functions in this section and discuss several important exponential models of growth and decay. The mathematics underlying the properties of these extraordinary functions and their relations to logarithms (next section) is beautiful and deep. We will investigate this in appropriate detail when we study the calculus of these functions in Chapters 3 and 6.

Exponential Growth

Table 8 shows the growth of \$100 invested in 1996 at an interest rate of 5.5%, compounded annually. After the first year, the value of the account is always 1.055 times its value in the previous year. After n years, the value is $y = 100 \cdot (1.055)^n$.

Compound interest provides an example of *exponential growth* and is modeled by a function of the form $y = P \cdot a^x$, where P is the initial investment and a is equal to 1 plus the interest rate expressed as a decimal.

Table 8 Savings account growth

Year	Amount (dollars)	Increase (dollars)
1996	100	
1997	$100(1.055) = 105.50$	5.50
1998	$100(1.055)^2 = 111.30$	5.80
1999	$100(1.055)^3 = 117.42$	6.12
2000	$100(1.055)^4 = 123.88$	6.46

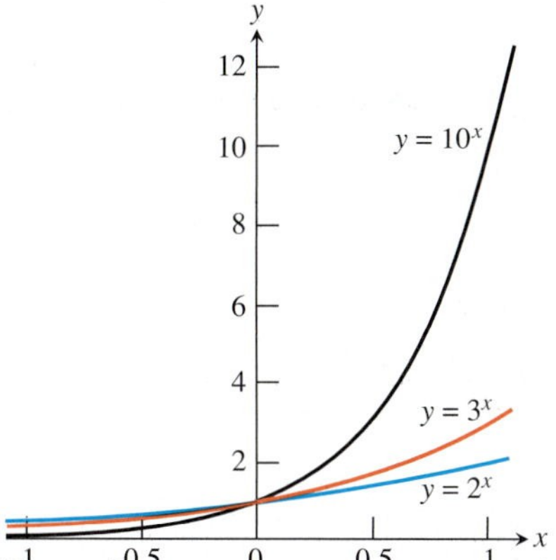

FIGURE 24 $y = 2^x, y = 3^x, y = 10^x$.

The equation $y = P \cdot a^x$, $a > 0$, $a \neq 1$, identifies a family of functions called *exponential functions*.

Example 1 Graphing $y = a^x$

Graph the functions $y = 2^x$, $y = 3^x$, and $y = 10^x$. For what values of x is it true that $2^x > 3^x > 10^x$?

Solution From the graph in Figure 24, the functions are increasing for all values of x. For $x < 0$, we have $2^x > 3^x > 10^x$. At $x = 0$, we have $2^x = 3^x = 10^x = 1$. For $x > 0$, we have $2^x < 3^x < 10^x$.

Definition Exponential Function

Let a be a positive real number other than 1. The function

$$f(x) = a^x$$

is the **exponential function with base a**.

The domain of $f(x) = a^x$ is $(-\infty, \infty)$ and the range is $(0, \infty)$. If $a > 1$, the graph of f looks like the graph of $y = 2^x$ in Figure 25a. If $0 < a < 1$, the graph of f looks like the graph of $y = (1/2)^x = 2^{-x}$ in Figure 25b.

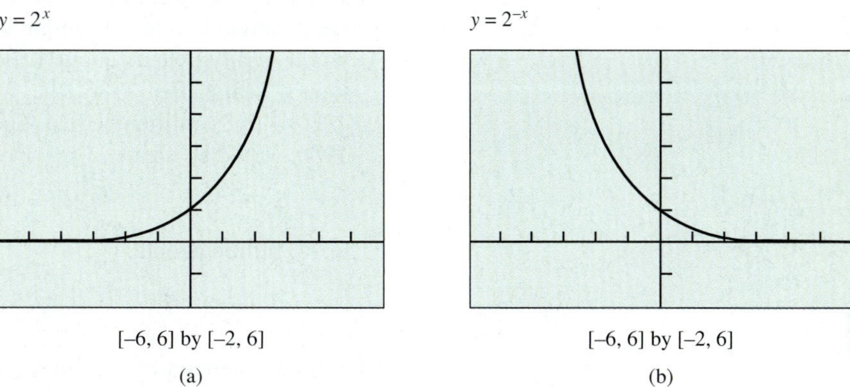

FIGURE 25 Graphs of (a) $y = 2^x$ and (b) $y = 2^{-x}$.

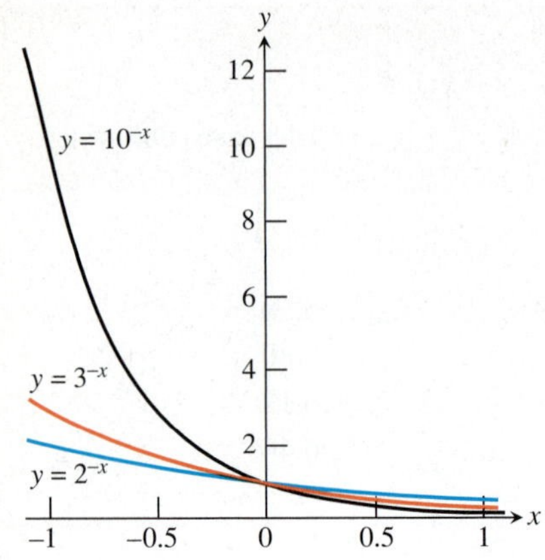

FIGURE 26 $y = 2^{-x}, y = 3^{-x}, y = 10^{-x}$.

Example 2 Graphing $y = a^{-x}$

Graph the functions $y = 2^{-x}$, $y = 3^{-x}$, and $y = 10^{-x}$. For what values of x is it true that $2^{-x} > 3^{-x} > 10^{-x}$?

Solution From the graph in Figure 26, the functions are decreasing for all values of x. For $x < 0$, we have $2^{-x} < 3^{-x} < 10^{-x}$. At $x = 0$, we have $2^{-x} = 3^{-x} = 10^{-x} = 1$. For $x > 0$, we have $2^{-x} > 3^{-x} > 10^{-x}$.

Exponential functions obey the rules for exponents.

Rules for Exponents

If $a > 0$ and $b > 0$, the following hold true for all real numbers x and y.

1. $a^x \cdot a^y = a^{x+y}$
2. $\dfrac{a^x}{a^y} = a^{x-y}$
3. $(a^x)^y = (a^y)^x = a^{xy}$
4. $a^x \cdot b^x = (ab)^x$
5. $\dfrac{a^x}{b^x} = \left(\dfrac{a}{b}\right)^x$

Population Growth

Population growth can sometimes be modeled with an exponential function. In Table 9, we give some values for the population of the world. We have also divided the population in one year by the population in the previous year to get an idea of how the population is growing. These ratios are in the third column.

Example 3 Predicting World Population

Use the data in Table 9 and an exponential model to predict the population of the world in 2010.

Solution Based on the third column in Table 9, although questionable given the variation in rates, we might be willing to conjecture that the population of the world in any year is about 1.018 times the population the year before. At any time t years after 1986, the world population would then be about $P(t) = 4936\,(1.018)^t$ million people. The population in 2010, which is $t = 24$ years after 1986, would be about

$$P(24) = 4936(1.018)^{24} \approx 7573.9$$

or 7.6 billion people.

The Exponential Function e^x

The most important exponential function for modeling natural, physical, and economic phenomena is the **natural exponential function,** whose base is the famous

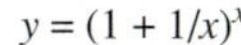

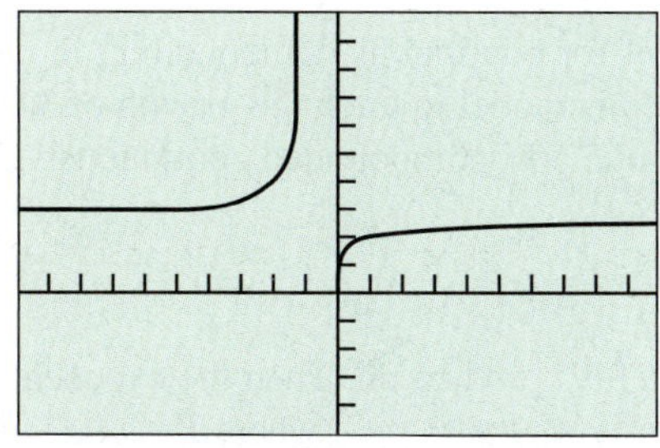

[–10, 10] by [–5, 10]

X	Y1	
1000	2.7169	
2000	2.7176	
3000	2.7178	
4000	2.7179	
5000	2.718	
6000	2.7181	
7000	2.7181	
Y1 = (1+1/X)^X		

FIGURE 27 A graph and table of values for $f(x) = (1 + 1/x)^x$ both suggest that as $x \to \infty$, $f(x) \to e \approx 2.718$.

Table 9 World population

Year	Population (millions)	Ratio
1986	4936	
1987	5023	$5023/4936 \approx 1.0176$
1988	5111	$5111/5023 \approx 1.0175$
1989	5201	$5201/5111 \approx 1.0176$
1990	5329	$5329/5201 \approx 1.0246$
1991	5422	$5422/5329 \approx 1.0175$

Source: Statistical Office of the United Nations, *Monthly Bulletin Statistics,* 1991.

number e, which is 2.718281828 to nine decimal places. We can define e to be the number that the function $f(x) = (1 + 1/x)^x$ approaches as x becomes large without bound. The graph and table in Figure 27 strongly suggest that such a number exists. You will learn more about the number e, and how it is obtained, in your study of calculus.

The exponential functions $y = e^{kx}$, where k is a nonzero constant, are frequently used as models of exponential growth or decay. For an example of exponential growth, interest **compounded continuously** uses the model $y = P \cdot e^{rt}$, where P is the initial investment, r is the interest rate as a decimal, and t is time in years. An example of exponential decay is the model $y = A \cdot e^{-1.2 \times 10^{-4} t}$, which represents how the radioactive element carbon-14 decays over time. Here A is the original amount of carbon-14 and t is the time in years. Carbon-14 decay is used to date the remains of dead organisms such as shells, seeds, and wooden artifacts.

Definitions Exponential Growth, Exponential Decay

The function $y = y_0 e^{kx}$ is a model for **exponential growth** if $k > 0$ and a model for **exponential decay** if $k < 0$.

Figure 28 shows graphs of exponential growth and exponential decay.

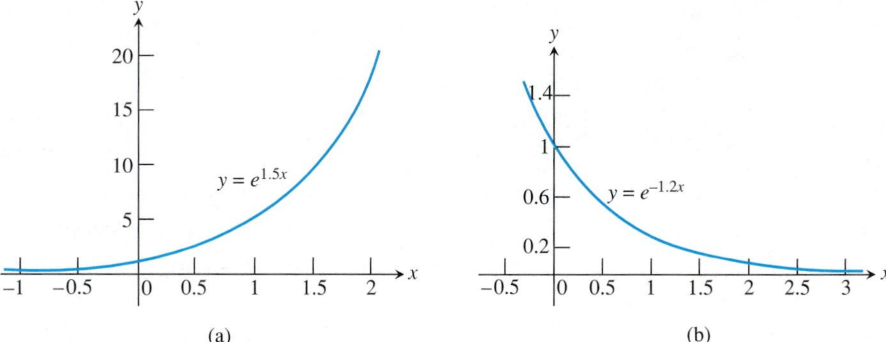

FIGURE 28 Graphs of (a) exponential growth, $k = 1.5 > 0$ and (b) exponential decay, $k = -1.2 < 0$.

Example 4 Savings Account Growth Revisited

Investment companies often use the model for continuous compounding in calculating the growth of an investment. Use this model to track the growth of \$100 invested in 1996 at an annual interest rate of 5.5%, compounded continuously.

Solution

Model

Let $x = 0$ represent 1996, $x = 1$ represent 1997, and so on. Then the exponential growth model for continuous compounding is $y(x) = P \cdot e^{rx}$, where $P = 100$ (the initial investment), $r = 0.055$ (the annual interest rate expressed as a decimal), and x is time in years. To predict the amount in the account in 2000, for example, we take $x = 4$ and calculate

$$\begin{aligned} y(4) &= 100 \cdot e^{0.055(4)} \\ &= 100 \cdot e^{0.22} \\ &= 124.61. \quad \text{Nearest cent} \end{aligned}$$

Comparing this result with the \$123.88 in the account when the interest is compounded annually (Table 10), we see that the investor earns more as the interest is compounded more frequently (in this case, *continuously*). In Table 10, we compare the values for the amount in the savings account for the years 1996 to 2000 when interest is compounded annually (Table 8) and continuously.

Table 10 Comparing savings account growth

Year	Amount (dollars), annual compounding	Amount (dollars), continuous compounding
1996	100.00	100.00
1997	105.50	105.65
1998	111.30	111.63
1999	117.42	117.94
2000	123.88	124.61

A bank might decide that it is worth the additional amount to be able to advertise, "We compound interest continuously," as a way to attract more customers.

Example 5 Modeling Radioactive Decay

Laboratory experiments indicate that some atoms emit a part of their mass as radiation, with the remainder of the atom re-forming to make an atom of some new element. For example, radioactive carbon-14 decays into nitrogen; radium eventually decays into lead. If y_0 is the number of radioactive nuclei present at time zero, the number still present at any later time t will be

$$y = y_0 e^{-rt}, \qquad r > 0.$$

The number r is called the **decay rate** of the radioactive substance. For carbon-14, the decay rate has been determined experimentally to be about $r = 1.2 \times 10^{-4}$

when t is measured in years. Predict the percent of carbon-14 present after 866 years have elapsed.

Solution If we start with an amount y_0 of carbon-14 nuclei, after 866 years we are left with the amount

$$y(866) = y_0 e^{(-1.2\times10^{-4})(866)}$$
$$\approx (0.901)y_0.$$

That is, after 866 years, we are left with about 90% of the original amount of carbon-14, so about 10% of the original nuclei have decayed. In Example 12 in the next section, you will see how to find the number of years required for half of the radioactive nuclei present in a sample to decay.

What Happened to a^x?

You may wonder why we use the family of functions $y = y_0 e^{kx}$ for different values of the constant k instead of the general exponential functions $y = Pa^x$. In the next section, we show that the exponential function a^x is the same as e^{kx} for an appropriate value of k. So the formula $y = y_0 e^{kx}$ covers the entire range of possibilities.

EXERCISES 3

In Exercises 1–6, match the function with the graphs from Figure 29. Try to do it without using your grapher.

1. $y = 2^x$

2. $y = 3^{-x}$

3. $y = -3^{-x}$

4. $y = -0.5^{-x}$

5. $y = 2^{-x} - 2$

6. $y = 1.5^x - 2$

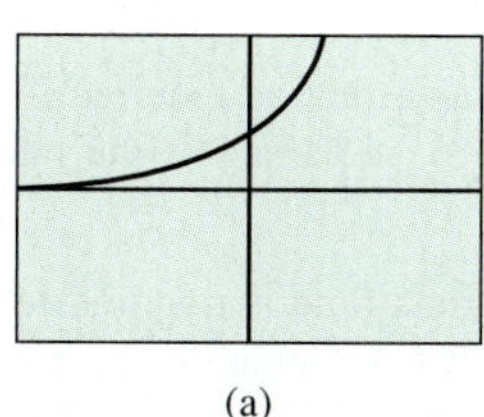

(a)

(b)

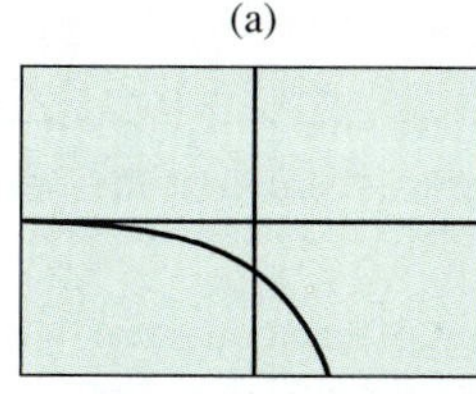

(c)

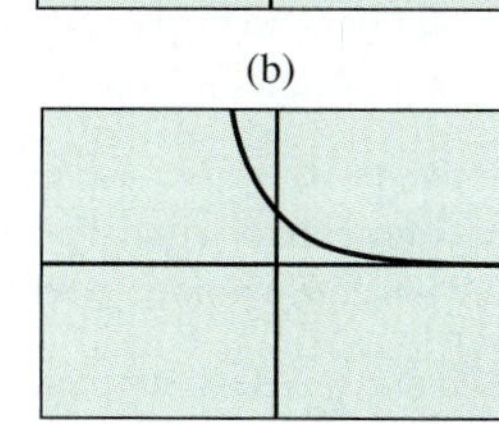

(d)

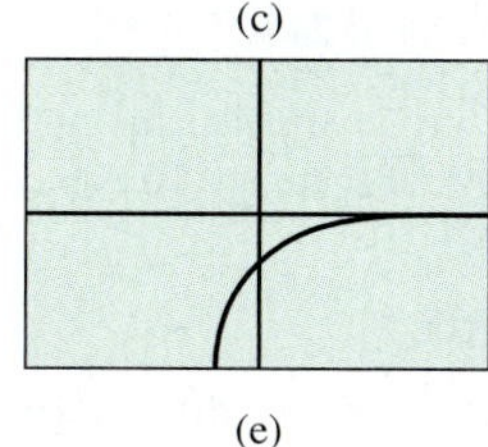

(e)

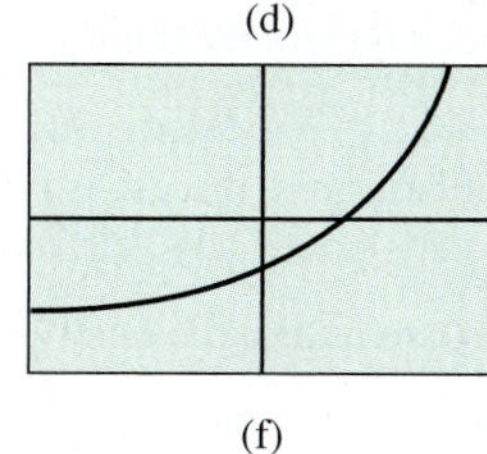

(f)

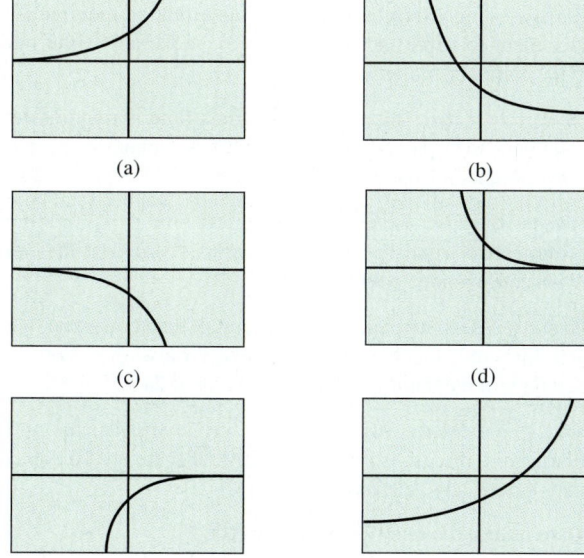

FIGURE 29 Graphs for Exercises 1–6.

In Exercises 7–10, graph the function. State its domain, range, and intercepts.

7. $y = -2^x + 3$

8. $y = e^x + 3$

9. $y = 3 \cdot e^{-x} - 2$

10. $y = -2^{-x} - 1$

In Exercises 11–14, rewrite the exponential expression to have the indicated base.

11. 9^{2x}, base 3

12. 16^{3x}, base 2

13. $(1/8)^{2x}$, base 2

14. $(1/27)^x$, base 3

In Exercises 15–18, copy and *work in groups of two or three* to complete the table for the function.

15. $y = 2x - 3$

x	y	Change (Δy)
1	?	
		?
2	?	
		?
3	?	
		?
4	?	

16. $y = -3x + 4$

x	y	Change (Δy)
1	?	
		?
2	?	
		?
3	?	
		?
4	?	

17. $y = x^2$

x	y	Change (Δy)
1	?	
		?
2	?	
		?
3	?	
		?
4	?	

18. $y = 3e^x$

x	y	Ratio (y_i/y_{i-1})
1	?	
		?
2	?	
		?
3	?	
		?
4	?	

19. *Writing to Learn* Explain how the change Δy is related to the slopes of the lines in Exercises 15 and 16. If the changes in x are constant for a linear function, what would you conclude about the corresponding changes in y?

20. *Writing to Learn* Describe how the change Δy from one x-value to the next in Exercise 17 is related to those x-values. What is the change Δy from $x = 1000$ to $x = 1001$? From $x = n$ to $x = n + 1$ for an arbitrary positive integer n?

Extending the Ideas

In Exercises 21 and 22, assume that the graph of the exponential function $f(x) = k \cdot a^x$ passes through the two points. Find the values of a and k.

21. $(1, 4.5), (-1, 0.5)$

22. $(1, 1.5), (-1, 6)$

Solving with Graphs

T In Exercises 23–26, use graphs to solve the equations.

23. $2^x = 5$

24. $e^x = 4$

25. $3^x - 0.5 = 0$

26. $3 - 2^{-x} = 0$

Theory and Examples

27. *World population* (*continuation of Example 3*) Use 1.018 and the population in 1991 to estimate the population of the world in 2010.

28. *Bacteria growth* The number of bacteria in a petri dish culture after t hours is

$$B = 100e^{0.693t}.$$

(a) What was the initial number of bacteria present?

(b) How many bacteria are present after 6 hours?

T **(c)** Approximately when will the number of bacteria be 200? Estimate the doubling time of the bacteria.

T In Exercises 29–40, use an exponential model and a graphing calculator to estimate the answer in each problem.

29. *Population growth* The population of Knoxville is 500,000 and is increasing at the rate of 3.75% each year. Approximately when will the population reach 1 million?

30. *Population growth* The population of Silver Run in the year 1890 was 6250. Assume the population increased at a rate of 2.75% per year.

(a) Estimate the population in 1915 and 1940.

(b) Approximately when did the population reach 50,000?

31. *Radioactive decay* The half-life of phosphorus-32 is about 14 days. There are 6.6 grams present initially.

(a) Express the amount of phosphorus-32 remaining as a function of time t.

(b) When will there be 1 gram remaining?

32. *Finding time* If John invests \$2300 in a savings account with a 6% interest rate compounded annually, how long will it take until John's account has a balance of \$4150?

33. *Doubling your money* Determine how much time is required for an investment to double in value if interest is earned at the rate of 6.25% compounded annually.

34. *Doubling your money* Determine how much time is required for an investment to double in value if interest is earned at the rate of 6.25% compounded monthly.

35. *Doubling your money* Determine how much time is required for an investment to double in value if interest is earned at the rate of 6.25% compounded continuously.

36. *Tripling your money* Determine how much time is required for an investment to triple in value if interest is earned at the rate of 5.75% compounded annually.

37. *Tripling your money* Determine how much time is required for an investment to triple in value if interest is earned at the rate of 5.75% compounded daily.

38. *Tripling your money* Determine how much time is required for an investment to triple in value if interest is earned at the rate of 5.75% compounded continuously.

39. *Cholera bacteria* Suppose that a colony of bacteria starts with 1 bacterium and doubles in number every half hour. How many bacteria will the colony contain at the end of 24 h?

40. *Eliminating a disease* Suppose that in any given year the number of cases of a disease is reduced by 20%. If there are 10,000 cases today, how many years will it take

(a) to reduce the number of cases to 1000?

(b) to eliminate the disease; that is, to reduce the number of cases to less than 1?

Regression Analysis: Exponential Models for Population

See page 5 for an introduction to regression analysis with a calculator.

T **41.** Table 11 gives data about the population of Mexico.

Table 11 Population of Mexico

Year	Population (millions)
1950	25.8
1960	34.9
1970	48.2
1980	66.8
1990	81.1

Source: The Statesman's Yearbook, 129th ed. (London: The Macmillan Press, Ltd., 1992).

(a) Let $x = 0$ represent 1900, $x = 1$ represent 1901, and so forth. Find an exponential regression equation for the data and superimpose its graph on a scatter plot of the data.

(b) Use the exponential regression equation to estimate the population of Mexico in 1900. How close is the estimate to the actual population in 1900 of 13,607,272?

(c) Use the exponential regression equation to estimate the annual rate of growth of the population of Mexico.

T **42.** Table 12 gives population data for South Africa.

Table 12 Population of South Africa

Year	Population (millions)
1904	5.2
1911	6.0
1921	6.9
1936	9.6
1946	11.4
1951	12.7
1960	16.0
1970	18.3
1980	20.6

Source: The Statesman's Yearbook, 129th ed. (London: The Macmillan Press, Ltd., 1992).

(a) Let $x = 0$ represent 1900, $x = 1$ represent 1901, and so forth. Find an exponential regression equation for the data and superimpose its graph on a scatter plot of the data.

(b) Use the exponential regression equation to estimate the population of South Africa in 1990.

(c) Use the exponential regression equation to estimate the annual rate of growth of the population of South Africa.

4 Inverse Functions and Logarithms

One-to-One Functions • Inverses • Finding Inverses • Logarithmic Functions • Properties of Logarithms • Applications

In this section, we define what it means for functions to be inverses of one another and look at what this says about the formulas and graphs of function-inverse pairs. We then study a logarithm function as the inverse of an exponential function with appropriate base, and present several important applications of logarithm functions.

One-to-One Functions

As you know, a function is a rule that assigns a single value in its range to each point in its domain. Some functions assign the same output to more than one input. For example, $f(x) = x^2$ assigns the output 4 to both 2 and -2. Other functions

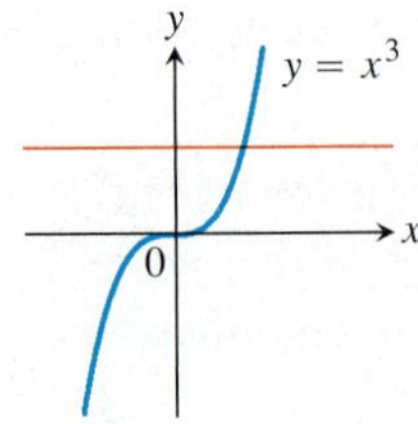

One-to-one: Graph meets each horizontal line once.

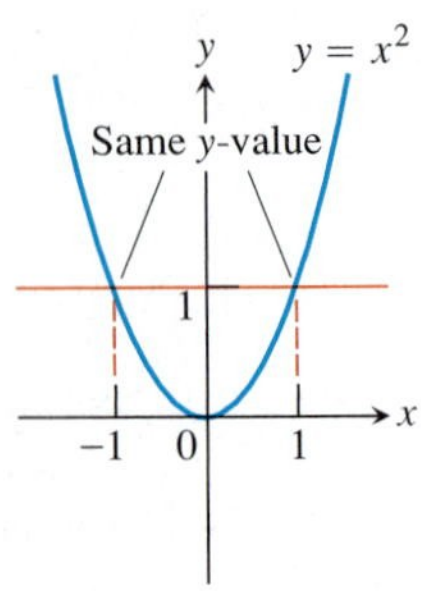

Not one-to-one: Graph meets some horizontal lines more than once.

FIGURE 30 Using the horizontal line test, we see that $y = x^3$ is one-to-one and $y = x^2$ is not.

never output a given value more than once. For example, the cubes of different numbers are always different.

If each output value of a function is associated with exactly one input value, the function is *one-to-one*.

Definition One-to-One Function

A function $f(x)$ is **one-to-one** on a domain D if $f(a) \neq f(b)$ whenever $a \neq b$.

The graph of a one-to-one function $y = f(x)$ can intersect any horizontal line at most once (the *horizontal line test*). If it intersects such a line more than once, it assumes the same y-value more than once and is not one-to-one (Figure 30).

Example 1 Using the Horizontal Line Test

Determine whether the functions are one-to-one.

(a) $f(x) = x^{2/3}$ **(b)** $g(x) = \sqrt{x}$

Solution As Figure 31a suggests, each horizontal line $y = c$, $c > 0$, intersects the graph of $f(x) = x^{2/3}$ twice so f is not one-to-one. As Figure 31b suggests, each horizontal line intersects the graph of $g(x) = \sqrt{x}$ either once or not at all. The function g is one-to-one.

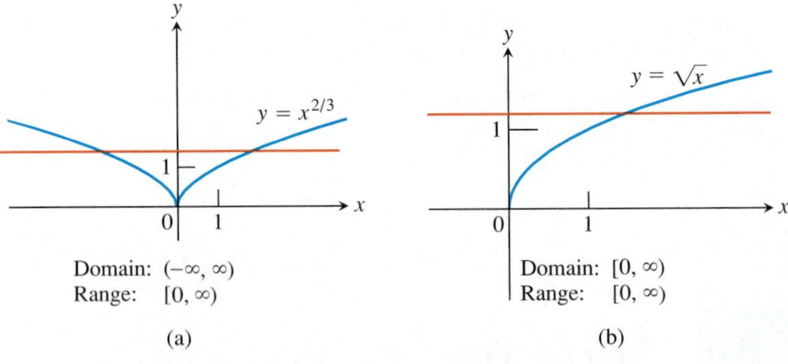

FIGURE 31 (a) The graph of $f(x) = x^{2/3}$ and a horizontal line. (b) The graph of $g(x) = \sqrt{x}$ and a horizontal line. (Example 1)

Inverses

Since each output of a one-to-one function comes from just one input, a one-to-one function can be reversed to send outputs back to the inputs from which they came. The function defined by reversing a one-to-one function f is the **inverse of f**. The functions in Tables 13 and 14 are inverses of one another. The symbol for the inverse of f is f^{-1}, read "f inverse." The -1 in f^{-1} is not an exponent; $f^{-1}(x)$ does not mean $1/f(x)$.

As Tables 13 and 14 suggest, composing a function with its inverse in either order sends each output back to the input from which it came. In other words, the result of composing a function and its inverse in either order is the **identity func-**

Table 13 Rental charge versus time

Time x (hours)	Charge y (dollars)
1	5.00
2	7.50
3	10.00
4	12.50
5	15.00
6	17.50

Table 14 Time versus rental charge

Charge x (dollars)	Time y (hours)
5.00	1
7.50	2
10.00	3
12.50	4
15.00	5
17.50	6

tion, the function that assigns each number to itself. Composing two functions f and g gives a way to test whether they are inverses of one another. Compute $f \circ g$ and $g \circ f$. If $(f \circ g)(x) = x$ and $(g \circ f)(x) = x$, then f and g are inverses of one another; otherwise, they are not. The functions $f(x) = x^3$ and $g(x) = x^{1/3}$ are inverses of one another because $(x^3)^{1/3} = x$ and $(x^{1/3})^3 = x$ for every number x.

A Test for Inverses

Functions f and g are an inverse pair if and only if

$$f(g(x)) = x \qquad \text{and} \qquad g(f(x)) = x.$$

In this case, $g = f^{-1}$ and $f = g^{-1}$.

Example 2 Testing for Inverses

(a) The functions

$$f(x) = 3x \qquad \text{and} \qquad g(x) = \frac{x}{3}$$

are an inverse pair because

$$f(g(x)) = f\left(\frac{x}{3}\right) = 3\left(\frac{x}{3}\right) = x \qquad \text{and} \qquad g(f(x)) = g(3x) = \frac{3x}{3} = x$$

for every number x.

(b) The functions

$$f(x) = x \qquad \text{and} \qquad g(x) = \frac{1}{x}$$

are *not* inverses because

$$f(g(x)) = f\left(\frac{1}{x}\right) = \frac{1}{x} \neq x.$$

Finding Inverses

How do we find the graph of the inverse of a function? Suppose, for example, that the function is the one pictured in Figure 32a. To read the graph, we start at the point x on the x-axis, go up to the graph, and then move over to the y-axis to read the value of y. If we start with y and want to find the x from which it came, we reverse the process (Figure 32b).

The graph of f is already the graph of f^{-1}, although the latter graph is not drawn in the usual way with the domain axis horizontal and the range axis vertical. For f^{-1}, the input-output pairs are reversed. To display the graph of f^{-1} in the usual way, we have to reverse the pairs by reflecting the graph in the 45° line $y = x$ (Figure 32c) and interchanging the letters x and y (Figure 32d). This step puts the independent variable, now called x, on the horizontal axis and the dependent variable, now called y, on the vertical axis.

That the graphs of f and f^{-1} are reflections of each other across the line $y = x$ is to be expected because the input-output pairs (a, b) of f have been reversed to produce the input-output pairs (b, a) of f^{-1}.

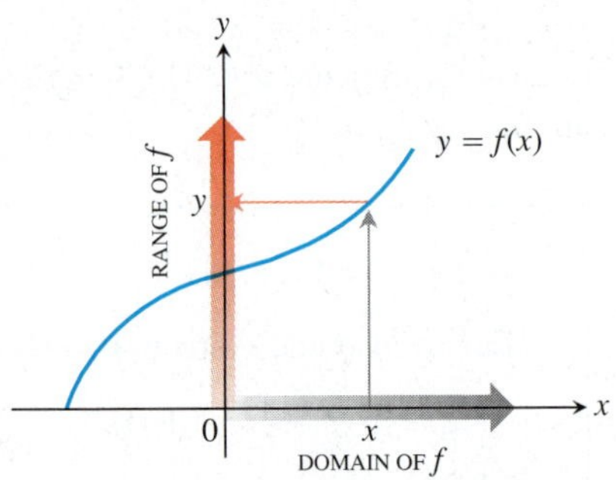

(a) To find the value of f at x, we start at x, go up to the curve, and then over to the y-axis.

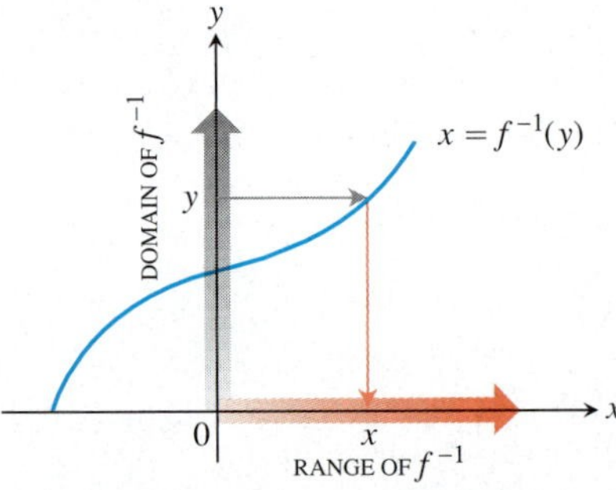

(b) The graph of f is already the graph of f^{-1}. To find the x that gave y, we start at y and go over to the curve and down to the x-axis. The domain of f^{-1} is the range of f. The range of f^{-1} is the domain of f.

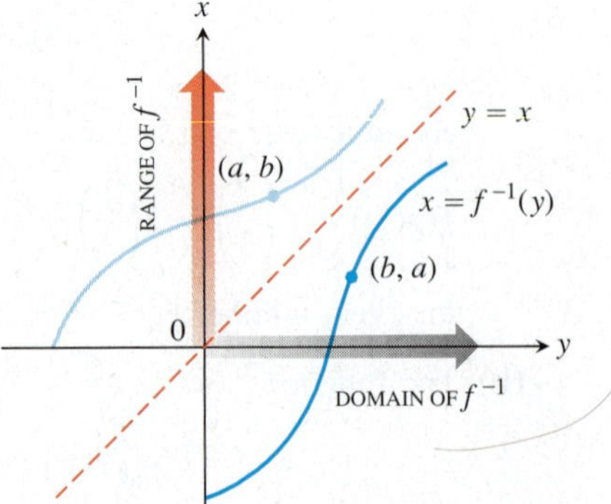

(c) To draw the graph of f^{-1} in the usual way, we reflect the system in the line $y = x$.

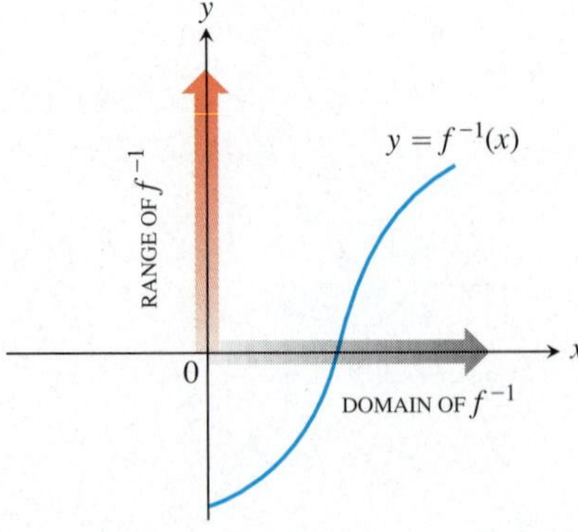

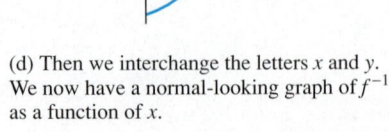

(d) Then we interchange the letters x and y. We now have a normal-looking graph of f^{-1} as a function of x.

FIGURE 32 The graph of $y = f^{-1}(x)$.

The pictures in Figure 32 tell how to express f^{-1} as a function of x algebraically.

Writing f^{-1} as a Function of x

Step 1. Solve the equation $y = f(x)$ for x in terms of y.

Step 2. Interchange x and y. The resulting formula will be $y = f^{-1}(x)$.

Example 3 Finding the Inverse Function

Find the inverse of $y = \frac{1}{2}x + 1$, expressed as a function of x.

Solution

Step 1: Solve for x in terms of y:

$$y = \frac{1}{2}x + 1$$
$$2y = x + 2$$
$$x = 2y - 2.$$

Step 2: Interchange x and y:

$$y = 2x - 2.$$

The inverse of the function $f(x) = (1/2)x + 1$ is the function $f^{-1}(x) = 2x - 2$.

Check: To check, we verify that both composites give the identity function:

$$f^{-1}(f(x)) = 2\left(\frac{1}{2}x + 1\right) - 2 = x + 2 - 2 = x$$

$$f(f^{-1}(x)) = \frac{1}{2}(2x - 2) + 1 = x - 1 + 1 = x.$$

See Figure 33.

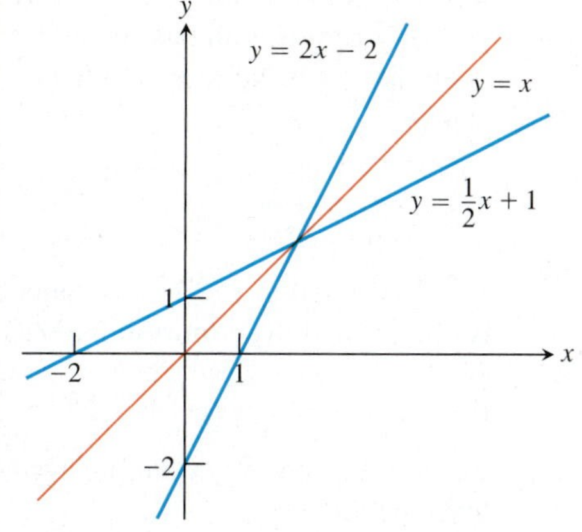

FIGURE 33 Graphing $f(x) = (1/2)x + 1$ and $f^{-1}(x) = 2x - 2$ together shows the graphs' symmetry with respect to the line $y = x$.

Example 4 Finding the Inverse Function

Find the inverse of the function $y = x^2, x \geq 0$, expressed as a function of x.

Solution

Step 1: Solve for x in terms of y:

$$y = x^2$$
$$\sqrt{y} = \sqrt{x^2} = |x| = x \qquad |x| = x \text{ because } x \geq 0$$

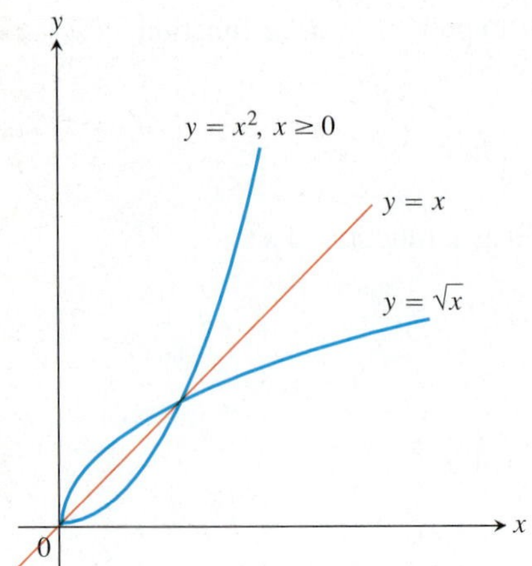

FIGURE 34 The functions $y = \sqrt{x}$ and $y = x^2, x \geq 0$, are inverses of one another. (Example 4)

Step 2: Interchange x and y:

$$y = \sqrt{x}.$$

The inverse of $y = x^2, x \geq 0$, is $y = \sqrt{x}$. See Figure 34.

Notice that, unlike the restricted function $y = x^2, x \geq 0$, the unrestricted function $y = x^2$ is not one-to-one and therefore has no inverse.

In Section 1.6, we show you an easy way to graph $y = f(x)$ and $y = f^{-1}(x)$ together on a graphing calculator or computer.

Logarithmic Functions

If a is any positive real number other than 1, the base a exponential function $f(x) = a^x$ is one-to-one. It therefore has an inverse. Its inverse is called the *base a logarithm function.*

Definition Base a Logarithm Function

The **base a logarithm function** $y = \log_a x$ is the inverse of the base a exponential function $y = a^x$ $(a > 0, a \neq 1)$.

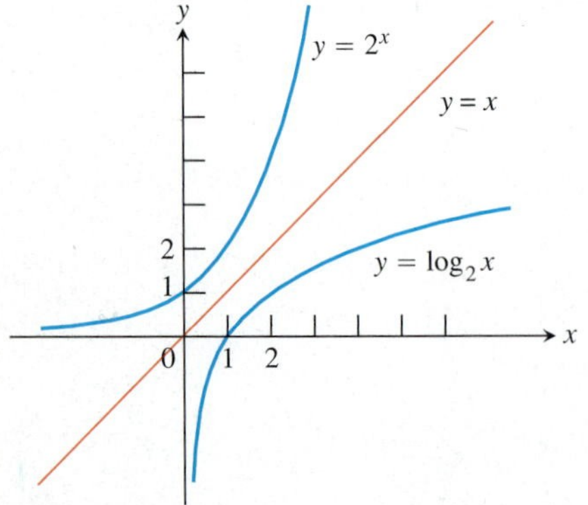

FIGURE 35 The graph of 2^x and its inverse, $\log_2 x$.

The domain of $\log_a x$ is $(0, \infty)$, the range of a^x. The range of $\log_a x$ is $(-\infty, \infty)$, the domain of a^x.

Because we have no technique for solving for x in terms of y in the equation $y = a^x$, we do not have an explicit formula for the logarithm function as a function of x. The graph of $y = \log_a x$, however, can be obtained by reflecting the graph of $y = a^x$ across the line $y = x$ (Figure 35).

Logarithms with base e and base 10 are so important in applications that calculators have special keys for them. They also have their own special notation and names:

$$\log_e x \quad \text{is written as} \quad \ln x.$$
$$\log_{10} x \quad \text{is written as} \quad \log x.$$

The function $y = \ln x$ is called the **natural logarithm function,** and $y = \log x$ is often called the **common logarithm function.**

Properties of Logarithms

Because a^x and $\log_a x$ are inverses, composing them in either order gives the identity function.

Inverse Properties for a^x and $\log_a x$

1. Base a: $a^{\log_a x} = x, \quad \log_a a^x = x, \quad a > 0, a \neq 1, x > 0$
2. Base e: $e^{\ln x} = x, \quad \ln e^x = x, \quad x > 0$

These properties help us solve equations that contain logarithms and exponential functions.

Example 5 Using the Inverse Properties

Solve for x: **(a)** $\ln x = 3t + 5$ **(b)** $e^{2x} = 10$

Solution

(a) $\ln x = 3t + 5$

$e^{\ln x} = e^{3t+5}$ Exponentiate both sides.

$x = e^{3t+5}$ Inverse Property

(b) $e^{2x} = 10$

$\ln e^{2x} = \ln 10$ Take logarithms of both sides.

$2x = \ln 10$ Inverse Property

$$x = \frac{1}{2}\ln 10 \approx 1.15$$

The logarithm function has the following arithmetic properties.

> **Properties of Logarithms**
> For any real numbers $x > 0$ and $y > 0$,
>
> **1.** *Product Rule:* $\log_a xy = \log_a x + \log_a y$
>
> **2.** *Quotient Rule:* $\log_a \frac{x}{y} = \log_a x - \log_a y$
>
> **3.** *Power Rule:* $\log_a x^y = y \log_a x$

Substituting a^x for x in the equation $x = e^{\ln x}$ enables us to rewrite a^x as a power of e:

$a^x = e^{\ln(a^x)}$ Substitute a^x for x in $x = e^{\ln x}$.

$= e^{x \ln a}$ Power Rule for logs

$= e^{(\ln a)x}$. Exponent rearranged

The exponential function a^x is the same as e^{kx} for $k = \ln a$. The usual formulation is the one without parentheses.

> Every exponential function is a power of the natural exponential function.
>
> $$a^x = e^{x \ln a}$$
>
> That is, a^x is the same as e^x raised to the power $\ln a$.

Example 6 Writing Exponentials as Powers of e

$$2^x = e^{(\ln 2)x} = e^{x \ln 2}$$
$$5^{-3x} = e^{(\ln 5)(-3x)} = e^{-3x \ln 5}$$

Example 7 Solving Equations with Logarithms

Solve for x:

$$3^{\log_3 (7)} - 4^{\log_4 (2)} = 5^{(\log_5 x - \log_5 x^2)}$$

Solution:

$$
\begin{aligned}
3^{\log_3 (7)} - 4^{\log_4 (2)} &= 5^{(\log_5 x - \log_5 x^2)} \\
3^{\log_3 (7)} - 4^{\log_4 (2)} &= 5^{\log_5 (x/x^2)} && \text{Quotient Rule} \\
7 - 2 &= \frac{x}{x^2} && \text{Inverse property} \\
5 &= \frac{1}{x} && \text{Cancellation, } x \neq 0 \\
\frac{1}{5} &= x
\end{aligned}
$$

Returning once more to the properties of a^x and $\log_a x$, we have

$$
\begin{aligned}
\ln x &= \ln a^{\log_a x} && \text{Inverse Property for } a^x \text{ and } \log_a x \\
&= (\log_a x)(\ln a). && \text{Power Rule for logarithms, with } y = \log_a x
\end{aligned}
$$

Rewriting this equation as $\log_a x = (\ln x)/(\ln a)$ shows that every logarithmic function is a constant multiple of $\ln x$.

Change of Base Formula
Every logarithmic function is a constant multiple of the natural logarithm.

$$\log_a x = \frac{\ln x}{\ln a} \qquad (a > 0, a \neq 1)$$

Example 8 Graphing a Base *a* Logarithm Function

Graph $f(x) = \log_2 x$.

Solution We use the change of base formula to rewrite $f(x)$.

$$f(x) = \log_2 x = \frac{\ln x}{\ln 2}$$

Figure 36 gives the graph.

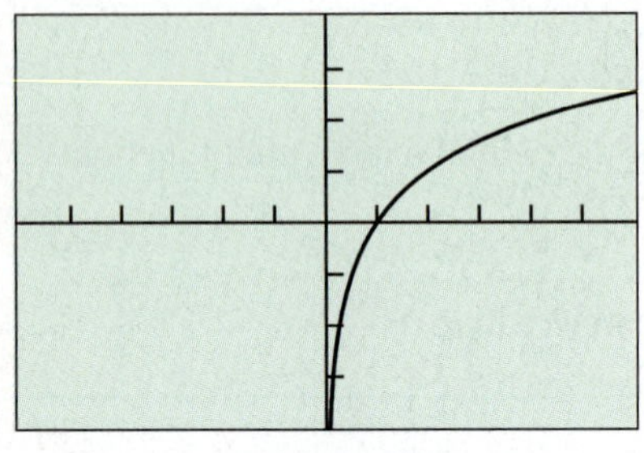

FIGURE 36 The graph of $f(x) = \log_2 x$ using $f(x) = (\ln x)/(\ln 2)$. (Example 8)

Applications

In Section 3, we used graphical methods to solve exponential growth and decay problems. Now we can use the properties of logarithms to solve the same problems algebraically.

Example 9 Finding Time

Sarah invests \$1000 in an account that earns 5.25% interest compounded annually. How long will it take the account to reach \$2500?

Solution

Model

The amount in the account at any time t in years is $1000(1.0525)^t$, so we need to solve the equation

$$1000(1.0525)^t = 2500\,.$$

Solve Algebraically

$$\begin{aligned} (1.0525)^t &= 2.5 && \text{Divide by 1000.}\\ \ln (1.0525)^t &= \ln 2.5 && \text{Take logarithms of both sides.}\\ t \ln 1.0525 &= \ln 2.5 && \text{Power Rule}\\ t &= \frac{\ln 2.5}{\ln 1.0525} \approx 17.9 \end{aligned}$$

Interpret

The amount in Sarah's account will be \$2500 in about 17.9 years, or about 17 years and 11 months.

Example 10 Earthquake Intensity

Earthquake intensity is often reported on the logarithmic Richter scale. Here the formula is

$$\text{Magnitude } R = \log\left(\frac{a}{T}\right) + B\,,$$

where a is the amplitude of the ground motion in microns at the receiving station, T is the period of the seismic wave in seconds, and B is an empirical factor that allows for the weakening of the seismic wave with increasing distance from the epicenter of the quake. For an earthquake 10,000 km from the receiving station, $B = 6.8$. If the recorded vertical ground motion is $a = 10$ microns and the period is $T = 1$ sec, the earthquake's magnitude is

$$R = \log\left(\frac{10}{1}\right) + 6.8 = 1 + 6.8 = 7.8\,.$$

An earthquake of this magnitude does great damage near its epicenter.

Typical Sound Levels

Threshold of hearing	0 db
Rustle of leaves	10 db
Average whisper	20 db
Quiet automobile	50 db
Ordinary conversation	65 db
Pneumatic drill 10 feet away	90 db
Threshold of pain	120 db

Example 11 Intensity of Sound

Another example of the use of common logarithms is the **decibel** or **db** ("dee bee") **scale** for measuring loudness. If I is the **intensity** of sound in watts per square meter, the decibel level of the sound is

$$\textbf{Sound level} = 10 \log (I \times 10^{12}) \text{ db}\,. \tag{1}$$

If you ever wondered why doubling the power of your audio amplifier increases the sound level by only a few decibels, Equation (1) provides the answer.

Doubling I in Equation (1) adds only about 3 db:

$$\begin{aligned}
\text{Sound level with } I \text{ doubled} &= 10\ \log\,(2I \times 10^{12}) && \text{Eq. (1) with } 2I \text{ for } I \\
&= 10 \log\,(2 \cdot I \times 10^{12}) \\
&= 10 \log 2 + 10 \log\,(I \times 10^{12}) \\
&= \text{original sound level} + 10 \log 2 \\
&\approx \text{original sound level} + 3 \cdot \log 2 \approx 0.30
\end{aligned}$$

Example 12 Half-Life of Polonium-210

The **half-life** of a radioactive element is the time required for half of the radioactive nuclei present in a sample to decay. It is a remarkable fact that the half-life is a constant that does not depend on the number of radioactive nuclei initially present in the sample, but only on the radioactive substance.

To see why, let y_0 be the number of radioactive nuclei initially present in the sample. Then the number y present at any later time t will be $y = y_0 e^{-kt}$. We seek the value of t at which the number of radioactive nuclei present equals half the original number:

$$\begin{aligned}
y_0 e^{-kt} &= \frac{1}{2} y_0 \\
e^{-kt} &= \frac{1}{2} \\
-kt &= \ln \frac{1}{2} = -\ln 2 && \text{Reciprocal Rule for logarithms} \\
t &= \frac{\ln 2}{k}. && (2)
\end{aligned}$$

This value of t is the half-life of the element. It depends only on the value of k; the number y_0 does not enter in.

The effective radioactive lifetime of polonium-210 is so short that we measure it in days rather than years. The number of radioactive atoms remaining after t days in a sample that starts with y_0 radioactive atoms is

$$y = y_0 e^{-5 \times 10^{-3} t}.$$

The element's half-life is

$$\begin{aligned}
\text{Half-life} &= \frac{\ln 2}{k} && \text{Eq. (2)} \\
&= \frac{\ln 2}{5 \times 10^{-3}} && \text{The } k \text{ from polonium's decay equation} \\
&\approx 139 \text{ days.}
\end{aligned}$$

EXERCISES 4

Identifying One-to-One Functions Graphically

Which of the functions graphed in Exercises 1–6 are one-to-one and which are not?

1.

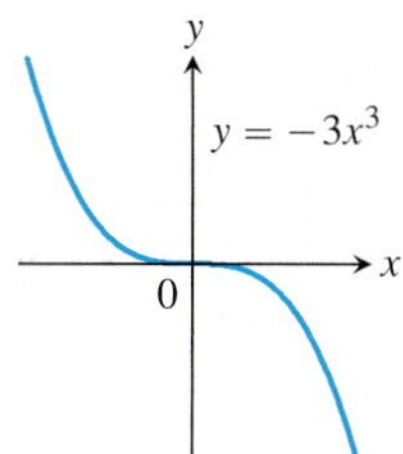

2.

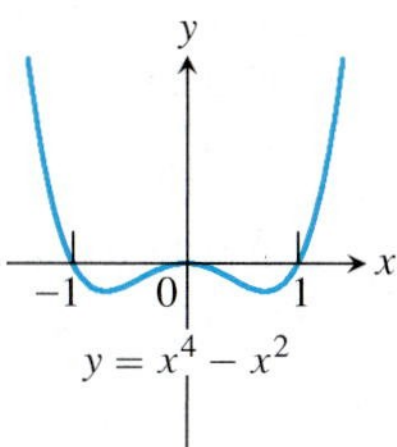

3.

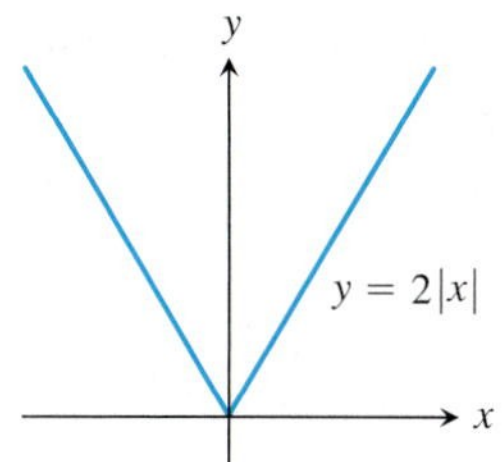

4.

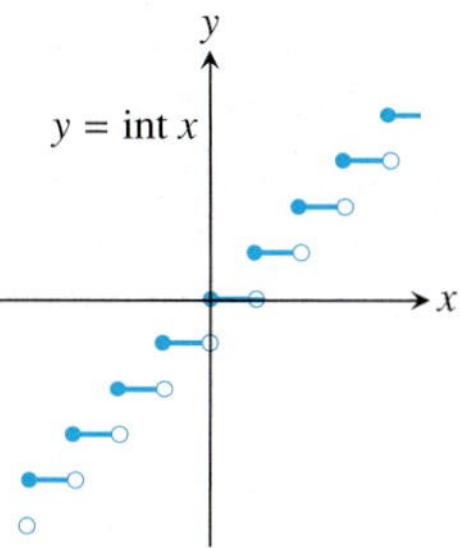

5.

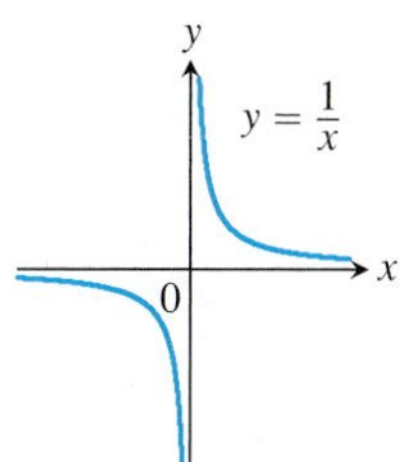

6.

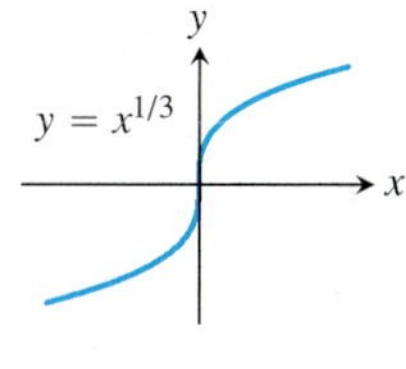

Graphing Inverse Functions

Each of Exercises 7–10 shows the graph of a function $y = f(x)$. Copy the graph and draw in the line $y = x$. Then use symmetry with respect to the line $y = x$ to add the graph of f^{-1} to your sketch. (It is not necessary to find a formula for f^{-1}.) Identify the domain and range of f^{-1}.

7.

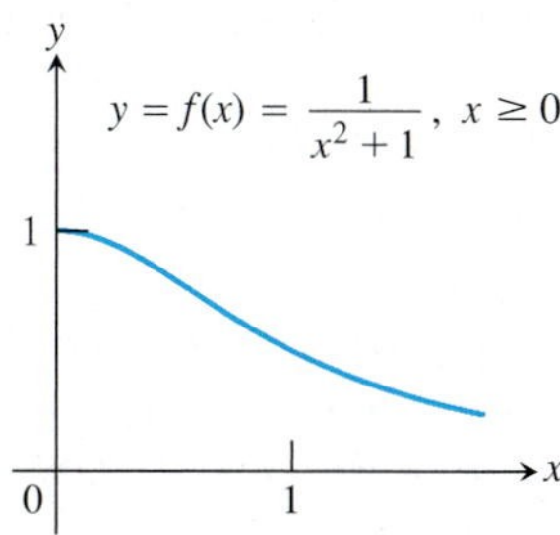

8.

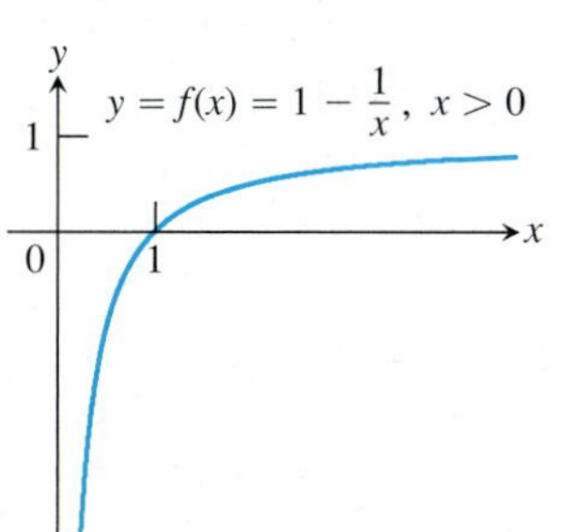

9.

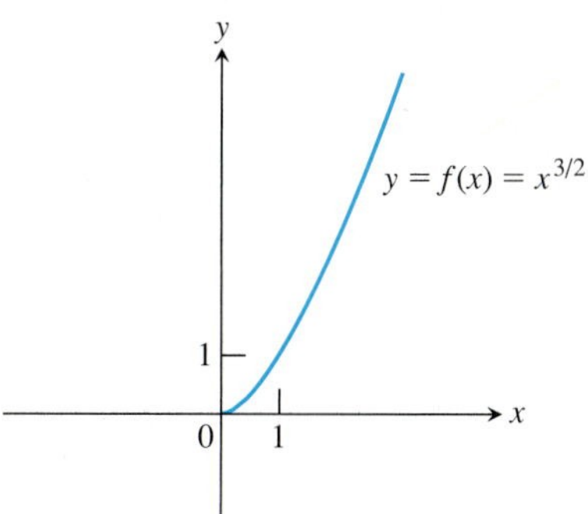

10.

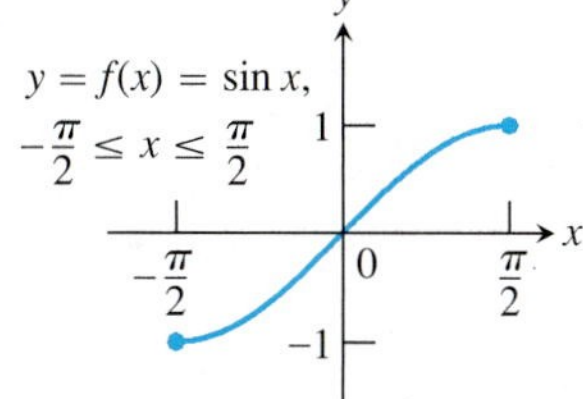

Formulas for Inverse Functions

Each of Exercises 11–16 gives a formula for a function $y = f(x)$ and shows the graphs of f and f^{-1}. Find a formula for f^{-1} in each case.

11. $f(x) = x^2 + 1, \quad x \geq 0$

12. $f(x) = x^2, \quad x \leq 0$

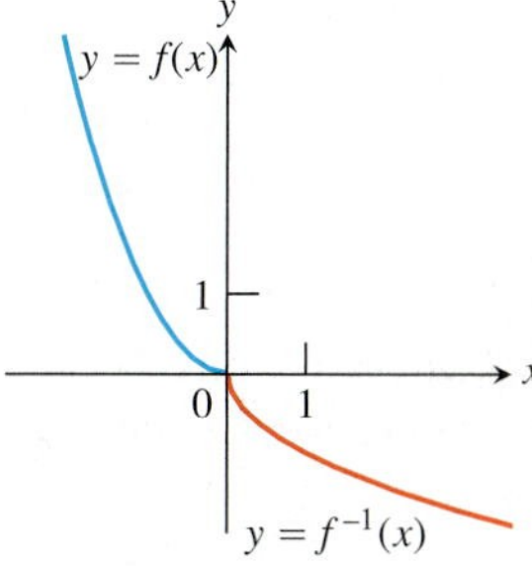

13. $f(x) = x^3 - 1$

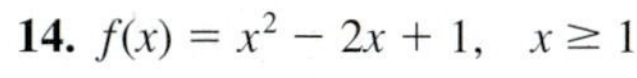

14. $f(x) = x^2 - 2x + 1, \quad x \geq 1$

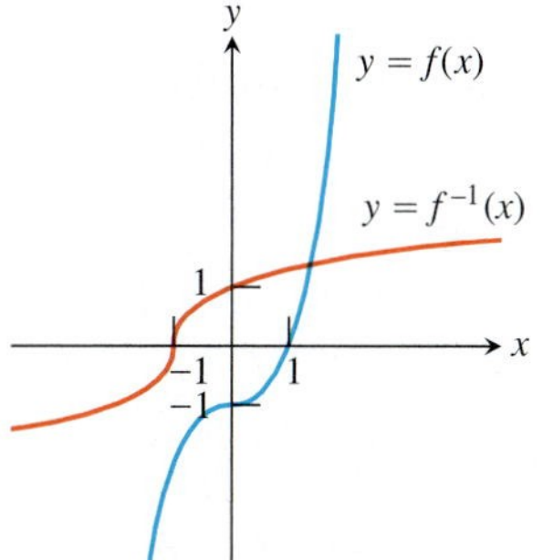

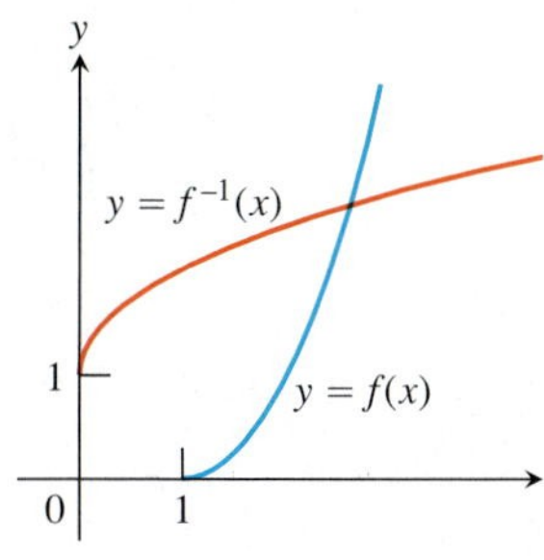

15. $f(x) = (x + 1)^2, \quad x \geq -1$

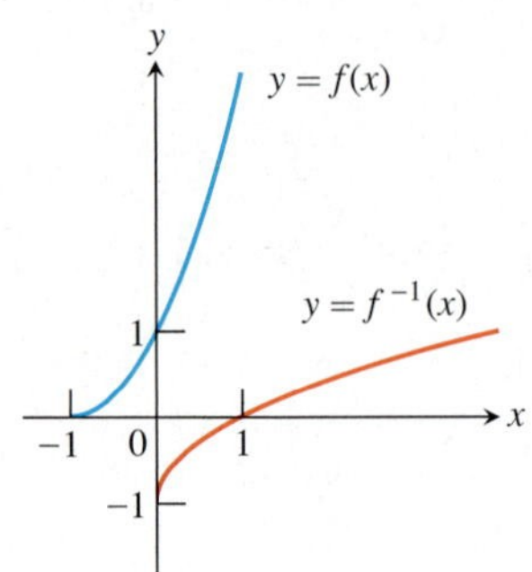

16. $f(x) = x^{2/3}, \quad x \geq 0$

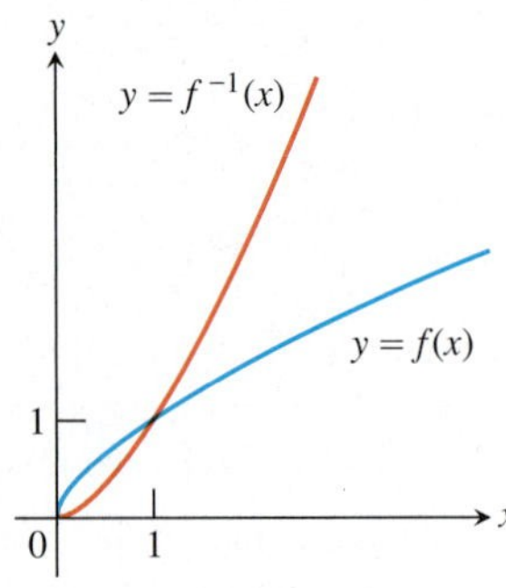

Finding Inverses

In Exercises 17–28, find f^{-1} and verify that

$$(f \circ f^{-1})(x) = (f^{-1} \circ f)(x) = x.$$

17. $f(x) = 2x + 3$

18. $f(x) = 5 - 4x$

19. $f(x) = x^3 - 1$

20. $f(x) = x^2 + 1, \quad x \geq 0$

21. $f(x) = x^2, \quad x \leq 0$

22. $f(x) = x^{2/3}, \quad x \geq 0$

23. $f(x) = -(x - 2)^2, \quad x \leq 2$

24. $f(x) = x^2 + 2x + 1, \quad x \geq -1$

25. $f(x) = \dfrac{1}{x^2}, \quad x > 0$

26. $f(x) = \dfrac{1}{x^3}$

27. $f(x) = \dfrac{2x + 1}{x + 3}$

28. $f(x) = \dfrac{x + 3}{x - 2}$

"Naturalizing" Exponential and Logarithmic Functions

In Exercises 29 and 30, express the exponential function as a power of e. Find the **(a)** domain and **(b)** range.

29. $y = 3^x - 1$

30. $y = 4^{x+1}$

In Exercises 31 and 32, express the function in terms of the natural logarithm. Find the **(a)** domain and **(b)** range. **(c)** Sketch the graph.

31. $y = 1 - (\ln 3) \log_3 x$

32. $y = (\ln 10) \log (x + 2)$

Solving Equations for Exponents

In Exercises 33–36, solve the equation algebraically. If you have a graphing calculator or computer grapher, support your solution graphically.

33. $(1.045)^t = 2$

34. $e^{0.05t} = 3$

35. $e^x + e^{-x} = 3$

36. $2^x + 2^{-x} = 5$

Solving Equations Containing Logarithmic Terms

In Exercises 37 and 38, solve for y.

37. $\ln y = 2t + 4$

38. $\ln (y - 1) - \ln 2 = x + \ln x$

Theory and Applications

39. Find a formula for f^{-1} and verify that $(f \circ f^{-1})(x) = (f^{-1} \circ f)(x) = x$.

(a) $f(x) = \dfrac{100}{1 + 2^{-x}}$ **(b)** $f(x) = \dfrac{50}{1 + 1.1^{-x}}$

40. *Inverse functions* We suggest that students *work in groups of two or three.*

Let $y = f(x) = mx + b, \quad m \neq 0.$

(a) *Writing to Learn* Give a convincing argument that f is a one-to-one function.

(b) Find a formula for the inverse of f. How are the slopes of the graphs of f and f^{-1} related?

(c) If the graphs of two functions are parallel lines with a nonzero slope, what can you say about the graphs of the inverses of the functions?

(d) If the graphs of two functions are perpendicular lines with a nonzero slope, what can you say about the graphs of the inverses of the functions?

41. *Radioactive decay* The half-life of a certain radioactive substance is 12 hours. There are 8 grams present initially.

(a) Express the amount of substance remaining as a function of time t.

(b) When will there be 1 gram remaining?

42. *Doubling your money* Determine how much time is required for a \$500 investment to double in value if interest is earned at the rate of 4.75% compounded annually.

43. *Population growth* The population of Glenbrook is 375,000 and is increasing at the rate of 2.25% per year. Predict when the population will be 1 million.

44. *Audio amplifiers* By what factor k do you have to multiply the intensity of I of the sound from your audio amplifier to add 10 db to the sound level?

45. *Audio amplifiers* You multiplied the intensity of the sound of your audio system by a factor of 10. By how many decibels did this increase the sound level?

46. *Radon-222* The decay equation for radon-222 gas is known to be $y = y_0 e^{-0.18t}$, with t in days. About how long will it take the radon in a sealed sample of air to fall to 90% of its original value?

Solving Equations and Comparing Functions

T In Exercises 47–50, use a graphing calculator to find the points of intersection of the two curves. Round your answers to 2 decimal places.

47. $y = 2x - 3$, $y = 5$

48. $y = -3x + 5$, $y = -3$

49. (a) $y = 2^x$, $y = 3$ (b) $y = 2^x$, $y = -1$

50. (a) $y = e^{-x}$, $y = 4$ (b) $y = e^{-x}$, $y = -1$

T For each of the function pairs in Exercises 51–54:

(a) Graph f and g together in a square window.

(b) Graph $f \circ g$.

(c) Graph $g \circ f$.

What can you conclude from the graphs?

51. $f(x) = x^3$, $g(x) = x^{1/3}$

52. $f(x) = x$, $g(x) = 1/x$

53. $f(x) = 3x$, $g(x) = x/3$

54. $f(x) = e^x$, $g(x) = \ln x$

T **55.** *Supporting the product rule* Let $y_1 = \ln(ax)$, $y_2 = \ln x$, and $y_3 = y_1 - y_2$.

(a) Graph y_1 and y_2 for $a = 2, 3, 4$, and 5. How do the graphs of y_1 and y_2 appear to be related?

(b) Support your finding by graphing y_3.

(c) Confirm your finding algebraically.

T **56.** *Supporting the quotient rule* Let $y_1 = \ln(x/a)$, $y_2 = \ln x$, $y_3 = y_2 - y_1$, and $y_4 = e^{y_3}$.

(a) Graph y_1 and y_2 for $a = 2, 3, 4$, and 5. How are the graphs of y_1 and y_2 related?

(b) Graph y_3 for $a = 2, 3, 4$, and 5. Describe the graphs.

(c) Graph y_4 for $a = 2, 3, 4$, and 5. Compare the graphs to the graph of $y = a$.

(d) Use $e^{y_3} = e^{y_2 - y_1} = a$ to solve for y_1.

T **57.** The equation $x^2 = 2^x$ has three solutions: $x = 2$, $x = 4$, and one other. Estimate the third solution as accurately as you can by graphing.

T **58.** Could $x^{\ln 2}$ possibly be the same as $2^{\ln x}$ for $x > 0$? Graph the two functions and explain what you see.

Logarithmic Regression Analysis: Oil Production

See page 5 for an introduction to regression analysis with a calculator.

T **59.** *Oil production in Indonesia* Table 15 shows the number of metric tons of oil produced by Indonesia for three different years.

(a) Use a calculator or computer to find the natural logarithm regression equation $y = a + b \ln x$ for the data in Table 15 and use it to estimate the number of metric tons of oil produced by Indonesia in 1982 and 2000. Let $x = 60$ represent 1960, $x = 70$ represent 1970, and so forth.

(b) Superimpose the graph of the logarithmic regression equation on a scatter plot of the data.

(c) Use the graph of the regression equation to predict the metric tons of oil production in 1982 and 2000.

Table 15 Indonesia's oil production

Year	Metric tons (millions)
1960	20.56
1970	42.10
1990	70.10

Source: Statesman's Yearbook, 129th ed. (London: The Macmillan Press, Ltd., 1992).

T **60.** *Oil production in Saudi Arabia*

(a) Find a natural logarithm regression equation for the data in Table 16.

(b) Estimate the number of metric tons of oil produced by Saudi Arabia in 1975.

(c) Predict when Saudi Arabian oil production will reach 400 million metric tons.

Table 16 Saudi Arabian oil production

Year	Metric tons (millions)
1960	61.09
1970	176.85
1990	321.93

Source: The Statesman's Yearbook, 129th ed. (London: The Macmillan Press, Ltd., 1992).

5 Trigonometric Functions and Their Inverses

Radian Measure • Graphs of Trigonometric Functions • Values of Trigonometric Functions • Periodicity • Even and Odd Trigonometric Functions • Transformations of Trigonometric Graphs • Identities • The Law of Cosines • Inverse Trigonometric Functions • Identities Involving Arc Sine and Arc Cosine

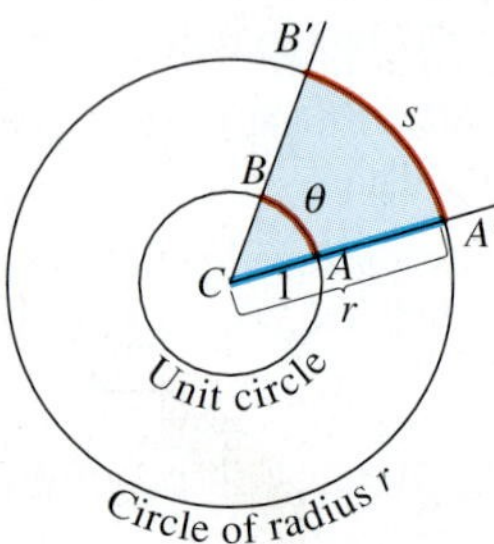

FIGURE 37 The radian measure of angle ACB is the length θ of arc AB on the unit circle centered at C. The value of θ can be found from any other circle, however, as the ratio s/r.

This section reviews the basic trigonometric functions and their inverses. The trigonometric functions are important because they are periodic, or repeating. They can therefore model many naturally occurring periodic processes such as daily temperature fluctuations in Earth's atmosphere, the wave behavior of musical notes, blood pressure in a heart, and the water level in a tidal basin.

Inverse trigonometric functions arise when we want to calculate angles from side measurements in triangles. You will see their usefulness in calculus in Chapters 6 and 7.

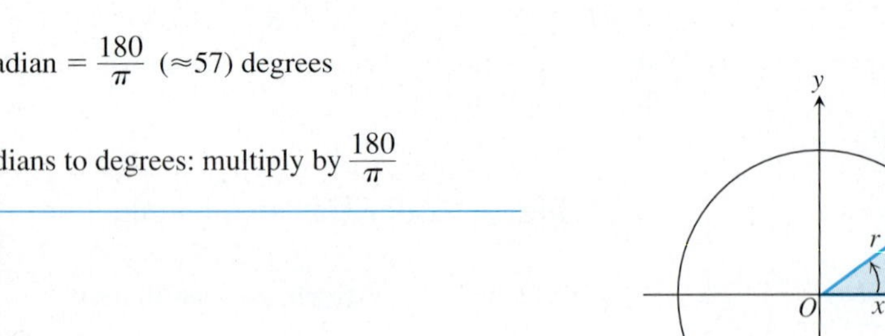

Conversion Formulas

1 degree $= \frac{\pi}{180}$ ($\approx$0.02) radians

Degrees to radians: multiply by $\frac{\pi}{180}$

1 radian $= \frac{180}{\pi}$ ($\approx$57) degrees

Radians to degrees: multiply by $\frac{180}{\pi}$

Radian Measure

The **radian measure** of the angle ACB at the center of the unit circle (Figure 37) equals the length of the arc that ACB cuts from the unit circle.

When an angle of measure θ is placed in *standard position* at the center of a circle of radius r (Figure 38), the six basic trigonometric functions of θ are defined as follows:

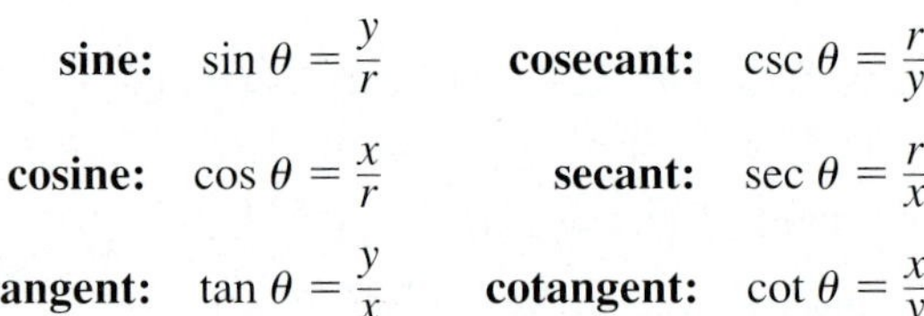

sine: $\sin\theta = \frac{y}{r}$ **cosecant:** $\csc\theta = \frac{r}{y}$

cosine: $\cos\theta = \frac{x}{r}$ **secant:** $\sec\theta = \frac{r}{x}$

tangent: $\tan\theta = \frac{y}{x}$ **cotangent:** $\cot\theta = \frac{x}{y}$

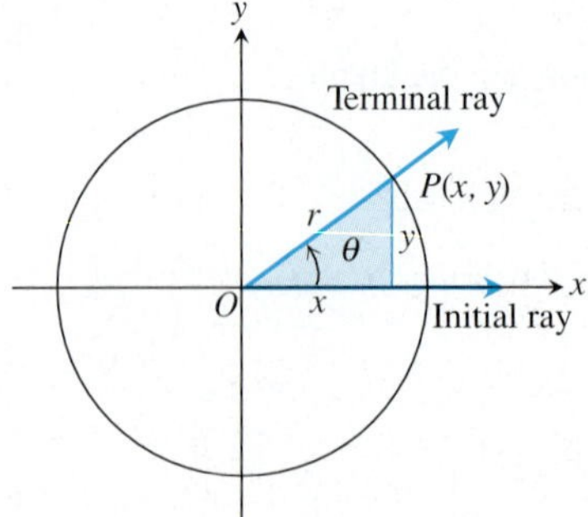

FIGURE 38 An angle θ in standard position.

Graphs of Trigonometric Functions

When we graph trigonometric functions in the coordinate plane, we usually denote the independent variable (radians) by x instead of θ (Figure 39).

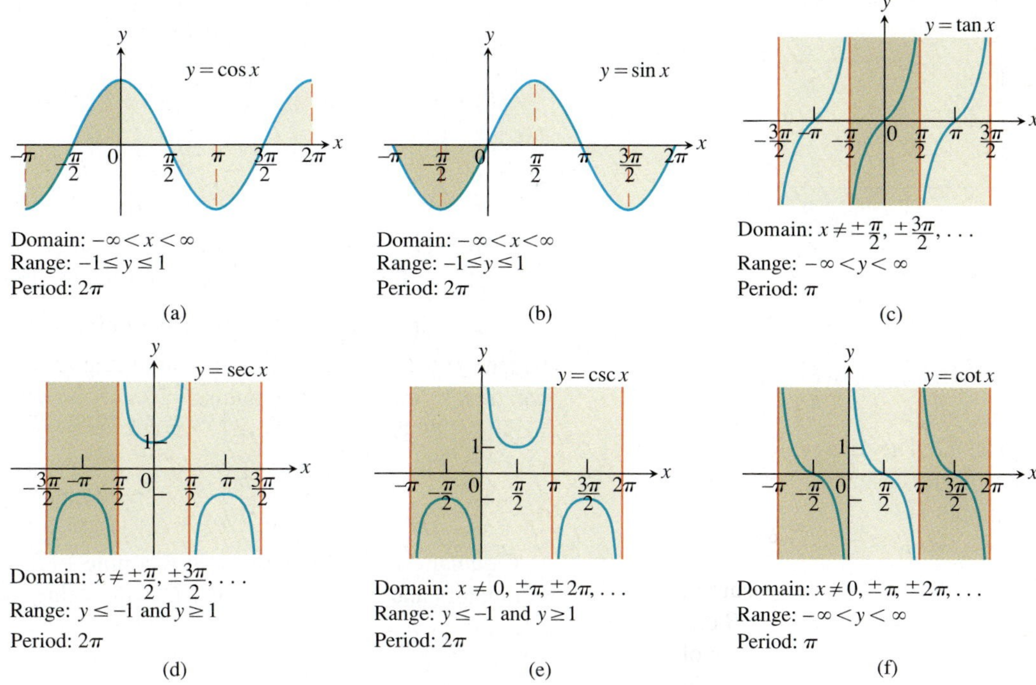

FIGURE 39 Graphs of the (a) cosine, (b) sine, (c) tangent, (d) secant, (e) cosecant, and (f) cotangent functions using radian measure.

Values of Trigonometric Functions

If the circle in Figure 40 has radius $r = 1$, the equations defining $\sin\theta$ and $\cos\theta$ become

$$\cos\theta = x, \qquad \sin\theta = y.$$

We can then calculate the values of the cosine and sine directly from the coordinates of P, if we happen to know them, or indirectly from the acute reference triangle made by dropping a perpendicular from P to the x-axis (Figure 41). We read the

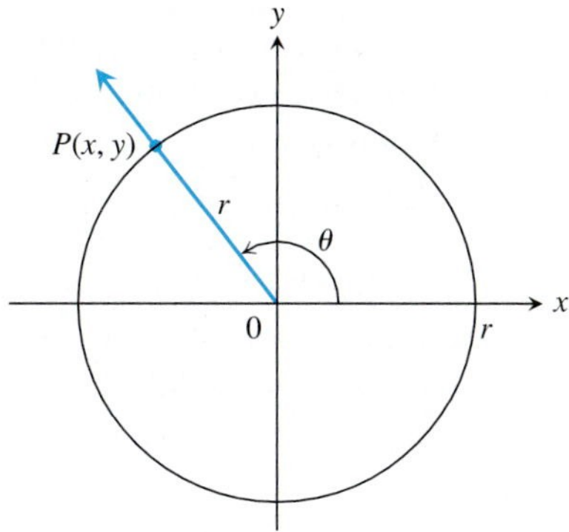

FIGURE 40 The trigonometric functions of a general angle θ are defined in terms of x, y, and r.

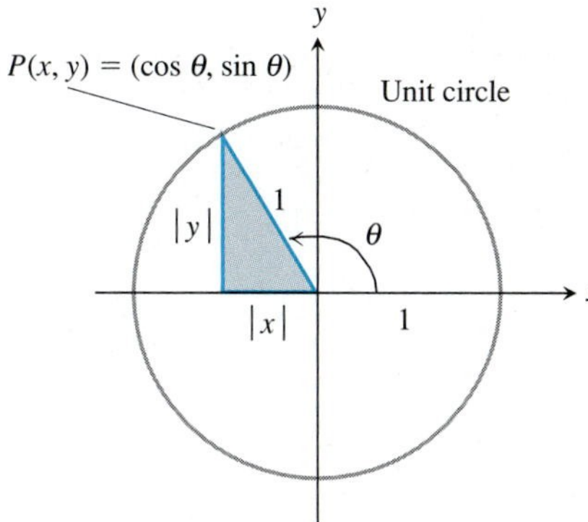

FIGURE 41 The acute reference triangle for an angle θ.

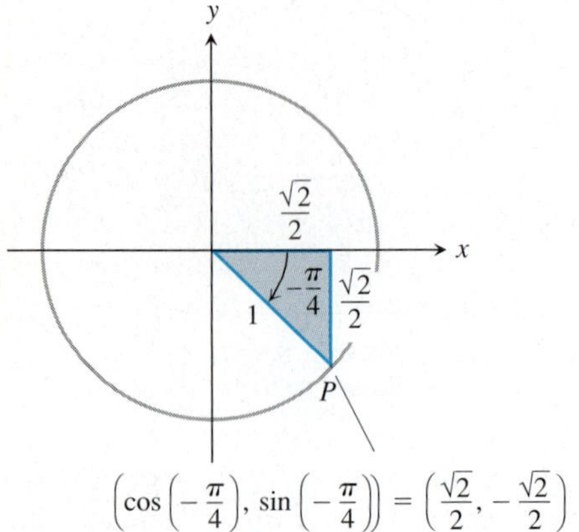

FIGURE 42 The triangle for calculating the sine and cosine of $-\pi/4$ radians. (Example 1)

magnitudes of x and y from the triangle's sides. The signs of x and y are determined by the quadrant in which the triangle lies.

Example 1 Finding Sine and Cosine Values

Find the sine and cosine of $-\pi/4$ radians.

Solution

Step 1: Draw the angle in standard position in the unit circle and write in the lengths of the sides of the reference triangle (Figure 42).

Step 2: Find the coordinates of the point P where the angle's terminal ray cuts the circle:

$$\cos\left(-\frac{\pi}{4}\right) = x\text{-coordinate of } P = \frac{\sqrt{2}}{2},$$

$$\sin\left(-\frac{\pi}{4}\right) = y\text{-coordinate of } P = -\frac{\sqrt{2}}{2}.$$

Calculations similar to those in Example 1 allow us to fill in Table 17. Most calculators and computers readily provide values of the trigonometric functions for angles given in either radians or degrees.

Table 17 Values of $\sin\theta$, $\cos\theta$, and $\tan\theta$ for selected values of θ

Degrees	**−180**	**−135**	**−90**	**−45**	**0**	**30**	**45**	**60**	**90**	**135**	**180**
θ (radians)	$-\pi$	$-3\pi/4$	$-\pi/2$	$-\pi/4$	**0**	$\pi/6$	$\pi/4$	$\pi/3$	$\pi/2$	$3\pi/4$	π
sin θ	0	$-\sqrt{2}/2$	-1	$-\sqrt{2}/2$	0	$1/2$	$\sqrt{2}/2$	$\sqrt{3}/2$	1	$\sqrt{2}/2$	0
cos θ	-1	$-\sqrt{2}/2$	0	$\sqrt{2}/2$	1	$\sqrt{3}/2$	$\sqrt{2}/2$	$1/2$	0	$-\sqrt{2}/2$	-1
tan θ	0	1		-1	0	$\sqrt{3}/3$	1	$\sqrt{3}$		-1	0

Periodicity

When an angle of measure θ and an angle of measure $\theta + 2\pi$ are in standard position, their terminal rays coincide. The two angles therefore have the same trigonometric function values:

$$\begin{aligned} \cos(\theta + 2\pi) &= \cos\theta \qquad \sin(\theta + 2\pi) = \sin\theta \qquad \tan(\theta + 2\pi) = \tan\theta \\ \sec(\theta + 2\pi) &= \sec\theta \qquad \csc(\theta + 2\pi) = \csc\theta \qquad \cot(\theta + 2\pi) = \cot\theta \end{aligned} \tag{1}$$

Similarly, $\cos(\theta - 2\pi) = \cos\theta$, $\sin(\theta - 2\pi) = \sin\theta$, and so on.

We see that the values of the trigonometric functions repeat at regular intervals. We describe this behavior by saying that the six basic trigonometric functions are *periodic*.

Definition Periodic Function, Period

A function $f(x)$ is **periodic** if there is a positive number p such that $f(x + p) = f(x)$ for every value of x. The smallest such value of p is the **period** of f.

Periods of Trigonometric Functions

Period π: $\tan(x + \pi) = \tan x$
$\cot(x + \pi) = \cot x$

Period 2π: $\sin(x + 2\pi) = \sin x$
$\cos(x + 2\pi) = \cos x$
$\sec(x + 2\pi) = \sec x$
$\csc(x + 2\pi) = \csc x$

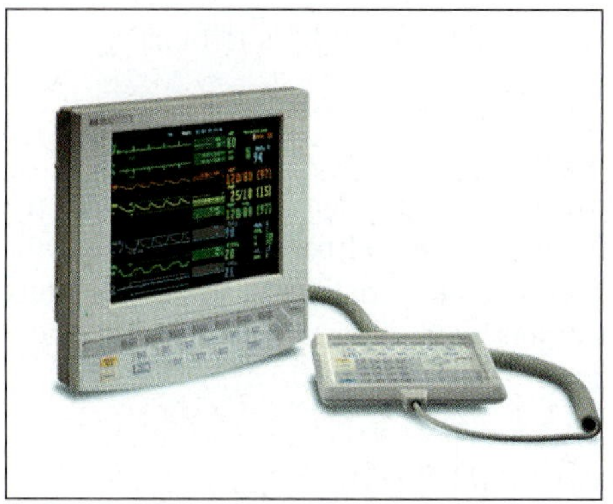

FIGURE 43 This compact patient monitor shows several periodic functions associated with the human body. This device dynamically monitors electrocardiogram (ECG) and respiration, and blood pressure.

As we can see in Figure 39, the functions $\cos x$, $\sin x$, $\sec x$, and $\csc x$ are periodic with period 2π. The functions $\tan x$ and $\cot x$ are periodic with period π.

Periodic functions are important because much of the behavior we study in science is periodic (Figure 43). Brain waves and heartbeats are periodic, as are household voltage and electric current. The electromagnetic field that heats food in a microwave oven is periodic, as are cash flows in seasonal businesses and the behavior of rotational machinery. The seasons are periodic, as is the weather. The phases of the moon are periodic, as are the motions of the planets. There is strong evidence that the ice ages are periodic, with a period of 90,000 to 100,000 years.

Why are trigonometric functions so important in the study of things periodic? The answer lies in a surprising and beautiful theorem from advanced calculus that says that every periodic function we want to use in mathematical modeling can be written as an algebraic combination of sines and cosines. Once we learn the calculus of sines and cosines, we can model the mathematical behavior of most periodic phenomena.

Even and Odd Trigonometric Functions

The graphs in Figure 39 suggest that $\cos x$ and $\sec x$ are even functions because their graphs are symmetric about the y-axis. The other four basic trigonometric functions are odd.

Example 2 Confirming Even and Odd

Show that the cosine is an even function and the sine is odd.

Solution From Figure 44, it follows that

$$\cos(-\theta) = \frac{x}{r} = \cos\theta, \qquad \sin(-\theta) = \frac{-y}{r} = -\sin\theta,$$

so the cosine is an even function and the sine is odd.

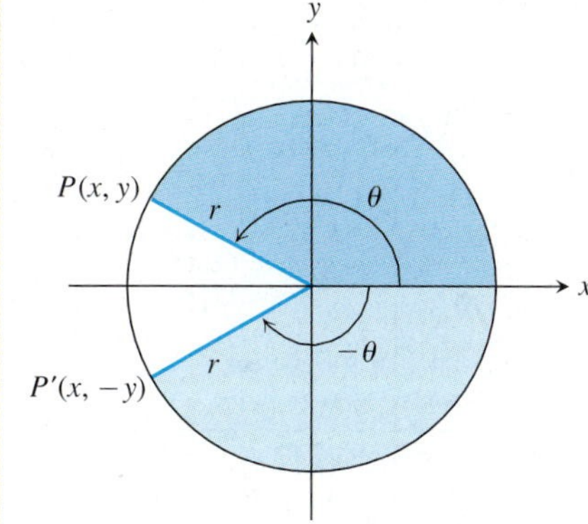

FIGURE 44 Angles of opposite sign. (Example 2)

We can use the results of Example 2 to establish the *parity* of the other four basic trigonometric functions. For example,

$$\sec(-\theta) = \frac{1}{\cos(-\theta)} = \frac{1}{\cos\theta} = \sec\theta,$$

$$\tan(-\theta) = \frac{\sin(-\theta)}{\cos(-\theta)} = \frac{-\sin\theta}{\cos\theta} = -\tan\theta,$$

so secant is an even function and tangent is odd. Similar steps will show that cosecant and cotangent are odd functions.

Transformations of Trigonometric Graphs

The rules for shifting, stretching, shrinking, and reflecting the graph of a function apply to the trigonometric functions. The following diagram will remind you of the controlling parameters.

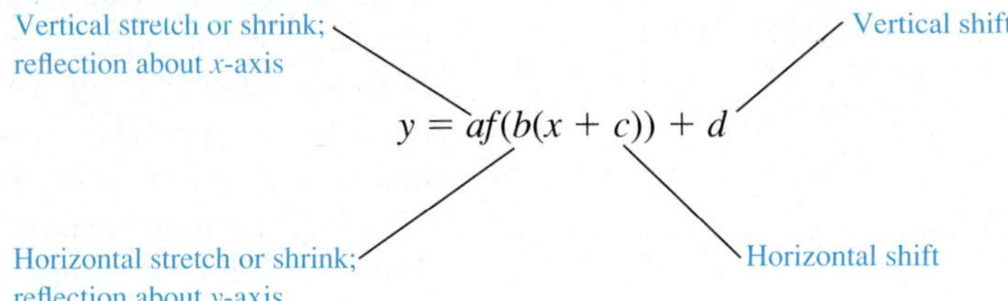

Example 3 Modeling Temperature in Alaska

The builders of the Trans-Alaska Pipeline used insulated pads to keep the pipeline heat from melting the permanently frozen soil beneath. To design the pads, it was necessary to take into account the variation in air temperature throughout the year. The variation was represented in the calculations by a general sine function or **sinusoid** of the form

$$f(x) = A \sin\left[\frac{2\pi}{B}(x - C)\right] + D, \tag{2}$$

where $|A|$ is the *amplitude*, $|B|$ is the *period*, C is the *horizontal shift*, and D is the *vertical shift* (Figure 45).

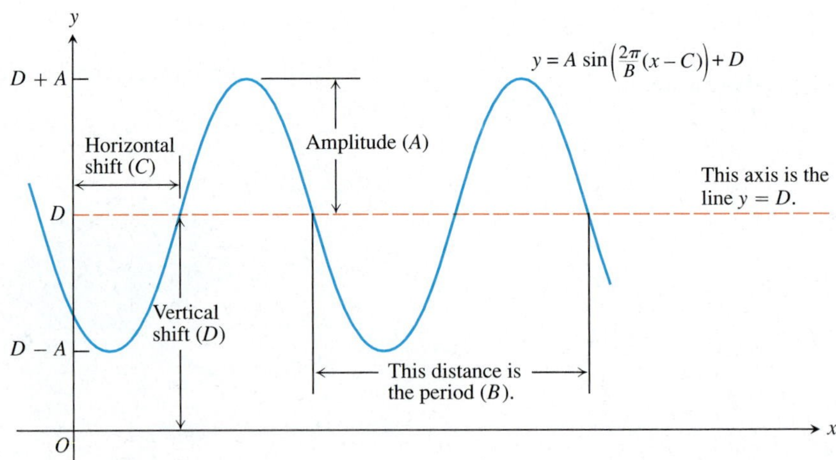

FIGURE 45 The general sine curve $y = A \sin[(2\pi/B)(x - C)] + D$, shown for A, B, C, and D positive. (Example 3)

Figure 46 shows how to use such a function to represent temperature data. The data points in the figure are plots of the mean daily air temperatures for Fairbanks, Alaska, based on records of the National Weather Service from 1941 to 1970. The sine function used to fit the data is

$$f(x) = 37 \sin\left[\frac{2\pi}{365}(x - 101)\right] + 25,$$

where f is temperature in degrees Fahrenheit and x is the number of the day counting from the beginning of the year. The fit is remarkably good.

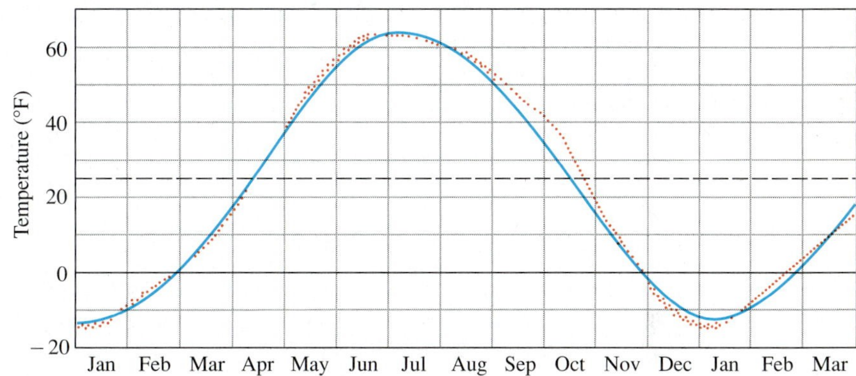

Source: "Is the Curve of Temperature Variation a Sine Curve?" by B. M. Lando and C. A. Lando, *The Mathematics Teacher,* Vol. 7, No. 6 (September 1977), Fig. 2, p. 53.

FIGURE 46 Normal mean air temperatures for Fairbanks, Alaska, plotted as data points (red). The approximating sine function (blue) is

$$f(x) = 37 \sin [(2\pi/365)(x - 101)] + 25 .$$

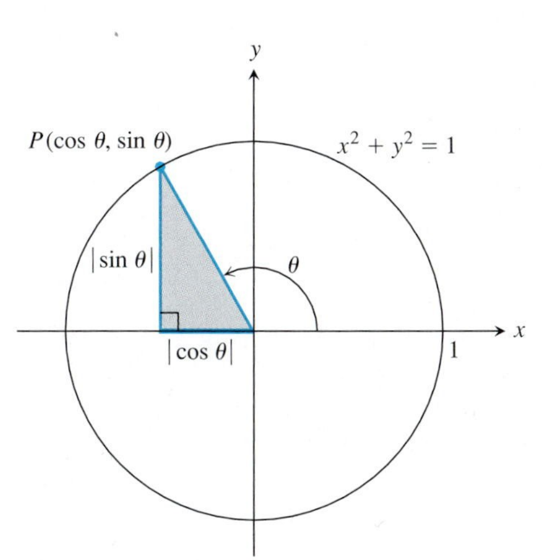

FIGURE 47 The reference triangle for a general angle θ.

Identities

Applying the Pythagorean Theorem to the reference right triangle we obtain by dropping a perpendicular from the point $P(\cos\theta, \sin\theta)$ on the unit circle to the x-axis (Figure 47) gives

$$\cos^2\theta + \sin^2\theta = 1 . \qquad (3)$$

This equation, true for all values of θ, is the most frequently used identity in trigonometry. Dividing this identity in turn by $\cos^2\theta$ and $\sin^2\theta$ gives

$$1 + \tan^2\theta = \sec^2\theta ,$$
$$1 + \cot^2\theta = \csc^2\theta .$$

The following formulas hold for all angles A and B.

All the trigonometric identities you will need in this book derive from Equations (3) and (4).

Angle Sum Formulas

$$\cos(A + B) = \cos A \cos B - \sin A \sin B$$
$$\sin(A + B) = \sin A \cos B + \cos A \sin B \qquad (4)$$

Substituting θ for both A and B in the angle sum formulas gives two more useful identities.

Instead of memorizing Equations (5) you might find it helpful to remember Equations (4) and then recall where they came from.

Double-Angle Formulas

$$\cos 2\theta = \cos^2\theta - \sin^2\theta$$
$$\sin 2\theta = 2 \sin\theta \cos\theta \tag{5}$$

The Law of Cosines

If a, b, and c are sides of a triangle ABC and if θ is the angle opposite c, then

$$c^2 = a^2 + b^2 - 2ab\cos\theta . \tag{6}$$

This equation is called the **law of cosines.**

We can see why the law holds if we introduce coordinate axes with the origin at C and the positive x-axis along one side of the triangle, as in Figure 48. The coordinates of A are $(b, 0)$; the coordinates of B are $(a\cos\theta, a\sin\theta)$. The square of the distance between A and B is therefore

$$\begin{aligned} c^2 &= (a\cos\theta - b)^2 + (a\sin\theta)^2 \\ &= a^2\underbrace{(\cos^2\theta + \sin^2\theta)}_{1} + b^2 - 2ab\cos\theta \\ &= a^2 + b^2 - 2ab\cos\theta . \end{aligned}$$

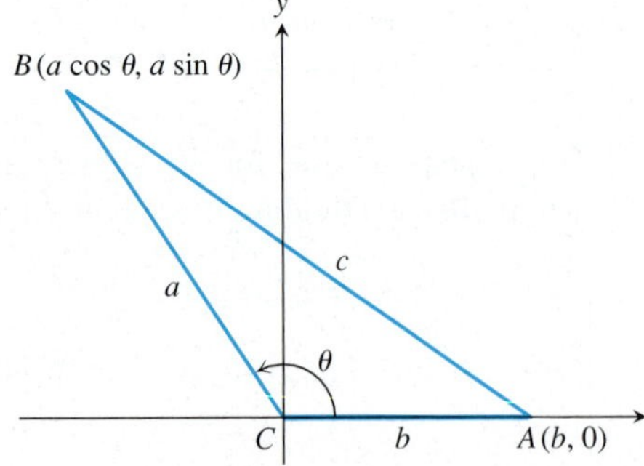

FIGURE 48 The square of the distance between A and B gives the law of cosines.

The law of cosines generalizes the Pythagorean Theorem. If $\theta = \pi/2$, then $\cos\theta = 0$ and $c^2 = a^2 + b^2$.

Inverse Trigonometric Functions

None of the six basic trigonometric functions graphed in Figure 39 is one-to-one. These functions do not have inverses. In each case, however, the domain can be restricted to produce a new function that does have an inverse, as illustrated in Example 4.

Example 4 Restricting the Domain of the Sine

Show that the function $y = \sin x$, $-\pi/2 \le x \le \pi/2$, is one-to-one, and graph its inverse.

Solution Figure 49a shows the graph of this restricted sine function. This function is one-to-one because it does not repeat any output values. It therefore has an inverse, which we graph in Figure 49b by interchanging the ordered pairs as in Section 4.

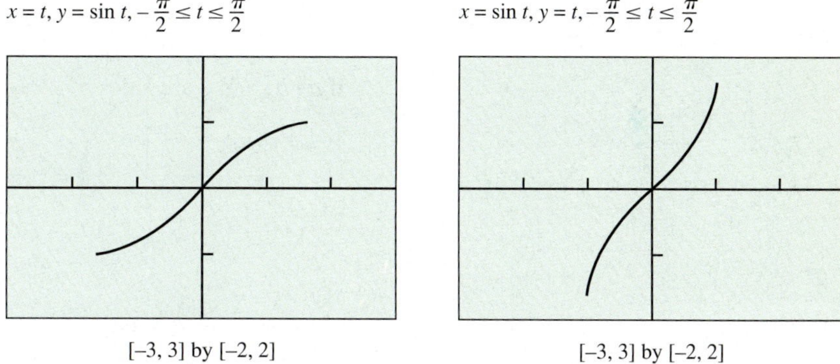

FIGURE 49 (a) A restricted sine function and (b) its inverse. The graphs were generated by a graphing calculator in parametric mode. See Section 6 for a review of parametric equations. (Example 4)

The inverse of the restricted sine function of Example 4 is called the *inverse sine function*. The inverse sine of x is the angle in the interval $[-\pi/2, \pi/2]$ whose sine is x. It is denoted by $\sin^{-1} x$ or $\arcsin x$. Either notation is read "arcsine of x" or "the inverse sine of x."

The domains of the other basic trigonometric functions can also be restricted to produce a function with an inverse. The domains and ranges of the resulting inverse functions become parts of their definitions.

Definitions Inverse Trigonometric Functions

Function	Domain	Range
$y = \cos^{-1} x$	$-1 \le x \le 1$	$0 \le y \le \pi$
$y = \sin^{-1} x$	$-1 \le x \le 1$	$-\frac{\pi}{2} \le y \le \frac{\pi}{2}$
$y = \tan^{-1} x$	$-\infty < x < \infty$	$-\frac{\pi}{2} < y < \frac{\pi}{2}$
$y = \sec^{-1} x$	$\lvert x \rvert \ge 1$	$0 \le y \le \pi, \quad y \ne \frac{\pi}{2}$
$y = \csc^{-1} x$	$\lvert x \rvert \ge 1$	$-\frac{\pi}{2} \le y \le \frac{\pi}{2}, \quad y \ne 0$
$y = \cot^{-1} x$	$-\infty < x < \infty$	$0 < y < \pi$

The graphs of the six inverse trigonometric functions are shown in Figure 50.

Domain: $-1 \leq x \leq 1$
Range: $0 \leq y \leq \pi$

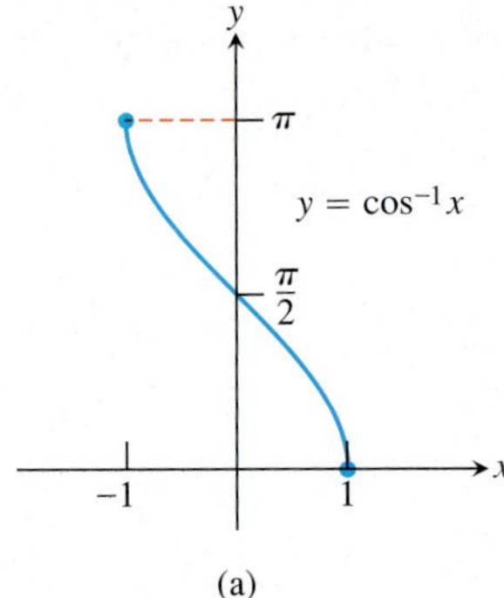

Domain: $-1 \leq x \leq 1$
Range: $-\frac{\pi}{2} \leq y \leq \frac{\pi}{2}$

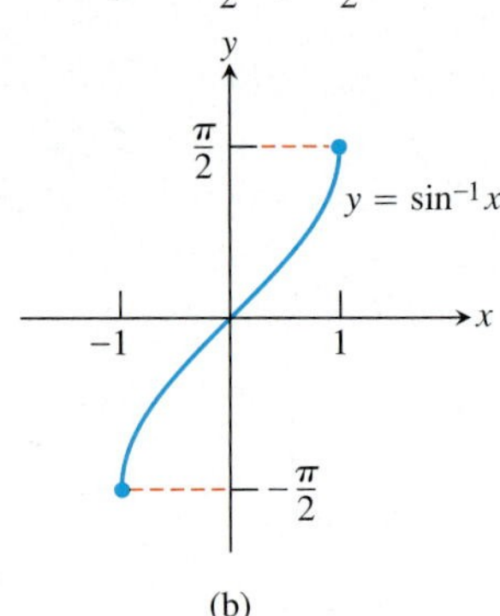

Domain: $-\infty < x < \infty$
Range: $-\frac{\pi}{2} < y < \frac{\pi}{2}$

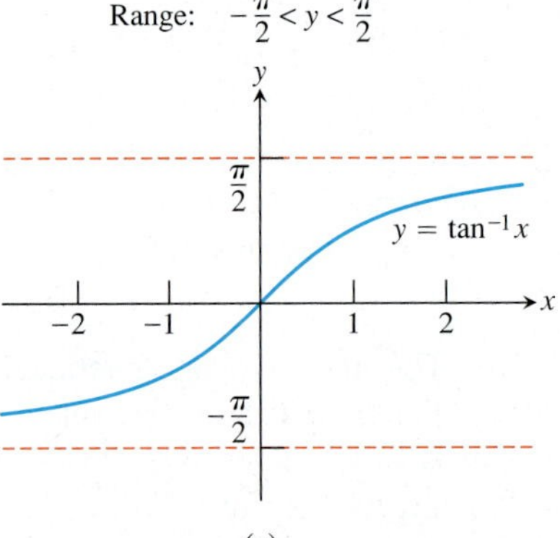

Domain: $x \leq -1$ or $x \geq 1$
Range: $0 \leq y \leq \pi,\ y \neq \frac{\pi}{2}$

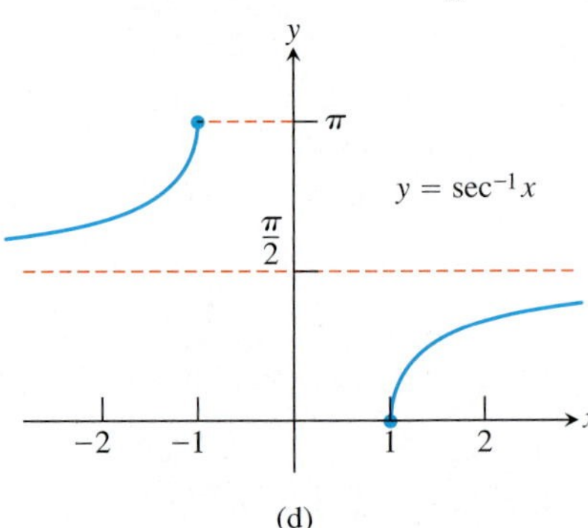

Domain: $x \leq -1$ or $x \geq 1$
Range: $-\frac{\pi}{2} \leq y \leq \frac{\pi}{2},\ y \neq 0$

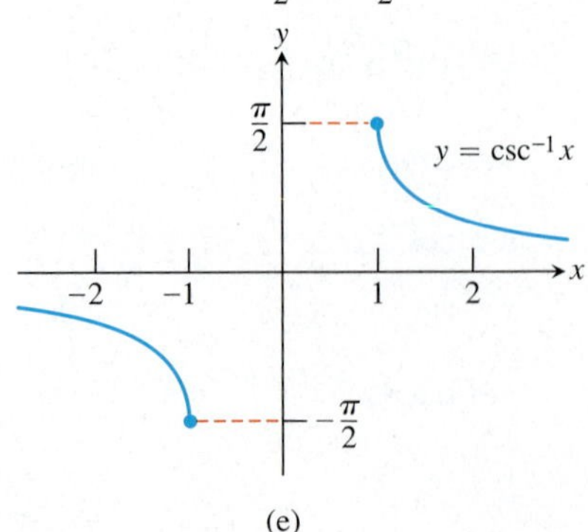

Domain: $-\infty < x < \infty$
Range: $0 < y < \pi$

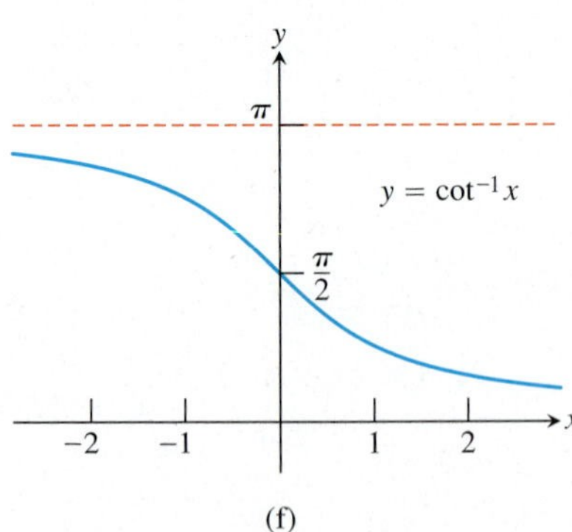

FIGURE 50 Graphs of (a) $y = \cos^{-1} x$, (b) $y = \sin^{-1} x$, (c) $y = \tan^{-1} x$, (d) $y = \sec^{-1} x$, (e) $y = \csc^{-1} x$, and (f) $y = \cot^{-1} x$.

The "Arc" in Arc Sine and Arc Cosine

The accompanying figure gives a geometric interpretation of $y = \sin^{-1} x$ and $y = \cos^{-1} x$ for radian angles in the first quadrant. For a unit circle, the equation $s = r\theta$ becomes $s = \theta$, so central angles and the arcs they subtend have the same measure. If $x = \sin y$, then, in addition to being the angle whose sine is x, y is also the length of arc on the unit circle that subtends an angle whose sine is x. So we call y "the arc whose sine is x."

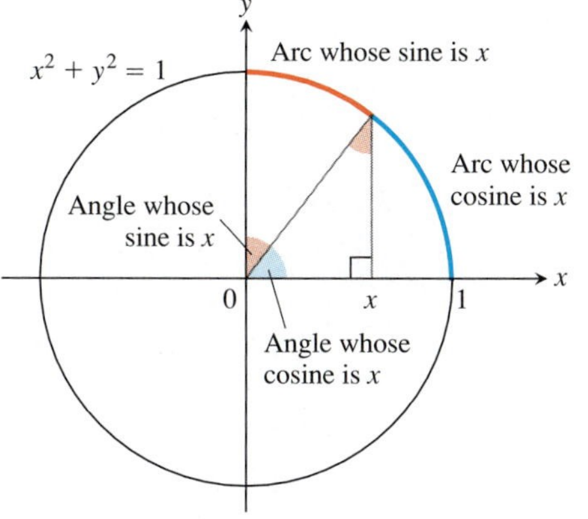

The domain and ranges (where appropriate) of the inverse functions are chosen so that the functions will have the following relationships:

$$\sec^{-1} x = \cos^{-1}(1/x),$$
$$\csc^{-1} x = \sin^{-1}(1/x),$$
$$\cot^{-1} x = \pi/2 - \tan^{-1} x.$$

We use these relationships to find values of $\sec^{-1} x$, $\csc^{-1} x$, and $\cot^{-1} x$ on calculators that give only $\cos^{-1} x$, $\sin^{-1} x$, and $\tan^{-1} x$.

Example 5 Common Values of $\sin^{-1} x$

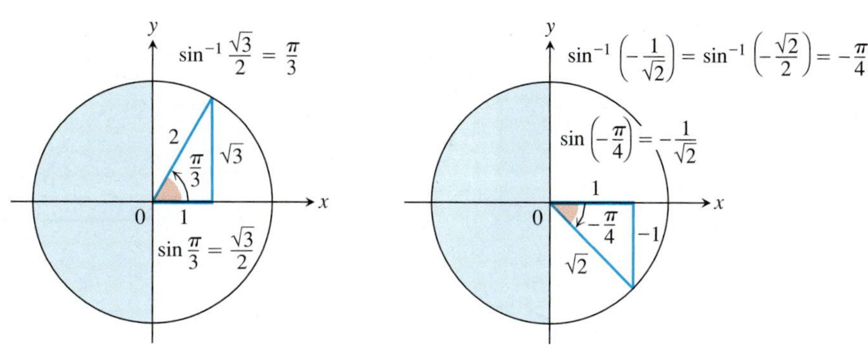

The angles come from the first and fourth quadrants because the range of $\sin^{-1} x$ is $[-\pi/2, \pi/2]$.

Example 6 Common Values of $\cos^{-1} x$

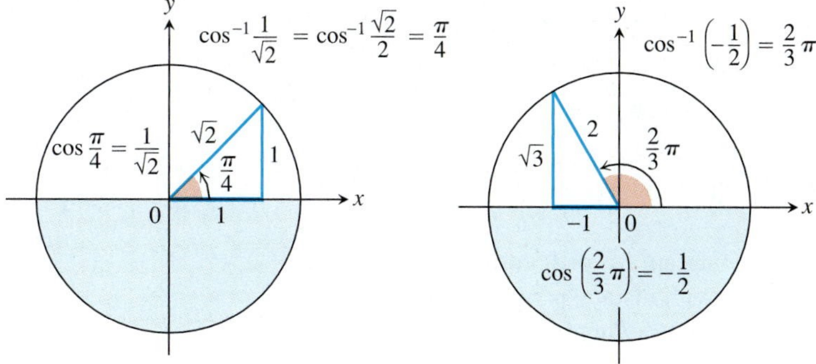

The angles come from the first and second quadrants because the range of $\cos^{-1} x$ is $[0, \pi]$.

x	$\sin^{-1} x$
$\sqrt{3}/2$	$\pi/3$
$\sqrt{2}/2$	$\pi/4$
$1/2$	$\pi/6$
$-1/2$	$-\pi/6$
$-\sqrt{2}/2$	$-\pi/4$
$-\sqrt{3}/2$	$-\pi/3$

x	$\cos^{-1} x$
$\sqrt{3}/2$	$\pi/6$
$\sqrt{2}/2$	$\pi/4$
$1/2$	$\pi/3$
$-1/2$	$2\pi/3$
$-\sqrt{2}/2$	$3\pi/4$
$-\sqrt{3}/2$	$5\pi/6$

x	$\tan^{-1} x$
$\sqrt{3}$	$\pi/3$
1	$\pi/4$
$\sqrt{3}/3$	$\pi/6$
$-\sqrt{3}/3$	$-\pi/6$
1	$-\pi/4$
$-\sqrt{3}$	$-\pi/3$

Example 7 Common Values of $\tan^{-1} x$

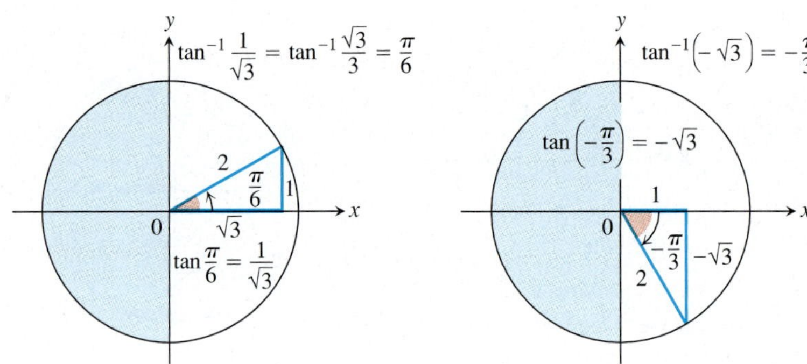

The angles come from the first and fourth quadrants because the range of $\tan^{-1} x$ is $(-\pi/2, \pi/2)$.

Example 8 Drift Correction

During an airplane flight from Chicago to St. Louis, the navigator determines that the plane is 12 mi off course, as shown in Figure 51. Find the angle a for a course parallel to the original, correct course, the angle b, and the correction angle $c = a + b$.

FIGURE 51 Diagram for drift correction (Example 8), with distances rounded to the nearest mile (drawing not to scale).

Solution

$$a = \sin^{-1} \frac{12}{180} \approx 0.067 \text{ radians} \approx 3.8°$$

$$b = \sin^{-1} \frac{12}{62} \approx 0.195 \text{ radians} \approx 11.2°$$

$$c = a + b \approx 15°.$$

Identities Involving Arc Sine and Arc Cosine

The graph of $y = \sin^{-1} x$ is symmetric about the origin as shown in Figure 50b. The arc sine is therefore an odd function:

$$\sin^{-1}(-x) = -\sin^{-1} x. \tag{7}$$

The graph of $y = \cos^{-1} x$ has no such symmetry. Instead, we can see from Figure 52 that the arc cosine of x satisfies the identity

$$\cos^{-1} x + \cos^{-1}(-x) = \pi, \tag{8}$$

or

$$\cos^{-1}(-x) = \pi - \cos^{-1} x. \tag{9}$$

And we can see from the triangle in Figure 53 that for $x > 0$,

$$\sin^{-1} x + \cos^{-1} x = \pi/2. \tag{10}$$

Equation (10) holds for the other values of x in $[-1, 1]$ as well.

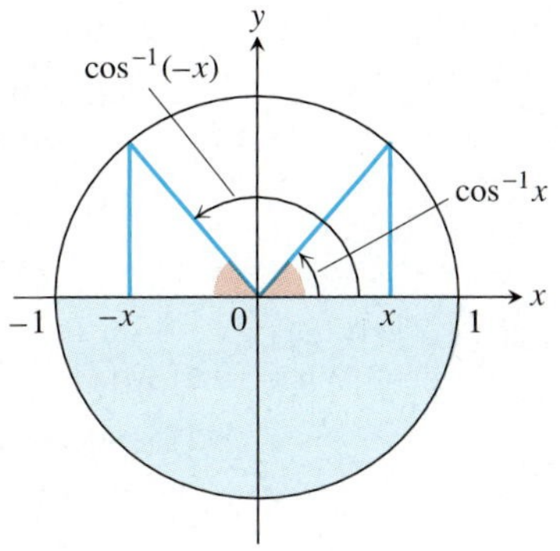

FIGURE 52 $\cos^{-1} x + \cos^{-1}(-x) = \pi$.

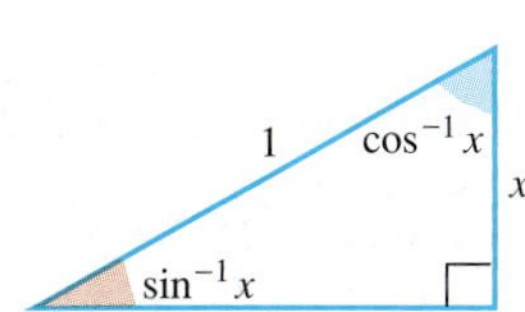

FIGURE 53 In this figure, $\sin^{-1} x + \cos^{-1} x = \pi/2$.

EXERCISES 5

Radians, Degrees, and Circular Arcs

1. On a circle of radius 10 m, how long is an arc that subtends a central angle of **(a)** $4\pi/5$ radians? **(b)** $110°$?

2. A central angle in a circle of radius 8 is subtended by an arc of length 10π. Find the angle's radian and degree measures.

Evaluating Trigonometric Functions

3. Copy and complete the following table of function values. If the function is undefined at a given angle, enter "UND." Do not use a calculator or tables.

θ	$-\pi$	$-2\pi/3$	0	$\pi/2$	$3\pi/4$
$\sin\theta$					
$\cos\theta$					
$\tan\theta$					
$\cot\theta$					
$\sec\theta$					
$\csc\theta$					

4. Copy and complete the following table of function values. If the function is undefined at a given angle, enter "UND." Do not use a calculator or tables.

θ	$-3\pi/2$	$-\pi/3$	$-\pi/6$	$\pi/4$	$5\pi/6$
$\sin\theta$					
$\cos\theta$					
$\tan\theta$					
$\cot\theta$					
$\sec\theta$					
$\csc\theta$					

In Exercises 5 and 6, the value of one of $\sin x$, $\cos x$, and $\tan x$ is given. Find the values of the other two in the specified interval.

5. **(a)** $\sin x = \dfrac{3}{5}, \quad x \text{ in } \left[\dfrac{\pi}{2}, \pi\right]$

(b) $\cos x = \dfrac{1}{3}, \quad x \text{ in } \left[-\dfrac{\pi}{2}, 0\right]$

6. **(a)** $\tan x = \frac{1}{2}, \quad x \text{ in } \left[\pi, \frac{3\pi}{2}\right]$

(b) $\sin x = -\frac{1}{2}, \quad x \text{ in } \left[\pi, \frac{3\pi}{2}\right]$

Graphing Trigonometric Functions

Graph the functions in Exercises 7–10. What is the period of each function?

7. **(a)** $\sin 2x$ **(b)** $\cos \pi x$

8. **(a)** $-\sin \frac{\pi x}{3}$ **(b)** $-\cos 2\pi x$

9. **(a)** $\cos\left(x - \frac{\pi}{2}\right)$ **(b)** $\sin\left(x + \frac{\pi}{2}\right)$

10. **(a)** $\sin\left(x - \frac{\pi}{4}\right) + 1$ **(b)** $\cos\left(x + \frac{\pi}{4}\right) - 1$

Graph the functions in Exercises 11 and 12 in the ts-plane (t-axis horizontal, s-axis vertical). What is the period of each function? What symmetries do the graphs have?

11. $s = \cot 2t$ **12.** $s = \sec\left(\frac{\pi t}{2}\right)$

Using the Angle Sum Formulas

In Exercises 13 and 14, express the given quantity in terms of $\sin x$ and $\cos x$.

13. **(a)** $\cos(\pi + x)$ **(b)** $\sin(2\pi - x)$

14. **(a)** $\sin\left(\frac{3\pi}{2} - x\right)$ **(b)** $\cos\left(\frac{3\pi}{2} + x\right)$

Use the angle sum formulas to derive the identities in Exercises 15 and 16.

15. **(a)** $\cos\left(x - \frac{\pi}{2}\right) = \sin x$

(b) $\cos(A - B) = \cos A \cos B + \sin A \sin B$

16. **(a)** $\sin\left(x + \frac{\pi}{2}\right) = \cos x$

(b) $\sin(A - B) = \sin A \cos B - \cos A \sin B$

17. What happens if you take $B = A$ in the identity $\cos(A - B) = \cos A \cos B + \sin A \sin B$? Does the result agree with something you already know?

18. What happens if you take $B = 2\pi$ in the angle sum formulas? Do the results agree with something you already know?

General Sine Curves

Identify A, B, C, and D in Equation (2) for the sine functions in Exercises 19 and 20 and sketch their graphs.

19. **(a)** $y = 2\sin(x + \pi) - 1$

(b) $y = \frac{1}{2}\sin(\pi x - \pi) + \frac{1}{2}$

20. **(a)** $y = -\frac{2}{\pi}\sin\left(\frac{\pi}{-2}t\right) + \frac{1}{\pi}$

(b) $y = \frac{L}{2\pi}\sin\frac{2\pi t}{L}, \quad L > 0$

21. *Temperature in Fairbanks, Alaska* Find the **(a)** amplitude, **(b)** period, **(c)** horizontal shift, and **(d)** vertical shift of the general sine function

$$f(x) = 37\sin\left(\frac{2\pi}{365}(x - 101)\right) + 25.$$

22. *Temperature in Fairbanks, Alaska* Use the equation in Exercise 21 to approximate the answers to the following questions about the temperature in Fairbanks, Alaska, shown in Figure 46. Assume that the year has 365 days.

(a) What are the highest and lowest mean daily temperatures shown?

(b) What is the average of the highest and lowest mean daily temperatures shown? Why is this average the vertical shift of the function?

Common Values of Inverse Trigonometric Functions

Use reference triangles like those in Examples 5–7 to find the angles in Exercises 23–26.

23. **(a)** $\tan^{-1} 1$ **(b)** $\tan^{-1}(-\sqrt{3})$ **(c)** $\tan^{-1}\left(\frac{1}{\sqrt{3}}\right)$

24. **(a)** $\sin^{-1}\left(\frac{-1}{2}\right)$ **(b)** $\sin^{-1}\left(\frac{1}{\sqrt{2}}\right)$ **(c)** $\sin^{-1}\left(\frac{-\sqrt{3}}{2}\right)$

25. **(a)** $\cos^{-1}\left(\frac{1}{2}\right)$ **(b)** $\cos^{-1}\left(\frac{-1}{\sqrt{2}}\right)$ **(c)** $\cos^{-1}\left(\frac{\sqrt{3}}{2}\right)$

26. **(a)** $\sec^{-1}(-\sqrt{2})$ **(b)** $\sec^{-1}\left(\frac{2}{\sqrt{3}}\right)$ **(c)** $\sec^{-1}(-2)$

Applications and Theory

27. You are sitting in a classroom next to the wall looking at the blackboard at the front of the room. The blackboard is 12 ft long and starts 3 ft from the wall you are sitting next to. Show that your viewing angle is

$$\alpha = \cot^{-1}\frac{x}{15} - \cot^{-1}\frac{x}{3}$$

if you are x ft from the front wall.

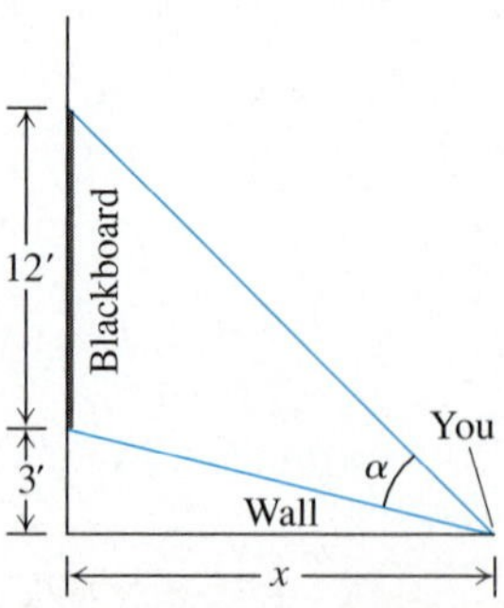

28. Find the angle α.

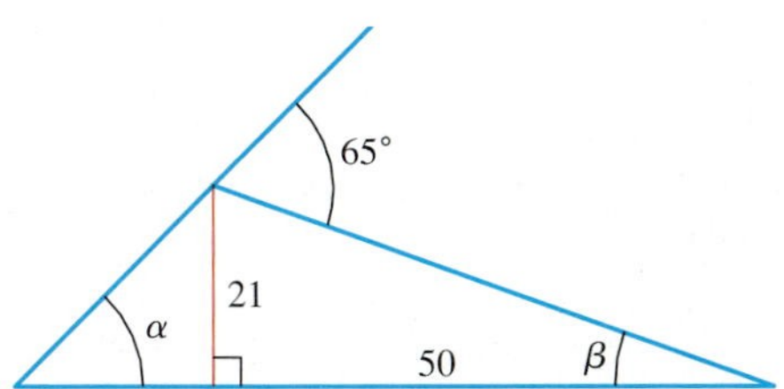

29. Apply the law of cosines to the triangle in the accompanying figure to derive a formula for $\cos(A - B)$.

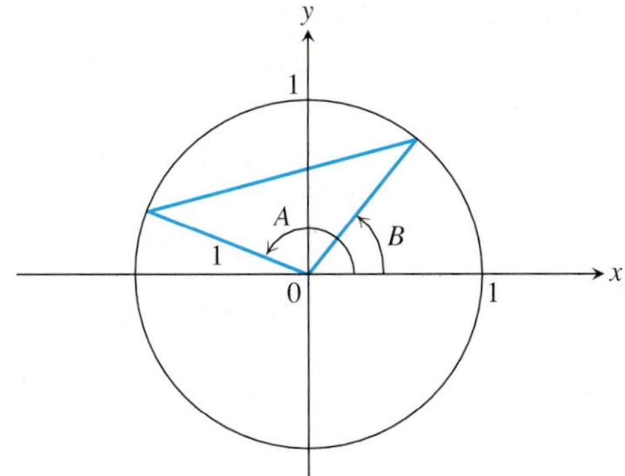

30. When applied to a figure similar to the one in Exercise 29, the law of cosines leads directly to the formula for $\cos(A + B)$. What is that figure, and how does the derivation go?

31. Here is an informal proof that $\tan^{-1} 1 + \tan^{-1} 2 + \tan^{-1} 3 = \pi$. Explain what is going on.

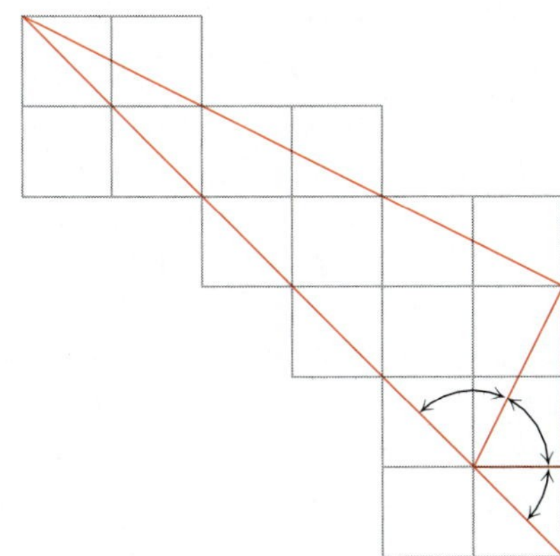

32. *Two derivations of the identity* $\sec^{-1}(-x) = \pi - \sec^{-1} x$

(a) (*Geometric*) Here is a pictorial proof that $\sec^{-1}(-x) = \pi - \sec^{-1} x$. See if you can tell what is going on.

(b) (*Algebraic*) Derive the identity $\sec^{-1}(-x) = \pi - \sec^{-1} x$ by combining the following two equations:

$$\cos^{-1}(-x) = \pi - \cos^{-1} x, \qquad \text{Eq. (9)}$$
$$\sec^{-1} x = \cos^{-1}(1/x).$$

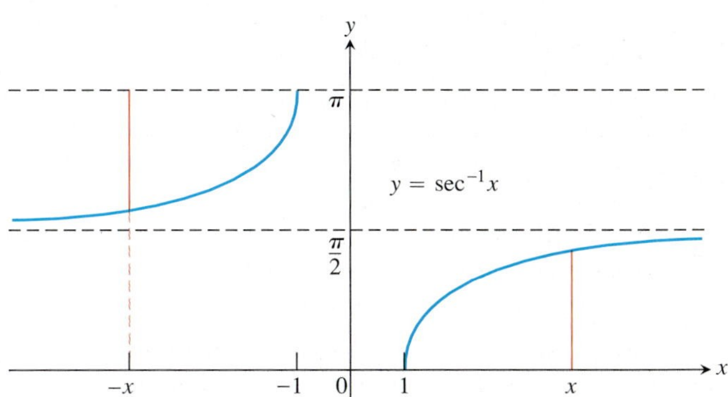

33. *The identity* $\sin^{-1} x + \cos^{-1} x = \pi/2$ Figure 53 establishes the identity for $0 < x < 1$. To establish it for the rest of $[-1, 1]$, verify by direct calculation that it holds for $x = 1, 0$, and -1. Then, for values of x in $(-1, 0)$, let $x = -a$, $a > 0$, and apply Equations (7) and (9) to the sum $\sin^{-1}(-a) + \cos^{-1}(-a)$.

34. Show that the sum $\tan^{-1} x + \tan^{-1}(1/x)$ is constant.

35. *The law of sines* The law of sines says that if a, b, and c are the sides opposite the angles A, B, and C in a triangle, then

$$\frac{\sin A}{a} = \frac{\sin B}{b} = \frac{\sin C}{c}.$$

Use the accompanying figures and the identity $\sin(\pi - \theta) = \sin\theta$, as required, to derive the law.

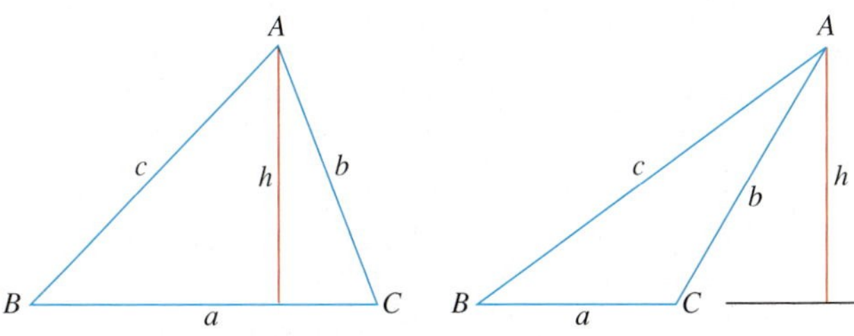

36. *The tangent sum formula* The standard formula for the tangent of the sum of two angles is

$$\tan(A + B) = \frac{\tan A + \tan B}{1 - \tan A \tan B}.$$

Derive the formula.

Solving Triangles and Comparing Functions

37. *Solving triangles*

(a) A triangle has sides $a = 2$ and $b = 3$ and angle $C = 60°$. Find the length of side c.

(b) A triangle has sides $a = 2$ and $b = 3$ and angle $C = 40°$. Find the length of side c.

38. *Solving triangles*

(a) A triangle has sides $a = 2$ and $b = 3$ and angle $C = 60°$ (as in Exercise 37, part (a). Find the sine of angle B using the law of sines from Exercise 35.

(b) A triangle has side $c = 2$ and angles $A = \pi/4$ and $B = \pi/3$. Find the length a of the side opposite A.

T **39.** *The approximation $\sin x \approx x$* It is often useful to know that, when x is measured in radians, $\sin x \approx x$ for numerically small values of x. In Section 3.6, we will see why the approximation holds. The approximation error is less than 1 in 500 if $|x| < 0.1$.

(a) With your grapher in radian mode, graph $y = \sin x$ and $y = x$ together in a viewing window about the origin. What do you see happening as x nears the origin?

(b) With your grapher in degree mode, graph $y = \sin x$ and $y = x$ together about the origin again. How is the picture different from the one obtained with radian mode?

(c) *A quick radian mode check* Is your calculator in radian mode? Evaluate $\sin x$ at a value of x near the origin, say $x = 0.1$. If $\sin x \approx x$, the calculator is in radian mode; if not, it isn't. Try it.

T **40.** *Functions and their reciprocals*

(a) Graph $y = \cos x$ and $y = \sec x$ together for $-3\pi/2 \le x \le 3\pi/2$. Comment on the behavior of $\sec x$ in relation to the signs and values of $\cos x$.

(b) Graph $y = \sin x$ and $y = \csc x$ together for $-\pi \le x \le 2\pi$. Comment on the behavior of $\csc x$ in relation to the signs and values of $\sin x$.

T In Exercises 41 and 42, find the domain and range of each composite function. Then graph the composites on separate screens. Do the graphs make sense in each case? Give reasons for your answers. Comment on any differences you see.

41. **(a)** $y = \tan^{-1}(\tan x)$ **(b)** $y = \tan(\tan^{-1} x)$

42. **(a)** $y = \sin^{-1}(\sin x)$ **(b)** $y = \sin(\sin^{-1} x)$

In Exercises 43–46, solve the equation in the specified interval.

43. $\tan x = 2.5, \quad 0 \le x < 2\pi$

44. $\cos x = -0.7, \quad 2\pi \le x < 4\pi$

45. $\sec x = -3, \quad -\pi \le x < \pi$

46. $\sin x = -0.5, \quad -\infty < x < \infty$

T **47.** *Trigonometric identities* Let $f(x) = \sin x + \cos x$.

(a) Graph $y = f(x)$. Describe the graph.

(b) Use the graph to identify the amplitude, period, horizontal shift, and vertical shift.

(c) Use the formula

$$\sin\alpha\cos\beta + \cos\alpha\sin\beta = \sin(\alpha + \beta)$$

for the sine of the sum of two angles to confirm your answers.

T **48.** *Newton's serpentine* Graph Newton's serpentine, $y = 4x/(x^2 + 1)$. Then graph $y = 2\sin(2\tan^{-1} x)$ in the same graphing window. What do you see? Explain.

Sinusoidal Regression Analysis: Musical Notes and Temperature

See page 5 for an introduction to regression analysis with a calculator. A sinusoidal regression equation is a general sine curve; see Equation (2). Many calculators and computers produce these regression equations for given data sets.

T **49.** *Finding the frequency of a musical note* Musical notes are pressure waves in the air. The wave behavior can be modeled with great accuracy by general sine curves. Devices called Calculator Based Laboratory™ (CBL) systems can record these waves with a microphone. The data in Table 18 give pressure displacement versus time in seconds of a musical note produced by a tuning fork and recorded with a CBL system.

(a) Find a sinusoidal regression equation (general sine curve) for the data and superimpose its graph on a scatter plot of the data.

(b) The *frequency* of a musical note, or wave, is measured in cycles per second, or hertz (1 Hz = 1 cycle per second). The frequency is the reciprocal of the *period* of the wave, which is measured in seconds per cycle. Estimate the frequency of the note produced by the tuning fork.

Table 18 Tuning fork data

Time	Pressure	Time	Pressure
0.00091	−0.080	0.00362	0.217
0.00108	0.200	0.00379	0.480
0.00125	0.480	0.00398	0.681
0.00144	0.693	0.00416	0.810
0.00162	0.816	0.00435	0.827
0.00180	0.844	0.00453	0.749
0.00198	0.771	0.00471	0.581
0.00216	0.603	0.00489	0.346
0.00234	0.368	0.00507	0.077
0.00253	0.099	0.00525	−0.164
0.00271	−0.141	0.00543	−0.320
0.00289	−0.309	0.00562	−0.354
0.00307	−0.348	0.00579	−0.248
0.00325	−0.248	0.00598	−0.035
0.00344	−0.041		

T **50.** *Temperature data* Table 19 gives the average monthly temperatures for St. Louis for a 12-month period starting with January. Model the monthly temperature with an equation of the form

$$y = a\sin(b(t - h)) + k,$$

y in degrees Fahrenheit, t in months, as follows:

(a) Find the value of b assuming that the period is 12 months.

(b) How is the amplitude a related to the difference $80° - 30°$?

(c) Use the information in part (b) to find k.

(d) Find h and write an equation for y.

(e) Superimpose a graph of y on a scatter plot of the data.

Table 19 Temperature data for St. Louis

Time (months)	Temperature (°F)
1	34
2	30
3	39
4	44
5	58
6	67
7	78
8	80
9	72
10	63
11	51
12	40

T **51.** *Sinusoidal regression* Table 20 gives the values of the function

$$f(x) = a \sin(bx + c) + d$$

accurate to two decimals.

(a) Find a sinusoidal regression equation for the data.

(b) Rewrite the equation with a, b, c, and d rounded to the nearest integer.

Table 20 Values of a function

x	$f(x)$
1	3.42
2	0.73
3	0.12
4	2.16
5	4.97
6	5.97

T **52.** We suggest that students *work in groups of two or three.* A musical note like that produced with a tuning fork or pitch meter is a series of pressure waves. Table 21 gives frequencies (in hertz) of musical notes on the tempered scale. The pressure versus time tuning fork data in Table 22 were collected using a CBL system and a microphone.

(a) Find a sinusoidal regression equation for the data in Table 22 and superimpose its graph on a scatter plot of the data.

(b) Determine the frequency of and identify the musical note produced by the tuning fork.

Table 21 Frequencies of notes

Note	Frequency (Hz)
C	262
C♯ or D♭	277
D	294
D♯ or E♭	311
E	330
F	349
F♯ or G♭	370
G	392
G♯ or A♭	415
A	440
A♯ or B♭	466
B	494
C (next octave)	523

Source: CBL™ System Experimental Workbook, Texas Instruments, Inc., 1994.

Table 22 Tuning fork data

Time (s)	Pressure	Time (s)	Pressure
0.0002368	1.29021	0.0049024	−1.06632
0.0005664	1.50851	0.0051520	0.09235
0.0008256	1.51971	0.0054112	1.44694
0.0010752	1.51411	0.0056608	1.51411
0.0013344	1.47493	0.0059200	1.51971
0.0015840	0.45619	0.0061696	1.51411
0.0018432	−0.89280	0.0064288	1.43015
0.0020928	−1.51412	0.0066784	0.19871
0.0023520	−1.15588	0.0069408	−1.06072
0.0026016	−0.04758	0.0071904	−1.51412
0.0028640	1.36858	0.0074496	−0.97116
0.0031136	1.50851	0.0076992	0.23229
0.0033728	1.51971	0.0079584	1.46933
0.0036224	1.51411	0.0082080	1.51411
0.0038816	1.45813	0.0084672	1.51971
0.0041312	0.32185	0.0087168	1.50851
0.0043904	−0.97676	0.0089792	1.36298
0.0046400	−1.51971		

6 Parametric Equations

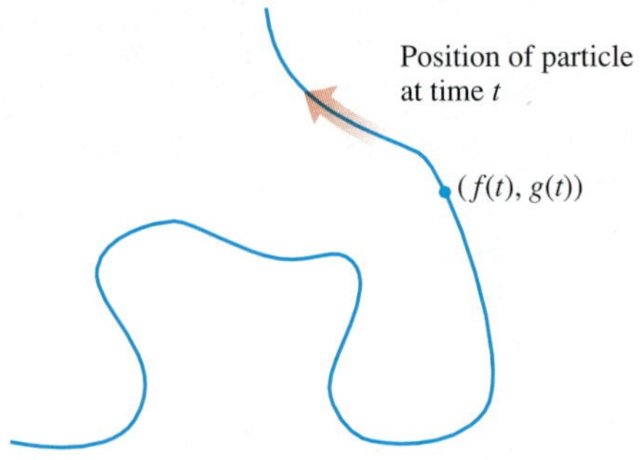

FIGURE 54 The path traced by a particle moving in the xy-plane is not always the graph of a function of x or a function of y.

When the path of a particle moving in the plane looks like the curve in Figure 54, we cannot describe it by an equation of the form $y = f(x)$ because there are vertical lines that intersect the curve more than once (see Exercise 25 in Section 2). Likewise, we cannot describe the curve by expressing x directly as a function of y. In this section, you learn another way to describe curves in terms of a third variable called a *parameter*. This powerful approach can also be used to describe ordinary function curves, like the ones we have been studying so far, and their inverses when they exist.

Parametrizations of Plane Curves

When the path of a particle moving in the plane looks like the curve in Figure 54, we express each of the particle's coordinates as a function of a third variable t and describe the path with a pair of equations, $x = f(t)$ and $y = g(t)$. For studying motion, t usually denotes time. Equations like these are better than a Cartesian formula because they tell us the particle's position $(x, y) = (f(t), g(t))$ at any time t.

Example 1 Moving Along a Parabola

The position $P(x, y)$ of a particle moving in the xy-plane is given by the equations and parameter interval

$$x = \sqrt{t}, \qquad y = t, \qquad t \geq 0.$$

Identify the path traced by the particle and describe the motion.

Solution We try to identify the path by eliminating t between the equations $x = \sqrt{t}$ and $y = t$. With any luck, this will produce a recognizable algebraic relation between x and y. We find that

$$y = t = (\sqrt{t})^2 = x^2.$$

Thus, the particle's position coordinates satisfy the equation $y = x^2$, so the particle moves along the parabola $y = x^2$.

It would be a mistake, however, to conclude that the particle's path is the entire parabola $y = x^2$; it is only half the parabola. The particle's x-coordinate is never negative. The particle starts at (0, 0) when $t = 0$ and rises into the first quadrant as t increases (Figure 55).

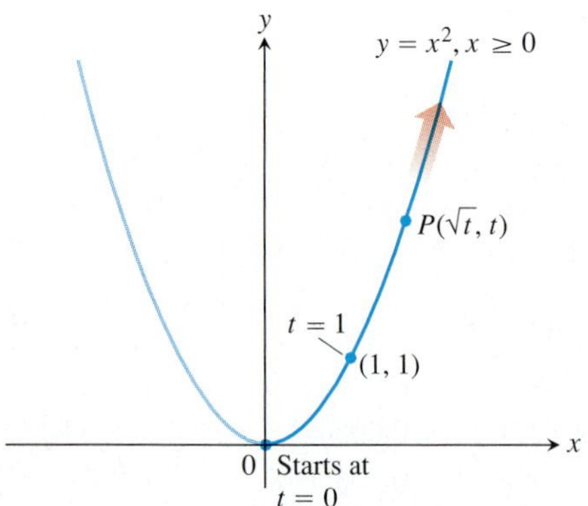

FIGURE 55 The equations $x = \sqrt{t}$ and $y = t$ and the interval $t \geq 0$ describe the motion of a particle that traces the right-hand half of the parabola $y = x^2$. (Example 1)

Definitions Parametric Curve, Parametric Equations

If x and y are given as functions

$$x = f(t), \qquad y = g(t)$$

over an interval of t-values, then the set of points $(x, y) = (f(t), g(t))$ defined by these equations is a **parametric curve.** The equations are **parametric equations** for the curve.

The variable t is a **parameter** for the curve, and its domain I is the **parameter interval.** If I is a closed interval, $a \le t \le b$, the point $(f(a), g(a))$ is the **initial point** of the curve. The point $(f(b), g(b))$ is the **terminal point.** When we give parametric equations and a parameter interval for a curve, we say that we have **parametrized** the curve. The equations and interval together constitute a **parametrization** of the curve.

In Example 1, the parameter interval is $[0, \infty)$, so $(0, 0)$ is the initial point. There is no terminal point.

A grapher can draw a parametrized curve only over a closed interval, so the portion it draws has endpoints even when the curve being graphed does not. Keep this in mind when you graph on a graphing calculator or computer.

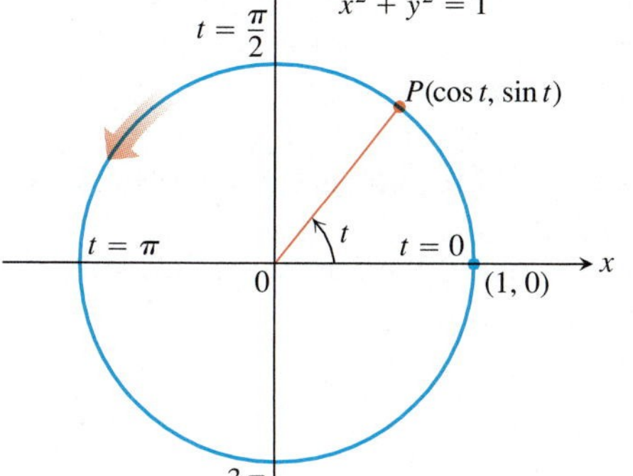

FIGURE 56 The equations $x = \cos t$ and $y = \sin t$ describe motion on the circle $x^2 + y^2 = 1$. The arrow shows the direction of increasing t. (Example 2)

Example 2 Moving Counterclockwise on a Circle

Graph the parametric curves

(a) $x = \cos t, \quad y = \sin t, \quad 0 \le t \le 2\pi$

(b) $x = a\cos t, \quad y = a\sin t, \quad 0 \le t \le 2\pi.$

Solution

(a) Since $x^2 + y^2 = \cos^2 t + \sin^2 t = 1$, the parametric curve lies along the unit circle $x^2 + y^2 = 1$. As t increases from 0 to 2π, the point $(x, y) = (\cos t, \sin t)$ starts at $(1, 0)$ and traces the entire circle once counterclockwise (Figure 56).

(b) For $x = a\cos t$, $y = a\sin t$, $0 \le t \le 2\pi$, we have $x^2 + y^2 = a^2\cos^2 t + a^2\sin^2 t = a^2$. The parametrization describes a motion that begins at the point $(a, 0)$ and traverses the circle $x^2 + y^2 = a^2$ once counterclockwise, returning to $(a, 0)$ at $t = 2\pi$.

FIGURE 57 The point $P(\cos t, -\sin t)$ moves clockwise as t increases from 0 to π. (Example 3)

Example 3 Traversing a Semicircle Clockwise

Graph the parametric curve

$$x = \cos t, \quad y = -\sin t, \quad 0 \le t \le \pi.$$

Find a Cartesian equation for a curve that contains the parametric curve. What portion of the graph of the Cartesian equation is traced by the parametric curve? Describe the motion.

Solution The point $(x, y) = (\cos t, -\sin t)$ moves on the circle $x^2 + y^2 = 1$. In contrast to Example 2, the motion is now clockwise. As t increases from 0 toward π, y becomes negative and x decreases. The point (x, y) moves around the bottom half of the circle, first falling to $(0, -1)$ and then rising to $(-1, 0)$. The motion stops at $t = \pi$, leaving only the bottom half of the circle covered (Figure 57).

Lines and Other Curves

Many other curves, including lines and line segments, can be defined parametrically.

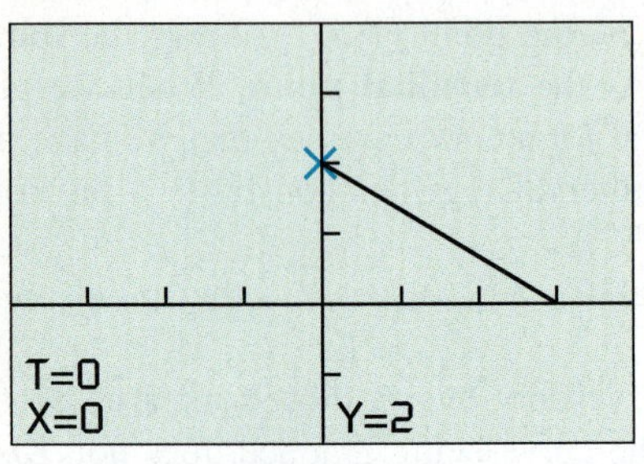

FIGURE 58 The graph of the line segment $x = 3t$, $y = 2 - 2t$, $0 \leq t \leq 1$, with trace on the initial point (0, 2). (Example 4)

Example 4 Moving Along a Straight Line

Graph and identify the parametric curve

$$x = 3t, \quad y = 2 - 2t, \quad 0 \leq t \leq 1.$$

What happens if the restriction on t is removed?

Solution When $t = 0$, the equations give $x = 0$ and $y = 2$. When $t = 1$, they give $x = 3$ and $y = 0$. If we substitute $t = x/3$ into the y equation, we obtain

$$y = 2 - 2\left(\frac{x}{3}\right) = -\frac{2}{3}x + 2.$$

Thus, the parametric curve traces the segment of the line $y = -(2/3)x + 2$ from (0, 2) to (3, 0) (Figure 58).

If we remove the restriction on t, changing the parameter interval from [0, 1] to $(-\infty, \infty)$, the parametrization traces the entire line $y = -(2/3)x + 2$.

Example 5 Parametrizing a Line Segment

Find a parametrization for the line segment with endpoints $(-2, 1)$ and $(3, 5)$.

Solution Using $(-2, 1)$ we create the parametric equations

$$x = -2 + at, \quad y = 1 + bt.$$

These represent a line, as we can see by solving each equation for t and equating to obtain

$$\frac{x + 2}{a} = \frac{y - 1}{b}.$$

This line goes through the point $(-2, 1)$ when $t = 0$. We determine a and b so that the line goes through (3, 5) when $t = 1$.

$$3 = -2 + a \Rightarrow a = 5 \qquad x = 3 \text{ when } t = 1.$$
$$5 = 1 + b \Rightarrow b = 4 \qquad y = 5 \text{ when } t = 1.$$

Therefore,

$$x = -2 + 5t, \quad y = 1 + 4t, \quad 0 \leq t \leq 1$$

is a parametrization of the line segment with initial point $(-2, 1)$ and terminal point (3, 5).

Example 6 Moving Along the Ellipse $x^2/a^2 + y^2/b^2 = 1$

Describe the motion of a particle whose position $P(x, y)$ at time t is given by

$$x = a\cos t, \quad y = b\sin t, \quad 0 \leq t \leq 2\pi.$$

Solution We find a Cartesian equation for the particle's coordinates by eliminating t between the equations

$$\cos t = \frac{x}{a}, \qquad \sin t = \frac{y}{b}.$$

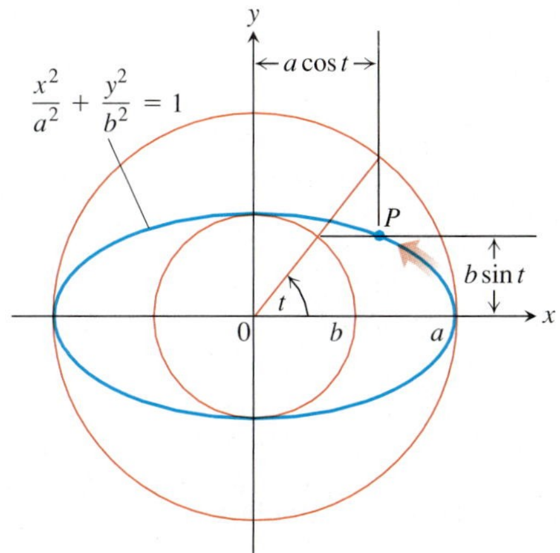

FIGURE 59 The ellipse in Example 6, drawn for $a > b > 0$. The coordinates of P are $x = a\cos t$, $y = b\sin t$.

The identity $\cos^2 t + \sin^2 t = 1$, yields

$$\left(\frac{x}{a}\right)^2 + \left(\frac{y}{b}\right)^2 = 1, \qquad \text{or} \qquad \frac{x^2}{a^2} + \frac{y^2}{b^2} = 1.$$

The particle's coordinates (x, y) satisfy the equation $(x^2/a^2) + (y^2/b^2) = 1$, so the particle moves along this ellipse. When $t = 0$, the particle's coordinates are

$$x = a\cos(0) = a, \qquad y = b\sin(0) = 0,$$

so the motion starts at $(a, 0)$. As t increases, the particle rises and moves toward the left, moving counterclockwise. It traverses the ellipse once, returning to its starting position $(a, 0)$ at $t = 2\pi$ (Figure 59).

Parametrizing Inverse Functions

We can graph or represent any function $y = f(x)$ parametrically as

$$x = t \qquad \text{and} \qquad y = f(t).$$

Interchanging t and $f(t)$ produces parametric equations for the inverse:

$$x = f(t) \qquad \text{and} \qquad y = t$$

(see Section 4).

For example, to graph the one-to-one function $f(x) = x^2$, $x \geq 0$, on a grapher together with its inverse and the line $y = x$, $x \geq 0$, use the parametric graphing option with

$$\begin{aligned} &\text{Graph of } f: && x_1 = t, && y_1 = t^2, && t \geq 0 \\ &\text{Graph of } f^{-1}: && x_2 = t^2, && y_2 = t \\ &\text{Graph of } y = x: && x_3 = t, && y_3 = t \end{aligned}$$

Figure 60 shows the three graphs.

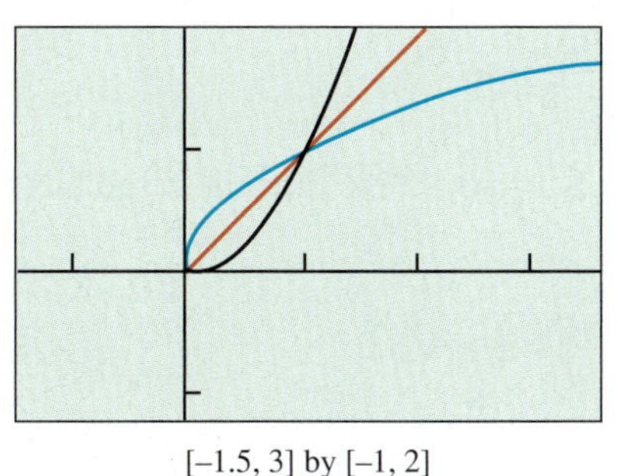

[−1.5, 3] by [−1, 2]

FIGURE 60 Parametric graphs of the function $f(x) = x^2$, $x \geq 0$, its inverse, and the line $y = x$.

An Application

Example 7 Dropping Emergency Supplies

A Red Cross aircaft is dropping emergency food and medical supplies into a disaster area. If the aircraft releases the supplies immediately above the edge of an open field 700 ft long and if the cargo moves along the path

$$x = 120t \qquad \text{and} \qquad y = -16t^2 + 500, \qquad t \geq 0$$

does the cargo land in the field? The coordinates x and y are measured in feet, and the parameter t (time since release) in seconds. Find a Cartesian equation for the path of the falling cargo (Figure 61).

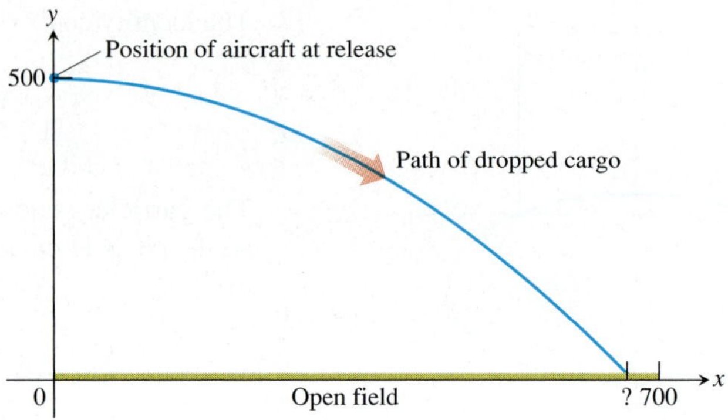

FIGURE 61 The path of the dropped cargo of supplies in Example 7.

Solution The cargo hits the ground when $y = 0$, which occurs at time t when

$$-16t^2 + 500 = 0 \qquad \text{Set } y = 0.$$

$$t^2 = \frac{500}{16} \qquad \text{Solve for } t.$$

$$t = \frac{5\sqrt{5}}{2} \text{ sec}. \qquad t \geq 0$$

The x-coordinate at the time of the release is $x = 0$. At the time the cargo hits the ground, the x-coordinate is

$$x = 120t = 120\left(\frac{5\sqrt{5}}{2}\right) = 300\sqrt{5} \text{ ft}.$$

Since $300\sqrt{5} \approx 670.8 < 700$, the cargo does land in the field.

We find a Cartesian equation for the cargo's coordinates by eliminating t between the parametric equations:

$$y = -16t^2 + 500 \qquad \text{Parametric equation for } y$$

$$= -16\left(\frac{x}{120}\right)^2 + 500 \qquad \text{Substitute for } t \text{ from the equation } x = 120t.$$

$$= -\frac{16}{14400}x^2 + 500 \qquad \text{Simplify.}$$

or

$$y = -\frac{1}{900}x^2 + 500.$$

Thus, the cargo moves along the parabola

$$y = -\frac{x^2}{900} + 500.$$

Standard Parametrizations

CIRCLE $x^2 + y^2 = a^2$:

$x = a \cos t$

$y = a \sin t$

$0 \le t \le 2\pi$

ELLIPSE $\dfrac{x^2}{a^2} + \dfrac{y^2}{b^2} = 1$:

$x = a \cos t$

$y = b \sin t$

$0 \le t \le 2\pi$

FUNCTION $y = f(x)$:

$x = t$

$y = f(t)$

INVERSE OF $y = f(x)$:

$x = f(t)$

$y = t$

EXERCISES 6

Finding Cartesian Equations from Parametric Equations

Exercises 1–18 give parametric equations and parameter intervals for the motion of a particle in the xy-plane. Identify the particle's path by finding a Cartesian equation for it. Graph the Cartesian equation. (The graphs will vary with the equation used.) Indicate the portion of the graph traced by the particle and the direction of motion.

1. $x = \cos t, \quad y = \sin t, \quad 0 \le t \le \pi$

2. $x = \cos 2t, \quad y = \sin 2t, \quad 0 \le t \le \pi$

3. $x = \sin(2\pi t), \quad y = \cos(2\pi t), \quad 0 \le t \le 1$

4. $x = \cos(\pi - t), \quad y = \sin(\pi - t), \quad 0 \le t \le \pi$

5. $x = 4 \cos t, \quad y = 2 \sin t, \quad 0 \le t \le 2\pi$

6. $x = 4 \sin t, \quad y = 5 \cos t, \quad 0 \le t \le 2\pi$

7. $x = 3t, \quad y = 9t^2, \quad -\infty < t < \infty$

8. $x = -\sqrt{t}, \quad y = t, \quad t \ge 0$

9. $x = t, \quad y = \sqrt{t}, \quad t \ge 0$

10. $x = \sec^2 t - 1, \quad y = \tan t, \quad -\pi/2 < t < \pi/2$

11. $x = -\sec t, \quad y = \tan t, \quad -\pi/2 < t < \pi/2$

12. $x = 2t - 5, \quad y = 4t - 7, \quad -\infty < t < \infty$

13. $x = 1 - t, \quad y = 1 + t, \quad -\infty < t < \infty$

14. $x = 3 - 3t, \quad y = 2t, \quad 0 \le t \le 1$

15. $x = t, \quad y = \sqrt{1 - t^2}, \quad -1 \le t \le 0$

16. $x = \sqrt{t + 1}, \quad y = \sqrt{t}, \quad t \ge 0$

17. $x = e^t + e^{-t}, \quad y = e^t - e^{-t}, \quad -\infty < t < \infty$

18. $x = \cos(e^t), \quad y = 2 \sin(e^t), \quad -\infty < t < \infty$

Determining Parametric Equations

19. Find parametric equations and a parameter interval for the motion of a particle that starts at $(a, 0)$ and traces the circle $x^2 + y^2 = a^2$

(a) once clockwise. **(b)** once counterclockwise.

(c) twice clockwise. **(d)** twice counterclockwise.

(There are many ways to do these, so your answers may not be the same as the ones in the back of the book.)

20. Find parametric equations and a parameter interval for the motion of a particle that starts at $(a, 0)$ and traces the ellipse $(x^2/a^2) + (y^2/b^2) = 1$

(a) once clockwise. **(b)** once counterclockwise.

(c) twice clockwise. **(d)** twice counterclockwise.

(As in Exercise 19, there are many correct answers.)

In Exercises 21–26, find a parametrization for the curve.

21. the line segment with endpoints $(-1, -3)$ and $(4, 1)$

22. the line segment with endpoints $(-1, 3)$ and $(3, -2)$

23. the lower half of the parabola $x - 1 = y^2$

24. the left half of the parabola $y = x^2 + 2x$

25. the ray (half line) with initial point $(2, 3)$ that passes through the point $(-1, -1)$

26. the ray (half line) with initial point $(-1, 2)$ that passes through the point $(0, 0)$

Parametric Graphing

T In Exercises 27–30, match the parametric equations with their graph. State the approximate dimensions of the viewing window. Give a parameter interval that traces the curve exactly once.

27. $x = 3\sin(2t)$, $y = 1.5\cos t$

28. $x = \sin^3 t$, $y = \cos^3 t$

29. $x = 7\sin t - \sin(7t)$, $y = 7\cos t - \cos(7t)$

30. $x = 12\sin t - 3\sin(6t)$, $y = 12\cos t + 3\cos(6t)$

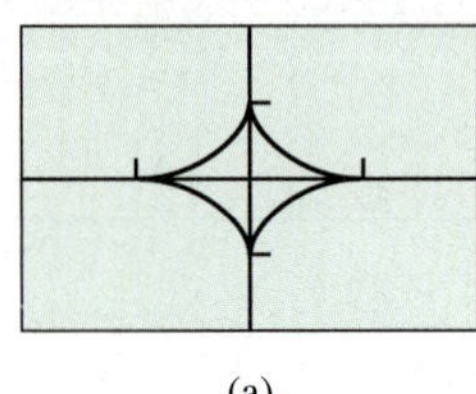

(a)

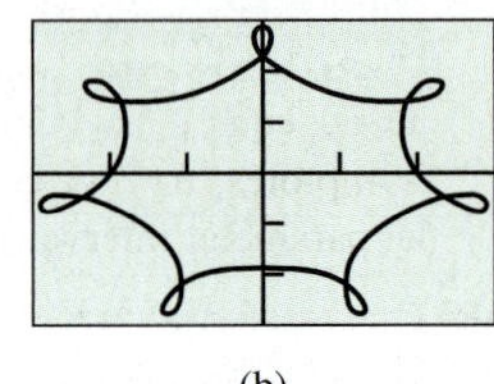

(b)

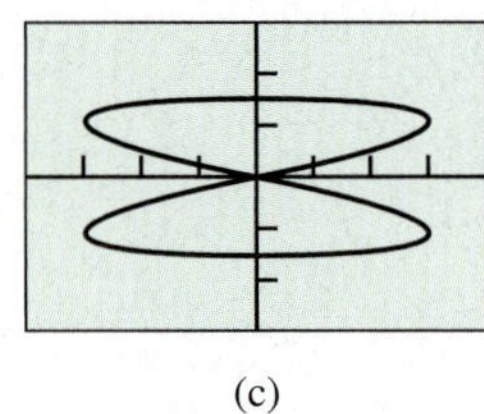

(c)

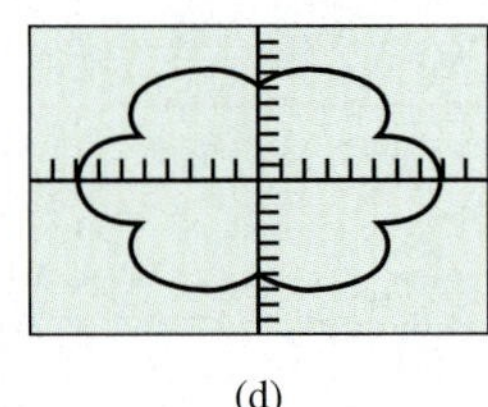

(d)

T In Exercises 31–38, use parametric graphing to graph f, f^{-1}, and $y = x$.

31. $f(x) = e^x$

32. $f(x) = 3^x$

33. $f(x) = 2^{-x}$

34. $f(x) = 3^{-x}$

35. $f(x) = \ln x$

36. $f(x) = \log x$

37. $f(x) = \sin^{-1} x$

38. $f(x) = \tan^{-1} x$

In Exercises 39–42, refer to the graph of

$$x = 3 - |t|, \qquad y = t - 1, \qquad -5 \le t \le 5,$$

shown in the accompanying figure. *Work in groups of two or three* to find the values of t that produce the graph in the given quadrant.

39. Quadrant I

40. Quadrant II

41. Quadrant III

42. Quadrant IV

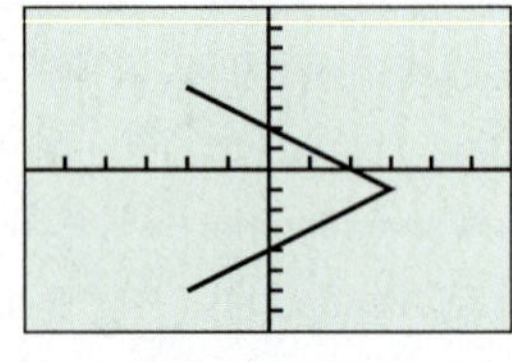

[–6, 6] by [–8, 8]

T In Exercises 43–48, graph the equations over the given intervals.

43. *Ellipse* $x = 4\cos t$, $y = 2\sin t$, over

(a) $0 \le t \le 2\pi$ **(b)** $0 \le t \le \pi$

(c) $-\pi/2 \le t \le \pi/2$.

44. *Hyperbola branch* $x = \sec t$ (enter as $1/\cos(t)$), $y = \tan t$ (enter as $\sin(t)/\cos(t)$), over

(a) $-1.5 \le t \le 1.5$ **(b)** $-0.5 \le t \le 0.5$

(c) $-0.1 \le t \le 0.1$.

45. *Parabola* $x = 2t + 3$, $y = t^2 - 1$, $-2 \le t \le 2$

46. *A nice curve (a deltoid)*

$$x = 2\cos t + \cos 2t, \qquad y = 2\sin t - \sin 2t, \qquad 0 \le t \le 2\pi$$

What happens if you replace 2 with -2 in the equations for x and y? Graph the new equations to find out.

47. *An even nicer curve*

$$x = 3\cos t + \cos 3t, \qquad y = 3\sin t - \sin 3t, \qquad 0 \le t \le 2\pi$$

What happens if you replace 3 with -3 in the equations for x and y? Graph the new equations to find out.

48. *Cycloid* $x = t - \sin t$, $y = 1 - \cos t$, over

(a) $0 \le t \le 2\pi$ **(b)** $0 \le t \le 4\pi$

(c) $\pi \le t \le 3\pi$.

Extending the Ideas

49. *The witch of Agnesi* The bell-shaped witch of Agnesi can be constructed as follows. Start with the circle of radius 1, centered at the point (0, 1), as shown in the accompanying figure.

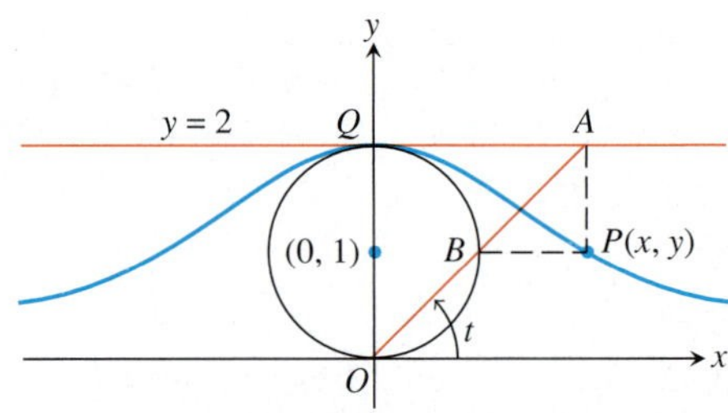

Choose a point A on the line $y = 2$, and connect it to the origin with a line segment. Call the point where the segment crosses the circle B. Let P be the point where the vertical line through A crosses the horizontal line through B. The witch is the curve traced by P as A moves along the line $y = 2$.

Find a parametrization for the witch by expressing the coordinates of P in terms of t, the radian measure of the angle that segment OA makes with the positive x-axis. The following equalities (which you may assume) will help:

(i) $x = AQ$.

(ii) $y = 2 - AB\sin t$.

(iii) $AB \cdot AO = (AQ)^2$.

50. *Parametrizing lines and segments*

(a) Show that the equations and parameter interval

$$x = x_0 + (x_1 - x_0)t, \qquad y = y_0 + (y_1 - y_0)t, \qquad -\infty < t < \infty,$$

describe the line through the points (x_0, y_0) and (x_1, y_1) (Figure 62).

(b) Using the same parameter interval, write parametric equations for the line through a point (x_1, y_1) and the origin.

(c) Using the same parameter interval, write parametric equations for the line through $(-1, 0)$ and $(0, 1)$.

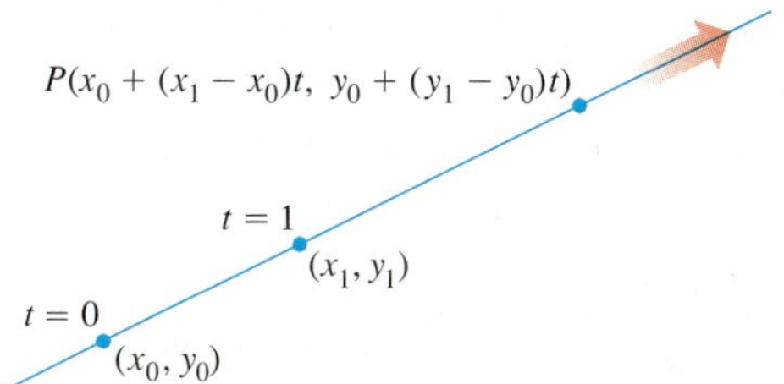

FIGURE 62 The line in Exercise 50(a). The arrow shows the direction of increasing t.

T **51.** *Graphing the witch of Agnesi* The witch of Agnesi is the curve

$$x = 2 \cot t, \quad y = 2 \sin^2 t, \quad 0 < t < \pi.$$

(a) Draw the curve using the window in the accompanying figure. What did you choose as a closed parameter interval for your grapher? In what direction is the curve traced? How far to the left and right of the origin do you think the curve extends?

$x = 2 \cot t,\ y = 2 \sin^2 t$

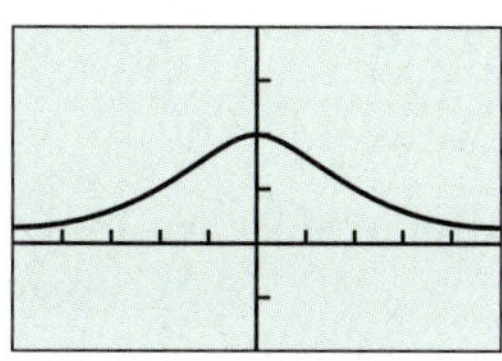

[–5, 5] by [–2, 4]

(b) Graph the same parametric equations using the parameter intervals $(-\pi/2, \pi/2)$, $(0, \pi/2)$, and $(\pi/2, \pi)$. In each case, describe the curve you see and the direction in which it is traced by your grapher.

(c) What happens if you replace $x = 2 \cot t$ by $x = -2 \cot t$ in the original parametrization? What happens if you use $x = 2 \cot (\pi - t)$?

T **52.** *Hyperboloids* Let $x = a \sec t$ and $y = b \tan t$.

(a) *Writing to Learn* Let $a = 1$, 2, or 3, and $b = 1$, 2, or 3, and graph using the parameter interval $(-\pi/2, \pi/2)$. Explain what you see and describe the role of a and b in these parametric equations. (Caution: If you get what appear to be asymptotes, try using the approximation $[-1.57, 1.57]$ for the parameter interval.)

(b) Let $a = 2$ and $b = 3$, and graph in the parameter interval $(\pi/2, 3\pi/2)$. Explain what you see.

(c) *Writing to Learn* Let $a = 2$ and $b = 3$, and graph using the parameter interval $(-\pi/2, 3\pi/2)$. Explain why you must be careful about graphing in this interval or any interval that contains $\pm\pi/2$.

(d) Use algebra to explain why

$$\left(\frac{x}{a}\right)^2 - \left(\frac{y}{b}\right)^2 = 1.$$

(e) Let $x = a \tan t$ and $y = b \sec t$. Repeat parts (a), (b), and (d) using an appropriate version of part (d).

Modeling Change

Mathematical Models • Simplification • Verifying a Model • A Model Construction Process • Empirical Modeling: Capturing the Trend of Collected Data • Using Calculus in Modeling

CD-ROM
WEBsite

To help us better understand our world, we often describe a particular phenomenon mathematically (by means of a function or an equation, for instance). Such a **mathematical model** is an idealization of the real-world phenomenon and never a completely accurate representation. Although any model has its limitations, a good one can provide valuable results and conclusions. In this section, we look at the modeling process and several illustrative examples.

Mathematical Models

In modeling our world, we are often interested in predicting the value of a variable at some time in the future. Perhaps it is a population, a real estate value, or the number of people with a communicative disease. Often, a mathematical model can help us understand a behavior better or aid us in planning for the future. Let's think of a mathematical model as a mathematical construct designed to study a particular real-world system or behavior of interest. The model allows us to reach mathematical conclusions about the behavior, as illustrated in Figure 63. These conclusions can be interpreted to help a decision maker plan for the future.

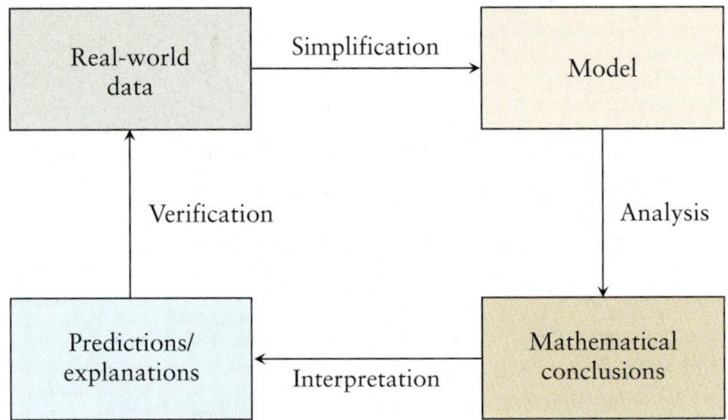

FIGURE 63 A flow of the modeling process beginning with an examination of real-world data.

Simplification

Most models simplify reality. Generally, models can only *approximate* real-world behavior. One very powerful simplifying relationship is **proportionality**.

Definition Proportionality

Two variables y and x are **proportional** (to one another) if one is always a constant multiple of the other; that is, if

$$y = kx$$

for some nonzero constant k.

The definition means that the graph of y versus x lies along a straight line through the origin. This graphical observation is useful in testing whether a given data collection reasonably assumes a proportionality relationship. If a proportionality is reasonable, a plot of one variable against the other should approximate a straight line through the origin. Here is an example.

Example 1 Testing for Proportionality in Driver Reaction Distance

During a panic stop, the driver of a car must react to an emergency, apply the brakes, and bring the car to a stop. What is a *safe following distance* for motorists? To answer this question, it would be helpful to know how far a vehicle travels at a given speed before the brakes are applied (*driver reaction distance*). The U.S. Bureau of Public Roads collected data for reaction and braking distances for a large number of motorists. (*Braking distance* is how far a vehicle travels after the brakes are applied until the vehicle stops.) In Table 23, x is the speed of an automobile in miles per hour (mph) and y is the distance in feet (ft) the automobile travels before the brakes are applied.

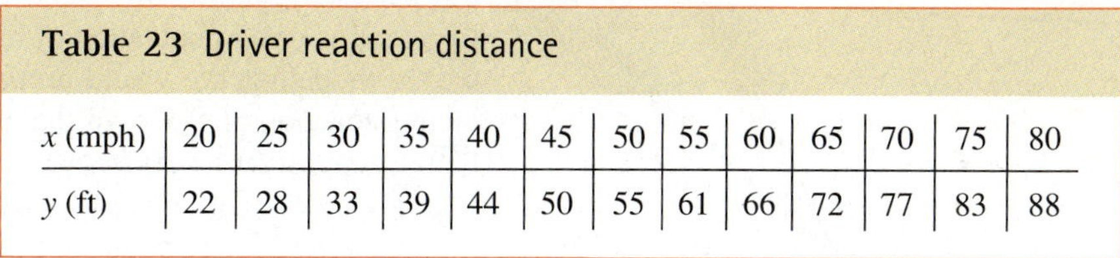

Table 23 Driver reaction distance

x (mph)	20	25	30	35	40	45	50	55	60	65	70	75	80
y (ft)	22	28	33	39	44	50	55	61	66	72	77	83	88

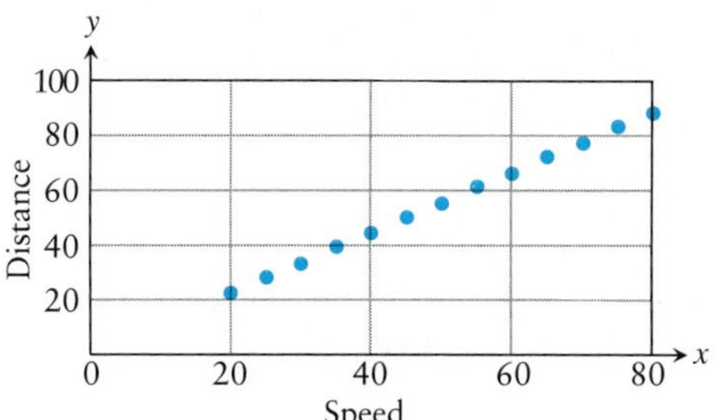

FIGURE 64 A plot of reaction distance versus speed.

For a collection of data representing many motorists, we might assume the time for a typical driver to react to an emergency to be approximately constant (independent of the speed). Then, the distance traveled while reacting is proportional to the speed. Let's test this proportionality assumption by plotting the distance traveled versus the speed. Figure 64 shows that the plotted points lie reasonably well along a straight line passing through the origin. Thus, the proportionality assumption appears valid.

We can even estimate the constant of proportionality from the graph. Using the first and last data points we find that a line approximating the data has a slope $rise/run = (88 - 22)/(80 - 20) = 1.1$. The proportionality model predicts driver reaction distance to be

$$y = 1.1x\,. \tag{1}$$

Verifying a Model

We can test to see how well Equation (1) fits the data by superimposing its graph on the scatter plot. Another procedure is to examine the errors or *residuals* (Table 24):

$$\text{Residuals} = \text{observations} - \text{predictions}\,.$$

Table 24 Computing the residuals

Speed (mph) x	**Observation (ft)**	**Prediction (ft)** $y = 1.1x$	**Residuals (ft)**
20	22	22.0	0.0
25	28	27.5	0.5
30	33	33.0	0.0
35	39	38.5	0.5
40	44	44.0	0.0
45	50	49.5	0.5
50	55	55.0	0.0
55	61	60.5	0.5
60	66	66.0	0.0
65	72	71.5	0.5
70	77	77.0	0.0
75	83	82.5	0.5
80	88	88.0	0.0

The residuals in Table 26 are relatively small (the largest is 0.5 ft, compared with a range of 22 to 88 ft) and do not have a bothersome pattern. Notice that at 60 mph, or 88 ft/sec, the average driver travels 66 ft while reacting. A typical driver requires (66 ft)/(88 ft/sec) = 0.75 sec to stop. This reaction time seems reasonable for a large and varying population of drivers. Given the imprecise nature of the problem we are modeling, we would probably accept this simple model as suitable for predicting reaction distance. In the exercises, we ask you to analyze *braking distance* before suggesting a *safe following distance* and devising a simple rule for motorists to follow in estimating a safe distance.

A powerful technique for judging a model's adequacy, and for gaining insight on how to improve the model, is to plot the residuals versus the independent variable. Then observe the relative magnitude of the errors. A pattern indicates that the model could be improved by capturing and incorporating the trend of this pattern.

A Model Construction Process

In learning to build models, the following process has been found to be useful. In constructing Equation (1), we roughly completed the following steps:

Steps for Model Construction

Step 1. *Identify the problem.* To estimate a safe following distance, we first decided to estimate the driver reaction distance.

Step 2. *Make assumptions on which variables to include and the relationships among the included variables.* Reaction distance depends on many factors, including speed, visibility, weather, and the driver's age. For simplicity, we assumed that reaction distance depends only on speed. We further assumed that reaction distance is proportional to the speed.

Step 3. *Find a function or graph that satisfies the relationships.* We tested the proportionality assumption by determining if the plot of reaction distance versus speed lies approximately along a line through the origin. Since it does, we could calculate the slope, which is the constant of proportionality.

Step 4. *Verify the model.* Analyze the residuals for size and pattern.

Empirical Modeling: Capturing the Trend of Collected Data

In Example 1, we hypothesized a relationship between a dependent variable and an independent variable. Another method for constructing models is to collect data and find a model that captures the trend of the data. This empirical approach has both advantages and disadvantages.

Example 2 Finding a Curve to Predict Population Levels

We may want to predict the future size of a population, such as the number of trout or catfish living in a fish farm. Figure 65 shows a scatter plot of the data collected by R. Pearl for a collection of yeast cells (measured as **biomass**) growing over time (measured in hours) in a nutrient.

Time (h) x	Biomass y
0	9.6
1	18.3
2	29
3	47.2
4	71.1
5	119.1
6	174.6
7	257.3

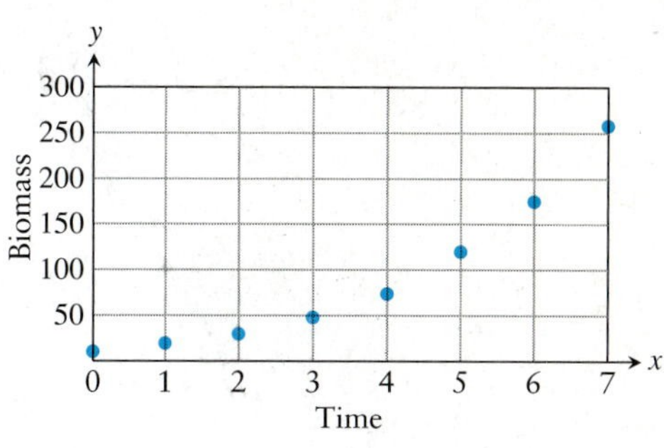

(Data from R. Pearl, "The Growth of Population," *Quart. Rev. Biol.*, Vol. 2 (1927), pp. 532–548.)

FIGURE 65 Biomass of a yeast culture versus elapsed time.

The plot of points appears to be reasonably smooth with an upward curving trend. We might attempt to capture this trend by fitting a polynomial (for example, a quadratic $y = ax^2 + bx + c$), a power curve ($y = ax^b$), or an exponential curve ($y = ae^{bx}$). Figure 66 shows the result of using a calculator to fit a quadratic model.

The quadratic model $y = 6.10x^2 - 9.28x + 16.43$ appears to fit the collected data reasonably well (Figure 66b). Using this model, we predict the population after 17 hours as $y(17) = 1622.65$. Let us examine more of Pearl's data to see if our quadratic model continues to be a good one.

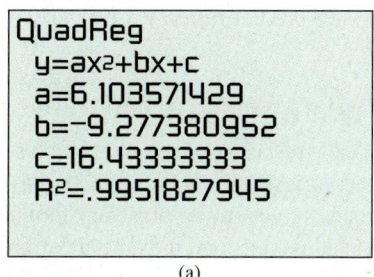

QuadReg
y=ax²+bx+c
a=6.103571429
b=−9.277380952
c=16.43333333
R²=.9951827945

(a)

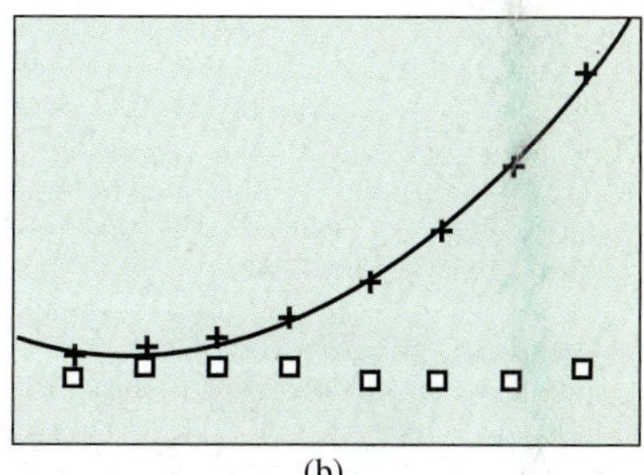

(b)

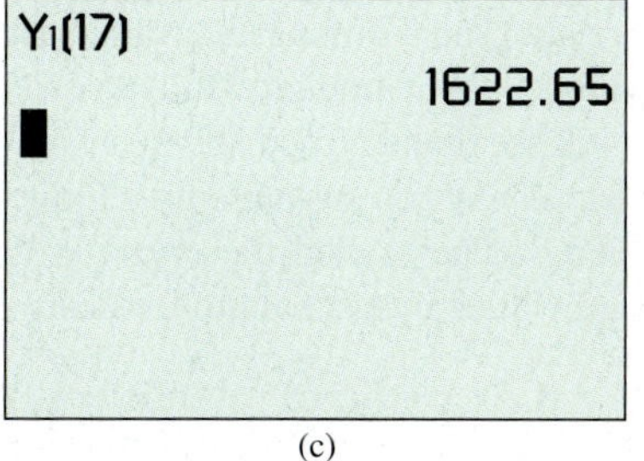

(c)

FIGURE 66 Using a calculator to (a) fit a quadratic; (b) overlay the data, model, and residuals; and (c) predict $y(17)$.

In Figure 67, we display all Pearl's data. Now you see that the prediction of $y(17) = 1622.65$ grossly overestimates the observed population of 659.6. Why did the quadratic model fail to predict a more accurate value?

Time (h) x	Observed y	Predicted y
0	9.6	16.4
1	18.3	13.3
2	29.0	22.3
3	47.2	43.5
4	71.1	77.0
5	119.1	122.6
6	174.6	180.5
7	257.3	250.6
8	350.7	332.8
9	441.0	427.3
10	513.3	534.0
11	559.7	652.9
12	594.8	784.0
13	629.4	927.3
14	640.8	1082.9
15	651.1	1250.6
16	655.9	1430.5
17	659.6	1622.7
18	661.8	1827.0

FIGURE 67 The rest of Pearl's data.

The problem lies in the danger of predicting beyond the range of data used to build the empirical model. (The range of data creating our model was $0 \leq x \leq 7$.) Such *extrapolation* is especially dangerous when the model selected is not supported by some underlying rationale suggesting the form of the model. In our yeast example, why would we expect a quadratic function as underlying population growth? Why not an exponential function? In the face of this, how then do we predict future values? Often, calculus can help.

Using Calculus in Modeling

CD-ROM
WEBsite

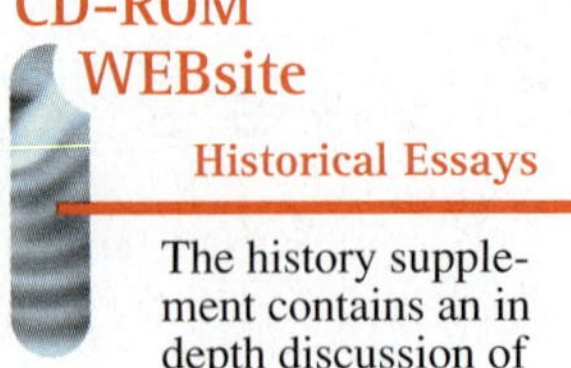

Historical Essays

The history supplement contains an in depth discussion of the development of calculus.

The application of calculus involves the study of *change*. The origins of calculus lie in our curiosity about, and need to develop, a deeper understanding of motion. The search for laws governing planetary motion, the study of the pendulum and its application to clock making, and the laws governing the flight of a cannonball were the kinds of problems stimulating the minds of mathematicians and scientists in the sixteenth and seventeenth centuries. In many cases, we observe how change is taking place and hypothesize relationships among the variables, much as we did in Example 1. In Chapter 6, we model population growth with calculus. In the case of the growth of a yeast culture, you will see that the food source available to the yeast constrains its growth. That is, the environment can sustain only a limited population. As the population approaches this limiting value (called the *carrying capacity*), growth slows. The model underlying the growth of a yeast culture for Pearl's data will turn out to be the *logistic* function

$$P = \frac{665}{1 + 73.8e^{-0.55t}}. \tag{2}$$

The graph of Equation (2) superimposed on a scatter plot of Pearl's data is displayed in Figure 68. You will see how Equation (2) comes about in Chapter 6.

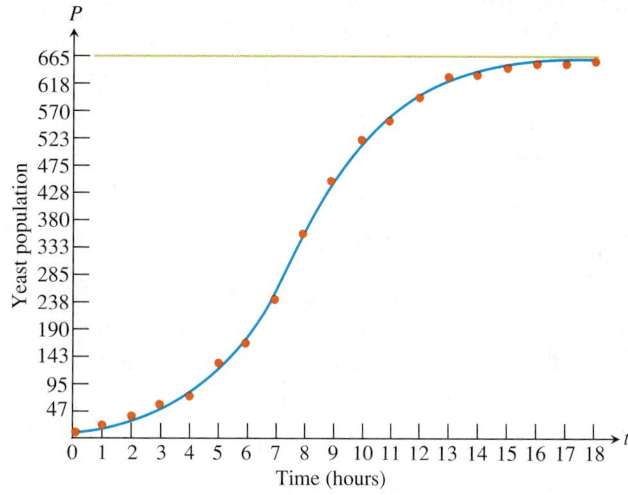

FIGURE 68 The logistic curve obtained from Equation (2) superimposed on a scatter plot of Pearl's observed data in Figure 67.

EXERCISES 7

1. *Constants of proportionality* Determine whether the following data support the stated proportionality assumption. If the assumption seems reasonable, estimate the constant of proportionality.

(a) y is proportional to x

y	1	2	3	4	5	6	7	8
x	5.9	12.1	17.9	23.9	29.9	36.2	41.8	48.2

(b) y is proportional to $x^{1/2}$

y	3.5	5	6	7	8
x	3	6	9	12	15

(c) y is proportional to 3^x

y	5	15	45	135	405	1215	3645	10,935
x	0	1	2	3	4	5	6	7

(d) y is proportional to $\ln x$

y	2	4.8	5.3	6.5	8.0	10.5	14.4	15.0
x	2.0	5.0	6.0	9.0	14.0	35.0	120.0	150.0

Constructing Models

2. *Spring elongation* The response of a spring to various loads must be modeled to design a vehicle such as a tank, dump truck, utility vehicle, or a luxury car that responds to road conditions in a desired way. We conducted an experiment to measure the stretch y of a spring in inches as a function of the number x of units of mass placed on the spring.

x (number of units of mass)	0	1	2	3	4	5
y (elongation in inches)	0	0.875	1.721	2.641	3.531	4.391

x (number of units of mass)	6	7	8	9	10
y (elongation in inches)	5.241	6.120	6.992	7.869	8.741

(a) Build a model relating the elongation of a spring to the number of units of mass.

(b) How well does your model fit the data?

(b) Predict the elongation of the spring for 13 units of mass. How comfortable are you with this prediction?

3. *Braking distance* How far does an automobile travel once the brakes have been applied? Consider the data below, where x is the speed of an automobile in miles per hour and y is the distance in feet required to stop the automobile once the brakes have been applied.

x (mph)	20	25	30	35	40	45	50	55	60	65	70	75
y (ft)	32	47	65	87	112	140	171	204	241	282	325	376

Build and test a model relating braking distance to speed.

4. *Safe following distance* Use Equation (1) for reaction distance and the model you built in Exercise 3 for braking distance to build a model for total stopping distance (reaction plus braking distances). A rule often given for safe following distance is to allow 2 seconds between your car and the car in front of you. Is this rule consistent with your model for total stopping distance? If not, suggest a better rule.

5. *Heart disease* Digoxin is used in the treatment of heart disease. Doctors must prescribe an amount that keeps the concentration of digoxin in the bloodstream above an *effective level* without exceeding a *safe level*. Begin by considering the rate of decay of digoxin in the bloodstream. Suppose that an initial dose of 0.5 mg is in the bloodstream. In the following table, x represents the number of days after taking the initial dose and y represents the amount of digoxin remaining in the bloodstream for a particular patient.

x	0	1	2	3	4	5	6	7	8
y	0.5000	0.345	0.238	0.164	0.113	0.078	0.054	0.037	0.026

(a) Build a model relating the amount of digoxin in the bloodstream to the number of days elapsed.

(b) How well does your model fit the data?

(c) Predict the amount of digoxin in the bloodstream after 12 days.

6. *Radioactivity* A radioactive dye is injected into a patient's veins to facilitate an X-ray procedure. Measuring the radioactivity in counts per minute (cpm) over the course of several minutes yielded the following table of values.

x time (min)	0	1	2	3	4	5
y radioactivity (cpm)	10,023	8174	6693	5500	4489	3683

x time (min)	6	7	8	9	10
y radioactivity (cpm)	3061	2479	2045	1645	1326

(a) Build a mathematical model relating radioactivity level to time elapsed.

(b) Compare the observations with the predictions.

(c) Use your model to predict when the radioactivity level will be below 500 cpm.

7. *Drug levels* As time passes, the blood concentration of a drug administered to laboratory animals decreases. The concentrations in parts per million (ppm) appear in the table below.

Concentration (ppm)	853	587	390	274	189	130
Time (days)	0	1	2	3	4	5

Concentration (ppm)	97	67	50	40	31
Time (days)	6	7	8	9	10

(a) Build a mathematical model relating the concentration level to the time elapsed.

(b) Compare the observations with the predictions.

(c) Use your model to predict when the concentration level will be below 10 parts per million (ppm).

CD-ROM WEBsite **8.** *Ponderosa pines* In the table, x represents the girth (distance around) of a pine tree measured in inches (in.) at shoulder height; y represents the board feet (bf) of lumber finally obtained.

x (in.)	17	19	20	23	25	28	32	38	39	41
y (bf)	19	25	32	57	71	113	123	252	259	294

Formulate and test the following two models: that usable board feet is proportional to **(a)** the square of the girth and **(b)** the cube of the girth. Which is better? Does one model provide a better "explanation" than the other?

CD-ROM WEBsite **9.** *Black bass* The following data represent the weight w in ounces of New York black bass for various lengths l in inches.

l (in.)	12.50	12.63	12.63	14.13	14.5	14.5	17.25	17.75
w (oz)	17	16	17	23	26	27	41	49

Build and test a model that assumes that weight is proportional to l^3. How well does the model fit?

CD-ROM WEBsite **10.** *Heart rate of mammals* The following data relate the weight in grams (g) of some mammals to their heart rate in beats per minute (bpm). Plot the data. Is there a trend? If so, find a function that captures the trend of the data. (*Hint:* Try models of the form $y = x^{-(1/n)}$ for n an integer.)

Mammal	**Body weight x (g)**	**Pulse rate y (bpm)**
Vesperugo pipistrellus (a really small bat)	4	660
Mouse	25	670
Rat	200	420
Guinea pig	300	300
Rabbit	2,000	205
Little dog	5,000	120
Big dog	30,000	85
Sheep	50,000	70
Human	70,000	72
Horse	450,000	38
Ox	500,000	40
Elephant	3,000,000	48

Relating Graphs to Behaviors

For Exercises 11–14, select the graph (or suggest your own) that best describes the behavior qualitatively. Explain your choice. Possible answers are (a) line (b) concave up increasing (c) concave up decreasing (d) concave down increasing (e) concave down decreasing (f) logistic

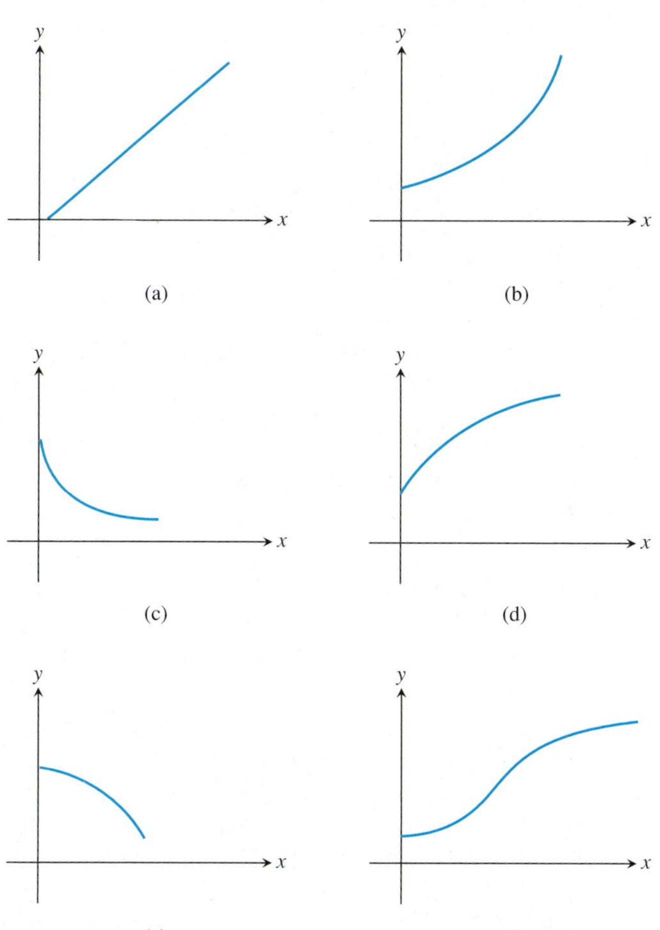

11. The concentration of a prescription drug in the bloodstream versus time

12. Your proficiency in a subject versus time spent studying it

13. The amount of carbon-14 remaining in a work of art versus time

14. Water is draining from an outlet at the bottom of a tank. Relate

(a) the velocity at which water is discharged with time elapsed.

(b) the depth of the water in the tank to the time elapsed.

15. Suggest a behavior that is qualitatively described by each of the graphs above.

16. Layers of plastic tinting are used to reduce the intensity of light. Relate the intensity of light transmitted with the number of layers of tinting.

In Exercises 17–20, sketch a graph that qualitatively describes each of the behaviors.

17. A golf ball is dropped from a height of 6 feet onto a concrete walkway. Relate the ball's height to time elapsed.

18. A marble is dropped into a bucket of oil. Relate

(a) the marble's speed with time.

(b) distance fallen with time.

19. A skydiver jumps from an airplane. After 4 seconds of free fall, the chute opens. Relate

(a) the skydiver's velocity with time.

(b) distance fallen with time.

20. An animal preserve is capable of sustaining a population level of 500 deer. Relate the deer population to time if the preserve is initially stocked with

(a) 300 deer.

(b) 500 deer.

(c) 600 deer.

21. Describe a behavior that is represented qualitatively by each of the following graphs.

(a)

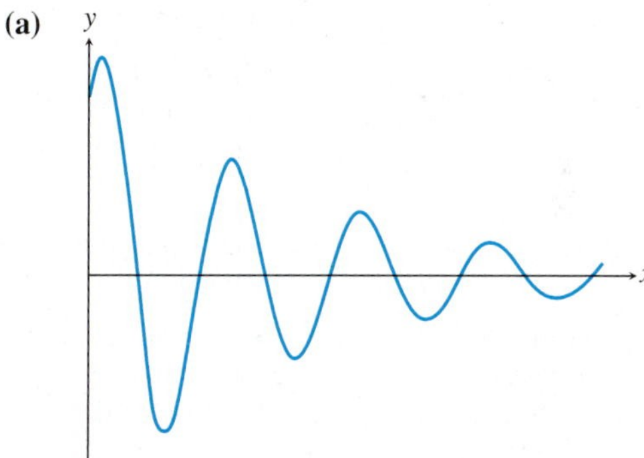

(b)

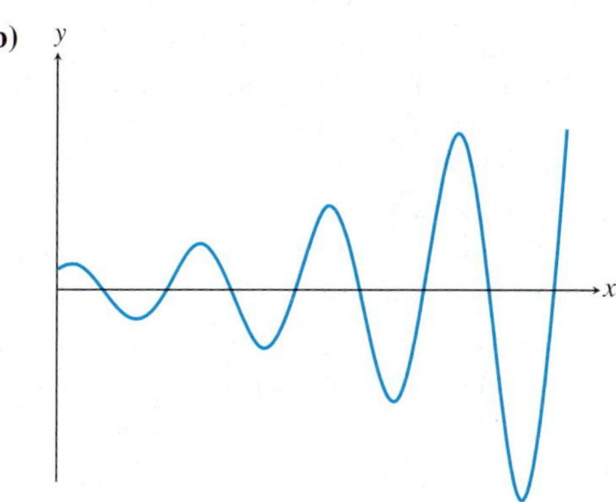

22. *Identifying a problem* From each of the following vaguely stated scenarios, identify a problem you would like to study. What

variables affect the behavior you have identified? Which variables are the most important?

(a) The population growth of a single species.

(b) An object is dropped from a great height. When and how hard will it hit the ground?

(c) How fast can a skier ski down a mountain slope?

(d) A physicist is interested in studying properties of light. She wants to understand the path of a ray of light as it travels through the air into a smooth lake, particularly at the interface of the two different media.

(e) The U.S. Food and Drug Administration is interested in knowing whether a new drug is effective in the control of a certain disease in the population.

(f) A retail store intends to construct a new parking lot. How should the lot be illuminated?

Questions to Guide Your Review

1. How can you write an equation for a line if you know the coordinates of two points on the line? The line's slope and the coordinates of one point on the line? The line's slope and y-intercept? Give examples.
2. What are the standard equations for lines perpendicular to the coordinate axes?
3. How are the slopes of *mutually* perpendicular lines related? What about parallel lines? Give examples.
4. What is a function? Give examples. How do you graph a real valued function whose domain and range are real numbers? What is an increasing function? A decreasing function?
5. What is an even function? An odd function? What symmetry properties do the graphs of such functions have? What advantage can we make of this? Give an example of a function that is neither even nor odd.
6. What is a piecewise defined function? Give examples. Define the absolute value function and draw its graph.
7. When is it possible to compose one function with another? Give examples of composites and their values at various points. Does the order in which functions are composed ever matter?
8. How do you change the equation $y = f(x)$ to shift its graph up or down? To the left or right? Give examples.
9. What is an exponential function? Give examples. What laws of exponents does it obey? How does it differ from a simple power function like $f(x) = x^n$? What kind of real-world phenomena are modeled by exponential functions?
10. What is the number e, and how is it defined? What are the domain and range of $f(x) = e^x$? What does its graph look like? How do the values of e^x relate to x^2, x^3, and so on?
11. What functions have inverses? How do you know if two functions f and g are inverses of one another? Give examples of functions that are (are not) inverses of one another.
12. How are the domains, ranges, and graphs of functions and their inverses related? Give an example.
13. What procedure can you sometimes use to express the inverse of a function of x as a function of x? How can you graph a function $y = f(x)$ together with its inverse $y = f^{-1}(x)$ parametrically on a graphing calculator or computer?
14. What is a logarithm function? What properties does it satisfy? What is the natural logarithm function? What are the domain and range of $y = \ln x$? What does its graph look like?
15. How is the graph of $\log_a x$ related to the graph of $\ln x$? What truth is there in the statement that there is really only one exponential function and one logarithmic function?
16. What is radian measure? How do you convert from radians to degrees? Degrees to radians?
17. Graph the six basic trigonometric functions. What symmetries do the graphs have?
18. How can you sometimes find values of trigonometric functions from triangles? Give examples.
19. What is a periodic function? Give examples. What are the periods of the six basic trigonometric functions?
20. How does the formula for the general sine function $f(x) = A \sin((2\pi/B)(x - C)) + D$ relate to the shifting, stretching, shrinking and reflection of its graph? Give examples. Graph the general sine curve and identify the constants A, B, C, and D.
21. Starting with the identity $\cos^2\theta + \sin^2\theta = 1$ and the formulas for $\cos(A + B)$ and $\sin(A + B)$, show how a variety of other trigonometric identities may be derived.
22. How are the inverse trigonometric functions defined? How can you sometimes use right triangles to find values of these functions? Give examples.
23. How can you find values of $\sec^{-1} x$, $\csc^{-1} x$, and $\cot^{-1} x$ using a calculator's keys for $\cos^{-1} x$, $\sin^{-1} x$, and $\tan^{-1} x$?
24. What is a parametrized curve in the xy plane? What is the initial point of the curve? The terminal point? If you find a Cartesian equation for the path of a particle whose motion in the plane is described parametrically, what kind of match can you expect between the Cartesian equation's graph and the path of motion? Give examples.
25. What is the standard parametrization for the circle $x^2 + y^2 = a^2$? For the ellipse $(x^2/a^2) + (y^2/b^2) = 1$? The graph of a function $y = f(x)$? The inverse of a function $y = f(x)$?

Practice Exercises

Lines

In Exercises 1–12, write an equation for the specified line.

1. through $(1, -6)$ with slope 3
2. through $(-1, 2)$ with slope $-1/2$
3. the vertical line through $(0, -3)$
4. through $(-3, 6)$ and $(1, -2)$
5. the horizontal line through $(0, 2)$
6. through $(3, 3)$ and $(-2, 5)$
7. with slope -3 and y-intercept 3
8. through $(3, 1)$ and parallel to $2x - y = -2$
9. through $(4, -12)$ and parallel to $4x + 3y = 12$
10. through $(-2, -3)$ and perpendicular to $3x - 5y = 1$
11. through $(-1, 2)$ and perpendicular to $(1/2)x + (1/3)y = 1$
12. with x-intercept 3 and y-intercept -5

Functions and Graphs

13. Express the area and circumference of a circle as functions of the circle's radius. Then express the area as a function of the circumference.
14. Express the radius of a sphere as a function of the sphere's surface area. Then express the surface area as a function of the volume.
15. A point P in the first quadrant lies on the parabola $y = x^2$. Express the coordinates of P as functions of the angle of inclination of the line joining P to the origin.
16. A hot-air balloon rising straight up from a level field is tracked by a range finder located 500 ft from the point of liftoff. Express the balloon's height as a function of the angle the line from the range finder to the balloon makes with the ground.

In Exercises 17–20, determine whether the graph of the function is symmetric about the y-axis, the origin, or neither.

17. $y = x^{1/5}$
18. $y = x^{2/5}$
19. $y = x^2 - 2x - 1$
20. $y = e^{-x^2}$

In Exercises 21–28, determine whether the function is even, odd, or neither.

21. $y = x^2 + 1$
22. $y = x^5 - x^3 - x$
23. $y = 1 - \cos x$
24. $y = \sec x \tan x$
25. $y = \dfrac{x^4 + 1}{x^3 - 2x}$
26. $y = 1 - \sin x$
27. $y = x + \cos x$
28. $y = \sqrt{x^4 - 1}$

In Exercises 29–38, find the **(a)** domain and **(b)** range.

29. $y = |x| - 2$
30. $y = -2 + \sqrt{1 - x}$
31. $y = \sqrt{16 - x^2}$
32. $y = 3^{2-x} + 1$
33. $y = 2e^{-x} - 3$
34. $y = \tan(2x - \pi)$
35. $y = 2\sin(3x + \pi) - 1$
36. $y = x^{2/5}$
37. $y = \ln(x - 3) + 1$
38. $y = -1 + \sqrt[3]{2 - x}$

Piecewise-Defined Functions

In Exercises 39 and 40, find the **(a)** domain and **(b)** range.

39. $y = \begin{cases} \sqrt{-x}, & -4 \le x \le 0 \\ \sqrt{x}, & 0 < x \le 4 \end{cases}$

40. $y = \begin{cases} -x - 2, & -2 \le x \le -1 \\ x, & -1 < x \le 1 \\ -x + 2, & 1 < x \le 2 \end{cases}$

In Exercises 41 and 42, write a piecewise formula for the function.

41.

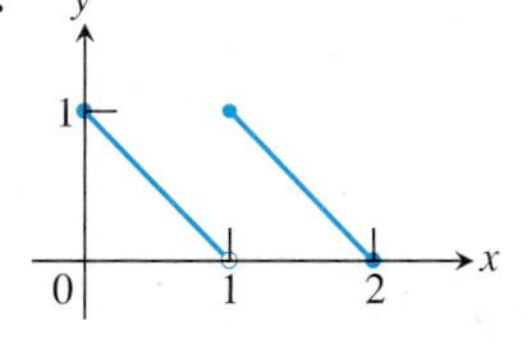

42.

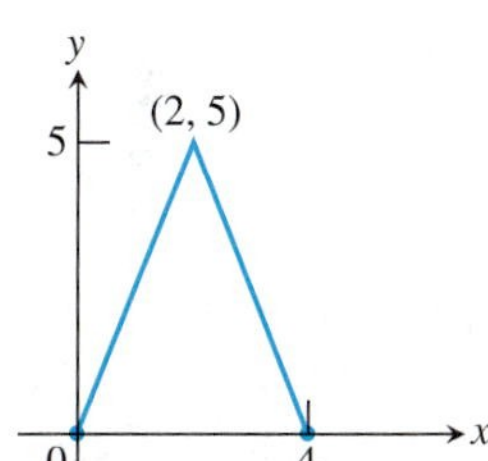

Composition of Functions

In Exercises 43 and 44, find

(a) $(f \circ g)(-1)$ **(b)** $(g \circ f)(2)$

(c) $(f \circ f)(x)$ **(d)** $(g \circ g)(x)$

43. $f(x) = \dfrac{1}{x}, \quad g(x) = \dfrac{1}{\sqrt{x + 2}}$
44. $f(x) = 2 - x, \quad g(x) = \sqrt[3]{x + 1}$

In Exercises 45 and 46, **(a)** write a formula for $f \circ g$ and $g \circ f$ and find the **(b)** domain and **(c)** range of each.

45. $f(x) = 2 - x^2, \quad g(x) = \sqrt{x + 2}$
46. $f(x) = \sqrt{x}, \quad g(x) = \sqrt{1 - x}$

Composition with absolute values In Exercises 47–52, graph f_1 and f_2 together. Then describe how applying the absolute value function before applying f_1 affects the graph.

$f_1(x)$	$f_2(x) = f_1(\|x\|)$
47. x	$\|x\|$
48. x^3	$\|x\|^3$
49. x^2	$\|x\|^2$
50. $\frac{1}{x}$	$\frac{1}{\|x\|}$
51. $\sqrt{x}$	$\sqrt{\|x\|}$
52. $\sin x$	$\sin \|x\|$

Composition with absolute values In Exercises 53–56, graph g_1 and g_2 together. Then describe how taking absolute values after applying g_1 affects the graph.

$g_1(x)$	$g_2(x) = \|g_1(x)\|$
53. x^3	$\|x^3\|$
54. $\sqrt{x}$	$\|\sqrt{x}\|$
55. $4 - x^2$	$\|4 - x^2\|$
56. $x^2 + x$	$\|x^2 + x\|$

Inverse Functions

57. **(a)** Graph the function $f(x) = \sqrt{1 - x^2}$, $0 \le x \le 1$. What symmetry does the graph have?

(b) Show that f is its own inverse. (Remember that $\sqrt{x^2} = x$ if $x \ge 0$.)

58. **(a)** Graph the function $f(x) = 1/x$. What symmetry does the graph have?

(b) Show that f is its own inverse.

In Exercises 59 and 60,

(a) find f^{-1} and show that $(f \circ f^{-1})(x) = (f^{-1} \circ f)(x) = x$.

(b) graph f and f^{-1} together.

59. $f(x) = 2 - 3x$

60. $f(x) = (x + 2)^2, \quad x \ge -2$

61. **(a)** Show that $f(x) = x^3$ and $g(x) = \sqrt[3]{x}$ are inverses of one another.

T **(b)** Graph f and g over an x-interval large enough to show the graphs intersecting at $(1, 1)$ and $(-1, -1)$. Be sure the picture shows the required symmetry in the line $y = x$.

62. **(a)** Show that $h(x) = x^3/4$ and $k(x) = (4x)^{1/3}$ are inverses of one another.

T **(b)** Graph h and k over an x-interval large enough to show the graphs intersecting at $(2, 2)$ and $(-2, -2)$. Be sure the picture shows the required symmetry about the line $y = x$.

63. **(a)** Find the inverse of $f(x) = x + 1$. Graph f and its inverse together. Add the line $y = x$ to your sketch, drawing it with dashes or dots for contrast.

(b) Find the inverse of $f(x) = x + b$ (b constant). How is the graph of f^{-1} related to the graph of f?

(c) What can you conclude about the inverses of functions whose graphs are lines parallel to the line $y = x$?

64. **(a)** Find the inverse of $f(x) = -x + 1$. Graph the line $y = -x + 1$ together with the line $y = x$. At what angle do the lines intersect?

(b) Find the inverse of $f(x) = -x + b$ (b constant). What angle does the line $y = -x + b$ make with the line $y = x$?

(c) What can you conclude about the inverses of functions whose graphs are lines perpendicular to the line $y = x$?

T **65.** *A decimal representation of e* Find e to as many decimal places as your calculator allows by solving the equation $\ln x = 1$.

T **66.** *The inverse relation between e^x and $\ln x$* Find out how good your calculator is at evaluating the composites

$$e^{\ln x} \quad \text{and} \quad \ln(e^x).$$

Algebraic Calculations with the Exponential and Logarithm

Find simpler expressions for the quantities in Exercises 67–70.

67. **(a)** $e^{\ln 7.2}$ **(b)** $e^{-\ln x^2}$ **(c)** $e^{\ln x - \ln y}$

68. **(a)** $e^{\ln(x^2+y^2)}$ **(b)** $e^{-\ln 0.3}$ **(c)** $e^{\ln \pi x - \ln 2}$

69. **(a)** $2 \ln \sqrt{e}$ **(b)** $\ln(\ln e^e)$ **(c)** $\ln(e^{-x^2-y^2})$

70. **(a)** $\ln(e^{\sec \theta})$ **(b)** $\ln(e^{e^x})$ **(c)** $\ln(e^{2 \ln x})$

Trigonometry

In Exercises 71 and 72, find the measure of the angle in radians and degrees.

71. $\sin^{-1}(0.6)$ **72.** $\tan^{-1}(-2.3)$

73. Find the six trigonometric values of $\theta = \cos^{-1}(3/7)$. Give exact answers.

74. Solve the equation $\sin x = -0.2$ in the following intervals.

(a) $0 \le x < 2\pi$ **(b)** $-\infty < x < \infty$

In Exercises 75 and 76, sketch the graph of the given function. What is the period of the function?

75. $y = \sin \frac{x}{2}$

76. $y = \cos \frac{\pi x}{2}$

77. Sketch the graph $y = 2\cos\left(x - \dfrac{\pi}{3}\right)$.

78. Sketch the graph $y = 1 + \sin\left(x + \dfrac{\pi}{4}\right)$.

In Exercises 79–82, ABC is a right triangle with the right angle at C. The sides opposite angles A, B, and C are a, b, and c, respectively.

79. (a) Find a and b if $c = 2$, $B = \pi/3$.

(b) Find a and c if $b = 2$, $B = \pi/3$.

80. (a) Express a in terms of A and c.

(b) Express a in terms of A and b.

81. (a) Express a in terms of B and b.

(b) Express c in terms of A and a.

82. (a) Express $\sin A$ in terms of a and c.

(b) Express $\sin A$ in terms of b and c.

In Exercises 83 and 84, show that the function is periodic and find its period.

83. $y = \sin^3 x$ 84. $y = |\tan x|$

Use the angle sum formulas to derive the identities in Exercises 85 and 86.

85. $\cos\left(x + \dfrac{\pi}{2}\right) = -\sin x$

86. $\sin\left(x - \dfrac{\pi}{2}\right) = -\cos x$

87. Evaluate $\sin\dfrac{7\pi}{12}$ as $\sin\left(\dfrac{\pi}{4} + \dfrac{\pi}{3}\right)$.

88. Evaluate $\cos\dfrac{11\pi}{12}$ as $\cos\left(\dfrac{\pi}{4} + \dfrac{2\pi}{3}\right)$.

Use reference triangles to find the angles in Exercises 89–92.

89. (a) $\sin^{-1}\left(\dfrac{1}{2}\right)$ (b) $\sin^{-1}\left(\dfrac{-1}{\sqrt{2}}\right)$ (c) $\sin^{-1}\left(\dfrac{\sqrt{3}}{2}\right)$

90. (a) $\cos^{-1}\left(\dfrac{-1}{2}\right)$ (b) $\cos^{-1}\left(\dfrac{1}{\sqrt{2}}\right)$ (c) $\cos^{-1}\left(\dfrac{-\sqrt{3}}{2}\right)$

91. (a) $\sec^{-1}\sqrt{2}$ (b) $\sec^{-1}\left(\dfrac{-2}{\sqrt{3}}\right)$ (c) $\sec^{-1} 2$

92. (a) $\cot^{-1} 1$ (b) $\cot^{-1}(-\sqrt{3})$ (c) $\cot^{-1}\left(\dfrac{1}{\sqrt{3}}\right)$

Evaluating Trigonometric and Inverse Trigonometric Functions

Find the values in Exercises 93–96.

93. $\sec\left(\cos^{-1}\dfrac{1}{2}\right)$

94. $\cot\left(\sin^{-1}\left(-\dfrac{\sqrt{3}}{2}\right)\right)$

95. $\tan(\sec^{-1} 1) + \sin(\csc^{-1}(-2))$

96. $\sec(\tan^{-1} 1 + \csc^{-1} 1)$

Evaluating Trigonometric Expressions

Evaluate the expressions in Exercises 97–100.

97. $\sec(\tan^{-1} 2x)$

98. $\tan\left(\sec^{-1}\dfrac{y}{5}\right)$

99. $\tan(\cos^{-1} x)$

100. $\sin\left(\tan^{-1}\dfrac{x}{\sqrt{x^2+1}}\right)$

Which of the expressions in Exercises 101–104 are defined and which are not? Give reasons for your answers.

101. (a) $\tan^{-1} 2$ (b) $\cos^{-1} 2$

102. (a) $\csc^{-1}\dfrac{1}{2}$ (b) $\csc^{-1} 2$

103. (a) $\sec^{-1} 0$ (b) $\sin^{-1}\sqrt{2}$

104. (a) $\cot^{-1}\left(\dfrac{1}{-2}\right)$ (b) $\cos^{-1}(-5)$

105. *Height of a pole* Two wires stretch from the top T of a vertical pole to points B and C on the ground, where C is 10 m closer to the base of the pole ,than is B. If wire BT makes an angle of 35° with the horizontal and wire CT makes an angle of 50° with the horizontal, how high is the pole?

106. *Height of a weather balloon* Observers at positions A and B 2 km apart simultaneously measure the angle of elevation of a weather balloon to be 40° and 70°, respectively. If the balloon is directly above a point on the line segment between A and B, find the height of the balloon.

T 107. (a) Graph the function $f(x) = \sin x + \cos(x/2)$.

(b) What appears to be the period of this function?

(c) Confirm your finding in part (b) algebraically.

T 108. (a) Graph $f(x) = \sin(1/x)$.

(b) What are the domain and range of f?

(c) Is f periodic? Give reasons for your answer.

Parametrizations

In Exercises 109–112, a parametrization is given for a curve.

(a) Find a Cartesian equation for a curve that contains the parametrized curve. What portion of the graph of the Cartesian equation is traced by the parametrized curve?

(b) Graph the curve. Identify the initial and terminal points, if any. Indicate the direction in which the curve is traced.

109. $x = 5\cos t, \quad y = 2\sin t, \quad 0 \le t \le 2\pi$

110. $x = 4\cos t, \quad y = 4\sin t, \quad \pi/2 \le t < 3\pi/2$

111. $x = 2 - t, \quad y = 11 - 2t, \quad -2 \le t \le 4$

112. $x = 1 + t, \quad y = (t - 1)^2, \quad t \le 1$

In Exercises 113–116, give a parametrization for the curve.

113. the line segment with endpoints $(-2, 5)$ and $(4, 3)$

114. the line through $(-3, -2)$ and $(4, -1)$

115. the ray with initial point $(2, 5)$ that passes through $(-1, 0)$

116. $y = x(x - 4), \quad x \le 2$

Additional Exercises: Theory, Examples, Applications

1. The graph of f is shown. Draw the graph of each function.

(a) $y = f(-x)$ **(b)** $y = -f(x)$

(c) $y = -2f(x + 1) + 1$ **(d)** $y = 3f(x - 2) - 2$

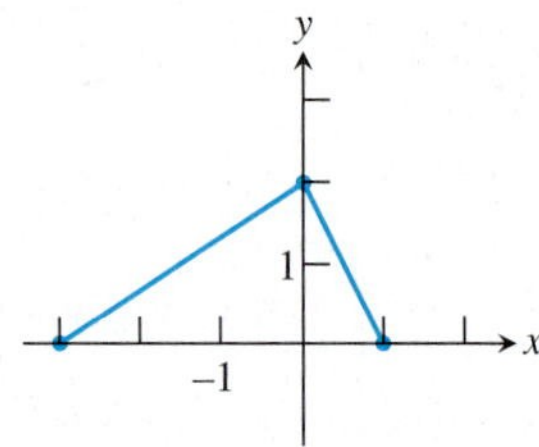

2. A portion of the graph of a function defined on $[-3, 3]$ is shown. Complete the graph assuming that the function is

(a) even. **(b)** odd.

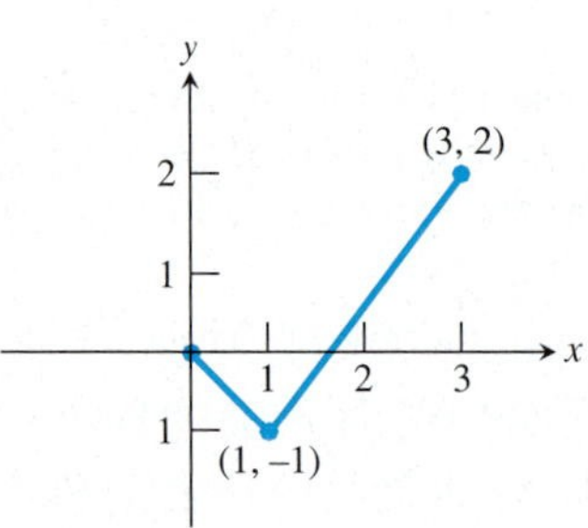

3. *Depreciation* Smith Hauling purchased an 18-wheel truck for \$100,000. The truck depreciates at the constant rate of \$10,000 per year for 10 years.

(a) Write an expression that gives the value y after x years.

(b) When is the value of the truck \$55,000?

4. *Drug absorption* A drug is administered intravenously for pain. The function

$$f(t) = 90 - 52\ln(1 + t), \qquad 0 \le t \le 4$$

gives the number of units of the drug remaining in the body after t hours.

(a) What was the initial number of units of the drug administered?

(b) How much is present after 2 hours?

(c) Draw the graph of f.

5. *Finding time* If Juanita invests \$1500 in a retirement account that earns 8% compounded annually, how long will it take this single payment to grow to \$5000?

6. Explain the following "proof without words" of the law of cosines. (Source: Sidney H. Kung, "Proof without Words: The Law of Cosines," *Mathematics Magazine*, Vol. 63, No. 5, Dec. 1990, p. 342.)

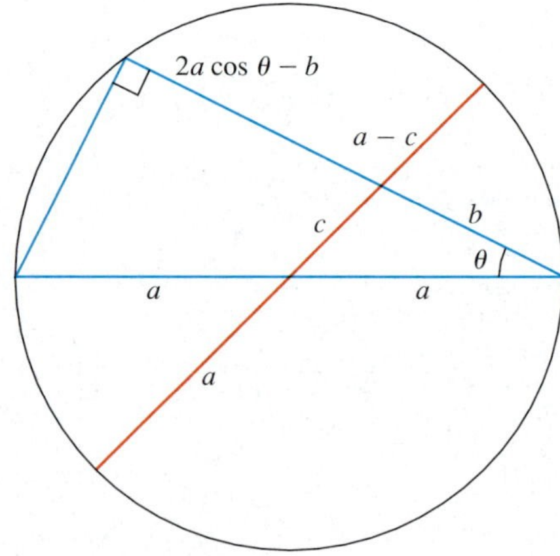

7. Show that the area of triangle ABC is given by $(1/2)ab\sin C = (1/2)bc\sin A = (1/2)ca\sin B$.

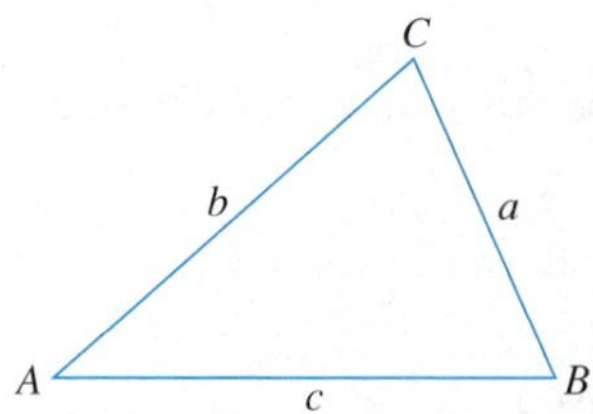

8. (a) Find the slope of the line from the origin to the midpoint P of side AB in the triangle in the accompanying figure $(a, b > 0)$.

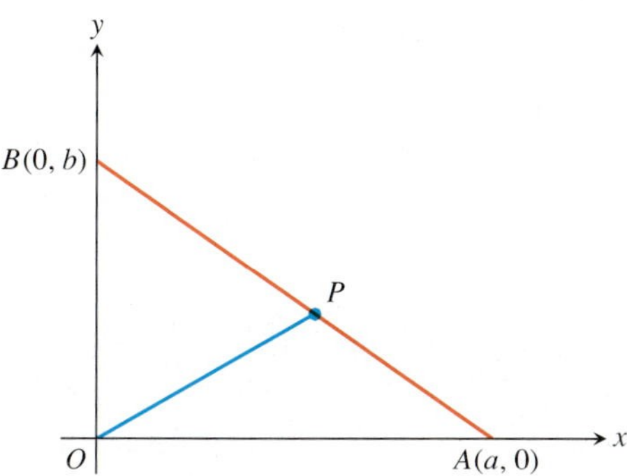

(b) When is OP perpendicular to AB?

9. The figure here shows an informal proof that

$$\tan^{-1}\frac{1}{2} + \tan^{-1}\frac{1}{3} = \frac{\pi}{4}.$$

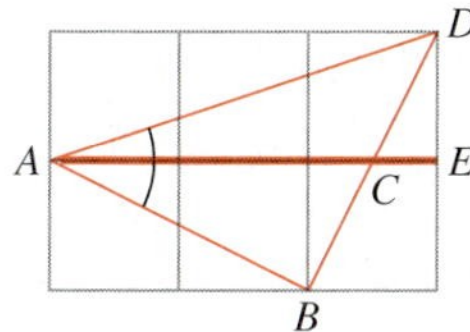

How does the argument go? (Source: Edward M. Harris, "Behold! Sums of Arctan," *College Mathematics Journal*, Vol. 18, No. 2, March 1987, p. 141.)

10. *Writing to Learn* For what $x > 0$, does $x^{(x^x)} = (x^x)^x$? Give reasons for your answer.

11. *Composition with an odd function*

(a) Let $h = g \circ f$ where g is an even function. Is h always an even function? Give reasons for your answer.

(b) Let $h = g \circ f$ where g is an odd function. Is h always an odd function? What if f is odd? What if f is even? Give reasons for your answer.

12. *The rule of 70* If you use the approximation $\ln 2 \approx 0.70$ (in place of $0.69314\ldots$), you can derive a rule of thumb that says, "To estimate how many years it will take an amount of money to double when invested at r percent compounded continuously, divide r into 70." For instance, an amount of money invested at 5% will double in about $70/5 = 14$ years. If you want it to double in 10 years instead, you have to invest it at $70/10 = 7\%$. Show how the rule of 70 is derived. (A similar "rule of 72" uses 72 instead of 70, because 72 has more integer factors.)

Functions and Graphs

13. Are there two functions f and g such that $f \circ g = g \circ f$? Give reasons for your answer.

14. Are there two functions f and g with the following property? The graphs of f and g are not straight lines but the graph of $f \circ g$ is a straight line. Give reasons for your answer.

15. If $f(x)$ is odd, can anything be said of $g(x) = f(x) - 2$? What if f is even instead? Give reasons for your answer.

16. If $g(x)$ is an odd function defined for all values of x, can anything be said about $g(0)$? Give reasons for your answer.

17. Graph the equation $|x| + |y| = 1 + x$.

18. Graph the equation $y + |y| = x + |x|$.

19. Show that if f is both even and odd, then $f(x) = 0$ for every x in the domain of f.

20. (a) *Even-odd decompositions* Let f be a function whose domain is symmetric about the origin; that is, $-x$ belongs to the domain whenever x does. Show that f is the sum of an even function and an odd function:

$$f(x) = E(x) + O(x),$$

where E is an even function and O is an odd function. (*Hint:* Let $E(x) = [f(x) + f(-x)]/2$. Show that $E(-x) = E(x)$, so that E is even. Then show that $O(x) = f(x) - E(x)$ is odd.)

(b) *Uniqueness* Show that there is only one way to write f as the sum of an even and an odd function. (*Hint:* One way is given in part (a). If also $f(x) = E_1(x) + O_1(x)$, where E_1 is even and O_1 is odd, show that $E - E_1 = O_1 - O$. Then use Exercise 19 to show that $E = E_1$ and $O = O_1$.)

21. *One-to-one functions* If f is a one-to-one function, prove that $g(x) = -f(x)$ is also one-to-one.

22. *One-to-one functions* If f is a one-to-one function and $f(x)$ is never zero, prove that $g(x) = 1/f(x)$ is also one-to-one.

23. *Domain and range* Suppose that $a \neq 0$, $b \neq 1$, and $b > 0$. Determine the domain and range of the function.

(a) $y = a(b^{c-x}) + d$ **(b)** $y = a\log_b(x - c) + d$

24. *Inverse functions* Let

$$f(x) = \frac{ax + b}{cx + d}, \quad c \neq 0, \quad ad - bc \neq 0.$$

(a) *Writing to Learn* Give a convincing argument that f is one-to-one.

(b) Find a formula for the inverse of f.

(c) Find the horizontal and vertical asymptotes of f.

(d) Find the horizontal and vertical asymptotes of f^{-1}. How are they related to those of f?

Modeling

25. *Constants of proportionality* Determine whether the following data support the stated proportionality assumption. If the assumption seems reasonable, estimate the constant of proportionality.

(a) y is proportional to x^2

y	6	13	24	39	58	81	108	139
x	0	1	2	3	4	5	6	7

(b) y is proportional to 4^x

y	0.6	2.4	9.6	38.4	153.6	614.4	2457.6	9830.4
x	0	1	2	3	4	5	6	7

26. *Cell count* A study of certain bacterial growth gives the cell counts shown.

x time (hr)	0	2	4	6	8	10
y cell count	597	893	1339	1995	2976	4433

x time (hr)	12	14	16	18	20
y cell count	6612	9865	14,719	21,956	32,763

(a) Build a mathematical model relating cell count to time elapsed.

(b) Compare the observations with the predictions.

(c) Use your model to predict when the cell count will reach 50,000.

27. *Spring elongation* The following table gives the elongation e in inches per inch (in./in.) for a given stress S on a steel wire measured in pounds per square in (lb/in.^{2}). Test the model $e = c_1 S$ by plotting the data. Estimate c_1 graphically.

$S \times 10^{-3}$	5	10	20	30	40	50
$e \times 10^5$	0	19	57	94	134	173

$S \times 10^{-3}$	60	70	80	90	100
$e \times 10^5$	216	256	297	343	390

(a) Build a model relating the elongation of a spring to the number of units of mass.

(b) How well does your model fit the data?

(c) Predict the elongation of the spring for a stress S of 200×10^{-3} lb/in.^{2}. How comfortable are you with this prediction?

28. *Vacuum pump* A mechanical vacuum pump, often called a "roughing pump," is used to evacuate a chamber of air. A pressure gauge measures the pressure in atmospheres (Pa). The data are recorded below.

Pressure (Pa)	100,000	36,788	13,537	4986	1837	671
Time (min)	0	1	2	3	4	5

(a) Build a model relating the pressure in the chamber to time elapsed.

(b) How well does your model fit the data?

(c) Predict when the pressure in the chamber will reach 200 Pa.

Regression Analysis

T **29.** *Doctoral degrees* Table 25 shows the number of doctoral degrees earned in the given academic year by Hispanic students. Let $x = 0$ represent 1970–71, $x = 1$ represent 1971–72, and so forth.

Table 25 Doctorates earned by Hispanic Americans

Year	Number of Degrees
1976–77	520
1980–81	460
1984–85	680
1988–89	630
1990–91	730
1991–92	810
1992–93	830

Source: U.S. Department of Education, as reported in the *Chronicle of Higher Education*, April 28, 1995.

(a) Find a linear regression equation for the data and superimpose its graph on a scatter plot of the data.

(b) Use the regression equation to predict the number of doctoral degrees that will be earned by Hispanic Americans in the academic year 2000–01.

(c) *Writing to Learn* Find the slope of the regression line. What does the slope represent?

T **30.** *Estimating population growth* Use the data in Table 26 about the population of New York State. Let $x = 60$ represent 1960, $x = 70$ represent 1970, an so forth.

Table 26 Population of New York State

Year	Population (millions)
1960	16.78
1980	17.56
1990	17.99

Source: The Statesman's Yearbook, 129th ed. (London: The Macmillan Press, Ltd., 1992).

(a) Find an exponential regression equation for the data.

(b) Use the regression equation to predict when the population will be 25 million.

(c) What annual rate of growth can we infer from the regression equation?

T **31.** *Sinusoidal regression* Table 27 gives the values of the function

$$f(x) = a \sin(bx + c) + d$$

accurate to two decimals.

(a) Find a sinusoidal regression equation for the data.

(b) Rewrite the equation with a, b, c, and d rounded to the nearest integer.

Table 27 Values of a function

x	$f(x)$
1	5.82
2	2.08
3	5.98
4	2.00
5	5.98
6	2.08

T **32.** *Oil production*

(a) Find a natural logarithm regression equation for the data in Table 28.

(b) Estimate the number of metric tons of oil produced by Canada in 1985.

(c) Predict when Canadian oil production will reach 120 metric tons.

Table 28 Canadian oil production

Year	Metric tons (millions)
1960	27.48
1970	69.95
1990	92.24

Source: The Statesman's Yearbook, 129th ed. (London: The Macmillan Press, Ltd., 1992).

T **33.** Table 29 gives some hypothetical data about energy consumption.

(a) Let $x = 0$ represent 1900, $x = 1$ represent 1910, and so forth. Find an exponential regression equation of the form $Q = ae^{bx}$ for the data and superimpose its graph on a scatter plot of the data.

(b) Use the exponential regression equation to estimate the energy consumption in 1996. What is the annual growth rate of energy consumption during the twentieth century?

Table 29 Energy consumption

Year	Consumption Q
1900	1.00
1910	2.01
1920	4.06
1930	8.17
1940	16.44
1950	33.12
1960	66.69
1970	134.29
1980	270.43
1990	544.57
2000	1096.63

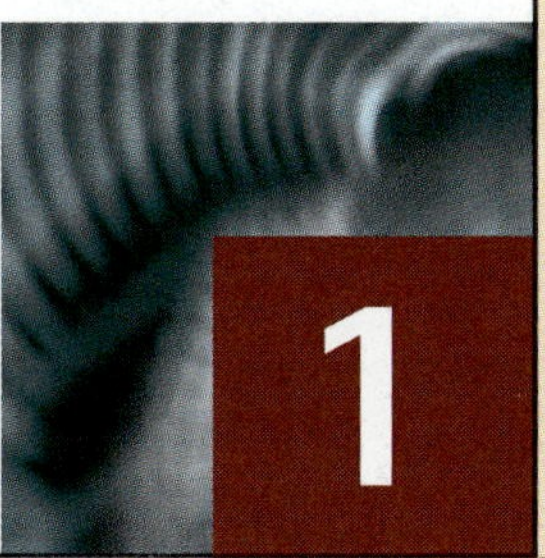

1 Limits and Continuity

OVERVIEW The concept of limit is one of the ideas that distinguish calculus from algebra and trigonometry. In this chapter, we show how to define and calculate limits of function values. The calculation rules are straightforward, and most of the limits we need can be found by substitution, graphical investigation, numerical approximation, algebra, or some combination of these.

The outputs of some functions vary continuously as their inputs vary—the smaller the change in the input, the smaller the change in the output. The values of other functions may jump or vary erratically no matter how carefully we control the inputs. The notion of limit gives a precise way to distinguish between these behaviors. We also use limits to define tangent lines to graphs of functions. This geometric application leads at once to the important concept of derivative of a function. The derivative, which we investigate thoroughly in Chapter 2, gives us a way to measure numerically the rate at which a function's values are changing at any instant.

1.1 Rates of Change and Limits

Average and Instantaneous Speed • Average Rates of Change and Secant Lines • Limits of Functions • Informal Definition of Limit • Precise Definition of Limit

CD-ROM WEBsite

In this section, we introduce average and instantaneous rates of change. These lead to the main idea of the section, the idea of limit.

CD-ROM WEBsite

Historical Biography

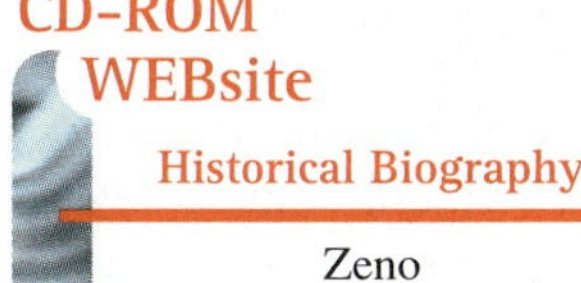

Zeno (490BC — 430BC)

Average and Instantaneous Speed

A moving body's **average speed** during an interval of time is found by dividing the distance covered by the time elapsed. The unit of measure is length per unit time: kilometers per hour, feet per second, or whatever is appropriate to the problem at hand.

Example 1 Finding an Average Speed

A rock breaks loose from the top of a tall cliff. What is its average speed during the first 2 sec of fall?

Solution Experiments show that a dense solid object dropped from rest to fall freely near the surface of the earth will fall

$$y = 16t^2$$

Free Fall

Near the surface of the earth, all bodies fall with the same constant acceleration. The distance a body falls after it is released from rest is a constant multiple of the square of the time fallen. At least, that is what happens when a body falls in a vacuum, where there is no air to slow it down. The square-of-time rule also holds for dense, heavy objects like rocks, ball bearings, and steel tools during the first few seconds of fall through air, before the velocity builds up to where air resistance begins to matter. When air resistance is absent or insignificant and the only force acting on a falling body is the force of gravity, we call the way the body falls *free fall.*

feet in the first t sec. The average speed of the rock over any given time interval is the distance traveled, Δy, divided by the length of the interval Δt. For the first 2 sec of fall, from $t = 0$ to $t = 2$, we have

$$\frac{\Delta y}{\Delta t} = \frac{16(2)^2 - 16(0)^2}{2 - 0} = 32 \text{ ft/sec}.$$

Example 2 Finding an Instantaneous Speed

Find the speed of the rock in Example 1 at the instant $t = 2$.

Solution

Solve Numerically

We can calculate the average speed of the rock over the interval from time $t = 2$ to any slightly later time $t = 2 + h$, $h > 0$, as

$$\frac{\Delta y}{\Delta t} = \frac{16(2 + h)^2 - 16(2)^2}{h}. \qquad (1)$$

We cannot use this formula to calculate the speed at the exact instant $t = 2$ because that would require taking $h = 0$, and $0/0$ is undefined. We can, however, get a good idea of what is happening at $t = 2$ by evaluating the formula at values of h *close* to 0. When we do, we see a clear pattern (Table 1.1).

As h approaches 0, the average speed approaches the limiting value 64 ft/sec.

Confirm Algebraically

If we expand the numerator of Equation (1) and simplify, we find that

$$\frac{\Delta y}{\Delta t} = \frac{16(2 + h)^2 - 16(2)^2}{h} = \frac{16(4 + 4h + h^2) - 64}{h}$$

$$= \frac{64h + 16h^2}{h} = 64 + 16h.$$

For values of h different from 0, the expressions on the right and left are equivalent and the average speed is $64 + 16h$ ft/sec. We can now see why the average speed has the limiting value $64 + 16(0) = 64$ ft/sec as h approaches 0.

Table 1.1 Average speeds over short time intervals starting at $t = 2$

$$\frac{\Delta y}{\Delta t} = \frac{16(2 + h)^2 - 16(2)^2}{h}$$

Length of time interval, h (sec)	Average speed for interval $\Delta y/\Delta t$ (ft/sec)
1	80
0.1	65.6
0.01	64.16
0.001	64.016
0.0001	64.0016
0.00001	64.00016

Average Rates of Change and Secant Lines

Given an arbitrary function $y = f(x)$, we calculate the average rate of change of y with respect to x over the interval $[x_1, x_2]$ by dividing the change in the value of y, $\Delta y = f(x_2) - f(x_1)$, by the length $\Delta x = x_2 - x_1 = h$ of the interval over which the change occurs.

Definition Average Rate of Change

The **average rate of change** of $y = f(x)$ with respect to x over the interval $[x_1, x_2]$ is

$$\frac{\Delta y}{\Delta x} = \frac{f(x_2) - f(x_1)}{x_2 - x_1} = \frac{f(x_1 + h) - f(x_1)}{h}, \qquad h \neq 0.$$

Geometrically, an average rate of change is a secant slope.

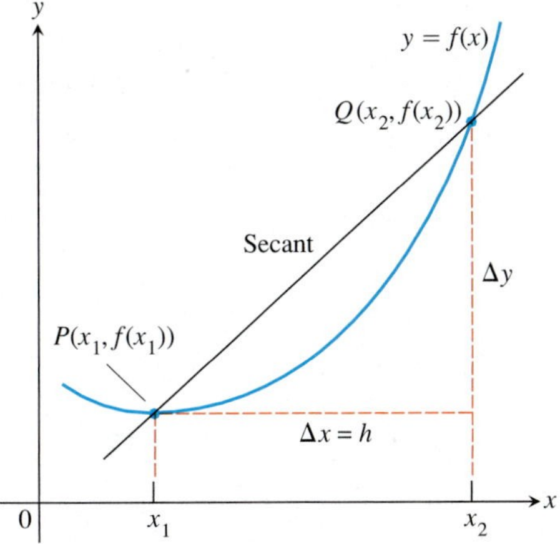

FIGURE 1.1 A secant to the graph $y = f(x)$. Its slope is $\Delta y/\Delta x$, the average rate of change of f over the interval $[x_1, x_2]$.

Notice that the rate of change of f over $[x_1, x_2]$ is the slope of the line through the points $P(x_1, f(x_1))$ and $Q(x_2, f(x_2))$ (Figure 1.1). In geometry, a line joining two points of a curve is a **secant** to the curve. Thus, the average rate of change of f from x_1 to x_2 is identical with the slope of secant PQ.

Engineers often want to know the rates at which temperatures change in materials to determine if cracks or other breaks will occur.

Example 3 The Change in Temperature of a Heat Shield

A mechanical engineer is designing a heat shield 1 in. thick for a reentry space vehicle. She has determined what the temperature u would be at each depth level x of the shield, as shown in Figure 1.2 (where the temperatures have been normalized to lie in the interval $0 \leq u \leq 1$). She must determine the maximum change in temperature per unit depth that the material would endure.

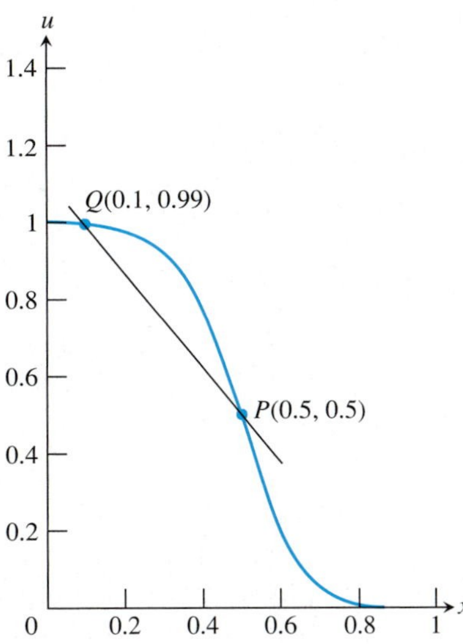

FIGURE 1.2 Temperature of a heat shield as a function of depth below the shield surface shortly after entering Earth's atmosphere.

Solution From the graph of the temperature function (Figure 1.2), the engineer sees that the slope of the curve is steepest at the point P where the depth is 0.5 in. The average rate of change of the temperature from a depth of 0.1 in. at point Q to the depth of 0.5 in. is

$$\text{Average rate of change:} \qquad \frac{\Delta u}{\Delta x} = \frac{0.99 - 0.5}{0.1 - 0.5} \approx -1.23 \text{ deg/in.}$$

This average is the slope of the secant through the points P and Q on the graph in Figure 1.2. This average, however, does not tell us how fast the temperature is changing at the point P itself. For that we need to examine the average rates of change over increasingly short depth intervals ending (or starting) at depth $x = 0.5$ in. In geometric terms, we find these rates by calculating the slopes of secants from P to Q, for a sequence of points Q approaching P along the curve (Figure 1.3).

The values in the table in Figure 1.3 show that the secant slopes go from -1.23 to -2.6 as the x-coordinate of Q increases from 0.1 to 0.4. Geometrically, the secants rotate clockwise about P and seem to approach a line that

Q	Slope of $PQ = \Delta u/\Delta x$
(0.1, 0.99)	$\frac{0.99 - 0.5}{0.1 - 0.5} \approx -1.23$
(0.2, 0.98)	$\frac{0.98 - 0.5}{0.2 - 0.5} \approx -1.60$
(0.3, 0.92)	$\frac{0.92 - 0.5}{0.3 - 0.5} \approx -2.10$
(0.4, 0.76)	$\frac{0.76 - 0.5}{0.4 - 0.5} \approx -2.60$

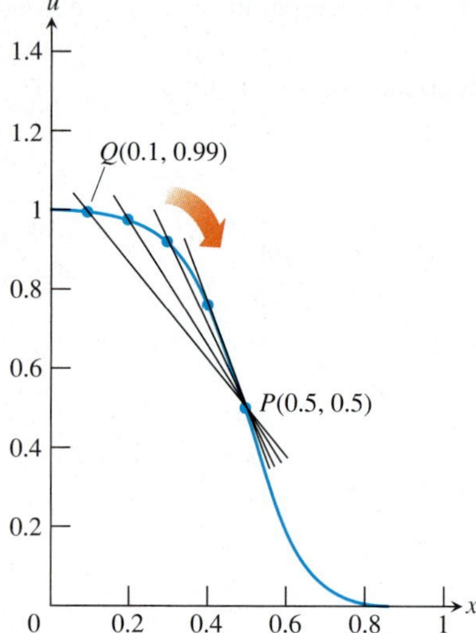

FIGURE 1.3 The positions and slopes of four secants through point P on the heat shield graph.

goes through P with the same steepness (slope) that the curve has at P. We will see that this line is called the *tangent* to the curve at P (Figure 1.4). Since the line appears to pass through the points $A(0.32, 1)$ and $B(0.68, 0)$, it has slope

$$\frac{1 - 0}{0.32 - 0.68} \approx -2.78 \text{ deg/in.}$$

At P, where the depth is 0.5 in., the temperature was changing at a rate of about -2.78 deg/in.

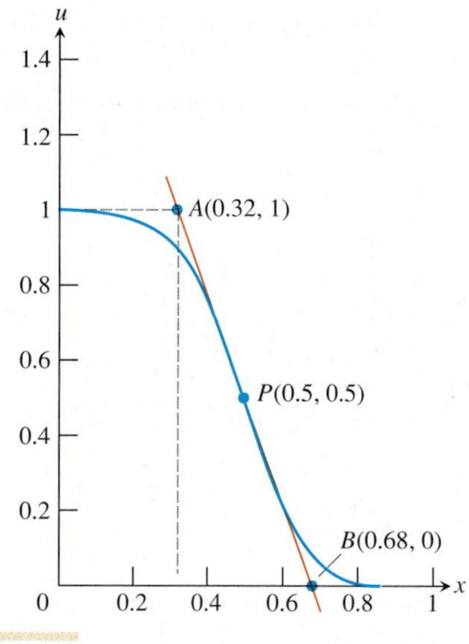

FIGURE 1.4 The tangent line at point P has the same steepness (slope) that the curve has at P.

The rate at which the rock in Example 2 was falling at the instant $t = 2$ and the rate at which the temperature in Example 3 was changing at depth 0.5 in. are called *instantaneous rates of change*. As the examples suggest, we find instantaneous rates as limiting values of average rates. In Example 3, we also pictured the tangent line to the temperature curve at depth 0.5 as a limiting position of secant lines. Instantaneous rates and tangent lines, intimately connected, appear in many other contexts. To talk about the two constructively and to understand the connection further, we need to investigate the process by which we determine limiting values, or *limits,* as we will soon call them.

Limits of Functions

Before we give a definition of limit, let us look at one more example.

CD-ROM WEBsite

Example 4 Behavior of a Function Near a Point

How does the function

$$f(x) = \frac{x^2 - 1}{x - 1}$$

behave near $x = 1$?

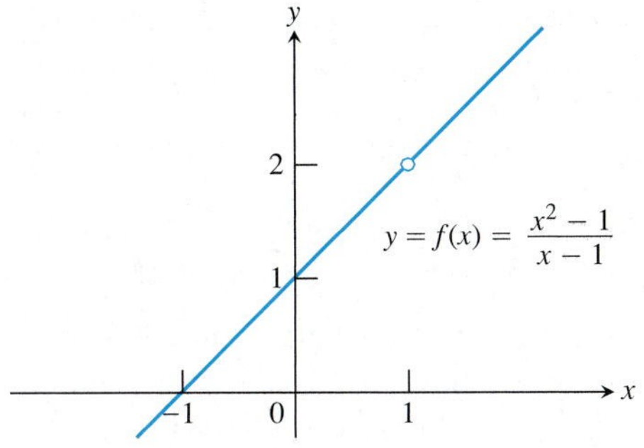

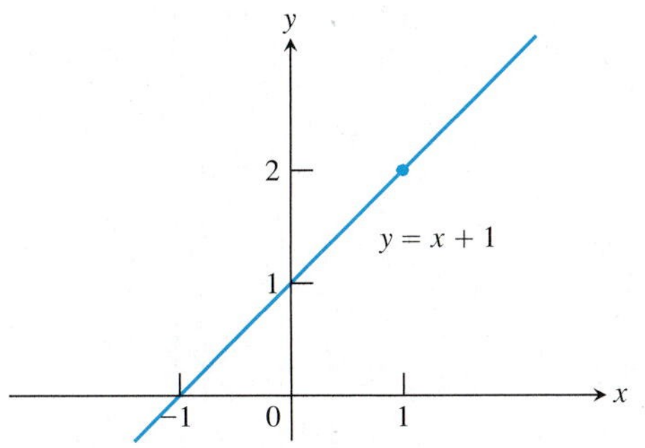

FIGURE 1.5 The graph of f is identical with the line $y = x + 1$ except at $x = 1$, where f is not defined.

Solution The given formula defines f for all real numbers x except $x = 1$ (we cannot divide by zero). For any $x \neq 1$, we can simplify the formula by factoring the numerator and canceling common factors:

$$f(x) = \frac{(x - 1)(x + 1)}{x - 1} = x + 1 \quad \text{for} \quad x \neq 1 .$$

The graph of f is thus the line $y = x + 1$ with the point (1, 2) *removed*. This removed point is shown as a "hole" in Figure 1.5. Even though $f(1)$ is not defined, it is clear that we can make the value of $f(x)$ *as close as we want* to 2 by choosing *x close enough* to 1 (Table 1.2).

Table 1.2 The closer x gets to 1, the closer $f(x) = (x^2 - 1)/(x - 1)$ seems to get to 2

Values of x below and above 1	$f(x) = \frac{x^2 - 1}{x - 1} = x + 1, \quad x \neq 1$
0.9	1.9
1.1	2.1
0.99	1.99
1.01	2.01
0.999	1.999
1.001	2.001
0.999999	1.999999
1.000001	2.000001

We say that $f(x)$ approaches arbitrarily close to 2 as x approaches 1, or, more simply, that $f(x)$ approaches the *limit* 2 as x approaches 1. We write this as

$$\lim_{x \to 1} f(x) = 2, \qquad \text{or} \qquad \lim_{x \to 1} \frac{x^2 - 1}{x - 1} = 2 .$$

Informal Definition of Limit

Let $f(x)$ be defined on an open interval about x_0, *except possibly at* x_0 *itself*. If $f(x)$ gets arbitrarily close to L for all x sufficiently close to x_0, we say that f approaches the **limit** L as x approaches x_0, and we write

$$\lim_{x \to x_0} f(x) = L .$$

This definition is "informal" because phrases like *arbitrarily close* and *sufficiently close* are imprecise; their meaning depends on the context. To a machinist manufacturing a piston, *close* may mean *within a few thousandths of an inch*. To an astronomer studying distant galaxies, *close* may mean *within a few thousand light-years*. The definition is clear enough, however, to enable us to recognize and evaluate limits of a number of specific functions.

Example 5 The Limit Value Does Not Depend On How the Function is Defined at x_0

The function f in Figure 1.6 has limit 2 as $x \to 1$ even though f is not defined at $x = 1$. The function g has limit 2 as $x \to 1$ even though $2 \neq g(1)$. The function h is the only one whose limit as $x \to 1$ equals its value at $x = 1$. For h, we have $\lim_{x\to 1} h(x) = h(1)$. This equality of limit and function value is special, and we return to it in Section 1.4.

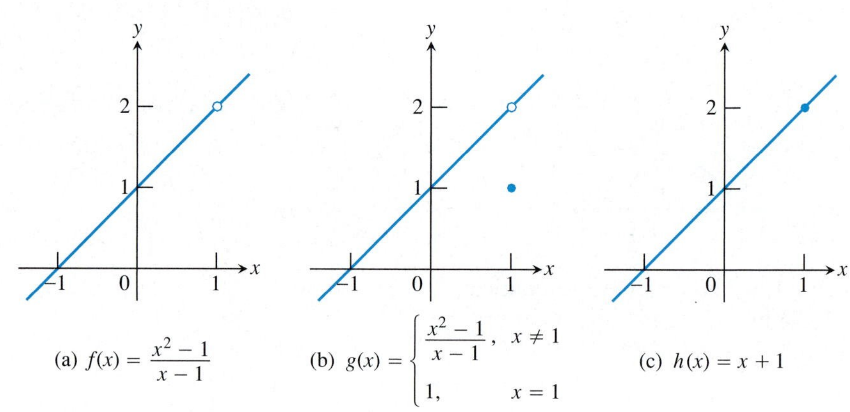

FIGURE 1.6 $\lim_{x\to 1} f(x) = \lim_{x\to 1} g(x) = \lim_{x\to 1} h(x) = 2$.

Example 6 Two Functions That Have Limits at Every Point

(a) If f is the **identity function** $f(x) = x$, then for any value of x_0 (Figure 1.7a),

$$\lim_{x\to x_0} f(x) = \lim_{x\to x_0} x = x_0 .$$

(b) If f is the **constant function** $f(x) = k$ (function with the constant value k), then for any value of x_0 (Figure 1.7b),

$$\lim_{x\to x_0} f(x) = \lim_{x\to x_0} k = k .$$

For instance,

$$\lim_{x\to 3} x = 3 \quad \text{and} \quad \lim_{x\to -7} (4) = \lim_{x\to 2} (4) = 4 .$$

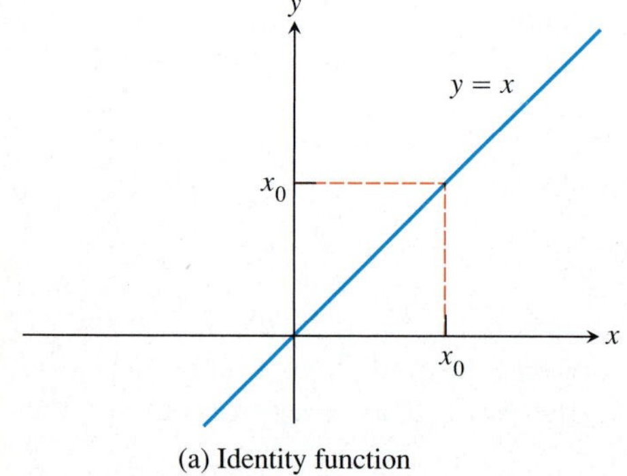

(a) Identity function

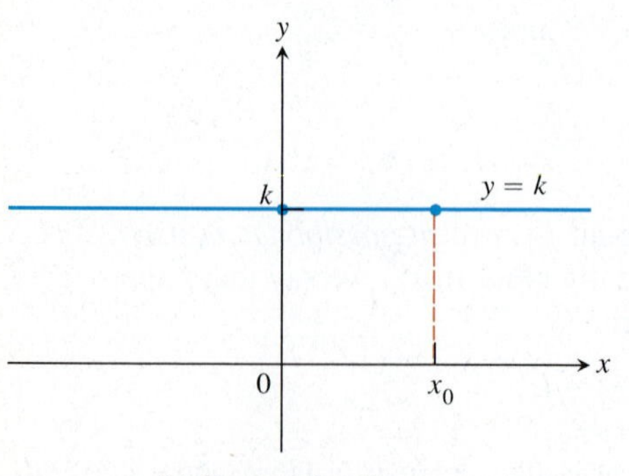

(b) Constant function

FIGURE 1.7 The functions in Example 6.

Some ways that limits can fail to exist are illustrated in Figure 1.8 and described in the next example.

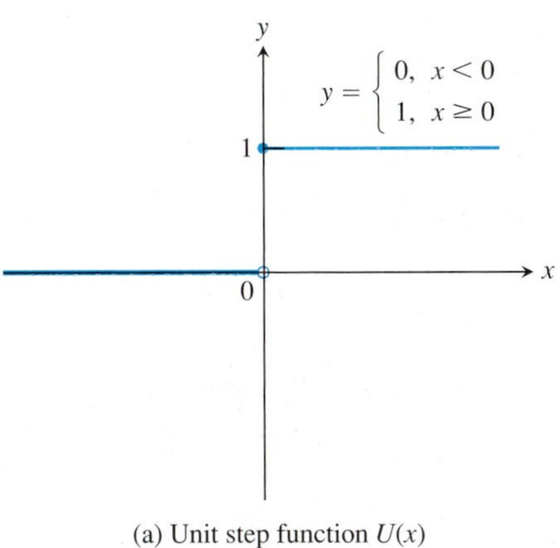

(a) Unit step function $U(x)$

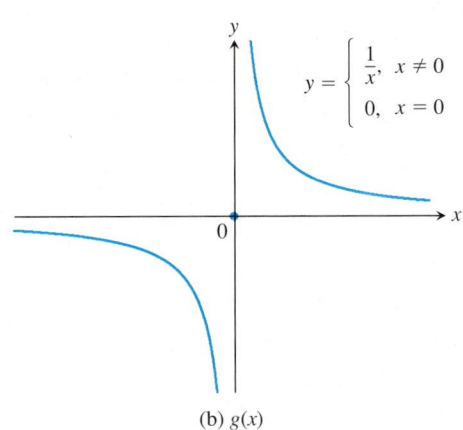

$y = \begin{cases} \frac{1}{x}, & x \neq 0 \\ 0, & x = 0 \end{cases}$

(b) $g(x)$

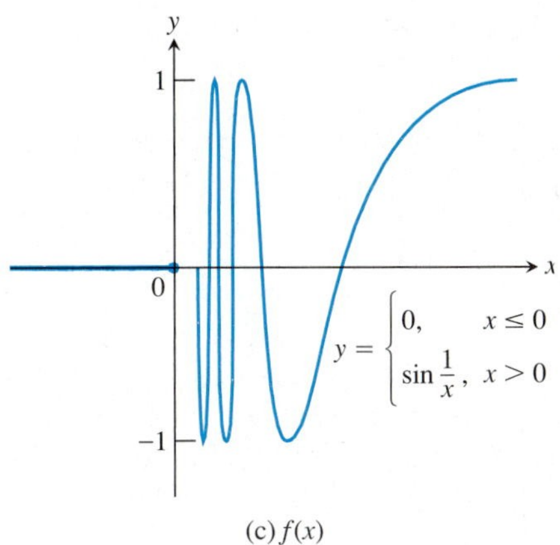

(c) $f(x)$

FIGURE 1.8 The functions in Example 7.

Example 7 Limits May Fail to Exist

Discuss the behavior of the following functions as $x \to 0$.

(a) $U(x) = \begin{cases} 0, & x < 0 \\ 1, & x \geq 0 \end{cases}$

(b) $g(x) = \begin{cases} \frac{1}{x}, & x \neq 0 \\ 0, & x = 0 \end{cases}$

(c) $f(x) = \begin{cases} 0, & x \leq 0 \\ \sin\frac{1}{x}, & x > 0 \end{cases}$

Solution

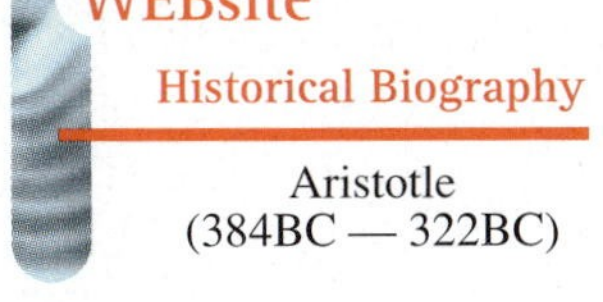

(a) It *jumps*: The **unit step function** $U(x)$ has no limit as $x \to 0$ because its values jump at $x = 0$. For negative values of x arbitrarily close to zero, $U(x) = 0$. For positive values of x arbitrarily close to zero, $U(x) = 1$. There is no *single* value L approached by $U(x)$ as $x \to 0$ (Figure 1.8a).

(b) It *grows too large to have a limit*: $g(x)$ has no limit as $x \to 0$ because the values of g grow arbitrarily large in absolute value as $x \to 0$ and do not stay close to *any* real number (Figure 1.8b).

(c) It *oscillates too much to have a limit*: $f(x)$ has no limit as $x \to 0$ because the function's values oscillate between $+1$ and -1 in every open interval containing 0. The values do not stay close to any one number as $x \to 0$ (Figure 1.8c).

Precise Definition of Limit

To show that the limit of $f(x)$ as $x \to x_0$ equals the number L, we need to show that the gap between $f(x)$ and L can be made "as small as we choose" if x is kept "close enough" to x_0. Let us see what this would require if we specified the size of the gap between $f(x)$ and L.

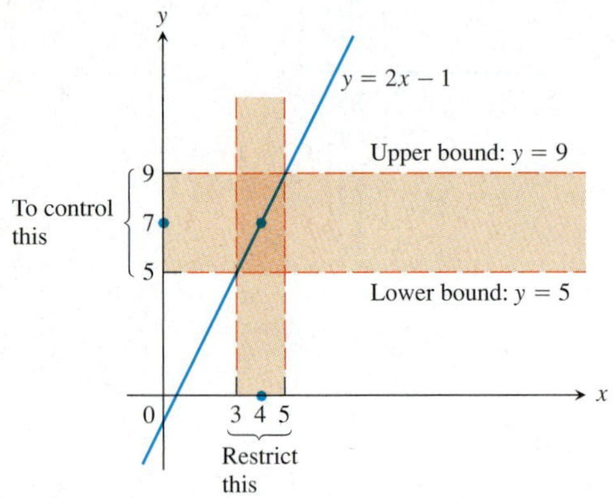

FIGURE 1.9 Keeping x within 1 unit of $x_0 = 4$ will keep y within 2 units of $y_0 = 7$.

Example 8 Controlling a Linear Function

How close to $x_0 = 4$ must we hold the input x to be sure that the output $y = 2x - 1$ lies within 2 units of $y_0 = 7$?

Solution We are asked: For what values of x is $|y - 7| < 2$? To find the answer we first express $|y - 7|$ in terms of x:

$$|y - 7| = |(2x - 1) - 7| = |2x - 8|.$$

The question then becomes: what values of x satisfy the inequality $|2x - 8| < 2$? To find out, we solve the inequality:

$$\begin{aligned} |2x - 8| &< 2 \\ -2 < 2x - 8 &< 2 \\ 6 < 2x &< 10 \\ 3 < x &< 5 \\ -1 < x - 4 &< 1. \end{aligned}$$

Keeping x within 1 unit of $x_0 = 4$ will keep y within 2 units of $y_0 = 7$ (Figure 1.9).

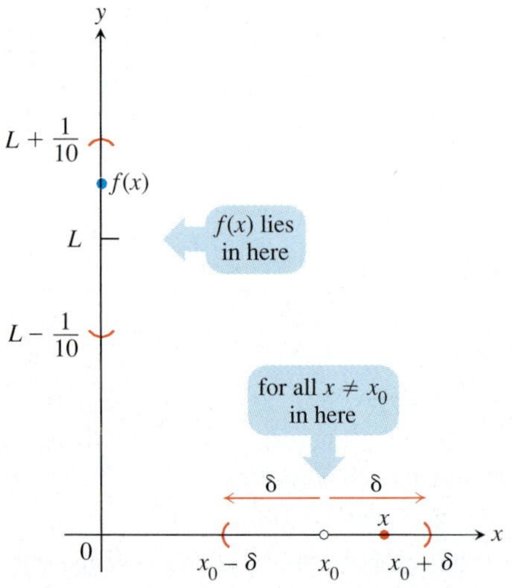

FIGURE 1.10 A preliminary stage in the development of the definition of limit.

In Example 8, we determined how close to hold a variable x to a particular value x_0 to ensure that the outputs $f(x)$ lie within a prescribed interval about the limit L. To show that the limit of $f(x)$ as $x \to x_0$ actually equals L, we must be able to show that the gap between $f(x)$ and L can be made less than *any prescribed error,* no matter how small, by holding x close enough to x_0.

Suppose that we are watching the values of a function $f(x)$ as x approaches x_0 (without taking on the value of x_0 itself). Certainly we want to be able to say that $f(x)$ stays within one-tenth of a unit of L as soon as x stays within some distance δ of x_0 (Figure 1.10). That in itself is not enough, however, because as x continues on its course toward x_0, what is to prevent $f(x)$ from jittering about within the interval from $L - 1/10$ to $L + 1/10$ without tending toward L? We can ensure that $f(x)$ is tending toward L by showing that no matter how tightly we narrow the gap between $f(x)$ and L, keeping x close enough to x_0 will keep $f(x)$ within that tolerance of L. This wording is the mathematical way to say that the closer x gets to x_0, the closer $y = f(x)$ gets to L.

Definition Formal Definition of Limit

Let $f(x)$ be defined on an open interval about x_0, except possibly at x_0 itself. We say that $f(x)$ approaches the **limit** L as x approaches x_0 and write

$$\lim_{x \to x_0} f(x) = L,$$

if, for every number $\epsilon > 0$, there exists a corresponding number $\delta > 0$ such that for all x,

$$0 < |x - x_0| < \delta \Rightarrow |f(x) - L| < \epsilon.$$

The limit definition is illustrated in Figure 1.11.

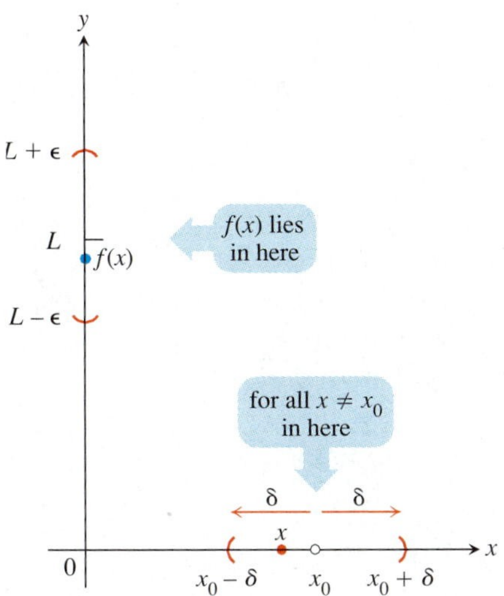

FIGURE 1.11 The relation of δ and ϵ in the definition of limit.

Example 9 Testing the Definition

Show that $\lim_{x \to 1} (5x - 3) = 2$.

Solution Set $x_0 = 1, f(x) = 5x - 3$, and $L = 2$ in the definition of limit. For any given $\epsilon > 0$, we have to find a suitable $\delta > 0$ so that if $x \neq 1$ and x is within distance δ of $x_0 = 1$, that is, if

$$0 < |x - 1| < \delta,$$

then $f(x)$ is within distance ϵ of $L = 2$, that is

$$|f(x) - 2| < \epsilon.$$

We find δ by working backwards from the ϵ-inequality:

$$\begin{aligned} |(5x - 3) - 2| = |5x - 5| &< \epsilon \\ 5|x - 1| &< \epsilon \\ |x - 1| &< \epsilon/5. \end{aligned}$$

Thus, we can take $\delta = \epsilon/5$ (Figure 1.12). If $0 < |x - 1| < \delta = \epsilon/5$, then

$$|(5x - 3) - 2| = |5x - 5| = 5|x - 1| < 5(\epsilon/5) = \epsilon,$$

which proves that $\lim_{x \to 1} (5x - 3) = 2$.

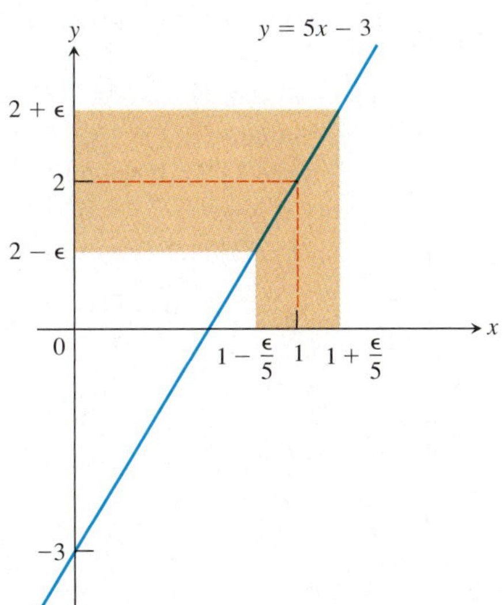

FIGURE 1.12 If $f(x) = 5x - 3$, then $0 < |x - 1| < \epsilon/5$ guarantees that $|f(x) - 2| < \epsilon$. (Example 9)

The value of $\delta = \epsilon/5$ is not the only value that will make $0 < |x - 1| < \delta$ imply $|5x - 5| < \epsilon$. Any smaller positive δ will do as well. The definition does not ask for a "best" positive δ, just one that will work.

In Example 9, the interval of values about x_0 for which $|f(x) - L|$ was less than ϵ was symmetric about x_0, and we could take δ to be half the length of the interval. When such symmetry is absent, as it usually is, we can take δ to be the distance from x_0 to the interval's nearer endpoint. This choice for δ is illustrated in the next example.

Example 10 Finding Delta Algebraically for a Given Epsilon

For the limit $\lim_{x \to 5} \sqrt{x-1} = 2$, find a $\delta > 0$ that works for $\epsilon = 1$. That is, find a $\delta > 0$ such that for all x

$$0 < |x - 5| < \delta \Rightarrow |\sqrt{x-1} - 2| < 1.$$

Solution We organize the search into two steps. First, we solve the inequality $|\sqrt{x-1} - 2| < 1$ to find an interval (a, b) about $x_0 = 5$ on which the inequality holds for all $x \neq x_0$. Then we find a value of $\delta > 0$ that places the interval $5 - \delta < x < 5 + \delta$ (centered at $x_0 = 5$) inside the interval (a, b).

Step 1: *Solve the inequality* $|\sqrt{x-1} - 2| < 1$ *to find an interval about* $x_0 = 5$ *on which the inequality holds for all* $x \neq x_0$.

$$\begin{aligned}|\sqrt{x-1} - 2| &< 1\\ -1 < \sqrt{x-1} - 2 &< 1\\ 1 < \sqrt{x-1} &< 3\\ 1 < x - 1 &< 9\\ 2 < x &< 10\end{aligned}$$

The inequality holds for all x in the open interval (2, 10), so it holds for all $x \neq 5$ in this interval as well.

Step 2: *Find a value of* $\delta > 0$ *to place the centered interval* $5 - \delta < x < 5 + \delta$ *inside the interval* (2, 10). The distance from 5 to the nearer endpoint of (2, 10) is 3 (Figure 1.13). If we take $\delta = 3$ or any smaller positive number, then the inequality $0 < |x - 5| < \delta$ will automatically place x between 2 and 10 to make $|\sqrt{x-1} - 2| < 1$ (Figure 1.14):

$$0 < |x - 5| < 3 \Rightarrow |\sqrt{x-1} - 2| < 1.$$

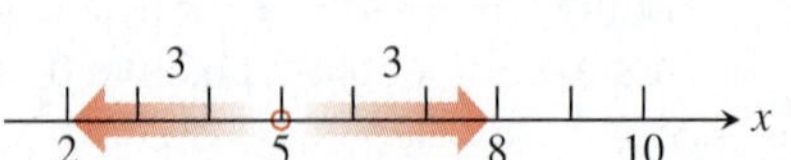

FIGURE 1.13 An open interval of radius 3 about $x_0 = 5$ will lie inside the open interval (2, 10).

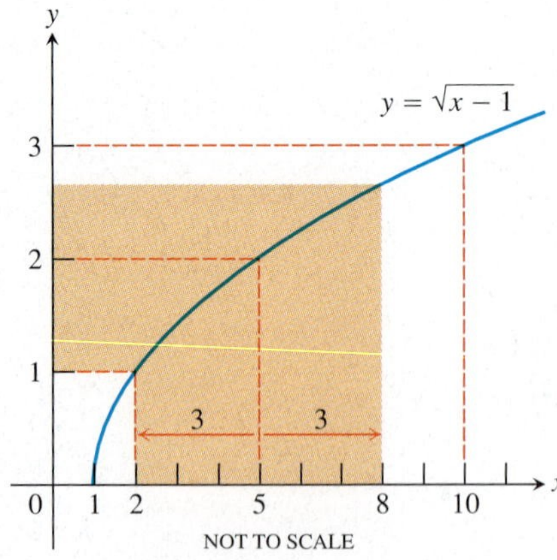

FIGURE 1.14 The function and intervals in Example 10.

EXERCISES 1.1

Average Rates of Change

In Exercises 1–4, find the average rate of change of the function over the given interval or intervals.

1. $f(x) = x^3 + 1$

   (a) $[2, 3]$ **(b)** $[-1, 1]$

2. $R(\theta) = \sqrt{4\theta + 1}$, $[0, 2]$

3. $h(t) = \cot t$

   (a) $[\pi/4, 3\pi/4]$ **(b)** $[\pi/6, \pi/2]$

4. $g(t) = 2 + \cos t$

   (a) $[0, \pi]$ **(b)** $[-\pi, \pi]$

5. *A Ford Mustang Cobra's speed* The accompanying figure shows the time-to-distance graph for a 1994 Ford Mustang Cobra accelerating from a standstill.

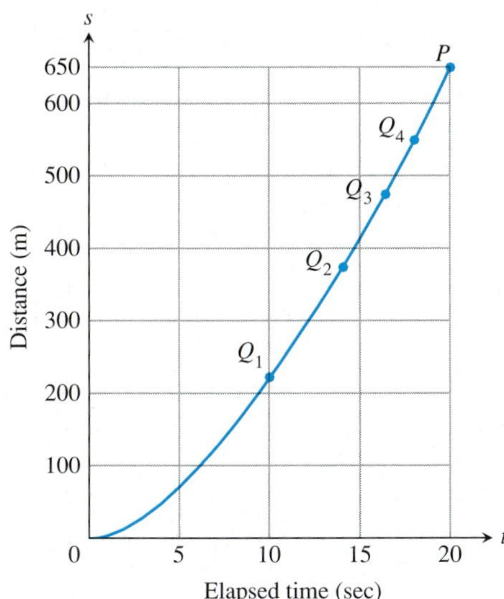

   (a) Estimate the slopes of secants PQ_1, PQ_2, PQ_3, and PQ_4, arranging them in order in a table. What are the appropriate units for these slopes?

   (b) Then estimate the Cobra's speed at time $t = 20$ sec.

6. *Speed of a falling wrench* The accompanying figure shows the plot of distance fallen versus time for a wrench that fell from the top of a communications mast on the moon to the station roof 80 m below.

   (a) Estimate the slopes of the secants PQ_1, PQ_2, PQ_3, and PQ_4, arranging them in a table like the one in Figure 1.3.

   (b) About how fast was the wrench going when it hit the roof?

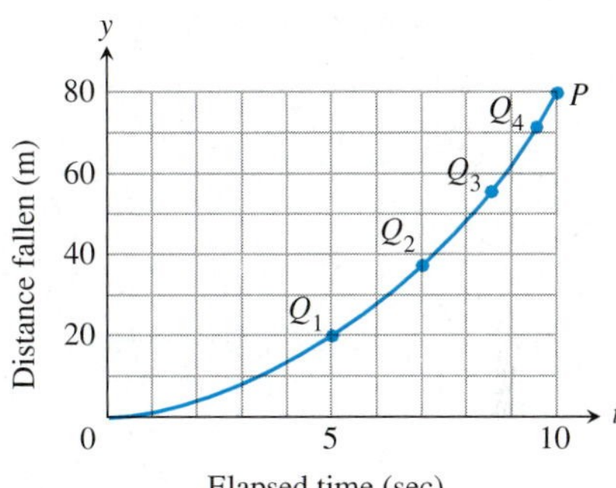

7. *Velocity of a ball* The accompanying data represent the distance a ball has traveled down an inclined plane. Estimate the instantaneous velocity at $t = 1$ by finding an upper and lower bound, and averaging them. That is, find $a \le v(1) \le b$ and estimate $v(1)$ as $(a + b)/2$.

Time t (sec)	**Distance traveled (ft)**
0	0
0.2	0.52
0.4	2.10
0.6	4.72
0.8	8.39
1.0	13.10
1.2	18.87
1.4	25.68

8. *Distance traveled by a train* A train accelerates from a standstill to a maximum cruising speed, then slows down to pass through a town at constant speed. After passing through the town the

train accelerates to cruising speed. Finally, the train decelerates smoothly to a stop as it reaches its destination. Sketch a possible graph of the distance the train has traveled as a function of time.

Limits From Graphs

9. For the function $g(x)$ graphed here, find the following limits or explain why they do not exist.

(a) $\lim_{x\to 1} g(x)$ **(b)** $\lim_{x\to 2} g(x)$ **(c)** $\lim_{x\to 3} g(x)$

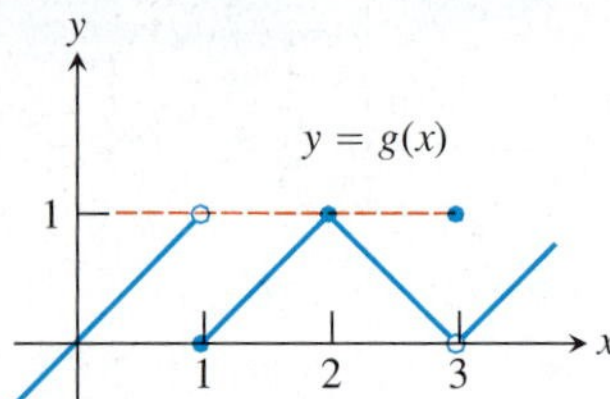

10. For the function $f(t)$ graphed here, find the following limits or explain why they do not exist.

(a) $\lim_{t\to -2} f(t)$ **(b)** $\lim_{t\to -1} f(t)$ **(c)** $\lim_{t\to 0} f(t)$

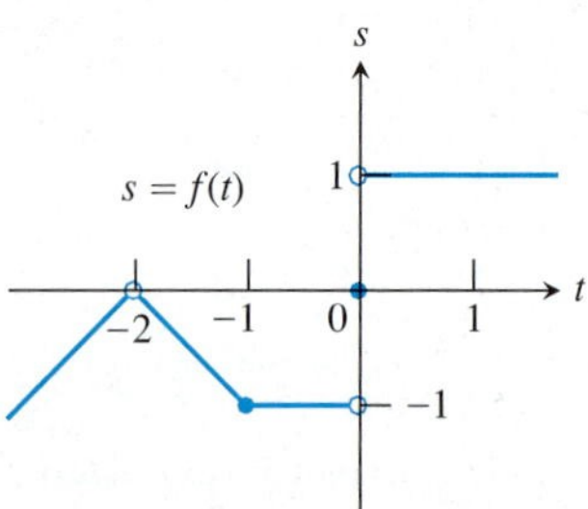

11. Which of the following statements about the function $y = f(x)$ graphed here are true, and which are false?

(a) $\lim_{x\to 0} f(x)$ exists.

(b) $\lim_{x\to 0} f(x) = 0$.

(c) $\lim_{x\to 0} f(x) = 1$.

(d) $\lim_{x\to 1} f(x) = 1$.

(e) $\lim_{x\to 1} f(x) = 0$.

(f) $\lim_{x\to x_0} f(x)$ exists at every point x_0 in $(-1, 1)$.

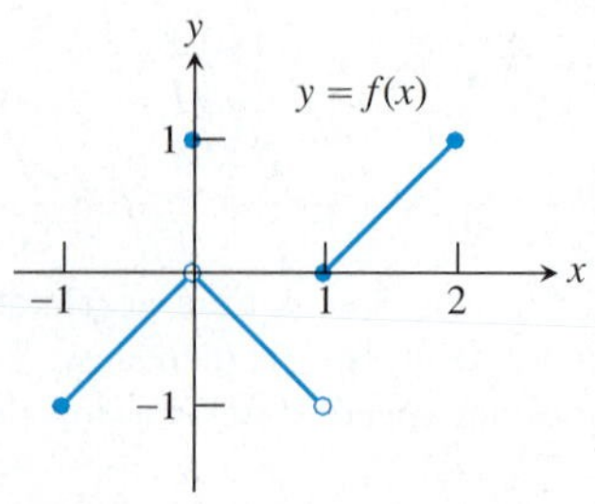

12. Which of the following statements about the function $y = f(x)$ graphed here are true, and which are false?

(a) $\lim_{x\to 2} f(x)$ does not exist.

(b) $\lim_{x\to 2} f(x) = 2$.

(c) $\lim_{x\to 1} f(x)$ does not exist.

(d) $\lim_{x\to x_0} f(x)$ exists at every point x_0 in $(-1, 1)$.

(e) $\lim_{x\to x_0} f(x)$ exists at every point x_0 in $(1, 3)$.

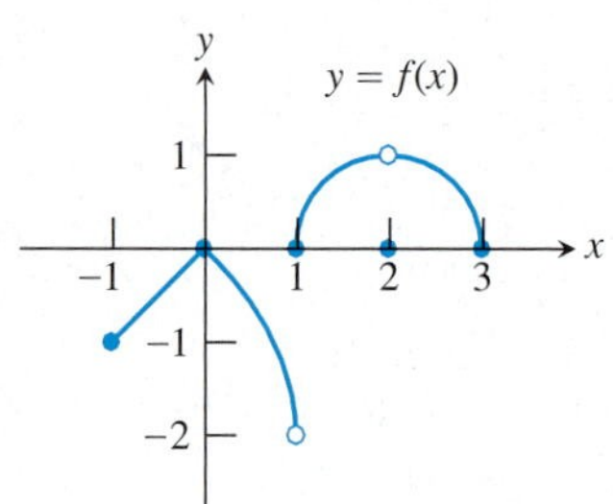

Existence of Limits

In Exercises 13 and 14, explain why the limits do not exist.

13. $\lim_{x\to 0} \dfrac{x}{|x|}$

14. $\lim_{x\to 1} \dfrac{1}{x-1}$

15. *Writing to Learn* Suppose that a function $f(x)$ is defined for all real values of x except $x = x_0$. Can anything be said about the existence of $\lim_{x\to x_0} f(x)$? Give reasons for your answer.

16. *Writing to Learn* Suppose that a function $f(x)$ is defined for all x in $[-1, 1]$. Can anything be said about the existence of $\lim_{x\to 0} f(x)$? Give reasons for your answer.

17. *Writing to Learn* If $\lim_{x\to 1} f(x) = 5$, must f be defined at $x = 1$? If it is, must $f(1) = 5$? Can we conclude *anything* about the values of f at $x = 1$? Explain.

18. *Writing to Learn* If $f(1) = 5$, must $\lim_{x\to 1} f(x)$ exist? If it does, then must $\lim_{x\to 1} f(x) = 5$? Can we conclude *anything* about $\lim_{x\to 1} f(x)$? Explain.

Estimating Limits

T You will find a graphing calculator useful for Exercises 19–26.

19. Let $f(x) = (x^2 - 9)/(x + 3)$.

(a) Make a table of the values of f at the points $x = -3.1$, -3.01, -3.001, and so on as far as your calculator can go. Then estimate $\lim_{x\to -3} f(x)$. What estimate do you arrive at if you evaluate f at $x = -2.9, -2.99, -2.999, \ldots$ instead?

(b) Support your conclusions in part (a) by graphing f near $x_0 = -3$ and using Zoom and Trace to estimate y-values on the graph as $x \to -3$.

(c) Find $\lim_{x\to -3} f(x)$ algebraically.

20. Let $g(x) = (x^2 - 2)/(x - \sqrt{2})$.

(a) Make a table of the values of g at the points $x = 1.4$, 1.41, 1.414, and so on through successive decimal approximations of $\sqrt{2}$. Estimate $\lim_{x \to \sqrt{2}} g(x)$.

(b) Support your conclusion in part (a) by graphing g near $x_0 = \sqrt{2}$ and using Zoom and Trace to estimate y-values on the graph as $x \to \sqrt{2}$.

(c) Find $\lim_{x \to \sqrt{2}} g(x)$ algebraically.

21. Let $G(x) = (x + 6)/(x^2 + 4x - 12)$.

(a) Make a table of the values of G at $x = -5.9$, -5.99, -5.999, and so on. Then estimate $\lim_{x \to -6} G(x)$. What estimate do you arrive at if you evaluate G at $x = -6.1$, -6.01, -6.001, . . . instead?

(b) Support your conclusions in part (a) by graphing G and using Zoom and Trace to estimate y-values on the graph as $x \to -6$.

(c) Find $\lim_{x \to -6} G(x)$ algebraically.

22. Let $h(x) = (x^2 - 2x - 3)/(x^2 - 4x + 3)$.

(a) Make a table of the values of h at $x = 2.9$, 2.99, 2.999, and so on. Then estimate $\lim_{x \to 3} h(x)$. What estimate do you arrive at if you evaluate h at $x = 3.1$, 3.01, 3.001, . . . instead?

(b) Support your conclusions in part (a) by graphing h near $x_0 = 3$ and using Zoom and Trace to estimate y-values on the graph as $x \to 3$.

(c) Find $\lim_{x \to 3} h(x)$ algebraically.

23. Let $g(\theta) = (\sin \theta)/\theta$.

(a) Make a table of the values of g at values of θ that approach $\theta_0 = 0$ from above and below. Then estimate $\lim_{\theta \to 0} g(\theta)$.

(b) Support your conclusion in part (a) by graphing g near $\theta_0 = 0$.

24. Let $G(t) = (1 - \cos t)/t^2$.

(a) Make tables of values of G at values of t that approach $t_0 = 0$ from above and below. Then estimate $\lim_{t \to 0} G(t)$.

(b) Support your conclusion in part (a) by graphing G near $t_0 = 0$.

25. Let $f(x) = x^{1/(1-x)}$.

(a) Make tables of values of f at values of x that approach $x_0 = 1$ from above and below. Does f appear to have a limit as $x \to 1$? If so, what is it? If not, why not?

(b) Support your conclusions in part (a) by graphing f near $x_0 = 1$.

26. Let $f(x) = (3^x - 1)/x$.

(a) Make tables of values of f at values of x that approach $x_0 = 0$ from above and below. Does f appear to have a limit as $x \to 0$? If so, what is it? If not, why not?

(b) Support your conclusions in part (a) by graphing f near $x_0 = 0$.

CD-ROM WEBsite

Finding Deltas Graphically

In Exercises 27–30, use the graphs to find a $\delta > 0$ such that for all x,

$$0 < |x - x_0| < \delta \Rightarrow |f(x) - L| < \epsilon.$$

27.

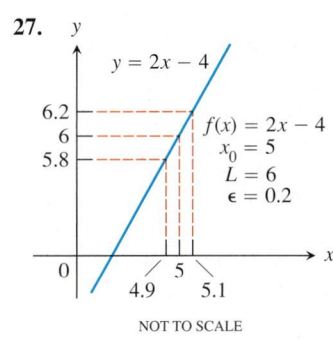

$y = 2x - 4$

$f(x) = 2x - 4$, $x_0 = 5$, $L = 6$, $\epsilon = 0.2$

6.2, 6, 5.8; 4.9, 5, 5.1

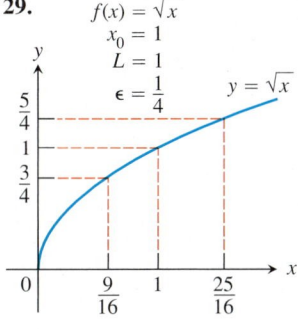

NOT TO SCALE

28.

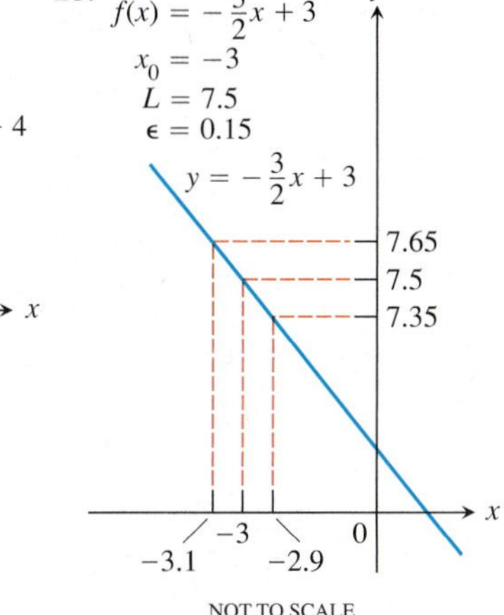

29. **30.**

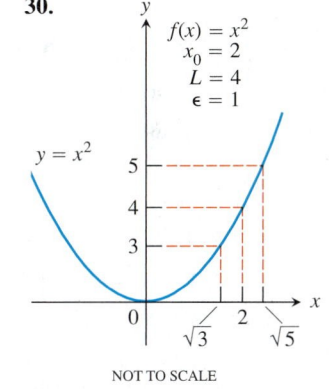

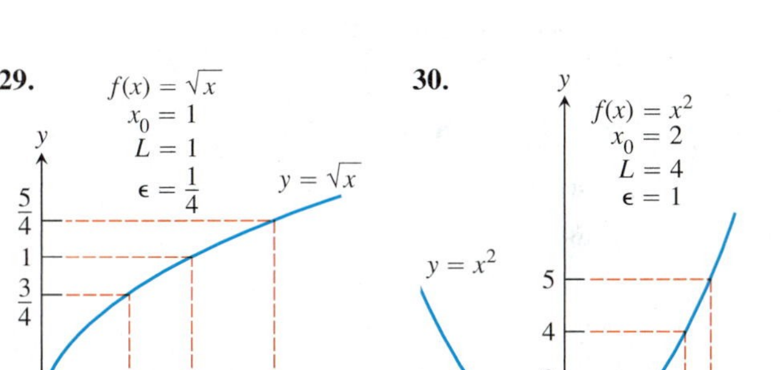

Finding Deltas Algebraically

Each of Exercises 31–36 gives a function $f(x)$ and numbers L, x_0, and $\epsilon > 0$. In each case, find an open interval about x_0 on which the inequality $|f(x) - L| < \epsilon$ holds. Then give a value for $\delta > 0$ such that for all x satisfying $0 < |x - x_0| < \delta$ the inequality $|f(x) - L| < \epsilon$ holds.

31. $f(x) = x + 1$, $L = 5$, $x_0 = 4$, $\epsilon = 0.01$

32. $f(x) = 2x - 2$, $L = -6$, $x_0 = -2$, $\epsilon = 0.02$

33. $f(x) = \sqrt{x + 1}$, $L = 1$, $x_0 = 0$, $\epsilon = 0.1$

34. $f(x) = \sqrt{19 - x}$, $L = 3$, $x_0 = 10$, $\epsilon = 1$

35. $f(x) = 1/x$, $L = 1/4$, $x_0 = 4$, $\epsilon = 0.05$

36. $f(x) = x^2$, $L = 3$, $x_0 = \sqrt{3}$, $\epsilon = 0.1$

Theory and Examples

37. *Grinding engine cylinders* Before contracting to grind engine cylinders to a cross-section area of 9 in.^{2}, you need to know how much deviation from the ideal cylinder diameter of $x_0 = 3.385$ in. you can allow and still have the area come within 0.01 in.^{2} of the required 9 in.^{2}. To find out, you let $A = \pi(x/2)^2$ and look for the interval in which you must hold x to make $|A - 9| \le 0.01$. What interval do you find?

38. *Manufacturing electrical resistors* Ohm's law for electrical circuits like the one shown in the accompanying figure states that $V = RI$. In this equation, V is a constant voltage, I is the current in amperes, and R is the resistance in ohms. Your firm has been asked to supply the resistors for a circuit in which V will be 120 volts and I is to be 5 ± 0.1 amp. In what interval does R have to lie for I to be within 0.1 amp of the target value $I_0 = 5$?

T 39. *Controlling outputs* Let $f(x) = \sqrt{3x - 2}$.

(a) Show that $\lim_{x \to 2} f(x) = 2 = f(2)$.

(b) Use a graph to estimate values for a and b so that $1.8 < f(x) < 2.2$ provided $a < x < b$.

(c) Use a graph to estimate values for a and b so that $1.99 < f(x) < 2.01$ provided $a < x < b$.

T 40. *Controlling outputs* Let $f(x) = \sin x$.

(a) Find $f(\pi/6)$.

(b) Use a graph to estimate an interval (a, b) about $x = \pi/6$ so that $0.3 < f(x) < 0.7$ provided $a < x < b$.

(c) Use a graph to estimate an interval (a, b) about $x = \pi/6$ so that $0.49 < f(x) < 0.51$ provided $a < x < b$.

41. *Free fall* A water balloon dropped from a window high above the ground falls $y = 4.9t^2$ m in t sec. Find the balloon's

(a) average speed during the first 3 sec. of fall.

(b) speed at the instant $t = 3$.

42. *Free fall on a small airless planet* A rock released from rest to fall on a small airless planet falls $y = gt^2$ m in t sec, g a constant. Suppose that the rock falls to the bottom of a crevasse 20 m below the release point and reaches the bottom in 4 sec.

(a) Find the value of g.

(b) Find the average speed for the fall.

(c) With what speed did the rock hit the bottom?

In Exercises 43–46, complete the following tables and state what you believe $\lim_{x \to 0} f(x)$ to be.

(a)

x	-0.1	-0.01	-0.001	-0.0001	. . .
$f(x)$	?	?	?	?	

(b)

x	0.1	0.01	0.001	0.0001	. . .
$f(x)$	?	?	?	?	

43. $f(x) = x \sin \frac{1}{x}$ **44.** $f(x) = \sin \frac{1}{x}$

45. $f(x) = \dfrac{10^x - 1}{x}$ **46.** $f(x) = x \sin(\ln|x|)$

COMPUTER EXPLORATIONS

Graphical Estimates of Limits

In Exercises 47–50, use a CAS to perform the following steps:

(a) Plot the function near the point x_0 being approached.

(b) From your plot guess the value of the limit.

(c) Evaluate the limit symbolically. How close was your guess?

47. $\lim_{x \to 2} \dfrac{x^4 - 16}{x - 2}$ **48.** $\lim_{x \to 0} \dfrac{1 - \cos x}{x \sin x}$

49. $\lim_{x \to -1} \dfrac{x^3 - x^2 - 5x - 3}{(x + 1)^2}$ **50.** $\lim_{x \to 3} \dfrac{x^2 - 9}{\sqrt{x^2 + 7} - 4}$

Finding Deltas Graphically

In Exercises 51–54, you will further explore finding deltas graphically. Use a CAS to perform the following steps:

(a) Plot the function $y = f(x)$ near the point x_0 being approached.

(b) Guess the value of the limit L and then evaluate the limit symbolically to see if you guessed correctly.

(c) Using the value $\epsilon = 0.2$, graph the bounding lines $y_1 = L - \epsilon$ and $y_2 = L + \epsilon$ together with the function f near x_0.

(d) From your graph in part (c), estimate a $\delta > 0$ such that for all x

$$0 < |x - x_0| < \delta \quad \Rightarrow \quad |f(x) - L| < \epsilon.$$

Test your estimate by plotting f, y_1, and y_2 over the interval $0 < |x - x_0| < \delta$. For your viewing window, use

$$x_0 - 2\delta \le x \le x_0 + 2\delta \quad \text{and} \quad L - 2\epsilon \le y \le L + 2\epsilon.$$

If any function values lie outside the interval $[L - \epsilon, L + \epsilon]$, your choice of δ was too large. Try again with a smaller estimate.

(e) Repeat parts (c) and (d) successively for $\epsilon = 0.1$, 0.05, and 0.001.

51. $f(x) = \dfrac{x^4 - 81}{x - 3}, \quad x_0 = 3$

52. $f(x) = \dfrac{5x^3 + 9x^2}{2x^5 + 3x^2}, \quad x_0 = 0$

53. $f(x) = \dfrac{\sin 2x}{3x}, \quad x_0 = 0$

CD-ROM WEBsite **54.** $f(x) = \dfrac{x(1 - \cos x)}{x - \sin x}, \quad x_0 = 0$

1.2 Finding Limits and One-Sided Limits

Properties of Limits • Eliminating Zero Denominators Algebraically • Sandwich Theorem • One-Sided Limits • Limits Involving $(\sin \theta)/\theta$

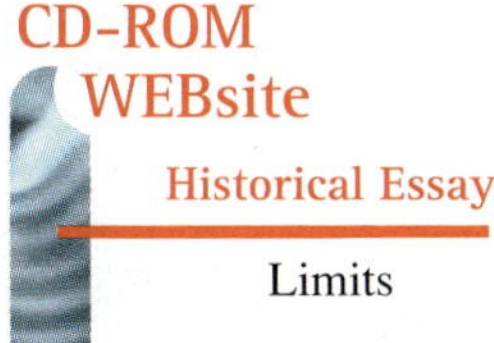

In the preceding section, we evaluated limits by examining graphs and numerical patterns. In this section, we see that many limits can be evaluated algebraically, using arithmetic and a few basic rules.

Properties of Limits

The next theorem tells how to calculate limits of functions that are arithmetic combinations of functions whose limits we already know. The rules are proved in Appendix 2.

Theorem 1 Limit Rules

If L, M, c, and k are real numbers and

$$\lim_{x \to c} f(x) = L \quad \text{and} \quad \lim_{x \to c} g(x) = M, \quad \text{then}$$

1. *Sum Rule:* $\lim_{x \to c} (f(x) + g(x)) = L + M$

The limit of the sum of two functions is the sum of their limits.

2. *Difference Rule:* $\lim_{x \to c} (f(x) - g(x)) = L - M$

The limit of the difference of two functions is the difference of their limits.

3. *Product Rule:* $\lim_{x \to c} (f(x) \cdot g(x)) = L \cdot M$

The limit of a product of two functions is the product of their limits.

4. *Constant Multiple Rule:* $\lim_{x \to c} (k \cdot f(x)) = k \cdot L$

The limit of a constant times a function is the constant times the limit of the function.

5. *Quotient Rule:* $\lim_{x \to c} \dfrac{f(x)}{g(x)} = \dfrac{L}{M}, \quad M \neq 0$

The limit of a quotient of two functions is the quotient of their limits, provided the limit of the denominator is not zero.

6. *Root rule:* If n is a positive integer, then

$$\lim_{x \to c} (f(x))^{1/n} = L^{1/n}$$

(If n is even, we assume that $L > 0$.)

The limit of the nth root is the nth root of the limit.

Here are some examples of how Theorem 1 can be used to find limits of polynomial and rational functions.

Example 1 Using Limit Rules

Use the observations $\lim_{x\to c} k = k$ and $\lim_{x\to c} x = c$ and the properties of limits to find the following limits.

(a) $\lim_{x\to c}(x^3 + 4x^2 - 3)$ **(b)** $\lim_{x\to c}\dfrac{x^4 + x^2 - 1}{x^2 + 5}$ **(c)** $\lim_{x\to -2}\sqrt{4x^2 - 3}$

Solution

(a)
$$\begin{aligned}\lim_{x\to c}(x^3 + 4x^2 - 3) &= \lim_{x\to c} x^3 + \lim_{x\to c} 4x^2 - \lim_{x\to c} 3 && \text{Sum and Difference Rules}\\ &= c^3 + 4c^2 - 3 && \text{Product and Multiple Rules}\end{aligned}$$

(b)
$$\begin{aligned}\lim_{x\to c}\frac{x^4 + x^2 - 1}{x^2 + 5} &= \frac{\lim_{x\to c}(x^4 + x^2 - 1)}{\lim_{x\to c}(x^2 + 5)} && \text{Quotient Rule}\\ &= \frac{\lim_{x\to c} x^4 + \lim_{x\to c} x^2 - \lim_{x\to c} 1}{\lim_{x\to c} x^2 + \lim_{x\to c} 5} && \text{Sum and Difference Rules}\\ &= \frac{c^4 + c^2 - 1}{c^2 + 5} && \text{Product Rule}\end{aligned}$$

(c)
$$\begin{aligned}\lim_{x\to -2}\sqrt{4x^2 - 3} &= \sqrt{\lim_{x\to -2}(4x^2 - 3)} && \text{Root Rule with } n = 2\\ &= \sqrt{\lim_{x\to -2} 4x^2 - \lim_{x\to -2} 3} && \text{Sum and Difference Rules}\\ &= \sqrt{4(-2)^2 - 3} && \text{Product and Multiple Rules}\\ &= \sqrt{16 - 3}\\ &= \sqrt{13}\end{aligned}$$

As suggested by Example 1, the formulas in Theorem 1 lead us to conclude that limits of polynomial functions can be found by substitution. The same is true for rational functions if their denominators are different from zero at the point of evaluation.

Theorem 2 Limits of Polynomials Can Be Found by Substitution

If $P(x) = a_n x^n + a_{n-1}x^{n-1} + \cdots + a_0$, then

$$\lim_{x\to c} P(x) = P(c) = a_n c^n + a_{n-1}c^{n-1} + \cdots + a_0.$$

Theorem 3 Limits of Rational Functions Can Be Found by Substitution If the Limit of the Denominator Is Not Zero

If $P(x)$ and $Q(x)$ are polynomials and $Q(c) \neq 0$, then

$$\lim_{x\to c}\frac{P(x)}{Q(x)} = \frac{P(c)}{Q(c)}.$$

Example 2 Limit of a Rational Function

$$\lim_{x \to -1} \frac{x^3 + 4x^2 - 3}{x^2 + 5} = \frac{(-1)^3 + 4(-1)^2 - 3}{(-1)^2 + 5} = \frac{0}{6} = 0$$

This result is similar to the second limit in Example 1 with $c = -1$, now done in one step.

Eliminating Zero Denominators Algebraically

Theorem 3 applies only if the denominator of the rational function is not zero at the limit point c. If the denominator is zero, canceling common factors in the numerator and denominator may reduce the fraction to one whose denominator is no longer zero at c. If this happens, we can find the limit by substitution in the simplified fraction.

Identifying Common Factors

It can be shown that if $Q(x)$ is a polynomial and $Q(c) = 0$, then $(x - c)$ is a factor of $Q(x)$. Thus, if the numerator and denominator of a rational function of x are both zero at $x = c$, they have $(x - c)$ as a common factor.

Example 3 Canceling a Common Factor

Evaluate

$$\lim_{x \to 1} \frac{x^2 + x - 2}{x^2 - x}.$$

Solution We cannot substitute $x = 1$ because it makes the denominator zero. We test the numerator to see if it, too, is zero at $x = 1$. It is, so it has a factor of $(x - 1)$ in common with the denominator. Canceling the $(x - 1)$'s gives a simpler fraction with the same values as the original for $x \neq 1$:

$$\frac{x^2 + x - 2}{x^2 - x} = \frac{(x - 1)(x + 2)}{x(x - 1)} = \frac{x + 2}{x}, \qquad \text{if } x \neq 1.$$

Using the simpler fraction, we find the limit of these values as $x \to 1$ by substitution:

$$\lim_{x \to 1} \frac{x^2 + x - 2}{x^2 - x} = \lim_{x \to 1} \frac{x + 2}{x} = \frac{1 + 2}{1} = 3.$$

See Figure 1.15.

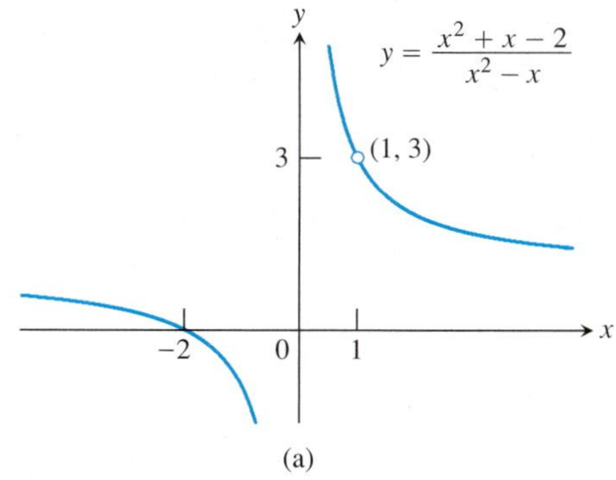

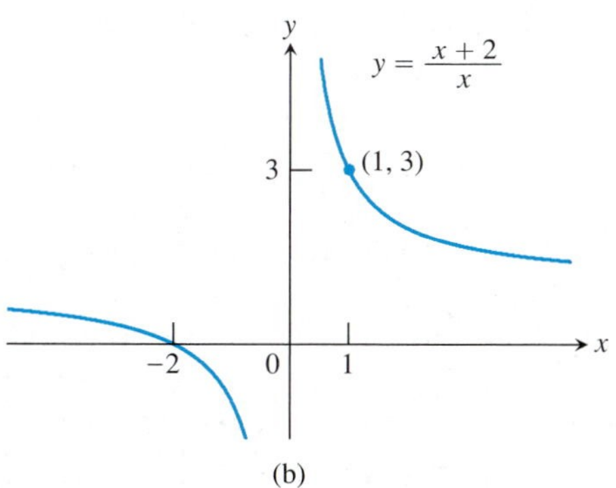

FIGURE 1.15 The graph of $f(x) = (x^2 + x - 2)/(x^2 - x)$ in (a) is the same as the graph of $g(x) = (x + 2)/x$ in (b) except at $x = 1$, where f is undefined. The functions have the same limit as $x \to 1$.

Example 4 Creating and Canceling a Common Factor

Evaluate

$$\lim_{h \to 0} \frac{\sqrt{2 + h} - \sqrt{2}}{h}.$$

Solution We cannot substitute $h = 0$, and the numerator and denominator have no obvious factors. We can, however, create a common factor by multiplying the numerator and denominator by the *conjugate expression* $\sqrt{2 + h} + \sqrt{2}$, obtained by changing the sign between the square roots:

$$\begin{aligned}
\frac{\sqrt{2 + h} - \sqrt{2}}{h} &= \frac{\sqrt{2 + h} - \sqrt{2}}{h} \cdot \frac{\sqrt{2 + h} + \sqrt{2}}{\sqrt{2 + h} + \sqrt{2}} \\
&= \frac{2 + h - 2}{h(\sqrt{2 + h} + \sqrt{2})} \\
&= \frac{h}{h(\sqrt{2 + h} + \sqrt{2})} && \text{Common factor of } h \\
&= \frac{1}{\sqrt{2 + h} + \sqrt{2}}. && \text{Cancel } h \text{ for } h \neq 0.
\end{aligned}$$

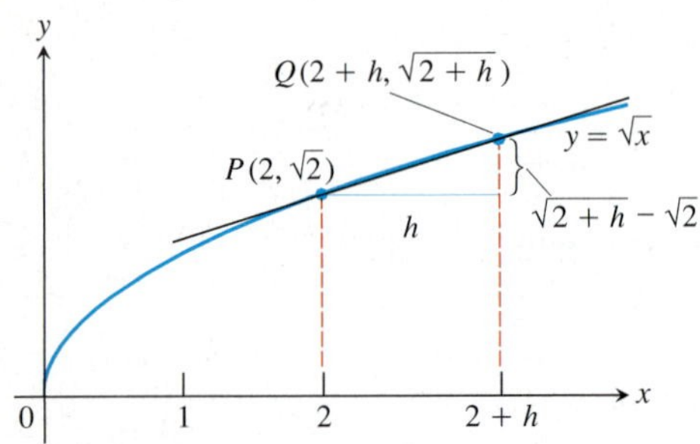

FIGURE 1.16 The limit of the slope $(\sqrt{2+h}-\sqrt{2})/h$ of secant PQ as $Q \to P$ along the curve is $1/(2\sqrt{2})$. (Example 4)

Therefore,

$$\lim_{h\to 0} \frac{\sqrt{2+h}-\sqrt{2}}{h} = \lim_{h\to 0} \frac{1}{\sqrt{2+h}+\sqrt{2}}$$

$$= \frac{1}{\sqrt{2+0}+\sqrt{2}}$$ Denominator not 0 at $h = 0$; substitute

$$= \frac{1}{2\sqrt{2}}.$$

Notice that $(\sqrt{2+h}-\sqrt{2})/h$ is the slope of the secant through the points $P(2, \sqrt{2})$ and $Q(2+h, \sqrt{2+h})$ on the curve $y = \sqrt{x}$. Figure 1.16 shows the secant for $h > 0$. Our calculation shows that $1/(2\sqrt{2})$ is the limiting value of this slope as $Q \to P$ along the curve from either side.

Sandwich Theorem

If we cannot find a limit directly, we may be able to find it indirectly with the Sandwich Theorem. The theorem refers to a function f whose values are sandwiched between the values of two other functions, g and h. If g and h have the same limit as $x \to c$, then f has this limit, too (Figure 1.17).

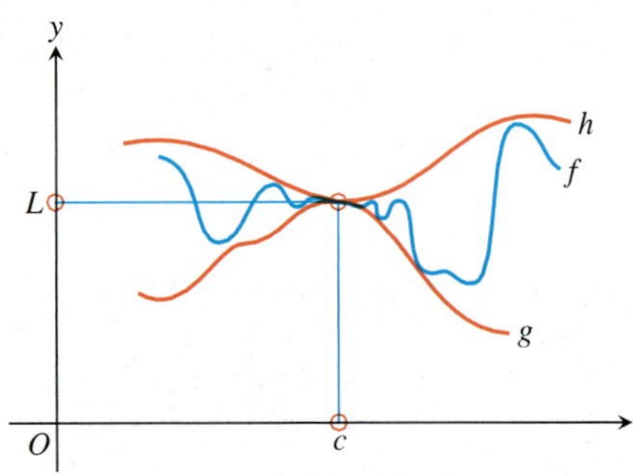

FIGURE 1.17 The graph of f is sandwiched between the graphs of g and h.

The Sandwich Theorem is sometimes called the Squeeze Theorem or the Pinching Theorem.

Theorem 4 Sandwich Theorem

Suppose that $g(x) \le f(x) \le h(x)$ for all x in some open interval containing c, except possibly at $x = c$ itself. Suppose also that

$$\lim_{x\to c} g(x) = \lim_{x\to c} h(x) = L.$$

Then $\lim_{x\to c} f(x) = L$.

You will find a proof of Theorem 4 in Appendix 2.

Example 5 Applying the Sandwich Theorem

Given that

$$1 - \frac{x^2}{4} \le u(x) \le 1 + \frac{x^2}{2} \qquad \text{for all } x \ne 0,$$

find $\lim_{x\to 0} u(x)$.

Solution Since

$$\lim_{x \to 0} (1 - (x^2/4)) = 1 \qquad \text{and} \qquad \lim_{x \to 0} (1 + (x^2/2)) = 1,$$

the Sandwich Theorem implies that $\lim_{x \to 0} u(x) = 1$ (Figure 1.18).

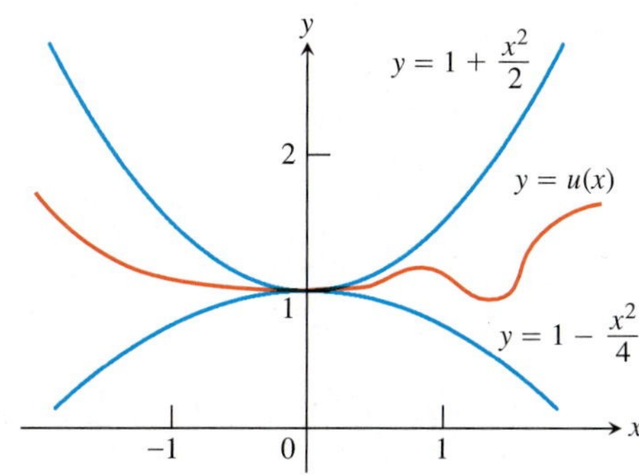

FIGURE 1.18 Any function $u(x)$ whose graph lies in the region between $y = 1 + (x^2/2)$ and $y = 1 - (x^2/4)$ has limit 1 as $x \to 0$.

CD-ROM WEBsite
Historical Biography
Euclid
(Ca. 365BC)

Example 6 Another Application of the Sandwich Theorem

(a) (Figure 1.19a) Since $-|\theta| \le \sin\theta \le |\theta|$ for all θ and $\lim_{\theta \to 0} (-|\theta|) = \lim_{\theta \to 0} |\theta| = 0$, we have

$$\lim_{\theta \to 0} \sin\theta = 0.$$

(b) (Figure 1.19b) Since $0 \le 1 - \cos\theta \le |\theta|$ for all θ, we have $\lim_{\theta \to 0} (1 - \cos\theta) = 0$ or

$$\lim_{\theta \to 0} \cos\theta = 1.$$

(c) For any function $f(x)$, if $\lim_{x \to c} |f(x)| = 0$, then $\lim_{x \to c} f(x) = 0$. The argument: $-|f(x)| \le f(x) \le |f(x)|$ and $-|f(x)|$ and $|f(x)|$ have limit 0 as $x \to c$.

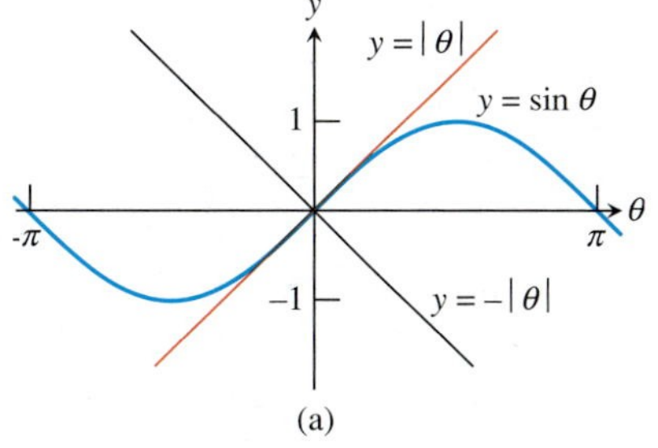

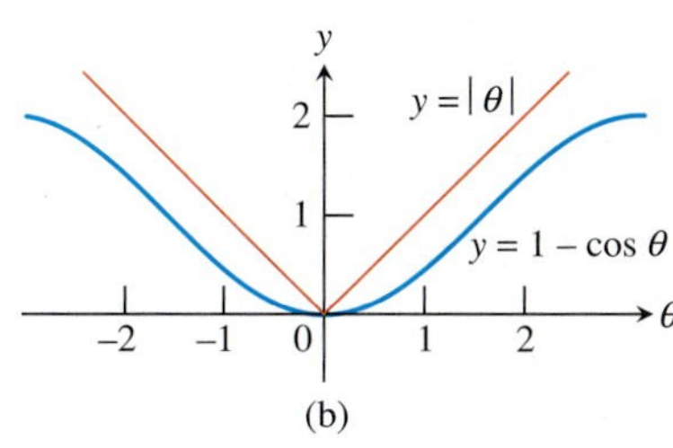

FIGURE 1.19 The Sandwich Theorem confirms that (a) $\lim_{\theta \to 0} \sin\theta = 0$ and (b) $\lim_{\theta \to 0} (1 - \cos\theta) = 0$.

CD-ROM WEBsite

One-Sided Limits

To have a limit L as x approaches a, a function f must be defined on *both sides* of a and its values $f(x)$ must approach L as x approaches a from either side. Because of this, ordinary limits are **two-sided.**

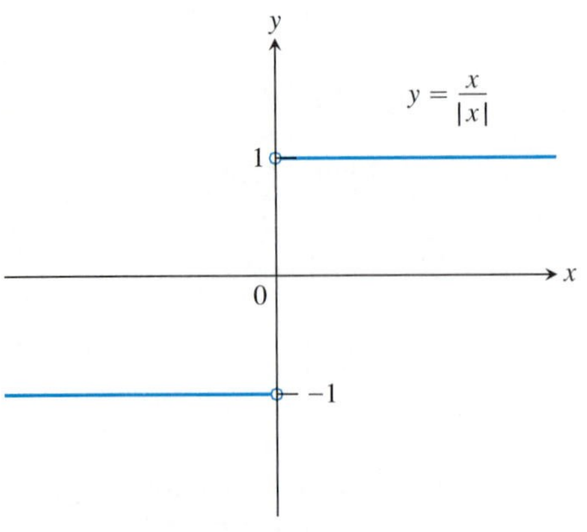

FIGURE 1.20 Different right-hand and left-hand limits at the origin.

If f fails to have a two-sided limit at a, it may still have a one-sided limit, that is, a limit if the approach is only from one side. If the approach is from the right, the limit is a **right-hand** limit. From the left, it is a **left-hand** limit.

The function $f(x) = x/|x|$ (Figure 1.20) has limit 1 as x approaches 0 from the right, and limit -1 as x approaches 0 from the left.

Definitions Right-Hand and Left-Hand Limits

Let $f(x)$ be defined on an interval (a, b), where $a < b$. If $f(x)$ approaches arbitrarily close to L as x approaches a from within that interval, then we say that f has **right-hand limit** L at a, and we write

$$\lim_{x \to a^+} f(x) = L.$$

Let $f(x)$ be defined on an interval (c, a), where $c < a$. If $f(x)$ approaches arbitrarily close to M as x approaches a from within the interval (c, a), then we say that f has **left-hand limit** M at a, and we write

$$\lim_{x \to a^-} f(x) = M.$$

The "+" and "−"

The significance of the signs in the notation for one-sided limits is that:

$\boldsymbol{x \to a^-}$ means x approaches a from the negative side of a, through values less than a.

$\boldsymbol{x \to a^+}$ means x approaches a from the positive side of a, through values greater than a.

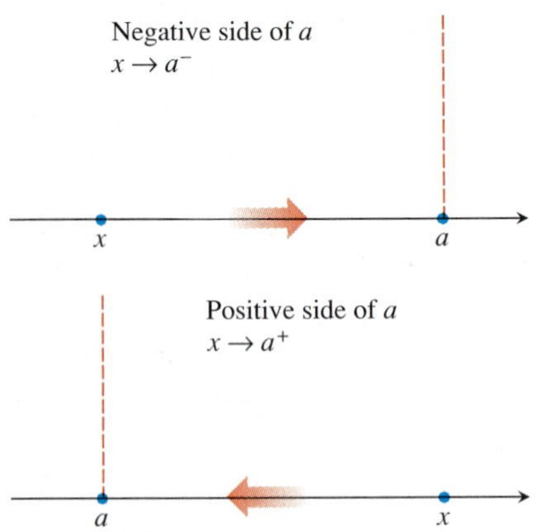

For the function $f(x) = x/|x|$ in Figure 1.20, we have

$$\lim_{x \to 0^+} f(x) = 1 \qquad \text{and} \qquad \lim_{x \to 0^-} f(x) = -1.$$

Example 7 One-Sided Limits for a Semicircle

The domain of $f(x) = \sqrt{4 - x^2}$ is $[-2, 2]$; its graph is the semicircle in Figure 1.21. We have

$$\lim_{x \to -2^+} \sqrt{4 - x^2} = 0 \qquad \text{and} \qquad \lim_{x \to 2^-} \sqrt{4 - x^2} = 0.$$

The function does not have a left-hand limit at $x = -2$ or a right-hand limit at $x = 2$. It does not have ordinary two-sided limits at either -2 or 2.

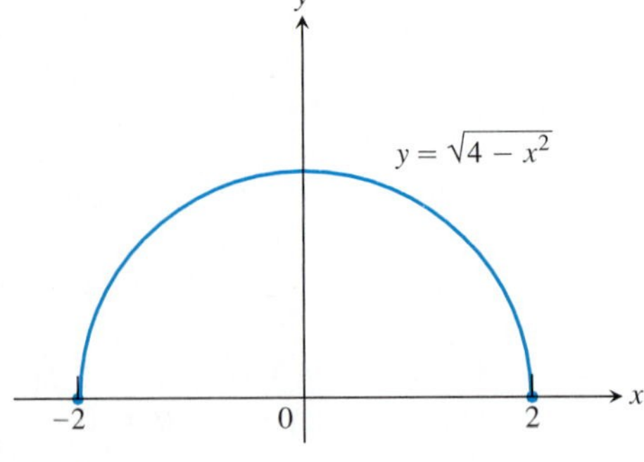

FIGURE 1.21 $\lim_{x \to 2^-} \sqrt{4 - x^2} = 0$ and $\lim_{x \to -2^+} \sqrt{4 - x^2} = 0$.

One-sided limits have all the properties listed in Theorem 1. The right-hand limit of the sum of two functions is the sum of their right-hand limits, and so on. The theorems for limits of polynomials and rational functions hold with one-sided limits, as does the Sandwich Theorem.

One-sided and two-sided limits are related in the following way:

The symbol $\Leftrightarrow$

The symbol $\Leftrightarrow$ is read "if and only if." It is a combination of the symbols $\Rightarrow$ (implies) and $\Leftarrow$ (is implied by).

Theorem 5 Relation Between One-Sided and Two-Sided Limits

A function $f(x)$ has a limit as x approaches c if and only if it has left-hand and right-hand limits there and these one-sided limits are equal:

$$\lim_{x \to c} f(x) = L \quad \Leftrightarrow \quad \lim_{x \to c^-} f(x) = L \quad \text{and} \quad \lim_{x \to c^+} f(x) = L\,.$$

Example 8 Limits of the Function Graphed in Figure 1.22

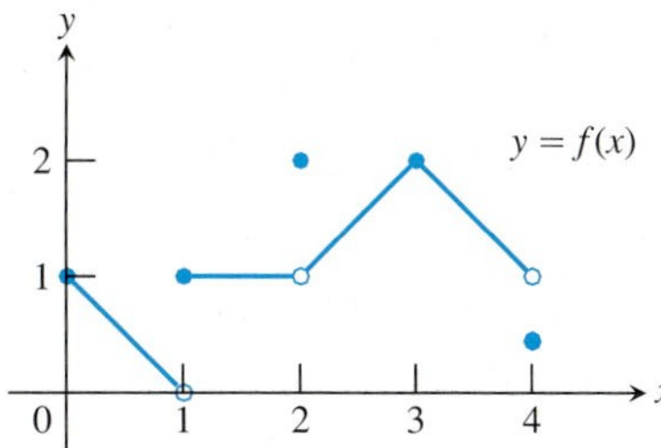

FIGURE 1.22 Graph of the function in Example 8.

At $x = 0$: $\lim_{x\to 0^+} f(x) = 1$,
$\lim_{x\to 0^-} f(x)$ and $\lim_{x\to 0} f(x)$ do not exist. The function is not defined to the left of $x = 0$.

At $x = 1$: $\lim_{x\to 1^-} f(x) = 0$ even though $f(1) = 1$,
$\lim_{x\to 1^+} f(x) = 1$,
$\lim_{x\to 1} f(x)$ does not exist. The right- and left-hand limits are not equal.

At $x = 2$: $\lim_{x\to 2^-} f(x) = 1$,
$\lim_{x\to 2^+} f(x) = 1$,
$\lim_{x\to 2} f(x) = 1$ even though $f(2) = 2$.

At $x = 3$: $\lim_{x\to 3^-} f(x) = \lim_{x\to 3^+} f(x) = \lim_{x\to 3} f(x) = f(3) = 2$

At $x = 4$: $\lim_{x\to 4^-} f(x) = 1$ even though $f(4) \neq 1$,
$\lim_{x\to 4^+} f(x)$ and $\lim_{x\to 4} f(x)$ do not exist. The function is not defined to the right of $x = 4$.

At every other point a in $[0, 4]$, $f(x)$ has limit $f(a)$.

The functions examined so far have had some kind of limit at each point of interest. In general, that need not be the case.

Example 9 A Function Oscillating Too Much

Show that $y = \sin(1/x)$ has no limit as x approaches zero from either side (Figure 1.23).

Solution As x approaches zero, its reciprocal, $1/x$, grows without bound and the values of $\sin(1/x)$ cycle repeatedly from -1 to 1. There is no single number L that the function's values stay increasingly close to as x approaches zero, which is true even if we restrict x to positive values or to negative values. The function has neither a right-hand limit nor a left-hand limit at $x = 0$.

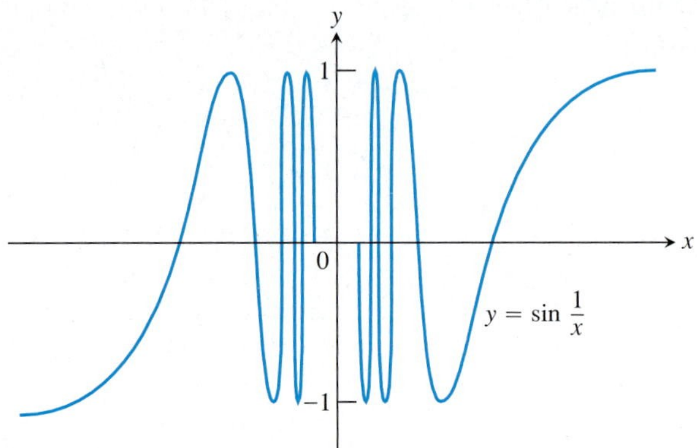

FIGURE 1.23 The function $y = \sin(1/x)$ has neither a right-hand nor a left-hand limit as x approaches zero. (Example 9)

Limits Involving $(\sin\theta)/\theta$

A central fact about $(\sin\theta)/\theta$ is that in radian measure its limit as $\theta \to 0$ is 1. We can see this in Figure 1.24 and confirm it algebraically using the Sandwich Theorem.

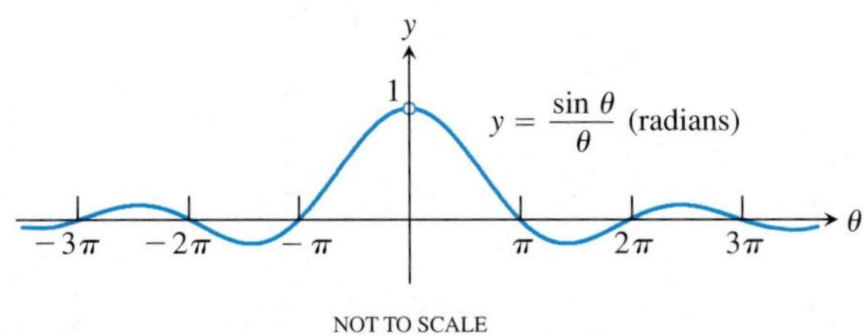

FIGURE 1.24 The graph of $f(\theta) = (\sin\theta)/\theta$.

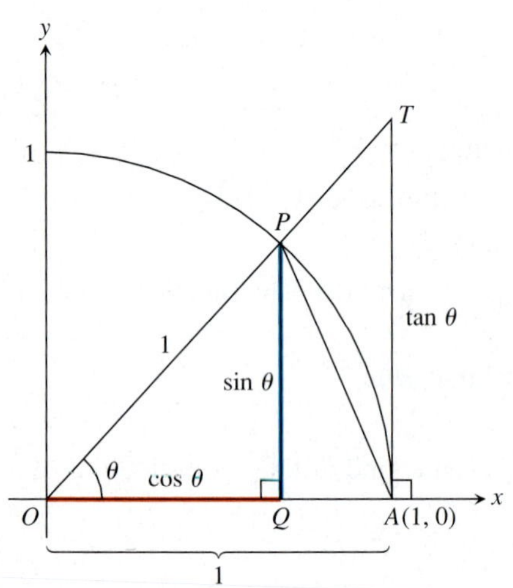

FIGURE 1.25 The figure for the proof of Theorem 6. $TA/OA = \tan\theta$, but $OA = 1$, so $TA = \tan\theta$.

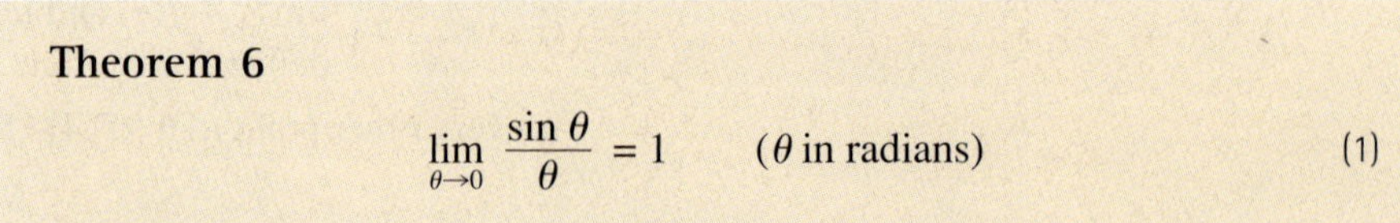

Theorem 6

$$\lim_{\theta\to 0}\frac{\sin\theta}{\theta} = 1 \qquad (\theta \text{ in radians}) \qquad (1)$$

Proof The plan is to show that the right-hand and left-hand limits are both 1. Then we will know that the two-sided limit is 1 as well.

To show that the right-hand limit is 1, we begin with positive values of θ less than $\pi/2$ (Figure 1.25). Notice that

$$\text{Area } \Delta OAP < \text{area sector } OAP < \text{area } \Delta OAT.$$

We can express these areas in terms of θ as follows:

Equation (2) is where radian measure comes in: The area of sector OAP is $\theta/2$ only if θ is measured in radians.

$$\text{Area } \Delta OAP = \frac{1}{2}\text{ base} \times \text{height} = \frac{1}{2}(1)(\sin\theta) = \frac{1}{2}\sin\theta$$

$$\text{Area sector } OAP = \frac{1}{2}r^2\theta = \frac{1}{2}(1)^2\theta = \frac{\theta}{2} \tag{2}$$

$$\text{Area } \Delta OAT = \frac{1}{2}\text{ base} \times \text{height} = \frac{1}{2}(1)(\tan\theta) = \frac{1}{2}\tan\theta .$$

Thus,

$$\frac{1}{2}\sin\theta < \frac{1}{2}\theta < \frac{1}{2}\tan\theta .$$

This last inequality goes the same way if we divide all three terms by the positive number $(1/2)\sin\theta$:

$$1 < \frac{\theta}{\sin\theta} < \frac{1}{\cos\theta} .$$

Taking reciprocals reverses the inequalities:

$$1 > \frac{\sin\theta}{\theta} > \cos\theta .$$

Since $\lim_{\theta\to 0^+} \cos\theta = 1$, the Sandwich Theorem gives

$$\lim_{\theta\to 0^+} \frac{\sin\theta}{\theta} = 1 .$$

Recall that $\sin\theta$ and θ are both *odd functions* (Preliminary Section 3). Therefore, $f(\theta) = (\sin\theta)/\theta$ is an *even function,* with a graph symmetric about the y-axis (see Figure 1.24). This symmetry implies that the left-hand limit at 0 exists and has the same value as the right-hand limit:

$$\lim_{\theta\to 0^-} \frac{\sin\theta}{\theta} = 1 = \lim_{\theta\to 0^+} \frac{\sin\theta}{\theta} ,$$

so $\lim_{\theta\to 0} (\sin\theta)/\theta = 1$ by Theorem 4.

CD-ROM WEBsite

Example 10 Using $\lim_{\theta\to 0} \frac{\sin\theta}{\theta} = 1$

Show that (a) $\lim_{h\to 0} \frac{\cos h - 1}{h} = 0$ and (b) $\lim_{x\to 0} \frac{\sin 2x}{5x} = \frac{2}{5}$.

Solution

(a) Using the half-angle formula $\cos h = 1 - 2\sin^2(h/2)$, we calculate

$$\begin{aligned}\lim_{h\to 0} \frac{\cos h - 1}{h} &= \lim_{h\to 0} -\frac{2\sin^2(h/2)}{h} \\ &= -\lim_{\theta\to 0} \frac{\sin\theta}{\theta}\sin\theta \qquad \text{Let } \theta = h/2. \\ &= -(1)(0) = 0 .\end{aligned}$$

(b) Equation (1) does not apply to the original fraction. We need a $2x$ in the denominator, not a $5x$. We produce it by multiplying numerator and denominator by $2/5$:

$$\begin{aligned}\lim_{x\to 0}\frac{\sin 2x}{5x} &= \lim_{x\to 0}\frac{(2/5)\cdot\sin 2x}{(2/5)\cdot 5x}\\ &= \frac{2}{5}\lim_{x\to 0}\frac{\sin 2x}{2x} \qquad \text{Now, Eq. (1) applies with } \theta = 2x.\\ &= \frac{2}{5}(1) = \frac{2}{5}\end{aligned}$$

EXERCISES 1.2

Estimating Limits Graphically

In Exercises 1–6, use the graph to estimate the limits and value of the function, or explain why the limits do not exist.

1.

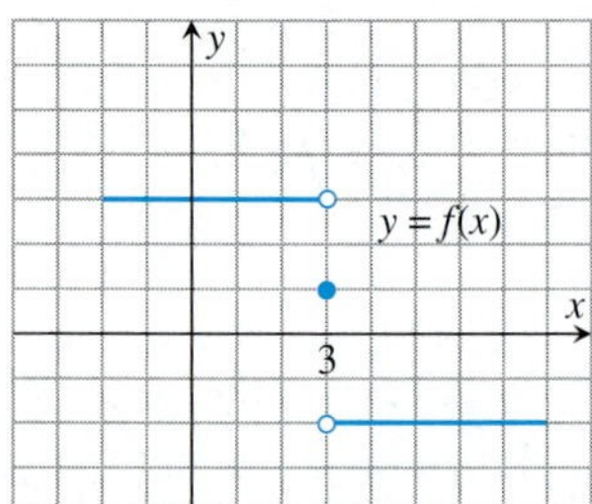

(a) $\lim_{x\to 3^-} f(x)$ (b) $\lim_{x\to 3^+} f(x)$ (c) $\lim_{x\to 3} f(x)$ (d) $f(3)$

2.

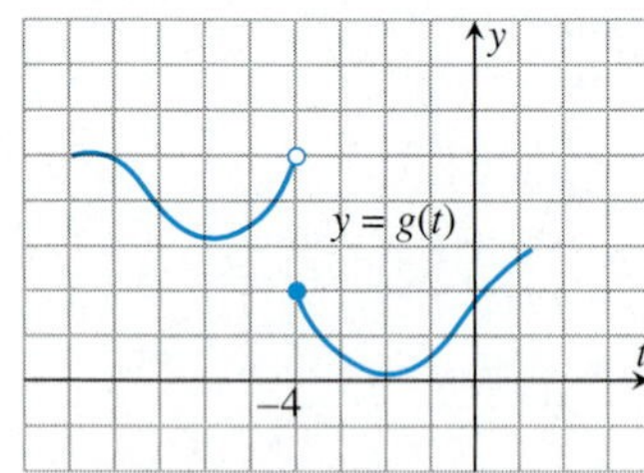

(a) $\lim_{t\to -4^-} g(t)$ (b) $\lim_{t\to -4^+} g(t)$ (c) $\lim_{t\to -4} g(t)$ (d) $g(-4)$

3.

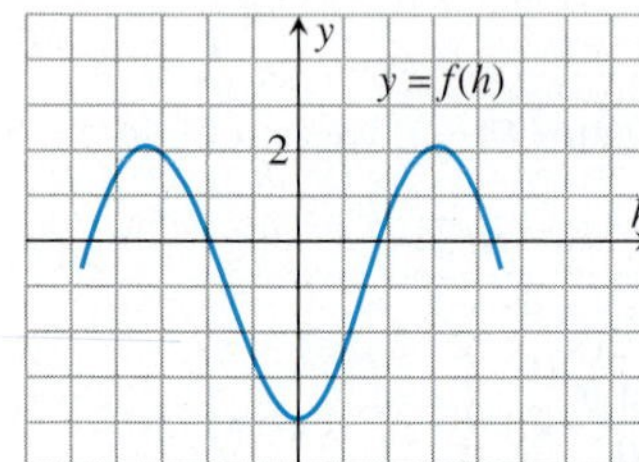

(a) $\lim_{h\to 0^-} f(h)$ (b) $\lim_{h\to 0^+} f(h)$ (c) $\lim_{h\to 0} f(h)$ (d) $f(0)$

4.

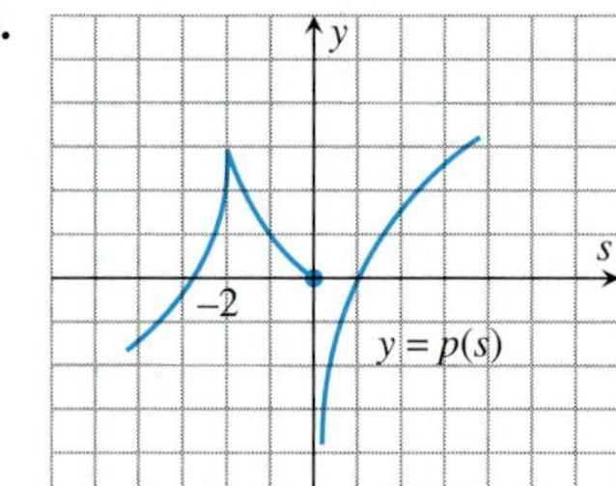

(a) $\lim_{s\to -2^-} p(s)$ (b) $\lim_{s\to -2^+} p(s)$ (c) $\lim_{s\to -2} p(s)$ (d) $p(-2)$

5.

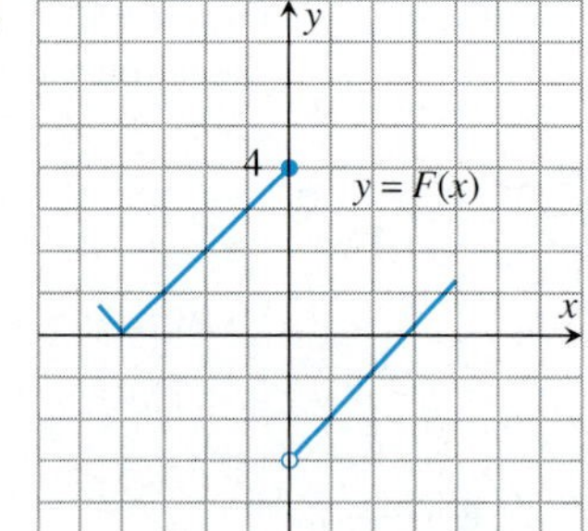

(a) $\lim_{x\to 0^-} F(x)$ (b) $\lim_{x\to 0^+} F(x)$ (c) $\lim_{x\to 0} F(x)$ (d) $F(0)$

6.

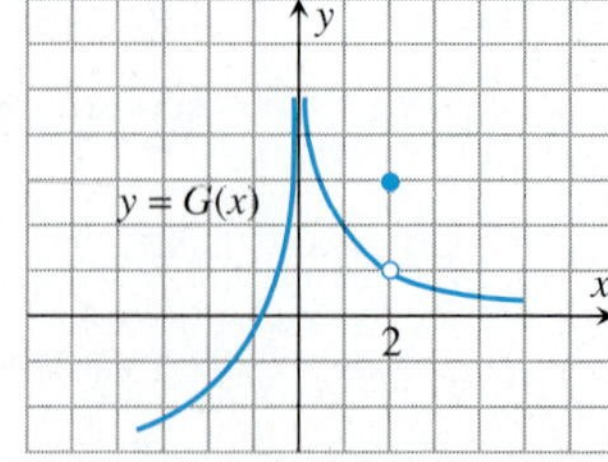

(a) $\lim_{x\to 2^-} G(x)$ (b) $\lim_{x\to 2^+} G(x)$ (c) $\lim_{x\to 2} G(x)$ (d) $G(2)$

Using Limit Rules

7. Suppose that $\lim_{x\to 0} f(x) = 1$ and $\lim_{x\to 0} g(x) = -5$. Name the rules in Theorem 1 that are used to accomplish steps (a), (b), and (c) of the following calculation.

$$\lim_{x\to 0} \frac{2f(x) - g(x)}{(f(x)+7)^{2/3}}$$

$$= \frac{\lim_{x\to 0} (2f(x) - g(x))}{\lim_{x\to 0} (f(x)+7)^{2/3}} \qquad \text{(a)}$$

$$= \frac{\lim_{x\to 0} 2f(x) - \lim_{x\to 0} g(x)}{\left(\lim_{x\to 0} (f(x)+7)\right)^{2/3}} \qquad \text{(b)}$$

$$= \frac{2\lim_{x\to 0} f(x) - \lim_{x\to 0} g(x)}{\left(\lim_{x\to 0} f(x) + \lim_{x\to 0} 7\right)^{2/3}} \qquad \text{(c)}$$

$$= \frac{(2)(1) - (-5)}{(1+7)^{2/3}} = \frac{7}{4}$$

8. Let $\lim_{x\to 1} h(x) = 5$, $\lim_{x\to 1} p(x) = 1$, and $\lim_{x\to 1} r(x) = -2$. Name the rules in Theorem 1 that are used to accomplish steps (a), (b), and (c) of the following calculation.

$$\lim_{x\to 1} \frac{\sqrt{5h(x)}}{p(x)(4 - r(x))}$$

$$= \frac{\lim_{x\to 1} \sqrt{5h(x)}}{\lim_{x\to 1} (p(x)(4 - r(x)))} \qquad \text{(a)}$$

$$= \frac{\sqrt{\lim_{x\to 1} 5h(x)}}{\left(\lim_{x\to 1} p(x)\right)\left(\lim_{x\to 1} (4 - r(x))\right)} \qquad \text{(b)}$$

$$= \frac{\sqrt{5\lim_{x\to 1} h(x)}}{\left(\lim_{x\to 1} p(x)\right)\left(\lim_{x\to 1} 4 - \lim_{x\to 1} r(x)\right)} \qquad \text{(c)}$$

$$= \frac{\sqrt{(5)(5)}}{(1)(4-2)} = \frac{5}{2}$$

9. Suppose that $\lim_{x\to c} f(x) = 5$ and $\lim_{x\to c} g(x) = -2$. Find

(a) $\lim_{x\to c} f(x)g(x)$ **(b)** $\lim_{x\to c} 2f(x)g(x)$

(c) $\lim_{x\to c} (f(x) + 3g(x))$ **(d)** $\lim_{x\to c} \dfrac{f(x)}{f(x) - g(x)}$

10. Suppose that $\lim_{x\to 4} f(x) = 0$ and $\lim_{x\to 4} g(x) = -3$. Find

(a) $\lim_{x\to 4} (g(x) + 3)$ **(b)** $\lim_{x\to 4} xf(x)$

(c) $\lim_{x\to 4} (g(x))^2$ **(d)** $\lim_{x\to 4} \dfrac{g(x)}{f(x) - 1}$

Limit Calculations

Find the limits in Exercises 11–14.

11. (a) $\lim_{x\to -7} (2x+5)$ **(b)** $\lim_{t\to 6} 8(t-5)(t-7)$

(c) $\lim_{y\to 2} \dfrac{y+2}{y^2+5y+6}$ **(d)** $\lim_{h\to 0} \dfrac{3}{\sqrt{3h+1}+1}$

12. (a) $\lim_{r\to -2} (r^3 - 2r^2 + 4r + 8)$ **(b)** $\lim_{x\to 2} \dfrac{x+3}{x+6}$

(c) $\lim_{y\to -3} (5-y)^{4/3}$ **(d)** $\lim_{\theta\to 5} \dfrac{\theta - 5}{\theta^2 - 25}$

13. (a) $\lim_{t\to -5} \dfrac{t^2+3t-10}{t+5}$ **(b)** $\lim_{x\to -2} \dfrac{-2x-4}{x^3+2x^2}$

(c) $\lim_{y\to 1} \dfrac{y-1}{\sqrt{y+3}-2}$

14. (a) $\lim_{x\to -1} \dfrac{\sqrt{x^2+8}-3}{x+1}$ **(b)** $\lim_{\theta\to 1} \dfrac{\theta^4 - 1}{\theta^3 - 1}$

(c) $\lim_{t\to 9} \dfrac{3-\sqrt{t}}{9-t}$

Using the Sandwich Theorem

15. *Writing to Learn* **(a)** It can be shown that the inequalities

$$1 - \frac{x^2}{6} < \frac{x \sin x}{2 - 2\cos x} < 1$$

hold for all values of x close to zero. What, if anything, does this tell you about

$$\lim_{x\to 0} \frac{x \sin x}{2 - 2\cos x}$$

Give reasons for your answer.

T **(b)** Graph $y = 1 - (x^2/6)$, $y = (x \sin x)/(2 - 2\cos x)$, and $y = 1$ together for $-2 \le x \le 2$. Comment on the behavior of the graphs as $x \to 0$.

16. *Writing to Learn* **(a)** The inequalities

$$\frac{1}{2} - \frac{x^2}{24} < \frac{1 - \cos x}{x^2} < \frac{1}{2}$$

hold for values of x close to zero. What, if anything, does this tell you about

$$\lim_{x\to 0} \frac{1 - \cos x}{x^2}?$$

Give reasons for your answer.

T **(b)** Graph $y = (1/2) - (x^2/24)$, $y = (1 - \cos x)/x^2$, and $y = 1/2$ together for $-2 \le x \le 2$. Comment on the behavior of the graphs as $x \to 0$.

Limits of Average Rates of Change

Because of their connections with secant lines, tangents, and instantaneous rates, limits of the form

$$\lim_{h\to 0} \frac{f(x_0 + h) - f(x_0)}{h}$$

occur frequently in calculus. In Exercises 17–20, evaluate this limit for the given x_0 and function f.

17. $f(x) = x^2, \quad x_0 = 1$

18. $f(x) = 3x - 4, \quad x_0 = 2$

19. $f(x) = 1/x, \quad x_0 = -2$

20. $f(x) = \sqrt{x}, \quad x_0 = 7$

Finding Limits Graphically

21. Which of the following statements about the function $y = f(x)$ graphed here are true and which are false? Give reasons for your answers.

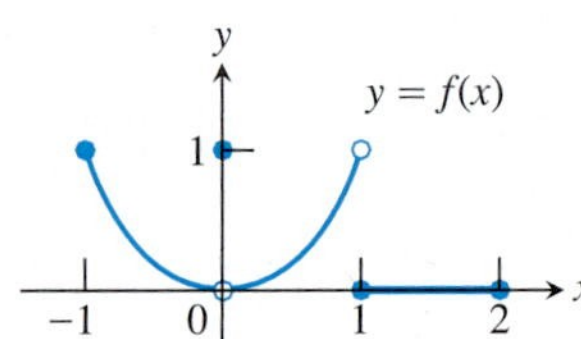

(a) $\lim_{x \to -1^+} f(x) = 1$. **(b)** $\lim_{x \to 0^-} f(x) = 0$.

(c) $\lim_{x \to 0^-} f(x) = 1$. **(d)** $\lim_{x \to 0^-} f(x) = \lim_{x \to 0^+} f(x)$.

(e) $\lim_{x \to 0} f(x)$ exists. **(f)** $\lim_{x \to 0} f(x) = 0$.

(g) $\lim_{x \to 0} f(x) = 1$. **(h)** $\lim_{x \to 1} f(x) = 1$.

(i) $\lim_{x \to 1} f(x) = 0$. **(j)** $\lim_{x \to 2^-} f(x) = 2$.

(k) $\lim_{x \to -1^-} f(x)$ does not exist. **(l)** $\lim_{x \to 2^+} f(x) = 0$.

22. Let

$$f(x) = \begin{cases} 3 - x, & x < 2 \\ \dfrac{x}{2} + 1, & x > 2. \end{cases}$$

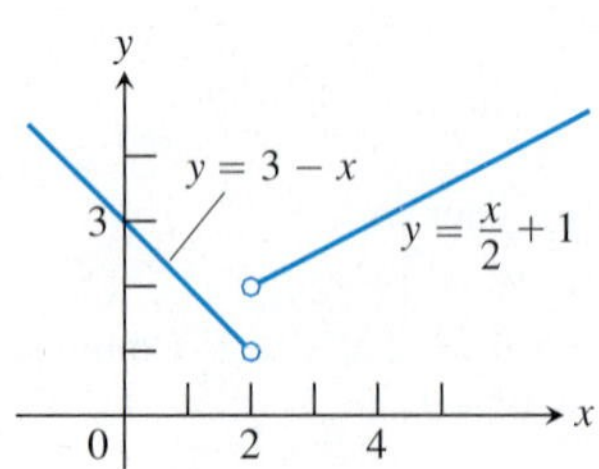

(a) Find $\lim_{x \to 2^+} f(x)$ and $\lim_{x \to 2^-} f(x)$.

(b) Does $\lim_{x \to 2} f(x)$ exist? If so, what is it? If not, why not?

(c) Find $\lim_{x \to 4^-} f(x)$ and $\lim_{x \to 4^+} f(x)$.

(d) Does $\lim_{x \to 4} f(x)$ exist? If so, what is it? If not, why not?

23. Let $f(x) = \begin{cases} 0, & x \le 0 \\ \sin \dfrac{1}{x}, & x > 0. \end{cases}$

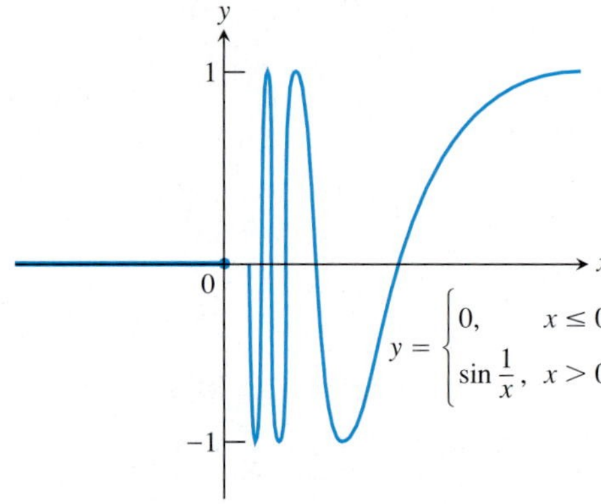

(a) Does $\lim_{x \to 0^+} f(x)$ exist? If so, what is it? If not, why not?

(b) Does $\lim_{x \to 0^-} f(x)$ exist? If so, what is it? If not, why not?

(c) Does $\lim_{x \to 0} f(x)$ exist? If so, what is it? If not, why not?

24. Let $g(x) = \sqrt{x} \sin(1/x)$.

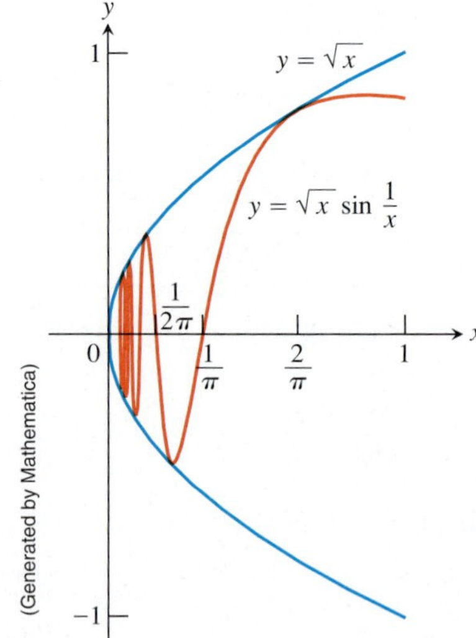

(a) Does $\lim_{x \to 0^+} g(x)$ exist? If so, what is it? If not, why not?

(b) Does $\lim_{x \to 0^-} g(x)$ exist? If so, what is it? If not, why not?

(c) Does $\lim_{x \to 0} g(x)$ exist? If so, what is it? If not, why not?

Graph the functions in Exercises 25 and 26. Then answer these questions.

(a) What are the domain and range of f?

(b) At what point c, if any, does $\lim_{x\to c} f(x)$ exist?

(c) At what point does only the left-hand limit exist?

(d) At what point does only the right-hand limit exist?

25. $f(x) = \begin{cases} \sqrt{1-x^2} & \text{if } 0 \le x < 1 \\ 1 & \text{if } 1 \le x < 2 \\ 2 & \text{if } x = 2 \end{cases}$

26. $f(x) = \begin{cases} x & \text{if } -1 \le x < 0 \quad \text{or} \quad 0 < x \le 1 \\ 1 & \text{if } x = 0 \\ 0 & \text{if } x < -1 \quad \text{or} \quad x > 1 \end{cases}$

Finding One-Sided Limits Algebraically

Find the limits in Exercises 27–32.

27. $\lim_{x\to -0.5^-} \sqrt{\dfrac{x+2}{x+1}}$

28. $\lim_{x\to -2^+} \left(\dfrac{x}{x+1}\right)\left(\dfrac{2x+5}{x^2+x}\right)$

29. $\lim_{h\to 0^+} \dfrac{\sqrt{h^2+4h+5}-\sqrt{5}}{h}$

30. $\lim_{h\to 0^-} \dfrac{\sqrt{6}-\sqrt{5h^2+11h+6}}{h}$

31. **(a)** $\lim_{x\to -2^+} (x+3)\dfrac{|x+2|}{x+2}$ **(b)** $\lim_{x\to -2^-} (x+3)\dfrac{|x+2|}{x+2}$

32. **(a)** $\lim_{x\to 1^+} \dfrac{\sqrt{2x}\,(x-1)}{|x-1|}$ **(b)** $\lim_{x\to 1^-} \dfrac{\sqrt{2x}\,(x-1)}{|x-1|}$

Theory and Examples

33. *Writing to Learn* If $x^4 \le f(x) \le x^2$ for x in $[-1, 1]$ and $x^2 \le f(x) \le x^4$ for $x < -1$ and $x > 1$, at what points c do you automatically know $\lim_{x\to c} f(x)$? What can you say about the value of the limit at these points?

34. *Writing to Learn* Suppose that $g(x) \le f(x) \le h(x)$ for all $x \ne 2$ and suppose that

$$\lim_{x\to 2} g(x) = \lim_{x\to 2} h(x) = -5.$$

Can we conclude anything about the values of f, g, and h at $x = 2$? Could $f(2) = 0$? Could $\lim_{x\to 2} f(x) = 0$? Give reasons for your answers.

35. *Deducing the value of a limit* If $\lim_{x\to -2} \dfrac{f(x)}{x^2} = 1$, find

(a) $\lim_{x\to -2} f(x)$ **(b)** $\lim_{x\to -2} \dfrac{f(x)}{x}$

36. *Deducing the value of a limit*

(a) If $\lim_{x\to 2} \dfrac{f(x)-5}{x-2} = 3$, find $\lim_{x\to 2} f(x)$.

(b) If $\lim_{x\to 2} \dfrac{f(x)-5}{x-2} = 4$, find $\lim_{x\to 2} f(x)$.

37. *Writing to Learn* Once you know $\lim_{x\to a^+} f(x)$ and $\lim_{x\to a^-} f(x)$ at an interior point of the domain of f, do you then know $\lim_{x\to a} f(x)$? Give reasons for your answer.

38. *Writing to Learn* If you know that $\lim_{x\to c} f(x)$ exists, can you find its value by calculating $\lim_{x\to c^+} f(x)$? Give reasons for your answer.

39. *Finding delta* Given $\epsilon > 0$, find an interval $I = (5, 5+\delta)$, $\delta > 0$, such that if x lies in I, then $\sqrt{x-5} < \epsilon$. What limit is being verified, and what is its value?

40. *Finding delta* Given $\epsilon > 0$, find an interval $I = (4-\delta, 4)$, $\delta > 0$, such that if x lies in I, then $\sqrt{4-x} < \epsilon$. What limit is being verified, and what is its value.

Even and Odd Functions

Recall that a function $y = f(x)$ defined on a domain D that is symmetric about the origin is **even** if $f(-x) = f(x)$ for all x in D and **odd** if $f(-x) = -f(x)$ for all x in D.

41. *Writing to Learn* Suppose that f is an odd function of x. Does knowing that $\lim_{x\to 0^+} f(x) = 3$ tell you anything about $\lim_{x\to 0^-} f(x)$? Give reasons for your answer.

42. *Writing to Learn* Suppose that f is an even function of x. Does knowing that $\lim_{x\to 2^-} f(x) = 7$ tell you anything about either $\lim_{x\to -2^-} f(x)$ or $\lim_{x\to -2^+} f(x)$? Give reasons for your answer.

COMPUTER EXPLORATIONS

43. **(a)** Graph $g(x) = x \sin(1/x)$ to estimate $\lim_{x\to 0} g(x)$, zooming in on the origin as necessary.

(b) *Writing to Learn* Now graph $k(x) = \sin(1/x)$. Compare the behaviors of g and k near the origin. What is the same? What is different?

44. **(a)** Graph $h(x) = x^2 \cos(1/x)$ to estimate $\lim_{x\to 0} h(x)$, zooming in on the origin as necessary.

(b) *Writing to Learn* Now graph $k(x) = \cos(1/x)$. Compare the behaviors of h and k near the origin. What is the same? What is different?

1.3 Limits Involving Infinity

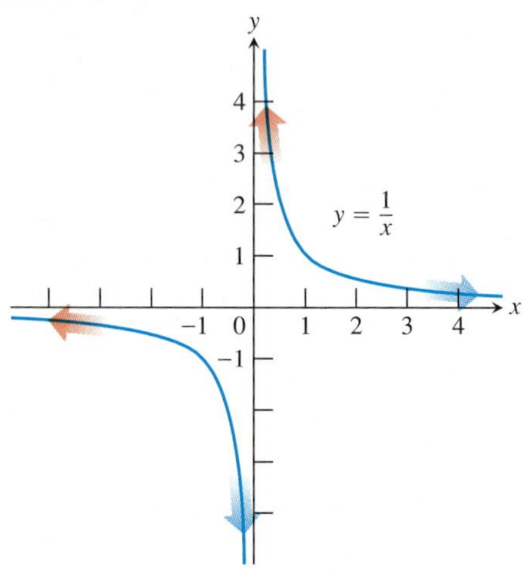

FIGURE 1.26 The graph of $y = 1/x$.

Finite Limits as $x \to \pm\infty$ • Limits of Rational Functions as $x \to \pm\infty$ • Horizontal and Vertical Asymptotes: Infinite Limits • Sandwich Theorem Revisited • Precise Definitions of Infinite Limits • End Behavior Models and Oblique Asymptotes

We analyze the graphs of rational functions (quotients of polynomial functions), as well as other functions with interesting limit behavior as $x \to \pm\infty$. Among the tools we use are horizontal and vertical asymptotes.

Finite Limits as x → ±∞

The symbol for infinity (∞) does not represent a real number. We use ∞ to describe the behavior of a function when the values in its domain or range outgrow all finite bounds. For example, the function $f(x) = 1/x$ is defined for all $x \neq 0$ (Figure 1.26). When x is positive and becomes increasingly large, $1/x$ becomes increasingly small. When x is negative and its magnitude becomes increasingly large, $1/x$ again becomes small. We summarize these observations by saying that $f(x) = 1/x$ has limit 0 as $x \to \pm\infty$.

CD-ROM
WEBsite

Definitions Limits as $x \to \pm\infty$

1. We say that $f(x)$ has the **limit L as x approaches infinity** and write

$$\lim_{x\to\infty} f(x) = L$$

if, as x moves increasingly far from the origin in the positive direction, $f(x)$ gets arbitrarily close to L.

2. We say that $f(x)$ has the **limit L as x approaches minus infinity** and write

$$\lim_{x\to-\infty} f(x) = L$$

if, as x moves increasingly far from the origin in the negative direction, $f(x)$ gets arbitrarily close to L.

The strategy for calculating limits of functions as $x \to \pm\infty$ is similar to the one for finite limits in Section 1.2. There we first found the limits of the constant and identity functions $y = k$ and $y = x$. We then extended these results to other functions by applying a theorem about limits of algebraic combinations. Here we do the same thing, except that the starting functions are $y = k$ and $y = 1/x$ instead of $y = k$ and $y = x$.

The basic facts to be verified as $x \to \pm\infty$ are given in the next example.

The Symbol Infinity (∞)

As always, the symbol ∞ does not represent a real number and we cannot use it in arithmetic in the usual way. Also, the symbol ∞ means $+\infty$. The two are used interchangeably.

Example 1 Limits of $1/x$ and k as $x \to \pm\infty$

Show that

(a) $\lim_{x\to\infty} \frac{1}{x} = \lim_{x\to-\infty} \frac{1}{x} = 0$

(b) $\lim_{x\to\infty} k = \lim_{x\to-\infty} k = k.$

Solution

(a) From Figure 1.26, we see that $y = 1/x$ gets closer and closer to zero as x moves increasingly far from the origin in either the positive or negative direction.

(b) No matter how far x is from the origin, the constant function $y = k$ always has exactly the value k.

Limits at infinity have properties similar to those of finite limits.

Theorem 7 Rules for Limits as $x \to \pm\infty$

If L, M, and k, are real numbers and

$$\lim_{x\to\pm\infty} f(x) = L \quad \text{and} \quad \lim_{x\to\pm\infty} g(x) = M, \quad \text{then}$$

1. *Sum Rule:* $\lim_{x\to\pm\infty} (f(x) + g(x)) = L + M$
2. *Difference Rule:* $\lim_{x\to\pm\infty} (f(x) - g(x)) = L - M$
3. *Product Rule:* $\lim_{x\to\pm\infty} (f(x) \cdot g(x)) = L \cdot M$
4. *Constant Multiple Rule:* $\lim_{x\to\pm\infty} (k \cdot f(x)) = k \cdot L$
5. *Quotient Rule:* $\lim_{x\to\pm\infty} \dfrac{f(x)}{g(x)} = \dfrac{L}{M}, \quad M \neq 0$
6. *Root Rule:* If n is a positive integer, then

$$\lim_{x\to\pm\infty} (f(x))^{1/n} = L^{1/n}$$

(If n is even, we assume that $L > 0$.)

These properties are just like the properties in Theorem 1, Section 1.2, and we use them the same way.

Example 2 Using Theorem 7

(a) $$\lim_{x\to\infty}\left(5 + \frac{1}{x}\right) = \lim_{x\to\infty} 5 + \lim_{x\to\infty} \frac{1}{x} \qquad \text{Sum Rule}$$

$$= 5 + 0 = 5 \qquad \text{Known limits}$$

(b) $$\lim_{x\to-\infty} \frac{\pi\sqrt{3}}{x^2} = \lim_{x\to-\infty} \pi\sqrt{3} \cdot \frac{1}{x} \cdot \frac{1}{x}$$

$$= \lim_{x\to-\infty} \pi\sqrt{3} \cdot \lim_{x\to-\infty} \frac{1}{x} \cdot \lim_{x\to-\infty} \frac{1}{x} \qquad \text{Product rule}$$

$$= \pi\sqrt{3} \cdot 0 \cdot 0 = 0 \qquad \text{Known limits}$$

Limits of Rational Functions as $x \to \pm\infty$

To determine the limit of a rational function as $x \to \pm\infty$, we can divide the numerator and denominator by the highest power of x in the denominator. What happens then depends on the degrees of the polynomials involved.

The **degree** of the polynomial $a_nx^n + a_{n-1}x^{n-1} + \cdots + a_1x + a_0$, $a_n \neq 0$, is n, the largest exponent.

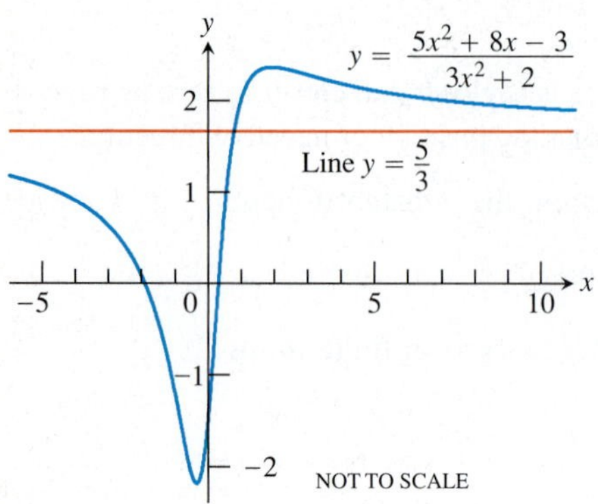

FIGURE 1.27 The function in Example 3.

Example 3 Numerator and Denominator of Same Degree

$$\lim_{x\to\infty} \frac{5x^2 + 8x - 3}{3x^2 + 2} = \lim_{x\to\infty} \frac{5 + (8/x) - (3/x^2)}{3 + (2/x^2)}$$

Divide numerator and denominator by x^2.

$$= \frac{5 + 0 - 0}{3 + 0} = \frac{5}{3}$$

See Fig. 1.27.

Example 4 Degree of Numerator Less Than Degree of Denominator

$$\lim_{x\to-\infty} \frac{11x + 2}{2x^3 - 1} = \lim_{x\to-\infty} \frac{(11/x^2) + (2/x^3)}{2 - (1/x^3)}$$

Divide numerator and denominator by x^3.

$$= \frac{0 + 0}{2 - 0} = 0$$

See Fig. 1.28.

Example 5 Degree of Numerator Greater Than Degree of Denominator

(a) $$\lim_{x\to-\infty} \frac{2x^2 - 3}{7x + 4} = \lim_{x\to-\infty} \frac{2x - (3/x)}{7 + (4/x)}$$

Divide numerator and denominator by x.

$$= -\infty$$

The numerator now approaches $-\infty$ while the denominator approaches 7, so the ratio $\to -\infty$. See Fig. 1.29.

(b) $$\lim_{x\to-\infty} \frac{-4x^3 + 7x}{2x^2 - 3x - 10} = \lim_{x\to-\infty} \frac{-4x + (7/x)}{2 - (3/x) - (10/x^2)}$$

Divide numerator and denominator by x^2.

$$= \infty$$

Numerator $\to \infty$, denominator $\to 2$; ratio $\to \infty$.

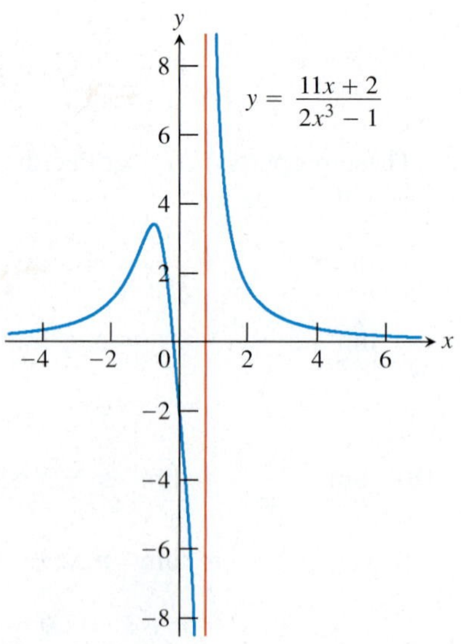

FIGURE 1.28 The graph of the function in Example 4. The graph approaches the x-axis as $|x|$ increases.

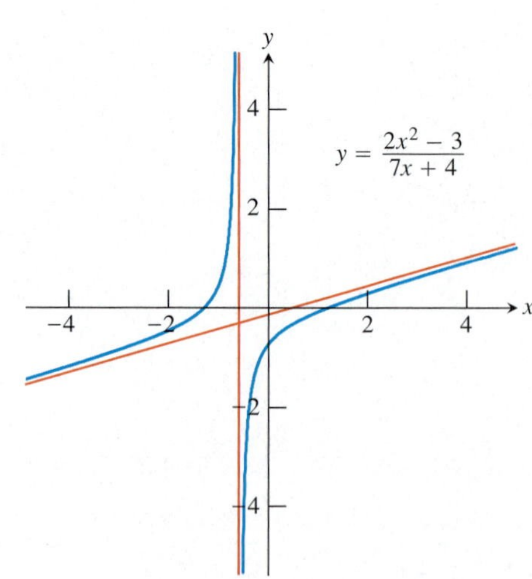

FIGURE 1.29 The function in Example 5(a).

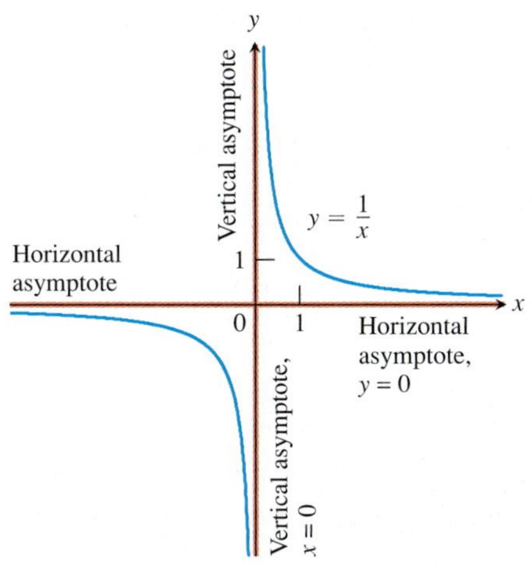

FIGURE 1.30 The coordinate axes are asymptotes of both branches of the hyperbola $y = 1/x$.

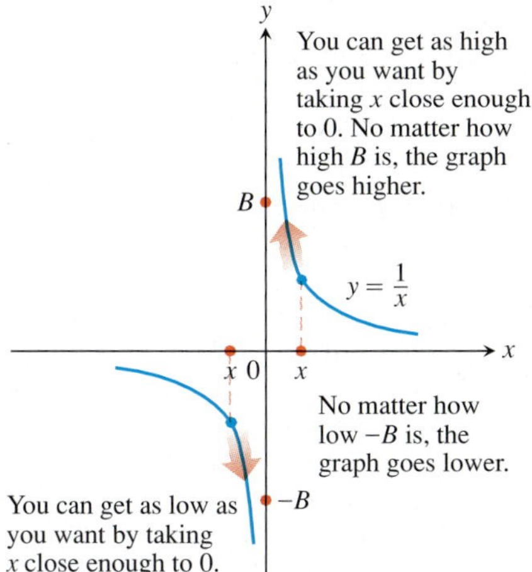

FIGURE 1.31 One-sided infinite limits:

$$\lim_{x\to 0^+} \frac{1}{x} = \infty \quad \text{and} \quad \lim_{x\to 0^-} \frac{1}{x} = -\infty.$$

Horizontal and Vertical Asymptotes: Infinite Limits

Looking at $f(x) = 1/x$ (Figure 1.30), we observe the following:

(a) As $x \to \infty$, $(1/x) \to 0$, and we write $\lim_{x\to\infty} (1/x) = 0$.

(b) As $x \to -\infty$, $(1/x) \to 0$, and we write $\lim_{x\to-\infty} (1/x) = 0$.

We say that the line $y = 0$ is a *horizontal asymptote* of the graph of f.

If the distance between the graph of a function and some fixed line approaches zero as the graph moves increasingly far from the origin, we say that the graph approaches the line asymptotically and that the line is an *asymptote* of the graph.

Let us look closely at the function $f(x) = 1/x$ in Figure 1.31. As $x \to 0^+$, the values of f grows without bound, eventually reaching and surpassing every positive real number. That is, given any positive real number B, however large, the values of f become larger still (Figure 1.31). Thus, f has no limit as $x \to 0^+$. It is nevertheless convenient to describe the behavior of f by saying that $f(x)$ approaches ∞ as $x \to 0^+$. We write

$$\lim_{x\to 0^+} f(x) = \lim_{x\to 0^+} \frac{1}{x} = \infty.$$

As $x \to 0^-$, the values of $f(x) = 1/x$ become arbitrarily large and negative. Given any negative real number $-B$, the values of f eventually lie below $-B$. (See Figure 1.31.) We write

$$\lim_{x\to 0^-} f(x) = \lim_{x\to 0^-} \frac{1}{x} = -\infty.$$

Notice that the denominator is zero at $x = 0$ and the function is undefined.

Definitions Horizontal and Vertical Asymptotes

A line $y = b$ is a **horizontal asymptote** of the graph of a function $y = f(x)$ if either

$$\lim_{x\to\infty} f(x) = b \qquad \text{or} \qquad \lim_{x\to-\infty} f(x) = b.$$

A line $x = a$ is a **vertical asymptote** of the graph if either

$$\lim_{x\to a^+} f(x) = \pm\infty \qquad \text{or} \qquad \lim_{x\to a^-} f(x) = \pm\infty.$$

Example 6 Looking for Asymptotes

Find the asymptotes of the curve

$$y = \frac{x+3}{x+2}.$$

Solution We are interested in the behavior as $x \to \pm\infty$ and as $x \to -2$, where the denominator is zero.

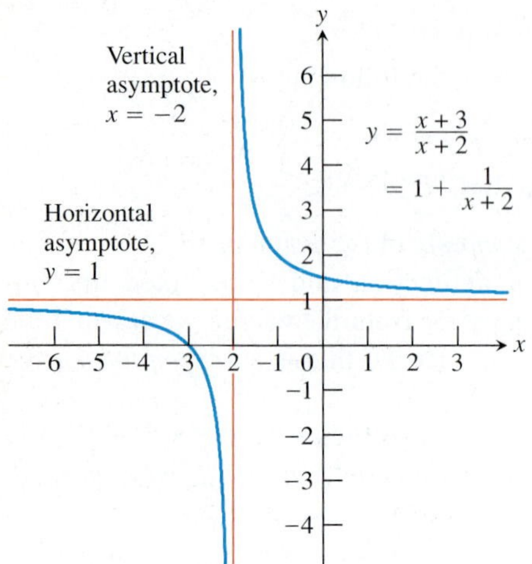

FIGURE 1.32 The lines $y = 1$ and $x = -2$ are asymptotes of the curve $y = (x + 3)/(x + 2)$. (Example 6)

The asymptotes are quickly revealed if we recast the rational function as a polynomial with a remainder, by dividing $(x + 2)$ into $(x + 3)$.

$$\begin{array}{r} 1 \\ x+2 \overline{)\,x+3} \\ \underline{x+2} \\ 1 \end{array}$$

This result enables us to rewrite y:

$$y = 1 + \frac{1}{x+2}.$$

We now see that the curve in question is the graph of $y = 1/x$ shifted 1 unit up and 2 units left (Figure 1.32). The asymptotes, instead of being the coordinate axes, are now the lines $y = 1$ and $x = -2$.

Example 7 Asymptotes Need Not Be Two-Sided

Find the asymptotes of the graph of

$$f(x) = -\frac{8}{x^2 - 4}.$$

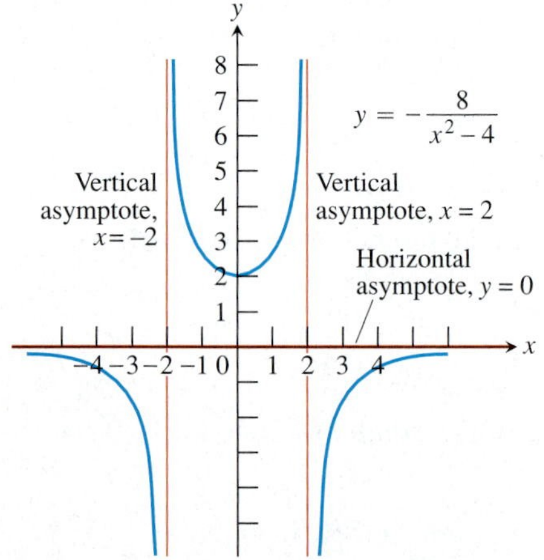

FIGURE 1.33 Graph of $y = -8/(x^2 - 4)$. Notice that the curve approaches the x-axis from only one side. Asymptotes do not have to be two-sided. (Example 7)

Solution We are interested in the behavior as $x \to \pm\infty$ and as $x \to \pm 2$, where the denominator is zero. Notice that f is an even function of x, so its graph is symmetric with respect to the y-axis.

The behavior as $x \to \pm\infty$. Since $\lim_{x\to\infty} f(x) = 0$, the line $y = 0$ is an asymptote of the graph to the right. By symmetry it is an asymptote to the left as well (Figure 1.33).

The behavior as $x \to \pm 2$. Since

$$\lim_{x\to 2^+} f(x) = -\infty \qquad \text{and} \qquad \lim_{x\to 2^-} f(x) = \infty,$$

the line $x = 2$ is a vertical asymptote both from the right and from the left. By symmetry, the same holds for the line $x = -2$.

There are no other asymptotes because f has a finite limit at every other point.

Example 8 Curves with Infinitely Many Asymptotes

The curves

$$y = \sec x = \frac{1}{\cos x} \qquad \text{and} \qquad y = \tan x = \frac{\sin x}{\cos x}$$

both have vertical asymptotes at odd-integer multiples of $\pi/2$, where $\cos x = 0$ (Figure 1.34).

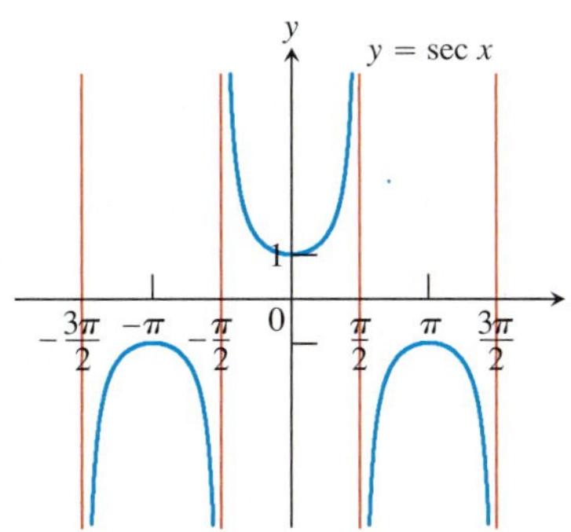

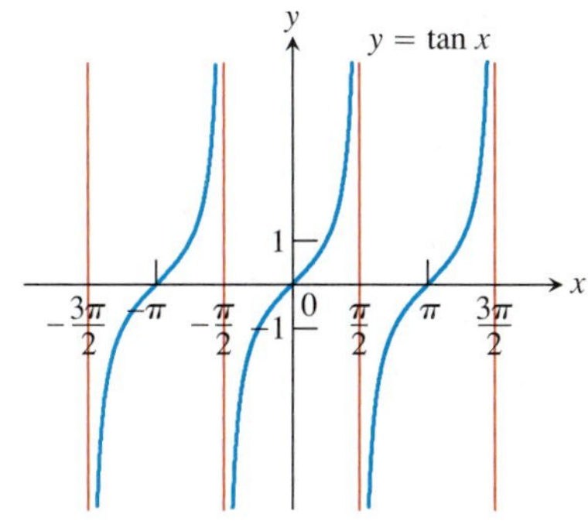

FIGURE 1.34 The graphs of sec x and tan x. (Example 8)

The graphs of

$$y = \csc x = \frac{1}{\sin x} \qquad \text{and} \qquad y = \cot x = \frac{\cos x}{\sin x}$$

have vertical asymptotes at integer multiples of π, where $\sin x = 0$ (Figure 1.35).

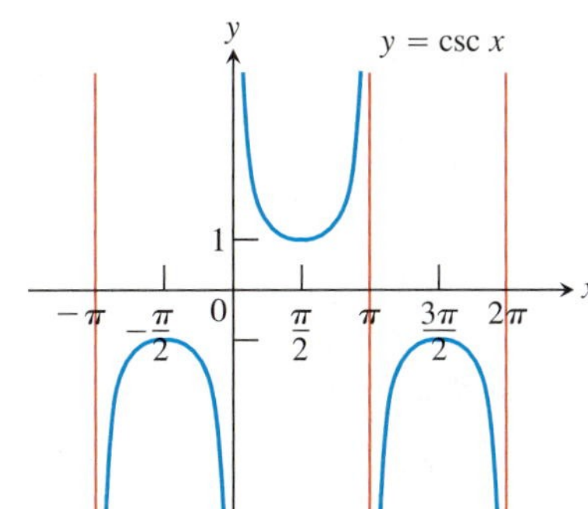

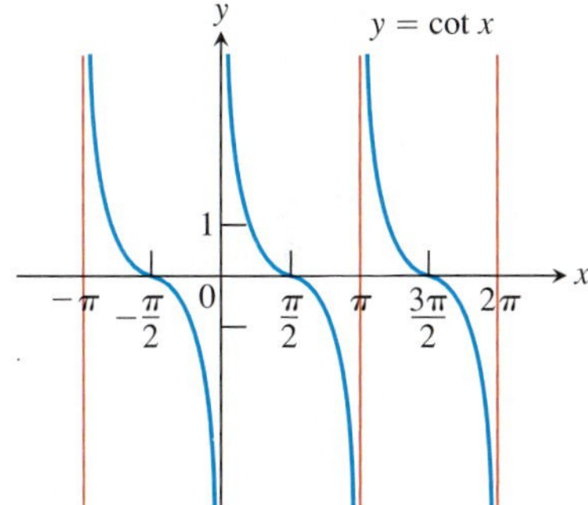

FIGURE 1.35 The graphs of csc x and cot x. (Example 8)

x	e^x
0	1.00000
−1	0.36788
−2	0.13534
−3	0.04979
−5	0.00674
−8	0.00034
−10	0.00005

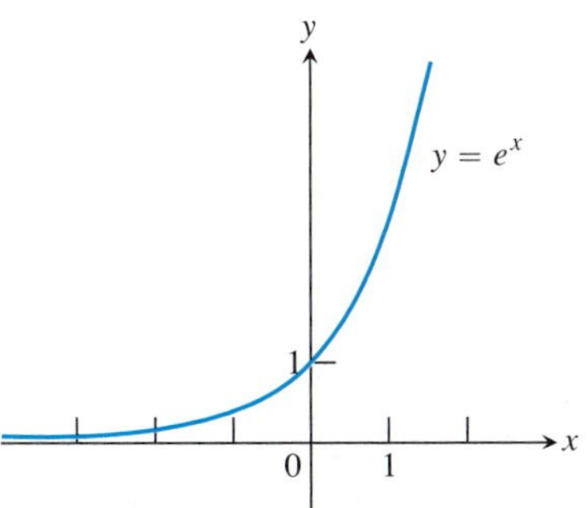

FIGURE 1.36 The line $y = 0$ is a horizontal asymptote of the graph $y = e^x$.

Example 9 Horizontal Asymptote of $y = e^x$

The curve

$$y = e^x$$

has the line $y = 0$ (the x-axis) as a horizontal asymptote. We see this from the graph in Figure 1.36 and its accompanying table of values. We write

$$\lim_{x \to -\infty} e^x = 0 .$$

Notice that the values of e^x approach 0 quite rapidly.

We can investigate the behavior of $y = f(x)$ as $x \to \pm\infty$ by investigating the limit of $y = f(1/x)$ as $x \to 0$.

Example 10 Substituting a New Variable

Find $\lim_{x \to \infty} \sin(1/x)$.

Solution We introduce the new variable $t = 1/x$. From Figure 1.30, we know that $t \to 0^+$ as $x \to \infty$. Therefore,

$$\lim_{x \to \infty} \sin \frac{1}{x} = \lim_{t \to 0^+} \sin t = 0.$$

Likewise, we can investigate the behavior of $y = f(1/x)$ as $x \to 0$ by investigating $y = f(x)$ as $x \to \pm\infty$.

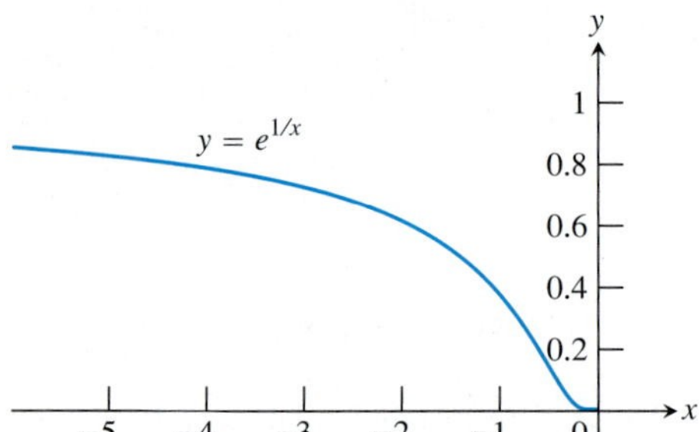

FIGURE 1.37 The graph of $y = e^{1/x}$ for $x < 0$ shows $\lim_{x \to 0^-} e^{1/x} = 0$. (Example 11)

Example 11 Using Substitution

Find $\lim_{x \to 0^-} e^{1/x}$.

Solution We let $t = 1/x$. From Figure 1.31, we know that $t \to -\infty$ as $x \to 0^-$. Therefore,

$$\lim_{x \to 0^-} e^{1/x} = \lim_{t \to -\infty} e^t = 0 \qquad \text{Example 9}$$

(Figure 1.37).

Sandwich Theorem Revisited

The Sandwich Theorem also holds for limits as $x \to \pm\infty$.

Example 12 Finding a Limit as x Approaches 0 or $\pm\infty$

Using the Sandwich Theorem, find the asymptotes of the curve

$$y = 2 + \frac{\sin x}{x}.$$

Solution We are interested in the behavior as $x \to \pm\infty$ and as $x \to 0$, where the denominator is zero.

The behavior as $x \to 0$. We know that $\lim_{x \to 0} (\sin x)/x = 1$, so there is no asymptote at the origin.

The behavior as $x \to \pm\infty$. Since

$$0 \le \left| \frac{\sin x}{x} \right| \le \left| \frac{1}{x} \right|$$

and $\lim_{x \to \pm\infty} |1/x| = 0$, we have $\lim_{x \to \pm\infty} (\sin x)/x = 0$ by the Sandwich Theorem. Hence,

$$\lim_{x \to \pm\infty} \left(2 + \frac{\sin x}{x} \right) = 2 + 0 = 2,$$

and the line $y = 2$ is an asymptote of the curve on both left and right (Figure 1.38).

This example illustrates that a curve may cross one of its horizontal asymptotes, perhaps many times.

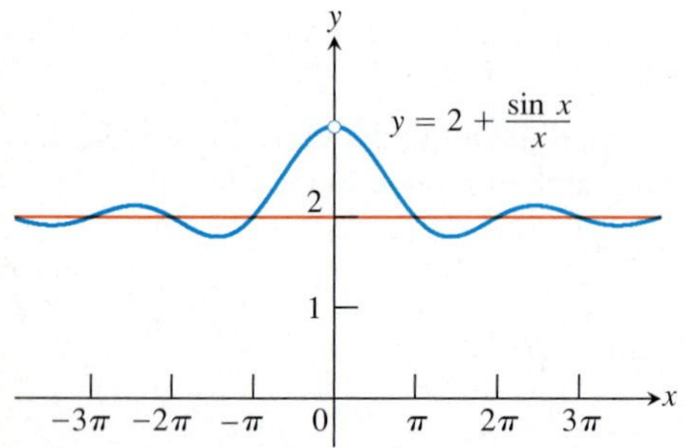

FIGURE 1.38 A curve may cross one of its asymptotes infinitely often. (Example 12)

Precise Definitions of Infinite Limits

Instead of requiring $f(x)$ to lie arbitrarily close to a finite number L for all x sufficiently close to x_0, the definitions of infinite limits require $f(x)$ to lie arbitrarily far from the origin. Except for this change, the language is identical with what we have seen before. Figures 1.39 and 1.40 accompany these definitions.

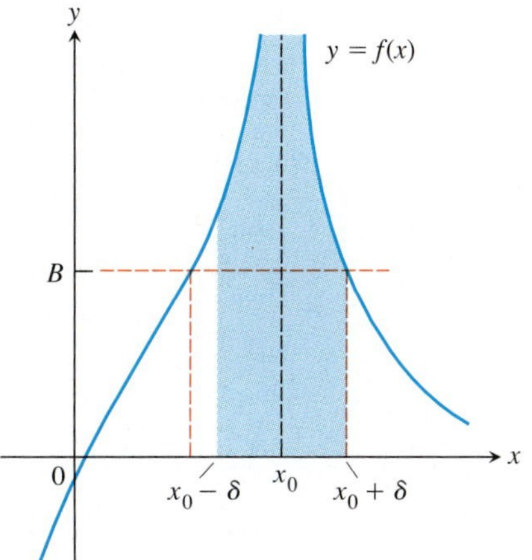

FIGURE 1.39 $\lim_{x\to x_0} f(x) = \infty$.

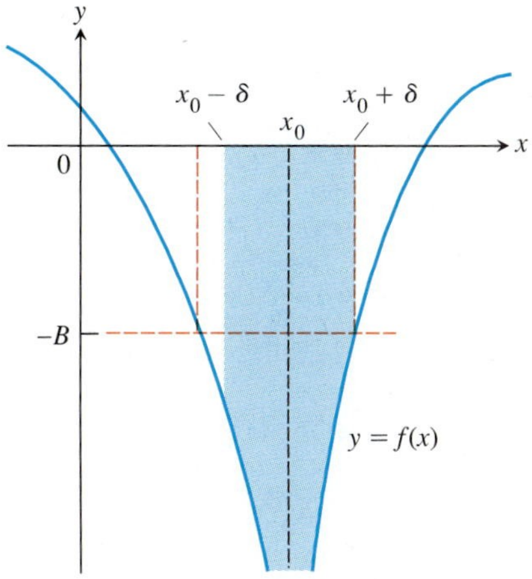

FIGURE 1.40 $\lim_{x\to x_0} f(x) = -\infty$.

Definition Infinite Limits

1. We say that $f(x)$ **approaches infinity as x approaches x_0**, and we write

$$\lim_{x\to x_0} f(x) = \infty,$$

if for every positive real number B there exists a corresponding $\delta > 0$ such that for all x

$$0 < |x - x_0| < \delta \Rightarrow f(x) > B.$$

2. We say that $f(x)$ **approaches minus infinity as x approaches x_0**, and write

$$\lim_{x\to x_0} f(x) = -\infty,$$

if for every negative real number $-B$ there exists a corresponding $\delta > 0$ such that for all x

$$0 < |x - x_0| < \delta \Rightarrow f(x) < -B.$$

The precise definitions of one-sided infinite limits at x_0 are similar.

CD-ROM WEBsite

End Behavior Models and Oblique Asymptotes

For numerically large values of x, we can sometimes model the behavior of a complicated function by a simpler one that acts virtually in the same way.

Example 13 Modeling Functions for $|x|$ Large

Let $f(x) = 3x^4 - 2x^3 + 3x^2 - 5x + 6$ and $g(x) = 3x^4$. Show that although f and g are quite different for numerically small values of x, they are virtually identical for $|x|$ large.

Solution

Solve Graphically

The graphs of f and g (Figure 1.41a), quite different near the origin, are virtually identical on a larger scale (Figure 1.41b).

$y = 3x^4 - 2x^3 + 3x^2 - 5x + 6$

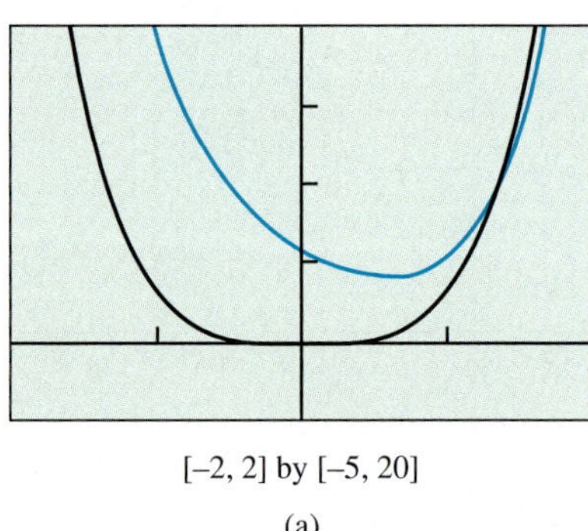

[–2, 2] by [–5, 20]

(a)

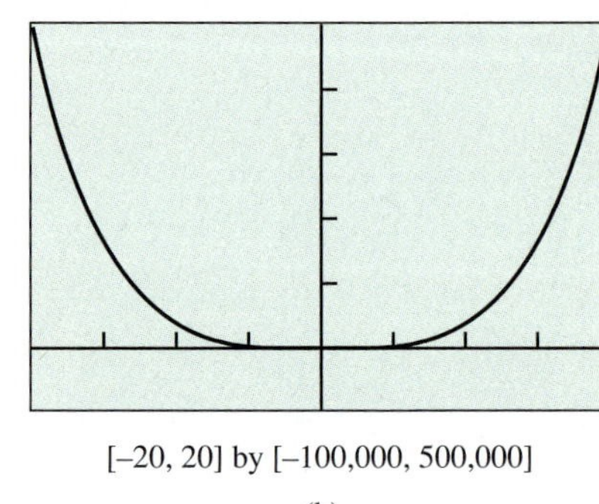

[–20, 20] by [–100,000, 500,000]

(b)

FIGURE 1.41 The graphs of f (upper curve) and g, (a) are distinct for $|x|$ small, and (b) nearly identical for $|x|$ large. (Example 13)

Confirm Analytically

We can test the claim that g models f for numerically large values of x by examining the ratio of the two functions as $x \to \pm\infty$. We find that

$$\begin{aligned}\lim_{x\to\pm\infty} \frac{f(x)}{g(x)} &= \lim_{x\to\pm\infty} \frac{3x^4 - 2x^3 + 3x^2 - 5x + 6}{3x^4} \\ &= \lim_{x\to\pm\infty} \left(1 - \frac{2}{3x} + \frac{1}{x^2} - \frac{5}{3x^3} + \frac{2}{x^4}\right) \\ &= 1,\end{aligned}$$

convincing evidence that f and g behave alike for $|x|$ large.

Definition End Behavior Model

The function g is

(a) a **right end behavior model** for f if and only if

$$\lim_{x\to\infty} \frac{f(x)}{g(x)} = 1$$

(b) a **left end behavior model** for f if and only if

$$\lim_{x\to-\infty} \frac{f(x)}{g(x)} = 1.$$

A function's right and left end behavior models need not be the same function.

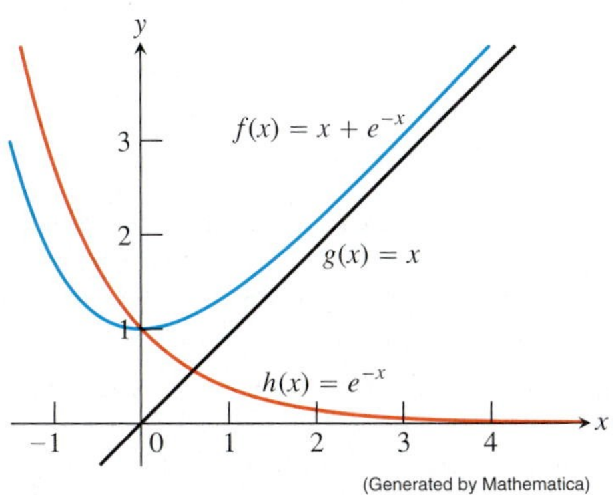

FIGURE 1.42 The graph of $f(x) = x + e^{-x}$ looks like the graph of $g(x) = x$ to the right of the y-axis and like the graph of $h(x) = e^{-x}$ to the left of the y-axis. (Example 14)

Example 14 Finding End Behavior Models

Let $f(x) = x + e^{-x}$. Show that $g(x) = x$ is a right end behavior model for f while $h(x) = e^{-x}$ is a left end behavior model for f.

Solution On the right,

$$\lim_{x\to\infty} \frac{f(x)}{g(x)} = \lim_{x\to\infty} \frac{x + e^{-x}}{x} = \lim_{x\to\infty} \left(1 + \frac{e^{-x}}{x}\right) = 1 \text{ because } \lim_{x\to\infty} \frac{e^{-x}}{x} = 0.$$

On the left,

$$\lim_{x\to-\infty} \frac{f(x)}{h(x)} = \lim_{x\to-\infty} \frac{x + e^{-x}}{e^{-x}} = \lim_{x\to-\infty} \left(\frac{x}{e^{-x}} + 1\right) = 1 \text{ because } \lim_{x\to-\infty} \frac{x}{e^{-x}} = 0.$$

(See Exercise 51). The graph of f in Figure 1.42 supports these end behavior conclusions.

In some instances we can find end behavior models for rational functions. If the degree of the numerator is one greater than the degree of the denominator, the graph of the rational function $f(x)$ has an **oblique (slanted) asymptote,** as in Figure 1.29. We find an equation for the asymptote by dividing numerator by denominator to express f as a linear function plus a remainder that goes to zero as $x \to \pm\infty$. Here's an example.

Example 15 Finding an Oblique Asymptote

Find the oblique asymptote for the graph of

$$f(x) = \frac{2x^2 - 3}{7x + 4}$$

in Figure 1.29.

Solution By long division, we find

$$f(x) = \frac{2x^2 - 3}{7x + 4} = \underbrace{\left(\frac{2}{7}x - \frac{8}{49}\right)}_{\text{linear function } g(x)} + \underbrace{\frac{-115}{49(7x + 4)}}_{\text{remainder}}.$$

As $x \to \pm\infty$, the remainder, whose magnitude gives the vertical distance between the graphs of f and g, goes to zero, making the (slanted) line

$$g(x) = \frac{2}{7}x - \frac{8}{49}$$

an asymptote of the graph of f (Figure 1.29). The function g is both a right and a left hand behavior model for f.

EXERCISES 1.3

Calculating Limits as $x \to \pm\infty$

In Exercises 1–4, find the limit of each function (**a**) as $x \to \infty$ and (**b**) as $x \to -\infty$. (You may wish to visualize your answer with a grapher.)

1. $f(x) = \pi - \dfrac{2}{x^2}$

2. $g(x) = \dfrac{1}{2 + (1/x)}$

3. $h(x) = \dfrac{-5 + (7/x)}{3 - (1/x^2)}$

4. $h(x) = \dfrac{3 - (2/x)}{4 + (\sqrt{2}/x^2)}$

Find the limits in Exercises 5 and 6.

5. $\lim\limits_{x\to\infty} \dfrac{\sin 2x}{x}$

6. $\lim\limits_{t\to-\infty} \dfrac{2 - t + \sin t}{t + \cos t}$

Limits of Rational Functions

In Exercises 7–14, find the limit of each function (**a**) as $x \to \infty$ and (**b**) as $x \to -\infty$.

7. $f(x) = \dfrac{2x + 3}{5x + 7}$

8. $f(x) = \dfrac{x + 1}{x^2 + 3}$

9. $f(x) = \dfrac{1 - 12x^3}{4x^2 + 12}$

10. $h(x) = \dfrac{7x^3}{x^3 - 3x^2 + 6x}$

11. $g(x) = \dfrac{3x^2 - 6x}{4x - 8}$

12. $f(x) = \dfrac{2x^5 + 3}{-x^2 + x}$

13. $h(x) = \dfrac{-2x^3 - 2x + 3}{3x^3 + 3x^2 - 5x}$

14. $h(x) = \dfrac{-x^4}{x^4 - 7x^3 + 7x^2 + 9}$

Limits with Noninteger or Negative Powers

The process by which we determine limits of rational functions applies equally well to ratios containing noninteger or negative powers of x: Divide numerator and denominator by the highest power of x in the denominator and proceed from there. Find the limits in Exercises 15–20.

15. $\lim\limits_{x\to\infty} \dfrac{2\sqrt{x} + x^{-1}}{3x - 7}$

16. $\lim\limits_{x\to\infty} \dfrac{2 + \sqrt{x}}{2 - \sqrt{x}}$

17. $\lim\limits_{x\to-\infty} \dfrac{\sqrt[3]{x} - \sqrt[5]{x}}{\sqrt[3]{x} + \sqrt[5]{x}}$

18. $\lim\limits_{x\to\infty} \dfrac{x^{-1} + x^{-4}}{x^{-2} - x^{-3}}$

19. $\lim\limits_{x\to\infty} \dfrac{2x^{5/3} - x^{1/3} + 7}{x^{8/5} + 3x + \sqrt{x}}$

20. $\lim\limits_{x\to-\infty} \dfrac{\sqrt[3]{x} - 5x + 3}{2x + x^{2/3} - 4}$

Inventing Graphs from Values and Limits

In Exercises 21 and 22, sketch the graph of a function $y = f(x)$ that satisfies the given conditions. No formulas are required; just label the coordinate axes and sketch an appropriate graph. (The answers are not unique, so your graphs may not be exactly like those in the answer section.)

21. $f(0) = 0$, $f(1) = 2$, $f(-1) = -2$, $\lim\limits_{x\to-\infty} f(x) = -1$, and $\lim\limits_{x\to\infty} f(x) = 1$

22. $f(0) = 0$, $\lim\limits_{x\to\pm\infty} f(x) = 0$, $\lim\limits_{x\to1^-} f(x) = \lim\limits_{x\to-1^+} f(x) = \infty$, $\lim\limits_{x\to1^+} f(x) = -\infty$, and $\lim\limits_{x\to-1^-} f(x) = -\infty$

Inventing Functions

In Exercises 23 and 24, find a function that satisfies the given conditions and sketch its graph. (The answers here are not unique. Any function that satisfies the conditions is acceptable. Feel free to use formulas defined in pieces if that will help.)

23. $\lim\limits_{x\to\pm\infty} f(x) = 0$, $\lim\limits_{x\to2^-} f(x) = \infty$, and $\lim\limits_{x\to2^+} f(x) = \infty$

24. $\lim\limits_{x\to-\infty} h(x) = -1$, $\lim\limits_{x\to\infty} h(x) = 1$, $\lim\limits_{x\to0^-} h(x) = -1$, and $\lim\limits_{x\to0^+} h(x) = 1$

Graphing Rational Functions

T Graph the rational functions in Exercises 25–34. Include the graphs and equations of the asymptotes.

25. $y = \dfrac{1}{x - 1}$

26. $y = \dfrac{x + 1}{x + 2}$

27. $y = \dfrac{2x^2 + x - 1}{x^2 - 1}$

28. $y = \dfrac{x^2 - 1}{x}$

29. $y = \dfrac{x^4 + 1}{x^2}$

30. $y = \dfrac{x^2 - 4}{x - 1}$

31. $y = \dfrac{x^2 - x + 1}{x - 1}$

32. $y = \dfrac{x}{x^2 - 1}$

33. $y = \dfrac{8}{x^2 + 4}$ (Agnesi's witch)

34. $y = \dfrac{4x}{x^2 + 4}$ (Newton's serpentine)

End Behavior Models

In Exercises 35–38, match the function with the graph of its end behavior model.

35. $y = \dfrac{2x^3 - 3x^2 + 1}{x + 3}$

36. $y = \dfrac{x^5 - x^4 + x + 1}{2x^2 + x - 3}$

37. $y = \dfrac{2x^4 - x^3 + x^2 - 1}{2 - x}$

38. $y = \dfrac{x^4 - 3x^3 + x^2 - 1}{1 - x^2}$

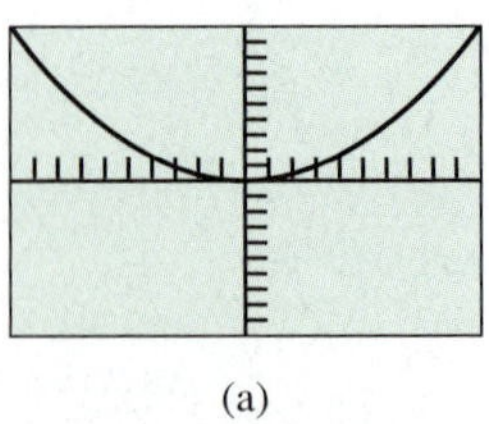

(a)

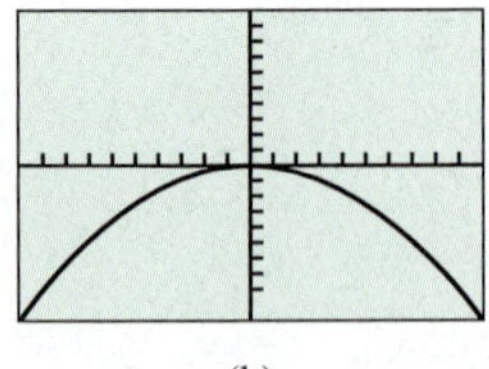

(b)

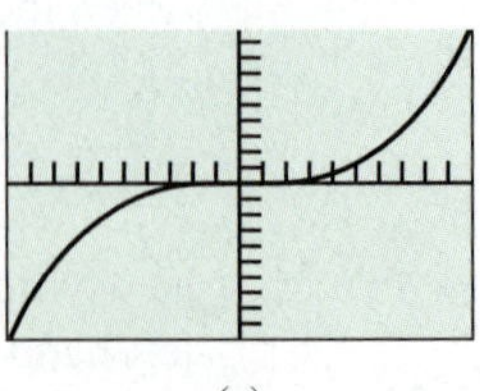

(c)

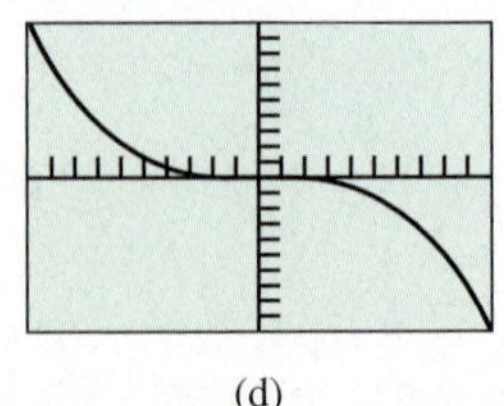

(d)

In Exercises 39–42, find (**a**) a simple basic function as a right end behavior model and (**b**) a simple basic function as a left end behavior model for the function.

39. $y = e^x - 2x$

40. $y = x^2 + e^{-x}$

41. $y = x + \ln|x|$

42. $y = x^2 + \sin x$

Theory and Examples

T **43.** (**a**) Estimate the value of

$$\lim_{x\to\infty} \left(\sqrt{x^2 + x + 1} - x\right)$$

by graphing the function

$$f(x) = \sqrt{x^2 + x + 1} - x.$$

(**b**) From a table of values of $f(x)$, guess the value of the limit in part (a). Then prove your guess is correct.

T **44.** Find $\lim_{x\to\infty} \left(\sqrt{x^2 + x} - \sqrt{x^2 - x}\right)$ graphically and confirm it algebraically.

45. *Writing to Learn* How many horizontal asymptotes can the graph of a given rational function have? Give reasons for your answer.

46. *Writing to Learn* How many vertical asymptotes can the graph of a given rational function have? Give reasons for your answer.

COMPUTER EXPLORATIONS

Comparing Graphs With Formulas

Graph the curves in Exercises 47–50. Explain the relation between the curve's formula and what you see.

47. $y = \dfrac{x}{\sqrt{4 - x^2}}$

48. $y = \dfrac{-1}{\sqrt{4 - x^2}}$

49. $y = x^{2/3} + \dfrac{1}{x^{1/3}}$

50. $y = \sin\left(\dfrac{\pi}{x^2 + 1}\right)$

Substituting 1/x

In Exercises 51–54, use the graph of $y = f(1/x)$ to find $\lim_{x\to\infty} f(x)$ and $\lim_{x\to-\infty} f(x)$.

51. $f(x) = xe^x$

52. $f(x) = x^2e^{-x}$

53. $f(x) = \dfrac{\ln|x|}{x}$

54. $f(x) = x \sin \dfrac{1}{x}$

55. $\displaystyle\lim_{x\to-\infty} \frac{\cos(1/x)}{1 + (1/x)}$

56. $\displaystyle\lim_{x\to\infty} \left(\frac{1}{x}\right)^{1/x}$

57. $\displaystyle\lim_{x\to\pm\infty} \left(3 + \frac{2}{x}\right)\left(\cos\frac{1}{x}\right)$

58. $\displaystyle\lim_{x\to\infty} \left(\frac{3}{x^2} - \cos\frac{1}{x}\right)\left(1 + \sin\frac{1}{x}\right)$

Finding Asymptotes

Graph the functions in Exercises 59–62. What asymptotes do the graphs have? Why are the asymptotes located where they are?

59. $y = -\dfrac{x^2 - 4}{x + 1}$

60. $y = \dfrac{x^3 - x^2 - 1}{x^2 - 1}$

61. $y = x^3 + \dfrac{3}{x}$

62. $y = 2\sin x + \dfrac{1}{x}$

Graph the functions in Exercises 63 and 64. Then answer the following questions.

(**a**) How does the graph behave as $x \to 0^+$?

(**b**) How does the graph behave as $x \to \pm\infty$?

(**c**) How does the graph behave at $x = 1$ and $x = -1$?

Give reasons for your answers.

63. $y = \dfrac{3}{2}\left(x - \dfrac{1}{x}\right)^{2/3}$

64. $y = \dfrac{3}{2}\left(\dfrac{x}{x - 1}\right)^{2/3}$

1.4 Continuity

Continuity at a Point • Continuous Functions • Algebraic Combinations • Composites • Intermediate Value Theorem for Continuous Functions

CD-ROM
WEBsite

When we plot function values generated in the laboratory or collected in the field, we often connect the plotted points with an unbroken curve to show what the function's values are likely to have been at the times we did not measure (Figure 1.43). In doing so, we are assuming that we are working with a *continuous function*, a function whose outputs vary continuously with the inputs and do not jump from one value to another without taking on the values in between.

Any function $y = f(x)$ whose graph can be sketched over its domain in one continuous motion without lifting the pencil is an example of a continuous function. We study the idea of continuity in this section.

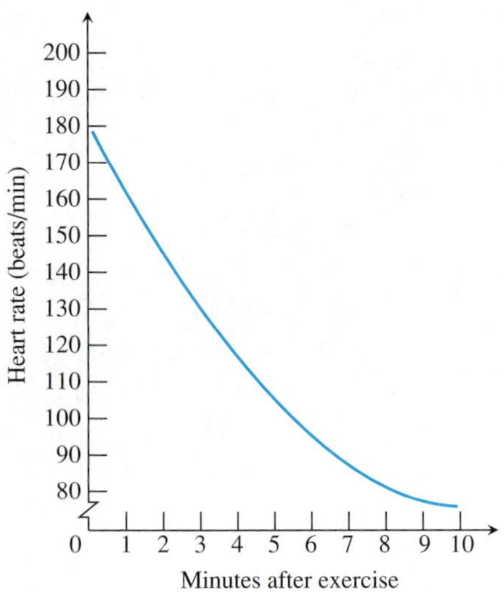

FIGURE 1.43 How the heart rate returns to a normal rate after running.

Continuity at a Point

Continuous functions are the functions we use to find a planet's closest point of approach to the sun or the peak concentration of antibodies in blood plasma. They are also the functions we use to describe how a body moves through space or how the speed of a chemical reaction changes with time. In fact, so many physical processes proceed continuously that throughout the eighteenth and nineteenth centuries it rarely occurred to anyone to look for any other kind of behavior. It came as a surprise when the physicists of the 1920's discovered that light comes in particles and that heated atoms emit light at discrete frequencies (Figure 1.44). As a result of these and other discoveries and because of the heavy use of discontinuous functions in computer science, statistics, and mathematical modeling, the issue of continuity has become one of practical as well as theoretical importance.

To understand continuity, we need to consider a function like the one in Figure 1.45, whose limits we investigated in Example 8, Section 1.2.

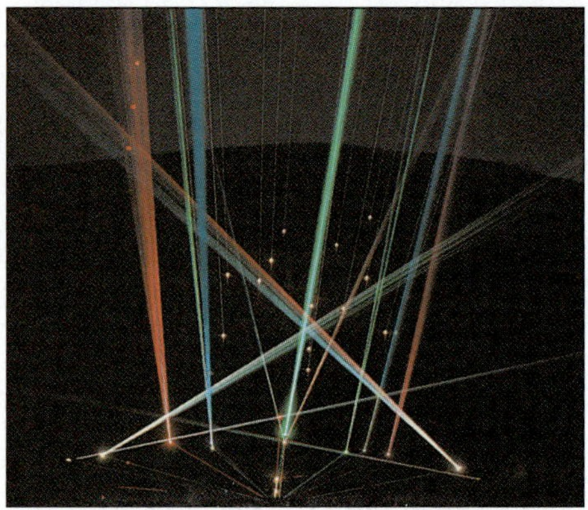

FIGURE 1.44 The laser was developed as a result of an understanding of the nature of the atom.

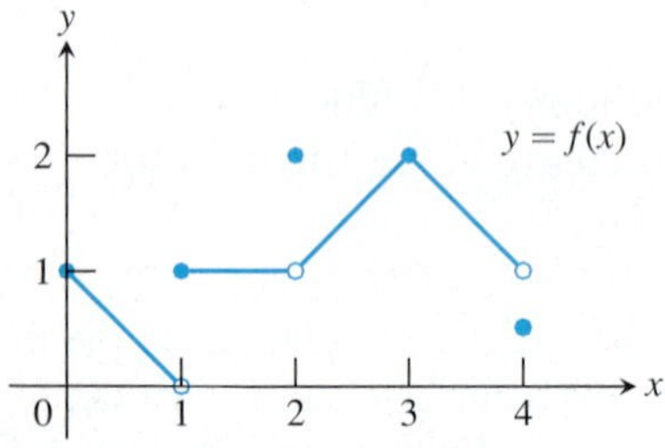

FIGURE 1.45 The function is continuous on $[0, 4]$ except at $x = 1$, $x = 2$, and $x = 4$. (Example 1)

Example 1 Investigating Continuity

Find the points at which the function f in Figure 1.45 is continuous and the points at which f is discontinuous.

Solution The function f is continuous at every point in its domain $[0, 4]$ except at $x = 1$, $x = 2$, and $x = 4$. At these points, there are breaks in the graph. Note the relationship between the limit of f and the value of f at each point of the function's domain.

Points at which f is continuous:

At $x = 0$, $\quad \lim_{x\to 0^+} f(x) = f(0)$.

At $x = 3$, $\quad \lim_{x\to 3} f(x) = f(3)$.

At $0 < c < 4$, $c \neq 1, 2$, $\quad \lim_{x\to c} f(x) = f(c)$.

Points at which f is discontinuous:

At $x = 1$, $\lim_{x\to 1} f(x)$ does not exist.

At $x = 2$, $\lim_{x\to 2} f(x) = 1$, but $1 \neq f(2)$.

At $x = 4$, $\lim_{x\to 4} f(x) = 1$, but $1 \neq f(4)$.

At $c < 0$, $c > 4$, these points are not in the domain of f.

CD-ROM WEBsite

Historical Biography

Johann van Waveren Hudde (1628 — 1704)

To define continuity at a point in a function's domain, we need to define continuity at an interior point (which involves a two-sided limit) and continuity at an endpoint (which involves a one-sided limit) (Figure 1.46).

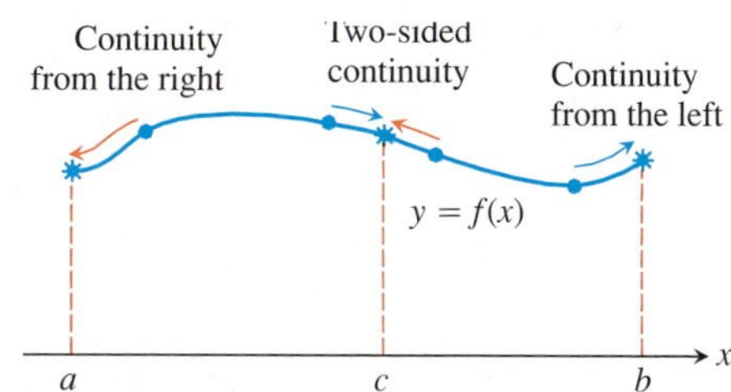

FIGURE 1.46 Continuity at points a, b, and c.

Definition Continuity at a Point

Interior point: A function $y = f(x)$ is **continuous at an interior point c** of its domain if

$$\lim_{x\to c} f(x) = f(c).$$

Endpoint: A function $y = f(x)$ is **continuous at a left endpoint a** or is **continuous at a right endpoint b** of its domain if

$$\lim_{x\to a^+} f(x) = f(a) \quad \text{or} \quad \lim_{x\to b^-} f(x) = f(b), \quad \text{respectively.}$$

If a function f is not continuous at a point c, we say that f is **discontinuous** at c and c is a **point of discontinuity** of f. Note that c need not be in the domain of f.

A function f is **right-continuous (continuous from the right)** at a point $x = c$ in its domain if $\lim_{x\to c^+} f(x) = f(c)$. It is **left-continuous (continuous from the left)** at c if $\lim_{x\to c^-} f(x) = f(c)$. Thus, a function is continuous at a left endpoint a of its domain if it is right-continuous at a and continuous at a right endpoint b of its domain if it is left-continuous at b. A function is continuous at an interior point c of its domain if and only if it is both right-continuous and left-continuous at c (Figure 1.46).

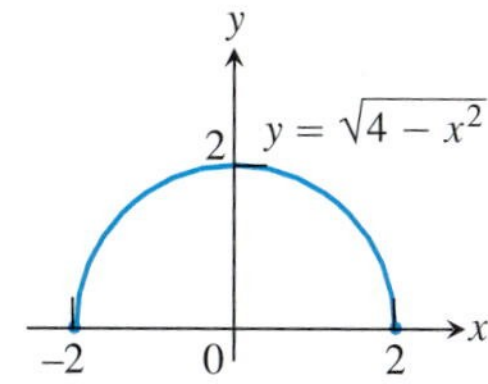

FIGURE 1.47 Continuous at every domain point.

Example 2 A Function Continuous Throughout Its Domain

The function $f(x) = \sqrt{4 - x^2}$ is continuous at every point of its domain, $[-2, 2]$ (Figure 1.47), including $x = -2$, where f is right-continuous, and $x = 2$, where f is left-continuous.

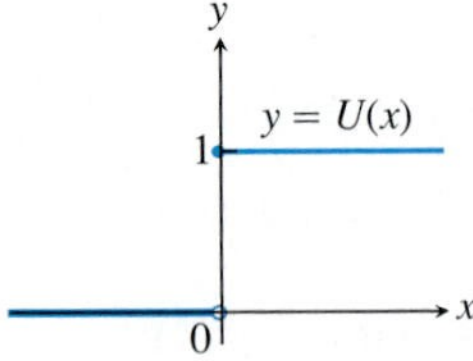

FIGURE 1.48 Right-continuous at the origin.

Example 3 A Function with a Jump Discontinuity

The unit step function $U(x)$, graphed in (Figure 1.48), is right-continuous at $x = 0$, but is neither left-continuous nor continuous there. It has a jump discontinuity at $x = 0$.

We summarize continuity at a point in the form of a test.

Continuity Test

A function $f(x)$ is continuous at $x = c$ if and only if it meets the following three conditions.

1. $f(c)$ exists (c lies in the domain of f)
2. $\lim_{x \to c} f(x)$ exists (f has a limit as $x \to c$)
3. $\lim_{x \to c} f(x) = f(c)$ (the limit equals the function value)

For one-sided continuity and continuity at an endpoint, the limits in parts 2 and 3 of the test should be replaced by the appropriate one-sided limits.

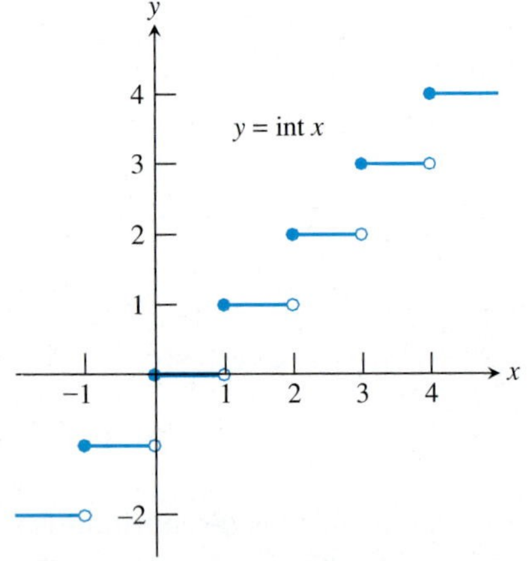

FIGURE 1.49 The function int x is continuous at every noninteger point. It is right-continuous, but not left-continuous, at every integer point. (Example 4)

Example 4 Finding Points of Continuity and Discontinuity

Find the points of continuity and the points of discontinuity of the greatest integer function $y = \text{int}\, x$ (Figure 1.49).

Solution For the function to be continuous at $x = c$, the limit as $x \to c$ must exist and must equal the value of the function at $x = c$. The greatest integer function is discontinuous at every integer. For example,

$$\lim_{x \to 3^-} \text{int}\, x = 2 \quad \text{and} \quad \lim_{x \to 3^+} \text{int}\, x = 3,$$

so the limit as $x \to 3$ does not exist. Notice that $\text{int}\, 3 = 3$, so the greatest integer function is right-continuous at $x = 3$. In general, if n is any integer,

$$\lim_{x \to n^-} \text{int}\, x = n - 1 \quad \text{and} \quad \lim_{x \to n^+} \text{int}\, x = n,$$

so the limit as $x \to n$ does not exist. Since $\text{int}\, n = n$, the greatest integer function is right-continuous at every integer n (but not left-continuous).

The greatest integer function is continuous at every real number other than the integers. For example,

$$\lim_{x \to 1.5} \text{int}\, x = 1 = \text{int}\, 1.5.$$

In general, if $n - 1 < c < n$, n an integer, then

$$\lim_{x \to c} \text{int}\, x = n - 1 = \text{int}\, c.$$

Figure 1.50 is a catalog of discontinuity types. The function in Figure 1.50a is continuous at $x = 0$. The function in Figure 1.50b would be continuous if it had

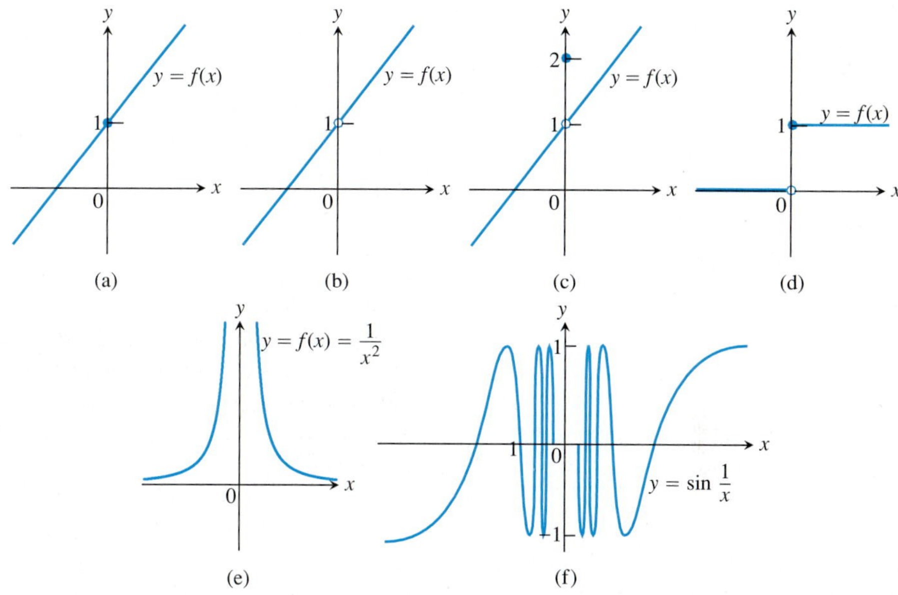

CD-ROM WEBsite

FIGURE 1.50 The function in (a) is continuous at $x = 0$; the functions in (b) through (f) are not.

$f(0) = 1$. The function in Figure 1.50c would be continuous if $f(0)$ were 1 instead of 2. The discontinuities in Figure 1.50b and c are **removable.** Each function has a limit as $x \to 0$, and we can remove the discontinuity by setting $f(0)$ equal to this limit.

The discontinuities in Figure 1.50d through f are more serious: $\lim_{x\to 0} f(x)$ does not exist, and there is no way to improve the situation by changing f at 0. The step function in Figure 1.50d has a **jump discontinuity:** The one-sided limits exist but have different values. The function $f(x) = 1/x^2$ in Figure 1.50e has an **infinite discontinuity.** The function in Figure 1.50f has an **oscillating discontinuity:** It oscillates too much to have a limit as $x \to 0$.

Continuous Functions

A function is **continuous on an interval** if and only if it is continuous at every point of the interval. A **continuous function** is one that is continuous at every point of its domain. A continuous function need not be continuous on every interval. For example, $y = 1/x$ is not continuous on $[-1, 1]$ (Figure 1.51).

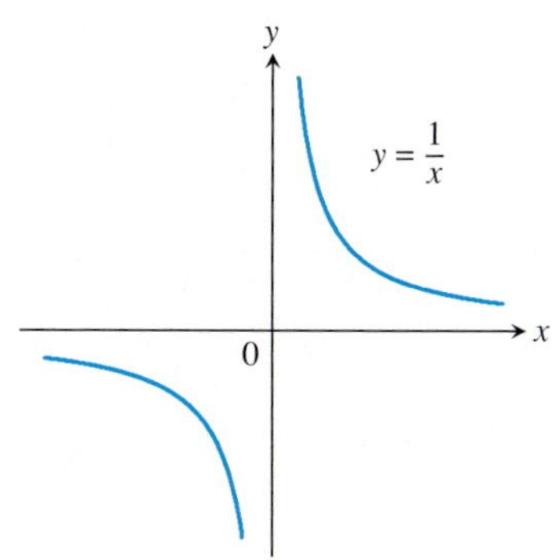

FIGURE 1.51 The function $y = 1/x$ is continuous at every value of x except $x = 0$. It has a point of discontinuity at $x = 0$. (Example 5)

Example 5 Identifying Continuous Functions

The function $y = 1/x$ (Figure 1.51) is a continuous function because it is continuous at every point of its domain. It has a point of discontinuity at $x = 0$, however, because it is not defined there.

The following types of functions are continuous at every point in their domains:

- polynomials
- rational functions

- root functions ($y = \sqrt[n]{x}$, n a positive integer greater than 1)
- trigonometric functions
- inverse trigonometric functions
- exponential functions
- logarithmic functions.

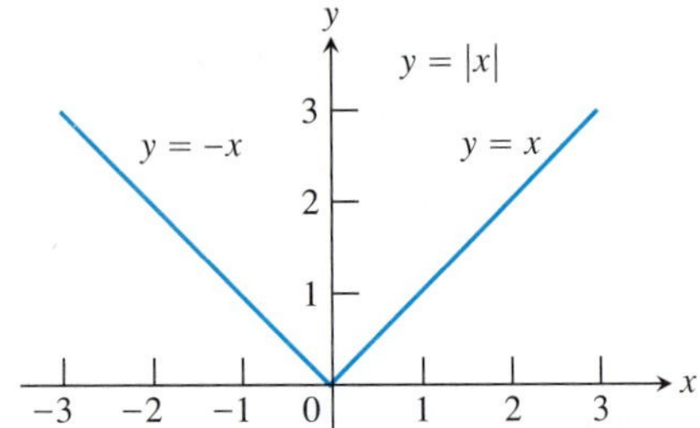

FIGURE 1.52 The sharp corner does not prevent the function from being continuous at the origin.

Polynomial functions f are continuous at every number c because $\lim_{x\to c} f(x) = f(c)$. Rational functions are continuous at every point of their domains. They have points of discontinuity at the zeros of their denominators. From their graphs, we are not surprised that the sine and cosine functions are continuous.

The inverse function of any continuous function is continuous. We see this because the graph of a continuous function f has no breaks, and the graph of f^{-1} is obtained by reflecting the graph of f about the line $y = x$ (so the graph of f^{-1} has no breaks either).

The exponential function $y = a^x$ was defined to be continuous, and therefore its inverse $y = \log_a x$ is also continuous over its domain.

The function $f(x) = |x|$ is continuous at every value of x (Figure 1.52). If $x > 0$, we have $f(x) = x$, a polynomial. If $x < 0$, we have $f(x) = -x$, another polynomial. Finally, at the origin, $\lim_{x\to 0} |x| = 0 = |0|$.

Algebraic Combinations

As you may have guessed, algebraic combinations of continuous functions are continuous wherever they are defined.

Theorem 8 Properties of Continuous Functions

If the functions f and g are continuous at $x = c$, then the following combinations are continuous at $x = c$.

1. *Sums*: $f + g$
2. *Differences*: $f - g$
3. *Products*: $f \cdot g$
4. *Constant multiples*: $k \cdot f$, for any number k
5. *Quotients*: f/g, provided $g(c) \neq 0$

Most of the results in Theorem 8 are easily proved from the limit rules in Theorem 1.

Composites

All composites of continuous functions are continuous. Thus, composites like

$$y = \sin(x^2) \qquad \text{and} \qquad y = |\cos x|$$

are continuous at every point at which they are defined. The idea is that if $f(x)$ is continuous at $x = c$ and $g(x)$ is continuous at $x = f(c)$, then $g \circ f$ is continuous at $x = c$ (Figure 1.53). In this case, the limit as $x \to c$ is $g(f(c))$.

Theorem 9 Composite of Continuous Functions

If f is continuous at c and g is continuous at $f(c)$, then the composite $g \circ f$ is continuous at c.

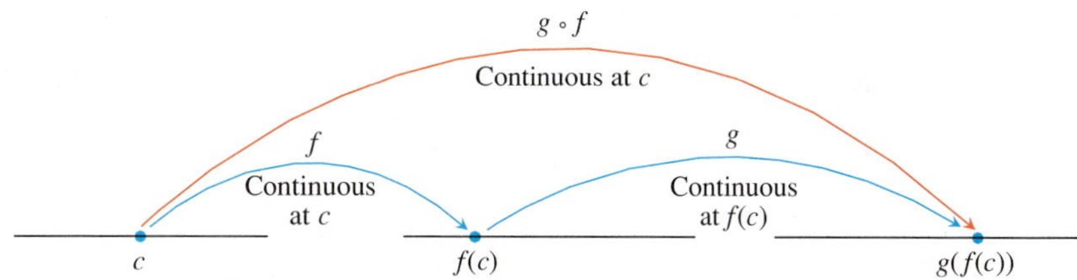

FIGURE 1.53 Composites of continuous functions are continuous.

Intuitively, Theorem 9 is reasonable because if x is close to c, then $f(x)$ is close to $f(c)$, and since g is continuous at $f(c)$, it follows that $g(f(x))$ is close to $g(f(c))$.

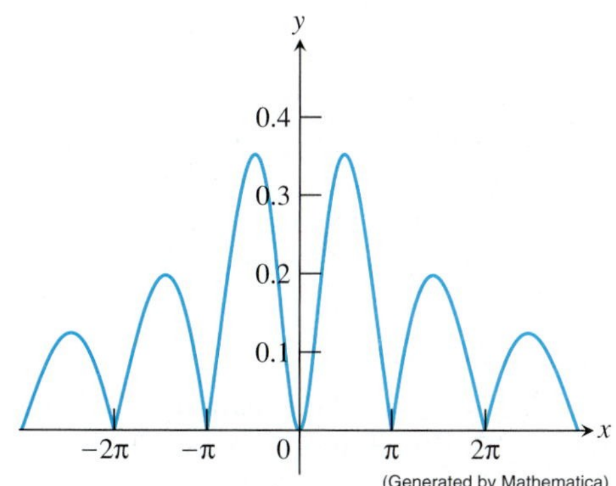

FIGURE 1.54 The graph suggests that $y = |(x \sin x)/(x^2 + 2)|$ is continuous. (Example 6)

Example 6 Using Theorem 9

Show that

$$y = \left| \frac{x \sin x}{x^2 + 2} \right|$$

is continuous.

Solution The graph (Figure 1.54) of $y = |(x \sin x)/(x^2 + 2)|$ suggests that the function is continuous at every value of x. By letting

$$g(x) = |x| \qquad \text{and} \qquad f(x) = \frac{x \sin x}{x^2 + 2},$$

we see that y is the composite $g \circ f$.

We know that the absolute value function g is continuous. The function f is continuous by Theorem 8. Their composite is continuous by Theorem 9.

Intermediate Value Theorem for Continuous Functions

Functions that are continuous on intervals have properties that make them particularly useful in mathematics and its applications. One of these is the *Intermediate Value Property*. A function is said to have the **Intermediate Value Property** if it never takes on two values without taking on all the values in between.

Theorem 10 The Intermediate Value Theorem for Continuous Functions

A function $y = f(x)$ that is continuous on a closed interval $[a, b]$ takes on every value between $f(a)$ and $f(b)$. In other words, if y_0 is any value between $f(a)$ and $f(b)$, then $y_0 = f(c)$ for some c in $[a, b]$.

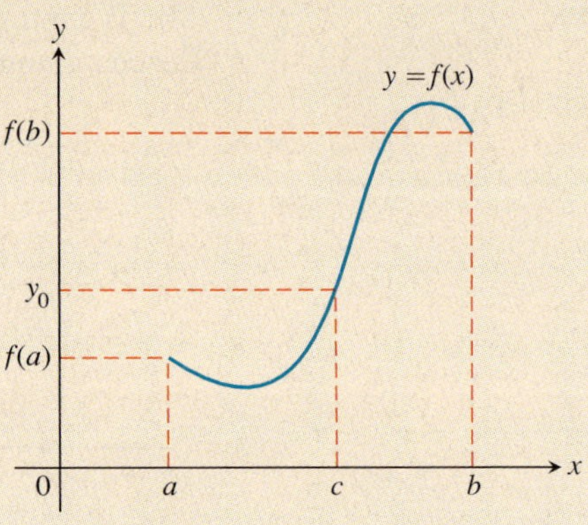

Geometrically, the Intermediate Value Theorem says that any horizontal line $y = y_0$ crossing the y-axis between the numbers $f(a)$ and $f(b)$ will cross the curve $y = f(x)$ at least once over the interval $[a, b]$.

The continuity of f on the interval is essential to Theorem 10. If f is discontinuous at even one point of the interval, the theorem's conclusion may fail, as it does for the function graphed in Figure 1.55.

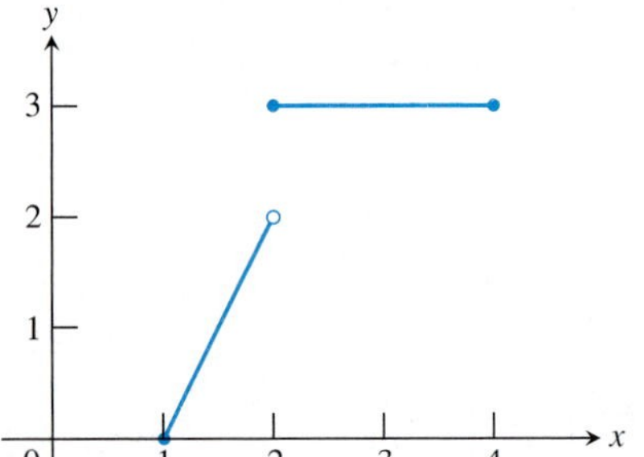

FIGURE 1.55 The function

$$f(x) = \begin{cases} 2x - 2, & 1 \le x < 2 \\ 3, & 2 \le x \le 4 \end{cases}$$

does not take on all values between $f(1) = 0$ and $f(4) = 3$; it misses all the values between 2 and 3.

A Consequence for Graphing: Connectivity Theorem 10 is the reason the graph of a function continuous on an interval I cannot have any breaks over the interval. It will be **connected,** a single, unbroken curve, like the graph of $\sin x$. It will not have jumps like the graph of the greatest integer function $\text{int } x$ or separate branches like the graph of $1/x$.

A Consequence for Root Finding We call a solution of the equation $f(x) = 0$ a **root** of the equation or **zero** of the function f. The Intermediate Value Theorem tells us that if f is continuous, then any interval on which f changes sign contains a zero of the function.

In practical terms, when we see the graph of a continuous function cross the horizontal axis on a computer screen, we know it is not stepping across. There really is a point where the function's value is zero. This consequence leads to a procedure for estimating the zeros of any continuous function we can graph:

A graphical procedure for root-finding.

1. Graph the function over a large interval to see roughly where the zeros are.
2. Zoom in on each zero to estimate its x-coordinate value.

You can practice this procedure on your graphing calculator or computer in some of the exercises.

USING TECHNOLOGY

Deceptive Pictures A graphing utility (calculator or Computer Algebra System) plots a graph much as you do when plotting by hand: by plotting points, or *pixels*, and then connecting them in succession. The resulting picture may be misleading when points on opposite sides of a point of discontinuity in the graph are incorrectly connected. To avoid incorrect connections, some systems allow you to use a "dot mode," which plots only the points. Dot mode, however, may not reveal enough information to portray the true behavior of the graph. Try the following four functions on your graphing device. If you can, plot them in both "connected" and "dot" modes.

$y_1 = x \operatorname{int} x$	at $x = 2$	jump discontinuity
$y_2 = \sin \dfrac{1}{x}$	at $x = 0$	oscillating discontinuity
$y_3 = \dfrac{1}{x - 2}$	at $x = 2$	infinite discontinuity
$y_4 = \dfrac{x^2 - 2}{x - \sqrt{2}}$	at $x = \sqrt{2}$	removable discontinuity

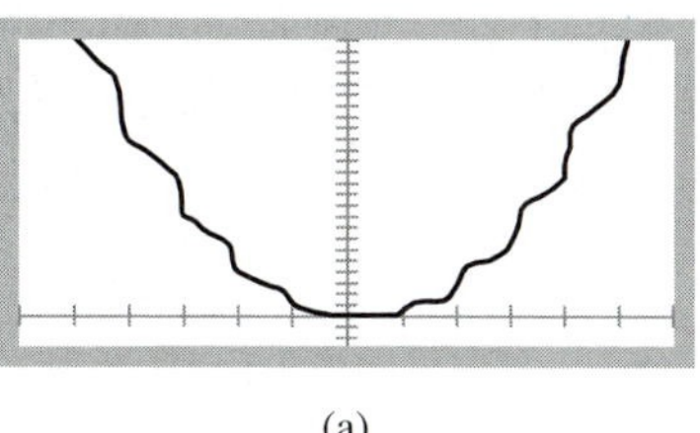

(a)

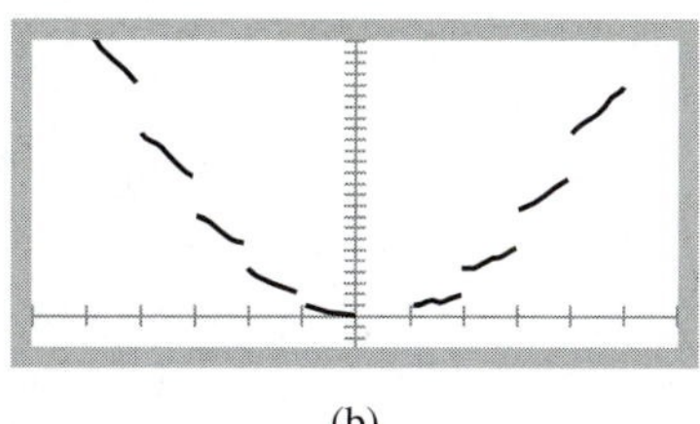

(b)

(a) $y_1 = x \times \operatorname{int} x$ incorrectly graphed in connected mode. (b) $y_1 = x \times \operatorname{int} x$ correctly graphed in dot mode.

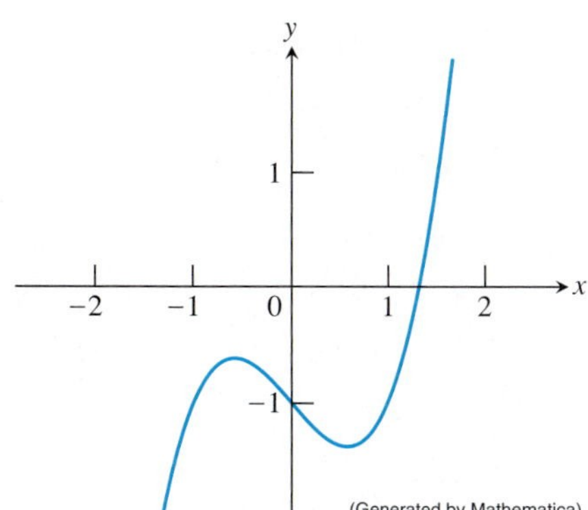

FIGURE 1.56 The graph of $f(x) = x^3 - x - 1$. (Example 7)

Example 7 Using the Intermediate Value Theorem

Is any real number exactly 1 less than its cube?

Solution We answer this question by applying the Intermediate Value Theorem in the following way. Any such number must satisfy the equation $x = x^3 - 1$ or, equivalently, $x^3 - x - 1 = 0$. Hence, we are looking for a zero value of the continuous function $f(x) = x^3 - x - 1$ (Figure 1.56). The function changes sign between 1 and 2, so there must be a point c between 1 and 2 where $f(c) = 0$.

EXERCISES 1.4

Continuity from Graphs

In Exercises 1–4, say whether the function graphed is continuous on $[-1, 3]$. If not, where does it fail to be continuous and why?

1.

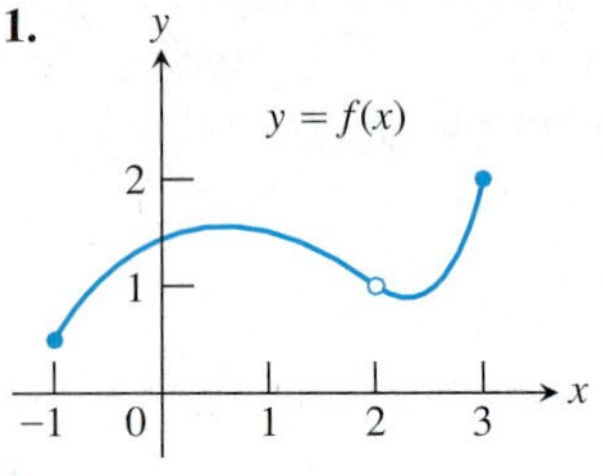

$y = f(x)$

2.

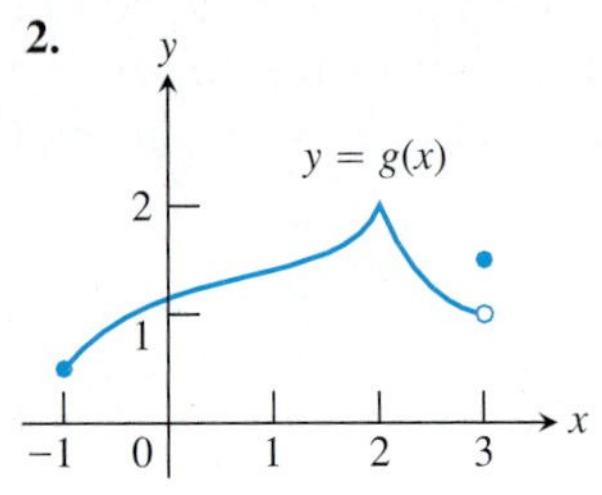

3.

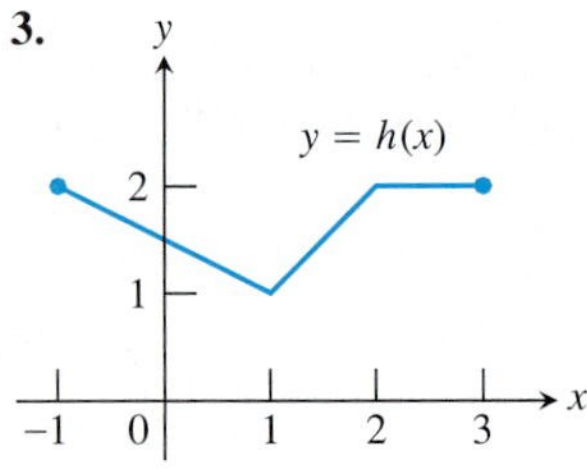

4.

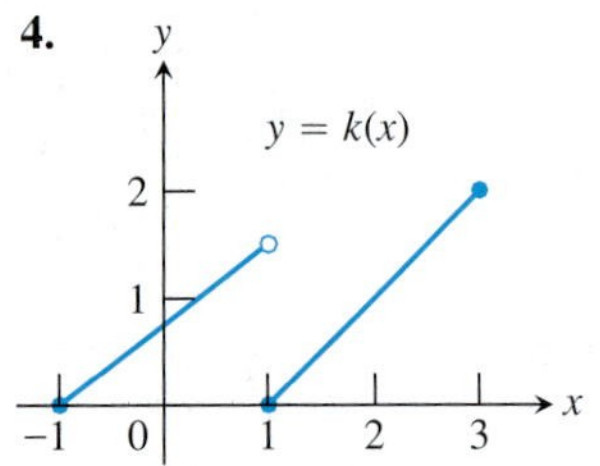

Exercises 5–10 are about the function

$$f(x) = \begin{cases} x^2 - 1, & -1 \le x < 0 \\ 2x, & 0 < x < 1 \\ 1, & x = 1 \\ -2x + 4, & 1 < x < 2 \\ 0, & 2 < x < 3 \end{cases}$$

graphed in Figure 1.57.

5. **(a)** Does $f(-1)$ exist?

(b) Does $\lim_{x \to -1^+} f(x)$ exist?

(c) Does $\lim_{x \to -1^+} f(x) = f(-1)$?

(d) Is f continuous at $x = -1$?

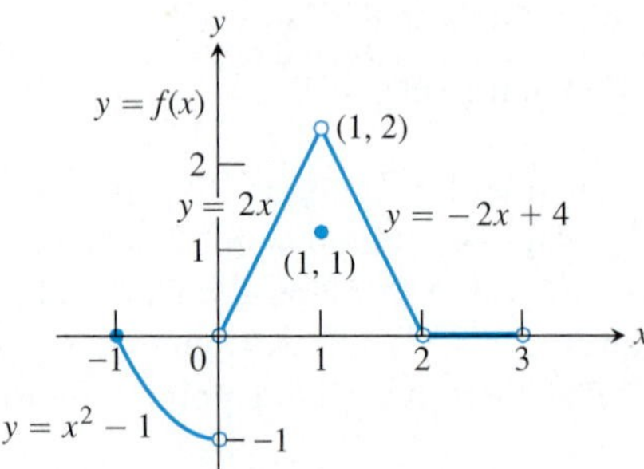

FIGURE 1.57 The graph for Exercises 5–10.

6. **(a)** Does $f(1)$ exist?

(b) Does $\lim_{x \to 1} f(x)$ exist?

(c) Does $\lim_{x \to 1} f(x) = f(1)$?

(d) Is f continuous at $x = 1$?

7. **(a)** Is f defined at $x = 2$? (Look at the definition of f.)

(b) Is f continuous at $x = 2$?

8. At what values of x is f continuous?

9. What value should be assigned to $f(2)$ to make the extended function continuous at $x = 2$?

10. To what new value should $f(1)$ be changed to remove the discontinuity?

Applying the Continuity Test

At which points do the functions in Exercises 11 and 12 fail to be continuous? At which points, if any, are the discontinuities removable? Not removable? Give reasons for your answers.

11. Exercise 11, Section 1.1

12. Exercise 12, Section 1.1

On what intervals are the functions in Exercises 13–20 continuous?

13. $y = \dfrac{1}{x - 2} - 3x$

14. $y = \dfrac{1}{(x + 2)^2} + 4$

15. $z = \dfrac{t + 1}{t^2 - 4t + 3}$

16. $u = \dfrac{1}{|t| + 1} - \dfrac{t^2}{2}$

17. $r = \dfrac{\cos \theta}{\theta}$

18. $y = \tan \dfrac{\pi\theta}{2}$

19. $s = \sqrt{2v + 3}$

20. $y = \sqrt[4]{3x - 1}$

Composite Functions

Find the limits in Exercises 21–24. Are the functions continuous at the point being approached?

21. $\lim\limits_{x \to \pi} \sin (x - \sin x)$

22. $\lim\limits_{t \to 0} \sin \left(\dfrac{\pi}{2} \cos (\tan t) \right)$

23. $\lim\limits_{y \to 1} \sec (y \sec^2 y - \tan^2 y - 1)$

24. $\lim\limits_{\theta \to 0} \tan \left(\dfrac{\pi}{4} \cos (\sin \theta^{1/3}) \right)$

Theory and Examples

25. *Writing to Learn* A function $y = f(x)$, continuous on $[0, 1]$, is known to be negative at $x = 0$ and positive at $x = 1$. What, if anything, does this say about the equation $f(x) = 0$? Illustrate with a sketch.

26. *Writing to Learn* Why does the equation $\cos x = x$ have at least one solution?

27. *Writing to Learn* Explain why the following five statements ask for the same information.

(a) Find the zeros of $f(x) = x^3 - 3x - 1$.

(b) Find the x-coordinate of the points where the curve $y = x^3$ crosses the line $y = 3x + 1$.

(c) Find all the values of x for which $x^3 - 3x = 1$.

(d) Find the x-coordinates of the points where the cubic curve $y = x^3 - 3x$ crosses the line $y = 1$.

(e) Solve the equation $x^3 - 3x - 1 = 0$.

28. *Solving an equation* If $f(x) = x^3 - 8x + 10$, show that there is at least one value of c for which $f(c)$ equals

(a) π

(b) $-\sqrt{3}$

(c) 5,000,000.

29. *Removable discontinuity* Give an example of a function $f(x)$ that is continuous for all values of x except $x = 2$, where it has a removable discontinuity. Explain how you know that f is discontinuous at $x = 2$ and how you know the discontinuity is removable.

30. *Nonremovable discontinuity* Give an example of a function $g(x)$ that is continuous for all values of x except $x = -1$, where it has a nonremovable discontinuity. Explain how you know that g is discontinuous there and why the discontinuity is not removable.

31. *Factoring a polynominal* Find rounded three-place values for r_1 through r_5 in the factorization

$$x^5 - x^4 - 5x^3 = (x - r_1)(x - r_2)(x - r_3)(x - r_4)(x - r_5).$$

T 32. *Factoring a polynomial* You want to rewrite the polynomial $x^3 - 3x - 1$ in the form $(x - r)q(x)$, where $q(x)$ is a quadratic polynomial. Rounded to three decimal places, what are your choices for r?

33. *A function discontinuous at every point*

(a) Use the fact that every nonempty interval of real numbers contains both rational and irrational numbers to show that the function

$$f(x) = \begin{cases} 1 & \text{if } x \text{ is rational} \\ 0 & \text{if } x \text{ is irrational} \end{cases}$$

is discontinuous at every point.

(b) Is f right-continuous or left-continuous at any point?

34. *Writing to Learn* If functions $f(x)$ and $g(x)$ are continuous for $0 \le x \le 1$, could $f(x)/g(x)$ possibly be discontinuous at a point of $[0, 1]$? Give reasons for your answer.

35. *Writing to Learn* Is it true that a continuous function that is never zero on an interval never changes sign on that interval? Give reasons for your answer.

36. *Stretching a rubber band* Is it true that if you stretch a rubber band by moving one end to the right and the other to the left, some point of the band will end up in its original position? Give reasons for your answer.

37. *A fixed point theorem* Suppose that a function f is continuous on the closed interval $[0, 1]$ and that $0 \le f(x) \le 1$ for every x in $[0, 1]$. Show that there must exist a number c in $[0, 1]$ such that $f(c) = c$ (c is called a **fixed point** of f).

38. *The sign-preserving property of continuous functions* Let f be defined on an interval (a, b) and suppose that $f(c) \ne 0$ at some c where f is continuous. Show that there is an interval $(c - \delta, c + \delta)$ about c where f has the same sign as $f(c)$. Notice how remarkable this conclusion is. Although f is defined throughout (a, b), it is not required to be continuous at any point except c. That and the condition $f(c) \ne 0$ are enough to make f different from zero (positive or negative) throughout an entire (small) interval.

T 39. *Salary negotiation* A welder's contract promises a 3.5% salary increase each year for 4 years, and Luisa has an initial salary of \$36,500.

(a) Show that Luisa's salary is given by

$$y = 36{,}500(1.035)^{\text{int } t},$$

where t is the time, measured in years, since Luisa signed the contract.

(b) Graph Luisa's salary function. At what values of t is it continuous?

T 40. *Airport parking* Valuepark charges \$1.10 per hour or fraction of an hour for airport parking. The maximum charge per day is \$7.25.

(a) Write a formula that gives the charge for x hours with $0 \le x \le 24$. (*Hint*: See Exercise 39.)

(b) Graph the function in part (a). At what values of x is it continuous?

COMPUTER EXPLORATIONS

Continuous Extension to a Point

As we saw in Section 1.2, a rational function may have a limit even at a point where its denominator is zero. If $f(c)$ is not defined but $\lim_{x \to c} f(x) = L$ exists, we can define a new function $F(x)$ by the rule

$$F(x) = \begin{cases} f(x) & \text{if } x \text{ is in the domain of } f \\ L & \text{if } x = c. \end{cases}$$

The function F is continuous at $x = c$. It is called the **continuous extension** of f to $x = c$. For rational functions f, continuous extensions are usually found by canceling common factors.

In Exercises 41–44, graph the function f to see whether it appears to have a continuous extension to the origin. If it does, use Trace and Zoom to find a good candidate for the extended function's value at $x = 0$. If the function does not appear to have a continuous extension, can it be extended to be continuous at the origin from the right or from the left? If so, what do you think the extended function's value(s) should be?

41. $f(x) = \dfrac{10^x - 1}{x}$

42. $f(x) = \dfrac{10^{|x|} - 1}{x}$

43. $f(x) = \dfrac{\sin x}{|x|}$

44. $f(x) = (1 + 2x)^{1/x}$

Solving Equations Graphically

Use a graphing calculator or computer grapher to solve the equations in Exercises 45–52. Round each solution to four decimal places.

45. $x^3 - 3x - 1 = 0$

46. $2x^3 - 2x^2 - 2x + 1 = 0$

47. $x(x - 1)^2 = 1$ (one root)

48. $x^x = 2$

49. $\sqrt{x} + \sqrt{1 + x} = 4$

50. $x^3 - 15x + 1 = 0$ (three roots)

51. $\cos x = x$ (one root). Be sure you are using radian mode.

52. $2 \sin x = x$ (three roots). Be sure you are using radian mode.

1.5 Tangent Lines

What *Is* a Tangent to a Curve? • Finding a Tangent to the Graph of a Function • Rates of Change: Derivative at a Point

This section continues the discussion of secants and tangents begun in Section 1.1. We calculate limits of secant slopes to find tangents to curves.

What Is a Tangent to a Curve?

For circles, tangency is straightforward. A line L is tangent to a circle at a point P if L passes through P perpendicular to the radius at P (Figure 1.58). Such a line just *touches* the circle. But what does it mean to say that a line L is tangent to some other curve C at a point P? Generalizing from the geometry of the circle, we might say that it means one of the following.

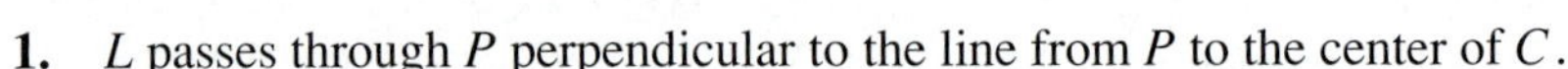

1. L passes through P perpendicular to the line from P to the center of C.
2. L passes through only one point of C, namely P.
3. L passes through P and lies on one side of C only.

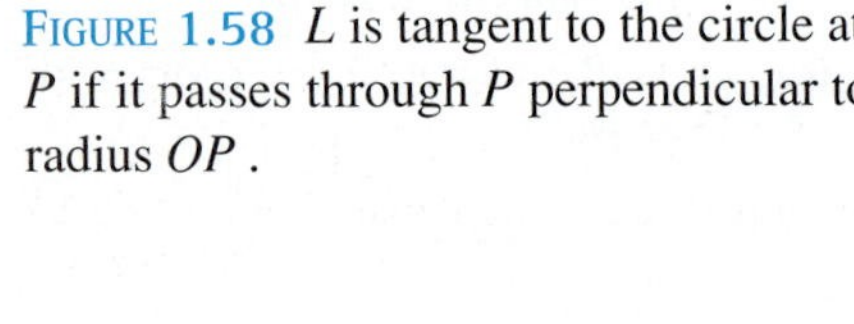

FIGURE 1.58 L is tangent to the circle at P if it passes through P perpendicular to radius OP.

Although these statements are valid if C is a circle, none of them works consistently for more general curves. Most curves do not have centers, and a line we may want to call tangent may intersect C at other points or cross C at the point of tangency (Figure 1.59).

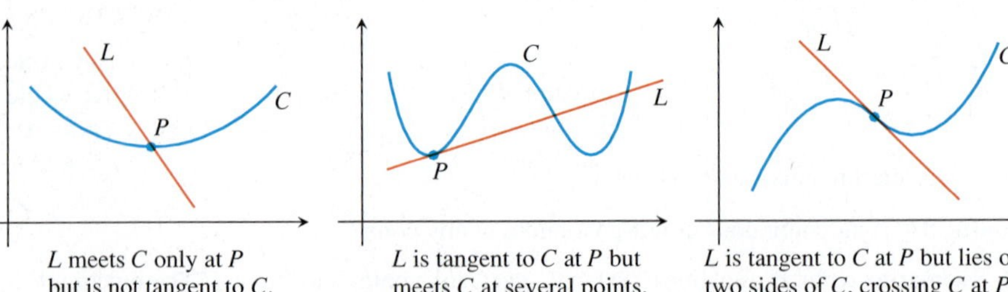

L
P
C
L meets C only at P but is not tangent to C.
C
L
P
L is tangent to C at P but meets C at several points.
L
C
P
L is tangent to C at P but lies on two sides of C, crossing C at P.

FIGURE 1.59 Exploding myths about tangent lines.

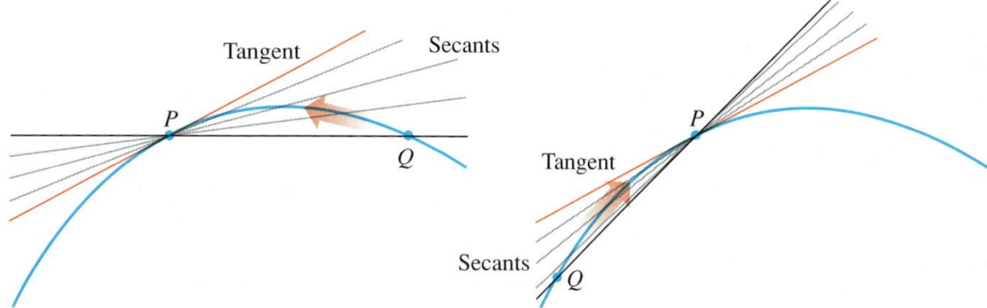

FIGURE 1.60 The dynamic approach to tangency. The tangent to the curve at P is the line through P whose slope is the limit of the secant slopes as $Q \to P$ from either side.

CD-ROM
WEBsite
Historical Biography

Pierre de Fermat
(1601 — 1665)

To define tangency for general curves, we need a *dynamic* approach that takes into account the behavior of the secants through P and nearby points Q as Q moves toward P along the curve (Figure 1.60). It goes like this:

1. We start with what we *can* calculate, namely the slope of the secant PQ.
2. Investigate the limit of the secant slope as Q approaches P along the curve.
3. If the limit exists, take it to be the slope of the curve at P and define the tangent to the curve at P to be the line through P with this slope.

This approach is what we were doing in the falling-rock and heat-shield examples in Section 1.1.

How do you find a tangent to a curve?

This problem was the dominant mathematical question of the early seventeenth century, and it is hard to overestimate how badly the scientists of the day wanted to know the answer. In optics, the tangent determined the angle at which a ray of light entered a curved lens. In mechanics, the tangent determined the direction of a body's motion at every point along its path. In geometry, the tangents to two curves at a point of intersection determined the angle at which the curves intersected. René Descartes went so far as to say that the problem of finding a tangent to a curve was "the most useful and most general problem not only that I know but even that I have any desire to know."

Example 1 Tangent Line to a Parabola

Find the slope of the parabola $y = x^2$ at the point $P(2, 4)$. Write an equation for the tangent to the parabola at this point.

Solution We begin with a secant line through $P(2, 4)$ and $Q(2 + h, (2 + h)^2)$ nearby. We then write an expression for the slope of the secant PQ and investigate what happens to the slope as Q approaches P along the curve:

$$\text{Secant slope} = \frac{\Delta y}{\Delta x} = \frac{(2+h)^2 - 2^2}{h} = \frac{h^2 + 4h + 4 - 4}{h}$$

$$= \frac{h^2 + 4h}{h} = h + 4.$$

If $h > 0$, then Q lies above and to the right of P, as in Figure 1.61. If $h < 0$, then Q lies to the left of P (not shown). In either case, as Q approaches P along the curve, h approaches zero and the secant slope approaches 4:

$$\lim_{h \to 0} (h + 4) = 4.$$

We take 4 to be the parabola's slope at P.

The tangent to the parabola at P is the line through P with slope 4:

$$y = 4 + 4(x - 2) \qquad \text{Point-slope equation}$$

$$y = 4x - 4.$$

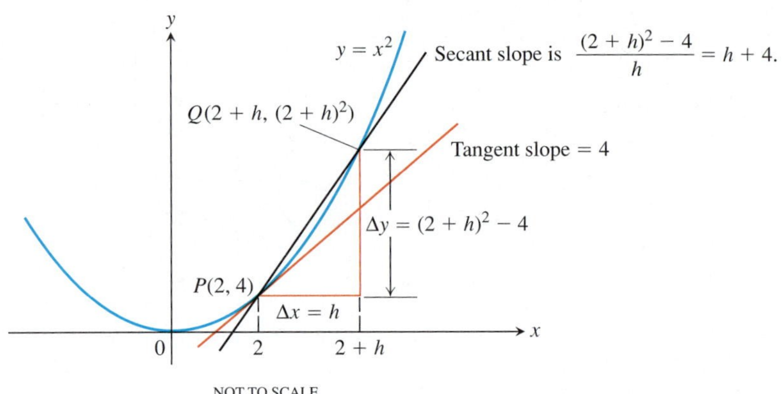

FIGURE 1.61 Diagram for finding the slope of the parabola $y = x^2$ at the point $P(2, 4)$. (Example 1)

Finding a Tangent to the Graph of a Function

To find a tangent to an arbitrary curve $y = f(x)$ at a point $P(x_0, f(x_0))$, we use the same dynamic procedure. We calculate the slope of the secant through P and a point $Q(x_0 + h, f(x_0 + h))$. We then investigate the limit of the slope as $h \to 0$ (Figure 1.62). If the limit exists, we call it the slope of the curve at P and define the tangent at P to be the line through P having this slope.

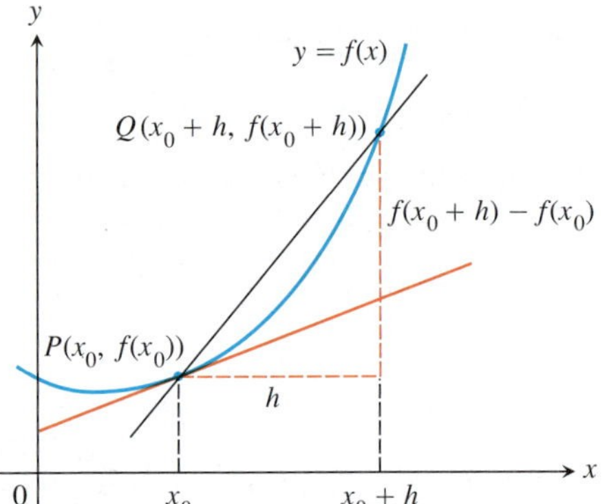

FIGURE 1.62 The tangent slope is

$$\lim_{h\to 0} \frac{f(x_0 + h) - f(x_0)}{h}$$

Definitions Slope and Tangent Line

The **slope of the curve** $y = f(x)$ at the point $P(x_0, f(x_0))$ is the number

$$m = \lim_{h\to 0} \frac{f(x_0 + h) - f(x_0)}{h} \qquad \text{(provided the limit exists).}$$

The **tangent line** to the curve at P is the line through P with this slope.

Whenever we make a new definition, it is a good idea to try it on familiar objects to be sure it gives the results we want in familiar cases. Example 2 shows that the new definition of slope agrees with the old definition when we apply it to nonvertical lines.

Example 2 Testing the Definition

Show that the line $y = mx + b$ is its own tangent at any point $(x_0, mx_0 + b)$.

Solution We let $f(x) = mx + b$ and organize the work into three steps.

Step 1: *Find* $f(x_0)$ *and* $f(x_0 + h)$.

$$f(x_0) = mx_0 + b$$
$$f(x_0 + h) = m(x_0 + h) + b = mx_0 + mh + b$$

Step 2: *Find the slope* $\lim_{h\to 0} (f(x_0+h)-f(x_0))/h$.

$$\lim_{h\to 0}\frac{f(x_0+h)-f(x_0)}{h} = \lim_{h\to 0}\frac{(mx_0+mh+b)-(mx_0+b)}{h}$$
$$= \lim_{h\to 0}\frac{mh}{h} = m$$

Step 3: *Find the tangent line using the point-slope equation.* The tangent line at the point (x_0, mx_0+b) is

$$y = (mx_0+b) + m(x-x_0)$$
$$y = mx_0 + b + mx - mx_0$$
$$y = mx + b.$$

How to Find the Tangent to the Curve $y = f(x)$ at (x_0, y_0)

1. Calculate $f(x_0)$ and $f(x_0+h)$.
2. Calculate the slope
$$m = \lim_{h\to 0}\frac{f(x_0+h)-f(x_0)}{h}.$$
3. If the limit exists, find the tangent line as $y = y_0 + m(x-x_0)$.

Example 3 Slope and Tangent to $y = 1/x$

(a) Find the slope of the curve $y = 1/x$ at $x = a$.

(b) Where does the slope equal $-1/4$?

(c) What happens to the tangent to the curve at the point $(a, 1/a)$ as a changes?

Solution

(a) Here $f(x) = 1/x$. The slope at $(a, 1/a)$ is

$$\lim_{h\to 0}\frac{f(a+h)-f(a)}{h} = \lim_{h\to 0}\frac{\dfrac{1}{a+h}-\dfrac{1}{a}}{h}$$
$$= \lim_{h\to 0}\frac{1}{h}\,\frac{a-(a+h)}{a(a+h)}$$
$$= \lim_{h\to 0}\frac{-h}{ha(a+h)}$$
$$= \lim_{h\to 0}\frac{-1}{a(a+h)} = -\frac{1}{a^2}.$$

Notice how we had to keep writing "$\lim_{h\to 0}$" before each fraction until the stage where we could evaluate the limit by substituting $h = 0$.

(b) The slope of $y = 1/x$ at the point where $x = a$ is $-1/a^2$. It will be $-1/4$ provided that

$$-\frac{1}{a^2} = -\frac{1}{4}.$$

This equation is equivalent to $a^2 = 4$, so $a = 2$ or $a = -2$. The curve has slope $-1/4$ at the two points $(2, 1/2)$ and $(-2, -1/2)$ (Figure 1.63).

(c) Notice that the slope $-1/a^2$ is always negative. As $a \to 0^+$, the slope approaches $-\infty$ and the tangent becomes increasingly steep (Figure 1.64). We see this situation again as $a \to 0^-$. As a moves away from the origin in either direction, the slope approaches 0^- and the tangent levels off.

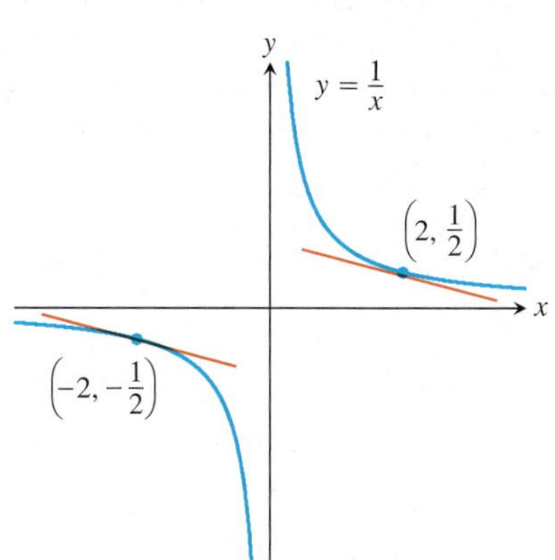

FIGURE 1.63 The two tangent lines to $y = 1/x$ having slope $-1/4$.

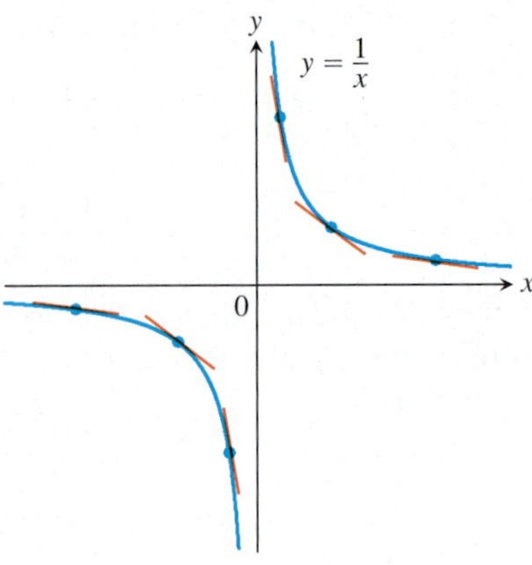

FIGURE 1.64 The tangent slopes, steep near the origin, become more gradual as the point of tangency moves away.

CD-ROM
WEBsite
Historical Biography

René François de Sluse (1622 — 1685)

Rates of Change: Derivative at a Point

The expression

$$\frac{f(x_0 + h) - f(x_0)}{h}$$

is called the **difference quotient of f at x_0 with increment h.** If the difference quotient has a limit as h approaches zero, that limit is called the **derivative of f at x_0.** If we interpret the difference quotient as a secant slope, the derivative gives the slope of the curve and tangent at the point where $x = x_0$. If we interpret the difference quotient as an average rate of change, as we did in Section 1.1, the derivative gives the function's rate of change with respect to x at the point $x = x_0$. The derivative is one of the two most important mathematical objects considered in calculus. We begin a thorough study of it in Chapter 2. The other important object is the integral, and we initiate its study in Chapter 4.

All these refer to the same thing

1. The slope of $y = f(x)$ at $x = x_0$
2. The slope of the tangent to the curve $y = f(x)$ at $x = x_0$
3. The rate of change of $f(x)$ with respect to x at $x = x_0$
4. The derivative of f at $x = x_0$
5. $\lim_{h \to 0} \frac{f(x_0 + h) - f(x_0)}{h}$

Example 4 Instantaneous Speed (Continuation of Section 1.1, Examples 1 and 2)

In Examples 1 and 2 in Section 1.1, we studied the speed of a rock falling freely from rest near the surface of the earth. We knew that the rock fell $y = 16t^2$ feet during the first t sec, and we used a sequence of average rates over increasingly short intervals to estimate the rock's speed at the instant $t = 2$. Exactly what *was* the rock's speed at this time?

Solution We let $f(t) = 16t^2$. The average speed of the rock over the interval between $t = 2$ and $t = 2 + h$ sec was

$$\frac{f(2 + h) - f(2)}{h} = \frac{16(2 + h)^2 - 16(2)^2}{h} = \frac{16(h^2 + 4h)}{h} = 16(h + 4).$$

The rock's speed at the instant $t = 2$ was

$$\lim_{h \to 0} 16(h + 4) = 16(0 + 4) = 64 \text{ ft/sec}.$$

Our original estimate of 64 ft/sec was right.

EXERCISES 1.5

Slopes and Tangent Lines

In Exercises 1–4, use the grid and a straight edge to make a rough estimate of the slope of the curve (in y-units per x-unit) at the points P_1 and P_2. Graphs can shift during a press run, so your estimates may be somewhat different from those in the back of the book.

1.

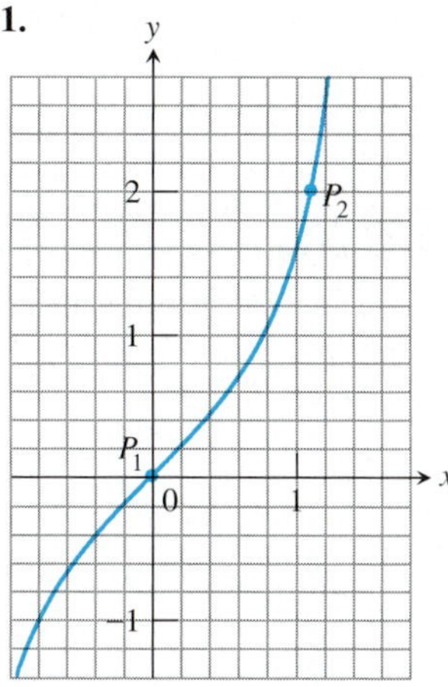

2.

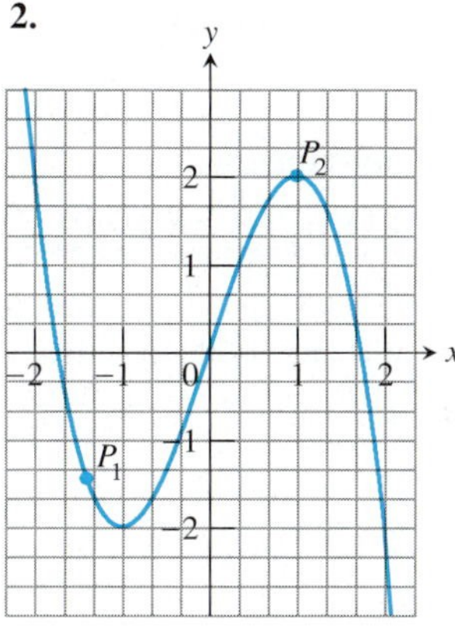

3.

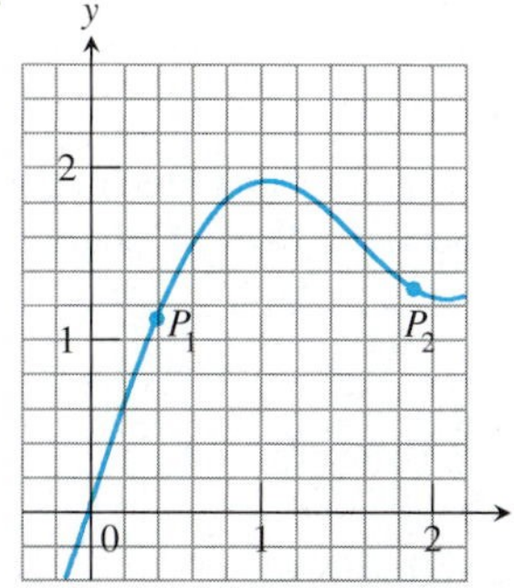

4.

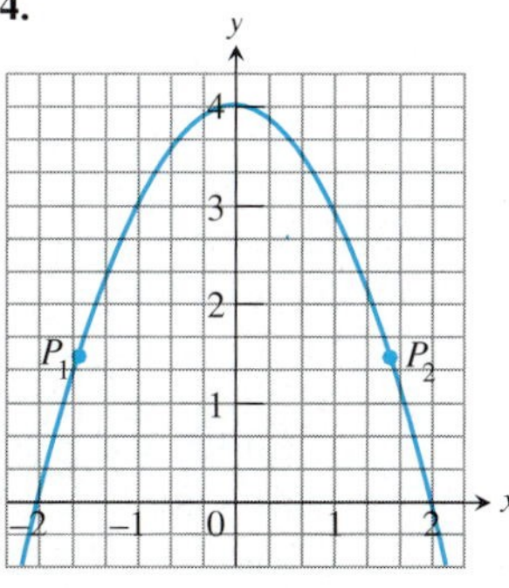

In Exercises 5–8, find an equation for the tangent to the curve at the given point. Sketch the curve and tangent together.

5. $y = 4 - x^2, \quad (-1, 3)$

6. $y = 2\sqrt{x}, \quad (1, 2)$

7. $y = x^3, \quad (-2, -8)$

8. $y = \dfrac{1}{x^3}, \quad (-2, -1/8)$

In Exercises 9–12, find the slope of the function's graph at the given point. Then find an equation for the tangent to the graph there.

9. $f(x) = x - 2x^2, \quad (1, -1)$

10. $h(t) = t^3 + 3t, \quad (1, 4)$

11. $g(u) = \dfrac{u}{u - 2}, \quad (3, 3)$

12. $f(x) = \sqrt{x + 1}, \quad (8, 3)$

In Exercises 13 and 14, find the slope of the curve at the indicated value of x.

13. $y = \dfrac{1}{x - 1}, \quad x = 3$

14. $y = \dfrac{x - 1}{x + 1}, \quad x = 0$

Tangent Lines with Specified Slopes

At what points do the graphs of the functions in Exercises 15 and 16 have horizontal tangents?

15. $f(x) = x^2 + 4x - 1$

16. $g(x) = x^3 - 3x$

17. Find equations for all tangents to the curve $y = 1/(x - 1)$ that have slope -1.

18. Find an equation for the tangent to the curve $y = \sqrt{x}$ that has slope $1/4$.

Rates of Change

19. *Object dropped from a tower* An object is dropped from the top of a 100-m-high tower. Its height above ground after t sec is $100 - 4.9t^2$ m. How fast is it falling 2 sec after it is dropped?

20. *Speed of a rocket* At t sec after liftoff, the height of a rocket is $3t^2$ ft. How fast is the rocket climbing 10 sec after liftoff?

21. *Circle's changing area* What is the rate of change of the area of a circle ($A = \pi r^2$) with respect to the radius when the radius is $r = 3$?

22. *Ball's changing volume* What is the rate of change of the volume of a ball ($V = (4/3)\pi r^3$) with respect to the radius when the radius is $r = 2$?

23. *Free fall on Mars* The equation for free fall at the surface of Mars is $s = 1.86t^2$ m, with t in seconds. Assume that a rock is dropped from the top of a 200-m cliff. Find the speed of the rock at $t = 1$ sec.

24. *Free fall on Jupiter* The equation for free fall at the surface of Jupiter is $s = 11.44t^2$ m with t in seconds. Assume that a rock is dropped from the top of a 500-m cliff. Find the speed of the rock at $t = 2$ sec.

Testing for Tangents

25. *Writing to Learn* Does the graph of

$$f(x) = \begin{cases} x^2 \sin(1/x), & x \neq 0 \\ 0, & x = 0 \end{cases}$$

have a tangent at the origin? Give reasons for your answer.

26. *Writing to Learn* Does the graph of

$$g(x) = \begin{cases} x \sin(1/x), & x \neq 0 \\ 0, & x = 0 \end{cases}$$

have a tangent at the origin? Give reasons for your answer.

Vertical Tangents

We say that the curve $y = f(x)$ has a **vertical tangent** at the point where $x = x_0$ if $\lim_{h \to 0} (f(x_0 + h) - f(x_0))/h = \infty$ or $-\infty$.

Vertical tangent at $x = 0$:

$$\lim_{h \to 0} \frac{f(0+h) - f(0)}{h} = \lim_{h \to 0} \frac{h^{1/3} - 0}{h} = \lim_{h \to 0} \frac{1}{h^{2/3}} = \infty$$

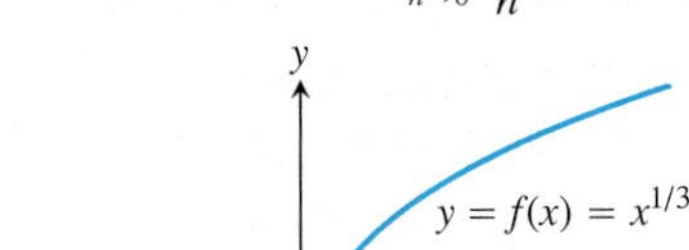

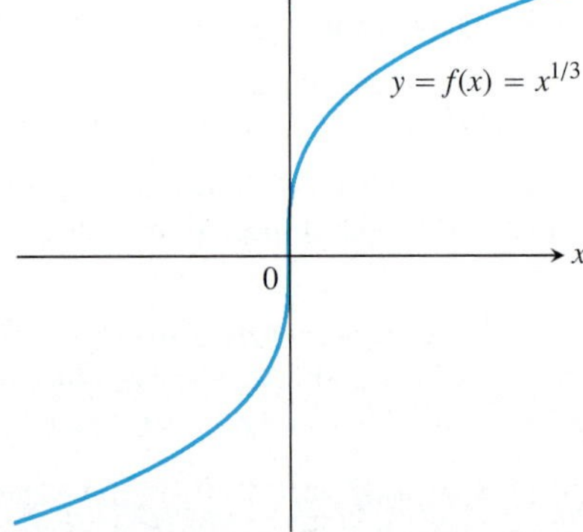

VERTICAL TANGENT AT ORIGIN

No vertical tangent at $x = 0$:

$$\lim_{h \to 0} \frac{g(0+h) - g(0)}{h} = \lim_{h \to 0} \frac{h^{2/3} - 0}{h} = \lim_{h \to 0} \frac{1}{h^{1/3}}$$

does not exist, because the limit is ∞ from the right and $-\infty$ from the left.

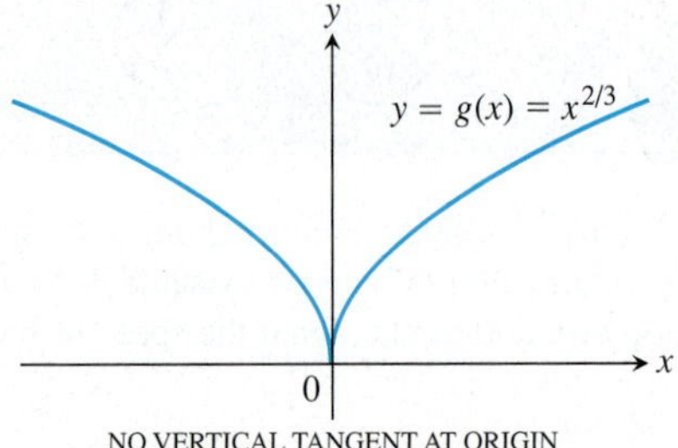

NO VERTICAL TANGENT AT ORIGIN

27. *Writing to Learn* Does the graph of

$$f(x) = \begin{cases} -1, & x < 0 \\ 0, & x = 0 \\ 1, & x > 0 \end{cases}$$

have a vertical tangent at the origin? Give reasons for your answer.

28. *Writing to Learn* Does the graph of

$$U(x) = \begin{cases} 0, & x < 0 \\ 1, & x \geq 0 \end{cases}$$

have a vertical tangent at the point (0, 1)? Give reasons for your answer.

COMPUTER EXPLORATIONS

In Exercises 29–32, use a calculator or computer to find the average rate of change of the function over each interval.

29. $f(x) = e^x$

(a) $[-2, 0]$ **(b)** $[1, 3]$

30. $f(x) = \ln x$

(a) $[1, 4]$ **(b)** $[100, 103]$

31. $f(t) = \cot t$

(a) $[\pi/4, 3\pi/4]$ **(b)** $[\pi/6, \pi/2]$

32. $f(t) = 2 + \cos t$

(a) $[0, \pi]$ **(b)** $[-\pi, \pi]$

33. *INS Funding* Table 1.3 gives the amount of federal funding for the Immigration and Naturalization Service (INS) in the United States for several years.

Table 1.3 Federal INS funding

Year	Funding ($ billions)
1993	1.5
1994	1.6
1995	2.1
1996	2.6
1997	3.1

Source: Immigration and Naturalization Service as reported by Bob Laird in *USA Today*, February 18, 1997.

(a) Find the average rate of change in funding from 1993 to 1995.

(b) Find the average rate of change from 1995 to 1997.

(c) Let $x = 0$ represent 1990, $x = 1$ represent 1991, and so forth. Find a quadratic regression equation for the data and superimpose its graph on a scatter plot of the data. (See page 5 for an introduction to regression analysis with a calculator.)

(d) Compute the average rates of change in parts (a) and (b) using the regression equation.

(e) Use the regression equation to find how fast the funding was growing in 1997.

34. *Academic funding by congress* Table 1.4 gives the amounts of money earmarked by the U.S. Congress for collegiate academic programs for several years.

Table 1.4 Congressional academic funding in the United States

Year	Funding ($ millions)
1988	225
1989	289
1990	270
1991	493
1992	684
1993	763
1994	651
1995	600
1996	296
1997	440

Source: The Chronicle of Higher Education, March 28, 1997.

(a) Let $x = 0$ represent 1980, $x = 1$ represent 1981, and so forth. Make a scatter plot of the data.

(b) Let P represent the point corresponding to 1997 and Q the point for any one of the previous years. Make a table of the slopes possible for the secant line PQ.

(c) *Writing to Learn* Based on the computations, explain why someone might be hesitant to make a prediction about the rate of change of congressional funding in 1997.

Grapher Explorations: Vertical Tangents

(a) Graph the curves in Exercises 35–44. Where do the graphs appear to have vertical tangents?

(b) Confirm your findings in part (a) with limit calculations. But before you do, read the introduction to Exercises 27 and 28.

35. $y = x^{2/5}$

36. $y = x^{4/5}$

37. $y = x^{1/5}$

38. $y = x^{3/5}$

39. $y = 4x^{2/5} - 2x$

40. $y = x^{5/3} - 5x^{2/3}$

41. $y = x^{2/3} - (x - 1)^{1/3}$

42. $y = x^{1/3} + (x - 1)^{1/3}$

43. $y = \begin{cases} -\sqrt{|x|}, & x \le 0 \\ \sqrt{x}, & x > 0 \end{cases}$

44. $y = \sqrt{|4 - x|}$

Graphing Secant and Tangent Lines

Use a CAS to perform the following steps for the functions in Exercises 45–48.

(a) Plot $y = f(x)$ over the interval $x_0 - 1/2 \le x \le x_0 + 3$.

(b) Holding x_0 fixed, the difference quotient

$$q(h) = \frac{f(x_0 + h) - f(x_0)}{h}$$

at x_0 becomes a function of the step size h. Enter this function into your CAS workspace.

(c) Find the limit of q as $h \to 0$.

(d) Define the secant lines $y = f(x_0) + q * (x - x_0)$ for $h = 3, 2$, and 1. Graph them together with f and the tangent line over the interval in part (a).

45. $f(x) = x^3 + 2x, \quad x_0 = 0$

46. $f(x) = x + \dfrac{5}{x}, \quad x_0 = 1$

47. $f(x) = x + \sin(2x), \quad x_0 = \pi/2$

48. $f(x) = \cos x + 4\sin(2x), \quad x_0 = \pi$

Questions to Guide Your Review

1. What is the average rate of change of the function $g(t)$ over the interval from $t = a$ to $t = b$? How is it related to a secant line?

2. What limit must be calculated to find the rate of change of a function $g(t)$ at $t = t_0$?

3. What is an informal definition of the limit

$$\lim_{x \to x_0} f(x) = L?$$

Why is the definition "informal"? Give examples. What exactly does $\lim_{x \to x_0} f(x) = L$ mean?

4. Does the existence and value of the limit of a function $f(x)$ as x approaches x_0 ever depend on what happens at $x = x_0$? Explain and give examples.

5. What function behaviors might occur for which the limit may fail to exist? Give examples.

6. What theorems are available for calculating limits? Give examples of how the theorems are used.
7. How are one-sided limits related to limits? How can this relationship sometimes be used to calculate a limit or prove it does not exist? Give examples.
8. What is the value of $\lim_{\theta\to 0}((\sin\theta)/\theta)$? Does it matter whether θ is measured in degrees or radians? Explain.
9. What do $\lim_{x\to\infty} f(x) = L$ and $\lim_{x\to-\infty} f(x) = L$ mean? Give examples.
10. What are $\lim_{x\to\pm\infty} k$ (k a constant) and $\lim_{x\to\pm\infty} (1/x)$? How do you extend these results to other functions? Give examples.
11. How do you find the limit of a rational function as $x \to \pm\infty$? Give examples.
12. What are horizontal, vertical, and oblique asymptotes? Give examples.
13. What conditions must be satisfied by a function if it is to be continuous at an interior point of its domain? At an endpoint?
14. How can looking at the graph of a function help you tell where the function is continuous?
15. What does it mean for a function to be right-continuous at a point? Left-continuous? How are continuity and one-sided continuity related?
16. What can be said about the continuity of polynomials? Of rational functions? Of trigonometric functions? Of exponential functions? Of logarithmic functions? Of rational powers and algebraic combinations of functions? Of composites of functions? Of absolute values of functions? Of inverses of functions?
17. What does it mean for a function to be continuous on an interval?
18. What does it mean for a function to be continuous? Give examples to illustrate that a function that is not continuous on its entire domain may still be continuous on selected intervals within the domain.
19. What are the basic types of discontinuity? Give an example of each. What is a removable discontinuity? Give an example.
20. What does it mean for a function to have the Intermediate Value Property? What conditions guarantee that a function has this property over an interval? What are the consequences for graphing and solving the equation $f(x) = 0$?
21. It is often said that a function is continuous if you can draw its graph without having to lift your pen from the paper. Why is that?
22. What does it mean for a line to be tangent to a curve C at a point P ?
23. What is the significance of the formula

$$\lim_{h\to 0} \frac{f(x+h)-f(x)}{h} ?$$

Interpret the formula geometrically and physically.

24. How do you find the tangent to the curve $y = f(x)$ a point (x_0, y_0) on the curve?
25. How does the slope of the curve $y = f(x)$ at $x = x_0$ relate to the function's rate of change with respect to x at $x = x_0$? To the derivative of f at x_0 ?

Practice Exercises

Limits and Continuity

1. Graph the function

$$f(x) = \begin{cases} 1, & x \le -1 \\ -x, & -1 < x < 0 \\ 1, & x = 0 \\ -x, & 0 < x < 1 \\ 1, & x \ge 1. \end{cases}$$

Then discuss, in detail, limits, one-sided limits, continuity, and one-sided continuity of f at $x = -1, 0$, and 1. Are any of the discontinuities removable? Explain.

2. Repeat the instructions of Exercise 1 for

$$f(x) = \begin{cases} 0, & x \le -1 \\ 1/x, & 0 < |x| < 1 \\ 0, & x = 1 \\ 1, & x > 1. \end{cases}$$

3. Suppose that $f(t)$ and $g(t)$ are defined for all t and that $\lim_{t\to t_0} f(t) = -7$ and $\lim_{t\to t_0} g(t) = 0$. Find the limit as $t \to t_0$ of the following functions.

(a) $3f(t)$ **(b)** $(f(t))^2$

(c) $f(t)\cdot g(t)$ **(d)** $\dfrac{f(t)}{g(t)-7}$

(e) $\cos(g(t))$ **(f)** $|f(t)|$

(g) $f(t)+g(t)$ **(h)** $1/f(t)$

4. Suppose that $f(x)$ and $g(x)$ are defined for all x and that $\lim_{x\to 0} f(x) = 1/2$ and $\lim_{x\to 0} g(x) = \sqrt{2}$. Find the limits as $x \to 0$ of the following functions.

(a) $-g(x)$ **(b)** $g(x)\cdot f(x)$

(c) $f(x)+g(x)$ **(d)** $1/f(x)$

(e) $x + f(x)$ **(f)** $\dfrac{f(x)\cdot\cos x}{x-1}$

In Exercises 5 and 6, find the value that $\lim_{x\to 0} g(x)$ must have if the given limit statements hold.

5. $\lim_{x\to 0}\left(\frac{4-g(x)}{x}\right)=1$

6. $\lim_{x\to -4}\left(x\lim_{x\to 0} g(x)\right)=2$

7. On what intervals are the following functions continuous?

(a) $f(x)=x^{1/3}$ **(b)** $g(x)=x^{3/4}$

(c) $h(x)=x^{-2/3}$ **(d)** $k(x)=x^{-1/6}$

8. On what intervals are the following functions continuous?

(a) $f(x)=\tan x$ **(b)** $g(x)=\csc x$

(c) $h(x)=e^{-x}$ **(d)** $k(x)=\dfrac{\sin x}{x}$

Finding Limits

In Exercises 9–16, find the limit or explain why it does not exist.

9. $\lim \dfrac{x^2-4x+4}{x^3+5x^2-14x}$

(a) as $x\to 0$ **(b)** as $x\to 2$

10. $\lim \dfrac{x^2+x}{x^5+2x^4+x^3}$

(a) as $x\to 0$ **(b)** as $x\to -1$

11. $\lim_{x\to 1}\dfrac{1-\sqrt{x}}{1-x}$

12. $\lim_{x\to a}\dfrac{x^2-a^2}{x^4-a^4}$

13. $\lim_{h\to 0}\dfrac{(x+h)^2-x^2}{h}$

14. $\lim_{x\to 0}\dfrac{(x+h)^2-x^2}{h}$

15. $\lim_{x\to 0}\dfrac{\frac{1}{2+x}-\frac{1}{2}}{x}$

16. $\lim_{x\to 0}\dfrac{(2+x)^3-8}{x}$

Find the limits in Exercises 17–28.

17. $\lim_{x\to\infty}\dfrac{2x+3}{5x+7}$

18. $\lim_{x\to -\infty}\dfrac{2x^2+3}{5x^2+7}$

19. $\lim_{x\to -\infty}\dfrac{x^2-4x+8}{3x^3}$

20. $\lim_{x\to\infty}\dfrac{1}{x^2-7x+1}$

21. $\lim_{x\to -\infty}\dfrac{x^2-7x}{x+1}$

22. $\lim_{x\to\infty}\dfrac{x^4+x^3}{12x^3+128}$

23. $\lim_{x\to\infty}\dfrac{\sin x}{\operatorname{int} x}$ (If you have a grapher, try graphing the function for $-5\le x\le 5$.)

24. $\lim_{\theta\to\infty}\dfrac{\cos\theta-1}{\theta}$ (If you have a grapher, try graphing $f(x)=x(\cos(1/x)-1)$ near the origin to "see" the limit at infinity.)

25. $\lim_{x\to\infty}\dfrac{x+\sin x+2\sqrt{x}}{x+\sin x}$

26. $\lim_{x\to\infty}\dfrac{x^{2/3}+x^{-1}}{x^{2/3}+\cos^2 x}$

27. $\lim_{x\to\infty} e^{-x^2}$

28. $\lim_{x\to -\infty} e^{1/x}$

T 29. Let $f(x)=x^3-x-1$.

(a) Show that f has a zero between -1 and 2.

(b) Solve the equation $f(x)=0$ graphically with an error of magnitude at most 10^{-8}.

(c) It can be shown that the exact value of the solution in part (b) is

$$\left(\frac{1}{2}+\frac{\sqrt{69}}{18}\right)^{1/3}+\left(\frac{1}{2}-\frac{\sqrt{69}}{18}\right)^{1/3}$$

Evaluate this exact answer and compare it with the value you found in part (b).

T 30. Let $f(\theta)=\theta^3-2\theta+2$.

(a) Show that f has a zero between -2 and 0.

(b) Solve the equation $f(\theta)=0$ graphically with an error of magnitude at most 10^{-4}.

(c) It can be shown that the exact value of the solution in part (b) is

$$\left(\sqrt{\frac{19}{27}}-1\right)^{1/3}-\left(\sqrt{\frac{19}{27}}+1\right)^{1/3}$$

Evaluate this exact answer and compare it with the value you found in part (b).

Additional Exercises: Theory, Examples, Applications

T 1. *Assigning a value to 0^0* The rules of exponents tell us that $a^0=1$ if a is any number different from zero. They also tell us that $0^n=0$ if n is any positive number.

If we tried to extend these rules to include the case 0^0, we would get conflicting results. The first rule would say $0^0=1$, whereas the second would say $0^0=0$.

We are not dealing with a question of right or wrong here. Neither rule applies as it stands, so there is no contradiction. We could, in fact, define 0^0 to have any value we wanted as long as we could persuade others to agree.

What value would you like 0^0 to have? Here is an example that might help you to decide. (See Exercise 2 below for another example.)

(a) Calculate x^x for $x=0.1$, 0.01, 0.001, and so on as far as your calculator can go. Record the values you get. What pattern do you see?

(b) Graph the function $y=x^x$ for $0<x\le 1$. Even though the function is not defined for $x\le 0$, the graph will approach the y-axis from the right. Toward what y-value does it seem to be headed? Zoom in to further support your idea.

T **2.** *A reason you might want 0^0 to be something other than 0 or 1* As the number x increases through positive values, the numbers $1/x$ and $1/(\ln x)$ both approach zero. What happens to the number

$$f(x) = \left(\frac{1}{x}\right)^{1/(\ln x)}$$

as x increases? Here are two ways to find out.

(a) Evaluate f for $x = 10, 100, 1000$, and so on as far as your calculator can reasonably go. What pattern do you see?

(b) Graph f in a variety of graphing windows, including windows that contain the origin. What do you see? Trace the y-values along the graph. What do you find?

3. *Lorentz contraction* In relativity theory, the length of an object, say a rocket, appears to an observer to depend on the speed at which the object is traveling with respect to the observer. If the observer measures the rocket's length as L_0 at rest, then at speed v the length will appear to be

$$L = L_0\sqrt{1 - \frac{v^2}{c^2}}.$$

This equation is the Lorentz contraction formula. Here, c is the speed of light in a vacuum, about 3×10^8 m/sec. What happens to L as v increases? Find $\lim_{v\to c^-} L$. Why was the left-hand limit needed?

4. *Controlling the flow from a draining tank* Torricelli's law says that if you drain a tank like the one in the figure shown, the rate y at which water runs out is a constant times the square root of the water's depth x. The constant depends on the size and shape of the exit valve.

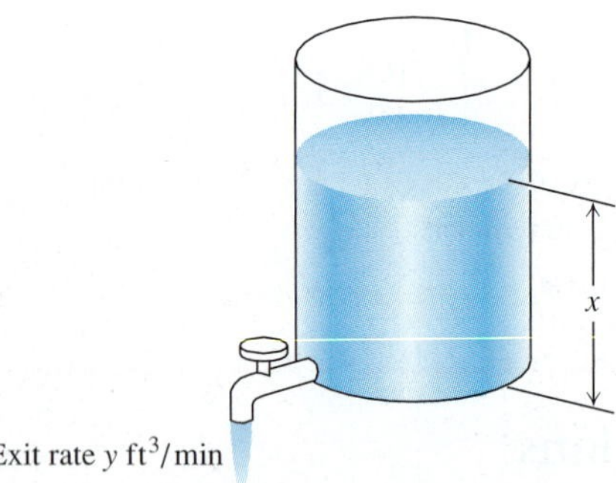

Suppose that $y = \sqrt{x}/2$ for a certain tank. You are trying to maintain a fairly constant exit rate by adding water to the tank with a hose from time to time. How deep must you keep the water if you want to maintain the exit rate

(a) within 0.2 ft^3/min of the rate $y_0 = 1$ ft^3/min?

(b) within 0.1 ft^3/min of the rate $y_0 = 1$ ft^3/min?

5. *Thermal expansion in precise equipment* As you may know, most metals expand when heated and contract when cooled. The dimensions of a piece of laboratory equipment are sometimes so critical that the shop where the equipment is made must be held at the same temperature as the laboratory where the equipment is to be used. A typical aluminum bar that is 10 cm wide at 70°F will be

$$y = 10 + (t - 70) \times 10^{-4}$$

centimeters wide at a nearby temperature t. Suppose that you are using a bar like this in a gravity wave detector, where its width must stay within 0.0005 cm of the ideal 10 cm. How close to $t_0 = 70$°F must you maintain the temperature to ensure that this tolerance is not exceeded?

6. *Writing to Learn: Antipodal points* Is there any reason to believe that there is always a pair of antipodal (diametrically opposite) points on Earth's equator where the temperatures are the same? Explain.

T **7.** *Roots of a quadratic equation that is almost linear* The equation $ax^2 + 2x - 1 = 0$, where a is a constant, has two roots if $a > -1$ and $a \neq 0$, one positive and one negative:

$$r_+(a) = \frac{-1 + \sqrt{1+a}}{a}, \qquad r_-(a) = \frac{-1 - \sqrt{1+a}}{a}.$$

(a) What happpens to $r_+(a)$ as $a \to 0$? As $a \to -1^+$?

(b) What happens to $r_-(a)$ as $a \to 0$? As $a \to -1^+$?

(c) Support your conclusions by graphing $r_+(a)$ and $r_-(a)$ as functions of a. Describe what you see.

(d) For added support, graph $f(x) = ax^2 + 2x - 1$ simultaneously for $a = 1, 0.5, 0.2, 0.1$, and 0.05.

8. *One-sided limits* If $\lim_{x\to 0^+} f(x) = A$ and $\lim_{x\to 0^-} f(x) = B$, find

(a) $\lim_{x\to 0^+} f(x^3 - x)$

(b) $\lim_{x\to 0^-} f(x^3 - x)$

(c) $\lim_{x\to 0^+} f(x^2 - x^4)$

(d) $\lim_{x\to 0^-} f(x^2 - x^4)$

9. *Limits and continuity* Which of the following statements are true, and which are false? If true, say why; if false, give a counterexample (that is, an example confirming the falsehood).

(a) If $\lim_{x\to a} f(x)$ exists but $\lim_{x\to a} g(x)$ does not exist, then $\lim_{x\to a} (f(x) + g(x))$ does not exist.

(b) If neither $\lim_{x\to a} (f(x)$ nor $\lim_{x\to a} g(x)$ exists, then $\lim_{x\to a} f(x) + g(x))$ does not exist.

(c) If f is continuous at x, then so is $|f|$.

(d) If $|f|$ is continuous at a, then so is f.

10. *Root of an equation* Show that the equation $x + 2\cos x = 0$ has at least one solution.

Formal Definition of Limit

In Exercises 11–14, use the formal definition of limit to prove that the function is continuous at x_0.

11. $f(x) = x^2 - 7, \quad x_0 = 1$

12. $g(x) = 1/(2x), \quad x_0 = 1/4$

13. $h(x) = \sqrt{2x - 3}, \quad x_0 = 2$

14. $F(x) = \sqrt{9 - x}, \quad x_0 = 5$

15. *A function continuous at only one point* Let

$$f(x) = \begin{cases} x & \text{if } x \text{ is rational} \\ 0 & \text{if } x \text{ is irrational.} \end{cases}$$

(a) Show that f is continuous at $x = 0$.

(b) Use the fact that every nonempty open interval of real numbers contains both rational and irrational numbers to show that f is not continuous at any nonzero value of x.

16. *The Dirichlet ruler function* If x is a rational number, then x can be written in a unique way as a quotient of integers m/n where $n > 0$ and m and n have no common factors greater than 1. (We say that such a fraction is in *lowest terms*. For example, $6/4$ written in lowest terms is $3/2$.) Let $f(x)$ be defined for all x in the interval $[0, 1]$ by

$$f(x) = \begin{cases} 1/n & \text{if } x = m/n \text{ is a rational number in lowest terms} \\ 0 & \text{if } x \text{ is irrational.} \end{cases}$$

For instance, $f(0) = f(1) = 1$, $f(1/2) = 1/2$, $f(1/3) = f(2/3) = 1/3$, $f(1/4) = f(3/4) = 1/4$, and so on.

(a) Show that f is discontinuous at every rational number in $[0, 1]$.

(b) Show that f is continuous at every irrational number in $[0, 1]$. (*Hint:* If ϵ is a given positive number, show that there are only finitely many rational numbers r in $[0, 1]$ such that $f(r) \geq \epsilon$.)

(c) Sketch the graph of f. Why do you think f is called the "ruler function"?

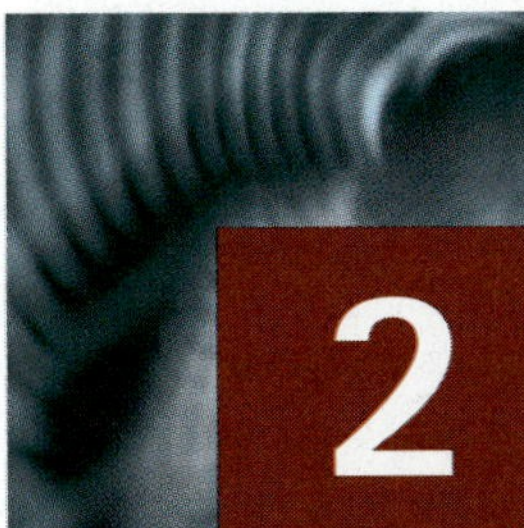

2 Derivatives

OVERVIEW In Chapter 1, we defined the slope of a curve at a point as the limit of secant slopes. This limit, called a derivative, measures the rate at which a function changes, and it is one of the most important ideas in calculus. Derivatives are used widely in engineering, science, economics, medicine, and computer science to calculate velocity and acceleration, to explain the behavior of machinery, to estimate the drop in water level as water is pumped out of a tank, and to predict the consequences of making errors in measurements. Finding derivatives by evaluating limits can be lengthy and difficult. In this chapter, we develop techniques to make calculating derivatives easier.

2.1 The Derivative as a Function

Definition of Derivative • Notation • Derivatives of Constants, Powers, Multiples, and Sums • Differentiable on an Interval; One-Sided Derivatives • Graphing f' from Estimated Values • Differentiable Functions are Continuous • Intermediate Value Property of Derivatives • Second- and Higher-Order Derivatives

CD-ROM
WEBsite
Historical Essay
The Derivative

At the end of Chapter 1, we defined the slope of a curve $y = f(x)$ at the point where $x = x_0$ to be

$$\lim_{h \to 0} \frac{f(x_0 + h) - f(x_0)}{h}.$$

We called this limit, when it existed, the derivative of f at x_0. We now investigate the derivative as a *function* derived from f by considering the limit at each point of the domain of f.

CD-ROM
WEBsite

Definition of Derivative

Definition Derivative Function

The **derivative** of the function $f(x)$ with respect to the variable x is the function f' whose value at x is

$$f'(x) = \lim_{h \to 0} \frac{f(x + h) - f(x)}{h},$$

provided the limit exists.

Calculating $f'(x)$ from the Definition of Derivative

Step 1. Write expressions for $f(x)$ and $f(x+h)$.

Step 2. Expand and simplify the difference quotient

$$\frac{f(x+h)-f(x)}{h}.$$

Step 3. Using the simplified quotient, find $f'(x)$ by evaluating the limit

$$f'(x)=\lim_{h\to 0}\frac{f(x+h)-f(x)}{h}.$$

The domain of f' is the set of points in the domain of f for which the limit exists. Its domain may be the same as the domain of f or it may be smaller. If f' exists at a particular x, we say that f is **differentiable (has a derivative)** at x. If f' exists at every point of the domain of f, we call f **differentiable.**

CD-ROM WEBsite

Example 1 Applying the Definition

(a) Find the derivative of $y=\sqrt{x}$ for $x>0$.

(b) Find the tangent line to the curve $y=\sqrt{x}$ at $x=4$.

Solution

(a)

Step 1: $f(x)=\sqrt{x}$ and $f(x+h)=\sqrt{x+h}$

Step 2:
$$\frac{f(x+h)-f(x)}{h}=\frac{\sqrt{x+h}-\sqrt{x}}{h}$$
$$=\frac{(x+h)-x}{h(\sqrt{x+h}+\sqrt{x})}\qquad \text{Multiply by } \frac{\sqrt{x+h}+\sqrt{x}}{\sqrt{x+h}+\sqrt{x}}.$$
$$=\frac{1}{\sqrt{x+h}+\sqrt{x}}$$

Step 3: $f'(x)=\lim_{h\to 0}\dfrac{1}{\sqrt{x+h}+\sqrt{x}}=\dfrac{1}{2\sqrt{x}}$

See Figure 2.1.

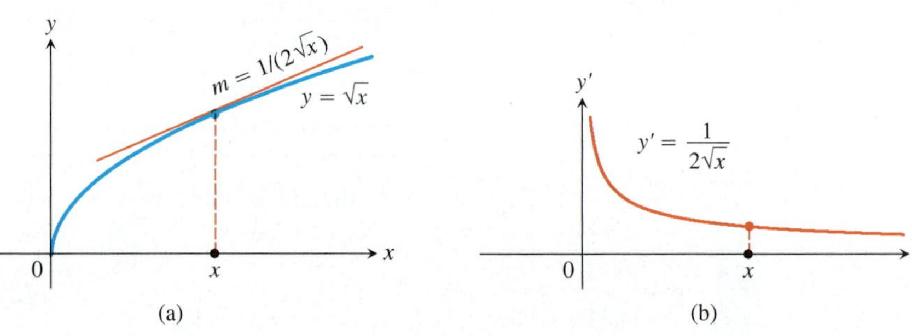

FIGURE 2.1 The graphs of (a) $y=\sqrt{x}$ and (b) $y'=1/(2\sqrt{x})$, $x>0$. The function is defined at $x=0$, but its derivative is not. (Example 1)

(b) The slope of the curve at $x=4$ is

$$f'(4)=\frac{1}{2\sqrt{4}}=\frac{1}{4}.$$

The tangent is the line through the point (4, 2) with slope 1/4 (Figure 2.2).

$$y=2+\frac{1}{4}(x-4)$$
$$y=\frac{1}{4}x+1$$

FIGURE 2.2 The curve $y=\sqrt{x}$ and its tangent at (4, 2). The tangent's slope is found by evaluating y' at $x=4$. (Example 1)

Notation

There are many ways to denote the derivative of a function $y = f(x)$. Besides $f'(x)$, the most common notations are these:

y'	"y prime"	Nice and brief but does not name the independent variable
$\frac{dy}{dx}$	"$dy\ dx$"	Names the variables and uses d for derivative
$\frac{df}{dx}$	"$df\ dx$"	Emphasizes the function's name
$\frac{d}{dx}f(x)$	"ddx of $f(x)$"	Emphasizes the idea that differentiation is an operation performed on f (Figure 2.3)

The "prime" notation comes from Newton's work, the d/dx notation from that of Leibniz.

We also read dy/dx as "the derivative of y with respect to x" and df/dx and $(d/dx)f(x)$ as "the derivative of f with respect to x."

The value

$$f'(a) = \lim_{h \to 0} \frac{f(a+h) - f(a)}{h}$$

of the derivative of $y = f(x)$ with respect to x at $x = a$ can also be denoted as

$$y'|_{x=a} \quad \text{or} \quad \left.\frac{dy}{dx}\right|_{x=a} \quad \text{or} \quad \left.\frac{d}{dx}f(x)\right|_{x=a}.$$

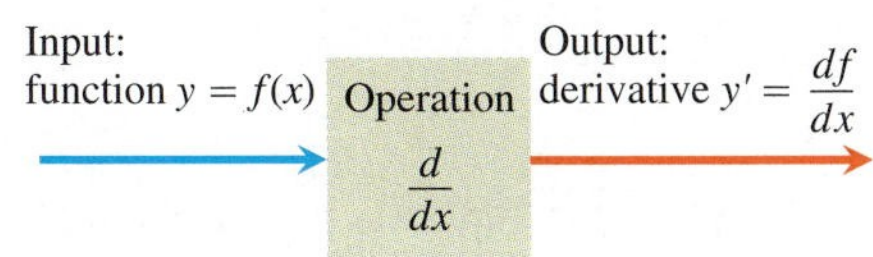

FIGURE 2.3 Flow diagram for the operation of taking a derivative with respect to x.

The symbol $|_{x=a}$, called an **evaluation symbol,** tells us to evaluate the expression to its left at $x = a$.

The process of calculating a derivative is called **differentiation.** Example 1 illustrates the process for the function $y = \sqrt{x}$. We now show how to differentiate functions without having to apply the definition each time.

Derivatives of Constants, Powers, Multiples, and Sums

The first rule is that the derivative of every constant function is the zero function.

Rule 1 Derivative of a Constant Function

If f has the constant value $f(x) = c$, then

$$\frac{df}{dx} = \frac{d}{dx}(c) = 0.$$

Example 2 Using Rule 1

If f has the constant value $f(x) = 8$, then

$$\frac{df}{dx} = \frac{d}{dx}(8) = 0.$$

Similarly,

$$\frac{d}{dx}\left(-\frac{\pi}{2}\right) = 0 \quad \text{and} \quad \frac{d}{dx}\left(\sqrt{3}\right) = 0.$$

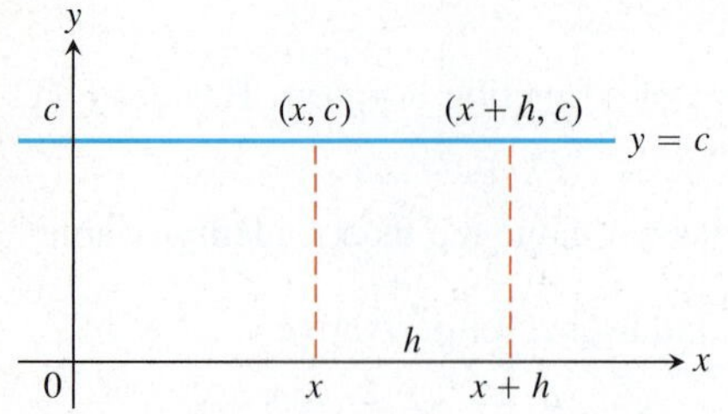

FIGURE 2.4 The rule $(d/dx)(c) = 0$ is another way to say that the values of constant functions never change and that the slope of a horizontal line is zero at every point.

Proof of Rule 1 We apply the definition of derivative to $f(x) = c$, the function whose outputs have the constant value c (Figure 2.4). At every value of x, we find that

$$f'(x) = \lim_{h \to 0} \frac{f(x+h) - f(x)}{h} = \lim_{h \to 0} \frac{c - c}{h} = \lim_{h \to 0} 0 = 0.$$

The second rule tells how to differentiate x^n if n is a positive integer.

Rule 2 Power Rule for Positive Integers

If n is a positive integer, then

$$\frac{d}{dx} x^n = nx^{n-1}.$$

To apply the Power Rule, we subtract 1 from the original exponent (n) and multiply the result by n.

Example 3 Interpreting Rule 2

f	x	x^2	x^3	x^4	...
f'	1	$2x$	$3x^2$	$4x^3$	...

Proof of Rule 2 If $f(x) = x^n$, then $f(x+h) = (x+h)^n$. Since n is a positive integer, we can use the fact that

$$a^n - b^n = (a - b)(a^{n-1} + a^{n-2}b + \cdots + ab^{n-2} + b^{n-1})$$

to simplify the difference quotient for f. Taking $x + h = a$ and $x = b$, we have $a - b = h$. Thus,

$$\begin{aligned} \frac{f(x+h) - f(x)}{h} &= \frac{(x+h)^n - x^n}{h} \\ &= \frac{(h)[(x+h)^{n-1} + (x+h)^{n-2}x + \cdots + (x+h)x^{n-2} + x^{n-1}]}{h} \\ &= \underbrace{(x+h)^{n-1} + (x+h)^{n-2}x + \cdots + (x+h)x^{n-2} + x^{n-1}}_{n \text{ terms, each with limit } x^{n-1} \text{ as } h \to 0}. \end{aligned}$$

Hence,

$$\frac{d}{dx} x^n = \lim_{h \to 0} \frac{f(x+h) - f(x)}{h} = nx^{n-1}.$$

The third rule says that when a differentiable function is multiplied by a constant, its derivative is multiplied by the same constant.

Rule 3 Constant Multiple Rule

If u is differentiable function of x, and c is a constant, then

$$\frac{d}{dx}(cu) = c\frac{du}{dx}.$$

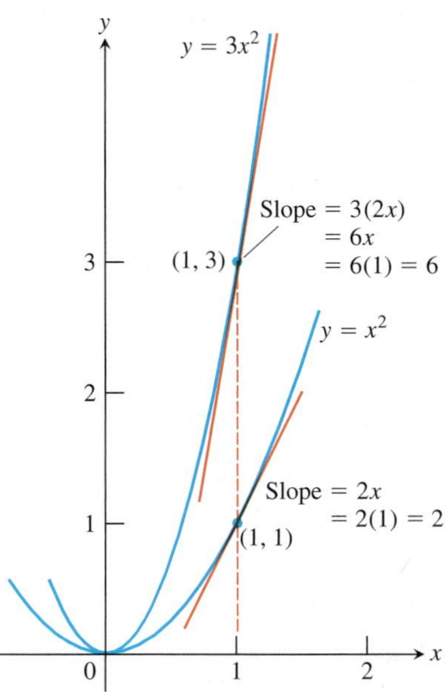

FIGURE 2.5 The graphs of $y = x^2$ and $y = 3x^2$. Tripling the y-coordinates triples the slope. (Example 4)

Denoting functions by u and v

The functions we are working with when we need a differentiation formula are likely to be denoted by letters like f and g. When we apply the formula, we do not want to find it using these same letters in some other way. To guard against this problem, we denote the functions in differentiation rules by letters like u and v that are not likely to be already in use.

Example 4 Using Rule 3

(a) $\dfrac{d}{dx}(3x^2) = 3 \cdot 2x = 6x$

Interpretation: Rescaling the graph of $y = x^2$ by multiplying each y-coordinate by 3 multiplies the slope at each point by 3 (Figure 2.5).

(b) *A useful special case:* The derivative of the negative of a differentiable function is the negative of the function's derivative. Rule 3 with $c = -1$ gives

$$\frac{d}{dx}(-u) = \frac{d}{dx}(-1 \cdot u) = -1 \cdot \frac{d}{dx}(u) = -\frac{du}{dx}.$$

Proof of Rule 3

$$\begin{aligned} \frac{d}{dx} cu &= \lim_{h \to 0} \frac{cu(x+h) - cu(x)}{h} && \text{Derivative definition with } f(x) = cu(x) \\ &= c \lim_{h \to 0} \frac{u(x+h) - u(x)}{h} && \text{Limit property} \\ &= c \frac{du}{dx} && u \text{ is differentiable.} \end{aligned}$$

The next rule says that the derivative of the sum of two differentiable functions is the sum of their derivatives.

Rule 4 Derivative Sum Rule

If u and v are differentiable functions of x, then their sum $u + v$ is differentiable at every point where u and v are both differentiable. At such points,

$$\frac{d}{dx}(u + v) = \frac{du}{dx} + \frac{dv}{dx}.$$

Example 5 Derivative of a Sum

$$\begin{aligned} y &= x^4 + 12x \\ \frac{dy}{dx} &= \frac{d}{dx}(x^4) + \frac{d}{dx}(12x) \\ &= 4x^3 + 12 \end{aligned}$$

Proof of Rule 4 We apply the definition of derivative to $f(x) = u(x) + v(x)$:

$$\begin{aligned} \frac{d}{dx}[u(x) + v(x)] &= \lim_{h \to 0} \frac{[u(x+h) + v(x+h)] - [u(x) + v(x)]}{h} \\ &= \lim_{h \to 0} \left[\frac{u(x+h) - u(x)}{h} + \frac{v(x+h) - v(x)}{h} \right] \\ &= \lim_{h \to 0} \frac{u(x+h) - u(x)}{h} + \lim_{h \to 0} \frac{v(x+h) - v(x)}{h} = \frac{du}{dx} + \frac{dv}{dx}. \end{aligned}$$

Combining the Sum Rule with the Constant Multiple rule gives the equivalent **Difference Rule,** which says that the derivative of a *difference* of differentiable functions is the difference of their derivatives.

$$\frac{d}{dx}(u - v) = \frac{d}{dx}[u + (-1)v] = \frac{du}{dx} + (-1)\frac{dv}{dx} = \frac{du}{dx} - \frac{dv}{dx}.$$

The Sum Rule also extends to sums of more than two functions, as long as there are only finitely many functions in the sum. If $u_1, u_2, \ldots, u_n$ are differentiable at x, then so is $u_1 + u_2 + \cdots + u_n$, and

$$\frac{d}{dx}(u_1 + u_2 + \cdots + u_n) = \frac{du_1}{dx} + \frac{du_2}{dx} + \cdots + \frac{du_n}{dx}.$$

Example 6 Derivative of a Polynomial

$$\begin{aligned} y &= x^3 + \frac{4}{3}x^2 - 5x + 1 \\ \frac{dy}{dx} &= \frac{d}{dx}x^3 + \frac{d}{dx}\left(\frac{4}{3}x^2\right) - \frac{d}{dx}(5x) + \frac{d}{dx}(1) \\ &= 3x^2 + \frac{4}{3}\cdot 2x - 5 + 0 \\ &= 3x^2 + \frac{8}{3}x - 5 \end{aligned}$$

All polynomials are differentiable.

Notice that we can differentiate any polynomial term by term, the way we differentiated the polynomials in Example 6.

Example 7 Finding Horizontal Tangents

Does the curve $y = x^4 - 2x^2 + 2$ have any horizontal tangents? If so, where?

Solution The horizontal tangents, if any, occur where the slope dy/dx is zero. To find these points, we follow these steps.

1. Calculate dy/dx:

$$\frac{dy}{dx} = \frac{d}{dx}(x^4 - 2x^2 + 2) = 4x^3 - 4x.$$

2. Solve the equation $\dfrac{dy}{dx} = 0$ for x:

$$\begin{aligned} 4x^3 - 4x &= 0 \\ 4x(x^2 - 1) &= 0 \\ x &= 0, 1, -1. \end{aligned}$$

The curve $y = x^4 - 2x^2 + 2$ has horizontal tangents at $x = 0$, 1, and -1. The corresponding points on the curve are (0, 2), (1, 1) and $(-1, 1)$. See Figure 2.6.

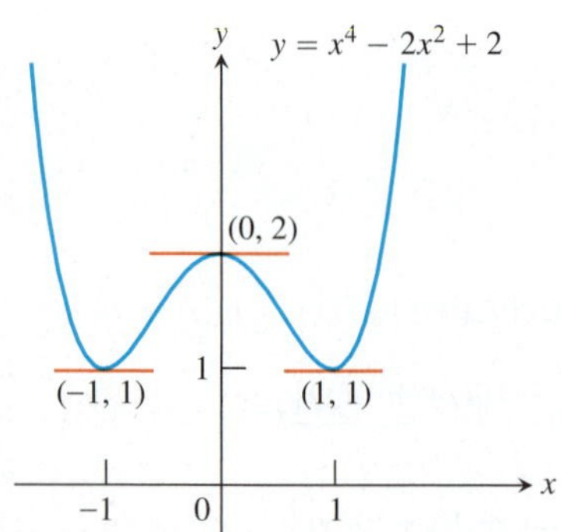

FIGURE 2.6 The curve $y = x^4 - 2x^2 + 2$ and its horizontal tangents. (Example 7)

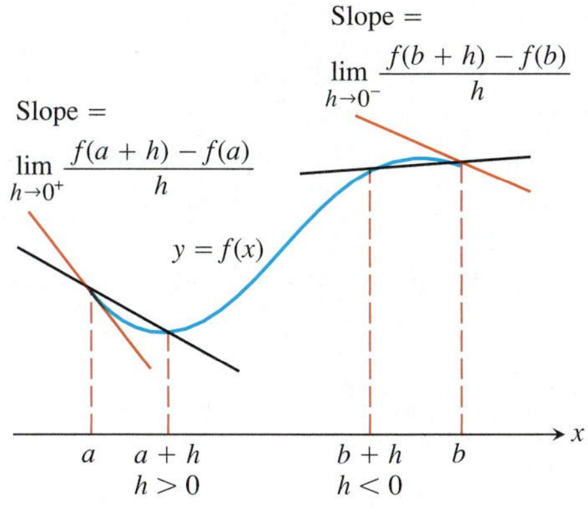

FIGURE 2.7 Derivatives at endpoints are one-sided limits.

Differentiable on an Interval; One-Sided Derivatives

A function $y = f(x)$ is **differentiable** on an open interval (finite or infinite) if it has a derivative at each point of the interval. It is differentiable on a closed interval $[a, b]$ if it is differentiable on the interior (a, b) and if the limits

$$\lim_{h\to 0^+} \frac{f(a+h)-f(a)}{h} \qquad \textbf{Right-hand derivative at } a$$

$$\lim_{h\to 0^-} \frac{f(b+h)-f(b)}{h} \qquad \textbf{Left-hand derivative at } b$$

exist at the endpoints (Figure 2.7).

Right-hand and left-hand derivatives may be defined at any point of a function's domain. The usual relation between one-sided and two-sided limits holds for these derivatives. Because of Theorem 5, Section 1.2, a function has a derivative at a point if and only if it has left-hand and right-hand derivatives there, and these one-sided derivatives are equal.

CD-ROM WEBsite

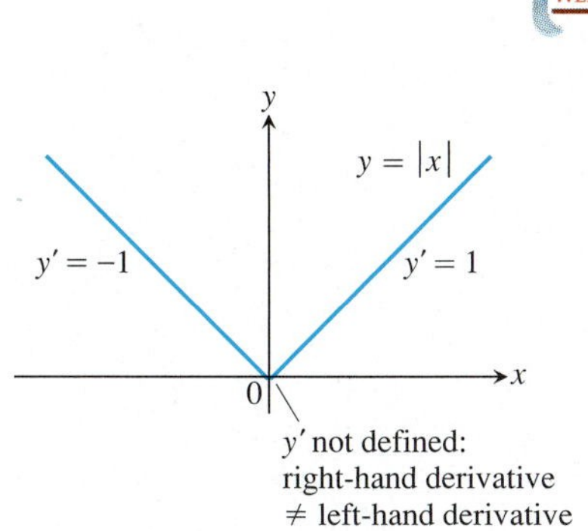

FIGURE 2.8 The function $y = |x|$ is not differentiable at the origin where the graph has a "corner."

Example 8 $y = |x|$ is Not Differentiable at the Origin

Show that the function $y = |x|$ is differentiable on $(-\infty, 0)$ and $(0, \infty)$ but has no derivative at $x = 0$.

Solution To the right of the origin,

$$\frac{d}{dx}(|x|) = \frac{d}{dx}(x) = \frac{d}{dx}(1 \cdot x) = 1. \qquad \frac{d}{dx}(mx+b) = m,\ |x| = x$$

To the left,

$$\frac{d}{dx}(|x|) = \frac{d}{dx}(-x) = \frac{d}{dx}(-1 \cdot x) = -1 \qquad |x| = -x$$

(Figure 2.8). There can be no derivative at the origin because the one-sided derivatives differ there:

$$\begin{aligned}
\text{Right-hand derivative of } |x| \text{ at zero} &= \lim_{h\to 0^+} \frac{|0+h|-|0|}{h} = \lim_{h\to 0^+} \frac{|h|}{h} \\
&= \lim_{h\to 0^+} \frac{h}{h} \qquad |h| = h \text{ when } h > 0. \\
&= \lim_{h\to 0^+} 1 = 1 \\
\text{Left-hand derivative of } |x| \text{ at zero} &= \lim_{h\to 0^-} \frac{|0+h|-|0|}{h} = \lim_{h\to 0^-} \frac{|h|}{h} \\
&= \lim_{h\to 0^-} \frac{-h}{h} \qquad |h| = -h \text{ when } h < 0. \\
&= \lim_{h\to 0^-} -1 = -1.
\end{aligned}$$

In general, if the graph of a function has a "corner," then there is no tangent at this point and f is not differentiable there. Thus, differentiability is a "smoothness" condition.

Graphing f′ from Estimated Values

When we measure the values of a function $y = f(x)$ in the laboratory or in the field (pressure versus temperature, say, or population versus time), we usually connect the data points with lines or curves to picture the graph of f. We can often make a reasonable plot of f' by estimating slopes on this graph. The following example shows how this is done and what can be learned from the process.

Example 9 Graphing a Derivative

Graph the derivative of the function $y = f(x)$ in Figure 2.9a.

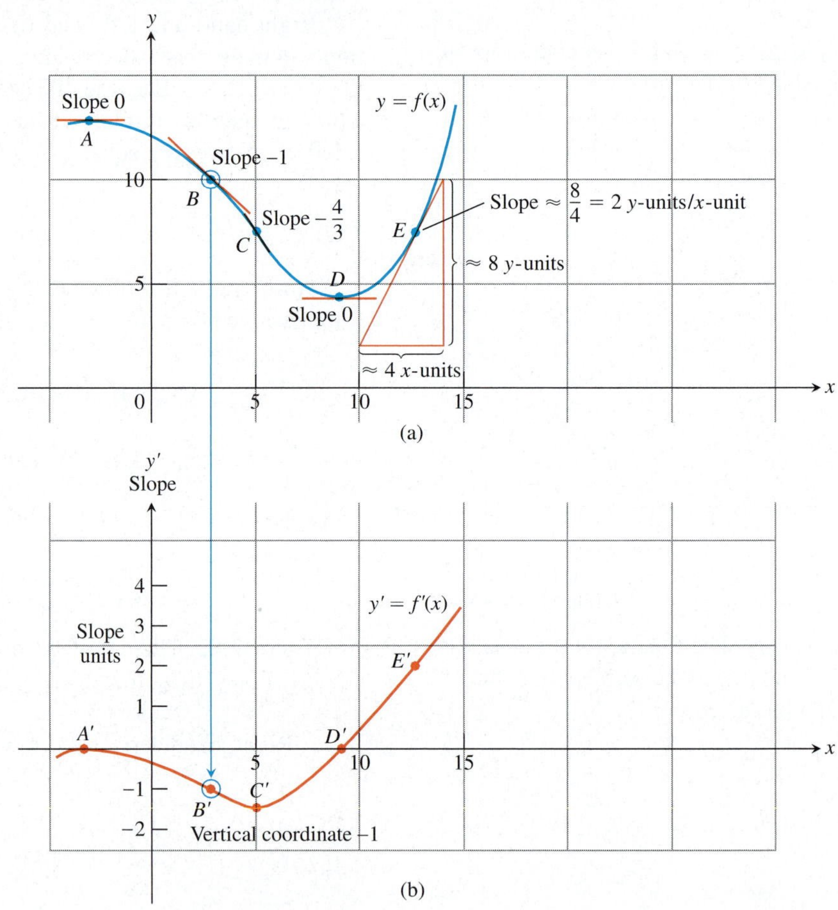

FIGURE 2.9 We made the graph of $y' = f'(x)$ in (b) by plotting slopes from the graph of $y = f(x)$ in (a). The vertical coordinate of B' is the slope at B and so on. The graph of $y' = f'(x)$ is a visual record of how the slope of f changes with x.

Solution We draw a pair of axes, marking the horizontal axis in x-units and the vertical axis in y'-units (Figure 2.9b). Next we sketch tangents to the graph of f at frequent intervals and use their slopes to estimate the values of $y' = f'(x)$ at these points. We plot the corresponding (x, y') pairs and connect them with a smooth curve.

From the graph of $y' = f'(x)$ we see at a glance

1. where the rate of change of f is positive, negative, or zero
2. the rough size of the growth rate at any x and its size in relation to the size of $f(x)$
3. where the rate of change itself is increasing or decreasing.

You can experiment further with these ideas in Exercises 15 through 20.

Differentiable Functions Are Continuous

A function is continuous at every point where it has a derivative.

CAUTION The converse of Theorem 1 is false. A function need not have a derivative at a point where it is continuous, as we saw in Example 8.

Theorem 1 Differentiability Implies Continuity
If f has a derivative at $x = c$, then f is continuous at $x = c$.

Proof Given that $f'(c)$ exists, we must show that $\lim_{x \to c} f(x) = f(c)$, or equivalently, that $\lim_{h \to 0} f(c+h) = f(c)$. If $h \neq 0$, then

$$\begin{aligned} f(c+h) &= f(c) + (f(c+h) - f(c)) \\ &= f(c) + \frac{f(c+h) - f(c)}{h} \cdot h. \end{aligned}$$

Now take limits as $h \to 0$. By Theorem 1 of Section 1.2,

$$\begin{aligned} \lim_{h \to 0} f(c+h) &= \lim_{h \to 0} f(c) + \lim_{h \to 0} \frac{f(c+h) - f(c)}{h} \cdot \lim_{h \to 0} h \\ &= f(c) + f'(c) \cdot 0 \\ &= f(c) + 0 \\ &= f(c). \end{aligned}$$

Similar arguments with one-sided limits show that if f has a derivative from one side (right or left) at $x = c$ then f is continuous from that side at $x = c$.

Theorem 1 gives another reason why a function may fail to have a derivative. If the function has a discontinuity at a point (for instance, a jump discontinuity), then it cannot be differentiable there. For instance, the greatest integer function $y = \text{int}\, x$ fails to be differentiable at every integer $x = n$ (Example 4, Section 1.4).

Intermediate Value Property of Derivatives

Not every function can be some function's derivative, as we see from the following theorem.

Theorem 2 Intermediate Value Property of Derivatives
If a and b are any two points in an interval on which f is differentiable, then f' takes on every value between $f'(a)$ and $f'(b)$.

Theorem 2 (which we will not prove) says that a function cannot *be* a derivative on an interval unless it has the Intermediate Value Property there (Figure 2.10). The question of when a function *is* a derivative is one of the central questions in all calculus, and Newton's and Leibniz's answer to this question revolutionized the world of mathematics. We see what their answer was in Chapter 4.

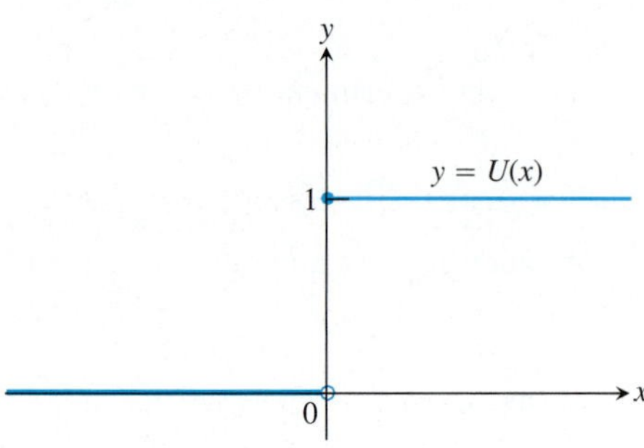

FIGURE 2.10 The unit step function does not have the Intermediate Value Property and cannot be the derivative of a function on the real line.

Second- and Higher-Order Derivatives

Notice that

$$\frac{d}{dx}\left(\frac{dy}{dx}\right)$$

does not mean multiplication. It means "the derivative of the derivative."

The derivative $y' = dy/dx$ is the **first (first-order) derivative** of y with respect to x. This derivative may itself be a differentiable function of x; if so, its derivative

$$y'' = \frac{dy'}{dx} = \frac{d}{dx}\left(\frac{dy}{dx}\right) = \frac{d^2y}{dx^2}$$

is called the **second (second-order) derivative** of y with respect to x.

If y'' is differentiable, its derivative, $y''' = dy''/dx = d^3y/dx^3$ is the **third (third-order) derivative** of y with respect to x. The names continue as you imagine, with

$$y^{(n)} = \frac{d}{dx}y^{(n-1)}$$

denoting the ***n*th (*n*th-order) derivative** of y with respect to x for any positive integer n.

We can interpret the second derivative as the rate of change of the slope of the tangent to the curve $y = f(x)$ at each point. You will see in the next chapter that the second derivative reveals whether a curve bends upward or downward from the tangent line as we move off the point of tangency. In the next section, we interpret both the second and third derivatives in terms of motion along a straight line.

How to read the symbols for derivatives

y'	"y prime"
y''	"y double prime"
$\frac{d^2y}{dx^2}$	"d squared y dx squared"
y'''	"y triple prime"
$y^{(n)}$	"y super n"
$\frac{d^ny}{dx^n}$	"d to the n of y by dx to the n"

Example 10 Finding Higher Derivatives

The first four derivatives of $y = x^3 - 3x^2 + 2$ are

First derivative:	$y' = 3x^2 - 6x$
Second derivative:	$y'' = 6x - 6$
Third derivative:	$y''' = 6$
Fourth derivative:	$y^{(4)} = 0.$

The function has derivatives of all orders, the fifth and later derivatives all being zero.

EXERCISES 2.1

Finding and Evaluating Derivative Functions

In Exercises 1–6, use the definition to find the function's derivative. Then evaluate the derivative at the indicated point or points.

1. $f(x) = 4 - x^2, \quad f'(-3), \quad f'(0)$

2. $g(t) = 1/t^2; \quad g'(-1), \quad g'(2)$

3. $\left.\dfrac{ds}{dt}\right|_{t=-1}$ if $s = t^3 - t^2$

4. $f(x) = x + \dfrac{9}{x}, \quad x = -3$

5. $p(\theta) = \sqrt{3\theta}, \quad \theta = 0.25$

6. $\left.\dfrac{dr}{d\theta}\right|_{\theta=0}$ if $r = \dfrac{2}{\sqrt{4 - \theta}}$

Derivative Calculations

In Exercises 7–10, find the first and second derivatives.

7. $y = x^2 + x + 8$

8. $s = 5t^3 - 3t^5$

9. $y = \dfrac{4x^3}{3} - 4$

10. $y = \dfrac{x^3 + 7}{x}$

Find the derivatives of all orders of the functions in Exercises 11 and 12.

11. $y = \dfrac{x^4}{2} - \dfrac{3}{2}x^2 - x$

12. $y = \dfrac{x^5}{120}$

Slopes and Tangents

13. (a) Find an equation for the tangent to the curve $y = x^3 - 4x + 1$ at the point (2, 1).

(b) What is the range of values of the curve's slope?

(c) Find equations for the tangents to the curve at the points where the slope of the curve is 8.

14. (a) Find an equation for the horizontal tangent to the curve $y = x - 3\sqrt{x}$.

(b) What is the range of values of the curve's slope?

Graphs

Match the functions graphed in Exercise 15–18 with the derivatives graphed in Figure 2.11.

15.

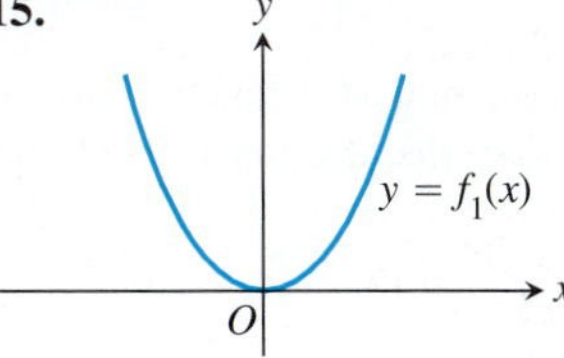

16.

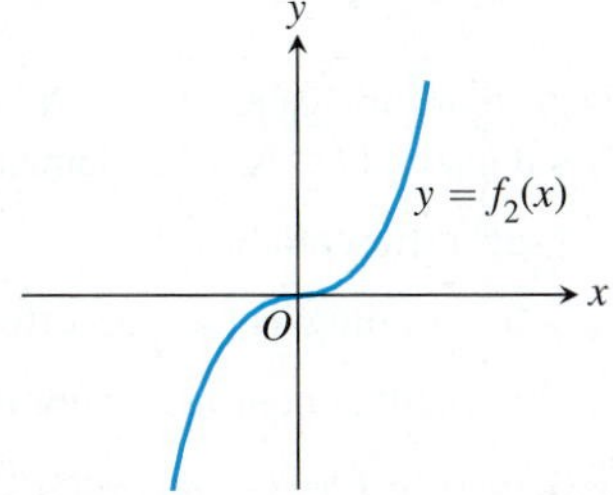

17. **18.**

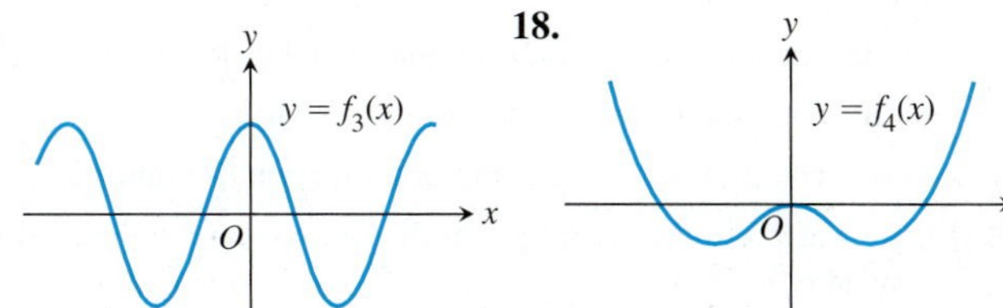

(a)

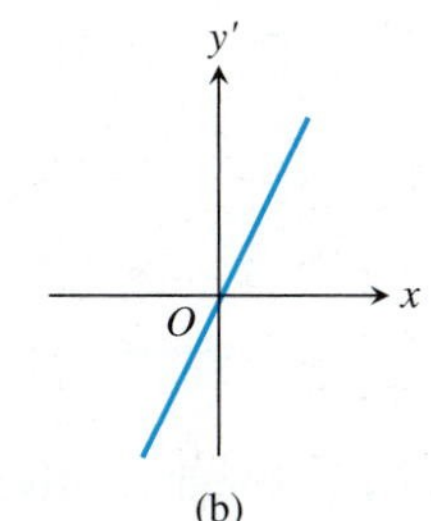

(b)

(c)

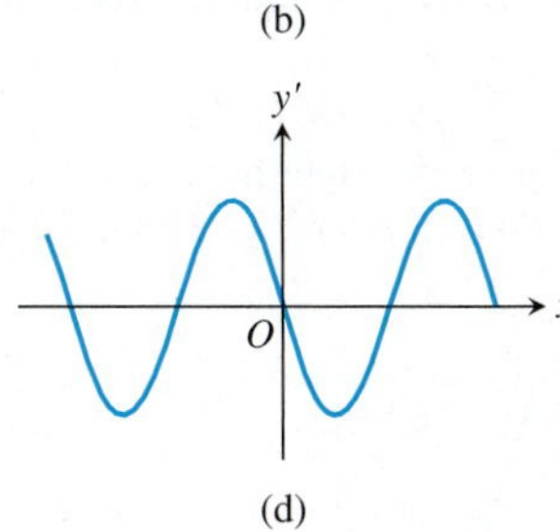

(d)

FIGURE 2.11 The derivative graphs for Exercises 15–18.

19. *Writing to Learn*

(a) The graph in Figure 2.12 is made of line segments joined end to end. At which points of the interval $[-4, 6]$ is f' not defined? Give reasons for your answer.

(b) Graph the derivative of f. Call the vertical axis the y'-axis. The graph should show a step function.

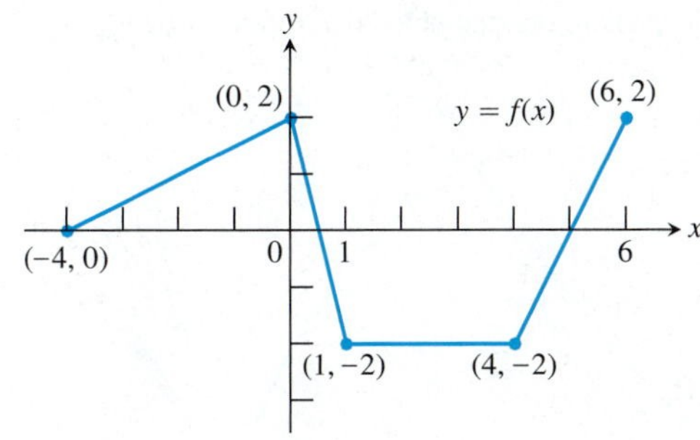

FIGURE 2.12 The graph for Exercise 19.

Recovering a Function from Its Derivative

20. (a) Use the following information to graph the function f over the closed interval $[-2, 5]$.

i. The graph of f is made of line segments joined end to end.

ii. The graph starts at the point $(-2, 3)$.

iii. The derivative of f is the step function in Figure 2.13.

(b) Repeat part (a) assuming that the graph starts at $(-2, 0)$ instead of $(-2, 3)$.

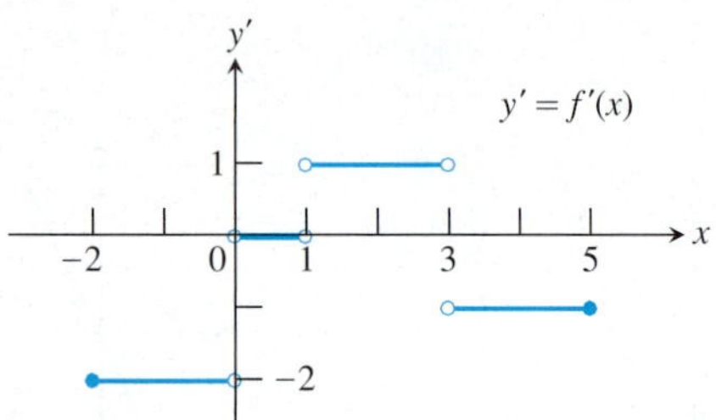

FIGURE 2.13 The derivative graph for Exercise 20.

One-Sided Derivatives

Compare the right-hand and left-hand derivatives to show that the functions in Exercises 21 and 22 are not differentiable at the point P.

21. **22.**

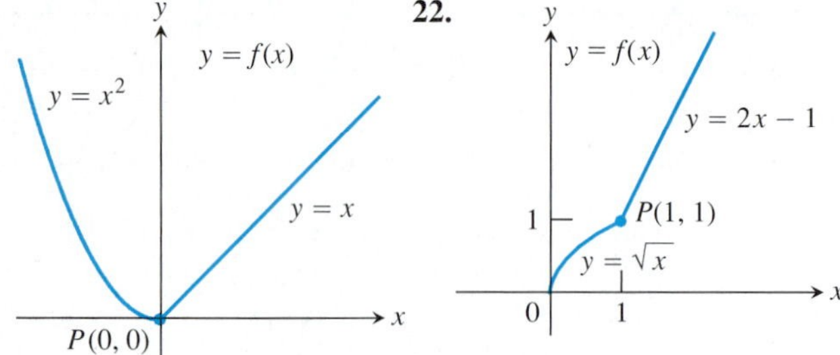

When Does a Function Not Have a Derivative at a Point?

A function has a derivative at a point x_0 if the slopes of the secant lines through $P(x_0, f(x_0))$ and a nearby point Q on the graph approach a limit as Q approaches P. Whenever the secants fail to take up a single limiting position or become vertical as Q approaches P, the derivative does not exist. A function whose graph is otherwise smooth will fail to have a derivative at a point where the graph has

(i) a *corner*, where the one-sided derivatives differ

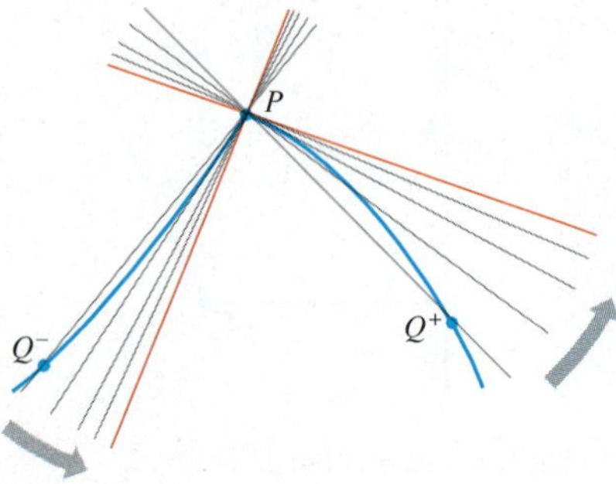

(ii) a *cusp*, where the slope of PQ approaches ∞ from one side and $-\infty$ from the other

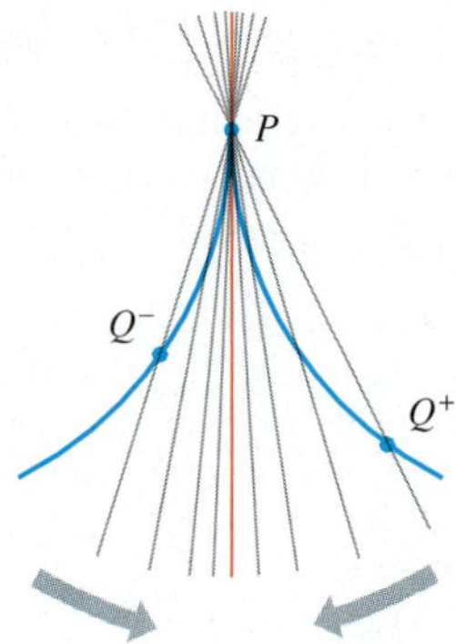

(iii) a *vertical tangent*, were the slope of PQ approaches ∞ from both sides or approaches $-\infty$ from both sides (here, $-\infty$)

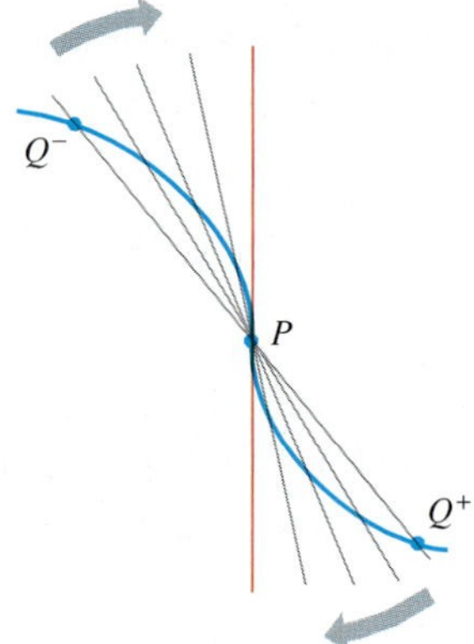

(iv) a *discontinuity*.

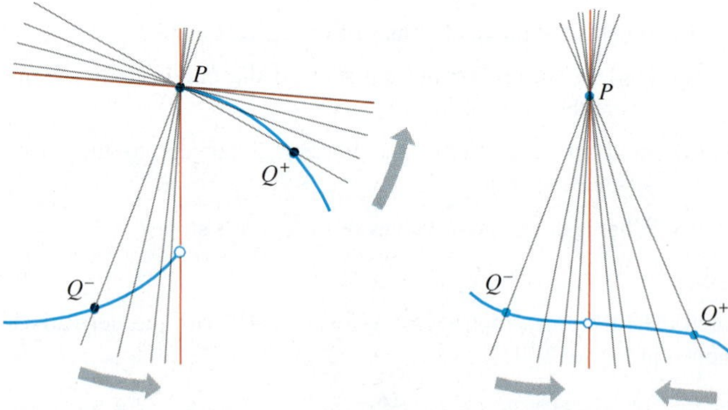

Each figure in Exercises 23–26 shows the graph of a function over a closed interval D. At what domain points does the function appear to be

(a) differentiable?

(b) continuous but not differentiable?

(c) neither continuous nor differentiable?

Give reasons for your answers.

23.

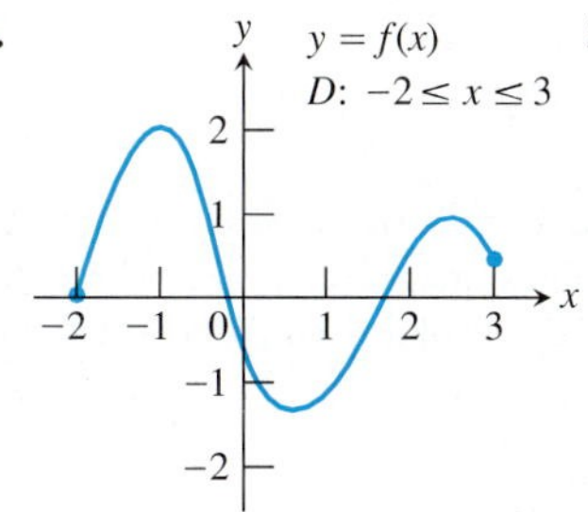

24.

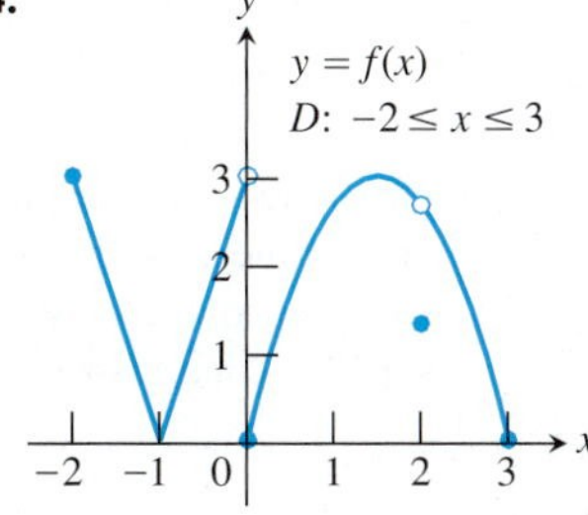

25.

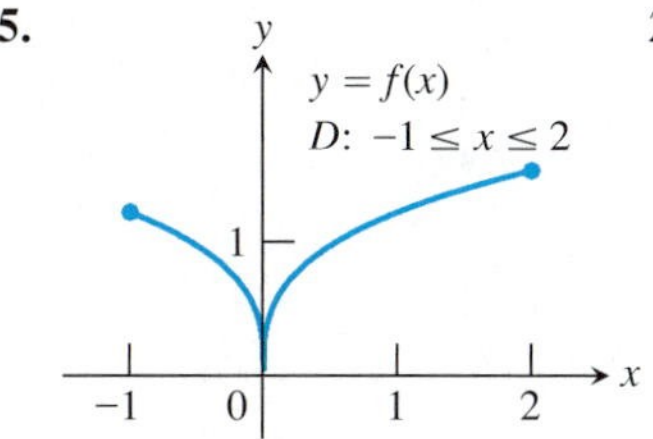

26.

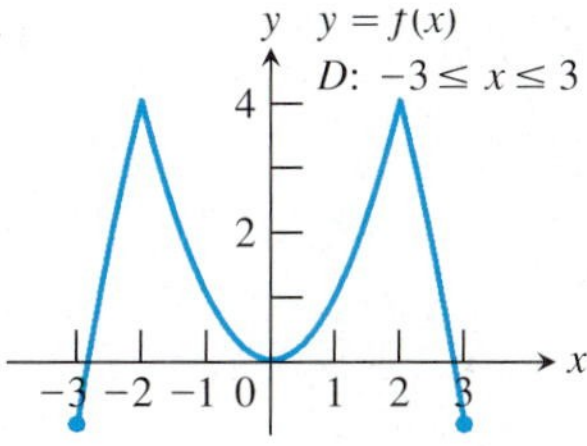

Interpreting Derivatives

In Exercises 27–30,

(a) Find the derivative $y' = f'(x)$ of the given function $y = f(x)$.

(b) Graph $y = f(x)$ and $y' = f'(x)$ side by side using separate sets of coordinate axes and answer the following questions.

(c) For what values of x, if any, is y' positive? Zero? Negative?

(d) Over what intervals of x-values, if any, does the function $y = f(x)$ increase as x increases? Decrease as x increases? How is this related to what you found in part (c)? (We say more about this relationship in Chapter 3.)

27. $y = -x^2$

28. $y = -1/x$

29. $y = x^3/3$

30. $y = x^4/4$

31. *Writing to Learn* Does the curve $y = x^3$ ever have a negative slope? If so, where? Give reasons for your answer.

32. *Writing to Learn* Does the curve $y = 2\sqrt{x}$ have any horizontal tangents? If so, where? Give reasons for your answer.

Theory and Examples

33. *Tangent to a parabola* Does the parabola $y = 2x^2 - 13x + 5$ have a tangent whose slope is -1? If so, find an equation for the line and the point of tangency. If not, why not?

34. *Tangent to $y = \sqrt{x}$* Does any tangent to the curve $y = \sqrt{x}$ cross the x-axis at $x = -1$? If so, find an equation for the line and the point of tangency. If not, why not?

35. *Greatest integer in x* Does any function differentiable on $(-\infty, \infty)$ have $y = \text{int } x$, the greatest integer in x (see Figure 1.49) as its derivative? Give reasons for your answer.

36. *Derivative of $y = |x|$* Graph the derivative of $f(x) = |x|$. Then graph $y = (|x| - 0)/(x - 0) = |x|/x$. What can you conclude?

37. *Derivative of $-f$* Does knowing that a function $f(x)$ is differentiable at $x = x_0$ tell you anything about the differentiability of the function $-f$ at $x = x_0$? Give reasons for your answer.

38. *Derivative of multiples* Does knowing that a function $g(t)$ is differentiable at $t = 7$ tell you anything about the differentiability of the function $3g$ at $t = 7$? Give reasons for your answer.

39. *Limit of a quotient* Suppose that functions $g(t)$ and $h(t)$ are defined for all values of t and $g(0) = h(0) = 0$. Can $\lim_{t\to 0} (g(t))/(h(t))$ exist? If it does exist, must it equal zero? Give reasons for your answers.

40. **(a)** Let $f(x)$ be a function satisfying $|f(x)| \le x^2$ for $-1 \le x \le 1$. Show that f is differentiable at $x = 0$ and find $f'(0)$.

(b) Show that

$$f(x) = \begin{cases} x^2 \sin \dfrac{1}{x}, & x \ne 0 \\ 0, & x = 0 \end{cases}$$

is differentiable at $x = 0$ and find $f'(0)$.

T **41.** *Writing to Learn* Graph $y = 1/(2\sqrt{x})$ in a window that has $0 \le x \le 2$. Then, on the same screen, graph

$$y = \frac{\sqrt{x+h} - \sqrt{x}}{h}$$

for $h = 1, 0.5, 0.1$. Then try $h = -1, -0.5, -0.1$. Explain what is going on.

T **42.** *Writing to Learn* Graph $y = 3x^2$ in a window that has $-2 \le x \le 2$, $0 \le y \le 3$. Then, on the same screen, graph

$$y = \frac{(x+h)^3 - x^3}{h}$$

for $h = 2, 1, 0.2$. Then try $h = -2, -1, -0.2$. Explain what is going on.

T **43.** *Weierstrass's nowhere differentiable continuous function* The sum of the first eight terms of the Weierstrass function $f(x) = \sum_{n=0}^{\infty} (2/3)^n \cos(9^n \pi x)$ is

$$g(x) = \cos(\pi x) + (2/3)^1 \cos(9\pi x) + (2/3)^2 \cos(9^2 \pi x) + (2/3)^3 \cos(9^3 \pi x) + \cdots + (2/3)^7 \cos(9^7 \pi x).$$

Graph this sum. Zoom in several times. How wiggly and bumpy is this graph? Specify a viewing window in which the displayed portion of the graph is smooth.

T **44.** *Writing to Learn* Use a graphing utility to graph the two functions $f(x) = 1 + \ln(x+1)$ and $g(x) = |x| + 1$ in the same viewing rectangle. Use the Zoom and Trace features to analyze the graphs near the point $(0, 1)$. What do you observe? Which function appears to be differentiable at this point? Give reasons for your answers.

COMPUTER EXPLORATIONS

Use a CAS to perform the following steps for the functions in Exercises 45–50.

(a) Plot $y = f(x)$ to see that function's global behavior.

(b) Define the difference quotient q at a general point x, with general stepsize h.

(c) Take the limit as $h \to 0$. What formula does this give?

(d) Substitute the value $x = x_0$ and plot the function $y = f(x)$ together with its tangent line at that point.

(e) Substitute various values for x larger and smaller than x_0 into the formula obtained in part (c). Do the numbers make sense with your picture?

(f) *Writing to Learn* Graph the formula obtained in part (c). What does it mean when its values are negative? Zero? Positive? Does this make sense with your plot from part (a)? Give reasons for your answer.

45. $f(x) = x^3 + x^2 - x, \quad x_0 = 1$

46. $f(x) = x^{1/3} + x^{2/3}, \quad x_0 = 1$

47. $f(x) = \dfrac{4x}{x^2 + 1}, \quad x_0 = 2$

48. $f(x) = \dfrac{x - 1}{3x^2 + 1}, \quad x_0 = -1$

49. $f(x) = \sin 2x, \quad x_0 = \pi/2$

50. $f(x) = x^2 \cos x, \quad x_0 = \pi/4$

2.2 The Derivative as a Rate of Change

Instantaneous Rates of Change • Motion Along a Line: Displacement, Velocity, Speed, Acceleration, and Jerk • Sensitivity to Change • Derivatives in Economics

In Section 1.1, we initiated the study of average and instantaneous rates of change. In this section, we continue our investigations of applications in which derivatives are used to model the rates at which things change in the world around us. We revisit the study of motion along a line and examine other applications.

It is natural to think of change as change with respect to time, but other variables can be treated in the same way. For example, a physician may want to know how change in dosage affects the body's response to a drug. An economist may want to study how the cost of producing steel varies with the number of tons produced.

Instantaneous Rates of Change

If we interpret the difference quotient $(f(x + h) - f(x))/h$ as the average rate of change in f over the interval from x to $x + h$, we can interpret its limit as $h \to 0$ as the rate at which f is changing at the point x.

Instantaneous rates are limits of average rates.

Definition Instantaneous Rate of Change

The **instantaneous rate of change** of f with respect to x at x_0 is the derivative

$$f'(x_0) = \lim_{h \to 0} \frac{f(x_0 + h) - f(x_0)}{h},$$

provided the limit exists.

It is conventional to use the word *instantaneous* even when x does not represent time. The word is, however, frequently omitted. When we say *rate of change*, we mean *instantaneous rate of change*.

Example 1 How a Circle's Area Changes with Its Diameter

The area A of a circle is related to its diameter by the equation

$$A = \frac{\pi}{4} D^2 .$$

How fast does the area change with respect to the diameter when the diameter is 10 m?

Solution The rate of change of the area with respect to the diameter is

$$\frac{dA}{dD} = \frac{\pi}{4} \cdot 2D = \frac{\pi D}{2} .$$

When $D = 10$ m, the area is changing at rate $(\pi/2)10 = 5\pi\ \mathrm{m}^2/\mathrm{m}$.

Motion Along a Line: Displacement, Velocity, Speed, Acceleration, and Jerk

Suppose that an object is moving along a coordinate line (say an s-axis) so that we know its position s on that line as a function of time t:

$$s = f(t) .$$

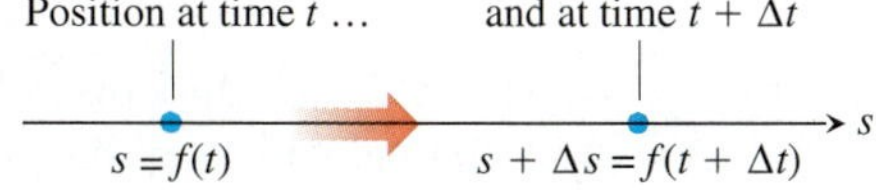

FIGURE 2.14 The positions of a body moving along a coordinate line at time t and shortly later at time $t + \Delta t$.

The **displacement** of the object over the time interval from t to $t + \Delta t$ (Figure 2.14) is

$$\Delta s = f(t + \Delta t) - f(t) ,$$

and the **average velocity** of the object over that time interval is

$$v_{av} = \frac{\text{displacement}}{\text{travel time}} = \frac{\Delta s}{\Delta t} = \frac{f(t + \Delta t) - f(t)}{\Delta t} .$$

To find the body's velocity at the exact instant t, we take the limit of the average velocity over the interval from t to $t + \Delta t$ as Δt shrinks to zero. This limit is the derivative of f with respect to t.

CD-ROM WEBsite

Definition (Instantaneous) Velocity

Velocity (instantaneous velocity) is the derivative of position with respect to time. If a body's position at time t is $s = f(t)$, then the body's velocity at time t is

$$v(t) = \frac{ds}{dt} = \lim_{\Delta t \to 0} \frac{f(t + \Delta t) - f(t)}{\Delta t} .$$

Example 2 Finding the Velocity of a Race Car

Figure 2.15 shows the time-to-distance graph of a 1996 Riley & Scott Mk III-Olds WSC race car. The slope of the secant PQ is the average velocity for the 3-sec interval from $t = 2$ to $t = 5$ sec; in this case, it is about 100 ft/sec or 68 mph.

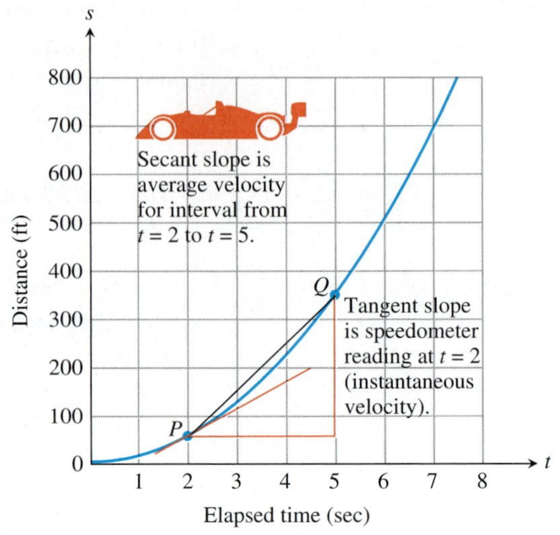

FIGURE 2.15 The time-to-distance graph for Example 2.

The slope of the tangent at P is the speedometer reading at $t = 2$ sec, about 57 ft/sec or 39 mph. The acceleration for the period shown is a nearly constant 28.5 ft/sec^2 during each second, which is about $0.89g$, where g is the acceleration due to gravity. The race car's top speed is an estimated 190 mph. (*Source: Road and Track*, March 1997.)

Besides telling how fast an object is moving, its velocity tells the direction of motion. When the object is moving forward (s increasing), the velocity is positive; when the body is moving backward (s decreasing), the velocity is negative.

If we drive to a friend's house and back at 30 mph, say, the speedometer will show 30 on the way over but it will not show -30 on the way back, even though our distance from home is decreasing. The speedometer always shows speed, which is the absolute value of velocity. Speed measures the rate of progress regardless of direction.

Definition Speed

Speed is the absolute value of velocity.

$$\text{Speed} = |v(t)| = \left|\frac{ds}{dt}\right|$$

Example 3 Horizontal Motion

Figure 2.16 shows the velocity $v = f'(t)$ of a particle moving on a coordinate line. The particle moves forward for the first 3 sec, moves backward for the next 2 sec, stands still for a second, and moves forward again. The particle achieves its greatest speed at time $t = 4$, while moving backward.

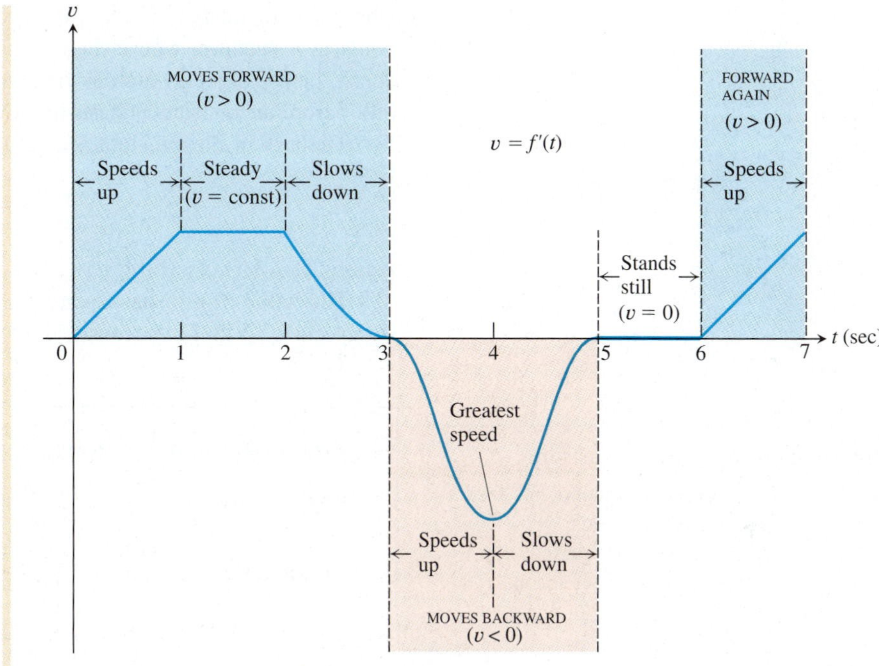

FIGURE 2.16 The velocity graph for Example 3.

The rate at which a body's velocity changes is the body's acceleration. The acceleration measures how quickly the body picks up or loses speed.

A sudden change in acceleration is called a "jerk." When a ride in a car or a bus is jerky, it is not that the accelerations involved are necessarily large but that the changes in acceleration are abrupt. Jerk is what spills our soft drink.

Definitions Acceleration, Jerk

Acceleration is the derivative of velocity with respect to time. If a body's position at time t is $s = f(t)$, then the body's acceleration at time t is

$$a(t) = \frac{dv}{dt} = \frac{d^2s}{dt^2}.$$

Jerk is the derivative of acceleration with respect to time:

$$j(t) = \frac{da}{dt} = \frac{d^3s}{dt^3}.$$

Near the surface of Earth all bodies fall with the same constant acceleration. Galileo's experiments with free fall revealed that the distance a body released from rest falls in time t is proportional to the square of the amount of time it has fallen. Today, we express this by saying that

$$s = \frac{1}{2}gt^2,$$

where s is distance and g is the acceleration due to Earth's gravity. This equation holds in a vacuum, where there is no air resistance, and closely models the fall of dense, heavy objects, such as rocks or steel tools, for the first few seconds of their fall, before air resistance starts to slow them down.

The jerk of the constant acceleration of gravity ($g = 32$ ft/sec^2) is zero:

$$j = \frac{d}{dt}(g) = 0.$$

An object does not exhibit jerkiness during free fall.

The value of g in the equation $s = (1/2)gt^2$ depends on the units used to measure t and s. With t in seconds (the usual unit), we have the following values:

Abbreviations for Units of Measure

ft/sec^2 "feet per second squared" or "feet per second per second"

m/sec^2 "meters per second squared" or "meters per second per second"

Free-Fall Equations (Earth)

English units: $g = 32 \dfrac{\text{ft}}{\text{sec}^2}$, $s = \dfrac{1}{2}(32)t^2 = 16t^2$ (s in feet)

Metric units: $g = 9.8 \dfrac{\text{m}}{\text{sec}^2}$, $s = \dfrac{1}{2}(9.8)t^2 = 4.9t^2$ (s in meters)

Example 4 Modeling Free Fall

Figure 2.17 shows the free fall of a heavy ball bearing released from rest at time $t = 0$ sec.

(a) How many meters does the ball fall in the first 2 sec?

(b) What is its velocity, speed, and acceleration then?

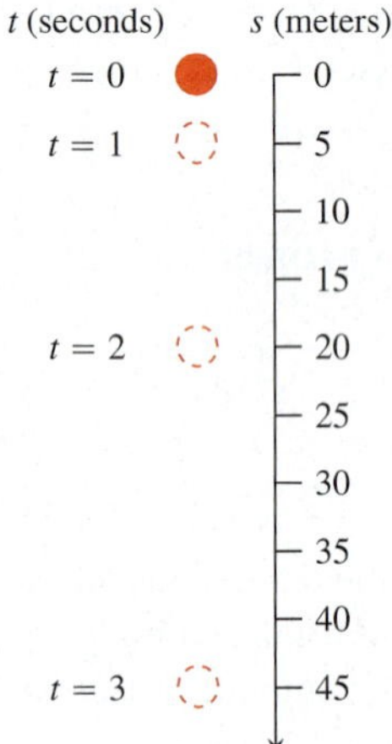

FIGURE 2.17 A ball bearing falling from rest. (Example 4)

Solution

(a) The metric free-fall equation is $s = 4.9t^2$. During the first 2 sec, the ball falls

$$s(2) = 4.9(2)^2 = 19.6 \text{ m}.$$

(b) At any time t, *velocity* is the derivative of position:

$$v(t) = s'(t) = \frac{d}{dt}(4.9t^2) = 9.8t.$$

At $t = 2$, the velocity is

$$v(2) = 19.6 \text{ m/sec}$$

in the downward (increasing s) direction. The *speed* at $t = 2$ is

$$\text{Speed} = |v(2)| = 19.6 \text{ m/sec}.$$

The *acceleration* at any time t is

$$a(t) = v'(t) = s''(t) = 9.8 \text{ m/sec}^2.$$

At $t = 2$, the acceleration is 9.8 m/sec^2.

CD-ROM WEBsite

Example 5 Modeling Vertical Motion

A dynamite blast blows a heavy rock straight up with a launch velocity of 160 ft/sec (about 109 mph) (Figure 2.18a). It reaches a height of $s = 160t - 16t^2$ ft after t sec.

(a) How high does the rock go?

(b) What are the velocity and speed of the rock when it is 256 ft above the ground on the way up? On the way down?

(c) What is the acceleration of the rock at any time t during its flight (after the blast)?

(d) When does the rock hit the ground again?

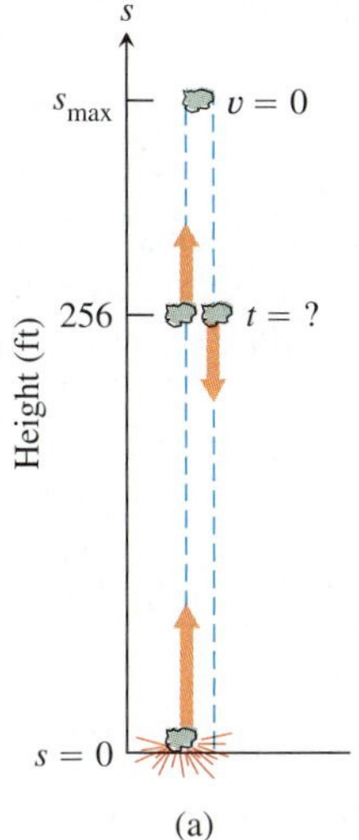

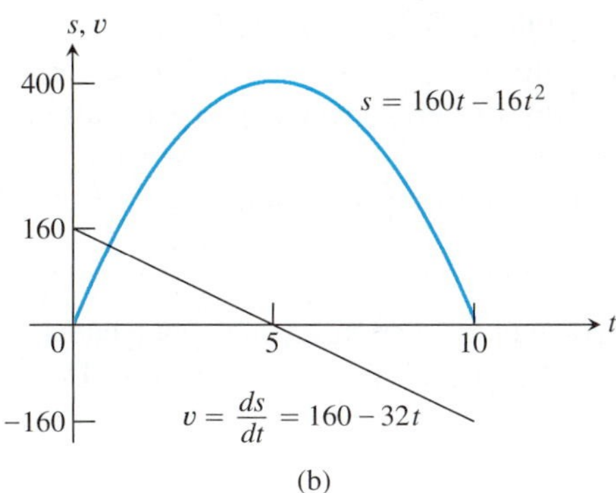

FIGURE 2.18 (a) The rock in Example 5. (b) The graphs of s and v as functions of time; s is largest when $v = ds/dt = 0$. The graph of s is *not* the path of the rock: It is a plot of height versus time. The slope of the plot is the rock's velocity, graphed here as a straight line.

Solution

(a) In the coordinate system we have chosen, s measures height from the ground up, so the velocity is positive on the way up and negative on the way down. The instant the rock is at its highest point is the one instant during the flight when the velocity is 0. To find the maximum height, all we need to do is to find when $v = 0$ and evaluate s at this time.

At any time t, the velocity is

$$v = \frac{ds}{dt} = \frac{d}{dt}(160t - 16t^2) = 160 - 32t \text{ ft/sec}.$$

The velocity is zero when

$$160 - 32t = 0 \qquad \text{or} \qquad t = 5 \text{ sec}.$$

The rock's height at $t = 5$ sec is

$$s_{max} = s(5) = 160(5) - 16(5)^2 = 800 - 400 = 400 \text{ ft}.$$

See Figure 2.18b.

(b) To find the rock's velocity at 256 ft on the way up and again on the way down, we first find the two values of t for which

$$s(t) = 160t - 16t^2 = 256.$$

To solve this equation, we write

$$\begin{aligned} 16t^2 - 160t + 256 &= 0 \\ 16(t^2 - 10t + 16) &= 0 \\ (t - 2)(t - 8) &= 0 \\ t = 2 \text{ sec}, t &= 8 \text{ sec}. \end{aligned}$$

The rock is 256 ft above the ground 2 sec after the explosion and again 8 sec after the explosion. The rock's velocities at these times are

$$v(2) = 160 - 32(2) = 160 - 64 = 96 \text{ ft/sec}$$
$$v(8) = 160 - 32(8) = 160 - 256 = -96 \text{ ft/sec}.$$

At both instants, the rock's speed is 96 ft/sec.

(c) At any time during its flight following the explosion, the rock's acceleration is a constant

$$a = \frac{dv}{dt} = \frac{d}{dt}(160 - 32t) = -32 \text{ ft/sec}^2 .$$

The acceleration is always downward. As the rock rises, it slows down; as it falls, it speeds up.

(d) The rock hits the ground at the positive time t for which $s = 0$. The equation $160t - 16t^2 = 0$ factors to give $16t(10 - t) = 0$, so it has solutions $t = 0$ and $t = 10$. At $t = 0$, the blast occurred and the rock was thrown upward. It returned to the ground 10 sec later.

USING TECHNOLOGY

Simulation of motion on a vertical line The parametric equations

$$x(t) = c \qquad y(t) = f(t)$$

will illuminate pixels along the vertical line $x = c$. If $f(t)$ denotes the height of a moving body at time t, graphing $(x(t), y(t)) = (c, f(t))$ will simulate the actual motion. Try it for the rock in Example 5 with $x(t) = 2$, say, and $y(t) = 160t - 16t^2$, in dot mode with t Step $= 0.1$. Why does the spacing of the dots vary? Why does the grapher seem to stop after it reaches the top? (Try the plots for $0 \leq t \leq 5$ and $5 \leq t \leq 10$ separately.)

For a second experiment, plot the parametric equations

$$x(t) = t, \qquad y(t) = 160t - 16t^2$$

together with the vertical line simulation of the motion, again in dot mode. Use what you know about the behavior of the rock from the calculations of Example 5 to select a window size that will display all the interesting behavior.

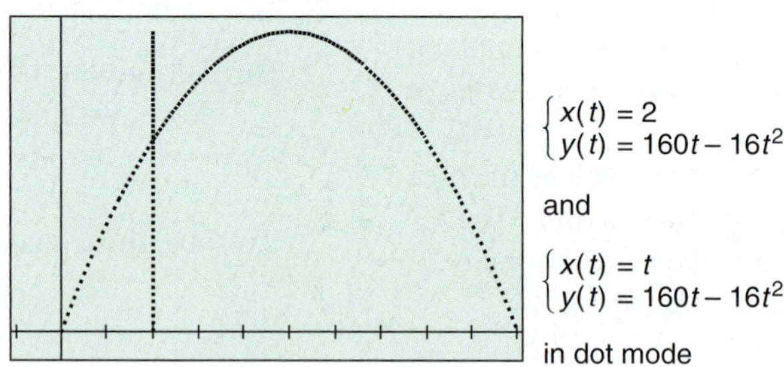

$\begin{cases} x(t) = 2 \\ y(t) = 160t - 16t^2 \end{cases}$ and $\begin{cases} x(t) = t \\ y(t) = 160t - 16t^2 \end{cases}$ in dot mode

Sensitivity to Change

When a small change in x produces a large change in the value of a function $f(x)$, we say that the function is relatively **sensitive** to changes in x. The derivative $f'(x)$ is a measure of this sensitivity.

Example 6 Genetic Data and Sensitivity to Change

The Austrian monk Gregor Johann Mendel (1822–1884), working with garden peas and other plants, provided the first scientific explanation of hybridization.

His careful records showed that if p (a number between 0 and 1) is the frequency of the gene for smooth skin in peas (dominant) and $(1 - p)$ is the frequency of the gene for wrinkled skin in peas, then the proportion of smooth-skinned peas in the next generation will be

$$y = 2p(1 - p) + p^2 = 2p - p^2.$$

The graph of y versus p in Figure 2.19a suggests that the value of y is more sensitive to a change in p when p is small than when p is large. Indeed, this fact is

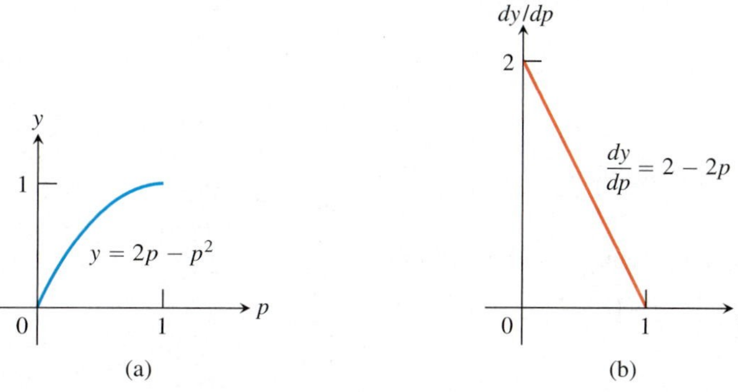

FIGURE 2.19 (a) The graph of $y = 2p - p^2$, describing the proportion of smooth-skinned peas. (b) The graph of dy/dp.

borne out by the derivative graph in Figure 2.19b, which shows that dy/dp is close to 2 when p is near 0 and close to 0 when p is near 1.

The implication for genetics is that introducing a few more dominant genes into a highly recessive population (where the frequency of wrinkled skin peas is small) will have a more dramatic effect on later generations than will a similar increase in a highly dominant population.

Derivatives in Economics

Engineers use the terms *velocity* and *acceleration* to refer to the derivatives of functions describing motion. Economists, too, have a specialized vocabulary for rates of change and derivatives. They call them *marginals*.

In a manufacturing operation, the *cost of production* $c(x)$ is a function of x, the number of units produced. The *marginal cost of production* is the rate of change of cost with respect to level of production, so it is dc/dx.

Suppose that $c(x)$ represents the dollars needed to produce x tons of steel in one week. It costs more to produce $x + h$ units per week, and the cost difference, divided by h, is the average cost of producing each additional ton:

$$\frac{c(x + h) - c(x)}{h} = \text{average cost of each of the additional } h \text{ tons of steel produced}.$$

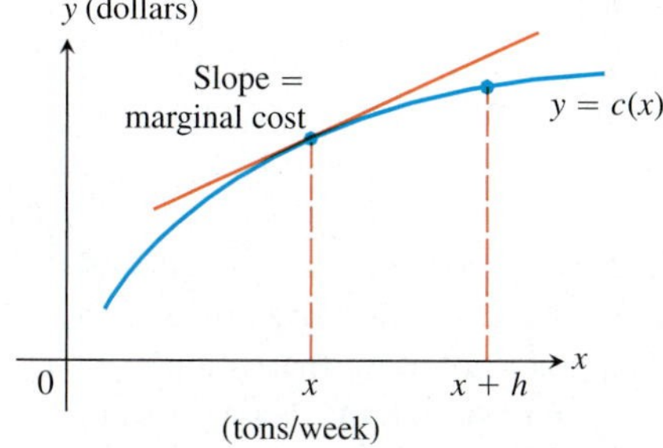

FIGURE 2.20 Weekly steel production: $c(x)$ is the cost of producing x tons per week. The cost of producing an additional h tons is $c(x + h) - c(x)$.

The limit of this ratio as $h \to 0$ is the *marginal cost* of producing more steel per week when the current weekly production is x tons (Figure 2.20).

$$\frac{dc}{dx} = \lim_{h \to 0} \frac{c(x + h) - c(x)}{h} = \text{marginal cost of production}.$$

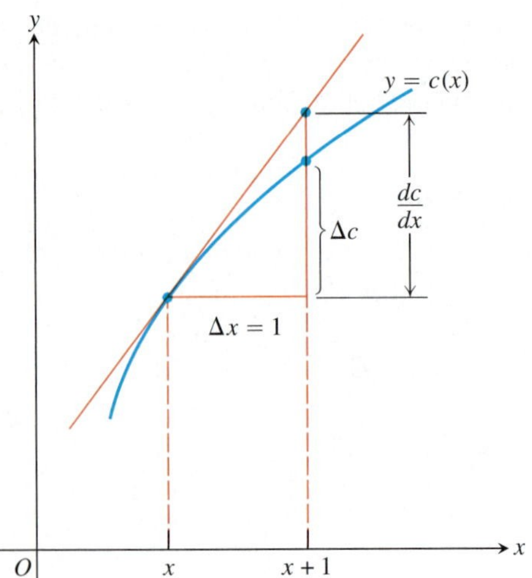

FIGURE 2.21 The marginal cost dc/dx is approximately the extra cost Δc of producing $\Delta x = 1$ more unit.

Sometimes the marginal cost of production is loosely defined to be the extra cost of producing one unit:

$$\frac{\Delta c}{\Delta x} = \frac{c(x + 1) - c(x)}{1},$$

which is approximated by the value of dc/dx at x. This approximation is acceptable if the slope of the graph of c does not change quickly near x. Then the difference quotient will be close to its limit dc/dx, which is the rise in the tangent line if $\Delta x = 1$ (Figure 2.21). The approximation works best for large values of x.

Choosing functions to illustrate economics

In case you are wondering why economists use polynomials of low degree to illustrate complicated phenomena like cost and revenue, here is the rationale: Although formulas for real phenomena are rarely available in any given instance, the theory of economics can still provide valuable guidance. The functions about which theory speaks can often be illustrated with low-degree polynomials on relevant intervals. Cubic polynomials provide a good balance between being easy to work with and being complicated enough to capture the behavior.

Example 7 Marginal Cost and Marginal Revenue

Suppose that it costs

$$c(x) = x^3 - 6x^2 + 15x$$

dollars to produce x radiators when 8 to 30 radiators are produced and that

$$r(x) = x^3 - 3x^2 + 12x$$

gives the dollar revenue from selling x radiators. Your shop currently produces 10 radiators a day. About how much extra will it cost to produce one more radiator a day, and what is your estimated increase in revenue for selling 11 radiators a day?

Solution The cost of producing one more radiator a day when 10 are produced is about $c'(10)$:

$$c'(x) = \frac{d}{dx}(x^3 - 6x^2 + 15x) = 3x^2 - 12x + 15$$

$$c'(10) = 3(100) - 12(10) + 15 = 195.$$

The additional cost will be about \$195. The marginal revenue is

$$r'(x) = \frac{d}{dx}(x^3 - 3x^2 + 12x) = 3x^2 - 6x + 12.$$

The marginal revenue function estimates the increase in revenue that will result from selling one additional unit. If you currently sell 10 radiators a day, you can expect your revenue to increase by about

$$r'(10) = 3(100) - 6(10) + 12 = \$252$$

if you increase sales to 11 radiators a day.

Example 8 Marginal Tax Rate

To get some feel for the language of marginal rates, consider marginal tax rates. If your marginal income tax rate is 28% and your income increases by \$1000, you can expect to pay an extra \$280 in taxes. This does not mean that you pay 28% of your entire income in taxes. It just means that at your current income level I, the rate of increase of taxes T with respect to income is $dT/dI = 0.28$. You will pay \$0.28 out of every extra dollar you earn in taxes. Of course, if you earn a lot more, you may land in a higher tax bracket and your marginal rate will increase.

EXERCISES 2.2

Motion Along a Coordinate Line

CD-ROM WEBsite

Exercises 1–4 give the positions $s = f(t)$ of a body moving on a coordinate line, with s in meters and t in seconds.

(a) Find the body's displacement and average velocity for the given time interval.

(b) Find the body's speed and acceleration at the endpoints of the interval.

(c) When if ever during the interval does the body change direction?

1. $s = t^2 - 3t + 2, \quad 0 \le t \le 2$

2. $s = 6t - t^2, \quad 0 \le t \le 6$

3. $s = -t^3 + 3t^2 - 3t, \quad 0 \le t \le 3$

4. $s = (t^4/4) - t^3 + t^2, \quad 0 \le t \le 3$

5. *Particle motion* At time t, the position of a body moving along the s-axis is $s = t^3 - 6t^2 + 9t$ m.

(a) Find the body's acceleration each time the velocity is zero.

(b) Find the body's speed each time the acceleration is zero.

(c) Find the total distance traveled by the body from $t = 0$ to $t = 2$.

6. *Particle motion* At time $t \ge 0$, the velocity of a body moving along the s-axis is $v = t^2 - 4t + 3$.

(a) Find the body's acceleration each time the velocity is zero.

(b) When is the body moving forward? Backward?

(c) When is the body's velocity increasing? Decreasing?

Free-Fall Applications

7. *Free fall on Mars and Jupiter* The equations for free fall at the surfaces of Mars and Jupiter (s in meters, t in seconds) are $s = 1.86t^2$ on Mars and $s = 11.44t^2$ on Jupiter. How long does it take a rock falling from rest to reach a velocity of 27.8 m/sec (about 100 km/h) on each planet?

8. *Lunar projectile motion* A rock thrown vertically upward from the surface of the moon at a velocity of 24 m/sec (about 86 km/h) reaches a height of $s = 24t - 0.8t^2$ meters in t sec.

(a) Find the rock's velocity and acceleration at time t. (The acceleration in this case is the acceleration of gravity on the moon.)

(b) How long does it take the rock to reach its highest point?

(c) How high does the rock go?

(d) How long does it take the rock to reach half its maximum height?

(e) How long is the rock aloft?

9. *Finding g on a small airless planet* Explorers on a small airless planet used a spring gun to launch a ball bearing vertically upward from the surface at a launch velocity of 15 m/sec. Because the acceleration of gravity at the planet's surface was g_s m/sec^2, the explorers expected the ball bearing to reach a height of $s = 15t - (1/2)g_s t^2$ meters t sec later. The ball bearing reached its maximum height 20 sec after being launched. What was the value of g_s?

CD-ROM WEBsite

10. *Speeding bullet* A 45-caliber bullet fired straight up from the surface of the moon would reach a height of $s = 832t - 2.6t^2$ feet after t sec. On Earth, in the absence of air, its height would be $s = 832t - 16t^2$ ft after t sec. How long will the bullet be aloft in each case? How high will the bullet go?

CD-ROM WEBsite

11. *Free fall from the Tower of Pisa* Had Galileo dropped a cannonball from the Tower of Pisa, 179 ft above the ground, the ball's height above ground t sec into the fall would have been $s = 179 - 16t^2$.

(a) What would have been the ball's velocity, speed, and acceleration at time t?

(b) About how long would it have taken the ball to hit the ground?

(c) What would have been the ball's velocity at the moment of impact?

12. *Galileo's free-fall formula* Galileo developed a formula for a body's velocity during free fall by rolling balls from rest down increasingly steep inclined planks and looking for a limiting formula that would predict a ball's behavior when the plank was vertical and the ball fell freely; see part (a) of the accompanying figure. He found that, for any given angle of the plank, the ball's velocity t sec into motion was a constant multiple of t. That is, the velocity was given by a formula of the form $v = kt$. The value of the constant k depended on the inclination of the plank.

In modern notation—part (b) of the figure—with distance in meters and time in seconds, what Galileo determined by experiment was that, for any given angle θ, the ball's velocity t sec into the roll was

$$v = 9.8(\sin\theta)t \text{ m/sec}.$$

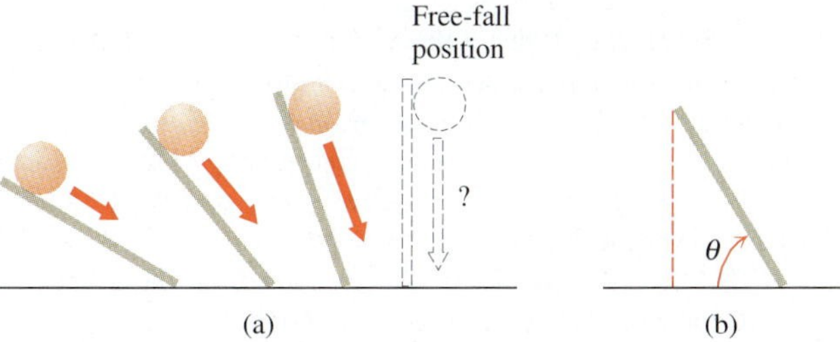

(a) What is the equation for the ball's velocity during free fall?

(b) Building on your work in part (a), what constant acceleration does a freely falling body experience near the surface of Earth?

Conclusions About Motion from Graphs

13. The accompanying figure shows the velocity $v = ds/dt = f(t)$ (m/sec) of a body moving along a coordinate line.

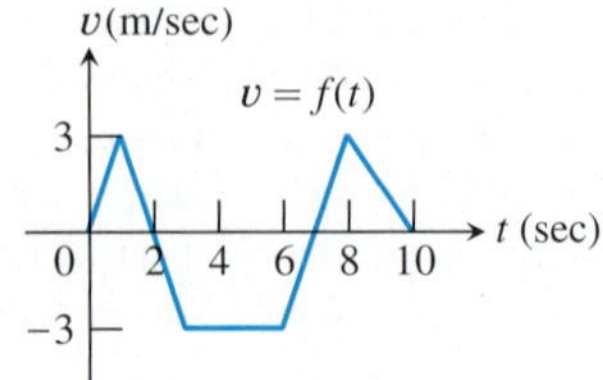

(a) When does the body reverse direction?

(b) When (approximately) is the body moving at a constant speed?

(c) Graph the body's speed for $0 \le t \le 10$.

(d) Graph the acceleration, where defined.

14. A particle P moves on the number line shown in part (a) of the accompanying figure. Part (b) shows the position of P as a function of time t.

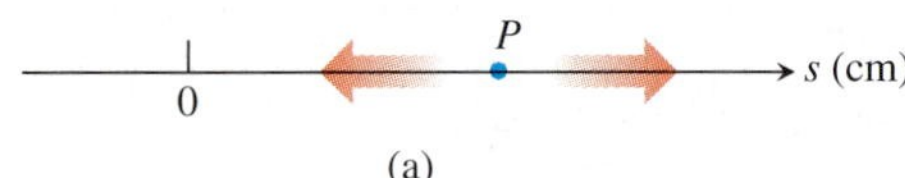

(a)

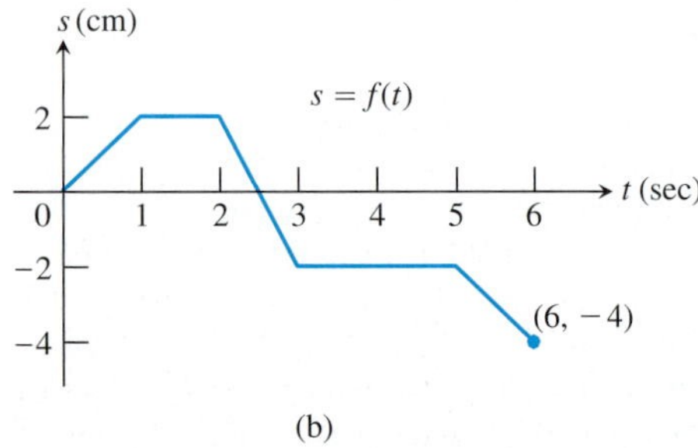

(b)

(a) When is P moving to the left? Moving to the right? Standing still?

(b) Graph the particle's velocity and speed (where defined).

15. *Launching a rocket* When a model rocket is launched, the propellant burns for a few seconds, accelerating the rocket upward. After burnout, the rocket coasts upward for a while and then begins to fall. A small explosive charge pops out a parachute shortly after the rocket starts down. The parachute slows the rocket to keep it from breaking when it lands.

The figure here shows velocity data from the flight of the model rocket. Use the data to answer the following.

(a) How fast was the rocket climbing when the engine stopped?

(b) For how many seconds did the engine burn?

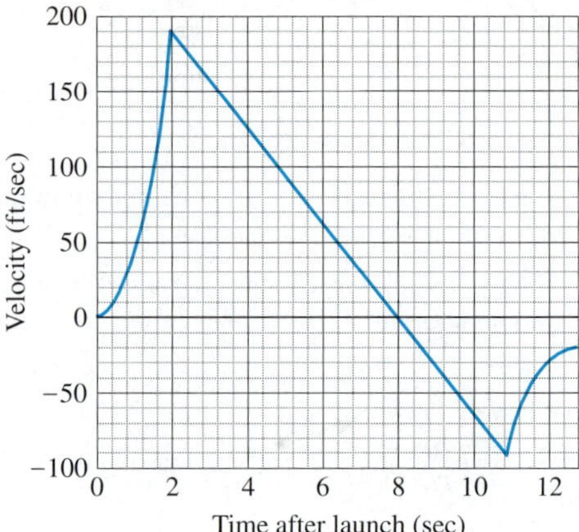

(c) When did the rocket reach its highest point? What was its velocity then?

(d) When did the parachute pop out? How fast was the rocket falling then?

(e) How long did the rocket fall before the parachute opened?

(f) When was the rocket's acceleration greatest?

(g) When was the acceleration constant? What was its value then (to the nearest integer)?

16. *A traveling truck* The accompanying graph shows the position s of a truck traveling on a highway. The truck starts at $t = 0$ and returns 15 h later at $t = 15$.

(a) Use the technique described in Section 2.1, Example 9, to graph the truck's velocity $v = ds/dt$ for $0 \le t \le 15$. Then repeat the process, with the velocity curve, to graph the truck's acceleration dv/dt.

(b) Suppose that $s = 15t^2 - t^3$. Graph ds/dt and d^2s/dt^2 and compare your graphs with those in part (a).

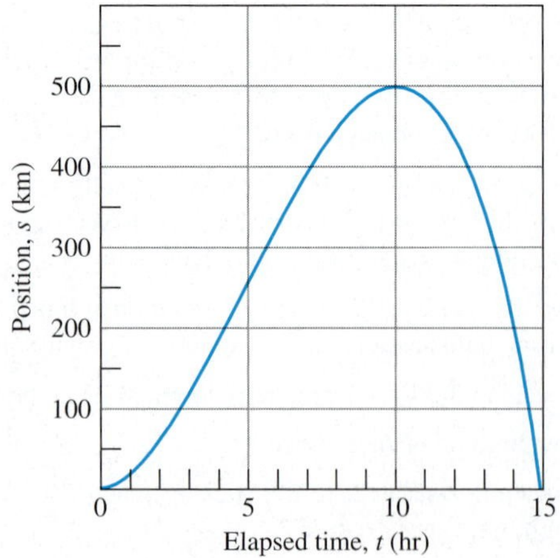

17. *Two falling balls* The multiflash photograph in the accompanying figure shows two balls falling from rest. The vertical rulers are marked in centimeters. Use the equation $s = 490t^2$ (the free-fall equation for s in centimeters and t in seconds) to answer the following questions.

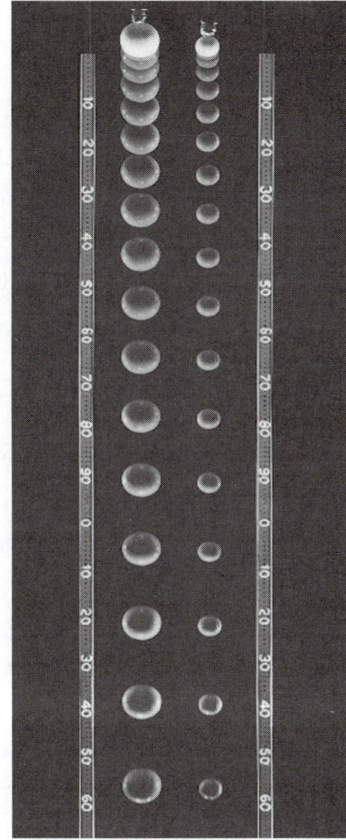

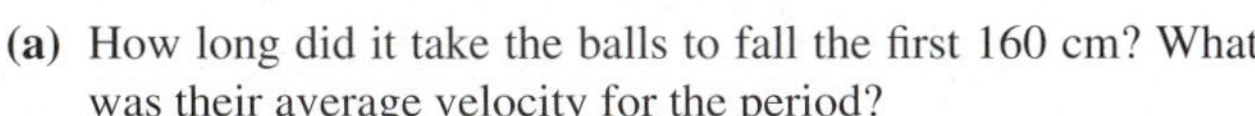

(a) How long did it take the balls to fall the first 160 cm? What was their average velocity for the period?

(b) How fast were the balls falling when they reached the 160-cm mark? What was their acceleration then?

(c) About how fast was the light flashing (flashes per second)?

18. *Writing to Learn* The graphs in Figure 2.22 show the position s, velocity $v = ds/dt$, and acceleration $a = d^2s/dt^2$ of a body moving along a coordinate line as functions of time t. Which graph is which? Give reasons for your answers.

19. *Writing to Learn* The graphs in Figure 2.23 show the position s, the velocity $v = ds/dt$, and the acceleration $a = d^2s/dt^2$ of a body moving along the coordinate line as functions of time t. Which graph is which? Give reasons for your answers.

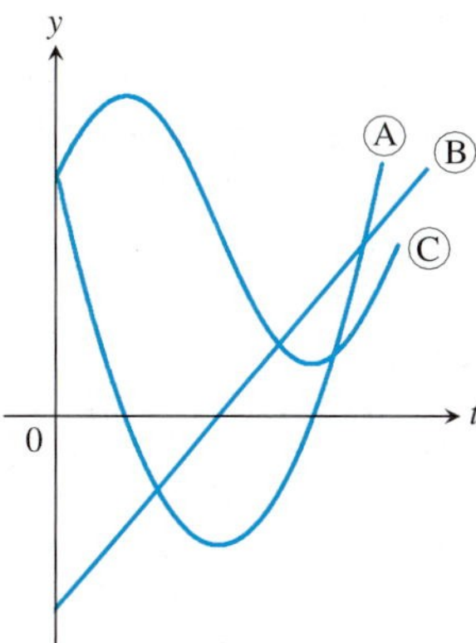

FIGURE 2.22 The graphs for Exercise 18.

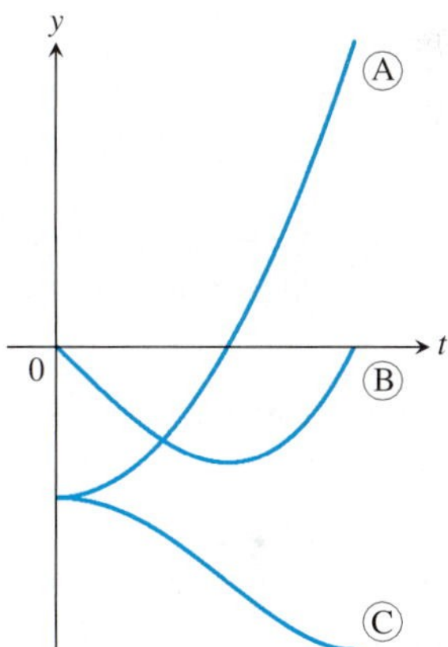

FIGURE 2.23 The graphs for Exercise 19.

Economics

20. *Marginal cost* Suppose that the dollar cost of producing x washing machines is $c(x) = 2000 + 100x - 0.1x^2$.

(a) Find the average cost per machine of producing the first 100 washing machines.

(b) Find the marginal cost when 100 washing machines are produced.

(c) Show that the marginal cost when 100 washing machines are produced is approximately the cost of producing one more washing machine after the first 100 have been made, by calculating the latter cost directly.

21. *Marginal revenue* Suppose that the revenue from selling x washing machines is

$$r(x) = 20{,}000\left(1 - \frac{1}{x}\right)$$

dollars.

(a) Find the marginal revenue when 100 machines are produced.

(b) Use the function $r'(x)$ to estimate the increase in revenue that will result from increasing production from 100 machines a week to 101 machines a week.

(c) Find the limit of $r'(x)$ as $x \to \infty$. How would you interpret this number?

Additional Applications

22. *Bacterium population* When a bactericide was added to a nutrient broth in which bacteria were growing, the bacterium population continued to grow for a while, but then stopped growing and began to decline. The size of the population at time t (hours) was $b = 10^6 + 10^4t - 10^3t^2$. Find the growth rates at

(a) $t = 0$ h.

(b) $t = 5$ h.

(c) $t = 10$ h.

23. *Draining a tank* The number of gallons of water in a tank t minutes after the tank has started to drain is $Q(t) = 200(30 - t)^2$. How fast is the water running out at the end of 10 min? What is the average rate at which the water flows out during the first 10 min?

T **24.** *Draining a tank* It takes 12 h to drain a storage tank by opening the valve at the bottom. The depth y of fluid in the tank t h after the valve is opened is given by the formula

$$y = 6\left(1 - \frac{t}{12}\right)^2 \text{ m}.$$

(a) Find the rate dy/dt (m/h) at which the tank is draining at time t.

(b) When is the fluid level in the tank falling fastest? Slowest? What are the values of dy/dt at these times?

(c) Graph y and dy/dt together and discuss the behavior of y in relation to the signs and values of dy/dt.

25. *Inflating a balloon* The volume $V = (4/3)\pi r^3$ of a spherical balloon changes with the radius.

(a) At what rate (ft^3/ft) does the volume change with respect to the radius when $r = 2$ ft?

(b) By approximately how much does the volume increase when the radius changes from 2 to 2.2 ft?

26. *Airplane takeoff* Suppose that the distance an aircraft travels along a runway before takeoff is given by $D = (10/9)t^2$, where D is measured in meters from the starting point and t is measured in seconds from the time the brakes are released. The aircraft will become airborne when its speed reaches 200 km/h. How long will it take to become airborne, and what distance will it travel in that time?

27. *Volcanic lava fountains* Although the November 1959 Kilauea Iki eruption on the island of Hawaii began with a line of fountains along the wall of the crater, activity was later confined to a single vent in the crater's floor, which at one point shot lava 1900 ft straight into the air (a world record). What was the lava's exit velocity in feet per second? In miles per hour? (*Hint*: If v_0 is the exit velocity of a particle of lava, its height t sec later will be $s = v_0t - 16t^2$ ft. Begin by finding the time at which $ds/dt = 0$. Neglect air resistance.)

T Exercises 28–31 give the position function $s = f(t)$ of a body moving along the s-axis as a function of time t. Graph f together with the velocity function $v(t) = ds/dt = f'(t)$ and the acceleration function $a(t) = d^2s/dt^2 = f''(t)$. Comment on the body's behavior in relation to the signs and values v and a. Include in your commentary such topics as the following:

(a) When is the body momentarily at rest?

(b) When does it move to the left (down) or to the right (up)?

(c) When does it change direction?

(d) When does it speed up and slow down?

(e) When is it moving fastest (highest speed)? Slowest?

(f) When is it farthest from the axis origin?

28. $s = 200t - 16t^2, \quad 0 \le t \le 12.5$ (a heavy object fired straight up from Earth's surface at 200 ft/sec)

29. $s = t^2 - 3t + 2, \quad 0 \le t \le 5$

30. $s = t^3 - 6t^2 + 7t, \quad 0 \le t \le 4$

31. $s = 4 - 7t + 6t^2 - t^3, \quad 0 \le t \le 4$

32. *Thoroughbred racing* A racehorse is running a 10-furlong race. (A furlong is 220 yards, although we will use furlongs and seconds as our units in this exercise.) As the horse passes each furlong marker (F), a steward records the time elapsed (t) since the beginning of the race, as shown in the table:

F	0	1	2	3	4	5	6	7	8	9	10
t	0	20	33	46	59	73	86	100	112	124	135

(a) How long does it take the horse to finish the race?

(b) What is the average speed of the horse over the first 5 furlongs?

(c) What is the approximate speed of the horse as it passes the 3-furlong marker?

(d) During which portion of the race is the horse running the fastest?

(e) During which portion of the race is the horse accelerating the fastest?

2.3 Derivatives of Products, Quotients, and Negative Powers

Products • Quotients • Negative Integer Powers of x

This section continues the discussion on how to differentiate functions without having to apply the definition each time.

Products

Although the derivative of the sum of two functions is the sum of their derivatives, the derivative of the product of two functions is *not* the product of their derivatives. For instance,

$$\frac{d}{dx}(x \cdot x) = \frac{d}{dx}(x^2) = 2x, \qquad \text{whereas} \qquad \frac{d}{dx}(x) \cdot \frac{d}{dx}(x) = 1 \cdot 1 = 1.$$

The derivative of a product of two functions is the sum of *two* products, as we now explain.

Rule 5 Derivative Product Rule

If u and v are differentiable at x, then so is their product uv, and

$$\frac{d}{dx}(uv) = u\frac{dv}{dx} + v\frac{du}{dx}.$$

The derivative of the product uv is u times the derivative of v plus v times the derivative of u. In prime notation, $(uv)' = uv' + vu'$.

Example 1 Using the Product Rule

Find the derivative of

$$y = \frac{1}{x}\left(x^2 + \frac{1}{x}\right).$$

Solution We apply the product rule with $u = 1/x$ and $v = x^2 + (1/x)$:

$$\frac{d}{dx}\left[\frac{1}{x}\left(x^2 + \frac{1}{x}\right)\right] = \frac{1}{x}\left(2x - \frac{1}{x^2}\right) + \left(x^2 + \frac{1}{x}\right)\left(-\frac{1}{x^2}\right)$$

$$= 2 - \frac{1}{x^3} - 1 - \frac{1}{x^3} \qquad \frac{d}{dx}(uv) = u\frac{dv}{dx} + v\frac{du}{dx}, \text{ and}$$

$$= 1 - \frac{2}{x^3}. \qquad \frac{d}{dx}\left(\frac{1}{x}\right) = -\frac{1}{x^2} \text{ by}$$

Example 3, Section 1.5.

Proof of Rule 5

$$\frac{d}{dx}(uv) = \lim_{h \to 0} \frac{u(x+h)v(x+h) - u(x)v(x)}{h}$$

Picturing the product rule

If $u(x)$ and $v(x)$ are positive and increase when x increases, and if $h > 0$,

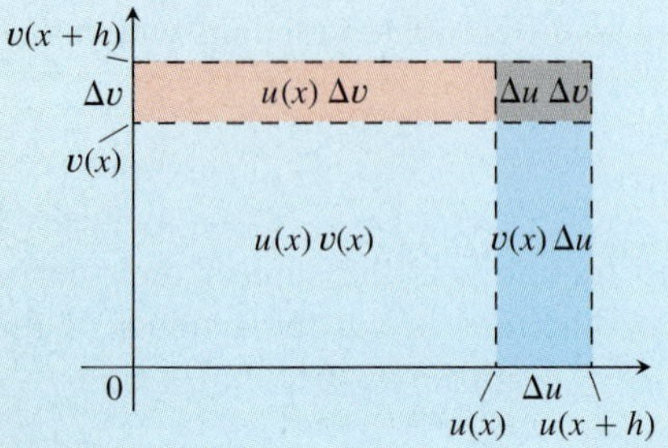

then the total shaded area in the picture is

$$\begin{aligned}&u(x+h)v(x+h) - u(x)v(x)\\ &= u(x+h)\,\Delta v + v(x+h)\,\Delta u\\ &- \Delta u\,\Delta v\,.\end{aligned}$$

Dividing both sides of this equation by h gives

$$\begin{aligned}&\frac{u(x+h)v(x+h) - u(x)v(x)}{h}\\ &\quad = u(x+h)\frac{\Delta v}{h} + v(x+h)\frac{\Delta u}{h}\\ &\quad - \Delta u\frac{\Delta v}{h}.\end{aligned}$$

As $h \to 0^+$,

$$\Delta u \cdot \frac{\Delta v}{h} \to 0 \cdot \frac{dv}{dx} = 0,$$

leaving

$$\frac{d}{dx}(uv) = u\frac{dv}{dx} + v\frac{du}{dx}.$$

To change this fraction into an equivalent one that contains difference quotients for the derivatives of u and v, we subtract and add $u(x+h)v(x)$ in the numerator:

$$\begin{aligned}\frac{d}{dx}(uv) &= \lim_{h\to 0}\frac{u(x+h)v(x+h) - u(x+h)v(x) + u(x+h)v(x) - u(x)v(x)}{h}\\ &= \lim_{h\to 0}\left[u(x+h)\frac{v(x+h)-v(x)}{h} + v(x)\frac{u(x+h)-u(x)}{h}\right]\\ &= \lim_{h\to 0} u(x+h)\cdot\lim_{h\to 0}\frac{v(x+h)-v(x)}{h} + v(x)\cdot\lim_{h\to 0}\frac{u(x+h)-u(x)}{h}.\end{aligned}$$

As h approaches zero, $u(x+h)$ approaches $u(x)$ because u, being differentiable at x, is continuous at x. The two fractions approach the values of dv/dx at x and du/dx at x. In short,

$$\frac{d}{dx}(uv) = u\frac{dv}{dx} + v\frac{du}{dx}.$$

In the following example, we have only numerical values with which to work.

Example 2 Derivative from Numerical Values

Let $y = uv$ be the product of the functions u and v. Find $y'(2)$ if

$$u(2) = 3, \qquad u'(2) = -4, \qquad v(2) = 1, \qquad \text{and} \qquad v'(2) = 2.$$

Solution From the Product Rule, in the form

$$y' = (uv)' = uv' + vu',$$

we have

$$\begin{aligned}y'(2) &= u(2)v'(2) + v(2)u'(2)\\ &= (3)(2) + (1)(-4) = 6 - 4 = 2.\end{aligned}$$

Quotients

Just as the derivative of the product of two differentiable functions is not the product of their derivatives, the derivative of the quotient of two functions is not the quotient of their derivatives. What happens instead is the Quotient Rule.

Rule 6 Derivative Quotient Rule

If u and v are differentiable at x and if $v(x) \neq 0$, then the quotient u/v is differentiable at x, and

$$\frac{d}{dx}\left(\frac{u}{v}\right) = \frac{v\dfrac{du}{dx} - u\dfrac{dv}{dx}}{v^2}.$$

Example 3 Using the Quotient Rule

Find the derivative of

$$y = \frac{t^2 - 1}{t^2 + 1}.$$

Solution We apply the Quotient Rule with $u = t^2 - 1$ and $v = t^2 + 1$:

$$\frac{dy}{dt} = \frac{(t^2+1)\cdot 2t - (t^2-1)\cdot 2t}{(t^2+1)^2} \qquad \frac{d}{dt}\left(\frac{u}{v}\right) = \frac{v(du/dt) - u(dv/dt)}{v^2}$$

$$= \frac{2t^3 + 2t - 2t^3 + 2t}{(t^2+1)^2}$$

$$= \frac{4t}{(t^2+1)^2}.$$

Proof of Rule 6

$$\frac{d}{dx}\left(\frac{u}{v}\right) = \lim_{h\to 0} \frac{\dfrac{u(x+h)}{v(x+h)} - \dfrac{u(x)}{v(x)}}{h}$$

$$= \lim_{h\to 0} \frac{v(x)u(x+h) - u(x)v(x+h)}{h\,v(x+h)v(x)}$$

To change the last fraction into an equivalent one that contains the difference quotients for the derivatives of u and v, we subtract and add $v(x)u(x)$ in the numerator. We then get

$$\frac{d}{dx}\left(\frac{u}{v}\right) = \lim_{h\to 0} \frac{v(x)u(x+h) - v(x)u(x) + v(x)u(x) - u(x)v(x+h)}{h\,v(x+h)v(x)}$$

$$= \lim_{h\to 0} \frac{v(x)\dfrac{u(x+h)-u(x)}{h} - u(x)\dfrac{v(x+h)-v(x)}{h}}{v(x+h)v(x)}.$$

Taking the limit in the numerator and denominator now gives the Quotient Rule.

Negative Integer Powers of x

The Power Rule for negative integers is the same as the rule for positive integers.

Rule 7 Power Rule for Negative Integers

If n is a negative integer and $x \neq 0$, then

$$\frac{d}{dx}(x^n) = nx^{n-1}.$$

Example 4 Using Rule 7

(a) $\dfrac{d}{dx}\left(\dfrac{1}{x}\right) = \dfrac{d}{dx}(x^{-1}) = (-1)x^{-2} = -\dfrac{1}{x^2}$

(b) $\dfrac{d}{dx}\left(\dfrac{4}{x^3}\right) = 4\dfrac{d}{dx}(x^{-3}) = 4(-3)x^{-4} = -\dfrac{12}{x^4}$

CD-ROM WEBsite
Historical Biography
René Descartes (1596 — 1650)

Proof of Rule 7 The proof uses the Quotient Rule in a clever way. If n is a negative integer, then $n = -m$, where m is a positive integer. Hence, $x^n = x^{-m} = 1/x^m$, and

$$\begin{aligned}
\frac{d}{dx}(x^n) &= \frac{d}{dx}\left(\frac{1}{x^m}\right) \\
&= \frac{x^m \cdot \dfrac{d}{dx}(1) - 1 \cdot \dfrac{d}{dx}(x^m)}{(x^m)^2} && \text{Quotient Rule with } u = 1 \text{ and } v = x^m \\
&= \frac{0 - mx^{m-1}}{x^{2m}} && \text{Since } m > 0,\ \frac{d}{dx}(x^m) = mx^{m-1}. \\
&= -mx^{-m-1} \\
&= nx^{n-1}. && \text{Since } -m = n
\end{aligned}$$

Example 5 Tangent to a Curve

Find an equation for the tangent to the curve

$$y = x + \frac{2}{x}$$

at the point (1, 3) (Figure 2.24).

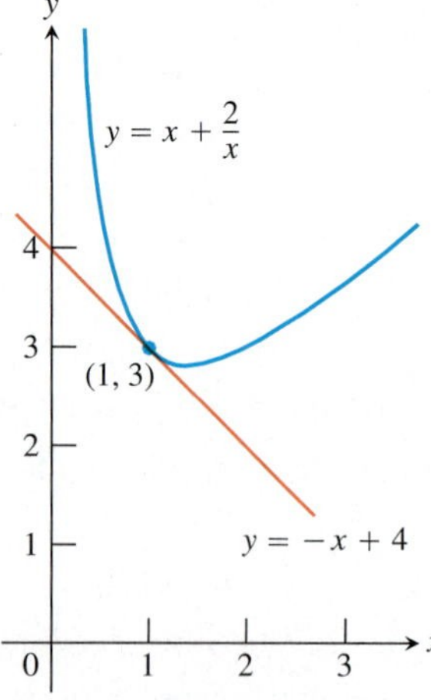

FIGURE 2.24 The tangent to the curve $y = x + (2/x)$ at $(1, 3)$. The curve has a third-quadrant portion not shown here. We see how to graph functions like this one in Chapter 3.

Solution The slope of the curve is

$$\frac{dy}{dx} = \frac{d}{dx}(x) + 2\frac{d}{dx}\left(\frac{1}{x}\right) = 1 + 2\left(-\frac{1}{x^2}\right) = 1 - \frac{2}{x^2}.$$

The slope at $x = 1$ is

$$\left.\frac{dy}{dx}\right|_{x=1} = \left[1 - \frac{2}{x^2}\right]_{x=1} = 1 - 2 = -1.$$

The line through (1, 3) with slope $m = -1$ is

$$\begin{aligned} y - 3 &= (-1)(x - 1) && \text{Point-slope equation} \\ y &= -x + 1 + 3 \\ y &= -x + 4. \end{aligned}$$

The choice of which rules to use in solving a differentiation problem can make a difference in how much work you have to do. Here is an example.

Example 6 Choosing Which Rule to Use

Rather than using the Quotient Rule to find the derivative of

$$y = \frac{(x - 1)(x^2 - 2x)}{x^4},$$

expand the numerator and divide by x^4:

$$y = \frac{(x - 1)(x^2 - 2x)}{x^4} = \frac{x^3 - 3x^2 + 2x}{x^4} = x^{-1} - 3x^{-2} + 2x^{-3}.$$

Then use the Sum and Power Rules:

$$\begin{aligned} \frac{dy}{dx} &= -x^{-2} - 3(-2)x^{-3} + 2(-3)x^{-4} \\ &= -\frac{1}{x^2} + \frac{6}{x^3} - \frac{6}{x^4}. \end{aligned}$$

Example 7 The Body's Reaction to Medicine

The reaction of the body to a dose of medicine is sometimes represented by an equation of the form

$$R = M^2\left(\frac{C}{2} - \frac{M}{3}\right),$$

where C is a positive constant and M is the amount of medicine absorbed in the blood. If the reaction is a change in blood pressure, R is measured in millimeters of mercury; if the reaction is a change in temperature, R is measured in degrees; and so on.

Find dR/dM. This derivative, as a function of M, is called the *sensitivity* of the body to the medicine.

Solution

$$\begin{aligned} \frac{dR}{dM} &= 2M\left(\frac{C}{2} - \frac{M}{3}\right) + M^2(-1/3) && \text{Product, Power, and Constant Multiple Rules} \\ &= MC - M^2 \end{aligned}$$

EXERCISES 2.3

Derivative Calculations

In Exercises 1–4, find the first and second derivatives.

1. $y = 6x^2 - 10x - 5x^{-2}$
2. $w = 3z^{-3} - \frac{1}{z}$
3. $r = \frac{1}{3s^2} - \frac{5}{2s}$
4. $r = \frac{12}{\theta} - \frac{4}{\theta^3} + \frac{1}{\theta^4}$

In Exercises 5 and 6, find y' **(a)** by applying the Product Rule and **(b)** by multiplying the factors to produce a sum of simpler terms to differentiate.

5. $y = (3 - x^2)(x^3 - x + 1)$
6. $y = \left(x + \frac{1}{x}\right)\left(x - \frac{1}{x} + 1\right)$

Find the derivatives of the functions in Exercises 7–14.

7. $y = \frac{2x + 5}{3x - 2}$
8. $g(x) = \frac{x^2 - 4}{x + 0.5}$
9. $f(t) = \frac{t^2 - 1}{t^2 + t - 2}$
10. $v = (1 - t)(1 + t^2)^{-1}$
11. $f(s) = \frac{\sqrt{s} - 1}{\sqrt{s} + 1}$
12. $r = 2\left(\frac{1}{\sqrt{\theta}} + \sqrt{\theta}\right)$
13. $y = \frac{1}{(x^2 - 1)(x^2 + x + 1)}$
14. $y = \frac{(x + 1)(x + 2)}{(x - 1)(x - 2)}$

Find the first and second derivatives of the functions in Exercises 15–18.

15. $s = \frac{t^2 + 5t - 1}{t^2}$
16. $r = \frac{(\theta - 1)(\theta^2 + \theta + 1)}{\theta^3}$
17. $w = \left(\frac{1 + 3z}{3z}\right)(3 - z)$
18. $p = \left(\frac{q^2 + 3}{12q}\right)\left(\frac{q^4 - 1}{q^3}\right)$

Using Numerical Values

19. Suppose that u and v are functions of x that are differentiable at $x = 0$ and that

$$u(0) = 5, \quad u'(0) = 3$$
$$v(0) = -1, \quad v'(0) = 2.$$

Find the values of the following derivatives at $x = 0$.

(a) $\frac{d}{dx}(uv)$ **(b)** $\frac{d}{dx}\left(\frac{u}{v}\right)$

(c) $\frac{d}{dx}\left(\frac{v}{u}\right)$ **(d)** $\frac{d}{dx}(7v - 2u)$

20. Suppose that u and v are differentiable functions of x and that

$$u(1) = 2, \quad u'(1) = 0$$
$$v(1) = 5, \quad v'(1) = -1.$$

Find the values of the following derivatives at $x = 1$.

(a) $\frac{d}{dx}(uv)$ **(b)** $\frac{d}{dx}\left(\frac{u}{v}\right)$

(c) $\frac{d}{dx}\left(\frac{v}{u}\right)$ **(d)** $\frac{d}{dx}(7v - 2u)$

Slopes and Tangents

21. Find the tangents to *Newton's serpentine* (graphed here) at the origin and the point (1, 2).

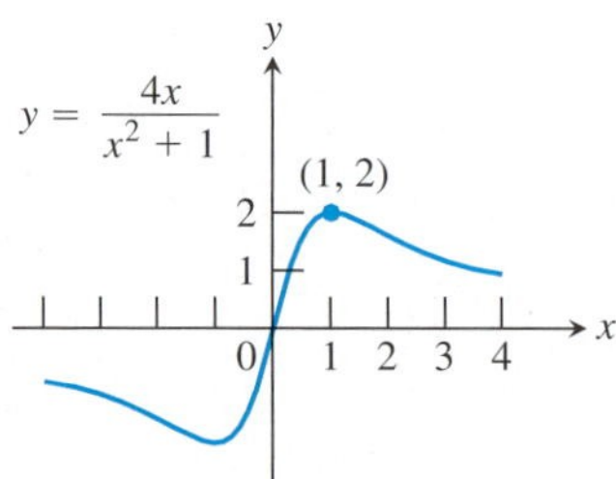

22. Find the tangent to the *witch of Agnesi* (graphed here) at the point (2, 1).

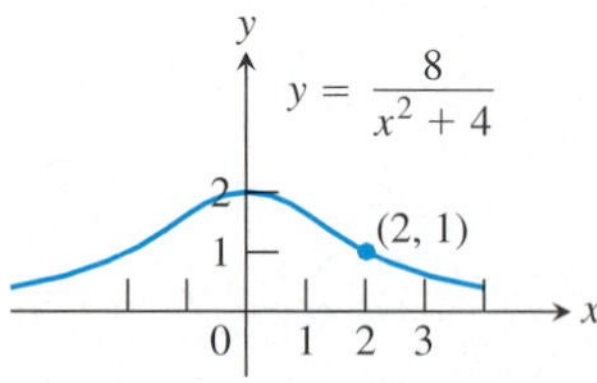

23. The curve $y = ax^2 + bx + c$ passes through the point (1, 2) and is tangent to the line $y = x$ at the origin. Find a, b, and c.

24. The curves $y = x^2 + ax + b$ and $y = cx - x^2$ have a common tangent at the point (1, 0). Find a, b, and c.

Theory and Examples

25. *Writing to Learn* Suppose that the function v in the Product Rule has a constant value c. What does the Product Rule then say? What does this say about the Constant Multiple Rule?

26. *The Reciprocal Rule*

(a) The Reciprocal Rule says that at any point where the function $v(x)$ is differentiable and different from zero,

$$\frac{d}{dx}\left(\frac{1}{v}\right) = -\frac{1}{v^2}\frac{dv}{dx}.$$

Show that the Reciprocal Rule is a special case of the Quotient Rule.

(b) Show that the Reciprocal Rule and the Product Rule together imply the Quotient Rule.

27. *Generalizing the Product Rule* The Product Rule gives the formula

$$\frac{d}{dx}(uv) = u\frac{dv}{dx} + v\frac{du}{dx}$$

for the derivative of the product uv of two differentiable functions of x.

(a) What is the analogous formula for the derivative of the product uvw of *three* differentiable functions of x?

(b) What is the formula for the derivative of the product $u_1u_2u_3u_4$ of *four* differentiable functions of x?

(c) What is the formula for the derivative of a product $u_1u_2u_3 \cdots u_n$ of a finite number n of differentiable functions of x?

28. *Rational powers*

(a) Find $\frac{d}{dx}(x^{3/2})$ by writing $x^{3/2}$ as $x \cdot x^{1/2}$ and using the Product Rule. Express your answer as a rational number times a rational power of x. Work parts (b) and (c) by a similar method.

(b) Find $\frac{d}{dx}(x^{5/2})$.

(c) Find $\frac{d}{dx}(x^{7/2})$.

(d) What patterns do you see in your answers to parts (a), (b), and (c)? Rational powers are explored in Section 2.6.

29. *Cylinder pressure* If gas in a cylinder is maintained at a constant temperature T, the pressure P is related to the volume V by a formula of the form

$$P = \frac{nRT}{V - nb} - \frac{an^2}{V^2},$$

in which a, b, n, and R are constants. Find dP/dV.

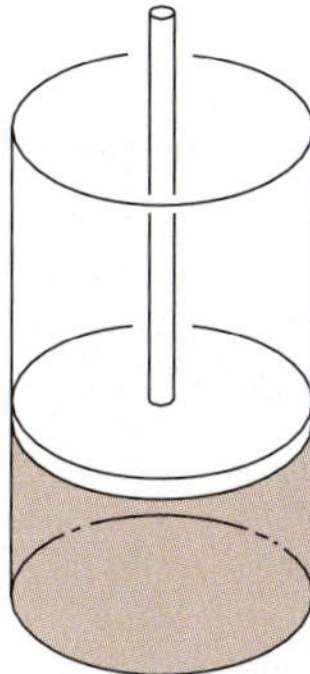

30. *The best quantity to order* One of the formulas for inventory management says that the average weekly cost of ordering, paying for, and holding merchandise is

$$A(q) = \frac{km}{q} + cm + \frac{hq}{2},$$

where q is the quantity you order when things run low (shoes, radios, brooms, or whatever the item might be); k is the cost of placing an order (the same, no matter how often you order); c is the cost of one item (a constant); m is the number of items sold each week (a constant); and h is the weekly holding cost per item (a constant that takes into account things such as space, utilities, insurance, and security). Find dA/dq and d^2A/dq^2.

2.4 Derivatives of Trigonometric Functions

Derivative of the Sine Function • Derivative of the Cosine Function • Simple Harmonic Motion • Derivatives of the Other Basic Trigonometric Functions • Continuity of Trigonometric Functions

Many of the phenomena we want information about are periodic (electromagnetic fields, heart rhythms, tides, weather). A surprising and beautiful theorem from advanced calculus says that every periodic function we are likely to use in mathematical modeling can be expressed in terms of sines and cosines. Thus, the derivatives of sines and cosines play a key role in describing periodic changes. This section shows how to differentiate the six basic trigonometric functions.

Derivative of the Sine Function

To calculate the derivative of $y = \sin x$, we combine the limits in Example 10(a) and Theorem 6 in Section 1.2 with the angle sum identity:

$$\sin(x + h) = \sin x \cos h + \cos x \sin h. \tag{1}$$

We have

$$\begin{aligned}
\frac{dy}{dx} &= \lim_{h \to 0} \frac{\sin(x+h) - \sin x}{h} && \text{Derivative definition} \\
&= \lim_{h \to 0} \frac{(\sin x \cos h + \cos x \sin h) - \sin x}{h} && \text{Eq. (1)} \\
&= \lim_{h \to 0} \frac{\sin x(\cos h - 1) + \cos x \sin h}{h} \\
&= \lim_{h \to 0} \left(\sin x \cdot \frac{\cos h - 1}{h}\right) + \lim_{h \to 0} \left(\cos x \cdot \frac{\sin h}{h}\right) \\
&= \sin x \cdot \lim_{h \to 0} \frac{\cos h - 1}{h} + \cos x \cdot \lim_{h \to 0} \frac{\sin h}{h} \\
&= \sin x \cdot 0 + \cos x \cdot 1 && \text{Example 10a and Theorem 6, Section 1.2} \\
&= \cos x.
\end{aligned}$$

Radian measure in calculus

In case you are wondering why calculus uses radian measure when the rest of the world seems to use degrees, the answer lies in the argument that the derivative of the sine is the cosine. The derivative of $\sin x$ is $\cos x$ only if x is measured in radians, because

$$\lim_{h \to 0} \frac{\sin h}{h} = 1$$

only for radian measure.

The derivative of the sine function is the cosine function.

$$\frac{d}{dx}(\sin x) = \cos x$$

Example 1 Derivatives Involving the Sine

(a) $y = x^2 - \sin x$:

$$\begin{aligned}
\frac{dy}{dx} &= 2x - \frac{d}{dx}(\sin x) && \text{Difference Rule} \\
&= 2x - \cos x.
\end{aligned}$$

(b) $y = \dfrac{\sin x}{x}$:

$$\begin{aligned}
\frac{dy}{dx} &= \frac{x \cdot \frac{d}{dx}(\sin x) - \sin x \cdot 1}{x^2} && \text{Quotient Rule} \\
&= \frac{x \cos x - \sin x}{x^2}.
\end{aligned}$$

CD-ROM WEBsite

Derivative of the Cosine Function

With the help of the angle sum formula,

$$\cos(x+h) = \cos x \cos h - \sin x \sin h, \tag{2}$$

we have

$$\begin{aligned}
\frac{d}{dx}(\cos x) &= \lim_{h\to 0}\frac{\cos(x+h)-\cos x}{h} && \text{Derivative definition}\\
&= \lim_{h\to 0}\frac{(\cos x\cos h - \sin x \sin h)-\cos x}{h} && \text{Eq. (2)}\\
&= \lim_{h\to 0}\frac{\cos x(\cos h - 1) - \sin x \sin h}{h}\\
&= \lim_{h\to 0}\cos x\cdot\frac{\cos h - 1}{h} - \lim_{h\to 0}\sin x\cdot\frac{\sin h}{h}\\
&= \cos x\cdot\lim_{h\to 0}\frac{\cos h - 1}{h} - \sin x\cdot\lim_{h\to 0}\frac{\sin h}{h}\\
&= \cos x\cdot 0 - \sin x\cdot 1 && \text{Example 10(a) and Theorem 6, Section 1.2}\\
&= -\sin x.
\end{aligned}$$

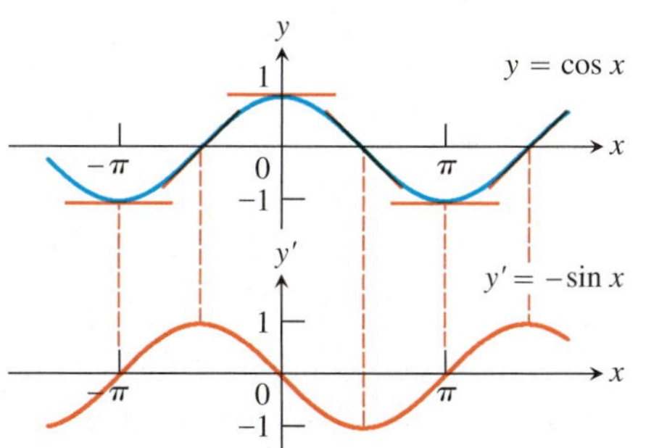

FIGURE 2.25 The curve $y' = -\sin x$ as the graph of the slopes of the tangents to the curve $y = \cos x$.

The derivative of the cosine function is the negative of the sine function.

$$\frac{d}{dx}(\cos x) = -\sin x$$

Figure 2.25 shows another way to visualize this result.

Example 2 Revisiting the Differentiation Rules

(a) $y = \sin x \cos x$:

$$\begin{aligned}
\frac{dy}{dx} &= \sin x\frac{d}{dx}(\cos x) + \cos x\frac{d}{dx}(\sin x) && \text{Product Rule}\\
&= \sin x(-\sin x) + \cos x(\cos x)\\
&= \cos^2 x - \sin^2 x.
\end{aligned}$$

(b) $y = \dfrac{\cos x}{1-\sin x}$:

$$\begin{aligned}
\frac{dy}{dx} &= \frac{(1-\sin x)\frac{d}{dx}(\cos x) - \cos x\frac{d}{dx}(1-\sin x)}{(1-\sin x)^2} && \text{Quotient Rule}\\
&= \frac{(1-\sin x)(-\sin x) - \cos x(0-\cos x)}{(1-\sin x)^2}\\
&= \frac{1-\sin x}{(1-\sin x)^2} && \sin^2 x + \cos^2 x = 1\\
&= \frac{1}{1-\sin x}.
\end{aligned}$$

Simple Harmonic Motion

The motion of a body bobbing freely up and down on the end of a spring or bungee cord is an example of *simple harmonic motion*. The next example describes a case in which there are no opposing forces such as friction or buoyancy to slow the motion down.

Example 3 Motion on a Spring

A body hanging from a spring (Figure 2.26) is stretched 5 units beyond its rest position and released at time $t = 0$ to bob up and down. Its position at any later time t is

$$s = 5 \cos t .$$

What are its velocity and acceleration at time t?

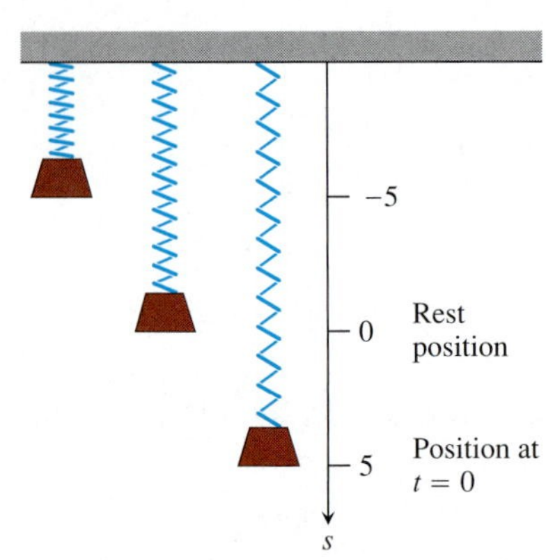

FIGURE 2.26 The body in Example 3.

Solution We have

Position: $s = 5 \cos t$

Velocity: $v = \dfrac{ds}{dt} = \dfrac{d}{dt}(5 \cos t) = -5 \sin t$

Acceleration: $a = \dfrac{dv}{dt} = \dfrac{d}{dt}(-5 \sin t) = -5 \cos t .$

Notice how much we can learn from these equations:

1. As time passes, the weight moves down and up between $s = -5$ and $s = 5$ on the s-axis. The amplitude of the motion is 5. The period of the motion is 2π.
2. The velocity $v = -5 \sin t$ attains its greatest magnitude, 5, when $\cos t = 0$, as the graphs show in Figure 2.27. Hence, the speed of the weight, $|v| = 5 |\sin t|$, is greatest when $\cos t = 0$, that is, when $s = 0$ (the rest position). The speed of the weight is zero when $\sin t = 0$. This occurs when $s = 5 \cos t = \pm 5$, at the endpoints of the interval of motion.
3. The acceleration value is always the exact opposite of the position value. When the weight is above the rest position, gravity is pulling it back down; when the weight is below the rest position, the spring is pulling it back up.
4. The acceleration, $a = -5 \cos t$, is zero only at the rest position, where $\cos t = 0$ and the force of gravity and the force from the spring offset each other. When the weight is anywhere else, the two forces are unequal and acceleration is nonzero. The acceleration is greatest in magnitude at the points farthest from the rest position, where $\cos t = \pm 1$.

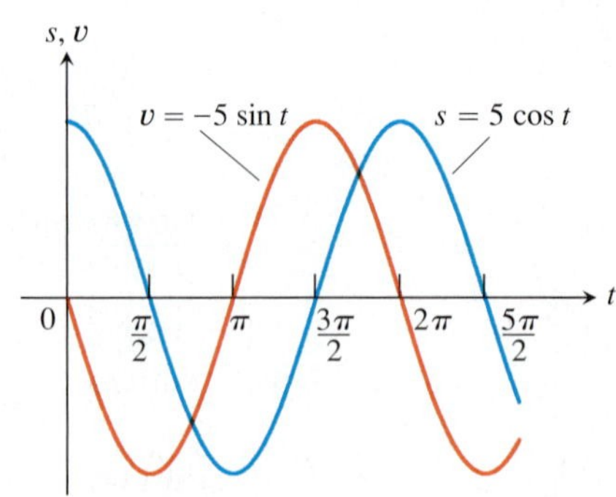

FIGURE 2.27 The graphs of the position and velocity of the body in Example 3.

Example 4 Jerk

The jerk of the simple harmonic motion in Example 3 is

$$j = \frac{da}{dt} = \frac{d}{dt}(-5 \cos t) = 5 \sin t .$$

It has its greatest magnitude when $\sin t = \pm 1$, not at the extremes of the displacement but at the rest position, where the acceleration changes direction and sign.

Derivatives of the Other Basic Trigonometric Functions

Because $\sin x$ and $\cos x$ are differentiable functions of x, the related functions

$$\tan x = \frac{\sin x}{\cos x}$$

$$\cot x = \frac{\cos x}{\sin x}$$

$$\sec x = \frac{1}{\cos x}$$

$$\csc x = \frac{1}{\sin x}$$

are differentiable at every value of x at which they are defined. Their derivatives, calculated from the Quotient Rule, are given by the following formulas.

Notice the minus signs in the derivative formulas for the cofunctions.

$$\frac{d}{dx}(\tan x) = \sec^2 x \tag{3}$$

$$\frac{d}{dx}(\sec x) = \sec x \tan x \tag{4}$$

$$\frac{d}{dx}(\cot x) = -\csc^2 x \tag{5}$$

$$\frac{d}{dx}(\csc x) = -\csc x \cot x \tag{6}$$

To show a typical calculation, we derive Equation (3). The other derivations are left to Exercise 44.

Example 5 Derivative of the Tangent Function

Find $d(\tan x)/dx$.

Solution

$$\frac{d}{dx}(\tan x) = \frac{d}{dx}\left(\frac{\sin x}{\cos x}\right) = \frac{\cos x \frac{d}{dx}(\sin x) - \sin x \frac{d}{dx}(\cos x)}{\cos^2 x} \quad \text{Quotient Rule}$$

$$= \frac{\cos x \cos x - \sin x (-\sin x)}{\cos^2 x}$$

$$= \frac{\cos^2 x + \sin^2 x}{\cos^2 x}$$

$$= \frac{1}{\cos^2 x} = \sec^2 x$$

Example 6 A Trigonometric Second Derivative

Find y'' if $y = \sec x$.

Solution

$$\begin{aligned} y &= \sec x \\ y' &= \sec x \tan x && \text{Eq. (4)} \\ y'' &= \frac{d}{dx}(\sec x \tan x) \\ &= \sec x \frac{d}{dx}(\tan x) + \tan x \frac{d}{dx}(\sec x) && \text{Product Rule} \\ &= \sec x (\sec^2 x) + \tan x (\sec x \tan x) \\ &= \sec^3 x + \sec x \tan^2 x \end{aligned}$$

Continuity of Trigonometric Functions

Since the six basic trigonometric functions are differentiable throughout their domains, they are continuous throughout their domains, by Theorem 1, Section 2.1. Hence, $\sin x$ and $\cos x$ are continuous for all x, supporting our observations in Section 1.4. Also, $\sec x$ and $\tan x$ are continuous except when x is a nonzero integer multiple of $\pi/2$, and $\csc x$ and $\cot x$ are continuous except at integer multiples of π. For each function, $\lim_{x \to c} f(x) = f(c)$ whenever $f(c)$ is defined. As a result, we can calculate the limits of many algebraic combinations and composites of trigonometric functions by direct substitution.

Example 7 Finding a Trigonometric Limit

$$\lim_{x \to 0} \frac{\sqrt{2 + \sec x}}{\cos(\pi - \tan x)} = \frac{\sqrt{2 + \sec 0}}{\cos(\pi - \tan 0)} = \frac{\sqrt{2 + 1}}{\cos(\pi - 0)} = \frac{\sqrt{3}}{-1} = -\sqrt{3}$$

EXERCISES 2.4

Derivatives

In Exercises 1–12, find dy/dx.

1. $y = -10x + 3 \cos x$
2. $y = \frac{3}{x} + 5 \sin x$
3. $y = \csc x - 4\sqrt{x} + 7$
4. $y = x^2 \cot x - \frac{1}{x^2}$
5. $y = (\sec x + \tan x)(\sec x - \tan x)$
6. $y = (\sin x + \cos x) \sec x$
7. $y = \frac{\cot x}{1 + \cot x}$
8. $y = \frac{\cos x}{1 + \sin x}$
9. $y = \frac{4}{\cos x} + \frac{1}{\tan x}$
10. $y = \frac{\cos x}{x} + \frac{x}{\cos x}$
11. $y = x^2 \sin x + 2x \cos x - 2 \sin x$
12. $y = x^2 \cos x - 2x \sin x - 2 \cos x$

In Exercises 13–16, find ds/dt.

13. $s = \tan t - t$
14. $s = t^2 - \sec t + 1$
15. $s = \frac{1 + \csc t}{1 - \csc t}$
16. $s = \frac{\sin t}{1 - \cos t}$

In Exercises 17–20, find $dr/d\theta$.

17. $r = 4 - \theta^2 \sin \theta$
18. $r = \theta \sin \theta + \cos \theta$
19. $r = \sec \theta \csc \theta$
20. $r = (1 + \sec \theta) \sin \theta$

In Exercises 21–24, find dp/dq.

21. $p = 5 + \frac{1}{\cot q}$
22. $p = (1 + \csc q) \cos q$
23. $p = \frac{\sin q + \cos q}{\cos q}$
24. $p = \frac{\tan q}{1 + \tan q}$
25. Find y'' if
    (a) $y = \csc x$.
    (b) $y = \sec x$.

26. Find $y^{(4)} = d^4 y/dx^4$ if

(a) $y = -2 \sin x$. (b) $y = 9 \cos x$.

Tangent Lines

In Exercises 27–30, graph the curves over the given intervals, together with their tangents at the given values of x. Label each curve and tangent with its equation.

27. $y = \sin x, \quad -3\pi/2 \le x \le 2\pi$

$x = -\pi, 0, 3\pi/2$

28. $y = \tan x, \quad -\pi/2 < x < \pi/2$

$x = -\pi/3, 0, \pi/3$

29. $y = \sec x, \quad -\pi/2 < x < \pi/2$

$x = -\pi/3, \pi/4$

30. $y = 1 + \cos x, \quad -3\pi/2 \le x \le 2\pi$

$x = -\pi/3, 3\pi/2$

T Do the graphs of the functions in Exercises 31–34 have any horizontal tangents in the interval $0 \le x \le 2\pi$? If so, where? If not, why not? Visualize your findings by graphing the functions with a grapher.

31. $y = x + \sin x$ 32. $y = 2x + \sin x$

33. $y = x - \cot x$ 34. $y = x + 2 \cos x$

35. Find all points on the curve $y = \tan x$, $-\pi/2 < x < \pi/2$, where the tangent line is parallel to the line $y = 2x$. Sketch the curve and tangent(s) together, labeling each with its equation.

36. Find all points on the curve $y = \cot x$, $0 < x < \pi$, where the tangent line is parallel to the line $y = -x$. Sketch the curve and tangent(s) together, labeling each with its equation.

In Exercises 37 and 38, find an equation for **(a)** the tangent to the curve at P and **(b)** the horizontal tangent to the curve at Q.

37.

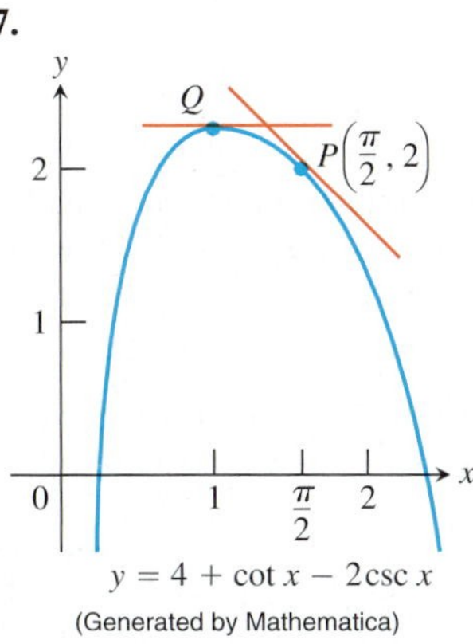

(Generated by Mathematica)

38.

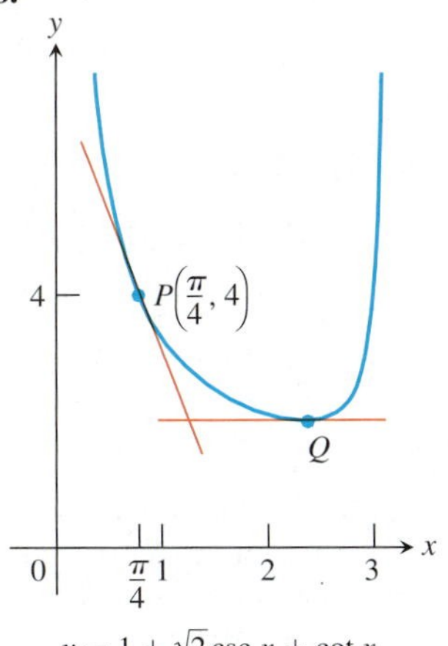

(Generated by Mathematica)

Simple Harmonic Motion

The equations in Exercises 39 and 40 give the position $s = f(t)$ of a body moving on a coordinate line (s in meters, t in seconds). Find the body's velocity, speed, acceleration, and jerk at time $t = \pi/4$ sec.

39. $s = 2 - 2 \sin t$ 40. $s = \sin t + \cos t$

Theory and Examples

41. *Writing to Learn* Is there a value of c that will make

$$f(x) = \begin{cases} \dfrac{\sin^2 3x}{x^2}, & x \ne 0 \\ c, & x = 0 \end{cases}$$

continuous at $x = 0$? Give reasons for your answer.

42. *Writing to Learn* Is there a value of b that will make

$$g(x) = \begin{cases} x + b, & x < 0 \\ \cos x, & x \ge 0 \end{cases}$$

continuous at $x = 0$? Differentiable at $x = 0$? Give reasons for your answers.

43. Find $d^{999}/dx^{999} (\cos x)$.

44. Derive the formula for the derivative with respect to x of

(a) $\sec x$.

(b) $\csc x$.

(c) $\cot x$.

T 45. Graph $y = \cos x$ for $-\pi \le x \le 2\pi$. On the same screen, graph

$$y = \frac{\sin (x + h) - \sin x}{h}$$

for $h = 1, 0.5, 0.3$, and 0.1. Then, in a new window, try $h = -1$, -0.5, and -0.3. What happens as $h \to 0^+$? As $h \to 0^-$? What phenomenon is being illustrated here?

T 46. Graph $y = -\sin x$ for $-\pi \le x \le 2\pi$. On the same screen, graph

$$y = \frac{\cos (x + h) - \cos x}{h}$$

for $h = 1, 0.5, 0.3$, and 0.1. Then, in a new window, try $h = -1, -0.5$, and -0.3. What happens as $h \to 0^+$? As $h \to 0^-$? What phenomenon is being illustrated here?

T 47. *Centered difference quotients* The **centered difference quotient**

$$\frac{f(x + h) - f(x - h)}{2h}$$

is used to approximate $f'(x)$ in numerical work because (1) its limit as $h \to 0$ equals $f'(x)$ when $f'(x)$ exists, and (2) it usually gives a better approximation of $f'(x)$ for a given value of h than Fermat's difference quotient

$$\frac{f(x+h)-f(x)}{h}.$$

See the accompanying figure.

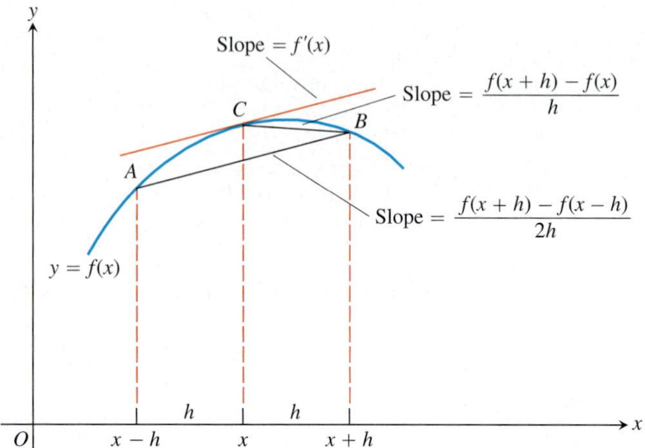

(a) To see how rapidly the centered difference quotient for $f(x) = \sin x$ converges to $f'(x) = \cos x$, graph $y = \cos x$ together with

$$y = \frac{\sin(x+h) - \sin(x-h)}{2h}$$

over the interval $[-\pi, 2\pi]$ for $h = 1, 0.5$, and 0.3. Compare the results with those obtained in Exercise 45 for the same values of h.

(b) To see how rapidly the centered difference quotient for $f(x) = \cos x$ converges to $f'(x) = -\sin x$, graph $y = -\sin x$ together with

$$y = \frac{\cos(x+h) - \cos(x-h)}{2h}$$

over the interval $[-\pi, 2\pi]$ for $h = 1, 0.5$, and 0.3. Compare the results with those obtained in Exercise 46 for the same values of h.

48. *A caution about centered difference quotients* *(Continuation of Exercise 47)* The quotient

$$\frac{f(x+h) - f(x-h)}{2h}$$

may have a limit as $h \to 0$ when f has no derivative at x. As a case in point, take $f(x) = |x|$ and calculate

$$\lim_{h\to 0} \frac{|0+h| - |0-h|}{2h}.$$

As you will see, the limit exists even though $f(x) = |x|$ has no derivative at $x = 0$. *Moral:* Before using a centered difference quotient, be sure the derivative exists.

T **49.** *Writing to Learn: Slopes on the graph of the tangent function* Graph $y = \tan x$ and its derivative together on $(-\pi/2, \pi/2)$. Does the graph of the tangent function appear to have a smallest slope? a largest slope? Is the slope ever negative? Give reasons for your answers.

T **50.** *Writing to Learn: Slopes on the graph of the cotangent function* Graph $y = \cot x$ and its derivative together for $0 < x < \pi$. Does the graph of the cotangent function appear to have a smallest slope? A largest slope? Is the slope ever positive? Give reasons for your answers.

T **51.** *Writing to Learn: Exploring (sin kx)/x* Graph $y = (\sin x)/x$, $y = (\sin 2x)/x$, and $y = (\sin 4x)/x$ together over the interval $-2 \le x \le 2$. Where does each graph appear to cross the y-axis? Do the graphs really intersect the axis? What would you expect the graphs of $y = (\sin 5x)/x$ and $y = (\sin(-3x))/x$ to do as $x \to 0$? Why? What about the graph of $y = (\sin kx)/x$ for other values of k? Give reasons for your answers.

T **52.** *Radians versus degrees: degree mode derivatives* What happens to the derivatives of $\sin x$ and $\cos x$ if x is measured in degrees instead of radians? To find out, take the following steps.

(a) With your graphing calculator or computer grapher in *degree mode*, graph

$$f(h) = \frac{\sin h}{h}$$

and estimate $\lim_{h\to 0} f(h)$. Compare your estimate with $\pi/180$. Is there any reason to believe the limit *should* be $\pi/180$?

(b) With your grapher still in degree mode, estimate

$$\lim_{h\to 0} \frac{\cos h - 1}{h}.$$

(c) Now go back to the derivation of the formula for the derivative of $\sin x$ in the text and carry out the steps of the derivation using degree-mode limits. What formula do you obtain for the derivative?

(d) Work through the derivation of the formula for the derivative of $\cos x$ using degree-mode limits. What formula do you obtain for the derivative?

(e) The disadvantages of the degree-mode formulas become apparent as you start taking derivatives of higher order. Try it. What are the second and third degree-mode derivatives of $\sin x$ and $\cos x$?

2.5 The Chain Rule and Parametric Equations

Derivative of a Composite Function • "Outside-Inside" Rule • Repeated Use of the Chain Rule • Slopes of Parametrized Curves • Power Chain Rule • Melting Ice Cubes

As we see in Chapters 3 and 4, many engineering applications of calculus involve finding a function with a given derivative. Sometimes we identify such a function right away. We know that $\cos x$ is the derivative of $\sin x$, for example, that $2x$ is the derivative of x^2, and that the sum $\cos x + 2x$ is the derivative of $\sin x + x^2$. But what if we're starting with the *product* of two derivatives instead of their sum? We know it does not come from the product of the differentiated functions because the derivative of their product is not the product of their derivatives.

So, where does a product of derivatives come from? The answer lies in a rule for differentiating composite functions called the Chain Rule. This section describes the rule and how to use it.

Derivative of a Composite Function

We begin with examples.

Example 1 Relating Derivatives

The function $y = 6x - 10 = 2(3x - 5)$ is the composite of the functions $y = 2u$ and $u = 3x - 5$. How are the derivatives of these functions related?

Solution We have

$$\frac{dy}{dx} = 6, \qquad \frac{dy}{du} = 2, \qquad \text{and} \qquad \frac{du}{dx} = 3.$$

Since $6 = 2 \cdot 3$, we see that

$$\frac{dy}{dx} = \frac{dy}{du} \cdot \frac{du}{dx}.$$

Is it an accident that

$$\frac{dy}{dx} = \frac{dy}{du} \cdot \frac{du}{dx}?$$

If we think of the derivative as a rate of change, our intuition allows us to see that this relationship is reasonable. If $y = f(u)$ changes twice as fast as u and $u = g(x)$ changes three times as fast as x, then we expect y to change six times as fast as x. This effect is much like that of a multiple gear train (Figure 2.28).

FIGURE 2.28 When gear A makes x turns, gear B makes u turns and gear C makes y turns. By comparing circumferences or counting teeth, we see that $y = u/2$ and $u = 3x$, so $y = 3x/2$. Thus, $dy/du = 1/2$, $du/dx = 3$, and $dy/dx = 3/2 = (dy/du)(du/dx)$.

Example 2 Relating Derivatives

The function

$$y = 9x^4 + 6x^2 + 1 = (3x^2 + 1)^2$$

is the composite of $y = u^2$ and $u = 3x^2 + 1$. Calculating derivatives, we see that

$$\begin{aligned} \frac{dy}{du} \cdot \frac{du}{dx} &= 2u \cdot 6x \\ &= 2(3x^2 + 1) \cdot 6x \qquad u = 3x^2 + 1 \\ &= 36x^3 + 12x \end{aligned}$$

and

$$\frac{dy}{dx} = \frac{d}{dx}(9x^4 + 6x^2 + 1) = 36x^3 + 12x .$$

Once again,

$$\frac{dy}{du} \cdot \frac{du}{dx} = \frac{dy}{dx} .$$

The derivative of the composite $f(g(x))$ at x is the derivative of f at $g(x)$ times the derivative of g at x. This observation is known as the Chain Rule (Figure 2.29).

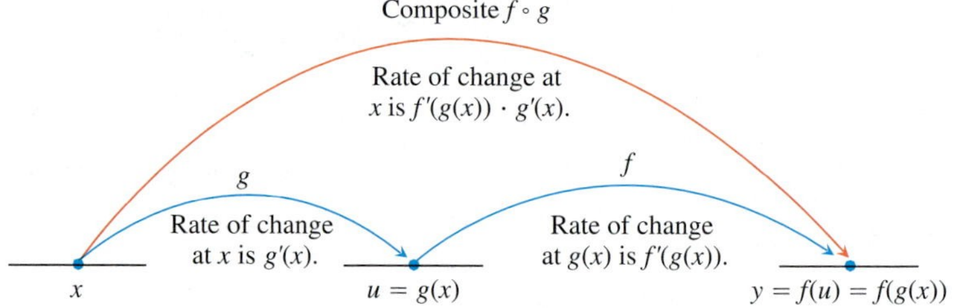

FIGURE 2.29 Rates of change multiply: The derivative of $f \circ g$ at x is the derivative of f at the point $g(x)$ times the derivative of g at the point x.

Theorem 3 The Chain Rule
If $f(u)$ is differentiable at the point $u = g(x)$ and $g(x)$ is differentiable at x, then the composite function $(f \circ g)(x) = f(g(x))$ is differentiable at x, and

$$(f \circ g)'(x) = f'(g(x)) \cdot g'(x) . \tag{1}$$

In Leibniz's notation, if $y = f(u)$ and $u = g(x)$, then

$$\frac{dy}{dx} = \frac{dy}{du} \cdot \frac{du}{dx}, \tag{2}$$

where dy/du is evaluated at $u = g(x)$.

It would be tempting to try to prove the Chain Rule by writing

$$\frac{\Delta y}{\Delta x} = \frac{\Delta y}{\Delta u} \cdot \frac{\Delta u}{\Delta x}$$

and taking the limit as $\Delta x \to 0$. This method would work if we knew that Δu, the change in u, was nonzero, but we do not know this. A small change in x could conceivably produce no change in u. The proof requires a different approach, using ideas in Section 3.6. (See Appendix 3.)

Example 3 Applying the Chain Rule

An object moves along the x-axis so that its position at any time $t \geq 0$ is given by $x(t) = \cos(t^2 + 1)$. Find the velocity of the object as a function of t.

Solution We know that the velocity is dx/dt. In this instance, x is a composite function: $x = \cos(u)$ and $u = t^2 + 1$. We have

$$\frac{dx}{du} = -\sin(u) \qquad x = \cos(u)$$

$$\frac{du}{dt} = 2t. \qquad u = t^2 + 1$$

By the Chain Rule,

$$\begin{aligned}\frac{dx}{dt} &= \frac{dx}{du} \cdot \frac{du}{dt} \\ &= -\sin(u) \cdot 2t \\ &= -\sin(t^2 + 1) \cdot 2t \\ &= -2t \sin(t^2 + 1).\end{aligned}$$

"Outside-Inside" Rule

It sometimes helps to think about the Chain Rule this way: If $y = f(g(x))$, then

$$\frac{dy}{dx} = f'(g(x)) \cdot g'(x). \tag{3}$$

In words, differentiate the "outside" function f and evaluate it at the "inside" function $g(x)$ left alone; then multiply by the derivative of the "inside function."

Example 4 Differentiating from the Outside In

Differentiate $\sin(x^2 + x)$ with respect to x.

Solution

$$\frac{d}{dx} \sin \underbrace{(x^2 + x)}_{\text{inside}} = \cos \underbrace{(x^2 + x)}_{\substack{\text{inside} \\ \text{left alone}}} \cdot \underbrace{(2x + 1)}_{\substack{\text{derivative of} \\ \text{the inside}}}$$

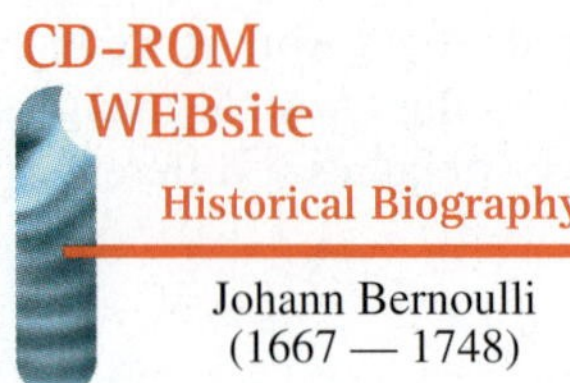

Repeated Use of the Chain Rule

We sometimes have to use the Chain Rule two or more times to find a derivative. Here is an example.

Example 5 A Three-Link "Chain"

Find the derivative of $g(t) = \tan(5 - \sin 2t)$.

Solution Notice here that the tangent is a function of $5 - \sin 2t$, whereas the sine is a function of $2t$, which is itself a function of t. Therefore, by the Chain Rule,

$$\begin{aligned} g'(t) &= \frac{d}{dt}(\tan(5 - \sin 2t)) \\ &= \sec^2(5 - \sin 2t) \cdot \frac{d}{dt}(5 - \sin 2t) && \text{Derivative of } \tan u \text{ with } u = 5 - \sin 2t \\ &= \sec^2(5 - \sin 2t) \cdot \left(0 - \cos 2t \cdot \frac{d}{dt}(2t)\right) && \text{Derivative of } 5 - \sin u \text{ with } u = 2t \\ &= \sec^2(5 - \sin 2t) \cdot (-\cos 2t) \cdot 2 \\ &= -2(\cos 2t)\sec^2(5 - \sin 2t). \end{aligned}$$

Slopes of Parametrized Curves

A parametrized curve $(x(t), y(t))$ is **differentiable** at t if x and y are differentiable at t. At a point on a differentiable parametrized curve where y is also a differentiable function of x, the derivatives dy/dt, dx/dt, and dy/dx are related by the Chain Rule:

$$\frac{dy}{dt} = \frac{dy}{dx} \cdot \frac{dx}{dt}.$$

If $dx/dt \neq 0$, we may divide both sides of this equation by dx/dt to solve for dy/dx.

Parametric Formula for dy/dx

If all three derivatives exist and $dx/dt \neq 0$,

$$\frac{dy}{dx} = \frac{dy/dt}{dx/dt}. \tag{4}$$

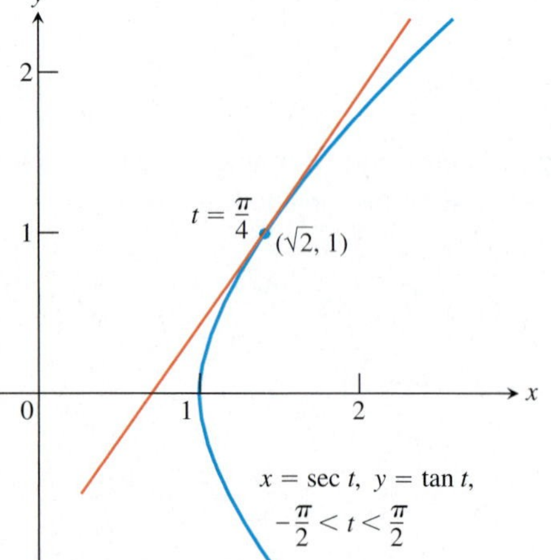

FIGURE 2.30 The hyperbola branch in Example 6. Equation (4) applies for every point on the graph except $(1, 0)$.

Example 6 Differentiating with a Parameter

Find the line tangent to the right-hand hyperbola branch defined parametrically by

$$x = \sec t, \qquad y = \tan t, \qquad -\frac{\pi}{2} < t < \frac{\pi}{2}$$

at the point $(\sqrt{2}, 1)$, where $t = \pi/4$ (Figure 2.30).

Solution All three of the derivatives in Equation (4) exist, and $dx/dt = \sec t \tan t \neq 0$ at the indicated point. Therefore, Equation (4) applies, and

$$\begin{aligned}\frac{dy}{dx} &= \frac{dy/dt}{dx/dt} \\ &= \frac{\sec^2 t}{\sec t \tan t} \\ &= \frac{\sec t}{\tan t} \\ &= \csc t .\end{aligned}$$

Setting $t = \pi/4$ gives

$$\left.\frac{dy}{dx}\right|_{t=\pi/4} = \csc(\pi/4) = \sqrt{2}.$$

The equation of the tangent line is

$$\begin{aligned}y - 1 &= \sqrt{2}(x - \sqrt{2}) \\ y &= \sqrt{2}x - 2 + 1 \\ y &= \sqrt{2}x - 1.\end{aligned}$$

If parametric equations define y as a twice-differentiable function of x, we can apply Equation (4) to the function $dy/dx = y'$ to calculate d^2y/dx^2 as a function of t:

$$\frac{d^2y}{dx^2} = \frac{d}{dx}(y') = \frac{dy'/dt}{dx/dt}. \qquad \text{Eq. (4) with } y' \text{ in place of } y$$

Parametric Formula for d^2y/dx^2

If the equations $x = f(t)$, $y = g(t)$ define y as a twice-differentiable function of x, then at any point where $dx/dt \neq 0$,

$$\frac{d^2y}{dx^2} = \frac{dy'/dt}{dx/dt}.$$

Example 7 Finding d^2y/dx^2 for a Parametrized Curve

Find d^2y/dx^2 as a function of t if $x = t - t^2$, $y = t - t^3$.

Finding d^2y/dx^2 in terms of t

Step 1. Express $y' = dy/dx$ in terms of t.

Step 2. Find dy'/dt.

Step 3. Divide dy'/dt by dx/dt.

Solution

Step 1: Express $y' = dy/dx$ in terms of t.

$$y' = \frac{dy}{dx} = \frac{dy/dt}{dx/dt} = \frac{1 - 3t^2}{1 - 2t}$$

Step 2: Differentiate y' with respect to t.

$$\frac{dy'}{dt} = \frac{d}{dt}\left(\frac{1 - 3t^2}{1 - 2t}\right) = \frac{2 - 6t + 6t^2}{(1 - 2t)^2}. \qquad \text{Quotient Rule}$$

Step 3: Divide dy'/dt by dx/dt.

$$\frac{d^2y}{dx^2} = \frac{dy'/dt}{dx/dt} = \frac{(2 - 6t + 6t^2)/(1 - 2t)^2}{1 - 2t} = \frac{2 - 6t + 6t^2}{(1 - 2t)^3}$$

Power Chain Rule

If f is a differentiable function of u and if u is a differentiable function of x, then substituting $y = f(u)$ into the Chain Rule formula

$$\frac{dy}{dx} = \frac{dy}{du} \cdot \frac{du}{dx}$$

leads to the formula

$$\frac{d}{dx} f(u) = f'(u) \frac{du}{dx}.$$

Here's an example of how it works: If n is an integer and $f(u) = u^n$, the Power Rules (Rules 2 and 7) tell us that $f'(u) = nu^{n-1}$. If u is a differentiable function of x, then we can use the Chain Rule to extend this to the **Power Chain Rule:**

$$\frac{d}{dx} u^n = nu^{n-1} \frac{du}{dx}. \qquad \frac{d}{du}(u^n) = nu^{n-1} \tag{5}$$

Example 8 Finding Tangent Slopes

$\sin^n x$ is short for $(\sin x)^n$, $n \neq -1$.

(a) Find the slope of the line tangent to the curve $y = \sin^5 x$ at the point where $x = \pi/3$.

(b) Show that the slope of every line tangent to the curve $y = 1/(1 - 2x)^3$ is positive.

Solution

(a) $$\frac{dy}{dx} = 5 \sin^4 x \cdot \frac{d}{dx} \sin x \qquad \text{Power Chain Rule with } u = \sin x,\ n = 5$$

$$= 5 \sin^4 x \cos x$$

The tangent line has slope

$$\left.\frac{dy}{dx}\right|_{x=\pi/3} = 5\left(\frac{\sqrt{3}}{2}\right)^4\left(\frac{1}{2}\right) = \frac{45}{32}.$$

(b) $$\frac{dy}{dx} = \frac{d}{dx}(1 - 2x)^{-3}$$

$$= -3(1 - 2x)^{-4} \cdot \frac{d}{dx}(1 - 2x) \qquad \text{Power Chain Rule with } u = (1 - 2x),\ n = -3$$

$$= -3(1 - 2x)^{-4} \cdot (-2)$$

$$= \frac{6}{(1 - 2x)^4}$$

At any point (x, y) on the curve, $x \neq 1/2$ and the slope of the tangent line is

$$\frac{dy}{dx} = \frac{6}{(1 - 2x)^4},$$

the quotient of two positive numbers.

Example 9 Radians Versus Degrees

It is important to remember that the formulas for the derivatives of both $\sin x$ and $\cos x$ were obtained under the assumption that x is measured in radians, *not* degrees. The Chain Rule gives us new insight into the difference between the two. Since $180° = \pi$ radians, $x° = \pi x/180$ radians where $x°$ means the angle x measured in degrees.

By the Chain Rule,

$$\frac{d}{dx}\sin(x°) = \frac{d}{dx}\sin\left(\frac{\pi x}{180}\right) = \frac{\pi}{180}\cos\left(\frac{\pi x}{180}\right) = \frac{\pi}{180}\cos(x°).$$

See Figure 2.31. Similarly, the derivative of $\cos(x°)$ is $-(\pi/180)\sin(x°)$.

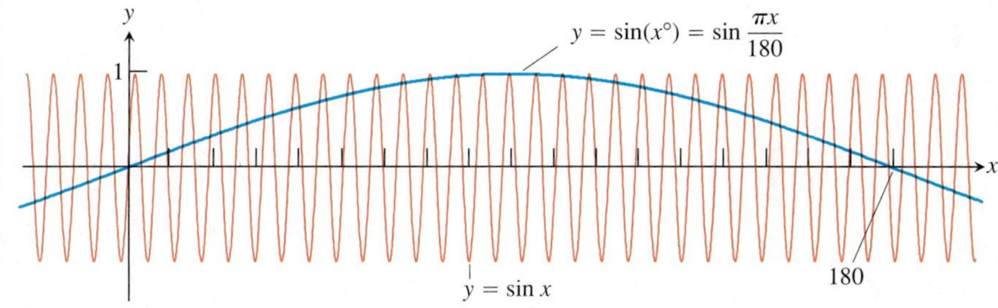

FIGURE 2.31 $\sin(x°)$ oscillates only $\pi/180$ times as often as $\sin x$ oscillates. Its maximum slope is $\pi/180$. (Example 9)

The factor $\pi/180$, annoying in the first derivative, would compound with repeated differentiation. We see at a glance the compelling reason for the use of radian measure.

Melting Ice Cubes

The state of California has serious droughts, and new sources of water are always under consideration. One of the proposals is to tow icebergs from polar waters to offshore locations near southern California, where the melting ice could provide fresh water. As a first approximation in analyzing this proposal, we might imagine the iceberg to be a large cube (or some other regularly shaped solid, like a rectangular solid or a pyramid).

Example 10 The Melting Ice Cube

How long will it take an ice cube to melt?

Solution We start with a mathematical model. We assume that the cube retains its cubical shape as it melts. If we call its edge length s, its volume is $V = s^3$ and its surface area is $6s^2$. We assume that V and s are differentiable functions of time t. We assume also that the cube's volume decreases at a rate that is proportional to its surface area. This latter assumption seems reasonable enough when we think that the melting takes place at the surface: Changing the amount of surface changes the amount of ice exposed to melt. In mathematical terms,

$$\frac{dV}{dt} = -k(6s^2), \qquad k > 0.$$

The minus sign indicates that the volume is decreasing. We assume that the proportionality factor k is constant. It probably depends on many things, such as the relative humidity of the surrounding air, the air temperature, and the incidence or absence of sunlight, to name only a few.

Finally, we need at least one more piece of information: How long will it take a specific percentage of the ice cube to melt? We have nothing to guide us unless we make one or more observations, but now let us assume a particular set of conditions in which the cube lost 1/4 of its volume during the first hour. (You could use letters instead of particular numbers, say $n\%$ in r hours. Then your answer would be in terms of n and r.) Mathematically, we now have the following problem.

Given: $V = s^3$ and $\frac{dV}{dt} = -k(6s^2)$

$V = V_0$ when $t = 0$

$V = (3/4)V_0$ when $t = 1$ h.

Find: The value of t when $V = 0$.

We apply the Chain Rule to differentiate $V = s^3$ with respect to t:

$$\frac{dV}{dt} = 3s^2 \frac{ds}{dt}.$$

We set this derivative equal to the given rate, $-k(6s^2)$, to get

$$3s^2 \frac{ds}{dt} = -6ks^2$$

$$\frac{ds}{dt} = -2k.$$

The side length is *decreasing* at the constant rate of $2k$ units per hour. Thus, if the initial length of the cube's side is s_0, the length of its side 1 hour later is $s_1 = s_0 - 2k$. This equation tells us that

$$2k = s_0 - s_1.$$

The melting time is the value of t that makes $2kt = s_0$. Hence,

$$t_{\text{melt}} = \frac{s_0}{2k} = \frac{s_0}{s_0 - s_1} = \frac{1}{1 - (s_1/s_0)},$$

but

$$\frac{s_1}{s_0} = \frac{\left(\frac{3}{4}V_0\right)^{1/3}}{(V_0)^{1/3}} = \left(\frac{3}{4}\right)^{1/3} \approx 0.91.$$

Therefore,

$$t_{\text{melt}} = \frac{1}{1 - 0.91} \approx 11 \text{ h}.$$

If 1/4 of the cube melts in 1 h, it will take about 10 h more for the rest of it to melt.

Of course the kinds of questions we ultimately need to answer are, How much of the ice will be lost in transit? And how long will it take the ice to turn into usable water? If we were interested in pursuing the matter further, our next step would be to test the model by running experiments and then to refine it on the basis of what we learned.

EXERCISES 2.5

Derivative Calculations

In Exercises 1–6, given $y = f(u)$ and $u = g(x)$, find $dy/dx = f'(g(x))g'(x)$.

1. $y = 6u - 9, \quad u = (1/2)x^4$

2. $y = 2u^3, \quad u = 8x - 1$

3. $y = \sin u, \quad u = 3x + 1$

4. $y = \cos u, \quad u = \sin x$

5. $y = \tan u, \quad u = 10x - 5$

6. $y = -\sec u, \quad u = x^2 + 7x$

In Exercises 7–12, write the function in the form $y = f(u)$ and $u = g(x)$. Then find dy/dx as a function of x.

7. $y = (4 - 3x)^9$

8. $y = \left(1 - \frac{x}{7}\right)^{-7}$

9. $y = \left(\frac{x^2}{8} + x - \frac{1}{x}\right)^4$

10. $y = \sec(\tan x)$

11. $y = \cot\left(\pi - \frac{1}{x}\right)$

12. $y = \sin^3 x$

Find the derivatives of the functions in Exercises 13–26.

13. $q = \sqrt{2r - r^2}$

14. $s = \sin\left(\frac{3\pi t}{2}\right) + \cos\left(\frac{3\pi t}{2}\right)$

15. $r = (\csc\theta + \cot\theta)^{-1}$

16. $r = -(\sec\theta + \tan\theta)^{-1}$

17. $y = x^2 \sin^4 x + x\cos^{-2} x$

18. $y = \frac{1}{x}\sin^{-5} x - \frac{x}{3}\cos^3 x$

19. $y = \frac{1}{21}(3x - 2)^7 + \left(4 - \frac{1}{2x^2}\right)^{-1}$

20. $y = (4x + 3)^4(x + 1)^{-3}$

21. $h(x) = x\tan(2\sqrt{x}) + 7$

22. $k(x) = x^2 \sec\left(\frac{1}{x}\right)$

23. $f(\theta) = \left(\frac{\sin\theta}{1 + \cos\theta}\right)^2$

24. $r = \sin(\theta^2)\cos(2\theta)$

25. $r = \sec\sqrt{\theta}\tan\left(\frac{1}{\theta}\right)$

26. $q = \sin\left(\frac{t}{\sqrt{t + 1}}\right)$

In Exercises 27–32, find dy/dt.

27. $y = \sin^2(\pi t - 2)$

28. $y = (1 + \cos 2t)^{-4}$

29. $y = (1 + \cot(t/2))^{-2}$

30. $y = \sin(\cos(2t - 5))$

31. $y = \left(1 + \tan^4\left(\frac{t}{12}\right)\right)^3$

32. $y = \sqrt{1 + \cos(t^2)}$

Tangents to Parametrized Curves

In Exercises 33–40, find an equation for the line tangent to the curve at the point defined by the given value of t. Also, find the value of d^2y/dx^2 at this point.

33. $x = 2\cos t, \quad y = 2\sin t, \quad t = \pi/4$

34. $x = \cos t, \quad y = \sqrt{3}\cos t, \quad t = 2\pi/3$

35. $x = t, \quad y = \sqrt{t}, \quad t = 1/4$

36. $x = -\sqrt{t + 1}, \quad y = \sqrt{3t}, \quad t = 3$

37. $x = 2t^2 + 3, \quad y = t^4, \quad t = -1$

38. $x = t - \sin t, \quad y = 1 - \cos t, \quad t = \pi/3$

39. $x = \cos t, \quad y = 1 + \sin t, \quad t = \pi/2$

40. $x = \sec^2 t - 1, \quad y = \tan t, \quad t = -\pi/4$

Second Derivatives

Find y'' in Exercises 41–44.

41. $y = \left(1 + \frac{1}{x}\right)^3$

42. $y = (1 - \sqrt{x})^{-1}$

43. $y = \frac{1}{9}\cot(3x - 1)$

44. $y = 9\tan\left(\frac{x}{3}\right)$

Finding Numerical Values of Derivatives

In Exercises 45–50, find the value of $(f \circ g)'$ at the given value of x.

45. $f(u) = u^5 + 1, \quad u = g(x) = \sqrt{x}, \quad x = 1$

46. $f(u) = 1 - \frac{1}{u}, \quad u = g(x) = \frac{1}{1 - x}, \quad x = -1$

47. $f(u) = \cot\frac{\pi u}{10}, \quad u = g(x) = 5\sqrt{x}, \quad x = 1$

48. $f(u) = u + \frac{1}{\cos^2 u}, \quad u = g(x) = \pi x, \quad x = 1/4$

49. $f(u) = \frac{2u}{u^2+1}, \quad u = g(x) = 10x^2 + x + 1, \quad x = 0$

50. $f(u) = \left(\frac{u-1}{u+1}\right)^2, \quad u = g(x) = \frac{1}{x^2} - 1, \quad x = -1$

51. Suppose that functions f and g and their derivatives with respect to x have the following values at $x = 2$ and $x = 3$.

x	$f(x)$	$g(x)$	$f'(x)$	$g'(x)$
2	8	2	1/3	−3
3	3	−4	2π	5

Find the derivatives with respect to x of the following combinations at the given value of x.

(a) $2f(x), \quad x = 2$
(b) $f(x) + g(x), \quad x = 3$
(c) $f(x) \cdot g(x), \quad x = 3$
(d) $f(x)/g(x), \quad x = 2$
(e) $f(g(x)), \quad x = 2$
(f) $\sqrt{f(x)}, \quad x = 2$
(g) $1/g^2(x), \quad x = 3$
(h) $\sqrt{f^2(x) + g^2(x)}, \quad x = 2$

52. Suppose that the functions f and g and their derivatives with respect to x have the following values at $x = 0$ and $x = 1$.

x	$f(x)$	$g(x)$	$f'(x)$	$g'(x)$
0	1	1	5	1/3
1	3	−4	−1/3	−8/3

Find the derivatives with respect to x of the following combinations at the given value of x.

(a) $5f(x) - g(x), \quad x = 1$
(b) $f(x)g^3(x), \quad x = 0$
(c) $\frac{f(x)}{g(x)+1}, \quad x = 1$
(d) $f(g(x)), \quad x = 0$
(e) $g(f(x)), \quad x = 0$
(f) $(x^{11} + f(x))^{-2}, \quad x = 1$
(g) $f(x + g(x)), \quad x = 0$

53. Find ds/dt when $\theta = 3\pi/2$ if $s = \cos\theta$ and $d\theta/dt = 5$.

54. Find dy/dt when $x = 1$ if $y = x^2 + 7x - 5$ and $dx/dt = 1/3$.

Choices in Composition

What happens if you can write a function as a composite in different ways? Do you get the same derivative each time? The Chain Rule says you should. Try it with the functions in Exercises 55 and 56.

55. Find dy/dx if $y = x$ by using the Chain Rule with y as a composite of

(a) $y = (u/5) + 7$ and $u = 5x - 35$

(b) $y = 1 + (1/u)$ and $u = 1/(x-1)$.

56. Find dy/dx if $y = x^{3/2}$ by using the Chain Rule with y as a composite of

(a) $y = u^3$ and $u = \sqrt{x}$
(b) $y = \sqrt{u}$ and $u = x^3$.

Tangents and Slopes

57. **(a)** Find the tangent to the curve $y = 2\tan(\pi x/4)$ at $x = 1$.

(b) *Writing to Learn: Slopes on a tangent curve* What is the smallest value the slope of the curve can ever have on the interval $-2 < x < 2$? Give reasons for your answer.

58. *Writing to Learn: Slopes on sine curves*

(a) Find equations for the tangents to the curves $y = \sin 2x$ and $y = -\sin(x/2)$ at the origin. Is there anything special about how the tangents are related? Give reasons for your answer.

(b) Can anything be said about the tangents to the curves $y = \sin mx$ and $y = -\sin(x/m)$ at the origin (m a constant $\neq 0$)? Give reasons for your answer.

(c) For a given m, what are the largest values the slopes of the curves $y = \sin mx$ and $y = -\sin(x/m)$ can ever have? Give reasons for your answer.

(d) The function $y = \sin x$ completes one period on the interval $[0, 2\pi]$, the function $y = \sin 2x$ completes two periods, the function $y = \sin(x/2)$ completes half a period, and so on. Is there any relation between the number of periods $y = \sin mx$ completes on $[0, 2\pi]$ and the slope of the curve $y = \sin mx$ at the origin? Give reasons for your answer.

Theory, Examples, and Applications

59. *Running machinery too fast* Suppose that a piston is moving straight up and down and that its position at time t sec is

$$s = A\cos(2\pi bt),$$

with A and b positive. The value of A is the amplitude of the motion, and b is the frequency (number of times the piston moves up and down each second). What effect does doubling the frequency have on the piston's velocity, acceleration, and jerk? (Once you find out, you will know why machinery breaks when you run it too fast.) See Figure 2.32.

FIGURE 2.32 The internal forces in the engine get so large that they tear the engine apart when the velocity is too great.

60. *Temperatures in Fairbanks, Alaska* The graph in Figure 2.33 shows the average Fahrenheit temperature in Fairbanks, Alaska, during a typical 365-day year. The equation that approximates the temperature on day x is

$$y = 37 \sin\left[\frac{2\pi}{365}(x - 101)\right] + 25.$$

(a) On what day is the temperature increasing the fastest?

(b) About how many degrees per day is the temperature increasing when it is increasing at its fastest?

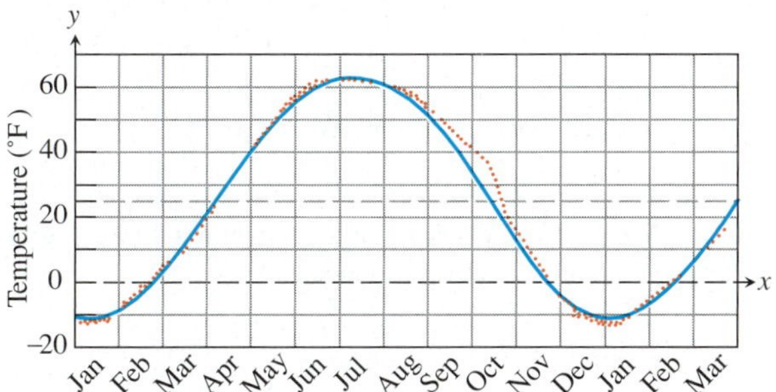

FIGURE 2.33 Normal mean air temperatures at Fairbanks, Alaska, plotted as data points, and the approximating sine function. (Exercise 60)

61. *Particle motion* The position of a particle moving along a coordinate line is $s = \sqrt{1 + 4t}$, with s in meters and t in seconds. Find the particle's velocity and acceleration at $t = 6$ sec.

62. *Constant acceleration* Suppose that the velocity of a falling body is $v = k\sqrt{s}$ m/sec (k a constant) at the instant the body has fallen s m from its starting point. Show that the body's acceleration is constant.

63. *Falling meteorite* The velocity of a heavy meteorite entering Earth's atmosphere is inversely proportional to $\sqrt{s}$ when it is s km from Earth's center. Show that the meteorite's acceleration is inversely proportional to s^2.

64. *Particle acceleration* A particle moves along the x-axis with velocity $dx/dt = f(x)$. Show that the particle's acceleration is $f(x)f'(x)$.

65. *Temperature and the period of a pendulum* For oscillations of small amplitude (short swings), we may safely model the relationship between the period T and the length L of a simple pendulum with the equation

$$T = 2\pi\sqrt{\frac{L}{g}},$$

where g is the constant acceleration of gravity at the pendulum's location. If we measure g in centimeters per second squared, we measure L in centimeters and T in seconds. If the pendulum is made of metal, its length will vary with temperature, either increasing or decreasing at a rate that is roughly proportional to L. In symbols, with u being temperature and k the proportionality constant,

$$\frac{dL}{du} = kL.$$

Assuming this to be the case, show that the rate at which the period changes with respect to temperature is $kT/2$.

66. *Writing to Learn: Chain Rule* Suppose that $f(x) = x^2$ and $g(x) = |x|$. Then the composites

$$(f \circ g)(x) = |x|^2 = x^2 \quad \text{and} \quad (g \circ f)(x) = |x^2| = x^2$$

are both differentiable at $x = 0$ even though g itself is not differentiable at $x = 0$. Does this contradict the Chain Rule? Explain.

67. *Writing to Learn: Tangents* Suppose that $u = g(x)$ is differentiable at $x = 1$ and that $y = f(u)$ is differentiable at $u = g(1)$. If the graph of $y = f(g(x))$ has a horizontal tangent at $x = 1$, can we conclude anything about the tangent to the graph of g at $x = 1$ or the tangent to the graph of f at $u = g(1)$? Give reasons for your answer.

68. *Writing to Learn* Suppose that $u = g(x)$ is differentiable at $x = -5$, $y = f(u)$ is differentiable at $u = g(-5)$, and $(f \circ g)'(-5)$ is negative. What, if anything, can be said about the values of $g'(-5)$ and $f'(g(-5))$?

T **69.** *The derivative of sin 2x* Graph the function $y = 2\cos 2x$ for $-2 \le x \le 3.5$. Then, on the same screen, graph

$$y = \frac{\sin 2(x + h) - \sin 2x}{h}$$

for $h = 1.0$, 0.5, and 0.2. Experiment with other values of h, including negative values. What do you see happening as $h \to 0$? Explain this behavior.

T **70.** *The derivative of cos (x^2)* Graph $y = -2x \sin(x^2)$ for $-2 \le x \le 3$. Then, on the same screen, graph

$$y = \frac{\cos((x + h)^2) - \cos(x^2)}{h}$$

for $h = 1.0$, 0.7, and 0.3. Experiment with other values of h. What do you see happening as $h \to 0$? Explain this behavior.

T The curves in Exercises 71 and 72 are called *Bowditch curves* or *Lissajous figures*. In each case, find the point in the interior of the first quadrant where the tangent to the curve is horizontal, and find the equations of the two tangents at the origin.

71.

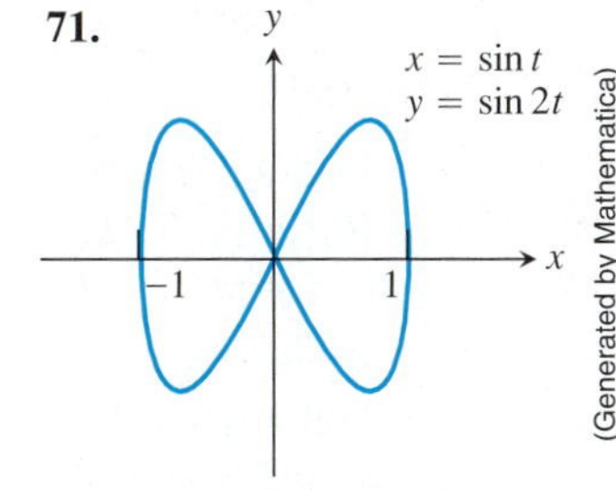

72.

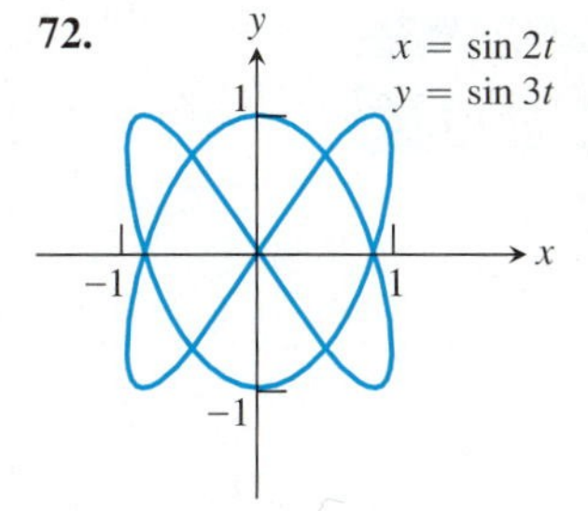

COMPUTER EXPLORATIONS

Trigonometric Polynomials

73. As Figure 2.34 shows, the trigonometric "polynomial"

$$s = f(t) = 0.78540 - 0.63662\cos 2t - 0.07074\cos 6t$$
$$- 0.02546\cos 10t - 0.01299\cos 14t$$

gives a good approximation of the sawtooth function $s = g(t)$ on the interval $[-\pi, \pi]$. How well does the derivative of f approximate the derivative of g at the points where dg/dt is defined? To find out, carry out the following steps.

(a) Graph dg/dt (where defined) over $[-\pi, \pi]$.

(b) Find df/dt.

(c) Graph df/dt. Where does the approximation of dg/dt by df/dt seem to be best? Least good? Approximations by trigonometric polynomials are important in the theories of heat and oscillation, but we must not expect too much of them, as we see in the next exercise.

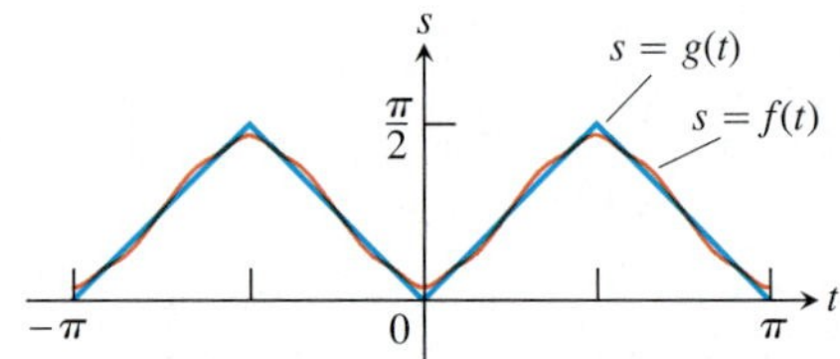

FIGURE 2.34 The approximation of a sawtooth function by a trigonometric "polynomial." (Exercise 73)

74. *(Continuation of Exercise 73)* In Exercise 73, the trigonometric polynomial $f(t)$ that approximated the sawtooth function $g(t)$ on $[-\pi, \pi]$ had a derivative that approximated the derivative of the sawtooth function. It is possible, however, for a trigonometric polynomial to approximate a function in a reasonable way without its derivative approximating the function's derivative at all well. As a case in point, the "polynomial"

$$s = h(t) = 1.2732\sin 2t + 0.4244\sin 6t + 0.25465\sin 10t$$
$$+ 0.18189\sin 14t + 0.14147\sin 18t$$

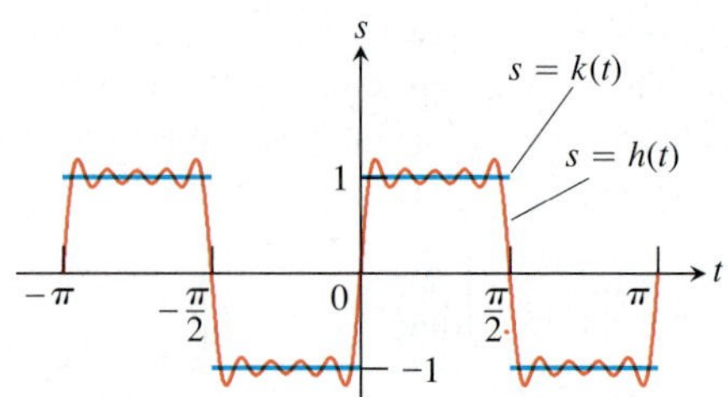

FIGURE 2.35 The approximation of a step function by a trigonometric "polynomial." (Exercise 74)

graphed in Figure 2.35 approximates the step function $s = k(t)$ shown there. Yet the derivative of h is nothing like the derivative of k.

(a) Graph dk/dt (where defined) over $[-\pi, \pi]$.

(b) Find dh/dt.

(c) Graph dh/dt to see how badly the graph fits the graph of dk/dt. Comment on what you see.

Parametrized Curves

Use a CAS to perform the following steps on the parametrized curves in Exercises 75–80.

(a) Plot the curve for the given interval of t values.

(b) Find dy/dx and d^2y/dx^2 at the point t_0.

(c) Find an equation for the tangent line to the curve at the point defined by the given value t_0. Plot the curve together with the tangent line on a single graph.

75. $x = \frac{1}{3}t^3,\quad y = \frac{1}{2}t^2,\quad 0 \le t \le 1,\quad t_0 = 1/2$

76. $x = 2t^3 - 16t^2 + 25t + 5,\quad y = t^2 + t - 3,\quad 0 \le t \le 6,\quad t_0 = 3/2$

77. $x = e^t - t^2,\quad y = t + e^{-t},\quad -1 \le t \le 2,\quad t_0 = 1$

78. $x = t - \cos t,\quad y = 1 + \sin t,\quad -\pi \le t \le \pi,\quad t_0 = \pi/4$

79. $x = e^t + \sin 2t,\quad y = e^t + \cos(t^2),\quad -\sqrt{2\pi} \le t \le \pi/4,\quad t_0 = -\pi/4$

80. $x = e^t\cos t,\quad y = e^t\sin t,\quad 0 \le t \le \pi,\quad t_0 = \pi/2$

2.6 Implicit Differentiation

Implicitly Defined Functions • Derivatives of Higher Order • Rational Powers of Differentiable Functions

In the law that describes how light changes direction as it enters a lens, the important angles are the angles the light makes with the line perpendicular to the surface of the lens at the point of entry (angles A and B in Figure 2.36). This line is called the *normal* to the surface at the point of entry. In a profile view of a lens like the one

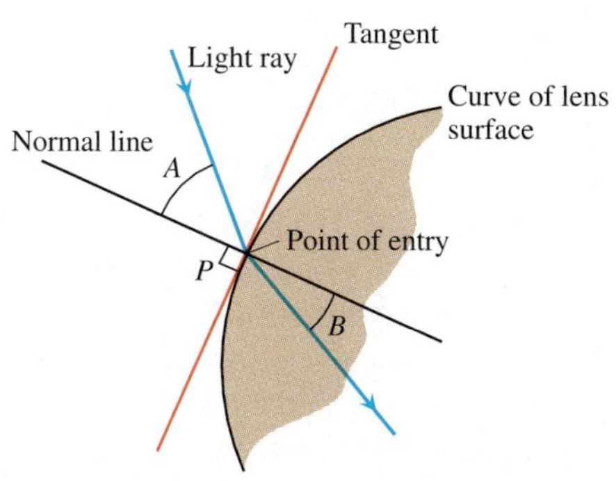

FIGURE 2.36 The profile of a lens, showing the bending (refraction) of a ray of light as it passes through the lens surface.

in Figure 2.36, the normal is the line perpendicular to the tangent to the profile curve at the point of entry.

Profiles of lenses are often described by equations of the form $F(x, y) = 0$. To find the normal to a point on the profile curve, we first need to find the slope of the tangent there by evaluating dy/dx. But what if we cannot put the equation in the form $y = f(x)$ to differentiate? In such cases, we may still be able to find dy/dx by a process called *implicit differentiation*. This section describes the technique and uses it to extend the Power Rule for differentiation to include rational exponents.

When Are the Functions Defined by $F(x, y) = 0$ Differentiable?

To justify implicit differentiation, we must know that the derivatives we want really do exist. That is, we need to know when we can rely on the functions defined by an expression $F(x, y) = 0$ to be differentiable. A theorem in advanced calculus guarantees this to be the case under certain conditions on the function F. All of the functions you will encounter in this section meet these conditions.

Implicitly Defined Functions

The graph of the equation $x^3 + y^3 - 9xy = 0$ (Figure 2.37) has a well-defined slope at nearly every point because it is the union of the graphs of the functions $y = f_1(x)$, $y = f_2(x)$, and $y = f_3(x)$, which are differentiable except at the origin and A. But how do we find the slope when we cannot conveniently solve the equation to find the functions? The answer is to treat y as a differentiable function of x and differentiate both sides of the equation with respect to x, using the differentiation rules for sums, products, and quotients and the Chain Rule. Then solve for dy/dx in terms of x and y *together* to obtain a formula that calculates the slope at any point (x, y) on the graph from the values of x and y. This process for finding dy/dx is called **implicit differentiation,** so named because the equation $x^3 + y^3 - 9xy = 0$ defines the functions f_1, f_2, and f_3 *implicitly* (i.e., hidden inside the equation), without giving us *explicit* formulas with which to work.

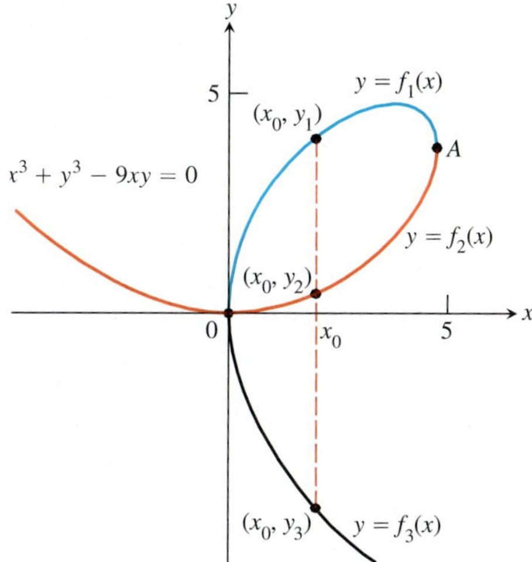

FIGURE 2.37 The curve $x^3 + y^3 - 9xy = 0$ is not the graph of any one function of x. The curve can, however, be divided into separate arcs that *are* the graphs of functions of x. This particular curve, called a *folium*, dates to Descartes in 1638.

Example 1 Differentiating Implicitly

Find dy/dx if $y^2 = x$.

Solution The equation $y^2 = x$ defines two differentiable functions of x that we can actually find, namely $y_1 = \sqrt{x}$ and $y_2 = -\sqrt{x}$ (Figure 2.38). We know how to calculate the derivative of each of these for $x > 0$:

$$\frac{dy_1}{dx} = \frac{1}{2\sqrt{x}} \quad \text{and} \quad \frac{dy_2}{dx} = -\frac{1}{2\sqrt{x}}.$$

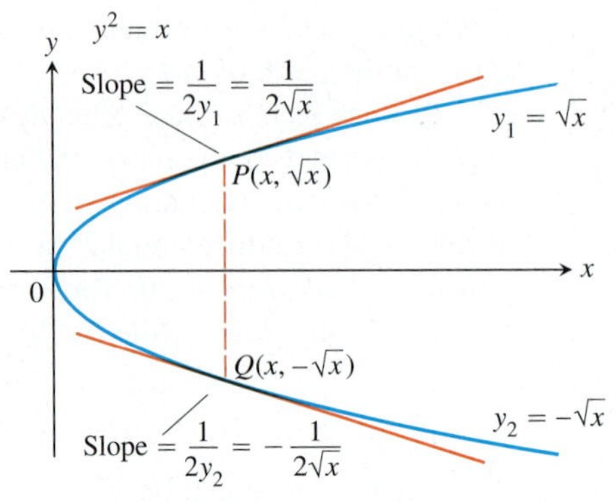

FIGURE 2.38 The equation $y^2 - x = 0$, or $y^2 = x$ as it is usually written, defines two differentiable functions of x on the interval $x \geq 0$. Example 1 shows how to find the derivatives of these functions without solving the equation $y^2 = x$ for y.

But suppose that we knew only that the equation $y^2 = x$ defined y as one or more differentiable functions of x for $x > 0$ without knowing exactly what these functions were. Could we still find dy/dx?

The answer is yes. To find dy/dx, we simply differentiate both sides of the equation $y^2 = x$ with respect to x, treating $y = f(x)$ as a differentiable function of x:

$$y^2 = x$$

$$2y\frac{dy}{dx} = 1 \qquad \text{The Chain Rule gives } \frac{d}{dx}y^2 = \frac{d}{dx}[f(x)]^2 = 2f(x)f'(x) = 2y\frac{dy}{dx}.$$

$$\frac{dy}{dx} = \frac{1}{2y}.$$

This one formula gives the derivatives we calculated for *both* explicit solutions $y_1 = \sqrt{x}$ and $y_2 = -\sqrt{x}$:

$$\frac{dy_1}{dx} = \frac{1}{2y_1} = \frac{1}{2\sqrt{x}} \quad \text{and} \quad \frac{dy_2}{dx} = \frac{1}{2y_2} = \frac{1}{2(-\sqrt{x})} = -\frac{1}{2\sqrt{x}}.$$

To calculate the derivatives of other implicitly defined functions, we proceed as in Example 1. We treat y as a differentiable implicit function of x and apply the usual rules to differentiate both sides of the defining equation.

Example 2 Differentiating Implicitly

Find dy/dx if $y^2 = x^2 + \sin xy$ (Figure 2.39).

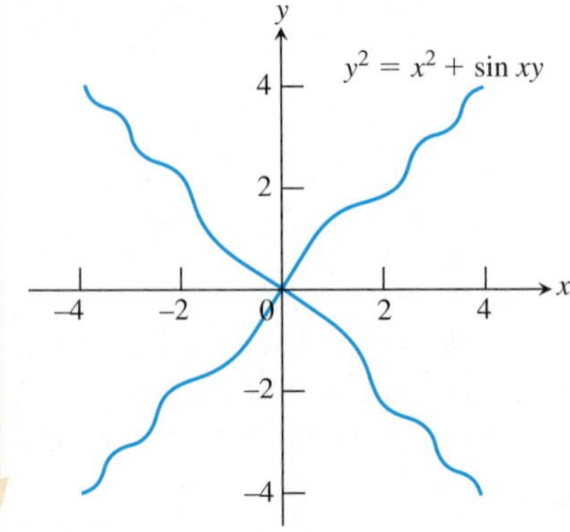

FIGURE 2.39 The graph of $y^2 = x^2 + \sin xy$ in Example 2. The example shows how to find slopes on this implicitly defined curve.

Implicit Differentiation Takes Four Steps

Step 1. Differentiate both sides of the equation with respect to x, treating y as a differentiable function of x.

Step 2. Collect the terms with dy/dx on one side of the equation.

Step 3. Factor out dy/dx.

Step 4. Solve for dy/dx.

Solution

$$y^2 = x^2 + \sin xy$$

$$\frac{d}{dx}(y^2) = \frac{d}{dx}(x^2) + \frac{d}{dx}(\sin xy) \qquad \text{Differentiate both sides with respect to } x \ldots$$

$$2y\frac{dy}{dx} = 2x + (\cos xy)\frac{d}{dx}(xy) \qquad \ldots \text{treating } y \text{ as a function of } x \text{ and using the Chain Rule.}$$

$$2y\frac{dy}{dx} = 2x + (\cos xy)\left(y + x\frac{dy}{dx}\right)$$

$$2y\frac{dy}{dx} - (\cos xy)\left(x\frac{dy}{dx}\right) = 2x + (\cos xy)\,y \qquad \text{Collect terms with } dy/dx \ldots$$

$$(2y - x\cos xy)\frac{dy}{dx} = 2x + y\cos xy \qquad \ldots \text{and factor out } dy/dx.$$

$$\frac{dy}{dx} = \frac{2x + y\cos xy}{2y - x\cos xy} \qquad \text{Solve for } dy/dx \text{ by dividing.}$$

Notice that the formula for dy/dx applies everywhere that the implicitly defined curve has a slope. Notice also that the derivative involves *both* variables x and y, not just the independent variable x.

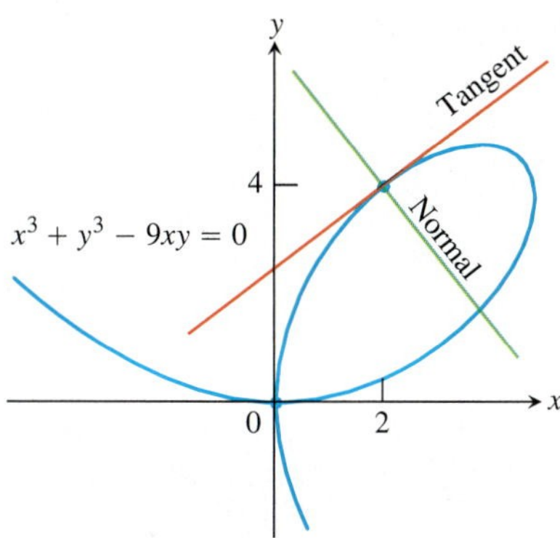

FIGURE 2.40 Example 3 shows how to find equations for the tangent and normal to the curve at $(2, 4)$.

Example 3 Tangent and Normal to the Folium of Descartes

Show that the point $(2, 4)$ lies on the curve $x^3 + y^3 - 9xy = 0$. Then find the tangent and normal to the curve there (Figure 2.40).

Solution The point $(2, 4)$ lies on the curve because its coordinates satisfy the equation given for the curve: $2^3 + 4^3 - 9(2)(4) = 8 + 64 - 72 = 0$.

To find the slope of the curve at $(2, 4)$, we first use implicit differentiation to find a formula for dy/dx:

$$x^3 + y^3 - 9xy = 0$$

$$\frac{d}{dx}(x^3) + \frac{d}{dx}(y^3) - \frac{d}{dx}(9xy) = \frac{d}{dx}(0) \qquad \text{Differentiate both sides with respect to } x.$$

$$3x^2 + 3y^2\frac{dy}{dx} - 9\left(x\frac{dy}{dx} + y\frac{dx}{dx}\right) = 0 \qquad \text{Treat } xy \text{ as a product and } y \text{ as a function of } x.$$

$$(3y^2 - 9x)\frac{dy}{dx} + 3x^2 - 9y = 0$$

$$3(y^2 - 3x)\frac{dy}{dx} = 9y - 3x^2 \qquad \text{Collect terms.}$$

$$\frac{dy}{dx} = \frac{3y - x^2}{y^2 - 3x}. \qquad \text{Solve for } dy/dx.$$

We then evaluate the derivative at $(x, y) = (2, 4)$:

$$\left.\frac{dy}{dx}\right|_{(2,4)} = \left.\frac{3y - x^2}{y^2 - 3x}\right|_{(2,4)} = \frac{3(4) - 2^2}{4^2 - 3(2)} = \frac{8}{10} = \frac{4}{5}.$$

The tangent at (2, 4) is the line through (2, 4) with slope 4/5:

$$y = 4 + \frac{4}{5}(x - 2)$$
$$y = \frac{4}{5}x + \frac{12}{5}.$$

The normal to the curve at (2, 4) is the line perpendicular to the tangent there, the line through (2, 4) with slope $-5/4$:

$$y = 4 - \frac{5}{4}(x - 2)$$
$$y = -\frac{5}{4}x + \frac{13}{2}.$$

The Word Normal

When analytic geometry was developed in the seventeenth century, European scientists wrote about their work and ideas in Latin, a language that all educated Europeans could be counted on to read and understand. The word *normalis*, which scholars used for "perpendicular" in Latin, became *normal* when they discussed geometry in English.

Derivatives of Higher Order

Implicit differentiation can also be used to find derivatives of higher order. Here is an example.

Example 4 Finding a Second Derivative Implicitly

Find d^2y/dx^2 if $2x^3 - 3y^2 = 8$.

Solution To start, we differentiate both sides of the equation with respect to x in order to find $y' = dy/dx$.

$$\frac{d}{dx}(2x^3 - 3y^2) = \frac{d}{dx}(8)$$
$$6x^2 - 6yy' = 0$$
$$x^2 - yy' = 0$$
$$y' = \frac{x^2}{y}, \qquad \text{when } y \neq 0$$

We now apply the Quotient Rule to find y''.

$$y'' = \frac{d}{dx}\left(\frac{x^2}{y}\right) = \frac{2xy - x^2y'}{y^2} = \frac{2x}{y} - \frac{x^2}{y^2} \cdot y'$$

Finally, we substitute $y' = x^2/y$ to express y'' in terms of x and y.

$$y'' = \frac{2x}{y} - \frac{x^2}{y^2}\left(\frac{x^2}{y}\right) = \frac{2x}{y} - \frac{x^4}{y^3}, \qquad \text{when } y \neq 0$$

Rational Powers of Differentiable Functions

We know that the rule

$$\frac{d}{dx}x^n = nx^{n-1}$$

holds when n is an integer. Using implicit differentiation we can show that it holds when n is any rational number.

Theorem 4 Power Rule for Rational Powers

If n is a rational number, then x^n is differentiable at every interior point of the domain of x^{n-1}, and

$$\frac{d}{dx} x^n = nx^{n-1}. \tag{1}$$

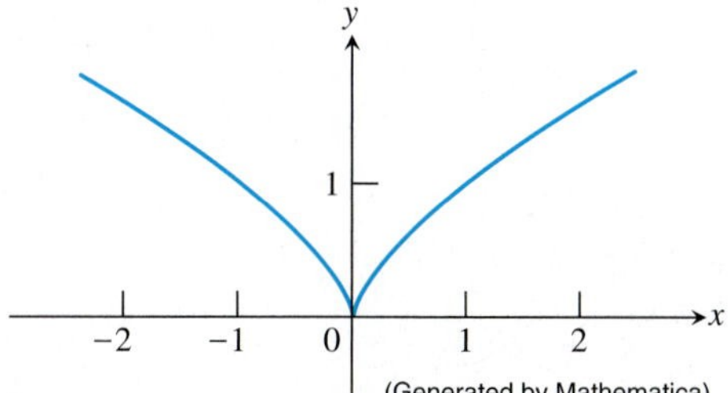

FIGURE 2.41 The graph of $y = x^{2/3}$ has a cusp at $x = 0$. (Example 5)

Example 5 Using the Rational Power Rule

(a) $\frac{d}{dx}(\sqrt{x}) = \frac{d}{dx}(x^{1/2}) = \frac{1}{2}x^{-1/2} = \frac{1}{2\sqrt{x}}$

Notice that $\sqrt{x}$ is defined at $x = 0$, but $1/(2\sqrt{x})$ is not.

(b) $\frac{d}{dx}(x^{2/3}) = \frac{2}{3}(x^{-1/3}) = \frac{2}{3x^{1/3}}$

The original function is defined for all real numbers, but the derivative is undefined at $x = 0$. Its graph has a *cusp* at $x = 0$ (Figure 2.41).

Proof of Theorem 4 Let p and q be integers with $q > 0$ and suppose that $y = \sqrt[q]{x^p} = x^{p/q}$. Then

$$y^q = x^p.$$

Since p and q are integers (for which we already have the Power Rule), we can differentiate both sides of the equation with respect to x and obtain

$$qy^{q-1}\frac{dy}{dx} = px^{p-1}.$$

If $y \neq 0$, we can divide both sides of the equation by qy^{q-1} to solve for dy/dx, obtaining

$$\begin{aligned}
\frac{dy}{dx} &= \frac{px^{p-1}}{qy^{q-1}} \\
&= \frac{p}{q}\cdot\frac{x^{p-1}}{(x^{p/q})^{q-1}} && y = x^{p/q} \\
&= \frac{p}{q}\cdot\frac{x^{p-1}}{x^{p-p/q}} && \frac{p}{q}(q-1) = p - \frac{p}{q} \\
&= \frac{p}{q}\cdot x^{(p-1)-(p-p/q)} && \text{A law of exponents} \\
&= \frac{p}{q}\cdot x^{(p/q)-1},
\end{aligned}$$

which proves the rule.

By combining this result with the Chain Rule, we get an extension of the Power Chain Rule to rational powers of u: If n is a rational number and u is a differentiable function of x, then u^n is a differentiable function of x and

$$\frac{d}{dx}u^n = nu^{n-1}\frac{du}{dx}, \tag{2}$$

provided that $u \neq 0$ if $n < 1$.

The restriction that $u \neq 0$ when $n < 1$ is necessary because 0 might be in the domain of u^n but not in the domain of u^{n-1}, as we see in the next example.

Example 6 Using the Rational Power and Chain Rules

(a) $$\frac{d}{dx}\overbrace{(1-x^2)^{1/4}}^{\text{function defined on } [-1,1]} = \frac{1}{4}(1-x^2)^{-3/4}(-2x) \qquad \text{Eq. (2) with } u = 1 - x^2 \text{ and } n = 1/4$$
$$= \underbrace{\frac{-x}{2(1-x^2)^{3/4}}}_{\text{derivative defined only on } (-1,\,1)}$$

(b) $$\frac{d}{dx}(\cos x)^{-1/5} = -\frac{1}{5}(\cos x)^{-6/5}\frac{d}{dx}(\cos x)$$
$$= -\frac{1}{5}(\cos x)^{-6/5}(-\sin x)$$
$$= \frac{1}{5}(\sin x)(\cos x)^{-6/5}$$

EXERCISES 2.6

Derivatives of Rational Powers

Find dy/dx in Exercises 1–6.

1. $y = x^{9/4}$

2. $y = \sqrt[3]{2x}$

3. $y = 7\sqrt{x+6}$

4. $y = (1-6x)^{2/3}$

5. $y = x(x^2+1)^{1/2}$

6. $y = x(x^2+1)^{-1/2}$

Find the first derivatives of the functions in Exercises 7–12.

7. $s = \sqrt[7]{t^2}$

8. $r = \sqrt[4]{\theta^{-3}}$

9. $y = \sin\left((2t+5)^{-2/3}\right)$

10. $f(x) = \sqrt{1-\sqrt{x}}$

11. $g(x) = 2(2x^{-1/2}+1)^{-1/3}$

12. $h(\theta) = \sqrt[3]{1+\cos(2\theta)}$

Differentiating Implicitly

Use implicit differentiation to find dy/dx in Exercises 13–22.

13. $x^2y + xy^2 = 6$

14. $2xy + y^2 = x + y$

15. $x^3 - xy + y^3 = 1$

16. $x^2(x-y)^2 = x^2 - y^2$

17. $y^2 = \dfrac{x-1}{x+1}$

18. $x^2 = \dfrac{x-y}{x+y}$

19. $x = \tan y$

20. $x + \sin y = xy$

21. $y \sin\left(\dfrac{1}{y}\right) = 1 - xy$

22. $y^2 \cos\left(\dfrac{1}{y}\right) = 2x + 2y$

Find $dr/d\theta$ in Exercises 23–26.

23. $\theta^{1/2} + r^{1/2} = 1$

24. $r - 2\sqrt{\theta} = \dfrac{3}{2}\theta^{2/3} + \dfrac{4}{3}\theta^{3/4}$

25. $\sin(r\theta) = \dfrac{1}{2}$

26. $\cos r + \cos\theta = r\theta$

Higher Derivatives

In Exercises 27–30, use implicit differentiation to find dy/dx and then d^2y/dx^2.

27. $x^{2/3} + y^{2/3} = 1$

28. $y^2 = x^2 + 2x$

29. $2\sqrt{y} = x - y$

30. $xy + y^2 = 1$

31. If $x^3 + y^3 = 16$, find the value of d^2y/dx^2 at the point $(2, 2)$.

32. If $xy + y^2 = 1$, find the value of d^2y/dx^2 at the point $(0, -1)$.

Implicitly Defined Parametrizations

Assuming that the equations in Exercises 33–36 define x and y implicitly as differentiable functions $x = f(t)$, $y = g(t)$, find the slope of the curve $x = f(t)$, $y = g(t)$ at the given value of t.

33. $x^2 - 2tx + 2t^2 = 4, \quad 2y^3 - 3t^2 = 4, \quad t = 2$

34. $x = \sqrt{5 - \sqrt{t}}, \quad y(t-1) = \ln y, \quad t = 1$

35. $x + 2x^{3/2} = t^2 + t, \quad y\sqrt{t+1} + 2t\sqrt{y} = 4, \quad t = 0$

36. $x \sin t + 2x = t, \quad t \sin t - 2t = y, \quad t = \pi$

Slopes, Tangents, and Normals

In Exercises 37 and 38, find the slope of the curve at the given points.

37. $y^2 + x^2 = y^4 - 2x$ at $(-2, 1)$ and $(-2, -1)$

38. $(x^2 + y^2)^2 = (x - y)^2$ at $(1, 0)$ and $(1, -1)$

In Exercises 39–46, verify that the given point is on the curve and find the lines that are **(a)** tangent and **(b)** normal to the curve at the given point.

39. $x^2 + xy - y^2 = 1$, $(2, 3)$

40. $x^2y^2 = 9$, $(-1, 3)$

41. $y^2 - 2x - 4y - 1 = 0$, $(-2, 1)$

42. $6x^2 + 3xy + 2y^2 + 17y - 6 = 0$, $(-1, 0)$

43. $2xy + \pi \sin y = 2\pi$, $(1, \pi/2)$

44. $x \sin 2y = y \cos 2x$, $(\pi/4, \pi/2)$

45. $y = 2 \sin (\pi x - y)$, $(1, 0)$

46. $x^2 \cos^2 y - \sin y = 0$, $(0, \pi)$

47. *Parallel tangents* Find the two points where the curve $x^2 + xy + y^2 = 7$ crosses the x-axis and show that the tangents to the curve at these points are parallel. What is the common slope of these tangents?

48. *Tangents parallel to coordinate axes* Find points on the curve $x^2 + xy + y^2 = 7$ **(a)** where the tangent is parallel to the x-axis and **(b)** where the tangent is parallel to the y-axis. In the latter case, dy/dx is not defined, but dx/dy is. What value does dx/dy have at these points?

49. *The eight curve* Find the slopes of the curve $y^4 = y^2 - x^2$ at the two points shown here.

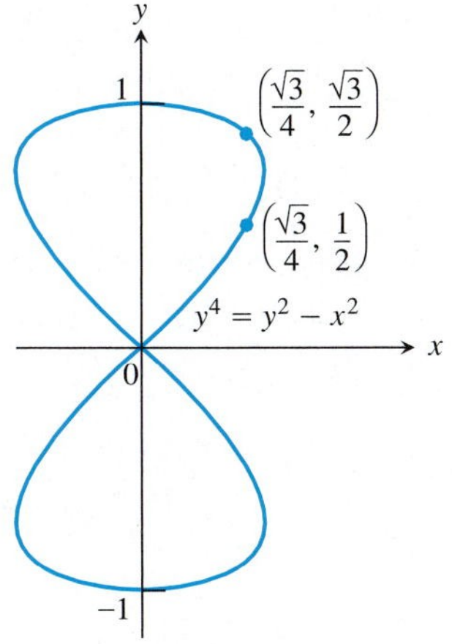

50. *The cissoid of Diocles (from about 200 B.C.)* Find equations for the tangent and normal to the cissoid of Diocles $y^2(2 - x) = x^3$ at $(1, 1)$.

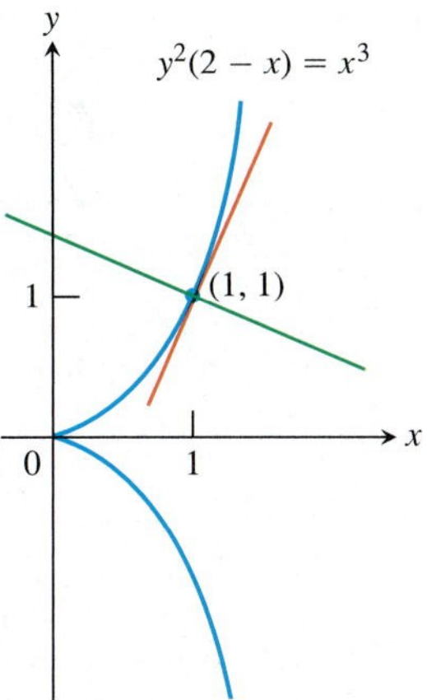

51. *The devil's curve (Gabriel Cramer, the Cramer of Cramer's rule, 1750)* Find the slopes of the devil's curve $y^4 - 4y^2 = x^4 - 9x^2$ at the four indicated points.

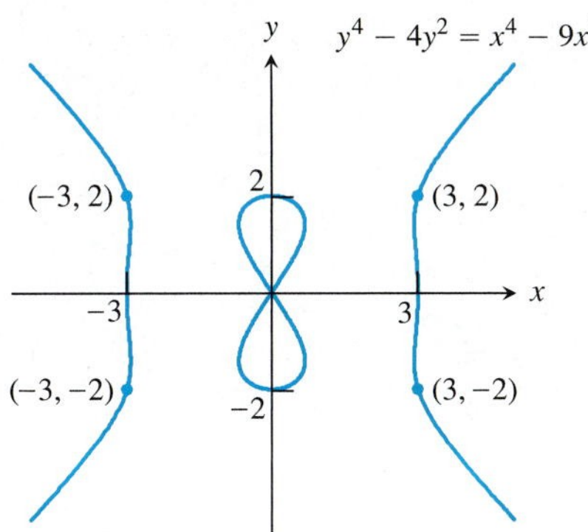

52. *The folium of Descartes* (See Figure 2.37)

(a) Find the slope of the folium of Descartes, $x^3 + y^3 - 9xy = 0$ at the points (4, 2) and (2, 4).

(b) At what point other than the origin does the folium have a horizontal tangent?

(c) Find the coordinates of the point A in Figure 2.37, where the folium has a vertical tangent.

Theory and Examples

53. Which of the following could be true if $f''(x) = x^{-1/3}$?

(a) $f(x) = \frac{3}{2}x^{2/3} - 3$ **(b)** $f(x) = \frac{9}{10}x^{5/3} - 7$

(c) $f'''(x) = -\frac{1}{3}x^{-4/3}$ **(d)** $f'(x) = \frac{3}{2}x^{2/3} + 6$

54. *Writing to Learn* Is there anything special about the tangents to the curves $2x^2 + 3y^2 = 5$ and $y^2 = x^3$ at the points $(1, \pm 1)$? Give reasons for your answer.

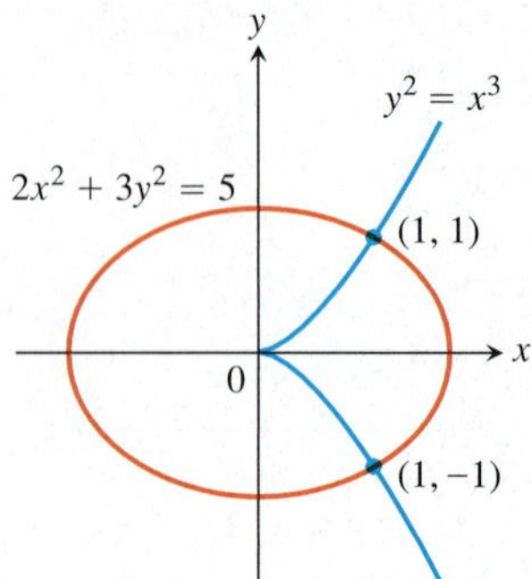

55. *Intersecting normal* The line that is normal to the curve $x^2 + 2xy - 3y^2 = 0$ at $(1, 1)$ intersects the curve at what other point?

56. *Normals parallel to a line* Find the normals to the curve $xy + 2x - y = 0$ that are parallel to the line $2x + y = 0$.

57. *Normals to a parabola* Show that if it is possible to draw three normals from the point $(a, 0)$ to the parabola $x = y^2$ shown here, then a must be greater than $1/2$. One of the normals is the x-axis. For what value of a are the other two normals perpendicular?

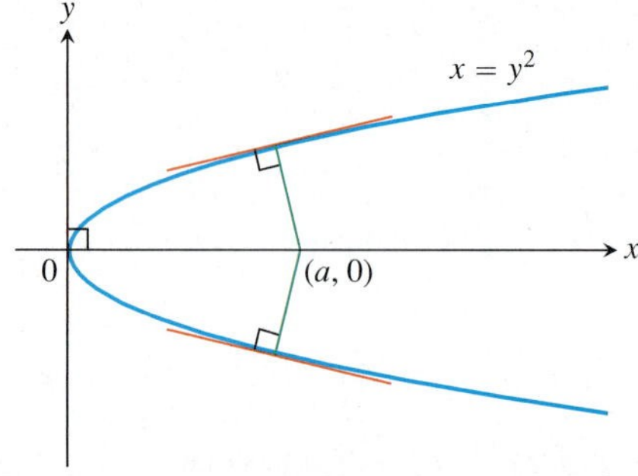

58. *Writing to Learn* What is the geometry behind the restrictions on the domains of the derivatives in Example 5 and Example 6(a)?

T In Exercises 59 and 60, find both dy/dx (treating y as a differential function of x) and dx/dy (treating x as a differentiable function of y). How do dy/dx and dx/dy seem to be related? Explain the relationship geometrically in terms of the graphs.

59. $xy^3 + x^2y = 6$

60. $x^3 + y^2 = \sin^2 y$

COMPUTER EXPLORATIONS

61. (a) Given that $x^4 + 4y^2 = 1$, find dy/dx two ways: (1) by solving for y and differentiating the resulting functions in the usual way and (2) by implicit differentiation. Do you get the same result each way?

(b) *Writing to Learn* Solve the equation $x^4 + 4y^2 = 1$ for y and graph the resulting functions together to produce a complete graph of the equation $x^4 + 4y^2 = 1$. Then add the graphs of the first derivatives of these functions to your display. Could you have predicted the general behavior of the derivative graphs from looking at the graph of $x^4 + 4y^2 = 1$? Could you have predicted the general behavior of the graph of $x^4 + 4y^2 = 1$ by looking at the derivative graphs? Give reasons for your answers.

62. (a) Given that $(x - 2)^2 + y^2 = 4$, find dy/dx two ways: (1) by solving for y and differentiating the resulting functions with respect to x and (2) by implicit differentiation. Do you get the same result each way?

(b) *Writing to Learn* Solve the equation $(x - 2)^2 + y^2 = 4$ for y and graph the resulting functions together to produce a complete graph of the equation $(x - 2)^2 + y^2 = 4$. Then add the graphs of the functions' first derivatives to your picture. Could you have predicted the general behavior of the derivative graphs from looking at the graph of $(x - 2)^2 + y^2 = 4$? Could you have predicted the general behavior of the graph of $(x - 2)^2 + y^2 = 4$ by looking at the derivative graphs? Give reasons for your answers.

Use a CAS to perform the following steps in Exercises 63–70.

(a) Plot the equation with the implicit plotter of a CAS. Check to see that the given point P satisfies the equation.

(b) Using implicit differentiation find a formula for the derivative dy/dx and evaluate it at the given point P.

(c) Use the slope found in part (b) to find an equation for the tangent line to the curve at P. Then plot the implicit curve and tangent line together on a single graph.

63. $x^3 - xy + y^3 = 7, \quad P(2, 1)$

64. $x^5 + y^3x + yx^2 + y^4 = 4, \quad P(1, 1)$

65. $y^2 + y = \dfrac{2 + x}{1 - x}, \quad P(0, 1)$

66. $y^3 + \cos xy = x^2, \quad P(1, 0)$

67. $x + \tan\left(\dfrac{y}{x}\right) = 2, \quad P\left(1, \dfrac{\pi}{4}\right)$

68. $xy^3 + \tan(x + y) = 1, \quad P\left(\dfrac{\pi}{4}, 0\right)$

69. $2y^2 + (xy)^{1/3} = x^2 + 2, \quad P(1, 1)$

70. $x\sqrt{1 + 2y} + y = x^2, \quad P(1, 0)$

2.7 Related Rates

Related Rate Equations • Solution Strategy

Suppose you want to measure the rate of ascent of a rocket during a vertical liftoff from the ground. With some elaborate and expensive precautions you could probably put an instrument of some kind on the ground under the rocket and take readings from it. But it would be safer and a lot cheaper just to stand some distance d away and measure the rate of change of the rocket's angle of elevation θ. With a little trigonometry you could then express the rocket's height h in terms of the angle and distance as $h = d \tan \theta$. Differentiating both sides of this equation with respect to time t would then express dh/dt, the rate you want, in terms of $d\theta/dt$, a rate you could easily measure. The problem of finding a rate you cannot measure easily from some other rate that you can is called a *related rate problem*. Related rate problems are the subject of this section.

Related Rate Equations

Suppose that a particle $P(x, y)$ is moving along a curve C in the plane so that its coordinates x and y are differentiable functions of time t. If D is the distance from the origin to P, then using the Chain Rule we can find an equation that relates dD/dt, dx/dt, and dy/dt.

$$D = \sqrt{x^2 + y^2}$$

$$\frac{dD}{dt} = \frac{1}{2}(x^2 + y^2)^{-1/2}\left(2x\frac{dx}{dt} + 2y\frac{dy}{dt}\right)$$

Any equation involving two or more variables that are differentiable functions of time t can be used to find an equation that relates their corresponding rates.

Example 1 Finding Related Rate Equations

Assume that the radius r and height h of a cone are differentiable functions of t and let V be the volume of the cone. Find an equation that relates dV/dt, dr/dt, and dh/dt.

Solution

$$V = \frac{\pi}{3}r^2h \qquad \text{Cone volume formula}$$

$$\frac{dV}{dt} = \frac{\pi}{3}\left(r^2 \cdot \frac{dh}{dt} + 2r\frac{dr}{dt} \cdot h\right) = \frac{\pi}{3}\left(r^2\frac{dh}{dt} + 2rh\frac{dr}{dt}\right)$$

Solution Strategy

How rapidly will the fluid level inside a vertical cylindrical storage tank drop if we pump the fluid out at some specified rate? A question like this one asks us to calculate a rate that we cannot measure directly from a rate that we can. To do so, we write an equation that relates the variables involved and differentiate it to get an equation that relates the rate we seek to the rate we know.

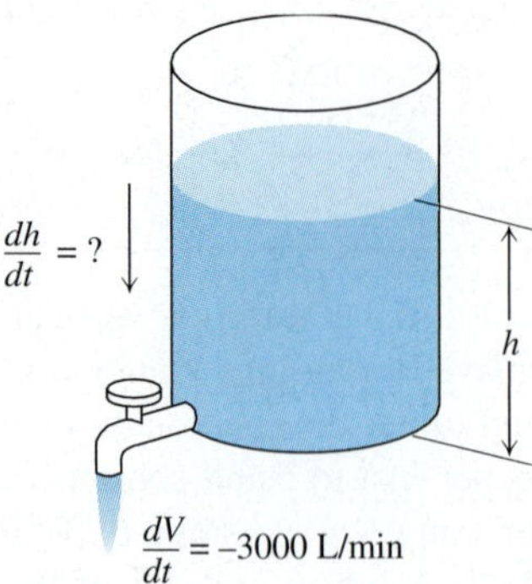

FIGURE 2.42 The cylindrical tank in Example 2.

Example 2 Pumping Out a Tank

How rapidly will the fluid level inside a vertical cylindrical tank drop if we pump the fluid out at the rate of 3000 L/min?

Solution We draw a picture of a partially filled vertical cylindrical tank, calling its radius r and the height of the fluid h (Figure 2.42). Call the volume of the fluid V.

As time passes, the radius remains constant, but V and h change. We think of V and h as differentiable functions of time and use t to represent time. We are told that

$$\frac{dV}{dt} = -3000.$$

We pump out at the rate of 3000 L/min. The rate is negative because the volume is decreasing.

We are asked to find

$$\frac{dh}{dt}.$$

How fast will the fluid level drop?

To find dh/dt, we first write an equation that relates h to V. The equation depends on the units chosen for V, r, and h. With V in liters and r and h in meters, the appropriate equation for the cylinder's volume is

$$V = 1000\pi r^2 h$$

because a cubic meter contains 1000 L.

Since V and h are differentiable functions of t, we can differentiate both sides of the equation $V = 1000\pi r^2 h$ with respect to t to get an equation that relates dh/dt to dV/dt:

$$\frac{dV}{dt} = 1000\pi r^2 \frac{dh}{dt}.$$

r is a constant.

We substitute the known value $dV/dt = -3000$ and solve for dh/dt:

$$\frac{dh}{dt} = \frac{-3000}{1000\pi r^2} = -\frac{3}{\pi r^2}.$$

The fluid level will drop at the rate of $3/(\pi r^2)$ m/min.

Interpret

The equation $dh/dt = -3/\pi r^2$ shows how the rate at which the fluid level drops depends on the tank's radius. If r is small, dh/dt will be large; if r is large, dh/dt will be small.

If $r = 1$ m: $\quad \frac{dh}{dt} = -\frac{3}{\pi} \approx -0.95 \text{ m/min} = -95 \text{ cm/min}.$

If $r = 10$ m: $\quad \frac{dh}{dt} = -\frac{3}{100\pi} \approx -0.0095 \text{ m/min} = -0.95 \text{ cm/min}.$

Related Rate Problem Strategy

Step 1. *Draw a picture and name the variables and constants.* Use t for time. Assume that all variables are differentiable functions of t.

Step 2. *Write down the numerical information* (in terms of the symbols you have chosen).

Step 3. *Write down what you are asked to find* (usually a rate, expressed as a derivative).

Step 4. *Write an equation that relates the variables.* You may have to combine two or more equations to get a single equation that relates the variable whose rate you want to the variables whose rates you know.

Step 5. *Differentiate with respect to t.* Then express the rate you want in terms of the rate and variables whose values you know.

Step 6. *Evaluate.* Use known values to find the unknown rate.

Example 3 A Rising Balloon

A hot-air balloon rising straight up from a level field is tracked by a range finder 500 ft from the liftoff point. At the moment the range finder's elevation angle is $\pi/4$, the angle is increasing at the rate of 0.14 rad/min. How fast is the balloon rising at that moment?

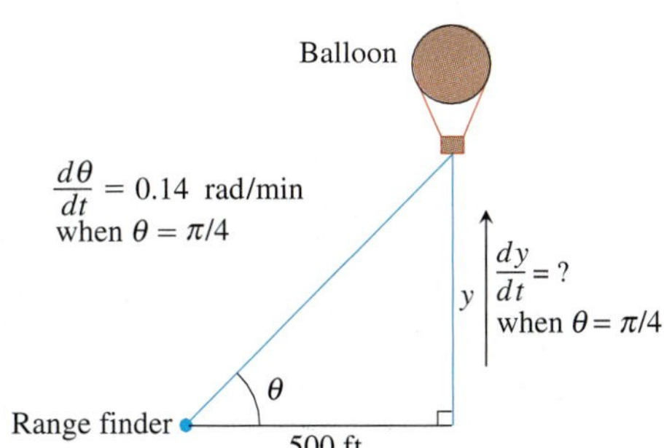

FIGURE 2.43 The balloon in Example 3.

Solution We answer the question in six steps.

Step 1: *Draw a picture and name the variables and constants* (Figure 2.43). The variables in the picture are

$\theta =$ the angle in radians the range finder makes with the ground

$y =$ the height in feet of the balloon.

We let t represent time in minutes and assume that θ and y are differentiable functions of t.

The one constant in the picture is the distance from the range finder to the liftoff point (500 ft). There is no need to give it a special symbol.

Step 2: *Write down the additional numerical information.*

$$\frac{d\theta}{dt} = 0.14 \text{ rad/min} \qquad \text{when} \qquad \theta = \frac{\pi}{4}$$

Step 3: *Write down what we are to find.* We want dy/dt when $\theta = \pi/4$.

Step 4: *Write an equation that relates the variables y and θ.*

$$\frac{y}{500} = \tan\theta \qquad \text{or} \qquad y = 500\tan\theta$$

Step 5: *Differentiate with respect to t using the Chain Rule.* The result tells how dy/dt (which we want) is related to $d\theta/dt$ (which we know).

$$\frac{dy}{dt} = 500(\sec^2\theta)\frac{d\theta}{dt}$$

Step 6: *Evaluate with $\theta = \pi/4$ and $d\theta/dt = 0.14$ to find dy/dt.*

$$\frac{dy}{dt} = 500(\sqrt{2})^2(0.14) = 140 \qquad \sec\frac{\pi}{4} = \sqrt{2}$$

Interpret

At the moment in question, the balloon is rising at the rate of 140 ft/min.

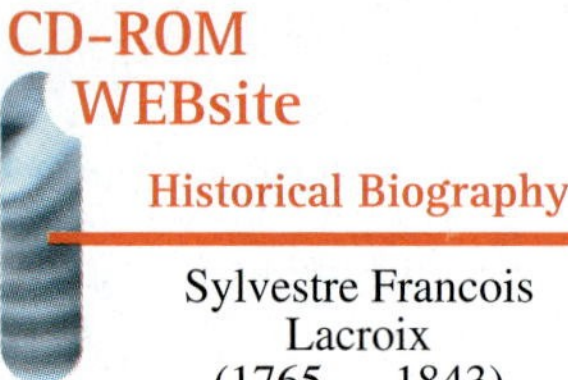

Historical Biography

Sylvestre Francois Lacroix (1765 — 1843)

Example 4 A Highway Chase

A police cruiser, approaching a right-angled intersection from the north, is chasing a speeding car that has turned the corner and is now moving straight east. When the cruiser is 0.6 mi north of the intersection and the car is 0.8 mi to the east, the police determine with radar that the distance between them and the car is increasing at 20 mph. If the cruiser is moving at 60 mph at the instant of measurement, what is the speed of the car?

Solution We carry out the steps of the strategy.

Step 1: *Picture and variables.* We picture the car and cruiser in the coordinate plane, using the positive x-axis as the eastbound highway and the positive y-axis as the southbound highway (Figure 2.44). We let t represent time and set

x = position of car at time t

y = position of cruiser at time t

s = distance between car and cruiser at time t.

We assume that x, y, and s are differentiable functions of t.

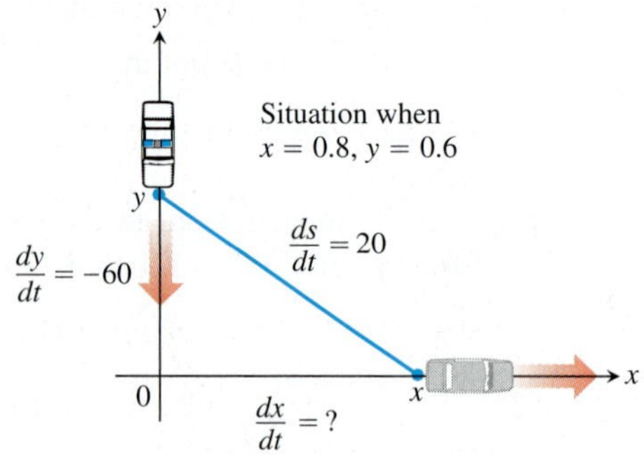

FIGURE 2.44 Figure for Example 4.

Step 2: *Numerical information.* At the instant in question,

$$x = 0.8 \text{ mi}, \qquad y = 0.6 \text{ mi}, \qquad \frac{dy}{dt} = -60 \text{ mph}, \qquad \frac{ds}{dt} = 20 \text{ mph}.$$

Note that dy/dt is negative because y is decreasing.

Step 3: *To find: dx/dt.*

Step 4: *How the variables are related:*

$$s^2 = x^2 + y^2$$

(We could also use $s = \sqrt{x^2 + y^2}$.)

Step 5: *Differentiate with respect to t.*

$$2s\frac{ds}{dt} = 2x\frac{dx}{dt} + 2y\frac{dy}{dt}$$

$$\frac{ds}{dt} = \frac{1}{s}\left(x\frac{dx}{dt} + y\frac{dy}{dt}\right)$$

$$= \frac{1}{\sqrt{x^2 + y^2}}\left(x\frac{dx}{dt} + y\frac{dy}{dt}\right)$$

Step 6: *Evaluate.* Use $x = 0.8$, $y = 0.6$, $dy/dt = -60$, $ds/dt = 20$, and solve for dx/dt.

$$20 = \frac{1}{\sqrt{(0.8)^2 + (0.6)^2}}\left(0.8\frac{dx}{dt} + (0.6)(-60)\right)$$

$$\frac{dx}{dt} = \frac{20\sqrt{(0.8)^2 + (0.6)^2} + (0.6)(60)}{0.8} = 70$$

Interpret

At the moment in question, the car's speed is 70 mph.

Example 5 Filling a Conical Tank

Water runs into a conical tank at the rate of $9\ \text{ft}^3/\text{min}$. The tank stands point down and has a height of 10 ft and a base radius of 5 ft. How fast is the water level rising when the water is 6 ft deep?

Solution We carry out the steps of the strategy.

Step 1: *Picture and variables.* Figure 2.45 shows a partially filled conical tank. The variables in the problem are

V = volume (ft^3) of the water in the tank at time t (min)

x = radius (ft) of the surface of the water at time t

y = depth (ft) of water in tank at time t.

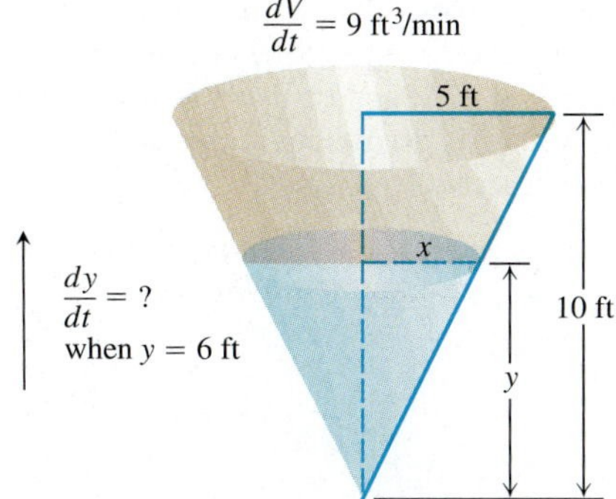

FIGURE 2.45 The conical tank in Example 5.

We assume that V, x, and y are differentiable functions of t. The constants are the dimensions of the tank.

Step 2: *Numerical information.* At the time in question,

$$y = 6\ \text{ft} \qquad \text{and} \qquad \frac{dV}{dt} = 9\ \text{ft}^3/\text{min}.$$

Step 3: *To find: dy/dt.*

Step 4: *How the variables are related:* The water forms a cone with volume

$$V = \frac{1}{3}\pi x^2 y.$$

This equation involves x as well as V and y. Because no information is given about x and dx/dt at the time in question, we need to eliminate x. The similar triangles in Figure 2.45 give us a way to express x in terms of y:

$$\frac{x}{y} = \frac{5}{10} \qquad \text{or} \qquad x = \frac{y}{2}.$$

Therefore,

$$V = \frac{1}{3}\pi\left(\frac{y}{2}\right)^2 y = \frac{\pi}{12}y^3.$$

Step 5: *Differentiate with respect to t.*

$$\frac{dV}{dt} = \frac{\pi}{12}\cdot 3y^2\frac{dy}{dt} = \frac{\pi}{4}y^2\frac{dy}{dt}$$

Step 6: *Evaluate.* Use $y = 6$ and $dV/dt = 9$ to solve for dy/dt.

$$9 = \frac{\pi}{4}(6)^2\frac{dy}{dt}$$

$$\frac{dy}{dt} = \frac{1}{\pi} \approx 0.32 \text{ ft/min}$$

Interpret

At the moment in question, the water level is rising at about 0.32 ft/min.

EXERCISES 2.7

1. *Area* Suppose that the radius r and area $A = \pi r^2$ of a circle are differentiable functions of t. Write an equation that relates dA/dt to dr/dt.

2. *Surface area* Suppose that the radius r and surface area $S = 4\pi r^2$ of a sphere are differentiable functions of t. Write an equation that relates dS/dt to dr/dt.

3. *Volume* The radius r and height h of a right circular cylinder are related to the cylinder's volume V by the formula $V = \pi r^2 h$.
   (a) How is dV/dt related to dh/dt if r is constant?
   (b) How is dV/dt related to dr/dt if h is constant?
   (c) How is dV/dt related to dr/dt and dh/dt if neither r nor h is constant?

4. *Volume* The radius r and height h of a right circular cone are related to the cone's volume V by the equation $V = (1/3)\pi r^2 h$.
   (a) How is dV/dt related to dh/dt if r is constant?
   (b) How is dV/dt related to dr/dt if h is constant?
   (c) How is dV/dt related to dr/dt and dh/dt if neither r nor h is constant?

5. *Changing voltage* The voltage V (volts), current I (amperes), and resistance R (ohms) of an electric circuit like the one shown here are related by the equation $V = IR$. Suppose that V is increasing at the rate of 1 volt/sec while I is decreasing at the rate of 1/3 amp/sec. Let t denote time in seconds.

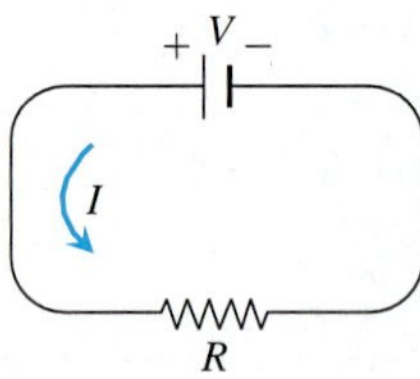

   (a) What is the value of dV/dt?
   (b) What is the value of dI/dt?
   (c) What equation relates dR/dt to dV/dt and dI/dt?

(d) Find the rate at which R is changing when $V = 12$ volts and $I = 2$ amp. Is R increasing, or decreasing?

6. *Electrical power* The power P (watts) of an electric circuit is related to the circuit's resistance R (ohms) and current i (amperes) by the equation $P = Ri^2$.

(a) How are dP/dt, dR/dt, and di/dt related if none of P, R, and i are constant?

(b) How is dR/dt related to di/dt if P is constant?

7. *Distance* Let x and y be differentiable functions of t and let $s = \sqrt{x^2 + y^2}$ be the distance between the points $(x, 0)$ and $(0, y)$ in the xy-plane.

(a) How is ds/dt related to dx/dt if y is constant?

(b) How is ds/dt related to dx/dt and dy/dt if neither x nor y is constant?

(c) How is dx/dt related to dy/dt if s is constant?

8. *Diagonals* If x, y, and z are lengths of the edges of a rectangular box, the common length of the box's diagonals is $s = \sqrt{x^2 + y^2 + z^2}$.

(a) Assuming that x, y, and z are differentiable functions of t, how is ds/dt related to dx/dt, dy/dt, and dz/dt?

(b) How is ds/dt related to dy/dt and dz/dt if x is constant?

(c) How are dx/dt, dy/dt, and dz/dt related if s is constant?

9. *Area* The area A of a triangle with sides of lengths a and b enclosing an angle of measure θ is

$$A = \frac{1}{2} ab \sin \theta .$$

(a) How is dA/dt related to $d\theta/dt$ if a and b are constant?

(b) How is dA/dt related to $d\theta/dt$ and da/dt if only b is constant?

(c) How is dA/dt related to $d\theta/dt$, da/dt, and db/dt if none of a, b, and θ are constant?

10. *Heating a plate* When a circular plate of metal is heated in an oven, its radius increases at the rate of 0.01 cm/min. At what rate is the plate's area increasing when the radius is 50 cm?

11. *Changing dimensions in a rectangle* The length l of a rectangle is decreasing at the rate of 2 cm/sec while the width w is increasing at the rate of 2 cm/sec. When $l = 12$ cm and $w = 5$ cm, find the rates of change of **(a)** the area, **(b)** the perimeter, and **(c)** the lengths of the diagonals of the rectangle. Which of these quantities are decreasing, and which are increasing?

12. *Changing dimensions in a rectangular box* Suppose that the edge lengths x, y, and z of a closed rectangular box are changing at the following rates:

$$\frac{dx}{dt} = 1 \text{ m/sec}, \quad \frac{dy}{dt} = -2 \text{ m/sec}, \quad \frac{dz}{dt} = 1 \text{ m/sec}.$$

Find the rates at which the box's **(a)** volume, **(b)** surface area, and **(c)** diagonal length $s = \sqrt{x^2 + y^2 + z^2}$ are changing at the instant when $x = 4$, $y = 3$, and $z = 2$.

13. *A sliding ladder* A 13-ft ladder is leaning against a house when its base starts to slide away. By the time the base is 12 ft from the house, the base is moving at the rate of 5 ft/sec.

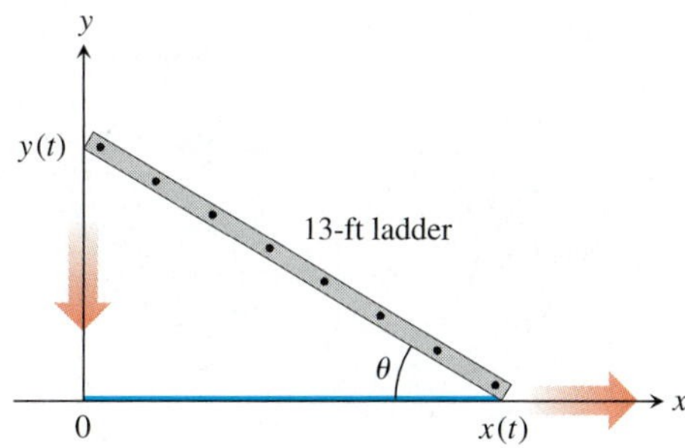

(a) How fast is the top of the ladder sliding down the wall then?

(b) At what rate is the area of the triangle formed by the ladder, wall, and ground changing then?

(c) At what rate is the angle θ between the ladder and the ground changing then?

14. *Commercial air traffic* Two commercial airplanes are flying at 40,000 ft along straight-line courses that intersect at right angles. Plane A is approaching the intersection point at a speed of 442 knots (nautical miles per hour; a nautical mile is 2000 yd). Plane B is approaching the intersection at 481 knots. At what rate is the distance between the planes changing when A is 5 nautical miles from the intersection point and B is 12 nautical miles from the intersection point?

15. *Flying a kite* A girl flies a kite at a height of 300 ft, the wind carrying the kite horizontally away from her at a rate of 25 ft/sec. How fast must she let out the string when the kite is 500 ft away from her?

16. *Boring a cylinder* The mechanics at Lincoln Automotive are reboring a 6-in.-deep cylinder to fit a new piston. The machine they are using increases the cylinder's radius one-thousandth of an inch every 3 min. How rapidly is the cylinder volume increasing when the bore (diameter) is 3.800 in.?

17. *A growing sand pile* Sand falls from a conveyor belt at the rate of 10 m^3/min onto the top of a conical pile. The height of the pile is always three-eighths of the base diameter. How fast are the **(a)** height and **(b)** radius changing when the pile is 4 m high? Answer in cm/min.

18. *A draining conical reservoir* Water is flowing at the rate of 50 m^3/min from a shallow concrete conical reservoir (vertex down) of base radius 45 m and height 6 m.

(a) How fast (cm/min) is the water level falling when the water is 5 m deep?

(b) How fast is the radius of the water's surface changing then? Answer in centimeters per minute.

19. *A draining hemispherical reservoir* Water is flowing at the rate of 6 m^3/min from a reservoir shaped like a hemispherical bowl of radius 13 m, shown here in profile. Answer the following questions, given that the volume of water in a hemispherical bowl of radius R is $V = (\pi/3)y^2(3R - y)$ when the water is y meters deep.

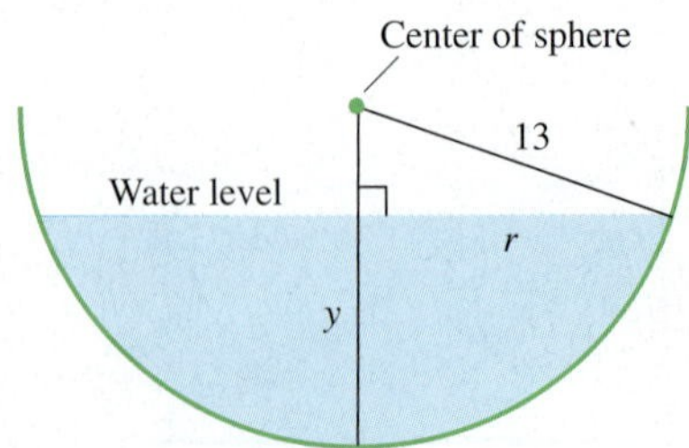

(a) At what rate is the water level changing when the water is 8 m deep?

(b) What is the radius r of the water's surface when the water is y m deep?

(c) At what rate is the radius r changing when the water is 8 m deep?

20. *A growing raindrop* Suppose that a drop of mist is a perfect sphere and that, through condensation, the drop picks up moisture at a rate proportional to its surface area. Show that under these circumstances the drop's radius increases at a constant rate.

21. *The radius of an inflating balloon* A spherical balloon is inflated with helium at the rate of 100π ft^3/min. How fast is the balloon's radius increasing at the instant the radius is 5 ft? How fast is the surface area increasing?

22. *Hauling in a dinghy* A dinghy is pulled toward a dock by a rope from the bow through a ring on the dock 6 ft above the bow. The rope is hauled in at the rate of 2 ft/sec.

(a) How fast is the boat approaching the dock when 10 ft of rope are out?

(b) At what rate is the angle θ changing then (see the figure)?

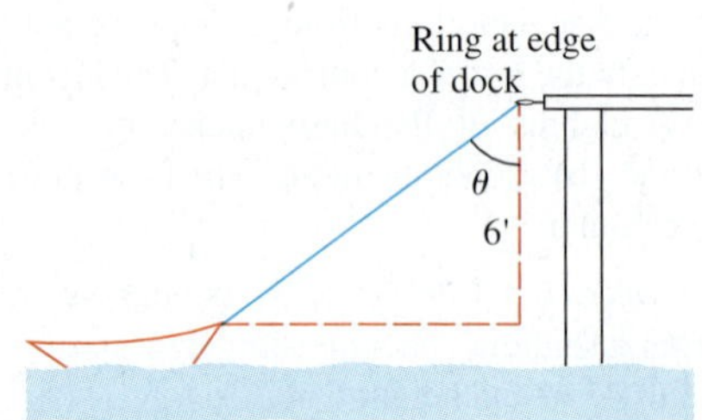

23. *A balloon and a bicycle* A balloon is rising vertically above a level, straight road at a constant rate of 1 ft/sec. Just when the balloon is 65 ft above the ground, a bicycle moving at a constant rate of 17 ft/sec passes under it. How fast is the distance $s(t)$ between the bicycle and balloon increasing 3 sec later?

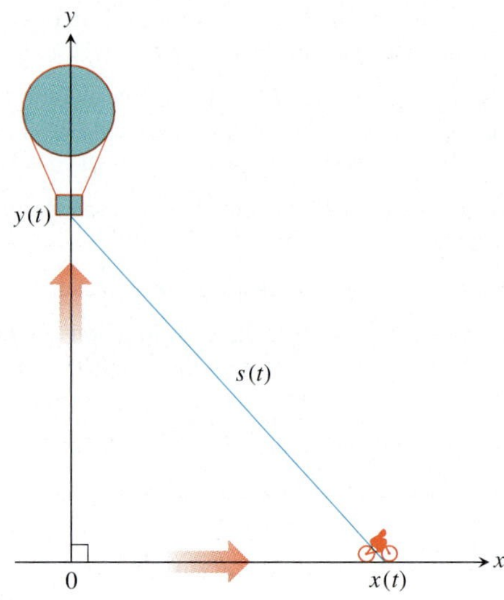

24. *Making coffee* Coffee is draining from a conical filter into a cylindrical coffeepot at the rate of 10 in^3/min.

(a) How fast is the level in the pot rising when the coffee in the cone is 5 in. deep?

(b) How fast is the level in the cone falling then?

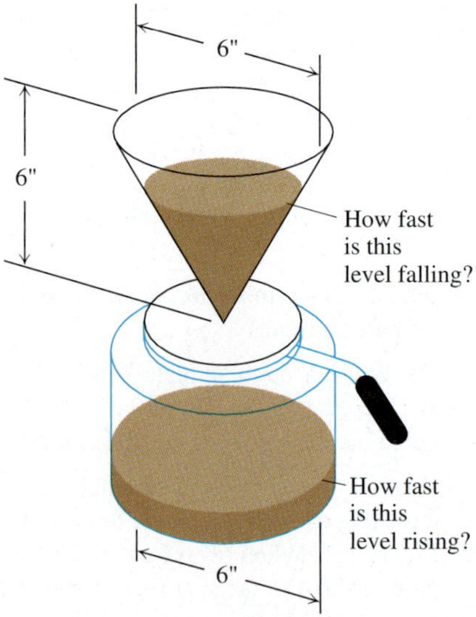

25. *Cardiac output* In the late 1860s, Adolf Fick, a professor of physiology in the Faculty of Medicine in Würtzberg, Germany, developed one of the methods we use today for measuring how much blood your heart pumps in a minute. Your cardiac output as you read this sentence is probably about 7 L/min. At rest it is likely to be a bit under 6 L/min. If you are a trained marathon runner running a marathon, your cardiac output can be as high as 30 L/min.

Your cardiac output can be calculated with the formula

$$y = \frac{Q}{D},$$

where Q is the number of milliliters of CO_2 you exhale in a minute and D is the difference between the CO_2 concentration (ml/L) in the blood pumped to the lungs and the CO_2 concentration in the blood returning from the lungs. With $Q = 233$ ml/min and $D = 97 - 56 = 41$ ml/L,

$$y = \frac{233 \text{ ml/min}}{41 \text{ ml/L}} \approx 5.68 \text{ L/min},$$

fairly close to the 6 L/min that most people have at basal (resting) conditions. (Data courtesy of J. Kenneth Herd, M.D., Quillan College of Medicine, East Tennessee State University.)

Suppose that when $Q = 233$ and $D = 41$, we also know that D is decreasing at the rate of 2 units a minute but that Q remains unchanged. What is happening to the cardiac output?

26. *Cost, revenue, and profit* A company can manufacture x items at a cost of $c(x)$ thousand dollars, a sales revenue of $r(x)$ thousand dollars, and a profit of $p(x) = r(x) - c(x)$ thousand dollars. Find dc/dt, dr/dt, and dp/dt for the following values of x and dx/dt.

(a) $r(x) = 9x$, $c(x) = x^3 - 6x^2 + 15x$, and $dx/dt = 0.1$ when $x = 2$

(b) $r(x) = 70x$, $c(x) = x^3 - 6x^2 + 45/x$, and $dx/dt = 0.05$ when $x = 1.5$

27. *Moving along a parabola* A particle moves along the parabola $y = x^2$ in the first quadrant in such a way that its x-coordinate (measured in meters) increases at a steady 10 m/sec. How fast is the angle of inclination θ of the line joining the particle to the origin changing when $x = 3$ m?

28. *Moving along another parabola* A particle moves from right to left along the parabolic curve $y = \sqrt{-x}$ in such a way that its x-coordinate (measured in meters) decreases at the rate of 8 m/sec. How fast is the angle of inclination θ of the line joining the particle to the origin changing when $x = -4$?

29. *Motion in the plane* The coordinates of a particle in the metric xy-plane are differentiable functions of time t with $dx/dt = -1$ m/sec and $dy/dt = -5$ m/sec. How fast is the particle's distance from the origin changing as it passes through the point $(5, 12)$?

30. *A moving shadow* A man 6 ft tall walks at the rate of 5 ft/sec toward a streetlight that is 16 ft above the ground. At what rate is the tip of his shadow moving? At what rate is the length of his shadow changing when he is 10 ft from the base of the light?

31. *Another moving shadow* A light shines from the top of a pole 50 ft high. A ball is dropped from the same height from a point 30 ft away from the light. (See accompanying figure.) How fast is the shadow of the ball moving along the ground 1/2 sec later? (Assume the ball falls a distance $s = 16t^2$ ft in t sec.)

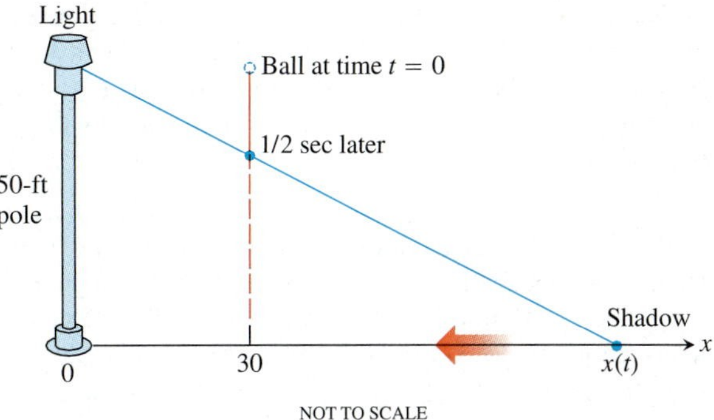

32. *Videotaping a moving car* You are videotaping a race from a stand 132 ft from the track, following a car that is moving at 180 mi/h (264 ft/sec). How fast will your camera angle θ be changing when the car is right in front of you? A half second later?

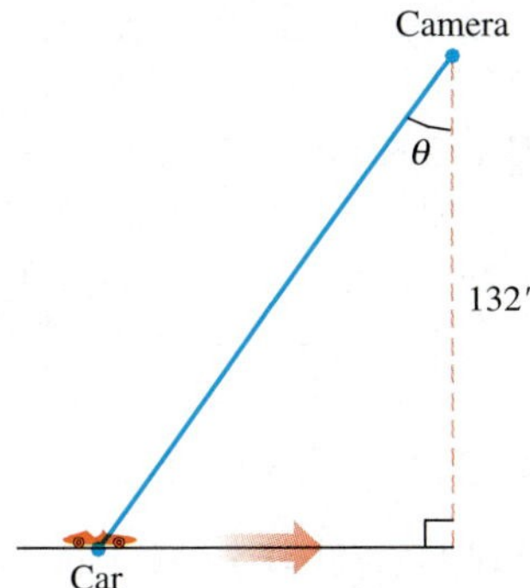

33. *A melting ice layer* A spherical iron ball 8 in. in diameter is coated with a layer of ice of uniform thickness. If the ice melts at the rate of 10 in^3/min, how fast is the thickness of the ice decreasing when it is 2 in. thick? How fast is the outer surface area of ice decreasing?

34. *Highway patrol* A highway patrol plane flies 3 mi above a level, straight road at a steady 120 mi/h. The pilot sees an oncoming car and with radar determines that at the instant the line-of-sight distance from plane to car is 5 mi, the line-of-sight distance is decreasing at the rate of 160 mi/h. Find the car's speed along the highway.

35. *A building's shadow* On a morning of a day when the sun will pass directly overhead, the shadow of an 80-ft building on level

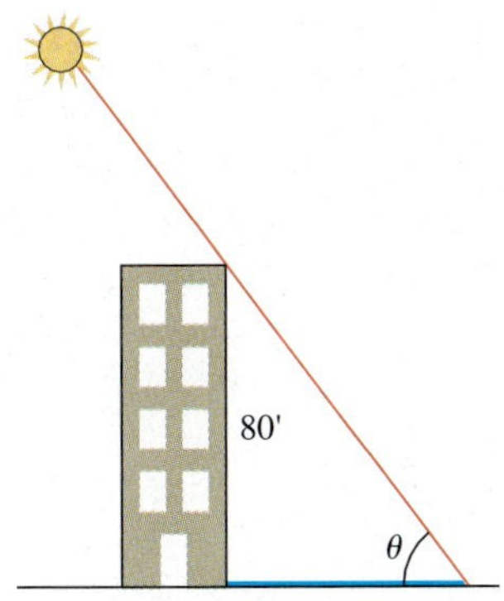

ground is 60 ft long. At the moment in question, the angle θ the sun makes with the ground is increasing at the rate of $0.27°/\text{min}$. At what rate is the shadow decreasing? (Remember to use radians. Express your answer in inches per minute, to the nearest tenth.)

36. *Walkers* A and B are walking on straight streets that meet at right angles. A approaches the intersection at 2 m/sec; B moves away from the intersection 1 m/sec. At what rate is the angle θ changing when A is 10 m from the intersection and B is 20 m from the intersection? Express your answer in degrees per second to the nearest degree.

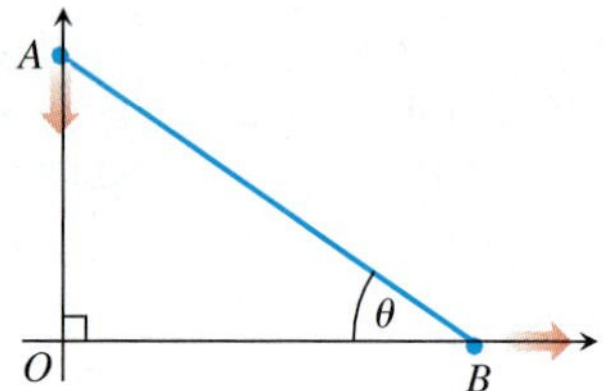

37. *Baseball players* A baseball diamond is a square 90 ft on a side. A player runs from first base to second at a rate of 16 ft/sec.

(a) At what rate is the player's distance from third base changing when the player is 30 ft from first base?

(b) At what rates are angles θ_1 and θ_2 (see the figure) changing at that time?

(c) The player slides into second base at the rate of 15 ft/sec. At what rates are angles θ_1 and θ_2 changing as the player touches base?

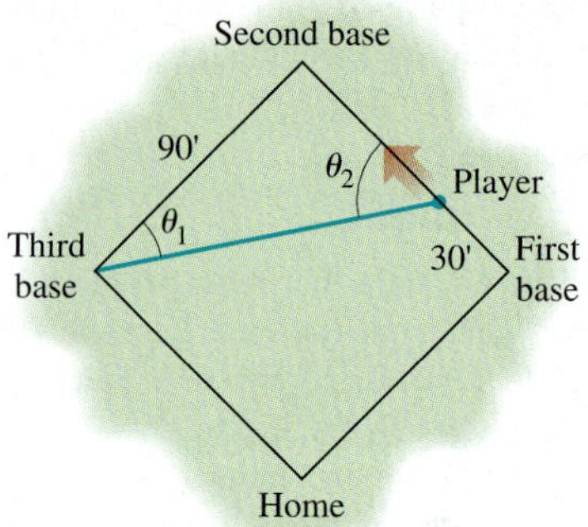

38. *Ships* Two ships are steaming straight away from a point O along routes that make a $120°$ angle. Ship A moves at 14 knots (nautical miles per hour; a nautical mile is 2000 yd). Ship B moves at 21 knots. How fast are the ships moving apart when $OA = 5$ and $OB = 3$ nautical miles?

2.8 Derivatives of Inverse Trigonometric Functions

Derivatives of Inverse Functions • Derivative of the Arcsine • Derivative of the Arctangent • Derivative of the Arcsecant • Derivatives of the Other Three

In this section, we learn when a differentiable function has an inverse function that is also differentiable. We then use that result (Theorem 5) to find formulas for the derivatives of the inverse trigonometric functions reviewed in Preliminary Section 5.

Derivatives of Inverse Functions

In Preliminary Section 4, we learned that the graph of the inverse of a function f can be obtained by reflecting the graph of f across the line $y = x$. If we combine that with our understanding of what makes a function differentiable, we can gain insight into the differentiability of inverse functions.

Example 1 Relating Slopes of Reflected Lines

If we calculate the derivatives of $f(x) = (1/2)x + 1$ and its inverse $f^{-1}(x) = 2x - 2$ from Example 3 in Preliminary Section 4, we see that

$$\frac{d}{dx} f(x) = \frac{d}{dx}\left(\frac{1}{2}x + 1\right) = \frac{1}{2}$$

$$\frac{d}{dx} f^{-1}(x) = \frac{d}{dx}(2x - 2) = 2.$$

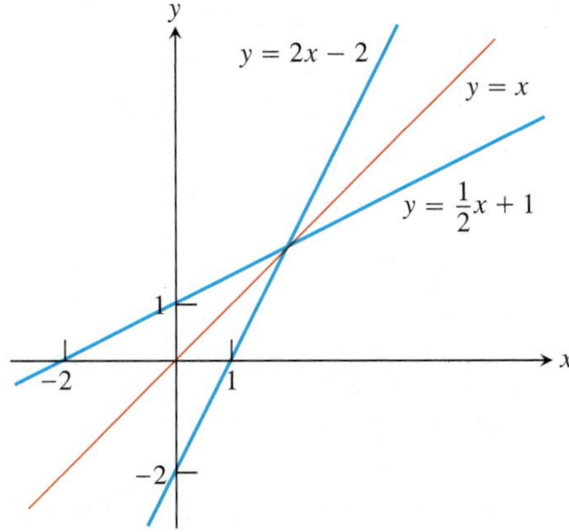

FIGURE 2.46 Graphing $f(x) = (1/2)x + 1$ and $f^{-1}(x) = 2x - 2$ together shows the graphs' symmetry with respect to the line $y = x$. Their slopes are reciprocals of one another.

The derivatives are reciprocals of one another. The graph of f is the line $y = (1/2)x + 1$, and the graph of f^{-1} is the line $y = 2x - 2$ (Figure 2.46). Their slopes are reciprocals of one another.

This case is not special. Reflecting any nonhorizontal or nonvertical line across the line $y = x$ inverts the line's slope. If the original line has slope $m \neq 0$ (Figure 2.47), the reflected line has slope $1/m$ (Exercise 38).

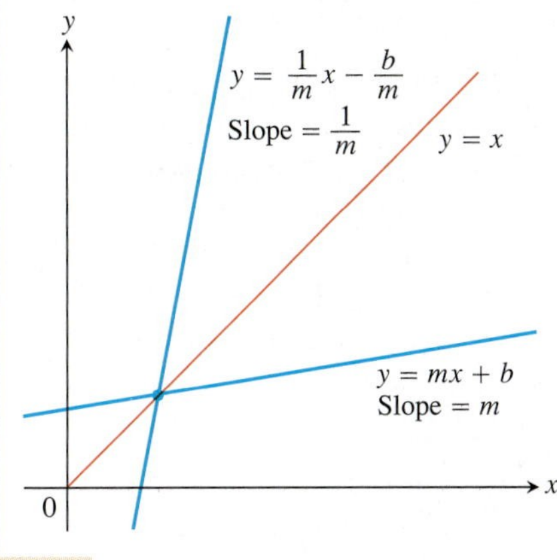

FIGURE 2.47 The slopes of nonvertical lines reflected across the line $y = x$ are reciprocals of one another.

As Figure 2.48 suggests, the reflection of a continuous curve with no cusps or corners will be another continuous curve with no cusps or corners. Indeed, if there is a tangent line to the graph of f at the point $(a, f(a))$, that line will reflect across $y = x$ to become a line tangent to the graph of f^{-1} at the point $(f(a), a)$. And, as we have just seen, the *slope* of the reflected tangent line, when it exists, will be the *reciprocal* of the slope of the original tangent line.

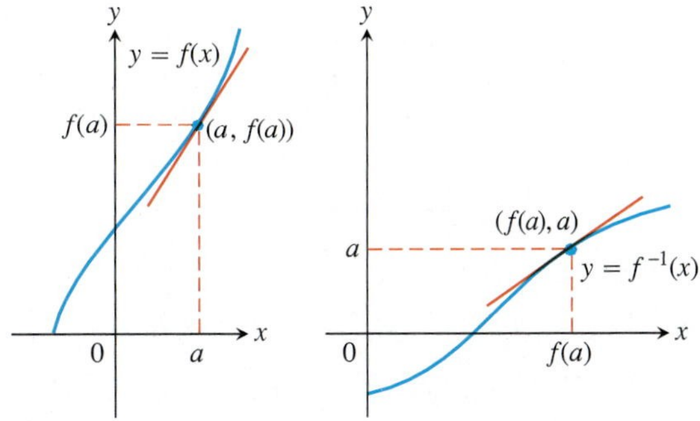

The slopes are reciprocal: $\left.\dfrac{df^{-1}}{dx}\right|_{f(a)} = \dfrac{1}{\left.\dfrac{df}{dx}\right|_{a}}$.

FIGURE 2.48 The graphs of inverse functions have reciprocal slopes at corresponding points.

All this serves as an introduction to the following theorem, which we will assume as we proceed to find derivatives of inverse functions. Although the essentials of the proof are illustrated in the geometry of Figure 2.48, a careful analytic proof is more appropriate for an advanced calculus text and is omitted here.

Theorem 5 Derivatives of Inverse Functions

If f is differentiable at every point of an interval I and df/dx is never zero on I, then f has an inverse and f^{-1} is differentiable at every point of the interval $f(I)$.

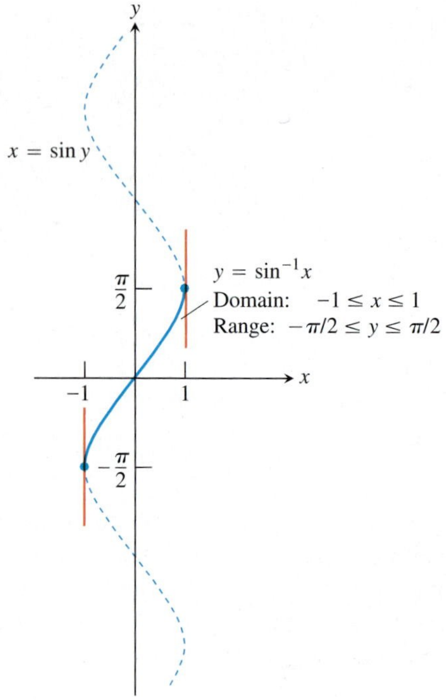

Derivative of the Arcsine

We know that the function $x = \sin y$ is differentiable in the open interval $-\pi/2 < y < \pi/2$ and that its derivative, the cosine, is positive there. Theorem 5 therefore assures us that the inverse function $y = \sin^{-1}(x)$ (the *arcsine* of x) is differentiable throughout the interval $-1 < x < 1$. We cannot expect the inverse to be differentiable at $x = -1$ or $x = 1$, however, because the tangents to the graph are vertical at these points (Figure 2.49).

We find the derivative of $y = \sin^{-1}(x)$ as follows:

$$
\begin{aligned}
y &= \sin^{-1} x \\
\sin y &= x && \text{Inverse function relationship} \\
\frac{d}{dx}(\sin y) &= \frac{d}{dx} x && \text{Differentiate both sides.} \\
\cos y \frac{dy}{dx} &= 1 && \text{Implicit differentiation.} \\
\frac{dy}{dx} &= \frac{1}{\cos y}.
\end{aligned}
$$

The division in the last step is safe because $\cos y \neq 0$ for $-\pi/2 < y < \pi/2$. In fact, $\cos y$ is *positive* for $-\pi/2 < y < \pi/2$, so we can replace $\cos y$ with $\sqrt{1-(\sin y)^2}$, which is $\sqrt{1-x^2}$. Thus,

$$\frac{d}{dx}(\sin^{-1} x) = \frac{1}{\sqrt{1-x^2}}.$$

If u is a differentiable function of x with $|u| < 1$, we apply the Chain Rule to get

$$\frac{d}{dx}\sin^{-1} u = \frac{1}{\sqrt{1-u^2}}\frac{du}{dx}, \qquad |u| < 1.$$

x = sin y
y = sin⁻¹x
Domain: $-1 \le x \le 1$
Range: $-\pi/2 \le y \le \pi/2$

FIGURE 2.49 The graph of $y = \sin^{-1} x$ has vertical tangents $x = -1$ and $x = 1$.

Example 2 Applying the Formula

$$\frac{d}{dx}(\sin^{-1} x^2) = \frac{1}{\sqrt{1-(x^2)^2}} \cdot \frac{d}{dx}(x^2) = \frac{2x}{\sqrt{1-x^4}}$$

Derivative of the Arctangent

Although the function $y = \sin^{-1}(x)$ has a rather narrow domain of $[-1, 1]$, the function $y = \tan^{-1} x$ is defined for all real numbers. It is also differentiable for all real numbers, as we will now see. The differentiation proceeds exactly as with the arcsine function above.

$$
\begin{aligned}
y &= \tan^{-1} x \\
\tan y &= x && \text{Inverse function relationship} \\
\frac{d}{dx}(\tan y) &= \frac{d}{dx} x \\
\sec^2 y \frac{dy}{dx} &= 1 && \text{Implicit differentiation} \\
\frac{dy}{dx} &= \frac{1}{\sec^2 y} \\
&= \frac{1}{1 + (\tan y)^2} && \text{Trig identity: } \sec^2 y = 1 + \tan^2 y \\
&= \frac{1}{1 + x^2}
\end{aligned}
$$

The derivative is defined for all real numbers. If u is a differentiable function of x, we get the Chain Rule form:

$$\frac{d}{dx} \tan^{-1} u = \frac{1}{1 + u^2} \frac{du}{dx}.$$

Example 3 Finding the Velocity of a Moving Particle

A particle moves along the x-axis so that its position at any time $t \geq 0$ is $x(t) = \tan^{-1} \sqrt{t}$. What is the velocity of the particle when $t = 16$?

Solution

$$v(t) = \frac{d}{dt} \tan^{-1} \sqrt{t} = \frac{1}{1 + (\sqrt{t})^2} \cdot \frac{d}{dt} \sqrt{t} = \frac{1}{1 + t} \cdot \frac{1}{2\sqrt{t}}$$

When $t = 16$, the velocity is

$$v(16) = \frac{1}{1 + 16} \cdot \frac{1}{2\sqrt{16}} = \frac{1}{136}.$$

Derivative of the Arcsecant

We find the derivative of $y = \sec^{-1} x$, $|x| > 1$, beginning as we did with the other inverse trigonometric functions.

$$
\begin{aligned}
y &= \sec^{-1} x \\
\sec y &= x && \text{Inverse function relationship} \\
\frac{d}{dx}(\sec y) &= \frac{d}{dx} x \\
\sec y \tan y \frac{dy}{dx} &= 1 \\
\frac{dy}{dx} &= \frac{1}{\sec y \tan y} && \text{Since } |x| > 1,\ y \text{ lies in } (0, \pi/2) \cup (\pi/2, \pi) \text{ and } \sec y \tan y \neq 0.
\end{aligned}
$$

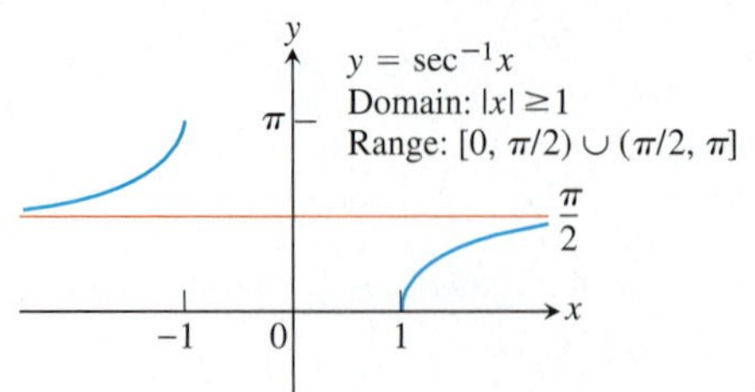

FIGURE 2.50 The slope of the curve $y = \sec^{-1} x$ is positive for both $x < -1$ and $x > 1$.

To express the result in terms of x, we use the relationships

$$\sec y = x \qquad \text{and} \qquad \tan y = \pm \sqrt{\sec^2 y - 1} = \pm \sqrt{x^2 - 1}$$

to get

$$\frac{dy}{dx} = \pm \frac{1}{x\sqrt{x^2 - 1}}.$$

Can we do anything about the $\pm$ sign? A glance at Figure 2.50 shows that the slope of the graph $y = \sec^{-1} x$ is always positive. That means that

$$\frac{d}{dx}\sec^{-1} x = \begin{cases} +\dfrac{1}{x\sqrt{x^2 - 1}} & \text{if } x > 1 \\ -\dfrac{1}{x\sqrt{x^2 - 1}} & \text{if } x < -1. \end{cases}$$

With the absolute value symbol we can write

$$\frac{d}{dx}\sec^{-1} x = \frac{1}{|x|\sqrt{x^2 - 1}}.$$

No $\pm$ is required in this new formula. If u is a differentiable function of x with $|u| > 1$, we have

$$\frac{d}{dx}\sec^{-1} u = \frac{1}{|u|\sqrt{u^2 - 1}}\frac{du}{dx}, \qquad |u| > 1$$

Example 4 Using the Formula

$$\begin{aligned} \frac{d}{dx}\sec^{-1}(5x^4) &= \frac{1}{|5x^4|\sqrt{(5x^4)^2 - 1}}\frac{d}{dx}(5x^4) \\ &= \frac{1}{5x^4\sqrt{25x^8 - 1}}(20x^3) \\ &= \frac{4}{x\sqrt{25x^8 - 1}} \end{aligned}$$

Derivatives of the Other Three

We could use the same technique to find the derivatives of the other three inverse trigonometric functions: arccosine, arccotangent, and arccosecant, but that is unnecessary, thanks to the following identities.

Inverse Function—Inverse Cofunction Identities

$$\cos^{-1} x = \pi/2 - \sin^{-1} x$$
$$\cot^{-1} x = \pi/2 - \tan^{-1} x$$
$$\csc^{-1} x = \pi/2 - \sec^{-1} x$$

It follows from these identities that the derivatives of the inverse cofunctions are the negatives of the derivatives of the corresponding inverse functions (see Exercises 23–25).

Example 5 A Tangent Line to the Arccotangent Curve

Find an equation for the line tangent to the graph of $y = \cot^{-1} x$ at $x = -1$.

Solution First, we note that

$$\cot^{-1}(-1) = \pi/2 - \tan^{-1}(-1) = \pi/2 - (-\pi/4) = 3\pi/4.$$

The slope of the tangent line is

$$\left.\frac{dy}{dx}\right|_{x=-1} = -\left.\frac{1}{1+x^2}\right|_{x=-1} = -\frac{1}{1+(-1)^2} = -\frac{1}{2},$$

so the tangent line has equation $y - 3\pi/4 = (-1/2)(x+1)$.

EXERCISES 2.8

Finding Derviatives

In Exercises 1–18, find the derivative of y with respect to the appropriate variable.

1. $y = \cos^{-1}(x^2)$

2. $y = \cos^{-1}(1/x)$

3. $y = \sin^{-1}\sqrt{2}t$

4. $y = \sin^{-1}(1-t)$

5. $y = \sec^{-1}(2s+1)$

6. $y = \sec^{-1} 5s$

7. $y = \csc^{-1}(x^2+1), \quad x > 0$

8. $y = \csc^{-1} x/2$

9. $y = \sec^{-1}\frac{1}{t}, \quad 0 < t < 1$

10. $y = \sin^{-1}\frac{3}{t^2}$

11. $y = \cot^{-1}\sqrt{t}$

12. $y = \cot^{-1}\sqrt{t-1}$

13. $y = s\sqrt{1-s^2} + \cos^{-1} s$

14. $y = \sqrt{s^2-1} - \sec^{-1} s$

15. $y = \tan^{-1}\sqrt{x^2-1} + \csc^{-1} x, \quad x > 1$

16. $y = \cot^{-1}\frac{1}{x} - \tan^{-1} x$

17. $y = x\sin^{-1} x + \sqrt{1-x^2}$

18. $y = \dfrac{1}{\sin^{-1}(2x)}$

Theory and Examples

19. **(a)** Find an equation for the line tangent to the graph of $y = \tan x$ at the point $(\pi/4, 1)$.

(b) Find an equation for the line tangent to the graph of $y = \tan^{-1} x$ at the point $(1, \pi/4)$.

20. Let $f(x) = x^5 + 2x^3 + x - 1$.

(a) Find $f(1)$ and $f'(1)$.

(b) Find $f^{-1}(3)$ and $(f^{-1})'(3)$.

21. Let $f(x) = \cos x + 3x$.

(a) Show that f has a differentiable inverse.

(b) Find $f(0)$ and $f'(0)$.

(c) Find $f^{-1}(1)$ and $(f^{-1})'(1)$.

22. *Motion of a particle* A particle moves along the x-axis so that its position at any time $t \geq 0$ is $x = \arctan t$.

(a) Prove that the particle is always moving to the right.

(b) Prove that the particle is always decelerating.

(c) What is the limiting position of the particle as t approaches infinity?

In Exercises 23–25, use the inverse function–inverse cofunction identities to derive the formula for the derivative of the function.

23. arccosine

24. arccotangent

25. arccosecant

26. *Identities* Confirm the following identities for $x > 0$.

(a) $\cos^{-1} x + \sin^{-1} x = \pi/2$

(b) $\tan^{-1} x + \cot^{-1} x = \pi/2$

(c) $\sec^{-1} x + \csc^{-1} x = \pi/2$

End Behavior Models

In Exercises 27–30, find **(a)** a right end behavior model, **(b)** a left end behavior model, and **(c)** any horizontal tangents for the function if they exist.

27. $y = \tan^{-1} x$

28. $y = \cot^{-1} x$

29. $y = \sec^{-1} x$

30. $y = \csc^{-1} x$

Derivatives of Inverse Functions

In Exercises 31–34:

(a) Find $f^{-1}(x)$.

(b) Graph f and f^{-1} together.

(c) Evaluate df/dx at $x = a$ and df^{-1}/dx at $x = f(a)$ to show that at these points $df^{-1}/dx = 1/(df/dx)$.

31. $f(x) = 2x + 3, \quad a = -1$

32. $f(x) = (1/5)x + 7, \quad a = -1$

33. $f(x) = 5 - 4x, \quad a = 1/2$

34. $f(x) = 2x^2, \quad x \geq 0, \quad a = 5$

35. (a) Show that $f(x) = x^3$ and $g(x) = \sqrt[3]{x}$ are inverses of one another.

T (b) Graph f and g over an x-interval large enough to show the graphs intersecting at $(1, 1)$ and $(-1, -1)$. Be sure the picture shows the required symmetry in the line $y = x$.

(c) Find the slopes of the tangents to the graphs of f and g at $(1, 1)$ and $(-1, -1)$ (four tangents in all).

(d) What lines are tangent to the curves at the origin?

36. (a) Show that $h(x) = x^3/4$ and $k(x) = (4x)^{1/3}$ are inverses of one another.

T (b) Graph h and k over an x-interval large enough to show the graphs intersecting at $(2, 2)$ and $(-2, -2)$. Be sure the picture shows the required symmetry about the line $y = x$.

(c) Find the slopes of the tangents to the graphs at h and k at $(2, 2)$ and $(-2, -2)$.

(d) What lines are tangent to the curves at the origin?

37. (a) Find the inverse of the function $f(x) = mx$, where m is a constant different from zero.

(b) What can you conclude about the inverse of a function $y = f(x)$ whose graph is a line through the origin with a nonzero slope m?

38. Show that the graph of the inverse of $f(x) = mx + b$, where m and b are constants and $m \neq 0$, is a line with slope $1/m$ and y-intercept $-b/m$.

COMPUTER EXPLORATIONS

Inverses and Derivatives

In Exercises 39–46, you will explore some functions and their inverses together with their derivatives and tangent lines at specified points. Perform the following steps using a CAS:

(a) Plot the function $y = f(x)$ together with its derivative over the given interval. Explain why you know that f is one-to-one over the interval.

(b) Solve the equation $y = f(x)$ for x as a function of y and name the resulting inverse function g.

(c) Find the equation for the line tangent to the graph of f at the specified point $(x_0, f(x_0))$.

(d) If we write $g(x)$ for the inverse $f^{-1}(x)$, the derivative of the inverse can be written as

$$g'(f(a)) = \frac{1}{f'(a)}, \qquad \text{or} \qquad g'(f(a)) \cdot f'(a) = 1.$$

Find the equation for the line tangent to the graph of g at the point $(f(x_0), x_0)$ located symmetrically across the 45° line $y = x$ (which is the graph of the identity function). Use the formula above to find the slope of this tangent line.

(e) Plot the functions f and g, the identity, the two tangent lines, and the line segment joining the points $(x_0, f(x_0))$ and $(f(x_0), x_0)$. Discuss the symmetries you see across the main diagonal.

39. $y = \sqrt{3x - 2}, \quad 2/3 \leq x \leq 4, \quad x_0 = 3$

40. $y = \dfrac{3x + 2}{2x - 11}, \quad -2 \leq x \leq 2, \quad x_0 = 1/2$

41. $y = \dfrac{4x}{x^2 + 1}, \quad -1 \leq x \leq 1, \quad x_0 = 1/2$

42. $y = \dfrac{x^3}{x^2 + 1}, \quad -1 \leq x \leq 1, \quad x_0 = 1/2$

43. $y = x^3 - 3x^2 - 1, \quad 2 \leq x \leq 5, \quad x_0 = 27/10$

44. $y = 2 - x - x^3, \quad -2 \leq x \leq 2, \quad x_0 = 3/2$

45. $y = e^x, \quad -3 \leq x \leq 5, \quad x_0 = 1$

46. $y = \sin x, \quad -\pi/2 \leq x \leq \pi/2, \quad x_0 = 1$

In Exercises 47 and 48, repeat the steps above to solve for the functions $y = f(x)$ and $x = f^{-1}(y)$ defined implicitly by the given equations over the interval.

47. $y^{1/3} - 1 = (x + 2)^3, \quad -5 \leq x \leq 5, \quad x_0 = -3/2$

48. $\cos y = x^{1/5}, \quad 0 \leq x \leq 1, \quad x_0 = 1/2$

2.9 Derivatives of Exponential and Logarithmic Functions

Derivative of e^x • Exponential Change • Derivative of a^x • Geometric Significance of $\lim\limits_{h \to 0} \dfrac{a^h - 1}{h}$ • Derivative of $\ln x$ • Derivative of $\log_a x$ • Power Rule for Arbitrary Real Powers

The most important function–inverse pair in mathematics and science is the pair consisting of the natural logarithm function $\ln x$ and the exponential function e^x. In this section, we learn how to differentiate these functions. We also begin to explore the properties that account for the amazing frequency with which the exponential function appears in science and engineering applications. We look at many of these applications in later chapters.

Derivative of e^x

At the end of the brief review of exponential functions in Preliminary Section 3, we mentioned that the function $y = e^{kx}$ was a particularly important function for modeling exponential growth. The number e was defined in that section to be the limit of $(1 + 1/x)^x$ as $x \to \infty$. This intriguing number shows up in many interesting applications, because the exponential function $y = e^x$ has the property that it is its own derivative. Let's see why this is true.

When we apply the definition of the derivative to $f(x) = a^x$, we see that the derivative is a constant multiple of a^x itself:

$$\begin{aligned}
\frac{d}{dx}(a^x) &= \lim_{h\to 0} \frac{a^{x+h} - a^x}{h} && \text{Derivative definition}\\
&= \lim_{h\to 0} \frac{a^x \cdot a^h - a^x}{h} && a^{x+h} = a^x \cdot a^h\\
&= \lim_{h\to 0} a^x \cdot \frac{a^h - 1}{h} && \text{Factoring out } a^x\\
&= a^x \cdot \lim_{h\to 0} \frac{a^h - 1}{h} && a^x \text{ is constant as } h \to 0.\\
&= \underbrace{\left(\lim_{h\to 0} \frac{a^h - 1}{h}\right)}_{\text{a fixed number } L} \cdot a^x. && (1)
\end{aligned}$$

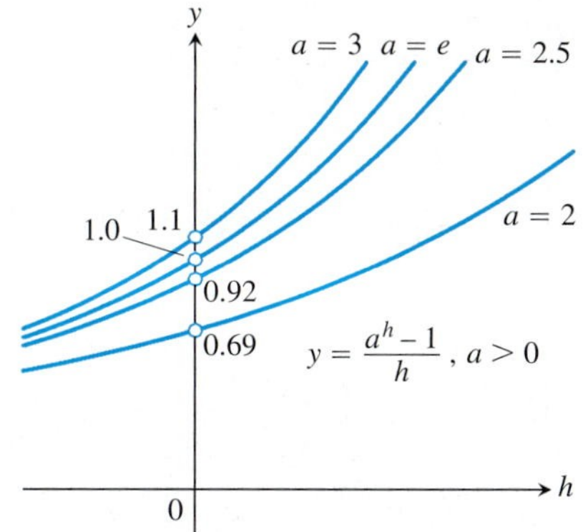

FIGURE 2.51 The position of the curve $y = (a^h - 1)/h$, $a > 0$, varies continuously with a.

It can be proved that the limit L exists. To estimate its value, we graph $y = (a^h - 1)/h$ as a function of h to see where the graph approaches the y-axis as $h \to 0$. Figure 2.51 shows the graphs for $a = 2$, 2.5, and 3 and for one other value we will discuss in a moment. The value of L is approximately 0.69 if $a = 2$, 0.92 if $a = 2.5$, and 1.10 if $a = 3$. Accordingly,

$$\frac{d}{dx}(2^x) \approx (0.69)2^x$$

$$\frac{d}{dx}(2.5^x) \approx (0.92)2.5^x$$

$$\frac{d}{dx}(3^x) \approx (1.10)3^x.$$

The good news is that we now have some idea of the values of these derivatives. The bad news is that the multipliers accumulate if we continue to differentiate. The calculations get messy. For instance,

$$\begin{aligned}
\frac{d}{dx}(2^x) &\approx (0.69)2^x\\
\frac{d^2}{dx^2}(2^x) &\approx (0.69)\frac{d}{dx}(2^x) \approx (0.69)^2 \cdot 2^x\\
&\vdots\\
\frac{d^n}{dx^n}(2^x) &\approx (0.69)^n \cdot 2^x.
\end{aligned}$$

Another look at Figure 2.51 suggests a way out. For $a = 2.5$, the graph of $y = (a^h - 1)/h$ steps across the y-axis at about 0.92. For $a = 3$, the graph steps across at about 1.1. Thus, at some value of a between 2.5 and 3, the graph should step across the axis at $y = 1$. The particular value of a for which this is true is the number e. That is, the number e gives the limit

$$\lim_{h \to 0} \frac{e^h - 1}{h} = 1. \tag{2}$$

That the limit is 1 creates a remarkable relationship between the function e^x and its derivative:

$$\begin{aligned} \frac{d}{dx}(e^x) &= \lim_{h \to 0} \left(\frac{e^h - 1}{h} \right) \cdot e^x && \text{Eq. (1) with } a = e \\ &= 1 \cdot e^x && \text{Eq. (2)} \\ &= e^x. \end{aligned}$$

In other words, the derivative of this particular function is itself!

$$\frac{d}{dx}(e^x) = e^x$$

If u is a differentiable function of x, then we have

$$\frac{d}{dx} e^u = e^u \frac{du}{dx}. \tag{3}$$

We make extensive use of this formula when we study exponential growth and decay in Chapter 4.

Is any other function its own derivative?

The zero function is also its own derivative, but this hardly seems worth mentioning. (Its value is always 0 and its slope is always 0.) In addition to e^x, however, we can also see that any constant *multiple* of e^x is its own derivative:

$$\frac{d}{dx}(c \cdot e^x) = c \cdot e^x.$$

The next obvious question is whether there are still *other* functions that are their own derivatives, and this time the answer is no. The only functions that satisfy the condition $dy/dx = y$ are functions of the form $y = ke^x$ (and notice that the zero function can be included in this category). We prove this significant fact in Chapter 4.

Example 1 Differentiating Exponentials

(a) $\frac{d}{dx}(e^{5x}) = e^{5x} \cdot \frac{d}{dx}(5x) = 5e^{5x}$ — Eq. (3) with $u = 5x$

(b) $\frac{d}{dx}(e^{kx}) = e^{kx} \cdot \frac{d}{dx}(kx) = ke^{kx}$ — Eq. (3) with $u = kx$, any number k

(c) $\frac{d}{dx}(e^{-x}) = -e^{-x}$ — Eq. (3) with $u = -1 \cdot x$

(d) $\frac{d}{dx}(e^{x^2}) = e^{x^2} \cdot \frac{d}{dx}(x^2) = 2xe^{x^2}$ — Eq. (3) with $u = x^2$

(e) $\frac{d}{dx} e^{\sin x} = e^{\sin x} \cdot \frac{d}{dx}(\sin x) = e^{\sin x} \cdot \cos x$ — Eq. (3) with $u = \sin x$

Exponential Change

In many instances in science, some positive quantity increases or decreases at a rate that at any given time t is proportional to the amount that is present at time t. In Section 5.4, we show that in these instances the amount can be represented by an equation of the form $y = y_0 e^{kt}$, where y_0 is the amount initially present at time $t = 0$.

> The equation
>
> $$y = y_0 e^{kt} \tag{4}$$
>
> is called the **law of exponential change.**

Example 2 Forecasting the Incidence of a Disease

One model for the way diseases spread assumes that the rate at which the number y of infected people changes is proportional to y itself. The more infected people there are, the faster the disease will spread. The fewer there are, the slower it will spread. If y_0 is the number of infected people at time $t = 0$, then the number of infected people at any time in the near future will be about

$$y = y_0 e^{kt}.$$

Let's use this equation to answer questions about the course of a particular disease. To be specific, suppose that a worldwide eradication program is reducing the number y of cases at the rate of 20% a year. There are 10,000 recorded cases today, and we want to know how many years it will take to lower the number to 1000.

Solution

Starting with the equation $y = y_0 e^{kt}$, there are three things to find:

1. the value of y_0
2. the value of k
3. the value of t that makes $y = 1000$.

The value of y_0

If we start counting time from today, then $y = 10{,}000$ when $t = 0$, so $y_0 = 10{,}000$. Our equation is

$$y = 10{,}000\, e^{kt}. \tag{5}$$

The value of k

The available information tells us that when $t = 1$ (i.e., when one year has passed), the number of cases will be 80% of its present value, or 8000. Hence,

$$\begin{aligned} 10{,}000\, e^{k(1)} &= 8000 && \text{Eq. (5) with } t = 1,\ y = 8000 \\ e^k &= 0.8 \\ \ln e^k &= \ln 0.8 && \text{Logarithm of both sides} \\ k &\approx -0.22 && \text{Calculator result} \end{aligned}$$

The value of t that makes $y = 1000$

We find t by solving the following equation for t:

$$\begin{aligned} 10{,}000\, e^{-0.22t} &= 1000 && \text{Eq. (5) with } k = -0.22,\ y = 1000 \\ e^{-0.22t} &= 0.1 \\ \ln e^{-0.22t} &= \ln 0.1 && \text{Logarithm of both sides} \\ -0.22t &\approx -2.3 \\ t &\approx 10.5. \end{aligned}$$

The number of cases will drop from the original 10,000 to 1000 in about 10.5 years.

Derivative of a^x

What about an exponential function with a base other than e? We will assume that the base is positive and different from 1, since negative numbers to arbitrary real powers are not always real numbers, and $y = 1^x$ is a constant function.

If $a > 0$ and $a \neq 1$, we can write a^x in terms of e^x. The formula for doing so is

$$a^x = e^{x \ln a},$$

reviewed in Preliminary Section 4. We can then find the derivative of a^x with the Chain Rule:

$$\frac{d}{dx} a^x = \frac{d}{dx} e^{x \ln a} = e^{x \ln a} \cdot \frac{d}{dx}(x \ln a) = e^{x \ln a} \cdot \ln a = a^x \ln a.$$

Thus, if u is a differentiable function of x, we get the following rule.

For $a > 0$ and $a \neq 1$,

$$\frac{d}{dx}(a^u) = a^u \ln a \frac{du}{dx}. \tag{6}$$

Geometric Significance of $\lim_{h \to 0} \frac{a^h - 1}{h}$

You may have noticed that the graph of each of the functions $f(x) = a^x$ passes through the point (0, 1) with a different slope. What is this slope? It is none other than

$$f'(0) = \lim_{h \to 0} \frac{a^{0+h} - a^0}{h} = \lim_{h \to 0} \frac{a^h - 1}{h}. \tag{7}$$

Since $f'(x) = a^x \ln a$, we see that $f'(0) = a^0 \ln a = \ln a$. Equation (7) then gives

$$\lim_{h \to 0} \frac{a^h - 1}{h} = \ln a \tag{8}$$

Example 3 Finding a Point on a Curve with a Specified Slope

At what point on the graph of the function $y = 2^t - 3$ does the tangent line have slope 21?

Solution The slope is the derivative:

$$\frac{d}{dt}(2^t - 3) = 2^t \cdot \ln 2 - 0 = 2^t \ln 2.$$

We want the value of t for which $2^t \ln 2 = 21$:

$$2^t \ln 2 = 21$$

$$2^t = \frac{21}{\ln 2}$$

$$\ln 2^t = \ln\left(\frac{21}{\ln 2}\right) \qquad \text{Logarithm of both sides}$$

$$t \cdot \ln 2 = \ln 21 - \ln(\ln 2) \qquad \text{Properties of logarithms}$$

$$t = \frac{\ln 21 - \ln(\ln 2)}{\ln 2}$$

$$t \approx 4.921 \qquad \text{Calculator result}$$

$$y = 2^t - 3 \approx 27.297. \qquad \text{Using the value of } t$$

The point is approximately (4.9, 27.3).

Derivative of ln x

Now that we know the derivative of e^x, it is relatively easy to find the derivative of its inverse function, $\ln x$.

$$y = \ln x$$

$$e^y = x \qquad \text{Inverse function relationship}$$

$$\frac{d}{dx}(e^y) = \frac{d}{dx}(x) \qquad \text{Differentiate implicitly.}$$

$$e^y \frac{dy}{dx} = 1$$

$$\frac{dy}{dx} = \frac{1}{e^y} = \frac{1}{x}$$

This equation answers what was once a perplexing problem: Is there a function with derivative x^{-1}? All the other power functions follow the Power Rule,

$$\frac{d}{dx}x^n = nx^{n-1}.$$

This formula, however, is not much help if one is looking for a function with x^{-1} as its derivative! Now we know why: the function we should be looking for is not a power function at all; it is the natural logarithm function.

If u is a differentiable function of x and $u > 0$,

$$\frac{d}{dx}\ln u = \frac{1}{u}\frac{du}{dx}. \qquad (9)$$

Example 4 Finding a Tangent Line Through the Origin

A line with slope m passes through the origin and is tangent to the graph of $y = \ln x$. What is the value of m?

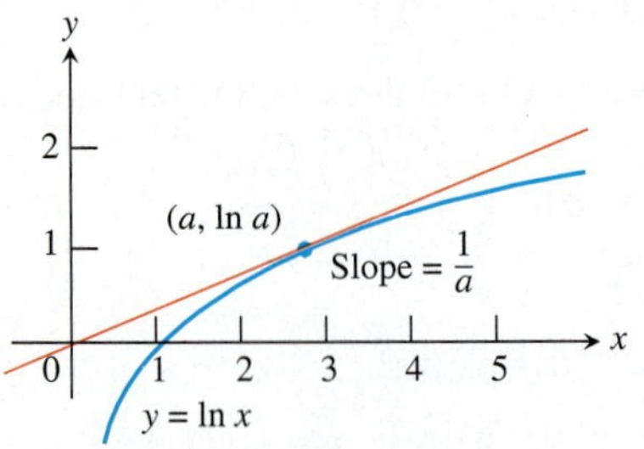

FIGURE 2.52 The tangent line intersects the curve at some point $(a, \ln a)$, where the slope of the curve is $1/a$. (Example 4)

Solution This problem is a little harder than it looks, since we do not know the point of tangency. We do, however, know two important facts about that point:

1. It has coordinates $(a, \ln a)$ for some positive a, and
2. The tangent line there has slope $m = 1/a$ (Figure 2.52).

Since the tangent line passes through the origin, its slope is

$$m = \frac{\ln a - 0}{a - 0} = \frac{\ln a}{a}.$$

Setting these two formulas for m equal to each other, we have

$$\begin{aligned} \frac{\ln a}{a} &= \frac{1}{a} \\ \ln a &= 1 \\ e^{\ln a} &= e^1 \\ a &= e \\ m &= \frac{1}{e}. \end{aligned}$$

Derivative of $\log_a x$

To find the derivative of $\log_a x$ for an arbitrary base $(a > 0,\ a \neq 1)$, we use the change-of-base formula for logarithms (reviewed in Preliminary Section 4) to express $\log_a x$ in terms of natural logarithms, as follows:

$$\log_a x = \frac{\ln x}{\ln a}.$$

The rest is easy:

$$\begin{aligned} \frac{d}{dx}\log_a x &= \frac{d}{dx}\left(\frac{\ln x}{\ln a}\right) \\ &= \frac{1}{\ln a} \cdot \frac{d}{dx}\ln x \qquad \text{Since } \ln a \text{ is a constant} \\ &= \frac{1}{\ln a} \cdot \frac{1}{x} \\ &= \frac{1}{x \ln a}. \end{aligned}$$

So, if u is a differentiable function of x and $u > 0$, the formula is as follows.

For $a > 0$ and $a \neq 1$,

$$\frac{d}{dx}\log_a u = \frac{1}{u \ln a}\frac{du}{dx}. \tag{10}$$

Example 5 Going the Long Way with the Chain Rule

Find dy/dx if $y = \log_a a^{\sin x}$.

Solution Carefully working from the outside in, we apply the Chain Rule to get:

$$\begin{aligned}
\frac{d}{dx}(\log_a a^{\sin x}) &= \frac{1}{a^{\sin x}\ln a}\cdot\frac{d}{dx}(a^{\sin x}) && \log_a u,\quad u = a^{\sin x}\\
&= \frac{1}{a^{\sin x}\ln a}\cdot a^{\sin x}\ln a\cdot\frac{d}{dx}(\sin x) && a^u,\quad u = \sin x\\
&= \frac{a^{\sin x}\ln a}{a^{\sin x}\ln a}\cdot\cos x\\
&= \cos x.
\end{aligned}$$

We could have saved ourselves a lot of work in Example 5 if we had noticed at the beginning that $\log_a a^{\sin x}$, being the composite of inverse functions, is equal to $\sin x$. It is always a good idea to simplify functions *before* differentiating, wherever possible. On the other hand, it is comforting to know that all these rules do work if applied correctly.

Power Rule for Arbitrary Real Powers

We are now ready to prove the Power Rule in its final form. As long as $x > 0$, we can write any real power of x as a power of e, specifically

$$x^n = e^{n\ln x}.$$

Doing so enables us to differentiate x^n for any real power n, as follows:

$$\begin{aligned}
\frac{d}{dx}(x^n) &= \frac{d}{dx}(e^{n\ln x})\\
&= e^{n\ln x}\cdot\frac{d}{dx}(n\ln x) && e^u,\quad u = n\ln x\\
&= e^{n\ln x}\cdot\frac{n}{x}\\
&= x^n\cdot\frac{n}{x}\\
&= nx^{n-1}.
\end{aligned}$$

The Chain Rule extends this result to the Power Rule's final form.

Rule 8 Power Rule for Arbitrary Real Powers

If u is a positive differentiable function of x and n is any real number, then u^n is a differentiable function of x, and

$$\frac{d}{dx}u^n = nu^{n-1}\frac{du}{dx}. \tag{11}$$

Example 6 Using the Power Rule in All Its Power

(a) If $y = x^{\sqrt{2}}$, then

$$\frac{dy}{dx} = \sqrt{2}x^{(\sqrt{2}-1)}.$$

(b) If $y = (2 + \sin 3x)^{\pi}$, then

$$\begin{aligned}\frac{d}{dx}(2 + \sin 3x)^{\pi} &= \pi(2 + \sin 3x)^{\pi-1}(\cos 3x) \cdot 3\\ &= 3\pi(2 + \sin 3x)^{\pi-1}(\cos 3x).\end{aligned}$$

Example 7 Finding a Derivative's Domain

If $f(x) = \ln(x - 3)$, find $f'(x)$. Identify the domain of f'.

Solution The domain of f is $(3, \infty)$, and

$$f'(x) = \frac{1}{x - 3}.$$

The domain of f' appears to be all $x \neq 3$. Since f is not defined for $x < 3$, however, neither is f'. Thus,

$$f'(x) = \frac{1}{x - 3}, \qquad x > 3.$$

That is, the domain of f' is $(3, \infty)$.

Sometimes the properties of logarithms can be used to simplify the differentiation process, even if we must introduce the logarithms ourselves as a step in the process. Example 8 shows a clever way to differentiate $y = x^x$ for $x > 0$.

Example 8 Logarithmic Differentiation

Find dy/dx for $y = x^x$, $x > 0$.

Solution

$$\begin{aligned}y &= x^x\\ \ln y &= \ln x^x && \text{Logarithm of both sides}\\ \ln y &= x \ln x && \text{Property of logs}\\ \frac{d}{dx}(\ln y) &= \frac{d}{dx}(x \ln x) && \text{Differentiate implicitly}\\ \frac{1}{y}\frac{dy}{dx} &= 1 \cdot \ln x + x \cdot \frac{1}{x}\\ \frac{dy}{dx} &= y(\ln x + 1)\\ \frac{dy}{dx} &= x^x(\ln x + 1)\end{aligned}$$

EXERCISES 2.9

Differentiating Exponentials

In Exercises 1–20, find dy/dx.

1. $y = 2e^x$
2. $y = e^{x+\sqrt{2}}$
3. $y = e^{-3x/2}$
4. $y = e^{-5x}$
5. $y = e^{2x/3}$
6. $y = e^{-x/4}$
7. $y = xe^x - e^x$
8. $y = x^2e^x - xe^x$
9. $y = e^{\sqrt{x}}$
10. $y = e^{(x^3)}$
11. $y = x^{\pi}$
12. $y = x^{1+\sqrt{2}}$
13. $y = x^{-\sqrt{2}}$
14. $y = x^{1-e}$
15. $y = 8^x$
16. $y = 9^{-x}$
17. $y = 3^{\csc x}$
18. $y = 3^{\cot x}$
19. $y = \dfrac{e^x}{e^{-x} + 1}$
20. $y = \dfrac{e^{-x}}{e^x + 1}$

Differentiating Logarithms

In Exercises 21–40, find dy/dx.

21. $y = \ln(x^2)$
22. $y = (\ln x)^2$
23. $y = \ln(1/x)$
24. $y = \ln(10/x)$
25. $y = \ln(x + 2)$
26. $y = \ln(2x + 2)$
27. $y = \ln(2 - \cos x)$
28. $y = \ln(x^2 + 1)$
29. $y = \ln(\ln x)$
30. $y = x \ln x - x$
31. $y = \log_4 x^2$
32. $y = \log_5 \sqrt{x}$
33. $y = \log_2(3x + 1)$
34. $y = \log_{10} \sqrt{x + 1}$
35. $y = \log_2(1/x)$
36. $y = 1/\log_2 x$
37. $y = \ln 2 \cdot \log_2 x$
38. $y = \log_3(1 + x \ln 3)$
39. $y = \log_{10} e^x$
40. $y = \ln 10^x$

Logarithmic Differentiation

In Exercises 41–46, use the technique of logarithmic differentiation to find dy/dx.

41. $y = x^{\ln x}$
42. $y = x^{(1/\ln x)}$
43. $y = (\sin x)^x, \quad 0 < x < \pi/2$
44. $y = x^{\tan x}, \quad x > 0$
45. $y = \sqrt[5]{\dfrac{(x-3)^4(x^2+1)}{(2x+5)^3}}$
46. $y = \dfrac{x\sqrt{x^2+1}}{(x+1)^{2/3}}$

Theory and Applications

47. *Tangent line* Find an equation for a line that is tangent to the graph of $y = e^x$ and goes through the origin.

48. *Normal line* Find an equation for a line that is normal to the graph of $y = xe^x$ and goes through the origin.

49. *Radioactive decay* The amount A (in grams) of radioactive plutonium remaining in a 20-g sample after t days is given by the formula

$$A = 20 \cdot (1/2)^{t/140}.$$

At what rate is the plutonium decaying when $t = 2$ days? Answer in appropriate units.

50. *Human evolution continues* The analysis of tooth shrinkage by C. Loring Brace and colleagues at the University of Michigan's Museum of Anthropology indicates that human tooth size is continuing to decrease and that the evolutionary process did not come to a halt some 30,000 years ago as many scientists contend. In northern Europeans, for example, tooth size reduction now has a rate of 1% per 1000 years.

    (a) If t represents time in years and y represents tooth size, use the condition that $y = 0.99y_0$ when $t = 1000$ to find the value of k in the equation $y = y_0e^{kt}$. Then use this value of k to answer the following questions.

    (b) In about how many years will human teeth be 90% of their present size?

    (c) What will be our descendants' tooth size 20,000 years from now (as a percentage of our present tooth size)?

    (*Source: LSA Magazine*, Vol. 12, No. 2 (Spring 1989), p. 19.)

51. *The U.S. population* The Museum of Science in Boston displays a running total of the U.S. population. On May 11, 1993, the total was increasing at the rate of 1 person every 14 sec. The displayed population figure for 3:45 P.M. that day was 257,313,431.

    (a) Assuming exponential growth at a constant rate, find the rate constant for the population's growth (people per 365-day year).

    (b) At this rate, what will the U.S. population be at 3:45 P.M. Boston time on May 11, 2001?

52. *The mean life of a radioactive nucleus* Physicists using the radioactivity equation $y = y_0e^{-kt}$ call the number $1/k$ the *mean life* of a radioactive nucleus. The mean life of a radon nucleus is about $1/0.18 \approx 5.6$ days. The mean life of a carbon-14 nucleus is more than 8000 years. Show that 95% of the radioactive nuclei originally present in a sample will disintegrate within three mean lifetimes, that is, by time $t = 3/k$. Thus, the mean life of a nucleus gives a quick way to estimate how long the radioactivity of a sample will last.

53. *Writing to Learn: The derivative of $g(x) = (0.5)^x$*

    (a) How are the derivative of $g(x) = (0.5)^x$ and the number

    $$L = \lim_{h \to 0} \frac{(0.5)^h - 1}{h}$$

    related? Give reasons for your answer.

    (b) What is the approximate value of L?

54. *Orthogonal families of curves* Prove that all curves in the family

$$y = -\frac{1}{2}x^2 + k$$

(k any constant) are perpendicular to all curves in the family $y = \ln x + c$ (c any constant) at their points of intersection. (See the accompanying figure.)

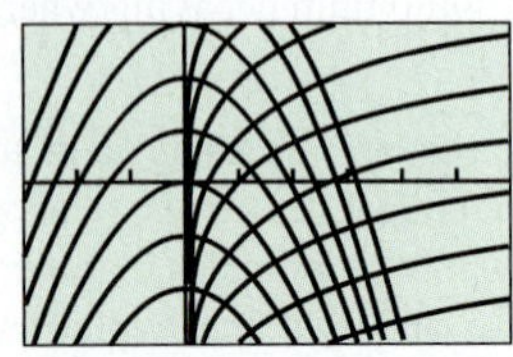

[–3, 6] by [–3, 3]

T **55.** *Writing to Learn: Why the graph of y = ln x has no horizontal asymptote* The graph of $y = \ln x$ looks as though it might be approaching a horizontal asymptote. Try it. Write an argument based on the graph of $y = e^x$ to explain why it does not.

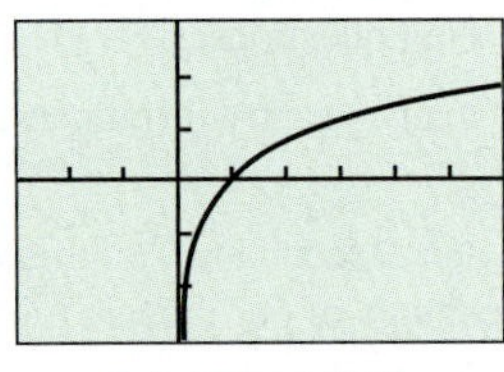

[–3, 6] by [–3, 3]

T **56.** *Solving $x^2 = 2^x$* How many solutions does the equation $x^2 = 2^x$ have? What are they?

57. *Which is bigger, π^e or e^π?* Calculators have taken some of the mystery out of this once-challenging question. (Go ahead and check; you will see that it is a surprisingly close call.) You can answer the question without a calculator, though, by using the result from Example 4 of this section.

Recall from that example that the line through the origin tangent to the graph of $y = \ln x$ has slope $1/e$.

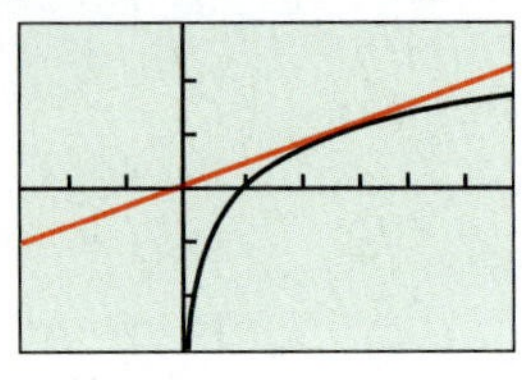

[–3, 6] by [–3, 3]

(a) Find an equation for this tangent line.

(b) Give an argument based on the graphs of $y = \ln x$ and the tangent line to explain why $\ln x < x/e$ for all positive $x \neq e$.

(c) Show that $\ln(x^e) < x$ for all positive $x \neq e$.

(d) Conclude that $x^e < e^x$ for all positive $x \neq e$.

(e) So which is bigger, π^e or e^π?

T **58.** *$x^{\ln 2}$ versus $2^{\ln x}$* Could $x^{\ln 2}$ possibly be the same as $2^{\ln x}$ for $x > 0$? Graph the two functions and explain what you see.

Questions To Guide Your Review

1. What is the derivative of a function f? How is its domain related to the domain of f? Give examples.
2. What role does the derivative play in defining slopes, tangents, and rates of change?
3. How can you sometimes graph the derivative of a function when all you have is a table of the function's values?
4. What does it mean for a function to be differentiable on an open interval? On a closed interval?
5. How are derivatives and one-sided derivatives related?
6. Describe geometrically when a function typically does *not* have a derivative at a point.
7. How is a function's differentiability at a point related to its continuity there if at all?
8. Could the unit step function

$$U(x) = \begin{cases} 0, & x < 0 \\ 1, & x \geq 0 \end{cases}$$

possibly be the derivative of some other function on $[-1, 1]$? Explain.

9. What rules do you know for calculating derivatives? Give some examples.
10. Explain how the three formulas

(a) $\dfrac{d}{dx}(x^n) = nx^{n-1}$

(b) $\dfrac{d}{dx}(cu) = c\dfrac{du}{dx}$

(c) $\dfrac{d}{dx}(u_1 + u_2 + \cdots + u_n) = \dfrac{du_1}{dx} + \dfrac{du_2}{dx} + \cdots + \dfrac{du_n}{dx}$

enable us to differentiate any polynomial.

11. What formula do we need, in addition to the three listed in Question 10, to differentiate rational functions?
12. What is a second derivative? A third derivative? How many derivatives do the functions you know have? Give examples.
13. What is the relationship between a function's average and instantaneous rates of change? Give an example.
14. How do derivatives arise in the study of motion? What can you learn about a body's motion along a line by examining the derivatives of the body's position function? Give examples.

15. How can derivatives arise in economics?

16. Give examples of still other applications of derivatives.

17. What do the limits $\lim_{h\to 0}(\sin h)/h$ and $\lim_{h\to 0}((\cos h - 1)/h)$ have to do with the derivatives of the sine and cosine functions? What *are* the derivatives of these functions?

18. Once you know the derivatives of $\sin x$ and $\cos x$, how can you find the derivatives of $\tan x$, $\cot x$, $\sec x$, and $\csc x$? What *are* the derivatives of these functions?

19. At what points are the six basic trigonometric functions continuous? How do you know?

20. What is the rule for calculating the derivative of a composite of two differentiable functions? How is such a derivative evaluated? Give examples.

21. What is the formula for the slope dy/dx of a parametrized curve $x = f(t)$, $y = g(t)$? When does the formula apply? When can you expect to be able to find d^2y/dx^2 as well? Give examples.

22. What is implicit differentiation? When do you need it? Give examples.

23. How do related rate problems arise? Give examples.

24. Outline a strategy for solving related rate problems. Illustrate with an example.

25. What are the derivatives of the inverse trigonometric functions? How do the domains of the derivatives compare with the domains of the functions?

26. What are the derivatives of the exponential function e^x and natural logarithm function $\ln x$? How do the domains of the derivatives compare with the domains of the functions?

27. What is the law of exponential change? What are some applications of the law?

28. What is the derivative of the exponential function a^x, $a > 0$ and $a \neq 1$? What is the geometric significance of the limit of $(a^h - 1)/h$ as $h \to 0$? What is the limit when a is the number e?

29. What is the derivative of $\log_a x$? Are there any restrictions on a?

30. What is logarithmic differentiation? Give an example.

31. How can you write any real power of x as a power of e? Are there any restrictions on x? How does this lead to the power rule for differentiating arbitrary real powers?

Practice Exercises

Derivatives of Functions

Find the derivatives of the functions in Exercises 1–42.

1. $y = x^5 - 0.125x^2 + 0.25x$
2. $y = x^3 - 3(x^2 + \pi^2)$
3. $y = x^7 + \sqrt{7x} - \dfrac{1}{\pi + 1}$
4. $y = (2x - 5)(4 - x)^{-1}$
5. $y = (\theta^2 + \sec\theta + 1)^3$
6. $s = \dfrac{\sqrt{t}}{1 + \sqrt{t}}$
7. $s = \dfrac{1}{\sqrt{t} - 1}$
8. $y = 2\tan^2 x - \sec^2 x$
9. $y = \dfrac{1}{\sin^2 x} - \dfrac{2}{\sin x}$
10. $s = \cos^4(1 - 2t)$
11. $s = \cot^3\left(\dfrac{2}{t}\right)$
12. $s = (\sec t + \tan t)^5$
13. $r = \sqrt{2\theta\sin\theta}$
14. $r = \sin(\theta + \sqrt{\theta + 1})$
15. $y = \dfrac{1}{2}x^2\csc\dfrac{2}{x}$
16. $y = x^{-1/2}\sec(2x)^2$
17. $y = 5\cot x^2$
18. $y = x^2\sin^2(2x^2)$
19. $s = \left(\dfrac{4t}{t + 1}\right)^{-2}$
20. $y = \left(\dfrac{\sqrt{x}}{1 + x}\right)^2$
21. $y = 4x\sqrt{x + \sqrt{x}}$
22. $r = \left(\dfrac{\sin\theta}{\cos\theta - 1}\right)^2$
23. $y = 20(3x - 4)^{1/4}(3x - 4)^{-1/5}$
24. $y = \dfrac{3}{(5x^2 + \sin 2x)^{3/2}}$
25. $y = \dfrac{1}{4}xe^{4x} - \dfrac{1}{16}e^{4x}$
26. $y = x^2e^{-2/x}$
27. $y = \ln(\sin^2\theta)$
28. $y = \log_2(x^2/2)$
29. $y = \log_5(3x - 7)$
30. $y = 8^{-t}$
31. $y = 5x^{3.6}$
32. $y = \sqrt{2}x^{-\sqrt{2}}$
33. $y = (x + 2)^{x+2}$
34. $y = 2(\ln x)^{x/2}$
35. $y = \sin^{-1}\sqrt{1 - u^2}$, $0 < u < 1$
36. $y = \ln\cos^{-1} x$
37. $y = z\cos^{-1} z - \sqrt{1 - z^2}$
38. $y = t\tan^{-1} t - \dfrac{1}{2}\ln t$
39. $y = (1 + t^2)\cot^{-1} 2t$
40. $y = z\sec^{-1} z - \sqrt{z^2 - 1}$, $z > 1$
41. $y = \csc^{-1}(\sec\theta)$, $0 < \theta < \pi/2$
42. $y = (1 + x^2)e^{\tan^{-1} x}$

Implicit Differentiation

In Exercises 43–54, find dy/dx.

43. $xy + 2x + 3y = 1$

44. $x^2 + xy + y^2 - 5x = 2$

45. $x^3 + 4xy - 3y^{4/3} = 2x$

46. $5x^{4/5} + 10y^{6/5} = 15$

47. $\sqrt{xy} = 1$

48. $x^2y^2 = 1$

49. $e^{x+2y} = 1$

50. $y^2 = 2e^{-1/x}$

51. $\ln(x/y) = 1$

52. $x \sin^{-1} y = 1 + x^2$

53. $ye^{\tan^{-1} x} = 2$

54. $x^y = \sqrt{2}$

In Exercises 55 and 56, find dr/ds.

55. $r \cos 2s + \sin^2 s = \pi$

56. $2rs - r - s + s^2 = -3$

57. Find d^2y/dx^2 by implicit differentiation:

(a) $x^3 + y^3 = 1$ (b) $y^2 = 1 - \dfrac{2}{x}$

58. (a) By differentiating $x^2 - y^2 = 1$ implicitly, show that $dy/dx = x/y$.

(b) Then show that $d^2y/dx^2 = -1/y^3$.

Numerical Values of Derivatives

59. Suppose that functions $f(x)$ and $g(x)$ and their first derivatives have the following values at $x = 0$ and $x = 1$.

x	$f(x)$	$g(x)$	$f'(x)$	$g'(x)$
0	1	1	−3	1/2
1	3	5	1/2	−4

Find the first derivatives of the following combinations at the given value of x.

(a) $6f(x) - g(x)$, $x = 1$ (b) $f(x)g^2(x)$, $x = 0$

(c) $\dfrac{f(x)}{g(x) + 1}$, $x = 1$ (d) $f(g(x))$, $x = 0$

(e) $g(f(x))$, $x = 0$ (f) $(x + f(x))^{3/2}$, $x = 1$

(g) $f(x + g(x))$, $x = 0$

60. Suppose that the function $f(x)$ and its first derivative have the following values at $x = 0$ and $x = 1$.

x	$f(x)$	$f'(x)$
0	9	−2
1	−3	1/5

Find the first derivatives of the following combinations at the given value of x.

(a) $\sqrt{x}f(x)$, $x = 1$ (b) $\sqrt{f(x)}$, $x = 0$

(c) $f(\sqrt{x})$, $x = 1$ (d) $f(1 - 5\tan x)$, $x = 0$

(e) $\dfrac{f(x)}{2 + \cos x}$, $x = 0$ (f) $10 \sin\left(\dfrac{\pi x}{2}\right) f^2(x)$, $x = 1$

61. Find the value of dy/dt at $t = 0$ if $y = 3 \sin 2x$ and $x = t^2 + \pi$.

62. Find the value of ds/du at $u = 2$ if $s = t^2 + 5t$ and $t = (u^2 + 2u)^{1/3}$.

63. Find the value of dw/ds at $s = 0$ if $w = \sin(e^{\sqrt{r}})$ and $r = 3\sin(s + \pi/6)$.

64. Find the value of dr/dt at $t = 0$ if $r = (\theta^2 + 7)^{1/3}$ and $\theta^2 e^t + \theta = 1$.

65. If $y^3 + y = 2\cos x$, find the value of d^2y/dx^2 at the point $(0, 1)$.

66. If $x^{1/3} + y^{1/3} = 4$, find d^2y/dx^2 at the point $(8, 8)$.

Derivative Definition

In Exercises 67 and 68, find the derivative using the definition.

67. $f(t) = \dfrac{1}{2t + 1}$

68. $g(x) = 2x^2 + 1$

69. *Writing to Learn*

(a) Graph the function

$$f(x) = \begin{cases} x^2, & -1 \le x < 0 \\ -x^2, & 0 \le x \le 1. \end{cases}$$

(b) Is f continuous at $x = 0$?

(c) Is f differentiable at $x = 0$?

Give reasons for your answers.

70. *Writing to Learn*

(a) Graph the function

$$f(x) = \begin{cases} x, & -1 \le x < 0 \\ \tan x, & 0 \le x \le \pi/4. \end{cases}$$

(b) Is f continuous at $x = 0$?

(c) Is f differentiable at $x = 0$?

Give reasons for your answers.

71. *Writing to Learn*

(a) Graph the function

$$f(x) = \begin{cases} x, & 0 \le x \le 1 \\ 2 - x, & 1 < x \le 2. \end{cases}$$

(b) Is f continuous at $x = 1$?

(c) Is f differentiable at $x = 1$?

Give reasons for your answers.

72. *Writing to Learn* For what value or values of the constant m, if any, is

$$f(x) = \begin{cases} \sin 2x, & x \leq 0 \\ mx, & x > 0 \end{cases}$$

(a) continuous at $x = 0$?

(b) differentiable at $x = 0$?

Give reasons for your answers.

Slopes, Tangents, and Normals

73. *Tangents with specified slope* Are there any points on the curve $y = (x/2) + 1/(2x - 4)$ where the slope is $-3/2$? If so, find them.

74. *Tangents with specified slope* Are there any points on the curve $y = x - e^{-x}$ where the slope is 2? If so, find them.

75. *Tangents intercepts* Find the x- and y-intercepts of the line that is tangent to the curve $y = x^3$ at the point $(-2, -8)$.

76. *Tangents perpendicular or parallel to lines* Find the points on the curve $y = 2x^3 - 3x^2 - 12x + 20$ where the tangent is

(a) perpendicular to the line $y = 1 - (x/24)$

(b) parallel to the line $y = \sqrt{2} - 12x$.

77. *Intersecting tangents* Show that the tangents to the curve $y = (\pi \sin x)/x$ at $x = \pi$ and $x = -\pi$ intersect at right angles.

78. *Tangent and normal lines* Find equations for the tangent and normal to the curve $y = 1 + \cos x$ at the point $(\pi/2, 1)$. Sketch the curve, tangent, and normal together, labeling each with its equation.

79. *Tangent parabola* The parabola $y = x^2 + C$ is to be tangent to the line $y = x$. Find C.

80. *Normal to a circle* Show that the normal line at any point of the circle $x^2 + y^2 = a^2$ passes through the origin.

Tangents and Normals to Implicitly Defined Curves

In Exercises 81–86, find equations for the lines that are tangent and normal to the curve at the given point.

81. $x^2 + 2y^2 = 9, \quad (1, 2)$

82. $e^x + y^2 = 2, \quad (0, 1)$

83. $xy + 2x - 5y = 2, \quad (3, 2)$

84. $(y - x)^2 = 2x + 4, \quad (6, 2)$

85. $x + \sqrt{xy} = 6, \quad (4, 1)$

86. $x^{3/2} + 2y^{3/2} = 17, \quad (1, 4)$

87. Find the slope of the curve $x^3y^3 + y^2 = x + y$ at the points $(1, 1)$ and $(1, -1)$.

88. *Writing to Learn* The graph shown suggests that the curve $y = \sin(x - \sin x)$ might have horizontal tangents at the x-axis. Does it? Give reasons for your answer.

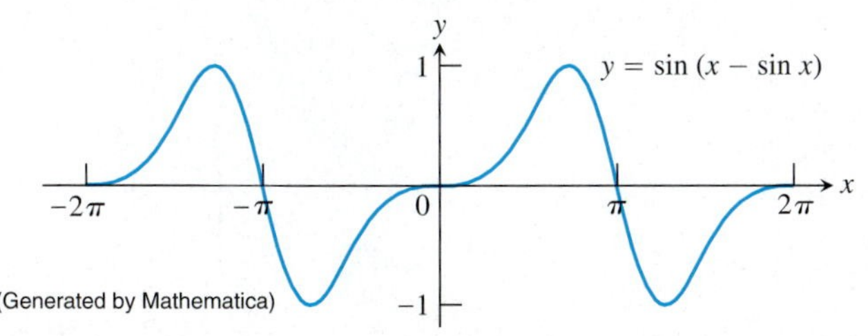

Analyzing Graphs

Each of the figures in Exercises 89 and 90 shows two graphs, the graph of a function $y = f(x)$ together with the graph of its derivative $f'(x)$. Which graph is which? How do you know?

89.

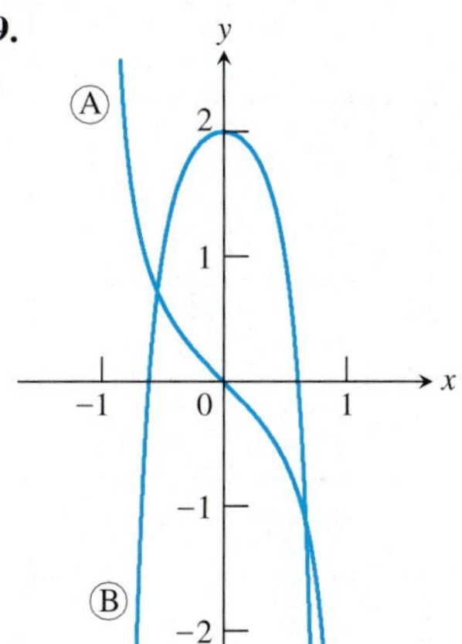

90.

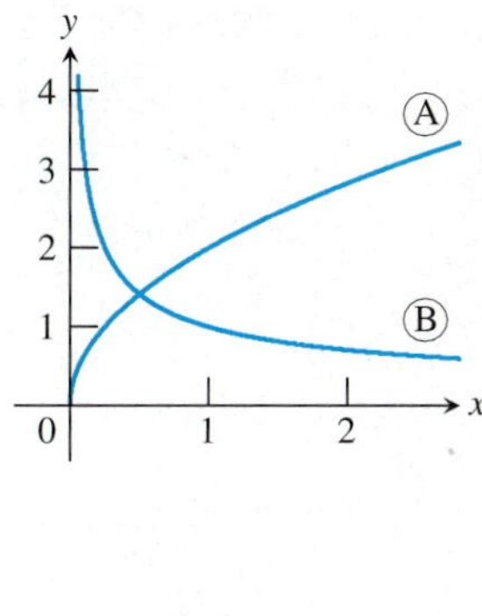

91. Use the following information to graph the function $y = f(x)$ for $-1 \leq x \leq 6$.

i. The graph of f is made of line segments joined end to end.

ii. The graph starts at the point $(-1, 2)$.

iii. The derivative of f, where defined, agrees with the step function shown here.

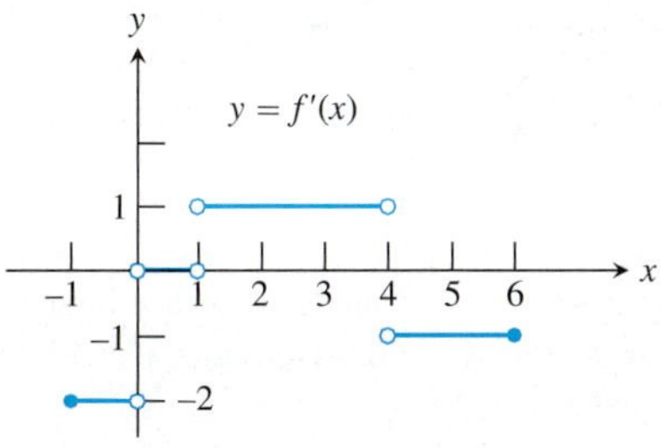

92. Repeat Exercise 91, supposing that the graph starts at $(-1, 0)$ instead of $(-1, 2)$.

Exercises 93 and 94 are about the graphs in Figure 2.53. The graphs in part (a) show the numbers of rabbits and foxes in a small arctic population. They are plotted as functions of time for 200 days. The number of rabbits increases at first, as the rabbits reproduce. But the foxes prey on the rabbits and, as the number of foxes increases, the rabbit population levels off and then drops. Figure 2.53b shows the graph of the derivative of the rabbit population. We made it by plotting slopes.

93. (a) What is the value of the derivative of the rabbit population in Figure 2.53 when the number of rabbits is largest? Smallest?

(b) What is the size of the rabbit population in Figure 2.53 when its derivative is largest? Smallest?

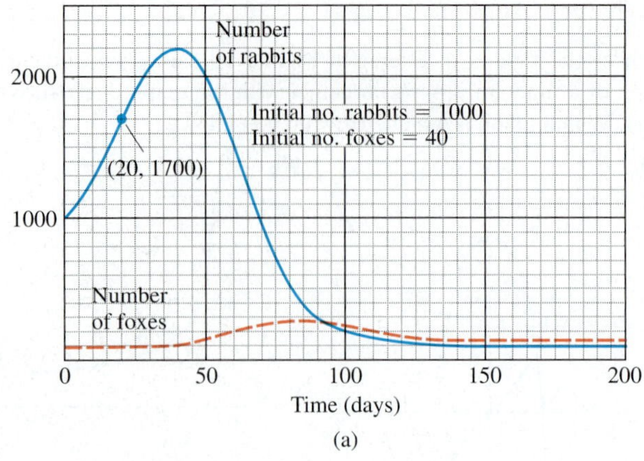

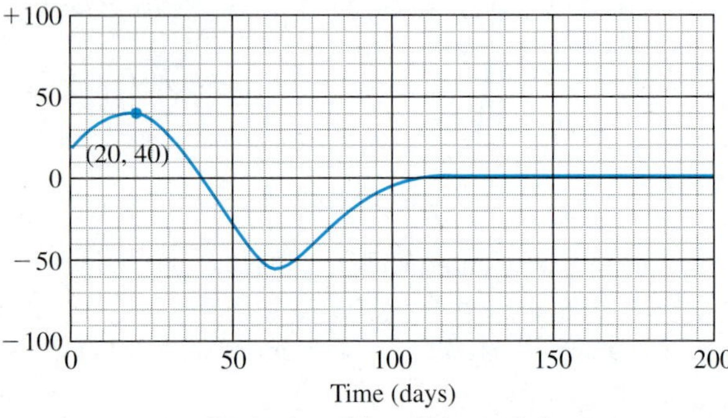

FIGURE 2.53 Rabbits and foxes in an arctic predator-prey food chain.

94. In what units should the slopes of the rabbit and fox population curves be measured?

Related Rates

95. *Right circular cylinder* The total surface area S of a right circular cylinder is related to the base radius r and height h by the equation $S = 2\pi r^2 + 2\pi rh$.

(a) How is dS/dt related to dr/dt if h is constant?

(b) How is dS/dt related to dh/dt if r is constant?

(c) How is dS/dt related to dr/dt and dh/dt if neither r nor h is constant?

(d) How is dr/dt related to dh/dt if S is constant?

96. *Right circular cone* The lateral surface area S of a right circular cone is related to the base radius r and height h by the equation $S = \pi r\sqrt{r^2 + h^2}$.

(a) How is dS/dt related to dr/dt if h is constant?

(b) How is dS/dt related to dh/dt if r is constant?

(c) How is dS/dt related to dr/dt and dh/dt if neither r nor h is constant?

97. *Circle's changing area* The radius of a circle is changing at the rate of $-2/\pi$ m/sec. At what rate is the circle's area changing when $r = 10$ m?

98. *Cube's changing edges* The volume of a cube is increasing at the rate of 1200 cm^3/min at the instant its edges are 20 cm long. At what rate are the edges changing at that instant?

99. *Resistors connected in parallel* If two resistors of R_1 and R_2 ohms are connected in parallel in an electric circuit to make an R-ohm resistor, the value of R can be found from the equation

$$\frac{1}{R} = \frac{1}{R_1} + \frac{1}{R_2}.$$

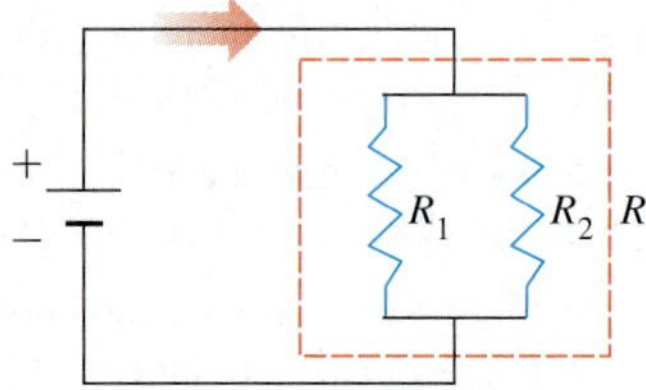

If R_1 is decreasing at the rate of 1 ohm/sec and R_2 is increasing at the rate of 0.5 ohm/sec, at what rate is R changing when $R_1 = 75$ ohms and $R_2 = 50$ ohms?

100. *Impedance in a series circuit* The impedance Z (ohms) in a series circuit is related to the resistance R (ohms) and reactance X (ohms) by the equation $Z = \sqrt{R^2 + X^2}$. If R is increasing at 3 ohms/sec and X is decreasing at 2 ohms/sec, at what rate is Z changing when $R = 10$ ohms and $X = 20$ ohms?

101. *Speed of moving particle* The coordinates of a particle moving in the metric xy-plane are differentiable functions of time t with $dx/dt = 10$ m/sec and $dy/dt = 5$ m/sec. How fast is the particle moving away from the origin as it passes through the point $(3, -4)$?

102. *Motion of a particle* A particle moves along the curve $y = x^{3/2}$ in the first quadrant in such a way that its distance from the origin increases at the rate of 11 units per second. Find dx/dt when $x = 3$.

103. *Draining a tank* Water drains from the conical tank shown in Figure 2.54 at the rate of 5 ft^3/min.

(a) What is the relation between the variables h and r in the figure?

(b) How fast is the water level dropping when $h = 6$ ft?

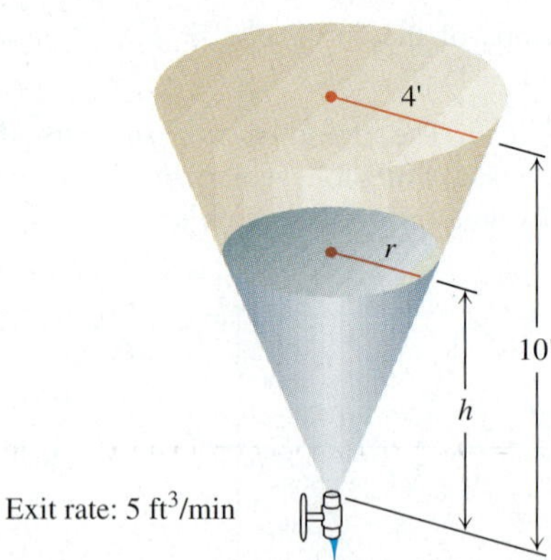

FIGURE 2.54 The conical tank in Exercise 103.

104. *Rotating spool* As television cable is pulled from a large spool to be strung from the telephone poles along a street, it unwinds from the spool in layers of constant radius (see Figure 2.55). If the truck pulling the cable moves at a steady 6 ft/sec (a touch over 4 mph), use the equation $s = r\theta$ to find how fast (in radians per second) the spool is turning when the layer of radius 1.2 ft is being unwound.

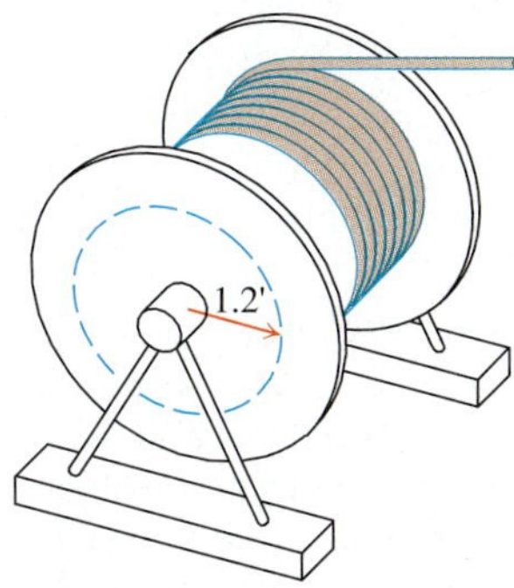

FIGURE 2.55 The television cable in Exercise 104.

105. *Moving searchlight beam* The figure shows a boat 1 km offshore, sweeping the shore with a searchlight. The light turns at a constant rate, $d\theta/dt = -0.6$ rad/sec.

(a) How fast is the light moving along the shore when it reaches point A?

(b) How many revolutions per minute is 0.6 rad/sec?

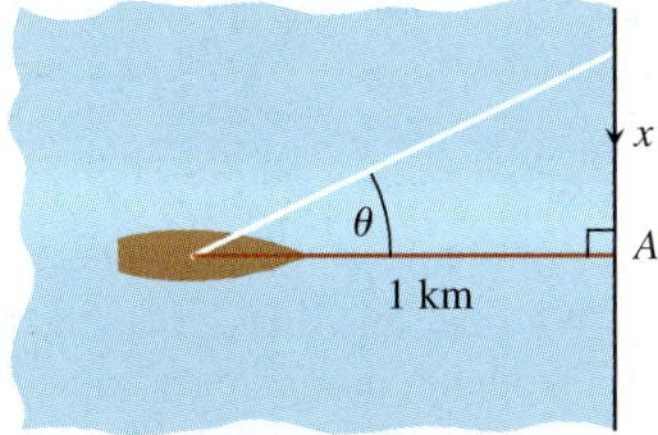

106. *Points moving on coordinate axes* Points A and B move along the x- and y-axes, respectively, in such a way that the distance r (meters) along the perpendicular from the origin to line AB remains constant. How fast is OA changing, and is it increasing, or decreasing, when $OB = 2r$ and B is moving toward O at the rate of $0.3r$ m/sec?

107. *Motion of a particle* A particle is traveling upward and to the right along the curve $y = \ln x$. Its x-coordinate is increasing at the rate $(dx/dt) = \sqrt{x}$ m/sec. At what rate is the y-coordinate changing at the point $(e^2, 2)$?

108. *Sliding down a slide* A girl is sliding down a slide shaped like the curve $y = 9e^{-x/3}$. Her y-coordinate is changing at the rate $dy/dt = (-1/4)\sqrt{9 - y}$ ft/sec. At approximately what rate is her x-coordinate changing when she reaches the bottom of the slide at $x = 9$ ft? (Take e^3 to be 20 and round your answer to the nearest foot per second.)

Logarithms

109. *Writing to Learn* The functions $f(x) = \ln 5x$ and $g(x) = \ln 3x$ differ by a constant. What constant? Give reasons for your answer.

110. *Writing to Learn*

(a) If $(\ln x)/x = (\ln 2)/2$, must $x = 2$?

(b) If $(\ln x)/x = -2 \ln 2$, must $x = 1/2$?

Give reasons for your answers.

111. *Writing to Learn* The quotient $(\log_4 x)/(\log_2 x)$ has a constant value. What value?
Give reasons for your answer.

112. *$\log_x (2)$ versus $\log_2 (x)$* How does $f(x) = \log_x (2)$ compare with $g(x) = \log_2 (x)$? Here is one way to find out.

(a) Use the equation $\log_a b = (\ln b)/(\ln a)$ to express $f(x)$ and $g(x)$ in terms of natural logarithms.

T **(b)** Graph f and g together. Comment on the behavior of f in relation to the signs and values of g.

113. *Fundamental frequency of a vibrating piano string* We measure the frequencies at which wires vibrate in cycles (trips back and forth) per second. The unit of measure is a *hertz*:1 cycle/sec. Middle A on a piano has a frequency 440 hertz. For any given wire, the fundamental frequency y is a function of four variables:

r: the radius of the wire

l: the length

d: the density of the wire

T: the tension (force) holding the wire taut.

With r and l in centimeters, d in grams per centimeter, and T in dynes (it takes about 100,000 dynes to lift an apple), the fundamental frequency of the wire is

$$y = \frac{1}{2rl}\sqrt{\frac{T}{\pi d}}.$$

If we keep all the variables fixed except one, then y can be alternatively thought of as four different functions of one variable, $y(r)$, $y(l)$, $y(d)$, and $y(T)$. How would changing each variable affect the string's fundamental frequency? To find out, calculate $y'(r)$, $y'(l)$, $y'(d)$, and $y'(T)$.

114. *Spread of measles* The spread of measles in a certain school is given by

$$P(t) = \frac{200}{1 + e^{5-t}},$$

where t is the number of days since the measles first appeared and $P(t)$ is the total number of students who have caught the measles to date.

(a) Estimate the initial number of students infected with measles.

(b) About how many students in all will get the measles?

(c) When will the rate of spread of measles be greatest? What is the rate?

115. Graph the function $f(x) = \tan^{-1}(\tan 2x)$ in the window $[-\pi, \pi]$ by $[-4, 4]$. Then answer the following questions.

T

(a) What is the domain of f?

(b) What is the range of f?

(c) At which points is f not differentiable?

(d) Describe the graph of f'.

116. If $x^2 - y^2 = 1$, find d^2y/dx^2 at the point $(2, \sqrt{3})$.

Additional Exercises: Theory, Examples, Applications

1. An equation like $\sin^2\theta + \cos^2\theta = 1$ is called an **identity** because it holds for all values of θ. An equation like $\theta = 0.5$ is not an identity because it holds only for selected values of θ, not all. If you differentiate both sides of a trigonometric identity in θ with respect to θ, the resulting new equation will also be an identity.

Differentiate the following to show that the resulting equations hold for all θ.

(a) $\sin 2\theta = 2\sin\theta\cos\theta$

(b) $\cos 2\theta = \cos^2\theta - \sin^2\theta$

2. *Writing to Learn* If the identity $\sin(x + a) = \sin x \cos a + \cos x \sin a$ is differentiated with respect to x, is the resulting equation also an identity? Does this principle apply to the equation $x^2 - 2x - 8 = 0$? Explain.

3. (a) Find values for the constants a, b, and c that will make

$$f(x) = \cos x \quad \text{and} \quad g(x) = a + bx + cx^2$$

satisfy the conditions

$$f(0) = g(0), \quad f'(0) = g'(0), \quad \text{and} \quad f''(0) = g''(0).$$

(b) Find values for b and c that will make

$$f(x) = \sin(x + a) \quad \text{and} \quad g(x) = b\sin x + c\cos x$$

satisfy the conditions

$$f(0) = g(0) \quad \text{and} \quad f'(0) = g'(0).$$

(c) For the determined values of a, b, and c, what happens for the third and fourth derivatives of f and g in each of parts (a) and (b)?

4. *Solutions to differential equations*

(a) Show that $y = \sin x$, $y = \cos x$, and $y = a\cos x + b\sin x$ (a and b constants) all satisfy the equation

$$y'' + y = 0.$$

(b) How would you modify the functions in part (a) to satisfy the equation

$$y'' + 4y = 0?$$

Generalize this result.

5. *An osculating circle* Find the values of h, k, and a that make the circle $(x - h)^2 + (y - k)^2 = a^2$ tangent to the parabola $y = x^2 + 1$ at the point $(1, 2)$ and that also make the second derivatives d^2y/dx^2 have the same value on both curves there. Circles like this one that are tangent to a curve and have the same second derivative as the curve at the point of tangency are called *osculating circles* (from the Latin *osculari*, meaning "to kiss"). We encounter them again in Chapter 10.

6. *Marginal revenue* A bus will hold 60 people. The number x of people per trip who use the bus is related to the fare charged (p dollars) by the law $p = [3 - (x/40)]^2$. Write an expression for the total revenue $r(x)$ per trip received by the bus company. What number of people per trip will make the marginal revenue dr/dx equal to zero? What is the corresponding fare? (This fare is the one that maximizes the revenue, so the bus company should probably rethink its fare policy.)

7. *Industrial production*

(a) Economists often use the expression "rate of growth" in relative rather than absolute terms. For example, let $u = f(t)$ be the number of people in the labor force at time t in a given industry. (We treat this function as though it were differentiable even though it is an integer-valued step function.)

Let $v = g(t)$ be the average production per person in the labor force at time t. The total production is then $y = uv$. If the labor force is growing at the rate of 4% per year ($du/dt = 0.04u$) and the production per worker is growing at the rate of 5% per year ($dv/dt = 0.05v$), find the rate of growth of the total production, y.

(b) Suppose that the labor force in part (a) is decreasing at the rate of 2% per year while the production per person is increasing at the rate of 3% per year. Is the total production increasing, or is it decreasing, and at what rate?

8. *Designing a gondola* The designer of a 30-ft-diameter spherical hot-air balloon wants to suspend the gondola 8 ft below the bottom of the balloon with cables tangent to the surface of the balloon, as shown in the accompanying figure. Two of the cables are shown running from the top edges of the gondola to their points of tangency, $(-12, -9)$ and $(12, -9)$. How wide should the gondola be?

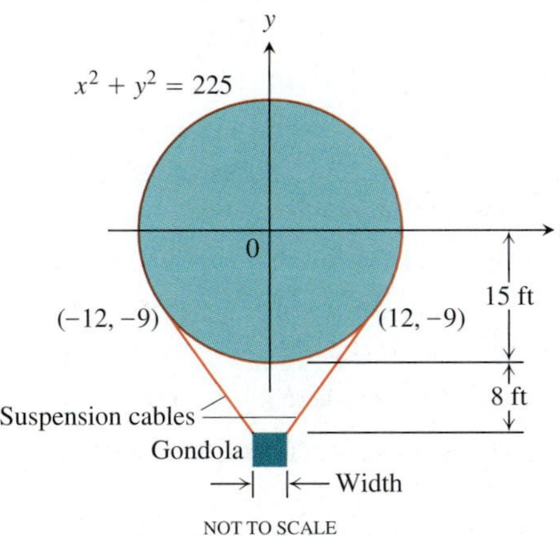

9. *Pisa by parachute* The photograph shows Mike McCarthy parachuting from the top of the Tower of Pisa on August 6, 1988. Make a rough sketch to show the shape of the graph of his speed during the jump.

Mike McCarthy of London jumped from the Tower of Pisa and then opened his parachute in what he said was a world record low-level parachute jump of 179 ft. (*Source: Boston Globe*, Aug. 6, 1988.)

10. *Motion of a particle* The position at time $t \geq 0$ of a particle moving along a coordinate line is

$$s = 10 \cos (t + \pi/4).$$

(a) What is the particle's starting position ($t = 0$)?

(b) What are the points farthest to the left and right of the origin reached by the particle?

(c) Find the particle's velocity and acceleration at the points in part (b).

(d) When does the particle first reach the origin? What are its velocity, speed, and acceleration then?

11. *Shooting a paper clip* On Earth, you can easily shoot a paper clip 64 ft straight up into the air with a rubber band. In t seconds after firing, the paper clip is $s = 64t - 16t^2$ ft above your hand.

(a) How long does it take the paper clip to reach its maximum height? With what velocity does it leave your hand?

(b) On the moon, the same acceleration will send the paper clip to a height of $s = 64t - 2.6t^2$ ft in t sec. About how long will it take the paper clip to reach its maximum height and how high will it go?

12. *Velocities of two particles* At time t sec, the positions of two particles on a coordinate line are $s_1 = 3t^3 - 12t^2 + 18t + 5$ m and $s_2 = -t^3 + 9t^2 - 12t$ m. When do the particles have the same velocities?

13. *Velocity of a particle* A particle of constant mass m moves along the x-axis. Its velocity v and position x satisfy the equation

$$\frac{1}{2} m (v^2 - v_0^2) = \frac{1}{2} k (x_0^2 - x^2),$$

where k, v_0, and x_0 are constants. Show that whenever $v \neq 0$,

$$m \frac{dv}{dt} = -kx.$$

14. *Average and instantaneous velocity*

(a) Show that if the position x of a moving point is given by a quadratic function of time t, $x = At^2 + Bt + C$, then the average velocity over any time interval $[t_1, t_2]$ is equal to the instantaneous velocity at the midpoint of the time interval.

(b) What is the geometric significance of the result in part (a)?

15. Find all values of the constants m and b for which the function

$$y = \begin{cases} \sin x & \text{for } x < \pi \\ mx + b & \text{for } x \geq \pi \end{cases}$$

is

(a) continuous at $x = \pi$.

(b) differentiable at $x = \pi$.

16. *Writing to Learn* Does the function

$$f(x) = \begin{cases} \dfrac{1 - \cos x}{x} & \text{for } x \neq 0 \\ 0 & \text{for } x = 0 \end{cases}$$

have a derivative at $x = 0$? Explain.

17. **(a)** For what values of a and b will

$$f(x) = \begin{cases} ax, & x < 2 \\ ax^2 - bx + 3, & x \geq 2 \end{cases}$$

be differentiable for all values of x?

(b) *Writing to Learn* Discuss the geometry of the resulting graph of f.

18. **(a)** For what values of a and b will

$$g(x) = \begin{cases} ax + b, & x \leq -1 \\ ax^3 + x + 2b, & x > -1 \end{cases}$$

be differentiable for all values of x?

(b) *Writing to Learn* Discuss the geometry of the resulting graph of g.

19. *Odd differentiable functions* Is there anything special about the derivative of an odd differentiable function of x? Give reasons for your answer.

20. *Even differentiable functions* Is there anything special about the derivative of an even differentiable function of x? Give reasons for your answer.

21. *A surprising result* Suppose that the functions f and g are defined throughout an open interval containing the point x_0, that f is differentiable at x_0, that $f(x_0) = 0$, and that g is continuous at x_0. Show that the product fg is differentiable at x_0. This process shows, for example, that although $|x|$ is not differentiable at $x = 0$, the product $x|x|$ is differentiable at $x = 0$.

22. *(Continuation of Exercise 21)* Use the result of Exercise 21 to show that the following functions are differentiable at $x = 0$.

(a) $|x| \sin x$ **(b)** $x^{2/3} \sin x$ **(c)** $\sqrt[3]{x}(1 - \cos x)$

(d) $h(x) = \begin{cases} x^2 \sin(1/x), & x \neq 0 \\ 0, & x = 0 \end{cases}$

23. *Writing to Learn* Is the derivative of

$$h(x) = \begin{cases} x^2 \sin(1/x), & x \neq 0 \\ 0, & x = 0 \end{cases}$$

continuous at $x = 0$? How about the derivative of $k(x) = xh(x)$? Give reasons for your answers.

24. Suppose that a function f satisfies the following conditions for all real values of x and y:

i. $f(x + y) = f(x) \cdot f(y)$.

ii. $f(x) = 1 + xg(x)$, where $\lim_{x \to 0} g(x) = 1$.

Show that the derivative $f'(x)$ exists at every value of x and that $f'(x) = f(x)$.

25. *The generalized product rule* Use mathematical induction (Appendix 1) to prove that if $y = u_1 u_2 \cdots u_n$ is a finite product of differentiable functions, then y is differentiable on their common domain and

$$\frac{dy}{dx} = \frac{du_1}{dx} u_2 \cdots u_n + u_1 \frac{du_2}{dx} \cdots u_n + \cdots + u_1 u_2 \cdots u_{n-1} \frac{du_n}{dx}.$$

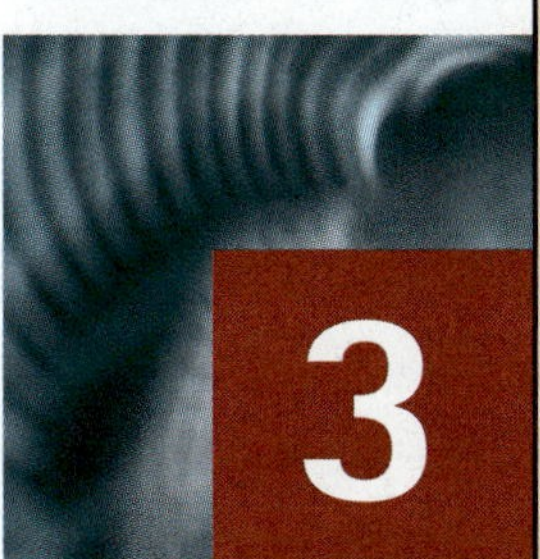

3 Applications of Derivatives

OVERVIEW In this chapter, we show how to use derivatives to find maximum and minimum values of functions, to predict and analyze the shapes of graphs, and to draw conclusions about the behavior of functions that solve differential equations. We also see how a tangent line captures the shape of a curve near the point of tangency and how it can be used to find the zeros of functions numerically. The key to many of these accomplishments is the Mean Value Theorem, a theorem whose corollaries provide the gateway to integral calculus in Chapter 4.

3.1 Extreme Values of Functions

The Drilling-Rig Problem • Absolute (Global) Extreme Values • Local (Relative) Extreme Values • Finding Extreme Values

One of the most useful things to learn from a function's derivative is whether the function assumes any maximum or minimum values on a given interval and where these values are located if it does. Once we can do this, we can solve problems like the drilling-rig problem.

The Drilling-Rig Problem

Example 1 Piping Oil from a Drilling Rig to a Refinery

A drilling rig 12 mi offshore is to be connected by pipe to a refinery onshore, 20 mi straight down the coast from the rig. If underwater pipe costs \$50,000 per mile and land-based pipe costs \$30,000 per mile, what combination of the two will give the least expensive connection?

Preliminary analysis:

We try a few possibilities to get a feel for the problem:

(a) *Smallest amount of underwater pipe*

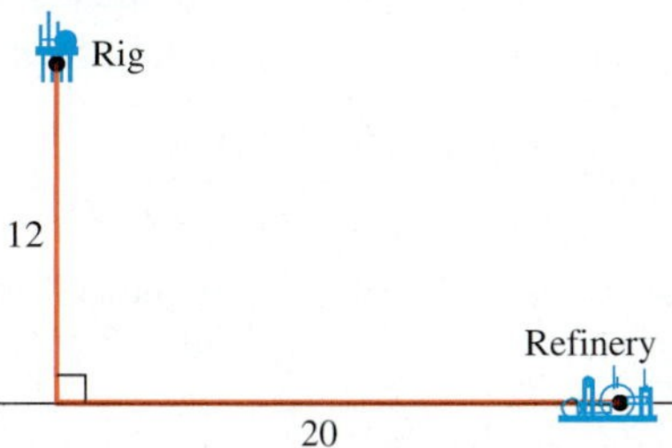

Underwater pipe is the more expensive, so we use as little as we can. We run straight to shore (12 mi) and use land pipe for the 20 mi to the refinery.

$$\begin{aligned}\text{Dollar cost} &= 12(50{,}000) + 20(30{,}000)\\ &= 1{,}200{,}000.\end{aligned}$$

(b) *All pipe under water (most direct route)*

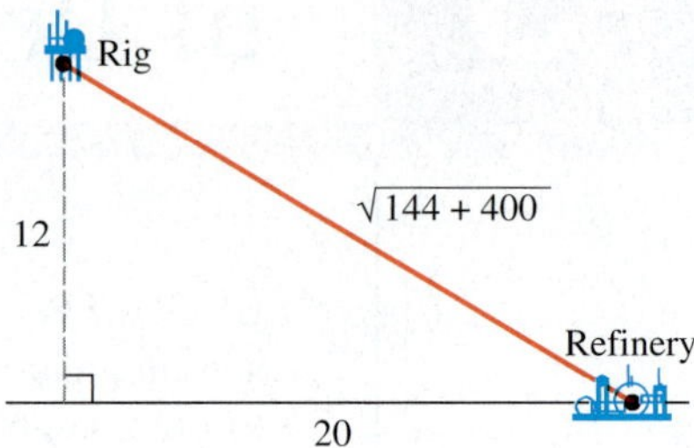

We go straight to the refinery under water.

$$\begin{aligned}\text{Dollar cost} &= \sqrt{544}(50{,}000)\\ &= 1{,}166{,}190\end{aligned}$$

This is less expensive than plan (a).

(c) *Something in between*

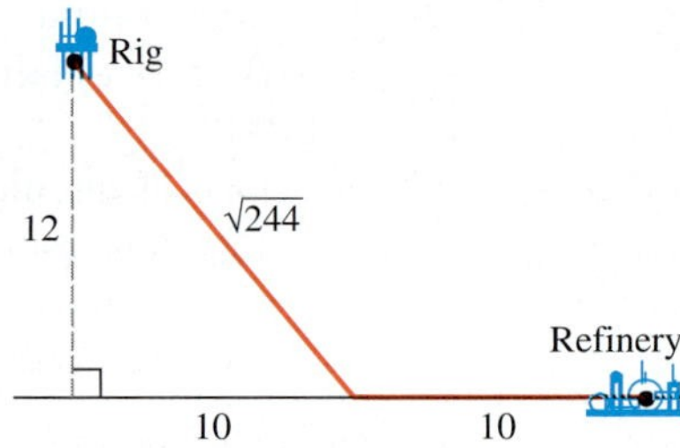

We run pipe underwater to the halfway point 10 mi from the refinery and go by land from there.

$$\begin{aligned}\text{Dollar cost} &= \sqrt{244}(50{,}000) + 10(30{,}000)\\ &\approx 1{,}081{,}025\end{aligned}$$

Neither extreme (shortest underwater segment or all pipe underwater) gives the best solution. Something in between is better.

The 10 mi point was an arbitrary choice. Would another choice be even better? If so, how do we find it? What's the best we can do? We find out using the mathematics we are about to develop, and we return to solve this problem at the end of the section.

Absolute (Global) Extreme Values

The largest and smallest values a function takes on, both locally and globally (see Figure 3.1), are always of great interest.

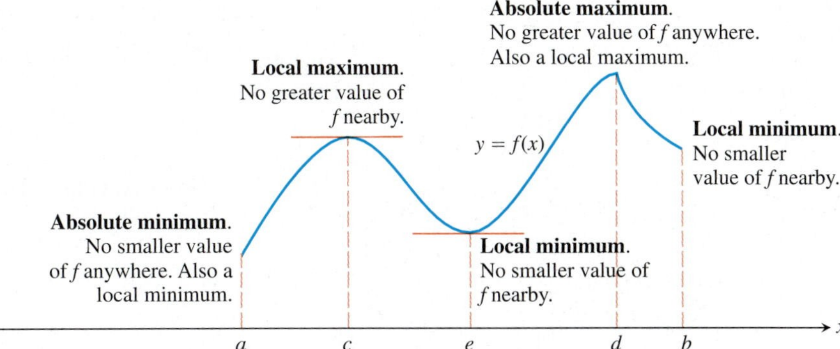

FIGURE 3.1 How to classify maxima and minima.

Definition Absolute Extreme Values

Let f be a function with domain D. Then $f(c)$ is the

(a) absolute maximum value on D if and only if $f(x) \leq f(c)$ for all x in D

(b) absolute minimum value on D if and only if $f(x) \geq f(c)$ for all x in D.

Absolute (or **global**) maximum and minimum values are also called **absolute extrema** (plural of the Latin *extremum*). We often omit the term "absolute" or "global" and just say maximum and minimum.

Example 2 shows that extreme values can occur at interior points or endpoints of intervals.

Example 2 Exploring Extreme Values

On $[-\pi/2, \pi/2]$, $f(x) = \cos x$ takes on a maximum value of 1 (once) and a minimum value of 0 (twice). The function $g(x) = \sin x$ takes on a maximum value of 1 and a minimum value of -1 (Figure 3.2).

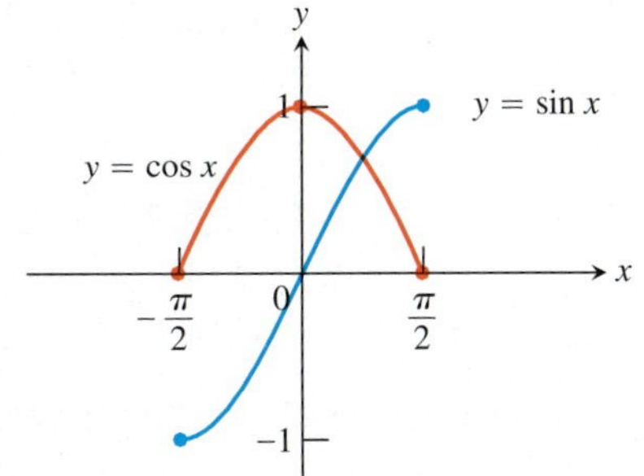

FIGURE 3.2 The graphs for Example 2.

Functions with the same defining rule can have different extrema, depending on the domain.

Example 3 Exploring Absolute Extrema

The absolute extrema of the following functions on their domains can be seen in Figure 3.3.

	Function rule	Domain D	Absolute extrema on D
(a)	$y = x^2$	$(-\infty, \infty)$	No absolute maximum. Absolute minimum of 0 at $x = 0$.
(b)	$y = x^2$	$[0, 2]$	Absolute maximum of 4 at $x = 2$. Absolute minimum of 0 at $x = 0$.
(c)	$y = x^2$	$(0, 2]$	Absolute maximum of 4 at $x = 2$. No absolute minimum.
(d)	$y = x^2$	$(0, 2)$	No absolute extrema.

CD-ROM WEBsite
Historical Biography
Daniel Bernoulli (1700 — 1789)

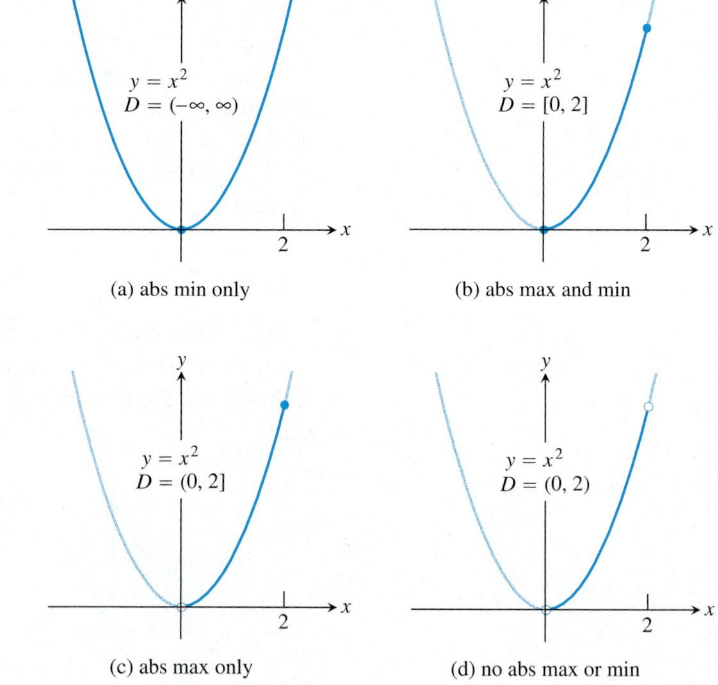

FIGURE 3.3 Graphs for Example 3.

Example 3 shows that a function may fail to have a maximum or minimum value. This cannot happen, however, with a continuous function on a finite closed interval.

Theorem 1 **The Extreme-Value Theorem for Continuous Functions**
If f is continuous at every point of a closed interval I, then f assumes both an absolute maximum value M and an absolute minimum value m somewhere in I. That is, there are numbers x_1 and x_2 in I with $f(x_1) = m$, $f(x_2) = M$, and $m \le f(x) \le M$ for every other x in I (Figure 3.4).

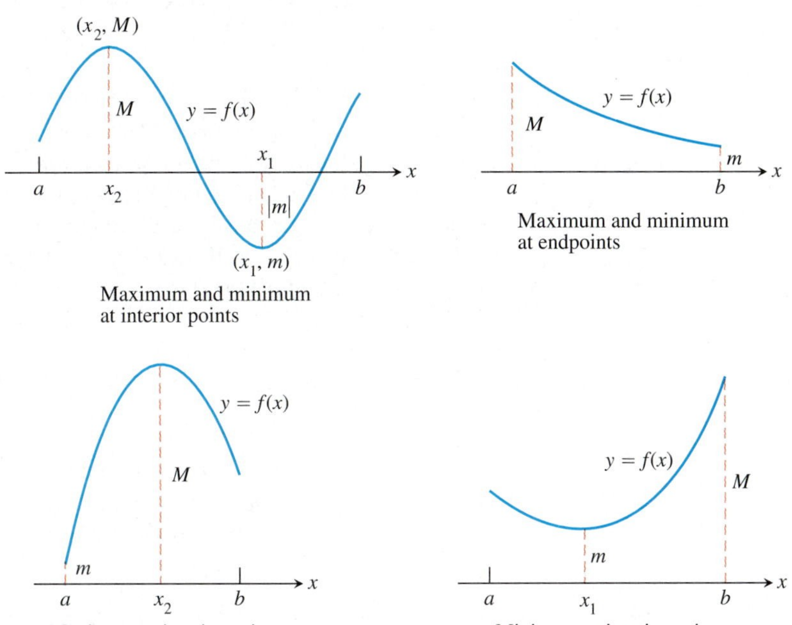

FIGURE 3.4 Some possibilities for a continuous function's maximum and minimum on a closed interval $[a, b]$.

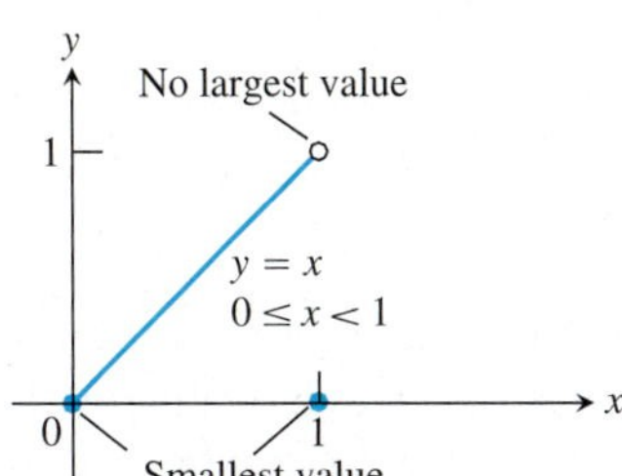

FIGURE 3.5 Even a single point of discontinuity can keep a function from having either a maximum or minimum value on a closed interval. The function

$$y = \begin{cases} x, & 0 \le x < 1 \\ 0, & x = 1 \end{cases}$$

is continuous at every point of [0, 1] except $x = 1$, yet its graph over [0, 1] does not have a highest point.

The proof of Theorem 1 requires a detailed knowledge of the real number system, and we do not give it here.

Example 4 **Extrema Depending on Continuity**

As Figure 3.5 shows, the requirements in Theorem 1 that the interval be closed and the function continuous are key ingredients. Without them, the conclusion of the theorem need not hold.

Local (Relative) Extreme Values

Figure 3.1 shows a graph with five points where a function has extreme values on its domain $[a, b]$. The function's absolute minimum occurs at a even though at e the function's value is smaller than at any other point *nearby*. The curve rises to the left and falls to the right around c, making $f(c)$ a maximum locally. The function attains its absolute maximum at d.

Definition Local Extreme Values

Let c be an interior point of the domain of the function f. Then $f(c)$ is a

(a) **local maximum value** at c if and only if $f(x) \leq f(c)$ for all x in some open interval containing c

(b) **local minimum value** at c if and only if $f(x) \geq f(c)$ for all x in some open interval containing c.

Local extrema are also called **relative extrema.** We can extend the definitions of local extrema to endpoints of intervals. A function f has a local maximum or local minimum *at an endpoint c* if the appropriate inequality holds for all x in some half-open interval containing c.

An absolute extremum is also a local extremum, because being an extreme value overall makes it an extreme value in its immediate neighborhood. Hence, *a list of all local extrema will automatically include the absolute extrema if there are any.*

Finding Extreme Values

The interior domain points where the function in Figure 3.1 has local extreme values are points where either f' is zero or f' does not exist. This is generally the case, as we see from the following theorem.

Theorem 2 Local Extreme Values

If a function f has a local maximum value or a local minimum value at an interior point c of its domain, and if f' exists at c, then

$$f'(c) = 0.$$

Theorem 2 says that a function's first derivative is always zero at an interior point where the function has a local extreme value and the derivative is defined. Hence the only places where a function f can possibly have an extreme value (local or global) are

1. interior points where $f' = 0$
2. interior points where f' is undefined
3. endpoints of the domain of f.

Most quests for extreme values call for finding the absolute extrema of a continuous function on a closed interval. Theorem 1 assures us that such values exist; Theorem 2 tells us where to look. The following definition helps us summarize these findings.

Definition Critical Point

A point in the domain of a function f at which $f' = 0$ or f' does not exist is a **critical point** of f.

Thus, in summary, extreme values occur only at critical points and endpoints.

Example 5 Finding Absolute Extrema on a Closed Interval

Find the absolute maximum and minimum values of $f(x) = 10x(2 - \ln x)$ on the interval $[1, e^2]$.

Solution Figure 3.6 suggests that f has its absolute maximum value near $x = 3$ and its absolute minimum value of 0 at $x = e^2$.

CD-ROM WEBsite

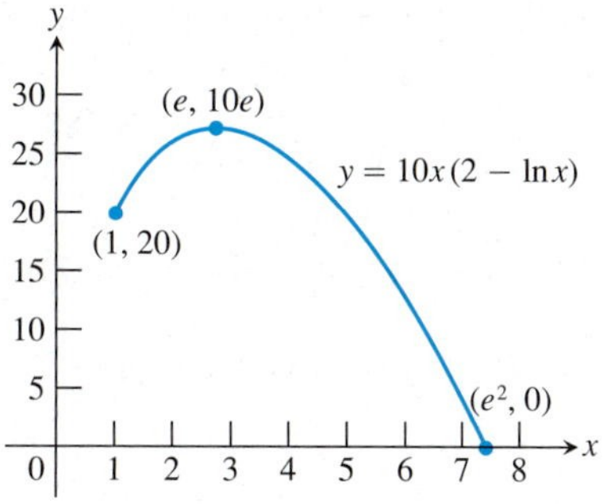

FIGURE 3.6 The extreme values of $f(x) = 10x\,(2 - \ln x)$ occur at $x = e$ and $x = e^2$. (Example 5)

We evaluate the function at the critical points and endpoints and take the largest and smallest of the resulting values.

The first derivative

$$f'(x) = 10(2 - \ln x) - 10x\left(\frac{1}{x}\right) = 10(1 - \ln x)$$

The only critical point in the domain $[1, e^2]$ is the point $x = e$, where $\ln x = 1$. The values of f at this one critical point and at the endpoints are

Critical point value: $f(e) = 10e$

Endpoint values: $f(1) = 10(2 - \ln 1) = 20$

$f(e^2) = 10e^2(2 - 2\ln e) = 0.$

We can see from this list that the function's absolute maximum value is $10e \approx 27.2$; it occurs at the critical interior point $x = e$. The absolute minimum value is 0 and occurs at the right endpoint $x = e^2$.

How to Find the Absolute Extrema of a Continuous Function f on a Closed Interval

Step 1. Evalute f at all critical points and endpoints.

Step 2. Take the largest and smallest of these values.

Proof of Theorem 2 To show that $f'(c)$ is zero at an interior local extremum, we show first that $f'(c)$ cannot be positive and second that $f'(c)$ cannot be negative. The only number that is neither positive nor negative is zero, so that is what $f'(c)$ must be.

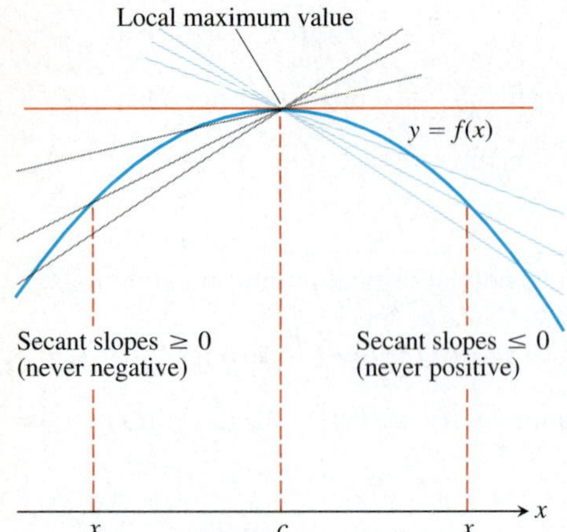

FIGURE 3.7 A curve with a local maximum value. The slope at c, simultaneously the limit of nonpositive numbers and nonnegative numbers, is zero.

To begin, suppose that f has a local maximum value at $x = c$ (Figure 3.7) so that $f(x) - f(c) \leq 0$ for all values of x near enough to c. Since c is an interior point of the domain of f, $f'(c)$ is defined by the two-sided limit

$$\lim_{x \to c} \frac{f(x) - f(c)}{x - c}.$$

Hence, the right-hand and left-hand limits both exist at $x = c$ and equal $f'(c)$. When we examine these limits separately, we find that

$$f'(c) = \lim_{x \to c^+} \frac{f(x) - f(c)}{x - c} \leq 0. \qquad \text{Because } (x - c) > 0 \text{ and } f(x) \leq f(c) \tag{1}$$

Similarly,

$$f'(c) = \lim_{x \to c^-} \frac{f(x) - f(c)}{x - c} \geq 0. \qquad \text{Because } (x - c) < 0 \text{ and } f(x) \leq f(c) \tag{2}$$

Together, (1) and (2) imply $f'(c) = 0$.

This proves the theorem for local maximum values. To prove it for local minimum values, we simply use $f(x) \geq f(c)$, which reverses the inequalities in (1) and (2).

In Example 6, we investigate the function whose graph was drawn in Example 3 of Preliminary Section 2.

Example 6 Finding Extreme Values

Find the extreme values of

$$f(x) = \frac{1}{\sqrt{4 - x^2}}.$$

Solution Figure 3.8 suggests that f has an absolute minimum of about 0.5 at $x = 0$. There also appear to be local maxima at $x = -2$ and $x = 2$. At these points, however, f is not defined, and there do not appear to be maxima anywhere else.

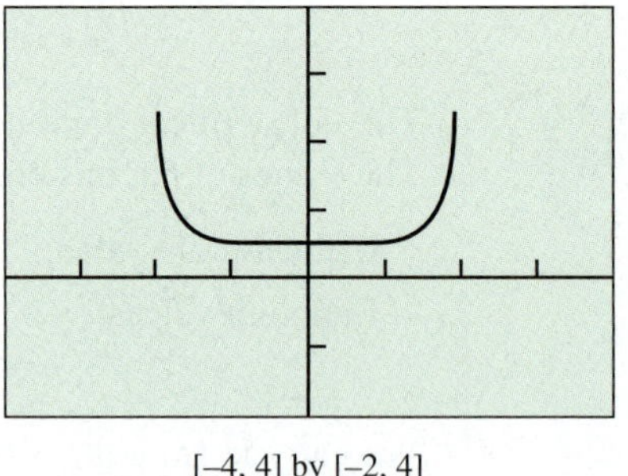

FIGURE 3.8 The graph of

$$f(x) = \frac{1}{\sqrt{4 - x^2}}.$$

(Example 6)

Let's confirm these graphical observations. The function f is defined only for $4 - x^2 > 0$, so its domain is the open interval $(-2, 2)$. The domain has no endpoints, so all the extreme values must occur at critical points. We rewrite the formula for f to find f':

$$f(x) = \frac{1}{\sqrt{4 - x^2}} = (4 - x^2)^{-1/2}.$$

Thus,

$$f'(x) = -\frac{1}{2}(4 - x^2)^{-3/2}(-2x) = \frac{x}{(4 - x^2)^{3/2}}.$$

The only critical point in the domain $(-2, 2)$ is $x = 0$. The value

$$f(0) = \frac{1}{\sqrt{4 - 0^2}} = \frac{1}{2}$$

is therefore the sole candidate for an extreme value.

To determine whether $1/2$ is an extreme value of f, we examine the formula

$$f(x) = \frac{1}{\sqrt{4 - x^2}}.$$

As x moves away from 0 on either side, the denominator gets smaller, the values of f increase, and the graph rises. We have a minimum value at $x = 0$, and the minimum is absolute.

The function has no maxima, either local or absolute. This does not violate Theorem 1 (the Extreme Value Theorem) because here f is defined on an *open* interval. To guarantee extreme points, Theorem 1 requires the interval be closed.

Example 7 Critical Points Need Not Give Extreme Values

Although a function's extrema can occur only at critical points and endpoints, not every critical point or endpoint signals the presence of an extreme value. Figure 3.9 illustrates this for interior points. Exercise 62 describes a function that fails to assume an extreme value at an endpoint of its domain.

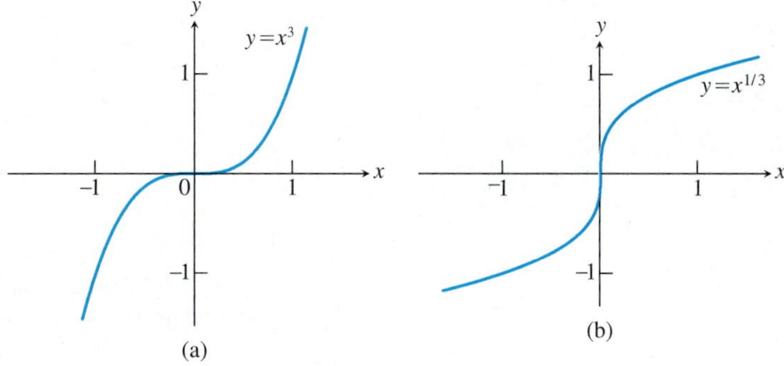

FIGURE 3.9 Critical points without extreme values. (a) $y' = 3x^2$ is 0 at $x = 0$, but $y = x^3$ has no extremum there. (b) $y' = (1/3)x^{-2/3}$ is undefined at $x = 0$, but $y = x^{1/3}$ has no extremum there.

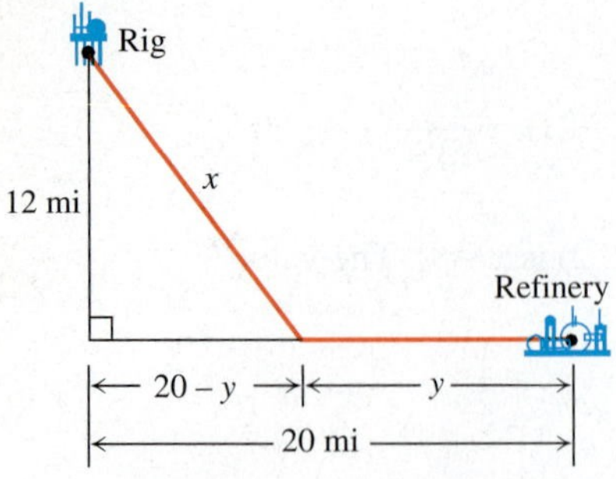

FIGURE 3.10 The figure for solving the drilling-rig problem. (Example 8)

Example 8 Solution to the Drilling-Rig Problem

As in the preliminary analysis at the beginning of the section, we draw a picture and sketch the relevant dimensions. This time, though, we also add the length x of underwater pipe and the length y of land-based pipe as variables (Figure 3.10).

The right angle opposite the rig is the key to expressing the relationship between x and y, for the Pythagorean Theorem gives

$$x^2 = 12^2 + (20 - y)^2$$
$$x = \sqrt{144 + (20 - y)^2}. \tag{3}$$

Only the positive root has meaning in this model.

The dollar cost of the pipeline is

$$c = 50{,}000x + 30{,}000y.$$

To express c as a function of a single variable, we can substitute for x, using Equation (3):

$$c = 50{,}000\sqrt{144 + (20 - y)^2} + 30{,}000y.$$

Our goal now is to find the minimum value of $c(y)$ on the interval $0 \le y \le 20$. The first derivative of c with respect to y is

$$\begin{aligned} c' &= 50{,}000 \cdot \frac{1}{2} \cdot \frac{2(20 - y)(-1)}{\sqrt{144 + (20 - y)^2}} + 30{,}000 \\ &= -50{,}000 \frac{20 - y}{\sqrt{144 + (20 - y)^2}} + 30{,}000. \end{aligned}$$

Setting c' equal to zero gives

$$\begin{aligned} 50{,}000\,(20 - y) &= 30{,}000\sqrt{144 + (20 - y)^2} \\ \frac{5}{3}(20 - y) &= \sqrt{144 + (20 - y)^2} \\ \frac{25}{9}(20 - y)^2 &= 144 + (20 - y)^2 \\ \frac{16}{9}(20 - y)^2 &= 144 \\ (20 - y) &= \pm\frac{3}{4} \cdot 12 = \pm 9 \\ y &= 20 \pm 9 \\ y &= 11 \quad \text{or} \quad y = 29. \end{aligned}$$

Only $y = 11$ lies in the interval of interest. The values of c at this one critical point and at the endpoints are

$$\begin{aligned} c(11) &= 1{,}080{,}000 \\ c(0) &= 1{,}166{,}190 \\ c(20) &= 1{,}200{,}000. \end{aligned}$$

The least expensive connection costs \$1,080,000, and we achieve it by running the line underwater to the point on shore 11 mi from the refinery.

EXERCISES 3.1

Finding Extrema from Graphs

In Exercises 1–6, determine from the graph whether the function has any absolute extreme values on $[a, b]$. Then explain how your answer is consistent with Theorem 1.

1.

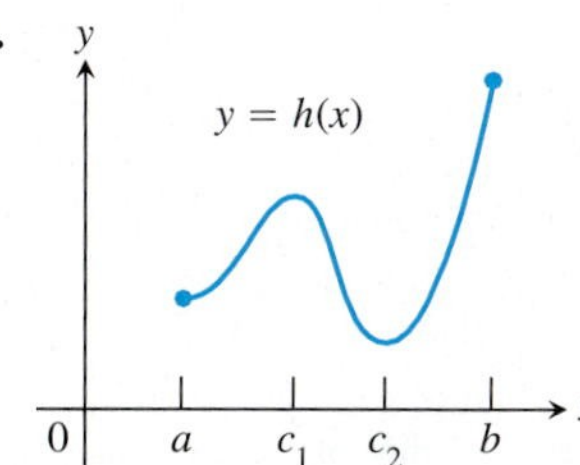

2.

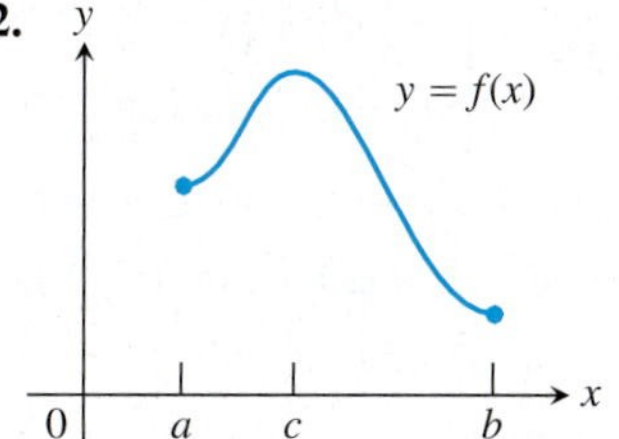

3.

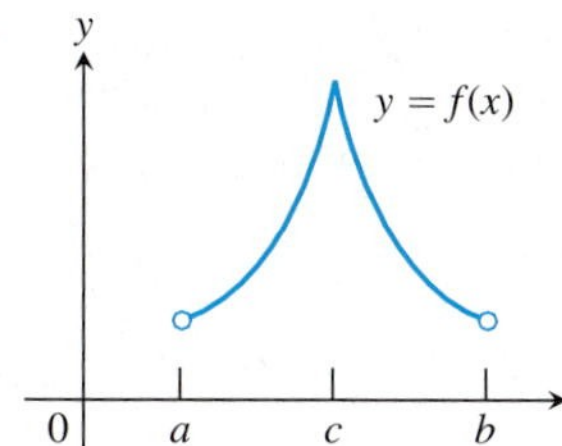

4.

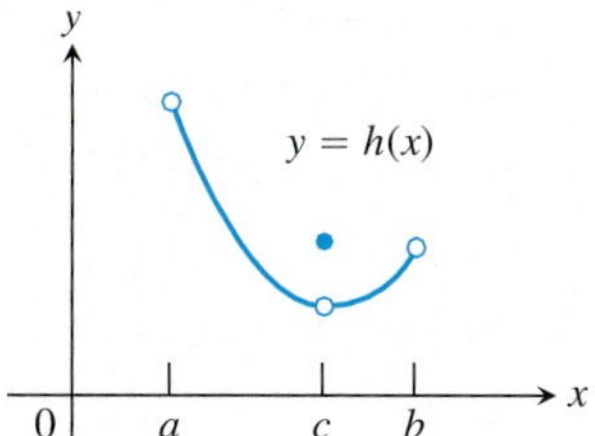

5.

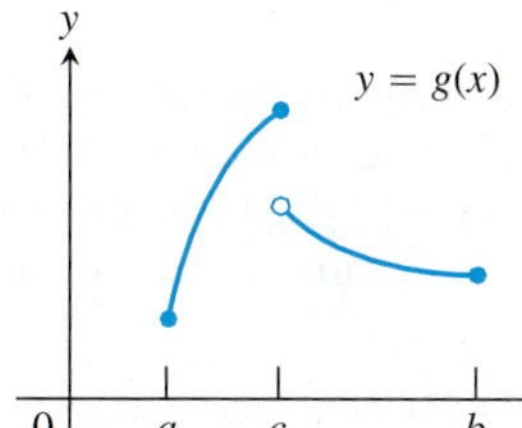

6.

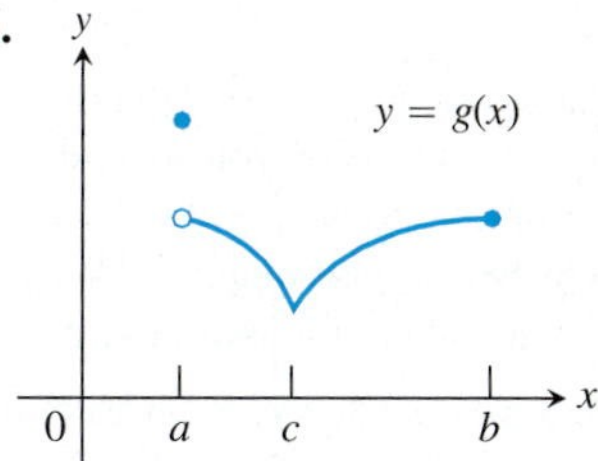

In Exercises 7–10, find the extreme values and where they occur.

7.

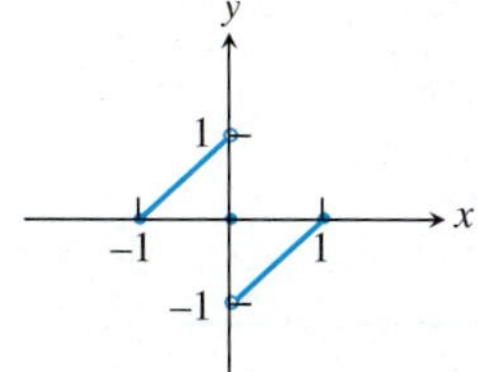

8.

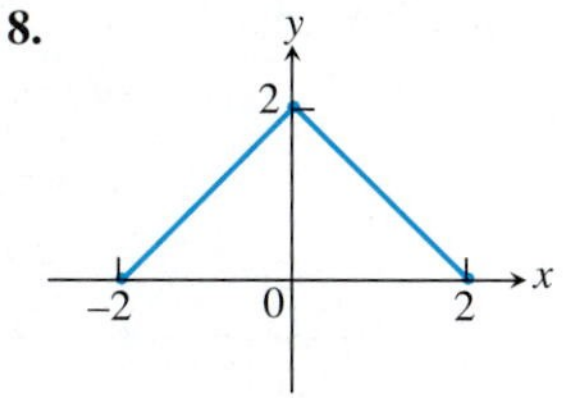

9.

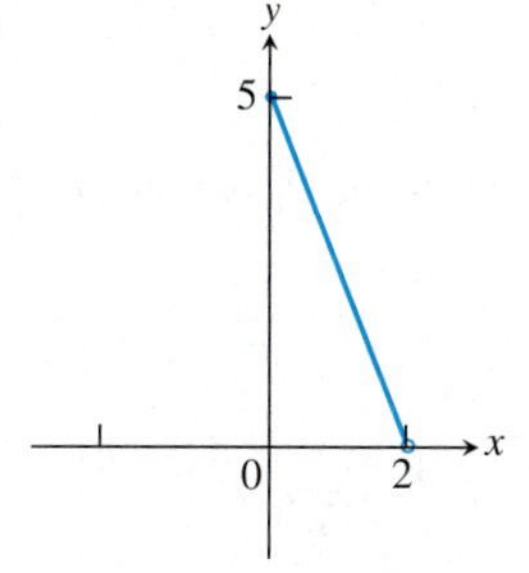

10.

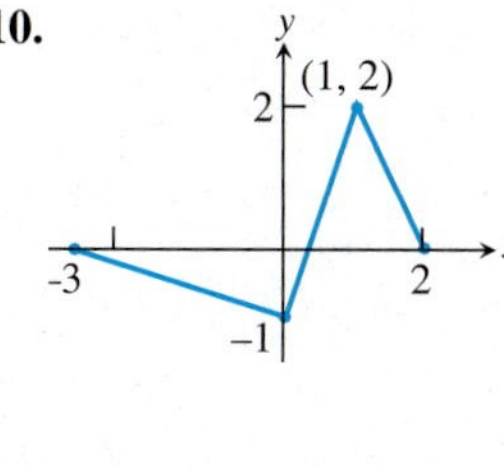

In Exercises 11–14, match the table with a graph.

11.

x	$f'(x)$
a	0
b	0
c	5

12.

x	$f'(x)$
a	0
b	0
c	-5

13.

x	$f'(x)$
a	does not exist
b	0
c	-2

14.

x	$f'(x)$
a	does not exist
b	does not exist
c	-1.7

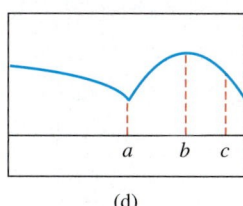

a b c

(a)

a b c

(b)

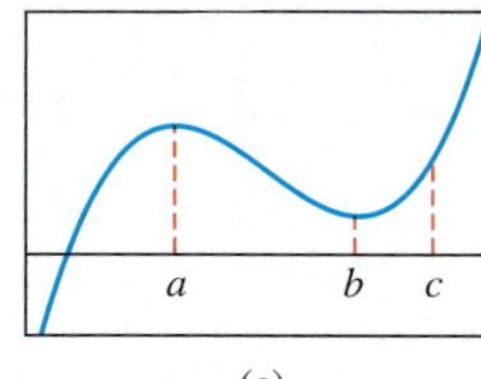

(c)

a b c

(d)

Absolute Extrema on Intervals

In Exercises 15–26, find the absolute extreme values of each function on the interval. Then graph the function. Identify the points on the graph where the absolute extrema occur and include their coordinates.

15. $f(x) = \frac{2}{3}x - 5, \quad -2 \le x \le 3$

16. $f(x) = x^2 - 6x + 9, \quad 0 \le x \le 5$

17. $f(x) = 4 - x^2, \quad -3 \le x \le 1$

18. $f(x) = \sqrt{3 + 2x - x^2}, \quad -0.5 \le x \le 3$

(*Hint:* Sketch the parabola first, then take the square root.)

19. $g(x) = \sin\left(x + \frac{\pi}{4}\right), \quad 0 \le x \le \frac{7\pi}{4}$

20. $g(x) = \sec x, \quad -\frac{\pi}{2} < x < \frac{3\pi}{2}$

21. $F(x) = -\frac{1}{x^2}, \quad 0.5 \le x \le 2$

22. $h(x) = \sqrt[3]{x}, \quad -1 \le x \le 8$

23. $f(x) = \frac{1}{x} + \ln x, \quad 0.5 \le x \le 4$

24. $g(x) = e^{-x}, \quad -1 \le x \le 1$

25. $h(x) = \ln(x + 1), \quad 0 \le x \le 3$

T **26.** $k(x) = e^{-x^2}, \quad -\infty < x < \infty$

Finding Extreme Values

In Exercises 27–36, find the extreme values of the function and where they occur.

27. $y = 2x^2 - 8x + 9$

28. $y = |3 + 2x - x^2|$

(*Hint:* Sketch the parabola first, then take absolute values.)

29. $y = \frac{1}{\sqrt{1 - x^2}}$

30. $y = \frac{1}{x^2 - 1}$

T **31.** $y = \frac{\ln x}{x}$

T **32.** $y = 4 + ex^2 - e^{-x^2}$

T **33.** $y = \frac{x}{x^2 + 1}$

T **34.** $y = \frac{x + 1}{x^2 + 2x + 2}$

35. $y = x^3 + x^2 - 8x + 5$

T **36.** $y = xe^{2x}$

Local Extrema and Critical Points

In Exercises 37–44, find the derivative at each critical point and determine the local extreme values.

37. $y = x^{2/3}(x + 2)$

38. $y = x^{2/3}(x^2 - 4)$

39. $y = x\sqrt{4 - x^2}$

40. $y = x^2\sqrt{3 - x}$

41. $y = \begin{cases} 4 - 2x, & x \le 1 \\ x + 1, & x > 1 \end{cases}$

42. $y = \begin{cases} 3 - x, & x < 0 \\ 3 + 2x - x^2, & x \ge 0 \end{cases}$

43. $y = \begin{cases} -x^2 - 2x + 4, & x \le 1 \\ -x^2 + 6x - 4, & x > 1 \end{cases}$

44. $y = \begin{cases} -\frac{1}{4}x^2 - \frac{1}{2}x + \frac{15}{4}, & x \le 1 \\ x^3 - 6x^2 + 8x, & x > 1 \end{cases}$

In Exercises 45 and 46, give reasons for your answers.

45. *Writing to Learn* Let $f(x) = (x - 2)^{2/3}$.

(a) Does $f'(2)$ exist?

(b) Show that the only local extreme value of f occurs at $x = 2$.

(c) Does the result in part (b) contradict the Extreme Value Theorem?

(d) Repeat parts (a) and (b) for $f(x) = (x - a)^{2/3}$, replacing 2 by a.

46. *Writing to Learn* Let $f(x) = |x^3 - 9x|$.

(a) Does $f'(0)$ exist?

(b) Does $f'(3)$ exist?

(c) Does $f'(-3)$ exist?

(d) Determine all extrema of f.

Optimization Applications

Whenever you are maximizing or minimizing a function of a single variable, we urge you to graph the function over the domain that is appropriate to the problem you are solving. The graph will provide insight before you begin to calculate and will furnish a visual context for understanding your answer.

47. *Constructing a pipeline* Supertankers off-load oil at a docking facility 4 mi offshore. The nearest refinery is 9 mi east of the shore point nearest the docking facility. A pipeline must be constructed connecting the docking facility with the refinery. The pipeline costs \$300,000 per mile if constructed underwater and \$200,000 per mile if overland.

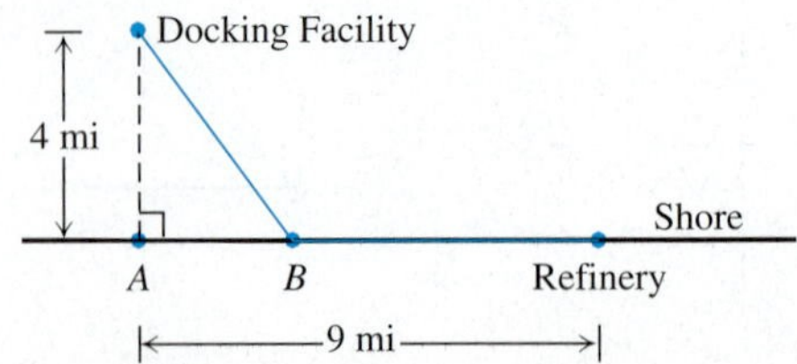

(a) Locate Point B to minimize the cost of the construction.

(b) The cost of underwater construction is expected to increase, whereas the cost of overland construction is expected to stay constant. At what cost does it become optimal to construct the pipeline directly to Point A?

48. *Upgrading a highway* A highway must be constructed to connect Village A with Village B. There is a rudimentary roadway that can be upgraded 50 mi south of the line connecting the two villages. The cost of upgrading the existing roadway is \$300,000 per mile, whereas the cost of constructing a new highway is \$500,000 per mile. Find the combination of upgrading and new construction that minimizes the cost of connecting the two villages. Clearly define the location of the proposed highway.

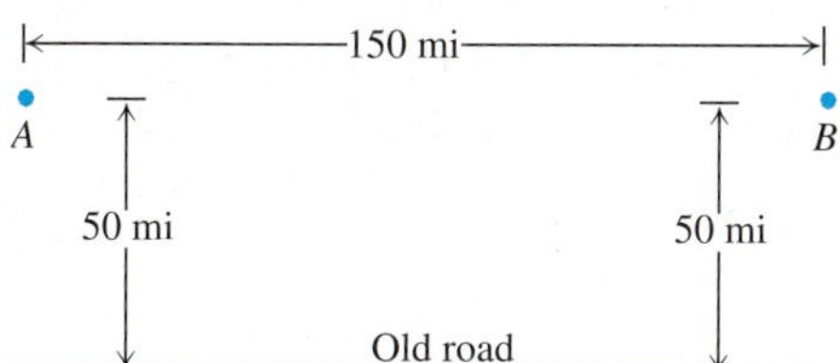

49. *Locating a pumping station* Two towns exist on the south side of a river. A pumping station is to be located to serve the two towns. A pipeline will be constructed from the pumping station to each of the towns along the line connecting the town and the pumping station. Locate the pumping station to minimize the amount of pipeline that must be constructed.

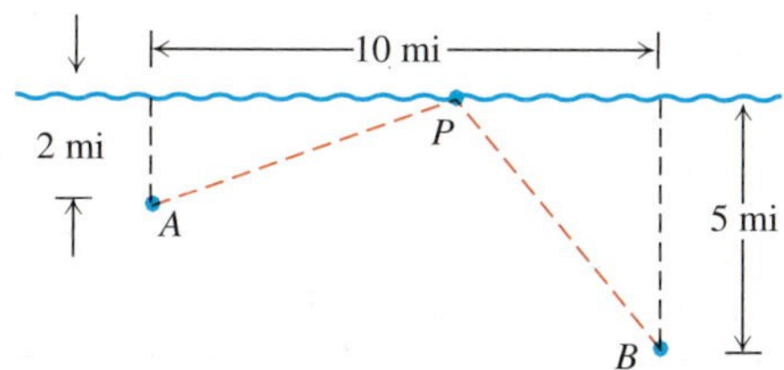

50. *Length of a guy wire* One tower is 50 ft high and another tower is 30 ft high. The towers are 150 ft apart. A guy wire is to run from Point A to the top of each tower.

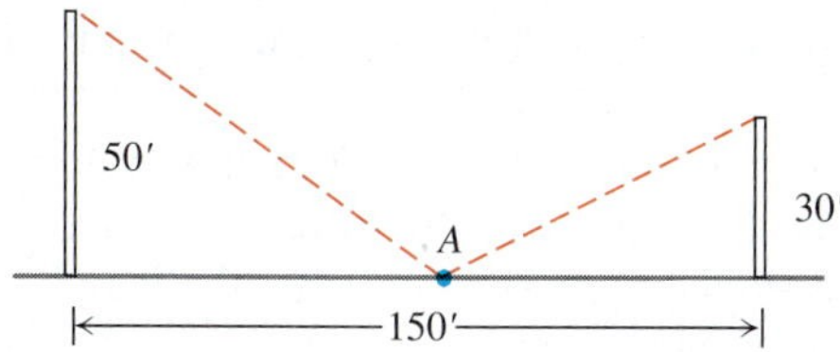

(a) Locate Point A so that the total length of guy wire is minimal.

(b) Show in general that regardless of the height of the tower, the length of guy wire is minimized if the angles at A are equal.

51. *Writing to Learn* The function

$$V(x) = x(10 - 2x)(16 - 2x), \qquad 0 < x < 5,$$

models the volume of a box.

(a) Find the extreme values of V.

(b) Interpret any values found in part (a) in terms of volume of the box.

52. *Writing to Learn* The function

$$P(x) = 2x + \frac{200}{x}, \qquad 0 < x < \infty,$$

models the perimeter of a rectangle of dimensions x by $100/x$.

(a) Find any extreme values of P.

(b) Give an interpretation in terms of perimeter of the rectangle for any values found in part (a).

53. *Area of a right triangle* What is the largest possible area for a right triangle whose hypotenuse is 5 cm long?

54. *Area of an athletic field* An athletic field is to be built in the shape of a rectangle x units long capped by semicircular regions of radius r at the two ends. The field is to be bounded by a 400 m racetrack.

(a) Express the area of the rectangular portion of the field as a function of x alone or r alone (your choice).

(b) What values of x and r give the rectangular portion the largest possible area?

55. *Maximum height of a vertically moving body* The height of a body moving vertically is given by

$$s = -\frac{1}{2}gt^2 + v_0 t + s_0, \qquad g > 0,$$

with s in meters and t in seconds. Find the body's maximum height.

56. *Peak alternating current* Suppose that at any given time t (in seconds) the current i (in amperes) in an alternating current circuit is $i = 2\cos t + 2\sin t$. What is the peak current for this circuit (largest magnitude)?

Theory and Examples

57. *A minimum with no derivatives* The function $f(x) = |x|$ has an absolute minimum value at $x = 0$ even though f is not differentiable at $x = 0$. Is this consistent with Theorem 2? Give reasons for your answer.

58. *Even functions* If an even function $f(x)$ has a local maximum value at $x = c$, can anything be said about the value of f at $x = -c$? Give reasons for your answer.

59. *Odd functions* If an odd function $g(x)$ has a local minimum value at $x = c$, can anything be said about the value of g at $x = -c$? Give reasons for your answer.

60. *Writing to Learn* We know how to find the extreme values of a continuous function $f(x)$ by investigating its values at critical points and endpoints. But what if there are no critical points or endpoints? What happens then? Do such functions really exist? Give reasons for your answers.

61. *Cubic functions* Consider the cubic function

$$f(x) = ax^3 + bx^2 + cx + d.$$

(a) Show that f can have 0, 1, or 2 critical points. Give examples and graphs to support your argument.

(b) How many local extreme values can f have?

T **62.** *Functions with no extreme values at endpoints*

(a) Graph the function

$$f(x) = \begin{cases} \sin \dfrac{1}{x}, & x > 0 \\ 0, & x = 0. \end{cases}$$

Explain why $f(0) = 0$ is not a local extreme value of f.

(b) Construct a function of your own that fails to have an extreme value at a domain endpoint.

T Graph the functions in Exercises 63–66. Then find the extreme values of the function on the interval and where they occur.

63. $f(x) = |x - 2| + |x + 3|, \quad -5 \le x \le 5$

64. $g(x) = |x - 1| - |x - 5|, \quad -2 \le x \le 7$

65. $h(x) = |x + 2| - |x - 3|, \quad -\infty < x < \infty$

66. $k(x) = |x + 1| + |x - 3|, \quad -\infty < x < \infty$

COMPUTER EXPLORATIONS

In Exercises 67–74, you will use a CAS to help find the absolute extrema of the given function over the specified closed interval. Perform the following steps.

(a) Plot the function over the interval to see its general behavior there.

(b) Find the interior points where $f' = 0$. (In some exercises, you may have to use the numerical equation solver to approximate a solution.) You may want to plot f' as well.

(c) Find the interior points where f' does not exist.

(d) Evaluate the function at all points found in parts (b) and (c) and at the endpoints of the interval.

(e) Find the function's absolute extreme values on the interval and identify where they occur.

67. $f(x) = x^4 - 8x^2 + 4x + 2, \quad [-20/25, 64/25]$

68. $f(x) = -x^4 + 4x^3 - 4x + 1, \quad [-3/4, 3]$

69. $f(x) = x^{2/3}(3 - x), \quad [-2, 2]$

70. $f(x) = 2 + 2x - 3x^{2/3}, \quad [-1, 10/3]$

71. $f(x) = \sqrt{x} + \cos x, \quad [0, 2\pi]$

72. $f(x) = x^{3/4} - \sin x + \dfrac{1}{2}, \quad [0, 2\pi]$

73. $f(x) = \pi x^2 e^{-3x/2}, \quad [0, 5]$

74. $f(x) = \ln(2x + x \sin x), \quad [1, 15]$

3.2 The Mean Value Theorem and Differential Equations

Rolle's Theorem • Mean Value Theorem • A Physical Interpretation • Mathematical Consequences • Finding Velocity and Position from Acceleration • Differential Equations and the Height of a Projectile

We have seen how to find the position of a body falling freely from rest as a function of time and from it how to derive the body's velocity and acceleration functions. But suppose that we had started with knowing only the body's acceleration, that is, with knowing the effect of gravity on the body and nothing else. Could we have worked backward to find the body's velocity and position functions?

The underlying mathematical question here is, What functions can have another function as derivative? What velocity functions can have a given acceleration function? What position functions can have a given velocity function? The corollaries of the Mean Value Theorem provide the answers.

The Mean Value Theorem itself connects the average rate of change of a function over an interval with the instantaneous rate of change of the function at a point within the interval.

Rolle's Theorem

There is strong geometric evidence that between any two points where a differentiable curve crosses the x-axis there is a point on the curve where the tangent is horizontal. A 300-year-old theorem of Michel Rolle assures us this is always the case.

Theorem 3 Rolle's Theorem

Suppose that $y = f(x)$ is continuous at every point of $[a, b]$ and differentiable at every point of (a, b). If

$$f(a) = f(b) = 0,$$

then there is at least one number c in (a, b) at which $f'(c) = 0$ (Figure 3.11).

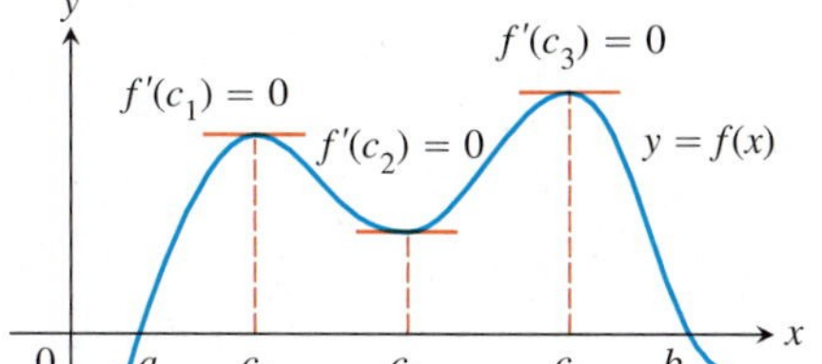

FIGURE 3.11 Rolle's Theorem says that a differentiable curve has at least one horizontal tangent between any two points where it crosses the x-axis. The curve here has three.

Proof Being continuous, f assumes absolute maximum and minimum values on $[a, b]$. These can occur only

1. at interior points where f' is zero
2. at interior points where f' does not exist
3. at the endpoints of the function's domain, in this case, a and b.

By hypothesis, f has a derivative at every interior point of $[a, b]$. That rules out option 2, leaving us with interior points where $f' = 0$ and with the two endpoints a and b.

If either the maximum or the minimum occurs at a point c inside the interval, then $f'(c) = 0$ by Theorem 2 in Section 3.1, and we have found a point for Rolle's Theorem.

If both maximum and minimum are at a or b, then the maximum and minimum values of f are both 0. Thus, f has the constant value 0, so $f' = 0$ throughout (a, b) and c can be taken anywhere in the interval. This now completes the proof.

The hypotheses of Theorem 3 are essential. If they fail at even one point, the graph may not have a horizontal tangent (Figure 3.12).

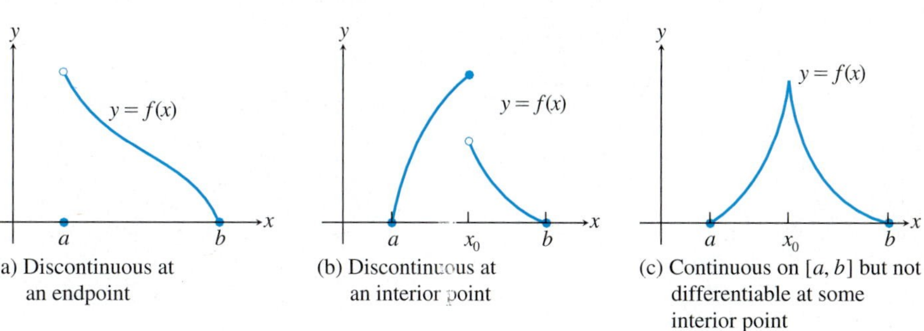

FIGURE 3.12 No horizontal tangent.

Mean Value Theorem

The Mean Value Theorem is Rolle's Theorem on a slant.

Theorem 4 The Mean Value Theorem

Suppose that $y = f(x)$ is continuous on a closed interval $[a, b]$ and differentiable on the interval's interior (a, b). Then there is at least one point c in (a, b) at which

$$\frac{f(b) - f(a)}{b - a} = f'(c). \tag{1}$$

Proof We picture the graph of f as a curve in the plane and draw a line through the points $A(a, f(a))$ and $B(b, f(b))$ (see Figure 3.13). The line is the graph of the function

$$g(x) = f(a) + \frac{f(b) - f(a)}{b - a}(x - a) \tag{2}$$

(point-slope equation). The vertical difference between the graphs of f and g at x is

$$\begin{aligned} h(x) &= f(x) - g(x) \\ &= f(x) - f(a) - \frac{f(b) - f(a)}{b - a}(x - a). \end{aligned} \tag{3}$$

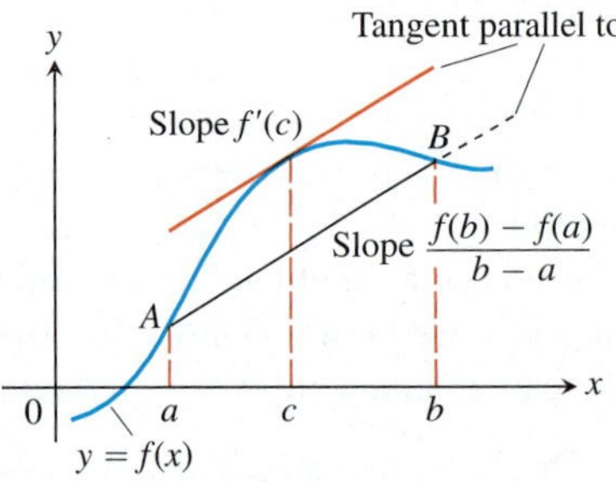

FIGURE 3.13 Geometrically, the Mean Value Theorem says that somewhere between A and B the curve has at least one tangent parallel to chord AB.

Figure 3.14 shows the graphs of f, g, and h together.

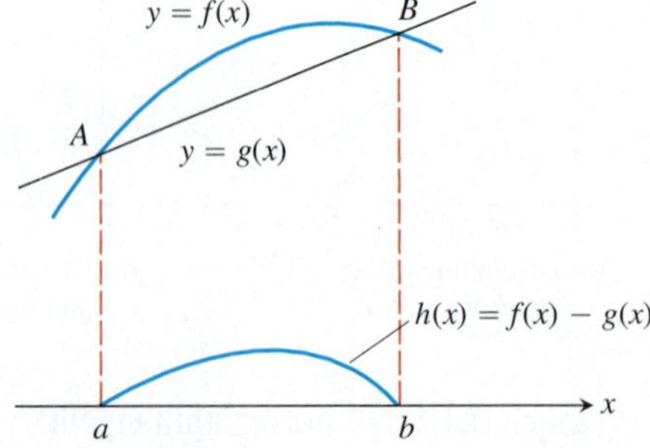

FIGURE 3.14 The chord AB is the graph of the function $g(x)$. The function $h(x) = f(x) - g(x)$ gives the vertical distance between the graphs of f and g at x.

The function h satisfies the hypotheses of Rolle's Theorem on $[a, b]$. It is continuous on $[a, b]$ and differentiable on (a, b) because f and g are. Also, $h(a) = h(b) = 0$ because the graphs of f and g both pass through A and B. Therefore, $h' = 0$ at some point c in (a, b). This is the point we want for Equation (1).

To verify Equation (1), we differentiate both sides of Equation (3) with respect to x and then set $x = c$:

$$h'(x) = f'(x) - \frac{f(b) - f(a)}{b - a} \qquad \text{Derivative of Eq. (3) . . .}$$

$$h'(c) = f'(c) - \frac{f(b) - f(a)}{b - a} \qquad \text{. . . with } x = c$$

$$0 = f'(c) - \frac{f(b) - f(a)}{b - a} \qquad h'(c) = 0$$

$$f'(c) = \frac{f(b) - f(a)}{b - a}, \qquad \text{Rearranged}$$

which is what we set out to prove.

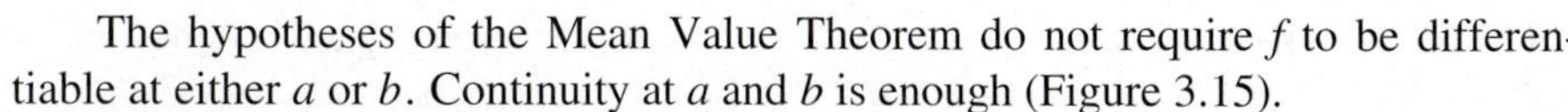

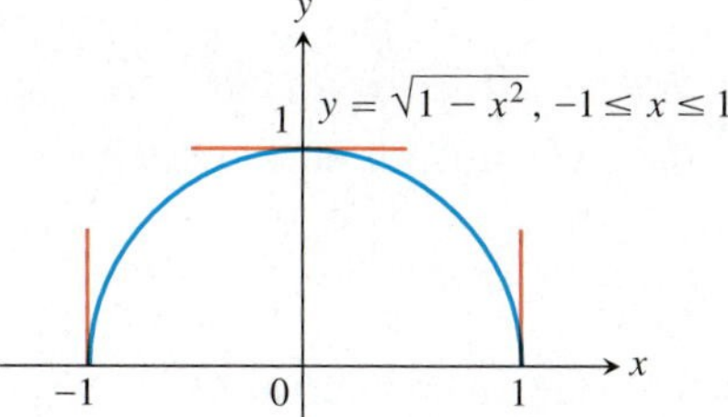

FIGURE 3.15 The function $f(x) = \sqrt{1 - x^2}$ satisfies the hypotheses (and conclusion) of the Mean Value Theorem on $[-1, 1]$ even though f is not differentiable at -1 and 1.

The hypotheses of the Mean Value Theorem do not require f to be differentiable at either a or b. Continuity at a and b is enough (Figure 3.15).

We usually do not know any more about c than the theorem says, which is that c exists. In a few cases, we can satisfy our curiosity about the identity of c, as in the next example. Our ability to identify c is the exception rather than the rule, however, and the importance of the theorem lies elsewhere.

Example 1 Exploring the Mean Value Theorem

The function $f(x) = x^2$ (Figure 3.16) is continuous for $0 \le x \le 2$ and differentiable for $0 < x < 2$. Since $f(0) = 0$ and $f(2) = 4$, the Mean Value Theorem says that at some point c in the interval the derivative $f'(x) = 2x$ must have the value $(4 - 0)/(2 - 0) = 2$. In this (exceptional) case, we can identify c by solving the equation $2c = 2$ to get $c = 1$.

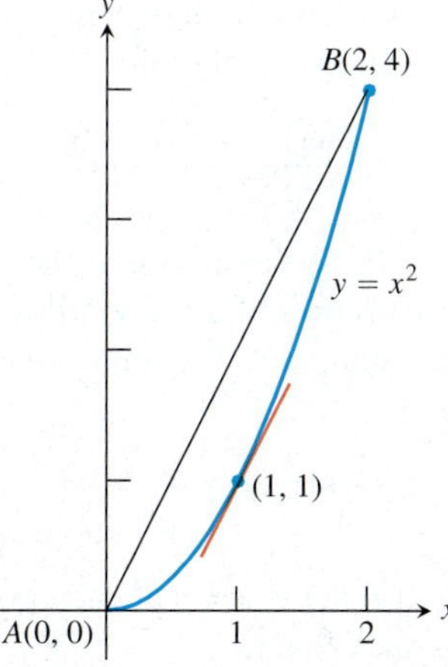

FIGURE 3.16 As we find in Example 1, $c = 1$ is where the tangent is parallel to the chord.

A Physical Interpretation

Think of the number $(f(b) - f(a))/(b - a)$ as the average change in f over $[a, b]$ and $f'(c)$ as an instantaneous change. The Mean Value Theorem says that the instantaneous change at some interior point must equal the average change over the entire interval.

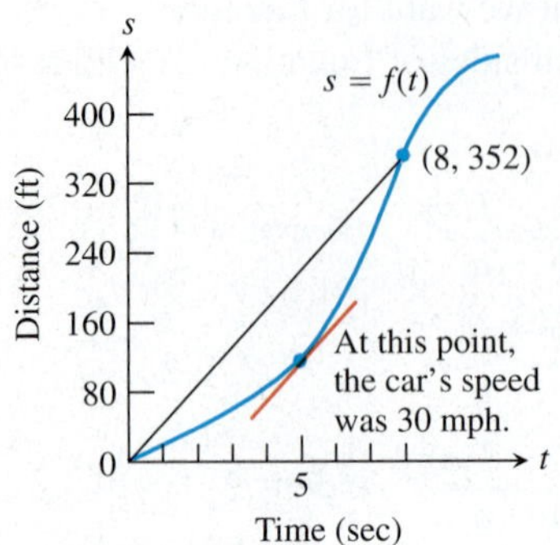

FIGURE 3.17 Distance versus elapsed time for the car in Example 2.

Example 2 Interpreting the Mean Value Theorem

If a car accelerating from zero takes 8 sec to go 352 ft, its average velocity for the 8-sec interval is $352/8 = 44$ ft/sec. At some point during the acceleration, the speedometer must read exactly 30 mph (44 ft/sec) (Figure 3.17).

Mathematical Consequences

The first corollary of the Mean Value Theorem tells us what kind of function has a zero derivative.

Corollary 1 Functions with Zero Derivatives Are Constant Functions

If $f'(x) = 0$ at each point of an interval I, then $f(x) = C$ for all x in I, where C is a constant.

We know that if a function f has a constant value on an interval I, then f is differentiable on I and $f'(x) = 0$ for all x in I. Corollary 1 provides the converse.

CD-ROM
WEBsite
Historical Biography
Bernard le Bouyer Fontenelle (1657 — 1757)

Proof of Corollary 1 We want to show f has a constant value on I. We do so by showing that if x_1 and x_2 are any two points in I, then $f(x_1) = f(x_2)$.

Suppose that x_1 and x_2 are two points in I, numbered from left to right so that $x_1 < x_2$. Then f satisfies the hypotheses of the Mean Value Theorem on $[x_1, x_2]$: It is differentiable at every point of $[x_1, x_2]$ and hence continuous at every point as well. Therefore,

$$\frac{f(x_2) - f(x_1)}{x_2 - x_1} = f'(c)$$

at some point c between x_1 and x_2. Since $f' = 0$ throughout I, this equation translates successively into

$$\frac{f(x_2) - f(x_1)}{x_2 - x_1} = 0, \qquad f(x_2) - f(x_1) = 0, \qquad \text{and} \qquad f(x_1) = f(x_2).$$

At the beginning of the section, we asked if we could work backward from the acceleration of a body falling freely from rest to find the body's velocity and position functions. The answer is yes, as a consequence of the next corollary.

Corollary 2 Functions with the Same Derivative Function on an Interval Differ by a Constant

If $f'(x) = g'(x)$ at each point of an interval I, then there exists a constant C such that $f(x) = g(x) + C$ for all x in I.

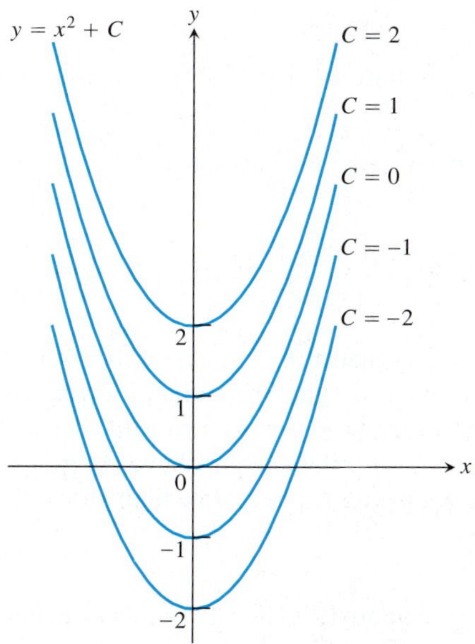

FIGURE 3.18 From a geometric point of view, Corollary 2 of the Mean Value Theorem says that the graphs of functions with identical derivatives on an interval can differ only by a vertical shift there. The graphs of the functions with derivative $2x$ are the parabolas $y = x^2 + C$, shown here for selected values of C.

Proof At each point x in I, the derivative of the difference function $h = f - g$ is

$$h'(x) = f'(x) - g'(x) = 0.$$

Thus, $h(x) = C$ on I (Corollary 1). That is, $f(x) - g(x) = C$ on I, so $f(x) = g(x) + C$.

Corollary 2 says that functions have identical derivatives on an interval only if their values on the interval have a constant difference. We know, for instance, that the derivative of $f(x) = x^2$ on $(-\infty, \infty)$ is $2x$. Any other function with derivative $2x$ on $(-\infty, \infty)$ must equal $x^2 + C$ for some value of C (Figure 3.18).

Example 3 Applying Corollary 2

Find the function $f(x)$ whose derivative is $\sin x$ and whose graph passes through the point $(0, 2)$.

Solution Since f has the same derivative as $g(x) = -\cos x$, we know that $f(x) = -\cos x + C$ for some constant C. The value of C can be determined from the condition that $f(0) = 2$ (the graph of f passes through $(0, 2)$):

$$f(0) = -\cos(0) + C = 2, \qquad \text{so } C = 3.$$

The formula for f is $f(x) = -\cos x + 3$.

Finding Velocity and Position from Acceleration

Here is how to find the velocity $v(t)$ and position $s(t)$ of a body falling freely from rest with acceleration 9.8 m/sec^2, the metric equivalent of 32 ft/sec^2.

We know $v(t)$ is a function whose derivative is 9.8. We also know the derivative of $g(t) = 9.8t$ is 9.8. By Corollary 2,

$$v(t) = 9.8t + C$$

for some constant C. Since the body falls from rest, $v(0) = 0$. This determines C:

$$9.8(0) + C = 0, \qquad \text{so } C = 0.$$

The velocity function must be $v(t) = 9.8t$. How about the position function $s(t)$?

We know $s(t)$ is a function whose derivative is $9.8t$. We also know the derivative of $h(t) = 4.9t^2$ is $9.8t$. By Corollary 2,

$$s(t) = 4.9t^2 + C$$

for some constant C. Since $s(0) = 0$,

$$4.9(0)^2 + C = 0 \qquad \text{and} \qquad C = 0.$$

The position function must be $s(t) = 4.9t^2$.

Differential Equations and the Height of a Projectile

A **differential equation** is an equation relating an unknown function and one or more of its derivatives. A function whose derivatives satisfy a differential equation is called a **solution** to the differential equation.

Example 4 Solutions to Differential Equations

(a) The function $s(t) = 4.9t^2$ is a solution of the differential equation $d^2s/dt^2 = 9.8\ \text{m/sec}^2$.

(b) The function $y = -\cos x + 3$ solves the differential equation $dy/dx = \sin x$.

Example 5 Finding a Projectile's Height from Its Acceleration, Initial Velocity, and Initial Position

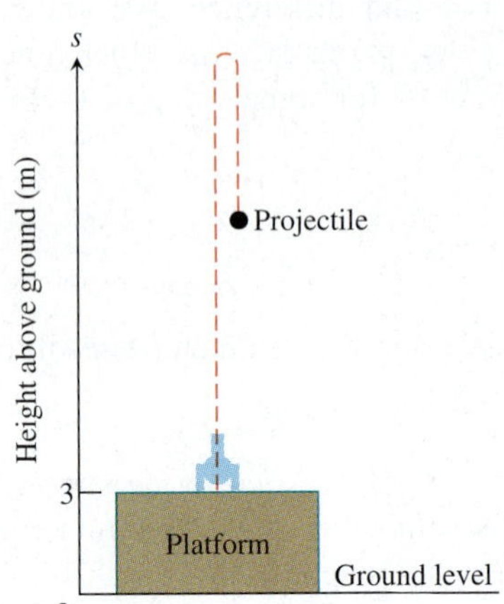

FIGURE 3.19 The sketch for modeling the projectile motion in Example 5.

A heavy projectile is fired straight up from a platform 3 m above the ground, with an initial velocity of 160 m /sec. Assume that the only force affecting the projectile during its flight is from gravity, which produces a downward acceleration of $9.8\ \text{m/sec}^2$. Find an equation for the projectile's height above the ground as a function of time t if $t = 0$ when the projectile is fired. How high above the ground is the projectile 3 sec after firing?

Solution To model the motion, we draw a figure (Figure 3.19) and let s denote the projectile's height above the ground at time t. We assume s to be a twice-differentiable function of t and represent the projectile's velocity and acceleration with the derivatives

$$v = \frac{ds}{dt} \quad \text{and} \quad a = \frac{dv}{dt} = \frac{d^2s}{dt^2}.$$

Since gravity acts in the direction of *decreasing* s in our model, the problem is to solve the differential equation

$$\frac{d^2s}{dt^2} = -9.8$$

knowing also that

$$v(0) = 160 \quad \text{and} \quad s(0) = 3.$$

Since the derivative of $g(t) = -9.8t$ is -9.8, Corollary 2 gives

$$v(t) = -9.8t + C$$

for some constant C. We can find the value of C from the initial condition $v(0) = 160$:

$$\begin{aligned} v(0) &= 160 \\ -9.8(0) + C &= 160 \\ C &= 160. \end{aligned}$$

This completes the formula for ds/dt:

$$\frac{ds}{dt} = -9.8t + 160.$$

We know $s(t)$ is a function whose derivative is $-9.8t + 160$. We also know the derivative of $h(t) = -4.9t^2 + 160t$ is $-9.8t + 160$. By Corollary 2,

$$s = -4.9t^2 + 160t + C.$$

We apply the second condition to find this new C:

$$s(0) = 3$$
$$-4.9(0)^2 + 160(0) + C = 3$$
$$C = 3.$$

This completes the formula for s as a function of t:

$$s = -4.9t^2 + 160t + 3.$$

To find the projectile's height 3 sec into the flight, we set $t = 3$ in the formula for s. The height is

$$s = -4.9(3)^2 + 160(3) + 3 = 438.9 \text{ m}.$$

We say more about solving differential equations in Chapters 4 and 6.

EXERCISES 3.2

Checking and Using Hypotheses

In Exercises 1–4, **(a)** show that the function f satisfies the hypotheses of the Mean Value Theorem on the given interval $[a, b]$ and **(b)** find each value of c in (a, b) that satisfies the equation

$$f'(c) = \frac{f(b) - f(a)}{b - a}.$$

1. $f(x) = x^2 + 2x - 1, \quad [0, 1]$

2. $f(x) = x^{2/3}, \quad [0, 1]$

3. $f(x) = \sin^{-1} x, \quad [-1, 1]$

4. $f(x) = \ln(x - 1), \quad [2, 4]$

5. *Writing to Learn* The function

$$f(x) = \begin{cases} x, & 0 \le x < 1 \\ 0, & x = 1 \end{cases}$$

is zero at $x = 0$ and $x = 1$ and differentiable on $(0, 1)$, but its derivative on $(0, 1)$ is never zero. How can this be? Doesn't Rolle's Theorem say the derivative has to be zero somewhere in $(0, 1)$? Give reasons for your answer.

6. For what values of a, m, and b does the function

$$f(x) = \begin{cases} 3, & x = 0 \\ -x^2 + 3x + a, & 0 < x < 1 \\ mx + b, & 1 \le x \le 2 \end{cases}$$

satisfy the hypotheses of the Mean Value Theorem on the interval $[0, 2]$?

Differential Equations

7. *Writing to Learn* Suppose that $f(-1) = 3$ and that $f'(x) = 0$ for all x. Must $f(x) = 3$ for all x? Give reasons for your answer.

8. *Writing to Learn* Suppose that $g(0) = 5$ and that $g'(t) = 2$ for all t. Must $g(t) = 2t + 5$ for all t? Give reasons for your answer.

In Exercises 9–12, find all possible functions with the given derivative.

9. **(a)** $y' = x$
(b) $y' = x^2$
(c) $y' = x^3$

10. **(a)** $y' = 2x$
(b) $y' = 2x - 1$
(c) $y' = 3x^2 + 2x - 1$

11. **(a)** $r' = \dfrac{1}{\theta}$
(b) $r' = 1 - \dfrac{1}{\theta}$
(c) $r' = 5 + \dfrac{1}{\theta}$

12. **(a)** $y' = \dfrac{1}{2\sqrt{t}}$
(b) $y' = \dfrac{1}{\sqrt{t}}$
(c) $y' = 4t - \dfrac{1}{\sqrt{t}}$

In Exercises 13–16, find the function with the given derivative whose graph passes through the point P.

13. $f'(x) = 2x - 1, \quad P(0, 0)$

14. $g'(x) = \dfrac{1}{x} + 2x, \quad P(1, -1)$

15. $f'(x) = e^{2x}, \quad P\left(0, \dfrac{3}{2}\right)$

16. $r'(t) = \sec t \tan t - 1, \quad P(0, 0)$

Finding Position from Velocity

Exercises 17–20 give the velocity $v = ds/dt$ and initial position of a body moving along a coordinate line. Find the body's position at time t.

17. $v = 9.8t + 5, \quad s(0) = 10$

18. $v = 32t - 2, \quad s(0.5) = 4$

19. $v = \sin \pi t, \quad s(0) = 0$

20. $v = \dfrac{1}{t+2}, \quad t > -2, \quad s(-1) = \dfrac{1}{2}$

Finding Position from Acceleration

Exercises 21–24 give the acceleration $a = d^2s/dt^2$, initial velocity, and initial position of a body moving on a coordinate line. Find the body's position at time t.

21. $a = e^t; \quad v(0) = 20, \quad s(0) = 5$

22. $a = 9.8; \quad v(0) = -3, \quad s(0) = 0$

23. $a = -4 \sin 2t; \quad v(0) = 2, \quad s(0) = -3$

24. $a = \dfrac{9}{\pi^2} \cos \dfrac{3t}{\pi}; \quad v(0) = 0, \quad s(0) = -1$

Gravitational Acceleration

25. *Free fall on the moon* On our moon, the acceleration of gravity is 1.6 m/sec^2. If a rock is dropped into a crevasse, how fast will it be going just before it hits bottom 30 sec later?

26. *Speed of a rocket* A rocket lifts off the surface of Earth with a constant acceleration of 20 m/sec^2. How fast will the rocket be going 1 min later?

27. *Diving off a platform* With approximately what velocity do you enter the water if you dive from a 10 m platform? (Use $g = 9.8 \text{ m/sec}^2$.)

28. *Projectile height on Mars* The acceleration of gravity near the surface of Mars is 3.72 m/sec^2. If a rock is blasted straight up from the surface with an initial velocity of 93 m/sec (about 208 mph), how high does it go? (*Hint:* When is the velocity zero?)

Linear Motion

29. *Motion along a coordinate line* A particle moves on a coordinate line with acceleration $a = d^2s/dt^2 = 15\sqrt{t} - (3/\sqrt{t})$, subject to the conditions that $ds/dt = 4$ and $s = 0$ when $t = 1$. Find

(a) the velocity $v = ds/dt$ in terms of t

(b) the position s in terms of t.

30. *Finding displacement from velocity* **(a)** Suppose that the velocity of a body moving along the s-axis is

$$\frac{ds}{dt} = v = 9.8t - 3.$$

i. Find the body's displacement over the time interval from $t = 1$ to $t = 3$ given that $s = 5$ when $t = 0$.

ii. Find the body's displacement from $t = 1$ to $t = 3$ given that $s = -2$ when $t = 0$.

iii. Now find the body's displacement from $t = 1$ to $t = 3$ given that $s = s_0$ when $t = 0$.

(b) *Writing to Learn* Suppose that the position s of a body moving along a coordinate line is a differentiable function of time t. Is it true that once you know a function whose derivative is the velocity function ds/dt you can find the body's displacement from $t = a$ to $t = b$ even if you do not know the body's exact position at either of those times? Give reasons for your answer.

Applications

31. *Temperature change* It took 14 sec for a mercury thermometer to rise from −19°C to 100°C when it was taken from a freezer and placed in boiling water. Show that somewhere along the way the mercury was rising at the rate of 8.5°C/sec.

32. *Speeding* A trucker handed in a ticket at a toll booth showing that in 2 h she had covered 159 mi on a toll road with speed limit 65 mph. The trucker was cited for speeding. Why?

33. *Triremes* Classical accounts tell us that a 170-oar trireme (ancient Greek or Roman warship) once covered 184 sea miles in 24 h. Explain why at some point during this feat the trireme's speed exceeded 7.5 knots (sea miles per hour).

34. *Running a marathon* A marathoner ran the 26.2 mi New York City Marathon in 2.2 h. Show that at least twice the marathoner was running at exactly 11 mph.

Theory and Examples

35. *The geometric mean of a and b* The **geometric mean** of two positive numbers a and b is the number $\sqrt{ab}$. Show that the value of c in the conclusion of the Mean Value Theorem for $f(x) = 1/x$ on an interval of positive numbers $[a, b]$ is $c = \sqrt{ab}$.

36. *The arithmetic mean of a and b* The **arithmetic mean** of two numbers a and b is the number $(a + b)/2$. Show that the value of c in the conclusion of the Mean Value Theorem for $f(x) = x^2$ on any interval $[a, b]$ is $c = (a + b)/2$.

T **37.** *Writing to Learn: A surprising graph* Graph the function

$$f(x) = \sin x \sin(x + 2) - \sin^2(x + 1).$$

What does the graph do? Why does the function behave this way? Give reasons for your answers.

T **38.** *Rolle's Theorem*

(a) Construct a polynomial $f(x)$ that has zeros at $x = -2, -1, 0, 1$, and 2.

(b) Graph f and its derivative f' together. How is what you see related to Rolle's theorem?

(c) Do $g(x) = \sin x$ and its derivative g' illustrate the same phenomenon?

39. *Unique solution* Assume that f is continuous on $[a, b]$ and differentiable on (a, b). Also assume that $f(a)$ and $f(b)$ have opposite signs and that $f' \neq 0$ between a and b. Show that $f(x) = 0$ exactly once between a and b.

40. *Parallel tangents* Assume that f and g are differentiable on $[a, b]$ and that $f(a) = g(a)$ and $f(b) = g(b)$. Show that there is at least one point between a and b where the tangents to the graphs of f and g are parallel or the same line. Illustrate with a sketch.

41. *Writing to Learn: Identical graphs* If the graphs of two differentiable functions $f(x)$ and $g(x)$ start at the same point in the plane and the functions have the same rate of change at every point, do the graphs have to be identical? Give reasons for your answer.

42. *Upper bounds* Show that for any numbers a and b, the inequality $|\sin b - \sin a| \le |b - a|$ is true.

43. *Sign of f'* Assume that f is differentiable on $a \le x \le b$ and that $f(b) < f(a)$. Show that f' is negative at some point between a and b.

44. Let f be a function defined on an interval $[a, b]$. What conditions could you place on f to guarantee that

$$\min f' \le \frac{f(b) - f(a)}{b - a} \le \max f',$$

where $\min f'$ and $\max f'$ refer to the minimum and maximum values of f' on $[a, b]$? Give reasons for your answers.

45. Use the inequalities in Exercise 44 to estimate $f(0.1)$ if $f'(x) = 1/(1 + x^4 \cos x)$ for $0 \le x \le 0.1$ and $f(0) = 1$.

46. Use the inequalities in Exercise 44 to estimate $f(0.1)$ if $f'(x) = 1/(1 - x^4)$ for $0 \le x \le 0.1$ and $f(0) = 2$.

COMPUTER EXPLORATIONS

Graphing Solutions of Differential Equations

Use a CAS to explore graphically solutions to each of the differential equations in Exercises 47–50. Perform the following steps to help with your explorations.

(a) Find the solution containing the arbitrary constant C using your CAS differential equation solver.

(b) Graph the solutions for $C = -2, -1, 0, 1, 2$ together.

(c) Find and graph the solution that passes through the specified point $P(x_0, y_0)$ over the given interval $[a, b]$.

47. $y' = x\sqrt{1 - x}, \quad [0, 1], \quad P(1/2, 1)$

48. $y' = \frac{1}{x}, \quad [1, 4], \quad P(2, -1)$

49. $y' = x \sin x, \quad [-4, 4], \quad P(\pi, -1)$

50. $y' = \frac{1}{1 + \sin x}, \quad [-\pi/2, \pi/2], \quad P(0, 1)$

3.3 The Shape of a Graph

First Derivative Test for Increasing Functions and Decreasing Functions • The First Derivative Test for Local Extrema • Concavity • Points of Inflection • Second Derivative Test for Local Extrema • Learning about Functions from Derivatives

What do we need to know to determine the shape of a graph? We need to know how it rises or falls as it goes along and how it bends. These distinguishing features are displayed in Figure 3.20. In this section, we see how the first and second derivatives of a function provide the information we need to determine a graph's shape. We begin by defining formally what it means for a function to be increasing or decreasing over an interval.

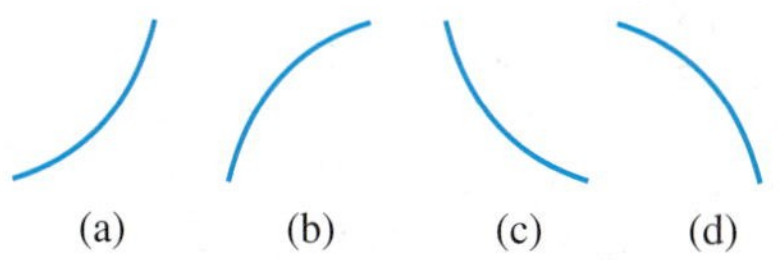

(a) (b) (c) (d)

FIGURE 3.20 The graph in (a) rises and bends upward. The graph in (b) rises and bends downward. The graph in (c) falls and bends upward. The graph in (d) falls and bends downward.

First Derivative Test for Increasing Functions and Decreasing Functions

What kinds of functions have positive derivatives or negative derivatives? The answer, provided by the Mean Value Theorem's third corollary, is this: The only functions with positive derivatives are increasing functions; the only functions with negative derivatives are decreasing functions.

Definitions Increasing Function, Decreasing Function

Let f be a function defined on an interval I. Then

1. f **increases** on I if for all points x_1 and x_2 in I, $x_1 < x_2 \Rightarrow f(x_1) < f(x_2)$.
2. f **decreases** on I if for all points x_1 and x_2 in I, $x_1 < x_2 \Rightarrow f(x_2) < f(x_1)$.

Corollary 3 The First Derivative Test for Increasing and Decreasing

Suppose that f is continuous on $[a, b]$ and differentiable on (a, b).

If $f' > 0$ at each point of (a, b), then f increases on $[a, b]$.

If $f' < 0$ at each point of (a, b), then f decreases on $[a, b]$.

Proof Let x_1 and x_2 be two points in $[a, b]$ with $x_1 < x_2$. The Mean Value Theorem applied to f on $[x_1, x_2]$ says that

$$f(x_2) - f(x_1) = f'(c)(x_2 - x_1)$$

for some c between x_1 and x_2. The sign of the right-hand side of this equation is the same as the sign of $f'(c)$ because $x_2 - x_1$ is positive. Therefore, $f(x_2) > f(x_1)$ if f' is positive on (a, b) and $f(x_2) < f(x_1)$ if f' is negative on (a, b).

Here is how we apply the first derivative test to find where a function is increasing and decreasing. The critical points of a function f partition the x-axis into intervals on which f' is either positive or negative. We determine the sign of f' in each interval by evaluating f' for one value of x in the interval. Then we apply Corollary 3.

Example 1 Using the First Derivative Test for Increasing and Decreasing

Find the critical points of $f(x) = x^3 - 12x - 5$ and identify the intervals on which f is increasing and decreasing.

Solution Figure 3.21 suggests that f has two critical points. Since f is continuous and differentiable for all real numbers, the critical points occur only at the zeros of f'.

$$\begin{aligned} f'(x) &= 3x^2 - 12 = 3(x^2 - 4) \\ &= 3(x + 2)(x - 2) \end{aligned}$$

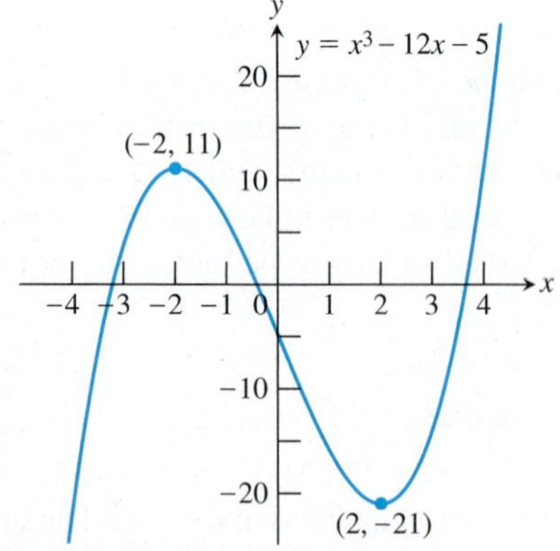

FIGURE 3.21 The graph of $f(x) = x^3 - 12x - 5$. (Example 1)

The zeros of f' are $x = -2$ and $x = 2$. They partition the x-axis into intervals as follows.

Intervals	$-\infty < x < -2$	$-2 < x < 2$	$2 < x < \infty$
Sign of f'	+	−	+
Behavior of f	increasing	decreasing	increasing

To determine the sign of f' on each interval, we found the sign of each factor on the interval and "multiplied" the signs of the factors to get the sign of f'. We then applied Corollary 3 to see that f increases on $(-\infty, -2)$, decreases on $(-2, 2)$, and increases on $(2, \infty)$.

Knowing where a function increases and decreases tells us how to test for the nature of local extreme values.

The First Derivative Test for Local Extrema

In Figure 3.22, at the points where f has a minimum value, $f' < 0$ immediately to the left and $f' > 0$ immediately to the right. (If the point is an endpoint, there is only one side to consider.) Thus, the curve is falling (values decreasing) on the left of the minimum value and rising (values increasing) on its right. Similarly, at the points where f has a maximum value, $f' > 0$ immediately to the left and $f' < 0$ immediately to the right. Thus, the curve is rising (values increasing) on the left of the maximum value and falling (values decreasing) on its right.

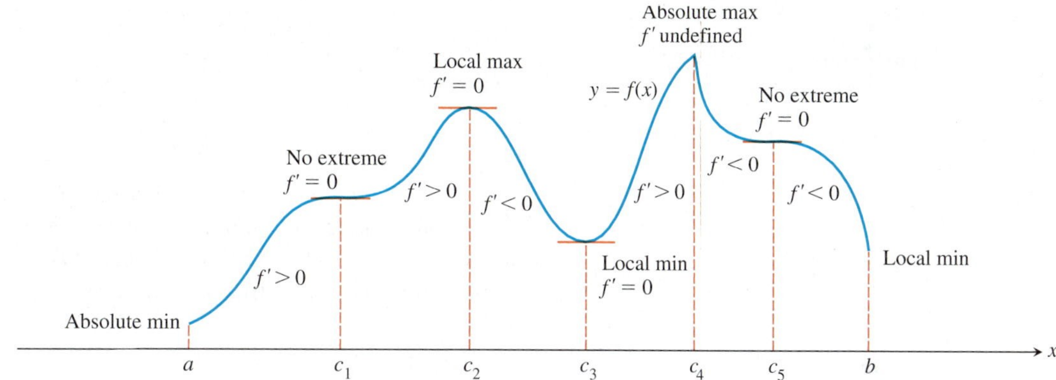

FIGURE 3.22 A function's first derivative tells how the graph rises and falls.

These observations lead to a test for the presence and nature of local extreme values of differentiable functions.

First Derivative Test for Local Extrema

At a critical point $x = c$,

1. f has a *local minimum* if f' changes from negative to positive at c
2. f has a *local maximum* if f' changes from positive to negative at c
3. f has *no local extreme* if f' has the same sign on both sides of c.

The test for local extrema at endpoints is similar but there is only one side to consider.

Example 2 Using the First Derivative Test for Local Extrema

Find the critical points of

$$f(x) = (x^2 - 3)\, e^x.$$

Identify the intervals on which f is increasing and decreasing. Find the function's local and absolute extreme values.

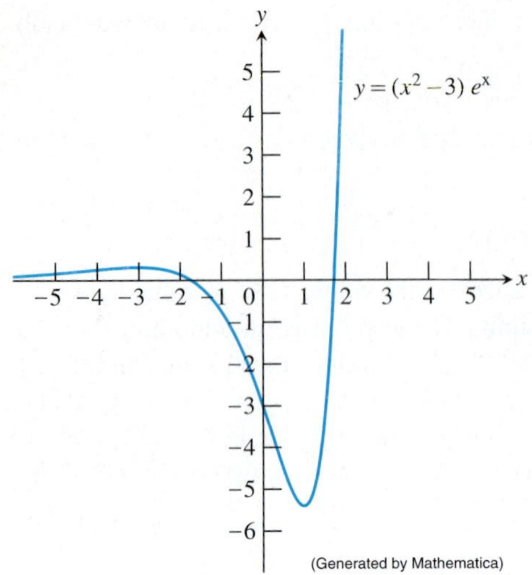

FIGURE 3.23 The graph of $f(x) = (x^2 - 3)e^x$. (Example 2)

Solution This time it is a little harder to see one of the extrema (Figure 3.23). The function f is continuous and differentiable for all real numbers, so the critical points occur only at the zeros of f'.

Using the Product Rule, we find

$$\begin{aligned} f'(x) &= (x^2 - 3) \cdot \frac{d}{dx} e^x + \frac{d}{dx}(x^2 - 3) \cdot e^x \\ &= (x^2 - 3) \cdot e^x + (2x) \cdot e^x \\ &= (x^2 + 2x - 3)e^x. \end{aligned}$$

Since e^x is never zero, the first derivative is zero if and only if

$$\begin{aligned} x^2 + 2x - 3 &= 0 \\ (x + 3)(x - 1) &= 0. \end{aligned}$$

The zeros $x = -3$ and $x = 1$ partition the x-axis into intervals as follows.

Intervals	$x < -3$	$-3 < x < 1$	$1 < x$
Sign of f'	+	−	+
Behavior of f	increasing	decreasing	increasing

We can see from the table that there is a local maximum (about 0.299) at $x = -3$ and a local minimum (about -5.437) at $x = 1$. The local minimum value is also an absolute minimum because $f(x) > 0$ for $|x| > \sqrt{3}$. There is no absolute maximum. The function increases on $(-\infty, -3)$ and $(1, \infty)$ and decreases on $(-3, 1)$.

We now show how to determine the way the graph of a function $y = f(x)$ bends. We know the information must be contained in y', but how do we find it? The answer, for functions that are twice differentiable except perhaps at isolated points, is to differentiate y'. Together y' and y'' tell us the shape of the function's graph. We see in the next section how this enables us to sketch solutions of certain differential equations.

Concavity

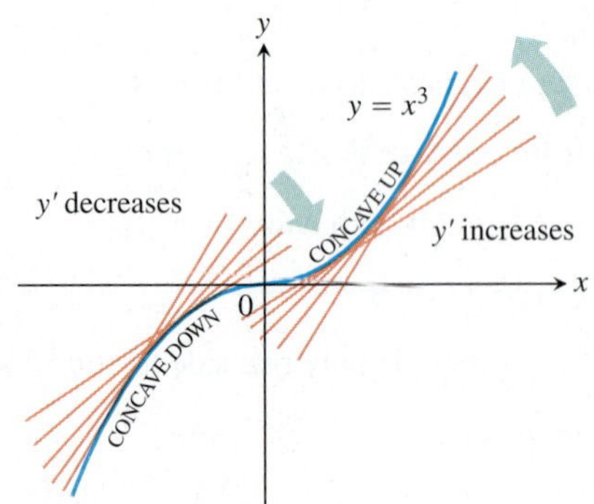

FIGURE 3.24 The graph of $f(x) = x^3$ is concave down on $(-\infty, 0)$ and concave up on $(0, \infty)$.

As you can see in Figure 3.24, the function $y = x^3$ rises as x increases, but the portions defined on the intervals $(-\infty, 0)$ and $(0, \infty)$ *turn* in different ways. Looking at tangents as we scan from left to right, we see that the slope y' of the curve decreases on the interval $(-\infty, 0)$ and then increases on the interval $(0, \infty)$. The curve $y = x^3$ is *concave down* on $(-\infty, 0)$ and *concave up* on $(0, \infty)$. The curve lies below the tangents where it is concave down and above the tangents where it is concave up.

Definition Concavity

The graph of a differentiable function $y = f(x)$ is

(a) concave up on an open interval I if y' is increasing on I

(b) concave down on an open interval I if y' is decreasing on I.

If a function $y = f(x)$ has a second derivative, then we can conclude that y' increases if $y'' > 0$ and y' decreases if $y'' < 0$.

Second Derivative Test for Concavity

The graph of a twice-differentiable function $y = f(x)$ is

(a) concave up on any interval where $y'' > 0$

(b) concave down on any interval where $y'' < 0$.

Example 3 Applying the Concavity Test

The curve $y = x^2$ (Figure 3.25) is concave up on $(-\infty, \infty)$ because its second derivative $y'' = 2$ is always positive.

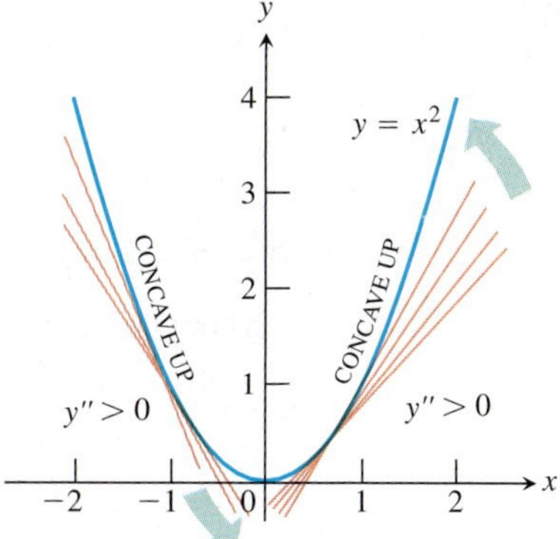

FIGURE 3.25 The graph of $f(x) = x^2$ is concave up on every interval.

Example 4 Determining Concavity

Determine the concavity of $y = 3 + \sin x$ on $[0, 2\pi]$.

Solution The graph of $y = 3 + \sin x$ is concave down on $(0, \pi)$, where $y'' = -\sin x$ is negative. It is concave up on $(\pi, 2\pi)$, where $y'' = -\sin x$ is positive (Figure 3.26).

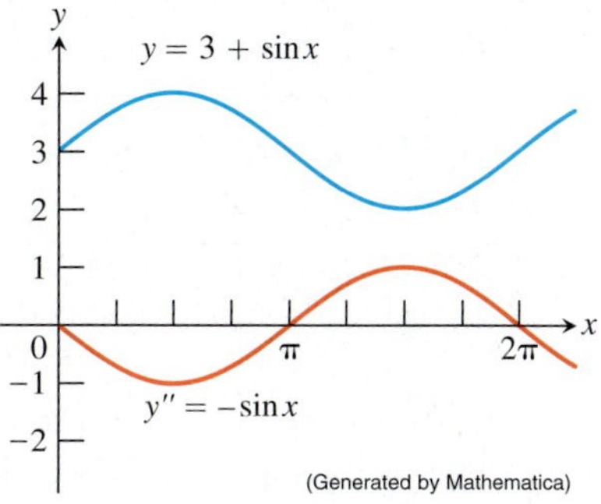

FIGURE 3.26 Using the graph of y'' to determine the concavity of y. (Example 4)

Points of Inflection

The curve $y = 3 + \sin x$ in Example 4 changes concavity at the point $(\pi, 3)$. We call $(\pi, 3)$ a *point of inflection* of the curve.

Definition Point of Inflection

A point where the graph of a function has a tangent line and where the concavity changes is a **point of inflection.**

A point on a curve where y'' is positive on one side and negative on the other is a point of inflection. At such a point, y'' is either zero (because derivatives have the Intermediate Value Property) or undefined. If y is a twice-differentiable function, $y'' = 0$ at a point of inflection and y' has a local maximum or minimum.

To study the motion of a body moving along a line as a function of time, we often are interested in knowing when the body's acceleration, given by the second derivative, is positive or negative. The points of inflection on the graph of the body's position function reveal where the acceleration changes sign.

Example 5 Studying Motion Along a Line

A particle is moving along a horizontal line with position function

$$s(t) = 2t^3 - 14t^2 + 22t - 5, \qquad t \geq 0.$$

Find the velocity and acceleration, and describe the motion of the particle.

Solution The velocity is

$$v(t) = s'(t) = 6t^2 - 28t + 22 = 2(t - 1)(3t - 11),$$

and the acceleration is

$$a(t) = v'(t) = s''(t) = 12t - 28 = 4(3t - 7).$$

When the function $s(t)$ is increasing, the particle is moving to the right; when $s(t)$ is decreasing, the particle is moving to the left.

Notice that the first derivative $(v = s')$ is zero when $t = 1$ and $t = 11/3$.

Intervals	$0 < t < 1$	$1 < t < 11/3$	$11/3 < t$
Sign of $v = s'$	$+$	$-$	$+$
Behavior of s	increasing	decreasing	increasing
Particle motion	right	left	right

The particle is moving to the right in the time intervals $[0, 1)$ and $(11/3, \infty)$, and moving to the left in $(1, 11/3)$.

The acceleration $a(t) = s''(t) = 4(3t - 7)$ is zero when $t = 7/3$.

Intervals	$0 < t < 7/3$	$7/3 < t$
Sign of $a = s''$	$-$	$+$
Graph of s	concave down	concave up

The accelerating force is directed toward the left during the time interval $[0, 7/3)$, is momentarily zero at $t = 7/3$, and is directed toward the right thereafter.

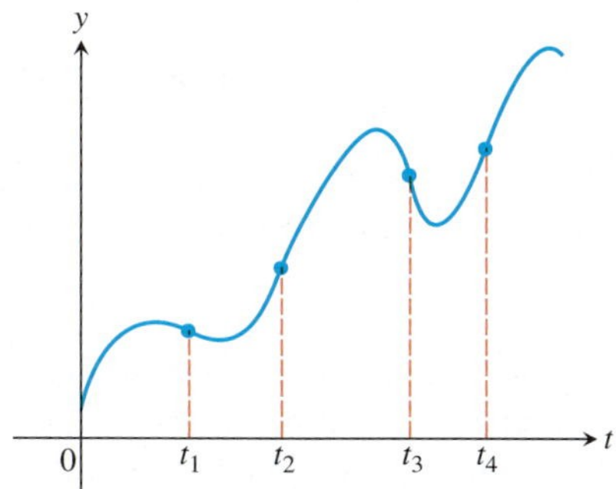

FIGURE 3.27 A hypothetical version of the Dow Jones Industrial Average for Example 6.

Example 6 Inflection Points and the Stock Market

The graph in Figure 3.27 shows a hypothetical version of the Dow Jones Industrial Average. The Dow Jones Industrial Average is a stock market *index*, capturing the overall increase of the stock market along with the local dips and rises.

One way to invest in the stock market is to buy shares of an index fund, which in turn buys a number of different stocks with the goal of tracking the index. The goal of an index fund director would certainly be to buy low (at local minima) and sell high (at local maxima). Such market timing is elusive, however, since it is impossible to predict the extreme values of the market. By the time an investor realizes that the market is truly on the upswing, the minimum has already passed.

Inflection points provide a way for an investor to predict a reversing trend before it happens, since inflection points signal fundamental changes in the *growth rate* of a function. Buying at (or close to) inflection points allows an investor to stay with the historical upward trend (the inflection points warn of a change in trend). This way an investor dampens the fluctuations of the market and, over the long run, captures the upward trend.

Example 7 Finding No Inflection Point Where $y'' = 0$

The curve $y = x^4$ has no inflection point at $x = 0$ (Figure 3.28). Even though $y'' = 12x^2$ is zero there, it does not change sign.

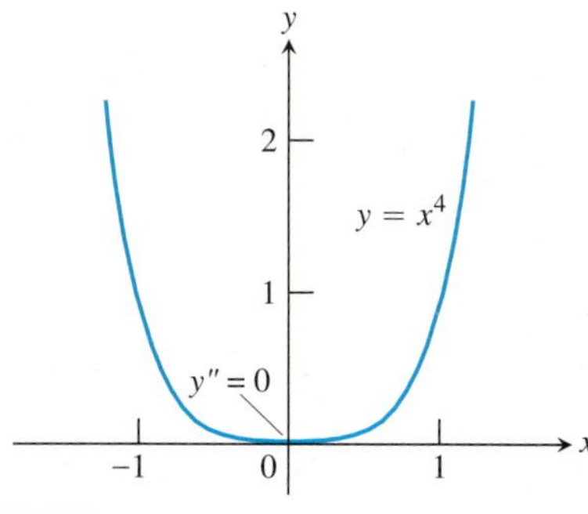

FIGURE 3.28 The graph of $y = x^4$ has no inflection point at the origin, even though $y'' = 0$ there.

Example 8 Finding an Inflection Point Where y'' Does Not Exist

The curve $y = x^{1/3}$ has a point of inflection at $x = 0$ (Figure 3.29), but y'' does not exist there.

$$y'' = \frac{d^2}{dx^2}\left(x^{1/3}\right) = \frac{d}{dx}\left(\frac{1}{3}x^{-2/3}\right) = -\frac{2}{9}x^{-5/3}$$

FIGURE 3.29 A point where y'' fails to exist can be a point of inflection.

We see from Example 7 that a zero second derivative does not always produce a point of inflection. From Example 8, we see that inflection points can also occur where there *is no* second derivative.

Second Derivative Test for Local Extrema

Instead of looking for sign changes in y' at critical points, we can sometimes use the following test to determine the presence of local extrema.

Theorem 5 Second Derivative Test for Local Extrema

1. If $f'(c) = 0$ and $f''(c) < 0$, then f has a local maximum at $x = c$.

2. If $f'(c) = 0$ and $f''(c) > 0$, then f has a local minimum at $x = c$.

This test requires us to know f'' *only at c itself* and not in an interval about c. This makes the test easy to apply. That's the good news. The bad news is that the test fails if $f''(c) = 0$ or if $f''(c)$ fails to exist. When this happens, go back to the first derivative test for local extreme values.

In Example 9, we apply the second derivative test to the function in Example 1.

Example 9 Using the Second Derivative Test

Find the extreme values of $f(x) = x^3 - 12x - 5$.

Solution We have

$$f'(x) = 3x^2 - 12 = 3(x^2 - 4)$$
$$f''(x) = 6x.$$

Testing the critical points $x = \pm 2$ (there are no endpoints), we find

$$f''(-2) = -12 < 0 \Rightarrow f \text{ has a local maximum at } x = -2$$

and

$$f''(2) = 12 > 0 \Rightarrow f \text{ has a local minimum at } x = 2.$$

Example 10 Using f' and f'' to Graph f

Sketch a graph of the function

$$f(x) = x^4 - 4x^3 + 10$$

using the following steps.

(a) Identify where the extrema of f occur.

(b) Find the intervals on which f is increasing and the intervals on which f is decreasing.

(c) Find where the graph of f is concave up and where it is concave down.

(d) Sketch a possible graph for f.

Solution f is continuous since $f'(x) = 4x^3 - 12x^2$ exists. The domain of f' is $(-\infty, \infty)$, so the domain of f is also $(-\infty, \infty)$. Thus, the critical points of f occur only at the zeros of f'. Since

$$f'(x) = 4x^3 - 12x^2 = 4x^2(x - 3)$$

the first derivative is zero at $x = 0$ and $x = 3$.

Intervals	$x < 0$	$0 < x < 3$	$3 < x$
Sign of f'	$-$	$-$	$+$
Behavior of f	decreasing	decreasing	increasing

(a) Using the first derivative test for local extrema and the table above, we see that there is no extremum at $x = 0$ and a local minimum at $x = 3$.

(b) Using the table above, we see that f is decreasing in $(-\infty, 0]$ and $[0, 3]$, and increasing in $[3, \infty)$.

(c) $f''(x) = 12x^2 - 24x = 12x(x - 2)$ is zero at $x = 0$ and $x = 2$.

Intervals	$x < 0$	$0 < x < 2$	$2 < x$
Sign of f''	$+$	$-$	$+$
Behavior of f	concave up	concave down	concave up

We see that f is concave up on the intervals $(-\infty, 0)$ and $(2, \infty)$, and concave down on $(0, 2)$.

(d) Summarizing the information in the two tables above, we obtain

$x < 0$	$0 < x < 2$	$2 < x < 3$	$3 < x$
decreasing	decreasing	decreasing	increasing
concave up	concave down	concave up	concave up

Figure 3.30 shows the graph of f.

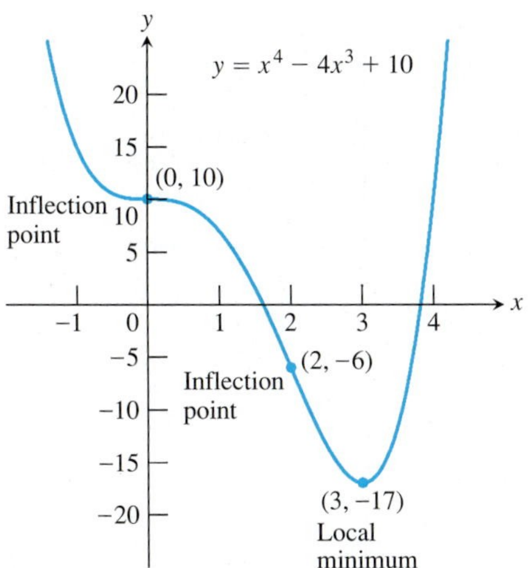

FIGURE 3.30 The graph of $f(x) = x^4 - 4x^3 + 10$. (Example 10)

Procedure for Graphing $y = f(x)$ by Hand

Step 1. Find y' and y''.

Step 2. Find the rise and fall of the curve.

Step 3. Determine the concavity of the curve.

Step 4. Make a summary and show the curve's general shape.

Step 5. Plot specific points and sketch the curve.

The steps in Example 10 give a general procedure for graphing by hand.

Learning about Functions from Derivatives

As we saw in Example 10, we can learn almost everything we need to know about a twice-differentiable function $y = f(x)$ by examining its first derivative. We can find where the function's graph rises and falls and where any local extrema are assumed. We can differentiate y' to learn how the graph bends as it passes over the intervals of rise and fall. We can determine the shape of the function's graph. The only information we cannot get from the derivative is how to place the graph in the xy-plane. But, as we discovered in Section 3.2, the only additional information we need to position the graph is the value of f at one point.

$y = f(x)$ Differentiable $\Rightarrow$ smooth, connected; graph may rise and fall	$y = f(x)$ $y' > 0 \Rightarrow$ rises from left to right; may be wavy	$y = f(x)$ $y' < 0 \Rightarrow$ falls from left to right; may be wavy
or $y'' > 0 \Rightarrow$ concave up throughout; no waves; graph may rise or fall	or $y'' < 0 \Rightarrow$ concave down throughout; no waves; graph may rise or fall	y'' changes sign Inflection point (if f is twice differentiable)
+ − or − + y' changes sign $\Rightarrow$ graph has local maximum or local minimum	$y' = 0$ and $y'' < 0$ at a point; graph has local maximum	$y' = 0$ and $y'' > 0$ at a point; graph has local minimum

EXERCISES 3.3

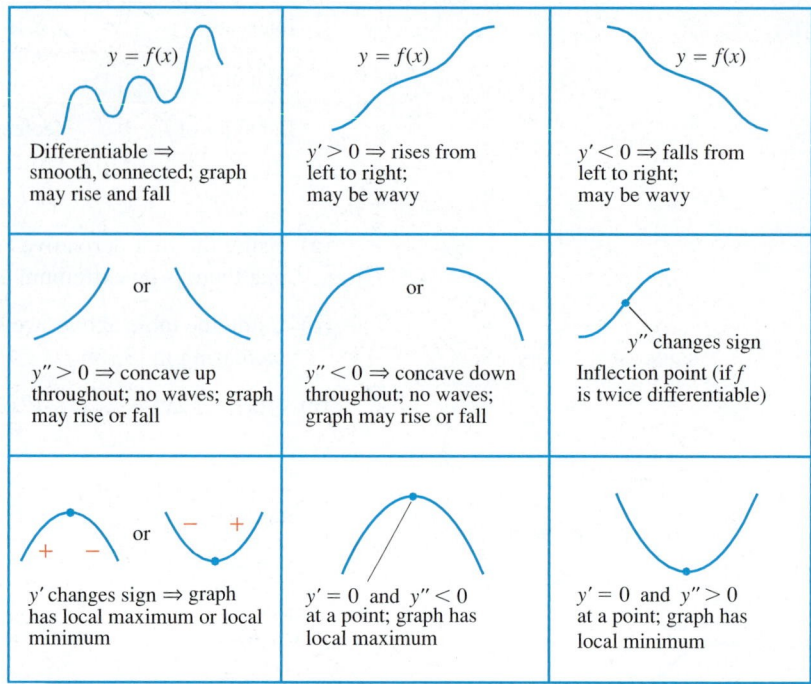

Sketching y from Graphs of y′ and y″

Each of Exercises 1–4 shows the graphs of the first and second derivatives of a function $y = f(x)$. Copy the picture and add to it a sketch of the approximate graph of f, given that the graph passes through the point P.

1. **2.**

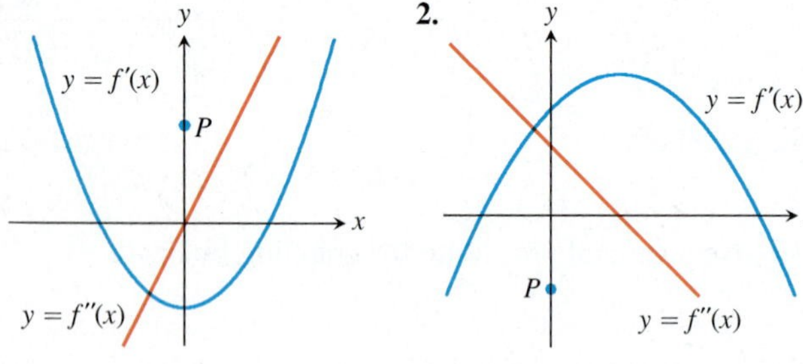

3.

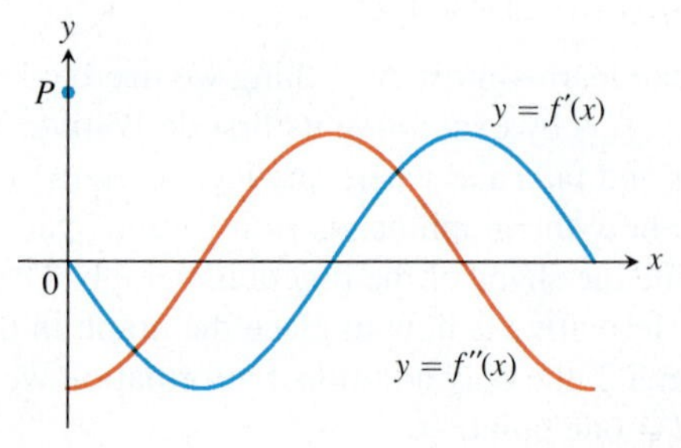

4.

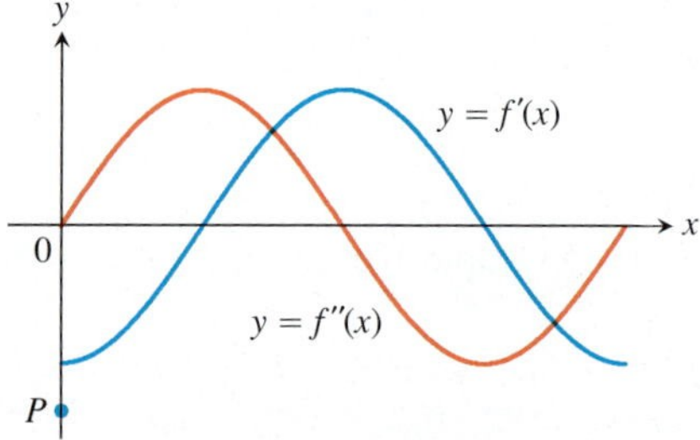

Sketching y from Signs of y′ and y″

5. Sketch the graph of a twice-differentiable function $y = f(x)$ with the following properties. Label coordinates where possible.

x	y	Derivatives
$x < 2$		$y' < 0,\ y'' > 0$
2	1	$y' = 0,\ y'' > 0$
$2 < x < 4$		$y' < 0,\ y'' > 0$
4	4	$y' > 0,\ y'' = 0$
$4 < x < 6$		$y' > 0,\ y'' < 0$
6	7	$y' = 0,\ y'' < 0$
$x > 6$		$y' < 0,\ y'' < 0$

6. Sketch the graph of a twice-differentiable function $y = f(x)$ that passes through the points $(-2, 2)$, $(-1, 1)$, $(0, 0)$, $(1, 1)$ and $(2, 2)$ and whose first two derivatives have the following sign patterns:

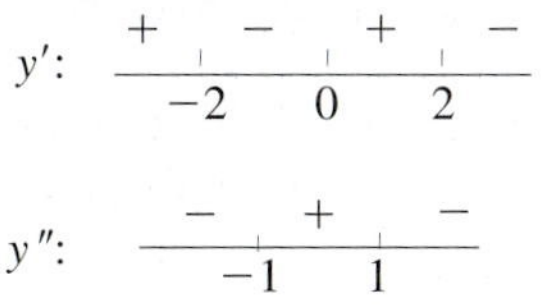

y': + on $(-\infty,-2)$, − on $(-2,0)$, + on $(0,2)$, − on $(2,\infty)$

y'': − on $(-\infty,-1)$, + on $(-1,1)$, − on $(1,\infty)$

Using Graphs to Analyze Functions

In Exercises 7 and 8, use the graph of the function f to estimate where **(a)** f' and **(b)** f'' are 0, positive, and negative.

7.

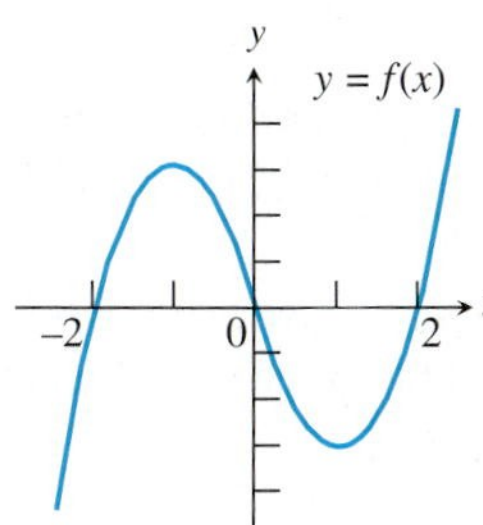

8.

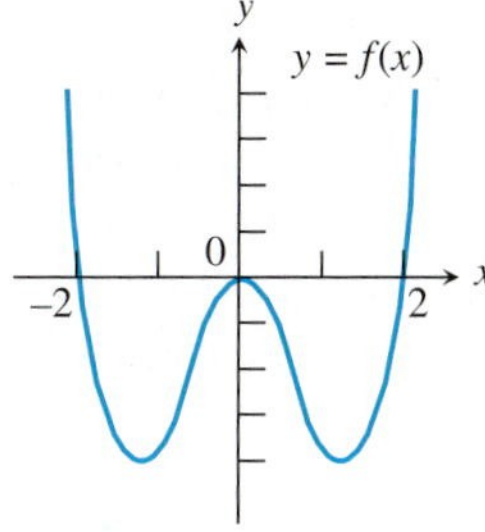

In Exercises 9–12, use the graph of f' to estimate the intervals on which the function f is **(a)** increasing or **(b)** decreasing. **(c)** Estimate where f has local extreme values.

9.

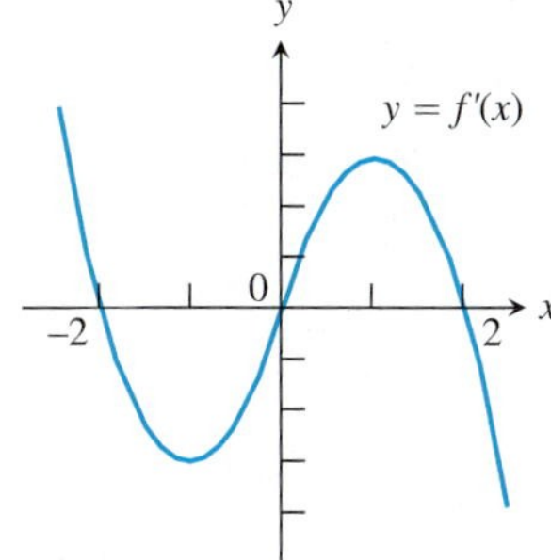

10.

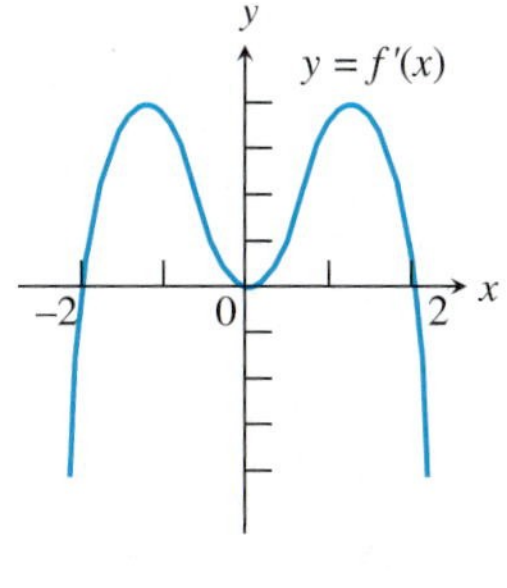

11. The domain of f' is $[0, 4) \cup (4, 6]$.

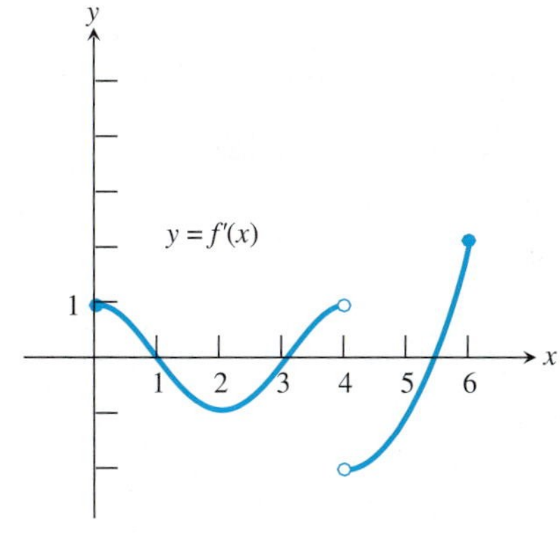

12. The domain of f' is $[0, 1) \cup (1, 2) \cup (2, 3]$.

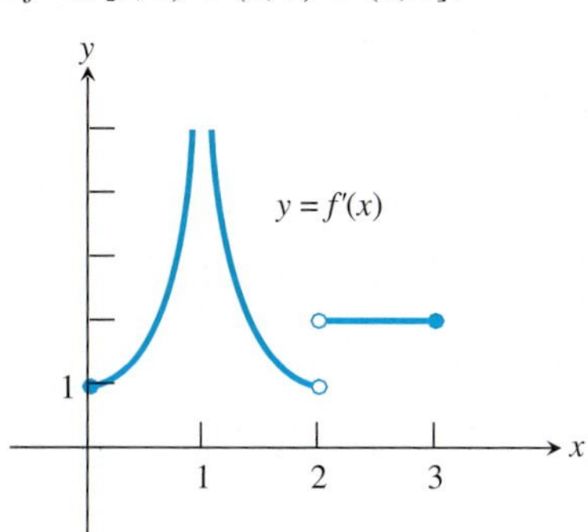

Analyzing f Given f′

Answer the following questions about the functions whose derivatives are given in Exercises 13–16.

(a) What are the critical points of f?

(b) On what intervals is f increasing or decreasing?

(c) At what points, if any, does f assume local maximum and minimum values?

13. $f'(x) = (x - 1)(x + 2)$

14. $f'(x) = (x - 1)^2(x + 2)$

15. $f'(x) = (x - 1)e^{-x}$

16. $f'(x) = x^{-1/3}(x + 2)$

Shape of a Graph

In Exercises 17–22, use analytic methods to find the intervals on which the function is

(a) increasing

(b) decreasing

(c) concave up

(d) concave down.

Then locate and identify any

(e) local extreme values

(f) inflection points.

17. $y = x^2 - x - 1$

18. $y = -2x^3 + 6x^2 - 3$

19. $y = 2x^4 - 4x^2 + 1$

20. $y = xe^{1/x}$

21. $y = x\sqrt{8 - x^2}$

22. $y = \begin{cases} 3 - x^2, & x < 0 \\ x^2 + 1, & x \geq 0 \end{cases}$

T In Exercises 23–32, graph the function to find the intervals on which the function is

(a) increasing

(b) decreasing

(c) concave up

(d) concave down.

Then locate and identify any

(e) local extreme values

(f) inflection points.

23. $y = 4x^3 + 21x^2 + 36x - 20$

24. $y = -x^4 + 4x^3 - 4x + 1$

25. $y = 2x^{1/5} + 3$

26. $y = 5 - x^{1/3}$

27. $y = x^{1/3}(x - 4)$

28. $y = x^2\sqrt{9 - x^2}$

29. $y = xe^{1/x^2}$

30. $y = \dfrac{\ln x}{x}$

31. $y = x^{1/4}(x + 3)$

32. $y = \dfrac{x}{x^2 + 1}$

Extrema and Points of Inflection

In Exercises 33 and 34, use the derivative of the function $y = f(x)$ to find the points at which f has a

(a) local maximum

(b) local minimum

(c) point of inflection.

33. $y' = (x - 1)^2(x - 2)$

34. $y' = (x - 1)^2(x - 2)(x - 4)$

In Exercises 35 and 36, *work in groups of two or three.*

(a) Find the absolute extrema of f and where they occur.

(b) Find any points of inflection.

(c) Sketch a possible graph of f.

35. f is continuous on [0, 3] and satisfies the following.

x	0	1	2	3
f	0	2	0	−2
f'	3	0	does not exist	−3
f''	0	−1	does not exist	0

x	$0 < x < 1$	$1 < x < 2$	$2 < x < 3$
f	+	+	−
f'	+	−	−
f''	−	−	−

36. f is an even function, continuous on [−3, 3], and satisfies the following.

x	0	1	2
f	2	0	−1
f'	does not exist	0	does not exist
f''	does not exist	0	does not exist

x	$0 < x < 1$	$1 < x < 2$	$2 < x < 3$
f	+	−	−
f'	−	−	+
f''	+	−	−

T In Exercises 37–40, find the inflection points (if any) on the graph of the function and the coordinates of the points on the graph where the function has a local maximum or local minimum value. Then graph the function in a region large enough to show all these points simultaneously. Add to your picture the graphs of the function's first and second derivatives. How are the values at which these graphs intersect the x-axis related to the graph of the function? In what other ways are the graphs of the derivatives related to the graph of the function?

37. $y = x^5 - 5x^4 - 240$

38. $y = x^3 - 12x^2$

39. $y = \dfrac{4}{5}x^5 + 16x^2 - 25$

40. $y = \dfrac{x^4}{4} - \dfrac{x^3}{3} - 4x^2 + 12x + 20$

T **41.** Graph $f(x) = 2x^4 - 4x^2 + 1$ and its first two derivatives together. Comment on the behavior of f in relation to the signs and values of f' and f''.

T **42.** Graph $f(x) = x \cos x$ and its second derivative together for $0 \le x \le 2\pi$. Comment on the behavior of the graph of f in relation to the signs and values of f''.

Motion Along a Line

In Exercises 43–46, a particle is moving along a line with position function $s(t)$. Find the **(a)** velocity and **(b)** acceleration, and **(c)** describe the motion of the particle for $t \ge 0$.

43. $s(t) = t^2 - 4t + 3$

44. $s(t) = 6 - 2t - t^2$

45. $s(t) = t^3 - 3t + 3$

46. $s(t) = 3t^2 - 2t^3$

In Exercises 47 and 48, the graph of the position function $y = s(t)$ of a particle moving along a line is given. At approximately what times is the particle's

(a) velocity equal to zero?

(b) acceleration equal to zero?

47.

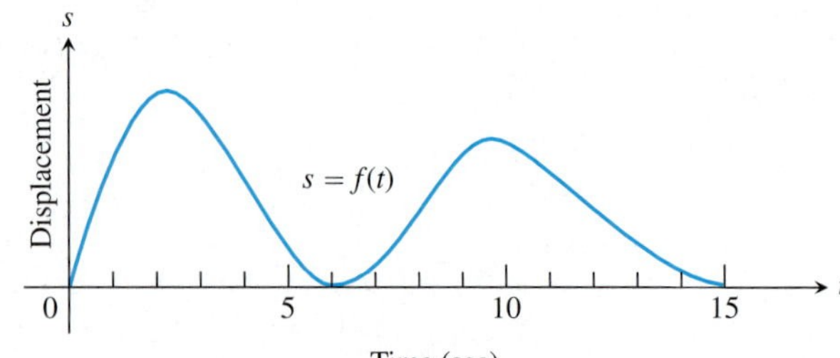

48.

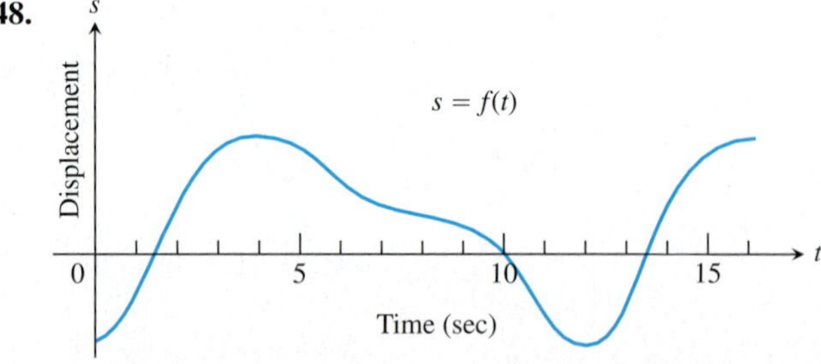

Theory and Examples

49. *Writing to Learn* If $f(x)$ is a differentiable function and $f'(c) = 0$ at an interior point c of f's domain, must f have a local maximum or minimum at $x = c$? Explain.

50. *Writing to Learn* If $f(x)$ is a twice-differentiable function and $f''(c) = 0$ at an interior point c of f's domain, must f have an inflection point at $x = c$? Explain.

51. *Connecting f and f′* Sketch a smooth curve $y = f(x)$ through the origin with the properties that $f'(x) < 0$ for $x < 0$ and $f'(x) > 0$ for $x > 0$.

52. *Connecting f and f″* Sketch a smooth curve $y = f(x)$ through the origin with the properties that $f''(x) < 0$ for $x < 0$ and $f''(x) > 0$ for $x > 0$.

53. *Connecting f, f′, and f″* Sketch a continuous curve $y = f(x)$ with the following properties. Label coordinates where possible.

$f(-2) = 8$	$f'(x) > 0$ for $\lvert x \rvert > 2$
$f(0) = 4$	$f'(x) < 0$ for $\lvert x \rvert < 2$
$f(2) = 0$	$f''(x) < 0$ for $x < 0$
$f'(2) = f'(-2) = 0$	$f''(x) > 0$ for $x > 0$

54. *Writing to Learn: Horizontal tangents* Does the curve $y = x^2 + 3 \sin 2x$ have a horizontal tangent near $x = -3$? Give reasons for your answer.

55. *Writing to Learn* For $x > 0$, sketch a curve $y = f(x)$ that has $f(1) = 0$ and $f'(x) = 1/x$. Can anything be said about the concavity of such a curve? Give reasons for your answer.

56. *Writing to Learn* Can anything be said about the graph of a function $y = f(x)$ that has a continuous second derivative that is never zero? Give reasons for your answer.

57. *Quadratic curves* What can you say about the inflection points of a quadratic curve $y = ax^2 + bx + c$, $a \neq 0$? Give reasons for your answer.

58. *Cubic curves* What can you say about the inflection points of a cubic curve $y = ax^3 + bx^2 + cx + d$, $a \neq 0$? Give reasons for your answer.

Counting Zeros

When we solve an equation $f(x) = 0$ numerically, we usually want to know beforehand how many solutions to look for in a given interval. With the help of Corollary 3, we can sometimes find out.

Suppose that

1. f is continuous on $[a, b]$ and differentiable on (a, b)
2. $f(a)$ and $f(b)$ have opposite signs
3. $f' > 0$ on (a, b) or $f' < 0$ on (a, b).

Then f has exactly one zero between a and b: It cannot have more than one because it is either increasing on $[a, b]$ or decreasing on $[a, b]$. Yet it has at least one, by the Intermediate Value Theorem (Section 1.4). For example, $f(x) = x^3 + 3x + 1$ has exactly one zero on $[-1, 1]$ because f is differentiable on $[-1, 1]$, $f(-1) = -3$ and $f(1) = 5$ have opposite signs, and $f'(x) = 3x^2 + 3 > 0$ for all x (see the accompanying figure.)

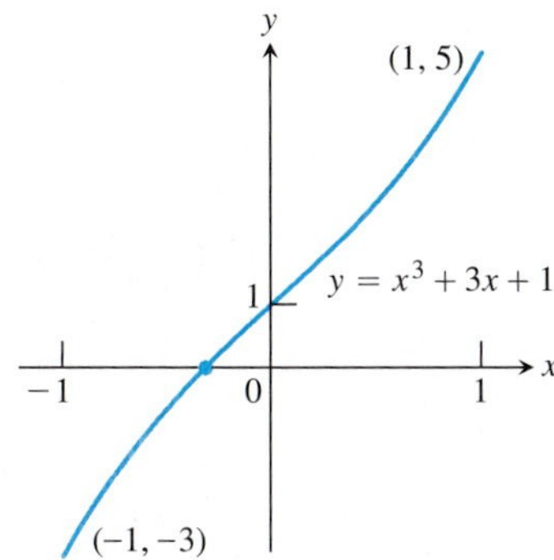

Show that the functions in Exercises 59–62 have exactly one zero in the given interval.

59. $f(x) = x^4 + 3x + 1, \quad [-2, -1]$

60. $g(t) = \sqrt{t} + \sqrt{1 + t} - 4, \quad (0, \infty)$

61. $r(\theta) = \theta + \sin^2\left(\dfrac{\theta}{3}\right) - 8, \quad (-\infty, \infty)$

62. $r(\theta) = \tan \theta - \cot \theta - \theta, \quad (0, \pi/2)$

COMPUTER EXPLORATIONS

63. *Writing to Learn: A family of cubic functions*

(a) On a common screen, graph $f(x) = x^3 + kx$ for $k = 0$ and nearby positive and negative values of k. How does the value of k seem to affect the shape of the graph?

(b) Find $f'(x)$. As you will see, $f'(x)$ is a quadratic function of x. Find the discriminant of the quadratic (the discriminant of $ax^2 + bx + c$ is $b^2 - 4ac$). For what values of k is the discriminant positive? Zero? Negative? For what values of k does f' have two zeros? One or no zeros? Now explain what the value of k has to do with the shape of the graph of f.

(c) Experiment with other values of k. What appears to happen as $k \to -\infty$? As $k \to \infty$?

64. *Writing to Learn: A family of quartic functions*

(a) On a common screen, graph $f(x) = x^4 + kx^3 + 6x^2$, $-1 \leq x \leq 4$ for $k = -4$, and some nearby values of k. How does the value of k seem to affect the shape of the graph?

(b) Find $f''(x)$. As you will see, $f''(x)$ is a quadratic function of x. What is the discriminant of this quadratic (see Exercise 63b)? For what values of k is the discriminant positive? Zero? Negative? For what values of k does $f''(x)$ have two zeros? One or no zeros? Now explain what the value of k has to do with the shape of the graph of f.

In Exercises 65 and 66, use a CAS to solve the problem.

65. *Logistic functions* Let $f(x) = c/(1 + ae^{-bx})$ with $a > 0$, $abc \neq 0$.

(a) Show that f is increasing on the interval $(-\infty, \infty)$ if $abc > 0$ and decreasing if $abc < 0$.

(b) Show that the point of inflection of f occurs at $x = (\ln a)/b$.

66. *Quartic polynomial functions* Let $f(x) = ax^4 + bx^3 + cx^2 + dx + e$ with $a \neq 0$.

(a) Show that the graph of f has 0 or 2 points of inflection.

(b) Write a condition that must be satisfied by the coefficients if the graph of f has 0 or 2 points of inflection.

3.4 Graphical Solutions of Autonomous Differential Equations

Equilibrium Values and Phase Lines • Stable and Unstable Equilibria • Cooling, a Falling Body Encountering Resistance, and Logistic Growth

We can build on our knowledge of how derivatives determine the shape of a graph to solve differential equations graphically. The starting points for doing so are the notions of *phase line* and *equilibrium value*. We arrive at these notions by investigating what happens when the derivative of a differentiable function is zero from a new point of view.

Equilibrium Values and Phase Lines

We have seen the important role the critical points play in determining how a function behaves and in finding its extreme points. Let's investigate what happens when the derivative of a function is zero from a slightly different point of view. In this case, the derivative dy/dx will be a function of y only (the dependent variable). For example, differentiating the equation

$$y^2 = x + 1$$

implicitly gives

$$2y\frac{dy}{dx} = 1 \qquad \text{or} \qquad \frac{dy}{dx} = \frac{1}{2y}.$$

A differential equation for which dy/dx is a function of y is called an **autonomous** differential equation.

Definition Equilibrium Values or Rest Points

If $dy/dx = g(y)$ is an autonomous differential equation, then the values of y for which $dy/dx = 0$ are called **equilibrium values** or **rest points.**

Thus, equilibrium values are those for which no change occurs in the dependent variable, so y is at *rest*. The emphasis is on the value of y where $dy/dx = 0$, not the value of x as in the preceding section.

Example 1 Finding Equilibrium Values

The equilibrium values for the autonomous differential equation

$$\frac{dy}{dx} = (y + 1)(y - 2)$$

are $y = -1$ and $y = 2$.

To construct a graphical solution to an autonomous differential equation like the one in Example 1, we first make a **phase line** for the equation, a plot on the y-axis that shows the equation's equilibrium values along the intervals where dy/dx and d^2y/dx^2 are positive and negative. Then we know where the solutions are increasing and decreasing, and the concavity of the solution curves. These are the essential features we found in the previous section, so we can determine the shapes of the solution curves without having to find formulas for them.

Example 2 Drawing a Phase Line and Sketching Solution Curves

Draw a phase line for the differential equation

$$\frac{dy}{dx} = (y + 1)(y - 2)$$

and use it to sketch solutions to the equation.

Solution

Step 1. *Draw a number line for* y *and mark the equilibrium values* $y = -1$ *and* $y = 2$, *where* $dy/dx = 0$.

Step 2. *Identify and label the intervals where* $y' > 0$ *and* $y' < 0$. This step resembles what we did in the preceding section, only now we are marking the y-axis instead of the x-axis.

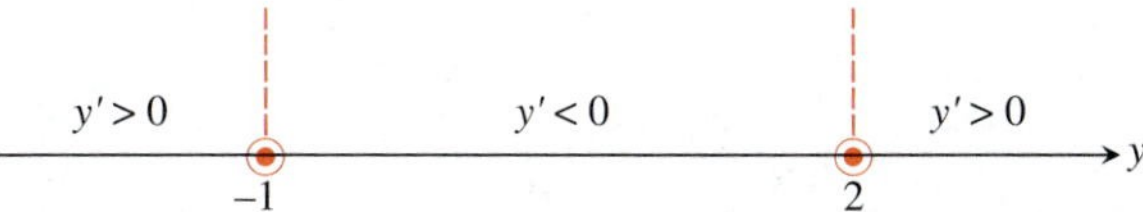

We can encapsulate the information about the sign of y' on the phase line itself. Since $y' > 0$ on the interval to the left of $y = -1$, a solution of the differential equation with a y-value less than -1 will increase from there toward $y = -1$. We display this information by drawing an arrow on the interval pointing to -1.

Similarly, $y' < 0$ between $y = -1$ and $y = 2$, so any solution with a value in this interval will decrease toward $y = -1$.

For $y > 2$, we have $y' > 0$, so a solution with a y-value greater than 2 will increase from there without bound.

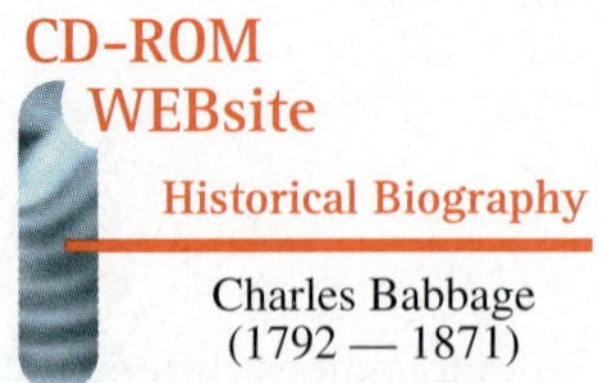

In short, solution curves below the horizontal line $y = -1$ in the xy-plane rise toward $y = -1$. Solution curves between the lines $y = -1$ and $y = 2$ fall away from $y = 2$ toward $y = -1$. Solution curves above $y = 2$ rise away from $y = 2$ and keep going up.

Step 3. *Calculate y'' and mark the intervals where $y'' > 0$ and $y'' < 0$.* To find y'', we differentiate y' *with respect to x*, using implicit differentiation.

$$y' = (y + 1)(y - 2) = y^2 - y - 2 \qquad \text{Formula for } y' \ldots$$

$$\begin{aligned} y'' &= \frac{d}{dx}(y') = \frac{d}{dx}(y^2 - y - 2) \qquad \ldots \text{differentiated implicitly with respect to } x. \\ &= 2yy' - y' \\ &= (2y - 1)y' \\ &= (2y - 1)(y + 1)(y - 2). \end{aligned}$$

From this formula, we see that y'' changes sign at $y = -1$, $y = 1/2$, and $y = 2$. We add the sign information to the phase line.

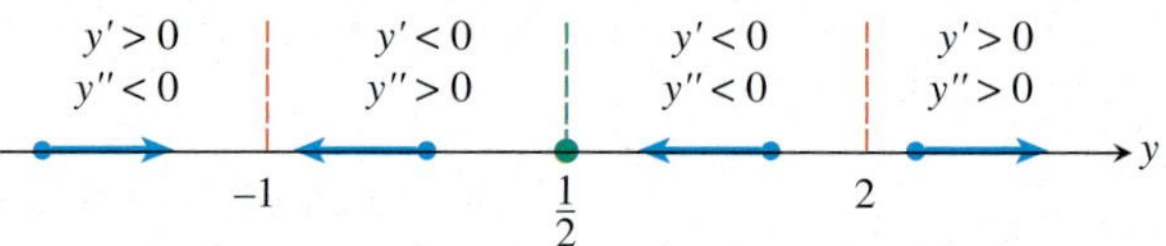

Step 4. *Sketch an assortment of solution curves in the xy-plane.* The horizontal lines $y = -1$, $y = 1/2$, and $y = 2$ partition the plane into horizontal bands in which we know the signs of y' and y''. In each band, this information tells us whether the solution curves rise or fall and how they bend as x increases (Figure 3.31).

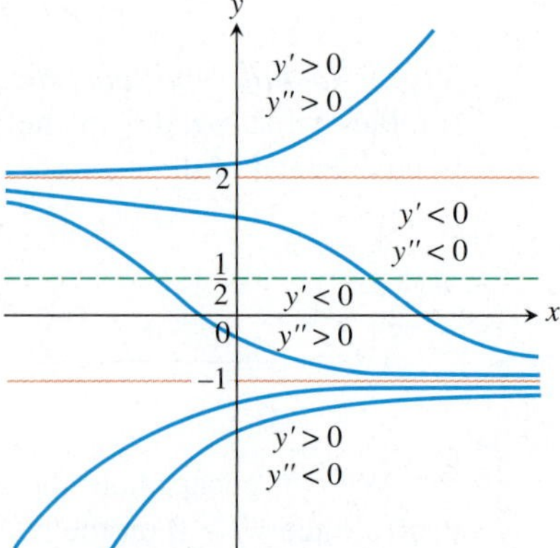

FIGURE 3.31 Graphical solutions from Example 2.

The "equilibrium lines" $y = -1$ and $y = 2$ are also solution curves. (The constant functions $y = -1$ and $y = 2$ satisfy the differential equation.) Solution curves that cross the line $y = 1/2$ have an inflection point there. The concavity changes from concave down (above the line) to concave up (below the line).

As predicted in step 2, solutions in the middle and lower bands approach the equilibrium value $y = -1$ as x increases. Solutions in the upper band rise steadily away from the value $y = 2$.

Stable and Unstable Equilibria

Look at Figure 3.31 once more, in particular at the behavior of the solution curves near the equilibrium values. Once a solution curve has a value near $y = -1$, it tends steadily toward that value; $y = -1$ is a **stable equilibrium.** The behavior near $y = 2$ is just the opposite: all solutions except the equilibrium solution $y = 2$ itself move *away* from it as x increases. We call $y = 2$ an **unstable equilibrium.** If the solution is *at* that value, it stays, but if it is off by any amount, no matter how small, it moves away. (Sometimes an equilibrium value is unstable because a solution moves away from it only on one side of the point.)

Now that we know what to look for, we can already see this behavior on the initial phase line. The arrows lead away from $y = 2$ and, once to the left of $y = 2$, toward $y = -1$.

Cooling, a Falling Body Encountering Resistance, and Logistic Growth

In the next example, we use a phase line to understand the behavior of a physical model. Isaac Newton postulated that the rate of change in the temperature of a cooled or heated object is proportional to the difference in temperature between the object and its surrounding medium. We can use this idea to describe how the object's temperature will change over time.

This postulate is called *Newton's Law of Cooling* (even though it applies equally well to heating).

Example 3 Cooling Soup

What happens to the temperature of the soup when a cup of hot soup is placed on a table in a room? We know the soup cools down, but what does a typical temperature curve look like as a function of time?

Solution We assume that the soup's Celsius temperature H is a differentiable function of time t, choose a suitable unit for t—minutes, say—and start measuring time at $t = 0$. We also assume that the volume of the surrounding medium is large enough so that the heat of the soup has a negligible effect on its surrounding temperature.

Suppose that the surrounding medium has a constant temperature of 15°C. We can then express the difference in temperature as $H(t) - 15$. According to Newton's Law of Cooling, there is a constant of proportionality $k > 0$ such that

$$\frac{dH}{dt} = -k(H - 15) \tag{1}$$

(*minus* k to give a negative derivative when $H > 15$).

Since $dH/dt = 0$ at $H = 15$, the temperature 15°C is an equilibrium value. If $H > 15$, Equation (1) tells us that $(H - 15) > 0$ and $dH/dt < 0$. If the object is hotter than the room, it will get cooler. Similarly, if $H < 15$, then $(H - 15) < 0$ and $dH/dt > 0$. An object cooler than the room will warm up. Thus, the behavior described by Equation (1) agrees with our intuition of how temperature should behave. These observations are captured in the initial phase line diagram in Figure 3.32.

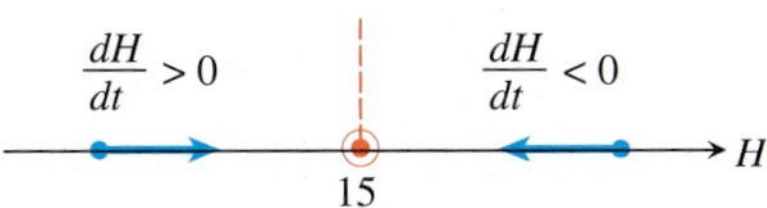

FIGURE 3.32 First step in constructing the phase line for Newton's Law of Cooling in Example 3. The temperature tends towards the equilibrium (surrounding-medium) value in the long run.

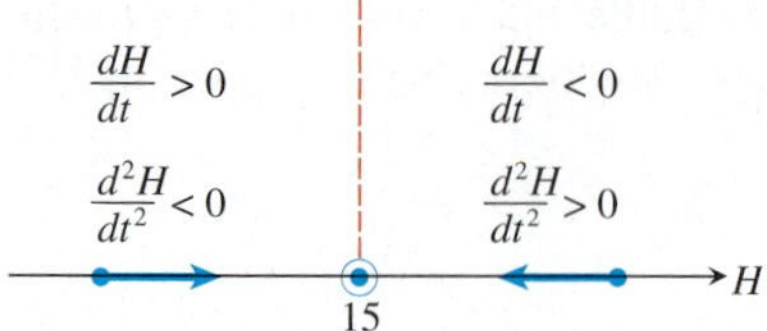

FIGURE 3.33 The complete phase line for Example 3.

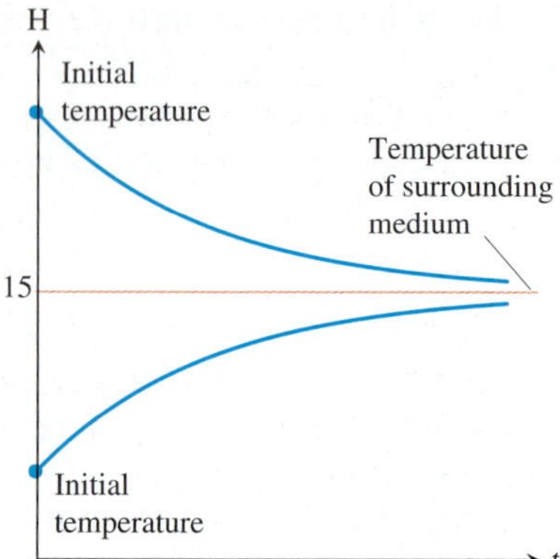

FIGURE 3.34 Temperature versus time. Regardless of initial temperature, the object's temperature $H(t)$ tends toward 15°C, the temperature of the surrounding medium.

We determine the concavity of the solution curves by differentiating both sides of Equation (1) with respect to t:

$$\frac{d}{dt}\left(\frac{dH}{dt}\right) = \frac{d}{dt}(-k(H - 15))$$

$$\frac{d^2H}{dt^2} = -k\,\frac{dH}{dt}.$$

Since $-k$ is negative, we see that d^2H/dt^2 is positive when $dH/dt < 0$ and negative when $dH/dt > 0$. Figure 3.33 adds this information to the phase line.

The completed phase line shows that if the temperature of the object is above the equilibrium value of 15°C, the graph of $H(t)$ will be decreasing and concave upward. If the temperature is below 15°C (the temperature of the surrounding medium), the graph of $H(t)$ will be increasing and concave downward. We use this information to sketch typical solution curves (Figure 3.34).

From the upper solution curve in Figure 3.34, we see that as the object cools down, the rate at which it cools slows down because dH/dt approaches zero. This observation is implicit in Newton's Law of Cooling and contained in the differential equation, but the flattening of the graph as time advances gives an immediate visual representation of the phenomenon. The ability to discern physical behavior from graphs is a powerful tool in understanding real-world systems.

Example 4 Analyzing the Fall of a Body Encountering a Resistive Force

Galileo and Newton both observed that the rate of change in momentum encountered by a moving object is equal to the net force applied to it. In mathematical terms,

$$F = \frac{d}{dt}(mv), \tag{2}$$

where F is the force and m and v the object's mass and velocity. If m varies with time, as it will if the object is a rocket burning fuel, the right-hand side of Equation (2) expands to

$$m\frac{dv}{dt} + v\frac{dm}{dt}$$

using the product rule. In many situations, however, m is constant, $dm/dt = 0$, and Equation (2) takes the simpler form

$$F = m\frac{dv}{dt} \qquad \text{or} \qquad F = ma, \tag{3}$$

known as *Newton's second law of motion.*

In free fall, the constant acceleration due to gravity is denoted by g and the one force acting downward on the falling body is

$$F_p = mg,$$

the propulsion due to gravity. If, however, we think of a real body falling through the air—say, a penny from a great height or a parachutist from an even greater height—we know that at some point air resistance is a factor in the speed of the fall. A more realistic model of free fall would include air resistance, shown as a force F_r in the schematic diagram in Figure 3.35.

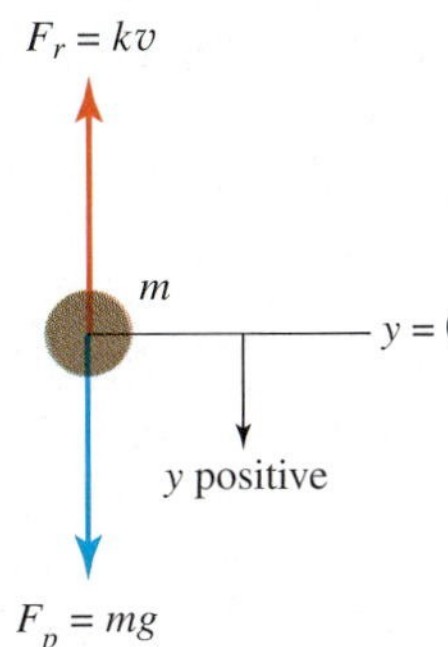

FIGURE 3.35 An object falling under the influence of gravity with a resistive force assumed to be proportional to the velocity.

For speeds well below the speed of sound, physical experiments have shown that F_r is approximately proportional to the body's velocity. The net force on the falling body is therefore

$$F = F_p - F_r$$

giving

$$ma = mg - kv$$

$$\frac{dv}{dt} = g - \frac{k}{m}v. \tag{4}$$

We can use a phase line to analyze the velocity functions that solve this differential equation.

The equilibrium point, obtained by setting the right-hand side of Equation (4) equal to zero, is

$$v = \frac{mg}{k}.$$

If the body is initially moving faster than this, dv/dt is negative and the body slows down. If the body is moving at a velocity below mg/k, $dv/dt > 0$ and the body speeds up. These observations are captured in the initial phase line diagram in Figure 3.36.

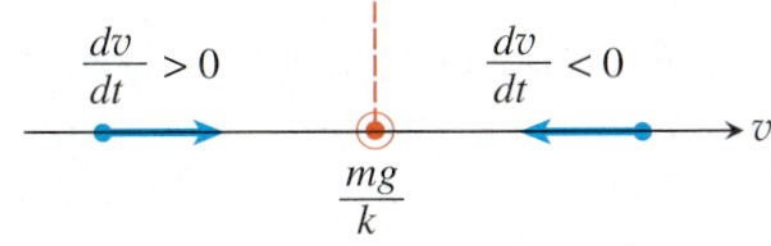

FIGURE 3.36 Initial phase line for Example 4.

We determine the concavity of the solution curves by differentiating both sides of Equation (4) with respect to t:

$$\frac{d^2v}{dt^2} = \frac{d}{dt}\left(g - \frac{k}{m}v\right) = -\frac{k}{m}\frac{dv}{dt}.$$

We see that $d^2v/dt^2 < 0$ when $v < mg/k$ and $d^2v/dt^2 > 0$ when $v > mg/k$. Figure 3.37 adds this information to the phase line. Notice the similarity to the phase

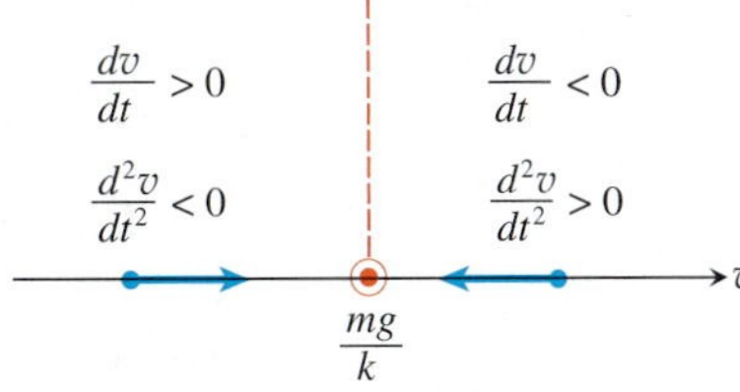

FIGURE 3.37 The completed phase line for Example 4.

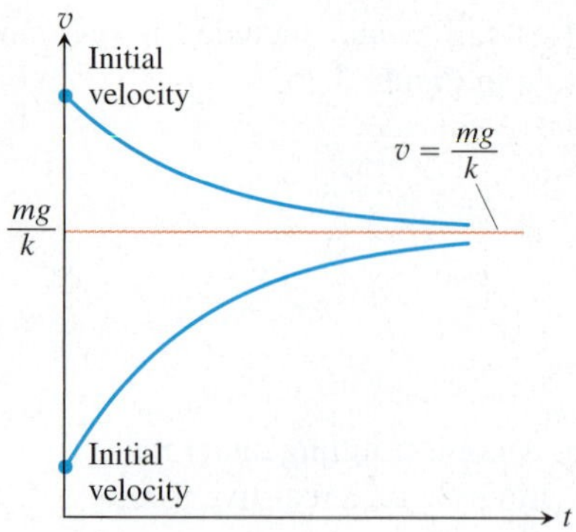

FIGURE 3.38 Typical velocity curves in Example 4. The value $v = mg/k$ is the terminal velocity.

Skydivers can vary their terminal velocity from 95 mph to 180 mph by changing the amount of body area opposing the fall.

line for Newton's Law of Cooling (Figure 3.33). The solution curves are similar as well (Figure 3.38).

Figure 3.38 shows two typical solution curves. Regardless of the initial velocity, we see the body's velocity tending toward the limiting value $v = mg/k$. This value, a stable equilibrium point, is called the body's **terminal velocity.**

Example 5 Analyzing Population Growth in a Limiting Environment

Suppose that $P = P(t)$ represents the number of individuals in a particular population at time t. For instance, P might be the number of yeast cells in a nutrient fluid or the number of white owls in the western United States. Assume that during a small increment of time Δt, a certain percentage of the population is born and another percentage dies. Then the average rate of change of P over the interval $[t, t + \Delta t]$ is

$$\frac{\Delta P}{\Delta t} = kP(t), \tag{5}$$

where $k > 0$ is the birth rate minus the death rate per individual per unit time.

Because the natural environment has only a limited number of resources to sustain life, it is reasonable to assume that only a maximum population M can be accommodated. As the population approaches this **limiting population** or **carrying capacity,** resources become less abundant and the growth rate k decreases. A simple relationship exhibiting this behavior is

$$k = r(M - P),$$

where $r > 0$ is a constant. Notice that k decreases as P increases toward M and that k is negative if P is greater than M. Substituting $r(M - P)$ for k in Equation (5) and taking the limit as $\Delta t \to 0$ gives the differential equation

$$\frac{dP}{dt} = r(M - P)P = rMP - rP^2. \tag{6}$$

The model (6) is referred to as **logistic growth.**

We can forecast the behavior of the population over time by analyzing the phase line for Equation (6). The equilibrium values are $P = M$ and $P = 0$, and we can see that $dP/dt > 0$ if $0 < P < M$ and $dP/dt < 0$ if $P > M$. These observations are recorded on the phase line in Figure 3.39.

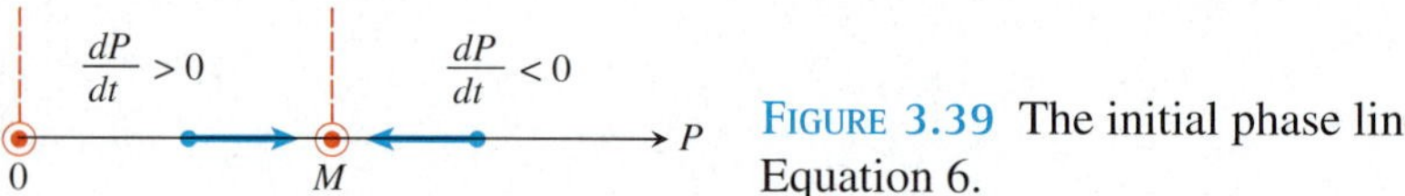

FIGURE 3.39 The initial phase line for Equation 6.

We determine the concavity of the population curves by differentiating both sides of Equation (6) with respect to t:

$$\begin{aligned}\frac{d^2P}{dt^2} &= \frac{d}{dt}(rMP - rP^2)\\ &= rM\frac{dP}{dt} - 2rP\frac{dP}{dt}\\ &= r(M - 2P)\frac{dP}{dt}.\end{aligned} \tag{7}$$

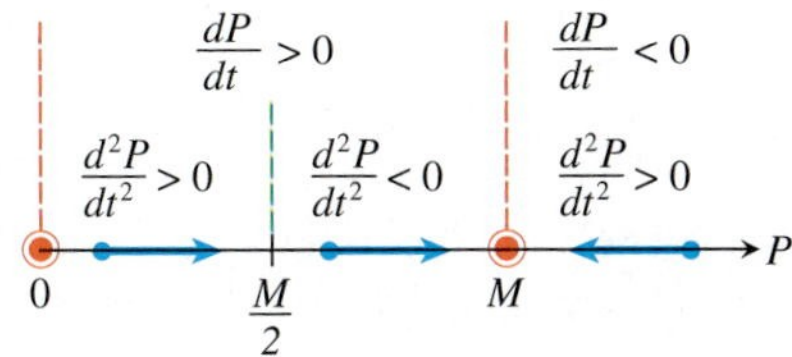

FIGURE 3.40 The completed phase line for logistic growth (Equation 6).

If $P = M/2$, then $d^2P/dt^2 = 0$. If $P < M/2$, then $(M - 2P)$ and dP/dt are positive and $d^2P/dt^2 > 0$. If $M/2 < P < M$, then $(M - 2P) < 0$, $dP/dt > 0$, and $d^2P/dt^2 < 0$. If $P > M$, then $(M - 2P)$ and dP/dt are both negative and $d^2P/dt^2 > 0$. We add this information to the phase line (Figure 3.40).

The lines $P = M/2$ and $P = M$ divide the first quadrant of the tP-plane into horizontal bands in which we know the signs of both dP/dt and d^2P/dt^2. In each band, we know how the solution curves rise and fall and how the bend behaves as time passes. The equilibrium lines $P = 0$ and $P = M$ are both population curves. Population curves crossing the line $P = M/2$ have an inflection point there, giving them a **sigmoid** shape (curved in two directions like a letter S). Figure 3.41 displays typical population curves. Figure 3.42 shows the growth of a laboratory yeast culture based on experimental data.

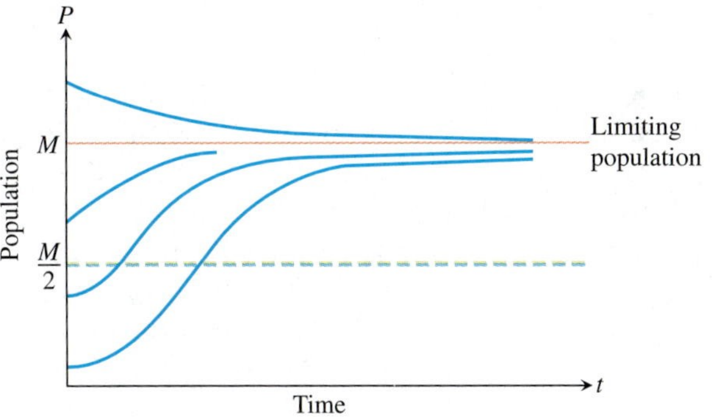

FIGURE 3.41 Population curves in Example 5.

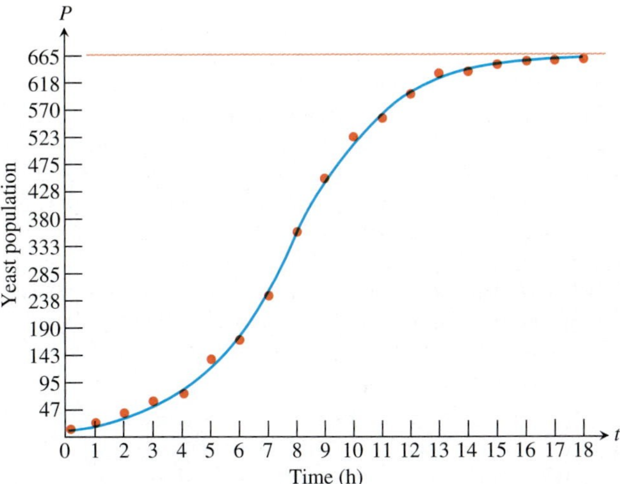

FIGURE 3.42 Logistic curve showing the growth of yeast in a culture. The dots indicate observed values. (Data from R. Pearl, "Growth of Population," *Quart. Rev. Biol.* 2 (1927): 532–548.)

EXERCISES 3.4

Phase Lines and Solution Curves

In Exercises 1–8,

(a) Identify the equilibrium values. Which are stable and which are unstable?

(b) Construct a phase line. Identify the signs of y' and y''.

(c) Sketch several solution curves.

1. $\dfrac{dy}{dx} = (y+2)(y-3)$

2. $\dfrac{dy}{dx} = y^2 - 4$

3. $\dfrac{dy}{dx} = y^3 - y$

4. $\dfrac{dy}{dx} = y^2 - 2y$

5. $y' = \sqrt{y}, \quad y > 0$

6. $y' = y - \sqrt{y}, \quad y > 0$

7. $y' = (y-1)(y-2)(y-3)$

8. $y' = y^3 - y^2$

Models of Population Growth

The autonomous differential equations in Exercises 9–12 represent models for population growth. For each exercise, use a phase line analysis to sketch solution curves for $P(t)$ selecting different starting values $P(0)$ (as in Example 5). Which equilibria are stable, and which are unstable?

9. $\dfrac{dP}{dt} = 1 - 2P$

10. $\dfrac{dP}{dt} = P(1 - 2P)$

11. $\dfrac{dP}{dt} = 2P(P - 3)$

12. $\dfrac{dP}{dt} = 3P(1 - P)\left(P - \dfrac{1}{2}\right)$

13. *Catastrophic continuation of Example 5* Suppose that a healthy population of some species is growing in a limited environment and that the current population P_0 is fairly close to the carrying capacity M_0. You might imagine a population of fish living in a freshwater lake in a wilderness area. Suddenly a catastrophe such as the Mount St. Helens volcanic eruption contaminates the lake and destroys a significant part of the food and oxygen on which the fish depend. The result is a new environment with a carrying capacity M_1 considerably less than M_0 and, in fact, less than the current population P_0. Starting at some time before the catastrophe, sketch a "before-and-after" curve that shows how the fish population responds to the change in environment.

14. *Controlling a population* The fish and game department in a certain state is planning to issue hunting permits to control the deer population (one deer per permit). It is known that if the deer population falls below a certain level m, the deer will become extinct. It is also known that if the deer population goes above the maximum carrying capacity M, the population will decrease back to M through disease and malnutrition.

(a) Discuss the reasonableness of the following model for the growth rate of the deer population as a function of time:

$$\frac{dP}{dt} = rP(M - P)(P - m),$$

where P is the population of the deer and r is a positive constant of proportionality. Include a phase line.

(b) Explain how this model differs from the logistic model $dP/dt = rP(M - P)$. Is it better or worse than the logistic model?

(c) Show that if $P > M$ for all t, then $\lim_{t\to\infty} P(t) = M$.

(d) What happens if $P < m$ for all t?

(e) Discuss the solutions to the differential equation. What are the equilibrium points of the model? Explain the dependence of the steady-state value of P on the initial values of P. About how many permits should be issued?

Applications and Examples

15. *Skydiving* If a body of mass m falling from rest under the action of gravity encounters an air resistance proportional to the square of velocity, then the body's velocity t seconds into the fall satisfies the equation

$$m\frac{dv}{dt} = mg - kv^2, \qquad k > 0$$

where k is a constant that depends on the body's aerodynamic properties and the density of the air. (We assume that the fall is too short to be affected by changes in the air's density.)

(a) Draw a phase line for the equation.

(b) Sketch a typical velocity curve.

(c) For a 160-lb skydiver ($mg = 160$) and with time in seconds and distance in feet, a typical value of k is 0.005. What is the diver's terminal velocity?

16. *Resistance proportional to $\sqrt{v}$* A body of mass m is projected vertically downward with initial velocity v_0. Assume that the resisting force is proportional to the square root of the velocity and find the terminal velocity from a graphical analysis.

17. *Sailing* A sailboat is running along a straight course with the wind providing a constant forward force of 50 lb. The only other force acting on the boat is resistance as the boat moves through the water. The resisting force is numerically equal to five times the boat's speed, and the initial velocity is 1 ft /sec. What is the maximum velocity in feet per second of the boat under this wind?

18. *The spread of information* Sociologists recognize a phenomenon called **social diffusion,** which is the spreading of a piece of information, technological innovation, or cultural fad among a population. The members of the population can be divided into two classes: those who have the information and those who do not. In a fixed population whose size is known, it is reasonable to assume that the rate of diffusion is proportional to the number who have the information times the number yet to receive it. If X denotes the number of individuals who have the information in a population of N people, then a mathematical model for social diffusion is given by

$$\frac{dX}{dt} = kX(N - X),$$

where t represents time and k is a positive constant.

(a) Discuss the reasonableness of the model.

(b) Construct a phase line identifying the sign of X' and X''.

(c) Sketch representative solution curves.

(d) Predict the value of X for which the information is spreading most rapidly. How many people eventually receive the information?

19. *Current in a circuit* The accompanying diagram represents an electrical circuit whose total resistance is a constant R ohms and whose self-inductance, shown as a coil, is L henries, also a constant. There is a switch whose terminals at a and b can be closed to connect a constant electrical source of V volts.

Ohm's law, $V = Ri$, has to be modified for such a circuit. The modified form is

$$L\frac{di}{dt} + Ri = V,$$

where i is the intensity of the current in amperes and t is the time in seconds. By solving this equation, we can predict how the current will flow after the switch is closed.

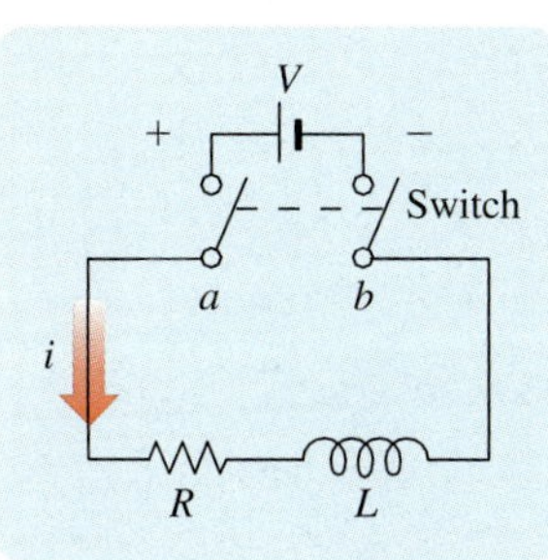

Use a phase line analysis to sketch the solution curve assuming that the switch in the RL circuit is closed at time $t = 0$. What happens to the current as $t \to \infty$? This value is called the **steady-state solution.**

20. *A pearl in shampoo* Suppose that a pearl is sinking in a thick fluid, like shampoo, subject to a frictional force opposing its fall and proportional to its velocity. Suppose that there is also a resistive buoyant force exerted by the shampoo. According to **Archimedes' principle,** the buoyant force equals the weight of the fluid displaced by the pearl. If m is the mass of the pearl and P is the mass of the shampoo displaced by the pearl as it descends, complete the following steps.

(a) Draw a schematic diagram showing the forces acting on the pearl as it sinks, as in Figure 3.35.

(b) Using $v(t)$ for the pearl's velocity as a function of time t, write a differential equation modeling the velocity of the pearl as a falling body.

(c) Construct a phase line identifying the signs of v' and v''.

(d) Sketch typical solution curves.

(e) What is the terminal velocity of the pearl?

3.5 Modeling and Optimization

Examples from Business and Industry • Examples from Mathematics and Physics • Fermat's Principle and Snell's Law • Examples from Economics • Modeling Discrete Phenomena with Differentiable Functions

To optimize something means to maximize or minimize some aspect of it. What is the size of the most profitable production run? What is the least expensive shape for an oil can? What is the stiffest beam we can cut from a 12 in. log? In the mathematical models in which we use functions to describe the things that interest us, we usually answer such questions by finding the greatest or smallest value of a differentiable function.

To illustrate, suppose that a cabinet maker uses exotic materials to produce custom furnishings. She has a contract to produce 5 units per day, which is also her capacity. For each of the raw materials she uses, she wants to determine *how much* and *how often* to have the material delivered. She is charged a delivery cost independent of the amount delivered, and she can rent as much storage as she likes. She reasons that if the delivery cost of a particular material is high and the storage cost low, she should deliver infrequently and store much. On the other hand, if the delivery cost is low and the storage cost high, she should deliver frequently and store little. But what is the minimum combined total cost of delivery and storage for each item? We answer this question later in this section.

Examples from Business and Industry

Example 1 Fabricating a Box

An open-top box is to be made by cutting small congruent squares from the corners of a 12-in.-by-12-in. sheet of tin and bending up the sides. How large should the squares cut from the corners be to make the box hold as much as possible?

Solution We start with a picture (Figure 3.43). In the figure, the corner squares are x in. on a side. The volume of the box is a function of this variable:

$$V(x) = x(12 - 2x)^2 = 144x - 48x^2 + 4x^3. \qquad V = hlw$$

Since the sides of the sheet of tin are only 12 in. long, $x \leq 6$ and the domain of V is the interval $0 \leq x \leq 6$.

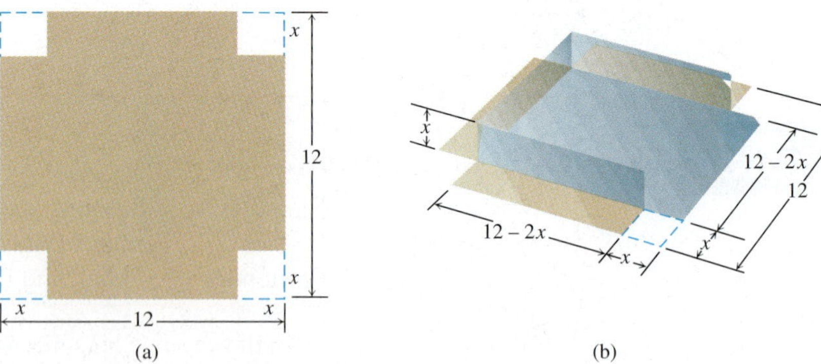

FIGURE 3.43 An open box made by cutting the corners from a square sheet of tin. (Example 1)

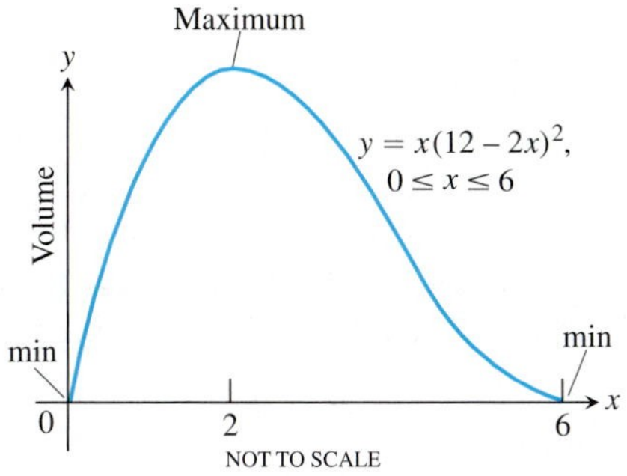

FIGURE 3.44 The volume of the box in Figure 3.43 graphed as a function of x.

A graph of V (Figure 3.44) suggests a minimum value of 0 at $x = 0$ and $x = 6$ and a maximum near $x = 2$. To learn more, we examine the first derivative of V with respect to x:

$$\frac{dV}{dx} = 144 - 96x + 12x^2 = 12(12 - 8x + x^2) = 12(2 - x)(6 - x).$$

Of the two zeros, $x = 2$ and $x = 6$, only $x = 2$ lies in the interior of the function's domain and makes the critical-point list. The values of V at this one critical point and two endpoints are

Critical-point value: $V(2) = 128$

Endpoint values: $V(0) = 0, \quad V(6) = 0.$

The maximum volume is 128 in^3. The cutout squares should be 2 in. on a side.

Example 2 Designing an Efficient Oil Can

You have been asked to design a 1 L oil can shaped like a right circular cylinder (Figure 3.45). What dimensions will use the least material?

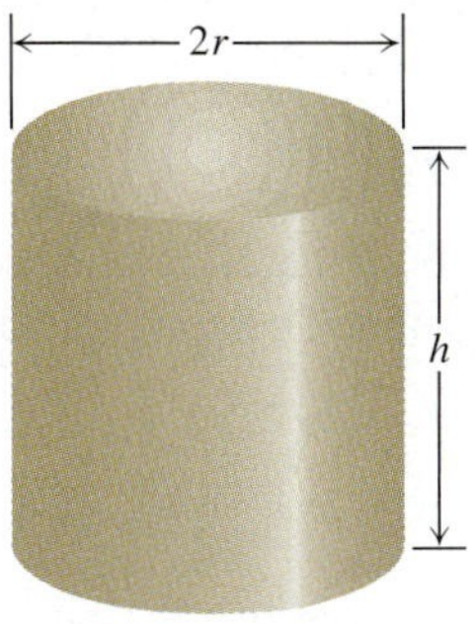

FIGURE 3.45 This 1 L can uses the least material when $h = 2r$. (Example 2)

Solution

Volume of can: If r and h are measured in centimeters, then the volume of the can in cubic centimeters is

$$\pi r^2 h = 1000. \qquad \text{1 liter} = 1000 \text{ cm}^3$$

Surface area of can: $A = \underbrace{2\pi r^2}_{\text{circular ends}} + \underbrace{2\pi rh}_{\text{cylinder wall}}$

How can we interpret the phrase "least material"? One possibility is to ignore the thickness of the material and the waste in manufacturing. Then we ask for dimensions r and h that make the total surface area as small as possible while satisfying the constraint $\pi r^2 h = 1000$. (Exercise 15 describes one way to take waste into account.)

Model

To express the surface area as a function of one variable, we solve for one of the variables in $\pi r^2 h = 1000$ and substitute that expression into the surface area formula. Solving for h is easier:

$$h = \frac{1000}{\pi r^2}.$$

Thus,

$$\begin{aligned} A &= 2\pi r^2 + 2\pi rh \\ &= 2\pi r^2 + 2\pi r\left(\frac{1000}{\pi r^2}\right) \\ &= 2\pi r^2 + \frac{2000}{r}. \end{aligned}$$

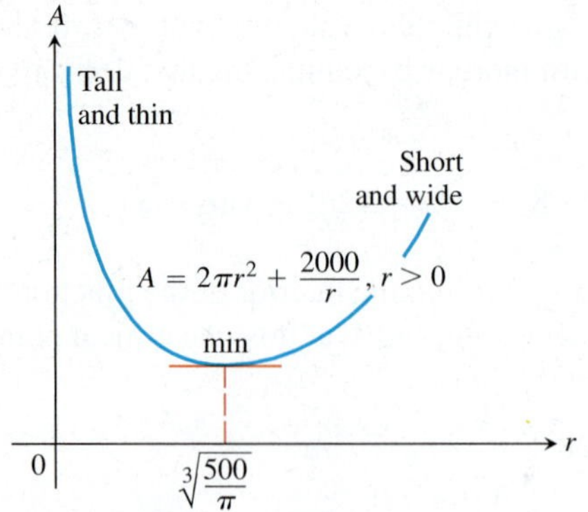

FIGURE 3.46 The graph of $A = 2\pi r^2 + 2000/r$ is concave up.

Solve Analytically

Our goal is to find a value of $r > 0$ that minimizes the value of A. Figure 3.46 suggests that such a value exists.

Notice from the graph that for small r (a tall thin container, like a piece of pipe), the term $2000/r$ dominates and A is large. For large r (a short wide container, like a pizza pan), the term $2\pi r^2$ dominates and A again is large. Since A is differentiable on $r > 0$, an interval with no endpoints, it can have a minimum value only where its first derivative is zero.

$$\frac{dA}{dr} = 4\pi r - \frac{2000}{r^2}$$

$$0 = 4\pi r - \frac{2000}{r^2} \qquad \text{Set } dA/dr = 0.$$

$$4\pi r^3 = 2000 \qquad \text{Multiply by } r^2.$$

$$r = \sqrt[3]{\frac{500}{\pi}} \approx 5.42 \qquad \text{Solve for } r.$$

Something happens at $r = \sqrt[3]{500/\pi}$, but what?

If the domain of A were a closed interval, we could find out by evaluating A at this critical point and the endpoints and comparing the results. But the domain is an open interval, so we must learn what is happening at $r = \sqrt[3]{500/\pi}$ by referring to the shape of A's graph. The second derivative

$$\frac{d^2A}{dr^2} = 4\pi + \frac{4000}{r^3}$$

is positive throughout the domain of A. The graph is therefore concave up and the value of A at $r = \sqrt[3]{500/\pi}$ an absolute minimum. The corresponding value of h (after a little algebra) is

$$h = \frac{1000}{\pi r^2} = 2\sqrt[3]{\frac{500}{\pi}} = 2r.$$

Interpret

The 1 L can that uses the least material has height equal to the diameter, with $r \approx 5.42$ cm and $h \approx 10.84$ cm.

Strategy for Solving Max-Min Problems

Step 1: *Understand the Problem* Read the problem carefully. Identify the information you need to solve the problem. What is unknown? What is given? What is sought?

Step 2:. *Develop a Mathematical Model of the Problem* Draw pictures and label the parts that are important to the problem. Introduce a variable to represent the quantity to be maximized or minimized. Using the variable, write a function whose extreme value gives the information sought.

Step 3: *Find the Domain of the Function* Determine what values of the variable make sense in the problem. Graph the function if possible.

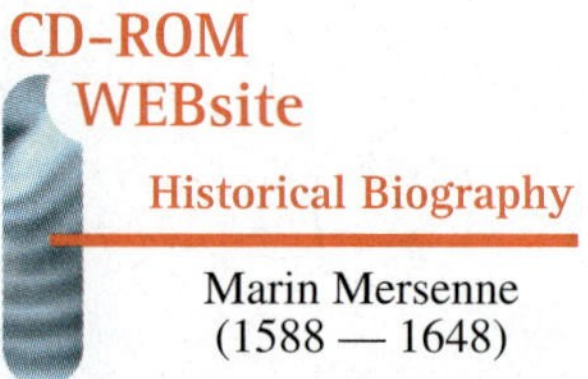

Step 4: *Identify the Critical Points and Endpoints* Find where the derivative is zero or fails to exist. Use what you know about the shape of the function's graph and the physics of the problem. Use the first and second derivatives to identify and classify critical points (where $f' = 0$ or does not exist).

Step 5: *Solve the Mathematical Model* If unsure of the result, support or confirm your solution with another method.

Step 6: *Interpret the Solution* Translate your mathematical result into the problem setting and decide whether the result makes sense.

Examples from Mathematics and Physics

Example 3 Inscribing Rectangles

A rectangle is to be inscribed in a semicircle of radius 2. What is the largest area the rectangle can have, and what are its dimensions?

Solution

Model

Let $(x, \sqrt{4 - x^2})$ be the coordinates of the corner of the rectangle obtained by placing the circle and rectangle in the coordinate plane (Figure 3.47). The length, height, and area of the rectangle can then be expressed in terms of the position x of the lower right-hand corner:

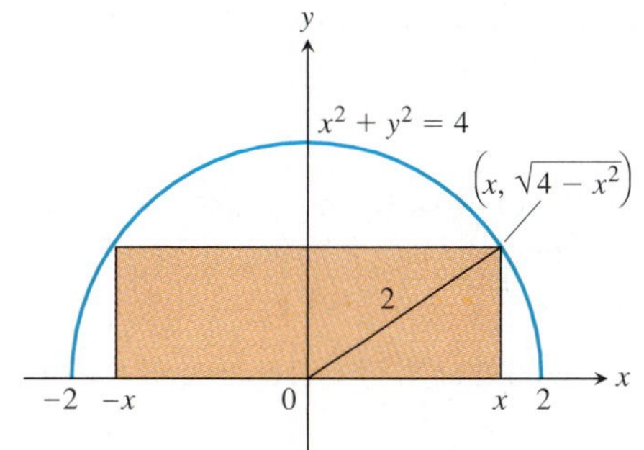

FIGURE 3.47 The rectangle and semicircle in Example 3.

$$\text{Length: } 2x, \qquad \text{height: } \sqrt{4 - x^2}, \qquad \text{area: } 2x \cdot \sqrt{4 - x^2}.$$

Notice that the values of x are to be found in the interval $0 \le x \le 2$, where the selected corner of the rectangle lies.

Our mathematical goal is now to find the absolute maximum value of the continuous function

$$A(x) = 2x\sqrt{4 - x^2}$$

on the domain $[0, 2]$.

Identify the Critical and Endpoints

The derivative

$$\frac{dA}{dx} = \frac{-2x^2}{\sqrt{4 - x^2}} + 2\sqrt{4 - x^2}$$

is not defined when $x = 2$ and is equal to zero when

$$\frac{-2x^2}{\sqrt{4 - x^2}} + 2\sqrt{4 - x^2} = 0$$

$$-2x^2 + 2(4 - x^2) = 0 \qquad \text{Multiply both sides by } \sqrt{4 - x^2}.$$

$$8 - 4x^2 = 0$$

$$x^2 = 2$$

$$x = \pm\sqrt{2}$$

Of the two zeros, $x = \sqrt{2}$ and $x = -\sqrt{2}$, only $x = \sqrt{2}$ lies in the interior of A's domain and makes the critical-point list. The values of A at the endpoints and at this one critical point are

Critical-point value: $\quad A(\sqrt{2}) = 2\sqrt{2}\sqrt{4-2} = 4$

Endpoint values: $\quad A(0) = 0, \qquad A(2) = 0.$

Interpret

The area has a maximum value of 4 when the rectangle is $\sqrt{4 - x^2} = \sqrt{2}$ units high and $2x = 2\sqrt{2}$ units long.

Fermat's Principle and Snell's Law

The speed of light depends on the medium through which it travels and tends to be slower in denser media. In a vacuum, it travels at the speed $c = 3 \times 10^8$ m/sec, but in Earth's atmosphere it travels slightly slower than that, and in glass slower still (about two-thirds as fast).

Fermat's principle in optics states that light always travels from one point to another along the quickest route (shortest time). This observation enables us to predict the path light will take when it travels from a point in one medium (air, say) to a point in another medium (say, glass or water).

Example 4 Finding the Path of a Light Ray

Find the path that a ray of light will follow in going from a point A in a medium where the speed of light is c_1 across a straight boundary to a point B in a medium where the speed of light is c_2.

Solution Since light traveling from A to B will do so by the quickest route, we look for a path that will minimize the travel time.

Model

We assume that A and B lie in the xy-plane and that the line separating the two media is the x-axis (Figure 3.48).

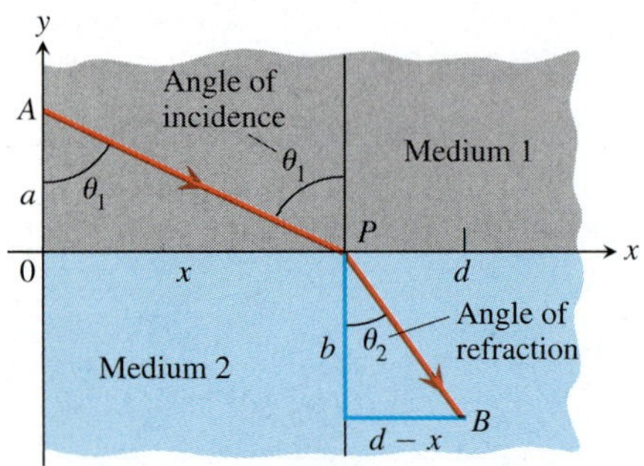

FIGURE 3.48 A light ray refracted (deflected from its path) as it passes from one medium to another. (Example 4)

In a uniform medium, where the speed of light remains constant, "shortest time" means "shortest path," and the ray of light will follow a straight line. Hence, the path from A to B will consist of a line segment from A to a boundary point P, followed by another line segment from P to B. From the formula distance equals rate times time, we have

$$\text{Time} = \frac{\text{distance}}{\text{rate}}.$$

The time required for light to travel from A to P is therefore

$$t_1 = \frac{AP}{c_1} = \frac{\sqrt{a^2 + x^2}}{c_1}.$$

From P to B, the time is

$$t_2 = \frac{PB}{c_2} = \frac{\sqrt{b^2 + (d-x)^2}}{c_2}.$$

The time from A to B is the sum of these:

$$t = t_1 + t_2 = \frac{\sqrt{a^2 + x^2}}{c_1} + \frac{\sqrt{b^2 + (d-x)^2}}{c_2}.$$

This equation expresses t as a differentiable function of x whose domain is $[0, d]$, and we want to find the absolute minimum value of t on this closed interval.

Identify the Critical and Endpoints

We find

$$\frac{dt}{dx} = \frac{x}{c_1\sqrt{a^2 + x^2}} - \frac{(d - x)}{c_2\sqrt{b^2 + (d - x)^2}}. \tag{1}$$

In terms of the angles θ_1 and θ_2 in Figure 3.48,

$$\frac{dt}{dx} = \frac{\sin\theta_1}{c_1} - \frac{\sin\theta_2}{c_2}.$$

dt/dx negative — dt/dx zero — dt/dx positive

– – – – – + + + + + + + + +

0 — x_0 — d — x

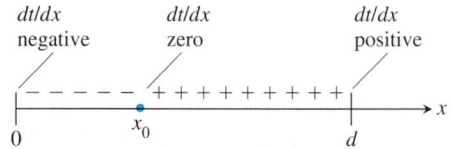

FIGURE 3.49 The sign pattern of dt/dx in Example 4.

We can see from Equation (1) that $dt/dx < 0$ at $x = 0$ and $dt/dx > 0$ at $x = d$. Hence, $dt/dx = 0$ at some point x_0 in between (Figure 3.49). There is only one such point because dt/dx is an increasing function of x (Exercise 58). At this point,

$$\frac{\sin\theta_1}{c_1} = \frac{\sin\theta_2}{c_2}.$$

This equation is **Snell's Law** or the **Law of Refraction.**

Interpret

We conclude that the path the ray of light follows is the one described by Snell's Law. Figure 3.50 shows how this works for air and water.

FIGURE 3.50 For air and water at room temperature, the light velocity ratio is 1.33, and Snell's Law becomes $\sin\theta_1 = 1.33\sin\theta_2$. In this laboratory photograph, $\theta_1 = 35.5°$, $\theta_2 = 26°$ and $(\sin 35.5°/\sin 26°) \approx 0.581/0.438 \approx 1.33$, as predicted.

This photograph also illustrates that angle of reflection = angle of incidence. (Exercise 42)

Examples from Economics

Here we want to point out two more places where calculus makes a contribution to economic theory. The first has to do with maximizing profit. The second has to do with minimizing average cost.

Suppose that

$$\begin{aligned} r(x) &= \text{the revenue from selling } x \text{ items} \\ c(x) &= \text{the cost of producing the } x \text{ items} \\ p(x) = r(x) - c(x) &= \text{the profit from selling } x \text{ items.} \end{aligned}$$

The **marginal revenue, marginal cost,** and **marginal profit** at this production level (x items) are

$$\frac{dr}{dx} = \text{marginal revenue}$$

$$\frac{dc}{dx} = \text{marginal cost}$$

$$\frac{dp}{dx} = \text{marginal profit.}$$

The first observation is about the relationship of p to these derivatives.

Theorem 6 Maximum Profit

At a production level yielding maximum profit, marginal revenue equals marginal cost.

Proof We assume that $r(x)$ and $c(x)$ are differentiable for all $x > 0$, so if $p(x) = r(x) - c(x)$ has a maximum value, it occurs at a production level at which $p'(x) = 0$. Since $p'(x) = r'(x) - c'(x)$, $p'(x) = 0$ implies that

$$r'(x) - c'(x) = 0 \qquad \text{or} \qquad r'(x) = c'(x).$$

Figure 3.51 gives more information about this situation.

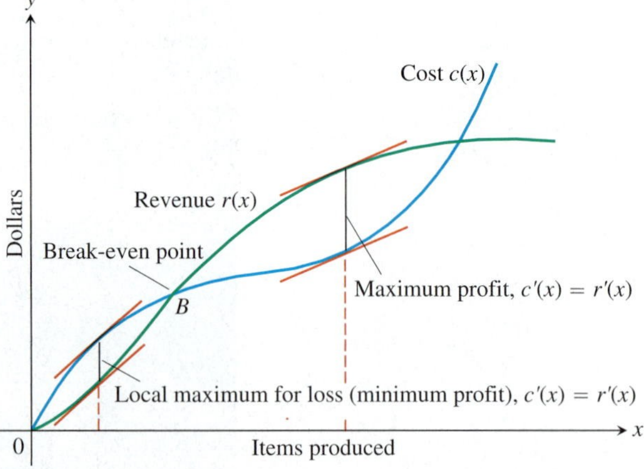

FIGURE 3.51 The graph of a typical cost function starts concave down and later turns concave up. It crosses the revenue curve at the break-even point B. To the left of B, the company operates at a loss. To the right, the company operates at a profit, with the maximum profit occurring where $c'(x) = r'(x)$. Farther to the right, cost exceeds revenue (perhaps because of a combination of rising labor and material costs and market saturation), and production levels become unprofitable again.

What guidance do we get from this observation? We know that a production level at which $p'(x) = 0$ need not be a level of maximum profit. It might be a level of minimum profit, for example. If we are making financial projections for our company, however, we should look for production levels at which marginal cost seems to equal marginal revenue. If there is a most profitable production level, it will be one of these.

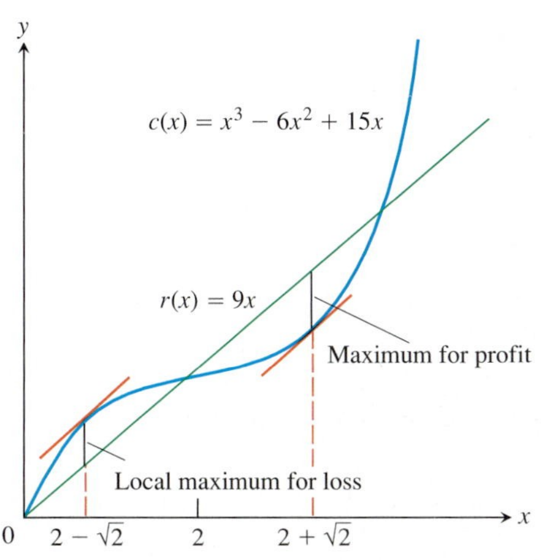

FIGURE 3.52 The cost and revenue curves for Example 5.

Example 5 Maximizing Profit

Suppose that $r(x) = 9x$ and $c(x) = x^3 - 6x^2 + 15x$, where x represents thousands of units. Is there a production level that maximizes profit? If so, what is it?

Solution Notice that $r'(x) = 9$ and $c'(x) = 3x^2 - 12x + 15$.

$$3x^2 - 12x + 15 = 9 \qquad \text{Set } c'(x) = r'(x).$$

$$3x^2 - 12x + 6 = 0$$

The two solutions of the quadratic equation are

$$x_1 = \frac{12 - \sqrt{72}}{6} = 2 - \sqrt{2} \approx 0.586 \qquad \text{and}$$

$$x_2 = \frac{12 + \sqrt{72}}{6} = 2 + \sqrt{2} \approx 3.414.$$

The possible production levels for maximum profit are $x \approx 0.586$ thousand units or $x \approx 3.414$ thousand units. The graphs in Figure 3.52 show that maximum profit occurs at about $x = 3.414$ (where revenue exceeds cost) and maximum loss occurs at about $x = 0.586$.

Example 6 Minimizing Costs

In the introduction to the chapter, we considered a cabinet maker using raw materials to produce 5 furnishings each day. Suppose that each delivery of a particular exotic material is \$5000, whereas the storage of that material is \$10 per day per unit stored, where a unit is the amount of material needed by her to produce one furnishing. How much material should be ordered each time and how often should the material be delivered to minimize her average daily cost in the production cycle between deliveries?

Solution

Model

If she asks for a delivery every x days, then she must order $5x$ units to have enough material for that delivery cycle. The *average* amount in storage is approximately one-half of the delivery amount, or $5x/2$. Thus, the cost of delivery and storage for each cycle is approximately

$$\text{Cost per cycle} = \text{delivery cost} + \text{storage costs}$$

$$\text{Cost per cycle} = \underbrace{5000}_{\substack{\text{delivery}\\ \text{cost}}} + \underbrace{\left(\frac{5x}{2}\right)}_{\substack{\text{average amount}\\ \text{stored}}} \cdot \underbrace{x}_{\substack{\text{number of}\\ \text{days stored}}} \cdot \underbrace{10}_{\substack{\text{storage cost}\\ \text{per day}}}$$

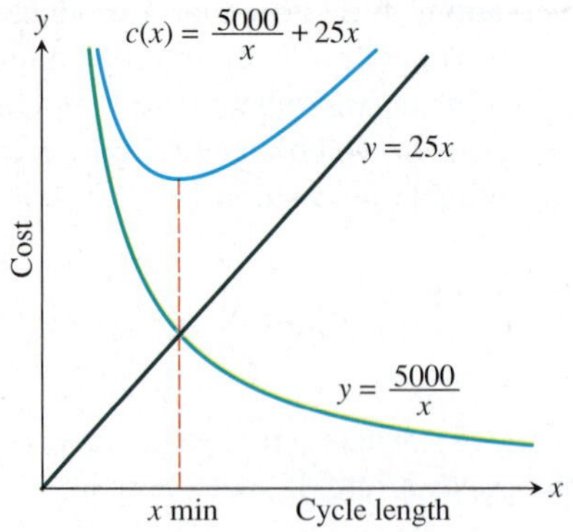

FIGURE 3.53 The average daily cost $c(x)$ is the sum of a hyperbola and a linear function. (Example 6)

We compute the *average daily cost* $c(x)$ by dividing the cost per cycle by the number of days x in the cycle (see Figure 3.53)

$$c(x) = \frac{5000}{x} + 25x, \quad x > 0.$$

As $x \to 0$ and as $x \to \infty$, the average daily cost becomes large. So we expect a minimum to exist, but where? Our goal is to determine the number of days x between deliveries that provides the absolute minimum cost.

Identify the Critical Points

We find the critical points by determining where the derivative is equal to zero:

$$c'(x) = -\frac{5000}{x^2} + 25 = 0$$
$$x = \pm\sqrt{200} \approx \pm 14.14.$$

Of the two critical points, only $\sqrt{200}$ lies in the domain of $c(x)$. The critical-point value of the average daily cost is

$$c(\sqrt{200}) = \frac{5000}{\sqrt{200}} + 25\sqrt{200} = 500\sqrt{2} \approx \$707.11.$$

We note that $c(x)$ is defined over the open interval $(0, \infty)$ with $c''(x) = 10000/x^3 > 0$. Thus, an absolute minimum exists at $x = \sqrt{200} \approx 14.14$ days.

Interpret

The cabinet maker should schedule a delivery of $5(14) = 70$ units of the exotic wood every 14 days.

Modeling Discrete Phenomena with Differentiable Functions

In case you are wondering how we can use differentiable functions $c(x)$ and $r(x)$ to describe the cost and revenue that comes from producing a number of items x that can only be an integer, here is the rationale.

When x is large, we can reasonably fit the cost and revenue data with smooth curves $c(x)$ and $r(x)$ that are defined not only at integer values of x but at the values in between. Once we have these differentiable functions, which are supposed to behave like the real cost and revenue when x is an integer, we can apply calculus to draw conclusions about their values. We then translate these mathematical conclusions into inferences about the real world that we hope will have predictive value. When they do, as is the case with the economic theory here, we say that the functions give a good model of reality.

What do we do when our calculus tells us that the best solution is a value of x that isn't an integer? In Example 6, if the number of days between deliveries must be an integer and certain materials come in lot sizes, we must round our answers. Should we round up or round down? That is, in the vicinity of the optimal time between deliveries, how *sensitive* is the change in cost to an increase or decrease in the time between deliveries?

Example 7 Sensitivity of the Minimum Cost

Should we round the number of days between deliveries up or down for the best solution in Example 6?

Solution

The average daily cost will increase by about \$0.03 if we round down from 14.14 to 14 days:

$$c(14) = \frac{5000}{14} + 25(14) = \$707.14$$

and

$$c(14) - c(14.14) = \$707.14 - \$707.11 = \$0.03.$$

On the other hand, $c(15) = \$708.33$, and our cost would increase by $\$708.33 - \$707.11 = \$1.22$ if we round up. Thus, it is better that we round x down to 14 days. In Exercise 51, you are asked to derive a general formula to obtain the optimal time between deliveries given the delivery and storage costs for any material.

EXERCISES 3.5

Whenever you are maximizing or minimizing a function of a single variable, we urge you to graph it over the domain that is appropriate to the problem you are solving. The graph will provide insight before you calculate and will furnish a visual context for understanding your answer.

Applications in Geometry

1. *Minimizing perimeter* What is the smallest perimeter possible for a rectangle whose area is 16 in.^{2}, and what are its dimensions?

2. *Finding area* Show that among all rectangles with an 8-m perimeter, the one with largest area is a square.

3. *Inscribing rectangles* The figure at right shows a rectangle inscribed in an isosceles right triangle whose hypotenuse is 2 units long.

   (a) Express the y-coordinate of P in terms of x. (*Hint:* Write an equation for the line AB.)

   (b) Express the area of the rectangle in terms of x.

   (c) What is the largest area the rectangle can have, and what are its dimensions?

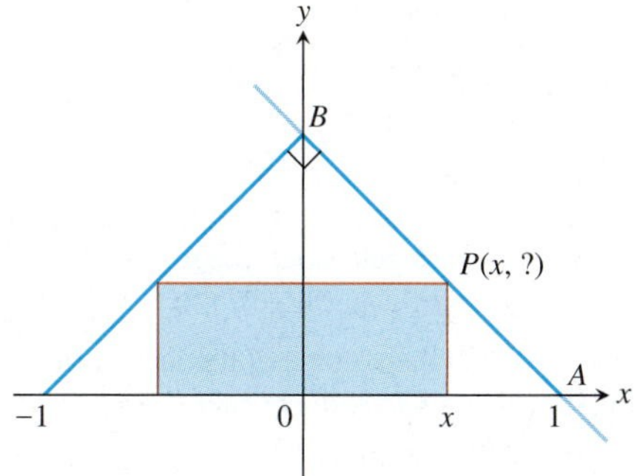

4. *Largest rectangle* A rectangle has its base on the x-axis and its upper two vertices on the parabola $y = 12 - x^2$. What is the largest area the rectangle can have, and what are its dimensions?

5. *Optimal dimensions* You are planning to make an open rectangular box from an 8-in.-by-15-in. piece of cardboard by cutting congruent squares from the corners and folding up the sides. What are the dimensions of the box of largest volume you can make this way, and what is its volume?

6. *Closing off the first quadrant* You are planning to close off a corner of the first quadrant with a line segment 20 units long running from $(a, 0)$ to $(0, b)$. Show that the area of the triangle enclosed by the segment is largest when $a = b$.

7. *The best fencing plan* A rectangular plot of farmland will be bounded on one side by a river and on the other three sides by a single-strand electric fence. With 800 m of wire at your disposal, what is the largest area you can enclose, and what are its dimensions?

8. *The shortest fence* A 216 m^2 rectangular pea patch is to be enclosed by a fence and divided into two equal parts by another fence parallel to one of the sides. What dimensions for the outer rectangle will require the smallest total length of fence? How much fence will be needed?

9. *Designing a tank* Your iron works has contracted to design and build a 500 ft^3, square-based, open-top, rectangular steel holding tank for a paper company. The tank is to be made by welding thin stainless steel plates together along their edges. As the production engineer, your job is to find dimensions for the base and height that will make the tank weigh as little as possible.

(a) What dimensions do you tell the shop to use?

(b) *Writing to Learn* Briefly describe how you took weight into account.

10. *Catching rainwater* A 1125 ft^3 open-top rectangular tank with a square base x ft on a side and y ft deep is to be built with its top flush with the ground to catch runoff water. The costs associated with the tank involve not only the material from which the tank is made but also an excavation charge proportional to the product xy.

(a) If the total cost is

$$c = 5(x^2 + 4xy) + 10xy,$$

what values of x and y will minimize it?

(b) *Writing to Learn* Give a possible scenario for the cost function in part (a).

11. *Designing a poster* You are designing a rectangular poster to contain 50 in.^{2} of printing with a 4-in. margin at the top and bottom and a 2-in. margin at each side. What overall dimensions will minimize the amount of paper used?

12. *Inscribing a cone* Find the volume of the largest right circular cone that can be inscribed in a sphere of radius 3.

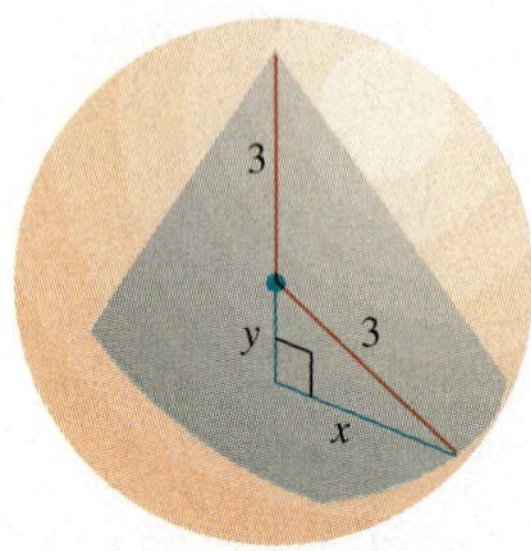

13. *Finding an angle* Two sides of a triangle have lengths a and b, and the angle between them is θ. What value of θ will maximize the triangle's area? (*Hint:* $A = (1/2)ab \sin \theta$.)

14. *Designing a can* What are the dimensions of the lightest open-top right circular cylindrical can that will hold a volume of 1000 cm^3? Compare the result here with the result in Example 2.

15. *Designing a can* You are designing a 1000 cm^3 right circular cylindrical can whose manufacture will take waste into account. There is no waste in cutting the aluminum for the side, but the top and bottom of radius r will be cut from squares that measure $2r$ units on a side. The total amount of aluminum used up by the can will therefore be

$$A = 8r^2 + 2\pi rh$$

rather than the $A = 2\pi r^2 + 2\pi rh$ in Example 2. In Example 2, the ratio of h to r for the most economical can was 2 to 1. What is the ratio now?

16. *Designing a box with a lid* A piece of cardboard measures 10 in. by 15 in. Two equal squares are removed from the corners of a 10 in. side as shown in the figure. Two equal rectangles are removed from the other corners so that the tabs can be folded to form a rectangular box with lid.

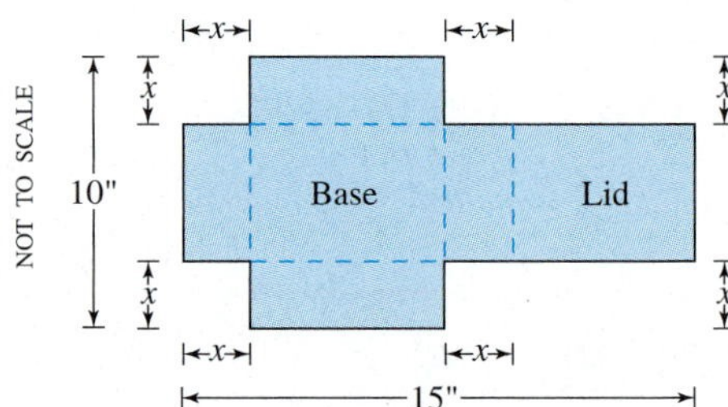

(a) Write a formula $V(x)$ for the volume of the box.

(b) Find the domain of V for the problem situation and graph V over this domain.

(c) Use a graphical method to find the maximum volume and the value of x that gives it.

(d) Confirm your result in part (c) analytically.

T **17.** *Designing a suitcase* A 24-in.-by-36-in. sheet of cardboard is folded in half to form a 24-in-by-18-in. rectangle as shown in the accompanying figure. Then four congruent squares of side length x are cut from the corners of the folded rectangle. The sheet is unfolded, and the six tabs are folded up to form a box with sides and a lid.

(a) Write a formula $V(x)$ for the volume of the box.

(b) Find the domain of V for the problem situation and graph V over this domain.

(c) Use a graphical method to find the maximum volume and the value of x that gives it.

(d) Confirm your result in (c) analytically.

(e) Find a value of x that yields a volume of 1120 in.^{3}.

(f) *Writing to Learn* Write a paragraph describing the issues that arise in part (b).

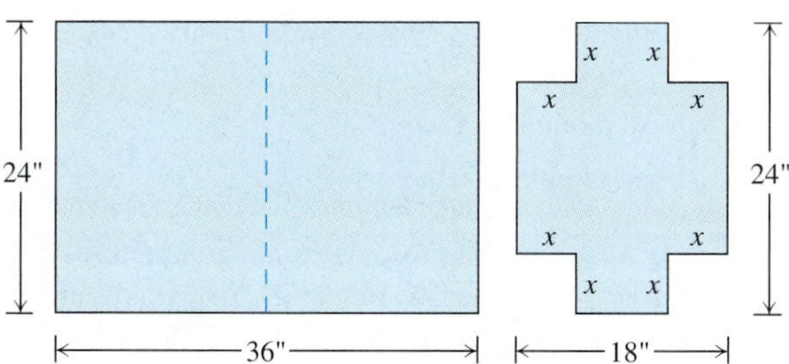

The sheet is then unfolded.

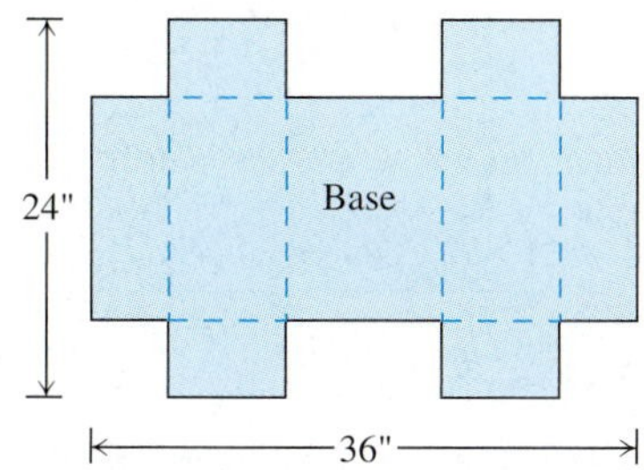

18. *Inscribing rectangles* A rectangle is to be inscribed under the arch of the curve $y = 4 \cos (0.5x)$ from $x = -\pi$ to $x = \pi$. What are the dimensions of the rectangle with largest area, and what is the largest area?

19. *Maximizing volume* Find the dimensions of a right circular cylinder of maximum volume that can be inscribed in a sphere of radius 10 cm. What is the maximum volume?

20. (a) The U.S. Postal Service will accept a box for domestic shipment only if the sum of its length and girth (distance around) does not exceed 108-in. What dimensions will give a box with a square end the largest possible volume?

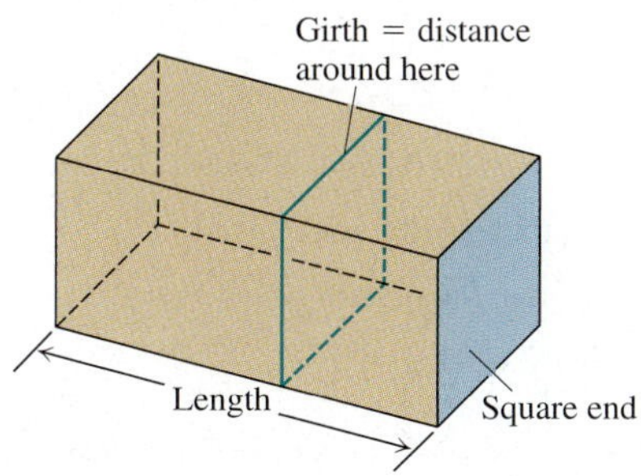

T **(b)** Graph the volume of a 108-in. box (length plus girth equals 108 in.) as a function of its length and compare what you see with your answer in part (a).

21. *(Continuation of Exercise 20.)*

(a) Suppose that instead of having a box with square ends you have a box with square sides so that its dimensions are h by h by w and the girth is $2h + 2w$. What dimensions will give the box its largest volume now?

T **(b)** Graph the volume as a function of h and compare what you see with your answer in part (a).

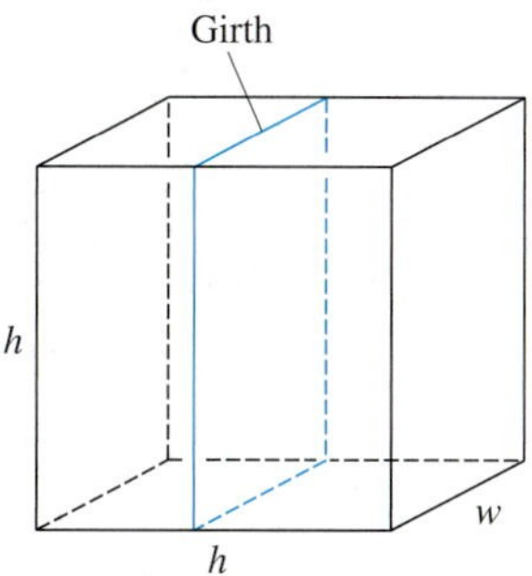

22. *Designing a window* A window is in the form of a rectangle surmounted by a semicircle. The rectangle is of clear glass, whereas the semicircle is of tinted glass that transmits only half as much light per unit area as clear glass does. The total perimeter is fixed. Find the proportions of the window that will admit the most light. Neglect the thickness of the frame.

23. *Constructing a silo* A silo (base not included) is to be constructed in the form of a cylinder surmounted by a hemisphere. The cost of construction per square unit of surface area is twice as great for the hemisphere as it is for the cylindrical sidewall. Determine the dimensions to be used if the volume is fixed and the cost of construction is to be kept to a minimum. Neglect the thickness of the silo and waste in construction.

24. *Finding an angle* The trough in the figure is to be made to the dimensions shown. Only the angle θ can be varied. What value of θ will maximize the trough's volume?

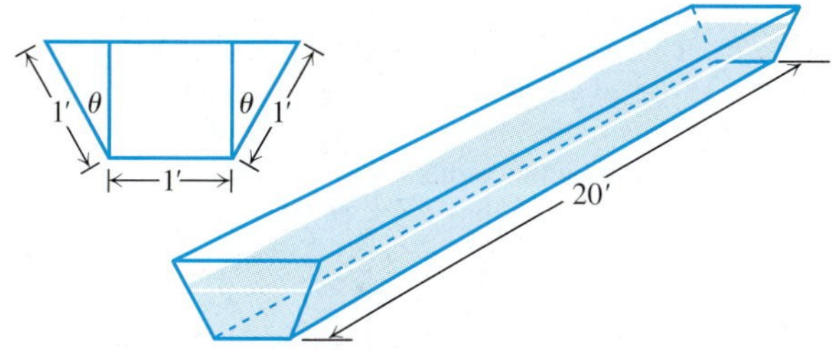

25. *Paper folding (Work in groups of two or three.)* A rectangular sheet of 8.5-in.-by-11-in. paper is placed on a flat surface. One of the corners is placed on the opposite longer edge, as shown in the figure, and held there as the paper is smoothed flat. The problem is to make the length of the crease as small as possible. Call the length L. Try it with paper.

(a) Show that $L^2 = 2x^3/(2x - 8.5)$.

(b) What value of x minimizes L^2?

(c) What is the minimum value of L?

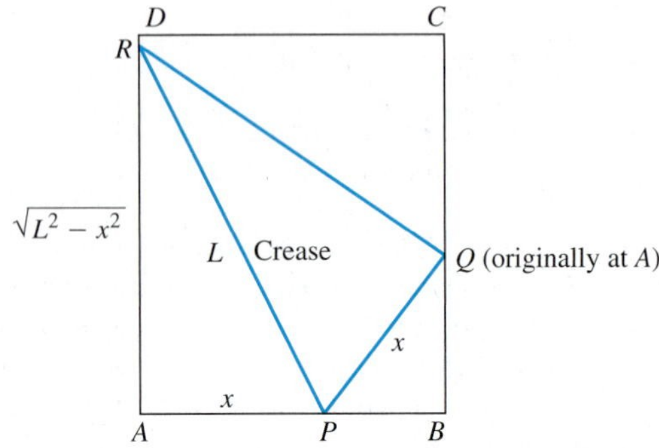

26. *Constructing cylinders* Compare the answers to the following two construction problems.

(a) A rectangular sheet of perimeter 36 cm and dimensions x cm by y cm is to be rolled into a cylinder as shown in part (a) of the figure. What values of x and y give the largest volume?

(b) The same sheet is to be revolved about one of the sides of length y to sweep out the cylinder as shown in part (b) of the figure. What values of x and y give the largest volume?

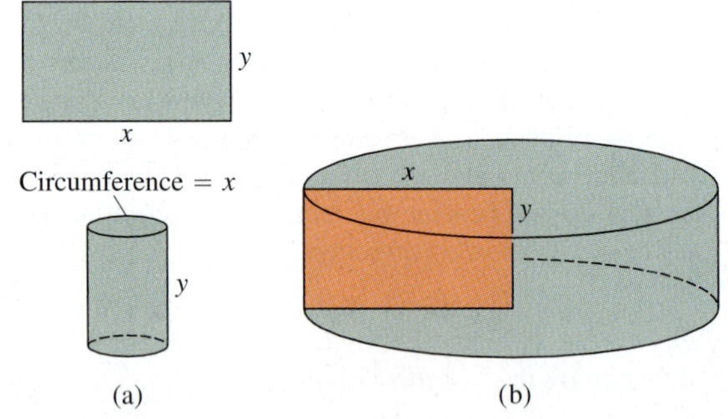

27. *Constructing cones* A right triangle whose hypotenuse is $\sqrt{3}$ m long is revolved about one of its legs to generate a right circular cone. Find the radius, height, and volume of the cone of greatest volume that can be made this way.

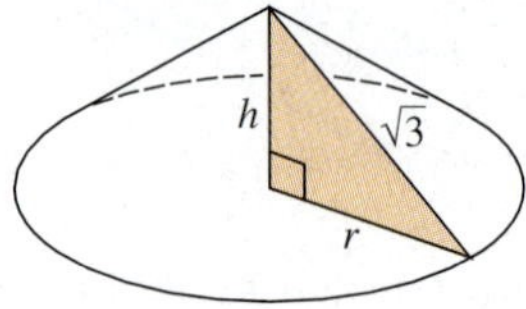

28. *Finding parameter values* What value of a makes $f(x) = x^2 + (a/x)$ have

(a) a local minimum at $x = 2$?

(b) a point of inflection at $x = 1$?

29. *Inscribing rectangles* The rectangle shown here has one side on the positive y-axis, one side on the positive x-axis, and its upper right-hand vertex on the curve $y = e^{-x^2}$. What dimensions give the rectangle its largest area, and what is that area?

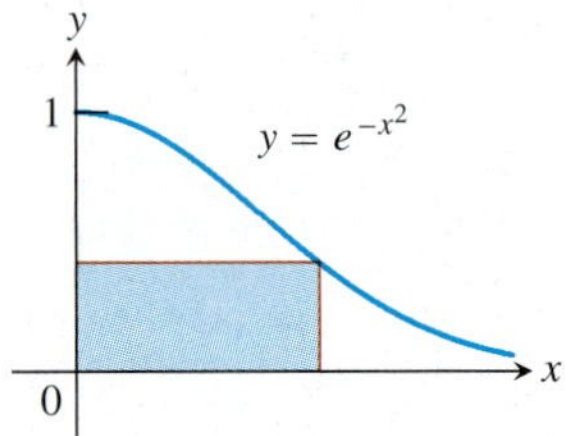

30. *Inscribing rectangles* The rectangle shown here has one side on the positive y-axis, one side on the positive x-axis, and its upper right-hand vertex on the curve $y = (\ln x)/x^2$. What dimensions give the rectangle its largest area, and what is that area?

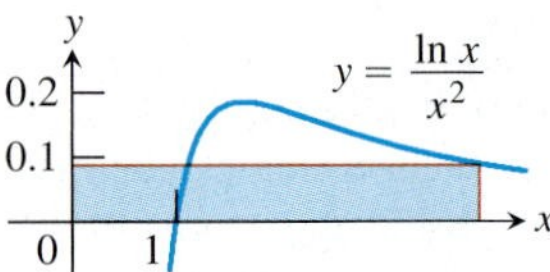

Physical Applications

31. *Vertical motion* The height of an object moving vertically is given by

$$s = -16t^2 + 96t + 112,$$

with s in feet and t in seconds. Find

(a) the object's velocity when $t = 0$

(b) its maximum height and when it occurs

(c) its velocity when $s = 0$.

32. *Quickest route* Jane is 2 mi offshore in a boat and wishes to reach a coastal village 6 mi down a straight shoreline from the point nearest the boat. She can row 2 mph and can walk 5 mph. Where should she land her boat to reach the village in the least amount of time?

33. *Shortest beam* The 8 ft wall shown in the accompanying figure stands 27 ft from the building. Find the length of the shortest straight beam that will reach to the side of the building from the ground outside the wall.

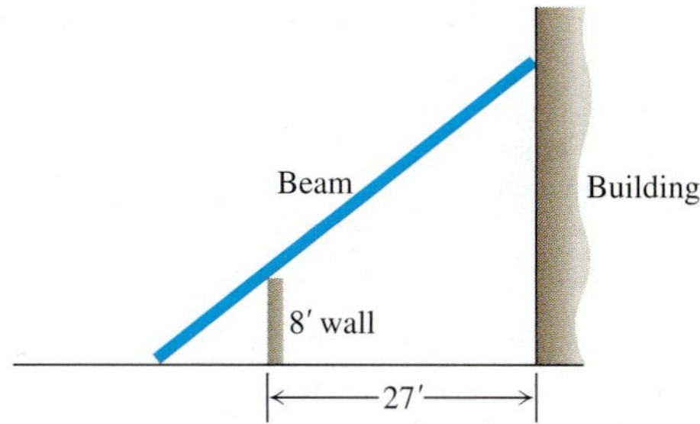

T **34.** *Strength of a beam* The strength S of a rectangular wooden beam is proportional to its width times the square of its depth. (See the accompanying figure.)

CD-ROM WEBsite

(a) Find the dimensions of the strongest beam that can be cut from a 12-in.-diameter cylindrical log.

(b) *Writing to Learn* Graph S as a function of the beam's width w, assuming the proportionality constant to be $k = 1$. Reconcile what you see with your answer in part (a).

(c) *Writing to Learn* On the same screen, graph S as a function of the beam's depth d, again taking $k = 1$. Compare the graphs with one another and with your answer in part (a). What would be the effect of changing to some other value of k? Try it.

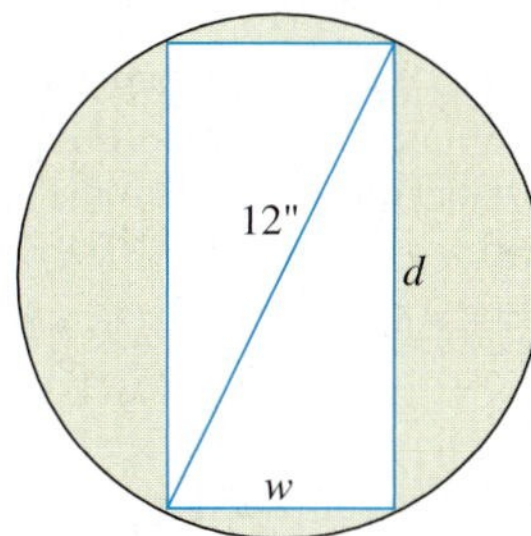

T **35.** *Stiffness of a beam* The stiffness S of a rectangular beam is proportional to its width times the cube of its depth.

CD-ROM WEBsite

(a) Find the dimensions of the stiffest beam that can be cut from a 12-in.-diameter cylindrical log.

(b) *Writing to Learn* Graph S as a function of the beam's width w, assuming the proportionality constant to be $k = 1$. Reconcile what you see with your answer in part (a).

(c) *Writing to Learn* On the same screen, graph S as a function of the beam's depth d, again taking $k = 1$. Compare the graphs with one another and with your answer in part (a). What would be the effect of changing to some other value of k? Try it.

36. *Locating a solar station* You are under contract to build a solar station at ground level on the east–west line between the two buildings shown here. How far from the taller building should you place the station to maximize the number of hours it will be in the sun on a day when the sun passes directly overhead? Begin by observing that

$$\theta = \pi - \cot^{-1}\frac{x}{60} - \cot^{-1}\frac{50 - x}{30}.$$

Then find the value of x that maximizes θ.

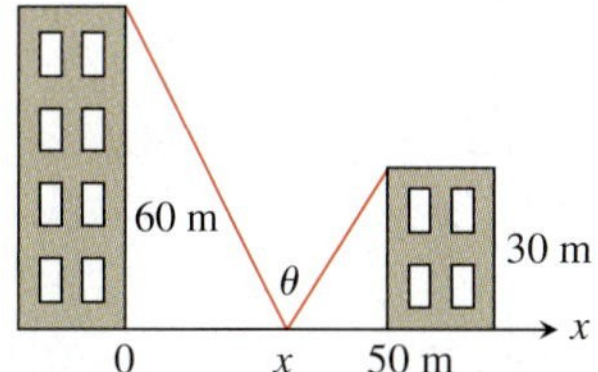

37. *Motion on a line* The positions of two particles on the s-axis are $s_1 = \sin t$ and $s_2 = \sin(t + \pi/3)$, with s_1 and s_2 in meters and t in seconds.

(a) At what time(s) in the interval $0 \le t \le 2\pi$ do the particles meet?

(b) What is the farthest apart that the particles ever get?

(c) When in the interval $0 \le t \le 2\pi$ is the distance between the particles changing the fastest?

38. *Electrical current* Suppose that at any time t(seconds) the current i (amp) in an alternating current circuit is $i = 2\cos t + 2\sin t$. What is the peak (largest magnitude) current for this circuit?

39. *Frictionless cart* A small frictionless cart, attached to the wall by a spring, is pulled 10 cm from its rest position and released at time $t = 0$ to roll back and forth for 4 sec. Its position at time t is $s = 10\cos \pi t$.

(a) What is the cart's maximum speed? When is the cart moving that fast? Where is it then? What is the magnitude of the acceleration then?

(b) Where is the cart when the magnitude of the acceleration is greatest? What is the cart's speed then?

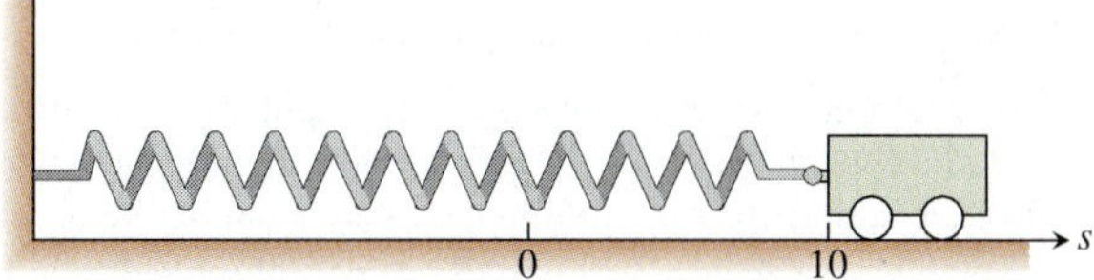

40. *Side-by-side hanging masses* Two masses hanging side by side from springs have positions $s_1 = 2\sin t$ and $s_2 = \sin 2t$, respectively.

(a) At what times in the interval $0 < t$ do the masses pass each other? (*Hint:* $\sin 2t = 2\sin t\cos t$.)

(b) When in the interval $0 \le t \le 2\pi$ is the vertical distance between the masses the greatest? What is this distance? (*Hint:* $\cos 2t = 2\cos^2 t - 1$.)

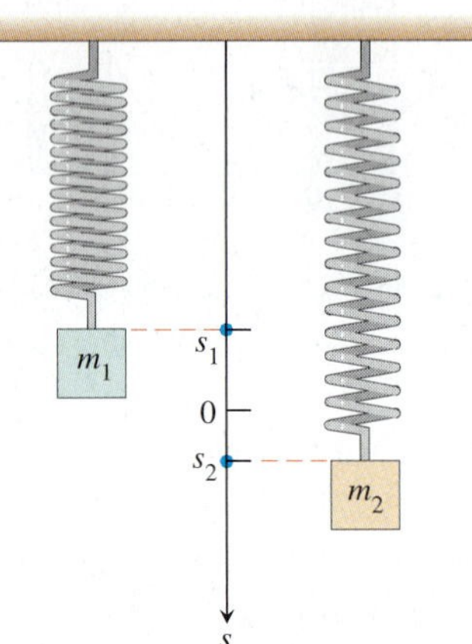

41. *Distance between two ships* At noon, ship A was 12 nautical miles due north of a ship B. Ship A was sailing south at 12 knots (nautical miles per hour; a nautical mile is 2000 yd) and continued to do so all day. Ship B was sailing east at 8 knots and continued to do so all day.

(a) Start counting time with $t = 0$ at noon and express the distance s between the ships as a function of t.

(b) How rapidly was the distance between the ships changing at noon? One hour later?

(c) The visibility that day was 5 nautical miles. Did the ships ever sight each other?

T **(d)** Graph s and ds/dt together as functions of t for $-1 \le t \le 3$, using different colors if possible. Compare the graphs and reconcile what you see with your answers in (b) and (c).

(e) The graph of ds/dt looks as if it might have a horizontal asymptote in the first quadrant. This in turn suggests that ds/dt approaches a limiting value as $t \to \infty$. What is this value? What is its relation to the ships' individual speeds?

42. *Fermat's principle in optics* Fermat's principle in optics states that light always travels from one point to another along a path that minimizes the travel time. Light from a source A is reflected by a plane mirror to a receiver at point B, as shown in the figure. Show that for the light to obey Fermat's principle, the angle of incidence must equal the angle of reflection, both measured from the line normal to the reflecting surface. (This result can also be derived without calculus. There is a purely geometric argument, which you may prefer.)

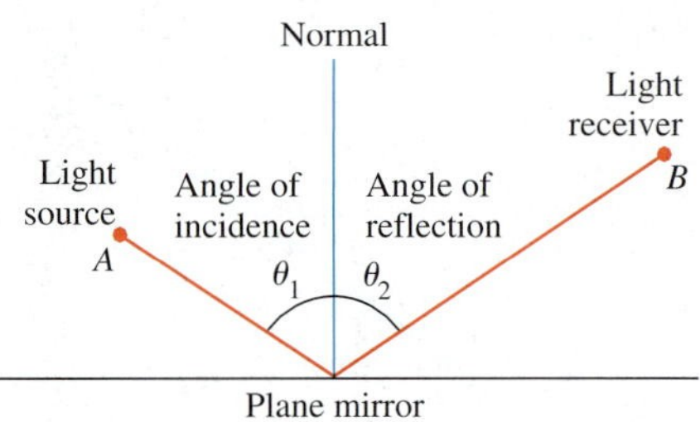

43. *Tin pest* When metallic tin is kept below 13.2°C, it slowly becomes brittle and crumbles to a gray powder. Tin objects eventually crumble to this gray powder spontaneously if kept in a cold climate for years. The Europeans who saw tin organ pipes in their churches crumble away years ago called the change *tin pest* because it seemed to be contagious, and indeed it was, for the gray powder is a catalyst for its own formation.

A *catalyst* for a chemical reaction is a substance that controls the rate of reaction without undergoing any permanent change in itself. An *autocatalytic reaction* is one whose product is a catalyst for its own formation. Such a reaction may proceed slowly at first if the amount of catalyst present is small and slowly again at the end, when most of the original substance is used up. But in between, when both the substance and its catalyst product are abundant, the reaction proceeds at a faster pace.

In some cases, it is reasonable to assume that the rate $v = dx/dt$ of the reaction is proportional both to the amount of the original substance present and to the amount of product. That is, v may be considered to be a function of x alone, and

$$v = kx(a - x) = kax - kx^2,$$

where

$x =$ the amount of product
$a =$ the amount of substance at the beginning
$k =$ a positive constant.

At what value of x does the rate v have a maximum? What is the maximum value of v?

44. *Airplane landing path* An airplane is flying at altitude H when it begins its descent to an airport runway that is at horizontal ground distance L from the airplane, as shown in the figure. Assume that the landing path of the airplane is the graph of a cubic polynomial function $y = ax^3 + bx^2 + cx + d$, where $y(-L) = H$ and $y(0) = 0$.

(a) What is dy/dx at $x = 0$?

(b) What is dy/dx at $x = -L$?

(c) Use the values for dy/dx at $x = 0$ and $x = -L$ together with $y(0) = 0$ and $y(-L) = H$ to show that

$$y(x) = H\left[2\left(\frac{x}{L}\right)^3 + 3\left(\frac{x}{L}\right)^2\right].$$

Landing path

H = cruising altitude

Airport

L

Business and Economics

45. *Selling backpacks* It costs you c dollars each to manufacture and distribute backpacks. If the backpacks sell at x dollars each, the number sold is given by

$$n = \frac{a}{x - c} + b(100 - x),$$

where a and b are certain positive constants. What selling price will bring a maximum profit?

46. *Tour service* You operate a tour service that offers the following rates:

$200 per person if 50 people (the minimum number to book the tour) go on the tour

For each additional person, up to a maximum of 80 people total, the rate per person is reduced by $2.

It costs $6000 (a fixed cost) plus $32 per person to conduct the tour. How many people does it take to maximize your profit?

47. *Wilson lot size formula* One of the formulas for inventory management says that the average weekly cost of ordering, paying for, and holding merchandise is

$$A(q) = \frac{km}{q} + cm + \frac{hq}{2},$$

where q is the quantity you order when things run low (shoes, radios, brooms, or whatever the item might be), k is the cost of placing an order (the same, no matter how often you order), c is the cost of one item (a constant), m is the number of items sold each week (a constant), and h is the weekly holding cost per item (a constant that takes into account things such as space, utilities, insurance, and security).

(a) Your job, as the inventory manager for your store, is to find the quantity that will minimize $A(q)$. What is it? (The formula you get for the answer is called the *Wilson lot size formula.*)

(b) Shipping costs sometimes depend on order size. When they do, it is more realistic to replace k by $k + bq$, the sum of k and a constant multiple of q. What is the most economical quantity to order now?

48. *Production level* Prove that the production level (if any) at which average cost is smallest is a level at which the average cost equals marginal cost.

49. *Production level* Show that if $r(x) = 6x$ and $c(x) = x^3 - 6x^2 + 15x$ are your revenue and cost functions, then the best you can do is break even (have revenue equal cost).

50. *Production level* Suppose that $c(x) = x^3 - 20x^2 + 20{,}000x$ is the cost of manufacturing x items. Find a production level that will minimize the average cost of making x items.

51. *Average daily cost* In Example 6, assume for any material that a cost of d is incurred per delivery, the storage cost is s per unit stored per day, and the production rate is p units per day.

(a) How much should be delivered every x days?

(b) Show that

$$\text{cost per cycle} = d + \frac{px}{2}sx.$$

(c) Find the time between deliveries x^* and the amount to deliver that minimizes the *average daily cost* of delivery and storage.

(d) Show that x^* occurs at the intersection of the hyperbola $y = d/x$ and the line $y = psx/2$.

52. *Minimizing average cost* Suppose that $c(x) = 2000 + 96x + 4x^{3/2}$ where x represents thousands of units. Is there a production level that minimizes average cost? If so, what is it?

Medicine

53. *Sensitivity to medicine (Continuation of Example 7, Section 2.3)* Find the amount of medicine to which the body is most sensitive by finding the value of M that maximizes the derivative dR/dM, where

$$R = M^2\left(\frac{C}{2} - \frac{M}{3}\right)$$

and C is a constant.

54. *How we cough*

(a) When we cough, the trachea (windpipe) contracts to increase the velocity of the air going out. This raises the questions of how much it should contract to maximize the velocity and whether it really contracts that much when we cough.

Under reasonable assumptions about the elasticity of the tracheal wall and about how the air near the wall is slowed by friction, the average flow velocity v can be modeled by the equation

$$v = c(r_0 - r)r^2 \text{ cm/sec}, \qquad \frac{r_0}{2} \le r \le r_0,$$

where r_0 is the rest radius of the trachea in centimeters and c is a positive constant whose value depends in part on the length of the trachea.

Show that v is greatest when $r = (2/3)r_0$, that is, when the trachea is about 33% contracted. The remarkable fact is that X-ray photographs confirm that the trachea contracts about this much during a cough.

T **(b)** Take r_0 to be 0.5 and c to be 1 and graph v over the interval $0 \le r \le 0.5$. Compare what you see with the claim that v is at a maximum when $r = (2/3)r_0$.

Theory and Examples

55. *An inequality for positive integers* Show that if a, b, c, and d are positive integers, then

$$\frac{(a^2+1)(b^2+1)(c^2+1)(d^2+1)}{abcd} \ge 16.$$

56. *The derivative dt / dx in Example 4*

(a) Show that

$$f(x) = \frac{x}{\sqrt{a^2+x^2}}$$

is an increasing function of x.

(b) Show that

$$g(x) = \frac{d-x}{\sqrt{b^2+(d-x)^2}}$$

is a decreasing function of x.

(c) Show that

$$\frac{dt}{dx} = \frac{x}{c_1\sqrt{a^2+x^2}} - \frac{d-x}{c_2\sqrt{b^2+(d-x)^2}}$$

is an increasing function of x.

57. *Writing to Learn* Let $f(x)$ and $g(x)$ be the differentiable functions graphed here. Point c is the point where the vertical distance between the curves is the greatest. Is there anything special about the tangents to the two curves at c? Give reasons for your answer.

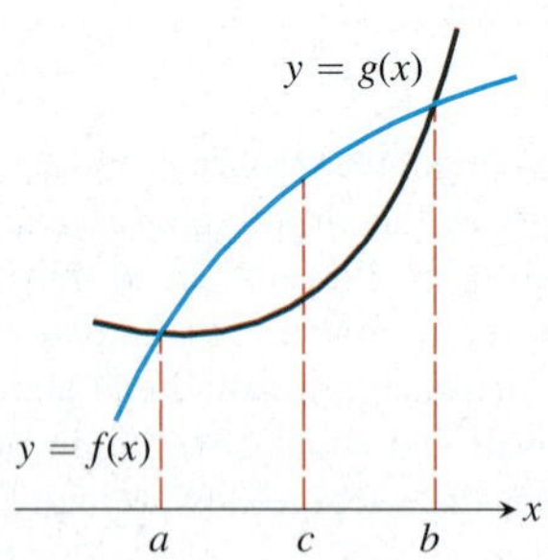

58. *Writing to Learn* You have been asked to determine whether the function $f(x) = 3 + 4\cos x + \cos 2x$ is ever negative.

(a) Explain why you need consider values of x only in the interval $[0, 2\pi]$.

(b) Is f ever negative? Explain.

59. *Absolute maximum*

(a) The function $y = \cot x - \sqrt{2}\csc x$ has an absolute maximum value on the interval $0 < x < \pi$. Find it.

T **(b)** Graph the function and compare what you see with your answer in part (a).

60. *Absolute minimum*

(a) The function $y = \tan x + 3\cot x$ has an absolute minimum value on the interval $0 < x < \pi/2$. Find it.

T **(b)** Graph the function and compare what you see with your answer in part (a).

61. *Calculus and geometry*

(a) How close does the curve $y = \sqrt{x}$ come to the point $(3/2, 0)$? (*Hint:* If you minimize the *square* of the distance, you can avoid square roots.)

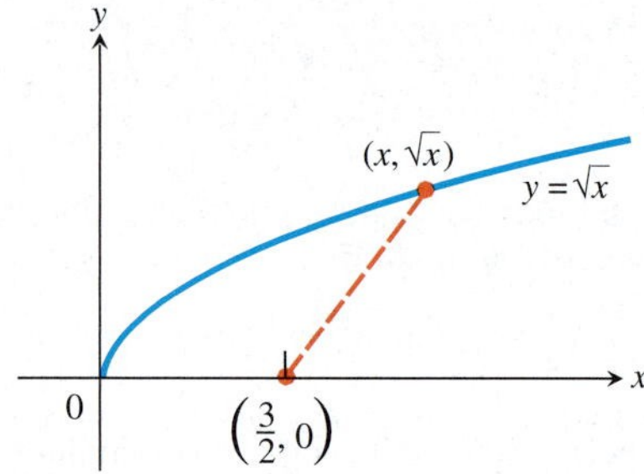

T **(b)** Graph the distance function and $y = \sqrt{x}$ together and reconcile what you see with your answer in part (a).

62. *Calculus and geometry*

(a) How close does the semicircle $y = \sqrt{16 - x^2}$ come to the point $(1, \sqrt{3})$?

T **(b)** Graph the distance function and $y = \sqrt{16 - x^2}$ together and reconcile what you see with your answer in part (a).

COMPUTER EXPLORATIONS

In Exercises 63 and 64, you may find it helpful to use a CAS.

63. *Generalized cone problem* A cone of height h and radius r is constructed from a flat, circular disk of radius a in. by removing a sector AOC of arc length x in. and then connecting the edges OA and OC.

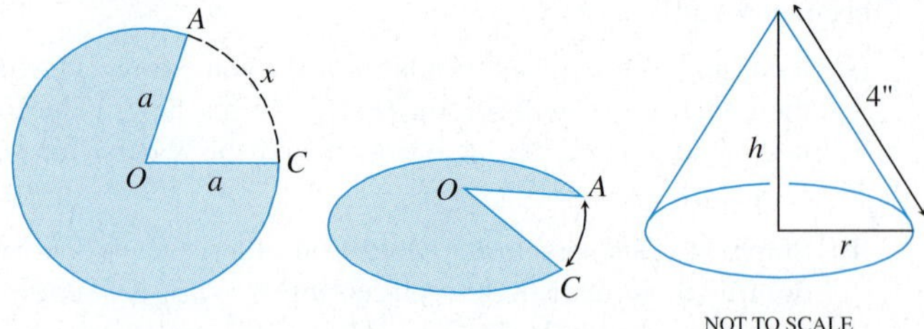

(a) Find a formula for the volume V of the cone in terms of x and a.

(b) Find r and h in the cone of maximum volume for $a = 4, 5, 6, 8$.

(c) *Writing to Learn* Find a simple relationship between r and h that is independent of a for the cone of maximum volume. Explain how you arrived at your relationship.

64. *Circumscribing an ellipse* Let $P(x, a)$ and $Q(-x, a)$ be two points on the upper half of the ellipse

$$\frac{x^2}{100} + \frac{(y-5)^2}{25} = 1$$

centered at (0, 5). A triangle RST is formed by using the tangent lines to the ellipse at Q and P as shown in the figure.

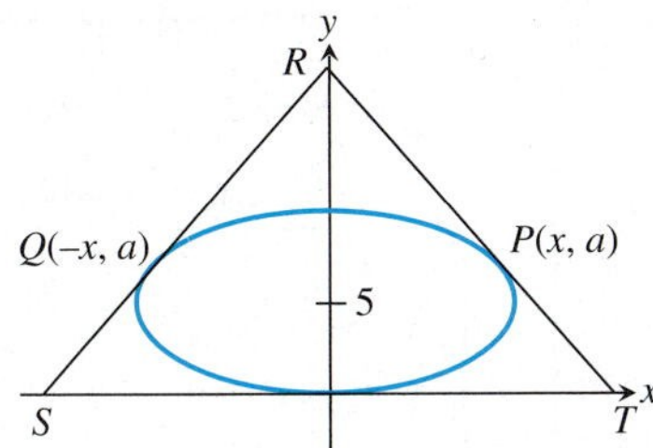

(a) Show that the area of the triangle is

$$A(x) = -f'(x)\left[x - \frac{f(x)}{f'(x)}\right]^2,$$

where $y = f(x)$ is the function representing the upper half of the ellipse.

(b) What is the domain of A? Draw the graph of A. How are the asymptotes of the graph related to the problem situation?

(c) Determine the height of the triangle with minimum area. How is it related to the y-coordinate of the center of the ellipse?

(d) Repeat parts (a) through (c) for the ellipse

$$\frac{x^2}{C^2} + \frac{(y-B)^2}{B^2} = 1$$

centered at $(0, B)$. Show that the triangle has minimum area when its height is $3B$.

3.6 Linearization and Differentials

Linearization • Differentials • Estimating Change with Differentials • Absolute, Relative, and Percentage Change • Sensitivity to Change • Error in Differential Approximation • Converting Mass to Energy

Sometimes we can approximate complicated functions with simpler ones that give the accuracy we want for specific applications and are easier to work with. The approximating functions discussed in this section are called *linearizations*, and they are based on tangent lines. Other approximating functions are discussed in Chapter 8.

We introduce new variables dx and dy and define them in a way that gives new meaning to the Leibniz notation dy/dx. We use dy to estimate error in measurement and sensitivity of a function to change.

Linearization

As you can see in Figure 3.54, the tangent to the curve $y = x^2$ lies close to the curve near the point of tangency. For a brief interval to either side, the y-values along the tangent line give a good approximation to the y-values on the curve. We observe this phenomenon by zooming in on the two graphs at the point of tangency, or by looking at tables of values for the difference between $f(x)$ and its tangent line near the x-coordinate of the point of tangency. Locally, every curve behaves like a straight line.

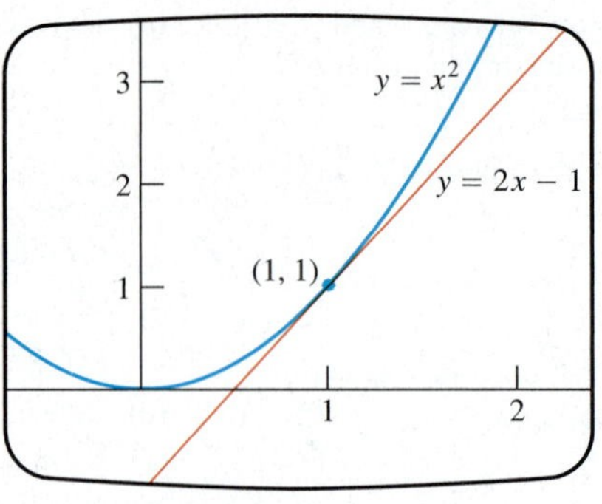

$y = x^2$ and its tangent $y = 2x - 1$ at (1, 1).

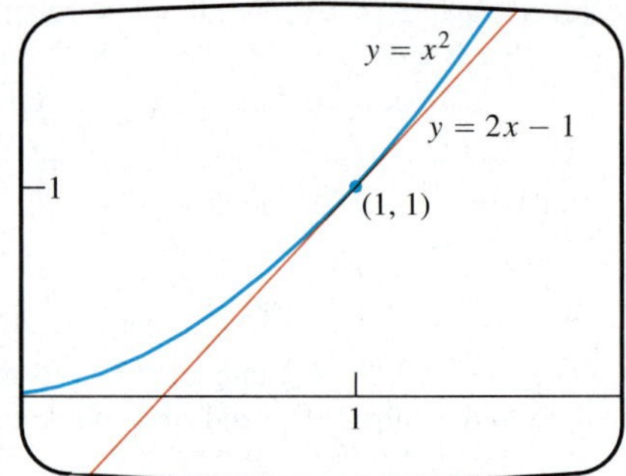

Tangent and curve very close near (1, 1).

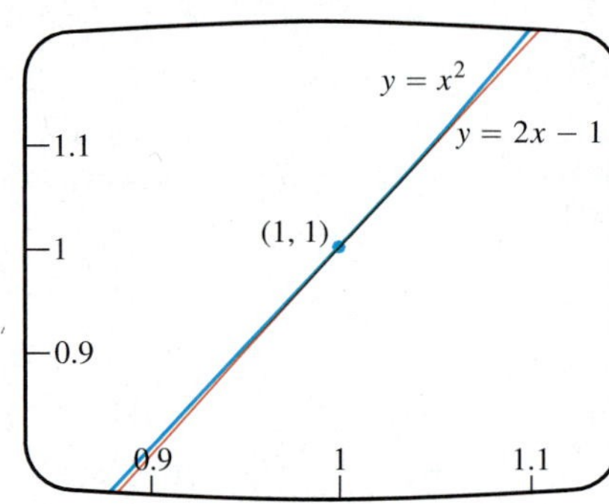

Tangent and curve very close throughout entire x-interval shown.

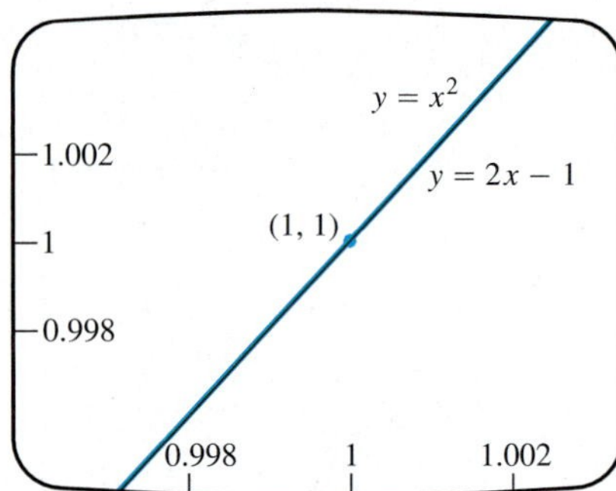

Tangent and curve closer still. Computer screen cannot distinguish tangent from curve on this x-interval.

FIGURE 3.54 The more we magnify the graph of a function near a point where the function is differentiable, the flatter the graph becomes and the more it resembles its tangent.

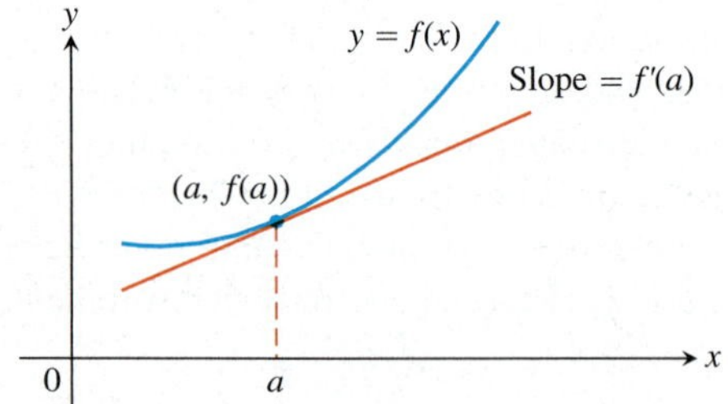

FIGURE 3.55 The tangent to the curve $y = f(x)$ at $x = a$ is the line $y = f(a) + f'(a)(x - a)$.

CD-ROM
WEBsite

In general, the tangent to $y = f(x)$ at a point $x = a$, where f is differentiable (Figure 3.55) passes through the point $(a, f(a))$, so its point-slope equation is

$$y = f(a) + f'(a)(x - a).$$

Thus, the tangent line is the graph of the linear function

$$L(x) = f(a) + f'(a)(x - a).$$

For as long as the line remains close to the graph of f, $L(x)$ gives a good approximation to $f(x)$.

Definition Linearization

If f is differentiable at $x = a$, then the approximating function

$$L(x) = f(a) + f'(a)(x - a) \tag{1}$$

is the **linearization** of f at a.

The approximation $f(x) \approx L(x)$ is the **standard linear approximation** of f at a. The point $x = a$ is the **center** of the approximation.

Example 1 Finding a Linearization

Find the linearization of $f(x) = \sqrt{1 + x}$ at $x = 0$ (Figure 3.56).

Solution Since

$$f'(x) = \frac{1}{2}(1 + x)^{-1/2},$$

we have $f(0) = 1, f'(0) = 1/2$, and

$$L(x) = f(a) + f'(a)(x - a) = 1 + \frac{1}{2}(x - 0) = 1 + \frac{x}{2}.$$

See Figure 3.56.

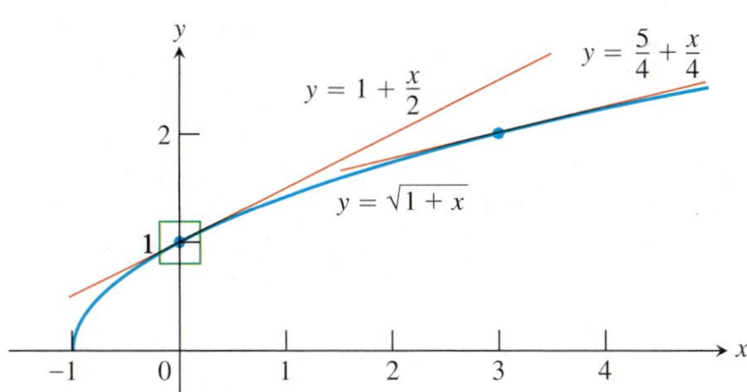

FIGURE 3.56 The graph of $y = \sqrt{1 + x}$ and its linearizations at $x = 0$ and $x = 3$. Figure 3.57 shows a magnified view of the small window about 1 on the y-axis.

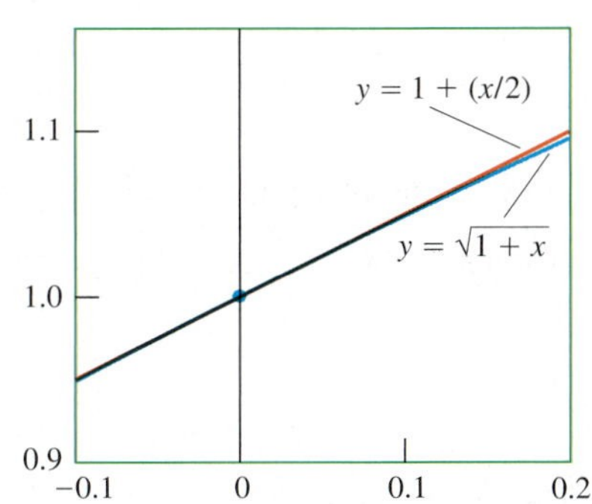

FIGURE 3.57 Magnified view of the window in Figure 3.56.

Look at how accurate the approximation $\sqrt{1 + x} \approx 1 + (x/2)$ is for values of x near 0.

Approximation	|True value − approximation|
$\sqrt{1.2} \approx 1 + \frac{0.2}{2} = 1.10$	$<10^{-2}$
$\sqrt{1.05} \approx 1 + \frac{0.05}{2} = 1.025$	$<10^{-3}$
$\sqrt{1.005} \approx 1 + \frac{0.005}{2} = 1.00250$	$<10^{-5}$

As we move away from zero, we lose accuracy. For example, for $x = 2$, the linearization gives 2 as the approximation for $\sqrt{3}$, which is not even accurate to one decimal place.

Do not be misled by the preceding calculations into thinking that whatever we do with a linearization is better done with a calculator. In practice, we would never use a linearization to find a particular square root. The utility of a linearization is its ability to replace a complicated formula by a simpler one over an entire interval of values. If we have to work with $\sqrt{1 + x}$ for x close to 0 and can tolerate the small amount of error involved, we can work with $1 + (x/2)$ instead. Of course, we then need to know how much error there is. We have a full story on error in Chapter 8.

A linear approximation normally loses accuracy away from its center. As Figure 3.56 suggests, the approximation $\sqrt{1+x} \approx 1 + (x/2)$ will probably be too crude to be useful near $x = 3$. There, we need the linearization at $x = 3$.

CD-ROM WEBsite

Historical Biography

François Viete (1540 — 1603)

Example 2 Finding a Second Linearization

Find the linearization of $f(x) = \sqrt{1+x}$ at $x = 3$.

Solution We evaluate Equation (1) for f at $a = 3$. With

$$f(3) = 2, \qquad f'(3) = \frac{1}{2}(1+x)^{-1/2}\bigg|_{x=3} = \frac{1}{4},$$

we have

$$L(x) = 2 + \frac{1}{4}(x-3) = \frac{5}{4} + \frac{x}{4}.$$

At $x = 3.2$, the linearization in Example 2 gives

$$\sqrt{1+x} = \sqrt{1+3.2} \approx \frac{5}{4} + \frac{3.2}{4} = 1.250 + 0.800 = 2.050,$$

which differs from the true value $\sqrt{4.2} \approx 2.04939$ by less than one one-thousandth. The linearization in Example 1 gives

$$\sqrt{1+x} = \sqrt{1+3.2} \approx 1 + \frac{3.2}{2} = 1 + 1.6 = 2.6,$$

a result that is off by more than 25%.

Example 3 Finding Roots and Powers

The most important linear approximation for roots and powers is

$$(1+x)^k \approx 1 + kx \qquad (x \text{ near } 0;\ \text{any number } k) \tag{2}$$

(Exercise 7). This approximation, good for values of x sufficiently close to zero, has broad application.

Common Linear Approximations, $x \approx 0$

$$\sin x \approx x$$
$$\cos x \approx 1$$
$$\tan x \approx x$$
$$e^x \approx 1 + x$$
$$\ln(1+x) \approx x$$
$$(1+x)^k \approx 1 + kx$$

(Exercises 6 and 7)

Example 4 Applying Example 3

The following approximations are consequences of Example 3.

$$\sqrt{1+x} \approx 1 + \frac{1}{2}x \qquad k = 1/2$$

$$\frac{1}{1-x} = (1-x)^{-1} \approx 1 + (-1)(-x) = 1 + x \qquad k = -1;\ \text{replace } x \text{ by } -x.$$

$$\sqrt[3]{1+5x^4} = (1+5x^4)^{1/3} \approx 1 + \frac{1}{3}(5x^4) = 1 + \frac{5}{3}x^4 \qquad k = 1/3;\ \text{replace } x \text{ by } 5x^4.$$

$$\frac{1}{\sqrt{1-x^2}} = (1-x^2)^{-1/2} \approx 1 + \left(-\frac{1}{2}\right)(-x^2) = 1 + \frac{1}{2}x^2 \qquad k = -1/2;\ \text{replace } x \text{ by } -x^2.$$

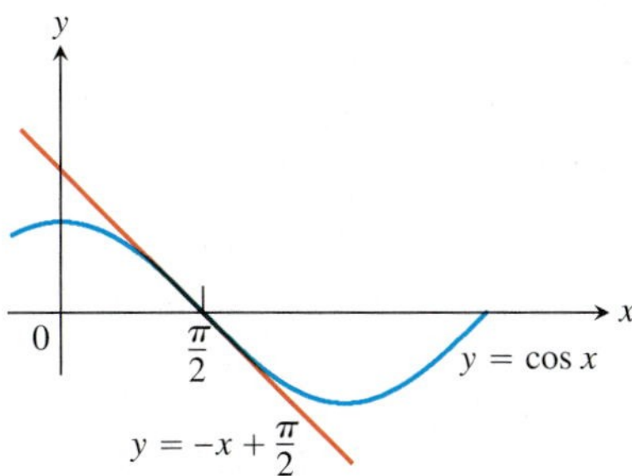

FIGURE 3.58 The graph of $f(x) = \cos x$ and its linearization at $x = \pi/2$. Near $x = \pi/2$, $\cos x \approx -x + (\pi/2)$. (Example 5)

Example 5 Finding a Linearization

Find the linearization of $f(x) = \cos x$ at $x = \pi/2$ (Figure 3.58).

Solution Since $f(\pi/2) = \cos(\pi/2) = 0$, $f'(x) = -\sin x$, and $f'(\pi/2) = -\sin(\pi/2) = -1$, we have

$$L(x) = f(a) + f'(a)(x - a)$$
$$= 0 + (-1)\left(x - \frac{\pi}{2}\right)$$
$$= -x + \frac{\pi}{2}.$$

Differentials

We sometimes use the notation dy/dx to represent the derivative y' of y with respect to x. Contrary to its appearance, it is not a ratio. We now introduce two new variables dx and dy with the property that if their ratio exists, it will be equal to the derivative.

Definition Differentials

Let $y = f(x)$ be a differentiable function. The **differential dx** is an independent variable. The **differential dy** is

$$dy = f'(x)\,dx.$$

Unlike the independent variable dx, the variable dy is always a dependent variable. It depends on both x and dx.

Example 6 Finding the Differential dy

Find dy if

(a) $y = x^5 + 37x$

(b) $y = \sin 3x$.

Solution

(a) $dy = (5x^4 + 37)\,dx$ **(b)** $dy = (3\cos 3x)\,dx$

The Meaning of dx and dy

In most contexts, the differential dx of the independent variable is its change Δx, but we do not impose this restriction on the definition.

Unlike the independent variable dx, the variable dy is always a dependent variable. It depends on both x and dx.

If $dx \neq 0$, then the quotient of the differential dy by the differential dx is equal to the derivative $f'(x)$ because

$$\frac{dy}{dx} = \frac{f'(x)\,dx}{dx} = f'(x).$$

We sometimes write

$$df = f'(x)\,dx$$

in place of $dy = f'(x)\,dx$, calling df the **differential of *f*.** For instance, if $f(x) = 3x^2 - 6$, then

$$df = d(3x^2 - 6) = 6x\,dx.$$

Every differentiation formula like

$$\frac{d(u+v)}{dx} = \frac{du}{dx} + \frac{dv}{dx} \quad \text{or} \quad \frac{d(\sin u)}{dx} = \cos u \frac{du}{dx}$$

has a corresponding differential form like

$$d(u+v) = du + dv \quad \text{or} \quad d(\sin u) = \cos u\,du.$$

Example 7 Finding Differentials of Functions

(a) $d(\tan 2x) = \sec^2(2x)\,d(2x) = 2\sec^2 2x\,dx$

(b) $d\left(\frac{x}{x+1}\right) = \frac{(x+1)\,dx - x\,d(x+1)}{(x+1)^2} = \frac{x\,dx + dx - x\,dx}{(x+1)^2} = \frac{dx}{(x+1)^2}$

Estimating Change with Differentials

Suppose that we know the value of a differentiable function $f(x)$ at a point a and we want to predict how much this value will change if we move to a nearby point $a + dx$. If dx is small, f and its linearization L at a will change by nearly the same amount (Figure 3.59). Since the values of L are simple to calculate, calculating the change in L offers a practical way to estimate the change in f.

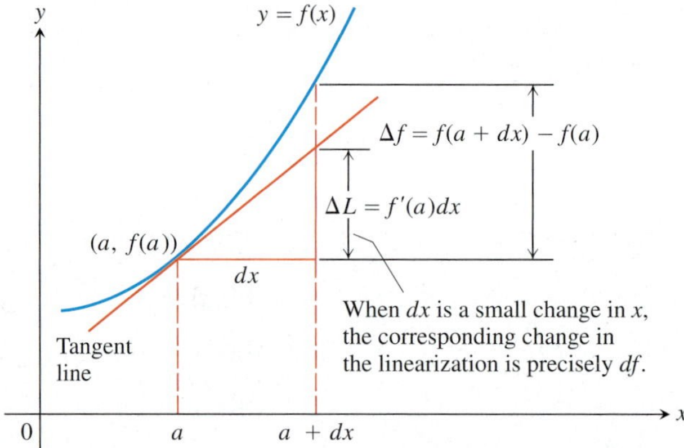

FIGURE 3.59 Approximating the change in the function f by the change in the linearization of f.

In the notation of Figure 3.59, the change in f is

$$\Delta f = f(a + dx) - f(a).$$

The corresponding change in L is

$$\begin{aligned}\Delta L &= L(a + dx) - L(a)\\ &= \underbrace{f(a) + f'(a)[(a + dx) - a]}_{L(a+dx)} - \underbrace{f(a)}_{L(a)}\\ &= f'(a)\,dx.\end{aligned}$$

Thus, the differential $df = f'(x)dx$ has a geometric interpretation: The value of df at $x = a$ is ΔL, the change in the linearization of f corresponding to the change dx.

Differential Estimate of Change

Let $f(x)$ be differentiable at $x = a$. The approximate change in the value of f when x changes from a to $a + dx$ is

$$df = f'(a)\,dx.$$

Example 8 Estimating Change with Differentials

The radius r of a circle increases from $a = 10$ m to 10.1 m (Figure 3.60). Use dA to estimate the increase in the circle's area A. Compare this estimate with the true change ΔA.

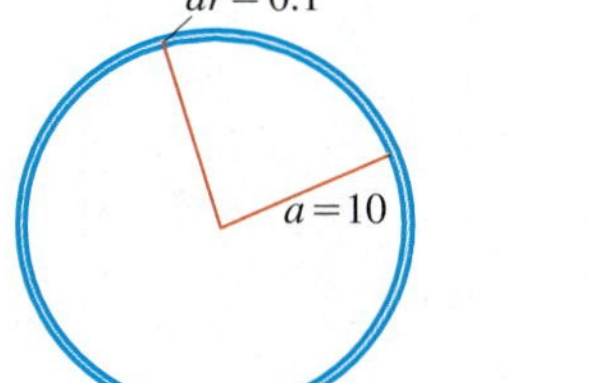

FIGURE 3.60 When dr is small compared with a, as it is when $dr = 0.1$ and $a = 10$, the differential $dA = 2\pi a\,dr$ gives a good estimate of ΔA. (Example 8)

Solution Since $A = \pi r^2$, the estimated increase is

$$dA = A'(a)\,dr = 2\pi a\,dr = 2\pi(10)(0.1) = 2\pi \text{ m}^2.$$

The true change is

$$\Delta A = \pi(10.1)^2 - \pi(10)^2 = (102.01 - 100)\pi = (\underbrace{2\pi}_{dA} + \underbrace{0.01\pi}_{\text{error}}) \text{ m}^2.$$

Absolute, Relative, and Percentage Change

As we move from a to a nearby point $a + dx$, we can describe the change in f in three ways:

	True	**Estimated**
Absolute change	$\Delta f = f(a + dx) - f(a)$	$df = f'(a)\,dx$
Relative change	$\dfrac{\Delta f}{f(a)}$	$\dfrac{df}{f(a)}$
Percentage change	$\dfrac{\Delta f}{f(a)} \times 100$	$\dfrac{df}{f(a)} \times 100$

Example 9 Computing Percentage Change

The estimated percentage change in the area of the circle in Example 8 is

$$\frac{dA}{A(a)} \times 100 = \frac{2\pi}{100\pi} \times 100 = 2\%.$$

The true percentage change is

$$\frac{\Delta A}{A(a)} \times 100 = \frac{2.01\pi}{100\pi} \times 100 = 2.01\%.$$

Example 10 Unclogging Arteries

In the late 1830s, French physiologist Jean Poiseuille ("pwa-ZOY") discovered the formula we use today to predict how much the radius of a partially clogged artery has to be expanded to restore normal flow. His formula,

$$V = kr^4,$$

says that the volume V of fluid flowing through a small pipe or tube in a unit of time at a fixed pressure is a constant times the fourth power of the tube's radius r. How will a 10% increase in r affect V?

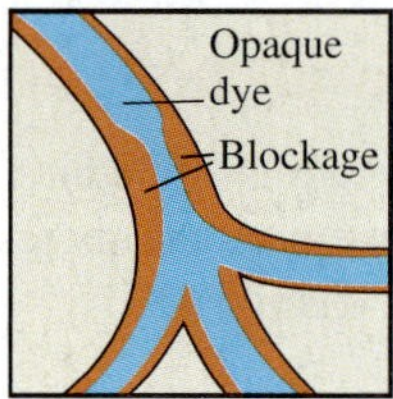

Angiography

An opaque dye is injected into a partially blocked artery to make the inside visible under X-rays. This reveals the location and severity of the blockage.

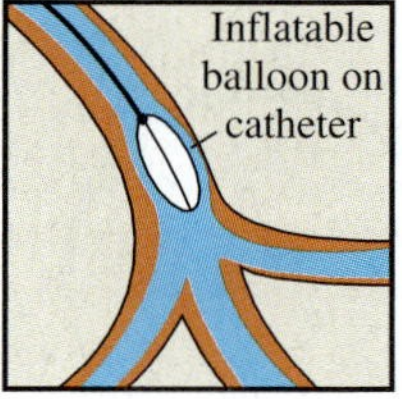

Angioplasty

A balloon-tipped catheter is inflated inside the artery to widen it at the blockage site.

Solution The differentials of r and V are related by the equation

$$dV = \frac{dV}{dr}\,dr = 4kr^3\,dr.$$

The relative change in V is

$$\frac{dV}{V} = \frac{4kr^3\,dr}{kr^4} = 4\frac{dr}{r}.$$

The relative change in V is 4 times the relative change in r, so a 10% increase in r will produce a 40% increase in the flow.

Sensitivity to Change

The equation $df = f'(x)\,dx$ tells how *sensitive* the output of f is to a change in input at different values of x. The larger the value of f' at x, the greater the effect of a given change dx.

Example 11 Finding the Depth of a Well

You want to calculate the depth of a well from the equation $s = 16t^2$ by timing how long it takes a heavy stone you drop to splash into the water below. How sensitive will your calculations be to a 0.1 sec error in measuring the time?

Solution The size of ds in the equation

$$ds = 32t\,dt$$

depends on how big t is. If $t = 2$ sec, the error caused by $dt = 0.1$ is only

$$ds = 32(2)(0.1) = 6.4 \text{ ft.}$$

Three seconds later at $t = 5$ sec, the error caused by the same dt is

$$ds = 32(5)(0.1) = 16 \text{ ft.}$$

Error in Differential Approximation

Let $f(x)$ be differentiable at $x = a$ and suppose that Δx is an increment of x. We have two ways to describe the change in f as x changes from a to $a + \Delta x$:

The true change: $\Delta f = f(a + \Delta x) - f(a)$

The differential estimate: $df = f'(a)\Delta x.$

How well does df approximate Δf?

We measure the approximation error by subtracting df from Δf:

$$\begin{aligned}\text{Approximation error} &= \Delta f - df \\ &= \Delta f - f'(a)\,\Delta x \\ &= \underbrace{f(a + \Delta x) - f(a)}_{\Delta f} - f'(a)\Delta x \\ &= \underbrace{\left(\frac{f(a + \Delta x) - f(a)}{\Delta x} - f'(a)\right)}_{\text{Call this part } \epsilon} \Delta x \\ &= \epsilon \cdot \Delta x.\end{aligned}$$

As $\Delta x \to 0$, the difference quotient

$$\frac{f(a + \Delta x) - f(a)}{\Delta x}$$

approaches $f'(a)$ (remember the definition of $f'(a)$), so the quantity in parentheses becomes a very small number (which is why we called it ϵ). In fact, $\epsilon \to 0$ as $\Delta x \to 0$. When Δx is small, the approximation error $\epsilon\,\Delta x$ is smaller still.

$$\underbrace{\Delta f}_{\substack{\text{true}\\\text{change}}} = \underbrace{f'(a)\Delta x}_{\substack{\text{estimated}\\\text{change}}} + \underbrace{\epsilon\,\Delta x}_{\text{error}}$$

Although we do not know exactly how small the error is and will not be able to make much progress on this front until Chapter 8, there is something worth noting here, namely the *form* taken by the equation.

Change in $y = f(x)$ near $x = a$

If $y = f(x)$ is differentiable at $x = a$, and x changes from a to $a + \Delta x$, the change Δy in f is given by an equation of the form

$$\Delta y = f'(a)\,\Delta x + \epsilon\,\Delta x \tag{3}$$

in which $\epsilon \to 0$ as $\Delta x \to 0$.

Converting Mass to Energy

Example 12 Using Approximations in Einstein's Physics

Newton's second law,

$$F = \frac{d}{dt}(mv) = m\frac{dv}{dt} = ma,$$

is stated with the assumption that mass is constant, but we know this is not strictly true because the mass of a body increases with velocity. In Einstein's corrected formula, mass has the value

$$m = \frac{m_0}{\sqrt{1 - v^2/c^2}},$$

where the "rest mass" m_0 represents the mass of a body that is not moving and c is the speed of light, which is about 300,000 km/sec. Use the approximation

$$\frac{1}{\sqrt{1 - x^2}} \approx 1 + \frac{1}{2}x^2 \tag{4}$$

from Example 4 to estimate the increase Δm in mass resulting from the added velocity v.

Solution

When v is very small compared with c, v^2/c^2 is close to zero and it is safe to use the approximation

$$\frac{1}{\sqrt{1 - v^2/c^2}} \approx 1 + \frac{1}{2}\left(\frac{v^2}{c^2}\right)$$

(Equation (4) with $x = v/c$) to write

$$m = \frac{m_0}{\sqrt{1 - v^2/c^2}} \approx m_0\left[1 + \frac{1}{2}\left(\frac{v^2}{c^2}\right)\right] = m_0 + \frac{1}{2}m_0v^2\left(\frac{1}{c^2}\right).$$

or

$$m \approx m_0 + \frac{1}{2}m_0v^2\left(\frac{1}{c^2}\right). \tag{5}$$

Equation (5) expresses the increase in mass that results from the added velocity v.

Energy Interpretation

In Newtonian physics, $(1/2)\,m_0v^2$ is the kinetic energy (KE) of the body, and if we rewrite Equation (5) in the form

$$(m - m_0)c^2 \approx \frac{1}{2}m_0v^2,$$

we see that

$$(m - m_0)c^2 \approx \frac{1}{2}m_0v^2 = \frac{1}{2}m_0v^2 - \frac{1}{2}m_0(0)^2 = \Delta(\text{KE}),$$

or

$$(\Delta m)c^2 \approx \Delta(\text{KE}). \tag{6}$$

In other words, the change in kinetic energy $\Delta(\text{KE})$ in going from velocity 0 to velocity v is approximately equal to $(\Delta m)c^2$.

With c equal to 3×10^8 m/sec, Equation (6) becomes

$$\Delta(\text{KE}) \approx 90{,}000{,}000{,}000{,}000{,}000\,\Delta m \text{ joules} \qquad \text{Mass in kilograms}$$

and we see that a small change in mass can create a large change in energy. The energy released by exploding a 20-kiloton atomic bomb, for instance, is the result of converting only 1 g of mass to energy. The products of the explosion weigh only 1 g less than the material exploded. A U.S. penny weighs about 3 g.

EXERCISES 3.6

CD-ROM WEBsite

Finding Linearizations

In Exercises 1–5, find the linearization $L(x)$ of $f(x)$ at $x = a$.

1. $f(x) = x^3 - 2x + 3, \quad a = 2$
2. $f(x) = \sqrt{x^2 + 9}, \quad a = -4$
3. $f(x) = x + \dfrac{1}{x}, \quad a = 1$
4. $f(x) = \sqrt[3]{x}, \quad a = -8$
5. $f(x) = \tan x, \quad a = \pi$
6. *Common linear approximations at $x = 0$* Find the linearizations of the following functions at $x = 0$.

   (a) $\sin x$ **(b)** $\cos x$ **(c)** $\tan x$

   (d) e^x **(e)** $\ln(1 + x)$

Linearizations for Powers and Roots

CD-ROM WEBsite

7. Show that the linearization of $f(x) = (1 + x)^k$ at $x = 0$ is $L(x) = 1 + kx$.

CD-ROM WEBsite

8. Use the linear approximation $(1 + x)^k \approx 1 + kx$ to find an approximation for the function $f(x)$ for values of x near zero.

   (a) $f(x) = (1 - x)^6$

   (b) $f(x) = \dfrac{2}{1 - x}$

   (c) $f(x) = \dfrac{1}{\sqrt{1 + x}}$

   (d) $f(x) = \sqrt{2 + x^2}$

   (e) $f(x) = (4 + 3x)^{1/3}$

   (f) $f(x) = \sqrt[3]{\left(1 - \dfrac{1}{2 + x}\right)^2}$

Linearization for Approximation

In Exercises 9–12, choose a linearization with center not at $x = a$ but at a nearby value at which the function and its derivative are easy to evaluate. State the linearization and the center.

9. $f(x) = 2x^2 + 4x - 3, \quad a = -0.9$
10. $f(x) = \sqrt[3]{x}, \quad a = 8.5$
11. $f(x) = \dfrac{x}{x + 1}, \quad a = 1.3$
12. $f(x) = \cos x, \quad a = 1.7$
13. *Faster than a calculator* Use the approximation $(1 + x)^k \approx 1 + kx$ to estimate the following.

    (a) $(1.0002)^{50}$ **(b)** $\sqrt[3]{1.009}$.

14. *Writing to Learn* Find the linearization of $f(x) = \sqrt{x + 1} + \sin x$ at $x = 0$. How is it related to the individual linearizations of $\sqrt{x + 1}$ and $\sin x$ at $x = 0$?

Derivatives in Differential Form

In Exercises 15–24, find dy.

15. $y = x^3 - 3\sqrt{x}$
16. $y = x\sqrt{1 - x^2}$
17. $y = x^2 \ln x$
18. $y = \dfrac{2\sqrt{x}}{3(1 + \sqrt{x})}$
19. $2y^{3/2} + xy - x = 0$
20. $xy^2 - 4x^{3/2} - y = 0$
21. $y = e^{\sin x}$
22. $y = \cos(x^2)$
23. $y = xe^x$
24. $y = \sec(x^2 - 1)$

Approximation Error

In Exercises 25–28, the function f changes value when x changes from a to $a + dx$. Find

(a) the absolute change $\Delta f = f(a + dx) - f(a)$

(b) the estimated change $df = f'(a)\,dx$

(c) the approximation error $|\Delta f - df|$.

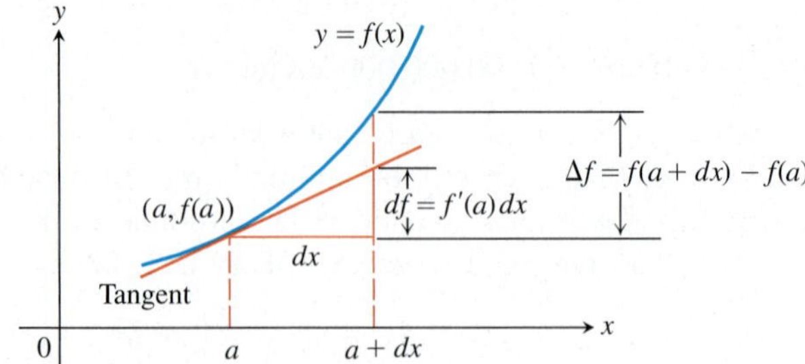

25. $f(x) = x^2 + 2x, \quad a = 0, \quad dx = 0.1$

26. $f(x) = x^3 - x, \quad a = 1, \quad dx = 0.1$

27. $f(x) = x^{-1}, \quad a = 0.5, \quad dx = 0.05$

28. $f(x) = x^4, \quad a = 1, \quad dx = 0.01$

Differential Estimates of Change

In Exercises 29–32, write a differential formula that estimates the given change in volume or surface area.

29. *Volume* The change in the volume $V = (4/3)\pi r^3$ of a sphere when the radius changes from a to $a + dr$

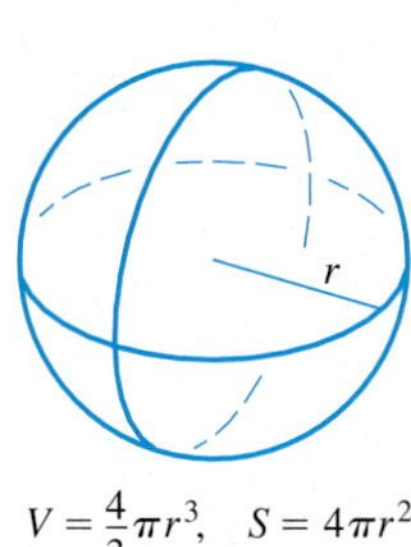

$V = \frac{4}{3}\pi r^3, \quad S = 4\pi r^2$

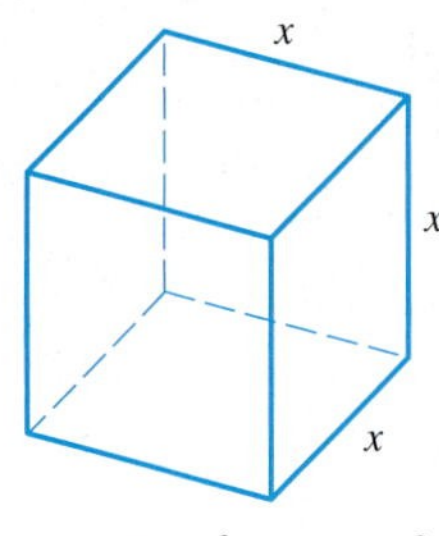

$V = x^3, \quad S = 6x^2$

30. *Surface area* The change in the surface area $S = 4\pi r^2$ of a sphere when the radius changes from a to $a + dr$

31. *Volume* The change in the volume $V = x^3$ of a cube when the edge lengths change from a to $a + dx$

32. *Surface area* The change in the surface area $S = 6x^2$ of a cube when the edge lengths change from a to $a + dx$

Theory and Examples

33. *Expanding circle* The radius of a circle is increased from 2.00 to 2.02 m.

(a) Estimate the resulting change in area.

(b) Express the estimate as a percentage of the circle's original area.

34. *Growing tree* The diameter of a tree was 10 in. During the following year, the circumference increased 2 in. About how much did the tree's diameter increase? The tree's cross-section area?

35. *Estimating volume* Estimate the volume of material in a cylindrical shell with height 30 in., radius 6 in., and shell thickness 0.5 in.

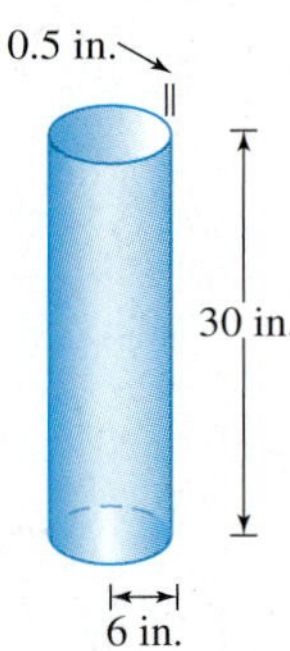

36. *Estimating height* A surveyor, standing 30 ft from the base of a building, measures the angle of elevation to the top of the building to be 75°. How accurately must the angle be measured for the percentage error in estimating the height of the building to be less than 4%?

37. *Tolerance* The height and radius of a right circular cylinder are equal, so the cylinder's volume is $V = \pi h^3$. The volume is to be calculated with an error of no more than 1% of the true value. Find approximately the greatest error that can be tolerated in the measurement of h, expressed as a percentage of h.

38. *Tolerance*

(a) About how accurately must the interior diameter of a 10-m high cylindrical storage tank be measured to calculate the tank's volume to within 1% of its true value?

(b) About how accurately must the tank's exterior diameter be measured to calculate the amount of paint it will take to paint the side of the tank to within 5% of the true amount?

39. *Minting coins* A manufacturer contracts to mint coins for the federal government. How much variation dr in the radius of the coins can be tolerated if the coins are to weigh within 1/1000 of their ideal weight? Assume that the thickness does not vary.

40. *Profit* The profit P for a certain manufacturer selling x items is

$$P(x) = 200xe^{-x/400}.$$

Estimate the change and percent change in P as sales change from $x = 145$ to $x = 150$ items.

41. *The effect of flight maneuvers on the heart* The amount of work done by the heart's main pumping chamber, the left ventricle, is given by the equation

$$W = PV + \frac{V\delta v^2}{2g},$$

where W is the work per unit time, P is the average blood pressure, V is the volume of blood pumped out during the unit of time, δ ("delta") is the weight density of the blood, v is the average velocity of the exiting blood, and g is the acceleration of gravity.

When P, V, δ, and v remain constant, W becomes a function of g, and the equation takes the form

$$W = a + \frac{b}{g} \qquad (a, b \text{ constant}).$$

As a member of NASA's medical team, you want to know how sensitive W is to apparent changes in g caused by flight maneuvers, and this depends on the initial value of g. As part of your investigation, you decide to compare the effect on W of a given change dg on the moon, where $g = 5.2$ ft/sec^2, with the effect the same change dg would have on Earth, where $g = 32$ ft/sec^2. Use the simplified equation above to find the ratio of dW_{moon} to dW_{Earth}.

42. *Measuring acceleration of gravity* When the length L of a clock pendulum is held constant by controlling its temperature, the pendulum's period T depends on the acceleration of gravity g. The period will therefore vary slightly as the clock is moved from place to place on the earth's surface, depending on the change in g. By keeping track of ΔT, we can estimate the variation in g from the equation $T = 2\pi(L/g)^{1/2}$ that relates T, g, and L.

(a) With L held constant and g as the independent variable, calculate dT and use it to answer parts (b) and (c).

(b) *Writing to Learn* If g increases, will T increase or decrease? Will a pendulum clock speed up or slow down? Explain.

(c) A clock with a 100 cm pendulum is moved from a location where $g = 980$ cm/sec^2 to a new location. This increases the period by $dT = 0.001$ sec. Find dg and estimate the value of g at the new location.

T **43.** *Zooming in to "see" differentiability* Is either of these functions differentiable at $x = 0$?

$$f(x) = |x| + 1, \qquad g(x) = \sqrt{x^2 + 0.0001} + 0.99$$

(a) We already know that f is not differentiable at $x = 0$; its graph has a corner there. Graph f and zoom in at the point (0, 1) several times. Does the corner show signs of straightening out?

(b) Now do the same thing with g. Does the graph of g show signs of straightening out? We know g *is* differentiable at $x = 0$ and, in fact, has a horizontal tangent there.

(c) How many zooms does it take before the graph of g looks exactly like a horizontal line?

(d) Now graph f and g *together* in a standard square viewing window. They appear to be identical until you start zooming in. The differentiable function eventually straightens out, whereas the nondifferentiable function remains impressively unchanged.

T **44.** *Reading derivatives from graphs* The idea that differentiable curves flatten out when magnified can be used to estimate the values of the derivatives of functions at particular points. We magnify the curve until the portion we see looks like a straight line through the point in question, and then we use the screen's coordinate grid to read the slope of the curve as the slope of the line it resembles.

(a) To see how the process works, try it first with the function $y = x^2$ at $x = 1$. The slope you read should be 2.

(b) Then try it with the curve $y = e^x$ at $x = 1$, $x = 0$, and $x = -1$. In each case, compare your estimate of the derivative with the value of e^x at the point. What pattern do you see? Test it with other values of x.

45. *The linearization is the best linear approximation* (This is why we use the linearization.) Suppose that $y = f(x)$ is differentiable at $x = a$ and that $g(x) = m(x - a) + c$ is a linear function in which m and c are constants. If the error $E(x) = f(x) - g(x)$ were small enough near $x = a$, we might think of using g as a linear approximation of f instead of the linearization $L(x) = f(a) + f'(a)(x - a)$. Show that if we impose on g the conditions

1. $E(a) = 0$ — The approximation error is zero at $x = a$.

2. $\lim_{x \to a} \frac{E(x)}{x - a} = 0$ — The error is negligible when compared with $x - a$.

then $g(x) = f(a) + f'(a)(x - a)$. Thus, the linearization $L(x)$ gives the only linear approximation whose error is both zero at $x = a$ and negligible in comparison with $x - a$.

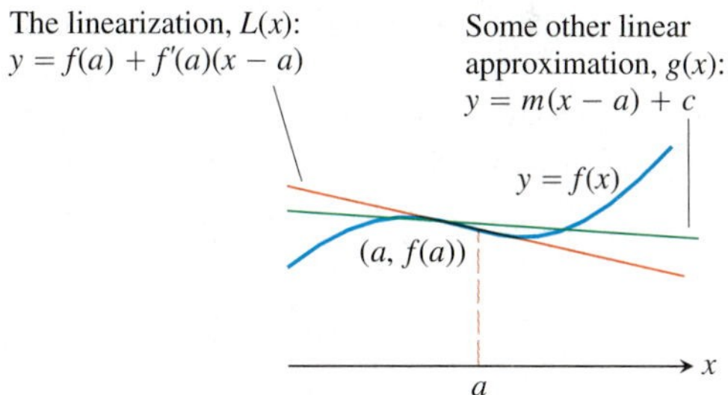

46. *Quadratic approximations*

(a) Let $Q(x) = b_0 + b_1(x - a) + b_2(x - a)^2$ be a quadratic approximation to $f(x)$ at $x = a$ with the properties:

i. $Q(a) = f(a)$

ii. $Q'(a) = f'(a)$

iii. $Q''(a) = f''(a)$.

Determine the coefficients b_0, b_1, and b_2.

(b) Find the quadratic approximation to $f(x) = 1/(1 - x)$ at $x = 0$.

T **(c)** Graph $f(x) = 1/(1 - x)$ and its quadratic approximation at $x = 0$. Then zoom in on the two graphs at the point (0, 1). Comment on what you see.

T **(d)** Find the quadratic approximation to $g(x) = 1/x$ at $x = 1$. Graph g and its quadratic approximation together. Comment on what you see.

T **(e)** Find the quadratic approximation to $h(x) = \sqrt{1 + x}$ at $x = 0$. Graph h and its quadratic approximation together. Comment on what you see.

(f) What are the linearizations of f, g, and h at the respective points in parts (b), (d), and (e)?

47. *The linearization of 2^x*

(a) Find the linearization of $f(x) = 2^x$ at $x = 0$. Then round its coefficients to two decimal places.

T **(b)** Graph the linearization and function together for $-3 \leq x \leq 3$ and $-1 \leq x \leq 1$.

48. *The linearization of $\log_3 x$*

(a) Find the linearization of $f(x) = \log_3 x$ at $x = 3$. Then round its coefficients to two decimal places.

T **(b)** Graph the linearization and function together in the window $0 \leq x \leq 8$ and $2 \leq x \leq 4$.

49. *Linearizations at inflection points* As Figure 3.58 suggests, linearizations fit particularly well at inflection points. You will understand why in Chapter 8. As another example, graph *Newton's serpentine*, $f(x) = 4x/(x^2 + 1)$, together with its linearizations at $x = 0$ and $x = \sqrt{3}$.

T **50.** *Writing to Learn: Repeated root-taking*

(a) Enter 2 in your calculator and take successive square roots by pressing the square root key repeatedly (or raising the displayed number repeatedly to the 0.5 power). What pattern do you see emerging? Explain what is going on. What happens if you take successive tenth roots instead?

(b) Repeat the procedure with 0.5 in place of 2 as the original entry. What happens now? Can you use any positive number x in place of 2? Explain what is going on.

COMPUTER EXPLORATIONS

In Exercises 51–54, you will use a CAS to estimate the magnitude of the error in using the linearization in place of the function over a specified interval I. Perform the following steps.

(a) Plot the function f over I.

(b) Find the linearization L of the function at the point a.

(c) Plot f and L together on a single graph.

(d) Plot the absolute error $|f(x) - L(x)|$ over I and find its maximum value.

(e) From your graph in part (d), estimate as large a $\delta > 0$ as you can, satisfying

$$|x - a| < \delta \Rightarrow |f(x) - L(x)| < \epsilon$$

for $\epsilon = 0.5, 0.1,$ and 0.01. Then check graphically to see if your δ-estimate holds true.

51. $f(x) = x^3 + x^2 - 2x, \quad [-1, 2], \quad a = 1$

52. $f(x) = \dfrac{x - 1}{4x^2 + 1}, \quad [-3/4, 1], \quad a = \dfrac{1}{2}$

53. $f(x) = x^{2/3}(x - 2), \quad [-2, 3], \quad a = 2$

54. $f(x) = \sqrt{x} - \sin x, \quad [0, 2\pi], \quad a = 2$

3.7 Newton's Method

Procedure for Newton's Method • The Practice • Convergence Is Usually Assured • But Things Can Go Wrong • Fractal Basins and Newton's Method

CD-ROM WEBsite
Historical Biography
Neils Henrik Abel (1802 — 1829)

We know simple formulas for solving linear and quadratic equations, and there are somewhat more complicated formulas for cubic and quartic equations (equations of degree three and four). At one time it was hoped that similar formulas might be found for quintic and higher-degree equations, but Norwegian mathematician Neils Henrik Abel showed that no formulas like these are possible for polynomial equations of degree greater than four.

When exact formulas for solving an equation $f(x) = 0$ are not available, we can turn to numerical techniques from calculus to approximate the solutions we seek. One of these techniques is *Newton's method* or, as it is more accurately called, the *Newton–Raphson method*. It is based on the idea of using tangent lines to replace the graph of $y = f(x)$ near the points where f is zero. Once again, linearization is the key to solving a practical problem.

Procedure for Newton's Method

Newton's method is a numerical technique for approximating a zero of a function with zeros of its linearizations. Under favorable circumstances, the zeros of the linearizations *converge* rapidly to an accurate approximation. Moreover the method applies to a wide range of functions and usually gets results in only a few steps. Here is how it works.

The initial estimate, x_0, may be found by graphing or just plain guessing. The method then uses the tangent to the curve $y = f(x)$ at $(x_0, f(x_0))$ to approximate the curve, calling the point where the tangent meets the x-axis x_1 (Figure 3.61). The number x_1 is usually a better approximation to the solution than is x_0. The point x_2 where the tangent to the curve at $(x_1, f(x_1))$ crosses the x-axis is the next approximation in the sequence. We continue on, using each approximation to generate the next, until we are close enough to the root to stop.

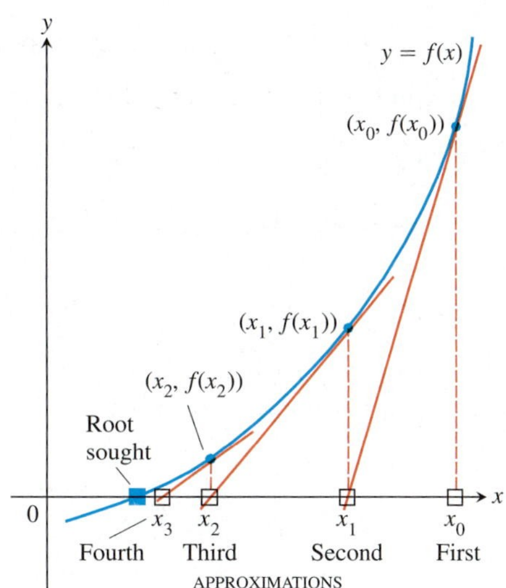

FIGURE 3.61 Newton's method starts with an initial guess x_0 and (under favorable circumstances) improves the guess one step at a time.

There is a formula for finding the $(n + 1)$st approximation x_{n+1} from the nth approximation x_n. The point-slope equation for the tangent to the curve at $(x_n, f(x_n))$ is

$$y - f(x_n) = f'(x_n)(x - x_n).$$

We can find where it crosses the x-axis by setting $y = 0$ (Figure 3.62).

$$\begin{aligned} 0 - f(x_n) &= f'(x_n)(x - x_n) \\ -f(x_n) &= f'(x_n) \cdot x - f'(x_n) \cdot x_n \\ f'(x_n) \cdot x &= f'(x_n) \cdot x_n - f(x_n) \\ x &= x_n - \frac{f(x_n)}{f'(x_n)} \qquad \text{If } f'(x_n) \neq 0 \end{aligned}$$

This value of x is the next approximation x_{n+1}. Here is a summary of Newton's method.

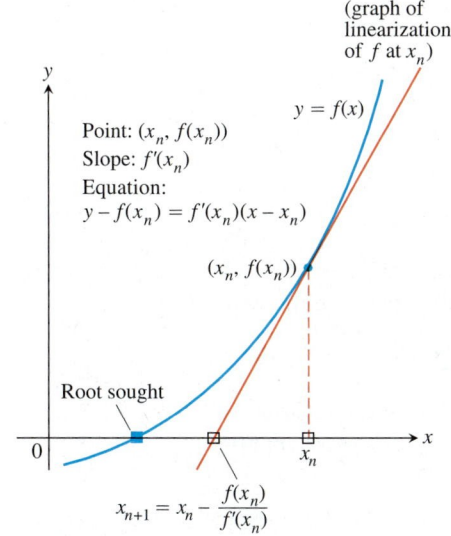

FIGURE 3.62 The geometry of the successive steps of Newton's method. From x_n we go up to the curve and follow the tangent line down to find x_{n+1}.

Procedure for Newton's Method

1. Guess a first approximation to a solution of the equation $f(x) = 0$. A graph of $y = f(x)$ may help.
2. Use the first approximation to get a second, the second to get a third, and so on, using the formula

$$x_{n+1} = x_n - \frac{f(x_n)}{f'(x_n)}. \tag{1}$$

The Practice

In our first example, we find decimal approximations to $\sqrt{2}$ by estimating the positive root of the equation $f(x) = x^2 - 2 = 0$.

Algorithm and Iteration

It is customary to call a specified sequence of computational steps like the one in Newton's method an *algorithm*. When an algorithm proceeds by repeating a given set of steps over and over, using the answer from the previous step as the input for the next, the algorithm is called *iterative* and each repetition is called an *iteration*. Newton's method is one of the really fast iterative techniques for finding roots.

Example 1 Finding the Square Root of 2

Find the positive root of the equation

$$f(x) = x^2 - 2 = 0.$$

Solution With $f(x) = x^2 - 2$ and $f'(x) = 2x$, Equation (1) becomes

$$x_{n+1} = x_n - \frac{x_n^2 - 2}{2x_n}.$$

To use our calculator efficiently, we rewrite this equation in a form that uses fewer arithmetic operations:

$$\begin{aligned} x_{n+1} &= x_n - \frac{x_n}{2} + \frac{1}{x_n} \\ &= \frac{x_n}{2} + \frac{1}{x_n}. \end{aligned}$$

The equation

$$x_{n+1} = \frac{x_n}{2} + \frac{1}{x_n}$$

enables us to go from each approximation to the next with just a few keystrokes. With the starting value $x_0 = 1$, we get the results in the first column of the following table. (To five decimal places, $\sqrt{2} = 1.41421$.)

	Error	Number of correct figures
$x_0 = 1$	-0.41421	1
$x_1 = 1.5$	0.08579	1
$x_2 = 1.41667$	0.00246	3
$x_3 = 1.41422$	0.00001	5

Newton's method is the method used by most calculators to calculate roots because it converges so fast (more about this later). If the arithmetic in the table in Example 1 had been carried to 13 decimal places instead of 5, then going one step further would have given $\sqrt{2}$ correctly to more than 10 decimal places.

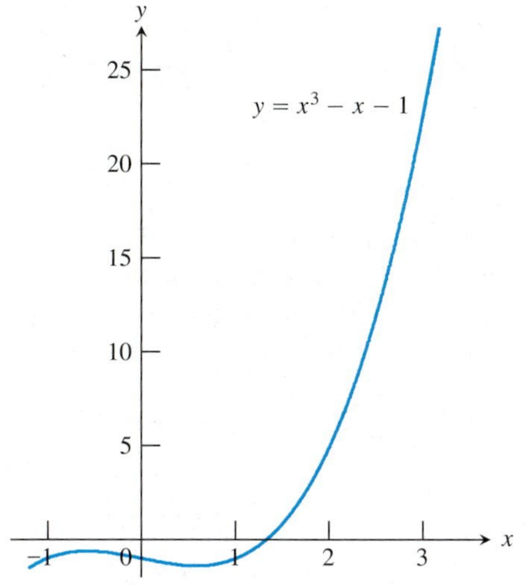

FIGURE 3.63 The graph of $f(x) = x^3 - x - 1$. (Example 2)

Example 2 Using Newton's Method

Find the x-coordinate of the point where the curve $y = x^3 - x$ crosses the horizontal line $y = 1$.

Solution The curve crosses the line when $x^3 - x = 1$ or $x^3 - x - 1 = 0$. When does $f(x) = x^3 - x - 1$ equal zero? The graph of f (Figure 3.63) shows a single root, located between $x = 1$ and $x = 2$. We apply Newton's method to f with the starting value $x_0 = 1$. The results are displayed in Table 3.1 and Figure 3.64.

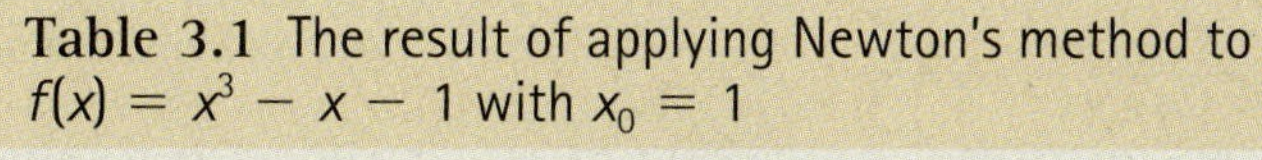

Table 3.1 The result of applying Newton's method to $f(x) = x^3 - x - 1$ with $x_0 = 1$

n	x_n	$f(x_n)$	$f'(x_n)$	$x_{n+1} = x_n - \dfrac{f(x_n)}{f'(x_n)}$
0	1	−1	2	1.5
1	1.5	0.875	5.75	1.3478 26087
2	1.3478 26087	0.1006 82173	4.4499 05482	1.3252 00399
3	1.3252 00399	0.0020 58362	4.2684 68292	1.3247 18174
4	1.3247 18174	0.0000 00924	4.2646 34722	1.3247 17957
5	1.3247 17957	1.8672E−13	4.2646 32999	1.3247 17957

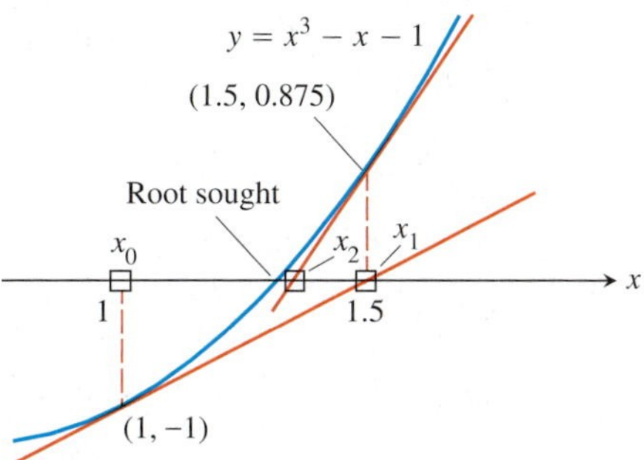

FIGURE 3.64 The first three x-values in Table 3.1.

At $n = 5$, we come to the result $x_6 = x_5 = 1.3247\ 17957$. When $x_{n+1} = x_n$, Equation (1) shows that $f(x_n) = 0$. We have found a solution of $f(x) = 0$ to nine decimals.

In Figure 3.65, we have indicated that the process in Example 2 might have started at the point $B_0(3, 23)$ on the curve, with $x_0 = 3$. Point B_0 is quite far from the x-axis, but the tangent at B_0 crosses the x-axis at about $(2.12, 0)$, so x_1 is still an improvement over x_0. If we use Equation (1) repeatedly as before, with $f(x) = x^3 - x - 1$ and $f'(x) = 3x^2 - 1$, we confirm the nine-place solution $x_7 = x_6 = 1.3247\ 17957$ in seven steps.

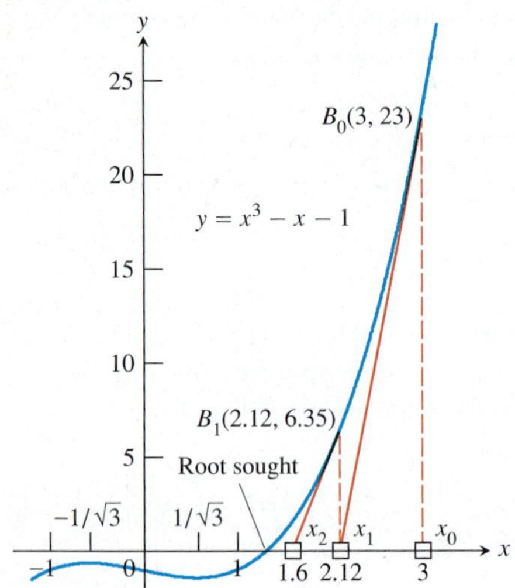

FIGURE 3.65 Any starting value x_0 to the right of $x = 1/\sqrt{3}$ will lead to the root.

The curve in Figure 3.65 has a local maximum at $x = -1/\sqrt{3}$ and a local minimum at $x = +1/\sqrt{3}$. We would not expect good results from Newton's method if we were to start with x_0 between these points, but we can start any place to the right of $x = 1/\sqrt{3}$ and get the answer. It would not be very clever to do so, but we could even begin far to the right of B_0, for example with $x_0 = 10$. It takes a bit longer, but the process still converges to the same answer as before.

Convergence Is Usually Assured

In practice, Newton's method usually converges with impressive speed, but this is not guaranteed. One way to test convergence is to begin by graphing the function to estimate a good starting value for x_0. You can test that you are getting closer to a zero of the function by evaluating $|f(x_n)|$ and check that the method is converging by evaluating $|x_n - x_{n+1}|$.

Theory does provide some help. A theorem from advanced calculus says that if

$$\left|\frac{f(x)f''(x)}{[f'(x)]^2}\right| < 1 \tag{2}$$

for all x in an interval about a root r, then the method will converge to r for any starting value x_0 in that interval.

Newton's method always converges if the curve $y = f(x)$ is convex ("bulges") toward the x-axis in the interval between x_0 and the root sought. (See Figure 3.66.)

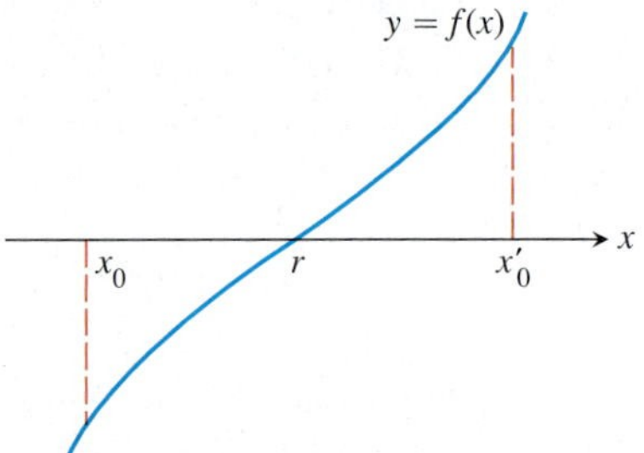

FIGURE 3.66 Newton's method will converge to r from either starting point.

Under favorable circumstances, the speed with which Newton's method converges to r is expressed by the advanced calculus formula

$$\underbrace{|x_{n+1} - r|}_{\text{error } e_{n+1}} \leq \frac{\max|f''|}{2\min|f'|}|x_n - r|^2 = \text{constant} \cdot \underbrace{|x_n - r|^2}_{\text{error } e_n}, \tag{3}$$

where max and min refer to the maximum and minimum values in an interval surrounding r. The formula says that the error in step $n + 1$ is no greater than a constant times the square of the error in step n. This may not seem like much, but think of what it says. If the constant is less than or equal to 1 and $|x_n - r| < 10^{-3}$, then $|x_{n+1} - r| < 10^{-6}$. *In a single step,* the method moves from three decimal places of accuracy to six!

The results in Equations (2) and (3) both assume that f is "nice." Hence, in the case of Equation (3), this means that f has only a single root at r, so that $f'(r) \neq 0$. If f has a multiple root at r, the convergence may be slower.

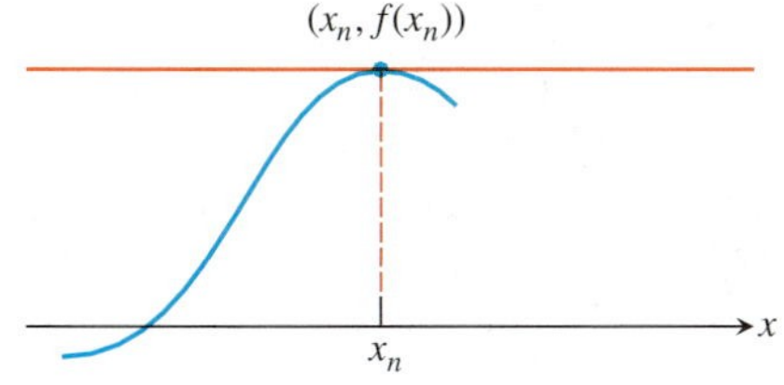

FIGURE 3.67 If $f'(x_n) = 0$, there is no intersection point to define x_{n+1}.

But Things Can Go Wrong

Newton's method stops if $f'(x_n) = 0$ (Figure 3.67). In that case, try a new starting point. Of course, f and f' may have a common root. To detect whether this is so, you could first find the solutions of $f'(x) = 0$ and check f at those values, or you can graph f and f' together.

Newton's method does not always converge. For instance, if

$$f(x) = \begin{cases} -\sqrt{r - x}, & x < r \\ \sqrt{x - r}, & x \geq r, \end{cases}$$

the graph will be like the one in Figure 3.68. If we begin with $x_0 = r - h$, we get $x_1 = r + h$, and successive approximations go back and forth between these two values. No amount of iteration brings us closer to the root than our first guess.

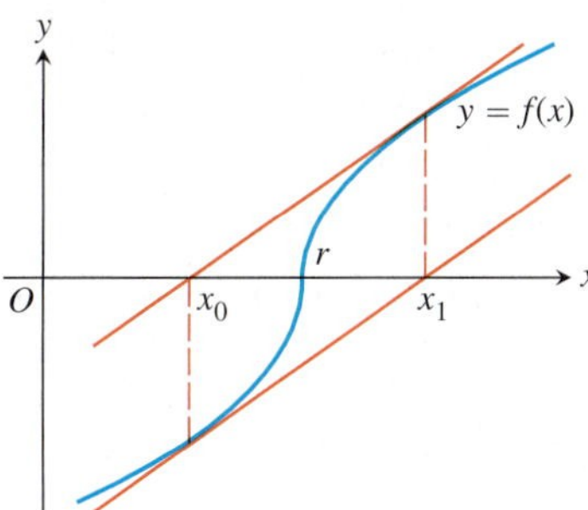

FIGURE 3.68 Newton's method fails to converge. You go from x_0 to x_1 and back to x_0, never getting any closer to r.

If Newton's method does converge, it converges to a root. Be careful, however. There are situations in which the method appears to converge but there is no root there. Fortunately, such situations are rare.

When Newton's method converges to a root, it may not be the root you have in mind. Figure 3.69 shows two ways this can happen.

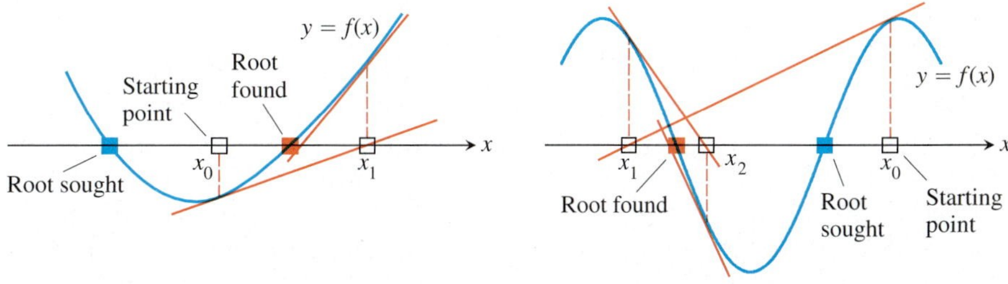

FIGURE 3.69 If you start too far away, Newton's method may miss the root you want.

Fractal Basins and Newton's Method

The process of finding roots by Newton's method can be uncertain in the sense that for some equations the final outcome can be extremely sensitive to the starting value's location.

The equation $4x^4 - 4x^2 = 0$ is a case in point (Figure 3.70a). Starting values in the blue zone on the x-axis lead to root A. Starting values in the black lead to root B, and starting values in the red zone lead to root C. The points $\pm\sqrt{2}/2$ give

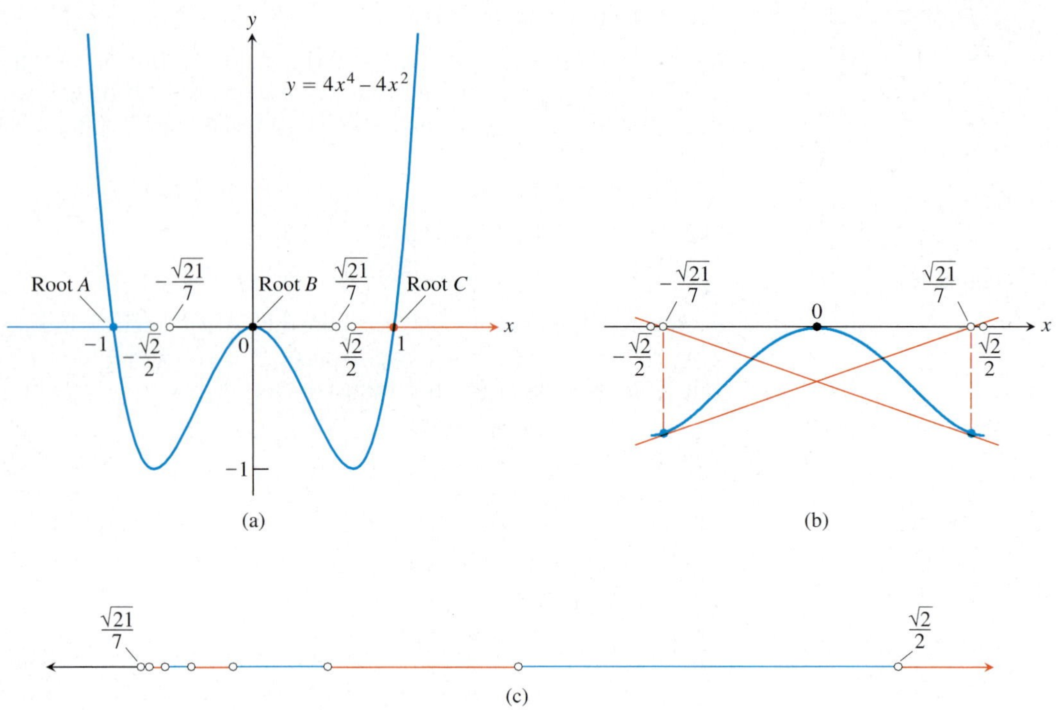

FIGURE 3.70 (a) Starting values in $(-\infty, -\sqrt{2}/2)$, $(-\sqrt{21}/7, \sqrt{21}/7)$, and $(\sqrt{2}/2, \infty)$ lead respectively to roots A, B, and C. (b) The values $x = \pm\sqrt{21}/7$ lead only to each other. (c) Between $\sqrt{21}/7$ and $\sqrt{2}/2$, there are infinitely many open intervals of points attracted to A alternating with open intervals of points attracted to C. This behavior is mirrored in the interval $(-\sqrt{2}/2, -\sqrt{21}/7)$.

horizontal tangents. The points $\pm\sqrt{21}/7$ "cycle," each leading to the other, and back (Figure 3.70b).

The interval between $\sqrt{21}/7$ and $\sqrt{2}/2$ contains infinitely many open intervals of points leading to root A, alternating with intervals of points leading to root C (Figure 3.70c). The boundary points separating consecutive intervals (there are infinitely many) do not lead to roots, but cycle back and forth from one to another. Moreover, as we select points that approach $\sqrt{21}/7$ from the right, it becomes increasingly difficult to distinguish which lead to root A and which to root C. On the same side of $\sqrt{21}/7$, we find arbitrarily close together points whose ultimate destinations are far apart.

If we think of the roots as "attractors" of other points, the coloring in Figure 3.70 shows the intervals of the points they attract (the "intervals of attraction"). You might think that points between roots A and B would be attracted to either A or B, but, as we see, that is not the case. Between A and B there are infinitely many intervals of points attracted to C. Similarly, between B and C lie infinitely many intervals of points attracted to A.

We encounter an even more dramatic example of such behavior when we apply Newton's method to solve the complex-number equation $z^6 - 1 = 0$. It has six solutions: 1, -1, and the four numbers $\pm(1/2) \pm (\sqrt{3}/2)i$. As Figure 3.71 suggests, each of the six roots has infinitely many "basins" of attraction in the complex plane (Appendix 3). Starting points in red basins are attracted to the root 1, those in the green basin to the root $(1/2) + (\sqrt{3}/2)i$, and so on. Each basin has a boundary whose complicated pattern repeats without end under successive magnifications. These basins are called **fractal basins.**

FIGURE 3.71 This computer-generated initial value portrait uses color to show where different points in the complex plane end up when they are used as starting values in applying Newton's method to solve the equation $z^6 - 1 = 0$. Red points go to 1, green points to $(1/2) + (\sqrt{3}/2)i$, dark blue points to $(-1/2) + (\sqrt{3}/2)i$, and so on. Starting values that generate sequences that do not arrive within 0.1 units of a root after 32 steps are colored black.

EXERCISES 3.7

Root-Finding

CD-ROM WEBsite

1. Use Newton's method to estimate the solutions of the equation $x^2 + x - 1 = 0$. Start with $x_0 = -1$ for the left-hand solution and with $x_0 = 1$ for the solution on the right. Then, in each case, find x_2.

2. Use Newton's method to estimate the one real solution of $x^3 + 3x + 1 = 0$. Start with $x_0 = 0$ and then find x_2.

3. Use Newton's method to estimate the two zeros of the function $f(x) = x^4 + x - 3$. Start with $x_0 = -1$ for the left-hand zero and with $x_0 = 1$ for the zero on the right. Then, in each case, find x_2.

4. Use Newton's method to estimate the two zeros of the function $f(x) = 2x - x^2 + 1$. Start with $x_0 = 0$ for the left-hand zero and with $x_0 = 2$ for the zero on the right. Then, in each case, find x_2.

In Exercises 5 and 6, use Newton's method to find all roots of the equation correct to six decimal places.

5. $e^{-x} = 2x + 1$

6. $\tan^{-1} x = 1 - 2x$

Theory, Examples, and Applications

7. *Guessing a root* Suppose that your first guess is lucky, in the sense that x_0 is a root of $f(x) = 0$. Assuming that $f'(x_0)$ is defined and not 0, what happens to x_1 and later approximations?

8. *Writing to Learn: Estimating pi* You plan to estimate $\pi/2$ to five decimal places by using Newton's method to solve the equation $\cos x = 0$. Does it matter what your starting value is? Give reasons for your answer.

9. *Oscillation* Show that if $h > 0$, applying Newton's method to

$$f(x) = \begin{cases} \sqrt{x}, & x \geq 0 \\ \sqrt{-x}, & x < 0 \end{cases}$$

leads to $x_1 = -h$ if $x_0 = h$ and to $x_1 = h$ if $x_0 = -h$. Draw a picture that shows what is going on.

10. *Approximations that get worse and worse* Apply Newton's method to $f(x) = x^{1/3}$ with $x_0 = 1$ and calculate x_1, x_2, x_3, and x_4. Find a formula for $|x_n|$. What happens to $|x_n|$ as $n \to \infty$? Draw a picture that shows what is going on.

11. *Writing to Learn* Explain why the following four statements ask for the same information:

    i. Find the roots of $f(x) = x^3 - 3x - 1$.

    ii. Find the x-coordinates of the intersections of the curve $y = x^3$ with the line $y = 3x + 1$.

    iii. Find the x-coordinates of the points where the curve $y = x^3 - 3x$ crosses the horizontal line $y = 1$.

    iv. Find the values of x where the derivative of $g(x) = (1/4)x^4 - (3/2)x^2 - x + 5$ equals zero.

CD-ROM WEBsite **12.** *Locating a planet* To calculate a planet's space coordinates, we have to solve equations like $x = 1 + 0.5 \sin x$. Graphing the function $f(x) = x - 1 - 0.5 \sin x$ suggests that the function has a root near $x = 1.5$. Use one application of Newton's method to improve this estimate. That is, start with $x_0 = 1.5$ and find x_1. (The value of the root is 1.49870 to five decimal places.) Remember to use radians.

T **13.** *A program for using Newton's method on a grapher* Let $f(x) = x^3 + 3x + 1$. Here is a home screen program to perform the computations in Newton's method.

(a) Let $y_0 = f(x)$ and $y_1 = \text{NDER}\, f(x)$.

(b) Store $x_0 = -0.3$ into x.

(c) Then store $x - (y_0/y_1)$ into x and press the "Enter" key over and over. Watch as the numbers converge to the zero of f.

(d) Use different values for x_0 and repeat steps (b) and (c).

(e) Write your own equation and use this approach to solve it using Newton's method. Compare your answer with the answer given by the built-in feature of your calculator that gives zeros of functions.

T **14.** *(Continuation of Exercise 11)*

(a) Use Newton's method to find the two negative zeros of $f(x) = x^3 - 3x - 1$ to five decimal places.

(b) Graph $f(x) = x^3 - 3x - 1$ for $-2 \le x \le 2.5$. Use the zoom and trace features to estimate the zeros of f to five decimal places.

(c) Graph $g(x) = 0.25x^4 - 1.5x^2 - x + 5$. Use the zoom and trace features with appropriate rescaling to find, to five decimal places, the values of x where the graph has horizontal tangents.

T CD-ROM WEBsite **15.** *Intersecting curves* The curve $y = \tan x$ crosses the line $y = 2x$ between $x = 0$ and $x = \pi/2$. Use Newton's method to find where.

T **16.** *Real solutions of a quartic* Use Newton's method to find the two real solutions of the equation $x^4 - 2x^3 - x^2 - 2x + 2 = 0$.

T **17.** *Finding solutions*

(a) How many solutions does the equation $\sin 3x = 0.99 - x^2$ have?

(b) Use Newton's method to find them.

T **18.** *Intersection of curves*

CD-ROM WEBsite **(a)** Does $\cos 3x$ ever equal x? Give reasons for your answer.

(b) Use Newton's method to find where.

T CD-ROM WEBsite **19.** *Multiple zeros* Find the four real zeros of the function $f(x) = 2x^4 - 4x^2 + 1$.

T CD-ROM WEBsite **20.** *Estimating pi* Estimate π to as many decimal places as your calculator will display by using Newton's method to solve the equation $\tan x = 0$ with $x_0 = 3$.

CD-ROM WEBsite **21.** *Intersection of curves* At what value(s) of x does $e^{-x^2} = x^2 - x + 1$?

22. *Intersection of curves* At what value(s) of x does $\ln(1 - x^2) = x - 1$?

23. *Finding a root* Use the Intermediate Value Theorem from Section 1.4 to show that $f(x) = x^3 + 2x - 4$ has a root between $x = 1$ and $x = 2$. Then find the root to five decimal places.

24. *Factoring a quartic* Find the approximate values of r_1 through r_4 in the factorization
CD-ROM WEBsite

$$8x^4 - 14x^3 - 9x^2 + 11x - 1 = 8(x - r_1)(x - r_2)(x - r_3)(x - r_4).$$

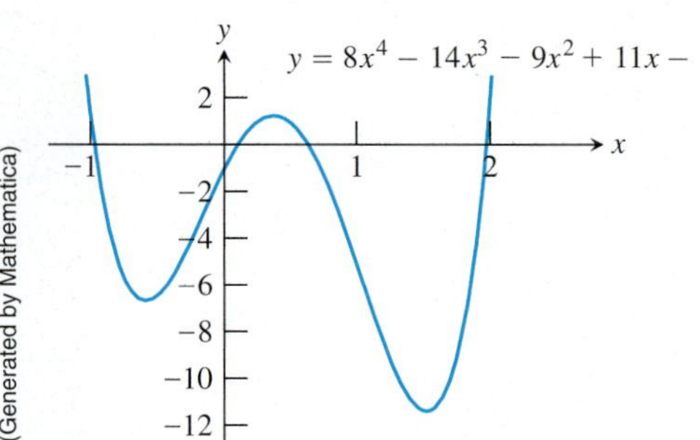

T CD-ROM WEBsite **25.** *Converging to different zeros* Use Newton's method to find the zeros of $f(x) = 4x^4 - 4x^2$ using the given starting values (Figure 3.70).

(a) $x_0 = -2$ and $x_0 = -0.8$, lying in $(-\infty, -\sqrt{2}/2)$

(b) $x_0 = -0.5$ and $x_0 = 0.25$, lying in $(-\sqrt{21}/7, \sqrt{21}/7)$

(c) $x_0 = 0.8$ and $x_0 = 2$, lying in $(\sqrt{2}/2, \infty)$

(d) $x_0 = -\sqrt{21}/7$ and $x_0 = \sqrt{21}/7$

26. *The sonobuoy problem* In submarine location problems, it is often necessary to find a submarine's closest point of approach (CPA) to a sonobuoy (sound detector) in the water. Suppose that the submarine travels on a parabolic path $y = x^2$ and that the buoy is located at the point $(2, -1/2)$.

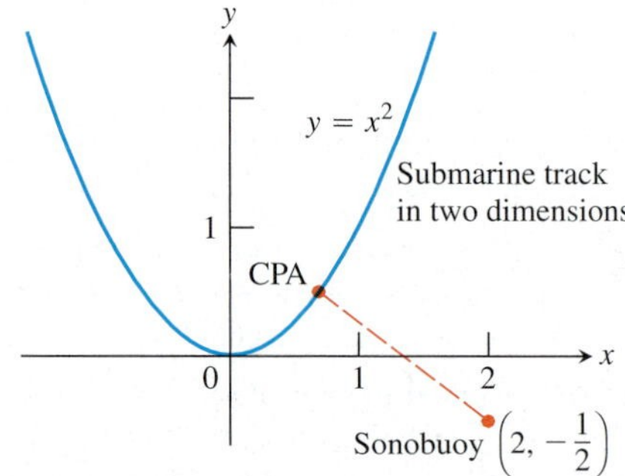

(*Source: The Contraction Mapping Principle*, by C. O. Wilde, UMAP Unit 326, Arlington, MA, COMAP, Inc.)

(a) Show that the value of x that minimizes the distance between the submarine and the buoy is a solution of the equation $x = 1/(x^2 + 1)$.

(b) Solve the equation $x = 1/(x^2 + 1)$ with Newton's method.

27. *Curves that are nearly flat at the root* Some curves are so flat that, in practice, Newton's method stops too far from the root to give a useful estimate. Try Newton's method on $f(x) = (x - 1)^{40}$ with a starting value of $x_0 = 2$ to see how close your machine comes to the root $x = 1$.

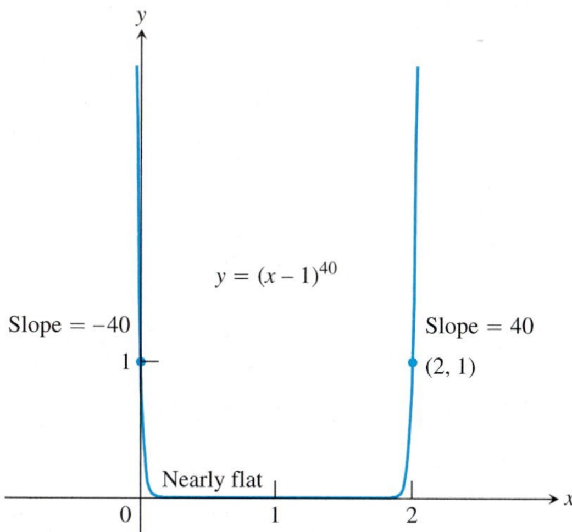

28. *Finding a root different from the one sought* All three roots of $f(x) = 4x^4 - 4x^2$ can be found by starting Newton's method near $x = \sqrt{21}/7$. Try it. See Figure 3.70.

29. *Finding an ion concentration* While trying to find the acidity of a saturated solution of magnesium hydroxide in hydrochloric acid, you derive the equation

$$\frac{3.64 \times 10^{-11}}{[H_3O^+]^2} = [H_3O^+] + 3.6 \times 10^{-4}$$

for the hydronium ion concentration $[H_3O^+]$. To find the value of $[H_3O^+]$, you set $x = 10^4[H_3O^+]$ and convert the equation to

$$x^3 + 3.6x^2 - 36.4 = 0.$$

You then solve this by Newton's method. What do you get for x? (Make it good to two decimal places.) For $[H_3O^+]$?

T **30.** *Complex roots* If you have a computer or a calculator that can be programmed to do complex-number arithmetic, experiment with Newton's method to solve the equation $z^6 - 1 = 0$. The recursion relation to use is

$$z_{n+1} = z_n - \frac{z_n^{\ 6} - 1}{6z_n^{\ 5}} \quad \text{or} \quad z_{n+1} = \frac{5}{6}z_n + \frac{1}{6z_n^{\ 5}}.$$

Try these starting values (among others): $2, i, \sqrt{3} + i$.

Questions to Guide Your Review

1. What can be said about the values of a function that is continuous on a closed interval?
2. What does it mean for a function to have a local extreme value on its domain? An absolute extreme value? How are local and absolute extreme values related, if at all? Give examples.
3. What is true of f' at an interior point where a local extremum occurs? How does this fact lead to a procedure for finding a function's local extreme values?
4. How do you find the absolute extrema of a continuous function on a closed interval? Give examples.
5. What are the hypotheses and conclusion of Rolle's Theorem? Are the hypotheses really necessary? Explain.
6. What are the hypotheses and conclusion of the Mean Value Theorem? What physical interpretations might the theorem have?
7. State the Mean Value Theorem's three corollaries.
8. How can you sometimes identify a function $f(x)$ by knowing f' and knowing the value of f at a point $x = x_0$? Give an example.
9. What is a differential equation? What is a solution to a differential equation? Give examples.
10. What is the First Derivative Test for Increasing and Decreasing? How can it be used to test for local extrema?
11. How do you test a twice-differentiable function to determine where its graph is concave up or concave down? Give examples.
12. What is an inflection point? Give an example. What physical significance do inflection points sometimes have?
13. What is the Second Derivative Test for Local Extrema? Give examples of how it is applied.
14. What do the derivatives of a function tell you about the shape of its graph?
15. What is an autonomous differential equation? What are its equilibrium values? How do they differ from critical points? What is a stable equilibrium value? Unstable?
16. How do you construct the phase line for an autonomous differential equation? How does the phase line help you produce a graph which qualitatively depicts a solution to the differential equation?
17. Outline a general strategy for solving max-min problems. Give examples.
18. What is the linearization $L(x)$ of a function $f(x)$ at a point $x = a$? What is required of f at a for the linearization to exist? How are linearizations used? Give examples.
19. If x moves from a to a nearby value $a + dx$, how do you estimate the corresponding change in the value of a differentiable function $f(x)$? How do you estimate the relative change? The percentage change? Give an example.
20. Describe Newton's method for solving equations. Give an example. What are some of the things to watch out for when you use the method?

Practice Exercises

Conclusions from Graphs

In Exercises 1–4, use the graph to answer the questions.

1. Identify any global extreme values of f and the values of x at which they occur.

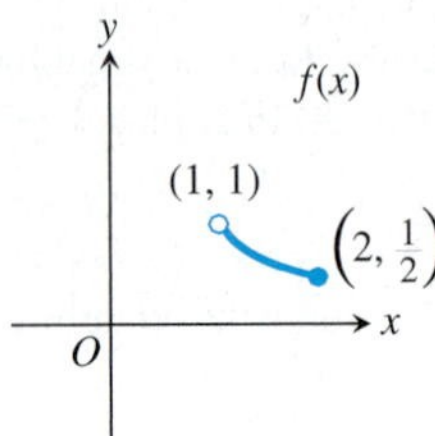

2. At which of the five points on the graph of $y = f(x)$ shown here

(a) are y' and y'' both negative?

(b) is y' negative and y'' positive?

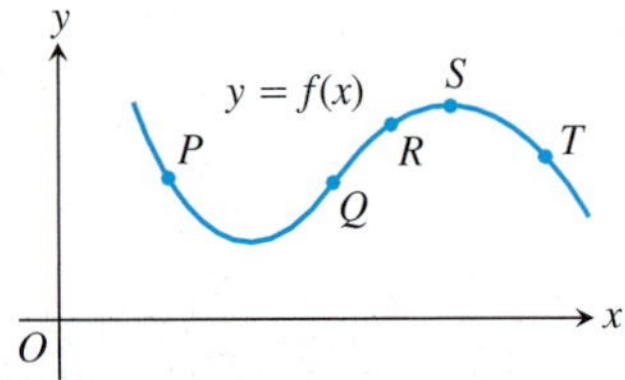

3. Estimate the intervals on which the function $y = f(x)$ is

(a) increasing

(b) decreasing.

(c) Use the given graph of f' to indicate where any local extreme values of the function occur, and whether each extreme is a relative maximum or minimum.

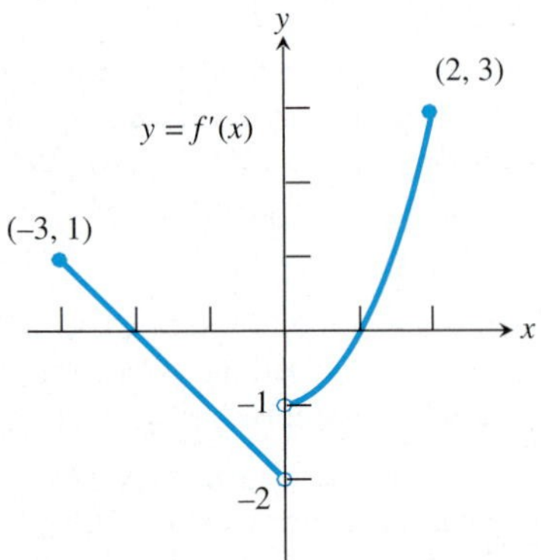

4. Here is the graph of a fruit fly population. On approximately what day did the population's growth rate change from increasing to decreasing?

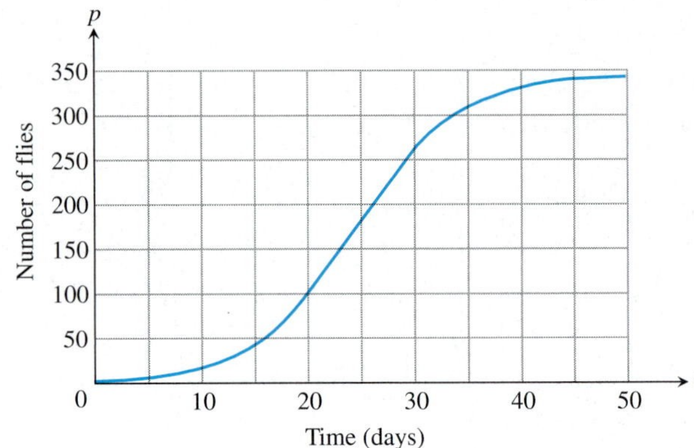

Existence of Extreme Values

5. *Writing to Learn: Local extrema* Does $f(x) = x^3 + 2x + \tan x$ have any local maximum or minimum values? Gives reasons for your answer.

6. *Writing to Learn: Local maxima* Does $g(x) = \csc x + 2 \cot x$ have any local maximum values? Give reasons for your answer.

7. *Writing to Learn: Extreme values* Does $f(x) = (7 + x)(11 - 3x)^{1/3}$ have an absolute minimum value? An absolute maximum? If so, find them or give reasons why they fail to exist. List all critical points of f.

8. *Writing to Learn: Local extrema* Find values of a and b such that the function

$$f(x) = \frac{ax + b}{x^2 - 1}$$

has a local extreme value of 1 at $x = 3$. Is this extreme value a local maximum, or a local minimum? Give reasons for your answer.

9. *Writing to Learn* The greatest integer function $f(x) = \text{int } x$, defined for all values of x, assumes a local maximum value of 0 at each point of $[0, 1)$. Could any of these local maximum values also be local minimum values of f? Give reasons for your answer.

10. (a) Give an example of a differentiable function f whose first derivative is zero at some point c even though f has neither a local maximum nor a local minimum at c.

(b) *Writing to Learn* How is this consistent with Theorem 2 in Section 3.1? Give reasons for your answer.

11. *Absolute extrema* The function $y = 1/x$ does not take on either a maximum or a minimum on the interval $0 < x < 1$ even though the function is continuous on this interval. Does this contradict the Extreme Value Theorem for Continuous Functions? Why?

12. *Absolute extrema* What are the maximum and minimum values of the function $y = |x|$ on the interval $-1 \le x < 1$? Notice that the interval is not closed. Is this consistent with the Extreme Value Theorem for Continuous Functions? Why?

The Mean Value Theorem

13. **(a)** Show that $g(t) = \sin^2 t - 3t$ decreases on every interval in its domain.

(b) *Writing to Learn* How many solutions does the equation $\sin^2 t - 3t = 5$ have? Give reasons for your answer.

14. **(a)** Show that $y = \tan \theta$ increases on every interval in its domain.

(b) *Writing to Learn* If the conclusion in (a) is really correct, how do you explain the fact that $\tan \pi = 0$ is less than $\tan(\pi/4) = 1$?

15. **(a)** Show that the equation $x^4 + 2x^2 - 2 = 0$ has exactly one solution on $[0, 1]$.

T **(b)** Using a calculator, find the solution to as many decimal places as you can.

16. **(a)** *An increasing function* Show that $f(x) = x/(x + 1)$ increases on every interval in its domain.

(b) *A function with no local extrema* Show that $f(x) = x^3 + 2x$ has no local maximum or minimum values.

17. *Water in a reservoir* As a result of a heavy rain, the volume of water in a reservoir increased by 1400 acre-ft in 24 h. Show that at some instant during that period, the reservoir's volume was increasing at a rate in excess of 225,000 gal/min. (An acre-foot is 43,560 ft^3, the volume that would cover one acre to the depth of one foot. A cubic foot holds 7.48 gal.)

18. *Writing to Learn* The formula $F(x) = 3x + C$ gives a different function for each value of C. All these functions, however, have the same derivative with respect to x, namely $F'(x) = 3$. Are these the only differentiable functions whose derivative is 3? Could there be any others? Give reasons for your answers.

19. *Writing to Learn* Show that

$$\frac{d}{dx}\left(\frac{x}{x+1}\right) = \frac{d}{dx}\left(-\frac{1}{x+1}\right)$$

even though

$$\frac{x}{x+1} \neq -\frac{1}{x+1}.$$

Doesn't this contradict Corollary 2 of the Mean Value Theorem? Give reasons for your answer.

20. *Comparing derivatives* Calculate the first derivatives of $f(x) = x^2/(x^2 + 1)$ and $g(x) = -1/(x^2 + 1)$. What can you conclude about the graphs of these functions?

Graphs and Sketching

Sketch the curves in Exercises 21–26.

21. $y = x^2 - (x^3/6)$

22. $y = -x^3 + 6x^2 - 9x + 3$

23. $y = (1/8)(x^3 + 3x^2 - 9x - 27)$

24. $y = x^3(8 - x)$

25. $y = (x - 3)^2 e^x$

26. $y = x\sqrt{3 - x}$

Each of Exercises 27 and 28 gives the first derivative of a function $y = f(x)$.

(a) At what points, if any, does the graph of f have a local maximum, local minimum, or inflection point?

(b) Sketch the general shape of the graph.

27. $y' = 16 - x^2$

28. $y' = 6x(x + 1)(x - 2)$

Finding Extreme Values

In Exercises 29 and 30, find the absolute maxima and minima of the functions and say where they are assumed.

29. $f(x) = e^{x/\sqrt{x^4+1}}$

30. $g(x) = e^{\sqrt{3-2x-x^2}}$

Motion

Drawing conclusions about motion from graphs Each of the graphs in Exercises 31 and 32 is the graph of the position function $s = f(t)$ of a body moving on a coordinate line (t represents time). At approximately what times (if any) is each body's

(a) velocity equal to zero?

(b) acceleration equal to zero?

During approximately what time intervals does the body move

(c) forward?

(d) backward?

31.

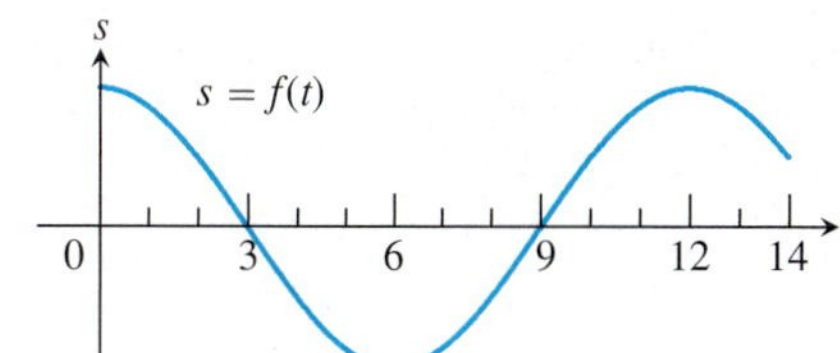

32.

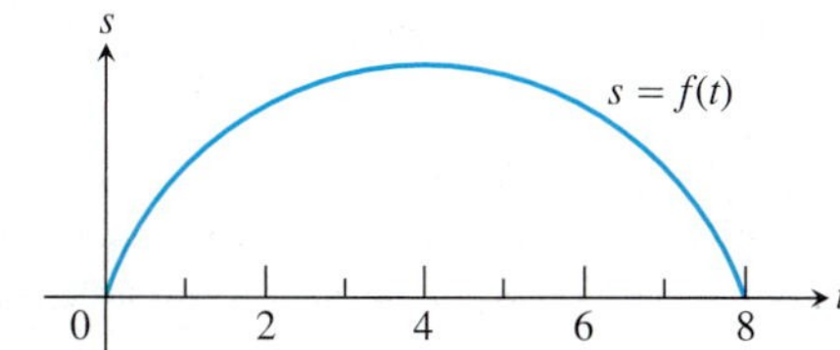

33. *Motion along a line* A particle is moving along a line with position function $s(t) = 3 + 4t - 3t^2 - t^3$. Find the

(a) velocity

(b) acceleration.

(c) Describe the motion of the particle for $t \geq 0$.

34. *Motion along a line* A particle is moving along a line with position function $s(t) = (1/2)t^4 - 4t^3 + 6t^2$, $t \geq 0$. During what time intervals does the particle move forward? Move backward?

Differential Equations

In Exercises 35–38, find all possible functions with the given derivative.

35. $f'(x) = x^{-5} + e^{-x}$

36. $f'(x) = \sec x \tan x$

37. $f'(x) = \dfrac{2}{x^2} + x^2 + 1, x > 0$

38. $f'(x) = \sqrt{x} + \dfrac{1}{\sqrt{x}}$

In Exercises 39 and 40, the velocity v or acceleration a of a particle is given along with its initial position. Find the particle's position s at time t.

39. $v = 9.8t + 5, \quad s = 10 \quad \text{when} \quad t = 0$

40. $a = 32, \quad v = 20 \quad \text{and} \quad s = 5 \quad \text{when} \quad t = 0$

Autonomous Differential Equations and Phase Lines

In Exercises 41 and 42,

(a) Identify the equilibrium values. Which are stable and which are unstable?

(b) Construct a phase line. Identify the signs of y' and y''.

(c) Sketch a representative selection of solution curves.

41. $\dfrac{dy}{dx} = y^2 - 1$

42. $\dfrac{dy}{dx} = y - y^2$

Optimization

43. *Area of sector* If the perimeter of the circular sector shown here is fixed at 100 ft, what values of r and s will give the sector the greatest area?

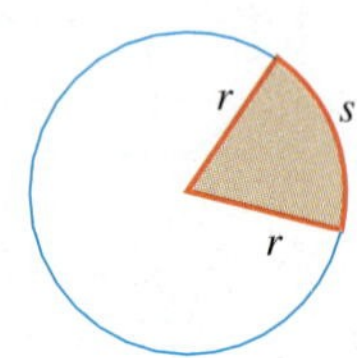

44. *Area of triangle* An isosceles triangle has its vertex at the origin and its base parallel to the x-axis with the vertices above the axis on the curve $y = 27 - x^2$. Find the largest area the triangle can have.

45. *Inscribing a cylinder* Find the height and radius of the largest right circular cylinder that can be put into a sphere of radius $\sqrt{3}$ as described in the figure.

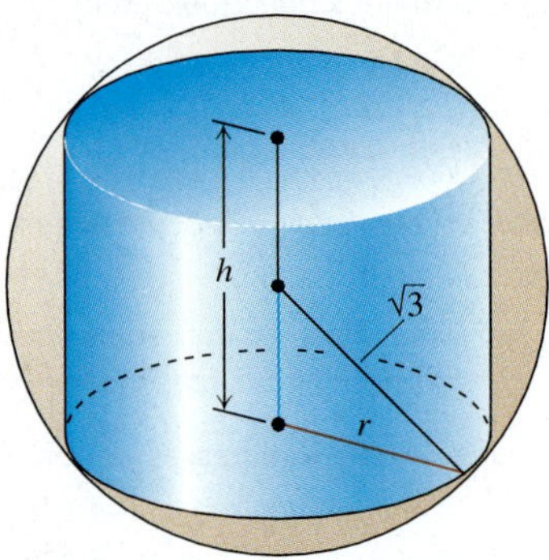

46. *Cone in a cone* The figure here shows two right circular cones, one upside down inside the other. The two bases are parallel, and the vertex of the smaller cone lies at the center of the larger cone's base. What values of r and h will give the smaller cone the largest possible volume?

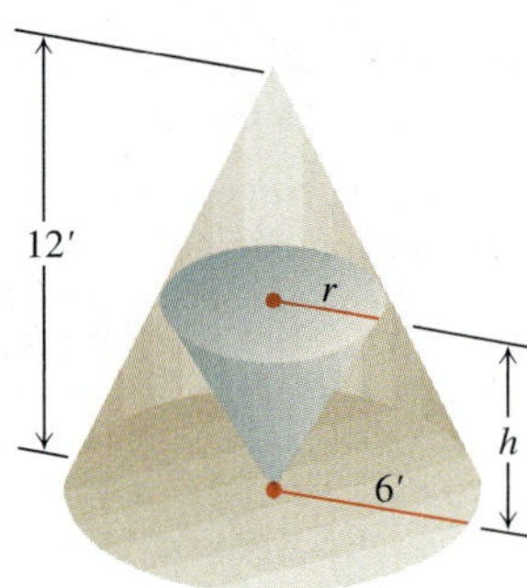

47. *Manufacturing tires* Your company can manufacture x hundred grade A tires and y hundred grade B tires a day, where $0 \leq x \leq 4$ and

$$y = \frac{40 - 10x}{5 - x}.$$

Your profit on a grade A tire is twice your profit on a grade B tire. What is the most profitable number of each kind to make?

48. *Particle motion* The positions of two particles on the s-axis are $s_1 = \cos t$ and $s_2 = \cos(t + \pi/4)$.

(a) What is the farthest apart the particles ever get?

(b) When do the particles collide?

T **49.** *Open-top box* An open-top rectangular box is constructed from a 10-in.-by-16.-in piece of cardboard by cutting squares of equal side length from the corners and folding up the sides. Find analytically the dimensions of the box of largest volume and the maximum volume. Support your answers graphically.

50. *Designing a vat* You are to design an open-top rectangular stainless-steel vat. It is to have a square base and a volume of 32 ft^3, to be welded from quarter-inch plate, and weigh no more than necessary. What dimensions do you recommend?

Linearization

51. Find the linearizations of

(a) $\tan x$ at $x = -\pi/4$ **(b)** $\sec x$ at $x = -\pi/4$.

Graph the curves and linearizations together.

52. We can obtain a useful linear approximation of the function $f(x) = 1/(1 + \tan x)$ at $x = 0$ by combining the approximations

$$\frac{1}{1+x} \approx 1 - x \qquad \text{and} \qquad \tan x \approx x$$

to get

$$\frac{1}{1 + \tan x} \approx 1 - x.$$

Show that this result is the standard linear approximation of $1/(1 + \tan x)$ at $x = 0$.

53. Find the linearization of $f(x) = e^x + \sin x - 0.5$ at $x = 0$.

54. Find the linearization of $f(x) = 2/(1 - x) + \sqrt{1 + x} - 3.1$ at $x = 0$.

Differential Estimates of Change

55. *Volume of a cone* Write a formula that estimates the change that occurs in the volume of a right circular cone when the radius changes from r_0 to $r_0 + dr$ and the height does not change.

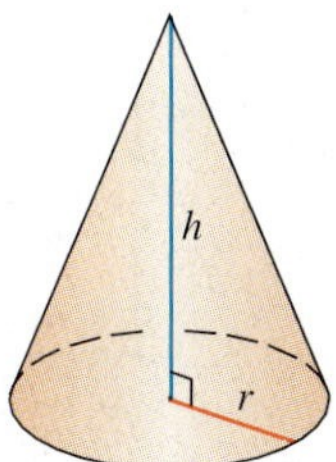

$V = \frac{1}{3}\pi r^2 h$

$S = \pi r\sqrt{r^2 + h^2}$

(Lateral surface area)

56. *Controlling error*

(a) How accurately should you measure the edge of a cube to be reasonably sure of calculating the cube's surface area with an error of no more than 2%?

(b) Suppose that the edge is measured with the accuracy required in part (a). About how accurately can the cube's volume be calculated from the edge measurement? To find out, estimate the percentage error in the volume calculation that might result from using the edge measurement.

57. *Compounding error* The circumference of the equator of a sphere is measured as 10 cm with a possible error of 0.4 cm. This measurement is then used to calculate the radius. The radius is then used to calculate the surface area and volume of the sphere. Estimate the percentage errors in the calculated values of

(a) the radius

(b) the surface area

(c) the volume.

58. *Finding height* To find the height of a lamppost (see figure), you stand a 6 ft pole 20 ft from the lamp and measure the length a of its shadow, finding it to be 15 ft, give or take an inch. Calculate the height of the lamppost using the value $a = 15$ and estimate the possible error in the result.

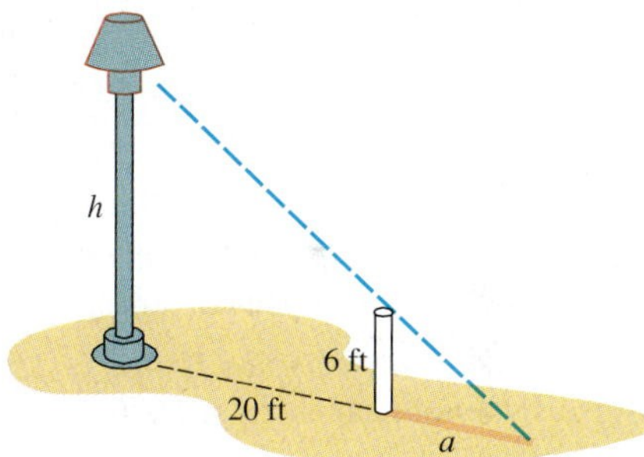

Newton's Method

T In Exercises 59–62, use Newton's method to estimate the zeros of the given function. Use a calculator and state your answers accurate to six decimal places.

59. $f(x) = 3x - x^3, \quad 1 \le x \le 2$

60. $f(x) = x^3 + \dfrac{4}{x^2} + 7, \quad x < 0$

61. $g(t) = 2\cos t - \sqrt{1 - t}, \quad -\infty < t \le 1$

62. $g(t) = \sqrt{t} + \sqrt{1 + t} - 4, \quad t > 0$

Additional Exercises: Theory, Examples, Applications

1. *Writing to Learn* What can you say about a function whose maximum and minimum values on an interval are equal? Give reasons for your answer.

2. *Writing to Learn* Is it true that a discontinuous function cannot have both an absolute maximum and an absolute minimum value on a closed interval? Give reasons for your answer.

3. *Writing to Learn* Can you conclude anything about the extreme values of a continuous function on an open interval? On a half-open interval? Give reasons for your answer.

4. *Local extrema* Use the sign pattern for the derivative

$$\frac{df}{dx} = 6(x-1)(x-2)^2(x-3)^3(x-4)^4$$

to identify the points where f has local maximum and minimum values.

5. *Local extrema*

   (a) Suppose that the first derivative of $y = f(x)$ is

   $$y' = 6(x+1)(x-2)^2.$$

   At what points, if any, does the graph of f have a local maximum, local minimum, or point of inflection?

   (b) Suppose that the first derivative of $y = f(x)$ is

   $$y' = 6x(x+1)(x-2).$$

   At what points, if any, does the graph of f have a local maximum, local minimum, or point of inflection?

6. *Writing to Learn: Bounding a function* If $f'(x) \le 2$ for all x, what is the most the values of f can increase on $[0, 6]$? Give reasons for your answer.

7. *Bounding a function* Suppose that f is continuous on $[a, b]$ and that c is an interior point of the interval. Show that if $f'(x) \le 0$ on $[a, c)$ and $f'(x) \ge 0$ on $(c, b]$, then $f(x)$ is never less than $f(c)$ on $[a, b]$.

8. *An inequality*

   (a) Show that $-1/2 \le x/(1+x^2) \le 1/2$ for every value of x.

   (b) Suppose that f is a function whose derivative is $f'(x) = x/(1+x^2)$. Use the result in part (a) to show that

   $$|f(b) - f(a)| \le \frac{1}{2}|b - a|$$

   for any a and b.

9. *Writing to Learn* The derivative of $f(x) = x^2$ is zero at $x = 0$, but f is not a constant function. Doesn't this contradict the corollary of the Mean Value Theorem that says that functions with zero derivatives are constant? Give reasons for your answer.

10. *Extrema and inflection points* Let $h = fg$ be the product of two differentiable functions of x.

    (a) If f and g are positive, with local maxima at $x = a$, and if f' and g' change sign at a, does h have a local maximum at a?

    (b) If the graphs of f and g have inflection points at $x = a$, does the graph of h have an inflection point at a?

    In either case, if the answer is yes, give a proof. If the answer is no, give a counterexample.

11. *Finding a function* Use the following information to find the values of a, b, and c in the formula $f(x) = (x+a)/(bx^2 + cx + 2)$.

    i. The values of a, b, and c are either 0 or 1.

    ii. The graph of f passes through the point $(-1, 0)$.

    iii. The line $y = 1$ is an asymptote of the graph of f.

12. *Horizontal tangent* For what value or values of the constant k will the curve $y = x^3 + kx^2 + 3x - 4$ have exactly one horizontal tangent?

13. *Largest inscribed triangle* Points A and B lie at the ends of a diameter of a unit circle and point C lies on the circumference. Is it true that the perimeter of triangle ABC is largest when the triangle is isosceles? How do you know?

14. *The ladder problem* What is the approximate length (in feet) of the longest ladder you can carry horizontally around the corner of the corridor shown here? Round your answer down to the nearest foot.

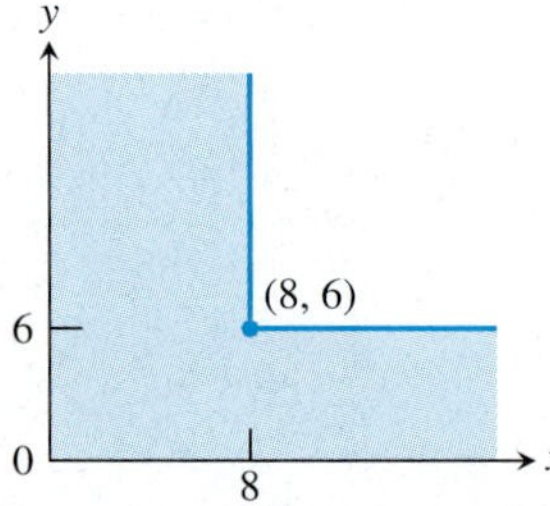

15. *Hole in a water tank* You want to bore a hole in the side of the tank shown here at a height that will make the stream of water coming out hit the ground as far from the tank as possible. If you drill the hole near the top, where the pressure is low, the water

will exit slowly but spend a relatively long time in the air. If you drill the hole near the bottom, the water will exit at a higher velocity but have only a short time to fall. Where is the best place, if any, for the hole? (*Hint:* How long will it take an exiting particle of water to fall from height y to the ground?)

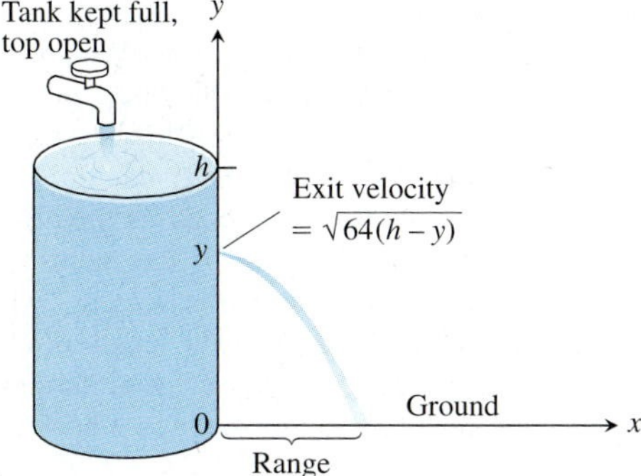

16. *Kicking a field goal* An American football player wants to kick a field goal with the ball being on a right hash mark. Assume that the goal posts are b feet apart and that the hash mark line is a distance $a > 0$ feet from the right goal post. (See the accompanying figure.) Find the distance h from the goal post line that gives the kicker his largest angle β. Assume that the football field is flat.

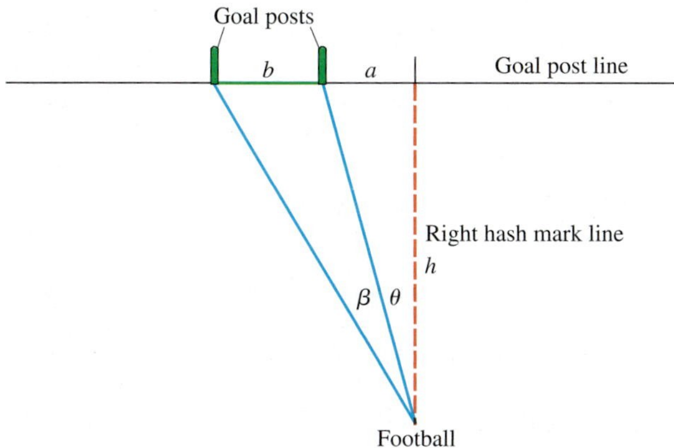

17. *A max-min problem with a variable answer* Sometimes the solution of a max-min problem depends on the proportions of the shapes involved. As a case in point, suppose that a right circular cylinder of radius r and height h is inscribed in a right circular cone of radius R and height H, as shown here. Find the value of r (in terms of R and H) that maximizes the total surface area of the cylinder (including top and bottom). As you will see, the solution depends on whether $H \le 2R$ or $H > 2R$.

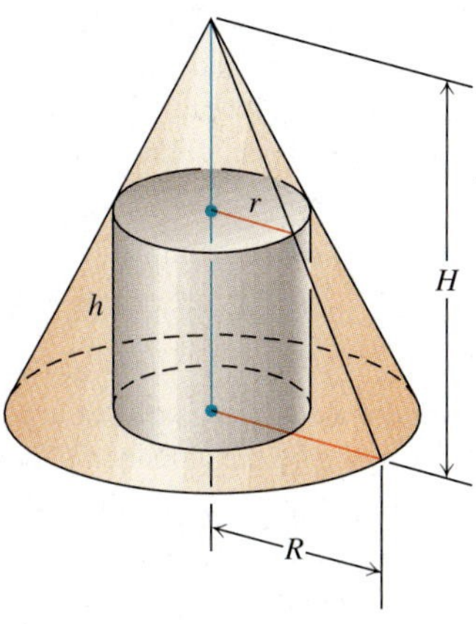

18. *Minimizing a parameter* Find the smallest value of the positive constant m that will make $mx - 1 + (1/x)$ greater than or equal to zero for all positive values of x.

19. *The second derivative test* The second derivative test for local maxima and minima (Section 3.3) says:

(a) f has a local maximum value at $x = c$ if $f'(c) = 0$ and $f''(c) < 0$

(b) f has a local minimum value at $x = c$ if $f'(c) = 0$ and $f''(c) > 0$.

To prove statement (a), let $\epsilon = (1/2)|f''(c)|$. Then use the fact that

$$f''(c) = \lim_{h \to 0} \frac{f'(c+h) - f'(c)}{h} = \lim_{h \to 0} \frac{f'(c+h)}{h}$$

to conclude that for some $\delta > 0$,

$$0 < |h| < \delta \Rightarrow \frac{f'(c+h)}{h} < f''(c) + \epsilon < 0.$$

Thus, $f'(c+h)$ is positive for $-\delta < h < 0$ and negative for $0 < h < \delta$. Prove statement (b) in a similar way.

20. *Schwarz's inequality*

(a) Show that if $a > 0$, then $f(x) = ax^2 + 2bx + c \ge 0$ for all (real) x if and only if $b^2 \le ac$.

(b) Derive **Schwarz's inequality,**

$$(a_1b_1 + a_2b_2 + \cdots + a_nb_n)^2 \le (a_1{}^2 + a_2{}^2 + \cdots + a_n{}^2)(b_1{}^2 + b_2{}^2 + \cdots + b_n{}^2),$$

by applying what you learned in part (a) to the sum

$$(a_1x + b_1)^2 + (a_2x + b_2)^2 + \cdots + (a_nx + b_n)^2.$$

(c) Show that equality holds in Schwarz's inequality only if there exists a real number x that makes a_ix equal $-b_i$ for every value of i from 1 to n.

21. *The period of a clock pendulum* The period T of a clock pendulum (time for one full swing and back) is given by the formula $T^2 = 4\pi^2 L/g$, where T is measured in seconds, $g = 32.2$ ft/sec^2, and L, the length of the pendulum, is measured in feet. Find approximately

(a) the length of a clock pendulum whose period is $T = 1$ sec

(b) the change dT in T if the pendulum in part (a) is lengthened 0.01 ft

(c) the amount the clock gains or loses in a day as a result of the period's changing by the amount dT found in part (b).

22. *Estimating reciprocals without division* You can estimate the value of the reciprocal of a number a without ever dividing by a if you apply Newton's method to the function $f(x) = (1/x) - a$. For example, if $a = 3$, the function involved is $f(x) = (1/x) - 3$.

(a) Graph $y = (1/x) - 3$. Where does the graph cross the x-axis?

(b) Show that the recursion formula in this case is

$$x_{n+1} = x_n(2 - 3x_n),$$

so there is no need for division.

T **23.** *Free fall in the fourteenth century* In the middle of the fourteenth century, Albert of Saxony (1316–1390) proposed a model of free fall that assumed that the velocity of a falling body was proportional to the distance fallen. It seemed reasonable to think that a body that had fallen 20 ft might be moving twice as fast as a body that had fallen 10 ft. And besides, none of the instruments in use at the time were accurate enough to prove otherwise. Today, we can see just how far off Albert of Saxony's model was by solving the initial value problem implicit in his model. Solve the problem and compare your solution graphically with the equation $s = 16t^2$. You will see that it describes a motion that starts too slowly at first and then becomes too fast too soon to be realistic.

24. *Group blood testing* During World War II, it was necessary to administer blood tests to large numbers of recruits. There are two standard ways to administer a blood test to N people. In method 1, each person is tested separately. In method 2, the blood samples of x people are pooled and tested as one large sample. If the test is negative, this one test is enough for all x people. If the test is positive, then each of the x people is tested separately, requiring a total of $x + 1$ tests. Using the second method and some probability theory, it can be shown that, on the average, the total number of tests y will be

$$y = N\left(1 - q^x + \frac{1}{x}\right).$$

With $q = 0.99$ and $N = 1000$, find the integer value of x that minimizes y. Also find the integer value of x that maximizes y. (This second result is not important to the real-life situation.) The group testing method was used in World War II with a savings of 80% over the individual testing method, but not with the given value of q.

25. *The best branching angles for blood vessels and pipes* When a smaller pipe branches off from a larger one in a flow system, we may want it to run off at an angle that is best from some energy-saving point of view. We might require, for instance, that energy loss due to friction be minimized along the section AOB shown in the following figure. In this diagram, B is a given point to be reached by the smaller pipe, A is a point in the larger pipe upstream from B, and O is the point where the branching occurs. A law due to Poiseuille states that the loss of energy due to friction in nonturbulent flow is proportional to the length of the path and inversely proportional to the fourth power of the radius. Thus, the loss along AO is $(kd_1)/R^4$ and along OB is $(kd_2)/r^4$, where k is a constant, d_1 is the length of AO, d_2 is the length of OB, R is the radius of the larger pipe, and r is the radius of the smaller pipe. The angle θ is to be chosen to minimize the sum of these two losses:

$$L = k\frac{d_1}{R^4} + k\frac{d_2}{r^4}.$$

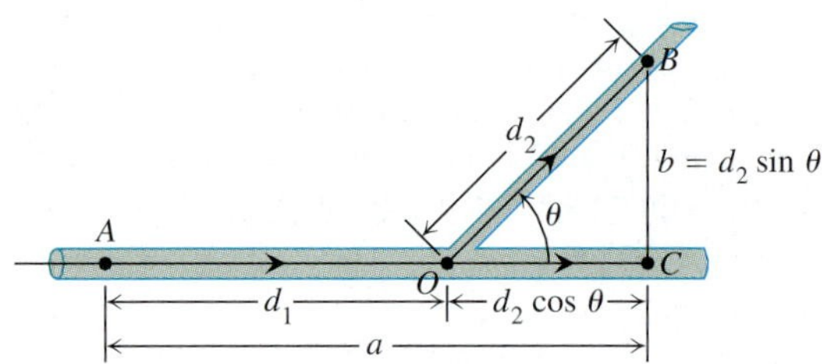

In our model, we assume that $AC = a$ and $BC = b$ are fixed. Thus, we have the relations

$$d_1 + d_2\cos\theta = a \qquad d_2\sin\theta = b$$

so that

$$d_2 = b\csc\theta, \qquad d_1 = a - d_2\cos\theta = a - b\cot\theta.$$

We can express the total loss L as a function of θ:

$$L = k\left(\frac{a - b\cot\theta}{R^4} + \frac{b\csc\theta}{r^4}\right).$$

(a) Show that the critical value of θ for which $dL/d\theta$ equals zero is

$$\theta_c = \cos^{-1}\frac{r^4}{R^4}.$$

(b) If the ratio of the pipe radii is $r/R = 5/6$, estimate to the nearest degree the optimal branching angle given in part (a).

The mathematical analysis described here is also used to explain the angles at which arteries branch in an animal's body. See *Introduction to Mathematics for Life Scientists*, 2nd ed., by E. Batschelet (New York: Springer-Verlag, 1976).

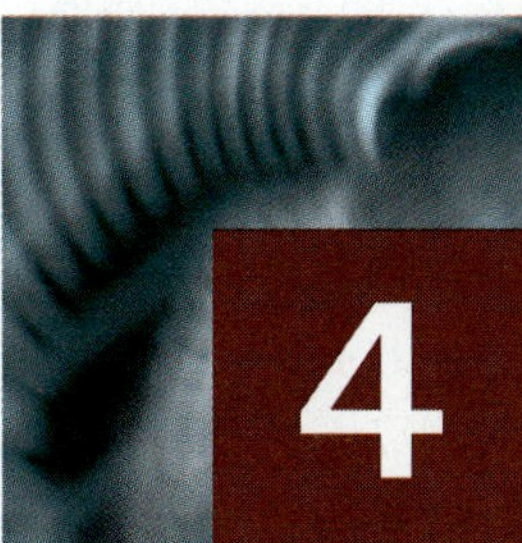

4 Integration

OVERVIEW We have seen how the need to calculate instantaneous rates of change led the discoverers of calculus to an investigation of the slopes of tangent lines and, ultimately, to the derivative, to what we call *differential* calculus. But they knew that derivatives revealed only half the story. In addition to a calculation method (a "calculus") to describe how functions were changing at a given instant, they also needed a method to describe how those instantaneous changes could accumulate over an interval to produce the function. That is, by studying how a behavior *changed*, they wanted to learn about the behavior itself. For example, from knowing the velocity of a moving object, they wanted to be able to determine its position as a function of time. This is why they were also investigating *areas under curves,* an investigation that ultimately led to the second main branch of calculus, called *integral* calculus.

Once they had the calculus for finding slopes of tangent lines and the calculus for finding areas under curves—two geometric operations that would seem to have nothing at all to do with each other—the challenge for Newton and Leibniz was to prove the connection that they know intuitively had to be there. The discovery of this connection (called the Fundamental Theorem of Calculus) brought differential and integral calculus together to become the single most powerful tool mathematicians ever acquired for understanding the universe.

4.1 Indefinite Integrals, Differential Equations, and Modeling

Finding Antiderivatives: Indefinite Integrals • Initial Value Problems • Mathematical Modeling

The process of determining a function $f(x)$ from one of its known values and its derivative $f'(x)$ has two steps. The first is to find a formula that gives all the functions that could possibly have f as a derivative. These functions are called the *antiderivatives* of f, and the formula that gives them all is called the *indefinite integral* of f. The second step is to use the known value to select the particular antiderivative we want from those in the indefinite integral.

Finding a formula that gives all a function's antiderivatives might seem like an impossible task, or at least to require a little magic, but this is not the case at all. If we can find even one antiderivative, we can find them all, because of the first two corollaries of the Mean Value Theorem of Section 3.2.

Finding Antiderivatives: Indefinite Integrals

We begin with a definition.

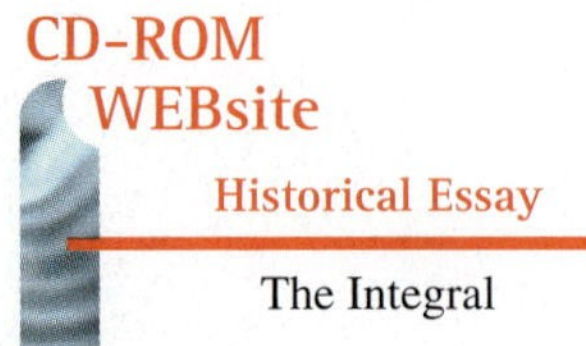

Definition Antiderivative of a Function

A function $F(x)$ is an **antiderivative** of a function $f(x)$ if

$$F'(x) = f(x)$$

for all x in the domain of f. The set of all antiderivatives of f is the **indefinite integral** of f with respect to x, denoted by

$$\int f(x)\,dx.$$

The symbol $\int$ is an **integral sign.** The function f is the **integrand** of the integral and x is the **variable of integration.**

According to Corollary 2 of the Mean Value Theorem (Section 3.2), once we have found one antiderivative F of a function f, the other antiderivatives differ from it by a constant. We indicate this in integral notation in the following way:

$$\int f(x)\,dx = F(x) + C \tag{1}$$

The constant C is the **constant of integration** or **arbitrary constant.** Equation (1) is read, "The indefinite integral of f with respect to x is $F(x) + C$." When we find $F(x) + C$, we say that we have **integrated** f and **evaluated** the integral.

Example 1 Finding An Indefinite Integral

Evaluate $\int e^{2x}\,dx$.

Solution

$$\int e^{2x}\,dx = \underbrace{\tfrac{1}{2}e^{2x}}_{\text{an antiderivative of } e^{2x}} + \underbrace{C}_{\text{the arbitrary constant}}$$

The formula $(1/2)e^{2x} + C$ generates all the antiderivatives of the function e^{2x}. The functions $(1/2)e^{2x} + 1$, $(1/2)e^{2x} - \pi$, and $(1/2)e^{2x} + \sqrt{2}$ are all antiderivatives of the function e^{2x}, as you can check by differentiation.

Many of the indefinite integrals needed in scientific work are found by reversing derivative formulas. You will see what we mean if you look at Table 4.1, which lists a number of standard integral forms side by side with their derivative-formula sources.

In case you are wondering why the integrals of the tangent, cotangent, secant, and cosecant do not appear in the table, the answer is that the usual formulas for them require logarithms. In Section 4.2, we find their antiderivatives.

Table 4.1 Integral formulas

Indefinite integral	Reversed derivative formula
1. $\int x^n\,dx = \dfrac{x^{n+1}}{n+1} + C$, $n \neq -1$, n rational	$\dfrac{d}{dx}\left(\dfrac{x^{n+1}}{n+1}\right) = x^n$
$\int dx = \int 1\,dx = x + C$ (special case)	$\dfrac{d}{dx}(x) = 1$
2. $\int \sin kx\,dx = -\dfrac{\cos kx}{k} + C$	$\dfrac{d}{dx}\left(-\dfrac{\cos kx}{k}\right) = \sin kx$
3. $\int \cos kx\,dx = \dfrac{\sin kx}{k} + C$	$\dfrac{d}{dx}\left(\dfrac{\sin kx}{k}\right) = \cos kx$
4. $\int \sec^2 x\,dx = \tan x + C$	$\dfrac{d}{dx}\tan x = \sec^2 x$
5. $\int \csc^2 x\,dx = -\cot x + C$	$\dfrac{d}{dx}(-\cot x) = \csc^2 x$
6. $\int \sec x \tan x\,dx = \sec x + C$	$\dfrac{d}{dx}\sec x = \sec x \tan x$
7. $\int \csc x \cot x\,dx = -\csc x + C$	$\dfrac{d}{dx}(-\csc x) = \csc x \cot x$
8. $\int e^{kx}\,dx = \dfrac{1}{k}e^{kx} + C$	$\dfrac{d}{dx}\left(\dfrac{1}{k}e^{kx}\right) = e^{kx}$
9. $\int \dfrac{1}{x}\,dx = \ln x + C, \quad x > 0$	$\dfrac{d}{dx}(\ln x) = \dfrac{1}{x}, \quad x > 0$
10. $\int \dfrac{1}{\sqrt{1-x^2}}\,dx = \sin^{-1} x + C$	$\dfrac{d}{dx}(\sin^{-1} x) = \dfrac{1}{\sqrt{1-x^2}}$
11. $\int \dfrac{1}{1+x^2}\,dx = \tan^{-1} x + C$	$\dfrac{d}{dx}(\tan^{-1} x) = \dfrac{1}{1+x^2}$
12. $\int \dfrac{1}{x\sqrt{x^2-1}}\,dx = \sec^{-1} x + C$, $x > 1$	$\dfrac{d}{dx}(\sec^{-1} x) = \dfrac{1}{x\sqrt{x^2-1}}, \quad x > 1$
13. $\int a^x\,dx = \left(\dfrac{1}{\ln a}\right)a^x$, $a > 0, \quad a \neq 1$	$\dfrac{d}{dx}\left(\dfrac{1}{\ln a}a^x\right) = a^x, \quad a > 0, a \neq 1$

Example 2 Selected Integrals from Table 4.1

(a) $\int x^5\,dx = \dfrac{x^6}{6} + C$ Formula 1 with $n = 5$

(b) $\int \dfrac{1}{\sqrt{x}}\,dx = \int x^{-1/2}\,dx = 2x^{1/2} + C = 2\sqrt{x} + C$ Formula 1 with $n = -1/2$

(c) $\int \sin 2x\, dx = -\frac{\cos 2x}{2} + C$ — Formula 2 with $k = 2$

(d) $\int e^{-3x}\, dx = -\frac{1}{3}e^{-3x} + C$ — Formula 8 with $k = -3$

(e) $\int 2^x\, dx = \left(\frac{1}{\ln 2}\right)2^x + C$ — Formula 13 with $a = 2$

(f)
$$\int \cos \frac{x}{2}\, dx = \int \cos\left(\frac{1}{2}x\right) dx$$
$$= \frac{\sin (1/2)x}{1/2} + C = 2 \sin \frac{x}{2} + C$$
Formula 3 with $k = 1/2$

Finding an integral formula can sometimes be difficult, but checking it, once found, is relatively easy: differentiate the right-hand side. The derivative should be the integrand.

Example 3 Checking Correctness of an Indefinite Integral

Right: $\int x \cos x\, dx = x \sin x + \cos x + C$

Reason: The derivative of the right-hand side is the integrand:

$$\frac{d}{dx}(x \sin x + \cos x + C) = x \cos x + \sin x - \sin x + 0 = x \cos x.$$

Wrong: $\int x \cos x\, dx = x \sin x + C$

Reason: The derivative of the right-hand side is not the integrand:

$$\frac{d}{dx}(x \sin x + C) = x \cos x + \sin x + 0 \neq x \cos x.$$

Do not worry about how to derive the correct integral formula in Example 3. We present a technique for doing so in Chapter 7.

Initial Value Problems

The problem of finding a function y of x when we know its derivative and its value y_0 at a particular point x_0 is called an **initial value problem.** We solve such problems in two steps, as demonstrated in Example 4.

Example 4 Finding a Curve from its Slope Function and a Point

Find the curve whose slope at the point (x, y) is $3x^2$ if the curve is required to pass through the point $(1, -1)$.

Solution In mathematical language, we are asked to solve the initial value problem that consists of the following.

The differential equation: $\frac{dy}{dx} = 3x^2$ — The curve's slope is $3x^2$.

The initial condition: $y(1) = -1$

1. Solve the differential equation:

$$\frac{dy}{dx} = 3x^2$$
$$\int \frac{dy}{dx}\,dx = \int 3x^2\,dx$$
$$y + C_1 = x^3 + C_2$$
$$y = x^3 + C \qquad \text{Constants of integration combined, giving the general solution}$$

This result tells us that y eqauls $x^3 + C$ for some value of C. We find that value from the condition $y(1) = -1$.

2. Evaluate C:

$$y = x^3 + C$$
$$-1 = (1)^3 + C \qquad \text{Initial condition } y(1) = -1$$
$$C = -2.$$

The curve we want is $y = x^3 - 2$ (Figure 4.1).

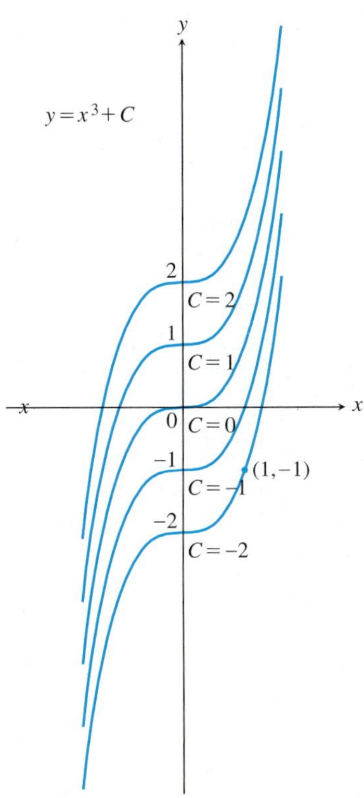

FIGURE 4.1 The curves $y = x^3 + C$ fill the coordinate plane without overlapping. In Example 4, we identify the curve $y = x^3 - 2$ as the one that passes through the given point $(1, -1)$.

The indefinite integral $F(x) + C$ of the function $f(x)$ gives the **general solution** $y = F(x) + C$ of the differential equation $dy/dx = f(x)$. The general solution gives all the solutions of the equation (there are infinitely many, one for each value of C). We **solve** the differential equation by finding its general solution. We then solve the initial value problem by finding the **particular solution** that satisfies the initial condition $y(x_0) = y_0$ (y has the value y_0 when $x = x_0$).

Solving initial value problems is important in mathematical modeling, the process by which we, as scientists and engineers, use mathematics to learn about the real world.

Mathematical Modeling

The development of a mathematical model usually takes four steps: First we observe something in the real world (for example, a ball bearing falling from rest or the trachea contracting during a cough) and construct a system of mathematical variables and relationships that imitate some of its important features. We build a mathematical metaphor for what we see. Next we apply mathematics to the variables and relationships to solve the model and draw conclusions about the variables. After that we translate the mathematical conclusions into information about the system under study. Finally, we check the information against observation to see if the model has predictive value. We also investigate the possibility that the model applies to other systems. The really good models are the ones that lead to conclusions that are consistent with observation, that have predictive value and broad application, and that are not too hard to use.

The natural cycle of mathematical imitation, deduction, interpretation, and confirmation is shown in the diagram for free fall.

The Modeling Process

Step 1. Observe real-world behavior.

Step 2. Make assumptions to identify variables and their relationships, creating a model.

Step 3. Solve the model to obtain mathematical solutions.

Step 4. Interpret the model and verify it is consistent with real-world observations.

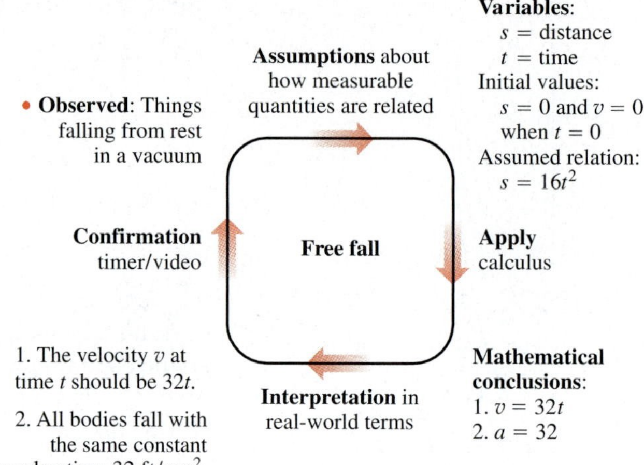

Observed: Things falling from rest in a vacuum
Assumptions about how measurable quantities are related
Variables:
s = distance
t = time
Initial values:
s = 0 and v = 0
when t = 0
Assumed relation:
s = 16t²
Confirmation timer/video
Free fall
Apply calculus
1. The velocity v at time t should be 32t.
2. All bodies fall with the same constant acceleration: 32 ft/sec².
Interpretation in real-world terms
Mathematical conclusions:
1. v = 32t
2. a = 32

Example 5 Dropping a Package from an Ascending Balloon

A balloon ascending at the rate of 12 ft/sec is at a height 80 ft above the ground when a package is dropped. How long does it take the package to reach the ground?

Solution Let $v(t)$ denote the velocity of the package at time t, and let $s(t)$ denote its height above the ground. The acceleration of gravity near the surface of the earth is 32 ft/sec^2. Assuming no other forces act on the dropped package, we have

$$\frac{dv}{dt} = -32. \qquad \text{Negative because gravity acts in the direction of decreasing } s.$$

This leads to the initial value problem

Differential equation: $\dfrac{dv}{dt} = -32$

Initial condition: $v(0) = 12,$

which is our mathematical model for the package's motion. We solve the initial value problem to obtain the velocity of the package.

1. Solve the differential equation:

$$\frac{dv}{dt} = -32$$

$$\int \frac{dv}{dt}\,dt = \int -32\,dt$$

$$v = -32t + C. \qquad \text{Constants combined as one}$$

Having found the general solution of the differential equation, we use the initial condition to find the particular solution that solves our problem.

2. Evaluate C:

$$12 = -32(0) + C \qquad \text{Initial condition } v(0) = 12$$
$$C = 12.$$

The solution of the initial value problem is

$$v = -32t + 12.$$

Since velocity is the derivative of height and the height of the package is 80 ft at the time $t = 0$ when it is dropped, we now have a second initial value problem.

Differential equation: $\dfrac{ds}{dt} = -32t + 12$ Set $v = ds/dt$ in the last equation.

Initial condition: $s(0) = 80$

We solve this initial value problem to find the height as a function of t.

Solve the differential equation:

$$\frac{ds}{dt} = -32t + 12$$
$$\int \frac{ds}{dt}\,dt = \int (-32t + 12)\,dt$$
$$s = -16t^2 + 12t + C. \qquad \text{Constants of integration combined, giving the general solution.}$$

Evaluate C:

$$80 = -16(0)^2 + 12(0) + C \qquad \text{Initial condition } s(0) = 80$$
$$C = 80.$$

The package's height above ground at time t is

$$s = -16t^2 + 12t + 80.$$

Use the solution: To find how long it takes the package to reach the ground, we set s equal to 0 and solve for t:

$$-16t^2 + 12t + 80 = 0$$
$$-4t^2 + 3t + 20 = 0$$
$$t = \frac{-3 \pm \sqrt{329}}{-8} \qquad \text{Quadratic formula}$$
$$t \approx -1.89, \qquad t \approx 2.64.$$

The package hits the ground about 2.64 sec after it is dropped from the balloon. (The negative root has no physical meaning.)

EXERCISES 4.1

Finding Antiderivatives

In Exercises 1–8, find an antiderivative for each function. Do as many as you can mentally. Check your answers by differentiation.

1. **(a)** $6x$ **(b)** x^7 **(c)** $x^7 - 6x + 8$
2. **(a)** $-3x^{-4}$ **(b)** x^{-4} **(c)** $x^{-4} + 2x + 3$
3. **(a)** $-\dfrac{2}{x^3}$ **(b)** $\dfrac{1}{2x^3}$ **(c)** $x^3 - \dfrac{1}{x^3}$
4. **(a)** $\dfrac{3}{2}\sqrt{x}$ **(b)** $\dfrac{1}{2\sqrt{x}}$ **(c)** $\sqrt{x} + \dfrac{1}{\sqrt{x}}$
5. **(a)** $\dfrac{2}{3}x^{-1/3}$ **(b)** $\dfrac{1}{3}x^{-2/3}$ **(c)** $-\dfrac{1}{3}x^{-4/3}$
6. **(a)** $-\pi \sin \pi x$ **(b)** $3 \sin x$ **(c)** $\sin \pi x - 3 \sin 3x$
7. **(a)** $\sec^2 x$ **(b)** $\dfrac{2}{3}\sec^2 \dfrac{x}{3}$ **(c)** $-\sec^2 \dfrac{3x}{2}$
8. **(a)** $\sec x \tan x$ **(b)** $4 \sec 3x \tan 3x$ **(c)** $\sec \dfrac{\pi x}{2} \tan \dfrac{\pi x}{2}$

Evaluating Integrals

Evaluate the integrals in Exercises 9–26. Check your answers by differentiation.

9. $\displaystyle\int (x + 1)\, dx$
10. $\displaystyle\int \left(3t^2 + \frac{t}{2}\right) dt$
11. $\displaystyle\int \left(\frac{1}{x} - \frac{5}{x^2 + 1}\right) dx$
12. $\displaystyle\int \left(\frac{1}{x^2} - x^2 - \frac{1}{3}\right) dx$
13. $\displaystyle\int \left(e^{-x} + 4^x\right) dx$
14. $\displaystyle\int (\sqrt{x} + \sqrt[3]{x})\, dx$
15. $\displaystyle\int \left(\frac{2}{\sqrt{1 - y^2}} - \frac{1}{y^{1/4}}\right) dy$
16. $\displaystyle\int \left(\frac{\sqrt{x}}{2} + \frac{2}{\sqrt{x}}\right) dx$
17. $\displaystyle\int \left(\frac{1}{7} - \frac{1}{y^{5/4}}\right) dy$
18. $\displaystyle\int 2x(1 - x^{-3})\, dx$
19. $\displaystyle\int \frac{t\sqrt{t} + \sqrt{t}}{t^2}\, dt$
20. $\displaystyle\int (-2 \cos t)\, dt$
21. $\displaystyle\int 7 \sin \frac{\theta}{3}\, d\theta$
22. $\displaystyle\int (-3 \csc^2 x)\, dx$
23. $\displaystyle\int (1 + \tan^2 \theta)\, d\theta$ (*Hint:* $1 + \tan^2 \theta = \sec^2 \theta$)
24. $\displaystyle\int \cot^2 x\, dx$ (*Hint:* $1 + \cot^2 x = \csc^2 x$)
25. $\displaystyle\int \cos \theta (\tan \theta + \sec \theta)\, d\theta$
26. $\displaystyle\int \frac{\csc \theta}{\csc \theta - \sin \theta}\, d\theta$

Checking Integration Formulas

Verify the integral formulas in Exercises 27–30 by differentiation. In Section 4.2, we will see where formulas like these come from.

27. $\displaystyle\int (7x - 2)^3\, dx = \frac{(7x - 2)^4}{28} + C$
28. $\displaystyle\int (3x + 5)^{-2}\, dx = -\frac{(3x + 5)^{-1}}{3} + C$
29. $\displaystyle\int \csc^2\left(\frac{x - 1}{3}\right) dx = -3 \cot\left(\frac{x - 1}{3}\right) + C$
30. $\displaystyle\int \frac{1}{x + 1}\, dx = \ln (x + 1) + C, \qquad x > -1$
31. Right or wrong? Say which for each formula and give a brief reason for each answer.

    (a) $\displaystyle\int x \sin x\, dx = \frac{x^2}{2} \sin x + C$

    (b) $\displaystyle\int x \sin x\, dx = -x \cos x + C$

    (c) $\displaystyle\int x \sin x\, dx = -x \cos x + \sin x + C$
32. Right or wrong? Say which for each formula and give a brief reason for each answer.

    (a) $\displaystyle\int (2x + 1)^2\, dx = \frac{(2x + 1)^3}{3} + C$

    (b) $\displaystyle\int 3(2x + 1)^2\, dx = (2x + 1)^3 + C$

    (c) $\displaystyle\int 6(2x + 1)^2\, dx = (2x + 1)^3 + C$

Initial Value Problems

33. *Writing to Learn* Which of the following graphs show the solution of the initial value problem

$$\frac{dy}{dx} = 2x, \qquad y = 4 \text{ when } x = 1?$$

(a) (b) (c)

Give reasons for your answer.

34. *Writing to Learn* Which of the following graphs shows the solution of the initial value problem

$$\frac{dy}{dx} = -x, \qquad y = 1 \text{ when } x = -1?$$

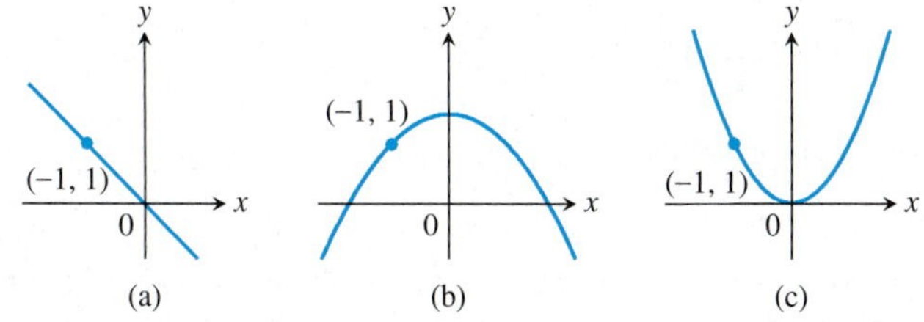

(a) (b) (c)

Give reasons for your answer.

Solve the initial value problems in Exercises 35–46.

35. $\dfrac{dy}{dx} = 2x - 7, \quad y(2) = 0$

36. $\dfrac{dy}{dx} = \dfrac{1}{x^2} + x, \quad x > 0; \quad y(2) = 1$

37. $\dfrac{dy}{dx} = 3x^{-2/3}, \quad y(-1) = -5$

38. $\dfrac{dy}{dx} = \dfrac{1}{2x}, \quad y(1) = -1$

39. $\dfrac{ds}{dt} = \cos t + \sin t, \quad s(\pi) = 1$

40. $\dfrac{dr}{d\theta} = -\pi \sin \pi\theta, \quad r(0) = 0$

41. $\dfrac{dv}{dt} = \dfrac{3}{t\sqrt{t^2 - 1}}, \quad t > 1, v(2) = 0$

42. $\dfrac{dv}{dt} = \dfrac{8}{1 + t^2} + \sec^2 t, \quad v(0) = 1$

43. $\dfrac{d^2y}{dx^2} = 2 - 6x; \quad y'(0) = 4, \quad y(0) = 1$

44. $\dfrac{d^2r}{dt^2} = \dfrac{2}{t^3}; \quad \left.\dfrac{dr}{dt}\right|_{t=1} = 1, \quad r(1) = 1$

45. $\dfrac{d^3y}{dx^3} = 6; \quad y''(0) = -8, \quad y'(0) = 0, \quad y(0) = 5$

46. $y^{(4)} = -\sin t + \cos t; \quad y'''(0) = 7, \quad y''(0) = y'(0) = -1, \quad y(0) = 0$

Finding Position from Velocity

Exercises 47 and 48 give the velocity $v = ds/dt$ and initial position of a body moving along a coordinate line. Find the body's position at time t.

47. $v = 9.8t + 5, \quad s(0) = 10$

48. $v = \dfrac{2}{\pi} \cos \dfrac{2t}{\pi}, \quad s(\pi^2) = 1$

Finding Position from Acceleration

Exercises 49 and 50 give the acceleration $a = d^2s/dt^2$, initial velocity, and initial position of a body moving on a coordinate line. Find the body's position at time t.

49. $a = 32; \quad v(0) = 20, \quad s(0) = 5$

50. $a = -4 \sin 2t; \quad v(0) = 2, \quad s(0) = -3$

Finding Curves

51. Find the curve $y = f(x)$ in the xy-plane that passes through the point (9, 4) and whose slope at each point is $3\sqrt{x}$.

52. **(a)** Find a curve $y = f(x)$ with the following properties:

i. $\dfrac{d^2y}{dx^2} = 6x$

ii. Its graph passes through the point (0, 1) and has a horizontal tangent there.

(b) *Writing to Learn* How many curves like this are there? How do you know?

Solution (Integral) Curves

Exercises 53–56 show solution curves of differential equations. In each exercise, find an equation for the curve through the labeled point.

53. **54.**

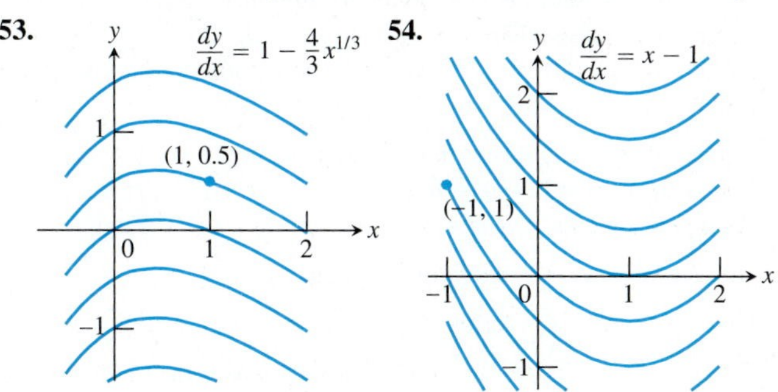

55.

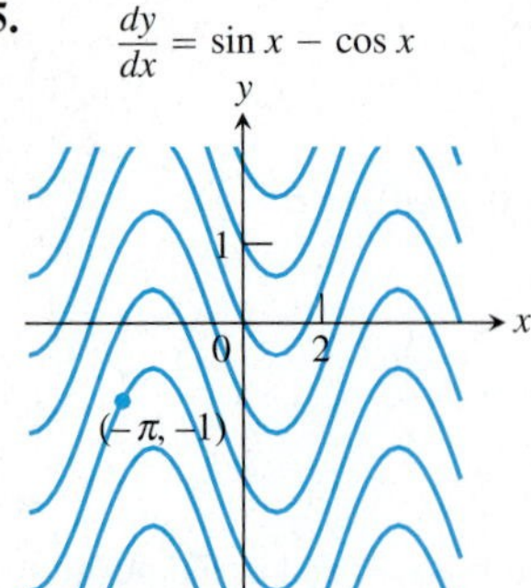

56.

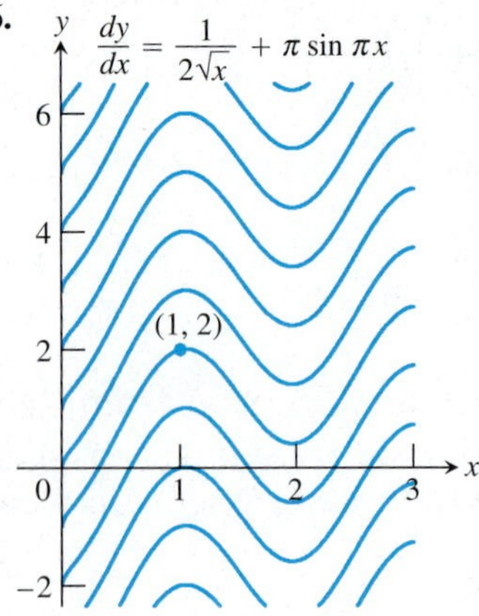

Applications

57. *Falling on the moon* On the moon, the acceleration due to gravity is 1.6 m/sec^2. If a rock is dropped into a crevasse, how fast will it be going just before it hits the bottom 30 sec later?

58. *Liftoff from Earth* A rocket lifts off the surface of Earth with a constant acceleration of 20 m/sec^2. How fast will the rocket be going 1 min later?

59. *Stopping a car in time* You are driving along a highway at a steady 60 mph (88 ft/sec) when you see an accident ahead and slam on the brakes. What constant deceleration is required to stop your car in 242 ft? To find out, carry out the following steps.

Step 1: Solve the initial value problem

Differential equation: $\dfrac{d^2s}{dt^2} = -k \quad (k \text{ constant})$

Initial conditions: $\dfrac{ds}{dt} = 88 \text{ and } s = 0 \text{ when } t = 0.$

Measuring time and distance from when the brakes are applied

Step 2: Find the value of t that makes $ds/dt = 0$. (The answer will involve k.)

Step 3: Find the value of k that makes $s = 242$ for the value of t you found in step 2.

60. *Stopping a motorcycle* The State of Illinois Cycle Rider Safety Program requires riders to be able to brake from 30 mph (44 ft/sec) to 0 in 45 ft. What constant deceleration does it take to do that?

61. *Motion along a coordinate line* A particle moves on a coordinate line with acceleration $a = d^2s/dt^2 = 15\sqrt{t} - (3/\sqrt{t})$, subject to the conditions that $ds/dt = 4$ and $s = 0$ when $t = 1$. Find

(a) the velocity $v = ds/dt$ in terms of t

(b) the position s in terms of t.

T **62.** *The hammer and the feather* When *Apollo 15* astronaut David Scott dropped a hammer and a feather on the moon to demonstrate that in a vacuum all bodies fall with the same (constant) acceleration, he dropped them from about 4 ft above the ground. The television footage of the event shows the hammer and feather falling more slowly than on Earth, where, in a vacuum, they would have taken only half a second to fall the 4 ft. How long did it take the hammer and feather to fall 4 ft on the moon? To find out, solve the following initial value problem for s as a function of t. Then find the value of t that makes s equal to 0.

Differential equation: $\dfrac{d^2s}{dt^2} = -5.2 \text{ ft/sec}^2$

Initial conditions: $\dfrac{ds}{dt} = 0 \text{ and } s = 4 \text{ when } t = 0$

63. *Motion with constant acceleration* The standard equation for the position s of a body moving with a constant acceleration a along a coordinate line is

$$s = \frac{a}{2}t^2 + v_0 t + s_0, \tag{2}$$

where v_0 and s_0 are the body's velocity and position at time $t = 0$. Derive this equation by solving the initial value problem

Differential equation: $\dfrac{d^2s}{dt^2} = a$

Initial conditions: $\dfrac{ds}{dt} = v_0 \text{ and } s = s_0 \text{ when } t = 0.$

64. *Writing to Learn: Free fall near the surface of a planet* *(Continuation of Exercise 63)* For free fall near the surface of a planet where the acceleration due to gravity has a constant magnitude of g length-units/sec^2, Equation (2) in Exercise 63 takes the form

$$s = -\frac{1}{2}gt^2 + v_0 t + s_0, \tag{3}$$

where s is the body's height above the surface. The equation has a minus sign because the acceleration acts downward, in the direction of decreasing s. The velocity v_0 is positive if the object is rising at time $t = 0$ and negative if the object is falling.

Instead of using the result of Exercise 63, you can derive Equation (3) directly by solving an appropriate initial value problem. What initial value problem? Solve it to be sure you have the right one, explaining the solution steps as you go along.

Theory and Examples

65. *Finding displacement from an antiderivative of velocity*

(a) Suppose that the velocity of a body moving along the s-axis is

$$\frac{ds}{dt} = v = 9.8t - 3.$$

i. Find the body's displacement over the time interval from $t = 1$ to $t = 3$ given that $s = 5$ when $t = 0$.

ii. Find the body's displacement from $t = 1$ to $t = 3$ given that $s = -2$ when $t = 0$.

iii. Now find the body's displacement from $t = 1$ to $t = 3$ given that $s = s_0$ when $t = 0$.

(b) Suppose that the position s of a body moving along a coordinate line is a differentiable function of time t. Is it true that once you know an antiderivative of the velocity function ds/dt you can find the body's displacement from $t = a$ to $t = b$ even if you do not know the body's exact position at either of those times? Give reasons for your answer.

66. *Uniqueness of solutions* If differentiable functions $y = F(x)$ and $y = G(x)$ both solve the initial value problem

$$\frac{dy}{dx} = f(x), \qquad y(x_0) = y_0,$$

on an interval I, must $F(x) = G(x)$ for every x in I? Give reasons for your answer.

COMPUTER EXPLORATIONS

Use a CAS to solve the initial value problems in Exercises 67–72. Plot the solution curves.

67. $y' = \cos^2 x + \sin x, \quad y(\pi) = 1$

68. $y' = xe^x, \quad y(\ln 2) = 0$

69. $y' = \ln x, \quad y(1) = 2$

70. $y' = \dfrac{1}{\sqrt{4 - x^2}}, \quad y(0) = 2$

71. $y'' = 3e^{x/2} + x, \quad y(0) = -1, \quad y'(0) = 4$

72. $y'' = \dfrac{2}{x} + \sqrt{x}, \quad y(1) = 0, \quad y'(1) = 0$

4.2 Integral Rules; Integration by Substitution

Rules of Algebra for Antiderivatives • The Integrals of $\sin^2 x$ and $\cos^2 x$ • The Power Rule in Integral Form • Substitution: Running the Chain Rule Backwards • The Integral $\int (1/u)\, du$ • The Integrals of tan x and cot x

Just as limits and derivatives obey algebraic rules, so do antiderivatives and indefinite integrals. In this section, we present and apply those rules to find antiderivatives for a variety of functions.

Rules of Algebra for Antiderivatives

We know the following from our previous study of the derivative.

1. A function is an antiderivative of a constant multiple kf of a function f if and only if it is k times an antiderivative of f.
2. In particular, a function is an antiderivative of $-f$ if and only if it is the negative of an antiderivative of f.
3. A function is an antiderivative of a sum or difference $f \pm g$ if and only if it is the sum or difference of an antiderivative of f and an antiderivative of g.

When we express these observations in integral notation, we get the standard arithmetic rules for indefinite integration (Table 4.2).

CD-ROM
WEBsite
Historical Biography
Jakob Bernoulli
(1654 — 1705)

Table 4.2 Rules for indefinite integration

1. *Constant Multiple Rule:* $\int kf(x)\, dx = k \int f(x)\, dx$
   (Does not work if k varies with x.)
2. *Rule for Negatives:* $\int -f(x)\, dx = -\int f(x)\, dx$
   (Rule 1 with $k = -1$)
3. *Sum and Difference Rule:* $\int [f(x) \pm g(x)]\, dx = \int f(x)\, dx \pm \int g(x)\, dx$

Example 1 Rewriting the Constant of Integration

$$\int 5 \sec x \tan x \, dx = 5 \int \sec x \tan x \, dx \qquad \text{Table 4.2, Rule 1}$$
$$= 5(\sec x + C) \qquad \text{Table 4.1, Formula 6}$$
$$= 5 \sec x + 5C \qquad \text{First form}$$
$$= 5 \sec x + C' \qquad \text{Shorter form, where } C' \text{ is } 5C$$
$$= 5 \sec x + C$$

Usual form, no prime. Since 5 times an arbitrary constant is an arbitrary constant, we rename C'.

What about the three different forms in Example 1? Each one gives all the antiderivatives of $f(x) = 5 \sec x \tan x$, so each answer is correct, but the least complicated of the three, and the usual choice, is the last one

$$\int 5 \sec x \tan x \, dx = 5 \sec x + C.$$

The Sum and Difference Rule for integration enables us to integrate expressions term by term. When we do so, we combine the individual constants of integration into a single arbitrary constant at the end.

Example 2 Term-by-Term Integration

Evaluate

$$\int (x^2 - 2x + 5) \, dx.$$

Solution If we recognize that $(x^3/3) - x^2 + 5x$ is an antiderivative of $x^2 - 2x + 5$, we can evaluate the integral as

$$\int (x^2 - 2x + 5) \, dx = \overbrace{\frac{x^3}{3} - x^2 + 5x}^{\text{antiderivative}} \underbrace{+ C}_{\text{arbitrary constant}}.$$

If we do not recognize the antiderivative right away, we can generate it term by term with the Sum and Difference Rule:

$$\int (x^2 - 2x + 5) \, dx = \int x^2 \, dx - \int 2x \, dx + \int 5 \, dx$$
$$= \frac{x^3}{3} + C_1 - x^2 + C_2 + 5x + C_3.$$

This formula is more complicated than it needs to be. If we combine C_1, C_2, and C_3 into a single constant $C = C_1 + C_2 + C_3$, the formula simplifies to

$$\frac{x^3}{3} - x^2 + 5x + C$$

and *still* gives all the antiderivatives there are. For this reason we recommend that you go right to the final form even if you elect to integrate term by term. Write

$$\int (x^2 - 2x + 5) \, dx = \int x^2 \, dx - \int 2x \, dx + \int 5 \, dx$$
$$= \frac{x^3}{3} - x^2 + 5x + C.$$

Find the simplest antiderivative you can for each part and add the constant of integration at the end.

The Integrals of $\sin^2 x$ and $\cos^2 x$

We can sometimes use trigonometric identities to transform integrals we do not know how to evaluate into integrals we do know how to evaluate. The integral formulas for $\sin^2 x$ and $\cos^2 x$ arise frequently in applications.

Example 3 Integrating $\sin^2 x$ and $\cos^2 x$

(a) $$\int \sin^2 x\, dx = \int \frac{1 - \cos 2x}{2}\, dx \qquad \sin^2 x = \frac{1 - \cos 2x}{2}$$

$$= \frac{1}{2}\int (1 - \cos 2x)\, dx = \frac{1}{2}\int dx - \frac{1}{2}\int \cos 2x\, dx$$

$$= \frac{1}{2}x - \frac{1}{2}\frac{\sin 2x}{2} + C = \frac{x}{2} - \frac{\sin 2x}{4} + C$$

(b) $$\int \cos^2 x\, dx = \int \frac{1 + \cos 2x}{2}\, dx \qquad \cos^2 x = \frac{1 + \cos 2x}{2}$$

$$= \frac{x}{2} + \frac{\sin 2x}{4} + C \qquad \text{As in part (a), but with a sign change}$$

CD-ROM
WEBsite
Historical Biography
Hippocrates of Chios
(ca. 440 BC)

The Power Rule in Integral Form

When u is a differentiable function of x and n is a rational number different from -1, the Chain Rule tells us that

$$\frac{d}{dx}\left(\frac{u^{n+1}}{n+1}\right) = u^n \frac{du}{dx}.$$

This same equation, from another point of view, says that $u^{n+1}/(n + 1)$ is one of the antiderivatives of the function $u^n\,(du/dx)$. Therefore,

$$\int \left(u^n \frac{du}{dx}\right) dx = \frac{u^{n+1}}{n+1} + C.$$

The integral on the left-hand side of this equation is usually written in the simpler "differential" form,

$$\int u^n\, du,$$

obtained by treating the dx's as differentials that cancel. We are thus led to the following rule.

Equation (1) actually holds for any real exponent $n \neq -1$, as we see in Chapter 6.

If u is any differentiable function, then

$$\int u^n\, du = \frac{u^{n+1}}{n+1} + C \qquad (n \neq -1,\ n \text{ rational}). \tag{1}$$

In deriving Equation (1), we assumed u to be a differentiable function of the variable x, but the name of the variable does not matter and does not appear in the final formula. We could have represented the variable with θ, t, y, or any other letter. Equation (1) says that whenever we can cast an integral in the form

$$\int u^n\,du \qquad (n \neq -1),$$

with u a differentiable function and du its differential, we can evaluate the integral as $[u^{n+1}/(n+1)] + C$.

Example 4 Using the Power Rule

$$\begin{aligned}
\int \sqrt{1+y^2}\cdot 2y\,dy &= \int u^{1/2}\,du && \text{Let } u = 1+y^2,\ du = 2y\,dy. \\
&= \frac{u^{(1/2)+1}}{(1/2)+1} + C && \text{Integrate, using Eq. (1) with } n = 1/2. \\
&= \frac{2}{3}u^{3/2} + C && \text{Simpler form} \\
&= \frac{2}{3}(1+y^2)^{3/2} + C && \text{Replace } u \text{ by } 1+y^2.
\end{aligned}$$

Example 5 Adjusting the Integrand by a Constant

$$\begin{aligned}
\int \sqrt{4t-1}\,dt &= \int u^{1/2}\cdot\frac{1}{4}\,du && \text{Let } u = 4t-1,\ du = 4\,dt,\ (1/4)\,du = dt. \\
&= \frac{1}{4}\int u^{1/2}\,du && \text{With the 1/4 out front, the integral is now in standard form.} \\
&= \frac{1}{4}\cdot\frac{u^{3/2}}{3/2} + C && \text{Integrate, using Eq. (1) with } n = 1/2. \\
&= \frac{1}{6}u^{3/2} + C && \text{Simpler form} \\
&= \frac{1}{6}(4t-1)^{3/2} + C && \text{Replace } u \text{ by } 4t-1.
\end{aligned}$$

Substitution: Running the Chain Rule Backwards

The substitutions in Examples 4 and 5 are all instances of the following general rule.

$$\begin{aligned}
\int f(g(x))\cdot g'(x)\,dx &= \int f(u)\,du && \text{1. Substitute } u = g(x),\ du = g'(x)\,dx. \\
&= F(u) + C && \text{2. Evaluate by finding an antiderivative } F(u) \text{ of } f(u). \text{ (Any one will do.)} \\
&= F(g(x)) + C && \text{3. Replace } u \text{ by } g(x).
\end{aligned}$$

These three steps are the steps of the substitution method of integration. The method works because $F(g(x))$ is an antiderivative of $f(g(x))\cdot g'(x)$ whenever F is an antiderivative of f:

$$\begin{aligned}
\frac{d}{dx}F(g(x)) &= F'(g(x))\cdot g'(x) && \text{Chain Rule} \\
&= f(g(x))\cdot g'(x). && \text{Because } F' = f
\end{aligned}$$

The Substitution Method of Integration

Take these steps to evaluate the integral

$$\int f(g(x))g'(x)\,dx,$$

when f and g' are continuous functions:

Step 1. Substitute $u = g(x)$ and $du = g'(x)\,dx$ to obtain the integral

$$\int f(u)\,du.$$

Step 2. Integrate with respect to u.

Step 3. Replace u by $g(x)$ in the result.

Example 6 Using Substitution

$$\int \cos(7\theta + 5)\, d\theta = \int \cos u \cdot \frac{1}{7}\, du$$

Let $u = 7\theta + 5$, $du = 7\, d\theta$, $(1/7)\, du = d\theta$.

$$= \frac{1}{7}\int \cos u\, du$$

With the $1/7$ out front, the integral is now in standard form.

$$= \frac{1}{7}\sin u + C$$

Integrate with respect to u.

$$= \frac{1}{7}\sin(7\theta + 5) + C$$

Replace u by $7\theta + 5$.

Example 7 Using Substitution

$$\int x^2 \sin(x^3)\, dx = \int \sin(x^3) \cdot x^2\, dx$$

$$= \int \sin u \cdot \frac{1}{3}\, du$$

Let $u = x^3$, $du = 3x^2\, dx$, $(1/3)\, du = x^2\, dx$,

$$= \frac{1}{3}\int \sin u\, du$$

$$= \frac{1}{3}(-\cos u) + C$$

Integrate with respect to u.

$$= -\frac{1}{3}\cos(x^3) + C$$

Replace u by x^3.

Example 8 Using Identities and Substitution

Evaluate

$$\int \frac{e^t\, dt}{\cos^2(e^t - 2)}.$$

Solution

$$\int \frac{e^t\, dt}{\cos^2(e^t - 2)} = \int \frac{du}{\cos^2 u}$$

$u = e^t - 2$, $du = e^t\, dt$.

$$= \int \sec^2 u\, du$$

$\frac{1}{\cos u} = \sec u$

$$= \tan u + C$$

$\frac{d}{du}\tan u = \sec^2 u$

$$= \tan(e^t - 2) + C$$

The success of the substitution method depends on finding a substitution that will change an integral we cannot evaluate directly into one that we can. If the first substitution fails, we can try to simplify the integrand further with an additional substitution or two. (You will see what we mean if you do Exercises 41 and 42). Alternatively, we can start afresh. There can be more than one good way to start, as in the next example.

Example 9 Using Different Substitutions

Evaluate

$$\int \frac{2z\, dz}{\sqrt[3]{z^2 + 1}}.$$

Solution We can use the substitution method of integration as an exploratory tool: Substitute for the most troublesome part of the integrand and see how things work out. For the integral here, we might try $u = z^2 + 1$ or we might even press our luck and take u to be the entire cube root. Here is what happens in each case.

Solution 1 Substitute $u = z^2 + 1$.

$$\int \frac{2z\,dz}{\sqrt[3]{z^2+1}} = \int \frac{du}{u^{1/3}} \qquad \text{Let } u = z^2 + 1,\ du = 2z\,dz.$$

$$= \int u^{-1/3}\,du \qquad \text{In the form } \textstyle\int u^n\,du$$

$$= \frac{u^{2/3}}{2/3} + C$$

$$= \frac{3}{2}u^{2/3} + C \qquad \text{Integrate with respect to } u.$$

$$= \frac{3}{2}(z^2+1)^{2/3} + C \qquad \text{Replace } u \text{ by } z^2 + 1.$$

Solution 2 Substitute $u = \sqrt[3]{z^2+1}$ instead.

$$\int \frac{2z\,dz}{\sqrt[3]{z^2+1}} = \int \frac{3u^2\,du}{u} \qquad \text{Let } u = \sqrt[3]{z^2+1},\ u^3 = z^2+1,\ 3u^2\,du = 2z\,dz.$$

$$= 3\int u\,du$$

$$= 3\cdot\frac{u^2}{2} + C \qquad \text{Integrate with respect to } u.$$

$$= \frac{3}{2}(z^2+1)^{2/3} + C \qquad \text{Replace } u \text{ by } (z^2+1)^{1/3}.$$

The Integral $\int (1/u)\,du$

The equation

$$\frac{d}{dx}\ln u = \frac{1}{u}\frac{du}{dx}, \qquad u > 0$$

leads to the integral formula

$$\int \frac{1}{u}\,du = \ln u + C \tag{2}$$

when u is a positive differentiable function, but what if u is negative? If u is negative, then $-u$ is positive and

$$\int \frac{1}{u}\,du = \int \frac{1}{(-u)}\,d(-u)$$

$$= \ln(-u) + C. \qquad \text{Eq. (2) with } u \text{ replaced by } -u \tag{3}$$

We can combine Equations (2) and (3) into a single formula by noticing that in each case the expression on the right is $\ln|u| + C$. In Equation (2), $\ln u = \ln|u|$ because $u > 0$; in Equation (3), $\ln(-u) = \ln|u|$ because $u < 0$. Whether u is positive or negative, the integral of $(1/u)\,du$ is $\ln|u| + C$.

$$\int \frac{1}{u}\,du = \ln|u| + C. \tag{4}$$

Formulas like Equation (4) with a single constant of integration are assumed to hold only over domains that are intervals.

Example 10 Using Equation 4

$$\int \frac{2x}{x^2-5}\,dx = \int \frac{du}{u} \qquad u = x^2-5,\ du = 2x\,dx$$
$$= \ln|u| + C \qquad \text{Eq. (4)}$$
$$= \ln|x^2-5| + C \qquad u = x^2-5$$

The Integrals of tan x and cot x

Equation (4) tells us at last how to integrate the tangent and cotangent functions. For the tangent,

$$\int \tan x\,dx = \int \frac{\sin x}{\cos x}\,dx = \int \frac{-du}{u} \qquad u = \cos x,\ du = -\sin x\,dx$$
$$= -\int \frac{du}{u} = -\ln|u| + C \qquad \text{Eq. (4)}$$
$$= -\ln|\cos x| + C = \ln\frac{1}{|\cos x|} + C \qquad \text{Reciprocal Rule}$$
$$= \ln|\sec x| + C.$$

For the cotangent,

$$\int \cot x\,dx = \int \frac{\cos x\,dx}{\sin x} = \int \frac{du}{u} \qquad u = \sin x,\ du = \cos x\,dx$$
$$= \ln|u| + C = \ln|\sin x| + C = -\ln|\csc x| + C.$$

$$\int \tan u\,du = -\ln|\cos u| + C = \ln|\sec u| + C$$
$$\int \cot u\,du = \ln|\sin u| + C = -\ln|\csc u| + C$$

EXERCISES 4.2

Evaluating Integrals

Evaluate the indefinite integrals in Exercises 1–10 by using the given substitutions to reduce the integrals to standard form.

1. $\int x\sin(2x^2)\,dx, \quad u = 2x^2$
2. $\int \left(1-\cos\frac{t}{2}\right)^2 \sin\frac{t}{2}\,dt, \quad u = 1-\cos\frac{t}{2}$
3. $\int 28(7x-2)^{-5}\,dx, \quad u = 7x-2$
4. $\int x^3(x^4-1)^2\,dx, \quad u = x^4-1$
5. $\int \frac{9r^2\,dr}{\sqrt{1-r^3}}, \quad u = 1-r^3$
6. $\int 12(y^4+4y^2+1)^2(y^3+2y)\,dy, \quad u = y^4+4y^2+1$

7. $\int \sqrt{x}\sin^2(x^{3/2}-1)\,dx, \quad u = x^{3/2}-1$

8. $\int \frac{1}{x^2}\cos^2\left(\frac{1}{x}\right)dx, \quad u = \frac{1}{x}$

9. $\int \csc^2 2\theta \cot 2\theta\, d\theta$

(a) Using $u = \cot 2\theta$ **(b)** Using $u = \csc 2\theta$

10. $\int \frac{dx}{\sqrt{5x+8}}$

(a) Using $u = 5x+8$ **(b)** Using $u = \sqrt{5x+8}$

Evaluate the integrals in Exercises 11–40.

11. $\int \sqrt{3-2s}\,ds$

12. $\int \frac{1}{\sqrt{5s+4}}\,ds$ **13.** $\int \frac{3\,dx}{(2-x)^2}$

14. $\int \theta\sqrt[4]{1-\theta^2}\,d\theta$

15. $\int 3y\sqrt{7-3y^2}\,dy$

16. $\int \frac{1}{\sqrt{x}(1+\sqrt{x})^2}\,dx$ **17.** $\int \frac{2e^{\sqrt{x}}}{\sqrt{x}}\,dx$

18. $\int \cos(3z+4)\,dz$

19. $\int \sec^2(3x+2)\,dx$

20. $\int \sin^5\frac{x}{3}\cos\frac{x}{3}\,dx$ **21.** $\int \frac{1}{t^2}e^{1/t}\,dt$

22. $\int r^2\left(\frac{r^3}{18}-1\right)^5 dr$

23. $\int x^{1/2}\sin(x^{3/2}+1)\,dx$

24. $\int \sec\left(v+\frac{\pi}{2}\right)\tan\left(v+\frac{\pi}{2}\right)dv$

25. $\int \frac{\sin(2t+1)}{\cos^2(2t+1)}\,dt$ **26.** $\int \frac{6\cos t}{(2+\sin t)^3}\,dt$

27. $\int \sqrt{\cot y}\csc^2 y\,dy$ **28.** $\int \frac{1}{t^2}\cos\left(\frac{1}{t}-1\right)dt$

29. $\int \frac{dx}{2x-1}$ **30.** $\int \frac{x\,dx}{x^2+4}$

31. $\int \frac{\sin t}{2-\cos t}\,dt$ **32.** $\int \frac{1}{\sin t\cos t}\,dt$

33. $\int \frac{dx}{\sqrt{1-4x^2}}$ **34.** $\int \frac{dx}{9+3x^2}$

35. $\int \frac{dx}{x\sqrt{25x^2-2}}$ **36.** $\int \frac{3\,dr}{\sqrt{1-4(r-1)^2}}$

37. $\int \frac{dx}{1+(3x+1)^2}$ **38.** $\int \frac{y\,dy}{\sqrt{1-y^4}}$

39. $\int \frac{e^x\,dx}{1+e^{2x}}$ **40.** $\int \frac{4\,dt}{t(1+\ln^2 t)}$

Simplifying Integrals Step by Step

If you do not know what substitution to make, try reducing the integral step by step, using a trial substitution to simplify the integral a bit and then another to simplify it some more. You will see what we mean if you try the sequences of substitutions in Exercises 41 and 42.

41. $\int \frac{18\tan^2 x\sec^2 x}{(2+\tan^3 x)^2}\,dx$

(a) $u = \tan x$, followed by $v = u^3$, then by $w = 2+v$

(b) $u = \tan^3 x$, followed by $v = 2+u$

(c) $u = 2+\tan^3 x$

42. $\int \sqrt{1+\sin^2(x-1)}\sin(x-1)\cos(x-1)\,dx$

(a) $u = x-1$, followed by $v = \sin u$, then by $w = 1+v^2$

(b) $u = \sin(x-1)$, followed by $v = 1+u^2$

(c) $u = 1+\sin^2(x-1)$

Evaluate the integrals in Exercises 43 and 44.

43. $\int \frac{(2r-1)\cos\sqrt{3(2r-1)^2+6}}{\sqrt{3(2r-1)^2+6}}\,dr$

44. $\int \frac{\sin\sqrt{\theta}}{\sqrt{\theta}\cos^3\sqrt{\theta}}\,d\theta$

Initial Value Problems

Solve the initial value problems in Exercises 45–58.

45. $\frac{ds}{dt} = 12t(3t^2-1)^3, \quad s(1) = 3$

46. $\frac{dy}{dx} = 4x(x^2+8)^{-1/3}, \quad y(0) = 0$

47. $\frac{ds}{dt} = 8\sin^2\left(t+\frac{\pi}{12}\right), \quad s(0) = 8$

48. $\frac{dy}{dx} = 1+\frac{1}{x}, \quad y(1) = 3$

49. $\frac{dy}{dt} = e^t\sin(e^t-2), \quad y(\ln 2) = 0$

50. $\frac{dy}{dt} = e^{-t}\sec^2(\pi e^{-t}), \quad y(\ln 4) = 2/\pi$

51. $\frac{d^2s}{dt^2} = -4\sin\left(2t-\frac{\pi}{2}\right), \quad s'(0) = 100, \quad s(0) = 0$

52. $\frac{d^2y}{dx^2} = \sec^2 x, \quad y(0) = 0 \text{ and } y'(0) = 1$

53. $\frac{d^2y}{dx^2} = 2e^{-x}, \quad y(0) = 1 \quad \text{and} \quad y'(0) = 0$

54. $\frac{d^2y}{dt^2} = 1 - e^{2t}, \quad y(1) = -1 \quad \text{and} \quad y'(1) = 0$

55. $\frac{dy}{dx} = \frac{1}{\sqrt{1-x^2}}, \quad y(0) = 0$

56. $\frac{dy}{dx} = \frac{1}{x^2+1} - 1, \quad y(0) = 1$

57. $\frac{dy}{dx} = \frac{1}{x\sqrt{x^2-1}}, \quad x > 1; \quad y(2) = \pi$

58. $\frac{dy}{dx} = \frac{1}{1+x^2} - \frac{2}{\sqrt{1-x^2}}, \quad y(0) = 2$

59. *Particle motion* The velocity of a particle moving back and forth on a line is $v = ds/dt = 6 \sin 2t$ m/sec for all t. If $s = 0$ when $t = 0$, find the value of s when $t = \pi/2$ sec.

60. *Particle motion* The acceleration of a particle moving back and forth on a line is $a = d^2s/dt^2 = \pi^2 \cos \pi t$ m/sec^2 for all t. If $s = 0$ and $v = 8$ m/sec when $t = 0$, find s when $t = 1$ sec.

Generalizing Integration Formulas

Formulas 10 and 11 in Table 4.1, Section 4.1, generalize as follows:

10.′ $\int \frac{1}{\sqrt{a^2-x^2}}\,dx = \sin^{-1}\frac{x}{a} + C$

11.′ $\int \frac{1}{a^2+x^2}\,dx = \frac{1}{a}\tan^{-1}\frac{x}{a} + C$

61. Verify formula 10.′

62. Verify formula 11.′

63. Evaluate

(a) $\int \frac{1}{\sqrt{9-x^2}}\,dx$ (b) $\int \frac{1}{3+x^2}\,dx$

64. Evaluate

(a) $\int \frac{1}{\sqrt{16-25x^2}}\,dx$ (b) $\int \frac{8}{1+4x^2}\,dx$

Theory and Examples

65. *Adjusting the integrand: Multiply it by a form of 1* Evaluate $\int \sec x\,dx$ (*Hint:* Multiply the integrand by

$$\frac{\sec x + \tan x}{\sec x + \tan x}$$

and then use a substitution to integrate the result.)

66. *Writing to Learn: Using different substitutions* It looks as if we can integrate $2 \sin x \cos x$ with respect to x in three different ways:

(a) $\int 2 \sin x \cos x\,dx = \int 2u\,du \qquad u = \sin x$
$= u^2 + C_1 = \sin^2 x + C_1$

(b) $\int 2 \sin x \cos x\,dx = \int -2u\,du \qquad u = \cos x$
$= -u^2 + C_2 = -\cos^2 x + C_2$

(c) $\int 2 \sin x \cos x\,dx = \int \sin 2x\,dx \qquad 2 \sin x \cos x = \sin 2x$
$= -\frac{\cos 2x}{2} + C_3.$

Can all three integrations be correct? Give reasons for your answer.

4.3 Estimating With Finite Sums

Area and Cardiac Output • Distance Traveled • Displacement versus Distance Traveled • Volume of a Sphere • Average Value of a Nonnegative Function • Conclusion

This section shows how practical questions can lead in natural ways to approximations by finite sums.

Area and Cardiac Output

The number of liters of blood your heart pumps in a fixed time interval is called your *cardiac output*. For a person at rest, the rate might be 5 or 6 L per minute. During strenuous exercise the rate might be as high as 30 L per minute. It might also be altered significantly by disease. How can a physician measure a patient's cardiac output without interrupting the flow of blood?

Table 4.3 Dye concentration data

Seconds after injection t	Dye concentration (adjusted for recirculation) c
5	0
7	3.8
9	8.0
11	6.1
13	3.6
15	2.3
17	1.45
19	0.91
21	0.57
23	0.36
25	0.23
27	0.14
29	0.09
31	0

One technique is to inject a dye into a main vein near the heart. The dye is drawn into the right side of the heart and pumped through the lungs and out the left side of the heart into the aorta, where its concentration can be measured every few seconds as the blood flows past. The data in Table 4.3 and the plot in Figure 4.2 (obtained from the data) show the response of a healthy, resting patient to an injection of 5.6 mg of dye.

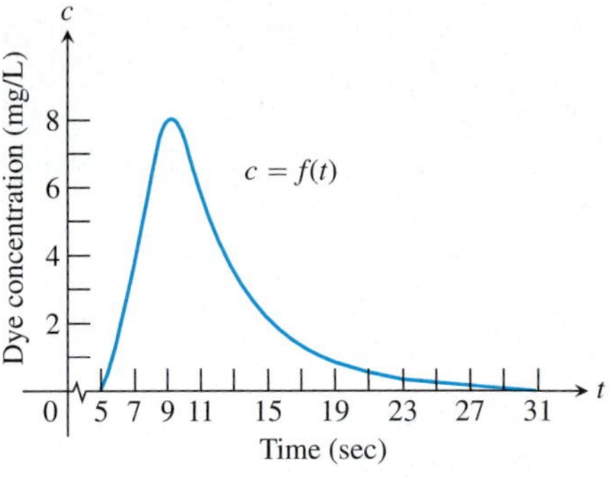

FIGURE 4.2 The dye concentrations from Table 4.3, plotted and fitted with a smooth curve. Time is measured with $t = 0$ at the time of injection. The dye concentrations are zero at the beginning, while the dye passes through the lungs. They then rise to a maximum at about $t = 9$ sec and taper to zero by $t = 31$ sec.

The graph shows dye concentration (measured in milligrams of dye per liter of blood) as a function of time (in seconds). How can we use this graph to obtain the cardiac output (measured in liters of blood per second)? The trick is to divide the *number of milligrams of dye* by the *area under the dye concentration curve*. You can see why this works if you consider what happens to the units:

$$\frac{\text{mg of dye}}{\text{units of area under curve}} = \frac{\text{mg of dye}}{\dfrac{\text{mg of dye}}{\text{L of blood}} \cdot \text{sec}}$$

$$= \frac{\text{mg of dye}}{\text{sec}} \cdot \frac{\text{L of blood}}{\text{mg of dye}}$$

$$= \frac{\text{L of blood}}{\text{sec}}.$$

So you are now ready to compute like a cardiologist.

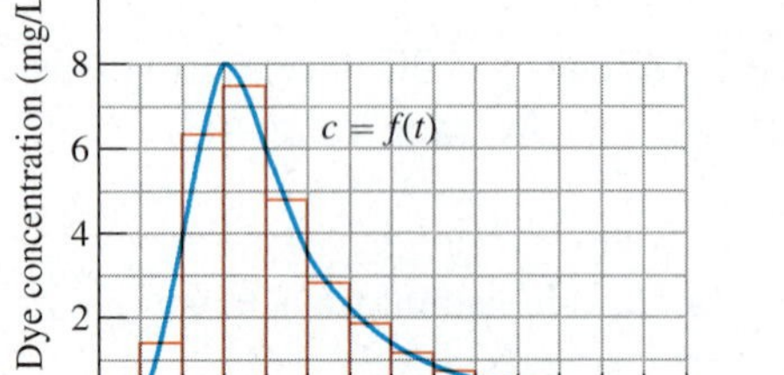

FIGURE 4.3 The region under the concentration curve of Figure 4.2 is approximated with rectangles. We ignore the portion from $t = 29$ to $t = 31$; its contribution is negligible.

Example 1 Computing Cardiac Output From Dye Concentration

Estimate the cardiac output of the patient whose data appear in Table 4.3 and Figure 4.2. Give the estimate in liters per minute.

Solution We have seen that we can obtain the cardiac output by dividing the amount of dye (5.6 mg for our patient) by the area under the curve in Figure 4.2. Now we need to find the area. None of the area formulas we know can be used for this irregularly shaped region, but we can get a good estimate of this area by approximating the region between the curve and the t-axis with rectangles and adding the areas of the rectangles (Figure 4.3). Each rectangle omits some of the area under the curve but includes area from outside the curve, which compensates. In Figure 4.3, each rectangle has a base 2 units long and a height that is equal to the height of the curve above the midpoint of the base. The rectangle's height acts as a sort of average value of the function over the time interval on which the rectangle stands. After reading rectangle heights from the curve, we multiply each rectangle's height and base to find its area and then get the following estimate:

$$\begin{aligned}\text{Area under curve} &\approx \text{sum of rectangle areas}\\ &\approx f(6)\cdot 2 + f(8)\cdot 2 + f(10)\cdot 2 + \cdots + f(28)\cdot 2\\ &\approx 2\cdot(1.4 + 6.3 + 7.5 + 4.8 + 2.8 + 1.9 + 1.1\\ &\quad + 0.7 + 0.5 + 0.3 + 0.2 + 0.1)\\ &\approx 2\cdot(27.6) = 55.2\ (\text{mg/L})\cdot\text{sec}.\end{aligned}$$

Dividing 5.6 mg by this figure gives an estimate for cardiac output in liters per second. Multiplying by 60 converts the estimate to liters per minute:

$$\frac{5.6\ \text{mg}}{55.2\ \text{mg}\cdot\text{sec/L}}\cdot\frac{60\ \text{sec}}{1\ \text{min}} \approx 6.09\ \text{L/min}.$$

USING TECHNOLOGY *Using a Grapher to Calculate Finite Sums* If your graphing utility has a method for evaluating sums, you might want to use it in this section. It is useful for approximating "definite" integrals later in the chapter. There will be other uses still later in your study of calculus.

Distance Traveled

Suppose that we know the velocity function $v = ds/dt = f(t)$ m/sec of a car moving down a highway and want to know how far the car will travel in the time interval $a \le t \le b$. If we know an antiderivative F of f, we can find the car's position function $s = F(t) + C$ and calculate the distance traveled as the difference between the car's positions at times $t = a$ and $t = b$ (as in Section 4.1, Exercise 65).

If we do not know an antiderivative of $v = f(t)$, we can approximate the answer with a sum in the following way. We partition $[a, b]$ into short time intervals *on each of which v is fairly constant.* Since velocity is the rate at which the car is traveling, we approximate the distance traveled on each time interval with the formula

$$\text{Distance} = \text{rate} \times \text{time} = f(t)\cdot\Delta t$$

and add the results across $[a, b]$. To be specific, suppose that the partitioned interval looks like

$|\leftarrow \Delta t \rightarrow|\leftarrow \Delta t \rightarrow|\leftarrow \Delta t \rightarrow|$

$a \quad t_1 \quad t_2 \quad t_3 \quad b \quad t$ (sec)

with the subintervals all of length Δt. Let t_1 be a point in the first subinterval. If the interval is short enough so that the rate is almost constant, the car will move about $f(t_1)\ \Delta t$ m during that interval. If t_2 is a point in the second interval, the car will move an additional $f(t_2)\ \Delta t$ m during that interval, and so on. The sum of these products approximates the total distance D traveled from $t = a$ to $t = b$. If we use n subintervals, then

$$D \approx f(t_1)\ \Delta t + f(t_2)\ \Delta t + \cdots + f(t_n)\ \Delta t. \tag{1}$$

Let's try this on the projectile in Example 5, Section 3.2. The projectile was fired straight into the air. Its velocity t sec into the flight was $v = f(t) = 160 - 9.8t$, and it rose 435.9 m from a height of 3 m to a height of 438.9 m during the first 3 sec of flight.

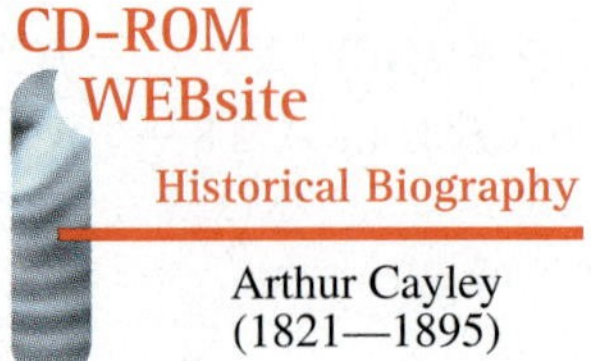

Example 2 Estimating the Height of a Projectile

The velocity function of a projectile fired straight into the air is $f(t) = 160 - 9.8t$. Use the summation techniques just described to estimate how far the projectile rises during the first 3 sec. How close do the sums come to the exact figure of 435.9 m?

Solution We explore the results for different numbers of intervals and different choices of evaluation points.

Three subintervals of length 1, *with f evaluated at left-hand endpoints*:

With f evaluated at $t = 0$, 1, and 2, we have

$$\begin{aligned} D &\approx f(t_1)\,\Delta t + f(t_2)\,\Delta t + f(t_3)\,\Delta t \qquad \text{Eq. (1)} \\ &\approx [160 - 9.8(0)](1) + [160 - 9.8(1)](1) + [160 - 9.8(2)](1) \\ &\approx 450.6. \end{aligned}$$

Three subintervals of length 1, *with f evaluated at right-hand endpoints*:

With f evaluated at $t = 1$, 2, and 3, we have

$$\begin{aligned} D &\approx f(t_1)\,\Delta t + f(t_2)\,\Delta t + f(t_3)\,\Delta t \qquad \text{Eq. (1)} \\ &\approx [160 - 9.8(1)](1) + [160 - 9.8(2)](1) + [160 - 9.8(3)](1) \\ &\approx 421.2. \end{aligned}$$

With six subintervals of length 1/2, *we get*

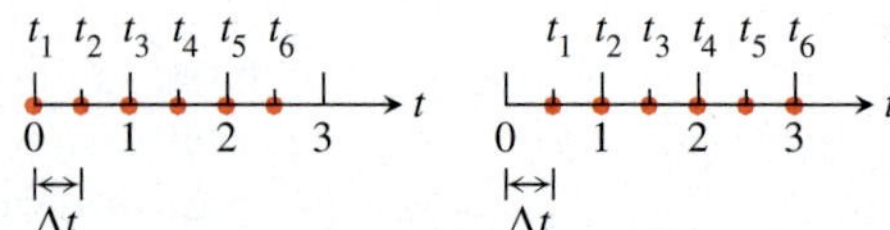

Using left-hand endpoints: $D \approx 443.25$.
Using right-hand endpoints: $D \approx 428.55$.

These six-interval estimates are somewhat closer than the three-interval estimates. The results improve as the subintervals get shorter.

As we can see in Table 4.4, the left-endpoint sums approach the true value 435.9 from above, whereas the right-endpoint sums approach it from below. The

Table 4.4 Travel-distance estimates

Number of subintervals	Length of each subinterval	Left-endpoint sum	Right-endpoint sum
3	1	450.6	421.2
6	0.5	443.25	428.55
12	0.25	439.58	432.23
24	0.125	437.74	434.06
48	0.0625	436.82	434.98
96	0.03125	436.36	435.44
192	0.015625	436.13	435.67

Error magnitude = | true value − calculated value |

true value lies between these upper and lower sums. The magnitude of the error in the closest entries is 0.23, a small percentage of the true value.

$$\text{Error percentage} = \frac{0.23}{435.9} \approx 0.05\%.$$

It would be safe to conclude from the table's last entities that the projectile rose about 436 m during its first 3 sec of flight.

Displacement versus Distance Traveled

If a body with position function $s(t)$ moves along a coordinate line without changing direction, we can calculate the total distance it travels from $t = a$ to $t = b$ by summing the distance traveled over small intervals, as in Example 2. If the body changes direction one or more times during the trip, then we need to use the body's *speed* $|v(t)|$, which is the absolute value of its velocity function, $v(t)$, to find the total distance traveled. Using the velocity itself, as in Example 2, only gives an estimate to the body's **displacement,** $s(b) - s(a)$, the difference between its initial and final positions.

To see why, partition the time interval $[a, b]$ into small enough equal subintervals Δt so that the body's velocity does not change very much from time t_{k-1} to t_k. Then $v(t_k)$ will give a good approximation of the velocity throughout the interval. Accordingly, the change in the body's position coordinate during the time interval will be about

$$v(t_k)\,\Delta t.$$

The change will be positive if $v(t_k)$ is positive and negative if $v(t_k)$ is negative.

In either case, the distance traveled during the subinterval will be about

$$|v(t_k)|\,\Delta t.$$

The **total trip distance** will be approximately the sum

$$|v(t_1)|\,\Delta t + |v(t_2)|\,\Delta t + \cdots + |v(t_n)|\,\Delta t. \tag{2}$$

Volume of a Sphere

Notice the mathematical similarity between Examples 1 and 2. In each case, we have a function f defined on a closed interval and estimate what we want to know with a sum of function values multiplied by interval lengths. We can use similar sums to estimate volumes.

Example 3 Estimating the Volume of a Sphere

CD-ROM WEBsite

Estimate the volume of a solid sphere of radius 4.

Solution We picture the sphere as if its surface were generated by revolving the graph of the function $f(x) = \sqrt{16 - x^2}$ about the x-axis (Figure 4.4a). We partition the interval $-4 \leq x \leq 4$ into n subintervals of equal length $\Delta x = 8/n$. We then slice the sphere with planes perpendicular to the x-axis at the partition points, cutting it like a round loaf of bread into n parallel slices of width Δx. When n is large, each slice can be approximated by a cylinder, a familiar geometric shape of known volume, $\pi r^2 h$. In our case, the cylinders lie on their sides and h is Δx, whereas r varies according to where we are on the x-axis. Let's choose $n = 8$ cylinders and pick the radius of each as the height $f(c_i) = \sqrt{16 - c_i^2}$ at its left-hand endpoint $x = c_i$ of the subinterval (Figure 4.4b). (The cylinder at $x = -4$ is degenerate because the cross section there is just a point.) We can then approximate the volume of the sphere by summing the cylinder volumes,

$$\pi r^2 h = \pi \left(\sqrt{16 - c_i^2}\right)^2 \Delta x.$$

The sum of the eight cylinders' volumes is

$$\begin{aligned} S_8 &= \pi[\sqrt{16 - c_1^2}\,]^2\,\Delta x + \pi[\sqrt{16 - c_2^2}\,]^2\,\Delta x + \pi[\sqrt{16 - c_3^2}\,]^2\,\Delta x \\ &\quad + \cdots + \pi[\sqrt{16 - c_8^2}\,]^2\,\Delta x \qquad \Delta x = \frac{8}{n} = 1 \\ &= \pi[(16 - (-4)^2) + (16 - (-3)^2) + (16 - (-2)^2) + \cdots + (16 - (3)^2)] \\ &= \pi[0 + 7 + 12 + 15 + 16 + 15 + 12 + 7] \\ &= 84\pi. \end{aligned}$$

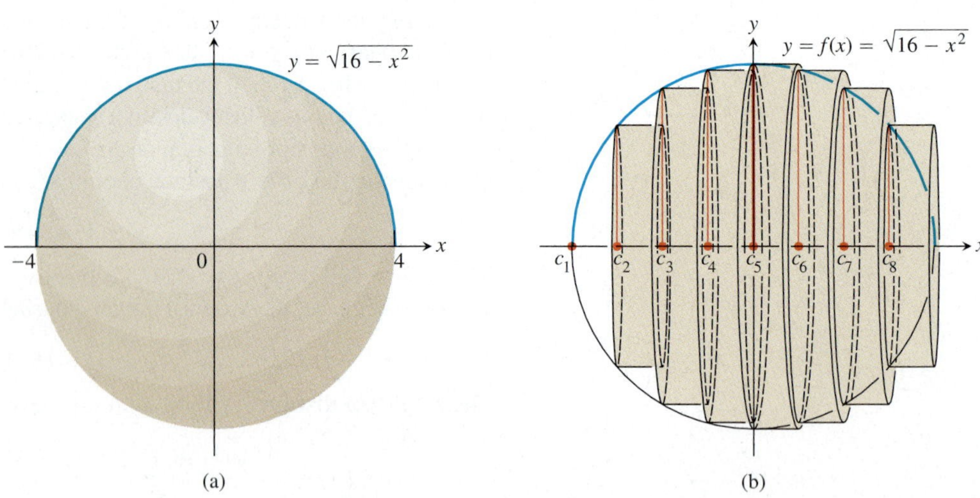

FIGURE 4.4 (a) The semicircle $y = \sqrt{16 - x^2}$ revolved about the x-axis to outline a sphere. (b) The solid sphere approximated with cross-section-based cylinders.

This result compares favorably with the sphere's true volume,

$$V = \frac{4}{3}\pi r^3 = \frac{4}{3}\pi(4)^3 = \frac{256\pi}{3}.$$

The difference between S_8 and V is a small percentage of V:

$$\text{Error percentage} = \frac{|V - S_8|}{V} = \frac{(256/3)\pi - 84\pi}{(256/3)\pi}$$

$$= \frac{256 - 252}{256} = \frac{1}{64} \approx 1.6\%.$$

With a finer partition (more subintervals), the approximation would be even better.

Average Value of a Nonnegative Function

To find the average of a finite set of values, we add them and divide by the number of values added. But what happens if we want to find the average of an infinite number of values? For example, what is the average value of the function $f(x) = x^2$ on the interval $[-1, 1]$? To see what this kind of "continuous" average might mean, imagine that we are pollsters sampling the function. We pick random x's between -1 and 1, square them, and average the squares. As we take larger samples, we expect this average to approach some number, which seems reasonable to call the *average of f over* $[-1, 1]$.

The graph in Figure 4.5a suggests that the average square should be less than $1/2$, because numbers with squares less than $1/2$ make up more than 70% of the interval $[-1, 1]$. If we had a computer to generate random numbers, we could carry

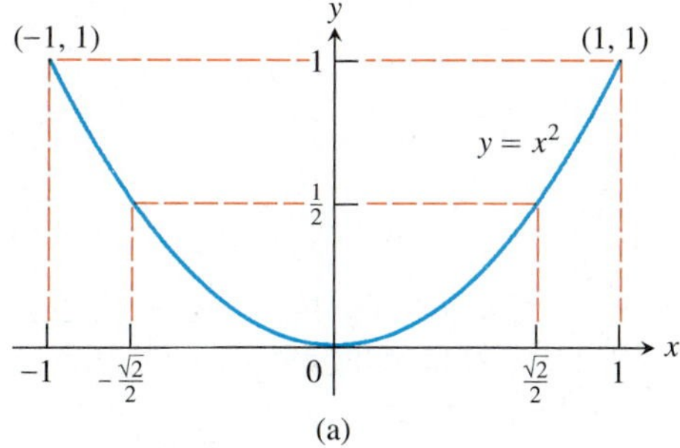

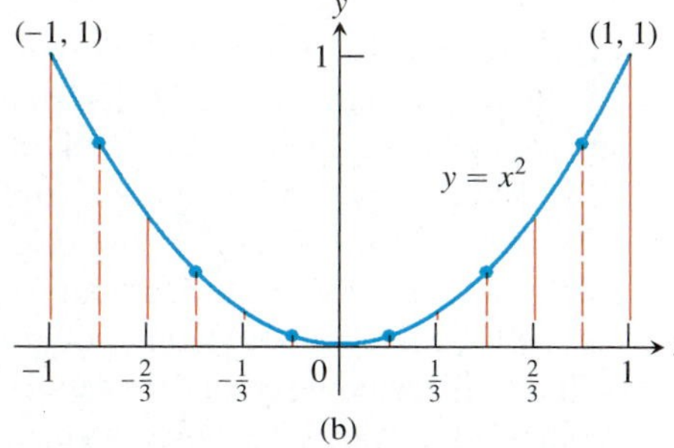

FIGURE 4.5 (a) The graph of $f(x) = x^2$, $-1 \le x \le 1$. (b) Values of f sampled at regular intervals.

out the sampling experiment described above, but it is much easier to estimate the average value with a finite sum.

Example 4 Estimating Average Value

Estimate the average value of the function $f(x) = x^2$ on the interval $[-1, 1]$.

Solution We look at the graph of $y = x^2$ and partition the interval $[-1, 1]$ into 6 subintervals of length $\Delta x = 1/3$ (Figure 4.5b).

It appears that a good estimate for the average square on each subinterval is the square of the midpoint of the subinterval. Since the subintervals have the same length, we can average these six estimates to get a final estimate for the average value over $[-1, 1]$.

$$\text{Average value} \approx \frac{\left(-\frac{5}{6}\right)^2 + \left(-\frac{3}{6}\right)^2 + \left(-\frac{1}{6}\right)^2 + \left(\frac{1}{6}\right)^2 + \left(\frac{3}{6}\right)^2 + \left(\frac{5}{6}\right)^2}{6}$$

$$\approx \frac{1}{6} \cdot \frac{25 + 9 + 1 + 1 + 9 + 25}{36} = \frac{70}{216} \approx 0.324$$

We will be able to show later that the average value is $1/3$.

Notice that

$$\frac{\left(-\frac{5}{6}\right)^2 + \left(-\frac{3}{6}\right)^2 + \left(-\frac{1}{6}\right)^2 + \left(\frac{1}{6}\right)^2 + \left(\frac{3}{6}\right)^2 + \left(\frac{5}{6}\right)^2}{6}$$

$$= \frac{1}{2}\left[\left(-\frac{5}{6}\right)^2 \cdot \frac{1}{3} + \left(-\frac{3}{6}\right)^2 \cdot \frac{1}{3} + \cdots + \left(\frac{5}{6}\right)^2 \cdot \frac{1}{3}\right]$$

$$= \frac{1}{\text{length of } [-1, 1]} \cdot \left[f\left(-\frac{5}{6}\right) \cdot \frac{1}{3} + f\left(-\frac{3}{6}\right) \cdot \frac{1}{3} + \cdots + f\left(\frac{5}{6}\right) \cdot \frac{1}{3}\right]$$

$$= \frac{1}{\text{length of } [-1, 1]} \cdot \left[\begin{array}{l}\text{a sum of function values} \\ \text{multiplied by interval lengths}\end{array}\right].$$

Once again our estimate has been achieved by multiplying function values by interval lengths and summing the results for all the intervals.

Conclusion

The examples in this section describe instances in which sums of function values multiplied by interval lengths provide approximations that are good enough to answer practical questions. You will find additional examples in the exercises.

The distance approximations in Example 2 improved as the intervals involved became shorter and more numerous. We knew this because we had already found the exact answer with antiderivatives in Section 3.2. If we had made our partitions of the time interval still finer, would the sums have approached the exact answer as a limit? Is the connection between the sums and the antiderivative in this case just a coincidence? Could we have calculated the area in Example 1, the volume in Example 3, and the average value in Example 4 with antiderivatives as well? As we will see, the answers are "Yes, they would have," "No, it is not a coincidence," and "Yes, we could have."

EXERCISES 4.3

Cardiac Output

1. The accompanying table gives dye concentrations for a dye-dilution cardiac-output determination like the one in Example 1. The amount of dye injected in this case was 5 mg instead of 5.6 mg. Use rectangles to estimate the area under the dye concentration curve and then go on to estimate the patient's cardiac output.

Seconds after injection	Dye concentration (adjusted for recirculation)
2	0
4	0.6
6	1.4
8	2.7
10	3.7
12	4.1
14	3.8
16	2.9
18	1.7
20	1.0
22	0.5
24	0

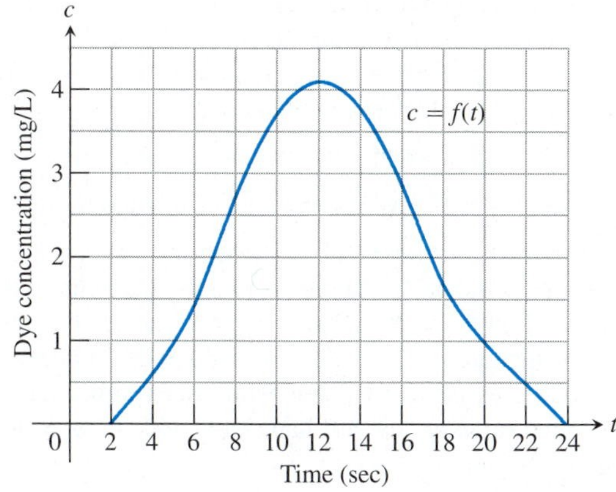

2. The accompanying table gives dye concentrations for a cardiac-output determination like the one in Example 1. The amount of dye injected in this case was 10 mg. Plot the data and connect the data points with a smooth curve. Estimate the area under the curve and calculate the cardiac output from this estimate.

Seconds after injection t	Dye concentration (adjusted for recirculation) c	Seconds after injection t	Dye concentration (adjusted for recirculation) c
0	0	16	7.9
2	0	18	7.8
4	0.1	20	6.1
6	0.6	22	4.7
8	2.0	24	3.5
10	4.2	26	2.1
12	6.3	28	0.7
14	7.5	30	0

Distance

3. *Distance traveled* The accompanying table shows the velocity of a model train engine moving along a track for 10 sec. Estimate the distance traveled by the engine using 10 subintervals of length 1 with

(a) left-endpoint values

(b) right-endpoint values.

Time (sec)	Velocity (in./sec)	Time (sec)	Velocity (in./sec)
0	0	6	11
1	12	7	6
2	22	8	2
3	10	9	6
4	5	10	0
5	13		

4. *Distance traveled upstream* You are sitting on the bank of a tidal river watching the incoming tide carry a bottle upstream. You record the velocity of the flow every five minutes for an hour, with the results shown in the table on the following page. About how far upstream did the bottle travel during that hour? Find an estimate using 12 subintervals of length 5 with

(a) left-endpoint values

(b) right-endpoint values.

Time (min)	Velocity (m /sec)	Time (min)	Velocity (m /sec)
0	1	35	1.2
5	1.2	40	1.0
10	1.7	45	1.8
15	2.0	50	1.5
20	1.8	55	1.2
25	1.6	60	0
30	1.4		

5. *Length of a road* You and a companion are about to drive a twisty stretch of dirt road in a car whose speedometer works but whose odometer (mileage counter) is broken. To find out how long this particular stretch of road is, you record the car's velocity at 10 sec intervals, with the results shown in the accompanying table. Estimate the length of the road using

(a) left-endpoint values

(b) right-endpoint values.

Time (sec)	Velocity (converted to ft/sec) (30 mi/h = 44 ft/sec)	Time (sec)	Velocity (converted to ft/sec) (30 mi/h = 44 ft/sec)
0	0	70	15
10	44	80	22
20	15	90	35
30	35	100	44
40	30	110	30
50	44	120	35
60	35		

6. *Distance from velocity data* The accompanying table gives data for the velocity of a vintage sports car accelerating from 0 to 142 mi/h in 36 sec (10 thousandths of an hour).

Time (h)	Velocity (mi/h)	Time (h)	Velocity (mi/h)
0.0	0	0.006	116
0.001	40	0.007	125
0.002	62	0.008	132
0.003	82	0.009	137
0.004	96	0.010	142
0.005	108		

(a) Use rectangles to estimate how far the car traveled during the 36 sec it took to reach 142 mi/h.

(b) Roughly how many seconds did it take the car to reach the halfway point? About how fast was the car going then?

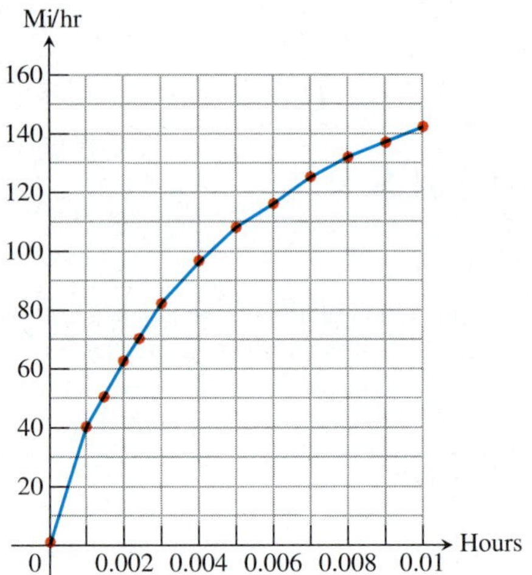

Volume

7. *Volume of a sphere* (*Continuation of Example 3*) Suppose that we approximate the volume V of the sphere in Example 3 by partitioning the interval $-4 \leq x \leq 4$ into four subintervals of length 2 and using cylinders based on the cross sections at the subintervals' left-hand endpoints. (As in Example 3, the leftmost cylinder will have a zero radius.)

CD-ROM WEBsite

(a) Find the sum S_4 of the volumes of the cylinders.

(b) Express $|V - S_4|$ as a percentage of V to the nearest percent.

8. *Volume of a solid sphere* To estimate the volume V of a solid sphere of radius 5, you partition its diameter into five subintervals of length 2. You then slice the sphere with planes perpendicular to the diameter at the subintervals' left-hand endpoints and add the volumes of cylinders of height 2 based on the cross sections of the sphere determined by these planes.

CD-ROM WEBsite

(a) Find the sum S_5 of the volumes of the cylinders.

(b) Express $|V - S_5|$ as a percentage of V to the nearest percent.

9. *Volume of a solid hemisphere* To estimate the volume V of a solid hemisphere of radius 4, imagine its axis of symmetry to be the interval $[0, 4]$ on the x-axis. Partition $[0, 4]$ into eight subintervals of equal length and approximate the solid with cylinders based on the circular cross sections of the hemisphere perpendicular to the x-axis at the subintervals' left-hand endpoints. (See the accompanying profile view.)

CD-ROM WEBsite

(a) *Writing to Learn* Find the sum S_8 of the volumes of the cylinders. Do you expect S_8 to overestimate V, or to underestimate V? Give reasons for your answer.

(b) Express $|V - S_8|$ as a percentage of V to the nearest percent.

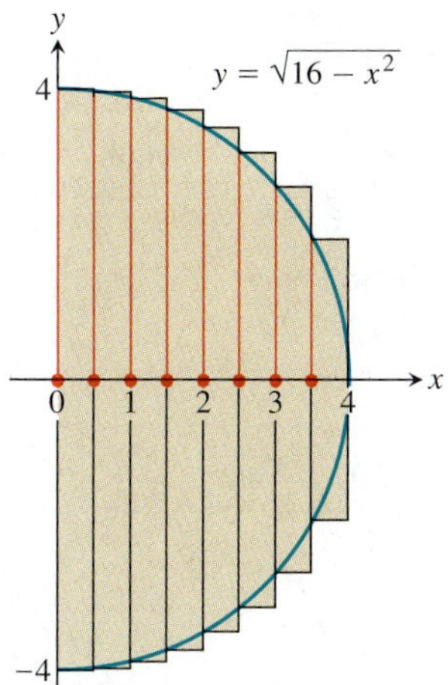

10. *Volume of a solid hemisphere* Repeat Exercise 9 using cylinders based on cross sections at the *right-hand* endpoints of the subintervals.

11. *Volume of water in a reservoir* A reservoir shaped like a hemispherical bowl of radius 8 m is filled with water to a depth of 4 m.

 (a) Find an estimate S of the water's volume by approximating the water with eight circumscribed solid cylinders.

 (b) As you will see in Section 4.5, Exercise 43, the water's volume is $V = 320\pi/3\ \text{m}^3$. Find the error $|V - S|$ as a percentage of V to the nearest percent.

12. *Volume of water in a swimming pool* A rectangular swimming pool is 30 ft wide and 50 ft long. The accompanying table shows the depth $h(x)$ of the water at 5 ft intervals from one end of the pool to the other. Estimate the volume of water in the pool using

 (a) left-endpoint values of h

 (b) right-endpoint values of h.

Position x **ft**	**Depth** $h(x)$ **ft**	**Position** x **ft**	**Depth** $h(x)$ **ft**
0	6.0	30	11.5
5	8.2	35	11.9
10	9.1	40	12.3
15	9.9	45	12.7
20	10.5	50	13.0
25	11.0		

13. *Volume of a nose cone* The nose "cone" of a rocket is a paraboloid obtained by revolving the curve $y = \sqrt{x}$, $0 \le x \le 5$, about the x-axis, where x is measured in feet. Estimate the volume V of the nose cone by partitioning the closed interval $[0, 5]$ into five subintervals of equal length, slicing the cone with planes perpendicular to the x-axis at the subintervals' left-hand endpoints, and construct cylinders of height 1 based on cross sections at these points. (See the accompanying figure.)

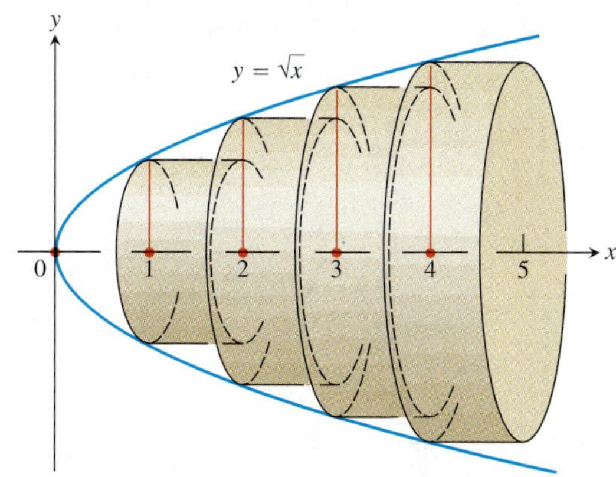

 (a) *Writing to Learn* Find the sum S_5 of the volumes of the cylinders. Do you expect S_5 to overestimate V or to underestimate V? Give reasons for your answer.

 (b) As you will see in Section 4.5, Exercise 44, the volume of the nose cone is $V = 25\pi/2\ \text{ft}^3$. Express $|V - S_5|$ as a percentage of V to the nearest percent.

14. *Volume of a nose cone* Repeat Exercise 13 using cylinders based on cross sections at the *right-hand* endpoints of the subintervals.

Velocity and Distance

15. *Free fall with air resistance* An object is dropped straight down from a helicopter. The object falls faster and faster but its acceleration (rate of change of its velocity) decreases over time because of air resistance. The acceleration is measured in feet per seconds squared and recorded every second after the drop for 5 sec, as shown in the table below.

t	0	1	2	3	4	5
a	32.00	19.41	11.77	7.14	4.33	2.63

 (a) Find an upper estimate for the speed when $t = 5$.

 (b) Find a lower estimate for the speed when $t = 5$.

 (c) Find an upper estimate for the distance fallen when $t = 3$.

16. *Distance traveled by a projectile* An object is shot straight upward from sea level with an initial velocity of 400 ft/sec.

 (a) Assuming that gravity is the only force acting on the object, give an upper estimate for its velocity after 5 sec have elapsed. Use $g = 32\ \text{ft/sec}^2$ for the gravitational constant.

 (b) Find a lower estimate for the height attained after 5 sec.

Average Value of a Function

In Exercises 17–20, use a finite sum to estimate the average value of f on the given interval by partitioning the interval into four subintervals of equal length and evaluating f at the subinterval midpoints.

17. $f(x) = x^3$ on $[0, 2]$

18. $f(x) = 1/x$ on $[1, 9]$

19. $f(t) = (1/2) + \sin^2 \pi t$ on $[0, 2]$

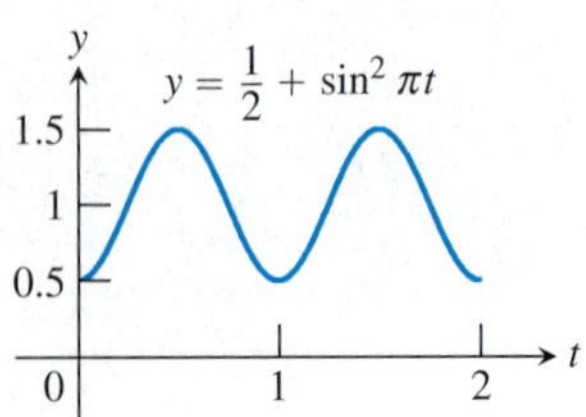

20. $f(t) = 1 - \left(\cos \frac{\pi t}{4}\right)^4$ on $[0, 4]$

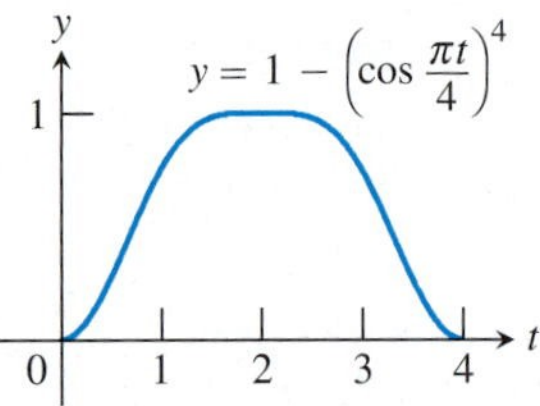

Pollution Control

21. *Water pollution* Oil is leaking out of a tanker damaged at sea. The damage to the tanker is worsening as evidenced by the increased leakage each hour, recorded in the following table.

Time (hours)	0	1	2	3	4
Leakage (gal/hr)	50	70	97	136	190

Time (hours)	5	6	7	8
Leakage (gal/hr)	265	369	516	720

(a) Give an upper and a lower estimate of the total quantity of oil that has escaped after 5 hours.

(b) Repeat part (a) for the quantity of oil that has escaped after 8 hours.

(c) The tanker continues to leak 720 gal/h after the first 8 hours. If the tanker originally contained 25,000 gal of oil, approximately how many more hours will elapse in the worst case before all the oil has spilled. In the best case?

22. *Air pollution* A power plant generates electricity by burning oil. Pollutants produced as a result of the burning process are removed by scrubbers in the smokestacks. Over time, the scrubbers become less efficient and eventually they must be replaced when the amount of pollution released exceeds government standards. Measurements are taken at the end of each month determining the rate at which pollutants are released into the atmosphere, recorded as follows.

Month	Jan	Feb	Mar	Apr	May	Jun
Pollutant Release rate (tons/day)	0.20	0.25	0.27	0.34	0.45	0.52

Month	Jul	Aug	Sep	Oct	Nov	Dec
Pollutant Release rate (tons/day)	0.63	0.70	0.81	0.85	0.89	0.95

(a) Assuming a 30-day month and that new scrubbers allow only 0.05 ton/day released, give an upper estimate of the total tonnage of pollutants released by the end of June. What is a lower estimate?

(b) In the best case, approximately when will a total of 125 tons of pollutants have been released into the atmosphere?

Area

23. *Area of a circle* Inscribe a regular n-sided polygon inside a circle of radius 1 and compute the area of the polygon for the following values of n.

(a) 4 (square) **(b)** 8 (octagon) **(c)** 16

(d) Compare the areas in parts (a), (b), and (c) with the area of the circle.

24. (*Continuation of Exercise 23*)

(a) Inscribe a regular n-sided polygon inside a circle of radius 1 and compute the area of one of the n congruent triangles formed by drawing radii to the vertices of the polygon.

(b) Compute the limit of the area of the inscribed polygon as $n \to \infty$.

(c) Repeat the computations in parts (a) and (b) for a circle of radius r.

COMPUTER EXPLORATIONS

Estimating Average Value

In Exercises 25–28, use a CAS to perform the following steps.

(a) Plot the functions over the given interval.

(b) Partition the interval into $n = 100$, 200, and 1000 subintervals of equal length, and evaluate the function at the midpoint of each subinterval.

(c) Compute the average value of the function values generated in part (b).

(d) Solve the equation $f(x) =$ (average value) for x using the average value calculated in part (c) for the $n = 1000$ partitioning.

25. $f(x) = \sin x$ on $[0, \pi]$

26. $f(x) = \sin^2 x$ on $[0, \pi]$

27. $f(x) = x \sin \frac{1}{x}$ on $\left[\frac{\pi}{4}, \pi\right]$

28. $f(x) = x \sin^2 \frac{1}{x}$ on $\left[\frac{\pi}{4}, \pi\right]$

4.4 Riemann Sums and Definite Integrals

Riemann Sums • Terminology and Notation of Integration • Area Under the Graph of a Nonnegative Function • Average Value of an Arbitrary Continuous Function • Properties of Definite Integrals

In the preceding section, we estimated distances, area, volumes, and average values with finite sums. The terms in the sums were obtained by multiplying selected function values by the lengths of intervals. In this section, we move beyond finite sums to see what happens in the limit, as the lengths of the intervals become infinitely small and their number infinitely large.

Riemann Sums

Sigma notation enables us to express a large sum in compact form:

$$\sum_{k=1}^{n} a_k = a_1 + a_2 + a_3 + \cdots + a_{n-1} + a_n.$$

The Greek capital letter Σ (sigma) stands for "sum." The index k tells us where to begin the sum (at the number below the Σ) and where to end (at the number above). If the symbol ∞ appears above the Σ, it indicates that the terms go on indefinitely.

Example 1 Using Sigma Notation

The sum in sigma notation	The sum written out, one term for each value of k	The value of the sum
$\sum_{k=1}^{5} k$	$1 + 2 + 3 + 4 + 5$	15
$\sum_{k=1}^{3} (-1)^k k$	$(-1)^1(1) + (-1)^2(2) + (-1)^3(3)$	$-1 + 2 - 3 = -2$
$\sum_{k=1}^{2} \frac{k}{k+1}$	$\frac{1}{1+1} + \frac{2}{2+1}$	$\frac{1}{2} + \frac{2}{3} = \frac{7}{6}$
$\sum_{k=4}^{5} \frac{k^2}{k-1}$	$\frac{4^2}{4-1} + \frac{5^2}{5-1}$	$\frac{16}{3} + \frac{25}{4} = \frac{139}{12}$

The lower limit of summation does not have to be 1; it can be any integer.

The sums in which we will be interested are called *Riemann* ("*ree*-mahn") *sums*, after Georg Friedrich Bernhard Riemann. Riemann sums are constructed in a particular way. We now describe that construction formally, in a more general context that does not confine us to nonnegative functions.

We begin with an arbitrary continuous function $f(x)$ defined on a closed interval $[a, b]$. Like the function graphed in Figure 4.6, it may have negative values as well as positive values.

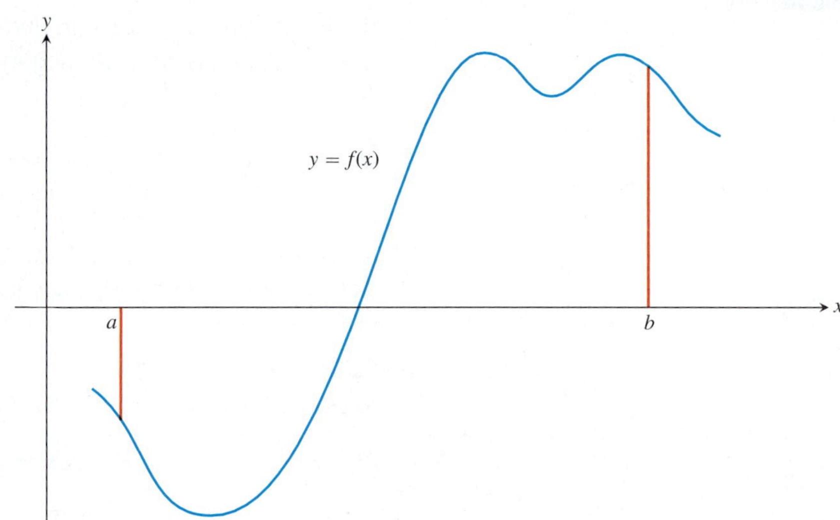

FIGURE 4.6 A typical continuous function $y = f(x)$ over a closed interval $[a, b]$.

We then partition the interval $[a, b]$ into n subintervals by choosing $n - 1$ points, say $x_1, x_2, \cdots, x_{n-1}$, between a and b subject only to the condition that

$$a < x_1 < x_2 < \cdots < x_{n-1} < b.$$

To make the notation consistent, we denote a by x_0 and b by x_n. The set

$$P = \{x_0, x_1, x_2, \cdots, x_n\}$$

is called a **partition** of $[a, b]$.

The partition P defines n closed **subintervals**

$$[x_0, x_1], [x_1, x_2], \ldots, [x_{n-1}, x_n].$$

The typical closed subinterval $[x_{k-1}, x_k]$ is called the ***k*th subinterval** of P.

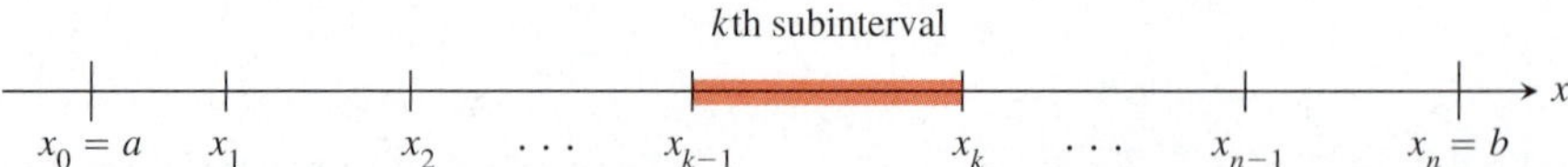

The length of the kth subinterval is $\Delta x_k = x_k - x_{k-1}$.

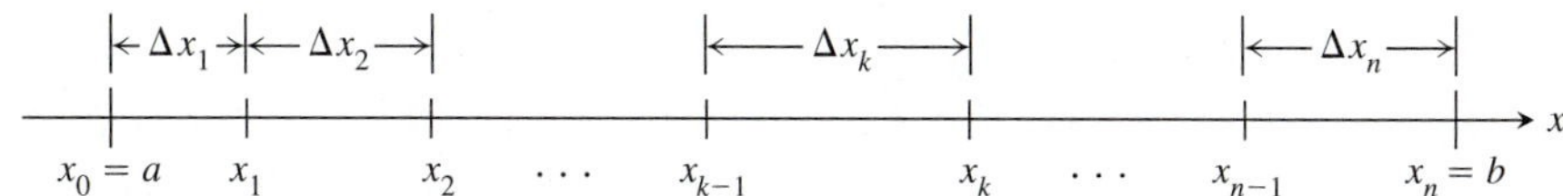

CD-ROM WEBsite

In each subinterval we select some number. Denote the number chosen from the kth subinterval by c_k.

Then, on each subinterval we stand a vertical rectangle that reaches from the x-axis to touch the curve at $(c_k, f(c_k))$. These rectangles could lie either above or below the x-axis (Figure 4.7).

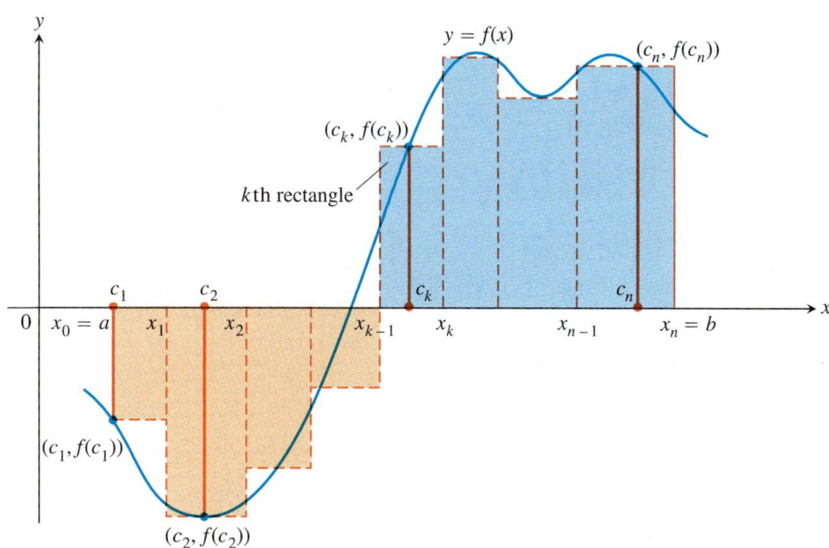

FIGURE 4.7 The rectangles approximate the region between the graph of the function $y = f(x)$ and the x-axis.

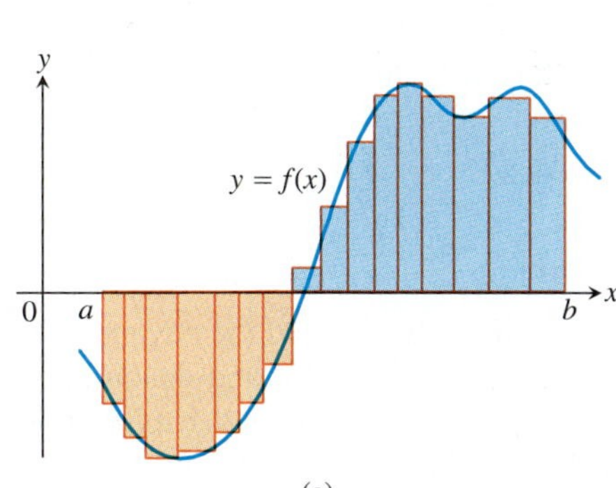

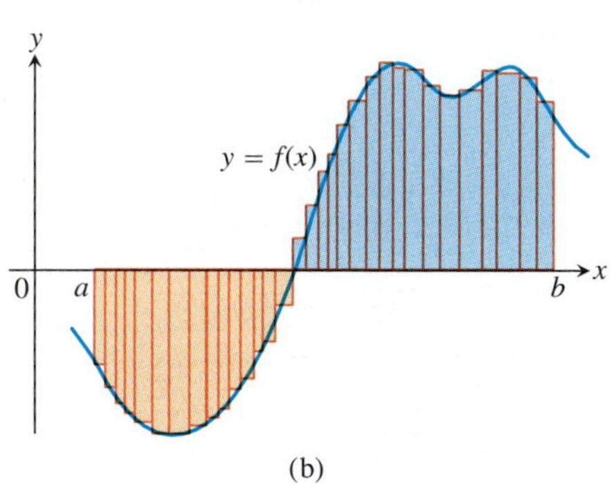

FIGURE 4.8 The curve of Figure 4.7 with rectangles from finer partitions of $[a, b]$. Finer partitions create more rectangles with shorter bases.

On each subinterval, we form the product $f(c_k) \cdot \Delta x_k$. This product can be positive, negative, or zero, depending on $f(c_k)$.

Finally, we take the sum of these products:

$$S_n = \sum_{k=1}^{n} f(c_k) \cdot \Delta x_k .$$

This sum, which depends on the partition P and the choice of the numbers c_k, is a **Riemann sum for f on the interval $[a, b]$.**

As the partitions of $[a, b]$ become finer and finer, we would expect the rectangles defined by the partitions to approximate the region between the x-axis and the graph of f with increasing accuracy (Figure 4.8). So we expect the associated Riemann sums to have a limiting value. Theorem 1 below assures us that they do, as long as the lengths of the subintervals all tend to zero. This latter condition is assured by requiring the longest subinterval length, called the **norm** of the partition and denoted by $\| P \|$, to tend to zero.

Despite the potential for variety in the sums $\Sigma f(c_k) \Delta x_k$ as the partitions change

Definition The Definite Integral as a Limit of Riemann Sums

Let f be a function defined on a closed interval $[a, b]$. For any partition P of $[a, b]$, let the numbers c_k be chosen arbitrarily in the subintervals $[x_{k-1}, x_k]$.

If there exists a number I such that

$$\lim_{\| P \| \to 0} \sum_{k=1}^{n} f(c_k) \Delta x_k = I$$

no matter how P and the c_k's are chosen, then f is **integrable** on $[a, b]$ and I is the **definite integral** of f over $[a, b]$.

and as the c_k's are chosen arbitrarily in the intervals of each partition, the sums always have the same limit as $\| P \| \to 0$ as long as f is *continuous* on $[a, b]$.

Theorem 1 The Existence of Definite Integrals

All continuous functions are integrable. That is, if a function f is continuous on an interval $[a, b]$, then its definite integral over $[a, b]$ exists.

Terminology and Notation of Integration

Leibniz's clever choice of notation for the derivative, dy/dx, had the advantage of retaining an identity as a "fraction" even though both numerator and denominator had tended to zero. Although not really fractions, derivatives can *behave* like fractions, so the notation makes profound results like the Chain Rule

$$\frac{dy}{dx} = \frac{dy}{du} \cdot \frac{du}{dx}$$

seem almost simple.

The notation that Leibniz introduced for the definite integral was equally inspired. In his derivative notation, the Greek letters ("Δ" for "difference") switch to Roman letters ("d" for "differential") in the limit,

$$\lim_{\Delta x \to 0} \frac{\Delta y}{\Delta x} = \frac{dy}{dx}.$$

In his definite integral notation, the Greek letters again become Roman letters in the limit,

$$\lim_{n \to \infty} \sum_{k=1}^{n} f(c_k) \Delta x = \int_a^b f(x)\, dx.$$

Notice that the difference Δx has again tended to zero, becoming a differential dx. The Greek "Σ" has become an elongated Roman "S," so that the integral can retain its identity as a "sum." The c_k's have become so crowded together in the limit that we no longer think of a choppy selection of x values between a and b, but rather of a continuous, unbroken sampling of x values from a to b. It is as if we were summing *all* products of the form $f(x)\,dx$ as x goes from a to b, so we can abandon the k and the n used in the finite sum expression.

The symbol

$$\int_a^b f(x)\,dx$$

is read as "the integral from a to b of f of x dee x" or sometimes as "the integral from a to b of f of x with respect to x." The component parts also have names:

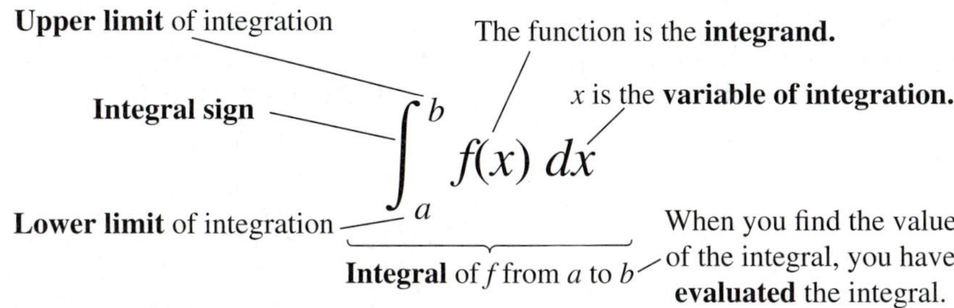
Upper limit of integration
The function is the **integrand.**
Integral sign
x is the **variable of integration.**
Lower limit of integration
When you find the value of the integral, you have **evaluated** the integral.
Integral of f from a to b

The value of the definite integral of a function over any particular interval depends on the function and not on the letter we choose to represent its independent variable. If we decide to use t or u instead of x, we simply write the integral as

$$\int_a^b f(t)\,dt \quad \text{or} \quad \int_a^b f(u)\,du \quad \text{instead of} \quad \int_a^b f(x)\,dx.$$

No matter how we represent the integral, it is the same *number*, defined as a limit of Riemann sums. Since it does not matter what letter we use to run from a to b, the variable of integration is called a **dummy variable.**

Example 2 Using the Notation

The interval $[-1, 3]$ is partitioned into n subintervals of equal length $\Delta x = 4/n$. Let m_k denote the midpoint of the kth subinterval. Express the limit

$$\lim_{n\to\infty} \sum_{k=1}^{n} (3(m_k)^2 - 2m_k + 5)\,\Delta x$$

as an integral.

Solution Since the midpoints m_k have been chosen from the subintervals of the partition, this expression is indeed a limit of Riemann sums. (The points chosen did not have to be midpoints; they could have been chosen from the subintervals in any arbitrary fashion.) The function being integrated is $f(x) = 3x^2 - 2x + 5$ over the interval $[-1, 3]$. Therefore,

$$\lim_{n\to\infty} \sum_{k=1}^{n} (3(m_k)^2 - 2m_k + 5)\,\Delta x = \int_{-1}^{3} (3x^2 - 2x + 5)\,dx.$$

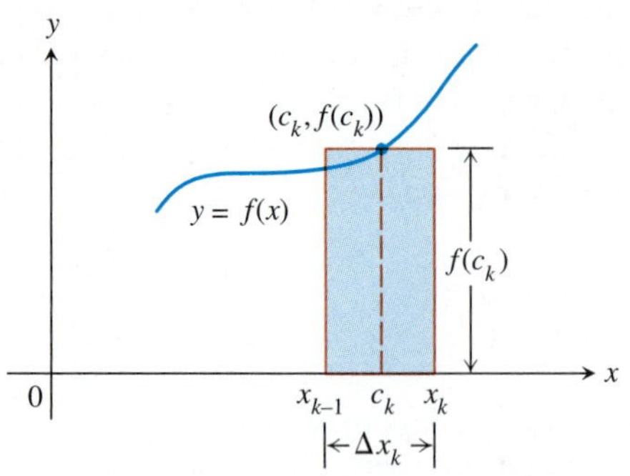

FIGURE 4.9 A term of a Riemann sum $\Sigma f(c_k)\,\Delta x_k$ for a nonnegative function f is either zero or the area of a rectangle such as the one shown.

Area Under the Graph of a Nonnegative Function

In Example 1, Section 4.3, we saw that we could approximate the area under the graph of a nonnegative continuous function $y = f(x)$ by summing the areas of finitely many rectangles with height equal to the height of the curve above the midpoint of the base subinterval. We now know why this is true. If an integrable function $y = f(x)$ is nonnegative throughout an interval $[a, b]$, each nonzero term $f(c_k)\,\Delta x_k$ is the area of a rectangle reaching from the x-axis up to the curve $y = f(x)$. (See Figure 4.9.)

The Riemann sum

$$\sum f(c_k)\,\Delta x_k,$$

which is the sum of the areas of these rectangles, gives an estimate of the area of the region between the curve and the x-axis from a to b. Since the rectangles give an increasingly good approximation of the region as we use partitions with smaller and smaller norms, we call the limiting value the area under the curve.

Definition Area Under a Curve (as a Definite Integral)

If $y = f(x)$ is nonnegative and integrable over a closed interval $[a, b]$, then the **area under the curve $y = f(x)$ from a to b** is the integral of f from a to b,

$$A = \int_a^b f(x)\,dx.$$

This definition works both ways: We can use integrals to calculate areas *and* we can use areas to calculate integrals.

Example 3 Area Under the Curve $f(x) = x$

Evaluate

$$\int_a^b x\,dx, \qquad 0 < a < b.$$

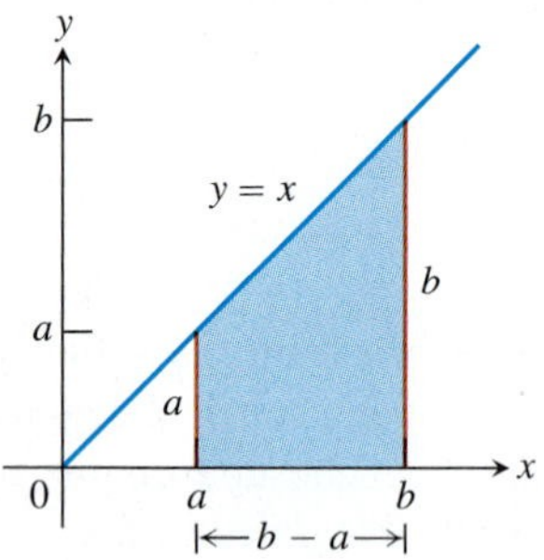

FIGURE 4.10 The region in Example 3.

Solution We sketch the region under the curve $y = x$, $a \le x \le b$ (Figure 4.10), and see that it is a trapezoid with height $(b - a)$ and bases a and b. The value of the integral is the area of this trapezoid:

$$\int_a^b x\,dx = (b - a)\cdot\frac{a + b}{2} = \frac{b^2}{2} - \frac{a^2}{2}.$$

Thus,

$$\int_1^{\sqrt{5}} x\,dx = \frac{(\sqrt{5})^2}{2} - \frac{(1)^2}{2} = 2$$

and so on.

Notice that $x^2/2$ is an antiderivative of x, further evidence of a connection between antiderivatives and summation.

Average Value of an Arbitrary Continuous Function

In Section 4.3, Example 4, we discussed the average value of a nonnegative continuous function. We are now ready to define average value without requiring f to be nonnegative and to show that every continuous function assumes its average value at least once.

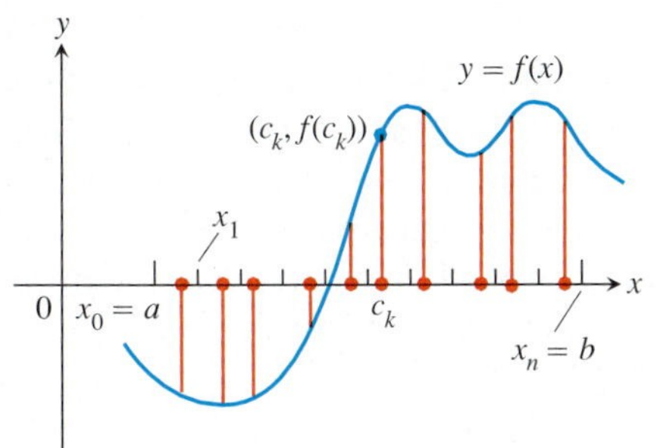

FIGURE 4.11 A sample of values of a function on an interval $[a, b]$.

We start once again with the idea from arithmetic that the average of n numbers is the sum of the numbers divided by n. For a continuous function f on a closed interval $[a, b]$, there may be infinitely many values to consider, but we can sample them in an orderly way. We partition $[a, b]$ into n subintervals of equal length (the length is $\Delta x = (b - a)/n$) and evaluate f at a point c_k in each subinterval (Figure 4.11). The average of the n sampled values is

$$\begin{aligned}
\frac{f(c_1) + f(c_2) + \cdots + f(c_n)}{n} &= \frac{1}{n} \cdot \sum_{k=1}^{n} f(c_k) && \text{the sum in sigma notation} \\
&= \frac{\Delta x}{b - a} \cdot \sum_{k=1}^{n} f(c_k) && \Delta x = \frac{b-a}{n} \\
&= \frac{1}{b - a} \cdot \underbrace{\sum_{k=1}^{n} f(c_k)\, \Delta x}_{\text{a Riemann sum for } f \text{ on } [a,\, b]}.
\end{aligned}$$

Thus, the average of the sampled values is always $1/(b - a)$ times a Riemann sum for f on $[a, b]$. As we increase the size of the sample and let the norm of the partition approach zero, the average must approach $(1/(b - a)) \int_a^b f(x)\, dx$. We are led by this remarkable fact to the following definition.

Definition Average (Mean) Value

If f is integrable on $[a, b]$, then its **average (mean) value** on $[a, b]$ is

$$\text{av}(f) = \frac{1}{b - a} \int_a^b f(x)\, dx.$$

Example 4 Finding an Average Value

Find the average value of $f(x) = \sqrt{4 - x^2}$ on $[-2, 2]$.

Solution We recognize $f(x) = \sqrt{4 - x^2}$ as a function whose graph is the upper semicircle of radius 2 centered at the origin.

The area between the semicircle and the x-axis from -2 to 2 can be computed using the geometry formula

$$\text{Area} = \frac{1}{2} \cdot \pi r^2 = \frac{1}{2} \cdot \pi (2)^2 = 2\pi.$$

Because the area is also the value of the integral of f from -2 to 2,

$$\int_{-2}^{2} \sqrt{4 - x^2}\, dx = 2\pi.$$

Therefore, the average value of f is

$$\text{av}(f) = \frac{1}{2-(-2)} \int_{-2}^{2} \sqrt{4-x^2}\, dx = \frac{1}{4}(2\pi) = \frac{\pi}{2}.$$

Properties of Definite Integrals

In defining $\int_a^b f(x)\, dx$ as a limit of sums $\Sigma f(c_k)\, \Delta x_k$, we moved from left to right across the interval $[a, b]$. What would happen if we integrated in the *opposite* direction? The integral would become $\int_b^a f(x)\, dx$—again a limit of sums of the form $\Sigma f(c_k)\, \Delta x_k$—but this time each of the Δx_k's would be negative as the x-values *decreased* from b to a. This would change the signs of all the terms in each Riemann sum and ultimately, the sign of the definite integral. This suggests the rule

$$\int_b^a f(x)\, dx = -\int_a^b f(x)\, dx.$$

Since the original definition did not apply to integrating backwards over an interval, we can treat this rule as a logical extension of the definition.

Although $[a, a]$ is technically not an interval, another logical extension of the definition is that $\int_a^a f(x)\, dx = 0$.

These are the first two rules in Table 4.5. The others are inherited from rules that hold for Riemann sums.

Table 4.5 Rules for definite integrals

1. *Order of Integration*: $\int_b^a f(x)\, dx = -\int_a^b f(x)\, dx$ — A definition
2. *Zero*: $\int_a^a f(x)\, dx = 0$ — Also a definition
3. *Constant Multiple*: $\int_a^b kf(x)\, dx = k\int_a^b f(x)\, dx$ — Any number k

   $\int_a^b -f(x)\, dx = -\int_a^b f(x)\, dx$ — $k = -1$
4. *Sum and Difference*: $\int_a^b (f(x) \pm g(x))\, dx = \int_a^b f(x)\, dx \pm \int_a^b g(x)\, dx$
5. *Additivity*: $\int_a^b f(x)\, dx + \int_b^c f(x)\, dx = \int_a^c f(x)\, dx$
6. *Max-Min Inequality*: If max f and min f are the maximum and minimum values of f on $[a, b]$, then

$$\min f \cdot (b-a) \le \int_a^b f(x)\, dx \le \max f \cdot (b-a).$$

7. *Domination*: $f(x) \ge g(x)$ on $[a, b] \Rightarrow \int_a^b f(x)\, dx \ge \int_a^b g(x)\, dx$

   $f(x) \ge 0$ on $[a, b] \Rightarrow \int_a^b f(x)\, dx \ge 0$ (special case)

Example 5 Using the Rules for Definite Integrals

Suppose that

$$\int_{-1}^{1} f(x)\,dx = 5, \qquad \int_{1}^{4} f(x)\,dx = -2, \qquad \text{and} \qquad \int_{-1}^{1} h(x)\,dx = 7.$$

Then

1. $\displaystyle \int_{4}^{1} f(x)\,dx = -\int_{1}^{4} f(x)\,dx = -(-2) = 2$ — Rule 2

2. $\displaystyle \int_{-1}^{1} [2f(x) + 3h(x)]\,dx = 2\int_{-1}^{1} f(x)\,dx + 3\int_{-1}^{1} h(x)\,dx$

   $\displaystyle = 2(5) + 3(7) = 31$ — Rules 3 and 4

3. $\displaystyle \int_{-1}^{4} f(x)\,dx = \int_{-1}^{1} f(x)\,dx + \int_{1}^{4} f(x)\,dx = 5 + (-2) = 3.$ — Rule 5

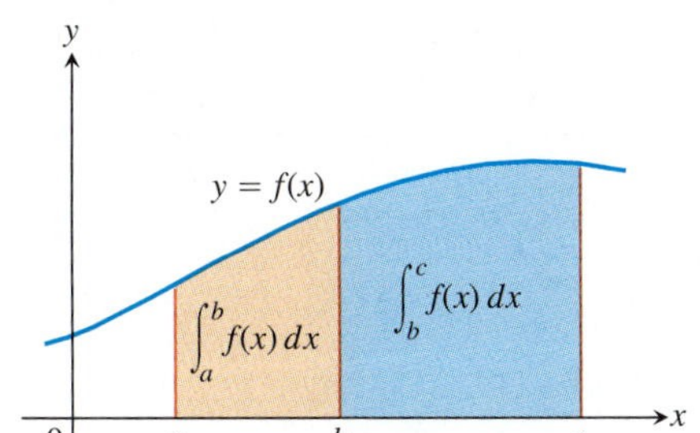

FIGURE 4.12 Additivity for definite integrals:

$$\int_a^b f(x)\,dx + \int_b^c f(x)\,dx = \int_a^c f(x)\,dx$$

$$\int_b^c f(x)\,dx = \int_a^c f(x)\,dx - \int_a^b f(x)\,dx.$$

Proof of Rule 3 Rule 3 says that the integral of k times a function is k times the integral of the function. This is true because

$$\begin{aligned}\int_a^b kf(x)\,dx &= \lim_{\|P\|\to 0} \sum_{i=1}^{n} kf(c_i)\,\Delta x_i \\ &= \lim_{\|P\|\to 0} k\sum_{i=1}^{n} f(c_i)\,\Delta x_i \\ &= k\lim_{\|P\|\to 0} \sum_{i=1}^{n} f(c_i)\,\Delta x_i = k\int_a^b f(x)\,dx.\end{aligned}$$

Figure 4.12 illustrates Rule 5 with a positive function, but the rule applies to any integrable function.

Proof of Rule 6 Rule 6 says that the integral of f over $[a, b]$ is never smaller than the minimum value of f times the length of the interval and never larger than the maximum value of f times the length of the interval. The reason is that for every partition of $[a, b]$ and for every choice of the points c_k,

$$\begin{aligned}\min f\cdot(b-a) &= \min f\cdot\sum_{k=1}^{n}\Delta x_k && \sum_{k=1}^{n}\Delta x_k = b-a \\ &= \sum_{k=1}^{n}\min f\cdot\Delta x_k \\ &\le \sum_{k=1}^{n} f(c_k)\,\Delta x_k && \min f \le f(c_k) \\ &\le \sum_{k=1}^{n}\max f\cdot\Delta x_k && f(c_k)\le \max f \\ &= \max f\cdot\sum_{k=1}^{n}\Delta x_k \\ &= \max f\cdot(b-a).\end{aligned}$$

In short, all Riemann sums for f on $[a, b]$ satisfy the inequality

$$\min f \cdot (b-a) \le \sum_{k=1}^{n} f(c_k)\,\Delta x_k \le \max f \cdot (b-a).$$

Hence, their limit, the integral, does too.

Example 6 Finding Bounds for an Integral

Show that the value of $\int_0^1 \sqrt{1 + \cos x}\,dx$ is less than $3/2$.

Solution The Max-Min Inequality for definite integrals (Rule 6) says that $\min f \cdot (b-a)$ is a *lower bound* for the value of $\int_a^b f(x)\,dx$ and that $\max f \cdot (b-a)$ is an *upper bound*. The maximum value of $\sqrt{1 + \cos x}$ on $[0, 1]$ is $\sqrt{1+1} = \sqrt{2}$, so

$$\int_0^1 \sqrt{1 + \cos x}\,dx \le \sqrt{2} \cdot (1 - 0) = \sqrt{2}.$$

Since $\int_0^1 \sqrt{1 + \cos x}\,dx$ is bounded from above by $\sqrt{2}$ (which is $1.414\ldots$), the integral is less than 3.2.

EXERCISES 4.4

Sigma Notation

Write the sums in Exercises 1–6 without sigma notation. Then evaluate them.

1. $\sum_{k=1}^{2} \frac{6k}{k+1}$

2. $\sum_{k=1}^{3} \frac{k-1}{k}$

3. $\sum_{k=1}^{4} \cos k\pi$

4. $\sum_{k=1}^{5} \sin k\pi$

5. $\sum_{k=1}^{3} (-1)^{k+1} \sin \frac{\pi}{k}$

6. $\sum_{k=1}^{4} (-1)^{k} \cos k\pi$

Rectangles for Riemann Sums

In Exercises 7–10, graph each function $f(x)$ over the given interval. Partition the interval into four subintervals of equal length. Then add to your sketch the rectangles associated with the Riemann sum $\Sigma_{k=1}^{4} f(c_k)\,\Delta x_k$, given that c_k is the **(a)** left-hand endpoint, **(b)** right-hand endpoint, and **(c)** midpoint of the kth subinterval. (Make a separate sketch for each set of rectangles.)

7. $f(x) = x^2 - 1, \quad [0, 2]$

8. $f(x) = -x^2, \quad [0, 1]$

9. $f(x) = \sin x, \quad [-\pi, \pi]$

10. $f(x) = \sin x + 1, \quad [-\pi, \pi]$

Expressing Limits as Integrals

Express the limits in Exercises 11–16 as definite integrals.

11. $\lim_{\|P\|\to 0} \sum_{k=1}^{n} c_k^{\,2}\,\Delta x_k$, where P is a partition of $[0, 2]$

12. $\lim_{\|P\|\to 0} \sum_{k=1}^{n} 2c_k^{\,3}\,\Delta x_k$, where P is a partition of $[-1, 0]$

13. $\lim_{\|P\|\to 0} \sum_{k=1}^{n} (c_k^{\,2} - 3c_k)\,\Delta x_k$, where P is a partition of $[-7, 5]$

14. $\lim_{\|P\|\to 0} \sum_{k=1}^{n} \frac{1}{1 - c_k}\,\Delta x_k$, where P is a partition of $[2, 3]$

15. $\lim_{\|P\|\to 0} \sum_{k=1}^{n} \sqrt{4 - c_k^{\,2}}\,\Delta x_k$, where P is a partition of $[0, 1]$

16. $\lim_{\|P\|\to 0} \sum_{k=1}^{n} (\sec c_k)\,\Delta x_k$, where P is a partition of $[-\pi/4, 0]$

Using Area to Evaluate Integrals

In Exercises 17–22, graph the integrands and use areas to evaluate the integrals.

17. $\int_{-2}^{4} \left(\frac{x}{2} + 3\right) dx$

18. $\int_{-3}^{3} \sqrt{9 - x^2}\,dx$

19. $\int_{-2}^{1} |x|\, dx$

20. $\int_{-1}^{1} (2 - |x|)\, dx$

21. $\int_{0}^{b} x\, dx, \ b > 0$

22. $\int_{a}^{b} 2s\, ds, \ 0 < a < b$

Average Value

In Exercises 23–26, find the average value of the function over the given interval, using the geometric method of Example 4.

23. $f(x) = 1 - x$ on $[0, 1]$

24. $f(x) = |x|$ on $[-1, 1]$

25. $f(x) = \sqrt{1 - x^2}$ on $[0, 1]$

26. $f(x) = \sqrt{1 - (x - 2)^2}$ on $[1, 2]$

Using Properties and Known Values to Find Other Integrals

27. Suppose that f and g are continuous and that

$$\int_{1}^{2} f(x)\, dx = -4, \quad \int_{1}^{5} f(x)\, dx = 6, \quad \int_{1}^{5} g(x)\, dx = 8.$$

Use the rules in Table 4.5 to find the following.

(a) $\int_{2}^{2} g(x)\, dx$ **(b)** $\int_{5}^{1} g(x)\, dx$

(c) $\int_{1}^{2} 3f(x)\, dx$ **(d)** $\int_{2}^{5} f(x)\, dx$

(e) $\int_{1}^{5} [f(x) - g(x)]\, dx$ **(f)** $\int_{1}^{5} [4f(x) - g(x)]\, dx$

28. Suppose that f and h are continuous and that

$$\int_{1}^{9} f(x)\, dx = -1, \quad \int_{7}^{9} f(x)\, dx = 5, \quad \int_{7}^{9} h(x)\, dx = 4.$$

Use the rules in Table 4.5 to find the following.

(a) $\int_{1}^{9} -2f(x)\, dx$ **(b)** $\int_{7}^{9} [f(x) + h(x)]\, dx$

(c) $\int_{7}^{9} [2f(x) - 3h(x)]\, dx$ **(d)** $\int_{9}^{1} f(x)\, dx$

(e) $\int_{1}^{7} f(x)\, dx$ **(f)** $\int_{9}^{7} [h(x) - f(x)]\, dx$

29. Suppose that $\int_1^2 f(x)\, dx = 5$. Find the following.

(a) $\int_{1}^{2} f(u)\, du$ **(b)** $\int_{1}^{2} \sqrt{3} f(z)\, dz$

(c) $\int_{2}^{1} f(t)\, dt$ **(d)** $\int_{1}^{2} [-f(x)]\, dx$

30. Suppose that $\int_{-3}^{0} g(t)\, dt = \sqrt{2}$. Find the following.

(a) $\int_{0}^{-3} g(t)\, dt$ **(b)** $\int_{-3}^{0} g(u)\, du$

(c) $\int_{-3}^{0} [-g(x)]\, dx$ **(d)** $\int_{-3}^{0} \frac{g(r)}{\sqrt{2}}\, dr$

31. Suppose that f is continuous and that $\int_0^3 f(z)\, dz = 3$ and $\int_0^4 f(z)\, dz = 7$. Find the following.

(a) $\int_{3}^{4} f(z)\, dz$ **(b)** $\int_{4}^{3} f(t)\, dt$

32. Suppose that h is continuous and that $\int_{-1}^{1} h(r)\, dr = 0$ and $\int_{-1}^{3} h(r)\, dr = 6$. Find the following.

(a) $\int_{1}^{3} h(r)\, dr$ **(b)** $-\int_{3}^{1} h(u)\, du$

Theory and Examples

33. *Maximizing an integral* What values of a and b maximize the value of

$$\int_{a}^{b} (x - x^2)\, dx\,?$$

34. *Minimizing an integral* What values of a and b minimize the value of

$$\int_{a}^{b} (x^4 - 2x^2)\, dx\,?$$

35. *Writing to Learn* Explain why the rule

$$\int_{a}^{b} k\, dx = k(b - a)$$

holds for *any* constant k.

36. *Integrals of nonnegative functions* Use the max-min inequality to show that if f is integrable then

$$f(x) \geq 0 \quad \text{on} \quad [a, b] \Rightarrow \int_{a}^{b} f(x)\, dx \geq 0.$$

37. *Upper and lower bounds* Use the max-min inequality to find upper and lower bounds for the value of

$$\int_{0}^{1} \frac{1}{1 + x^2}\, dx.$$

38. *Upper and lower bounds* (*Continuation of Exercise 37*) Use the max-min inequality to find upper and lower bounds for

$$\int_{0}^{0.5} \frac{1}{1 + x^2}\, dx \quad \text{and} \quad \int_{0.5}^{1} \frac{1}{1 + x^2}\, dx.$$

Add these to arrive at an improved estimate of

$$\int_{0}^{1} \frac{1}{1 + x^2}\, dx.$$

39. *Average speed on a trip* If you average 30 mph on a 150 mi trip and then return over the same 150 mi at the rate of 50 mph, what is your average speed for the round trip? Give reasons for your answer. (*Source*: David H. Pleacher, *The Mathematics Teacher*, Vol. 85, No. 6 (September 1992), pp 445–446.)

40. *Average rate of water release* A dam released 1000 m^3 of water at 10 m^3/min and then released another 1000 m^3 at 20 m^3/min. What was the average rate at which the water was released? Give reasons for your answer.

COMPUTER EXPLORATIONS

Finding Riemann Sums

If your CAS can draw rectangles associated with Riemann sums, use it to draw rectangles associated with Riemann sums that converge to the integrals in Exercises 41–46. Use $n = 4$, 10, 20, and 50 subintervals of equal length in each case.

41. $\int_0^1 (1 - x)\, dx = \frac{1}{2}$

42. $\int_0^1 (x^2 + 1)\, dx = \frac{4}{3}$

43. $\int_{-\pi}^{\pi} \cos x\, dx = 0$

44. $\int_0^{\pi/4} \sec^2 x\, dx = 1$

45. $\int_{-1}^{1} |x|\, dx = 1$

46. $\int_1^2 \frac{1}{x}\, dx$ (The integral's value is ln 2.)

4.5 The Mean Value and Fundamental Theorems

Mean Value Theorem for Definite Integrals • Fundamental Theorem, Part 1 • A Geometric Interpretation • Fundamental Theorem, Part 2 • Area Connection

This section presents two of the most important theorems in integral calculus. The Mean Value Theorem for Definite Integrals asserts that a continuous function on a closed interval assumes its average value at least once in the interval. The Fundamental Theorem connects integration and differentiation and comes in two parts. Its independent discovery by Leibniz and Newton started the mathematical developments that fueled the scientific revolution for the next 200 years and constitutes what is still regarded as the most important computational discovery in the history of the world.

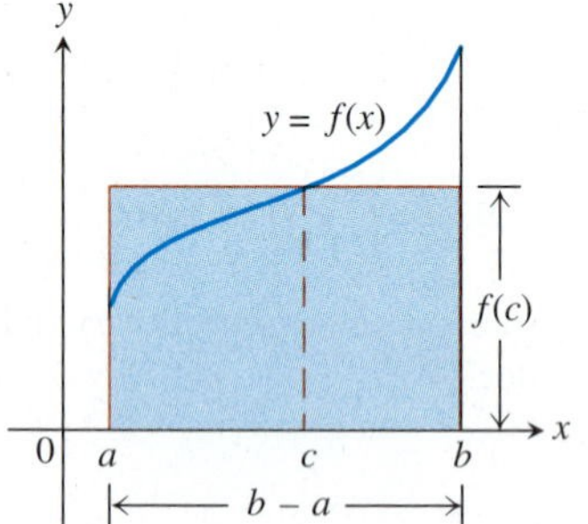

FIGURE 4.13 The value $f(c)$ in the Mean Value Theorem is, in a sense, the average (or *mean*) height of f on $[a, b]$. When $f \geq 0$, the area of the shaded rectangle is the area under the graph of f from a to b,

$$f(c)(b - a) = \int_a^b f(x)\, dx.$$

Mean Value Theorem for Definite Integrals

In the previous section, we defined the average value of a continuous function over a closed interval $[a, b]$ as the definite integral $\int_a^b f(x)\, dx$ divided by the length $b - a$ of the interval. The Mean Value Theorem for Definite Integrals asserts that this average value is *always* taken on at least once in the interval. This is no mere coincidence. Look at the graph in Figure 4.13 and imagine rectangles with base $(b - a)$ and heights ranging from the minimum of f (a rectangle too small to give the integral) to the maximum of f (a rectangle too large). Somewhere in between there is a "just right" rectangle, and its topside will intersect the graph of f if f is continuous.

Theorem 2 The Mean Value Theorem for Definite Integrals

If f is continuous on $[a, b]$, then at some point c in $[a, b]$,

$$f(c) = \frac{1}{b - a} \int_a^b f(x)\, dx.$$

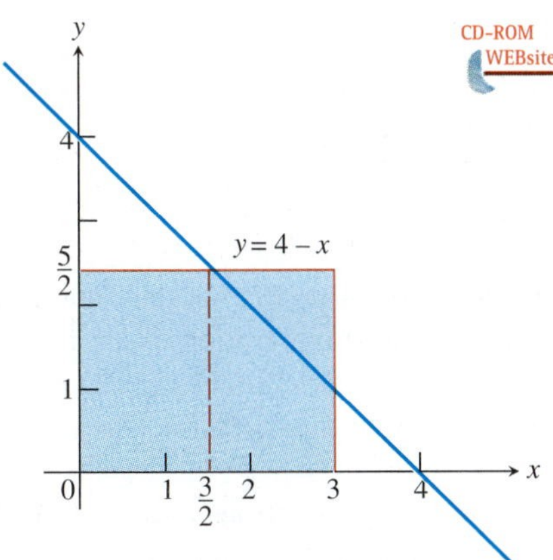

FIGURE 4.14 The rectangle with base [0, 3] and with height equal to 5/2 (the average value of the function $f(x) = 4 - x$) has area equal to the area between the graph of f and the x-axis from 0 to 3. (Example 1)

Example 1 Applying Theorem 2

Find the average value of $f(x) = 4 - x$ on [0, 3] and where f actually takes on this value at some point in the given domain.

Solution

$$\begin{aligned}
\text{av}(f) &= \frac{1}{b-a}\int_a^b f(x)\,dx \\
&= \frac{1}{3-0}\int_0^3 (4-x)\,dx = \frac{1}{3}\left(\int_0^3 4\,dx - \int_0^3 x\,dx\right) \\
&= \frac{1}{3}\left(4(3-0) - \left(\frac{3^2}{2} - \frac{0^2}{2}\right)\right) \qquad \text{Section 4.4, Example 3} \\
&= 4 - \frac{3}{2} = \frac{5}{2}.
\end{aligned}$$

The average value of $f(x) = 4 - x$ over [0, 3] is $5/2$. The function assumes this value when $4 - x = 5/2$ or $x = 3/2$ (Figure 4.14).

In Example 1, we found a point where f assumed its average value by setting $f(x)$ equal to the calculated average value and solving for x, but this does not prove that such a point will always exist. It proves only that it existed in Example 1. To prove Theorem 2, we need a more general argument.

Proof of Theorem 2 If we divide both sides of the max-min inequality (Table 4.5, Rule 6) by $(b - a)$, we obtain

$$\min f \le \frac{1}{b-a}\int_a^b f(x)\,dx \le \max f.$$

Since f is continuous, the Intermediate Value Theorem for Continuous Functions (Section 1.4) says that f must assume every value between $\min f$ and $\max f$. It must therefore assume the value $(1/(b-a))\int_a^b f(x)\,dx$ at some point c in $[a, b]$.

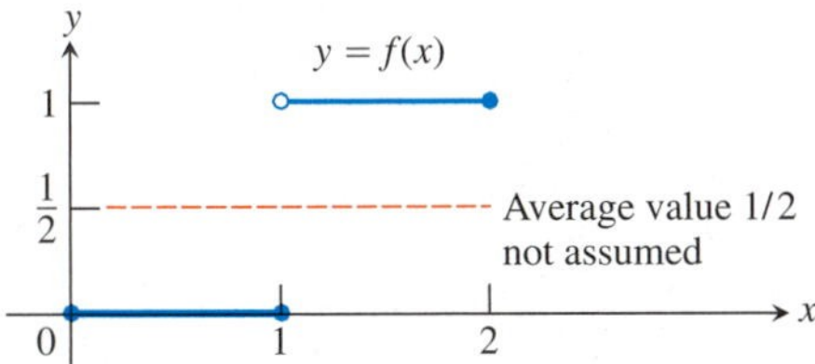

FIGURE 4.15 A discontinuous function need not assume its average value.

The continuity of f is important here. A discontinuous function can step over its average value (Figure 4.15).

What else can we learn from Theorem 2? Here is an example.

Example 2 Average Value of Zero

Show that if f is continuous on $[a, b]$, $a \neq b$, and if

$$\int_a^b f(x)\,dx = 0,$$

then $f(x) = 0$ at least once in $[a, b]$.

Solution The average value of f on $[a, b]$ is

$$\text{av}(f) = \frac{1}{b-a}\int_a^b f(x)\,dx = \frac{1}{b-a}\cdot 0 = 0.$$

By Theorem 2, f assumes this value at some point c in $[a, b]$.

Fundamental Theorem, Part 1

If $f(t)$ is an integrable function, the integral from any fixed number a to another number x defines a function F whose value at x is

$$F(x) = \int_a^x f(t)\, dt. \tag{1}$$

For example, if f is nonnegative and x lies to the right of a, $F(x)$ is the area under the graph from a to x. The variable x is the upper limit of integration of an integral, but F is just like any other real-valued function of a real variable. For each value of the input x, there is a well-defined numerical output, in this case the integral of f from a to x.

Equation (1) gives an important way to define new functions and to describe solutions of differential equations (more about this later). The reason for mentioning Equation (1) now, however, is the connection it makes between integrals and derivatives. If f is any continuous function whatever, then F is a differentiable function of x whose derivative is f itself. At every value of x,

$$\frac{d}{dx} F(x) = \frac{d}{dx} \int_a^x f(t)\, dt = f(x).$$

This idea is so important that it is the first part of the Fundamental Theorem of Calculus.

Theorem 3 The Fundamental Theorem of Calculus, Part 1

If f is continuous on $[a, b]$, then the function

$$F(x) = \int_a^x f(t)\, dt$$

has a derivative at every point x in $[a, b]$, and

$$\frac{dF}{dx} = \frac{d}{dx} \int_a^x f(t)\, dt = f(x). \tag{2}$$

This conclusion is beautiful, powerful, deep, and surprising, and Equation (2) may well be one of the most important equations in mathematics. It says that the differential equation $dF/dx = f$ has a solution for every continuous function f. It says that every continuous function f is the derivative of some other function, namely $\int_a^x f(t)\, dt$. It says that every continuous function has an antiderivative. And it says that the process of integration and differentiation are inverses of one another.

Example 3 Applying the Fundamental Theorem

Find

$$\frac{d}{dx} \int_{-\pi}^x \cos t\, dt \qquad \text{and} \qquad \frac{d}{dx} \int_0^x \frac{1}{1 + t^2}\, dt$$

by using the Fundamental Theorem.

Solution

$$\frac{d}{dx}\int_{-\pi}^{x} \cos t\,dt = \cos x \qquad \text{Eq. 2 with } f(t) = \cos t$$

$$\frac{d}{dx}\int_{0}^{x} \frac{1}{1+t^2}\,dt = \frac{1}{1+x^2}. \qquad \text{Eq. 2 with } f(t) = \frac{1}{1+t^2}.$$

Example 4 Applying the Fundamental Theorem With the Chain Rule

Find dy/dx if $y = \int_1^{x^2} \cos t\,dt$.

Solution The upper limit of integration is not x but x^2. This makes y a composite of

$$y = \int_1^u \cos t\,dt \qquad \text{and} \qquad u = x^2.$$

We must therefore apply the Chain Rule when finding dy/dx.

$$\begin{aligned}
\frac{dy}{dx} &= \frac{dy}{du} \cdot \frac{du}{dx} \\
&= \left(\frac{d}{du}\int_1^u \cos t\,dt\right) \cdot \frac{du}{dx} \\
&= \cos u \cdot \frac{du}{dx} \\
&= \cos(x^2) \cdot 2x \\
&= 2x \cos x^2
\end{aligned}$$

Example 5 Variable Lower Limits of Integration

Find dy/dx.

(a) $y = \displaystyle\int_x^5 3t \sin t\,dt$ **(b)** $y = \displaystyle\int_{1+3x^2}^4 \frac{1}{2+e^t}\,dt$

Solution Rule 1 for integrals in Section 4.4 sets these up for the Fundamental Theorem.

(a)

$$\begin{aligned}
\frac{d}{dx}\int_x^5 3t \sin t\,dt &= \frac{d}{dx}\left(-\int_5^x 3t \sin t\,dt\right) && \text{Rule 1} \\
&= -\frac{d}{dx}\int_5^x 3t \sin t\,dt \\
&= -3x \sin x
\end{aligned}$$

(b)

$$\begin{aligned}
\frac{d}{dx}\int_{1+3x^2}^4 \frac{1}{2+e^t}\,dt &= \frac{d}{dx}\left(-\int_4^{1+3x^2} \frac{1}{2+e^t}\,dt\right) && \text{Rule 1} \\
&= -\frac{d}{dx}\int_4^{1+3x^2} \frac{1}{2+e^t}\,dt \\
&= -\frac{1}{2+e^{(1+3x^2)}} \frac{d}{dx}(1+3x^2) && \text{Eq. (2) and the Chain Rule} \\
&= -\frac{6x}{2+e^{(1+3x^2)}}
\end{aligned}$$

Example 6 Constructing a Function With A Given Derivative and Value

Find a function $y = f(x)$ with derivative

$$\frac{dy}{dx} = \tan x$$

that satisfies the condition $f(3) = 5$.

Solution The Fundamental Theorem makes it easy to construct a function with derivative $\tan x$:

$$y = \int_3^x \tan t \, dt.$$

Since $y(3) = 0$, we have only to add 5 to this function to construct one with derivative $\tan x$ whose value at $x = 3$ is 5:

$$f(x) = \int_3^x \tan t \, dt + 5.$$

Although the solution to the problem in Example 6 satisfies the two required conditions, you might question whether it is in a useful form. Not many years ago, this form might have posed a computation problem. Indeed, for such problems much effort has been expended over the centuries trying to find solutions that do not involve integrals. Since $\ln |\cos t|$ is an antiderivative of $\tan t$, you will see momentarily that we can write the solution in Example 6 as

$$y = \ln \left| \frac{\cos 3}{\cos x} \right| + 5.$$

Now that computers and calculators are capable of evaluating integrals, however, the form given in Example 6 is not only useful, but in some ways preferable. It is certainly easier to find and is always available.

A Geometric Interpretation

If the values of f are positive, the equation

$$\frac{d}{dx} \int_a^x f(t)\, dt = f(x)$$

has a nice geometric interpretation: The integral of f from a to x is the area $A(x)$ of the region between the graph of f and the x-axis from a to x. Imagine the area swept out by a wiper blade clearing the raindrops from the windshield on a bus. As the blade moves past x, the rate at which the cleared area is being swept out is precisely the height of the vertical blade $f(x)$ (Figure 4.16).

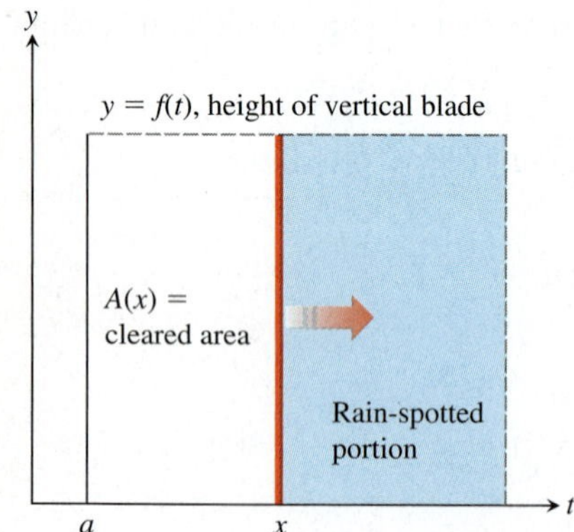

FIGURE 4.16 The rate at which the wiper blade on a bus clears the windshield of rain as the blade moves past x is the height of the blade. In symbols, $dA/dx = f(x)$.

Proof of Theorem 3 We prove Theorem 3 by applying the definition of derivative directly to the function $F(x)$. This means writing out the difference quotient

$$\frac{F(x+h) - F(x)}{h} \tag{3}$$

and showing that its limit as $h \to 0$ is the number $f(x)$.

When we replace $F(x + h)$ and $F(x)$ by their defining integrals, the numerator in Equation (3) becomes

$$F(x+h) - F(x) = \int_a^{x+h} f(t)\,dt - \int_a^x f(t)\,dt.$$

The Additivity Rule for integrals (Table 4.5, Rule 5) simplifies the right-hand side to

$$\int_x^{x+h} f(t)\,dt,$$

so that Equation (3) becomes

$$\begin{aligned} \frac{F(x+h) - F(x)}{h} &= \frac{1}{h}[F(x+h) - F(x)] \\ &= \frac{1}{h}\int_x^{x+h} f(t)\,dt. \end{aligned} \tag{4}$$

According to the Mean Value Theorem for Definite Integrals, the value of the last expression in Equation (4) is one of the values taken on by f in the interval joining x and $x + h$. That is, for some number c in this interval,

$$\frac{1}{h}\int_x^{x+h} f(t)\,dt = f(c). \tag{5}$$

We can therefore find out what happens to $(1/h)$ times the integral as $h \to 0$ by watching what happens to $f(c)$ as $h \to 0$.

What does happen to $f(c)$ as $h \to 0$? As $h \to 0$, the endpoint $x + h$ approaches x, pushing c ahead of it like a bead on a wire.

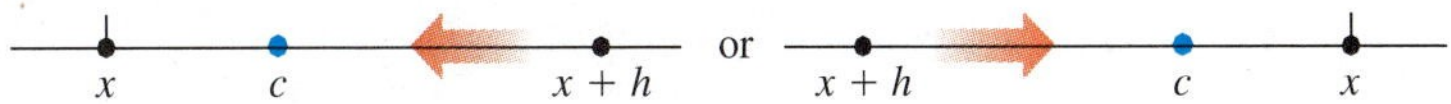

So c approaches x, and, since f is continuous at x, $f(c)$ approaches $f(x)$:

$$\lim_{h\to 0} f(c) = f(x). \tag{6}$$

Going back to the beginning, then, we have

$$\begin{aligned} \frac{dF}{dx} &= \lim_{h\to 0}\frac{F(x+h) - F(x)}{h} && \text{Definition of derivative} \\ &= \lim_{h\to 0}\frac{1}{h}\int_x^{x+h} f(t)\,dt && \text{Eq. (4)} \\ &= \lim_{h\to 0} f(c) && \text{Eq. (5)} \\ &= f(x). && \text{Eq. (6)} \end{aligned}$$

This concludes the proof.

Fundamental Theorem, Part 2

The second part of the Fundamental Theorem of Calculus shows how to evaluate definite integrals directly from antiderivatives.

Theorem 3 (continued) The Fundamental Theorem of Calculus, Part 2

If f is continuous at every point of $[a, b]$ and if F is any antiderivative of f on $[a, b]$, then

$$\int_a^b f(x)\,dx = F(b) - F(a).$$

This part of the Fundamental Theorem is also called the **Integral Evaluation Theorem.**

CD-ROM
WEBsite

Proof Part 1 of the Fundamental Theorem tells us that an antiderivative of f exists, namely

$$G(x) = \int_a^x f(t)\,dt.$$

Thus, if F is *any* antiderivative of f, then $F(x) = G(x) + C$ for some constant C (by Corollary 2 of the Mean Value Theorem for Derivatives, Section 3.2).

Evaluating $F(b) - F(a)$, we have

$$\begin{aligned} F(b) - F(a) &= [G(b) + C] - [G(a) + C] \\ &= G(b) - G(a) \\ &= \int_a^b f(t)\,dt - \int_a^a f(t)\,dt \\ &= \int_a^b f(t)\,dt - 0 \\ &= \int_a^b f(t)\,dt. \end{aligned}$$

At the risk of repeating ourselves: It is difficult to overestimate the power of the simple equation

$$\int_a^b f(x)\,dx = F(b) - F(a).$$

It says that any definite integral of any continuous function f can be calculated without taking limits, without calculating Riemann sums, and often without effort, as long as an antiderivative of f can be found. If you can imagine what it was like before this theorem (and before computing machines), when approximations by tedious sums were the only alternative for solving many real-world problems, then you can imagine what a miracle calculus was thought to be. If any equation deserves to be called the Fundamental Theorem of Calculus, this equation is surely the (second) one.

How to Evaluate $\int_a^b f(x)\,dx$

Step 1. Find an antiderivative F of f. Any antiderivative will do, so pick the simplest one you can.

Step 2. Calculate the number $F(b) - F(a)$.

This number will be $\int_a^b f(x)\,dx$.

Integral Evaluation Notation

The usual notation for $F(b) - F(a)$ is

$$F(x)\Big]_a^b \quad \text{or} \quad \Big[F(x)\Big]_a^b,$$

depending on whether F has one or more terms. This notation provides a compact "recipe" for the evaluation, allowing us to show the antiderivative in an intermediate step.

Example 7 Evaluating an Integral

Evaluate $\int_{-1}^{3} (x^3 + 1)\,dx$ using an antiderivative.

Solution

A simple antiderivative of $x^3 + 1$ is $(x^4/4) + x$. Therefore,

$$\int_{-1}^{3} (x^3 + 1)\,dx = \left[\frac{x^4}{4} + x\right]_{-1}^{3}$$

$$= \left(\frac{81}{4} + 3\right) - \left(\frac{1}{4} - 1\right)$$

$$= 24.$$

Area Connection

We can now compute areas using antiderivatives, but we must be careful to distinguish net area (in which area below the x-axis is counted as negative) from total area. The unmodified word "area" will be taken to mean *total area.*

Example 8 Finding Area Using Antiderivatives

Find the area of the region between the x-axis and the graph of $f(x) = x^3 - x^2 - 2x$, $-1 \le x \le 2$.

Solution First find the zeros of f. Since

$$f(x) = x^3 - x^2 - 2x = x(x^2 - x - 2) = x(x + 1)(x - 2),$$

the zeros are $x = 0$, -1, and 2 (Figure 4.17). The zeros partition $[-1, 2]$ into two subintervals: $[-1, 0]$, on which $f \ge 0$ and $[0, 2]$, on which $f \le 0$. We integrate f over each subinterval and add the absolute values of the calculated values.

Integral over $[-1, 0]$: $\displaystyle\int_{-1}^{0} (x^3 - x^2 - 2x)\,dx = \left[\frac{x^4}{4} - \frac{x^3}{3} - x^2\right]_{-1}^{0}$

$$= 0 - \left[\frac{1}{4} + \frac{1}{3} - 1\right] = \frac{5}{12}$$

Integral over $[0, 2]$: $\displaystyle\int_{0}^{2} (x^3 - x^2 - 2x)\,dx = \left[\frac{x^4}{4} - \frac{x^3}{3} - x^2\right]_{0}^{2}$

$$= \left[4 - \frac{8}{3} - 4\right] - 0 = -\frac{8}{3}$$

Enclosed area: Total enclosed area $= \dfrac{5}{12} + \left|-\dfrac{8}{3}\right| = \dfrac{37}{12}$

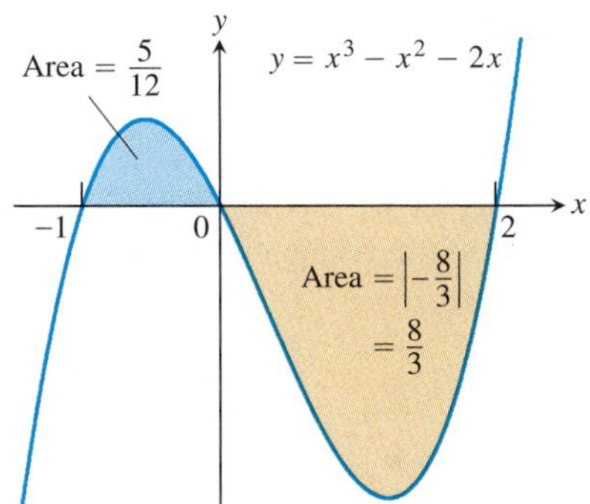

FIGURE 4.17 The region between the curve $y = x^3 - x^2 - 2x$ and the x-axis. (Example 8)

How to Find Total Area

To find the area between the graph of $y = f(x)$ and the x-axis over the interval $[a, b]$ analytically, do the following.

Step 1. Partition $[a, b]$ with the zeros of f.

Step 2. Integrate f over each subinterval.

Step 3. Add the absolute values of the integrals.

Example 9 Household Electricity

We model the voltage in our home wiring with the sine function

$$V = V_{max} \sin 120\pi t,$$

which expresses the voltage V in volts as a function of time t in seconds. The function runs through 60 cycles each second (its frequency is 60 hertz, or 60 Hz). The positive constant V_{max} ("vee max") is the **peak voltage.**

The average value of V over the half-cycle from 0 to 1/120 sec (see Figure 4.18) is

$$\begin{aligned} V_{av} &= \frac{1}{(1/120) - 0} \int_0^{1/120} V_{max} \sin 120\pi t \, dt \\ &= 120\, V_{max} \left[-\frac{1}{120\pi} \cos 120\pi t \right]_0^{1/120} \\ &= \frac{V_{max}}{\pi} [-\cos \pi + \cos 0] \\ &= \frac{2V_{max}}{\pi}. \end{aligned}$$

The average value of the voltage over a full cycle, as we can see from Figure 4.18, is zero. (Also see Exercise 52.) If we measured the voltage with a standard moving-coil galvanometer, the meter would read zero.

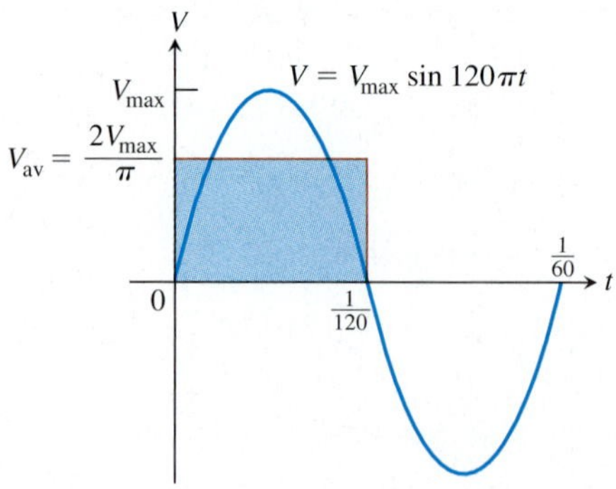

FIGURE 4.18 The graph of the voltage $V = V_{max} \sin 120\pi t$ over a full cycle. Its average value over a half-cycle is $2V_{max}/\pi$. Its average value over a full cycle is zero. (Example 9)

To measure the voltage effectively, we use an instrument that measures the square root of the average value of the square of the voltage, namely

$$V_{rms} = \sqrt{(V^2)_{av}}\,.$$

The subscript "rms" (read the letters separately) stands for "root mean square." Since the average value of $V^2 = (V_{\max})^2 \sin^2 120\pi t$ over a cycle is

$$(V^2)_{\text{av}} = \frac{1}{(1/60) - 0} \int_0^{1/60} (V_{\max})^2 \sin^2 120\pi t \, dt = \frac{(V_{\max})^2}{2} \tag{7}$$

(Exercise 52, part (c)), the rms voltage is

$$V_{\text{rms}} = \sqrt{\frac{(V_{\max})^2}{2}} = \frac{V_{\max}}{\sqrt{2}}. \tag{8}$$

The values given for household currents and voltages are always rms values. Thus, "115 volts ac" means that the rms voltage is 115. The peak voltage,

$$V_{\max} = \sqrt{2}\, V_{\text{rms}} = \sqrt{2} \cdot 115 \approx 163 \text{ volts},$$

obtained from Equation (8), is considerably higher.

EXERCISES 4.5

Evaluating Integrals

Evaluate the integrals in Exercises 1–14.

1. $\displaystyle\int_{-2}^{0} (2x + 5)\, dx$

2. $\displaystyle\int_{0}^{4} \left(3x - \frac{x^3}{4}\right) dx$

3. $\displaystyle\int_{0}^{1} (x^2 + \sqrt{x})\, dx$

4. $\displaystyle\int_{-2}^{-1} \frac{2}{x^2}\, dx$

5. $\displaystyle\int_{0}^{\pi} (1 + \cos x)\, dx$

6. $\displaystyle\int_{0}^{\pi/3} 2 \sec^2 x\, dx$

7. $\displaystyle\int_{\pi/4}^{3\pi/4} \csc\theta \cot\theta\, d\theta$

8. $\displaystyle\int_{0}^{\pi/2} \frac{1 + \cos 2t}{2}\, dt$

9. $\displaystyle\int_{-\pi/2}^{\pi/2} (8y^2 + \sin y)\, dy$

10. $\displaystyle\int_{-1}^{1} (r + 1)^2\, dr$

11. $\displaystyle\int_{1}^{\sqrt{2}} \left(\frac{u^2}{2} - \frac{1}{u^5}\right) du$

12. $\displaystyle\int_{4}^{9} \frac{1 - \sqrt{u}}{\sqrt{u}}\, du$

13. $\displaystyle\int_{0}^{1} xe^{x^2}\, dx$

14. $\displaystyle\int_{0}^{\ln 2} e^{3x}\, dx$

Derivatives of Integrals

Find the derivatives in Exercises 15–18

(a) by evaluating the integral and differentiating the result

(b) by differentiating the integral directly.

15. $\displaystyle\frac{d}{dx}\int_{0}^{\sqrt{x}} \cos t\, dt$

16. $\displaystyle\frac{d}{dx}\int_{1}^{\sin x} 3t^2\, dt$

17. $\displaystyle\frac{d}{dt}\int_{0}^{t^4} \sqrt{u}\, du$

18. $\displaystyle\frac{d}{d\theta}\int_{0}^{\tan\theta} \sec^2 y\, dy$

Find dy/dx in Exercises 19–24.

19. $\displaystyle y = \int_{0}^{x} \sqrt{1 + t^2}\, dt$

20. $\displaystyle y = \int_{1}^{x} \frac{1}{t}\, dt, \quad x > 0$

21. $\displaystyle y = \int_{\sqrt{x}}^{0} \sin(t^2)\, dt$

22. $\displaystyle y = \int_{0}^{x^2} \cos\sqrt{t}\, dt$

23. $\displaystyle y = \int_{1}^{x^{1/3}} e^{(t^3+1)}\, dt$

24. $\displaystyle y = \int_{e^x}^{e} \ln t\, dt, \quad x > 1$

Evaluating Integrals Using Substitutions

In Exercises 25–28, use a substitution to find an antiderivative and then apply the Fundamental Theorem to evaluate the integral.

25. $\displaystyle\int_{0}^{1} (1 - 2x)^3\, dx$

26. $\displaystyle\int_{0}^{1} t\sqrt{t^2 + 1}\, dt$

27. $\displaystyle\int_{0}^{\pi} \sin^2\left(1 + \frac{\theta}{2}\right) d\theta$

28. $\displaystyle\int_{0}^{\pi} \sin^2\frac{x}{4} \cos\frac{x}{4}\, dx$

Initial Value Problems

Solve the initial value problems in Exercises 29–32.

29. $\displaystyle\frac{dy}{dx} = \sec x, \quad y(2) = 3$

30. $\dfrac{dy}{dx} = x\sqrt{1 + x^2}, \quad y(1) = -2$

31. $\dfrac{dy}{dt} = e^t \sin(e^t - 2), \quad y(\ln 2) = 0$

32. $\dfrac{d^2y}{dt^2} = 1 - e^{2t}, \quad y(1) = -1 \text{ and } y'(1) = 0$

Area

In Exercises 33–36, find the total area of the region between the curve and the x-axis.

33. $y = -x^2 - 2x, \quad -3 \le x \le 2$

34. $y = x^3 - 3x^2 + 2x, \quad 0 \le x \le 2$

35. $y = x^3 - 4x, \quad -2 \le x \le 2$

36. $y = x^{1/3} - x, \quad -1 \le x \le 8$

Find the areas of the shaded regions in Exercises 37 and 38.

37.

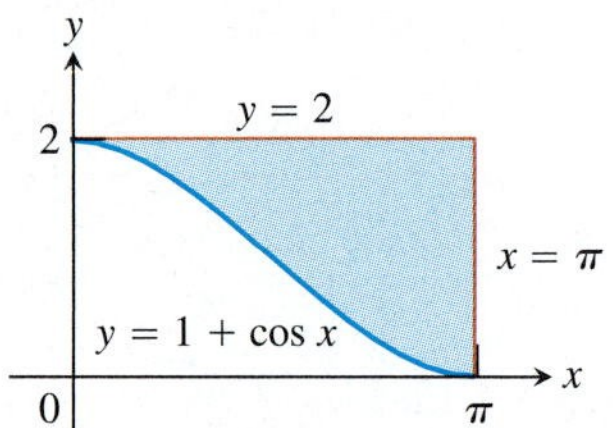

38.

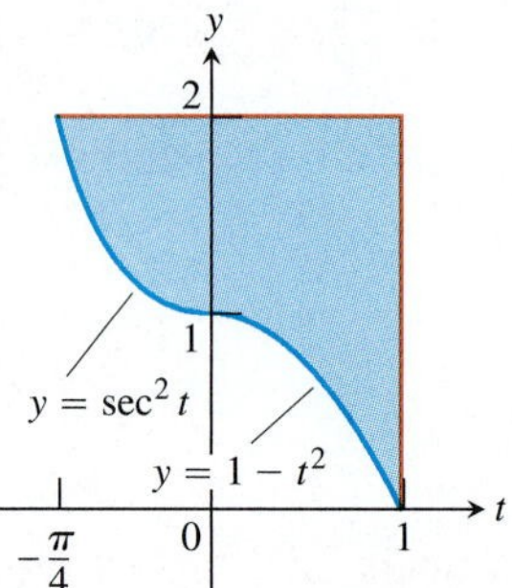

Applications

39. *Cost from marginal cost* The marginal cost of printing a poster when x posters have been printed is

$$\frac{dc}{dx} = \frac{1}{2\sqrt{x}}$$

dollars. Find

(a) $c(100) - c(1)$, the cost of printing posters 2–100

(b) $c(400) - c(100)$, the cost of printing posters 101–400.

40. *Revenue from marginal revenue* Suppose that a company's marginal revenue from the manufacture and sale of egg beaters is

$$\frac{dr}{dx} = 2 - \frac{2}{(x + 1)^2},$$

where r is measured in thousands of dollars and x in thousands of units. How much money should the company expect from a production run of $x = 3$ thousand egg beaters? To find out, integrate the marginal revenue from $x = 0$ to $x = 3$.

Drawing Conclusions about Motion from Graphs

41. *Writing to Learn* Suppose that f is the differentiable function shown in the accompanying graph and that the position at time t (seconds) of a particle moving along a coordinate axis is

$$s(t) = \int_0^t f(x)\,dx$$

meters. Use the graph to answer the following questions. Give reasons for your answers.

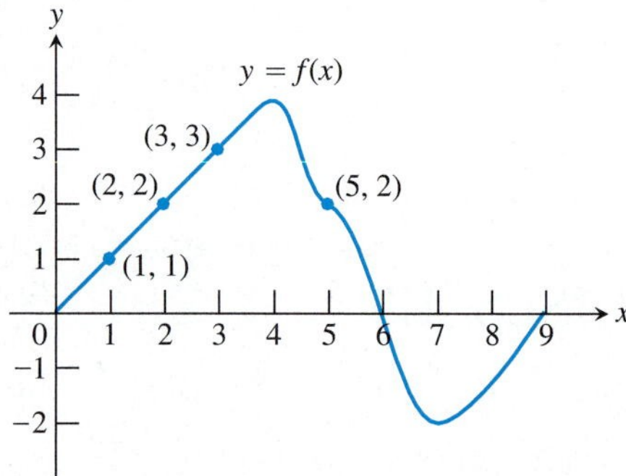

(a) What is the particle's velocity at time $t = 5$?

(b) Is the acceleration of the particle at time $t = 5$ positive or negative?

(c) What is the particle's position at time $t = 3$?

(d) At what time during the first 9 sec does s have its largest value?

(e) Approximately when is the acceleration zero?

(f) When is the particle moving toward the origin? Away from the origin?

(g) On which side of the origin does the particle lie at time $t = 9$?

42. *Writing to Learn* Suppose that g is the differentiable function shown in the accompanying graph and that the position at time t (sec) of a particle moving along a coordinate axis is

$$s(t) = \int_0^t g(x)\,dx$$

meters. Use the graph to answer the following questions. Give reasons for your answers.

(a) What is the particle's velocity at $t = 3$?

(b) Is the acceleration at time $t = 3$ positive or negative?

(c) What is the particle's position at time $t = 3$?

(d) When does the particle pass through the origin?

(e) When is the acceleration zero?

(f) When is the particle moving away from the origin? Toward the origin?

(g) On which side of the origin does the particle lie at $t = 9$?

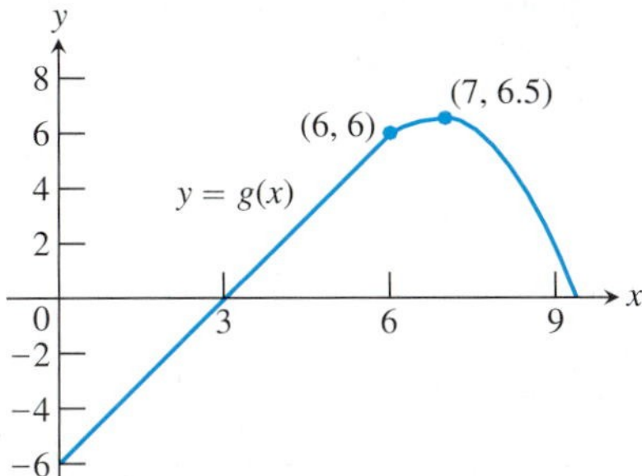

Volumes from Section 4.3

43. (*Continuation of Section 4.3, Exercise 11*) The approximating sums for the volume of water in Exercise 11, Section 4.3, are Riemann sums for an integral. What integral? Evaluate it to find the volume.

44. (*Continuation of Section 4.3, Exercise 13*) The approximating sums for the volume of the rocket nose cone in Exercise 13, Section 4.3, is a Riemann sum for an integral. What integral? Evaluate it to find the volume.

Theory and Examples

45. Suppose that $\int_1^x f(t)\,dt = x^2 - 2x + 1$. Find $f(x)$.

46. Find $f(4)$ if $\int_0^x f(t)\,dt = x \cos \pi x$.

47. *Linearization* Find the linearization of

$$f(x) = 2 - \int_2^{x+1} \frac{9}{1+t}\,dt$$

at $x = 1$.

48. *Linearization* Find the linearization of

$$g(x) = 3 + \int_1^{x^2} \sec(t-1)\,dt$$

at $x = -1$.

49. *Writing to Learn* Suppose that f has a positive derivative for all values of x and that $f(1) = 0$. Which of the following statements must be true of the function

$$g(x) = \int_0^x f(t)\,dt?$$

Give reasons for your answers.

(a) g is a differentiable function of x.

(b) g is a continuous function of x.

(c) The graph of g has a horizontal tangent at $x = 1$.

(d) g has a local maximum at $x = 1$.

(e) g has a local minimum at $x = 1$.

(f) The graph of g has an inflection point at $x = 1$.

(g) The graph of dg/dx crosses the x-axis at $x = 1$.

50. *Writing to Learn* Suppose that f has a negative derivative for all values of x and that $f(1) = 0$. Which of the following statements must be true of the function

$$h(x) = \int_0^x f(t)\,dt?$$

Give reasons for your answers.

(a) h is a twice-differentiable function of x.

(b) h and dh/dx are both continuous.

(c) The graph of h has a horizontal tangent at $x = 1$.

(d) h has a local maximum at $x = 1$.

(e) h has a local minimum at $x = 1$.

(f) The graph of h has an inflection point at $x = 1$.

(g) The graph of dh/dx crosses the x-axis at $x = 1$.

51. *Archimedes' area formula for parabolas* Archimedes (287–212 B.C.), inventor, military engineer, physicist, and the greatest mathematician of classical times in the western world, discovered that the area under a parabolic arch is two-thirds the base times the height.

(a) Use an integral to find the area under the arch

$$y = 6 - x - x^2, \quad -3 \le x \le 2.$$

(b) Find the height of the arch.

(c) Show that the area is two-thirds the base b times the height h.

(d) Sketch the parabolic arch $y = h - (4h/b^2)x^2$, $-b/2 \le x \le b/2$, assuming that h and b are positive. Then use calculus to find the area of the region enclosed between the arch and the x-axis.

52. (*Continuation of Example 9*) *Household electricity*

(a) Show by evaluating the integral in the expression

$$\frac{1}{(1/60) - 0} \int_0^{1/60} V_{max} \sin 120\pi t\,dt$$

that the average value of $V = V_{max} \sin 120\pi t$ over a full cycle is zero.

(b) The circuit that runs your electric stove is rated 240 volts rms. What is the peak value of the allowable voltage?

(c) Show that

$$\int_0^{1/60} (V_{max})^2 \sin^2 120\pi t\,dt = \frac{(V_{max})^2}{120}.$$

T **53.** *The Fundamental Theorem* If f is continuous, we expect

$$\lim_{h \to 0} \frac{1}{h} \int_x^{x+h} f(t)\,dt$$

to equal $f(x)$, as in the proof of Part 1 of the Fundamental Theorem. For instance, if $f(t) = \cos t$, then

$$\frac{1}{h} \int_x^{x+h} \cos t\,dt = \frac{\sin(x+h) - \sin x}{h}. \qquad (9)$$

The right-hand side of Equation (9) is the difference quotient for the derivative of the sine, and we expect its limit as $h \to 0$ to be $\cos x$.

Graph $\cos x$ for $-\pi \le x \le 2\pi$. Then, in a different color if possible, graph the right-hand side of Equation (9) as a function of x for $h = 2, 1, 0.5$, and 0.1. Watch how the latter curves converge to the graph of the cosine as $h \to 0$.

T **54.** Repeat Exercise 53 for $f(t) = 3t^2$. What is

$$\lim_{h \to 0} \frac{1}{h} \int_x^{x+h} 3t^2\,dt = \lim_{h \to 0} \frac{(x+h)^3 - x^3}{h}?$$

Graph $f(x) = 3x^2$ for $-1 \le x \le 1$. Then graph the quotient $((x+h)^3 - x^3)/h$ as a function of x for $h = 1, 0.5, 0.2$, and 0.1. Watch how the latter curves converge to the graph of $3x^2$ as $h \to 0$.

COMPUTER EXPLORATIONS

In Exercises 55–58, let $F(x) = \int_a^x f(t)\,dt$ for the specified function f and interval $[a, b]$. Use a CAS to perform the following steps and answer the questions posed.

(a) Plot the functions f and F together over $[a, b]$.

(b) Solve the equation $F'(x) = 0$. What can you see to be true about the graphs of f and F at points where $F'(x) = 0$? Is your observation borne out by Part 1 of the Fundamental Theorem coupled with information provided by the derivative? Explain your answer.

(c) Over what intervals (approximately) is the function F increasing and decreasing? What is true about f over those intervals?

(d) Calculate the derivative f' and plot it together with F. What can you see to be true about the graph of F at points where $f'(x) = 0$? Is your observation borne out by Part 1 of the Fundamental Theorem? Explain your answer.

55. $f(x) = x^3 - 4x^2 + 3x, \quad [0, 4]$

56. $f(x) = 2x^4 - 17x^3 + 46x^2 - 43x + 12, \quad [0, 9/2]$

57. $f(x) = \sin 2x \cos \frac{x}{3}, \quad [0, 2\pi]$

58. $f(x) = x \cos \pi x, \quad [0, 2\pi]$

In Exercises 59–64, let $F(x) = \int_a^{u(x)} f(t)\,dt$ for the specified a, u, and f. Use a CAS to perform the following steps and answer the questions posed.

(a) Find the domain of F.

(b) Calculate $F'(x)$ and determine its zeros. Over what intervals is F increasing? decreasing?

(c) Calculate $F''(x)$ and determine its zero. Identify the local extrema and the points of inflection of F.

(d) Using the information from parts (a) through (c), draw a rough sketch of $y = F(x)$ over its domain. Then graph $F(x)$ on your CAS to support your sketch.

59. $a = 1, \quad u(x) = x^2, \quad f(x) = \sqrt{1 - x^2}$

60. $a = 0, \quad u(x) = x^2, \quad f(x) = \sqrt{1 - x^2}$

61. $a = 0, \quad u(x) = 1 - x, \quad f(x) = x^2 - 2x - 3$

62. $a = 0, \quad u(x) = 1 - x^2, \quad f(x) = x^2 - 2x - 3$

63. Calculate $(d/dx) \int_a^{u(x)} f(t)\,dt$ and check your answer using a CAS.

64. Calculate $(d^2/dx^2) \int_a^{u(x)} f(t)dt$ and check your answer using a CAS.

4.6 Substitution in Definite Integrals

Substitution Formula • Area Between Curves • Boundaries with Changing Formulas

There are two methods for evaluating a definite integral by substitution, and they both work well. One is to find the corresponding indefinite integral by substitution and use one of the resulting antiderivatives to evaluate the definite integral by the Fundamental Theorem. The other is to use a formula we now study.

Substitution Formula

CD-ROM
WEBsite
Historical Biography

Isaac Barrow
(1630—1677)

Substitution in Definite Integrals

THE FORMULA

$$\int_a^b f(g(x)) \cdot g'(x)\, dx = \int_{g(a)}^{g(b)} f(u)\, du \qquad (1)$$

HOW TO USE IT

Substitute $u = g(x)$, $du = g'(x)\, dx$, and integrate from $g(a)$ to $g(b)$.

To use the formula , make the same u-substitution you would use to evaluate the corresponding indefinite integral. Then integrate with respect to u from the value u has at $x = a$ to the value u has at $x = b$.

To see why Equation (1) is true, we let F denote any antiderivative of f. Then

$$\begin{aligned}\int_a^b f(g(x)) \cdot g'(x)\, dx &= F(g(x))\Big]_{x=a}^{x=b} && \frac{d}{dx}F(g(x)) = F'(g(x))\,g'(x) = f(g(x))g'(x)\\ &= F(g(b)) - F(g(a))\\ &= F(u)\Big]_{u=g(a)}^{u=g(b)}\\ &= \int_{g(a)}^{g(b)} f(u)\, du. && \text{Fundamental Theorem, Part 2}\end{aligned}$$

Example 1 Using the Substitution Formula

Use Equation (1) to evaluate

$$\int_{-1}^{1} 3x^2\sqrt{x^3+1}\, dx$$

Solution

Transform the integral and evaluate the transformed integral with the transformed limits given by Equation (1).

$$\begin{aligned}\int_{-1}^{1} 3x^2\sqrt{x^3+1}\, dx &\\ &= \int_0^2 \sqrt{u}\, du && \text{Let } u = x^3+1,\ du = 3x^2\,dx. \text{ When } x=-1,\ u=(-1)^3+1=0. \text{ When } x=1,\ u=(1)^3+1=2.\\ &= \frac{2}{3}u^{2/3}\Big]_0^2 && \text{Evaluate the new definite integral.}\\ &= \frac{2}{3}[2^{3/2} - 0^{3/2}] = \frac{2}{3}[2\sqrt{2}] = \frac{4\sqrt{2}}{3}\end{aligned}$$

Instead of using The Substitution Formula we could use antiderivatives and the Fundamental Theorem.

Example 2 Evaluation without Using the Formula

Evaluate

$$\int_{-1}^{1} 3x^2\sqrt{x^3+1}\,dx$$

by transforming the integral as an indefinite integral, integrating, changing back to x, and using the original x-limits.

Solution

$$\int 3x^2\sqrt{x^3+1}\,dx = \int \sqrt{u}\,du \qquad \text{Let } u = x^3+1,\ du = 3x^2\,dx.$$

$$= \frac{2}{3}u^{3/2} + C \qquad \text{Integrate with respect to } u.$$

$$= \frac{2}{3}(x^3+1)^{3/2} + C \qquad \text{Replace } u \text{ by } x^3+1.$$

$$\int_{-1}^{1} 3x^2\sqrt{x^3+1}\,dx = \frac{2}{3}(x^3+1)^{3/2}\Big]_{-1}^{1} \qquad \text{Use the integral just found, with limits of integration for } x.$$

$$= \frac{2}{3}\left[((1)^3+1)^{3/2} - ((-1)^3+1)^{3/2}\right]$$

$$= \frac{2}{3}\left[2^{3/2} - 0^{3/2}\right] = \frac{2}{3}\left[2\sqrt{2}\right] = \frac{4\sqrt{2}}{3}$$

Which method is better, transforming the integral, integrating, and transforming back to use the original limits of integration or evaluating the transformed integral with transformed limits? For the integrand $3x^2\sqrt{x^3+1}$, using the Substitution Formula in Example 1 seems easier, but that is not always the case. As a rule, it is best to know both methods and to use whichever one seems better at the time.

Here is another example of evaluating a transformed integral with transformed limits.

Example 3 Using the Substitution Formula

$$\int_{\pi/4}^{\pi/2} \cot\theta\,\csc^2\theta\,d\theta = \int_{1}^{0} u\cdot(-du)$$

Let $u = \cot\theta$, $du = -\csc^2\theta\,d\theta$,
$-du = \csc^2\theta\,d\theta$.
When $\theta = \pi/4$, $u = \cot(\pi/4) = 1$.
When $\theta = \pi/2$, $u = \cot(\pi/2) = 0$.

$$= -\int_{1}^{0} u\,du$$

$$= -\frac{u^2}{2}\Big]_{1}^{0}$$

$$= -\left[\frac{(0)^2}{2} - \frac{(1)^2}{2}\right] = \frac{1}{2}$$

USING TECHNOLOGY *Visualizing Integrals with Elusive Antiderivatives*

CD-ROM WEBsite

Many integrable functions, such as the important

$$f(x) = e^{-x^2}$$

from probability theory, *do not* have antiderivatives that can be expressed in terms of elementary functions. Nevertheless, we know the antiderivative of f exists by Part 1 of the Fundamental Theorem of Calculus. Use your graphing utility to visualize the integral function

$$F(x) = \int_0^x e^{-t^2}\, dt.$$

What can you say about $F(x)$? Where is it increasing and decreasing? Where are its extreme values, if any? What can you say about the concavity of its graph?

Area Between Curves

We next determine how to find the areas of regions in the coordinate plane by integrating the functions that define the regions' boundaries.

Suppose that we want to find the area of a region that is bounded above by the curve $y = f(x)$, below by the curve $y = g(x)$, and on the left and right by the lines $x = a$ and $x = b$ (Figure 4.19). The region might accidentally have a shape whose area we could find with geometry, but if f and g are arbitrary continuous functions we usually have to find the area with an integral.

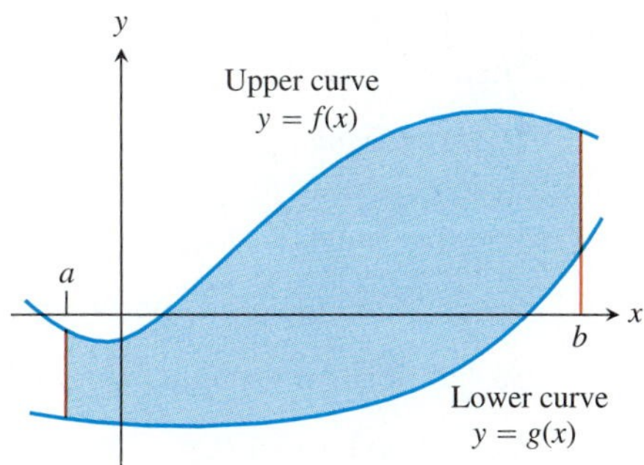

FIGURE 4.19 The region between the curves $y = f(x)$ and $y = g(x)$ and the lines $x = a$ and $x = b$.

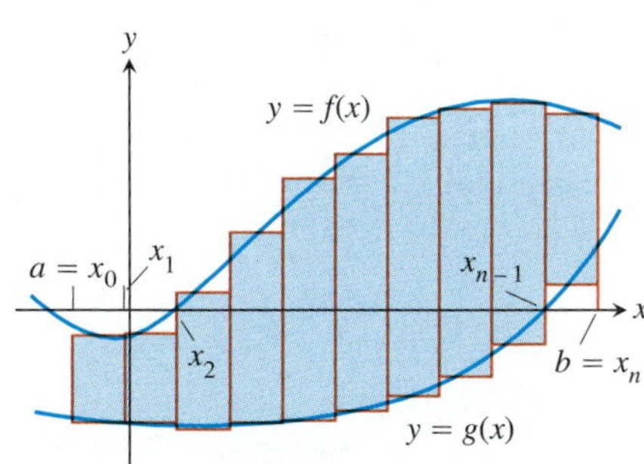

FIGURE 4.20 We approximate the region with rectangles perpendicular to the x-axis.

To see what the integral should be, we first approximate the region with n vertical rectangles based on a partition $P = \{x_0, x_1, \ldots, x_n\}$ of $[a, b]$ (Figure 4.20). The area of the kth rectangle (Figure 4.21) is

$$\Delta A_k = \text{height} \times \text{width} = [f(c_k) - g(c_k)]\, \Delta x_k.$$

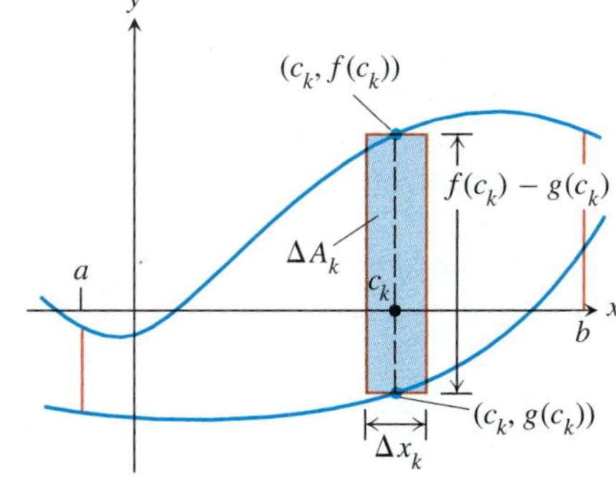

FIGURE 4.21 ΔA_k = area of kth rectangle, $f(c_k) - g(c_k)$ = height, and Δx_k = width.

We then approximate the area of the region by adding the areas of the n rectangles:

$$A \approx \sum_{k=1}^{n} \Delta A_k = \sum_{k=1}^{n} [f(c_k) - g(c_k)]\, \Delta x_k. \qquad \text{Riemann sum}$$

As $\| P \| \to 0$ the sums on the right approach the limit $\int_a^b [f(x) - g(x)]\, dx$ because f and g are continuous. We take the area of the region to be the value of this integral. That is,

$$A = \lim_{\| P \| \to 0} \sum_{k=1}^{n} [f(c_k) - g(c_k)]\, \Delta x_k = \int_a^b [f(x) - g(x)]\, dx.$$

Definition Area Between Curves

If f and g are continuous with $f(x) \geq g(x)$ throughout $[a, b]$, then the area of the region between the curves $y = f(x)$ and $y = g(x)$ from a to b is the integral of $[f - g]$ from a to b:

$$A = \int_a^b [f(x) - g(x)]\, dx. \tag{2}$$

To apply Equation (2) we take the following steps.

How to Find the Area Between Two Curves

Step 1. *Graph the curves and draw a representative rectangle.* This reveals which curve is f (upper curve) and which is g (lower curve). It also helps find the limits of integration if you do not already know them.

Step 2. *Find the limits of integration.*

Step 3. *Write a formula for* $f(x) - g(x)$. Simplify it if you can.

Step 4. *Integrate* $[f(x) - g(x)]$ *from* a *to* b. The number you get is the area.

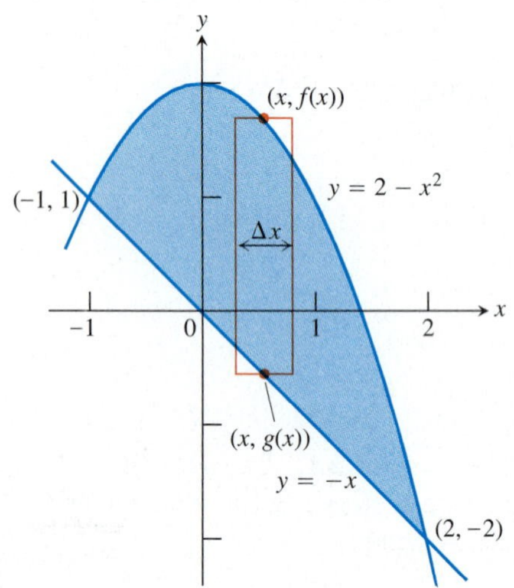

FIGURE 4.22 The region in Example 4 with a typical approximating rectangle.

Example 4 Area Between Intersecting Curves

Find the area of the region enclosed by the parabola $y = 2 - x^2$ and the line $y = -x$.

Solution

Step 1. *Sketch the curves and a vertical rectangle* (Figure 4.22). Identifying the upper and the lower curves, we take $f(x) = 2 - x^2$ and $g(x) = -x$. The x-coordinates of the intersection points are the limits of integration.

Step 2. *Find the limits of integration.* We find the limits of integration by solving $y = 2 - x^2$ and $y = -x$ simultaneously for x:

$$\begin{aligned} 2 - x^2 &= -x && \text{Equate } f(x) \text{ and } g(x). \\ x^2 - x - 2 &= 0 && \text{Rewrite.} \\ (x + 1)(x - 2) &= 0 && \text{Factor.} \\ x = -1, \quad x &= 2. && \text{Solve.} \end{aligned}$$

The region runs from $x = -1$ to $x = 2$. The limits of integration are $a = -1$, $b = 2$.

Step 3. *Simplify the formula for* $f(x) - g(x)$.

$$f(x) - g(x) = (2 - x^2) - (-x) = 2 - x^2 + x$$
$$= 2 + x - x^2$$

Rearrangement a matter of taste

Step 4. *Integrate* $[f(x) - g(x)]$ *from a to b.*

$$A = \int_a^b [f(x) - g(x)]\,dx = \int_{-1}^{2} (2 + x - x^2)\,dx = \left[2x + \frac{x^2}{2} - \frac{x^3}{3}\right]_{-1}^{2}$$
$$= \left(4 + \frac{4}{2} - \frac{8}{3}\right) - \left(-2 + \frac{1}{2} + \frac{1}{3}\right)$$
$$= 6 + \frac{3}{2} - \frac{9}{3} = \frac{9}{2}$$

Boundaries with Changing Formulas

If the formula for a bounding curve changes at one or more points, we partition the region into subregions that correspond to the formula changes and apply Equation (2) to each subregion.

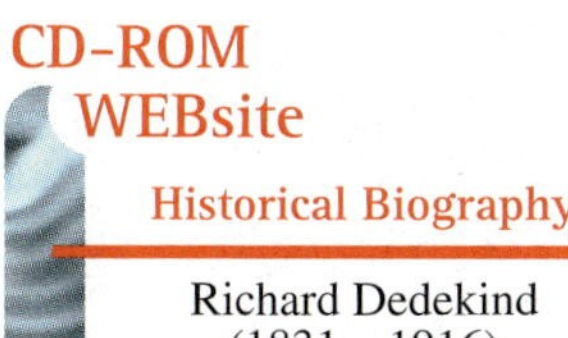

CD-ROM WEBsite

Historical Biography

Richard Dedekind (1831—1916)

Example 5 Changing the Integral to Match a Boundary Change

Find the area of the region in the first quadrant that is bounded above by $y = \sqrt{x}$ and below by the x-axis and the line $y = x - 2$.

Solution

Step 1. The sketch (Figure 4.23) shows that the region's upper boundary is the graph of $f(x) = \sqrt{x}$. The lower boundary changes from $g(x) = 0$ for $0 \le x \le 2$ to $g(x) = x - 2$ for $2 \le x \le 4$ (there is agreement at $x = 2$). We partition the region at $x = 2$ into subregions A and B and sketch a representative rectangle for each subregion.

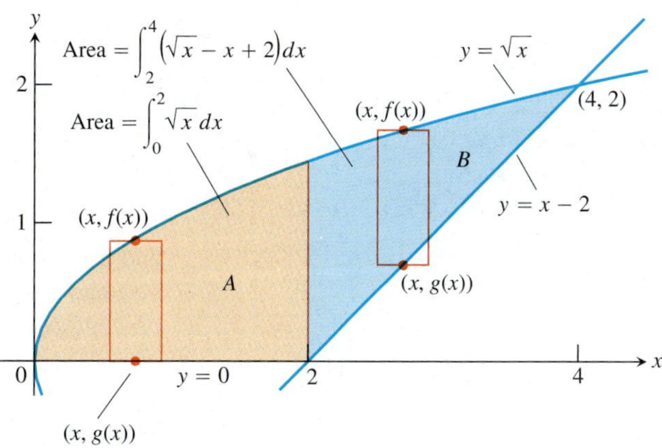

FIGURE 4.23 When the formula for a bounding curve changes, the area integral changes to match. (Example 5)

Step 2. The limits of integration for region A are $a = 0$ and $b = 2$. The left-hand limit for region B is $a = 2$. To find the right-hand limit, we solve the equations $y = \sqrt{x}$ and $y = x - 2$ simultaneously for x:

$$\sqrt{x} = x - 2 \qquad \text{Equate } f(x) \text{ and } g(x).$$
$$x = (x-2)^2 = x^2 - 4x + 4 \qquad \text{Square both sides.}$$
$$x^2 - 5x + 4 = 0 \qquad \text{Rewrite.}$$
$$(x-1)(x-4) = 0 \qquad \text{Factor.}$$
$$x = 1, \qquad x = 4. \qquad \text{Solve.}$$

Only the value $x = 4$ satisfies the equation $\sqrt{x} = x - 2$. The value $x = 1$ is an extraneous root introduced by squaring. The right-hand limit is $b = 4$.

Step 3. For $0 \le x \le 2$: $\quad f(x) - g(x) = \sqrt{x} - 0 = \sqrt{x}$

For $2 \le x \le 4$: $\quad f(x) - g(x) = \sqrt{x} - (x - 2) = \sqrt{x} - x + 2$

Step 4. We add the area of subregions A and B to find the total area:

$$\text{Total area} = \underbrace{\int_0^2 \sqrt{x}\,dx}_{\text{area of } A} + \underbrace{\int_2^4 (\sqrt{x} - x + 2)\,dx}_{\text{area of } B}$$

$$= \left[\frac{2}{3}x^{3/2}\right]_0^2 + \left[\frac{2}{3}x^{3/2} - \frac{x^2}{2} + 2x\right]_2^4$$

$$= \frac{2}{3}(2)^{3/2} - 0 + \left(\frac{2}{3}(4)^{3/2} - 8 + 8\right) - \left(\frac{2}{3}(2)^{3/2} - 2 + 4\right)$$

$$= \frac{2}{3}(8) - 2 = \frac{10}{3}.$$

USING TECHNOLOGY

The Intersection of Two Graphs One of the difficult and sometimes frustrating parts of integration applications is finding the limits of integration. To do this, you often have to find the zeroes of a function or the intersection points of two curves.

To solve the equation $f(x) = g(x)$ using a graphing utility, you enter

$$y_1 = f(x) \qquad \text{and} \qquad y_2 = g(x)$$

and use the grapher routine to find the points of intersection. Alternatively, you can solve the equation $f(x) - g(x) = 0$ with a root finder. Try both procedures with

$$f(x) = \ln x \qquad \text{and} \qquad g(x) = 3 - x.$$

When points of intersection are not clearly revealed or you suspect hidden behavior, additional work with the graphing utility or further use of calculus may be necessary.

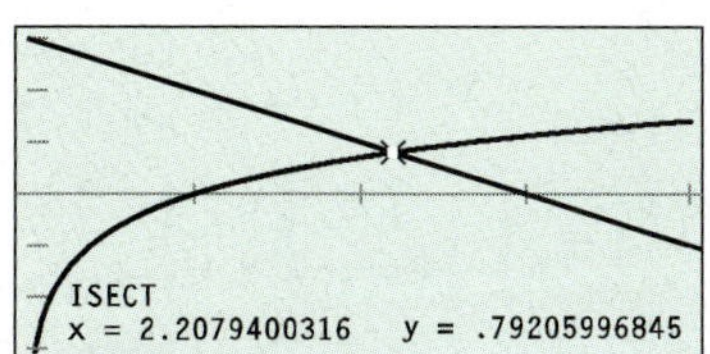

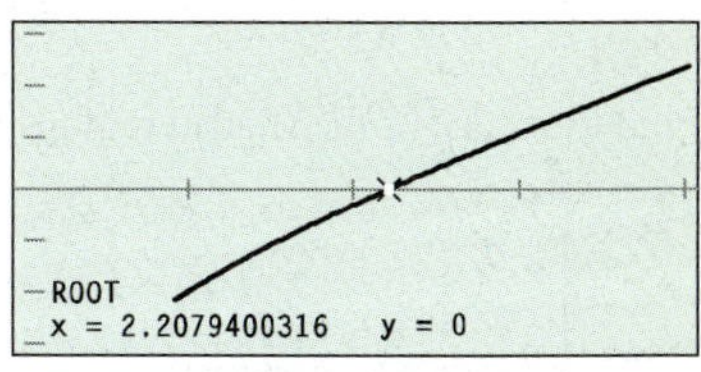

(a) The intersecting curves $y_1 = \ln x$ and $y_2 = 3 - x$, using a built-in function to find the intersection

(b) Using a built-in root finder to find the zero of $f(x) = \ln x - 3 + x$

EXERCISES 4.6

Use the substitution formula to evaluate the integrals in Exercises 1–16.

1. (a) $\displaystyle\int_0^3 \sqrt{y+1}\, dy$

(b) $\displaystyle\int_{-1}^0 \sqrt{y+1}\, dy$

2. (a) $\displaystyle\int_0^{\pi/4} \tan x \sec^2 x\, dx$

(b) $\displaystyle\int_{-\pi/4}^0 \tan x \sec^2 x\, dx$

3. (a) $\displaystyle\int_0^{\pi} 3\cos^2 x \sin x\, dx$

(b) $\displaystyle\int_{2\pi}^{3\pi} 3\cos^2 x \sin x\, dx$

4. (a) $\displaystyle\int_0^{\sqrt{7}} t(t^2+1)^{1/3}\, dt$

(b) $\displaystyle\int_{-\sqrt{7}}^0 t(t^2+1)^{1/3}\, dt$

5. (a) $\displaystyle\int_{-1}^1 \frac{5r}{(4+r^2)^2}\, dr$

(b) $\displaystyle\int_0^1 \frac{5r}{(4+r^2)^2}\, dr$

6. (a) $\displaystyle\int_0^{\sqrt{3}} \frac{4x}{\sqrt{x^2+1}}\, dx$

(b) $\displaystyle\int_{-\sqrt{3}}^{\sqrt{3}} \frac{4x}{\sqrt{x^2+1}}\, dx$

7. (a) $\displaystyle\int_0^{2\pi} \frac{\cos z}{\sqrt{4+3\sin z}}\, dz$

(b) $\displaystyle\int_{-\pi}^{\pi} \frac{\cos z}{\sqrt{4+3\sin z}}\, dz$

8. $\displaystyle\int_0^1 \sqrt{t^5+2t}\,(5t^4+2)\, dt$

9. $\displaystyle\int_1^4 \frac{dy}{2\sqrt{y}\,(1+\sqrt{y})^2}$

10. $\displaystyle\int_0^{\pi/6} \cos^{-3} 2\theta \sin 2\theta\, d\theta$

11. $\displaystyle\int_0^{\pi/4} (1-\sin 2t)^{3/2} \cos 2t\, dt$

12. $\displaystyle\int_0^1 (4y - y^2 + 4y^3 + 1)^{-2/3}(12y^2 - 2y + 4)\, dy$

13. $\displaystyle\int_0^{\pi/2} e^{\sin x} \cos x\, dx$

14. $\displaystyle\int_0^{\pi/4} (1 + e^{\tan\theta})\sec^2\theta\, d\theta$

15. $\displaystyle\int_{\ln(\pi/6)}^{\ln(\pi/2)} 2e^v \cos e^v\, dv$

16. $\displaystyle\int_0^{\sqrt{\ln \pi}} 2xe^{x^2} \cos(e^{x^2})\, dx$

Initial Value Problems

Solve the initial value problems in Exercises 17 and 18.

17. $\dfrac{dy}{dt} = e^{-t}\sec^2(\pi e^{-t}), \quad y(\ln 4) = 2/\pi$

18. $\dfrac{d^2y}{dx^2} = 2e^{-x}, \quad y(0) = 1 \text{ and } y'(0) = 0$

Area

Find the areas of the shaded regions in Exercises 19–24.

19.

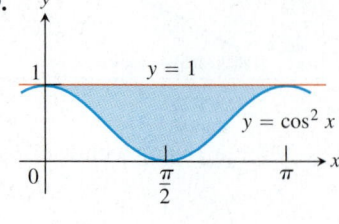

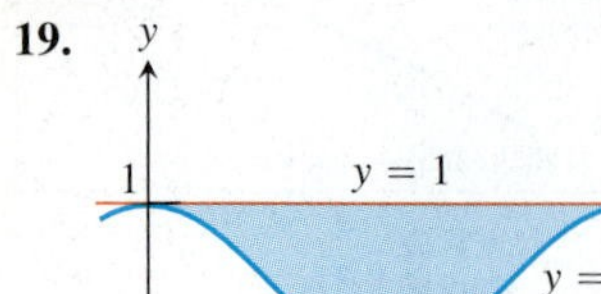

20.

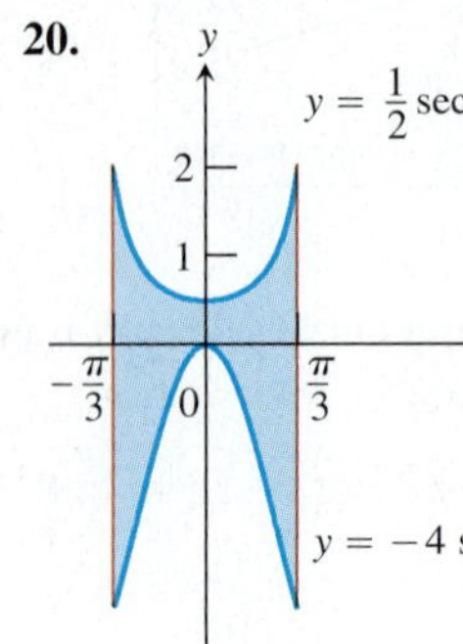

21.

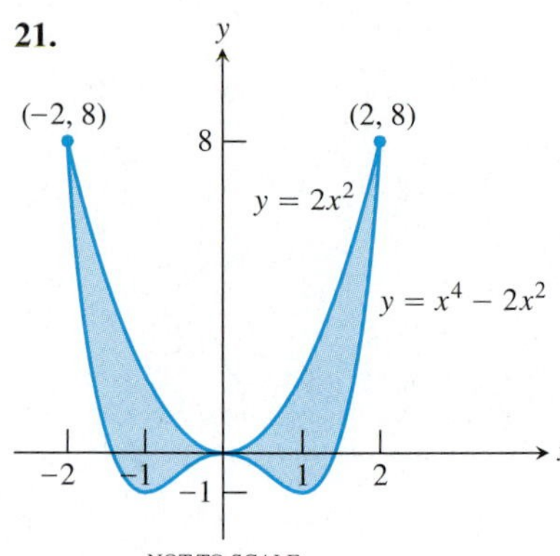

22.

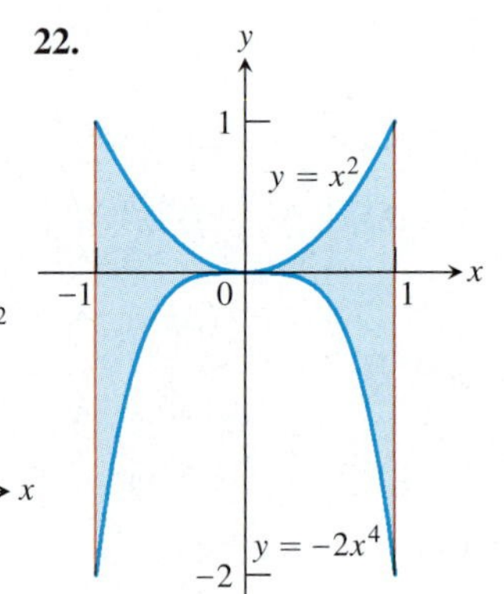

23.

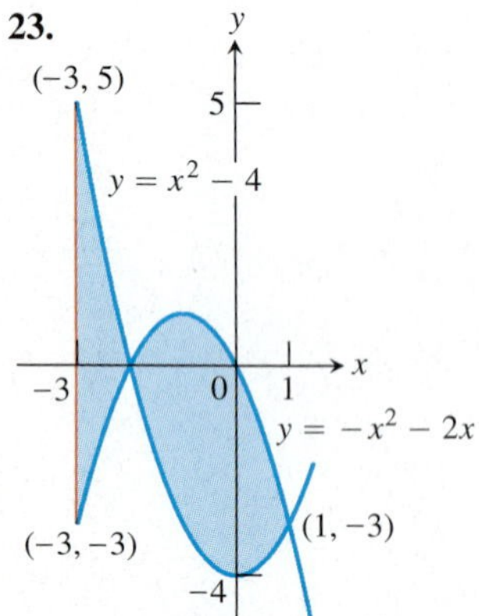

24.

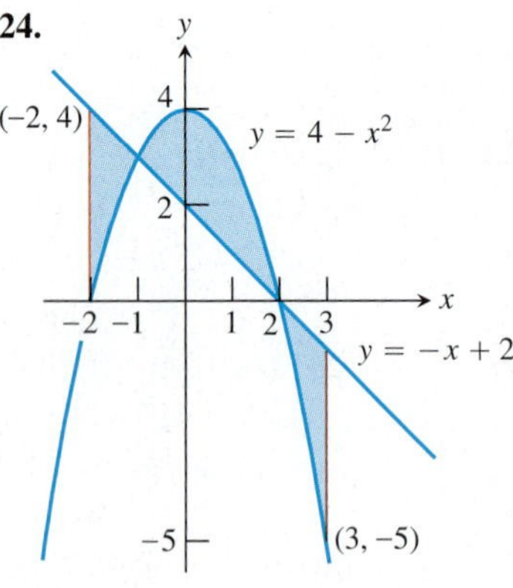

In Exercises 25–28, graph the function over the given interval. Then

(a) integrate the function over the interval

(b) find the area of the region between the graph and the x-axis.

25. $y = x^2 - 6x + 8, \quad [0, 3]$

26. $y = -x^2 + 5x - 4, \quad [0, 2]$

27. $y = 2x - x^2, \quad [0, 3]$

28. $y = x^2 - 4x, \quad [0, 5]$

Find the areas of the regions enclosed by the lines and curves in Exercises 29–38.

29. $y = x^2 - 2$ and $y = 2$

30. $y = -x^2 - 2x$ and $y = x$

31. $y = x^2$ and $y = -x^2 + 4x$

32. $y = 7 - 2x^2$ and $y = x^2 + 4$

33. $y = x^4 - 4x^2 + 4$ and $y = x^2$

34. $y = |x^2 - 4|$ and $y = (x^2/2) + 4$

35. $y = 2\sin x$ and $y = \sin 2x, \quad 0 \le x \le \pi$

36. $y = 8\cos x$ and $y = \sec^2 x, \quad -\pi/3 \le x \le \pi/3$

37. $y = \sin(\pi x/2)$ and $y = x$

38. $y = \sec^2 x$, $y = \tan^2 x$, $x = -\pi/4$, and $x = \pi/4$

39. Find the area of the region in the first quadrant bounded by the line $y = x$, the line $x = 2$, the curve $y = 1/x^2$, and the x-axis.

40. Find the area of the "triangular" region in the first quadrant bounded on the left by the y-axis and on the right by the curves $y = \sin x$ and $y = \cos x$.

41. Find the area of the region between the curve $y = 3 - x^2$ and the line $y = -1$.

42. Find the area of the region in the first quadrant bounded on the left by the y-axis, below by the line $y = x/4$, above left by the curve $y = 1 + \sqrt{x}$, and above right by the curve $y = 2/\sqrt{x}$.

COMPUTER EXPLORATIONS

In Exercises 43–46, you will find the area between curves in the plane when you cannot find their points of intersection using simple algebra. Use a CAS to perform the following steps.

(a) Plot the curves together to see what they look like and how many points of intersection they have.

(b) Use the numerical equation solver in your CAS to find all the points of intersection.

(c) Integrate $|f(x) - g(x)|$ over consecutive pairs of intersection values.

(d) Sum together the integrals found in part (c).

43. $f(x) = \dfrac{x^3}{3} - \dfrac{x^2}{2} - 2x + \dfrac{1}{3}, \quad g(x) = x - 1$

44. $f(x) = \dfrac{x^4}{2} - 3x^3 + 10, \quad g(x) = 8 - 12x$

45. $f(x) = x + \sin(2x), \quad g(x) = x^3$

46. $f(x) = x^2\cos x, \quad g(x) = x^3 - x$

4.7 Numerical Integration

Trapezoidal Approximations • Error in the Trapezoidal Approximation • Approximations Using Parabolas • Error in Simpson's Rule • Which Rule Gives Better Results? • Round-off Errors

As we have seen, the ideal way to evaluate a definite integral $\int_a^b f(x)\,dx$ is to find a formula $F(x)$ for one of the antiderivatives of $f(x)$ and calculate the number $F(b) - F(a)$. But some antiderivatives are hard to find, and still others, like the antiderivatives of $(\sin x)/x$ and $\sqrt{1 + x^4}$, have no elementary formulas. We do not mean merely that no one has yet succeeded in finding elementary formulas for the antiderivatives of $(\sin x)/x$ and $\sqrt{1 + x^4}$. We mean it has been proved that no such formulas exist.

Whatever the reason, when we cannot evaluate a definite integral with an antiderivative, we turn to numerical methods such as the Trapezoidal Rule and Simpson's Rule, described in this section.

Trapezoidal Approximations

When we cannot find a workable antiderivative for a function f that we have to integrate, we partition the interval of integration, replace f by a closely fitting polynomial on each subinterval, integrate the polynomials, and add the results to approximate the integral of f. We start with straight-line segments giving trapezoids.

As shown in Figure 4.24, if $[a, b]$ is partitioned into n subintervals of equal length $h = (b - a)/n$, the graph of f on $[a, b]$ can be approximated by a straight-line segment over each subinterval.

The region between the curve and the x-axis is then approximated by the trapezoids, the area of each trapezoid being the length of its horizontal "altitude" times

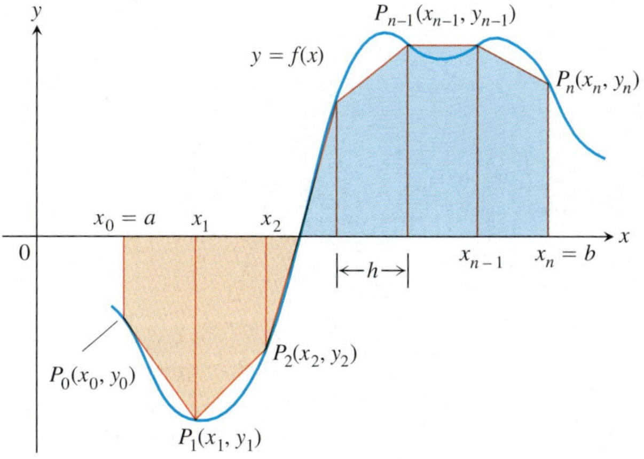

The length $h = (b - a)/n$ is called the **step size.** It is conventional to use h in this context instead of Δx.

FIGURE 4.24 The Trapezoidal Rule approximates short stretches of the curve $y = f(x)$ with line segments. To approximate the integral of f from a to b, we add the "signed" areas of the trapezoids made by joining the ends of the segments to the x-axis.

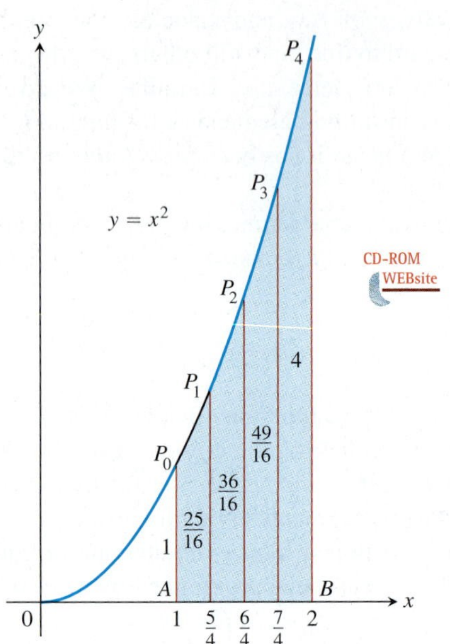

FIGURE 4.25 The trapezoidal approximation of the area under the graph of $y = x^2$ from $x = 1$ to $x = 2$ is a slight overestimate.

the average of its two vertical "bases." We add the areas of the trapezoids, counting area above the x-axis as positive and area below the axis as negative:

$$\begin{aligned} T &= \frac{1}{2}(y_0 + y_1)h + \frac{1}{2}(y_1 + y_2)h + \cdots + \frac{1}{2}(y_{n-2} + y_{n-1})h + \frac{1}{2}(y_{n-1} + y_n)h \\ &= h\left(\frac{1}{2}y_0 + y_1 + y_2 + \cdots + y_{n-1} + \frac{1}{2}y_n\right) \\ &= \frac{h}{2}(y_0 + 2y_1 + 2y_2 + \cdots + 2y_{n-1} + y_n), \end{aligned}$$

where

$$y_0 = f(a), \quad y_1 = f(x_1), \quad \ldots, \quad y_{n-1} = f(x_{n-1}), \quad y_n = f(b).$$

The Trapezoidal Rule says: Use T to estimate the integral of f from a to b.

CD-ROM WEBsite

The Trapezoidal Rule

To approximate $\int_a^b f(x)\,dx$, use

$$T = \frac{h}{2}(y_0 + 2y_1 + 2y_2 + \cdots + 2y_{n-1} + y_n),$$

The y's are the values of f at the partition points

$$x_0 = a, \quad x_1 = a + h, \quad x_2 = a + 2h, \quad \ldots, \quad x_{n-1} = a + (n-1)h, \quad x_n = b,$$

where $h = (b - a)/n$.

Table 4.6

x	$y = x^2$
1	1
$\frac{5}{4}$	$\frac{25}{16}$
$\frac{6}{4}$	$\frac{36}{16}$
$\frac{7}{4}$	$\frac{49}{16}$
2	4

Example 1 Applying the Trapezoidal Rule

Use the Trapezoidal Rule with $n = 4$ to estimate $\int_1^2 x^2\,dx$. Compare the estimate with the exact value.

Solution Partition [1, 2] into four subintervals of equal length (Figure 4.25). Then evaluate $y = x^2$ at each partition point (Table 4.6).

Using these y values, $n = 4$, and $h = (2 - 1)/4 = 1/4$ in the Trapezoidal Rule, we have

$$\begin{aligned} T &= \frac{h}{2}(y_0 + 2y_1 + 2y_2 + 2y_3 + y_4) \\ &= \frac{1}{8}\left(1 + 2\left(\frac{25}{16}\right) + 2\left(\frac{36}{16}\right) + 2\left(\frac{49}{16}\right) + 4\right) \\ &= \frac{75}{32} = 2.34375. \end{aligned}$$

The exact value of the integral is

$$\int_1^2 x^2\,dx = \frac{x^3}{3}\bigg]_1^2 = \frac{8}{3} - \frac{1}{3} = \frac{7}{3}.$$

The T approximation overestimates the integral by about half a percent of its true value of 7/3. The percentage error is $(2.34375 - 7/3)/(7/3) \approx 0.00446$, or 0.446%.

We could have predicted that the Trapezoidal Rule would overestimate the integral in Example 1 by considering the geometry of the graph in Figure 4.25. Since the parabola is concave *up,* the approximating segments lie above the curve, giving each trapezoid slightly more area than the corresponding strip under the curve. In Figure 4.24, we see that the straight segments lie *under* the curve on those intervals where the curve is concave *down,* causing the Trapezoidal Rule to *underestimate* the integral on those intervals. The interpretation of "area" changes where the curve lies below the x-axis, but it is still the case that the higher y-values give the greater signed area. Thus, we can always say that T overestimates the integral where the graph is concave up and underestimates the integral where the graph is concave down.

Example 2 Averaging Temperatures

An observer measures the outside temperature every hour from noon until midnight, recording the temperatures in the following table.

Time	N	1	2	3	4	5	6	7	8	9	10	11	M
Temp	63	65	66	68	70	69	68	68	65	64	62	58	55

What was the average temperature for the 12-hour period?

Solution We are looking for the average value of a continuous function (temperature) for which we know values at discrete times that are one unit apart. We need to find

$$\text{av}(f) = \frac{1}{b-a}\int_a^b f(x)\,dx,$$

without having a formula for $f(x)$. The integral, however, can be approximated by the Trapezoid Rule, using the temperatures in the table as function values at the points of a 12-subinterval partition of the 12-hour interval (making $h = 1$).

$$\begin{aligned} T &= \frac{h}{2}(y_0 + 2y_1 + 2y_2 + \cdots + 2y_{11} + y_{12}) \\ &= \frac{1}{2}(63 + 2\cdot 65 + 2\cdot 66 + \cdots + 2\cdot 58 + 55) \\ &= 782 \end{aligned}$$

Using T to approximate $\int_a^b f(x)\,dx$, we have

$$\text{av}(f) \approx \frac{1}{b-a}\cdot T = \frac{1}{12}\cdot 782 \approx 65.17.$$

Rounding to be consistent with the data given, we estimate the average temperature as 65 degrees.

Error in the Trapezoidal Approximation

Pictures suggest that the magnitude of the error

$$E_T = \int_a^b f(x)\,dx - T$$

in the trapezoidal approximation will decrease as the **step size** h decreases, because the trapezoids fit the curve better as their number increases. A theorem from advanced calculus assures us that this will be the case if f has a continuous second derivative.

> **Error Estimate for the Trapezoidal Rule**
>
> If f'' is continuous and M is any upper bound for the values of $|f''|$ on $[a, b]$, then
>
> $$|E_T| \le \frac{b-a}{12} h^2 M, \tag{1}$$
>
> where $h = (b-a)/n$.

Although theory tells us there will always be a smallest safe value of M, in practice we can hardly ever find it. Instead, we find the best value we can and go on from there to estimate $|E_T|$. This may seem sloppy, but it works. To make $|E_T|$ small for a given M, we make h small.

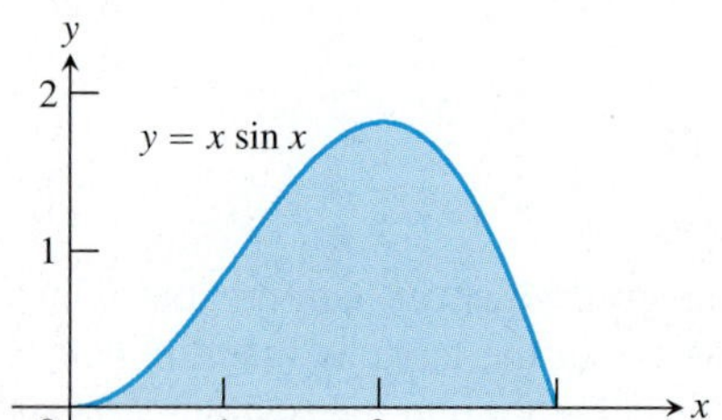

FIGURE 4.26 Graph of the integrand in Example 3.

Example 3 Bounding the Trapezoidal Rule Error

Find an upper bound for the error incurred in estimating

$$\int_0^\pi x \sin x\,dx$$

with the trapezoidal rule with $n = 10$ steps (Figure 4.26).

Solution With $a = 0$, $b = \pi$, and $h = (b-a)/n = \pi/10$, Equation (1) gives

$$|E_T| \le \frac{b-a}{12} h^2 M = \frac{\pi}{12}\left(\frac{\pi}{10}\right)^2 M = \frac{\pi^3}{1200} M.$$

The number M can be any upper bound for the magnitude of the second derivative of $f(x) = x \sin x$ on $[0, \pi]$. A routine calculation gives

$$f''(x) = 2\cos x - x \sin x,$$

so

$$\begin{aligned} |f''(x)| &= |2\cos x - x\sin x| \\ &\le 2|\cos x| + |x||\sin x| && \text{Triangle inequality: } |a+b| \le |a| + |b| \\ &\le 2 \cdot 1 + \pi \cdot 1 = 2 + \pi. && |\cos x| \text{ and } |\sin x| \text{ never exceed 1, and } 0 \le x \le \pi. \end{aligned}$$

We can safely take $M = 2 + \pi$. Therefore,

$$|E_T| \leq \frac{\pi^3}{1200} M = \frac{\pi^3(2 + \pi)}{1200} < 0.133. \qquad \text{Rounded up to be safe}$$

The absolute error is no greater than 0.133.

For greater accuracy, we would not try to improve M but would take more steps. With $n = 100$ steps, for example, $h = \pi/100$ and

$$|E_T| \leq \frac{\pi}{12}\left(\frac{\pi}{100}\right)^2 M = \frac{\pi^3(2 + \pi)}{120{,}000} < 0.00133 = 1.33 \times 10^{-3}.$$

Approximations Using Parabolas

Riemann sums and the Trapezoidal Rule both give reasonable approximations to the integral of a continuous function over a closed interval. The Trapezoidal Rule is more efficient, giving a better approximation for small values of n, which makes it a faster algorithm for numerical integration.

Indeed, the only shortcoming of the Trapezoidal Rule seems to be that it depends on approximating curved arcs with straight segments. You might think that an algorithm that approximates the curve with *curved* pieces would be even more efficient (and hence faster for machines), and you would be right. One such algorithm uses parabolas and is known as *Simpson's Rule.* Simpson's Rule for approximating $\int_a^b f(x)\,dx$ is based on approximating f with quadratic polynomials instead of linear polynomials. We approximate the graph with parabolic arcs instead of line segments (Figure 4.27).

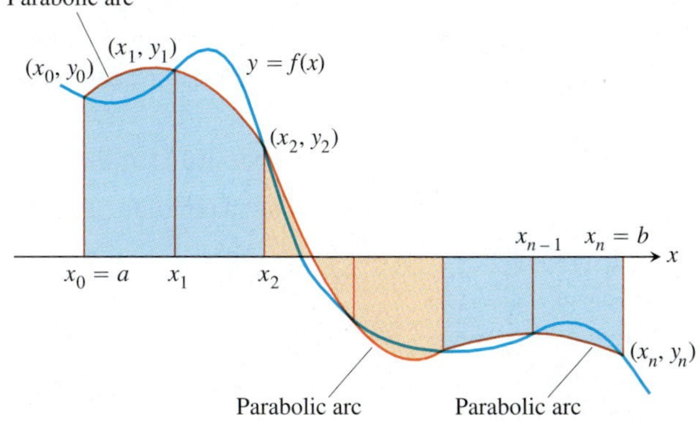

FIGURE 4.27 Simpson's Rule approximates short stretches of a curve with parabolic arcs.

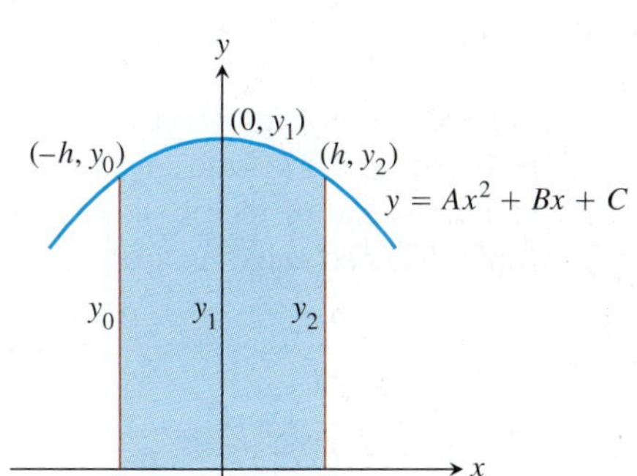

FIGURE 4.28 By integrating from $-h$ to h, we find the shaded area to be

$$\frac{h}{3}(y_0 + 4y_1 + y_2).$$

The integral of the quadratic polynomial $y = Ax^2 + Bx + C$ in Figure 4.28 from $x = -h$ to $x = h$ is

$$\int_{-h}^{h} (Ax^2 + Bx + C)\,dx = \frac{h}{3}(y_0 + 4y_1 + y_2) \tag{2}$$

(Appendix 4). Simpson's Rule follows from partitioning $[a, b]$ into an even number of subintervals of equal length h, applying Equation (2) to successive interval pairs, and adding the results.

Simpson's Rule

To approximate $\int_a^b f(x)\,dx$, use

$$S = \frac{h}{3}(y_0 + 4y_1 + 2y_2 + 4y_3 + \cdots + 2y_{n-2} + 4y_{n-1} + y_n).$$

The y's are the values of f at the partition points

$$x_0 = a, \quad x_1 = a + h, \quad x_2 = a + 2h, \quad \ldots, \quad x_{n-1} = a + (n-1)h, \quad x_n = b.$$

The number n is even, and $h = (b - a)/n$.

Table 4.7

x	$y = 5x^4$
0	0
$\frac{1}{2}$	$\frac{5}{16}$
1	5
$\frac{3}{2}$	$\frac{405}{16}$
2	80

Example 4 Applying Simpson's Rule

Use Simpson's Rule with $n = 4$ to approximate $\int_0^2 5x^4\,dx$.

Solution Partition [0, 2] into four subintervals and evaluate $y = 5x^4$ at the partition points (Table 4.7). Then apply Simpson's Rule with $n = 4$ and $h = 1/2$:

$$\begin{aligned} S &= \frac{h}{3}(y_0 + 4y_1 + 2y_2 + 4y_3 + y_4) \\ &= \frac{1}{6}\left(0 + 4\left(\frac{5}{16}\right) + 2(5) + 4\left(\frac{405}{16}\right) + 80\right) \\ &= 32\frac{1}{12}. \end{aligned}$$

This estimate differs from the exact value (32) by only 1/12, a percentage error of less than three-tenths of one percent, and this was with just four subintervals.

CD-ROM WEBsite

Error in Simpson's Rule

The magnitude of the Simpson's Rule error,

$$E_S = \int_a^b f(x)\,dx - S,$$

decreases with the step size, as we would expect from our experience with the Trapezoidal Rule. The inequality for controlling the Simpson's Rule error, however, assumes f to have a continuous fourth derivative instead of merely a continuous second derivative. The formula, once again from advanced calculus, is this.

Error Estimate for Simpson's Rule

If $f^{(4)}$ is continuous and M is an upper bound for the values of $|f^{(4)}|$ on $[a, b]$, then

$$|E_S| \le \frac{b-a}{180}h^4 M, \tag{3}$$

where $h = (b - a)/n$.

As with the Trapezoidal Rule, we can almost never find the smallest possible value of M. We just find the best value we can and go on from there.

Example 5 Bounding the Error in Simpson's Rule

What estimate does Equation (3) give for the error in Simpson's approximation in Example 4?

Solution To estimate the error, we first find an upper bound M for the magnitude of the fourth derivative of $f(x) = 5x^4$ on the interval $0 \le x \le 2$. Since the fourth derivative has the constant value $f^{(4)}(x) = 120$, we may safely take $M = 120$. With $b - a = 2$ and $h = 1/2$, Equation (3) gives

$$|E_S| \le \frac{b-a}{180} h^4 M = \frac{2}{180}\left(\frac{1}{2}\right)^4 (120) = \frac{1}{12}.$$

Which Rule Gives Better Results?

The answer lies in the error-control formulas

$$|E_T| \le \frac{b-a}{12} h^2 M, \qquad |E_S| \le \frac{b-a}{180} h^4 M.$$

The M's, of course, mean different things, the first being an upper bound on $|f''|$ and the second an upper bound on $|f^{(4)}|$. But there is more. The factor $(b - a)/180$ in the Simpson formula is one-fifteenth of the factor $(b - a)/12$ in the trapezoidal formula. More important still, the Simpson formula has an h^4, whereas the trapezoidal formula has only an h^2. If h is one-tenth, then h^2 is one-hundredth but h^4 is only one ten-thousandth. If both M's are 1, for example, and $b - a = 1$, then, with $h = 1/10$,

$$|E_T| \le \frac{1}{12}\left(\frac{1}{10}\right)^2 \cdot 1 = \frac{1}{1200},$$

whereas

$$|E_S| \le \frac{1}{180}\left(\frac{1}{10}\right)^4 \cdot 1 = \frac{1}{1{,}800{,}000} = \frac{1}{1500} \cdot \frac{1}{1200}.$$

For roughly the same amount of computational effort, we get better accuracy with Simpson's Rule, at least in this case.

The h^2 versus h^4 is the key. If h is less than 1, then h^4 can be significantly smaller than h^2. On the other hand, if h equals 1, there is no difference between h^2 and h^4. If h is greater than 1, the value of h^4 may be significantly larger than the value of h^2. In the latter two cases, the error formulas offer little help. We have to go back to the geometry of the curve $y = f(x)$ to see whether trapezoids or parabolas, if either, are going to give the results we want.

Trapezoidal versus Simpson

If Simpson's Rule is more accurate, why bother with the Trapezoidal Rule? There are two reasons. First, the Trapezoidal Rule is useful in a number of specific applications because it leads to much simpler expressions. Second, the Trapezoidal Rule is the basis for *Rhomberg integration,* one of the most satisfactory machine methods when high precision is required.

Example 6 Comparing the Trapezoidal Rule and Simpson's Rule Approximations

We know from the Fundamental Theorem that the value of ln 2 can be calculated from the integral

$$\ln 2 = \int_1^2 \frac{1}{x}\,dx.$$

Table 4.8 shows T and S values for approximations of $\int_1^2 (1/x)\,dx$ using various values of n. Notice how Simpson's Rule dramatically improves over the Trapezoidal Rule. In particular, notice that when we double the value of n (thereby halving the value of h), the T error is divided by 2 *squared*, whereas the S error is divided by 2 *to the fourth.*

Table 4.8 Trapezoidal Rule approximations (T_n) and Simpson's Rule approximations (S_n) of $\ln 2 = \int_1^2 (1/x)\,dx$

n	T_n	|Error| less than . . .	S_n	|Error| less than . . .
10	0.6937714032	0.0006242227	0.6931502307	0.0000030502
20	0.6933033818	0.0001562013	0.6931473747	0.0000001942
30	0.6932166154	0.0000694349	0.6931472190	0.0000000385
40	0.6931862400	0.0000390595	0.6931471927	0.0000000122
50	0.6931721793	0.0000249988	0.6931471856	0.0000000050
100	0.6931534305	0.0000062500	0.6931471809	0.0000000004

This has a dramatic effect as h gets very small. The Simpson approximation for $n = 50$ rounds accurately to seven places and for $n = 100$ agrees to nine decimal places (billionths)!

Example 7 Draining A Swamp

A town wants to drain and fill a small polluted swamp (Figure 4.29). The swamp averages 5 ft deep. About how many cubic yards of dirt will it take to fill the area after the swamp is drained?

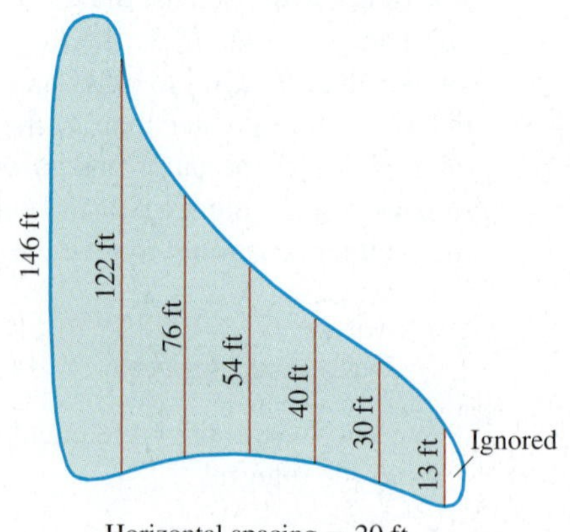

FIGURE 4.29 The swamp in Example 7.

Solution To calculate the volume of the swamp, we estimate the surface area and multiply by 5. To estimate the area, we use Simpson's Rule with $h = 20$ ft and the y's equal to the distances measured across the swamp, as shown in Figure 4.29.

$$S = \frac{h}{3}(y_0 + 4y_1 + 2y_2 + 4y_3 + 2y_4 + 4y_5 + y_6)$$

$$= \frac{20}{3}(146 + 488 + 152 + 216 + 80 + 120 + 13) = 8100$$

The volume is about $(8100)(5) = 40{,}500$ ft^3 or 1500 yd^3.

Round-off Errors

Although decreasing the step size h reduces the error in the Simpson and trapezoidal approximations in theory, it may fail to do so in practice. When h is very small, say $h = 10^{-5}$, computer or calculator round-off errors in the arithmetic required to evaluate S and T may accumulate to such an extent that the error formulas no longer describe what is going on. Shrinking h below a certain size can actually make things worse. Although this is not an issue in this book, you should consult a text on numerical analysis for alternative methods if you are having problems with round-off.

EXERCISES 4.7

Estimating Integrals

CD-ROM WEBsite

The instructions for the integrals in Exercises 1–10 have two parts, one for the Trapezoidal Rule and one for Simpson's Rule.

I. *Using the Trapezoidal Rule*

(a) Estimate the integral with $n = 4$ steps and use Equation (1) to find an upper bound for $|E_T|$.

(b) Evaluate the integral directly and find $|E_T|$.

(c) Use the formula $(|E_T|/(\text{true value})) \times 100$ to express $|E_T|$ as a percentage of the integral's true value.

II. *Using Simpson's Rule*

(a) Estimate the integral with $n = 4$ steps and use Equation (3) to find an upper bound for $|E_S|$.

(b) Evaluate the integral directly, and find $|E_S|$.

(c) Use the formula $(|E_S|/(\text{true value})) \times 100$ to express $|E_S|$ as a percentage of the integral's true value.

1. $\int_1^2 x\,dx$

2. $\int_1^3 (2x - 1)\,dx$

3. $\int_{-1}^1 (x^2 + 1)\,dx$

4. $\int_{-2}^0 (x^2 - 1)\,dx$

5. $\int_0^2 (t^3 + t)\,dt$

6. $\int_{-1}^1 (t^3 + 1)\,dt$

7. $\int_1^2 \frac{1}{s^2}\,ds$

8. $\int_2^4 \frac{1}{(s-1)^2}\,ds$

9. $\int_0^\pi \sin t\,dt$

10. $\int_0^1 \sin \pi t\,dt$

In Exercises 11–14, use the tabulated values of the integrand to estimate the integral with **(a)** the Trapezoidal Rule and **(b)** Simpson's Rule with $n = 8$ steps. Round your answers to five decimal places. Then **(c)** find the integral's exact value and the approximation error E_T or E_S, as appropriate.

11. $\int_0^1 x\sqrt{1 - x^2}\,dx$

x	$x\sqrt{1-x^2}$
0	0.0
0.125	0.12402
0.25	0.24206
0.375	0.34763
0.5	0.43301
0.625	0.48789
0.75	0.49608
0.875	0.42361
1.0	0

12. $\displaystyle\int_0^3 \frac{\theta}{\sqrt{16+\theta^2}}\,d\theta$

θ	$\theta/\sqrt{16+\theta^2}$
0	0.0
0.375	0.09334
0.75	0.18429
1.125	0.27075
1.5	0.35112
1.875	0.42443
2.25	0.49026
2.625	0.58466
3.0	0.6

13. $\displaystyle\int_{-\pi/2}^{\pi/2} \frac{3\cos t}{(2+\sin t)^2}\,dt$

t	$(3\cos t)/(2+\sin t)^2$
−1.57080	0.0
−1.17810	0.99138
−0.78540	1.26906
−0.39270	1.05961
0	0.75
0.39270	0.48821
0.78540	0.28946
1.17810	0.13429
1.57080	0

14. $\displaystyle\int_{\pi/4}^{\pi/2} (\csc^2 y)\sqrt{\cot y}\,dy$

y	$(\csc^2 y)\sqrt{\cot y}$
0.78540	2.0
0.88357	1.51606
0.98175	1.18237
1.07992	0.93998
1.17810	0.75402
1.27627	0.60145
1.37445	0.46364
1.47262	0.31688
1.57080	0

Applications

15. *Volume of water in a swimming pool* A rectangular swimming pool is 30 ft wide and 50 ft long. The table below shows the depth $h(x)$ of the water at 5 ft intervals from one end of the pool to the other. Estimate the volume of water in the pool using the Trapezoidal Rule with $n = 10$, applied to the integral.

$$V = \int_0^{50} 30 \cdot h(x)\,dx.$$

Position (ft) x	Depth (ft) $h(x)$	Position (ft) x	Depth (ft) $h(x)$
0	6.0	30	11.5
5	8.2	35	11.9
10	9.1	40	12.3
15	9.9	45	12.7
20	10.5	50	13.0
25	11.0		

16. *Stocking a fish pond* As the fish and game warden of your township, you are responsible for stocking the town pond with fish before the fishing season. The average depth of the pond is 20 ft. Using a scaled map, you measure distance across the pond at 200 ft intervals, as shown in the accompanying diagram.

(a) Use the Trapezoidal Rule to estimate the volume of the pond.

(b) You plan to start the season with one fish per 1000 cubic feet. You intend to have at least 25% of the opening day's fish population left at the end of the season. What is the maximum number of licenses the town can sell if the average seasonal catch is 20 fish per license?

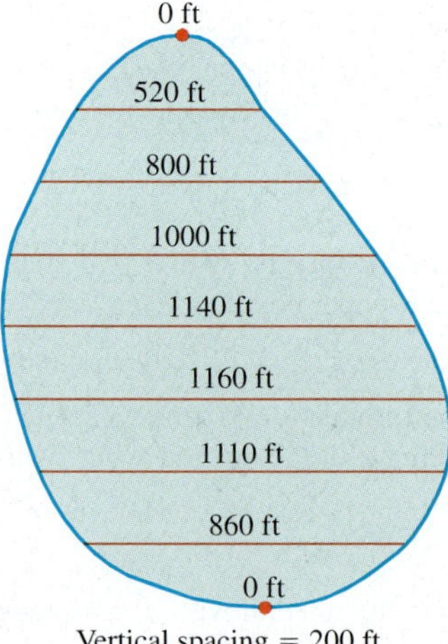

Vertical spacing = 200 ft

17. *Ford® Mustang Cobra™* The accompanying table shows time-to-speed data for a 1994 Ford Mustang Cobra accelerating from rest to 130 mph. How far had the Mustang traveled by the time it reached this speed? (Use trapezoids to estimate the area under the velocity curve, but be careful: The time intervals may vary in length.)

Speed change	Time (sec)
Zero to 30 mph	2.2
40 mph	3.2
50 mph	4.5
60 mph	5.9
70 mph	7.8
80 mph	10.2
90 mph	12.7
100 mph	16.0
110 mph	20.6
120 mph	26.2
130 mph	37.1

Source: Car and Driver, April 1994.

18. *Aerodynamic drag* A vehicle's aerodynamic drag is determined in part by its cross-section area, so, all other things being equal, engineers try to make this area as small as possible. Use Simpson's Rule to estimate the cross-section area of the body of James Worden's solar-powered Solectria® automobile at MIT from the diagram.

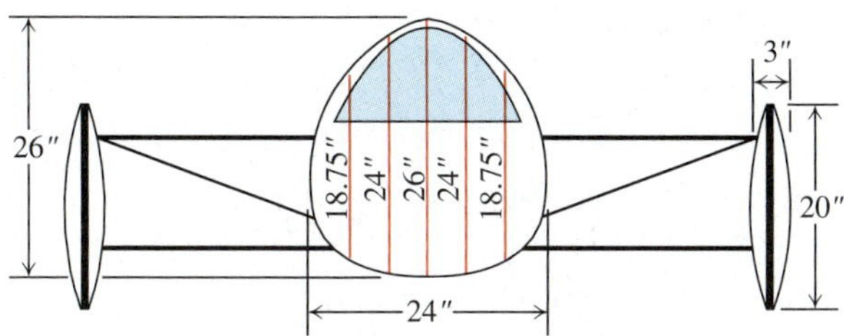

19. *Wing design* The design of a new airplane requires a gasoline tank of constant cross-section area in each wing. A scale drawing of a cross section is shown here. The tank must hold 5000 lb of gasoline, which has a density of 42 lb/ft^3. Estimate the length of the tank.

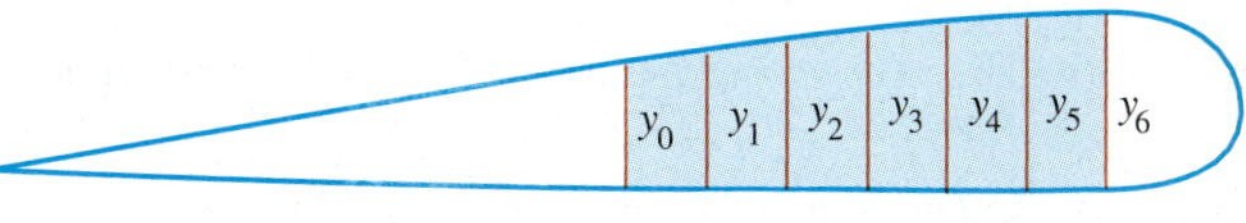

$y_0 = 1.5$ ft, $y_1 = 1.6$ ft, $y_2 = 1.8$ ft, $y_3 = 1.9$ ft,
$y_4 = 2.0$ ft, $y_5 = y_6 = 2.1$ ft Horizontal spacing = 1 ft

20. *Oil Consumption on Pathfinder Island* A diesel generator runs continuously, consuming oil at a gradually increasing rate until it must be temporarily shut down to have the filters replaced.

Use the Trapezoidal Rule to estimate the amount of oil consumed by the generator during that week.

Day	Oil consumption rate (liters/hour)
Sun	0.019
Mon	0.020
Tue	0.021
Wed	0.023
Thu	0.025
Fri	0.028
Sat	0.031
Sun	0.035

Theory and Examples

21. *Polynomials of low degree* The magnitude of the error in the trapezoidal approximation of $\int_a^b f(x)\,dx$ is

$$|E_T| = \frac{b-a}{12} h^2 |f''(c)|,$$

where c is some point (usually unidentified) in (a, b). If f is a linear function of x, then $f''(c) = 0$, so $E_T = 0$ and T gives the exact value of the integral for any value of h. This is no surprise, really, for if f is linear, the line segments approximating the graph of f fit the graph exactly. The surprise comes with Simpson's Rule. The magnitude of the error in Simpson's Rule is

$$|E_S| = \frac{b-a}{180} h^4 |f^{(4)}(c)|,$$

where once again c lies in (a, b). If f is a polynomial of degree less than 4, then $f^{(4)} = 0$ no matter what c is, so $E_S = 0$ and S

gives the integral's exact value, even if we use only two steps. As a case in point, use Simpson's Rule with $n = 2$ to estimate

$$\int_0^2 x^3\, dx.$$

Compare your answer with the integral's exact value.

22. *Usable values of the sine-integral function* **The sine-integral function,**

$$\text{Si}\,(x) = \int_0^x \frac{\sin t}{t}\, dt, \qquad \text{"Sine integral of } x\text{"}$$

is one of the many functions in engineering whose formulas cannot be simplified. There is no elementary formula for the antiderivative of $(\sin t)/t$. The values of Si (x), however, are readily estimated by numerical integration.

Although the notation does not show it explicitly, the function being integrated is

$$f(t) = \begin{cases} \dfrac{\sin t}{t}, & t \neq 0 \\ 1, & t = 0, \end{cases}$$

the continuous extension of $(\sin t)/t$ to the interval $[0, x]$. The function has derivatives of all orders at every point of its domain. Its graph is smooth and you can expect good results from Simpson's Rule.

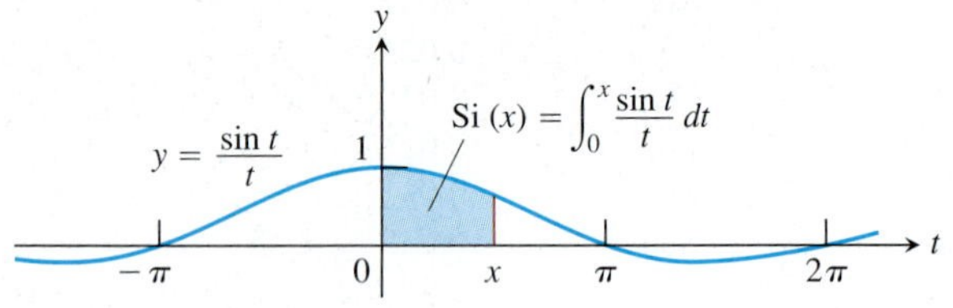

(a) Use the fact that $|f^{(4)}| \leq 1$ on $[0, \pi/2]$ to give an upper bound for the error that will occur if

$$\text{Si}\left(\frac{\pi}{2}\right) = \int_0^{\pi/2} \frac{\sin t}{t}\, dt$$

is estimated by Simpson's Rule with $n = 4$.

(b) Estimate Si $(\pi/2)$ by Simpson's Rule with $n = 4$.

(c) Express the error bound you found in part (a) as a percentage of the value you found in part (b).

23. *The error function* **The error function,**

$$\text{erf}\,(x) = \frac{2}{\sqrt{\pi}} \int_0^x e^{-t^2}\, dt,$$

important in probability and in the theories of heat flow and signal transmission, must be evaluated numerically because there is no elementary expression for the antiderivative of e^{-t^2}.

(a) Use Simpson's Rule with $n = 10$ to estimate erf (1).

(b) In the interval $[0, 1]$,

$$\left|\frac{d^4}{dt^4}\left(e^{-t^2}\right)\right| \leq 12.$$

Give an upper bound for the magnitude of the error of the estimate in part (a).

24. *Writing to Learn* In Example 2 (before rounding), we found the average temperature to be 65.17 degrees when we used the integral approximation, yet the average of the 13 discrete temperatures is only 64.69 degrees. Considering the shape of the temperature curve, explain why you would expect the average of the 13 discrete temperatures to be less than the average value of the temperature function on the entire interval.

In Exercises 25 and 26, *work in groups of two or three.*

T **25.** Consider the integral $\int_{-1}^{1} \sin\,(x^2)\, dx$.

(a) Find f'' for $f(x) = \sin\,(x^2)$.

(b) Graph $y = f''(x)$ in the viewing window $[-1, 1]$ by $[-3, 3]$.

(c) Explain why the graph in part (b) suggests that $|f''(x)| \leq 3$ for $-1 \leq x \leq 1$.

(d) Show that the error estimate for the Trapezoidal Rule in this case becomes

$$|E_T| \leq \frac{h^2}{2}.$$

(e) Show that the Trapezoidal Rule error will be less than or equal to 0.01 in magnitude if $h \leq 0.1$.

(f) How large must n be for $h \leq 0.1$?

T **26.** Consider the integral $\int_{-1}^{1} \sin\,(x^2)\, dx$.

(a) Find $f^{(4)}$ for $f(x) = \sin\,(x^2)$.

(b) Graph $y = f^{(4)}(x)$ in the viewing window $[-1, 1]$ by $[-30, 10]$.

(c) Explain why the graph in part (b) suggests that $|f^{(4)}(x)| \leq 30$ for $-1 \leq x \leq 1$.

(d) Show that the error estimate for Simpson's Rule in this case becomes

$$|E_S| \leq \frac{h^4}{3}.$$

(e) Show that the Simpson's Rule error will be less than or equal to 0.01 in magnitude if $h \leq 0.4$.

(f) How large must n be for $h \leq 0.4$?

COMPUTER EXPLORATIONS

As we mentioned at the beginning of the section, the definite integrals of many continuous functions cannot be evaluated with the Fundamental Theorem of Calculus because their antiderivatives lack elementary formulas. Numerical integration offers a practical way to estimate the values of these so-called *nonelementary integrals*. If your calculator or computer has a numerical integration routine, try it on the integrals in Exercises 27–30.

27. $\int_{-1}^{1} 2\sqrt{1-x^2}\,dx$ The exact value is π.

28. $\int_{0}^{1} \sqrt{1+x^4}\,dx$ An integral that came up in Newton's research

29. $\int_{0}^{\pi/2} \frac{\sin x}{x}\,dx$

30. $\int_{0}^{\pi/2} \sin(x^2)\,dx$ An integral associated with the diffraction of light

31. Consider the integral $\int_0^{\pi} \sin x\,dx$.

(a) Find the Trapezoidal Rule approximations for $n = 10, 100$, and 1000.

(b) Record the errors with as many decimal places of accuracy as you can.

(c) What pattern do you see?

(d) *Writing to Learn* Explain how the error bound for E_T accounts for the pattern in part (c).

32. (*Continuation of Exercise 31*) Repeat Exercise 31 with Simpson's Rule and E_S.

Questions To Guide Your Review

1. Can a function have more than one antiderivative? If so, how are the antiderivatives related? Explain.
2. What is an indefinite integral? How do you evaluate one? What general formulas do you know for evaluating indefinite integrals?
3. How can you sometimes solve a differential equation of the form $dy/dx = f(x)$?
4. What is an initial value problem? How do you solve one? Give an example.
5. How can you sometimes use a trigonometric identity to transform an unfamiliar integral into one you know how to evaluate?
6. If you know the acceleration of a body moving along a coordinate line as a function of time, what more do you need to know to find the body's position function? Give an example.
7. How is integration by substitution related to the Chain Rule?
8. How can you sometimes evaluate indefinite integrals by substitution? Give examples.
9. How can you sometimes estimate quantities like distance traveled, area, volume, and average value with finite sums? Why might you want to do so?
10. What is sigma notation? What advantage does it offer? Give examples.
11. What is a Riemann Sum? Why might you want to consider such a sum?
12. What is the norm of a partition of a closed interval?
13. What is the definite integral of a function f over a closed interval $[a, b]$? When can you be sure it exists?
14. What is the relation between definite integrals and area? Describe some other interpretations of definite integrals.
15. What is the average value of an integrable function over a closed interval? Must the function assume its average value? Explain.
16. What does a function's average value have to do with sampling a function's values?
17. Describe the rules for working with definite integrals (Table 4.5). Give examples.
18. What is the Fundamental Theorem of Calculus? Why is it so important? Illustrate each part of the theorem with an example.
19. How does the Fundamental Theorem provide a solution to the initial value problem $dy/dx = f(x)$, $y(x_0) = y_0$, when f is continuous?
20. How does the method of substitution work for definite integrals? Give examples.
21. How do you define and calculate the area of the region between the graphs of two continuous functions? Give an example.
22. You are collaborating to produce a short "how-to" manual for numerical integration, and you are writing about the Trapezoidal Rule.

    (a) What would you say about the rule itself and how to use it? How to achieve accuracy?

    (b) What would you say if you were writing about Simpson's Rule instead?
23. How would you compare the relative merits of Simpson's Rule and the Trapezoidal Rule?

Practice Exercises

Evaluating Indefinite Integrals

Evaluate the integrals in Exercises 1–20.

1. $\int (x^3 + 5x - 7)\, dx$

2. $\int \left(8t^3 - \frac{t^2}{2} + t\right) dt$

3. $\int \left(3\sqrt{t} + \frac{4}{t^2}\right) dt$

4. $\int \left(\frac{1}{2\sqrt{t}} - \frac{3}{t^4}\right) dt$

5. $\int \frac{r\, dr}{(r^2 + 5)^2}$

6. $\int \frac{6r^2\, dr}{(r^3 - \sqrt{2})^3}$

7. $\int 3\theta\sqrt{2 - \theta^2}\, d\theta$

8. $\int \frac{\theta^2}{9\sqrt{73 + \theta^3}}\, d\theta$

9. $\int e^x \sin (e^x)\, dx$

10. $\int e^t \cos (3e^t - 2)\, dt$

11. $\int e^x \sec^2 (e^x - 7)\, dx$

12. $\int e^y \csc (e^y + 1) \cot (e^y + 1)\, dy$

13. $\int \frac{\tan (\ln v)}{v}\, dv$

14. $\int \sec \frac{\theta}{3} \tan \frac{\theta}{3}\, d\theta$

15. $\int \sin^2 \frac{x}{4}\, dx$

16. $\int \frac{(\ln x)^{-3}}{x}\, dx$

17. $\int \frac{1}{r} \csc^2 (1 + \ln r)\, dr$

18. $\int \frac{\cos (1 - \ln v)}{v}\, dv$

19. $\int x3^{x^2}\, dx$

20. $\int 2^{\tan x} \sec^2 x\, dx$

Finite Sums and Estimates

21. *Flight of a model rocket* The accompanying figure shows the graph of the velocity (ft/sec) of a model rocket for the first 8 sec after launch. The rocket accelerated straight up for the first 2 sec and then coasted to reach its maximum height at $t = 8$ sec.

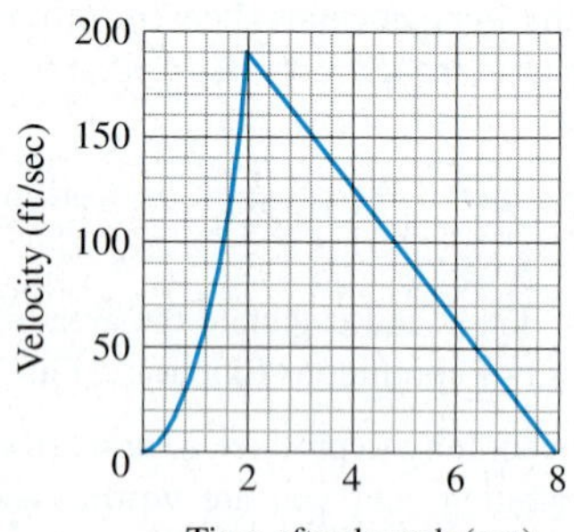

(a) Assuming that the rocket was launched from ground level, about how high did it go? (This is the rocket in Section 2.2, Exercise 15, but you do not need to do Exercise 15 to do the exercise here.)

(b) Sketch a graph of the rocket's height aboveground as a function of time for $0 \le t \le 8$.

22. *Analyzing linear motion*

(a) The accompanying figure shows the velocity (m/sec) of a body moving along the s-axis during the time interval from $t = 0$ to $t = 10$ sec. About how far did the body travel during those 10 sec?

(b) Sketch a graph of the body's position s as a function of t for $0 \le t \le 10$ assuming $s(0) = 0$.

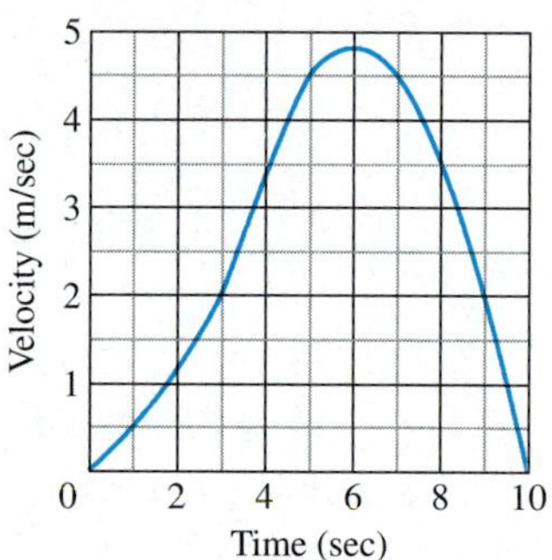

Definite Integrals

In Exercises 23–26, express each limit as a definite integral. Then evaluate the integral to find the value of the limit. In each case, P is a partition of the given interval and the numbers c_k are chosen from the subintervals of P.

23. $\lim_{\|P\|\to 0} \sum_{k=1}^{n} (2c_k - 1)^{-1/2}\, \Delta x_k$, where P is a partition of $[1, 5]$

24. $\lim_{\|P\|\to 0} \sum_{k=1}^{n} c_k(c_k^{\,2} - 1)^{1/3}\, \Delta x_k$, where P is a partition of $[1, 3]$

25. $\lim_{\|P\|\to 0} \sum_{k=1}^{n} \left(\cos \left(\frac{c_k}{2}\right)\right) \Delta x_k$, where P is a partition of $[-\pi, 0]$

26. $\lim_{\|P\|\to 0} \sum_{k=1}^{n} (\sin c_k)(\cos c_k)\, \Delta x_k$, where P is a partition of $[0, \pi/2]$

Using Properties and Known Values to Find Other Integrals

27. If $\int_{-2}^{2} 3f(x)\, dx = 12$, $\int_{-2}^{5} f(x)\, dx = 6$, and $\int_{-2}^{5} g(x)\, dx = 2$, find the values of the following.

(a) $\int_{-2}^{2} f(x)\, dx$

(b) $\int_{2}^{5} f(x)\, dx$

(c) $\int_{5}^{-2} g(x)\, dx$

(d) $\int_{-2}^{5} (-\pi g(x))\, dx$

(e) $\int_{-2}^{5} \left(\frac{f(x) + g(x)}{5}\right) dx$

28. If $\int_0^2 f(x)\,dx = \pi$, $\int_0^2 7g(x)\,dx = 7$, and $\int_0^1 g(x)\,dx = 2$, find the values of the following.

(a) $\int_0^2 g(x)\,dx$

(b) $\int_1^2 g(x)\,dx$

(c) $\int_2^0 f(x)\,dx$

(d) $\int_0^2 \sqrt{2}\, f(x)\,dx$

(e) $\int_0^2 (g(x) - 3f(x))\,dx$

Evaluating Definite Integrals

Evaluate the integrals in Exercises 29–52.

29. $\int_{-1}^{1} (3x^2 - 4x + 7)\,dx$

30. $\int_0^1 (8s^3 - 12s^2 + 5)\,ds$

31. $\int_1^4 \left(\frac{x}{8} + \frac{1}{2x}\right) dx$

32. $\int_1^8 \left(\frac{2}{3x} - \frac{8}{x^2}\right) dx$

33. $\int_1^4 \frac{dt}{t\sqrt{t}}$

34. $\int_1^4 \frac{(1 + \sqrt{u})^{1/2}}{\sqrt{u}}\,du$

35. $\int_0^1 \frac{36\,dx}{(2x + 1)^3}$

36. $\int_0^1 \frac{dr}{\sqrt[3]{(7 - 5r)^2}}$

37. $\int_0^{\ln 5} e^r(3e^r + 1)^{-3/2}\,dr$

38. $\int_0^{\ln 9} e^\theta(e^\theta - 1)^{1/2}\,d\theta$

39. $\int_e^{e^2} \frac{1}{x\sqrt{\ln x}}\,dx$

40. $\int_0^{\pi/4} \cos^2\left(4t - \frac{\pi}{4}\right) dt$

41. $\int_1^e \frac{1}{x}(1 + 7\ln x)^{-1/3}\,dx$

42. $\int_{\pi/4}^{3\pi/4} \csc^2 x\,dx$

43. $\int_{-2}^{2} \frac{3\,dt}{4 + 3t^2}$

44. $\int_1^3 \frac{(\ln(v + 1))^2}{v + 1}\,dv$

45. $\int_{-\pi/3}^{0} \sec x \tan x\,dx$

46. $\int_{\pi/4}^{3\pi/4} \csc z \cot z\,dz$

47. $\int_{\sqrt{2}/3}^{2/3} \frac{dy}{|y|\sqrt{9y^2 - 1}}$

48. $\int_{-1}^{1} 2x \sin(1 - x^2)\,dx$

49. $\int_0^{\pi/2} \frac{3 \sin x \cos x}{\sqrt{1 + 3\sin^2 x}}\,dx$

50. $\int_{-2}^{-1} \frac{2\,dv}{v^2 + 4v + 5}$

51. $\int_{-1}^{1} \frac{3\,dv}{4v^2 + 4v + 4}$

52. $\int_{\pi^2/36}^{\pi^2/4} \frac{\cos\sqrt{t}}{\sqrt{t}\sin\sqrt{t}}\,dt$

Area

In Exercises 53–56, find the total area of the region between the graph of f and the x-axis.

53. $f(x) = x^2 - 4x + 3, \quad 0 \le x \le 3$

54. $f(x) = 1 - (x^2/4), \quad -2 \le x \le 3$

55. $f(x) = 5 - 5x^{2/3}, \quad -1 \le x \le 8$

56. $f(x) = 1 - \sqrt{x}, \quad 0 \le x \le 4$

Find the areas of the regions enclosed by the curves and lines in Exercises 57–64.

57. $y = x, \quad y = 1/x^2, \quad x = 2$

58. $y = x, \quad y = 1/\sqrt{x}, \quad x = 2$

59. $\sqrt{x} + \sqrt{y} = 1, \quad x = 0, \quad y = 0$

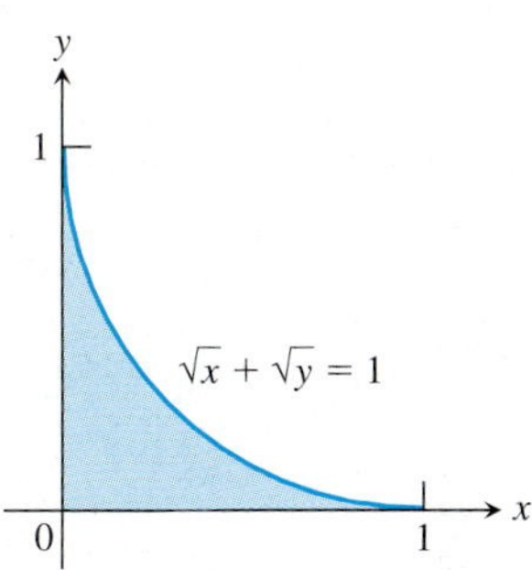

60. $x^3 + \sqrt{y} = 1, \quad x = 0, \quad y = 0, \quad \text{for } 0 \le x \le 1$

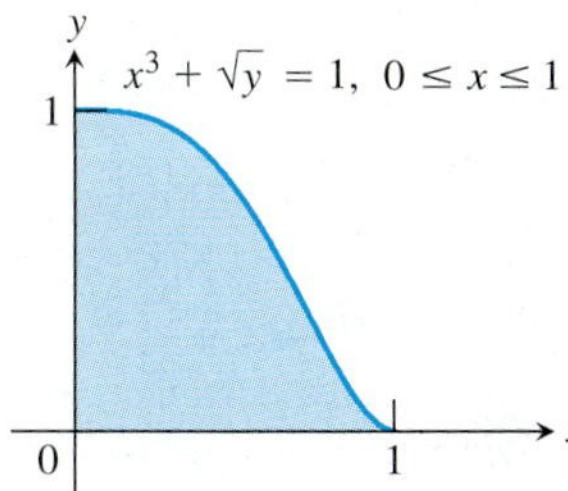

61. $y = \sin x, \quad y = x, \quad 0 \le x \le \pi/4$

62. $y = |\sin x|, \quad y = 1, \quad -\pi/2 \le x \le \pi/2$

63. $y = 2\sin x, \quad y = \sin 2x, \quad 0 \le x \le \pi$

64. $y = 8\cos x, \quad y = \sec^2 x, \quad -\pi/3 \le x \le \pi/3$

65. Find the extreme values of $f(x) = x^3 - 3x^2$ and find the area of the region enclosed by the graph of f and the x-axis.

66. Find the area of the region cut from the first quadrant by the curve $x^{1/3} + y^{1/3} = 1$.

Initial Value Problems

Solve the initial value problems in Exercises 67–70.

67. $\frac{dy}{dx} = \frac{x^2 + 1}{x^2}, \quad y(1) = -1$

68. $\frac{dy}{dx} = e^{-x-y-2}, \quad y(0) = -2$

69. $\frac{d^2r}{dt^2} = 15\sqrt{t} + \frac{3}{\sqrt{t}}; \quad r'(1) = 8, \quad r(1) = 0$

70. $\dfrac{d^3r}{dt^3} = -\cos t; \quad r''(0) = r'(0) = 0, \quad r(0) = -1$

71. Show that $y = x^2 + \int_1^x (1/t)\,dt$ solves the initial value problem

$$\frac{d^2y}{dx^2} = 2 - \frac{1}{x^2}; \qquad y'(1) = 3, \qquad y(1) = 1.$$

72. Show that $y = \int_0^x (1 + 2\sqrt{\sec t})\,dt$ solves the initial value problem

$$\frac{d^2y}{dx^2} = \sqrt{\sec x}\,\tan x; \qquad y'(0) = 3, \qquad y(0) = 0.$$

Express the solutions of the initial value problems in Exercises 73 and 74 in terms of integrals.

73. $\dfrac{dy}{dx} = \dfrac{\sin x}{x}, \quad y(5) = -3$

74. $\dfrac{dy}{dx} = \sqrt{2 - \sin^2 x}, \quad y(-1) = 2$

Average Values

75. Find the average value of $f(x) = mx + b$

(a) over $[-1, 1]$

(b) over $[-k, k]$.

76. Find the average value of

(a) $y = \sqrt{3x}$ over $[0, 3]$

(b) $y = \sqrt{ax}$ over $[0, a]$.

77. *Writing to Learn: Average and instantaneous rates of change* Let f be a function that is differentiable on $[a, b]$. In Chapter 1, we defined the average rate of change of f over $[a, b]$ to be

$$\frac{f(b) - f(a)}{b - a}$$

and the instantaneous rate of change of f at x to be $f'(x)$. In this chapter, we defined the average value of a function. For the new definition of average to be consistent with the old one, we should have

$$\frac{f(b) - f(a)}{b - a} = \text{average value of } f' \text{ on } [a, b].$$

Is this the case? Give reasons for your answer.

78. *Writing to Learn: Average value* Is it true that the average value of an integrable function over an interval of length 2 is half the function's integral over the interval? Give reasons for your answer.

Differentiating Integrals

In Exercises 79–82, find dy/dx.

79. $y = \displaystyle\int_2^x \sqrt{2 + \cos^3 t}\,dt$

80. $y = \displaystyle\int_2^{7x^2} \sqrt{2 + \cos^3 t}\,dt$

81. $y = \displaystyle\int_x^1 \frac{6}{3 + t^4}\,dt$

82. $y = \displaystyle\int_x^{e^{2x}} \frac{1}{t^2 + 1}\,dt$

Numerical Integration

83. A direct calculation from Example 3, Section 4.2 shows that

$$\int_0^{\pi} 2\sin^2 x\,dx = \pi.$$

How close do you come to this value by using the Trapezoidal Rule with $n = 6$? Simpson's Rule with $n = 6$? Try them and find out.

84. *Fuel efficiency* An automobile computer gives a digital readout of fuel consumption in gallons per hour. During a trip, a passenger recorded the fuel consumption every 5 min for a full hour of travel.

Time	Gal/h	Time	Gal/h
0	2.5	35	2.5
5	2.4	40	2.4
10	2.3	45	2.3
15	2.4	50	2.4
20	2.4	55	2.4
25	2.5	60	2.3
30	2.6		

(a) Use the Trapezoidal Rule to approximate the total fuel consumption during the hour.

(b) If the automobile covered 60 mi in the hour, what was its fuel efficiency (in miles per gallon) for that portion of the trip?

85. *Mean temperature* Compute the average value of the temperature function

$$f(x) = 37\sin\left(\frac{2\pi}{365}(x - 101)\right) + 25$$

for a 365-day year. This is one way to estimate the annual mean air temperature in Fairbanks, Alaska. The National Weather Service's official figure, a numerical average of the daily normal mean air temperatures for the year, is 25.7°F, which is slightly higher than the average value of $f(x)$. Figure 2.34 shows why.

86. *Specific heat of a gas* Heat capacity C_v is the amount of heat required to raise the temperature of a given mass of gas with constant volume by 1°C, measured in units of cal/deg-mol (calories per degree gram molecular weight). The heat capacity of oxygen depends on its temperature T and satisfies the formula

$$C_v = 8.27 + 10^{-5}(26T - 1.87T^2).$$

Find the average value of C_v for $20° \le T \le 675°\text{C}$ and the temperature at which it is attained.

87. *A new parking lot* To meet the demand for parking, your town has allocated the area shown here. As the town engineer, you have been asked by the town council to find out if the lot can be built for \$11,000. The cost to clear the land will be \$0.10 a square foot, and the lot will cost \$2.00 a square foot to pave. Can the job be done for \$11,000?

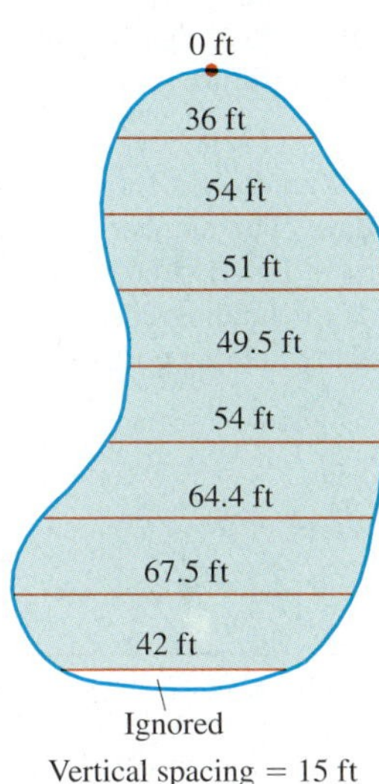

88. *Rubber-band-powered sled* A sled powered by a wound rubber band moves along a track until friction and the unwinding of the rubber band gradually slow it to a stop. A speedometer in the sled monitors its speed, which is recorded at 3 sec intervals during the 27 sec run.

Time (sec)	Speed (ft/sec)
0	5.30
3	5.25
6	5.04
9	4.71
12	4.25
15	3.66
18	2.94
21	2.09
24	1.11
27	0

(a) Give an upper estimate and a lower estimate for the distance traveled by the sled.

(b) Use the Trapezoidal Rule to estimate the distance traveled by the sled.

Theory and Examples

89. *Writing to Learn* Is it true that every function $y = f(x)$ that is differentiable on $[a, b]$ is itself the derivative of some function on $[a, b]$? Give reasons for your answer.

90. *Writing to Learn* Suppose that $F(x)$ is an antiderivative of $f(x) = \sqrt{1 + x^4}$. Express $\int_0^1 \sqrt{1 + x^4}\, dx$ in terms of F and give a reason for your answer.

91. *Expressing a solution as a definite integral* Express the function $y(x)$ with

$$\frac{dy}{dx} = \frac{\sin x}{x} \quad \text{and} \quad y(5) = 3$$

as a definite integral.

92. *A differential equation* Show that $y = \sin x + \int_x^{\pi} \cos 2t\, dt + 1$ satisfies both of the following conditions:

i. $y'' = -\sin x + 2 \sin 2x$

ii $y = 1$ and $y' = -2$ when $x = \pi$.

93. *A function defined by an integral* The graph of a function f consists of a semicircle and two line segments as shown below.

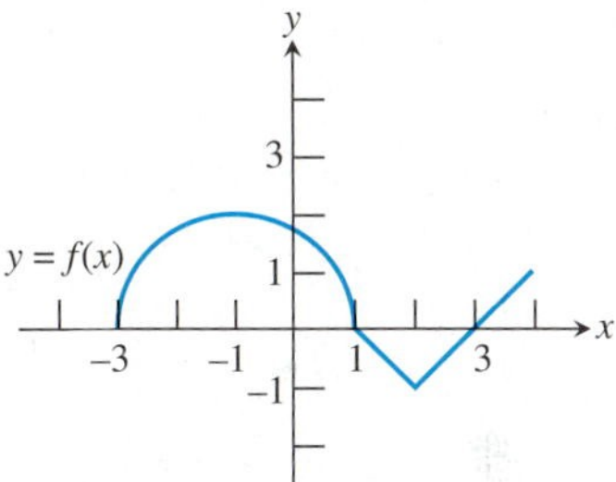

Let $g(x) = \int_1^x f(t)\, dt$.

(a) Find $g(1)$. **(b)** Find $g(3)$. **(c)** Find $g(-1)$.

(d) Find all the values of x on the open interval $(-3, 4)$ at which g has a relative maximum.

(e) Write an equation for the line tangent to the graph of g at $x = -1$.

(f) Find the x-coordinate of each point of inflection of the graph of g on the open interval $(-3, 4)$.

(g) Find the range of g.

94. *Skydiving* Skydivers A and B are in a helicopter hovering at 6400 ft. Skydiver A jumps and descends for 4 sec before opening her parachute. The helicopter then climbs to 7000 ft and hovers there. Forty-five seconds after A leaves the aircraft, B jumps and descends for 13 sec before opening her parachute. Both skydivers descend at 16 ft/sec with parachutes open. Assume that the skydivers fall freely (with acceleration -32 ft/sec^2) before their parachutes open.

(a) At what altitude does A's parachute open?

(b) At what altitude does B's parachute open?

(c) Which skydiver lands first?

Average Daily Inventory

Average value is used in economics to study such things as average daily inventory. If $I(t)$ is the number of radios, tires, shoes, or whatever product a firm has on hand on day t (we call I an **inventory function**), the average value of I over a time period $[0, T]$ is called the firm's average daily inventory for the period.

$$\textbf{Average daily inventory} = \text{av}(I) = \frac{1}{T}\int_0^T I(t)\,dt.$$

If h is the dollar cost of holding one item per day, the product $\text{av}(I) \cdot h$ is the **average daily holding cost** for the period.

95. As a wholesaler, Tracey Burr Distributors (TBD) receives a shipment of 1200 cases of chocolate bars every 30 days. TBD sells the chocolate to retailers at a steady rate, and t days after a shipment arrives, its inventory of cases on hand is $I(t) = 1200 - 40t$, $0 \le t \le 30$. What is TBD's average daily inventory for the 30-day period? What is its average daily holding cost if the cost of holding one case is 3¢ a day?

96. Rich Wholesale Foods, a manufacturer of cookies, stores its cases of cookies in an air-conditioned warehouse for shipment every 14 days. Rich tries to keep 600 cases on reserve to meet occasional peaks in demand, so a typical 14-day inventory function is $I(t) = 600 + 600t$, $0 \le t \le 14$. The daily holding cost for each case is 4¢ per day. Find Rich's average daily inventory and average daily holding cost.

97. Solon Container receives 450 drums of plastic pellets every 30 days. The inventory function (drums on hand as a function of days) is $I(t) = 450 - t^2/2$. Find the average daily inventory. If the holding cost for one drum is 2¢ per day, find the average daily holding cost.

98. Mitchell Mailorder receives a shipment of 600 cases of athletic socks every 60 days. The number of cases on hand t days after the shipment arrives is $I(t) = 600 - 20\sqrt{15t}$. Find the average daily inventory. If the holding cost for one case is 1/2¢ per day, find the average daily holding cost.

Additional Exercises: Theory, Examples, Applications

Theory and Examples

1. **(a)** If $\int_0^1 7f(x)\,dx = 7$, does $\int_0^1 f(x)\,dx = 1$?

(b) If $\int_0^1 f(x)\,dx = 4$ and $f(x) \ge 0$, does $\int_0^1 \sqrt{f(x)}\,dx = \sqrt{4} = 2$?

Give reasons for your answers.

2. Suppose that $\int_{-2}^2 f(x)\,dx = 4$, $\int_2^5 f(x)\,dx = 3$, $\int_{-2}^5 g(x)\,dx = 2$. Which, if any, of the following statements are true?

(a) $\int_5^2 f(x)\,dx = -3$ **(b)** $\int_{-2}^5 (f(x) + g(x))\,dx = 9$

(c) $f(x) \le g(x)$ on the interval $-2 \le x \le 5$

3. *Initial value problem* Show that

$$y = \frac{1}{a}\int_0^x f(t)\sin a(x - t)\,dt$$

solves the initial value problem

$$\frac{d^2y}{dx^2} + a^2y = f(x), \quad \frac{dy}{dx} = 0 \quad \text{and} \quad y = 0 \quad \text{when} \quad x = 0.$$

(*Hint:* $\sin(ax - at) = \sin ax \cos at - \cos ax \sin at$.)

4. *Proportionality* Suppose that x and y are related by the equation

$$x = \int_0^y \frac{1}{\sqrt{1 + 4t^2}}\,dt.$$

Show that d^2y/dx^2 is proportional to y and find the constant of proportionality.

5. Find $f(4)$ if

(a) $\displaystyle\int_0^{x^2} f(t)\,dt = x\cos \pi x$

(b) $\displaystyle\int_0^{f(x)} t^2\,dt = x\cos \pi x.$

6. Find $f(\pi/2)$ from the following information.

i. f is positive and continuous.

ii. The area under the curve $y = f(x)$ from $x = 0$ to $x = a$ is

$$\frac{a^2}{2} + \frac{a}{2}\sin a + \frac{\pi}{2}\cos a.$$

7. The area of the region in the xy-plane enclosed by the x-axis, the curve $y = f(x)$, $f(x) \ge 0$, and the lines $x = 1$ and $x = b$ is equal to $\sqrt{b^2 + 1} - \sqrt{2}$ for all $b > 1$. Find $f(x)$.

8. Prove that

$$\int_0^x\left(\int_0^u f(t)\,dt\right)du = \int_0^x f(u)(x - u)\,du.$$

(*Hint:* Express the integral on the right-hand side as the difference of two integrals. Then show that both sides of the equation have the same derivative with respect to x.)

9. *Finding a curve* Find the equation for the curve in the xy-plane that passes through the point $(1, -1)$ if its slope at x is always $3x^2 + 2$.

10. *Shoveling dirt* You sling a shovelfull of dirt up from the bottom of a hole with an initial velocity of 32 ft/sec. The dirt must rise 17 ft above the release point to clear the edge of the hole. Is that enough speed to get the dirt out, or had you better duck?

Piecewise Continuous Functions

Although we are mainly interested in continuous functions, many functions in applications are piecewise continuous. A function $f(x)$ is **piecewise continuous on a closed interval I** if f has only finitely many discontinuities in I, the limits

$$\lim_{x\to c^-} f(x) \quad \text{and} \quad \lim_{x\to c^+} f(x)$$

exist and are finite at every interior point of I, and the appropriate one-sided limits exist and are finite at the endpoints of I. All piecewise continuous functions are integrable. The points of discontinuity partition I into open and half-open subintervals on which f is continuous, and the limit criteria above guarantee that f has a continuous extension to the closure of each subinterval. To integrate a piecewise continuous function, we integrate the individual extensions and add the results. The integral of

$$f(x) = \begin{cases} 1 - x, & -1 \le x < 0 \\ x^2, & 0 \le x < 2 \\ -1, & 2 \le x \le 3 \end{cases}$$

(Figure 4.30) over $[-1, 3]$ is

$$\begin{aligned} \int_{-1}^{3} f(x)\,dx &= \int_{-1}^{0} (1 - x)\,dx + \int_{0}^{2} x^2\,dx + \int_{2}^{3} (-1)\,dx \\ &= \left[x - \frac{x^2}{2}\right]_{-1}^{0} + \left[\frac{x^3}{3}\right]_{0}^{2} + \left[-x\right]_{2}^{3} \\ &= \frac{3}{2} + \frac{8}{3} - 1 = \frac{19}{6}. \end{aligned}$$

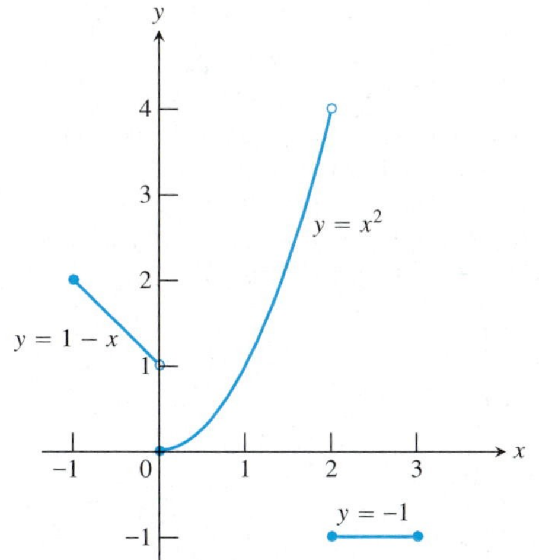

FIGURE 4.30 Piecewise continuous functions like this are integrated piece by piece.

The Fundamental Theorem applies to piecewise continuous functions with the restriction that $(d/dx)\int_a^x f(t)\,dt$ is expected to equal $f(x)$ only at values of x at which f is continuous. There is a similar restriction on Leibniz's Rule below.

Graph the functions in Exercises 11–16 and integrate them over their domains.

11. $f(x) = \begin{cases} x^{2/3}, & -8 \le x < 0 \\ -4, & 0 \le x \le 3, \end{cases}$

12. $f(x) = \begin{cases} \sqrt{-x}, & -4 \le x < 0 \\ x^2 - 4, & 0 \le x \le 3 \end{cases}$

13. $g(t) = \begin{cases} t, & 0 \le t < 1 \\ \sin \pi t, & 1 \le t \le 2 \end{cases}$

14. $h(z) = \begin{cases} \sqrt{1 - z}, & 0 \le z < 1 \\ (7z - 6)^{-1/3}, & 1 \le z \le 2 \end{cases}$

15. $f(x) = \begin{cases} 1, & -2 \le x < -1 \\ 1 - x^2, & -1 \le x < 1 \\ 2, & 1 \le x \le 2 \end{cases}$

16. $h(r) = \begin{cases} r, & -1 \le r < 0 \\ 1 - r^2, & 0 \le r < 1 \\ 1, & 1 \le r \le 2 \end{cases}$

17. Find the average value of the function graphed in Figure 4.31a.

18. Find the average value of the function graphed in Figure 4.31b.

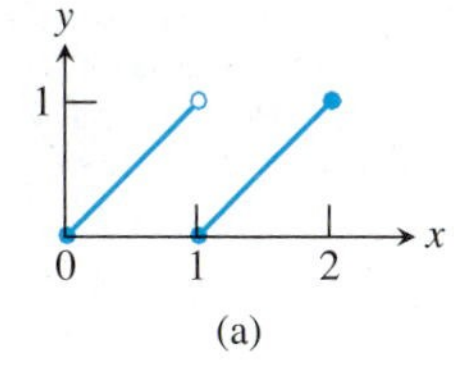

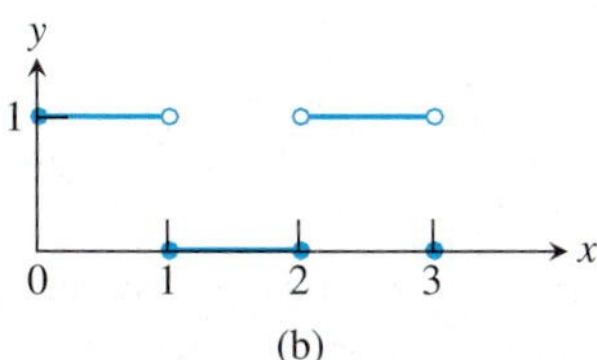

FIGURE 4.31 The graphs for Exercises 17 and 18.

Leibniz's Rule

In applications, we sometimes encounter functions like

$$f(x) = \int_{\sin x}^{x^2} (1 + t)\,dt \quad \text{and} \quad g(x) = \int_{\sqrt{x}}^{2\sqrt{x}} \sin t^2\,dt,$$

defined by integrals that have variable upper limits of integration and variable lower limits of integration at the same time. The first integral can be evaluated directly, but the second cannot. We may find the derivative of either integral, however, by a formula called **Leibniz's Rule.**

Leibniz's Rule

If f is continuous on $[a, b]$ and if $u(x)$ and $v(x)$ are differentiable functions of x whose values lie in $[a, b]$, then

$$\frac{d}{dx}\int_{u(x)}^{v(x)} f(t)\,dt = f(v(x))\frac{dv}{dx} - f(u(x))\frac{du}{dx}.$$

Figure 4.32 gives a geometric interpretation of Leibniz's Rule. It shows a carpet of variable width $f(t)$ that is being rolled up at the left at the same time x as it is being unrolled at the right. (In this interpretation, time is x, not t.) At time x, the floor is covered from $u(x)$ to $v(x)$. The rate du/dx at which the carpet is being rolled up need not be the same as the rate dv/dx at which the carpet is being laid down. At any given time x, the area covered by carpet is

$$A(x) = \int_{u(x)}^{v(x)} f(t)\,dt.$$

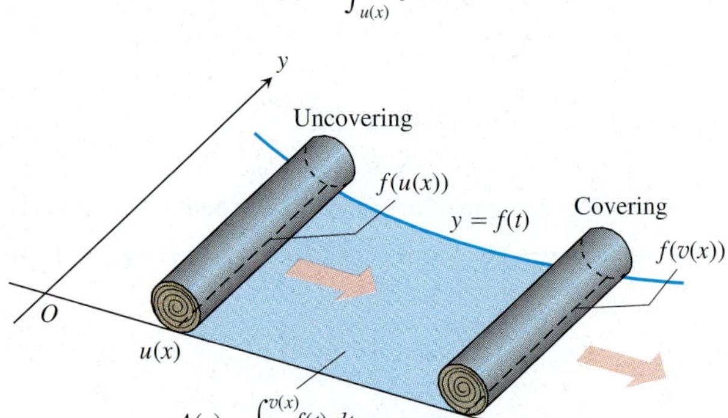

FIGURE 4.32 Rolling and unrolling a carpet; a geometric interpretation of Leibniz's Rule:

$$\frac{dA}{dx} = f(v(x))\frac{dv}{dx} - f(u(x))\frac{du}{dx}.$$

At what rate is the covered area changing? At the instant x, $A(x)$ is increasing by the width $f(v(x))$ of the unrolling carpet times the rate dv/dx at which the carpet is being unrolled. That is, $A(x)$ is being increased at the rate

$$f(v(x))\frac{dv}{dx}.$$

At the same time, A is being decreased at the rate

$$f(u(x))\frac{du}{dx},$$

the width at the end that is being rolled up times the rate du/dx. The net rate of change in A is

$$\frac{dA}{dx} = f(v(x))\frac{dv}{dx} - f(u(x))\frac{du}{dx},$$

which is precisely Leibniz's Rule.

To prove the rule, let F be an antiderivative of f *on* [a, b]. Then

$$\int_{u(x)}^{v(x)} f(t)\,dt = F(v(x)) - F(u(x)).$$

Differentiating both sides of this equation with respect to x gives the equation we want:

$$\begin{aligned}\frac{d}{dx}\int_{u(x)}^{v(x)} f(t)\,dt &= \frac{d}{dx}[F(v(x)) - F(u(x))]\\ &= F'(v(x))\frac{dv}{dx} - F'(u(x))\frac{du}{dx}\\ &= f(v(x))\frac{dv}{dx} - f(u(x))\frac{du}{dx}.\end{aligned}$$

You will see another way to derive the rule in Chapter 11, additional Exercise 3.

Use Leibniz's Rule to find the derivatives of the functions in Exercises 19–21.

19. $f(x) = \displaystyle\int_{1/x}^{x} \frac{1}{t}\,dt$

20. $f(x) = \displaystyle\int_{\cos x}^{\sin x} \frac{1}{1-t^2}\,dt$

21. $g(y) = \displaystyle\int_{\sqrt{y}}^{2\sqrt{y}} \sin t^2\,dt$

22. Use Leibniz's Rule to find the value of x that maximizes the value of the integral

$$\int_{x}^{x+3} t(5-t)\,dt.$$

Problems like this arise in the mathematical theory of political elections. See "The Entry Problem in a Political Race" by Steven J. Brams and Philip D. Straffin Jr., in *Political Equilibrium*, edited by Peter Ordeshook and Kenneth Shepfle (Boston: Kluwer-Nijhoff, 1982), pp. 181–195.

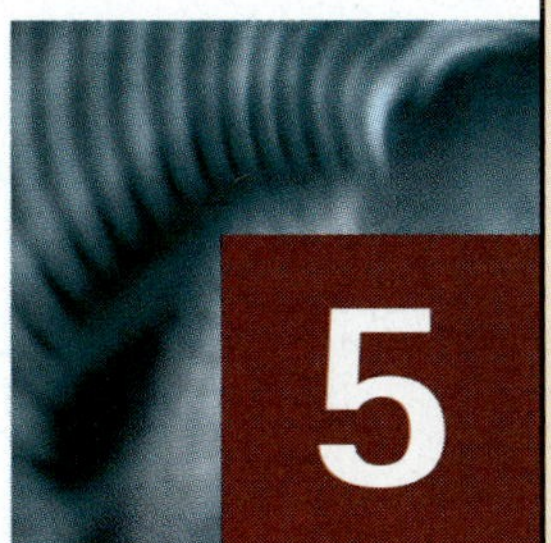

5 Applications of Integrals

OVERVIEW Many things we want to know can be calculated with integrals: the volumes of solids, the lengths of curves, the amount of work it takes to pump liquids from below ground, the forces against floodgates, the coordinates of the points where solid objects will balance. We define all these as limits of Riemann sums of continuous functions on closed intervals—that is, as integrals—and evaluate these limits with calculus.

5.1 Volumes by Slicing and Rotation About an Axis

Volumes by Slicing • Solids of Revolution: Circular Cross Sections • Solids of Revolution: Washer Cross Sections

In Section 4.3, Example 3, we estimated the volume of a sphere by partitioning it into thin slices that were nearly cylindrical and summing the cylinders' volumes in what turned out later to be a Riemann sum. Had we known how at the time, we could have continued on to express the volume of the sphere as a definite integral.

Starting the same way, we can now find the volumes of a great many solids by integration.

Volumes by Slicing

Suppose that we want to find the volume of a solid like the one in Figure 5.1. The cross section of the solid at each point x in the interval $[a, b]$ is a region $R(x)$ of area $A(x)$. If A is a continuous function of x, we can use it to define and calculate the volume of the solid as an integral in the following way.

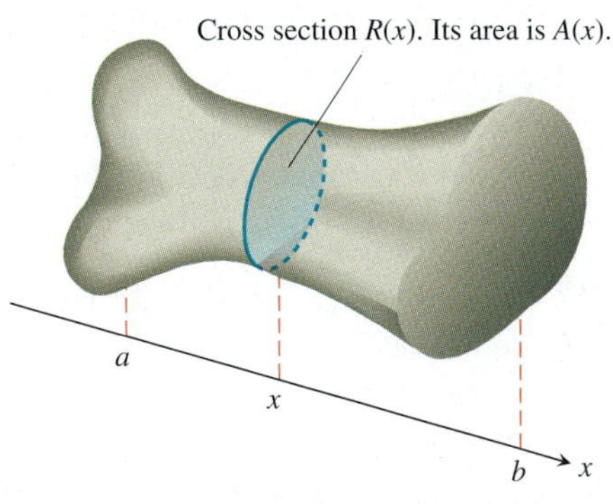

FIGURE 5.1 If the area $A(x)$ of the cross section $R(x)$ is a continuous function of x, we can find the volume of the solid by integrating $A(x)$ from a to b.

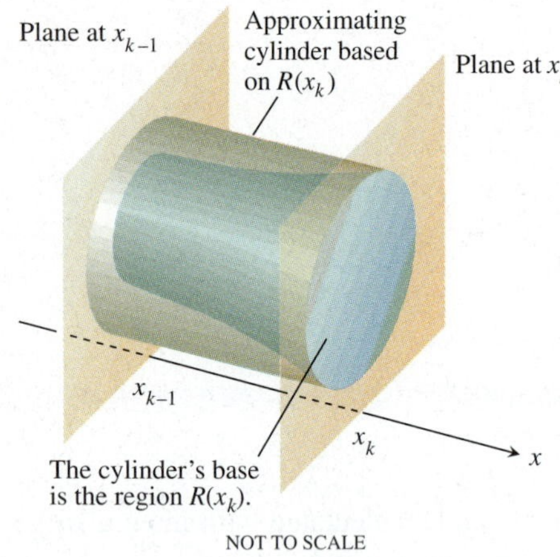

FIGURE 5.2 Enlarged view of the slice of the solid between the planes at x_{k-1} and x_k and its approximating cylinder.

We partition $[a, b]$ into subintervals of length Δx and slice the solid, as we would a loaf of bread, by planes perpendicular to the x-axis at the partition points. The kth slice, the one between the planes at x_{k-1} and x_k, has approximately the same volume as the cylinder between the two planes based on the region $R(x_k)$ (Figure 5.2).

The volume of the cylinder is

$$V_k = \text{base area} \times \text{height} = A(x_k) \times \Delta x.$$

The sum

$$\sum V_k = \sum A(x_k) \times \Delta x$$

approximates the volume of the solid.

This is a Riemann sum for $A(x)$ on $[a, b]$. We expect the approximations to improve as the norms of the partitions go to zero, so we define their limiting integral to be the *volume of the solid.*

Definition Volume of a Solid

The **volume** of a solid of known integrable cross-section area $A(x)$ from $x = a$ to $x = b$ is the integral of A from a to b,

$$V = \int_a^b A(x)\, dx.$$

To apply this formula, we proceed as follows.

How to Find Volume by the Method of Slicing

Step 1. *Sketch the solid and a typical cross section.*

Step 2. *Find a formula for* $A(x)$.

Step 3. *Find the limits of integration.*

Step 4. *Integrate* $A(x)$ *to find the volume.*

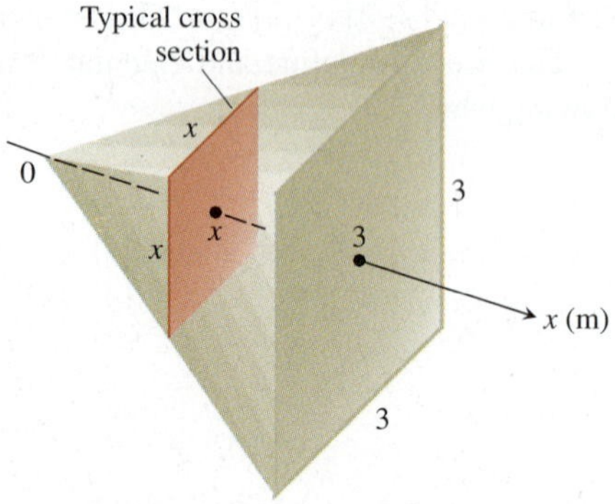

Example 1 Volume of a Pyramid

A pyramid 3 m high has a square base that is 3 m on a side. The cross section of the pyramid perpendicular to the altitude x m down from the vertex is a square x m on a side. Find the volume of the pyramid.

Solution

Step 1: *A sketch.* We draw the pyramid with its altitude along the x-axis and its vertex at the origin and include a typical cross section (Figure 5.3).

FIGURE 5.3 The cross sections of the pyramid in Example 1 are squares.

Step 2: *A formula for* $A(x)$. The cross section at x is a square x meters on a side, so its area is

$$A(x) = x^2.$$

Step 3: *The limits of integration.* The squares go from $x = 0$ to $x = 3$.

Step 4: *Integrate to find the volume.*

$$V = \int_0^3 A(x)\,dx = \int_0^3 x^2\,dx = \frac{x^3}{2}\bigg]_0^3 = 9\text{ m}^3$$

CD-ROM
WEBsite
Historical Biography

Bonaventura Cavalieri
(1598 — 1647)

Example 2 Cavalieri's Volume Theorem

Cavalieri's volume theorem says that solids with equal altitudes and identical cross-section areas at each height have the same volume (Figure 5.4). This follows immediately from the definition of volume, because the cross-section area function $A(x)$ and the interval $[a, b]$ are the same for both solids.

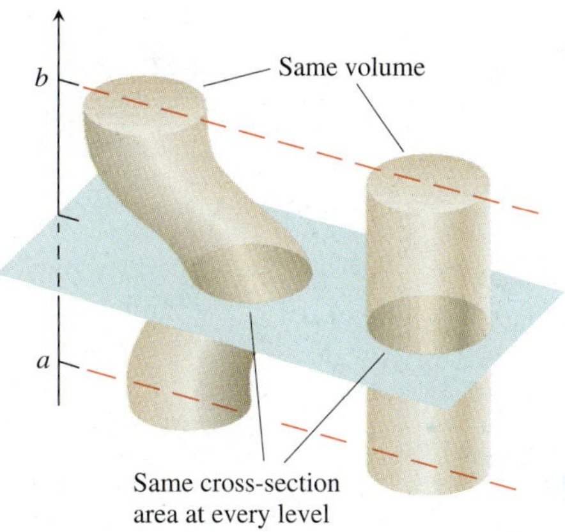

FIGURE 5.4 Cavalieri's Theorem: These solids have the same volume. You can illustrate this yourself with stacks of coins.

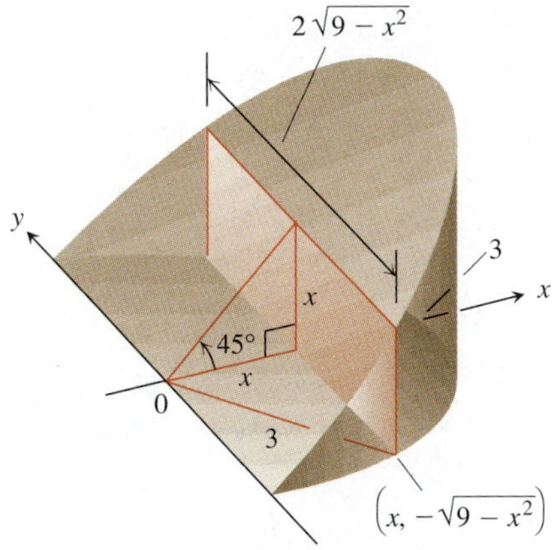

FIGURE 5.5 The wedge of Example 3, sliced perpendicular to the x-axis. The cross sections are rectangles.

Example 3 Volume of a Wedge

A curved wedge is cut from a cylinder of radius 3 by two planes. One plane is perpendicular to the axis of the cylinder. The second plane crosses the first plane at a 45° angle at the center of the cylinder. Find the volume of the wedge.

Solution

Step 1: *A sketch.* We draw the wedge and sketch a typical cross section perpendicular to the x-axis (Figure 5.5).

Step 2: *The formula for* $A(x)$. The cross section at x is a rectangle of area

$$A(x) = (\text{height})(\text{width}) = (x)\,(2\sqrt{9 - x^2})$$
$$= 2x\sqrt{9 - x^2},\text{ units squared.}$$

Step 3: *The limits of integration.* The rectangles run from $x = 0$ to $x = 3$.

Step 4: *Integrate to find the volume.*

$$V = \int_a^b A(x)\,dx = \int_0^3 2x\sqrt{9-x^2}\,dx$$

$$= -\frac{2}{3}(9-x^2)^{3/2}\Big]_0^3$$

$$= 0 + \frac{2}{3}(9)^{3/2} \qquad \text{Let } u = 9 - x^2,\ du = -2x\,dx, \text{ integrate, and substitute back.}$$

$$= 18, \text{ units cubed.}$$

Solids of Revolution: Circular Cross Sections

The most common application of the method of slicing is to solids of revolution. **Solids of revolution** are solids whose shapes can be generated by revolving plane regions about axes. The only thing that changes when the cross sections are circular is the formula for the area $A(x)$.

The typical cross section of the solid perpendicular to the axis of revolution is a disk of radius $R(x)$ and area

$$A(x) = \pi(\text{radius})^2 = \pi[R(x)]^2.$$

For this reason, the method is often called the **disk method.** Here are several examples.

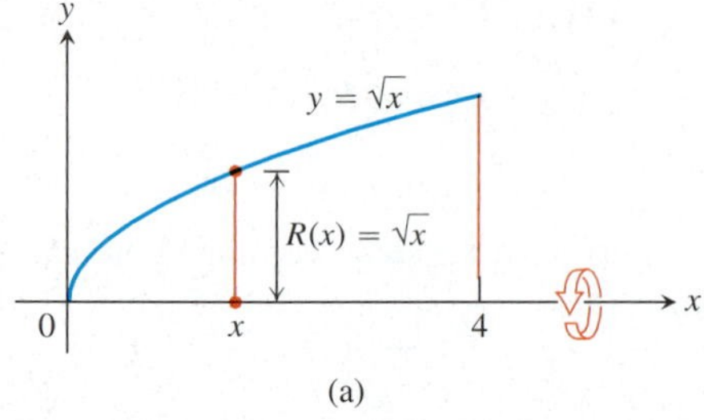

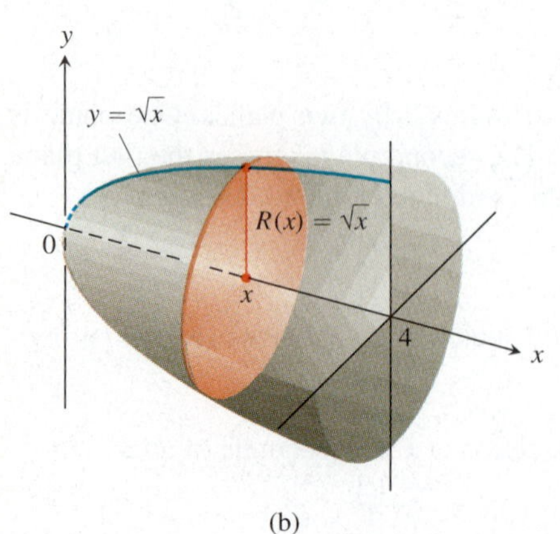

FIGURE 5.6 The region (a) and solid (b) in Example 4.

Example 4 A Solid of Revolution (Rotation About the x-axis)

The region between the curve $y = \sqrt{x}$, $0 \le x \le 4$, and the x-axis is revolved around the x-axis to generate a solid. Find its volume.

Solution We draw figures showing the region, a typical radius, and the generated solid (Figure 5.6). The volume is

$$V = \int_a^b \pi[R(x)]^2\,dx$$

$$= \int_0^4 \pi[\sqrt{x}]^2\,dx \qquad R(x) = \sqrt{x}$$

$$= \pi\int_0^4 x\,dx = \pi\,\frac{x^2}{2}\Big]_0^4 = \pi\,\frac{(4)^2}{2} = 8\pi \text{ units cubed.}$$

The axis of revolution in the next example is not the x-axis, but the rule for calculating the volume is the same: Integrate $\pi(\text{radius})^2$ between appropriate limits.

Example 5 A Solid of Revolution (Rotation About the Line $y = 1$)

Find the volume of the solid generated by revolving the region bounded by $y = \sqrt{x}$ and the lines $y = 1$, $x = 4$ about the line $y = 1$.

Solution We draw figures showing the region, a typical radius, and the generated solid (Figure 5.7). The volume is

$$
\begin{aligned}
V &= \int_1^4 \pi[R(x)]^2\,dx \\
&= \int_1^4 \pi[\sqrt{x} - 1]^2\,dx \qquad R(x) = \sqrt{x} - 1 \\
&= \pi \int_1^4 [x - 2\sqrt{x} + 1]\,dx \\
&= \pi \left[\frac{x^2}{2} - 2 \cdot \frac{2}{3}x^{3/2} + x\right]_1^4 = \frac{7\pi}{6} \text{ units cubed.}
\end{aligned}
$$

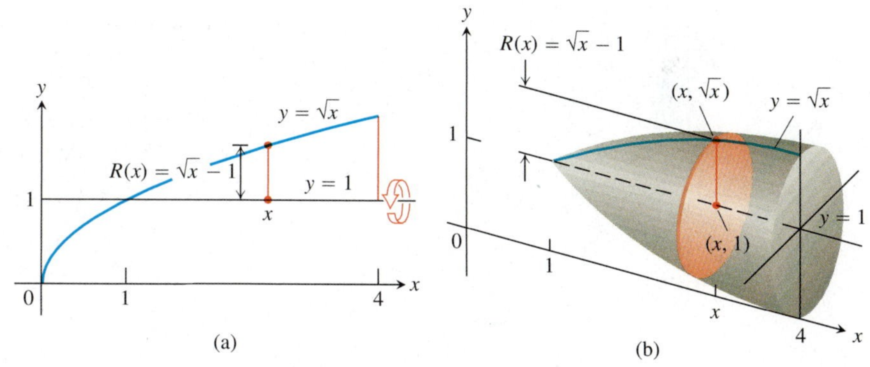

FIGURE 5.7 The region (a) and solid (b) in Example 5.

How to Find Volumes for Circular Cross Sections (Disk Method)

Step 1. *Draw the region and identify the radius function* $R(x)$.

Step 2. *Square* $R(x)$ *and multiply by* π.

Step 3. *Integrate* to find the volume.

To find the volume of a solid generated by revolving a region between the y-axis and a curve $x = R(y)$, $c \le y \le d$, about the y-axis, we use the same method with x replaced by y. In this case, the circular cross section is

$$A(y) = \pi\,[\text{radius}]^2 = \pi\,[R(y)]^2.$$

Example 6 Revolution About the y-axis

Find the volume of the solid generated by revolving the region between the y-axis and the curve $x = 2/y$, $1 \le y \le 4$, about the y-axis.

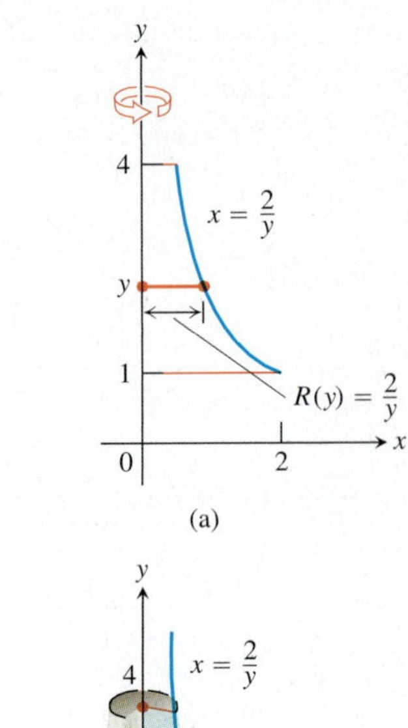

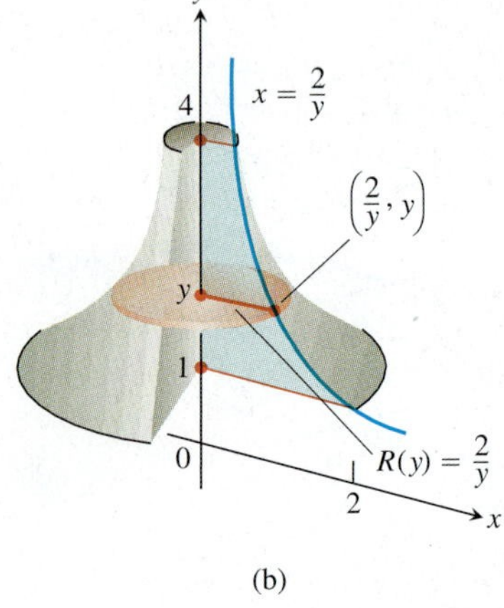

FIGURE 5.8 The region (a) and solid (b) in Example 6.

Solution We draw figures showing the region, a typical radius, and the generated solid (Figure 5.8). The volume is

$$\begin{aligned} V &= \int_1^4 \pi[R(y)]^2\, dy \\ &= \int_1^4 \pi\left(\frac{2}{y}\right)^2 dy \qquad R(y) = \frac{2}{y} \\ &= \pi\int_1^4 \frac{4}{y^2}\, dy = 4\pi\left[-\frac{1}{y}\right]_1^4 = 4\pi\left[\frac{3}{4}\right] \\ &= 3\pi \text{ units cubed.} \end{aligned}$$

Example 7 Revolution About a Vertical Axis

Find the volume of the solid generated by revolving the region between the parabola $x = y^2 + 1$ and the line $x = 3$ about the line $x = 3$.

Solution We draw figures showing the region, a typical radius, and the generated solid (Figure 5.9). The volume is

$$\begin{aligned} V &= \int_{-\sqrt{2}}^{\sqrt{2}} \pi\,[R(y)]^2\, dy \\ &= \int_{-\sqrt{2}}^{\sqrt{2}} \pi\,[2 - y^2]^2\, dy \qquad R(y) = 3 - (y^2 + 1) = 2 - y^2 \\ &= \pi\int_{-\sqrt{2}}^{\sqrt{2}} [4 - 4y^2 + y^4]\, dy \\ &= \pi\left[4y - \frac{4}{3}y^3 + \frac{y^5}{5}\right]_{-\sqrt{2}}^{\sqrt{2}} \\ &= \frac{64\pi\sqrt{2}}{15} \text{ units cubed.} \end{aligned}$$

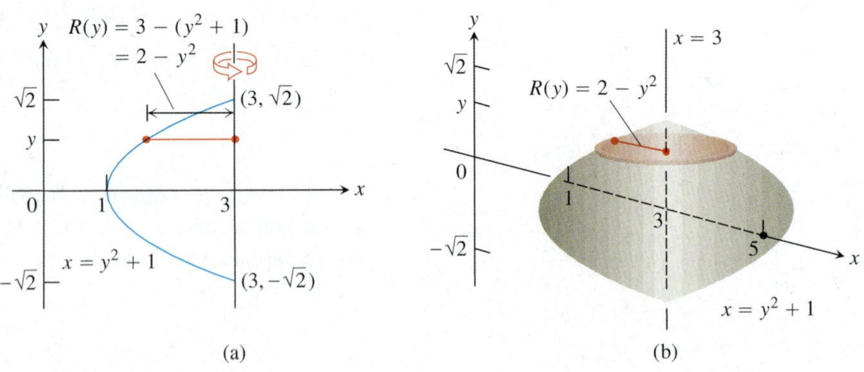

FIGURE 5.9 The region (a) and solid (b) in Example 7.

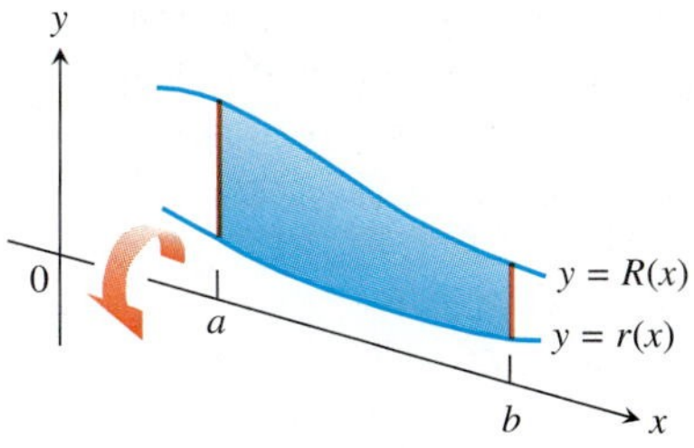

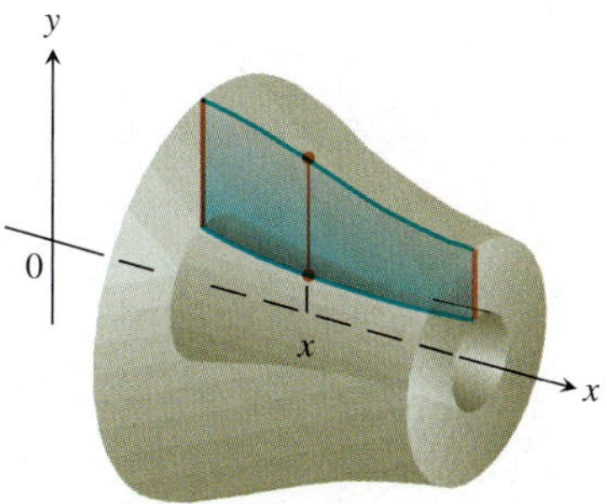

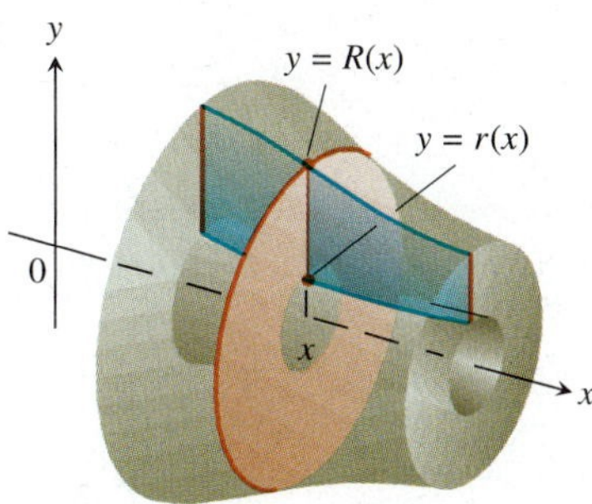

FIGURE 5.10 The cross sections of the solid of revolution generated here are washers, not disks, so the integral $\int_a^b A(x)\,dx$ leads to a slightly different formula.

Solids of Revolution: Washer Cross Sections

If the region we revolve to generate a solid does not border on or cross the axis of revolution, the solid has a hole in it (Figure 5.10). The cross-sections perpendicular to the axis of revolution are washers instead of disks. The dimensions of a typical washer are

Outer radius: $R(x)$
Inner radius: $r(x)$

The washer's area is

$$A(x) = \pi[R(x)]^2 - \pi[r(x)]^2 = \pi([R(x)]^2 - [r(x)]^2).$$

Example 8 A Washer Cross Section (Rotation About the x-axis)

The region bounded by the curve $y = x^2 + 1$ and the line $y = -x + 3$ is revolved about the x-axis to generate a solid. Find the volume of the solid.

Solution

Step 1: Draw the region and sketch a line segment across it perpendicular to the axis of revolution (the red segment in Figure 5.11).

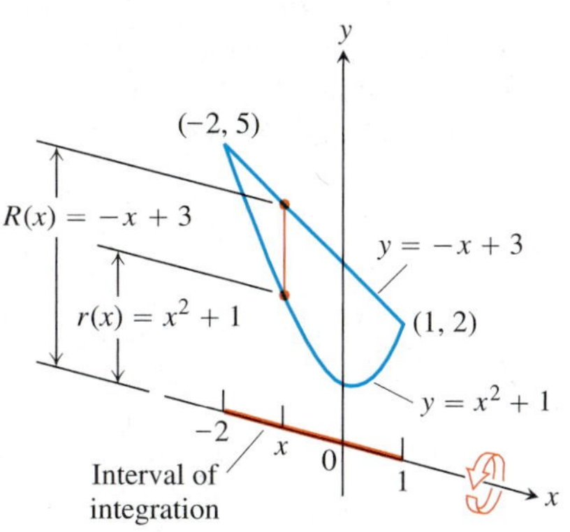

FIGURE 5.11. The region in Example 8 spanned by a line segment perpendicular to the axis of revolution. When the region is revolved about the x-axis, the line segment will generate a washer.

Step 2: Find the limits of integration by finding the x-coordinates of the intersection points of the curve and line in Figure 5.11.

$$\begin{aligned} x^2 + 1 &= -x + 3 \\ x^2 + x - 2 &= 0 \\ (x + 2)(x - 1) &= 0 \\ x = -2, \qquad x &= 1 \end{aligned}$$

Step 3: Find the outer and inner radii of the washer that would be swept out by the line segment if it were revolved about the x-axis along with the region. (We

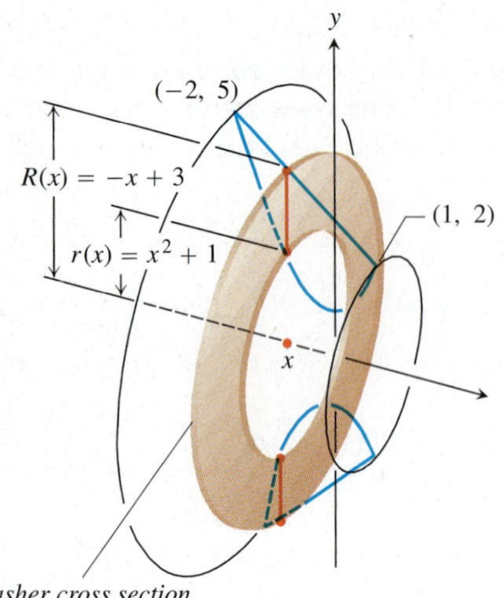

FIGURE 5.12 The inner and outer radii of the washer swept out by the line segment in Figure 5.11.

drew the washer in Figure 5.12, but in your own work you need not do that.) These radii are the distances of the ends of the line segment from the axis of revolution.

Outer radius: $R(x) = -x + 3$

Inner radius: $r(x) = x^2 + 1$

Step 4: Evaluate the volume integral.

$$V = \int_a^b \pi([R(x)]^2 - [r(x)]^2)\,dx$$

$$= \int_{-2}^{1} \pi((-x+3)^2 - (x^2+1)^2)\,dx \qquad \text{Values from steps 2 and 3}$$

$$= \int_{-2}^{1} \pi(8 - 6x - x^2 - x^4)\,dx \qquad \text{Expressions squared and combined}$$

$$= \pi\left[8x - 3x^2 - \frac{x^3}{3} - \frac{x^5}{5}\right]_{-2}^{1} = \frac{117\pi}{5} \text{ units cubed.}$$

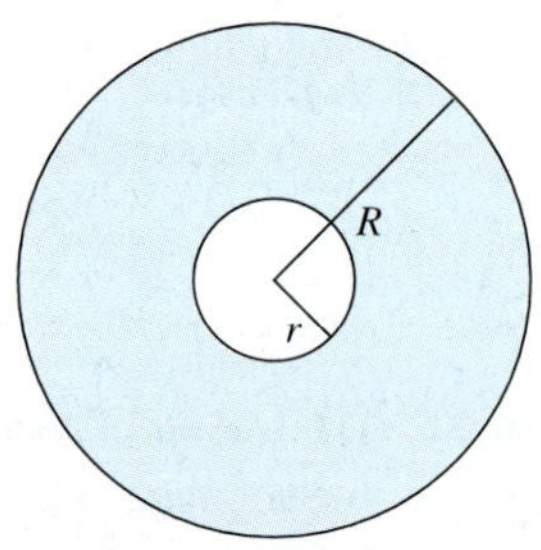

The area of a washer is $\pi R^2 - \pi r^2$.

How to Find Volumes for Washer Cross Sections

Step 1. *Draw the region and sketch a line segment across it perpendicular to the axis of revolution.* When the region is revolved, this segment will generate a typical washer cross section of the generated solid.

Step 2. *Find the limits of integration.*

Step 3. *Find the outer and inner radii* of the washer swept out by the line segment.

Step 4. *Integrate* to find the volume.

To find the volume of a solid generated by revolving a region about the y-axis, we use the steps listed above but integrate with respect to y instead of x.

Example 9 A Washer Cross Section (Rotation About the y-axis)

The region bounded by the parabola $y = x^2$ and the line $y = 2x$ in the first quadrant is revolved about the y-axis to generate a solid. Find the volume of a solid.

Solution

Step 1: Draw the region and sketch a line segment across it perpendicular to the axis of revolution, in this case the y-axis (Figure 5.13).

Step 2. The line and parabola intersect at $y = 0$ and $y = 4$, so the limits of integration are $c = 0$ and $d = 4$.

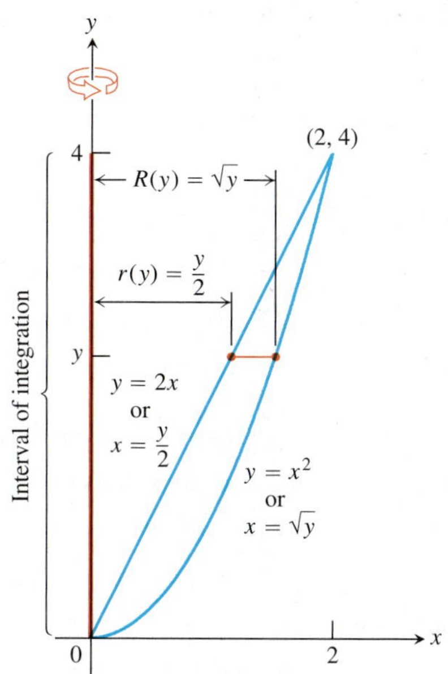

FIGURE 5.13 The region, limits of integration, and radii in Example 9.

Step 3. The radii of the washer swept out by the line segment are $R(y) = \sqrt{y}$, $r(y) = y/2$ (Figures 5.13 and 5.14).

Step 4. Integrate to find the volume:

$$V = \int_c^d \pi([R(y)]^2 - [r(y)]^2)\, dy$$

$$= \int_0^4 \pi\left(\left[\sqrt{y}\right]^2 - \left[\frac{y}{2}\right]^2\right) dy \qquad \text{Values from steps 2 and 3}$$

$$= \pi \int_0^4 \left(y - \frac{y^2}{4}\right) dy = \pi\left[\frac{y^2}{2} - \frac{y^3}{12}\right]_0^4 = \frac{8}{3}\pi \text{ units cubed.}$$

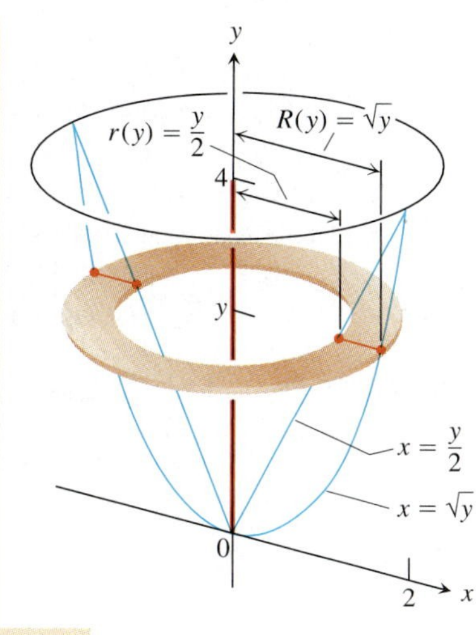

FIGURE 5.14 The washer swept out by the line segment in Figure 5.13.

EXERCISES 5.1

Cross-Section Areas

In Exercises 1 and 2, find a formula for the area $A(x)$ of the cross sections of the solid perpendicular to the x-axis.

1. The solid lies between planes perpendicular to the x-axis at $x = -1$ and $x = 1$. In each case, the cross sections perpendicular to the x-axis between these planes run from the semicircle $y = -\sqrt{1 - x^2}$ to the semicircle $y = \sqrt{1 - x^2}$.

   (a) The cross sections are circular disks with diameters in the xy-plane.

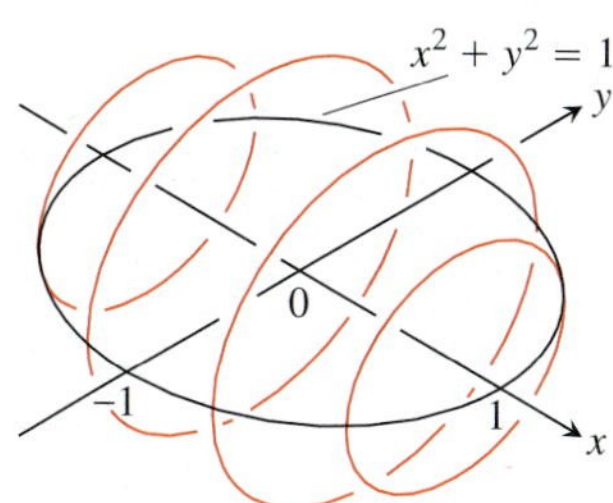

(b) The cross sections are squares with bases in the xy-plane.

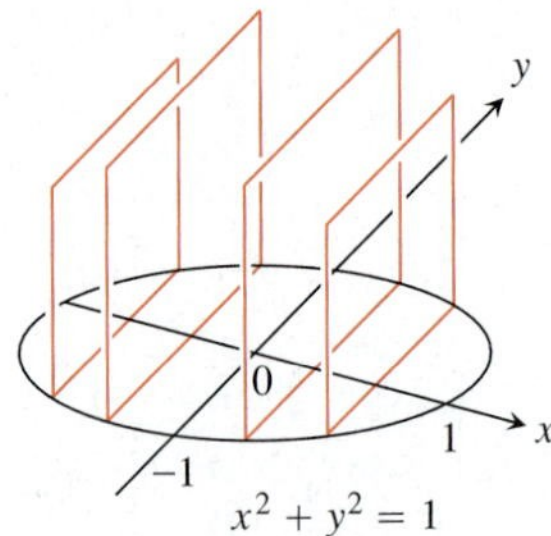

(c) The cross sections are squares with diagonals in the xy-plane. (The length of a square's diagonal is $\sqrt{2}$ times the length of its sides.)

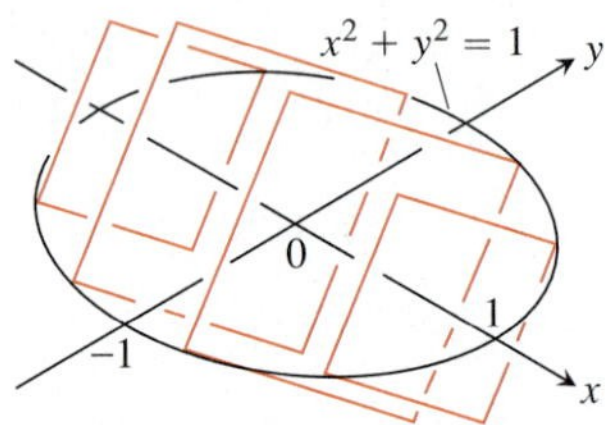

(d) The cross sections are equilateral triangles with bases in the xy-plane.

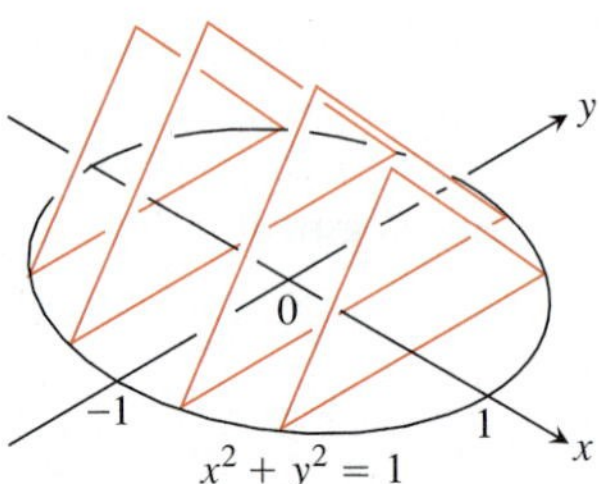

2. The solid lies between planes perpendicular to the x-axis at $x = 0$ and $x = 4$. The cross sections perpendicular to the x-axis between these planes run from the parabola $y = -\sqrt{x}$ to the parabola $y = \sqrt{x}$.

(a) The cross sections are circular disks with diameters in the xy-plane.

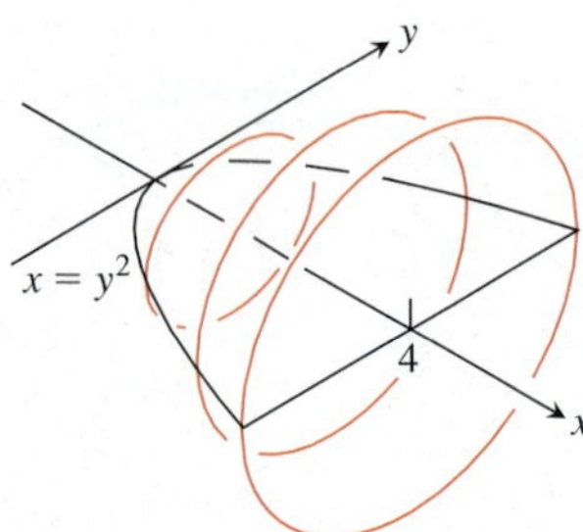

(b) The cross sections are squares with bases in the xy-plane.

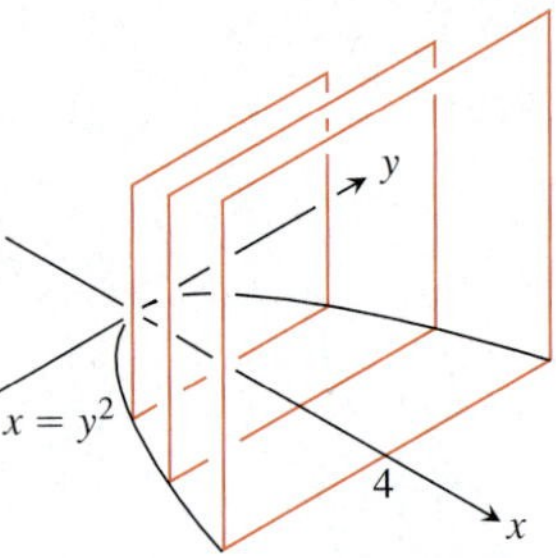

(c) The cross sections are squares with diagonals in the xy-plane.

(d) The cross sections are equilateral triangles with bases in the xy-plane.

Volumes by Slicing

Find the volumes of the solids in Exercises 3–10.

3. The solid lies between planes perpendicular to the x-axis at $x = 0$ and $x = 4$. The cross sections perpendicular to the x-axis on the interval $0 \le x \le 4$ are squares whose diagonals run from the parabola $y = -\sqrt{x}$ to the parabola $y = \sqrt{x}$.

4. The solid lies between planes perpendicular to the x-axis at $x = -1$ and $x = 1$. The cross sections perpendicular to the x-axis are circular disks whose diameters run from the parabola $y = x^2$ to the parabola $y = 2 - x^2$.

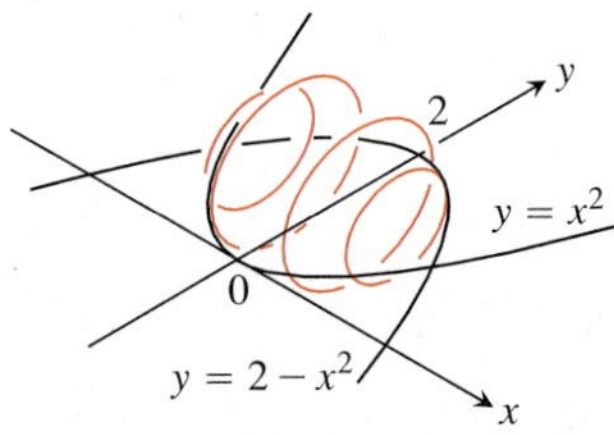

5. The solid lies between planes perpendicular to the x-axis at $x = -1$ and $x = 1$. The cross sections perpendicular to the x-axis are

(a) circles whose diameters stretch from the curve $y = -1/\sqrt{1 + x^2}$ to the curve $y = 1/\sqrt{1 + x^2}$

(b) vertical squares whose base edges run from the curve $y = 1/\sqrt{1 + x^2}$ to the curve $y = 1/\sqrt{1 + x^2}$.

6. The solid lies between planes perpendicular to the x-axis at $x = -\sqrt{2}/2$ and $x = \sqrt{2}/2$. The cross sections are

(a) circles whose diameters stretch from the x-axis to the curve $y = 2/\sqrt[4]{1 - x^2}$

(b) squares whose diagonals stretch from the x-axis to the curve $y = 2/\sqrt[4]{1 - x^2}$.

7. The base of a solid is the region between the curve $y = 2\sqrt{\sin x}$ and the interval $[0, \pi]$ on the x-axis. The cross sections perpendicular to the x-axis are

(a) equilateral triangles with bases running from the x-axis to the curve as shown in the figure

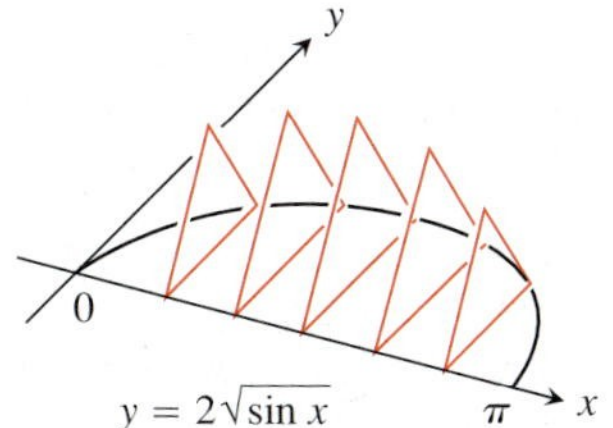

(b) squares with bases running from the x-axis to the curve.

8. The solid lies between planes perpendicular to the x-axis at $x = -\pi/3$ and $x = \pi/3$. The cross sections perpendicular to the x-axis are

(a) circular disks with diameters running from the curve $y = \tan x$ to the curve $y = \sec x$

(b) squares whose bases run from the curve $y = \tan x$ to the curve $y = \sec x$.

9. The solid lies between planes perpendicular to the y-axis at $y = 0$ and $y = 2$. The cross sections perpendicular to the y-axis are circular disks with diameters running from the y-axis to the parabola $x = \sqrt{5}y^2$.

10. The base of the solid is the disk $x^2 + y^2 \leq 1$. The cross sections by planes perpendicular to the y-axis between $y = -1$ and $y = 1$ are isosceles right triangles with one leg in the disk.

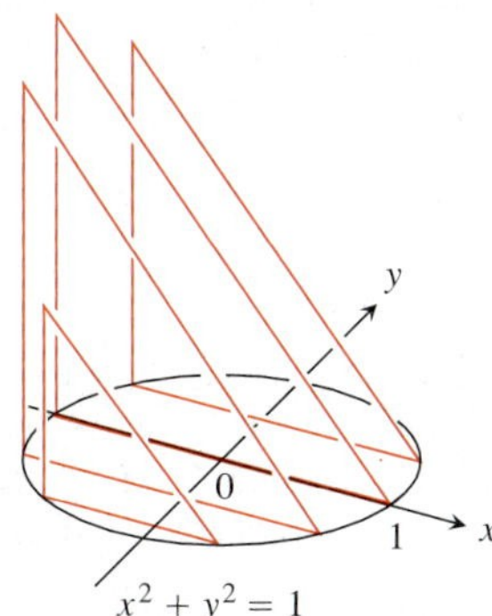

11. *A twisted solid* A square of side length s lies in a plane perpendicular to a line L. One vertex of the square lies on L. As this square moves a distance h along L, the square turns one revolution about L to generate a corkscrew-like column with square cross sections.

(a) Find the volume of the column.

(b) *Writing to Learn* What will the volume be if the square turns twice instead of once? Give reasons for your answer.

12. *Writing to Learn* A solid lies between planes perpendicular to the x-axis at $x = 0$ and $x = 12$. The cross sections cut by planes perpendicular to the x-axis are circular disks whose diameters run from the line $y = x/2$ to the line $y = x$ as shown in the accompanying figure. Explain why the solid has the same volume as a right circular cone with base radius 3 and height 12.

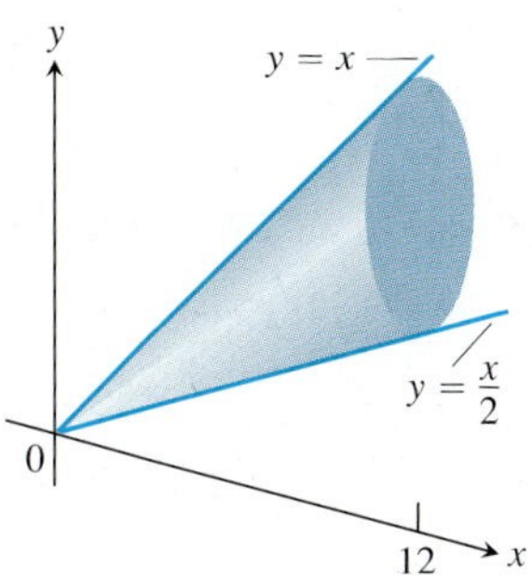

Solids of Revolution: Circular Cross Sections

In Exercises 13–16, find the volume of the solid generated by revolving the shaded region about the given axis.

13. About the x-axis

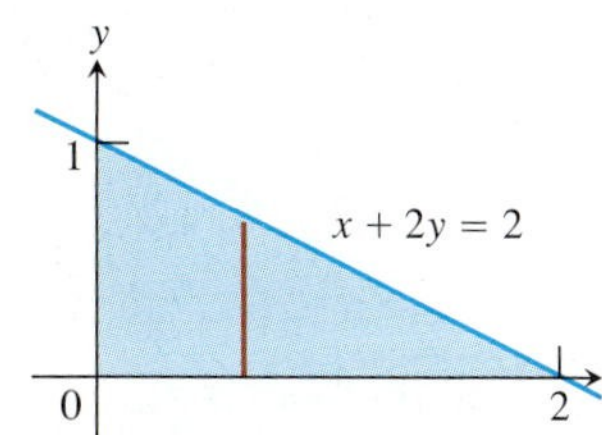

14. About the y-axis

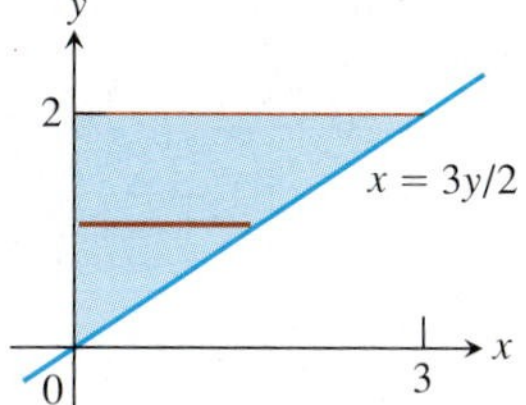

15. About the y-axis

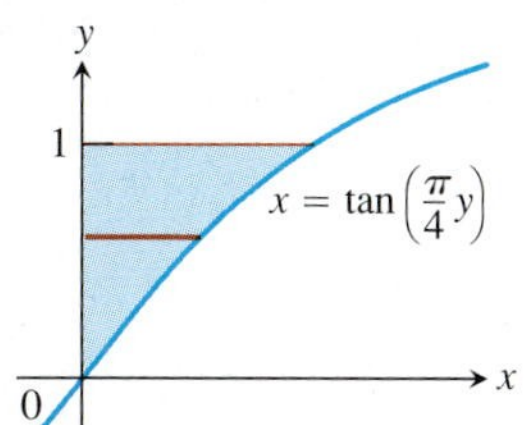

16. About the x-axis

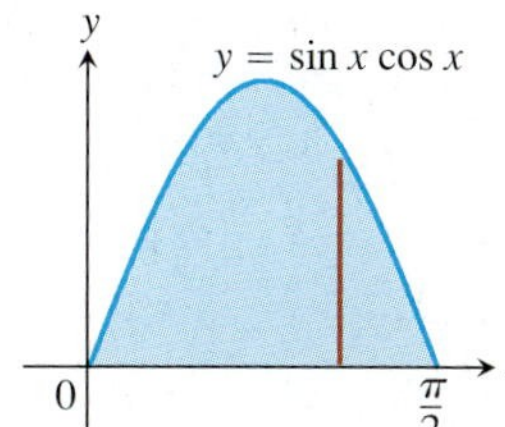

CD-ROM WEBsite

Find the volumes of the solids generated by revolving the regions bounded by the lines and curves in Exercises 17–22, about the x-axis.

17. $y = x^2, \quad y = 0, \quad x = 2$

18. $y = x^3, \quad y = 0, \quad x = 2$

19. $y = \sqrt{9 - x^2}, \quad y = 0$

20. $y = e^{-x}, \quad y = 0, \quad x = 0, \quad x = 1$

21. $y = \sqrt{\cos x}, \quad 0 \leq x \leq \pi/2, \quad y = 0, \quad x = 0$

22. $y = \sqrt{\cot x}, \quad y = 0, \quad x = \pi/6, \quad x = \pi/2$

In Exercises 23 and 24, find the volume of the solid generated by revolving the region about the given line.

23. The region in the first quadrant bounded above by the line $y = \sqrt{2}$, below by the curve $y = \sec x \tan x$, and on the left by the y-axis, about the line $y = \sqrt{2}$

24. The region in the first quadrant bounded above by the line $y = 2$, below by the curve $y = 2 \sin x$, $0 \leq x \leq \pi/2$, and on the left by the y-axis, about the line $y = 2$

Find the volumes of the solids generated by revolving the regions bounded by the lines and curves in Exercises 25–30 about the y-axis.

25. The region enclosed by $x = \sqrt{5}y^2, \quad x = 0, \quad y = -1, \quad y = 1$

26. The region enclosed by $x = y^{3/2}, \quad x = 0, \quad y = 2$

27. The region enclosed by $x = \sqrt{2 \sin 2y}, \quad 0 \le y \le \pi/2, \quad x = 0$

28. The region enclosed by $x = 2/\sqrt{y+1}, \quad y = 3, \quad x = 0, \quad y = 0$

29. $x = 2/(y+1), \quad x = 0, \quad y = 0, \quad y = 3$

30. $x = \sqrt{2y}/(y^2+1), \quad x = 0, \quad y = 1$

Solids of Revolution: Washer Cross Sections

Find the volumes of the solids generated by revolving the shaded regions in Exercises 31 and 32 about the indicated axes.

31. The x-axis

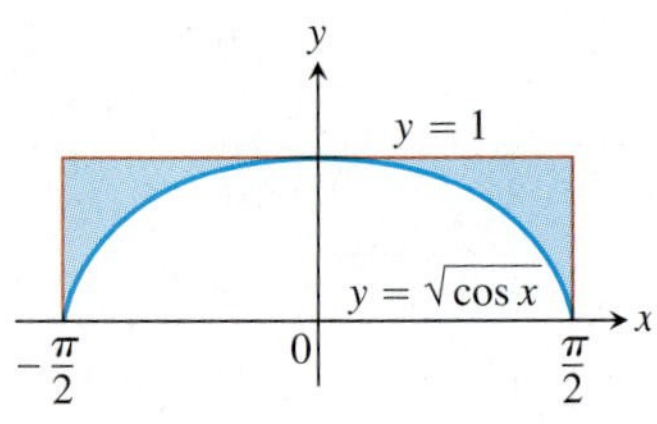

32. The y-axis

CD-ROM
WEBsite

Find the volumes of the solids generated by revolving the regions bounded by the lines and curves in Exercises 33–38 about the x-axis.

33. $y = x, \quad y = 1, \quad x = 0$

34. $y = 2\sqrt{x}, \quad y = 2, \quad x = 0$

35. $y = x^2 + 1, \quad y = x + 3$

36. $y = 4 - x^2, \quad y = 2 - x$

37. $y = \sec x, \quad y = \sqrt{2}, \quad -\pi/4 \le x \le \pi/4$

38. $y = \sec x, \quad y = \tan x, \quad x = 0, \quad x = 1$

In Exercises 39–42, find the volume of the solid generated by revolving each region about the y-axis.

39. The region enclosed by the triangle with vertices (1, 0), (2, 1), and (1, 1)

40. The region enclosed by the triangle with vertices (0, 1), (1, 0), and (1, 1)

41. The region in the first quadrant bounded above by the parabola $y = x^2$, below by the x-axis, and on the right by the line $x = 2$

42. The region in the first quadrant bounded on the left by the circle $x^2 + y^2 = 3$, on the right by the line $x = \sqrt{3}$, and above by the line $y = \sqrt{3}$

In Exercises 43 and 44, find the volume of the solid generated by revolving each region about the given axis.

43. The region in the first quadrant bounded above by the curve $y = x^2$, below by the x-axis, and on the right by the line $x = 1$, about the line $x = -1$

44. The region in the second quadrant bounded above by the curve $y = -x^3$, below by the x-axis, and on the left by the line $x = -1$, about the line $x = -2$

Volumes of Solids of Revolution

45. Find the volume of the solid generated by revolving the region bounded by $y = \sqrt{x}$ and the lines $y = 2$ and $x = 0$ about

(a) the x-axis

(b) the y-axis

(c) the line $y = 2$

(d) the line $x = 4$.

46. Find the volume of the solid generated by revolving the triangular region bounded by the lines $y = 2x$, $y = 0$, and $x = 1$ about

(a) the line $x = 1$

(b) the line $x = 2$.

47. Find the volume of the solid generated by revolving the region bounded by the parabola $y = x^2$ and the line $y = 1$ about

(a) the line $y = 1$

(b) the line $y = 2$

(c) the line $y = -1$.

48. By integration, find the volume of the solid generated by revolving the triangular region with vertices (0, 0), $(b, 0)$, $(0, h)$ about

(a) the x-axis

(b) the y-axis.

Theory and Applications

49. *The volume of a torus* The disk $x^2 + y^2 \le a^2$ is revolved about the line $x = b$ $(b > a)$ to generate a solid shaped like a doughnut and called a *torus*. Find its volume. (*Hint:* $\int_{-a}^{a} \sqrt{a^2 - y^2}\, dy = \pi a^2/2$, since it is the area of a semicircle of radius a.)

50, *Volume of a bowl* A bowl has a shape that can be generated by revolving the graph of $y = x^2/2$ between $y = 0$ and $y = 5$ about the y-axis.

(a) Find the volume of the bowl.

(b) *Related rates* If we fill the bowl with water at a constant rate of 3 cubic units per second, how fast will the water level in the bowl be rising when the water is 4 units deep?

51. *Volume of a bowl*

(a) A hemispherical bowl of radius a contains water to a depth h. Find the volume of water in the bowl.

(b) *Related rates* Water runs into a sunken concrete hemispherical bowl of radius 5 m at the rate of 0.2 m^3/sec. How fast is the water level in the bowl rising when the water is 4 m deep?

52. *Writing to Learn* Explain how you could estimate the volume of a solid of revolution by measuring the shadow cast on a table parallel to its axis of revolution by a light shining directly above it.

53. *Volume of a hemisphere* Derive the formula $V = (2/3)\,\pi R^3$ for the volume of a hemisphere of radius R by comparing its cross sections with the cross sections of a solid right circular cylinder of radius R and height R from which a solid right circular cone of base radius R and height R has been removed as suggested by the accompanying figure.

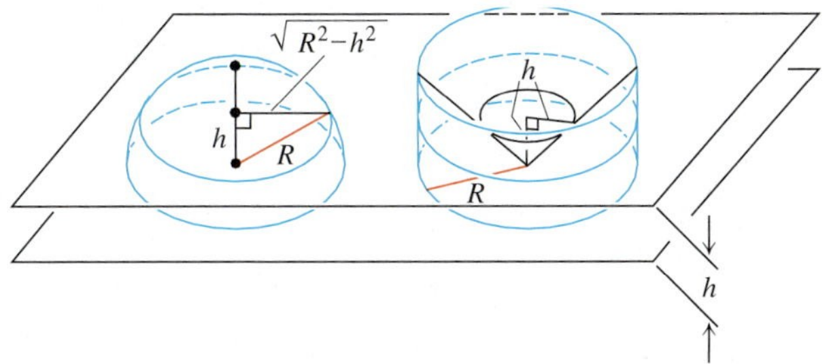

54. *Consistency of volume definitions* The volume formulas in calculus are consistent with the standard formulas from geometry in the sense that they agree on objects to which both apply.

(a) As a case in point, show that if you revolve the region enclosed by the semicircle $y = \sqrt{a^2 - x^2}$ and the x-axis about the x-axis to generate a solid sphere, the calculus formula for volume at the beginning of the section will give $(4/3)\pi a^3$ for the volume just as it should.

(b) Use calculus to the find the volume of a right circular cone of height h and base radius r.

55. *Designing a wok* You are designing a wok frying pan that will be shaped like a spherical bowl with handles. A bit of experimentation at home persuades you that you can get one that holds about 3 L if you make it 9 cm deep and give the sphere a radius of 16 cm. To be sure, you picture the wok as a solid of revolution, as shown here, and calculate its volume with an integral. To the nearest cubic centimeter, what volume do you really get? (1 L = 1000 cm^3.)

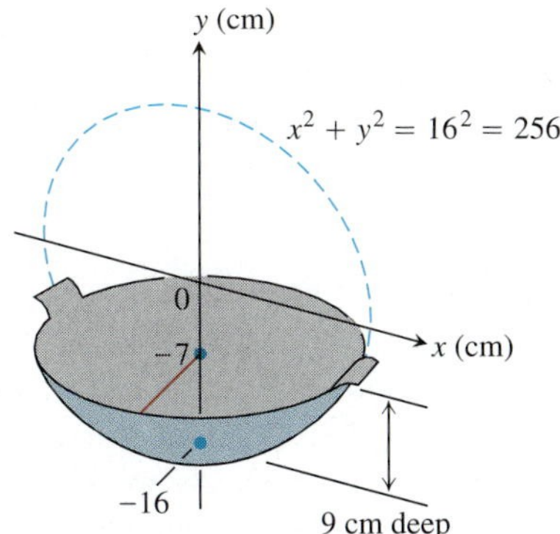

56. *Designing a plumb bob* Having been asked to design a brass plumb bob that will weigh in the neighborhood of 190 g, you decide to shape it like the solid of revolution shown here. Find the plumb bob's volume. If you specify a brass that weighs 8.5 g/cm^3, how much will the plumb bob weigh (to the nearest gram)?

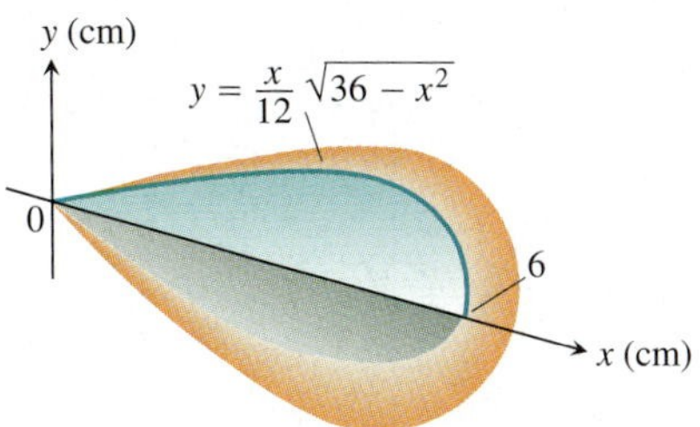

57. *Max-Min* The arch $y = \sin x,\ 0 \le x \le \pi$, is revolved about the line $y = c,\ 0 \le c \le 1$, to generate the solid in the figure.

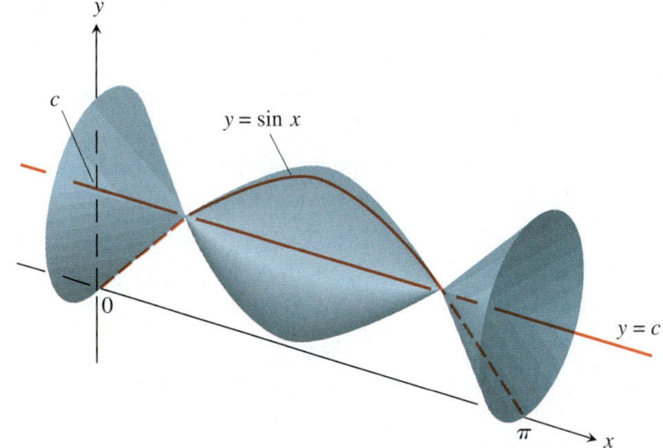

(a) Find the value of c that minimizes the volume of the solid. What is the minimum volume?

(b) What value of c in [0, 1] maximizes the volume of the solid?

T **(c)** *Writing to Learn* Graph the solid's volume as a function of c, first for $0 \le c \le 1$ and then on a larger domain. What happens to the volume of the solid as c moves away from [0, 1]? Does this make sense physically? Give reasons for your answers.

58. *An auxiliary fuel tank* You are designing an auxiliary fuel tank that will fit under a helicopter's fuselage to extend its range. After some experimentation at your drawing board, you decide to shape the tank like the surface generated by revolving the curve $y = 1 - (x^2/16),\ -4 \le x \le 4$, about the x-axis (dimensions in feet).

(a) How many cubic feet of fuel will the tank hold (to the nearest cubic foot)?

(b) A cubic foot holds 7.481 gal. If the helicopter gets 2 mi to the gallon, how many additional miles will the helicopter be able to fly once the tank is installed (to the nearest mile)?

59. *A vase* We wish to estimate the volume of a flower vase using only a calculator, a string, and a ruler. We measure the height of the vase to be 6 in. We then use the string and the ruler to find circumferences of the vase (in inches) at half-inch intervals. (We list them from the top down to correspond with the picture of the vase.)

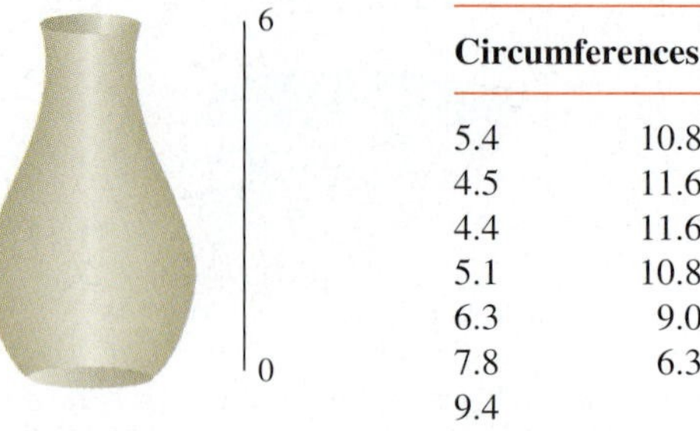

Circumferences	
5.4	10.8
4.5	11.6
4.4	11.6
5.1	10.8
6.3	9.0
7.8	6.3
9.4	

(a) Find the areas of the cross sections that correspond to the given circumferences.

(b) Express the volume of the vase as an integral with respect to y over the interval $[0, 6]$.

(c) Approximate the integral using the Trapezoidal Rule with $n = 12$.

(d) *Writing to Learn* Approximate the integral using Simpson's Rule with $n = 12$. Which result do you think is more accurate? Give reasons for your answer.

60. *A sailboat's displacement* To find the volume of water displaced by a sailboat, the common practice is to partition the waterline into 10 subintervals of equal length, measure the cross-section area $A(x)$ of the submerged portion of the hull at each partition point, and then use Simpson's Rule to estimate the integral of $A(x)$ from one end of the waterline to the other. The table here lists the area measurements at "Stations" 0 through 10, as the partition points are called, for the cruising sloop *Pipedream*, shown here. The common subinterval length (distance between consecutive stations) is $h = 2.54$ ft (about 2 ft 6-1/2 in., chosen for the convenience of the builder).

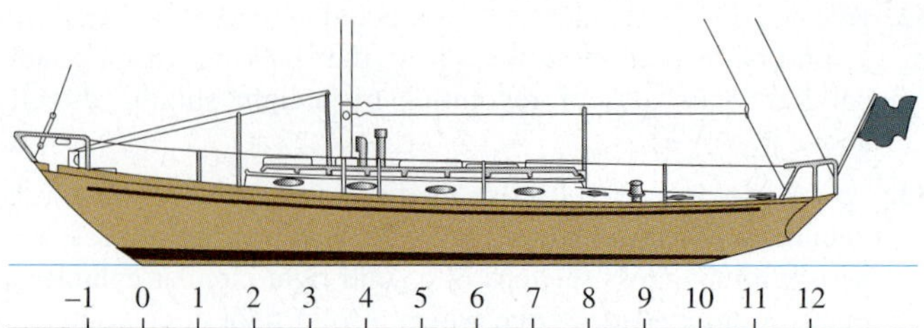

(a) Estimate *Pipedream*'s displacement volume to the nearest cubic foot.

Station	Submerged area (ft^2)
0	0
1	1.07
2	3.84
3	7.82
4	12.20
5	15.18
6	16.14
7	14.00
8	9.21
9	3.24
10	0

(b) The figures in the table are for seawater, which weighs 64 lb/ft^3. How many pounds of water does *Pipedream* displace? (Displacement is given in pounds for small craft and in long tons (1 long ton = 2240 lb) for larger vessels.)

(Data from *Skene's Elements of Yacht Design* by Francis S. Kinney, (Dodd, Mead, 1962.))

(c) *Prismatic coefficients* A boat's prismatic coefficient is the ratio of the displacement volume to the volume of a prism whose height equals the boat's waterline length and whose base equals the area of the boat's largest submerged cross section. The best sailboats have prismatic coefficients between 0.51 and 0.54. Find *Pipedream*'s prismatic coefficient, given a waterline length of 25.4 ft and a largest submerged cross section area of 16.14 ft^2 (at Station 6).

5.2 Modeling Volume Using Cylindrical Shells

Volume by Cylindrical Shells • The Shell Formula

There is another way to find volumes of solids of rotation that can be useful when the axis of revolution is perpendicular to the axis containing the natural interval of integration. Instead of summing volumes of thin slices, we sum volumes of thin cylindrical shells that grow outward from the axis of revolution like tree rings.

Volume by Cylindrical Shells

Here is an example of using thin cylindrical shells to find the volume of a solid.

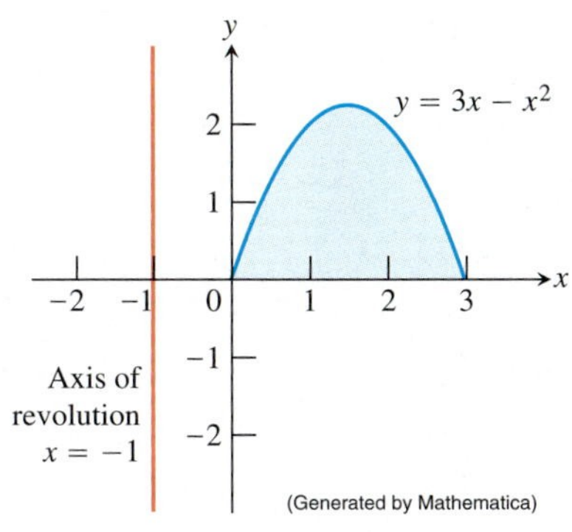

FIGURE 5.15 The graph of the region in Example 1, before revolution.

Example 1 Finding a Volume Using Shells

The region enclosed by the x-axis and the parabola $y = f(x) = 3x - x^2$ is revolved about the line $x = -1$ to generate the shape of a solid (Figures 5.15 and 5.16). What is the volume of the solid?

Solution Integrating with respect to y would be awkward here, as it is not easy to get the original parabola in terms of y. (Try finding the volume by washers and you will soon see what we mean.) To integrate with respect to x, you can do the problem by *cylindrical shells*, which requires that you cut the solid in a rather unusual way.

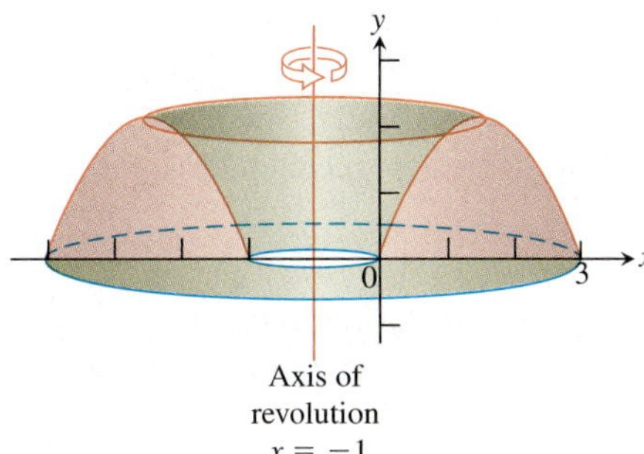

FIGURE 5.16 The region in Figure 5.15 is revolved about the line $x = -1$ to form a solid cake. The natural interval of integration is along the x-axis, perpendicular to the axis of revolution. (Example 1)

Step 1: Instead of cutting a wedge shape, cut a *cylindrical* slice by cutting straight down (parallel to the axis of revolution) all the way around close to the inside hole. Then cut another cylindrical slice around the enlarged hole, then another, and so on. The radii of the cylinders gradually increase, and the heights of the cylinders follow the contour of the parabola: smaller to larger, then back to smaller (Figure 5.17). Each slice is sitting over a subinterval of the x-axis of length Δx. Its radius is approximately $(1 + x_k)$, and its height is approximately $3x_k - x_k^2$.

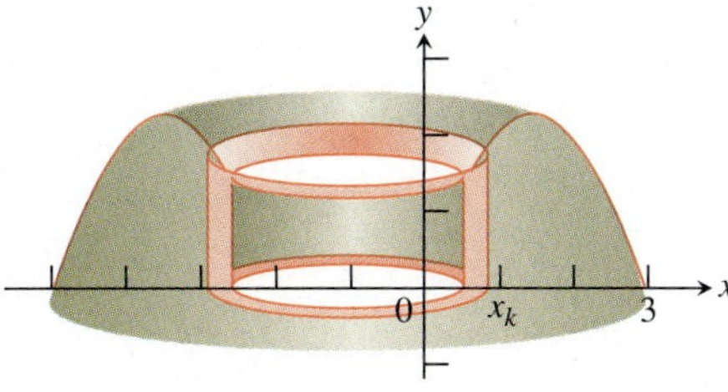

FIGURE 5.17 Cutting the solid into thin cylindrical slices, working from the inside out. Each slice occurs at some x_k between 0 and 3 and has thickness Δx. (Example 1)

Step 2: If you unroll the cylinder at x_k and flatten it out, it becomes (essentially) a rectangular slab with thickness Δx (Figure 5.18). The inner circumference of the

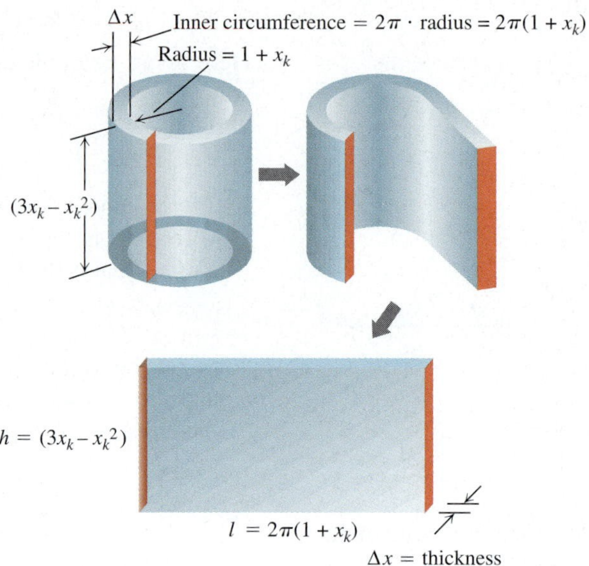

FIGURE 5.18 Imagine cutting and unrolling a cylindrical shell to get a flat (nearly) rectangular solid. (Example 1)

cylinder is $2\pi \cdot$ radius $= 2\pi\,(1 + x_k)$, and this is the length of the rolled out rectangular slab. Therefore, the volume of the flat (nearly) rectangular solid is

$$\begin{aligned}\Delta V &\approx \text{length} \times \text{height} \times \text{thickness}\\ &\approx 2\pi(1 + x_k) \cdot (3x_k - x_k^2) \cdot \Delta x\end{aligned}$$

Step 3: Summing together the volumes of the individual cylindrical shells over the interval $0 \le x \le 3$ gives the Riemann sum $\Sigma\ 2\pi(x_k + 1)(3x_k - x_k^2)\ \Delta x$. Taking the limit as the thickness $\Delta x \to 0$ gives the volume integral

$$\begin{aligned}V &= \int_0^3 2\pi(x + 1)(3x - x^2)\,dx\\ &= \int_0^3 2\pi(3x^2 + 3x - x^3 - x^2)\,dx\\ &= 2\pi\int_0^3 (2x^2 + 3x - x^3)\,dx\\ &= 2\pi\left[\frac{2}{3}x^3 + \frac{3}{2}x^2 - \frac{1}{4}x^4\right]_0^3\\ &= \frac{45\pi}{2}\ \text{units cubed.}\end{aligned}$$

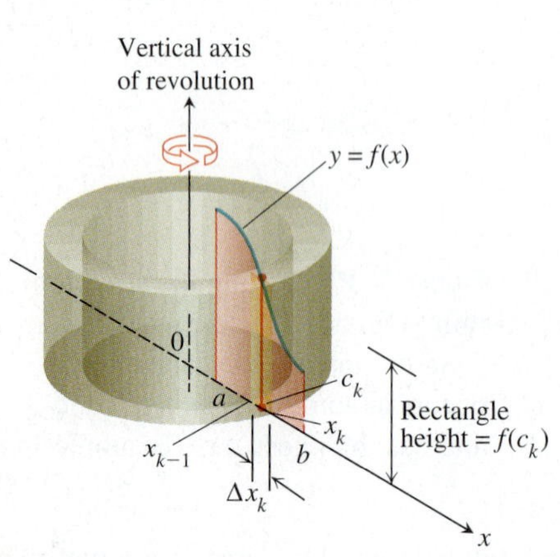

FIGURE 5.19 The shell swept out by the kth rectangle.

The Shell Formula

Suppose that we revolve the tinted region in Figure 5.19 about a vertical line to generate a solid. To estimate the volume of the solid, we can approximate the region with rectangles based on a partition P of the interval $[a, b]$ over which the region stands. The typical approximating rectangle is Δx_k units wide by $f(c_k)$ units high, where c_k is the midpoint of the rectangle's base. A formula from geometry tells us that the volume of the shell swept out by the rectangle is

$$\Delta V_k = 2\pi \times \text{average shell radius} \times \text{shell height} \times \text{thickness}.$$

We approximate the volume of the solid by adding the volumes of the shells swept out by the n rectangles based on P:

$$V \approx \sum_{k=1}^{n} \Delta V_k.$$

The limit of this sum as $\|P\| \to 0$ gives the volume of the solid:

$$\begin{aligned} V &= \lim_{\|P\| \to 0} \sum \Delta V_k \\ &= \int_a^b 2\pi \begin{pmatrix}\text{shell} \\ \text{radius}\end{pmatrix} \begin{pmatrix}\text{shell} \\ \text{height}\end{pmatrix} dx. \end{aligned}$$

Shell Formula for Revolution About a Vertical Line

The volume of the solid generated by revolving the region between the x-axis and the graph of a continuous function $y = f(x) \geq 0$, $0 \leq a \leq x \leq b$, about a vertical line is

$$V = \int_a^b 2\pi \begin{pmatrix}\text{shell} \\ \text{radius}\end{pmatrix} \begin{pmatrix}\text{shell} \\ \text{height}\end{pmatrix} dx$$

Example 2 Cylindrical Shells Revolving About the y-axis

The region bounded by the curve $y = \sqrt{x}$, the x-axis, and the line $x = 4$ is revolved about the y-axis to generate a solid. Find the volume of the solid.

Solution

Step 1: Sketch the region and draw a line segment across it *parallel* to the axis of revolution (Figure 5.20). Label the segment's height (shell height) and distance from the axis of revolution (shell radius). The width of the segment is the shell thickness dx. (We drew the shell in Figure 5.21, but you need not do that.)

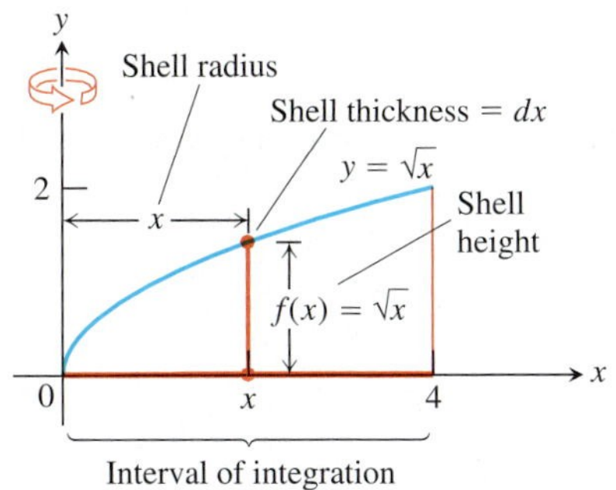

FIGURE 5.20 The region, shell dimensions, and interval of integration in Example 2.

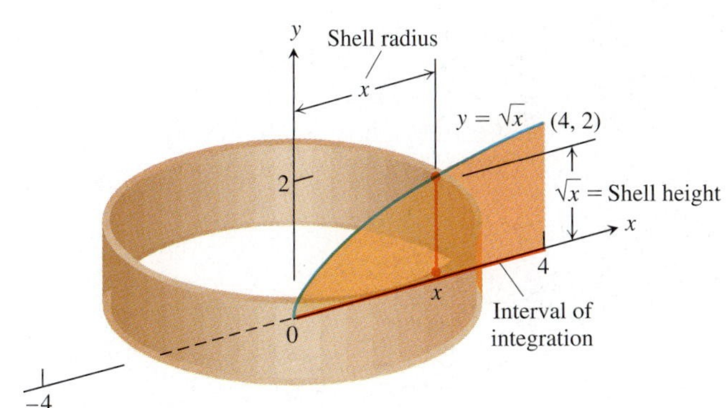

FIGURE 5.21 The shell swept out by the line segment in Figure 5.20.

Step 2: Find the limits of integration for the thickness variable (x runs from $a = 0$ to $b = 4$) and write the volume integral using the Shell Formula:

$$V = \int_a^b 2\pi \begin{pmatrix}\text{shell}\\ \text{radius}\end{pmatrix}\begin{pmatrix}\text{shell}\\ \text{height}\end{pmatrix} dx$$

$$= \int_0^4 2\pi(x)(\sqrt{x})\, dx.$$

Step 3: Integrate to find the volume:

$$V = \int_0^4 2\pi(x)(\sqrt{x})\, dx$$

$$= 2\pi \int_0^4 x^{3/2}\, dx = 2\pi \left[\frac{2}{5} x^{5/2}\right]_0^4 = \frac{128\pi}{5} \text{ units cubed.}$$

So far, we have used vertical axes of revolution. For horizontal axes, we replace the x's with y's.

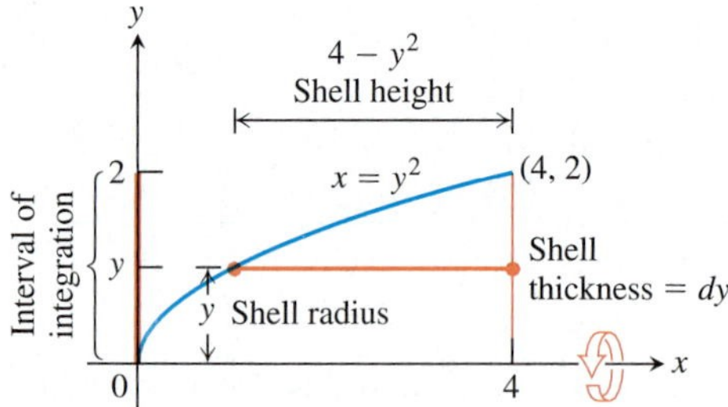

FIGURE 5.22 The region, shell dimensions, and interval of integration in Example 3.

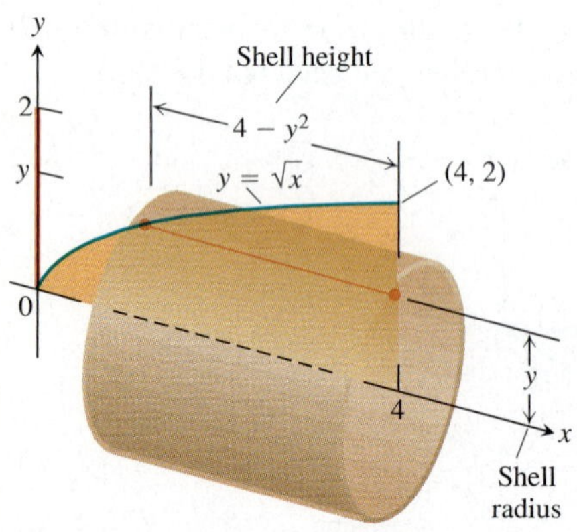

FIGURE 5.23 The shell swept out by the line segment in Figure 5.22.

Example 3 Cylindrical Shells Revolving About the x-axis

The region bounded by the curve $y = \sqrt{x}$, the x-axis, and the line $x = 4$ is revolved about the x-axis to generate a solid. Find the volume of the solid.

Solution

Step 1: Sketch the region and draw a line segment across it parallel to the axis of revolution (Figure 5.22). Label the segment's length (shell height) and distance from the axis of revolution (shell radius). The width of the segment is the shell thickness dy. (We drew the shell in Figure 5.23, but you need not do that.)

Step 2: Find the limits of integration for the thickness variable (y runs from 0 to 2) and write the volume integral using the Shell Formula:

$$V = \int_0^2 2\pi \begin{pmatrix}\text{shell}\\ \text{radius}\end{pmatrix}\begin{pmatrix}\text{shell}\\ \text{height}\end{pmatrix} dy$$

$$= \int_0^2 2\pi(y)(4 - y^2)\, dy.$$

Step 3: Integrate to find the volume:

$$V = \int_0^2 2\pi(y)(4 - y^2)\, dy$$

$$= 2\pi\left[2y^2 - \frac{y^4}{4}\right]_0^2 = 8\pi \text{ units cubed.}$$

How to Use the Shell Method

Regardless of the position of the axis of revolution (horizontal or vertical), the steps for implementing the shell method are these.

Step 1. *Draw the region and sketch a line segment* across it *parallel* to the axis of revolution. *Label* the segment's height or length (shell height), distance from the axis of revolution (shell radius), and width (shell thickness).

Step 2. *Find the limits of integration* for the thickness variable and write the volume integral.

Step 3. *Integrate* the product 2π (shell radius) (shell height) with respect to the thickness variable (x or y) to find the volume.

EXERCISES 5.2

In Exercises 1–6, use the shell method to find the volumes of the solids generated by revolving the shaded region about the indicated axis.

1.

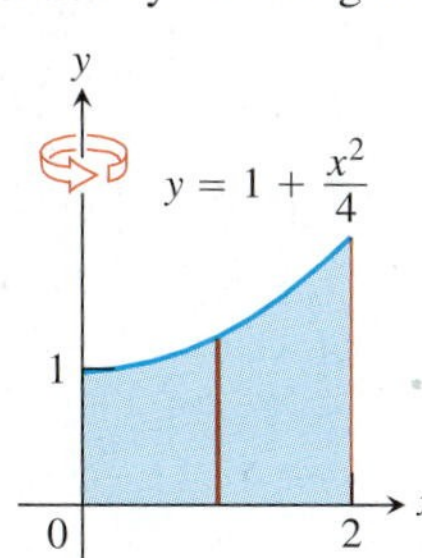

2.

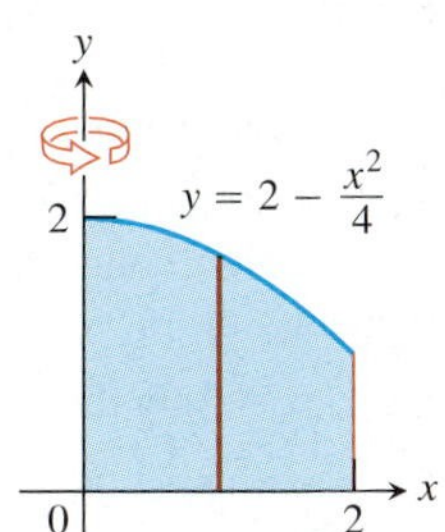

3.

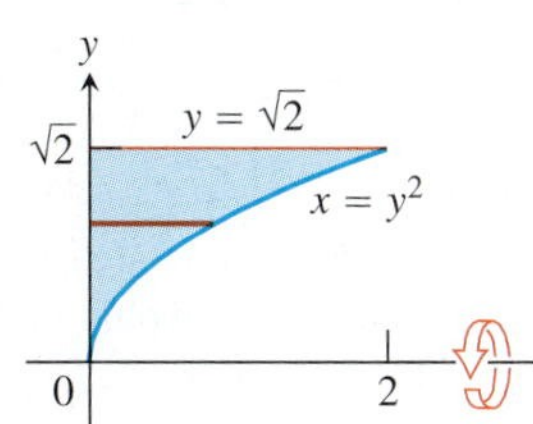

4.

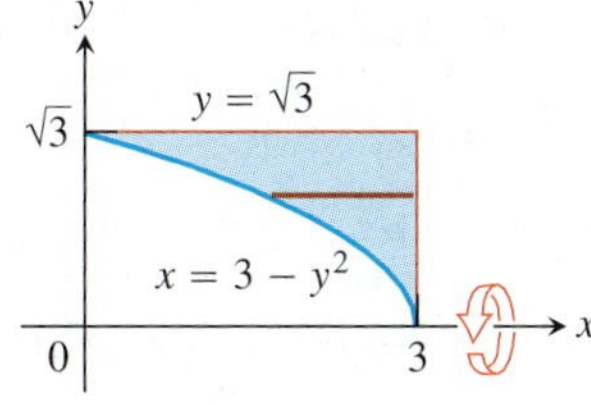

5. The y-axis

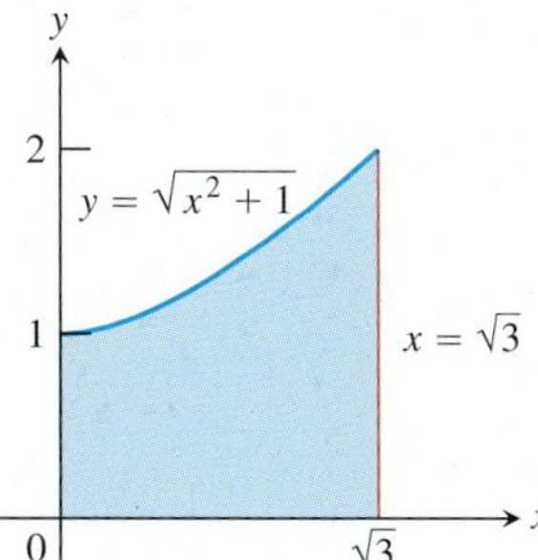

6. The y-axis

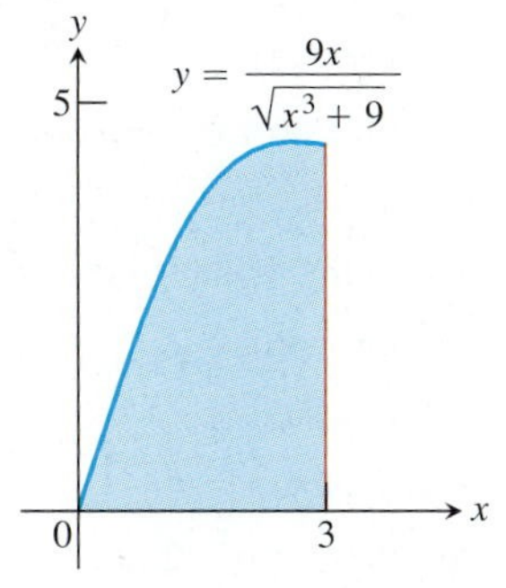

Revolution About the y-axis

Use the shell method to find the volumes of the solids generated by revolving the regions bounded by the curves and lines in Exercises 7–14 about the y-axis.

7. $y = x, \quad y = -x/2, \quad x = 2$

8. $y = 2x, \quad y = x/2, \quad x = 1$

9. $y = x^2, \quad y = 2 - x, \quad x = 0$, for $x \geq 0$

10. $y = 2 - x^2, \quad y = x^2, \quad x = 0$

11. $y = e^{-x^2}, \quad y = 0, \quad x = 0, \quad x = 1$

12. $y = 3/(2\sqrt{x}), \quad y = 0, \quad x = 1, \quad x = 4$

13. Let $f(x) = \begin{cases} (\sin x)/x, & 0 < x \leq \pi \\ 1, & x = 0 \end{cases}$

(a) Show that $xf(x) = \sin x, 0 \leq x \leq \pi$.

(b) Find the volume of the solid generated by revolving the shaded region about the y-axis.

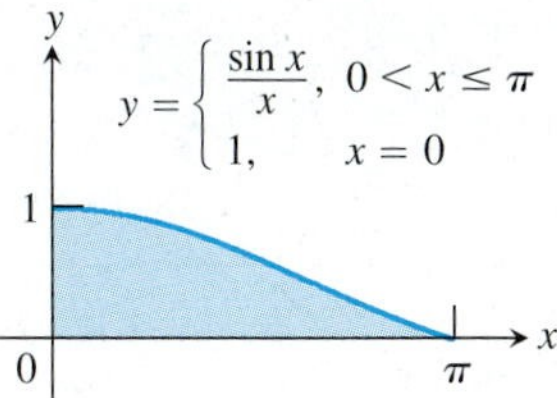

14. Let $g(x) = \begin{cases} (\tan x)^2/x, & 0 < x \leq \pi/4 \\ 0, & x = 0 \end{cases}$

(a) Show that $xg(x) = (\tan x)^2, 0 \leq x \leq \pi/4$.

(b) Find the volume of the solid generated by revolving the shaded region about the y-axis.

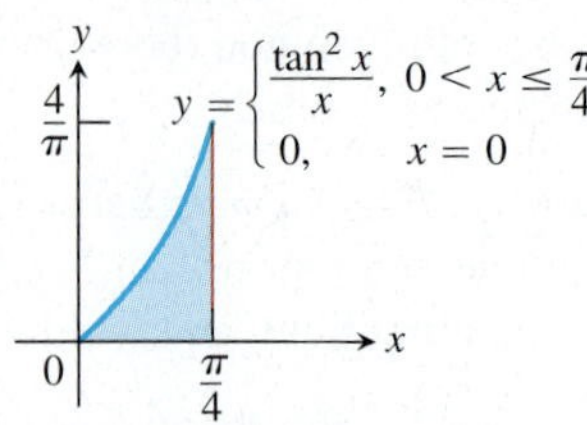

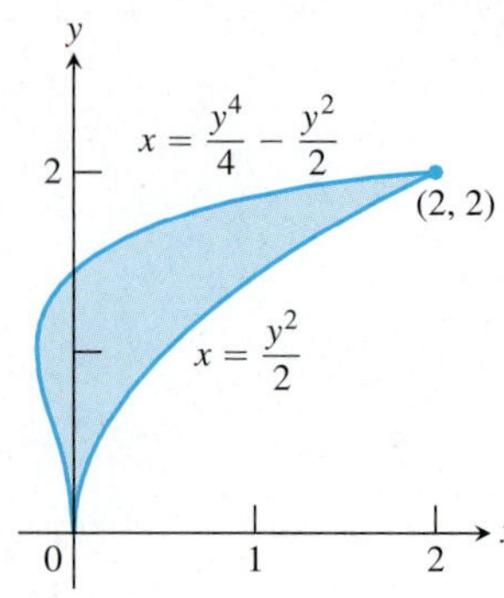

Revolution About the x-axis

Use the shell method to find the volumes of the solids generated by revolving the regions bounded by the curves and lines in Exercises 15–22 about the x-axis.

15. $x = \sqrt{y}, \quad x = -y, \quad y = 2$

16. $x = y^2, \quad x = -y, \quad y = 2, \quad y \geq 0$

17. $x = 2y - y^2, \quad x = 0$

18. $x = 2y - y^2, \quad x = y$

19. $y = |x|, \quad y = 1$

20. $x = e^{y^2}, \quad y = 0, \quad x = 0, \quad y = 1$

21. $y = \sqrt{x}, \quad y = 0, \quad y = x - 2$

22. $y = \sqrt{x}, \quad y = 0, \quad y = 2 - x$

Revolution About Horizontal Lines

In Exercises 23 and 24, use the shell method to find the volumes of the solids generated by revolving the shaded regions about the indicated axes.

23. **(a)** The x-axis

(b) The line $y = 1$

(c) The line $y = 8/5$

(d) The line $y = -2/5$

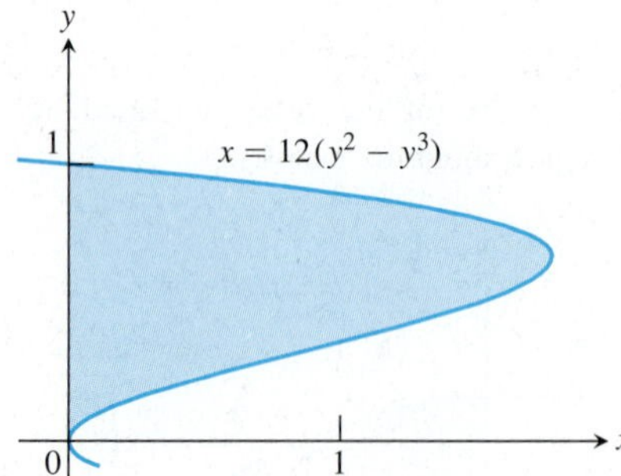

24. **(a)** The x-axis

(b) The line $y = 2$

(c) The line $y = 5$

(d) The line $y = -5/8$

Comparing the Washer and Shell Models

For some regions, both the washer and shell methods work well for the solid generated by revolving the region about the coordinate axes, but this is not always the case. When a region is revolved about the y-axis, for example, and washers are used, we must integrate with respect to y. It may not be possible, however, to express the integrand in terms of y. In such a case, the shell method allows us to integrate with respect to x instead, Exercises 25 and 26 provide some insight.

25. Compute the volume of the solid generated by revolving the region bounded by $y = x$ and $y = x^2$ about each coordinate axis using

(a) the shell method

(b) the washer method.

26. Compute the volume of the solid generated by revolving the triangular region bounded by the lines $2y = x + 4$, $y = x$, and $x = 0$ about

(a) the x-axis using the washer method

(b) the y-axis using the shell method

(c) the line $x = 4$ using the shell method

(d) The line $y = 8$ using the washer method.

Choosing Shells or Washers

In Exercises 27–34, find the volumes of the solids generated by revolving the regions about the given axes. If you think it would be better to use washers in any given instance, feel free to do so.

27. The triangle with vertices $(1, 1)$, $(1, 2)$, and $(2, 2)$ about

(a) the x-axis

(b) the y-axis

(c) the line $x = 10/3$

(d) the line $y = 1$

28. The region bounded by $y = \sqrt{x}$, $y = 2$, $x = 0$ about

(a) the x-axis

(b) the y-axis

(c) the line $x = 4$

(d) the line $y = 2$

29. The region in the first quadrant bounded by the curve $x = y - y^3$ and the y-axis about

(a) the x-axis

(b) the line $y = 1$

30. The region in the first quadrant bounded by $x = y - y^3$, $x = 1$, and $y = 1$ about

(a) the x-axis

(b) the y-axis

(c) the line $x = 1$

(d) the line $y = 1$

31. The region bounded by $y = \sqrt{x}$ and $y = x^2/8$ about

(a) the x-axis

(b) the y-axis

32. The region bounded by $y = 2x - x^2$ and $y = x$ about

(a) the y-axis

(b) the line $x = 1$

33. The region in the first quadrant that is bounded above by the curve $y = 1/x^{1/4}$, on the left by the line $x = 1/16$, and below by the line $y = 1$ is revolved about the x-axis to generate a solid. Find the volume of the solid by

(a) the washer method

(b) the shell method.

34. The region in the first quadrant that is bounded above by the curve $y = 1/\sqrt{x}$, on the left by the line $y = 1/4$, and below by the line $y = 1$ is revolved about the y-axis to generate a solid. Find the volume of the solid by

(a) the washer method **(b)** the shell method.

Choosing Disks, Washers, or Shells

35. The region shown here is to be revolved about the x-axis to generate a solid. Which of the methods (disk, washer, shell) could you use to find the volume of the solid? How many integrals would be required in each case? Explain.

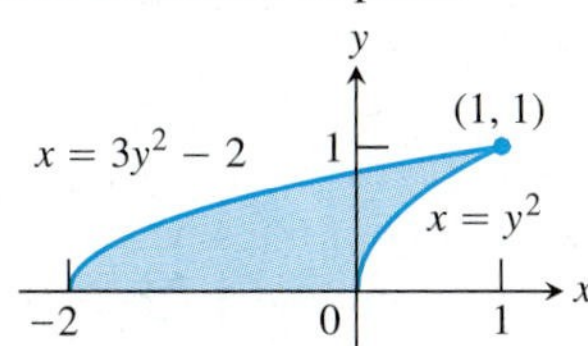

36. The region shown here is to be revolved about the y-axis to generate a solid. Which of the methods (disk, washer, shell) could you use to find the volume of the solid? How many integrals would be required in each case? Give reasons for your answers.

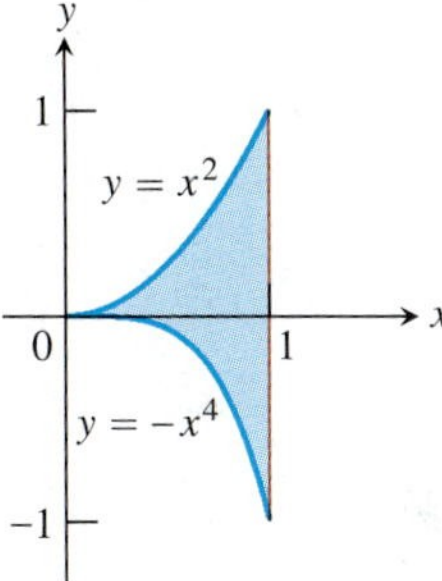

5.3 Lengths of Plane Curves

A Sine Wave • Length of a Smooth Curve • Dealing with Discontinuities in dy/dx • The Short Differential Formula • Parametric Arc Length Formula

If you are hiking along the rim of the Grand Canyon, how many miles do you walk? As a highway engineer, how do you estimate the cost of paving a curving mountainous road based on its total length? To answer questions like this, you need to know how to calculate the length of a curve.

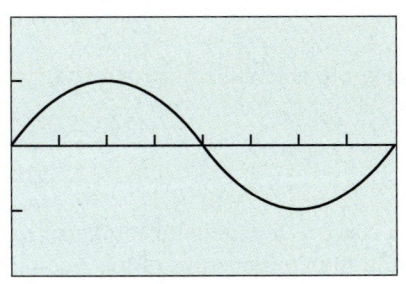

[0, 2π] by [–2, 2]

FIGURE 5.24 One wave of a sine curve has to be longer than 2π.

A Sine Wave

How long is a sine wave (Figure 5.24)?

The usual meaning of *wavelength* refers to the fundamental period, which for $y = \sin x$ is 2π. But how long is the curve itself? If you straightened it out like a piece of string along the positive x-axis with one end at 0, where would the other end be?

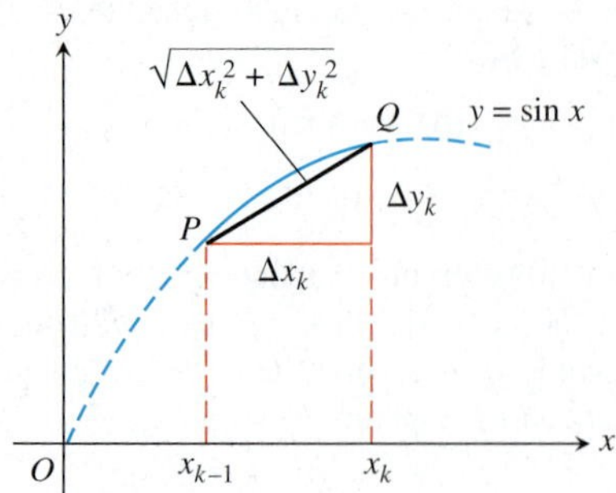

FIGURE 5.25 The line segment approximating the arc PQ of the sine curve above the subinterval $[x_{k-1}, x_k]$. (Example 1)

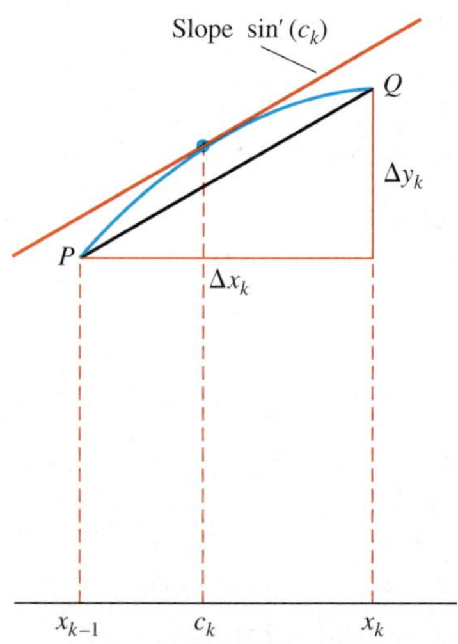

FIGURE 5.26 The portion of the sine curve above $[x_{k-1}, x_k]$. At some c_k in the interval, $\sin'(c_k) = \Delta y_k/\Delta x_k$, the slope of segment PQ. (Example 1)

Example 1 The Length of a Sine Wave

What is the length of the curve $y = \sin x$ from $x = 0$ to $x = 2\pi$?

Solution We answer this question with integration, following our usual plan of breaking the whole into measurable parts. We partition $[0, 2\pi]$ into intervals so short that the pieces of curve (call them "arcs") lying directly above the intervals are nearly straight. That way, each arc is nearly the same as the line segment joining its two ends, and we can take the length of the segment as an approximation to the length of the arc.

Figure 5.25 shows the segment approximating the arc above the subinterval $[x_{k-1}, x_k]$. The length of the segment is $\sqrt{\Delta x_k^{\,2} + \Delta y_k^{\,2}}$. The sum

$$\sum \sqrt{\Delta x_k^{\,2} + \Delta y_k^{\,2}}$$

over the entire partition approximates the length of the curve. All we need now is to find the limit of this sum as the norms of the partitions go to zero. That's the usual plan, but this time there is a problem. Do you see it?

The problem is that the sums as written are not Riemann sums. They do not have the form $\Sigma f(c_k)\,\Delta x$. We can rewrite them as Riemann sums if we multiply and divide each square root by Δx_k.

$$\sum \sqrt{\Delta x_k^{\,2} + \Delta y_k^{\,2}} = \sum \frac{\sqrt{(\Delta x_k)^2 + (\Delta y_k)^2}}{\Delta x_k}\,\Delta x_k$$

$$= \sum \sqrt{1 + \left(\frac{\Delta y_k}{\Delta x_k}\right)^2}\,\Delta x_k$$

This is better, but we still need to write the last square root as a function evaluated at some c_k in the kth subinterval. For this, we call on the Mean Value Theorem for differentiable functions (Section 3.2), which says that since $\sin x$ is continuous on $[x_{k-1}, x_k]$ and is differentiable on (x_{k-1}, x_k), there is a point c_k in $[x_{k-1}, x_k]$ at which $\Delta y_k/\Delta x_k = \sin' c_k$ (Figure 5.26). That gives us

$$\sum \sqrt{1 + (\sin' c_k)^2}\,\Delta x_k,$$

which *is* a Riemann sum.

Now we take the limit as the norms of the subdivisions go to zero and find that the length of one wave of the sine function is

$$\int_0^{2\pi} \sqrt{1 + (\sin' x)^2}\,dx = \int_0^{2\pi} \sqrt{1 + \cos^2 x}\,dx \approx 7.64.$$

We used a numerical integrator on a calculator to obtain the (approximate) value of the integral.

Length of a Smooth Curve

We are almost ready to define the length of a curve as a definite integral, using the procedure of Example 1. We first call attention to two properties of the sine function that came into play along the way.

We obviously used *differentiability* when we invoked the Mean Value Theorem to replace $\Delta y_k/\Delta x_k$ by $\sin'(c_k)$ for some c_k in the interval $[x_{k-1}, x_k]$. Less obviously, we used the continuity of the derivative of sine in passing from $\Sigma \sqrt{1 + (\sin' c_k)^2}\,\Delta x_k$ to the Riemann integral. The requirement for finding the

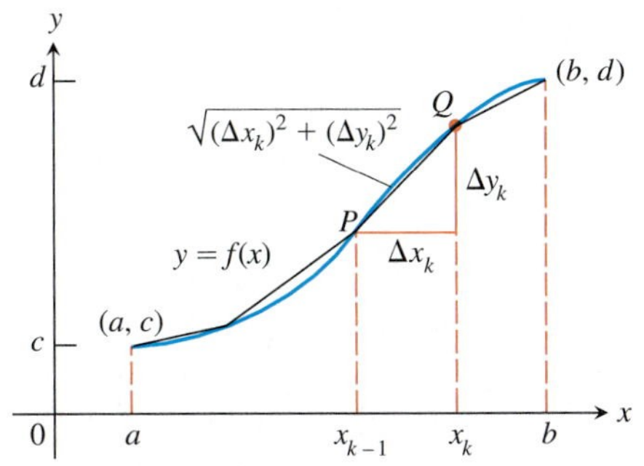

FIGURE 5.27 The graph of f, approximated by line segments.

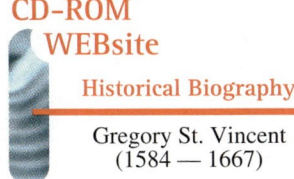

length of a curve by this method, then, is that the function have a continuous first derivative. We call this property **smoothness.** A function with a continuous first derivative is **smooth** and its graph is a **smooth curve.**

Let us review the process, this time with a general smooth function $f(x)$. Suppose that the graph of f begins at the point (a, c) and ends at (b, d), as shown in Figure 5.27. We partition the interval $a \le x \le b$ into subintervals so short that the arcs of the curve above them are nearly straight. The length of the segment approximating the arc above the subinterval $[x_{k-1}, x_k]$ is $\sqrt{\Delta x_k^2 + \Delta y_k^2}$. The sum $\Sigma \sqrt{\Delta x_k^2 + \Delta y_k^2}$ approximates the length of the curve. We apply the Mean Value Theorem to f on each subinterval to rewrite the sum as a Riemann sum,

$$\sum \sqrt{\Delta x_k^2 + \Delta y_k^2} = \sum \sqrt{1 + \left(\frac{\Delta y_k}{\Delta x_k}\right)^2}\,\Delta x_k$$

$$= \sum \sqrt{1 + (f'(c_k))^2}\,\Delta x_k. \qquad \text{For some point } c_k \text{ in } (x_{k-1}, x_k)$$

Passing to the limit as the norms of the subdivisions go to zero gives the length of the curve as

$$L = \int_a^b \sqrt{1 + (f'(x))^2}\,dx = \int_a^b \sqrt{1 + \left(\frac{dy}{dx}\right)^2}\,dx.$$

CD-ROM WEBsite Historical Biography
Gregory St. Vincent (1584 — 1667)

We could as easily have transformed $\Sigma \sqrt{\Delta x_k^2 + \Delta y_k^2}$ into a Riemann sum by dividing and multiplying by Δy_k, giving a formula that involves x as a function of y (say, $x = g(y)$) on the interval $[c, d]$:

$$L \approx \sum \frac{\sqrt{(\Delta x_k)^2 + (\Delta y_k)^2}}{\Delta y_k}\,\Delta y_k = \sum \sqrt{1 + \left(\frac{\Delta x_k}{\Delta y_k}\right)^2}\,\Delta y_k$$

$$= \sum \sqrt{1 + (g'(c_k))^2}\,\Delta y_k. \qquad \text{For some } c_k \text{ in } (y_{k-1}, y_k)$$

The limit of these sums, as the norms of the subdivisions go to zero, gives another reasonable way to calculate the curve's length,

$$L = \int_c^d \sqrt{1 + (g'(y))^2}\,dy = \int_c^d \sqrt{1 + \left(\frac{dx}{dy}\right)^2}\,dy.$$

Putting these two formulas together, we have the following definition for the length of a smooth curve.

CD-ROM WEBsite

Arc Length Formulas for Length of a Smooth Curve

If f is smooth on $[a, b]$, the **length** of the curve $y = f(x)$ from a to b is the number

$$L = \int_a^b \sqrt{1 + \left(\frac{dy}{dx}\right)^2}\,dx. \tag{1}$$

If g is smooth on $[c, d]$, the **length** of the curve $x = g(y)$ from c to d is the number

$$L = \int_c^d \sqrt{1 + \left(\frac{dx}{dy}\right)^2}\,dy \tag{2}$$

Example 2 Applying the Arc Length Formula

Find the *exact* length of the curve

$$y = \frac{x^2}{2} - \frac{\ln x}{4} \quad \text{for} \quad 2 \le x \le 4.$$

Solution

$$\frac{dy}{dx} = x - \frac{1}{4x},$$

which is continuous on $[2, 4]$. Therefore,

$$\begin{aligned}
L &= \int_2^4 \sqrt{1 + \left(\frac{dy}{dx}\right)^2}\, dx \\
&= \int_2^4 \sqrt{1 + \left(x^2 - \frac{1}{2} + \frac{1}{16x^2}\right)}\, dx && \text{Square } x - \frac{1}{4x} \text{ for } dy/dx \\
&= \int_2^4 \sqrt{x^2 + \frac{1}{2} + \frac{1}{16x^2}}\, dx && \text{Combine terms} \\
&= \int_2^4 \sqrt{\left(x + \frac{1}{4x}\right)^2}\, dx && \text{Factor} \\
&= \int_2^4 \left(x + \frac{1}{4x}\right) dx && x + (1/4x) \ge 0 \text{ on } [2, 4] \\
&= \left(\frac{x^2}{2} + \frac{\ln x}{4}\right)\Bigg]_2^4 && \text{Integrate} \\
&= 6 + \frac{\ln 4}{4} - \frac{\ln 2}{4} \\
&= 6 + \frac{\ln 2}{4} \approx 6.173.
\end{aligned}$$

Dealing with Discontinuities in dy/dx

At a point on a curve where dy/dx fails to exist, dx/dy may exist and we may be able to find the curve's length by expressing x as a function of y and applying Equation (2).

Example 3 Applying Equation (2)

Find the length of the curve $y = (x/2)^{2/3}$ from $x = 0$ to $x = 2$.

Solution The derivative

$$\frac{dy}{dx} = \frac{2}{3}\left(\frac{x}{2}\right)^{-1/3}\left(\frac{1}{2}\right) = \frac{1}{3}\left(\frac{2}{x}\right)^{1/3}$$

is not defined at $x = 0$, so we cannot find the curve's length with Equation (1).

We therefore rewrite the equation to express x in terms of y:

$$y = \left(\frac{x}{2}\right)^{2/3}$$

$$y^{3/2} = \frac{x}{2} \qquad \text{Raise both sides to the power 3/2.}$$

$$x = 2y^{3/2}. \qquad \text{Solve for } x.$$

From this, we see that the curve whose length we want is also the graph of $x = 2y^{3/2}$ from $y = 0$ to $y = 1$ (Figure 5.28).

The derivative

$$\frac{dx}{dy} = 2\left(\frac{3}{2}\right)y^{1/2} = 3y^{1/2}$$

is continuous on [0, 1]. We may therefore use Equation (2) to find the curve's length:

$$L = \int_c^d \sqrt{1 + \left(\frac{dx}{dy}\right)^2}\, dy = \int_0^1 \sqrt{1 + 9y}\, dy \qquad \text{Eq. (2) with } c = 0,\ d = 1$$

$$= \frac{1}{9} \cdot \frac{2}{3}(1 + 9y)^{3/2}\Big]_0^1 \qquad \text{Let } u = 1 + 9y \text{ and } du/9 = dy, \text{ integrate, and substitute back.}$$

$$= \frac{2}{27}(10\sqrt{10} - 1) \approx 2.27.$$

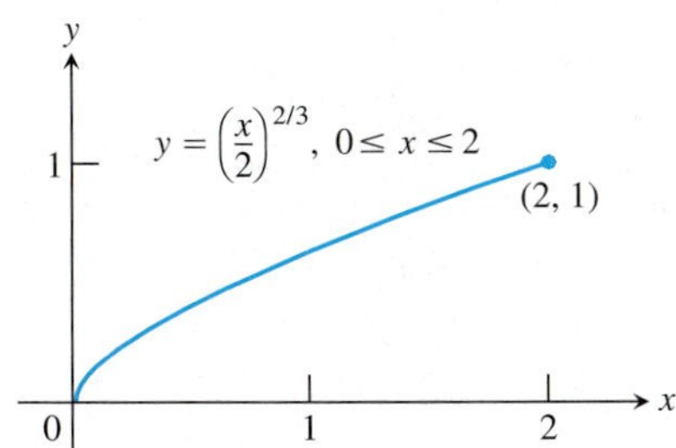

FIGURE 5.28 The graph of $y = (x/2)^{2/3}$ from $x = 0$ to $x = 2$ is also the graph of $x = 2y^{3/2}$ from $y = 0$ to $y = 1$.

The Short Differential Formula

Equations (1) and (2),

$$L = \int_a^b \sqrt{1 + \left(\frac{dy}{dx}\right)^2}\, dx \qquad \text{and} \qquad L = \int_c^d \sqrt{1 + \left(\frac{dx}{dy}\right)^2}\, dy,$$

are often written with differentials instead of derivatives. This is done formally by thinking of the derivatives as quotients of differentials and bringing the dx and dy inside the radicals to cancel the denominators. In the first integral, we have

$$\sqrt{1 + \left(\frac{dy}{dx}\right)^2}\, dx = \sqrt{1 + \frac{dy^2}{dx^2}}\, dx = \sqrt{dx^2 + \frac{dy^2}{dx^2}\, dx^2} = \sqrt{dx^2 + dy^2}.$$

In the second integral, we have

$$\sqrt{1 + \left(\frac{dx}{dy}\right)^2}\, dy = \sqrt{1 + \frac{dx^2}{dy^2}}\, dy = \sqrt{dy^2 + \frac{dx^2}{dy^2}\, dy^2} = \sqrt{dx^2 + dy^2}.$$

In either case, we end up with the same differential formula:

$$L = \int_\alpha^\beta \sqrt{dx^2 + dy^2}. \tag{3}$$

Of course, dx and dy must be expressed in terms of a common variable, and appropriate limits of integration α and β must be found before the integration in Equation (3) is performed.

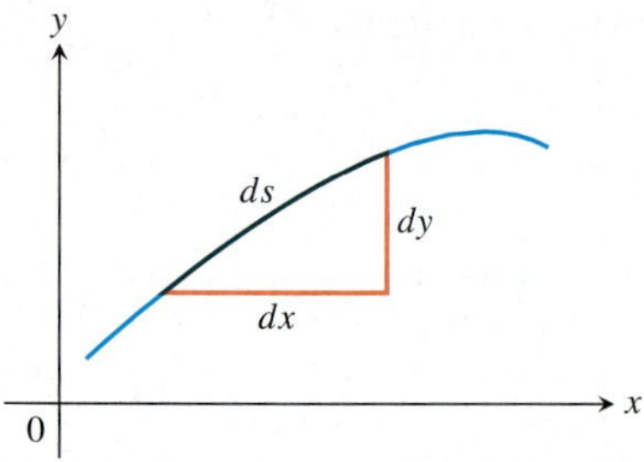

FIGURE 5.29 Think of dx and dy as two sides of a small triangle whose "hypotenuse" is $ds = \sqrt{dx^2 + dy^2}$.

We can shorten Equation (3) still further. Think of dx and dy as two sides of a small triangle whose "hypotenuse" is $ds = \sqrt{dx^2 + dy^2}$ (Figure 5.29). The differential ds is then regarded as a differential of arc length that can be integrated between appropriate limits to give the length of the curve. With $\sqrt{dx^2 + dy^2}$ set equal to ds, the integral in Equation (3) simply becomes the integral of ds.

Definition The Arc Length Differential and the Differential Formula for Arc Length

$$ds = \sqrt{dx^2 + dy^2} \qquad L = \int ds$$

arc length differential — differential formula for arc length

Parametric Arc Length Formula

Another way to deal with discontinuities in dy/dx is to use a parametrization for the plane curve.

CD-ROM WEBsite Historical Biography: James Gregory (1638 — 1675)

Suppose that a curve C can be described by the parametric equations $x = f(t)$ and $y = g(t)$, $\alpha \leq t \leq \beta$. The curve is **smooth** if f and g have continuous first derivatives that are not simultaneously zero. We find an integral for the length of a smooth curve $x = f(t)$, $y = g(t)$, $\alpha \leq t \leq \beta$, by rewriting the integral $L = \int \sqrt{dx^2 + dy^2}$ from Equation (3) in the following way:

$$\begin{aligned} L &= \int_{t=\alpha}^{t=\beta} \sqrt{dx^2 + dy^2} \\ &= \int_{\alpha}^{\beta} \sqrt{\left(\frac{(dx)^2}{(dt)^2} + \frac{(dy)^2}{(dt)^2}\right) dt^2} = \int_{\alpha}^{\beta} \sqrt{\left(\frac{dx}{dt}\right)^2 + \left(\frac{dy}{dt}\right)^2}\, dt. \end{aligned}$$

The only requirement besides the continuity of the integrand is that the point $P(x, y) = P(f(t), g(t))$ not trace any portion of the curve more than once as t moves from α to β.

Parametric Formula for Arc Length

If a curve C is described by the parametric equations $x = f(t)$, $y = g(t)$, $\alpha \leq t \leq \beta$, where f' and g' are continuous and not simultaneously zero on $[\alpha, \beta]$, and if C is traversed exactly once as t increases from α to β, then the length of C is

$$L = \int_{a}^{\beta} \sqrt{\left(\frac{dx}{dt}\right)^2 + \left(\frac{dy}{dt}\right)^2}\, dt. \tag{4}$$

What if there are two different parametrizations for a curve whose length we want to find; does it matter which one we use? The answer, from advanced calculus, is no, as long as the parametrization we choose meets the conditions preceding Equation (4).

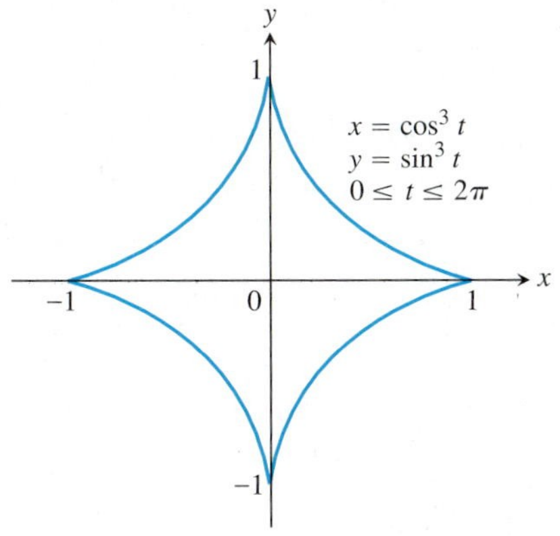

FIGURE 5.30 The astroid in Example 4.

Example 4 Applying the Parametric Formula

Find the length of the astroid (Figure 5.30)

$$x = \cos^3 t, \qquad y = \sin^3 t, \qquad 0 \le t \le 2\pi.$$

Solution Because of the curve's symmetry with respect to the coordinate axes, its length is four times the length of the first-quadrant portion. We have

$$x = \cos^3 t, \qquad y = \sin^3 t$$

$$\left(\frac{dx}{dt}\right)^2 = [3\cos^2 t\,(-\sin t)]^2 = 9\cos^4 t\sin^2 t$$

$$\left(\frac{dy}{dt}\right)^2 = [3\sin^2 t\,(\cos t)]^2 = 9\sin^4 t\cos^2 t$$

$$\sqrt{\left(\frac{dx}{dt}\right)^2 + \left(\frac{dy}{dt}\right)^2} = \sqrt{9\cos^2 t\sin^2 t\,\underbrace{(\cos^2 t + \sin^2 t)}_{1}}$$

$$= \sqrt{9\cos^2 t\sin^2 t}$$

$$= 3|\cos t\ \sin t|$$

$$= 3\ \cos t\ \sin t. \qquad \cos t \sin t \ge 0 \text{ for } 0 \le t \le \pi/2$$

Therefore,

$$\text{Length of first-quadrant portion} = \int_0^{\pi/2} 3\ \cos t\ \sin t\, dt$$

$$= \frac{3}{2}\int_0^{\pi/2} \sin 2t\, dt \qquad \cos t \sin t = (1/2)\sin 2t$$

$$= -\frac{3}{4}\cos 2t\Big]_0^{\pi/2} = \frac{3}{2}.$$

The length of the astroid is four times this: $4(3/2) = 6$.

EXERCISES 5.3

Finding Lengths of Curves

Find the lengths of the curves in Exercises 1–10. If you have a grapher, you may want to graph these curves to see what they look like.

CD-ROM WEBsite **1.** $y = (1/3)(x^2 + 2)^{3/2}$ from $x = 0$ to $x = 3$

CD-ROM WEBsite **2.** $y = x^{3/2}$ from $x = 0$ to $x = 4$

3. $x = (y^3/3) + 1/(4y)$ from $y = 1$ to $y = 3$

(*Hint:* $1 + (dx/dy)^2$ is a perfect square.)

4. $x = (y^{3/2}/3) - y^{1/2}$ from $y = 1$ to $y = 9$

(*Hint:* $1 + (dx/dy)^2$ is a perfect square.)

5. $x = (y^4/4) + 1/(8y^2)$ from $y = 1$ to $y = 2$

(*Hint:* $1 + (dx/dy)^2$ is a perfect square.)

6. $x = (y^3/6) + 1/(2y)$ from $y = 2$ to $y = 3$

(*Hint:* $1 + (dx/dy)^2$ is a perfect square.)

7. $y = (3/4)x^{4/3} - (3/8)x^{2/3} + 5, \quad 1 \le x \le 8$ CD-ROM WEBsite

CD-ROM WEBsite **8.** $y = (x^3/3) + x^2 + x + 1/(4x + 4), \quad 0 \le x \le 2$

9. $x = \int_0^y \sqrt{\sec^4 t - 1}\, dt, \quad -\pi/4 \le y \le \pi/4$

CD-ROM WEBsite **10.** $y = \int_{-2}^x \sqrt{3t^4 - 1}\, dt, \quad -2 \le x \le -1$

Lengths of Parametrized Curves

Find the lengths of the curves in Exercises 11–16.

11. $x = a\cos t, \quad y = a\sin t, \quad 0 \le t \le 2\pi$

12. $x = \cos t, \quad y = t + \sin t, \quad 0 \le t \le \pi$

13. $x = e^t - t, \quad y = 4e^{t/2}, \quad 0 \le t \le 3$

14. $x = t^2/2, \quad y = (2t + 1)^{3/2}/3, \quad 0 \le t \le 4$

15. $x = e^t \cos t, \quad y = e^t \sin t, \quad 0 \le t \le \pi$

16. $x = 8 \cos t + 8t \sin t, \quad y = 8 \sin t - 8t \cos t, \quad 0 \le t \le \pi/2$

Theory and Examples

17. *Writing to Learn* Is there a smooth curve $y = f(x)$ whose length over the interval $0 \le x \le a$ is always $\sqrt{2}a$? Give reasons for your answer.

18. *Using tangent fins to derive the length formula for curves* Assume that f is smooth on $[a, b]$ and partition the interval $[a, b]$ in the usual way. In each subinterval $[x_{k-1}, x_k]$, construct the *tangent fin* at the point $(x_{k-1}, f(x_{k-1}))$, as shown in the figure.

(a) Show that the length of the kth tangent fin over the interval $[x_{k-1}, x_k]$ equals $\sqrt{(\Delta x_k)^2 + (f'(x_{k-1})\, \Delta x_k)^2}$.

(b) Show that

$$\lim_{n\to\infty} \sum_{k=1}^{n} (\text{length of } k\text{th tangent fin}) = \int_a^b \sqrt{1 + (f'(x))^2}\, dx,$$

which is the length L of the curve $y = f(x)$ from a to b.

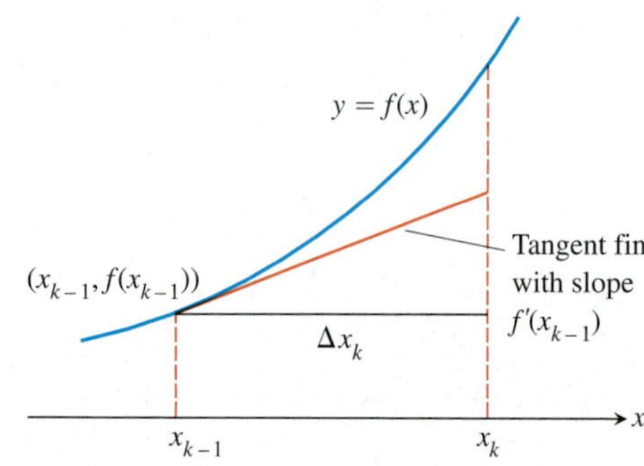

19. (a) Find a curve through the point $(1, 1)$ whose length integral is

$$L = \int_1^4 \sqrt{1 + \frac{1}{4x}}\, dx.$$

(b) *Writing to Learn* How many such curves are there? Give reasons for your answer.

20. (a) Find a curve through the point $(0, 1)$ whose length integral is

$$L = \int_1^2 \sqrt{1 + \frac{1}{y^4}}\, dy.$$

(b) *Writing to Learn* How many such curves are there? Give reasons for your answer.

Finding Integrals for Lengths of Curves

In Exercises 21–28, do the following.

(a) Set up an integral for the length of the curve.

T **(b)** Graph the curve to see what it looks like.

T **(c)** Use your grapher's or computer's integral evaluator to find the curve's length numerically.

CD-ROM WEBsite

CD-ROM WEBsite **21.** $y = x^2, \quad -1 \le x \le 2$

22. $y = \tan x, \quad -\pi/3 \le x \le 0$

23. $x = \sin y, \quad 0 \le y \le \pi$

24. $x = \sqrt{1 - y^2}, \quad -1/2 \le y \le 1/2$

25. $y^2 + 2y = 2x + 1$ from $(-1, -1)$ to $(7, 3)$

26. $y = \sin x - x \cos x, \quad 0 \le x \le \pi$

27. $y = \int_0^x \tan t\, dt, \quad 0 \le x \le \pi/6$

28. $x = \int_0^y \sqrt{\sec^2 t - 1}\, dt, \quad -\pi/3 \le y \le \pi/4$

29. *Fabricating metal sheets* Your metal fabrication company is bidding for a contract to make sheets of corrugated steel roofing like the one shown here. The cross sections of the corrugated sheets are to conform to the curve

$$y = \sin\left(\frac{3\pi}{20}x\right), \qquad 0 \le x \le 20 \text{ in.}$$

If the roofing is to be stamped from flat sheets by a process that does not stretch the material, how wide should the original material be? Give your answer to two decimal places.

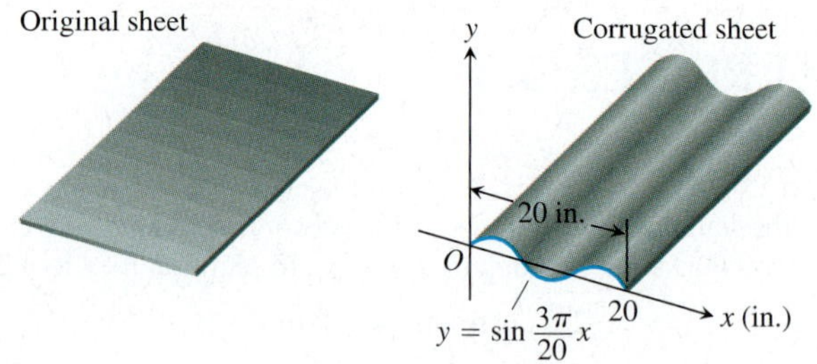

30. *Tunnel construction* Your engineering firm is bidding for the contract to construct the tunnel shown in the accompanying figure. The tunnel is 300 ft long and 50 ft wide at the base. The cross section is shaped like one arch of the curve $y = 25 \cos(\pi x/50)$. Upon completion, the tunnel's inside surface (excluding the roadway) will be treated with a waterproof sealer that costs \$1.75 per square foot to apply. How much will it cost to apply the sealer?

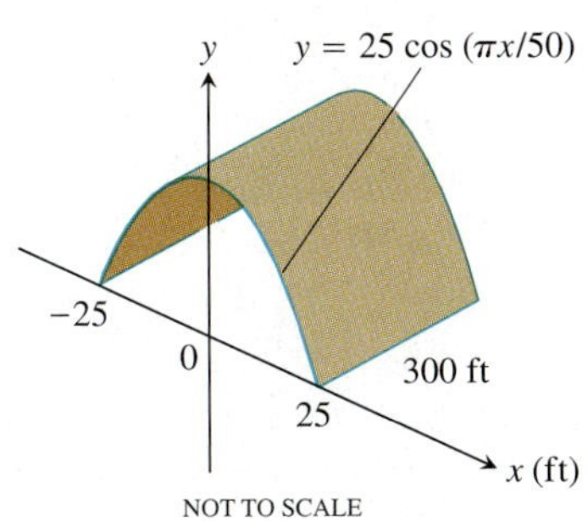

COMPUTER EXPLORATIONS

Polygonal Approximations

In Exercises 31–36, use a CAS to perform the following steps for the given curve over the closed interval.

(a) Plot the curve together with the polygonal path approximations for $n = 2, 4, 8$ partition points over the interval. (See Figure 5.27).

(b) Find the corresponding approximation to the length of the curve by summing the lengths of the line segments.

(c) Evaluate the length of the curve using an integral. Compare your approximations for $n = 2, 4, 8$ with the actual length given by the integral. How does the actual length compare with the approximations as n increases? Explain your answer.

31. $f(x) = \sqrt{1 - x^2}, \quad -1 \le x \le 1$

32. $f(x) = x^{1/3} + x^{2/3}, \quad 0 \le x \le 2$

33. $f(x) = \ln(1 - x^2), \quad 0 \le x \le 1/2$

34. $f(x) = x^2 \cos x, \quad 0 \le x \le \pi$

35. $f(x) = \dfrac{x - 1}{4x^2 + 1}, \quad -\dfrac{1}{2} \le x \le 1$

36. $f(x) = \ln\left(\dfrac{e^x + 1}{e^x - 1}\right), \quad 1 \le x \le 2$

5.4 First-Order Separable Differential Equations

General First-Order Differential Equations and Solutions • Separable Equations • Slope Fields: Viewing Solution Curves • Solving the Initial Value Problem $dy/dt = ky$, $y(0) = y_0$ • Heat Transfer: Newton's Law of Cooling Revisited • Resistance Proportional to Velocity • A Moving Body Coasting to a Stop • Torricelli's Law

CD-ROM
WEBsite
Historical Biography

Jules Henri Poincare
(1854 — 1912)

In calculating derivatives by implicit differentiation (Section 2.6), we found that the expression for the derivative dy/dx often contained both variables x and y, not just the independent variable x. In this section, we study initial value problems in which the derivative has the form $dy/dx = f(x, y)$.

General First-Order Differential Equations and Solutions

A **first-order differential equation** is a relation

$$\frac{dy}{dx} = f(x, y) \tag{1}$$

in which $f(x, y)$ is a function of two variables defined on a region in the xy-plane. A **solution** of Equation (1) is a differentiable function $y = y(x)$ defined on an interval of x-values (perhaps infinite) such that

$$\frac{d}{dx}\, y(x) = f(x, y(x))$$

on that interval. The initial condition that $y(x_0) = y_0$ amounts to requiring the solution curve $y = y(x)$ to pass through the point (x_0, y_0).

Example 1 Verifying that a Function is a Solution

Show that the function

$$y = \frac{1}{x} + \frac{x}{2}$$

solves the first-order initial value problem

$$\frac{dy}{dx} = 1 - \frac{y}{x}, \qquad y(2) = \frac{3}{2}.$$

Solution The equation

$$\frac{dy}{dx} = 1 - \frac{y}{x}$$

is a first-order differential equation in which $f(x, y) = 1 - (y/x)$.

The function $y = (1/x) + (x/2)$ solves the differential equation because the two sides of the equation agree when we substitute $(1/x) + (x/2)$ for y:

On the left: $$\frac{dy}{dx} = \frac{d}{dx}\left(\frac{1}{x} + \frac{x}{2}\right) = -\frac{1}{x^2} + \frac{1}{2}.$$

On the right: $$1 - \frac{y}{x} = 1 - \frac{1}{x}\left(\frac{1}{x} + \frac{x}{2}\right)$$
$$= 1 - \frac{1}{x^2} - \frac{1}{2} = -\frac{1}{x^2} + \frac{1}{2}.$$

The function satisfies the initial condition because

$$y(2) = \left(\frac{1}{x} + \frac{x}{2}\right)_{x=2} = \frac{1}{2} + \frac{2}{2} = \frac{3}{2}.$$

Separable Equations

We sometimes write $y' = f(x, y)$ for $dy/dx = f(x, y)$.

The equation $y' = f(x, y)$ is **separable** if f can be expressed as a product of a function of x and a function of y. The differential equation then has the form

$$\frac{dy}{dx} = g(x)\,h(y).$$

If $h(y) \neq 0$, we can **separate the variables** by dividing both sides by h, obtaining, in succession

$$\frac{1}{h(y)}\frac{dy}{dx} = g(x)$$

$$\int \frac{1}{h(y)}\frac{dy}{dx}\,dx = \int g(x)\,dx \qquad \text{Integrate with respect to } x.$$

$$\int \frac{1}{h(y)}\,dy = \int g(x)\,dx.$$

With x and y now separated, we simply integrate each side to get the solutions we seek by expressing y either explicitly or implicitly as a function of x, up to an arbitrary constant.

Example 2 Solving a Separable Equation

Solve the differential equation

$$\frac{dy}{dx} = (1 + y^2)e^x.$$

Solution Since $1 + y^2$ is never zero, we can solve the equation by separating the variables.

$$\frac{dy}{dx} = (1 + y^2)e^x$$

$$dy = (1 + y^2)e^x\,dx \qquad \text{Treat } dy/dx \text{ as a quotient of differentials and multiply both sides by } dx.$$

$$\frac{dy}{1 + y^2} = e^x\,dx \qquad \text{Divide by } (1 + y^2).$$

$$\int \frac{dy}{1 + y^2} = \int e^x\,dx \qquad \text{Integrate both sides.}$$

$$\tan^{-1} y = e^x + C \qquad C \text{ represents the combined constants of integration.}$$

The equation $\tan^{-1} y = e^x + C$ gives y as an implicit function of x. When $-\pi/2 < e^x + C < \pi/2$, we can solve for y as an explicit function of x by taking the tangent of both sides:

$$\tan\,(\tan^{-1} y) = \tan\,(e^x + C)$$

$$y = \tan\,(e^x + C).$$

Slope Fields: Viewing Solution Curves

CD-ROM WEBsite

Each time we specify an initial condition $y(x_0) = y_0$ for the solution of a differential equation $y' = f(x, y)$, the **solution curve** (graph of the solution) is required to pass through the point (x_0, y_0) and to have slope $f(x_0, y_0)$ there. We can picture these slopes graphically by drawing short line segments of slope $f(x, y)$ at selected points (x, y) in the region of the xy-plane that constitutes the domain of f. Each segment has the same slope as the solution curve through (x, y) and so is tangent to the curve there. We see how the curves behave by following these tangents (Figure 5.31).

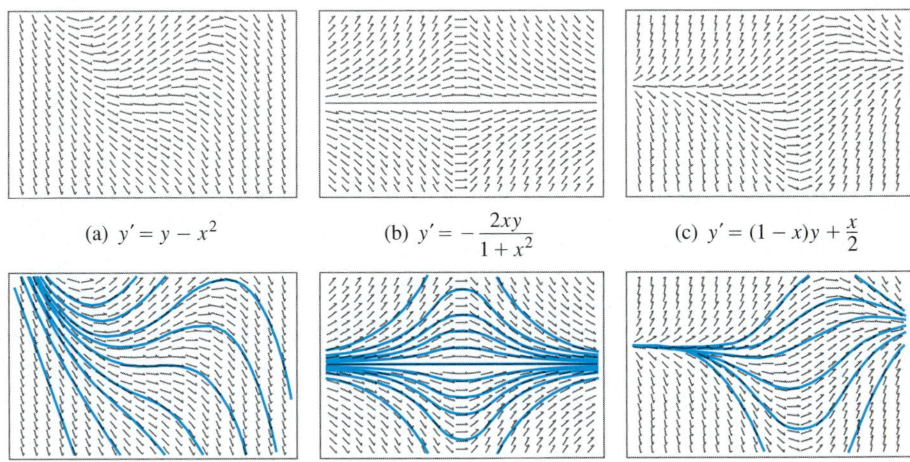

FIGURE 5.31 Slope fields (top row) and selected solution curves (bottom row). In computer renditions, slope segments are sometimes portrayed with arrows, as they are here. This is not to be taken as an indication that slopes have directions, however, because they do not.

Constructing a slope field with pencil and paper can be quite tedious. All our examples were generated by a computer.

Solving the Initial Value Problem $dy/dt = ky,\ y(0) = y_0$

We know that all the functions $y = Ae^{kt}$, A and k constant, satisfy the differential equation $dy/dt = ky$ because

$$\frac{d}{dt}(Ae^{kt}) = A\frac{d}{dt}(e^{kt}) = A(ke^{kt}) = ky.$$

Using the fact that $\int (1/y)\,dy = \ln|y| + C$, we can show that these are the *only* functions satisfying the equation.

We see right away that the constant function $y = 0$ is a solution of $dy/dt = ky$. To find the nonzero solutions we separate the variables and integrate with respect to t:

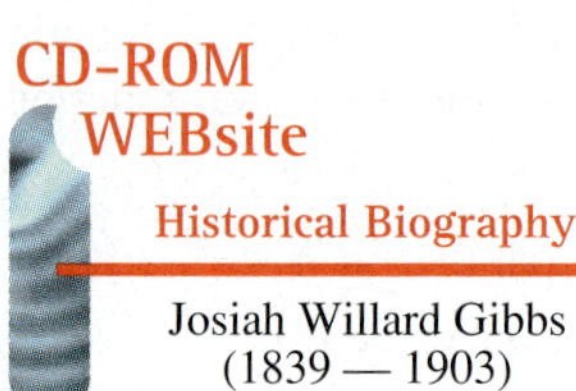

$$\begin{aligned}
\frac{dy}{dt} &= ky \\
\frac{1}{y}\frac{dy}{dt} &= k && y \neq 0 \\
\int \frac{1}{y}\frac{dy}{dt}\,dt &= \int k\,dt \\
\int \frac{1}{y}\,dy &= \int k\,dt && dy = (dy/dt)\,dt \\
\ln|y| &= kt + C && \text{Constants of integration combined} \\
|y| &= e^{kt+C} && \text{Exponentiate} \\
|y| &= e^{C}\cdot e^{kt} && \text{Law of exponents} \\
y &= \pm e^{C}e^{kt} && \text{If } |y| = r\text{, then } y = \pm r. \\
y &= Ae^{kt}. && \text{Writing } A \text{ for } \pm e^{C}
\end{aligned}$$

Thus, all solutions of the equation $dy/dt = ky$ have the form $y = Ae^{kt}$. If, further, we have the initial condition $y = y_0$ when $t = 0$, then

$$y_0 = Ae^{k(0)} = A$$

and the solution of the initial value problem is $y = y_0e^{kt}$.

We can summarize our work with this problem as follows.

The Initial Value Problem $dy/dt = ky,\ y(0) = y_0$

The unique solution of the problem is

$$y = y_0e^{kt}.$$

This equation is the **law of exponential change.** It describes growth if $k > 0$ and decay if $k < 0$. The number k is the **rate constant** of the equation.

CD-ROM
WEBsite
Historical Essay
Differential Equations

Heat Transfer: Newton's Law of Cooling Revisited

In Section 3.4, we obtained the shape of the temperature curve of a heated object immersed in a cooler surrounding medium, such as hot soup in a tin cup cooling to the temperature of the surrounding air. The differential equation describing this cooling process is based on the principle that the rate at which an object's temperature is changing at any given time is roughly proportional to the difference between its temperature and the temperature of the surrounding medium (*Newton's Law of Cooling*, which applies to warming as well).

If H is the temperature of the object at time t and if H_s is the constant surrounding temperature, then the differential equation is

$$\frac{dH}{dt} = -k(H - H_s). \tag{2}$$

If we substitute y for $(H - H_s)$, then

$$\begin{aligned}
\frac{dy}{dt} &= \frac{d}{dt}(H - H_s) = \frac{dH}{dt} - \frac{d}{dt}(H_s) \\
&= \frac{dH}{dt} - 0 && H_s \text{ is a constant.} \\
&= \frac{dH}{dt} \\
&= -k(H - H_s) && \text{Eq. (2)} \\
&= -ky. && H - H_s = y
\end{aligned}$$

Now we know that the solution of $dy/dt = -ky$ is $y = y_0 e^{-kt}$, where $y(0) = y_0$. Substituting $(H - H_s)$ for y gives

$$H - H_s = (H_0 - H_s)e^{-kt}, \tag{3}$$

where H_0 is the temperature at $t = 0$. This is the equation of Newton's Law of Cooling.

Example 3 Cooling a Hard-Boiled Egg

A hard-boiled egg at 98°C is put in a sink of 18°C water. After 5 min, the egg's temperature is 38°C. Assuming that the water has not warmed appreciably, how much longer will it take the egg to reach 20°C?

Solution We find how long it would take the egg to cool from 98°C to 20°C and subtract the 5 min that have already elapsed.

Step 1: *Solve the initial value problem.* Using Equation (3) with $H_s = 18$ and $H_0 = 98$, the egg's temperature t min after it is put in the sink is

$$H = 18 + (98 - 18)e^{-kt} = 18 + 80e^{-kt}.$$

To find k, we use the information that $H = 38$ when $t = 5$:

$$\begin{aligned}
38 &= 18 + 80e^{-5k} \\
e^{-5k} &= \frac{1}{4} \\
\ln\,(e^{-5k}) &= \ln\left(\frac{1}{4}\right) && \text{Natural logarithm of both sides} \\
-5k &= \ln\frac{1}{4} = -\ln 4 \\
k &= \frac{1}{5}\ln 4 = 0.2\ln 4.
\end{aligned}$$

The egg's temperature at time t is $H = 18 + 80e^{-(0.2\ln 4)t}$.

Step 2: *Find the time t when H = 20.*

$$\begin{aligned} 20 &= 18 + 80e^{-(0.2 \ln 4)t} \\ 80e^{-(0.2 \ln 4)t} &= 2 \\ e^{-(0.2 \ln 4)t} &= \frac{1}{40} \\ -(0.2 \ln 4)t &= \ln \frac{1}{40} = -\ln 40 \qquad \text{Logarithms of both sides} \\ t &= \frac{\ln 40}{0.2 \ln 4} \approx 13 \text{ min} \end{aligned}$$

Step 3: *Interpretation of the time required to reach $H = 20$.* The egg's temperature will reach 20°C about 13 min after it is put in water to cool. Since it took 5 min to reach 38°C, it will take about 8 min more to reach 20°C.

Resistance Proportional to Velocity

In some cases, it is reasonable to assume that, other forces being absent, the resistance encountered by a moving object, such as a car coasting to a stop, is proportional to the object's velocity. The slower the object moves, the less its forward progress is resisted by the air through which it passes. To describe this in mathematical terms, we picture the object as a mass m moving along a coordinate line with position function s and velocity v at time t. The resisting force opposing the motion is

$$\text{Force} = \text{mass} \times \text{acceleration} = m\frac{dv}{dt}.$$

We can express the assumption that the resisting force is proportional to velocity by writing

$$m\frac{dv}{dt} = -kv \qquad \text{or} \qquad \frac{dv}{dt} = -\frac{k}{m}v \qquad (k > 0).$$

This is a differential equation of exponential change. The solution to the differential equation with initial condition $v = v_0$ at $t = 0$ is

$$v = v_0 e^{-(k/m)t}. \tag{4}$$

A Moving Body Coasting to a Stop

What can we learn from Equation (4)? For one thing, we can see that if m is something large, like the mass of a 20,000-ton ore boat in Lake Erie, it will take a long time for the velocity to approach zero. For another, we can integrate the equation to find s as a function of t.

Suppose that a body is coasting to a stop and the only force acting on it is a resistance proportional to its speed. How far will it coast? To find out, we start with Equation (4) and solve the initial value problem

$$\frac{ds}{dt} = v_0 e^{-(k/m)t}, \qquad s(0) = 0.$$

Integrating with respect to t gives

$$s = -\frac{v_0 m}{k} e^{-(k/m)t} + C.$$

Substituting $s = 0$ when $t = 0$ gives

$$0 = -\frac{v_0 m}{k} + C \qquad \text{and} \qquad C = \frac{v_0 m}{k}.$$

The body's position at time t is therefore

$$s(t) = -\frac{v_0 m}{k} e^{-(k/m)t} + \frac{v_0 m}{k} = \frac{v_0 m}{k}(1 - e^{-(k/m)t}). \tag{5}$$

To find how far the body will coast, we find the limit of $s(t)$ as $t \to \infty$. Since $-(k/m) < 0$, we know that $e^{-(k/m)t} \to 0$ as $t \to \infty$, so that

$$\begin{aligned} \lim_{t\to\infty} s(t) &= \lim_{t\to\infty} \frac{v_0 m}{k}(1 - e^{-(k/m)t}) \\ &= \frac{v_0 m}{k}(1 - 0) = \frac{v_0 m}{k}. \end{aligned}$$

Thus,

$$\text{Distance coasted} = \frac{v_0 m}{k}. \tag{6}$$

This is an ideal figure, of course. Only in mathematics can time stretch to infinity. The number $v_0 m/k$ is only an upper bound (albeit a useful one). It is true to life in one respect, at least: if m is large it will take a lot of energy to stop the body. That is why ocean liners have to be docked by tugboats. Any liner of conventional design entering a slip with enough speed to steer would smash into the pier before it could stop.

Weight versus Mass

Weight is the force that results from gravity pulling on a mass. The two are related by the equation in Newton's second law,

$$\text{Weight} = \text{mass} \times \text{acceleration}.$$

To convert mass to weight, multiply by the acceleration of gravity. To convert weight to mass, divide by the acceleration of gravity. In the metric system,

$$\text{Newtons} = \text{kilograms} \times 9.8$$

and

$$\text{Newtons}/9.8 = \text{kilograms}.$$

In the English system, where weight is measured in pounds, mass is measured in slugs. Thus,

$$\text{Pounds} = \text{slugs} \times 32$$

and

$$\text{Pounds}/32 = \text{slugs}.$$

A skater weighing 192 lb has a mass of

$$192/32 = 6 \text{ slugs}.$$

Example 4 A Coasting Ice Skater

For a 192 lb skater, the k in Equation (4) is about 1/3 slug/sec and $m = 192/32 = 6$ slugs. How long will it take the skater to coast from 11 ft/sec (7.5 mph) to 1 ft/sec? How far will the skater coast before coming to a complete stop?

Solution We answer the first question by solving Equation (4) for t:

$$\begin{aligned} 11e^{-t/18} &= 1 \\ e^{-t/18} &= 1/11 \\ -t/18 &= \ln(1/11) = -\ln 11 \\ t &= 18 \ln 11 \approx 43 \text{ sec}. \end{aligned}$$

Eq. (4) with $k = 1/3$, $m = 6$, $v_0 = 11$, $v = 1$

We answer the second question with Equation (6):

$$\begin{aligned} \text{Distance coasted} &= \frac{v_0 m}{k} = \frac{11 \cdot 6}{1/3} \\ &= 198 \text{ ft}. \end{aligned}$$

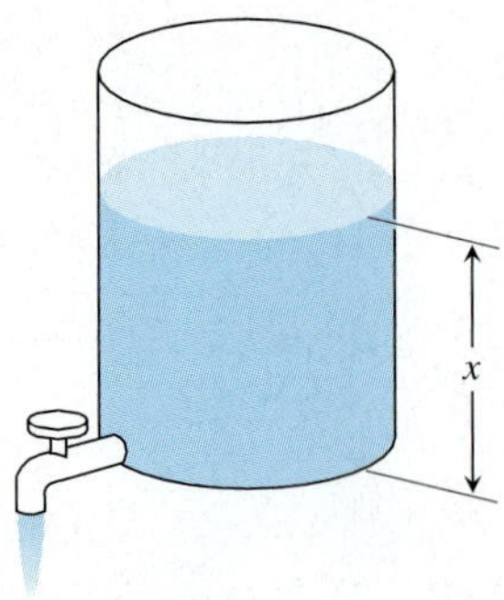

FIGURE 5.32 The rate at which water runs out is $0.5\sqrt{x}$ ft³/min. (Example 5)

Torricelli's Law

Torricelli's law says that if you drain a tank like the one in Figure 5.32, the rate at which the water runs out is a constant times the square root of the water's depth x. The constant depends on the size of the exit value. In Example 5, we assume that the constant is $1/2$.

Example 5 Draining a Tank

A right circular cylindrical tank with radius 5 ft and height 16 ft that was initially full of water is being drained at the rate of $0.5\sqrt{x}\ \text{ft}^3/\text{min}$. Find a formula for the depth and the amount of water in the tank at any time t. How long will it take the tank to empty?

Solution The volume of a right circular cylinder with radius r and height h is $V = \pi r^2 h$.

Model

The volume of water in the tank (Figure 5.32) is

$$V = \pi r^2 h = \pi(5)^2 x = 25\pi x.$$

Differential equation:

$$\frac{dV}{dt} = 25\pi \frac{dx}{dt}$$

$$-0.5\sqrt{x} = 25\pi \frac{dx}{dt} \qquad \text{Negative because } V \text{ is decreasing and } dx/dt < 0$$

$$\frac{dx}{dt} = -\frac{\sqrt{x}}{50\pi} \qquad \text{Torricelli's law}$$

Initial condition:

$$x(0) = 16 \qquad \text{The water is 16 ft deep when } t = 0.$$

Solve Analytically

We first solve the differential equation by separating the variables.

$$x^{-1/2}\,dx = -\frac{1}{50\pi}\,dt$$

$$\int x^{-1/2}\,dx = -\int \frac{1}{50\pi}\,dt \qquad \text{Integrate both sides.}$$

$$2x^{1/2} = -\frac{1}{50\pi}t + C \qquad \text{Constants combined}$$

The initial condition $x(0) = 16$ determines the value of C.

$$2(16)^{1/2} = -\frac{1}{50\pi}(0) + C$$

$$C = 8$$

With $C = 8$, we have

$$2x^{1/2} = -\frac{1}{50\pi}t + 8 \qquad \text{or} \qquad x^{1/2} = 4 - \frac{t}{100\pi}.$$

The formulas we seek are

$$x = \left(4 - \frac{t}{100\pi}\right)^2 \quad \text{and} \quad V = 25\pi x = 25\pi\left(4 - \frac{t}{100\pi}\right)^2.$$

Interpret

At any time t, the water in the tank is $(4 - t/(100\pi))^2$ ft deep and the amount of water is $25\pi(4 - t/(100\pi))^2$ ft^3. At $t = 0$, we have $x = 16$ ft and $V = 400\pi$ ft^3, as required. The tank will empty ($V = 0$) in $t = 400\pi$ minutes, which is about 21 h.

EXERCISES 5.4

Verifying Solutions

In Exercises 1 and 2, show that each function $y = f(x)$ is a solution of the accompanying differential equation.

1. $2y' + 3y = e^{-x}$

(a) $y = e^{-x}$

(b) $y = e^{-x} + e^{-(3/2)x}$

(c) $y = e^{-x} + Ce^{-(3/2)x}$

2. $y' = y^2$

(a) $y = -\frac{1}{x}$ **(b)** $y = -\frac{1}{x+3}$ **(c)** $y = -\frac{1}{x+C}$

In Exercises 3 and 4, show that each function is a solution of the given initial value problem.

3. Differential equation: $y' = e^{-x^2} - 2xy$

Initial condition: $y(2) = 0$

Solution candidate: $y = (x - 2)e^{-x^2}$

4. Differential equation: $xy' + y = -\sin x, \quad x > 0$

Initial condition: $y\left(\frac{\pi}{2}\right) = 0$

Solution candidate: $y = \frac{\cos x}{x}$

Separable Equations

Solve the differential equation in Exercises 5–14.

5. $2\sqrt{xy}\,\frac{dy}{dx} = 1, \quad x, y > 0$

6. $\frac{dy}{dx} = x^2\sqrt{y}, \quad y > 0$

7. $\frac{dy}{dx} = e^{x-y}$

8. $\frac{dy}{dx} = 3x^2e^{-y}$

9. $\frac{dy}{dx} = \sqrt{y}\cos^2\sqrt{y}$

10. $\sqrt{2xy}\,\frac{dy}{dx} = 1$

11. $\sqrt{x}\,\frac{dy}{dx} = e^{y+\sqrt{x}}, \quad x > 0$

12. $(\sec x)\frac{dy}{dx} = e^{y+\sin x}$

13. $\frac{dy}{dx} = 2x\sqrt{1 - y^2}, \quad -1 < y < 1$

14. $\frac{dy}{dx} = \frac{e^{2x-y}}{e^{x+y}}$

Applications

15. *Atmospheric pressure* Earth's atmospheric pressure p is often modeled by assuming that the rate dp/dh at which p changes with the altitude h above sea level is proportional to p. Suppose that the pressure at sea level is 1013 millibars (about 14.7 pounds per square inch) and that the pressure at an altitude of 20 km is 90 millibars.

(a) Solve the initial value problem:

Differential equation: $dp/dh = kp$ (k a constant)

Initial condition: $p = p_0$ when $h = 0$

to express p in terms of h. Determine the values of p_0 and k from the given altitude-pressure data.

(b) What is the atmospheric pressure at $h = 50$ km?

(c) At what altitude does the pressure equal 900 millibars?

16. *First-order chemical reactions* In some chemical reactions, the rate at which the amount of a substance changes with time is proportional to the amount present. For the change of δ-glucono lactone into gluconic acid, for example,

$$\frac{dy}{dt} = -0.6y$$

when t is measured in hours. If there are 100 g of δ-glucono lactone present when $t = 0$, how many grams will be left after the first hour?

17. *The inversion of sugar* The processing of raw sugar has a step called "inversion" that changes the sugar's molecular structure. Once the process has begun, the rate of change of the amount of raw sugar is proportional to the amount of raw sugar remaining. If 1000 kg of raw sugar reduces to 800 kg of raw sugar during the first 10 h, how much raw sugar will remain after another 14 h?

18. *Working underwater* The intensity $L(x)$ of light x ft beneath the surface of the ocean satisfies the differential equation

$$\frac{dL}{dx} = -kL.$$

As a diver, you know from experience that diving to 18 ft in the Caribbean Sea cuts the intensity in half. You cannot work without artificial light when the intensity falls below one-tenth of the surface value. About how deep can you expect to work without artificial light?

19. *Voltage in a discharging capacitor* Suppose that electricity is draining from a capacitor at a rate that is proportional to the voltage V across its terminals and that, if t is measured in seconds,

$$\frac{dV}{dt} = -\frac{1}{40}V.$$

Solve this equation for V, using V_0 to denote the value of V when $t = 0$. How long will it take the voltage to drop to 10% of its original value?

20. *Oil depletion* Suppose that the amount of oil pumped from one of the canyon wells in Whittier, California decreases at the continuous rate of 10% per year. When will the well's output fall to one-fifth of its present value?

CD-ROM WEBsite **21.** *Glucose fed intravenously* A hospital patient is fed glucose intravenously (directly to the bloodstream) at a rate of r units per minute. The body removes glucose from the bloodstream at a rate proportional to the amount $Q(t)$ present in the bloodstream at time t.

(a) Write a differential equation modeling the change of glucose in the bloodstream over time.

(b) Solve the differential equation in part (a) with the initial condition $Q(0) = Q_0$.

(c) Find the limit of $Q(t)$ as $t \to \infty$.

22. *Cooling soup* Suppose that a cup of soup cooled from 90°C to 60°C after 10 min in a room whose temperature was 20°C. Use Newton's Law of Cooling to answer the following questions.

(a) How much longer would it take the soup to cool to 35°C?

(b) Instead of being left to stand in the room, the cup of 90°C soup is put in a freezer whose temperature is −15°C. How long will it take the soup to cool from 90°C to 35°C?

23. *A beam of unknown temperature* An aluminum beam was brought from the outside cold into a machine shop where the temperature was held at 65°F. After 10 min, the beam warmed to 35°F, and after another 10 min, it was 50°F. Use Newton's Law of Cooling to estimate the beam's initial temperature.

24. *Surrounding medium of unknown temperature* A pan of warm water (46°C) was put in a refrigerator. Ten minutes later, the water's temperature was 39°C; 10 min after that, it was 33°C. Use Newton's Law of Cooling to estimate how cold the refrigerator was.

25. *Continuous price discounting* To encourage buyers to place 100-unit orders, your firm's sales department applies a continuous discount that makes the unit price a function $p(x)$ of the number of units x ordered. The discount decreases the price at the rate of \$0.01 per unit ordered. The price per unit for a 100-unit order is $p(100) = \$20.09$.

(a) Find $p(x)$ by solving the following initial value problem:

Differential equation: $\dfrac{dp}{dx} = -\dfrac{1}{100}p$

Initial condition: $p(100) = 20.09.$

(b) Find the unit price $p(10)$ for a 10-unit order and the unit price $p(90)$ for a 90-unit order.

(c) The sales department has asked you to find out if it is discounting so much that the firm's revenue, $r(x) = x \cdot p(x)$, will actually be less for a 100-unit order than, say, for a 90-unit order. Reassure them by showing that r has its maximum value at $x = 100$.

26. *World's population* The table gives estimates of the world population, in millions, over two centuries:

Year	1750	1800	1850	1900	1950
Population	728	906	1171	1608	2517

(a) Use the exponential model $dP/dt = kP$, $k > 0$, and the population figures for 1750 and 1800 to predict the world population in 1900 and 1950. Compare with the actual figures.

(b) Use the exponential model and the population figures for 1850 and 1900 to predict the world population in 1950. Compare with the actual population.

(c) *Writing to Learn* Use the exponential model and the population figures for 1900 and 1950 to predict the world population in 1999. Compare with the actual 1999 population of 6 billion and try to explain the discrepancy.

27. *Radioactive decay* Experiments have shown that the rate at which a radioactive element decays (as measured by the number of nuclei that change per unit time) is approximately proportional to the number $y(t)$ of radioactive nuclei present at time t. The constant of proportionality is called the **decay constant.** Write and solve a differential equation modeling radioactive decay assuming the initial condition $y(0) = y_0$.

28. *Half-life* (*Continuation of Exercise 27*) The **half-life** of a radioactive element is the time required for half of the radioactive nuclei present in a sample to decay (see Exercise 27).

(a) Show that the half-life of a radioactive element with decay constant k is

$$\text{Half-life} = \frac{\ln 2}{k}.$$

(b) *Polonium-210* The effective radioactive lifetime of polonium-210 is so short we measure it in days rather than years. Find the half-life of polonium-210 assuming that its (per day) decay constant is $k = 5 \times 10^{-3}$.

29. *Carbon-14 dating* People who do carbon-14 dating use a figure of 5700 years for its half-life (see Exercises 27 and 28). Find the age of a sample in which 10% of the radioactive nuclei originally present have decayed.

30. *The age of Crater Lake* The charcoal from a tree killed in the volcanic eruption that formed Crater Lake in Oregon contained 44.5% of the carbon-14 found in living matter. About how old is Crater Lake?

31. *The sensitivity of carbon-14 dating to measurement* To see the effect of a relatively small error in the estimate of the amount of carbon-14 in a sample being dated, consider this hypothetical situation.

(a) A fossilized bone found in central Illinois in the year A.D. 2000 contains 17% of its original carbon-14 content. Estimate the year the animal died.

(b) Repeat part (a) assuming 18% instead of 17%.

(c) Repeat part (a) assuming 16% instead of 17%.

32. *Art forgery* A painting attributed to Vermeer (1632–1675), which should contain no more than 96.2% of its original carbon-14, contains 99.5% instead. About how old is the forgery?

Resistance Proportional to Velocity

33. *A coasting cyclist* For a 160 lb cyclist on a 15 lb bicycle on level ground, the k in Equation (4) is about 1/5 slug/sec and $m = 160/32 = 5$ slugs. The cyclist starts coasting at 22 ft/sec (15 mph).

(a) About how far will the cyclist coast before reaching a complete stop?

(b) To the nearest second, about how long will it take the cyclist's speed to drop to 1 ft/sec?

34. *A coasting battleship* For a 56,000-ton Iowa class battleship, $m = 1{,}750{,}000$ slugs and the k in Equation (4) might be 3000 slugs/sec. Suppose that the battleship loses power when it is moving at a speed of 22 ft/sec (13.0 knots).

(a) About how far will the ship coast before it stops?

(b) About how long will it take the ship's speed to drop to 1 ft/sec?

COMPUTER EXPLORATIONS

Slope Fields and Solution Curves

In Exercises 35–40, obtain a slope field and graph the solution curves passing through the given points.

35. $y' = y$ with

(a) $(0, 1)$

(b) $(0, 2)$

(c) $(0, -1)$

36. $y' = 2(y - 4)$ with

(a) $(0, 1)$

(b) $(0, 4)$

(c) $(0, 5)$

37. $y' = y(2 - y)$ with

(a) $(0, 1/2)$

(b) $(0, 3/2)$

(c) $(0, 2)$

(d) $(0, 3)$

38. $y' = y^2$ with

(a) $(0, 1)$

(b) $(0, 2)$

(c) $(0, -1)$

(d) $(0, 0)$

39. $y' = \dfrac{3y}{x}$ with

(a) $(-3, 2)$

(b) $(1, 1)$

(c) $(2, 4)$

40. $y' = \dfrac{xy}{x^2 + 4}$ with

(a) $(0, 2)$

(b) $(0, -6)$

(c) $(-2\sqrt{3}, -4)$

5.5 Springs, Pumping and Lifting

Work Done by a Constant Force • Work Done by a Variable Force Along a Line • Hooke's Law for Springs: $F = kx$ • Pumping Liquids from Containers

Most dams are built with an overflow device known as a "glory hole" to provide an outlet for the water behind the dam when the water level exceeds a certain height. Unfortunately, the glory hole can get clogged by debris that can be cleared only after the water in the hole is pumped out. To acquire powerful enough pumps to do the job, it is necessary to find the amount of work required to pump the water from the hole.

In everyday life, *work* means an activity that requires muscular or mental effort. In science, the term refers specifically to a force acting on a body and the body's subsequent displacement. This section shows how to calculate work.

Work Done by a Constant Force

When a body moves a distance d along a straight line as a result of being acted on by a force of constant magnitude F in the direction of motion, we calculate the **work** W done by the force on the body with the formula

$$W = Fd \qquad \text{(constant-force formula for work).} \tag{1}$$

Right away, we can see a considerable difference between what we are used to calling work and what this formula says work is. If you push a car down the street, you will be doing work on the car, both by your own reckoning and by Equation (1). But if you push against the car and the car does not move, Equation (1) says you will do no work on the car, even if you push for an hour.

From Equation (1), we see that the unit of work in any system is the unit of force multiplied by the unit of distance. In SI units (SI stands for *Système International*, or International System), the unit of force is a newton, the unit of distance is a meter, and the unit of work is a newton-meter $(\mathrm{N} \cdot \mathrm{m})$. This combination appears so often it has a special name, the **joule.** In the British system, the unit of work is the foot-pound, a unit frequently used by engineers.

Joules

The joule, abbreviated **J** and pronounced "jewel," is named after the English physicist James Prescott Joule (1818–1889). The defining equation is

$$1 \text{ joule} = (1 \text{ newton})(1 \text{ meter}).$$

In symbols, $1\ \mathrm{J} = 1\ \mathrm{N} \cdot \mathrm{m}$.

It takes a force of about 1 N to lift an apple from a table. If you lift it 1 m, you have done about 1 J of work on the apple.

Example 1 Jacking Up a Car

If you jack up the side of a 2000 lb car 1.25 ft to change a tire (you have to apply a constant vertical force of about 1000 lb), you will perform $1000 \times 1.25 = 1250\ \mathrm{ft} \cdot \mathrm{lb}$ of work on the car. In SI units, you have applied a force of 4448 N through a distance of 0.381 m to do $4448 \times 0.381 \approx 1695$ J of work.

Work Done by a Variable Force Along a Line

If the force you apply varies along the way, as it will if you are lifting a leaking bucket or compressing a spring, the formula $W = Fd$ has to be replaced by an integral formula that takes the variation in F into account.

Suppose that the force performing the work acts along a line that we can model with the x-axis and that its magnitude F is a continuous function of the position. We want to find the work done over the interval from $x = a$ to $x = b$. We partition $[a, b]$ in the usual way and choose an arbitrary point c_k in each subinterval $[x_{k-1}, x_k]$. If the subinterval is short enough, F, being continuous, will not vary much from x_{k-1} to x_k. The amount of work done across the interval will be about $F(c_k)$ times the distance Δx_k, the same as it would be if F were constant and we could apply Equation (1). The total work done from a to b is therefore approximated by the Riemann sum

$$\sum_{k=1}^{n} F(c_k)\,\Delta x_k .$$

We expect the approximation to improve as the norm of the partition goes to zero, so we define the work done by the force from a to b to be the integral of F from a to b.

Definition Work

The **work** done by a variable force $F(x)$ directed along the x-axis from $x = a$ to $x = b$ is

$$W = \int_a^b F(x)\,dx. \tag{2}$$

The units of the integral are joules if F is in newtons and x is in meters, and foot-pounds if F is in pounds and x in feet.

Example 2 Applying the Definition of Work

The work done by a force of $F(x) = 1/x^2$ N along the x-axis from $x = 1$ m to $x = 10$ m is

$$W = \int_1^{10} \frac{1}{x^2}\,dx = -\frac{1}{x}\bigg]_1^{10} = -\frac{1}{10} + 1 = 0.9\text{ J}.$$

Example 3 Lifting a Leaky Bucket

A leaky 5 lb bucket is lifted from the ground into the air by pulling in 20 ft of rope at a constant speed (Figure 5.33). The rope weighs 0.08 lb/ft. The bucket starts with 2 gal of water (16 lb) and leaks at a constant rate. It finishes draining just as it reaches the top. How much work was spent

(a) lifting the water alone

(b) lifting the water and bucket together

(c) lifting the water, bucket, and rope?

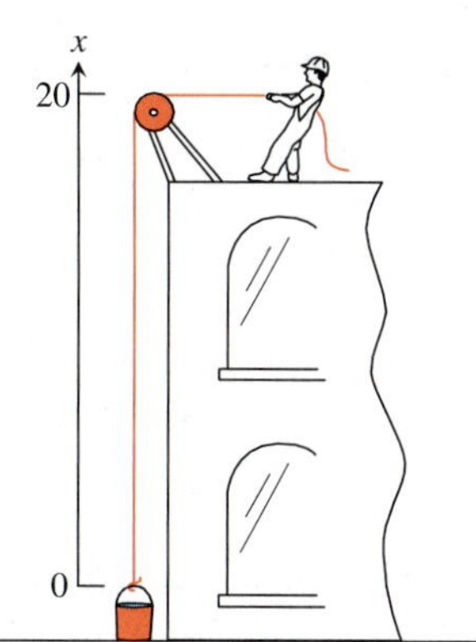

FIGURE 5.33 The leaky bucket in Example 3.

Solution

(a) *The water alone.* The force required to lift the water is equal to the water's weight, which varies steadily from 16 to 0 lb over the 20-ft lift. When the bucket is x ft off the ground, the water weighs

$$F(x) = \underbrace{16}_{\substack{\text{original weight}\\ \text{of water}}} \cdot \underbrace{((20 - x)/20)}_{\substack{\text{proportion left}\\ \text{at elevation } x}} = 16\left(1 - \frac{x}{20}\right) = 16 - \frac{4x}{5}\text{ lb}.$$

The work done is

$$\begin{aligned} W &= \int_a^b F(x)\,dx && \text{Use Eq. (2) for variable forces.} \\ &= \int_0^{20} \left(16 - \frac{4x}{5}\right) dx \\ &= \left[16x - \frac{2x^2}{5}\right]_0^{20} \\ &= 320 - 160 = 160\text{ ft}\cdot\text{lb} \end{aligned}$$

(b) *The water and bucket together.* According to Equation (1), it takes $5 \times 20 = 100$ ft · lb to lift a 5 lb weight 20 ft. Therefore,

$$160 + 100 = 260 \text{ ft} \cdot \text{lb}$$

of work were spent lifting the water and bucket together.

(c) *The water, bucket, and rope.* Now the total weight at level x is

$$F(x) = \underbrace{\left(16 - \frac{4x}{5}\right)}_{\substack{\text{variable weight}\\ \text{of water}}} + \underbrace{5}_{\substack{\text{constant weight}\\ \text{of bucket}}} + \underbrace{\overbrace{(0.08)}^{\text{lb/ft}}\overbrace{(20 - x)}^{\text{ft}}}_{\substack{\text{weight of rope}\\ \text{paid out at}\\ \text{elevation } x}}.$$

The work lifting the rope is

$$\begin{aligned}\text{Work on rope} &= \int_0^{20} (0.08)(20 - x)\, dx = \int_0^{20} (1.6 - 0.08x)\, dx \\ &= \Big[1.6x - 0.04x^2\Big]_0^{20} = 32 - 16 = 16 \text{ ft} \cdot \text{lb}.\end{aligned}$$

The total work for the water, bucket, and rope combined is

$$160 + 100 + 16 = 276 \text{ ft} \cdot \text{lb}.$$

Hooke's Law for Springs: F = kx

Hooke's law says that the force it takes to stretch or compress a spring x length units from its natural (unstressed) length is proportional to x. In symbols,

$$F = kx. \tag{3}$$

The constant k, measured in force units per unit length, is a characteristic of the spring, called the **force constant** (or spring constant) of the spring. Hooke's law Equation (3), gives good results as long as the force doesn't distort the metal in the spring. We assume that the forces in this section are too small to do that.

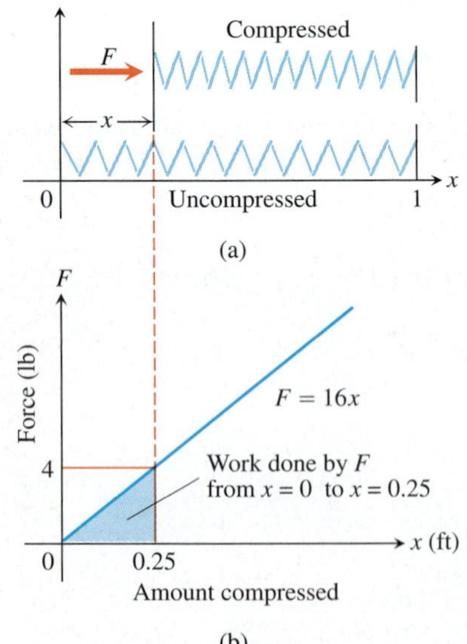

FIGURE 5.34 The force F needed to hold a spring under compression increases linearly as the spring is compressed.

Example 4 Compressing a Spring

Find the work required to compress a spring from its natural length of 1 ft to a length of 0.75 ft if the force constant is $k = 16$ lb/ft.

Solution We picture the uncompressed spring laid out along the x-axis with its movable end at the origin and its fixed end at $x = 1$ ft (Figure 5.34). This enables us to describe the force required to compress the spring from 0 to x with the formula $F = 16x$. To compress the spring from 0 to 0.25 ft, the force must increase from

$$F(0) = 16 \cdot 0 = 0 \text{ lb} \quad \text{to} \quad F(0.25) = 16 \cdot 0.25 = 4 \text{ lb}.$$

The work done by F over this interval is

$$W = \int_0^{0.25} 16x\, dx = 8x^2 \Big]_0^{0.25} = 0.5 \text{ ft} \cdot \text{lb}.$$

Eq. (2) with $a = 0$, $b = 0.25$, $F(x) = 16x$

Example 5 Stretching a Spring

A spring has a natural length of 1 m. A force of 24 N stretches the spring to a length of 1.8 m.

(a) Find the force constant k.

(b) How much work will it take to stretch the spring 2 m beyond its natural length?

(c) How far will a 45 N force stretch the spring?

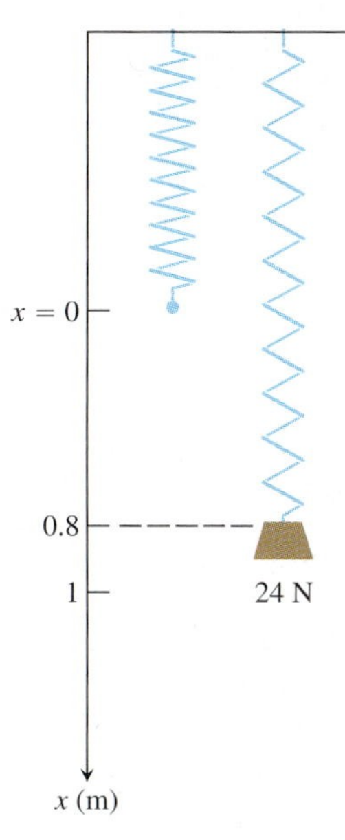

FIGURE 5.35 A 24 N weight stretches this spring 0.8 m beyond its unstressed length.

Solution

(a) *The force constant.* We find the force constant from Equation (3). A force of 24 N stretches the spring 0.8 m, so

$$24 = k(0.8) \qquad \text{Eq. (3) with } F = 24,\ x = 0.8$$

$$k = 24/0.8 = 30 \text{ N/m}.$$

(b) *The work to stretch the spring* 2 m. We imagine the unstressed spring hanging along the x-axis with its free end at $x = 0$ (Figure 5.35). The force required to stretch the spring x m beyond its natural length is the force required to pull the free end of the spring x units from the origin. Hooke's law with $k = 30$ says that this force is

$$F(x) = 30x.$$

The work done by F on the spring from $x = 0$ m to $x = 2$ m is

$$W = \int_0^2 30x\,dx = 15x^2\Big]_0^2 = 60 \text{ J}.$$

(c) *How far will a 45 N force stretch the spring?* We substitute $F = 45$ in the equation $F = 30x$ to find

$$45 = 30x, \qquad \text{or} \qquad x = 1.5 \text{ m}.$$

A 45 N force will stretch the spring 1.5 m. No calculus is required to find this.

Pumping Liquids from Containers

How much work does it take to pump all or part of the liquid from a container? To find out, we imagine lifting the liquid out one thin horizontal slab at a time and applying the equation $W = Fd$ to each slab. We then evaluate the integral this leads to as the slabs become thinner and more numerous. The integral we get each time depends on the weight of the liquid and the dimensions of the container, but the way we find the integral is always the same. The next examples show what to do.

Example 6 Pumping Water from a Cylindrical Tank

How much work does it take to pump the water from a full upright circular cylindrical tank of radius 5 m and height 10 m to a level of 4 m above the top of the tank?

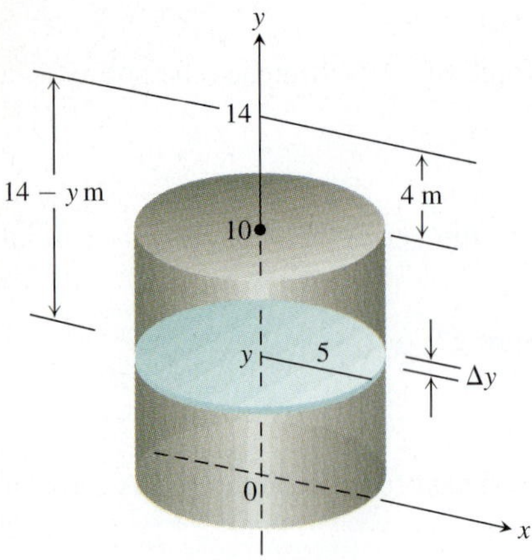

FIGURE 5.36 To find the work it takes to pump the water from a tank, think of lifting the water one thin slab at a time.

Solution We draw the tank (Figure 5.36), add coordinate axes, and imagine the water divided into thin horizontal slabs by planes perpendicular to the y-axis at the points of a partition P of the interval $[0, 10]$.

The typical slab between the planes at y and $y + \Delta y$ has a volume of

$$\Delta V = \pi(\text{radius})^2(\text{thickness}) = \pi(5)^2\,\Delta y = 25\pi\,\Delta y\ \text{m}^3.$$

The force $F(y)$ required to lift the slab is equal to its weight,

$$\begin{aligned} F(y) &= 9800\,\Delta V \qquad \text{Water weighs } 9800\ \text{N/m}^3. \\ &= 9800(25\pi\,\Delta y) = 245{,}000\pi\,\Delta y\ \text{N}. \end{aligned}$$

The distance through which F must act is about $(14 - y)$ m, so the work done lifting the slab is about

$$\Delta W = \text{force} \times \text{distance} = 245{,}000\pi(14 - y)\,\Delta y\ \text{J}.$$

The work it takes to lift all the water is approximately

$$W \approx \sum_{0}^{10} \Delta W = \sum_{0}^{10} 245{,}000\pi(14 - y)\,\Delta y\ \text{J}.$$

This is a Riemann sum for the function $245{,}000\pi(14 - y)$ over the interval $0 \le y \le 10$. The work of pumping the tank dry is the limit of these sums as $\|P\| \to 0$:

$$\begin{aligned} W &= \int_0^{10} 245{,}000\pi(14 - y)\,dy = 245{,}000\pi \int_0^{10} (14 - y)\,dy \\ &= 245{,}000\pi\left[14y - \frac{y^2}{2}\right]_0^{10} = 245{,}000\pi[90] \\ &\approx 69{,}272{,}118 \approx 69.3 \times 10^6\ \text{J}. \end{aligned}$$

A 1-horsepower output motor rated at 746 J/sec could empty the tank in a little less than 26 h.

Example 7 Pumping Oil from a Conical Tank

The conical tank in Figure 5.37 is filled to within 2 ft of the top with olive oil weighing 57 lb/ft³. How much work does it take to pump the oil to the rim of the tank?

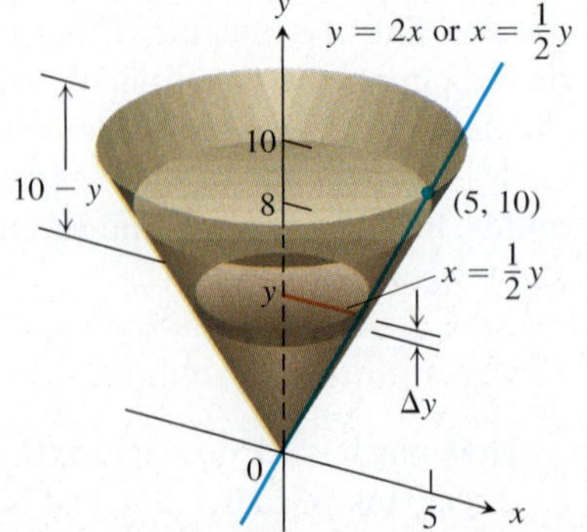

FIGURE 5.37 The olive oil in Example 7.

Solution We imagine the oil divided into thin slabs by planes perpendicular to the y-axis at the points of a partition of the interval $[0, 8]$.

The typical slab between the planes at y and $y + \Delta y$ has a volume of about

$$\Delta V = \pi(\text{radians})^2(\text{thickness}) = \pi\left(\frac{1}{2}y\right)^2 \Delta y = \frac{\pi}{4}y^2\,\Delta y \text{ ft}^3.$$

The force $F(y)$ required to lift this slab is equal to its weight,

$$F(y) = 57\,\Delta V = \frac{57\pi}{4}y^2\,\Delta y \text{ lb}$$

Weight = weight per unit volume × volume

The distance through which $F(y)$ must act to lift this slab to the level of the rim of the cone is about $(10 - y)$ ft, so the work done lifting the slab is about

$$\Delta W = \frac{57\pi}{4}(10 - y)y^2\,\Delta y \text{ ft} \cdot \text{lb}.$$

The work done lifting all the slabs from $y = 0$ to $y = 8$ to the rim is approximately

$$W \approx \sum_{0}^{8} \frac{57\pi}{4}(10 - y)y^2\,\Delta y \text{ ft} \cdot \text{lb}.$$

This is a Riemann sum for the function $(57\pi/4)(10 - y)y^2$ on the interval from $y = 0$ to $y = 8$. The work of pumping the oil to the rim is the limit of these sums as the norm of the partition goes to zero.

$$\begin{aligned} W &= \int_0^8 \frac{57\pi}{4}(10 - y)y^2\,dy \\ &= \frac{57\pi}{4}\int_0^8 (10y^2 - y^3)\,dy \\ &= \frac{57\pi}{4}\left[\frac{10y^3}{3} - \frac{y^4}{4}\right]_0^8 \approx 30{,}561 \text{ ft} \cdot \text{lb} \end{aligned}$$

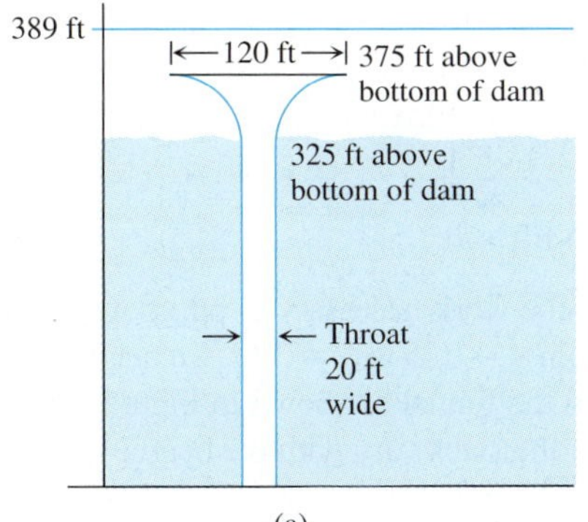

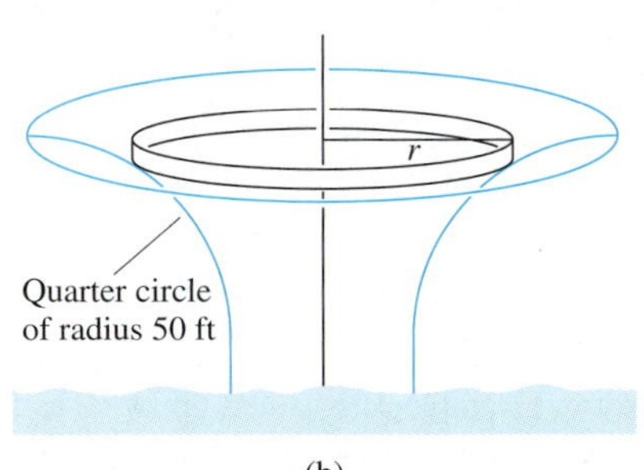

FIGURE 5.38 (a) Cross section of the glory hole for a dam and (b) the top of the glory hole.

Example 8 Pumping Water from a Glory Hole

A glory hole is a vertical drain pipe that keeps the water behind a dam from getting too high. The top of the glory hole for a dam is 14 ft below the top of the dam and 375 ft above the bottom (Figure 5.38). The hole needs to be pumped out from time to time to permit the removal of seasonal debris as mentioned in the opening remarks to this section.

From the cross section in Figure 5.38a, we see that the glory hole is a funnel-shaped drain. The throat of the funnel is 20 ft wide and the head is 120 ft across. The outside boundary of the head cross section are quarter circles formed with 50 ft radii, shown in Figure 5.38b. The glory hole is formed by rotating a cross section around its center. Consequently, all horizontal cross sections are circular disks throughout the entire glory hole. We calculate the work required to pump water from

(a) the throat of the hole

(b) the funnel portion.

Solution

(a) *Pumping from the throat.* A typical slab in the throat between the planes at y and $y + \Delta y$ has a volume of about

$$\Delta V = \pi(\text{radius})^2(\text{thickness}) = \pi(10)^2\,\Delta y \ \text{ft}^3.$$

The force $F(y)$ required to lift this slab is equal to its weight (about 62.4 lbs/ft^3 for water),

$$F(y) = 62.4\,\Delta V = 6{,}240\pi\,\Delta y \ \text{lb}.$$

The distance through which $F(y)$ must act to lift this slab to the top of the hole is $(375 - y)$ ft, so the work done lifting the slab is

$$\Delta W = 6240\pi(375 - y)\,\Delta y \ \text{ft}\cdot\text{lb}.$$

The work done in pumping the water from the throat is the integral of a differential slab element from $y = 0$ to $y = 325$,

$$\begin{aligned} W &= \int_0^{325} 6240\pi(375 - y)\,dy \\ &= 6240\pi\left[375y - \frac{y^2}{2}\right]_0^{325} \\ &\approx 1{,}353{,}869{,}354 \ \text{ft}\cdot\text{lb} \end{aligned}$$

(b) *Pumping from the funnel.* To compute the work necessary to pump water from the funnel portion of the glory hole, from $y = 325$ to $y = 375$, we need to compute ΔV for approximating elements in the funnel as shown in Figure 5.39. As can be seen from the figure, the radii of the slabs vary with the height y.

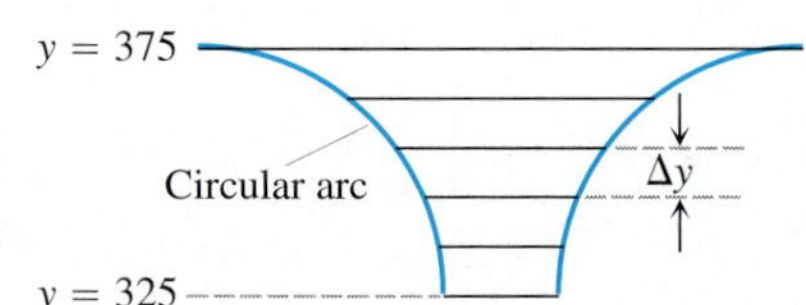

FIGURE 5.39 The glory hole funnel portion.

In Exercises 33 and 34, you are asked to complete the analysis to determine the total work required to pump the water and power of the pumps necessary to pump out the glory hole.

EXERCISES 5.5

Work Done by a Variable Force

1. *Leaky bucket* The workers in Example 3 changed to a larger bucket that held 5 gal (40 lb) of water, but the new bucket had an even larger leak so that it, too, was empty by the time it reached the top. Assuming that the water leaked out at a steady rate, how much work was done lifting the water? (Do not include the rope and bucket.)

2. *Leaky bucket* The bucket in Example 3 is hauled up twice as fast so that there is still 1 gal (8 lb) of water left when the bucket reaches the top. How much work is done lifting the water this time? (Do not include the rope and bucket.)

3. *Lifting a rope* A mountain climber is about to haul up a 50 m length of hanging rope. How much work will it take if the rope weighs 0.624 N/m?

4. *Leaky sand bag* A bag of sand originally weighing 144 lb was lifted at a constant rate. As it rose, sand also leaked out at a constant rate. The sand was half gone by the time the bag had been lifted 18 ft. How much work was done lifting the sand this far? (Neglect the weight of the bag and lifting equipment.)

5. *Lifting an elevator cable* An electric elevator with a motor at the top has a multistrand cable weighing 4.5 lb/ft. When the car is at the first floor, 180 ft of cable are paid out, and effectively 0 ft are out when the car is at the top floor. How much work does the motor do just lifting the cable when it takes the car from the first floor to the top?

6. *Force of attraction* When a particle of mass m is at $(x, 0)$, it is attracted toward the origin with a force whose magnitude is k/x^2. If the particle starts from rest at $x = b$ and is acted on by no other forces, find the work done on it by the time it reaches $x = a$, $0 < a < b$.

7. *Compressing gas* Suppose that the gas in a circular cylinder of cross-section area A is being compressed by a piston. If p is the pressure of the gas in pounds per square inch and V is the volume in cubic inches, show that the work done in compressing the gas from state (p_1, V_1) to state (p_2, V_2) is given by the equation

$$\text{Work} = \int_{(p_1,V_1)}^{(p_2,V_2)} p\,dV.$$

(*Hint:* In the coordinates suggested in the figure here, $dV = A\,dx$. The force against the piston is pA.)

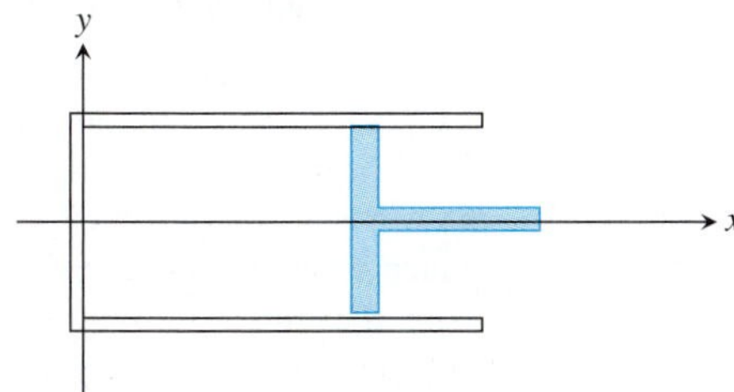

8. (*Continuation of Exercise 7*) Use the integral in Exercise 7 to find the work done in compressing the gas from $V_1 = 243$ in.^{3} to $V_2 = 32$ in.^{3} if $p_1 = 50$ lb/in^3 and p and V obey the gas law $pV^{1.4} =$ constant (for adiabatic processes).

Springs

9. *Spring constant* It took 1800 J of work to stretch a spring from its natural length of 2 m to a length of 5 m. Find the spring's force constant.

10. *Stretching a spring* A spring has a natural length of 10 in. An 800 lb force stretches the spring to 14 in.

    (a) Find the force constant.

    (b) How much work is done in stretching the spring from 10 in. to 12 in.?

    (c) How far beyond its natural length will a 1600 lb force stretch the spring?

11. *Stretching a rubber band* A force of 2 N will stretch a rubber band 2 cm (0.02 m). Assuming Hooke's law applies, how far will a 4 N force stretch the rubber band? How much work does it take to stretch the rubber band this far?

12. *Stretching a spring* If a force of 90 N stretches a spring 1 m beyond its natural length, how much work does it take to stretch the spring 5 m beyond its natural length?

13. *Subway car springs* It takes a force of 21,714 lb to compress a coil spring assembly on a New York City Transit Authority subway car from its free height of 8 in. to its fully compressed height of 5 in.

    (a) What is the assembly's force constant?

    (b) How much work does it take to compress the assembly the first half inch? the second half inch? Answer to the nearest in. · lb.

    (Data courtesy of Bombardier, Inc., Mass Transit Division, for spring assemblies in subway cars delivered to the New York City Transit Authority from 1985 to 1987.)

14. *Bathroom scale* A bathroom scale is compressed 1/16 in. when a 150 lb person stands on it. Assuming that the scale behaves like a spring that obeys Hooke's law, how much does someone who compresses the scale 1/8 in. weigh? How much work is done compressing the scale 1/8 in.?

Pumping Liquids from Containers

The Weight of Water

Because of Earth's rotation and variations in its gravitational field, the weight of a cubic foot of water at sea level can vary from about 62.26 lb at the equator to as much as 62.59 lb near the poles, a variation of about 0.5%. A cubic foot that weighs about 62.4 lb in Melbourne and New York City will weigh 62.5 lb in Juneau and Stockholm. Although 62.4 is a typical figure and a common textbook value, there is considerable variation.

15. *Pumping water* The rectangular tank shown in the accompanying figure, with its top at ground level, is used to catch runoff water. Assume that the water weighs 62.4 lb/ft^3.

    (a) How much work does it take to empty the tank by pumping the water back to ground level once the tank is full?

    (b) If the water is pumped to ground level with a (5/11)-horsepower (hp) motor (work output 250 ft · lb/sec), how long will it take to empty the full tank (to the nearest minute)?

    (c) Show that the pump in part (b) will lower the water level 10 ft (halfway) during the first 25 min of pumping.

(d) *The weight of water* What are the answers to parts (a) and (b) in a location where water weighs 62.26 lb/ft^3? 62.59 lb/ft^3?

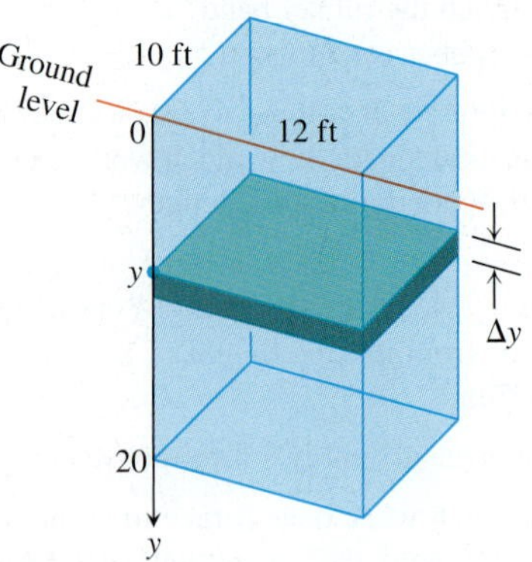

16. *Emptying a cistern* The rectangular cistern (storage tank for rainwater) shown below has its top 10 ft below ground level. The cistern, currently full, is to be emptied for inspection by pumping its contents to ground level.

(a) How much work will it take to empty the cistern?

(b) How long will it take a 1/2 hp pump, rated at 275 ft · lb/sec, to pump the tank dry?

(c) How long will it take the pump in part (b) to empty the tank halfway? (It will be less than half the time required to empty the tank completely.)

(d) *The weight of water* What are the answers to parts (a) through (c) in a location where water weighs 62.26 lb/ft^3? 62.59 lb/ft^3?

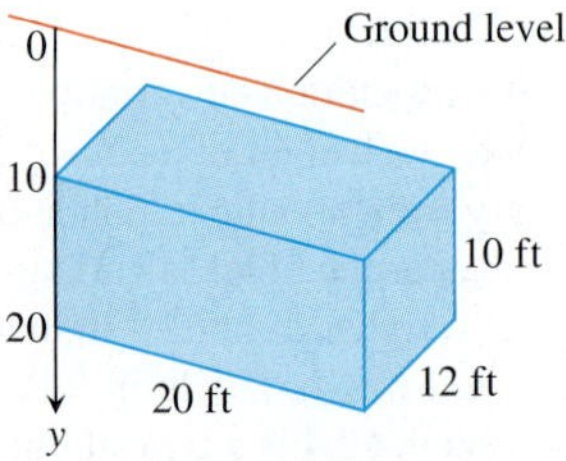

17. *Pumping water* How much work would it take to pump the water from the tank in Example 6 to the level of the top of the tank (instead of 4 m higher)?

18. *Pumping a half-full tank* Suppose that, instead of being full, the tank in Example 6 is only half full. How much work does it take to pump the remaining water to a level 4 m above the top of the tank?

19. *Emptying a tank* A vertical right circular cylindrical tank measures 30 ft high and 20 ft in diameter. It is full of kerosene weighing 51.2 lb/ft^3. How much work does it take to pump the kerosene to the level of the top of the tank?

20. *Writing to Learn* The cylindrical tank shown here can be filled by pumping water from a lake 15 ft below the bottom of the tank. There are two ways to go about it. One is to pump the water through a hose attached to a valve in the bottom of the tank. The other is to attach the hose to the rim of the tank and let the water pour in. Which way will be faster? Give reasons for your answer.

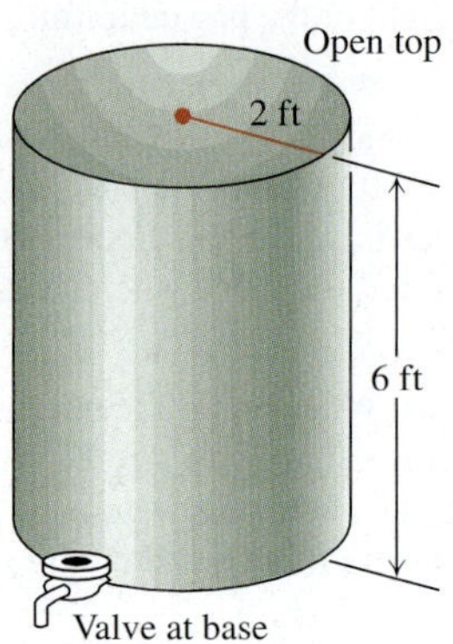

21. (a) *Pumping milk* Suppose that the conical container in Example 7 contains milk (weighing 64.5 lb/ft^3) instead of olive oil. How much work will it take to pump the contents to the rim?

(b) *Pumping oil* How much work will it take to pump the oil in Example 7 to a level 3 ft above the cone's rim?

22. *Pumping seawater* To design the interior surface of a huge stainless-steel tank, you revolve the curve $y = x^2$, $0 \le x \le 4$, about the y-axis. The container, with dimensions in meters, is to be filled with seawater, which weighs $10{,}000 \text{ N/m}^3$. How much work will it take to empty the tank by pumping the water to the tank's top?

23. *Emptying a water reservoir* We model pumping from spherical containers the way we do from other containers, with the axis of integration along the vertical axis of the sphere. Use the figure here to find how much work it takes to empty a full hemispherical water reservoir of radius 5 m by pumping the water to a height of 4 m above the top of the reservoir. Water weighs 9800 N/m^3.

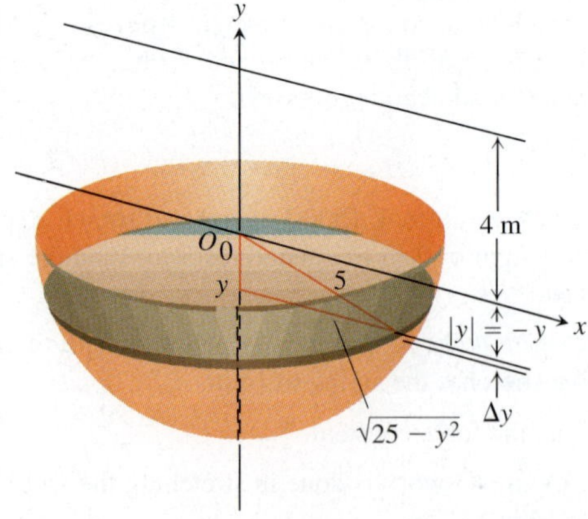

24. *Writing to Learn* You are in charge of the evacuation and repair of the storage tank shown here. The tank is a hemisphere of radius

10 ft and is full of benzene weighing 56 lb/ft^3. A firm you contacted says it can empty the tank for 1/2¢ per foot-pound of work. Find the work required to empty the tank by pumping the benzene to an outlet 2 ft above the top of the tank. If you have \$5000 budgeted for the job, can you afford to hire the firm?

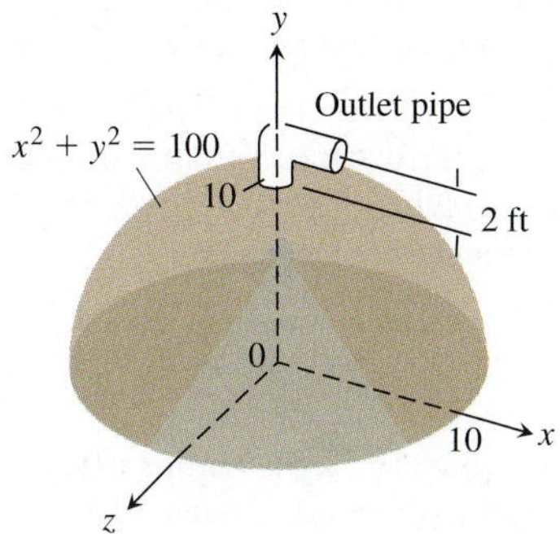

Work and Kinetic Energy

25. *Kinetic energy* If a variable force of magnitude $F(x)$ moves a body of mass m along the x-axis from x_1 to x_2, the body's velocity v can be written as dx/dt (where t represents time). Use Newton's Second Law of Motion $F = m(dv/dt)$ and the Chain Rule

$$\frac{dv}{dt} = \frac{dv}{dx}\frac{dx}{dt} = v\frac{dv}{dx}$$

to show that the net work done by the force in moving the body from x_1 to x_2 is

$$W = \int_{x_1}^{x_2} F(x)\,dx = \frac{1}{2}mv_2{}^2 - \frac{1}{2}mv_1{}^2,$$

where v_1 and v_2 are the body's velocities at x_1 and x_2. In physics, the expression $(1/2)mv^2$ is called the *kinetic energy* of the body moving with velocity v. Therefore, *the work done by the force equals the change in the body's kinetic energy*, and we can find the work by calculating this change.

In Exercises 26–32, use the result of Exercise 25.

26. *Tennis* A 2 oz tennis ball was served at 160 ft/sec (about 109 mph). How much work was done on the ball to make it go this fast? (To find the ball's mass from its weight, express the weight in pounds and divide by 32 ft/sec^2, the acceleration of gravity.)

27. *Baseball* How many foot-pounds of work does it take to throw a baseball 90 mph? A baseball weighs 5 oz, or 0.3125 lb.

28. *Golf* A 1.6 oz golf ball is driven off the tee at a speed of 280 ft/sec (about 191 mph). How many foot-pounds of work are done on the ball getting it into the air?

29. *Tennis* During the match in which Pete Sampras won the 1990 U.S. Open men's tennis championship, Sampras hit a serve that was clocked at a phenomenal 124 mph. How much work did Sampras have to do on the 2 oz ball to get it to that speed?

30. *Football* A quarterback threw a 14.5 oz football 88 ft/sec (60 mph). How many foot-pounds of work were done on the ball to get it to this speed?

31. *Softball* How much work has to be performed on a 6.5 oz softball to pitch it 132 ft/sec (90 mph)?

32. *A ball bearing* A 2 oz steel ball bearing is placed on a vertical spring whose force constant is $k = 18$ lb/ft. The spring is compressed 2 in. and released. About how high does the ball bearing go?

33. *(Continuation of Example 8) Pumping the funnel of the glory hole*

(a) Find the radius of the cross section (funnel portion) of the glory hole in Example 8 as a function of the height y above the floor of the dam (from $y = 325$ to $y = 375$).

(b) Find ΔV for the funnel section of the glory hole (from $y = 325$ to $y = 375$).

(c) Find the work necessary to pump out the funnel section by formulating and evaluating the appropriate definite integral.

34. *(Continuation of Exercise 33) Pumping water from a glory hole*

(a) Find the total work necessary to pump out the glory hole, by adding the work necessary to pump both the throat and funnel sections.

(b) Your answer to part (a) is in foot-pounds. A more useful form is horsepower-hours, since motors are rated in horsepower. To convert from foot-pounds to horsepower-hours, divide by 1.98×10^6. How many hours would it take a 1000 horsepower motor to pump out the glory hole, assuming that the motor was fully efficient?

Weight versus Mass

Weight is the force that results from gravity pulling on a mass. The two are related by the equation in Newton's second law,

$$\text{Weight} = \text{mass} \times \text{acceleration}.$$

Thus,

$$\text{Newtons} = \text{kilograms} \times \text{m/sec}^2,$$
$$\text{Pounds} = \text{slugs} \times \text{ft/sec}^2.$$

To convert mass to weight, multiply by the acceleration of gravity. To convert weight to mass, divide by the acceleration of gravity.

35. *Drinking a milkshake* The truncated conical container shown in the accompanying figure is full of strawberry milkshake that weighs 4/9 oz/in.^{3}. As you can see, the container is 7 in. deep, 2.5 in. across at the base, and 3.5 in. across at the top (a standard container size). The straw sticks up an inch above the top. About how much work does it take to suck up the milkshake through the straw (neglecting friction)? Answer in inch-ounces.

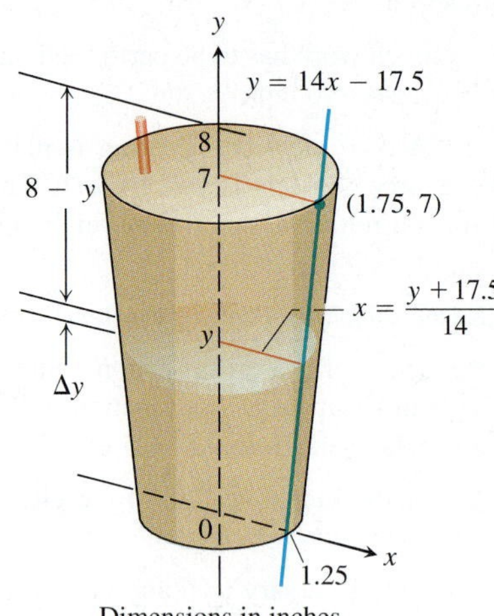

Dimensions in inches

36. *Water tower* Your town has decided to drill a well to increase its water supply. As the town engineer, you have determined that a water tower will be necessary to provide the pressure needed for distribution, and you have designed the system shown here. The water is to be pumped from a 300 ft well through a vertical 4 in. pipe into the base of a cylindrical tank 20 ft in diameter and 25 ft high. The base of the tank will be 60 ft above ground. The pump is a 3 hp pump, rated at 1650 ft · lb/sec. To the nearest hour, how long will it take to fill the tank the first time? (Include the time it takes to fill the pipe.) Assume that water weighs 62.4 lb/ft^3.

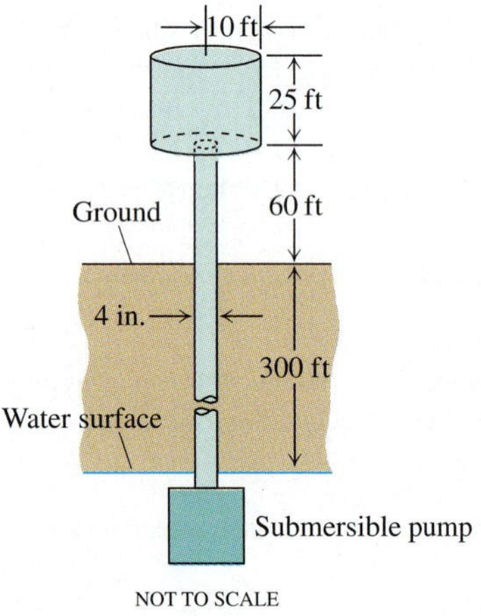

37. *Putting a satellite in orbit* The strength of Earth's gravitational field varies with the distance r from Earth's center, and the magnitude of the gravitational force experienced by a satellite of mass m during and after launch is

$$F(r) = \frac{mMG}{r^2}.$$

Here, $M = 5.975 \times 10^{24}$ kg is Earth's mass, $G = 6.6720 \times 10^{-11}$ N · m^{2}kg^{-2} is the universal gravitational constant, and r is measured in meters. The work it takes to lift a 1000 kg satellite from Earth's surface to a circular orbit 35,780 km above Earth's center is therefore given by the integral

$$\text{Work} = \int_{6,370,000}^{35,780,000} \frac{1000MG}{r^2}\,dr \text{ joules}.$$

Evaluate the integral. The lower limit of integration is Earth's radius in meters at the launch site. (This calculation does not take into account energy spent lifting the launch vehicle or energy spent bringing the satellite to orbit velocity.)

38. *Forcing electrons together* Two electrons r meters apart repel each other with a force of

$$F = \frac{23 \times 10^{-29}}{r^2} \text{ newtons}.$$

(a) Suppose that one electron is held fixed at the point (1, 0) on the x-axis (units in meters). How much work does it take to move a second electron along the x-axis from the point $(-1, 0)$ to the origin?

(b) Suppose that an electron is held fixed at each of the points $(-1, 0)$ and $(1, 0)$. How much work does it take to move a third electron along the x-axis from (5, 0) to (3, 0)?

5.6 Fluid Forces

The Constant-Depth Formula for Fluid Force • The Variable-Depth Formula

Engineers design dams to be thicker at the bottom than at the top (Figure 5.40) because the pressure against them increases with the depth. To generate hydroelectric power, gates (called *penstocks*) located near the base of the dam are opened to allow water under tremendous pressures to flow to turbine generators. Engineers must compute the total force against these gates at various depths to design the gates

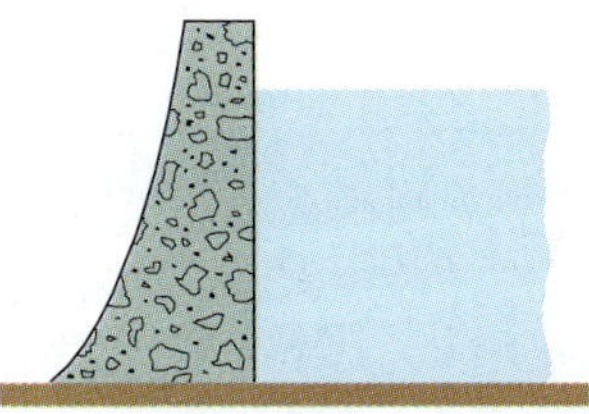

FIGURE 5.40 To withstand the increasing pressure, dams are built thicker as they go down.

Weight-Density

A fluid's weight-density is its weight per unit volume. Typical values (in pounds per cubic feet) are

Gasoline	42
Mercury	849
Milk	64.5
Molasses	100
Olive oil	57
Seawater	64
Water	62.4

themselves and the hydraulic systems that open and close them. It is remarkable that the pressure at any point on a dam depends only on how far below the surface the point is and not on how much the surface of the dam happens to be tilted at that point. The pressure in pounds per square foot at a point h ft below the surface is always $62.4h$. The number 62.4 is the weight-density of water in pounds per cubic foot.

The formula, pressure $= 62.4h$, makes sense when you think of the units involved:

$$\frac{\text{lb}}{\text{ft}^2} = \frac{\text{lb}}{\text{ft}^3} \times \text{ft}.$$

As you can see, this equation depends only on units and not on the fluid involved. The pressure h feet below the surface of any fluid is the fluid's weight-density times h.

The Pressure-Depth Equation

In a fluid that is standing still, the pressure p at depth h is the fluid's weight-density w times h:

$$p = wh. \qquad (1)$$

In this section, we use the equation $p = wh$ to derive a formula for the total force exerted by a fluid against all or part of a vertical or horizontal containing wall.

The Constant-Depth Formula for Fluid Force

In a container of fluid with a flat horizontal base, the total force exerted by the fluid against the base can be calculated by multiplying the area of the base by the pressure at the base. We can do this because total force equals force per unit area (pressure) times area. (See Figure 5.41.) If F, p, and A are the total force, pressure, and area, then

$$\begin{aligned} F &= \text{total force} = \text{force per unit area} \times \text{area} \\ &= \text{pressure} \times \text{area} = pA \\ &= whA. \qquad && p = wh \text{ from Eq. (1)} \end{aligned}$$

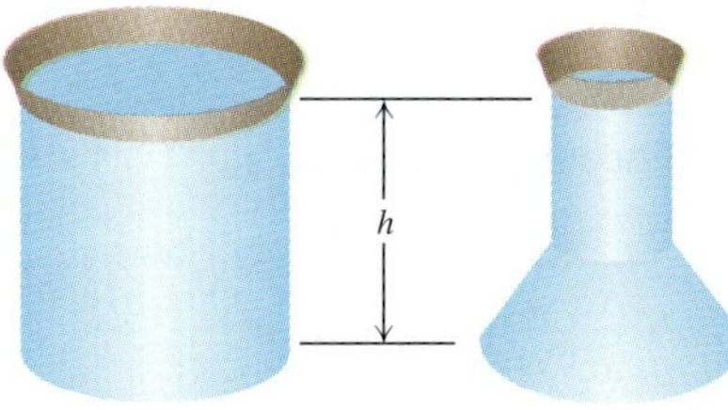

FIGURE 5.41 These containers are filled with water to the same depth and have the same base area. The total force is therefore the same on the bottom of each container. The containers' shapes do not matter here.

Fluid Force on a Constant-Depth Surface

$$F = pA = whA \qquad (2)$$

Example 1 The Great Molasses Flood of 1919

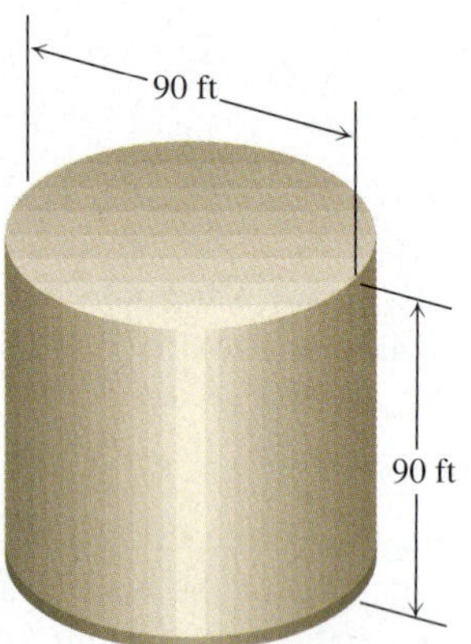

FIGURE 5.42 Schematic drawing of the molasses tank in Example 1.

At 1:00 P.M. on January 15, 1919 (an unseasonably warm day), a 90-ft-high, 90-ft-diameter cylindrical metal tank in which the Puritan Distilling Company stored molasses at the corner of Foster and Commercial streets in Boston's North End exploded. Molasses flooded the streets 30 ft deep, trapping pedestrians and horses, knocking down buildings, and oozing into homes. It was eventually tracked all over town and even made its way into the suburbs via trolley cars and people's shoes. It took weeks to clean up. Given that the tank was full of molasses weighing 100 lb/ft^3, what was the total force exerted by the molasses on the bottom of the tank at the time it ruptured (Figure 5.42)?

Solution At the bottom of the tank, the molasses exerted a constant pressure of

$$p = wh = \left(100 \, \frac{\text{lb}}{\text{ft}^3}\right)(90 \text{ ft}) = 9000 \, \frac{\text{lb}}{\text{ft}^2}.$$

Since the area of the base was $\pi(45)^2$, the total force on the base was

$$\left(9000 \, \frac{\text{lb}}{\text{ft}^2}\right)(2025 \, \pi \text{ ft}^2) \approx 57{,}255{,}526 \text{ lb}.$$

For a flat plate submerged *horizontally,* like the bottom of the molasses tank in Example 1, the downward force acting on its upper face due to liquid pressure is given by Equation (2). If the plate is submerged *vertically,* however, then the pressure against it will be different at different depths and Equation (2) no longer is usable in that form (because h varies). By dividing the plate into many narrow horizontal bands or strips, we can create a Riemann sum whose limit is the fluid force against the side of the submerged vertical plate. Here is the procedure.

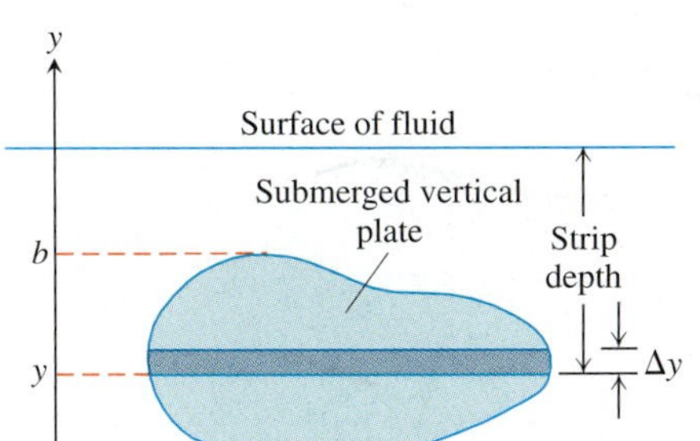

FIGURE 5.43 The force exerted by a fluid against one side of a thin horizontal strip is about ΔF = pressure × area = w × (strip depth) × $L(y)\,\Delta y$.

The Variable-Depth Formula

Suppose we want to know the force exerted by a fluid against one side of a vertical plate submerged in a fluid of weight-density w. To find it, we model the plate as a region extending from $y = a$ to $y = b$ in the xy-plane (Figure 5.43). We partition $[a, b]$ in the usual way and imagine the region to be cut into thin horizontal strips by planes perpendicular to the y-axis at the partition points. The typical strip from y to $y + \Delta y$ is Δy units wide by $L(y)$ units long. We assume $L(y)$ to be a continuous function of y.

The pressure varies across the strip from top to bottom. If the strip is narrow enough, however, the pressure will remain close to its bottom-edge value of w × (strip depth). The force exerted by the fluid against one side of the strip will be about

$$\begin{aligned}\Delta F &= (\text{pressure along bottom edge}) \times (\text{area}) \\ &= w \times (\text{strip depth}) \times L(y)\,\Delta y.\end{aligned}$$

The force against the entire plate will be about

$$\sum_a^b \Delta F = \sum_a^b (w \times (\text{strip depth}) \times L(y)\,\Delta y). \tag{3}$$

The sum in Equation (3) is a Riemann sum for a continuous function on $[a, b]$, and we expect the approximations to improve as the norm of the partition goes to zero. The force against the plate is the limit of these sums.

The Integral for Fluid Force Against a Vertical Flat Plate

Suppose that a plate submerged vertically in fluid of weight-density w runs from $y = a$ to $y = b$ on the y-axis. Let $L(y)$ be the length of the horizontal strip measured from left to right along the surface of the plate at level y. Then the force exerted by the fluid against one side of the plate is

$$F = \int_a^b w \cdot (\text{strip depth}) \cdot L(y)\,dy. \tag{4}$$

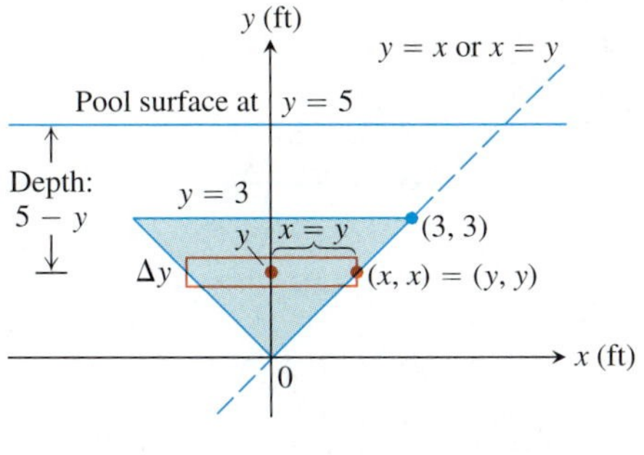

FIGURE 5.44 To find the force on one side of the submerged plate in Example 2, we can use a coordinate system like the one here.

Example 2 Applying the Integral for Fluid Force

A flat isosceles right triangular plate with base 6 ft and height 3 ft is submerged vertically, base up, 2 ft below the surface of a swimming pool. Find the force exerted by the water against one side of the plate.

Solution We establish a coordinate system to work in by placing the origin at the plate's bottom vertex and running the y-axis upward along the plate's axis of symmetry (Figure 5.44). The surface of the pool lies along the line $y = 5$ and the plate's top edge along the line $y = 3$. The plate's right-hand edge lies along the line $y = x$, with the upper right vertex at $(3, 3)$. The length of a thin strip at level y is

$$L(y) = 2x = 2y.$$

The depth of the strip beneath the surface is $(5 - y)$. The force exerted by the water against one side of the plate is therefore

$$F = \int_a^b w \times \begin{pmatrix}\text{strip}\\\text{depth}\end{pmatrix} \times L(y)\,dy \qquad \text{Eq. (4)}$$

$$= \int_0^3 62.4(5 - y)2y\,dy$$

$$= 124.8 \int_0^3 (5y - y^2)\,dy$$

$$= 124.8 \left[\frac{5}{2}y^2 - \frac{y^3}{3}\right]_0^3 = 1684.8 \text{ lb.}$$

How to Find Fluid Force

Whatever coordinate system you use, you can find the fluid force against one side of a submerged vertical plate or wall by taking these steps.

Step 1. *Find expressions* for the length and depth of a typical thin horizontal strip.

Step 2. *Multiply their product by the fluid's weight-density w and integrate* over the interval of depths occupied by the plate or wall.

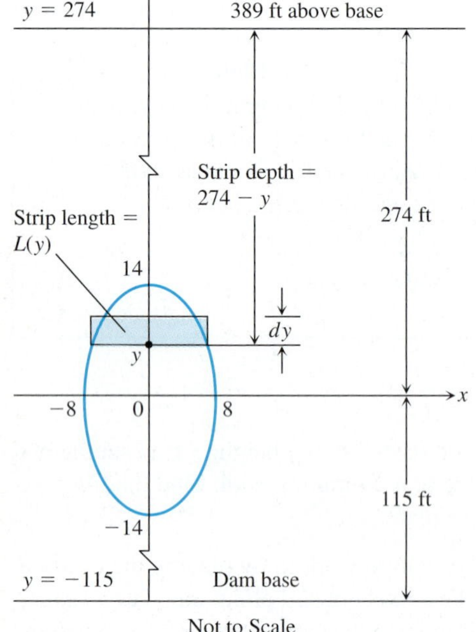

FIGURE 5.46 Dimensions and coordinates for calculating the maximum possible force against each penstock gate. (Example 3)

Example 3 The Snake River Dam and Hydroelectric Power

Water from the Snake River Dam is used to generate hydroelectric power at an adjacent electrical plant. Water under extreme pressure flows through three penstock intake gates located as shown in Figure 5.45a. The lowest point on the gate is 101 ft above the base of the dam, whereas the highest point is 129 ft above the base of the dam. The width of the gate at its center is 16 ft. Each gate is inset to be vertical. To design and build the gate, engineers must know the maximum total force against the gate.

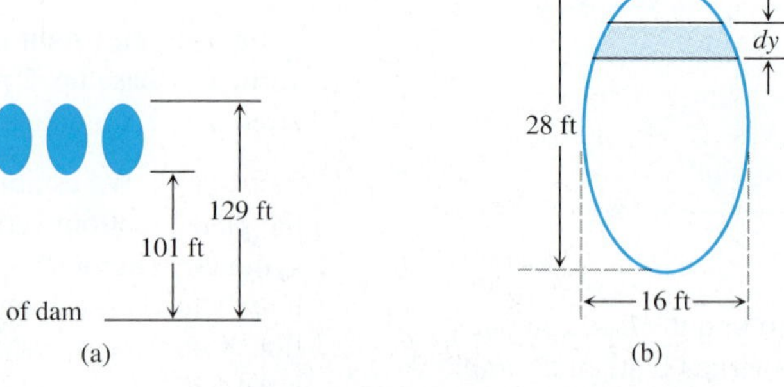

FIGURE 5.45 (a) Location of penstock gates from base of the dam. (b) Enlarged penstock gate subdivided into horizontal strips.

We now approximate the gate with horizontal rectangular strips of width dy and length $L(y)$. The force against the gate, from Equation (4), is

$$F = \int_a^b w \times \begin{pmatrix} \text{strip} \\ \text{depth} \end{pmatrix} \times L(y)\, dy$$

$$= \int_{-14}^{14} 61.4\,(274 - y)L(y)\, dy. \tag{5}$$

To complete the analysis, we use the equation for the gate's elliptical boundary to formulate $L(y)$ and perform the integration (Exercise 13). You will be astonished by the size of the force the gate must be designed to withstand. (Figure 5.46)

EXERCISES 5.6

The weight-densities of the fluids in the following exercises can be found in the table on page 465.

1. *Triangular plate* Calculate the fluid force on one side of the plate in Example 2 using the coordinate system shown here.

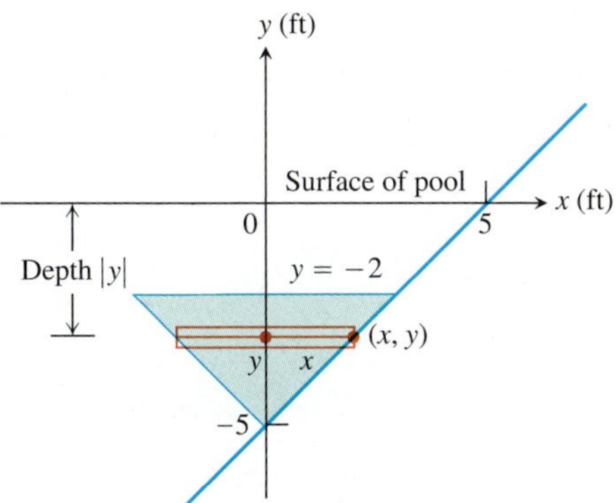

2. *Triangular plate* Calculate the fluid force on one side of the plate in Example 2 using the coordinate system shown here.

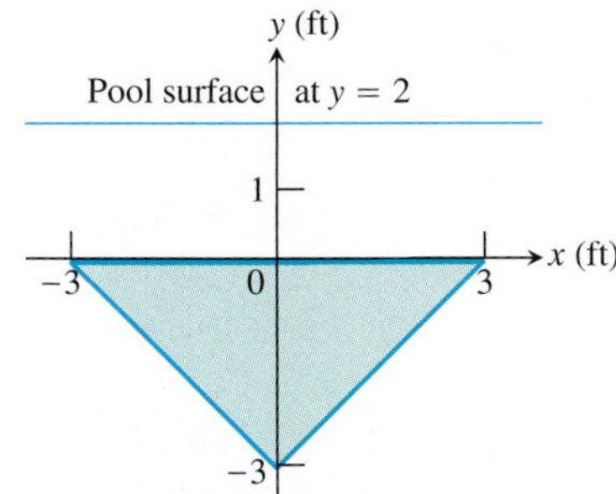

3. *Lowered triangular plate* The plate in Example 2 is lowered another 2 ft into the water. What is the fluid force on one side of the plate now?

4. *Raised triangular plate* The plate in Example 2 is raised to put its top edge at the surface of the pool. What is the fluid force on one side of the plate now?

5. *Triangular plate* The isosceles triangular plate shown here is submerged vertically 1 ft below the surface of a freshwater lake.

   (a) Find the fluid force against one face of the plate.

   (b) What would be the fluid force on one side of the plate if the water were seawater instead of freshwater?

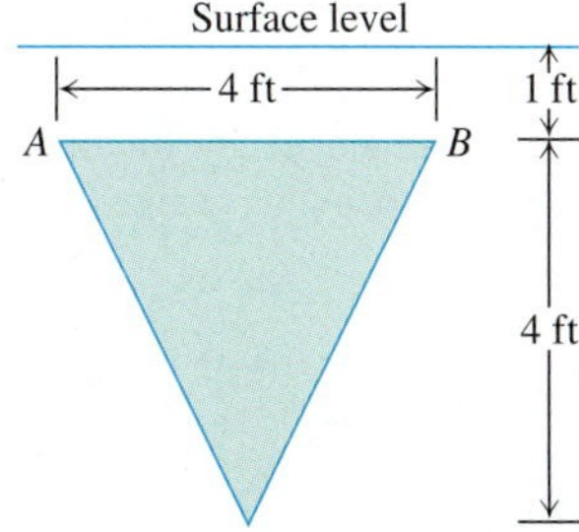

6. *Rotated triangular plate* The plate in Exercise 5 is revolved 180° about line AB so that part of the plate sticks out of the lake, as shown here. What force does the water exert on one face of the plate now?

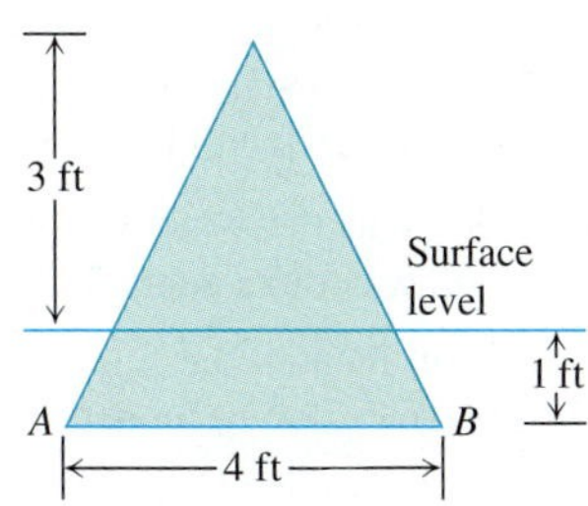

7. *New England Aquarium* The viewing portion of the rectangular glass window in a typical fish tank at the New England Aquarium in Boston is 63 in. wide and runs from 0.5 in. below the water's surface to 33.5 in. below the surface. Find the fluid force against this portion of the window. The weight-density of seawater is 64 lb/ft^3. (In case you were wondering, the glass is 3/4 in. thick and the tank walls extend 4 in. above the water to keep the fish from jumping out.)

8. *Fish tank* A horizontal rectangular freshwater fish tank with base 2×4 ft and height 2 ft (interior dimensions) is filled to within 2 in. of the top.

(a) Find the fluid force against each side and end of the tank.

(b) *Writing to Learn* If the tank is sealed and stood on end (without spilling), so that one of the square ends is the base, what does that do to the fluid forces on the rectangular sides?

9. *Semicircular plate* A semicircular plate 2 ft in diameter sticks straight down into fresh water with the diameter along the surface. Find the force exerted by the water on one side of the plate.

10. *Milk truck* A tank truck hauls milk in a 6-ft-diameter horizontal right circular cylindrical tank. How much force does the milk exert on each end of the tank when the tank is half full?

11. *Tank with parabolic gate* The cubical metal tank shown here has a parabolic gate, held in place by bolts and designed to withstand a fluid force of 160 lb without rupturing. The liquid you plan to store has a weight-density of 50 lb/ft^3.

(a) What is the fluid force on the gate when the liquid is 2 ft deep?

(b) What is the maximum height to which the container can be filled without exceeding its design limitation?

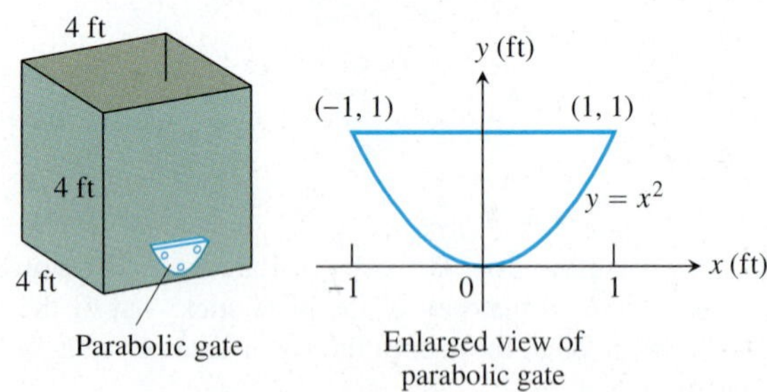

12. *Window in a tank* The rectangular tank shown here has a 1 ft × 1 ft square window 1 ft above the base. The window is designed to withstand a fluid force of 312 lb without cracking.

(a) What fluid force will the window have to withstand if the tank is filled with water to a depth of 3 ft?

(b) To what level can the tank be filled with water without exceeding the window's design limitation?

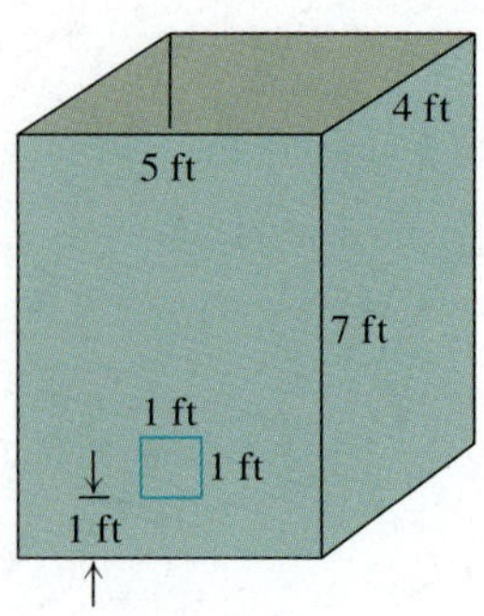

13. *Force against penstock gate (Example 3)*

(a) Find an equation for the elliptical gate boundary in Figure 5.46.

(b) Use the equation found in part (a) to write a formula for the strip length $L(y)$.

T **(c)** Use the formula found in part (b) to complete the integral in Equation (5) and evaluate the integral numerically.

14. *Swimming pool drain plate* Water is running into the rectangular swimming pool shown here at the rate of 1000 ft^3/h.

(a) Find the fluid force against the triangular drain plate after 9 h of filling.

(b) The drain plate is designed to withstand a fluid force of 520 lb. How high can you fill the pool without exceeding this limitation?

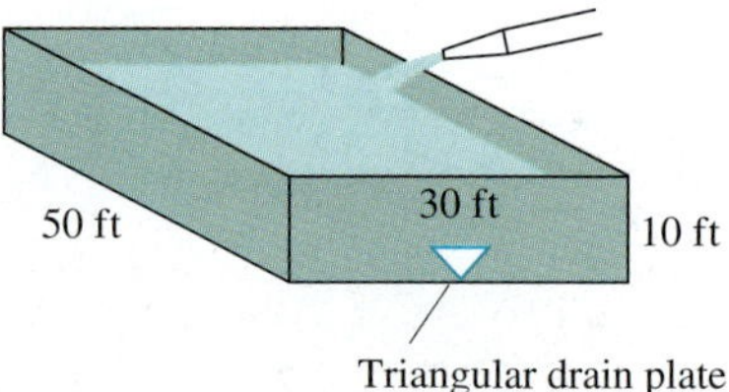

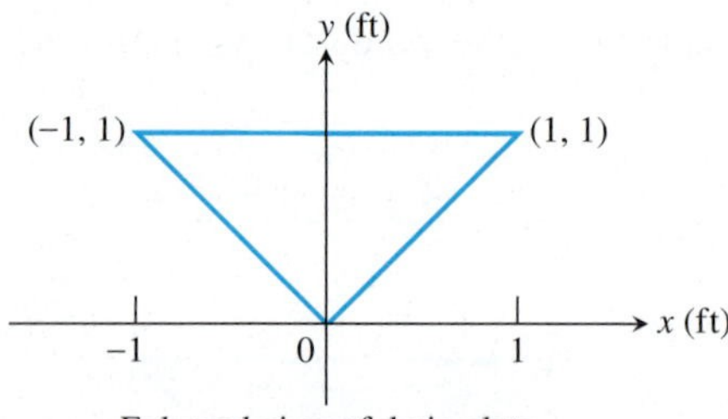

15. (a) *Average pressure* A vertical rectangular plate a units long by b units wide is submerged in a fluid of weight density w with its long edges parallel to the fluid's surface. Find the average value of the pressure along the vertical dimension of the plate.

(b) Show that the force exerted by the fluid on one side of the plate is the average value of the pressure times the area of the plate.

16. *Tank with a movable end* Water pours into the tank shown here at the rate of 4 ft^3/min. The tank's cross sections are 4-ft-diameter semicircles. One end of the tank is movable, but moving it to increase the volume compresses a spring. The spring constant is $k = 100$ lb/ft. If the end of the tank moves 5 ft against the spring, the water will drain out of a safety hole in the bottom at the rate of 5 ft^3/min. Will the movable end reach the hole before the tank overflows?

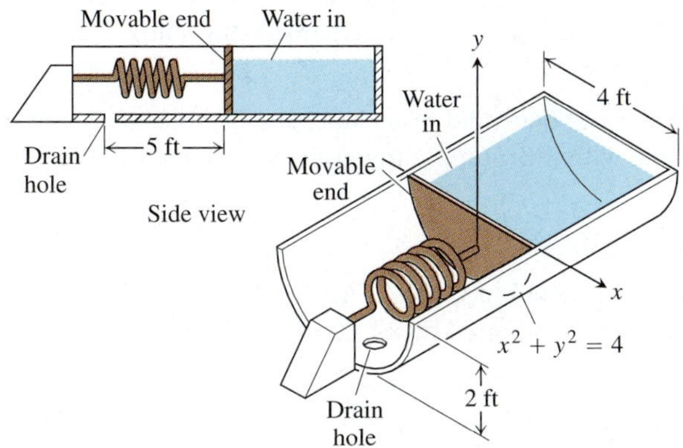

17. *Watering trough* The vertical ends of a watering trough are isosceles triangles like the one shown here (dimensions in feet).

(a) Find the fluid force against the ends when the trough is full.

(b) How many inches do you have to lower the water level in the trough to cut the fluid force on the ends in half? (Answer to the nearest half inch.)

(c) *Writing to Learn* Does it matter how long the trough is? Give reasons for your answer.

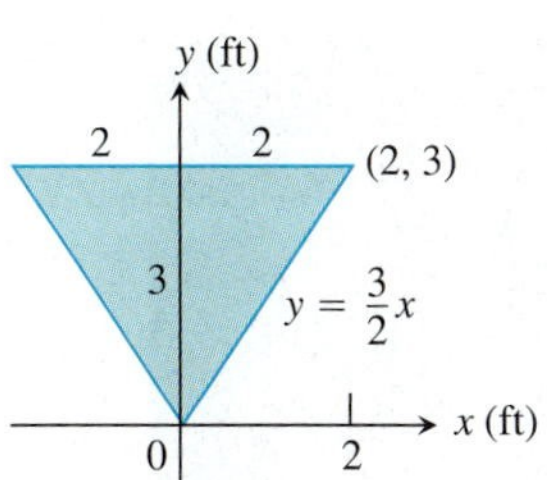

18. *Watering trough* The vertical ends of a watering trough are squares 3 ft on a side.

(a) Find the fluid force against the ends when the trough is full.

(b) How many inches do you have to lower the water level in the trough to reduce the fluid force by 25%?

19. *Milk carton* A rectangular milk carton measures 3.75 × 3.75 in. at the base and is 7.75 in. tall. Find the force of the milk on one side when the carton is full.

20. *Olive oil can* A standard olive oil can measures 5.75 by 3.5 in. at the base and is 10 in. tall. Find the fluid force against the base and each side when the can is full.

21. *Water in a trough* The end plates of the trough shown here were designed to withstand a fluid force of 6667 lb. How many cubic feet of water can the tank hold without exceeding this limitation? Round down to the nearest cubic foot.

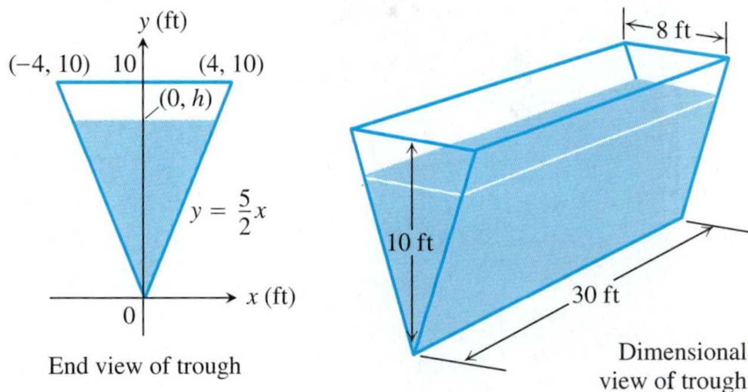

5.7 Moments and Centers of Mass

Masses Along a Line • Wires and Thin Rods • Masses Distributed over a Plane Region • Thin, Flat Plates • Centroids

Many structures and mechanical systems behave as if their masses were concentrated at a single point, called the center of mass (Figure 5.47). It is important to know how to locate this point, and doing so is basically a mathematical enterprise.

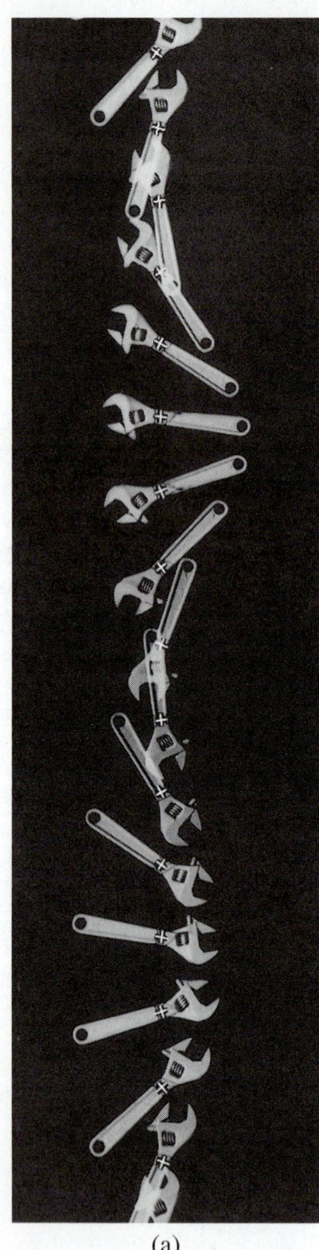

(a)

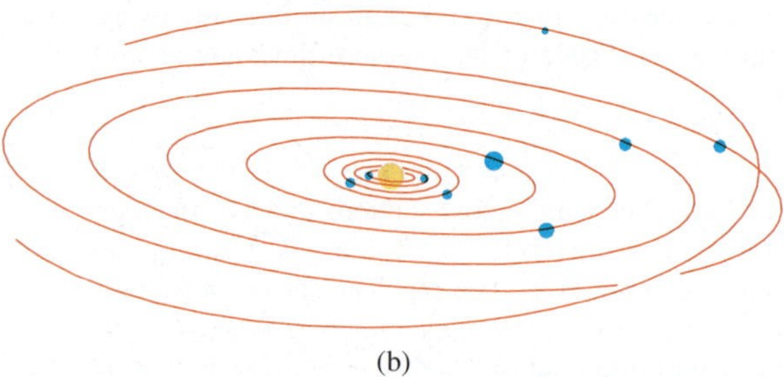

(b)

FIGURE 5.47 (a) The motion of this wrench gliding on ice seems haphazard until we notice that the wrench is simply turning about its center of mass as the center glides in a straight line. (b) The planets, asteroids, and comets of our solar system revolve about their collective center of mass. (It lies inside the sun.)

For the moment, we deal with one- and two-dimensional objects. Three-dimensional objects are best done with the multiple integrals of Chapter 12.*

Masses Along a Line

We develop our mathematical model in stages. The first stage is to imagine masses m_1, m_2, and m_3 on a rigid x-axis supported by a fulcrum at the origin.

x_1 0 x_2 x_3 x

m_1 m_2 m_3

Fulcrum at origin

The resulting system might balance, or it might not. It depends on how large the masses are and how they are arranged.

Each mass m_k exerts a downward force $m_k g$ equal to the magnitude of the mass times the acceleration of gravity. Each of these forces has a tendency to turn the axis about the origin, the way you turn a seesaw. This turning effect, called a **torque,** is measured by multiplying the force $m_k g$ by the signed distance x_k from the point of application to the origin. Masses to the left of the origin exert negative (counterclockwise) torque. Masses to the right of the origin exert positive (clockwise) torque.

The sum of the torques measures the tendency of a system to rotate about the origin. This sum is called the **system torque.**

$$\text{System torque} = m_1 g x_1 + m_2 g x_2 + m_3 g x_3 \tag{1}$$

The system will balance if and only if its torque is zero.

If we factor out the g in Equation (1), we see that the system torque is

$$\underbrace{g}_{\text{a feature of the environment}} \cdot \underbrace{(m_1 x_1 + m_2 x_2 + m_3 x_3)}_{\text{a feature of the system}}$$

Mass versus Weight

Weight is the force that results from gravity pulling on a mass. If an object of mass m is placed in a location where the acceleration of gravity is g, the object's weight there is

$$F = mg$$

(as in Newton's second law).

*The presentation in Chapter 12 is entirely self-contained. The present section may therefore be omitted, if desired. The essential ideas are repeated in Chapter 12.

Thus, the torque is the product of the gravitational acceleration g, which is a feature of the environment in which the system happens to reside, and the number $(m_1x_1 + m_2x_2 + m_3x_3)$, which is a feature of the system itself, a constant that stays the same no matter where the system is placed.

The number $(m_1x_1 + m_2x_2 + m_3x_3)$ is called the **moment of the system about the origin.** It is the sum of the **moments** m_1x_1, m_2x_2, m_3x_3 of the individual masses.

$$M_0 = \text{Moment of system about origin} = \sum m_k x_k$$

(We shift to sigma notation here to allow for sums with more terms. For $\sum m_k x_k$, read "summation $m_k\, x_k$.")

We usually want to know where to place the fulcrum to make the system balance, that is, at what point $\bar{x}$ to place it to make the torques add to zero.

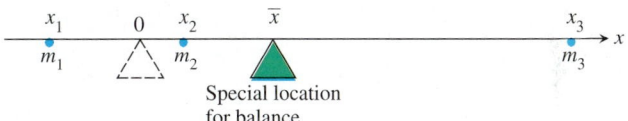

x_1 0 x_2 $\bar{x}$ x_3 x
m_1 m_2 m_3
Special location for balance

The torque of each mass about the fulcrum in this special location is

$$\begin{aligned}\text{Torque of } m_k \text{ about } \bar{x} &= \begin{pmatrix}\text{signed distance} \\ \text{of } m_k \text{ from } \bar{x}\end{pmatrix}\begin{pmatrix}\text{downward} \\ \text{force}\end{pmatrix} \\ &= (x_k - \bar{x})m_k g.\end{aligned}$$

When we write the equation that says that the sum of these torques is zero, we get an equation we can solve for $\bar{x}$:

$$\begin{aligned}
\sum (x_k - \bar{x})m_k g &= 0 && \text{Sum of the torques equals zero} \\
g\sum (x_k - \bar{x})m_k &= 0 && \text{Constant Multiple Rule for Sums} \\
\sum (m_k x_k - \bar{x}m_k) &= 0 && g \text{ divided out, } m_k \text{ distributed} \\
\sum m_k x_k - \sum \bar{x}m_k &= 0 && \text{Difference Rule for Sums} \\
\sum m_k x_k &= \bar{x}\sum m_k && \text{Rearranged, Constant Multiple Rule again} \\
\bar{x} &= \frac{\sum m_k x_k}{\sum m_k}. && \text{Solved for } \bar{x}
\end{aligned}$$

This last equation tells us to find $\bar{x}$ by dividing the system's moment about the origin by the system's total mass:

$$\bar{x} = \frac{\sum m_k x_k}{\sum m_k} = \frac{\text{system moment about origin}}{\text{system mass}}.$$

The point $\bar{x}$ is called the system's **center of mass.**

Wires and Thin Rods

In many applications, we want to know the center of mass of a rod or a thin strip of metal. In cases like these where we can model the distribution of mass with a continuous function, the summation signs in our formulas become integrals in a manner we now describe.

Imagine a long, thin strip lying along the x-axis from $x = a$ to $x = b$ and cut into small pieces of mass Δm_k by a partition of the interval $[a, b]$.

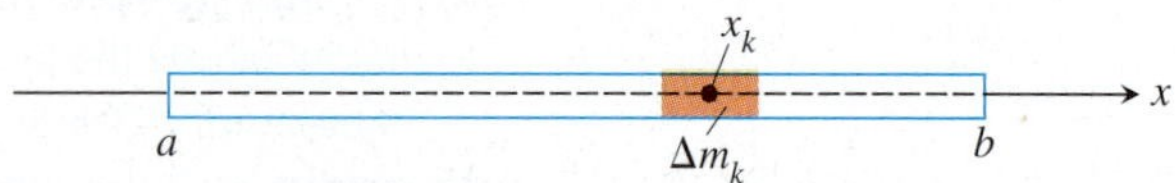

The kth piece is Δx_k units long and lies approximately x_k units from the origin. Now observe three things.

First, the strip's center of mass $\bar{x}$ is nearly the same as that of the system of point masses we would get by attaching each mass Δm_k to the point x_k:

$$\bar{x} \approx \frac{\text{system moment}}{\text{system mass}}.$$

Second, the moment of each piece of the strip about the origin is approximately $x_k \, \Delta m_k$, so the system moment is approximately the sum of the $x_k \, \Delta m_k$:

$$\text{System moment} \approx \sum x_k \, \Delta m_k.$$

Density

A material's density is its mass per unit volume. In practice, however, we tend to use units we can conveniently measure. For wires, rods, and narrow strips we use mass per unit length. For flat sheets and plates we use mass per unit area.

Third, if the density of the strip at x_k is $\delta(x_k)$, expressed in terms of mass per unit length and if δ is continuous, then Δm_k is approximately equal to $\delta(x_k) \, \Delta x_k$ (mass per unit length times length):

$$\Delta m_k \approx \delta(x_k) \, \Delta x_k.$$

Combining these three observations gives

$$\bar{x} \approx \frac{\text{system moment}}{\text{system mass}} \approx \frac{\sum x_k \, \Delta m_k}{\sum \Delta m_k} \approx \frac{\sum x_k \delta(x_k) \, \Delta x_k}{\sum \delta(x_k) \, \Delta x_k}. \tag{2}$$

The sum in the last numerator in Equation (2) is a Riemann sum for the continuous function $x\delta(x)$ over the closed interval $[a, b]$. The sum in the denominator is a Riemann sum for the function $\delta(x)$ over this interval. We expect the approximations in Equation (2) to improve as the strip is partitioned more finely, and we are led to the equation

$$\bar{x} = \frac{\int_a^b x\delta(x) \, dx}{\int_a^b \delta(x) \, dx}.$$

This is the formula we use to find $\bar{x}$.

Moment, Mass, and Center of Mass of a Thin Rod or Strip Along the x-axis with Density Function $\delta(x)$

Moment about the origin: $$M_0 = \int_a^b x\delta(x) \, dx \tag{3a}$$

Mass: $$M = \int_a^b \delta(x) \, dx \tag{3b}$$

Center of mass: $$\bar{x} = \frac{M_0}{M} \tag{3c}$$

To find a center of mass, divide moment by mass.

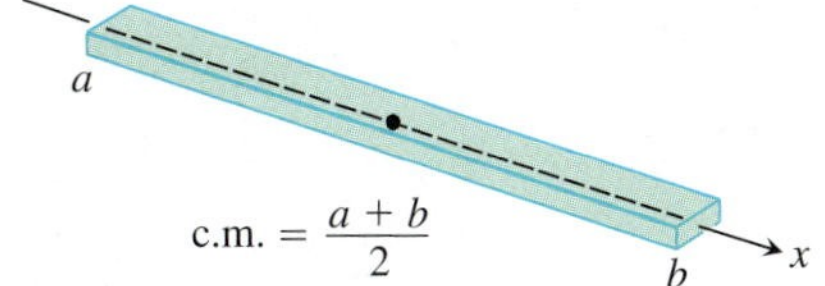

FIGURE 5.48 The center of mass of a straight, thin rod or strip of constant density lies halfway between its ends.

Example 1 Strips and Rods of Constant Density

Show that the center of mass of a straight, thin strip or rod of constant density lies halfway between its two ends.

Solution We model the strip as a portion of the x-axis from $x = a$ to $x = b$ (Figure 5.48). Our goal is to show that $\bar{x} = (a + b)/2$, the point halfway between a and b.

The key is the density's having a constant value. This enables us to regard the function $\delta(x)$ in the integrals in Equations (3) as a constant (call it δ), with the result that

$$M_0 = \int_a^b \delta x\, dx = \delta \int_a^b x\, dx = \delta\left[\frac{1}{2}x^2\right]_a^b = \frac{\delta}{2}(b^2 - a^2)$$

$$M = \int_a^b \delta\, dx = \delta \int_a^b dx = \delta[x]_a^b = \delta(b - a)$$

$$\bar{x} = \frac{M_0}{M} = \frac{\frac{\delta}{2}(b^2 - a^2)}{\delta(b - a)}$$

$$= \frac{a + b}{2}.$$

The δ's cancel in the formula for $\bar{x}$.

Example 2 Variable-Density Rod

The 10-m-long rod in Figure 5.49 thickens from left to right so that its density, instead of being constant, is $\delta(x) = 1 + (x/10)$ kg/m. Find the rod's center of mass.

Solution The rod's moment about the origin (Equation 3(a)) is

$$M_0 = \int_0^{10} x\delta(x)\, dx = \int_0^{10} x\left(1 + \frac{x}{10}\right) dx = \int_0^{10} \left(x + \frac{x^2}{10}\right) dx$$

$$= \left[\frac{x^2}{2} + \frac{x^3}{30}\right]_0^{10} = 50 + \frac{100}{3} = \frac{250}{3}\ \text{kg} \cdot \text{m}.$$

The units of a moment are mass × length.

The rod's mass (Equation 3(b)) is

$$M = \int_0^{10} \delta(x)\, dx = \int_0^{10} \left(1 + \frac{x}{10}\right) dx = \left[x + \frac{x^2}{20}\right]_0^{10} = 10 + 5 = 15\ \text{kg}.$$

The center of mass (Equation 3(c)) is located at the point

$$\bar{x} = \frac{M_0}{M} = \frac{250}{3} \cdot \frac{1}{15} = \frac{50}{9} \approx 5.56\ \text{m}.$$

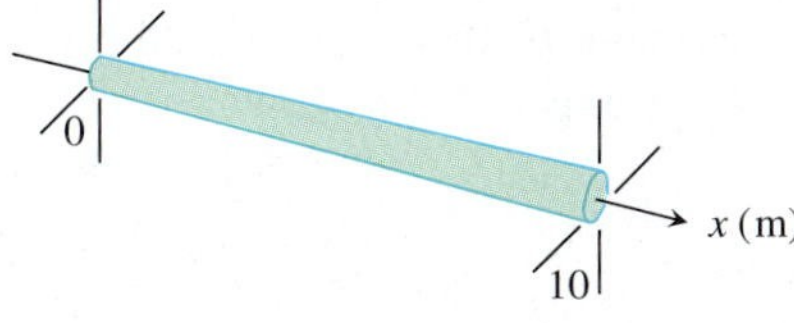

FIGURE 5.49 We can treat a rod of variable thickness as a rod of variable density. See Example 2.

Masses Distributed over a Plane Region

Suppose that we have a finite collection of masses located in the plane, with mass m_k at the point (x_k, y_k) (see Figure 5.50). The mass of the system is

$$\text{System mass:} \qquad M = \sum m_k.$$

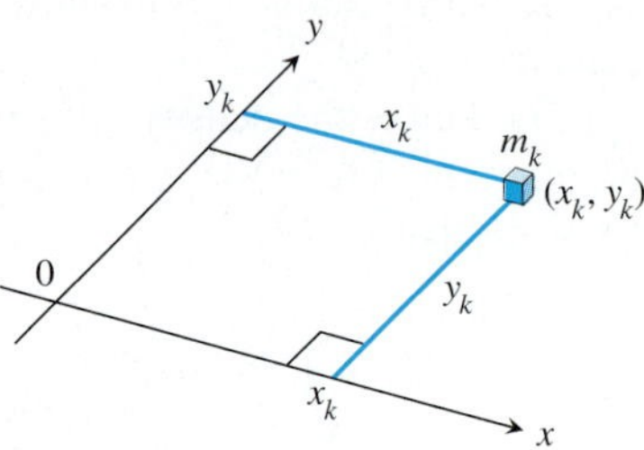

FIGURE 5.50 Each mass m_k has a moment about each axis.

Each mass m_k has a moment about each axis. Its moment about the x-axis is $m_k y_k$, and its moment about the y-axis is $m_k x_k$. The moments of the entire system about the two axes are

$$\text{Moment about } x\text{-axis:} \qquad M_x = \sum m_k y_k,$$

$$\text{Moment about } y\text{-axis:} \qquad M_y = \sum m_k x_k.$$

The x-coordinate of the system's center of mass is defined to be

$$\bar{x} = \frac{M_y}{M} = \frac{\sum m_k x_k}{\sum m_k}. \tag{4}$$

With this choice of $\bar{x}$, as in the one-dimensional case, the system balances about the line $x = \bar{x}$ (Figure 5.51).

The y-coordinate of the system's center of mass is defined to be

$$\bar{y} = \frac{M_x}{M} = \frac{\sum m_k y_k}{\sum m_k}. \tag{5}$$

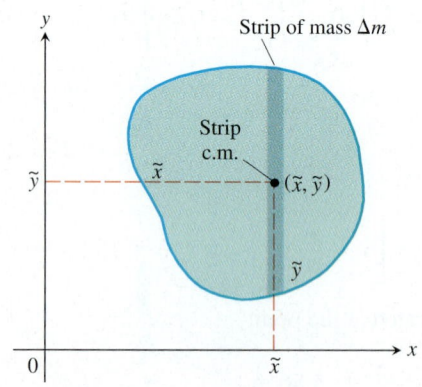

FIGURE 5.51 A two-dimensional array of masses balances on its center of mass.

With this choice of $\bar{y}$, the system balances about the line $y = \bar{y}$ as well. The torques exerted by the masses about the line $y = \bar{y}$ cancel out. Thus, as far as balance is concerned, the system behaves as if all its mass were at the single point $(\bar{x}, \bar{y})$. We call this point the system's *center of mass*.

Thin, Flat Plates

In many applications, we need to find the center of mass of a thin, flat plate: a disk of aluminum, say, or a triangular sheet of steel. In such cases, we assume the distribution of mass to be continuous, and the formulas we use to calculate $\bar{x}$ and $\bar{y}$ contain integrals instead of finite sums. The integrals arise in the following way.

Imagine the plate occupying a region in the xy-plane, cut into thin strips parallel to one of the axes (in Figure 5.52, the y-axis). The center of mass of a typical strip is $(\tilde{x}, \tilde{y})$. We treat the strip's mass Δm as if it were concentrated at $(\tilde{x}, \tilde{y})$. The moment of the strip about the y-axis is then $\tilde{x}\,\Delta m$. The moment of the strip about the x-axis is $\tilde{y}\,\Delta m$. Equations (4) and (5) then become

$$\bar{x} = \frac{M_y}{M} = \frac{\sum \tilde{x}\,\Delta m}{\sum \Delta m}, \qquad \bar{y} = \frac{M_x}{M} = \frac{\sum \tilde{y}\,\Delta m}{\sum \Delta m}.$$

FIGURE 5.52 A plate cut into thin strips parallel to the y-axis. The moment exerted by a typical strip about each axis is the moment its mass Δm would exert if concentrated at the strip's center of mass $(\tilde{x}, \tilde{y})$.

As in the one-dimensional case, the sums are Riemann sums for integrals and approach these integrals as limiting values as the strips into which the plate is cut become narrower and narrower. We write these integrals symbolically as

$$\bar{x} = \frac{\int \tilde{x}\, dm}{\int dm} \qquad \text{and} \qquad \bar{y} = \frac{\int \tilde{y}\, dm}{\int dm}.$$

Moments, Mass, and Center of Mass of a Thin Plate Covering a Region in the xy-plane

Moment about the x-axis: $\quad M_x = \int \tilde{y}\, dm$

Moment about the y-axis: $\quad M_y = \int \tilde{x}\, dm \qquad (6)$

Mass: $\quad M = \int dm$

Center of mass: $\quad \bar{x} = \frac{M_y}{M}, \qquad \bar{y} = \frac{M_x}{M}$

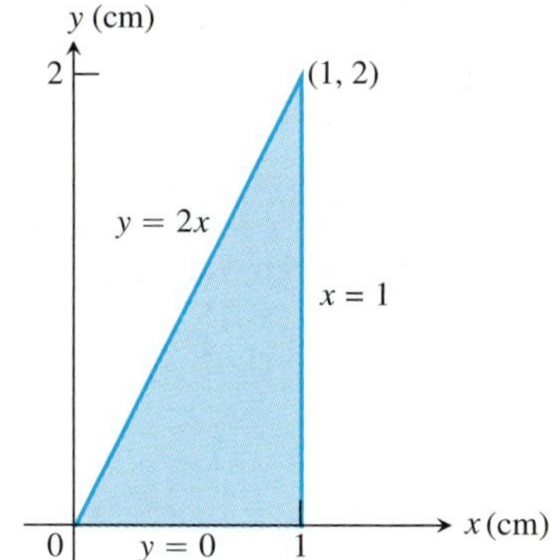

FIGURE 5.53 The plate in Example 3.

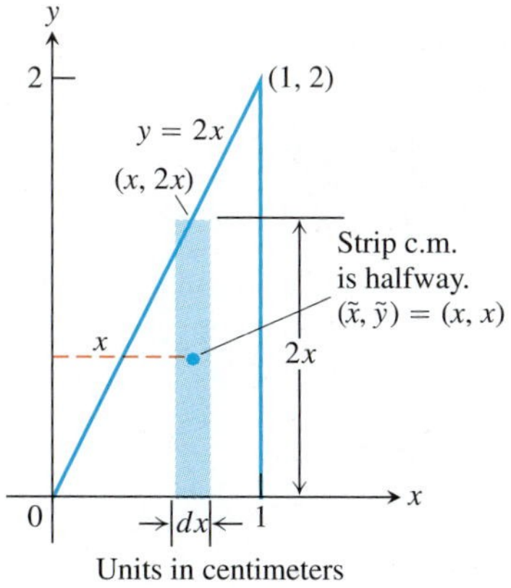

To evaluate these integrals, we picture the plate in the coordinate plane and sketch a strip of mass parallel to one of the coordinate axes. We then express the strip's mass dm and the coordinates $(\tilde{x}, \tilde{y})$ of the strip's center of mass in terms of x or y. Finally, we integrate $\tilde{y}\, dm$, $\tilde{x}\, dm$, and dm between limits of integration determined by the plate's location in the plane.

Example 3 Constant-Density Plate

The triangular plate shown in Figure 5.53 has a constant density of $\delta = 3\ \text{g/cm}^2$. Find

(a) the plate's moment M_y about the y-axis

(b) the plate's mass M

(c) the x-coordinate of the plate's center of mass (c.m.).

FIGURE 5.54 Modeling the plate in Example 3 with vertical strips.

Solution

Method 1: *Vertical strips* (Figure 5.54)

(a) The moment M_y: The typical vertical strip has

center of mass (c.m.): $\quad (\tilde{x}, \tilde{y}) = (x, x)$

length: $\quad 2x$

width: $\quad dx$

area: $\quad dA = 2x\, dx$

mass: $\quad dm = \delta\, dA = 3 \cdot 2x\, dx = 6x\, dx$

distance of c.m. from y-axis: $\quad \tilde{x} = x.$

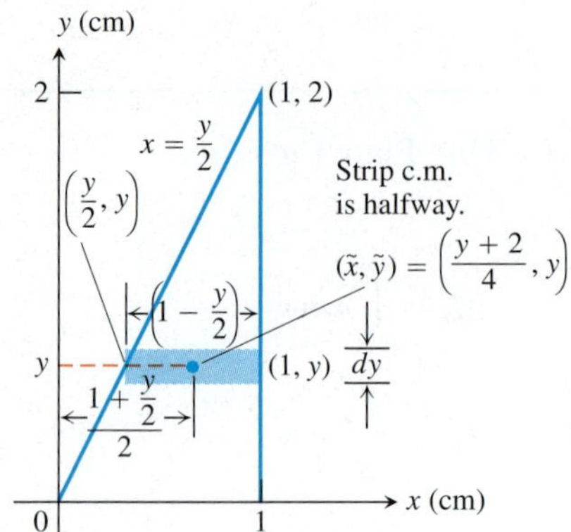

FIGURE 5.55 Modeling the plate in Example 3 with horizontal strips.

The moment of the strip about the y-axis is

$$\tilde{x}\, dm = x \cdot 6x\, dx = 6x^2\, dx.$$

The moment of the plate about the y-axis is therefore

$$M_y = \int \tilde{x}\, dm = \int_0^1 6x^2\, dx = 2x^3 \Big]_0^1 = 2 \text{ g} \cdot \text{cm}.$$

(b) The plate's mass:

$$M = \int dm = \int_0^1 6x\, dx = 3x^2 \Big]_0^1 = 3 \text{ g}.$$

(c) The x-coordinate of the plate's center of mass:

$$\bar{x} = \frac{M_y}{M} = \frac{2 \text{ g} \cdot \text{cm}}{3 \text{ g}} = \frac{2}{3} \text{ cm}.$$

By a similar computation, we could find M_x and $\bar{y} = M_x/M$.

Method 2: *Horizontal strips* (Figure 5.55).

(a) The moment M_y: The y-coordinate of the center of mass of a typical horizontal strip is y (see the figure), so

$$\tilde{y} = y.$$

The x-coordinate is the x-coordinate of the point halfway across the triangle. This makes it the average of $y/2$ (the strip's left-hand x-value) and 1 (the strip's right-hand x-value):

$$\tilde{x} = \frac{(y/2) + 1}{2} = \frac{y}{4} + \frac{1}{2} = \frac{y + 2}{4}.$$

We also have

length: $1 - \frac{y}{2} = \frac{2 - y}{2}$

width: dy

area: $dA = \frac{2 - y}{2}\, dy$

mass: $dm = \delta\, dA = 3 \cdot \frac{2 - y}{2}\, dy$

distance of c.m. to y-axis: $\tilde{x} = \frac{y + 2}{4}.$

The moment of the strip about the y-axis is

$$\tilde{x}\, dm = \frac{y + 2}{4} \cdot 3 \cdot \frac{2 - y}{2}\, dy = \frac{3}{8}(4 - y^2)\, dy.$$

The moment of the plate about the y-axis is

$$M_y = \int \tilde{x}\, dm = \int_0^2 \frac{3}{8}(4 - y^2)\, dy = \frac{3}{8}\left[4y - \frac{y^3}{3}\right]_0^2 = \frac{3}{8}\left(\frac{16}{3}\right) = 2 \text{ g} \cdot \text{cm}.$$

How to Find a Plate's Center of Mass

Step 1. Picture the plate in the xy-plane.

Step 2. Sketch a strip of mass parallel to one of the coordinate axes and find its dimensions.

Step 3. Find the strip's mass dm and center of mass $(\tilde{x}, \tilde{y})$.

Step 4. Integrate $\tilde{y}\, dm$, $\tilde{x}\, dm$, and dm to find M_x, M_y, and M.

Step 5. Divide the moments by the mass to calculate $\bar{x}$ and $\bar{y}$.

(b) The plate's mass:

$$M = \int dm = \int_0^2 \frac{3}{2}(2 - y)\,dy = \frac{3}{2}\left[2y - \frac{y^2}{2}\right]_0^2 = \frac{3}{2}(4 - 2) = 3 \text{ g}.$$

(c) The x-coordinate of the plate's center of mass:

$$\bar{x} = \frac{M_y}{M} = \frac{2 \text{ g} \cdot \text{cm}}{3 \text{ g}} = \frac{2}{3} \text{ cm}.$$

By a similar computation, we could find M_x and $\bar{y}$.

If the distribution of mass in a thin, flat plate has an axis of symmetry, the center of mass will lie on this axis. If there are two axes of symmetry, the center of mass will lie at their intersection. These facts often help to simplify our work.

Example 4 Constant-Density Plate

Find the center of mass of a thin plate of constant density δ covering the region bounded above by the parabola $y = 4 - x^2$ and below by the x-axis (Figure 5.56).

Solution Since the plate is symmetric about the y-axis and its density is constant, the distribution of mass is symmetric about the y-axis and the center of mass lies on the y-axis. Thus, $\bar{x} = 0$. It remains to find $\bar{y} = M_x/M$.

A trial calculation with horizontal strips (Figure 5.56a) leads to an inconvenient integration

$$M_x = \int_0^4 2\delta\, y\sqrt{4 - y}\,dy.$$

We therefore model the distribution of mass with vertical strips instead (Figure 5.56b). The typical vertical strip has

center of mass (c.m.): $(\tilde{x}, \tilde{y}) = \left(x, \dfrac{4 - x^2}{2}\right)$

length: $4 - x^2$

width: dx

area: $dA = (4 - x^2)\,dx$

mass: $dm = \delta\, dA = \delta(4 - x^2)\,dx$

distance from c.m. to x-axis: $\tilde{y} = \dfrac{4 - x^2}{2}$.

The moment of the strip about the x-axis is

$$\tilde{y}\,dm = \frac{4 - x^2}{2} \cdot \delta(4 - x^2)\,dx = \frac{\delta}{2}(4 - x^2)^2\,dx.$$

The moment of the plate about the x-axis is

$$M_x = \int \tilde{y}\,dm = \int_{-2}^{2} \frac{\delta}{2}(4 - x^2)^2\,dx$$

$$= \frac{\delta}{2}\int_{-2}^{2}(16 - 8x^2 + x^4)\,dx = \frac{256}{15}\delta. \qquad (7)$$

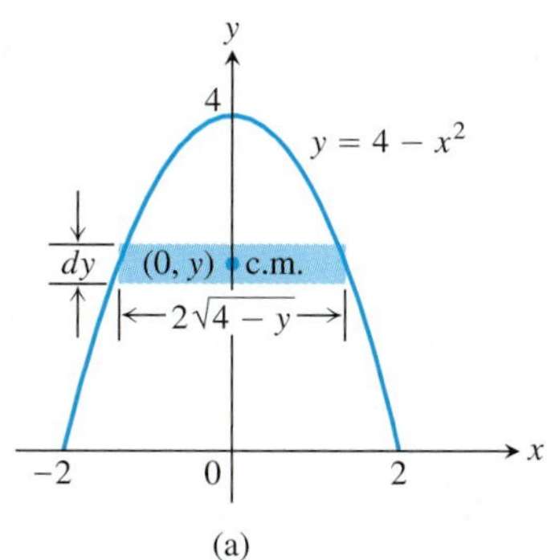

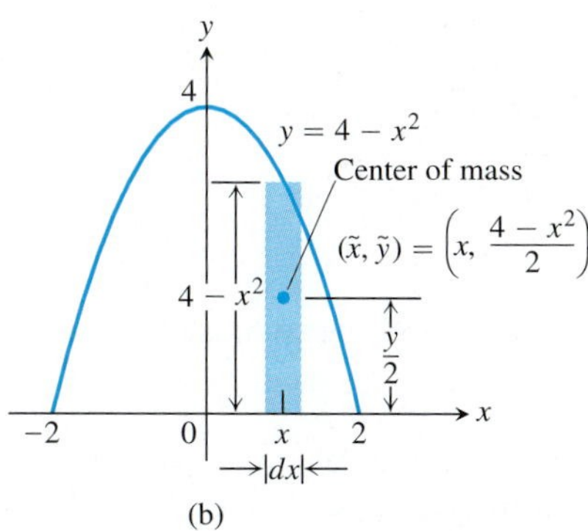

FIGURE 5.56 Modeling the plate in Example 4 with (a) horizontal strips leads to an inconvenient integration, so we model with (b) vertical strips instead.

The mass of the plate is

$$M = \int dm = \int_{-2}^{2} \delta(4 - x^2)\, dx = \frac{32}{3}\,\delta. \tag{8}$$

Therefore,

$$\bar{y} = \frac{M_x}{M} = \frac{(256/15)\delta}{(32/3)\delta} = \frac{8}{5}.$$

The plate's center of mass is the point

$$(\bar{x}, \bar{y}) = \left(0, \frac{8}{5}\right).$$

Example 5 Variable-Density Plate

Find the center of mass of the plate in Example 4 if the density at the point (x, y) is $\delta = 2x^2$, twice the square of the distance from the point to the y-axis.

Solution The mass distribution is still symmetric about the y-axis, so $\bar{x} = 0$. With $\delta = 2x^2$, Equations (7) and (8) become

$$\begin{aligned} M_x &= \int \tilde{y}\, dm = \int_{-2}^{2} \frac{\delta}{2}(4 - x^2)^2\, dx = \int_{-2}^{2} x^2(4 - x^2)^2\, dx \\ &= \int_{-2}^{2} (16x^2 - 8x^4 + x^6)\, dx = \frac{2048}{105} \end{aligned} \tag{7'}$$

$$\begin{aligned} M &= \int dm = \int_{-2}^{2} \delta(4 - x^2)\, dx = \int_{-2}^{2} 2x^2(4 - x^2)\, dx \\ &= \int_{-2}^{2} (8x^2 - 2x^4)\, dx = \frac{256}{15}. \end{aligned} \tag{8'}$$

Therefore,

$$\bar{y} = \frac{M_x}{M} = \frac{2048}{105} \cdot \frac{15}{256} = \frac{8}{7}.$$

The plate's new center of mass is

$$(\bar{x}, \bar{y}) = \left(0, \frac{8}{7}\right).$$

Example 6 Constant-Density Wire

Find the center of mass of a wire of constant density δ shaped like a semicircle of radius a.

Solution We model the wire with the semicircle $y = \sqrt{a^2 - x^2}$ (Figure 5.57). The distribution of mass is symmetric about the y-axis, so $\bar{x} = 0$. To find $\bar{y}$, we imagine the wire divided into short segments. The typical segment (Figure 5.57a) has

length: $ds = a\, d\theta$

mass: $dm = \delta\, ds = \delta a\, d\theta$ Mass per unit length times length

distance of c.m. to x-axis: $\tilde{y} = a \sin\theta$.

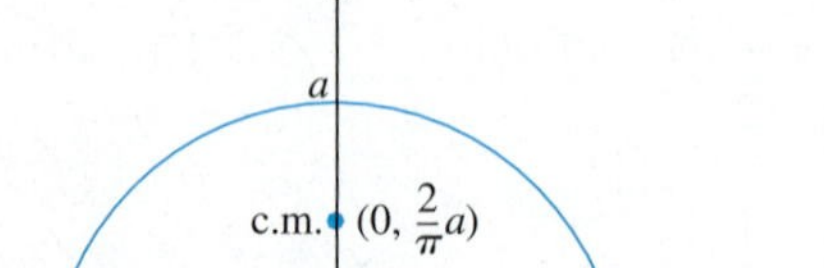

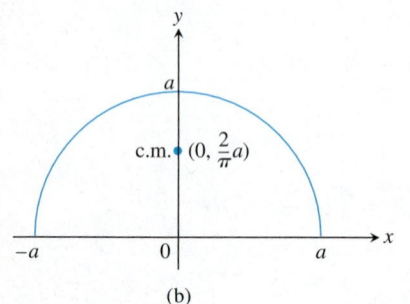

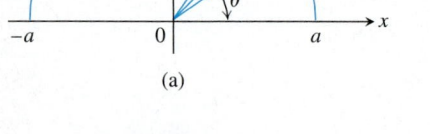

FIGURE 5.57 The semicircular wire in Example 6. (a) The dimensions and variables used in finding the center of mass. (b) The center of mass does not lie on the wire.

Hence,

$$\bar{y} = \frac{\int \tilde{y}\, dm}{\int dm} = \frac{\int_0^\pi a \sin\theta \cdot \delta a\, d\theta}{\int_0^\pi \delta a\, d\theta} = \frac{\delta a^2[-\cos\theta]_0^\pi}{\delta a \pi} = \frac{2}{\pi} a.$$

The center of mass lies on the axis of symmetry at the point $(0, 2a/\pi)$, about two-thirds of the way up from the origin (Figure 5.57b).

Centroids

When the density function is constant, it cancels out of the numerator and denominator of the formulas for $\bar{x}$ and $\bar{y}$. This happened in nearly every example in this section. As far as $\bar{x}$ and $\bar{y}$ were concerned, δ might as well have been 1. Thus, when the density is constant, the location of the center of mass is a feature of the geometry of the object and not of the material from which it is made. In such cases, engineers may call the center of mass the **centroid** of the shape, as in "Find the centroid of a triangle or a solid cone." To do so, just set δ equal to 1 and proceed to find $\bar{x}$ and $\bar{y}$ as before, by dividing moments by masses.

EXERCISES 5.7

Thin Rods

1. An 80 lb child and a 100 lb child are balancing on a seesaw. The 80 lb child is 5 ft from the fulcrum. How far from the fulcrum is the 100 lb child?

2. The ends of a log are placed on two scales. One scale reads 100 kg and the other 200 kg. Where is the log's center of mass?

3. The ends of two thin steel rods of equal length are welded together to make a right-angled frame. Locate the frame's center of mass. (*Hint:* Where is the center of mass of each rod?)

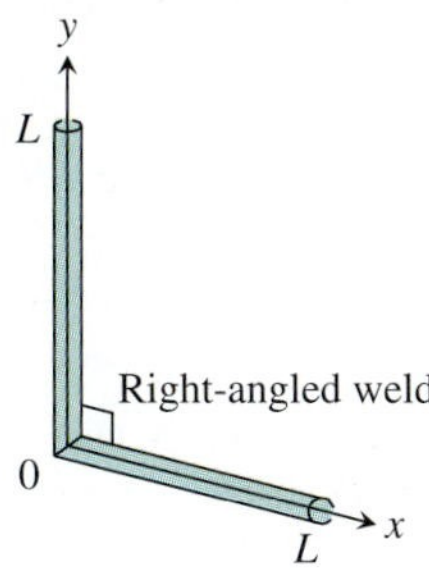

4. You weld the ends of two steel rods into a right-angled frame. One rod is twice the length of the other. Where is the frame's center of mass? (*Hint:* Where is the center of mass of each rod?)

Exercises 5–12 give density functions of thin rods lying along various intervals of the x-axis. Use Equations (3a) through (3c) to find each rod's moment about the origin, mass, and center of mass.

5. $\delta(x) = 4, \quad 0 \le x \le 2$

6. $\delta(x) = 4, \quad 1 \le x \le 3$

7. $\delta(x) = 1 + (x/3), \quad 0 \le x \le 3$

8. $\delta(x) = 2 - (x/4), \quad 0 \le x \le 4$

9. $\delta(x) = 1 + (1/\sqrt{x}), \quad 1 \le x \le 4$

10. $\delta(x) = 3(x^{-3/2} + x^{-5/2}), \quad 0.25 \le x \le 1$

11. $\delta(x) = \begin{cases} 2 - x, & 0 \le x < 1 \\ x, & 1 \le x \le 2 \end{cases}$

12. $\delta(x) = \begin{cases} x + 1, & 0 \le x < 1 \\ 2, & 1 \le x \le 2 \end{cases}$

Thin Plates with Constant Density

In Exercises 13–24, find the center of mass of a thin plate of constant density δ covering the given region.

13. The region bounded by the parabola $y = x^2$ and the line $y = 4$

14. The region bounded by the parabola $y = 25 - x^2$ and the x-axis

15. The region bounded by the parabola $y = x - x^2$ and the line $y = -x$

16. The region enclosed by the parabolas $y = x^2 - 3$ and $y = -2x^2$

17. The region bounded by the y-axis and the curve $x = y - y^3$, $0 \le y \le 1$

18. The region bounded by the parabola $x = y^2 - y$ and the line $y = x$

19. The region bounded by the x-axis and the curve $y = \cos x$, $-\pi/2 \le x \le \pi/2$

20. The region between the x-axis and the curve $y = \sec^2 x$, $-\pi/4 \le x \le \pi/4$

21. The region bounded by the parabolas $y = 2x^2 - 4x$ and $y = 2x - x^2$

22. **(a)** The region cut from the first quadrant by the circle $x^2 + y^2 = 9$

  (b) The region bounded by the x-axis and the semicircle $y = \sqrt{9 - x^2}$

  Compare your answer in part (b) with the answer in part (a).

23. The "triangular" region in the first quadrant between the circle $x^2 + y^2 = 9$ and the lines $x = 3$ and $y = 3$. (*Hint:* Use geometry to find the area.)

24. The region bounded above by the curve $y = 1/x^3$, below by the curve $y = -1/x^3$, and on the left and right by the lines $x = 1$ and $x = a > 1$. Also, find $\lim_{a\to\infty} \bar{x}$.

Thin Plates with Varying Density

25. Find the center of mass of a thin plate covering the region between the x-axis and the curve $y = 2/x^2$, $1 \le x \le 2$, if the plate's density at the point (x, y) is $\delta(x) = x^2$.

26. Find the center of a thin plate covering the region bounded below by the parabola $y = x^2$ and above by the line $y = x$ if the plate's density at the point (x, y) is $\delta(x) = 12x$.

27. The region bounded by the curves $y = \pm 4/\sqrt{x}$ and the lines $x = 1$ and $x = 4$ is revolved about the y-axis to generate a solid.

  (a) Find the volume of the solid.

  (b) Find the center of mass of a thin plate covering the region if the plate's density at the point (x, y) is $\delta(x) = 1/x$.

  (c) Sketch the plate and show the center of mass in your sketch.

28. The region between the curve $y = 2/x$ and the x-axis from $x = 1$ to $x = 4$ is revolved about the x-axis to generate a solid.

  (a) Find the volume of the solid.

  (b) Find the center of mass of a thin plate covering the region if the plate's density at the point (x, y) is $\delta(x) = \sqrt{x}$.

  (c) Sketch the plate and show the center of mass in your sketch.

Centroids of Triangles

29. *The centroid of a triangle lies at the intersection of the triangle's medians (Figure 5.58a)* You may recall that the point inside a triangle that lies one-third of the way from each side toward the opposite vertex is the point where the triangle's three medians intersect. Show that the centroid lies at the intersection of the medians by showing that it too lies one-third of the way from each side toward the opposite vertex. To do so, take the following steps.

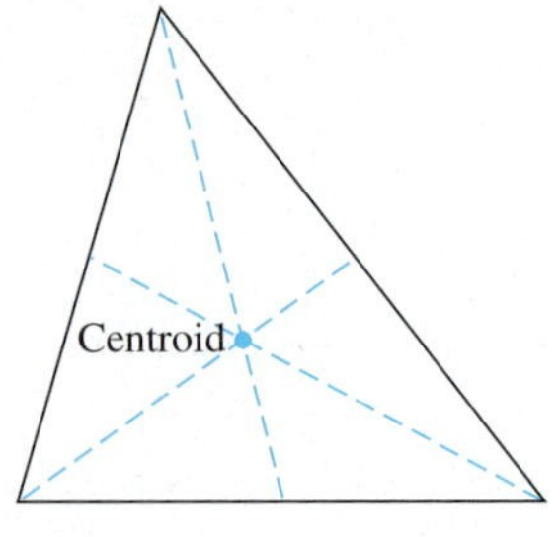

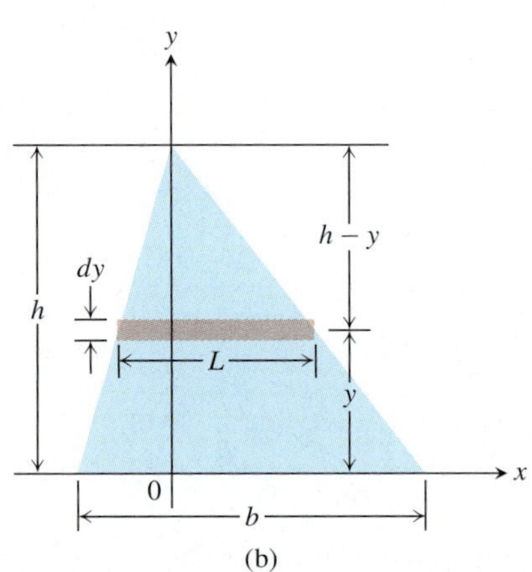

FIGURE 5.58 The triangle in Exercise 29. (a) The centroid. (b) The dimensions and variables to use in locating the center of mass.

  i. Stand one side of the triangle on the x-axis as in Figure 5.58b. Express dm in terms of L and dy.

  ii. Use similar triangles to show that $L = (b/h)(h - y)$. Substitute this expression for L in your formula for dm.

  iii. Show that $\bar{y} = h/3$.

  iv. Extend the argument to the other sides.

Use the result in Exercise 29 to find the centroids of the triangles whose vertices appear in Exercises 30–34. Assume $a, b > 0$.

30. $(-1, 0), (1, 0), (0, 3)$

31. $(0, 0), (1, 0), (0, 1)$

32. $(0, 0), (a, 0), (0, a)$

33. $(0, 0), (a, 0), (0, b)$

34. $(0, 0), (a, 0), (a/2, b)$

Thin Wires

35. *Constant density* Find the moment about the x-axis of a wire of constant density that lies along the curve $y = \sqrt{x}$ from $x = 0$ to $x = 2$.

36. *Constant density* Find the moment about the x-axis of a wire of constant density that lies along the curve $y = x^3$ from $x = 0$ to $x = 1$.

37. *Variable density* Suppose that the density of the wire in Example 6 is $\delta = k \sin \theta$ (k constant). Find the center of mass.

38. *Variable density* Suppose that the density of the wire in Example 6 is $\delta = 1 + k|\cos\theta|$ (k constant). Find the center of mass.

Engineering Formulas

Verify the statements and formulas in Exercises 39–42.

39. The coordinates of the centroid of a differentiable plane curve are

$$\bar{x} = \frac{\int x\,ds}{\text{length}}, \qquad \bar{y} = \frac{\int y\,ds}{\text{length}}.$$

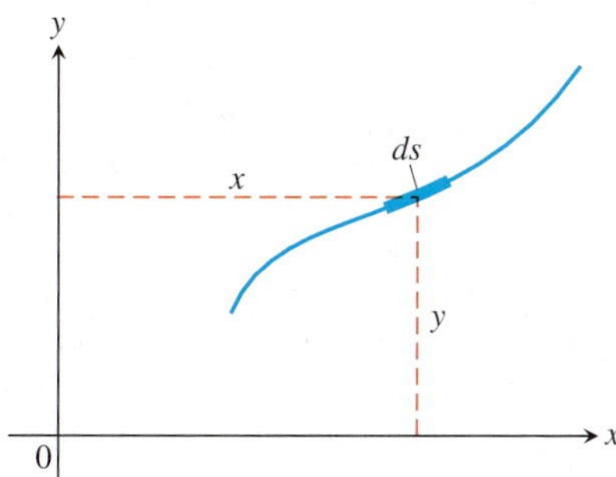

40. Whatever the value of $p > 0$ in the equation $y = x^2/(4p)$, the y-coordinate of the centroid of the parabolic segment shown here is $\bar{y} = (3/5)a$.

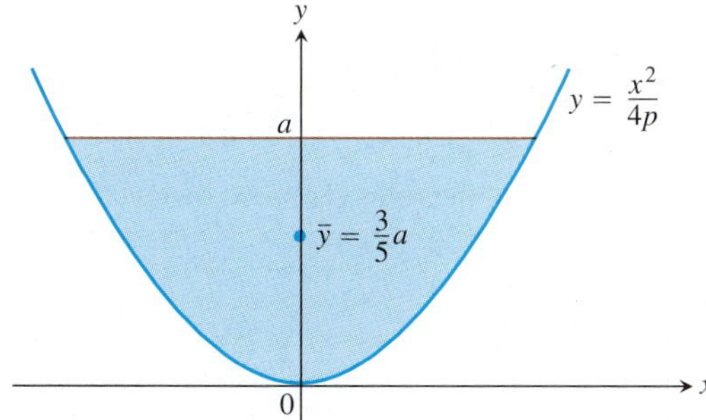

41. For wires and thin rods of constant density shaped like circular arcs centered at the origin and symmetric about the y-axis, the y-coordinate of the center of mass is

$$\bar{y} = \frac{a \sin \alpha}{\alpha} = \frac{ac}{s}.$$

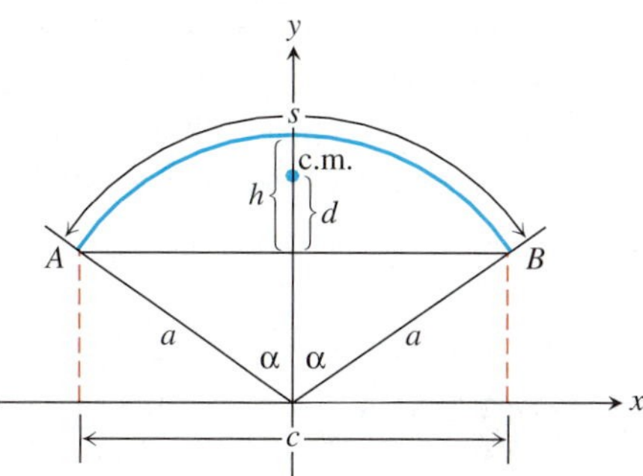

42. (*Continuation of Exercise 41*)

(a) Show that when α is small, the distance d from the centroid to chord AB is about $2h/3$ (in the notation of the figure here) by taking the following steps.

i. Show that

$$\frac{d}{h} = \frac{\sin\alpha - \alpha\cos\alpha}{\alpha - \alpha\cos\alpha}. \tag{9}$$

T **ii.** Graph

$$f(\alpha) = \frac{\sin\alpha - \alpha\cos\alpha}{\alpha - \alpha\cos\alpha}$$

and use the trace feature to show that $\lim_{\alpha\to 0^+} f(\alpha) \approx 2/3$.

(b) The error (difference between d and $2h/3$) is small even for angles greater than 45°. See for yourself by evaluating the right-hand side of Equation (9) for $\alpha = 0.2, 0.4, 0.6, 0.8$, and 1.0 rad.

Questions to Guide Your Review

1. How do you define and calculate the volumes of solids by the method of slicing? Give an example.
2. How are the disk and washer methods for calculating volumes derived from the method of slicing? Give examples of volume calculations by these methods.
3. Describe the method of cylindrical shells. Give an example.
4. How do you define and calculate the length of the graph of a smooth function $y = f(x)$ over a closed interval? Give an example.
5. How do you find the length of a smooth parametrized curve $x = f(t)$, $y = g(t)$, $a \le t \le b$? What does smoothness have to do with length? What else do you need to know about the parametrization to find the curve's length? Give examples.
6. What is a first-order differential equation? When is a function a solution of such an equation?
7. How do you solve separable first-order differential equations?
8. What is the slope field of a differential equation $y' = f(x, y)$? What can we learn from such fields?

9. What is an initial value problem? How do you solve one? Give an example.

10. What is the law of exponential change? How can it be derived from an initial value problem? What are some of the applications of the law?

11. How do you define and calculate the work done by a variable force directed along a portion of the x-axis? How do you calculate the work it takes to pump a liquid from a tank? Give examples.

12. What is Hooke's law for springs? When might Hooke's law give poor results in modeling the behavior of a spring? Give examples.

13. How do you calculate the force exerted by a liquid against a portion of a flat vertical wall? Give an example.

14. What is a center of mass?

15. How do you locate the center of mass of a straight, narrow rod or strip of material? Give an example. If the density of the material is constant, you can tell right away where the center of mass is. Where is it?

16. How do you locate the center of mass of a thin flat plate of material? Give an example.

Practice Exercises

Volumes

Find the volumes of the solids in Exercises 1–16.

1. The solid lies between planes perpendicular to the x-axis at $x = 0$ and $x = 1$. The cross sections perpendicular to the x-axis between these planes are circular disks whose diameters run from the parabola $y = x^2$ to the parabola $y = \sqrt{x}$.

2. The base of the solid is the region in the first quadrant between the line $y = x$ and the parabola $y = 2\sqrt{x}$. The cross sections of the solid perpendicular to the x-axis are equilateral triangles whose bases stretch from the line to the curve.

3. The solid lies between planes perpendicular to the x-axis at $x = \pi/4$ and $x = 5\pi/4$. The cross sections between these planes are circular disks whose diameters run from the curve $y = 2\cos x$ to the curve $y = 2\sin x$.

4. The solid lies between planes perpendicular to the x-axis at $x = 0$ and $x = 6$. The cross sections between these planes are squares whose bases run from the x-axis up to the curve $x^{1/2} + y^{1/2} = \sqrt{6}$.

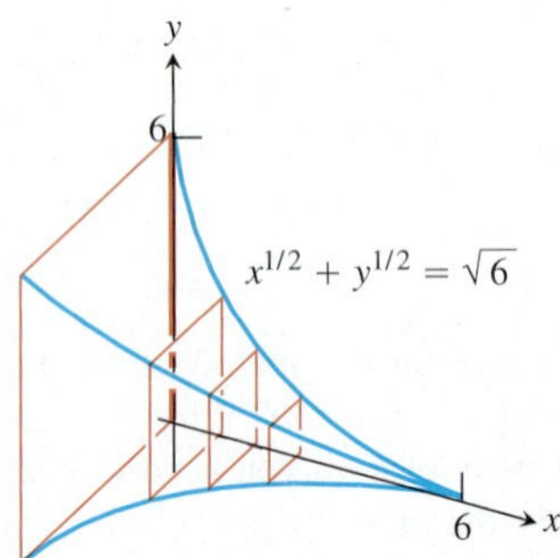

5. The solid lies between planes perpendicular to the x-axis at $x = 0$ and $x = 4$. The cross sections of the solid perpendicular to the x-axis between these planes are circular disks whose diameters run from the curve $x^2 = 4y$ to the curve $y^2 = 4x$.

6. The base of the solid is the region bounded by the parabola $y^2 = 4x$ and the line $x = 1$ in the xy-plane. Each cross section perpendicular to the x-axis is an equilateral triangle with one edge in the plane. (The triangles all lie on the same side of the plane.)

7. Find the volume of the solid generated by revolving the region bounded by the x-axis, the curve $y = 3x^4$, and the lines $x = 1$ and $x = -1$ about

(a) the x-axis

(b) the y-axis

(c) the line $x = 1$

(d) the line $y = 3$.

8. Find the volume of the solid generated by revolving the "triangular" region bounded by the curve $y = 4/x^3$ and the lines $x = 1$ and $y = 1/2$ about

(a) the x-axis

(b) the y-axis

(c) the line $x = 2$

(d) the line $y = 4$.

9. Find the volume of the solid generated by revolving the region bounded on the left by the parabola $x = y^2 + 1$ and on the right by the line $x = 5$ about

(a) the x-axis

(b) the y-axis

(c) the line $x = 5$.

10. Find the volume of the solid generated by revolving the region bounded by the parabola $y^2 = 4x$ and the line $y = x$ about

(a) the x-axis

(b) the y-axis

(c) the line $x = 4$

(d) the line $y = 4$.

11. Find the volume of the solid generated by revolving the region enclosed by the graphs of $y = e^{x/2}$, $y = 1$, and $x = \ln 3$ about the x-axis.

12. Find the volume of the solid generated by revolving the region bounded by the curve $y = \sin x$ and the lines $x = 0$, $x = \pi$, and $y = 2$ about the line $y = 2$.

13. Find the volume of the solid generated by revolving the region between the x-axis and the curve $y = x^2 - 2x$ about

(a) the x-axis

(b) the line $y = -1$

(c) the line $x = 2$

(d) the line $y = 2$.

14. Find the volume of the solid generated by revolving about the x-axis the region bounded by $y = 2 \tan x$, $y = 0$, $x = -\pi/4$, and $x = \pi/4$. (The region lies in the first and third quadrants and resembles a skewed bow tie.)

15. *Volume of a solid sphere with a hole* A round hole of radius $\sqrt{3}$ ft is bored through the center of a solid sphere of radius 2 ft. Find the volume of material removed from the sphere.

CD-ROM WEBsite **16.** *Volume of a football* The profile of a football resembles the ellipse shown here. Find the football's volume to the nearest cubic inch.

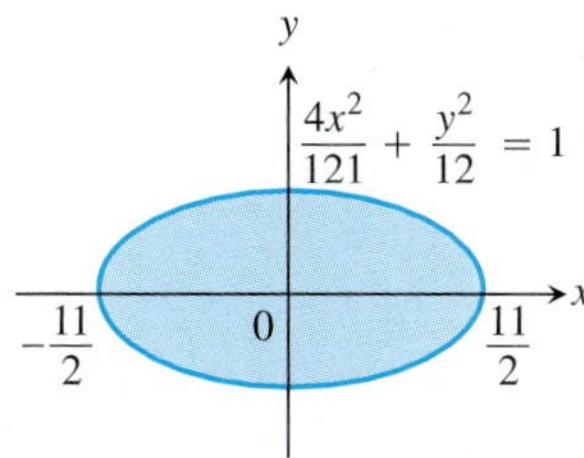

Lengths of Curves

Find the lengths of the curves in Exercises 17–22.

CD-ROM WEBsite **17.** $y = x^{1/2} - (1/3)x^{3/2}, \quad 1 \le x \le 4$

18. $x = y^{2/3}, \quad 1 \le y \le 8$

CD-ROM WEBsite **19.** $y = (5/12)x^{6/5} - (5/8)x^{4/5}, \quad 1 \le x \le 32$

20. $x = (y^3/12) + (1/y), \quad 1 \le y \le 2$

21. $x = 5 \cos t - \cos 5t, \quad y = 5 \sin t - \sin 5t, \quad 0 \le t \le \pi/2$

22. $x = t^2, \quad y = 2t, \quad 0 \le t \le 1$

23. *Finding a function* Find a function f that has a continuous derivative on $(0, \infty)$ and that has both of the following properties.

i. The graph of f goes through the point $(1, 1)$.

ii. The length L of the curve from $(1, 1)$ to any point $(x, f(x))$ is given by the formula $L = \ln x + f(x) - 1$.

24. Find the length of the enclosed loop $x = t^2$, $y = (t^3/3) - t$ shown here. The loop starts at $t = -\sqrt{3}$ and ends at $t = \sqrt{3}$.

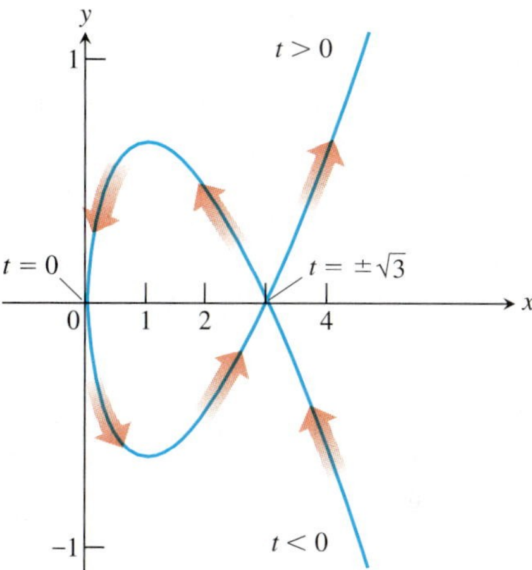

Work

25. *Lifting equipment* A rock climber is about to haul up 100 N (about 22.5 lb) of equipment that has been hanging beneath her on 40 m of rope that weighs 0.8 newton per meter. How much work will it take? (*Hint:* Solve for the rope and equipment separately, then add.)

26. *Leaky tank truck* You drove an 800 gal tank truck of water from the base of Mt. Washington to the summit and discovered on arrival that the tank was only half full. You started with a full tank, climbed at a steady rate, and accomplished the 4750 ft elevation change in 50 min. Assuming that the water leaked out at a steady rate, how much work was spent in carrying water to the top? Do not count the work done in getting yourself and the truck there. Water weighs 8 lb/U.S. gal.

27. *Stretching a spring* If a force of 20 lb is required to hold a spring 1 ft beyond its unstressed length, how much work does it take to stretch the spring this far? An additional foot?

28. *Garage door spring* A force of 200 N will stretch a garage door spring 0.8 m beyond its unstressed length. How far will a 300 N force stretch the spring? How much work does it take to stretch the spring this far from its unstressed length?

29. *Pumping a reservoir* A reservoir shaped like a right circular cone, point down, 20 ft across the top and 8 ft deep, is full of water. How much work does it take to pump the water to a level 6 ft above the top?

30. *Pumping a reservoir* (*Continuation of Exercise 29*) The reservoir is filled to a depth of 5 ft, and the water is to be pumped to the same level as the top. How much work does it take?

31. *Pumping a conical tank* A right circular conical tank, point down, with top radius 5 ft and height 10 ft is filled with a liquid whose weight-density is 60 lb/ft^3. How much work does it take to pump the liquid to a point 2 ft above the tank? If the pump is driven by a motor rated at 275 ft · lb/sec (1/2 hp), how long will it take to empty the tank?

32. *Pumping a cylindrical tank* A storage tank is a right circular cylinder 20 ft long and 8 ft in diameter with its axis horizontal. If the tank is half full of olive oil weighing 57 lb/ft^3, find the work done in emptying it through a pipe that runs from the bottom of the tank to an outlet that is 6 ft above the top of the tank.

Fluid Force

33. *Trough of water* The vertical triangular plate shown here is the end plate of a trough full of water ($w = 62.4$). What is the fluid force against the plate?

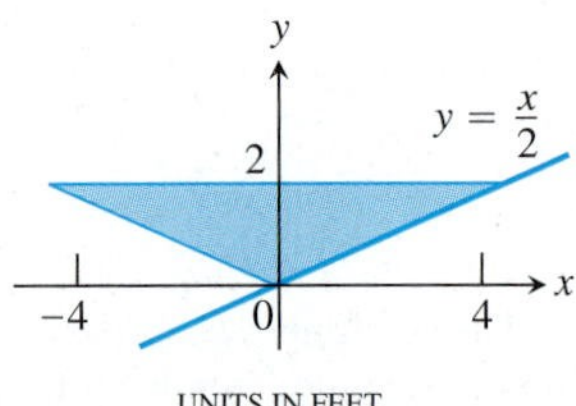

34. *Trough of maple syrup* The vertical trapezoidal plate shown here is the end plate of a trough full of maple syrup weighing 75 lb/ft^3. What is the force exerted by the syrup against the end plate of the trough when the syrup is 10 in. deep?

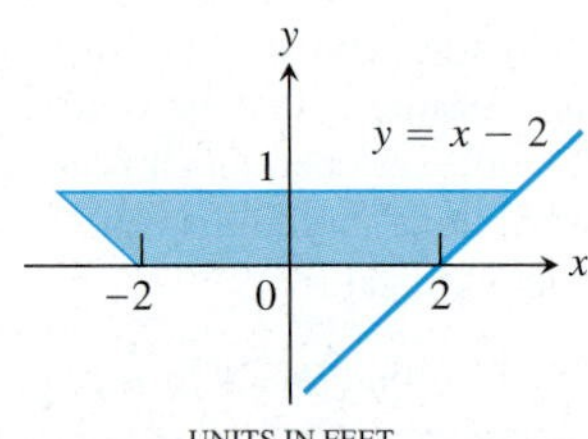

35. *Force on a parabolic gate* A flat vertical gate in the face of a dam is shaped like the parabolic region between the curve $y = 4x^2$ and the line $y = 4$, with measurements in feet. The top of the gate lies 5 ft below the surface of the water. Find the force exerted by the water against the gate ($w = 62.4$).

36. *Force on a trapezoid* The isosceles trapezoidal plate shown here is submerged vertically in water ($w = 62.4$) with its upper edge 4 ft below the surface. Find the fluid force on one side of the plate.

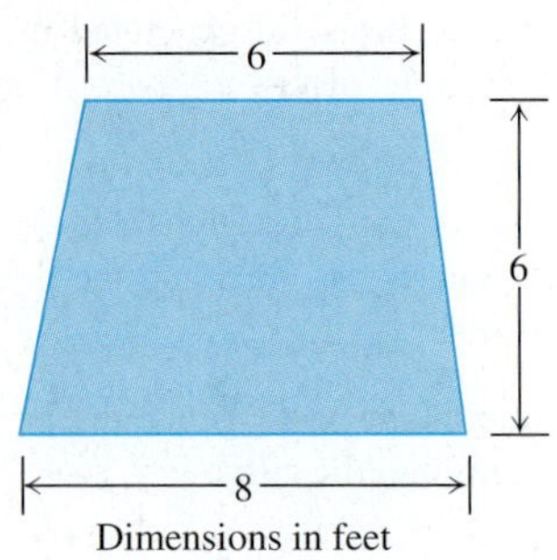

First-Order Separable Equations

Solve the differential equations in Exercises 37–42.

37. $\dfrac{dy}{dx} = x^2\sqrt{y}, \quad y > 0$

38. $\dfrac{dy}{dx} = e^{2x+3y}$

39. $x\dfrac{dy}{dx} = y \ln x$

40. $\csc x \dfrac{dy}{dx} = e^{\cos x - y}$

41. $(\sec^2 \sqrt{x})\dfrac{dx}{dt} = \sqrt{x}$

42. $\sin t - (x \cos^2 t)\dfrac{dx}{dt} = 0, \qquad -\pi/2 < t < \pi/2$

Centroids and Centers of Mass

43. Find the centroid of a thin, flat plate covering the region enclosed by the parabolas $y = 2x^2$ and $y = 3 - x^2$.

44. Find the centroid of a thin, flat plate covering the region enclosed by the x-axis, the lines $x = 2$ and $x = -2$, and the parabola $y = x^2$.

45. Find the centroid of a thin, flat plate covering the "triangular" region in the first quadrant bounded by the y-axis, the parabola $y = x^2/4$, and the line $y = 4$.

46. Find the centroid of a thin, flat plate covering the region enclosed by the parabola $y^2 = x$ and the line $x = 2y$.

47. *Variable density* Find the center of mass of a thin, flat plate covering the region enclosed by the parabola $y^2 = x$ and the line $x = 2y$ if the density function is $\delta(y) = 1 + y$. (Use horizontal strips.)

48. **(a)** *Constant density* Find the center of mass of a thin plate of constant density covering the region between the curve $y = 3/x^{3/2}$ and the x-axis from $x = 1$ to $x = 9$.

(b) *Variable density* Find the plate's center of mass if, instead of being constant, the density is $\delta(x) = x$. (Use vertical strips.)

Additional Exercises: Theory, Examples, Applications

Volume and Length

1. A solid is generated by revolving about the x-axis the region bounded by the graph of the continuous function $y = f(x)$, the x-axis, and the lines $x = 0$ and $x = a$. Its volume, for all $a > 0$, is $a^2 + a$. Find $f(x)$.

2. Suppose that the function $f(x)$ is nonnegative and continuous for $x \geq 0$. Suppose also that, for every positive number b, revolving the region enclosed by the graph of f, the coordinate axes, and the line $x = b$ about the y-axis generates a solid of volume $2\pi b^3$. Find $f(x)$.

3. Suppose that the increasing function $f(x)$ is smooth for $x \geq 0$ and that $f(0) = a$. Let $s(x)$ denote the length of the graph of f from $(0, a)$ to $(x, f(x))$, $x > 0$. Find $f(x)$ if $s(x) = Cx$ for some constant C. What are the allowable values for C?

4. **(a)** Show that for $0 < \alpha \leq \pi/2$,

$$\int_0^{\alpha} \sqrt{1 + \cos^2 \theta}\, d\theta > \sqrt{\alpha^2 + \sin^2 \alpha}.$$

   (b) Generalize the result in part (a).

Work and Fluid Force

5. *Work and kinetic energy* Suppose that a 1.6 oz golf ball is placed on a vertical spring with force constant $k = 2$ lb/in. The spring is compressed 6 in. and released. About how high does the ball go (measured from the spring's rest position)?

6. *Fluid force* A triangular plate ABC is submerged in water with its plane vertical. The side AB, 4 ft long, is 6 ft below the surface of the water, whereas the vertex C is 2 ft below the surface. Find the force exerted by the water on one side of the plate.

7. *Average pressure* A vertical rectangular plate is submerged in a fluid with its top edge parallel to the fluid's surface. Show that the force exerted by the fluid on one side of the plate equals the average value of the pressure up and down the plate times the area of the plate.

8. *Vertical square plate* The container profile shown here is filled with two nonmixing liquids of weight density w_1 and w_2. Find the fluid force on one side of the vertical square plate $ABCD$. The points B and D lie in the boundary layer and the square is $6\sqrt{2}$ ft on a side. (*Hint:* If y is measured downward from the fluid's surface, then the pressure is $p = w_1 y$ for $0 \leq y \leq 8$ and $p = 8w_1 + w_2(y - 8)$ for $y > 8$.)

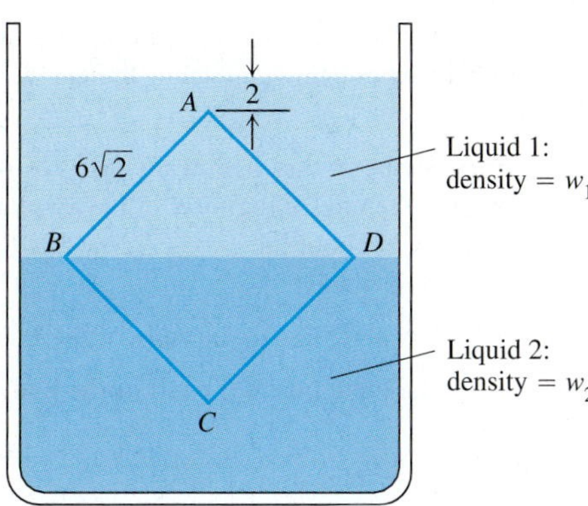

First-Order Differential Equations

9. *Radioactivity* A small amount A_0 of a radioactive substance is placed in a lead container. After 24 h, it is observed that 6/7 of the original amount is remaining. If the rate of decay is proportional to the amount of substance present at any time, what is the half-life of this substance? How long will it take for the substance to be reduced to 1/5 of its original amount?

10. *Gypsy moth population* The rate of increase of the gypsy moth caterpillar population is directly proportional to the number P of the caterpillars present at any time t. The results of a survey show that the caterpillars increased from 2 million in 1979 to 3 million in 1981. Predict the gypsy moth caterpillar population in 1985.

11. *Cooling broth* In a room at a constant temperature of 70°F, the temperature of a cup of broth changes from 200° at noon to 190° at 1 P.M. Predict the temperature of the broth at 3:30 P.M.

12. *A murder victim* The body of a murder victim was discovered at 11 P.M. one evening. The police doctor on call arrived at 11:30 P.M. and immediately took the temperature of the body, which was 94.6°F. He again took the temperature after 1 h, when it was 93.4°F, and he noted that the temperature of the room was a constant 70°F. Use Newton's Law of Cooling to estimate the time of death, assuming that the victim's normal body temperature was 98.6°F.

Moments and Centers of Mass

13. *Limiting position of a centroid* Find the centroid of the region bounded below by the x-axis and above by the curve $y = 1 - x^n$, n an even positive integer. What is the limiting position of the centroid as $n \to \infty$?

14. *Telephone pole* If you haul a telephone pole on a two-wheeled carriage behind a truck, you want the wheels to be 3 ft or so behind the pole's center of mass to provide an adequate "tongue" weight. NYNEX's class 1 40 ft wooden poles have a 27 in. circumference at the top and a 43.5 in. circumference at the base. About how far from the top is the center of mass?

15. *Constant density* Suppose that a thin metal plate of area A and constant density δ occupies a region R in the xy-plane, and let M_y be the plate's moment about the y-axis. Show that the plate's moment about the line $x = b$ is

(a) $M_y - b\delta A$ if the plate lies to the right of the line

(b) $b\delta A - M_y$ if the plate lies to the left of the line.

16. *Variable density* Find the center of mass of a thin plate covering the region bounded by the curve $y^2 = 4ax$ and the line $x = a$, $a =$ positive constant, if the density at (x, y) is directly proportional to (a) x, (b) $|y|$.

17. (a) *Concentric circles* Find the centroid of the region in the first quadrant bounded by two concentric circles and the coordinate axes, if the circles have radii a and b, $0 < a < b$, and their centers are at the origin.

(b) *Writing to Learn* Find the limits of the coordinates of the centroid as a approaches b and discuss the meaning of the result.

18. *Cutting a corner from a square* A triangular corner is cut from a square 1 ft on a side. The area of the triangle removed is 36 in^2. If the centroid of the remaining region is 7 in. from one side of the original square, how far is it from the remaining sides?

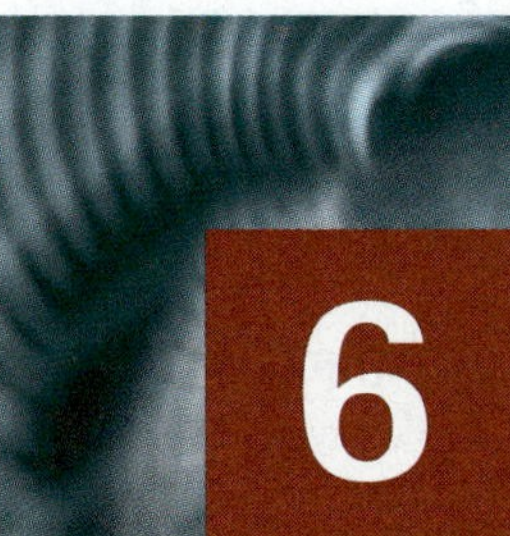

6 Transcendental Functions and Differential Equations

OVERVIEW The functions $\ln x$ and e^x are probably the best-known function–inverse pair. In this chapter, we take a new view of the calculus of these functions and broaden our acquaintance with the amazing range of problems they solve. We also introduce the hyperbolic functions and their inverses, with applications to integration, skydiving, trucking, and hanging cables.

6.1 Logarithms

In this section, we define the natural logarithm function as an integral, based on the Fundamental Theorem of Calculus. This is a different approach from starting with e^x and defining $\ln x$ as its inverse, as we did in the Preliminary chapter. We calculate the derivative of $\ln x$ based on its new definition, which differs from the way we made the calculation in Section 2.9. Of course, the results are exactly the same.

The importance of logarithms came at first from the improvement they brought to arithmetic. The revolutionary properties of logarithms made possible the calculations of the great seventeenth-century advances in offshore navigation and celestial mechanics. Nowadays we do complicated arithmetic with calculators, but the properties of logarithms remain as important as ever in calculus and modeling.

The Natural Logarithm Function

The natural logarithm of a positive number x, written as $\ln x$, is the value of an integral.

Definition The Natural Logarithm Function

$$\ln x = \int_1^x \frac{1}{t}\, dt, \qquad x > 0$$

If $x > 1$, then $\ln x$ is the area under the curve $y = 1/t$ from $t = 1$ to $t = x$ (Figure 6.1). For $0 < x < 1$, $\ln x$ gives the negative of the area under the curve from x to 1. The function is not defined for $x \leq 0$. We also have

$$\ln 1 = \int_1^1 \frac{1}{t}\, dt = 0. \qquad \text{Upper and lower limits equal}$$

Notice that we show the graph of $y = 1/x$ in Figure 6.1, but use $y = 1/t$ in the integral. Using x for everything would have us writing

$$\ln x = \int_1^x \frac{1}{x}\, dx,$$

with x meaning two different things. So we change the variable of integration to t.

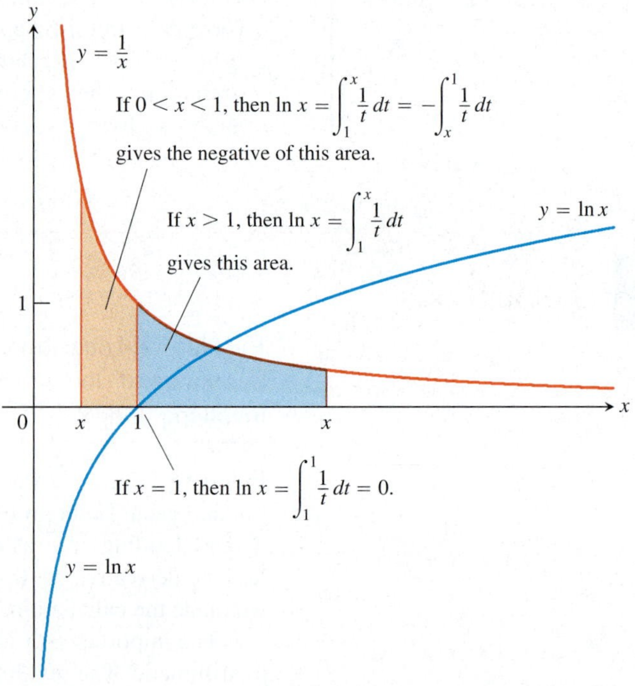

FIGURE 6.1 The graph of $y = \ln x$ and its relation to the function $y = 1/x$, $x > 0$. The graph of the logarithm rises above the x-axis as x moves from 1 to the right, and it falls below the axis as x moves from 1 to the left.

Derivative of y = ln x

By the first part of the Fundamental Theorem of Calculus (in Section 4.5),

$$\frac{d}{dx} \ln x = \frac{d}{dx} \int_1^x \frac{1}{t}\, dt = \frac{1}{x}.$$

CD-ROM WEBsite

For every positive value of x, therefore,

$$\frac{d}{dx} \ln x = \frac{1}{x}, \qquad x > 0$$

Since $x > 0$ in the definition of $\ln x$, we see that its derivative is always positive so $\ln x$ is increasing everywhere in its domain. The second derivative, $-1/x^2$, is negative, so the graph of $\ln x$ is everywhere concave down.

If u is a differentiable function of x whose values are positive, so that $\ln u$ is defined, then applying the Chain Rule

$$\frac{dy}{dx} = \frac{dy}{du}\frac{du}{dx}$$

to the function $y = \ln u$ gives

$$\frac{d}{dx}\ln u = \frac{d}{du}\ln u \cdot \frac{du}{dx} = \frac{1}{u}\frac{du}{dx}.$$

$$\frac{d}{dx}\ln u = \frac{1}{u}\frac{du}{dx}, \qquad u > 0 \tag{1}$$

Example 1 Derivatives of Natural Logarithms

(a) $\dfrac{d}{dx}\ln 2x = \dfrac{1}{2x}\dfrac{d}{dx}(2x) = \dfrac{1}{2x}(2) = \dfrac{1}{x}$

(b) Equation (1) with $u = x^2 + 3$ gives

$$\frac{d}{dx}\ln(x^2+3) = \frac{1}{x^2+3}\cdot\frac{d}{dx}(x^2+3) = \frac{1}{x^2+3}\cdot 2x = \frac{2x}{x^2+3}.$$

Notice the remarkable occurrence in Example 1(a). The function $y = \ln 2x$ has the same derivative as the function $y = \ln x$. This is true of $y = \ln ax$ for any number a:

$$\frac{d}{dx}\ln ax = \frac{1}{ax}\cdot\frac{d}{dx}(ax) = \frac{1}{ax}(a) = \frac{1}{x}. \tag{2}$$

Derivative of $\log_a u$

To find the derivative of a base a logarithm, we first convert it to a natural logarithm (see Preliminary Section 4). If u is a positive differentiable function of x, then

$$\frac{d}{dx}(\log_a u) = \frac{d}{dx}\left(\frac{\ln u}{\ln a}\right) = \frac{1}{\ln a}\frac{d}{dx}(\ln u) = \frac{1}{\ln a}\cdot\frac{1}{u}\frac{du}{dx}.$$

For $a > 0$ and $a \neq 1$,

$$\frac{d}{dx}(\log_a u) = \frac{1}{\ln a}\cdot\frac{1}{u}\frac{du}{dx}, \qquad u > 0 \tag{3}$$

Equation (3) is the same formula we derived in Section 2.9.

Example 2 Differentiating Base *a* Logarithms

$$\frac{d}{dx}\log_{10}(3x+1) = \frac{1}{\ln 10}\cdot\frac{1}{3x+1}\frac{d}{dx}(3x+1) = \frac{3}{(\ln 10)(3x+1)}$$

Laws of Logarithms

For any numbers $a > 0$ and $x > 0$,

1. $\ln ax = \ln a + \ln x$
2. $\ln \frac{a}{x} = \ln a - \ln x$
3. $\ln x^n = n \ln x$

Proof that ln ax = ln a + ln x The argument is unusual, and elegant. It starts by observing that $\ln ax$ and $\ln x$ have the same derivative; see Equation (2). According to Corollary 1 of the Mean Value Theorem, then, the functions must differ by a constant, which means that

$$\ln ax = \ln x + C \tag{4}$$

for some C. With this much accomplished, it remains only to show that C equals $\ln a$.

Equation (4) holds for all positive values of x, so it must hold for $x = 1$. Hence,

$$\begin{aligned} \ln(a\cdot 1) &= \ln 1 + C \\ \ln a &= 0 + C && \ln 1 = 0 \\ C &= \ln a. && \text{Rearranged} \end{aligned}$$

Substituting $C = \ln a$ in Equation (4) gives the equation we wanted to prove:

$$\ln ax = \ln a + \ln x. \tag{5}$$

Proof that ln (a/x) = ln a − ln x We get this from Equation (5) in two stages. Equation (5) with a replaced by $1/x$ gives

$$\begin{aligned} \ln\frac{1}{x} + \ln x &= \ln\left(\frac{1}{x}\cdot x\right) \\ &= \ln 1 = 0, \end{aligned}$$

so that

$$\ln\frac{1}{x} = -\ln x.$$

Equation (5) with x replaced by $1/x$ then gives

$$\ln \frac{a}{x} = \ln \left(a \cdot \frac{1}{x}\right) = \ln a + \ln \frac{1}{x}$$
$$= \ln a - \ln x.$$

Proof that $\ln x^n = n \ln x$ (assuming n rational) We use the same-derivative argument again. For all positive values of x,

$$\frac{d}{dx} \ln x^n = \frac{1}{x^n} \frac{d}{dx}(x^n) \qquad \text{Eq. (1) with } u = x^n$$
$$= \frac{1}{x^n} nx^{n-1}$$
$$= n \cdot \frac{1}{x} = \frac{d}{dx}(n \ln x).$$

Here is where we need n to be rational, at least for now. We have proved the Power Rule only for rational exponents.

Since $\ln x^n$ and $n \ln x$ have the same derivative,

$$\ln x^n = n \ln x + C$$

for some constant C. Taking x to be 1 identifies C as zero, and we're done.

As for using the rule $\ln x^n = n \ln x$ for irrational values of n, go right ahead and do so. It does hold for all n, and there is no need to pretend otherwise. Be aware that the rule is far from proved.

The Integral $\int (1/u)\, du$

Equation (1) leads to the integral formula

$$\int \frac{1}{u}\, du = \ln u + C$$

when u is a positive differentiable function. As we saw in Section 4.2 this formula can be generalized to include negative functions as well.

Whether $u < 0$ or $u > 0$,

$$\int \frac{1}{u}\, du = \ln |u| + C. \qquad (6)$$

We know that

$$\int u^n\, du = \frac{u^{n+1}}{n+1} + C, \qquad n \neq -1.$$

Equation (6) explains what to do when n equals -1.

Equation (6) says that integrals of a certain *form* lead to logarithms. That is,

$$\int \frac{f'(x)}{f(x)}\,dx = \ln|f(x)| + C$$

whenever $f(x)$ is a differentiable function that maintains a constant sign on the domain given for it.

Example 3 Using Substitution

$$\int_{-\pi/2}^{\pi/2} \frac{4\cos\theta}{3+2\sin\theta}\,d\theta = \int_1^5 \frac{2}{u}\,du \qquad u = 3+2\sin\theta,\ du = 2\cos\theta\,d\theta,\ u(-\pi/2)=1,\ u(\pi/2)=5$$

$$= 2\ln|u|\Big]_1^5 \qquad \text{Eq. (6)}$$

$$= 2\ln|5| - 2\ln|1| = 2\ln 5$$

Example 4 Integral of tan u

$$\int_0^{\pi/6} \tan 2x\,dx = \int_0^{\pi/3} \tan u \cdot \frac{du}{2} = \frac{1}{2}\int_0^{\pi/3} \tan u\,du \qquad \text{Substitute } u = 2x,\ dx = du/2,\ u(0)=0,\ u(\pi/6)=\pi/3.$$

$$= \frac{1}{2}\ln|\sec u|\Big]_0^{\pi/3} = \frac{1}{2}(\ln 2 - \ln 1) = \frac{1}{2}\ln 2$$

Integrals Involving $\log_a x$

To evaluate integrals involving base a logarithms, we convert them to natural logarithms.

Example 5 Using Substitution

$$\int \frac{\log_2 x}{x}\,dx = \frac{1}{\ln 2}\int \frac{\ln x}{x}\,dx \qquad \log_2 x = \frac{\ln x}{\ln 2}$$

$$= \frac{1}{\ln 2}\int u\,du \qquad u = \ln x,\ du = \frac{1}{x}\,dx$$

$$= \frac{1}{\ln 2}\frac{u^2}{2} + C = \frac{1}{\ln 2}\frac{(\ln x)^2}{2} + C = \frac{(\ln x)^2}{2\ln 2} + C$$

EXERCISES 6.1

Derivatives of Logarithms

In Exercises 1–28, find the derivative of y with respect to x, t, or θ, as appropriate.

1. $y = \ln 3x$
2. $y = \ln(\theta + 1)$
3. $y = \ln x^3$
4. $y = (\ln x)^3$
5. $y = t(\ln t)^2$
6. $y = t\sqrt{\ln t}$
7. $y = \dfrac{x^4}{4}\ln x - \dfrac{x^4}{16}$
8. $y = \dfrac{1 + \ln t}{t}$
9. $y = \dfrac{\ln t}{t}$
10. $y = \dfrac{x \ln x}{1 + \ln x}$
11. $y = \dfrac{\ln x}{1 + \ln x}$
12. $y = \ln(\ln(\ln x))$
13. $y = \theta(\sin(\ln\theta) + \cos(\ln\theta))$
14. $y = \ln(\sec\theta + \tan\theta)$
15. $y = \ln\dfrac{1}{x\sqrt{x+1}}$
16. $y = \sqrt{\ln\sqrt{t}}$
17. $y = \dfrac{1 + \ln t}{1 - \ln t}$
18. $y = \dfrac{1 + (\ln t)^2}{1 - (\ln t)^2}$
19. $y = \ln(\sec(\ln\theta))$
20. $y = \ln\left(\dfrac{(x^2 + 1)^5}{\sqrt{1 - x}}\right)$
21. $y = \log_2 5\theta$
22. $y = \log_4 x + \log_4 x^2$
23. $y = \log_2 r \cdot \log_4 r$
24. $y = \log_3\left(\left(\dfrac{x + 1}{x - 1}\right)^{\ln 3}\right)$
25. $y = \theta\sin(\log_7\theta)$
26. $y = 3\log_8(\log_2 t)$

In Exercises 27 and 28, use Leibniz's Rule in the Chapter 4 Additional Exercises to find the derivative of y with respect to x.

27. $y = \displaystyle\int_{x^2/2}^{x^2} \ln\sqrt{t}\,dt$
28. $y = \displaystyle\int_{\sqrt{x}}^{\sqrt[3]{x}} \ln t\,dt$

Integration

Evaluate the integrals in Exercises 29–46.

29. $\displaystyle\int_{-3}^{-2} \frac{dx}{x}$
30. $\displaystyle\int_{-1}^{0} \frac{3\,dx}{3x - 2}$
31. $\displaystyle\int \frac{2y\,dy}{y^2 - 25}$
32. $\displaystyle\int \frac{8r\,dr}{4r^2 - 5}$
33. $\displaystyle\int_0^{\pi} \frac{\sin t}{2 - \cos t}\,dt$
34. $\displaystyle\int_0^{\pi/3} \frac{4\sin\theta}{1 - 4\cos\theta}\,d\theta$
35. $\displaystyle\int_1^2 \frac{2\ln x}{x}\,dx$
36. $\displaystyle\int_2^4 \frac{dx}{x\ln x}$
37. $\displaystyle\int_2^4 \frac{dx}{x(\ln x)^2}$
38. $\displaystyle\int_2^{16} \frac{dx}{2x\sqrt{\ln x}}$
39. $\displaystyle\int \frac{3\sec^2 t}{6 + 3\tan t}\,dt$
40. $\displaystyle\int \frac{\sec y \tan y}{2 + \sec y}\,dy$
41. $\displaystyle\int_0^{\pi/2} \tan\frac{x}{2}\,dx$
42. $\displaystyle\int_{\pi/4}^{\pi/2} \cot t\,dt$
43. $\displaystyle\int_{\pi/2}^{\pi} 2\cot\frac{\theta}{3}\,d\theta$
44. $\displaystyle\int_0^{\pi/12} 6\tan 3x\,dx$
45. $\displaystyle\int \frac{dx}{2\sqrt{x} + 2x}$
46. $\displaystyle\int \frac{\sec x\,dx}{\sqrt{\ln(\sec x + \tan x)}}$

Theory and Applications

47. *Absolute extrema* Locate and identify the absolute extreme values of
   (a) $\ln(\cos x)$ on $[-\pi/4, \pi/3]$
   (b) $\cos(\ln x)$ on $[1/2, 2]$.

48. *$\ln x < x$ if $x > 1$*
   (a) Prove that $f(x) = x - \ln x$ is increasing for $x > 1$.
   (b) Using part (a), show that $\ln x < x$ if $x > 1$.

49. *Area* Find the area between the curves $y = \ln x$ and $y = \ln 2x$ from $x = 1$ to $x = 5$.

50. *Area* find the area between the curve $y = \tan x$ and the x-axis from $x = -\pi/4$ to $x = \pi/3$.

Initial Value Problems

Solve the initial value problems in Exercises 51 and 52.

51. $\dfrac{dy}{dx} = 1 + \dfrac{1}{x}, \quad y(1) = 3$

52. $\dfrac{d^2y}{dx^2} = \sec^2 x, \quad y(0) = 0 \quad \text{and} \quad y'(0) = 1$

Logarithms to Other Bases

Evaluate the integrals in Exercises 53–58.

53. $\displaystyle\int \frac{\log_{10} x}{x}\, dx$

54. $\displaystyle\int_1^4 \frac{\ln 2 \log_2 x}{x}\, dx$

55. $\displaystyle\int_0^2 \frac{\log_2 (x+2)}{x+2}\, dx$

56. $\displaystyle\int_0^9 \frac{2 \log_{10} (x+1)}{x+1}\, dx$

57. $\displaystyle\int \frac{dx}{x \log_{10} x}$

58. $\displaystyle\int \frac{dx}{x(\log_8 x)^2}$

CD-ROM WEBsite

59. *The linearization of ln (1 + x) at x = 0* Instead of approximating $\ln x$ near $x = 1$, we approximate $\ln (1 + x)$ near $x = 0$. We get a simpler formula this way.

(a) Derive the linearization $\ln (1 + x) \approx x$ at $x = 0$.

(b) Estimate to 5 decimal places the error involved in replacing $\ln (1 + x)$ by x on the interval $[0, 0.1]$.

(c) Graph $\ln (1 + x)$ and x together for $0 \le x \le 0.5$. Use different colors, if available. At what points does the approximation of $\ln (1 + x)$ seem best? Least good? By reading coordinates from the graphs, find as good an upper bound for the error as your grapher will allow.

T **60.** *Estimating values of ln x with Simpson's Rule* Although linearizations are good for replacing the logarithmic function over short intervals, Simpson's Rule is better for estimating *particular* values of $\ln x$.

As a case in point, the values of $\ln (1.2)$ and $\ln (0.8)$ to 5 places are

$$\ln (1.2) = 0.18232, \qquad \ln (0.8) = -0.22314.$$

Estimate $\ln (1.2)$ and $\ln (0.8)$ first with the formula $\ln (1 + x) \approx x$ and then use Simpson's Rule with $n = 2$. (Impressive, isn't it?)

6.2 Exponential Functions

The Inverse of $\ln x$ and the Number e • The Natural Exponential Function $y = e^x$ • Equations Involving $\ln x$ and e^x • The Derivative and Integral of e^x • The Number e Expressed as a Limit • The Derivative and Integral of a^u

Whenever we have a quantity y whose rate of change over time is proportional to the amount of y present, we have a function that satisfies the differential equation

$$\frac{dy}{dt} = ky.$$

If, in addition, $y = y_0$ when $t = 0$, the function is the exponential function $y = y_0 e^{kt}$. This section defines the exponential function as the inverse of $\ln x$ and explores the properties that account for the amazing frequency with which the function appears in mathematics and its applications. Our approach to the natural exponential function $y = e^x$ here is different from that taken in Section 2.9.

The Inverse of ln x and the Number e

The function $\ln x$, being an increasing function of x with domain $(0, \infty)$ and range $(-\infty, \infty)$, has an inverse $\ln^{-1} x$ with domain $(-\infty, \infty)$ and range $(0, \infty)$.

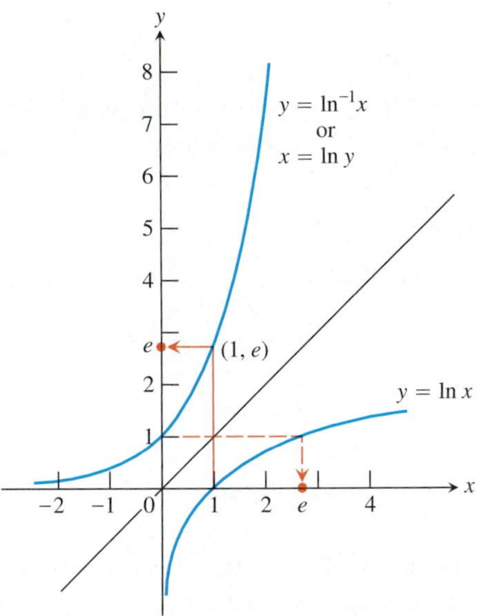

FIGURE 6.2 The graphs of $y = \ln x$ and $y = \ln^{-1} x$. The number e is $\ln^{-1} 1$.

The graph of $\ln^{-1} x$ is the graph of $\ln x$ reflected across the line $y = x$. As you can see from Figure 6.2,

$$\lim_{x\to\infty} \ln^{-1} x = \infty \qquad \text{and} \qquad \lim_{x\to-\infty} \ln^{-1} x = 0.$$

The number $\ln^{-1} 1$ is denoted by the letter e (Figure 6.2).

Definition The Number e

$$e = \ln^{-1} 1$$

Although e is not a rational number, later in this section we see one way to calculate it as a limit. In Equation (2) of Section 2.9, we used another limit to calculate e.

CD-ROM
WEBsite
Historical Biography
Charles Hermite
(1822 — 1901)

The Natural Exponential Function $y = e^x$

We can raise the number e to a rational power x in the usual way:

$$e^2 = e \cdot e, \qquad e^{-2} = \frac{1}{e^2}, \qquad e^{1/2} = \sqrt{e},$$

and so on. Since e is positive, e^x is positive too. Thus, e^x has a logarithm. When we take the logarithm, we find that

$$\ln e^x = x \ln e = x \cdot 1 = x. \tag{1}$$

Since $\ln x$ is one-to-one and $\ln(\ln^{-1} x) = x$, Equation (1) tells us that

$$e^x = \ln^{-1} x \qquad \text{for } x \text{ rational.} \tag{2}$$

Equation (2) provides a way to extend the definition of e^x to irrational values of x. The function $\ln^{-1} x$ is defined for all x, so we can use it to assign a value to e^x at every point where e^x had no previous value.

Definition Natural Exponential Function

For every real number x,

$$e^x = \ln^{-1} x.$$

Equations Involving ln x and e^x

Since $\ln x$ and e^x are inverses of one another, we have

Inverse Equations for e^x and $\ln x$

$$e^{\ln x} = x \qquad (\text{all } x > 0) \tag{3}$$

$$\ln(e^x) = x \qquad (\text{all } x) \tag{4}$$

The Derivative and Integral of e^x

We now calculate the derivative of e^x as the inverse of the natural logarithm. This is a different method than we used in Section 2.9. The exponential function is differentiable because it is the inverse of a differentiable function whose derivative is never zero. Starting with $y = e^x$, we have, in order,

$$
\begin{aligned}
y &= e^x \\
\ln y &= x && \text{Logarithms of both sides} \\
\frac{1}{y}\frac{dy}{dx} &= 1 && \text{Derivative of both sides with respect to } x \\
\frac{dy}{dx} &= y && \text{Multiply both sides by } y \\
\frac{dy}{dx} &= e^x && y \text{ replaced by } e^x
\end{aligned}
$$

As we saw in Section 2.9, the only functions that behave this way are constant multiples of e^x. The Chain Rule extends this result in the usual way to a more general form, just as we saw in Section 2.9. We repeat the rule here.

If u is any differentiable function of x, then

$$\frac{d}{dx}e^u = e^u\frac{du}{dx}. \tag{5}$$

The integral equivalent of Equation (5) is

$$\int e^u\,du = e^u + C.$$

Example 1 Integrating an Exponential

$$
\begin{aligned}
\int_0^{\ln 2} e^{3x}\,dx &= \int_0^{\ln 8} e^u \cdot \frac{1}{3}\,du && u = 3x,\ \frac{1}{3}du = dx,\ u(0) = 0, \\
& && u(\ln 2) = 3\ln 2 = \ln 2^3 = \ln 8 \\
&= \frac{1}{3}\int_0^{\ln 8} e^u\,du \\
&= \frac{1}{3}e^u\Big]_0^{\ln 8} \\
&= \frac{1}{3}[8 - 1] = \frac{7}{3}
\end{aligned}
$$

Example 2 Using Substitution

$$
\begin{aligned}
\int_0^{\pi/2} e^{\sin x}\cos x\,dx &= \int_0^1 e^u\,du && u = \sin x,\ du = \cos x\,dx \\
& && u(0) = 0,\ u(\pi/2) = 1 \\
&= e^u\Big]_0^1 \\
&= e^1 - e^0 = e - 1
\end{aligned}
$$

Example 3 Solving an Initial Value Problem

Solve the initial value problem

$$e^y \frac{dy}{dx} = 2x, \qquad x > \sqrt{3}, \qquad y(2) = 0.$$

Solution We integrate both sides of the differential equation with respect to x to obtain

$$e^y = x^2 + C.$$

We use the initial condition to determine C:

$$\begin{aligned} C &= e^0 - (2)^2 \\ &= 1 - 4 = -3. \end{aligned}$$

This completes the formula for e^y:

$$e^y = x^2 - 3. \tag{6}$$

To find y, we take logarithms of both sides:

$$\begin{aligned} \ln e^y &= \ln\,(x^2 - 3) \\ y &= \ln\,(x^2 - 3). \end{aligned} \tag{7}$$

Notice that the solution is valid for $x > \sqrt{3}$.

It is always a good idea to check a solution in the original equation. From Equations (6) and (7), we have

$$\begin{aligned} e^y \frac{dy}{dx} &= e^y \frac{d}{dx} \ln\,(x^2 - 3) && \text{Eq. (7)} \\ &= e^y \frac{2x}{x^2 - 3} && \text{Eq. (6)} \\ &= (x^2 - 3)\frac{2x}{x^2 - 3} \\ &= 2x. \end{aligned}$$

The solution checks.

Historical Biography
C. L. F. Lindemann (1852 — 1939)

Transcendental Numbers and Transcendental Functions

Numbers that are solutions of polynomial equations with rational coefficients are called **algebraic:** -2 is algebraic because it satisfies the equation $x + 2 = 0$, and $\sqrt{3}$ is algebraic because it satisfies the equation $x^2 - 3 = 0$. Numbers that are not algebraic are called **transcendental,** a term coined by Euler to describe numbers like e and π that appeared to "transcend the power of algebraic methods." Not until a hundred years after Euler's death (1873), however, did Charles Hermite prove the transcendence of e in the sense that we describe. A few years later (1882), C. L. F. Lindemann proved the transcendence of π.

Today, we call a function $y = f(x)$ algebraic if it satisfies an equation of the form

$$P_n y^n + \cdots + P_1 y + P_0 = 0$$

in which the P's are polynomials in x with rational coefficients. The function $y = 1/\sqrt{x+1}$ is algebraic because it satisfies the equation $(x + 1)y^2 - 1 = 0$. Here the polynomials are $P_2 = x + 1$, $P_1 = 0$, and $P_0 = -1$. Polynomials and rational functions with rational coefficients are algebraic, as are all sums, products, quotients, rational powers, and rational roots of algebraic functions.

Functions that are not algebraic are called transcendental. The six basic trigonometric functions are transcendental, as are the inverses of the trigonometric functions and the exponential and logarithmic functions that are the main subject of this chapter.

The Number e Expressed as a Limit

We have defined e as the number for which $\ln e = 1$. The next theorem shows one way to calculate e as a limit.

Theorem 1 The Number e as a Limit

$$\lim_{x \to 0} (1 + x)^{1/x} = e$$

Proof If $f(x) = \ln x$, then $f'(x) = 1/x$ and $f'(1) = 1$. By definition of the derivative,

$$\begin{aligned} f'(1) &= \lim_{h \to 0} \frac{f(1 + h) - f(1)}{h} = \lim_{x \to 0} \frac{f(1 + x) - f(1)}{x} \\ &= \lim_{x \to 0} \frac{\ln\,(1 + x) - \ln 1}{x} = \lim_{x \to 0} \frac{1}{x} \ln\,(1 + x) \\ &= \lim_{x \to 0} \ln\,(1 + x)^{1/x} = \ln\left[\lim_{x \to 0} (1 + x)^{1/x}\right] && \ln \text{ is continuous.} \end{aligned}$$

Because $f'(1) = 1$, we obtain

$$\ln \left[\lim_{x \to 0} (1 + x)^{1/x} \right] = 1$$

so that

$$\lim_{x \to 0} (1 + x)^{1/x} = e.$$

X	Y1
1.	2.
0.1	2.59374
0.01	2.70481
0.001	2.71692
0.0001	2.71815
0.00001	2.71827
0.000001	2.71828

Y1 = (1+X)^(1/X)

FIGURE 6.3 A table of values for $f(x) = (1 + x)^{1/x}$.

Using a calculator, we produced the table in Figure 6.3. To 15 places,

$$e = 2.7\ 1828\ 1828\ 45\ 90\ 45.$$

Since the derivative of the exponential function defined here as the inverse of $\ln x$ is the same as the derivative of the exponential function defined in Section 2.9, Corollary 2 of the Mean Value Theorem guarantees that the functions differ by a constant C. Both functions, however, have the same value at $x = 0$:

$$e^0 = 1.$$

Therefore, we obtain immediately that $C = 0$, and both functions are exactly the same. We simply took alternate ways in defining them.

The Derivative and Integral of a^u

We already know that the general exponential function is defined by

$$a^x = e^{x \ln a},$$

where $a > 0$ and $a \neq 1$. We also know (Section 2.9) that

$$\frac{d}{dx}(a^u) = a^u \ln a \frac{du}{dx}. \tag{8}$$

Since $a \neq 1$, so that $\ln a \neq 0$, we can divide both sides of Equation (8) by $\ln a$ to obtain

$$a^u \frac{du}{dx} = \frac{1}{\ln a} \frac{d}{dx}(a^u).$$

Integrating with respect to x then gives

$$\int a^u \frac{du}{dx}\, dx = \int \frac{1}{\ln a} \frac{d}{dx}(a^u)\, dx = \frac{1}{\ln a} \int \frac{d}{dx}(a^u)\, dx = \frac{1}{\ln a} a^u + C.$$

Writing the first integral in differential form gives

$$\int a^u\, du = \frac{a^u}{\ln a} + C. \tag{9}$$

Example 4 Integrating General Exponentials

(a) $\int 2^x\,dx = \dfrac{2^x}{\ln 2} + C$ Eq. (9) with $a = 2,\ u = x$

(b) $\int 2^{\sin x} \cos x\,dx$

$= \int 2^u\,du = \dfrac{2^u}{\ln 2} + C$

$= \dfrac{2^{\sin x}}{\ln 2} + C$ $u = \sin x$ in Eq. (9)

EXERCISES 6.2

Derivatives of Natural Exponentials

In Exercises 1–14, find the derivative of y with respect to x, t, or θ, as appropriate.

1. $y = e^{-2x/3}$

2. $y = e^{5-7x}$

3. $y = e^{(4\sqrt{x}+x^2)}$

4. $y = (1 + 3x)e^{-x}$

5. $y = (x^2 - 2x + 2)e^x$

6. $y = e^{\theta}(\sin\theta + \cos\theta)$

7. $y = \ln(3\theta e^{-\theta})$

8. $y = \cos(e^{-\theta^2})$

9. $y = \ln(2e^{-t}\sin t)$

10. $y = \ln\left(\dfrac{e^{\theta}}{1 + e^{\theta}}\right)$

11. $y = \ln\left(\dfrac{\sqrt{\theta}}{1 + \sqrt{\theta}}\right)$

12. $y = e^{\sin t}(\ln t^2 + 1)$

13. $y = \displaystyle\int_0^{\ln x} \sin e^t\,dt$

14. $y = \displaystyle\int_{e^{4\sqrt{x}}}^{e^{2x}} \ln t\,dt$

Implicit Differentiation

In Exercises 15–18, find dy/dx.

15. $\ln y = e^y \sin x$

16. $\ln xy = e^{x+y}$

17. $e^{2x} = \sin(x + 3y)$

18. $\tan y = e^x + \ln x$

Integrals of Natural Exponentials

Evaluate the integrals in Exercises 19–32.

19. $\int (e^{3x} + 5e^{-x})\,dx$

20. $\displaystyle\int_{\ln 2}^{\ln 3} e^x\,dx$

21. $\int 8e^{(x+1)}\,dx$

22. $\displaystyle\int_{\ln 4}^{\ln 9} e^{x/2}\,dx$

23. $\displaystyle\int \frac{e^{-\sqrt{r}}}{\sqrt{r}}\,dr$

24. $\int 2t\,e^{-t^2}\,dt$

25. $\displaystyle\int \frac{e^{1/x}}{x^2}\,dx$

26. $\displaystyle\int \frac{e^{-1/x^2}}{x^3}\,dx$

27. $\displaystyle\int_0^{\pi/4} (1 + e^{\tan\theta})\sec^2\theta\,d\theta$

28. $\displaystyle\int_{\pi/4}^{\pi/2} (1 + e^{\cot\theta})\csc^2\theta\,d\theta$

29. $\int e^{\sec \pi t} \sec \pi t \tan \pi t\,dt$

30. $\displaystyle\int_0^{\sqrt{\ln \pi}} 2xe^{x^2}\cos(e^{x^2})\,dx$

31. $\displaystyle\int \frac{e^r}{1 + e^r}\,dr$

32. $\displaystyle\int \frac{dx}{1 + e^x}$

Derivatives of General Exponentials

In Exercises 33–44, find the derivative of y with respect to the given independent variable.

33. $y = 2^x$

34. $y = 5^{\sqrt{s}}$

35. $y = x^{\pi}$

36. $y = (\cos\theta)^{\sqrt{2}}$

37. $y = 7^{\sec\theta}\ln 7$

38. $y = 2^{\sin 3t}$

39. $y = t^{1-e}$

40. $y = (\ln\theta)^{\pi}$

41. $y = \log_3\left(\left(\dfrac{x+1}{x-1}\right)^{\ln 3}\right)$

42. $y = \log_5\sqrt{\left(\dfrac{7x}{3x+2}\right)^{\ln 5}}$

43. $y = \log_7\left(\dfrac{\sin\theta\cos\theta}{e^{\theta}\,2^{\theta}}\right)$

44. $y = \log_2\left(\dfrac{x^2e^2}{2\sqrt{x+1}}\right)$

Logarithmic Differentiation

In Exercises 45–48, use logarithmic differentiation (Example 8, Section 2.9) to find the derivative of y with respect to the given independent variable.

45. $y = (x + 1)^x$

46. $y = t^{\sqrt{t}}$

47. $y = x^{\sin x}$

48. $y = (\ln x)^{\ln x}$

Integrals of General Exponentials

Evaluate the integrals in Exercises 49–56.

49. $\int_1^{\sqrt{2}} x 2^{(x^2)}\,dx$

50. $\int_0^{\pi/2} 7^{\cos t} \sin t\,dt$

51. $\int_1^2 \frac{2^{\ln x}}{x}\,dx$

52. $\int 3x^{\sqrt{3}}\,dx$

53. $\int x^{\sqrt{2}-1}\,dx$

54. $\int_0^3 (\sqrt{2}+1)x^{\sqrt{2}}\,dx$

55. $\int_1^e x^{(\ln 2)-1}\,dx$

56. $\int_1^{e^x} \frac{3^{\ln t}}{t}\,dt$

Initial Value Problems

Solve the initial value problems in Exercises 57–60.

57. $\frac{dy}{dt} = e^t \sin(e^t - 2), \quad y(\ln 2) = 0$

58. $\frac{dy}{dt} = e^{-t} \sec^2(\pi e^{-t}), \quad y(\ln 4) = 2/\pi$

59. $\frac{d^2y}{dx^2} = 2e^{-x}, \quad y(0) = 1 \quad \text{and} \quad y'(0) = 0$

60. $\frac{d^2y}{dt^2} = 1 - e^{2t}, \quad y(1) = -1 \quad \text{and} \quad y'(1) = 0$

Theory and Applications

61. *Absolute extrema* Find the absolute maximum and minimum values of $f(x) = e^x - 2x$ on $[0, 1]$.

62. *A periodic function* Where does the periodic function $f(x) = 2e^{\sin(x/2)}$ take on its extreme values, and what are these values?

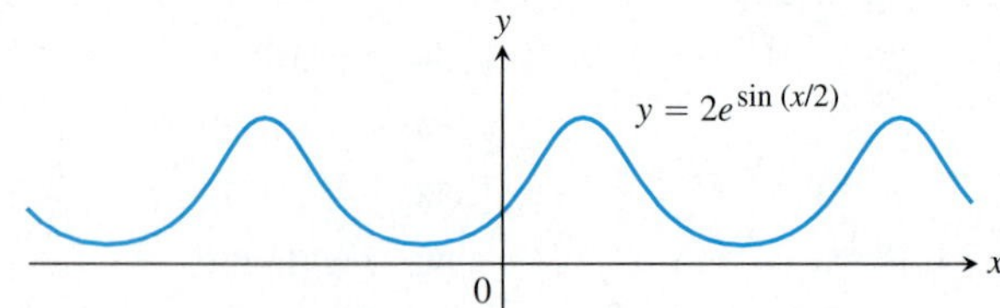

63. *Absolute maximum* Find the absolute maximum value of $f(x) = x^2 \ln(1/x)$ and say where it is assumed.

64. *Area* Find the area of the "triangular" region in the first quadrant that is bounded above by the curve $y = e^{2x}$, below by the curve $y = e^x$, and on the right by the line $x = \ln 3$.

65. *Exponential limit* Show that $\lim_{k\to\infty} (1 + (r/k))^k = e^r$.

66. *Length of a curve* Find a curve through the origin in the xy-plane whose length from $x = 0$ to $x = 1$ is

$$L = \int_0^1 \sqrt{1 + \frac{1}{4}e^x}\,dx.$$

67. Show that for any number $a > 1$,

$$\int_1^a \ln x\,dx + \int_0^{\ln a} e^y\,dy = a \ln a.$$

(See the accompanying figure.)

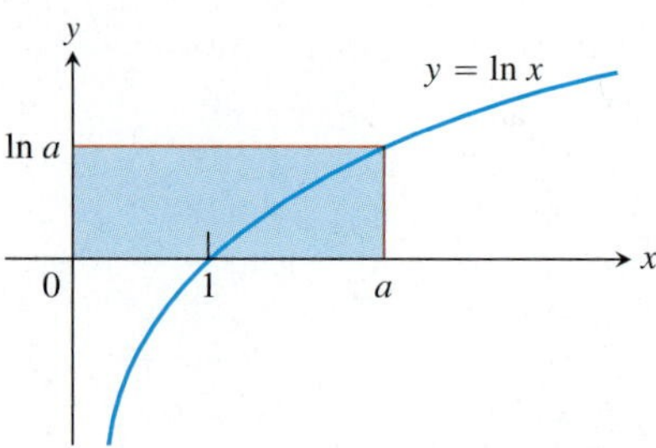

68. *The geometric, logarithmic, and arithmetic mean inequality*

(a) Show that the graph of e^x is concave up over every interval of x-values.

(b) Show, by reference to the accompanying figure, that if $0 < a < b$, then

$$e^{(\ln a + \ln b)/2} \cdot (\ln b - \ln a) < \int_{\ln a}^{\ln b} e^x\,dx < \frac{e^{\ln a} + e^{\ln b}}{2} \cdot (\ln b - \ln a).$$

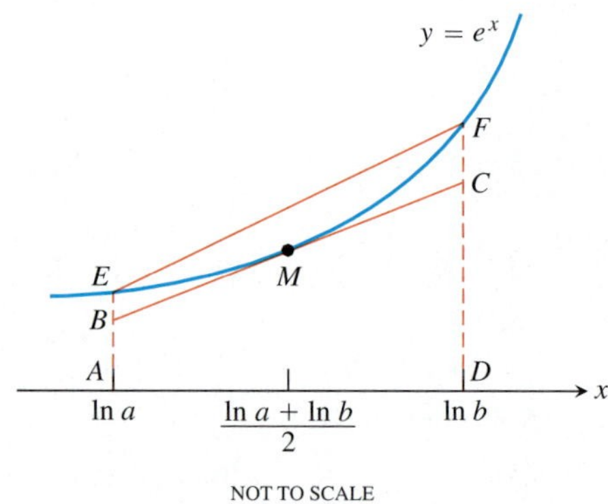

(c) Use the inequality in part (b) to conclude that

$$\sqrt{ab} < \frac{b - a}{\ln b - \ln a} < \frac{a + b}{2}.$$

This inequality says that the geometric mean of two positive numbers is less than their logarithmic mean, which in turn is less than their arithmetic mean.

(For more about this inequality, see "The Geometric, Logarithmic, and Arithmetic Mean Inequality" by Frank Burk, *American Mathematical Monthly*, Vol. 94, No. 6 (June–July 1987), pp. 527–528.)

T **69.** Graph $f(x) = (x - 3)^2 e^x$ and its first derivative together. Comment on the behavior of f in relation to the signs and values of f'. Identify significant points on the graphs with calculus, as necessary.

70. *The inverse relation between e^x and $\ln x$* Find out how good your calculator is at evaluating the composites

$$e^{\ln x} \qquad \text{and} \qquad \ln(e^x).$$

71. *The linearization of e^x at $x = 0$*

(a) Derive the linear approximation $e^x \approx 1 + x$ at $x = 0$.

T **(b)** Graph e^x and $1 + x$ together for $-2 \le x \le 2$. Use different colors, if available. On what intervals does the approximation appear to overestimate e^x? Underestimate e^x?

(c) Estimate to 5 decimal places the magnitude of the error involved in replacing e^x by $1 + x$ on the interval $[0, 0.2]$.

72. *A decimal representation of e* Find e to as many decimal places as your calculator allows by solving the equation $\ln x = 1$.

6.3 Linear First-Order Differential Equations

Linear First-Order Equations • Solving the Linear Equation • Mixture Problems • *RL* Circuits

In Section 5.4, we derived the law of exponential change, $y = y_0 e^{kt}$, as the solution of the initial value problem $dy/dt = ky$, $y(0) = y_0$. As we saw, this problem models radioactive decay, heat transfer, and a great many other phenomena. In this section, we study initial value problems based on the equation $dy/dx = f(x, y)$, in which f is a function of both the independent and dependent variables. The function f will have a particular form, called a *linear form*, and the associated differential equations have applications that are broader still.

Linear First-Order Equations

A first-order differential equation that can be written in the form

$$\frac{dy}{dx} + P(x)y = Q(x), \tag{1}$$

where P and Q are functions of x, is a **linear** first-order equation. Equation (1) is the equation's **standard form.**

Example 1 Finding the Standard Form

Put the following equation in standard form:

$$x\frac{dy}{dx} = x^2 + 3y, \qquad x > 0.$$

Solution

$$x\frac{dy}{dx} = x^2 + 3y$$

$$\frac{dy}{dx} = x + \frac{3}{x}y \qquad \text{Divide by } x.$$

$$\frac{dy}{dx} - \frac{3}{x}y = x \qquad \text{Standard form with } P(x) = -3/x \text{ and } Q(x) = x$$

Notice that $P(x)$ is $-3/x$, not $+3/x$. The standard form is $y' + P(x)y = Q(x)$, so the minus sign is part of the formula for $P(x)$.

Example 2 Linearity of the Exponential Growth Model

The equation

$$\frac{dy}{dx} = ky$$

with which we modeled resistance proportional to velocity and temperature change in Section 5.4 is a linear first-order equation. Its standard form is

$$\frac{dy}{dx} - ky = 0. \qquad P(x) = -k \text{ and } Q(x) = 0$$

Solving the Linear Equation

We solve the equation

$$\frac{dy}{dx} + P(x)y = Q(x) \tag{2}$$

by multiplying both sides by a *positive* function $v(x)$ that transforms the left-hand side into the derivative of the product $v(x) \cdot y$. We will show how to find v in a moment, but first we want to show how, once found, it provides the solution we seek.

Here is why multiplying by v works:

$$\frac{dy}{dx} + P(x)y = Q(x) \qquad \text{Original equation is in standard form.}$$

$$v(x)\frac{dy}{dx} + P(x)v(x)y = v(x)Q(x) \qquad \text{Multiply by positive } v(x).$$

$$\frac{d}{dx}(v(x) \cdot y) = v(x)Q(x) \qquad v(x) \text{ is chosen to make } v\frac{dy}{dx} + Pvy = \frac{d}{dx}(v \cdot y).$$

$$v(x) \cdot y = \int v(x)Q(x)\, dx \qquad \text{Integrate with respect to } x.$$

$$y = \frac{1}{v(x)}\int v(x)Q(x)\, dx. \qquad \text{Solve for } y. \tag{3}$$

We call $v(x)$ an **integrating factor** for Equation (2) because its presence makes the equation integrable.

Equation (3) expresses the solution of Equation (2) in terms of the functions $v(x)$ and $Q(x)$.

Why doesn't the formula for $P(x)$ appear in the solution as well? It does, but indirectly, in the construction of the positive function $v(x)$. We have

$$\frac{d}{dx}(vy) = v\frac{dy}{dx} + Pvy \qquad \text{Condition imposed on } v$$

$$v\frac{dy}{dx} + y\frac{dv}{dx} = v\frac{dy}{dx} + Pvy \qquad \text{Product Rule for derivatives}$$

$$y\frac{dv}{dx} = Pvy \qquad \text{The terms } v\frac{dy}{dx} \text{ cancel.}$$

This last equation will hold if

$$\frac{dv}{dx} = Pv \qquad \text{Variables separated}$$

$$\frac{dv}{v} = P\,dx \qquad \text{Integrate both sides.}$$

$$\int \frac{dv}{v} = \int P\,dx \qquad \text{Since } v > 0, \text{ we do not need absolute value signs in } \ln v.$$

$$\ln v = \int P\,dx \qquad \text{Exponentiate both sides to solve for } v.$$

$$e^{\ln v} = e^{\int P\,dx}$$

$$v = e^{\int P\,dx} \tag{4}$$

From this, we see that any function v that satisfies Equation (4) will enable us to solve Equation (2) with the formula in Equation (3). We do not need the most general possible v, only one that will work. Therefore, it will do no harm to simplify our lives by choosing the simplest possible antiderivative of P for $\int P\,dx$. Notice too that any function v satisfying Equation (4) is positive.

Theorem 2

The solution of the linear equation

$$\frac{dy}{dx} + P(x)\,y = Q(x)$$

is

$$y = \frac{1}{v(x)} \int v(x)\,Q(x)\,dx, \tag{5}$$

where

$$v(x) = e^{\int P(x)\,dx}. \tag{6}$$

In the formula for v, we do not need the most general antiderivative of $P(x)$. Any antiderivative will do.

Example 3 Applying Theorem 2

Solve the equation

$$x\frac{dy}{dx} = x^2 + 3y, \qquad x > 0.$$

Solution We solve the equation in four steps.

Step 1: *Put the equation in standard form to identify* P *and* Q.

$$\frac{dy}{dx} - \frac{3}{x}y = x, \qquad P(x) = -\frac{3}{x}, \qquad Q(x) = x. \qquad \text{Example 1}$$

Step 2: *Find an antiderivative of $P(x)$* (any one will do).

$$\int P(x)\,dx = \int -\frac{3}{x}\,dx = -3\int \frac{1}{x}\,dx = -3\ln|x| = -3\ln x \qquad x > 0$$

Step 3: *Find the integrating factor $v(x)$.*

$$v(x) = e^{\int P(x)\,dx} = e^{-3\ln x} = e^{\ln x^{-3}} = \frac{1}{x^3} \qquad \text{Eq. (6)}$$

Step 4: *Find the solution.*

$$\begin{aligned} y &= \frac{1}{v(x)}\int v(x)Q(x)\,dx && \text{Eq. (5)} \\ &= \frac{1}{(1/x^3)}\int\left(\frac{1}{x^3}\right)(x)\,dx && \text{Values from steps 1–3} \\ &= x^3 \cdot \int \frac{1}{x^2}\,dx \\ &= x^3\left(-\frac{1}{x} + C\right) && \text{Don't forget the } C\ldots \\ &= -x^2 + Cx^3 && \ldots\text{ it provides part of the answer.} \end{aligned}$$

The solution is $y = -x^2 + Cx^3,\ x > 0$.

How to Solve a Linear First-Order Equation

Step 1. Put it in standard form.

Step 2. Find an antiderivative of $P(x)$.

Step 3. Find $v(x) = e^{\int P(x)\,dx}$.

Step 4. Use Equation (5) to find y.

Example 4 Solving a Linear First-Order Initial Value Problem

Solve the equation

$$xy' = x^2 + 3y, \qquad x > 0,$$

given the initial condition $y(1) = 2$.

Solution We first solve the differential equation (Example 3), obtaining

$$y = -x^2 + Cx^3, \qquad x > 0.$$

We then use the initial condition to find the right value for C:

$$\begin{aligned} y &= -x^2 + Cx^3 \\ 2 &= -(1)^2 + C(1)^3 && y = 2 \text{ when } x = 1 \\ C &= 2 + (1)^2 = 3. \end{aligned}$$

The solution of the initial value problem is the function $y = -x^2 + 3x^3$.

CD-ROM
WEBsite
Historical Biography
Adrien Marie Legendre
(1752 — 1833)

Mixture Problems

A chemical in a liquid solution (or dispersed in a gas) runs into a container holding the liquid (or the gas) with, possibly, a specified amount of the chemical dissolved as well. The mixture is kept uniform by stirring and flows out of the container at a known rate. In this process, it is often important to know the concentration of the chemical in the container at any given time. The differential equation describing the process is based on the formula

$$\begin{matrix}\text{Rate of change} \\ \text{of amount} \\ \text{in container}\end{matrix} = \begin{pmatrix}\text{rate at which} \\ \text{chemical} \\ \text{arrives}\end{pmatrix} - \begin{pmatrix}\text{rate at which} \\ \text{chemical} \\ \text{departs.}\end{pmatrix} \tag{7}$$

If $y(t)$ is the amount of chemical in the container at time t and $V(t)$ is the total volume of liquid in the container at time t, then the departure rate of the chemical at time t is

$$\begin{aligned}\text{Departure rate} &= \frac{y(t)}{V(t)} \cdot (\text{outflow rate}) \\ &= \begin{pmatrix}\text{concentration in} \\ \text{container at time } t\end{pmatrix} \cdot (\text{outflow rate}).\end{aligned} \tag{8}$$

Accordingly, Equation (7) becomes

$$\frac{dy}{dt} = (\text{chemical's arrival rate}) - \frac{y(t)}{V(t)} \cdot (\text{outflow rate}). \tag{9}$$

If, say, y is measured in pounds, V in gallons, and t in minutes, the units in Equation (9) are

$$\frac{\text{pounds}}{\text{minutes}} = \frac{\text{pounds}}{\text{minutes}} - \frac{\text{pounds}}{\text{gallons}} \cdot \frac{\text{gallons}}{\text{minutes}}.$$

Here is an example.

Example 5 Oil Refinery Storage Tank

In an oil refinery, a storage tank contains 2000 gal of gasoline that initially has 100 lb of an additive dissolved in it. In preparation for winter weather, gasoline containing 2 lb of additive per gallon is pumped into the tank at a rate of 40 gal/min. The well-mixed solution is pumped out at a rate of 45 gal/min. How much of the additive is in the tank 20 min after the pumping process begins (Figure 6.4)?

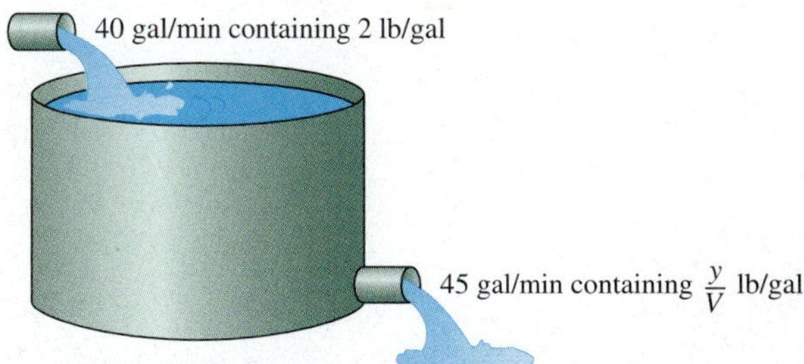

FIGURE 6.4 Storage tank in Example 5.

Solution

Differential Equation Model

Let y be the amount (in pounds) of additive in the tank at time t. We know that $y = 100$ when $t = 0$. The number of gallons of gasoline and additive in solution in the tank at any time t is

$$\begin{aligned}V(t) &= 2000 \text{ gal} + \left(40 \frac{\text{gal}}{\text{min}} - 45 \frac{\text{gal}}{\text{min}}\right)(t \text{ min}) \\ &= (2000 - 5t) \text{ gal}.\end{aligned}$$

Therefore,

$$\begin{aligned}\text{Rate out} &= \frac{y(t)}{V(t)} \cdot \text{outflow rate} && \text{Eq. (8)}\\ &= \left(\frac{y}{2000-5t}\right)45 && \text{Outflow rate is 45 gal/min. and } V = 2000-5t.\\ &= \frac{45y}{2000-5t}\,\frac{\text{lb}}{\text{min}}.\end{aligned}$$

Also,

$$\begin{aligned}\text{Rate in} &= \left(2\,\frac{\text{lb}}{\text{gal}}\right)\left(40\,\frac{\text{gal}}{\text{min}}\right)\\ &= 80\,\frac{\text{lb}}{\text{min}}.\end{aligned}$$

The differential equation modeling the mixture process is

$$\frac{dy}{dt} = 80 - \frac{45y}{2000-5t} \qquad \text{Eq. (9)}$$

in pounds per minute.

Analytical Solution

To solve this differential equation, we first write it in standard form:

$$\frac{dy}{dt} + \frac{45}{2000-5t}\,y = 80.$$

Thus, $P(t) = 45/(2000-5t)$ and $Q(t) = 80$.

An antiderivative for $P(t)$ is

$$\begin{aligned}\int P(t)\,dt &= \int \frac{45}{2000-5t}\,dt\\ &= -9\ln(2000-5t). && 2000-5t>0\end{aligned}$$

(Remember, any antiderivative for P will do.)

The integrating factor is

$$\begin{aligned}v(t) &= e^{\int P\,dt}\\ &= e^{-9\ln(2000-5t)}\\ &= (2000-5t)^{-9}.\end{aligned}$$

The general solution to the differential equation is

$$\begin{aligned}y &= \frac{1}{(2000-5t)^{-9}}\int (2000-5t)^{-9}(80)\,dt && \text{Eq. (5)}\\ &= \frac{80}{(2000-5t)^{-9}}\left(\frac{(2000-5t)^{-8}}{(-8)(-5)} + C\right) && \text{Don't forget the } C.\\ &= 2(2000-5t) + C(2000-5t)^9. && \text{Using } C \text{ again for } 80\,C\end{aligned}$$

Because $y = 100$ when $t = 0$, we can determine the value of C:

$$100 = 2(2000-0) + C(2000-0)^9$$

$$C = -\frac{3900}{(2000)^9}.$$

The solution of the initial value problem is

$$y = 2(2000 - 5t) - \frac{3900}{(2000)^9}(2000 - 5t)^9.$$

Interpretation

The amount of additive 20 minutes after the pumping begins is

$$y(20) = 2[2000 - 5(20)] - \frac{3900}{(2000)^9}[2000 - 5(20)]^9 \approx 1342.03 \text{ lb}.$$

RL Circuits

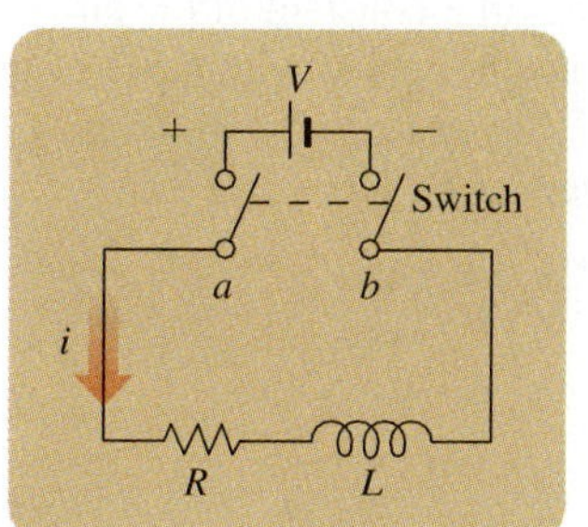

FIGURE 6.5 The *RL* circuit in Example 6.

The diagram in Figure 6.5 represents an electrical circuit whose total resistance is a constant R ohms and whose self-inductance, shown as a coil, is L henries, also a constant. There is a switch whose terminals at a and b can be closed to connect a constant electrical source of V volts.

Ohm's law, $V = RI$, has to be modified for such a circuit. The modified form is

$$L\frac{di}{dt} + Ri = V, \tag{10}$$

where i is the intensity of the current in amperes and t is the time in seconds. By solving this equation, we can predict how the current will flow after the switch is closed.

Example 6 Electric Current Flow

The switch in the RL circuit in Figure 6.5 is closed at time $t = 0$. How will the current flow as a function of time?

Solution Equation (10) is a linear first-order differential equation for i as a function of t. Its standard form is

$$\frac{di}{dt} + \frac{R}{L}i = \frac{V}{L}, \tag{11}$$

and the corresponding solution, from Theorem 2, given that $i = 0$ when $t = 0$, is

$$i = \frac{V}{R} - \frac{V}{R}e^{-(R/L)t} \tag{12}$$

(Exercise 32). Since R and L are positive, $-(R/L)$ is negative and $e^{-(R/L)t} \to 0$ as $t \to \infty$. Thus,

$$\lim_{t\to\infty} i = \lim_{t\to\infty}\left(\frac{V}{R} - \frac{V}{R}e^{-(R/L)t}\right) = \frac{V}{R} - \frac{V}{R}\cdot 0 = \frac{V}{R}.$$

At any given time, the current is theoretically less than V/R, but as time passes the current approaches the **steady-state value** V/R. According to the equation

$$L\frac{di}{dt} + Ri = V,$$

$I = V/R$ is the current that will flow in the circuit if either $L = 0$ (no inductance) or $di/dt = 0$ (steady current, $i =$ constant) (Figure 6.6).

Equation (12) expresses the solution of Equation (11) as the sum of two terms: a **steady-state solution** V/R and a **transient solution** $-(V/R)e^{-(R/L)t}$ that tends to zero as $t \to \infty$.

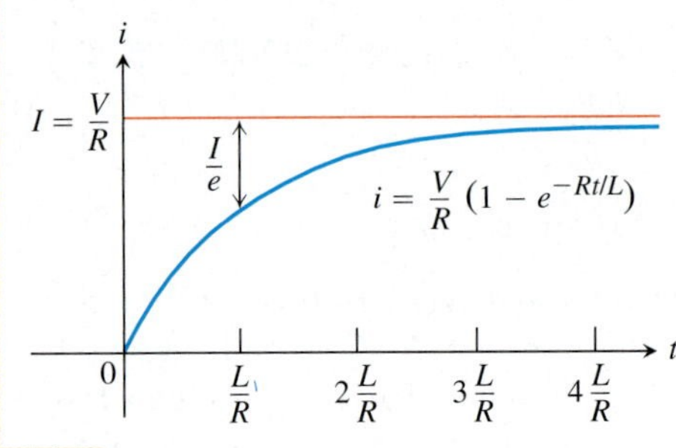

FIGURE 6.6 The growth of the current in the RL circuit in Example 6 is the current's steady-state value. The number $t = L/R$ is the time constant of the circuit. The current gets to within 5% of its steady-state value in 3 time constants. (Exercise 31)

EXERCISES 6.3

Linear First-Order Equations

Solve the differential equations in Exercises 1–14.

1. $x\dfrac{dy}{dx} + y = e^x, \quad x > 0$
2. $e^x\dfrac{dy}{dx} + 2e^x y = 1$
3. $xy' + 3y = \dfrac{\sin x}{x^2}, \quad x > 0$
4. $y' + (\tan x)y = \cos^2 x, \quad -\pi/2 < x < \pi/2$
5. $x\dfrac{dy}{dx} + 2y = 1 - \dfrac{1}{x}, \quad x > 0$
6. $(1 + x)y' + y = \sqrt{x}$
7. $2y' = e^{x/2} + y$
8. $e^{2x}y' + 2e^{2x}y = 2x$
9. $xy' - y = 2x \ln x$
10. $x\dfrac{dy}{dx} = \dfrac{\cos x}{x} - 2y, \quad x > 0$
11. $(t - 1)^3\dfrac{ds}{dt} + 4(t - 1)^2 s = t + 1, \quad t > 1$
12. $(t + 1)\dfrac{ds}{dt} + 2s = 3(t + 1) + \dfrac{1}{(t + 1)^2}, \quad t > -1$
13. $\sin\theta\dfrac{dr}{d\theta} + (\cos\theta)r = \tan\theta, \quad 0 < \theta < \pi/2$
14. $\tan\theta\dfrac{dr}{d\theta} + r = \sin^2\theta, \quad 0 < \theta < \pi/2$

Solving Initial Value Problems

Solve the initial value problems in Exercises 15–20.

	Differential equation	Initial condition
15.	$\dfrac{dy}{dt} + 2y = 3$	$y(0) = 1$
16.	$t\dfrac{dy}{dt} + 2y = t^3, \quad t > 0$	$y(2) = 1$
17.	$\theta\dfrac{dy}{d\theta} + y = \sin\theta, \quad \theta > 0$	$y(\pi/2) = 1$
18.	$\theta\dfrac{dy}{d\theta} - 2y = \theta^3 \sec\theta \tan\theta, \quad \theta > 0$	$y(\pi/3) = 2$
19.	$(x + 1)\dfrac{dy}{dx} - 2(x^2 + x)y = \dfrac{e^{x^2}}{x + 1}, \quad x > -1$	$y(0) = 5$
20.	$\dfrac{dy}{dx} + xy = x$	$y(0) = -6$

21. What do you get when you use Theorem 2 to solve the following initial value problem for y as a function of t?
$$\frac{dy}{dt} = ky \qquad (k \text{ constant}), \qquad y(0) = y_0$$

22. Use Theorem 2 to solve the following initial value problem for v as a function of t.
$$\frac{dv}{dt} + \frac{k}{m}v = 0 \qquad (k \text{ and } m \text{ positive constants}), \qquad v(0) = v_0$$

Theory and Examples

23. *Writing to Learn* Is either of the following equations correct? Give reasons for your answers.

(a) $x \int \frac{1}{x}\, dx = x \ln |x| + C$

(b) $x \int \frac{1}{x}\, dx = x \ln |x| + Cx$

24. *Continuous compounding* You have \$1000 with which to open an account and plan to add \$1000 per year. All funds in the account will earn 10% interest per year, compounded continuously. If the added deposits are also credited to your account continuously, the number of dollars x in your account at time t (years) will satisfy the initial value problem

$$\frac{dx}{dt} = 1000 + 0.10x, \qquad x(0) = 1000.$$

(a) Solve the initial value problem for x as a function of t.

(b) About how many years will it take for the amount in your account to reach \$100,000?

25. *Salt mixture* A tank initially contains 100 gal of brine in which 50 lb of salt are dissolved. A brine containing 2 lb/gal of salt runs into the tank at the rate of 5 gal/min. The mixture is kept uniform by stirring and flows out of the tank at the rate of 4 gal/min.

(a) At what rate (pounds per minute) does salt enter the tank at time t?

(b) What is the volume of brine in the tank at time t?

(c) At what rate (pounds per minute) does salt leave the tank at time t?

(d) Write down and solve the initial value problem describing the mixing process.

(e) Find the concentration of salt in the tank 25 min after the process starts.

26. *Mixture problem* A 200 gallon tank is half full of distilled water. Beginning at time $t = 0$, a solution containing 0.5 lb/gal of concentrate enters the tank at the rate of 5 gal/min, and the well-stirred mixture is withdrawn at the rate of 3 gal/min.

(a) At what time will the tank be full?

(b) At the time the tank is full, how many pounds of concentrate will it contain?

27. *Fertilizer mixture* A tank contains 100 gal of fresh water. A solution containing 1 lb/gal of soluble lawn fertilizer runs into the tank at the rate of 1 gal/min, and the mixture is pumped out of the tank at the rate of 3 gal/min. Find the maximum amount of fertilizer in the tank and the time required to reach the maximum.

28. *Carbon monoxide pollution* An executive conference room of a corporation contains 4500 ft^3 of air initially free of carbon monoxide. Starting at time $t = 0$, cigarette smoke containing 4% carbon monoxide is blown into the room at the rate of 0.3 ft^3/min. A ceiling fan keeps the air in the room well circulated and the air leaves the room at the same rate of 0.3 ft^3/min. Find the time when the concentration of carbon monoxide in the room reaches 0.01%.

29. *Current in a closed RL circuit* How many seconds after the switch in an RL circuit is closed will it take the current i to reach half of its steady-state value? Notice that the time depends on R and L and not on how much voltage is applied.

30. *Current in an open RL circuit* If the switch is thrown open after the current in an RL circuit has built up to its steady-state value $I = V/R$, the decaying current (graphed here) obeys the equation

$$L\frac{di}{dt} + Ri = 0,$$

which is Equation (10) with $V = 0$.

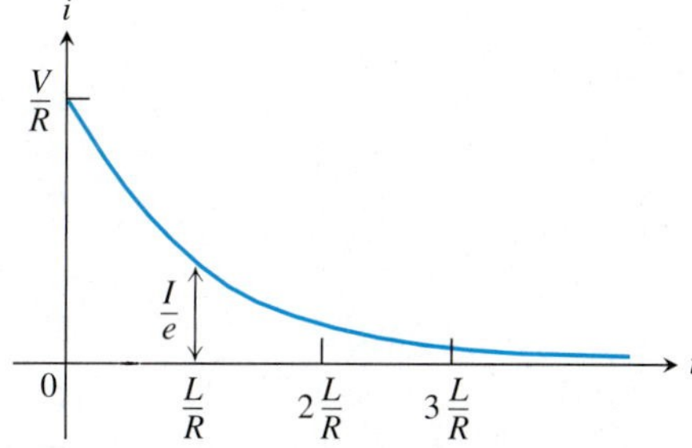

(a) Solve the equation above to express i as a function of t.

(b) How long after the switch is thrown will it take the current to fall to half its original value?

(c) Show that the value of the current when $t = L/R$ is I/e. (The significance of this time is explained in the next exercise.)

31. *Time constants* Engineers call the number L/R the *time constant* of the RL circuit in Figure 6.6. The significance of the time constant is that the current will reach 95% of its final value within 3 time constants of the time the switch is closed (Figure 6.6). Thus, the time constant gives a built-in measure of how rapidly an individual circuit will reach equilibrium.

(a) Find the value of i in Equation (12) that corresponds to $t = 3L/R$ and show that it is about 95% of the steady-state value $I = V/R$.

(b) Approximately what percentage of the steady-state current will be flowing in the circuit 2 time constants after the switch is closed (i.e., when $t = 2L/R$)?

32. *Derivation of Equation (12) in Example 6.*

(a) Use Theorem 2 to show that the solution of the equation

$$\frac{di}{dt} + \frac{R}{L}i = \frac{V}{L}$$

is

$$i = \frac{V}{R} + Ce^{-(R/L)t}.$$

(b) Then use the initial condition $i(0) = 0$ to determine the value of C. This will complete the derivation of Equation (12).

(c) Show that $i = V/R$ is a solution of Equation (11) and that $i = Ce^{-(R/L)t}$ satisfies the equation

$$\frac{di}{dt} + \frac{R}{L}i = 0.$$

6.4 Euler's Method: Population Models

Euler's Method • Graphical Solutions • Improved Euler's Method • Exponential Population Model • Logistic Population Model

If we do not require or cannot immediately find an *exact* solution for an initial value problem $y' = f(x, y)$, $y(x_0) = y_0$, we can probably use a computer to generate a table of approximate numerical values of y for values of x in an appropriate interval. Such a table is called a **numerical solution** of the problem and the method by which we generate the table is called a **numerical method.** Numerical methods are generally fast and accurate, and they are often the methods of choice when exact formulas are unnecessary, unavailable, or overly complicated. In this section, we study one such method, called Euler's method, upon which many other numerical methods are based.

Euler's Method

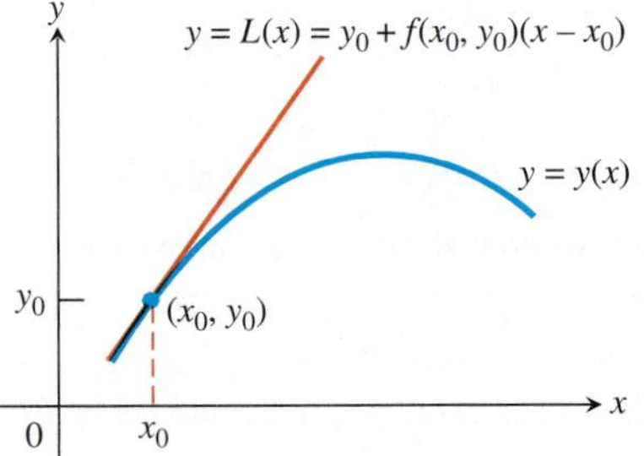

FIGURE 6.7 The linearization of $y = y(x)$ at $x = x_0$.

Given a differential equation $dy/dx = f(x, y)$ and an initial condition $y(x_0) = y_0$, we can approximate the solution $y = y(x)$ by its linearization

$$L(x) = y(x_0) + y'(x_0)(x - x_0) \qquad \text{or} \qquad L(x) = y_0 + f(x_0, y_0)(x - x_0).$$

The function $L(x)$ gives a good approximation to the solution $y(x)$ in a short interval about x_0 (Figure 6.7). The basis of Euler's method is to patch together a string of linearizations to approximate the curve over a longer stretch. Here is how the method works.

We know the point (x_0, y_0) lies on the solution curve. Suppose that we specify a new value for the independent variable to be $x_1 = x_0 + dx$. If the increment dx is small, then

$$y_1 = L(x_1) = y_0 + f(x_0, y_0)\,dx$$

is a good approximation to the exact solution value $y = y(x_1)$. So from the point (x_0, y_0), which lies *exactly* on the solution curve, we obtain the point (x_1, y_1), which lies very close to the point $(x_1, y(x_1))$ on the solution curve (Figure 6.8).

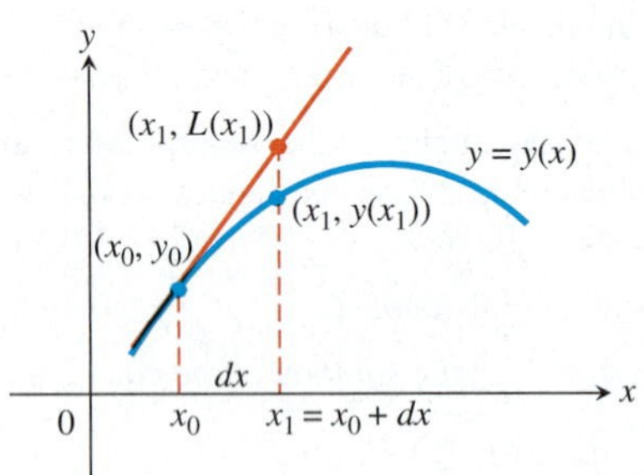

FIGURE 6.8 The first Euler step approximates $y(x_1)$ with $L(x_1)$.

Using the point (x_1, y_1) and the slope $f(x_1, y_1)$ of the solution curve through (x_1, y_1), we take a second step. Setting $x_2 = x_1 + dx$, we use the linearization of the solution curve through (x_1, y_1) to calculate

$$y_2 = y_1 + f(x_1, y_1)\,dx.$$

This gives the next approximation (x_2, y_2) to values along the solution curve $y = y(x)$ (Figure 6.9). Continuing in this fashion, take a third step from the point (x_2, y_2) with slope $f(x_2, y_2)$ to obtain the third approximation

$$y_3 = y_2 + f(x_2, y_2)\,dx,$$

and so on. We are literally building an approximation to one of the solutions by following the direction of the slope field of the differential equation.

The steps in Figure 6.9 are drawn large to illustrate the construction process, so the approximation looks crude. In practice, dx would be small enough to make the red curve hug the blue one and give a good approximation throughout.

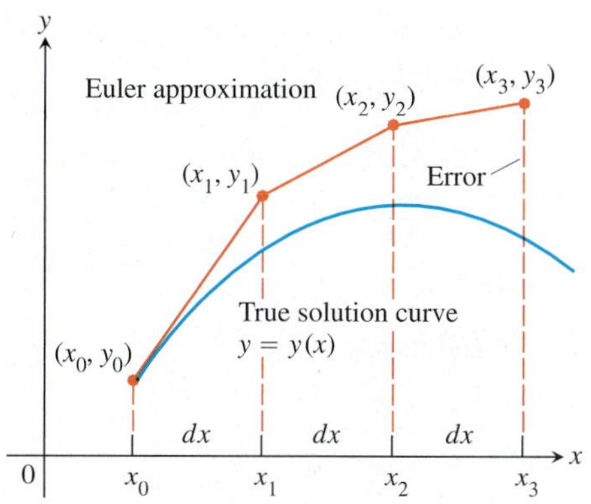

FIGURE 6.9 Three steps in the Euler approximation to the solution of the initial value problem $y' = f(x, y)$, $y(x_0) = y_0$. As we take more steps, the errors involved usually accumulate, but not in the exaggerated way shown here.

Example 1 Using Euler's Method

Find the first three approximations y_1, y_2, y_3 using Euler's method for the initial value problem

$$y' = 1 + y, \qquad y(0) = 1,$$

starting at $x_0 = 0$ with $dx = 0.1$.

Solution We have $x_0 = 0$, $y_0 = 1$, $x_1 = x_0 + dx = 0.1$, $x_2 = x_0 + 2dx = 0.2$, and $x_3 = x_0 + 3dx = 0.3$.

$$\begin{aligned}\textit{First:}\quad y_1 &= y_0 + f(x_0, y_0)\,dx\\ &= y_0 + (1 + y_0)\,dx\\ &= 1 + (1 + 1)(0.1) = 1.2\end{aligned}$$

$$\begin{aligned}\textit{Second:}\quad y_2 &= y_1 + f(x_1, y_1)\,dx\\ &= y_1 + (1 + y_1)\,dx\\ &= 1.2 + (1 + 1.2)(0.1) = 1.42\end{aligned}$$

$$\begin{aligned}\textit{Third:}\quad y_3 &= y_2 + f(x_2, y_2)\,dx\\ &= y_2 + (1 + y_2)\,dx\\ &= 1.42 + (1 + 1.42)(0.1) = 1.662\end{aligned}$$

The step-by-step process used in Example 1 can be continued easily. Using equally spaced values for the independent variable in the table and generating n of them, set

$$\begin{aligned}x_1 &= x_0 + dx\\ x_2 &= x_1 + dx\\ &\vdots\\ x_n &= x_{n-1} + dx.\end{aligned}$$

CD-ROM
WEBsite
Historical Biography
Leonhard Euler
(1703 — 1783)

Then calculate the approximations to the solution,

$$\begin{aligned}y_1 &= y_0 + f(x_0, y_0)\,dx\\ y_2 &= y_1 + f(x_1, y_1)\,dx\\ &\vdots\\ y_n &= y_{n-1} + f(x_{n-1}, y_{n-1})\,dx.\end{aligned}$$

The number of steps n can be as large as we like, but errors can accumulate if n is too large.

Euler's method is easy to program on a computer or programmable calculator. A program generates a table of numerical solutions to an initial value problem, allowing us to input x_0 and y_0, the number of steps n, and the step size dx. It then calculates the approximate solution values $y_1, y_2, \ldots, y_n$ in iterative fashion, as just described.

In Exercise 13, you will show that the exact solution to the initial problem of Example 1 is $y = 2e^x - 1$. We use this information in Example 2.

Example 2 Investigating the Accuracy of Euler's Method

Use Euler's method to solve

$$y' = 1 + y, \qquad y(0) = 1,$$

on the interval $0 \le x \le 1$, starting at $x_0 = 0$ and taking

(a) $dx = 0.1$

(b) $dx = 0.05$.

Compare the approximations with the values of the exact solution $y = 2e^x - 1$.

Solution

(a) We used a programmable calculator to generate the approximate values in Table 6.1. The "error" column is obtained by subtracting the unrounded Euler values from the unrounded values found using the exact solution. All entries are then rounded to 4 decimal places.

By the time we reach $x = 1$ (after 10 steps), the error is about 5.6% of the exact solution.

(b) One way to try to reduce the error is to decrease the step size. Table 6.2 shows the results and their comparisons with the exact solutions when we decrease the step size to 0.05, doubling the number of steps to 20. As in

Table 6.1 Euler solution of $y' = 1 + y$, $y(0) = 1$, step size $dx = 0.1$

x	y (Euler)	y (exact)	Error
0	1	1	0
0.1	1.2	1.2103	0.0103
0.2	1.42	1.4428	0.0228
0.3	1.662	1.6997	0.0377
0.4	1.9282	1.9836	0.0554
0.5	2.2210	2.2974	0.0764
0.6	2.5431	2.6442	0.1011
0.7	2.8974	3.0275	0.1301
0.8	3.2872	3.4511	0.1639
0.9	3.7159	3.9192	0.2033
1.0	4.1875	4.4366	0.2491

Table 6.1, all computations are performed before rounding. This time when we reach $x = 1$, the relative error is only about 2.9%.

Table 6.2 Euler solution of $y' = 1 + y$, $y(0) = 1$, step size $dx = 0.05$

x	y (Euler)	y (exact)	Error
0	1	1	0
0.05	1.1	1.1025	0.0025
0.10	1.205	1.2103	0.0053
0.15	1.3153	1.3237	0.0084
0.20	1.4310	1.4428	0.0118
0.25	1.5526	1.5681	0.0155
0.30	1.6802	1.6997	0.0195
0.35	1.8142	1.8381	0.0239
0.40	1.9549	1.9836	0.0287
0.45	2.1027	2.1366	0.0340
0.50	2.2578	2.2974	0.0397
0.55	2.4207	2.4665	0.0458
0.60	2.5917	2.6442	0.0525
0.65	2.7713	2.8311	0.0598
0.70	2.9599	3.0275	0.0676
0.75	3.1579	3.2340	0.0761
0.80	3.3657	3.4511	0.0853
0.85	3.5840	3.6793	0.0953
0.90	3.8132	3.9192	0.1060
0.95	4.0539	4.1714	0.1175
1.00	4.3066	4.4366	0.1300

It might be tempting to reduce the step size even further in Example 2 to obtain greater accuracy. Each additional calculation, however, not only requires additional calculator time but more importantly adds to the buildup of round-off errors due to the approximate representations of numbers inside the grapher.

The analysis of error and the investigation of methods to reduce it when making numerical calculations are important but are appropriate for a more advanced course. There are numerical methods more accurate than Euler's method, as you will see in your further study of differential equations. We study one improvement here.

Graphical Solutions

Investigating a graph is usually more instructive than analyzing a large data table. If we plot the data pairs in a numerical solution of an initial value problem, we obtain a **graphical solution,** as shown in Example 3.

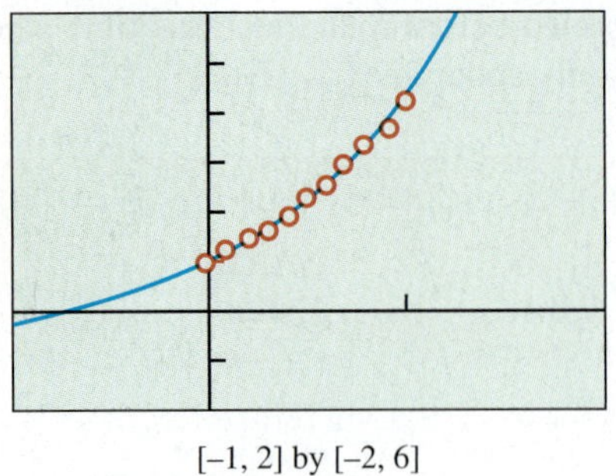

[−1, 2] by [−2, 6]

FIGURE 6.10 The graph of $y = 2e^x - 1$ superimposed on a scatter plot of the Euler approximations shown in Table 6.1. (Example 3)

Example 3 Visualizing Euler's Approximations

Figure 6.10 gives a visualization of the numerical solution shown in Table 6.1 by superimposing the graph of the exact solution on a scatter plot of the data points in the table.

Improved Euler's Method

We can improve on Euler's method by taking an average of two slopes. We first estimate y_n as in the original Euler method, but denote it by z_n. We then take the average of $f(x_{n-1}, y_{n-1})$ and $f(x_n, z_n)$ in place of $f(x_{n-1}, y_{n-1})$ in the next step. Thus, we calculate the next approximation y_n using

$$z_n = y_{n-1} + f(x_{n-1}, y_{n-1})\,dx$$

$$y_n = y_{n-1} + \left[\frac{f(x_{n-1}, y_{n-1}) + f(x_n, z_n)}{2}\right] dx.$$

Example 4 Investigating the Accuracy of the Improved Euler's Method

Use the improved Euler's method to solve

$$y' = 1 + y, \qquad y(0) = 1$$

on the interval $0 \le x \le 1$, starting at $x_0 = 0$ and taking $dx = 0.1$. Compare the approximations with the values of the exact solution $y = 2e^x - 1$.

CD-ROM
WEBsite
Historical Biography
Carl Runge
(1856 — 1927)

Solution We used a programmable calculator to generate the approximate values in Table 6.3. The "error" column is obtained by subtracting the unrounded improved Euler values from the unrounded values found using the exact solution. All entries are then rounded to 4 decimal places.

By the time we reach $x = 1$ (after 10 steps), the relative error is about 0.19%.

Table 6.3 Improved Euler solution of $y' = 1 + y$, $y(0) = 1$, step size $dx = 0.1$

x	y (improved Euler)	y (exact)	Error
0	1	1	0
0.1	1.21	1.2103	0.0003
0.2	1.4421	1.4428	0.0008
0.3	1.6985	1.6997	0.0013
0.4	1.9818	1.9836	0.0018
0.5	2.2949	2.2974	0.0025
0.6	2.6409	2.6442	0.0034
0.7	3.0231	3.0275	0.0044
0.8	3.4456	3.4511	0.0055
0.9	3.9124	3.9192	0.0068
1.0	4.4282	4.4366	0.0084

By comparing Tables 6.1 and 6.3, we see that the improved Euler's method is considerably more accurate than the regular Euler's method, at least for the initial value problem $y' = 1 + y,\ y(0) = 1$.

Exponential Population Model

Strictly speaking, the number of individuals in a population is a discontinuous function of time because it takes on only whole number values. One common way to model a population, however, is with a differentiable function P growing at a rate proportional to the size of the population. Thus, for some constant k,

$$\frac{dP}{dt} = kP.$$

Notice that

$$\frac{dP/dt}{P} = k \tag{1}$$

is constant. This rate is called the **relative growth rate.** As we saw in Section 5.4, we can represent the population by the model $P = P_0 e^{kt}$, where P_0 is the size of the population at time $t = 0$.

Table 6.4 gives the world population at midyear for the years 1980 to 1989. Taking $dt = 1$ and $dP \approx \Delta P$, we see from the table that the relative growth rate in Equation (1) is approximately the constant 0.017.

Table 6.4 World population (midyear)

Year	Population (millions)	$\Delta P/P$
1980	4454	$76/4454 \approx 0.0171$
1981	4530	$80/4530 \approx 0.0177$
1982	4610	$80/4610 \approx 0.0174$
1983	4690	$80/4690 \approx 0.0171$
1984	4770	$81/4770 \approx 0.0170$
1985	4851	$82/4851 \approx 0.0169$
1986	4933	$85/4933 \approx 0.0172$
1987	5018	$87/5018 \approx 0.0173$
1988	5105	$85/5105 \approx 0.0167$
1989	5190	

Source: U.S. Bureau of the Census (Sept., 1999): www.census.gov/ipc/www/worldpop.html.

Example 5 Predicting World Population

Find an initial value problem model for world population and use it to predict the population in midyear 1999. Graph the model and the data.

Solution We let $t = 0$ represent 1980, $t = 1$ represent 1981, and so forth. The year 1999 will be represented by $t = 19$. If we approximate the ratios in Table 6.4 by $k = 0.017$, we obtain the initial value problem

Differential equation: $\dfrac{dP}{dt} = 0.017P$

Initial condition: $P(0) = 4454$.

The solution to this initial value problem gives the population function $P = 4454e^{0.017t}$. Midyear 1999 is represented by $t = 19$, so

$$P(19) \approx 6152.$$

Figure 6.11 shows the graph of the model superimposed on a scatter plot of the data.

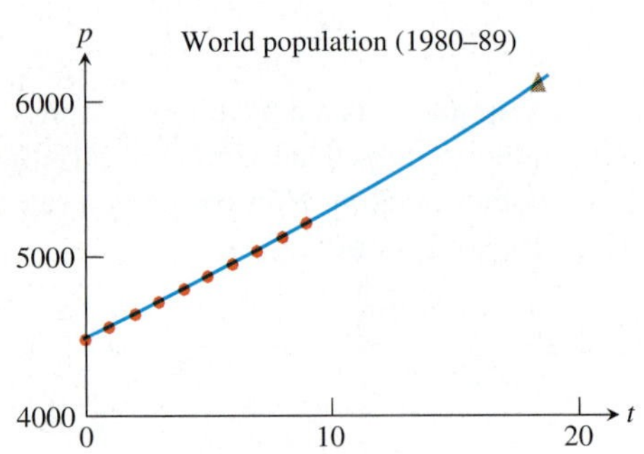

FIGURE 6.11 Notice that the value of the solution $P = 4454e^{0.017t}$ is 6152.16 when $t = 19$. (Example 5)

Interpret

This model predicts the world population in midyear 1999 to be about 6152 million, or 6.15 billion, which is more than the actual population of 5996 million given by the U.S. Bureau of the Census. In the next example, we examine more recent data to see if there has been a change in the growth rate.

Logistic Population Model

The exponential model for population growth assumes unlimited growth and assumes that the relative growth rate (Equation (1)) is constant. This may be reasonable for the years 1980 to 1989, but let's look at more recent data. Table 6.5 shows the world population for the years 1990 to 1999.

Table 6.5 Recent world population

Year	Population (millions)	$\Delta P/P$
1990	5277	$82/5277 \approx 0.0155$
1991	5359	$83/5359 \approx 0.0155$
1992	5442	$81/5442 \approx 0.0149$
1993	5523	$80/5523 \approx 0.0145$
1994	5603	$79/5603 \approx 0.0141$
1995	5682	$79/5682 \approx 0.0139$
1996	5761	$79/5761 \approx 0.0137$
1997	5840	$79/5840 \approx 0.0135$
1998	5919	$77/5919 \approx 0.0130$
1999	5996	

Source: U.S. Bureau of the Census (Sept., 1999): www.census.gov/ipc/www/worldpop.html.

From Tables 6.4 and 6.5, we see that the relative growth rate is positive but decreases as the population increases due to environmental, economic, and other factors. On average, the growth rate decreases by about 0.00036 per year over the years 1990 to 1999. That is, the graph of k in Equation (1) is closer to being a line with a negative slope $-r = -0.00036$.

In Example 5 of Section 3.4, we proposed the **logistic growth model**

$$\frac{dP}{dt} = r(M - P)P, \tag{2}$$

where M is the maximum population, or **carrying capacity,** that the environment is capable of sustaining in the long run. Comparing Equation (2) with the exponential model we see that $k = r(M - P)$ is indeed a linearly decreasing function of the population. The graphical solution curves to the logistic model (2) were obtained in Section 3.4 and are displayed (again) in Figure 6.12. Notice from the graphs that if $P < M$, the population grows toward M; if $P > M$, the growth rate will be negative (as $r > 0$, $M > 0$) and the population decreasing.

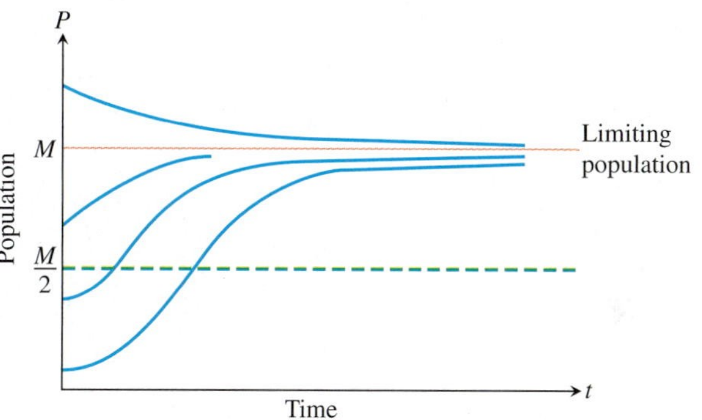

FIGURE 6.12 Solution curves to the logistic population model $dP/dt = r(M - P)P$.

CD-ROM
WEBsite

Example 6 Modeling a Bear Population

A national park is known to be capable of supporting 100 grizzly bears, but no more. Ten bears are in the park at present. We model the population with a logistic differential equation with $r = 0.001$.

(a) Draw and describe a slope field for the differential equation.

(b) Use Euler's method with step size $dt = 1$ to estimate the population size in 20 years.

(c) Find a logistic growth analytic solution $P(t)$ for the population and draw its graph.

(d) When will the bear population reach 50?

Solution

(a) *Slope field.* The carrying capacity is 100, so $M = 100$. The solution we seek is a solution to the following differential equation.

$$\frac{dP}{dt} = 0.001(100 - P)P$$

Figure 6.13 shows a slope field for this differential equation. There appears to be a horizontal asymptote at $P = 100$. The solution curves fall toward this level from above and rise toward it from below.

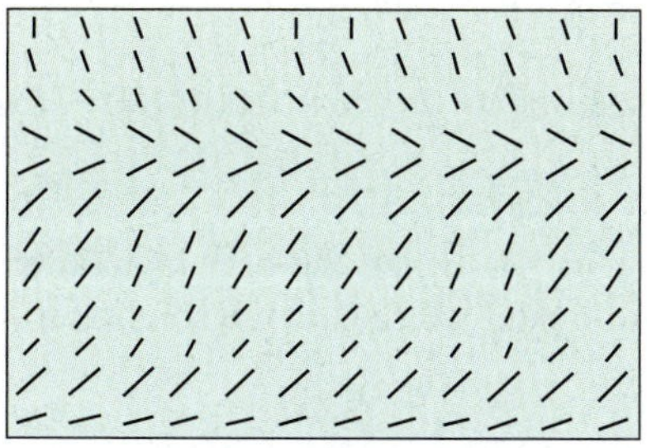

[0, 150] by [0, 150]

FIGURE 6.13 A slope field for the logistic differential equation

$$\frac{dP}{dt} = 0.001(100 - P)P.$$

(Example 6)

(b) *Euler's method.* With step size $dt = 1$, $t_0 = 0$, $P(0) = 10$, and

$$\frac{dP}{dt} = f(t, P) = 0.001(100 - P)P,$$

we obtain the approximations in Table 6.6, using the iteration formula

$$P_n = P_{n-1} + 0.001(100 - P_{n-1})P_{n-1}.$$

Table 6.6 Euler solution of $dP/dt = 0.001(100 - P)P$, $P(0) = 10$, step size $dt = 1$

t	P (Euler)	t	P (Euler)
0	10		
1	10.9	11	24.3629
2	11.8712	12	26.2056
3	12.9174	13	28.1395
4	14.0423	14	30.1616
5	15.2493	15	32.2680
6	16.5417	16	34.4536
7	17.9222	17	36.7119
8	19.3933	18	39.0353
9	20.9565	19	41.4151
10	22.6130	20	43.8414

There are approximately 44 grizzly bears after 20 years. Figure 6.14 shows a graph of the Euler approximation over the interval $0 \le t \le 150$ with step size $dt = 1$. It looks like the lower curves we sketched in Figure 6.12.

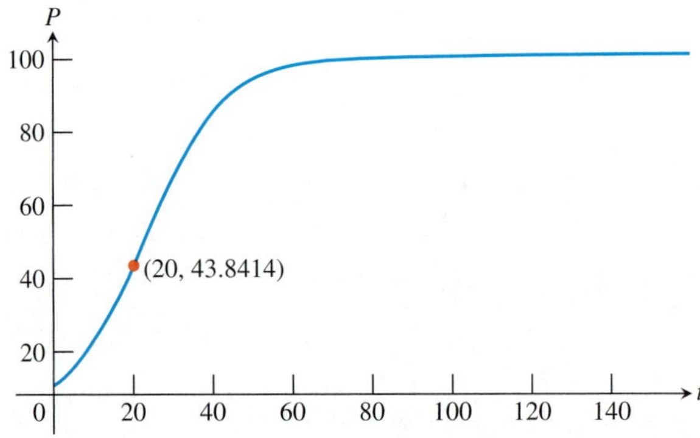

FIGURE 6.14 Euler approximations of the solution to $dP/dt = 0.001(100 - P)P$, $P(0) = 10$, step size $dt = 1$.

(c) *Analytic solution.* We can assume that $t = 0$ when the bear population is 10, so $P(0) = 10$. The logistic growth model we seek is the solution to the following initial value problem.

Differential equation: $\dfrac{dP}{dt} = 0.001(100 - P)P$

Initial condition: $P(0) = 10$

To prepare for integration, we rewrite the differential equation in the form

$$\frac{1}{P(100 - P)} \frac{dP}{dt} = 0.001.$$

The fraction $1/(P(100 - P))$ can be rewritten as

$$\frac{1}{P(100-P)} = \frac{1}{100}\left(\frac{1}{P} + \frac{1}{100-P}\right).$$

(You will learn a general method for obtaining such partial fraction decompositions in the next chapter.) Substituting this expression into the differential equation and multiplying both sides by 100 we obtain

$$\left(\frac{1}{P} + \frac{1}{100-P}\right)\frac{dP}{dt} = 0.1$$

$$\ln|P| - \ln|100 - P| = 0.1t + C \qquad \text{Integrate with respect to } t.$$

$$\ln\left|\frac{P}{100-P}\right| = 0.1t + C$$

$$\ln\left|\frac{100-P}{P}\right| = -0.1t - C \qquad \ln\frac{a}{b} = -\ln\frac{b}{a}$$

$$\left|\frac{100-P}{P}\right| = e^{-0.1t-C} \qquad \text{Exponentiate.}$$

$$\frac{100-P}{P} = (\pm e^{-C})e^{-0.1t}$$

$$\frac{100}{P} - 1 = Ae^{-0.1t} \qquad \text{Let } A = \pm e^{-C}.$$

$$P = \frac{100}{1 + Ae^{-0.1t}}. \qquad \text{Solve for } P.$$

This is the general solution to the differential equation. When $t = 0$, $P = 10$, and we obtain

$$10 = \frac{100}{1 + Ae^0}$$

$$1 + A = 10$$

$$A = 9.$$

Thus, the logistic growth model is

$$P = \frac{100}{1 + 9e^{-0.1t}}.$$

Its graph (Figure 6.15) is superimposed on the slope field from Figure 6.13.

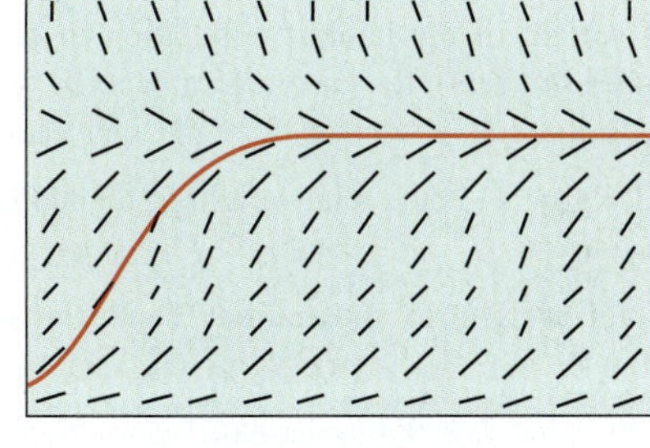

[0, 150] by [0, 150]

FIGURE 6.15 The graph of

$$P = \frac{100}{1 + 9e^{-0.1t}}$$

superimposed on a slope field for

$$\frac{dP}{dt} = 0.001(100 - P)P.$$

(Example 6)

(d) *Interpretation.* When will the bear population reach 50? For this model,

$$50 = \frac{100}{1 + 9e^{-0.1t}}$$

$$1 + 9e^{-0.1t} = 2$$

$$e^{-0.1t} = \frac{1}{9}$$

$$e^{0.1t} = 9$$

$$t = \frac{\ln 9}{0.1} \approx 22 \text{ years}.$$

The solution of the general logistic differential equation (Equation (2))

$$\frac{dP}{dt} = r(M - P)P$$

can be obtained as in Example 6. In Exercise 14, we ask you to show that the solution is

$$P = \frac{M}{1 + Ae^{-rMt}}.$$

The value of A is determined by an appropriate initial condition.

EXERCISES 6.4

Calculating Euler Approximations

In Exercises 1–4, use Euler's method to calculate the first three approximations to the given initial value problem for the specified increment size. Calculate the exact solution and compare the accuracy of your approximations with the exact solution. Round your results to 4 decimal places.

1. $y' = x(1 - y), \quad y(1) = 0, \quad dx = 0.2$
2. $y' = 1 - \frac{y}{x}, \quad y(2) = -1, \quad dx = 0.5$
3. $y' = 2xy + 2y, \quad y(0) = 3, \quad dx = 0.2$
4. $y' = y^2(1 + 2x), \quad y(-1) = 1, \quad dx = 0.5$
5. Use the Euler method with $dx = 0.2$ to estimate $y(1)$ if $y' = y$ and $y(0) = 1$. What is the exact value of $y(1)$?
6. Use the Euler method with $dx = 0.2$ to estimate $y(2)$ if $y' = y/x$ and $y(1) = 2$. What is the exact value of $y(2)$?

Improved Euler's Method

In Exercises 7 and 8, use the improved Euler's method to calculate the first three approximations to the given initial value problem. Compare the approximations with the values of the exact solution.

7. $y' = 2y(x + 1), \quad y(0) = 3, \quad dx = 0.2$
(See Exercise 3 for the exact solution.)
8. $y' = x(1 - y), \quad y(1) = 0, \quad dx = 0.2$
(See Exercise 1 for the exact solution.)
9. *Guppy population* A 2000 gal tank can support no more than 150 guppies. Six guppies are introduced into the tank. Assume that the rate of growth of the population is

$$\frac{dP}{dt} = 0.0015(150 - P)P,$$

where time t is in weeks.

(a) Find a formula for the guppy population in terms of t.

(b) How long will it take for the guppy population to be 100? 125?

10. *Gorilla population* A certain wild animal preserve can support no more than 250 lowland gorillas. Twenty-eight gorillas were known to be in the preserve in 1970. Assume that the rate of growth of the population is

$$\frac{dP}{dt} = 0.0004\,(250 - P)P,$$

where time t is in years.

(a) Find a formula for the gorilla population in terms of t.

(b) How long will it take for the gorilla population to reach the carrying capacity of the preserve?

11. *Pacific halibut fishery* The Pacific halibut fishery has been modeled by the logistic equation

$$\frac{dy}{dt} = r(M - y)y$$

where $y(t)$ is the total weight of the halibut population in kilograms at time t (measured in years), the carrying capacity is estimated to be $M = 8 \times 10^7$ kg, and $r = 0.08875 \times 10^{-7}$ per year.

(a) If $y(0) = 1.6 \times 10^7$ kg, what is the total weight of the halibut population after 1 year?

(b) When will the total weight in the halibut fishery reach 4×10^7 kg?

12. *Modified logistic model* Suppose that the logistic differential equation in Example 6 is modified to

$$\frac{dP}{dt} = 0.001(100 - P)P - c$$

for some constant c.

(a) *Writing to Learn* Explain the meaning of the constant c. What values for c might be realistic for the grizzly bear population?

T (b) Draw a direction field for the differential equation when $c = 1$. What are the equilibrium solutions (Section 3.4)?

(c) *Writing to Learn* Sketch several solution curves in your direction field from part (a). Describe what happens to the grizzly bear population for various initial populations.

13. *Exact solutions* Find the exact solutions to the following initial value problems.

(a) $y' = 1 + y, \quad y(0) = 1$

(b) $y' = 0.5(400 - y)y, \quad y(0) = 2$

14. *Logistic differential equation* Show that the solution of the differential equation

$$\frac{dP}{dt} = r(M - P)P$$

is

$$P = \frac{M}{1 + Ae^{-rMt}},$$

where A is an arbitrary constant.

15. *Catastrophic solution* Let k and P_0 be positive constants.

(a) Solve the initial value problem.

$$\frac{dP}{dt} = kP^2, \qquad P(0) = P_0$$

T **(b)** Show that the graph of the solution in part (a) has a vertical asymptote at a positive value of t. What is that value of t?

16. *Extinct populations* In Exercise 14, Section 3.4, we presented the population model

$$\frac{dP}{dt} = r(M - P)(P - m),$$

where $r > 0$, M is the maximum population, and m is the minimum population below which the species becomes extinct.

(a) Let $m = 100$ and $M = 1200$, and assume that $m < P < M$. Show that the differential equation can be rewritten in the form

$$\left[\frac{1}{1200 - P} + \frac{1}{P - 100}\right]\frac{dP}{dt} = 1100r.$$

Use a procedure similar to that used in Example 6 to solve this differential equation.

(b) Find the solution to part (a) that satisfies $P(0) = 300$.

(c) Solve the differential equation with the restriction $m < P < M$.

COMPUTER EXPLORATIONS

Euler's Method

In Exercises 17–20, use Euler's method with the specified step size to estimate the value of the solution at the given point x^*. Find the value of the exact solution at x^*.

17. $y' = 2xe^{x^2}, \quad y(0) = 2, \quad dx = 0.1, \quad x^* = 1$

18. $y' = y + e^x - 2, \quad y(0) = 2, \quad dx = 0.5, \quad x^* = 2$

19. $y' = y^2/\sqrt{x}, \quad y(1) = -1, \quad dx = 0.5, \quad x^* = 5$

20. $y' = y - e^{2x}, \quad y(0) = 1, \quad dx = 1/3, \quad x^* = 2$

In Exercises 21 and 22, **(a)** find the exact solution of the initial value problem. Then compare the accuracy of the approximation with $y(x^*)$ using Euler's method starting at x_0 with step size **(b)** 0.2, **(c)** 0.1, and **(d)** 0.05.

21. $y' = 2y^2(x - 1), \quad y(2) = -1/2, \quad x_0 = 2, \quad x^* = 3$

22. $y' = y - 1, \quad y(0) = 3, \quad x_0 = 0, \quad x^* = 1$

Improved Euler's Method

In Exercises 23 and 24, compare the accuracy of the approximation with $y(x^*)$ using the improved Euler's method starting at x_0 with step size

(a) 0.2

(b) 0.1

(c) 0.05.

(d) *Writing to Learn* Describe what happens to the error as the step size decreases.

23. $y' = 2y^2(x - 1), \quad y(2) = -1/2, \quad x_0 = 2, \quad x^* = 3$

(See Exercise 21 for the exact solution.)

24. $y' = y - 1, \quad y(0) = 3, \quad x_0 = 0, \quad x^* = 1$

(See Exercise 22 for the exact solution.)

Exploring Differential Equations Graphically

Use a CAS to explore graphically each of the differential equations in Exercises 25–28. Perform the following steps to help with your explorations.

(a) Plot a slope field for the differential equation in the given xy-window.

(b) Find the general solution for the differential equation using your CAS differential equation solver.

(c) Graph the solutions for the values of the arbitrary constant $C = -2, -1, 0, 1, 2$ superimposed on your slope field plot.

(d) Find and graph the solution that satisfies the specified initial condition over the interval $[0, b]$.

(e) Find the Euler numerical approximation to the solution of the initial value problem with 4 subintervals of the x-interval and plot the Euler approximation superimposed on the graph produced in part (d).

(f) Repeat part (e) for 8, 16, and 32 subintervals. Plot these three Euler approximations superimposed on the graph from part (e).

(g) Find the error (y (exact) $-$ y (Euler)) at the specified point $x = b$ for each of your four Euler approximations. Discuss the improvement in the percentage error.

25. $y' = x + y, \quad y(0) = -7/10; \quad -4 \le x \le 4, \quad -4 \le y \le 4; \quad b = 1$

26. $y' = -x/y, \quad y(0) = 2; \quad -3 \le x \le 3, \quad -3 \le y \le 3; \quad b = 2$

27. *A logistic equation* $y' = y(2 - y), \quad y(0) = 1/2; \quad 0 \le x \le 4, \quad 0 \le y \le 3; \quad b = 3$

28. $y' = (\sin x)(\sin y), \quad y(0) = 2; \quad -6 \le x \le 6, \quad -6 \le y \le 6;$ $b = 3\pi/2$

Exercises 29 and 30 have no explicit solution in terms of elementary functions. Use a CAS to explore graphically each of the differential equations, performing as many of the steps (a) through (g) above as possible.

CD-ROM WEBsite **29.** $y' = \cos(2x - y), \quad y(0) = 2; \quad 0 \le x \le 5, \quad 0 \le y \le 5; \quad y(2)$

CD-ROM WEBsite **30.** *A Gompertz equation* $y' = y(1/2 - \ln y), \quad y(0) = 1/3;$ $0 \le x \le 4, \quad 0 \le y \le 3; \quad y(3)$

CD-ROM 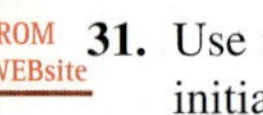 **31.** Use a CAS to find the solutions of $y' + y = f(x)$ subject to the initial condition $y(0) = 0$ if $f(x)$ is

(a) $2x$ **(b)** $\sin 2x$

(c) $3e^{x/2}$ **(d)** $2e^{-x/2}\cos 2x$.

Graph all four solutions over the interval $-2 \le x \le 6$ to compare the results.

32. CD-ROM WEBsite **(a)** Use a CAS to plot the slope field of the differential equation

$$y' = \frac{3x^2 + 4x + 2}{2(y - 1)}$$

over the region $-3 \le x \le 3$ and $-3 \le y \le 3$.

(b) Separate the variables and use a CAS integrator to find the general solution in implicit form.

(c) Using a CAS implicit function grapher, plot solution curves for the arbitrary constant values $C = -6, -4, -2, 0, 2, 4, 6$.

(d) Find and graph the solution that satisfies the initial condition $y(0) = -1$.

6.5 Hyperbolic Functions

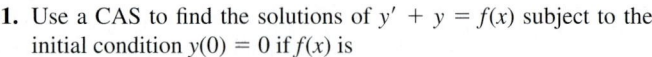

Definitions and Identities • Derivatives and Integrals • Inverse Hyperbolic Functions • Useful Identities • Derivatives and Integrals

Every function f that is defined on an interval centered at the origin can be written in a unique way as the sum of one even function and one odd function. The decomposition is

$$f(x) = \underbrace{\frac{f(x) + f(-x)}{2}}_{\text{even part}} + \underbrace{\frac{f(x) - f(-x)}{2}}_{\text{odd part}}.$$

If we write e^x this way, we get

$$e^x = \underbrace{\frac{e^x + e^{-x}}{2}}_{\text{even part}} + \underbrace{\frac{e^x - e^{-x}}{2}}_{\text{odd part}}.$$

The even and odd parts of e^x, called the hyperbolic cosine and hyperbolic sine of x, respectively, are useful in their own right. They describe the motions of waves in elastic solids, the shapes of hanging electric power lines, and the temperature distributions in metal cooling fins. The center line of the Gateway Arch to the West in St. Louis is a weighted hyperbolic cosine curve. In this section, we give a brief introduction to hyperbolic functions.

Definitions and Identities

The hyperbolic cosine and hyperbolic sine functions are defined by the first two equations in Table 6.7. The table also lists the definitions of the hyperbolic tangent, cotangent, secant, and cosecant. As we see, the hyperbolic functions bear a number of similarities to the trigonometric functions after which they are named. (See Exercise 84 as well.)

Hyperbolic functions satisfy the identities in Table 6.8. Except for differences in sign, these are identities we already know for trigonometric functions.

The notation $\cosh x$ is often read "kosh x," rhyming with "gosh x," and $\sinh x$ is pronounced as if spelled "cinch x."

Table 6.7 The six basic hyperbolic functions (see Figure 6.16 for graphs)

Hyperbolic cosine of x:	$\cosh x = \dfrac{e^x + e^{-x}}{2}$
Hyperbolic sine of x:	$\sinh x = \dfrac{e^x - e^{-x}}{2}$
Hyperbolic tangent:	$\tanh x = \dfrac{\sinh x}{\cosh x} = \dfrac{e^x - e^{-x}}{e^x + e^{-x}}$
Hyperbolic cotangent:	$\coth x = \dfrac{\cosh x}{\sinh x} = \dfrac{e^x + e^{-x}}{e^x - e^{-x}}$
Hyperbolic secant:	$\operatorname{sech} x = \dfrac{1}{\cosh x} = \dfrac{2}{e^x + e^{-x}}$
Hyperbolic cosecant:	$\operatorname{csch} x = \dfrac{1}{\sinh x} = \dfrac{2}{e^x - e^{-x}}$

Table 6.8 Identities for hyperbolic functions

$$\sinh 2x = 2 \sinh x \cosh x$$

$$\cosh 2x = \cosh^2 x + \sinh^2 x$$

$$\cosh^2 x = \frac{\cosh 2x + 1}{2}$$

$$\sinh^2 x = \frac{\cosh 2x - 1}{2}$$

$$\cosh^2 x - \sinh^2 x = 1$$

$$\tanh^2 x = 1 - \operatorname{sech}^2 x$$

$$\coth^2 x = 1 + \operatorname{csch}^2 x$$

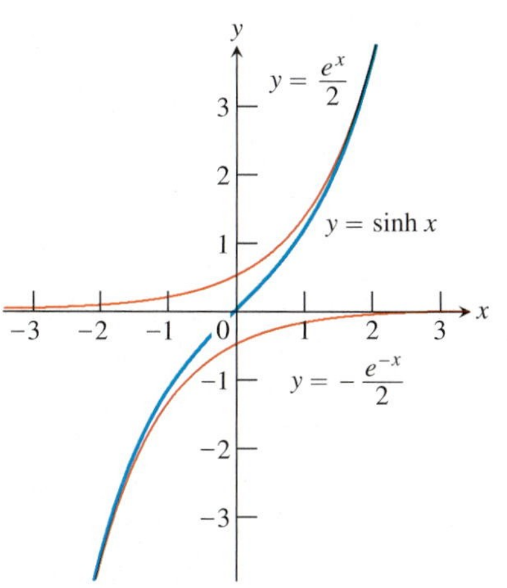

(a) The hyperbolic sine and its component exponentials

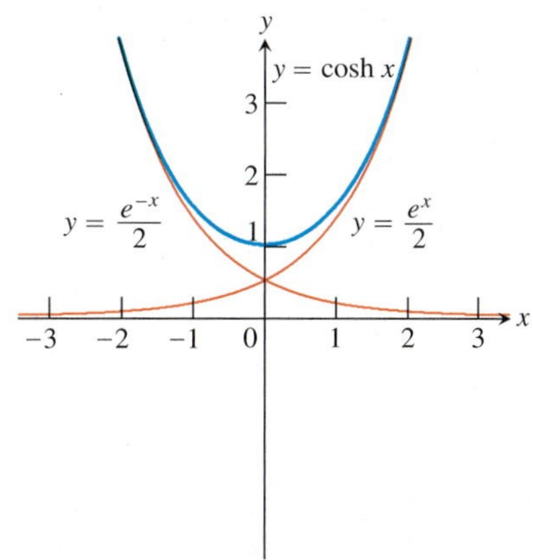

(b) The hyperbolic cosine and its component exponentials

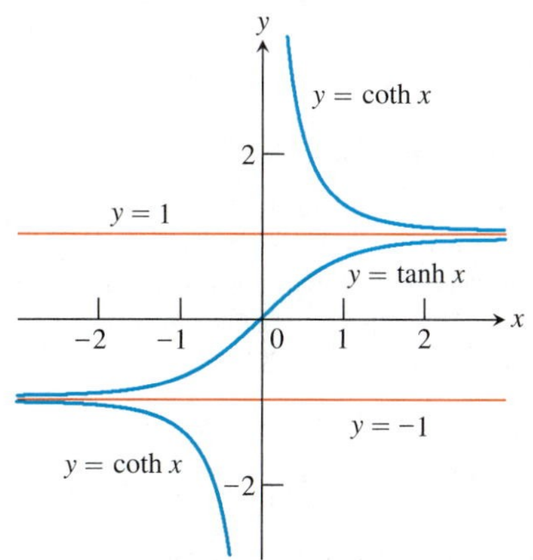

(c) The graphs of $y = \tanh x$ and $y = \coth x = 1/\tanh x$

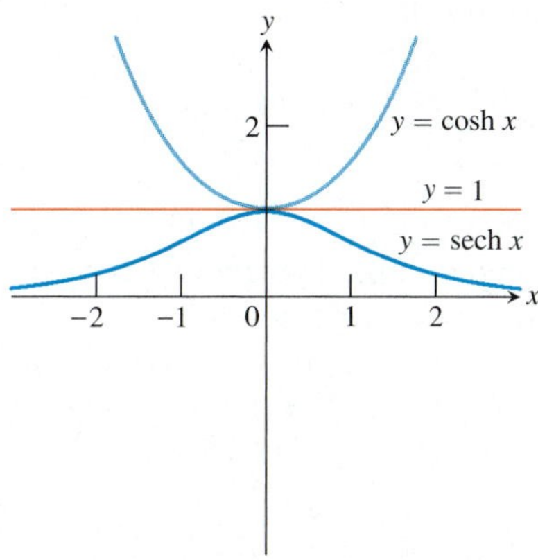

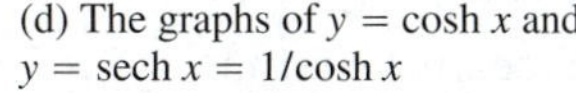

(d) The graphs of $y = \cosh x$ and $y = \operatorname{sech} x = 1/\cosh x$

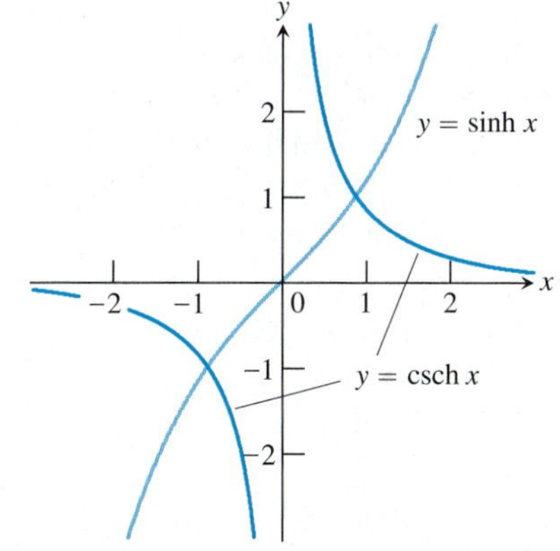

(e) The graphs of $y = \sinh x$ and $y = \operatorname{csch} x = 1/\sinh x$

FIGURE 6.16 Graphs of the six basic hyperbolic functions.

Table 6.9 Derivatives of hyperbolic functions

$$\frac{d}{dx}(\sinh u) = \cosh u \frac{du}{dx}$$

$$\frac{d}{dx}(\cosh u) = \sinh u \frac{du}{dx}$$

$$\frac{d}{dx}(\tanh u) = \operatorname{sech}^2 u \frac{du}{dx}$$

$$\frac{d}{dx}(\coth u) = -\operatorname{csch}^2 u \frac{du}{dx}$$

$$\frac{d}{dx}(\operatorname{sech} u) = -\operatorname{sech} u \tanh u \frac{du}{dx}$$

$$\frac{d}{dx}(\operatorname{csch} u) = -\operatorname{csch} u \coth u \frac{du}{dx}$$

Derivatives and Integrals

The six hyperbolic functions, being rational combinations of the differentiable functions e^x and e^{-x}, have derivatives at every point at which they are defined (Table 6.9).

Again, there are similarities with trigonometric functions. The derivative formulas in Table 6.9 lead to the integral formulas in Table 6.10.

Example 1 Finding Derivatives and Integrals

(a) $$\frac{d}{dt}(\tanh \sqrt{1+t^2}) = \operatorname{sech}^2 \sqrt{1+t^2} \cdot \frac{d}{dt}(\sqrt{1+t^2})$$

$$= \frac{t}{\sqrt{1+t^2}} \operatorname{sech}^2 \sqrt{1+t^2}$$

(b) $$\int \coth 5x\, dx = \int \frac{\cosh 5x}{\sinh 5x}\, dx = \frac{1}{5}\int \frac{du}{u} \qquad u = \sinh 5x,\ du = 5\cosh 5x\, dx$$

$$= \frac{1}{5}\ln|u| + C = \frac{1}{5}\ln|\sinh 5x| + C$$

(c) $$\int_0^1 \sinh^2 x\, dx = \int_0^1 \frac{\cosh 2x - 1}{2}\, dx \qquad \text{Table 6.8}$$

$$= \frac{1}{2}\int_0^1 (\cosh 2x - 1)\, dx = \frac{1}{2}\left[\frac{\sinh 2x}{2} - x\right]_0^1$$

$$= \frac{\sinh 2}{4} - \frac{1}{2} \approx 0.40672$$

(d) $$\int_0^{\ln 2} 4e^x \sinh x\, dx = \int_0^{\ln 2} 4e^x \frac{e^x - e^{-x}}{2}\, dx = \int_0^{\ln 2} (2e^{2x} - 2)\, dx$$

$$= \left[e^{2x} - 2x\right]_0^{\ln 2} = (e^{2\ln 2} - 2\ln 2) - (1 - 0)$$

$$= 4 - 2\ln 2 - 1$$

$$\approx 1.6137$$

Evaluating Hyperbolic Functions

Like many standard functions, hyperbolic functions and their inverses are easily evaluated with calculators, which have special keys or keystroke sequences for that purpose.

Table 6.10 Integral formulas for hyperbolic functions

$$\int \sinh u\, du = \cosh u + C$$

$$\int \cosh u\, du = \sinh u + C$$

$$\int \operatorname{sech}^2 u\, du = \tanh u + C$$

$$\int \operatorname{csch}^2 u\, du = -\coth u + C$$

$$\int \operatorname{sech} u \tanh u\, du = -\operatorname{sech} u + C$$

$$\int \operatorname{csch} u \coth u\, du = -\operatorname{csch} u + C$$

Inverse Hyperbolic Functions

We use the inverses of the six basic hyperbolic functions in integration. Since $d(\sinh x)/dx = \cosh x > 0$, the hyperbolic sine is an increasing function of x. We denote its inverse by

$$y = \sinh^{-1} x.$$

For every value of x in the interval $-\infty < x < \infty$, the value of $y = \sinh^{-1} x$ is the number whose hyperbolic sine is x. The graphs of $y = \sinh x$ and $y = \sinh^{-1} x$ are shown in Figure 6.17a.

The function $y = \cosh x$ is not one-to-one, as we can see from the graph in Figure 6.16b. The restricted function $y = \cosh x$, $x \geq 0$, however, is one-to-one and therefore has an inverse, denoted by

$$y = \cosh^{-1} x.$$

For every value of $x \geq 1$, $y = \cosh^{-1} x$ is the number in the interval $0 \leq y < \infty$ whose hyperbolic cosine is x. The graphs of $y = \cosh x$, $x \geq 0$, and $y = \cosh^{-1} x$ are shown in Figure 6.17b.

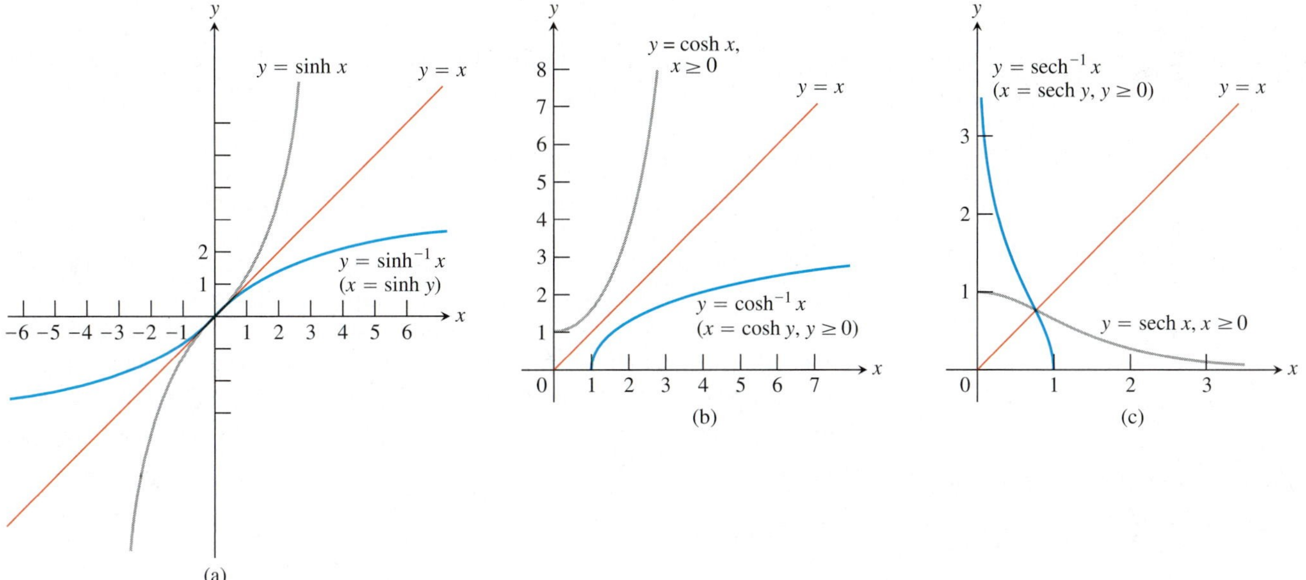

FIGURE 6.17 The graphs of the inverse hyperbolic sine, cosine, and secant of x. Notice the symmetries about the line $y = x$.

Like $y = \cosh x$, the function $y = \operatorname{sech} x = 1/\cosh x$ fails to be one-to-one, but its restriction to nonnegative values of x does have an inverse, denoted by

$$y = \operatorname{sech}^{-1} x.$$

For every value of x in the interval $(0, 1]$, $y = \operatorname{sech}^{-1} x$ is the nonnegative number whose hyperbolic secant is x. The graphs of $y = \operatorname{sech} x$, $x \geq 0$, and $y = \operatorname{sech}^{-1} x$ are shown in Figure 6.17c.

The hyperbolic tangent, cotangent, and cosecant are one-to-one on their domains and therefore have inverses, denoted by

$$y = \tanh^{-1} x, \qquad y = \coth^{-1} x, \qquad y = \operatorname{csch}^{-1} x.$$

These functions are graphed in Figure 6.18.

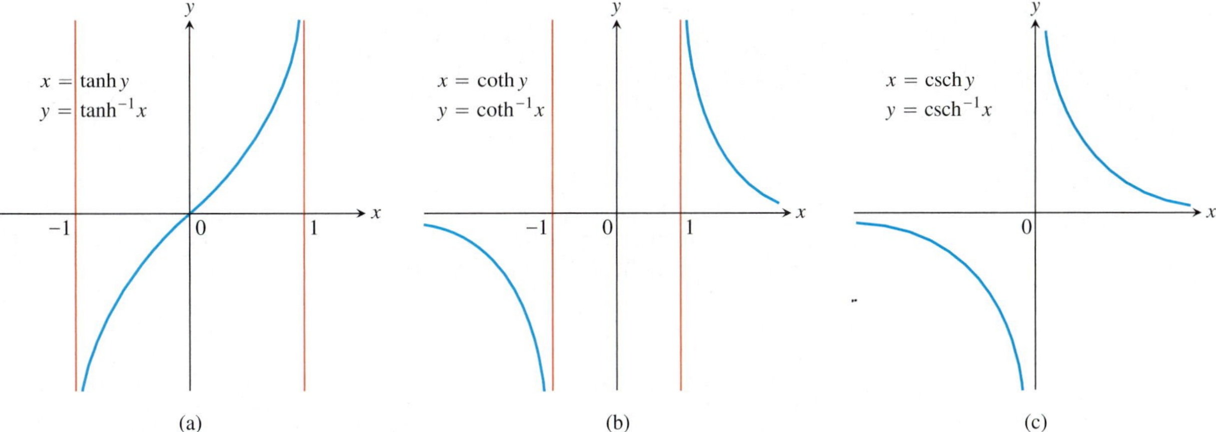

FIGURE 6.18 The graphs of the inverse hyperbolic tangent, cotangent, and cosecant of x.

Table 6.11 Identities for inverse hyperbolic functions

$$\operatorname{sech}^{-1} x = \cosh^{-1} \frac{1}{x}$$

$$\operatorname{csch}^{-1} x = \sinh^{-1} \frac{1}{x}$$

$$\coth^{-1} x = \tanh^{-1} \frac{1}{x}$$

Table 6.12 Derivatives of inverse hyperbolic functions

$$\frac{d(\sinh^{-1} u)}{dx} = \frac{1}{\sqrt{1+u^2}} \frac{du}{dx}$$

$$\frac{d(\cosh^{-1} u)}{dx} = \frac{1}{\sqrt{u^2-1}} \frac{du}{dx}, \quad u > 1$$

$$\frac{d(\tanh^{-1} u)}{dx} = \frac{1}{1-u^2} \frac{du}{dx}, \quad |u| < 1$$

$$\frac{d(\coth^{-1} u)}{dx} = \frac{1}{1-u^2} \frac{du}{dx}, \quad |u| > 1$$

$$\frac{d(\operatorname{sech}^{-1} u)}{dx} = \frac{-du/dx}{u\sqrt{1-u^2}}, \quad 0 < u < 1$$

$$\frac{d(\operatorname{csch}^{-1} u)}{dx} = \frac{-du/dx}{|u|\sqrt{1+u^2}}, \quad u \neq 0$$

Useful Identities

We use the identities in Table 6.11 to calculate the values of $\operatorname{sech}^{-1} x$, $\operatorname{csch}^{-1} x$, and $\coth^{-1} x$ on calculators that give only $\cosh^{-1} x$, $\sinh^{-1} x$, and $\tanh^{-1} x$.

Derivatives and Integrals

The chief use of inverse hyperbolic functions lies in integrations that reverse the derivative formulas in Table 6.12.

The restrictions $|u| < 1$ and $|u| > 1$ on the derivative formulas for $\tanh^{-1} u$ and $\coth^{-1} u$ come from the natural restrictions on the values of these functions. (See Figure 6.18a and b.) The distinction between $|u| < 1$ and $|u| > 1$ becomes important when we convert the derivative formulas into integral formulas. If $|u| < 1$, the integral of $1/(1-u^2)$ is $\tanh^{-1} u + C$. If $|u| > 1$, the appropriate integral is $\coth^{-1} u + C$.

Example 2 Derivative of the Inverse Hyperbolic Cosine

Show that if u is a differentiable function of x whose values are greater than 1, then

$$\frac{d}{dx}(\cosh^{-1} u) = \frac{1}{\sqrt{u^2-1}} \frac{du}{dx}.$$

Solution First we find the derivative of $y = \cosh^{-1} x$ for $x > 1$:

$$y = \cosh^{-1} x$$

$$x = \cosh y \qquad \text{Equivalent equation}$$

$$1 = \sinh y \frac{dy}{dx} \qquad \text{Differentiation with respect to } x$$

$$\frac{dy}{dx} = \frac{1}{\sinh y} = \frac{1}{\sqrt{\cosh^2 y - 1}} \qquad \text{Since } x > 1,\ y > 0 \text{ and } \sinh y > 0.$$

$$= \frac{1}{\sqrt{x^2-1}} \qquad \cosh y = x$$

In short,

$$\frac{d}{dx}(\cosh^{-1} x) = \frac{1}{\sqrt{x^2-1}}.$$

The Chain Rule gives the final result:

$$\frac{d}{dx}(\cosh^{-1} u) = \frac{1}{\sqrt{u^2-1}} \frac{du}{dx}.$$

With appropriate substitutions, the derivative formulas in Table 6.12 lead to the integration formulas in Table 6.13.

Example 3 Using Table 6.13

Evaluate

$$\int_0^1 \frac{2dx}{\sqrt{3+4x^2}}.$$

CD-ROM
WEBsite
Historical Biography

Sonya Kovalevsky
(1850 — 1891)

Table 6.13 Integrals leading to inverse hyperbolic functions

1. $$\int \frac{du}{\sqrt{a^2+u^2}} = \sinh^{-1}\left(\frac{u}{a}\right) + C, \qquad a > 0$$

2. $$\int \frac{du}{\sqrt{u^2-a^2}} = \cosh^{-1}\left(\frac{u}{a}\right) + C, \qquad u > a > 0$$

3. $$\int \frac{du}{a^2-u^2} = \begin{cases} \dfrac{1}{a}\tanh^{-1}\left(\dfrac{u}{a}\right) + C & \text{if } u^2 < a^2 \\ \dfrac{1}{a}\coth^{-1}\left(\dfrac{u}{a}\right) + C & \text{if } u^2 > a^2 \end{cases}$$

4. $$\int \frac{du}{u\sqrt{a^2-u^2}} = -\frac{1}{a}\operatorname{sech}^{-1}\left(\frac{u}{a}\right) + C, \qquad 0 < u < a$$

5. $$\int \frac{du}{u\sqrt{a^2+u^2}} = -\frac{1}{a}\operatorname{csch}^{-1}\left|\frac{u}{a}\right| + C, \qquad u \neq 0$$

Solution The indefinite integral is

$$\int \frac{2\,dx}{\sqrt{3+4x^2}} = \int \frac{du}{\sqrt{a^2+u^2}} \qquad u = 2x,\quad du = 2\,dx,\quad a = \sqrt{3}$$

$$= \sinh^{-1}\left(\frac{u}{a}\right) + C \qquad \text{Formula from Table 6.13}$$

$$= \sinh^{-1}\left(\frac{2x}{\sqrt{3}}\right) + C.$$

Therefore,

$$\int_0^1 \frac{2\,dx}{\sqrt{3+4x^2}} = \sinh^{-1}\left(\frac{2x}{\sqrt{3}}\right)\Bigg]_0^1 = \sinh^{-1}\left(\frac{2}{\sqrt{3}}\right) - \sinh^{-1}(0)$$

$$= \sinh^{-1}\left(\frac{2}{\sqrt{3}}\right) - 0 \approx 0.98665.$$

EXERCISES 6.5

Hyperbolic Function Values and Identities

Each of Exercises 1–4 gives a value of $\sinh x$ or $\cosh x$. Use the definitions and the identity $\cosh^2 x - \sinh^2 x = 1$ to find the values of the remaining five hyperbolic functions.

1. $\sinh x = -\dfrac{3}{4}$

2. $\sinh x = \dfrac{4}{3}$

3. $\cosh x = \dfrac{17}{15}, \quad x > 0$

4. $\cosh x = \dfrac{13}{5}, \quad x > 0$

Rewrite the expressions in Exercises 5–10 in terms of exponentials and simplify the results as much as you can.

5. $2\cosh(\ln x)$

6. $\sinh(2\ln x)$

7. $\cosh 5x + \sinh 5x$

8. $\cosh 3x - \sinh 3x$

9. $(\sinh x + \cosh x)^4$

10. $\ln(\cosh x + \sinh x) + \ln(\cosh x - \sinh x)$

11. Use the identities

$$\sinh(x+y) = \sinh x \cosh y + \cosh x \sinh y$$
$$\cosh(x+y) = \cosh x \cosh y + \sinh x \sinh y$$

to show that

(a) $\sinh 2x = 2 \sinh x \cosh x$

(b) $\cosh 2x = \cosh^2 x + \sinh^2 x$.

12. Use the definitions of $\cosh x$ and $\sinh x$ to show that

$$\cosh^2 x - \sinh^2 x = 1.$$

Derivatives

In Exercises 13–24, find the derivative of y with respect to the appropriate variable.

13. $y = 6 \sinh \frac{x}{3}$

14. $y = \frac{1}{2} \sinh(2x + 1)$

15. $y = 2\sqrt{t} \tanh \sqrt{t}$

16. $y = t^2 \tanh \frac{1}{t}$

17. $y = \ln(\sinh z)$

18. $y = \ln(\cosh z)$

19. $y = \operatorname{sech} \theta(1 - \ln \operatorname{sech} \theta)$

20. $y = \operatorname{csch} \theta(1 - \ln \operatorname{csch} \theta)$

21. $y = \ln \cosh v - \frac{1}{2} \tanh^2 v$

22. $y = \ln \sinh v - \frac{1}{2} \coth^2 v$

23. $y = (x^2 + 1) \operatorname{sech}(\ln x)$

(*Hint:* Before differentiating, express in terms of exponentials and simplify.)

24. $y = (4x^2 - 1) \operatorname{csch}(\ln 2x)$

In Exercises 25–36, find the derivative of y with respect to the appropriate variable.

25. $y = \sinh^{-1} \sqrt{x}$

26. $y = \cosh^{-1} 2\sqrt{x+1}$

27. $y = (1 - \theta) \tanh^{-1} \theta$

28. $y = (\theta^2 + 2\theta) \tanh^{-1}(\theta + 1)$

29. $y = (1 - t) \coth^{-1} \sqrt{t}$

30. $y = (1 - t^2) \coth^{-1} t$

31. $y = \cos^{-1} x - x \operatorname{sech}^{-1} x$

32. $y = \ln x + \sqrt{1 - x^2} \operatorname{sech}^{-1} x$

33. $y = \operatorname{csch}^{-1} \left(\frac{1}{2}\right)^{\theta}$

34. $y = \operatorname{csch}^{-1} 2^{\theta}$

35. $y = \sinh^{-1}(\tan x)$

36. $y = \cosh^{-1}(\sec x), \quad 0 < x < \pi/2$

Integration Formulas

Verify the integration formulas in Exercises 37–40.

37. **(a)** $\int \operatorname{sech} x \, dx = \tan^{-1}(\sinh x) + C$

(b) $\int \operatorname{sech} x \, dx = \sin^{-1}(\tanh x) + C$

38. $\int x \operatorname{sech}^{-1} x \, dx = \frac{x^2}{2} \operatorname{sech}^{-1} x - \frac{1}{2}\sqrt{1 - x^2} + C$

39. $\int x \coth^{-1} x \, dx = \frac{x^2 - 1}{2} \coth^{-1} x + \frac{x}{2} + C$

40. $\int \tanh^{-1} x \, dx = x \tanh^{-1} x + \frac{1}{2} \ln(1 - x^2) + C$

Indefinite Integrals

Evaluate the integrals in Exercises 41–50.

41. $\int \sinh 2x \, dx$

42. $\int \sinh \frac{x}{5} \, dx$

43. $\int 6 \cosh \left(\frac{x}{2} - \ln 3\right) dx$

44. $\int 4 \cosh(3x - \ln 2) \, dx$

45. $\int \tanh \frac{x}{7} \, dx$

46. $\int \coth \frac{\theta}{\sqrt{3}} \, d\theta$

47. $\int \operatorname{sech}^2 \left(x - \frac{1}{2}\right) dx$

48. $\operatorname{csch}^2(5 - x) \, dx$

49. $\int \frac{\operatorname{sech} \sqrt{t} \tanh \sqrt{t} \, dt}{\sqrt{t}}$

50. $\int \frac{\operatorname{csch}(\ln t) \coth(\ln t) \, dt}{t}$

Definite Integrals

Evaluate the integrals in Exercises 51–60.

51. $\int_{\ln 2}^{\ln 4} \coth x \, dx$

52. $\int_0^{\ln 2} \tanh 2x \, dx$

53. $\int_{-\ln 4}^{-\ln 2} 2e^{\theta} \cosh \theta \, d\theta$

54. $\int_0^{\ln 2} 4e^{-\theta} \sinh \theta \, d\theta$

55. $\int_{-\pi/4}^{\pi/4} \cosh(\tan \theta) \sec^2 \theta \, d\theta$

56. $\int_0^{\pi/2} 2 \sinh(\sin \theta) \cos \theta \, d\theta$

57. $\int_1^2 \frac{\cosh(\ln t)}{t} \, dt$

58. $\int_1^4 \frac{8 \cosh \sqrt{x}}{\sqrt{x}} \, dx$

59. $\int_{-\ln 2}^{0} \cosh^2 \left(\frac{x}{2}\right) dx$

60. $\int_0^{\ln 10} 4 \sinh^2 \left(\frac{x}{2}\right) dx$

Evaluating Inverse Hyperbolic Functions and Related Integrals

When hyperbolic function keys are not available on a calculator, it is still possible to evaluate the inverse hyperbolic functions by expressing them as logarithms, as shown here.

$$\sinh^{-1} x = \ln\left(x + \sqrt{x^2 + 1}\right), \quad -\infty < x < \infty$$

$$\cosh^{-1} x = \ln\left(x + \sqrt{x^2 - 1}\right), \quad x \geq 1$$

$$\tanh^{-1} x = \frac{1}{2} \ln \frac{1 + x}{1 - x}, \quad |x| < 1$$

$$\operatorname{sech}^{-1} x = \ln\left(\frac{1 + \sqrt{1 - x^2}}{x}\right), \quad 0 < x \leq 1$$

$$\operatorname{csch}^{-1} x = \ln\left(\frac{1}{x} + \frac{\sqrt{1 + x^2}}{|x|}\right), \quad x \neq 0$$

$$\coth^{-1} x = \frac{1}{2} \ln \frac{x + 1}{x - 1}, \quad |x| > 1$$

Use the formulas in the table above to express the numbers in Exercises 61–66 in terms of natural logarithms.

61. $\sinh^{-1}(-5/12)$ **62.** $\cosh^{-1}(5/3)$

63. $\tanh^{-1}(-1/2)$ **64.** $\coth^{-1}(5/4)$

65. $\operatorname{sech}^{-1}(3/5)$ **66.** $\operatorname{csch}^{-1}(-1/\sqrt{3})$

Evaluate the integrals in Exercises 67–74 in terms of

(a) inverse hyperbolic functions

(b) natural logarithms.

67. $\int_0^{2\sqrt{3}} \frac{dx}{\sqrt{4+x^2}}$ **68.** $\int_0^{1/3} \frac{6dx}{\sqrt{1+9x^2}}$

69. $\int_{5/4}^{2} \frac{dx}{1-x^2}$ **70.** $\int_0^{1/2} \frac{dx}{1-x^2}$

71. $\int_{1/5}^{3/13} \frac{dx}{x\sqrt{1-16x^2}}$ **72.** $\int_1^2 \frac{dx}{x\sqrt{4+x^2}}$

73. $\int_0^{\pi} \frac{\cos x\, dx}{\sqrt{1+\sin^2 x}}$ **74.** $\int_1^e \frac{dx}{x\sqrt{1+(\ln x)^2}}$

Applications and Theory

75. (a) Show that if a function f is defined on an interval symmetric about the origin (so that f is defined at $-x$ whenever it is defined at x), then

$$f(x) = \frac{f(x)+f(-x)}{2} + \frac{f(x)-f(-x)}{2}. \quad (1)$$

Then show that $(f(x) + f(-x))/2$ is even and that $(f(x) - f(-x))/2$ is odd.

(b) *Writing to Learn* Equation (1) simplifies considerably if f itself is (i) even or (ii) odd. What are the new equations? Give reasons for your answers.

76. *Writing to Learn* Derive the formula $\sinh^{-1} x = \ln(x + \sqrt{x^2+1})$, $-\infty < x < \infty$. Explain in your derivation why the plus sign is used with the square root instead of the minus sign.

77. *Skydiving* If a body of mass m falling from rest under the action of gravity encounters an air resistance proportional to the square of the velocity, then the body's velocity t sec into the fall satisfies the differential equation

$$m\frac{dv}{dt} = mg - kv^2,$$

where k is a constant that depends on the body's aerodynamic properties and the density of the air. (We assume that the fall is short enough so that the variation in the air's density will not affect the outcome significantly.)

(a) Show that

$$v = \sqrt{\frac{mg}{k}}\tanh\left(\sqrt{\frac{gk}{m}}\,t\right)$$

satisfies the differential equation and the initial condition that $v = 0$ when $t = 0$.

(b) Find the body's *limiting velocity*, $\lim_{t\to\infty} v$.

(c) For a 160 lb skydiver ($mg = 160$), with time in seconds and distance in feet, a typical value for k is 0.005. What is the diver's limiting velocity?

78. *Accelerations whose magnitudes are proportional to displacement* Suppose that the position of a body moving along a coordinate line at time t is

(a) $s = a\cos kt + b\sin kt$,

(b) $s = a\cosh kt + b\sinh kt$.

Show in both cases that the acceleration d^2s/dt^2 is proportional to s but that in the first case it is directed toward the origin, whereas in the second case it is directed away from the origin.

79. *Tractor trailers and the tractrix* When a tractor trailer turns into a cross street or driveway, its rear wheels follow a curve like the one shown here.

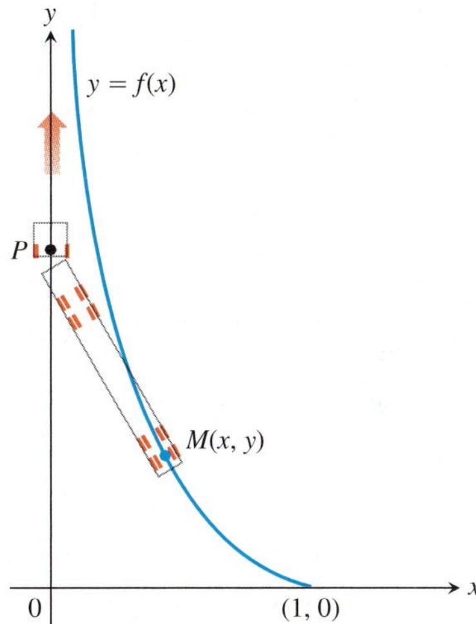

(This is why the rear wheels sometimes ride up over the curb.) We can find an equation for the curve if we picture the rear wheels as a mass M at the point (1, 0) on the x-axis attached by a rod of unit length to a point P representing the cab at the origin. As the point P moves up the y-axis, it drags M along behind it. The curve traced by M—called a *tractrix* from the Latin word *tractum*, for "drag"—can be shown to be the graph of the function $y = f(x)$ that solves the initial value problem

Differential equation: $\frac{dy}{dx} = -\frac{1}{x\sqrt{1-x^2}} + \frac{x}{\sqrt{1-x^2}}$

Initial condition: $y = 0$ when $x = 1$.

Solve the initial value problem to find an equation for the curve. (You need an inverse hyperbolic function.)

80. *Area* Show that the area of the region in the first quadrant enclosed by the curve $y = (1/a)\cosh ax$, the coordinate axes, and the line $x = b$ is the same as the area of a rectangle of height $1/a$

and length s, where s is the length of the curve from $x = 0$ to $x = b$.

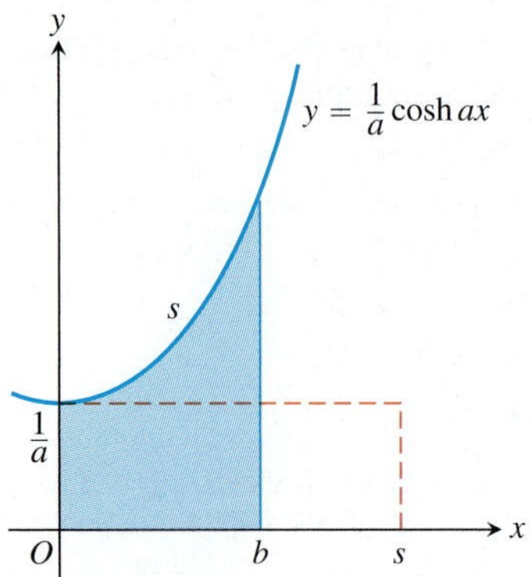

81. *Volume* A region in the first quadrant is bounded above by the curve $y = \cosh x$, below by the curve $y = \sinh x$, and on the left and right by the y-axis and the line $x = 2$, respectively. Find the volume of the solid generated by revolving the region about the x-axis.

82. *Volume* The region enclosed by the curve $y = \operatorname{sech} x$, the x-axis, and the lines $x = \pm\ln\sqrt{3}$ is revolved about the x-axis to generate a solid. Find the volume of the solid.

83. *Arc length* Find the length of the segment of the curve $y = (1/2)\cosh 2x$ from $x = 0$ to $x = \ln\sqrt{5}$.

84. *The hyperbolic in hyperbolic functions* In case you are wondering where the name *hyperbolic* comes from, here is the answer: Just as $x = \cos u$ and $y = \sin u$ are identified with points (x, y) on the unit circle, the functions $x = \cosh u$ and $y = \sinh u$ are identified with points (x, y) on the right-hand branch of the unit hyperbola, $x^2 - y^2 = 1$ (Figure 6.19).

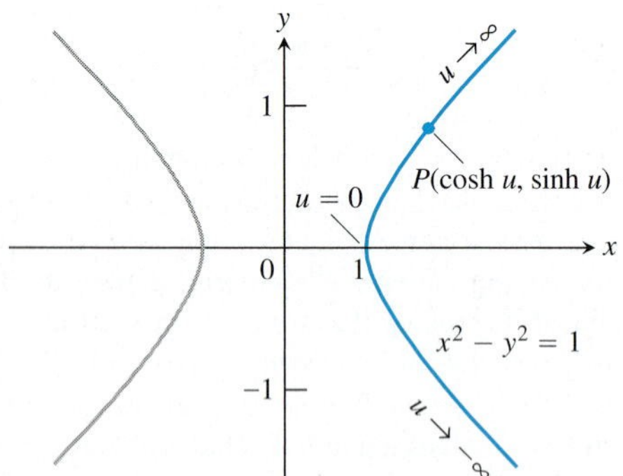

FIGURE 6.19 Since $\cosh^2 u - \sinh^2 u = 1$, the point $(\cosh u, \sinh u)$ lies on the right-hand branch of the hyperbola $x^2 - y^2 = 1$ for every value of u. (Exercise 84)

Another analogy between hyperbolic and circular functions is that the variable u in the coordinates $(\cosh u, \sinh u)$ for the points of the right-hand branch of the hyperbola $x^2 - y^2 = 1$ is twice the area of the sector AOP pictured in Figure 6.20. To see why this is so, carry out the following steps.

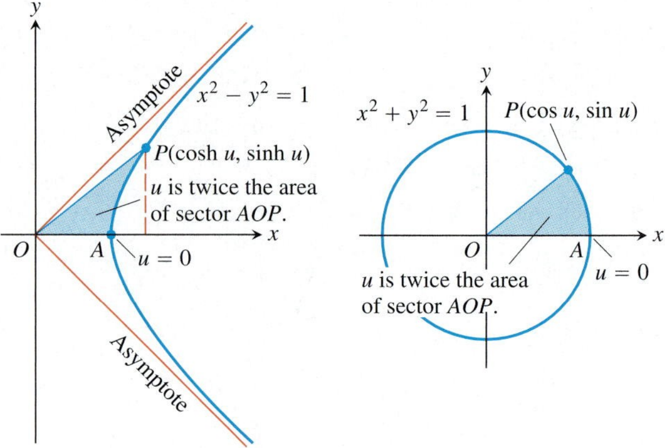

FIGURE 6.20 One of the analogies between hyperbolic and circular functions is revealed by these two diagrams. (Exercise 84)

(a) Show that the area $A(u)$ of sector AOP is

$$A(u) = \frac{1}{2}\cosh u \sinh u - \int_1^{\cosh u} \sqrt{x^2 - 1}\, dx.$$

(b) Differentiate both sides of the equation in part (a) with respect to u to show that

$$A'(u) = \frac{1}{2}.$$

(c) Solve this last equation for $A(u)$. What is the value of $A(0)$? What is the value of the constant of integration C in your solution? With C determined, what does your solution say about the relationship of u to $A(u)$?

Hanging Cables

85. Imagine a cable, like a telephone line or TV cable, strung from one support to another and hanging freely. The cable's weight per unit length is w and the horizontal tension at its lowest point is a vector of length H. If we choose a coordinate system for the plane of the cable in which the x-axis is horizontal, the force of gravity is straight down, the positive y-axis points straight up, and the lowest point of the cable lies at the point $y = H/w$ on the y-axis (Figure 6.21), then it can be shown that the cable lies along the graph of the hyperbolic cosine

$$y = \frac{H}{w}\cosh\frac{w}{H}x.$$

Such a curve is sometimes called a **chain curve** or a **catenary,** the latter deriving from the Latin *catena*, meaning "chain."

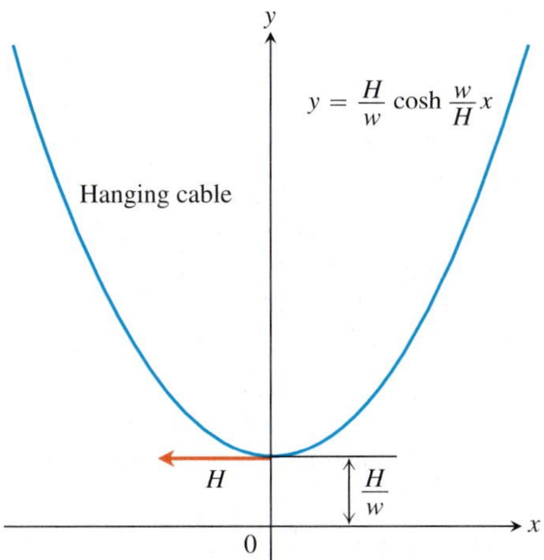

FIGURE 6.21 In a coordinate system chosen to match H and w in the manner shown, a hanging cable lies along the curve $y = (H/w)\cosh(wx/H)$.

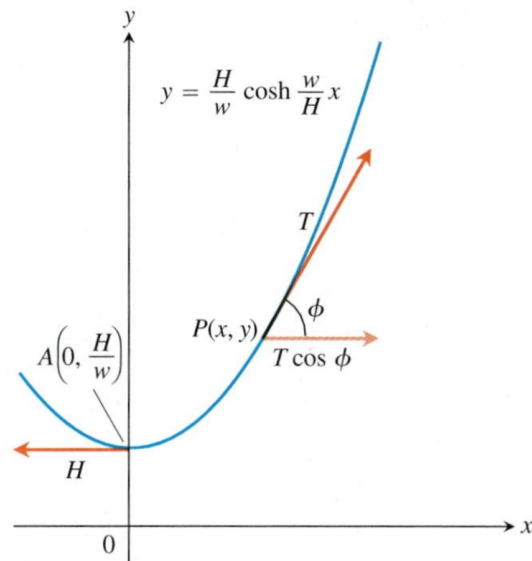

FIGURE 6.22 As discussed in Exercise 85, $T = wy$ in this coordinate system.

(a) Let $P(x, y)$ denote an arbitrary point on the cable. Figure 16.22 displays the tension at P as a vector of length (magnitude) T, as well as the tension H at the lowest point A. Show that the cable's slope at P is

$$\tan\phi = \frac{dy}{dx} = \sinh\frac{w}{H}x.$$

(b) Using the result from part (a) and the fact that the horizontal tension at P must equal H (the cable is not moving), show that $T = wy$. This means that the magnitude of the tension at $P(x, y)$ is exactly equal to the weight of y units of cable.

86. (*Continuation of Exercise 85*) The length of arc AP in Figure 6.22 is $s = (1/a)\sinh ax$, where $a = w/H$. Show that the coordinates of P may be expressed in terms of s as

$$x = \frac{1}{a}\sinh^{-1} as, \qquad y = \sqrt{s^2 + \frac{1}{a^2}}.$$

87. *The sag and horizontal tension in a cable* The ends of a cable 32 ft long and weighing 2 lb/ft are fastened at the same level to posts 30 ft apart.

(a) Model the cable with the equation

$$y = \frac{1}{a}\cosh ax, \qquad -15 \le x \le 15.$$

Use information from Exercise 86 to show that a satisfies the equation

$$16a = \sinh 15a. \tag{2}$$

T **(b)** Solve Equation (2) graphically by estimating the coordinates of the points where the graphs of the equations $y = 16a$ and $y = \sinh 15a$ intersect in the ay-plane.

T **(c)** Solve Equation (2) for a numerically. Compare your solution with the value you found in part (b).

(d) Estimate the horizontal tension in the cable at the cable's lowest point.

T **(e)** Using the value found for a in part (c), graph the catenary

$$y = \frac{1}{a}\cosh ax$$

over the interval $-15 \le x \le 15$. Estimate the sag in the cable at its center.

Questions to Guide Your Review

1. What is the natural logarithm function in terms of an integral? What are its domain, range, and derivative? What arithmetic properties does it have? Comment on its graph.
2. What integrals lead to logarithms? Give examples.
3. How is the exponential function e^x defined in terms of the natural logarithm? What are its domain, range, and derivative? What laws of exponents does it obey? Comment on its graph.
4. How is the number e defined? How can it be expressed as a limit?
5. What is the slope field of a differential equation $y' = f(x, y)$? What can we learn from such fields?
6. How do you solve linear first-order differential equations? What applications are modeled by these equations?
7. Describe Euler's method for solving the initial value problem $y' = f(x, y), y(x_0) = y_0$ numerically. Give an example. Comment on the method's accuracy. Why might you want to solve an initial value problem numerically?
8. Describe the improved Euler's method for solving the initial value problem $y' = f(x, y), y(x_0) = y_0$ numerically. How does it compare with Euler's method?
9. Why is the exponential model unrealistic for predicting long-term population growth? How does the logistic model correct for the deficiency in the exponential model for population growth? What is the logistic differential equation? What is the form of its solution? Describe the graph of the logistic solution.
10. What are the six basic hyperbolic functions? Comment on their domains, ranges, and graphs. What are some of the identities relating them?
11. What are the derivatives of the six basic hyperbolic functions? What are the corresponding integral formulas? What similarities do you see here with the six basic trigonometric functions?
12. How are the inverse hyperbolic functions defined? Comment on their domains, ranges, and graphs. How can you find values of $\operatorname{sech}^{-1} x$, $\operatorname{csch}^{-1} x$, and $\coth^{-1} x$ using a calculator's keys for $\cosh^{-1} x$, $\sinh^{-1} x$, and $\tanh^{-1} x$?
13. What integrals lead naturally to inverse hyperbolic functions?

Practice Exercises

Integration

Evaluate the integrals in Exercises 1–30.

1. $\int e^x \sin(e^x)\,dx$
2. $\int e^t \cos(3e^t - 2)\,dt$
3. $\int e^x \sec^2(e^x - 7)\,dx$
4. $\int e^y \csc(e^y + 1)\cot(e^y + 1)\,dy$
5. $\int \sec^2(x) e^{\tan x}\,dx$
6. $\int \csc^2(x) e^{\cot x}\,dx$
7. $\int_{-1}^{1} \frac{dx}{3x - 4}$
8. $\int_1^e \frac{\sqrt{\ln x}}{x}\,dx$
9. $\int_0^{\pi} \tan\frac{x}{3}\,dx$
10. $\int_{1/6}^{1/4} 2\cot \pi x\,dx$
11. $\int_0^4 \frac{2t}{t^2 - 25}\,dt$
12. $\int_{-\pi/2}^{\pi/6} \frac{\cos t}{1 - \sin t}\,dt$
13. $\int \frac{\tan(\ln v)}{v}\,dv$
14. $\int \frac{dv}{v \ln v}$
15. $\int \frac{(\ln x)^{-3}}{x}\,dx$
16. $\int \frac{\ln(x - 5)}{x - 5}\,dx$
17. $\int \frac{1}{r}\csc^2(1 + \ln r)\,dr$
18. $\int \frac{\cos(1 - \ln v)}{v}\,dv$
19. $\int x 3^{x^2}\,dx$
20. $\int 2^{\tan x} \sec^2 x\,dx$
21. $\int_1^7 \frac{3}{x}\,dx$
22. $\int_1^{32} \frac{1}{5x}\,dx$
23. $\int_{-2}^{-1} e^{-(x+1)}\,dx$
24. $\int_{-\ln 2}^{0} e^{2w}\,dw$
25. $\int_1^3 \frac{(\ln(v + 1))^2}{v + 1}\,dv$
26. $\int_2^4 (1 + \ln t)t \ln t\,dt$
27. $\int_1^8 \frac{\log_4 \theta}{\theta}\,d\theta$
28. $\int_1^e \frac{8 \ln 3 \log_3 \theta}{\theta}\,d\theta$
29. $\int \frac{2\,dy}{\sqrt{1 + 25y^2}}$
30. $\int \frac{3\,dy}{4 - 49y^2}$

Theory and Applications

In Exercises 31 and 32, find the absolute maximum and minimum values of each function on the given interval.

31. $y = x \ln 2x - x,\quad \left[\frac{1}{2e}, \frac{e}{2}\right]$
32. $y = 10x(2 - \ln x),\quad (0, e^2]$

33. *Area* Find the area between the curve $y = 2(\ln x)/x$ and the x-axis from $x = 1$ to $x = e$.

34. (a) *Area* Show that the area between the curve $y = 1/x$ and the x-axis from $x = 10$ to $x = 20$ is the same as the area between the curve and the x-axis from $x = 1$ to $x = 2$.

(b) *Area* Show that the area between the curve $y = 1/x$ and the x-axis from ka to kb is the same as the area between the curve and the x-axis from $x = a$ to $x = b$ $(0 < a < b,\ k > 0)$.

35. *Traveling particle* A particle is traveling upward and to the right along the curve $y = \ln x$. Its x-coordinate is increasing at the rate $(dx/dt) = \sqrt{x}$ m/sec. At what rate is the y-coordinate changing at the point $(e^2, 2)$?

36. *Volume* The region between the curve $y = 1/(2\sqrt{x})$ and the x-axis from $x = 1/4$ to $x = 4$ is revolved about the x-axis to generate a solid. Find the volume of the solid.

37. *Writing to Learn* The functions $f(x) = \ln 5x$ and $g(x) = \ln 3x$ differ by a constant. What constant? Give reasons for your answer.

38. *Writing to Learn*

(a) If $(\ln x)/x = (\ln 2)/2$, must $x = 2$?

(b) If $(\ln x)/x = -2\ln 2$, must $x = 1/2$?

Give reasons for your answers.

39. *Escape velocity* The gravitational attraction F exerted by an airless moon on a body of mass m at a distance s from the moon's center is given by the equation $F = -mgR^2s^{-2}$, where g is the acceleration of gravity at the moon's surface and R is the moon's radius (Figure 6.23). The force F is negative because it acts in the direction of decreasing s.

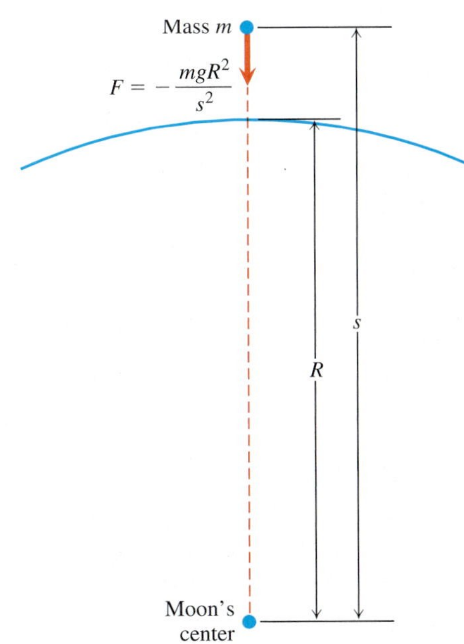

FIGURE 6.23 Diagram for Exercise 39.

If the body is projected vertically upward from the moon's surface with an initial velocity v_0 at time $t = 0$, use Newton's second law, $F = ma$, to show that the body's velocity at position s is given by the equation

$$v^2 = \frac{2gR^2}{s} + v_0^2 - 2gR.$$

Thus, the velocity remains positive as long as $v_0 \geq \sqrt{2gR}$. The velocity $v_0 = \sqrt{2gR}$ is the moon's **escape velocity.** A body projected upward with this velocity or a greater one will escape from the moon's gravitational pull.

40. (*Continuation of Exercise 39*) Show that if $v_0 = \sqrt{2gR}$, then

$$s = R\left(1 + \frac{3v_0}{2R}t\right)^{2/3}.$$

Initial Value Problems

Solve the initial value problems in Exercises 41–44.

Differential equation	Initial condition
41. $y' \cos x - y \sin x = \sin 2x$	$y(0) = 1$
42. $\dfrac{dy}{dx} = -\dfrac{y \ln y}{1 + x^2}$	$y(0) = e^2$
43. $(x + 1)\dfrac{dy}{dx} + 2y = x, \quad x > -1$	$y(0) = 1$
44. $x\dfrac{dy}{dx} + 2y = x^2 + 1, \quad x > 0$	$y(1) = 1$

T **45.** *Extreme values* Graph the following functions and use what you see to locate and estimate the extreme values, identify the coordinates of the inflection points, and identify the intervals on which the graphs are concave up and concave down. Then confirm your estimates by working with the functions' derivatives.

(a) $y = (\ln x)/\sqrt{x}$ **(b)** $y = e^{-x^2}$

(c) $y = (1 + x)e^{-x}$

T **46.** *Absolute minimum* Graph $f(x) = x \ln x$. Does the function appear to have an absolute minimum value? Confirm your answer with calculus.

47. *Age of charcoal* What is the age of a sample of charcoal in which 90% of the carbon-14 originally present has decayed?

48. *Cooling a pie* A deep-dish apple pie, whose internal temperature was 220°F when removed from the oven, was set out on a breezy 40°F porch to cool. Fifteen minutes later, the pie's internal temperature was 180°F. How long did it take the pie to cool from there to 70°F?

49. (a) Solve the differential equation

$$xy' + y = x \cos x, \qquad x > 0.$$

T **(b)** Graph several members of the family of solution curves. How does the curve vary as C changes?

50. *Confirming a solution* Show that

$$y = \int_0^x \sin(t^2)\,dt + x^3 + x + 2$$

is the solution of the initial value problem.

Differential equation: $y'' = 2x\cos(x^2) + 6x$

Initial conditions: $y'(0) = 1, \quad y(0) = 2$

Slope Fields and Euler's Method

51. *Sketching solutions* Draw a possible graph for the function $y = f(x)$ with slope field given in the figure that satisfies the initial condition $y(0) = 0$.

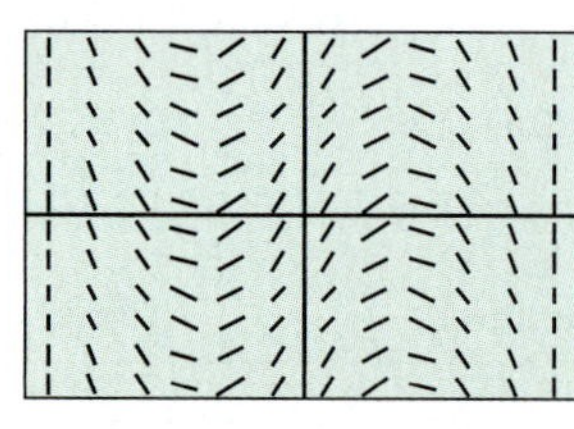

[−10, 10] by [−10, 10]

In Exercises 52 and 53, use the stated method to solve the initial value problem on the given interval starting at x_0 with $dx = 0.1$.

T **52.** *Euler;* $y' = y + \cos x, \quad y(0) = 0; \quad 0 \le x \le 2; \quad x_0 = 0$

T **53.** *Improved Euler;* $y' = (2 - y)(2x + 3), \quad y(-3) = 1;$ $-3 \le x \le -1; \quad x_0 = -3$

In Exercises 54 and 55, use the stated method with $dx = 0.05$ to estimate $y(c)$ where y is the solution to the given initial value problem.

T **54.** *Improved Euler;* $c = 3; \quad \dfrac{dy}{dx} = \dfrac{x - 2y}{x + 1}, \quad y(0) = 1$

T **55.** *Euler;* $c = 4; \quad \dfrac{dy}{dx} = \dfrac{x^2 - 2y + 1}{x}, \quad y(1) = 1$

In Exercises 56 and 57, use the stated method to solve the initial value problem graphically, starting at $x_0 = 0$ with

(a) $dx = 0.1$ **(b)** $dx = -0.1$.

T **56.** *Euler;* $\dfrac{dy}{dx} = \dfrac{1}{e^{x+y+2}}, \quad y(0) = -2$

T **57.** *Improved Euler;* $\dfrac{dy}{dx} = -\dfrac{x^2 + y}{e^y + x}, \quad y(0) = 0$

58. (a) *Finding an exact solution* Use analytic methods to find the exact solution to

$$\frac{dP}{dt} = 0.002P\left(1 - \frac{P}{800}\right), \qquad P(0) = 50.$$

T **(b)** *Numerical solution* Use Euler's method to solve the equation in part (a) for $0 \le t \le 20$ and $dt = 0.5$. Compare the approximation $P(20)$ with the exact solution value.

CD-ROM WEBsite

In Exercises 59–62, sketch part of the equation's slope field. Then add to your sketch the solution curve that passes through the point $P(1, -1)$. Use Euler's method with $x_0 = 1$ and $dx = 0.2$ to estimate $y(2)$. Round your answers to 4 decimal places. Find the exact value of $y(2)$ for comparison.

59. $y' = x$ **60.** $y' = 1/x$

61. $y' = xy$ **62.** $y' = 1/y$

Additional Exercises: Theory, Examples, Applications

1. *Area between curves* Find the areas between the curves $y = 2(\log_2 x)/x$ and $y = 2(\log_4 x)/x$ and the x-axis from $x = 1$ to $x = e$. What is the ratio of the larger area to the smaller?

2. *Derivative of integral* Find $f'(2)$ if

$$f(x) = e^{g(x)} \quad \text{and} \quad g(x) = \int_2^x \frac{t}{1 + t^4}\,dt.$$

3. *Derivative of an integral*

(a) Find df/dx if

$$f(x) = \int_1^{e^x} \frac{2\ln t}{t}\,dt.$$

(b) Find $f(0)$.

(c) *Writing to Learn* What can you conclude about the graph of f? Give reasons for your answer.

4. *Writing to Learn: The inequality* $\pi^e < e^\pi$

(a) Why does Figure 6.24 "prove" that $\pi^e < e^\pi$? (Source: "Proof without Words" by Fouad Nakhil, *Mathematics Magazine*, Vol. 60, No. 3 (June 1987), p. 165.)

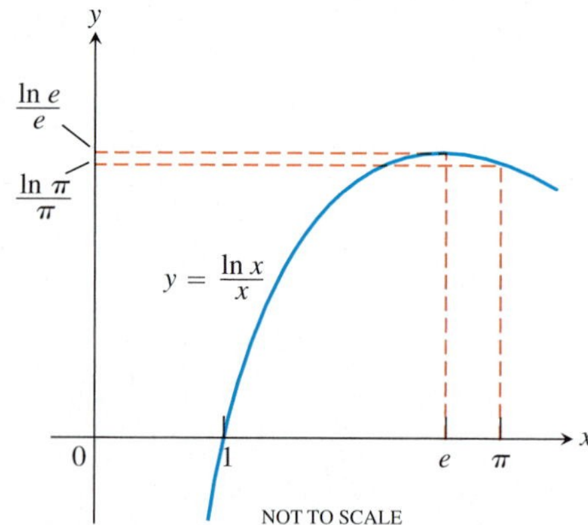

FIGURE 6.24 The figure for Exercise 4.

(b) Figure 6.24 assumes that $f(x) = (\ln x)/x$ has an absolute maximum value at $x = e$. How do you know it does?

5. *An unexpected equality* Use the accompanying figure to show that

$$\int_0^{\pi/2} \sin x\, dx = \frac{\pi}{2} - \int_0^1 \sin^{-1} x\, dx.$$

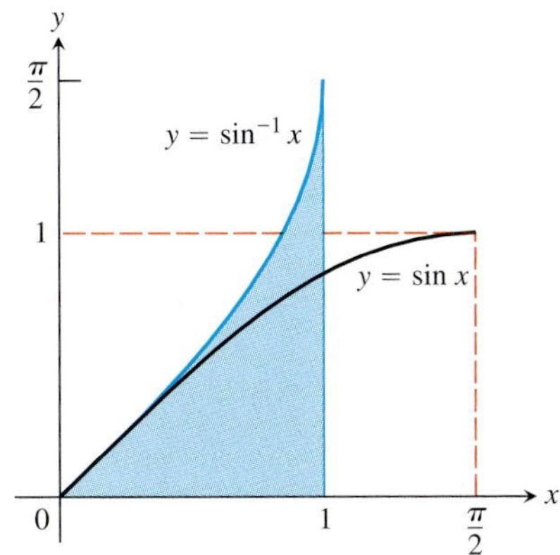

6. *Writing to Learn: Napier's inequality* Here are two pictorial proofs that

$$b > a > 0 \Rightarrow \frac{1}{b} < \frac{\ln b - \ln a}{b - a} < \frac{1}{a}.$$

Explain what is going on in each case.

(a)

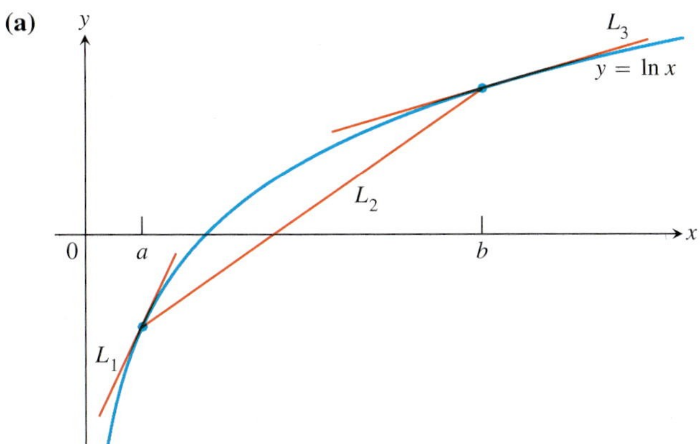

(b)

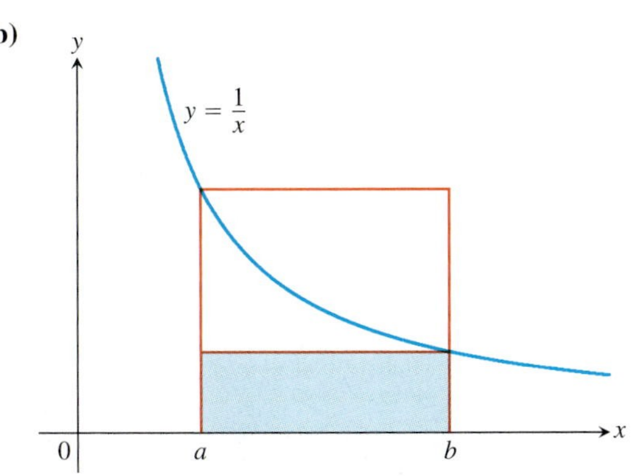

(*Source*: Roger B. Nelson, *College Mathematics Journal*, Vol. 24, No. 2 (March 1993), p. 165.)

7. *Supporting a solution* Give two ways to provide graphical support for the integral formula

$$\int x^2 \ln x\, dx = \frac{x^3}{3} \ln x - \frac{x^3}{9} + C.$$

8. *Writing to Learn* Graph $f(x) = (\sin x)^{\sin x}$ in the window $[0, 3\pi]$ by $[-2, 5]$. Explain what you see.

Applications

9. *Seasonal-growth model* In some population growth models, a periodic function of time is introduced as a multiplier of the population to account for seasonal variations in the rate of growth. For example, the population could be affected by seasonal changes in the food or water supplies.

(a) Solve the seasonal-growth model

T $$\frac{dP}{dt} = kP \cos(at - b), \qquad P(0) = P_0,$$

where k, a, and b are positive constants.

T **(b)** *Writing to Learn* Graph your solution for various values of k, a, and b. Explain the effects these values have on the solution. What happens to $P(t)$ in the long term (i.e., as $t \to \infty$)?

10. *Transport through a cell membrane* Under some conditions, the result of the movement of a dissolved substance across a cell's membrane is described by the equation

$$\frac{dy}{dt} = k\frac{A}{V}(c - y).$$

In this equation, y is the concentration of the substance inside the cell and dy/dt is the rate at which y changes over time. The letters k, A, V, and c stand for constants, k being the *permeability coefficient* (a property of the membrane), A the surface area of the membrane, V the cell's volume, and c the concentration of the substance outside the cell. The equation says that the rate at which the concentration changes within the cell is proportional to the difference between it and the outside concentration.

(a) Solve the equation for $y(t)$, using y_0 to denote $y(0)$.

(b) Find the steady state concentration, $\lim_{t\to\infty} y(t)$.
(Based on *Some Mathematical Models in Biology* edited by R. M. Thrall, J. A. Mortimer, K. R. Rebman, R. F. Baum, rev. ed., December 1967, PB-202 364, pp. 101–103; distributed by N.T.I.S., U.S. Department of Commerce.)

T **11.** *Writing to Learn* Graph $f(x) = \tan^{-1} x + \tan^{-1}(1/x)$ for $-5 \le x \le 5$. Then use calculus to explain what you see. How would you expect f to behave beyond the interval $[-5, 5]$? Give reasons for your answer.

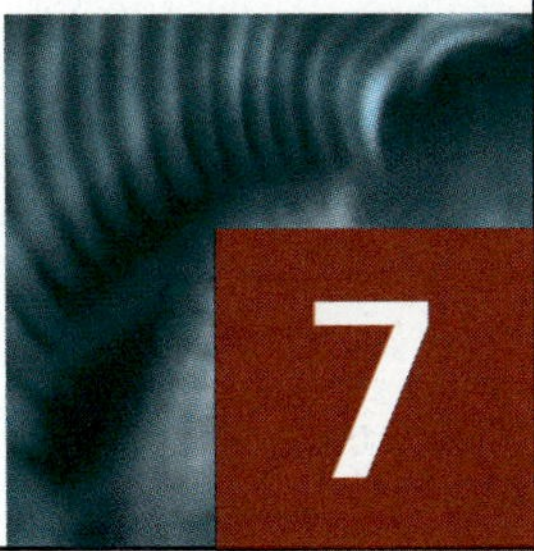

7 Integration Techniques, l'Hôpital's Rule, and Improper Integrals

OVERVIEW We have seen how integrals arise in modeling real phenomena and in measuring objects in the world around us, and we know in theory how integrals are evaluated with antiderivatives. The more sophisticated our models become, however, the more involved our integrals become. We need to know how to change these more involved integrals into forms we can work with. One goal of this chapter is to show how to change unfamiliar integrals into integrals we can recognize, find in a table, or evaluate with a computer.

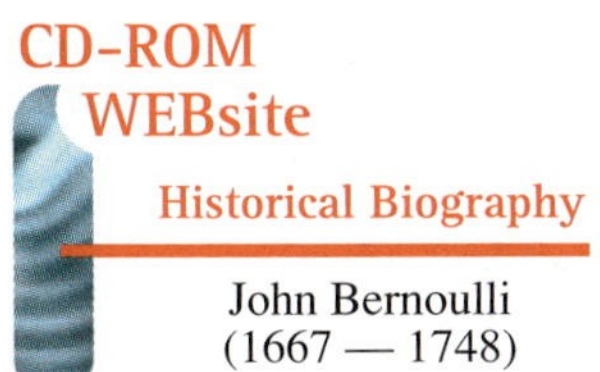

We already know two techniques for doing so: algebraic manipulation and substitution. Here, we carry these techniques a step further and introduce a powerful new technique called integration by parts. We also show how all rational functions can be integrated. Finally, we extend our ideas to integrals where one or both limits of integration are infinite and to integrals whose integrands become unbounded on the interval of integration. Just before doing so, we pause to present l'Hôpital's Rule for calculating limits of fractions whose numerators and denominators both approach zero. L'Hôpital's Rule was actually discovered by John Bernoulli but ended up being named for l'Hôpital after l'Hôpital popularized it in a calculus book he wrote.

7.1 Basic Integration Formulas

Algebraic Procedures

As we saw in Section 4.1, we evaluate an indefinite integral by finding an antiderivative of the integrand and adding an arbitrary constant. Table 7.1 shows the basic forms of the integrals we have evaluated so far. There is a more extensive table at the back of the book; we discuss it in Section 7.5.

Table 7.1 Basic integration formulas

1. $\int du = u + C$
2. $\int k\,du = ku + C \qquad \text{(any number } k\text{)}$
3. $\int (du + dv) = \int du + \int dv$
4. $\int u^n\,du = \frac{u^{n+1}}{n+1} + C \qquad (n \neq -1)$
5. $\int \frac{du}{u} = \ln |u| + C$
6. $\int \sin u\,du = -\cos u + C$
7. $\int \cos u\,du = \sin u + C$
8. $\int \sec^2 u\,du = \tan u + C$
9. $\int \csc^2 u\,du = -\cot u + C$
10. $\int \sec u \tan u\,du = \sec u + C$
11. $\int \csc u \cot u\,du = -\csc u + C$
12. $\int \tan u\,du = -\ln |\cos u| + C = \ln |\sec u| + C$
13. $\int \cot u\,du = \ln |\sin u| + C = -\ln |\csc u| + C$
14. $\int e^u\,du = e^u + C$
15. $\int a^u\,du = \frac{a^u}{\ln a} + C \qquad (a > 0, a \neq 1)$
16. $\int \sinh u\,du = \cosh u + C$
17. $\int \cosh u\,du = \sinh u + C$
18. $\int \frac{du}{\sqrt{a^2 - u^2}} = \sin^{-1}\left(\frac{u}{a}\right) + C$
19. $\int \frac{du}{a^2 + u^2} = \frac{1}{a}\tan^{-1}\left(\frac{u}{a}\right) + C$
20. $\int \frac{du}{u\sqrt{u^2 - a^2}} = \frac{1}{a}\sec^{-1}\left|\frac{u}{a}\right| + C$
21. $\int \frac{du}{\sqrt{a^2 + u^2}} = \sinh^{-1}\left(\frac{u}{a}\right) + C \qquad (a > 0)$
22. $\int \frac{du}{\sqrt{u^2 - a^2}} = \cosh^{-1}\left(\frac{u}{a}\right) + C \qquad (u > a > 0)$

Algebraic Procedures

We often have to rewrite an integral to match it to a standard formula.

Example 1 Making a Simplifying Substitution

Evaluate

$$\int \frac{2x - 9}{\sqrt{x^2 - 9x + 1}}\,dx.$$

Solution

$$\begin{aligned}
\int \frac{2x - 9}{\sqrt{x^2 - 9x + 1}}\,dx &= \int \frac{du}{\sqrt{u}} && u = x^2 - 9x + 1,\ du = (2x - 9)\,dx \\
&= \int u^{-1/2}\,du \\
&= \frac{u^{(-1/2)+1}}{(-1/2) + 1} + C && \text{Table 7.1, Formula 4, with } n = -1/2 \\
&= 2u^{1/2} + C \\
&= 2\sqrt{x^2 - 9x + 1} + C
\end{aligned}$$

Example 2 Completing the Square

Evaluate

$$\int \frac{dx}{\sqrt{8x - x^2}}.$$

Solution We complete the square to write the radicand as

$$\begin{aligned} 8x - x^2 &= -(x^2 - 8x) = -(x^2 - 8x + 16 - 16) \\ &= -(x^2 - 8x + 16) + 16 = 16 - (x - 4)^2. \end{aligned}$$

Then

$$\begin{aligned} \int \frac{dx}{\sqrt{8x - x^2}} &= \int \frac{dx}{\sqrt{16 - (x - 4)^2}} \\ &= \int \frac{du}{\sqrt{a^2 - u^2}} && a = 4, u = (x - 4), du = dx \\ &= \sin^{-1}\left(\frac{u}{a}\right) + C && \text{Table 7.1, Formula 18} \\ &= \sin^{-1}\left(\frac{x - 4}{4}\right) + C. \end{aligned}$$

Example 3 Expanding a Power and Using a Trigonometric Identity

Evaluate

$$\int (\sec x + \tan x)^2\, dx.$$

Solution We expand the integrand and get

$$(\sec x + \tan x)^2 = \sec^2 x + 2 \sec x \tan x + \tan^2 x.$$

The first two terms on the right-hand side of this equation are old friends; we can integrate them at once. How about $\tan^2 x$? There is an identity that connects it with $\sec^2 x$:

$$\tan^2 x + 1 = \sec^2 x, \qquad \tan^2 x = \sec^2 x - 1.$$

We replace $\tan^2 x$ by $\sec^2 x - 1$ and get

$$\begin{aligned} \int (\sec x + \tan x)^2\, dx &= \int (\sec^2 x + 2 \sec x \tan x + \sec^2 x - 1)\, dx \\ &= 2\int \sec^2 x\, dx + 2\int \sec x \tan x\, dx - \int 1\, dx \\ &= 2 \tan x + 2 \sec x - x + C. \end{aligned}$$

Example 4 Eliminating a Square Root

Evaluate

$$\int_0^{\pi/4} \sqrt{1 + \cos 4x}\, dx.$$

Solution We use the identity

$$\cos^2\theta = \frac{1 + \cos 2\theta}{2}, \qquad \text{or} \qquad 1 + \cos 2\theta = 2\cos^2\theta.$$

With $\theta = 2x$, this identity becomes

$$1 + \cos 4x = 2\cos^2 2x.$$

Hence,

$$\begin{aligned}
\int_0^{\pi/4} \sqrt{1 + \cos 4x}\, dx &= \int_0^{\pi/4} \sqrt{2}\sqrt{\cos^2 2x}\, dx \\
&= \sqrt{2}\int_0^{\pi/4} |\cos 2x|\, dx && \sqrt{u^2} = |u| \\
&= \sqrt{2}\int_0^{\pi/4} \cos 2x\, dx && \text{On } [0, \pi/4],\ \cos 2x \ge 0, \text{ so } |\cos 2x| = \cos 2x. \\
&= \sqrt{2}\left[\frac{\sin 2x}{2}\right]_0^{\pi/4} \\
&= \sqrt{2}\left[\frac{1}{2} - 0\right] = \frac{\sqrt{2}}{2}.
\end{aligned}$$

Example 5 Reducing an Improper Fraction

Evaluate

$$\int \frac{3x^2 - 7x}{3x + 2}\, dx.$$

Solution The integrand is an improper fraction (degree of numerator greater than or equal to degree of denominator). To integrate it, we divide first, getting a quotient plus a remainder that is a proper fraction:

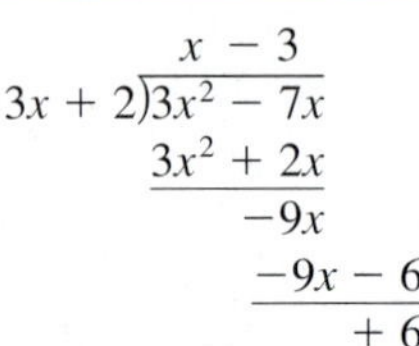

$$\frac{3x^2 - 7x}{3x + 2} = x - 3 + \frac{6}{3x + 2}.$$

Therefore,

$$\int \frac{3x^2 - 7x}{3x + 2}\, dx = \int \left(x - 3 + \frac{6}{3x + 2}\right) dx = \frac{x^2}{2} - 3x + 2\ln|3x + 2| + C.$$

Reducing an improper fraction by long division (Example 5) does not always lead to an expression we can integrate directly. We see what to do about that in Section 7.3.

Example 6 Separating a Fraction

Evaluate

$$\int \frac{3x + 2}{\sqrt{1 - x^2}}\, dx.$$

Solution We first separate the integrand to get

$$\int \frac{3x + 2}{\sqrt{1 - x^2}}\, dx = 3\int \frac{x\, dx}{\sqrt{1 - x^2}} + 2\int \frac{dx}{\sqrt{1 - x^2}}.$$

In the first of these new integrals, we substitute

$$u = 1 - x^2, \qquad du = -2x\,dx, \qquad \text{and} \qquad x\,dx = -\frac{1}{2}\,du.$$

$$3\int \frac{x\,dx}{\sqrt{1-x^2}} = 3\int \frac{(-1/2)\,du}{\sqrt{u}} = -\frac{3}{2}\int u^{-1/2}\,du$$

$$= -\frac{3}{2}\cdot\frac{u^{1/2}}{1/2} + C_1 = -3\sqrt{1-x^2} + C_1$$

The second of the new integrals is a standard form,

$$2\int \frac{dx}{\sqrt{1-x^2}} = 2\sin^{-1}x + C_2.$$

Combining these results and renaming $C_1 + C_2$ as C gives

$$\int \frac{3x+2}{\sqrt{1-x^2}}\,dx = -3\sqrt{1-x^2} + 2\sin^{-1}x + C.$$

Example 7 Multiplying by a Form of 1

Evaluate

$$\int \sec x\,dx.$$

Solution

$$\int \sec x\,dx = \int (\sec x)(1)\,dx = \int \sec x \cdot \frac{\sec x + \tan x}{\sec x + \tan x}\,dx$$

$$= \int \frac{\sec^2 x + \sec x\tan x}{\sec x + \tan x}\,dx$$

$$= \int \frac{du}{u} \qquad u = \tan x + \sec x,\quad du = (\sec^2 x + \sec x\tan x)\,dx$$

$$= \ln|u| + C = \ln|\sec x + \tan x| + C.$$

With cosecants and cotangents in place of secants and tangents, the method of Example 7 leads to a companion formula for the integral of the cosecant (see Exercise 93).

Table 7.2 The secant and cosecant integrals
1. $\int \sec u\,du = \ln\lvert\sec u + \tan u\rvert + C$
2. $\int \csc u\,du = -\ln\lvert\csc u + \cot u\rvert + C$

Procedures for Matching Integrals to Basic Formulas

PROCEDURE	EXAMPLE
Making a simplifying substitution	$\dfrac{2x-9}{\sqrt{x^2-9x+1}}\,dx = \dfrac{du}{\sqrt{u}}$
Completing the square	$\sqrt{8x-x^2} = \sqrt{16-(x-4)^2}$
Using a trigonometric identity	$(\sec x + \tan x)^2 = \sec^2 x + 2\sec x\tan x + \tan^2 x$ $= \sec^2 x + 2\sec x\tan x + (\sec^2 x - 1)$ $= 2\sec^2 x + 2\sec x\tan x - 1$
Eliminating a square root	$\sqrt{1+\cos 4x} = \sqrt{2\cos^2 2x} = \sqrt{2}\,\lvert\cos 2x\rvert$
Reducing an improper fraction	$\dfrac{3x^2-7x}{3x+2} = x - 3 + \dfrac{6}{3x+2}$
Separating a fraction	$\dfrac{3x+2}{\sqrt{1-x^2}} = \dfrac{3x}{\sqrt{1-x^2}} + \dfrac{2}{\sqrt{1-x^2}}$
Multiplying by a form of 1	$\sec x = \sec x \cdot \dfrac{\sec x + \tan x}{\sec x + \tan x}$ $= \dfrac{\sec^2 x + \sec x\tan x}{\sec x + \tan x}$

EXERCISES 7.1

Basic Substitutions

Evaluate each integral in Exercises 1–36 by using a substitution to reduce it to standard form.

1. $\displaystyle\int \frac{16x\,dx}{\sqrt{8x^2+1}}$

2. $\displaystyle\int \frac{3\cos x\,dx}{\sqrt{1+3\sin x}}$

3. $\displaystyle\int 3\sqrt{\sin v}\,\cos v\,dv$

4. $\displaystyle\int \cot^3 y\,\csc^2 y\,dy$

5. $\displaystyle\int_0^1 \frac{16x\,dx}{8x^2+2}$

6. $\displaystyle\int_{\pi/4}^{\pi/3} \frac{\sec^2 z}{\tan z}\,dz$

7. $\displaystyle\int \frac{dx}{\sqrt{x}(\sqrt{x}+1)}$

8. $\displaystyle\int \frac{dx}{x-\sqrt{x}}$

9. $\displaystyle\int \cot(3-7x)\,dx$

10. $\displaystyle\int \csc(\pi x - 1)\,dx$

11. $\displaystyle\int e^\theta \csc(e^\theta + 1)\,d\theta$

12. $\displaystyle\int \frac{\cot(3+\ln x)}{x}\,dx$

13. $\displaystyle\int \sec\frac{t}{3}\,dt$

14. $\displaystyle\int x\sec(x^2-5)\,dx$

15. $\displaystyle\int \csc(s-\pi)\,ds$

16. $\displaystyle\int \frac{1}{\theta^2}\csc\frac{1}{\theta}\,d\theta$

17. $\displaystyle\int_0^{\sqrt{\ln 2}} 2x\,e^{x^2}\,dx$

18. $\displaystyle\int_{\pi/2}^{\pi} (\sin y)\,e^{\cos y}\,dy$

19. $\displaystyle\int e^{\tan v}\sec^2 v\,dv$

20. $\displaystyle\int \frac{e^{\sqrt{t}}\,dt}{\sqrt{t}}$

21. $\displaystyle\int 3^{x+1}\,dx$

22. $\displaystyle\int \frac{2^{\ln x}}{x}\,dx$

23. $\displaystyle\int \frac{2^{\sqrt{w}}\,dw}{2\sqrt{w}}$

24. $\displaystyle\int 10^{2\theta}\,d\theta$

25. $\displaystyle\int \frac{9\,du}{1+9u^2}$

26. $\displaystyle\int \frac{4\,dx}{1+(2x+1)^2}$

27. $\displaystyle\int_0^{1/6} \frac{dx}{\sqrt{1-9x^2}}$

28. $\displaystyle\int_0^1 \frac{dt}{\sqrt{4-t^2}}$

29. $\displaystyle\int \frac{2s\,ds}{\sqrt{1-s^4}}$

30. $\displaystyle\int \frac{2\,dx}{x\sqrt{1-4\ln^2 x}}$

31. $\int \frac{6\,dx}{x\sqrt{25x^2-1}}$

32. $\int \frac{dr}{r\sqrt{r^2-9}}$

33. $\int \frac{dx}{e^x+e^{-x}}$

34. $\int \frac{dy}{\sqrt{e^{2y}-1}}$

35. $\int_1^{e^{\pi/3}} \frac{dx}{x\cos(\ln x)}$

36. $\int \frac{\ln x\,dx}{x+4x\ln^2 x}$

Completing the Square

Evaluate each integral in Exercises 37–42 by completing the square and using a substitution to reduce it to standard form.

37. $\int_1^2 \frac{8\,dx}{x^2-2x+2}$

38. $\int_2^4 \frac{2\,dx}{x^2-6x+10}$

39. $\int \frac{dt}{\sqrt{-t^2+4t-3}}$

40. $\int \frac{d\theta}{\sqrt{2\theta-\theta^2}}$

41. $\int \frac{dx}{(x+1)\sqrt{x^2+2x}}$

42. $\int \frac{dx}{(x-2)\sqrt{x^2-4x+3}}$

Trigonometric Identities

Evaluate the integrals in Exercises 43–46 by using trigonometric identities and substitutions to reduce it to standard form.

43. $\int (\sec x + \cot x)^2\,dx$

44. $\int (\csc x - \tan x)^2\,dx$

45. $\int \csc x \sin 3x\,dx$

46. $\int (\sin 3x \cos 2x - \cos 3x \sin 2x)\,dx$

Improper Fractions

Evaluate each integral in Exercises 47–52 by reducing the improper fraction and using a substitution (if necessary) to reduce it to standard form.

47. $\int \frac{x}{x+1}\,dx$

48. $\int \frac{x^2}{x^2+1}\,dx$

49. $\int_{\sqrt{2}}^3 \frac{2x^3}{x^2-1}\,dx$

50. $\int_{-1}^3 \frac{4x^2-7}{2x+3}\,dx$

51. $\int \frac{4t^3-t^2+16t}{t^2+4}\,dt$

52. $\int \frac{2\theta^3-7\theta^2+7\theta}{2\theta-5}\,d\theta$

Separating Fractions

Evaluate each integral in Exercises 53–56 by separating the fraction and using a substitution (if necessary) to reduce it to standard form.

53. $\int \frac{1-x}{\sqrt{1-x^2}}\,dx$

54. $\int \frac{x+2\sqrt{x-1}}{2x\sqrt{x-1}}\,dx$

55. $\int_0^{\pi/4} \frac{1+\sin x}{\cos^2 x}\,dx$

56. $\int_0^{1/2} \frac{2-8x}{1+4x^2}\,dx$

Multiplying by a Form of 1

Evaluate each integral in Exercises 57–62 by multiplying by a form of 1 and using a substitution (if necessary) to reduce it to standard form.

57. $\int \frac{1}{1+\sin x}\,dx$

58. $\int \frac{1}{1+\cos x}\,dx$

59. $\int \frac{1}{\sec\theta+\tan\theta}\,d\theta$

60. $\int \frac{1}{\csc\theta+\cot\theta}\,d\theta$

61. $\int \frac{1}{1-\sec x}\,dx$

62. $\int \frac{1}{1-\csc x}\,dx$

Eliminating Square Roots

Evaluate each integral in Exercises 63–70 by eliminating the square root.

63. $\int_0^{2\pi} \sqrt{\frac{1-\cos x}{2}}\,dx$

64. $\int_0^{\pi} \sqrt{1-\cos 2x}\,dx$

65. $\int_{\pi/2}^{\pi} \sqrt{1+\cos 2t}\,dt$

66. $\int_{-\pi}^{0} \sqrt{1+\cos t}\,dt$

67. $\int_{-\pi}^{0} \sqrt{1-\cos^2\theta}\,d\theta$

68. $\int_{\pi/2}^{\pi} \sqrt{1-\sin^2\theta}\,d\theta$

69. $\int_{-\pi/4}^{\pi/4} \sqrt{1+\tan^2 y}\,dy$

70. $\int_{-\pi/4}^{0} \sqrt{\sec^2 y-1}\,dy$

Assorted Integrations

Evaluate each integral in Exercises 71–82 by using any technique you think is appropriate.

71. $\int_{\pi/4}^{3\pi/4} (\csc x - \cot x)^2\,dx$

72. $\int_0^{\pi/4} (\sec x + 4\cos x)^2\,dx$

73. $\int \cos\theta \csc(\sin\theta)\,d\theta$

74. $\int \left(1+\frac{1}{x}\right)\cot(x+\ln x)\,dx$

75. $\int (\csc x - \sec x)(\sin x + \cos x)\,dx$

76. $\int 3\sinh\left(\frac{x}{2}+\ln 5\right)dx$

77. $\int \frac{6\,dy}{\sqrt{y}(1+y)}$

78. $\int \frac{dx}{x\sqrt{4x^2-1}}$

79. $\int \frac{7\,dx}{(x-1)\sqrt{x^2-2x-48}}$

80. $\int \frac{dx}{(2x+1)\sqrt{4x^2+4x}}$

81. $\int \sec^2 t \tan(\tan t)\,dt$

82. $\int \frac{dx}{x\sqrt{3+x^2}}$

Trigonometric Powers

83. (a) Evaluate $\int \cos^3\theta\,d\theta$. (*Hint:* $\cos^2\theta = 1-\sin^2\theta$.)

(b) Evaluate $\int \cos^5\theta\,d\theta$.

(c) Without actually evaluating the integral, explain how you would evaluate $\int \cos^9\theta\,d\theta$.

84. **(a)** Evaluate $\int \sin^3 \theta \, d\theta$. (*Hint:* $\sin^2 \theta = 1 - \cos^2 \theta$.)

(b) Evaluate $\int \sin^5 \theta \, d\theta$.

(c) Evaluate $\int \sin^7 \theta \, d\theta$.

(d) Without actually evaluating the integral, explain how you would evaluate $\int \sin^{13} \theta \, d\theta$.

85. **(a)** Express $\int \tan^3 \theta \, d\theta$ in terms of $\int \tan \theta \, d\theta$. Then evaluate $\int \tan^3 \theta \, d\theta$. (*Hint:* $\tan^2 \theta = \sec^2 \theta - 1$.)

(b) Express $\int \tan^5 \theta \, d\theta$ in terms of $\int \tan^3 \theta \, d\theta$.

(c) Express $\int \tan^7 \theta \, d\theta$ in terms of $\int \tan^5 \theta \, d\theta$.

(d) Express $\int \tan^{2k+1} \theta \, d\theta$, where k is a positive integer, in terms of $\int \tan^{2k-1} \theta \, d\theta$.

86. **(a)** Express $\int \cot^3 \theta \, d\theta$ in terms of $\int \cot \theta \, d\theta$. Then evaluate $\int \cot^3 \theta \, d\theta$. (*Hint:* $\cot^2 \theta = \csc^2 \theta - 1$.)

(b) Express $\int \cot^5 \theta \, d\theta$ in terms of $\int \cot^3 \theta \, d\theta$.

(c) Express $\int \cot^7 \theta \, d\theta$ in terms of $\int \cot^5 \theta \, d\theta$.

(d) Express $\int \cot^{2k+1} \theta \, d\theta$, where k is a positive integer, in terms of $\int \cot^{2k-1} \theta \, d\theta$.

Theory and Examples

87. *Area* Find the area of the region bounded above by $y = 2 \cos x$ and below by $y = \sec x$, $-\pi/4 \le x \le \pi/4$.

88. *Area* Find the area of the "triangular" region that is bounded from above and below by the curves $y = \csc x$ and $y = \sin x$, $\pi/6 \le x \le \pi/2$, and on the left by the line $x = \pi/6$.

89. *Volume* Find the volume of the solid generated by revolving the region in Exercise 87 about the x-axis.

90. *Volume* Find the volume of the solid generated by revolving the region in Exercise 88 about the x-axis.

91. *Arc length* Find the length of the curve $y = \ln(\cos x)$, $0 \le x \le \pi/3$.

92. *Arc length* Find the length of the curve $y = \ln(\sec x)$, $0 \le x \le \pi/4$.

93. *The integral of csc x* Repeat the derivation in Example 7, using cofunctions, to show that

$$\int \csc x \, dx = -\ln |\csc x + \cot x| + C.$$

94. *Using different substitutions* Show that the integral

$$\int ((x^2 - 1)(x + 1))^{-2/3} \, dx$$

can be evaluated with any of the following substitutions.

(a) $u = 1/(x + 1)$

(b) $u = ((x - 1)/(x + 1))^k$ for $k = 1, 1/2, 1/3, -1/3, -2/3$, and -1

(c) $u = \tan^{-1} x$

(d) $u = \tan^{-1} \sqrt{x}$

(e) $u = \tan^{-1}((x - 1)/2)$

(f) $u = \cos^{-1} x$

(g) $u = \cosh^{-1} x$

What is the value of the integral? (*Source:* "Problems and Solutions," *College Mathematics Journal,* Vol. 21, No. 5 (Nov. 1990), pp. 425–426.)

7.2 Integration by Parts

Product Rule in Integral Form • Repeated Use • Solving for the Unknown Integral • Tabular Integration

Since

$$\int x \, dx = \frac{1}{2}x^2 + C$$

and

$$\int x^2 \, dx = \frac{1}{3}x^3 + C,$$

it is apparent that

$$\int x \cdot x \, dx \neq \int x \, dx \cdot \int x \, dx.$$

In other words, the integral of a product is generally *not* the product of the individual integrals:

$$\int f(x)g(x)\,dx \neq \int f(x)\,dx \cdot \int g(x)\,dx.$$

Integration by parts is a technique for simplifying integrals of the form

$$\int f(x)g(x)\,dx$$

in which f can be differentiated repeatedly and g can be integrated repeatedly without difficulty. The integral

$$\int xe^x\,dx$$

is such an integral because $f(x) = x$ can be differentiated twice to become zero and $g(x) = e^x$ can be integrated repeatedly without difficulty. Integration by parts also applies to integrals like

$$\int e^x \sin x\,dx$$

in which each part of the integrand appears again after repeated differentiation or integration.

In this section, we describe integration by parts and show how to apply it.

Product Rule in Integral Form

When u and v are differentiable functions of x, the Product Rule for differentiation tells us that

$$\frac{d}{dx}(uv) = u\frac{dv}{dx} + v\frac{du}{dx}.$$

Integrating both sides with respect to x and rearranging leads to the integral equation

$$\begin{aligned}\int\left(u\frac{dv}{dx}\right)dx &= \int\left(\frac{d}{dx}(uv)\right)dx - \int\left(v\frac{du}{dx}\right)dx\\ &= uv - \int\left(v\frac{du}{dx}\right)dx.\end{aligned}$$

When this equation is written in the simpler differential notation, we obtain the following formula.

Integration by Parts Formula

$$\int u\,dv = uv - \int v\,du \qquad (1)$$

This formula expresses one integral, $\int u\,dv$, in terms of a second integral, $\int v\,du$. With a proper choice of u and v, the second integral may be easier to evaluate than the

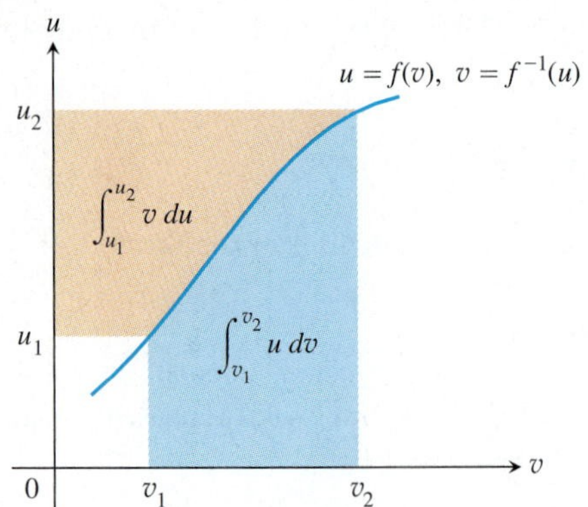

FIGURE 7.1 The area of the blue region, $\int_{v_1}^{v_2} u\,dv$, equals the area of the large rectangle, u_2v_2, minus the areas of the small rectangle, u_1v_1, and the sand-colored region, $\int_{u_1}^{u_2} v\,du$.

In symbols,

$\int_{v_1}^{v_2} u\,dv = (u_2v_2 - u_1v_1) - \int_{u_1}^{u_2} v\,du.$

first. This is the reason for the importance of the formula. When faced with an integral we cannot handle, we can replace it by one with which we might have more success.

The equivalent formula for definite integrals is

$$\int_{v_1}^{v_2} u\,dv = (u_2v_2 - u_1v_1) - \int_{u_1}^{u_2} v\,du. \tag{2}$$

Figure 7.1 shows how the different parts of the formula may be interpreted as areas.

Example 1 Using Integration by Parts

Evaluate

$$\int x\cos x\,dx.$$

Solution We use the formula

$$\int u\,dv = uv - \int v\,du$$

with

$$u = x, \qquad dv = \cos x\,dx.$$

To complete the formula, we take the differential of u and find the simplest antiderivative of $\cos x$.

$$du = dx, \qquad v = \sin x$$

Then,

$$\int x\cos x\,dx = x\sin x - \int \sin x\,dx = x\sin x + \cos x + C.$$

Let's examine the choices available for u and v in Example 1.

Example 2 Investigating Integration by Parts

What are the choices for u and dv when we apply integration by parts to

$$\int x\cos x\,dx = \int u\,dv?$$

Which choices lead to a successful evaluation of the original integral?

Solution There are four possible choices.

1. $u = 1$ and $dv = x\cos x\,dx$ **2.** $u = x$ and $dv = \cos x\,dx$

3. $u = x\cos x$ and $dv = dx$ **4.** $u = \cos x$ and $dv = x\,dx$

Choice 1 won't do because we still don't know how to integrate $dv = x\cos x\,dx$ to get v.

When and How to Use Integration by Parts

When: If substitution doesn't work, try integration by parts.

How: Start with an integral of the form

$$\int f(x)g(x)\,dx.$$

Match this with an integral of the form

$$\int u\,dv$$

by choosing dv to be part of the integrand including dx and possibly $f(x)$ or $g(x)$.

Guideline for choosing u and dv: The formula

$$\int u\,dv = uv - \int v\,du$$

gives a new integral on the right side of the equation. You must be able to readily integrate dv to get the right side. If the new integral is more complex than the original one, try a different choice for u and dv.

Choice 2 works well, as we saw in Example 1.

Choice 3 leads to

$$u = x\cos x, \qquad dv = dx,$$
$$du = (\cos x - x\sin x)\,dx, \qquad v = x,$$

and the new integral

$$\int v\,du = \int (x\cos x - x^2\sin x)\,dx.$$

This is worse than the integral we started with.

Choice 4 leads to

$$u = \cos x, \qquad dv = x\,dx,$$
$$du = -\sin x\,dx, \qquad v = \frac{x^2}{2},$$

and the new integral

$$\int v\,du = -\int \frac{x^2}{2}\sin x\,dx.$$

This, too, is worse.

The goal of integration by parts is to go from an integral $\int u\,dv$ that we don't see how to evaluate to an integral $\int v\,du$ that we can evaluate. Generally, you choose dv first to be as much of the integrand, including dx, as you can readily integrate; u is the leftover part. Keep in mind that integration by parts does not always work.

Example 3 Finding Area

Find the area of the region bounded by the curve $y = xe^{-x}$ and the x-axis from $x = 0$ to $x = 4$.

Solution The region is shaded in Figure 7.2. Its area is

$$\int_0^4 xe^{-x}\,dx.$$

We use the formula $\int u\,dv = uv - \int v\,du$ with

$$u = x, \qquad dv = e^{-x}\,dx,$$
$$du = dx, \qquad v = -e^{-x}.$$

Then

$$\int xe^{-x}\,dx = -xe^{-x} - \int(-e^{-x})\,dx$$
$$= -xe^{-x} + \int e^{-x}\,dx$$
$$= -xe^{-x} - e^{-x} + C.$$

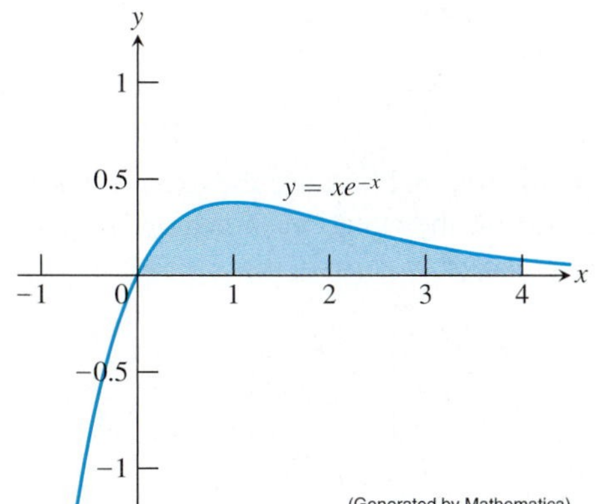

FIGURE 7.2 The region in Example 3.

Thus,

$$\int_0^4 xe^{-x}\,dx = [-xe^{-x} - e^{-x}]_0^4$$
$$= (-4e^{-4} - e^{-4}) - (-e^0) = 1 - 5e^{-4} \approx 0.91.$$

Example 4 Integral of the Natural Logarithm

Find

$$\int \ln x\,dx.$$

Solution Since $\int \ln x\,dx$ can be written as $\int \ln x \cdot 1\,dx$, we use the formula $\int u\,dv = uv - \int v\,du$ with

$u = \ln x$	Simplifies when differentiated	$dv = dx$	Easy to integrate
$du = \frac{1}{x}\,dx,$		$v = x.$	Simplest antiderivative

Then

$$\int \ln x\,dx = x\ln x - \int x \cdot \frac{1}{x}\,dx = x\ln x - \int dx = x\ln x - x + C.$$

Repeated Use

Sometimes we have to use integration by parts more than once.

Example 5 Repeated Use of Integration by Parts

Evaluate

$$\int x^2e^x\,dx.$$

Solution With $u = x^2$, $dv = e^x\,dx$, $du = 2x\,dx$, and $v = e^x$, we have

$$\int x^2e^x\,dx = x^2e^x - 2\int xe^x\,dx.$$

The new integral is less complicated than the original because the exponent on x is reduced by one. To evaluate the integral on the right, we integrate by parts again with $u = x$, $dv = e^x\,dx$. Then $du = dx$, $v = e^x$, and

$$\int xe^x\,dx = xe^x - \int e^x\,dx = xe^x - e^x + C.$$

Hence,

$$\int x^2e^x\,dx = x^2e^x - 2\int xe^x\,dx$$
$$= x^2e^x - 2xe^x + 2e^x + C.$$

The technique of Example 5 works for any integral $\int x^n e^x\,dx$ in which n is a positive integer, because differentiating x^n will eventually lead to zero and integrating e^x is easy. We say more on this later in this section when we discuss *tabular integration.*

Solving for the Unknown Integral

Integrals like the one in the next example occur in electrical engineering. Their evaluation requires two integrations by parts, followed by solving for the unknown integral.

Example 6 Solving for the Unknown Integral

Evaluate

$$\int e^x \cos x\,dx.$$

Solution Let $u = e^x$ and $dv = \cos x\,dx$. Then $du = e^x\,dx$, $v = \sin x$, and

$$\int e^x \cos x\,dx = e^x \sin x - \int e^x \sin x\,dx.$$

The second integral is like the first except that it has $\sin x$ in place of $\cos x$. To evaluate it, we use integration by parts with

$$u = e^x, \qquad dv = \sin x\,dx, \qquad v = -\cos x, \qquad du = e^x\,dx.$$

Then

$$\begin{aligned}\int e^x \cos x\,dx &= e^x \sin x - \left(-e^x \cos x - \int (-\cos x)(e^x\,dx)\right)\\ &= e^x \sin x + e^x \cos x - \int e^x \cos x\,dx.\end{aligned}$$

The unknown integral now appears on both sides of the equation. Adding the integral to both sides gives

$$2\int e^x \cos x\,dx = e^x \sin x + e^x \cos x + C.$$

Dividing by 2 and renaming the constant of integration gives

$$\int e^x \cos x\,dx = \frac{e^x \sin x + e^x \cos x}{2} + C.$$

When making repeated use of integration by parts in circumstances like Example 6, once a choice for u and dv is made, it is usually not a good idea to switch choices in the second stage of the problem. Doing so may result in undoing the work. For example, if we had switched to the substitution $u = \sin x$, $dv = e^x\,dx$ in the second integration, we would have obtained

$$\begin{aligned}\int e^x \cos x\,dx &= e^x \sin x - \left(e^x \sin x - \int e^x \cos x\,dx\right)\\ &= \int e^x \cos x\,dx,\end{aligned}$$

undoing the first integration by parts. The tabular integration technique introduced next prevents this mistake.

Tabular Integration

We have seen that integrals of the form $\int f(x)g(x)\,dx$, in which f can be differentiated repeatedly to become zero and g can be integrated repeatedly without difficulty, are natural candidates for integration by parts. If many repetitions are required, however, the calculations can be cumbersome. In situations like this, there is a way to organize the calculations that saves a great deal of work. It is **tabular integration,** as shown in Examples 7 and 8.

Example 7 Using Tabular Integration

Evaluate

$$\int x^2 e^x\,dx.$$

Solution With $f(x) = x^2$ and $g(x) = e^x$, we list:

$f(x)$ and its derivatives		$g(x)$ and its integrals
x^2	$(+)$	e^x
$2x$	$(-)$	e^x
2	$(+)$	e^x
0		e^x

We combine the products of the functions connected by the arrows according to the operation signs above the arrows to obtain

$$\int x^2 e^x\,dx = x^2e^x - 2xe^x + 2e^x + C.$$

Compare this with the result in Example 5.

Example 8 Using Tabular Integration

Evaluate

$$\int x^3 \sin x\,dx.$$

Solution With $f(x) = x^3$ and $g(x) = \sin x$, we list:

$f(x)$ and its derivatives		$g(x)$ and its integrals
x^3	$(+)$	$\sin x$
$3x^2$	$(-)$	$-\cos x$
$6x$	$(+)$	$-\sin x$
6	$(-)$	$\cos x$
0		$\sin x$

For more about tabular integration, see the Additional Exercises at the end of this chapter.

Again we combine the products of the functions connected by the arrows according to the operation signs above the arrows to obtain

$$\int x^3 \sin x\, dx = -x^3 \cos x + 3x^2 \sin x + 6x \cos x - 6 \sin x + C.$$

EXERCISES 7.2

Integration by Parts

Evaluate the integrals in Exercises 1–24.

1. $\int x \sin \frac{x}{2}\, dx$

2. $\int \theta \cos \pi\theta\, d\theta$

3. $\int t^2 \cos t\, dt$

4. $\int x^2 \sin x\, dx$

5. $\int_1^2 x \ln x\, dx$

6. $\int_1^e x^3 \ln x\, dx$

7. $\int \tan^{-1} y\, dy$

8. $\int \sin^{-1} y\, dy$

9. $\int x \sec^2 x\, dx$

10. $\int 4x \sec^2 2x\, dx$

11. $\int x^3 e^x\, dx$

12. $\int p^4 e^{-p}\, dp$

13. $\int (x^2 - 5x)e^x\, dx$

14. $\int (r^2 + r + 1)e^r\, dr$

15. $\int x^5 e^x\, dx$

16. $\int t^2 e^{4t}\, dt$

17. $\int_0^{\pi/2} \theta^2 \sin 2\theta\, d\theta$

18. $\int_0^{\pi/2} x^3 \cos 2x\, dx$

19. $\int_{2/\sqrt{3}}^{2} t \sec^{-1} t\, dt$

20. $\int_0^{1/\sqrt{2}} 2x \sin^{-1}(x^2)\, dx$

21. $\int e^\theta \sin \theta\, d\theta$

22. $\int e^{-y} \cos y\, dy$

23. $\int e^{2x} \cos 3x\, dx$

24. $\int e^{-2x} \sin 2x\, dx$

Substitution and Integration by Parts

Evaluate the integrals in Exercises 25–30 by using a substitution prior to integration by parts.

25. $\int e^{\sqrt{3s+9}}\, ds$

26. $\int_0^1 x\sqrt{1-x}\, dx$

27. $\int_0^{\pi/3} x \tan^2 x\, dx$

28. $\int \ln(x + x^2)\, dx$

29. $\int \sin(\ln x)\, dx$

30. $\int z(\ln z)^2\, dz$

Differential Equations

In Exercises 31–34, solve the differential equation.

31. $\frac{dy}{dx} = x^2 e^{4x}$

32. $\frac{dy}{dx} = x^2 \ln x$

33. $\frac{dy}{d\theta} = \sin \sqrt{\theta}$

34. $\frac{dy}{d\theta} = \theta \sec \theta \tan \theta$

Theory and Examples

35. *Finding area* Find the area of the region enclosed by the curve $y = x \sin x$ and the x-axis (see the accompanying figure) for

(a) $0 \le x \le \pi$

(b) $\pi \le x \le 2\pi$

(c) $2\pi \le x \le 3\pi$.

(d) What pattern do you see here? What is the area between the curve and the x-axis for $n\pi \le x \le (n + 1)\pi$, n an arbitrary nonnegative integer? Give reasons for your answer.

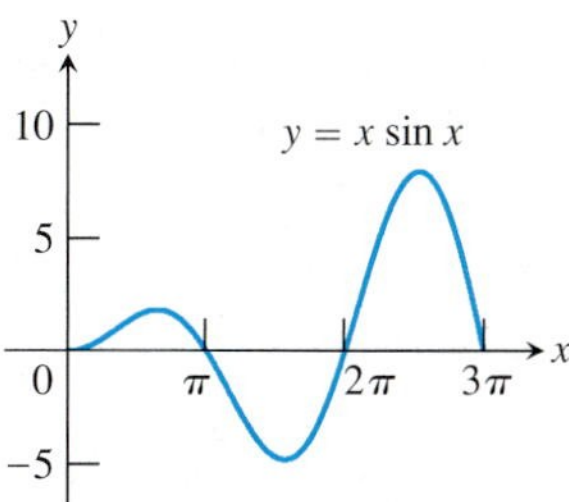

36. *Finding area* Find the area of the region enclosed by the curve $y = x \cos x$ and the x-axis (see the accompanying figure) for

(a) $\pi/2 \le x \le 3\pi/2$

(b) $3\pi/2 \le x \le 5\pi/2$

(c) $5\pi/2 \le x \le 7\pi/2$.

(d) What pattern do you see? What is the area between the curve and the x-axis for

$$\left(\frac{2n-1}{2}\right)\pi \le x \le \left(\frac{2n+1}{2}\right)\pi,$$

n an arbitrary positive integer? Give reasons for your answer.

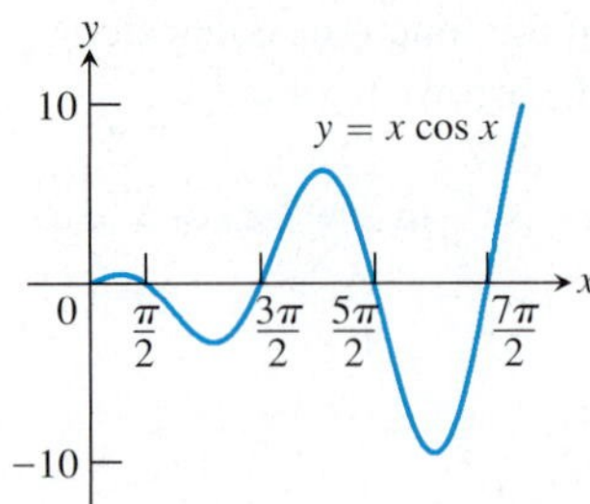

37. *Finding volume* Find the volume of the solid generated by revolving the region in the first quadrant bounded by the coordinate axes, the curve $y = e^x$, and the line $x = \ln 2$ about the line $x = \ln 2$.

38. *Finding volume* Find the volume of the solid generated by revolving the region in the first quadrant bounded by the coordinate axes, the curve $y = e^{-x}$, and the line $x = 1$

(a) about the y-axis,

(b) about the line $x = 1$.

39. *Finding volume* Find the volume of the solid generated by revolving the region in the first quadrant bounded by the coordinate axes and the curve $y = \cos x$, $0 \le x \le \pi/2$, about

(a) the y-axis

(b) the line $x = \pi/2$.

40. *Finding volume* Find the volume of the solid generated by revolving the region bounded by the x-axis and the curve $y = x \sin x$, $0 \le x \le \pi$, about

(a) the y-axis

(b) the line $x = \pi$.

(See Exercise 35 for a graph.)

41. *Average value* A retarding force, symbolized by the dashpot in the figure, slows the motion of the weighted spring so that the mass's position at time t is

$$y = 2e^{-t} \cos t, \qquad t \ge 0.$$

Find the average value of y over the interval $0 \le t \le 2\pi$.

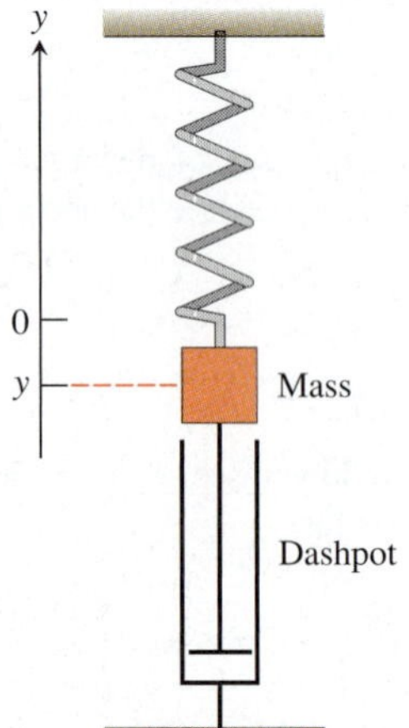

42. *Average value* In a mass-spring-dashpot system like the one in Exercise 41, the mass's position at time t is

$$y = 4e^{-t}(\sin t - \cos t), \qquad t \ge 0.$$

Find the average value of y over the interval $0 \le t \le 2\pi$.

Reduction Formulas

In Exercises 43–46, use integration by parts to establish the *reduction formula*.

43. $\int x^n \cos x\, dx = x^n \sin x - n \int x^{n-1} \sin x\, dx$

44. $\int x^n \sin x\, dx = -x^n \cos x + n \int x^{n-1} \cos x\, dx$

45. $\int x^n e^{ax}\, dx = \frac{x^n e^{ax}}{a} - \frac{n}{a} \int x^{n-1} e^{ax}\, dx, \quad a \ne 0$

46. $\int (\ln x)^n\, dx = x(\ln x)^n - n \int (\ln x)^{n-1}\, dx$

Integrating Inverses of Functions

47. *Integrating inverse functions* Assume that the function f has an inverse.

(a) Show that

$$\int f^{-1}(x)\, dx = \int y f'(y)\, dy.$$

(*Hint:* Use the substitution $y = f^{-1}(x)$.)

(b) Use integration by parts on the second integral in part (a) to show that

$$\int f^{-1}(x)\, dx = \int y f'(y)\, dy = x f^{-1}(x) - \int f(y)\, dy.$$

48. *Integrating inverse functions* Assume that the function f has an inverse. Use integration by parts directly to show that

$$\int f^{-1}(x)\, dx = x f^{-1}(x) - \int x \left(\frac{d}{dx} f^{-1}(x) \right) dx.$$

In Exercises 49–52, evaluate the integral using

(a) the technique of Exercise 47

(b) the technique of Exercise 48.

(c) Show that the expressions (with $C = 0$) obtained in parts (a) and (b) are the same.

49. $\int \sin^{-1} x\, dx$

50. $\int \tan^{-1} x\, dx$

51. $\int \cos^{-1} x\, dx$

52. $\int \log_2 x\, dx$

7.3 Partial Fractions

Partial Fractions • General Description of the Method • The Heaviside "Cover-up" Method for Linear Factors • Other Ways to Determine the Coefficients

In studying population modeling in Example 6 (Section 6.4 or 6.6), we solved the logistic differential equation

$$\frac{dP}{dt} = 0.001P(100 - P)$$

by rewriting it as

$$\frac{100}{P(100 - P)}\, dP = 0.1\, dt, \qquad \text{Variables separated}$$

expanding the fraction on the left into two basic fractions,

$$\frac{100}{P(100 - P)} = \frac{1}{P} + \frac{1}{100 - P},$$

and integrating both sides to find the solution

$$\ln|P| - \ln|100 - P| = 0.1t + C.$$

This expansion technique is the **method of partial fractions.** Any rational function can be written as a sum of basic fractions, called **partial fractions,** using the method of partial fractions. We can then integrate the rational function by integrating the sum of partial fractions instead.

Partial Fractions

We sum algebraic fractions by finding a common denominator, summing the resulting fractions, and finally simplifying. For instance,

$$\begin{aligned}\frac{2}{x+1} + \frac{3}{x-3} &= \frac{2(x-3)}{(x+1)(x-3)} + \frac{3(x+1)}{(x-3)(x+1)} \\ &= \frac{2x - 6 + 3x + 3}{x^2 - 2x - 3} \\ &= \frac{5x - 3}{x^2 - 2x - 3}.\end{aligned}$$

It is easy to find the integral

$$\int \frac{5x - 3}{x^2 - 2x - 3}\, dx$$

if we can "reverse" the process above and obtain

$$\begin{aligned}\int \frac{5x-3}{x^2 - 2x - 3}\, dx &= \int \frac{2}{x+1}\, dx + \int \frac{3}{x-3}\, dx \\ &= 2\ln|x+1| + 3\ln|x-3| + C.\end{aligned}$$

More generally, a theorem from advanced algebra (mentioned later in more detail) says that every rational function, no matter how complicated, can be rewritten as a sum of simpler fractions (that we can integrate with techniques we already know). Let's see how this simpler sum can be found by the method of partial fractions applied to our preceding example.

Example 1 Using Partial Fractions

Use the method of partial fractions to evaluate

$$\int \frac{5x - 3}{x^2 - 2x - 3}\,dx.$$

Undetermined Coefficients

The A and B in the partial fraction decomposition are referred to as **undetermined coefficients.**

Solution First we factor the denominator: $x^2 - 2x - 3 = (x + 1)(x - 3)$. Then we determine the values of A and B so that

$$\frac{5x - 3}{x^2 - 2x - 3} = \frac{A}{x + 1} + \frac{B}{x - 3}.$$

We proceed by clearing the fractions:

$$5x - 3 = A(x - 3) + B(x + 1) \qquad \text{Multiply both sides of the equation by } (x + 1)(x - 3).$$

$$= (A + B)x - 3A + B. \qquad \text{Combine terms.}$$

Now equate the coefficients to obtain the following system of linear equations.

$$A + B = 5$$
$$-3A + B = -3$$

Solving these equations simultaneously yields $A = 2$ and $B = 3$. Therefore,

$$\int \frac{5x - 3}{x^2 - 2x - 3}\,dx = \int \frac{2}{x + 1}\,dx + \int \frac{3}{x - 3}\,dx$$
$$= 2 \ln |x + 1| + 3 \ln |x - 3| + C.$$

General Description of the Method

Success in writing a rational function $f(x)/g(x)$ as a sum of partial fractions depends on two things:

- *The degree of $f(x)$ must be less than the degree of $g(x)$.* That is, the fraction must be *proper*. If it isn't, divide $f(x)$ by $g(x)$ and work with the remainder term. See Example 4 of this section.
- *We must know the factors of $g(x)$.* In theory, any polynomial with real coefficients can be written as a product of real linear factors and real quadratic factors. In practice, the factors may be hard to find.

Here is how we find the partial fractions of a proper fraction $f(x)/g(x)$ when the factors of g are known.

Method of Partial Fractions (f(x)/g(x) Proper)

Step 1: Let $x - r$ be a linear factor of $g(x)$. Suppose that $(x - r)^m$ is the highest power of $x - r$ that divides $g(x)$. Then, to this factor, assign the sum of the m partial fractions:

$$\frac{A_1}{x-r} + \frac{A_2}{(x-r)^2} + \cdots + \frac{A_m}{(x-r)^m}.$$

Do this for each distinct linear factor of $g(x)$.

Step 2: Let $x^2 + px + q$ be a quadratic factor of $g(x)$. Suppose that $(x^2 + px + q)^n$ is the highest power of this factor that divides $g(x)$. Then, to this factor, assign the sum of the n partial fractions:

$$\frac{B_1x + C_1}{x^2 + px + q} + \frac{B_2x + C_2}{(x^2 + px + q)^2} + \cdots + \frac{B_nx + C_n}{(x^2 + px + q)^n}.$$

Do this for each distinct quadratic factor of $g(x)$ that cannot be factored into linear factors with real coefficients.

Step 3: Set the original fraction $f(x)/g(x)$ equal to the sum of all these partial fractions. Clear the resulting equation of fractions and arrange the terms in decreasing powers of x.

Step 4: Equate the coefficients of corresponding powers of x and solve the resulting equations for the undetermined coefficients.

Example 2 Using a Repeated Linear Factor

Express as a sum of partial fractions:

$$\frac{6x+7}{(x+2)^2}.$$

Solution According to the description above, we must express the fraction as a sum of partial fractions with undetermined coefficients.

$$\frac{6x+7}{(x+2)^2} = \frac{A}{x+2} + \frac{B}{(x+2)^2}$$

$$6x + 7 = A(x+2) + B \qquad \text{Multiply both sides by } (x+2)^2.$$

$$= Ax + (2A + B) \qquad \text{Combine terms.}$$

Equating coefficients of corresponding powers of x gives

$$A = 6 \quad \text{and} \quad 2A + B = 12 + B = 7, \quad \text{or}$$

$$A = 6 \quad \text{and} \quad B = -5.$$

Therefore,

$$\frac{6x+7}{(x+2)^2} = \frac{6}{x+2} - \frac{5}{(x+2)^2}.$$

Example 3 Using Partial Fractions

Evaluate

$$\int \frac{6x+7}{(x+2)^2}\,dx.$$

Solution

$$\begin{aligned}\int \frac{6x+7}{(x+2)^2}\,dx &= \int \left(\frac{6}{x+2} - \frac{5}{(x+2)^2}\right) dx \qquad \text{Example 2}\\ &= 6\int \frac{dx}{x+2} - 5\int (x+2)^{-2}\,dx\\ &= 6\ln|x+2| + 5(x+2)^{-1} + C\end{aligned}$$

Example 4 Integrating an Improper Fraction

Evaluate

$$\int \frac{2x^3-4x^2-x-3}{x^2-2x-3}\,dx.$$

Solution First we divide the denominator into the numerator to get a polynomial plus a proper fraction.

$$\begin{array}{r} 2x \\ x^2-2x-3\,\overline{)\,2x^3-4x^2-x-3} \\ \underline{2x^3-4x^2-6x} \\ 5x-3 \end{array}$$

Then we write the improper fraction as a polynomial plus a proper fraction.

$$\frac{2x^3-4x^2-x-3}{x^2-2x-3} = 2x + \frac{5x-3}{x^2-2x-3}$$

Finally, using $\int 2x\,dx = x^2$ and Example 1 we obtain

$$\begin{aligned}\int \frac{2x^3-4x^2-x-3}{x^2-2x-3}\,dx &= \int 2x\,dx + \int \frac{5x-3}{x^2-2x-3}\,dx\\ &= x^2 + 2\ln|x+1| + 3\ln|x-3| + C.\end{aligned}$$

Example 5 Solving an Initial Value Problem

Find the solution to $dy/dx = 2xy(y^2+1)$ that satisfies $y(0) = 1$.

Solution Separating the variables, we rewrite the differential equation as

$$\frac{1}{y(y^2+1)}\,dy = 2x\,dx.$$

Integrating both sides gives

$$\int \frac{1}{y(y^2+1)}\,dy = \int 2x\,dx = x^2 + C_1.$$

We use partial fractions to rewrite the integrand on the left.

$$\frac{1}{y(y^2+1)} = \frac{A}{y} + \frac{By+C}{y^2+1}$$

Notice the numerator over $y^2 + 1$: For quadratic factors, we use first degree numerators, not constant numerators. Clearing the equations of fractions gives

$$\begin{aligned} 1 &= A(y^2+1) + (By+C)y && \text{Multiply by } y(y^2+1). \\ &= (A+B)y^2 + Cy + A \end{aligned}$$

Equating coefficients of like terms gives $A + B = 0$, $C = 0$, and $A = 1$. Solving these equations simultaneously, we find $A = 1$, $B = -1$, and $C = 0$. Accordingly,

$$\begin{aligned} \int \frac{1}{y(y^2+1)}\,dy &= \int \frac{1}{y}\,dy - \int \frac{y}{y^2+1}\,dy \\ &= \ln|y| - \frac{1}{2}\ln(y^2+1) + C_2. \end{aligned}$$

The solution to the differential equation is

$$\ln|y| - \frac{1}{2}\ln(y^2+1) = x^2 + C. \qquad C = C_1 - C_2$$

Substituting $x = 0$ and $y = 1$, we find

$$0 - \frac{1}{2}\ln 2 = C, \qquad \text{or} \qquad C = -\ln\sqrt{2}.$$

The solution to the initial value problem is

$$\ln|y| - \frac{1}{2}\ln(y^2+1) = x^2 - \ln\sqrt{2}.$$

A quadratic polynomial is **irreducible** if it cannot be written as the product of two linear factors with real coefficients.

Example 6 Integrating with an Irreducible Quadratic Factor in the Denominator

Evaluate

$$\int \frac{-2x+4}{(x^2+1)(x-1)^2}\,dx$$

using partial fractions.

Solution The denominator has an irreducible quadratic factor as well as a repeated linear factor, so we write

$$\frac{-2x+4}{(x^2+1)(x-1)^2} = \frac{Ax+B}{x^2+1} + \frac{C}{x-1} + \frac{D}{(x-1)^2}. \tag{1}$$

Clearing the equation of fractions gives

$$\begin{aligned} -2x+4 &= (Ax+B)(x-1)^2 + C(x-1)(x^2+1) + D(x^2+1) \\ &= (A+C)x^3 + (-2A+B-C+D)x^2 \\ &\quad + (A-2B+C)x + (B-C+D). \end{aligned}$$

Equating coefficients of like terms gives

Coefficients of x^3:	$0 = A + C$
Coefficients of x^2:	$0 = -2A + B - C + D$
Coefficients of x^1:	$-2 = A - 2B + C$
Coefficients of x^0:	$4 = B - C + D$

We solve these equations simultaneously to find the values of A, B, C, and D:

$-4 = -2A, \quad A = 2$	Subtract fourth equation from second.
$C = -A = -2$	From the first equation
$B = 1$	$A = 2$ and $C = -2$ in third equation.
$D = 4 - B + C = 1.$	From the fourth equation

We substitute these values into Equation (1), obtaining

$$\frac{-2x+4}{(x^2+1)(x-1)^2} = \frac{2x+1}{x^2+1} - \frac{2}{x-1} + \frac{1}{(x-1)^2}.$$

Finally, using the expansion above we can integrate:

$$\begin{aligned}\int \frac{-2x+4}{(x^2+1)(x-1)^2}\,dx &= \int \left(\frac{2x+1}{x^2+1} - \frac{2}{x-1} + \frac{1}{(x-1)^2}\right) dx \\ &= \int \left(\frac{2x}{x^2+1} + \frac{1}{x^2+1} - \frac{2}{x-1} + \frac{1}{(x-1)^2}\right) dx \\ &= \ln\,(x^2+1) + \tan^{-1} x - 2\ln|x-1| - \frac{1}{x-1} + C.\end{aligned}$$

Oliver Heaviside (1850 — 1925)

The Heaviside "Cover-up" Method for Linear Factors

When the degree of the polynomial $f(x)$ is less than the degree of $g(x)$ and

$$g(x) = (x - r_1)(x - r_2)\cdots(x - r_n)$$

is a product of n distinct linear factors, each raised to the first power, there is a quick way to expand $f(x)/g(x)$ by partial fractions.

Example 7 Using the Heaviside Method

Find A, B, and C in the partial-fraction expansion

$$\frac{x^2+1}{(x-1)(x-2)(x-3)} = \frac{A}{x-1} + \frac{B}{x-2} + \frac{C}{x-3}. \tag{2}$$

Solution If we multiply both sides of Equation (2) by $(x - 1)$ to get

$$\frac{x^2+1}{(x-2)(x-3)} = A + \frac{B(x-1)}{x-2} + \frac{C(x-1)}{x-3}$$

and set $x = 1$, the resulting equation gives the value of A:

$$\frac{(1)^2+1}{(1-2)(1-3)} = A + 0 + 0,$$

$$A = 1.$$

Thus, the value of A is the number we would have obtained if we had covered the factor $(x-1)$ in the denominator of the original fraction

$$\frac{x^2+1}{(x-1)(x-2)(x-3)} \tag{3}$$

and evaluated the rest at $x=1$:

$$A = \frac{(1)^2+1}{\underset{\substack{\Uparrow \\ \text{Cover}}}{\boxed{(x-1)}}\;(1-2)(1-3)} = \frac{2}{(-1)(-2)} = 1.$$

Similarly, we find the value of B in Equation (2) by covering the factor $(x-2)$ in Equation (3) and evaluating the rest at $x=2$:

$$B = \frac{(2)^2+1}{(2-1)\;\underset{\substack{\Uparrow \\ \text{Cover}}}{\boxed{(x-2)}}\;(2-3)} = \frac{5}{(1)(-1)} = -5.$$

Finally, C is found by covering the $(x-3)$ in Equation (3) and evaluating the rest at $x=3$:

$$C = \frac{(3)^2+1}{(3-1)(3-2)\;\underset{\substack{\Uparrow \\ \text{Cover}}}{\boxed{(x-3)}}} = \frac{10}{(2)(1)} = 5.$$

Heaviside Method

Step 1: *Write the quotient with $g(x)$ factored:*

$$\frac{f(x)}{g(x)} = \frac{f(x)}{(x-r_1)(x-r_2)\cdots(x-r_n)}.$$

Step 2: *Cover the factors $(x-r_i)$ of $g(x)$ one at a time,* each time replacing all the uncovered x's by the number r_i. This gives a number A_i for each root r_i:

$$A_1 = \frac{f(r_1)}{(r_1-r_2)\cdots(r_1-r_n)}$$

$$A_2 = \frac{f(r_2)}{(r_2-r_1)(r_2-r_3)\cdots(r_2-r_n)}$$

$$\vdots$$

$$A_n = \frac{f(r_n)}{(r_n-r_1)(r_n-r_2)\cdots(r_n-r_{n-1})}.$$

Step 3: *Write the partial-fraction expansion of $f(x)/g(x)$* as

$$\frac{f(x)}{g(x)} = \frac{A_1}{(x-r_1)} + \frac{A_2}{(x-r_2)} + \cdots + \frac{A_n}{(x-r_n)}.$$

Example 8 Integrating with the Heaviside Method

Evaluate

$$\int \frac{x + 4}{x^3 + 3x^2 - 10x}\,dx.$$

Solution The degree of $f(x) = x + 4$ is less than the degree of $g(x) = x^3 + 3x^2 - 10x$, and, with $g(x)$ factored,

$$\frac{x + 4}{x^3 + 3x^2 - 10x} = \frac{x + 4}{x(x - 2)(x + 5)}.$$

The roots of $g(x)$ are $r_1 = 0$, $r_2 = 2$, and $r_3 = -5$. We find

$$A_1 = \frac{0 + 4}{\boxed{x}\;(0 - 2)(0 + 5)} = \frac{4}{(-2)(5)} = -\frac{2}{5}$$

⇑ Cover

$$A_2 = \frac{2 + 4}{2\;\boxed{(x - 2)}\;(2 + 5)} = \frac{6}{(2)(7)} = \frac{3}{7}$$

⇑ Cover

$$A_3 = \frac{-5 + 4}{(-5)(-5 - 2)\;\boxed{(x + 5)}} = \frac{-1}{(-5)(-7)} = -\frac{1}{35}.$$

⇑ Cover

Therefore,

$$\frac{x + 4}{x(x - 2)(x + 5)} = -\frac{2}{5x} + \frac{3}{7(x - 2)} - \frac{1}{35(x + 5)},$$

and

$$\int \frac{x + 4}{x(x - 2)(x + 5)}\,dx = -\frac{2}{5}\ln|x| + \frac{3}{7}\ln|x - 2| - \frac{1}{35}\ln|x + 5| + C.$$

Other Ways to Determine the Coefficients

Another way to determine the constants that appear in partial fractions is to differentiate, as in the next example. Still another is to assign selected numerical values to x.

Example 9 Using Differentiation

Find A, B, and C in the equation

$$\frac{x - 1}{(x + 1)^3} = \frac{A}{x + 1} + \frac{B}{(x + 1)^2} + \frac{C}{(x + 1)^3}.$$

Solution We first clear fractions:

$$x - 1 = A(x + 1)^2 + B(x + 1) + C.$$

Substituting $x = -1$ shows $C = -2$. We then differentiate both sides with respect to x, obtaining

$$1 = 2A(x + 1) + B.$$

Substituting $x = -1$ shows $B = 1$. We differentiate again to get $0 = 2A$, which shows $A = 0$. Hence,

$$\frac{x - 1}{(x + 1)^3} = \frac{1}{(x + 1)^2} - \frac{2}{(x + 1)^3}.$$

In some problems, assigning small values to x such as $x = 0, \pm 1, \pm 2$, to get equations in A, B, and C provides a fast alternative to other methods.

Example 10 Assigning Numerical Values to x

Find A, B, and C in

$$\frac{x^2+1}{(x-1)(x-2)(x-3)} = \frac{A}{x-1} + \frac{B}{x-2} + \frac{C}{x-3}.$$

Solution Clear fractions to get

$$x^2 + 1 = A(x-2)(x-3) + B(x-1)(x-3) + C(x-1)(x-2).$$

Then let $x = 1, 2, 3$ successively to find A, B, and C:

$$\begin{aligned} x = 1: \quad (1)^2 + 1 &= A(-1)(-2) + B(0) + C(0) \\ 2 &= 2A \\ A &= 1 \end{aligned}$$

$$\begin{aligned} x = 2: \quad (2)^2 + 1 &= A(0) + B(1)(-1) + C(0) \\ 5 &= -B \\ B &= -5 \end{aligned}$$

$$\begin{aligned} x = 3: \quad (3)^2 + 1 &= A(0) + B(0) + C(2)(1) \\ 10 &= 2C \\ C &= 5. \end{aligned}$$

Conclusion:

$$\frac{x^2+1}{(x-1)(x-2)(x-3)} = \frac{1}{x-1} - \frac{5}{x-2} + \frac{5}{x-3}.$$

EXERCISES 7.3

Expanding Quotients into Partial Fractions

Expand the quotients in Exercises 1–8 by partial fractions.

1. $\dfrac{5x-13}{(x-3)(x-2)}$
2. $\dfrac{5x-7}{x^2-3x+2}$
3. $\dfrac{x+4}{(x+1)^2}$
4. $\dfrac{2x+2}{x^2-2x+1}$
5. $\dfrac{z+1}{z^2(z-1)}$
6. $\dfrac{z}{z^3-z^2-6z}$
7. $\dfrac{t^2+8}{t^2-5t+6}$
8. $\dfrac{t^4+9}{t^4+9t^2}$

Nonrepeated Linear Factors

In Exercises 9–16, express the integrands as a sum of partial fractions and evaluate the integrals.

9. $\displaystyle\int \frac{dx}{1-x^2}$
10. $\displaystyle\int \frac{dx}{x^2+2x}$
11. $\displaystyle\int \frac{x+4}{x^2+5x-6}\,dx$
12. $\displaystyle\int \frac{2x+1}{x^2-7x+12}\,dx$
13. $\displaystyle\int_4^8 \frac{y\,dy}{y^2-2y-3}$
14. $\displaystyle\int_{1/2}^1 \frac{y+4}{y^2+y}\,dy$
15. $\displaystyle\int \frac{dt}{t^3+t^2-2t}$
16. $\displaystyle\int \frac{x+3}{2x^3-8x}\,dx$

Repeated Linear Factors

In Exercises 17–20, express the integrands as a sum of partial fractions and evaluate the integrals.

17. $\int_0^1 \frac{x^3\,dx}{x^2+2x+1}$

18. $\int_{-1}^0 \frac{x^3\,dx}{x^2-2x+1}$

19. $\int \frac{dx}{(x^2-1)^2}$

20. $\int \frac{x^2\,dx}{(x-1)(x^2+2x+1)}$

Irreducible Quadratic Factors

In Exercises 21–28, express the integrands as a sum of partial fractions and evaluate the integrals.

21. $\int_0^1 \frac{dx}{(x+1)(x^2+1)}$

22. $\int_1^{\sqrt{3}} \frac{3t^2+t+4}{t^3+t}\,dt$

23. $\int \frac{y^2+2y+1}{(y^2+1)^2}\,dy$

24. $\int \frac{8x^2+8x+2}{(4x^2+1)^2}\,dx$

25. $\int \frac{2s+2}{(s^2+1)(s-1)^3}\,ds$

26. $\int \frac{s^4+81}{s(s^2+9)^2}\,ds$

27. $\int \frac{2\theta^3+5\theta^2+8\theta+4}{(\theta^2+2\theta+2)^2}\,d\theta$

28. $\int \frac{\theta^4-4\theta^3+2\theta^2-3\theta+1}{(\theta^2+1)^3}\,d\theta$

Improper Fractions

In Exercises 29–34, perform long division on the integrand, write the proper fraction as a sum of partial fractions, and then evaluate the integral.

29. $\int \frac{2x^3-2x^2+1}{x^2-x}\,dx$

30. $\int \frac{x^4}{x^2-1}\,dx$

31. $\int \frac{9x^3-3x+1}{x^3-x^2}\,dx$

32. $\int \frac{16x^3}{4x^2-4x+1}\,dx$

33. $\int \frac{y^4+y^2-1}{y^3+y}\,dy$

34. $\int \frac{2y^4}{y^3-y^2+y-1}\,dy$

Evaluating Integrals

Evaluate the integrals in Exercises 35–40.

35. $\int \frac{e^t\,dt}{e^{2t}+3e^t+2}$

36. $\int \frac{e^{4t}+2e^{2t}-e^t}{e^{2t}+1}\,dt$

37. $\int \frac{\cos y\,dy}{\sin^2 y+\sin y-6}$

38. $\int \frac{\sin\theta\,d\theta}{\cos^2\theta+\cos\theta-2}$

39. $\int \frac{(x-2)^2\tan^{-1}(2x)-12x^3-3x}{(4x^2+1)(x-2)^2}\,dx$

40. $\int \frac{(x+1)^2\tan^{-1}(3x)+9x^3+x}{(9x^2+1)(x+1)^2}\,dx$

Initial Value Problems

Solve the initial value problems in Exercises 41–48.

41. $(t^2-3t+2)\frac{dx}{dt}=1 \quad (t>2), \quad x(3)=0$

42. $(3t^4+4t^2+1)\frac{dx}{dt}=2\sqrt{3}, \quad x(1)=-\pi\sqrt{3}/4$

43. $(t^2+2t)\frac{dx}{dt}=2x+2 \quad (t,x>0), \quad x(1)=1$

44. $(t+1)\frac{dx}{dt}=x^2+1 \quad (t>-1), \quad x(0)=\pi/4$

45. $\frac{dy}{dx}=e^x(y^2-y), \quad y(0)=2$

46. $\frac{dy}{d\theta}=(y+1)^2\sin\theta \quad y(\pi/2)=0$

47. $\frac{dy}{dx}=\frac{1}{x^2-3x+2}, \quad y(3)=0$

48. $\frac{ds}{dt}=\frac{2s+2}{t^2+2t}, \quad s(1)=1$

Applications and Examples

In Exercises 49 and 50, find the volume of the solid generated by revolving the shaded region about the indicated axis.

49. The x-axis

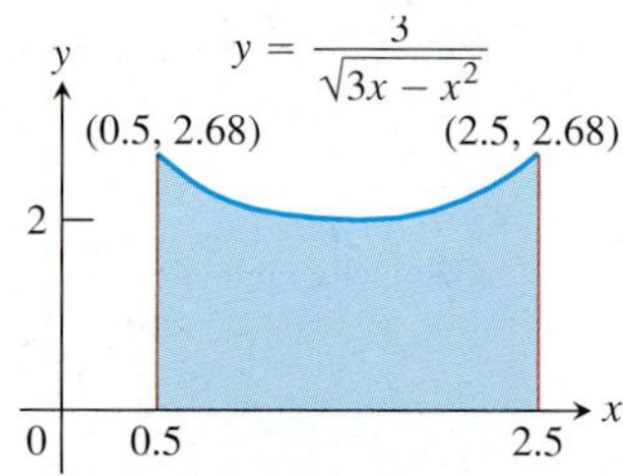

50. The y-axis

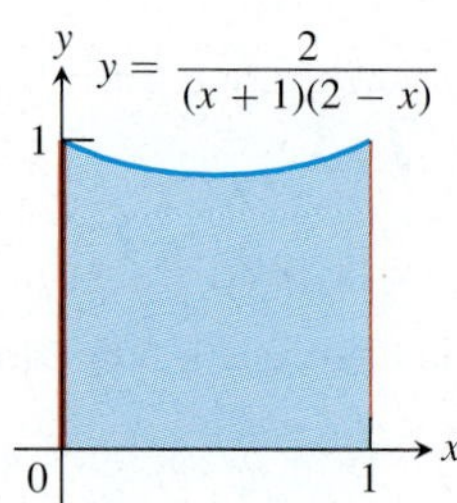

51. *Social diffusion* Sociologists sometimes use the phrase "social diffusion" to describe the way information spreads through a population. The information might be a rumor, a cultural fad, or news about a technical innovation. In a sufficiently large population, the number of people x who have the information is treated as a differentiable function of time t, and the rate of diffusion, dx/dt, is assumed to be proportional to the number of people who have the information times the number of people who do not. This leads to the equation

$$\frac{dx}{dt} = kx(N - x),$$

where N is the number of people in the population.

Suppose that t is in days, $k = 1/250$, and two people start a rumor at time $t = 0$ in a population of $N = 1000$ people.

(a) Find x as a function of t.

(b) When will half the population have heard the rumor? (This is when the rumor will be spreading the fastest.)

52. *Second-order chemical reactions* Many chemical reactions are the result of the interaction of two molecules that undergo a change to produce a new product. The rate of the reaction typically depends on the concentrations of the two kinds of molecules. If a is the amount of substance A and b is the amount of substance B at time $t = 0$, and if x is the amount of product at time t, then the rate of formation of x may be given by the differential equation

$$\frac{dx}{dt} = k(a - x)(b - x),$$

or

$$\frac{1}{(a - x)(b - x)} \frac{dx}{dt} = k,$$

where k is a constant for the reaction. Integrate both sides of this equation to obtain a relation between x and t

(a) if $a = b$

(b) if $a \neq b$.

Assume in each case that $x = 0$ when $t = 0$.

7.4 Trigonometric Substitutions

Three Basic Substitutions

Trigonometric substitutions enable us to replace the binomials $a^2 + x^2$, $a^2 - x^2$, and $x^2 - a^2$ by single squared terms and thereby transform a number of integrals containing square roots into integrals we can evaluate directly.

Three Basic Substitutions

The most common substitutions are $x = a \tan \theta$, $x = a \sin \theta$, and $x = a \sec \theta$. They come from the reference right triangles in Figure 7.3.

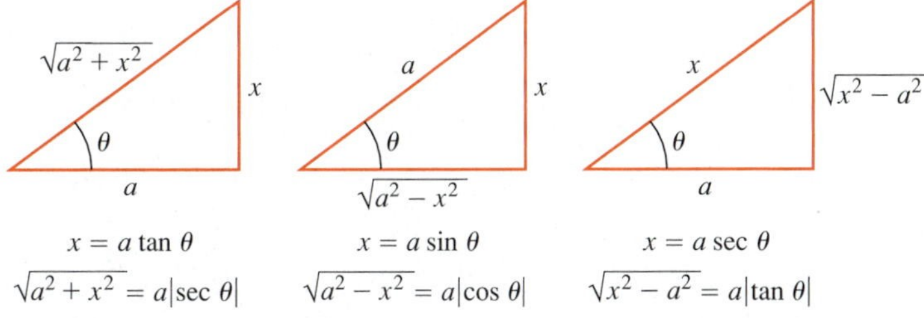

FIGURE 7.3 Reference triangles for trigonometric substitutions that change binomials into single squared terms.

With $x = a \tan \theta$,

$$a^2 + x^2 = a^2 + a^2 \tan^2 \theta = a^2(1 + \tan^2 \theta) = a^2 \sec^2 \theta.$$

With $x = a \sin \theta$,

$$a^2 - x^2 = a^2 - a^2 \sin^2 \theta = a^2(1 - \sin^2 \theta) = a^2 \cos^2 \theta.$$

With $x = a \sec \theta$,

$$x^2 - a^2 = a^2 \sec^2 \theta - a^2 = a^2(\sec^2 \theta - 1) = a^2 \tan^2 \theta.$$

Trigonometric Substitutions

1. $x = a \tan \theta$ replaces $a^2 + x^2$ by $a^2 \sec^2 \theta$.
2. $x = a \sin \theta$ replaces $a^2 - x^2$ by $a^2 \cos^2 \theta$.
3. $x = a \sec \theta$ replaces $x^2 - a^2$ by $a^2 \tan^2 \theta$.

We want any substitution we use in an integration to be reversible so that we can change back to the original variable afterward. For example, if $x = a \tan \theta$, we want to be able to set $\theta = \tan^{-1}(x/a)$ after the integration takes place. If $x = a \sin \theta$, we want to be able to set $\theta = \sin^{-1}(x/a)$ when we're done, and similarly for $x = a \sec \theta$.

For reversibility,

$$x = a \tan \theta \quad \text{requires} \quad \theta = \tan^{-1}\left(\frac{x}{a}\right) \quad \text{with} \quad -\frac{\pi}{2} < \theta < \frac{\pi}{2}$$

$$x = a \sin \theta \quad \text{requires} \quad \theta = \sin^{-1}\left(\frac{x}{a}\right) \quad \text{with} \quad -\frac{\pi}{2} \le \theta \le \frac{\pi}{2}$$

$$x = a \sec \theta \quad \text{requires} \quad \theta = \sec^{-1}\left(\frac{x}{a}\right) \quad \text{with} \quad \begin{cases} 0 \le \theta < \frac{\pi}{2}, & \frac{x}{a} \ge 1 \\ \frac{\pi}{2} < \theta \le \pi, & \frac{x}{a} \le -1. \end{cases}$$

Example 1 Using the Substitution $x = a \tan \theta$

Evaluate

$$\int \frac{dx}{\sqrt{4 + x^2}}.$$

Solution We set

$$x = 2 \tan \theta, \qquad dx = 2 \sec^2 \theta \, d\theta, \qquad -\frac{\pi}{2} < \theta < \frac{\pi}{2},$$

$$4 + x^2 = 4 + 4 \tan^2 \theta = 4(1 + \tan^2 \theta) = 4 \sec^2 \theta.$$

Then

$$\begin{aligned}
\int \frac{dx}{\sqrt{4 + x^2}} &= \int \frac{2 \sec^2 \theta \, d\theta}{\sqrt{4 \sec^2 \theta}} = \int \frac{\sec^2 \theta \, d\theta}{|\sec \theta|} && \sqrt{\sec^2 \theta} = |\sec \theta| \\
&= \int \sec \theta \, d\theta && \sec \theta > 0 \text{ for } -\frac{\pi}{2} < \theta < \frac{\pi}{2} \\
&= \ln |\sec \theta + \tan \theta| + C \\
&= \ln \left| \frac{\sqrt{4 + x^2}}{2} + \frac{x}{2} \right| + C && \text{From Fig. 7.4} \\
&= \ln |\sqrt{4 + x^2} + x| + C'. && \text{Taking } C' = C - \ln 2
\end{aligned}$$

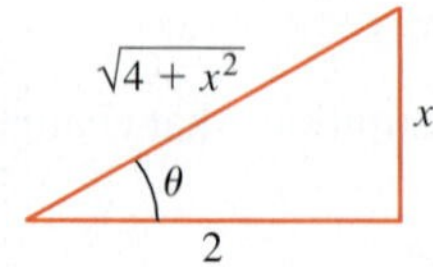

FIGURE 7.4 Reference triangle for $x = 2 \tan \theta$ (Example 1):

$$\tan \theta = \frac{x}{2}$$

and

$$\sec \theta = \frac{\sqrt{4 + x^2}}{2}.$$

Notice how we expressed $\ln|\sec\theta + \tan\theta|$ in terms of x: We drew a reference triangle for the original substitution $x = 2\tan\theta$ (Figure 7.4) and read the ratios from the triangle.

Example 2 Using the Substitution $x = a\sin\theta$

Evaluate

$$\int \frac{x^3\,dx}{\sqrt{9-x^2}}, \qquad -3 < x < 3.$$

Solution We set

$$x = 3\sin\theta, \qquad dx = 3\cos\theta\,d\theta, \qquad -\frac{\pi}{2} < \theta < \frac{\pi}{2}$$

$$9 - x^2 = 9 - 9\sin^2\theta = 9(1-\sin^2\theta) = 9\cos^2\theta.$$

Then

$$\begin{aligned}\int \frac{x^3\,dx}{\sqrt{9-x^2}} &= \int \frac{27\sin^3\theta \cdot 3\cos\theta\,d\theta}{|3\cos\theta|} \\ &= 27\int \sin^3\theta\,d\theta && \cos\theta > 0 \text{ for } -\frac{\pi}{2} < \theta < \frac{\pi}{2} \\ &= 27\int (1-\cos^2\theta)\sin\theta\,d\theta && \sin^2\theta = 1-\cos^2\theta \\ &= -27\cos\theta + 9\cos^3\theta + C \\ &= -27\cdot\frac{\sqrt{9-x^2}}{3} + 9\left(\frac{\sqrt{9-x^2}}{3}\right)^3 + C && \text{Figure 7.5, } a = 3 \\ &= -9\sqrt{9-x^2} + \frac{(9-x^2)^{3/2}}{3} + C.\end{aligned}$$

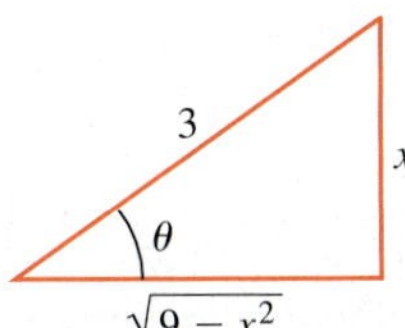

FIGURE 7.5 Reference triangle for $x = 3\sin\theta$ (Example 2):

$$\sin\theta = \frac{x}{3}$$

and

$$\cos\theta = \frac{\sqrt{9-x^2}}{3}.$$

Example 3 Using the Substitution $x = a\sec\theta$

Evaluate

$$\int \frac{dx}{\sqrt{25x^2-4}}, \qquad x > \frac{2}{5}.$$

Solution We first rewrite the radical as

$$\begin{aligned}\sqrt{25x^2-4} &= \sqrt{25\left(x^2 - \frac{4}{25}\right)} \\ &= 5\sqrt{x^2 - \left(\frac{2}{5}\right)^2}\end{aligned}$$

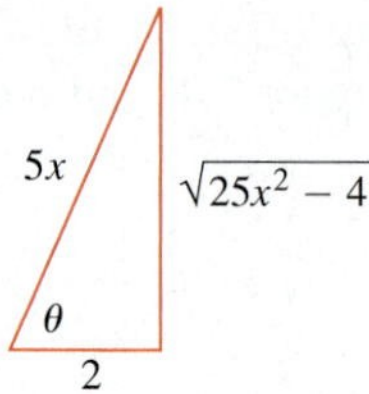

FIGURE 7.6 If $x = (2/5)\sec\theta, 0 \le \theta < \pi/2$, then $\theta = \sec^{-1}(5x/2)$, and we can read the values of the other trigonometric functions of θ from this right triangle.

to put the radicand in the form $x^2 - a^2$. We then substitute

$$x = \frac{2}{5}\sec\theta, \qquad dx = \frac{2}{5}\sec\theta\tan\theta\,d\theta, \qquad 0 < \theta < \frac{\pi}{2}$$

$$x^2 - \left(\frac{2}{5}\right)^2 = \frac{4}{25}\sec^2\theta - \frac{4}{25}$$

$$= \frac{4}{25}(\sec^2\theta - 1) = \frac{4}{25}\tan^2\theta$$

$$\sqrt{x^2 - \left(\frac{2}{5}\right)^2} = \frac{2}{5}|\tan\theta| = \frac{2}{5}\tan\theta. \qquad \text{tan } \theta > 0 \text{ for } 0 < \theta < \pi/2$$

With these substitutions, we have

$$\int \frac{dx}{\sqrt{25x^2 - 4}} = \int \frac{dx}{5\sqrt{x^2 - (4/25)}} = \int \frac{(2/5)\sec\theta\tan\theta\,d\theta}{5\cdot(2/5)\tan\theta}$$

$$= \frac{1}{5}\int \sec\theta\,d\theta = \frac{1}{5}\ln|\sec\theta + \tan\theta| + C$$

$$= \frac{1}{5}\ln\left|\frac{5x}{2} + \frac{\sqrt{25x^2 - 4}}{2}\right| + C \qquad \text{Fig. 7.6}$$

A trigonometric substitution can sometimes help us to evaluate an integral containing an integer power of a quadratic binomial, as in the next example.

Example 4 Finding the Volume of a Solid of Revolution

Find the volume of the solid generated by revolving about the x-axis the region bounded by the curve $y = 4/(x^2 + 4)$, the x-axis, and the lines $x = 0$ and $x = 2$.

Solution We sketch the region (Figure 7.7) and use the disk method:

$$V = \int_0^2 \pi[R(x)]^2\,dx = 16\pi\int_0^2 \frac{dx}{(x^2 + 4)^2}. \qquad R(x) = \frac{4}{x^2 + 4}$$

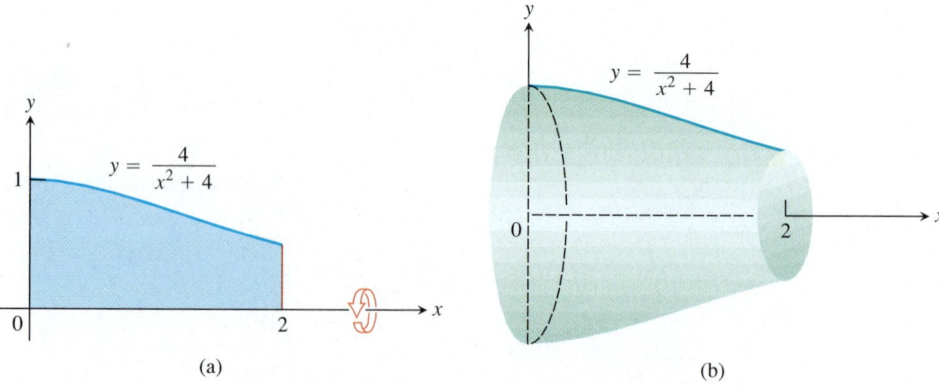

FIGURE 7.7 The region (a) and solid (b) in Example 4.

To evaluate the integral, we set

$$x = 2\tan\theta, \qquad dx = 2\sec^2\theta\, d\theta, \qquad \theta = \tan^{-1}\frac{x}{2},$$

$$x^2 + 4 = 4\tan^2\theta + 4 = 4(\tan^2\theta + 1) = 4\sec^2\theta$$

(Figure 7.8). With these substitutions,

$$\begin{aligned} V &= 16\pi \int_0^2 \frac{dx}{(x^2+4)^2} \\ &= 16\pi \int_0^{\pi/4} \frac{2\sec^2\theta\, d\theta}{(4\sec^2\theta)^2} \qquad \theta = 0 \text{ when } x = 0;\ \theta = \pi/4 \text{ when } x = 2 \\ &= 16\pi \int_0^{\pi/4} \frac{2\sec^2\theta\, d\theta}{16\sec^4\theta} = \pi \int_0^{\pi/4} 2\cos^2\theta\, d\theta \\ &= \pi \int_0^{\pi/4} (1 + \cos 2\theta)\, d\theta = \pi\left[\theta + \frac{\sin 2\theta}{2}\right]_0^{\pi/4} \qquad 2\cos^2\theta = 1 + \cos 2\theta \\ &= \pi\left[\frac{\pi}{4} + \frac{1}{2}\right] \approx 4.04. \end{aligned}$$

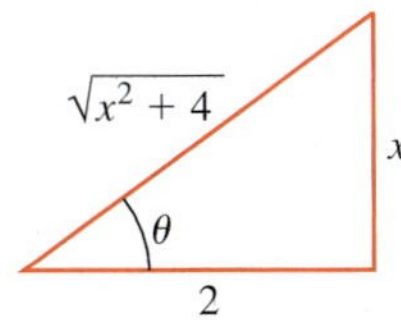

FIGURE 7.8 Reference triangle for $x = 2\tan\theta$. (Example 4)

EXERCISES 7.4

Basic Trigonometric Substitutions

Evaluate the integrals in Exercises 1–18.

1. $\int \frac{dy}{\sqrt{9 + y^2}}$

2. $\int \frac{3\, dy}{\sqrt{1 + 9y^2}}$

3. $\int \sqrt{25 - t^2}\, dt$

4. $\int \sqrt{1 - 9t^2}\, dt$

5. $\int \frac{dx}{\sqrt{4x^2 - 49}}, \quad x > \frac{7}{2}$

6. $\int \frac{5\, dx}{\sqrt{25x^2 - 9}}, \quad x > \frac{3}{5}$

7. $\int \frac{dx}{x^2\sqrt{x^2 - 1}}, \quad x > 1$

8. $\int \frac{2\, dx}{x^3\sqrt{x^2 - 1}}, \quad x > 1$

9. $\int \frac{x^3\, dx}{\sqrt{x^2 + 4}}$

10. $\int \frac{dx}{x^2\sqrt{x^2 + 1}}$

11. $\int \frac{8\, dw}{w^2\sqrt{4 - w^2}}$

12. $\int \frac{\sqrt{9 - w^2}}{w^2}\, dw$

13. $\int \frac{dx}{(x^2 - 1)^{3/2}}, \quad x > 1$

14. $\int \frac{x^2\, dx}{(x^2 - 1)^{5/2}}, \quad x > 1$

15. $\int \frac{(1 - x^2)^{3/2}}{x^6}\, dx$

16. $\int \frac{(1 - x^2)^{1/2}}{x^4}\, dx$

17. $\int \frac{8\, dx}{(4x^2 + 1)^2}$

18. $\int \frac{6\, dt}{(9t^2 + 1)^2}$

Combining Substitutions

In Exercises 19–26, use an appropriate substitution and then a trigonometric substitution to evaluate the integrals.

19. $\int_0^{\ln 4} \frac{e^t\, dt}{\sqrt{e^{2t} + 9}}$

20. $\int_{\ln(3/4)}^{\ln(4/3)} \frac{e^t\, dt}{(1 + e^{2t})^{3/2}}$

21. $\int_{1/12}^{1/4} \frac{2\, dt}{\sqrt{t} + 4t\sqrt{t}}$

22. $\int_1^e \frac{dy}{y\sqrt{1 + (\ln y)^2}}$

23. $\int \frac{dx}{x\sqrt{x^2 - 1}}$

24. $\int \frac{dx}{1 + x^2}$

25. $\int \frac{x\, dx}{\sqrt{x^2 - 1}}$

26. $\int \frac{dx}{\sqrt{1 - x^2}}$

Initial Value Problems

Solve the initial value problems in Exercises 27–30 for y as a function of x.

27. $x\frac{dy}{dx} = \sqrt{x^2 - 4}, \quad x \geq 2, \quad y(2) = 0$

28. $\sqrt{x^2 - 9}\,\frac{dy}{dx} = 1, \quad x > 3, \quad y(5) = \ln 3$

29. $(x^2 + 4)\frac{dy}{dx} = 3, \quad y(2) = 0$

30. $(x^2 + 1)^2\frac{dy}{dx} = \sqrt{x^2 + 1}, \quad y(0) = 1$

Applications

31. *Finding area* Find the area of the region in the first quadrant that is enclosed by the coordinate axes and the curve $y = \sqrt{9 - x^2}/3$.

32. *Finding volume* Find the volume of the solid generated by revolving about the x-axis the region in the first quadrant enclosd by the coordinate axes, the curve $y = 2/(1 + x^2)$, and the line $x = 1$.

The Substitution $z = \tan(x/2)$

33. The substitution

$$z = \tan \frac{x}{2}$$

(see figure) reduces the problem of integrating any rational function of $\sin x$ and $\cos x$ to a problem involving a rational function of z. This in turn can be solved by partial fractions. Show that each equation is true.

(a) $\tan \frac{x}{2} = \frac{\sin x}{1 + \cos x}$ **(b)** $\cos x = \frac{1 - z^2}{1 + z^2}$

(c) $\sin x = \frac{2z}{1 + z^2}$ **(d)** $dx = \frac{2\,dz}{1 + z^2}$

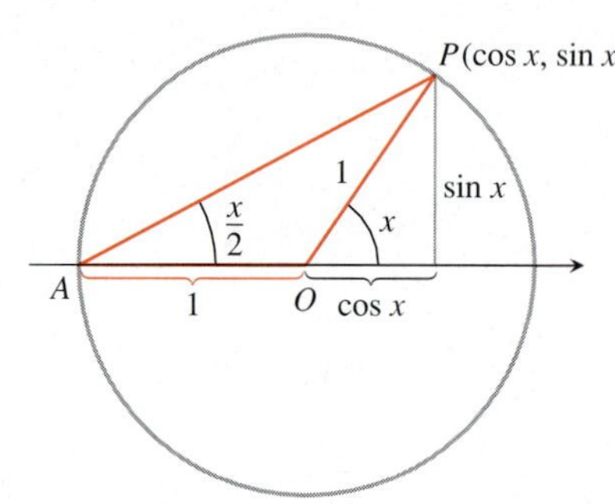

In Exercises 34–41, use the substitution $z = \tan(x/2)$ and the results in Exercise 33 to evaluate the integral.

34. $\int \frac{dx}{1 + \sin x}$

35. $\int \frac{dx}{1 - \cos x}$

36. $\int \frac{d\theta}{1 - \sin \theta}$

37. $\int \frac{dt}{1 + \sin t + \cos t}$

38. $\int_0^{\pi/2} \frac{d\theta}{2 + \cos \theta}$

39. $\int_{\pi/2}^{2\pi/3} \frac{\cos \theta \, d\theta}{\sin \theta \cos \theta + \sin \theta}$

40. $\int \frac{dt}{\sin t - \cos t}$

41. $\int \frac{\cos t \, dt}{1 - \cos t}$

7.5 Integral Tables, Computer Algebra Systems, and Monte Carlo Integration

Integral Tables • Integration with a CAS • Monte Carlo Numerical Integration

As you know, the basic techniques of integration are substitution and integration by parts. We apply these techniques to transform unfamiliar integrals into integrals whose forms we recognize or can find in a table. If a computer algebra system (CAS) is available, you can usually use it to evaluate an integral. Another technique for approximating the value of a definite integral is called *Monte Carlo Integration.* We investigate these methods for evaluating integrals in this section.

Integral Tables

A Brief Table of Integrals appears at the back of the book, after the index. The integration formulas are stated in terms of constants a, b, c, m, n, and so on. These constants can usually assume any real value and need not be integers. Occasional limitations on their values are stated with the formulas. Formula 5 requires $n \neq -1$, for example, and Formula 11 requires $n \neq -2$.

The formulas also assume that the constants do not take on values that require dividing by zero or taking even roots of negative numbers. For example, Formula 8 assumes that $a \neq 0$, and Formula 13(a) cannot be used unless b is negative.

Example 1 Using a Table of Integrals

Evaluate

$$\int \left(\frac{1}{x^2+1} + \frac{1}{x^2-2x+5} \right) dx.$$

Solution

$$\int \left(\frac{1}{x^2+1} + \frac{1}{x^2-2x+5} \right) dx = \int \frac{1}{x^2+1}\, dx + \int \frac{1}{x^2-2x+5}\, dx \qquad (1)$$

Both integrals on the right-hand side of Equation 1 can be evaluated using Formula 16 from the Brief Table of Integrals.

$$\textbf{16.} \int \frac{dx}{a^2+x^2} = \frac{1}{a} \tan^{-1} \frac{x}{a} + C$$

David Hilbert (1862 — 1943)

You probably remember that

$$\int \frac{dx}{1+x^2} = \tan^{-1} x + C,$$

but if you don't, you can obtain the result by setting $a = 1$ in Formula 16.

To find the value of the second integral, we need to complete the square:

$$x^2 - 2x + 5 = x^2 - 2x + 1 + 4 = (x-1)^2 + 4.$$

Then

$$\begin{aligned} \int \frac{dx}{x^2-2x+5} &= \int \frac{dx}{(x-1)^2+4} \\ &= \int \frac{du}{u^2+4} && u = x-1,\ du = dx \\ &= \frac{1}{2} \tan^{-1} \frac{u}{2} + C && \text{Formula 16 with } a = 2 \\ &= \frac{1}{2} \tan^{-1} \left(\frac{x-1}{2} \right) + C. \end{aligned}$$

Combining the two integrations gives

$$\int \left(\frac{1}{x^2+1} + \frac{1}{x^2-2x+5} \right) dx = \tan^{-1} x + \frac{1}{2} \tan^{-1} \left(\frac{x-1}{2} \right) + C.$$

The manipulation and substitution in Example 1 are typical of what is needed to use integral tables to evaluate integrals. Here's another example.

Example 2 Using a Table of Integrals

Find

$$\int \frac{dx}{x^2\sqrt{2x-4}}.$$

Solution We begin with Formula 15:

$$\textbf{15.}\quad \int \frac{dx}{x^2\sqrt{ax+b}} = -\frac{\sqrt{ax+b}}{bx} - \frac{a}{2b}\int \frac{dx}{x\sqrt{ax+b}} + C.$$

With $a = 2$ and $b = -4$, we have

$$\int \frac{dx}{x^2\sqrt{2x-4}} = -\frac{\sqrt{2x-4}}{-4x} + \frac{2}{2\cdot 4}\int \frac{dx}{x\sqrt{2x-4}} + C.$$

We then use Formula 13(a) to evaluate the integral on the right:

$$\textbf{13(a).}\quad \int \frac{dx}{x\sqrt{ax-b}} = \frac{2}{\sqrt{b}}\tan^{-1}\sqrt{\frac{ax-b}{b}} + C.$$

With $a = 2$ and $b = 4$, we have

$$\int \frac{dx}{x\sqrt{2x-4}} = \frac{2}{\sqrt{4}}\tan^{-1}\sqrt{\frac{2x-4}{4}} + C = \tan^{-1}\sqrt{\frac{x-2}{2}} + C.$$

Combining the two integrations gives

$$\int \frac{dx}{x^2\sqrt{2x-4}} = \frac{\sqrt{2x-4}}{4x} + \frac{1}{4}\tan^{-1}\sqrt{\frac{x-2}{2}} + C.$$

Integration with a CAS

A powerful capability of computer algebra systems is their facility to integrate symbolically. This is performed with the **integrate command** specified by the particular system (e.g., **int** in Maple, **Integrate** in Mathematica).

Example 3 Using a CAS with a Named Function

Suppose that you want to evaluate the indefinite integral of the function

$$f(x) = x^2\sqrt{a^2+x^2}.$$

Using Maple, you first define or name the function:

```
> f := x^2 * sqrt (a^2 + x^2);
```

Then you use the integrate command on f, identifying the variable of integration:

```
> int(f, x);
```

Maple returns the answer

$$\frac{1}{4}x(a^2+x^2)^{3/2} - \frac{1}{8}a^2x\sqrt{a^2+x^2} - \frac{1}{8}a^4\ln\left(x+\sqrt{a^2+x^2}\right).$$

If you want to see if the answer can be simplified, enter

```
> simplify(");
```

Maple returns

$$\frac{1}{8}a^2x\sqrt{a^2+x^2} + \frac{1}{4}x^3\sqrt{a^2+x^2} - \frac{1}{8}a^4\ln\left(x+\sqrt{a^2+x^2}\right).$$

If you want the definite integral for $0 \le x \le \pi/2$, you can use the format

```
> int(f, x = 0..Pi/2);
```

Maple (Version 3.0) will return the expression

$$\frac{1}{64}(4a^2+\pi^2)^{3/2}\pi - \frac{1}{8}a^4 \ln\left(\frac{1}{2}\pi + \frac{1}{2}\sqrt{4a^2+\pi^2}\right) - \frac{1}{32}a^2\sqrt{4a^2+\pi^2}\,\pi + \frac{1}{8}a^4 \ln(\sqrt{a^2}).$$

You can also find the definite integral for a particular value of the constant a:

```
> a := 1;
> int(f, x = 0..1);
```

Maple returns the numerical answer

$$\frac{3}{8}\sqrt{2} - \frac{1}{8}\ln(1+\sqrt{2}).$$

Example 4 Using a CAS Without Naming the Function

Use a CAS to find

$$\int \sin^2 x \cos^3 x \, dx.$$

Solution With Maple, we have the entry

```
> int((sin ^2)(x)*(cos ^3)(x), x);
```

with the immediate return

$$-\frac{1}{5}\sin(x)\cos(x)^4 + \frac{1}{15}\cos(x)^2\sin(x) + \frac{2}{15}\sin(x).$$

Example 5 A CAS May Not Return a Closed Form Solution

Use a CAS to find

$$\int (\cos^{-1} ax)^2 \, dx.$$

Solution Using Maple, we enter

```
> int((arccos(a*x))^2, x);
```

and Maple returns the expression

$$\int \arccos(ax)^2 \, dx,$$

indicating that it does not have a closed form solution. In the next chapter, you see how series expansion may help to evaluate such an integral.

Computer algebra systems vary in how they process integrations. We used Maple in Examples 3 through 5. Mathematica would have returned somewhat different results:

1. In Example 3, given

In [1]: = Integrate [x^2 * Sqrt [a^2 + x^2], x]

Mathematica returns

$$\text{Out [1]} = \text{Sqrt}\,[a^2 + x^2]\left(\frac{a^2x}{8} + \frac{x^3}{4}\right) - \frac{a^4\,\text{Log}\,[x + \text{Sqrt}\,[a^2 + x^2]]}{8}$$

without having to simplify an intermediate result. The answer is close to Formula 22 in the integral tables.

2. The Mathematica answer to the integral

In [2]: = Integrate [Sin [x]^2 * Cos [x]^3, x]

in Example 4 is

$$\text{Out [2]} = \frac{30\,\text{Sin}\,[x] - 5\,\text{Sin}\,[3x] - 3\,\text{Sin}\,[5x]}{240}$$

differing from the Maple answer.

3. Mathematica does give a result for the integration

In [3]: = Integrate [ArcCos [a * x]^2, x]

in Example 5:

$$\text{Out [3]} = -2x - \frac{2\,\text{Sqrt}\,[1 - a^2x^2]\,\text{ArcCos}\,[ax]}{a} + x\text{ArcCos}\,[ax]^2$$

Although a CAS is very powerful and can aid us in solving difficult problems, each CAS has its own limitations. There are even situations where a CAS may further complicate a problem (in the sense of producing an answer that is extremely difficult to use or interpret). On the other hand, a little mathematical thinking on your part may reduce the problem to one that is quite easy to handle. We provide an example in Exercise 49.

Monte Carlo Numerical Integration

In many applications, we encounter integrals such as $\int_1^{100} e^{x^2}\,dx$, which cannot be evaluated using the analytic techniques presented in this chapter. We can approximate a definite integral like that using numerical techniques such as the Trapezoidal Rule or Simpson's Rule (Section 4.7). Another numerical method for approximating definite integrals, called **Monte Carlo integration,** generalizes conveniently to multiple integrals (Chapter 12) as well. The method is typically accomplished with the aid of a computer. We illustrate how it works with a simple example.

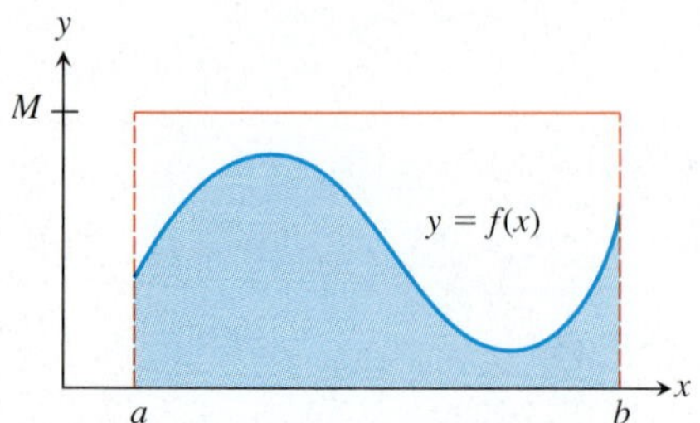

FIGURE 7.9 Imagine the rectangle as a dartboard.

Example 6 Monte Carlo Area Under a Nonnegative Curve

Imagine the area under the continuous curve $y = f(x)$, $0 \le f(x) \le M$, over the closed interval $a \le x \le b$ as a portion of the area of the total rectangle depicted in Figure 7.9. Randomly select a large number of points $P(x, y)$ within the rectangle and use them to estimate the area under the curve.

Solution To "randomly select" a point within the rectangle, we use a computer or calculator programmed with a random number generator. First we ask the computer to generate a random number x satisfying $a \leq x \leq b$. In theory, all numbers in the closed interval $[a, b]$ have an equal likelihood of being selected. Next we ask the computer to generate a second random number y satisfying $0 \leq y \leq M$. Again any number in $[0, M]$ has equal likelihood of being selected, at least in theory. The point $P(x, y)$ therefore lies somewhere inside the rectangle in Figure 7.9. Once the random point $P(x, y)$ is selected, ask yourself if it lies within the shaded region below the curve; that is, does the y-coordinate satisfy $0 \leq y < f(x)$? If the answer is yes, then count the point P by adding one to some counter. Two counters will be necessary; one to count the total points generated and a second to count those points that lie below the curve (see Figure 7.10). We pick a very large number of points P using the method just described, remembering to count those that lie below the curve. We can then approximate the area under the curve from the formula

$$\frac{\text{Area under curve}}{\text{area of rectangle}} \approx \frac{\text{number of points counted below curve}}{\text{total number of random points}}. \tag{2}$$

For example, Table 7.3 gives estimates to the area below the curve $y = \cos x$ and above the x-axis over the interval $-\pi/2 \leq x \leq \pi/2$. We chose the number

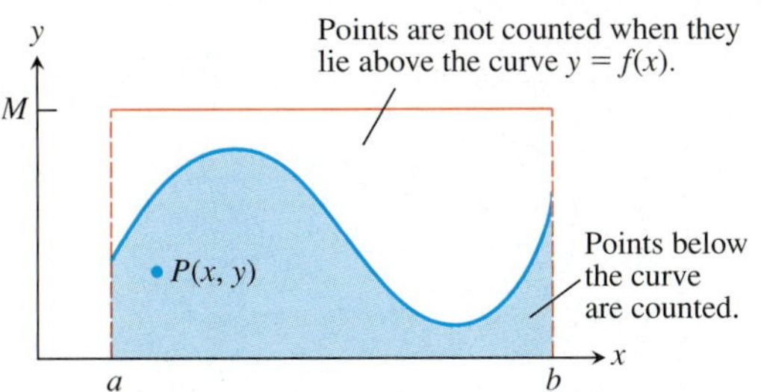

FIGURE 7.10 The area under the nonnegative curve $y = f(x)$, $a \leq x \leq b$, is contained within the rectangle of height M and base $b - a$.

Table 7.3 Monte Carlo approximation to the area under the curve $y = \cos x$ over the interval $-\pi/2 \leq x \leq \pi/2$

Number of points	Approximation to area	Number of points	Approximation to area
100	2.07345	2,000	1.94465
200	2.13628	3,000	1.97711
300	2.01064	4,000	1.99962
400	2.12058	5,000	2.01429
500	2.04832	6,000	2.02319
600	2.09440	8,000	2.00669
700	2.02857	10,000	2.00873
800	1.99491	15,000	2.00978
900	1.99666	20,000	2.01093
1,000	1.96664	30,000	2.01186

$M = 2$ as the height of the rectangle, and we estimated the area after randomly selecting 100, 200, . . . , 30,000 points in the rectangle and applying the estimation formula (2).

The actual area under the curve $y = \cos x$ over the given interval is 2 square units. Note that even with the relatively large number of points generated the error is significant. For functions of one variable, Monte Carlo integration is generally not competitive with the techniques you learned in Section 4.7. The lack of an error bound and the potential difficulty in finding an upper bound M are disadvantages as well. Nevertheless, the Monte Carlo technique can be extended to functions of several variables and becomes more practical in that situation.

EXERCISES 7.5

Using Integral Tables

Use the table of integrals at the back of the book to evaluate the integrals in Exercises 1–20.

1. $\int \frac{dx}{x\sqrt{x-3}}$

2. $\int \frac{x\,dx}{\sqrt{x-2}}$

3. $\int x\sqrt{2x-3}\,dx$

4. $\int \frac{\sqrt{9-4x}}{x^2}\,dx$

5. $\int x\sqrt{4x-x^2}\,dx$

6. $\int \frac{dx}{x\sqrt{7+x^2}}$

7. $\int \frac{\sqrt{4-x^2}}{x}\,dx$

8. $\int \sqrt{25-p^2}\,dp$

9. $\int \frac{r^2}{\sqrt{4-r^2}}\,dr$

10. $\int \frac{d\theta}{5+4\sin 2\theta}$

11. $\int e^{2t}\cos 3t\,dt$

12. $\int x\cos^{-1}x\,dx$

13. $\int \frac{ds}{(9-s^2)^2}$

14. $\int \frac{\sqrt{4x+9}}{x^2}\,dx$

15. $\int \frac{\sqrt{3t-4}}{t}\,dt$

16. $\int x^2\tan^{-1}x\,dx$

17. $\int \sin 3x\cos 2x\,dx$

18. $\int 8\sin 4t\sin\frac{t}{2}\,dt$

19. $\int \cos\frac{\theta}{3}\cos\frac{\theta}{4}\,d\theta$

20. $\int \cos\frac{\theta}{2}\cos 7\theta\,d\theta$

Substitution and Integral Tables

In Exercises 21–32, use a substitution to change the integral into one you can find in the table. Then evaluate the integral.

21. $\int \frac{x^3+x+1}{(x^2+1)^2}\,dx$

22. $\int \frac{x^2+6x}{(x^2+3)^2}\,dx$

23. $\int \sin^{-1}\sqrt{x}\,dx$

24. $\int \frac{\cos^{-1}\sqrt{x}}{\sqrt{x}}\,dx$

25. $\int \cot t\sqrt{1-\sin^2 t}\,dt, \quad 0 < t < \pi/2$

26. $\int \frac{dt}{\tan t\sqrt{4-\sin^2 t}}$

27. $\int \frac{dy}{y\sqrt{3+(\ln y)^2}}$

28. $\int \frac{\cos\theta\,d\theta}{\sqrt{5+\sin^2\theta}}$

29. $\int \frac{3\,dr}{\sqrt{9r^2-1}}$

30. $\int \frac{3\,dy}{\sqrt{1+9y^2}}$

31. $\int \cos^{-1}\sqrt{x}\,dx$

32. $\int \tan^{-1}\sqrt{y}\,dy$

Powers of x Times Exponentials

Evaluate the integrals in Exercises 33–36 using table Formulas 103–106. These integrals can also be evaluated using tabular integration (Section 7.2).

33. $\int xe^{3x}\,dx$

34. $\int x^3e^{x/2}\,dx$

35. $\int x^2 2^x\,dx$

36. $\int x\pi^x\,dx$

Hyperbolic Functions

Use the integral tables to evaluate the integrals in Exercises 37–40.

37. $\int \frac{1}{8}\sinh^5 3x\,dx$

38. $\int \frac{\cosh^4\sqrt{x}}{\sqrt{x}}\,dx$

39. $\int x^2\cosh 3x\,dx$

40. $\int x\sinh 5x\,dx$

Theory and Examples

Exercises 41–44 refer to formulas in the table of integrals at the back of the book.

41. Derive Formula 9 by using the substitution $u = ax + b$ to evaluate

$$\int \frac{x}{(ax+b)^2}\,dx.$$

42. Derive Formula 29 by using a trigonometric substitution to evaluate

$$\int \sqrt{a^2 - x^2}\, dx.$$

43. Derive Formula 110 by evaluating

$$\int x^n (\ln ax)^m\, dx$$

by integration by parts.

44. Derive Formula 99 by evaluating

$$\int x^n \sin^{-1} ax\, dx$$

by integration by parts.

45. *Finding volume* The head of your firm's accounting department has asked you to find a formula she can use in a computer program to calculate the year-end inventory of gasoline in the company's tanks. A typical tank is shaped like a right circular cylinder of radius r and length L, mounted horizontally, as shown here. The data come to the accounting office as depth measurements taken with a vertical measuring stick marked in centimeters.

(a) Show, in the notation of the figure here, that the volume of gasoline that fills the tank to a depth d is

$$V = 2L \int_{-r}^{-r+d} \sqrt{r^2 - y^2}\, dy.$$

(b) Evaluate the integral.

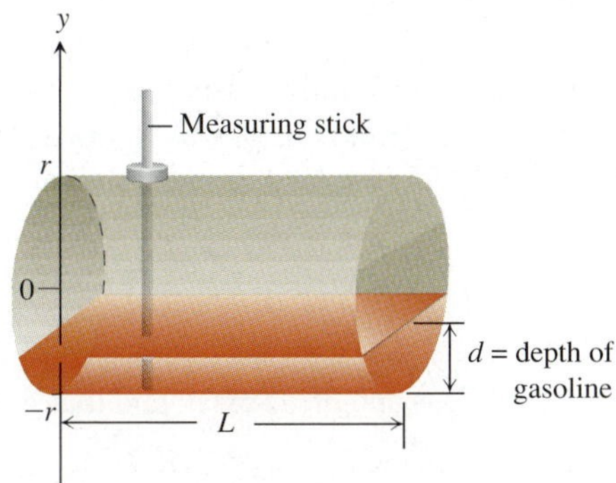

Measurements in centimeters

46. *Writing to Learn* What is the largest value

$$\int_a^b \sqrt{x - x^2}\, dx$$

can have for any a and b? Give reasons for your answer.

COMPUTER EXPLORATIONS

In Exercises 47 and 48, use a CAS to perform the integrations.

47. Evaluate the integrals in parts (a) through (c).

(a) $\int x \ln x\, dx$ **(b)** $\int x^2 \ln x\, dx$

(c) $\int x^3 \ln x\, dx$

(d) What pattern do you see? Predict a formula for $\int x^4 \ln x\, dx$ and then see if you are correct by evaluating it with a CAS.

(e) What is the formula for $\int x^n \ln x\, dx$, $n \geq 1$? Check your answer using a CAS.

48. Evaluate the integrals in parts (a) through (c).

(a) $\int \frac{\ln x}{x^2}\, dx$ **(b)** $\int \frac{\ln x}{x^3}\, dx$

(c) $\int \frac{\ln x}{x^4}\, dx$

(d) What pattern do you see? Predict the formula for

$$\int \frac{\ln x}{x^5}\, dx$$

and then see if you are correct by evaluating it with a CAS.

(e) What is the formula for

$$\int \frac{\ln x}{x^n}\, dx, \qquad n \geq 2?$$

Check your answer using a CAS.

49. (a) Use a CAS to evaluate

$$\int_0^{\pi/2} \frac{\sin^n x}{\sin^n x + \cos^n x}\, dx,$$

where n is an arbitrary positive integer. Does your CAS find the result?

(b) In succession, find the integral when $n = 1, 2, 3, 5, 7$. Comment on the complexity of the results.

(c) Now substitute $x = (\pi/2) - u$ and add the new and old integrals. What is the value of

$$\int_0^{\pi/2} \frac{\sin^n x}{\sin^n x + \cos^n x}\, dx?$$

This exercise illustrates how a little mathematical ingenuity solves a problem not immediately amenable to solution by a CAS.

Monte Carlo Integration

Use Monte Carlo integration to compute the integrals in Exercises 50–55. Compare your answers with exact answers obtained by using a computer algebra system or an analytic method previously presented.

50. $\int_0^1 xe^{-2x}\, dx$ **51.** $\int_{\pi/2}^{\pi} (\sin y)e^{\cos y}\, dy$

52. $\int_0^{1/\sqrt{2}} 2x \sin^{-1}(x^2)\, dx$ **53.** $\int_0^1 z\sqrt{1 - z}\, dz$

54. $\int_0^{1/2} \frac{t^3\, dt}{t^2 - 2t + 1}$ **55.** $\int_1^2 (\ln \theta)^3\, d\theta$

7.6 L'Hôpital's Rule

Indeterminate Form 0/0 • Indeterminate Forms ∞/∞, $\infty \cdot 0$, $\infty - \infty$ • Indeterminate Forms 1^∞, 0^0, ∞^0

CD-ROM
WEBsite
Historical Biography
Guillaume François Antoine de l'Hôpital (1661 — 1704)

As noted earlier, John Bernoulli discovered a rule for calculating limits of fractions whose numerators and denominators both approach zero. The rule is known today as **l'Hôpital's Rule,** after Guillaume François Antoine de l'Hôpital, Marquis de St. Mesme, a French nobleman who wrote the first introductory differential calculus text, where the rule first appeared in print.

Indeterminate Form 0/0

If the continuous functions $f(x)$ and $g(x)$ are both zero at $x = a$, then

$$\lim_{x\to a} \frac{f(x)}{g(x)}$$

cannot be found by substituting $x = a$. The substitution produces $0/0$, a meaningless expression known as an **indeterminate form.** Our experience so far has been that limits that lead to indeterminate forms may or may not be hard to find algebraically. It took a lot of analysis in Section 1.2 to find $\lim_{x\to 0} (\sin x)/x$. But we have had remarkable success with the limit

$$f'(a) = \lim_{x\to a} \frac{f(x) - f(a)}{x - a},$$

from which we calculate derivatives and which always produces the equivalent of $0/0$ when we substitute $x = a$. L'Hôpital's Rule enables us to draw on our success with derivatives to evaluate limits that otherwise lead to indeterminate forms.

Theorem 1 L'Hôpital's Rule (First Form)

Suppose that $f(a) = g(a) = 0$, that $f'(a)$ and $g'(a)$ exist, and that $g'(a) \neq 0$. Then

$$\lim_{x\to a} \frac{f(x)}{g(x)} = \frac{f'(a)}{g'(a)}.$$

Caution

To apply l'Hôpital's Rule to f/g, divide the derivative of f by the derivative of g. Do not fall into the trap of taking the derivative of f/g. The quotient to use is f'/g', not $(f/g)'$.

Example 1 Using L'Hôpital's Rule

(a) $$\lim_{x\to 0} \frac{3x - \sin x}{x} = \left.\frac{3 - \cos x}{1}\right|_{x=0} = 2$$

(b) $$\lim_{x\to 0} \frac{\sqrt{1+x} - 1}{x} = \left.\frac{\frac{1}{2\sqrt{1+x}}}{1}\right|_{x=0} = \frac{1}{2}$$

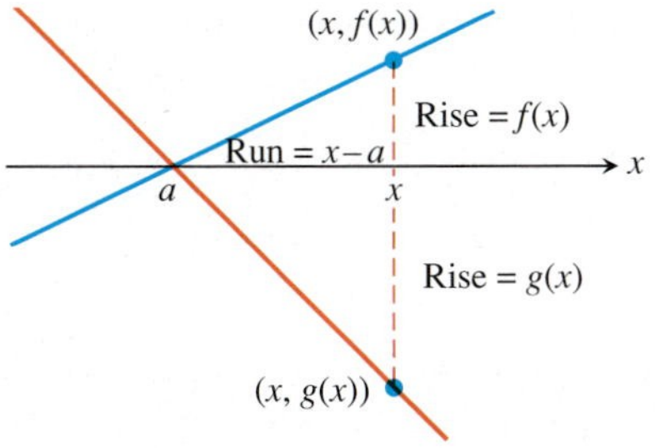

FIGURE 7.11 A zoom-in view of the graphs of the differentiable functions f and g at $x = a$. (Theorem 1)

Proof of Theorem 1

Graphical Argument

If we zoom in on the graphs of f and g at $(a, f(a)) = (a, g(a)) = (a, 0)$, the graphs (Figure 7.11) appear to be straight lines because differentiable functions are locally linear. Let m_1 and m_2 be the slopes of the lines for f and g, respectively. Then for x near a,

$$\frac{f(x)}{g(x)} = \frac{\dfrac{f(x)}{x-a}}{\dfrac{g(x)}{x-a}} = \frac{m_1}{m_2}.$$

As $x \to a$, m_1 and m_2 approach $f'(a)$ and $g'(a)$, respectively. Therefore,

$$\lim_{x\to a} \frac{f(x)}{g(x)} = \lim_{x\to a} \frac{m_1}{m_2} = \frac{f'(a)}{g'(a)}.$$

Confirm Analytically

Working backwards from $f'(a)$ and $g'(a)$, which are themselves limits, we have

$$\frac{f'(a)}{g'(a)} = \frac{\lim\limits_{x\to a} \dfrac{f(x)-f(a)}{x-a}}{\lim\limits_{x\to a} \dfrac{g(x)-g(a)}{x-a}} = \lim_{x\to a} \frac{\dfrac{f(x)-f(a)}{x-a}}{\dfrac{g(x)-g(a)}{x-a}}$$

$$= \lim_{x\to a} \frac{f(x)-f(a)}{g(x)-g(a)} = \lim_{x\to a} \frac{f(x)-0}{g(x)-0} = \lim_{x\to a} \frac{f(x)}{g(x)}.$$

Sometimes after differentiation, the new numerator and denominator both equal zero at $x = a$, as we see in Example 2. In these cases, we apply a stronger form of l'Hôpital's Rule.

Theorem 2 L'Hôpital's Rule (Stronger Form)

Suppose that $f(a) = g(a) = 0$, that f and g are differentiable on an open interval I containing a, and that $g'(x) \neq 0$ on I if $x \neq a$. Then

$$\lim_{x\to a} \frac{f(x)}{g(x)} = \lim_{x\to a} \frac{f'(x)}{g'(x)},$$

assuming that the limit on the right side exists.

A proof of the finite-limit case of Theorem 2 is given in Appendix 6.

Example 2 Applying Stronger Form of L'Hôpital's Rule

$$\lim_{x\to 0} \frac{\sqrt{1+x} - 1 - x/2}{x^2} \qquad \frac{0}{0}$$

$$= \lim_{x\to 0} \frac{(1/2)(1+x)^{-1/2} - 1/2}{2x} \qquad \text{Still } \frac{0}{0}\text{; differentiate again.}$$

$$= \lim_{x\to 0} \frac{-(1/4)(1+x)^{-3/2}}{2} = -\frac{1}{8} \qquad \text{Not } \frac{0}{0}\text{; limit is found.}$$

When you apply l'Hôpital's Rule, look for a change from 0/0 to something else. This is where the limit is revealed.

Example 3 Incorrectly Applying Stronger Form of L'Hôpital's Rule

$$\lim_{x\to 0} \frac{1-\cos x}{x+x^2} \qquad \frac{0}{0}$$

$$= \lim_{x\to 0} \frac{\sin x}{1+2x} = \frac{0}{1} = 0 \qquad \text{Not } \frac{0}{0}\text{; limit is found.}$$

If we continue to differentiate in an attempt to apply l'Hôpital's Rule once more, we get

$$\lim_{x\to 0} \frac{1-\cos x}{x+x^2} = \lim_{x\to 0} \frac{\sin x}{1+2x} = \lim_{x\to 0} \frac{\cos x}{2} = \frac{1}{2},$$

which is wrong.

L'Hôpital's Rule applies to one-sided limits as well.

Example 4 Using L'Hôpital's Rule with One-Sided Limits

Recall that ∞ and $+\infty$ mean the same thing.

(a) $$\lim_{x\to 0^+} \frac{\sin x}{x^2} \qquad \frac{0}{0}$$

$$= \lim_{x\to 0^+} \frac{\cos x}{2x} = \infty \qquad \frac{1}{0}$$

(b) $$\lim_{x\to 0^-} \frac{\sin x}{x^2} \qquad \frac{0}{0}$$

$$= \lim_{x\to 0^-} \frac{\cos x}{2x} = -\infty \qquad -\frac{1}{0}$$

When we reach a point where one of the derivatives approaches 0, as in Example 4, and the other does not, then the limit in question is 0 (if the numerator approaches 0) or infinity (if the denominator approaches 0).

Indeterminate Forms ∞/∞, $\infty \cdot 0$, $\infty - \infty$

A version of l'Hôpital's Rule also applies to quotients that lead to the indeterminant form ∞/∞. If $f(x)$ and $g(x)$ both approach infinity as $x \to a$, then

$$\lim_{x\to a} \frac{f(x)}{g(x)} = \lim_{x\to a} \frac{f'(x)}{g'(x)},$$

provided the latter limit exists. The a here (and in the indeterminate form 0/0) may itself be finite or infinite and may be an endpoint of the interval I of Theorem 2.

Example 5 Working with the Indeterminate Form ∞/∞

Find

(a) $\displaystyle\lim_{x\to \pi/2} \frac{\sec x}{1+\tan x}$

(b) $\displaystyle\lim_{x\to \infty} \frac{\ln x}{2\sqrt{x}}$.

Solution

(a) The numerator and denominator are discontinuous at $x = \pi/2$, so we investigate the one-sided limits there. To apply l'Hôpital's Rule, we can choose I to be any open interval with $x = \pi/2$ as an endpoint.

$$\lim_{x\to(\pi/2)^-} \frac{\sec x}{1 + \tan x} \qquad \frac{\infty}{\infty} \text{ from the left}$$

$$= \lim_{x\to(\pi/2)^-} \frac{\sec x \tan x}{\sec^2 x} = \lim_{x\to(\pi/2)^-} \sin x = 1$$

The right-hand limit is 1 also, with $(-\infty)/(-\infty)$ as the indeterminate form. Therefore, the two-sided limit is equal to 1.

(b) $\lim_{x\to\infty} \dfrac{\ln x}{2\sqrt{x}} = \lim_{x\to\infty} \dfrac{1/x}{1/\sqrt{x}} = \lim_{x\to\infty} \dfrac{1}{\sqrt{x}} = 0$

We can sometimes handle the indeterminate forms $\infty \cdot 0$ and $\infty - \infty$ by using algebra to get $0/0$ or ∞/∞ instead. Here again we do not mean to suggest that there is a number $\infty \cdot 0$ or $\infty - \infty$ any more than we mean to suggest that there is a number $0/0$ or ∞/∞. These forms are not numbers but are descriptions of function behavior.

Example 6 Working with the Indeterminate Form $\infty \cdot 0$

Find

(a) $\lim_{x\to\infty} \left(x \sin \dfrac{1}{x}\right)$

(b) $\lim_{x\to-\infty} \left(x \sin \dfrac{1}{x}\right)$.

Solution

(a)

$$\lim_{x\to\infty} \left(x \sin \frac{1}{x}\right) \qquad \infty \cdot 0$$

$$= \lim_{h\to0^+} \left(\frac{1}{h} \sin h\right) \qquad \text{Let } h = 1/x.$$

$$= 1$$

(b) Similarly,

$$\lim_{x\to-\infty} \left(x \sin \frac{1}{x}\right) = 1.$$

Example 7 Working with the Indeterminate Form $\infty - \infty$

Find

$$\lim_{x\to0} \left(\frac{1}{\sin x} - \frac{1}{x}\right).$$

Solution If $x \to 0^+$, then $\sin x \to 0^+$ and

$$\frac{1}{\sin x} - \frac{1}{x} \to \infty - \infty.$$

Similarly, if $x \to 0^-$, then $\sin x \to 0^-$ and

$$\frac{1}{\sin x} - \frac{1}{x} \to -\infty - (-\infty) = -\infty + \infty.$$

Neither form reveals what happens in the limit. To find out, we first combine the fractions:

$$\frac{1}{\sin x} - \frac{1}{x} = \frac{x - \sin x}{x \sin x} \qquad \text{Common denominator is } x \sin x.$$

Then apply l'Hôpital's Rule to the result:

$$\begin{aligned}
\lim_{x\to 0}\left(\frac{1}{\sin x} - \frac{1}{x}\right) &= \lim_{x\to 0}\frac{x - \sin x}{x \sin x} && \frac{0}{0}\\
&= \lim_{x\to 0}\frac{1 - \cos x}{\sin x + x\cos x} && \text{Still } \frac{0}{0}\\
&= \lim_{x\to 0}\frac{\sin x}{2\cos x - x\sin x} = \frac{0}{2} = 0.
\end{aligned}$$

Indeterminate Forms 1^∞, 0^0, ∞^0

Limits that lead to the indeterminate forms 1^∞, 0^0, and ∞^0 can sometimes be handled by taking logarithms first. We use l'Hôpital's Rule to find the limit of the logarithm and then exponentiate to reveal the original function's behavior.

Since $b = e^{\ln b}$ for every positive number b, we can write $f(x)$ as

$$f(x) = e^{\ln f(x)}$$

for any positive function $f(x)$.

$$\lim_{x\to a} \ln f(x) = L \quad \Rightarrow \quad \lim_{x\to a} f(x) = \lim_{x\to a} e^{\ln f(x)} = e^{\lim_{x\to a} \ln f(x)} = e^L$$

Here a can be finite or infinite.

In Preliminary Section 3, we used graphs and tables to investigate the values of $f(x) = (1 + 1/x)^x$ as $x \to \infty$. Now we find this limit with l'Hôpital's Rule.

CD-ROM
WEBsite
Historical Biography
Augustin-Louis Cauchy (1789 — 1857)

Example 8 Working with the Indeterminate Form 1^∞

Find

$$\lim_{x\to\infty}\left(1 + \frac{1}{x}\right)^x.$$

Solution Let $f(x) = (1 + 1/x)^x$. Then taking logarithms of both sides converts the indeterminate form 1^∞ to $0/0$, to which we can apply l'Hôpital's Rule.

$$\ln f(x) = \ln\left(1 + \frac{1}{x}\right)^x = x \ln\left(1 + \frac{1}{x}\right) = \frac{\ln\left(1 + \frac{1}{x}\right)}{\frac{1}{x}}$$

We apply l'Hôpital's Rule to the last expression above.

$$\lim_{x\to\infty} \ln f(x) = \lim_{x\to\infty} \frac{\ln\left(1+\dfrac{1}{x}\right)}{\dfrac{1}{x}} \qquad \frac{0}{0}$$

$$= \lim_{x\to\infty} \frac{\dfrac{1}{1+\dfrac{1}{x}}\left(-\dfrac{1}{x^2}\right)}{-\dfrac{1}{x^2}} \qquad \text{Differentiate numerator and denominator.}$$

$$= \lim_{x\to\infty} \frac{1}{1+\dfrac{1}{x}} = 1$$

Therefore,

$$\lim_{x\to\infty}\left(1+\frac{1}{x}\right)^x = \lim_{x\to\infty} f(x) = \lim_{x\to\infty} e^{\ln f(x)} = e^1 = e.$$

Example 9 Working with the Indeterminate Form 0^0

Determine whether $\lim_{x\to 0^+} x^x$ exists and find its value if it does.

Solution

The limit leads to the indeterminate form 0^0. To convert the problem to one involving $0/0$, we let $f(x) = x^x$ and take the logarithm of both sides.

$$\ln f(x) = x \ln x = \frac{\ln x}{1/x}$$

Applying l'Hôpital's Rule to $(\ln x)/(1/x)$, we obtain

$$\lim_{x\to 0^+} \ln f(x) = \lim_{x\to 0^+} \frac{\ln x}{1/x} \qquad \frac{-\infty}{\infty}$$

$$= \lim_{x\to 0^+} \frac{1/x}{-1/x^2} \qquad \text{Differentiate.}$$

$$= \lim_{x\to 0^+} (-x) = 0.$$

Therefore,

$$\lim_{x\to 0^+} x^x = \lim_{x\to 0^+} f(x) = \lim_{x\to 0^+} e^{\ln f(x)} = e^0 = 1.$$

Example 10 Working with the Indeterminate Form ∞^0

Find $\lim_{x\to\infty} x^{1/x}$.

Solution

Let $f(x) = x^{1/x}$. Then

$$\ln f(x) = \frac{\ln x}{x}.$$

Applying l'Hôpital's Rule to $\ln f(x)$, we obtain

$$\lim_{x\to\infty} \ln f(x) = \lim_{x\to\infty} \frac{\ln x}{x} \qquad \frac{\infty}{\infty}$$
$$= \lim_{x\to\infty} \frac{1/x}{1} \qquad \text{Differentiate.}$$
$$= \lim_{x\to\infty} \frac{1}{x} = 0.$$

Therefore,

$$\lim_{x\to\infty} x^{1/x} = \lim_{x\to\infty} f(x) = \lim_{x\to\infty} e^{\ln f(x)} = e^0 = 1.$$

EXERCISES 7.6

Finding Limits

In Exercises 1–6, use l'Hôpital's Rule to evaluate the limit. Then evaluate the limit using a method studied in Chapter 1.

1. $\lim_{x\to 2} \dfrac{x-2}{x^2-4}$

2. $\lim_{x\to 0} \dfrac{\sin 5x}{x}$

3. $\lim_{x\to\infty} \dfrac{5x^2-3x}{7x^2+1}$

4. $\lim_{x\to 1} \dfrac{x^3-1}{4x^3-x-3}$

5. $\lim_{x\to 0} \dfrac{1-\cos x}{x^2}$

6. $\lim_{x\to\infty} \dfrac{2x^2+3x}{x^3+x+1}$

Applying l'Hôpital's Rule

Use l'Hôpital's Rule to find the limits in Exercises 7–38.

7. $\lim_{\theta\to 0} \dfrac{\sin \theta^2}{\theta}$

8. $\lim_{\theta\to\pi/2} \dfrac{1-\sin\theta}{1+\cos 2\theta}$

9. $\lim_{t\to 0} \dfrac{\cos t - 1}{e^t - t - 1}$

10. $\lim_{t\to 1} \dfrac{t-1}{\ln t - \sin \pi t}$

11. $\lim_{x\to\infty} \dfrac{\ln(x+1)}{\log_2 x}$

12. $\lim_{x\to\infty} \dfrac{\log_2 x}{\log_3(x+3)}$

13. $\lim_{y\to 0^+} \dfrac{\ln(y^2+2y)}{\ln y}$

14. $\lim_{y\to\pi/2} \left(\dfrac{\pi}{2}-y\right)\tan y$

15. $\lim_{x\to 0^+} x\ln x$

16. $\lim_{x\to\infty} x\tan\dfrac{1}{x}$

17. $\lim_{x\to 0^+} (\csc x - \cot x + \cos x)$

18. $\lim_{x\to\infty} (\ln 2x - \ln(x+1))$

19. $\lim_{x\to 0^+} (\ln x - \ln \sin x)$

20. $\lim_{x\to 0^+} \left(\dfrac{1}{x}-\dfrac{1}{\sqrt{x}}\right)$

21. $\lim_{x\to 0} (e^x+x)^{1/x}$

22. $\lim_{x\to 0} \left(\dfrac{1}{x^2}\right)^x$

23. $\lim_{x\to\pm\infty} \dfrac{3x-5}{2x^2-x+2}$

24. $\lim_{x\to 0} \dfrac{\sin 7x}{\tan 11x}$

25. $\lim_{x\to\infty} (\ln x)^{1/x}$

26. $\lim_{x\to\infty} (1+2x)^{1/(2\ln x)}$

27. $\lim_{x\to 1} (x^2-2x+1)^{x-1}$

28. $\lim_{x\to(\pi/2)^-} (\cos x)^{\cos x}$

29. $\lim_{x\to 0^+} (1+x)^{1/x}$

30. $\lim_{x\to 1} x^{1/(x-1)}$

31. $\lim_{x\to 0^+} (\sin x)^x$

32. $\lim_{x\to 0^+} (\sin x)^{\tan x}$

33. $\lim_{x\to 1^+} x^{1/(1-x)}$

34. $\lim_{x\to\infty} x^2 e^{-x}$

35. $\lim_{x\to\infty} \displaystyle\int_x^{2x} \frac{1}{t}\,dt$

36. $\lim_{x\to\infty} \dfrac{1}{x\ln x}\displaystyle\int_1^x \ln t\,dt$

37. $\lim_{\theta\to 0} \dfrac{\cos\theta - 1}{e^\theta - \theta - 1}$

38. $\lim_{t\to\infty} \dfrac{e^t+t^2}{e^t-t}$

Theory and Applications

L'Hôpital's Rule does not help with the limits in Exercises 39–42. Try it; you just keep on cycling. Find the limits some other way.

39. $\lim_{x\to\infty} \dfrac{\sqrt{9x+1}}{\sqrt{x+1}}$

40. $\lim_{x\to 0^+} \dfrac{\sqrt{x}}{\sqrt{\sin x}}$

41. $\lim_{x\to(\pi/2)^-} \dfrac{\sec x}{\tan x}$

42. $\lim_{x\to 0^+} \dfrac{\cot x}{\csc x}$

43. *Writing to Learn* Which one is correct, and which one is wrong? Give reasons for your answers.

(a) $\lim_{x\to 3} \dfrac{x-3}{x^2-3} = \lim_{x\to 3} \dfrac{1}{2x} = \dfrac{1}{6}$

(b) $\lim_{x\to 3} \dfrac{x-3}{x^2-3} = \dfrac{0}{6} = 0$

44. *∞/∞ form* Give an example of two differentiable functions f and g with $\lim_{x\to\infty} f(x) = \lim_{x\to\infty} g(x) = \infty$ that satisfy the following.

(a) $\displaystyle\lim_{x\to\infty} \frac{f(x)}{g(x)} = 3$ **(b)** $\displaystyle\lim_{x\to\infty} \frac{f(x)}{g(x)} = 0$

(c) $\displaystyle\lim_{x\to\infty} \frac{f(x)}{g(x)} = \infty$

45. *Writing to Learn: Continuous extension* Find a value of c that makes the function

$$f(x) = \begin{cases} \dfrac{9x - 3\sin 3x}{5x^3}, & x \neq 0 \\ c, & x = 0 \end{cases}$$

continuous at $x = 0$. Explain why your value of c works.

46. *L'Hôpital's Rule* Let

$$f(x) = \begin{cases} x + 2, & x \neq 0 \\ 0, & x = 0 \end{cases} \quad \text{and} \quad g(x) = \begin{cases} x + 1, & x \neq 0 \\ 0, & x = 0. \end{cases}$$

(a) Show that

$$\lim_{x\to 0} \frac{f'(x)}{g'(x)} = 1 \quad \text{but} \quad \lim_{x\to 0} \frac{f(x)}{g(x)} = 2.$$

(b) *Writing to Learn* Explain why this does not contradict l'Hôpital's Rule.

47. *Interest compounded continuously*

(a) Show that

$$\lim_{k\to\infty} A_0\left(1 + \frac{r}{k}\right)^{kt} = A_0 e^{rt}.$$

(b) *Writing to Learn* Explain how the limit in part (a) connects interest compounded k times per year with interest compounded continuously.

T **48.** *0/0 form* Estimate the value of

$$\lim_{x\to 1} \frac{2x^2 - (3x + 1)\sqrt{x} + 2}{x - 1}$$

by graphing. Then confirm your estimate with l'Hôpital's Rule.

T **49.** *0/0 form*

(a) Estimate the value of

$$\lim_{x\to 1} \frac{(x - 1)^2}{x \ln x - x - \cos \pi x}$$

by graphing $f(x) = (x - 1)^2/(x \ln x - x - \cos \pi x)$ near $x = 1$. Then confirm your estimate with l'Hôpital's Rule.

(b) Graph f for $0 < x \leq 11$.

50. *Why 0^∞ and $0^{-\infty}$ are not indeterminate forms* Assume that $f(x)$ is nonnegative in an open interval containing c and $\lim_{x\to c} f(x) = 0$.

(a) If $\lim_{x\to c} g(x) = \infty$, show that $\lim_{x\to c} f(x)^{g(x)} = 0$.

(b) If $\lim_{x\to c} g(x) = -\infty$, show that $\lim_{x\to c} f(x)^{g(x)} = \infty$.

T **51.** *Grapher precision* Let

$$f(x) = \frac{1 - \cos x^6}{x^{12}}.$$

Explain why some graphs of f may give false information about $\lim_{x\to 0} f(x)$. (*Hint:* Try the window $[-1, 1]$ by $[-0.5, 1]$.)

T **52.** *$\infty - \infty$ form*

(a) Estimate the value of

$$\lim_{x\to\infty} (x - \sqrt{x^2 + x})$$

by graphing $f(x) = x - \sqrt{x^2 + x}$ over a suitably large interval of x-values.

(b) Now confirm your estimate by finding the limit with l'Hôpital's Rule. As the first step, multiply $f(x)$ by the fraction $(x + \sqrt{x^2 + x})/(x + \sqrt{x^2 + x})$ and simplify the new numerator.

53. *Exponential functions*

(a) Use the equation

$$a^x = e^{x \ln a}$$

to find the domain of

$$f(x) = \left(1 + \frac{1}{x}\right)^x.$$

(b) Find $\lim_{x\to -1^-} f(x)$.

(c) Find $\lim_{x\to -\infty} f(x)$.

54. *Extended exponentials* Given that $x > 0$, find the maximum value, if any, of

(a) $x^{1/x}$ **(b)** x^{1/x^2}

(c) x^{1/x^n} (n a positive integer)

(d) Show that $\lim_{x\to\infty} x^{1/x^n} = 1$ for every positive integer n.

T **55.** *The place of $\ln x$ among the powers of x* The natural logarithm

$$\ln x = \int_1^x \frac{1}{t}\, dt$$

fills the gap in the set of formulas

$$\int t^{k-1}\, dt = \frac{t^k}{k} + C, \qquad k \neq 0,$$

but the formulas themselves do not reveal how well the logarithm fits in. We can see the nice fit graphically if we select the specific antiderivatives

$$\int_1^x t^{k-1}\,dt = \frac{x^k - 1}{k}, \qquad x > 0,$$

and compare their graphs with the graph of $\ln x$.

(a) Graph the functions $f(x) = (x^k - 1)/k$ together with $\ln x$ on the interval $0 \le x \le 50$ for $k = \pm 1, \pm 0.5, \pm 0.1$, and ± 0.05.

(b) Show that

$$\lim_{k \to 0} \frac{x^k - 1}{k} = \ln x.$$

(*Source:* "The Place of ln x Among the Powers of x" by Henry C. Finlayson, *American Mathematical Monthly,* Vol. 94, No. 5 (May 1987), p. 450.)

T **56.** *The continuous extension of* $(\sin x)^x$ *to* $[0, \pi]$

(a) Graph $f(x) = (\sin x)^x$ on the interval $0 \le x \le \pi$. What value would you assign to f to make it continuous at $x = 0$?

(b) Verify your conclusion in part (a) by finding $\lim_{x \to 0^+} f(x)$ with l'Hôpital's Rule.

(c) Returning to the graph, estimate the maximum value of f on $[0, \pi]$. About where is max f taken on?

(d) Sharpen your estimate in part (c) by graphing f' in the same window to see where its graph crosses the x-axis. To simplify your work, you might want to delete the exponential factor from the expression for f' and graph just the factor that has a zero.

(e) Sharpen your estimate of the location of max f further still by solving the equation $f' = 0$ numerically.

(f) Estimate max f by evaluating f at the locations you found in parts (c), (d), and (e). What is your best value for max f?

7.7 Improper Integrals

Infinite Limits of Integration • The Integral $\int_1^\infty dx/x^p$ • Integrands with Infinite Discontinuities • Tests for Convergence and Divergence • Computer Algebra Systems

CD-ROM WEBsite

Up to now, we have required our definite integrals to have two properties: first, that the domain of integration, from a to b, be finite, and second, that the range of the integrand be finite on this domain. In practice, however, we frequently encounter problems that fail to meet one or both of these conditions. As an example of an infinite domain, we might consider the area under the curve $y = (\ln x)/x^2$ from $x = 1$ to $x = \infty$ (Figure 7.12a). As an example of an infinite range, we might

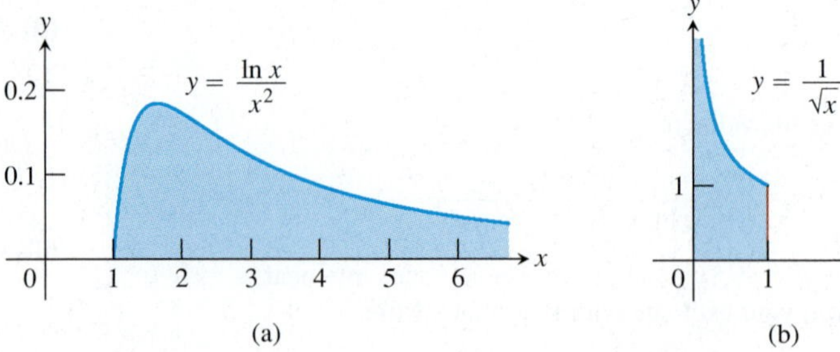

FIGURE 7.12 Are the areas under these infinite curves finite?

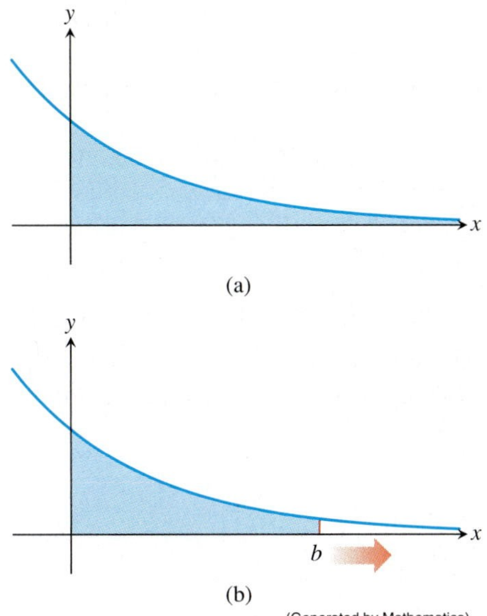

FIGURE 7.13 (a) The area in the first quadrant under the curve $y = e^{-x/2}$ is (b)

$$\lim_{b\to\infty} \int_0^b e^{-x/2}\,dx.$$

consider the area under the curve $y = 1/\sqrt{x}$ between $x = 0$ and $x = 1$ (Figure 7.12b). We treat both examples in the same reasonable way. We ask, "What is the integral when the domain is slightly less?" and examine the answer as the domain increases to the limit. We do the finite case and then see what happens as we approach infinity.

Infinite Limits of Integration

Consider the infinite region that lies under the curve $y = e^{-x/2}$ in the first quadrant (Figure 7.13a). You might think this region has infinite area, but we will see that the natural value to assign is finite. Here is how we assign a value to the area. First we find the area $A(b)$ of the portion of the region that is bounded on the right by $x = b$ (Figure 7.13b).

$$A(b) = \int_0^b e^{-x/2}\,dx = -2e^{-x/2}\Big]_0^b = -2e^{-b/2} + 2$$

Then we find the limit of $A(b)$ as $b \to \infty$.

$$\lim_{b\to\infty} A(b) = \lim_{b\to\infty} (-2e^{-b/2} + 2) = 2$$

The value we assign to the area under the curve from 0 to ∞ is

$$\int_0^\infty e^{-x/2}\,dx = \lim_{b\to\infty} \int_0^b e^{-x/2}\,dx = 2.$$

Definition Improper Integrals with Infinite Integration Limits

Integrals with infinite limits of integration are **improper integrals.**

1. If $f(x)$ is continuous on $[a, \infty)$, then

$$\int_a^\infty f(x)\,dx = \lim_{b\to\infty} \int_a^b f(x)\,dx.$$

2. If $f(x)$ is continuous on $(-\infty, b]$, then

$$\int_{-\infty}^b f(x)\,dx = \lim_{a\to-\infty} \int_a^b f(x)\,dx.$$

3. If $f(x)$ is continuous on $(-\infty, \infty)$, then

$$\int_{-\infty}^\infty f(x)\,dx = \int_{-\infty}^c f(x)\,dx + \int_c^\infty f(x)\,dx,$$

where c is any real number.

In parts 1 and 2, if the limit is finite, the improper integral **converges** and the limit is the **value** of the improper integral. If the limit fails to exist, the improper integral **diverges.** In part 3, the integral on the left-hand side of the equation

converges if both improper integrals on the right-hand side converge; otherwise it **diverges** and has no value. It can be shown that the choice of c in part 3 is unimportant. We can evaluate or determine the convergence or divergence of $\int_{-\infty}^{\infty} f(x)\,dx$ with any convenient choice.

Example 1 Evaluating an Improper Integral on $[1, \infty)$

Is the area under the curve $y = (\ln x)/x^2$ from $x = 1$ to $x = \infty$ finite? If so, what is it?

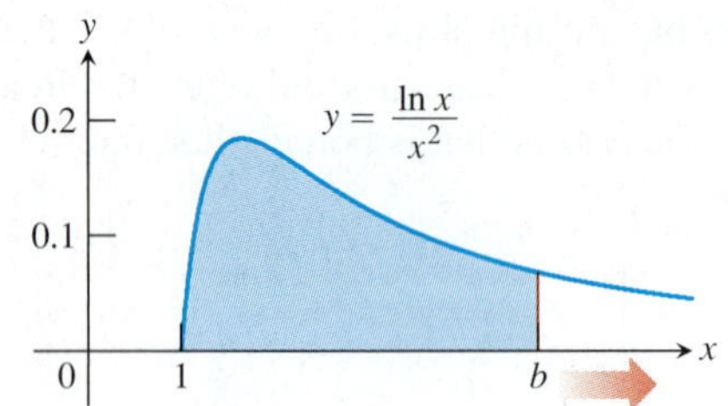

FIGURE 7.14 The area under this curve is

$$\lim_{b\to\infty}\int_1^b \frac{\ln x}{x^2}\,dx.$$

(Example 1)

Solution We find the area under the curve from $x = 1$ to $x = b$ and examine the limit as $b \to \infty$. If the limit is finite, we take it to be the area under the infinite curve (Figure 7.14). The area from 1 to b is

$$\int_1^b \frac{\ln x}{x^2}\,dx = \left[(\ln x)\left(-\frac{1}{x}\right)\right]_1^b - \int_1^b \left(-\frac{1}{x}\right)\left(\frac{1}{x}\right)dx$$

Integration by parts with $u = \ln x$, $dv = dx/x^2$, $du = dx/x$, $v = -1/x$

$$= -\frac{\ln b}{b} - \left[\frac{1}{x}\right]_1^b$$

$$= -\frac{\ln b}{b} - \frac{1}{b} + 1.$$

The limit of the area as $b \to \infty$ is

$$\int_1^\infty \frac{\ln x}{x^2}\,dx = \lim_{b\to\infty}\int_1^b \frac{\ln x}{x^2}\,dx$$

$$= \lim_{b\to\infty}\left[-\frac{\ln b}{b} - \frac{1}{b} + 1\right]$$

$$= -\left[\lim_{b\to\infty}\frac{\ln b}{b}\right] - 0 + 1$$

$$= -\left[\lim_{b\to\infty}\frac{1/b}{1}\right] + 1 = 0 + 1 = 1.$$

l'Hôpital's Rule

Thus, the improper integral converges and the area has finite value 1.

Example 2 Evaluating an Integral on $(-\infty, \infty)$

Evaluate

$$\int_{-\infty}^{\infty} \frac{dx}{1+x^2}.$$

Solution According to the definition (part 3), we can write

$$\int_{-\infty}^{\infty} \frac{dx}{1+x^2} = \int_{-\infty}^{0} \frac{dx}{1+x^2} + \int_{0}^{\infty} \frac{dx}{1+x^2}.$$

CD-ROM
WEBsite
Historical Biography
Lejeune Dirichlet
(1805 — 1859)

Next we evaluate each improper integral on the right-hand side of the equation above.

$$\int_{-\infty}^{0} \frac{dx}{1+x^2} = \lim_{a\to-\infty} \int_{a}^{0} \frac{dx}{1+x^2}$$

$$= \lim_{a\to-\infty} \tan^{-1} x\Big]_{a}^{0}$$

$$= \lim_{a\to-\infty} (\tan^{-1} 0 - \tan^{-1} a) = 0 - \left(-\frac{\pi}{2}\right) = \frac{\pi}{2}$$

$$\int_{0}^{\infty} \frac{dx}{1+x^2} = \lim_{b\to\infty} \int_{0}^{b} \frac{dx}{1+x^2}$$

$$= \lim_{b\to\infty} \tan^{-1} x\Big]_{0}^{b}$$

$$= \lim_{b\to\infty} (\tan^{-1} b - \tan^{-1} 0) = \frac{\pi}{2} - 0 = \frac{\pi}{2}$$

Thus,

$$\int_{-\infty}^{\infty} \frac{dx}{1+x^2} = \frac{\pi}{2} + \frac{\pi}{2} = \pi.$$

The Integral $\int_{1}^{\infty} \frac{dx}{x^p}$

The function $y = 1/x$ is the boundary between the convergent and divergent improper integrals with integrands of the form $y = 1/x^p$. Example 3 explains.

Example 3 Determining Convergence

For what values of p does the integral $\int_1^\infty dx/x^p$ converge? When the integral does converge, what is its value?

Solution If $p \neq 1$,

$$\int_{1}^{b} \frac{dx}{x^p} = \frac{x^{-p+1}}{-p+1}\Big]_{1}^{b} = \frac{1}{1-p}(b^{-p+1} - 1) = \frac{1}{1-p}\left(\frac{1}{b^{p-1}} - 1\right).$$

Thus,

$$\int_{1}^{\infty} \frac{dx}{x^p} = \lim_{b\to\infty} \int_{1}^{b} \frac{dx}{x^p}$$

$$= \lim_{b\to\infty} \left[\frac{1}{1-p}\left(\frac{1}{b^{p-1}} - 1\right)\right] = \begin{cases} \dfrac{1}{p-1}, & p > 1 \\ \infty, & p < 1 \end{cases}$$

because

$$\lim_{b\to\infty} \frac{1}{b^{p-1}} = \begin{cases} 0, & p > 1 \\ \infty, & p < 1. \end{cases}$$

Therefore, the integral converges to the value $1/(p-1)$ if $p > 1$ and it diverges if $p < 1$.

If $p = 1$,

$$\begin{aligned}\int_1^\infty \frac{dx}{x^p} &= \int_1^\infty \frac{dx}{x}\\ &= \lim_{b\to\infty} \int_1^b \frac{dx}{x}\\ &= \lim_{b\to\infty} \ln x\Big]_1^b\\ &= \lim_{b\to\infty} (\ln b - \ln 1) = \infty\end{aligned}$$

and the integral diverges.

Integrands with Infinite Discontinuities

Another type of improper integral arises when the integrand has a vertical asymptote—an infinite discontinuity—at a limit of integration or at some point between the limits of integration.

Consider the infinite region in the first quadrant that lies under the curve $y = 1/\sqrt{x}$ from $x = 0$ to $x = 1$ (Figure 7.12b). First we find the area of the portion from a to 1 (Figure 7.15).

$$\int_a^1 \frac{dx}{\sqrt{x}} = 2\sqrt{x}\Big]_a^1 = 2 - 2\sqrt{a}$$

Then we find the limit of this area as $a \to 0^+$.

$$\lim_{a\to 0^+} \int_a^1 \frac{dx}{\sqrt{x}} = \lim_{a\to 0^+} (2 - 2\sqrt{a}) = 2$$

The area under the curve from 0 to 1 is

$$\int_0^1 \frac{dx}{\sqrt{x}} = \lim_{a\to 0^+} \int_a^1 \frac{dx}{\sqrt{x}} = 2.$$

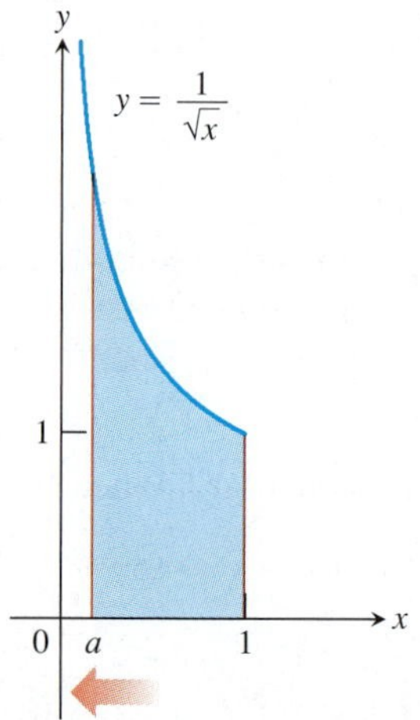

FIGURE 7.15 The area under this curve is $\lim_{a\to 0^+} \int_a^1 \left(\frac{1}{\sqrt{x}}\right) dx$.

Definition Improper Integrals with Infinite Discontinuities

Integrals of functions that become infinite at a point within the interval of integration are **improper integrals.**

1. If $f(x)$ is continuous on $(a, b]$, then
$$\int_a^b f(x)\,dx = \lim_{c\to a^+} \int_c^b f(x)\,dx.$$

2. If $f(x)$ is continuous on $[a, b)$, then
$$\int_a^b f(x)\,dx = \lim_{c\to b^-} \int_a^c f(x)\,dx.$$

3. If $f(x)$ is continuous on $[a, c) \cup (c, b]$, then

$$\int_a^b f(x)\,dx = \int_a^c f(x)\,dx + \int_c^b f(x)\,dx.$$

In parts 1 and 2, if the limit is finite, the improper integral **converges** and the limit is the **value** of the improper integral. If the limit fails to exist, the improper integral **diverges.** In part 3, the integral on the left-hand side of the equation **converges** if both integrals on the right-hand side have values; otherwise it **diverges.**

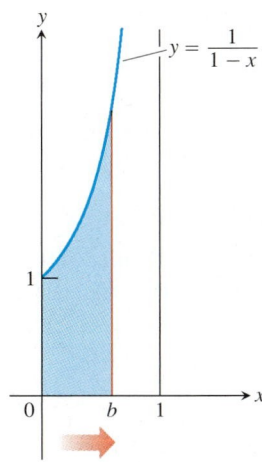

FIGURE 7.16 If the limit exists,

$$\int_0^1 \left(\frac{1}{1-x}\right) dx = \lim_{b\to 1^-} \int_0^b \frac{1}{1-x}\,dx.$$

(Example 4)

Example 4 A Divergent Improper Integral

Investigate the convergence of

$$\int_0^1 \frac{1}{1-x}\,dx.$$

Solution The integrand $f(x) = 1/(1-x)$ is continuous on $[0, 1)$ but becomes infinite as $x \to 1^-$ (Figure 7.16). We evaluate the integral as

$$\begin{aligned}\lim_{b\to 1^-} \int_0^b \frac{1}{1-x}\,dx &= \lim_{b\to 1^-} \left[-\ln|1-x|\right]_0^b \\ &= \lim_{b\to 1^-} \left[-\ln(1-b) + 0\right] = \infty.\end{aligned}$$

The limit is infinite, so the integral diverges.

Example 5 Infinite Discontinuity at an Interior Point

Evaluate

$$\int_0^3 \frac{dx}{(x-1)^{2/3}}.$$

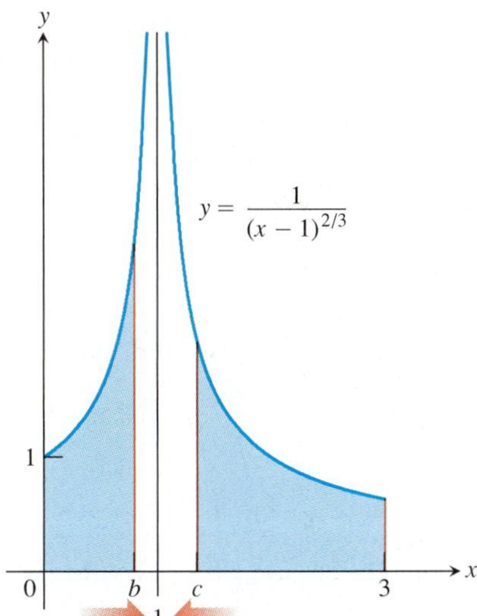

FIGURE 7.17 Example 5 investigates the convergence of

$$\int_0^3 \frac{1}{(x-1)^{2/3}}\,dx.$$

Solution The integrand has a vertical asymptote at $x = 1$ and is continuous on $[0, 1)$ and $(1, 3]$ (Figure 7.17). Thus, by part 3 of the definition above,

$$\int_0^3 \frac{dx}{(x-1)^{2/3}} = \int_0^1 \frac{dx}{(x-1)^{2/3}} + \int_1^3 \frac{dx}{(x-1)^{2/3}}.$$

Next, we evaluate each improper integral on the right-hand side of this equation.

$$\begin{aligned}\int_0^1 \frac{dx}{(x-1)^{2/3}} &= \lim_{b\to 1^-} \int_0^b \frac{dx}{(x-1)^{2/3}} \\ &= \lim_{b\to 1^-} 3(x-1)^{1/3}\Big]_0^b \\ &= \lim_{b\to 1^-} \left[3(b-1)^{1/3} + 3\right] = 3\end{aligned}$$

$$\begin{aligned}\int_1^3 \frac{dx}{(x-1)^{2/3}} &= \lim_{c\to 1^+} \int_c^3 \frac{dx}{(x-1)^{2/3}} \\ &= \lim_{c\to 1^+} 3(x-1)^{1/3}\Big]_c^3 \\ &= \lim_{c\to 1^+} \left[3(3-1)^{1/3} - 3(c-1)^{1/3}\right] = 3\sqrt[3]{2}\end{aligned}$$

We conclude that

$$\int_0^3 \frac{dx}{(x-1)^{2/3}} = 3 + 3\sqrt[3]{2}.$$

Example 6 Finding the Volume of an Infinite Solid

The cross sections of the solid horn in Figure 7.18 perpendicular to the x-axis are circular disks with diameters reaching from the x-axis to the curve $y = e^x$, $-\infty < x \leq \ln 2$. Find the volume of the horn.

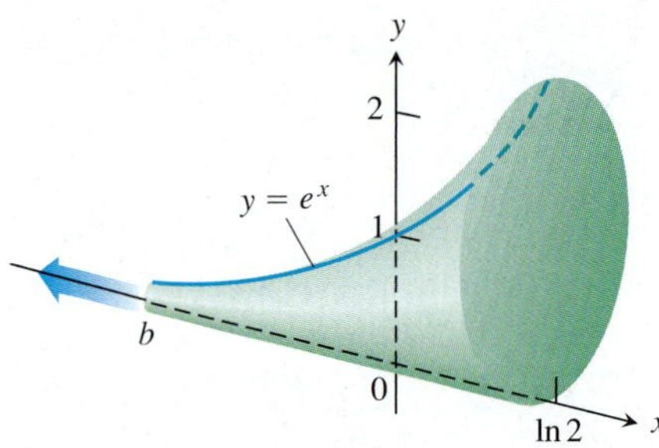

FIGURE 7.18 The calculation in Example 6 shows that this infinite horn has a finite volume.

Solution The area of a typical cross section is

$$A(x) = \pi(\text{radius})^2 = \pi\left(\frac{1}{2}y\right)^2 = \frac{\pi}{4}e^{2x}.$$

We define the volume of the horn to be the limit as $b \to -\infty$ of the volume of the portion from b to $\ln 2$. As in Section 5.1 (the method of slicing), the volume of this portion is

$$V = \int_b^{\ln 2} A(x)\,dx = \int_b^{\ln 2} \frac{\pi}{4}e^{2x}\,dx = \frac{\pi}{8}e^{2x}\bigg]_b^{\ln 2}$$

$$= \frac{\pi}{8}(e^{\ln 4} - e^{2b}) = \frac{\pi}{8}(4 - e^{2b}).$$

As $b \to -\infty$, $e^{2b} \to 0$ and $V \to (\pi/8)(4 - 0) = \pi/2$. The volume of the horn is $\pi/2$.

Example 7 Finding Circumference

Use the arc length formula (Section 5.3) to show that the circumference of the circle $x^2 + y^2 = 4$ is 4π.

Solution One fourth of this circle is given by $y = \sqrt{4 - x^2}$, $0 \leq x \leq 2$. Its arc length is

$$L = \int_0^2 \sqrt{1 + (y')^2}\,dx, \qquad \text{where} \qquad y' = -\frac{x}{\sqrt{4 - x^2}}.$$

The integral is improper because y' is not defined at $x = 2$. We evaluate it as a limit.

$$\begin{aligned} L = \int_0^2 \sqrt{1 + (y')^2}\, dx &= \int_0^2 \sqrt{1 + \frac{x^2}{4 - x^2}}\, dx \\ &= \int_0^2 \sqrt{\frac{4}{4 - x^2}}\, dx \\ &= \lim_{b \to 2^-} \int_0^b \sqrt{\frac{4}{4 - x^2}}\, dx \\ &= \lim_{b \to 2^-} \int_0^b \sqrt{\frac{1}{1 - (x/2)^2}}\, dx \\ &= \lim_{b \to 2^-} 2 \sin^{-1} \frac{x}{2} \Big]_0^b \\ &= \lim_{b \to 2^-} 2 \left[\sin^{-1} \frac{b}{2} - 0 \right] = \pi \end{aligned}$$

The circumference of the quarter circle is π; the circumference of the circle is 4π.

Tests for Convergence and Divergence

When we cannot evaluate an improper integral directly (often the case in practice), we first try to determine whether it converges or diverges. If the integral diverges, that's the end of the story. If it converges, we can then use numerical methods to approximate its value. The principal tests for convergence or divergence are the Direct Comparison Test and the Limit Comparison Test.

Bounded Monotonic Functions

It can be shown that a monotonic function $f(x)$ that is bounded on an infinite interval (a, ∞) must have a finite limit as $x \to \infty$. In Example 8, this is applied to the function

$$f(b) = \int_1^b e^{-x^2}\, dx$$

as $b \to \infty$.

Example 8 Investigating Convergence

Does the integral $\int_1^\infty e^{-x^2}\, dx$ converge?

Solution By definition,

$$\int_1^\infty e^{-x^2}\, dx = \lim_{b \to \infty} \int_1^b e^{-x^2}\, dx.$$

We cannot evaluate the latter integral directly because there is no simple formula for the antiderivative of e^{-x^2}. We must therefore determine its convergence or divergence some other way. Because $e^{-x^2} > 0$ for all x, $\int_1^b e^{-x^2}\, dx$ is an increasing function of b. Therefore, as $b \to \infty$, the integral either becomes infinite as $b \to \infty$ or it is bounded from above and is forced to converge (have a finite limit).

The two curves $y = e^{-x^2}$ and $y = e^{-x}$ intersect at $(1, e^{-1})$, and $0 < e^{-x^2} \le e^{-x}$ for $x \ge 1$ (Figure 7.19). Thus, for any $b > 1$,

$$0 < \int_1^b e^{-x^2}\, dx \le \int_1^b e^{-x}\, dx = -e^{-b} + e^{-1} < e^{-1} \approx 0.368.$$

Rounded up to be safe

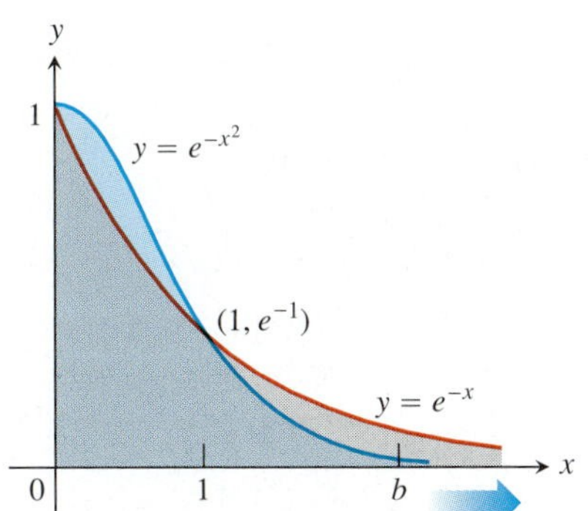

FIGURE 7.19 The graph of e^{-x^2} lies below the graph of e^{-x} for $x > 1$. (Example 8)

As an increasing function of b bounded above by 0.368, the integral $\int_1^b e^{-x^2}\,dx$ must converge as $b \to \infty$. This does not tell us much about the value of the improper integral, however, except that it is positive and less than 0.368.

The comparison of e^{-x^2} and e^{-x} in Example 8 is a special case of the following test.

CD-ROM
WEBsite
Historical Biography
Karl Weierstrass
(1815–1897)

Theorem 3 Direct Comparison Test

Let f and g be continuous on $[a, \infty)$ with $0 \le f(x) \le g(x)$ for all $x \ge a$. Then

1. $\int_a^\infty f(x)\,dx$ converges if $\int_a^\infty g(x)\,dx$ converges

2. $\int_a^\infty g(x)\,dx$ diverges if $\int_a^\infty f(x)\,dx$ diverges.

Example 9 Using the Direct Comparison Test

(a) $\int_1^\infty \frac{\sin^2 x}{x^2}\,dx$ converges because

$$0 \le \frac{\sin^2 x}{x^2} \le \frac{1}{x^2} \quad \text{on} \quad [1, \infty) \quad \text{and} \quad \int_1^\infty \frac{1}{x^2}\,dx \text{ converges.}$$

Example 3

(b) $\int_1^\infty \frac{1}{\sqrt{x^2 - 0.1}}\,dx$ diverges because

$$\frac{1}{\sqrt{x^2 - 0.1}} \ge \frac{1}{x} \quad \text{on} \quad [1, \infty) \quad \text{and} \quad \int_1^\infty \frac{1}{x}\,dx \text{ diverges.}$$

Example 3

Theorem 4 Limit Comparison Test

If the positive functions f and g are continuous on $[a, \infty)$ and if

$$\lim_{x \to \infty} \frac{f(x)}{g(x)} = L, \qquad 0 < L < \infty,$$

then

$$\int_a^\infty f(x)\,dx \quad \text{and} \quad \int_a^\infty g(x)\,dx$$

both converge or both diverge.

A proof of Theorem 4 is given in advanced calculus.

Although the improper integrals of two functions from a to ∞ may both converge, this does not mean that their integrals necessarily have the same value, as the next example shows.

Example 10 Using the Limit Comparison Test

Show that

$$\int_1^\infty \frac{dx}{1+x^2}$$

converges by comparison with $\int_1^\infty (1/x^2)\,dx$. Find and compare the two integral values.

Solution The functions $f(x) = 1/x^2$ and $g(x) = 1/(1+x^2)$ are positive and continuous on $[1, \infty)$. Also,

$$\begin{aligned}\lim_{x\to\infty}\frac{f(x)}{g(x)} &= \lim_{x\to\infty}\frac{1/x^2}{1/(1+x^2)} = \lim_{x\to\infty}\frac{1+x^2}{x^2}\\ &= \lim_{x\to\infty}\left(\frac{1}{x^2}+1\right) = 0+1 = 1,\end{aligned}$$

a positive finite limit (Figure 7.20). Therefore, $\int_1^\infty \frac{dx}{1+x^2}$ converges because $\int_1^\infty \frac{dx}{1+x^2}$ converges.

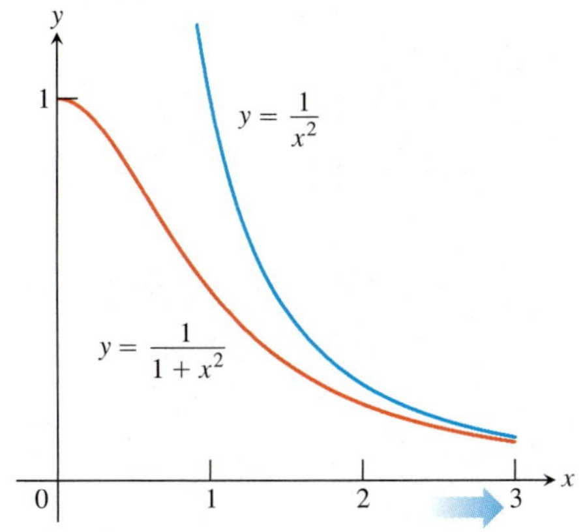

The integrals converge to different values, however.

$$\int_1^\infty \frac{dx}{x^2} = \frac{1}{2-1} = 1 \qquad \text{Example 3}$$

and

$$\begin{aligned}\int_1^\infty \frac{dx}{1+x^2} &= \lim_{b\to\infty}\int_1^b \frac{dx}{1+x^2}\\ &= \lim_{b\to\infty}\,[\tan^{-1} b - \tan^{-1} 1] = \frac{\pi}{2} - \frac{\pi}{4} = \frac{\pi}{4}\end{aligned}$$

FIGURE 7.20 The functions in Example 10.

Example 11 Using the Limit Comparison Test

Show that

$$\int_2^\infty \frac{3}{e^x - 5}\,dx$$

converges.

Solution From Example 8, it is easy to see that $\int_2^\infty e^{-x}\,dx = \int_2^\infty (1/e^x)\,dx$ converges. Because

$$\lim_{x\to\infty}\frac{1/e^x}{3/(e^x-5)} = \lim_{x\to\infty}\frac{e^x-5}{3e^x} = \lim_{x\to\infty}\left(\frac{1}{3} - \frac{5}{3e^x}\right) = \frac{1}{3},$$

the integral

$$\int_2^\infty \frac{3}{e^x - 5}\,dx$$

also converges.

Computer Algebra Systems

Computer algebra systems can evaluate many convergent improper integrals.

Example 12 Using a CAS

Evaluate the integral

$$\int_2^\infty \frac{x+3}{(x-1)(x^2+1)}\,dx.$$

Solution Using Maple, enter

```
> f: = (x + 3)/((x − 1)*(x^2 + 1));
```

Then use the integration command

```
> int(f, x=2..infinity);
```

Maple returns the answer

$$-\frac{1}{2}\pi + \ln(5) + \arctan(2).$$

To obtain a numerical result, use the evaluation command **evalf** and specify the number of digits, as follows:

```
> evalf(", 6);
```

The ditto symbol (") instructs the computer to evaluate the last expression on the screen, in this case $(-1/2)\pi + \ln(5) + \arctan(2)$. Maple returns 1.14579.

Using Mathematica, entering

```
In [1]: = Integrate [(x + 3)/((x − 1)(x^2 + 1)), {x, 2, Infinity}]
```

returns

$$\text{Out [1]} = \frac{-\text{Pi}}{2} + \text{ArcTan [2]} + \text{Log [5]}.$$

To obtain a numerical result with six digits, use the command "N[%, 6]"; it also yields 1.14579.

Types of Improper Integrals Discussed in This Section

INFINITE LIMITS OF INTEGRATION

1. Upper limit

$$\int_1^{\infty} \frac{\ln x}{x^2}\,dx = \lim_{b\to\infty} \int_1^b \frac{\ln x}{x^2}\,dx$$

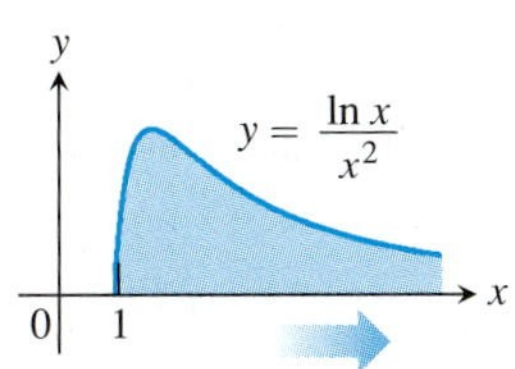

2. Lower limit

$$\int_{-\infty}^{0} \frac{dx}{1+x^2} = \lim_{a\to-\infty} \int_a^0 \frac{dx}{1+x^2}$$

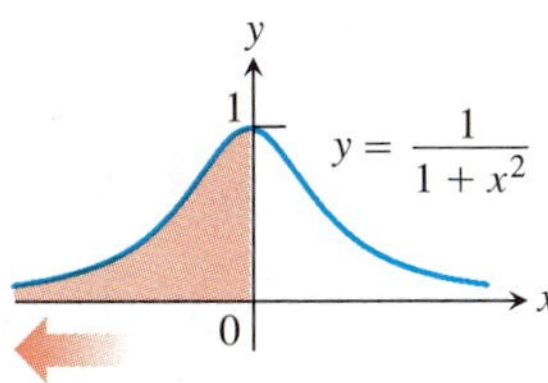

3. Both limits

$$\int_{-\infty}^{\infty} \frac{dx}{1+x^2} = \lim_{b\to-\infty} \int_b^0 \frac{dx}{1+x^2} + \lim_{c\to\infty} \int_0^c \frac{dx}{1+x^2}$$

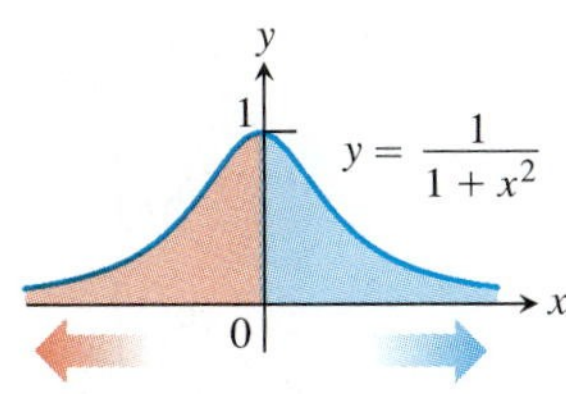

INTEGRAND BECOMES INFINITE

4. Upper endpoint

$$\int_0^1 \frac{dx}{(x-1)^{2/3}} = \lim_{b\to1^-} \int_0^b \frac{dx}{(x-1)^{2/3}}$$

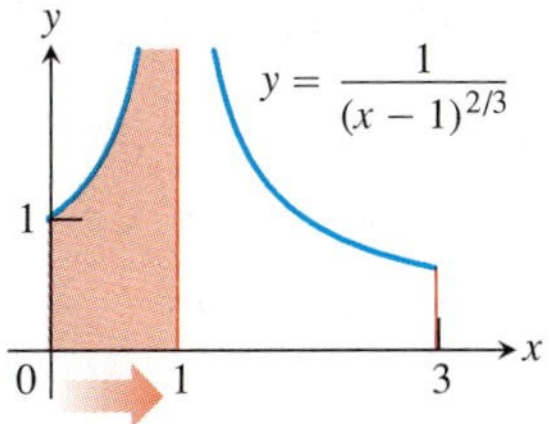

5. Lower endpoint

$$\int_1^3 \frac{dx}{(x-1)^{2/3}} = \lim_{d\to1^+} \int_d^3 \frac{dx}{(x-1)^{2/3}}$$

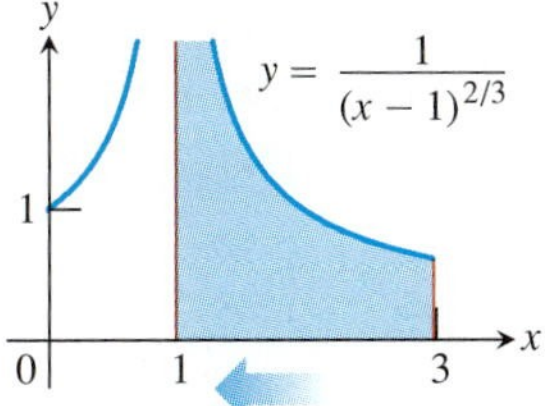

6. Interior point

$$\int_0^3 \frac{dx}{(x-1)^{2/3}} = \int_0^1 \frac{dx}{(x-1)^{2/3}} + \int_1^3 \frac{dx}{(x-1)^{2/3}}$$

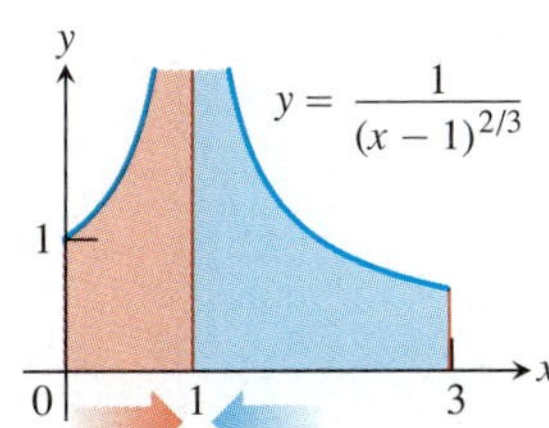

EXERCISES 7.7

Identifying Improper Integrals

In Exercises 1–6, do the following.

(a) State why the integral is improper or involves improper integrals.

(b) Determine whether the integral converges or diverges.

(c) Evaluate the integral if it converges.

1. $\displaystyle\int_0^\infty \frac{dx}{x^2+1}$

2. $\displaystyle\int_0^1 \frac{dx}{\sqrt{x}}$

3. $\displaystyle\int_{-8}^1 \frac{dx}{x^{1/3}}$

4. $\displaystyle\int_{-\infty}^\infty \frac{2x\,dx}{(x^2+1)^2}$

5. $\displaystyle\int_0^{\ln 2} x^{-2}e^{1/x}\,dx$

6. $\displaystyle\int_0^{\pi/2} \cot\theta\,d\theta$

Evaluating Improper Integrals

Evaluate each integral in Exercises 7–34 or state that it diverges.

7. $\displaystyle\int_1^\infty \frac{dx}{x^{1.001}}$

8. $\displaystyle\int_{-1}^1 \frac{dx}{x^{2/3}}$

9. $\displaystyle\int_0^4 \frac{dr}{\sqrt{4-r}}$

10. $\displaystyle\int_0^1 \frac{dr}{r^{0.999}}$

11. $\displaystyle\int_0^1 \frac{dx}{\sqrt{1-x^2}}$

12. $\displaystyle\int_{-\infty}^2 \frac{2\,dx}{x^2+4}$

13. $\displaystyle\int_{-\infty}^{-2} \frac{2\,dx}{x^2-1}$

14. $\displaystyle\int_2^\infty \frac{3\,dt}{t^2-t}$

15. $\displaystyle\int_0^1 \frac{\theta+1}{\sqrt{\theta^2+2\theta}}\,d\theta$

16. $\displaystyle\int_0^2 \frac{s+1}{\sqrt{4-s^2}}\,ds$

17. $\displaystyle\int_0^\infty \frac{dx}{(1+x)\sqrt{x}}$

18. $\displaystyle\int_1^\infty \frac{dx}{x\sqrt{x^2-1}}$

19. $\displaystyle\int_1^2 \frac{ds}{s\sqrt{s^2-1}}$

20. $\displaystyle\int_{-1}^\infty \frac{d\theta}{\theta^2+5\theta+6}$

21. $\displaystyle\int_2^\infty \frac{2}{v^2-v}\,dv$

22. $\displaystyle\int_2^\infty \frac{2\,dt}{t^2-1}$

23. $\displaystyle\int_0^2 \frac{ds}{\sqrt{4-s^2}}$

24. $\displaystyle\int_0^1 \frac{4r\,dr}{\sqrt{1-r^4}}$

25. $\displaystyle\int_0^\infty \frac{dv}{(1+v^2)(1+\tan^{-1}v)}$

26. $\displaystyle\int_0^\infty \frac{16\tan^{-1}x}{1+x^2}\,dx$

27. $\displaystyle\int_{-1}^4 \frac{dx}{\sqrt{|x|}}$

28. $\displaystyle\int_0^2 \frac{dx}{\sqrt{|x-1|}}$

29. $\displaystyle\int_{-\infty}^0 \theta e^\theta\,d\theta$

30. $\displaystyle\int_0^\infty 2e^{-\theta}\sin\theta\,d\theta$

31. $\displaystyle\int_{-\infty}^\infty e^{-|x|}\,dx$

32. $\displaystyle\int_{-\infty}^\infty 2xe^{-x^2}\,dx$

33. $\displaystyle\int_0^1 x\ln x\,dx$

34. $\displaystyle\int_0^1 (-\ln x)\,dx$

Testing for Convergence

In Exercises 35–64, use integration, the Direct Comparison Test, or the Limit Comparison Test to test the integrals for convergence. If more than one method applies, use whatever method you prefer.

35. $\displaystyle\int_0^{\pi/2} \tan\theta\,d\theta$

36. $\displaystyle\int_0^{\pi/2} \cot\theta\,d\theta$

37. $\displaystyle\int_0^\pi \frac{\sin\theta\,d\theta}{\sqrt{\pi-\theta}}$

38. $\displaystyle\int_{-\pi/2}^{\pi/2} \frac{\cos\theta\,d\theta}{(\pi-2\theta)^{1/3}}$

39. $\displaystyle\int_0^{\ln 2} x^{-2}e^{-1/x}\,dx$

40. $\displaystyle\int_0^1 \frac{e^{-\sqrt{x}}}{\sqrt{x}}\,dx$

41. $\displaystyle\int_0^\pi \frac{dt}{\sqrt{t}+\sin t}$

42. $\displaystyle\int_0^1 \frac{dt}{t-\sin t}$ (*Hint:* $t \ge \sin t$ for $t \ge 0$)

43. $\displaystyle\int_0^2 \frac{dx}{1-x^2}$

44. $\displaystyle\int_0^2 \frac{dx}{1-x}$

45. $\displaystyle\int_{-1}^1 \ln|x|\,dx$

46. $\displaystyle\int_{-1}^1 -x\ln|x|\,dx$

47. $\displaystyle\int_1^\infty \frac{dx}{x^3+1}$

48. $\displaystyle\int_4^\infty \frac{dx}{\sqrt{x}-1}$

49. $\displaystyle\int_2^\infty \frac{dv}{\sqrt{v-1}}$

50. $\displaystyle\int_4^\infty \frac{2\,dt}{t^{3/2}-1}$

51. $\displaystyle\int_0^\infty \frac{dx}{\sqrt{x^6+1}}$

52. $\displaystyle\int_2^\infty \frac{dx}{\sqrt{x^2-1}}$

53. $\displaystyle\int_1^\infty \frac{\sqrt{x+1}}{x^2}\,dx$

54. $\displaystyle\int_2^\infty \frac{x\,dx}{\sqrt{x^4-1}}$

55. $\displaystyle\int_\pi^\infty \frac{2+\cos x}{x}\,dx$

56. $\displaystyle\int_\pi^\infty \frac{1+\sin x}{x^2}\,dx$

57. $\displaystyle\int_0^\infty \frac{d\theta}{1+e^\theta}$

58. $\displaystyle\int_2^\infty \frac{1}{\ln x}\,dx$

59. $\displaystyle\int_1^\infty \frac{e^x}{x}\,dx$

60. $\displaystyle\int_{e^e}^\infty \ln(\ln x)\,dx$

61. $\displaystyle\int_1^\infty \frac{1}{\sqrt{e^x-x}}\,dx$

62. $\displaystyle\int_1^\infty \frac{1}{e^x-2^x}\,dx$

63. $\displaystyle\int_{-\infty}^\infty \frac{dx}{\sqrt{x^4+1}}$

64. $\displaystyle\int_{-\infty}^\infty \frac{dx}{e^x+e^{-x}}$

Theory and Examples

65. Find the values of p for which each integral converges.

(a) $\displaystyle\int_1^2 \frac{dx}{x(\ln x)^p}$ **(b)** $\displaystyle\int_2^\infty \frac{dx}{x(\ln x)^p}$

66. *$\int_{-\infty}^{\infty} f(x)\,dx$ may not equal $\lim_{b\to\infty}\int_{-b}^{b} f(x)\,dx$* Show that

$$\int_0^\infty \frac{2x\,dx}{x^2+1}$$

diverges and hence that

$$\int_{-\infty}^\infty \frac{2x\,dx}{x^2+1}$$

diverges. Then show that

$$\lim_{b\to\infty}\int_{-b}^{b} \frac{2x\,dx}{x^2+1} = 0.$$

Exercises 67–69 are about the infinite region in the first quadrant between the curve $y = e^{-x}$ and the x-axis.

67. *Area* Find the area of the region.

68. *Volume* Find the volume of the solid generated by revolving the region about the y-axis.

69. *Volume* Find the volume of the solid generated by revolving the region about the x-axis.

70. *Area* Find the area of the region that lies between the curves $y = \sec x$ and $y = \tan x$ from $x = 0$ to $x = \pi/2$.

71. *Writing to Learn* Here is an argument that $\ln 3$ equals $\infty - \infty$. Where does the argument go wrong? Give reasons for your answer.

$$\ln 3 = \ln 1 + \ln 3 = \ln 1 - \ln\frac{1}{3}$$

$$= \lim_{b\to\infty} \ln\left(\frac{b-2}{b}\right) - \ln\frac{1}{3}$$

$$= \lim_{b\to\infty}\left[\ln\frac{x-2}{x}\right]_3^b$$

$$= \lim_{b\to\infty}\left[\ln(x-2) - \ln x\right]_3^b$$

$$= \lim_{b\to\infty}\int_3^b\left(\frac{1}{x-2} - \frac{1}{x}\right)dx$$

$$= \int_3^\infty\left(\frac{1}{x-2} - \frac{1}{x}\right)dx$$

$$= \int_3^\infty \frac{1}{x-2}\,dx - \int_3^\infty \frac{1}{x}\,dx$$

$$= \lim_{b\to\infty}\left[\ln(x-2)\right]_3^b - \lim_{b\to\infty}\left[\ln x\right]_3^b$$

$$= \infty - \infty$$

72. *Comparing integrals* Show that if $f(x)$ is integrable on every interval of real numbers and a and b are real numbers with $a < b$, then

(a) $\int_{-\infty}^a f(x)\,dx$ and $\int_a^\infty f(x)\,dx$ both converge if and only if $\int_{-\infty}^b f(x)\,dx$ and $\int_b^\infty f(x)\,dx$ both converge

(b) $\int_{-\infty}^a f(x)\,dx + \int_a^\infty f(x)\,dx = \int_{-\infty}^b f(x)\,dx + \int_b^\infty f(x)\,dx$ when the integrals involved converge.

73. *Estimating the value of a convergent improper integral whose domain is infinite*

(a) Show that

$$\int_3^\infty e^{-3x}\,dx = \frac{1}{3}e^{-9} < 0.000042$$

and hence that $\int_3^\infty e^{-x^2}\,dx < 0.000042$. Explain why this means that $\int_0^\infty e^{-x^2}\,dx$ can be replaced by $\int_0^3 e^{-x^2}\,dx$ without introducing an error of magnitude greater than 0.000042.

T **(b)** *Numerical integrator* Evaluate $\int_0^3 e^{-x^2}\,dx$ numerically.

74. *Sine-integral function* The integral

$$\text{Si}(x) = \int_0^x \frac{\sin t}{t}\,dt,$$

called the **sine-integral function,** has important applications in optics.

T **(a)** Plot the integrand $(\sin t)/t$ for $t > 0$. Is the Si function everywhere increasing or decreasing? Do you think $\text{Si}(x) = 0$ for $x > 0$? Check your answers by graphing the function $\text{Si}(x)$ for $0 \le x \le 25$.

(b) Explore the convergence of

$$\int_0^\infty \frac{\sin t}{t}\,dt.$$

If it converges, what is its value?

75. *Error function* The function

$$\text{erf}(x) = \int_0^x \frac{2e^{-t^2}}{\sqrt{\pi}}\,dt,$$

called the **error function,** has important applications in probability and statistics.

T **(a)** Plot the error function for $0 \le x \le 25$.

(b) Explore the convergence of

$$\int_0^\infty \frac{2e^{-t^2}}{\sqrt{\pi}}\,dt.$$

If it converges, what appears to be its value? You will see how to confirm your estimate in Section 12.3, Exercise 37.

76. *Normal probability distribution function* The function

$$f(x) = \frac{1}{\sigma\sqrt{2\pi}}e^{-\frac{1}{2}\left(\frac{x-\mu}{\sigma}\right)^2},$$

is called the **normal probability density function** with mean μ and standard deviation σ. The number μ tells where the distribution is centered, and σ measures the "scatter" around the mean.

From the theory of probability, it is known that

$$\int_{-\infty}^{\infty} f(x)\, dx = 1.$$

In what follows, let $\mu = 0$ and $\sigma = 1$.

T **(a)** Draw the graph of f. Find the intervals on which f is increasing, the intervals on which f is decreasing, and any local extreme values and where they occur.

(b) Evaluate

$$\int_{-n}^{n} f(x)\, dx$$

for $n = 1, 2, 3$.

(c) Give a convincing argument that

$$\int_{-\infty}^{\infty} f(x)\, dx = 1.$$

(*Hint:* Show that $0 < f(x) < e^{-x/2}$ for $x > 1$, and for $b > 1$,

$$\int_{b}^{\infty} e^{-x/2}\, dx \to 0 \quad \text{as} \quad b \to \infty.)$$

COMPUTER EXPLORATIONS

Exploring Integrals of $x^p \ln x$

In Exercises 77–80, use a CAS to explore the integrals for various values of p (include noninteger values). For what values of p does the integral converge? What is the value of the integral when it does converge? Plot the integrand for various values of p.

77. $\displaystyle\int_0^e x^p \ln x\, dx$

78. $\displaystyle\int_e^\infty x^p \ln x\, dx$

79. $\displaystyle\int_0^\infty x^p \ln x\, dx$

80. $\displaystyle\int_{-\infty}^\infty x^p \ln |x|\, dx$

Questions to Guide Your Review

1. What basic integration formulas do you know?
2. What procedures do you know for matching integrals to basic formulas?
3. What is the formula for integration by parts? Where does it come from? Why might you want to use it?
4. When applying the formula for integration by parts, how do you choose the u and dv? How can you apply integration by parts to an integral of the form $\int f(x)\, dx$?
5. What is tabular integration? Give an example.
6. What is the goal of the method of partial fractions?
7. When the degree of a polynomial $f(x)$ is less than the degree of a polynomial $g(x)$, how do you write $f(x)/g(x)$ as a sum of partial fractions if $g(x)$
   (a) is a product of distinct linear factors
   (b) consists of a repeated linear factor
   (c) contains an irreducible quadratic factor?

   What do you do if the degree of f is *not* less than the degree of g?
8. What substitutions are sometimes used to change quadratic binomials into single squared terms? Why might you want to make such a change?
9. What restrictions can you place on the variables involved in the three basic trigonometric substitutions to make sure the substitutions are reversible (have inverses)?
10. How are integral tables typically used? What do you do if a particular integral you want to evaluate is not listed in the table?
11. Describe Monte Carlo integration for finding the area beneath a nonnegative curve and above the x-axis.
12. Describe l'Hôpital's Rule. How do you know when to use the rule and when to stop? Give an example.
13. How can you sometimes handle limits that lead to indeterminate forms 1^∞, 0^0, and ∞^0? Give examples.
14. What is an improper integral? How are the values of various types of improper integrals defined? Give examples.
15. What tests are available for determining the convergence and divergence of improper integrals that cannot be evaluated directly? Give examples of their use.

Practice Exercises

Integration Using Substitutions

Evaluate the integrals in Exercises 1–46. To transform each inegral into a recognizable basic form, it may be necessary to use one or more of the techniques of algebraic substitution, completing the square, separating fractions, long division, or trigonometric substitution.

1. $\int x\sqrt{4x^2 - 9}\,dx$
2. $\int x(2x + 1)^{1/2}\,dx$
3. $\int \frac{x\,dx}{\sqrt{8x^2 + 1}}$
4. $\int \frac{y\,dy}{25 + y^2}$
5. $\int \frac{t^3\,dt}{\sqrt{9 - 4t^4}}$
6. $\int z^{2/3}(z^{5/3} + 1)^{2/3}\,dz$
7. $\int \frac{\sin 2\theta\,d\theta}{(1 - \cos 2\theta)^2}$
8. $\int \frac{\cos 2t}{1 + \sin 2t}\,dt$
9. $\int \sin 2x\,e^{\cos 2x}\,dx$
10. $\int e^{\theta}\sec^2(e^{\theta})\,d\theta$
11. $\int 2^{x-1}\,dx$
12. $\int \frac{dv}{v\ln v}$
13. $\int \frac{dx}{(x^2 + 1)(2 + \tan^{-1} x)}$
14. $\int \frac{2\,dx}{\sqrt{1 - 4x^2}}$
15. $\int \frac{dt}{\sqrt{16 - 9t^2}}$
16. $\int \frac{dt}{9 + t^2}$
17. $\int \frac{4\,dx}{5x\sqrt{25x^2 - 16}}$
18. $\int \frac{dx}{\sqrt{4x - x^2 - 3}}$
19. $\int \frac{dy}{y^2 - 4y + 8}$
20. $\int \frac{dv}{(v + 1)\sqrt{v^2 + 2v}}$
21. $\int \cos^2 3x\,dx$
22. $\int \sin^3 \frac{\theta}{2}\,d\theta$
23. $\int \tan^3 2t\,dt$
24. $\int \frac{dx}{2\sin x\cos x}$
25. $\int \frac{2\,dx}{\cos^2 x - \sin^2 x}$
26. $\int_{\pi/4}^{\pi/2} \sqrt{\csc^2 y - 1}\,dy$
27. $\int_{\pi/4}^{3\pi/4} \sqrt{\cot^2 t + 1}\,dt$
28. $\int_0^{2\pi} \sqrt{1 - \sin^2 \frac{x}{2}}\,dx$
29. $\int_{-\pi/2}^{\pi/2} \sqrt{1 - \cos 2t}\,dt$
30. $\int_{\pi}^{2\pi} \sqrt{1 + \cos 2t}\,dt$
31. $\int \frac{x^2}{x^2 + 4}\,dx$
32. $\int \frac{x^3}{9 + x^2}\,dx$
33. $\int \frac{2y - 1}{y^2 + 4}\,dy$
34. $\int \frac{y + 4}{y^2 + 1}\,dy$
35. $\int \frac{t + 2}{\sqrt{4 - t^2}}\,dt$
36. $\int \frac{2t^2 + \sqrt{1 - t^2}}{t\sqrt{1 - t^2}}\,dt$
37. $\int \frac{\tan x\,dx}{\tan x + \sec x}$
38. $\int x\csc(x^2 + 3)\,dx$
39. $\int \cot\left(\frac{x}{4}\right)dx$
40. $\int x\sqrt{1 - x}\,dx$
41. $\int (16 + z^2)^{-3/2}\,dz$
42. $\int \frac{dy}{\sqrt{25 + y^2}}$
43. $\int \frac{dx}{x^2\sqrt{1 - x^2}}$
44. $\int \frac{x^2\,dx}{\sqrt{1 - x^2}}$
45. $\int \frac{dx}{\sqrt{x^2 - 9}}$
46. $\int \frac{12\,dx}{(x^2 - 1)^{3/2}}$

Integration by Parts

Evaluate the integrals in Exercises 47–54 using integration by parts.

47. $\int \ln(x + 1)\,dx$
48. $\int x^2 \ln x\,dx$
49. $\int \tan^{-1} 3x\,dx$
50. $\int \cos^{-1}\left(\frac{x}{2}\right)dx$
51. $\int (x + 1)^2 e^x\,dx$
52. $\int x^2 \sin(1 - x)\,dx$
53. $\int e^x \cos 2x\,dx$
54. $\int e^{-2x}\sin 3x\,dx$

Partial Fractions

Evaluate the integrals in Exercises 55–66. It may be necessary to use a substitution first.

55. $\int \frac{x\,dx}{x^2 - 3x + 2}$
56. $\int \frac{dx}{x(x + 1)^2}$
57. $\int \frac{\sin\theta\,d\theta}{\cos^2\theta + \cos\theta - 2}$
58. $\int \frac{3x^2 + 4x + 4}{x^3 + x}\,dx$
59. $\int \frac{v + 3}{2v^3 - 8v}\,dv$
60. $\int \frac{dt}{t^4 + 4t^2 + 3}$
61. $\int \frac{x^3 + x^2}{x^2 + x - 2}\,dx$
62. $\int \frac{x^3 + 4x^2}{x^2 + 4x + 3}\,dx$
63. $\int \frac{2x^3 + x^2 - 21x + 24}{x^2 + 2x - 8}\,dx$
64. $\int \frac{dx}{x(3\sqrt{x + 1})}$
65. $\int \frac{ds}{e^s - 1}$
66. $\int \frac{ds}{\sqrt{e^s + 1}}$

Trigonometric Substitutions

Evaluate the integrals in Exercises 67–70 (a) without using a trigonometric substitution and then (b) using a trigonometric substitution.

67. $\int \frac{y\,dy}{\sqrt{16-y^2}}$

68. $\int \frac{x\,dx}{\sqrt{4+x^2}}$

69. $\int \frac{x\,dx}{4-x^2}$

70. $\int \frac{t\,dt}{\sqrt{4t^2-1}}$

Quadratic Terms

Evaluate the integrals in Exercises 71–74.

71. $\int \frac{x\,dx}{9-x^2}$

72. $\int \frac{dx}{x(9-x^2)}$

73. $\int \frac{dx}{9-x^2}$

74. $\int \frac{dx}{\sqrt{9-x^2}}$

Assorted Integrations

Evaluate the integrals in Exercises 75–114. The integrals are listed in random order.

75. $\int \frac{x\,dx}{1+\sqrt{x}}$

76. $\int \frac{dx}{x(x^2+1)^2}$

77. $\int \frac{\cos\sqrt{x}}{\sqrt{x}}\,dx$

78. $\int \frac{dx}{\sqrt{-2x-x^2}}$

79. $\int \frac{du}{\sqrt{1+u^2}}$

80. $\int \frac{2-\cos x+\sin x}{\sin^2 x}\,dx$

81. $\int \frac{9\,dv}{81-v^4}$

82. $\int \theta\cos(2\theta+1)\,d\theta$

83. $\int \frac{x^3\,dx}{x^2-2x+1}$

84. $\int \frac{d\theta}{\sqrt{1+\sqrt{\theta}}}$

85. $\int \frac{2\sin\sqrt{x}\,dx}{\sqrt{x}\sec\sqrt{x}}$

86. $\int \frac{x^5\,dx}{x^4-16}$

87. $\int \frac{d\theta}{\theta^2-2\theta+4}$

88. $\int \frac{dr}{(r+1)\sqrt{r^2+2r}}$

89. $\int \frac{\sin 2\theta\,d\theta}{(1+\cos 2\theta)^2}$

90. $\int \frac{dx}{(x^2-1)^2}$

91. $\int \frac{x\,dx}{\sqrt{2-x}}$

92. $\int \frac{dy}{y^2-2y+2}$

93. $\int \ln\sqrt{x-1}\,dx$

94. $\int \frac{x\,dx}{\sqrt{8-2x^2-x^4}}$

95. $\int \frac{z+1}{z^2(z^2+4)}\,dz$

96. $\int x^3 e^{(x^2)}\,dx$

97. $\int \frac{\tan^{-1}x}{x^2}\,dx$

98. $\int \frac{e^t\,dt}{e^{2t}+3e^t+2}$

99. $\int \frac{1-\cos 2x}{1+\cos 2x}\,dx$

100. $\int \frac{\cos(\sin^{-1}x)}{\sqrt{1-x^2}}\,dx$

101. $\int \frac{\cos x\,dx}{\sin^3 x-\sin x}$

102. $\int \frac{e^t\,dt}{1+e^t}$

103. $\int_1^\infty \frac{\ln y}{y^3}\,dy$

104. $\int \frac{\cot v\,dv}{\ln\sin v}$

105. $\int \frac{dx}{(2x-1)\sqrt{x^2-x}}$

106. $\int e^{\ln\sqrt{x}}\,dx$

107. $\int e^\theta\sqrt{3+4e^\theta}\,d\theta$

108. $\int \frac{dv}{\sqrt{e^{2v}-1}}$

109. $\int (27)^{3\theta+1}\,d\theta$

110. $\int x^5\sin x\,dx$

111. $\int \frac{dr}{1+\sqrt{r}}$

112. $\int \frac{8\,dy}{y^3(y+2)}$

113. $\int \frac{8\,dm}{m\sqrt{49m^2-4}}$

114. $\int \frac{dt}{t(1+\ln t)\sqrt{(\ln t)(2+\ln t)}}$

Limits

In Exercises 115–128, find the limit.

115. $\lim_{t\to 0} \frac{t-\ln(1+2t)}{t^2}$

116. $\lim_{t\to 0} \frac{\tan 3t}{\tan 5t}$

117. $\lim_{x\to 0} \frac{x\sin x}{1-\cos x}$

118. $\lim_{x\to 1} x^{1/(1-x)}$

119. $\lim_{x\to\infty} x^{1/x}$

120. $\lim_{x\to\infty} \left(1+\frac{3}{x}\right)^x$

121. $\lim_{r\to\infty} \frac{\cos r}{\ln r}$

122. $\lim_{\theta\to\pi/2} \left(\theta-\frac{\pi}{2}\right)\sec\theta$

123. $\lim_{x\to 1} \left(\frac{1}{x-1}-\frac{1}{\ln x}\right)$

124. $\lim_{x\to 0^+} \left(1+\frac{1}{x}\right)^x$

125. $\lim_{\theta\to 0^+} (\tan\theta)^\theta$

126. $\lim_{\theta\to\infty} \theta^2\sin\left(\frac{1}{\theta}\right)$

127. $\lim_{x\to\infty} \frac{x^3-3x^2+1}{2x^2+x-3}$

128. $\lim_{x\to\infty} \frac{3x^2-x+1}{x^4-x^3+2}$

Improper Integrals

Evaluate the improper integrals in Exercises 129–138, or state that it diverges.

129. $\int_0^3 \frac{dx}{\sqrt{9-x^2}}$

130. $\int_0^1 \ln x\,dx$

131. $\int_{-1}^1 \frac{dy}{y^{2/3}}$

132. $\int_{-2}^0 \frac{d\theta}{(\theta+1)^{3/5}}$

133. $\int_3^\infty \frac{2\,du}{u^2-2u}$

134. $\int_1^\infty \frac{3v-1}{4v^3-v^2}\,dv$

135. $\int_0^\infty x^2 e^{-x}\,dx$

136. $\int_{-\infty}^0 xe^{3x}\,dx$

137. $\int_{-\infty}^\infty \frac{dx}{4x^2+9}$

138. $\int_{-\infty}^\infty \frac{4\,dx}{x^2+16}$

Convergence or Divergence

Which of the improper integrals in Exercises 139–144 converge, and which diverge? Give reasons for your answer.

139. $\int_6^\infty \frac{d\theta}{\sqrt{\theta^2+1}}$

140. $\int_0^\infty e^{-u}\cos u\,du$

141. $\int_1^\infty \frac{\ln z}{z}\,dz$

142. $\int_1^\infty \frac{e^{-t}}{\sqrt{t}}\,dt$

143. $\int_{-\infty}^\infty \frac{dx}{e^x+e^{-x}}$

144. $\int_{-\infty}^\infty \frac{dx}{x^2(1+e^x)}$

Initial Value Problems

In Exercises 145–148, solve the initial value problem.

145. $\frac{dy}{dx} = e^x(y^2 - y), \quad y(0) = 2$

146. $\frac{dy}{d\theta} = (y+1)^2 \sin\theta, \quad y(\pi/2) = 0$

147. $\frac{dy}{dx} = \frac{1}{x^2 - 3x + 2}, \quad y(3) = 0$

148. $\frac{ds}{dt} = \frac{2s+2}{t^2+2t}, \quad s(1) = 1$

Additional Exercises: Theory, Examples, Applications

Challenging Integrals

Evaluate the integrals in Exercises 1–10.

1. $\int (\sin^{-1} x)^2\,dx$

2. $\int \frac{dx}{x(x+1)(x+2)\cdots(x+m)}$

3. $\int x\sin^{-1} x\,dx$

4. $\int \sin^{-1}\sqrt{y}\,dy$

5. $\int \frac{d\theta}{1-\tan^2\theta}$

6. $\int \ln(\sqrt{x} + \sqrt{1+x})\,dx$

7. $\int \frac{dt}{t - \sqrt{1-t^2}}$

8. $\int \frac{(2e^{2x} - e^x)\,dx}{\sqrt{3e^{2x} - 6e^x - 1}}$

9. $\int \frac{dx}{x^4+4}$

10. $\int \frac{dx}{x^6-1}$

Limits

Find the limits in Exercises 11–16.

11. $\lim_{b\to 1^-} \int_0^b \frac{dx}{\sqrt{1-x^2}}$

12. $\lim_{x\to\infty} \frac{1}{x}\int_0^x \tan^{-1} t\,dt$

13. $\lim_{x\to 0^+} (\cos\sqrt{x})^{1/x}$

14. $\lim_{x\to\infty} (x+e^x)^{2/x}$

15. $\lim_{x\to\infty} \int_{-x}^x \sin t\,dt$

16. $\lim_{x\to 0^+} x\int_x^1 \frac{\cos t}{t^2}\,dt$

Theory and Applications

17. *Finding arc length* Find the length of the curve

$$y = \int_0^x \sqrt{\cos 2t}\,dt, \qquad 0 \le x \le \pi/4.$$

18. *Finding arc length* Find the length of the curve $y = \ln(1 - x^2)$, $0 \le x \le 1/2$.

19. *Finding volume* The region in the first quadrant that is enclosed by the x-axis and the curve $y = 3x\sqrt{1-x}$ is revolved about the y-axis to generate a solid. Find the volume of the solid.

20. *Finding volume* The region in the first quadrant that is enclosed by the x-axis, the curve $y = 5/(x\sqrt{5-x})$, and the lines $x = 1$ and $x = 4$ is revolved about the x-axis to generate a solid. Find the volume of the solid.

21. *Finding volume* The region in the first quadrant enclosed by the coordinate axes, the curve $y = e^x$, and the line $x = 1$ is revolved about the y-axis to generate a solid. Find the volume of the solid.

22. *Finding volume* The region in the first quadrant that is bounded above by the curve $y = e^x - 1$, below by the x-axis, and on the right by the line $x = \ln 2$ is revolved about the line $x = \ln 2$ to generate a solid. Find the volume of the solid.

23. *Finding volume* Let R be the "triangular" region in the first quadrant that is bounded above by the line $y = 1$, below by the curve $y = \ln x$, and on the left by the line $x = 1$. Find the volume of the solid generated by revolving R about

(a) the x-axis

(b) the line $y = 1$.

24. *Finding volume* (*Continuation of Exercise 23*) Find the volume of the solid generated by revolving the region R about

(a) the y-axis

(b) the line $x = 1$.

25. *Finding volume* The region between the curve

$$y = f(x) = \begin{cases} 0, & x = 0 \\ x \ln x, & 0 < x \le 2 \end{cases}$$

is revolved about the x-axis to generate the solid shown here.

(a) Show that f is continuous at $x = 0$.

(b) Find the volume of the solid.

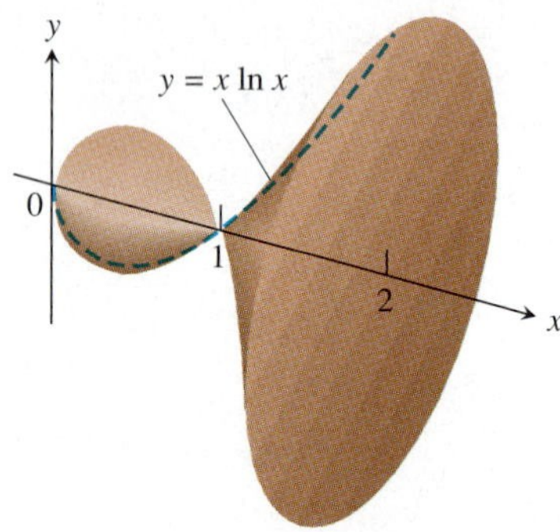

26. *Finding volume* The infinite region bounded by the coordinate axes and the curve $y = -\ln x$ in the first quadrant is revolved about the x-axis to generate a solid. Find the volume of the solid.

27. *Finding a limit* Find

$$\lim_{n\to\infty} \int_0^1 \frac{n y^{n-1}}{1 + y}\, dy.$$

28. *An integral formula* Derive the integral formula

$$\int x\left(\sqrt{x^2 - a^2}\right)^n dx = \frac{\left(\sqrt{x^2 - a^2}\right)^{n+2}}{n + 2} + C, \qquad n \ne -2.$$

29. *An inequality* Prove that

$$\frac{\pi}{6} < \int_0^1 \frac{dx}{\sqrt{4 - x^2 - x^3}} < \frac{\pi\sqrt{2}}{8}.$$

(*Hint:* Observe that for $0 < x < 1$, we have $4 - x^2 > 4 - x^2 - x^3 > 4 - 2x^2$, with the left-hand side becoming an equality for $x = 0$ and the right-hand side becoming an equality for $x = 1$.)

30. *Writing to Learn* For what value or values of a does

$$\int_1^\infty \left(\frac{ax}{x^2 + 1} - \frac{1}{2x}\right) dx$$

converge? Evaluate the corresponding integral(s).

31. *Evaluating an integral* Suppose that for a certain function f it is known that

$$f'(x) = \frac{\cos x}{x}, \qquad f(\pi/2) = a, \qquad \text{and} \qquad f(3\pi/2) = b.$$

Use integration by parts to evaluate

$$\int_{\pi/2}^{3\pi/2} f(x)\, dx.$$

32. *Finding equal areas* Find a positive number a satisfying

$$\int_0^a \frac{dx}{1 + x^2} = \int_a^\infty \frac{dx}{1 + x^2}.$$

33. *The length of an astroid* The graph of the equation $x^{2/3} + y^{2/3} = 1$ is one of a family of curves called *astroids* (not "asteroids") because of their starlike appearance (see accompanying figure). Find the length of this particular astroid.

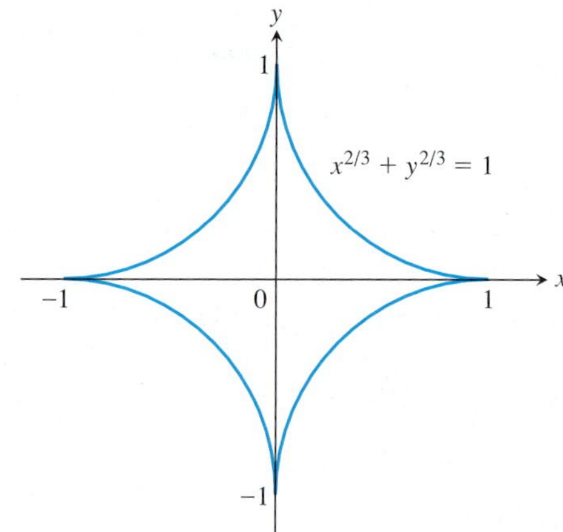

34. Find a curve through the origin whose length is

$$\int_0^4 \sqrt{1 + \frac{1}{4x}}\, dx.$$

35. *A rational function* Find the second-degree polynomial $P(x)$ such that $P(0) = 1$, $P'(0) = 0$, and

$$\int \frac{P(x)}{x^3(x - 1)^2}\, dx$$

is a rational function.

36. *Writing to Learn* Without evaluating either integral, explain why

$$2\int_{-1}^1 \sqrt{1 - x^2}\, dx = \int_{-1}^1 \frac{dx}{\sqrt{1 - x^2}}.$$

(*Source:* Peter A. Lindstrom, *Mathematics Magazine,* Vol. 45, No. 1 (January 1972), p. 47.)

37. *Infinite area and finite volume* What values of p have the following property? The area of the region between the curve $y = x^{-p}$, $1 \le x < \infty$, and the x-axis is infinite, but the volume of the solid generated by revolving the region about the x-axis is finite.

38. *Infinite area and finite volume* What values of p have the following property? The area of the region in the first quadrant enclosed by the curve $y = x^{-p}$, the y-axis, the line $x = 1$, and the interval $[0, 1]$ on the x-axis is infinite, but the volume of the solid generated by revolving the region about the x-axis is finite.

39. *Finite area*

T **(a)** Graph the function $f(x) = e^{(x-e^x)}$, $-5 \le x \le 3$.

(b) Show that

$$\int_{-\infty}^{\infty} f(x)\, dx$$

converges and find its value.

40. *An integral connecting π to the approximation 22/7*

(a) Evaluate

$$\int_0^1 \frac{x^4(x-1)^4}{x^2+1}\, dx.$$

(b) How good is the approximation $\pi \approx 22/7$? Find out by expressing $(\pi - 22/7)$ as a percentage of π.

T **(c)** Graph the function

$$y = \frac{x^4(x-1)^4}{x^2+1}$$

for $0 \le x \le 1$. Experiment with the range on the y-axis set between 0 and 1, then between 0 and 0.5, and then decreasing the range until the graph can be seen. What do you conclude about the area under the curve?

Tabular Integration

The technique of tabular integration also applies to integrals of the form $\int f(x)g(x)\, dx$ when neither function can be differentiated repeatedly to become zero. For example, to evaluate

$$\int e^{2x} \cos x\, dx,$$

we begin as before with a table listing successive derivatives of e^{2x} and integrals of $\cos x$:

e^{2x} and its derivatives		$\cos x$ and its integrals	
e^{2x}	+	$\cos x$	
$2e^{2x}$	−	$\sin x$	
$4e^{2x}$	+	$-\cos x$	← *Stop here:* Row is same as first row except for multiplicative constants (4 on the left, −1 on the right).

We stop differentiating and integrating as soon as we reach a row that is the same as the first row except for multiplicative constants. We interpret the table as saying

$$\int e^{2x} \cos x\, dx = +(e^{2x} \sin x) - (2e^{2x}(-\cos x)) + \int (4e^{2x})(-\cos x)\, dx.$$

We take signed products from the diagonal arrows and a signed integral for the last horizontal arrow. Transposing the integral on the right-hand side over to the left-hand side now gives

$$5 \int e^{2x} \cos x\, dx = e^{2x} \sin x + 2e^{2x} \cos x$$

or

$$\int e^{2x} \cos x\, dx = \frac{e^{2x} \sin x + 2e^{2x} \cos x}{5} + C,$$

after dividing by 5 and adding the constant of integration.

Use tabular integration to evaluate the integrals in Exercises 41–48.

41. $\int e^{2x} \cos 3x\, dx$

42. $\int e^{3x} \sin 4x\, dx$

43. $\int \sin 3x \sin x\, dx$

44. $\int \cos 5x \sin 4x\, dx$

45. $\int e^{ax} \sin bx\, dx$

46. $\int e^{ax} \cos bx\, dx$

47. $\int \ln (ax)\, dx$

48. $\int x^2 \ln (ax)\, dx$

The Gamma Function and Stirling's Formula

Euler's gamma function $\Gamma(x)$ ("gamma of x"; Γ is a Greek capital g) uses an integral to extend the factorial function from the nonnegative integers to other real values. The formula is

$$\Gamma(x) = \int_0^{\infty} t^{x-1} e^{-t}\, dt, \qquad x > 0.$$

For each positive x, the number $\Gamma(x)$ is the integral of $t^{x-1}e^{-t}$ with respect to t from 0 to ∞. Figure 7.21 shows the graph of Γ near the origin. You will see how to calculate $\Gamma(1/2)$ if you do Additional Exercise 31 in Chapter 12.

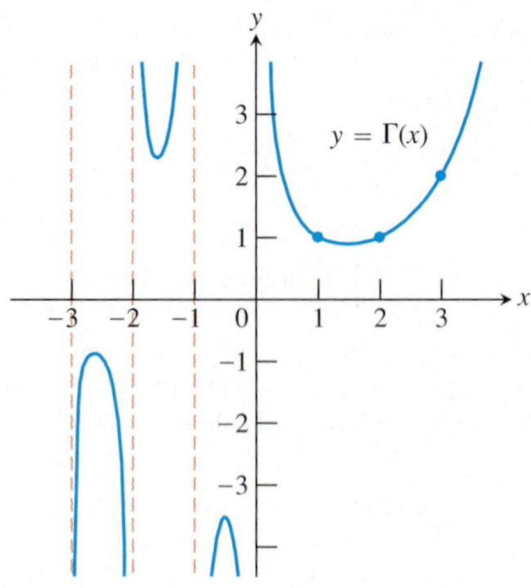

FIGURE 7.21 $\Gamma(x)$ is a continuous function of x whose value at each positive integer $n + 1$ is $n!$. The defining integral formula for Γ is valid only for $x > 0$, but we can extend Γ to negative noninteger values of x with the formula $\Gamma(x) = (\Gamma(x + 1))/x$, which is the subject of Exercise 49.

49. *If n is a nonnegative integer,* $\Gamma(n + 1) = n!$

(a) Show that $\Gamma(1) = 1$.

(b) Then apply integration by parts to the integral for $\Gamma(x + 1)$ to show that $\Gamma(x + 1) = x\Gamma(x)$. This gives

$$\begin{aligned}\Gamma(2) &= 1\Gamma(1) = 1\\ \Gamma(3) &= 2\Gamma(2) = 2\\ \Gamma(4) &= 3\Gamma(3) = 6\\ &\vdots\\ \Gamma(n+1) &= n\Gamma(n) = n! \qquad (1)\end{aligned}$$

(c) Use mathematical induction to verify Equation (1) for every nonnegative integer n.

50. *Stirling's formula* Scottish mathematician James Stirling (1692–1770) showed that

$$\lim_{x\to\infty}\left(\frac{e}{x}\right)^x\sqrt{\frac{x}{2\pi}}\,\Gamma(x) = 1,$$

so for large x,

$$\Gamma(x) = \left(\frac{x}{e}\right)^x\sqrt{\frac{2\pi}{x}}\,(1 + \epsilon(x)), \qquad \epsilon(x)\to 0 \text{ as } x\to\infty. \qquad (2)$$

Dropping $\epsilon(x)$ leads to the approximation

$$\Gamma(x) \approx \left(\frac{x}{e}\right)^x\sqrt{\frac{2\pi}{x}} \qquad \textbf{(Stirling's formula)}. \qquad (3)$$

(a) *Stirling's approximation for* $n!$ Use Equation (3) and the fact that $n! = n\Gamma(n)$ to show that

$$n! \approx \left(\frac{n}{e}\right)^n\sqrt{2n\pi} \qquad \textbf{(Stirling's approximation)}. \qquad (4)$$

As you will see if you do Exercise 64 in Section 8.1, Equation (4) leads to the approximation

$$\sqrt[n]{n!} \approx \frac{n}{e}. \qquad (5)$$

(b) Compare your calculator's value for $n!$ with the value given by Stirling's approximation for $n = 10, 20, 30, \ldots$, as far as your calculator can go.

(c) A refinement of Equation (2) gives

$$\Gamma(x) = \left(\frac{x}{e}\right)^x\sqrt{\frac{2\pi}{x}}\,e^{1/(12x)}(1 + \epsilon(x)),$$

or

$$\Gamma(x) \approx \left(\frac{x}{e}\right)^x\sqrt{\frac{2\pi}{x}}\,e^{1/(12x)},$$

which tells us that

$$n! \approx \left(\frac{n}{e}\right)^n\sqrt{2n\pi}\,e^{1/(12n)}. \qquad (6)$$

Compare the values given for 10! by your calculator, Stirling's approximation, and Equation (6).

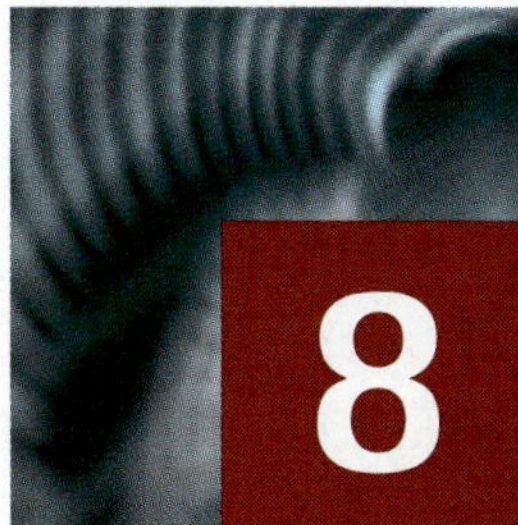

8 Infinite Series

OVERVIEW One infinite process that had puzzled mathematicians for centuries was the summing of infinite series. Sometimes an infinite series of terms added to a number, as in

$$\frac{1}{2}+\frac{1}{4}+\frac{1}{8}+\frac{1}{16}+\cdots=1.$$

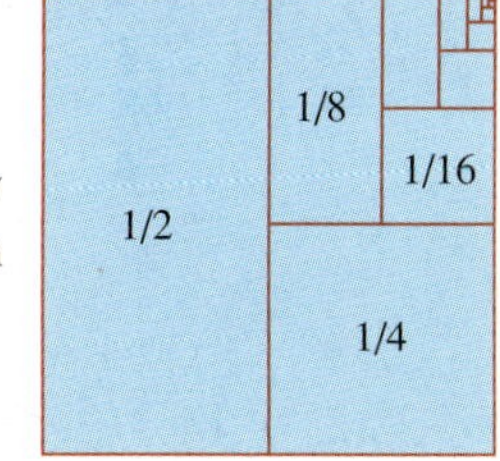

(You can see this by adding the areas in the "infinitely halved" unit square at the right.) Sometimes the infinite sum was infinite, however, as in

$$\frac{1}{1}+\frac{1}{2}+\frac{1}{3}+\frac{1}{4}+\frac{1}{5}+\cdots=\infty$$

(although this is far from obvious), and sometimes the infinite sum was impossible to pin down, as in

$$1-1+1-1+1-1+\cdots.$$

(Is it 0? Is it 1? Is it neither?)

Nonetheless, mathematicians like Gauss and Euler successfully used infinite series to derive previously inaccessible results. Laplace used infinite series to prove the stability of the solar system (although that does not stop some people from worrying about it today when they feel that "too many" planets have swung to the same side of the sun). It was years later that careful analysts like Cauchy developed the theoretical foundation for series computations, sending many mathematicians (including Laplace) back to their desks to verify their results.

Infinite series form the basis for a remarkable formula that enables us to express many functions as "infinite polynomials" and at the same time tells how much error we incur if we truncate those polynomials to make them finite. In addition to providing effective polynomial approximations of differentiable functions, these infinite polynomials (called power series) have many other uses. We also see how to use infinite sums of trigonometric terms, called Fourier series, to represent important functions used in science and engineering applications. Infinite series provide an efficient way to evaluate nonelementary integrals, and they solve differential equations that give insight into heat flow, vibration, chemical diffusion, and signal transmission. What you learn here sets the stage for the roles played by series of functions of all kinds in science and mathematics.

8.1 Limits of Sequences of Numbers

Definitions and Notation • Convergence and Divergence • Calculating Limits of Sequences • Using L'Hôpital's Rule • Limits That Arise Frequently

Informally, a sequence is an ordered list of things, but in this chapter, the things will usually be numbers. We have seen sequences before, such as the sequence $x_0, x_1, \ldots, x_n, \ldots$ of numbers generated by Newton's method. Later we consider sequences involving powers of x and others involving trigonometric terms like $\sin x$, $\cos x$, $\sin 2x$, $\cos 2x$, . . . , $\sin nx$, $\cos nx$, ... A central question is whether a sequence has a limit or not.

CD-ROM
WEBsite
Historical Essay
Sequences and Series

Definitions and Notation

We can list the integer multiples of 3 by assigning each multiple a position:

Domain:	1	2	3 . . .	n . . .
	↓	↓	↓	↓
Range:	3	6	9	$3n$

The first number is 3, the second 6, the third 9, and so on. The assignment is a function that assigns $3n$ to the nth place. That is the basic idea for constructing sequences. There is a function placing each number in the range in its correct ordered position.

Definition Sequence

An infinite **sequence** of numbers is a function whose domain is the set of integers greater than or equal to some integer n_0.

Usually, n_0 is 1 and the domain of the sequence is the set of positive integers. Sometimes, however, we want to start sequences elsewhere. We take $n_0 = 0$ when we begin Newton's method. We might take $n_0 = 3$ if we were defining a sequence of n-sided polygons.

Sequences are defined the same way as other functions, some typical rules being

$$a(n) = \sqrt{n}, \qquad a(n) = (-1)^{n+1}\frac{1}{n}, \qquad a(n) = \frac{n-1}{n}$$

(Example 1 and Figure 8.1). To indicate that the domains are sets of integers, we use a letter like n from the middle of the alphabet for the independent variable, instead of the x, y, z, and t used widely in other contexts. The formulas in the defining rules, however, like those above, are often valid for domains larger than the set of positive integers. This can be an advantage, as we will see. The number

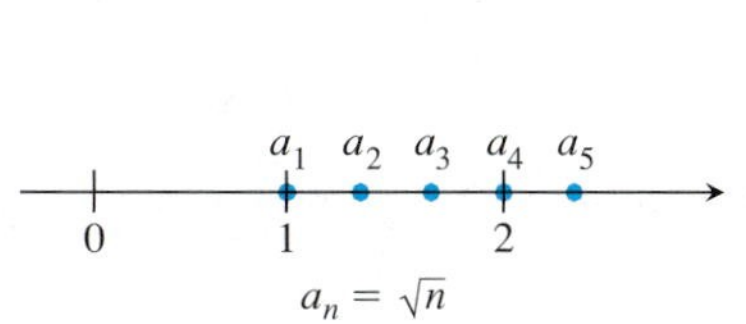

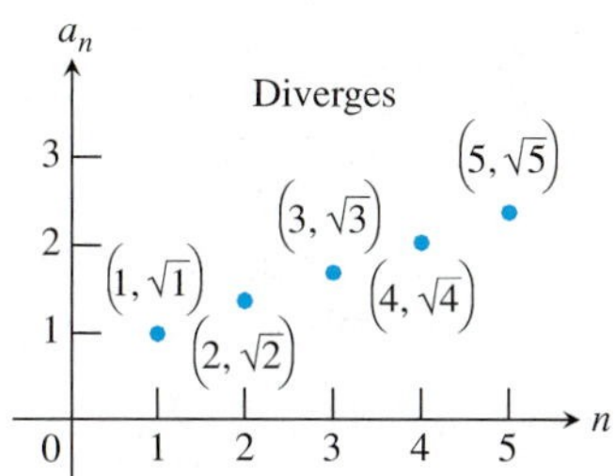

(a) The terms $a_n = \sqrt{n}$ eventually surpass every integer, so the sequence $\{a_n\}$ diverges, . . .

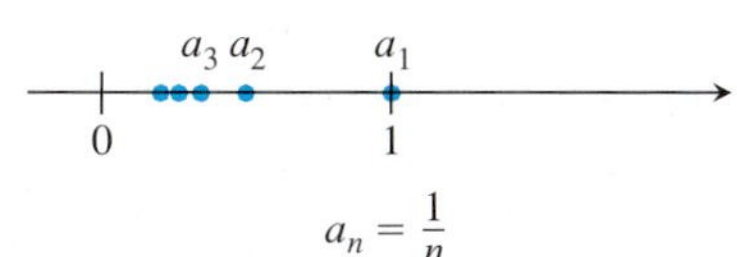

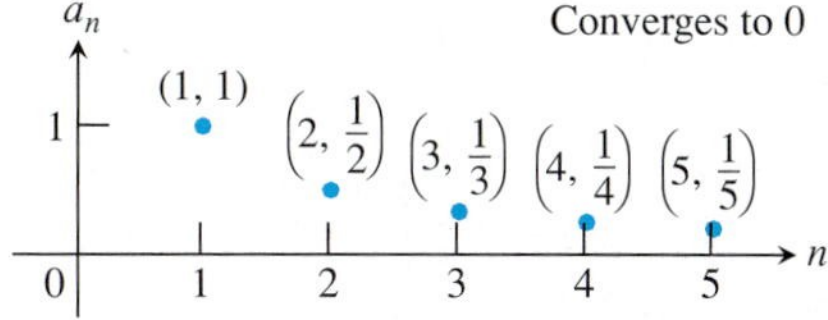

(b) . . . but the terms $a_n = 1/n$ decrease steadily and get arbitrarily close to 0 as n increases, so the sequence $\{a_n\}$ converges to 0.

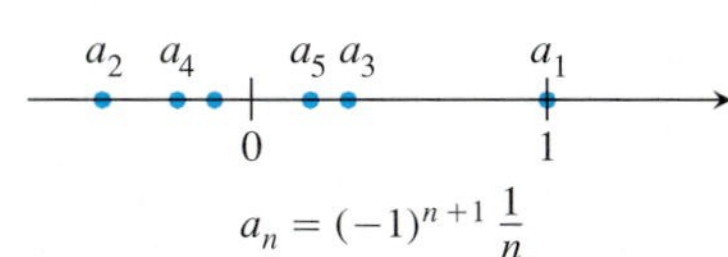

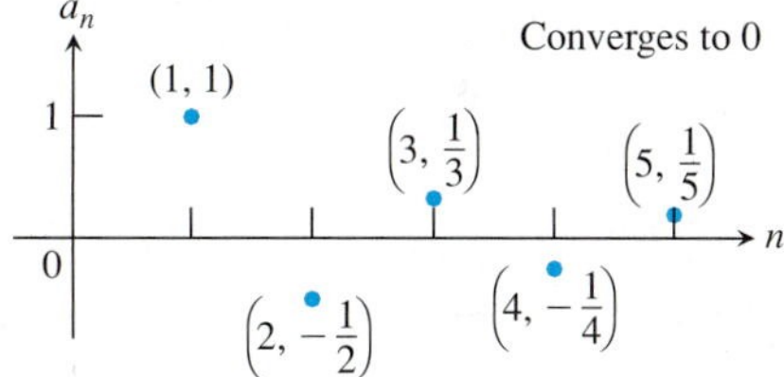

(c) The terms $a_n = (-1)^{n+1}(1/n)$ alternate in sign but still converge to 0.

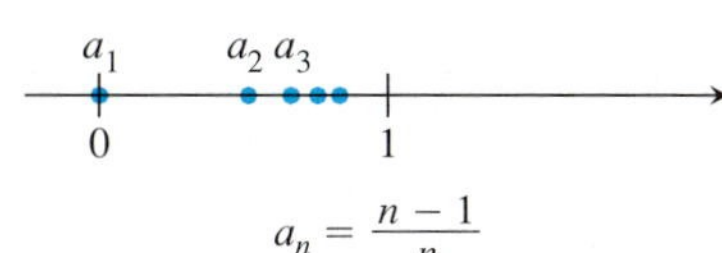

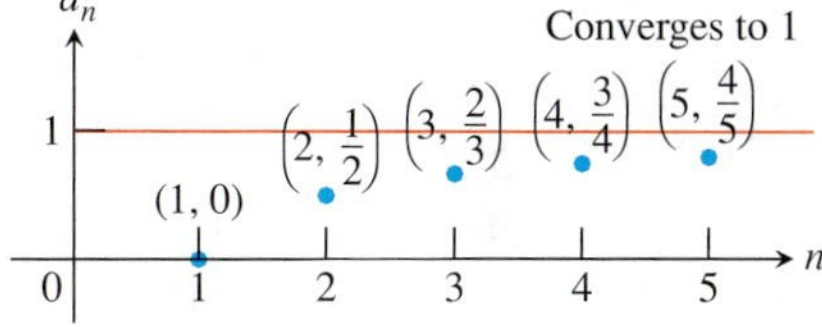

(d) The terms $a_n = (n-1)/n$ approach 1 steadily and get arbitrarily close as n increases, so the sequence $\{a_n\}$ converges to 1.

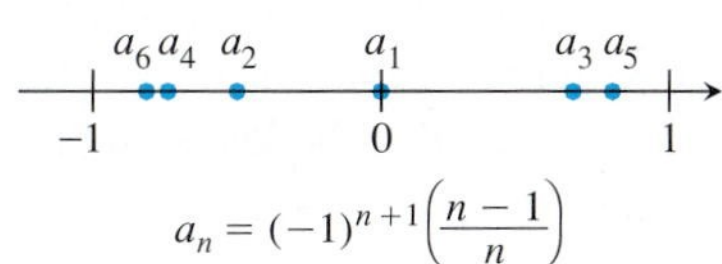

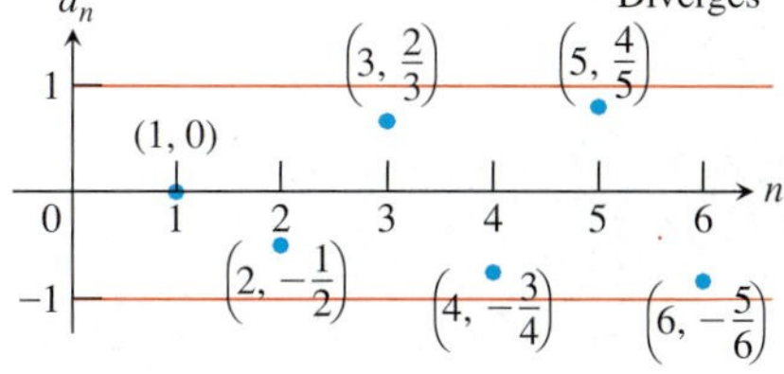

(e) The terms $a_n = (-1)^{n+1}[(n-1)/n]$ alternate in sign. The positive terms approach 1. But the negative terms approach -1 as n increases, so the sequence $\{a_n\}$ diverges.

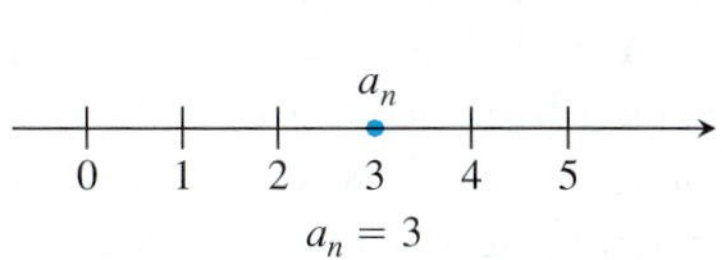

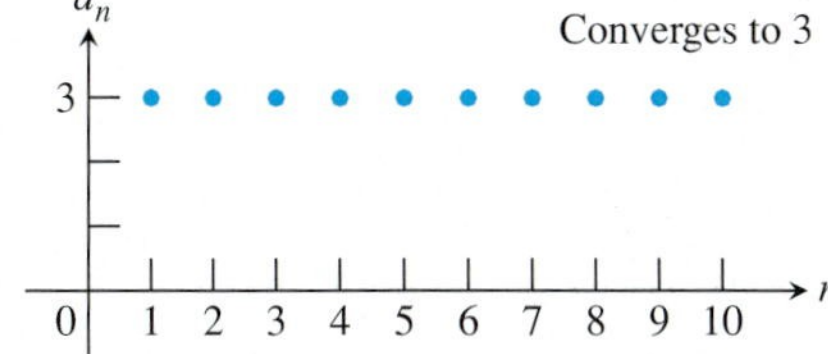

(f) The terms in the sequence of constants $a_n = 3$ have the same value regardless of n, so the sequence $\{a_n\}$ converges to 3.

FIGURE 8.1 The sequences of Example 1 are graphed here in two different ways: by plotting the numbers a_n on a horizontal axis and by plotting the points (n, a_n) in the coordinate plane.

$a(n)$ is the ***n*th term** of the sequence, or the **term with index *n*.** If $a(n) = (n-1)/n$, we have

First term	Second term	Third term		nth term
$a(1) = 0$	$a(2) = \frac{1}{2}$,	$a(3) = \frac{2}{3}$,	$\ldots,$	$a(n) = \frac{n-1}{n}$.

When we use the subscript notation a_n for $a(n)$, the sequence is written

$$a_1 = 0, \qquad a_2 = \frac{1}{2}, \qquad a_3 = \frac{2}{3}, \qquad \ldots, \qquad a_n = \frac{n-1}{n}.$$

To describe sequences, we often write the first few terms as well as a formula for the nth term.

Example 1 Describing Sequences

We write	For the sequence whose defining rule is
(a) $1, \sqrt{2}, \sqrt{3}, \sqrt{4}, \ldots, \sqrt{n}, \ldots$	$a_n = \sqrt{n}$
(b) $1, \frac{1}{2}, \frac{1}{3}, \ldots, \frac{1}{n}, \ldots$	$a_n = \frac{1}{n}$
(c) $1, -\frac{1}{2}, \frac{1}{3}, -\frac{1}{4}, \ldots, (-1)^{n+1}\frac{1}{n}, \ldots$	$a_n = (-1)^{n+1}\frac{1}{n}$
(d) $0, \frac{1}{2}, \frac{2}{3}, \frac{3}{4}, \ldots, \frac{n-1}{n}, \ldots$	$a_n = \frac{n-1}{n}$
(e) $0, -\frac{1}{2}, \frac{2}{3}, -\frac{3}{4}, \ldots, (-1)^{n+1}\left(\frac{n-1}{n}\right), \ldots$	$a_n = (-1)^{n+1}\left(\frac{n-1}{n}\right)$
(f) $3, 3, 3, \ldots, 3, \ldots$	$a_n = 3$

Notation We refer to the sequence whose nth term is a_n with the notation $\{a_n\}$ ("the sequence a sub n"). The second sequence in Example 1 is $\{1/n\}$ ("the sequence 1 over n"); the last sequence is $\{3\}$ ("the constant sequence 3").

Convergence and Divergence

As Figure 8.1 shows, the sequences of Example 1 do not behave the same way. The sequences $\{1/n\}$, $\{(-1)^{n+1}(1/n)\}$, and $\{(n-1)/n\}$ each seem to approach a single limiting value as n increases, and $\{3\}$ is at a limiting value from the very first. On the other hand, terms of $\{(-1)^{n+1}(n-1)/n\}$ seem to accumulate near two different values, -1 and 1, whereas the terms of $\{\sqrt{n}\}$ become increasingly large and do not accumulate anywhere.

The following definition distinguishes those sequences that approach a unique limiting value L, as n increases, from those that do not.

Definitions Converges, Diverges, Limit

The sequence $\{a_n\}$ **converges** to the number L if to every positive number ϵ there corresponds an integer N such that for all n,

$$n > N \Rightarrow |a_n - L| < \epsilon.$$

If no such number L exists, we say that $\{a_n\}$ **diverges.**

If $\{a_n\}$ converges to L, we write $\lim_{n\to\infty} a_n = L$, or simply $a_n \to L$, and call L the **limit** of the sequence (Figure 8.2).

CD-ROM
WEBsite
Historical Biography
Nicole Oresme
(ca. 1320 — 1382)

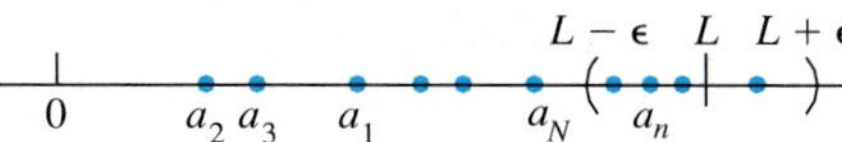

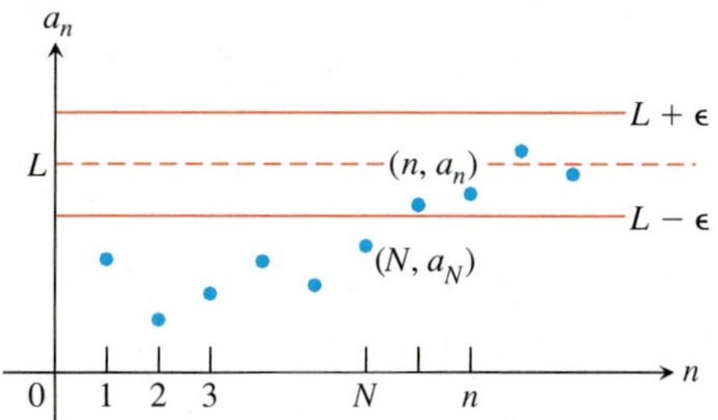

FIGURE 8.2 $a_n \to L$ if $y = L$ is a horizontal asymptote of the sequence of points $\{(n, a_n)\}$. In this figure, all the a_n's after a_N lie within ϵ of L.

Example 2 Testing the Definition

Show that

(a) $\lim_{n\to\infty} \dfrac{1}{n} = 0$

(b) $\lim_{n\to\infty} k = k$ (any constant k).

Solution

(a) Let $\epsilon > 0$ be given. We must show that there exists an integer N such that for all n,

$$n > N \Rightarrow \left|\frac{1}{n} - 0\right| < \epsilon.$$

This implication will hold if $(1/n) < \epsilon$ or $n > 1/\epsilon$. If N is any integer greater than $1/\epsilon$, the implication will hold for all $n > N$. This proves $\lim_{n\to\infty} (1/n) = 0$.

(b) Let $\epsilon > 0$ be given. We must show that there exists an integer N such that for all n,

$$n > N \Rightarrow |k - k| < \epsilon.$$

Since $k - k = 0$, we can use any positive integer for N and the implication will hold. This proves that $\lim_{n\to\infty} k = k$ for any constant k.

Example 3 A Divergent Sequence

Show that $\{(-1)^{n+1}[(n-1)/n]\}$ diverges.

Solution Take a positive ϵ smaller than 1 so that the bands shown in Figure 8.3 about the lines $y = 1$ and $y = -1$ do not overlap. Any $\epsilon < 1$ will do. Convergence to 1 would require every point of the graph beyond a certain index N to lie inside the upper band, but this will never happen. As soon as a point (n, a_n) lies in the upper band, every alternate point starting with $(n + 1, a_{n+1})$ will lie in the lower band. Hence, the sequence cannot converge to 1. Likewise, it cannot converge to -1. On the other hand, because the terms of the sequence get alternately closer to 1 and -1, they never accumulate near any other value. Therefore, the sequence diverges.

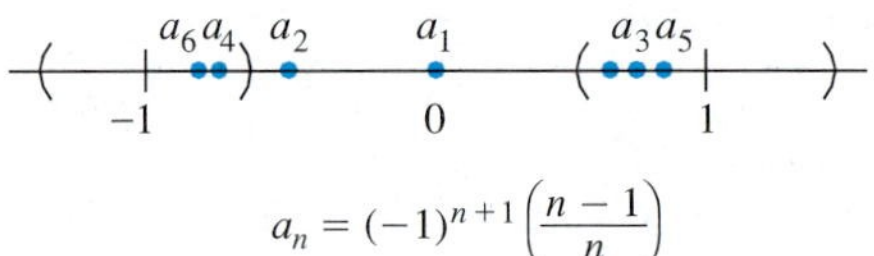

Neither the ϵ-interval about 1 nor the ϵ-interval about -1 contains all a_n satisfying $n \geq N$ for some N.

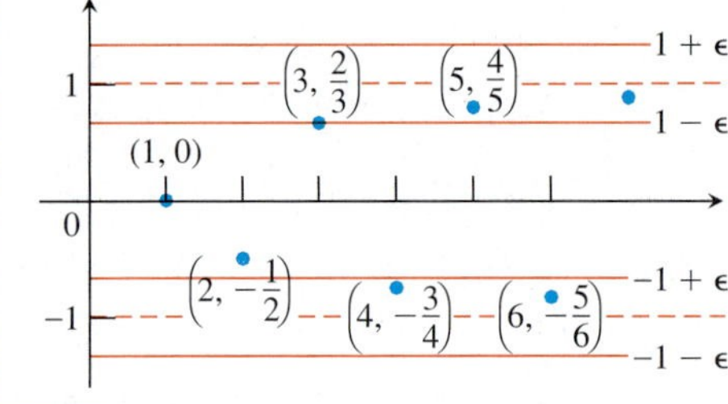

FIGURE 8.3 The sequence $\{(-1)^{n+1}[(n-1)/n]\}$ diverges.

The behavior of $\{(-1)^{n+1}[(n - 1)/n]\}$ is qualitatively different from that of $\{\sqrt{n}\}$, which diverges because it outgrows every real number L. To describe the behavior of $\{\sqrt{n}\}$, we write

$$\lim_{n\to\infty} (\sqrt{n}) = \infty.$$

In speaking of infinity as a limit of a sequence $\{a_n\}$, we do not mean that the difference between a_n and infinity becomes small as n increases. We mean that a_n becomes numerically large as n increases.

Calculating Limits of Sequences

The study of limits would be cumbersome if we had to answer every question about convergence by applying the definition. Fortunately, three theorems make this largely unnecessary. The first theorem is not surprising, based on our previous work with limits. We omit the proofs.

Theorem 1 Limit Laws for Sequences

Let $\{a_n\}$ and $\{b_n\}$ be sequences of real numbers and let A and B be real numbers. The following rules hold if $\lim_{n\to\infty} a_n = A$ and $\lim_{n\to\infty} b_n = B$.

1. *Sum Rule*: $\lim_{n\to\infty} (a_n + b_n) = A + B$
2. *Difference Rule*: $\lim_{n\to\infty} (a_n - b_n) = A - B$
3. *Product Rule*: $\lim_{n\to\infty} (a_n \cdot b_n) = A \cdot B$
4. *Constant Multiple Rule*: $\lim_{n\to\infty} (k \cdot b_n) = k \cdot B$ (any number k)
5. *Quotient Rule*: $\lim_{n\to\infty} \dfrac{a_n}{b_n} = \dfrac{A}{B}$ if $B \neq 0$

Example 4 Applying the Limit Laws

By combining Theorem 1 with the limit results in Example 2, we have

(a) $$\lim_{n\to\infty}\left(-\frac{1}{n}\right) = -1 \cdot \lim_{n\to\infty}\frac{1}{n} = -1 \cdot 0 = 0$$

(b) $$\lim_{n\to\infty}\left(\frac{n-1}{n}\right) = \lim_{n\to\infty}\left(1 - \frac{1}{n}\right) = \lim_{n\to\infty} 1 - \lim_{n\to\infty}\frac{1}{n} = 1 - 0 = 1$$

(c) $$\lim_{n\to\infty}\frac{5}{n^2} = 5 \cdot \lim_{n\to\infty}\frac{1}{n} \cdot \lim_{n\to\infty}\frac{1}{n} = 5 \cdot 0 \cdot 0 = 0$$

(d) $$\lim_{n\to\infty}\frac{4 - 7n^6}{n^6 + 3} = \lim_{n\to\infty}\frac{(4/n^6) - 7}{1 + (3/n^6)} = \frac{0 - 7}{1 + 0} = -7.$$

Example 5 Constant Multiples of Divergent Sequences Diverge

Every nonzero multiple of a divergent sequence $\{a_n\}$ diverges. Suppose, to the contrary, that $\{ca_n\}$ converges for some number $c \neq 0$. Then, by taking $k = 1/c$ in the Constant Multiple Rule in Theorem 1, we see that the sequence

$$\left\{\frac{1}{c} \cdot ca_n\right\} = \{a_n\}$$

converges. Thus, $\{ca_n\}$ cannot converge unless $\{a_n\}$ also converges. If $\{a_n\}$ does not converge, then $\{ca_n\}$ does not converge.

You are asked to prove the next Theorem in Exercise 69.

Theorem 2 The Sandwich Theorem for Sequences

Let $\{a_n\}$, $\{b_n\}$, and $\{c_n\}$ be sequences of real numbers. If $a_n \leq b_n \leq c_n$ holds for all n beyond some index N and if $\lim_{n\to\infty} a_n = \lim_{n\to\infty} c_n = L$, then $\lim_{n\to\infty} b_n = L$ also.

An immediate consequence of Theorem 2 is that if $|b_n| \le c_n$ and $c_n \to 0$, then $b_n \to 0$ because $-c_n \le b_n \le c_n$. We use this fact in the next example.

Example 6 Using the Sandwich Theorem

Since $1/n \to 0$, we know that

(a) $\dfrac{\cos n}{n} \to 0$ because $\left|\dfrac{\cos n}{n}\right| = \dfrac{|\cos n|}{n} \le \dfrac{1}{n}$

(b) $\dfrac{1}{2^n} \to 0$ because $\dfrac{1}{2^n} \le \dfrac{1}{n}$

(c) $(-1)^n \dfrac{1}{n} \to 0$ because $\left|(-1)^n \dfrac{1}{n}\right| \le \dfrac{1}{n}$.

The application of Theorems 1 and 2 is broadened by a theorem stating that applying a continuous function to a convergent sequence produces a convergent sequence. We state the theorem without proof (Exercise 70).

Theorem 3 The Continuous Function Theorem for Sequences

Let $\{a_n\}$ be a sequence of real numbers. If $a_n \to L$ and if f is a function that is continuous at L and defined at all a_n, then $f(a_n) \to f(L)$.

Example 7 Applying Theorem 3

Show that $\sqrt{(n+1)/n} \to 1$.

Solution We know that $(n+1)/n \to 1$. Taking $f(x) = \sqrt{x}$ and $L = 1$ in Theorem 3 gives $\sqrt{(n+1)/n} \to \sqrt{1} = 1$.

Example 8 The Sequence $\{2^{1/n}\}$

The sequence $\{1/n\}$ converges to 0. By taking $a_n = 1/n$, $f(x) = 2^x$, and $L = 0$ in Theorem 3, we see that $2^{1/n} = f(1/n) \to f(L) = 2^0 = 1$. The sequence $\{2^{1/n}\}$ converges to 1 (Figure 8.4).

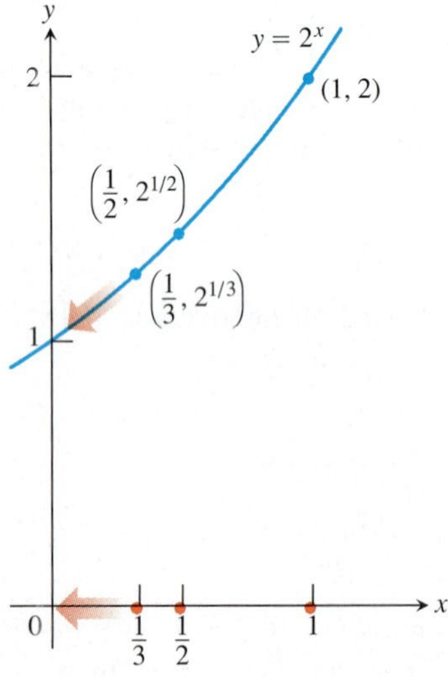

FIGURE 8.4 As $n \to \infty$, $1/n \to 0$ and $2^{1/n} \to 2^0$.

Using L'Hôpital's Rule

The next theorem enables us to use l'Hôpital's Rule to find the limits of some sequences. It matches values of a (usually differentiable) function with the values of a given sequence.

Theorem 4

Suppose that $f(x)$ is a function defined for all $x \ge n_0$ and that $\{a_n\}$ is a sequence of real numbers such that $a_n = f(n)$ for $n \ge n_0$. Then

$$\lim_{x\to\infty} f(x) = L \Rightarrow \lim_{n\to\infty} a_n = L.$$

Example 9 Applying L'Hôpital's Rule

Show that

$$\lim_{n\to\infty} \frac{\ln n}{n} = 0.$$

Solution The function $(\ln x)/x$ is defined for all $x \geq 1$ and agrees with the given sequence at positive integers. Therefore, by Theorem 4, $\lim_{n\to\infty} (\ln n)/n$ will equal $\lim_{x\to\infty} (\ln x)/x$ if the latter exists. A single application of l'Hôpital's Rule shows that

$$\lim_{x\to\infty} \frac{\ln x}{x} = \lim_{x\to\infty} \frac{1/x}{1} = \frac{0}{1} = 0.$$

We conclude that $\lim_{n\to\infty} (\ln n)/n = 0$.

When we use l'Hôpital's Rule to find the limit of a sequence, we often treat n as a continuous real variable and differentiate directly with respect to n. This saves us from having to rewrite the formula for a_n as we did in Example 9.

Example 10 Applying L'Hôpital's Rule

Find

$$\lim_{n\to\infty} \frac{2^n}{5n}.$$

Solution By l'Hôpital's Rule (differentiating with respect to n),

$$\begin{aligned} \lim_{n\to\infty} \frac{2^n}{5n} &= \lim_{n\to\infty} \frac{2^n \cdot \ln 2}{5} \\ &= \infty. \end{aligned}$$

Proof of Theorem 4 Suppose that $\lim_{x\to\infty} f(x) = L$. Then for each positive number ϵ there is a number M such that for all x,

$$x > M \Rightarrow |f(x) - L| < \epsilon.$$

Let N be an integer greater than M and greater than or equal to n_0. Then

$$n > N \Rightarrow a_n = f(n) \qquad \text{and} \qquad |a_n - L| = |f(n) - L| < \epsilon.$$

Example 11 Applying L'Hôpital's Rule to Determine Convergence

Does the sequence whose nth term is

$$a_n = \left(\frac{n+1}{n-1}\right)^n$$

converge? If so, find $\lim_{n\to\infty} a_n$.

Solution The limit leads to the indeterminate form 1^∞. We can apply l'Hôpital's Rule if we first change the form to $\infty \cdot 0$ by taking the natural logarithm of a_n:

$$\begin{aligned}\ln a_n &= \ln\left(\frac{n+1}{n-1}\right)^n \\ &= n \ln\left(\frac{n+1}{n-1}\right).\end{aligned}$$

Then,

$$\begin{aligned}\lim_{n\to\infty} \ln a_n &= \lim_{n\to\infty} n \ln\left(\frac{n+1}{n-1}\right) && \infty \cdot 0 \\ &= \lim_{n\to\infty} \frac{\ln\left(\frac{n+1}{n-1}\right)}{1/n} && \frac{0}{0} \\ &= \lim_{n\to\infty} \frac{-2/(n^2-1)}{-1/n^2} && \text{l'Hôpital's Rule} \\ &= \lim_{n\to\infty} \frac{2n^2}{n^2-1} = 2.\end{aligned}$$

Since $\ln a_n \to 2$ and $f(x) = e^x$ is continuous, Theorem 3 tells us that

$$a_n = e^{\ln a_n} \to e^2.$$

The sequence $\{a_n\}$ converges to e^2.

Table 8.1

1. $\lim_{n\to\infty} \frac{\ln n}{n} = 0$
2. $\lim_{n\to\infty} \sqrt[n]{n} = 1$
3. $\lim_{n\to\infty} x^{1/n} = 1 \qquad (x > 0)$
4. $\lim_{n\to\infty} x^n = 0 \qquad (|x| < 1)$
5. $\lim_{n\to\infty} \left(1 + \frac{x}{n}\right)^n = e^x \qquad (\text{any } x)$
6. $\lim_{n\to\infty} \frac{x^n}{n!} = 0 \qquad (\text{any } x)$

In formulas (3) through (6), x remains fixed as $n \to \infty$.

Limits That Arise Frequently

The limits in Table 8.1 arise frequently. The first limit is from Example 9. The next two can be proved by taking logarithms and applying Theorem 3 (Exercises 67 and 68). The remaining proofs can be found in Appendix 7.

Factorial Notation

The notation $n!$ ("n factorial") means the product $1 \cdot 2 \cdot 3 \cdots n$ of the integers from 1 to n. Notice that $(n+1)! = (n+1) \cdot n!$. Thus, $4! = 1 \cdot 2 \cdot 3 \cdot 4 = 24$ and $5! = 1 \cdot 2 \cdot 3 \cdot 4 \cdot 5 = 5 \cdot 4! = 120$. We define $0!$ to be 1. Factorials grow even faster than exponentials, as the following table suggests.

n	e^n(rounded)	$n!$
1	3	1
5	148	120
10	22,026	3,628,800
20	4.9×10^8	2.4×10^{18}

Example 12 Limits from Table 8.1

(a) $\frac{\ln(n^2)}{n} = \frac{2 \ln n}{n} \to 2 \cdot 0 = 0$ — Formula 1

(b) $\sqrt[n]{n^2} = n^{2/n} = (n^{1/n})^2 \to (1)^2 = 1$ — Formula 2

(c) $\sqrt[n]{3n} = 3^{1/n}(n^{1/n}) \to 1 \cdot 1 = 1$ — Formula 3 with $x = 3$ and Formula 2

(d) $\left(-\frac{1}{2}\right)^n \to 0$ — Formula 4 with $x = -\frac{1}{2}$

(e) $\left(\frac{n-2}{n}\right)^n = \left(1 + \frac{-2}{n}\right)^n \to e^{-2}$ — Formula 5 with $x = -2$

(f) $\frac{100^n}{n!} \to 0$ — Formula 6 with $x = 100$

EXERCISES 8.1

Finding Terms of a Sequence

Each of Exercises 1–4 gives a formula for the nth term a_n of a sequence $\{a_n\}$. Find the values of a_1, a_2, a_3, and a_4.

1. $a_n = \dfrac{1-n}{n^2}$

2. $a_n = \dfrac{1}{n!}$

3. $a_n = \dfrac{(-1)^{n+1}}{2n-1}$

4. $a_n = \dfrac{2^n}{2^{n+1}}$

Finding Formulas for Sequences

In Exercises 5–12, find a formula for the nth term of the sequence.

5. The sequence $1, -1, 1, -1, 1, \ldots$ 1's with alternating signs

6. The sequence $1, -4, 9, -16, 25, \ldots$ Squares of the positive integers, with alternating signs

7. The sequence $0, 3, 8, 15, 24, \ldots$ Squares of the positive integers diminished by 1

8. The sequence $-3, -2, -1, 0, 1, \ldots$ Integers beginning with -3

9. The sequence $1, 5, 9, 13, 17, \ldots$ Every other odd positive integer

10. The sequence $2, 6, 10, 14, 18, \ldots$ Every other even positive integer

11. The sequence $1, 0, 1, 0, 1, \ldots$ Alternating 1's and 0's

12. The sequence $0, 1, 1, 2, 2, 3, 3, 4, \ldots$ Each positive integer repeated

Finding Limits

Which of the sequences $\{a_n\}$ in Exercises 13–56 converge, and which diverge? Find the limit of each convergent sequence.

13. $a_n = 2 + (0.1)^n$

14. $a_n = \dfrac{n + (-1)^n}{n}$

15. $a_n = \dfrac{1-2n}{1+2n}$

16. $a_n = \dfrac{1-5n^4}{n^4+8n^3}$

17. $a_n = \dfrac{n^2-2n+1}{n-1}$

18. $a_n = \dfrac{n+3}{n^2+5n+6}$

19. $a_n = 1 + (-1)^n$

20. $a_n = (-1)^n\left(1 - \dfrac{1}{n}\right)$

21. $a_n = \left(\dfrac{n+1}{2n}\right)\left(1 - \dfrac{1}{n}\right)$

22. $a_n = \dfrac{(-1)^{n+1}}{2n-1}$

23. $a_n = \sqrt{\dfrac{2n}{n+1}}$

24. $a_n = \sin\left(\dfrac{\pi}{2} + \dfrac{1}{n}\right)$

25. $a_n = \dfrac{\sin n}{n}$

26. $a_n = \dfrac{\sin^2 n}{2^n}$

27. $a_n = \dfrac{n}{2^n}$

28. $a_n = \dfrac{\ln(n+1)}{\sqrt{n}}$

29. $a_n = \dfrac{\ln n}{n^{1/n}}$

30. $a_n = \ln n - \ln(n+1)$

31. $a_n = \left(1 + \dfrac{7}{n}\right)^n$

32. $a_n = \left(1 - \dfrac{1}{n}\right)^n$

33. $a_n = \sqrt[n]{10n}$

34. $a_n = \sqrt[n]{n^2}$

35. $a_n = \left(\dfrac{3}{n}\right)^{1/n}$

36. $a_n = (n+4)^{1/(n+4)}$

37. $a_n = \sqrt[n]{4^n n}$

38. $a_n = \sqrt[n]{3^{2n+1}}$

39. $a_n = \dfrac{n!}{n^n}$ (*Hint*: Compare with $1/n$.)

40. $a_n = \dfrac{(-4)^n}{n!}$

41. $a_n = \dfrac{n!}{10^{6n}}$

42. $a_n = \dfrac{n!}{2^n \cdot 3^n}$

43. $a_n = \left(\dfrac{1}{n}\right)^{1/(\ln n)}$

44. $a_n = \ln\left(1 + \dfrac{1}{n}\right)^n$

45. $a_n = \left(\dfrac{3n+1}{3n-1}\right)^n$

46. $a_n = \left(\dfrac{n}{n+1}\right)^n$

47. $a_n = \left(\dfrac{x^n}{2n+1}\right)^{1/n}, \quad x > 0$

48. $a_n = \left(1 - \dfrac{1}{n^2}\right)^n$

49. $a_n = \dfrac{3^n \cdot 6^n}{2^{-n} \cdot n!}$

50. $a_n = \dfrac{n^2}{2n-1}\sin\dfrac{1}{n}$

51. $a_n = \tan^{-1} n$

52. $a_n = \dfrac{1}{\sqrt{n}}\tan^{-1} n$

53. $a_n = \left(\dfrac{1}{3}\right)^n + \dfrac{1}{\sqrt{2^n}}$

54. $a_n = \sqrt[n]{n^2+n}$

55. $a_n = \dfrac{(\ln n)^5}{\sqrt{n}}$

56. $a_n = n - \sqrt{n^2-n}$

Calculator Explorations of Limits

In Exercises 57–60, experiment with a calculator to find a value of N that will make the inequality hold for all $n > N$. Assuming that the inequality is the one from the formal definition of the limit of a sequence, what sequence is being considered in each case, and what is its limit?

57. $|\sqrt[n]{0.5} - 1| < 10^{-3}$

58. $|\sqrt[n]{n} - 1| < 10^{-3}$

59. $(0.9)^n < 10^{-3}$

60. $(2^n/n!) < 10^{-7}$

Theory and Examples

61. A sequence of rational numbers is described as follows:

$$\frac{1}{1}, \frac{3}{2}, \frac{7}{5}, \frac{17}{12}, \ldots, \frac{a}{b}, \frac{a+2b}{a+b}, \ldots.$$

Here the numerators form one sequence, the denominators form a second sequence, and their ratios form a third sequence. Let x_n and y_n be, respectively, the numerator and the denominator of the nth fraction $r_n = x_n/y_n$.

(a) Verify that $x_1^2 - 2y_1^2 = -1$, $x_2^2 - 2y_2^2 = +1$, and, more generally, that if $a^2 - 2b^2 = -1$ or $+1$, then

$$(a + 2b)^2 - 2(a + b)^2 = +1 \quad \text{or} \quad -1,$$

respectively.

(b) The fractions $r_n = x_n/y_n$ approach a limit as n increases. What is that limit? (*Hint*: Use part (a) to show that $r_n^2 - 2 = \pm(1/y_n)^2$ and that y_n is not less than n.)

62. **(a)** Suppose that $f(x)$ is differentiable for all x in $[0, 1]$ and that $f(0) = 0$. Define the sequence $\{a_n\}$ by the rule $a_n = nf(1/n)$. Show that $\lim_{n\to\infty} a_n = f'(0)$.

Use the result in part (a) to find the limits of the following sequences $\{a_n\}$.

(b) $a_n = n \tan^{-1} \dfrac{1}{n}$ **(c)** $a_n = n(e^{1/n} - 1)$

(d) $a_n = n \ln\left(1 + \dfrac{2}{n}\right)$

63. *Pythagorean triples* A triple of positive integers a, b, and c is called a **Pythagorean triple** if $a^2 + b^2 = c^2$. Let a be an odd positive integer and let

$$b = \left\lfloor \frac{a^2}{2} \right\rfloor \quad \text{and} \quad c = \left\lceil \frac{a^2}{2} \right\rceil$$

be, respectively, the integer floor and ceiling for $a^2/2$.

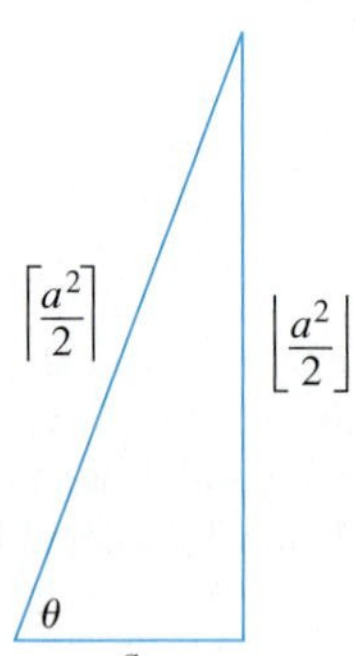

(a) Show that $a^2 + b^2 = c^2$. (*Hint*: Let $a = 2n + 1$ and express b and c in terms of n.)

(b) By direct calculation, or by appealing to the figure, find

$$\lim_{a\to\infty} \frac{\left\lfloor \dfrac{a^2}{2} \right\rfloor}{\left\lceil \dfrac{a^2}{2} \right\rceil}.$$

64. *The nth root of n!*

(a) Show that $\lim_{n\to\infty} (2n\pi)^{1/(2n)} = 1$ and hence, using Stirling's approximation (Chapter 7, Additional Exercise 50, part (a), that

$$\sqrt[n]{n!} \approx \frac{n}{e} \quad \text{for large values of } n.$$

(b) Test the approximation in part (a) for $n = 40, 50, 60, \ldots$ as far as your calculator will allow.

65. **(a)** Assuming that $\lim_{n\to\infty} (1/n^c) = 0$ if c is any positive constant, show that

$$\lim_{n\to\infty} \frac{\ln n}{n^c} = 0$$

if c is any positive constant.

(b) Prove that $\lim_{n\to\infty} (1/n^c) = 0$ if c is any positive constant. (*Hint*: If $\epsilon = 0.001$ and $c = 0.04$, how large should N be to ensure that $|1/n^c - 0| < \epsilon$ if $n > N$?)

66. *The Zipper Theorem* Prove the "Zipper Theorem" for sequences: If $\{a_n\}$ and $\{b_n\}$ both converge to L, then the sequence

$$a_1, b_1, a_2, b_2, \ldots, a_n, b_n, \ldots$$

converges to L.

67. Prove that $\lim_{n\to\infty} \sqrt[n]{n} = 1$.

68. Prove that $\lim_{n\to\infty} x^{1/n} = 1 \ (x > 0)$.

69. Prove Theorem 2.

70. Prove Theorem 3.

71. *Terms become arbitrarily close in convergent sequences* Prove that if $\{a_n\}$ is a convergent sequence, then to every positive number ϵ there corresponds an integer N such that for all m and n,

$$m > N \quad \text{and} \quad n > N \Rightarrow |a_m - a_n| < \epsilon.$$

72. *Uniqueness of limits* Prove that limits of sequences are unique. That is, show that if L_1 and L_2 are numbers such that $a_n \to L_1$ and $a_n \to L_2$, then $L_1 = L_2$.

73. *Convergence and absolute value* Prove that a sequence $\{a_n\}$ converges to 0 if and only if the sequence of absolute values $\{|a_n|\}$ converges to 0.

74. *Improving automobile production* According to a front-page article in the December 15, 1992 issue of the *Wall Street Journal*, Ford Motor Company now uses about $7\frac{1}{4}$ h of labor to produce stampings for the average vehicle, down from an estimated 15 h in 1980. The Japanese need only about $3\frac{1}{2}$ h.

Ford's improvement since 1980 represents an average decrease of 6% per year. If that rate continues, then n years from now Ford will use about

$$S_n = 7.25(0.94)^n$$

hours of labor to produce stampings for the average vehicle. Assuming that the Japanese continue to spend $3\frac{1}{2}$ h per vehicle, how many more years will it take Ford to catch up? Find out two ways:

(a) Find the first term of the sequence $\{S_n\}$ that is less than or equal to 3.5.

T (b) Graph $f(x) = 7.25(0.94)^x$ and use Trace to find where the graph crosses the line $y = 3.5$.

COMPUTER EXPLORATIONS

Looking for Signs of Convergence and Divergence

Use a CAS to perform the following steps for the sequences in Exercises 75–84.

(a) Calculate and then plot the first 25 terms of the sequence. Does the sequence appear to converge or diverge? If it does converge, what is the limit L?

(b) If the sequence converges, find an integer N such that $|a_n - L| \le 0.01$ for $n \ge N$. How far in the sequence do you have to get for the terms to lie within 0.0001 of L?

75. $a_n = \sqrt[n]{n}$

76. $a_n = \left(1 + \frac{0.5}{n}\right)^n$

77. $a_n = \sin n$

78. $a_n = n \sin \frac{1}{n}$

79. $a_n = \frac{\sin n}{n}$

80. $a_n = \frac{\ln n}{n}$

81. $a_n = (0.9999)^n$

82. $a_n = 123456^{1/n}$

83. $a_n = \frac{8^n}{n!}$

84. $a_n = \frac{n^{41}}{19^n}$

8.2 Subsequences, Bounded Sequences, and Picard's Method

Subsequences • Monotonic and Bounded Sequences • Recursively Defined Sequences • Picard's Method for Finding Roots

This section continues our study of the convergence or divergence of a sequence.

Subsequences

If the terms of one sequence appear in another sequence in their given order, we call the first sequence a **subsequence** of the second.

Example 1 Subsequences of the Sequence of Positive Integers

(a) The subsequence of even integers: $2, 4, 6, \ldots, 2n, \ldots$

(b) The subsequence of odd integers: $1, 3, 5, \ldots, 2n - 1, \ldots$

(c) The subsequence of primes: $2, 3, 5, 7, 11, \ldots$

Subsequences are important for two reasons:

1. If a sequence $\{a_n\}$ converges to L, then all its subsequences converge to L. If we know that a sequence converges, it may be quicker to find or estimate its limit by examining a particular subsequence.

2. If any subsequence of a sequence $\{a_n\}$ diverges or if two subsequences have different limits, then $\{a_n\}$ diverges. For example, the sequence $\{(-1)^n\}$ diverges because the subsequence $-1, -1, -1, \ldots$ of odd-numbered terms converge to -1, whereas the subsequence $1, 1, 1, \ldots$ of even-numbered terms converges to 1, a different limit.

The convergence or divergence of a sequence has nothing to do with how the sequence begins. It depends only on how the tails behave.

Subsequences also provide a new way to view convergence. A **tail** of a sequence is a subsequence that consists of all terms of the sequence from some index N on. In other words, a tail is one of the sets $\{a_n \mid n \geq N\}$. Another way to say that $a_n \to L$ is to say that every ϵ-interval about L contains a tail of the sequence.

Monotonic and Bounded Sequences

Definition Nondecreasing, Nonincreasing, Monotonic Sequence

A sequence $\{a_n\}$ with the property that $a_n \leq a_{n+1}$ for all n is called a **nondecreasing sequence**; that is, $a_1 \leq a_2 \leq a_3 \leq \ldots$.

It is called **nonincreasing** if $a_n \geq a_{n+1}$ for all n. A sequence is **monotonic** if it is either nondecreasing or nonincreasing.

CD-ROM
WEBsite
Historical Biography
Fibonacci
(1170 — 1240)

Example 2 Monotonic Sequences

(a) The sequence $1, 2, 3, \ldots, n, \ldots$ of natural numbers is nondecreasing.

(b) The sequence $\frac{1}{2}, \frac{2}{3}, \frac{3}{4}, \ldots, \frac{n}{n+1}, \ldots$ is nondecreasing.

(c) The sequence $\frac{3}{8}, \frac{3}{9}, \frac{3}{10}, \ldots, \frac{3}{n+7}, \ldots$ is nonincreasing.

(d) The constant sequence $\{3\}$ is both nondecreasing and nonincreasing.

Example 3 A Nondecreasing Sequence

Show that the sequence

$$a_n = \frac{n-1}{n+1}$$

is nondecreasing.

Solution

(a) We show that for all $n \geq 1$, $a_n \leq a_{n+1}$; that is,

$$\frac{n-1}{n+1} \leq \frac{(n+1)-1}{(n+1)+1}.$$

The inequality is equivalent to the one we get by cross multiplication:

$$\begin{aligned} \frac{n-1}{n+1} \leq \frac{(n+1)-1}{(n+1)+1} &\Leftrightarrow \frac{n-1}{n+1} \leq \frac{n}{n+2} \\ &\Leftrightarrow (n-1)(n+2) \leq n(n+1) \\ &\Leftrightarrow n^2+n-2 \leq n^2+n \\ &\Leftrightarrow -2 \leq 0. \end{aligned}$$

Since $-2 \leq 0$ is true, $a_n \leq a_{n+1}$ and the sequence $\{a_n\}$ is nondecreasing.

(b) Another way to show that $\{a_n\}$ is nondecreasing is by defining $f(n) = a_n$ and establishing that $f'(x) \geq 0$. In this example, $f(n) = (n-1)/(n+1)$ and

$$f'(x) = \frac{d}{dx}\left(\frac{x-1}{x+1}\right)$$

$$= \frac{(x+1)(1) - (x-1)(1)}{(x+1)^2} \qquad \text{Quotient Rule}$$

$$= \frac{2}{(x+1)^2} > 0.$$

Therefore, f is an increasing function, so $f(n+1) \geq f(n)$, or $a_{n+1} \geq a_n$.

Definition Bounded from Above, Upper Bound, Bounded from Below, Lower Bound, Bounded Sequence

A sequence $\{a_n\}$ is **bounded from above** if there exists a number M such that $a_n \leq M$ for all n. The number M is an **upper bound** for $\{a_n\}$. The sequence is **bounded from below** if there exists a number m such that $m \leq a_n$ for all n. The number m is a **lower bound** for $\{a_n\}$. If it is bounded from above and below, then $\{a_n\}$ is a **bounded sequence.**

Example 4 Applying The Definition For Boundedness

(a) The sequence $1, 2, 3, \ldots, n, \ldots$ has no upper bound, but it is bounded below by $m = 1$.

(b) The sequence $\frac{1}{2}, \frac{2}{3}, \frac{3}{4}, \ldots, \frac{n}{n+1}, \ldots$ is bounded above by $M = 1$ and below by $m = \frac{1}{2}$.

(c) The sequence $-1, 2, -3, 4, \ldots, (-1)^n n, \ldots$ is neither bounded from above nor from below.

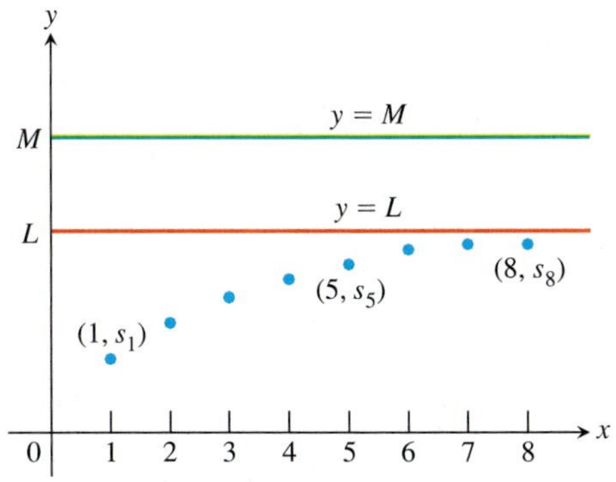

We know that not every bounded sequence converges because the sequence $a_n = (-1)^n$ is bounded $(-1 \leq a_n \leq 1)$ but is divergent. Also, not every monotonic sequence converges because the sequence $1, 2, 3, \ldots, n, \ldots$ of natural numbers is monotonic but diverges. If a sequence is *both* bounded and monotonic, however, then it must converge. This fact is summarized in the following theorem.

Theorem 5 Monotonic Sequence Theorem

Every bounded, monotonic sequence is convergent.

y
y = M
M
y = L
L
(8, s_8)
(5, s_5)
(1, s_1)
0 1 2 3 4 5 6 7 8 x

FIGURE 8.5 If the terms of a nondecreasing sequence have an upper bound M, they have a limit $L \leq M$.

Although we do not prove Theorem 5, Figure 8.5 helps us understand why it is true for a nondecreasing sequence that is bounded from above. Since the sequence is nondecreasing and cannot go above M, the terms are forced to bunch together around some number $L \leq M$.

Example 5 Applying Theorem 5

(a) The nondecreasing sequence $\left\{\dfrac{n}{n+1}\right\}$ is convergent because it is bounded above by $M = 1$. In fact,

$$\lim_{n\to\infty} \frac{n}{n+1} = \lim_{n\to\infty} \frac{1}{1+(1/n)} = \frac{1}{1+0} = 1,$$

and the sequence converges to $L = 1$.

(b) The nonincreasing sequence $\left\{\dfrac{1}{n+1}\right\}$ is bounded below by $m = 0$ and is therefore convergent. It converges to $L = 0$.

Recursively Defined Sequences

Recursion formulas arise regularly in computer programs and numerical routines for solving differential equations, such as Euler's method.

So far, the sequences we have studied calculated each a_n directly from the value of n. Sequences, however, are often defined **recursively** by giving

1. The value(s) of the initial term or terms and
2. A rule, called a **recursion formula,** for calculating any later term from terms that precede it.

Example 6 Sequences Constructed Recursively

(a) The statements $a_1 = 1$ and $a_n = a_{n-1} + 1$ define the sequence $1, 2, 3, \ldots, n, \ldots$ of positive integers. With $a_1 = 1$, we have $a_2 = a_1 + 1 = 2$, $a_3 = a_2 + 1 = 3$, and so on.

(b) The statements $a_1 = 1$ and $a_n = n \cdot a_{n-1}$ define the sequence $1, 2, 6, 24, \ldots, n!, \ldots$ of factorials. With $a_1 = 1$, we have $a_2 = 2 \cdot a_1 = 2$, $a_3 = 3 \cdot a_2 = 6$, $a_4 = 4 \cdot a_3 = 24$, and so on.

(c) The statements $a_1 = 1$, $a_2 = 1$, and $a_{n+1} = a_n + a_{n-1}$ define the sequence $1, 1, 2, 3, 5, \ldots$ of **Fibonacci numbers.** With $a_1 = 1$ and $a_2 = 1$, we have $a_3 = 1 + 1 = 2$, $a_4 = 2 + 1 = 3$, $a_5 = 3 + 2 = 5$, and so on.

(d) As we can see by applying Newton's method, the statements $x_0 = 1$ and $x_{n+1} = x_n - [(\sin x_n - x_n^2)/(\cos x_n - 2x_n)]$ define a sequence that converges to a solution of the equation $\sin x - x^2 = 0$.

Picard's Method for Finding Roots

CD-ROM WEBsite
Historical Biography
Charles Émile Picard (1856 — 1941)

The problem of solving the equation

$$f(x) = 0 \tag{1}$$

is equivalent to that of solving the equation

$$g(x) = f(x) + x = x,$$

obtained by adding x to both sides of Equation (1). By this simple change, we cast Equation (1) into a form that may render it solvable on a computer by a powerful method called **Picard's method.**

If the domain of g contains the range of g, we can start with a point x_0 in the domain and apply g repeatedly to get

$$x_1 = g(x_0), \qquad x_2 = g(x_1), \qquad x_3 = g(x_2), \ldots.$$

Under simple restrictions that we describe shortly, the sequence generated by the recursion formula $x_{n+1} = g(x_n)$ will converge to a point x for which $g(x) = x$. This point solves the equation $f(x) = 0$ because

$$f(x) = g(x) - x = x - x = 0.$$

A point x for which $g(x) = x$ is a **fixed point** of g. We see from the last equation that the fixed points of g are precisely the roots of f.

Example 7 Testing Picard's Method

Solve the equation

$$\frac{1}{4}x + 3 = x.$$

Solution By algebra, we know that the solution is $x = 4$. To apply Picard's method, we take

$$g(x) = \frac{1}{4}x + 3,$$

choose a starting point, say $x_0 = 1$, and calculate the initial terms of the sequence $x_{n+1} = g(x_n)$. Table 8.2 lists the results. In 10 steps, the solution of the original equation is found with an error of magnitude less than 3×10^{-6}.

Table 8.2 Successive iterates of $g(x) = (1/4)x + 3$, starting with $x_0 = 1$

x_n	$x_{n+1} = g(x_n) = (1/4)\,x_n + 3$
$x_0 = 1$	$x_1 = g(x_0) = (1/4)(1) + 3 = 3.25$
$x_1 = 3.25$	$x_2 = g(x_1) = (1/4)(3.25) + 3 = 3.8125$
$x_2 = 3.8125$	$x_3 = g(x_2) = 3.9531\ 25$
$x_3 = 3.9531\ 25$	$x_4 = 3.9882\ 8125$
.	$x_5 = 3.9970\ 70313$
.	$x_6 = 3.9992\ 67578$
.	$x_7 = 3.9998\ 16895$
	$x_8 = 3.9999\ 54224$
	$x_9 = 3.9999\ 88556$
	$x_{10} = 3.9999\ 97139$
	⋮

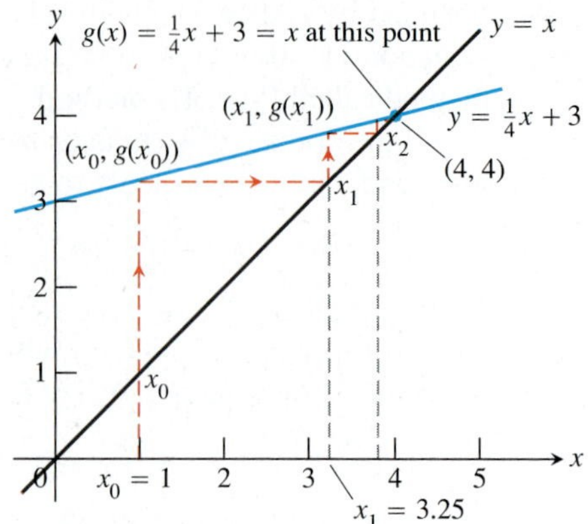

FIGURE 8.6 The Picard solution of the equation $g(x) = (1/4)x + 3 = x$. (Example 7)

Figure 8.6 shows the geometry of the solution. We start with $x_0 = 1$ and calculate the first value $g(x_0)$. This becomes the second x-value x_1. The second y-value $g(x_1)$ becomes the third x-value x_2, and so on. The process is shown as a path (called the *iteration path*) that starts at $x_0 = 1$, moves up to $(x_0, g(x_0)) = (x_0, x_1)$, over to (x_1, x_1), up to $(x_1, g(x_1))$, and so on. The path converges to the point where the graph of g meets the line $y = x$. This is the point where $g(x) = x$.

Example 8 Using Picard's Method

Solve the equation $\cos x = x$.

Solution We take $g(x) = \cos x$, choose $x_0 = 1$ as a starting value, and use the recursion formula $x_{n+1} = g(x_n)$ to find

$$x_0 = 1, \qquad x_1 = \cos 1, \qquad x_2 = \cos(x_1), \ldots .$$

We can approximate the first 50 terms or so on a calculator in radian mode by entering 1 and taking the cosine repeatedly. The display stops changing when $\cos x = x$ to the number of decimal places in the display.

Try it for yourself. As you continue to take the cosine, the successive approximations lie alternately above and below the fixed point $x = 0.739085133\ldots$.

Figure 8.7 shows that the values oscillate this way because the path of the procedure spirals around the fixed point.

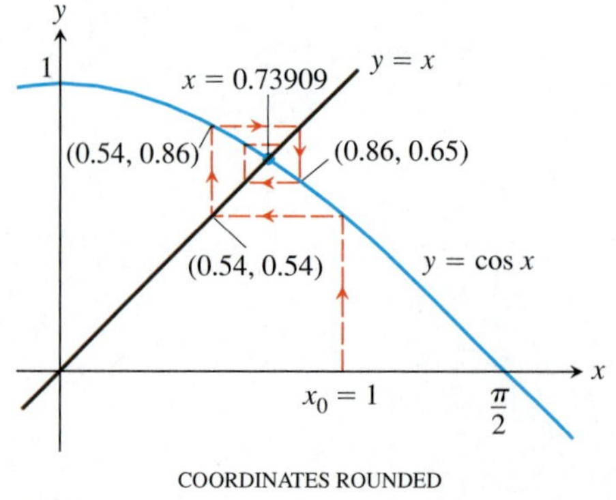

FIGURE 8.7 The solution of $\cos x = x$ by Picard's method starting at $x_0 = 1$. (Example 8)

Example 9 Picard's Method May Fail to Solve an Equation

Picard's method will not solve the equation

$$g(x) = 4x - 12 = x.$$

As Figure 8.8 shows, any choice of x_0 except $x_0 = 4$, the solution itself, generates a divergent sequence that moves away from the solution.

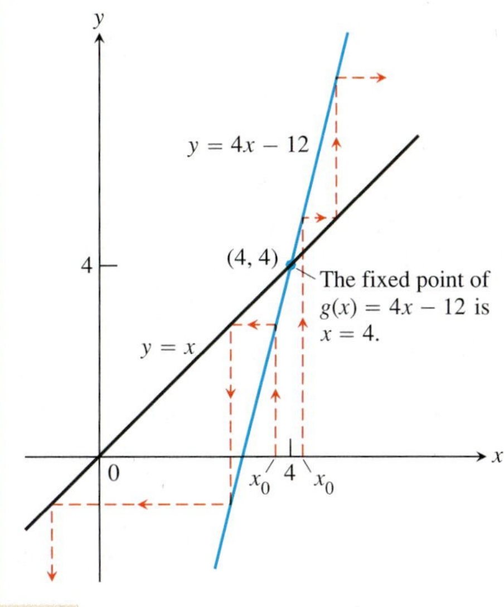

FIGURE 8.8 Applying the Picard method to $g(x) = 4x - 12$ will not find the fixed point unless x_0 is the fixed point 4 itself. (Example 9)

The difficulty in Example 9 can be traced to the slope of the line $y = 4x - 12$ exceeding 1, the slope of the line $y = x$. Conversely, the process worked in Example 7 because the slope of the line $y = (1/4)x + 3$ was numerically less than 1. A theorem from advanced calculus tells us that if $g'(x)$ is continuous on a closed interval I whose interior contains a solution of the equation $g(x) = x$ and if $|g'(x)| < 1$ on I, then any choice of x_0 in the interior of I will lead to the solution.

EXERCISES 8.2

Finding Terms of a Recursively Defined Sequence

Each of Exercises 1–6 gives the first term or two of a sequence along with a recursion formula for the remaining terms. Write out the first ten terms of the sequence.

1. $a_1 = 1, \quad a_{n+1} = a_n + (1/2^n)$

2. $a_1 = 1, \quad a_{n+1} = a_n/(n + 1)$

3. $a_1 = 2, \quad a_{n+1} = (-1)^{n+1} a_n/2$

4. $a_1 = -2, \quad a_{n+1} = na_n/(n + 1)$

5. $a_1 = a_2 = 1, \quad a_{n+2} = a_{n+1} + a_n$

6. $a_1 = 2, \quad a_2 = -1, \quad a_{n+2} = a_{n+1}/a_n$

T **7.** *Sequences generated by Newton's method* Newton's method, applied to a differentiable function $f(x)$, begins with a starting value x_0 and constructs from it a sequence of numbers $\{x_n\}$ that under favorable circumstances converges to a zero of f. The recursion formula for the sequence is

$$x_{n+1} = x_n - \frac{f(x_n)}{f'(x_n)}.$$

(a) Show that the recursion formula for $f(x) = x^2 - a$, $a > 0$, can be written as $x_{n+1} = (x_n + a/x_n)/2$.

(b) *Writing to Learn* Starting with $x_0 = 1$ and $a = 3$, calculate successive terms of the sequence until the display begins to repeat. What number is being approximated? Explain.

T **8.** (*Continuation of Exercise 7*) Repeat part (b) of Exercise 7 with $a = 2$ in place of $a = 3$.

T **9.** *Newton's method* The following sequences come from the recursion formula for Newton's method (see Exercise 7).

Do the sequences converge? If so, to what value? In each case, begin by identifying the function f that generates the sequence.

(a) $x_0 = 1, \quad x_{n+1} = x_n - \dfrac{x_n^2 - 2}{2x_n} = \dfrac{x_n}{2} + \dfrac{1}{x_n}$

(b) $x_0 = 1, \quad x_{n+1} = x_n - \dfrac{\tan x_n - 1}{\sec^2 x_n}$

(c) $x_0 = 1, \quad x_{n+1} = x_n - 1$

T **10.** *A recursive definition of $\pi/2$* If you start with $x_1 = 1$ and define the subsequent terms of $\{x_n\}$ by the rule $x_n = x_{n-1} + \cos x_{n-1}$, you generate a sequence that converges rapidly to $\pi/2$.

(a) Try it.

(b) Use the accompanying figure to explain why the convergence is so rapid.

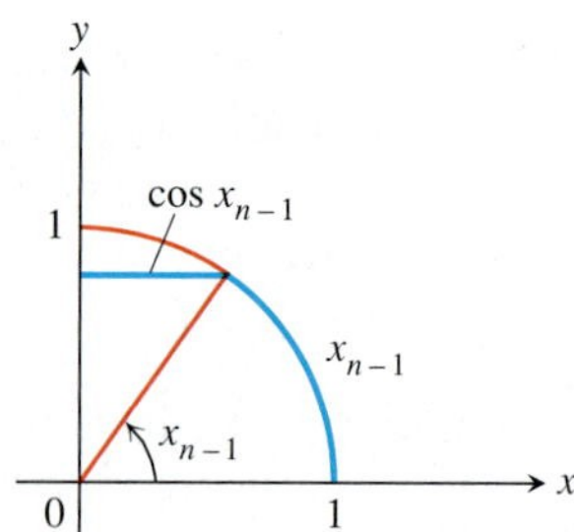

Theory and Examples

In Exercises 11–14, determine if the sequence is nondecreasing and if it is bounded from above.

11. $a_n = \dfrac{3n + 1}{n + 1}$

12. $a_n = \dfrac{(2n + 3)!}{(n + 1)!}$

13. $a_n = \dfrac{2^n 3^n}{n!}$

14. $a_n = 2 - \dfrac{2}{n} - \dfrac{1}{2^n}$

Which of the sequences in Exercises 15–24 converge, and which diverge? Give reasons for your answers.

15. $a_n = 1 - \dfrac{1}{n}$

16. $a_n = n - \dfrac{1}{n}$

17. $a_n = \dfrac{2^n - 1}{2^n}$

18. $a_n = \dfrac{2^n - 1}{3^n}$

19. $a_n = ((-1)^n + 1)\left(\dfrac{n + 1}{n}\right)$

20. The first term of a sequence is $x_1 = \cos(1)$. The next terms are $x_2 = x_1$ or $\cos(2)$, whichever is larger, and $x_3 = x_2$ or $\cos(3)$, whichever is larger (farther to the right). In general,

$$x_{n+1} = \max\{x_n, \cos(n + 1)\}.$$

21. $a_n = \dfrac{n + 1}{n}$

22. $a_n = \dfrac{1 + \sqrt{2n}}{\sqrt{n}}$

23. $a_n = \dfrac{1 - 4^n}{2^n}$

24. $a_n = \dfrac{4^{n+1} + 3^n}{4^n}$

25. *Limits and subsequences* Prove that if two subsequences of a sequence $\{a_n\}$ have different limits $L_1 \neq L_2$, then $\{a_n\}$ diverges.

26. *Even and odd indices* For a sequence $\{a_n\}$, the terms of even index are denoted by a_{2k} and the terms of odd index by a_{2k+1}. Prove that if $a_{2k} \to L$ and $a_{2k+1} \to L$, then $a_n \to L$.

Picard's Method

Use Picard's method to solve the equations in Exercises 27–32.

27. $\sqrt{x} = x$

28. $x^2 = x$

29. $\cos x + x = 0$

30. $\cos x = x + 1$

31. $x - \sin x = 0.1$

32. $\sqrt{x} = 4 - \sqrt{1 + x}$ (*Hint*: Square both sides first.)

33. Solving the equation $\sqrt{x} = x$ by Picard's method finds the solution $x = 1$ but not the solution $x = 0$. Why? (*Hint*: Graph $y = x$ and $y = \sqrt{x}$ together.)

34. Solving the equation $x^2 = x$ by Picard's method with $|x_0| \neq 1$ can find the solution $x = 0$ but not the solution $x = 1$. Why? (*Hint*: Graph $y = x^2$ and $y = x$ together.)

COMPUTER EXPLORATIONS

Looking for Convergence of Recursively Defined Sequences

Use a CAS to perform the following steps for the sequences in Exercises 35 and 36.

(a) Calculate and then plot the first 25 terms of the sequence. Does the sequence appear to be bounded from above or below? Does it appear to converge or diverge? If it does converge, what is the limit L?

(b) If the sequence converges, find an integer N such that $|a_n - L| \leq 0.01$ for $n \geq N$. How far in the sequence do you have to get for the terms to lie within 0.0001 of L?

35. $a_1 = 1, \quad a_{n+1} = a_n + \dfrac{1}{5^n}$

36. $a_1 = 1, \quad a_{n+1} = a_n + (-2)^n$

37. *Compound interest, deposits, and withdrawals* If you invest an amount of money A_0 at a fixed annual interest rate r compounded m times per year and if the constant amount b is added to the account at the end of each compounding period (or taken from the account if $b < 0$), then the amount you have after $n + 1$ compounding periods is

$$A_{n+1} = \left(1 + \frac{r}{m}\right)A_n + b. \tag{2}$$

(a) If $A_0 = 1000$, $r = 0.02015$, $m = 12$, and $b = 50$, calculate and plot the first 100 points (n, A_n). How much money is in your account at the end of 5 years? Does $\{A_n\}$ converge? Is $\{A_n\}$ bounded?

(b) Repeat part (a) with $A_0 = 5000$, $r = 0.0589$, $m = 12$, and $b = -50$.

(c) If you invest 5000 dollars in a certificate of deposit (CD) that pays 4.5% annually, compounded quarterly, and you make no further investments in the CD, approximately how many years will it take before you have 20,000 dollars? What if the CD earns 6.25%?

(d) It can be shown that for any $k \geq 0$, the sequence defined recursively by Equation (2) satisfies the relation

$$A_k = \left(1 + \frac{r}{m}\right)^k \left(A_0 + \frac{mb}{r}\right) - \frac{mb}{r}. \qquad (3)$$

For the values of the constants A_0, r, m, and b given in part (a), validate this assertion by comparing the values of the first 50 terms of both sequences. Then show by direct substitution that the terms in Equation (3) satisfy the recursion formula (2).

38. *Logistic difference equation and bifurcation* The recursive relation

$$a_{n+1} = ra_n(1 - a_n)$$

is called the **logistic difference equation,** and when the initial value a_0 is given, the equation defines the **logistic sequence** $\{a_n\}$. Throughout this exercise, we choose a_0 in the interval $0 < a_0 < 1$, say $a_0 = 0.3$.

(a) Choose $r = 3/4$. Calculate and plot the points (n, a_n) for the first 100 terms in the sequence. Does it appear to converge? What do you guess is the limit? Does the limit seem to depend on your choice of a_0?

(b) Choose several values of r in the interval $1 < r < 3$ and repeat the procedures in part (a). Be sure to choose some points near the endpoints of the interval. Describe the behavior of the sequences you observe in your plots.

(c) Now examine the behavior of the sequence for values of r near the endpoints of the interval $3 < r < 3.45$. The transition value $r = 3$ is called a **bifurcation value,** and the new behavior of the sequence in the interval is called an **attracting 2-cycle.** Explain why this reasonably describes the behavior.

(d) Next explore the behavior for r values near the endpoints of each of the intervals $3.45 < r < 3.54$ and $3.54 < r < 3.55$. Plot the first 200 terms of the sequences. Describe in your own words the behavior observed in your plots for each interval. Among how many values does the sequence appear to oscillate for each interval? The values $r = 3.45$ and $r = 3.54$ (rounded to 2 decimal places) are also called bifurcation values because the behavior of the sequence changes as r crosses over those values.

(e) The situation gets even more interesting. There is actually an increasing sequence of bifurcation values $3 < 3.45 < 3.54 < \ldots < c_n < c_{n+1} \ldots$ such that for $c_n < r < c_{n+1}$, the logistic sequence $\{a_n\}$ eventually oscillates steadily among 2^n values, called an **attracting 2^n-cycle.** Moreover, the bifurcation sequence $\{c_n\}$ is bounded above by 3.57 (so it converges). If you choose a value of $r < 3.57$, you will observe a 2^n-cycle of some sort. Choose $r = 3.5695$ and plot 300 points.

(f) Let us see what happens when $r > 3.57$. Choose $r = 3.65$ and calculate and plot the first 300 terms of $\{a_n\}$. Observe how the terms wander around in an unpredictable, chaotic fashion. You cannot predict the value of a_{n+1} from the value of a_n.

(g) For $r = 3.65$, choose two starting values of a_0 that are close together, say $a_0 = 0.3$ and $a_0 = 0.301$. Calculate and plot the first 300 values of the sequences determined by each starting value. Compare the behaviors observed in your plots. How far out do you go before the corresponding terms of your two sequences appear to depart from each other? Repeat the exploration for $r = 3.75$. Can you see how the plots look different depending on your choice of a_0? We say that the logistic sequence is **sensitive to the initial condition** a_0.

8.3 Infinite Series

Series and Partial Sums • Geometric Series • Divergent Series • *n*th-Term Test for Divergence • Adding or Deleting Terms • Reindexing • Combining Series

In mathematics and science, we often write functions as infinite polynomials, such as

$$\frac{1}{1 - x} = 1 + x + x^2 + x^3 + \cdots + x^n + \cdots, \qquad |x| < 1$$

(we see the importance of doing so as the chapter continues). For any allowable value of x, we evaluate the polynomial as an infinite sum of constants, a sum we call an *infinite series*. The goal of this section is to familiarize ourselves with infinite series.

Series and Partial Sums

The first thing to get straight about an infinite series is that it is not simply an example of addition. Addition of real numbers is a *binary* operation, meaning that we really add numbers two at a time. The only reason that $1 + 2 + 3$ makes sense as "addition" is that we can *group* the numbers and then add them two at a time. The associative property of addition guarantees that we get the same sum no matter how we group them:

$$1 + (2 + 3) = 1 + 5 = 6 \qquad \text{and} \qquad (1 + 2) + 3 = 3 + 3 = 6.$$

In short, a *finite sum* of real numbers always produces a real number (the result of a finite number of binary additions), but an *infinite sum* of real numbers is something else entirely. That is why we need a careful definition of infinite series.

We begin by asking how to assign meaning to an expression like

$$1 + \frac{1}{2} + \frac{1}{4} + \frac{1}{8} + \frac{1}{16} + \cdots.$$

The way to do so is not to try to add all the terms at once (we cannot) but rather to add the terms one at a time from the beginning and look for a pattern in how these "partial sums" grow.

Partial sum		**Value**
First:	$s_1 = 1$	$2 - 1$
Second:	$s_2 = 1 + \frac{1}{2}$	$2 - \frac{1}{2}$
Third:	$s_3 = 1 + \frac{1}{2} + \frac{1}{4}$	$2 - \frac{1}{4}$
$\vdots$	$\vdots$	$\vdots$
nth	$s_n = 1 + \frac{1}{2} + \frac{1}{4} + \cdots + \frac{1}{2^{n-1}}$	$2 - \frac{1}{2^{n-1}}$

Indeed there is a pattern. The partial sums form a sequence whose nth term is

$$s_n = 2 - \frac{1}{2^{n-1}}.$$

(We see why momentarily.) This sequence converges to 2 because $\lim_{n\to\infty} (1/2^n) = 0$. We say,

$$\text{"The sum of the infinite series } 1 + \frac{1}{2} + \frac{1}{4} + \cdots + \frac{1}{2^{n-1}} + \cdots \text{ is 2."}$$

Is the sum of any finite number of terms in this series equal to 2? No. Can we actually add an infinite number of terms one by one? No. We can, however, still define their sum by defining it to be the limit of the sequence of partial sums as $n \to \infty$, in this case 2 (Figure 8.9). Our knowledge of sequences and limits enables us to break away from the confines of finite sums to define this entirely new concept.

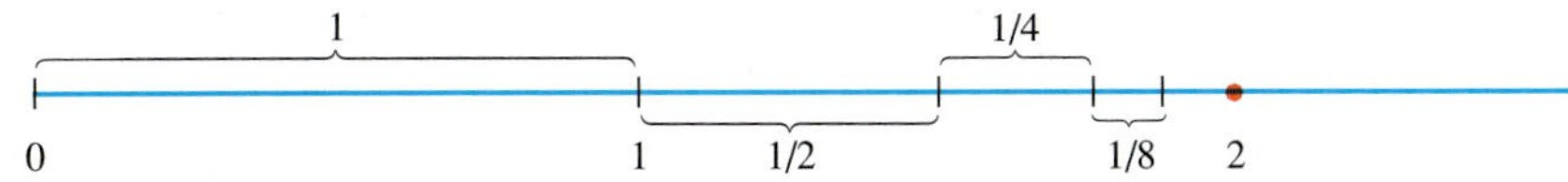

FIGURE 8.9 As the lengths 1, ½, ¼, ⅛, . . . are added one by one, the sum approaches 2.

CD-ROM
WEBsite
Historical Biography

Blaise Pascal
(1623 — 1662)

Definition Infinite Series

Given a sequence of numbers $\{a_n\}$, an expression of the form

$$a_1 + a_2 + a_3 + \cdots + a_n + \cdots$$

is an **infinite series.** The number a_n is the ***n*th term** of the series.

The **partial sums** of the series form a sequence

$$\begin{aligned} s_1 &= a_1 \\ s_2 &= a_1 + a_2 \\ s_3 &= a_1 + a_2 + a_3 \\ &\vdots \\ s_n &= \sum_{k=1}^{n} a_k \\ &\vdots \end{aligned}$$

of real numbers, each defined as a finite sum. If the sequence of partial sums has a limit S as $n \to \infty$, we say that the series **converges** to the sum S, and we write

$$a_1 + a_2 + a_3 + \cdots + a_n + \cdots = \sum_{k=1}^{\infty} a_k = S.$$

Otherwise, we say that the series **diverges.**

Example 1 Identifying a Convergent Series

Does the series

$$\frac{3}{10} + \frac{3}{100} + \frac{3}{1000} + \cdots + \frac{3}{10^n} + \cdots$$

converge?

Solution Here is the sequence of partial sums, written in decimal form.

$$0.3, 0.33, 0.333, 0.3333, \ldots$$

This sequence has a limit $0.\overline{3}$, which we recognize as the fraction $1/3$. The series converges to the sum $1/3$.

When we begin to study a given series $a_1 + a_2 + \cdots + a_n + \cdots$, we might not know whether it converges or diverges. In either case, it is convenient to use sigma notation to write the series as

CD-ROM WEBsite

$$\sum_{n=1}^{\infty} a_n, \qquad \sum_{k=1}^{\infty} a_k, \qquad \text{or} \qquad \sum a_n.$$

A useful shorthand when summation from 1 to ∞ is understood

Geometric Series

The series in Example 1 is a **geometric series** because each term is obtained from its preceding term by multiplying by the same number r, in this case, $r = 1/10$. (The series of areas for the infinitely halved square at the beginning of this chapter is also geometric.) The convergence of geometric series is one of the few infinite processes with which mathematicians were reasonably comfortable prior to calculus. Let's see why.

Geometric series are series of the form

$$a + ar + ar^2 + \cdots + ar^{n-1} + \cdots = \sum_{n=1}^{\infty} ar^{n-1}$$

in which a and r are fixed real numbers and $a \neq 0$. The **ratio** r can be positive, as in

$$1 + \frac{1}{2} + \frac{1}{4} + \cdots + \left(\frac{1}{2}\right)^{n-1} + \cdots,$$

or negative, as in

$$1 - \frac{1}{3} + \frac{1}{9} - \cdots + \left(-\frac{1}{3}\right)^{n-1} + \cdots.$$

If $|r| \neq 1$, we can determine the convergence or divergence of the series in the following way, starting with the nth partial sum:

$$s_n = a + ar + ar^2 + \cdots + ar^{n-1}$$

$$rs_n = ar + ar^2 + \cdots + ar^{n-1} + ar^n$$ Multiply s_n by r.

$$s_n - rs_n = a - ar^n$$ Subtract rs_n from s_n. Most of the terms on the right cancel.

$$s_n(1 - r) = a(1 - r^n)$$ Factor.

$$s_n = \frac{a(1 - r^n)}{1 - r}, \qquad (r \neq 1).$$ We can solve for s_n if $r \neq 1$.

If $|r| < 1$, then $r^n \to 0$ as $n \to \infty$ (Table 8.1, Formula 4) and $s_n \to a/(1 - r)$. If $|r| > 1$, then $|r^n| \to \infty$ and the series diverges.

If $r = 1$, the nth partial sum of the geometric series is

$$s_n = a + a(1) + a(1)^2 + \cdots + a(1)^{n-1} = na,$$

and the series diverges because $\lim_{n\to\infty} s_n = \pm\infty$, depending on the sign of a. If $r = -1$, the series diverges because the nth partial sums alternate between a and 0. Let's summarize our results.

The equation

$$\sum_{n=1}^{\infty} ar^{n-1} = \frac{a}{1 - r}, \qquad |r| < 1$$

holds only if the summation begins with $n = 1$.

The **geometric series**

$$a + ar + ar^2 + ar^3 + \cdots + ar^{n-1} + \cdots = \sum_{n=1}^{\infty} ar^{n-1}$$

converges to the sum $a/(1 - r)$ if $|r| < 1$ and diverges if $|r| \geq 1$.

a

ar

ar^2

ar^3

(a)

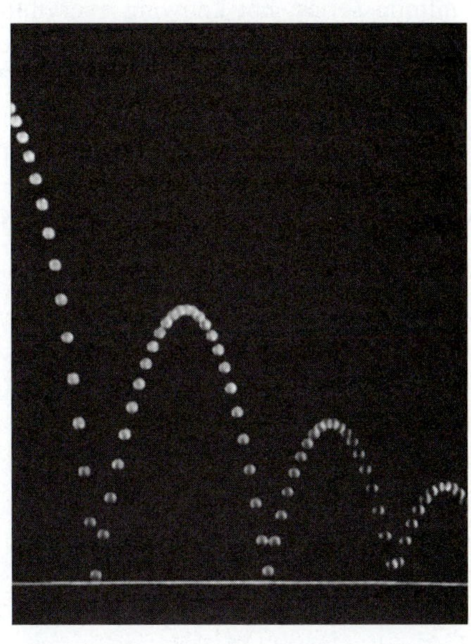

(b)

FIGURE 8.10 (a) Example 3 shows how to use a geometric series to calculate the total vertical distance traveled by a bouncing ball if the height of each rebound is reduced by the factor r. (b) A stroboscopic photo of a bouncing ball.

This completely settles the issue for geometric series. We know which ones converge and which ones diverge, and for the convergent ones, we know what the sums must be. The interval $-1 < r < 1$ is the **interval of convergence.**

Example 2 Analyzing Geometric Series

Tell whether each series converges or diverges. If it converges, give its sum.

(a) $\displaystyle\sum_{n=1}^{\infty} 3\left(\frac{1}{2}\right)^{n-1}$

(b) $1 - \dfrac{1}{2} + \dfrac{1}{4} - \dfrac{1}{8} + \cdots + \left(-\dfrac{1}{2}\right)^{n-1} + \cdots$

(c) $\displaystyle\sum_{k=0}^{\infty}\left(\frac{3}{5}\right)^{k} = \sum_{k=1}^{\infty}\left(\frac{3}{5}\right)^{k-1}$

(d) $\dfrac{\pi}{2} + \dfrac{\pi^2}{4} + \dfrac{\pi^3}{8} + \cdots$

Solution

(a) First term is $a = 3$ and $r = 1/2$. The series converges to

$$\frac{3}{1-(1/2)} = 6.$$

(b) First term is $a = 1$ and $r = -1/2$. The series converges to

$$\frac{1}{1-(-1/2)} = \frac{2}{3}.$$

(c) First term is $a = (3/5)^0 = 1$ and $r = 3/5$. The series converges to

$$\frac{1}{1-(3/5)} = \frac{5}{2}.$$

(d) In this series, $r = \pi/2 > 1$. The series diverges.

Example 3 A Bouncing Ball

You drop a ball from a meters above a flat surface. Each time the ball hits the surface after falling a distance h, it rebounds a distance rh, where r is positive but less than 1. Find the total distance the ball travels up and down (Figure 8.10).

Solution

The total distance is

$$s = a + \underbrace{2ar + 2ar^2 + 2ar^3 + \cdots}_{\text{This sum is } 2ar/(1-r).} = a + \frac{2ar}{1-r} = a\,\frac{1+r}{1-r}.$$

CD-ROM WEBsite

If $a = 6$ m and $r = 2/3$, for instance, the distance is

$$s = 6\,\frac{1+(2/3)}{1-(2/3)} = 6\left(\frac{5/3}{1/3}\right) = 30 \text{ m}.$$

Example 4 Repeating Decimals

Express the repeating decimal 5.23 23 23 . . . as the ratio of two integers.

Solution

$$\begin{aligned} 5.23\ 23\ 23\ldots &= 5 + \frac{23}{100} + \frac{23}{(100)^2} + \frac{23}{(100)^3} + \cdots \\ &= 5 + \frac{23}{100}\underbrace{\left(1 + \frac{1}{100} + \left(\frac{1}{100}\right)^2 + \cdots\right)}_{1/(1-0.01)} \qquad a = 1,\ r = 1/100 \\ &= 5 + \frac{23}{100}\left(\frac{1}{0.99}\right) = 5 + \frac{23}{99} = \frac{518}{99} \end{aligned}$$

We have hardly begun our study of infinite series, but knowing everything there is to know about the convergence and divergence of an *entire class* of series (geometric) is an impressive start. Like the Renaissance mathematicians, we are ready to explore where this might lead.

Unfortunately, formulas like the one for the sum of a convergent geometric series are rare, and we usually have to settle for an estimate of a series' sum (more about this later). The next example, however, is another case in which we can find the sum exactly.

Example 5 A Nongeometric but Telescoping Series

Find the sum of the series

$$\sum_{n=1}^{\infty} \frac{1}{n(n+1)}.$$

Solution We look for a pattern in the sequence of partial sums that might lead to a formula for s_k. The key is partial fractions. The observation that

$$\frac{1}{k(k+1)} = \frac{1}{k} - \frac{1}{k+1}$$

permits us to write the partial sum

$$\sum_{n=1}^{k} \frac{1}{n(n+1)} = \frac{1}{1\cdot 2} + \frac{1}{2\cdot 3} + \cdots + \frac{1}{k\cdot(k+1)}$$

as

$$s_k = \left(\frac{1}{1} - \frac{1}{2}\right) + \left(\frac{1}{2} - \frac{1}{3}\right) + \cdots + \left(\frac{1}{k} - \frac{1}{k+1}\right).$$

Removing parentheses and canceling the terms of opposite sign collapses the sum to

$$s_k = 1 - \frac{1}{k+1}.$$

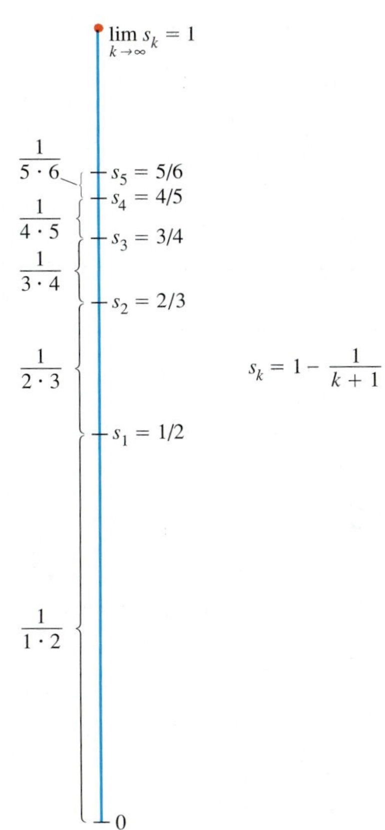

FIGURE 8.11 The partial sums of the series in Example 5.

We now see that $s_k \to 1$ as $k \to \infty$. The series converges, and its sum is 1 (Figure 8.11).

$$\sum_{n=1}^{\infty} \frac{1}{n(n+1)} = 1.$$

Divergent Series

Geometric series with $|r| \geq 1$ are not the only series to diverge.

Example 6 Identifying a Divergent Series

Does the series $1 - 1 + 1 - 1 + 1 - 1 + \cdots$ converge?

Solution You might be tempted to pair the terms as

$$(1-1) + (1-1) + (1-1) + \cdots.$$

That strategy, however, requires an *infinite* number of pairings, so it cannot be justified by the associative property of addition. This is an infinite series, not a finite sum, so if it has a sum it *has to be* the limit of its sequence of partial sums,

$$1, 0, 1, 0, 1, 0, 1, \ldots.$$

Since this sequence has no limit, the series has no sum. It diverges.

Example 7 Partial Sums Outgrowing Any Bound

(a) The series

$$\sum_{n=1}^{\infty} n^2 = 1 + 4 + 9 + \cdots + n^2 + \cdots$$

diverges because the partial sums grow beyond every number L. After $n = 1$, the partial sum $s_n = 1 + 4 + 9 + \cdots + n^2$ is greater than n^2.

(b) The series

$$\sum_{n=1}^{\infty} \frac{n+1}{n} = \frac{2}{1} + \frac{3}{2} + \frac{4}{3} + \cdots + \frac{n+1}{n} + \cdots$$

diverges because the partial sums eventually outgrow every preassigned number. Each term is greater than 1, so the sum of n terms is greater than n.

nth-Term Test for Divergence

Observe that $\lim_{n\to\infty} a_n$ must equal zero if the series $\sum_{n=1}^{\infty} a_n$ converges. To see why, let S represent the series' sum and $s_n = a_1 + a_2 + \cdots + a_n$ the nth partial sum. When n is large, both s_n and s_{n-1} are close to S, so their difference, a_n, is close to zero. More formally,

$$a_n = s_n - s_{n-1} \to S - S = 0. \qquad \text{Difference Rule for sequences}$$

CAUTION Theorem 6 *does not say* that $\sum_{n=1}^{\infty} a_n$ converges if $a_n \to 0$. It is possible for a series to diverge when $a_n \to 0$.

Theorem 6 **Limit of the nth Term of a Convergent Series**

If $\sum_{n=1}^{\infty} a_n$ converges, then $a_n \to 0$.

Theorem 6 leads to a test for detecting the kind of divergence that occurred in Examples 6 and 7.

nth-Term Test for Divergence

$\sum_{n=1}^{\infty} a_n$ diverges if $\lim_{n\to\infty} a_n$ fails to exist or is different from zero.

Example 8 Applying the nth-Term Test

(a) $\displaystyle\sum_{n=1}^{\infty} n^2$ diverges because $n^2 \to \infty$.

(b) $\displaystyle\sum_{n=1}^{\infty} \frac{n+1}{n}$ diverges because $\dfrac{n+1}{n} \to 1$.

(c) $\displaystyle\sum_{n=1}^{\infty} (-1)^{n+1}$ diverges because $\displaystyle\lim_{n\to\infty} (-1)^{n+1}$ does not exist.

(d) $\displaystyle\sum_{n=1}^{\infty} \frac{-n}{2n+5}$ diverges because $\displaystyle\lim_{n\to\infty} \left(\frac{-n}{2n+5}\right) = -\frac{1}{2} \neq 0$.

Example 9 $a_n \to 0$, but the Series Diverges

The series

$$1 + \underbrace{\frac{1}{2} + \frac{1}{2}}_{2 \text{ terms}} + \underbrace{\frac{1}{4} + \frac{1}{4} + \frac{1}{4} + \frac{1}{4}}_{4 \text{ terms}} + \cdots + \underbrace{\frac{1}{2^n} + \frac{1}{2^n} + \cdots + \frac{1}{2^n}}_{2^n \text{ terms}} + \cdots$$

$$= 1 + 1 + 1 + \cdots + 1 + \cdots$$

diverges even though its terms form a sequence that converges to 0.

Adding or Deleting Terms

We can always add a finite number of terms to a series or delete a finite number of terms without altering the series' convergence or divergence, although in the case of convergence, this will usually change the sum. If $\sum_{n=1}^{\infty} a_n$ converges, then $\sum_{n=k}^{\infty} a_n$ converges for any $k > 1$, and

$$\sum_{n=1}^{\infty} a_n = a_1 + a_2 + \cdots + a_{k-1} + \sum_{n=k}^{\infty} a_n.$$

Conversely, if $\sum_{n=k}^{\infty} a_n$ converges for any $k > 1$, then $\sum_{n=1}^{\infty} a_n$ converges. Thus,

$$\sum_{n=1}^{\infty} \frac{1}{5^n} = \frac{1}{5} + \frac{1}{25} + \frac{1}{125} + \sum_{n=4}^{\infty} \frac{1}{5^n}$$

and

$$\sum_{n=4}^{\infty} \frac{1}{5^n} = \left(\sum_{n=1}^{\infty} \frac{1}{5^n}\right) - \frac{1}{5} - \frac{1}{25} - \frac{1}{125}.$$

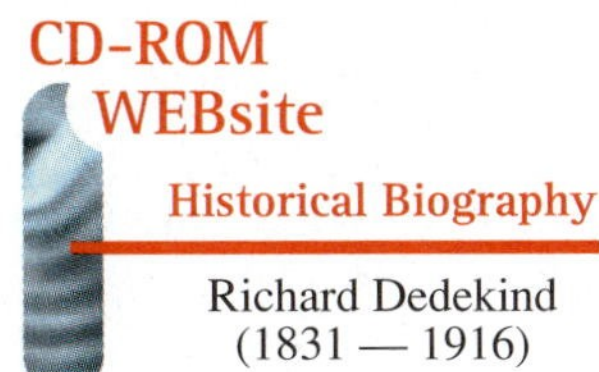

Richard Dedekind
(1831 — 1916)

Reindexing

As long as we preserve the order of its terms, we can reindex any series without altering its convergence (see Example 2c). To raise the starting value of the index h units, replace the n in the formula for a_n by $n - h$:

$$\sum_{n=1}^{\infty} a_n = \sum_{n=1+h}^{\infty} a_{n-h} = a_1 + a_2 + a_3 + \cdots.$$

To lower the starting value of the index h units, replace the n in the formula for a_n by $n + h$:

$$\sum_{n=1}^{\infty} a_n = \sum_{n=1-h}^{\infty} a_{n+h} = a_1 + a_2 + a_3 + \cdots.$$

It works like a horizontal shift.

Example 10 Reindexing A Geometric Series

We can write the geometric series that starts with

$$1 + \frac{1}{2} + \frac{1}{4} + \cdots$$

as

$$\sum_{n=0}^{\infty} \frac{1}{2^n}, \qquad \sum_{n=5}^{\infty} \frac{1}{2^{n-5}}, \qquad \text{or even} \qquad \sum_{n=-4}^{\infty} \frac{1}{2^{n+4}}.$$

The partial sums remain the same no matter what indexing we choose.

We usually give preference to indexings that lead to simple expressions.

Combining Series

Whenever we have two convergent series, we can add them term by term, subtract them term by term, or multiply them by constants to make new convergent series.

Theorem 7 Properties of Convergent Series

If $\Sigma a_n = A$ and $\Sigma b_n = B$ are convergent series, then

1. *Sum Rule:* $\Sigma (a_n + b_n) = \Sigma a_n + \Sigma b_n = A + B$
2. *Difference Rule:* $\Sigma (a_n - b_n) = \Sigma a_n - \Sigma b_n = A - B$
3. *Constant Multiple Rule:* $\Sigma ka_n = k \Sigma a_n = kA$ (any number k).

Example 11 Applying Theorem 7

Find the sums of the following series.

(a)
$$\sum_{n=1}^{\infty} \frac{3^{n-1} - 1}{6^{n-1}} = \sum_{n=1}^{\infty} \left(\frac{1}{2^{n-1}} - \frac{1}{6^{n-1}} \right)$$

$$= \sum_{n=1}^{\infty} \frac{1}{2^{n-1}} - \sum_{n=1}^{\infty} \frac{1}{6^{n-1}}$$ Difference Rule

$$= \frac{1}{1 - (1/2)} - \frac{1}{1 - (1/6)}$$ Geometric series with $a = 1$ and $r = 1/2, 1/6$

$$= 2 - \frac{6}{5}$$

$$= \frac{4}{5}$$

(b)
$$\sum_{n=1}^{\infty} \frac{4}{2^{n-1}} = 4 \sum_{n=1}^{\infty} \frac{1}{2^{n-1}}$$ Constant Multiple Rule

$$= 4\left(\frac{1}{1 - (1/2)} \right)$$ Geometric series with $a = 1$, $r = 1/2$

$$= 8$$

Proof of Theorem 7 The three rules for series follow from the analogous rules for sequences in Theorem 1, Section 8.1. To prove the Sum Rule for series, let

$$A_n = a_1 + a_2 + \cdots + a_n, \qquad B_n = b_1 + b_2 + \cdots + b_n.$$

Then the partial sums of $\Sigma\,(a_n + b_n)$ are

$$\begin{aligned} S_n &= (a_1 + b_1) + (a_2 + b_2) + \cdots + (a_n + b_n) \\ &= (a_1 + \cdots + a_n) + (b_1 + \cdots + b_n) \\ &= A_n + B_n. \end{aligned}$$

Since $A_n \to A$ and $B_n \to B$, we have $S_n \to A + B$ by the Sum Rule for sequences. The proof of the Difference Rule is similar.

To prove the Constant Multiple Rule for series, observe that the partial sums of $\Sigma\, ka_n$ form the sequence

$$S_n = ka_1 + ka_2 + \cdots + ka_n = k(a_1 + a_2 + \cdots + a_n) = kA_n,$$

which converges to kA by the Constant Multiple Rule for sequences.

Interpreting Theorem 7 for Divergence

1. Every nonzero constant multiple of a divergent series diverges.
2. If $\Sigma\, a_n$ converges and $\Sigma\, b_n$ diverges, then $\Sigma\,(a_n + b_n)$ and $\Sigma\,(a_n - b_n)$ both diverge.

We omit the proofs.

EXERCISES 8.3

Finding nth Partial Sums

In Exercises 1–6, find a formula for the nth partial sum of each series and use it to find the series' sum if the series converges.

1. $2 + \frac{2}{3} + \frac{2}{9} + \frac{2}{27} + \cdots + \frac{2}{3^{n-1}} + \cdots$

2. $\frac{9}{100} + \frac{9}{100^2} + \frac{9}{100^3} + \cdots + \frac{9}{100^n} + \cdots$

3. $1 - \frac{1}{2} + \frac{1}{4} - \frac{1}{8} + \cdots + (-1)^{n-1}\frac{1}{2^{n-1}} + \cdots$

4. $1 - 2 + 4 - 8 + \cdots + (-1)^{n-1}2^{n-1} + \cdots$

5. $\frac{1}{2 \cdot 3} + \frac{1}{3 \cdot 4} + \frac{1}{4 \cdot 5} + \cdots + \frac{1}{(n+1)(n+2)} + \cdots$

6. $\frac{5}{1 \cdot 2} + \frac{5}{2 \cdot 3} + \frac{5}{3 \cdot 4} + \cdots + \frac{5}{n(n+1)} + \cdots$

Series with Geometric Terms

In Exercises 7–12, write out the first few terms of each series to show how the series starts. Then find the sum of the series.

7. $\sum_{n=0}^{\infty} \frac{(-1)^n}{4^n}$

8. $\sum_{n=1}^{\infty} \frac{7}{4^n}$

9. $\sum_{n=0}^{\infty} \left(\frac{5}{2^n} + \frac{1}{3^n}\right)$

10. $\sum_{n=0}^{\infty} \left(\frac{5}{2^n} - \frac{1}{3^n}\right)$

11. $\sum_{n=0}^{\infty} \left(\frac{1}{2^n} + \frac{(-1)^n}{5^n}\right)$

12. $\sum_{n=0}^{\infty} \left(\frac{2^{n+1}}{5^n}\right)$

Telescoping Series

Use partial fractions to find the sum of each series in Exercises 13–16.

13. $\sum_{n=1}^{\infty} \frac{4}{(4n-3)(4n+1)}$

14. $\sum_{n=1}^{\infty} \frac{6}{(2n-1)(2n+1)}$

15. $\sum_{n=1}^{\infty} \frac{40n}{(2n-1)^2(2n+1)^2}$

16. $\sum_{n=1}^{\infty} \frac{2n+1}{n^2(n+1)^2}$

Find the sums of the series in Exercises 17 and 18.

17. $\sum_{n=1}^{\infty} \left(\frac{1}{\sqrt{n}} - \frac{1}{\sqrt{n+1}}\right)$

18. $\sum_{n=1}^{\infty} \left(\frac{1}{\ln(n+2)} - \frac{1}{\ln(n+1)}\right)$

Convergence or Divergence

Which series in Exercises 19–32 converge, and which diverge? Give reasons for your answers. If a series converges, find its sum.

19. $\sum_{n=0}^{\infty} \left(\frac{1}{\sqrt{2}}\right)^n$

20. $\sum_{n=0}^{\infty} (\sqrt{2})^n$

21. $\sum_{n=1}^{\infty} (-1)^{n+1}\frac{3}{2^n}$

22. $\sum_{n=0}^{\infty} \frac{\cos n\pi}{5^n}$

23. $\sum_{n=0}^{\infty} e^{-2n}$

24. $\sum_{n=1}^{\infty} \ln \frac{1}{n}$

25. $\sum_{n=0}^{\infty} \frac{1}{x^n}, \quad |x| > 1$

26. $\sum_{n=0}^{\infty} \frac{2^n - 1}{3^n}$

27. $\sum_{n=1}^{\infty} \left(1 - \frac{1}{n}\right)^n$

28. $\sum_{n=0}^{\infty} \left(\frac{e}{\pi}\right)^n$

29. $\sum_{n=1}^{\infty} \ln\left(\frac{n}{n+1}\right)$

30. $\sum_{n=0}^{\infty} \frac{e^{n\pi}}{\pi^{ne}}$

31. $\sum_{n=0}^{\infty} \frac{n!}{1000^n}$

32. $\sum_{n=1}^{\infty} \frac{n^n}{n!}$

Geometric Series

In each of the geometric series in Exercises 33–36, write out the first few terms of the series to find a and r and find the sum of the series. Then express the inequality $|r| < 1$ in terms of x and find the values of x for which the inequality holds and the series converges.

33. $\sum_{n=0}^{\infty} (-1)^n x^n$

34. $\sum_{n=0}^{\infty} (-1)^n x^{2n}$

35. $\sum_{n=0}^{\infty} 3\left(\frac{x-1}{2}\right)^n$

36. $\sum_{n=0}^{\infty} \frac{(-1)^n}{2}\left(\frac{1}{3+\sin x}\right)^n$

In Exercises 37–40, find the values of x for which the given geometric series converges. Also, find the sum of the series (as a function of x) for those values of x.

37. $\sum_{n=0}^{\infty} 2^n x^n$

38. $\sum_{n=0}^{\infty} (-1)^n x^{-2n}$

39. $\sum_{n=0}^{\infty} \left(-\frac{1}{2}\right)^n (x-3)^n$

40. $\sum_{n=0}^{\infty} (\ln x)^n$

Repeating Decimals

Express each of the numbers in Exercises 41–46 as the ratio of two integers.

41. $0.\overline{23} = 0.23\ 23\ 23\ldots$

42. $0.\overline{234} = 0.234\ 234\ 234\ldots$

43. $0.\overline{7} = 0.7777\ldots$

44. $1.\overline{414} = 1.414\ 414\ 414\ldots$

45. $1.24\overline{123} = 1.24\ 123\ 123\ 123\ldots$

46. $3.\overline{142857} = 3.142857\ 142857\ldots$

Theory and Examples

CD-ROM WEBsite 47. *Bouncing ball distance* A ball is dropped from a height of 4 m. Each time it strikes the pavement after falling from a height of h meters, it rebounds to a height of $0.75h$ m. Find the total distance the ball travels up and down.

CD-ROM WEBsite 48. *Total time for bouncing* Find the total number of seconds the ball in Exercise 47 is traveling. (*Hint:* The formula $s = 4.9t^2$ gives $t = \sqrt{s/4.9}$.)

49. *Summing areas* The accompanying figure shows the first five of a sequence of squares. The outermost square has an area of 4 m^2. Each of the other squares is obtained by joining the midpoints of the sides of the squares before it. Find the sum of the areas of all the squares.

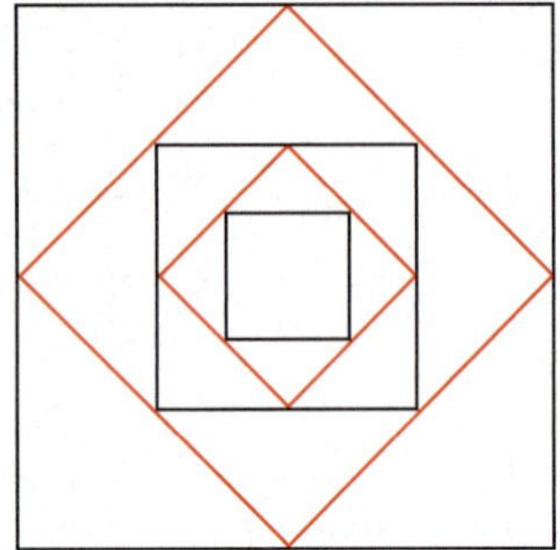

50. *Summing areas* The accompanying figure shows the first three rows and part of the fourth row of a sequence of rows of semicircles. There are 2^n semicircles in the nth row, each of radius $1/2^n$. Find the sum of the areas of all the semicircles.

51. *Helge von Koch's snowflake curve* Start with an equilateral triangle whose sides have length 1, calling it Curve 1. On the middle third of each side, build an equilateral triangle pointing outward. Then erase the interiors of the old middle thirds. Call the expanded curve Curve 2. Now put equilateral triangles, again pointing outward, on the middle thirds of the sides of Curve 2. Erase the interiors of the old middle thirds to make Curve 3. Repeat the process, as shown, to define an infinite sequence of plane curves. The limit curve of the sequence is Koch's snowflake curve.

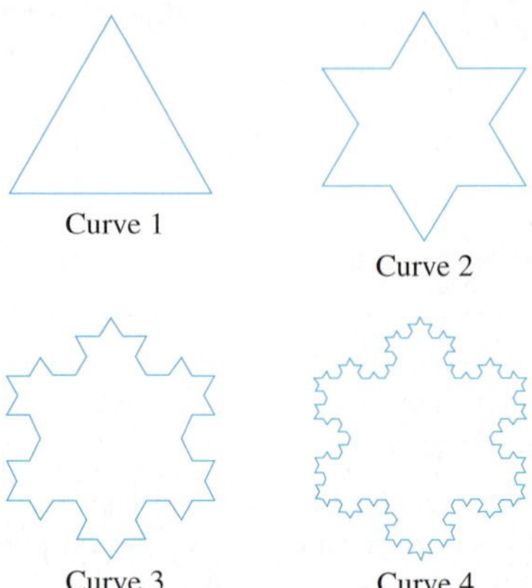

Here's how to show that the snowflake is a curve of infinite length enclosing a region of finite area.

(a) Find the length L_n of the nth curve C_n and show that $\lim_{n\to\infty} L_n = \infty$.

(b) Find the area A_n of the region enclosed by C_n and calculate $\lim_{n\to\infty} A_n$.

52. *Writing to Learn* The accompanying figure provides an informal proof that $\sum_{n=1}^{\infty} (1/n^2)$ is less than 2. Explain what is going on. (*Source*: "Convergence with Pictures" by P. J. Rippon, *American Mathematical Monthly*, Vol. 93, No. 6 (1986), pp. 476—478.)

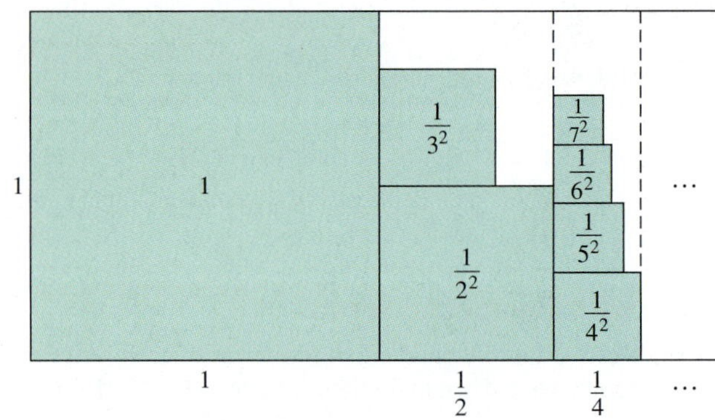

53. *Reindexing* The series in Exercise 5 can also be written as

$$\sum_{n=1}^{\infty} \frac{1}{(n+1)(n+2)} \quad \text{and} \quad \sum_{n=-1}^{\infty} \frac{1}{(n+3)(n+4)}.$$

Write it as a sum beginning with

(a) $n = -2$

(b) $n = 0$

(c) $n = 5$.

54. *Writing to Learn* Make up an infinite series of nonzero terms whose sum is

(a) 1

(b) -3

(c) 0.

Can you make an infinite series of nonzero terms that converges to any number you want? Explain.

55. *Geometric series* Find the value of b for which

$$1 + e^b + e^{2b} + e^{3b} + \cdots = 9.$$

56. *Modified geometric series* For what values of r does the infinite series

$$1 + 2r + r^2 + 2r^3 + r^4 + 2r^5 + r^6 + \cdots$$

converge? Find the sum of the series when it converges.

57. *Error using a partial sum* Show that the error $(L - s_n)$ obtained by replacing a convergent geometric series with one of its partial sums s_n is $ar^n/(1 - r)$.

58. *Term-by-term product* Find convergent geometric series $A = \Sigma\, a_n$ and $B = \Sigma\, b_n$ that illustrate that $\Sigma\, a_n b_n$ may converge without being equal to AB.

59. *Term-by-term quotient* Show by example that $\Sigma\, (a_n/b_n)$ may converge to something other than A/B even when $A = \Sigma\, a_n$, $B = \Sigma\, b_n \neq 0$, and no b_n equals 0.

60. *Term-by-term quotient* Show by example that $\Sigma\, (a_n/b_n)$ may diverge even though $\Sigma\, a_n$ and $\Sigma\, b_n$ converge and no b_n equals 0.

61. *Term-by-term reciprocals* If $\Sigma\, a_n$ converges and $a_n > 0$ for all n, can anything be said about $\Sigma\, (1/a_n)$? Give reasons for your answer.

62. *Adding or deleting terms* What happens if you add a finite number of terms to a divergent series or delete a finite number of terms from a divergent series? Give reasons for your answer.

63. *Summing convergent and divergent series* If $\Sigma\, a_n$ converges and $\Sigma\, b_n$ diverges, can anything be said about their term-by-term sum $\Sigma\, (a_n + b_n)$? Give reasons for your answer.

8.4 Series of Nonnegative Terms

Integral Test • Harmonic Series and p-Series • Comparison Tests • Ratio and Root Tests

Given a series $\Sigma\, a_n$, we have two questions.

1. Does the series converge?
2. If it converges, what is its sum?

In this section, we study series that do not have negative terms. The reason for this restriction is that the partial sums of these series form nondecreasing sequences, and nondecreasing sequences that are bounded from above always converge. The partial sums are nondecreasing because $s_{n+1} = s_n + a_n$ and $a_n \geq 0$:

$$s_1 \leq s_2 \leq s_3 \leq \cdots \leq s_n \leq s_{n+1} \leq \cdots .$$

From the Monotonic Sequence Theorem (Theorem 5, Section 8.2), the series will converge if $\{s_n\}$ is bounded from above.

Corollary of Theorem 5

A series $\sum_{n=1}^{\infty} a_n$ of nonnegative terms converges if its partial sums are bounded from above.

This result is the basis for the tests to establish convergence we study in this section.

Integral Test

We introduce the Integral Test with an example.

Example 1 Applying Corollary of Theorem 5

Show that the series

$$\sum_{n=1}^{\infty} \frac{1}{n^2} = 1 + \frac{1}{4} + \frac{1}{9} + \frac{1}{16} + \cdots + \frac{1}{n^2} + \cdots$$

converges.

Solution We determine the convergence of $\Sigma_{n=1}^{\infty} (1/n^2)$ by comparing it with $\int_1^{\infty} (1/x^2)\,dx$. To carry out the comparison, we think of the terms of the series as values of the function $f(x) = 1/x^2$ and interpret these values as the areas of rectangles under the curve $y = 1/x^2$.

As Figure 8.12 shows,

$$\begin{aligned} s_n &= \frac{1}{1^2} + \frac{1}{2^2} + \frac{1}{3^2} + \cdots + \frac{1}{n^2} \\ &= f(1) + f(2) + f(3) + \cdots + f(n) \\ &< f(1) + \int_1^n \frac{1}{x^2}\,dx \\ &< 1 + \int_1^{\infty} \frac{1}{x^2}\,dx \\ &< 1 + 1 = 2. \end{aligned}$$

As in Section 7.7. Example 3, with $p = 2$, $\int_1^{\infty} (1/x^2)\,dx = 1$.

Thus, the partial sums of $\Sigma_{n=1}^{\infty} (1/n^2)$ are bounded from above (by 2) and the series converges. The sum of the series is known to be $\pi^2/6 \approx 1.64493$.

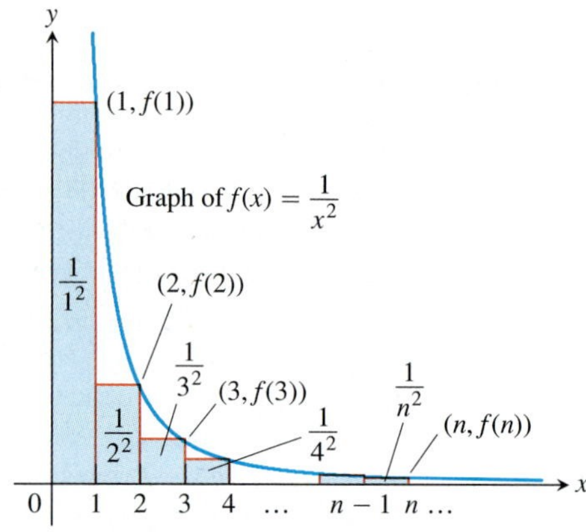

FIGURE 8.12 Figure for the area comparisons in Example 1.

The Integral Test

Let $\{a_n\}$ be a sequence of positive terms. Suppose that $a_n = f(n)$, where f is a continuous, positive, decreasing function of x for all $x \geq N$ (N a positive integer). Then the series $\sum_{n=N}^{\infty} a_n$ and the integral $\int_N^{\infty} f(x)\,dx$ both converge or both diverge.

Proof We establish the test for the case $N = 1$. The proof for general N is similar.

We start with the assumption that f is a decreasing function with $f(n) = a_n$ for every n. This leads us to observe that the rectangles in Figure 8.13a, which have areas $a_1, a_2, \ldots, a_n$, collectively enclose more area than that under the curve $y = f(x)$ from $x = 1$ to $x = n + 1$. That is,

$$\int_1^{n+1} f(x)\,dx \leq a_1 + a_2 + \cdots + a_n.$$

CAUTION The series and integral need not have the same value in the convergent case. In Example 1, $\sum_{n=1}^{\infty} (1/n^2) = \pi^2/6$, whereas $\int_1^{\infty} (1/x^2)\,dx = 1$.

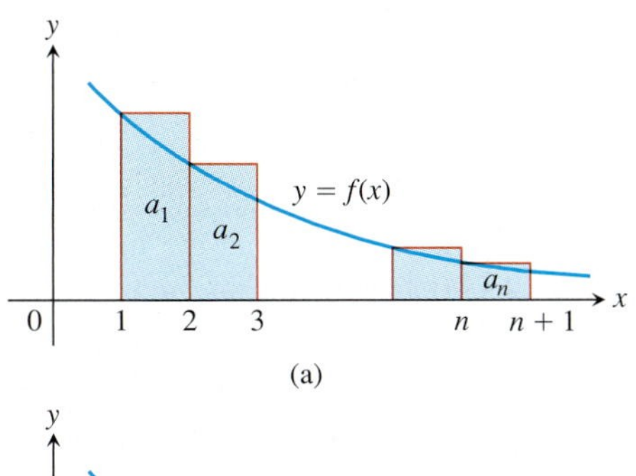

(a)

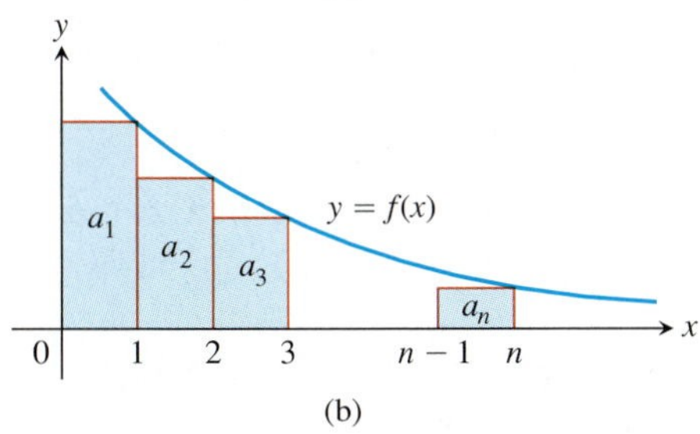

(b)

FIGURE 8.13 Subject to the conditions of the Integral Test, the series $\Sigma_{n=1}^{\infty} a_n$ and the integral $\int_1^\infty f(x)\,dx$ both converge or both diverge.

In Figure 8.13b the rectangles have been faced to the left instead of to the right. If we momentarily disregard the first rectangle, of area a_1, we see that

$$a_2 + a_3 + \cdots + a_n \le \int_1^n f(x)\,dx.$$

If we include a_1, we have

$$a_1 + a_2 + \cdots + a_n \le a_1 + \int_1^n f(x)\,dx.$$

Combining these results gives

$$\int_1^{n+1} f(x)\,dx \le a_1 + a_2 + \cdots + a_n \le a_1 + \int_1^n f(x)\,dx.$$

If $\int_1^\infty f(x)\,dx$ is finite, the right-hand inequality shows that $\Sigma\, a_n$ is finite. If $\int_1^\infty f(x)\,dx$ is infinite, the left-hand inequality shows that $\Sigma\, a_n$ is infinite.

Hence, the series and the integral are both finite or both infinite.

Example 2 Applying the Integral Test

Does $\displaystyle\sum_{n=1}^{\infty} \frac{1}{n\sqrt{n}}$ converge?

Solution The Integral Test applies because

$$f(x) = \frac{1}{x\sqrt{x}}$$

is a continuous, positive, decreasing function of x for $x > 1$.

We have

$$\begin{aligned}\int_1^\infty \frac{1}{x\sqrt{x}}\,dx &= \lim_{k\to\infty} \int_1^k x^{-3/2}\,dx \\ &= \lim_{k\to\infty} \left[-2x^{-1/2}\right]_1^k \\ &= \lim_{k\to\infty} \left(-\frac{2}{\sqrt{k}} + 2\right) \\ &= 2.\end{aligned}$$

Since the integral converges, so must the series.

Harmonic Series and p-Series

The Integral Test can be used to settle the question of convergence for any series of the form $\Sigma_{n=1}^{\infty} (1/n^p)$, p a real constant. (The series in Example 2 had this form, with $p = 3/2$.) Such series are called ***p*-series.**

The p-Series

$$\sum_{n=1}^{\infty} \frac{1}{n^p} = \frac{1}{1^p} + \frac{1}{2^p} + \frac{1}{3^p} + \cdots + \frac{1}{n^p} + \cdots$$

(p a real constant) converges if $p > 1$ and diverges if $p \le 1$.

Proof From Example 3 in Section 7.7, the integral $\int_1^\infty dx/x^p$ converges if $p > 1$ and diverges if $p \leq 1$. From the Integral Test, the same holds true of the p-series $\Sigma_{n=1}^{\infty}(1/n^p)$: It converges if $p > 1$ and diverges if $p \leq 1$.

The p-series with $p = 1$ is the **harmonic series**, and it is probably the most famous divergent series in mathematics. The p-Series Test shows that the harmonic series is just *barely* divergent; if we increase p to 1.000000001, for instance, the series converges!

The slowness with which the partial sums of the harmonic series approaches infinity is most impressive. Consider the following example.

What Is Harmonic About the Harmonic Series?

The terms in the harmonic series correspond to the nodes on a vibrating string that produce multiples of the fundamental frequency. For example, 1/2 produces the harmonic that is twice the fundamental frequency, 1/3 produces a frequency that is three times the fundamental frequency, and so on. The fundamental frequency is the lowest note or pitch we hear when a string is plucked.

Example 3 The Slow Divergence of the Harmonic Series

Approximately how many terms of the harmonic series are required to form a partial sum larger than 20?

Solution The graphs tell the story (Figure 8.14).

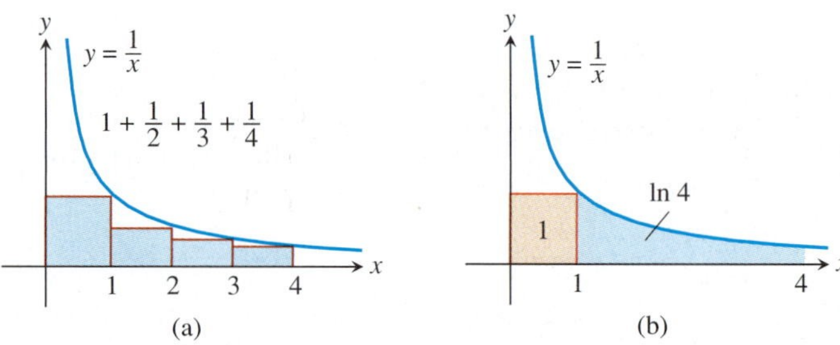

FIGURE 8.14 Finding an upper bound for one of the partial sums of the harmonic series. (Example 3)

Let H_n denote the nth partial sum of the harmonic series. Comparing the two graphs, we see that $H_4 < (1 + \ln 4)$ and (in general) that $H_n \leq (1 + \ln n)$. If we wish H_n to be greater than 20, then

$$\begin{aligned} 1 + \ln n > H_n &> 20 \\ 1 + \ln n &> 20 \\ \ln n &> 19 \\ n &> e^{19}. \end{aligned}$$

The exact value of e^{19} rounds up to 178,482,301. It will take *at least* that many terms of the harmonic series to move the partial sums beyond 20. It would take your calculator several weeks to compute a sum with this many terms. Nonetheless, the harmonic series really does diverge!

Comparison Tests

The p-Series Test tells everything there is to know about the convergence or divergence of series of the form $\Sigma\,(1/n^p)$. This is admittedly a rather narrow class of series, but we can test many other kinds (including those in which the nth term is any rational function of n) by *comparing* them with p-series.

CD-ROM

WEBsite

Historical Biography

Albert of Saxony
(ca. 1316 — 1390)

The Direct Comparison Test

Let $\Sigma\, a_n$ be a series with no negative terms.

(a) $\Sigma\, a_n$ converges if there is a convergent series $\Sigma\, c_n$ with $a_n \leq c_n$ for all $n > N$, for some integer N.

(b) $\Sigma\, a_n$ diverges if there is a divergent series of nonnegative terms $\Sigma\, d_n$ with $a_n \geq d_n$ for all $n > N$, for some integer N.

Proof In part (a), the partial sums of $\Sigma\, a_n$ are bounded above by

$$M = a_1 + a_2 + \cdots + a_n + \sum_{n=N+1}^{\infty} c_n.$$

They therefore form a nondecreasing sequence with a limit $L \leq M$.

In part (b), the partial sums of $\Sigma\, a_n$ are not bounded from above. If they were, the partial sums for $\Sigma\, d_n$ would be bounded by

$$M^* = d_1 + d_2 + \cdots + d_N + \sum_{n=N+1}^{\infty} a_n$$

and $\Sigma\, d_n$ would have to converge instead of diverge.

To apply the Direct Comparison Test to a series, we need not include the early terms of the series. We can start the test with any index N provided that we include all the terms of the series being tested from there on.

Example 4 Applying the Direct Comparison Test

Does the following series converge?

$$5 + \frac{2}{3} + 1 + \frac{1}{7} + \frac{1}{2} + \frac{1}{3!} + \frac{1}{4!} + \cdots + \frac{1}{k!} + \cdots$$

Solution We ignore the first four terms and compare the remaining terms with those of the convergent geometric series $\sum_{n=1}^{\infty} (1/2^n)$. We see that

$$\frac{1}{2} + \frac{1}{3!} + \frac{1}{4!} + \cdots \leq \frac{1}{2} + \frac{1}{4} + \frac{1}{8} + \cdots.$$

Therefore, the original series converges by the Direct Comparison Test.

To apply the Direct Comparison Test, we need to have on hand a list of series whose convergence or divergence we know. Here is what we know so far:

Convergent series	**Divergent series**
Geometric series with $\|r\| < 1$	Geometric series with $\|r\| \geq 1$
Telescoping series like $\sum_{n=1}^{\infty} \frac{1}{n(n+1)}$	The harmonic series $\sum_{n=1}^{\infty} \frac{1}{n}$
The series $\sum_{n=0}^{\infty} \frac{1}{n!}$	Any series $\sum a_n$ for which $\lim_{n\to\infty} a_n$ does not exist or $\lim_{n\to\infty} a_n \neq 0$
Any p-series $\sum_{n=1}^{\infty} \frac{1}{n^p}$ with $p > 1$	Any p-series $\sum_{n=1}^{\infty} \frac{1}{n^p}$ with $p \leq 1$

The Direct Comparison Test is one method of comparison; the *Limit Comparison Test* is another.

The Limit Comparison Test

Suppose that $a_n > 0$ and $b_n > 0$ for all $n \geq N$ (N a positive integer).

1. If $\lim_{n\to\infty} \frac{a_n}{b_n} = c, 0 < c < \infty$, then $\Sigma\, a_n$ and $\Sigma\, b_n$ both converge or both diverge.

2. If $\lim_{n\to\infty} \frac{a_n}{b_n} = 0$ and $\Sigma\, b_n$ converges, then $\Sigma\, a_n$ converges.

3. If $\lim_{n\to\infty} \frac{a_n}{b_n} = \infty$ and $\Sigma\, b_n$ diverges, then $\Sigma\, a_n$ diverges.

Proof We will prove part (1). Parts (2) and (3) are left as Exercise 67.

Since $c/2 > 0$, there exists an integer N such that for all n,

$$n > N \Rightarrow \left| \frac{a_n}{b_n} - c \right| < \frac{c}{2}.$$

Limit definition with $\epsilon = c/2$, $L = c$, and a_n replaced by a_n/b_n.

Thus, for $n > N$,

$$-\frac{c}{2} < \frac{a_n}{b_n} - c < \frac{c}{2},$$

$$\frac{c}{2} < \frac{a_n}{b_n} < \frac{3c}{2},$$

$$\left(\frac{c}{2}\right) b_n < a_n < \left(\frac{3c}{2}\right) b_n.$$

If $\Sigma\, b_n$ converges then $\Sigma\, (3c/2)b_n$ converges and $\Sigma\, a_n$ converges by the Direct Comparison Test. If $\Sigma\, b_n$ diverges, then $\Sigma\, (c/2)b_n$ diverges and $\Sigma\, a_n$ diverges by the Direct Comparison Test.

Example 5 Using the Limit Comparison Test

Determine whether the series converge or diverge.

(a) $\frac{3}{4}+\frac{5}{9}+\frac{7}{16}+\frac{9}{25}+\cdots=\sum_{n=1}^{\infty}\frac{2n+1}{(n+1)^2}=\sum_{n=1}^{\infty}\frac{2n+1}{n^2+2n+1}$

(b) $\frac{1}{1}+\frac{1}{3}+\frac{1}{7}+\frac{1}{15}+\cdots=\sum_{n=1}^{\infty}\frac{1}{2^n-1}$

(c) $\frac{1+2\ln 2}{9}+\frac{1+3\ln 3}{14}+\frac{1+4\ln 4}{21}+\cdots=\sum_{n=2}^{\infty}\frac{1+n\ln n}{n^2+5}$

Solution

We could just as well have taken $b_n = 2/n$, but $1/n$ is simpler.

(a) Let $a_n = (2n + 1)/(n^2 + 2n + 1)$. For n large, we expect a_n to behave like $2n/n^2 = 2/n$, so we let $b_n = 1/n$. Since

$$\sum_{n=1}^{\infty} b_n = \sum_{n=1}^{\infty}\frac{1}{n}$$

diverges and

$$\lim_{n\to\infty}\frac{a_n}{b_n}=\lim_{n\to\infty}\frac{2n^2+n}{n^2+2n+1}=2,$$

$\Sigma\, a_n$ diverges by part 1 of the Limit Comparison Test.

(b) Let $a_n = 1/(2^n - 1)$. For n large, we expect a_n to behave like $1/2^n$, so we let $b_n = 1/2^n$. Since

$$\sum_{n=1}^{\infty} b_n = \sum_{n=1}^{\infty}\frac{1}{2^n}$$

converges and

$$\begin{aligned}\lim_{n\to\infty}\frac{a_n}{b_n}&=\lim_{n\to\infty}\frac{2^n}{2^n-1}\\&=\lim_{n\to\infty}\frac{1}{1-(1/2^n)}\\&=1,\end{aligned}$$

$\Sigma\, a_n$ converges by part 1 of the Limit Comparison Test.

(c) Let $a_n = (1 + n\ln n)/(n^2 + 5)$. For n large, we expect a_n to behave like $(n\ln n)/n^2 = (\ln n)/n$, which is greater than $1/n$ for $n \geq 3$, so we take $b_n = 1/n$. Since

$$\sum_{n=2}^{\infty} b_n = \sum_{n=2}^{\infty}\frac{1}{n}$$

diverges and

$$\begin{aligned}\lim_{n\to\infty}\frac{a_n}{b_n}&=\lim_{n\to\infty}\frac{n+n^2\ln n}{n^2+5}\\&=\infty,\end{aligned}$$

$\Sigma\, a_n$ diverges by part 3 of the Limit Comparison Test.

Ratio and Root Tests

The Ratio Test measures the rate of growth (or decline) of a series by examining the ratio a_{n+1}/a_n. For a geometric series $\Sigma\, ar^n$, this rate is a constant $((ar^{n+1})/(ar^n) = r)$, and the series converges if and only if its ratio is less than 1 in absolute value. The Ratio Test is a powerful rule extending that result.

The Ratio Test

Let $\Sigma\, a_n$ be a series with positive terms and suppose that

$$\lim_{n\to\infty} \frac{a_{n+1}}{a_n} = \rho.$$

Then

(a) the series *converges* if $\rho < 1$
(b) the series *diverges* if $\rho > 1$ or ρ is infinite
(c) the test is *inconclusive* if $\rho = 1$.

Proof

(a) $\boldsymbol{\rho < 1}$**.** Let r be a number between ρ and 1. Then the number $\epsilon = r - \rho$ is positive. Since

$$\frac{a_{n+1}}{a_n} \to \rho,$$

a_{n+1}/a_n must lie within ϵ of ρ when n is large enough, say for all $n \geq N$. In particular

$$\frac{a_{n+1}}{a_n} < \rho + \epsilon = r, \qquad \text{when } n \geq N.$$

That is,

$$\begin{aligned}
a_{N+1} &< r\,a_N,\\
a_{N+2} &< r\,a_{N+1} < r^2 a_N,\\
a_{N+3} &< r\,a_{N+2} < r^3 a_N,\\
&\vdots\\
a_{N+m} &< r\,a_{N+m-1} < r^m a_N.
\end{aligned}$$

These inequalities show that the terms of our series, after the Nth term, approach zero more rapidly than the terms in a geometric series with ratio $r < 1$. More precisely, consider the series $\Sigma\, c_n$, where $c_n = a_n$ for $n = 1, 2, \ldots, N$ and $c_{N+1} = ra_N$, $c_{N+2} = r^2 a_N, \ldots, c_{N+m} = r^m a_N, \ldots$. Now $a_n \leq c_n$ for all n, and

$$\begin{aligned}
\sum_{n=1}^{\infty} c_n &= a_1 + a_2 + \cdots + a_{N-1} + a_N + ra_N + r^2 a_N + \cdots\\
&= a_1 + a_2 + \cdots + a_{N-1} + a_N(1 + r + r^2 + \cdots).
\end{aligned}$$

The geometric series $1 + r + r^2 + \cdots$ converges because $|r| < 1$, so $\Sigma\, c_n$ converges. Since $a_n \le c_n$, $\Sigma\, a_n$ also converges.

(b) $\mathbf{1 < \rho \le \infty}$**.** From some index M on,

$$\frac{a_{n+1}}{a_n} > 1 \qquad \text{and} \qquad a_M < a_{M+1} < a_{M+2} < \cdots.$$

The terms of the series do not approach zero as n becomes infinite, and the series diverges by the nth-Term Test.

(c) $\boldsymbol{\rho = 1}$**.** The two series

$$\sum_{n=1}^{\infty} \frac{1}{n} \qquad \text{and} \qquad \sum_{n=1}^{\infty} \frac{1}{n^2}$$

show that some other test for convergence must be used when $\rho = 1$.

For $\displaystyle\sum_{n=1}^{\infty} \frac{1}{n}$: $\quad \dfrac{a_{n+1}}{a_n} = \dfrac{1/(n+1)}{1/n} = \dfrac{n}{n+1} \to 1.$

For $\displaystyle\sum_{n=1}^{\infty} \frac{1}{n^2}$: $\quad \dfrac{a_{n+1}}{a_n} = \dfrac{1/(n+1)^2}{1/n^2} = \left(\dfrac{n}{n+1}\right)^2 \to 1^2 = 1.$

In both cases, $\rho = 1$, yet the first series diverges, whereas the second converges.

The Ratio Test is often effective when the terms of a series contain factorials of expressions involving n or expressions raised to the nth power.

Example 6 Applying the Ratio Test

Investigate the convergence of the following series.

(a) $\displaystyle\sum_{n=0}^{\infty} \frac{2^n + 5}{3^n}$ **(b)** $\displaystyle\sum_{n=1}^{\infty} \frac{(2n)!}{n!n!}$ **(c)** $\displaystyle\sum_{n=1}^{\infty} \frac{4^n n! n!}{(2n)!}$

Solution

(a) For the series $\sum_{n=0}^{\infty} (2^n + 5)/3^n$,

$$\frac{a_{n+1}}{a_n} = \frac{(2^{n+1} + 5)/3^{n+1}}{(2^n + 5)/3^n} = \frac{1}{3} \cdot \frac{2^{n+1} + 5}{2^n + 5} = \frac{1}{3} \cdot \left(\frac{2 + 5 \cdot 2^{-n}}{1 + 5 \cdot 2^{-n}}\right) \to \frac{1}{3} \cdot \frac{2}{1} = \frac{2}{3}.$$

The series converges because $\rho = 2/3$ is less than 1. This does *not* mean that $2/3$ is the sum of the series. In fact,

$$\sum_{n=0}^{\infty} \frac{2^n + 5}{3^n} = \sum_{n=0}^{\infty} \left(\frac{2}{3}\right)^n + \sum_{n=0}^{\infty} \frac{5}{3^n} = \frac{1}{1 - (2/3)} + \frac{5}{1 - (1/3)} = \frac{21}{2}.$$

(b) If $a_n = \dfrac{(2n)!}{n!n!}$, then $a_{n+1} = \dfrac{(2n+2)!}{(n+1)!(n+1)!}$ and

$$\frac{a_{n+1}}{a_n} = \frac{n!n!(2n+2)(2n+1)(2n)!}{(n+1)!(n+1)!(2n)!}$$

$$= \frac{(2n+2)(2n+1)}{(n+1)(n+1)} = \frac{4n+2}{n+1} \to 4.$$

The series diverges because $\rho = 4$ is greater than 1.

(c) If $a_n = 4^n n!n!/(2n)!$, then

$$\frac{a_{n+1}}{a_n} = \frac{4^{n+1}(n+1)!(n+1)!}{(2n+2)(2n+1)(2n)!} \cdot \frac{(2n)!}{4^n n!n!}$$
$$= \frac{4(n+1)(n+1)}{(2n+2)(2n+1)} = \frac{2(n+1)}{2n+1} \rightarrow 1.$$

Because the limit is $\rho = 1$, we cannot decide from the Ratio Test whether the series converges. When we notice that $a_{n+1}/a_n = (2n+2)/(2n+1)$, we conclude that a_{n+1} is always greater than a_n because $(2n+2)/(2n+1)$ is always greater than 1. Therefore, all terms are greater than or equal to $a_1 = 2$, and the nth term does not approach zero as $n \rightarrow \infty$. The series diverges.

The nth-Root Test is another useful tool for answering the question of convergence for series with nonnegative terms. We state the result here without proof.

The nth-Root Test

Let $\Sigma\, a_n$ be a series with $a_n \geq 0$ for $n \geq N$ and suppose that

$$\lim_{n\to\infty} \sqrt[n]{a_n} = \rho.$$

Then

(a) the series *converges* if $\rho < 1$
(b) the series *diverges* if $\rho > 1$ or ρ is infinite
(c) the test is *inconclusive* if $\rho = 1$.

Example 7 Applying the *n*th-Root Test

Let

$$a_n = \begin{cases} n/2^n, & n \text{ odd} \\ 1/2^n, & n \text{ even.} \end{cases}$$

Does $\Sigma\, a_n$ converge?

Solution We apply the nth-Root Test, finding that

$$\sqrt[n]{a_n} = \begin{cases} \sqrt[n]{n}/2, & n \text{ odd} \\ 1/2, & n \text{ even.} \end{cases}$$

Therefore,

$$\frac{1}{2} \leq \sqrt[n]{a_n} \leq \frac{\sqrt[n]{n}}{2}.$$

Since $\sqrt[n]{n} \rightarrow 1$ (Section 8.1, Table 8.1), we have $\lim_{n\to\infty} \sqrt[n]{a_n} = 1/2$ by the Sandwich Theorem. The limit is less than 1, so the series converges by the nth-Root Test.

Example 8 Applying the *n*th-Root Test

Which of the following series converge, and which diverge?

(a) $\sum_{n=1}^{\infty} \frac{n^2}{2^n}$ **(b)** $\sum_{n=1}^{\infty} \frac{2^n}{n^2}$

Solution

(a) $\sum_{n=1}^{\infty} \frac{n^2}{2^n}$ converges because $\sqrt[n]{\frac{n^2}{2^n}} = \frac{\sqrt[n]{n^2}}{\sqrt[n]{2^n}} = \frac{(\sqrt[n]{n})^2}{2} \to \frac{1}{2} < 1.$

(b) $\sum_{n=1}^{\infty} \frac{2^n}{n^2}$ diverges because $\sqrt[n]{\frac{2^n}{n^2}} = \frac{2}{(\sqrt[n]{n})^2} \to \frac{2}{1} > 1.$

EXERCISES 8.4

Integral Test

Use the Integral Test to determine which of the series in Exercises 1–8 converge and which diverge.

1. $\sum_{n=1}^{\infty} \frac{5}{n+1}$
2. $\sum_{n=1}^{\infty} \frac{1}{2n-1}$
3. $\sum_{n=2}^{\infty} \frac{\ln n}{n}$
4. $\sum_{n=2}^{\infty} \frac{\ln n}{\sqrt{n}}$
5. $\sum_{n=1}^{\infty} \frac{e^n}{1+e^{2n}}$
6. $\sum_{n=1}^{\infty} \frac{1}{\sqrt{n}(\sqrt{n}+1)}$
7. $\sum_{n=3}^{\infty} \frac{(1/n)}{(\ln n)\sqrt{\ln^2 n - 1}}$
8. $\sum_{n=1}^{\infty} \frac{1}{n(1+\ln^2 n)}$

Direct Comparison Test

Use the Direct Comparison Test to determine which of the series in Exercises 9–14 converge and which diverge.

9. $\sum_{n=1}^{\infty} \frac{1}{2\sqrt{n}+\sqrt[3]{n}}$
10. $\sum_{n=1}^{\infty} \frac{3}{n+\sqrt{n}}$
11. $\sum_{n=1}^{\infty} \frac{\sin^2 n}{2^n}$
12. $\sum_{n=1}^{\infty} \frac{1+\cos n}{n^2}$
13. $\sum_{n=1}^{\infty} \left(\frac{n}{3n+1}\right)^n$
14. $\sum_{n=3}^{\infty} \frac{1}{\ln(\ln n)}$

Limit Comparison Test

Use the Limit Comparison Test to determine which of the series in Exercises 15–20 converge and which diverge.

15. $\sum_{n=2}^{\infty} \frac{1}{(\ln n)^2}$
16. $\sum_{n=1}^{\infty} \frac{(\ln n)^2}{n^3}$
17. $\sum_{n=1}^{\infty} \frac{(\ln n)^3}{n^3}$
18. $\sum_{n=2}^{\infty} \frac{1}{\sqrt{n}\ln n}$
19. $\sum_{n=1}^{\infty} \frac{(\ln n)^2}{n^{3/2}}$
20. $\sum_{n=1}^{\infty} \frac{1}{1+\ln n}$

Ratio Test

Use the Ratio Test to determine which of the series in Exercises 21–28 converge and which diverge.

21. $\sum_{n=1}^{\infty} \frac{n^{\sqrt{2}}}{2^n}$
22. $\sum_{n=1}^{\infty} n^2 e^{-n}$
23. $\sum_{n=1}^{\infty} n! e^{-n}$
24. $\sum_{n=1}^{\infty} \frac{n!}{10^n}$
25. $\sum_{n=1}^{\infty} \frac{n^{10}}{10^n}$
26. $\sum_{n=1}^{\infty} \frac{n \ln n}{2^n}$
27. $\sum_{n=1}^{\infty} \frac{(n+1)(n+2)}{n!}$
28. $\sum_{n=1}^{\infty} e^{-n}(n^3)$

Root Test

Use the *n*th-Root Test to determine which of the series in Exercises 29–34 converge and which diverge.

29. $\sum_{n=1}^{\infty} \frac{(\ln n)^n}{n^n}$
30. $\sum_{n=1}^{\infty} \left(\frac{1}{n} - \frac{1}{n^2}\right)^n$
31. $\sum_{n=2}^{\infty} \frac{n}{(\ln n)^n}$
32. $\sum_{n=2}^{\infty} \frac{n}{(\ln n)^{(n/2)}}$
33. $\sum_{n=1}^{\infty} \frac{(n!)^n}{(n^n)^2}$
34. $\sum_{n=1}^{\infty} \frac{n^n}{(2^n)^2}$

Determining Convergence or Divergence

Which of the series in Exercises 35–60 converge, and which diverge? Give reasons for your answers. (When checking your answers, remember that there may be more than one way to determine a series' convergence or divergence.)

35. $\sum_{n=1}^{\infty} e^{-n}$

36. $\sum_{n=1}^{\infty} \frac{n}{n+1}$

37. $\sum_{n=1}^{\infty} \frac{3}{\sqrt{n}}$

38. $\sum_{n=1}^{\infty} \frac{-2}{n\sqrt{n}}$

39. $\sum_{n=1}^{\infty} \frac{1}{(1+\ln n)^2}$

40. $\sum_{n=2}^{\infty} \frac{\ln(n+1)}{n+1}$

41. $\sum_{n=2}^{\infty} \frac{1}{n\sqrt{n^2-1}}$

42. $\sum_{n=1}^{\infty} \left(\frac{n-2}{n}\right)^n$

43. $\sum_{n=1}^{\infty} \frac{(n+3)!}{3!n!3^n}$

44. $\sum_{n=1}^{\infty} \frac{n2^n(n+1)!}{3^n n!}$

45. $\sum_{n=1}^{\infty} \frac{n!}{(2n+1)!}$

46. $\sum_{n=1}^{\infty} \frac{n!}{n^n}$

47. $\sum_{n=1}^{\infty} \frac{8\tan^{-1} n}{1+n^2}$

48. $\sum_{n=1}^{\infty} \frac{n}{n^2+1}$

49. $\sum_{n=1}^{\infty} \operatorname{sech} n$

50. $\sum_{n=1}^{\infty} \operatorname{sech}^2 n$

51. $\sum_{n=1}^{\infty} \frac{2+(-1)^n}{1.25^n}$

52. $\sum_{n=1}^{\infty} \left(1-\frac{1}{3n}\right)^n$

53. $\sum_{n=1}^{\infty} \frac{\ln n}{n^3}$

54. $\sum_{n=1}^{\infty} \frac{\ln n}{n}$

55. $\sum_{n=1}^{\infty} \frac{10n+1}{n(n+1)(n+2)}$

56. $\sum_{n=3}^{\infty} \frac{5n^3-3n}{n^2(n-2)(n^2+5)}$

57. $\sum_{n=1}^{\infty} \frac{\tan^{-1} n}{n^{1.1}}$

58. $\sum_{n=1}^{\infty} \frac{\sec^{-1} n}{n^{1.3}}$

59. $\sum_{n=1}^{\infty} n \sin \frac{1}{n}$

60. $\sum_{n=1}^{\infty} \frac{2}{1+e^n}$

Terms Defined Recursively

Which of the series $\sum_{n=1}^{\infty} a_n$ defined by the formulas in Exercises 61–66 converge, and which diverge? Give reasons for your answers.

61. $a_1 = 2, \quad a_{n+1} = \frac{1+\sin n}{n} a_n$

62. $a_1 = 1, \quad a_{n+1} = \frac{1+\tan^{-1} n}{n} a_n$

63. $a_1 = \frac{1}{3}, \quad a_{n+1} = \frac{3n-1}{2n+5} a_n$

64. $a_1 = 3, \quad a_{n+1} = \frac{n}{n+1} a_n$

65. $a_1 = \frac{1}{3}, \quad a_{n+1} = \sqrt[n]{a_n}$

66. $a_1 = \frac{1}{2}, \quad a_{n+1} = (a_n)^{n+1}$

Theory and Examples

67. Prove

(a) Part 2 of the Limit Comparison Test

(b) Part 3 of the Limit Comparison Test.

68. *Writing to Learn* If $\sum_{n=1}^{\infty} a_n$ is a convergent series of nonnegative numbers, can anything be said about $\sum_{n=1}^{\infty} (a_n/n)$? Explain.

69. *Writing to Learn* Suppose that $a_n > 0$ and $b_n > 0$ for $n \ge N$ (N an integer). If $\lim_{n\to\infty} (a_n/b_n) = \infty$ and Σa_n converges, can anything be said about $\Sigma\, b_n$? Give reasons for your answer.

70. *Term-by-term squaring* Prove that if $\Sigma\, a_n$ is a convergent series of nonnegative terms, then $\Sigma\, a_n^2$ converges.

For what values of a, if any, do the series in Exercises 71 and 72 converge?

71. $\sum_{n=1}^{\infty} \left(\frac{a}{n+2} - \frac{1}{n+4}\right)$

72. $\sum_{n=3}^{\infty} \left(\frac{1}{n-1} - \frac{2a}{n+1}\right)$

73. *The Cauchy Condensation Test* The Cauchy Condensation Test says: Let $\{a_n\}$ be a nonincreasing sequence ($a_n \ge a_{n+1}$ for all n) of positive terms that converges to 0. Then $\Sigma\, a_n$ converges if and only if $\Sigma\, 2^n a_{2^n}$ converges. For example, $\Sigma\,(1/n)$ diverges because $\Sigma 2^n \cdot (1/2^n) = \Sigma 1$ diverges. Show why the test works.

74. Use the Cauchy Condensation Test from Exercise 73 to show that

(a) $\sum_{n=2}^{\infty} \frac{1}{n \ln n}$ diverges

(b) $\sum_{n=1}^{\infty} \frac{1}{n^p}$ converges if $p > 1$ and diverges if $p \le 1$.

75. *Logarithmic p-series*

(a) Show that

$$\int_2^{\infty} \frac{dx}{x(\ln x)^p} \qquad (p \text{ a positive constant})$$

converges if and only if $p > 1$.

(b) What implications does the fact in part (a) have for the convergence of the series

$$\sum_{n=2}^{\infty} \frac{1}{n(\ln n)^p}?$$

Give reasons for your answer.

76. (Continuation of Exercise 75) Use the result in Exercise 75 to determine which of the following series converge and which diverge. Support your answer in each case.

(a) $\sum_{n=2}^{\infty} \frac{1}{n(\ln n)}$

(b) $\sum_{n=2}^{\infty} \frac{1}{n(\ln n)^{1.01}}$

(c) $\sum_{n=2}^{\infty} \frac{1}{n \ln(n^3)}$

(d) $\sum_{n=2}^{\infty} \frac{1}{n(\ln n)^3}$

77. *Another logarithmic p-series* Show that neither the Ratio Test nor the nth-Root Test provides information about the convergence of

$$\sum_{n=2}^{\infty} \frac{1}{(\ln n)^p} \qquad (p \text{ constant}).$$

78. Let

$$a_n = \begin{cases} n/2^n & \text{if } n \text{ is a prime number} \\ 1/2^n & \text{otherwise.} \end{cases}$$

Does $\Sigma\, a_n$ converge? Give reasons for your answer.

79. *p-Series* Neither the Ratio Test nor the nth-Root Test helps with p-series. Try these tests on

$$\sum_{n=1}^{\infty} \frac{1}{n^p}$$

and show that both fail to provide information about convergence.

COMPUTER EXPLORATION

A Current Mystery

80. It is not yet known whether the series

$$\sum_{n=1}^{\infty} \frac{1}{n^3 \sin^2 n}$$

converges or diverges. Use a CAS to explore the behavior of the series by performing the following steps.

(a) Define the sequence of partial sums

$$s_k = \sum_{n=1}^{k} \frac{1}{n^3 \sin^2 n}.$$

What happens when you try to find the limit of s_k as $k \to \infty$? Does your CAS find a closed-form answer for this limit?

(b) Plot the first 100 points (k, s_k) for the sequence of partial sums. Do they appear to converge? What would you estimate the limit to be?

(c) Next plot the first 200 points (k, s_k). Discuss the behavior in your own words.

(d) Plot the first 400 points (k, s_k). What happens when $k = 355$? Calculate the number $355/113$. Explain from your calculation what happened at $k = 355$. For what values of k would you guess this behavior might occur again?

You will find an interesting discussion of this series in Chapter 72 of *Mazes for the Mind* by Clifford A. Pickover (New York: St. Martin's Press, 1992).

8.5 Alternating Series, Absolute and Conditional Convergence

Alternating Series • Absolute Convergence • Rearranging Series • Procedure for Determining Convergence

The convergence tests investigated so far apply only to series with nonnegative terms. In this section, we learn how to deal with series that may have negative terms. An important example is the *alternating series,* whose terms alternate in sign. We also learn which convergent series can have their terms rearranged (that is, changing the order in which they appear) without changing their sum.

Alternating Series

A series in which the terms are alternately positive and negative is an **alternating series.**

Here are three examples.

$$1 - \frac{1}{2} + \frac{1}{3} - \frac{1}{4} + \frac{1}{5} - \cdots + \frac{(-1)^{n+1}}{n} + \cdots \tag{1}$$

$$-2 + 1 - \frac{1}{2} + \frac{1}{4} - \frac{1}{8} + \cdots + \frac{(-1)^n 4}{2^n} + \cdots \tag{2}$$

$$1 - 2 + 3 - 4 + 5 - 6 + \cdots + (-1)^{n+1} n + \cdots \tag{3}$$

Series (1), called the **alternating harmonic series,** converges, as we see shortly. Series (2), a geometric series with $a = -2$ and $r = -1/2$, converges to $-2/[1 + (1/2)] = -4/3$. Series (3) diverges by the nth-Term Test.

We prove the convergence of the alternating harmonic series by applying the following test.

Theorem 8 **The Alternating Series Test (Leibniz's Theorem)**

The series

$$\sum_{n=1}^{\infty} (-1)^{n+1} u_n = u_1 - u_2 + u_3 - u_4 + \cdots$$

converges if all three of the following conditions are satisfied.

1. The u_n's are all positive.
2. $u_n \geq u_{n+1}$ for all $n \geq N$, for some integer N.
3. $u_n \to 0$.

Proof If n is an even integer, say $n = 2m$, then the sum of the first n term is

$$\begin{aligned} s_{2m} &= (u_1 - u_2) + (u_3 - u_4) + \cdots + (u_{2m-1} - u_{2m}) \\ &= u_1 - (u_2 - u_3) - (u_4 - u_5) - \cdots - (u_{2m-2} - u_{2m-1}) - u_{2m}. \end{aligned}$$

The first equality shows that s_{2m} is the sum of m nonnegative terms, since each term in parentheses is positive or zero. Hence, $s_{2m+2} \geq s_{2m}$, and the sequence $\{s_{2m}\}$ is nondecreasing. The second equality shows that $s_{2m} \leq u_1$. Since $\{s_{2m}\}$ is nondecreasing and bounded from above, it has a limit, say

$$\lim_{m\to\infty} s_{2m} = L. \tag{4}$$

If n is an odd integer, say $n = 2m + 1$, then the sum of the first n terms is $s_{2m+1} = s_{2m} + u_{2m+1}$. Since $u_n \to 0$,

$$\lim_{m\to\infty} u_{2m+1} = 0$$

and, as $m \to \infty$,

$$s_{2m+1} = s_{2m} + u_{2m+1} \to L + 0 = L. \tag{5}$$

Combining the results of Equations (4) and (5) gives $\lim_{n\to\infty} s_n = L$ (Section 8.2, Exercise 26). Figure 8.15 illustrates the convergence of the partial sums to their limit L.

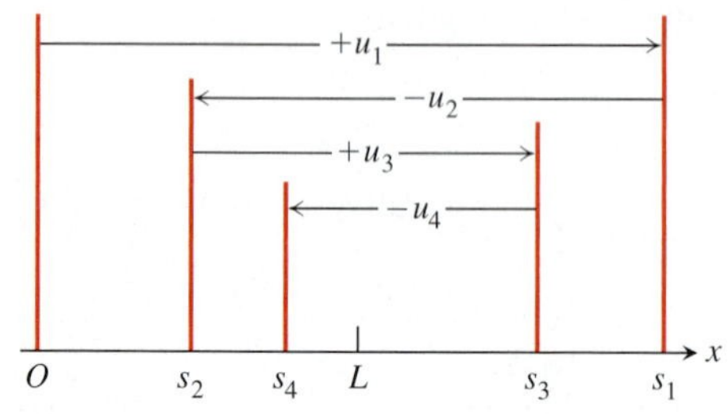

FIGURE 8.15 The partial sums of an alternating series that satisfies the hypotheses of Theorem 8 for $N = 1$ straddle the limit from the beginning.

Figure 8.15 actually shows more than the *fact* of convergence; it also shows the *way* that an alternating series converges when it satisfies the conditions of the test. The partial sums keep "overshooting" the limit as they go back and forth on the number line, gradually closing in as the terms tend to zero. If we stop at the nth partial sum, we know that the next term $(\pm u_{n+1})$ will again cause us to overshoot the limit in the positive direction or negative direction, depending on the sign carried by u_{n+1}. This gives us a convenient bound for the **truncation error**, which we state as another theorem.

Theorem 9 The Alternating Series Estimation Theorem

If the alternating series $\sum_{n=1}^{\infty} (-1)^{n+1} u_n$ satisfies the conditions of Theorem 8, then the truncation error for the nth partial sum is less than u_{n+1} and has the same sign as the unused term.

Example 1 The Alternating Harmonic Series

Prove that the alternating harmonic series is convergent, but that the corresponding series of absolute values is not convergent. Find a bound for the truncation error after 99 terms.

Solution The terms are strictly alternating in sign and decrease in absolute value from the start:

$$1 > \frac{1}{2} > \frac{1}{3} > \cdots.$$

Also,

$$\frac{1}{n} \to 0.$$

By the Alternating Series Test,

$$\sum_{n=1}^{\infty} \frac{(-1)^{n+1}}{n}$$

converges.

On the other hand, the series $\sum_{n=1}^{\infty} (1/n)$ of absolute values is the harmonic series, which diverges.

The Alternating Series Estimation Theorem guarantees that the truncation error after 99 terms is less than $u_{99+1} = 1/(99+1) = 1/100$.

A Note on the Error Bound

Theorem 9 does not give a *formula* for the truncation error but a *bound* for the truncation error. The bound might be fairly conservative. For example, the first 99 terms of the alternating harmonic series add to about 0.6981721793, whereas the series itself has a sum of $\ln 2 \approx 0.6931471806$. That makes the actual truncation error very close to 0.005, about half the size of the bound of 0.01 given by Theorem 9.

Example 2 Applying The Estimation Theorem

We try Theorem 9 on a series whose sum we know:

$$\sum_{n=0}^{\infty} (-1)^n \frac{1}{2^n} = 1 - \frac{1}{2} + \frac{1}{4} - \frac{1}{8} + \frac{1}{16} - \frac{1}{32} + \frac{1}{64} - \frac{1}{128} \;\vdots\; + \frac{1}{256} - \cdots.$$

The theorem says that if we truncate the series after the eighth term, we throw away a total that is positive and less than $1/256$. The sum of the first eight terms is 0.6640 625. The sum of the series is

$$\frac{1}{1-(-1/2)} = \frac{1}{3/2} = \frac{2}{3}.$$

The difference, $(2/3) - 0.6640\,625 = 0.0026\,0416\,6\ldots$, is positive and less than $(1/256) = 0.0039\,0625$.

Absolute Convergence

Definition Absolute Convergence

A series $\Sigma\, a_n$ **converges absolutely** (is **absolutely convergent**) if the corresponding series of absolute values, $\Sigma\, |a_n|$, converges.

Niccolo Tartaglia
(1499 — 1557)

The geometric series

$$1 - \frac{1}{2} + \frac{1}{4} - \frac{1}{8} + \cdots$$

converges absolutely because the corresponding series of absolute values

$$1 + \frac{1}{2} + \frac{1}{4} + \frac{1}{8} + \cdots$$

converges. The alternating harmonic series (Example 1) does not converge absolutely. The corresponding series of absolute values is the (divergent) harmonic series.

Definition Conditional Convergence

A series that converges but does not converge absolutely **converges conditionally.**

The alternating harmonic series converges conditionally.

Example 3 Absolute and Conditional Convergence

Determine which of the following series is absolutely or conditionally convergent, or divergent.

(a) $\displaystyle\sum_{n=1}^{\infty} (-1)^{n+1} \frac{1}{\sqrt{n}} = 1 - \frac{1}{\sqrt{2}} + \frac{1}{\sqrt{3}} - \frac{1}{\sqrt{4}} + \cdots$

(b) $\displaystyle\sum_{n=2}^{\infty} (-1)^n \left(1 - \frac{1}{n}\right)^n = \left(\frac{1}{2}\right)^2 - \left(\frac{2}{3}\right)^3 + \left(\frac{3}{4}\right)^4 - \cdots$

(c) $\displaystyle\sum_{n=1}^{\infty} (-1)^{n(n+1)/2} \frac{1}{2^n} = -\frac{1}{2} - \frac{1}{4} + \frac{1}{8} + \frac{1}{16} - \cdots$

Solution

(a) The series converges by the Alternating Series Test since $(1/\sqrt{n}) > (1/\sqrt{n+1})$ and $(1/\sqrt{n}) \to 0$. The series $\sum_{n=1}^{\infty} (1/\sqrt{n})$ of absolute values diverges, however, because it is a p-series with $p = (1/2) < 1$. Therefore, the given series is *conditionally convergent*.

(b) The series *diverges* by the nth-Term Test since $\lim_{n\to\infty}\ (1 - (1/n))^n = e^{-1} \neq 0$ (Table 8.1, Formula 5).

(c) This is *not* an alternating series. However,

$$\sum_{n=1}^{\infty} \left| (-1)^{n(n+1)/2} \frac{1}{2^n} \right| = \sum_{n=1}^{\infty} \frac{1}{2^n}$$

is a convergent geometric series, and we conclude that the given series is *absolutely convergent.*

Absolute convergence is important for two reasons. First, we have good tests for convergence of series of positive terms. Second, if a series converges absolutely, then it converges. That is the thrust of the next theorem.

CAUTION We can rephrase Theorem 10 to say that *every absolutely convergent series converges*. The converse statement is false, however: Many convergent series do not converge absolutely.

Theorem 10 The Absolute Convergence Test

If $\sum_{n=1}^{\infty} |a_n|$ converges, then $\sum_{n=1}^{\infty} a_n$ converges.

Proof For each n,

$$-|a_n| \leq a_n \leq |a_n|, \qquad \text{so} \qquad 0 \leq a_n + |a_n| \leq 2|a_n|.$$

If $\sum_{n=1}^{\infty} |a_n|$ converges, then $\sum_{n=1}^{\infty} 2|a_n|$ converges and, by the Direct Comparison Test, the nonnegative series $\sum_{n=1}^{\infty} (a_n + |a_n|)$ converges. The equality $a_n = (a_n + |a_n|) - |a_n|$ now lets us express $\sum_{n=1}^{\infty} a_n$ as the difference of two convergent series:

$$\sum_{n=1}^{\infty} a_n = \sum_{n=1}^{\infty} (a_n + |a_n| - |a_n|) = \sum_{n=1}^{\infty} (a_n + |a_n|) - \sum_{n=1}^{\infty} |a_n|.$$

Therefore, $\sum_{n=1}^{\infty} a_n$ converges.

Example 4 Applying the Absolute Convergence Test

For

$$\sum_{n=1}^{\infty} (-1)^{n+1} \frac{1}{n^2} = 1 - \frac{1}{4} + \frac{1}{9} - \frac{1}{16} + \cdots,$$

the corresponding series of absolute values is the convergent series

$$\sum_{n=1}^{\infty} \frac{1}{n^2} = 1 + \frac{1}{4} + \frac{1}{9} + \frac{1}{16} + \cdots.$$

The original alternating series converges because it converges absolutely.

Example 5 Applying the Absolute Convergence Test

For

$$\sum_{n=1}^{\infty} \frac{\sin n}{n^2} = \frac{\sin 1}{1} + \frac{\sin 2}{4} + \frac{\sin 3}{9} + \cdots,$$

the corresponding series of absolute values is

$$\sum_{n=1}^{\infty}\left|\frac{\sin n}{n^2}\right| = \frac{|\sin 1|}{1} + \frac{|\sin 2|}{4} + \cdots,$$

which converges by comparison with $\sum_{n=1}^{\infty}(1/n^2)$ because $|\sin n| \le 1$ for every n. The original series converges absolutely; therefore, it converges.

CD-ROM WEBsite
Historical Biography
Georg Cantor
(1845 — 1918)

Example 6 Alternating p-Series

If p is a positive constant, the sequence $\{1/n^p\}$ is a decreasing sequence with limit zero. Therefore, the alternating p-series

$$\sum_{n=1}^{\infty}\frac{(-1)^{n-1}}{n^p} = 1 - \frac{1}{2^p} + \frac{1}{3^p} - \frac{1}{4^p} + \cdots, \qquad p > 0,$$

converges.

If $p > 1$, the series converges absolutely. If $0 < p \le 1$, the series converges conditionally.

Conditional convergence: $1 - \frac{1}{\sqrt{2}} + \frac{1}{\sqrt{3}} - \frac{1}{\sqrt{4}} + \cdots$

Absolute convergence: $1 - \frac{1}{2^{3/2}} + \frac{1}{3^{3/2}} - \frac{1}{4^{3/2}} + \cdots$

Rearranging Series

Theorem 11 The Rearrangement Theorem for Absolutely Convergent Series

If $\sum_{n=1}^{\infty} a_n$ converges absolutely and $b_1, b_2, \ldots, b_n, \ldots$, is any arrangement of the sequence $\{a_n\}$, then $\sum b_n$ converges absolutely and

$$\sum_{n=1}^{\infty} b_n = \sum_{n=1}^{\infty} a_n.$$

(For an outline of the proof, see Exercise 60.)

Example 7 Applying the Rearrangement Theorem

As we saw in Example 4, the series

$$1 - \frac{1}{4} + \frac{1}{9} - \frac{1}{16} + \cdots + (-1)^{n-1}\frac{1}{n^2} + \cdots$$

converges absolutely. A possible rearrangement of the terms of the series might start with a positive term, then two negative terms, then three positive terms, then four negative terms, and so on: After k terms of one sign, take $k + 1$ terms of the opposite sign. The first 10 terms of such a series look like this:

$$1 - \frac{1}{4} - \frac{1}{16} + \frac{1}{9} + \frac{1}{25} + \frac{1}{49} - \frac{1}{36} - \frac{1}{64} - \frac{1}{100} - \frac{1}{144} + \cdots.$$

The Rearrangement Theorem says that both series converge to the same value. In this example, if we had the second series to begin with, we would probably be glad to exchange it for the first, if we knew that we could. We can do even better: The sum of either series is also equal to

$$\sum_{n=1}^{\infty} \frac{1}{(2n-1)^2} - \sum_{n=1}^{\infty} \frac{1}{(2n)^2}.$$

(See Exercise 61.)

CAUTION If we rearrange infinitely many terms of a conditionally convergent series, we can get results that are far different from the sum of the original series.

The kind of behavior illustrated by this example is typical of what can happen with any conditionally convergent series. Moral: Add the terms of a conditionally convergent series in the order given.

Example 8 Rearranging the Alternating Harmonic Series

The alternating harmonic series

$$\frac{1}{1} - \frac{1}{2} + \frac{1}{3} - \frac{1}{4} + \frac{1}{5} - \frac{1}{6} + \frac{1}{7} - \frac{1}{8} + \frac{1}{9} - \frac{1}{10} + \frac{1}{11} - \cdots$$

can be rearranged to diverge or to reach any preassigned sum.

(a) *Rearranging* $\sum_{n=1}^{\infty} (-1)^{n+1}/n$ *to diverge.* The series of terms $\Sigma\ [1/(2n-1)]$ diverges to $+\infty$, and the series of terms $\Sigma\ (-1/2n)$ diverges to $-\infty$. No matter how far out in the sequence of odd-numbered terms we begin, we can always add enough positive terms to get an arbitrarily large sum. Similarly, with the negative terms, no matter how far out we start, we can add enough consecutive even-numbered terms to get a negative sum of arbitrarily large absolute value. If we wished to do so, we could start adding odd-numbered terms until we had a sum greater than $+3$, say, and then follow that with enough consecutive negative terms to make the new total less than -4. We could then add enough positive terms to make the total greater than $+5$ and follow with consecutive unused negative terms to make a new total less than -6, and so on. In this way, we could make the swings arbitrarily large in either direction.

(b) *Rearranging* $\sum_{n=1}^{\infty} (-1)^{n+1}/n$ *to converge to* 1. Another possibility is to focus on a particular limit. Suppose that we try to get sums that converge to 1. We start with the first term, $1/1$, and then subtract $1/2$. Next we add $1/3$ and $1/5$, which brings the total back to 1 or above. Then we add consecutive negative terms until the total is less than 1. We continue in this manner: When the sum is less than 1, add positive terms until the total is 1 or more, then subtract (add negative) terms until the total is again less than 1. This process can be continued indefinitely. Because both the odd-numbered terms and the even-numbered terms of the original series approach zero as $n \to \infty$, the amount by which our partial sums exceed 1 or fall below it approaches zero. So the new series converges to 1. The rearranged series starts like this:

$$\frac{1}{1} - \frac{1}{2} + \frac{1}{3} + \frac{1}{5} - \frac{1}{4} + \frac{1}{7} + \frac{1}{9} - \frac{1}{6} + \frac{1}{11} + \frac{1}{13} - \frac{1}{8} + \frac{1}{15} + \frac{1}{17} - \frac{1}{10}$$
$$+ \frac{1}{19} + \frac{1}{21} - \frac{1}{12} + \frac{1}{23} + \frac{1}{25} - \frac{1}{14} + \frac{1}{27} - \frac{1}{16} + \cdots.$$

Procedure For Determining Convergence

The following flowchart is often useful for deciding whether a given infinite series converges or diverges.

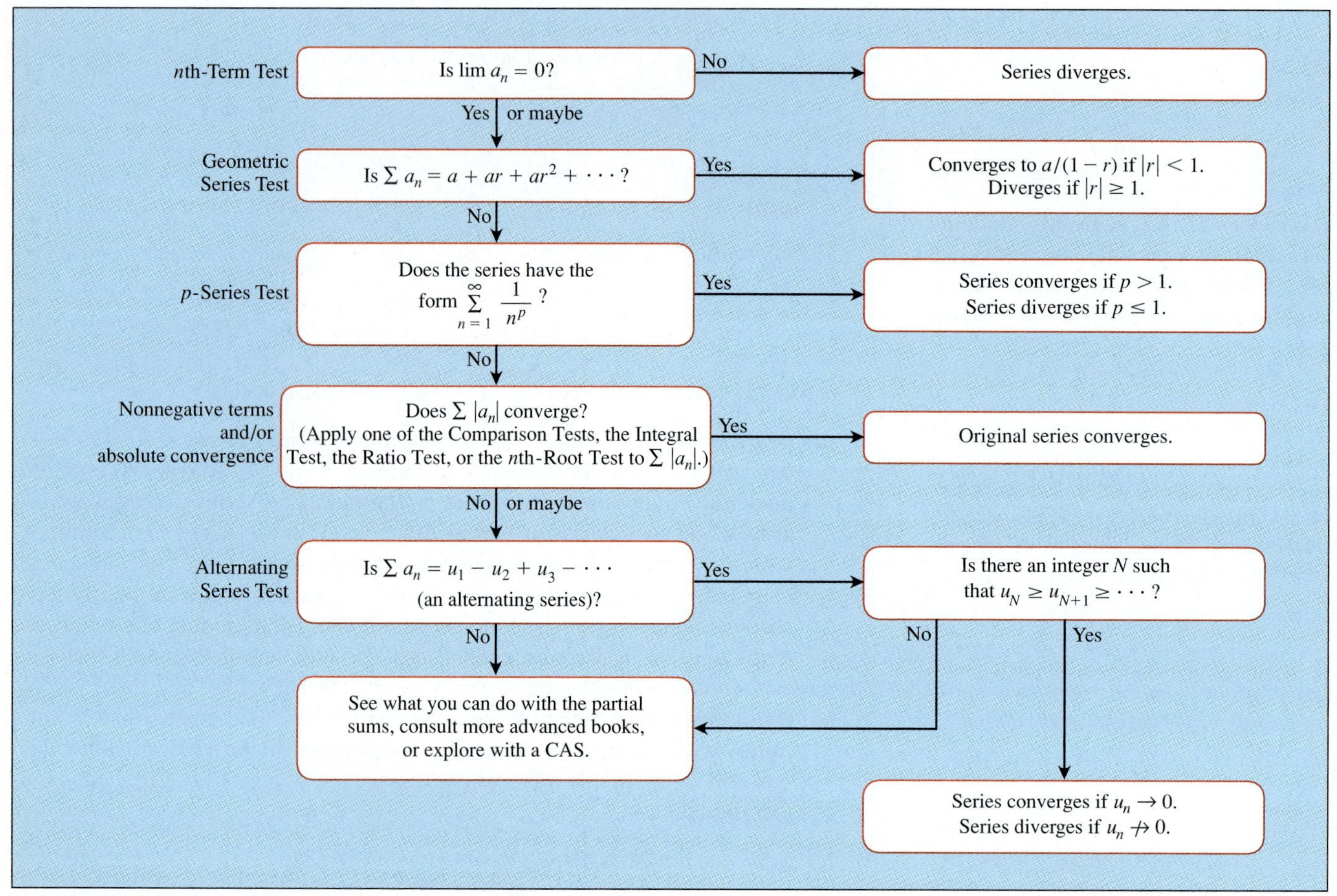

FLOWCHART 8.1 Procedure for determining convergence.

EXERCISES 8.5

Determining Convergence or Divergence

Which of the alternating series in Exercises 1–10 converge, and which diverge? Give reasons for your answers.

1. $\sum_{n=1}^{\infty} (-1)^{n+1} \frac{1}{n^2}$

2. $\sum_{n=1}^{\infty} (-1)^{n+1} \frac{1}{n^{3/2}}$

3. $\sum_{n=1}^{\infty} (-1)^{n+1} \left(\frac{n}{10}\right)^n$

4. $\sum_{n=1}^{\infty} (-1)^{n+1} \frac{10^n}{n^{10}}$

5. $\sum_{n=2}^{\infty} (-1)^{n+1} \frac{1}{\ln n}$

6. $\sum_{n=1}^{\infty} (-1)^{n+1} \frac{\ln n}{n}$

7. $\sum_{n=2}^{\infty} (-1)^{n+1} \frac{\ln n}{\ln n^2}$

8. $\sum_{n=1}^{\infty} (-1)^n \ln\left(1 + \frac{1}{n}\right)$

9. $\sum_{n=1}^{\infty} (-1)^{n+1} \frac{\sqrt{n}+1}{n+1}$

10. $\sum_{n=1}^{\infty} (-1)^{n+1} \frac{3\sqrt{n+1}}{\sqrt{n}+1}$

Absolute Versus Conditional Convergence

Which of the series in Exercises 11–44 converge absolutely, which converge conditionally, and which diverge? Give reasons for your answers.

11. $\sum_{n=1}^{\infty} (-1)^{n+1} (0.1)^n$

12. $\sum_{n=1}^{\infty} (-1)^{n+1} \frac{(0.1)^n}{n}$

13. $\sum_{n=1}^{\infty} (-1)^n \frac{1}{\sqrt{n+1}}$

14. $\sum_{n=1}^{\infty} \frac{(-1)^n}{1+\sqrt{n}}$

15. $\sum_{n=1}^{\infty} (-1)^{n+1} \frac{n}{n^3+1}$

16. $\sum_{n=1}^{\infty} (-1)^{n+1} \frac{n!}{2^n}$

17. $\sum_{n=1}^{\infty} (-1)^n \frac{1}{n+3}$

18. $\sum_{n=1}^{\infty} (-1)^n \frac{\sin n}{n^2}$

19. $\sum_{n=1}^{\infty} (-1)^{n+1} \frac{3+n}{5+n}$

20. $\sum_{n=2}^{\infty} (-1)^n \frac{1}{\ln (n^3)}$

21. $\sum_{n=1}^{\infty} (-1)^{n+1} \frac{1+n}{n^2}$

22. $\sum_{n=1}^{\infty} \frac{(-2)^{n+1}}{n+5^n}$

23. $\sum_{n=1}^{\infty} (-1)^n n^2 (2/3)^n$

24. $\sum_{n=1}^{\infty} (-1)^{n+1} (\sqrt[n]{10})$

25. $\sum_{n=1}^{\infty} (-1)^n \frac{\tan^{-1} n}{n^2+1}$

26. $\sum_{n=2}^{\infty} (-1)^{n+1} \frac{1}{n \ln n}$

27. $\sum_{n=1}^{\infty} (-1)^n \frac{n}{n+1}$

28. $\sum_{n=1}^{\infty} (-1)^n \frac{\ln n}{n - \ln n}$

29. $\sum_{n=1}^{\infty} \frac{(-100)^n}{n!}$

30. $\sum_{n=1}^{\infty} (-5)^{-n}$

31. $\sum_{n=1}^{\infty} \frac{(-1)^{n-1}}{n^2+2n+1}$

32. $\sum_{n=2}^{\infty} (-1)^n \left(\frac{\ln n}{\ln n^2}\right)^n$

33. $\sum_{n=1}^{\infty} \frac{\cos n\pi}{n\sqrt{n}}$

34. $\sum_{n=1}^{\infty} \frac{\cos n\pi}{n}$

35. $\sum_{n=1}^{\infty} \frac{(-1)^n (n+1)^n}{(2n)^n}$

36. $\sum_{n=1}^{\infty} \frac{(-1)^{n+1}(n!)^2}{(2n)!}$

37. $\sum_{n=1}^{\infty} (-1)^n \frac{(2n)!}{2^n n!\, n}$

38. $\sum_{n=1}^{\infty} (-1)^n \frac{(n!)^2 3^n}{(2n+1)!}$

39. $\sum_{n=1}^{\infty} (-1)^n (\sqrt{n+1} - \sqrt{n})$

40. $\sum_{n=1}^{\infty} (-1)^n (\sqrt{n^2+n} - n)$

41. $\sum_{n=1}^{\infty} (-1)^n (\sqrt{n+\sqrt{n}} - \sqrt{n})$

42. $\sum_{n=1}^{\infty} \frac{(-1)^n}{\sqrt{n}+\sqrt{n+1}}$

43. $\sum_{n=1}^{\infty} (-1)^n \operatorname{sech} n$

44. $\sum_{n=1}^{\infty} (-1)^n \operatorname{csch} n$

Error Estimation

In Exercises 45–48, estimate the magnitude of the error involved in using the sum of the first four terms to approximate the sum of the entire series.

45. $\sum_{n=1}^{\infty} (-1)^{n+1} \frac{1}{n}$ It can be shown that the sum is ln 2.

46. $\sum_{n=1}^{\infty} (-1)^{n+1} \frac{1}{10^n}$

47. $\sum_{n=1}^{\infty} (-1)^{n+1} \frac{(0.01)^n}{n}$ As you will see in Section 8.6 the sum is ln (1.01).

48. $\frac{1}{1+t} = \sum_{n=0}^{\infty} (-1)^n t^n, \quad 0 < t < 1$

Approximate the sums in Exercises 49 and 50 with an error of magnitude less than 5×10^{-6}.

49. $\sum_{n=0}^{\infty} (-1)^n \frac{1}{(2n)!}$ As you will see in Section 8.7, the sum is cos 1, the cosine of 1 radian.

50. $\sum_{n=0}^{\infty} (-1)^n \frac{1}{n!}$ As you will see in Section 8.7, the sum is e^{-1}.

Theory and Examples

51. **(a)** *Writing to Learn* The series

$$\frac{1}{3} - \frac{1}{2} + \frac{1}{9} - \frac{1}{4} + \frac{1}{27} - \frac{1}{8} + \cdots + \frac{1}{3^n} - \frac{1}{2^n} + \cdots$$

does not meet one of the conditions of Theorem 8. Which one?

(b) Find the sum of the series in part (a).

52. The limit L of an alternating series that satisfies the conditions of Theorem 8 lies between the values of any two consecutive partial sums. This suggests using the average

$$\frac{s_n + s_{n+1}}{2} = s_n + \frac{1}{2}(-1)^{n+2} a_{n+1}$$

to estimate L. Compute

$$s_{20} + \frac{1}{2} \cdot \frac{1}{21}$$

as an approximation to the sum of the alternating harmonic series. The exact sum is $\ln 2 = 0.6931\ldots$.

53. *The sign of the remainder of an alternating series that satisfies the conditions of Theorem 8* Prove the assertion in Theorem 9 that whenever an alternating series satisfying the conditions of Theorem 8 is approximated with one of its partial sums, then the remainder (sum of the unused terms) has the same sign as the first unused term. (*Hint:* Group the remainder's terms in consecutive pairs.)

54. *Writing to Learn* Show that the sum of the first $2n$ terms of the series

$$1 - \frac{1}{2} + \frac{1}{2} - \frac{1}{3} + \frac{1}{3} - \frac{1}{4} + \frac{1}{4} - \frac{1}{5} + \frac{1}{5} - \frac{1}{6} + \cdots$$

is the same as the sum of the first n terms of the series

$$\frac{1}{1\cdot 2} + \frac{1}{2\cdot 3} + \frac{1}{3\cdot 4} + \frac{1}{4\cdot 5} + \frac{1}{5\cdot 6} + \cdots.$$

Do these series converge? What is the sum of the first $2n+1$ terms of the first series? If the series converge, what is their sum?

55. *Divergence* Show that if $\sum_{n=1}^{\infty} a_n$ diverges, then $\sum_{n=1}^{\infty} |a_n|$ diverges.

56. Show that if $\sum_{n=1}^{\infty} a_n$ converges absolutely, then

$$\left|\sum_{n=1}^{\infty} a_n\right| \le \sum_{n=1}^{\infty} |a_n|.$$

57. *Rules of absolute convergence* Show that if $\sum_{n=1}^{\infty} a_n$ and $\sum_{n=1}^{\infty} b_n$ both converge absolutely, then so does

(a) $\sum_{n=1}^{\infty} (a_n + b_n)$ **(b)** $\sum_{n=1}^{\infty} (a_n - b_n)$

(c) $\sum_{n=1}^{\infty} ka_n$ (k any number)

58. *Term-by-term products* Show by example that $\sum_{n=1}^{\infty} a_n b_n$ may diverge even if $\sum_{n=1}^{\infty} a_n$ and $\sum_{n=1}^{\infty} b_n$ both converge.

T **59.** *Rearrangement* In Example 8, suppose that the goal is to arrange the terms to get a new series that converges to $-1/2$. Start the new arrangement with the first negative term, which is $-1/2$. Whenever you have a sum that is less than or equal to $-1/2$, start introducing positive terms, taken in order, until the new total is greater than $-1/2$. Then add negative terms until the total is less than or equal to $-1/2$ again. Continue this process until your partial sums have been above the target at least three times and finish at or below it. If s_n is the sum of the first n terms of your new series, plot the points (n, s_n) to illustrate how the sums are behaving.

60. *Outline of the proof of the Rearrangement Theorem (Theorem 11)*

(a) Let ϵ be a positive real number, let $L = \sum_{n=1}^{\infty} a_n$, and let $s_k = \sum_{n=1}^{k} a_n$. Show that for some index N_1 and for some index $N_2 \geq N_1$,

$$\sum_{n=N_1}^{\infty} |a_n| < \frac{\epsilon}{2} \quad \text{and} \quad |s_{N_2} - L| < \frac{\epsilon}{2}.$$

Since all the terms $a_1, a_2, \ldots, a_{N_2}$ appear somewhere in the sequence $\{b_n\}$, there is an index $N_3 \geq N_2$ such that if $n \geq N_3$, then $\left(\sum_{n=1}^{k} b_k\right) - s_{N_2}$ is at most a sum of terms a_m with $m \geq N_1$. Therefore, if $n \geq N_3$,

$$\left|\sum_{k=1}^{n} b_k - L\right| \leq \left|\sum_{k=1}^{n} b_k - s_{N_2}\right| + |s_{N_2} - L|$$

$$\leq \sum_{k=N_1}^{\infty} |a_k| + |s_{N_2} - L| < \epsilon.$$

(b) The argument in part (a) shows that if $\sum_{n=1}^{\infty} a_n$ converges absolutely, then $\sum_{n=1}^{\infty} b_n$ converges and $\sum_{n=1}^{\infty} b_n = \sum_{n=1}^{\infty} a_n$. Now show that because $\sum_{n=1}^{\infty} |a_n|$ converges, $\sum_{n=1}^{\infty} |b_n|$ converges to $\sum_{n=1}^{\infty} |a_n|$.

61. *Unzipping absolutely convergent series*

(a) Show that if $\sum_{n=1}^{\infty} |a_n|$ converges and

$$b_n = \begin{cases} a_n & \text{if } a_n \geq 0 \\ 0 & \text{if } a_n < 0, \end{cases}$$

then $\sum_{n=1}^{\infty} b_n$ converges.

(b) Use the results in part (a) to show likewise that if $\sum_{n=1}^{\infty} |a_n|$ converges and

$$c_n = \begin{cases} 0 & \text{if } a_n \geq 0 \\ a_n & \text{if } a_n < 0, \end{cases}$$

then $\sum_{n=1}^{\infty} c_n$ converges.

In other words, if a series converges absolutely, its positive terms form a convergent series, and so do its negative terms. Furthermore,

$$\sum_{n=1}^{\infty} a_n = \sum_{n=1}^{\infty} b_n + \sum_{n=1}^{\infty} c_n$$

because $b_n = (a_n + |a_n|)/2$ and $c_n = (a_n - |a_n|)/2$.

62. *Alternating harmonic series revisited* What is wrong here: Multiply both sides of the alternating harmonic series.

$$S = 1 - \frac{1}{2} + \frac{1}{3} - \frac{1}{4} + \frac{1}{5} - \frac{1}{6} + \frac{1}{7} - \frac{1}{8} + \frac{1}{9} - \frac{1}{10} + \frac{1}{11} - \frac{1}{12} + \cdots$$

by 2 to get

$$2S = 2 - 1 + \frac{2}{3} - \frac{1}{2} + \frac{2}{5} - \frac{1}{3} + \frac{2}{7} - \frac{1}{4} + \frac{2}{9} - \frac{1}{5} + \frac{2}{11} - \frac{1}{6} + \cdots.$$

Collect terms with the same denominator, as the arrows indicate, to arrive at

$$2S = 1 - \frac{1}{2} + \frac{1}{3} - \frac{1}{4} + \frac{1}{5} - \frac{1}{6} + \cdots.$$

The series on the right-hand side of this equation is the series we started with. Therefore, $2S = S$, and dividing by S gives $2 = 1$. (*Source*: "Riemann's Rearrangement Theorem" by Stewart Galanor, *Mathematics Teacher*, Vol. 80, No. 8 (1987), pp. 675–681.)

63. Draw a figure similar to Figure 8.15 to illustrate the convergence of the series in Theorem 8 when $N > 1$.

8.6 Power Series

Power Series and Convergence • The Radius and Interval of Convergence • Term-by-Term Differentiation • Term-by-Term Integration • Multiplication of Power Series

If $|x| < 1$, then the geometric series formula assures us that

$$1 + x + x^2 + x^3 + \cdots + x^n + \cdots = \frac{1}{1 - x}.$$

Consider this statement for a moment. The expression on the right defines a function whose domain is the set of all numbers $x \neq 1$. The expression on the left defines a function whose domain is the interval of convergence, $|x| < 1$. The equality is understood to hold only on this latter domain, where both sides of the equation are defined. On this domain, the series *represents* the function $1/(1 - x)$.

In this section, we study the "infinite polynomials" like $\sum_{n=0}^{\infty} x^n$, and in the next section, we take up the question of representing a particular function with such an infinite polynomial (called a power series).

Power Series and Convergence

The expression $\sum_{n=0}^{\infty} c_n x^n$ is like a polynomial in that it is a sum of coefficients times powers of x, but polynomials have *finite* degrees and do not suffer from divergence for the wrong values of x. Just as an infinite series of numbers is not a mere sum, this series of powers of x is not a mere polynomial.

When we set $x = 0$ in the expression

$$\sum_{n=0}^{\infty} c_n x^n = c_0 + c_1 x + c_2 x^2 + \cdots + c_n x^n + \cdots,$$

we get c_0 on the right but $c_0 \cdot 0^0$ on the left. Since 0^0 is not a number, this is a slight flaw in the notation, which we agree to overlook. The same situation arises when we set

$$x = a \quad \text{in} \quad \sum_{n=0}^{\infty} c_n (x - a)^n.$$

In either case, we agree that the expression will equal c_0. (It really *should* equal c_0, so we are not compromising the mathematics; we are clarifying the notation we use to convey the mathematics.)

Definition Power Series

An expression of the form

$$\sum_{n=0}^{\infty} c_n x^n = c_0 + c_1 x + c_2 x^2 + \cdots + c_n x^n + \cdots$$

is a **power series centered at $x = 0$.** An expression of the form

$$\sum_{n=0}^{\infty} c_n (x - a)^n = c_0 + c_1(x - a) + c_2(x - a)^2 + \cdots + c_n(x - a)^n + \cdots$$

is a **power series centered at $x = a$.** The term $c_n(x - a)^n$ is the ***n*th term;** the number a is the **center.**

Example 1 The Geometric Series

The geometric series

$$\sum_{n=0}^{\infty} x^n = 1 + x + x^2 + \cdots + x^n + \cdots$$

is a power series centered at $x = 0$. It converges to $1/(1 - x)$ on the interval $-1 < x < 1$, also centered at $x = 0$. (Figure 8.16). This is typical behavior, as we soon see. A power series converges for all x, converges on a finite interval with the same center as the series, or converges only at the center itself.

Up to now, we have used the equation

$$\frac{1}{1 - x} = 1 + x + x^2 + \cdots + x^n + \cdots, \qquad -1 < x < 1$$

as a formula for the sum of the series on the right.

We now change the focus: We think of the partial sums of the series on the right as polynomials $P_n(x)$ that approximate the function on the left. For values of x near zero, we need take only a few terms of the series to get a good approximation. As

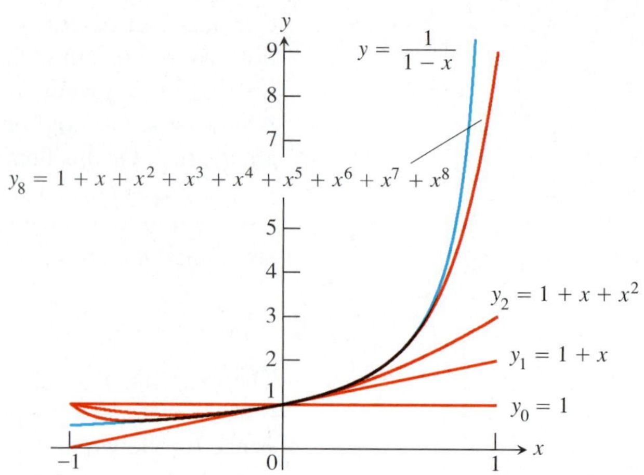

FIGURE 8.16 The graphs of $f(x) = 1/(1-x)$ and four of its polynomial approximations. (Example 1)

we move toward $x = 1$, or -1, we must take more terms. Figure 8.16 shows the graphs of $f(x) = 1/(1-x)$ and the approximating polynomials $y_n = P_n(x)$ for $n = 0, 1, 2$, and 8.

Example 2 Applying the Definition

The power series

$$1 - \frac{1}{2}(x-2) + \frac{1}{4}(x-2)^2 + \cdots + \left(-\frac{1}{2}\right)^n (x-2)^n + \cdots \tag{1}$$

is centered at $a = 2$ with coefficients $c_0 = 1$, $c_1 = -1/2$, $c_2 = 1/4, \ldots, c_n = (-1/2)^n$. This is a geometric series with first term 1 and ratio $r = -\frac{x-2}{2}$. The series converges for $\left|\frac{x-2}{2}\right| < 1$ or $0 < x < 4$. The sum is

$$\frac{1}{1-r} = \frac{1}{1 + \frac{x-2}{2}} = \frac{2}{x},$$

so

$$\frac{2}{x} = 1 - \frac{(x-2)}{2} + \frac{(x-2)^2}{4} - \cdots + \left(-\frac{1}{2}\right)^n (x-2)^n + \cdots, \qquad 0 < x < 4.$$

Series (1) generates useful polynomial approximations of $f(x) = 2/x$ for values of x near 2:

$$P_0(x) = 1$$

$$P_1(x) = 1 - \frac{1}{2}(x-2) = 2 - \frac{x}{2}$$

$$P_2(x) = 1 - \frac{1}{2}(x-2) + \frac{1}{4}(x-2)^2 = 3 - \frac{3x}{2} + \frac{x^2}{4},$$

and so on (Figure 8.17).

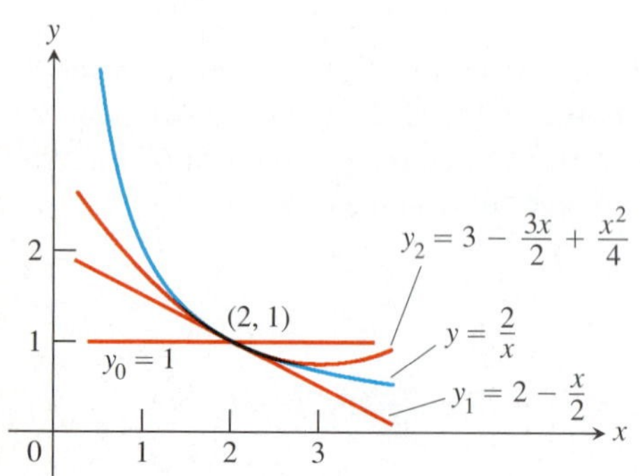

FIGURE 8.17 The graphs of $f(x) = 2/x$ and its first three polynomial approximations. (Example 2)

The Radius and Interval of Convergence

The power series in Examples 1 and 2 happen to be geometric, so we could find the intervals for which they converge. For nongeometric series, we begin by noting that any power series of the form $\sum_{n=0}^{\infty} c_n(x-a)^n$ always converges at $x = a$, thus assuring us of at least one coordinate on the real number line where the series must converge. We have encountered power series like the series in Examples 1 and 2 that converge only on a finite interval centered at a. Some power series converge for all real numbers. A useful fact about power series is that those are the only possibilities, as the following theorem attests.

Theorem 12 The Convergence Theorem for Power Series

There are three possibilities for $\sum_{n=0}^{\infty} c_n(x-a)^n$ with respect to convergence.

1. There is a positive number R such that the series diverges for $|x-a| > R$ but converges for $|x-a| < R$. The series may or may not converge at either of the endpoints $x = a - R$ and $x = a + R$.
2. The series converges for every x $(R = \infty)$.
3. The series converges at $x = a$ and diverges elsewhere $(R = 0)$.

The number R is the **radius of convergence,** and the set of all values of x for which the series converges is the **interval of convergence.** The radius of convergence completely determines the interval of convergence if R is either zero or infinite. For $0 < R < \infty$, however, there remains the question of what happens at the endpoints of the interval. The next example illustrates how to find the interval of convergence.

Example 3 Finding the Interval of Convergence Using the Ratio Test

For what values of x do the following power series converge?

(a) $\displaystyle\sum_{n=1}^{\infty} (-1)^{n-1}\frac{x^n}{n} = x - \frac{x^2}{2} + \frac{x^3}{3} - \cdots$

(b) $\displaystyle\sum_{n=1}^{\infty} (-1)^{n-1}\frac{x^{2n-1}}{2n-1} = x - \frac{x^3}{3} + \frac{x^5}{5} - \cdots$

(c) $\displaystyle\sum_{n=0}^{\infty} \frac{x^n}{n!} = 1 + x + \frac{x^2}{2!} + \frac{x^3}{3!} + \cdots$

(d) $\displaystyle\sum_{n=0}^{\infty} n!\,x^n = 1 + x + 2!\,x^2 + 3!\,x^3 + \cdots$

Solution Apply the Ratio Test to the series $\Sigma\,|u_n|$, where u_n is the nth term of the series in question.

(a) $$\left|\frac{u_{n+1}}{u_n}\right| = \frac{n}{n+1}|x| \to |x|.$$

The series converges absolutely for $|x| < 1$. It diverges if $|x| > 1$ because the nth term does not converge to zero. At $x = 1$, we get the alternating harmonic series $1 - 1/2 + 1/3 - 1/4 + \cdots$, which converges. At $x = -1$, we get $-1 - 1/2 - 1/3 - 1/4 - \cdots$, the negative of the harmonic series; it diverges. Series (a) converges for $-1 < x \le 1$ and diverges elsewhere.

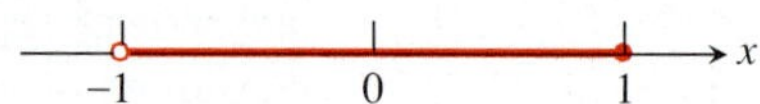

(b) $\left|\dfrac{u_{n+1}}{u_n}\right| = \dfrac{2n-1}{2n+1}x^2 \to x^2.$

The series converges absolutely for $x^2 < 1$. It diverges for $x^2 > 1$ because the nth term does not converge to zero. At $x = 1$, the series becomes $1 - 1/3 + 1/5 - 1/7 + \cdots$, which converges by the Alternating Series Theorem. It also converges at $x = -1$ because it is again an alternating series that satisfies the conditions for convergence. The value at $x = -1$ is the negative of the value at $x = 1$. Series (b) converges for $-1 \le x \le 1$ and diverges elsewhere.

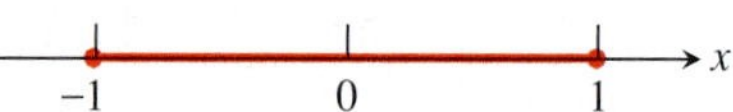

(c) $\left|\dfrac{u_{n+1}}{u_n}\right| = \left|\dfrac{x^{n+1}}{(n+1)!}\cdot\dfrac{n!}{x^n}\right| = \dfrac{|x|}{n+1} \to 0$ for every x.

The series converges absolutely for all x.

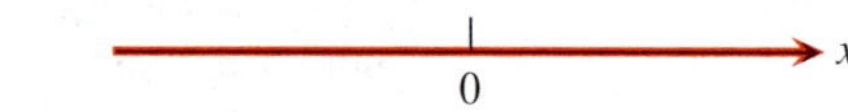

(d) $\left|\dfrac{u_{n+1}}{u_n}\right| = \left|\dfrac{(n+1)!x^{n+1}}{n!x^n}\right| = (n+1)|x| \to \infty$ unless $x = 0$.

The series diverges for all values of x except $x = 0$.

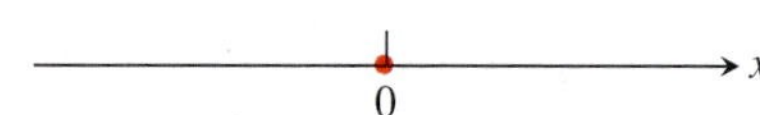

Here's a summary of steps for finding the interval of convergence of a power series.

Finding the Interval of Convergence

Step 1: *Use the Ratio Test (or nth-Root Test) to find the interval where the series converges absolutely.* Ordinarily, this is an open interval

$$|x - a| < R \qquad \text{or} \qquad a - R < x < a + R.$$

Step 2: *If the interval of absolute convergence is finite, test for convergence or divergence at each endpoint,* as in Examples 3(a) and (b). Use a Comparison Test, the Integral Test, or the Alternating Series Test.

Step 3: *If the interval of absolute convergence is $a - R < x < a + R$, the series diverges for $|x - a| > R$* (it does not even converge conditionally), because the nth term does not approach zero for those values of x.

The convergence of a power series is absolute at every point in the interior of the interval. If a power series converges absolutely for all values of x, we say that its **radius of convergence is infinite.** If it converges only at $x = a$, the **radius of convergence is zero.**

Term-by-Term Differentiation

A theorem from advanced calculus says that a power series can be differentiated term by term at each interior point of its interval of convergence.

A WORD OF CAUTION Term-by-term differentiation might not work for other kinds of series. For example, the trigonometric series

$$\sum_{n=1}^{\infty} \frac{\sin(n!x)}{n^2}$$

converges for all x. But if we differentiate term by term, we get the series

$$\sum_{n=1}^{\infty} \frac{n!\cos(n!x)}{n^2}$$

which diverges for all x.

Theorem 13 The Term-by-Term Differentiation Theorem

If $\Sigma\, c_n(x-a)^n$ converges for $a - R < x < a + R$ for some $R > 0$, it defines a function f:

$$f(x) = \sum_{n=0}^{\infty} c_n(x-a)^n, \qquad a - R < x < a + R.$$

Such a function f has derivatives of all orders inside the interval of convergence. We can obtain the derivatives by differentiating the original series term by term:

$$f'(x) = \sum_{n=1}^{\infty} nc_n(x-a)^{n-1}$$

$$f''(x) = \sum_{n=2}^{\infty} n(n-1)c_n(x-a)^{n-2},$$

and so on. Each of these derived series converges at every interior point of the interval of convergence of the original series.

Example 4 Applying Term-by-Term Differentiation

Find series for $f'(x)$ and $f''(x)$ if

$$f(x) = \frac{1}{1-x} = 1 + x + x^2 + x^3 + x^4 + \cdots + x^n + \cdots$$

$$= \sum_{n=0}^{\infty} x^n, \qquad -1 < x < 1.$$

Solution

$$f'(x) = \frac{1}{(1-x)^2} = 1 + 2x + 3x^2 + 4x^3 + \cdots + nx^{n-1} + \cdots$$

$$= \sum_{n=1}^{\infty} nx^{n-1}, \qquad -1 < x < 1$$

$$f''(x) = \frac{2}{(1-x)^3} = 2 + 6x + 12x^2 + \cdots + n(n-1)x^{n-2} + \cdots$$

$$= \sum_{n=2}^{\infty} n(n-1)\,x^{n-2}, \qquad -1 < x < 1$$

Term-by-Term Integration

Another advanced theorem states that a power series can be integrated term by term throughout its interval of convergence.

Theorem 14 The Term-by-Term Integration Theorem

Suppose that

$$f(x) = \sum_{n=0}^{\infty} c_n(x-a)^n$$

converges for $a - R < x < a + R$ $(R > 0)$. Then

$$\sum_{n=0}^{\infty} c_n \frac{(x-a)^{n+1}}{n+1}$$

converges for $a - R < x < a + R$ and

$$\int f(x)\,dx = \sum_{n=0}^{\infty} c_n \frac{(x-a)^{n+1}}{n+1} + C$$

for $a - R < x < a + R$.

Example 5 A Series for $\tan^{-1} x$, $-1 \le x \le 1$

Identify the function

$$f(x) = x - \frac{x^3}{3} + \frac{x^5}{5} - \cdots, \qquad -1 \le x \le 1.$$

Solution We differentiate the original series term by term and get

$$f'(x) = 1 - x^2 + x^4 - x^6 + \cdots, \qquad -1 < x < 1.$$

This is a geometric series with first term 1 and ratio $-x^2$, so

$$f'(x) = \frac{1}{1-(-x^2)} = \frac{1}{1+x^2}.$$

We can now integrate $f'(x) = 1/(1+x^2)$ to get

$$\int f'(x)\,dx = \int \frac{dx}{1+x^2} = \tan^{-1} x + C.$$

The series for $f(x)$ is zero when $x = 0$, so $C = 0$. Hence,

$$f(x) = x - \frac{x^3}{3} + \frac{x^5}{5} - \frac{x^7}{7} + \cdots = \tan^{-1} x, \qquad -1 < x < 1.$$

In Section 8.8, we see that the series also converges to $\tan^{-1} x$ at $x = \pm 1$.

Notice that the original series in Example 5 converges at both endpoints of the original interval of convergence, but Theorem 13 can guarantee the convergence of the differentiated series only inside the interval.

Example 6 A Series for $\ln(1 + x)$, $-1 < x \le 1$

The series

$$\frac{1}{1+t} = 1 - t + t^2 - t^3 + \cdots$$

converges on the open interval $-1 < t < 1$. Therefore,

$$\ln(1+x) = \int_0^x \frac{1}{1+t}\,dt = t - \frac{t^2}{2} + \frac{t^3}{3} - \frac{t^4}{4} + \cdots \bigg]_0^x$$

$$= x - \frac{x^2}{2} + \frac{x^3}{3} - \frac{x^4}{4} + \cdots, \qquad -1 < x < 1.$$

It can also be shown that the series converges at $x = 1$ to the number $\ln 2$, but that was not guaranteed by the theorem.

Multiplication of Power Series

Still another advanced theorem states that absolutely converging power series can be multiplied the way we multiply polynomials to produce new absolutely convergent series.

Theorem 15 The Series Multiplication Theorem for Power Series

If $A(x) = \sum_{n=0}^{\infty} a_n x^n$ and $B(x) = \sum_{n=0}^{\infty} b_n x^n$ converge absolutely for $|x| < R$ and if

$$c_n = a_0 b_n + a_1 b_{n-1} + a_2 b_{n-2} + \cdots + a_{n-1} b_1 + a_n b_0 = \sum_{k=0}^{n} a_k b_{n-k},$$

then $\sum_{n=0}^{\infty} c_n x^n$ converges absolutely to $A(x)\,B(x)$ for $|x| < R$:

$$\left(\sum_{n=0}^{\infty} a_n x^n\right) \cdot \left(\sum_{n=0}^{\infty} b_n x^n\right) = \sum_{n=0}^{\infty} c_n x^n.$$

Example 7 Applying the Multiplication Theorem

Multiply the geometric series

$$\sum_{n=0}^{\infty} x^n = 1 + x + x^2 + \cdots + x^n + \cdots = \frac{1}{1-x} \qquad \text{for } |x| < 1,$$

by itself to get a power series for $1/(1-x)^2$, for $|x| < 1$.

Solution Let

$$A(x) = \sum_{n=0}^{\infty} a_n x^n = 1 + x + x^2 + \cdots + x^n + \cdots = 1/(1-x)$$

$$B(x) = \sum_{n=0}^{\infty} b_n x^n = 1 + x + x^2 + \cdots + x^n + \cdots = 1/(1-x)$$

and

$$c_n = \underbrace{a_0 b_n + a_1 b_{n-1} + \cdots + a_k b_{n-k} + \cdots + a_n b_0}_{n+1 \text{ terms}}$$

$$= \underbrace{1 + 1 + \cdots + 1}_{n+1 \text{ ones}} = n + 1.$$

Then, by the Series Multiplication Theorem,

$$A(x) \cdot B(x) = \sum_{n=0}^{\infty} c_n x^n = \sum_{n=0}^{\infty} (n+1)x^n$$

$$= 1 + 2x + 3x^2 + 4x^3 + \cdots + (n+1)x^n + \cdots$$

is the series for $1/(1-x)^2$. The series all converge absolutely for $|x| < 1$.

Example 4 gives the same answer because

$$\frac{d}{dx}\left(\frac{1}{1-x}\right) = \frac{1}{(1-x)^2}.$$

EXERCISES 8.6

Intervals of Convergence

In Exercises 1–32, (a) find the series' radius and interval of convergence. For what values of x does the series converge (b) absolutely and (c) conditionally?

1. $\sum_{n=0}^{\infty} x^n$

2. $\sum_{n=0}^{\infty} (x+5)^n$

3. $\sum_{n=0}^{\infty} (-1)^n(4x+1)^n$

4. $\sum_{n=1}^{\infty} \frac{(3x-2)^n}{n}$

5. $\sum_{n=0}^{\infty} \frac{(x-2)^n}{10^n}$

6. $\sum_{n=0}^{\infty} (2x)^n$

7. $\sum_{n=0}^{\infty} \frac{nx^n}{n+2}$

8. $\sum_{n=1}^{\infty} \frac{(-1)^n(x+2)^n}{n}$

9. $\sum_{n=1}^{\infty} \frac{x^n}{n\sqrt{n}3^n}$

10. $\sum_{n=1}^{\infty} \frac{(x-1)^n}{\sqrt{n}}$

11. $\sum_{n=0}^{\infty} \frac{(-1)^n x^n}{n!}$

12. $\sum_{n=0}^{\infty} \frac{3^n x^n}{n!}$

13. $\sum_{n=0}^{\infty} \frac{x^{2n+1}}{n!}$

14. $\sum_{n=0}^{\infty} \frac{(2x+3)^{2n+1}}{n!}$

15. $\sum_{n=0}^{\infty} \frac{x^n}{\sqrt{n^2+3}}$

16. $\sum_{n=0}^{\infty} \frac{(-1)^n x^n}{\sqrt{n^2+3}}$

17. $\sum_{n=0}^{\infty} \frac{n(x+3)^n}{5^n}$

18. $\sum_{n=0}^{\infty} \frac{nx^n}{4^n(n^2+1)}$

19. $\sum_{n=0}^{\infty} \frac{\sqrt{n}\,x^n}{3^n}$

20. $\sum_{n=1}^{\infty} \sqrt[n]{n}(2x+5)^n$

21. $\sum_{n=1}^{\infty} \left(1+\frac{1}{n}\right)^n x^n$

22. $\sum_{n=1}^{\infty} (\ln n)\, x^n$

23. $\sum_{n=1}^{\infty} n^n x^n$

24. $\sum_{n=0}^{\infty} n!\,(x-4)^n$

25. $\sum_{n=1}^{\infty} \frac{(-1)^{n+1}(x+2)^n}{n2^n}$

26. $\sum_{n=0}^{\infty} (-2)^n(n+1)(x-1)^n$

27. $\sum_{n=2}^{\infty} \frac{x^n}{n(\ln n)^2}$ (Get the information you need about $\sum 1/(n(\ln n)2)$ from Section 8.4, Exercise 75).

28. $\sum_{n=2}^{\infty} \frac{x^n}{n \ln n}$ (Get the information you need about $\sum 1/(n(\ln n)2)$ from Section 8.4, Exercise 75).

29. $\sum_{n=1}^{\infty} \frac{(4x-5)^{2n+1}}{n^{3/2}}$

30. $\sum_{n=1}^{\infty} \frac{(3x+1)^{n+1}}{2n+2}$

31. $\sum_{n=1}^{\infty} \frac{(x+\pi)^n}{\sqrt{n}}$

32. $\sum_{n=0}^{\infty} \frac{(x-\sqrt{2})^{2n+1}}{2^n}$

Geometric Series in x

In Exercises 33–38, find the series' interval of convergence and, within this interval, the sum of the series as a function of x.

33. $\sum_{n=0}^{\infty} \frac{(x-1)^{2n}}{4^n}$

34. $\sum_{n=0}^{\infty} \frac{(x+1)^{2n}}{9^n}$

35. $\sum_{n=0}^{\infty} \left(\frac{\sqrt{x}}{2}-1\right)^n$

36. $\sum_{n=0}^{\infty} (\ln x)^n$

37. $\sum_{n=0}^{\infty} \left(\frac{x^2+1}{3}\right)^n$

38. $\sum_{n=0}^{\infty} \left(\frac{x^2-1}{2}\right)^n$

Theory and Examples

39. *Term-by-term differentiation* For what values of x does the series

$$1 - \frac{1}{2}(x-3) + \frac{1}{4}(x-3)^2 + \cdots + \left(-\frac{1}{2}\right)^n (x-3)^n + \cdots$$

converge? What is its sum? What series do you get if you differentiate the given series term by term? For what values of x does the new series converge? What is its sum?

40. *Term-by-term integration* If you integrate the series in Exercise 39 term by term, what new series do you get? For what values of x does the new series converge, and what is another name for its sum?

41. *Power series for* $\sin x$ The series

$$\sin x = x - \frac{x^3}{3!} + \frac{x^5}{5!} - \frac{x^7}{7!} + \frac{x^9}{9!} - \frac{x^{11}}{11!} + \cdots$$

converges to $\sin x$ for all x.

(a) Find the first six terms of a series for $\cos x$. For what values of x should the series converge?

(b) By replacing x by $2x$ in the series for $\sin x$, find a series that converges to $\sin 2x$ for all x.

(c) Using the result in part (a) and series multiplication, calculate the first six terms of a series for $2 \sin x \cos x$. Compare your answer with the answer in part (b).

42. *Power series for* e^x The series

$$e^x = 1 + x + \frac{x^2}{2!} + \frac{x^3}{3!} + \frac{x^4}{4!} + \frac{x^5}{5!} + \cdots$$

converges to e^x for all x.

(a) Find a series for $(d/dx)e^x$. Do you get the series for e^x? Explain your answer.

(b) Find a series for $\int e^x\,dx$. Do you get the series for e^x? Explain your answer.

(c) Replace x by $-x$ in the series for e^x to find a series that converges to e^{-x} for all x. Then multiply the series for e^x and e^{-x} to find the first six terms of a series for $e^{-x} \cdot e^x$.

43. *Power series for* $\tan x$ The series

$$\tan x = x + \frac{x^3}{3} + \frac{2x^5}{15} + \frac{17x^7}{315} + \frac{62x^9}{2835} + \cdots$$

converges to $\tan x$ for $-\pi/2 < x < \pi/2$.

(a) Find the first five terms of the series for $\ln|\sec x|$. For what values of x should the series converge?

(b) Find the first five terms of the series for $\sec^2 x$. For what values of x should this series converge?

(c) Check your result in part (b) by squaring the series given for $\sec x$ in Exercise 44.

44. *Power series for* $\sec x$ The series for

$$\sec x = 1 + \frac{x^2}{2} + \frac{5}{24}x^4 + \frac{61}{720}x^6 + \frac{277}{8064}x^8 + \cdots$$

converges to $\sec x$ for $-\pi/2 < x < \pi/2$.

(a) Find the first five terms of a power series for the function $\ln|\sec x + \tan x|$. For what values of x should the series converge?

(b) Find the first four terms of a series for $\sec x \tan x$. For what values of x should the series converge?

(c) Check your result in part (b) by multiplying the series for $\sec x$ by the series given for $\tan x$ in Exercise 43.

45. *Uniqueness of convergent power series*

(a) Show that if two power series $\sum_{n=0}^{\infty} a_n x^n$ and $\sum_{n=0}^{\infty} b_n x^n$ are convergent and equal for all values of x in an open interval $(-c, c)$, then $a_n = b_n$ for every n. (*Hint:* Let $f(x) = \sum_{n=0}^{\infty} a_n x^n = \sum_{n=0}^{\infty} b_n x^n$. Differentiate term by term to show that a_n and b_n both equal $f^{(n)}(0)/(n!)$.)

(b) Show that if $\sum_{n=0}^{\infty} a_n x^n = 0$ for all x in an open interval $(-c, c)$, then $a_n = 0$ for every n.

46. *The sum of the series* $\sum_{n=0}^{\infty} (n^2/2^n)$ To find the sum of this series, express $1/(1 - x)$ as a geometric series, differentiate both sides of the resulting equation with respect to x, multiply both sides of the result by x, differentiate again, multiply by x again, and set x equal to $1/2$. What do you get? (*Source*: David E. Dobbs's letter to the editor, *Illinois Mathematics Teacher*, Vol. 33, Issue 4 (1982), p. 27.)

47. *Convergence at endpoints* Show by examples that the convergence of a power series at an endpoint of its interval of convergence may be either conditional or absolute.

48. *Intervals of convergence* Make up a power series whose interval of convergence is

(a) $(-3, 3)$ **(b)** $(-2, 0)$ **(c)** $(1, 5)$.

8.7 Taylor and Maclaurin Series

Constructing a Series • Taylor and Maclaurin Series • Taylor Polynomials • Remainder of a Taylor Polynomial • Estimating the Remainder • Truncation Error • Table of Maclaurin Series • Combining Taylor Series

A comprehensive understanding of geometric series served us well in the last section, enabling us to find power series to represent certain functions and functions that are equivalent to certain power series (all these equivalencies being subject to

the condition of convergence). In this section, we learn a more general technique for constructing power series, one that makes good use of the tools of calculus. In many cases, these series can provide useful polynomial approximations of the generating functions.

Constructing a Series

We know that within its interval of convergence, the sum of a power series is a continuous function with derivatives of all orders, but what about the other way around? If a function $f(x)$ has derivatives of all orders on an interval I, can it be expressed as a power series on I? If it can, what will its coefficients be?

We can answer the last question readily if we assume that $f(x)$ is the sum of a power series

$$\begin{aligned} f(x) &= \sum_{n=0}^{\infty} a_n(x-a)^n \\ &= a_0 + a_1(x-a) + a_2(x-a)^2 + \cdots + a_n(x-a)^n + \cdots \end{aligned}$$

with a positive radius of convergence. By repeated term-by-term differentiation within the interval of convergence I, we obtain

$$\begin{aligned} f'(x) &= a_1 + 2a_2(x-a) + 3a_3(x-a)^2 + \cdots + na_n(x-a)^{n-1} + \cdots \\ f''(x) &= 1 \cdot 2a_2 + 2 \cdot 3a_3(x-a) + 3 \cdot 4a_4(x-a)^2 + \cdots \\ f'''(x) &= 1 \cdot 2 \cdot 3a_3 + 2 \cdot 3 \cdot 4a_4(x-a) + 3 \cdot 4 \cdot 5a_5(x-a)^2 + \cdots, \end{aligned}$$

with the nth derivative, for all n, being

$$f^{(n)}(x) = n!a_n + \text{a sum of terms with } (x-a) \text{ as a factor.}$$

Since these equations all hold at $x = a$, we have

$$\begin{aligned} f'(a) &= a_1, \\ f''(a) &= 1 \cdot 2a_2, \\ f'''(a) &= 1 \cdot 2 \cdot 3a_3, \end{aligned}$$

and, in general,

$$f^{(n)}(a) = n!a_n.$$

These formulas reveal a marvelous pattern in the coefficients of any power series $\sum_{n=0}^{\infty} a_n(x-a)^n$ that converges to the values of f on I ("represents f on I," we say). If there *is* such a series (still an open question), then there is only one such series and its nth coefficient is

$$a_n = \frac{f^{(n)}(a)}{n!}.$$

If f has a series representation, then the series must be

$$f(x) = f(a) + f'(a)(x-a) + \frac{f''(a)}{2!}(x-a)^2 + \cdots + \frac{f^{(n)}(a)}{n!}(x-a)^n + \cdots. \tag{1}$$

If we start with an arbitrary function f that is infinitely differentiable on an interval I centered at $x = a$ and use it to generate the series in Equation (1), however, will the series then converge to $f(x)$ at each x in the interior of I? The answer is maybe; for some functions it will, but for other functions it will not, as we will see.

Taylor and Maclaurin Series

CD-ROM
WEBsite

Historical Biographies

Brook Taylor
(1685 — 1731)

Colin Maclaurin
(1698 — 1746)

Definitions Taylor Series, Maclaurin Series

Let f be a function with derivatives of all orders throughout some interval containing a as an interior point. Then the **Taylor series generated by f at $x = a$ is**

$$\sum_{k=0}^{\infty} \frac{f^{(k)}(a)}{k!}(x-a)^k = f(a) + f'(a)(x-a) + \frac{f''(a)}{2!}(x-a)^2 + \cdots + \frac{f^{(n)}(a)}{n!}(x-a)^n + \cdots.$$

The **Maclaurin series generated by f** is

$$\sum_{k=0}^{\infty} \frac{f^{(k)}(0)}{k!}x^k = f(0) + f'(0)x + \frac{f''(0)}{2!}x^2 + \cdots + \frac{f^{(n)}(0)}{n!}x^n + \cdots,$$

the Taylor series generated by f at $x = 0$.

Example 1 Finding a Taylor Series

Find the Taylor series generated by $f(x) = 1/x$ at $a = 2$. Where, if anywhere, does the series converge to $1/x$?

Solution We need to find $f(2), f'(2), f''(2), \ldots$. Taking derivatives, we get

$$\begin{aligned}
f(x) &= x^{-1}, & f(2) &= 2^{-1} = \frac{1}{2}, \\
f'(x) &= -x^{-2}, & f'(2) &= -\frac{1}{2^2}, \\
f''(x) &= 2!x^{-3}, & \frac{f''(2)}{2!} &= 2^{-3} = \frac{1}{2^3}, \\
f'''(x) &= -3!\,x^{-4}, & \frac{f'''(2)}{3!} &= -\frac{1}{2^4}, \\
&\vdots & &\vdots \\
f^{(n)}(x) &= (-1)^n n! x^{-(n+1)}, & \frac{f^{(n)}(2)}{n!} &= \frac{(-1)^n}{2^{n+1}}.
\end{aligned}$$

The Taylor series is

$$f(2) + f'(2)(x-2) + \frac{f''(2)}{2!}(x-2)^2 + \cdots + \frac{f^{(n)}}{n!}(x-2)^n + \cdots$$
$$= \frac{1}{2} - \frac{(x-2)}{2^2} + \frac{(x-2)^2}{2^3} - \cdots + (-1)^n\frac{(x-2)^n}{2^{n+1}} + \cdots.$$

This is a geometric series with first term $1/2$ and ratio $r = -(x-2)/2$. It converges absolutely for $|x-2| < 2$, and its sum is

$$\frac{1/2}{1+(x-2)/2} = \frac{1}{2+(x-2)} = \frac{1}{x}.$$

In this example, the Taylor series generated by $f(x) = 1/x$ at $a = 2$ converges to $1/x$ for $|x-2| < 2$ or $0 < x < 4$.

Taylor Polynomials

The linearization of a differentiable function f at a point a is the polynomial

$$P_1(x) = f(a) + f'(a)(x-a).$$

If f has derivatives of higher order at a, then it has higher-order polynomial approximations as well, one for each available derivative. These polynomials are called the Taylor polynomials of f.

We speak of a Taylor polynomial of *order n* rather than *degree n* because $f^{(n)}(a)$ may be zero. The first two Taylor polynomials of $\cos x$ at $x = 0$, for example, are $P_0(x) = 1$ and $P_1(x) = 1$. The first-order polynomial has degree zero, not one.

Definition Taylor Polynomial of Order n

Let f be a function with derivatives of order k for $k = 1, 2, \ldots, N$ in some interval containing a as an interior point. Then for any integer n from 0 through N, the **Taylor polynomial of order *n*** generated by f at $x = a$ is the polynomial

$$P_n(x) = f(a) + f'(a)(x-a) + \frac{f''(a)}{2!}(x-a)^2 + \cdots + \frac{f^{(k)}(a)}{k!}(x-a)^k + \cdots + \frac{f^{(n)}(a)}{n!}(x-a)^n.$$

Just as the linearization of f at $x = a$ provides the best linear approximation of f in the neighborhood of a, the higher-order Taylor polynomials provide the best polynomial approximations of their respective degrees. (See Exercise 58.)

CD-ROM WEBsite

Example 2 Finding Taylor Polynomials for e^x

Find the Taylor series and the Taylor polynomials generated by $f(x) = e^x$ at $x = 0$.

Solution Since

$$f(x) = e^x, \quad f'(x) = e^x, \quad \ldots, \quad f^{(n)}(x) = e^x, \ldots,$$

we have

$$f(0) = e^0 = 1, \quad f'(0) = 1, \quad \ldots, \quad f^{(n)}(0) = 1, \ldots.$$

The Taylor series generated by f at $x = 0$ is

$$f(0) + f'(0)x + \frac{f''(0)}{2!}x^2 + \cdots + \frac{f^{(n)}(0)}{n!}x^n + \cdots = 1 + x + \frac{x^2}{2} + \cdots + \frac{x^n}{n!} + \cdots$$
$$= \sum_{k=0}^{\infty} \frac{x^k}{k!}.$$

By definition, this is also the Maclaurin series for e^x. We soon see that the series converges to e^x at every x.

The Taylor polynomial of order n at $x = 0$ is

$$P_n(x) = 1 + x + \frac{x^2}{2} + \cdots + \frac{x^n}{n!}.$$

See Figure 8.18.

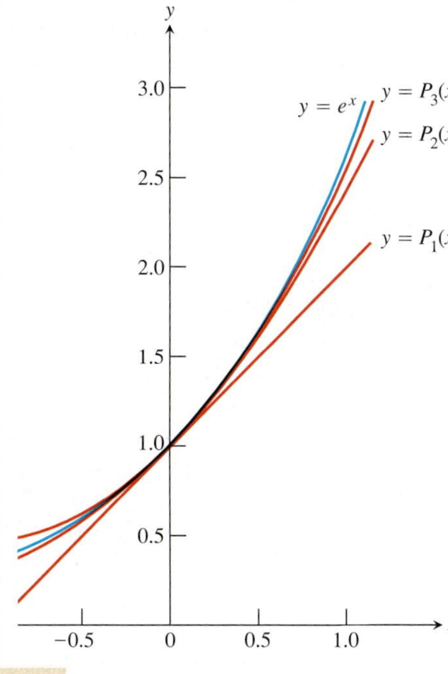

FIGURE 8.18 The graph of $f(x) = e^x$ and its Taylor polynomials

$P_1(x) = 1 + x$

$P_2(x) = 1 + x + (x^2/2!)$

$P_3(x) = 1 + x + (x^2/2!) + (x^3/3!)$.

Notice the very close agreement near the center $x = 0$.

CD-ROM WEBsite

Example 3 Finding Taylor Polynomials For cos x

Find the Taylor series and Taylor polynomials generated by $f(x) = \cos x$ at $x = 0$.

Solution The cosine and its derivatives are

$$\begin{aligned} f(x) &= \cos x, & f'(x) &= -\sin x, \\ f''(x) &= -\cos x, & f^{(3)}(x) &= \sin x, \\ &\vdots & &\vdots \\ f^{(2n)}(x) &= (-1)^n \cos x, & f^{(2n+1)}(x) &= (-1)^{n+1} \sin x. \end{aligned}$$

At $x = 0$, the cosines are 1 and the sines are 0, so

$$f^{(2n)}(0) = (-1)^n, \qquad f^{(2n+1)}(0) = 0.$$

The Taylor series generated by f at 0 is

$$\begin{aligned} &f(0) + f'(0)x + \frac{f''(0)}{2!}x^2 + \frac{f'''(0)}{3!}x^3 + \cdots + \frac{f^{(n)}(0)}{n!}x^n + \cdots \\ &\quad = 1 + 0 \cdot x - \frac{x^2}{2!} + 0 \cdot x^3 + \frac{x^4}{4!} + \cdots + (-1)^n \frac{x^{2n}}{(2n)!} + \cdots = \sum_{n=0}^{\infty} \frac{(-1)^n x^{2n}}{(2n)!}. \end{aligned}$$

By definition, this is also the Maclaurin series for $\cos x$. We see later that the series converges to $\cos x$ at every x.

Because $f^{(2n+1)}(0) = 0$, the Taylor polynomials of orders $2n$ and $2n + 1$ are identical:

$$P_{2n}(x) = P_{2n+1}(x) = 1 - \frac{x^2}{2!} + \frac{x^4}{4!} - \cdots + (-1)^n \frac{x^{2n}}{(2n)!}.$$

Figure 8.19 shows how well these polynomials approximate $f(x) = \cos x$ near $x = 0$. Only the right-hand portions of the graphs are given because the graphs are symmetric about the y-axis.

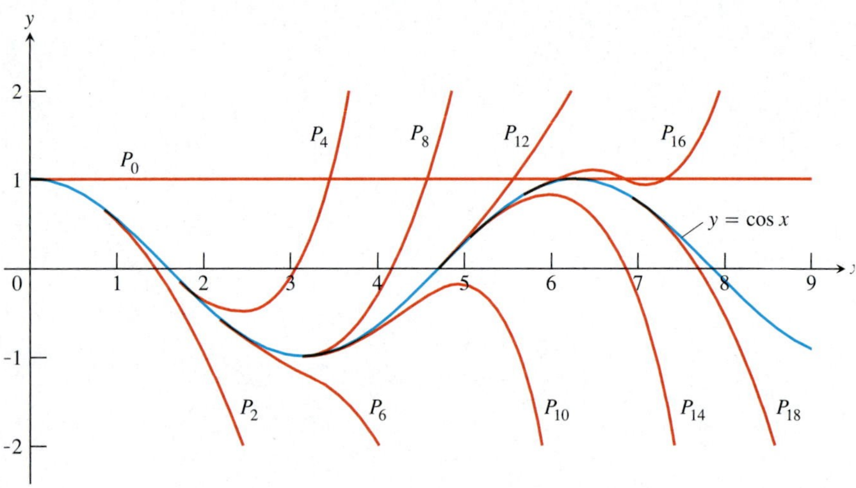

FIGURE 8.19 The polynomials

$$P_{2n}(x) = \sum_{k=0}^{n} \frac{(-1)^k x^{2k}}{(2k)!}$$

converge to $\cos x$ as $n \to \infty$. We can deduce the behavior of $\cos x$ arbitrarily far away solely from knowing the values of the cosine and its derivatives at $x = 0$.

Infinitely differentiable functions that are represented by their Taylor series only at isolated points are, in practice, quite rare.

Example 4 A Function f Whose Taylor Series Converges at Every x But Converges to $f(x)$ Only at $x = 0$

It can be shown (although not easily) that

$$f(x) = \begin{cases} 0, & x = 0 \\ e^{-1/x^2}, & x \neq 0 \end{cases}$$

(Figure 8.20) has derivatives of all orders at $x = 0$ and that $f^{(n)}(0) = 0$ for all n. Hence, the Taylor series generated by f at $x = 0$ is

$$\begin{aligned} f(0) + f'(0)x + \frac{f''(0)}{2!}x^2 + \cdots + \frac{f^{(n)}(0)}{n!}x^n + \cdots \\ = 0 + 0 \cdot x + 0 \cdot x^2 + \cdots + 0 \cdot x^n + \cdots \\ = 0 + 0 + \cdots + 0 + \cdots. \end{aligned}$$

The series converges for every x (its sum is 0) but converges to $f(x)$ only at $x = 0$.

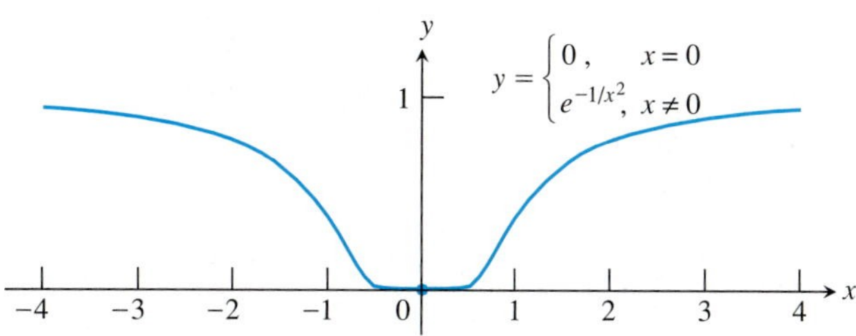

FIGURE 8.20 The graph of the continuous extension of $y = e^{-1/x^2}$ is so flat at the origin that all of its derivatives there are zero. (Example 4)

Two questions still remain.

1. For what values of x can we normally expect a Taylor series to converge to its generating function?
2. How accurately do a function's Taylor polynomials approximate the function on a given interval?

We answer these questions next.

Remainder of a Taylor Polynomial

We need a measure of the accuracy in approximating a function value $f(x)$ by its Taylor polynomial $P_n(x)$. We can use the idea of a **remainder** $R_n(x)$ defined by

$$f(x) = P_n(x) + R_n(x).$$

Exact value — Approximate value — Remainder

The absolute value $|R_n(x)| = |f(x) - P_n(x)|$ is called the **error** associated with the approximation.

The next theorem gives a way to estimate the remainder associated with a Taylor polynomial.

Theorem 16 Taylor's Theorem

If f is differentiable through order $n + 1$ in an open interval I containing a, then for each x in I, there exists a number c between x and a such that

$$f(x) = f(a) + f'(a)(x - a) + \frac{f''(a)}{2!}(x - a)^2 + \cdots + \frac{f^{(n)}(a)}{n!}(x - a)^n + R_n(x),$$

where

$$R_n(x) = \frac{f^{(n+1)}(c)}{(n + 1)!}(x - a)^{n+1}.$$

Taylor's Theorem is a generalization of the Mean Value Theorem (Exercise 49). The proof is lengthy and given in Appendix 8.

If $R_n(x) \to 0$ as $n \to \infty$ for all x in I, we say that the Taylor series generated by f at $x = a$ **converges** to f on I, and we write

$$f(x) = \sum_{k=0}^{\infty} \frac{f^{(k)}(a)}{k!}(x-a)^k.$$

Example 5 The Maclaurin Series for e^x Revisited

Show that the Taylor series generated by $f(x) = e^x$ at $x = 0$ converges to $f(x)$ for every real value of x.

Solution The function has derivatives of all orders throughout the interval $I = (-\infty, \infty)$, and from Example 2,

$$e^x = 1 + x + \frac{x^2}{2!} + \cdots + \frac{x^n}{n!} + R_n(x),$$

where

$$R_n(x) = \frac{e^c}{(n+1)!}x^{n+1} \qquad \text{for some } c \text{ between 0 and } x.$$

Since e^x is an increasing function of x, e^c lies between $e^0 = 1$ and e^x. When x is negative, so is c, and $e^c < 1$. When x is zero, $e^x = 1$ and $R_n(x) = 0$. When x is positive, so is c, and $e^c < e^x$. Thus,

$$|R_n(x)| \le \frac{|x|^{n+1}}{(n+1)!} \qquad \text{when } x \le 0,$$

and

$$|R_n(x)| < e^x \frac{x^{n+1}}{(n+1)!} \qquad \text{when } x > 0.$$

Finally, because

$$\lim_{n\to\infty} \frac{x^{n+1}}{(n+1)!} = 0 \qquad \text{for every } x, \qquad \text{Table 8.1, Formula 6}$$

$\lim_{n\to\infty} R_n(x) = 0$, and the series converges to e^x for every x.

Estimating the Remainder

It is often possible to estimate $R_n(x)$ as we did in Example 5. This method of estimation is so convenient that we state it as a theorem for future reference.

Theorem 17 The Remainder Estimation Theorem

If there are positive constants M and r such that $|f^{(n+1)}(t)| \le Mr^{n+1}$ for all t between a and x, inclusive, then the remainder term $R_n(x)$ in Taylor's Theorem satisfies the inequality

$$|R_n(x)| \le M\frac{r^{n+1}|x-a|^{n+1}}{(n+1)!}.$$

If these conditions hold for every n and all the other conditions of Taylor's Theorem are satisfied by f, then the series converges to $f(x)$.

In the simplest examples, we can take $r = 1$ provided f and all its derivatives are bounded in magnitude by some constant M. In other cases, we may need to consider r. For example, if $f(x) = 2\cos(3x)$, each time we differentiate we get a factor of 3 and r needs to be greater than 1. In this particular case, we can take $r = 3$ along with $M = 2$.

We are now ready to look at some examples of how the Remainder Estimation Theorem and Taylor's Theorem can be used together to settle questions of convergence. As you will see, they can also be used to determine the accuracy with which a function is approximated by one of its Taylor polynomials.

Example 6 The Maclaurin Series for sin *x*

Show that the Maclaurin series for $\sin x$ converges to $\sin x$ for all x.

Solution The function and its derivatives are

$$\begin{aligned} f(x) &= \sin x, & f'(x) &= \cos x, \\ f''(x) &= -\sin x, & f'''(x) &= -\cos x, \\ &\vdots & &\vdots \\ f^{(2k)}(x) &= (-1)^k \sin x, & f^{(2k+1)}(x) &= (-1)^k \cos x, \end{aligned}$$

so

$$f^{(2k)}(0) = 0 \qquad \text{and} \qquad f^{(2k+1)}(0) = (-1)^k.$$

The series has only odd-powered terms and, for $n = 2k + 1$, Taylor's Theorem gives

$$\sin x = x - \frac{x^3}{3!} + \frac{x^5}{5!} - \cdots + \frac{(-1)^k x^{2k+1}}{(2k+1)!} + R_{2k+1}(x).$$

All the derivatives of $\sin x$ have absolute values less than or equal to 1, so we can apply the Remainder Estimation Theorem with $M = 1$ and $r = 1$ to obtain

$$|R_{2k+1}(x)| \le 1 \cdot \frac{|x|^{2k+2}}{(2k+2)!}.$$

Since $(|x|^{2k+2}/(2k+2)!) \to 0$ as $k \to \infty$, whatever the value of x, $R_{2k+1}(x) \to 0$, and the Maclaurin series for $\sin x$ converges to $\sin x$ for every x.

Example 7 The Maclaurin Series for cos *x* Revisited

Show that the Maclaurin series for $\cos x$ converges to $\cos x$ for every value of x.

Solution We add the remainder term to the Taylor polynomial for $\cos x$ in Example 3 to obtain Taylor's formula for $\cos x$ with $n = 2k$:

$$\cos x = 1 - \frac{x^2}{2!} + \frac{x^4}{4!} - \cdots + (-1)^k \frac{x^{2k}}{(2k)!} + R_{2k}(x).$$

Because the derivatives of the cosine have absolute value less than or equal to 1, the Remainder Estimation Theorem with $M = 1$ and $r = 1$ gives

$$|R_{2k}(x)| \le 1 \cdot \frac{|x|^{2k+1}}{(2k+1)!}.$$

For every value of x, $R_{2k} \to 0$ as $k \to \infty$. Therefore, the series converges to $\cos x$ for every value of x.

Truncation Error

The Maclaurin series for e^x converges to e^x for all x, but we still need to decide how many terms to use to approximate e^x to a given degree of accuracy. We get this information from the Remainder Estimation Theorem.

Example 8 Calculating the Number e

Calculate e with an error of less than 10^{-6}.

Solution We can use the result of Example 2 with $x = 1$ to write

$$e = 1 + 1 + \frac{1}{2!} + \cdots + \frac{1}{n!} + R_n(1),$$

with

$$R_n(1) = e^c \frac{1}{(n+1)!} \qquad \text{for some } c \text{ between 0 and 1.}$$

For the purposes of this example, we assume that we know that $e < 3$. Hence, we are certain that

$$\frac{1}{(n+1)!} < R_n(1) < \frac{3}{(n+1)!}$$

because $1 < e^c < 3$ for $0 < c < 1$.

By experiment, we find that $1/9! > 10^{-6}$, whereas $3/10! < 10^{-6}$. Thus, we should take $(n + 1)$ to be at least 10 or n to be at least 9. With an error of less than 10^{-6},

$$e = 1 + 1 + \frac{1}{2} + \frac{1}{3!} + \cdots + \frac{1}{9!} \approx 2.7182\,82.$$

Example 9 Sine Function as a Polynomial of Degree 3

For what values of x can we replace $\sin x$ by $x - (x^3/3!)$ with an error of magnitude no greater than 3×10^{-4}?

Solution Using the result of Example 6, $x - (x^2/3!) = 0 + x + 0x^2 - (x^3/3!) + 0x^4$ is the Taylor polynomial of order 4 as well as of order 3 for $\sin x$. Then,

$$\sin x = x - \frac{x^3}{3!} + 0 + R_4,$$

and the Remainder Estimation Theorem with $M = r = 1$ gives

$$|R_4| \le 1 \cdot \frac{|x|^5}{5!} = \frac{|x|^5}{120}.$$

Therefore, the error will be less than or equal to 3×10^{-4} if

$$\frac{|x|^5}{120} < 3 \times 10^{-4} \qquad \text{or} \qquad |x| < \sqrt[5]{360 \times 10^{-4}} \approx 0.514.$$

Rounded down, to be safe

The Alternating Series Estimation Theorem tells us something that the Remainder Estimation Theorem does not: namely, that the estimate $x - (x^3/3!)$ for $\sin x$ is an underestimate when x is positive because then $x^5/120$ is positive.

Figure 8.21 shows the graph of $\sin x$, along with the graphs of a number of its approximating Taylor polynomials. The graph of $P_3(x) = x - (x^3/3!)$ is almost indistinguishable from the sine curve when $-1 \le x \le 1$.

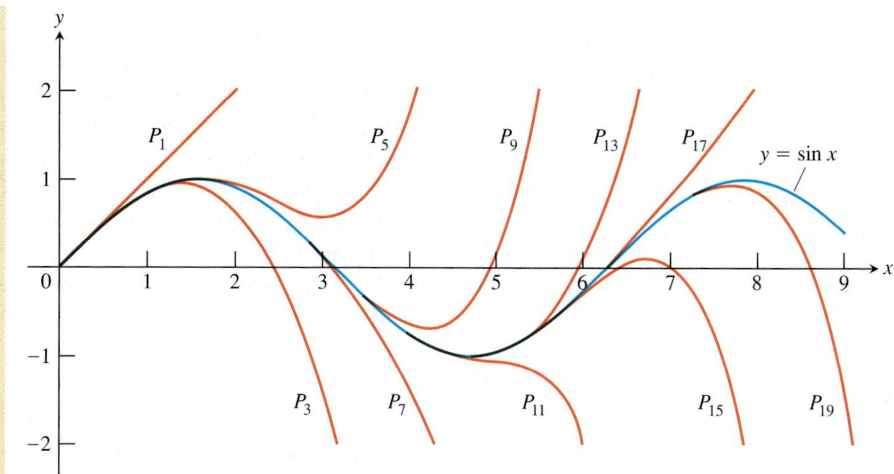

FIGURE 8.21 The polynomials

$$P_{2n+1}(x) = \sum_{k=0}^{n} \frac{(-1)^k x^{2k+1}}{(2k+1)!}$$

converge to $\sin x$ as $n \to \infty$.

CD-ROM
WEBsite

Table of Maclaurin Series

Here we list some of the most useful Maclaurin series, which have all been derived in one way or another in this chapter. The exercises will ask you to use these series as basic building blocks for constructing other series (e.g., $\tan^{-1} x^2$ or $7xe^x$). We also list the intervals of convergence.

Maclaurin series

1. $\frac{1}{1-x} = 1 + x + x^2 + \cdots + x^n + \cdots = \sum_{n=0}^{\infty} x^n \qquad (|x| < 1)$

2. $\frac{1}{1+x} = 1 - x + x^2 - \cdots + (-x)^n + \cdots = \sum_{n=0}^{\infty} (-1)^n x^n \qquad (|x| < 1)$

3. $e^x = 1 + x + \frac{x^2}{2!} + \cdots + \frac{x^n}{n!} + \cdots = \sum_{n=0}^{\infty} \frac{x^n}{n!} \qquad$ (all real x)

4. $\sin x = x - \frac{x^3}{3!} + \frac{x^5}{5!} - \cdots + (-1)^n \frac{x^{2n+1}}{(2n+1)!} + \cdots = \sum_{n=0}^{\infty} (-1)^n \frac{x^{2n+1}}{(2n+1)!} \qquad$ (all real x)

5. $\cos x = 1 - \frac{x^2}{2!} + \frac{x^4}{4!} - \cdots + (-1)^n \frac{x^{2n}}{(2n)!} + \cdots = \sum_{n=0}^{\infty} (-1)^n \frac{x^{2n}}{(2n)!} \qquad$ (all real x)

6. $\ln(1+x) = x - \frac{x^2}{2} + \frac{x^3}{3} - \cdots + (-1)^{n-1} \frac{x^n}{n} + \cdots = \sum_{n=1}^{\infty} (-1)^{n-1} \frac{x^n}{n} \qquad (-1 < x \le 1)$

7. $\tan^{-1} x = x - \frac{x^3}{3} + \frac{x^5}{5} - \cdots + (-1)^n \frac{x^{2n+1}}{2n+1} + \cdots = \sum_{n=0}^{\infty} (-1)^n \frac{x^{2n+1}}{2n+1} \qquad (|x| \le 1)$

Combining Taylor Series

On the intersection of their intervals of convergence, Taylor series can be added, subtracted, and multiplied by constants and powers of x, and the results are once again Taylor series. The Taylor series for $f(x) + g(x)$ is the sum of the Taylor series for $f(x)$ and the Taylor series for $g(x)$ because the nth derivative of $f + g$ is $f^{(n)} + g^{(n)}$, and so on. We can obtain the Maclaurin series for $(1 + \cos 2x)/2$ by substituting $2x$ in the Maclaurin series for $\cos x$, adding 1, and dividing the result by 2. The Maclaurin series for $\sin x + \cos x$ is the term-by-term sum of the series for $\sin x$ and $\cos x$. We obtain the Maclaurin series for $x \sin x$ by multiplying all the terms of the Maclaurin series for $\sin x$ by x.

Example 10 Finding a Maclaurin Series by Substitution

Find the Maclaurin series for $\cos 2x$.

Solution We can find the Maclaurin series for $\cos 2x$ by substituting $2x$ for x in the Maclaurin series for $\cos x$:

$$\begin{aligned}\cos 2x &= \sum_{k=0}^{\infty} \frac{(-1)^k (2x)^{2k}}{(2k)!} = 1 - \frac{(2x)^2}{2!} + \frac{(2x)^4}{4!} - \frac{(2x)^6}{6!} + \cdots \qquad \text{Eq. (5) with } 2x \text{ for } x\\ &= 1 - \frac{2^2x^2}{2!} + \frac{2^4x^4}{4!} - \frac{2^6x^6}{6!} + \cdots\\ &= \sum_{k=0}^{\infty} (-1)^k \frac{2^{2k}x^{2k}}{(2k)!}.\end{aligned}$$

Equation (5) holds for $-\infty < x < \infty$, implying that it holds for $-\infty < 2x < \infty$, so the newly created series converges for all x. Exercise 54 explains why the series is in fact the Maclaurin series for $\cos 2x$.

Example 11 Finding a Maclaurin Series by Multiplication

Find the Maclaurin series for $x \sin x$.

Solution We can find the Maclaurin series for $x \sin x$ by multiplying the Maclaurin series for $\sin x$ (Equation 4) by x:

$$\begin{aligned}x \sin x &= x\left(x - \frac{x^3}{3!} + \frac{x^5}{5!} - \frac{x^7}{7!} + \cdots\right)\\ &= x^2 - \frac{x^4}{3!} + \frac{x^6}{5!} - \frac{x^8}{7!} + \cdots.\end{aligned}$$

The new series converges for all x because the series for $\sin x$ converges for all x. Exercise 54 explains why the series is the Maclaurin series for $x \sin x$.

EXERCISES 8.7

Finding Taylor Polynomials

CD-ROM WEBsite

In Exercises 1–6 find the Taylor polynomials of orders 0, 1, 2, and 3 generated by f at a.

1. $f(x) = \ln x, \quad a = 1$
2. $f(x) = \ln(1 + x), \quad a = 0$
3. $f(x) = \dfrac{1}{(x + 2)}, \quad a = 0$
4. $f(x) = \sin x, \quad a = \pi/4$
5. $f(x) = \cos x, \quad a = \pi/4$
6. $f(x) = \sqrt{x}, \quad a = 4$

Finding Maclaurin Series

Find the Maclaurin series for the functions in Exercises 7–14.

7. e^{-x}
8. $\dfrac{1}{1 + x}$
9. $\sin 3x$
10. $7 \cos(-x)$
11. $\cosh x = \dfrac{e^x + e^{-x}}{2}$
12. $\sinh x = \dfrac{e^x - e^{-x}}{2}$
13. $x^4 - 2x^3 - 5x + 4$
14. $(x + 1)^2$

Finding Taylor Series

In Exercises 15–20, find the Taylor series generated by f at $x = a$.

15. $f(x) = x^3 - 2x + 4, \quad a = 2$
16. $f(x) = 3x^5 - x^4 + 2x^3 + x^2 - 2, \quad a = -1$
17. $f(x) = 1/x^2, \quad a = 1$
18. $f(x) = x/(1 - x), \quad a = 0$
19. $f(x) = e^x, \quad a = 2$
20. $f(x) = 2^x, \quad a = 1$

Maclaurin Series by Substitution

Use substitution as in Example 10 to find the Maclaurin series of the functions in Exercises 21–24.

21. e^{-5x}
22. $e^{-x/2}$
23. $\sin\left(\dfrac{\pi x}{2}\right)$
24. $\cos\sqrt{x}$

More Maclaurin Series

Using the series in the Maclaurin series table as basic building blocks, combine series expressions to find Maclaurin series for the functions in Exercises 25–34.

25. xe^x
26. $x^2 \sin x$
27. $\dfrac{x^2}{2} - 1 + \cos x$
28. $\sin x - x + \dfrac{x^3}{3!}$
29. $x \cos \pi x$
30. $\cos^2 x$ (*Hint*: $\cos^2 x = (1 + \cos 2x)/2$.)
31. $\sin^2 x$
32. $\dfrac{x^2}{1 - 2x}$
33. $x \ln(1 + 2x)$
34. $\dfrac{1}{(1 - x)^2}$

Error Estimates

35. *Writing to Learn* For approximately what values of x can you replace $\sin x$ by $x - (x^3/6)$ with an error of magnitude no greater than 5×10^{-4}? Give reasons for your answer.
36. *Writing to Learn* If $\cos x$ is replaced by $1 - (x^2/2)$ and $|x| < 0.5$, what estimate can be made of the error? Does $1 - (x^2/2)$ tend to be too large or too small? Give reasons for your answer.
37. *Linear approximation for* $\sin x$ How close is the approximation $\sin x = x$ when $|x| < 10^{-3}$? For which of these values of x is $x < \sin x$?
38. *Linear approximation for* $\sqrt{1 + x}$ The estimate $\sqrt{1 + x} = 1 + (x/2)$ is used when x is small. Estimate the error when $|x| < 0.01$.
39. *Quadratic approximation for* e^x
    (a) The approximation $e^x = 1 + x + (x^2/2)$ is used when x is small. Use the Remainder Estimation Theorem to estimate the error when $|x| < 0.1$.
    (b) When $x < 0$, the series for e^x is an alternating series. Use the Alternating Series Estimation Theorem to estimate the error that results from replacing e^x by $1 + x + (x^2/2)$ when $-0.1 < x < 0$. Compare your estimate with the one you obtained in part (a).
40. *Cubic approximation for* $\sinh x$ Estimate the error in the approximation $\sinh x = x + (x^3/3!)$ when $|x| < 0.5$. (*Hint*: Use R_4, not R_3.)
41. *Linear approximation for* e^h When $0 \le h \le 0.01$, show that e^h may be replaced by $1 + h$ with an error of magnitude no greater than 0.6% of h. Use $e^{0.01} = 1.01$.
42. *Approximating* $\ln(1 + x)$ *by* x For what positive values of x can you replace $\ln(1 + x)$ by x with an error of magnitude no greater than 1% of the value of x?

43. *Estimating $\pi/4$* You plan to estimate $\pi/4$ by evaluating the Maclaurin series for $\tan^{-1} x$ at $x = 1$. Use the Alternating Series Estimation Theorem to determine how many terms of the series you would have to add to be sure the estimate is good to 2 decimal places.

44. *Bounding $y = (\sin x)/x$*

(a) Use the Maclaurin series for $\sin x$ and the Alternating Series Estimation Theorem to show that

$$1 - \frac{x^2}{6} < \frac{\sin x}{x} < 1, \qquad x \neq 0.$$

T **(b)** *Writing to Learn* Graph $f(x) = (\sin x)/x$ together with the functions $y = 1 - (x^2/6)$ and $y = 1$ for $-5 \leq x \leq 5$. Comment on the relationships among the graphs.

Quadratic Approximations

CD-ROM WEBsite The Taylor polynomial of order 2 generated by a twice-differentiable function $f(x)$ at $x = a$ is called the **quadratic approximation** of f at $x = a$. In Exercises 45–48, find the

(a) linearization (Taylor polynomial of order 1) at $x = 0$

(b) quadratic approximation of f at $x = 0$.

45. $f(x) = \ln(\cos x)$

46. $f(x) = e^{\sin x}$

47. $f(x) = 1/\sqrt{1 - x^2}$

48. $f(x) = \cosh x$

Theory and Examples

49. *Taylor's Theorem and the Mean Value Theorem* Explain how the Mean Value Theorem (Section 3.2, Theorem 4) is a special case of Taylor's Theorem.

50. *Linearizations at inflection points* (*Continuation of Section 3.6, Exercise 49*) Show that if the graph of a twice-differentiable function $f(x)$ has an inflection point at $x = a$, then the linearization of f at $x = a$ is also the quadratic approximation of f at $x = a$. This explains why tangent lines fit so well at inflection points.

51. *The (Second) Second Derivative Test* Use the equation

$$f(x) = f(a) + f'(a)(x - a) + \frac{f''(c_2)}{2}(x - a)^2$$

to establish the following test.

Let f have continuous first and second derivatives and suppose that $f'(a) = 0$. Then

(a) f has a local maximum at a if $f'' \leq 0$ throughout an interval whose interior contains a

(b) f has a local minimum at a if $f'' \geq 0$ throughout an interval whose interior contains a.

52. *A cubic approximation* Use Taylor's formula with $a = 0$ and $n = 3$ to find the standard cubic approximation of $f(x) = 1/(1 - x)$ at $x = 0$. Give an upper bound for the magnitude of the error in the approximation when $|x| \leq 0.1$.

53. *Improving approximations to π*

(a) Let P be an approximation of π accurate to n decimals. Show that $P + \sin P$ gives an approximation correct to $3n$ decimals. (*Hint*: Let $P = \pi + x$.)

(b) Try it with a calculator.

54. *The Maclaurin series generated by $f(x) = \sum_{n=0}^{\infty} a_n x^n$ is $\sum_{n=0}^{\infty} a_n x^n$* A function defined by a power series $\sum_{n=0}^{\infty} a_n x^n$ with a radius of convergence $c > 0$ has a Maclaurin series that converges to the function at every point of $(-c, c)$. Show this by showing that the Maclaurin series generated by $f(x) = \sum_{n=0}^{\infty} a_n x^n$ is the series $\sum_{n=0}^{\infty} a_n x^n$ itself.

An immediate consequence of this is that series like

$$x \sin x = x^2 - \frac{x^4}{3!} + \frac{x^6}{5!} - \frac{x^8}{7!} + \cdots$$

and

$$x^2 e^x = x^2 + x^3 + \frac{x^4}{2!} + \frac{x^5}{3!} + \cdots,$$

obtained by multiplying Maclaurin series by powers of x, as well as series obtained by integration and differentiation of convergent power series are themselves the Maclaurin series generated by the functions they represent.

55. *Maclaurin series for even functions and odd functions* Suppose that $f(x) = \sum_{n=0}^{\infty} a_n x^n$ converges for all x in an open interval $(-c, c)$.

(a) Show that if f is even, then $a_1 = a_3 = a_5 = \cdots = 0$; that is, the series for f contains only even powers of x.

(b) Show that if f is odd, then $a_0 = a_2 = a_4 = \cdots = 0$; that is, the series for f contains only odd powers of x.

56. *Taylor polynomials of periodic functions*

(a) Show that every continuous periodic function $f(x)$, $-\infty < x < \infty$, is bounded in magnitude by showing that there exists a positive constant M such that $|f(x)| \leq M$ for all x.

(b) Show that the graph of every Taylor polynomial of positive degree generated by $f(x) = \cos x$ must eventually move away from the graph of $\cos x$ as $|x|$ increases. You can see this in Figure 8.19. The Taylor polynomials of $\sin x$ behave in a similar way (Figure 8.21).

T **57. (a)** *Two graphs* Graph the curves $y = (1/3) - (x^2)/5$ and $y = (x - \tan^{-1} x)/x^3$ together with the line $y = 1/3$.

(b) Use a Maclaurin series to explain what you see. What is

$$\lim_{x \to 0} \frac{x - \tan^{-1} x}{x^3}?$$

58. *Of all polynomials of degree $\leq n$, the Taylor polynomial of n gives the best approximation* Suppose that $f(x)$ is differentiable on an interval centered at $x = a$ and that $g(x) = b_0 + b_1(x - a) + \cdots + b_n(x - a)^n$ is a polynomial of degree n with constant coefficients $b_0, \ldots, b_n$. Let $E(x) = f(x) - g(x)$. Show that if we impose on g the conditions

(a) $E(a) = 0$ — The approximation error is zero at $x = a$.

(b) $\lim_{x \to a} \frac{E(x)}{(x-a)^n} = 0,$ — The error is negligible when compared to $(x-a)^n$.

then

$$g(x) = f(a) + f'(a)(x-a) + \frac{f''(a)}{2!}(x-a)^2 + \cdots + \frac{f^{(n)}(a)}{n!}(x-a)^n.$$

Thus, the Taylor polynomial $P_n(x)$ is the only polynomial of degree less than or equal to n whose error is both zero at $x = a$ and negligible when compared with $(x-a)^n$.

COMPUTER EXPLORATIONS

Linear, Quadratic, and Cubic Approximations

Taylor's formula with $n = 1$ and $a = 0$ gives the linearization of a function at $x = 0$. With $n = 2$ and $n = 3$, we obtain the standard quadratic and cubic approximations. In these exercises, we explore the errors associated with these approximations. We seek answers to two questions:

(a) For what values of x can the function be replaced by each approximation with an error less than 10^{-2}?

(b) What is the maximum error we could expect if we replace the function by each approximation over the specified interval?

Using a CAS, perform the following steps to aid in answering questions (a) and (b) for the functions and intervals in Exercises 59–64.

Step 1: Plot the function over the specified interval.

Step 2: Find the Taylor polynomials $P_1(x)$, $P_2(x)$, and $P_3(x)$ at $x = 0$.

Step 3: Calculate the $(n+1)$st derivative $f^{(n+1)}(c)$ associated with the remainder term for each Taylor polynomial. Plot the derivative as a function of c over the specified interval and estimate its maximum absolute value, M.

Step 4: Calculate the remainder $R_n(x)$ for each polynomial. Using the estimate M from step 3 in place of $f^{(n+1)}(c)$, plot $R_n(x)$ over the specified interval. Then estimate the values of x that answer question (a).

Step 5: Compare your estimated error with the actual error $E_n(x) = |f(x) - P_n(x)|$ by plotting $E_n(x)$ over the specified interval. This will help answer question (b).

Step 6: Graph the function and its three Taylor approximations together. Discuss the graphs in relation to the information discovered in steps 4 and 5.

59. $f(x) = \frac{1}{\sqrt{1+x}}, \quad |x| \le \frac{3}{4}$

60. $f(x) = (1+x)^{3/2}, \quad -\frac{1}{2} \le x \le 2$

61. $f(x) = \frac{x}{x^2+1}, \quad |x| \le 2$

62. $f(x) = (\cos x)(\sin 2x), \quad |x| \le 2$

63. $f(x) = e^{-x}\cos 2x, \quad |x| \le 1$

64. $f(x) = e^{x/3}\sin 2x, \quad |x| \le 2$

8.8 Applications of Power Series

Binomial Series for Powers and Roots • Series Solutions of Differential Equations • Evaluating Indeterminate Forms • Arctangents

This section shows how power series are used by scientists and engineers in a variety of applications.

Binomial Series for Powers and Roots

The Maclaurin series generated by $f(x) = (1+x)^m$, when m is constant, is

$$1 + mx + \frac{m(m-1)}{2!}x^2 + \frac{m(m-1)(m-2)}{3!}x^3 + \cdots + \frac{m(m-1)(m-2)\cdots(m-k+1)}{k!}x^k + \cdots.$$

This series, called the **binomial series,** converges absolutely for $|x| < 1$. To derive the series, we first list the function and its derivatives:

$$\begin{aligned}
f(x) &= (1 + x)^m \\
f'(x) &= m(1 + x)^{m-1} \\
f''(x) &= m(m - 1)(1 + x)^{m-2} \\
f'''(x) &= m(m - 1)(m - 2)(1 + x)^{m-3} \\
&\vdots \\
f^{(k)}(x) &= m(m - 1)(m - 2) \cdots (m - k + 1)(1 + x)^{m-k}.
\end{aligned}$$

We then evaluate these at $x = 0$ and substitute into the Maclaurin series formula to obtain the binomial series.

If m is an integer greater than or equal to zero, the series stops after $(m + 1)$ terms because the coefficients from $k = m + 1$ on are zero.

If m is not a positive integer or zero, the series is infinite and converges for $|x| < 1$. To see why, let u_k be the term involving x^k. Then apply the Ratio Test for absolute convergence to see that

$$\left|\frac{u_{k+1}}{u_k}\right| = \left|\frac{m - k}{k + 1}x\right| \to |x| \qquad \text{as } k \to \infty.$$

Our derivation of the binomial series shows only that it is generated by $(1 + x)^m$ and converges for $|x| < 1$. The derivation does not show that the series converges to $(1 + x)^m$. It does, but we assume that part without proof.

Binomial Series

For $-1 < x < 1$,

$$(1 + x)^m = 1 + \sum_{k=1}^{\infty} \binom{m}{k} x^k,$$

where we define

$$\binom{m}{1} = m, \qquad \binom{m}{2} = \frac{m(m - 1)}{2!},$$

and

$$\binom{m}{k} = \frac{m(m - 1)(m - 2) \cdots (m - k + 1)}{k!} \qquad \text{for } k \geq 3.$$

Example 1 Using the Binomial Series

If $m = -1$,

$$\binom{-1}{1} = -1, \qquad \binom{-1}{2} = \frac{-1(-2)}{2!} = 1,$$

and

$$\binom{-1}{k} = \frac{-1(-2)(-3) \cdots (-1 - k + 1)}{k!} = (-1)^k \left(\frac{k!}{k!}\right) = (-1)^k.$$

With these coefficient values, the binomial series formula gives the familiar geometric series

$$(1+x)^{-1} = 1 + \sum_{k=1}^{\infty} (-1)^k x^k = 1 - x + x^2 - x^3 + \cdots + (-1)^k x^k + \cdots .$$

Example 2 Using the Binomial Series

We know from Section 3.6, Example 1, that $\sqrt{1+x} \approx 1 + (x/2)$ for $|x|$ small. With $m = 1/2$, the binomial series gives quadratic and higher-order approximations as well, along with error estimates that come from the Alternating Series Estimation Theorem:

$$\begin{aligned}(1+x)^{1/2} &= 1 + \frac{x}{2} + \frac{\left(\frac{1}{2}\right)\left(-\frac{1}{2}\right)}{2!}x^2 + \frac{\left(\frac{1}{2}\right)\left(-\frac{1}{2}\right)\left(-\frac{3}{2}\right)}{3!}x^3 \\ &\quad + \frac{\left(\frac{1}{2}\right)\left(-\frac{1}{2}\right)\left(-\frac{3}{2}\right)\left(-\frac{5}{2}\right)}{4!}x^4 + \cdots \\ &= 1 + \frac{x}{2} - \frac{x^2}{8} + \frac{x^3}{16} - \frac{5x^4}{128} + \cdots .\end{aligned}$$

Substitution for x gives still other approximations. For example,

$$\sqrt{1-x^2} \approx 1 - \frac{x^2}{2} - \frac{x^4}{8} \qquad \text{for } |x^2| \text{ small}$$

$$\sqrt{1-\frac{1}{x}} \approx 1 - \frac{1}{2x} - \frac{1}{8x^2} \qquad \text{for } \left|\frac{1}{x}\right| \text{ small, that is, } |x| \text{ large.}$$

Series Solutions of Differential Equations

When we cannot find a relatively simple expression for the solution of an initial value problem or differential equation, we try to get information about the solution in other ways. One way is to try to find a power series representation for the solution. If we can do so, we immediately have a source of polynomial approximations of the solution, which may be all that we really need. The first example (Example 3) deals with a first-order linear differential equation that could be solved as a linear equation using the method studied earlier. The example shows how, not knowing this, we can solve the equation with power series. The second example (Example 4) deals with an equation that cannot be solved by previous methods.

Example 3 Series Solution of an Initial Value Problem

Solve the initial value problem

$$y' - y = x, \qquad y(0) = 1.$$

Solution We assume that there is a solution of the form

$$y = a_0 + a_1x + a_2x^2 + \cdots + a_{n-1}x^{n-1} + a_nx^n + \cdots . \tag{1}$$

Our goal is to find values for the coefficients a_k that make the series and its first derivative

$$y' = a_1 + 2a_2x + 3a_3x^2 + \cdots + na_nx^{n-1} + \cdots \tag{2}$$

satisfy the given differential equation and initial condition. The series $y' - y$ is the difference of the series in Equations (1) and (2):

$$\begin{aligned} y' - y = (a_1 + a_0) + (2a_2 - a_1)x + (3a_3 - a_2)x^2 + \cdots \\ + (na_n - a_{n-1})x^{n-1} + \cdots. \end{aligned} \tag{3}$$

CD-ROM
WEBsite
Historical Biography
John Van Neumann
(1903 — 1957)

If y is to satisfy the equation $y' - y = x$, the series in Equation (3) must equal x. Since power series representations are unique, as you saw if you did Exercise 45 in Section 8.6, the coefficients in Equation (3) must satisfy the equations

$$\begin{aligned} a_1 - a_0 &= 0 && \text{Constant terms} \\ 2a_2 - a_1 &= 1 && \text{Coefficients of } x \\ 3a_3 - a_2 &= 0 && \text{Coefficients of } x^2 \\ &\vdots && \vdots \\ na_n - a_{n-1} &= 0 && \text{Coefficients of } x^{n-1} \\ &\vdots && \vdots \end{aligned}$$

We can also see from Equation (1) that $y = a_0$ when $x = 0$, so that $a_0 = 1$ (this being the initial condition). Putting it all together, we have

$$a_0 = 1, \qquad a_1 = a_0 = 1, \qquad a_2 = \frac{1 + a_1}{2} = \frac{1+1}{2} = \frac{2}{2},$$
$$a_3 = \frac{a_2}{3} = \frac{2}{3 \cdot 2} = \frac{2}{3!}, \qquad \ldots, \qquad a_n = \frac{a_{n-1}}{n} = \frac{2}{n!}, \quad \ldots.$$

Substituting these coefficient values into the equation for y (Equation (1)) gives

$$\begin{aligned} y &= 1 + x + 2 \cdot \frac{x^2}{2!} + 2 \cdot \frac{x^3}{3!} + \cdots + 2 \cdot \frac{x^n}{n!} + \cdots \\ &= 1 + x + 2\underbrace{\left(\frac{x^2}{2!} + \frac{x^3}{3!} + \cdots + \frac{x^n}{n!} + \cdots\right)}_{\text{The Maclaurin series for } e^x - 1 - x} \\ &= 1 + x + 2(e^x - 1 - x) = 2e^x - 1 - x. \end{aligned}$$

The solution of the initial value problem is $y = 2e^x - 1 - x$.

As a check, we see that

$$y(0) = 2e^0 - 1 - 0 = 2 - 1 = 1$$

and

$$y' - y = (2e^x - 1) - (2e^x - 1 - x) = x.$$

Example 4 Solving a Differential Equation

Find a power series solution for

$$y'' + x^2 y = 0. \tag{4}$$

Solution We assume that there is a solution of the form

$$y = a_0 + a_1 x + a_2 x^2 + \cdots + a_n x^n + \cdots \tag{5}$$

and find what the coefficients a_k have to be to make the series and its second derivative

$$y'' = 2a_2 + 3 \cdot 2a_3 x + \cdots + n(n-1)a_n x^{n-2} + \cdots \tag{6}$$

satisfy Equation (4). The series for $x^2 y$ is x^2 times the right-hand side of Equation (5):

$$x^2 y = a_0 x^2 + a_1 x^3 + a_2 x^4 + \cdots + a_n x^{n+2} + \cdots. \tag{7}$$

The series for $y'' + x^2 y$ is the sum of the series in Equations (6) and (7):

$$\begin{aligned} y'' + x^2 y = {} & 2a_2 + 6a_3 x + (12a_4 + a_0)x^2 + (20a_5 + a_1)x^3 \\ & + \cdots + (n(n-1)a_n + a_{n-4})x^{n-2} + \cdots. \end{aligned} \tag{8}$$

Notice that the coefficient of x^{n-2} in Equation (7) is a_{n-4}. If y and its second derivative y'' are to satisfy Equation (4), the coefficients of the individual powers of x on the right-hand side of Equation (8) must all be zero:

$$2a_2 = 0, \quad 6a_3 = 0, \quad 12a_4 + a_0 = 0, \quad 20a_5 + a_1 = 0, \tag{9}$$

and for all $n \geq 4$,

$$n(n-1)a_n + a_{n-4} = 0. \tag{10}$$

We can see from Equation (5) that

$$a_0 = y(0), \qquad a_1 = y'(0).$$

In other words, the first two coefficients of the series are the values of y and y' at $x = 0$. The equations in Equation (9) and the recursion formula in Equation (10) enable us to evaluate all the other coefficients in terms of a_0 and a_1.

The first two of Equations (9) give

$$a_2 = 0, \qquad a_3 = 0.$$

Equation (10) shows that if $a_{n-4} = 0$, then $a_n = 0$, so we conclude that

$$a_6 = 0, \quad a_7 = 0, \quad a_{10} = 0, \quad a_{11} = 0,$$

and whenever $n = 4k + 2$ or $4k + 3$, a_n is zero. For the other coefficients, we have

$$a_n = \frac{-a_{n-4}}{n(n-1)}$$

so that

$$a_4 = \frac{-a_0}{4 \cdot 3}, \qquad a_8 = \frac{-a_4}{8 \cdot 7} = \frac{a_0}{3 \cdot 4 \cdot 7 \cdot 8},$$

$$a_{12} = \frac{-a_8}{11 \cdot 12} = \frac{-a_0}{3 \cdot 4 \cdot 7 \cdot 8 \cdot 11 \cdot 12}$$

and

$$a_5 = \frac{-a_1}{5 \cdot 4}, \qquad a_9 = \frac{-a_5}{9 \cdot 8} = \frac{a_1}{4 \cdot 5 \cdot 8 \cdot 9},$$

$$a_{13} = \frac{-a_9}{12 \cdot 13} = \frac{-a_1}{4 \cdot 5 \cdot 8 \cdot 9 \cdot 12 \cdot 13}.$$

The answer is best expressed as the sum of two separate series, one multiplied by a_0, the other by a_1:

$$y = a_0\left(1 - \frac{x^4}{3 \cdot 4} + \frac{x^8}{3 \cdot 4 \cdot 7 \cdot 8} - \frac{x^{12}}{3 \cdot 4 \cdot 7 \cdot 8 \cdot 11 \cdot 12} + \cdots\right)$$

$$+ a_1\left(x - \frac{x^5}{4 \cdot 5} + \frac{x^9}{4 \cdot 5 \cdot 8 \cdot 9} - \frac{x^{13}}{4 \cdot 5 \cdot 8 \cdot 9 \cdot 12 \cdot 13} + \cdots\right).$$

Both series converge absolutely for all x, as is readily seen by the Ratio Test.

Evaluating Indeterminate Forms

We can sometimes evaluate indeterminate forms by expressing the functions involved as Taylor series.

Example 5 Limits Using Power Series

Evaluate

$$\lim_{x \to 0} \frac{\sin x - \tan x}{x^3}.$$

Solution The Maclaurin series for $\sin x$ and $\tan x$, to terms in x^5, are

$$\sin x = x - \frac{x^3}{3!} + \frac{x^5}{5!} - \cdots, \qquad \tan x = x + \frac{x^3}{3} + \frac{2x^5}{15} + \cdots.$$

Hence,

$$\sin x - \tan x = -\frac{x^3}{2} - \frac{x^5}{8} - \cdots = x^3\left(-\frac{1}{2} - \frac{x^2}{8} - \cdots\right)$$

and

$$\lim_{x \to 0} \frac{\sin x - \tan x}{x^3} = \lim_{x \to 0}\left(-\frac{1}{2} - \frac{x^2}{8} - \cdots\right)$$

$$= -\frac{1}{2}.$$

If we apply series to calculate $\lim_{n \to 0}\left((1/\sin x) - (1/x)\right)$, we not only find the limit successfully but also discover an approximation formula for $\csc x$.

Example 6 Limits Using Power Series

Find

$$\lim_{x \to 0}\left(\frac{1}{\sin x} - \frac{1}{x}\right).$$

Solution

$$\frac{1}{\sin x} - \frac{1}{x} = \frac{x - \sin x}{x \sin x} = \frac{x - \left(x - \dfrac{x^3}{3!} + \dfrac{x^5}{5!} - \cdots\right)}{x \cdot \left(x - \dfrac{x^3}{3!} + \dfrac{x^5}{5!} - \cdots\right)}$$

$$= \frac{x^3\left(\dfrac{1}{3!} - \dfrac{x^2}{5!} + \cdots\right)}{x^2\left(1 - \dfrac{x^2}{3!} + \cdots\right)} = x\,\frac{\dfrac{1}{3!} - \dfrac{x^2}{5!} + \cdots}{1 - \dfrac{x^2}{3!} + \cdots}.$$

Therefore,

$$\lim_{x\to 0}\left(\frac{1}{\sin x} - \frac{1}{x}\right) = \lim_{x\to 0}\left(x\,\frac{\dfrac{1}{3!} - \dfrac{x^2}{5!} + \cdots}{1 - \dfrac{x^2}{3!} + \cdots}\right) = 0.$$

From the quotient on the right, we can see that if $|x|$ is small, then

$$\frac{1}{\sin x} - \frac{1}{x} \approx x \cdot \frac{1}{3!} = \frac{x}{6} \qquad \text{or} \qquad \csc x \approx \frac{1}{x} + \frac{x}{6}.$$

Arctangents

In Section 8.6, Example 5, we found a series for $\tan^{-1} x$ by differentiating to get

$$\frac{d}{dx}\tan^{-1} x = \frac{1}{1 + x^2} = 1 - x^2 + x^4 - x^6 + \cdots$$

and integrating to get

$$\tan^{-1} x = x - \frac{x^3}{3} + \frac{x^5}{5} - \frac{x^7}{7} + \cdots.$$

We did not, however, prove the term-by-term integration theorem on which this conclusion depended. We now derive the series again by integrating both sides of the finite formula

$$\frac{1}{1 + t^2} = 1 - t^2 + t^4 - t^6 + \cdots + (-1)^n t^{2n} + \frac{(-1)^{n+1} t^{2n+2}}{1 + t^2},$$

in which the last term comes from adding the remaining terms as a geometric series with first term $a = (-1)^{n+1} t^{2n+2}$ and ratio $r = -t^2$. Integrating both sides of the last equation from $t = 0$ to $t = x$ gives

$$\tan^{-1} x = x - \frac{x^3}{3} + \frac{x^5}{5} - \frac{x^7}{7} + \cdots + (-1)^n \frac{x^{2n+1}}{2n + 1} + R(n, x),$$

where

$$R(n, x) = \int_0^x \frac{(-1)^{n+1} t^{2n+2}}{1 + t^2}\, dt.$$

The denominator of the integrand is greater than or equal to 1; hence,

$$|R(n, x)| \le \int_0^{|x|} t^{2n+2}\, dt = \frac{|x|^{2n+3}}{2n+3}.$$

We take this route instead of finding the Maclaurin series directly because the formulas for the higher-order derivatives of $\tan^{-1} x$ are unmanageable.

If $|x| \le 1$, the right side of this inequality approaches zero as $n \to \infty$. Therefore, $\lim_{n\to\infty} R(n, x) = 0$ if $|x| \le 1$ and

$$\tan^{-1} x = \sum_{n=0}^{\infty} \frac{(-1)^n x^{2n+1}}{2n+1}, \qquad |x| \le 1.$$

When we put $x = 1$ in the series for $\tan^{-1} x$ we get **Leibniz's formula:**

$$\frac{\pi}{4} = 1 - \frac{1}{3} + \frac{1}{5} - \frac{1}{7} + \frac{1}{9} - \cdots + \frac{(-1)^n}{2n+1} + \cdots.$$

This series converges too slowly to be a useful source of decimal approximations of π. It is better to use a formula like

$$\pi = 48 \tan^{-1} \frac{1}{18} + 32 \tan^{-1} \frac{1}{57} - 20 \tan^{-1} \frac{1}{239},$$

which uses values of x closer to zero.

EXERCISES 8.8

Binomial Series

Find the first four terms of the binomial series for the functions in Exercises 1–10.

1. $(1 + x)^{1/2}$
2. $(1 + x)^{1/3}$
3. $(1 - x)^{-1/2}$
4. $(1 - 2x)^{1/2}$
5. $\left(1 + \frac{x}{2}\right)^{-2}$
6. $\left(1 - \frac{x}{2}\right)^{-2}$
7. $(1 + x^3)^{-1/2}$
8. $(1 + x^2)^{-1/3}$
9. $\left(1 + \frac{1}{x}\right)^{1/2}$
10. $\left(1 - \frac{2}{x}\right)^{1/3}$

Find the binomial series for the functions in Exercises 11–14.

11. $(1 + x)^4$
12. $(1 + x^2)^3$
13. $(1 - 2x)^3$
14. $\left(1 - \frac{x}{2}\right)^4$

Initial Value Problems

Find series solutions for the initial value problems in Exercises 15–32.

15. $y' + y = 0, \quad y(0) = 1$
16. $y' - 2y = 0, \quad y(0) = 1$
17. $y' - y = 1, \quad y(0) = 0$
18. $y' + y = 1, \quad y(0) = 2$
19. $y' - y = x, \quad y(0) = 0$
20. $y' + y = 2x, \quad y(0) = -1$
21. $y' - xy = 0, \quad y(0) = 1$
22. $y' - x^2 y = 0, \quad y(0) = 1$
23. $(1 - x)y' - y = 0, \quad y(0) = 2$
24. $(1 + x^2)y' + 2xy = 0, \quad y(0) = 3$
25. $y'' - y = 0, \quad y'(0) = 1$ and $y(0) = 0$
26. $y'' + y = 0, \quad y'(0) = 0$ and $y(0) = 1$
27. $y'' + y = x, \quad y'(0) = 1$ and $y(0) = 2$
28. $y'' - y = x, \quad y'(0) = 2$ and $y(0) = -1$
29. $y'' - y = -x, \quad y'(2) = -2$ and $y(2) = 0$
30. $y'' - x^2 y = 0, \quad y'(0) = b$ and $y(0) = a$
31. $y'' + x^2 y = x, \quad y'(0) = b$ and $y(0) = a$
32. $y'' - 2y' + y = 0, \quad y'(0) = 1$ and $y(0) = 0$

Approximating Integral Functions by Polynomials

In Exercises 33–36, find a polynomial that will approximate $F(x)$ throughout the given interval with an error of magnitude less than 10^{-3}.

33. $F(x) = \int_0^x \sin t^2\, dt, \quad [0, 1]$
34. $F(x) = \int_0^x t^2 e^{-t^2}\, dt, \quad [0, 1]$
35. $F(x) = \int_0^x \tan^{-1} t\, dt, \quad$ (a) $[0, 0.5]$ (b) $[0, 1]$
36. $F(x) = \int_0^x \frac{\ln (1+t)}{t}\, dt, \quad$ (a) $[0, 0.5]$ (b) $[0, 1]$

Indeterminate Forms

Use series to evaluate the limits in Exercises 37–42.

37. $\lim_{x \to 0} \dfrac{e^x - (1+x)}{x^2}$

38. $\lim_{t \to 0} \dfrac{1 - \cos t - (t^2/2)}{t^4}$

39. $\lim_{x \to \infty} x^2(e^{-1/x^2} - 1)$

40. $\lim_{y \to 0} \dfrac{\tan^{-1} y - \sin y}{y^3 \cos y}$

41. $\lim_{x \to 0} \dfrac{\ln(1 + x^2)}{1 - \cos x}$

42. $\lim_{x \to \infty} (x+1) \sin \dfrac{1}{x+1}$

Theory and Examples

43. *Series for* $\ln(1-x), |x| < 1$ Replace x by $-x$ in the Maclaurin series for $\ln(1+x)$ to obtain a series for $\ln(1-x)$. Then subtract this from the Maclaurin series for $\ln(1+x)$ to show that for $|x| < 1$,

$$\ln \frac{1+x}{1-x} = 2\left(x + \frac{x^3}{3} + \frac{x^5}{5} + \cdots\right).$$

44. *Writing to Learn* How many terms of the Maclaurin series for $\ln(1+x)$ should you add to be sure of calculating $\ln(1.1)$ with an error of magnitude less than 10^{-8}? Give reasons for your answer.

45. *Writing to Learn* According to the Alternating Series Estimation Theorem, how many terms of the Maclaurin series for $\tan^{-1} 1$ would you have to add to be sure of finding $\pi/4$ with an error of magnitude less than 10^{-3}? Give reasons for your answer.

46. *Maclaurin series for* $\tan^{-1} x$ Show that the Maclaurin series for $f(x) = \tan^{-1} x$ diverges for $|x| > 1$.

47. *Taylor polynomial for* $\sin^{-1} x$

(a) Use the binomial series and that

$$\frac{d}{dx} \sin^{-1} x = (1 - x^2)^{-1/2}$$

to generate the first four nonzero terms of the Maclaurin series for $\sin^{-1} x$. What is the radius of convergence?

(b) *Taylor polynomial for* $\cos^{-1} x$ Use your result in part (a) to find the first five nonzero terms of the Maclaurin series for $\cos^{-1} x$.

48. *Maclaurin series for* $\sin^{-1} x$ Integrate the binomial series for the function $(1 - x^2)^{-1/2}$ to show that for $|x| < 1$,

$$\sin^{-1} x = x + \sum_{n=1}^{\infty} \frac{1 \cdot 3 \cdot 5 \cdot \cdots \cdot (2n-1)}{2 \cdot 4 \cdot 6 \cdot \cdots \cdot (2n)} \frac{x^{2n+1}}{2n+1}.$$

49. *Series for* $\tan^{-1} x$ *for* $|x| > 1$ Derive the series

$$\tan^{-1} x = \frac{\pi}{2} - \frac{1}{x} + \frac{1}{3x^3} - \frac{1}{5x^5} + \cdots, \quad x > 1$$

$$\tan^{-1} x = -\frac{\pi}{2} - \frac{1}{x} + \frac{1}{3x^3} - \frac{1}{5x^5} + \cdots, \quad x < -1$$

by integrating the series

$$\frac{1}{1+t^2} = \frac{1}{t^2} \cdot \frac{1}{1 + (1/t^2)} = \frac{1}{t^2} - \frac{1}{t^4} + \frac{1}{t^6} - \frac{1}{t^8} + \cdots$$

in the first case from x to ∞ and in the second case from $-\infty$ to x.

50. *The value of* $\sum_{n=0}^{\infty} \tan^{-1}(2/n^2)$

(a) Use the formula for the tangent of the difference of two angles to show that

$$\tan(\tan^{-1}(n+1) - \tan^{-1}(n-1)) = \frac{2}{n^2}$$

and hence that

$$\tan^{-1} \frac{2}{n^2} = \tan^{-1}(n+1) - \tan^{-1}(n-1).$$

(b) Show that

$$\sum_{n=1}^{N} \tan^{-1} \frac{2}{n^2} = \tan^{-1}(N+1) + \tan^{-1} N - \frac{\pi}{4}.$$

(c) Find the value of $\sum_{n=1}^{\infty} \tan^{-1}(2/n^2)$.

8.9 Fourier Series

Coefficients in the Fourier Series Expansion • Convergence of the Fourier Series • Periodic Extension

CD-ROM
WEBsite
Historical Biography
Jean-Baptiste Joseph Fourier (1766 — 1830)

When investigating the problem of heat conduction in a long thin insulated rod, French mathematician Jean-Baptiste Joseph Fourier needed to express a function $f(x)$ as a trigonometric series. Generally, if $f(x)$ is defined on the interval $-L < x < L$, we need to know the coefficients a_0, a_n, and b_n $(n \geq 1)$ for which

$$f(x) = \frac{a_0}{2} + \sum_{n=1}^{\infty} \left(a_n \cos \frac{n\pi x}{L} + b_n \sin \frac{n\pi x}{L}\right). \tag{1}$$

Notice that the interval $-L < x < L$ is *symmetric* about the origin. Equation (1) is called a **Fourier series** for f on the interval $(-L, L)$. These series have a wide range of science and engineering applications in the study of heat conduction, wave phenomena, concentrations of chemicals and pollutants, and other models of the physical world. In this section, we introduce these important trigonometric series representations of a given function f.

Coefficients in the Fourier Series Expansion

Suppose that f is a function defined over the *symmetric* interval $-L < x < L$. Assume that f is expressible as the trigonometric series given by Equation (1). We want to find a way to calculate the coefficients $a_0, a_1, a_2, \ldots, b_1, b_2, \ldots$. The key to the calculations is the definite integral, based on the results in Table 8.3.

Table 8.3 Trigonometric Integrals

If m and n are positive integers, then

1. $\displaystyle\int_{-L}^{L} \cos\frac{n\pi x}{L}\,dx = 0$

2. $\displaystyle\int_{-L}^{L} \sin\frac{n\pi x}{L}\,dx = 0$

3. $\displaystyle\int_{-L}^{L} \cos\frac{n\pi x}{L}\cos\frac{m\pi x}{L}\,dx = \begin{cases} 0, & m \neq n, \\ L, & m = n \end{cases}$

4. $\displaystyle\int_{-L}^{L} \sin\frac{n\pi x}{L}\cos\frac{m\pi x}{L}\,dx = 0$

5. $\displaystyle\int_{-L}^{L} \sin\frac{n\pi x}{L}\sin\frac{m\pi x}{L}\,dx = \begin{cases} 0, & m \neq n, \\ L, & m = n. \end{cases}$

(We ask you to evaluate these trigonometric integrals in Exercises 17 through 21.)

The term $a_0/2$ in Equation (1) keeps consistency with the formulas calculating the Fourier coefficients.

Calculation of a_0 We integrate both sides of Equation (1) from $-L$ to L and assume that the operations for integration and summation can be interchanged to obtain

$$\int_{-L}^{L} f(x)\,dx = \frac{a_0}{2}\int_{-L}^{L} dx + \sum_{n=1}^{\infty} a_n \int_{-L}^{L} \cos\frac{n\pi x}{L}\,dx + \sum_{n=1}^{\infty} b_n \int_{-L}^{L} \sin\frac{n\pi x}{L}\,dx. \tag{2}$$

For every positive integer n, the last two integrals on the right-hand side of Equation (2) are zero (Formulas 1 and 2 in Table 8.3).Therefore,

$$\int_{-L}^{L} f(x)\,dx = \frac{a_0}{2}\int_{-L}^{L} dx = \frac{a_0 x}{2}\bigg]_{-L}^{L} = La_0.$$

Solving for a_0 yields

$$a_0 = \frac{1}{L}\int_{-L}^{L} f(x)\,dx. \tag{3}$$

Calculation of a_m We multiply both sides of Equation (1) by cos $(m\pi x/L)$, $m > 0$, and integrate the result from $-L$ to L:

$$\begin{aligned}\int_{-L}^{L} f(x) \cos \frac{m\pi x}{L}\, dx &= \frac{a_0}{2} \int_{-L}^{L} \cos \frac{m\pi x}{L}\, dx \\ &+ \sum_{n=1}^{\infty} a_n \int_{-L}^{L} \cos \frac{n\pi x}{L} \cos \frac{m\pi x}{L}\, dx \\ &+ \sum_{n=1}^{\infty} b_n \int_{-L}^{L} \sin \frac{n\pi x}{L} \cos \frac{m\pi x}{L}\, dx.\end{aligned} \tag{4}$$

The first integral on the right-hand side of Equation (4) is zero (Formula 1 in Table 8.3). Formulas 3 and 4 in Table 8.3, further reduce the equation to

$$\int_{-L}^{L} f(x) \cos \frac{m\pi x}{L}\, dx = a_m \int_{-L}^{L} \cos \frac{m\pi x}{L} \cos \frac{m\pi x}{L}\, dx = La_m.$$

Therefore,

$$a_m = \frac{1}{L} \int_{-L}^{L} f(x) \cos \frac{m\pi x}{L}\, dx. \tag{5}$$

Calculation of b_m We multiply both sides of Equation (1) by sin $(m\pi x/L)$, $m > 0$, and integrate the result from $-L$ to L:

$$\begin{aligned}\int_{-L}^{L} f(x) \sin \frac{m\pi x}{L}\, dx &= \frac{a_0}{2} \int_{-L}^{L} \sin \frac{m\pi x}{L}\, dx \\ &+ \sum_{n=1}^{\infty} a_n \int_{-L}^{L} \cos \frac{n\pi x}{L} \sin \frac{m\pi x}{L}\, dx \\ &+ \sum_{n=1}^{\infty} b_n \int_{-L}^{L} \sin \frac{n\pi x}{L} \sin \frac{m\pi x}{L}\, dx.\end{aligned}$$

From Formulas 2, 4, and 5 in Table 8.3, we obtain

$$\int_{-L}^{L} f(x) \sin \frac{m\pi x}{L}\, dx = b_m \int_{-L}^{L} \sin \frac{m\pi x}{L} \sin \frac{m\pi x}{L}\, dx = Lb_m.$$

Therefore,

$$b_m = \frac{1}{L} \int_{-L}^{L} f(x) \sin \frac{m\pi x}{L}\, dx. \tag{6}$$

The trigonometric series (1), whose coefficients a_0, a_n, b_n are determined by Equations (3), (5), and (6), respectively (with m replaced by n), is called the **Fourier series expansion** of the function f over the interval $-L < x < L$. The constants a_0, a_n, and b_n are the **Fourier coefficients** of f.

Example 1 Finding a Fourier Series Expansion

Find the Fourier series expansion of the function

$$f(x) = \begin{cases} 1, & -\pi < x < 0, \\ x, & 0 < x < \pi, \end{cases}$$

(Figure 8.22).

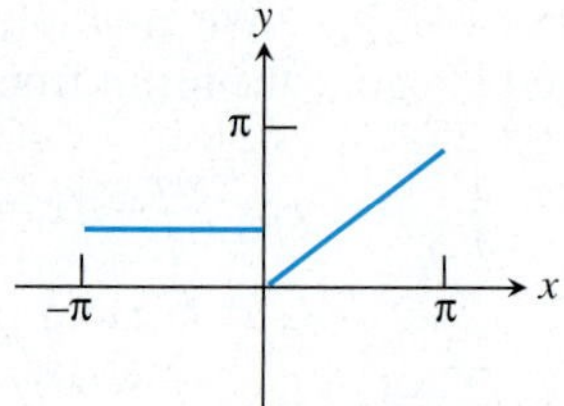

FIGURE 8.22 The piecewise continuous function in Example 1.

Solution Notice from Figure 8.22 that $L = \pi$. Thus, from Equation (3) we have

$$\begin{aligned} a_0 &= \frac{1}{\pi}\int_{-\pi}^{\pi} f(x)\,dx \\ &= \frac{1}{\pi}\int_{-\pi}^{0} dx + \frac{1}{\pi}\int_{0}^{\pi} x\,dx \\ &= 1 + \frac{\pi}{2}. \end{aligned}$$

To find a_n, we use Equation (5) with m replaced by n:

$$\begin{aligned} a_n &= \frac{1}{\pi}\int_{-\pi}^{\pi} f(x)\cos nx\,dx \\ &= \frac{1}{\pi}\int_{-\pi}^{0} \cos nx\,dx + \frac{1}{\pi}\int_{0}^{\pi} x\cos nx\,dx \\ &= \frac{1}{n\pi}\sin nx\Big]_{-\pi}^{0} + \frac{1}{\pi}\left[\frac{x}{n}\sin nx\right]_{0}^{\pi} - \frac{1}{\pi n}\int_{0}^{\pi}\sin nx\,dx \\ &= \frac{1}{\pi n^2}\cos nx\Big]_{0}^{\pi} \\ &= \frac{1}{\pi n^2}(\cos n\pi - 1) \\ &= \frac{(-1)^n - 1}{\pi n^2}. \qquad \cos n\pi = (-1)^n \end{aligned}$$

In a similar manner, from Equation (6) with m replaced by n:

$$\begin{aligned} b_n &= \frac{1}{\pi}\int_{-\pi}^{\pi} f(x)\sin nx\,dx \\ &= \frac{1}{\pi}\int_{-\pi}^{0} \sin nx\,dx + \frac{1}{\pi}\int_{0}^{\pi} x\sin nx\,dx \\ &= \frac{(-1)^n(1-\pi) - 1}{n\pi}. \end{aligned}$$

Therefore, the Fourier expansion is

$$f(x) = \frac{1}{2} + \frac{\pi}{4} + \sum_{n=1}^{\infty}\frac{(-1)^n - 1}{\pi n^2}\cos nx + \sum_{n=1}^{\infty}\frac{(-1)^n(1-\pi) - 1}{\pi n}\sin nx.$$

A graph of the Fourier series approximations as n varies up to 1, 5, and 20 terms is given in Figure 8.23. Notice how the approximations get closer and closer to the graph of the function at all points of continuity as n increases. At the point $x = 0$, where f is discontinuous, the Fourier approximations approach the value

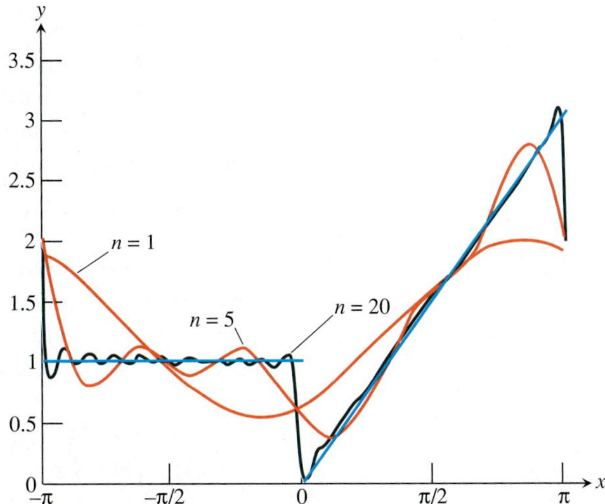

FIGURE 8.23 Fourier series approximations of the function in Example 1 as n varies up to 1, 5, and 20 terms in the infinite series. As n increases, the Fourier approximations approach the actual $f(x)$ values.

0.5 halfway between the jump. These results are consistent with the theorem on Fourier convergence stated below.

In calculating the coefficients a_0, a_n, and b_n, we assumed that f was integrable on the interval $(-L, L)$. We also assumed that the trigonometric series on the right side of Equation (1), as well as the series obtained when we multiply it by $\cos(m\pi x/L)$ or $\sin(m\pi x/L)$, converge in a way to permit term-by-term integration. These convergence issues are studied in advanced calculus, as is the question of when the Fourier series actually equals $f(x)$ for $-L < x < L$. Most of the functions you will encounter in applications guarantee both convergence of the series and its equality with f. We say a little more about this in a moment, but first we summarize our results.

Definition Fourier Series

The **Fourier series** of a function $f(x)$ defined on the interval $-L < x < L$ is

$$f(x) = \frac{a_0}{2} + \sum_{n=1}^{\infty}\left[a_n \cos\frac{n\pi x}{L} + b_n \sin\frac{n\pi x}{L}\right], \tag{7}$$

where

$$a_0 = \frac{1}{L}\int_{-L}^{L} f(x)\,dx, \tag{8}$$

$$a_n = \frac{1}{L}\int_{-L}^{L} f(x)\cos\frac{n\pi x}{L}\,dx, \tag{9}$$

$$b_n = \frac{1}{L}\int_{-L}^{L} f(x)\sin\frac{n\pi x}{L}\,dx. \tag{10}$$

Convergence of the Fourier Series

We now state without proof the result concerning the convergence of the Fourier series expansion for a wide class of functions commonly encountered in simplified models of several physical behaviors. Recall that a function f is *piecewise continuous* over an interval I if both limits

$$\lim_{x\to c^+} f(x) = f(c^+) \qquad \text{and} \qquad \lim_{x\to c^-} f(x) = f(c^-)$$

exist at every interior point c in I and, moreover, the appropriate one-sided limits exist at the endpoints of I, and f has at most finitely many discontinuities in I. Notice that a piecewise continuous function over a closed interval must be bounded (so it cannot tend toward infinity).

Theorem 18 Convergence of Fourier Series

If the function f and its derivative f' are piecewise continuous over the interval $-L < x < L$, then f equals its Fourier series at all points of continuity. At a point c where a jump discontinuity occurs in f, the Fourier series converges to the average

$$\frac{f(c^+) + f(c^-)}{2},$$

where $f(c^+)$ and $f(c^-)$ denote the right and left limits of f at c, respectively.

Example 2 Convergence Values

The function in Example 1 satisfies the conditions of Theorem 18. For every $x \neq 0$ in the interval $-\pi < x < \pi$, the Fourier series converges to $f(x)$. At $x = 0$, the function has a jump discontinuity and the Fourier series converges to the average value

$$\frac{f(0^+) + f(0^-)}{2} = \frac{0+1}{2} = \frac{1}{2}$$

(Figure 8.23).

Periodic Extension

The trigonometric terms $\sin\,(n\pi x/L)$ and $\cos\,(n\pi x/L)$ in the Fourier series are periodic with period $2L$:

$$\begin{aligned}\sin\frac{n\pi(x+2L)}{L} &= \sin\frac{n\pi x}{L}\cos 2n\pi + \cos\frac{n\pi x}{L}\sin 2n\pi \\ &= \sin\frac{n\pi x}{L}\end{aligned}$$

and

$$\begin{aligned}\cos\frac{n\pi(x+2L)}{L} &= \cos\frac{n\pi x}{L}\cos 2n\pi - \sin\frac{n\pi x}{L}\sin 2n\pi \\ &= \cos\frac{n\pi x}{L}.\end{aligned}$$

It follows that the Fourier series is also periodic with period $2L$. Thus, the Fourier series not only represents the function f over the interval $-L < x < L$, but it also produces the **periodic extension** of f over the entire real-number line. From Theorem 18, the series converges to the average value $[f(L^-) + f(-L^+)]/2$ at the endpoints of the interval, as well as to this value extended periodically to $\pm 3L, \pm 5L, \pm 7L$, and so forth.

Example 3 Convergence and Periodic Extension

The Fourier series for $f(x) = x$ on $-\pi < x < \pi$ is

$$f(x) = \sum_{n=1}^{\infty} \frac{2(-1)^{n+1}}{n} \sin nx.$$

(You are asked to find this series in Exercise 3.) The series converges to the periodic extension of $f(x) = x$ on the entire x-axis. The solid dots in Figure 8.24 represent the value

$$\frac{f(\pi^+) + f(\pi^-)}{2} = \frac{\pi + (-\pi)}{2} = 0.$$

The series converges to 0 at the interval endpoints $\pm\pi, \pm 3\pi, \pm 5\pi, \ldots$.

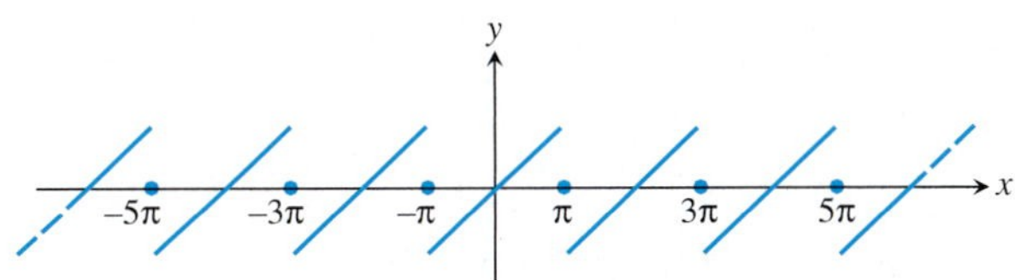

FIGURE 8.24 The Fourier series of $f(x) = x$ converges to f over the interval $-\pi < x < \pi$ and to its periodic extension along the real axis (Theorem 18).

EXERCISES 8.9

Finding Fourier Series

In Exercises 1–14, find the Fourier series expansion for the functions over the specified intervals.

1. $f(x) = 1, \quad -\pi < x < \pi$

2. $f(x) = \begin{cases} -1, & -\pi < x < 0 \\ 1, & 0 < x < \pi \end{cases}$

3. $f(x) = x, \quad -\pi < x < \pi$

4. $f(x) = 1 - x, \quad -\pi < x < \pi$

-ROM WEBsite 5. $f(x) = \dfrac{x^2}{4}, \quad -\pi < x < \pi$

6. $f(x) = \begin{cases} 0, & -\pi < x < 0 \\ x^2, & 0 < x < \pi \end{cases}$

7. $f(x) = e^x, \quad -\pi < x < \pi$

8. $f(x) = \begin{cases} 0, & -\pi < x < 0 \\ e^x, & 0 < x < \pi \end{cases}$

9. $f(x) = \begin{cases} 0, & -\pi < x < 0 \\ \cos x, & 0 < x < \pi \end{cases}$

10. $f(x) = \begin{cases} -x, & -2 < x < 0 \\ 2, & 0 < x < 2 \end{cases}$

11. $f(x) = \begin{cases} 0, & -\pi < x < -\dfrac{\pi}{2} \\ 1, & -\dfrac{\pi}{2} < x < \dfrac{\pi}{2} \\ 0, & \dfrac{\pi}{2} < x < \pi \end{cases}$

12. $f(x) = |x|, \quad -1 < x < 1$

13. $f(x) = |2x - 1|, \quad -1 < x < 1$

14. $f(x) = x|x|, \quad -\pi < x < \pi$

Theory and Examples

15. Use the Fourier series in Exercise 5 to show that

CD-ROM WEBsite

$$1 + \frac{1}{4} + \frac{1}{9} + \frac{1}{16} + \frac{1}{25} + \cdots = \frac{\pi^2}{6}.$$

16. Use the Fourier series in Exercise 6 to show that

$$1 - \frac{1}{4} + \frac{1}{9} - \frac{1}{16} + \cdots = \frac{\pi^2}{12}.$$

Establish the results in Exercises 17–21, where m and n are positive integers.

17. $\displaystyle\int_{-L}^{L} \cos \frac{m\pi x}{L}\, dx = 0 \quad \text{for all } m.$

18. $\displaystyle\int_{-L}^{L} \sin \frac{m\pi x}{L}\, dx = 0 \quad \text{for all } m.$

19. $\displaystyle\int_{-L}^{L} \cos \frac{n\pi x}{L} \cos \frac{m\pi x}{L}\, dx = \begin{cases} 0, & m \neq n \\ L, & m = n \end{cases}$

(*Hint*: $\cos A \cos B = (1/2)\,[\cos (A + B) + \cos (A - B)]$.)

20. $\displaystyle\int_{-L}^{L} \sin \frac{n\pi x}{L} \sin \frac{m\pi x}{L}\, dx = \begin{cases} 0, & m \neq n \\ L, & m = n \end{cases}$

(*Hint*: $\sin A \sin B = (1/2)\,[\cos (A - B) - \cos (A + B)]$.)

21. $\displaystyle\int_{-L}^{L} \sin \frac{n\pi x}{L} \cos \frac{m\pi x}{L}\, dx = 0 \quad \text{for all } m \text{ and } n.$

(*Hint*: $\sin A \cos B = (1/2)\,[\sin (A + B) + \sin (A - B)]$.)

22. *Writing to Learn: Fourier series of sums of functions* If $f(x)$ and $g(x)$ both satisfy the conditions of Theorem 18, is the Fourier series of $f(x) + g(x)$ on $(-L, L)$ the sum of the Fourier series of $f(x)$ and $g(x)$ on the interval? Give reasons for your answers.

23. *Term-by-term differentiation*

(a) Use Theorem 18 to verify that the Fourier series for $f(x) = x$ in Exercise 3 converges to $f(x)$ for $-\pi < x < \pi$.

(b) Although $f'(x) = 1$, show that the series obtained from term-by-term differentiation of the Fourier series in part (a) diverges.

(c) *Writing to Learn* What do you conclude from part (b)? Give reasons for your answer.

24. *Term-by-term integration* In advanced calculus, it is proved that the Fourier series of a piecewise continuous function on $[-L, L]$ can be integrated term by term. Use this fact to show that if $f(x)$ is piecewise continuous on $-\pi < x < \pi$, then

$$\int_{-\pi}^{x} f(s)\, ds = \frac{1}{2} a_0(x + \pi)$$

$$+ \sum_{n=1}^{\infty} \frac{1}{n}(a_n \sin nx - b_n(\cos nx - \cos n\pi)) \quad \text{for } -\pi \leq x \leq \pi,$$

where a_0, a_n, and b_n are the Fourier coefficients of f.

8.10 Fourier Cosine and Sine Series

Integrals of Even and Odd Functions • Even Extension: Fourier Cosine Series • Odd Extension: Fourier Sine Series • Gibbs Phenomenon

In modeling heat conduction in a long, thin insulated rod or wire, we assume that the x-axis is aligned with the length L of the rod and that $0 < x < L$. The temperature $u(x, t)$ along the length of the rod generally varies with both position x and time t. (In Chapter 12, you will study functions like this, which depend on two or more independent variables for their values.) The problem is to determine $u(x, t)$ given the initial temperature $u(x, 0) = f(x)$ along the rod. For example, it might be hot at one end and cooler at the other end, so heat will flow from the hot to the cool end and we might want to know what the temperature distribution will look like in an hour's time. One method used to solve this problem requires the expansion

$$f(x) = \sum_{n=1}^{\infty} b_n \sin \frac{n\pi x}{L}$$

over the *nonsymmetric* interval $0 < x < L$. How then can we calculate a Fourier series expansion for f? To do so, we extend the function so that it is defined over the symmetric interval $-L < x < L$. How, though, do we define the extension of f for $-L < x < 0$? The answer is that we can define the extension to be *any function* over $-L < x < 0$ we choose as long as the extension and its derivative are piecewise continuous (in order to satisfy the hypothesis of Theorem 18). No matter what piecewise continuous function we define as the extension over $-L < x < 0$, the resulting Fourier series is guaranteed to equal $f(x)$ at all points of continuity over the original domain $0 < x < L$. Of course, the Fourier series also converges to whatever extension function we have chosen for $-L < x < 0$. Nevertheless, there are two special extensions that are particularly useful and whose Fourier coefficients are especially easy to calculate; these are the even and odd extensions of f.

Integrals of Even and Odd Functions

Recall (Preliminaries Chapter, Section 2) that a function $g(x)$ is an even function of x if $g(-x) = g(x)$ for all x in the domain of g. If $g(-x) = -g(x)$ instead, g is said to be an odd function of x. The function $\cos x$ is even, the function $\sin x$ is odd. The graph of an even function is symmetric about the y-axis, whereas the graph of an odd function is symmetric about the origin (Figure 8.25).

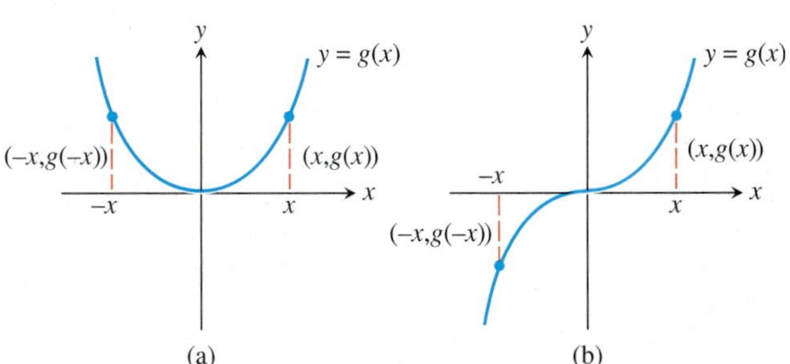

FIGURE 8.25 (a) The graph of an even function is symmetric about the y-axis. (b) The graph of an odd function is symmetric about the origin.

This observation can make the integrals of even and odd functions over intervals symmetric about the origin relatively easy to calculate. For instance, if we consider the "appropriately signed" portions of the graphs in Figure 8.25, we obtain the following:

Odd function:

$$\int_{-L}^{L} g(x)\,dx = 0. \tag{1}$$

Even function:

$$\int_{-L}^{L} g(x)\,dx = 2\int_{0}^{L} g(x)\,dx. \tag{2}$$

Because of rules (1) and (2), even and odd extensions of a function are convenient to use. The following results also hold for even and odd functions.

1. The product of two even functions is even.
2. The product of an even function with an odd function is odd.
3. The product of two odd functions is even.

CD-ROM WEBsite
Historical Biography
John William Strutt Rayleigh (1842 — 1909)

Even Extension: Fourier Cosine Series

Suppose that the function $y = f(x)$ is specified for the interval $0 < x < L$. We define the **even extension of** f by requiring that

$$f(-x) = f(x), \qquad -L < x < L.$$

Graphically, we obtain the even extension by reflecting $y = f(x)$ about the y-axis. The even extension of a function is illustrated in Figure 8.26. Therefore, if we use the even extension for a function f, we obtain the Fourier coefficients

$$a_0 = \frac{1}{L}\int_{-L}^{L} f(x)\,dx = \frac{2}{L}\int_{0}^{L} f(x)\,dx,$$

$$a_n = \frac{1}{L}\int_{-L}^{L} \underbrace{f(x)\cos\frac{n\pi x}{L}}_{\text{Even}}\,dx = \frac{2}{L}\int_{0}^{L} f(x)\cos\frac{n\pi x}{L}\,dx,$$

$$b_n = \frac{1}{L}\int_{-L}^{L} \underbrace{f(x)\sin\frac{n\pi x}{L}}_{\text{Odd}}\,dx = 0.$$

The Fourier series of f is

$$f(x) = \frac{a_0}{2} + \sum_{n=1}^{\infty} a_n \cos\frac{n\pi x}{L}.$$

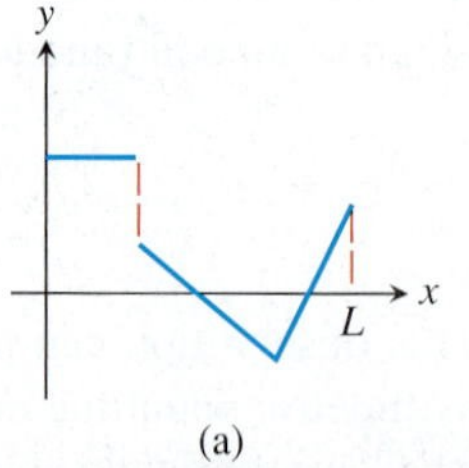

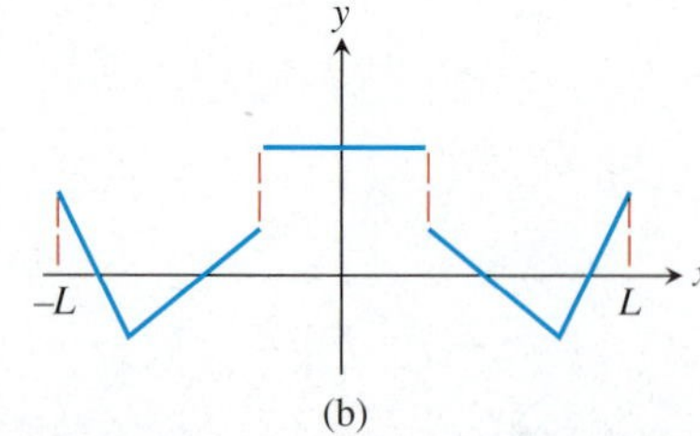

FIGURE 8.26 (a) The original piecewise continuous function f defined over nonsymmetric interval $0 < x < L$. (b) The even extension of f over $-L < x < L$.

Because the Fourier coefficients b_n are all zero, no sine terms appear in the Fourier series expansion, and the series is called the **Fourier cosine series** of the function f. It converges to the original function f over the interval $0 < x < L$ and to the even extension over the interval $-L < x < 0$ (assuming the piecewise continuity of f and f'). We summarize this result.

CD-ROM WEBsite

Fourier Cosine Series

The Fourier series of an even function on the interval $-L < x < L$ is the **cosine series**

$$f(x) = \frac{a_0}{2} + \sum_{n=1}^{\infty} a_n \cos \frac{n\pi x}{L}, \tag{3}$$

where

$$a_0 = \frac{2}{L}\int_0^L f(x)\,dx, \tag{4}$$

$$a_n = \frac{2}{L}\int_0^L f(x) \cos \frac{n\pi x}{L}\,dx. \tag{5}$$

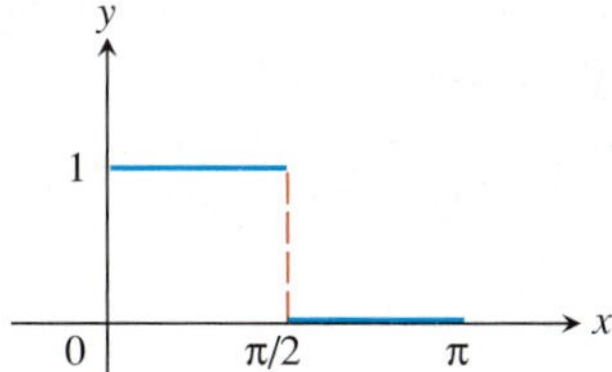

FIGURE 8.27 The function given in Example 1.

Example 1 Finding a Fourier Cosine Series

Find the Fourier cosine series for the function

$$f(x) = \begin{cases} 1, & 0 < x < \frac{\pi}{2}, \\ 0, & \frac{\pi}{2} < x < \pi, \end{cases}$$

depicted in Figure 8.27.

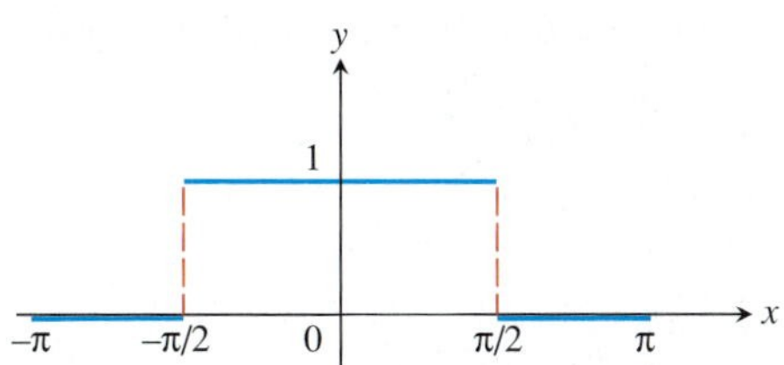

FIGURE 8.28 The even extension of the function in Example 1.

Solution For the Fourier cosine series, we select the even extension of the function over $-\pi < x < \pi$ as shown in Figure 8.28. The Fourier coefficients are

$$a_0 = \frac{2}{\pi}\int_0^{\pi} f(x)\,dx \qquad \text{Eq. (4) with } L = \pi$$

$$= \frac{2}{\pi}\int_0^{\pi/2} dx \qquad f(x) = 0 \text{ for } \pi/2 < x < \pi$$

$$= \frac{2x}{\pi}\Bigg]_0^{\pi/2} = 1$$

$$a_n = \frac{2}{\pi}\int_0^{\pi} f(x) \cos \frac{n\pi x}{\pi}\,dx \qquad \text{Eq. (5) with } L = \pi$$

$$= \frac{2}{\pi}\int_0^{\pi/2} \cos nx\,dx$$

$$= \frac{2}{n\pi} \sin \frac{n\pi}{2}.$$

Therefore, we have the Fourier cosine expansion

$$f(x) = \frac{1}{2} + \sum_{n=1}^{\infty} \frac{2}{n\pi} \sin \frac{n\pi}{2} \cos nx.$$

The Fourier cosine series equals exactly the values of $f(x)$ for $x \neq \pi/2$; at the point $x = \pi/2$, the value of the Fourier cosine series is $1/2$. Plots of Fourier cosine approximations for $f(x)$ as n varies up to 1, 5, and 20 terms are given in Figure 8.29.

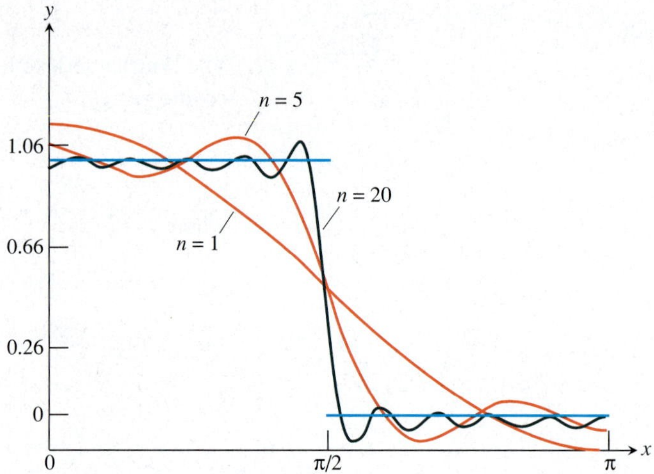

FIGURE 8.29 Fourier cosine series approximations of the function in Example 1 as n varies up to 1, 5, and 20 terms in the infinite series. As n increases, the Fourier cosine approximations approach the actual values of function $f(x)$. Each Fourier cosine approximation passes through the value $y = 0.5$, the midvalue of the jump, at the point of discontinuity $x = \pi/2$.

Odd Extension: Fourier Sine Series

Consider again a function $y = f(x)$ specified for the interval $0 < x < L$. We define the **odd extension of f** by requiring that

$$f(-x) = -f(x), \qquad -L < x < L.$$

Graphically, we obtain the odd extension by reflecting $y = f(x)$ about the origin. The odd extension of a function is illustrated in Figure 8.30. For the odd extension of f, we obtain

$$a_0 = \frac{1}{L} \int_{-L}^{L} f(x)\, dx = 0$$

$$a_n = \frac{1}{L} \int_{-L}^{L} \underbrace{f(x) \cos \frac{n\pi x}{L}}_{\text{Odd}}\, dx = 0$$

$$b_n = \frac{1}{L} \int_{-L}^{L} \underbrace{f(x) \sin \frac{n\pi x}{L}}_{\text{Even}}\, dx = \frac{2}{L} \int_{0}^{L} f(x) \sin \frac{n\pi x}{L}\, dx.$$

The Fourier series of f is

$$f(x) = \sum_{n=1}^{\infty} b_n \sin \frac{n\pi x}{L}.$$

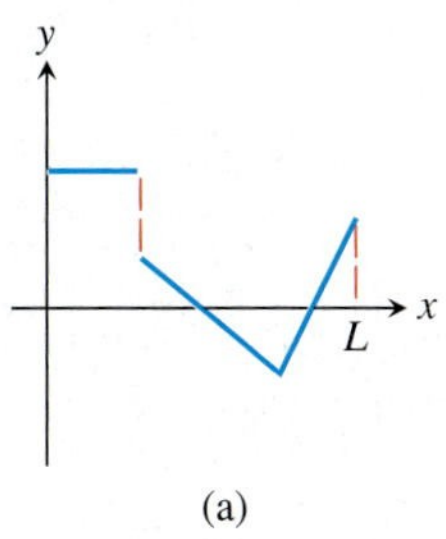

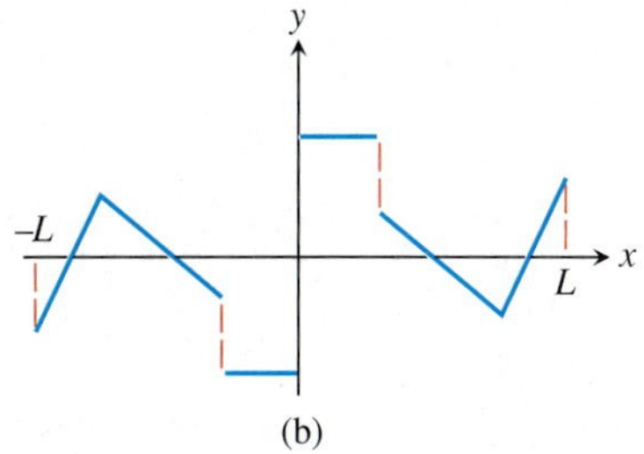

FIGURE 8.30 (a) The original piecewise continuous function f defined over nonsymmetric interval $0 < x < L$. (b) The odd extension of f over $-L < x < L$.

Because the Fourier coefficients a_0 and a_n are all zero, no cosine terms appear in the Fourier series expansion, and the series is called the **Fourier sine series** of the function f. This series converges to the original function f over the interval $0 < x < L$, and to the *odd* extension over the interval $-L < x < 0$ (assuming the piecewise continuity of f and f'). We summarize this result.

CD-ROM WEBsite

Fourier Sine Series

The Fourier series of an odd function on the interval $-L < x < L$ is the **sine series**

$$f(x) = \sum_{n=1}^{\infty} b_n \sin \frac{n\pi x}{L}, \tag{6}$$

where

$$b_n = \frac{2}{L} \int_0^L f(x) \sin \frac{n\pi x}{L}\, dx. \tag{7}$$

Example 2 Finding a Fourier Sine Series

Find the Fourier sine series for the function

$$f(x) = \begin{cases} 1, & 0 < x < \dfrac{\pi}{2}, \\ 0, & \dfrac{\pi}{2} < x < \pi, \end{cases}$$

in Example 1.

Solution We select the *odd* extension of the function $f(x)$. The Fourier coefficients are

$$\begin{aligned} b_n &= \frac{2}{\pi} \int_0^{\pi} f(x) \sin \frac{n\pi x}{\pi}\, dx && \text{Eq. (7) with } L = \pi \\ &= \frac{2}{\pi} \int_0^{\pi/2} \sin nx\, dx && f(x) = 0 \text{ for } \pi/2 < x < \pi \\ &= -\frac{2}{n\pi} \cos nx \Big]_0^{\pi/2} = \frac{2}{n\pi}\left(1 - \cos \frac{n\pi}{2}\right). \end{aligned}$$

Therefore, we have the Fourier sine expansion

$$f(x) = \sum_{n=1}^{\infty} \frac{2}{n\pi}\left(1 - \cos\frac{n\pi}{2}\right) \sin nx.$$

A graph of the Fourier sine approximations for $f(x)$ for $n = 1, 5$, and 20 terms is shown in Figure 8.31.

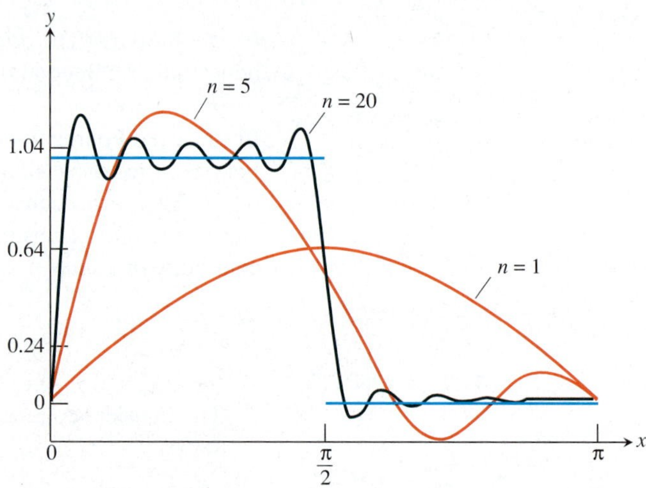

FIGURE 8.31 Fourier sine approximations of the function in Example 2 for $n = 1, 5$, and 20 terms in the infinite series. As n increases, the Fourier sine approximations approach the values of $f(x)$. The approximations converge to the midpoint of the jump at the point of discontinuity, $x = \pi/2$.

CD-ROM
WEBsite
Historical Biography
Josiah Willard Gibbs
(1839 — 1903)

Gibbs Phenomenon

In Figures 8.29 and 8.31, the **overshoot** at $x = \pi/2^-$ and the **undershoot** at $x = \pi/2^+$ are characteristic of Fourier series expansions near points of discontinuity. Known as the **Gibbs phenomenon,** after American mathematical physicist Josiah Willard Gibbs, this characteristic persists even when a large number of terms are summed. The combined overshoot and undershoot amount to about 18% of the difference between the function's values at the point of discontinuity. Figure 8.32 reveals the phenomenon for

$$f(x) = \begin{cases} 1, & 0 \le x < 1 \\ 0, & x = 1 \end{cases}$$

with $n = 2, 4, 8, 16, 32$, and 64 terms. With more terms, the highest peak moves closer to the discontinuity at $x = 1$, but the overshoot remains approximately 1.09.

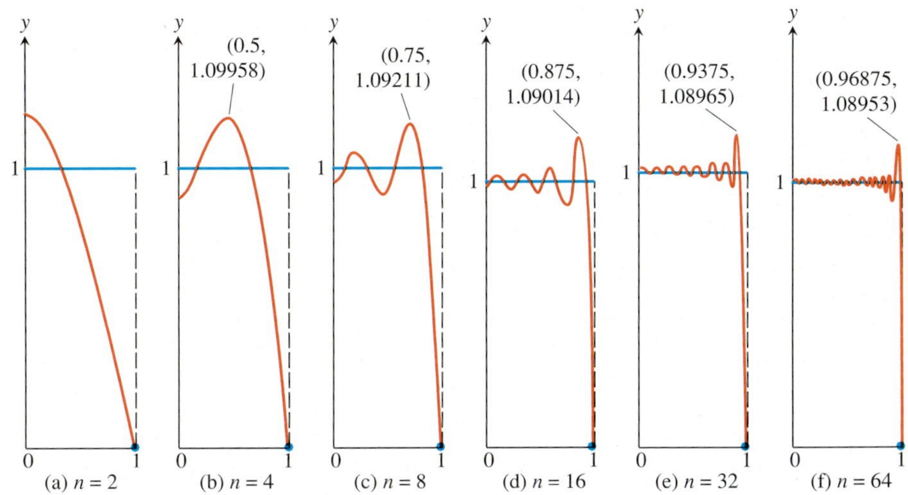

FIGURE 8.32 The Gibbs phenomenon for $n = 2, 4, 8, 16, 32$, and 64 terms. The highest peak moves from 0.5 to 0.75, then to 0.875, and so forth, getting closer to the discontinuity at $x = 1$. The overshoot is always close to 1.09, or about 9% of the distance between $y = 0$ and $y = 1$ at the discontinuity, $x = 1$.

EXERCISES 8.10

Finding Fourier Cosine Series

Each of Exercises 1–8 gives a function $f(x)$ defined on an interval $(0, L)$. Graph f and its even extension to $(-L, L)$. Then find the Fourier cosine series expansion for f.

1. $f(x) = x, \quad 0 < x < \pi$

2. $f(x) = \sin x, \quad 0 < x < \pi$

3. $f(x) = e^x, \quad 0 < x < 1$

4. $f(x) = \cos x, \quad 0 < x < \pi$

5. $f(x) = \begin{cases} 1, & 0 < x < 1 \\ -x, & 1 < x < 2 \end{cases}$

6. $f(x) = \begin{cases} -1, & 0 < x < 0.5 \\ 1, & 0.5 < x < 1 \end{cases}$

CD-ROM WEBsite

7. $f(x) = |2x - 1|, \quad 0 < x < 1$

8. $f(x) = |2x - \pi|, \quad 0 < x < \pi$

Finding Fourier Sine Series

Each of Exercises 9–16 gives a function $f(x)$ defined on an interval $(0, L)$. Graph f and its odd extension to $(-L, L)$. Then find the Fourier sine series expansion for f.

9. $f(x) = -x, \quad 0 < x < 1$

10. $f(x) = x^2, \quad 0 < x < \pi$

11. $f(x) = \cos x, \quad 0 < x < \pi$

12. $f(x) = e^x, \quad 0 < x < 1$

13. $f(x) = \sin x, \quad 0 < x < \pi$

14. $f(x) = \begin{cases} x, & 0 < x < 1 \\ 1, & 1 < x < 2 \end{cases}$

15. $f(x) = \begin{cases} 1 - x, & 0 < x < 1 \\ 0, & 1 < x < 2 \end{cases}$

16. $f(x) = |2x - \pi|, \quad 0 < x < \pi$

CD-ROM WEBsite

Theory and Examples

17. *A series for $\pi/4$*

(a) Find the Fourier sine series for

$$f(x) = \begin{cases} 1, & 0 < x < \pi \\ 0, & x = 0 \text{ and } x = \pi. \end{cases}$$

(b) Use the result of part (a) to show that

$$\frac{\pi}{4} = 1 - \frac{1}{3} + \frac{1}{5} - \frac{1}{7} + \cdots.$$

18. *A function with a triangular graph*

(a) Graph the triangular function

$$f(x) = \begin{cases} 1 - x, & 0 < x < 1 \\ x - 1, & 1 < x < 2 \end{cases}$$

(b) Find a Fourier series expansion for $f(x)$.

(c) Find a Fourier cosine series expansion for $f(x)$.

19. *Evaluating a series* Use the result of Exercise 2 to find the value of

$$\sum_{n=1}^{\infty} \frac{(-1)^n}{4n^2 - 1}.$$

20. *Fourier sine series* Given the function

$$f(x) = 2 - x, \qquad 0 < x < 2,$$

define a function whose Fourier sine series representation will converge to $f(x)$ for all values of x. (*Note*: The answer is not unique.)

Questions to Guide Your Review

1. What is an infinite sequence? What does it mean for such a sequence to converge? To diverge? Give examples.
2. What theorems are available for calculating limits of sequences? Give examples.
3. What theorem sometimes enables us to use l'Hôpital's Rule to calculate the limit of a sequence? Give an example.
4. What six sequence limits are likely to arise when you work with sequences and series?
5. What is a subsequence? Why are subsequences important? What uses can be found for subsequences? Give examples.
6. What is a nondecreasing sequence? A nonincreasing sequence? A monotonic sequence? Under what circumstances do these sequences have a limit? Give examples.
7. What is Picard's method for solving the equation $f(x) = 0$? Give an example.
8. What is an infinite series? What does it mean for such a series to converge? To diverge? Give examples.
9. What is a geometric series? When does such a series converge? Diverge? When it does converge, what is its sum? Give examples.
10. Besides geometric series, what other convergent and divergent series do you know?
11. What is the nth-Term Test for Divergence? What is the idea behind the test?
12. What can be said about term-by-term sums and differences of convergent series? About constant multiples of convergent and divergent series?
13. What happens if you add a finite number of terms to a convergent series? A divergent series? What happens if you delete a finite number of terms from a convergent series? A divergent series?
14. Under what circumstances will an infinite series of nonnegative terms converge? Diverge? Why study series of nonnegative terms?
15. What is the Integral Test? What is the reasoning behind it? Give an example of its use.
16. When do p-series converge? Diverge? How do you know? Give examples of convergent and divergent p-series.
17. What are the Direct Comparison Test and the Limit Comparison Test? What is the reasoning behind these tests? Give examples of their use.
18. What are the Ratio and Root Tests? Do they always give you the information you need to determine convergence or divergence? Give examples.
19. What is an alternating series? What theorem is available for determining the convergence of such a series?
20. How can you estimate the error involved in approximating the sum of an alternating series with one of the series' partial sums? What is the reasoning behind the estimate?
21. What is absolute convergence? Conditional convergence? How are the two related?
22. What do you know about rearranging the terms of an absolutely convergent series? Of a conditionally convergent series? Give examples.
23. What is a power series? How do you test a power series for convergence? What are the possible outcomes?
24. What are the basic facts about
    (a) term-by-term differentiation of power series?
    (b) term-by-term integration of power series?
    (c) multiplication of power series?
    Give examples.
25. What is the Taylor series generated by a function $f(x)$ at a point $x = a$? What information do you need about f to construct the series? Give an example.
26. What is a Maclaurin series?
27. Does a Taylor series always converge to its generating function? Explain.
28. What are Taylor polynomials? Of what use are they?
29. What is Taylor's Theorem? What does it say about the errors involved in using Taylor polynomials to approximate functions? In particular, what does the Remainder Estimation Theorem say about the error in a linearization? A quadratic approximation?
30. What is the binomial series? On what interval does it converge? How is it used?
31. What are the Maclaurin series for $1/(1 - x)$, $1/(1 + x)$, e^x, $\sin x$, $\cos x$, $\ln(1 + x)$, and $\tan^{-1} x$?
32. What is a Fourier series? How do you calculate the Fourier coefficients for a function $f(x)$ defined over the interval $-L < x < L$? Under what conditions does a Fourier series converge to its generating function? What happens at a point of discontinuity?
33. What is the periodic extension to the entire real line of a function $f(x)$ defined on $-L < x < L$?
34. What is the even extension to $-L < x < 0$ of a function $f(x)$ defined on $0 < x < L$? What is a Fourier cosine series? How do you calculate its coefficients?
35. What is the odd extension to $-L < x < 0$ of a function $f(x)$ defined on $0 < x < L$? What is a Fourier sine series? How do you calculate its coefficients?
36. What is the Gibbs phenomenon? How is it affected when you sum more and more terms in the Fourier series?

Practice Exercises

Convergent or Divergent Sequences

Which of the sequences whose nth terms appear in Exercises 1–18 converge, and which diverge? Find the limit of each convergent sequence.

1. $a_n = 1 + \frac{(-1)^n}{n}$
2. $a_n = \frac{1-(-1)^n}{\sqrt{n}}$
3. $a_n = \frac{1-2^n}{2^n}$
4. $a_n = 1 + (0.9)^n$
5. $a_n = \sin \frac{n\pi}{2}$
6. $a_n = \sin n\pi$
7. $a_n = \frac{\ln (n^2)}{n}$
8. $a_n = \frac{\ln (2n+1)}{n}$
9. $a_n = \frac{n + \ln n}{n}$
10. $a_n = \frac{\ln (2n^3+1)}{n}$
11. $a_n = \left(\frac{n-5}{n}\right)^n$
12. $a_n = \left(1 + \frac{1}{n}\right)^{-n}$
13. $a_n = \sqrt[n]{\frac{3^n}{n}}$
14. $a_n = \left(\frac{3}{n}\right)^{1/n}$
15. $a_n = n(2^{1/n} - 1)$
16. $a_n = \sqrt[n]{2n+1}$
17. $a_n = \frac{(n+1)!}{n!}$
18. $a_n = \frac{(-4)^n}{n!}$

Convergent Series

Find the sums of the series in Exercises 19–24.

19. $\sum_{n=3}^{\infty} \frac{1}{(2n-3)(2n-1)}$
20. $\sum_{n=2}^{\infty} \frac{-2}{n(n+1)}$
21. $\sum_{n=1}^{\infty} \frac{9}{(3n-1)(3n+2)}$
22. $\sum_{n=3}^{\infty} \frac{-8}{(4n-3)(4n+1)}$
23. $\sum_{n=0}^{\infty} e^{-n}$
24. $\sum_{n=1}^{\infty} (-1)^n \frac{3}{4^n}$

Convergent or Divergent Series

Which of the series in Exercises 25–40 converge absolutely, which converge conditionally, and which diverge? Give reasons for your answers.

25. $\sum_{n=1}^{\infty} \frac{1}{\sqrt{n}}$
26. $\sum_{n=1}^{\infty} \frac{-5}{n}$
27. $\sum_{n=1}^{\infty} \frac{(-1)^n}{\sqrt{n}}$
28. $\sum_{n=1}^{\infty} \frac{1}{2n^3}$
29. $\sum_{n=1}^{\infty} \frac{(-1)^n}{\ln (n+1)}$
30. $\sum_{n=2}^{\infty} \frac{1}{n(\ln n)^2}$
31. $\sum_{n=1}^{\infty} \frac{\ln n}{n^3}$
32. $\sum_{n=3}^{\infty} \frac{\ln n}{\ln (\ln n)}$
33. $\sum_{n=1}^{\infty} \frac{(-1)^n}{n\sqrt{n^2+1}}$
34. $\sum_{n=1}^{\infty} \frac{(-1)^n 3n^2}{n^3+1}$
35. $\sum_{n=1}^{\infty} \frac{n+1}{n!}$
36. $\sum_{n=1}^{\infty} \frac{(-1)^n(n^2+1)}{2n^2+n-1}$
37. $\sum_{n=1}^{\infty} \frac{(-3)^n}{n!}$
38. $\sum_{n=1}^{\infty} \frac{2^n 3^n}{n^n}$
39. $\sum_{n=1}^{\infty} \frac{1}{\sqrt{n(n+1)(n+2)}}$
40. $\sum_{n=2}^{\infty} \frac{1}{n\sqrt{n^2-1}}$

Power Series

In Exercises 41–50, (a) find the series' radius and interval of convergence. Then identify the values of x for which the series converges (b) absolutely and (c) conditionally.

41. $\sum_{n=1}^{\infty} \frac{(x+4)^n}{n3^n}$
42. $\sum_{n=1}^{\infty} \frac{(x-1)^{2n-2}}{(2n-1)!}$
43. $\sum_{n=1}^{\infty} \frac{(-1)^{n-1}(3x-1)^n}{n^2}$
44. $\sum_{n=0}^{\infty} \frac{(n+1)(2x+1)^n}{(2n+1)2^n}$
45. $\sum_{n=1}^{\infty} \frac{x^n}{n^n}$
46. $\sum_{n=1}^{\infty} \frac{x^n}{\sqrt{n}}$
47. $\sum_{n=0}^{\infty} \frac{(n+1)\,x^{2n-1}}{3^n}$
48. $\sum_{n=0}^{\infty} \frac{(-1)^n(x-1)^{2n+1}}{2n+1}$
49. $\sum_{n=1}^{\infty} (\operatorname{csch} n)\, x^n$
50. $\sum_{n=1}^{\infty} (\coth n)\, x^n$

Maclaurin Series

Each of the series in Exercises 51–56 is the value of the Maclaurin series of a function $f(x)$ at a particular point. What function and what point? What is the sum of the series?

51. $1 - \frac{1}{4} + \frac{1}{16} - \cdots + (-1)^n \frac{1}{4^n} + \cdots$
52. $\frac{2}{3} - \frac{4}{18} + \frac{8}{81} - \cdots + (-1)^{n-1} \frac{2^n}{n3^n} + \cdots$
53. $\pi - \frac{\pi^3}{3!} + \frac{\pi^5}{5!} - \cdots + (-1)^n \frac{\pi^{2n+1}}{(2n+1)!} + \cdots$
54. $1 - \frac{\pi^2}{9 \cdot 2!} + \frac{\pi^4}{81 \cdot 4!} - \cdots + (-1)^n \frac{\pi^{2n}}{3^{2n}(2n)!} + \cdots$
55. $1 + \ln 2 + \frac{(\ln 2)^2}{2!} + \cdots + \frac{(\ln 2)^n}{n!} + \cdots$
56. $\frac{1}{\sqrt{3}} - \frac{1}{9\sqrt{3}} + \frac{1}{45\sqrt{3}} - \cdots + (-1)^{n-1} \frac{1}{(2n-1)(\sqrt{3})^{2n-1}} + \cdots$

Find Maclaurin series for the functions in Exercises 57–64.

57. $\dfrac{1}{1-2x}$

58. $\dfrac{1}{1+x^3}$

59. $\sin \pi x$

60. $\sin \dfrac{2x}{3}$

61. $\cos (x^{5/2})$

62. $\cos \sqrt{5x}$

63. $e^{(\pi x/2)}$

64. e^{-x^2}

Taylor Series

In Exercises 65–68, find the first four nonzero terms of the Taylor series generated by f at $x = a$.

65. $f(x) = \sqrt{3 + x^2}$ at $x = -1$

66. $f(x) = 1/(1 - x)$ at $x = 2$

67. $f(x) = 1/(x + 1)$ at $x = 3$

68. $f(x) = 1/x$ at $x = a > 0$

Initial Value Problems

Use power series to solve the initial value problems in Exercises 69–76.

69. $y' + y = 0, \quad y(0) = -1$

70. $y' - y = 0, \quad y(0) = -3$

71. $y' + 2y = 0, \quad y(0) = 3$

72. $y' + y = 1, \quad y(0) = 0$

73. $y' - y = 3x, \quad y(0) = -1$

74. $y' + y = x, \quad y(0) = 0$

75. $y' - y = x, \quad y(0) = 1$

76. $y' - y = -x, \quad y(0) = 2$

Indeterminate Forms

In Exercises 77–82:

(a) Use power series to evaluate the limit.

T **(b)** Then use a grapher to support your calculation.

77. $\lim\limits_{x\to 0} \dfrac{7 \sin x}{e^{2x} - 1}$

78. $\lim\limits_{\theta\to 0} \dfrac{e^{\theta} - e^{-\theta} - 2\theta}{\theta - \sin \theta}$

79. $\lim\limits_{t\to 0} \left(\dfrac{1}{2 - 2\cos t} - \dfrac{1}{t^2}\right)$

80. $\lim\limits_{h\to 0} \dfrac{(\sin h)/h - \cos h}{h^2}$

81. $\lim\limits_{z\to 0} \dfrac{1 - \cos^2 z}{\ln (1 - z) + \sin z}$

82. $\lim\limits_{y\to 0} \dfrac{y^2}{\cos y - \cosh y}$

83. Use a series representation of $\sin 3x$ to find values of r and s for which

$$\lim_{x\to 0}\left(\frac{\sin 3x}{x^3} + \frac{r}{x^2} + s\right) = 0.$$

84. **(a)** Show that the approximation $\csc x \approx 1/x + x/6$ in Section 8.8, Example 6, leads to the approximation $\sin x \approx 6x/(6 + x^2)$.

T **(b)** *Writing to Learn* Compare the accuracies of the approximations $\sin x \approx x$ and $\sin x \approx 6x/(6 + x^2)$ by comparing the graphs of $f(x) = \sin x - x$ and $g(x) = \sin x - (6x/(6 + x^2))$. Describe what you find.

Fourier Series

In Exercises 85–90, find the Fourier series of f on the given interval.

85. $f(x) = \begin{cases} -1, & -\pi < x < 0 \\ 2, & 0 < x < \pi \end{cases}$

86. $f(x) = \begin{cases} 0, & -1 < x < 0 \\ x, & 0 < x < 1 \end{cases}$

87. $f(x) = x + \pi, \quad -\pi < x < \pi$

88. $f(x) = \begin{cases} 0, & -\pi < x < 0 \\ \sin x, & 0 < x < \pi \end{cases}$

89. $f(x) = \begin{cases} 1, & -2 < x < 0 \\ 1 + x, & 0 < x < 2 \end{cases}$

90. $f(x) = \begin{cases} 0, & -2 < x < 0 \\ x, & 0 < x < 1 \\ 1, & 1 < x < 2 \end{cases}$

Fourier Cosine and Sine Series

In Exercises 91–96, find

(a) the Fourier cosine series

(b) the Fourier sine series of f on the given interval.

91. $f(x) = \begin{cases} 1, & 0 < x < 1/2 \\ 0, & 1/2 < x < 1 \end{cases}$

92. $f(x) = \begin{cases} 0, & 0 < x < 1 \\ x, & 1 < x < 2 \end{cases}$

93. $f(x) = \sin \pi x, \quad 0 < x < 1$

94. $f(x) = \cos x, \quad 0 < x < \pi/2$

95. $f(x) = 2x + x^2, \quad 0 < x < 3$

96. $f(x) = e^{-x}, \quad 0 < x < 2$

Theory and Examples

97. *A convergent series*

(a) Show that the series

$$\sum_{n=1}^{\infty}\left(\sin \frac{1}{2n} - \sin \frac{1}{2n+1}\right)$$

converges.

(b) *Writing to Learn* Estimate the magnitude of the error involved in using the sum of the sines through $n = 20$ to approximate the sum of the series. Is the approximation too large, or too small? Give reasons for your answer.

98. **(a)** *A convergent series* Show that the series

$$\sum_{n=1}^{\infty}\left(\tan\frac{1}{2n}-\tan\frac{1}{2n+1}\right)$$

converges.

(b) *Writing to Learn* Estimate the magnitude of the error in using the sum of the tangents through $-\tan(1/41)$ to approximate the sum of the series. Is the approximation too large or too small? Give reasons for your answer.

99. *Radius of convergence* Find the radius of convergence of the series

$$\sum_{n=1}^{\infty}\frac{2\cdot 5\cdot 8\cdot\cdots\cdot(3n-1)}{2\cdot 4\cdot 6\cdot\cdots\cdot(2n)}x^n.$$

100. *Radius of convergence* Find the radius of convergence of the series

$$\sum_{n=1}^{\infty}\frac{3\cdot 5\cdot 7\cdot\cdots\cdot(2n-1)}{4\cdot 9\cdot 14\cdot\cdots\cdot(5n-1)}(x-1)^n.$$

101. *nth partial sum* Find a closed-form formula for the nth partial sum of the series $\sum_{n=2}^{\infty}\ln(1-(1/n^2))$ and use it to determine the convergence or divergence of the series.

102. *nth partial sum* Evaluate $\sum_{k=2}^{\infty}(1/(k^2-1))$ by finding the limit as $n\to\infty$ of the series' nth partial sum.

103. **(a)** *Interval of convergence* Find the interval of convergence of the series

$$y=1+\frac{1}{6}x^3+\frac{4}{720}x^6+\cdots+\frac{1\cdot 4\cdot 7\cdot\cdots\cdot(3n-2)}{(3n)!}x^{3n}+\cdots.$$

(b) *Differential equation* Show that the function defined by the series satisfies a differential equation of the form

$$\frac{d^2y}{dx^2}=x^a y+b$$

and find the values of the constants a and b.

104. **(a)** *Maclaurin series* Find the Maclaurin series for the function $x^2/(1+x)$.

(b) Does the series converge at $x=1$? Explain.

105. *Writing to Learn* If $\sum_{n=1}^{\infty}a_n$ and $\sum_{n=1}^{\infty}b_n$ are convergent series of nonnegative numbers, can anything be said about $\sum_{n=1}^{\infty}a_nb_n$? Give reasons for your answer.

106. *Writing to Learn* If $\sum_{n=1}^{\infty}a_n$ and $\sum_{n=1}^{\infty}b_n$ are divergent series of nonnegative numbers, can anything be said about $\sum_{n=1}^{\infty}a_nb_n$? Give reasons for your answer.

107. *Sequence and series* Prove that the sequence $\{x_n\}$ and the series $\sum_{k=1}^{\infty}(x_{k+1}-x_k)$ both converge or both diverge.

108. *Convergence* Prove that $\sum_{n=1}^{\infty}(a_n/(1+a_n))$ converges if $a_n>0$ for all n and $\sum_{n=1}^{\infty}a_n$ converges.

109. **(a)** *Divergence* Suppose that $a_1, a_2, a_3, \ldots, a_n$ are positive numbers satisfying the following conditions:

i. $a_1\ge a_2\ge a_3\ge\cdots$

ii. the series $a_2+a_4+a_8+a_{16}+\cdots$ diverges.

Show that the series

$$\frac{a_1}{1}+\frac{a_2}{2}+\frac{a_3}{3}+\cdots$$

diverges.

(b) Use the result in part (a) to show that

$$1+\sum_{n=2}^{\infty}\frac{1}{n\ln n}$$

diverges.

110. *Estimating an integral* Suppose that you wish to obtain a quick estimate for the value of $\int_0^1 x^2e^x\,dx$. There are several ways to do this.

(a) Use the Trapezoidal Rule with $n=2$ to estimate $\int_0^1 x^2e^x\,dx$.

(b) Write out the first three nonzero terms of the Maclaurin series for x^2e^x to obtain the fourth Maclaurin polynomial $P(x)$ for x^2e^x. Use $\int_0^1 P(x)\,dx$ to obtain another estimate for $\int_0^1 x^2e^x\,dx$.

(c) *Writing to Learn* The second derivative of $f(x)=x^2e^x$ is positive for all $x>0$. Explain why this enables you to conclude that the trapezoidal rule estimate obtained in part (a) is too large.

(d) *Writing to Learn* All the derivatives of $f(x)=x^2e^x$ are positive for $x>0$. Explain why this enables you to conclude that all Maclaurin polynomial approximations to $f(x)$ for x in $[0,1]$ will be too small. (*Hint*: $f(x)=P_n(x)+R_n(x)$.)

(e) Use integration by parts to evaluate $\int_0^1 x^2e^x\,dx$.

111. *Series for* $\tan^{-1}x$

(a) Integrate from $t=0$ to $t=x$ both sides of the equation

$$\frac{1}{1+t^2}=1-t^2+t^4-t^6+\cdots+(-1)^nt^{2n}+\frac{(-1)^{n+1}t^{2n+2}}{1+t^2}$$

in which the last term comes from adding the remaining terms as a geometric series with first term $a=(-1)^{n+1}t^{2n+2}$ and ratio $r=-t^2$.

(b) Show that the remainder term from part (a) is

$$R_n(x)=\int_0^x\frac{(-1)^{n+1}t^{2n+2}}{1+t^2}\,dt$$

and find $\lim_{n\to\infty}R_n(x)$ if $|x|\le 1$.

(c) Find a power series for $\tan^{-1}x$ based on the result in part (b).

(d) Set $x = 1$ in the series for $\tan^{-1} x$ to obtain **Leibniz's formula**

$$\frac{\pi}{4} = 1 - \frac{1}{3} + \frac{1}{5} - \frac{1}{7} + \frac{1}{9} - \cdots + \frac{(-1)^n}{2n+1} + \cdots.$$

112. *Evaluating nonelementary integrals* As you know, Maclaurin series can be used to express nonelementary integrals in terms of series.

(a) Express $\int_0^x \sin t^2\, dt$ as a power series.

(b) According to the Alternating Series Estimation Theorem, how many terms of the series in part (a) should you use to estimate $\int_0^1 \sin x^2\, dx$ with an error of less than 0.001?

113. *Picard's method for slope greater than 1* Example 9 in Section 8.2 showed that we cannot apply Picard's method to find a fixed point of $g(x) = 4x - 12$, but we can apply the method to find a fixed point of $g^{-1}(x) = (1/4)x + 3$ because the derivative of g^{-1} is 1/4, whose value is less than 1 in magnitude on any interval. In Example 7 of Section 8.2, we found the fixed point of g^{-1} to be $x = 4$. Now notice that 4 is also a fixed point of g, since

$$g(4) = 4(4) - 12 = 4.$$

In finding the fixed point of g^{-1}, we found the fixed point of g.

A function and its inverse always have the same fixed points. The graphs of the functions are symmetric about the line $y = x$ and therefore intersect the line at the same points.

We now see that the application of Picard's method is quite broad. Suppose g is one-to-one, with a continuous first derivative whose magnitude is greater than 1 on a closed interval I whose interior contains a fixed point of g. Then the derivative of g^{-1}, being the reciprocal of g', has magnitude less than 1 on I. Picard's method applied to g^{-1} on I will find the fixed point of g. As cases in point, find the fixed points of the following functions.

(a) $g(x) = 2x + 3$

(b) $g(x) = 1 - 4x$

Additional Exercises: Theory, Examples, Applications

Convergence or Divergence

Which of the series $\sum_{n=1}^{\infty} a_n$ defined by the formulas in Exercises 1–4 converge, and which diverge? Give reasons for your answers.

1. $\displaystyle\sum_{n=1}^{\infty} \frac{1}{(3n-2)^{n+(1/2)}}$

2. $\displaystyle\sum_{n=1}^{\infty} \frac{(\tan^{-1} n)^2}{n^2+1}$

3. $\displaystyle\sum_{n=1}^{\infty} (-1)^n \tanh n$

4. $\displaystyle\sum_{n=2}^{\infty} \frac{\log_n (n!)}{n^3}$

Which of the series $\sum_{n=1}^{\infty} a_n$ defined by the formulas in Exercises 5–8 converge, and which diverge? Give reasons for your answers.

5. $a_1 = 1, \quad a_{n+1} = \dfrac{n(n+1)}{(n+2)(n+3)}\, a_n$ (*Hint*: Write out several terms, see which factors cancel, and then generalize.)

6. $a_1 = a_2 = 7, \quad a_{n+1} = \dfrac{n}{(n-1)(n+1)}\, a_n$ if $n \geq 2$

7. $a_1 = a_2 = 1, \quad a_{n+1} = \dfrac{1}{1+a_n}$ if $n \geq 2$

8. $a_n = 1/3^n$ if n is odd, $\quad a_n = n/3^n$ if n is even

Choosing Centers for Taylor Series

Taylor's formula

$$f(x) = f(a) + f'(a)(x-a) + \frac{f''(a)}{2!}(x-a)^2 + \cdots + \frac{f^{(n)}(a)}{n!}(x-a)^n + \frac{f^{(n+1)}(c)}{(n+1)!}(x-a)^{n+1}$$

expresses the value of f at x in terms of the values of f and its derivatives at $x = a$. In numerical computations, we therefore need a to be a point where we know the values of f and its derivatives. We also need a to be close enough to the values of f we are interested in to make $(x-a)^{n+1}$ so small we can neglect the remainder.

In Exercises 9–14, what Taylor series would you choose to represent the function near the given value of x? (There may be more than one good answer.) Write out the first four nonzero terms of the series you choose.

9. $\cos x$ near $x = 1$

10. $\sin x$ near $x = 6.3$

11. e^x near $x = 0.4$

12. $\ln x$ near $x = 1.3$

13. $\cos x$ near $x = 69$

14. $\tan^{-1} x$ near $x = 2$

Theory and Examples

15. *nth root of $a^n + b^n$* Let a and b be constants with $0 < a < b$. Does the sequence $\{(a^n + b^n)^{1/n}\}$ converge? If it does converge, what is the limit?

16. *Repeating decimal* Find the sum of the infinite series

$$1 + \frac{2}{10} + \frac{3}{10^2} + \frac{7}{10^3} + \frac{2}{10^4} + \frac{3}{10^5} + \frac{7}{10^6} + \frac{2}{10^7} + \frac{3}{10^8} + \frac{7}{10^9} + \cdots.$$

17. *Summing integrals* Evaluate

$$\sum_{n=0}^{\infty} \int_n^{n+1} \frac{1}{1+x^2}\, dx.$$

18. *Absolute convergence* Find all values of x for which

$$\sum_{n=1}^{\infty} \frac{nx^n}{(n+1)(2x+1)^n}$$

converges absolutely.

19. *Euler's constant* Graphs like those in Figure 8.13, suggest that as n increases, there is little change in the difference between the sum

$$1 + \frac{1}{2} + \cdots + \frac{1}{n}$$

and the integral

$$\ln n = \int_1^n \frac{1}{x}\, dx.$$

To explore this idea, carry out the following steps.

(a) By taking $f(x) = 1/x$ in Figure 8.13, show that

$$\ln(n+1) \le 1 + \frac{1}{2} + \cdots + \frac{1}{n} \le 1 + \ln n$$

or

$$0 < \ln(n+1) - \ln n \le 1 + \frac{1}{2} + \cdots + \frac{1}{n} - \ln n \le 1.$$

Thus, the sequence

$$a_n = 1 + \frac{1}{2} + \cdots + \frac{1}{n} - \ln n$$

is bounded from below and from above.

(b) Show that

$$\frac{1}{n+1} < \int_n^{n+1} \frac{1}{x}\, dx = \ln(n+1) - \ln n$$

and use this result to show that the sequence $\{a_n\}$ in part (a) is nonincreasing.

Since a nonincreasing sequence that is bounded from below converges, the numbers a_n defined in part (a) converge:

$$1 + \frac{1}{2} + \cdots + \frac{1}{n} - \ln n \to \gamma.$$

The number γ, whose value is 0.5772 . . . , is called *Euler's constant.* In contrast to other special numbers like π and e, no other expression with a simple law of formulation has ever been found for γ.

20. *Generalizing Euler's constant* The figure below shows the graph of a positive twice-differentiable decreasing functizon f whose second derivative is positive on $(0, \infty)$. For each n, the number A_n is the area of the lunar region between the curve and the line segment joining the points $(n, f(n))$ and $(n+1, f(n+1))$.

(a) Use the figure to show that $\sum_{n=1}^{\infty} A_n < (1/2)(f(1) - f(2))$.

(b) Next show the existence of

$$\lim_{n\to\infty}\left[\sum_{k=1}^{n} f(k) - \frac{1}{2}(f(1) + f(n)) - \int_1^n f(x)\, dx\right].$$

(c) Then show the existence of

$$\lim_{n\to\infty}\left[\sum_{k=1}^{n} f(k) - \int_1^n f(x)\, dx\right].$$

If $f(x) = 1/x$, the limit in part (c) is Euler's constant. (*Source*: "Convergence with Pictures" by P. J. Rippon, *American Mathematical Monthly*, Vol. 93, No. 6 (1986), pp. 476–478.)

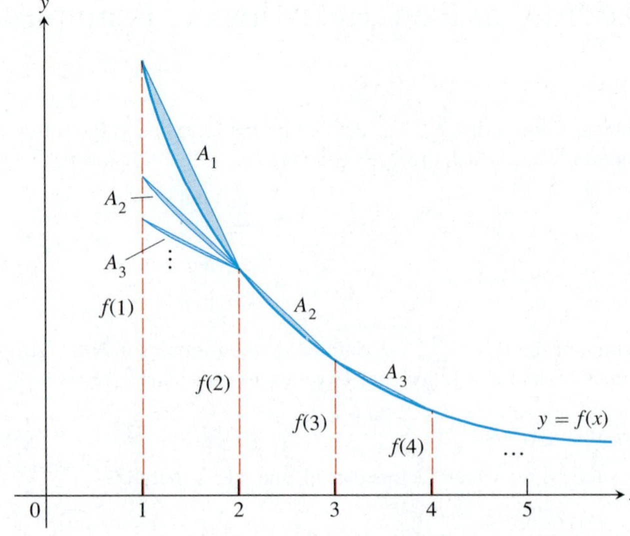

21. *Punching out triangles* This exercise refers to the "right side up" equilateral triangle with sides of length $2b$ in the accompanying figure. "Upside down" equilateral triangles are removed from the original triangle as the sequence of pictures suggests. The sum of the areas removed from the original triangle forms an infinite series.

(a) Find this infinite series.

(b) Find the sum of this infinite series and hence find the total area removed from the original triangle.

(c) Is every point on the original triangle removed? Explain why or why not.

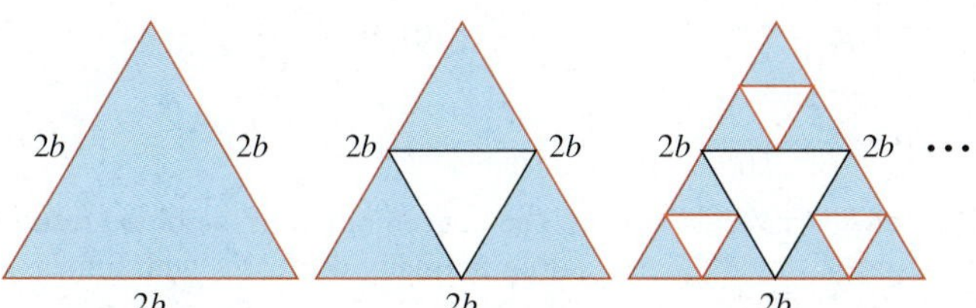

22. *A fast estimate of $\pi/2$* As you saw if you did Exercise 10 in Section 8.2, the sequence generated by starting with $x_0 = 1$ and applying the recursion formula $x_{n+1} = x_n + \cos x_n$ converges rapidly to $\pi/2$. To explain the speed of the convergence, let $\epsilon_n = (\pi/2) - x_n$. (See the accompanying figure.) Then

$$\begin{aligned}\epsilon_{n+1} &= \frac{\pi}{2} - x_n - \cos x_n \\ &= \epsilon_n - \cos\left(\frac{\pi}{2} - \epsilon_n\right) \\ &= \epsilon_n - \sin \epsilon_n \\ &= \frac{1}{3!}(\epsilon_n)^3 - \frac{1}{5!}(\epsilon_n)^5 + \cdots.\end{aligned}$$

Use this equality to show that

$$0 < \epsilon_{n+1} < \frac{1}{6}(\epsilon_n)^3.$$

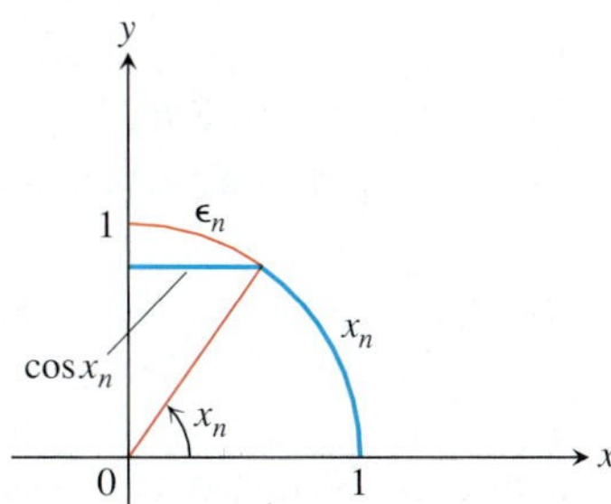

23. *Computer exploration*

(a) *Writing to Learn* Does the value of

$$\lim_{n\to\infty}\left(1 - \frac{\cos(a/n)}{n}\right)^n, \qquad a \text{ constant},$$

appear to depend on the value of a? If so, how?

(b) *Writing to Learn* Does the value of

$$\lim_{n\to\infty}\left(1 - \frac{\cos(a/n)}{bn}\right)^n, \qquad a \text{ and } b \text{ constant}, b \neq 0,$$

appear to depend on the value of b? If so, how?

(c) Use calculus to confirm your findings in parts (a) and (b).

24. Show that if $\sum_{n=1}^{\infty} a_n$ converges, then

$$\sum_{n=1}^{\infty}\left(\frac{1 + \sin(a_n)}{2}\right)^n$$

converges.

25. *Radius of convergence* Find a value for the constant b that will make the radius of convergence of the power series

$$\sum_{n=2}^{\infty}\frac{b^n x^n}{\ln n}$$

equal to 5.

26. *Writing to Learn: Transcendental functions* How do you know that the functions $\sin x$, $\ln x$, and e^x are not polynomials? Give reasons for your answer.

27. *Raabe's (or Gauss's) Test* The following test, which we state without proof, is an extension of the Ratio Test.

Raabe's Test: If $\sum_{n=1}^{\infty} u_n$ is a series of positive constants and there exist constants C, K, and N such that

$$\frac{u_n}{u_{n+1}} = 1 + \frac{C}{n} + \frac{f(n)}{n^2},$$

where $|f(n)| < K$ for $n \geq N$, then $\sum_{n=1}^{\infty} u_n$ converges if $C > 1$ and diverges if $C \leq 1$.

Show that the results of Raabe's Test agree with what you know about the series $\sum_{n=1}^{\infty}(1/n^2)$ and $\sum_{n=1}^{\infty}(1/n)$.

28. *Using Raabe's Test* Suppose that the terms of $\sum_{n=1}^{\infty} u_n$ are defined recursively by the formulas

$$u_1 = 1, \qquad u_{n+1} = \frac{(2n-1)^2}{(2n)(2n+1)}u_n.$$

Apply Raabe's Test to determine whether the series converges.

29. Assume that $\sum_{n=1}^{\infty} a_n$ converges, $a_n \neq 1$, and $a_n > 0$ for all n.

(a) *Squaring terms* Show that $\sum_{n=1}^{\infty} a_n^2$ converges.

(b) *Writing to Learn* Does $\sum_{n=1}^{\infty} a_n/(1 - a_n)$ converge? Explain.

30. (*Continuation of Exercise 29*) If $\sum_{n=1}^{\infty} a_n$ converges and if $1 > a_n > 0$ for all n, show that $\sum_{n=1}^{\infty} \ln(1 - a_n)$ converges. (*Hint*: First show that $|\ln(1 - a_n)| \le a_n/(1 - a_n)$.)

31. Nicole Oresme's Theorem Prove Nicole Oresme's Theorem that

$$1 + \frac{1}{2} \cdot 2 + \frac{1}{4} \cdot 3 + \cdots + \frac{n}{2^{n-1}} + \cdots = 4.$$

(*Hint*: differentiate both sides of the equation $1/(1 - x) = 1 + \sum_{n=1}^{\infty} x^n$.)

32. **(a)** Term-by-term differentiation Show that

$$\sum_{n=1}^{\infty} \frac{n(n+1)}{x^n} = \frac{2x^2}{(x-1)^3}$$

for $|x| > 1$ by differentiating the identity

$$\sum_{n=1}^{\infty} x^{n+1} = \frac{x^2}{1-x}$$

twice, multiplying the result by x, and then replacing x by $1/x$.

(b) Use part (a) to find the real solution greater than 1 of the equation

$$x = \sum_{n=1}^{\infty} \frac{n(n+1)}{x^n}.$$

33. Summing exponential powers Use the integral test to show that

$$\sum_{n=0}^{\infty} e^{-n^2}$$

converges.

34. Writing to Learn If $\sum_{n=1}^{\infty} a_n$ is a convergent series of positive numbers, can anything be said about the convergence of $\sum_{n=1}^{\infty} \ln(1 + a_n)$? Give reasons for your answer.

35. Quality control

(a) Differentiate the series

$$\frac{1}{1-x} = 1 + x + x^2 + \cdots + x^n + \cdots$$

to obtain a series for $1/(1 - x)^2$.

(b) Rolling dice In one throw of two dice, the probability of getting a roll of 7 is $p = 1/6$. If you throw the dice repeatedly, the probability that a 7 will appear for the first time at the nth throw is $q^{n-1}p$, where $q = 1 - p = 5/6$. The expected number of throws until a 7 first appears is $\sum_{n=1}^{\infty} nq^{n-1}p$. Find the sum of this series.

(c) As an engineer applying statistical control to an industrial operation, you inspect items taken at random from the assembly line. You classify each sampled item as either "good" or "bad." If the probability of an item's being good is p and of an item's being bad is $q = 1 - p$, the probability that the first bad item found is the nth one inspected is $p^{n-1}q$. The average number inspected up to and including the first bad item found is $\sum_{n=1}^{\infty} np^{n-1}q$. Evaluate this sum, assuming $0 < p < 1$.

36. Expected value Suppose that a random variable X may assume the values 1, 2, 3, . . . , with probabilities $p_1, p_2, p_3, \ldots$, where p_k is the probability that X equals k ($k = 1, 2, 3, \ldots$). Suppose also that $p_k \ge 0$ and $\sum_{k=1}^{\infty} p_k = 1$. The **expected value** of X, denoted by $E(X)$, is the number $\sum_{k=1}^{\infty} kp_k$, provided the series converges. In each of the following cases, show that $\sum_{k=1}^{\infty} p_k = 1$ and find $E(X)$ if it exists. (*Hint*: See Exercise 35.)

(a) $p_k = 2^{-k}$ **(b)** $p_k = \dfrac{5^{k-1}}{6^k}$

(c) $p_k = \dfrac{1}{k(k+1)} = \dfrac{1}{k} - \dfrac{1}{k+1}$

37. Safe and effective dosage The concentration in the blood resulting from a single dose of a drug normally decreases with time as the drug is eliminated from the body. Doses may therefore need to be repeated periodically to keep the concentration from dropping below some particular level. One model for the effect of repeated doses gives the residual concentration just before the $(n + 1)$st dose as

$$R_n = C_0e^{-kt_0} + C_0e^{-2kt_0} + \cdots + C_0e^{-nkt_0},$$

where C_0 = the change in concentration achievable by a single dose (milligrams per milliliter), k = the *elimination constant* (per hour), and t_0 = time between doses (hours). See the accompanying figure.

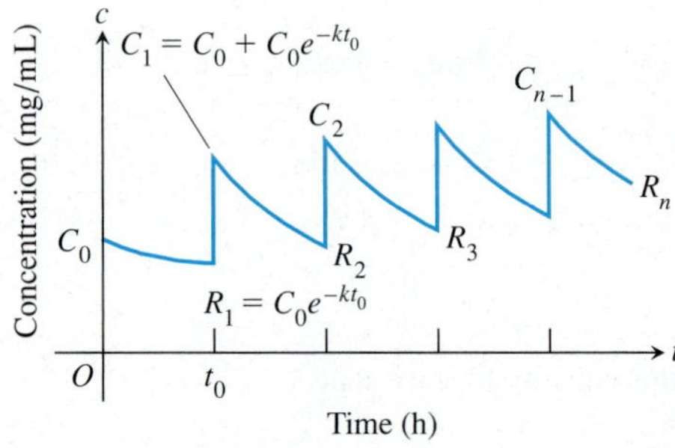

(a) Write R_n in closed form as a single fraction and find $R = \lim_{n\to\infty} R_n$.

(b) Calculate R_1 and R_{10} for $C_0 = 1$ mg/mL, $k = 0.1\ \text{h}^{-1}$, and $t_0 = 10$ h. How good an estimate of R is R_{10}?

(c) If $k = 0.01\ \text{h}^{-1}$ and $t_0 = 10$ h, find the smallest n such that $R_n > (1/2)R$.

(*Source*: *Prescribing Safe and Effective Dosage* by B. Horelick and S. Koont (Lexington, MA: COMAP, Inc., 1979).)

38. Time between drug doses (continuation of Exercise 37) If a drug is known to be ineffective below a concentration C_L and harmful above some higher concentration C_H, one needs to find values of

C_0 and t_0 that will produce a concentration that is safe (not above C_H) but effective (not below C_L). See the accompanying figure.

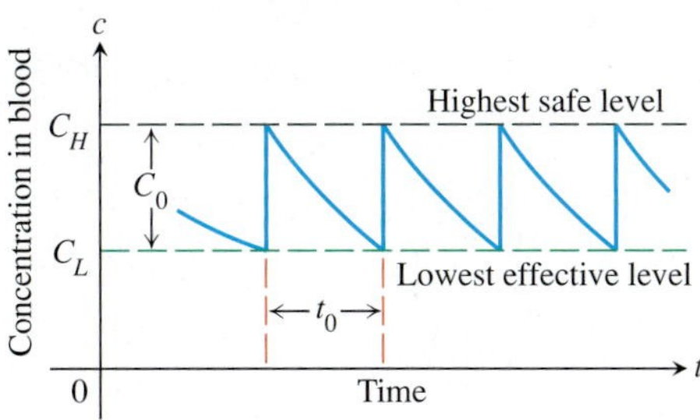

We therefore want to find values for C_0 and t_0 for which

$$R = C_L \qquad \text{and} \qquad C_0 + R = C_H.$$

Thus, $C_0 = C_H - C_L$. When these values are substituted in the equation for R obtained in part (a) of Exercise 37, the resulting equation simplifies to

$$t_0 = \frac{1}{k} \ln \frac{C_H}{C_L}.$$

To reach an effective level rapidly, one might administer a "loading" dose that would produce a concentration of C_H milligrams per milliliter. This could be followed every t_0 hours by a dose that raises the concentration by $C_0 = C_H - C_L$ milligrams per milliliter.

(a) Verify the preceding equation for t_0.

(b) If $k = 0.05\ \text{h}^{-1}$ and the highest safe concentration is e times the lowest effective concentration, find the length of time between doses that will assure safe and effective concentrations.

(c) *Writing to Learn* Given $C_H = 2$ mg/mL, $C_L = 0.5$ mg/mL, and $k = 0.02\ \text{h}^{-1}$, determine a scheme for administering the drug.

(d) Suppose that $k = 0.2\ \text{h}^{-1}$ and that the smallest effective concentration is 0.03 mg/mL. A single dose that produces a concentration of 0.1 mg/mL is administered. About how long will the drug remain effective?

39. *An infinite product* The infinite product

$$\prod_{n=1}^{\infty} (1 + a_n) = (1 + a_1)(1 + a_2)(1 + a_3) \cdots$$

is said to converge if the series

$$\sum_{n=1}^{\infty} \ln (1 + a_n),$$

obtained by taking the natural logarithm of the product, converges. Prove that the product converges if $a_n > -1$ for every n and if $\sum_{n=1}^{\infty} |a_n|$ converges. (*Hint*: Show that

$$|\ln (1 + a_n)| \le \frac{|a_n|}{1 - |a_n|} \le 2|a_n|$$

when $|a_n| < 1/2$.)

40. *Extended logarithmic p-series* If p is a constant, show that the series

$$1 + \sum_{n=3}^{\infty} \frac{1}{n \cdot \ln n \cdot [\ln (\ln n)]^p}$$

(a) converges if $p > 1$

(b) diverges if $p \le 1$.

In general, if $f_1(x) = x, f_{n+1}(x) = \ln (f_n(x))$ and n takes on the values 1, 2, 3, . . . , we find that $f_2(x) = \ln x, f_3(x) = \ln (\ln x)$, and so on. If $f_n(a) > 1$, then

$$\int_a^{\infty} \frac{dx}{f_1(x) f_2(x) \cdots f_n(x) (f_{n+1}(x))^p}$$

converges if $p > 1$ and diverges if $p \le 1$.

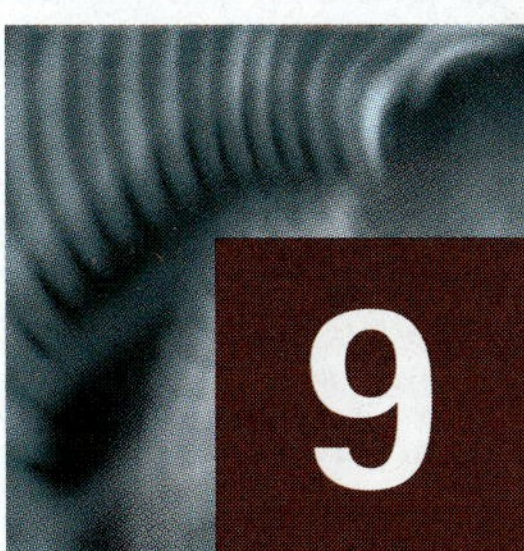

9 Vectors in the Plane and Polar Functions

OVERVIEW When a body travels in the xy-plane, the parametric equations $x = f(t)$ and $y = g(t)$ can be used to model the body's motion and path. In this chapter, we introduce the *vector* form of parametric equations, which allows us to track the positions of moving bodies with vectors, calculate the directions and magnitudes of their velocities and accelerations, and predict the effects of the forces we see working on them.

One of the principal applications of vector functions is the analysis of motion in space. Planetary motion is best described with polar coordinates (another of Newton's inventions, although Jakob Bernoulli usually gets the credit because he published first), so we investigate curves, derivatives, and integrals in this new coordinate system.

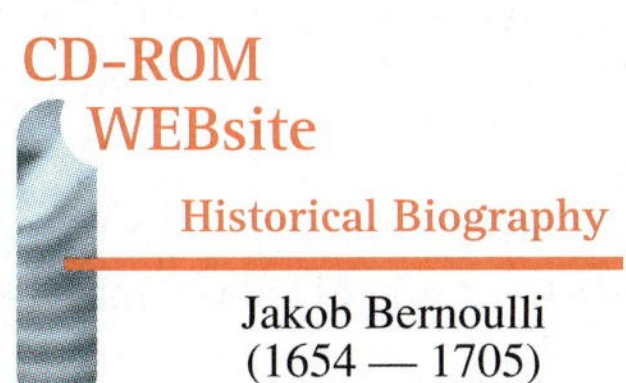

CD-ROM
WEBsite
Historical Biography
Jakob Bernoulli
(1654 — 1705)

9.1 Vectors in the Plane

Component Form • Zero Vector and Unit Vectors • Vector Algebra Operations • Standard Unit Vectors • Length and Direction • Tangents and Normals

Some of the things we measure are determined by their magnitudes. To record mass, length, or time, for example, we need only write down a number and name an appropriate unit of measure. These are *scalar quantities*, and the associated real numbers are **scalars.** We need more information to describe a force, displacement, or velocity. To describe a force, we need to record the direction in which it acts as well as how large it is. To describe a body's displacement, we have to say in what direction it moved as well as how far. To describe a body's velocity, we have to know where the body is headed as well as how fast it is going.

Component Form

A quantity such as force, displacement, or velocity is represented by a **directed line segment** (Figure 9.1). The arrow points in the direction of the action and its length gives the magnitude of the action in terms of a suitably chosen unit. For example, a force vector points in the direction in which it is applied and its length is a measure of its strength; a velocity vector points in the direction of motion and its length is the speed of the moving object. (We say more about force and velocity in Sections 9.3 and 9.4.)

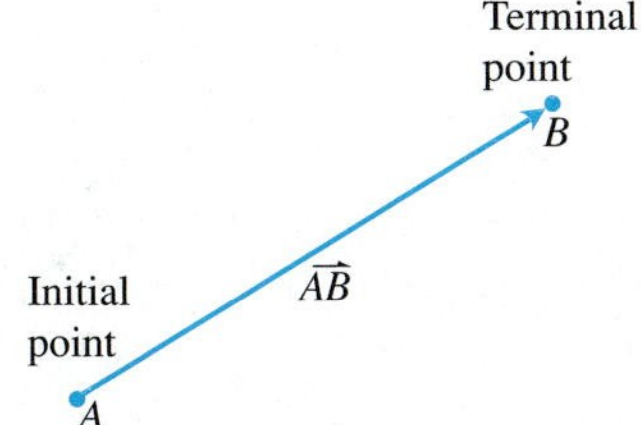

FIGURE 9.1 The directed line segment $\overrightarrow{AB}$.

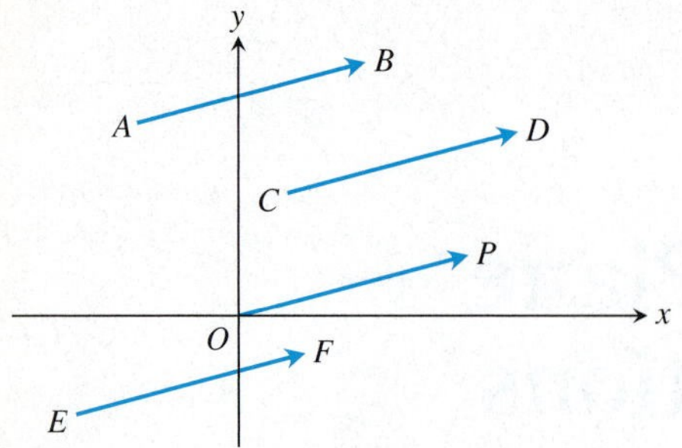

FIGURE 9.2 The four arrows (directed line segments) shown here have the same length and direction. They therefore represent the same vector, and we write $\overrightarrow{AB} = \overrightarrow{CD} = \overrightarrow{OP} = \overrightarrow{EF}$.

The directed line segment $\overrightarrow{AB}$ has **initial point** A and **terminal point** B; its **length** is denoted by $|\overrightarrow{AB}|$. Directed line segments that have the same length and direction are **equivalent.**

Definitions Vector, Equal Vectors

A **vector** in the plane is a directed line segment. Two vectors are **equal** (or **the same**) if they have the same length and direction.

Thus, the arrows we use when we draw vectors are understood to represent the same vector if they have the same length, are parallel, and point in the same direction (Figure 9.2).

In textbooks, vectors are usually written in lowercase, boldface letters, for example **u**, **v**, and **w**. Sometimes we use uppercase boldface letters, such as **F**, to denote a force vector. In handwritten form, it is customary to draw small arrows above the letters, for example $\vec{u}$, $\vec{v}$, $\vec{w}$, and $\vec{F}$.

Example 1 Showing Vectors are Equal

Let $A = (0, 0)$, $B = (3, 4)$, $C = (-4, 2)$, and $D = (-1, 6)$. Show that the vectors $\mathbf{u} = \overrightarrow{AB}$ and $\mathbf{v} = \overrightarrow{CD}$ are equal.

CD-ROM WEBsite

Solution We need to show that **u** and **v** have the same length and direction (Figure 9.3). We use the distance formula to find their lengths.

$$|\mathbf{u}| = |\overrightarrow{AB}| = \sqrt{(3-0)^2 + (4-0)^2} = 5$$
$$|\mathbf{v}| = |\overrightarrow{CD}| = \sqrt{(-1-(-4))^2 + (6-2)^2} = 5$$

Next we calculate the slopes of the two line segments.

$$\text{Slope of } \overrightarrow{AB} = \frac{4-0}{3-0} = \frac{4}{3}, \qquad \text{slope of } \overrightarrow{CD} = \frac{6-2}{-1-(-4)} = \frac{4}{3}$$

The line segments have the same direction because they are parallel and directed toward the upper right. Therefore, $\mathbf{u} = \mathbf{v}$ because they have the same length and direction.

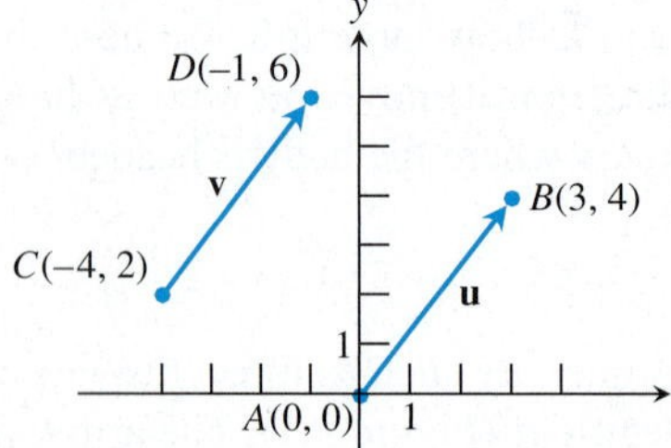

FIGURE 9.3 Two equal vectors. They have the same length and direction. (Example 1)

Let $\mathbf{v} = \overrightarrow{PQ}$. There is one directed line segment equivalent to $\overrightarrow{PQ}$ whose initial point is the origin (Figure 9.4). It is the representative of **v** in **standard position** and is the vector we normally use to represent **v**.

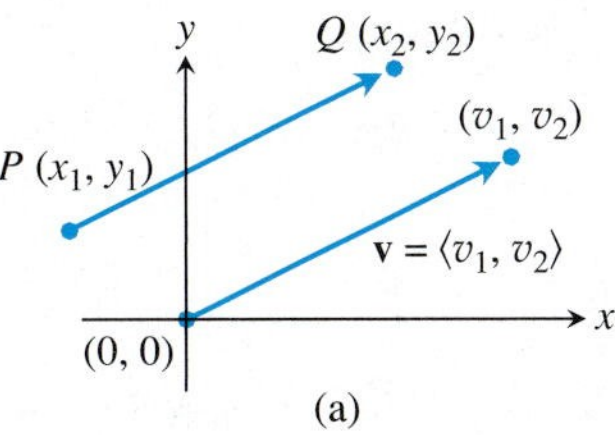

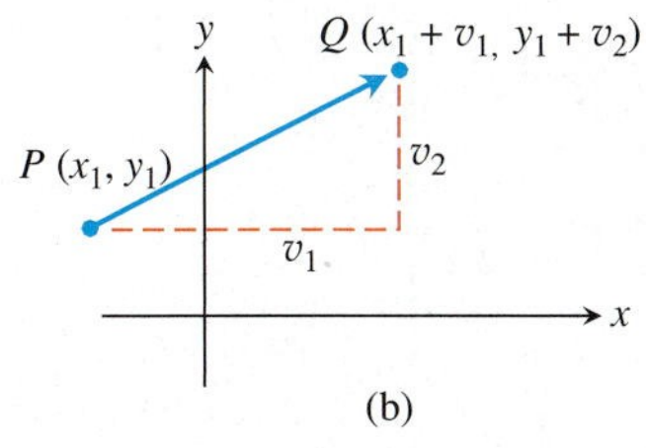

FIGURE 9.4 (a) The standard position of a vector is where the initial point is the origin. (b) The coordinates of Q satisfy $x_2 = x_1 + v_1$ and $y_2 = y_1 + v_2$.

CD-ROM
WEBsite
Historical Biography
William Rowan Hamilton
(1805 — 1865)

Definition Component Form of a Vector

If **v** is a vector in the plane equal to the vector with initial point (0, 0) and terminal point (v_1, v_2), then the **component form** of **v** is

$$\mathbf{v} = \langle v_1, v_2 \rangle.$$

Thus, a vector in the plane is also an ordered pair $\langle v_1, v_2 \rangle$ of real numbers. The numbers v_1 and v_2 are the **components** of **v**. The vector $\langle v_1, v_2 \rangle$ is called the **position vector** of the point (v_1, v_2).

Observe that if $\mathbf{v} = \langle v_1, v_2 \rangle$ is represented by the directed line segment $\overrightarrow{PQ}$, where the initial point is $P(x_1, y_1)$ and the terminal point is $Q(x_2, y_2)$, then $x_1 + v_1 = x_2$ and $y_1 + v_2 = y_2$ (Figure 9.4) so that $v_1 = x_2 - x_1$ and $v_2 = y_2 - y_1$ are the components of $\overrightarrow{PQ}$. In summary,

Given the points $P(x_1, y_1)$ and $Q(x_2, y_2)$, the position vector $\mathbf{v} = \langle v_1, v_2 \rangle$ equivalent to $\overrightarrow{PQ}$ is

$$\mathbf{v} = \langle x_2 - x_1, y_2 - y_1 \rangle.$$

Two vectors $\langle a, b \rangle$ and $\langle c, d \rangle$ are equal if and only if $a = c$ and $b = d$, so that $\langle v_1, v_2 \rangle = \langle x_2 - x_1, y_2 - y_1 \rangle$.

The **magnitude** or **length** of the vector $\overrightarrow{PQ}$ is the length of any of its equivalent directed line segment representations. In particular, if $\mathbf{v} = \langle x_2 - x_1, y_2 - y_1 \rangle$ is the position vector for $\overrightarrow{PQ}$ (Figure 9.4), then the distance formula gives the magnitude or length of **v**, denoted by the symbol $|\mathbf{v}|$ or $\|\mathbf{v}\|$.

The **magnitude** or **length** of the vector $\mathbf{v} = \overrightarrow{PQ}$ is

$$|\mathbf{v}| = \sqrt{{v_1}^2 + {v_2}^2} = \sqrt{(x_2 - x_1)^2 + (y_2 - y_1)^2}$$

(Figure 9.4).

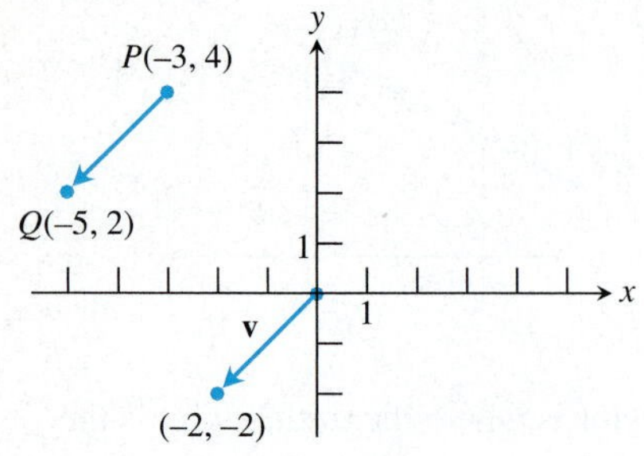

FIGURE 9.5 The vector $\overrightarrow{PQ}$ equals the position vector $v = \langle -2, -2 \rangle$. (Example 2)

Example 2 Finding the Component Form and Length of a Vector

Find the **(a)** component form and **(b)** length of the vector with initial point $P = (-3, 4)$ and terminal point $Q = (-5, 2)$.

Solution

(a) The position vector **v** representing $\overrightarrow{PQ}$ has components $v_1 = x_2 - x_1 = (-5) - (-3) = -2$ and $v_2 = y_2 - y_1 = 2 - 4 = -2$ (Figure 9.5). The component form of $\overrightarrow{PQ}$ is

$$\mathbf{v} = \langle -2, -2 \rangle.$$

(b) The length of $\mathbf{v} = \overrightarrow{PQ}$ is

$$|\mathbf{v}| = \sqrt{(-2)^2 + (-2)^2} = 2\sqrt{2}.$$

Example 3 Force Moving a Cart

A small cart is being pulled along a smooth horizontal floor with a 20 lb force **F** making a 45° angle to the floor (Figure 9.6). What is the effective force moving the cart forward?

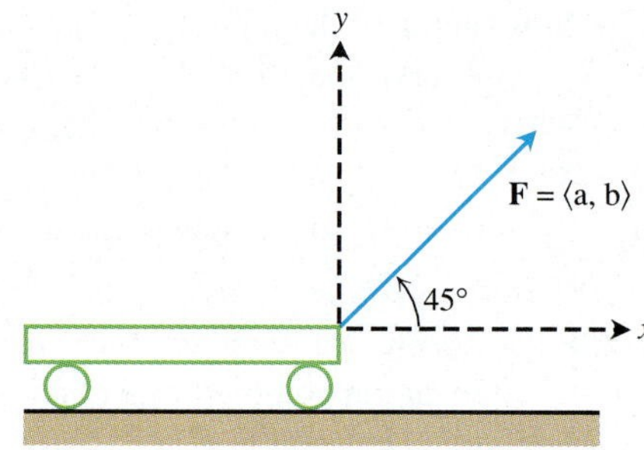

FIGURE 9.6 The force pulling the cart forward is represented by the vector **F** of length 20 (pounds) making an angle of 45° with the horizontal ground (positive x-axis). (Example 3)

Solution The effective force is the horizontal component of $\mathbf{F} = \langle a, b \rangle$, given by

$$a = |\mathbf{F}| \cos 45° = (20)\left(\frac{\sqrt{2}}{2}\right) \approx 14.14 \text{ lb}.$$

Zero Vector and Unit Vectors

The only vector with length 0 is the **zero vector**

$$\mathbf{0} = \langle 0, 0 \rangle.$$

The zero vector is also the only vector with no specific direction.

Any vector **v** of length 1 is a **unit vector.** If $\mathbf{v} = \langle v_1, v_2 \rangle$ makes an angle θ with the positive x-axis, then

$$v_1 = |\mathbf{v}| \cos \theta = \cos \theta \qquad |\mathbf{v}| = 1$$
$$v_2 = |\mathbf{v}| \sin \theta = \sin \theta$$

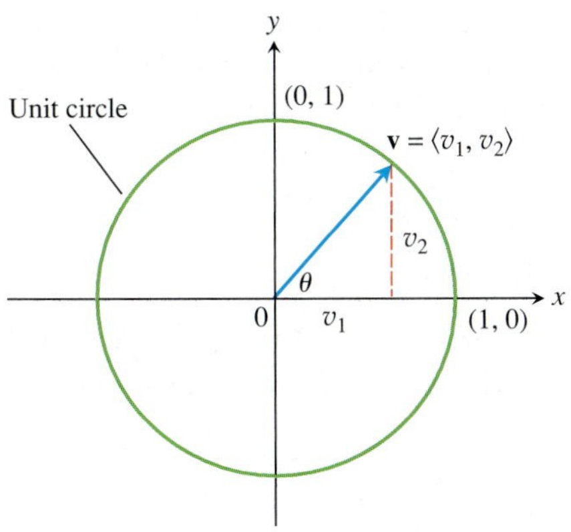

FIGURE 9.7 The unit vector $\mathbf{v} = \langle v_1, v_2 \rangle$ has length 1, so $v_1 = \cos\theta$ and $v_2 = \sin\theta$, where θ is the angle $\mathbf{v}$ makes with the positive x-axis. As θ varies from 0 to 2π, the terminal point of $\mathbf{v}$ traces the unit circle.

(Figure 9.7). In summary,

> A unit vector $\mathbf{v}$ in the plane having angle θ with the positive x-axis is represented by
>
> $$\mathbf{v} = \langle \cos\theta, \sin\theta \rangle.$$

As θ varies from 0 to 2π, the terminal point of a unit vector $\mathbf{v}$ traces the unit circle counterclockwise, taking into account all possible directions.

Vector Algebra Operations

Two principal operations involving vectors are *vector addition* and *scalar multiplication*.

> **Definitions** Vector Addition and Multiplication of a Vector by a Scalar
>
> Let $\mathbf{u} = \langle u_1, u_2 \rangle$, $\mathbf{v} = \langle v_1, v_2 \rangle$ be vectors with k a scalar (real number).
>
> **Addition:** $\mathbf{u} + \mathbf{v} = \langle u_1, u_2 \rangle + \langle v_1, v_2 \rangle = \langle u_1 + v_1, u_2 + v_2 \rangle$
>
> **Scalar multiplication:** $k\mathbf{u} = \langle ku_1, ku_2 \rangle$

The definition of vector addition is illustrated geometrically in Figure 9.8a, where the initial point of one vector is placed at the terminal point of the other. Another interpretation is shown in Figure 9.8b (called the **parallelogram law** of addition), where the sum, called the **resultant vector,** is the diagonal of the parallelogram. In physics, forces add vectorially as do velocities, accelerations, and so on.

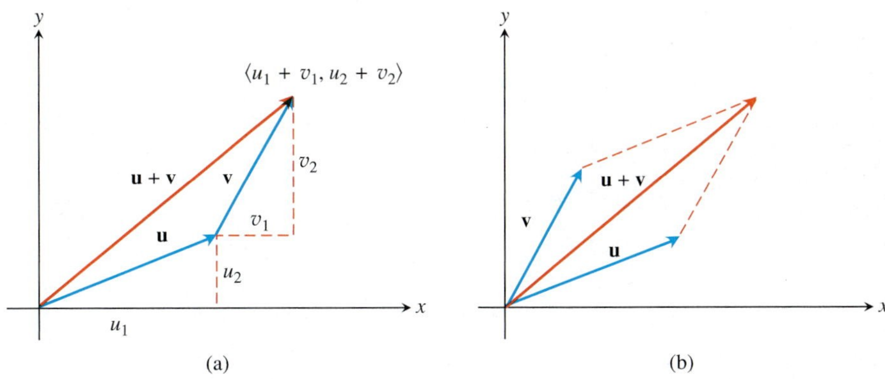

FIGURE 9.8 (a) Geometric interpretation of the vector sum. (b) The parallelogram law of vector addition.

A geometric interpretation of the product $k\mathbf{u}$ of the scalar k and vector $\mathbf{u}$ is displayed in Figure 9.9. First, if $k > 0$, then $k\mathbf{u}$ has the same direction as $\mathbf{u}$; if $k < 0$, then the direction of $k\mathbf{u}$ is opposite to that of $\mathbf{u}$. Comparing the lengths of $\mathbf{u} = \langle u_1, u_2 \rangle$ and $k\mathbf{u}$, we see that

$$\begin{aligned} |k\mathbf{u}| &= \sqrt{(ku_1)^2 + (ku_2)^2} = \sqrt{k^2(u_1{}^2 + u_2{}^2)} \\ &= \sqrt{k^2}\sqrt{u_1{}^2 + u_2{}^2} = |k||\mathbf{u}|. \end{aligned}$$

That is, the length of $k\mathbf{u}$ is the absolute value of the scalar k times the length of $\mathbf{u}$. In particular, the vector $(-1)\,\mathbf{u} = -\mathbf{u}$ has the same length as $\mathbf{u}$ but points in the opposite direction.

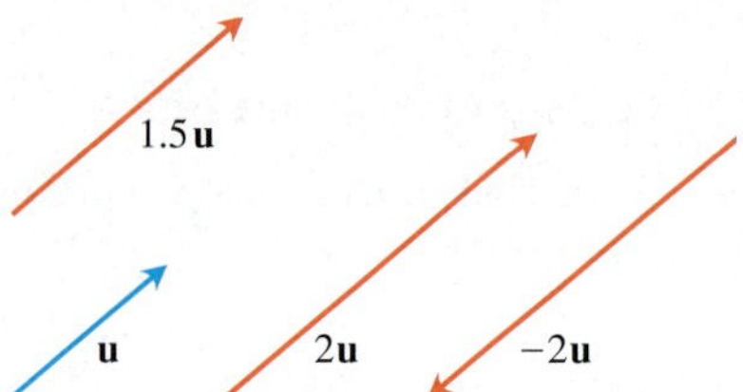

FIGURE 9.9 Scalar multiples of **u**.

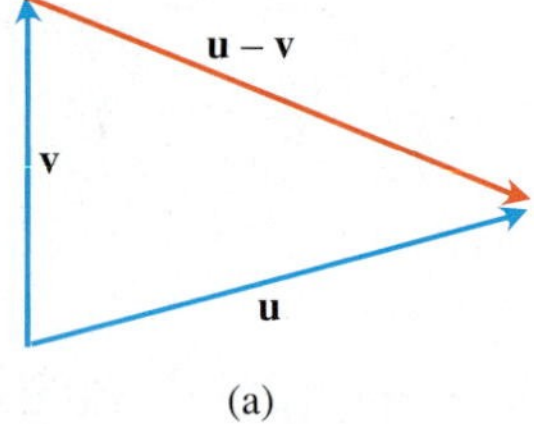

(a)

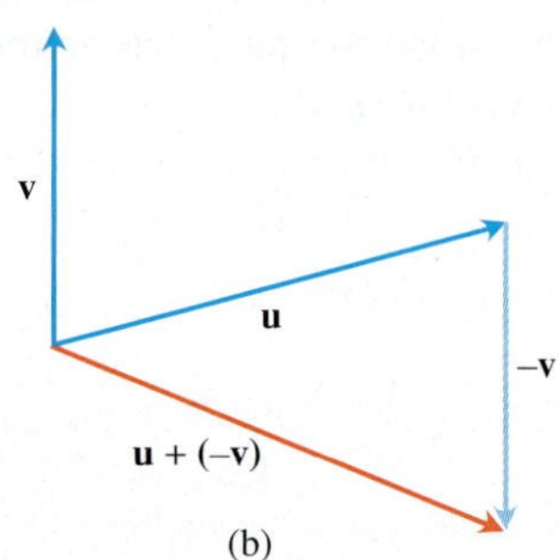

(b)

FIGURE 9.10 (a) The vector $\mathbf{u} - \mathbf{v}$, when added to $\mathbf{v}$, gives $\mathbf{u}$. (b) $\mathbf{u} - \mathbf{v} = \mathbf{u} + (-\mathbf{v})$.

By the **difference** $\mathbf{u} - \mathbf{v}$ of two vectors, we mean

$$\mathbf{u} - \mathbf{v} = \mathbf{u} + (-\mathbf{v}).$$

If $\mathbf{u} = \langle u_1, u_2 \rangle$ and $\mathbf{v} = \langle v_1, v_2 \rangle$, then

$$\mathbf{u} - \mathbf{v} = \langle u_1 - v_1, u_2 - v_2 \rangle.$$

Note that $(\mathbf{u} - \mathbf{v}) + \mathbf{v} = \mathbf{u}$, so adding the vector $(\mathbf{u} - \mathbf{v})$ to $\mathbf{v}$ gives $\mathbf{u}$ (Figure 9.10a). Figure 9.10b shows the difference $\mathbf{u} - \mathbf{v}$ as the sum $\mathbf{u} + (-\mathbf{v})$.

Example 4 Performing Operations on Vectors

Let $\mathbf{u} = \langle -1, 3 \rangle$ and $\mathbf{v} = \langle 4, 7 \rangle$. Find

(a) $2\mathbf{u} + 3\mathbf{v}$ **(b)** $\mathbf{u} - \mathbf{v}$ **(c)** $\left|\frac{1}{2}\mathbf{u}\right|$.

Solution

(a) $$\begin{aligned} 2\mathbf{u} + 3\mathbf{v} &= 2\langle -1, 3 \rangle + 3\langle 4, 7 \rangle \\ &= \langle 2(-1) + 3(4),\ 2(3) + 3(7) \rangle = \langle 10, 27 \rangle \end{aligned}$$

(b) $$\begin{aligned} \mathbf{u} - \mathbf{v} &= \langle -1, 3 \rangle - \langle 4, 7 \rangle \\ &= \langle -1 - 4,\ 3 - 7 \rangle = \langle -5, -4 \rangle \end{aligned}$$

(c) $$\left|\frac{1}{2}\mathbf{u}\right| = \left|\left\langle -\frac{1}{2}, \frac{3}{2} \right\rangle\right| = \sqrt{\left(-\frac{1}{2}\right)^2 + \left(\frac{3}{2}\right)^2} = \frac{1}{2}\sqrt{10}$$

Vector operations have many of the properties of ordinary arithmetic. These properties are readily verified using the definitions of vector addition and multiplication by a scalar.

Properties of Vector Operations

Let $\mathbf{u}, \mathbf{v}, \mathbf{w}$ be vectors and a, b be scalars.

1. $\mathbf{u} + \mathbf{v} = \mathbf{v} + \mathbf{u}$
2. $(\mathbf{u} + \mathbf{v}) + \mathbf{w} = \mathbf{u} + (\mathbf{v} + \mathbf{w})$
3. $\mathbf{u} + \mathbf{0} = \mathbf{u}$
4. $\mathbf{u} + (-\mathbf{u}) = \mathbf{0}$
5. $0\mathbf{u} = \mathbf{0}$
6. $1\mathbf{u} = \mathbf{u}$
7. $a(b\mathbf{u}) = (ab)\mathbf{u}$
8. $a(\mathbf{u} + \mathbf{v}) = a\mathbf{u} + a\mathbf{v}$
9. $(a + b)\mathbf{u} = a\mathbf{u} + b\mathbf{u}$

An important application of vectors occurs in navigation.

Example 5 Finding Ground Speed and Direction

A Boeing® 727® airplane, flying due east at 500 mph in still air, encounters a 70 mph tailwind acting in the direction 60° north of east. The airplane holds its compass heading due east but, because of the wind, acquires a new ground speed and direction. What are they?

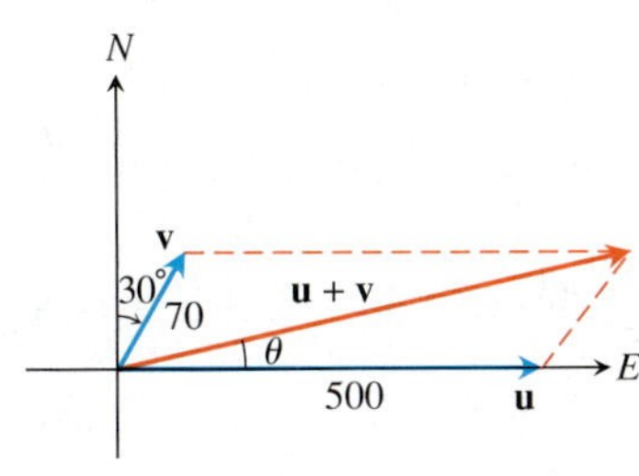

FIGURE 9.11 Vectors representing the velocities of the airplane and tailwind in Example 5.

Solution If $\mathbf{u}$ = the velocity of the airplane alone and $\mathbf{v}$ = the velocity of the tailwind, then $|\mathbf{u}| = 500$ and $|\mathbf{v}| = 70$ (Figure 9.11). We need to find the magnitude and direction of the *resultant vector* $\mathbf{u} + \mathbf{v}$. If we let the positive x-axis represent east and the positive y-axis represent north, then the component forms of $\mathbf{u}$ and $\mathbf{v}$ are

$$\mathbf{u} = \langle 500, 0 \rangle \quad \text{and} \quad \mathbf{v} = \langle 70 \cos 60°, 70 \sin 60° \rangle = \langle 35, 35\sqrt{3} \rangle.$$

Therefore,

$$\mathbf{u} + \mathbf{v} = \langle 535, 35\sqrt{3} \rangle$$
$$|\mathbf{u} + \mathbf{v}| = \sqrt{535^2 + (35\sqrt{3})^2} \approx 538.4$$

and

$$\theta = \tan^{-1} \frac{35\sqrt{3}}{535} \approx 6.5°. \qquad \text{Fig. 9.11}$$

Interpret

The new ground speed of the airplane is about 538.4 mph, and its new direction is about 6.5° north of east.

Standard Unit Vectors

Any vector $\mathbf{v} = \langle a, b \rangle$ in the plane can be written as a *linear combination* of the two **standard unit vectors**

$$\mathbf{i} = \langle 1, 0 \rangle \quad \text{and} \quad \mathbf{j} = \langle 0, 1 \rangle$$

as follows:

$$\mathbf{v} = \langle a, b \rangle = \langle a, 0 \rangle + \langle 0, b \rangle = a\langle 1, 0 \rangle + b\langle 0, 1 \rangle$$
$$= a\mathbf{i} + b\mathbf{j}.$$

The vector $\mathbf{v}$ is a **linear combination** of the vectors $\mathbf{i}$ and $\mathbf{j}$; the scalar a is the **horizontal** or **i-component** of $\mathbf{v}$ and the scalar b is the **vertical** or **j-component** of $\mathbf{v}$. The **slope** of a nonvertical vector $\mathbf{v} = \langle a, b \rangle$ is the slope shared by the lines parallel to it. Thus, if $a \neq 0$, the vector $\mathbf{v}$ has slope b/a (Figure 9.12).

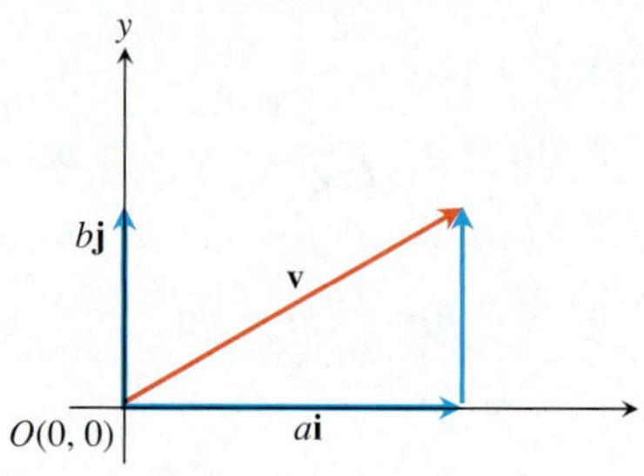

FIGURE 9.12 $\mathbf{v}$ is a linear combination of $\mathbf{i}$ and $\mathbf{j}$.

Example 6 Expressing Vectors as Linear Combinations of i and j

Let $P = (-1, 5)$ and $Q = (3, 2)$. Write the vector $\mathbf{v} = \overrightarrow{PQ}$ as a linear combination of $\mathbf{i}$ and $\mathbf{j}$ and find its slope.

Solution The component form of $\mathbf{v}$ is $\langle 3 - (-1), 2 - 5 \rangle = \langle 4, -3 \rangle$. Thus,

$$\mathbf{v} = \langle 4, -3 \rangle = 4\mathbf{i} + (-3)\mathbf{j} = 4\mathbf{i} - 3\mathbf{j}.$$

The slope of $\mathbf{v}$ is $-3/4$.

CD-ROM
WEBsite

Historical Biography

Hermann Grassmann
(1809 — 1877)

Length and Direction

In studying motion, we often want to know the direction an object is headed and how fast it is going.

Example 7 Expressing Velocity as Speed Times Direction

If $\mathbf{v} = 3\mathbf{i} - 4\mathbf{j}$ is a velocity vector, express $\mathbf{v}$ as a product of its speed times a unit vector in the direction of motion.

Solution Speed is the magnitude (length) of $\mathbf{v}$:

$$|\mathbf{v}| = \sqrt{(3)^2 + (-4)^2} = \sqrt{9 + 16} = 5.$$

The vector $\mathbf{v}/|\mathbf{v}|$ has the same direction as $\mathbf{v}$:

$$\frac{\mathbf{v}}{|\mathbf{v}|} = \frac{3\mathbf{i} - 4\mathbf{j}}{5} = \frac{3}{5}\mathbf{i} - \frac{4}{5}\mathbf{j}.$$

Moreover, $\mathbf{v}/|\mathbf{v}|$ is a unit vector:

$$\left|\frac{\mathbf{v}}{|\mathbf{v}|}\right| = \sqrt{\left(\frac{3}{5}\right)^2 + \left(-\frac{4}{5}\right)^2} = \sqrt{\frac{9}{25} + \frac{6}{25}} = 1.$$

Thus,

$$\mathbf{v} = 3\mathbf{i} - 4\mathbf{j} = \underbrace{5}_{\text{Length (speed)}}\underbrace{\left(\frac{3}{5}\mathbf{i} - \frac{4}{5}\mathbf{j}\right)}_{\text{Direction of motion}}.$$

Generally, if $\mathbf{v} \neq \mathbf{0}$, then its length $|\mathbf{v}|$ is not zero and

$$\left|\frac{1}{|\mathbf{v}|}\mathbf{v}\right| = \frac{1}{|\mathbf{v}|}|\mathbf{v}| = 1.$$

That is, $\mathbf{v}/|\mathbf{v}|$ is a unit vector in the direction of $\mathbf{v}$. We can therefore express $\mathbf{v}$ in terms of its two important features, length and direction, by writing $\mathbf{v} = |\mathbf{v}|(\mathbf{v}/|\mathbf{v}|)$.

If $\mathbf{v} \neq \mathbf{0}$, then

1. $\dfrac{\mathbf{v}}{|\mathbf{v}|}$ is a unit vector in the direction of $\mathbf{v}$;

2. the equation $\mathbf{v} = |\mathbf{v}|\dfrac{\mathbf{v}}{|\mathbf{v}|}$ expresses $\mathbf{v}$ in terms of its length and direction.

The unit vector $\mathbf{v}/|\mathbf{v}|$ is called the **direction** of $\mathbf{v}$. Thus, $\mathbf{v} = 5((3/5)\mathbf{i} - (4/5)\mathbf{j})$ expresses the velocity vector in Example 7 as a *product* of its length and direction.

Tangents and Normals

When an object is moving along a path in the plane (or in space), its velocity is a vector tangent to the path. Moreover, if the object is speeding up or slowing down, forces are acting in the tangent direction and perpendicular (or normal) to it. (We investigate motion along a path in the plane in Section 9.3.)

A vector is **tangent** or **normal** to a curve at a point P if it is parallel or normal, respectively, to the line that is tangent to the curve at P. Example 8 shows how to find such vectors for a differentiable curve $y = f(x)$ in the plane.

Example 8 Finding Vectors Tangent and Normal to a Curve

An object is moving along the curve

$$y = \frac{x^3}{2} + \frac{1}{2}.$$

Find unit vectors tangent and normal to the curve at the point $(1, 1)$.

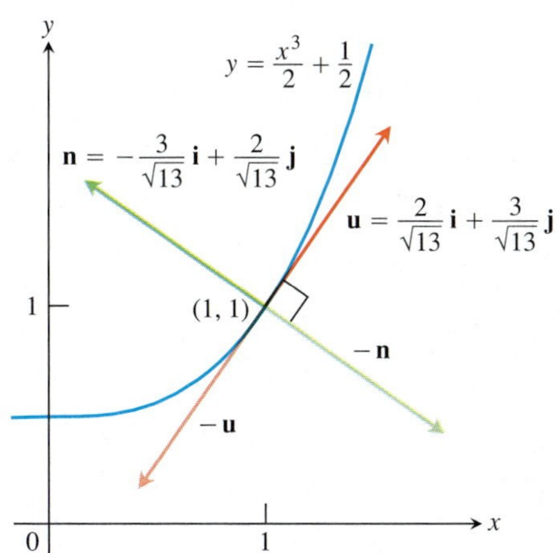

FIGURE 9.13 The unit tangent and normal vectors at the point $(1, 1)$ on the curve $y = (x^3/2) + 1/2$. (Example 8)

Solution We find the unit vectors that are parallel and normal to the curve's tangent line at $(1, 1)$ (Figure 9.13).

The slope of the line tangent to the curve at $(1, 1)$ is

$$y' = \left.\frac{3x^2}{2}\right|_{x=1} = \frac{3}{2}.$$

We look for a unit vector with this slope. The vector $\mathbf{v} = 2\mathbf{i} + 3\mathbf{j}$ has slope $3/2$, as does every nonzero multiple of $\mathbf{v}$. To find a multiple of $\mathbf{v}$ that is a unit vector, we divide $\mathbf{v}$ by

$$|\mathbf{v}| = \sqrt{2^2 + 3^2} = \sqrt{13},$$

obtaining

$$\mathbf{u} = \frac{\mathbf{v}}{|\mathbf{v}|} = \frac{2}{\sqrt{13}}\mathbf{i} + \frac{3}{\sqrt{13}}\mathbf{j}.$$

The vector **u** is tangent to the curve at (1, 1) because it has the same direction as **v**. Of course,

$$-\mathbf{u} = -\frac{2}{\sqrt{13}}\mathbf{i} - \frac{3}{\sqrt{13}}\mathbf{j},$$

which points in the opposite direction, is also tangent to the curve at (1, 1). Without some additional requirement (such as specifying the direction of motion), there is no reason to prefer one of these vectors to the other.

To find unit vectors normal to the curve at (1, 1), we look for unit vectors whose slopes are the negative reciprocal of the slope of **u**. This is quickly done by interchanging the scalar components of **u** and changing the sign of one of them. We obtain

$$\mathbf{n} = -\frac{3}{\sqrt{13}}\mathbf{i} + \frac{2}{\sqrt{13}}\mathbf{j} \quad \text{and} \quad -\mathbf{n} = \frac{3}{\sqrt{13}}\mathbf{i} - \frac{2}{\sqrt{13}}\mathbf{j}.$$

Again, either one will do. The vectors have opposite directions but both are normal to the curve at (1, 1). (See Figure 9.13.)

EXERCISES 9.1

Component Form

In Exercises 1–8, let $\mathbf{u} = \langle 3, -2\rangle$ and $\mathbf{v} = \langle -2, 5\rangle$. Find the **(a)** component form and **(b)** magnitude (length) of the vector.

1. $3\mathbf{u}$

2. $-2\mathbf{v}$

3. $\mathbf{u} + \mathbf{v}$

4. $\mathbf{u} - \mathbf{v}$

5. $2\mathbf{u} - 3\mathbf{v}$

6. $-2\mathbf{u} + 5\mathbf{v}$

7. $\frac{3}{5}\mathbf{u} + \frac{4}{5}\mathbf{v}$

8. $-\frac{5}{13}\mathbf{u} + \frac{12}{13}\mathbf{v}$

In Exercises 9–16, find the component form of the vector.

9. The vector $\overrightarrow{PQ}$, where $P = (1, 3)$ and $Q = (2, -1)$

10. The vector $\overrightarrow{OP}$ where O is the origin and P is the midpoint of segment RS, where $R = (2, -1)$ and $S = (-4, 3)$

11. The vector from the point $A = (2, 3)$ to the origin

12. The sum of $\overrightarrow{AB}$ and $\overrightarrow{CD}$, where $A = (1, -1)$, $B = (2, 0)$, $C = (-1, 3)$, and $D = (-2, 2)$

13. The unit vector that makes an angle $\theta = 2\pi/3$ with the positive x-axis

14. The unit vector that makes an angle $\theta = -3\pi/4$ with the positive x-axis

15. The unit vector obtained by rotating the vector $\langle 0, 1\rangle$ 120° counterclockwise about the origin

16. The unit vector obtained by rotating the vector $\langle 1, 0\rangle$ 135° counterclockwise about the origin

Geometry and Calculation

In Exercises 17 and 18, copy vectors **u, v,** and **w** head to tail as needed to sketch the indicated vector.

17.

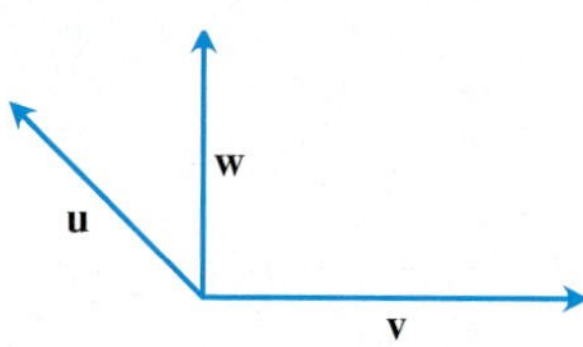

(a) $\mathbf{u} + \mathbf{v}$ **(b)** $\mathbf{u} + \mathbf{v} + \mathbf{w}$

(c) $\mathbf{u} - \mathbf{v}$ **(d)** $\mathbf{u} - \mathbf{w}$

18.

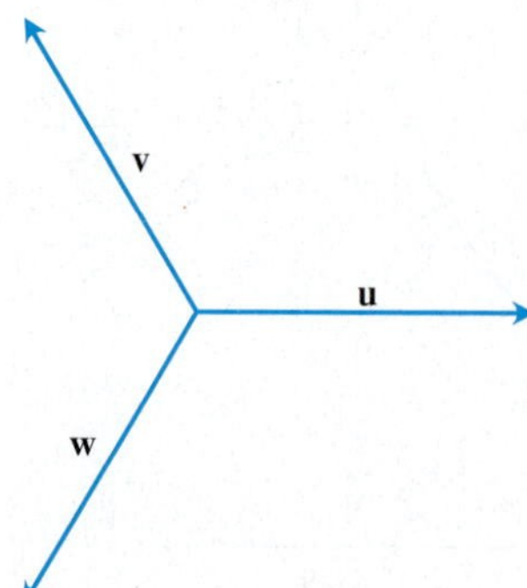

(a) $\mathbf{u} - \mathbf{v}$ **(b)** $\mathbf{u} - \mathbf{v} + \mathbf{w}$

(c) $2\mathbf{u} - \mathbf{v}$ **(d)** $\mathbf{u} + \mathbf{v} + \mathbf{w}$

Using Linear Combinations

Express the vectors in Exercises 19–24 in the form $a\mathbf{i} + b\mathbf{j}$ and sketch them as arrows in the coordinate plane beginning at the origin.

19. $\overrightarrow{P_1P_2}$ if P_1 is the point (5, 7) and P_2 is the point (2, 9)

20. $\overrightarrow{P_1P_2}$ if P_1 is the point (1, 2) and P_2 is the point (−3, 5)

21. $\overrightarrow{AB}$ if A is the point (−5, 3) and B is the point (−10, 8)

22. $\overrightarrow{AB}$ if A is the point (−7, −8) and B is the point (6, 11)

23. $\overrightarrow{P_1P_2}$ if P_1 is the point (1, 3) and P_2 is the point (2, −1)

24. $\overrightarrow{P_3P_4}$ if P_3 is the point (1, 3) and P_4 is the midpoint of the line segment P_1P_2 joining $P_1(2, -1)$ and $P_2(-4, 3)$

Unit Vectors

Sketch the vectors in Exercises 25–28 and express each vector in the form $a\mathbf{i} + b\mathbf{j}$.

25. The unit vectors $\mathbf{u} = (\cos\theta)\mathbf{i} + (\sin\theta)\mathbf{j}$ for $\theta = \pi/6$ and $\theta = 2\pi/3$. Include the circle $x^2 + y^2 = 1$ in your sketch.

26. The unit vectors $\mathbf{u} = (\cos\theta)\mathbf{i} + (\sin\theta)\mathbf{j}$ for $\theta = -\pi/4$ and $\theta = -3\pi/4$. Include the circle $x^2 + y^2 = 1$ in your sketch.

27. The unit vector obtained by rotating $\mathbf{j}$ counterclockwise $3\pi/4$ rad about the origin

28. The unit vector obtained by rotating $\mathbf{j}$ clockwise $2\pi/3$ rad about the origin

In Exercises 29–32, find a unit vector in the direction of the given vector.

29. $\langle 3, 4\rangle$ **30.** $\langle 4, -3\rangle$

31. $\langle -15, 8\rangle$ **32.** $\langle -5, -2\rangle$

For the vectors in Exercises 33 and 34, find unit vectors $\mathbf{u} = (\cos\theta)\mathbf{i} + (\sin\theta)\mathbf{j}$ in the same direction.

33. $6\mathbf{i} - 8\mathbf{j}$ **34.** $-\mathbf{i} + 3\mathbf{j}$

Length and Direction

In Exercises 35 and 36, express each vector as a product of its length and direction.

35. $5\mathbf{i} + 12\mathbf{j}$ **36.** $2\mathbf{i} - 3\mathbf{j}$

37. Find the unit vectors that are parallel to the vector $3\mathbf{i} - 4\mathbf{j}$ (two vectors in all).

38. Find a vector of length 2 whose direction is the opposite of the direction of the vector $-\mathbf{i} + 2\mathbf{j}$. How many such vectors are there?

Tangent and Normal Vectors

In Exercises 39–42, find the unit vectors that are tangent and normal to the curve at the given point (four vectors in all). Then sketch the vectors and curve together.

39. $y = x^2$, (2, 4)

40. $x^2 + 2y^2 = 6$, (2, 1)

41. $y = \tan^{-1} x$, $(1, \pi/4)$

42. $y = \sum_{n=0}^{\infty} \frac{x^n}{n!}$, (0, 1)

In Exercises 43–46, find the unit vectors that are tangent and normal to the curve at the given point (four vectors in all).

43. $3x^2 + 8xy + 2y^2 - 3 = 0$, (1, 0)

44. $x^2 - 6xy + 8y^2 - 2x - 1 = 0$, (1, 1)

45. $y = \int_0^x \sqrt{3 + t^4}\, dt$, (0, 0)

46. $y = \int_e^x \ln(\ln t)\, dt$, $(e, 0)$

Theory and Applications

47. *Linear combination* Let $\mathbf{u} = 2\mathbf{i} + \mathbf{j}$, $\mathbf{v} = \mathbf{i} + \mathbf{j}$, and $\mathbf{w} = \mathbf{i} - \mathbf{j}$. Find scalars a and b such that $\mathbf{u} = a\mathbf{v} + b\mathbf{w}$.

48. *Linear combination* Let $\mathbf{u} = \mathbf{i} - 2\mathbf{j}$, $\mathbf{v} = 2\mathbf{i} + 3\mathbf{j}$, and $\mathbf{w} = \mathbf{i} + \mathbf{j}$. Write $\mathbf{u} = \mathbf{u}_1 + \mathbf{u}_2$, where $\mathbf{u}_1$ is parallel to $\mathbf{v}$ and $\mathbf{u}_2$ is parallel to $\mathbf{w}$. (See Exercise 47.)

49. *Force vector* You are pulling on a suitcase with a force $\mathbf{F}$ (pictured here) whose magnitude is $|\mathbf{F}| = 10$ lb. Find the $\mathbf{i}$- and $\mathbf{j}$-components of $\mathbf{F}$.

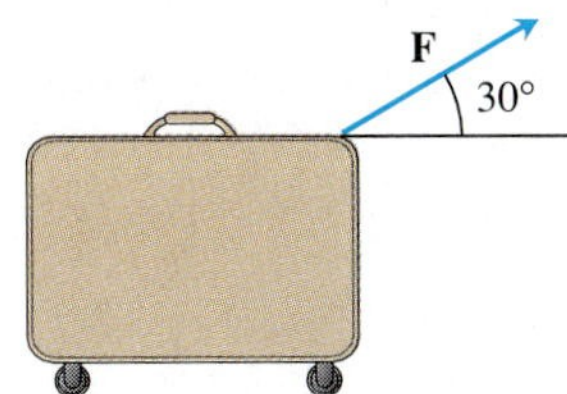

50. *Force vector* A kite string exerts a 12 lb pull ($|\mathbf{F}| = 12$) on a kite and makes a 45° angle with the horizontal. Find the horizontal and vertical components of $\mathbf{F}$.

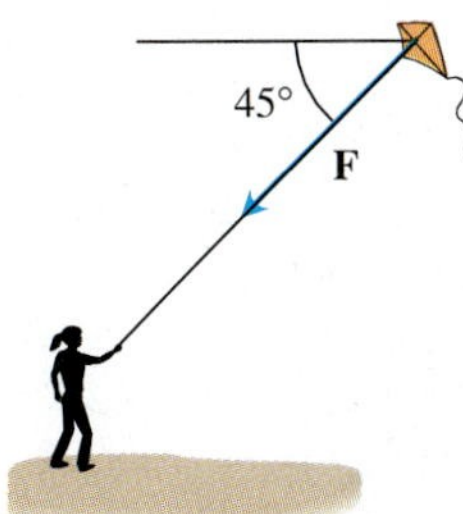

51. *Velocity* An airplane is flying in the direction 25° west of north at 800 km/h. Find the component form of the velocity of the airplane, assuming that the positive x-axis represents due east and the positive y-axis represents due north.

52. *Velocity* An airplane is flying in the direction 10° east of south at 600 km/h. Find the component form of the velocity of the airplane, assuming that the positive x-axis represents due east and the positive y-axis represents due north.

53. *Location* A bird flies from its nest 5 km in the direction 60° north of east, where it stops to rest on a tree. It then flies 10 km in the direction due southeast and lands atop a telephone pole. Place an xy-coordinate system so that the origin is the bird's nest, the x-axis points east, and the y-axis points north.

(a) At what point is the tree located?

(b) At what point is the telephone pole?

54. *Location* A bird flies from its nest 7 km in the direction northeast, where it stops to rest on a tree. It then flies 8 km in the direction 30° south of west and lands atop a telephone pole. Place an xy-coordinate system so that the origin is the bird's nest, the x-axis points east, and the y-axis points north.

(a) At what point is the tree located?

(b) At what point is the telephone pole?

9.2 Dot Products

Angle Between Vectors • Laws of the Dot Product • Perpendicular (Orthogonal) Vectors • Vector Projections • Work • Writing a Vector as a Sum of Orthogonal Vectors

If a force $\mathbf{F}$ is applied to a particle moving along a path, we often need to know the magnitude of the force in the direction of motion. If $\mathbf{v}$ is parallel to the tangent line to the path at the point where $\mathbf{F}$ is applied, then we want the magnitude of $\mathbf{F}$ in the direction of $\mathbf{v}$. Figure 9.14 shows that the scalar quantity we seek is the length $|\mathbf{F}| \cos \theta$, where θ is the angle between the two vectors $\mathbf{F}$ and $\mathbf{v}$.

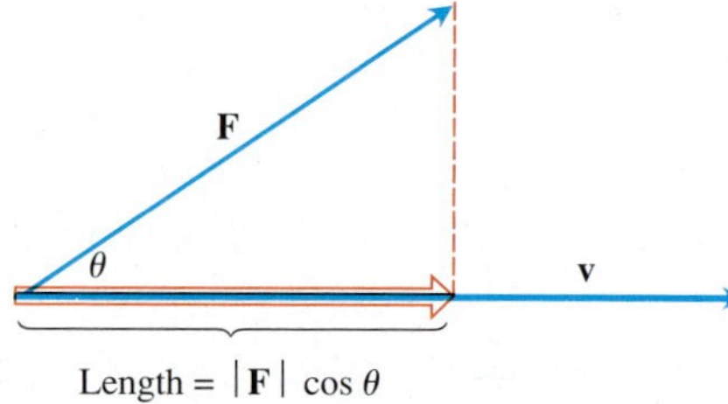

FIGURE 9.14 The magnitude of the force $\mathbf{F}$ in the direction of vector $\mathbf{v}$ is the length $|\mathbf{F}| \cos \theta$ of the projection of $\mathbf{F}$ onto $\mathbf{v}$.

In this section, you learn how to calculate easily the angle between two vectors directly from their components. A key part of the calculation is an expression called the *dot product*. Dot products are also called *scalar* products because the product results in a scalar, not a vector. After investigating the dot product, we apply it to finding the projection of one vector onto another (as displayed in Figure 9.14) and to finding the work done by a constant force acting through a displacement.

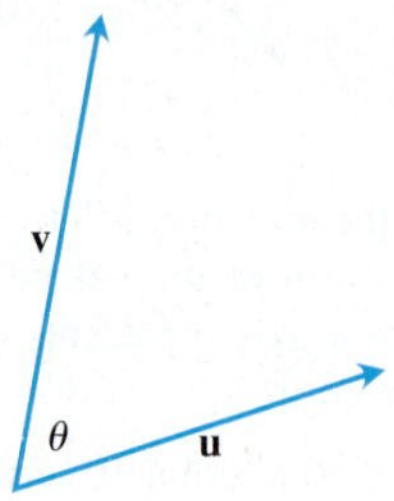

Angle Between Vectors

When two nonzero vectors $\mathbf{u}$ and $\mathbf{v}$ are placed so their initial points coincide, they form an angle θ of measure $0 \le \theta \le \pi$ (Figure 9.15). This angle is the **angle between $\mathbf{u}$ and $\mathbf{v}$.**

FIGURE 9.15 The angle between $\mathbf{u}$ and $\mathbf{v}$.

Theorem 1 gives a formula we can use to determine the angle between two vectors.

Theorem 1 Angle Between Two Vectors

The angle θ between two nonzero vectors $\mathbf{u} = \langle u_1, u_2 \rangle$ and $\mathbf{v} = \langle v_1, v_2 \rangle$ is given by

$$\theta = \cos^{-1} \frac{u_1 v_1 + u_2 v_2}{|\mathbf{u}||\mathbf{v}|}.$$

Before proving Theorem 1 (which is a consequence of the law of cosines), let's focus attention on the expression $u_1 v_1 + u_2 v_2$ in the calculation for θ.

Definition Dot Product (Inner Product)

The **dot product** (or **inner product**) $\mathbf{u} \cdot \mathbf{v}$ ("**u** dot **v**") of vectors $\mathbf{u} = \langle u_1, u_2 \rangle$ and $\mathbf{v} = \langle v_1, v_2 \rangle$ is the number

$$\mathbf{u} \cdot \mathbf{v} = u_1 v_1 + u_2 v_2.$$

Example 1 Finding Dot Products

(a) $\langle 1, -2 \rangle \cdot \langle -6, 2 \rangle = (1)(-6) + (-2)(2)$
$= -6 - 4 = -10$

(b) $\left(\frac{1}{2}\mathbf{i} + 3\mathbf{j}\right) \cdot (4\mathbf{i} - \mathbf{j}) = \left(\frac{1}{2}\right)(4) + (3)(-1)$
$= 2 - 3 = -1$

We can use the dot product to rewrite the formula in Theorem 1 for finding the angle between two vectors.

Corollary Angle Between Two Vectors

The angle between nonzero vectors **u** and **v** is

$$\theta = \cos^{-1}\left(\frac{\mathbf{u} \cdot \mathbf{v}}{|\mathbf{u}||\mathbf{v}|}\right).$$

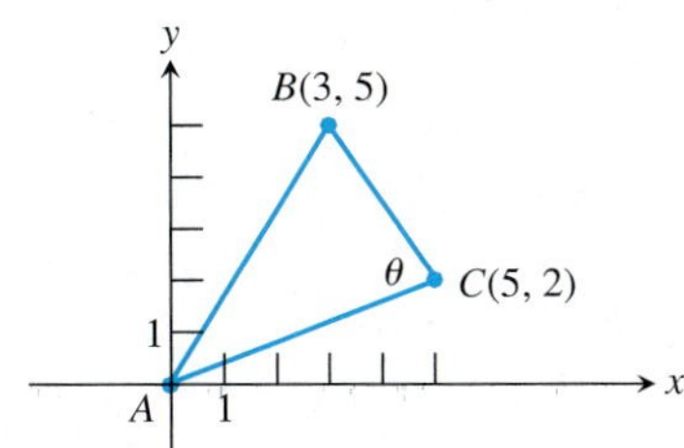

FIGURE 9.16 The triangle in Example 2.

Example 2 Finding an Angle of a Triangle

Find the angle θ in the triangle ABC determined by the vertices $A = (0, 0)$, $B = (3, 5)$, and $C = (5, 2)$ (Figure 9.16).

Solution The angle θ is the angle between the vectors $\overrightarrow{CA}$ and $\overrightarrow{CB}$. The component forms of these two vectors are

$$\overrightarrow{CA} = \langle -5, -2 \rangle \quad \text{and} \quad \overrightarrow{CB} = \langle -2, 3 \rangle.$$

First we calculate the dot product and magnitudes of these two vectors.

$$\overrightarrow{CA} \cdot \overrightarrow{CB} = (-5)(-2) + (-2)(3) = 4$$
$$|\overrightarrow{CA}| = \sqrt{(-5)^2 + (-2)^2} = \sqrt{29}$$
$$|\overrightarrow{CB}| = \sqrt{(-2)^2 + (3)^2} = \sqrt{13}$$

Then applying the corollary to Theorem 1, we have

$$\theta = \cos^{-1}\left(\frac{\overrightarrow{CA} \cdot \overrightarrow{CB}}{|\overrightarrow{CA}||\overrightarrow{CB}|}\right)$$
$$= \cos^{-1}\left(\frac{4}{(\sqrt{29})(\sqrt{13})}\right)$$
$$\approx 78.1° \quad \text{or} \quad 1.36 \text{ radians.}$$

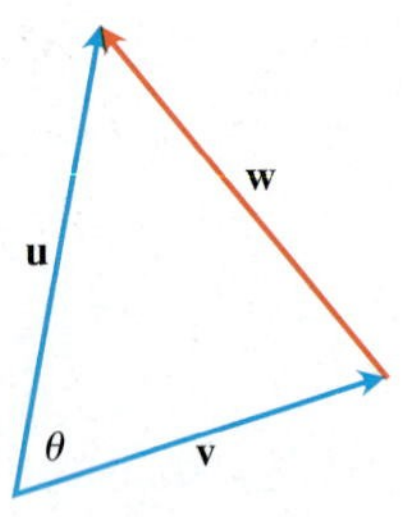

FIGURE 9.17 The parallelogram law of addition of vectors gives $\mathbf{w} = \mathbf{u} - \mathbf{v}$.

Proof of Theorem 1 Applying the law of cosines to the triangle in Figure 9.17, we find that

$$|\mathbf{w}|^2 = |\mathbf{u}|^2 + |\mathbf{v}|^2 - 2|\mathbf{u}||\mathbf{v}|\cos\theta$$
$$2|\mathbf{u}||\mathbf{v}|\cos\theta = |\mathbf{u}|^2 + |\mathbf{v}|^2 - |\mathbf{w}|^2.$$

Because $\mathbf{w} = \mathbf{u} - \mathbf{v}$, the component form of $\mathbf{w}$ is $\langle u_1 - v_1, u_2 - v_2 \rangle$. Thus,

$$|\mathbf{u}|^2 = (\sqrt{u_1^2 + u_2^2})^2 = u_1^2 + u_2^2$$
$$|\mathbf{v}|^2 = (\sqrt{v_1^2 + v_2^2})^2 = v_1^2 + v_2^2$$
$$|\mathbf{w}|^2 = (\sqrt{(u_1 - v_1)^2 + (u_2 - v_2)^2})^2 = (u_1 - v_1)^2 + (u_2 - v_2)^2$$
$$= (u_1^2 - 2u_1v_1 + v_1^2) + (u_2^2 - 2u_2v_2 + v_2^2)$$

and

$$|\mathbf{u}|^2 + |\mathbf{v}|^2 - |\mathbf{w}|^2 = 2(u_1v_1 + u_2v_2).$$

Therefore,

$$2|\mathbf{u}||\mathbf{v}|\cos\theta = |\mathbf{u}|^2 + |\mathbf{v}|^2 - |\mathbf{w}|^2 = 2(u_1v_1 + u_2v_2)$$
$$|\mathbf{u}||\mathbf{v}|\cos\theta = u_1v_1 + u_2v_2$$
$$\cos\theta = \frac{u_1v_1 + u_2v_2}{|\mathbf{u}||\mathbf{v}|}.$$

So

$$\theta = \cos^{-1}\left(\frac{u_1v_1 + u_2v_2}{|\mathbf{u}||\mathbf{v}|}\right).$$

Laws of the Dot Product

The dot product obeys many of the laws that hold for ordinary products of real numbers (scalars).

Properties of the Dot Product

If $\mathbf{u}$, $\mathbf{v}$, and $\mathbf{w}$ are any vectors and c is a scalar, then

1. $\mathbf{u} \cdot \mathbf{v} = \mathbf{v} \cdot \mathbf{u}$
2. $(c\mathbf{u}) \cdot \mathbf{v} = \mathbf{u} \cdot (c\mathbf{v}) = c(\mathbf{u} \cdot \mathbf{v})$
3. $\mathbf{u} \cdot (\mathbf{v} + \mathbf{w}) = \mathbf{u} \cdot \mathbf{v} + \mathbf{u} \cdot \mathbf{w}$
4. $\mathbf{u} \cdot \mathbf{u} = |\mathbf{u}|^2$
5. $\mathbf{0} \cdot \mathbf{u} = 0.$

CD-ROM
WEBsite
Historical Biography
Carl Friedrich Gauss
(1777 — 1855)

The properties are easy to prove using the definition. For instance, here are the proofs of Properties 1 and 3.

1. $\mathbf{u} \cdot \mathbf{v} = u_1v_1 + u_2v_2 = v_1u_1 + v_2u_2 = \mathbf{v} \cdot \mathbf{u}$

3. $$\begin{aligned} \mathbf{u} \cdot (\mathbf{v} + \mathbf{w}) &= \langle u_1, u_2 \rangle \cdot \langle v_1 + w_1, v_2 + w_2 \rangle \\ &= u_1 (v_1 + w_1) + u_2(v_2 + w_2) \\ &= u_1v_1 + u_1w_1 + u_2v_2 + u_2w_2 \\ &= (u_1v_1 + u_2v_2) + (u_1w_1 + u_2w_2) \\ &= \mathbf{u} \cdot \mathbf{v} + \mathbf{u} \cdot \mathbf{w} \end{aligned}$$

Perpendicular (Orthogonal) Vectors

Two nonzero vectors $\mathbf{u}$ and $\mathbf{v}$ are perpendicular or **orthogonal** if the angle between them is $\pi/2$. For such vectors, we automatically have $\mathbf{u} \cdot \mathbf{v} = 0$ because $\cos(\pi/2) = 0$. The converse is also true. If $\mathbf{u}$ and $\mathbf{v}$ are nonzero vectors with $\mathbf{u} \cdot \mathbf{v} = |\mathbf{u}||\mathbf{v}| \cos \theta = 0$, then $\cos \theta = 0$ and $\theta = \cos^{-1} 0 = \pi/2$.

Definition Orthogonal Vectors

Vectors $\mathbf{u}$ and $\mathbf{v}$ are **orthogonal (perpendicular)** if and only if $\mathbf{u} \cdot \mathbf{v} = 0$.

Example 3 Applying the Definition of Orthogonality

(a) $\mathbf{u} = \langle 3, -2 \rangle$ and $\mathbf{v} = \langle 4, 6 \rangle$ are orthogonal because $\mathbf{u} \cdot \mathbf{v} = (3)(4) + (-2)(6) = 0$.

(b) $\mathbf{u} = \mathbf{i} + 2\mathbf{j}$ is orthogonal to $\mathbf{v} = -10\mathbf{i} + 5\mathbf{j}$ because $\mathbf{u} \cdot \mathbf{v} = (1)(-10) + (2)(5) = 0$.

(c) $\mathbf{0}$ is orthogonal to every vector $\mathbf{u}$ since $\mathbf{0} \cdot \mathbf{u} = 0$ from Property 5.

We now return to the problem of projecting one vector onto another, posed in the opening to this section.

Vector Projections

The **vector projection** of $\mathbf{u} = \overrightarrow{PQ}$ onto a nonzero vector $\mathbf{v} = \overrightarrow{PS}$ (Figure 9.18) is the vector $\overrightarrow{PR}$ determined by dropping a perpendicular from Q to the line PS. The notation for this vector is

$$\text{proj}_{\mathbf{v}}\,\mathbf{u} \qquad \text{("the vector projection of } \mathbf{u} \text{ onto } \mathbf{v}\text{")}.$$

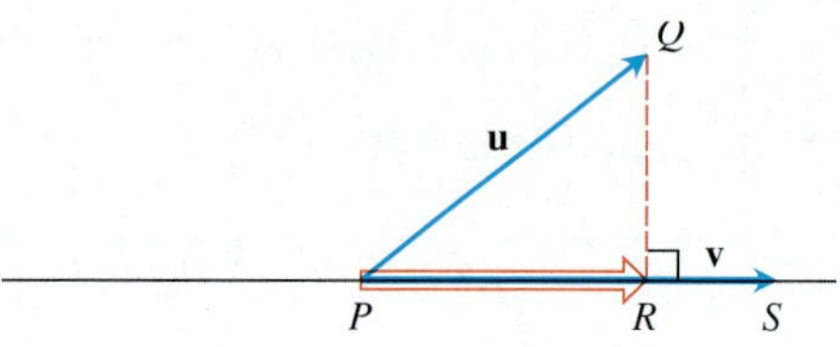

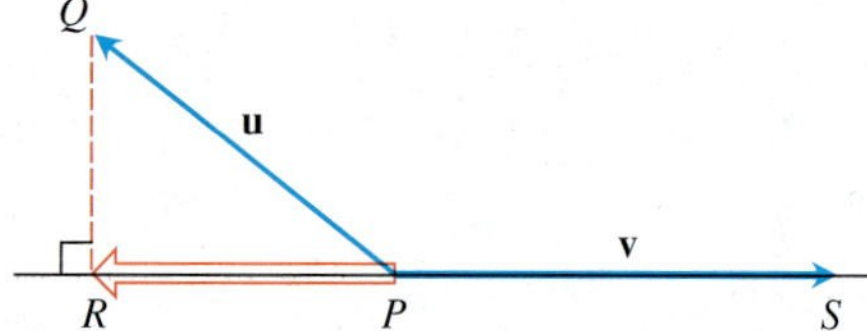

FIGURE 9.18 The vector projection of **u** onto **v**.

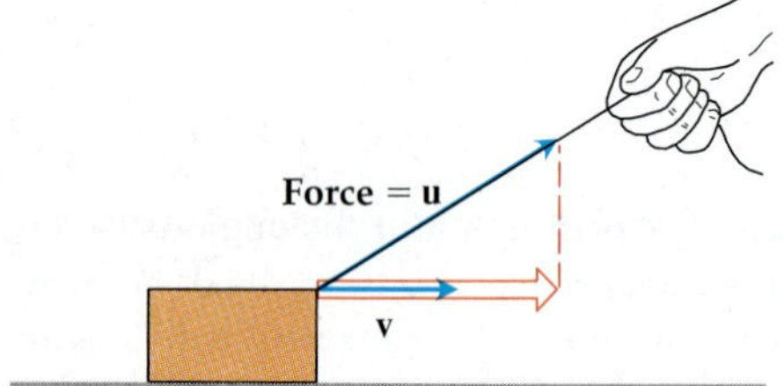

FIGURE 9.19 If we pull on the box with force **u**, the effective force moving the box forward in the direction **v** is the projection of **u** onto **v**.

If **u** represents a force, then $\text{proj}_{\mathbf{v}}\,\mathbf{u}$ represents the effective force in the direction of **v** (Figure 9.19).

If the angle θ between **u** and **v** is acute, $\text{proj}_{\mathbf{v}}\,\mathbf{u}$ has length $|\mathbf{u}|\cos\theta$ and direction $\mathbf{v}/|\mathbf{v}|$ (Figure 9.20). If θ is obtuse, $\cos\theta < 0$ and $\text{proj}_{\mathbf{v}}\,\mathbf{u}$ has length $-|\mathbf{u}|\cos\theta$ and direction $-\mathbf{v}/|\mathbf{v}|$. In any case,

$$\begin{aligned}
\text{proj}_{\mathbf{v}}\,\mathbf{u} &= (|\mathbf{u}|\cos\theta)\frac{\mathbf{v}}{|\mathbf{v}|} \\
&= \left(\frac{\mathbf{u}\cdot\mathbf{v}}{|\mathbf{v}|}\right)\frac{\mathbf{v}}{|\mathbf{v}|} && |\mathbf{u}|\cos\theta = \frac{|\mathbf{u}||\mathbf{v}|\cos\theta}{|\mathbf{v}|} \\
&= \left(\frac{\mathbf{u}\cdot\mathbf{v}}{|\mathbf{v}|^2}\right)\mathbf{v}. && = \frac{\mathbf{u}\cdot\mathbf{v}}{|\mathbf{v}|}
\end{aligned}$$

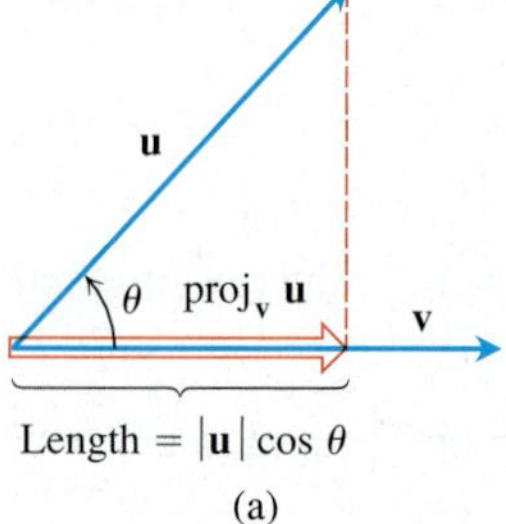

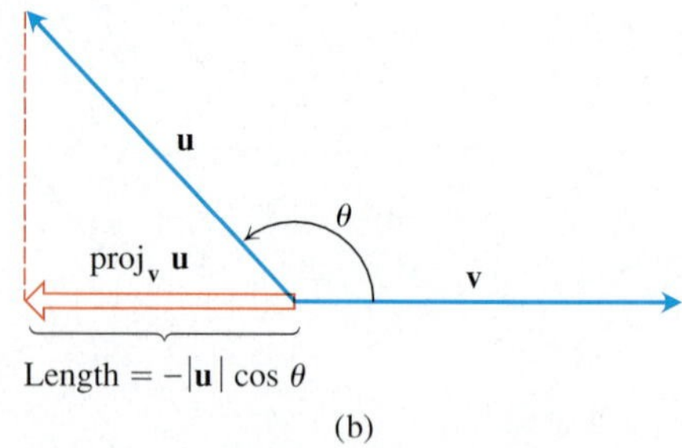

FIGURE 9.20 The length of $\text{proj}_{\mathbf{v}}\,\mathbf{u}$ is (a) $|\mathbf{u}|\cos\theta$ if $\cos\theta \geq 0$ and (b) $-|\mathbf{u}|\cos\theta$ if $\cos\theta < 0$.

The number $|\mathbf{u}| \cos \theta$ is called the **scalar component of u in the direction of v.** To summarize,

Vector projection of **u** onto **v**:

$$\text{proj}_{\mathbf{v}} \mathbf{u} = \left(\frac{\mathbf{u} \cdot \mathbf{v}}{|\mathbf{v}|^2} \right) \mathbf{v}$$

Scalar component of **u** in the direction of **v**:

$$|\mathbf{u}| \cos \theta = \frac{\mathbf{u} \cdot \mathbf{v}}{|\mathbf{v}|} = \mathbf{u} \cdot \frac{\mathbf{v}}{|\mathbf{v}|}$$

Example 4 Finding Vector Projections and Scalar Components

Find the vector projection of a force $\mathbf{F} = 5\mathbf{i} + 2\mathbf{j}$ onto $\mathbf{v} = \mathbf{i} - 3\mathbf{j}$ and the scalar component of **F** in the direction of **v**.

Solution The vector projection is

$$\begin{aligned}
\text{proj}_{\mathbf{v}} \mathbf{F} &= \left(\frac{\mathbf{F} \cdot \mathbf{v}}{|\mathbf{v}|^2} \right) \mathbf{v} \\
&= \frac{5 - 6}{1 + 9} (\mathbf{i} - 3\mathbf{j}) = -\frac{1}{10} (\mathbf{i} - 3\mathbf{j}) \\
&= -\frac{1}{10} \mathbf{i} + \frac{3}{10} \mathbf{j}.
\end{aligned}$$

The scalar component of **F** in the direction of **v** is

$$|\mathbf{F}| \cos \theta = \frac{\mathbf{F} \cdot \mathbf{v}}{|\mathbf{v}|} = \frac{5 - 6}{\sqrt{1 + 9}} = -\frac{1}{\sqrt{10}}.$$

Work

In Chapter 5, we calculated the work done by a constant force of magnitude F in moving an object through a distance d as $W = Fd$. That formula holds only if the force is directed along the line of motion. If a force **F** moving an object through a displacement $\mathbf{D} = \overrightarrow{PQ}$ has some other direction, the work is performed by the component of **F** in the direction of **D**. If θ is the angle between **F** and **D** (Figure 9.21), then

$$\begin{aligned}
\text{Work} &= \begin{pmatrix} \text{scalar component of } \mathbf{F} \\ \text{in the direction of } \mathbf{D} \end{pmatrix} (\text{length of } \mathbf{D}) \\
&= (|\mathbf{F}| \cos \theta) |\mathbf{D}| \\
&= \mathbf{F} \cdot \mathbf{D}.
\end{aligned}$$

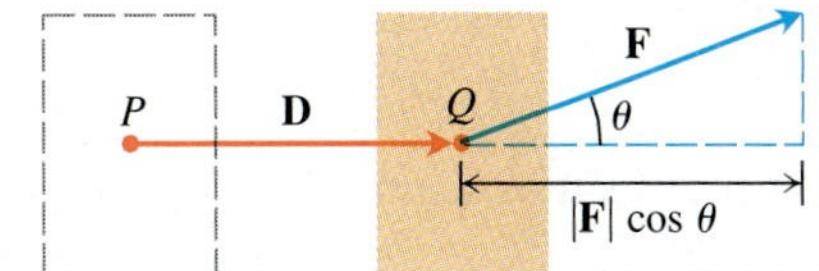

FIGURE 9.21 The work done by a constant force **F** during a displacement **D** is $(|\mathbf{F}| \cos \theta) |\mathbf{D}|$.

Definition Work Done by a Constant Force

The **work** done by a constant force $\mathbf{F}$ acting through a displacement $\mathbf{D} = \overrightarrow{PQ}$ is

$$W = \mathbf{F} \cdot \mathbf{D} = |\mathbf{F}||\mathbf{D}| \cos \theta,$$

where θ is the angle between $\mathbf{F}$ and $\mathbf{D}$.

Work

The standard units of work are the foot-pound and the newton-meter, both force-distance units. The newton-meter is usually called a *joule*.

Example 5 Applying the Definition of Work

If $|\mathbf{F}| = 40$ N (newtons), $|\mathbf{D}| = 3$ m, and $\theta = 60°$, the work done by $\mathbf{F}$ in acting from P to Q is

$$\begin{aligned} \text{Work} &= |\mathbf{F}||\mathbf{D}| \cos \theta && \text{Definition} \\ &= (40)(3) \cos 60° && \text{Given values} \\ &= (120)(1/2) \\ &= 60 \text{ J (joules)}. \end{aligned}$$

We encounter more interesting work problems in Chapter 13 when we learn to find the work done by a variable force along a *path* in space.

Writing a Vector as a Sum of Orthogonal Vectors

We know one way to write a vector $\mathbf{u} = \langle a, b \rangle$ as a sum of orthogonal vectors:

$$\mathbf{u} = \langle a, b \rangle = a\mathbf{i} + b\mathbf{j}$$

(since $\mathbf{i} \cdot \mathbf{j} = 0$). Sometimes, however, it is more informative to express $\mathbf{u}$ as a different sum. In mechanics, for instance, we often need to write a vector $\mathbf{u}$ as a sum of a vector parallel to a given vector $\mathbf{v}$ and a vector orthogonal to $\mathbf{v}$. As an example, in studying the motion of a particle moving along a path in the plane (or space), it is desirable to know the components of the acceleration vector in the direction of the tangent to the path (at a point) and of the normal to the path. (These *tangential* and *normal components* of acceleration are investigated in Section 10.6.) The acceleration vector can then be expressed as the sum of its (vector) tangential and normal components (which reflect important geometric properties about the nature of the path itself, such as *curvature*). Velocity and acceleration vectors are studied in the next section.

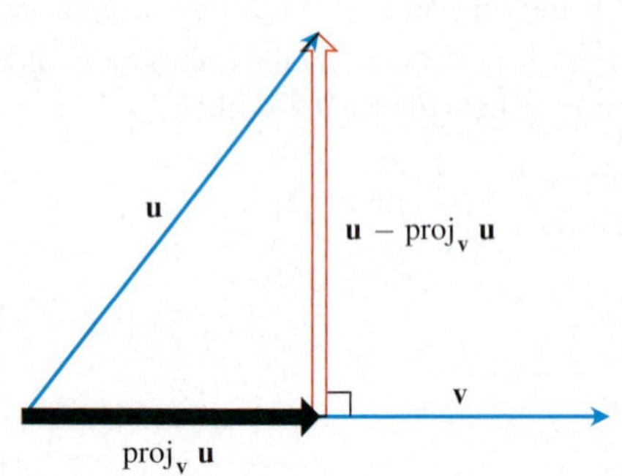

FIGURE 9.22 Writing $\mathbf{u}$ as the sum of vectors parallel and orthogonal to $\mathbf{v}$.

Generally, for vectors $\mathbf{u}$ and $\mathbf{v}$, it is easy to see from Figure 9.22 that the vector

$$\mathbf{u} - \text{proj}_{\mathbf{v}}\, \mathbf{u}$$

is orthogonal to the projection vector $\text{proj}_{\mathbf{v}}\, \mathbf{u}$ (which has the same direction as $\mathbf{v}$). Thus,

$$\mathbf{u} = \text{proj}_{\mathbf{v}}\, \mathbf{u} + (\mathbf{u} - \text{proj}_{\mathbf{v}}\, \mathbf{u})$$

expresses $\mathbf{u}$ as a sum of orthogonal vectors.

How to Write u as a Vector Parallel to v Plus a Vector Orthogonal to v

$$\mathbf{u} = \text{proj}_{\mathbf{v}}\,\mathbf{u} + (\mathbf{u} - \text{proj}_{\mathbf{v}}\,\mathbf{u})$$

$$= \underbrace{\left(\frac{\mathbf{u}\cdot\mathbf{v}}{|\mathbf{v}|^2}\right)\mathbf{v}}_{\text{Parallel to } \mathbf{v}} + \underbrace{\left(\mathbf{u} - \left(\frac{\mathbf{u}\cdot\mathbf{v}}{|\mathbf{v}|^2}\right)\mathbf{v}\right)}_{\text{Orthogonal to } \mathbf{v}}$$

Example 6 Sum of Orthogonal Vectors

In Example 8, Section 9.1, we found the vector $\mathbf{v} = 2\mathbf{i} + 3\mathbf{j}$ to be tangent to the path

$$y = \frac{x^3}{2} + \frac{1}{2}$$

at the point $(1, 1)$. If $\mathbf{u} = 4\mathbf{i} - \mathbf{j}$ is the acceleration at the point of a particle moving along the path, express $\mathbf{u}$ as the sum of a vector parallel to $\mathbf{v}$ and a vector orthogonal to $\mathbf{v}$.

Solution With $\mathbf{u}\cdot\mathbf{v} = 8 - 3 = 5$ and $|\mathbf{v}|^2 = \mathbf{v}\cdot\mathbf{v} = 4 + 9 = 13$, we have

$$\begin{aligned}\mathbf{u} &= \left(\frac{\mathbf{u}\cdot\mathbf{v}}{|\mathbf{v}|^2}\right)\mathbf{v} + \left(\mathbf{u} - \left(\frac{\mathbf{u}\cdot\mathbf{v}}{|\mathbf{v}|^2}\right)\mathbf{v}\right)\\ &= \frac{5}{13}(2\mathbf{i} + 3\mathbf{j}) + \left(4\mathbf{i} - \mathbf{j} - \frac{5}{13}(2\mathbf{i} + 3\mathbf{j})\right)\\ &= \left(\frac{10}{13}\mathbf{i} + \frac{15}{13}\mathbf{j}\right) + \left(\frac{42}{13}\mathbf{i} - \frac{28}{13}\mathbf{j}\right).\end{aligned}$$

Check: The first vector in the sum is parallel to $\mathbf{v}$ because it is $(5/13)\mathbf{v}$. The second vector is orthogonal to $\mathbf{v}$ because

$$\left(\frac{42}{13}\mathbf{i} - \frac{28}{13}\mathbf{j}\right)\cdot(2\mathbf{i} + 3\mathbf{j}) = \frac{84}{13} - \frac{84}{13} = 0.$$

In the next chapter, we learn how to write velocity and acceleration vectors as linear combinations of other mutually orthogonal vectors.

EXERCISES 9.2

Calculations

In Exercises 1–6, find

(a) $\mathbf{v}\cdot\mathbf{u}$, $|\mathbf{v}|$, $|\mathbf{u}|$

(b) the cosine of the angle between $\mathbf{v}$ and $\mathbf{u}$

(c) the scalar component of $\mathbf{u}$ in the direction of $\mathbf{v}$

(d) the vector $\text{proj}_{\mathbf{v}}\,\mathbf{u}$.

1. $\mathbf{v} = 2\mathbf{i} - 4\mathbf{j}$, $\mathbf{u} = 2\mathbf{i} + 4\mathbf{j}$

2. $\mathbf{v} = 2\mathbf{i} + 10\mathbf{j}$, $\mathbf{u} = 2\mathbf{i} + 2\mathbf{j}$

3. $\mathbf{v} = -\mathbf{i} + \mathbf{j}$, $\mathbf{u} = \sqrt{2}\mathbf{i} + \sqrt{3}\mathbf{j}$

4. $\mathbf{v} = 5\mathbf{i} + \mathbf{j}$, $\mathbf{u} = 2\mathbf{i} + \sqrt{17}\mathbf{j}$

5. $\mathbf{v} = \left\langle \frac{1}{\sqrt{2}}, \frac{1}{\sqrt{3}} \right\rangle, \quad \mathbf{u} = \left\langle \frac{1}{\sqrt{2}}, -\frac{1}{\sqrt{3}} \right\rangle$

6. $\mathbf{v} = \left\langle \frac{1}{\sqrt{2}}, \frac{1}{\sqrt{2}} \right\rangle, \quad \mathbf{u} = \left\langle -\frac{1}{\sqrt{2}}, -\frac{1}{\sqrt{2}} \right\rangle$

Angles Between Vectors

Find the angles between the vectors in Exercises 7–10 to the nearest hundredth of a radian.

7. $\mathbf{v} = 2\mathbf{i} + \mathbf{j}, \quad \mathbf{u} = \mathbf{i} + 2\mathbf{j}$

8. $\mathbf{v} = 2\mathbf{i} - 2\mathbf{j}, \quad \mathbf{u} = 3\mathbf{i}$

9. $\mathbf{v} = \sqrt{3}\mathbf{i} - 7\mathbf{j}, \quad \mathbf{u} = \sqrt{3}\mathbf{i} + \mathbf{j}$

10. $\mathbf{v} = \mathbf{i} + \sqrt{2}\mathbf{j}, \quad \mathbf{u} = -\mathbf{i} + \mathbf{j}$

11. *Triangle* Find the measures of the angles of the triangle whose vertices are $A = (-1, 0)$, $B = (2, 1)$, and $C = (1, -2)$.

12. *Rectangle* Find the measures of the angles between the diagonals of the rectangle whose vertices are $A = (1, 0)$, $B = (0, 3)$, $C = (3, 4)$, and $D = (4, 1)$.

Geometry and Examples

13. *Writing to Learn: Sums and differences* In the accompanying figure, it looks as if $\mathbf{v}_1 + \mathbf{v}_2$ and $\mathbf{v}_1 - \mathbf{v}_2$ are orthogonal. Is this mere coincidence, or are there circumstances under which we may expect the sum of two vectors to be orthogonal to their difference? Give reasons for your answer.

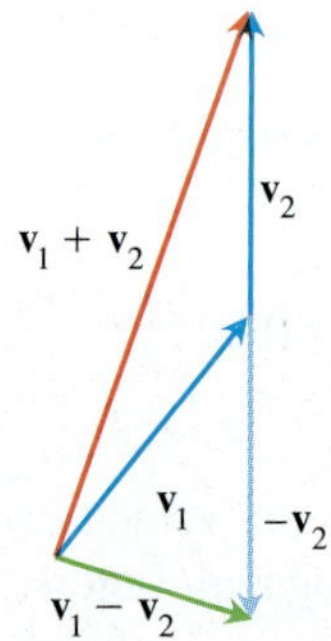

14. *Orthogonality on a circle* Suppose that AB is the diameter of a circle with center O and that C is a point on one of the two arcs joining A and B. Show that $\overrightarrow{CA}$ and $\overrightarrow{CB}$ are orthogonal.

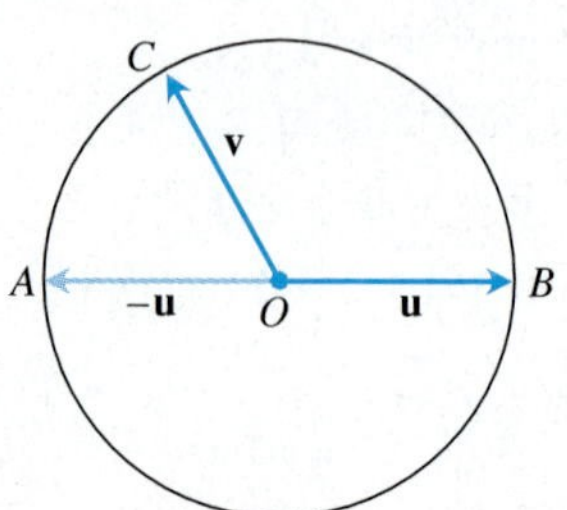

15. *Diagonals of a rhombus* Show that the diagonals of a rhombus (parallelogram with sides of equal length) are perpendicular.

16. *Perpendicular diagonals* Show that squares are the only rectangles with perpendicular diagonals.

17. *When parallelograms are rectangles* Prove that a parallelogram is a rectangle if and only if its diagonals are equal in length. (This fact is often exploited by carpenters.)

18. *Diagonal of parallelogram* Show that the indicated diagonal of the parallelogram determined by vectors $\mathbf{u}$ and $\mathbf{v}$ bisects the angle between $\mathbf{u}$ and $\mathbf{v}$ if $|\mathbf{u}| = |\mathbf{v}|$.

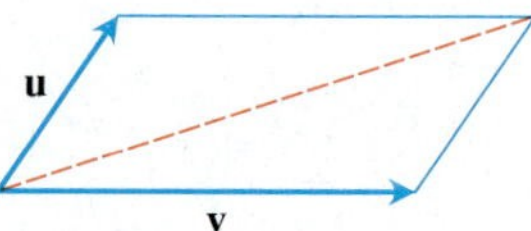

19. *Projectile motion* A gun with muzzle velocity of 1200 ft/sec is fired at an angle of 8° above the horizontal. Find the horizontal and vertical components of the velocity.

20. *Inclined plane* Suppose that a box is being towed up an inclined plane as shown in the figure. Find the force $\mathbf{w}$ needed to make the component of the force parallel to the inclined plane equal to 2.5 lb.

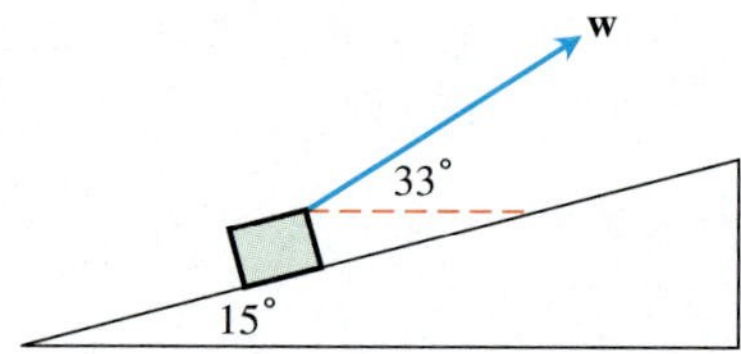

Theory and Examples

21. (a) *Cauchy-Schwartz inequality* Use the fact that $\mathbf{u} \cdot \mathbf{v} = |\mathbf{u}||\mathbf{v}| \cos \theta$ to show that the inequality $|\mathbf{u} \cdot \mathbf{v}| \le |\mathbf{u}||\mathbf{v}|$ holds for any vectors $\mathbf{u}$ and $\mathbf{v}$.

    (b) *Writing to Learn* Under what circumstances, if any, does $|\mathbf{u} \cdot \mathbf{v}|$ equal $|\mathbf{u}||\mathbf{v}|$? Give reasons for your answer.

22. *Writing to Learn* Copy the axes and vector shown here. Then shade in the points (x, y) for which $(x\mathbf{i} + y\mathbf{j}) \cdot \mathbf{v} \le 0$. Justify your answer.

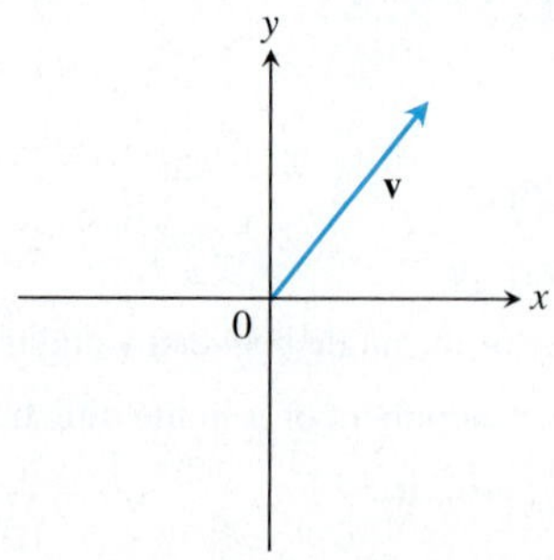

23. *Orthogonal unit vectors* If $\mathbf{u}_1$ and $\mathbf{u}_2$ are orthogonal unit vectors and $\mathbf{v} = a\mathbf{u}_1 + b\mathbf{u}_2$, find $\mathbf{v} \cdot \mathbf{u}_1$.

24. *Writing to Learn: Cancellation in dot products* In real-number multiplication, if $uv_1 = uv_2$ and $u \neq 0$, we can cancel the u and conclude that $v_1 = v_2$. Does the same rule hold for the dot product: If $\mathbf{u} \cdot \mathbf{v}_1 = \mathbf{u} \cdot \mathbf{v}_2$ and $\mathbf{u} \neq \mathbf{0}$, can you conclude that $\mathbf{v}_1 = \mathbf{v}_2$? Give reasons for your answer.

Equations for Lines in the Plane

25. *Line perpendicular to a vector* Show that the vector $\mathbf{v} = a\mathbf{i} + b\mathbf{j}$ is perpendicular to the line $ax + by = c$ by establishing that the slope of $\mathbf{v}$ is the negative reciprocal of the slope of the given line.

26. *Line parallel to a vector* Show that the vector $\mathbf{v} = a\mathbf{i} + b\mathbf{j}$ is parallel to the line $bx - ay = c$ by establishing that the slope of the line segment representing $\mathbf{v}$ is the same as the slope of the given line.

In Exercises 27–30, use the result of Exercise 25 to find an equation for the line through P perpendicular to $\mathbf{v}$. Then sketch the line. Include $\mathbf{v}$ in your sketch *as a vector starting at the origin.*

27. $P(2, 1)$, $\mathbf{v} = \mathbf{i} + 2\mathbf{j}$

28. $P(-1, 2)$, $\mathbf{v} = -2\mathbf{i} - \mathbf{j}$

29. $P(-2, -7)$, $\mathbf{v} = -2\mathbf{i} + \mathbf{j}$

30. $P(11, 10)$, $\mathbf{v} = 2\mathbf{i} - 3\mathbf{j}$

In Exercises 31–34, use the result of Exercise 26 to find an equation for the line through P parallel to $\mathbf{v}$. Then sketch the line. Include $\mathbf{v}$ in your sketch *as a vector starting at the origin.*

31. $P(-2, 1)$, $\mathbf{v} = \mathbf{i} - \mathbf{j}$

32. $P(0, -2)$, $\mathbf{v} = 2\mathbf{i} + 3\mathbf{j}$

33. $P(1, 2)$, $\mathbf{v} = -\mathbf{i} - 2\mathbf{j}$

34. $P(1, 3)$, $\mathbf{v} = 3\mathbf{i} - 2\mathbf{j}$

Work

35. *Work along a line* Find the work done by a force $\mathbf{F} = 5\,\mathbf{i}$ (magnitude 5 N) in moving an object along the line from the origin to the point $(1, 1)$ (distance in meters).

36. *Locomotive* The union Pacific's *Big Boy* locomotive could pull 6000-ton trains with a tractive effort (pull) of 602,148 N (135,375 lb). At this level of effort, about how much work did *Big Boy* do on the (approximately straight) 605 km journey from San Francisco to Los Angeles?

37. *Inclined plane* How much work does it take to slide a crate 20 m along a loading dock by pulling on it with a 200 N force at an angle of 30° from the horizontal?

38. *Sailboat* The wind passing over a boat's sail exerted a 1000 lb magnitude force $\mathbf{F}$ as shown here. How much work did the wind perform in moving the boat forward 1 mi? Answer in foot-pounds.

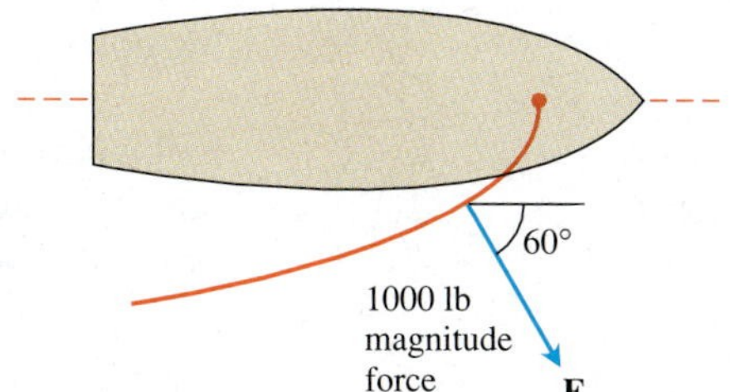

Angles Between Lines in the Plane

The acute angle between intersecting lines that do not cross at right angles is the same as the angle determined by vectors normal to the lines or by the vectors parallel to the lines.

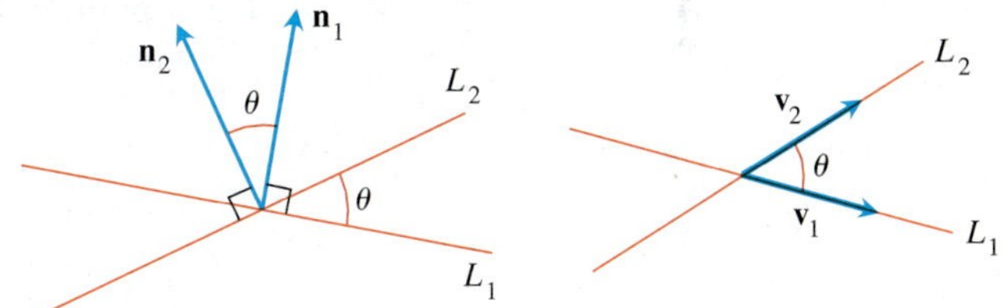

Use this fact and the results of Exercise 25 or 26 to find the acute angles between the lines in Exercises 39–44.

39. $3x + y = 5$, $2x - y = 4$

40. $y = \sqrt{3}x - 1$, $y = -\sqrt{3}x + 2$

41. $\sqrt{3}x - y = -2$, $x - \sqrt{3}y = 1$

42. $x + \sqrt{3}y = 1$, $(1 - \sqrt{3})x + (1 + \sqrt{3})y = 8$

43. $3x - 4y = 3$, $x - y = 7$

44. $12x + 5y = 1$, $2x - 2y = 3$

Angles Between Differentiable Curves

The angles between two differentiable curves at a point of intersection are the angles between the curves' tangent lines at these points. Find the angles between the curves in Exercises 45–48.

45. $y = (3/2) - x^2$, $y = x^2$ (two points of intersection)

46. $x = (3/4) - y^2$, $x = y^2 - (3/4)$ (two points of intersection)

47. $y = x^3$, $x = y^2$ (two points of intersection)

48. $y = -x^2$, $y = \sqrt[3]{x}$ (two points of intersection)

9.3 Vector-Valued Functions

Planar Curves • Limits and Continuity • Derivatives • Motion • Integrals

In this section, we show how to use the calculus of vectors to study the paths, velocities, and accelerations of bodies moving in the plane. This allows us to answer questions later about the motions of projectiles.

CD-ROM
WEBsite
Historical Biography

James Clerk Maxwell
(1831 — 1879)

Planar Curves

When a particle moves through the plane during a time interval I, we think of the particle's coordinates as functions defined on I:

$$x = f(t), \qquad y = g(t), \qquad t \in I. \tag{1}$$

The points $(x, y) = (f(t), g(t))$, $t \in I$, make up the curve in the plane that is the particle's **path.** The equations and interval in Equation (1) parametrize the curve. The vector

$$\mathbf{r}(t) = \overrightarrow{OP} = \langle f(t), g(t) \rangle = f(t)\mathbf{i} + g(t)\mathbf{j} \tag{2}$$

from the origin to the particle's **position** $P(f(t), g(t))$ at time t is the particle's **position vector.** The functions f and g are the **component functions (components)** of the position vector. We think of the particle's path as the **curve traced by r** during the time interval I (Figure 9.23).

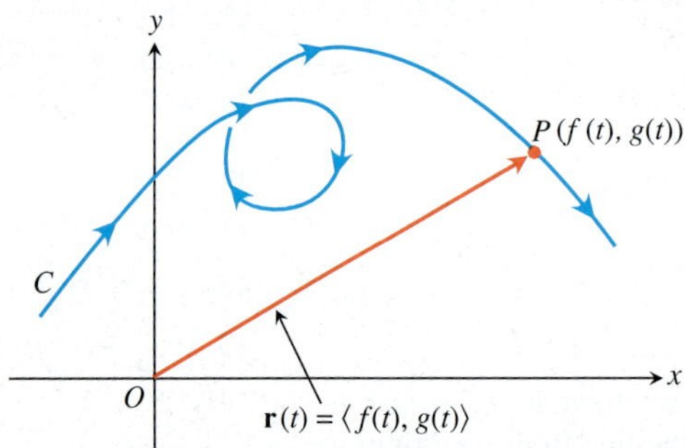

FIGURE 9.23 The path (curve) C is traced by the position vector $\mathbf{r}(t)$ during the time interval I.

Equation (2) defines $\mathbf{r}$ as a *vector function* of the real variable t on the interval I. More generally, a **vector function** or **vector-valued function** on a domain D is a rule that assigns a vector in the plane to each element in D. The curve traced by a vector function is its **graph.**

We refer to real-valued functions as **scalar functions** to distinguish them from vector functions. The components of $\mathbf{r}$ are scalar functions of t. When we define a vector-valued function by giving its component functions, we assume that the vector function's domain to be the common domain of the components.

Example 1 Graphing an Archimedes Spiral

Graph the vector function

$$\mathbf{r}(t) = (t \cos t)\mathbf{i} + (t \sin t)\mathbf{j}, \qquad t \geq 0.$$

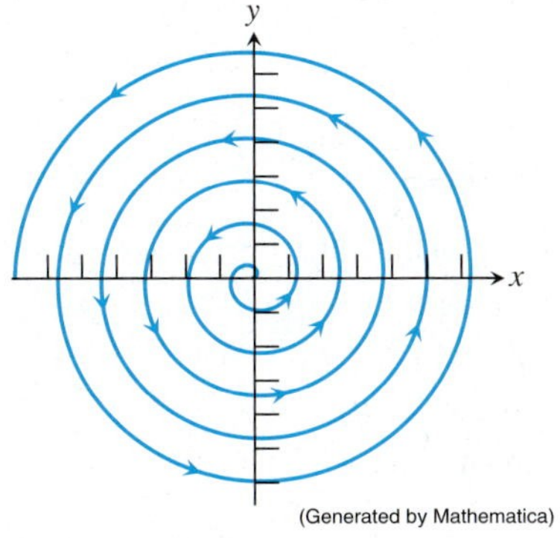

FIGURE 9.24 The graph of $\mathbf{r}(t) = (t \cos t)\mathbf{i} + (t \sin t)\mathbf{j}$, $t \geq 0$, is the curve $x = t\cos t$, $y = t \sin t$, $t \geq 0$. (Example 1)

Solution We can graph the vector function parametrically on a graphing calculator or computer using

$$x = t\cos t, \qquad y = t\sin t, \qquad t \geq 0.$$

As t increases from 0 to 2π, the point (x, y) starts at the origin (0, 0) and then winds once around the origin, getting farther away from the origin as t increases. This spiral continues winding around the origin, getting farther and farther away as t increases beyond 2π. The spiral is shown in Figure 9.24.

Limits and Continuity

We define limits of vector functions in terms of their scalar components.

Definition Limit

Let $\mathbf{r}(t) = f(t)\mathbf{i} + g(t)\mathbf{j}$. If

$$\lim_{t\to c} f(t) = L_1 \qquad \text{and} \qquad \lim_{t\to c} g(t) = L_2,$$

then the **limit** of $\mathbf{r}(t)$ as t approaches c is

$$\lim_{t\to c} \mathbf{r}(t) = \mathbf{L} = L_1\mathbf{i} + L_2\mathbf{j}.$$

Example 2 Finding a Limit of a Vector Function

If $\mathbf{r}(t) = (\cos t)\mathbf{i} + (\sin t)\mathbf{j}$, then

$$\lim_{t\to\pi/4} \mathbf{r}(t) = \left(\lim_{t\to\pi/4} \cos t\right)\mathbf{i} + \left(\lim_{t\to\pi/4} \sin t\right)\mathbf{j} = \frac{\sqrt{2}}{2}\mathbf{i} + \frac{\sqrt{2}}{2}\mathbf{j}.$$

We define continuity for vector functions in the same way we define continuity for scalar functions.

Definition Continuity at a Point

A vector function $\mathbf{r}(t)$ is **continuous at a point** $t = c$ in its domain if

$$\lim_{t\to c} \mathbf{r}(t) = \mathbf{r}(c).$$

A vector function $\mathbf{r}(t)$ is **continuous** if it is continuous at every point in its domain. Since limits of vector functions are defined in terms of components, we have the following test for continuity.

Component Test for Continuity at a Point

The vector function $\mathbf{r}(t) = f(t)\mathbf{i} + g(t)\mathbf{j}$ is continuous at $t = c$ if and only if f and g are continuous at $t = c$.

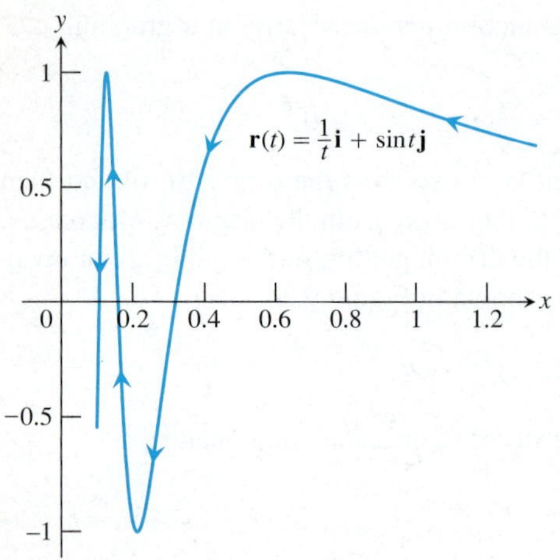

FIGURE 9.25 As $t > 0$ increases, the path of $\mathbf{r}(t)$ oscillates between $y = -1$ and $y = 1$, approaching the y-axis as the **i**-component approaches 0. As $t \to 0^+$, the **i**-component approaches ∞ and the **j**-component approaches 0 through positive values. (The portion of the graph to the left of the y-axis is not shown.)

Example 3 Finding Points of Continuity and Discontinuity

(a) The vector function

$$\mathbf{r}(t) = (t \cos t)\mathbf{i} + (t \sin t)\mathbf{j}$$

is continuous everywhere because the component functions, $t \cos t$, and $t \sin t$ are continuous everywhere.

(b) The vector function

$$\mathbf{r}(t) = \frac{1}{t}\mathbf{i} + (\sin t)\mathbf{j}$$

is not continuous at $t = 0$ because the first component is not continuous at $t = 0$. It is a continuous vector function, however, because it is continuous on its domain, the set of all nonzero real numbers (Figure 9.25).

Derivatives

Suppose that $\mathbf{r}(t) = f(t)\mathbf{i} + g(t)\mathbf{j}$ is the position vector of a particle moving along a curve in the plane and that f and g are differentiable functions of t. Then (see Figure 9.26) the difference between the particle's positions at time $t + \Delta t$ and time t is $\Delta\mathbf{r} = \mathbf{r}(t + \Delta t) - \mathbf{r}(t)$. In terms of components,

$$\begin{aligned}\Delta\mathbf{r} &= \mathbf{r}(t + \Delta t) - \mathbf{r}(t)\\ &= [f(t + \Delta t)\mathbf{i} + g(t + \Delta t)\mathbf{j}] - [f(t)\mathbf{i} + g(t)\mathbf{j}]\\ &= [f(t + \Delta t) - f(t)]\mathbf{i} + [g(t + \Delta t) - g(t)]\mathbf{j}.\end{aligned}$$

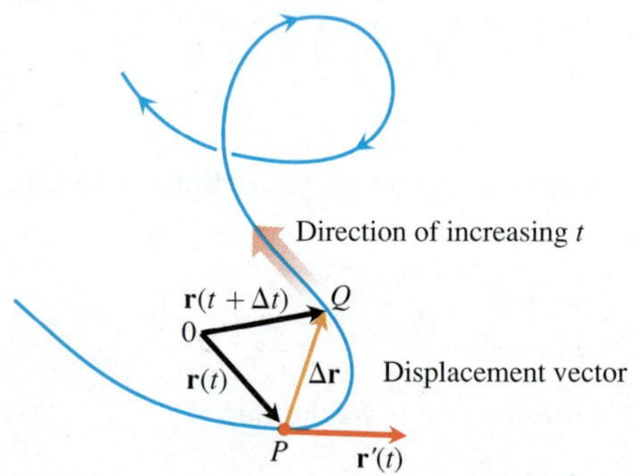

FIGURE 9.26 Between time t and time $t + \Delta t$, the particle moving along the path shown here undergoes the displacement $\overrightarrow{PQ} = \Delta\mathbf{r}$. The vector sum $\mathbf{r}(t) + \Delta\mathbf{r}$ gives the new position, $\mathbf{r}(t + \Delta t)$. As $\Delta t \to 0$, the point Q approaches P along the curve and the vector $\Delta\mathbf{r}/\Delta t$ approaches the limiting tangent position $\mathbf{r}'(t)$.

As Δt approaches zero, three things seem to happen simultaneously. First, Q approaches P along the curve. Second, the secant line PQ seems to approach a limiting position tangent to the curve at P. Third, the quotient $\Delta\mathbf{r}/\Delta t$ approaches the limit

$$\begin{aligned}\lim_{\Delta t \to 0} \frac{\Delta\mathbf{r}}{\Delta t} &= \left[\lim_{\Delta t \to 0} \frac{f(t + \Delta t) - f(t)}{\Delta t}\right]\mathbf{i} + \left[\lim_{\Delta t \to 0} \frac{g(t + \Delta t) - g(t)}{\Delta t}\right]\mathbf{j}\\ &= \left[\frac{df}{dt}\right]\mathbf{i} + \left[\frac{dg}{dt}\right]\mathbf{j}.\end{aligned}$$

We are therefore led by past experience to the following definition.

Definition Derivative at a Point

The vector function $\mathbf{r}(t) = f(t)\mathbf{i} + g(t)\mathbf{j}$ has a **derivative (is differentiable) at t** if f and g have derivatives at t. The derivative is the vector function

$$\mathbf{r}'(t) = \frac{d\mathbf{r}}{dt} = \lim_{\Delta t \to 0} \frac{\mathbf{r}(t + \Delta t) - \mathbf{r}(t)}{\Delta t} = \frac{df}{dt}\mathbf{i} + \frac{dg}{dt}\mathbf{j}.$$

A vector function **r** is **differentiable** if it is differentiable at every point of its domain. The curve traced by **r** is **smooth** if $d\mathbf{r}/dt$ is continuous and never **0**, that is,

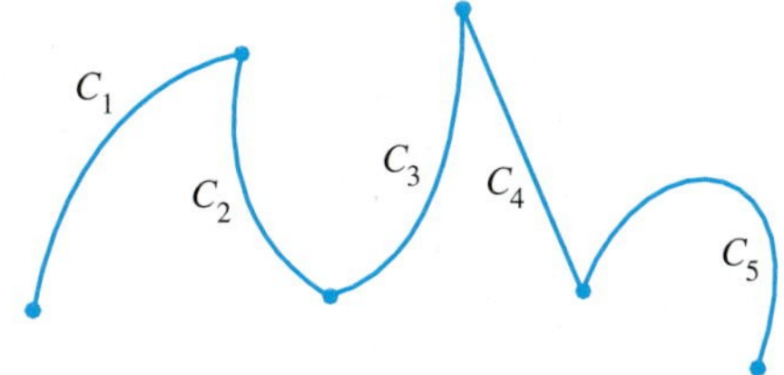

FIGURE 9.27 A piecewise smooth curve made by connecting five smooth curves end to end in continuous fashion.

if f and g have continuous first derivatives that are not simultaneously 0. On a smooth curve, there are no sharp corners or cusps.

The vector $d\mathbf{r}/dt$, when different from $\mathbf{0}$, is also a vector *tangent* to the curve at each point where it exists. The **tangent line** to the curve at a point $P = (f(a), g(a))$ is defined to be the line through P parallel to $d\mathbf{r}/dt$ at $t = a$ (Figure 9.26).

A curve that is made up of a finite number of smooth curves pieced together in a continuous fashion is **piecewise smooth** (Figure 9.27).

Example 4 Finding Derivatives

Find the derivative $d\mathbf{r}/dt$ of the vector function

$$\mathbf{r}(t) = (t \cos t)\mathbf{i} + (t \sin t)\mathbf{j}.$$

Solution

$$\begin{aligned}\mathbf{r}'(t) = \frac{d\mathbf{r}}{dt} &= \frac{d}{dt}(t \cos t)\mathbf{i} + \frac{d}{dt}(t \sin t)\mathbf{j} \\ &= (\cos t - t \sin t)\mathbf{i} + (\sin t + t \cos t)\mathbf{j}\end{aligned}$$

Because the derivatives of vector functions are computed component by component, the rules for differentiating vector functions have the same form as the rules for differentiating scalar functions.

Differentiation Rules for Vector Functions

Let $\mathbf{u}$ and $\mathbf{v}$ be differentiable vector functions of t, $\mathbf{C}$ a constant vector, c any scalar, and f any differentiable scalar function.

1. *Constant Function Rule*: $\dfrac{d}{dt}\mathbf{C} = \mathbf{0}$
2. *Scalar Multiple Rules*: $\dfrac{d}{dt}[c\mathbf{u}(t)] = c\mathbf{u}'(t)$

   $\dfrac{d}{dt}[f(t)\mathbf{u}(t)] = f'(t)\mathbf{u}(t) + f(t)\mathbf{u}'(t)$
3. *Sum Rule*: $\dfrac{d}{dt}[\mathbf{u}(t) + \mathbf{v}(t)] = \mathbf{u}'(t) + \mathbf{v}'(t)$
4. *Difference Rule*: $\dfrac{d}{dt}[\mathbf{u}(t) - \mathbf{v}(t)] = \mathbf{u}'(t) - \mathbf{v}'(t)$
5. *Dot Product Rule*: $\dfrac{d}{dt}[\mathbf{u}(t) \cdot \mathbf{v}(t)] = \mathbf{u}'(t) \cdot \mathbf{v}(t) + \mathbf{u}(t) \cdot \mathbf{v}'(t)$
6. *Chain Rule*: $\dfrac{d}{dt}[\mathbf{u}(f(t))] = f'(t)\mathbf{u}'(f(t))$

We will prove the Dot Product Rule but leave the rest for Exercises 37–40.

Proof of Rule 5 Suppose that

$$\mathbf{u} = u_1(t)\mathbf{i} + u_2(t)\mathbf{j} \quad \text{and} \quad \mathbf{v} = v_1(t)\mathbf{i} + v_2(t)\mathbf{j}.$$

Then

$$\begin{aligned}\frac{d}{dt}(\mathbf{u}\cdot\mathbf{v}) &= \frac{d}{dt}(u_1v_1 + u_2v_2)\\ &= u_1'v_1 + u_2'v_2 + u_1v_1' + u_2v_2'\\ &= \quad \mathbf{u}'\cdot\mathbf{v} \quad + \quad \mathbf{u}\cdot\mathbf{v}'.\end{aligned}$$

Example 5 Applying the Differentiation Rules

For the functions given by $\mathbf{u}(t) = 2t^3\mathbf{i} - t^2\mathbf{j}$, $\mathbf{v}(t) = (1/t)\mathbf{i} + (\sin t)\mathbf{j}$, and $f(t) = e^{-t}$, find

(a) $\dfrac{d}{dt}[f(t)\,\mathbf{u}(t)]$ **(b)** $\dfrac{d}{dt}[\mathbf{u}(t) + \mathbf{v}(t)]$

(c) $\dfrac{d}{dt}[\mathbf{u}(t)\cdot\mathbf{v}(t)]$.

Solution

(a) Because $f'(t) = -e^{-t}$ and $\mathbf{u}'(t) = 6t^2\mathbf{i} - 2t\mathbf{j}$, we have

$$\begin{aligned}\frac{d}{dt}[f(t)\,\mathbf{u}(t)] &= f'(t)\,\mathbf{u}(t) + f(t)\,\mathbf{u}'(t)\\ &= (-e^{-t})(2t^3\mathbf{i} - t^2\mathbf{j}) + e^{-t}(6t^2\mathbf{i} - 2t\mathbf{j})\\ &= e^{-t}(6t^2 - 2t^3)\mathbf{i} + e^{-t}(t^2 - 2t)\mathbf{j}\\ &= 2t^2e^{-t}(3 - t)\mathbf{i} + te^{-t}(t - 2)\mathbf{j}.\end{aligned}$$

(b) $\mathbf{u}'(t) = 6t^2\mathbf{i} - 2t\mathbf{j}$ and $\mathbf{v}'(t) = -\dfrac{1}{t^2}\mathbf{i} + (\cos t)\mathbf{j}$ so that

$$\begin{aligned}\frac{d}{dt}[\mathbf{u}(t) + \mathbf{v}(t)] &= \mathbf{u}'(t) + \mathbf{v}'(t)\\ &= (6t^2\mathbf{i} - 2t\mathbf{j}) + \left(-\frac{1}{t^2}\mathbf{i} + (\cos t)\mathbf{j}\right)\\ &= \left(6t^2 - \frac{1}{t^2}\right)\mathbf{i} + (\cos t - 2t)\mathbf{j}.\end{aligned}$$

(c) Using the derivatives in part (b) and the Dot Product Rule, we have

$$\begin{aligned}\frac{d}{dt}[\mathbf{u}(t)\cdot\mathbf{v}(t)] &= \mathbf{u}'(t)\cdot\mathbf{v}(t) + \mathbf{u}(t)\cdot\mathbf{v}'(t)\\ &= (6t^2\mathbf{i} - 2t\mathbf{j})\cdot\left(\frac{1}{t}\mathbf{i} + (\sin t)\mathbf{j}\right) + (2t^3\mathbf{i} - t^2\mathbf{j})\cdot\left(-\frac{1}{t^2}\mathbf{i} + (\cos t)\mathbf{j}\right)\\ &= (6t^2)\left(\frac{1}{t}\right) + (-2t)(\sin t) + (2t^3)\left(-\frac{1}{t^2}\right) + (-t^2)(\cos t)\\ &= 6t - 2t\sin t - 2t - t^2\cos t\\ &= 4t - 2t\sin t - t^2\cos t.\end{aligned}$$

Observe that the derivative of the dot product of vector functions is a scalar function.

Check:

$$\frac{d}{dt}[\mathbf{u}(t)\cdot\mathbf{v}(t)] = \frac{d}{dt}(2t^2 - t^2 \sin t)$$
$$= 4t - 2t \sin t - t^2 \cos t.$$

Motion

Look once again at Figure 9.26. We drew the figure for Δt positive, so $\Delta \mathbf{r}$ points forward, in the direction of the motion. The vector $\Delta \mathbf{r}/\Delta t$ (not shown), having the same direction as $\Delta \mathbf{r}$, points forward also. Had Δt been negative, $\Delta \mathbf{r}$ would have pointed backwards, against the direction of motion. The quotient $\Delta \mathbf{r}/\Delta t$, however, being a negative scalar multiple of $\Delta \mathbf{r}$, would have once again pointed forward. No matter how $\Delta \mathbf{r}$ points, $\Delta \mathbf{r}/\Delta t$ points forward and we expect the vector $d\mathbf{r}/dt = \lim_{\Delta t \to 0} \Delta \mathbf{r}/\Delta t$, when different from $\mathbf{0}$, to do the same. This means that the derivative $d\mathbf{r}/dt$ is just what we want for modeling a particle's velocity. It points in the direction of motion and gives the rate of change of position with respect to time. For a smooth curve, the velocity is never zero; the particle does not stop or reverse direction.

Definitions Velocity, Speed, Acceleration, Direction of Motion

If $\mathbf{r}$ is the position vector of a particle moving along a smooth curve in the plane, then at any time t,

1. $\mathbf{v}(t) = \dfrac{d\mathbf{r}}{dt}$ is the particle's **velocity vector** and is tangent to the curve
2. $|\mathbf{v}(t)|$, the magnitude of $\mathbf{v}$, is the particle's **speed**
3. $\mathbf{a}(t) = \dfrac{d\mathbf{v}}{dt} = \dfrac{d^2\mathbf{r}}{dt^2}$, the derivative of velocity and the second derivative of position, is the particle's **acceleration vector**
4. $\dfrac{\mathbf{v}}{|\mathbf{v}|}$, a unit vector, is the **direction of motion.**

We can express the velocity of a moving particle as the product of its speed and direction (Example 7, Section 9.1).

$$\text{Velocity} = |\mathbf{v}|\left(\frac{\mathbf{v}}{|\mathbf{v}|}\right) = (\text{speed})(\text{direction})$$

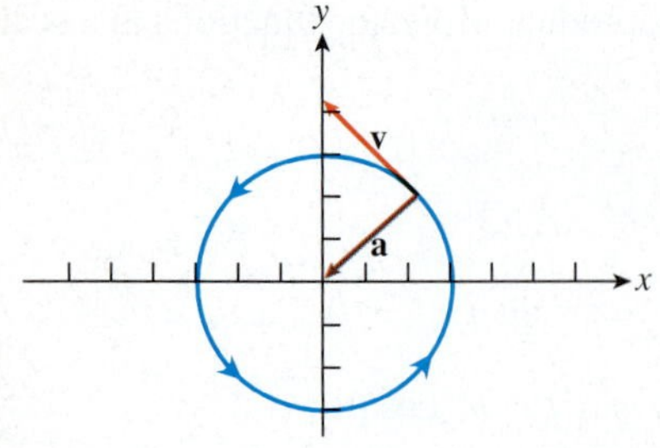

FIGURE 9.28 At $t = \pi/4$, the velocity vector $-(3/\sqrt{2})\mathbf{i} + (3/\sqrt{2})\mathbf{j}$ is tangent to the circle and the acceleration vector $-(3/\sqrt{2})\mathbf{i} - (3/\sqrt{2})\mathbf{j}$ is perpendicular to the tangent, pointing toward the center of the circle. (Example 6)

Example 6 Studying Motion on a Circle

The vector $\mathbf{r}(t) = (3 \cos t)\mathbf{i} + (3 \sin t)\mathbf{j}$ gives the position of a particle at time t moving counterclockwise on the circle of radius 3 centered at the origin (Figure 9.28). Find

(a) the velocity and acceleration vectors

(b) the velocity, acceleration, speed, and direction of motion at $t = \pi/4$

(c) $\mathbf{v} \cdot \mathbf{a}$. Interpret this result geometrically.

Solution

(a) $\mathbf{v} = \dfrac{d\mathbf{r}}{dt} = (-3 \sin t)\mathbf{i} + (3 \cos t)\mathbf{j}$

$\mathbf{a} = \dfrac{d\mathbf{v}}{dt} = (-3 \cos t)\mathbf{i} - (3 \sin t)\mathbf{j}$

(b) At $t = \pi/4$, the particle's velocity and acceleration are

Velocity: $$\mathbf{v}\left(\frac{\pi}{4}\right) = \left(-3 \sin \frac{\pi}{4}\right)\mathbf{i} + \left(3 \cos \frac{\pi}{4}\right)\mathbf{j} = -\frac{3}{\sqrt{2}}\mathbf{i} + \frac{3}{\sqrt{2}}\mathbf{j}$$

Acceleration: $$\mathbf{a}\left(\frac{\pi}{4}\right) = \left(-3 \cos \frac{\pi}{4}\right)\mathbf{i} - \left(3 \sin \frac{\pi}{4}\right)\mathbf{j} = -\frac{3}{\sqrt{2}}\mathbf{i} - \frac{3}{\sqrt{2}}\mathbf{j}.$$

Its speed and direction are

Speed: $$\left|\mathbf{v}\left(\frac{\pi}{4}\right)\right| = \sqrt{\left(\frac{-3}{\sqrt{2}}\right)^2 + \left(\frac{3}{\sqrt{2}}\right)^2} = 3$$

Direction: $$\frac{\mathbf{v}(\pi/4)}{|\mathbf{v}(\pi/4)|} = \frac{-3/\sqrt{2}}{3}\mathbf{i} + \frac{3/\sqrt{2}}{3}\mathbf{j} = -\frac{1}{\sqrt{2}}\mathbf{i} + \frac{1}{\sqrt{2}}\mathbf{j}.$$

(c) $\mathbf{v} \cdot \mathbf{a} = 9 \sin t \cos t - 9 \sin t \cos t = 0$

Thus, in this example, **v** and **a** are perpendicular for all values of t.

Figure 9.28 shows the path and the velocity and acceleration vectors at $t = \pi/4$.

Example 7 Studying Motion

The vector $\mathbf{r}(t) = (2t^3 - 3t^2)\mathbf{i} + (t^3 - 12t)\mathbf{j}$ gives the position of a moving particle at time t.

(a) Write an equation for the line tangent to the path of the particle at the point where $t = -1$.

(b) Find the coordinates of each point on the path where the horizontal component of the velocity is 0.

Solution

(a) $\mathbf{v}(t) = \dfrac{d\mathbf{r}}{dt} = (6t^2 - 6t)\mathbf{i} + (3t^2 - 12)\mathbf{j}$

At $t = -1$, $\mathbf{r}(-1) = -5\mathbf{i} + 11\mathbf{j}$ and $\mathbf{v}(-1) = 12\mathbf{i} - 9\mathbf{j}$. Thus, we want the equation of the line through $(-5, 11)$ with slope $-9/12 = -3/4$.

$$y - 11 = -\frac{3}{4}(x + 5) \qquad \text{or} \qquad y = -\frac{3}{4}x + \frac{29}{4}$$

(b) The horizontal component of the velocity is $6t^2 - 6t$. It equals 0 when $t = 0$ and $t = 1$. The point corresponding to $t = 0$ is the origin $(0, 0)$; the point corresponding to $t = 1$ is $(-1, -11)$.

Integrals

A differentiable vector function $\mathbf{R}(t)$ is an **antiderivative** of a vector function $\mathbf{r}(t)$ on an interval I if $d\mathbf{R}/dt = \mathbf{r}$ at each point t of I. If $\mathbf{R}$ is an antiderivative of $\mathbf{r}$ on I, it can be shown, working one component at a time, that every antiderivative of $\mathbf{r}$ on I has the form $\mathbf{R} + \mathbf{C}$ for some constant vector $\mathbf{C}$ (Exercise 35).

Definition Indefinite Integral

The **indefinite integral** of $\mathbf{r}$ with respect to t is the set of all antiderivatives of $\mathbf{r}$, denoted by $\int \mathbf{r}(t)\, dt$. If $\mathbf{R}$ is any antiderivative of $\mathbf{r}$, then

$$\int \mathbf{r}(t)\, dt = \mathbf{R}(t) + \mathbf{C}.$$

The usual arithmetic rules for indefinite integrals apply.

Example 8 Finding Antiderivatives

$$\int ((\cos t)\mathbf{i} - 2t\mathbf{j})\, dt = \left(\int \cos t\, dt\right)\mathbf{i} - \left(\int 2t\, dt\right)\mathbf{j} \tag{3}$$

$$= (\sin t + C_1)\mathbf{i} - (t^2 + C_2)\mathbf{j} \tag{4}$$

$$= (\sin t)\mathbf{i} - t^2\mathbf{j} + \mathbf{C} \qquad \mathbf{C} = C_1\mathbf{i} + C_2\mathbf{j}$$

As with integration of scalar functions, we recommend that you skip the steps in Equations 3 and 4 and go directly to the final form. Find an antiderivative for each component and add a constant vector at the end.

As with derivatives and indefinite integrals, definite integrals of vector functions are calculated component by component.

Definition Definite Integral

If the components of $\mathbf{r}(t) = f(t)\mathbf{i} + g(t)\mathbf{j}$ are integrable on $[a, b]$, then so is $\mathbf{r}$, and the **definite integral** of $\mathbf{r}$ from a to b is

$$\int_a^b \mathbf{r}(t)\, dt = \left(\int_a^b f(t)\, dt\right)\mathbf{i} + \left(\int_a^b g(t)\, dt\right)\mathbf{j}.$$

Example 9 Evaluating Definite Integrals

$$\int_0^{\pi} ((\cos t)\mathbf{i} - 2t\mathbf{j})\,dt = \left(\int_0^{\pi} \cos t\,dt\right)\mathbf{i} - \left(\int_0^{\pi} 2t\,dt\right)\mathbf{j}$$

$$= \left(\sin t\Big]_0^{\pi}\right)\mathbf{i} - \left(t^2\Big]_0^{\pi}\right)\mathbf{j} = 0\mathbf{i} - \pi^2\mathbf{j} = -\pi^2\mathbf{j}$$

Example 10 Finding a Path

The velocity vector of a particle moving in the plane (scaled in meters) is

$$\frac{d\mathbf{r}}{dt} = \frac{1}{t+1}\mathbf{i} + 2t\mathbf{j}, \qquad t \geq 0.$$

(a) Find the particle's position as a vector function of t if $\mathbf{r} = (\ln 2)\mathbf{i}$ when $t = 1$.

(b) Find the distance the particle travels from $t = 0$ to $t = 2$.

Solution

(a) $\mathbf{r} = \left(\int \frac{dt}{t+1}\right)\mathbf{i} + \left(\int 2t\,dt\right)\mathbf{j} = (\ln\,(t+1))\mathbf{i} + t^2\mathbf{j} + \mathbf{C}$

$\mathbf{r}(1) = (\ln\,2)\mathbf{i} + \mathbf{j} + \mathbf{C} = (\ln\,2)\mathbf{i}$

Thus, $\mathbf{C} = -\mathbf{j}$ and

$$\mathbf{r} = (\ln\,(t+1))\mathbf{i} + (t^2 - 1)\mathbf{j}.$$

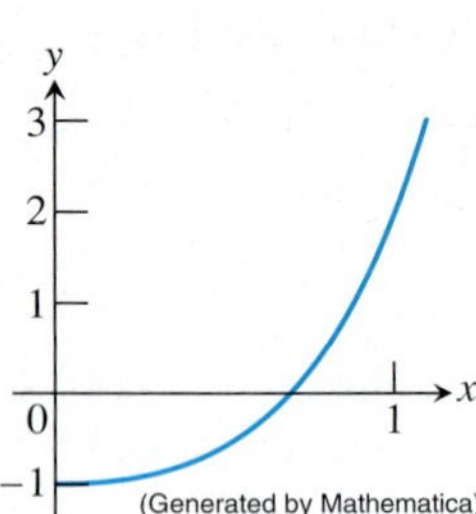

FIGURE 9.29 The path of the particle in Example 10 for $0 \leq t \leq 2$.

(b) The parametrization

$$x = \ln\,(t+1), \qquad y = t^2 - 1, \qquad 0 \leq t \leq 2$$

is smooth, and because x and y are increasing functions of t, the path is traversed exactly once as t increases from 0 to 2 (Figure 9.29). The length is

$$L = \int_0^2 \sqrt{\left(\frac{dx}{dt}\right)^2 + \left(\frac{dy}{dt}\right)^2}\,dt = \int_0^2 \sqrt{\left(\frac{1}{t+1}\right)^2 + (2t)^2}\,dt \approx 4.34\text{ m}.$$

EXERCISES 9.3

Studying Motion

In Exercises 1–4, $\mathbf{r}(t)$ is the position vector of a particle in the plane at time t.

T **(a)** Draw the graph of the path of the particle.

(b) Find the velocity and acceleration vectors.

(c) Find the particle's speed and direction of motion at the given value of t.

(d) Write the particle's velocity at that time as the product of its speed and direction.

1. $\mathbf{r}(t) = (2\cos t)\mathbf{i} + (3\sin t)\mathbf{j}, \quad t = \pi/2$

2. $\mathbf{r}(t) = (\cos 2t)\mathbf{i} + (2\sin t)\mathbf{j}, \quad t = 0$

3. $\mathbf{r}(t) = (\sec t)\mathbf{i} + (\tan t)\mathbf{j}, \quad t = \pi/6$

4. $\mathbf{r}(t) = (2\ln\,(t+1))\mathbf{i} + (t^2)\mathbf{j}, \quad t = 1$

In Exercises 5–8, $\mathbf{r}(t)$ is the position vector of a particle in the plane at time t. Find the time, or times, in the given time interval when the velocity and acceleration vectors are perpendicular.

5. $\mathbf{r}(t) = (t - \sin t)\mathbf{i} + (1 - \cos t)\mathbf{j}, \quad 0 \leq t \leq 2\pi$

6. $\mathbf{r}(t) = (\sin t)\mathbf{i} + t\mathbf{j}, \quad t \geq 0$

7. $\mathbf{r}(t) = (3 \cos t)\mathbf{i} + (4 \sin t)\mathbf{j}, \quad t \geq 0$

8. $\mathbf{r}(t) = (5 \cos t)\mathbf{i} + (5 \sin t)\mathbf{j}, \quad t \geq 0$

In Exercises 9 and 10, $\mathbf{r}(t)$ is the position vector of a particle in the plane at time t. Find the angle between the velocity and acceleration vectors at the given value of t.

9. $\mathbf{r}(t) = (2 \cos t)\mathbf{i} + (\sin t)\mathbf{j}, \quad t = \pi/4$

10. $\mathbf{r}(t) = (3t + 1)\mathbf{i} + (t^2)\mathbf{j}, \quad t = 0$

Limits and Continuity

In Exercises 11 and 12, **(a)** evaluate the limit, and **(b)** find the values of t for which the vector function is continuous and **(c)** discontinuous.

11. $\lim_{t \to 3} \left[t\mathbf{i} + \dfrac{t^2 - 9}{t^2 + 3t}\mathbf{j} \right]$

12. $\lim_{t \to 0} \left[\dfrac{\sin 2t}{t}\mathbf{i} + (\ln (t + 1))\mathbf{j} \right]$

Tangents and Normals

In Exercises 13 and 14, find an equation for the line that is **(a)** tangent and **(b)** normal to the curve $\mathbf{r}(t)$ at the point determined by the given value of t.

13. $\mathbf{r}(t) = (\sin t)\mathbf{i} + (t^2 - \cos t)\mathbf{j}, \quad t = 0$

14. $\mathbf{r}(t) = (2 \cos t - 3)\mathbf{i} + (3 \sin t + 1)\mathbf{j}, \quad t = \pi/4$

Integration

In Exercises 15–18, evaluate the integral.

15. $\int_1^2 [(6 - 6t)\mathbf{i} + 3\sqrt{t}\mathbf{j}]\, dt$

16. $\int_{-\pi/4}^{\pi/4} [(\sin t)\mathbf{i} + (1 + \cos t)\mathbf{j}]\, dt$

17. $\int [(\sec t \tan t)\mathbf{i} + (\tan t)\mathbf{j}]\, dt$

18. $\int \left[\dfrac{1}{t}\mathbf{i} + \dfrac{1}{5 - t}\mathbf{j} \right] dt$

Initial Value Problems

In Exercises 19–22, solve the initial value problem for $\mathbf{r}$ as a vector function of t.

19. $\dfrac{d\mathbf{r}}{dt} = \dfrac{3}{2}(t + 1)^{1/2}\mathbf{i} + e^{-t}\mathbf{j}, \quad \mathbf{r}(0) = \mathbf{0}$

20. $\dfrac{d\mathbf{r}}{dt} = (t^3 + 4t)\mathbf{i} + t\mathbf{j}, \quad \mathbf{r}(0) = \mathbf{i} + \mathbf{j}$

21. $\dfrac{d^2\mathbf{r}}{dt^2} = -32\mathbf{j}, \quad \mathbf{r}(0) = 100\mathbf{i}, \quad \left.\dfrac{d\mathbf{r}}{dt}\right|_{t=0} = 8\mathbf{i} + 8\mathbf{j}$

22. $\dfrac{d^2\mathbf{r}}{dt^2} = -\mathbf{i} - \mathbf{j}, \quad \mathbf{r}(0) = 10\mathbf{i} + 10\mathbf{j}, \quad \left.\dfrac{d\mathbf{r}}{dt}\right|_{t=0} = \mathbf{0}$

Paths and Motion

23. *Finding distance traveled* The position of a particle in the plane at time t is given by

$$\mathbf{r}(t) = (1 - \cos t)\mathbf{i} + (t - \sin t)\mathbf{j}.$$

Find the distance the particle travels along the path from $t = 0$ to $t = 2\pi/3$.

24. *Length of a path* Let C be the path traced by

$$\mathbf{r}(t) = \left(\frac{1}{4}e^{4t} - t \right)\mathbf{i} + (e^{2t})\mathbf{j}, \qquad 0 \leq t \leq 2.$$

(a) Find the initial and terminal points of C.

(b) Find the length of C.

25. *Velocity on a path* The position of a particle is given by

$$\mathbf{r}(t) = (\sin t)\mathbf{i} + (\cos 2t)\mathbf{j}.$$

(a) Find the velocity vector for the particle.

(b) For what values of t in the interval $0 \leq t \leq 2\pi$ is $d\mathbf{r}/dt$ equal to $\mathbf{0}$?

(c) *Writing to Learn* Find a Cartesian equation for a curve that contains the particle's path. What portion of the graph of the Cartesian equation is traced by the particle? Describe the motion as t increases from 0 to 2π.

26. *Revisiting Example 7* The position of a particle is given by $\mathbf{r}(t) = (2t^3 - 3t^2)\mathbf{i} + (t^3 - 12t)\mathbf{j}$.

(a) Find dy/dx in terms of t.

(b) *Writing to Learn* Find the x- and y-coordinates for each critical point of the path (point where dy/dx is zero or does not exist). Does the path have a vertical or horizontal tangent at the critical point? Explain.

27. *Finding a position vector* At time $t = 0$, a particle is located at the point $(1, 2)$. It travels in a straight line to the point $(4, 1)$, has speed 2 at $(1, 2)$, and constant acceleration $3\mathbf{i} - \mathbf{j}$. Find an equation for the position vector $\mathbf{r}(t)$ of the particle at time t.

28. *Studying a motion* The path of a particle for $t > 0$ is given by

$$\mathbf{r}(t) = \left(t + \frac{2}{t} \right)\mathbf{i} + (3t^2)\mathbf{j}.$$

(a) Find the coordinates of each point on the path where the horizontal component of the velocity of the particle is zero.

(b) Find dy/dx when $t = 1$.

(c) Find d^2y/dx^2 when $y = 12$.

29. *Motion on circular paths* Each of equations (a) through (e) describes the motion of a particle having the same path, namely, the unit circle $x^2 + y^2 = 1$. Although the path of each particle in (a) through (e) is the same, the behavior, or "dynamics," of each particle is different. For each particle, answer the following questions.

i. Does the particle have constant speed? If so, what is its constant speed?

ii. Is the particle's acceleration vector always orthogonal to its velocity vector?

iii. Does the particle move clockwise or counterclockwise around the circle?

iv. Does the particle begin at the point (1, 0)?

(a) $\mathbf{r}(t) = (\cos t)\mathbf{i} + (\sin t)\mathbf{j}, \quad t \geq 0$

(b) $\mathbf{r}(t) = (\cos 2t)\mathbf{i} + (\sin 2t)\mathbf{j}, \quad t \geq 0$

(c) $\mathbf{r}(t) = \cos(t - \pi/2)\mathbf{i} + \sin(t - \pi/2)\mathbf{j}, \quad t \geq 0$

(d) $\mathbf{r}(t) = (\cos t)\mathbf{i} - (\sin t)\mathbf{j}, \quad t \geq 0$

(e) $\mathbf{r}(t) = \cos(t^2)\mathbf{i} + \sin(t^2)\mathbf{j}, \quad t \geq 0$

30. *Motion on a parabola* A particle moves along the top of the parabola $y^2 = 2x$ from left to right at a constant speed of 5 units per second. Find the velocity of the particle as it moves through the point (2, 2).

Applications

31. *Flying a kite* The position of a kite is given by

$$\mathbf{r}(t) = \frac{t}{8}\mathbf{i} - \frac{3}{64}t(t - 160)\mathbf{j},$$

where $t \geq 0$ is measured in seconds and distance is measured in meters.

(a) How long is the kite above ground?

(b) How high is the kite at $t = 40$ sec?

(c) At what rate is the kite's altitude increasing at $t = 40$ sec?

(d) At what time does the kite start to lose altitude?

32. *Colliding particles* The paths of two particles for $t \geq 0$ are given by

$$\mathbf{r}_1(t) = (t - 3)\mathbf{i} + (t - 3)^2\mathbf{j},$$

$$\mathbf{r}_2(t) = \left(\frac{3t}{2} - 4\right)\mathbf{i} + \left(\frac{3t}{2} - 2\right)\mathbf{j}.$$

(a) Determine the exact time(s) at which the particles collide.

(b) Find the direction of motion of each particle at the time(s) of collision.

33. *A satellite in circular orbit* A satellite of mass m is moving at a constant speed v around a planet of mass M in a circular orbit of radius r_0, as measured from the planet's center of mass. Determine the satellite's orbital period T (the time to complete one full orbit), as follows.

(a) Coordinatize the orbital plane by placing the origin at the planet's center of mass, with the satellite on the x-axis at $t = 0$ and moving counterclockwise, as in the accompanying figure.

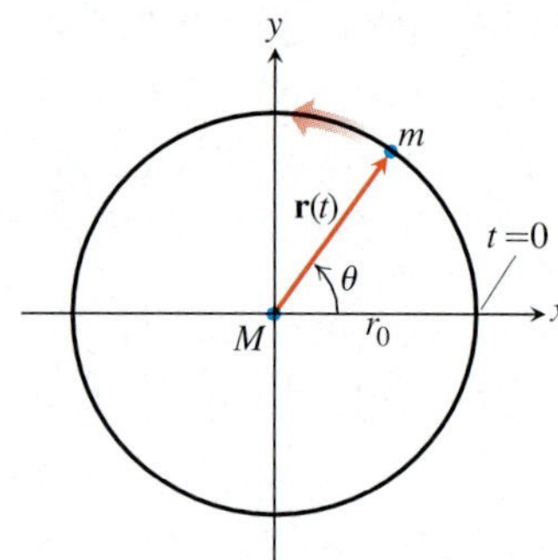

Let $\mathbf{r}(t)$ be the satellite's position vector at time t. Show that $\theta = vt/r_0$ and hence that

$$\mathbf{r}(t) = \left(r_0 \cos \frac{vt}{r_0}\right)\mathbf{i} + \left(r_0 \sin \frac{vt}{r_0}\right)\mathbf{j}.$$

(b) Find the acceleration of the satellite.

(c) According to Newton's law of gravitation, the gravitational force exerted on the satellite by the planet is directed toward the origin and is given by

$$\mathbf{F} = \left(-\frac{GmM}{r_0^{\,2}}\right)\frac{\mathbf{r}}{r_0},$$

where G is the universal constant of gravitation. Using Newton's second law, $\mathbf{F} = m\mathbf{a}$, show that $v^2 = GM/r_0$.

(d) Show that the orbital period T satisfies $vT = 2\pi r_0$.

(e) From parts (c) and (d), deduce that

$$T^2 = \frac{4\pi^2}{GM}r_0^{\,3};$$

that is, the square of the period of a satellite in circular orbit is proportional to the cube of the radius from the orbital center.

34. *Rowing across a river* A straight river is 100 m wide. A rowboat leaves the far shore at time $t = 0$. The person in the boat rows at a rate of 20 m/min always toward the near shore. The velocity of the river at (x, y) is

$$\mathbf{v} = \left(-\frac{1}{250}(y - 50)^2 + 10\right)\mathbf{i} \quad \text{m/min}, \qquad 0 < y < 100.$$

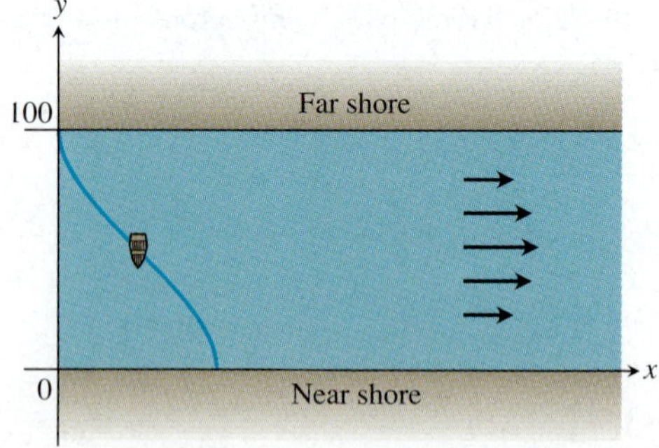

(a) Given that $\mathbf{r}(0) = 0\mathbf{i} + 100\mathbf{j}$, what is the position of the boat at time t?

(b) How far downstream will the boat land on the near shore?

Theory and Examples

35. *Antiderivatives of vector functions*

(a) Use Corollary 2 in Section 3.2 (a consequence of the Mean Value Theorem for scalar functions) to show that two vector functions $\mathbf{R}_1(t)$ and $\mathbf{R}_2(t)$ that have identical derivatives on an interval I differ by a constant vector value throughout I.

(b) Use the result in part (a) to show that if $\mathbf{R}(t)$ is any antiderivative of $\mathbf{r}(t)$ on I, then every other antiderivative of $\mathbf{r}(t)$ on I equals $\mathbf{R}(t) + \mathbf{C}$ for some constant vector $\mathbf{C}$.

36. *Constant-length vector functions* Let $\mathbf{v}$ be a differentiable vector function of t. Show that if $\mathbf{v} \cdot (d\mathbf{v}/dt) = 0$ for all t, then $|\mathbf{v}|$ is constant.

37. *Constant function rule* Prove that if $\mathbf{u}$ is the vector function with the constant value $\mathbf{C}$, then $d\mathbf{u}/dt = \mathbf{0}$.

38. *Scalar multiple rules*

(a) Prove that if $\mathbf{u}$ is a differentiable function of t and c is any real number, then

$$\frac{d(c\mathbf{u})}{dt} = c\frac{d\mathbf{u}}{dt}.$$

(b) Prove that if $\mathbf{u}$ is a differentiable function of t and f is a differentiable scalar function of t, then

$$\frac{d(f\mathbf{u})}{dt} = \frac{df}{dt}\mathbf{u} + f\frac{d\mathbf{u}}{dt}.$$

39. *Sum and difference rules* Prove that if $\mathbf{u}$ and $\mathbf{v}$ are differentiable functions of t, then

(a) $\dfrac{d}{dt}(\mathbf{u} + \mathbf{v}) = \dfrac{d\mathbf{u}}{dt} + \dfrac{d\mathbf{v}}{dt}.$

(b) $\dfrac{d}{dt}(\mathbf{u} - \mathbf{v}) = \dfrac{d\mathbf{u}}{dt} - \dfrac{d\mathbf{v}}{dt}.$

40. *Chain rule* Prove that if $\mathbf{u}$ is a differentiable vector function of s and $s = f(t)$ is a differentiable scalar function of t, then

$$\frac{d}{dt}[\mathbf{u}(f(t))] = f'(t)\mathbf{u}'(f(t)).$$

41. *Differentiable vector functions are continuous* Show that if $\mathbf{r}(t) = f(t)\mathbf{i} + g(t)\mathbf{j}$ is differentiable at $t = c$, then $\mathbf{r}$ is continuous at c as well.

42. *Integration properties* Establish the following properties of integrable vector functions.

(a) *Constant Scalar Multiple Rule:*

$$\int_a^b k\mathbf{r}(t)\,dt = k\int_a^b \mathbf{r}(t)\,dt$$

for any scalar constant k.

(b) *Sum and Difference Rules:*

$$\int_a^b (\mathbf{r}_1(t) \pm \mathbf{r}_2(t))\,dt = \int_a^b \mathbf{r}_1(t)\,dt \pm \int_a^b \mathbf{r}_2(t)\,dt$$

(c) *Constant Vector Multiple Rule:*

$$\int_a^b \mathbf{C} \cdot \mathbf{r}(t)\,dt = \mathbf{C} \cdot \int_a^b \mathbf{r}(t)\,dt$$

for any constant vector $\mathbf{C}$.

43. *Fundamental theorem of calculus* The Fundamental Theorem of Calculus for scalar functions of a real variable holds for vector functions of a real variable as well.

(a) Prove this by using the theorem for scalar functions to show that if a vector function $\mathbf{r}(t)$ is continuous for $a \le t \le b$, then

$$\frac{d}{dt}\int_a^t \mathbf{r}(q)\,dq = \mathbf{r}(t)$$

at every point t of $[a, b]$.

(b) Use the conclusion in part (b) of Exercise 35 to show that if $\mathbf{R}$ is any antiderivative of $\mathbf{r}$ on $[a, b]$, then

$$\int_a^b \mathbf{r}(t)\,dt = \mathbf{R}(b) - \mathbf{R}(a).$$

9.4 Modeling Projectile Motion

CD-ROM

Ideal Projectile Motion • Height, Flight Time, and Range • Ideal Trajectories Are Parabolic • Firing from (x_0, y_0) • Projectile Motion with Wind Gusts

When we shoot a projectile into the air, we usually want to know beforehand how far it will go (will it reach the target?), how high it will rise (will it clear the hill?), and when it will land (when do we get results?) We get this information from the di-

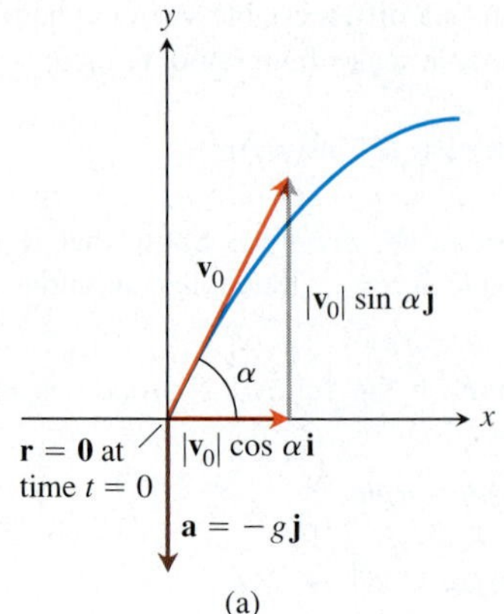

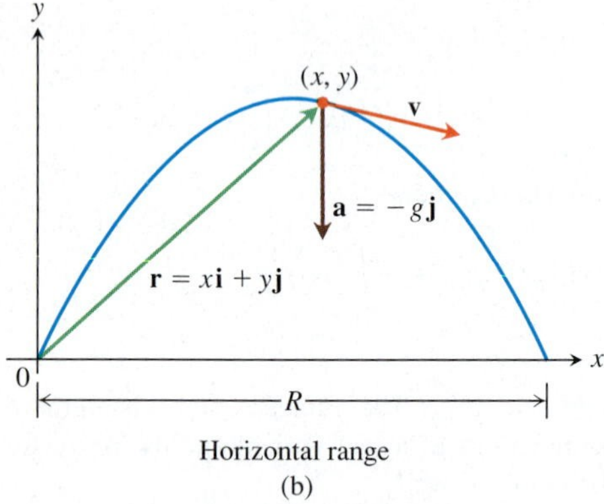

FIGURE 9.30 (a) Position, velocity, acceleration, and launch angle at $t = 0$. (b) Position, velocity, and acceleration at a later time t

rection and magnitude of the projectile's initial velocity vector, using Newton's second law of motion.

Ideal Projectile Motion

We are going to model *ideal* projectile motion. This assumes that the projectile behaves like a particle moving in a vertical coordinate plane and that the only force acting on the projectile during its flight (close to Earth's surface) is the constant force of gravity, always pointing straight down.

We assume that the projectile is launched from the origin at time $t = 0$ into the first quadrant with an initial velocity $\mathbf{v}_0$ (Figure 9.30). If $\mathbf{v}_0$ makes an angle α with the horizontal, then

$$\mathbf{v}_0 = (|\mathbf{v}_0| \cos \alpha)\mathbf{i} + (|\mathbf{v}_0| \sin \alpha)\mathbf{j}.$$

If we use the simpler notation v_0 for the initial speed $|\mathbf{v}_0|$, then

$$\mathbf{v}_0 = (v_0 \cos \alpha)\mathbf{i} + (v_0 \sin \alpha)\mathbf{j}. \tag{1}$$

The projectile's initial position is

$$\mathbf{r}_0 = 0\mathbf{i} + 0\mathbf{j} = \mathbf{0}. \tag{2}$$

Newton's second law of motion says that the force acting on the projectile is equal to the projectile's mass m times its acceleration, or $m(d^2\mathbf{r}/dt^2)$ if $\mathbf{r}$ is the projectile's position vector and t is time. If the force is solely the gravitational force $-mg\mathbf{j}$, then

$$m\frac{d^2\mathbf{r}}{dt^2} = -mg\mathbf{j} \qquad \text{and} \qquad \frac{d^2\mathbf{r}}{dt^2} = -g\mathbf{j}.$$

We find $\mathbf{r}$ as a function of t by solving the following initial value problem.

Differential equation: $\dfrac{d^2\mathbf{r}}{dt^2} = -g\mathbf{j}$

Initial conditions: $\mathbf{r} = \mathbf{r}_0$ and $\dfrac{d\mathbf{r}}{dt} = \mathbf{v}_0$ when $t = 0$

The first integration gives

$$\frac{d\mathbf{r}}{dt} = -(gt)\mathbf{j} + \mathbf{v}_0.$$

A second integration gives

$$\mathbf{r} = -\frac{1}{2}gt^2\mathbf{j} + \mathbf{v}_0 t + \mathbf{r}_0.$$

Substituting the values of $\mathbf{v}_0$ and $\mathbf{r}_0$ from Equations 1 and 2 gives

$$\mathbf{r} = -\frac{1}{2}gt^2\mathbf{j} + \underbrace{(v_0 \cos \alpha)t\mathbf{i} + (v_0 \sin \alpha)t\mathbf{j}}_{\mathbf{v}_0 t} + \mathbf{0}$$

or

$$\mathbf{r} = (v_0 \cos \alpha)\,t\mathbf{i} + \left((v_0 \sin \alpha)\,t - \frac{1}{2}gt^2\right)\mathbf{j}. \tag{3}$$

Equation (3) is the **vector equation** for ideal projectile motion. The angle α is the projectile's **launch angle (firing angle, angle of elevation),** and v_0, as we said before, is the projectile's **initial speed.**

The components of **r** give

$$x = (v_0 \cos \alpha)t \qquad \text{and} \qquad y = (v_0 \sin \alpha)t - \frac{1}{2}gt^2, \tag{4}$$

where x is the distance downrange and y is the height of the projectile at time $t \geq 0$.

CD-ROM
WEBsite
Historical Biography
Joseph Louis Lagrange
(1736 — 1813)

Example 1 Firing an Ideal Projectile

A projectile is fired from the origin over horizontal ground at an initial speed of 500 m/sec and a launch angle of 60°. Where will the projectile be 10 sec later?

Solution We use Equation (3) with $v_0 = 500$, $\alpha = 60°$, $g = 9.8$, and $t = 10$ to find the projectile's components 10 sec after firing.

$$\begin{aligned}
\mathbf{r} &= (v_0 \cos \alpha)t\mathbf{i} + \left((v_0 \sin \alpha)t - \frac{1}{2}gt^2\right)\mathbf{j} \\
&= (500)\left(\frac{1}{2}\right)(10)\mathbf{i} + \left((500)\left(\frac{\sqrt{3}}{2}\right)10 - \left(\frac{1}{2}\right)(9.8)(100)\right)\mathbf{j} \\
&\approx 2500\mathbf{i} + 3840\mathbf{j}.
\end{aligned}$$

Interpret

Ten seconds after firing, the projectile is about 3840 m in the air and 2500 m downrange.

Height, Flight Time, and Range

Equation (3) enables us to answer most questions about the ideal motion for a projectile fired from the origin.

The projectile reaches its highest point when its vertical velocity component is zero, that is, when

$$\frac{dy}{dt} = v_0 \sin \alpha - gt = 0, \qquad \text{or} \qquad t = \frac{v_0 \sin \alpha}{g}.$$

For this value of t, the value of y is

$$y_{\max} = (v_0 \sin \alpha)\left(\frac{v_0 \sin \alpha}{g}\right) - \frac{1}{2}g\left(\frac{v_0 \sin \alpha}{g}\right)^2 = \frac{(v_0 \sin \alpha)^2}{2g}.$$

To find when the projectile lands when fired over horizontal ground, we set the vertical component equal to zero in Equation (3) and solve for t.

$$\begin{aligned}
(v_0 \sin \alpha)t - \frac{1}{2}gt^2 &= 0 \\
t\left(v_0 \sin \alpha - \frac{1}{2}gt\right) &= 0 \\
t = 0, \qquad t &= \frac{2v_0 \sin \alpha}{g}
\end{aligned}$$

Since 0 is the time the projectile is fired, $(2v_0 \sin \alpha)/g$ must be the time when the projectile strikes the ground.

To find the projectile's **range** R, the distance from the origin to the point of impact on horizontal ground, we find the value of the horizontal component when $t = (2v_0 \sin \alpha)/g$.

$$x = (v_0 \cos \alpha)t$$

$$R = (v_0 \cos \alpha)\left(\frac{2v_0 \sin \alpha}{g}\right) = \frac{v_0^2}{g}(2 \sin \alpha \cos \alpha) = \frac{v_0^2}{g} \sin 2\alpha$$

The range is largest when $\sin 2\alpha = 1$ or $\alpha = 45°$.

Height, Flight Time, and Range for Ideal Projectile Motion

For ideal projectile motion when an object is launched from the origin over a horizontal surface with initial speed v_0 and launch angle α:

Maximum height: $$y_{max} = \frac{(v_0 \sin \alpha)^2}{2g}$$

Flight time: $$t = \frac{2v_0 \sin \alpha}{g}$$

Range: $$R = \frac{v_0^2}{g} \sin 2\alpha.$$

Example 2 Investigating Ideal Projectile Motion

Find the maximum height, flight time, and range of a projectile fired from the origin over horizontal ground at an initial speed of 500 m/sec and a launch angle of 60° (same projectile as Example 1).

Solution

Maximum height: $$y_{max} = \frac{(v_0 \sin \alpha)^2}{2g} = \frac{(500 \sin 60°)^2}{2(9.8)} \approx 9566.33 \text{ m}$$

Flight time: $$t = \frac{2v_0 \sin \alpha}{g} = \frac{2(500) \sin 60°}{9.8} \approx 88.37 \text{ sec}$$

Range: $$R = \frac{v_0^2}{g} \sin 2\alpha = \frac{(500)^2 \sin 120°}{9.8} \approx 22{,}092.48 \text{ m}$$

From Equation (3), the position vector of the projectile is

$$\begin{aligned}\mathbf{r} &= (v_0 \cos \alpha)t\mathbf{i} + \left((v_0 \sin \alpha)t - \frac{1}{2}gt^2\right)\mathbf{j} \\ &= (500 \cos 60°)t\mathbf{i} + \left((500 \sin 60°)t - \frac{1}{2}(9.8)t^2\right)\mathbf{j} \\ &= 250t\mathbf{i} + ((250\sqrt{3})t - 4.9t^2)\mathbf{j}.\end{aligned}$$

A graph of the projectile's path is shown in Figure 9.31.

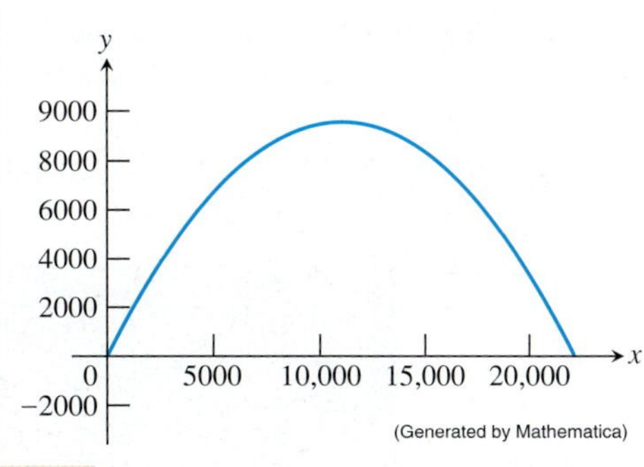

FIGURE 9.31 The graph of the parametric equations $x = 250t$, $y = (250\sqrt{3})t - 4.9t^2$ for $0 \le t \le 88.4$. (Example 2)

Ideal Trajectories Are Parabolic

CD-ROM
WEBsite
Historical Biography
Henry Briggs
(1561 — 1631)

It is often claimed that water from a hose traces a parabola in the air, but any one who looks closely enough will see this is not so. The air slows the water down, and its forward progress is too slow at the end to keep pace with the rate at which it falls.

What is really being claimed is that ideal projectiles move along parabolas, and this we can see from Equation (4). If we substitute $t = x/(v_0 \cos \alpha)$ from the first equation into the second, we obtain the Cartesian-coordinate equation

$$y = -\left(\frac{g}{2{v_0}^2 \cos^2 \alpha}\right) x^2 + (\tan \alpha)\, x.$$

This equation has the form $y = ax^2 + bx$, so its graph is a parabola.

Firing from (x_0, y_0)

If we fire our ideal projectile from the point (x_0, y_0) instead of the origin (Figure 9.32), the position vector for the path of motion is

$$\mathbf{v} = (x_0 + (v_0 \cos \alpha)\, t)\mathbf{i} + \left(y_0 + (v_0 \sin \alpha)\, t - \frac{1}{2}\, g t^2\right)\mathbf{j}, \tag{5}$$

as you are asked to show in Exercise 19.

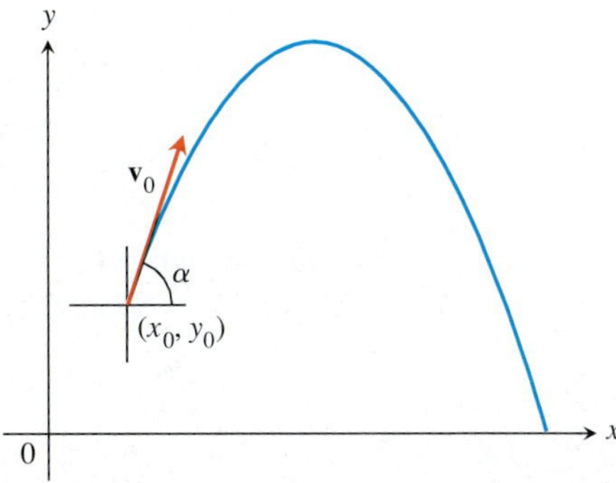

FIGURE 9.32 The path of a projectile fired from (x_0, y_0) with an initial velocity $\mathbf{v}_0$ at an angle of α degrees with the horizontal.

Example 3 Firing a Flaming Arrow

To open the 1992 Summer Olympics in Barcelona, bronze medalist archer Antonio Rebollo lit the Olympic torch with a flaming arrow (Figure 9.33). Suppose that Rebollo shot the arrow at a height of 6 ft above ground level 90 ft from the 70-ft-high cauldron, and he wanted the arrow to reach maximum height exactly 4 ft above the center of the cauldron (Figure 9.34).

FIGURE 9.33 Spanish archer Antonio Rebollo lights the Olympic torch in Barcelona with a flaming arrow.

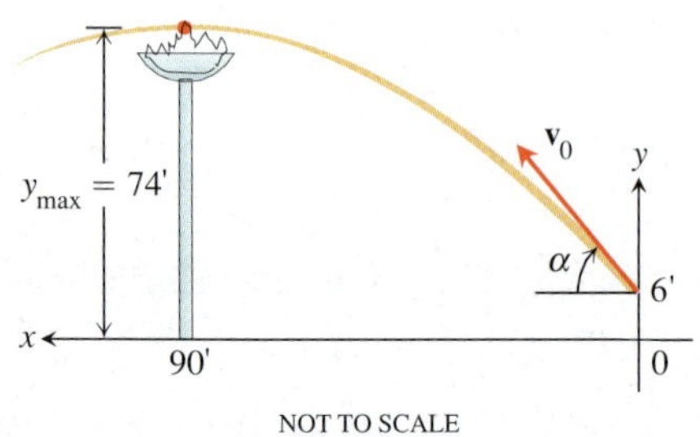

FIGURE 9.34 Ideal path of the arrow that lit the Olympic torch.

(a) Express y_{max} in terms of the initial speed v_0 and firing angle α.

(b) Use $y_{max} = 74$ ft (Figure 9.34) and the result from part (a) to find the value of $v_0 \sin \alpha$.

(c) Find the value of $v_0 \cos \alpha$.

(d) Find the initial firing angle of the arrow.

Solution

(a) We use a coordinate system in which the positive x-axis lies along the ground toward the left (to match the second photograph in Figure 9.33) and the coordinates of the flaming arrow at $t = 0$ are $x_0 = 0$ and $y_0 = 6$ (Figure 9.34). We have

$$y = y_0 + (v_0 \sin \alpha)t - \frac{1}{2}gt^2 \qquad \text{Equation (5), } \mathbf{j}\text{-component}$$

$$= 6 + (v_0 \sin \alpha)t - \frac{1}{2}gt^2. \qquad y_0 = 6$$

We find the time when the arrow reaches its highest point by setting $dy/dt = 0$ and solving for t, obtaining

$$t = \frac{v_0 \sin \alpha}{g}.$$

For this value of t, the value of y is

$$y_{max} = 6 + (v_0 \sin \alpha)\left(\frac{v_0 \sin \alpha}{g}\right) - \frac{1}{2}g\left(\frac{v_0 \sin \alpha}{g}\right)^2$$

$$= 6 + \frac{(v_0 \sin \alpha)^2}{2g}.$$

(b) Using $y_{max} = 74$ and g $= 32$, we see from the preceeding equation in part (a) that

$$74 = 6 + \frac{(v_0 \sin \alpha)^2}{2(32)}$$

or

$$v_0 \sin \alpha = \sqrt{(68)(64)}.$$

(c) When the arrow reaches y_{max}, the horizontal distance traveled to the center of the cauldron is $x = 90$ ft. We substitute the time to reach y_{max} from part (a) and the horizontal distance $x = 90$ ft into the **i**-component of Equation (5) to obtain

$$x = x_0 + (v_0 \cos \alpha)t \qquad \text{Equation (5), } \mathbf{i}\text{-component}$$

$$90 = 0 + (v_0 \cos \alpha)t \qquad x = 90,\ x_0 = 0$$

$$= (v_0 \cos \alpha)\left(\frac{v_0 \sin \alpha}{g}\right). \qquad t = (v_0 \sin \alpha)/g$$

Solving this equation for $v_0 \cos \alpha$ and using $g = 32$ and the result from part (b), we have

$$v_0 \cos \alpha = \frac{90g}{v_0 \sin \alpha} = \frac{(90)(32)}{\sqrt{(68)(64)}}.$$

(d) Parts (b) and (c) together tell us that

$$\tan \alpha = \frac{v_0 \sin \alpha}{v_0 \cos \alpha} = \frac{(\sqrt{(68)(64)}\,)^2}{(90)(32)} = \frac{68}{45}$$

or

$$\alpha = \tan^{-1}\left(\frac{68}{45}\right) \approx 56.5°.$$

This is Rebollo's firing angle.

Projectile Motion with Wind Gusts

The next example shows how to account for another force acting on a projectile. We also assume that the path of the baseball in Example 4 lies in a vertical plane.

Example 4 Hitting a Baseball

A baseball is hit when it is 3 ft above the ground. It leaves the bat with initial speed of 152 ft/sec, making an angle of 20° with the horizontal. At the instant the ball is hit, an instantaneous gust of wind blows in the horizontal direction directly opposite the direction the ball is taking toward the outfield, adding a component of $-8.8\mathbf{i}$ (ft/sec) to the ball's initial velocity (8.8 ft/sec = 6 mph).

(a) Find a vector equation (position vector) for the path of the baseball.

(b) How high does the baseball go, and when does it reach maximum height?

(c) Assuming that the ball is not caught, find its range and flight time.

Solution

(a) Using Equation (1) and accounting for the gust of wind, the initial velocity of the baseball is

$$\begin{aligned}\mathbf{v}_0 &= (v_0 \cos \alpha)\mathbf{i} + (v_0 \sin \alpha)\mathbf{j} - 8.8\mathbf{i} \\ &= (152 \cos 20°)\mathbf{i} + (152 \sin 20°)\mathbf{j} - (8.8)\mathbf{i} \\ &= (152 \cos 20° - 8.8)\mathbf{i} + (152 \sin 20°)\mathbf{j}.\end{aligned}$$

The initial position is $\mathbf{r}_0 = 0\mathbf{i} + 3\mathbf{j}$. Integration of $d^2\mathbf{r}/dt^2 = -g\mathbf{j}$ gives

$$\frac{d\mathbf{r}}{dt} = (gt)\mathbf{j} + \mathbf{v}_0.$$

A second integration gives

$$\mathbf{r} = -\frac{1}{2}gt^2\mathbf{j} + \mathbf{v}_0 t + \mathbf{r}_0.$$

Substituting the values of $\mathbf{v}_0$ and $\mathbf{r}_0$ into the last equation gives the position vector of the baseball.

$$\begin{aligned}\mathbf{r} &= -\frac{1}{2}gt^2\mathbf{j} + \mathbf{v}_0 t + \mathbf{r}_0 \\ &= -16t^2\mathbf{j} + (152 \cos 20° - 8.8)t\mathbf{i} + (152 \sin 20°)t\mathbf{j} + 3\mathbf{j} \\ &= (152 \cos 20° - 8.8)t\mathbf{i} + (3 + (152 \sin 20°)t - 16t^2)\mathbf{j}.\end{aligned}$$

(b) The baseball reaches its highest point when the vertical component of velocity is zero, or

$$\frac{dy}{dt} = 152 \sin 20° - 32t = 0.$$

Solving for t we find

$$t = \frac{152 \sin 20°}{32} \approx 1.62 \text{ sec}.$$

Substituting this time into the vertical component for $\mathbf{r}$ gives the maximum height

$$\begin{aligned}y_{\max} &= 3 + (152 \sin 20°)(1.62) - 16(1.62)^2 \\ &\approx 45.2 \text{ ft}.\end{aligned}$$

That is, the maximum height of the baseball is about 45.2 ft, reached about 1.6 sec after leaving the bat.

(c) To find when the baseball lands, we set the vertical component for **r** equal to 0 and solve for t:

$$3 + (152 \sin 20°)t - 16t^2 = 0$$
$$3 + (51.99)t - 16t^2 = 0.$$

The solution values are about $t = 3.3$ sec and $t = -0.06$ sec. Substituting the positive time into the horizontal component for **r**, we find the range

$$R = (152 \cos 20° - 8.8)(3.3)$$
$$\approx 442.3 \text{ ft}.$$

Thus, the horizontal range is about 442.3 ft, and the flight time is about 3.3 sec.

In Exercises 29 through 31, we consider projectile motion when there is air resistance slowing down the flight.

EXERCISES 9.4

Projectile flights in the following exercises are to be treated as ideal unless stated otherwise. All launch angles are assumed to be measured from the horizontal. All projectiles are assumed to be launched from the origin over a horizontal surface unless stated otherwise.

1. *Travel time* A projectile is fired at a speed of 840 m/sec at an angle of 60°. How long will it take to get 21 km downrange?

2. *Finding muzzle speed* Find the muzzle speed of a gun whose maximum range is 24.5 km.

3. *Flight time and height* A projectile is fired with an initial speed of 500 m/sec at an angle of elevation of 45°.
   (a) When and how far away will the projectile strike?
   (b) How high overhead will the projectile be when it is 5 km downrange?
   (c) What is the greatest height reached by the projectile?

4. *Throwing a baseball* A baseball is thrown from the stands 32 ft above the field at an angle of 30° up from the horizontal. When and how far away will the ball strike the ground if its initial speed is 32 ft/sec?

5. *Shot put* An athlete puts a 16 lb shot at an angle of 45° to the horizontal from 6.5 ft above the ground at an initial speed of 44 ft/sec as suggested in the accompanying figure. How long after launch and how far from the inner edge of the stopboard does the shot land?

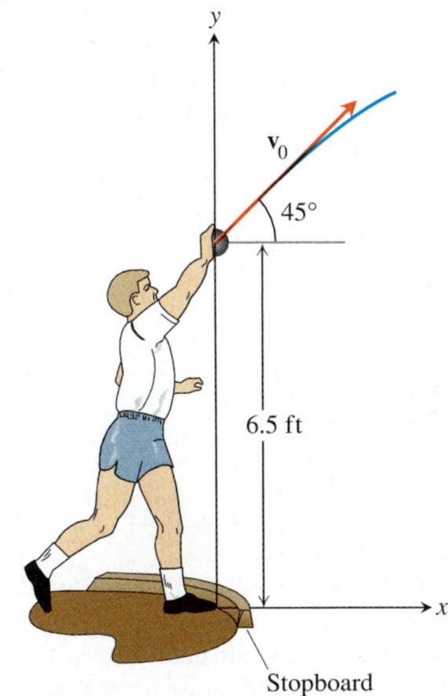

6. *(Continuation of Exercise 5)* Because of its initial elevation, the shot in Exercise 5 would have gone slightly farther if it had been launched at a 40° angle. How much farther? Answer in inches.

7. *Firing golf balls* A spring gun at ground level fires a golf ball at an angle of 45°. The ball lands 10 m away.

(a) What was the ball's initial speed?

(b) For the same initial speed, find the two firing angles that make the range 6 m.

8. *Beaming electrons* An electron in a TV tube is beamed horizontally at a speed of 5×10^6 m/sec toward the face of the tube 40 cm away. About how far will the electron drop before it hits?

9. *Finding golf ball speed* Laboratory tests designed to find how far golf balls of different hardness go when hit with a driver showed that a 100-compression ball hit with a club-head speed of 100 mph at a launch angle of 9° carried 248.8 yd. What was the launch speed of the ball? (It was more than 100 mph. At the same time the club head was moving forward, the compressed ball was kicking away from the club face, adding to the ball's forward speed.)

10. *Writing to Learn:* A *human cannonball* is to be fired with an initial speed of $v_0 = 80\sqrt{10}/3$ ft/sec. The circus performer (of the right caliber, naturally) hopes to land on a special cushion located 200 ft downrange at the same height as the muzzle of the cannon. The circus is being held in a large room with a flat ceiling 75 ft higher than the muzzle. Can the performer be fired to the cushion without striking the ceiling? If so, what should the cannon's angle of elevation be?

11. *Writing to Learn* A golf ball leaves the ground at a 30° angle at a speed of 90 ft/sec. Will it clear the top of a 30 ft tree that is in the way, 135 ft down the fairway? Explain.

12. *Elevated green* A golf ball is hit with an initial speed of 116 ft/sec at an angle of elevation of 45° from the tee to a green that is elevated 45 ft above the tee as shown in the diagram. Assuming that the pin, 369 ft downrange, does not get in the way, where will the ball land in relation to the pin?

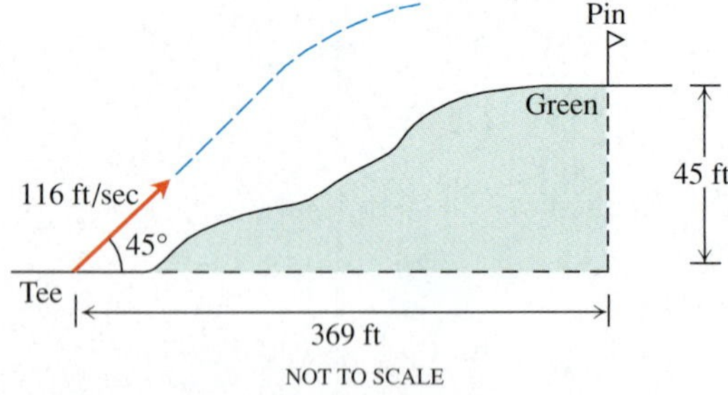

13. *The Green Monster* A baseball hit by a Boston Red Sox player at a 20° angle from 3 ft above the ground just cleared the left end of the "Green Monster," the left-field wall in Fenway Park (retired after the 2002 season). This wall is 37 ft high and 315 ft from home plate (see the accompanying figure).

(a) What was the initial speed of the ball?

(b) How long did it take the ball to reach the wall?

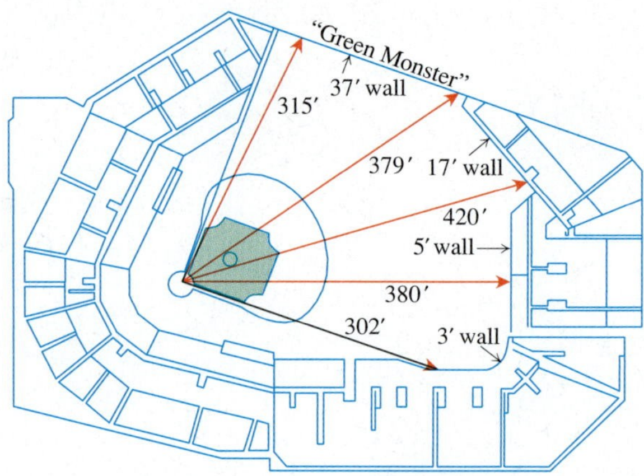

14. *Equal-range firing angles* Show that a projectile fired at an angle of α degrees, $0 < \alpha < 90$, has the same range as a projectile fired at the same speed at an angle of $(90 - \alpha)$ degrees. (In models that take air resistance into account, this symmetry is lost.)

15. *Equal-range firing angles* What two angles of elevation will enable a projectile to reach a target 16 km downrange on the same level as the gun if the projectile's initial speed is 400 m/sec?

16. *Range and height versus speed*

(a) Show that doubling a projectile's initial speed at a given launch angle multiplies its range by 4.

(b) By about what percentage should you increase the initial speed to double the height and range?

17. *Shot put* In Moscow in 1987, Natalya Lisouskaya set a women's world record by putting an 8 lb 13 oz shot 73 ft 10 in. Assuming that she launched the shot at a 40° angle to the horizontal from 6.5 ft above the ground, what was the shot's initial speed?

18. *Height versus time* Show that a projectile attains three-quarters of its maximum height in half the time it takes to reach the maximum height.

19. *Firing from* (x_0, y_0) Derive the equations

$$x = x_0 + (v_0 \cos \alpha)t,$$

$$y = y_0 + (v_0 \sin \alpha)t - \frac{1}{2}gt^2,$$

(see Equation 5 in the text) by solving the following initial value problem for a vector $\mathbf{r}$ in the plane.

Differential equation: $\dfrac{d^2\mathbf{r}}{dt^2} = -g\mathbf{j}$

Initial conditions: $\mathbf{r}(0) = x_0\mathbf{i} + y_0\mathbf{j}$

$$\frac{d\mathbf{r}}{dt}(0) = (v_0 \cos \alpha)\mathbf{i} + (v_0 \sin \alpha)\mathbf{j}$$

20. *Flaming arrow* Using the firing angle found in Example 3, find the speed at which the flaming arrow left Rebollo's bow. See Figure 9.34.

21. *Flaming arrow* The cauldron in Example 3 is 12 ft in diameter. Using Equation 5 and Example 3c, find how long it takes the flaming arrow to cover the horizontal distance to the rim. How high is the arrow at this time?

22. *Writing to Learn* Describe the path of a projectile given by Equations 4 when $\alpha = 90°$.

23. *Model train* The accompanying multiflash photograph shows a model train engine moving at a constant speed on a straight horizontal track. As the engine moved along, a marble was fired into the air by a spring in the engine's smokestack. The marble, which continued to move with the same forward speed as the engine, rejoined the engine 1 sec after it was fired. Measure the angle the marble's path made with the horizontal and use the information to find how high the marble went and how fast the engine was moving.

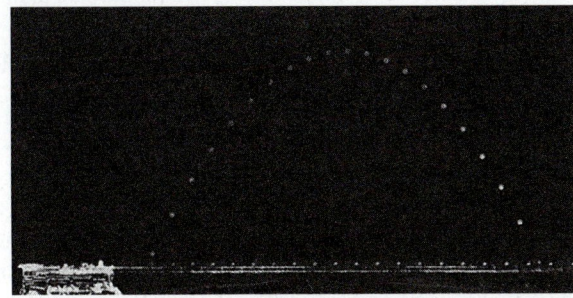

24. *Writing to Learn: Colliding marbles* The figure shows an experiment with two marbles. Marble A was launched toward marble B with launch angle α and initial speed v_0. At the same instant, marble B was released to fall from rest at $R \tan \alpha$ units directly above a spot R units downrange from A. The marbles were found to collide regardless of the value of v_0. Was this mere coincidence, or must this happen? Give reasons for your answer.

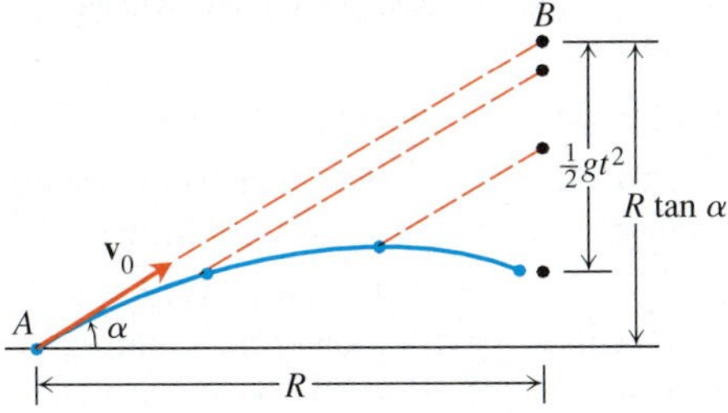

25. *Launching downhill* An ideal projectile is launched straight down an inclined plane as shown in the accompanying figure.

(a) Show that the greatest downhill range is achieved when the initial velocity vector bisects angle AOR.

(b) *Writing to Learn* If the projectile were fired uphill instead of down, what launch angle would maximize its range? Give reasons for your answer.

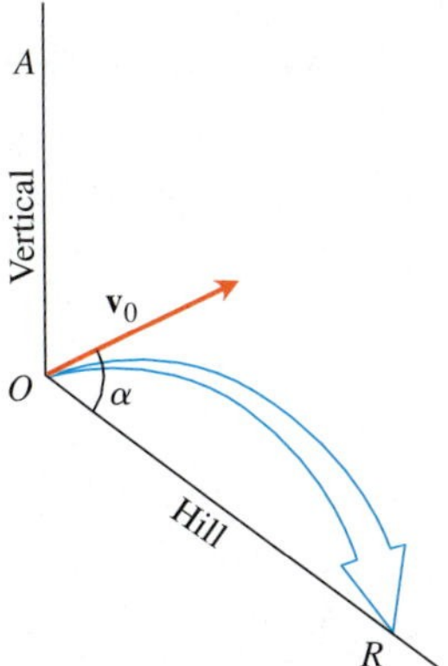

26. *Hitting a baseball under a wind gust* A baseball is hit when it is 2.5 ft above the ground. It leaves the bat with an initial velocity of 145 ft/sec at a launch angle of 23°. At the instant the ball is hit, an instantaneous gust of wind blows against the ball, adding a component of $-14\mathbf{i}$ (ft/sec) to the ball's initial velocity. A 15-ft-high fence lies 300 ft from home plate in the direction of the flight.

(a) Find a vector equation for the path of the baseball.

(b) How high does the baseball go, and when does it reach maximum height?

(c) Find the range and flight time of the baseball, assuming that the ball is not caught.

(d) When is the baseball 20 ft high? How far (ground distance) is the baseball from home plate at that height?

(e) *Writing to Learn* Has the batter hit a home run? Explain.

27. *Volleyball* A volleyball is hit when it is 4 ft above the ground and 12 ft from a 6-ft-high net. It leaves the point of impact with an initial velocity of 35 ft/sec at an angle of 27° and slips by the opposing team untouched.

(a) Find a vector equation for the path of the volleyball.

(b) How high does the volleyball go, and when does it reach maximum height?

(c) Find its range and flight time.

(d) When is the volleyball 7 ft above the ground? How far (ground distance) is the volleyball from where it will land?

(e) *Writing to Learn* Suppose that the net is raised to 8 ft. Does this change things? Explain.

28. *Where trajectories crest* For a projectile fired from the ground at launch angle α with initial speed v_0, consider α as a variable and v_0 as a fixed constant. For each α, $0 < \alpha < \pi/2$, we obtain a parabolic trajectory as shown in the accompanying figure. Show

that the points in the plane that give the maximum heights of these parabolic trajectories all lie on the ellipse

$$x^2 + 4\left(y - \frac{v_0^{\,2}}{4g}\right)^2 = \frac{v_0^{\,4}}{4g^2},$$

where $x \geq 0$.

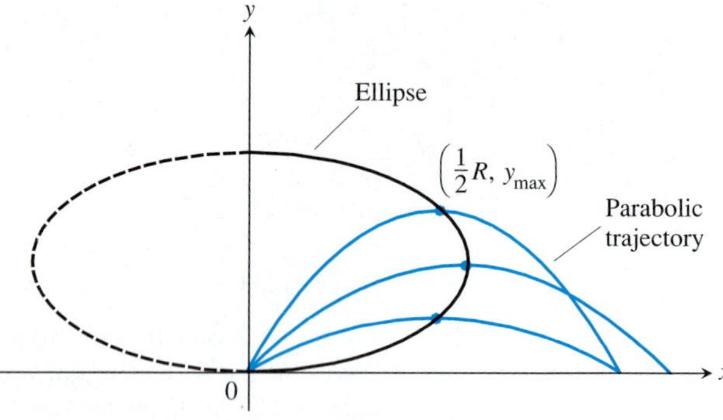

Projectile Motion with Linear Drag

The main force affecting the motion of a projectile, other than gravity, is air resistance. This slowing down force is **drag force,** and it acts in a direction *opposite* to the velocity of the projectile (see accompanying figure). For projectiles moving through the air at relatively low speeds, however, the drag force is (very nearly) proportional to the speed (to the first power) and so is called **linear.**

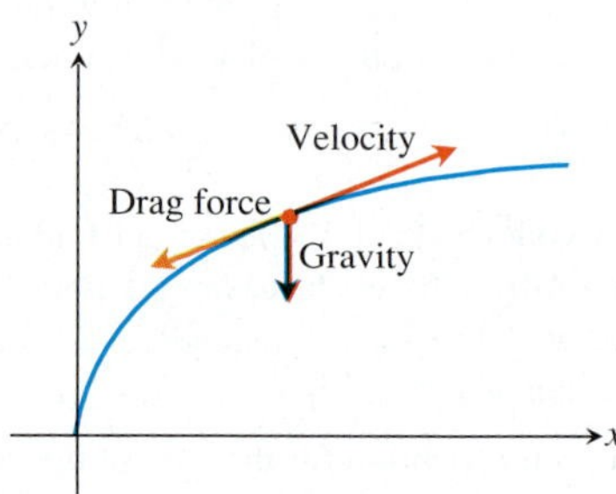

29. *Linear drag* Derive the equations

$$x = \frac{v_0}{k}(1 - e^{-kt})\cos\alpha$$

$$y = \frac{v_0}{k}(1 - e^{-kt})(\sin\alpha) + \frac{g}{k^2}(1 - kt - e^{-kt})$$

by solving the following initial value problem for a vector **r** in the plane.

Differential equation: $\frac{d^2\mathbf{r}}{dt^2} = -g\mathbf{j} - k\mathbf{v} = -g\mathbf{j} - k\frac{d\mathbf{r}}{dt}$

Initial conditions: $\mathbf{r}(0) = \mathbf{0}$

$$\left.\frac{d\mathbf{r}}{dt}\right|_{t=0} = \mathbf{v}_0 = (v_0 \cos\alpha)\mathbf{i} + (v_0 \sin\alpha)\mathbf{j}$$

The **drag coefficient** k is a positive constant representing resistance due to air density, v_0 and α are the projectile's initial speed and launch angle, and g is the acceleration of gravity.

30. *Hitting a baseball with linear drag* Consider the baseball problem in Example 4 when there is linear drag (see Exercise 29). Assume a drag coefficient $k = 0.12$, but no gust of wind.

(a) From Exercise 29, find a vector form for the path of the baseball.

(b) How high does the baseball go, and when does it reach maximum height?

(c) Find the range and flight time of the baseball.

(d) When is the baseball 30 ft high? How far (ground distance) is the baseball from home plate at that height?

(e) *Writing to Learn* A 10-ft-high outfield fence is 340 ft from home plate in the direction of the flight of the baseball. The outfielder can jump and catch any ball up to 11 ft off the ground to stop it from going over the fence. Has the batter hit a home run?

31. *Hitting a baseball with linear drag under a wind gust* Consider again the baseball problem in Example 4. This time assume a drag coefficient of 0.08 *and* an instantaneous gust of wind that adds a component of $-17.6\mathbf{i}$ (ft/sec) to the initial velocity at the instant the baseball is hit.

(a) Find a vector equation for the path of the baseball.

(b) How high does the baseball go, and when does it reach maximum height?

(c) Find the range and flight time of the baseball?

(d) When is the baseball 35 ft high? How far (ground distance) is the baseball from home plate at that height?

(e) *Writing to Learn* A 20-ft-high outfield fence is 380 ft from home plate in the direction of the flight of the baseball. Has the batter hit a home run? If "yes," what change in the horizontal component of the ball's initial velocity would have kept the ball in the park? If "no," what change would have allowed it to be a home run?

9.5 Polar Coordinates and Graphs

Polar Coordinates • Polar Graphing • Symmetry • Relating Polar and Cartesian Coordinates • Finding Points Where Polar Graphs Intersect

In radar tracking, an operator is interested in the bearing or angle the tracked object makes with some fixed ray (e.g., a directed line pointing due east) and how far away the object is currently located. In this section, we study a coordinate system devised by Newton, called the **polar coordinate system,** which is practical to use for such purposes.

Polar Coordinates

To define polar coordinates, we first choose a point in the plane called the **pole** (or **origin**) and labeled O. Then we draw an **initial ray** (or **polar axis**) starting at O. This ray is usually drawn horizontally and pointing to the right, corresponding to the positive x-axis in Cartesian coordinates (Figure 9.35). Then each point P can be located by assigning to it a **polar coordinate pair** (r, θ) in which r gives the directed distance from O to P and θ gives the directed angle from the initial ray to ray OP.

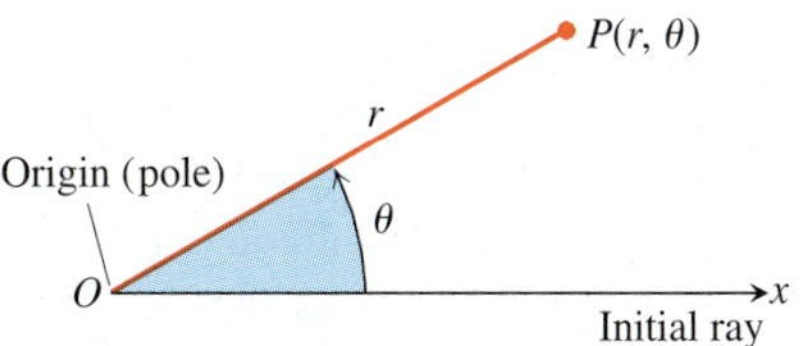

FIGURE 9.35 To define polar coordinates for the plane, we start with an origin, called the pole, and an initial ray.

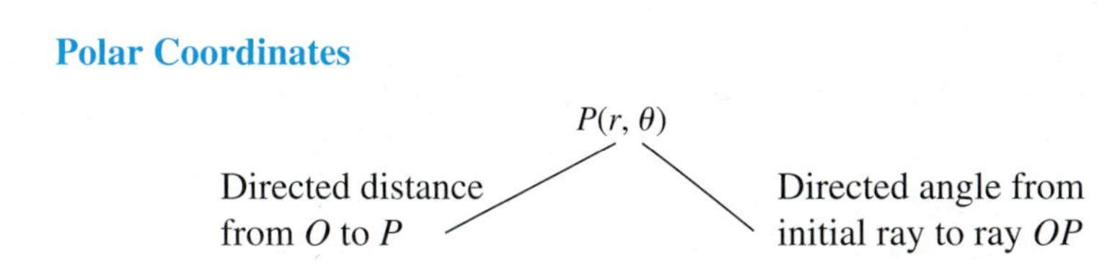

As in trigonometry, θ is positive when measured counterclockwise and negative when measured clockwise. The angle associated with a given point is not unique. For instance, the point 2 units from the origin along the ray $\theta = \pi/6$ has polar coordinates $r = 2$, $\theta = \pi/6$. It also has coordinates $r = 2$, $\theta = -11\pi/6$ (Figure 9.36).

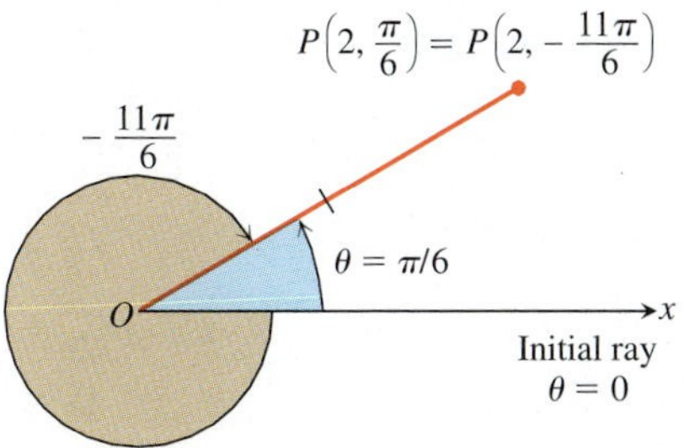

FIGURE 9.36 Polar coordinates are not unique.

There are also occasions when we wish to allow r to be negative. This is why we use *directed* distance in defining $P(r, \theta)$. The point $P(2, 7\pi/6)$ can be reached by turning $7\pi/6$ radians counterclockwise from the initial ray and going forward 2 units (Figure 9.37). It can also be reached by turning $\pi/6$ radians counterclockwise from the initial ray and going *backwards* 2 units. So the point also has polar coordinates $r = -2$, $\theta = \pi/6$.

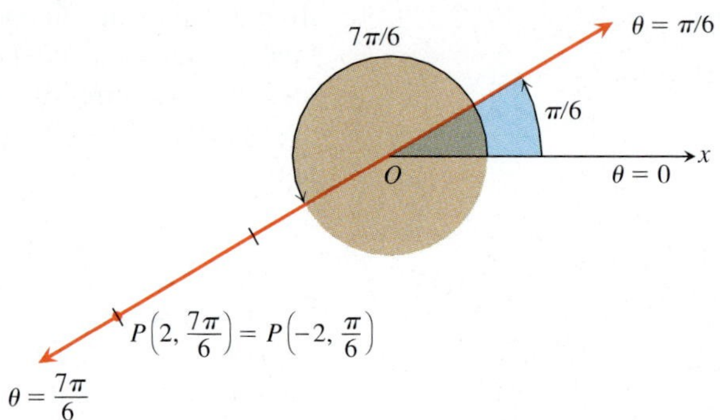

FIGURE 9.37 Polar coordinates can have negative r-values.

Polar Graphing

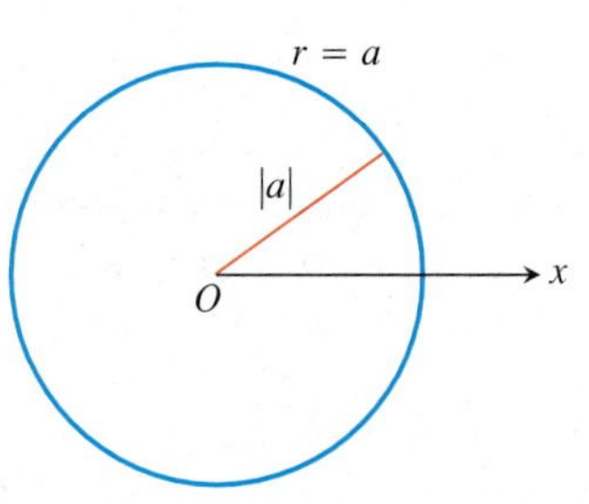

FIGURE 9.38 The polar equation for this circle is $r = a$.

If we hold r fixed at a constant value $a \neq 0$, the point $P(r, \theta)$ will lie $|a|$ units from the origin O. As θ varies over any interval of length 2π, P traces a circle of radius $|a|$ centered at O (Figure 9.38).

If we hold θ fixed at a constant value $\theta = \alpha$ and let r vary between $-\infty$ and ∞, the point $P(r, \theta)$ traces the line through O that makes an angle of measure α with the initial ray.

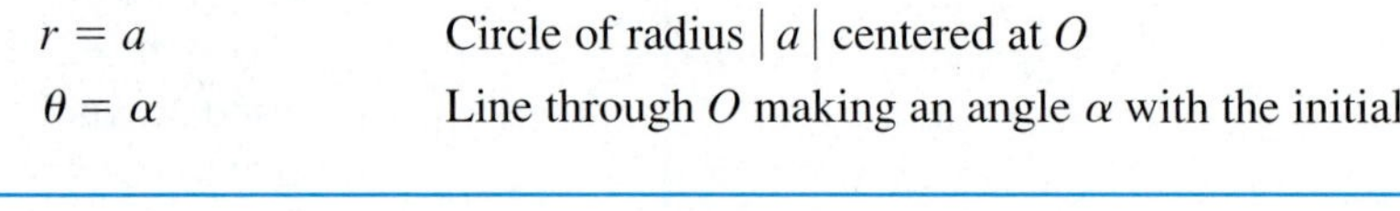

Equation	Polar Graph
$r = a$	Circle of radius $\lvert a \rvert$ centered at O
$\theta = \alpha$	Line through O making an angle α with the initial ray

Example 1 Finding Polar Equations for Graphs

(a) $r = 1$ and $r = -1$ are equations for the circle of radius 1 centered at O.

(b) $\theta = \pi/6$, $\theta = 7\pi/6$, and $\theta = -5\pi/6$ are equations for the line in Figure 9.37.

Equations of the form $r = a$ and $\theta = \alpha$ can be combined to define regions, segments, and rays.

Example 2 Graphing Equations and Inequalities

Graph the set of points whose polar coordinates satisfy the given conditions.

(a) $1 \leq r \leq 2$ and $0 \leq \theta \leq \dfrac{\pi}{2}$

(b) $-3 \leq r \leq 2$ and $\theta = \dfrac{\pi}{4}$

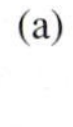

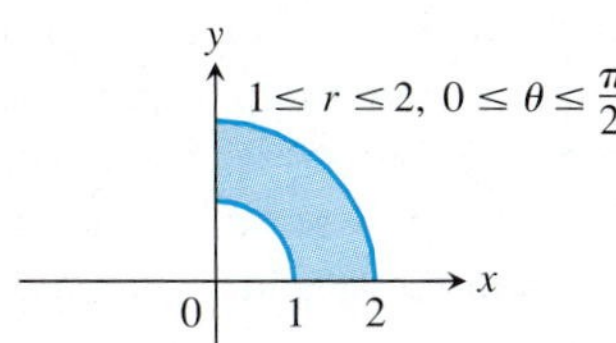

(b)

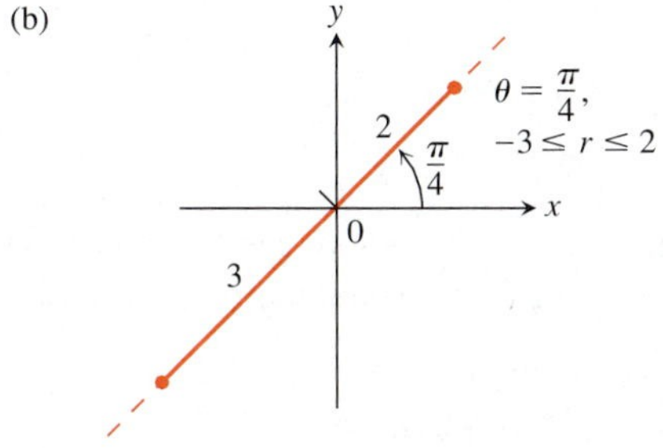

(c)

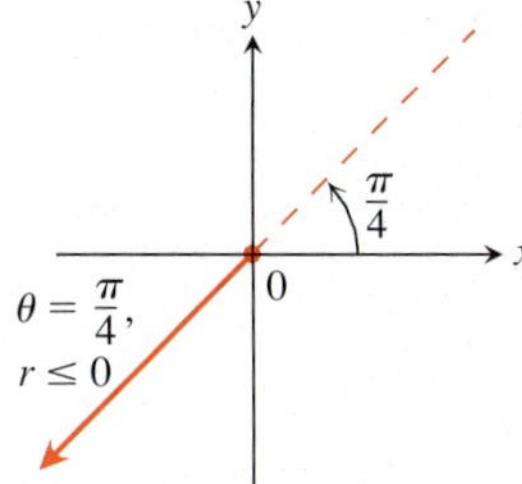

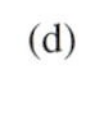

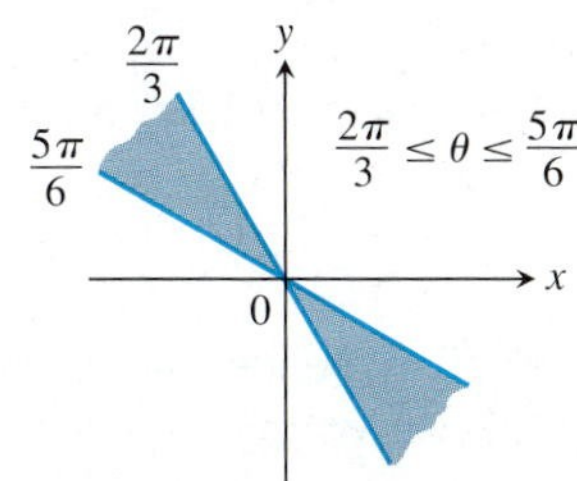

FIGURE 9.39 The graphs of typical inequalities in r and θ. (Example 2)

(c) $r \le 0$ and $\theta = \frac{\pi}{4}$

(d) $\frac{2\pi}{3} \le \theta \le \frac{5\pi}{6}$ (no restriction on r)

Solution The graphs are shown in Figure 9.39.

Symmetry

Figure 9.40 illustrates the standard polar coordinate tests for symmetry.

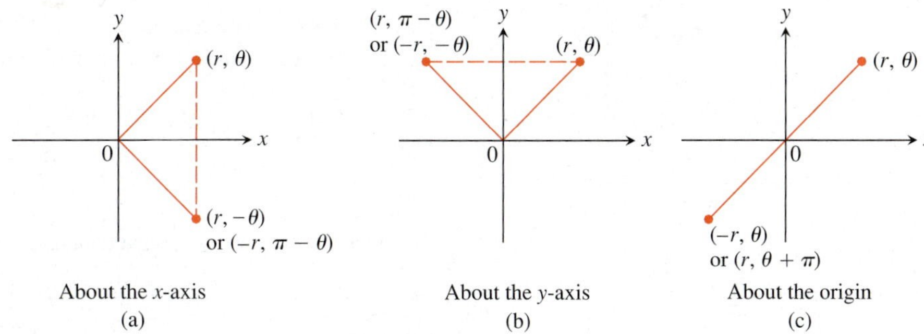

FIGURE 9.40 Three tests for symmetry.

Symmetry Tests for Polar Graphs

1. **Symmetry about the x-axis:** If the point (r, θ) lies on the graph, the point $(r, -\theta)$ or $(-r, \pi - \theta)$ lies on the graph (Figure 9.40a).
2. **Symmetry about the y-axis:** If the point (r, θ) lies on the graph, the point $(r, \pi - \theta)$ or $(-r, -\theta)$ lies on the graph (Figure 9.40b).
3. **Symmetry about the origin:** If the point (r, θ) lies on the graph, the point $(-r, \theta)$ or $(r, \theta + \pi)$ lies on the graph (Figure 9.40c).

Example 3 A cardioid

Graph the curve $r = 1 - \cos\theta$.

Solution The curve is symmetric about the x-axis because

$$\begin{aligned} (r, \theta) \text{ on the graph} &\Rightarrow r = 1 - \cos\theta \\ &\Rightarrow r = 1 - \cos(-\theta) \qquad \cos\theta = \cos(-\theta) \\ &\Rightarrow (r, -\theta) \text{ on the graph.} \end{aligned}$$

As θ increases from 0 to π, $\cos\theta$ decreases from 1 to -1 and $r = 1 - \cos\theta$ increases from a minimum value of 0 to a maximum value of 2. As θ continues on from π to 2π, $\cos\theta$ increases from -1 back to 1 and r decreases from 2 back to 0. The curve starts to repeat when $\theta = 2\pi$ because the cosine has period 2π.

The curve leaves the origin with slope $\tan(0) = 0$ and returns to the origin with slope $\tan(2\pi) = 0$.

θ	$r = 1 - \cos\theta$
0	0
$\frac{\pi}{3}$	$\frac{1}{2}$
$\frac{\pi}{2}$	1
$\frac{2\pi}{3}$	$\frac{3}{2}$
π	2

(a)

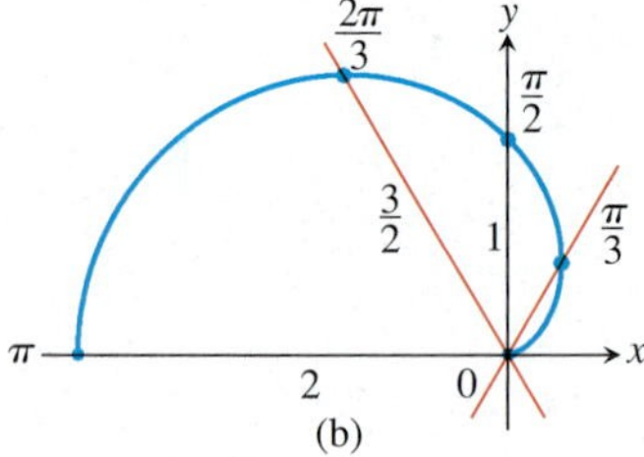

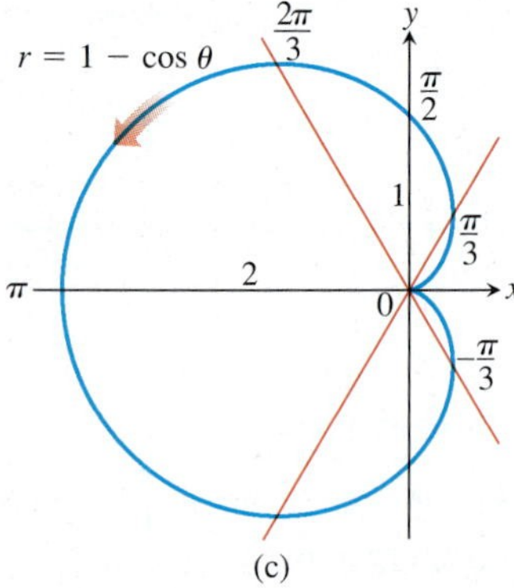

FIGURE 9.41 The steps in graphing the cardioid $r = 1 - \cos\theta$ (Example 3). The arrow shows the direction of increasing θ.

We make a table of values from $\theta = 0$ to $\theta = \pi$, plot the points, draw a smooth curve through them with a horizontal tangent at the origin, and reflect the curve across the x-axis to complete the graph (Figure 9.41). The curve is called a *cardioid* because of its heart shape. Cardioid shapes appear in the cams that direct the even layering of thread on bobbins and reels, and in the signal-strength patterns of certain radio antennae.

Example 4 Polar Graphing

Graph the curve $r^2 = 4\cos\theta$.

Solution The equation $r^2 = 4\cos\theta$ requires $\cos\theta \geq 0$, so we get the entire graph by running θ from $-\pi/2$ to $\pi/2$. The curve is symmetric about the x-axis because

$$(r, \theta) \text{ on the graph} \Rightarrow r^2 = 4\cos\theta$$
$$\Rightarrow r^2 = 4\cos(-\theta) \qquad \cos\theta = \cos(-\theta)$$
$$\Rightarrow (r, -\theta) \text{ on the graph.}$$

The curve is also symmetric about the origin because

$$(r, \theta) \text{ on the graph} \Rightarrow r^2 = 4\cos\theta$$
$$\Rightarrow (-r)^2 = 4\cos\theta$$
$$\Rightarrow (-r, \theta) \text{ on the graph.}$$

Together, these two symmetries imply symmetry about the y-axis.

The curve passes through the origin when $\theta = -\pi/2$ and $\theta = \pi/2$. It has a vertical tangent both times because $\tan\theta$ is infinite.

For each value of θ in the interval between $-\pi/2$ and $\pi/2$, the formula $r^2 = 4\cos\theta$ gives two values of r:

$$r = \pm 2\sqrt{\cos\theta}.$$

We make a short table of values, plot the corresponding points, and use information about symmetry and tangents to guide us in connecting the points with a smooth curve (Figure 9.42).

θ	$\cos\theta$	$r = \pm 2\sqrt{\cos\theta}$
0	1	± 2
$\pm\frac{\pi}{6}$	$\frac{\sqrt{3}}{2}$	± 1.9
$\pm\frac{\pi}{4}$	$\frac{1}{\sqrt{2}}$	± 1.7
$\pm\frac{\pi}{3}$	$\frac{1}{2}$	± 1.4
$\pm\frac{\pi}{2}$	0	0

(a)

$r^2 = 4\cos\theta$

Loop for $r = -2\sqrt{\cos\theta}$, $-\frac{\pi}{2} \leq \theta \leq \frac{\pi}{2}$

Loop for $r = 2\sqrt{\cos\theta}$, $-\frac{\pi}{2} \leq \theta \leq \frac{\pi}{2}$

(b)

FIGURE 9.42 The graph of $r^2 = 4\cos\theta$. The arrows show the direction of increasing θ. The values of r in the table are rounded. (Example 4)

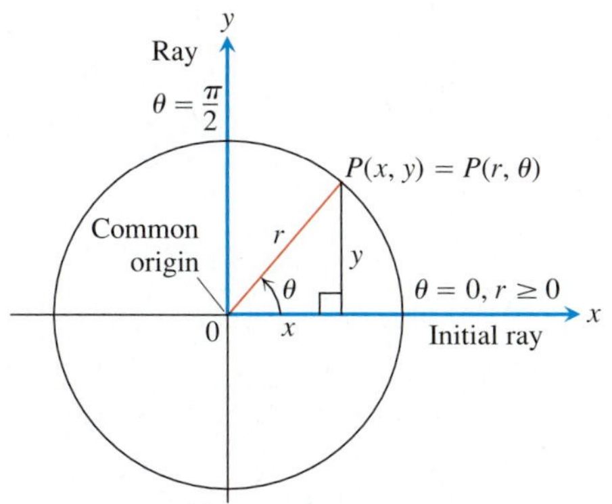

FIGURE 9.43 The usual way to relate polar and Cartesian coordinates.

Relating Polar and Cartesian Coordinates

When we use both polar and Cartesian coordinates in a plane, we place the two origins together and take the polar initial ray as the positive x-axis. The ray $\theta = \pi/2$, $r > 0$, becomes the positive y-axis (Figure 9.43). The two coordinate systems are then related by the following equations.

Equations Relating Polar and Cartesian Coordinates

$$x = r\cos\theta, \qquad y = r\sin\theta, \qquad x^2 + y^2 = r^2, \qquad \frac{y}{x} = \tan\theta$$

We use these equations and algebra (sometimes a lot of it!) to rewrite polar equations in Cartesian form and vice versa.

Example 5 Equivalent Equations

Polar equation	Cartesian equivalent
$r\cos\theta = 2$	$x = 2$
$r^2\cos\theta\sin\theta = 4$	$xy = 4$
$r^2\cos^2\theta - r^2\sin^2\theta = 1$	$x^2 - y^2 = 1$
$r = 1 + 2r\cos\theta$	$y^2 - 3x^2 - 4x - 1 = 0$
$r = 1 - \cos\theta$	$x^4 + y^4 + 2x^2y^2 + 2x^3 + 2xy^2 - y^2 = 0$

With some curves, we are better off with polar coordinates; with others, we aren't.

Example 6 Converting Cartesian to Polar

Find a polar equation for the circle $x^2 + (y - 3)^2 = 9$ (Figure 9.44). Support graphically.

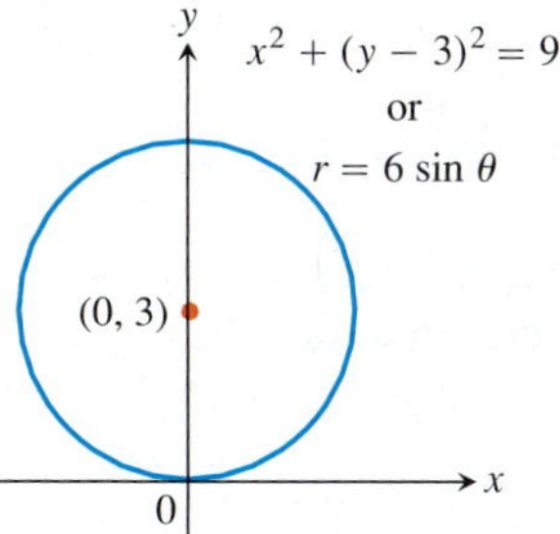

FIGURE 9.44 The circle in Example 6.

Solution

$$\begin{aligned} x^2 + y^2 - 6y + 9 &= 9 && \text{Expand } (y-3)^2. \\ x^2 + y^2 - 6y &= 0 \\ r^2 - 6r\sin\theta &= 0 && x^2 + y^2 = r^2,\quad y = r\sin\theta \\ r = 0 \text{ or } r - 6\sin\theta &= 0 \end{aligned}$$

The equation $r = 6\sin\theta$ includes the possibility that $r = 0$.

Example 7 Converting Polar to Cartesian

Find a Cartesian equivalent for the polar equation. Identify the graph.

(a) $r^2 = 4r\cos\theta$ **(b)** $r = \dfrac{4}{2\cos\theta - \sin\theta}$

Solution

(a)

$$r^2 = 4r \cos \theta$$

$$x^2 + y^2 = 4x \qquad r^2 = x^2 + y^2,\ r \cos \theta = x$$

$$x^2 - 4x + y^2 = 0$$

$$x^2 - 4x + 4 + y^2 = 4 \qquad \text{Completing the square}$$

$$(x - 2)^2 + y^2 = 4$$

The graph of the equivalent Cartesian equation $(x - 2)^2 + y^2 = 4$ is a circle with radius 2 and center $(2, 0)$.

(b)

$$r = \frac{4}{2 \cos \theta - \sin \theta}$$

$$r(2 \cos \theta - \sin \theta) = 4$$

$$2r \cos \theta - r \sin \theta = 4$$

$$2x - y = 4 \qquad r \cos \theta = x,\ r \sin \theta = y$$

$$y = 2x - 4$$

The graph of the equivalent Cartesian equation $y = 2x - 4$ is a line with slope 2 and y-intercept -4.

Finding Points Where Polar Graphs Intersect

That we can represent a point in different ways in polar coordinates makes extra care necessary in deciding when a point lies on the graph of a polar equation and in determining the points in which polar graphs intersect. The problem is that a point of intersection may satisfy the equation of one curve with polar coordinates that are different from the ones with which it satisfies the equation of another curve. Thus, solving the equations of two curves simultaneously may not identify all their points of intersection. The only sure way to identify all the points of intersection is to graph the equations.

Example 8 Deceptive Coordinates

Show that the point $(2, \pi/2)$ lies on the curve $r = 2 \cos 2\theta$.

Solution It may seem at first that the point $(2, \pi/2)$ does not lie on the curve because substituting the given coordinates into the equation gives

$$2 = 2 \cos 2\left(\frac{\pi}{2}\right) = 2 \cos \pi = -2,$$

which is not a true equality. The magnitude is right, but the sign is wrong. This suggests looking for a pair of coordinates for the given point in which r is negative, for example, $(-2, -(\pi/2))$. If we try these in the equation $r = 2 \cos 2\theta$, we find

$$-2 = 2 \cos 2\left(-\frac{\pi}{2}\right) = 2(-1) = -2,$$

and the equation is satisfied. The point $(2, \pi/2)$ does lie on the curve.

Example 9 Elusive Intersection Points

Find the points of intersection of the curves

$$r^2 = 4 \cos \theta \qquad \text{and} \qquad r = 1 - \cos \theta.$$

Solution In Cartesian coordinates, we can always find the points where two curves cross by solving their equations simultaneously. In polar coordinates, the story is different. Simultaneous solution may reveal some intersection points without revealing others. In this example, simultaneous solution reveals only two of the four intersection points. The others are found by graphing. (Also, see Exercise 79.)

If we substitute $\cos \theta = r^2/4$ in the equation $r = 1 - \cos \theta$, we get

$$r = 1 - \cos \theta = 1 - \frac{r^2}{4}$$
$$4r = 4 - r^2$$
$$r^2 + 4r - 4 = 0$$
$$r = -2 \pm 2\sqrt{2}. \qquad \text{Quadratic formula}$$

CD-ROM
WEBsite
Historical Biography
Johannes Kepler
(1571 — 1630)

The value $r = -2 - 2\sqrt{2}$ has too large an absolute value to belong to either curve. The values of θ corresponding to $r = -2 + 2\sqrt{2}$ are

$$\begin{aligned} \theta &= \cos^{-1}(1 - r) && \text{From } r = 1 - \cos\theta \\ &= \cos^{-1}(1 - (2\sqrt{2} - 2)) && \text{Set } r = 2\sqrt{2} - 2. \\ &= \cos^{-1}(3 - 2\sqrt{2}) \\ &= \pm 80^\circ. && \text{Rounded to the nearest degree} \end{aligned}$$

We have thus identified two intersection points: $(r, \theta) = (2\sqrt{2} - 2, \pm 80^\circ)$.

If we graph the equations $r^2 = 4 \cos \theta$ and $r = 1 - \cos \theta$ together (Figure 9.45), as we can now do by combining the graphs in Figures 9.41 and 9.42, we see that the curves also intersect at the point $(2, \pi)$ and the origin. Why weren't the r-values of these points revealed by the simultaneous solution? The answer is that the points $(0, 0)$ and $(2, \pi)$ are not on the curves "simultaneously." They are not reached at the same value of θ. On the curve $r = 1 - \cos \theta$, the point $(2, \pi)$ is reached when $\theta = \pi$. On the curve $r^2 = 4 \cos \theta$, it is reached when $\theta = 0$, where it is identified not by the coordinates $(2, \pi)$, which do not satisfy the equation, but by the coordinates $(-2, 0)$, which do. Similarly, the cardioid reaches the origin when $\theta = 0$, but the curve $r^2 = 4 \cos \theta$ reaches the origin when $\theta = \pi/2$.

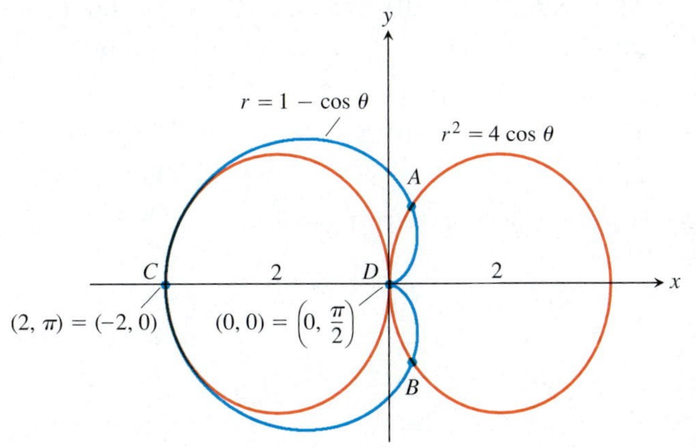

FIGURE 9.45 The four points of intersection of the curves $r = 1 - \cos \theta$ and $r^2 = 4 \cos \theta$ (Example 9). Only A and B were found by simultaneous solution. The other two were disclosed by graphing.

USING TECHNOLOGY

Finding Intersections The *simultaneous mode* of a graphing utility gives new meaning to the *simultaneous solution* of a pair of polar coordinate equations. A simultaneous solution occurs only where the two graphs "collide" while they are being drawn simultaneously and not where one graph intersects the other at a point that had been illuminated earlier. The distinction is particularly important in the areas of traffic control or missile defense. For example, in traffic control, the only issue is whether two aircraft are in the same place at the same time. The question of whether the curves the craft follow intersect is unimportant.

To illustrate, graph the polar equations

$$r = \cos 2\theta \qquad \text{and} \qquad r = \sin 2\theta$$

in simultaneous mode with $0 \leq \theta < 2\pi$, θ Step $= 0.1$ and view dimensions [xmin, xmax] $= [-1, 1]$ by [ymin, ymax] $= [-1, 1]$. *While the graphs are being drawn on the screen*, count the number of times the two graphs illuminate a single pixel simultaneously. Explain why these points of intersection of the two graphs correspond to simultaneous solutions of the equations. (You may find it helpful to slow down the graphing by making θ Step smaller, say 0.05, for example.) In how many points total do the graphs actually intersect?

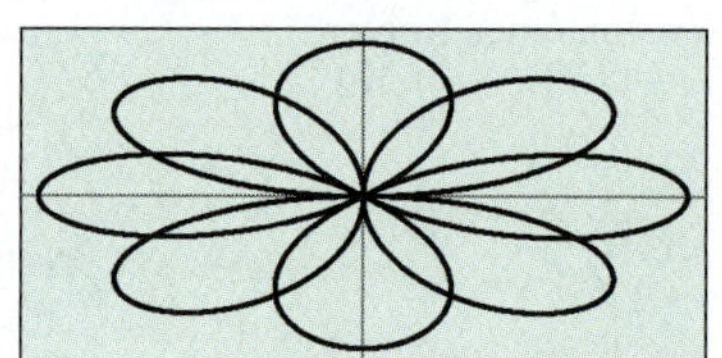

EXERCISES 9.5

Polar Coordinate Pairs

In Exercises 1 and 2, determine which polar coordinate pairs name the same point.

1. **(a)** $(3, 0)$ **(b)** $(-3, 0)$ **(c)** $(2, 2\pi/3)$
(d) $(2, 7\pi/3)$ **(e)** $(-3, \pi)$ **(f)** $(2, \pi/3)$
(g) $(-3, 2\pi)$ **(h)** $(-2, -\pi/3)$

2. **(a)** $(-2, \pi/3)$ **(b)** $(2, -\pi/3)$ **(c)** (r, θ)
(d) $(r, \theta + \pi)$ **(e)** $(-r, \theta)$ **(f)** $(2, -2\pi/3)$
(g) $(-r, \theta + \pi)$ **(h)** $(-2, 2\pi/3)$

In Exercises 3 and 4, plot the points with the given polar coordinates and find their Cartesian coordinates.

3. **(a)** $(\sqrt{2}, \pi/4)$ **(b)** $(1, 0)$
(c) $(0, \pi/2)$ **(d)** $(-\sqrt{2}, \pi/4)$

4. **(a)** $(-3, 5\pi/6)$ **(b)** $(5, \tan^{-1}(4/3))$
(c) $(-1, 7\pi)$ **(d)** $(2\sqrt{3}, 2\pi/3)$

In Exercises 5 and 6, plot the points with the given Cartesian coordinates and find two sets of polar coordinates for each.

5. **(a)** $(-1, 1)$ **(b)** $(1, -\sqrt{3})$
(c) $(0, 3)$ **(d)** $(-1, 0)$

6. **(a)** $(-\sqrt{3}, -1)$ **(b)** $(3, 4)$
(c) $(0, -2)$ **(d)** $(2, 0)$

Graphing Polar Equations and Inequalities

In Exercises 7–18, graph the set of points whose polar coordinates satisfy the given equations and inequalities.

7. $r = 2$

8. $0 \leq r \leq 2$

9. $r \geq 1$

10. $0 \leq \theta \leq \pi/6, \quad r \geq 0$

11. $\theta = 2\pi/3, \quad r \leq -2$

12. $\theta = \pi/3, \quad -1 \leq r \leq 3$

13. $0 \leq \theta \leq \pi, \quad r = 1$

14. $0 \leq \theta \leq \pi, \quad r = -1$

15. $\theta = \pi/2, \quad r \leq 0$

16. $\pi/4 \le \theta \le 3\pi/4, \quad 0 \le r \le 1$

17. $-\pi/4 \le \theta \le \pi/4, \quad -1 \le r \le 1$

18. $0 \le \theta \le \pi/2, \quad 1 \le |r| \le 2$

Polar to Cartesian Equations

In Exercises 19–36, replace the polar equation by an equivalent Cartesian equation. Then identify or describe the graph.

19. $r \sin \theta = 0$

20. $r \cos \theta = 0$

21. $r = 4 \csc \theta$

22. $r = -3 \sec \theta$

23. $r \cos \theta + r \sin \theta = 1$

24. $r^2 = 1$

25. $r^2 = 4r \sin \theta$

26. $r = \dfrac{5}{\sin \theta - 2 \cos \theta}$

27. $r^2 \sin 2\theta = 2$

28. $r = \cot \theta \csc \theta$

29. $r = (\csc \theta)\, e^{r \cos \theta}$

30. $\cos^2 \theta = \sin^2 \theta$

31. $r \sin \theta = \ln r + \ln \cos \theta$

32. $r^2 + 2r^2 \cos \theta \sin \theta = 1$

33. $r^2 = -4r \cos \theta$

34. $r = 8 \sin \theta$

35. $r = 2 \cos \theta + 2 \sin \theta$

36. $r \sin\left(\theta + \dfrac{\pi}{6}\right) = 2$

Cartesian to Polar Equations

In Exercises 37–48, replace the Cartesian equation by an equivalent polar equation.

37. $x = 7$

38. $y = 1$

39. $x = y$

40. $x - y = 3$

41. $x^2 + y^2 = 4$

42. $x^2 - y^2 = 1$

43. $\dfrac{x^2}{9} + \dfrac{y^2}{4} = 1$

44. $xy = 2$

45. $y^2 = 4x$

46. $x^2 + xy + y^2 = 1$

47. $x^2 + (y - 2)^2 = 4$

48. $(x - 3)^2 + (y + 1)^2 = 4$

Symmetries and Polar Graphs

In Exercises 49–58, **(a)** graph the polar curve. **(b)** What is the shortest length a θ-interval can have and still produce the graph?

49. $r = 1 + \cos \theta$

50. $r = 2 - 2 \cos \theta$

51. $r^2 = -\sin 2\theta$

52. $r = 1 - \sin \theta$

53. $r = 1 - 2 \sin 3\theta$

54. $r = \sin(\theta/2)$

55. $r = \theta$

56. $r = 1 + \sin \theta$

57. $r = 2 \cos 3\theta$

58. $r = 1 + 2 \sin \theta$

In Exercises 59–62, determine the symmetries of the curve.

59. $r^2 = 4 \cos 2\theta$

60. $r^2 = 4 \sin 2\theta$

61. $r = 2 + \sin \theta$

62. $r^2 = -\cos 2\theta$

63. *Writing to Learn: Vertical and horizontal lines*

(a) Explain why every vertical line in the plane has a polar equation of the form $r = a \sec \theta$.

(b) Find an analogous polar equation for horizontal lines. Give reasons for your answer.

64. *Writing to Learn: Do two symmetries imply three?* If a curve has any two of the symmetries listed at the beginning of the section, can anything be said about its having or not having the third symmetry? Give reasons for your answer.

Intersections

65. Show that the point $(2, 3\pi/4)$ lies on the curve $r = 2 \sin 2\theta$.

66. Show that $(1/2, 3\pi/2)$ lies on the curve $r = -\sin(\theta/3)$.

Find the points of intersection of the pairs of curves in Exercises 67–70.

67. $r = 1 + \cos \theta, \quad r = 1 - \cos \theta$

68. $r = 2 \sin \theta, \quad r = 2 \sin 2\theta$

69. $r = \cos \theta, \quad r = 1 - \cos \theta$

70. $r = 1, \quad r^2 = 2 \sin 2\theta$

T Find the points of intersection of the pairs of curves in Exercises 71–74.

71. $r^2 = \sin 2\theta, \quad r^2 = \cos 2\theta$

72. $r = 1 + \cos \dfrac{\theta}{2}, \quad r = 1 - \sin \dfrac{\theta}{2}$

73. $r = 1, \quad r = 2 \sin 2\theta$

74. $r = 1, \quad r^2 = 2 \sin 2\theta$

T **75.** Which of the following has the same graph as $r = 1 - \cos \theta$?

(a) $r = -1 - \cos \theta$

(b) $r = 1 + \cos \theta$

Confirm your answer with alegbra.

T **76.** *Rose curves* Let $r = 2 \sin n\theta$.

(a) Graph $r = 2 \sin n\theta$ for $n = \pm 2, \pm 4, \pm 6$. Describe the curves.

(b) What is the smallest length a θ-interval can have and still produce the graphs in part (a)?

(c) Based on your observations in part (a), describe the graph of $r = 2 \sin n\theta$ when n is a nonzero even integer.

(d) Graph $r = 2 \sin n\theta$ for $n = \pm 3, \pm 5, \pm 7$. Describe the curves.

(e) What is the smallest length a θ-interval can have and still produce the graphs in part (d)?

(f) Based on your observations in part (d), describe the graph of $r = 2 \sin n\theta$ when n is an odd integer different from ± 1.

T **77.** *A rose within a rose* Graph the equation $r = 1 - 2 \sin 3\theta$.

T **78.** *The nephroid of Freeth* Graph the nephroid of Freeth:

$$r = 1 + 2\sin\frac{\theta}{2}.$$

Theory and Examples

79. (*Continuation of Example 9*) The simultaneous solution of the equations

$$r^2 = 4\cos\theta \qquad (1)$$

$$r = 1 - \cos\theta \qquad (2)$$

in the text did not reveal the points (0, 0) and $(2, \pi)$ in which their graphs intersected.

(a) We could have found the point $(2, \pi)$, however, by replacing the (r, θ) in Equation (1) by the equivalent $(-r, \theta + \pi)$ to obtain

$$\begin{aligned} r^2 &= 4\cos\theta \\ (-r)^2 &= 4\cos(\theta + \pi) \qquad (3) \\ r^2 &= -4\cos\theta. \end{aligned}$$

Solve Equations (2) and (3) simultaneously to show that $(2, \pi)$ is a common solution. (This will still not reveal that the graphs intersect at (0, 0).)

(b) The origin is still a special case. (It often is.) Here is one way to handle it: Set $r = 0$ in Equations (1) and (2) and solve each equation for a corresponding value of θ. Since $(0, \theta)$ is the origin for *any* θ, this will show that both curves pass through the origin even if they do so for different θ-values.

CD-ROM WEBsite

80. *Relating polar equations to parametric equations* Let $r = f(\theta)$ be a polar curve.

(a) *Writing to Learn* Explain why

$$x = f(t)\cos t, \qquad y = f(t)\sin t$$

are parametric equations for the curve.

T **(b)** Use part (a) to write parametric equations for the circle $r = 3$. Support your answer by graphing the parametric equations.

T **(c)** Repeat part (b) with $r = 1 - \cos\theta$.

T **(d)** Repeat part (b) with $r = 3\sin 2\theta$.

81. *Distance formula* Show that the distance between points (r_1, θ_1), and (r_2, θ_2) in polar coordinates is

$$d = \sqrt{r_1{}^2 + r_2{}^2 - 2r_1r_2\cos(\theta_1 - \theta_2)}.$$

82. *Height of a cardioid* Find the maximum height above the x-axis of the cardioid $r = 2(1 + \cos\theta)$.

9.6 Calculus of Polar Curves

Slope • Area in the Plane • Length of a Curve

In this section, you will see how to find slopes, areas, and lengths of polar curves $r = f(\theta)$.

Slope

The slope of a polar curve $r = f(\theta)$ is given by dy/dx, not by $r' = df/d\theta$. To see why, think of the graph of f as the graph of the parametric equations

$$x = r\cos\theta = f(\theta)\cos\theta, \qquad y = r\sin\theta = f(\theta)\sin\theta.$$

If f is a differentiable function of θ, then so are x and y, and when $dx/d\theta \neq 0$, we can calculate dy/dx from the parametric formula

$$\frac{dy}{dx} = \frac{dy/d\theta}{dx/d\theta} \qquad \text{Section 2.5, Equation (4) with } t = \theta$$

$$= \frac{\dfrac{d}{d\theta}(f(\theta)\sin\theta)}{\dfrac{d}{d\theta}(f(\theta)\cos\theta)}$$

$$= \frac{\dfrac{df}{d\theta}\sin\theta + f(\theta)\cos\theta}{\dfrac{df}{d\theta}\cos\theta - f(\theta)\sin\theta} \qquad \text{Product Rule for Derivatives}$$

Slope of the Polar Curve $r = f(\theta)$

$$\left.\frac{dy}{dx}\right|_{(r,\theta)} = \frac{f'(\theta)\sin\theta + f(\theta)\cos\theta}{f'(\theta)\cos\theta - f(\theta)\sin\theta}, \tag{1}$$

provided $dx/d\theta \neq 0$ at (r, θ).

We can see from Equation 1 and its derivation that the curve $r = f(\theta)$ has a

1. Horizontal tangent at a point where $dy/d\theta = 0$ and $dx/d\theta \neq 0$
2. Vertical tangent at a point where $dx/d\theta = 0$ and $dy/d\theta \neq 0$.

If both derivatives are zero, no conclusion can be drawn without further investigation, as illustrated in Example 1.

Example 1 Finding Horizontal and Vertical Tangents

Find the horizontal and vertical tangents to the graph of the cardioid $r = 1 - \cos\theta$, $0 \leq \theta \leq 2\pi$.

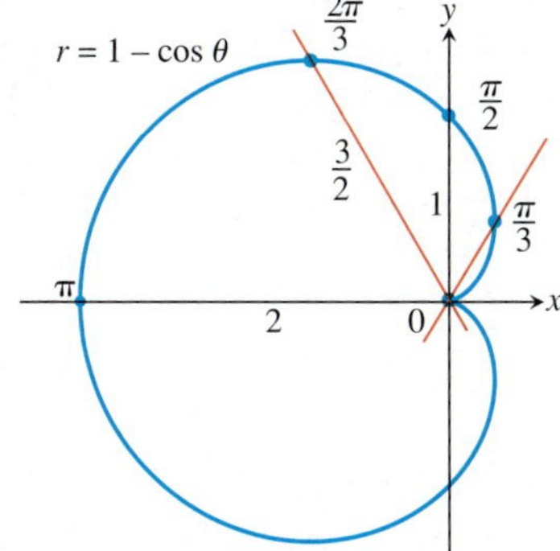

FIGURE 9.46 Where are the horizontal and vertical tangents to this cardioid? (Example 1)

Solution The graph in Figure 9.46 suggests that there are at least two horizontal and three vertical tangents.

The parametric form of the equation is

$$x = r\cos\theta = (1 - \cos\theta)\cos\theta = \cos\theta - \cos^2\theta,$$
$$y = r\sin\theta = (1 - \cos\theta)\sin\theta = \sin\theta - \cos\theta\sin\theta.$$

We need to find the zeros of $dy/d\theta$ and $dx/d\theta$.

(a) Zeros of $dy/d\theta$ in $[0, 2\pi]$:

$$\frac{dy}{d\theta} = \cos\theta + \sin^2\theta - \cos^2\theta = \cos\theta + (1 - \cos^2\theta) - \cos^2\theta$$
$$= 1 + \cos\theta - 2\cos^2\theta = (1 + 2\cos\theta)(1 - \cos\theta).$$

Now,

$$1 - \cos\theta = 0 \Rightarrow \theta = 0, 2\pi$$
$$1 + 2\cos\theta = 0 \Rightarrow \theta = 2\pi/3, 4\pi/3.$$

Thus, $dy/d\theta = 0$ in $0 \leq \theta \leq 2\pi$ if $\theta = 0, 2\pi/3, 4\pi/3$, or 2π.

(b) Zeros of $dx/d\theta$ in $[0, 2\pi]$:

$$\frac{dx}{d\theta} = -\sin\theta + 2\cos\theta\sin\theta = (2\cos\theta - 1)\sin\theta.$$

Now,

$$2\cos\theta - 1 = 0 \Rightarrow \theta = \pi/3, 5\pi/3$$
$$\sin\theta = 0 \Rightarrow \theta = 0, \pi, 2\pi.$$

Thus, $dx/d\theta = 0$ in $0 \leq \theta \leq 2\pi$ if $\theta = 0, \pi/3, \pi, 5\pi/3$, or 2π.

We can now see that there are horizontal tangents ($dy/d\theta = 0$, $dx/d\theta \neq 0$) at the points where $\theta = 2\pi/3$ and $4\pi/3$, and vertical tangents ($dx/d\theta = 0$, $dy/d\theta \neq 0$) at the points where $\theta = \pi/3$, π, and $5\pi/3$.

At the points where $\theta = 0$ or 2π, the right side of Equation (1) takes the form $0/0$. We can use l'Hôpital's Rule (Section 7.6) to see that

$$\lim_{\theta \to 0,\, 2\pi} \frac{dy/d\theta}{dx/d\theta} = \lim_{\theta \to 0, 2\pi} \frac{1 + \cos\theta - 2\cos^2\theta}{2\cos\theta\sin\theta - \sin\theta}$$

$$= \lim_{\theta \to 0, 2\pi} \frac{-\sin\theta + 4\cos\theta\sin\theta}{2\cos^2\theta - 2\sin^2\theta - \cos\theta} = \frac{0}{1} = 0.$$

The curve has a horizontal tangent at the point where $\theta = 0$ or 2π. Summarizing, we have

Horizontal tangents at	$(0, 0) = (0, 2\pi)$, $(1.5, 2\pi/3)$, $(1.5, 4\pi/3)$
Vertical tangents at	$(0.5, \pi/3)$, $(2, \pi)$, $(0.5, 5\pi/3)$.

If the curve $r = f(\theta)$ passes through the origin at $\theta = \theta_0$, then $f(\theta_0) = 0$, and Equation (1) gives

$$\left.\frac{dy}{dx}\right|_{(0,\,\theta_0)} = \frac{f'(\theta_0)\sin\theta_0}{f'(\theta_0)\cos\theta_0} = \tan\theta_0,$$

provided $f'(\theta_0) \neq 0$ (not the case in Example 1). The reason we say "slope at $(0, \theta_0)$" and not just "slope at the origin" is that a polar curve may pass through the origin more than once, with different slopes at different θ-values.

Example 2 Finding Tangent Lines at the Pole (Origin)

Find the lines tangent to the rose curve

$$r = f(\theta) = 2\sin 3\theta, \qquad 0 \leq \theta \leq \pi,$$

at the pole.

Solution $f(\theta)$ is zero when $\theta = 0$, $\pi/3$, $2\pi/3$, and π. The derivative $f'(\theta) = 6\cos 3\theta$ is not zero at these four values of θ. Thus, this curve has tangent lines at the pole (Figure 9.47) with slopes $\tan 0 = \tan \pi = 0$, $\tan(\pi/3) = \sqrt{3}$, and $\tan(2\pi/3) = -\sqrt{3}$. The three corresponding tangent lines are $y = 0$, $y = \sqrt{3}x$, and $y = -\sqrt{3}x$.

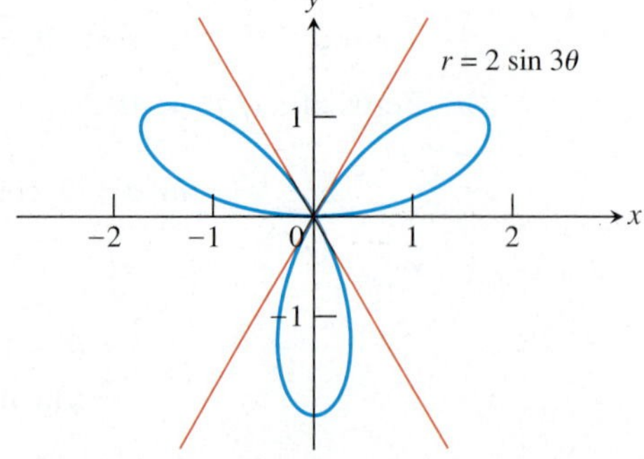

FIGURE 9.47 The three tangent lines to $r = f(\theta) = 2\sin 3\theta$, $0 \leq \theta \leq \pi$, are $y = 0$, $y = \sqrt{3}x$, and $y = -\sqrt{3}x$. (Example 2)

Area in the Plane

The region OTS in Figure 9.48 is bounded by the rays $\theta = \alpha$ and $\theta = \beta$ and the curve $r = f(\theta)$. We approximate the region with n nonoverlapping circular sectors based on a partition P of angle TOS. The typical sector has radius $r_k = f(\theta_k)$ and central angle of radian measure $\Delta\theta_k$. Its area is

$$A_k = \frac{1}{2} r_k^{\,2}\, \Delta\theta_k = \frac{1}{2} (f(\theta_k))^2\, \Delta\theta_k .$$

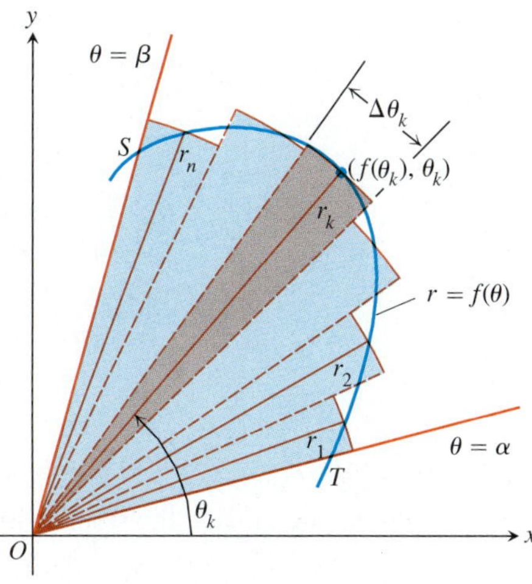

FIGURE 9.48 To derive a formula for the area of region OTS, we approximate the region with fan-shaped circular sectors.

The area of the region OTS is approximately

$$\sum_{k=1}^{n} A_k = \sum_{k=1}^{n} \frac{1}{2} (f(\theta_k))^2\, \Delta\theta_k .$$

If f is continuous, we expect the approximations to improve as $\|P\| \to 0$, and we are led to the following formula for the region's area:

$$\begin{aligned} A &= \lim_{\|P\| \to 0} \sum_{k=1}^{n} \frac{1}{2} (f(\theta_k))^2\, \Delta\theta_k \\ &= \int_{\alpha}^{\beta} \frac{1}{2} (f(\theta))^2\, d\theta . \end{aligned}$$

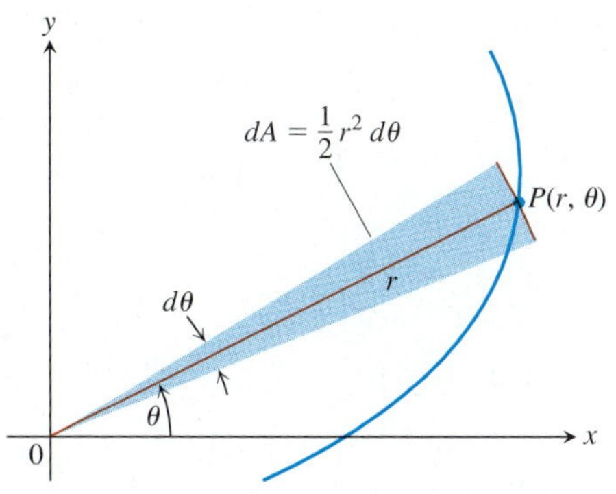

FIGURE 9.49 The area differential dA.

Area in Polar Coordinates

The **area** of the region **between the origin and the curve** $r = f(\theta)$, $\alpha \le \theta \le \beta$, is

$$A = \int_{\alpha}^{\beta} \frac{1}{2} r^2\, d\theta .$$

This is the integral of the **area differential** (Figure 9.49),

$$dA = \frac{1}{2} r^2\, d\theta .$$

Example 3 Finding Area

Find the area of the region in the plane enclosed by the cardioid $r = 2(1 + \cos\theta)$.

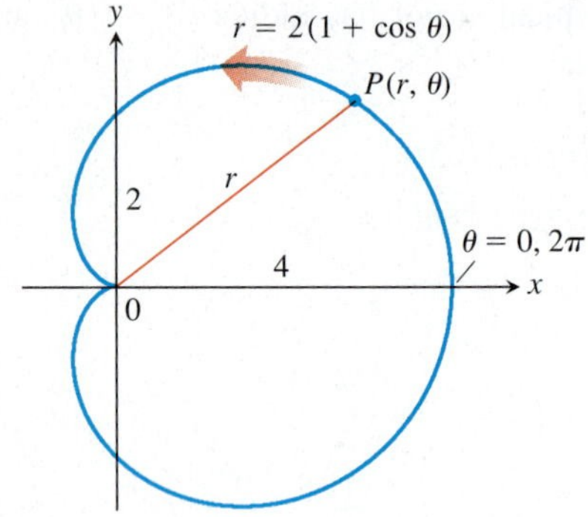

FIGURE 9.50 The cardioid in Example 3.

Solution We graph the cardioid (Figure 9.50) and determine that the *radius* r sweeps out the region exactly once as θ runs from 0 to 2π. The area is therefore

$$\begin{aligned}
\int_{\theta=0}^{\theta=2\pi} \frac{1}{2} r^2\, d\theta &= \int_0^{2\pi} \frac{1}{2} \cdot 4(1 + \cos\theta)^2\, d\theta \\
&= \int_0^{2\pi} 2(1 + 2\cos\theta + \cos^2\theta)\, d\theta \\
&= \int_0^{2\pi} \left(2 + 4\cos\theta + 2\,\frac{1 + \cos 2\theta}{2}\right) d\theta \\
&= \int_0^{2\pi} (3 + 4\cos\theta + \cos 2\theta)\, d\theta \\
&= \left[3\theta + 4\sin\theta + \frac{\sin 2\theta}{2}\right]_0^{2\pi} = 6\pi - 0 = 6\pi.
\end{aligned}$$

Example 4 Finding Area

Find the area inside the smaller loop of the limaçon $r = 2\cos\theta + 1$.

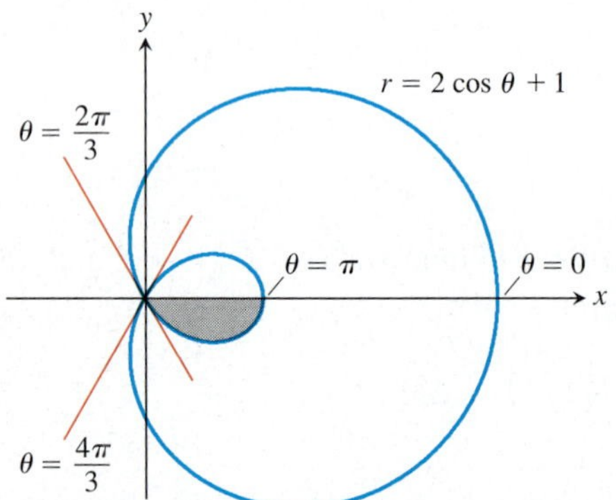

FIGURE 9.51 The limaçon in Example 4. Limaçon (pronounced LEE-ma-sahn) is an old French word for *snail*.

Solution After sketching the curve (Figure 9.51), we see that the smaller loop is traced out by the point (r, θ) as θ increases from $\theta = 2\pi/3$ to $\theta = 4\pi/3$. Since the curve is symmetric about the x-axis (the equation is unaltered when we replace θ by $-\theta$), we may calculate the area of the shaded half of the inner loop by integrating from $\theta = 2\pi/3$ to $\theta = \pi$. The area we seek will be twice the resulting integral:

$$A = 2\int_{2\pi/3}^{\pi} \frac{1}{2} r^2\, d\theta = \int_{2\pi/3}^{\pi} r^2\, d\theta.$$

Since

$$\begin{aligned}
r^2 &= (2\cos\theta + 1)^2 = 4\cos^2\theta + 4\cos\theta + 1 \\
&= 4 \cdot \frac{1 + \cos 2\theta}{2} + 4\cos\theta + 1 \\
&= 2 + 2\cos 2\theta + 4\cos\theta + 1 \\
&= 3 + 2\cos 2\theta + 4\cos\theta,
\end{aligned}$$

we have

$$\begin{aligned}
A &= \int_{2\pi/3}^{\pi} (3 + 2\cos 2\theta + 4\cos\theta)\, d\theta \\
&= \Big[3\theta + \sin 2\theta + 4\sin\theta\Big]_{2\pi/3}^{\pi} \\
&= (3\pi) - \left(2\pi - \frac{\sqrt{3}}{2} + 4 \cdot \frac{\sqrt{3}}{2}\right) \\
&= \pi - \frac{3\sqrt{3}}{2}.
\end{aligned}$$

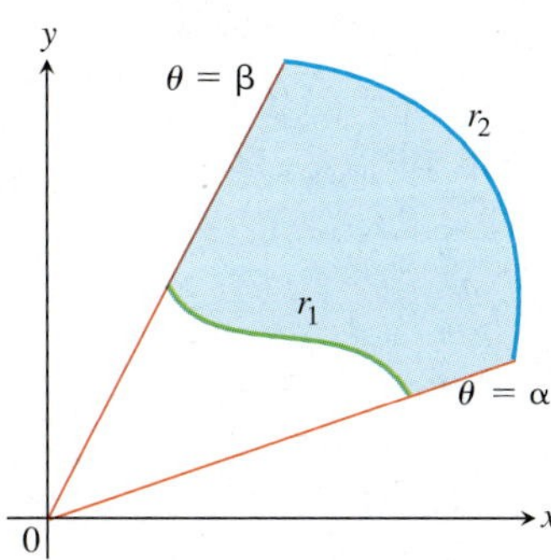

FIGURE 9.52 The area of the shaded region is calculated by subtracting the area of the region between r_1 and the origin from the area of the region between r_2 and the origin.

To find the area of a region like the one in Figure 9.52 which lies between two polar curves $r_1 = r_1(\theta)$ and $r_2 = r_2(\theta)$ from $\theta = \alpha$ to $\theta = \beta$, we subtract the integral of $(1/2)r_1^2$ from the integral of $(1/2)r_2^2$. This leads to the following formula.

Area Between Polar Curves

The area of the region $0 \le r_1(\theta) \le r_2(\theta)$, $\alpha \le \theta \le \beta$, is

$$A = \int_{\alpha}^{\beta} \frac{1}{2} r_2^2 \, d\theta - \int_{\alpha}^{\beta} \frac{1}{2} r_1^2 \, d\theta = \int_{\alpha}^{\beta} \frac{1}{2} (r_2^2 - r_1^2) \, d\theta. \tag{2}$$

Example 5 Finding Area Between Curves

Find the area of the region that lies inside the circle $r = 1$ and outside the cardioid $r = 1 - \cos\theta$.

Solution The region is shown in Figure 9.53. The outer curve is $r_2 = 1$, the inner curve is $r_1 = 1 - \cos\theta$, and θ runs from $-\pi/2$ to $\pi/2$. The area, from Equation (2), is

$$\begin{aligned} A &= \int_{-\pi/2}^{\pi/2} \frac{1}{2}(r_2^2 - r_1^2)\, d\theta \\ &= 2\int_{0}^{\pi/2} \frac{1}{2}(r_2^2 - r_1^2)\, d\theta \qquad \text{Symmetry} \\ &= \int_{0}^{\pi/2} (1 - (1 - 2\cos\theta + \cos^2\theta))\, d\theta \\ &= \int_{0}^{\pi/2} (2\cos\theta - \cos^2\theta)\, d\theta = \int_{0}^{\pi/2} \left(2\cos\theta - \frac{1 + \cos 2\theta}{2}\right) d\theta \\ &= \left[2\sin\theta - \frac{\theta}{2} - \frac{\sin 2\theta}{4}\right]_{0}^{\pi/2} = 2 - \frac{\pi}{4}. \end{aligned}$$

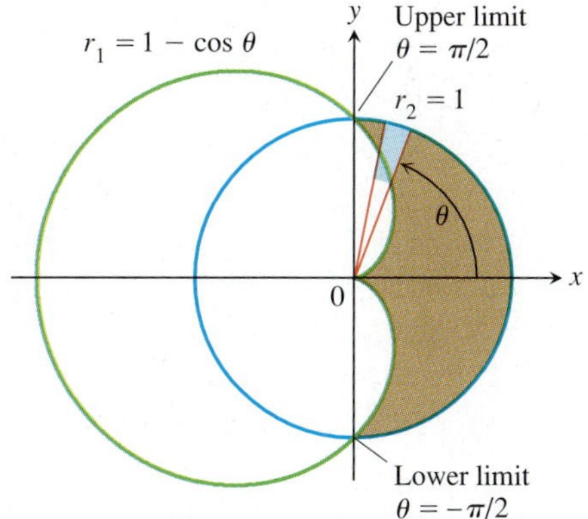

FIGURE 9.53 The region and limits of integration in Example 5.

Length of a Curve

We can obtain a polar coordinate formula for the length of a curve $r = f(\theta)$, $\alpha \le \theta \le \beta$, by parametrizing the curve as

$$x = r\cos\theta = f(\theta)\cos\theta, \qquad y = r\sin\theta = f(\theta)\sin\theta, \qquad \alpha \le \theta \le \beta. \tag{3}$$

The parametric length formula from Section 5.3 then gives the length as

$$L = \int_{\alpha}^{\beta} \sqrt{\left(\frac{dx}{d\theta}\right)^2 + \left(\frac{dy}{d\theta}\right)^2}\, d\theta.$$

This equation becomes

$$L = \int_{\alpha}^{\beta} \sqrt{r^2 + \left(\frac{dr}{d\theta}\right)^2}\, d\theta$$

when Equations (3) are substituted for x and y (Exercise 41).

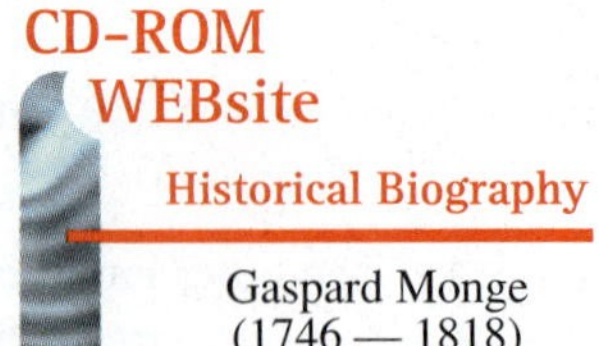

CD-ROM
WEBsite
Historical Biography
Gaspard Monge
(1746 — 1818)

Length of a Polar Curve

If $r = f(\theta)$ has a continuous first derivative for $\alpha \le \theta \le \beta$ and if the point $P(r, \theta)$ traces the curve $r = f(\theta)$ exactly once as θ runs from α to β, then the length of the curve is

$$L = \int_{\alpha}^{\beta} \sqrt{r^2 + \left(\frac{dr}{d\theta}\right)^2}\, d\theta. \qquad (4)$$

Example 6 Finding the Length of a Cardioid

Find the length of the cardioid $r = 1 - \cos\theta$.

Solution The graph is shown in Figure 9.54. The point $P(r, \theta)$ traces the curve once counterclockwise as θ runs from 0 to 2π, so these are the values we take for the limits of integration.

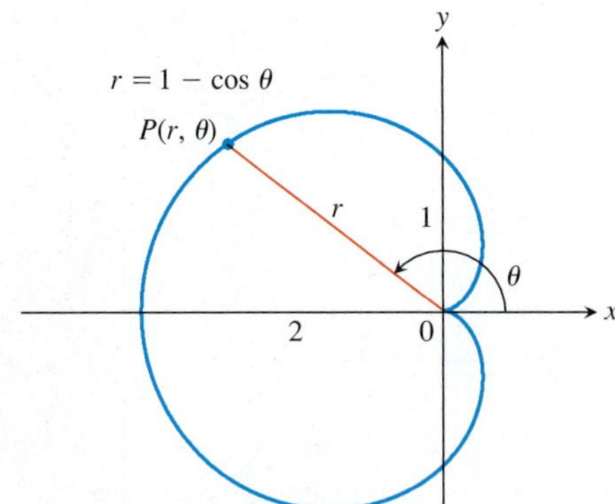

FIGURE 9.54 Example 6 calculates the length of this cardioid.

Since $r = 1 - \cos\theta$, $dr/d\theta = \sin\theta$, and we have

$$r^2 + \left(\frac{dr}{d\theta}\right)^2 = (1 - \cos\theta)^2 + (\sin\theta)^2$$

$$= 1 - 2\cos\theta + \underbrace{\cos^2\theta + \sin^2\theta}_{1} = 2 - 2\cos\theta.$$

Therefore,

$$\begin{aligned} L &= \int_\alpha^\beta \sqrt{r^2 + \left(\frac{dr}{d\theta}\right)^2}\, d\theta && \text{Equation (4)} \\ &= \int_0^{2\pi} \sqrt{2 - 2\cos\theta}\, d\theta \\ &= \int_0^{2\pi} \sqrt{4\sin^2\frac{\theta}{2}}\, d\theta && 1 - \cos\theta = 2\sin^2\frac{\theta}{2} \\ &= \int_0^{2\pi} \left|2\sin\frac{\theta}{2}\right| d\theta \\ &= \int_0^{2\pi} 2\sin\frac{\theta}{2}\, d\theta && \sin\frac{\theta}{2} \ge 0 \text{ for } 0 \le \theta \le 2\pi \\ &= \left[-4\cos\frac{\theta}{2}\right]_0^{2\pi} = 4 + 4 = 8. \end{aligned}$$

EXERCISES 9.6

Slopes of Polar Curves

In Exercises 1–4, find the slope of the curve at each indicated point.

1. $r = -1 + \sin\theta, \quad \theta = 0, \pi$

2. $r = \cos 2\theta, \quad \theta = 0, \pm\pi/2, \pi$

3. $r = 2 - 3\sin\theta$

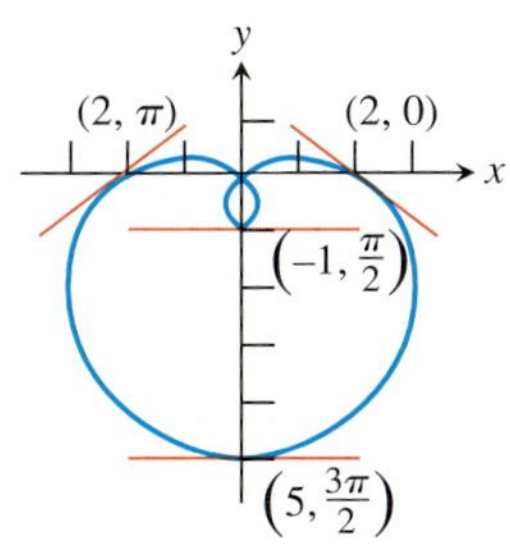

4. $r = 3(1 - \cos\theta)$

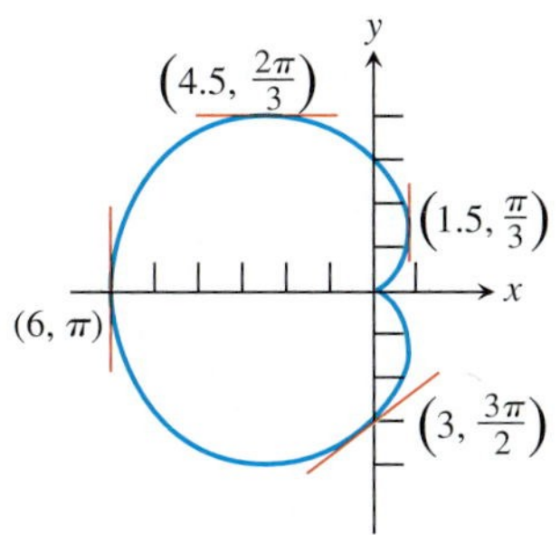

In Exercises 5–8, find the tangent lines at the pole.

5. $r = 3\cos\theta, \quad 0 \le \theta \le 2\pi$

6. $r = 2\cos 3\theta, \quad 0 \le \theta \le \pi$

7. $r = \sin 5\theta, \quad 0 \le \theta \le \pi$

8. $r = 2\sin 2\theta, \quad 0 \le \theta \le 2\pi$

In Exercises 9–12, find equations for the horizontal and vertical tangent lines to the curve.

9. $r = -1 + \sin\theta, \quad 0 \le \theta \le 2\pi$

10. $r = 1 + \cos\theta, \quad 0 \le \theta \le 2\pi$

11. $r = 2\sin\theta, \quad 0 \le \theta \le \pi$

12. $r = 3 - 4\cos\theta, \quad 0 \le \theta \le 2\pi$

Areas Inside Polar Curves

Find the areas of the regions in Exercises 13–18.

13. Inside the oval limaçon $r = 4 + 2\cos\theta$

14. Inside the cardioid $r = a(1 + \cos\theta), \quad a > 0$

15. Inside one leaf of the four-leaved rose $r = \cos 2\theta$

16. Inside the lemniscate $r^2 = 2a^2\cos 2\theta, \quad a > 0$

17. Inside one loop of the lemniscate $r^2 = 4\sin 2\theta$

18. Inside the six-leaved rose $r^2 = 2\sin 3\theta$

Areas Shared by Polar Regions

Find the areas of the regions in Exercises 19–28.

19. Shared by the circles $r = 2\cos\theta$ and $r = 2\sin\theta$

20. Shared by the circles $r = 1$ and $r = 2\sin\theta$

21. Shared by the circle $r = 2$ and the cardioid $r = 2(1 - \cos\theta)$

22. Shared by the cardioids $r = 2(1 + \cos\theta)$ and $r = 2(1 - \cos\theta)$

23. Inside the lemniscate $r^2 = 6\cos 2\theta$ and outside the circle $r = \sqrt{3}$

24. Inside the circle $r = 3a\cos\theta$ and outside the cardioid $r = a(1 + \cos\theta)$, $a > 0$

25. Inside the circle $r = -2\cos\theta$ and outside the circle $r = 1$

26. **(a)** Inside the outer loop of the limaçon $r = 1 + 2\cos\theta$ (See Figure 9.51.)

(b) Inside the outer loop and outside the inner loop of the limaçon $r = 1 + 2\cos\theta$

27. Inside the circle $r = 6$ above the line $r = 3\csc\theta$

28. Inside the lemniscate $r^2 = 6\cos 2\theta$ to the right of the line $r = (3/2)\sec\theta$

29. **(a)** Find the area of the shaded region in the accompanying figure.

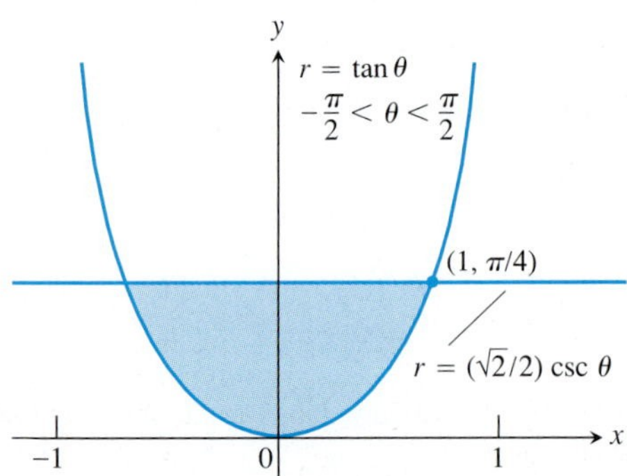

(b) *Writing to Learn* It looks as if the graph of $r = \tan\theta$, $-\pi/2 < \theta < \pi/2$, could be asymptotic to the lines $x = 1$ and $x = -1$. Is it? Give reasons for your answer.

30. *Writing to Learn* The area of the region that lies inside the cardioid curve $r = 1 + \cos\theta$ and outside the circle $r = \cos\theta$ is not

$$\frac{1}{2}\int_0^{2\pi} [(1 + \cos\theta)^2 - \cos^2\theta]\, d\theta = \pi.$$

Why not? What *is* the area? Give reasons for your answers.

Lengths of Polar Curves

Find the lengths of the curves in Exercises 31–39.

31. The spiral $r = \theta^2$, $0 \le \theta \le \sqrt{5}$

32. The spiral $r = e^{\theta}/\sqrt{2}$, $0 \le \theta \le \pi$

33. The cardioid $r = 1 + \cos\theta$

34. The curve $r = a\sin^2(\theta/2)$, $0 \le \theta \le \pi$, $a > 0$

35. The parabolic segment $r = 6/(1 + \cos\theta)$, $0 \le \theta \le \pi/2$

36. The parabolic segment $r = 2/(1 - \cos\theta)$, $\pi/2 \le \theta \le \pi$

37. The curve $r = \cos^3(\theta/3)$, $0 \le \theta \le \pi/4$

38. The curve $r = \sqrt{1 + \sin 2\theta}$, $0 \le \theta \le \pi\sqrt{2}$

39. The curve $r = \sqrt{1 + \cos 2\theta}$, $0 \le \theta \le \pi\sqrt{2}$

40. *Circumferences of circles* As usual, when faced with a new formula, it is a good idea to try it on familiar objects to be sure it gives results consistent with past experience. Use the length formula in Equation (4) to calculate the circumferences of the following circles ($a > 0$):

(a) $r = a$ **(b)** $r = a\cos\theta$ **(c)** $r = a\sin\theta$.

Theory and Examples

41. *Length of a polar curve* Assuming that the necessary derivatives are continuous, show how the substitutions

$$x = f(\theta)\cos\theta, \qquad y = f(\theta)\sin\theta$$

(Equations (3) in the text) transform

$$L = \int_{\alpha}^{\beta} \sqrt{\left(\frac{dx}{d\theta}\right)^2 + \left(\frac{dy}{d\theta}\right)^2}\, d\theta$$

into

$$L = \int_{\alpha}^{\beta} \sqrt{r^2 + \left(\frac{dr}{d\theta}\right)^2}\, d\theta.$$

42. *Average value* If f is continuous, the average value of the polar coordinate r over the curve $r = f(\theta)$, $\alpha \le \theta \le \beta$, with respect to θ is

$$r_{\text{av}} = \frac{1}{\beta - \alpha}\int_{\alpha}^{\beta} f(\theta)\, d\theta.$$

Use this formula to find the average value of r with respect to θ over the following curves ($a > 0$).

(a) The cardioid $r = a(1 - \cos\theta)$

(b) The circle $r = a$

(c) The circle $r = a\cos\theta$, $-\pi/2 \le \theta \le \pi/2$

43. *Writing to Learn* Can anything be said about the relative lengths of the curves

$$r = f(\theta), \qquad \alpha \le \theta \le \beta,$$

and

$$r = 2f(\theta), \qquad \alpha \le \theta \le \beta?$$

Give reasons for your answer.

44. *Videocassette tape length* The length of a tape wound onto a take-up reel as shown in the figure is

$$L = \int_0^\alpha \sqrt{r^2 + \left(\frac{b}{2\pi}\right)^2}\, d\theta,$$

where b is the tape thickness and

$$r = r_0 + \left(\frac{\alpha}{2\pi}\right) b$$

is the radius of the tape on the take-up reel. The initial radius of the tape on the take-up reel is r_0 and α is the angle in radians through which the wheel has turned.

(a) Find a spiral that models the tape accumulating on the take-up reel using polar graphing with $r_0 = 1.75$ cm and $b = 0.06$ cm.

(b) Confirm the formula for L analytically.

(c) Determine the length of tape on the take-up reel if the reel has turned through an angle of 80π with $r_0 = 1.75$ cm and $b = 0.06$ cm.

(d) Assume that b is very small in comparison to r at any time. Show analytically that

$$L_a = \int_0^\alpha r\, d\theta$$

is an excellent approximation to the exact value of L.

(e) For the values given in part (c), compare L_a with L.

45. (*Continuation of Exercise 44*) Let n be the number of complete turns the take-up reel has made.

(a) Find a formula for n in terms of L, the tape length.

(b) When a VCR operates, the tape moves past the heads at a constant speed. Describe the speed of the take-up reel as time progresses.

(c) Suppose that the VCR tape counter is the number n of complete turns of the take-up reel. Describe the counter values as a function of time t.

Questions to Guide Your Review

1. When do directed line segments in the plane represent the same vector?
2. How are vectors added and subtracted geometrically? Algebraically?
3. How do you find a vector's magnitude and direction?
4. If a vector is multiplied by a positive scalar, how is the result related to the original vector? What if the scalar is zero? Negative?
5. Define the *dot product* (*scalar product*) of two vectors. Which algebraic laws are satisfied by dot products? Give examples. When is the dot product of two vectors equal to zero?
6. What geometric interpretation does the dot product have? Give examples.
7. What is the vector projection of a vector **u** onto a vector **v**? How do you write **u** as the sum of a vector parallel to **v** and a vector orthogonal to **v**?
8. State the rules for differentiating and integrating vector functions. Give examples.
9. How do you define and calculate the velocity, speed, direction of motion, and acceleration of a body moving along a sufficiently differentiable plane curve? Give examples.
10. What are the vector and parametric equations for ideal projectile motion? How do you find an ideal projectile's maximum height, flight time, and range? Give examples.
11. What are polar coordinates? What equations relate polar coordinates to Cartesian coordinates? Why might you want to change from one coordinate system to the other?
12. What consequence does the lack of uniqueness of polar coordinates have for graphing? Give an example.
13. How do you graph equations in polar coordinates? Include in your discussion symmetry and slope. Give examples.
14. How do you find the area of a region $0 \le r_1(\theta) \le r \le r_2(\theta)$, $\alpha \le \theta \le \beta$, in the polar coordinate plane? Give examples.
15. Under what conditions can you find the length of a curve $r = f(\theta)$, $\alpha \le \theta \le \beta$, in the polar coordinate plane? Give an example of a typical calculation.

Practice Exercises

Vector Calculations

In Exercises 1–4, let $\mathbf{u} = \langle -3, 4\rangle$ and $\mathbf{v} = \langle 2, -5\rangle$. Find **(a)** the component form of the vector and **(b)** its magnitude.

1. $3\mathbf{u} - 4\mathbf{v}$
2. $\mathbf{u} + \mathbf{v}$
3. $-2\mathbf{u}$
4. $5\mathbf{v}$

In Exercises 5–8, find the component form of the vector.

5. The vector obtained by rotating $\langle 0, 1\rangle$ through an angle of $2\pi/3$ radians
6. The unit vector that makes an angle of $\pi/6$ radian with the positive x-axis
7. The vector 2 units long in the direction $4\mathbf{i} - \mathbf{j}$
8. The vector 5 units long in the direction opposite to the direction of $(3/5)\mathbf{i} + (4/5)\mathbf{j}$

Length and Direction

Express the vectors in Exercises 9–12 in terms of their lengths and directions.

9. $\sqrt{2}\mathbf{i} + \sqrt{2}\mathbf{j}$
10. $-\mathbf{i} - \mathbf{j}$
11. Velocity vector to $\mathbf{r} = (2\cos t)\mathbf{i} + (2\sin t)\mathbf{j}$ at the point $(0, 2)$.
12. Velocity vector to $\mathbf{r} = (e^t \cos t)\mathbf{i} + (e^t \sin t)\mathbf{j}$ when $t = \ln 2$.

Tangent and Normal Vectors

In Exercises 13 and 14, find the unit vectors that are tangent and normal to the curve at point P.

13. $y = \tan x, \quad P(\pi/4, 1)$

14. $x^2 + y^2 = 25, \quad P(3, 4)$

Vector Projections

15. Copy the vectors **u** and **v** and sketch the vector projection of **v** onto **u**.

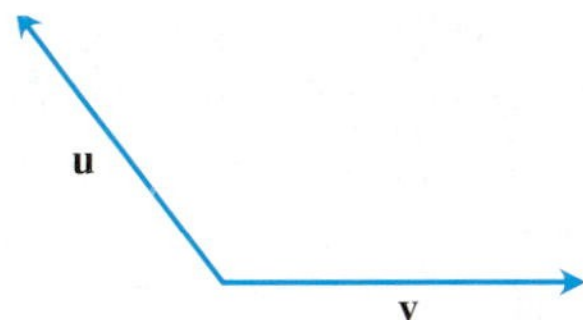

16. Express vectors **a**, **b**, and **c** in terms of **u** and **v**.

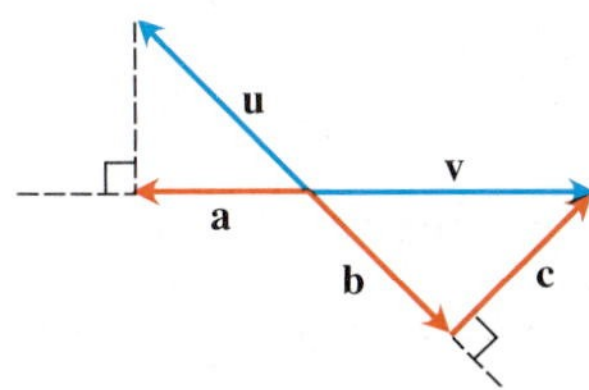

In Exercises 17 and 18, find $|\mathbf{v}|$, $|\mathbf{u}|$, $\mathbf{v} \cdot \mathbf{u}$, $\mathbf{u} \cdot \mathbf{v}$, the angle between **v** and **u**, the scalar component of **u** in the direction of **v**, and the vector projection of **u** onto **v**.

17. $\mathbf{v} = \mathbf{i} + \mathbf{j}$

$\mathbf{u} = 2\mathbf{i} + \mathbf{j}$

18. $\mathbf{v} = \mathbf{i} + \mathbf{j}$

$\mathbf{u} = -\mathbf{i} - 3\mathbf{j}$

In Exercises 19 and 20, write **u** as the sum of a vector parallel to **v** and a vector orthogonal to **v**.

19. $\mathbf{v} = 2\mathbf{i} - \mathbf{j}$

$\mathbf{u} = \mathbf{i} + \mathbf{j}$

20. $\mathbf{v} = \mathbf{i} - 2\mathbf{j}$

$\mathbf{u} = -\mathbf{i} + \mathbf{j}$

Velocity and Acceleration Vectors

In Exercises 21 and 22, $\mathbf{r}(t)$ is the position vector of a particle in the plane at time t.

(a) Find the velocity and acceleration vectors.

(b) Find the speed at the given value of t.

(c) Find the angle between the velocity and acceleration vectors at the given value of t.

21. $\mathbf{r}(t) = (4 \cos t)\mathbf{i} + (\sqrt{2} \sin t)\mathbf{j}, \quad t = \pi/4$

22. $\mathbf{r}(t) = (\sqrt{3} \sec t)\mathbf{i} + (\sqrt{3} \tan t)\mathbf{j}, \quad t = 0$

23. *Maximum speed* The position of a particle in the plane at time t is

$$\mathbf{r} = \frac{1}{\sqrt{1 + t^2}}\mathbf{i} + \frac{t}{\sqrt{1 + t^2}}\mathbf{j}.$$

Find the particle's greatest speed.

24. *Writing to Learn: Minimum speed* The position of a particle in the plane at time $t \geq 0$ is

$$\mathbf{r}(t) = (e^t \cos t)\mathbf{i} + (e^t \sin t)\mathbf{j}.$$

Find the particle's minimum speed. Does it have a maximum speed? Give reasons for your answer.

Integrals and Initial Value Problems

In Exercises 25 and 26, evaluate the integral.

25. $\displaystyle\int_0^1 [(3 + 6t)\mathbf{i} + (6\pi \cos \pi t)\mathbf{j}]\, dt$

26. $\displaystyle\int_e^{e^2} \left[\left(\frac{2 \ln t}{t}\right)\mathbf{i} + \left(\frac{1}{t \ln t}\right)\mathbf{j}\right] dt$

In Exercises 27–30, solve the initial value problem.

27. $\dfrac{d\mathbf{r}}{dt} = -(\sin t)\mathbf{i} + (\cos t)\mathbf{j}, \quad \mathbf{r}(0) = \mathbf{j}$

28. $\dfrac{d\mathbf{r}}{dt} = \dfrac{1}{t^2 + 1}\mathbf{i} + \dfrac{t}{\sqrt{t^2 + 1}}\mathbf{j}, \quad \mathbf{r}(0) = \mathbf{i} + \mathbf{j}$

29. $\dfrac{d^2\mathbf{r}}{dt^2} = 2\mathbf{j}, \quad \left.\dfrac{d\mathbf{r}}{dt}\right|_{t=0} = \mathbf{0}, \quad \mathbf{r}(0) = \mathbf{i}$

30. $\dfrac{d^2\mathbf{r}}{dt^2} = -2\mathbf{i} - 2\mathbf{j}, \quad \left.\dfrac{d\mathbf{r}}{dt}\right|_{t=1} = 4\mathbf{i}, \quad \mathbf{r}(1) = 3\mathbf{i} + 3\mathbf{j}$

Graphs in the Polar Plane

Sketch the regions defined by the polar coordinate inequalities in Exercises 31 and 32.

31. $0 \leq r \leq 6 \cos \theta$

32. $-4 \sin \theta \leq r \leq 0$

Match each graph in Exercises 33–40 with the appropriate equation (a) through (l). There are more equations than graphs, so some equations will not be matched.

(a) $r = \cos 2\theta$

(b) $r \cos \theta = 1$

(c) $r = \dfrac{6}{1 - 2 \cos \theta}$

(d) $r = \sin 2\theta$

(e) $r = \theta$

(f) $r^2 = \cos 2\theta$

(g) $r = 1 + \cos\theta$ **(h)** $r = 1 - \sin\theta$

(i) $r = \dfrac{2}{1 - \cos\theta}$ **(j)** $r^2 = \sin 2\theta$

(k) $r = -\sin\theta$ **(l)** $r = 2\cos\theta + 1$

33. Four-leaved rose

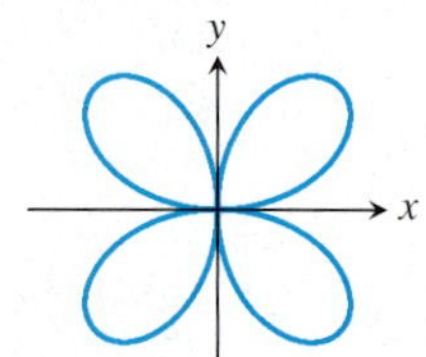

34. Spiral

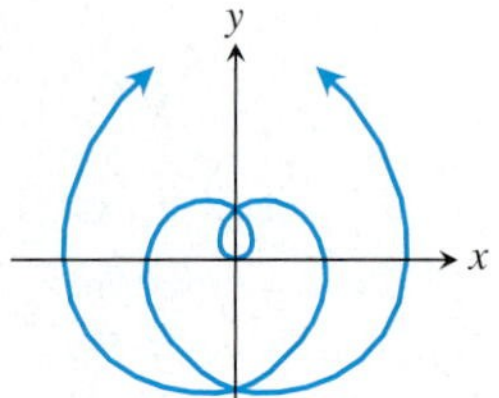

35. Limaçon

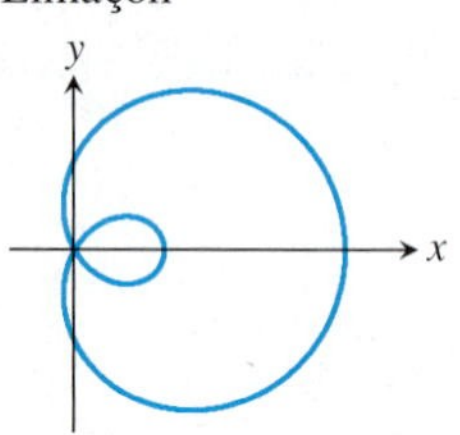

36. Lemniscate

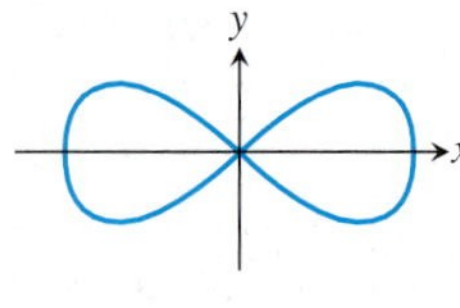

37. Circle

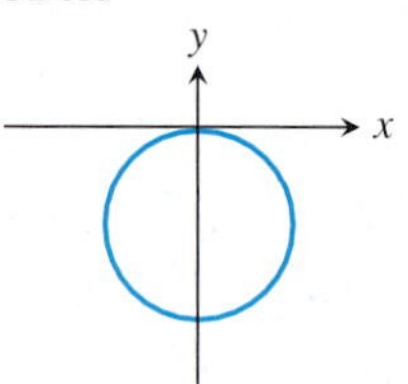

38. Cardioid

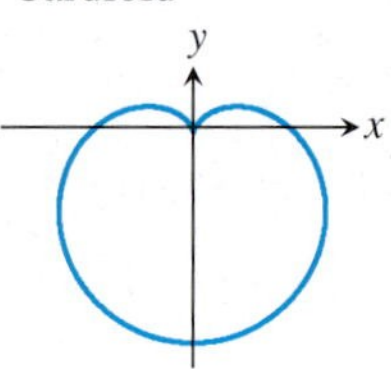

39. Parabola

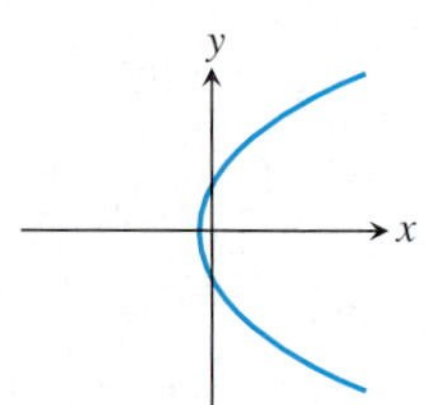

40. Lemniscate

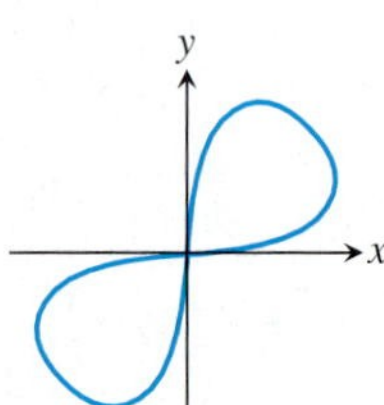

In Exercises 41–44, (**a**) graph the polar curve. (**b**) What is the smallest length θ-interval that will produce the graph?

41. $r = \cos 2\theta$

42. $r\cos\theta = 1$

43. $r^2 = \sin 2\theta$

44. $r = -\sin\theta$

Tangent Lines to Polar Graphs

In Exercises 45 and 46, find the tangent lines at the pole.

45. $r = \cos 2\theta, \quad 0 \le \theta \le 2\pi$

46. $r = 1 + \cos 2\theta, \quad 0 \le \theta \le 2\pi$

In Exercises 47 and 48, find equations for the horizontal and vertical tangent lines to the curve.

47. $r = 1 - \cos(\theta/2), \quad 0 \le \theta \le 4\pi$

48. $r = 2(1 - \sin\theta), \quad 0 \le \theta \le 2\pi$

49. Find equations for the lines that are tangent to the tips of the petals of the four-leaved rose $r = \sin 2\theta$.

50. Find equations for the lines that are tangent to the graph of the cardioid $r = 1 + \sin\theta$ at the points where it crosses the x-axis.

Polar to Cartesian Equations

In Exercises 51–56, replace the polar equation by an equivalent Cartesian equation. Then identify or describe the graph.

51. $r\cos\theta = r\sin\theta$

52. $r = 3\cos\theta$

53. $r = 4\tan\theta\sec\theta$

54. $r\cos(\theta + \pi/3) = 2\sqrt{3}$

55. $r = 2\sec\theta$

56. $r = -(3/2)\csc\theta$

Cartesian to Polar Equations

In Exercises 57–60, replace the Cartesian equation by an equivalent polar equation.

57. $x^2 + y^2 + 5y = 0$

58. $x^2 + y^2 - 2y = 0$

59. $x^2 + 4y^2 = 16$

60. $(x + 2)^2 + (y - 5)^2 = 16$

Area in the Polar Plane

Find the areas of the regions in the polar coordinate plane described in Exercises 61–64.

61. Enclosed by the limaçon $r = 2 - \cos\theta$

62. Enclosed by one leaf of the three-leaved rose $r = \sin 3\theta$

63. Inside the "figure eight" $r = 1 + \cos 2\theta$ and outside the circle $r = 1$

64. Inside the cardioid $r = 2(1 + \sin\theta)$ and outside the circle $r = 2\sin\theta$

Lengths of Polar Curves

Find the lengths of the curves given by the polar coordinate equations in Exercises 65–68.

65. $r = -1 + \cos\theta$

66. $r = 2\sin\theta + 2\cos\theta, \quad 0 \le \theta \le \pi/2$

67. $r = 8 \sin^3 (\theta/3), \quad 0 \le \theta \le \pi/4$

68. $r = \sqrt{1 + \cos 2\theta}, \quad -\pi/2 \le \theta \le \pi/2$

Theory and Examples

69. *Navigation* An airplane, flying in the direction 80° east of north at 540 mph in still air, encounters a 55 mph tailwind acting in the direction 100° east of north. The airplane holds its compass heading but, because of the wind, acquires a different ground speed and direction. What are they?

70. *Combining forces* A force of 120 lb pulls up on an object at an angle of 20° with the horizontal. A second force of 300 lb pulls down on the object at an angle of −5°. Find the direction and length of the resultant force vector.

71. *Shot put* A shot leaves the thrower's hand 6.5 ft above the ground at a 45° angle at 44 ft/sec. Where is it 3 sec later?

72. *Javelin* A javelin leaves the thrower's hand 7 ft above the ground at a 45° angle at 80 ft/sec. How high does it go?

73. *Rolling wheel* A circular wheel with radius 1 ft and center C rolls to the right along the x-axis at a half-turn per second (see figure). At time t seconds, the position vector of the point P on the wheel's circumference is

$$\mathbf{r}(t) = (\pi t - \sin \pi t)\mathbf{i} + (1 - \cos \pi t)\mathbf{j}.$$

T **(a)** Graph the curve traced by P during the interval $0 \le t \le 3$.

(b) Find velocity and acceleration vectors $\mathbf{v}$ and $\mathbf{a}$ at $t = 0, 1, 2,$ and 3.

(c) *Writing to Learn* At any given time, what is the forward speed of the topmost point of the wheel? Of C? Give reasons for your answers.

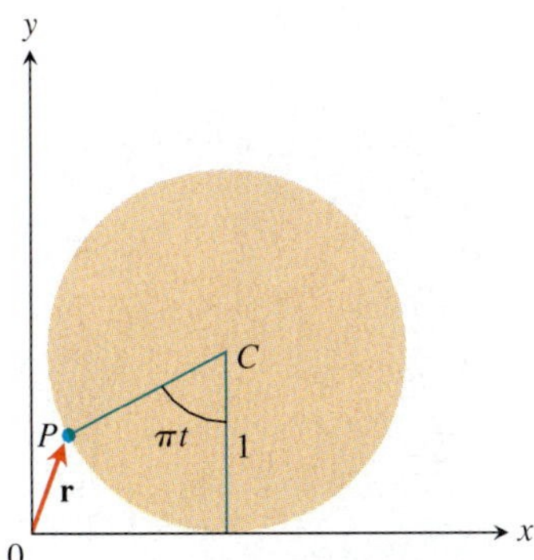

74. *The dictator* The Civil War mortar Dictator weighed so much (17,120 lb) that it had to be mounted on a railroad car. It had a 13 in. bore and used a 20 lb powder charge to fire a 200 lb shell. The mortar was made by Mr. Charles Knapp in his ironworks in Pittsburgh, Pennsylvania, and was used by the Union army in 1864 in the siege of Petersburg, Virginia. How far did it shoot? Here we have a difference of opinion. The ordnance manual claimed 4325 yd, whereas field officers claimed 4752 yd. Assuming a 45° firing angle, what muzzle speeds are involved here?

75. *World's record for popping a champagne cork*

(a) Until 1988, the world's record for popping a champagne cork, 109 ft 6 in., was held by Captain Michael Hill of the British Royal Artillery. Assuming that Captain Hill held the bottle at ground level at a 45° angle and the cork behaved like an ideal projectile, how fast was the cork going as it left the bottle?

(b) A new world record, 177 ft 9 in., was set on June 5, 1988, by Prof. Emeritus Heinrich of Rensselaer Polytechnic Institute, firing from 4 ft above ground level at a 45° angle at the Woodbury Vineyards Winery, New York. Assuming an ideal trajectory, what was the cork's initial speed?

76. *Javelin* In Potsdam in 1988, Petra Felke of (then) East Germany set a women's world record by throwing a javelin 262 ft 5 in.

(a) Assuming that Felke launched the javelin at a 40° angle to the horizontal from 6.5 ft above the ground, what was the javelin's initial speed?

(b) How high did the javelin go?

77. *Synchronous curves* By eliminating α from the ideal projectile equations

$$x = (v_0 \cos \alpha)t, \qquad y = (v_0 \sin \alpha)t - \frac{1}{2}gt^2,$$

show that $x^2 + (y + gt^2/2)^2 = v_0^2t^2$. This shows that projectiles launched simultaneously from the origin at the same initial speed will, at any given instant, all lie on the circle of radius v_0t centered at $(0, -gt^2/2)$, regardless of their launch angle. These circles are the *synchronous curves* of the launching.

78. *Hitting a baseball under a wind gust* A baseball is hit when it is 4 ft above the ground. It leaves the bat with an intial velocity of 155 ft/sec, making an angle of 18° with the horizontal. At the instant the ball is hit, an instantaneous 11.7 ft/sec gust of wind blows in the horizontal direction against the ball, adding a component of

$-11.7\mathbf{i}$ to the ball's initial velocity. A 10-foot-high fence is 380 ft from home plate in the direction of the flight.

(a) Find vector and parametric equations for the path of the baseball.

(b) How high does the baseball go, and when does it reach maximum height?

(c) Find the range and flight time of the baseball.

(d) When is the baseball 25 ft high? How far (ground distance) is the baseball from home plate at that height?

(e) *Writing to Learn* Has the batter hit a home run? Explain.

79. *Linear drag* (*Continuation of Exercise 78*) Consider the baseball problem of Exercise 78 again. This time, assume a linear drag model with a drag coefficient of 0.09.

(a) Find vector and parametric equations for the path of the baseball.

(b) How high does the baseball go, and when does it reach maximum height?

(c) Find the range and flight time of the baseball.

(d) When is the baseball 30 ft high? How far (ground distance) is the baseball from home plate at that height?

(e) Has the batter hit a home run? If "yes," find a drag coefficient that would have prevented a home run. If "no," find a drag coefficient that would have allowed the hit to be a home run.

80. *Parallelogram* The accompanying figure shows parallelogram $ABCD$ and the midpoint P of diagonal BD.

(a) Express $\overrightarrow{BD}$ in terms of $\overrightarrow{AB}$ and $\overrightarrow{AD}$.

(b) Express $\overrightarrow{AP}$ in terms of $\overrightarrow{AB}$ and $\overrightarrow{AD}$.

(c) Prove that P is also the midpoint of diagonal AC.

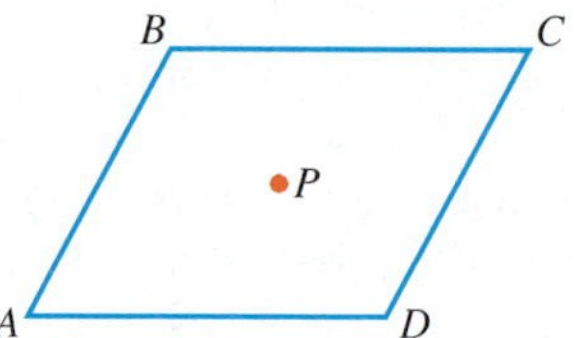

81. *Archimedes spirals* The graph of an equation of the form $r = a\theta$, where a is a nonzero constant, is called an Archimedes spiral. Is there anything special about the widths between the successive turns of such a spiral?

Additional Exercises: Theory, Examples, Applications

1. *Rowing across a river* A straight river is 20 m wide. The velocity of the river at (x, y) is

$$\mathbf{v} = -\frac{3x(20-x)}{100}\mathbf{j} \text{ m/min}, \qquad 0 \le x \le 20.$$

A boat leaves the shore at (0, 0) and travels through the water with a constant velocity. It arrives at the opposite shore at (20, 0). The speed of the boat is always $\sqrt{20}$ m/min.

(a) Find the velocity of the boat.

(b) Find the location of the boat at time t.

(c) Sketch the path of the boat.

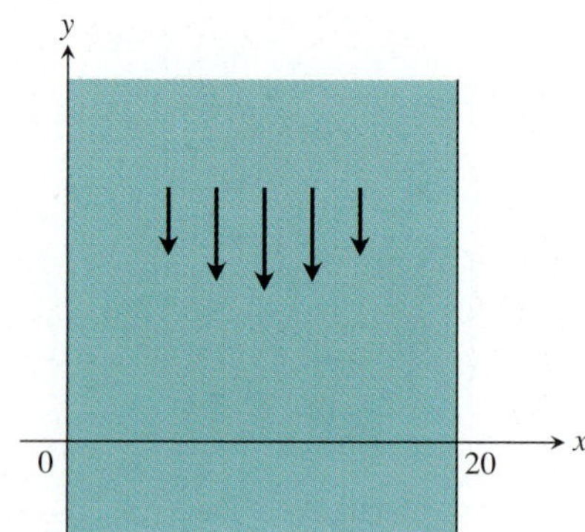

2. *Motion on a circle* A particle moves in the plane so that its velocity and position vectors are always orthogonal. Show that the particle moves in a circle centered at the origin.

3. *Angle between position and acceleration vectors* Suppose that $\mathbf{r}(t) = (e^t \cos t)\mathbf{i} + (e^t \sin t)\mathbf{j}$. Show that the angle between $\mathbf{r}$ and $\mathbf{a}$ never changes. What *is* the angle?

4. *Motion on a circle* A particle moves around the unit circle in the xy-plane. Its position at time t is $\mathbf{r} = x\,\mathbf{i} + y\,\mathbf{j}$, where x and y are differentiable functions of t. Find dy/dt if $\mathbf{v} \cdot \mathbf{i} = y$. Is the motion clockwise, or counterclockwise?

5. *Motion on a cubic* You send a message through a pneumatic tube that follows the curve $9y = x^3$ (distance in meters). At the point $(3, 3)$, $\mathbf{v} \cdot \mathbf{i} = 4$ and $\mathbf{a} \cdot \mathbf{i} = -2$. Find the values of $\mathbf{v} \cdot \mathbf{j}$ and $\mathbf{a} \cdot \mathbf{j}$ at $(3, 3)$.

6. *Angle bisection* Show that $\mathbf{w} = |\mathbf{v}|\mathbf{u} + |\mathbf{u}|\mathbf{v}$ bisects the angle between $\mathbf{u}$ and $\mathbf{v}$.

Polar Coordinates

7. **(a)** *Finding a polar equation* Find an equation in polar coordinates for the curve

$$x = e^{2t} \cos t, \qquad y = e^{2t} \sin t, \qquad -\infty < t < \infty.$$

(b) *Length of curve* Find the length of the curve from $t = 0$ to $t = 2\pi$.

8. *Length of curve* Find the length of the curve $r = 2 \sin^3 (\theta/3)$, $0 \le \theta \le 3\pi$, in the polar coordinate plane.

9. *Polar area* Sketch the regions enclosed by the curves $r = 2a \cos^2 (\theta/2)$ and $r = 2a \sin^2 (\theta/2)$, $a > 0$, in the polar coordinate plane and find the area of the portion of the plane they have in common.

T 10. Graph the curve $r = \cos 5\theta + n \cos \theta$, $0 \le \theta \le \pi$ for integers $n = -5$ (heart) to $n = 5$ (bell). (*Source*: *The College Mathematics Journal*, Vol. 25, No. 1 (Jan. 1994).

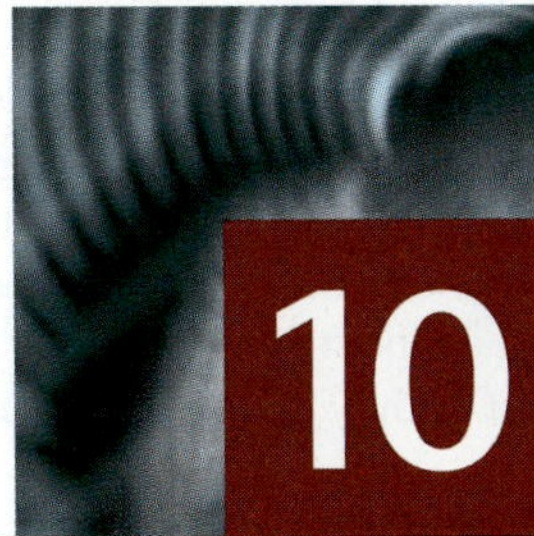

10 Vectors and Motion in Space

OVERVIEW This chapter introduces vectors within a three-dimensional coordinate system. Just as the coordinate plane is the natural place to study functions of a single variable, coordinate space is the place to study functions of two variables (or more). We establish coordinates in space by adding a third axis that measures distance above and below the xy-plane. This axis is designated the z-axis, and the standard unit vector parallel to it pointing in the positive direction is denoted by $\mathbf{k}$.

When a body travels through space, the equations $x = f(t)$, $y = g(t)$, and $z = h(t)$ that give the body's coordinates as functions of time serve as parametric equations for the body's motion and path. With vector notation, we can condense these into a single equation $\mathbf{r}(t) = f(t)\mathbf{i} + g(t)\mathbf{j} + h(t)\mathbf{k}$ that gives the body's position as a vector function of time.

In this chapter, we show how to use calculus to study the paths, velocities, and accelerations of moving bodies. As we go along, we see how our work answers the standard questions about the paths and motions of planets and satellites. In the final section, we use our new vector calculus to derive Kepler's laws of planetary motion from Newton's laws of motion and gravitation.

10.1 Cartesian (Rectangular) Coordinates and Vectors in Space

Cartesian Coordinates • Vectors in Space • Magnitude • The Zero Vector • Unit Vectors • Length and Direction • Distance and Spheres in Space • Midpoints

Our goal now is to describe the three-dimensional Cartesian coordinate system. Then we can define and study vectors in space.

Cartesian Coordinates

To locate points in space, we use three mutually perpendicular coordinate axes, arranged as in Figure 10.1. The axes shown there make a *right-handed* coordinate frame. When you hold your right hand so that the fingers curl from the positive x-axis toward the positive y-axis, your thumb points along the positive z-axis.

The Cartesian coordinates (x, y, z) of a point P in space are the numbers at which the planes through P perpendicular to the axes cut the axes. Cartesian

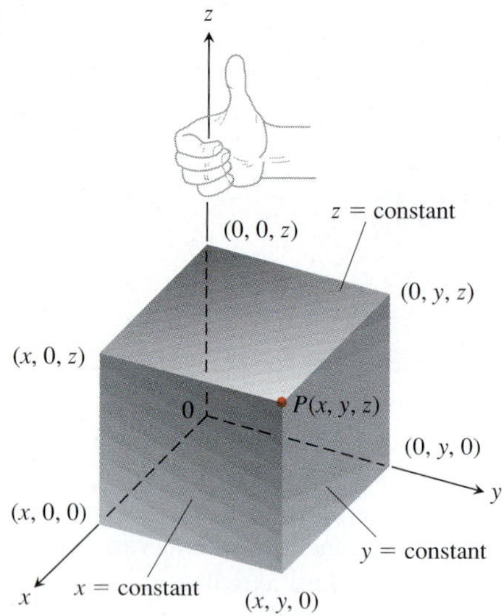

FIGURE 10.1 The Cartesian coordinate system is right-handed.

coordinates for space are also called **rectangular coordinates** because the axes that define them meet at right angles.

Points on the x-axis have y- and z-coordinates equal to zero. That is, they have coordinates of the form $(x, 0, 0)$. Similarly, points on the y-axis have coordinates of the form $(0, y, 0)$. Points on the z-axis have coordinates of the form $(0, 0, z)$.

The planes determined by the coordinates axes are the ***xy*-plane,** whose standard equation is $z = 0$; the ***yz*-plane,** whose standard equation is $x = 0$; and the ***xz*-plane,** whose standard equation is $y = 0$. They meet at the **origin** $(0, 0, 0)$ (Figure 10.2).

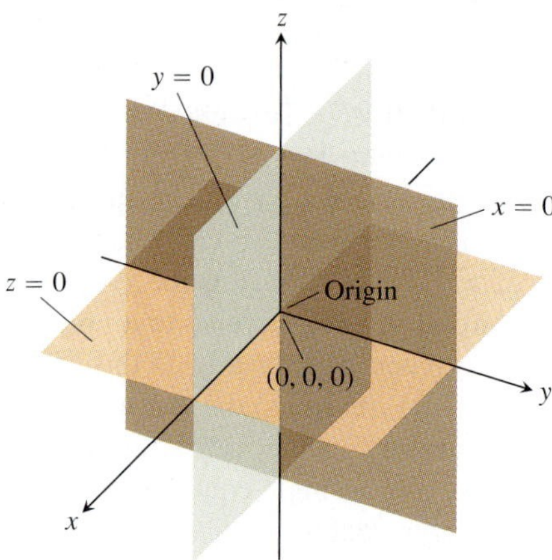

FIGURE 10.2 The planes $x = 0$, $y = 0$, and $z = 0$ divide space into eight octants.

The three **coordinate planes** $x = 0$, $y = 0$, and $z = 0$ divide space into eight cells called **octants.** The octant in which the point coordinates are all positive is called the **first octant;** there is no conventional numbering for the other seven octants.

In the following examples, we match coordinate equations and inequalities with the sets of points they define in space.

Example 1 Interpreting Equations and Inequalities Geometrically

(a) $z \geq 0$ — The half-space consisting of the points on and above the xy-plane.

(b) $x = -3$ — The plane perpendicular to the x-axis at $x = -3$. This plane lies parallel to the yz-plane and 3 units behind it.

(c) $z = 0,\ x \leq 0,\ y \geq 0$ — The second quadrant of the xy-plane.

(d) $x \geq 0,\ y \geq 0,\ z \geq 0$ — The first octant.

(e) $-1 \leq y \leq 1$ — The slab between the planes $y = -1$ and $y = 1$ (planes included).

(f) $y = -2,\ z = 2$ — The line in which the planes $y = -2$ and $z = 2$ intersect. Alternatively, the line through the point $(0, -2, 2)$ parallel to the x-axis.

Example 2 Graphing Equations

What points $P(x, y, z)$ satisfy the equations

$$x^2 + y^2 = 4 \qquad \text{and} \qquad z = 3?$$

Solution The points lie in the horizontal plane $z = 3$ and, in this plane, make up the circle $x^2 + y^2 = 4$. We call this set of points "the circle $x^2 + y^2 = 4$ in the plane $z = 3$" or, more simply, "the circle $x^2 + y^2 = 4$, $z = 3$" (Figure 10.3).

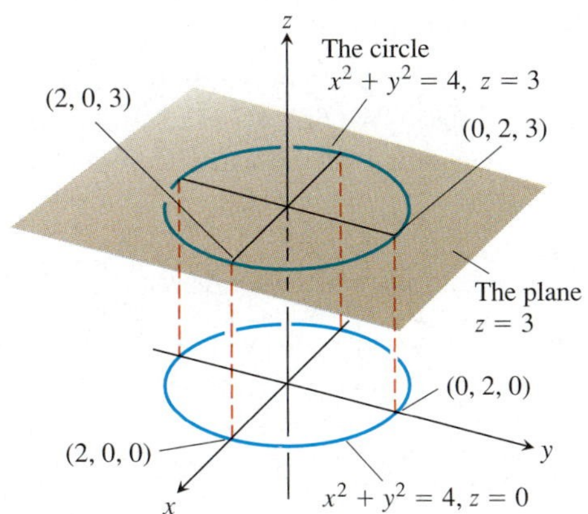

FIGURE 10.3 The circle $x^2 + y^2 = 4$, $z = 3$.

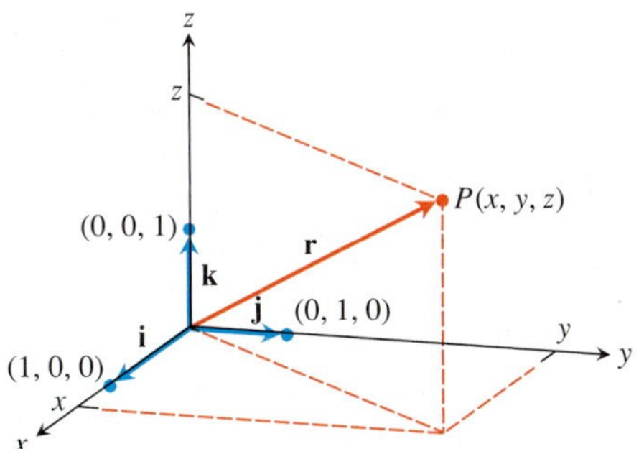

FIGURE 10.4 The position vector of a point in space.

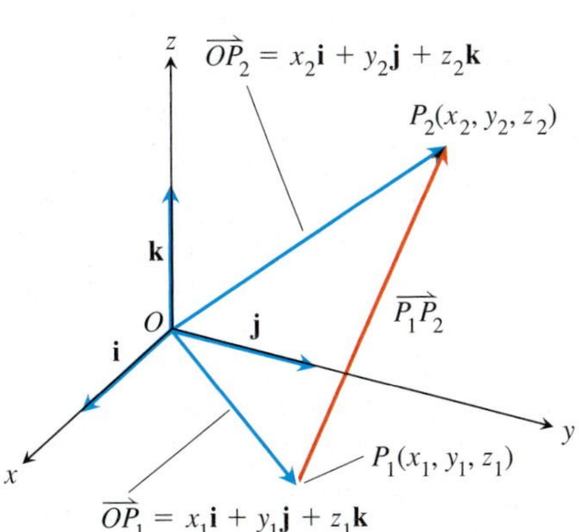

FIGURE 10.5 The vector from P_1 to P_2 is $\overrightarrow{P_1P_2} = (x_2 - x_1)\mathbf{i} + (y_2 - y_1)\mathbf{j} + (z_2 - z_1)\mathbf{k}$.

Vectors in Space

Vectors in space are like vectors in the plane except there is a third component. Just as in the plane (Section 9.1), **vectors** in space are directed line segments. Two such vectors are **equal** if they have the same length and direction. Vectors are used to represent forces, displacements, velocities, and accelerations in space. In this section, we summarize the properties of vectors in space (which are identical to the properties studied in Section 9.1 for planar vectors).

If $\mathbf{v}$ is a vector in space equal to a vector with initial point $(0, 0, 0)$ and terminal point (v_1, v_2, v_3), then the **component form** of $\mathbf{v}$ is $\mathbf{v} = \langle v_1, v_2, v_3 \rangle$. As in the plane, this is also the **position vector** of the point (v_1, v_2, v_3). The vector from the initial point $P_1(x_1, y_1, z_1)$ to the terminal point $P_2(x_2, y_2, z_2)$ is $\mathbf{v} = \overrightarrow{P_1P_2} = \langle x_2 - x_1, y_2 - y_1, z_2 - z_1 \rangle$.

The vectors represented by the directed line segments from the origin to the points $(1, 0, 0)$, $(0, 1, 0)$, and $(0, 0, 1)$ are the **standard unit vectors** and are denoted by $\mathbf{i}$, $\mathbf{j}$, and $\mathbf{k}$ (Figure 10.4). The position vector $\mathbf{r}$ from the origin to the typical point $P(x, y, z)$ can then be written as

$$\mathbf{r} = \overrightarrow{OP} = x\mathbf{i} + y\mathbf{j} + z\mathbf{k}.$$

Thus, the vector $\overrightarrow{P_1P_2} = \langle x_2 - x_1, y_2 - y_1, z_2 - z_1 \rangle$ can be expressed as

$$\overrightarrow{P_1P_2} = (x_2 - x_1)\mathbf{i} + (y_2 - y_1)\mathbf{j} + (z_2 - z_1)\mathbf{k}.$$

(See Figure 10.5.)We use this as our primary notation for vectors in space.

The definitions of addition, subtraction, and scalar multiplication are the same as in the plane. They also satisfy the same properties and interpretations.

Definitions Vector Operations

Let $\mathbf{u} = u_1\mathbf{i} + u_2\mathbf{j} + u_3\mathbf{k}$ and $\mathbf{v} = v_1\mathbf{i} + v_2\mathbf{j} + v_3\mathbf{k}$ be vectors with k a scalar (real number).

Addition: $\mathbf{u} + \mathbf{v} = (u_1 + v_1)\mathbf{i} + (u_2 + v_2)\mathbf{j} + (u_3 + v_3)\mathbf{k}$

Subtraction: $\mathbf{u} - \mathbf{v} = (u_1 - v_1)\mathbf{i} + (u_2 - v_2)\mathbf{j} + (u_3 - v_3)\mathbf{k}$

Scalar Multiplication: $k\mathbf{u} = (ku_1)\mathbf{i} + (ku_2)\mathbf{j} + (ku_3)\mathbf{k}$

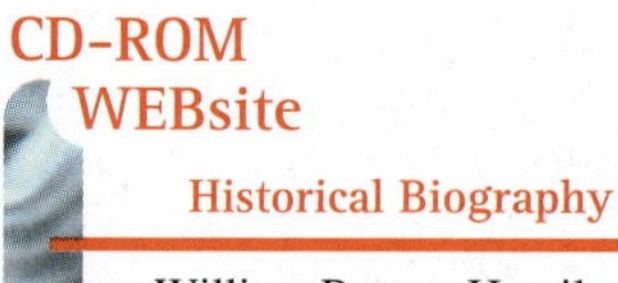

CD-ROM
WEBsite
Historical Biography
William Rowan Hamilton
(1805 — 1865)

Magnitude

As for planar vectors, magnitude and direction are important features of a vector in space. We find a formula for the magnitude (length) of $\mathbf{v} = v_1\mathbf{i} + v_2\mathbf{j} + v_3\mathbf{k}$ by applying the Pythagorean Theorem to the right triangles in Figure 10.6. From triangle ABC,

$$|\overrightarrow{AC}| = \sqrt{v_1^2 + v_2^2},$$

and from triangle ACD,

$$\begin{aligned} |\mathbf{v}| = |v_1\mathbf{i} + v_2\mathbf{j} + v_3\mathbf{k}| = |\overrightarrow{AD}| &= \sqrt{|\overrightarrow{AC}|^2 + |\overrightarrow{CD}|^2} \\ &= \sqrt{v_1^2 + v_2^2 + v_3^2}. \end{aligned}$$

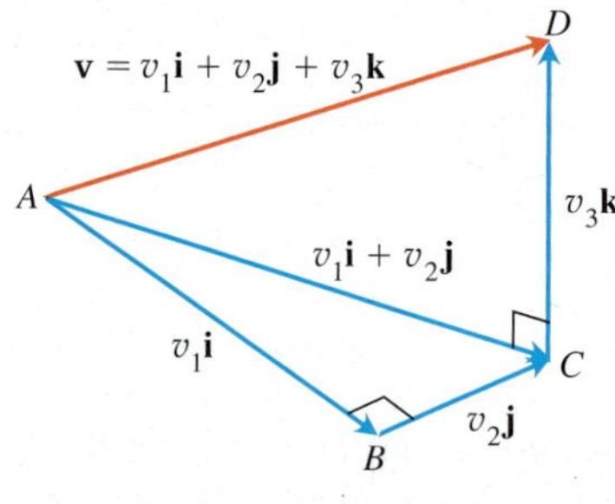

FIGURE 10.6 We find the length of $\mathbf{v} = \overrightarrow{AD}$ by applying the Pythagorean Theorem to the right triangles ABC and ACD.

Definition Magnitude (Length) of a Space Vector

The **magnitude (length)** of $\mathbf{v} = v_1\mathbf{i} + v_2\mathbf{j} + v_3\mathbf{k}$ is

$$|\mathbf{v}| = |v_1\mathbf{i} + v_2\mathbf{j} + v_3\mathbf{k}| = \sqrt{v_1^2 + v_2^2 + v_3^2}.$$

The Zero Vector

The **zero vector** in space is the vector $\mathbf{0} = \langle 0, 0, 0\rangle = 0\mathbf{i} + 0\mathbf{j} + 0\mathbf{k}$. As in the plane, $\mathbf{0}$ has zero length and no direction.

Unit Vectors

A **unit vector** in space is a vector of length 1. The lengths of the standard unit vectors are

$$\begin{aligned} |\mathbf{i}| &= |1\mathbf{i} + 0\mathbf{j} + 0\mathbf{k}| = \sqrt{1^2 + 0^2 + 0^2} = 1 \\ |\mathbf{j}| &= |0\mathbf{i} + 1\mathbf{j} + 0\mathbf{k}| = \sqrt{0^2 + 1^2 + 0^2} = 1 \\ |\mathbf{k}| &= |0\mathbf{i} + 0\mathbf{j} + 1\mathbf{k}| = \sqrt{0^2 + 0^2 + 1^2} = 1 \end{aligned}$$

confirming that the standard unit vectors are indeed unit vectors.

If $\mathbf{v} \neq \mathbf{0}$, then $\mathbf{v}/|\mathbf{v}|$ is a unit vector in the direction of $\mathbf{v}$.

Example 3 Finding a Unit Vector

Find a unit vector $\mathbf{u}$ in the direction of the vector from $P_1(1, 0, 1)$ to $P_2(3, 2, 0)$.

Solution We divide $\overrightarrow{P_1P_2}$ by its length:

$$\overrightarrow{P_1P_2} = (3 - 1)\mathbf{i} + (2 - 0)\mathbf{j} + (0 - 1)\mathbf{k} = 2\mathbf{i} + 2\mathbf{j} - \mathbf{k}$$

$$|\overrightarrow{P_1P_2}| = \sqrt{(2)^2 + (2)^2 + (-1)^2} = \sqrt{4 + 4 + 1} = \sqrt{9} = 3$$

$$\mathbf{u} = \frac{\overrightarrow{P_1P_2}}{|\overrightarrow{P_1P_2}|} = \frac{2\mathbf{i} + 2\mathbf{j} - \mathbf{k}}{3} = \frac{2}{3}\mathbf{i} + \frac{2}{3}\mathbf{j} - \frac{1}{3}\mathbf{k}.$$

Length and Direction

As in the plane, if $\mathbf{v} \neq \mathbf{0}$ is a nonzero vector in space, then $\mathbf{v}/|\mathbf{v}|$ is a unit vector in the direction of $\mathbf{v}$. The equation

$$\mathbf{v} = \underbrace{|\mathbf{v}|}_{\text{Length}} \underbrace{\frac{\mathbf{v}}{|\mathbf{v}|}}_{\text{Direction}}$$

expresses $\mathbf{v}$ as a product of its length and direction.

Example 4 Velocity as Speed Times Direction

Express the velocity vector $\mathbf{v} = \mathbf{i} - 2\mathbf{j} + 3\mathbf{k}$ of a projectile as a product of its speed and direction.

Solution As in the plane, speed is the magnitude of the velocity vector, and we have

$$|\mathbf{v}| = \sqrt{1^2 + (-2)^2 + 3^2} = \sqrt{14}.$$

Then,

$$\begin{aligned}\mathbf{v} &= |\mathbf{v}|\frac{\mathbf{v}}{|\mathbf{v}|}\\ &= \sqrt{14} \cdot \frac{\mathbf{i} - 2\mathbf{j} + 3\mathbf{k}}{\sqrt{14}}\\ &= \sqrt{14}\left(\frac{1}{\sqrt{14}}\mathbf{i} - \frac{2}{\sqrt{14}}\mathbf{j} + \frac{3}{\sqrt{14}}\mathbf{k}\right) = (\text{length of } \mathbf{v}) \cdot (\text{direction of } \mathbf{v}).\end{aligned}$$

Interpret If distance is measured in feet and time in seconds, then the speed of the object is $\sqrt{14}$ ft/sec and it is moving in the direction of the unit vector $(1/\sqrt{14})\mathbf{i} - (2/\sqrt{14})\mathbf{j} + (3/\sqrt{14})\mathbf{k}$.

Example 5 A Force Vector

A force of 6 N is applied in the direction of the vector $\mathbf{v} = 2\mathbf{i} + 2\mathbf{j} - \mathbf{k}$. Express the force $\mathbf{F}$ as a product of its length and direction.

Solution The force vector is

$$\mathbf{F} = 6\frac{\mathbf{v}}{|\mathbf{v}|} = 6\frac{2\mathbf{i} + 2\mathbf{j} - \mathbf{k}}{\sqrt{2^2 + 2^2 + (-1)^2}} = 6\frac{2\mathbf{i} + 2\mathbf{j} - \mathbf{k}}{3}$$

$$= 6\left(\frac{2}{3}\mathbf{i} + \frac{2}{3}\mathbf{j} - \frac{1}{3}\mathbf{k}\right).$$

Distance and Spheres in Space

The distance between two points P_1 and P_2 in space is the length of $\overrightarrow{P_1P_2}$.

The Distance Between $P_1(x_1, y_1, z_1)$ and $P_2(x_2, y_2, z_2)$

$$|\overrightarrow{P_1P_2}| = \sqrt{(x_2 - x_1)^2 + (y_2 - y_1)^2 + (z_2 - z_1)^2}$$

Example 6 Finding the Distance Between Two Points

The distance between $P_1(2, 1, 5)$ and $P_2(-2, 3, 0)$ is

$$|\overrightarrow{P_1P_2}| = \sqrt{(-2 - 2)^2 + (3 - 1)^2 + (0 - 5)^2}$$

$$= \sqrt{16 + 4 + 25}$$

$$= \sqrt{45} \approx 6.708.$$

We use the distance formula to write equations for spheres in space (Figure 10.7). A point $P(x, y, z)$ lies on the sphere of radius a centered at $P_0(x_0, y_0, z_0)$ precisely when $|\overrightarrow{P_0P}| = a$ or

$$(x - x_0)^2 + (y - y_0)^2 + (z - z_0)^2 = a^2.$$

The Standard Equation for the Sphere of Radius a and Center (x_0, y_0, z_0)

$$(x - x_0)^2 + (y - y_0)^2 + (z - z_0)^2 = a^2$$

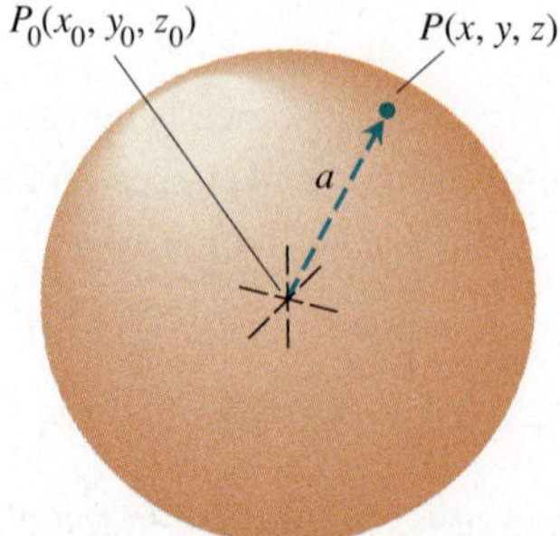

FIGURE 10.7 The sphere $(x - x_0)^2 + (y - y_0)^2 + (z - z_0)^2 = a^2$.

Example 7 Finding the Center and Radius of a Sphere

Find the center and radius of the sphere

$$x^2 + y^2 + z^2 + 3x - 4z + 1 = 0.$$

Solution We find the center and radius of a sphere the way we find the center and radius of a circle: Complete the squares on the x-, y-, and z-terms as necessary and write each quadratic as a squared linear expression. Then, from the equation in standard form, read off the center and radius. For the sphere here, we have

$$x^2 + y^2 + z^2 + 3x - 4z + 1 = 0$$

$$(x^2 + 3x) + y^2 + (z^2 - 4z) = -1$$

$$\left(x^2 + 3x + \left(\frac{3}{2}\right)^2\right) + y^2 + \left(z^2 - 4z + \left(\frac{-4}{2}\right)^2\right) = -1 + \left(\frac{3}{2}\right)^2 + \left(\frac{-4}{2}\right)^2$$

$$\left(x + \frac{3}{2}\right)^2 + y^2 + (z - 2)^2 = -1 + \frac{9}{4} + 4 = \frac{21}{4}.$$

From this standard form, we read that $x_0 = -3/2$, $y_0 = 0$, $z_0 = 2$, and $a = \sqrt{21}/2$. The center is $(-3/2, 0, 2)$. The radius is $\sqrt{21}/2$.

Example 8 Interpreting Equations and Inequalities

(a) $x^2 + y^2 + z^2 < 4$	The interior of the sphere $x^2 + y^2 + z^2 = 4$.
(b) $x^2 + y^2 + z^2 \le 4$	The solid ball bounded by the sphere $x^2 + y^2 + z^2 = 4$. Alternatively, the sphere $x^2 + y^2 + z^2 = 4$ together with its interior.
(c) $x^2 + y^2 + z^2 > 4$	The exterior of the sphere $x^2 + y^2 + z^2 = 4$.
(d) $x^2 + y^2 + z^2 = 4$, $z \le 0$	The lower hemisphere cut from the sphere $x^2 + y^2 + z^2 = 4$ by the xy-plane (the plane $z = 0$).

Midpoints

The coordinates of the midpoint of a line segment are found by averaging.

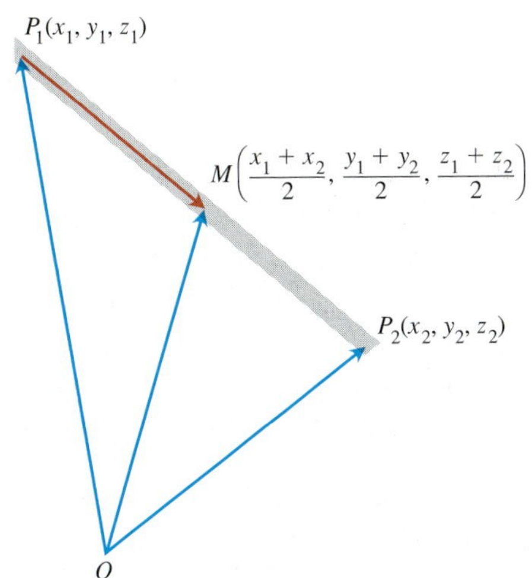

FIGURE 10.8 The coordinates of the midpoint are the averages of the coordinates of P_1 and P_2.

> The **midpoint** M of the line segment joining points $P_1(x_1, y_1, z_1)$ and $P_2(x_2, y_2, z_2)$ is the point
>
> $$\left(\frac{x_1 + x_2}{2}, \frac{y_1 + y_2}{2}, \frac{z_1 + z_2}{2}\right).$$

To see why, observe (Figure 10.8) that

$$\begin{aligned}\overrightarrow{OM} &= \overrightarrow{OP_1} + \frac{1}{2}(\overrightarrow{P_1P_2}) = \overrightarrow{OP_1} + \frac{1}{2}(\overrightarrow{OP_2} - \overrightarrow{OP_1})\\ &= \frac{1}{2}(\overrightarrow{OP_1} + \overrightarrow{OP_2})\\ &= \frac{x_1 + x_2}{2}\mathbf{i} + \frac{y_1 + y_2}{2}\mathbf{j} + \frac{z_1 + z_2}{2}\mathbf{k}.\end{aligned}$$

Example 9 Finding Midpoints

The midpoint of the segment joining $P_1(3, -2, 0)$ and $P_2(7, 4, 4)$

$$\left(\frac{3+7}{2}, \frac{-2+4}{2}, \frac{0+4}{2}\right) = (5, 1, 2).$$

EXERCISES 10.1

Sets, Equations, and Inequalities

In Exercises 1–10, give a geometric description of the set of points in space whose coordinates satisfy the given pairs of equations.

1. $x = 2, \quad y = 3$

2. $x = -1, \quad z = 0$

3. $y = 0, \quad z = 0$

4. $x = 1, \quad y = 0$

5. $x^2 + y^2 = 4, \quad z = -2$

6. $x^2 + z^2 = 4, \quad y = 0$

7. $x^2 + y^2 + z^2 = 1, \quad x = 0$

8. $x^2 + y^2 + z^2 = 25, \quad y = -4$

9. $x^2 + y^2 + (z + 3)^2 = 25, \quad z = 0$

10. $x^2 + (y - 1)^2 + z^2 = 4, \quad y = 0$

In Exercises 11–16, describe the sets of points in space whose coordinates satisfy the given inequalities or combinations of equations and inequalities.

11. **(a)** $x \ge 0, \quad y \ge 0, \quad z = 0$

(b) $x \ge 0, \quad y \le 0, \quad z = 0$

12. **(a)** $0 \le x \le 1$

(b) $0 \le x \le 1, \quad 0 \le y \le 1$

(c) $0 \le x \le 1, \quad 0 \le y \le 1, \quad 0 \le z \le 1$

13. **(a)** $x^2 + y^2 + z^2 \le 1$

(b) $x^2 + y^2 + z^2 > 1$

14. **(a)** $x^2 + y^2 \le 1, \quad z = 0$

(b) $x^2 + y^2 \le 1, \quad z = 3$

(c) $x^2 + y^2 \le 1$, no restriction on z

15. **(a)** $x^2 + y^2 + z^2 = 1, \quad z \ge 0$

(b) $x^2 + y^2 + z^2 \le 1, \quad z \ge 0$

16. **(a)** $x = y, \quad z = 0$

(b) $x = y$, no restriction on z

In Exercises 17–26, describe the given set with a single equation or with a pair of equations.

17. The plane perpendicular to the

(a) x-axis at $(3, 0, 0)$

(b) y-axis at $(0, -1, 0)$

(c) z-axis at $(0, 0, -2)$

18. The plane through the point $(3, -1, 2)$ perpendicular to the

(a) x-axis **(b)** y-axis **(c)** z-axis

19. The plane through the point $(3, -1, 1)$ parallel to the

(a) xy-plane **(b)** yz-plane **(c)** xz-plane

20. The circle of radius 2 centered at $(0, 0, 0)$ and lying in the

(a) xy-plane **(b)** yz-plane **(c)** xz-plane

21. The circle of radius 2 centered at $(0, 2, 0)$ and lying in the

(a) xy-plane **(b)** yz-plane **(c)** plane $y = 2$

22. The circle of radius 1 centered at $(-3, 4, 1)$ and lying in a plane parallel to the

(a) xy-plane **(b)** yz-plane **(c)** xz-plane

23. The line through the point $(1, 3, -1)$ parallel to the

(a) x-axis **(b)** y-axis **(c)** z-axis

24. The set of points in space equidistant from the origin and the point $(0, 2, 0)$

25. The circle in which the plane through the point $(1, 1, 3)$ perpendicular to the z-axis meets the sphere of radius 5 centered at the origin

26. The set of points in space that lie 2 units from the point $(0, 0, 1)$ and, at the same time, 2 units from the point $(0, 0, -1)$

Write inequalities to describe the sets in Exercises 27–32.

27. The slab bounded by the planes $z = 0$ and $z = 1$ (planes included)

28. The solid cube in the first octant bounded by the coordinate planes and the planes $x = 2$, $y = 2$, and $z = 2$

29. The half-space consisting of the points on and below the xy-plane

30. The upper hemisphere of the sphere of radius 1 centered at the origin

31. The **(a)** interior and **(b)** exterior of the sphere of radius 1 centered at the point $(1, 1, 1)$

32. The closed region bounded by the spheres of radius 1 and radius 2 centered at the origin. (*Closed* means the spheres are to be included. Had we wanted the spheres left out, we would have asked for the *open* region bounded by the spheres. This is analogous to the way we use *closed* and *open* to describe intervals: *closed* means endpoints included, *open* means endpoints left out. Closed sets include boundaries; open sets leave them out.)

Length and Direction

In Exercises 33–38, express each vector as a product of its length and direction.

33. $2\mathbf{i} + \mathbf{j} - 2\mathbf{k}$

34. $9\mathbf{i} - 2\mathbf{j} + 6\mathbf{k}$

35. $5\mathbf{k}$

36. $\frac{3}{5}\mathbf{i} + \frac{4}{5}\mathbf{k}$

37. $\frac{1}{\sqrt{6}}\mathbf{i} - \frac{1}{\sqrt{6}}\mathbf{j} - \frac{1}{\sqrt{6}}\mathbf{k}$

38. $\frac{\mathbf{i}}{\sqrt{3}} + \frac{\mathbf{j}}{\sqrt{3}} + \frac{\mathbf{k}}{\sqrt{3}}$

39. Find the vectors whose lengths and directions are given. Try to do the calculations without writing.

Length	Direction
(a) 2	$\mathbf{i}$
(b) $\sqrt{3}$	$-\mathbf{k}$
(c) $\frac{1}{2}$	$\frac{3}{5}\mathbf{j} + \frac{4}{5}\mathbf{k}$
(d) 7	$\frac{6}{7}\mathbf{i} - \frac{2}{7}\mathbf{j} + \frac{3}{7}\mathbf{k}$

40. Find the vectors whose lengths and directions are given. Try to do the calculations without writing.

Length	Direction
(a) 7	$-\mathbf{j}$
(b) $\sqrt{2}$	$-\frac{3}{5}\mathbf{i} - \frac{4}{5}\mathbf{k}$
(c) $\frac{13}{12}$	$\frac{3}{13}\mathbf{i} - \frac{4}{13}\mathbf{j} - \frac{12}{13}\mathbf{k}$
(d) $a > 0$	$\frac{1}{\sqrt{2}}\mathbf{i} + \frac{1}{\sqrt{3}}\mathbf{j} - \frac{1}{\sqrt{6}}\mathbf{k}$

41. Find a vector of magnitude 7 in the direction of $\mathbf{v} = 12\mathbf{i} - 5\mathbf{k}$.

42. Find a vector of magnitude 3 in the direction opposite to the direction of $\mathbf{v} = (1/2)\mathbf{i} - (1/2)\mathbf{j} - (1/2)\mathbf{k}$.

Vectors Determined by Points; Midpoints and Distance

In Exercises 43–46, find

(a) the distance between points P_1 and P_2

(b) the direction of $\overrightarrow{P_1P_2}$

(c) the midpoint of line segment P_1P_2.

43. $P_1(-1, 1, 5)$, $P_2(2, 5, 0)$

44. $P_1(1, 4, 5)$, $P_2(4, -2, 7)$

45. $P_1(3, 4, 5)$, $P_2(2, 3, 4)$

46. $P_1(0, 0, 0)$, $P_2(2, -2, -2)$

47. If $\overrightarrow{AB} = \mathbf{i} + 4\mathbf{j} - 2\mathbf{k}$ and B is the point $(5, 1, 3)$, find A.

48. If $\overrightarrow{AB} = -7\mathbf{i} + 3\mathbf{j} + 8\mathbf{k}$ and A is the point $(-2, -3, 6)$, find B.

Spheres

In Exercises 49 and 50, find equations for the spheres with given center and radius.

49. Center: $(1, 2, 3)$, radius: $\sqrt{14}$

50. Center: $(0, -1, 5)$, radius: 2

Find the centers and radii of the spheres in Exercises 51–56.

51. $(x + 2)^2 + y^2 + (z - 2)^2 = 8$

52. $\left(x + \frac{1}{2}\right)^2 + \left(y + \frac{1}{2}\right)^2 + \left(z + \frac{1}{2}\right)^2 = \frac{21}{4}$

53. $x^2 + y^2 + z^2 + 4x - 4z = 0$

54. $x^2 + y^2 + z^2 - 6y + 8z = 0$

55. $2x^2 + 2y^2 + 2z^2 + x + y + z = 9$

56. $3x^2 + 3y^2 + 3z^2 + 2y - 2z = 9$

Theory and Examples

57. *Distance to coordinate axes* Find a formula for the distance from the point $P(x, y, z)$ to the

(a) x-axis **(b)** y-axis **(c)** z-axis.

58. *Distance to coordinate planes* Find a formula for the distance from the point $P(x, y, z)$ to the

(a) xy-plane **(b)** yz-plane **(c)** xz-plane.

59. *Medians of a triangle* Suppose that A, B, and C are the corner points of the thin triangular plate of constant density shown in the accompanying figure.

(a) Find the vector from C to the midpoint M of side AB.

(b) Find the vector from C to the point that lies two-thirds of the way from C to M on the median CM.

(c) Find the coordinates of the point in which the medians of ΔABC intersect. According to Exercise 29, Section 5.6, this point is the plate's center of mass.

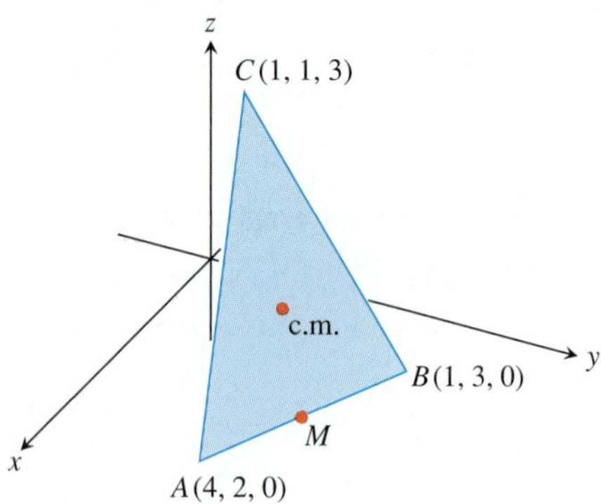

60. *Geometry and vectors* Find the vector from the origin to the point of intersection of the medians of the triangle whose vertices are

$$A(1, -1, 2), \quad B(2, 1, 3), \quad \text{and} \quad C(-1, 2, -1).$$

61. *Quadrilateral* Let $ABCD$ be a general, not necessarily planar, quadrilateral in space. Show that the two segments joining the midpoints of opposite sides of $ABCD$ bisect each other. (*Hint:* Show that the segments have the same midpoint.)

62. *Writing to Learn* Vectors are drawn from the center of a regular n-sided polygon in the plane to the vertices of the polygon. Show that the sum of the vectors is zero. (*Hint:* What happens to the sum if you rotate the polygon about its center?)

63. *Triangle* Suppose that A, B, and C are vertices of a triangle and that a, b, and c are, respectively, the midpoints of the opposite sides. Show that $\overrightarrow{Aa} + \overrightarrow{Bb} + \overrightarrow{Cc} = 0$.

10.2 Dot and Cross Products

Dot Products • Properties of the Dot Product • Perpendicular (Orthogonal) Vectors and Projections • The Cross Product of Two Vectors in Space • Properties of the Cross Product • Determinant Formula for $\mathbf{u} \times \mathbf{v}$ • Torque • Triple Scalar or Box Product

In this section, we extend the definition of the dot product studied in Section 9.2 to vectors in space. Then we introduce a new product, the *cross product*, for vectors in space. The cross product is useful for the geometry of space such as describing how a plane is tilting by identifying a vector that is perpendicular to the plane. The direction of this vector tells us the "inclination" of the plane, just as the slope or angle of inclination describes how lines in the plane are tilting.

Dot Products

The dot product (or inner product) of two vectors in space is defined in the same way as for vectors in the plane (see Section 9.2). When two nonzero vectors $\mathbf{u}$ and $\mathbf{v}$ are placed so their initial points coincide, they form an angle θ of measure $0 \le \theta \le \pi$.

Definition Dot Product (Inner Product)

The **dot product (inner product)** $\mathbf{u} \cdot \mathbf{v}$ ("$\mathbf{u}$ dot $\mathbf{v}$") of vectors $\mathbf{u}$ and $\mathbf{v}$ is the number

$$\mathbf{u} \cdot \mathbf{v} = |\mathbf{u}||\mathbf{v}| \cos \theta,$$

where θ is the angle between $\mathbf{u}$ and $\mathbf{v}$.

In Section 9.2, we proved (see Theorem 1) that the dot product can be expressed using the components of the vectors. The same proof yields the following formula.

Computing the Dot Product

If $\mathbf{u} = u_1\mathbf{i} + u_2\mathbf{j} + u_3\mathbf{k}$ and $\mathbf{v} = v_1\mathbf{i} + v_2\mathbf{j} + v_3\mathbf{k}$, then

$$\mathbf{u} \cdot \mathbf{v} = u_1v_1 + u_2v_2 + u_3v_3.$$

Thus, to find the dot product of two given vectors, we multiply their corresponding **i-**, **j-**, and **k-**components and add the results. This is the same procedure we used for planar vectors except there were only two components.

Solving for θ in the definition of dot product gives a formula for finding angles between vectors in space.

Angle Between Nonzero Vectors

The angle between two nonzero vectors **u** and **v** is

$$\theta = \cos^{-1}\left(\frac{\mathbf{u} \cdot \mathbf{v}}{|\mathbf{u}||\mathbf{v}|}\right).$$

Example 1 Finding the Angle Between Two Vectors in Space

Find the angle between $\mathbf{u} = \mathbf{i} - 2\mathbf{j} - 2\mathbf{k}$ and $\mathbf{v} = 6\mathbf{i} + 3\mathbf{j} + 2\mathbf{k}$.

Solution We use the formula above:

$$\mathbf{u} \cdot \mathbf{v} = (1)(6) + (-2)(3) + (-2)(2) = 6 - 6 - 4 = -4$$

$$|\mathbf{u}| = \sqrt{(1)^2 + (-2)^2 + (-2)^2} = \sqrt{9} = 3$$

$$|\mathbf{v}| = \sqrt{(6)^2 + (3)^2 + (2)^2} = \sqrt{49} = 7$$

$$\theta = \cos^{-1}\left(\frac{\mathbf{u} \cdot \mathbf{v}}{|\mathbf{u}||\mathbf{v}|}\right)$$

$$= \cos^{-1}\left(\frac{-4}{(3)(7)}\right) \approx 1.76 \text{ rad.}$$

Properties of the Dot Product

We can use the component form of the dot product to establish the following properties (which are the same as for vectors in the plane studied in Section 9.2).

> **Properties of Dot Products**
>
> If $\mathbf{u}$, $\mathbf{v}$, and $\mathbf{w}$ are any vectors and c is a scalar, then
>
> 1. $\mathbf{u} \cdot \mathbf{v} = \mathbf{v} \cdot \mathbf{u}$
> 2. $(c\mathbf{u}) \cdot \mathbf{v} = \mathbf{u} \cdot (c\mathbf{v}) = c(\mathbf{u} \cdot \mathbf{v})$
> 3. $\mathbf{u} \cdot (\mathbf{v} + \mathbf{w}) = \mathbf{u} \cdot \mathbf{v} + \mathbf{u} \cdot \mathbf{w}$
> 4. $\mathbf{u} \cdot \mathbf{u} = |\mathbf{u}|^2$
> 5. $\mathbf{0} \cdot \mathbf{u} = 0$

Perpendicular (Orthogonal) Vectors and Projections

Just as for vectors in the plane, two nonzero vectors $\mathbf{u}$ and $\mathbf{v}$ are perpendicular or **orthogonal** if and only if $\mathbf{u} \cdot \mathbf{v} = 0$. The **vector projection** of $\mathbf{u} = \overrightarrow{PQ}$ onto a nonzero vector $\mathbf{v} = \overrightarrow{PS}$ (Figure 10.9) is the vector $\overrightarrow{PR}$ determined by dropping a perpendicular from Q to the line PS. This is exactly the same definition as for vectors in the plane (Section 9.2). The notation for this vector is

$$\text{proj}_{\mathbf{v}}\,\mathbf{u} \qquad (\text{"the vector projection of } \mathbf{u} \text{ onto } \mathbf{v}\text{"}).$$

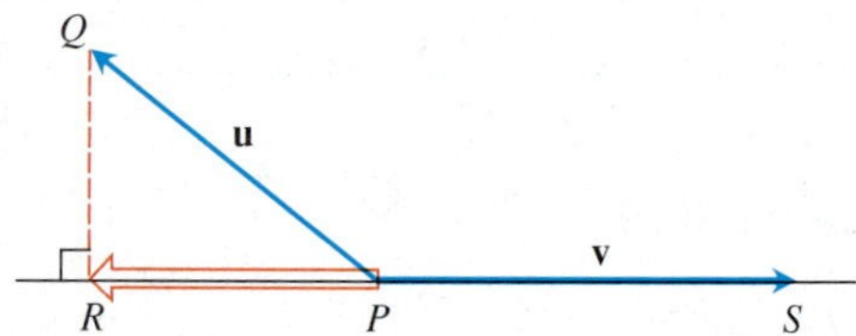

FIGURE 10.9 The vector projection of $\mathbf{u}$ onto $\mathbf{v}$.

If $\mathbf{u}$ represents a force, then $\text{proj}_{\mathbf{v}}\,\mathbf{u}$ represents the effective force in the direction of $\mathbf{v}$ (Figure 10.10).

Calculation of $\text{proj}_{\mathbf{v}}\,\mathbf{u}$ is the same as before.

> **Vector projection of $\mathbf{u}$ onto $\mathbf{v}$:**
>
> $$\text{proj}_{\mathbf{v}}\,\mathbf{u} = \left(\frac{\mathbf{u} \cdot \mathbf{v}}{|\mathbf{v}|^2}\right)\mathbf{v} \tag{1}$$

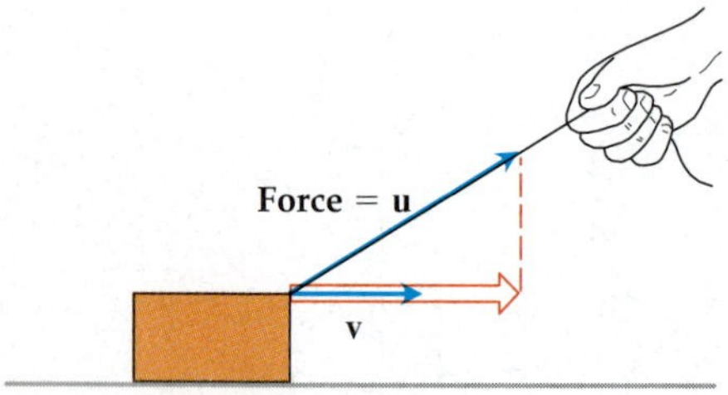

FIGURE 10.10 If we pull on a box with force $\mathbf{v}$, the effective force in the direction of $\mathbf{u}$ is the vector projection of $\mathbf{v}$ onto $\mathbf{u}$.

The number $|\mathbf{u}| \cos\theta$ is called the **scalar component of $\mathbf{u}$ in the direction of $\mathbf{v}$.** Since

$$|\mathbf{u}| \cos\theta = \frac{\mathbf{u} \cdot \mathbf{v}}{|\mathbf{v}|} = \mathbf{u} \cdot \frac{\mathbf{v}}{|\mathbf{v}|}, \tag{2}$$

we can find the scalar component by "dotting" $\mathbf{u}$ with the direction of $\mathbf{v}$.

Example 2 Finding the Vector Projection

Find the vector projection of $\mathbf{u} = 6\mathbf{i} + 3\mathbf{j} + 2\mathbf{k}$ onto $\mathbf{v} = \mathbf{i} - 2\mathbf{j} - 2\mathbf{k}$ and the scalar component of $\mathbf{u}$ in the direction of $\mathbf{v}$.

Solution We find $\text{proj}_{\mathbf{v}}\,\mathbf{u}$ from Equation (1):

$$\begin{aligned} \text{proj}_{\mathbf{v}}\,\mathbf{u} &= \frac{\mathbf{u} \cdot \mathbf{v}}{\mathbf{v} \cdot \mathbf{v}}\,\mathbf{v} = \frac{6 - 6 - 4}{1 + 4 + 4}(\mathbf{i} - 2\mathbf{j} - 2\mathbf{k}) \\ &= -\frac{4}{9}(\mathbf{i} - 2\mathbf{j} - 2\mathbf{k}) = -\frac{4}{9}\mathbf{i} + \frac{8}{9}\mathbf{j} + \frac{8}{9}\mathbf{k}. \end{aligned}$$

We find the scalar component of $\mathbf{u}$ in the direction of $\mathbf{v}$ from Equation (2):

$$\begin{aligned}|\mathbf{u}| \cos\theta = \mathbf{u} \cdot \frac{\mathbf{v}}{|\mathbf{v}|} &= (6\mathbf{i} + 3\mathbf{j} + 2\mathbf{k}) \cdot \left(\frac{1}{3}\mathbf{i} - \frac{2}{3}\mathbf{j} - \frac{2}{3}\mathbf{k}\right) \\ &= 2 - 2 - \frac{4}{3} = -\frac{4}{3}.\end{aligned}$$

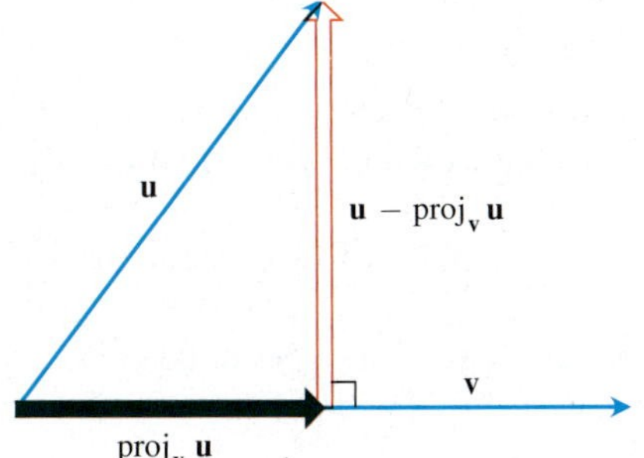

FIGURE 10.11 Writing $\mathbf{u}$ as the sum of vectors parallel and orthogonal to $\mathbf{v}$.

As with vectors in the plane, we can express a vector $\mathbf{u}$ as a sum of a vector parallel to a vector $\mathbf{v}$ and a vector orthogonal to $\mathbf{v}$. We accomplish this with the equation

$$\mathbf{u} = \operatorname{proj}_{\mathbf{v}} \mathbf{u} + (\mathbf{u} - \operatorname{proj}_{\mathbf{v}} \mathbf{u}),$$

shown in Figure 10.11.

Example 3 Force on a Spacecraft

A force $\mathbf{F} = 2\mathbf{i} + \mathbf{j} - 3\mathbf{k}$ N is applied to a spacecraft with velocity vector $\mathbf{v} = 3\mathbf{i} - \mathbf{j}$. Express $\mathbf{F}$ as a sum of a vector parallel to $\mathbf{v}$ and a vector orthogonal to $\mathbf{v}$.

Solution

$$\begin{aligned}\mathbf{F} &= \operatorname{proj}_{\mathbf{v}} \mathbf{F} + (\mathbf{F} - \operatorname{proj}_{\mathbf{v}} \mathbf{F}) \\ &= \frac{\mathbf{F}\cdot\mathbf{v}}{\mathbf{v}\cdot\mathbf{v}}\mathbf{v} + \left(\mathbf{F} - \frac{\mathbf{F}\cdot\mathbf{v}}{\mathbf{v}\cdot\mathbf{v}}\mathbf{v}\right) \\ &= \left(\frac{6-1}{9+1}\right)\mathbf{v} + \left(\mathbf{F} - \left(\frac{6-1}{9+1}\right)\mathbf{v}\right) \\ &= \frac{5}{10}(3\mathbf{i} - \mathbf{j}) + \left(2\mathbf{i} + \mathbf{j} - 3\mathbf{k} - \frac{5}{10}(3\mathbf{i} - \mathbf{j})\right) \\ &= \left(\frac{3}{2}\mathbf{i} - \frac{1}{2}\mathbf{j}\right) + \left(\frac{1}{2}\mathbf{i} + \frac{3}{2}\mathbf{j} - 3\mathbf{k}\right).\end{aligned}$$

Interpret The force $(3/2)\mathbf{i} - (1/2)\mathbf{j}$ is the effective force parallel to the velocity $\mathbf{v}$. The force $(1/2)\mathbf{i} + (3/2)\mathbf{j} - 3\mathbf{k}$ is orthogonal to $\mathbf{v}$. To check that this vector is orthogonal to $\mathbf{v}$, we find the dot product:

$$\left(\frac{1}{2}\mathbf{i} + \frac{3}{2}\mathbf{j} - 3\mathbf{k}\right) \cdot (3\mathbf{i} - \mathbf{j}) = \frac{3}{2} - \frac{3}{2} = 0.$$

The Cross Product of Two Vectors in Space

We start with two nonzero vectors $\mathbf{u}$ and $\mathbf{v}$ in space. If $\mathbf{u}$ and $\mathbf{v}$ are not parallel, they determine a plane. We select a unit vector $\mathbf{n}$ perpendicular to the plane by the **right-hand rule.** This means that we choose $\mathbf{n}$ to be the unit (normal) vector that points the way your right thumb points when your fingers curl through the angle

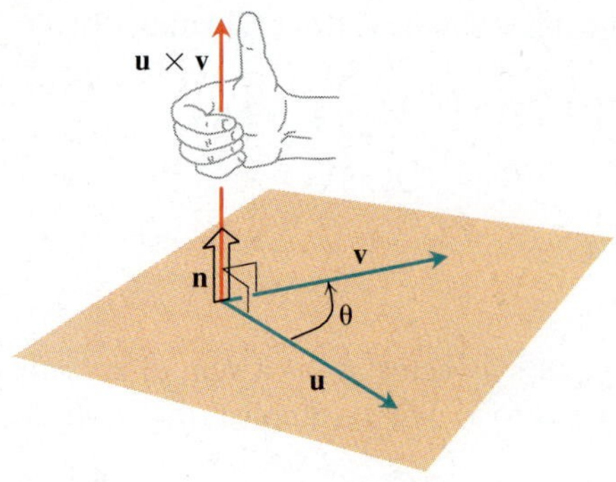

FIGURE 10.12 The construction of $\mathbf{u} \times \mathbf{v}$.

θ from $\mathbf{u}$ to $\mathbf{v}$ (Figure 10.12). Then the **vector product** $\mathbf{u} \times \mathbf{v}$ ("**u** cross **v**") is the *vector* defined as follows.

Definition Vector (Cross) Product

$$\mathbf{u} \times \mathbf{v} = (|\mathbf{u}||\mathbf{v}| \sin \theta)\,\mathbf{n}$$

The vector $\mathbf{u} \times \mathbf{v}$ is orthogonal to both $\mathbf{u}$ and $\mathbf{v}$ because it is a scalar multiple of $\mathbf{n}$. The vector product of $\mathbf{u}$ and $\mathbf{v}$ is often called the **cross product** of $\mathbf{u}$ and $\mathbf{v}$ because of the cross in the notation $\mathbf{u} \times \mathbf{v}$.

Since the sines of 0 and π are both zero, it makes sense to define the cross product of two parallel nonzero vectors to be $\mathbf{0}$.

If one or both of $\mathbf{u}$ and $\mathbf{v}$ are zero, we also define $\mathbf{u} \times \mathbf{v}$ to be zero. This way, the cross product of two vectors $\mathbf{u}$ and $\mathbf{v}$ is zero if and only if $\mathbf{u}$ and $\mathbf{v}$ are parallel or one or both of them are zero.

Parallel Vectors
Nonzero vectors $\mathbf{u}$ and $\mathbf{v}$ are parallel if and only if $\mathbf{u} \times \mathbf{v} = \mathbf{0}$.

Properties of the Cross Product

The cross product obeys the following laws.

Properties of the Cross Product
If $\mathbf{u}$, $\mathbf{v}$, and $\mathbf{w}$ are any vectors and r, s are scalars, then

1. $(r\mathbf{u}) \times (s\mathbf{v}) = (rs)(\mathbf{u} \times \mathbf{v})$
2. $\mathbf{u} \times (\mathbf{v} + \mathbf{w}) = \mathbf{u} \times \mathbf{v} + \mathbf{u} \times \mathbf{w}$
3. $(\mathbf{v} + \mathbf{w}) \times \mathbf{u} = \mathbf{v} \times \mathbf{u} + \mathbf{w} \times \mathbf{u}$
4. $\mathbf{v} \times \mathbf{u} = -(\mathbf{u} \times \mathbf{v})$
5. $\mathbf{0} \times \mathbf{u} = \mathbf{0}$

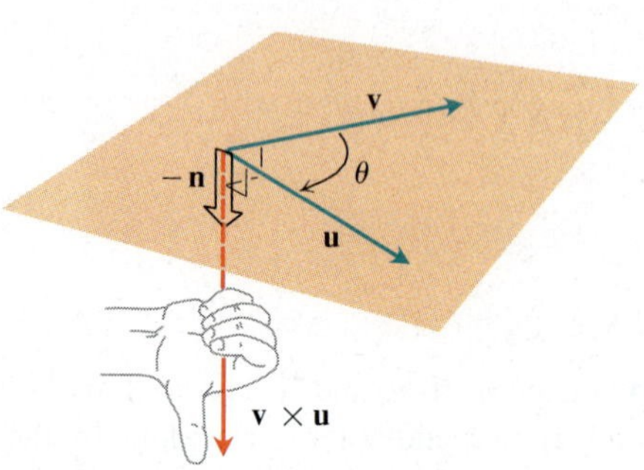

FIGURE 10.13 The construction of $\mathbf{v} \times \mathbf{u}$.

To visualize Property 4, for example, we notice that when the fingers of our right hand curl through the angle θ from $\mathbf{v}$ to $\mathbf{u}$, our thumb points the opposite way and the unit vector we choose in forming $\mathbf{v} \times \mathbf{u}$ is the negative of the one we choose in forming $\mathbf{u} \times \mathbf{v}$ (Figure 10.13).

Property 1 can be verified by applying the definition of cross product to both sides of the equation and comparing the results. Property 2 is proved in Appendix 9. Property 3 follows by multiplying both sides of the equation in Property 2 by -1 and reversing the order of the products using Property 4. Property 5 is by definition. As a rule, cross-product multiplication is *not associative* because $(\mathbf{u} \times \mathbf{v}) \times \mathbf{w}$ lies in the plane of $\mathbf{u}$ and $\mathbf{v}$ whereas $\mathbf{u} \times (\mathbf{v} \times \mathbf{w})$ lies in the plane of $\mathbf{v}$ and $\mathbf{w}$.

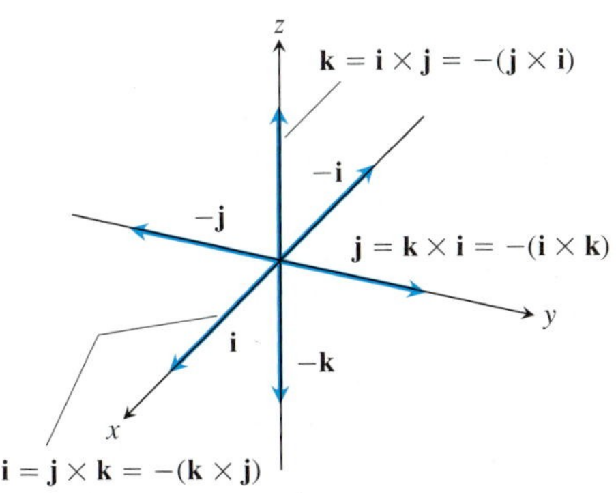

FIGURE 10.14 The pairwise cross products of **i**, **j**, and **k**.

When we apply the definition to calculate the pairwise cross products of **i**, **j**, and **k**, we find (Figure 10.14)

$$\mathbf{i} \times \mathbf{j} = -(\mathbf{j} \times \mathbf{i}) = \mathbf{k}$$
$$\mathbf{j} \times \mathbf{k} = -(\mathbf{k} \times \mathbf{j}) = \mathbf{i}$$
$$\mathbf{k} \times \mathbf{i} = -(\mathbf{i} \times \mathbf{k}) = \mathbf{j}$$

Diagram for recalling these products

and

$$\mathbf{i} \times \mathbf{i} = \mathbf{j} \times \mathbf{j} = \mathbf{k} \times \mathbf{k} = \mathbf{0}.$$

$|\mathbf{u} \times \mathbf{v}|$ Is the Area of a Parallelogram

Because **n** is a unit vector, the magnitude of $\mathbf{u} \times \mathbf{v}$ is

$$|\mathbf{u} \times \mathbf{v}| = |\mathbf{u}||\mathbf{v}|\,|\sin\theta|\,|\mathbf{n}| = |\mathbf{u}||\mathbf{v}|\,\sin\theta.$$

This is the area of the parallelogram determined by **u** and **v** (Figure 10.15), $|\mathbf{u}|$ being the base of the parallelogram and $|\mathbf{v}||\sin\theta|$ the height.

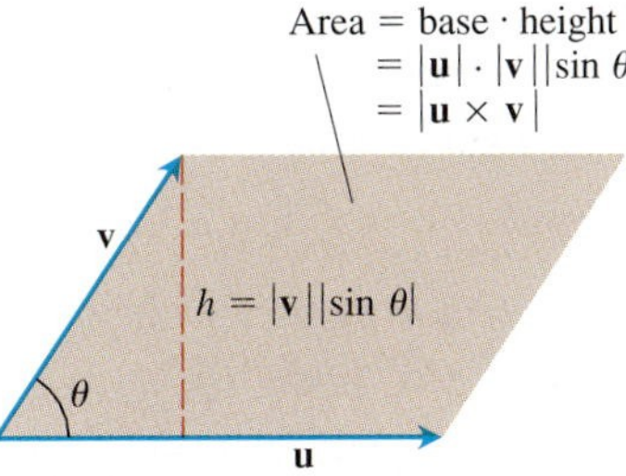

FIGURE 10.15 The parallelogram determined by **u** and **v**.

Determinant Formula for $\mathbf{u} \times \mathbf{v}$

Our next objective is to calculate $\mathbf{u} \times \mathbf{v}$ from the components of **u** and **v** relative to a Cartesian coordinate system.

Suppose that

$$\mathbf{u} = u_1\mathbf{i} + u_2\mathbf{j} + u_3\mathbf{k}, \qquad \mathbf{v} = v_1\mathbf{i} + v_2\mathbf{j} + v_3\mathbf{k}.$$

Then the distributive laws and the rules for multiplying **i**, **j**, and **k** tell us that

$$\begin{aligned}
\mathbf{u} \times \mathbf{v} &= (u_1\mathbf{i} + u_2\mathbf{j} + u_3\mathbf{k}) \times (v_1\mathbf{i} + v_2\mathbf{j} + v_3\mathbf{k}) \\
&= u_1v_1\mathbf{i} \times \mathbf{i} + u_1v_2\mathbf{i} \times \mathbf{j} + u_1v_3\mathbf{i} \times \mathbf{k} \\
&\quad + u_2v_1\mathbf{j} \times \mathbf{i} + u_2v_2\mathbf{j} \times \mathbf{j} + u_2v_3\mathbf{j} \times \mathbf{k} \\
&\quad + u_3v_1\mathbf{k} \times \mathbf{i} + u_3v_2\mathbf{k} \times \mathbf{j} + u_3v_3\mathbf{k} \times \mathbf{k} \\
&= (u_2v_3 - u_3v_2)\mathbf{i} - (u_1v_3 - u_3v_1)\mathbf{j} + (u_1v_2 - u_2v_1)\mathbf{k}.
\end{aligned}$$

The terms in the last line are the same as the terms in the expansion of the symbolic determinant

$$\begin{vmatrix} \mathbf{i} & \mathbf{j} & \mathbf{k} \\ u_1 & u_2 & u_3 \\ v_1 & v_2 & v_3 \end{vmatrix}.$$

Determinants

(For more information, see Appendix 10.)

$$\begin{vmatrix} a & b \\ c & d \end{vmatrix} = ad - bc$$

EXAMPLE

$$\begin{aligned}
\begin{vmatrix} 2 & 1 \\ -4 & 3 \end{vmatrix} &= (2)(3) - (1)(-4) \\
&= 6 + 4 = 10
\end{aligned}$$

$$\begin{vmatrix} a_1 & a_2 & a_3 \\ b_1 & b_2 & b_3 \\ c_1 & c_2 & c_3 \end{vmatrix}$$
$$= a_1\begin{vmatrix} b_2 & b_3 \\ c_2 & c_3 \end{vmatrix} - a_2\begin{vmatrix} b_1 & b_3 \\ c_1 & c_3 \end{vmatrix} + a_3\begin{vmatrix} b_1 & b_2 \\ c_1 & c_2 \end{vmatrix}$$

EXAMPLE

$$\begin{vmatrix} -5 & 3 & 1 \\ 2 & 1 & 1 \\ -4 & 3 & 1 \end{vmatrix}$$
$$= (-5)\begin{vmatrix} 1 & 1 \\ 3 & 1 \end{vmatrix} - (3)\begin{vmatrix} 2 & 1 \\ -4 & 1 \end{vmatrix} + (1)\begin{vmatrix} 2 & 1 \\ -4 & 3 \end{vmatrix}$$
$$= -5(1 - 3) - 3(2 + 4) + 1(6 + 4)$$
$$= 10 - 18 + 10 = 2$$

We therefore have the following rule.

Calculating Cross Products Using Determinants

If $\mathbf{u} = u_1\mathbf{i} + u_2\mathbf{j} + u_3\mathbf{k}$ and $\mathbf{v} = v_1\mathbf{i} + v_2\mathbf{j} + v_3\mathbf{k}$, then

$$\mathbf{u} \times \mathbf{v} = \begin{vmatrix} \mathbf{i} & \mathbf{j} & \mathbf{k} \\ u_1 & u_2 & u_3 \\ v_1 & v_2 & v_3 \end{vmatrix}.$$

Example 4 Calculating Cross Products with Determinants

Find $\mathbf{u} \times \mathbf{v}$ and $\mathbf{v} \times \mathbf{u}$ if $\mathbf{u} = 2\mathbf{i} + \mathbf{j} + \mathbf{k}$ and $\mathbf{v} = -4\mathbf{i} + 3\mathbf{j} + \mathbf{k}$.

Solution

$$\mathbf{u} \times \mathbf{v} = \begin{vmatrix} \mathbf{i} & \mathbf{j} & \mathbf{k} \\ 2 & 1 & 1 \\ -4 & 3 & 1 \end{vmatrix} = \begin{vmatrix} 1 & 1 \\ 3 & 1 \end{vmatrix}\mathbf{i} - \begin{vmatrix} 2 & 1 \\ -4 & 1 \end{vmatrix}\mathbf{j} + \begin{vmatrix} 2 & 1 \\ -4 & 3 \end{vmatrix}\mathbf{k}$$

$$= -2\mathbf{i} - 6\mathbf{j} + 10\mathbf{k}$$

$$\mathbf{v} \times \mathbf{u} = -(\mathbf{u} \times \mathbf{v}) = 2\mathbf{i} + 6\mathbf{j} - 10\mathbf{k}$$

Example 5 Finding Vectors Perpendicular to a Plane

Find a vector perpendicular to the plane of $P(1, -1, 0)$, $Q(2, 1, -1)$, and $R(-1, 1, 2)$ (Figure 10.16).

Solution The vector $\overrightarrow{PQ} \times \overrightarrow{PR}$ is perpendicular to the plane because it is perpendicular to both vectors. In terms of components,

$$\overrightarrow{PQ} = (2 - 1)\mathbf{i} + (1 + 1)\mathbf{j} + (-1 - 0)\mathbf{k} = \mathbf{i} + 2\mathbf{j} - \mathbf{k}$$

$$\overrightarrow{PR} = (-1 - 1)\mathbf{i} + (1 + 1)\mathbf{j} + (2 - 0)\mathbf{k} = -2\mathbf{i} + 2\mathbf{j} + 2\mathbf{k}$$

$$\overrightarrow{PQ} \times \overrightarrow{PR} = \begin{vmatrix} \mathbf{i} & \mathbf{j} & \mathbf{k} \\ 1 & 2 & -1 \\ -2 & 2 & 2 \end{vmatrix} = \begin{vmatrix} 2 & -1 \\ 2 & 2 \end{vmatrix}\mathbf{i} - \begin{vmatrix} 1 & -1 \\ -2 & 2 \end{vmatrix}\mathbf{j} + \begin{vmatrix} 1 & 2 \\ -2 & 2 \end{vmatrix}\mathbf{k}$$

$$= 6\mathbf{i} + 6\mathbf{k}.$$

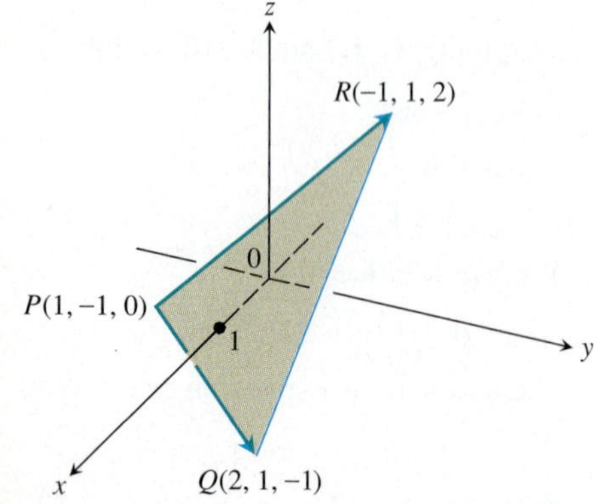

FIGURE 10.16 The area of triangle PQR is half of $|\overrightarrow{PQ} \times \overrightarrow{PR}|$. (Example 6)

Example 6 Finding the Area of a Triangle

Find the area of the triangle with vertices $P(1, -1, 0)$, $Q(2, 1, -1)$, and $R(-1, 1, 2)$ (Figure 10.16).

Solution The area of the parallelogram determined by P, Q, and R is

$$|\overrightarrow{PQ} \times \overrightarrow{PR}| = |6\mathbf{i} + 6\mathbf{k}| \quad \text{Values from Example 5.}$$

$$= \sqrt{(6)^2 + (6)^2} = \sqrt{2 \cdot 36} = 6\sqrt{2}.$$

The triangle's area is half of this, or $3\sqrt{2}$.

Example 7 Finding a Unit Normal to a Plane

Find a unit vector perpendicular to the plane of $P(1, -1, 0)$, $Q(2, 1, -1)$, and $R(-1, 1, 2)$.

Solution Since $\overrightarrow{PQ} \times \overrightarrow{PR}$ is perpendicular to the plane, its direction $\mathbf{n}$ is a unit vector perpendicular to the plane. Taking values from Examples 5 and 6, we have

$$\mathbf{n} = \frac{\overrightarrow{PQ} \times \overrightarrow{PR}}{|\overrightarrow{PQ} \times \overrightarrow{PR}|} = \frac{6\mathbf{i} + 6\mathbf{k}}{6\sqrt{2}} = \frac{1}{\sqrt{2}}\mathbf{i} + \frac{1}{\sqrt{2}}\mathbf{k}.$$

Torque

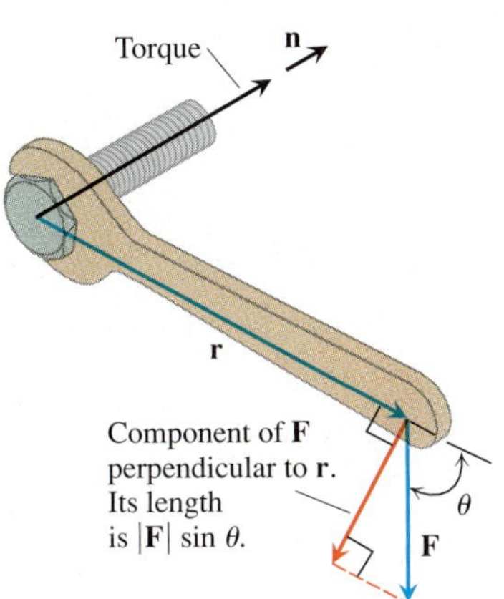

FIGURE 10.17 The torque vector describes the tendency of the force $\mathbf{F}$ to drive the bolt forward.

When we turn a bolt by applying a force $\mathbf{F}$ to a wrench (Figure 10.17), the torque we produce acts along the axis of the bolt to drive the bolt forward. The magnitude of the torque depends on how far out on the wrench the force is applied and on how much of the force is perpendicular to the wrench at the point of application. The number we use to measure the torque's magnitude is the product of the length of the lever arm $\mathbf{r}$ and the scalar component of $\mathbf{F}$ perpendicular to $\mathbf{r}$. In the notation of Figure 10.17,

$$\text{Magnitude of torque vector} = |\mathbf{r}||\mathbf{F}| \sin \theta,$$

or $|\mathbf{r} \times \mathbf{F}|$. If we let $\mathbf{n}$ be a unit vector along the axis of the bolt in the direction of the torque, then a complete description of the torque vector is $\mathbf{r} \times \mathbf{F}$, or

$$\text{Torque vector} = (|\mathbf{r}||\mathbf{F}| \sin \theta)\,\mathbf{n}.$$

Recall that we defined $\mathbf{u} \times \mathbf{v}$ to be $\mathbf{0}$ when $\mathbf{u}$ and $\mathbf{v}$ are parallel. This is consistent with the torque interpretation as well. If the force $\mathbf{F}$ in Figure 10.17 is parallel to the wrench, meaning that we are trying to turn the bolt by pushing or pulling along the line of the wrench's handle, the torque produced is zero.

Example 8 Finding the Magnitude of a Torque

The magnitude of the torque generated by force $\mathbf{F}$ at the pivot point P in Figure 10.18 is

$$\begin{aligned} |\overrightarrow{PQ} \times \mathbf{F}| &= |\overrightarrow{PQ}||\mathbf{F}| \sin 70^\circ \\ &\approx (3)(20)(0.94) \\ &\approx 56.4 \text{ ft-lb}. \end{aligned}$$

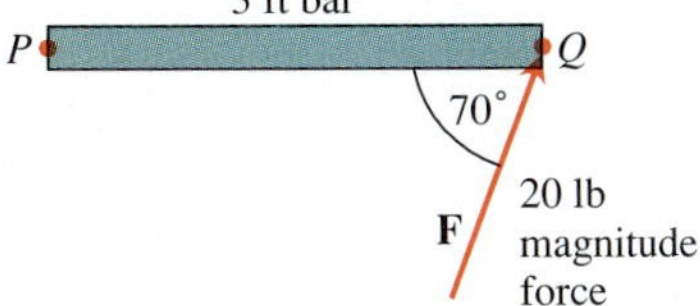

FIGURE 10.18 The magnitude of the torque exerted by $\mathbf{F}$ at P is about 56.4 ft-lb. (Example 8)

Triple Scalar or Box Product

The product $(\mathbf{u} \times \mathbf{v}) \cdot \mathbf{w}$ is called the **triple scalar product** of $\mathbf{u}$, $\mathbf{v}$, and $\mathbf{w}$ (in that order). As you can see from the formula

$$|(\mathbf{u} \times \mathbf{v}) \cdot \mathbf{w}| = |\mathbf{u} \times \mathbf{v}||\mathbf{w}||\cos \theta|,$$

the absolute value of the product is the volume of the parallelepiped (parallelogram-sided box) determined by $\mathbf{u}$, $\mathbf{v}$, and $\mathbf{w}$ (Figure 10.19). The number $|\mathbf{u} \times \mathbf{v}|$ is the area of the base parallelogram. The number $|\mathbf{w}||\cos\theta|$ is the parallelepiped's height. Because of this geometry, $(\mathbf{u} \times \mathbf{v}) \cdot \mathbf{w}$ is also called the **box product** of $\mathbf{u}$, $\mathbf{v}$, and $\mathbf{w}$.

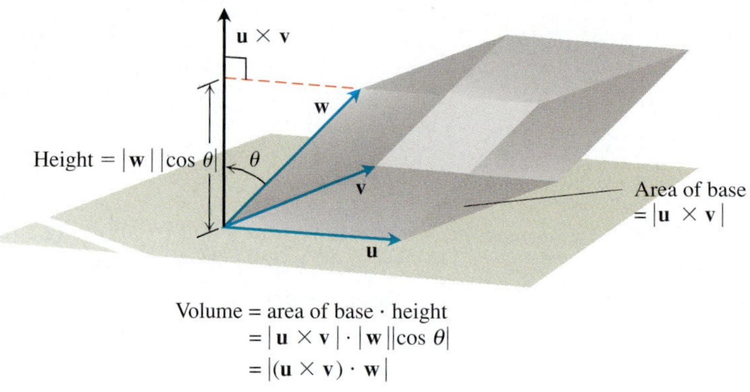

FIGURE 10.19 The number $|(\mathbf{u} \times \mathbf{v}) \cdot \mathbf{w}|$ is the volume of a parallelepiped.

The dot and cross may be interchanged in a triple scalar product without altering its value.

By treating the planes of $\mathbf{v}$ and $\mathbf{w}$ and of $\mathbf{w}$ and $\mathbf{u}$ as the base planes of the parallelepiped determined by $\mathbf{u}$, $\mathbf{v}$, and $\mathbf{w}$, we see that

$$(\mathbf{u} \times \mathbf{v}) \cdot \mathbf{w} = (\mathbf{v} \times \mathbf{w}) \cdot \mathbf{u} = (\mathbf{w} \times \mathbf{u}) \cdot \mathbf{v}.$$

Since the dot product is commutative, we also have

$$(\mathbf{u} \times \mathbf{v}) \cdot \mathbf{w} = \mathbf{u} \cdot (\mathbf{v} \times \mathbf{w}).$$

The triple scalar product can be evaluated as a determinant:

$$\begin{aligned}(\mathbf{u} \times \mathbf{v}) \cdot \mathbf{w} &= \left[\begin{vmatrix} u_2 & u_3 \\ v_2 & v_3 \end{vmatrix}\mathbf{i} - \begin{vmatrix} u_1 & u_3 \\ v_1 & v_3 \end{vmatrix}\mathbf{j} + \begin{vmatrix} u_1 & u_2 \\ v_1 & v_2 \end{vmatrix}\mathbf{k}\right] \cdot \mathbf{w} \\ &= w_1\begin{vmatrix} u_2 & u_3 \\ v_2 & v_3 \end{vmatrix} - w_2\begin{vmatrix} u_1 & u_3 \\ v_1 & v_3 \end{vmatrix} + w_3\begin{vmatrix} u_1 & u_2 \\ v_1 & v_2 \end{vmatrix} \\ &= \begin{vmatrix} u_1 & u_2 & u_3 \\ v_1 & v_2 & v_3 \\ w_1 & w_2 & w_3 \end{vmatrix}.\end{aligned}$$

Triple Scalar Product

$$(\mathbf{u} \times \mathbf{v}) \cdot \mathbf{w} = \begin{vmatrix} u_1 & u_2 & u_3 \\ v_1 & v_2 & v_3 \\ w_1 & w_2 & w_3 \end{vmatrix}$$

Example 9 Finding the Volume of a Parallelepiped

Find the volume of the box (parallelepiped) determined by $\mathbf{u} = \mathbf{i} + 2\mathbf{j} - \mathbf{k}$, $\mathbf{v} = -2\mathbf{i} + 3\mathbf{k}$, and $\mathbf{w} = 7\mathbf{j} - 4\mathbf{k}$.

Solution Using a calculator, we find

$$(\mathbf{u} \times \mathbf{v}) \cdot \mathbf{w} = \begin{vmatrix} 1 & 2 & -1 \\ -2 & 0 & 3 \\ 0 & 7 & -4 \end{vmatrix} = -23.$$

The volume is $|(\mathbf{u} \times \mathbf{v}) \cdot \mathbf{w}| = 23$ units cubed.

EXERCISES 10.2

Dot Product and Projections

In Exercises 1–6, find

(a) $\mathbf{v} \cdot \mathbf{u}, |\mathbf{v}|, |\mathbf{u}|$

(b) the cosine of the angle between **v** and **u**

(c) the scalar component of **u** in the direction of **v**

(d) the vector $\text{proj}_{\mathbf{v}}\, \mathbf{u}$.

1. $\mathbf{v} = 2\mathbf{i} - 4\mathbf{j} + \sqrt{5}\mathbf{k}, \quad \mathbf{u} = -2\mathbf{i} + 4\mathbf{j} - \sqrt{5}\mathbf{k}$

2. $\mathbf{v} = (3/5)\mathbf{i} + (4/5)\mathbf{k}, \quad \mathbf{u} = 5\mathbf{i} + 12\mathbf{j}$

3. $\mathbf{v} = 10\mathbf{i} + 11\mathbf{j} - 2\mathbf{k}, \quad \mathbf{u} = 3\mathbf{j} + 4\mathbf{k}$

4. $\mathbf{v} = 2\mathbf{i} + 10\mathbf{j} - 11\mathbf{k}, \quad \mathbf{u} = 2\mathbf{i} + 2\mathbf{j} + \mathbf{k}$

5. $\mathbf{v} = 5\mathbf{j} - 3\mathbf{k}, \quad \mathbf{u} = \mathbf{i} + \mathbf{j} + \mathbf{k}$

6. $\mathbf{v} = -\mathbf{i} + \mathbf{j}, \quad \mathbf{u} = \sqrt{2}\mathbf{i} + \sqrt{3}\mathbf{j} + 2\mathbf{k}$

Decomposing Vectors

In Exercises 7–9, write **u** as the sum of a vector parallel to **v** and a vector orthogonal to **v**.

7. $\mathbf{u} = 3\mathbf{j} + 4\mathbf{k}, \quad \mathbf{v} = \mathbf{i} + \mathbf{j}$

8. $\mathbf{u} = \mathbf{j} + \mathbf{k}, \quad \mathbf{v} = \mathbf{i} + \mathbf{j}$

9. $\mathbf{u} = 8\mathbf{i} + 4\mathbf{j} - 12\mathbf{k}, \quad \mathbf{v} = \mathbf{i} + 2\mathbf{j} - \mathbf{k}$

10. *Sum of vectors* $\mathbf{u} = \mathbf{i} + (\mathbf{j} + \mathbf{k})$ is already the sum of a vector parallel to **i** and a vector orthogonal to **i**. If you use $\mathbf{v} = \mathbf{i}$, in the decomposition $\mathbf{u} = \text{proj}_{\mathbf{v}}\, \mathbf{u} + (\mathbf{u} - \text{proj}_{\mathbf{v}}\, \mathbf{u})$, do you get $\text{proj}_{\mathbf{v}}\, \mathbf{u} = \mathbf{i}$ and $(\mathbf{u} - \text{proj}_{\mathbf{v}}\, \mathbf{u}) = \mathbf{j} + \mathbf{k}$? Try it and find out.

Angles Between Vectors

Find the angles between the vectors in Exercises 11–14 to the nearest hundredth of a radian.

11. $\mathbf{u} = 2\mathbf{i} + \mathbf{j}, \quad \mathbf{v} = \mathbf{i} + 2\mathbf{j} - \mathbf{k}$

12. $\mathbf{u} = 2\mathbf{i} - 2\mathbf{j} + \mathbf{k}, \quad \mathbf{v} = 3\mathbf{i} + 4\mathbf{k}$

13. $\mathbf{u} = \sqrt{3}\mathbf{i} - 7\mathbf{j}, \quad \mathbf{v} = \sqrt{3}\mathbf{i} + \mathbf{j} - 2\mathbf{k}$

14. $\mathbf{u} = \mathbf{i} + \sqrt{2}\mathbf{j} - \sqrt{2}\mathbf{k}, \mathbf{v} = -\mathbf{i} + \mathbf{j} + \mathbf{k}$

15. *Direction angles and direction cosines* The **direction angles** α, β, and γ of a vector $\mathbf{v} = a\mathbf{i} + b\mathbf{j} + c\mathbf{k}$ are defined as follows:

α is the angle between **v** and the positive x-axis $(0 \le \alpha \le \pi)$

β is the angle between **v** and the positive y-axis $(0 \le \beta \le \pi)$

γ is the angle between **v** and the positive z-axis $(0 \le \gamma \le \pi)$.

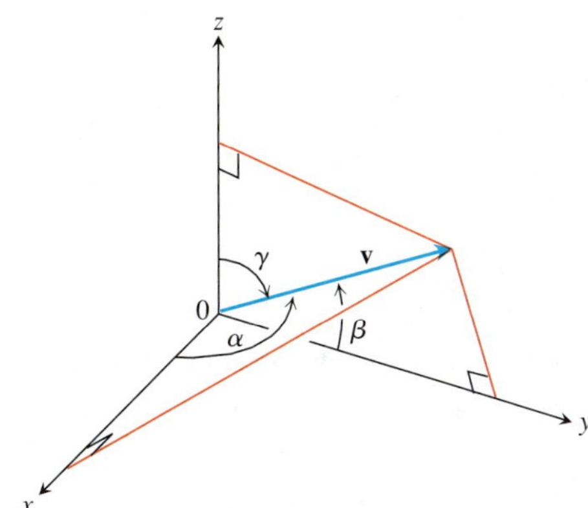

(a) Show that

$$\cos \alpha = \frac{a}{|\mathbf{v}|}, \qquad \cos \beta = \frac{b}{|\mathbf{v}|}, \qquad \cos \gamma = \frac{c}{|\mathbf{v}|},$$

and $\cos^2 \alpha + \cos^2 \beta + \cos^2 \gamma = 1$. These cosines are called the **direction cosines** of **v**.

(b) *Unit vectors are built from direction cosines* Show that if $\mathbf{v} = a\mathbf{i} + b\mathbf{j} + c\mathbf{k}$ is a unit vector, then a, b, and c *are* the direction cosines of **v**.

16. *Water main construction* A water main is to be constructed with a 20% grade in the north direction and a 10% grade in the east direction. Determine the angle θ required in the water main for the turn from north to east.

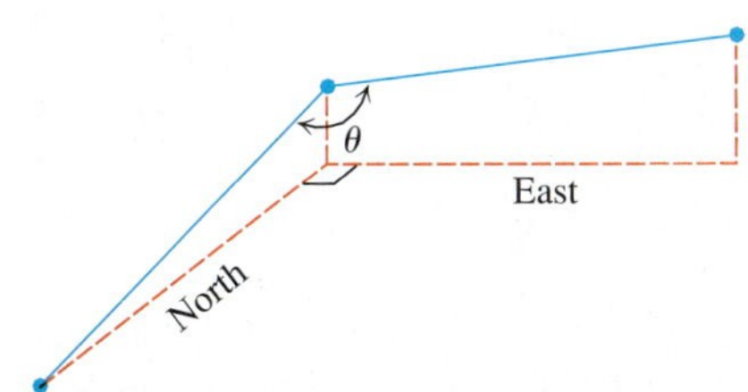

Cross-Product Calculations

In Exercises 17–24, find the length and direction (when defined) of $\mathbf{u} \times \mathbf{v}$ and $\mathbf{v} \times \mathbf{u}$.

17. $\mathbf{u} = 2\mathbf{i} - 2\mathbf{j} - \mathbf{k}, \quad \mathbf{v} = \mathbf{i} - \mathbf{k}$

18. $\mathbf{u} = 2\mathbf{i} + 3\mathbf{j}, \quad \mathbf{v} = -\mathbf{i} + \mathbf{j}$

19. $\mathbf{u} = 2\mathbf{i} - 2\mathbf{j} + 4\mathbf{k}, \quad \mathbf{v} = -\mathbf{i} + \mathbf{j} - 2\mathbf{k}$

20. $\mathbf{u} = \mathbf{i} + \mathbf{j} - \mathbf{k}, \quad \mathbf{v} = \mathbf{0}$

21. $\mathbf{u} = 2\mathbf{i}, \quad \mathbf{v} = -3\mathbf{j}$

22. $\mathbf{u} = \mathbf{i} \times \mathbf{j}, \quad \mathbf{v} = \mathbf{j} \times \mathbf{k}$

23. $\mathbf{u} = -8\mathbf{i} - 2\mathbf{j} - 4\mathbf{k}, \quad \mathbf{v} = 2\mathbf{i} + 2\mathbf{j} + \mathbf{k}$

24. $\mathbf{u} = \frac{3}{2}\mathbf{i} - \frac{1}{2}\mathbf{j} + \mathbf{k}, \quad \mathbf{v} = \mathbf{i} + \mathbf{j} + 2\mathbf{k}$

In Exercises 25–28, sketch the coordinate axes and then include the vectors $\mathbf{u}$, $\mathbf{v}$, and $\mathbf{u} \times \mathbf{v}$ as vectors starting at the origin.

25. $\mathbf{u} = \mathbf{i} - \mathbf{k}, \quad \mathbf{v} = \mathbf{j}$

26. $\mathbf{u} = \mathbf{i} - \mathbf{k}, \quad \mathbf{v} = \mathbf{j} + \mathbf{k}$

27. $\mathbf{u} = \mathbf{i} + \mathbf{j}, \quad \mathbf{v} = \mathbf{i} - \mathbf{j}$

28. $\mathbf{u} = \mathbf{j} + 2\mathbf{k}, \quad \mathbf{v} = \mathbf{i}$

Triangles in Space

In Exercises 29–32:

(a) Find the area of the triangle determined by the points P, Q, and R.

(b) Find a unit vector perpendicular to plane PQR.

29. $P(1, -1, 2), \quad Q(2, 0, -1), \quad R(0, 2, 1)$

30. $P(1, 1, 1), \quad Q(2, 1, 3), \quad R(3, -1, 1)$

31. $P(2, -2, 1), \quad Q(3, -1, 2), \quad R(3, -1, 1)$

32. $P(-2, 2, 0), \quad Q(0, 1, -1), \quad R(-1, 2, -2)$

Triple Scalar Products

In Exercises 33–36, verify that $(\mathbf{u} \times \mathbf{v}) \cdot \mathbf{w} = (\mathbf{v} \times \mathbf{w}) \cdot \mathbf{u} = (\mathbf{w} \times \mathbf{u}) \cdot \mathbf{v}$ and find the volume of the parallelepiped (box) determined by $\mathbf{u}$, $\mathbf{v}$, and $\mathbf{w}$.

	u	v	w
33.	$2\mathbf{i}$	$2\mathbf{j}$	$2\mathbf{k}$
34.	$\mathbf{i} - \mathbf{j} + \mathbf{k}$	$2\mathbf{i} + \mathbf{j} - 2\mathbf{k}$	$-\mathbf{i} + 2\mathbf{j} - \mathbf{k}$
35.	$2\mathbf{i} + \mathbf{j}$	$2\mathbf{i} - \mathbf{j} + \mathbf{k}$	$\mathbf{i} + 2\mathbf{k}$
36.	$\mathbf{i} + \mathbf{j} - 2\mathbf{k}$	$-\mathbf{i} - \mathbf{k}$	$2\mathbf{i} + 4\mathbf{j} - 2\mathbf{k}$

Theory and Examples

37. *Writing to Learn: Parallel and perpendicular vectors* Let $\mathbf{u} = 5\mathbf{i} - \mathbf{j} + \mathbf{k}$, $\mathbf{v} = \mathbf{j} - 5\mathbf{k}$, $\mathbf{w} = -15\mathbf{i} + 3\mathbf{j} - 3\mathbf{k}$. Which vectors, if any, are (a) perpendicular? (b) Parallel? Give reasons for your answers.

38. *Writing to Learn: Parallel and perpendicular vectors* Let $\mathbf{u} = \mathbf{i} + 2\mathbf{j} - \mathbf{k}$, $\mathbf{v} = -\mathbf{i} + \mathbf{j} + \mathbf{k}$, $\mathbf{w} = \mathbf{i} + \mathbf{k}$, $\mathbf{r} = -(\pi/2)\mathbf{i} - \pi\mathbf{j} + (\pi/2)\mathbf{k}$. Which vectors, if any, are (a) perpendicular? (b) Parallel? Give reasons for your answers.

In Exercises 39 and 40, find the magnitude of the torque exerted by $\mathbf{F}$ on the bolt at P if $|\overrightarrow{PQ}| = 8$ in. and $|\mathbf{F}| = 30$ lb. Answer in foot-pounds.

39.

F
60°
Q
P

40.

135°
F
Q
P

41. Which of the following are *always true*, and which are *not always true*? Give reasons for your answers.

(a) $|\mathbf{u}| = \sqrt{\mathbf{u} \cdot \mathbf{u}}$ **(b)** $\mathbf{u} \cdot \mathbf{u} = |\mathbf{u}|$

(c) $\mathbf{u} \times \mathbf{0} = \mathbf{0} \times \mathbf{u} = \mathbf{0}$ **(d)** $\mathbf{u} \times (-\mathbf{u}) = \mathbf{0}$

(e) $\mathbf{u} \times \mathbf{v} = \mathbf{v} \times \mathbf{u}$

(f) $\mathbf{u} \times (\mathbf{v} + \mathbf{w}) = \mathbf{u} \times \mathbf{v} + \mathbf{u} \times \mathbf{w}$

(g) $(\mathbf{u} \times \mathbf{v}) \cdot \mathbf{v} = 0$

(h) $(\mathbf{u} \times \mathbf{v}) \cdot \mathbf{w} = \mathbf{u} \cdot (\mathbf{v} \times \mathbf{w})$

42. Which of the following are *always true*, and which are *not always true*? Give reasons for your answers.

(a) $\mathbf{u} \cdot \mathbf{v} = \mathbf{v} \cdot \mathbf{u}$ **(b)** $\mathbf{u} \times \mathbf{v} = -(\mathbf{v} \times \mathbf{u})$

(c) $(-\mathbf{u}) \times \mathbf{v} = -(\mathbf{u} \times \mathbf{v})$

(d) $(c\mathbf{u}) \cdot \mathbf{v} = \mathbf{u} \cdot (c\mathbf{v}) = c(\mathbf{u} \cdot \mathbf{v})$ (any number c)

(e) $c(\mathbf{u} \times \mathbf{v}) = (c\mathbf{u}) \times \mathbf{v} = \mathbf{u} \times (c\mathbf{v})$ (any number c)

(f) $\mathbf{u} \cdot \mathbf{u} = |\mathbf{u}|^2$ **(g)** $(\mathbf{u} \times \mathbf{u}) \cdot \mathbf{u} = 0$

(h) $(\mathbf{u} \times \mathbf{v}) \cdot \mathbf{u} = \mathbf{v} \cdot (\mathbf{u} \times \mathbf{v})$

43. Given nonzero vectors $\mathbf{u}$, $\mathbf{v}$, and $\mathbf{w}$, use dot-product and cross-product notation, as appropriate, to describe the following.

(a) The vector projection of $\mathbf{u}$ onto $\mathbf{v}$

(b) A vector orthogonal to $\mathbf{u}$ and $\mathbf{v}$

(c) A vector orthogonal to $\mathbf{u} \times \mathbf{v}$ and $\mathbf{w}$

(d) The volume of the parallelepiped determined by $\mathbf{u}$, $\mathbf{v}$, and $\mathbf{w}$

44. Given nonzero vectors $\mathbf{u}$, $\mathbf{v}$, and $\mathbf{w}$, use dot-product and cross-product notation to describe the following.

(a) A vector orthogonal to $\mathbf{u} \times \mathbf{v}$ and $\mathbf{u} \times \mathbf{w}$

(b) A vector orthogonal to $\mathbf{u} + \mathbf{v}$ and $\mathbf{u} - \mathbf{v}$

(c) A vector of length $|\mathbf{u}|$ in the direction of $\mathbf{v}$

(d) The area of the parallelogram determined by $\mathbf{u}$ and $\mathbf{w}$

45. *Writing to Learn* Let $\mathbf{u}$, $\mathbf{v}$, and $\mathbf{w}$ be vectors. Which of the following make sense, and which do not? Give reasons for your answers.

(a) $(\mathbf{u} \times \mathbf{v}) \cdot \mathbf{w}$

(b) $\mathbf{u} \times (\mathbf{v} \cdot \mathbf{w})$

(c) $\mathbf{u} \times (\mathbf{v} \times \mathbf{w})$

(d) $\mathbf{u} \cdot (\mathbf{v} \cdot \mathbf{w})$

46. *Writing to Learn: Cross products of three vectors* Show that except in degenerate cases, $(\mathbf{u} \times \mathbf{v}) \times \mathbf{w}$ lies in the plane of $\mathbf{u}$ and $\mathbf{v}$, whereas $\mathbf{u} \times (\mathbf{v} \times \mathbf{w})$ lies in the plane of $\mathbf{v}$ and $\mathbf{w}$. What *are* the degenerate cases?

47. *Cancellation in cross products* If $\mathbf{u} \times \mathbf{v} = \mathbf{u} \times \mathbf{w}$ and $\mathbf{u} \neq \mathbf{0}$, then does $\mathbf{v} = \mathbf{w}$? Give reasons for your answer.

48. *Double cancellation* If $\mathbf{u} \neq \mathbf{0}$ and if $\mathbf{u} \times \mathbf{v} = \mathbf{u} \times \mathbf{w}$ and $\mathbf{u} \cdot \mathbf{v} = \mathbf{u} \cdot \mathbf{w}$, then does $\mathbf{v} = \mathbf{w}$? Give reasons for your answer.

Area in the Plane

Find the areas of the parallelograms whose vertices are given in Exercises 49–52.

49. $A(1, 0)$, $B(0, 1)$, $C(-1, 0)$, $D(0, -1)$

50. $A(0, 0)$, $B(7, 3)$, $C(9, 8)$, $D(2, 5)$

51. $A(-1, 2)$, $B(2, 0)$, $C(7, 1)$, $D(4, 3)$

52. $A(-6, 0)$, $B(1, -4)$, $C(3, 1)$, $D(-4, 5)$

Find the areas of the triangles whose vertices are given in Exercises 53–56.

53. $A(0, 0)$, $B(-2, 3)$, $C(3, 1)$

54. $A(-1, -1)$, $B(3, 3)$, $C(2, 1)$

55. $A(-5, 3)$, $B(1, -2)$, $C(6, -2)$

56. $A(-6, 0)$, $B(10, -5)$, $C(-2, 4)$

57. *Writing to Learn: Triangle area* Find a formula for the area of the triangle in the xy-plane with vertices at $(0, 0)$, (a_1, a_2), and (b_1, b_2). Explain your work.

58. *Triangle area* Find a concise formula for the area of a triangle with vertices (a_1, a_2), (b_1, b_2), and (c_1, c_2).

10.3 Lines and Planes in Space

Lines and Line Segments in Space • Equations for Planes in Space • Lines of Intersection

In the calculus of functions of a single variable, we began with lines and used our knowledge of lines to study curves in the plane. We investigated tangents and found that, when highly magnified, differentiable curves were effectively linear.

To study the calculus of functions of more than one variable in the next chapter, we begin much the same way. We start with planes and use our knowledge of planes to study the surfaces that are the graphs of functions in space.

This section shows how to use scalar and vector products to write equations for lines, line segments, and planes in space.

Lines and Line Segments in Space

In the plane, a line is determined by a point and a number giving the slope of the line. Analogously, in space a line is determined by a point and a *vector* giving the direction of the line.

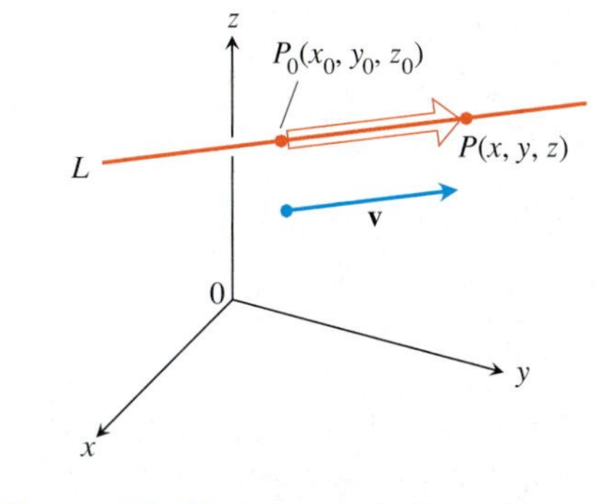

Suppose that L is a line in space passing through a point $P_0(x_0, y_0, z_0)$ parallel to a vector $\mathbf{v} = v_1\mathbf{i} + v_2\mathbf{j} + v_3\mathbf{k}$. Then L is the set of all points $P(x, y, z)$ for which $\overrightarrow{P_0P}$ is parallel to $\mathbf{v}$ (Figure 10.20). Thus, $\overrightarrow{P_0P} = t\mathbf{v}$ for some scalar parameter t. The value of t depends on the location of the point P along the line, and the domain of t is $(-\infty, \infty)$. The expanded form of the equation $\overrightarrow{P_0P} = t\mathbf{v}$ is

$$(x - x_0)\mathbf{i} + (y - y_0)\mathbf{j} + (z - z_0)\mathbf{k} = t(v_1\mathbf{i} + v_2\mathbf{j} + v_3\mathbf{k}),$$

and this last equation can be rewritten as

$$x\mathbf{i} + y\mathbf{j} + z\mathbf{k} = x_0\mathbf{i} + y_0\mathbf{j} + z_0\mathbf{k} + t(v_1\mathbf{i} + v_2\mathbf{j} + v_3\mathbf{k}). \tag{1}$$

FIGURE 10.20 A point P lies on L through P_0 parallel to $\mathbf{v}$ if and only if $\overrightarrow{P_0P}$ is a scalar multiple of $\mathbf{v}$.

If $\mathbf{r}(t)$ is the position vector of a point $P(x, y, z)$ on the line and $\mathbf{r}_0$ is the position vector of the point $P_0(x_0, y_0, z_0)$, then Equation (1) gives the following vector form for the equation of a line in space.

Vector Equation for a Line

A vector equation for the line L through $P_0(x_0, y_0, z_0)$ parallel to $\mathbf{v}$ is

$$\mathbf{r}(t) = \mathbf{r}_0 + t\mathbf{v}, \qquad -\infty < t < \infty, \tag{2}$$

where $\mathbf{r}$ is the position vector of a point $P(x, y, z)$ on L and $\mathbf{r}_0$ is the position vector of $P_0(x_0, y_0, z_0)$.

Equating the corresponding components of the two sides of Equation (1) gives three scalar equations involving the parameter t:

$$x = x_0 + tv_1, \qquad y = y_0 + tv_2, \qquad z = z_0 + tv_3.$$

These equations give us the standard parametrization of the line for the parameter interval $-\infty < t < \infty$.

Parametric Equations for a Line

The standard parametrization of the line through $P_0(x_0, y_0, z_0)$ parallel to $\mathbf{v} = v_1\mathbf{i} + v_2\mathbf{j} + v_3\mathbf{k}$ is

$$x = x_0 + tv_1, \qquad y = y_0 + tv_2, \qquad z = z_0 + tv_3, \qquad -\infty < t < \infty \tag{3}$$

Example 1 Parametrizing a Line Through a Point Parallel to a Vector

Find parametric equations for the line through $(-2, 0, 4)$ parallel to $\mathbf{v} = 2\mathbf{i} + 4\mathbf{j} - 2\mathbf{k}$ (Figure 10.21).

Solution With $P_0(x_0, y_0, z_0)$ equal to $(-2, 0, 4)$ and $v_1\mathbf{i} + v_2\mathbf{j} + v_3\mathbf{k}$ equal to $2\mathbf{i} + 4\mathbf{j} - 2\mathbf{k}$, Equations (3) become

$$x = -2 + 2t, \qquad y = 4t, \qquad z = 4 - 2t.$$

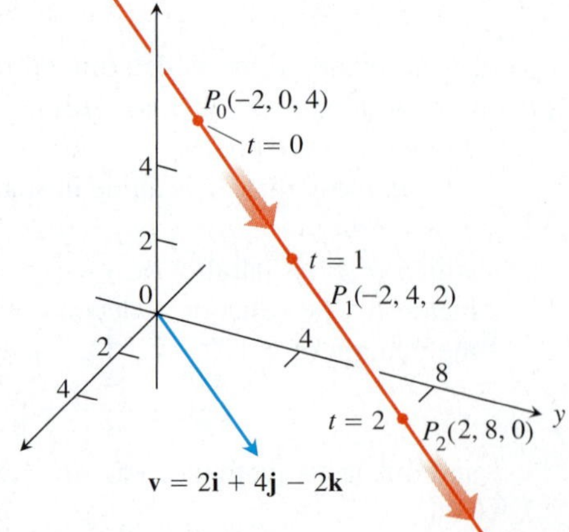

FIGURE 10.21 Selected points and parameter values on the line $x = -2 + 2t$, $y = 4t$, $z = 4 - 2t$. The arrows show the direction of increasing t. (Example 1)

Example 2 Parametrizing a Line Through Two Points

Find parametric equations for the line through $P(-3, 2, -3)$ and $Q(1, -1, 4)$.

Solution The vector

$$\begin{aligned}\overrightarrow{PQ} &= (1-(-3))\mathbf{i} + (-1-2)\mathbf{j} + (4-(-3))\mathbf{k} \\ &= 4\mathbf{i} - 3\mathbf{j} + 7\mathbf{k}\end{aligned}$$

is parallel to the line, and Equations (3) with $(x_0, y_0, z_0) = (-3, 2, -3)$ give

$$x = -3 + 4t, \qquad y = 2 - 3t, \qquad z = -3 + 7t.$$

We could have chosen $Q(1, -1, 4)$ as the "base point" and written

$$x = 1 + 4t, \qquad y = -1 - 3t, \qquad z = 4 + 7t.$$

These equations serve as well as the first; they simply place you at a different point on the line for a given value of t.

Notice that parametrizations are not unique. Not only can the "base point" change, but so can the parameter. The equations $x = -3 + 4t^3$, $y = 2 - 3t^3$, and $z = -3 + 7t^3$ also parametrize the line in Example 2.

To parametrize a line segment joining two points, we first parametrize the line through the points. We then find the t-values for the endpoints and restrict t to lie in the closed interval bounded by these values. The line equations together with this added restriction parametrize the segment.

Example 3 Parametrizing a Line Segment

Parametrize the line segment joining the points $P(-3, 2, -3)$ and $Q(1, -1, 4)$ (Figure 10.22).

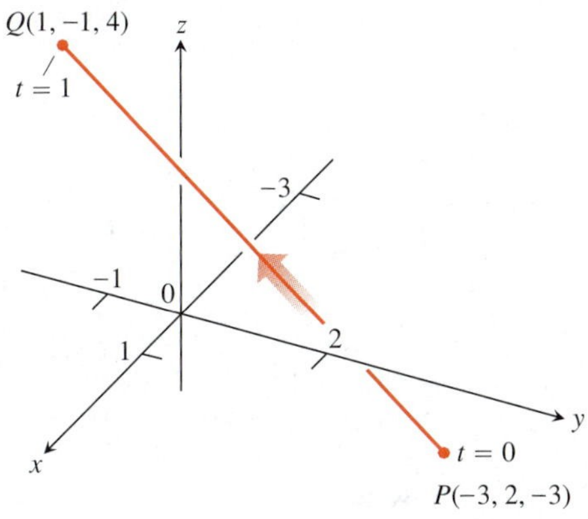

FIGURE 10.22 Example 3 derives a parametrization of line segment PQ. The arrow shows the direction of increasing t.

Solution We begin with equations for the line through P and Q, taking them, in this case, from Example 2:

$$x = -3 + 4t, \qquad y = 2 - 3t, \qquad z = -3 + 7t.$$

We observe that the point

$$(x, y, z) = (-3 + 4t, 2 - 3t, -3 + 7t)$$

passes through $P(-3, 2, -3)$ at $t = 0$ and $Q(1, -1, 4)$ at $t = 1$. We add the restriction $0 \le t \le 1$ to parametrize the segment:

$$x = -3 + 4t, \qquad y = 2 - 3t, \qquad z = -3 + 7t, \qquad 0 \le t \le 1.$$

The vector form (Equation (2)) for a line in space is more revealing if we think of a line as the path of a particle starting at position $P_0(x_0, y_0, z_0)$ and moving in the direction of vector $\mathbf{v}$. Rewriting Equation (2), we have

$$\begin{aligned}\mathbf{r}(t) &= \mathbf{r}_0 + t\mathbf{v} \\ &= \underbrace{\mathbf{r}_0}_{\text{Initial position}} + \underbrace{t}_{\text{Time}}\underbrace{|\mathbf{v}|}_{\text{Speed}}\underbrace{\frac{\mathbf{v}}{|\mathbf{v}|}}_{\text{Direction}}. \end{aligned} \tag{4}$$

In other words, the position of the particle at time t is its initial position plus its rate $\times$ time (distance moved) in the direction $\mathbf{v}/|\mathbf{v}|$ of its straight-line motion.

Example 4 Flight of a Helicopter

A helicopter is to fly directly from a helipad at the origin toward the point $(1, 1, 1)$ at a speed of 60 ft/sec. What is the position of the helicopter after 10 sec?

Solution We place the origin at the starting position (helipad) of the helicopter. Then the unit vector

$$\mathbf{u} = \frac{1}{\sqrt{3}}\mathbf{i} + \frac{1}{\sqrt{3}}\mathbf{j} + \frac{1}{\sqrt{3}}\mathbf{k}$$

gives the flight direction of the helicopter. From Equation (4), the position of the helicopter at any time t is

$$\begin{aligned}\mathbf{r}(t) &= \mathbf{r}_0 + t\,(\text{speed})\mathbf{u} \\ &= \mathbf{0} + t(60)\left(\frac{1}{\sqrt{3}}\mathbf{i} + \frac{1}{\sqrt{3}}\mathbf{j} + \frac{1}{\sqrt{3}}\mathbf{k}\right) \\ &= 20\sqrt{3}\,t\,(\mathbf{i} + \mathbf{j} + \mathbf{k}).\end{aligned}$$

When $t = 10$ sec,

$$\begin{aligned}\mathbf{r}(10) &= 200\sqrt{3}(\mathbf{i} + \mathbf{j} + \mathbf{k}) \\ &= \langle 200\sqrt{3}, 200\sqrt{3}, 200\sqrt{3}\rangle.\end{aligned}$$

Interpret After 10 sec of flight from the origin toward $(1, 1, 1)$, the helicopter is located at the point $(200\sqrt{3}, 200\sqrt{3}, 200\sqrt{3})$ in space. It has traveled a distance of $(60 \text{ ft/sec})(10 \text{ sec}) = 600$ ft, which is the length of the vector $\mathbf{r}(10)$.

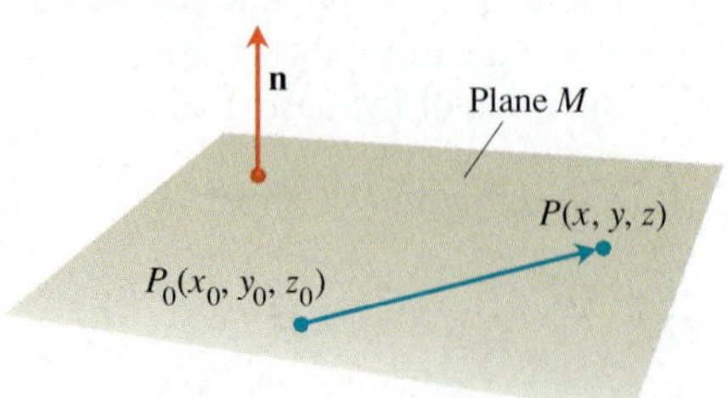

FIGURE 10.23 The standard equation for a plane in space is defined in terms of a vector normal to the plane: A point P lies in the plane through P_0 normal to $\mathbf{n}$ if and only if $\mathbf{n} \cdot \overrightarrow{P_0P} = 0$.

Equations for Planes in Space

A plane in space is determined by knowing a point on the plane and its "tilt" or orientation. This "tilt" is defined by specifying a vector that is perpendicular or normal to the plane.

Suppose that plane M passes through a point $P_0(x_0, y_0, z_0)$ and is normal (perpendicular) to the nonzero vector $\mathbf{n} = A\mathbf{i} + B\mathbf{j} + C\mathbf{k}$. Then M is the set of all points $P(x, y, z)$ for which $\overrightarrow{P_0P}$ is orthogonal to $\mathbf{n}$ (Figure 10.23). Thus, the dot product $\mathbf{n} \cdot \overrightarrow{P_0P} = 0$. This equation is equivalent to

$$(A\mathbf{i} + B\mathbf{j} + C\mathbf{k}) \cdot [(x - x_0)\mathbf{i} + (y - y_0)\mathbf{j} + (z - z_0)\mathbf{k}] = 0$$

or

$$A(x - x_0) + B(y - y_0) + C(z - z_0) = 0.$$

Equation for a Plane

The **plane through $P_0(x_0, y_0, z_0)$ normal to $\mathbf{n} = A\mathbf{i} + B\mathbf{j} + C\mathbf{k}$** has

Vector equation: $\mathbf{n} \cdot \overrightarrow{P_0P} = 0$

Component equation: $A(x - x_0) + B(y - y_0) + C(z - z_0) = 0$

Component equation simplified: $Ax + By + Cz = D$, where $D = Ax_0 + By_0 + Cz_0$

Example 5 Finding an Equation for a Plane

Find an equation for the plane through $P_0(-3, 0, 7)$ perpendicular to $\mathbf{n} = 5\mathbf{i} + 2\mathbf{j} - \mathbf{k}$.

Solution The component equation is

$$5(x - (-3)) + 2(y - 0) + (-1)(z - 7) = 0.$$

Simplifying, we obtain

$$5x + 15 + 2y - z + 7 = 0$$
$$5x + 2y - z = -22.$$

$A\mathbf{i} + B\mathbf{j} + C\mathbf{k}$ is normal to the plane $Ax + By + Cz = D$.

Notice in Example 5 how the components of $\mathbf{n} = 5\mathbf{i} + 2\mathbf{j} - \mathbf{k}$ became the coefficients of x, y, and z in the equation $5x + 2y - z = -22$.

Example 6 Finding an Equation for a Plane Through Three Points

Find an equation for the plane through $A(0, 0, 1)$, $B(2, 0, 0)$, and $C(0, 3, 0)$.

Solution We find a vector normal to the plane and use it with one of the points (it does not matter which) to write an equation for the plane.

The cross product

$$\overrightarrow{AB} \times \overrightarrow{AC} = \begin{vmatrix} \mathbf{i} & \mathbf{j} & \mathbf{k} \\ 2 & 0 & -1 \\ 0 & 3 & -1 \end{vmatrix} = 3\mathbf{i} + 2\mathbf{j} + 6\mathbf{k}$$

is normal to the plane. We substitute the components of this vector and the coordinates of A(0, 0, 1) into the component form of the equation to obtain

$$3(x - 0) + 2(y - 0) + 6(z - 1) = 0$$
$$3x + 2y + 6z = 6.$$

Just as lines are parallel if and only if they have the same direction, two planes are **parallel** if and only if their normals are parallel, or $\mathbf{n}_1 = k\mathbf{n}_2$ for some scalar k.

Lines of Intersection

Two planes that are not parallel intersect in a line.

Example 7 Finding a Vector Parallel to the Line of Intersection of Two Planes

Find a vector parallel to the line of intersection of the planes $3x - 6y - 2z = 15$ and $2x + y - 2z = 5$.

Solution The line of intersection of two planes is perpendicular to the planes' normal vectors $\mathbf{n}_1$ and $\mathbf{n}_2$ (Figure 10.24) and therefore parallel to $\mathbf{n}_1 \times \mathbf{n}_2$. Turning this around, $\mathbf{n}_1 \times \mathbf{n}_2$ is a vector parallel to the planes' line of intersection. In our case,

$$\mathbf{n}_1 \times \mathbf{n}_2 = \begin{vmatrix} \mathbf{i} & \mathbf{j} & \mathbf{k} \\ 3 & -6 & -2 \\ 2 & 1 & -2 \end{vmatrix} = 14\mathbf{i} + 2\mathbf{j} + 15\mathbf{k}.$$

Any nonzero scalar multiple of $\mathbf{n}_1 \times \mathbf{n}_2$ will do as well.

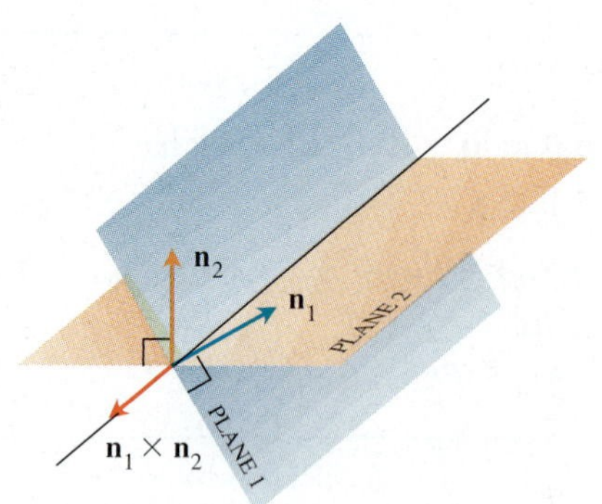

FIGURE 10.24 How the line of intersection of two planes is related to the planes' normal vectors. (Example 7)

Example 8 Parametrizing the Line of Intersection of Two Planes

Find parametric equations for the line in which the planes $3x - 6y - 2z = 15$ and $2x + y - 2z = 5$ intersect.

Solution We find a vector parallel to the line and a point on the line and use Equations (3).

Example 7 identifies $\mathbf{v} = 14\mathbf{i} + 2\mathbf{j} + 15\mathbf{k}$ as a vector parallel to the line. To find a point on the line, we can take any point common to the two planes. Substituting $z = 0$ in the plane equations and solving for x and y simultaneously identifies one of these points as $(3, -1, 0)$. The line is

$$x = 3 + 14t, \qquad y = -1 + 2t, \qquad z = 15t.$$

Sometimes we want to know where a line and a plane intersect. For example, if we are looking at a flat plate and a line segment passes through it, we may be interested in knowing what portion of the line segment is hidden from our view by the plate. This application is used in computer graphics (Exercise 62).

Example 9 Finding the Intersection of a Line and a Plane

Find the point where the line

$$x = \frac{8}{3} + 2t, \qquad y = -2t, \qquad z = 1 + t$$

intersects the plane $3x + 2y + 6z = 6$.

Solution The point

$$\left(\frac{8}{3} + 2t, -2t, 1 + t\right)$$

lies in the plane if its coordinates satisfy the equation of the plane, that is, if

$$\begin{aligned} 3\left(\frac{8}{3} + 2t\right) + 2(-2t) + 6(1 + t) &= 6 \\ 8 + 6t - 4t + 6 + 6t &= 6 \\ 8t &= -8 \\ t &= -1. \end{aligned}$$

The point of intersection is

$$(x, y, z)\big|_{t=-1} = \left(\frac{8}{3} - 2, 2, 1 - 1\right) = \left(\frac{2}{3}, 2, 0\right).$$

EXERCISES 10.3

Lines and Line Segments

Find vector and parametric equations for the lines in Exercises 1–10.

1. The line through the point $P(3, -4, -1)$ parallel to the vector $\mathbf{i} + \mathbf{j} + \mathbf{k}$

2. The line through $P(1, 2, -1)$ and $Q(-1, 0, 1)$

3. The line through $P(-2, 0, 3)$ and $Q(3, 5, -2)$

4. The line through the origin parallel to the vector $2\mathbf{j} + \mathbf{k}$

5. The line through the point $(3, -2, 1)$ parallel to the line $x = 1 + 2t$, $y = 2 - t$, $z = 3t$

6. The line through $(1, 1, 1)$ parallel to the z-axis

7. The line through $(2, 4, 5)$ perpendicular to the plane $3x + 7y - 5z = 21$

8. The line through $(0, -7, 0)$ perpendicular to the plane $x + 2y + 2z = 13$

9. The line through $(2, 3, 0)$ perpendicular to the vectors $\mathbf{u} = \mathbf{i} + 2\mathbf{j} + 3\mathbf{k}$ and $\mathbf{v} = 3\mathbf{i} + 4\mathbf{j} + 5\mathbf{k}$

10. The x-axis

Find parametrizations for the line segments joining the points in Exercises 11–14. Draw coordinate axes and sketch each segment, indicating the direction of increasing t for your parametrization.

11. $(0, 0, 0)$, $(1, 1, 3/2)$

12. $(1, 0, 0)$, $(1, 1, 0)$

13. $(0, 1, 1)$, $(0, -1, 1)$

14. $(1, 0, -1)$, $(0, 3, 0)$

Planes

Find equations for the planes in Exercises 15–20.

15. The plane through $P_0(0, 2, -1)$ normal to $\mathbf{n} = 3\mathbf{i} - 2\mathbf{j} - \mathbf{k}$

16. The plane through $(1, -1, 3)$ parallel to the plane $3x + y + z = 7$

17. The plane through $(1, 1, -1)$, $(2, 0, 2)$, and $(0, -2, 1)$

18. The plane through $(2, 4, 5)$, $(1, 5, 7)$, and $(-1, 6, 8)$

19. The plane through $P_0(2, 4, 5)$ perpendicular to the line

$$x = 5 + t, \qquad y = 1 + 3t, \qquad z = 4t$$

20. The plane through $A(1, -2, 1)$ perpendicular to the vector from the origin to A

21. Find the point of intersection of the lines $x = 2t + 1$, $y = 3t + 2$, $z = 4t + 3$, and $x = s + 2$, $y = 2s + 4$, $z = -4s - 1$, and then find the plane determined by these lines.

22. Find the point of intersection of the lines $x = t$, $y = -t + 2$, $z = t + 1$, and $x = 2s + 2$, $y = s + 3$, $z = 5s + 6$, and then find the plane determined by these lines.

In Exercises 23 and 24, find the plane determined by the intersecting lines.

23. $L1$: $x = -1 + t$, $y = 2 + t$, $z = 1 - t$, $-\infty < t < \infty$

$L2$: $x = 1 - 4s$, $y = 1 + 2s$, $z = 2 - 2s$, $-\infty < s < \infty$

24. $L1$: $x = t$, $y = 3 - 3t$, $z = -2 - t$, $-\infty < t < \infty$

$L2$: $x = 1 + s$, $y = 4 + s$, $z = -1 + s$, $-\infty < s < \infty$

25. Find a plane through $P_0(2, 1, -1)$ and perpendicular to the line of intersection of the planes $2x + y - z = 3$, $x + 2y + z = 2$.

26. Find a plane through the points $P_1(1, 2, 3)$, $P_2(3, 2, 1)$ and perpendicular to the plane $4x - y + 2z = 7$.

Distance from a Point to a Line

27. Follow these steps to find the distance from a point S to a line that passes through a point P parallel to a vector $\mathbf{v}$ as shown in the figure.

(a) Show that the length of the component of $\overrightarrow{PS}$ normal to the line is $|\overrightarrow{PS}| \sin\theta$.

(b) *Distance formula* Show that the distance d from S to the line through P parallel to $\mathbf{v}$ is

$$d = \frac{|\overrightarrow{PS} \times \mathbf{v}|}{|\mathbf{v}|}.$$

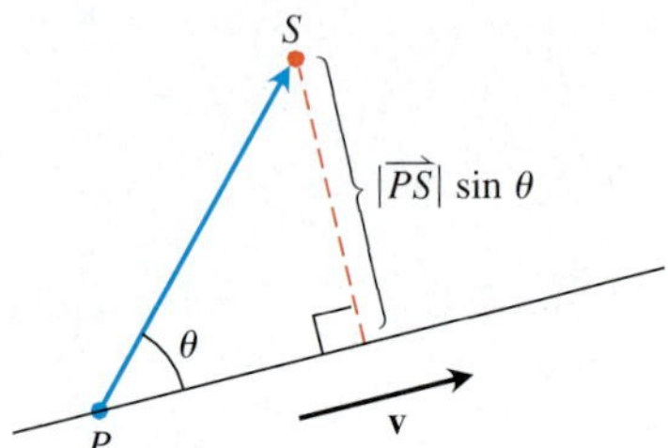

In Exercises 28–30, use the result of Exercise 27 to find the distance from the point to the line.

28. $(0, 0, 0)$; $x = 5 + 3t$, $y = 5 + 4t$, $z = -3 - 5t$

29. $(2, 1, 3)$; $x = 2 + 2t$, $y = 1 + 6t$, $z = 3$

30. $(3, -1, 4)$; $x = 4 - t$, $y = 3 + 2t$, $z = -5 + 3t$

Distance from a Point to a Plane

31. Follow these steps to find the distance d from a point S to a plane $Ax + By + Cz = D$.

(a) Find a point P on the plane.

(b) Find $\overrightarrow{PS}$.

(c) *Distance formula* Show that the distance is

$$d = \left| \overrightarrow{PS} \cdot \frac{\mathbf{n}}{|\mathbf{n}|} \right|,$$

where $\mathbf{n} = A\mathbf{i} + B\mathbf{j} + C\mathbf{k}$.

The accompanying figure shows the situation for finding the distance from $S(1, 1, 3)$ to the plane $3x + 2y + 6z = 6$. In Exercises 32–34, use the result of Exercise 31 to find the distance from the point to the plane.

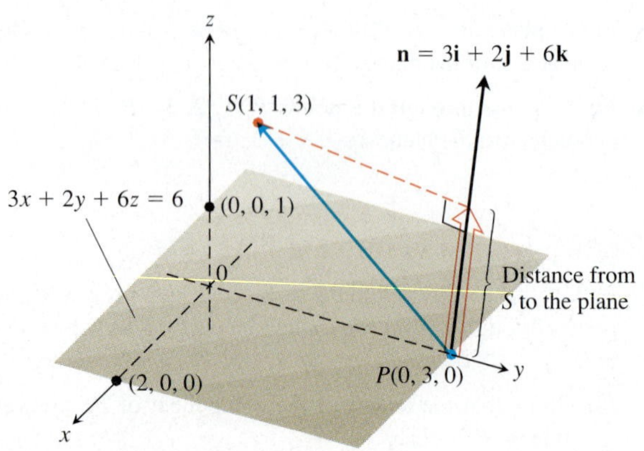

32. $(2, -3, 4)$, $x + 2y + 2z = 13$

33. $(0, 1, 1)$, $4y + 3z = -12$

34. $(0, -1, 0)$, $2x + y + z = 4$

35. Find the distance from the plane $x + 2y + 6z = 1$ to the plane $x + 2y + 6z = 10$.

36. Find the distance from the line $x = 2 + t$, $y = 1 + t$, $z = -1/2 - (1/2)t$ to the plane $x + 2y + 6z = 10$.

Angle Between Planes

37. The **angle between two intersecting planes** is defined to be the (acute) angle determined by the normal vectors as shown in the figure.

(a) *Angle formula* If $\mathbf{n}_1$ and $\mathbf{n}_2$ are the normals to two planes, show that the angle between the planes is

$$\theta = \cos^{-1}\left(\frac{\mathbf{n}_1 \cdot \mathbf{n}_2}{|\mathbf{n}_1||\mathbf{n}_2|}\right).$$

(b) *Finding an angle* Show that the angle between the planes $3x - 6y - 2z = 15$ and $2x + y - 2z = 5$ is about 1.38 radians.

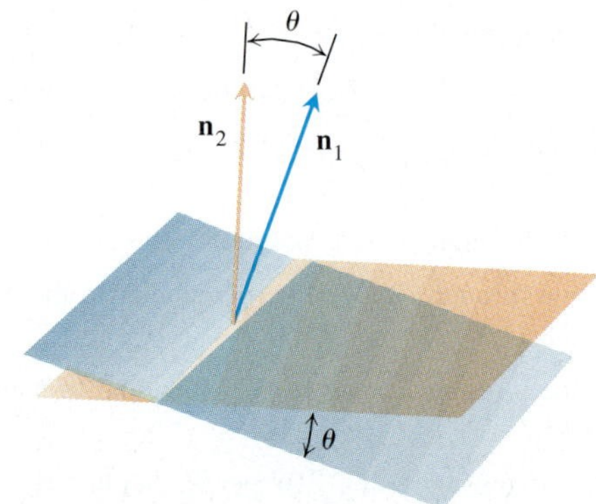

In Exercises 38–40, use the result of Exercise 37 to find the acute angle between the planes.

38. $x + y + z = 1$, $z = 0$

39. $2x + 2y - z = 3$, $x + 2y + z = 2$

40. $4y + 3z = -12$, $3x + 2y + 6z = 6$

Intersecting Lines and Planes

In Exercises 41–44, find the point in which the line meets the plane.

41. $x = 1 - t$, $y = 3t$, $z = 1 + t$; $2x - y + 3z = 6$

42. $x = 2$, $y = 3 + 2t$, $z = -2 - 2t$; $6x + 3y - 4z = -12$

43. $x = 1 + 2t$, $y = 1 + 5t$, $z = 3t$; $x + y + z = 2$

44. $x = -1 + 3t$, $y = -2$, $z = 5t$; $2x - 3z = 7$

Find parametrizations for the lines in which the planes in Exercises 45–48 intersect.

45. $x + y + z = 1, \quad x + y = 2$

46. $3x - 6y - 2z = 3, \quad 2x + y - 2z = 2$

47. $x - 2y + 4z = 2, \quad x + y - 2z = 5$

48. $5x - 2y = 11, \quad 4y - 5z = -17$

Given two lines in space, they are parallel, or they intersect, or they are skew (imagine, for example, the flight paths of two planes in the sky). Exercises 49 and 50 each give three lines. In each exercise, determine whether the lines, taken two at a time, are parallel, intersect, or are skew. If they intersect, find the point of intersection.

49. $L1$: $x = 3 + 2t,\ y = -1 + 4t,\ z = 2 - t,\ -\infty < t < \infty$

$L2$: $x = 1 + 4s,\ y = 1 + 2s,\ z = -3 + 4s,\ -\infty < s < \infty$

$L3$: $x = 3 + 2r,\ y = 2 + r,\ z = -2 + 2r,\ -\infty < r < \infty$

50. $L1$: $x = 1 + 2t,\ y = -1 - t,\ z = 3t,\ -\infty < t < \infty$

$L2$: $x = 2 - s,\ y = 3s,\ z = 1 + s,\ -\infty < s < \infty$

$L3$: $x = 5 + 2r,\ y = 1 - r,\ z = 8 + 3r,\ -\infty < r < \infty$

Theory and Examples

51. *Finding a line* Use Equations (3) to generate a parametrization of the line through $P(2, -4, 7)$ parallel to $\mathbf{v}_1 = 2\mathbf{i} - \mathbf{j} + 3\mathbf{k}$. Then generate another parametrization of the line using the point $P_2(-2, -2, 1)$ and the vector $\mathbf{v}_2 = -\mathbf{i} + (1/2)\mathbf{j} - (3/2)\mathbf{k}$.

52. *Finding a plane* Use the component form to generate an equation for the plane through $P_1(4, 1, 5)$ normal to $\mathbf{n}_1 = \mathbf{i} - 2\mathbf{j} + \mathbf{k}$. Then generate another equation for the same plane using the point $P_2(3, -2, 0)$ and the normal vector

$$\mathbf{n}_2 = -\sqrt{2}\mathbf{i} + 2\sqrt{2}\mathbf{j} - \sqrt{2}\mathbf{k}.$$

53. *Writing to Learn* Find the points in which the line $x = 1 + 2t$, $y = -1 - t$, $z = 3t$ meets the coordinate planes. Describe the reasoning behind your answer.

54. *Writing to Learn* Find equations for the line in the plane $z = 3$ that makes an angle of $\pi/6$ rad with $\mathbf{i}$ and an angle of $\pi/3$ rad with $\mathbf{j}$. Describe the reasoning behind your answer.

55. *Writing to Learn* Is the line $x = 1 - 2t$, $y = 2 + 5t$, $z = -3t$ parallel to the plane $2x + y - z = 8$? Give reasons for your answer.

56. *Writing to Learn* How can you tell when two planes $A_1x + B_1y + C_1z = D_1$ and $A_2x + B_2y + C_2z = D_2$ are parallel? Perpendicular? Give reasons for your answer.

57. *Writing to Learn: Planes intersecting in a given line* Find two different planes whose intersection is the line $x = 1 + t$, $y = 2 - t$, $z = 3 + 2t$. Write an equation for each plane in the form $Ax + By + Cz = D$. Explain how you found your planes.

58. *Writing to Learn* Find a plane through the origin that meets the plane M: $2x + 3y + z = 12$ in a right angle. How do you know that your plane is perpendicular to M?

59. *Writing to Learn* For any nonzero numbers a, b, and c, the graph of $(x/a) + (y/b) + (z/c) = 1$ is a plane. Which planes have an equation of this form? Give reasons for your answer.

60. *Writing to learn* Suppose that L_1 and L_2 are disjoint (nonintersecting) nonparallel lines. Is it possible for a nonzero vector to be perpendicular to both L_1 and L_2? Give reasons for your answer.

Computer Graphics

CD-ROM WEBsite

61. *Perspective in computer graphics* In computer graphics and perspective drawing, we need to represent objects seen by the eye in space as images on a two-dimensional plane. Suppose that the eye is at $E(x_0, 0, 0)$ as shown here and that we want to represent a point $P_1(x_1, y_1, z_1)$ as a point on the yz-plane. We do this by projecting P_1 onto the plane with a ray from E. The point P_1 will be portrayed as the point $P(0, y, z)$. The problem for us as graphics designers is to find y and z given E and P_1.

(a) Write a vector equation that holds between $\overrightarrow{EP}$ and $\overrightarrow{EP_1}$. Use the equation to express y and z in terms of x_0, x_1, y_1, and z_1.

(b) Test the formulas obtained for y and z in part (a) by investigating their behavior at $x_1 = 0$ and $x_1 = x_0$ and by seeing what happens as $x_0 \to \infty$. What do you find?

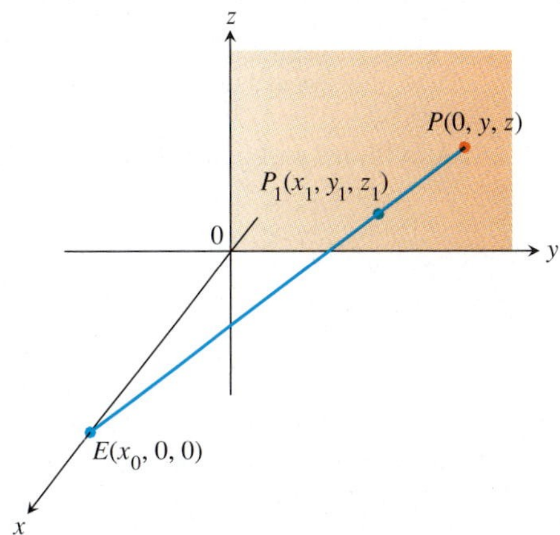

62. *Hidden lines* Here is another typical problem in computer graphics. Your eye is at (4, 0, 0). You are looking at a triangular plate whose vertices are at (1, 0, 1), (1, 1, 0), and (−2, 2, 2). The line segment from (1, 0, 0) to (0, 2, 2) passes through the plate. What portion of the line segment is hidden from your view by the plate? (This is an exercise in finding intersections of lines and planes.)

10.4 Cylinders and Quadric Surfaces

Cylinders • Quadric Surfaces

Up to now, we have studied two special types of surfaces necessary to understanding vector calculus and the calculus of space, namely spheres and planes in space. In this section, we extend our inventory to include a variety of cylinders and quadric surfaces. Quadric surfaces are surfaces defined by second-degree equations in x, y, and z. Spheres are quadric surfaces, but there are others of equal interest.

Cylinders

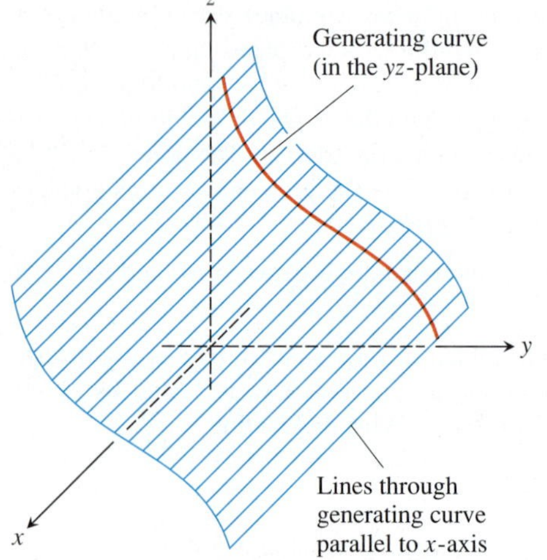

FIGURE 10.25 A cylinder and generating curve.

A **cylinder** is the surface composed of all the lines that (1) lie parallel to a given line in space and (2) pass through a given plane curve. The curve is a **generating curve** for the cylinder (Figure 10.25). In solid geometry, where *cylinder* means *circular cylinder*, the generating curves are circles, but now we allow generating curves of any kind. The cylinder in our first example is generated by a parabola.

When graphing a cylinder or other surface by hand or analyzing one generated by a computer, it helps to look at the curves formed by intersecting the surface with planes parallel to the coordinate planes. These curves are called **cross sections** or **traces.**

Example 1 The Parabolic Cylinder $y = x^2$

Find an equation for the cylinder made by the lines parallel to the z-axis that pass through the parabola $y = x^2$, $z = 0$ (Figure 10.26).

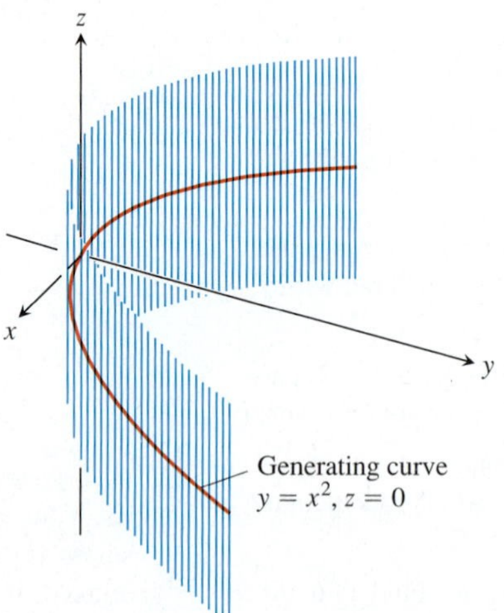

FIGURE 10.26 The cylinder of lines passing through the parabola $y = x^2$ in the xy-plane parallel to the z-axis. (Example 1)

Solution Suppose that the point $P_0(x_0, x_0^2, 0)$ lies on the parabola $y = x^2$ in the xy-plane. Then, for any value of z, the point $Q(x_0, x_0^2, z)$ will lie on the cylinder because it lies on the line $x = x_0, y = x_0^2$ through P_0 parallel to the z-axis. Conversely, any point $Q(x_0, x_0^2, z)$ whose y-coordinate is the square of its x-coordinate lies on the cylinder because it lies on the line $x = x_0, y = x_0^2$ through P_0 parallel to the z-axis (Figure 10.27).

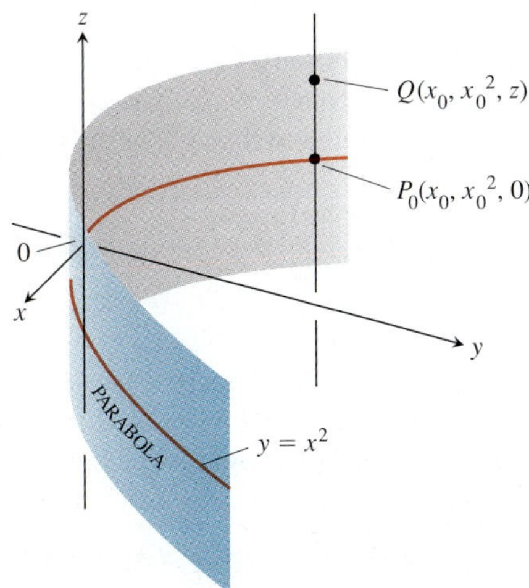

FIGURE 10.27 Every point of the cylinder in Figure 10.26 has coordinates of the form (x_0, x_0^2, z). We call the cylinder "the cylinder $y = x^2$."

Regardless of the value of z, therefore, the points on the surface are the points whose coordinates satisfy the equation $y = x^2$. This makes $y = x^2$ an equation for the cylinder. Because of this, we call the cylinder "the cylinder $y = x^2$."

As Example 1 suggests, any curve $f(x, y) = c$ in the xy-plane defines a cylinder parallel to the z-axis whose equation is also $f(x, y) = c$. The equation $x^2 + y^2 = 1$ defines the circular cylinder made by the lines parallel to the z-axis that pass through the circle $x^2 + y^2 = 1$ in the xy-plane. The equation $x^2 + 4y^2 = 9$ defines the elliptical cylinder made by the lines parallel to the z-axis that pass through the ellipse $x^2 + 4y^2 = 9$ in the xy-plane.

In a similar way, any curve $g(x, z) = c$ in the xz-plane defines a cylinder parallel to the y-axis whose space equation is also $g(x, z) = c$ (Figure 10.28). Any curve $h(y, z) = c$ defines a cylinder parallel to the x-axis whose space equation is also $h(y, z) = c$ (Figure 10.29).

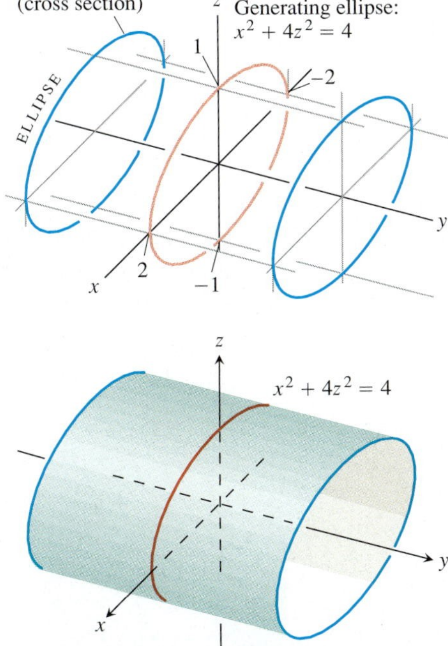

FIGURE 10.28 The elliptic cylinder $x^2 + 4z^2 = 4$ is made of lines parallel to the y-axis and passing through the ellipse $x^2 + 4z^2 = 4$ in the xz-plane. The cross sections or "traces" of the cylinder in planes perpendicular to the y-axis are ellipses congruent to the generating ellipse. The cylinder extends along the entire y-axis.

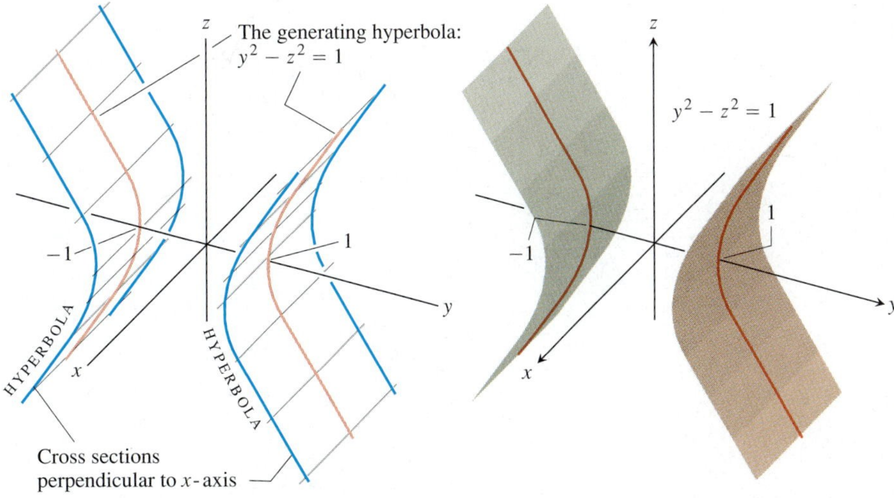

FIGURE 10.29 The hyperbolic cylinder $y^2 - z^2 = 1$ is made of lines parallel to the x-axis and passing through the hyperbola $y^2 - z^2 = 1$ in the yz-plane. The cross sections of the cylinder in planes perpendicular to the x-axis are hyperbolas congruent to the generating hyperbola.

> **Equation of a Cylinder**
> An equation in any two of the three Cartesian coordinates defines a cylinder parallel to the axis of the third coordinate.

The axis of a cylinder need not be parallel to a coordinate axis, however.

Quadric Surfaces

The next type of surface we study is a *quadric* surface. These surfaces are the three-dimensional analogues of ellipses, parabolas, and hyperbolas.

A **quadric surface** is the graph in space of a second-degree equation in x, y, and z. The most general form is

$$Ax^2 + By^2 + Cz^2 + Dxy + Eyz + Fxz + Gx + Hy + Jz + K = 0,$$

where A, B, C, and so on are constants, but the equation can be simplified by translation and rotation, as in the two-dimensional case. We will study only the simpler equations. Although the definition did not require it, the cylinders in Figures 10.27 through 10.29 were also examples of quadric surfaces. The basic quadric surfaces are **ellipsoids, paraboloids, elliptic cones,** and **hyperboloids.** (We can think of spheres as special ellipsoids.) We now present examples of each type.

Example 2 Graphing Ellipsoids

The **ellipsoid**

$$\frac{x^2}{a^2} + \frac{y^2}{b^2} + \frac{z^2}{c^2} = 1 \tag{1}$$

(Figure 10.30) cuts the coordinate axes at $(\pm a, 0, 0)$, $(0, \pm b, 0)$, and $(0, 0, \pm c)$. It lies within the rectangular box defined by the inequalities $|x| \leq a$, $|y| \leq b$, and $|z| \leq c$. The surface is symmetric with respect to each of the coordinate planes because the variables in the defining equation are squared.

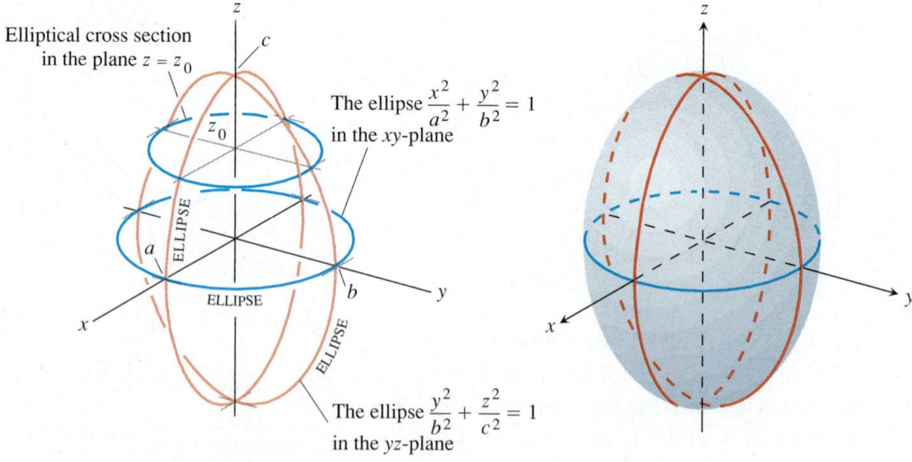

FIGURE 10.30 The ellipsoid

$$\frac{x^2}{a^2} + \frac{y^2}{b^2} + \frac{z^2}{c^2} = 1$$

in Example 2.

The curves in which the three coordinate planes cut the surface are ellipses. For example,

$$\frac{x^2}{a^2} + \frac{y^2}{b^2} = 1 \qquad \text{when} \qquad z = 0.$$

The section cut from the surface by the plane $z = z_0, |z_0| < c$, is the ellipse

$$\frac{x^2}{a^2(1 - (z_0/c)^2)} + \frac{y^2}{b^2(1 - (z_0/c)^2)} = 1.$$

If any two of the semiaxes a, b, and c are equal, the surface is an **ellipsoid of revolution.** If all three are equal, the surface is a sphere.

Example 3 Graphing Paraboloids

The **elliptic paraboloid**

$$\frac{x^2}{a^2} + \frac{y^2}{b^2} = \frac{z}{c} \tag{2}$$

is symmetric with respect to the planes $x = 0$ and $y = 0$ (Figure 10.31). The only intercept on the axes is the origin. Except for this point, the surface lies above (if $c > 0$) or entirely below (if $c < 0$) the xy-plane, depending on the sign of c. The sections cut by the coordinate planes are

$x = 0$: the parabola $z = \dfrac{c}{b^2} y^2$

$y = 0$: the parabola $z = \dfrac{c}{a^2} x^2$

$z = 0$: the point $(0, 0, 0)$.

Each plane $z = z_0$ above the xy-plane cuts the surface in the ellipse

$$\frac{x^2}{a^2} + \frac{y^2}{b^2} = \frac{z_0}{c}.$$

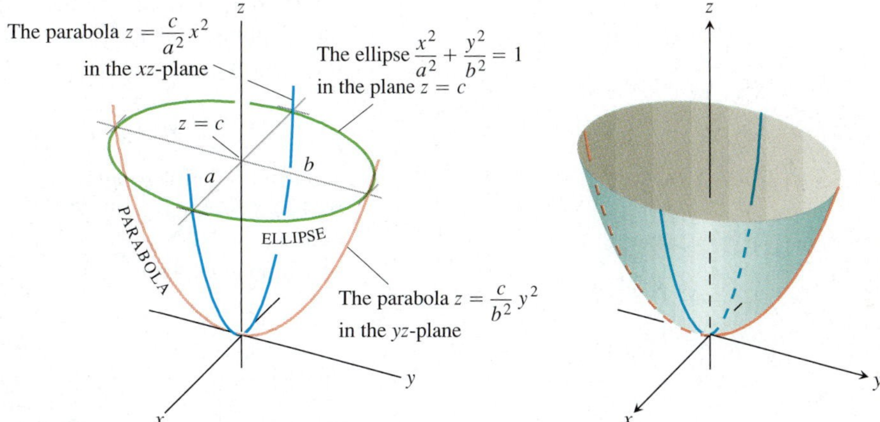

FIGURE 10.31 The elliptic paraboloid $(x^2/a^2) + (y^2/b^2) = z/c$ in Example 3, shown for $c > 0$. The cross sections perpendicular to the z-axis above the xy-plane are ellipses. The cross sections in the planes that contain the z-axis are parabolas.

Example 4 Graphing Cones

The **elliptic cone**

$$\frac{x^2}{a^2} + \frac{y^2}{b^2} = \frac{z^2}{c^2} \tag{3}$$

is symmetric with respect to the three coordinate planes (Figure 10.32). The sections cut by the coordinate planes are

$$x = 0: \quad \text{the lines } z = \pm \frac{c}{b} y$$

$$y = 0: \quad \text{the lines } z = \pm \frac{c}{a} x$$

$$z = 0: \quad \text{the point } (0, 0, 0).$$

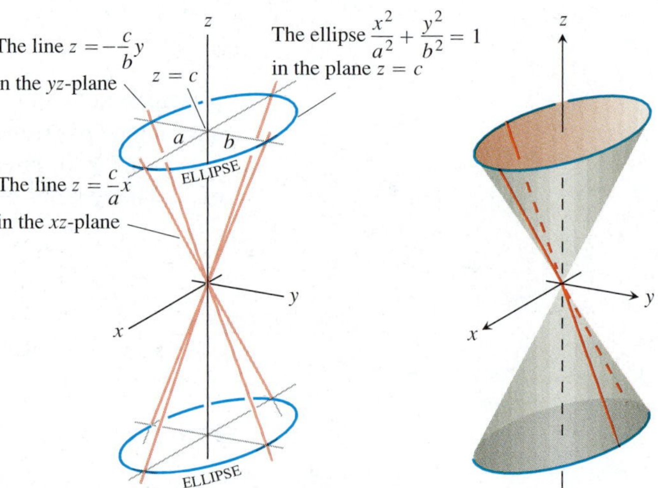

FIGURE 10.32 The elliptic cone $(x^2/a^2) + (y^2/b^2) = (z^2/c^2)$ in Example 4. Planes perpendicular to the z-axis cut the cone in ellipses above and below the xy-plane. Vertical planes that contain the z-axis cut it in pairs of intersecting lines.

The sections cut by planes $z = z_0$ above and below the xy-plane are ellipses whose centers lie on the z-axis and whose vertices lie on the lines given above.

If $a = b$, the cone is a right circular cone.

Example 5 Graphing Hyperboloids

The **hyperboloid of one sheet**

$$\frac{x^2}{a^2} + \frac{y^2}{b^2} - \frac{z^2}{c^2} = 1 \tag{4}$$

is symmetric with respect to each of the three coordinate planes (Figure 10.33).

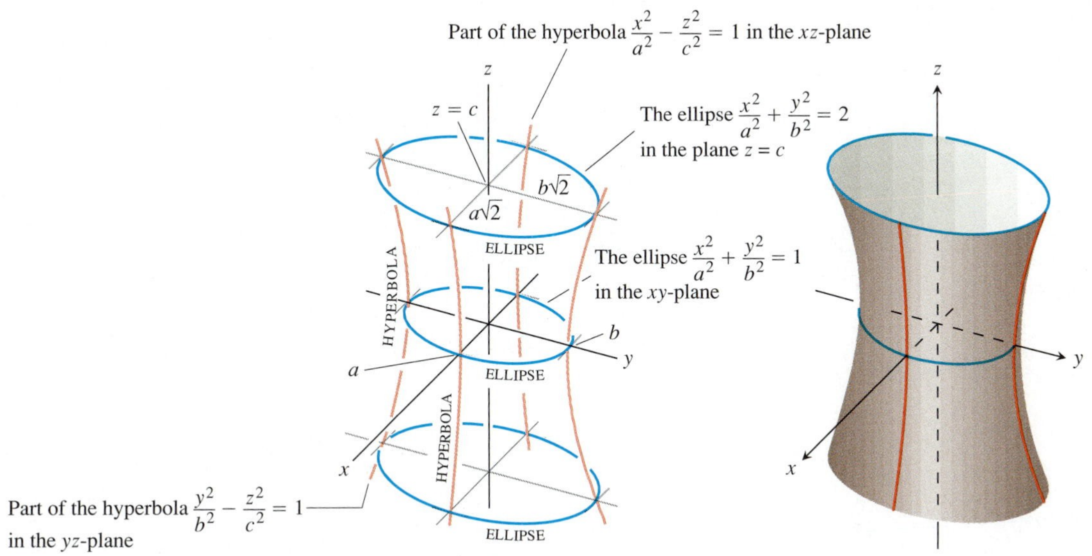

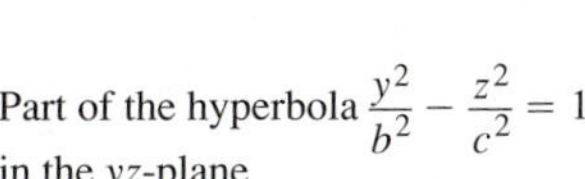

FIGURE 10.33 The hyperboloid $(x^2/a^2) + (y^2/b^2) - (z^2/c^2) = 1$ in Example 5. Planes perpendicular to the z-axis cut it in ellipses. Vertical planes containing the z-axis cut it in hyperbolas.

The sections cut out by the coordinate planes are

$$x = 0: \quad \text{the hyperbola } \frac{y^2}{b^2} - \frac{z^2}{c^2} = 1$$

$$y = 0: \quad \text{the hyperbola } \frac{x^2}{a^2} - \frac{z^2}{c^2} = 1$$

$$z = 0: \quad \text{the ellipse } \frac{x^2}{a^2} + \frac{y^2}{b^2} = 1.$$

The plane $z = z_0$ cuts the surface in an ellipse with center on the z-axis and vertices on one of the hyperbolic sections above.

The surface is connected, meaning that it is possible to travel from one point on it to any other without leaving the surface. For this reason, it is said to have *one* sheet, in contrast to the hyperboloid in the next example, which has two sheets.

If $a = b$, the hyperboloid is a surface of revolution.

Example 6 Graphing Hyperboloids

The **hyperboloid of two sheets**

$$\frac{z^2}{c^2} - \frac{x^2}{a^2} - \frac{y^2}{b^2} = 1 \tag{5}$$

is symmetric with respect to the three coordinate planes (Figure 10.34). The plane $z = 0$ does not intersect the surface; in fact, for a horizontal plane to intersect the surface, we must have $|z| \geq c$. The hyperbolic sections

$$x = 0: \quad \frac{z^2}{c^2} - \frac{y^2}{b^2} = 1$$

$$y = 0: \quad \frac{z^2}{c^2} - \frac{x^2}{a^2} = 1$$

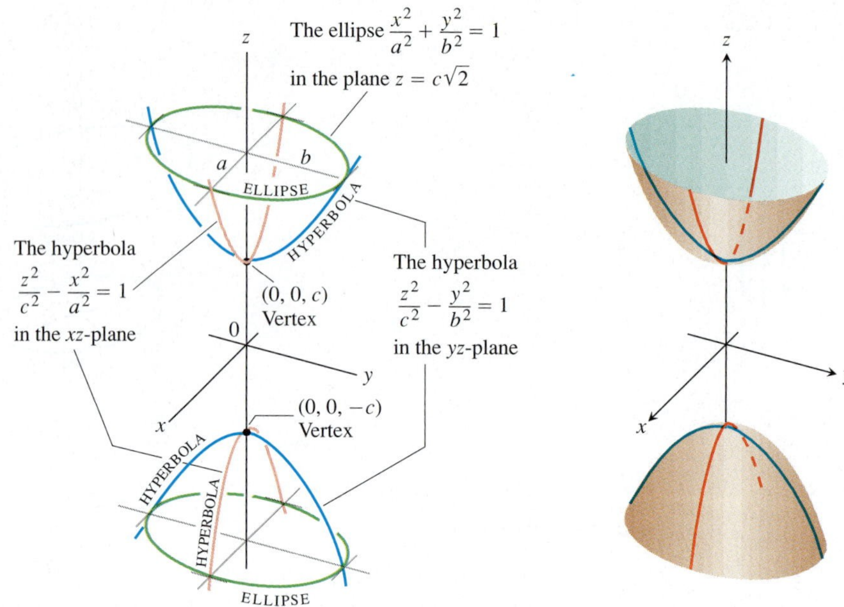

FIGURE 10.34 The hyperboloid $(z^2/c^2) - (x^2/a^2) - (y^2/b^2) = 1$ in Example 6. Planes perpendicular to the z-axis above and below the vertices cut it in ellipses. Vertical planes containing the z-axis cut it in hyperbolas.

have their vertices and foci on the z-axis. The surface is separated into two portions, one above the plane $z = c$ and the other below the plane $z = -c$. This accounts for its name.

Equations (4) and (5) have different numbers of negative terms. The number in each case is the same as the number of sheets of the hyperboloid. If we replace the 1 on the right side of either Equation (4) or Equation (5) by 0, we obtain the equation

$$\frac{x^2}{a^2} + \frac{y^2}{b^2} = \frac{z^2}{c^2}$$

for an elliptic cone (Equation (3)). The hyperboloids are asymptotic to this cone (Figure 10.35) in the same way that the hyperbolas

$$\frac{x^2}{a^2} - \frac{y^2}{b^2} = \pm 1$$

are asymptotic to the lines

$$\frac{x^2}{a^2} - \frac{y^2}{b^2} = 0$$

in the xy-plane.

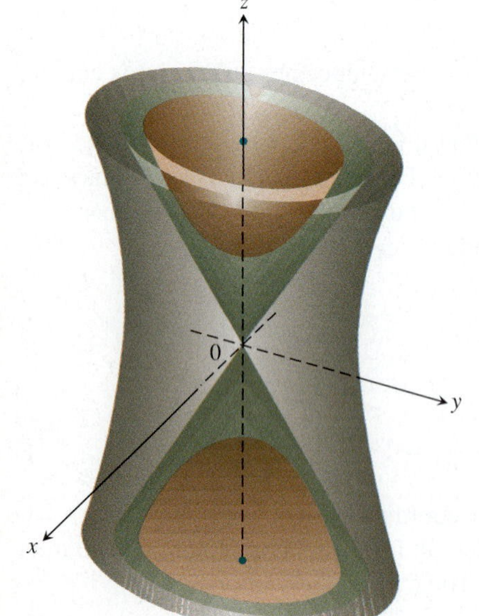

FIGURE 10.35 Both hyperboloids are asymptotic to the cone. (Example 6)

Example 7 Graphing a Saddle

The **hyperbolic paraboloid**

$$\frac{y^2}{b^2} - \frac{x^2}{a^2} = \frac{z}{c}, \qquad c > 0 \tag{6}$$

has symmetry with respect to the planes $x = 0$ and $y = 0$ (Figure 10.36). The sections in these planes are

$$x = 0: \quad \text{the parabola } z = \frac{c}{b^2} y^2 \tag{7}$$

$$y = 0: \quad \text{the parabola } z = -\frac{c}{a^2} x^2. \tag{8}$$

In the plane $x = 0$, the parabola opens upward from the origin. The parabola in the plane $y = 0$ opens downward.

If we cut the surface by a plane $z = z_0 > 0$, the section is a hyperbola,

$$\frac{y^2}{b^2} - \frac{x^2}{a^2} = \frac{z_0}{c},$$

with its focal axis parallel to the y-axis and its vertices on the parabola in Equation (7). If z_0 is negative, the focal axis is parallel to the x-axis and the vertices lie on the parabola in Equation (8).

Near the origin, the surface is shaped like a saddle. To a person traveling along the surface in the yz-plane, the origin looks like a minimum. To a person traveling in the xz-plane, the origin looks like a maximum. Such a point is called a **minimax** or **saddle point** of a surface.

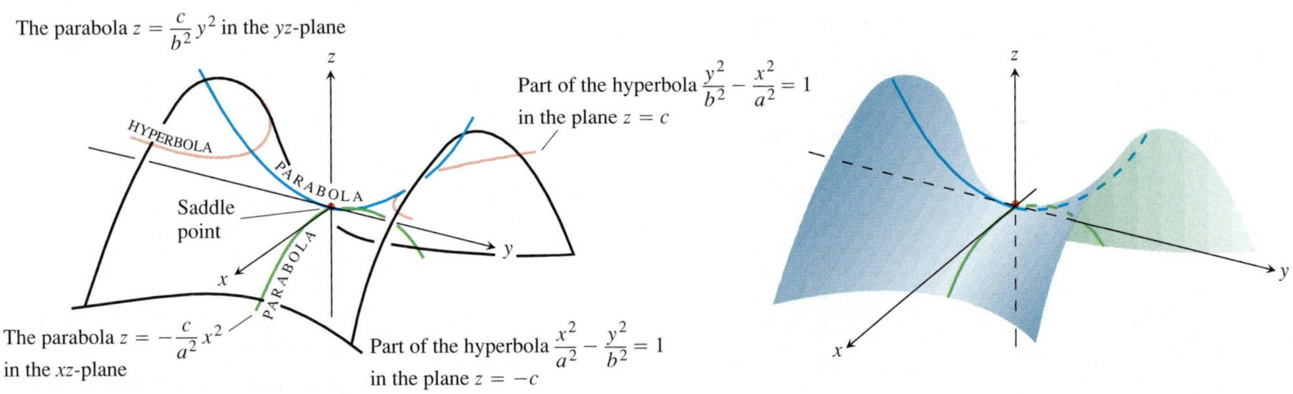

FIGURE 10.36 The hyperbolic paraboloid $(y^2/b^2) - (x^2/a^2) = z/c$, $c > 0$. The cross sections in planes perpendicular to the z-axis above and below the xy-plane are hyperbolas. The cross sections in planes perpendicular to the other axes are parabolas.

USING TECHNOLOGY

Visualizing in space A computer algebra system (CAS) or other computer graphing utility can help in visualizing surfaces in space. It can draw traces in different planes with far more patience than most people can muster. Many computer graphing systems can rotate a figure so you can see it as if it were a physical model you could turn in your hand. Hidden-line algorithms (see Exercise 62, Section 10.3) are used to block out portions of the surface that you would not see from your current viewing angle. Often a CAS will require surfaces to be entered in parametric form, as discussed in Section 13.6 (see also CAS Exercises 57 through 60 in Section 11.1). Sometimes you may have to manipulate the grid mesh to see all portions of a surface.

EXERCISES 10.4

Matching Equations with Surfaces

In Exercises 1–12, match the equation with the surface it defines. Also, identify each surface by type (paraboloid, ellipsoid, etc.). The surfaces are labeled (a) through (l).

1. $x^2 + y^2 + 4z^2 = 10$

2. $z^2 + 4y^2 - 4x^2 = 4$

3. $9y^2 + z^2 = 16$

4. $y^2 + z^2 = x^2$

5. $x = y^2 - z^2$

6. $x = -y^2 - z^2$

7. $x^2 + 2z^2 = 8$

8. $z^2 + x^2 - y^2 = 1$

9. $x = z^2 - y^2$

10. $z = -4x^2 - y^2$

11. $x^2 + 4z^2 = y^2$

12. $9x^2 + 4y^2 + 2z^2 = 36$

(a)

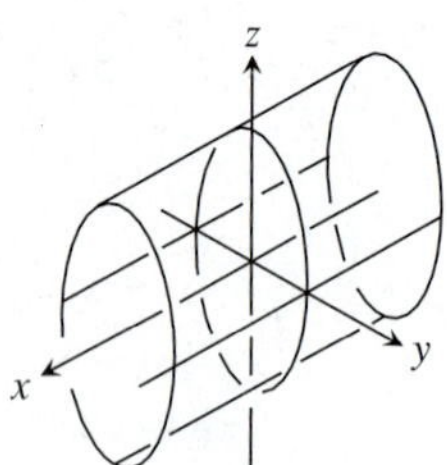

(b)

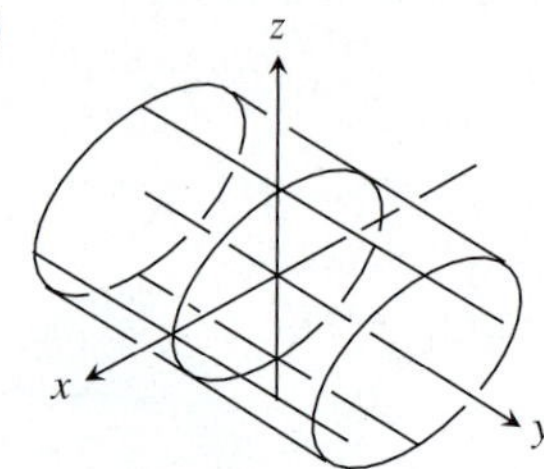

(c)

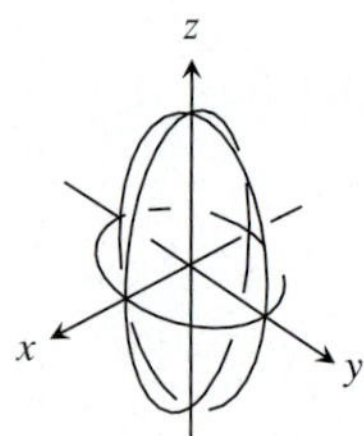

(d)

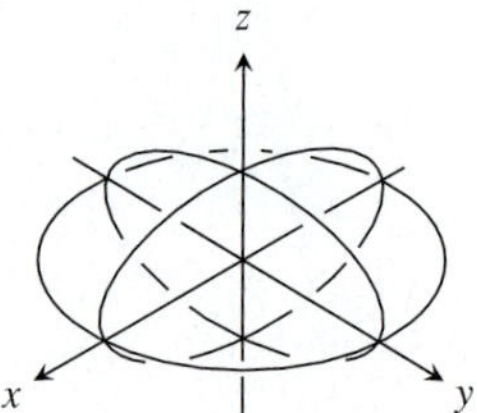

(e)

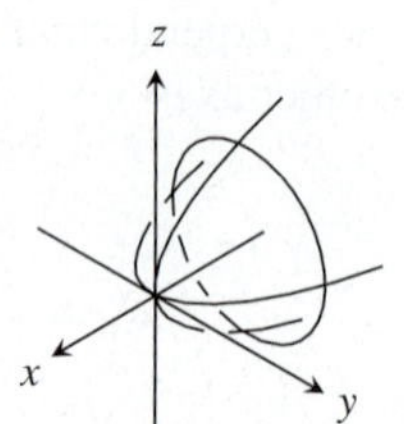

(f)

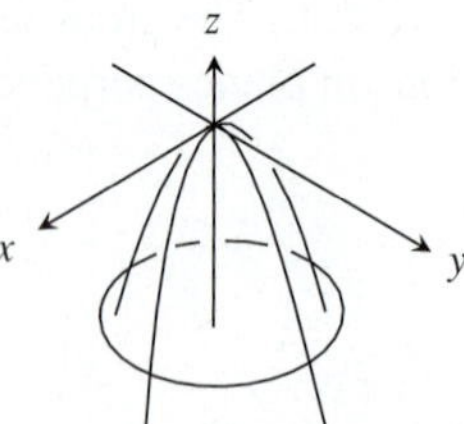

(g)

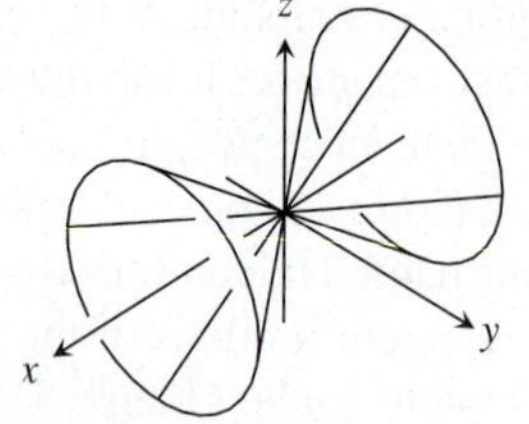

(h)

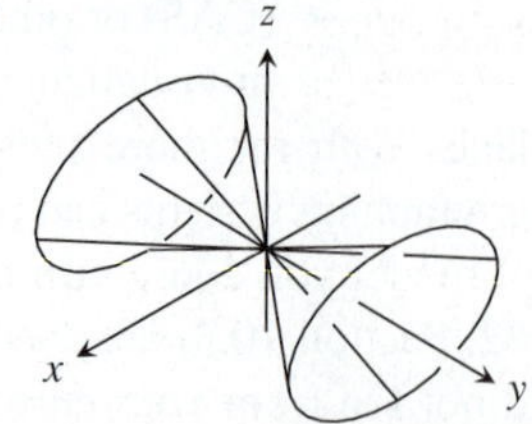

(i)

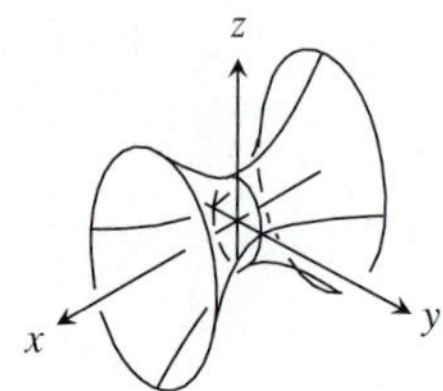

(j)

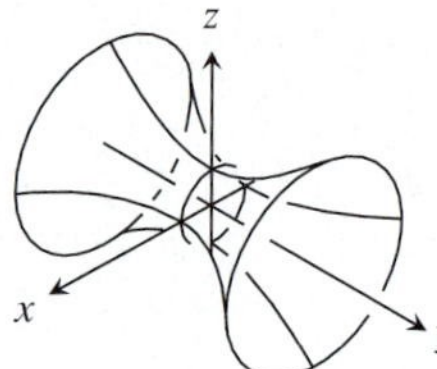

(k)

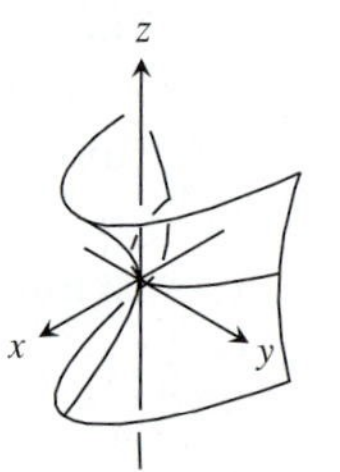

(l)

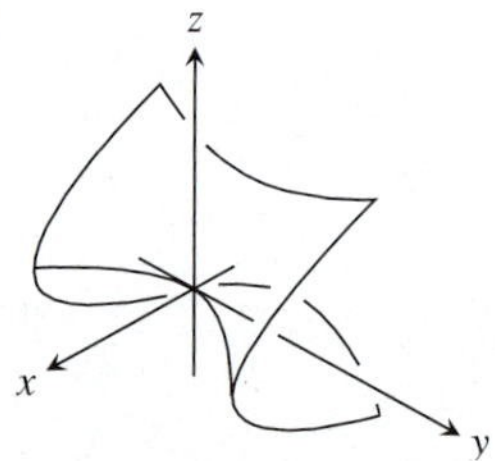

Theory and Examples

13. *Area and volume* Express the area A of the cross section cut from the ellipsoid

$$x^2 + \frac{y^2}{4} + \frac{z^2}{9} = 1$$

by the plane $z = c$ as a function of c. (The area of an ellipse with semiaxes a and b is πab.)

(b) Use slices perpendicular to the z-axis to find the volume of the ellipsoid in part (a).

(c) Now find the volume of the ellipsoid

$$\frac{x^2}{a^2} + \frac{y^2}{b^2} + \frac{z^2}{c^2} = 1.$$

Does your formula give the volume of a sphere of radius a if $a = b = c$?

14. *Volume of a barrel* The barrel shown here is shaped like an ellipsoid with equal pieces cut from the ends by planes perpendicular to the z-axis. The cross sections perpendicular to the z-axis are circular. The barrel is $2h$ units high, its midsection radius is R, and its end radii are both r. Find a formula for the barrel's volume. Then check two things. First, suppose that the sides of the barrel are straightened to turn the barrel into a cylinder of radius R and height $2h$. Does your formula give the cylinder's volume? Second, suppose that $r = 0$ and $h = R$

so the barrel is a sphere. Does your formula give the sphere's volume?

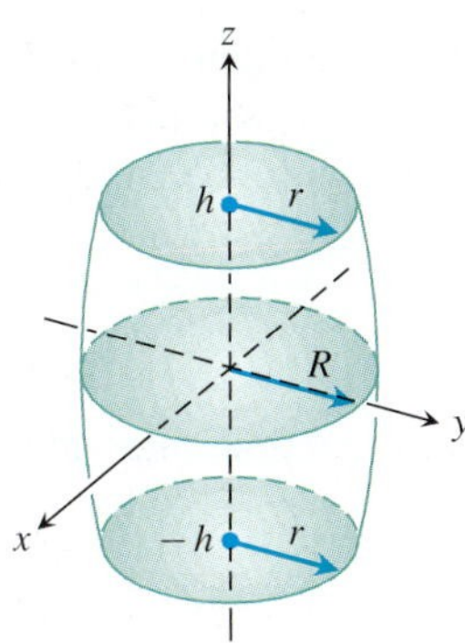

15. *Volume of paraboloid* Show that the volume of the segment cut from the paraboloid

$$\frac{x^2}{a^2} + \frac{y^2}{b^2} = \frac{z}{c}$$

by the plane $z = h$ equals half the segment's base times its altitude. (Figure 10.31 shows the segment for the special case $h = c$.)

16. *Volume of hyperboloid*

(a) Find the volume of the solid bounded by the hyperboloid

$$\frac{x^2}{a^2} + \frac{y^2}{b^2} - \frac{z^2}{c^2} = 1$$

and the planes $z = 0$ and $z = h$, $h > 0$.

(b) Express your answer in part (a) in terms of h and the areas A_0 and A_h of the regions cut by the hyperboloid from the planes $z = 0$ and $z = h$.

(c) Show that the volume in part (a) is also given by the formula

$$V = \frac{h}{6}(A_0 + 4A_m + A_h),$$

where A_m is the area of the region cut by the hyperboloid from the plane $z = h/2$.

Graphing Surfaces

T Plot the surfaces in Exercises 17–20 over the indicated domains. If you can, rotate the surface into different viewing positions.

17. $z = y^2, \quad -2 \le x \le 2, \quad -0.5 \le y \le 2$

18. $z = 1 - y^2, \quad -2 \le x \le 2, \quad -2 \le y \le 2$

19. $z = x^2 + y^2, \quad -3 \le x \le 3, \quad -3 \le y \le 3$

20. $z = x^2 + 2y^2$ over

(a) $-3 \le x \le 3, \quad -3 \le y \le 3$

(b) $-1 \le x \le 1, \quad -2 \le y \le 3$

(c) $-2 \le x \le 2, \quad -2 \le y \le 2$

(d) $-2 \le x \le 2, \quad -1 \le y \le 1$

COMPUTER EXPLORATIONS

Surface Plots

CD-ROM WEBsite

Use a CAS to plot the surfaces in Exercises 21–26. Identify the type of quadric surface from your graph.

21. $\dfrac{x^2}{9} + \dfrac{y^2}{36} = 1 - \dfrac{z^2}{25}$

22. $\dfrac{x^2}{9} - \dfrac{z^2}{9} = 1 - \dfrac{y^2}{16}$

23. $5x^2 = z^2 - 3y^2$

24. $\dfrac{y^2}{16} = 1 - \dfrac{x^2}{9} + z$

25. $\dfrac{x^2}{9} - 1 = \dfrac{y^2}{16} + \dfrac{z^2}{2}$

26. $y - \sqrt{4 - z^2} = 0$

10.5 Vector-Valued Functions and Space Curves

Space Curves • Limits and Continuity • Derivatives and Motion • Differentiation Rules • Vector Functions of Constant Length • Integrals of Vector Functions

Just as we did for planar curves in Section 9.3, to track a particle moving in space, we run a vector **r** from the origin to the particle (Figure 10.37) and study the

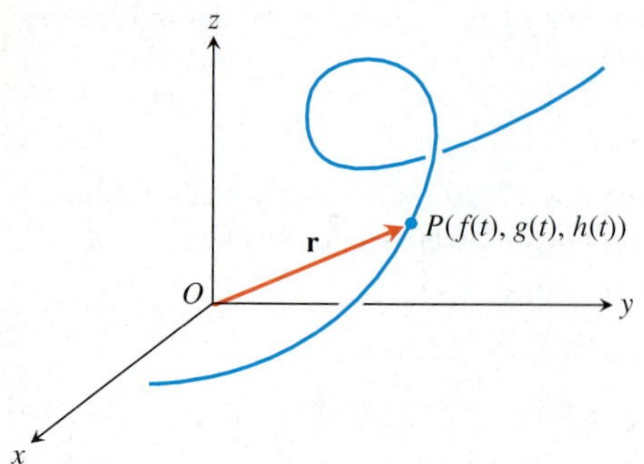

FIGURE 10.37 The position vector $\mathbf{r} = \overrightarrow{OP}$ of a particle moving through space is a function of time.

changes in $\mathbf{r}$. If the particle's position coordinates are twice-differentiable functions of time, then so is $\mathbf{r}$, and we can find the particle's velocity and acceleration vectors at any time by differentiating $\mathbf{r}$. Conversely, if we know either the particle's velocity vector or acceleration vector as a continuous function of time and if we have enough information about the particle's initial velocity and position, we can find $\mathbf{r}$ as a function of time by integration. We study space curves in the remainder of this chapter.

Space Curves

When a particle moves through space during a time interval I, we think of the particle's coordinates as functions defined on I:

$$x = f(t), \qquad y = g(t), \qquad z = h(t), \qquad t \in I. \tag{1}$$

The points $(x, y, z) = (f(t), g(t), h(t))$, $t \in I$, make up the **curve** in space that we call the particle's **path.** The equations and interval in Equation (1) **parametrize** the curve. A curve in space can also be represented in vector form. The vector

$$\mathbf{r}(t) = \overrightarrow{OP} = f(t)\mathbf{i} + g(t)\mathbf{j} + h(t)\mathbf{k} \tag{2}$$

from the origin to the particle's **position** $P(f(t), g(t), h(t))$ at time t is the particle's **position vector.** The functions f, g, and h are the **component functions (components)** of the position vector. We think of the particle's path as the **curve traced by r** during the time interval I. Figure 10.38 displays several space curves generated by a computer graphing program. It would not be easy to plot these curves by hand.

CD-ROM WEBsite

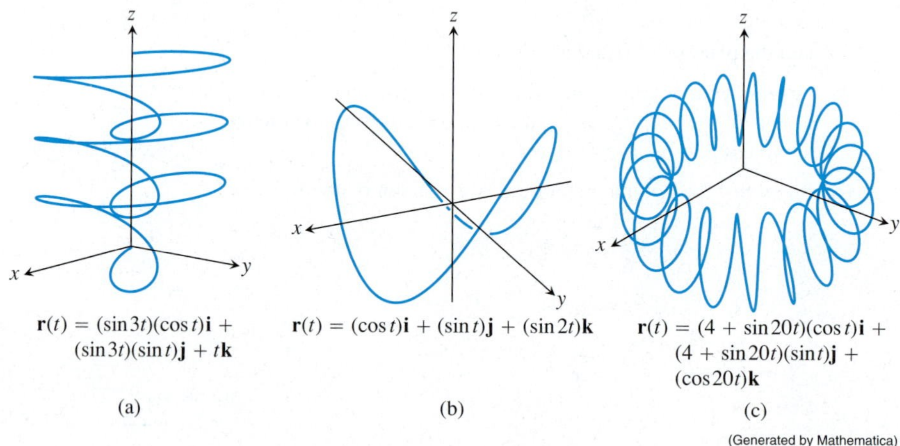

FIGURE 10.38 Computer-generated space curves are defined by the position vectors $\mathbf{r}(t)$.

Equation (2) defines $\mathbf{r}$ as a vector function of the real variable t on the interval I. More generally, a **vector function** or **vector-valued function** on a domain set D is a rule that assigns a vector in space to each element in D. For now, the domains will be intervals of real numbers resulting in a space curve. Later, in Chapter 13, the domains will be regions in the plane. Vector functions will then represent surfaces in space. Vector functions on a domain in the plane or space also give rise to "vector fields," which are important to the study of the flow of a fluid, gravitational fields, and electromagnetic phenomena. We investigate vector fields and their applications in Chapter 13.

As in Chapter 9, we refer to real-valued functions as **scalar functions** to distinguish them from vector functions. The components of $\mathbf{r}$ are scalar functions of t. When we define a vector-valued function by giving its component functions, we assume the vector function's domain to be the common domain of the components. You should find this material to be very similar to the material on planar curves in Section 9.3.

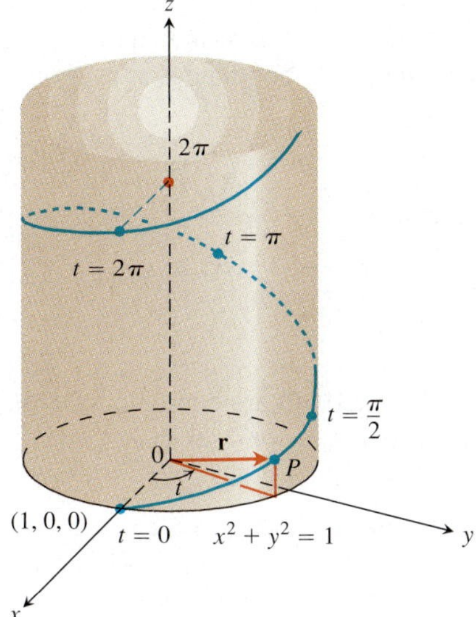

FIGURE 10.39 The upper half of the helix $\mathbf{r}(t) = (\cos t)\mathbf{i} + (\sin t)\mathbf{j} + t\mathbf{k}$. (Example 1)

Example 1 Graphing a Helix

Graph the vector function

$$\mathbf{r}(t) = (\cos t)\mathbf{i} + (\sin t)\mathbf{j} + t\mathbf{k}.$$

Solution

The vector function

$$\mathbf{r}(t) = (\cos t)\mathbf{i} + (\sin t)\mathbf{j} + t\mathbf{k}$$

is defined for all real values of t. The curve traced by $\mathbf{r}$ is a helix (from an old Greek word for "spiral") that winds around the circular cylinder $x^2 + y^2 = 1$ (Figure 10.39). The curve lies on the cylinder because the **i**- and **j**-components of $\mathbf{r}$, being the x- and y-coordinates of the tip of $\mathbf{r}$, satisfy the cylinder's equation:

$$x^2 + y^2 = (\cos t)^2 + (\sin t)^2 = 1.$$

The curve rises as the **k**-component $z = t$ increases. Each time t increases by 2π, the curve completes one turn around the cylinder. The equations

$$x = \cos t, \qquad y = \sin t, \qquad z = t$$

parametrize the helix, the interval $-\infty < t < \infty$ being understood. You will find more helices in Figure 10.40.

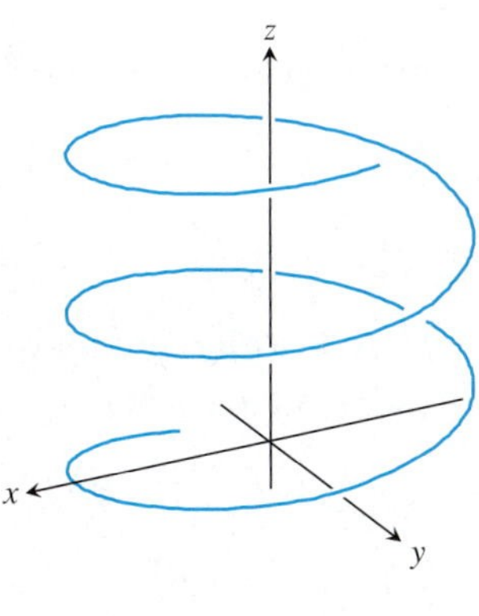

$\mathbf{r}(t) = (\cos t)\mathbf{i} + (\sin t)\mathbf{j} + t\mathbf{k}$

(Generated by Mathematica)

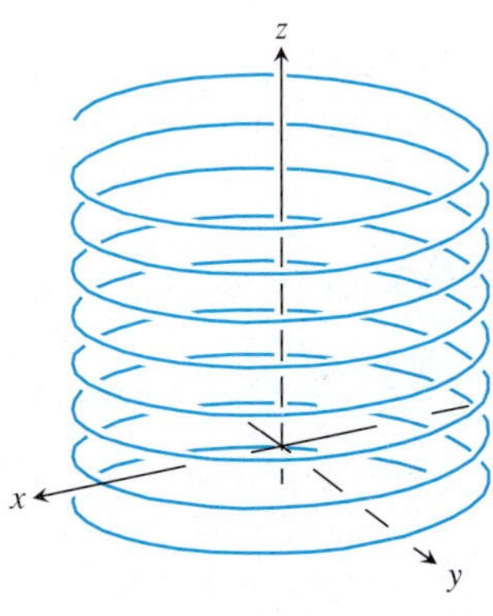

$\mathbf{r}(t) = (\cos t)\mathbf{i} + (\sin t)\mathbf{j} + 0.3t\mathbf{k}$

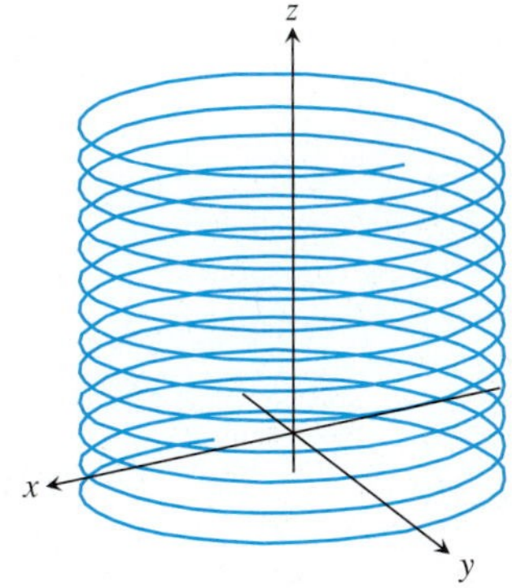

$\mathbf{r}(t) = (\cos 5t)\mathbf{i} + (\sin 5t)\mathbf{j} + t\mathbf{k}$

FIGURE 10.40 Helices drawn by computer.

Limits and Continuity

We define limits and continuity of vector-valued functions for space the same way we define limits of vector-valued functions for the plane.

Definition Limit and Continuity

If $\mathbf{r}(t) = f(t)\mathbf{i} + g(t)\mathbf{j} + h(t)\mathbf{k}$, then

$$\lim_{t\to t_0} \mathbf{r}(t) = \left(\lim_{t\to t_0} f(t)\right)\mathbf{i} + \left(\lim_{t\to t_0} g(t)\right)\mathbf{j} + \left(\lim_{t\to t_0} h(t)\right)\mathbf{k}. \tag{3}$$

A vector function $\mathbf{r}(t)$ is **continuous at a point** $t = t_0$ in its domain if $\lim_{t\to t_0} \mathbf{r}(t) = \mathbf{r}(t_0)$. The function is **continuous** if it is continuous at every point in its domain.

From Equation (3), we see that $\mathbf{r}(t)$ is continuous at $t = t_0$ if and only if each component function is continuous there.

Example 2 Continuity of Space Curves

All the space curves shown in Figures 10.38 and 10.40 are continuous because their component functions are continuous at every value of t in $(-\infty, \infty)$.

Example 3 Finding Limits of Vector Functions

If $\mathbf{r}(t) = (\cos t)\mathbf{i} + (\sin t)\mathbf{j} + t\mathbf{k}$, then

$$\begin{aligned}\lim_{t\to\pi/4} \mathbf{r}(t) &= \left(\lim_{t\to\pi/4} \cos t\right)\mathbf{i} + \left(\lim_{t\to\pi/4} \sin t\right)\mathbf{j} + \left(\lim_{t\to\pi/4} t\right)\mathbf{k} \\ &= \frac{\sqrt{2}}{2}\mathbf{i} + \frac{\sqrt{2}}{2}\mathbf{j} + \frac{\pi}{4}\mathbf{k}.\end{aligned}$$

Derivatives and Motion

The derivative of a vector function in space is defined in the same way as for planar functions, but with one more component.

Definition Derivative at a Point

The vector function $\mathbf{r}(t) = f(t)\mathbf{i} + g(t)\mathbf{j} + h(t)\mathbf{k}$ is **differentiable at** $t = t_0$ if f, g, and h are differentiable at t_0. The **derivative** is the vector

$$\mathbf{r}'(t) = \frac{d\mathbf{r}}{dt} = \lim_{\Delta t\to 0} \frac{\mathbf{r}(t + \Delta t) - \mathbf{r}(t)}{\Delta t} = \frac{df}{dt}\mathbf{i} + \frac{dg}{dt}\mathbf{j} + \frac{dh}{dt}\mathbf{k}.$$

A vector function $\mathbf{r}$ is **differentiable** if it is differentiable at every point of its domain. The curve traced by $\mathbf{r}$ is **smooth** if $\mathbf{r}'(t)$ is continuous and never $\mathbf{0}$, that is, if f, g, and h have continuous first derivatives that are not simultaneously 0.

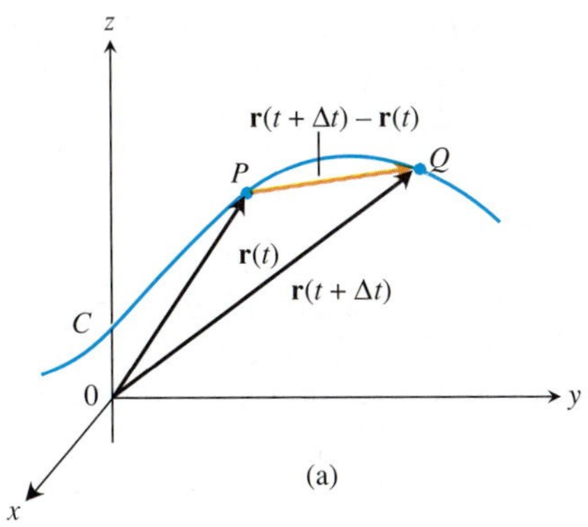

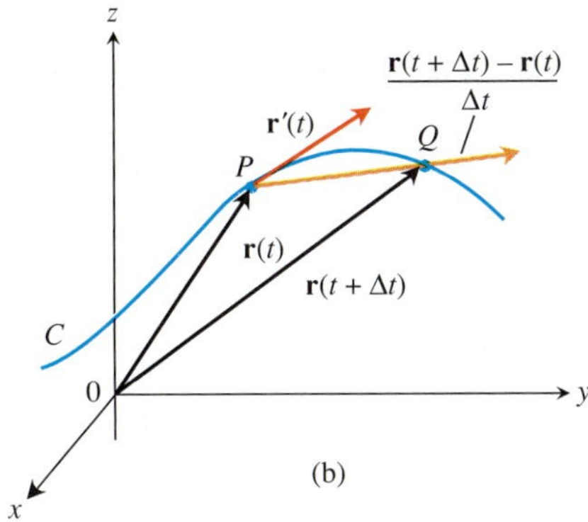

FIGURE 10.41 As $\Delta t \to 0$, the point Q approaches the point P along the curve C, and the vector $\overrightarrow{PQ}/\Delta t$ becomes the tangent vector $\mathbf{r}'(t)$ in the limit.

The geometric significance of the definition of derivative is the same as for planar curves and is shown in Figure 10.41. The points P and Q have position vectors $\mathbf{r}(t)$ and $\mathbf{r}(t + \Delta t)$, and the vector $\overrightarrow{PQ}$ is represented by $\mathbf{r}(t + \Delta t) - \mathbf{r}(t)$. For $\Delta t > 0$, the scalar multiple $(1/\Delta t)(\mathbf{r}(t + \Delta t) - \mathbf{r}(t))$ points in the same direction as the vector $\overrightarrow{PQ}$. As $\Delta t \to 0$, this vector approaches a vector that is tangent to the curve at P (Figure 10.41b). We define $\mathbf{r}'(t)$, when different from $\mathbf{0}$, to be the vector **tangent** to the curve at P. The **tangent line** to the curve at a point $(f(t_0), g(t_0), h(t_0))$ is defined to be the line through the point parallel to $\mathbf{r}'(t_0)$. We require $d\mathbf{r}/dt \neq \mathbf{0}$ for a smooth curve to make sure the curve has a continuously turning tangent at each point. On a smooth curve, there are no sharp corners or cusps.

A curve that is made up of a finite number of smooth curves pieced together in a continuous fashion is called **piecewise smooth** (Figure 10.42).

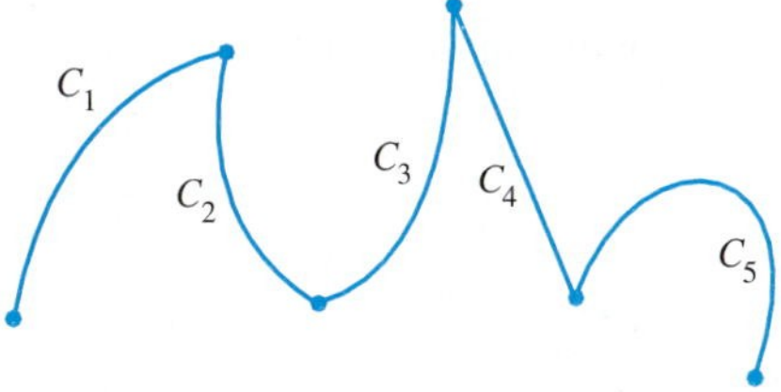

FIGURE 10.42 A piecewise smooth curve made up of five smooth curves connected end to end in continuous fashion.

As we found for vector functions in the plane, when $d\mathbf{r}/dt$ is different from $\mathbf{0}$, the derivative models a particle's velocity as it moves along the space curve defined by $\mathbf{r}$. The derivative points in the direction of motion and gives the rate of change of position with respect to time. For a smooth curve, the velocity is never zero; the particle does not stop or reverse direction.

CD-ROM WEBsite

Definitions Velocity, Speed, Acceleration, Direction of Motion

If $\mathbf{r}$ is the position vector of a particle moving along a smooth curve in space, then at any time t, the following definitions apply.

1. $\mathbf{v}(t) = \dfrac{d\mathbf{r}}{dt}$, the derivative of position, is the particle's **velocity vector** and is tangent to the curve.
2. $|\mathbf{v}(t)|$, the magnitude of $\mathbf{v}$, is the particle's **speed.**
3. $\mathbf{a}(t) = \dfrac{d\mathbf{v}}{dt} = \dfrac{d^2\mathbf{r}}{dt^2}$, the derivative of velocity and the second derivative of position, is the particle's **acceleration vector.**
4. $\dfrac{\mathbf{v}}{|\mathbf{v}|}$, a unit vector, is the **direction of motion.**

As for plane curves, we can express the velocity of a moving particle as the product of its speed and direction:

$$\text{Velocity} = |\mathbf{v}|\left(\frac{\mathbf{v}}{|\mathbf{v}|}\right) = (\text{speed})(\text{direction}).$$

In Section 10.3, we found this expression for velocity useful in locating, for example, the position of a helicopter moving along a straight line in space. Now let's look at an example of an object moving along a (nonlinear) space curve.

CD-ROM
WEBsite
Historical Biography

James Clerk Maxwell
(1831 — 1879)

Example 4 Flight of a Hang Glider

A person on a hang glider is spiraling upward due to rapidly rising air on a path having position vector $\mathbf{r}(t) = (3 \cos t)\mathbf{i} + (3 \sin t)\mathbf{j} + t^2\mathbf{k}$. The path is similar to that of a helix (although it's *not* a helix, as you will see in Section 10.7) and is shown in Figure 10.43 for $0 \le t \le 4\pi$. Find

(a) the velocity and acceleration vectors

(b) the glider's speed at any time t

(c) the times, if any, when the glider's acceleration is orthogonal to its velocity.

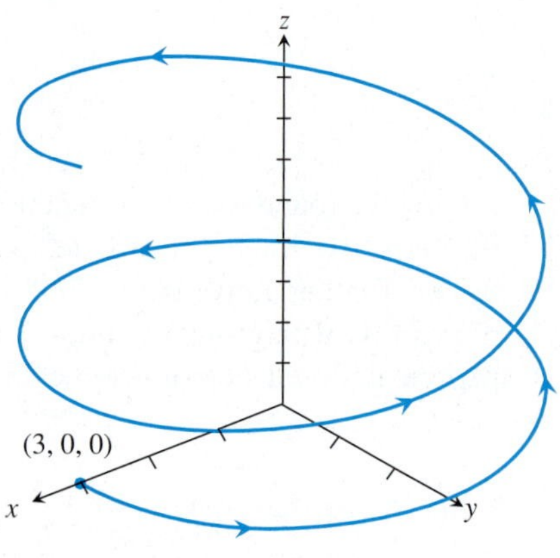

FIGURE 10.43 The path of a hang glider with position vector $\mathbf{r}(t) = (3 \cos t)\mathbf{i} + (3 \sin t)\mathbf{j} + t^2\mathbf{k}$. (Example 4)

Solution

(a) $\mathbf{r} = (3 \cos t)\mathbf{i} + (3 \sin t)\mathbf{j} + t^2\mathbf{k}$

$$\mathbf{v} = \frac{d\mathbf{r}}{dt} = -(3 \sin t)\mathbf{i} + (3 \cos t)\mathbf{j} + 2t\mathbf{k}$$

$$\mathbf{a} = \frac{d^2\mathbf{r}}{dt^2} = -(3 \cos t)\mathbf{i} - (3 \sin t)\mathbf{j} + 2\mathbf{k}$$

(b) Speed is the magnitude of $\mathbf{v}$:

$$\begin{aligned} |\mathbf{v}(t)| &= \sqrt{(-3 \sin t)^2 + (3 \cos t)^2 + (2t)^2} \\ &= \sqrt{9 \sin^2 t + 9 \cos^2 t + 4t^2} \\ &= \sqrt{9 + 4t^2}. \end{aligned}$$

The glider is moving faster and faster as it rises along its path.

(c) To find the times when $\mathbf{v}$ and $\mathbf{a}$ are orthogonal, we look for values of t for which

$$\mathbf{v} \cdot \mathbf{a} = 9 \sin t \cos t - 9 \cos t \sin t + 4t = 4t = 0.$$

Thus, the only time the acceleration vector is orthogonal to $\mathbf{v}$ is when $t = 0$. We study acceleration for motions along paths in more detail in Section 10.7. There we discover how the acceleration vector reveals the curving nature and tendency of the path to "twist" out of a plane containing the velocity vector.

Differentiation Rules

Because the derivatives of vector functions may be computed component by component, the rules for differentiating vector functions have the same form as the rules for differentiating scalar functions.

Differentiation Rules for Vector Functions

Let $\mathbf{u}$ and $\mathbf{v}$ be differentiable vector functions of t, $\mathbf{C}$ a constant vector, c any scalar, and f any differentiable scalar function.

1. *Constant Function Rule:* $\dfrac{d}{dt}\mathbf{C} = \mathbf{0}$
2. *Scalar Multiple Rules:* $\dfrac{d}{dt}[c\mathbf{u}(t)] = c\mathbf{u}'(t)$

   $\dfrac{d}{dt}[f(t)\mathbf{u}(t)] = f'(t)\mathbf{u}(t) + f(t)\mathbf{u}'(t)$
3. *Sum Rule:* $\dfrac{d}{dt}[\mathbf{u}(t) + \mathbf{v}(t)] = \mathbf{u}'(t) + \mathbf{v}'(t)$
4. *Difference Rule:* $\dfrac{d}{dt}[\mathbf{u}(t) - \mathbf{v}(t)] = \mathbf{u}'(t) - \mathbf{v}'(t)$
5. *Dot Product Rule:* $\dfrac{d}{dt}[\mathbf{u}(t) \cdot \mathbf{v}(t)] = \mathbf{u}'(t) \cdot \mathbf{v}(t) + \mathbf{u}(t) \cdot \mathbf{v}'(t)$
6. *Cross Product Rule:* $\dfrac{d}{dt}[\mathbf{u}(t) \times \mathbf{v}(t)] = \mathbf{u}'(t) \times \mathbf{v}(t) + \mathbf{u}(t) \times \mathbf{v}'(t)$
7. *Chain Rule:* $\dfrac{d}{dt}[\mathbf{u}(f(t))] = f'(t)\mathbf{u}'(f(t))$

When you use the Cross Product Rule, remember to preserve the order of the factors. If $\mathbf{u}$ comes first on the left side of the equation, it must also come first on the right or the signs will be wrong.

Applying the differentiation rules is the same as for planar vector functions (Example 5, Section 9.3) except that now we have a third component. As before, the rules can be proved by applying the definition or the corresponding differentiation formulas for scalar functions to the components of the vector functions. For example, here's how to prove the Cross Product and Chain Rules.

Proof of the Cross Product Rule We model the proof after the proof of the product rule for scalar functions. According to the definition of derivative,

$$\frac{d}{dt}(\mathbf{u} \times \mathbf{v}) = \lim_{h\to 0} \frac{\mathbf{u}(t+h) \times \mathbf{v}(t+h) - \mathbf{u}(t) \times \mathbf{v}(t)}{h}.$$

To change this fraction into an equivalent one that contains the difference quotients for the derivatives of $\mathbf{u}$ and $\mathbf{v}$, we subtract and add $\mathbf{u}(t) \times \mathbf{v}(t+h)$ in the numerator.

Then

$$\begin{aligned}\frac{d}{dt}(\mathbf{u}\times\mathbf{v}) &= \lim_{h\to 0}\frac{\mathbf{u}(t+h)\times\mathbf{v}(t+h)-\mathbf{u}(t)\times\mathbf{v}(t+h)+\mathbf{u}(t)\times\mathbf{v}(t+h)-\mathbf{u}(t)\times\mathbf{v}(t)}{h}\\ &= \lim_{h\to 0}\left[\frac{\mathbf{u}(t+h)-\mathbf{u}(t)}{h}\times\mathbf{v}(t+h)+\mathbf{u}(t)\times\frac{\mathbf{v}(t+h)-\mathbf{v}(t)}{h}\right]\\ &= \lim_{h\to 0}\frac{\mathbf{u}(t+h)-\mathbf{u}(t)}{h}\times\lim_{h\to 0}\mathbf{v}(t+h)+\lim_{h\to 0}\mathbf{u}(t)\times\lim_{h\to 0}\frac{\mathbf{v}(t+h)-\mathbf{v}(t)}{h}.\end{aligned}$$

The last of these equalities holds because the limit of the cross product of two vector functions is the cross product of their limits if the latter exist (Exercise 39). As h approaches zero, $\mathbf{v}(t+h)$ approaches $\mathbf{v}(t)$ because $\mathbf{v}$, being differentiable at t, is continuous at t (Exercise 40). The two fractions approach the values of $d\mathbf{u}/dt$ and $d\mathbf{v}/dt$ at t. In short,

$$\frac{d}{dt}(\mathbf{u}\times\mathbf{v}) = \frac{d\mathbf{u}}{dt}\times\mathbf{v}+\mathbf{u}\times\frac{d\mathbf{v}}{dt}.$$

As an algebraic convenience, we sometimes write the product of a scalar c and a vector $\mathbf{v}$ as $\mathbf{v}c$ instead of $c\mathbf{v}$. This permits us, for instance, to write the Chain Rule in a familiar form:

$$\frac{du}{dt} = \frac{d\mathbf{u}}{ds}\frac{ds}{dt},$$

where $s = f(t)$.

Proof of the Chain Rule Suppose that $\mathbf{u}(s) = a(s)\mathbf{i}+b(s)\mathbf{j}+c(s)\mathbf{k}$ is a differentiable vector function of s and that $s = f(t)$ is a differentiable scalar function of t. Then a, b, and c are differentiable functions of t, and the Chain Rule for differentiable real-valued functions gives

$$\begin{aligned}\frac{d}{dt}[\mathbf{u}(s)] &= \frac{da}{dt}\mathbf{i}+\frac{db}{dt}\mathbf{j}+\frac{dc}{dt}\mathbf{k}\\ &= \frac{da}{ds}\frac{ds}{dt}\mathbf{i}+\frac{db}{ds}\frac{ds}{dt}\mathbf{j}+\frac{dc}{ds}\frac{ds}{dt}\mathbf{k}\\ &= \frac{ds}{dt}\left(\frac{da}{ds}\mathbf{i}+\frac{db}{ds}\mathbf{j}+\frac{dc}{ds}\mathbf{k}\right)\\ &= \frac{ds}{dt}\frac{d\mathbf{u}}{ds}\\ &= f'(t)\mathbf{u}'(f(t)). \qquad s = f(t)\end{aligned}$$

Vector Functions of Constant Length

When we track a particle moving on a sphere centered at the origin (Figure 10.44), the position vector has a constant length equal to the radius of the sphere. The velocity vector $d\mathbf{r}/dt$, tangent to the path of motion, is tangent to the sphere and hence perpendicular to $\mathbf{r}$. This is always the case for a differentiable vector function of constant length: The vector and its first derivative are orthogonal. With the length constant, the change in the function is a change in direction only, and direction changes take place at right angles. We can also obtain this result by direct calculation:

$$\begin{aligned}\mathbf{r}(t)\cdot\mathbf{r}(t) &= c^2 && |\mathbf{r}(t)| = c \text{ is constant.}\\ \frac{d}{dt}[\mathbf{r}(t)\cdot\mathbf{r}(t)] &= 0 && \text{Differentiate both sides.}\\ \mathbf{r}'(t)\cdot\mathbf{r}(t)+\mathbf{r}(t)\cdot\mathbf{r}'(t) &= 0 && \text{Rule 5 with } \mathbf{r}(t)=\mathbf{u}(t)=\mathbf{v}(t)\\ 2\mathbf{r}'(t)\cdot\mathbf{r}(t) &= 0.\end{aligned}$$

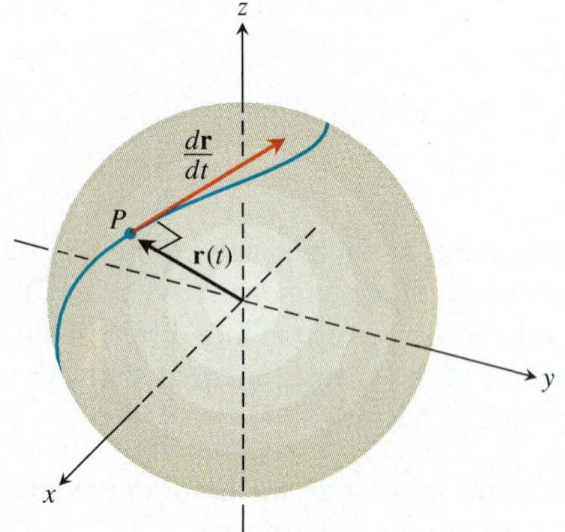

FIGURE 10.44 If a particle moves on a sphere in such a way that its position $\mathbf{r}$ is a differentiable function of time, then $\mathbf{r}\cdot(d\mathbf{r}/dt) = 0$.

The vectors $\mathbf{r}'(t)$ and $\mathbf{r}(t)$ are orthogonal because their dot product is 0. In summary,

We will use this observation repeatedly in Section 10.7.

If $\mathbf{r}$ is a differentiable vector function of t of constant length, then

$$\mathbf{r} \cdot \frac{d\mathbf{r}}{dt} = 0. \tag{4}$$

Example 5 Supporting Equation (4)

Show that $\mathbf{r}(t) = (\sin t)\mathbf{i} + (\cos t)\mathbf{j} + \sqrt{3}\mathbf{k}$ has constant length and is orthogonal to its derivative.

Solution

$$\mathbf{r}(t) = (\sin t)\mathbf{i} + (\cos t)\mathbf{j} + \sqrt{3}\mathbf{k}$$

$$|\mathbf{r}(t)| = \sqrt{(\sin t)^2 + (\cos t)^2 + (\sqrt{3})^2} = \sqrt{1+3} = 2$$

$$\frac{d\mathbf{r}}{dt} = (\cos t)\mathbf{i} - (\sin t)\mathbf{j}$$

$$\mathbf{r} \cdot \frac{d\mathbf{r}}{dt} = \sin t \cos t - \sin t \cos t = 0$$

Integrals of Vector Functions

A differentiable vector function $\mathbf{R}(t)$ is an **antiderivative** of a vector function $\mathbf{r}(t)$ on an interval I if $d\mathbf{R}/dt = \mathbf{r}$ at each point of I. If $\mathbf{R}$ is an antiderivative of $\mathbf{r}$ on I, it can be shown, working one component at a time, that every antiderivative of $\mathbf{r}$ on I has the form $\mathbf{R} + \mathbf{C}$ for some constant vector $\mathbf{C}$ (Exercise 45). The set of all antiderivatives of $\mathbf{r}$ on I is the **indefinite integral** of $\mathbf{r}$ on I.

Definition Indefinite Integral

The **indefinite integral** of $\mathbf{r}$ with respect to t is the set of all antiderivatives of $\mathbf{r}$, denoted by $\int \mathbf{r}(t)\,dt$. If $\mathbf{R}$ is any antiderivative of $\mathbf{r}$, then

$$\int \mathbf{r}(t)\,dt = \mathbf{R}(t) + \mathbf{C}.$$

The usual arithmetic rules for indefinite integrals apply.

Example 6 Finding Antiderivatives

$$\int ((\cos t)\mathbf{i} + \mathbf{j} - 2t\mathbf{k})\,dt = \left(\int \cos t\,dt\right)\mathbf{i} + \left(\int dt\right)\mathbf{j} - \left(\int 2t\,dt\right)\mathbf{k} \tag{5}$$

$$= (\sin t + C_1)\mathbf{i} + (t + C_2)\mathbf{j} - (t^2 + C_3)\mathbf{k} \tag{6}$$

$$= (\sin t)\mathbf{i} + t\mathbf{j} - t^2\mathbf{k} + \mathbf{C} \qquad \mathbf{C} = C_1\mathbf{i} + C_2\mathbf{j} - C_3\mathbf{k}$$

As in the integration of scalar functions, we recommend that you skip the steps in Equations (5) and (6) and go directly to the final form. Find an antiderivative for each component and add a constant vector at the end.

Definite integrals of vector functions are defined in terms of components.

Definition Definite Integral

If the components of $\mathbf{r}(t) = f(t)\mathbf{i} + g(t)\mathbf{j} + h(t)\mathbf{k}$ are integrable over $[a, b]$, then so is $\mathbf{r}$, and the **definite integral** of $\mathbf{r}$ from a to b is

$$\int_a^b \mathbf{r}(t)\,dt = \left(\int_a^b f(t)\,dt\right)\mathbf{i} + \left(\int_a^b g(t)\,dt\right)\mathbf{j} + \left(\int_a^b h(t)\,dt\right)\mathbf{k}.$$

Example 7 Evaluating Definite Integrals

$$\begin{aligned}\int_0^\pi ((\cos t)\mathbf{i} + \mathbf{j} - 2t\mathbf{k})\,dt &= \left(\int_0^\pi \cos t\,dt\right)\mathbf{i} + \left(\int_0^\pi dt\right)\mathbf{j} - \left(\int_0^\pi 2t\,dt\right)\mathbf{k}\\ &= [\sin t]_0^\pi\,\mathbf{i} + [t]_0^\pi\,\mathbf{j} - [t^2]_0^\pi\,\mathbf{k}\\ &= [0-0]\mathbf{i} + [\pi - 0]\mathbf{j} - [\pi^2 - 0^2]\mathbf{k}\\ &= \pi\mathbf{j} - \pi^2\mathbf{k}\end{aligned}$$

Example 8 Revisiting the Flight of a Glider

Suppose that we did not know the path of the glider in Example 4, but only its acceleration vector $\mathbf{a}(t) = -(3\cos t)\mathbf{i} - (3\sin t)\mathbf{j} + 2\mathbf{k}$. We also know that initially (at time $t = 0$), the glider departed from the point $(3, 0, 0)$ with velocity $\mathbf{v}(0) = 3\mathbf{j}$. Find the glider's position as a function of t.

Solution Our goal is to find $\mathbf{r}(t)$ knowing

The differential equation: $\mathbf{a} = \dfrac{d^2\mathbf{r}}{dt^2} = -(3\cos t)\mathbf{i} - (3\sin t)\mathbf{j} + 2\mathbf{k}$

The initial conditions: $\mathbf{v}(0) = 3\mathbf{j}$ and $\mathbf{r}(0) = 3\mathbf{i} + 0\mathbf{j} + 0\mathbf{k}$.

Integrating both sides of the differential equation with respect to t gives

$$\mathbf{v}(t) = -(3\sin t)\mathbf{i} + (3\cos t)\mathbf{j} + 2t\mathbf{k} + \mathbf{C}_1.$$

We use $\mathbf{v}(0) = 3\mathbf{j}$ to find $\mathbf{C}_1$:

$$\begin{aligned}3\mathbf{j} &= -(3\sin 0)\mathbf{i} + (3\cos 0)\mathbf{j} + (0)\mathbf{k} + \mathbf{C}_1\\ 3\mathbf{j} &= 3\mathbf{j} + \mathbf{C}_1\\ \mathbf{C}_1 &= \mathbf{0}.\end{aligned}$$

The glider's velocity as a function of time is

$$\frac{d\mathbf{r}}{dt} = \mathbf{v}(t) = -(3\sin t)\mathbf{i} + (3\cos t)\mathbf{j} + 2t\mathbf{k}.$$

Integrating both sides of this last differential equation gives

$$\mathbf{r}(t) = (3\cos t)\mathbf{i} + (3\sin t)\mathbf{j} + t^2\mathbf{k} + \mathbf{C}_2.$$

We then use the initial condition $\mathbf{r}(0) = 3\mathbf{i}$ to find $\mathbf{C}_2$:

$$\begin{aligned} 3\mathbf{i} &= (3 \cos 0)\mathbf{i} + (3 \sin 0)\mathbf{j} + (0^2)\mathbf{k} + \mathbf{C}_2 \\ 3\mathbf{i} &= 3\mathbf{i} + (0)\mathbf{j} + (0)\mathbf{k} + \mathbf{C}_2 \\ \mathbf{C}_2 &= \mathbf{0}. \end{aligned}$$

The glider's position as a function of t is

$$\mathbf{r}(t) = (3 \cos t)\mathbf{i} + (3 \sin t)\mathbf{j} + t^2\mathbf{k}.$$

This is the path of the glider we know from Example 4 and is shown in Figure 10.43.

Note: It was peculiar to this example that both of the constant vectors of integration, $\mathbf{C}_1$ and $\mathbf{C}_2$, turned out to be $\mathbf{0}$. Exercises 23 and 24 give different results.

EXERCISES 10.5

Velocity and Acceleration in Space

In Exercises 1–6, $\mathbf{r}(t)$ is the position of a particle in space at time t. Find the particle's velocity and acceleration vectors. Then find the particle's speed and direction of motion at the given value of t. Write the particle's velocity at that time as the product of its speed and direction.

1. $\mathbf{r}(t) = (t + 1)\mathbf{i} + (t^2 - 1)\mathbf{j} + 2t\mathbf{k}, \quad t = 1$

2. $\mathbf{r}(t) = (1 + t)\mathbf{i} + \dfrac{t^2}{\sqrt{2}}\mathbf{j} + \dfrac{t^3}{3}\mathbf{k}, \quad t = 1$

3. $\mathbf{r}(t) = (2 \cos t)\mathbf{i} + (3 \sin t)\mathbf{j} + 4t\mathbf{k}, \quad t = \pi/2$

4. $\mathbf{r}(t) = (\sec t)\mathbf{i} + (\tan t)\mathbf{j} + \dfrac{4}{3}t\mathbf{k}, \quad t = \pi/6$

5. $\mathbf{r}(t) = (2 \ln (t + 1))\mathbf{i} + t^2\mathbf{j} + \dfrac{t^2}{2}\mathbf{k}, \quad t = 1$

6. $\mathbf{r}(t) = (e^{-t})\mathbf{i} + (2 \cos 3t)\mathbf{j} + (2 \sin 3t)\mathbf{k}, \quad t = 0$

In Exercises 7–10, $\mathbf{r}(t)$ is the position of a particle in space at time t. Find the angle between the velocity and acceleration vectors at time $t = 0$.

7. $\mathbf{r}(t) = (3t + 1)\mathbf{i} + \sqrt{3}t\mathbf{j} + t^2\mathbf{k}$

8. $\mathbf{r}(t) = \left(\dfrac{\sqrt{2}}{2}t\right)\mathbf{i} + \left(\dfrac{\sqrt{2}}{2}t - 16t^2\right)\mathbf{j}$

9. $\mathbf{r}(t) = (\ln (t^2 + 1))\mathbf{i} + (\tan^{-1} t)\mathbf{j} + \sqrt{t^2 + 1}\,\mathbf{k}$

10. $\mathbf{r}(t) = \dfrac{4}{9}(1 + t)^{3/2}\mathbf{i} + \dfrac{4}{9}(1 - t)^{3/2}\mathbf{j} + \dfrac{1}{3}t\mathbf{k}$

In Exercises 11 and 12, $\mathbf{r}(t)$ is the position vector of a particle in space at time t. Find the time or times in the given time interval when the velocity and acceleration vectors are orthogonal.

11. $\mathbf{r}(t) = (t - \sin t)\mathbf{i} + (1 - \cos t)\mathbf{j}, \quad 0 \le t \le 2\pi$

12. $\mathbf{r}(t) = (\sin t)\mathbf{i} + t\mathbf{j} + (\cos t)\mathbf{k}, \quad t \ge 0$

Integrating Vector-Valued Functions

Evaluate the integrals in Exercises 13–18.

13. $\displaystyle\int_0^1 [t^3\mathbf{i} + 7\mathbf{j} + (t + 1)\mathbf{k}]\, dt$

14. $\displaystyle\int_1^2 \left[(6 - 6t)\mathbf{i} + 3\sqrt{t}\mathbf{j} + \left(\frac{4}{t^2}\right)\mathbf{k}\right] dt$

15. $\displaystyle\int_{-\pi/4}^{\pi/4} [(\sin t)\mathbf{i} + (1 + \cos t)\mathbf{j} + (\sec^2 t)\mathbf{k}]\, dt$

16. $\displaystyle\int_0^{\pi/3} [(\sec t \tan t)\mathbf{i} + (\tan t)\mathbf{j} + (2 \sin t \cos t)\mathbf{k}]\, dt$

17. $\displaystyle\int_1^4 \left[\frac{1}{t}\mathbf{i} + \frac{1}{5 - t}\mathbf{j} + \frac{1}{2t}\mathbf{k}\right] dt$

18. $\displaystyle\int_0^1 \left[\frac{2}{\sqrt{1 - t^2}}\mathbf{i} + \frac{\sqrt{3}}{1 + t^2}\mathbf{k}\right] dt$

Initial Value Problems for Vector-Valued Functions

Solve the initial value problems in Exercises 19–24 for $\mathbf{r}$ as a vector function of t.

19. Differential equation: $\dfrac{d\mathbf{r}}{dt} = -t\mathbf{i} - t\mathbf{j} - t\mathbf{k}$

Initial condition: $\mathbf{r}(0) = \mathbf{i} + 2\mathbf{j} + 3\mathbf{k}$

20. Differential equation: $\dfrac{d\mathbf{r}}{dt} = (180t)\mathbf{i} + (180t - 16t^2)\mathbf{j}$

Initial condition: $\mathbf{r}(0) = 100\mathbf{j}$

21. Differential equation: $\dfrac{d\mathbf{r}}{dt} = \dfrac{3}{2}(t + 1)^{1/2}\mathbf{i} + e^{-t}\mathbf{j} + \dfrac{1}{t + 1}\mathbf{k}$

Initial condition: $\mathbf{r}(0) = \mathbf{k}$

22. Differential equation: $\frac{d\mathbf{r}}{dt} = (t^3 + 4t)\mathbf{i} + t\mathbf{j} + 2t^2\mathbf{k}$

Initial condition: $\mathbf{r}(0) = \mathbf{i} + \mathbf{j}$

23. Differential equation: $\frac{d^2\mathbf{r}}{dt^2} = -32\mathbf{k}$

Initial conditions: $\mathbf{r}(0) = 100\mathbf{k}$ and

$$\left.\frac{d\mathbf{r}}{dt}\right|_{t=0} = 8\mathbf{i} + 8\mathbf{j}$$

24. Differential equation: $\frac{d^2\mathbf{r}}{dt^2} = -(\mathbf{i} + \mathbf{j} + \mathbf{k})$

Initial conditions: $\mathbf{r}(0) = 10\mathbf{i} + 10\mathbf{j} + 10\mathbf{k}$ and

$$\left.\frac{d\mathbf{r}}{dt}\right|_{t=0} = \mathbf{0}$$

Tangent Lines to Smooth Curves

As mentioned in the text, the tangent line to a smooth curve $\mathbf{r}(t) = f(t)\mathbf{i} + g(t)\mathbf{j} + h(t)\mathbf{k}$ at $t = t_0$ is the line that passes through the point $(f(t_0), g(t_0), h(t_0))$ parallel to $\mathbf{v}(t_0)$, the curve's velocity vector at t_0. In Exercises 25–28, find parametric equations for the line that is tangent to the given curve at the given parameter value $t = t_0$.

25. $\mathbf{r}(t) = (\sin t)\mathbf{i} + (t^2 - \cos t)\mathbf{j} + e^t\mathbf{k}, \quad t_0 = 0$

26. $\mathbf{r}(t) = (2 \sin t)\mathbf{i} + (2 \cos t)\mathbf{j} + 5t\mathbf{k}, \quad t_0 = 4\pi$

27. $\mathbf{r}(t) = (a \sin t)\mathbf{i} + (a \cos t)\mathbf{j} + bt\mathbf{k}, \quad t_0 = 2\pi$

28. $\mathbf{r}(t) = (\cos t)\mathbf{i} + (\sin t)\mathbf{j} + (\sin 2t)\mathbf{k}, \quad t_0 = \frac{\pi}{2}$

Motion Along a Straight Line

29. At time $t = 0$, a particle is located at the point $(1, 2, 3)$. It travels in a straight line to the point $(4, 1, 4)$, has speed 2 at $(1, 2, 3)$ and constant acceleration $3\mathbf{i} - \mathbf{j} + \mathbf{k}$. Find an equation for the position vector $\mathbf{r}(t)$ of the particle at time t.

30. A particle traveling in a straight line is located at the point $(1, -1, 2)$ and has speed 2 at time $t = 0$. The particle moves toward the point $(3, 0, 3)$ with constant acceleration $2\mathbf{i} + \mathbf{j} + \mathbf{k}$. Find its position vector $\mathbf{r}(t)$ at time t.

Theory and Examples

31. *Motion along a cycloid* A particle moves in the xy-plane in such a way that its position at time t is

$$\mathbf{r}(t) = (t - \sin t)\mathbf{i} + (1 - \cos t)\mathbf{j}.$$

T **(a)** Graph $\mathbf{r}(t)$. The resulting curve is called a cycloid.

(b) Find the maximum and minimum values of $|\mathbf{v}|$ and $|\mathbf{a}|$. (*Hint:* Find the extreme values of $|\mathbf{v}|^2$ and $|\mathbf{a}|^2$ first and take square roots later.)

32. *Motion along a circle* Show that the vector-valued function

$$\mathbf{r}(t) = (2\mathbf{i} + 2\mathbf{j} + \mathbf{k}) + (\cos t)\left(\frac{1}{\sqrt{2}}\mathbf{i} - \frac{1}{\sqrt{2}}\mathbf{j}\right) + (\sin t)\left(\frac{1}{\sqrt{3}}\mathbf{i} + \frac{1}{\sqrt{3}}\mathbf{j} + \frac{1}{\sqrt{3}}\mathbf{k}\right)$$

describes the motion of a particle moving on the circle of radius 1 centered at the point $(2, 2, 1)$ and lying in the plane $x + y - 2z = 2$.

33. *Motion along an ellipse* A particle moves around the ellipse $(y/3)^2 + (z/2)^2 = 1$ in the yz-plane in such a way that its position at time t is

$$\mathbf{r}(t) = (3 \cos t)\mathbf{j} + (2 \sin t)\mathbf{k}$$

Find the maximum and minimum values of $|\mathbf{v}|$ and $|\mathbf{a}|$. (*Hint:* Find the extreme values of $|\mathbf{v}|^2$ and $|\mathbf{a}|^2$ first and take square roots later.)

34. *Constant magnitude* Let $\mathbf{v}$ be a differentiable vector function of t. Show that if $\mathbf{v} \cdot (d\mathbf{v}/dt) = 0$ for all t, then $|\mathbf{v}|$ is constant.

35. *Constant function rule* Prove that if $\mathbf{u}$ is the vector function with the constant value $\mathbf{C}$, then $d\mathbf{u}/dt = \mathbf{0}$.

36. *Scalar multiple rules*

(a) Prove that if $\mathbf{u}$ is a differentiable function of t and c is any real number, then

$$\frac{d(c\mathbf{u})}{dt} = c\frac{d\mathbf{u}}{dt}.$$

(b) Prove that if $\mathbf{u}$ is a differentiable function of t and f is a differentiable scalar function of t, then

$$\frac{d}{dt}(f\mathbf{u}) = \frac{df}{dt}\mathbf{u} + f\frac{d\mathbf{u}}{dt}.$$

37. *Sum and difference rules* Prove that if $\mathbf{u}$ and $\mathbf{v}$ are differentiable functions of t, then

$$\frac{d}{dt}(\mathbf{u} + \mathbf{v}) = \frac{d\mathbf{u}}{dt} + \frac{d\mathbf{v}}{dt}$$

and

$$\frac{d}{dt}(\mathbf{u} - \mathbf{v}) = \frac{d\mathbf{u}}{dt} - \frac{d\mathbf{v}}{dt}.$$

38. *Component test for continuity at a point* Show that the vector function $\mathbf{r}$ defined by the rule $\mathbf{r}(t) = f(t)\mathbf{i} + g(t)\mathbf{j} + h(t)\mathbf{k}$ is continuous at $t = t_0$ if and only if f, g, and h are continuous at t_0.

39. *Limits of cross products of vector functions* Suppose that $\mathbf{r}_1(t) = f_1(t)\mathbf{i} + f_2(t)\mathbf{j} + f_3(t)\mathbf{k}$, $\mathbf{r}_2(t) = g_1(t)\mathbf{i} + g_2(t)\mathbf{j} + g_3(t)\mathbf{k}$, $\lim_{t\to t_0} \mathbf{r}_1(t) = \mathbf{u}$, and $\lim_{t\to t_0} \mathbf{r}_2(t) = \mathbf{v}$. Use the determinant formula for cross products and the Limit Product Rule for scalar functions to show that

$$\lim_{t\to t_0} (\mathbf{r}_1(t) \times \mathbf{r}_2(t)) = \mathbf{u} \times \mathbf{v}.$$

40. *Differentiable vector functions are continuous* Show that if $\mathbf{r}(t) = f(t)\mathbf{i} + g(t)\mathbf{j} + h(t)\mathbf{k}$ is differentiable at $t = t_0$, then it is continuous at t_0 as well.

41. *Derivatives of triple scalar products*

(a) Show that if $\mathbf{u}$, $\mathbf{v}$, and $\mathbf{w}$ are differentiable vector functions of t, then

$$\frac{d}{dt}(\mathbf{u}\cdot\mathbf{v}\times\mathbf{w}) = \frac{d\mathbf{u}}{dt}\cdot\mathbf{v}\times\mathbf{w} + \mathbf{u}\cdot\frac{d\mathbf{v}}{dt}\times\mathbf{w} + \mathbf{u}\cdot\mathbf{v}\times\frac{d\mathbf{w}}{dt}. \quad (7)$$

(b) Show that Equation (7) is equivalent to

$$\frac{d}{dt}\begin{vmatrix} u_1 & u_2 & u_3 \\ v_1 & v_2 & v_3 \\ w_1 & w_2 & w_3 \end{vmatrix} = \begin{vmatrix} \frac{du_1}{dt} & \frac{du_2}{dt} & \frac{du_3}{dt} \\ v_1 & v_2 & v_3 \\ w_1 & w_2 & w_3 \end{vmatrix} + \begin{vmatrix} u_1 & u_2 & u_3 \\ \frac{dv_1}{dt} & \frac{dv_2}{dt} & \frac{dv_3}{dt} \\ w_1 & w_2 & w_3 \end{vmatrix} + \begin{vmatrix} u_1 & u_2 & u_3 \\ v_1 & v_2 & v_3 \\ \frac{dw_1}{dt} & \frac{dw_2}{dt} & \frac{dw_3}{dt} \end{vmatrix}. \quad (8)$$

Equation (8) says that the derivative of a 3 by 3 determinant of differentiable functions is the sum of the three determinants obtained from the original by differentiating one row at a time. The result extends to determinants of any order.

42. *(Continuation of Exercise 41.)* Suppose that $\mathbf{r}(t) = f(t)\mathbf{i} + g(t)\mathbf{j} + h(t)\mathbf{k}$ and that f, g, and h have derivatives through order three. Use Equation (7) or (8) to show that

$$\frac{d}{dt}\left(\mathbf{r}\cdot\frac{d\mathbf{r}}{dt}\times\frac{d^2\mathbf{r}}{dt^2}\right) = \mathbf{r}\cdot\left(\frac{d\mathbf{r}}{dt}\times\frac{d^3\mathbf{r}}{dt^3}\right). \quad (9)$$

(*Hint:* Differentiate on the left and look for vectors whose products are zero.)

43. *Properties of integrable vector functions* Establish the following properties of integrable vector functions.

(a) The *Constant Scalar Multiple Rule:*

$$\int_a^b k\mathbf{r}(t)\,dt = k\int_a^b \mathbf{r}(t)\,dt \qquad \text{(any scalar } k\text{)}$$

The *Rule for Negatives,*

$$\int_a^b (-\mathbf{r}(t))\,dt = -\int_a^b \mathbf{r}(t)\,dt,$$

is obtained by taking $k = -1$.

(b) The *Sum and Difference Rules:*

$$\int_a^b (\mathbf{r}_1(t) \pm \mathbf{r}_2(t))\,dt = \int_a^b \mathbf{r}_1(t)\,dt \pm \int_a^b \mathbf{r}_2(t)\,dt$$

(c) The *Constant Vector Multiple Rules:*

$$\int_a^b \mathbf{C}\cdot\mathbf{r}(t)\,dt = \mathbf{C}\cdot\int_a^b \mathbf{r}(t)\,dt \qquad \text{(any constant vector } \mathbf{C}\text{)}$$

and

$$\int_a^b \mathbf{C}\times\mathbf{r}(t)\,dt = \mathbf{C}\times\int_a^b \mathbf{r}(t)\,dt \qquad \text{(any constant vector } \mathbf{C}\text{)}$$

44. *Products of scalar and vector functions* Suppose that the scalar function $u(t)$ and the vector function $\mathbf{r}(t)$ are both defined for $a \le t \le b$.

(a) Show that $u\mathbf{r}$ is continuous on $[a, b]$ if u and $\mathbf{r}$ are continuous on $[a, b]$.

(b) If u and $\mathbf{r}$ are both differentiable on $[a, b]$, show that $u\mathbf{r}$ is differentiable on $[a, b]$ and that

$$\frac{d}{dt}(u\mathbf{r}) = u\frac{d\mathbf{r}}{dt} + \mathbf{r}\frac{du}{dt}.$$

45. *Antiderivatives of vector functions*

(a) Use Corollary 2 of the Mean Value Theorem for scalar functions to show that if two vector functions $\mathbf{R}_1(t)$ and $\mathbf{R}_2(t)$ have identical derivatives on an interval I, then the functions differ by a constant vector value throughout I.

(b) Use the result in part (a) to show that if $\mathbf{R}(t)$ is any antiderivative of $\mathbf{r}(t)$ on I, then every other antiderivative of $\mathbf{r}$ on I equals $\mathbf{R}(t) + \mathbf{C}$ for some constant vector $\mathbf{C}$.

46. *The Fundamental Theorem of Calculus* The Fundamental Theorem of Calculus for scalar functions of a real variable holds for vector functions of a real variable as well. Prove this by using the theorem for scalar functions to show first that if a vector function $\mathbf{r}(t)$ is continuous for $a \le t \le b$, then

$$\frac{d}{dt}\int_a^t \mathbf{r}(\tau)\,d\tau = \mathbf{r}(t)$$

at every point t of $[a, b]$. Then use the conclusion in part (b) of Exercise 45 to show that if $\mathbf{R}$ is any antiderivative of $\mathbf{r}$ on $[a, b]$, then

$$\int_a^b \mathbf{r}(t)\,dt = \mathbf{R}(b) - \mathbf{R}(a).$$

COMPUTER EXPLORATIONS

Drawing Tangents to Space Curves

Use a CAS to perform the following steps in Exercises 47–50.

(a) Plot the space curve traced out by the position vector $\mathbf{r}$.

(b) Find the components of the velocity vector $d\mathbf{r}/dt$.

(c) Evaluate $d\mathbf{r}/dt$ at the given point t_0 and find an equation for the tangent line to the curve at $\mathbf{r}(t_0)$.

(d) Plot the tangent line together with the curve over the given interval.

47. $\mathbf{r}(t) = (\sin t - t\cos t)\mathbf{i} + (\cos t + t\sin t)\mathbf{j} + t^2\mathbf{k}, \quad 0 \le t \le 6\pi,$ $t_0 = 3\pi/2$

48. $\mathbf{r}(t) = \sqrt{2}t\mathbf{i} + e^t\mathbf{j} + e^{-t}\mathbf{k}, \quad -2 \le t \le 3, \quad t_0 = 1$

49. $\mathbf{r}(t) = (\sin 2t)\mathbf{i} + (\ln(1+t))\mathbf{j} + t\mathbf{k}, \quad 0 \le t \le 4\pi, \quad t_0 = \pi/4$

50. $\mathbf{r}(t) = (\ln(t^2+2))\mathbf{i} + (\tan^{-1} 3t)\mathbf{j} + \sqrt{t^2+1}\,\mathbf{k}, \quad -3 \le t \le 5,$ $t_0 = 3$

Exploring Helices

In Exercises 51 and 52, you will explore graphically the behavior of the helix

$$\mathbf{r}(t) = (\cos at)\mathbf{i} + (\sin at)\mathbf{j} + bt\mathbf{k}$$

as you change the values of the constants a and b. Use a CAS to perform the steps in each exercise.

51. Set $b = 1$. Plot the helix $\mathbf{r}(t)$ together with the tangent line to the curve at $t = 3\pi/2$ for $a = 1, 2, 4$, and 6 over the interval $0 \le t \le 4\pi$. Describe in your own words what happens to the graph of the helix and the position of the tangent line as a increases through these positive values.

52. Set $a = 1$. Plot the helix $\mathbf{r}(t)$ together with the tangent line to the curve at $t = 3\pi/2$ for $b = 1/4, 1/2, 2$, and 4 over the interval $0 \le t \le 4\pi$. Describe in your own words what happens to the graph of the helix and the position of the tangent line as b increases through these positive values.

10.6 Arc Length and the Unit Tangent Vector T

Arc Length Along a Curve • Speed on a Smooth Curve • Unit Tangent Vector T • Curvature and the Principal Unit Normal for Plane Curves • Circle of Curvature and Radius of Curvature

Imagine the motions you might experience traveling at high speeds along a path through the air or space. Specifically, imagine the motions of turning to your left or right and the up-and-down motions tending to lift you from, or pin you down to, your seat. Pilots flying through the atmosphere, turning and twisting in flight acrobatics, certainly experience these motions. Modern roller-coaster rides try to capture them for thrill seekers who are more earthbound. The intensity of the experience is heightened by the "tightness" of the turning, the strength of the "lift" perpendicular to your seat, and the overall speed along the path. Turns that are too tight, descents or climbs that are too steep, or either one coupled with high and increasing speed can cause an aircraft to spin out of control, possibly even to break up in midair, and crash to Earth.

In this section and the next, we study the features of the curve's shape that describe mathematically the sharpness of turning and the twisting perpendicular to the forward motion. We further see how this geometry of the curve is actually carried numerically in the velocity and acceleration vectors defining the motion (just as speed is intrinsic to the velocity vector itself).

Arc Length Along a Curve

Early in our calculus studies, we learned that speed is the derivative of distance with respect to time. Up to now, we considered motion occurring primarily along a straight line (although we have also looked at projectile motion along a parabolic arc). To study motion along other smooth space curves, we need to have a measurable length along the curve. This enables us to locate points along these curves by giving their directed distance s along the curve from some **base point,** the way we locate points on coordinate axes by giving their directed distance from the origin (Figure 10.45). Time is the natural parameter for describing a moving body's velocity and acceleration, but s is the natural parameter for studying a curve's shape. Both parameters are useful in studying space curves, as we soon see.

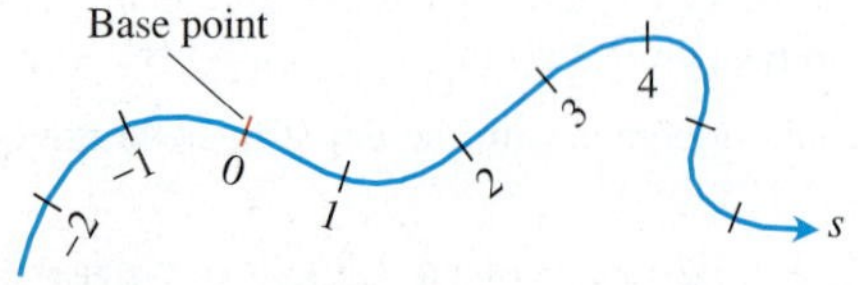

FIGURE 10.45 Smooth curves can be scaled like number lines, the coordinate of each point being its directed distance from a preselected base point.

The following formula defines how to measure distance along a smooth curve in space. It is the three-dimensional form of the parametric formula we obtained for planar curves in Section 5.3, and it should come as no surprise.

Definition Arc Length: Length of a Smooth Curve

The **length** of a smooth curve $\mathbf{r}(t) = f(t)\mathbf{i} + g(t)\mathbf{j} + h(t)\mathbf{k}$, $a \le t \le b$, that is traced exactly once as t increases from $t = a$ to $t = b$ is

$$L = \int_a^b \sqrt{\left(\frac{df}{dt}\right)^2 + \left(\frac{dg}{dt}\right)^2 + \left(\frac{dh}{dt}\right)^2}\, dt$$

$$= \int_a^b \sqrt{\left(\frac{dx}{dt}\right)^2 + \left(\frac{dy}{dt}\right)^2 + \left(\frac{dz}{dt}\right)^2}\, dt. \qquad (1)$$

Just as for plane curves, we can calculate the length of a curve in space from any convenient parametrization that meets the stated conditions. We omit the proof.

The square root in Equation (1) is $|\mathbf{v}|$, the length of a velocity vector $d\mathbf{r}/dt$. This enables us to write the formula for length a shorter way.

CD-ROM WEBsite

Arc Length Formula (Short Form)

$$L = \int_a^b |\mathbf{v}|\, dt \qquad (2)$$

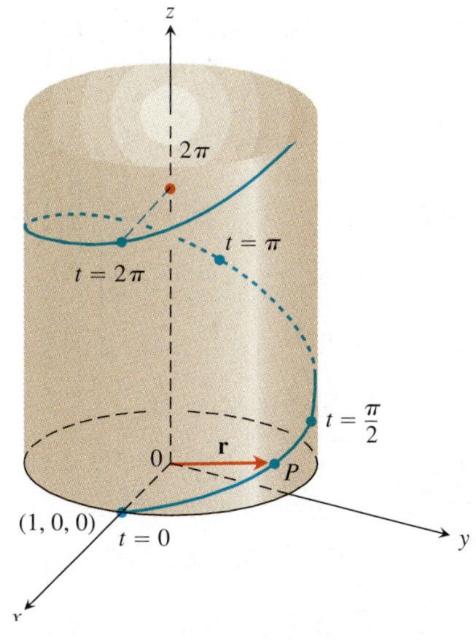

FIGURE 10.46 The helix $\mathbf{r}(t) = (\cos t)\mathbf{i} + (\sin t)\mathbf{j} + t\mathbf{k}$ in Example 1.

Example 1 Distance Traveled by a Glider

A glider is soaring upward along the helix $\mathbf{r}(t) = (\cos t)\mathbf{i} + (\sin t)\mathbf{j} + t\mathbf{k}$. How far does the glider travel along its path from $t = 0$ to $t = 2\pi \approx 6.28$ sec?

Solution The path segment during this time corresponds to one full turn of the helix (Figure 10.46). The length of this portion of the curve is

$$L = \int_a^b |\mathbf{v}|\, dt = \int_0^{2\pi} \sqrt{(-\sin t)^2 + (\cos t)^2 + (1)^2}\, dt$$

$$= \int_0^{2\pi} \sqrt{2}\, dt = 2\pi\sqrt{2} \text{ units of length.}$$

This is $\sqrt{2}$ times the length of the circle in the xy-plane over which the helix stands.

If we choose a base point $P(t_0)$ on a smooth curve C parametrized by t, each value of t determines a point $P(t) = (x(t), y(t), z(t))$ on C and a "directed distance"

$$s(t) = \int_{t_0}^{t} |\mathbf{v}(\tau)|\, d\tau,$$

measured along C from the base point (Figure 10.47). If $t > t_0$, $s(t)$ is the distance from $P(t_0)$ to $P(t)$. If $t < t_0$, $s(t)$ is the negative of the distance. Each value of s determines a point on C and this parametrizes C with respect to s. We call s an **arc length parameter** for the curve. The parameter's value increases in the direction of increasing t. The arc length parameter is particularly effective for investigating the turning and twisting nature of a space curve.

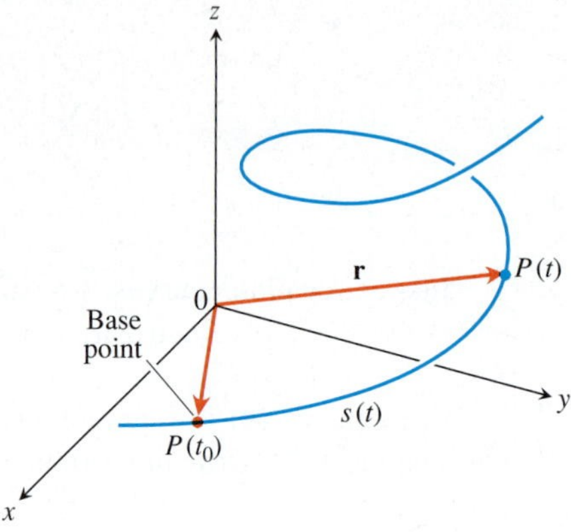

FIGURE 10.47 The directed distance along the curve from $P(t_0)$ to any point $P(t)$ is

$$s(t) = \int_{t_0}^{t} |\mathbf{v}(\tau)|\, d\tau.$$

We use the Greek letter τ ("tau") as the variable of integration in Equation (3) because the letter t is already in use as the upper limit.

Arc Length Parameter with Base Point $P(t_0)$

$$s(t) = \int_{t_0}^{t} \sqrt{[x'(\tau)]^2 + [y'(\tau)]^2 + [z'(\tau)]^2}\, d\tau = \int_{t_0}^{t} |\mathbf{v}(\tau)|\, d\tau \tag{3}$$

If a curve $\mathbf{r}(t)$ is already given in terms of some parameter t and $s(t)$ is the arc length function given by Equation (3), then we may be able to solve for t as a function of s: $t = t(s)$. Then the curve can be reparametrized in terms of s by substituting for t: $\mathbf{r} = \mathbf{r}(t(s))$. Here's a simple example.

Example 2 Finding an Arc Length Parametrization

If $t_0 = 0$, the arc length parameter along the helix

$$\mathbf{r}(t) = (\cos t)\mathbf{i} + (\sin t)\mathbf{j} + t\mathbf{k}$$

from t_0 to t is

$$\begin{aligned} s(t) &= \int_{t_0}^{t} |\mathbf{v}(\tau)|\, d\tau && \text{Eq. (3)} \\ &= \int_{0}^{t} \sqrt{2}\, d\tau && \text{Value from Example 1} \\ &= \sqrt{2}\, t. \end{aligned}$$

Solving this equation for t gives $t = s/\sqrt{2}$. Substituting into the position vector $\mathbf{r}$ gives the following arc length parametrization for the helix:

$$\mathbf{r}(t(s)) = \left(\cos \frac{s}{\sqrt{2}}\right)\mathbf{i} + \left(\sin \frac{s}{\sqrt{2}}\right)\mathbf{j} + \frac{s}{\sqrt{2}}\mathbf{k}.$$

Unlike Example 2, the arc length parametrization is generally difficult to find analytically for a curve already given in terms of some other parameter t. Fortunately, however, we rarely need an exact formula for $s(t)$ or its inverse $t(s)$.

Speed on a Smooth Curve

Since the derivatives beneath the radical in Equation (3) are continuous (the curve is smooth), the Fundamental Theorem of Calculus tells us that s is a differentiable function of t with derivative

$$\frac{ds}{dt} = |\mathbf{v}(t)|. \tag{4}$$

As we already knew, the speed with which a particle moves along its path is the magnitude of $\mathbf{v}$.

Notice that although the base point $P(t_0)$ plays a role in defining s in Equation (3), it plays no role in Equation (4). The rate at which a moving particle covers distance along its path has nothing to do with how far away the base point is.

Notice also that $ds/dt > 0$ since, by definition, $|\mathbf{v}|$ is never zero for a smooth curve. We see once again that s is an increasing function of t.

Unit Tangent Vector T

We already know the velocity vector $\mathbf{v} = d\mathbf{r}/dt$ is tangent to the curve and that the vector

$$\mathbf{T} = \frac{\mathbf{v}}{|\mathbf{v}|}$$

is therefore a unit vector tangent to the (smooth) curve. There is more to the story, however, when we consider the arc length parameter. Since $ds/dt > 0$ for the curves we are considering, s is one-to-one and has an inverse that gives t as a differentiable function of s (Section 6.2). The derivative of the inverse is

$$\frac{dt}{ds} = \frac{1}{ds/dt} = \frac{1}{|\mathbf{v}|}.$$

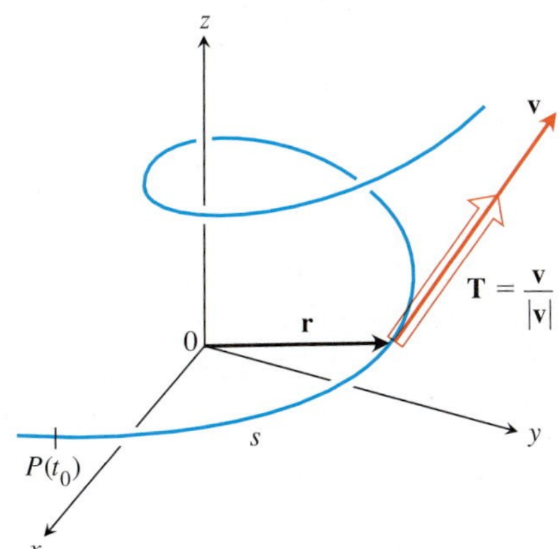

FIGURE 10.48 We find the unit tangent vector $\mathbf{T}$ by dividing $\mathbf{v}$ by $|\mathbf{v}|$.

This makes $\mathbf{r}$ a differentiable function of s whose derivative can be calculated with the Chain Rule to be

$$\frac{d\mathbf{r}}{ds} = \frac{d\mathbf{r}}{dt}\frac{dt}{ds} = \mathbf{v}\frac{1}{|\mathbf{v}|} = \frac{\mathbf{v}}{|\mathbf{v}|} = \mathbf{T}.$$

This equation says that $d\mathbf{r}/ds$ is the unit tangent vector in the direction of the velocity vector $\mathbf{v}$ (Figure 10.48).

Definition Unit Tangent Vector

The **unit tangent vector** of a differentiable curve $\mathbf{r}(t)$ is

$$\mathbf{T} = \frac{d\mathbf{r}}{ds} = \frac{d\mathbf{r}/dt}{ds/dt} = \frac{\mathbf{v}}{|\mathbf{v}|}. \tag{5}$$

The unit tangent vector $\mathbf{T}$ is a differentiable function of t whenever $\mathbf{v}$ is a differentiable function of t. As we see in the next section, $\mathbf{T}$ is one of three unit vectors in a traveling reference frame that is used to describe the motion of space vehicles and other bodies moving in three dimensions.

Example 3 Finding the Unit Tangent Vector T

Find the unit tangent vector of the curve

$$\mathbf{r}(t) = (3\cos t)\mathbf{i} + (3\sin t)\mathbf{j} + t^2\mathbf{k}$$

representing the path of the glider in Example 4, Section 10.5.

Solution In that example, we found

$$\mathbf{v} = \frac{d\mathbf{r}}{dt} = -(3\sin t)\mathbf{i} + (3\cos t)\mathbf{j} + 2t\mathbf{k}$$

and

$$|\mathbf{v}| = \sqrt{9 + 4t^2}.$$

Thus,

$$\mathbf{T} = \frac{\mathbf{v}}{|\mathbf{v}|} = -\frac{3\sin t}{\sqrt{9+4t^2}}\mathbf{i} + \frac{3\cos t}{\sqrt{9+4t^2}}\mathbf{j} + \frac{2t}{\sqrt{9+4t^2}}\mathbf{k}.$$

Curvature and the Principal Unit Normal Vector for Plane Curves

To understand how a curve is "turning" as opposed to "twisting," it is easiest to begin with curves in the plane (turning, but no twisting out of the plane). Based on that understanding, we will take the next step and study space curves in Section 10.7.

As a particle moves along a smooth curve in the plane, $\mathbf{T} = d\mathbf{r}/ds$ turns as the curve bends. Since $\mathbf{T}$ is a unit vector, its length remains constant and only its direction changes as the particle moves along the curve. The rate at which $\mathbf{T}$ turns per unit of length along the curve is called the *curvature* (Figure 10.49). The traditional symbol for the curvature function is the Greek letter κ ("kappa").

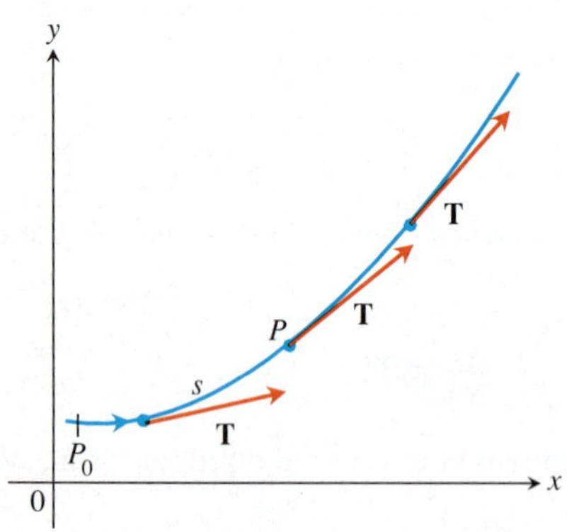

FIGURE 10.49 As P moves along the curve in the direction of increasing arc length, the unit tangent vector turns. The value of $|d\mathbf{T}/ds|$ at P is called the *curvature* of the curve at P.

Definition Curvature

If $\mathbf{T}$ is the unit vector of a smooth curve, the **curvature** function of the curve is

$$\kappa = \left|\frac{d\mathbf{T}}{ds}\right|.$$

If $|d\mathbf{T}/ds|$ is large, $\mathbf{T}$ turns sharply as the particle passes through P, and the curvature at P is large. If $|d\mathbf{T}/ds|$ is close to zero, $\mathbf{T}$ turns more slowly and the curvature at P is smaller.

If a smooth curve $\mathbf{r}(t)$ is already given in terms of some parameter t other than the arc length parameter s, we can calculate the curvature as

$$\begin{aligned}\kappa &= \left|\frac{d\mathbf{T}}{ds}\right| = \left|\frac{d\mathbf{T}}{dt}\frac{dt}{ds}\right| \\ &= \frac{1}{|ds/dt|}\left|\frac{d\mathbf{T}}{dt}\right| \\ &= \frac{1}{|\mathbf{v}|}\left|\frac{d\mathbf{T}}{dt}\right|. \qquad \frac{ds}{dt} = |\mathbf{v}|\end{aligned}$$

Formula for Calculating Curvature

If $\mathbf{r}(t)$ is a smooth curve, then the curvature is

$$\kappa = \frac{1}{|\mathbf{v}|}\left|\frac{d\mathbf{T}}{dt}\right|, \tag{6}$$

where $\mathbf{T} = \mathbf{v}/|\mathbf{v}|$ is the unit tangent vector.

Testing the definition, we see in Examples 4 and 5 that the curvature is constant for straight lines and circles.

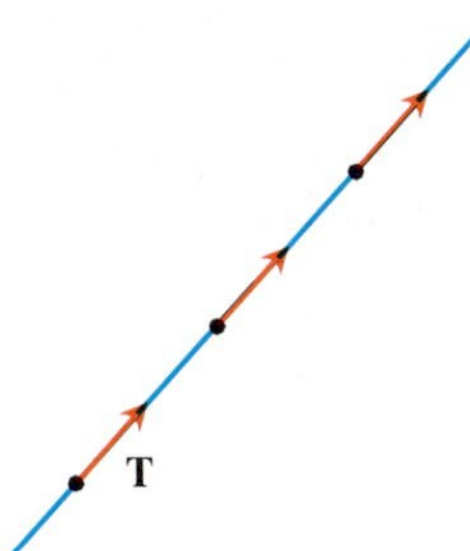

FIGURE 10.50 Along a straight line, $\mathbf{T}$ always points in the same direction. The curvature, $|d\mathbf{T}/ds|$, is zero. (Example 4)

Example 4 The Curvature of a Straight Line is Zero

On a straight line, the unit tangent vector $\mathbf{T}$ always points in the same direction, so its components are constants. Therefore, $|d\mathbf{T}/ds| = |\mathbf{0}| = 0$ (Figure 10.50).

Example 5 The Curvature of a Circle of Radius a is $1/a$

To see why, we begin with the parametrization

$$\mathbf{r}(t) = (a\cos t)\mathbf{i} + (a\sin t)\mathbf{j}$$

of a circle of radius a. Then,

$$\begin{aligned}\mathbf{v} &= \frac{d\mathbf{r}}{dt} = -(a\sin t)\mathbf{i} + (a\cos t)\mathbf{j} \\ |\mathbf{v}| &= \sqrt{(-a\sin t)^2 + (a\cos t)^2} = \sqrt{a^2} = |a| = a. \qquad \text{Since } a > 0,\ |a| = a.\end{aligned}$$

From this we find

$$\begin{aligned}\mathbf{T} &= \frac{\mathbf{v}}{|\mathbf{v}|} = -(\sin t)\mathbf{i} + (\cos t)\mathbf{j} \\ \frac{d\mathbf{T}}{dt} &= -(\cos t)\mathbf{i} - (\sin t)\mathbf{j} \\ \left|\frac{d\mathbf{T}}{dt}\right| &= \sqrt{\cos^2 t + \sin^2 t} = 1.\end{aligned}$$

Hence, for any value of the parameter t,

$$\kappa = \frac{1}{|\mathbf{v}|}\left|\frac{d\mathbf{T}}{dt}\right| = \frac{1}{a}(1) = \frac{1}{a}.$$

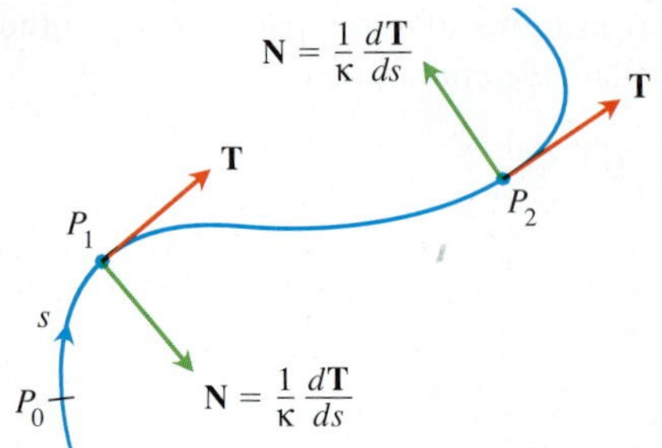

FIGURE 10.51 The vector $d\mathbf{T}/ds$, normal to the curve, always points in the direction in which $\mathbf{T}$ is turning. The vector $\mathbf{N}$ is the direction of $d\mathbf{T}/ds$.

Although the formula for calculating κ in Equation (6) is also valid for space curves, in the next section we find a computational formula that is usually more convenient to apply.

Among the vectors orthogonal to the unit tangent vector $\mathbf{T}$ is one of particular significance because it points in the direction in which the curve is turning. Since $\mathbf{T}$ has constant length (namely, 1), the derivative $d\mathbf{T}/ds$ is orthogonal to $\mathbf{T}$ (Section 10.5). Therefore, if we divide $d\mathbf{T}/ds$ by its length κ, we obtain a *unit* vector $\mathbf{N}$ orthogonal to $\mathbf{T}$ (Figure 10.51).

Definition Principal Unit Normal

At a point where $\kappa \neq 0$, the **principal unit normal** vector for a curve in the plane is

$$\mathbf{N} = \frac{1}{\kappa}\frac{d\mathbf{T}}{ds}.$$

The vector $d\mathbf{T}/ds$ points in the direction in which $\mathbf{T}$ turns as the curve bends. Therefore, if we face in the direction of increasing arc length, the vector $d\mathbf{T}/ds$ points toward the right if $\mathbf{T}$ turns clockwise and toward the left if $\mathbf{T}$ turns counterclockwise. In other words, the principal normal vector $\mathbf{N}$ will point toward the concave side of the curve (Figure 10.51).

If a smooth curve $\mathbf{r}(t)$ is already given in terms of some parameter t other than the arc length parameter s, we can use the Chain Rule to calculate $\mathbf{N}$ directly:

$$\begin{aligned}\mathbf{N} &= \frac{d\mathbf{T}/ds}{|d\mathbf{T}/ds|}\\ &= \frac{(d\mathbf{T}/dt)(dt/ds)}{|d\mathbf{T}/dt||dt/ds|}\\ &= \frac{d\mathbf{T}/dt}{|d\mathbf{T}/dt|}.\end{aligned}$$

This formula enables us to find $\mathbf{N}$ without having to find κ and s first.

Formula for Calculating N

If $\mathbf{r}(t)$ is a smooth curve, then the principal unit normal is

$$\mathbf{N} = \frac{d\mathbf{T}/dt}{|d\mathbf{T}/dt|}, \qquad (7)$$

where $\mathbf{T} = \mathbf{v}/|\mathbf{v}|$ is the unit tangent vector.

Example 6 Finding T and N

Find $\mathbf{T}$ and $\mathbf{N}$ for the circular motion

$$\mathbf{r}(t) = (\cos 2t)\mathbf{i} + (\sin 2t)\mathbf{j}.$$

Solution We first find $\mathbf{T}$:

$$\mathbf{v} = -(2 \sin 2t)\mathbf{i} + (2 \cos 2t)\mathbf{j}$$

$$|\mathbf{v}| = \sqrt{4 \sin^2 2t + 4 \cos^2 2t} = 2$$

$$\mathbf{T} = \frac{\mathbf{v}}{|\mathbf{v}|} = -(\sin 2t)\mathbf{i} + (\cos 2t)\mathbf{j}.$$

From this we find

$$\frac{d\mathbf{T}}{dt} = -(2 \cos 2t)\mathbf{i} - (2 \sin 2t)\mathbf{j}$$

$$\left|\frac{d\mathbf{T}}{dt}\right| = \sqrt{4 \cos^2 2t + 4 \sin^2 2t} = 2$$

and

$$\mathbf{N} = \frac{d\mathbf{T}/dt}{|d\mathbf{T}/dt|}$$

$$= -(\cos 2t)\mathbf{i} - (\sin 2t)\mathbf{j}. \qquad \text{Eq. (7)}$$

Notice that $\mathbf{T} \cdot \mathbf{N} = 0$, verifying that $\mathbf{N}$ is orthogonal to $\mathbf{T}$.

Circle of Curvature and Radius of Curvature

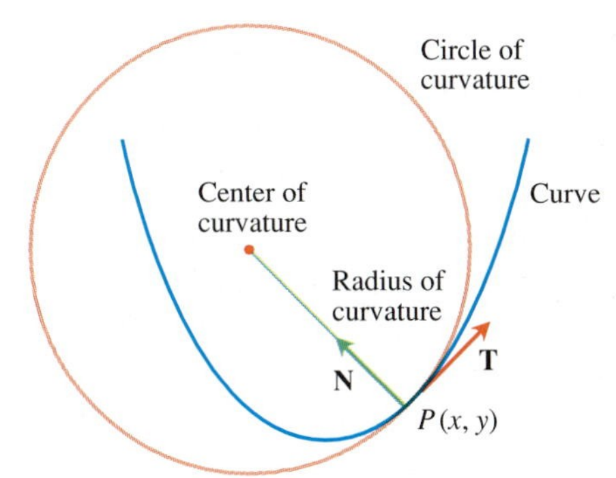

FIGURE 10.52 The osculating circle at $P(x, y)$ lies toward the inner side of the curve.

The **circle of curvature** or **osculating circle** at a point P on a plane curve where $\kappa \neq 0$ is the circle in the plane of the curve that

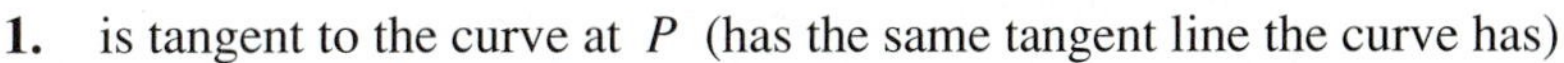

1. is tangent to the curve at P (has the same tangent line the curve has)
2. has the same curvature the curve has at P
3. lies toward the concave or inner side of the curve (as in Figure 10.52).

The **radius of curvature** of the curve at P is the radius of the circle of curvature, which, according to Example 5, is

$$\text{Radius of curvature} = \rho = \frac{1}{\kappa}.$$

To find ρ, we find κ and take the reciprocal. The **center of curvature** of the curve at P is the center of the circle of curvature.

Example 7 Finding the Osculating Circle for a Parabola

Find and graph the osculating circle of the parabola $y = x^2$ at the origin.

Solution We parametrize the parabola using the parameter $t = x$ (Preliminary Section 6),

$$\mathbf{r}(t) = t\mathbf{i} + t^2\mathbf{j}.$$

First we find the curvature of the parabola at the origin, using Equation (6):

$$\mathbf{v} = \frac{d\mathbf{r}}{dt} = \mathbf{i} + 2t\mathbf{j}$$

$$|\mathbf{v}| = \sqrt{1 + 4t^2}$$

so that

$$\mathbf{T} = \frac{\mathbf{v}}{|\mathbf{v}|} = (1 + 4t^2)^{-1/2}\mathbf{i} + 2t(1 + 4t^2)^{-1/2}\mathbf{j}.$$

From this we find

$$\frac{d\mathbf{T}}{dt} = -4t(1 + 4t^2)^{-3/2}\mathbf{i} + [2(1 + 4t^2)^{-1/2} - 8t^2(1 + 4t^2)^{-3/2}]\mathbf{j}.$$

At the origin, $t = 0$, so the curvature is

$$\begin{aligned} \kappa(0) &= \frac{1}{|\mathbf{v}(0)|}\left|\frac{d\mathbf{T}}{dt}(0)\right| \qquad \text{Eq. (6)} \\ &= \frac{1}{\sqrt{1}}|0\mathbf{i} + 2\mathbf{j}| \\ &= (1)\sqrt{0^2 + 2^2} = 2. \end{aligned}$$

Therefore, the radius of curvature is $1/\kappa = \dfrac{1}{2}$ and the center of the circle is $\left(0, \dfrac{1}{2}\right)$ (see Figure 10.53). The equation of the osculating circle is

$$(x - 0)^2 + \left(y - \frac{1}{2}\right)^2 = \left(\frac{1}{2}\right)^2$$

or

$$x^2 + \left(y - \frac{1}{2}\right)^2 = \frac{1}{4}.$$

You can see from Figure 10.53 that the osculating circle is a better approximation to the parabola at the origin than is the tangent line approximation $y = 0$.

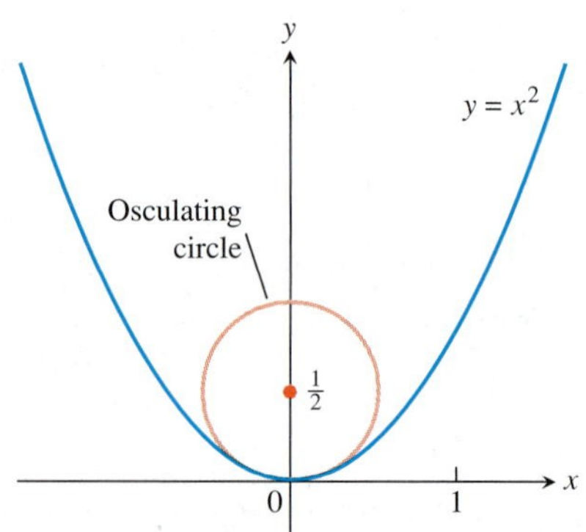

FIGURE 10.53 The osculating circle for the parabola $y = x^2$ at the origin is $x^2 + \left(y - \frac{1}{2}\right)^2 = \frac{1}{4}$. (Example 7)

EXERCISES 10.6

Finding Unit Tangent Vectors and Lengths of Curves in Space

In Exercises 1–8, find the curve's unit tangent vector. Also, find the length of the indicated portion of the curve.

1. $\mathbf{r}(t) = (2\cos t)\mathbf{i} + (2\sin t)\mathbf{j} + \sqrt{5}t\mathbf{k}, \quad 0 \le t \le \pi$
2. $\mathbf{r}(t) = (6\sin 2t)\mathbf{i} + (6\cos 2t)\mathbf{j} + 5t\mathbf{k}, \quad 0 \le t \le \pi$
3. $\mathbf{r}(t) = t\mathbf{i} + (2/3)t^{3/2}\mathbf{k}, \quad 0 \le t \le 8$
4. $\mathbf{r}(t) = (2 + t)\mathbf{i} - (t + 1)\mathbf{j} + t\mathbf{k}, \quad 0 \le t \le 3$
5. $\mathbf{r}(t) = (\cos^3 t)\mathbf{j} + (\sin^3 t)\mathbf{k}, \quad 0 \le t \le \pi/2$
6. $\mathbf{r}(t) = 6t^3\mathbf{i} - 2t^3\mathbf{j} - 3t^3\mathbf{k}, \quad 1 \le t \le 2$
7. $\mathbf{r}(t) = (t\cos t)\mathbf{i} + (t\sin t)\mathbf{j} + (2\sqrt{2}/3)t^{3/2}\mathbf{k}, \quad 0 \le t \le \pi$
8. $\mathbf{r}(t) = (t\sin t + \cos t)\mathbf{i} + (t\cos t - \sin t)\mathbf{j}, \quad \sqrt{2} \le t \le 2$
9. Find the point on the curve
$$\mathbf{r}(t) = (5\sin t)\mathbf{i} + (5\cos t)\mathbf{j} + 12t\mathbf{k}$$
at a distance 26π units along the curve from the point $(0, 5, 0)$ when $t = 0$ in the direction of increasing arc length.
10. Find the point on the curve
$$\mathbf{r}(t) = (12\sin t)\mathbf{i} - (12\cos t)\mathbf{j} + 5t\mathbf{k}$$
at a distance 13π units along the curve from the point $(0, -12, 0)$ when $t = 0$ in the direction opposite to the direction of increasing arc length.

Arc Length Parameter

In Exercises 11–14, find the arc length parameter along the curve from the point where $t = 0$ by evaluating the integral

$$s = \int_0^t |\mathbf{v}(\tau)|\, d\tau$$

from Equation (3). Then find the length of the indicated portion of the curve.

11. $\mathbf{r}(t) = (4 \cos t)\mathbf{i} + (4 \sin t)\mathbf{j} + 3t\mathbf{k}, \quad 0 \le t \le \pi/2$

12. $\mathbf{r}(t) = (\cos t + t \sin t)\mathbf{i} + (\sin t - t \cos t)\mathbf{j}, \quad \pi/2 \le t \le \pi$

13. $\mathbf{r}(t) = (e^t \cos t)\mathbf{i} + (e^t \sin t)\mathbf{j} + e^t\mathbf{k}, \quad -\ln 4 \le t \le 0$

14. $\mathbf{r}(t) = (1 + 2t)\mathbf{i} + (1 + 3t)\mathbf{j} + (6 - 6t)\mathbf{k}, \quad -1 \le t \le 0$

Plane Curves

Find $\mathbf{T}$, $\mathbf{N}$, and κ for the plane curves in Exercises 15–18.

15. $\mathbf{r}(t) = t\mathbf{i} + (\ln \cos t)\mathbf{j}, \quad -\pi/2 < t < \pi/2$

16. $\mathbf{r}(t) = (\ln \sec t)\mathbf{i} + t\mathbf{j}, \quad -\pi/2 < t < \pi/2$

17. $\mathbf{r}(t) = (2t + 3)\mathbf{i} + (5 - t^2)\mathbf{j}$

18. $\mathbf{r}(t) = (\cos t + t \sin t)\mathbf{i} + (\sin t - t \cos t)\mathbf{j}, \quad t > 0$

Theory and Examples

19. *Arc length* Find the length of the curve

$$\mathbf{r}(t) = (\sqrt{2}t)\mathbf{i} + (\sqrt{2}t)\mathbf{j} + (1 - t^2)\mathbf{k}$$

from $(0, 0, 1)$ to $(\sqrt{2}, \sqrt{2}, 0)$.

20. *Length of helix* The length $2\pi\sqrt{2}$ of the turn of the helix in Example 1 is also the length of the diagonal of a square 2π units on a side. Show how to obtain this square by cutting away and flattening a portion of the cylinder around which the helix winds.

21. *Ellipse*

(a) Show that the curve $\mathbf{r}(t) = (\cos t)\mathbf{i} + (\sin t)\mathbf{j} + (1 - \cos t)\mathbf{k}$, $0 \le t \le 2\pi$, is an ellipse by showing that it is the intersection of a right circular cylinder and a plane. Find equations for the cylinder and plane.

(b) Sketch the ellipse on the cylinder. Add to your sketch the unit tangent vectors at $t = 0, \pi/2, \pi$, and $3\pi/2$.

(c) Show that the acceleration vector always lies parallel to the plane (orthogonal to a vector normal to the plane). Thus, if you draw the acceleration as a vector attached to the ellipse, it will lie in the plane of the ellipse. Add the acceleration vectors for $t = 0, \pi/2, \pi$, and $3\pi/2$ to your sketch.

(d) Write an integral for the length of the ellipse. Do not try to evaluate the integral; it is nonelementary.

T **(e)** *Numerical integrator* Estimate the length of the ellipse to two decimal places.

22. *Length is independent of parametrization* To illustrate that the length of a smooth space curve does not depend on the parametrization you use to compute it, calculate the length of one turn of the helix in Example 1 with the following parametrizations.

(a) $\mathbf{r}(t) = (\cos 4t)\mathbf{i} + (\sin 4t)\mathbf{j} + 4t\mathbf{k}, \quad 0 \le t \le \pi/2$

(b) $\mathbf{r}(t) = [\cos (t/2)]\mathbf{i} + [\sin (t/2)]\mathbf{j} + (t/2)\mathbf{k}, \quad 0 \le t \le 4\pi$

(c) $\mathbf{r}(t) = (\cos t)\mathbf{i} - (\sin t)\mathbf{j} - t\mathbf{k}, \quad -2\pi \le t \le 0$

23. *Circle of curvature* Find an equation for the circle of curvature of the curve $\mathbf{r}(t) = t\mathbf{i} + (\sin t)\mathbf{j}$ at the point $(\pi/2, 1)$. (The curve parametrizes the graph of $y = \sin x$ in the xy-plane.)

24. *Circle of curvature* Find an equation for the circle of curvature of the curve $\mathbf{r}(t) = (2 \ln t)\mathbf{i} - [t + (1/t)]\mathbf{j}$, $e^{-2} \le t \le e^2$, at the point $(0, -2)$, where $t = 1$.

10.7 The TNB Frame; Tangential and Normal Components of Acceleration

Curvature and Normal Vectors for Space Curves • Torsion and the Binormal Vector • Tangential and Normal Components of Acceleration • Formulas for Computing Curvature and Torsion

If you are traveling along a space curve, the Cartesian $\mathbf{i}$, $\mathbf{j}$, and $\mathbf{k}$ coordinate system for representing the vectors describing your motion are not truly relevant to you. What is meaningful instead are the vectors representative of your forward direction (the unit tangent vector $\mathbf{T}$), the direction in which your path is turning (the unit normal vector $\mathbf{N}$), and the tendency of your motion to "twist" out of the plane created by these vectors in the direction perpendicular to this plane (defined by the *unit binormal vector* $\mathbf{B} = \mathbf{T} \times \mathbf{N}$). Expressing the acceleration vector along the curve as a linear combination of this **TNB** frame of mutually orthogonal unit vectors traveling with the motion (Figure 10.54) is particularly revealing of the nature of the path and motion along it.

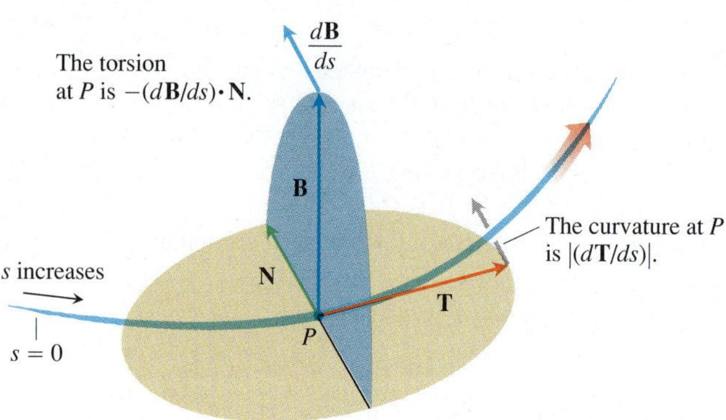

FIGURE 10.54 Every moving body travels with a **TNB** frame that characterizes the geometry of its path of motion.

For example, $|d\mathbf{T}/ds|$ tells how much a vehicle's path turns to the left or right as it moves along; it is called the *curvature* of the vehicle's path. The number $-(d\mathbf{B}/ds)\cdot\mathbf{N}$ tells how much a vehicle's path rotates or twists out of its plane of motion as the vehicle moves along; it is called the *torsion* of the vehicle's path. Look at Figure 10.54 again. If P is a train climbing up a curved track, the rate at which the headlight turns from side to side per unit distance is the curvature of the track. The rate at which the engine tends to twist out of the plane formed by $\mathbf{T}$ and $\mathbf{N}$ is the torsion.

Curvature and Normal Vectors for Space Curves

The unit tangent vector $\mathbf{T}$ for space curves is defined the same way as for planar curves. If the smooth curve is specified by the position vector $\mathbf{r}(t)$ as a function of some parameter t, and if s is the arc length parameter of the curve, then $\mathbf{T} = d\mathbf{r}/ds = \mathbf{v}/|\mathbf{v}|$. The **curvature** in space is then defined to be

$$\kappa = \left|\frac{d\mathbf{T}}{ds}\right| = \frac{1}{|\mathbf{v}|}\left|\frac{d\mathbf{T}}{dt}\right| \tag{1}$$

just as for plane curves (Section 10.6, Equation 6). The vector $d\mathbf{T}/ds$ is orthogonal to $\mathbf{T}$, and we define the **principal unit normal** to be

$$\mathbf{N} = \frac{1}{\kappa}\frac{d\mathbf{T}}{ds} = \frac{d\mathbf{T}/dt}{|d\mathbf{T}/dt|}. \tag{2}$$

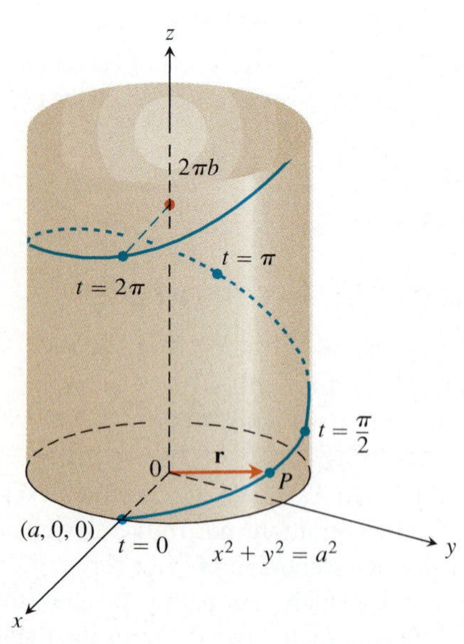

FIGURE 10.55 The helix

$$\mathbf{r}(t) = (a\cos t)\mathbf{i} + (a\sin t)\mathbf{j} + bt\mathbf{k},$$

drawn with a and b positive and $t \geq 0$. (Example 1)

Example 1 Finding Curvature

Find the curvature for the helix (Figure 10.55)

$$\mathbf{r}(t) = (a\cos t)\mathbf{i} + (a\sin t)\mathbf{j} + bt\mathbf{k}, \qquad a, b \geq 0, \qquad a^2 + b^2 \neq 0.$$

Solution We calculate $\mathbf{T}$ from the velocity vector $\mathbf{v}$:

$$\mathbf{v} = -(a\sin t)\mathbf{i} + (a\cos t)\mathbf{j} + b\mathbf{k}$$

$$|\mathbf{v}| = \sqrt{a^2\sin^2 t + a^2\cos^2 t + b^2} = \sqrt{a^2 + b^2}$$

$$\mathbf{T} = \frac{\mathbf{v}}{|\mathbf{v}|} = \frac{1}{\sqrt{a^2 + b^2}}[-(a\sin t)\mathbf{i} + (a\cos t)\mathbf{j} + b\mathbf{k}].$$

Then using Equation (1),

$$
\begin{aligned}
\kappa &= \frac{1}{|\mathbf{v}|}\left|\frac{d\mathbf{T}}{dt}\right| \\
&= \frac{1}{\sqrt{a^2+b^2}}\left|\frac{1}{\sqrt{a^2+b^2}}\left[-(a\cos t)\mathbf{i}-(a\sin t)\mathbf{j}\right]\right| \\
&= \frac{a}{a^2+b^2}\left|-(\cos t)\mathbf{i}-(\sin t)\mathbf{j}\right| \\
&= \frac{a}{a^2+b^2}\sqrt{(\cos t)^2+(\sin t)^2} = \frac{a}{a^2+b^2}.
\end{aligned}
$$

From this equation, we see that increasing b for a fixed a decreases the curvature. Decreasing a for a fixed b eventually decreases the curvature as well. Stretching a spring tends to straighten it.

If $b = 0$, the helix reduces to a circle of radius a and its curvature reduces to $1/a$, as it should. If $a = 0$, the helix becomes the z-axis, and its curvature reduces to 0, again as it should.

Example 2 Finding the Principal Unit Normal Vector N

Find **N** for the helix in Example 1.

Solution We have

$$\frac{d\mathbf{T}}{dt} = -\frac{1}{\sqrt{a^2+b^2}}\left[(a\cos t)\mathbf{i}+(a\sin t)\mathbf{j}\right] \qquad \text{Example 1}$$

$$\left|\frac{d\mathbf{T}}{dt}\right| = \frac{1}{\sqrt{a^2+b^2}}\sqrt{a^2\cos^2 t + a^2\sin^2 t} = \frac{a}{\sqrt{a^2+b^2}}$$

$$
\begin{aligned}
\mathbf{N} &= \frac{d\mathbf{T}/dt}{|d\mathbf{T}/dt|} \qquad \text{Eq. (2)} \\
&= -\frac{\sqrt{a^2+b^2}}{a}\cdot\frac{1}{\sqrt{a^2+b^2}}\left[(a\cos t)\mathbf{i}+(a\sin t)\mathbf{j}\right] \\
&= -(\cos t)\mathbf{i}-(\sin t)\mathbf{j}.
\end{aligned}
$$

Torsion and the Binormal Vector

The **binormal vector** of a curve in space is $\mathbf{B} = \mathbf{T}\times\mathbf{N}$, a unit vector orthogonal to both $\mathbf{T}$ and $\mathbf{N}$ (Figure 10.56). Together $\mathbf{T}$, $\mathbf{N}$, and $\mathbf{B}$ define a moving right-handed vector frame that plays a significant role in calculating the paths of particles moving through space.

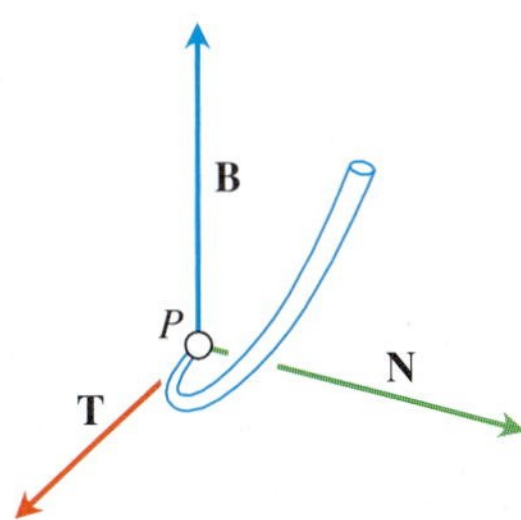

FIGURE 10.56 The vectors $\mathbf{T}$, $\mathbf{N}$, and $\mathbf{B}$ (in that order) make a right-handed frame of mutually orthogonal unit vectors in space. You can call it the **Frenet** ("fre-*nay*") **frame** (after Jean-Frédéric Frenet, 1816–1900), or you can call it the **TNB frame.**

How does $d\mathbf{B}/ds$ behave in relation to $\mathbf{T}$, $\mathbf{N}$, and $\mathbf{B}$? From the rule for differentiating a cross product, we have

$$\frac{d\mathbf{B}}{ds} = \frac{d\mathbf{T}}{ds}\times\mathbf{N} + \mathbf{T}\times\frac{d\mathbf{N}}{ds}.$$

Since $\mathbf{N}$ is the direction of $d\mathbf{T}/ds$, $(d\mathbf{T}/ds)\times\mathbf{N} = \mathbf{0}$ and

$$\frac{d\mathbf{B}}{ds} = \mathbf{0} + \mathbf{T}\times\frac{d\mathbf{N}}{ds} = \mathbf{T}\times\frac{d\mathbf{N}}{ds}.$$

From this we see that $d\mathbf{B}/ds$ is orthogonal to $\mathbf{T}$ since a cross product is orthogonal to its factors.

Since $d\mathbf{B}/ds$ is also orthogonal to $\mathbf{B}$ (the latter has constant length), it follows that $d\mathbf{B}/ds$ is orthogonal to the plane of $\mathbf{B}$ and $\mathbf{T}$. In other words, $d\mathbf{B}/ds$ is parallel to $\mathbf{N}$, so $d\mathbf{B}/ds$ is a scalar multiple of $\mathbf{N}$. In symbols,

$$\frac{d\mathbf{B}}{ds} = -\tau\mathbf{N}.$$

The minus sign in this equation is traditional. The scalar τ is called the torsion along the curve. Notice that

$$\frac{d\mathbf{B}}{ds} \cdot \mathbf{N} = -\tau\mathbf{N} \cdot \mathbf{N} = -\tau(1) = -\tau,$$

so that

$$\tau = -\frac{d\mathbf{B}}{ds} \cdot \mathbf{N}.$$

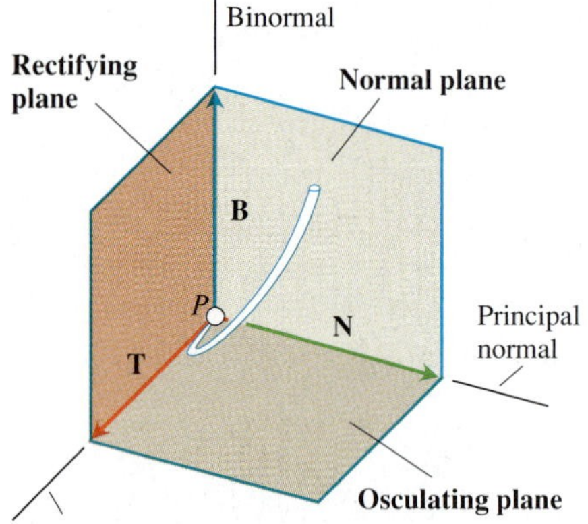

FIGURE 10.57 The names of the three planes determined by $\mathbf{T}$, $\mathbf{N}$, and $\mathbf{B}$.

Definition Torsion

Let $\mathbf{B} = \mathbf{T} \times \mathbf{N}$. The **torsion** function of a smooth curve is

$$\tau = -\frac{d\mathbf{B}}{ds} \cdot \mathbf{N}.$$

Unlike the curvature κ, which is never negative, the torsion τ may be positive, negative, or zero.

The three planes determined by $\mathbf{T}$, $\mathbf{N}$, and $\mathbf{B}$ are shown in Figure 10.57. The curvature $\kappa = |d\mathbf{T}/ds|$ can be thought of as the rate at which the normal plane turns as the point P moves along its path. Similarly, the torsion $\tau = -(d\mathbf{B}/ds) \cdot \mathbf{N}$ is the rate at which the osculating plane turns about $\mathbf{T}$ as P moves along the curve. Torsion measures how the curve twists.

Tangential and Normal Components of Acceleration

When a body is accelerated by gravity, brakes, a combination of rocket motors, or whatever, we usually want to know how much of the acceleration acts in the direction of motion, in the tangential direction $\mathbf{T}$. We can find out if we use the Chain Rule to rewrite $\mathbf{v}$ as

$$\mathbf{v} = \frac{d\mathbf{r}}{dt} = \frac{d\mathbf{r}}{ds}\frac{ds}{dt} = \mathbf{T}\frac{ds}{dt}$$

and differentiate both ends of this string of equalities to get

$$\begin{aligned}
\mathbf{a} &= \frac{d\mathbf{v}}{dt} = \frac{d}{dt}\left(\mathbf{T}\frac{ds}{dt}\right) = \frac{d^2s}{dt^2}\mathbf{T} + \frac{ds}{dt}\frac{d\mathbf{T}}{dt} \\
&= \frac{d^2s}{dt^2}\mathbf{T} + \frac{ds}{dt}\left(\frac{d\mathbf{T}}{ds}\frac{ds}{dt}\right) = \frac{d^2s}{dt^2}\mathbf{T} + \frac{ds}{dt}\left(\kappa\mathbf{N}\frac{ds}{dt}\right) \qquad \frac{d\mathbf{T}}{ds} = \kappa\mathbf{N} \text{ from Eq. (2)} \\
&= \frac{d^2s}{dt^2}\mathbf{T} + \kappa\left(\frac{ds}{dt}\right)^2\mathbf{N}.
\end{aligned}$$

Definition Tangential and Normal Components of Acceleration

$$\mathbf{a} = a_{\mathrm{T}}\mathbf{T} + a_{\mathrm{N}}\mathbf{N}, \tag{3}$$

where

$$a_{\mathrm{T}} = \frac{d^2s}{dt^2} = \frac{d}{dt}|\mathbf{v}| \quad \text{and} \quad a_{\mathrm{N}} = \kappa\left(\frac{ds}{dt}\right)^2 = \kappa|\mathbf{v}|^2 \tag{4}$$

are the **tangential** and **normal** scalar components of acceleration.

Equation (3) is remarkable in that $\mathbf{B}$ does not appear. No matter how the path of the moving body we are watching may appear to twist and turn in space, the acceleration $\mathbf{a}$ *always lies in the plane of* $\mathbf{T}$ and $\mathbf{N}$ orthogonal to $\mathbf{B}$. The equation also tells us exactly how much of the acceleration takes place tangent to the motion (d^2s/dt^2) and how much takes place normal to the motion $[\kappa(ds/dt)^2]$ (Figure 10.58).

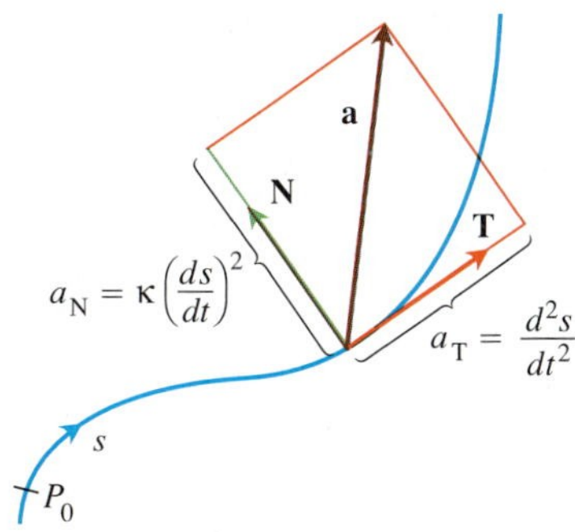

FIGURE 10.58 The tangential and normal components of acceleration. The acceleration $\mathbf{a}$ always lies in the plane of $\mathbf{T}$ and $\mathbf{N}$, orthogonal to $\mathbf{B}$.

What information can we read from Equations (4)? By definition, acceleration $\mathbf{a}$ is the rate of change of velocity $\mathbf{v}$, and in general, both the length and direction of $\mathbf{v}$ change as a body moves along its path. The tangential component of acceleration a_{T} measures the rate of change of the *length* of $\mathbf{v}$ (that is, the change in the speed). The normal component of acceleration a_{N} measures the rate of change of the *direction* of $\mathbf{v}$.

Notice that the normal scalar component of the acceleration is the curvature times the *square* of the speed. This explains why you have to hold on when your car makes a sharp (large κ), high-speed (large $|\mathbf{v}|$) turn. If you double the speed of your car, you will experience four times the normal component of acceleration for the same curvature.

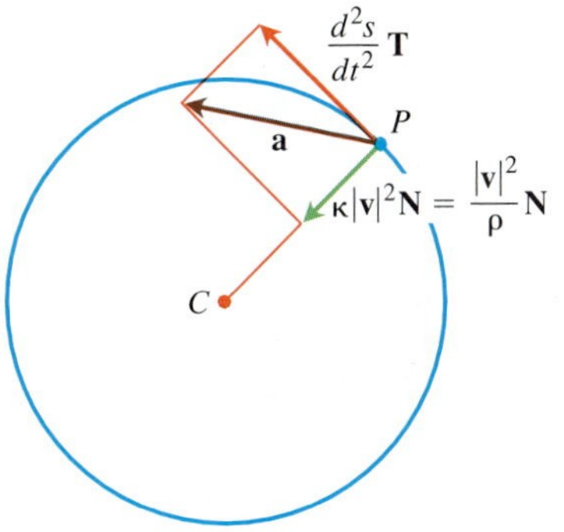

If a body moves in a circle at a constant speed, d^2s/dt^2 is zero and all the acceleration points along $\mathbf{N}$ toward the circle's center. If the body is speeding up or slowing down, $\mathbf{a}$ has a nonzero tangential component (Figure 10.59).

FIGURE 10.59 The tangential and normal components of the acceleration of a body that is speeding up as it moves counterclockwise around a circle of radius ρ.

To calculate a_{N}, we usually use the formula $a_{\mathrm{N}} = \sqrt{|\mathbf{a}|^2 - a_{\mathrm{T}}^2}$, which comes from solving the equation $|\mathbf{a}|^2 = \mathbf{a} \cdot \mathbf{a} = a_{\mathrm{T}}^2 + a_{\mathrm{N}}^2$ for a_{N}. With this formula, we can find a_{N} without having to calculate κ first.

Formula for Calculating the Normal Component of Acceleration

$$a_{\mathrm{N}} = \sqrt{|\mathbf{a}|^2 - a_{\mathrm{T}}^2} \tag{5}$$

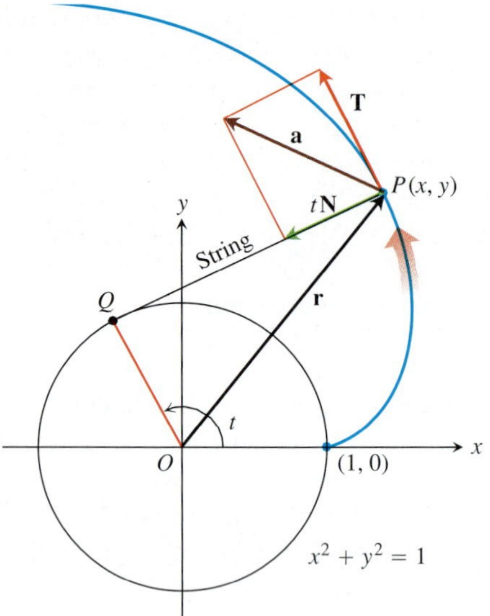

FIGURE 10.60 The tangential and normal components of the acceleration of the motion $\mathbf{r}(t) = (\cos t + t \sin t)\mathbf{i} + (\sin t - t \cos t)\mathbf{j}$, for $t > 0$. If a string wound around a fixed circle is unwound while held taught in the plane of the circle, its end P traces an involute of the circle. (Example 3)

Example 3 Finding the Acceleration Scalar Components a_T, a_N

Without finding $\mathbf{T}$ and $\mathbf{N}$, write the acceleration of the motion

$$\mathbf{r}(t) = (\cos t + t \sin t)\mathbf{i} + (\sin t - t \cos t)\mathbf{j}, \qquad t > 0$$

in the form $\mathbf{a} = a_{\mathrm{T}}\mathbf{T} + a_{\mathrm{N}}\mathbf{N}$. (The path of the motion is the involute of the circle in Figure 10.60.

Solution We use the first of Equations (4) to find a_{T}:

$$\mathbf{v} = \frac{d\mathbf{r}}{dt} = (-\sin t + \sin t + t \cos t)\mathbf{i} + (\cos t - \cos t + t \sin t)\mathbf{j}$$
$$= (t \cos t)\mathbf{i} + (t \sin t)\mathbf{j}$$
$$|\mathbf{v}| = \sqrt{t^2 \cos^2 t + t^2 \sin^2 t} = \sqrt{t^2} = |t| = t \qquad t > 0$$
$$a_{\mathrm{T}} = \frac{d}{dt}|\mathbf{v}| = \frac{d}{dt}(t) = 1. \qquad \text{Eq. (4)}$$

Knowing a_{T}, we use Equation (5) to find a_{N}:

$$\mathbf{a} = (\cos t - t \sin t)\mathbf{i} + (\sin t + t \cos t)\mathbf{j}$$
$$|\mathbf{a}|^2 = t^2 + 1 \qquad \text{After some algebra}$$
$$a_{\mathrm{N}} = \sqrt{|\mathbf{a}|^2 - a_{\mathrm{T}}^2}$$
$$= \sqrt{(t^2 + 1) - (1)} = \sqrt{t^2} = t.$$

We then use Equation (3) to find $\mathbf{a}$:

$$a = a_{\mathrm{T}}\mathbf{T} + a_{\mathrm{N}}\mathbf{N} = (1)\mathbf{T} + (t)\,\mathbf{N} = \mathbf{T} + t\mathbf{N}.$$

See Figure 10.60.

Formulas for Computing Curvature and Torsion

We now give some easy-to-use formulas for computing the curvature and torsion of a smooth curve. From Equation (3), we have

$$\mathbf{v} \times \mathbf{a} = \left(\frac{ds}{dt}\mathbf{T}\right) \times \left[\frac{d^2s}{dt^2}\mathbf{T} + \kappa\left(\frac{ds}{dt}\right)^2\mathbf{N}\right] \qquad \text{From Section 10.6, Eq. (5), } \mathbf{v} = d\mathbf{r}/dt = (ds/dt)\mathbf{T}$$
$$= \left(\frac{ds}{dt}\frac{d^2s}{dt^2}\right)(\mathbf{T} \times \mathbf{T}) + \kappa\left(\frac{ds}{dt}\right)^3(\mathbf{T} \times \mathbf{N})$$
$$= \kappa\left(\frac{ds}{dt}\right)^3\mathbf{B}. \qquad \mathbf{T} \times \mathbf{T} = \mathbf{0} \text{ and } \mathbf{T} \times \mathbf{N} = \mathbf{B}$$

It follows that

$$|\mathbf{v} \times \mathbf{a}| = \kappa\left|\frac{ds}{dt}\right|^3|\mathbf{B}| = \kappa|\mathbf{v}|^3. \qquad \frac{ds}{dt} = |\mathbf{v}| \text{ and } |\mathbf{B}| = 1$$

Solving for κ gives the following formula.

Vector Formula for Curvature

$$\kappa = \frac{|\mathbf{v} \times \mathbf{a}|}{|\mathbf{v}|^3} \qquad (6)$$

Newton's Dot Notation for Derivatives

The dots in Equation (7) denote differentiation with respect to t, one derivative for each dot. Thus, $\dot{x}$ ("x dot") means dx/dt, $\ddot{x}$ ("x double dot") means d^2x/dt^2, and $\dddot{x}$ ("x triple dot") means d^3x/dt^3. Similarly, $\dot{y} = dy/dt$ and so on.

Equation (6) calculates the curvature, a geometric property of the curve, from the velocity and acceleration of any vector representation of the curve in which $|\mathbf{v}|$ is different from zero. Take a moment to think about how remarkable this really is: From any formula for motion along a curve, no matter how variable the motion may be (as long as $\mathbf{v}$ is never zero), we can calculate a physical property of the curve that seems to have nothing to do with the way the curve is traversed.

The most widely used formula for torsion, derived in more advanced texts, is

$$\tau = \frac{\begin{vmatrix} \dot{x} & \dot{y} & \dot{z} \\ \ddot{x} & \ddot{y} & \ddot{z} \\ \dddot{x} & \dddot{y} & \dddot{z} \end{vmatrix}}{|\mathbf{v} \times \mathbf{a}|^2} \qquad (\text{if } \mathbf{v} \times \mathbf{a} \neq \mathbf{0}). \tag{7}$$

This formula calculates the torsion directly from the derivatives of the component functions $x = f(t)$, $y = g(t)$, $z = h(t)$ that make up $\mathbf{r}$. The determinant's first row comes from $\mathbf{v}$, the second row comes from $\mathbf{a}$, and the third row comes from $\dot{\mathbf{a}} = d\mathbf{a}/dt$.

Example 4 Finding Curvature and Torsion

Use Equations (6) and (7) to find κ and τ for the helix

$$\mathbf{r}(t) = (a \cos t)\mathbf{i} + (a \sin t)\mathbf{j} + bt\mathbf{k}, \qquad a, b \geq 0, \qquad a^2 + b^2 \neq 0.$$

Solution We calculate the curvature with Equation (6):

$$\mathbf{v} = -(a \sin t)\mathbf{i} + (a \cos t)\mathbf{j} + b\mathbf{k}$$

$$\mathbf{a} = -(a \cos t)\mathbf{i} - (a \sin t)\mathbf{j}$$

$$\mathbf{v} \times \mathbf{a} = \begin{vmatrix} \mathbf{i} & \mathbf{j} & \mathbf{k} \\ -a \sin t & a \cos t & b \\ -a \cos t & -a \sin t & 0 \end{vmatrix}$$

$$= (ab \sin t)\mathbf{i} - (ab \cos t)\mathbf{j} + a^2\mathbf{k}$$

$$\kappa = \frac{|\mathbf{v} \times \mathbf{a}|}{|\mathbf{v}|^3} = \frac{\sqrt{a^2b^2 + a^4}}{(a^2 + b^2)^{3/2}} = \frac{a\sqrt{a^2 + b^2}}{(a^2 + b^2)^{3/2}} = \frac{a}{a^2 + b^2}. \tag{8}$$

Notice that Equation (8) agrees with the result in Example 1, where we calculated the curvature directly from its definition.

To evaluate Equation (7) for the torsion, we find the entries in the determinant by differentiating $\mathbf{r}$ with respect to t. We already have $\mathbf{v}$ and $\mathbf{a}$, and

$$\dot{\mathbf{a}} = \frac{d\mathbf{a}}{dt} = (a \sin t)\mathbf{i} - (a \cos t)\mathbf{j}.$$

Hence,

$$\tau = \frac{\begin{vmatrix} \dot{x} & \dot{y} & \dot{z} \\ \ddot{x} & \ddot{y} & \ddot{z} \\ \dddot{x} & \dddot{y} & \dddot{z} \end{vmatrix}}{|\mathbf{v} \times \mathbf{a}|^2} = \frac{\begin{vmatrix} -a \sin t & a \cos t & b \\ -a \cos t & -a \sin t & 0 \\ a \sin t & -a \cos t & 0 \end{vmatrix}}{\left(a\sqrt{a^2 + b^2}\right)^2}$$

Value of $|\mathbf{v} \times \mathbf{a}|$ from Eq. (8)

$$= \frac{b(a^2 \cos^2 t + a^2 \sin^2 t)}{a^2(a^2 + b^2)}$$

$$= \frac{b}{a^2 + b^2}. \tag{9}$$

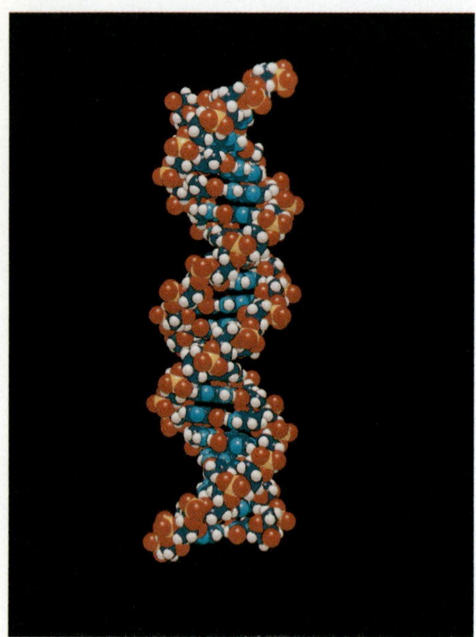

FIGURE 10.61 The helical shape of a DNA molecule is characterized by its constant curvature and torsion.

From Equation (9), we see that the torsion of a helix about a circular cylinder is constant. In fact, constant curvature and constant torsion characterize the helix among all curves in space.

The DNA molecule, the basic building block of life forms, is designed in the form of two helices winding around each other, a little like the rungs and sides of a twisted rope ladder (Figure 10.61). Not only is the space occupied by the DNA molecule very much smaller than it would be if it were unraveled, but when the molecule is damaged, the imperfect piece can be snipped out by a kind of molecular scissors (because the curvature and torsion functions are constant) and the DNA made right again.

Formulas for Curves in Space

Unit tangent vector: $\mathbf{T} = \dfrac{\mathbf{v}}{|\mathbf{v}|}$

Principal unit normal vector: $\mathbf{N} = \dfrac{d\mathbf{T}/dt}{|d\mathbf{T}/dt|}$

Binormal vector: $\mathbf{B} = \mathbf{T} \times \mathbf{N}$

Curvature: $\kappa = \left|\dfrac{d\mathbf{T}}{ds}\right| = \dfrac{|\mathbf{v} \times \mathbf{a}|}{|\mathbf{v}|^3}$

Torsion: $\tau = -\dfrac{d\mathbf{B}}{ds} \cdot \mathbf{N} = \dfrac{\begin{vmatrix} \dot{x} & \dot{y} & \dot{z} \\ \ddot{x} & \ddot{y} & \ddot{z} \\ \dddot{x} & \dddot{y} & \dddot{z} \end{vmatrix}}{|\mathbf{v} \times \mathbf{a}|^2}$

Tangential and normal scalar components of acceleration:

$$\mathbf{a} = a_{\mathrm{T}}\mathbf{T} + a_{\mathrm{N}}\mathbf{N}$$

$$a_{\mathrm{T}} = \frac{d}{dt}|\mathbf{v}|$$

$$a_{\mathrm{N}} = \kappa|\mathbf{v}|^2 = \sqrt{|\mathbf{a}|^2 - a_{\mathrm{T}}^2}$$

EXERCISES 10.7

Space Curves

Find $\mathbf{T}, \mathbf{N}, \mathbf{B}, \kappa$, and τ for the space curves in Exercises 1–8.

1. $\mathbf{r}(t) = (3 \sin t)\mathbf{i} + (3 \cos t)\mathbf{j} + 4t\mathbf{k}$
2. $\mathbf{r}(t) = (\cos t + t \sin t)\mathbf{i} + (\sin t - t \cos t)\mathbf{j} + 3\mathbf{k}$
3. $\mathbf{r}(t) = (e^t \cos t)\mathbf{i} + (e^t \sin t)\mathbf{j} + 2\mathbf{k}$
4. $\mathbf{r}(t) = (6 \sin 2t)\mathbf{i} + (6 \cos 2t)\mathbf{j} + 5t\mathbf{k}$
5. $\mathbf{r}(t) = (t^3/3)\mathbf{i} + (t^2/2)\mathbf{j}, \quad t > 0$
6. $\mathbf{r}(t) = (\cos^3 t)\mathbf{i} + (\sin^3 t)\mathbf{j}, \quad 0 < t < \pi/2$
7. $\mathbf{r}(t) = t\mathbf{i} + (a \cosh (t/a))\mathbf{j}, \quad a > 0$
8. $\mathbf{r}(t) = (\cosh t)\mathbf{i} - (\sinh t)\mathbf{j} + t\mathbf{k}$

In Exercises 9 and 10, write $\mathbf{a}$ in the form $a_{\mathrm{T}}\mathbf{T} + a_{\mathrm{N}}\mathbf{N}$ without finding $\mathbf{T}$ and $\mathbf{N}$.

9. $\mathbf{r}(t) = (a \cos t)\mathbf{i} + (a \sin t)\mathbf{j} + bt\mathbf{k}$
10. $\mathbf{r}(t) = (1 + 3t)\mathbf{i} + (t - 2)\mathbf{j} - 3t\mathbf{k}$

In Exercises 11–14, write $\mathbf{a}$ in the form $\mathbf{a} = a_{\mathrm{T}}\mathbf{T} + a_{\mathrm{N}}\mathbf{N}$ at the given value of t without finding $\mathbf{T}$ and $\mathbf{N}$.

11. $\mathbf{r}(t) = (t + 1)\mathbf{i} + 2t\mathbf{j} + t^2\mathbf{k}, \quad t = 1$
12. $\mathbf{r}(t) = (t \cos t)\mathbf{i} + (t \sin t)\mathbf{j} + t^2\mathbf{k}, \quad t = 0$
13. $\mathbf{r}(t) = t^2\mathbf{i} + (t + (1/3)t^3)\mathbf{j} + (t - (1/3)t^3)\mathbf{k}, \quad t = 0$
14. $\mathbf{r}(t) = (e^t \cos t)\mathbf{i} + (e^t \sin t)\mathbf{j} + \sqrt{2}e^t\mathbf{k}, \quad t = 0$

In Exercises 15 and 16, find **r**, **T**, **N**, and **B** at the given value of t. Then find equations for the osculating, normal, and rectifying planes (Figure 10.57) at that value of t.

15. $\mathbf{r}(t) = (\cos t)\mathbf{i} + (\sin t)\mathbf{j} - \mathbf{k}, \quad t = \pi/4$

16. $\mathbf{r}(t) = (\cos t)\mathbf{i} + (\sin t)\mathbf{j} + t\mathbf{k}, \quad t = 0$

Physical Applications

17. *Writing to Learn* The speedometer on your car reads a steady 35 mph. Could you be accelerating? Explain.

18. *Writing to Learn* Can anything be said about the acceleration of a particle that is moving in space at a constant speed? Give reasons for your answer.

19. *Writing to Learn* Can anything be said about the speed of a particle whose acceleration is always orthogonal to its velocity? Give reasons for your answer.

20. *Motion along a parabola* An object of mass m travels along the parabola $y = x^2$ with a constant speed of 10 units/sec. What is the force on the object due to its acceleration at (0, 0)? At ($\sqrt{2}$, 2)? Write your answers in terms of **i** and **j**. (Remember Newton's law, $\mathbf{F} = m\mathbf{a}$.)

21. *Writing to Learn* The following is a quotation from an article in the *American Mathematical Monthly,* titled "Curvature in the Eighties" by Robert Osserman (October 1990, p. 731):

> Curvature also plays a key role in physics. The magnitude of a force required to move an object at constant speed along a curved path is, according to Newton's laws, a constant multiple of the curvature of the trajectories.

Explain mathematically why the second sentence of the quotation is true.

22. *What happens if $a_N = 0$?* Show that a moving particle will move in a straight line if the normal component of its acceleration is zero.

Curvature and Torsion

CD-ROM WEBsite

23. *A formula for the curvature of the graph of a function in the xy-plane*

(a) The graph $y = f(x)$ in the xy-plane automatically has the parametrization $x = x$, $y = f(x)$, and the vector formula $\mathbf{r}(x) = x\mathbf{i} + f(x)\mathbf{j}$. Use this formula to show that if f is a twice-differentiable function of x, then

$$\kappa(x) = \frac{|f''(x)|}{[1 + (f'(x))^2]^{3/2}}.$$

(b) Use the formula for κ in part (a) to find the curvature of $y = \ln(\cos x)$, $-\pi/2 < x < \pi/2$.

(c) Show that the curvature is zero at a point of inflection.

24. *A formula for the curvature of a parametrized plane curve*

(a) Show that the curvature of a smooth curve $\mathbf{r}(t) = f(t)\mathbf{i} + g(t)\mathbf{j}$ defined by twice-differentiable functions $x = f(t)$ and $y = g(t)$ is given by the formula

$$\kappa = \frac{|\dot{x}\ddot{y} - \dot{y}\ddot{x}|}{(\dot{x}^2 + \dot{y}^2)^{3/2}}.$$

Apply the formula to find the curvatures of the following curves.

(b) $\mathbf{r}(t) = t\mathbf{i} + (\ln \sin t)\mathbf{j}, \quad 0 < t < \pi$

(c) $\mathbf{r}(t) = [\tan^{-1}(\sinh t)]\mathbf{i} + (\ln \cosh t)\mathbf{j}$

25. *Normals to plane curves*

(a) Show that $\mathbf{n}(t) = -g'(t)\mathbf{i} + f'(t)\mathbf{j}$ and $-\mathbf{n}(t) = g'(t)\mathbf{i} - f'(t)\mathbf{j}$ are both normal to the curve $\mathbf{r}(t) = f(t)\mathbf{i} + g(t)\mathbf{j}$ at the point $(f(t), g(t))$.

To obtain **N** for a particular plane curve, we can choose the one of **n** or $-\mathbf{n}$ from part (a) that points toward the concave side of the curve, and make it into a unit vector. (See Figure 10.51.) Apply this method to find **N** for the following curves.

(b) $\mathbf{r}(t) = t\mathbf{i} + e^{2t}\mathbf{j}$

(c) $\mathbf{r}(t) = \sqrt{4 - t^2}\mathbf{i} + t\mathbf{j}, \quad -2 \le t \le 2$

26. *(Continuation of Exercise 25)*

(a) Use the method of Exercise 25 to find **N** for the curve $\mathbf{r}(t) = t\mathbf{i} + (1/3)t^3\mathbf{j}$ when $t < 0$ and when $t > 0$.

(b) *Writing to Learn* Calculate

$$\mathbf{N} = \frac{d\mathbf{T}/dt}{|d\mathbf{T}/dt|}, \qquad t \ne 0,$$

for the curve in part (a). Does **N** exist at $t = 0$? Graph the curve and explain what is happening to **N** as t passes from negative to positive values.

27. *Curvature extremes* Show that the parabola $y = ax^2$, $a \ne 0$, has its largest curvature at its vertex and has no minimum curvature. (*Note:* Since the curvature of a curve remains the same if the curve is translated or rotated, this result is true for any parabola.)

28. *Curvature extremes* Show that the ellipse $x = a\cos t$, $y = b\sin t$, $a > b > 0$, has its largest curvature on its major axis and its smallest curvature on its minor axis. (As in Exercise 27, the same is true for any ellipse.)

29. *Writing to Learn: Maximizing the curvature of a helix* In Example 1, we found the curvature of the helix $\mathbf{r}(t) = (a\cos t)\mathbf{i} + (a\sin t)\mathbf{j} + bt\mathbf{k}$ $(a, b \ge 0)$ to be $\kappa = a/(a^2 + b^2)$. What is the largest value κ can have for a given value of b? Give reasons for your answer.

30. *A sometime shortcut to curvature* If you already know $|a_N|$ and $|\mathbf{v}|$, then the formula $a_N = \kappa|\mathbf{v}|^2$ gives a convenient way to find the curvature. Use it to find the curvature and radius of curvature of the curve

$$\mathbf{r}(t) = (\cos t + t\sin t)\mathbf{i} + (\sin t - t\cos t)\mathbf{j}, \qquad t > 0.$$

(Take a_N and $|\mathbf{v}|$ from Example 3.)

31. *Curvature and torsion for a line* Show that κ and τ are both zero for the line

$$\mathbf{r}(t) = (x_0 + At)\mathbf{i} + (y_0 + Bt)\mathbf{j} + (z_0 + Ct)\mathbf{k}.$$

32. *Total curvature* We find the **total curvature** of the portion of a smooth curve that runs from $s = s_0$ to $s = s_1 > s_0$ by integrating κ from s_0 to s_1. If the curve has some other parameter, say t, then the total curvature is

$$K = \int_{s_0}^{s_1} \kappa\, ds = \int_{t_0}^{t_1} \kappa \frac{ds}{dt}\, dt = \int_{t_0}^{t_1} \kappa |\mathbf{v}|\, dt,$$

where t_0 and t_1 correspond to s_0 and s_1. Find the total curvature of the portion of the helix $\mathbf{r}(t) = (3 \cos t)\mathbf{i} + (3 \sin t)\mathbf{j} + t\mathbf{k}$, $0 \le t \le 4\pi$.

33. (*Continuation of Exercise 32.*) Find the total curvatures of the following curves.

(a) The involute of the unit circle: $\mathbf{r}(t) = (\cos t + t \sin t)\mathbf{i} + (\sin t - t \cos t)\mathbf{j}$, $a \le t \le b$ $(a > 0)$. (Exercise 30 gives a convenient way to find κ. Use values from Example 3.)

(b) The parabola $y = x^2$, $-\infty < x < \infty$.

34. *Writing to Learn: The torsion of a helix* In Example 4, we found the torsion of the helix

$$\mathbf{r}(t) = (a \cos t)\mathbf{i} + (a \sin t)\mathbf{j} + bt\mathbf{k}, \qquad a, b \ge 0$$

to be $\tau = b/(a^2 + b^2)$. What is the largest value τ can have for a given value of a? Give reasons for your answer.

35. *Differentiable curves with zero torsion lie in planes* That a sufficiently differentiable curve with zero torsion lies in a plane is a special case of a particle whose velocity remains perpendicular to a fixed vector $\mathbf{C}$ moving in a plane perpendicular to $\mathbf{C}$. This, in turn, can be viewed as the solution of the following problem in calculus.

Suppose that $\mathbf{r}(t) = f(t)\mathbf{i} + g(t)\mathbf{j} + h(t)\mathbf{k}$ is twice differentiable for all t in an interval $[a, b]$, that $\mathbf{r} = 0$ when $t = a$, and that $\mathbf{v} \cdot \mathbf{k} = 0$ for all t in $[a, b]$. Then $h(t) = 0$ for all t in $[a, b]$.

Solve this problem. (*Hint:* Start with $\mathbf{a} = d^2\mathbf{r}/dt^2$ and apply the initial conditions in reverse order.)

36. *A formula that calculates* τ *from* $\mathbf{B}$ *and* $\mathbf{v}$ If we start with the definition $\tau = -(d\mathbf{B}/ds) \cdot \mathbf{N}$ and apply the Chain Rule to rewrite $d\mathbf{B}/ds$ as

$$\frac{d\mathbf{B}}{ds} = \frac{d\mathbf{B}}{dt}\frac{dt}{ds} = \frac{d\mathbf{B}}{dt}\frac{1}{|\mathbf{v}|},$$

we arrive at the formula

$$\tau = -\frac{1}{|\mathbf{v}|}\left(\frac{d\mathbf{B}}{dt} \cdot \mathbf{N}\right).$$

The advantage of this formula over Equation (7) is that it is easier to derive and state. The disadvantage is that it can take a lot of work to evaluate without a computer. Use the new formula to find the torsion of the helix in Example 4.

The formula

$$\kappa(x) = \frac{|f''(x)|}{[1 + (f'(x))^2]^{3/2}},$$

derived in Exercise 23, expresses the curvature $\kappa(x)$ of a twice-differentiable plane curve $y = f(x)$ as a function of x. Find the curvature function of each of the curves in Exercises 37–40. Then graph $f(x)$ together with $\kappa(x)$ over the given interval. You will find some surprises.

T **37.** $y = x^2$, $-2 \le x \le 2$

T **38.** $y = x^4/4$, $-2 \le x \le 2$

T **39.** $y = \sin x$, $0 \le x \le 2\pi$

T **40.** $y = e^x$, $-1 \le x \le 2$

COMPUTER EXPLORATIONS

Circles of Curvature

In Exercises 41–48, you will use a CAS to explore the osculating circle at a point P on a plane curve where $\kappa \ne 0$. Use a CAS to perform the following steps:

(a) Plot the plane curve given in parametric or function form over the specified interval to see what it looks like.

(b) Calculate the curvature κ of the curve at the given value t_0 using the appropriate formula from Exercise 23 or 24. Use the parametrization $x = t$ and $y = f(t)$ if the curve is given as a function $y = f(x)$.

(c) Find the unit normal vector $\mathbf{N}$ at t_0. Notice that the signs of the components of $\mathbf{N}$ depend on whether the unit tangent vector $\mathbf{T}$ is turning clockwise or counterclockwise at $t = t_0$. (See Exercise 25.)

(d) If $\mathbf{C} = a\mathbf{i} + b\mathbf{j}$ is the vector from the origin to the center (a, b) of the osculating circle (see Section 10.6), find the center $\mathbf{C}$ from the vector equation

$$\mathbf{C} = \mathbf{r}(t_0) + \frac{1}{\kappa(t_0)}\mathbf{N}(t_0).$$

The point $P(x_0, y_0)$ on the curve is given by the position vector $\mathbf{r}(t_0)$.

(e) Plot implicitly the equation $(x - a)^2 + (y - b)^2 = 1/\kappa^2$ of the osculating circle. Then plot the curve and osculating circle together. You may need to experiment with the size of the viewing window, but be sure it is square.

41. $\mathbf{r}(t) = (3 \cos t)\mathbf{i} + (5 \sin t)\mathbf{j}$, $0 \le t \le 2\pi$, $t_0 = \pi/4$

42. $\mathbf{r}(t) = (\cos^3 t)\mathbf{i} + (\sin^3 t)\mathbf{j}$, $0 \le t \le 2\pi$, $t_0 = \pi/4$

43. $\mathbf{r}(t) = t^2\mathbf{i} + (t^3 - 3t)\mathbf{j}$, $-4 \le t \le 4$, $t_0 = 3/5$

44. $\mathbf{r}(t) = (t^3 - 2t^2 - t)\mathbf{i} + \dfrac{3t}{\sqrt{1 + t^2}}\mathbf{j}$, $-2 \le t \le 5$, $t_0 = 1$

45. $\mathbf{r}(t) = (2t - \sin t)\mathbf{i} + (2 - 2 \cos t)\mathbf{j}$, $0 \le t \le 3\pi$, $t_0 = 3\pi/2$

46. $\mathbf{r}(t) = (e^{-t} \cos t)\mathbf{i} + (e^{-t} \sin t)\mathbf{j}$, $0 \le t \le 6\pi$, $t_0 = \pi/4$

47. $y = x^2 - x$, $-2 \le x \le 5$, $x_0 = 1$

48. $y = x(1 - x)^{2/5}$, $-1 \le x \le 2$, $x_0 = 1/2$

10.8 Planetary Motion and Satellites

In this section, we derive Kepler's laws of planetary motion from Newton's laws of motion and gravitation and discuss the orbits of Earth satellites. The derivation of Kepler's laws from Newton's is one of the triumphs of calculus. It draws on almost everything we have studied so far, including the algebra and geometry of vectors in space, the calculus of vector functions, the solutions of differential equations and initial value problems, and polar coordinates.

Motion in Polar and Cylindrical Coordinates

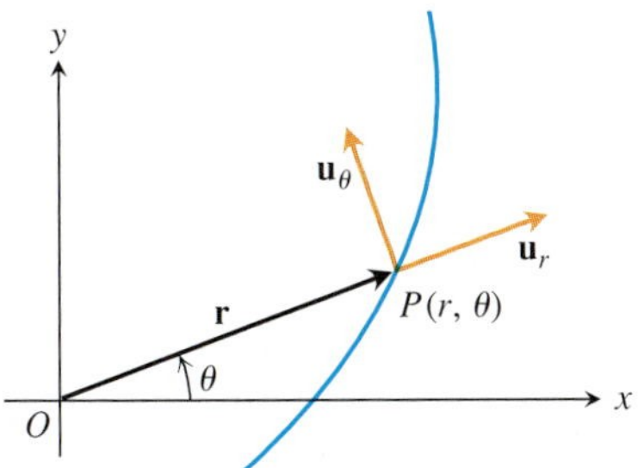

FIGURE 10.62 The length of $\mathbf{r}$ is the positive polar coordinate r of the point P. Thus, $\mathbf{u}_r$, which is $\mathbf{r}/|\mathbf{r}|$, is also $\mathbf{r}/r$. Equations (1) express $\mathbf{u}_r$ and $\mathbf{u}_\theta$ in terms of $\mathbf{i}$ and $\mathbf{j}$.

When a particle moves along a curve in the polar coordinate plane, we express its position, velocity, and acceleration in terms of the moving unit vectors

$$\mathbf{u}_r = (\cos\theta)\mathbf{i} + (\sin\theta)\mathbf{j}, \qquad \mathbf{u}_\theta = -(\sin\theta)\mathbf{i} + (\cos\theta)\mathbf{j}, \tag{1}$$

shown in Figure 10.62. The vector $\mathbf{u}_r$ points along the position vector $\overrightarrow{OP}$, so $\mathbf{r} = r\mathbf{u}_r$. The vector $\mathbf{u}_\theta$, orthogonal to $\mathbf{u}_r$, points in the direction of increasing θ.

We find from Equations (1) that

$$\begin{aligned} \frac{d\mathbf{u}_r}{d\theta} &= -(\sin\theta)\mathbf{i} + (\cos\theta)\mathbf{j} = \mathbf{u}_\theta \\ \frac{d\mathbf{u}_\theta}{d\theta} &= -(\cos\theta)\mathbf{i} - (\sin\theta)\mathbf{j} = -\mathbf{u}_r. \end{aligned} \tag{2}$$

When we differentiate $\mathbf{u}_r$ and $\mathbf{u}_\theta$ with respect to t to find how they change with time, the Chain Rule gives

$$\dot{\mathbf{u}}_r = \frac{d\mathbf{u}_r}{d\theta}\dot{\theta} = \dot{\theta}\mathbf{u}_\theta, \qquad \dot{\mathbf{u}}_\theta = \frac{d\mathbf{u}_\theta}{d\theta}\dot{\theta} = -\dot{\theta}\mathbf{u}_r. \tag{3}$$

Hence,

$$\mathbf{v} = \dot{\mathbf{r}} = \frac{d}{dt}(r\mathbf{u}_r) = \dot{r}\mathbf{u}_r + r\dot{\mathbf{u}}_r = \dot{r}\mathbf{u}_r + r\dot{\theta}\mathbf{u}_\theta. \tag{4}$$

See Figure 10.63.

As in the previous section, we use Newton's dot notation for time derivatives to keep the formulas as simple as we can: $\dot{\mathbf{u}}_r$ means $d\mathbf{u}_r/dt$, $\dot{\theta}$ means $d\theta/dt$, and so on.

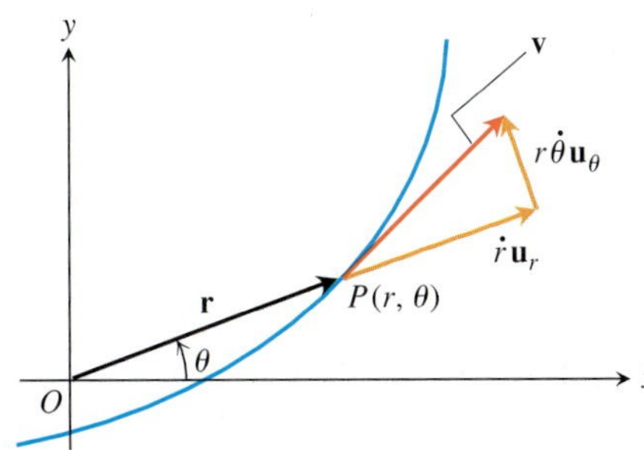

FIGURE 10.63 In polar coordinates, the velocity vector is

$$\mathbf{v} = \dot{r}\mathbf{u}_r + r\dot{\theta}\mathbf{u}_\theta$$

The acceleration is

$$\mathbf{a} = \dot{\mathbf{v}} = (\ddot{r}\mathbf{u}_r + \dot{r}\dot{\mathbf{u}}_r) + (\dot{r}\dot{\theta}\mathbf{u}_\theta + r\ddot{\theta}\mathbf{u}_\theta + r\dot{\theta}\dot{\mathbf{u}}_\theta). \tag{5}$$

When Equations (3) are used to evaluate $\dot{\mathbf{u}}_r$ and $\dot{\mathbf{u}}_\theta$ and the components are separated, the equation for acceleration becomes

$$\mathbf{a} = (\ddot{r} - r\dot{\theta}^2)\mathbf{u}_r + (r\ddot{\theta} + 2\dot{r}\dot{\theta})\mathbf{u}_\theta. \tag{6}$$

To extend these equations of motion to space, we add $z\mathbf{k}$ to the right-hand side of the equation $\mathbf{r} = r\mathbf{u}_r$. Then, in cylindrical coordinates,

Notice that $|\mathbf{r}| \neq r$ if $z \neq 0$.

$$\begin{aligned}\mathbf{r} &= r\mathbf{u}_r + z\mathbf{k} \\ \mathbf{v} &= \dot{r}\mathbf{u}_r + r\dot{\theta}\mathbf{u}_\theta + \dot{z}\mathbf{k} \\ \mathbf{a} &= (\ddot{r} - r\dot{\theta}^2)\mathbf{u}_r + (r\ddot{\theta} + 2\dot{r}\dot{\theta})\mathbf{u}_\theta + \ddot{z}\mathbf{k}.\end{aligned} \tag{7}$$

The vectors $\mathbf{u}_r$, $\mathbf{u}_\theta$, and $\mathbf{k}$ make a right-handed frame (Figure 10.64) in which

$$\mathbf{u}_r \times \mathbf{u}_\theta = \mathbf{k}, \qquad \mathbf{u}_\theta \times \mathbf{k} = \mathbf{u}_r, \qquad \mathbf{k} \times \mathbf{u}_r = \mathbf{u}_\theta. \tag{8}$$

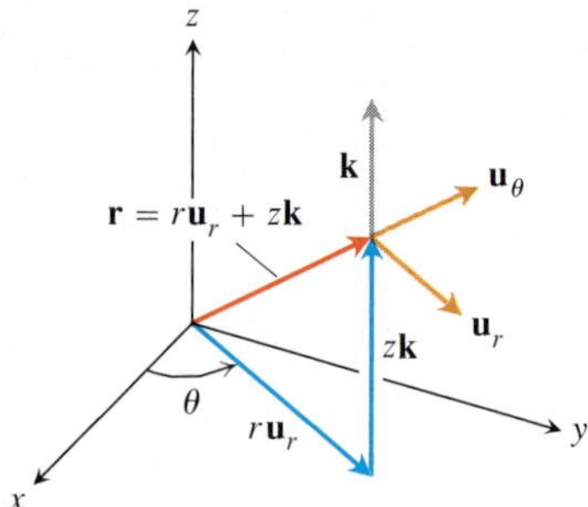

FIGURE 10.64 Position vector and basic unit vectors in cylindrical coordinates.

Planets Move in Planes

Newton's Law of Gravitation says that if $\mathbf{r}$ is the radius vector from the center of a sun of mass M to the center of a planet of mass m, then the force $\mathbf{F}$ of the gravitational attraction between the planet and sun is

$$\mathbf{F} = -\frac{GmM}{|\mathbf{r}|^2}\frac{\mathbf{r}}{|\mathbf{r}|} \tag{9}$$

(Figure 10.65). The number G is the **universal gravitational constant.** If we measure mass in kilograms, force in newtons, and distance in meters, G is about $6.6726 \times 10^{-11}\ \mathrm{Nm^2kg^{-2}}$.

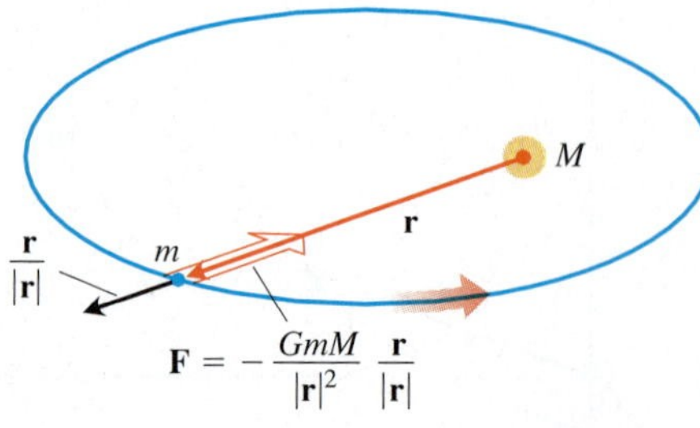

FIGURE 10.65 The force of gravity is directed along the line joining the centers of mass.

Combining Equation (9) with Newton's second law $\mathbf{F} = m\ddot{\mathbf{r}}$ for the force acting on the planet, gives

$$m\ddot{\mathbf{r}} = -\frac{GmM}{|\mathbf{r}|^2}\frac{\mathbf{r}}{|\mathbf{r}|}$$

$$\ddot{\mathbf{r}} = -\frac{GM}{|\mathbf{r}|^2}\frac{\mathbf{r}}{|\mathbf{r}|}. \tag{10}$$

The planet is accelerated toward the sun's center at all times.

Equation (10) says $\ddot{\mathbf{r}}$ is a scalar multiple of $\mathbf{r}$, so that

$$\mathbf{r} \times \ddot{\mathbf{r}} = \mathbf{0}. \tag{11}$$

A routine calculation shows $\mathbf{r} \times \ddot{\mathbf{r}}$ to be the derivative of $\mathbf{r} \times \dot{\mathbf{r}}$:

$$\frac{d}{dt}(\mathbf{r} \times \dot{\mathbf{r}}) = \underbrace{\dot{\mathbf{r}} \times \dot{\mathbf{r}}}_{\mathbf{0}} + \mathbf{r} \times \ddot{\mathbf{r}} = \mathbf{r} \times \ddot{\mathbf{r}}. \tag{12}$$

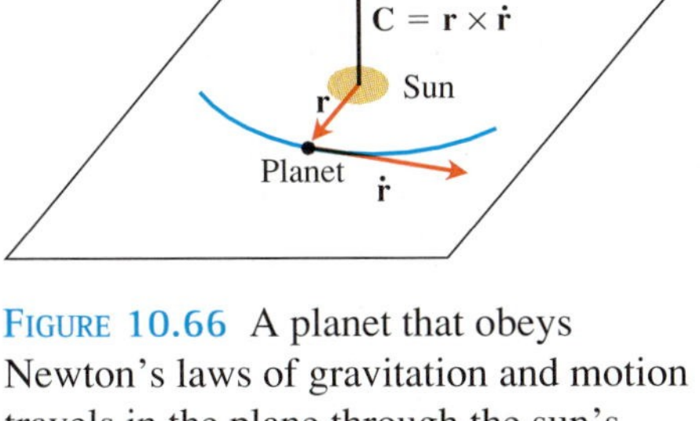

FIGURE 10.66 A planet that obeys Newton's laws of gravitation and motion travels in the plane through the sun's center of mass perpendicular to $\mathbf{C} = \mathbf{r} \times \dot{\mathbf{r}}$.

Hence, Equation (11) is equivalent to

$$\frac{d}{dt}(\mathbf{r} \times \dot{\mathbf{r}}) = \mathbf{0}, \tag{13}$$

which integrates to

$$\mathbf{r} \times \dot{\mathbf{r}} = \mathbf{C} \tag{14}$$

for some constant vector $\mathbf{C}$.

Equation (14) tells us that $\mathbf{r}$ and $\dot{\mathbf{r}}$ always lie in a plane perpendicular to $\mathbf{C}$. Hence, the planet moves in a fixed plane through the center of its sun (Figure 10.66).

Coordinates and Initial Conditions

We now introduce coordinates in a way that places the origin at the sun's center of mass and makes the plane of the planet's motion the polar coordinate plane. This makes $\mathbf{r}$ the planet's polar coordinate position vector and makes $|\mathbf{r}|$ equal to r and $\mathbf{r}/|\mathbf{r}|$ equal to $\mathbf{u}_r$. We also position the z-axis in a way that makes $\mathbf{k}$ the direction of $\mathbf{C}$. Thus, $\mathbf{k}$ has the same right-hand relation to $\mathbf{r} \times \dot{\mathbf{r}}$ that $\mathbf{C}$ does, and the planet's motion is counterclockwise when viewed from the positive z-axis. This makes θ increase with t, so that $\dot{\theta} > 0$ for all t. Finally, we rotate the polar coordinate plane about the z-axis, if necessary, to make the initial ray coincide with the direction $\mathbf{r}$ has when the planet is closest to the sun. This runs the ray through the planet's **perihelion** position (Figure 10.67).

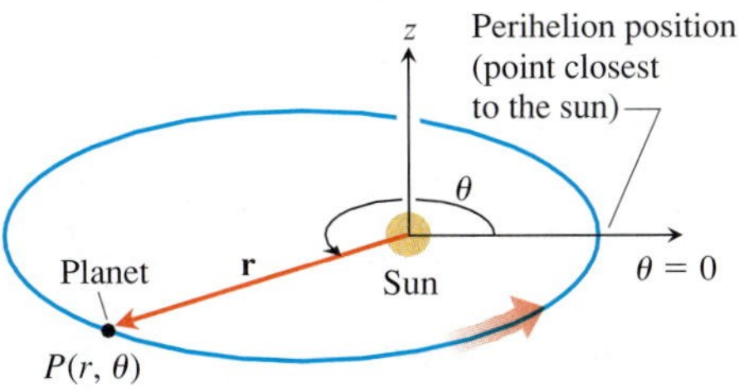

FIGURE 10.67 The coordinate system for planetary motion. The motion is counterclockwise when viewed from above, as it is here, and $\dot{\theta} > 0$.

If we measure time so that $t = 0$ at perihelion, we have the following initial conditions for the planet's motion.

1. $r = r_0$, the minimum radius, when $t = 0$
2. $\dot{r} = 0$ when $t = 0$ (because r has a minimum value then)
3. $\theta = 0$ when $t = 0$
4. $|\mathbf{v}| = v_0$ when $t = 0$

Since

$$\begin{aligned} v_0 &= |\mathbf{v}|_{t=0} \\ &= |\dot{r}\mathbf{u}_r + r\dot{\theta}\mathbf{u}_\theta|_{t=0} && \text{Eq. (4)} \\ &= |r\dot{\theta}\mathbf{u}_\theta|_{t=0} && \dot{r} = 0 \text{ when } t = 0 \\ &= (|r\dot{\theta}||\mathbf{u}_\theta|)_{t=0} \\ &= |r\dot{\theta}|_{t=0} && |\mathbf{u}_\theta| = 1 \\ &= (r\dot{\theta})_{t=0}, && r \text{ and } \dot{\theta} \text{ both positive} \end{aligned}$$

we also know that

5. $r\dot{\theta} = v_0$ when $t = 0$.

Kepler's First Law (The Conic Section Law)

Kepler's first law says that a planet's path is a conic section with the sun at one focus. The eccentricity of the conic is

$$e = \frac{r_0 {v_0}^2}{GM} - 1 \tag{15}$$

and the polar equation is

$$r = \frac{(1 + e)r_0}{1 + e \cos \theta}. \tag{16}$$

The derivation uses Kepler's second law, so we will state and prove the second law before proving the first law.

Kepler's Second Law (The Equal Area Law)

Kepler's second law says that the radius vector from the sun to a planet (the vector $\mathbf{r}$ in our model) sweeps out equal areas in equal times (Figure 10.68). To derive the law, we use Equation (4) to evaluate the cross product $\mathbf{C} = \mathbf{r} \times \dot{\mathbf{r}}$ from Equation (14):

$$\begin{aligned} \mathbf{C} &= \mathbf{r} \times \dot{\mathbf{r}} = \mathbf{r} \times \mathbf{v} \\ &= r\mathbf{u}_r \times (\dot{r}\,\mathbf{u}_r + r\dot{\theta}\mathbf{u}_\theta) && \text{Eq. (4)} \\ &= r\dot{r}\underbrace{(\mathbf{u}_r \times \mathbf{u}_r)}_{\mathbf{0}} + r(r\dot{\theta})\underbrace{(\mathbf{u}_r \times \mathbf{u}_\theta)}_{\mathbf{k}} && (17) \\ &= r(r\dot{\theta})\mathbf{k}. \end{aligned}$$

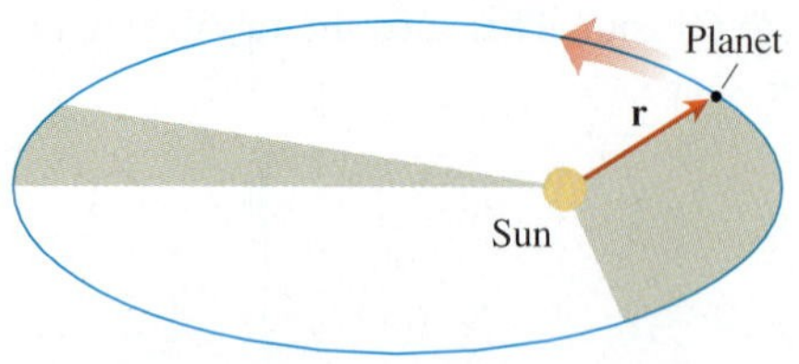

FIGURE 10.68 The line joining a planet to its sun sweeps over equal areas in equal times.

Setting t equal to zero shows that

$$\mathbf{C} = [r(r\dot{\theta})]_{t=0}\mathbf{k} = r_0 v_0 \mathbf{k}. \tag{18}$$

Substituting this value for $\mathbf{C}$ in Equation (17) gives

$$r_0 v_0 \mathbf{k} = r^2 \theta \mathbf{k}, \qquad \text{or} \qquad r^2 \dot{\theta} = r_0 v_0. \tag{19}$$

This is where the area comes in. The area differential in polar coordinates is

$$dA = \frac{1}{2} r^2 \, d\theta$$

(Section 9.6). Accordingly, dA/dt has the constant value

$$\frac{dA}{dt} = \frac{1}{2} r^2 \dot{\theta} = \frac{1}{2} r_0 v_0, \tag{20}$$

which is Kepler's second law.

For Earth, r_0 is about 150,000,000 km, v_0 is about 30 km/sec, and dA/dt is about 2,250,000,000 km^2/sec. Every time your heart beats, Earth advances 30 km along its orbit, and the radius joining Earth to the sun sweeps out 2,250,000,000 km^2 of area.

Proof of Kepler's First Law

To prove that a planet moves along a conic section with one focus at its sun, we need to express the planet's radius r as a function of θ. This requires a long sequence of calculations and some substitutions that are not altogether obvious.

We begin with the equation that comes from equating the coefficients of $\mathbf{u}_r = \mathbf{r}/|\mathbf{r}|$ in Equations (6) and (10):

$$\ddot{r} - r\dot{\theta}^2 = -\frac{GM}{r^2}. \tag{21}$$

We eliminate $\dot{\theta}$ temporarily by replacing it with $r_0 v_0 / r^2$ from Equation (19) and rearrange the resulting equation to get

$$\ddot{r} = \frac{r_0^2 v_0^2}{r^3} - \frac{GM}{r^2}. \tag{22}$$

We change this into a first order equation by a change of variable. With

$$p = \frac{dr}{dt}, \qquad \frac{d^2 r}{dt^2} = \frac{dp}{dt} = \frac{dp}{dr}\frac{dr}{dt} = p\frac{dp}{dr}, \qquad \text{Chain Rule}$$

Equation (22) becomes

$$p\frac{dp}{dr} = \frac{r_0^2 v_0^2}{r^3} - \frac{GM}{r^2}. \tag{23}$$

Multiplying through by 2 and integrating with respect to r gives

$$p^2 = (\dot{r})^2 = -\frac{r_0^2 v_0^2}{r^2} + \frac{2GM}{r} + C_1. \tag{24}$$

The initial conditions that $r = r_0$ and $\dot{r} = 0$ when $t = 0$ determine the value of C_1 to be

$$C_1 = v_0^2 - \frac{2GM}{r_0}.$$

Accordingly, Equation (24), after a suitable rearrangement, becomes

$$\dot{r}^2 = v_0^2\left(1 - \frac{r_0^2}{r^2}\right) + 2GM\left(\frac{1}{r} - \frac{1}{r_0}\right). \tag{25}$$

The effect of going from Equation (21) to Equation (25) has been to replace a second-order differential equation in r by a first order differential equation in r. Our goal is still to express r in terms of θ, so we now bring θ back into the picture. To accomplish this, we divide both sides of Equation (25) by the squares of the corresponding sides of the equation $r^2\dot{\theta} = r_0 v_0$ (Equation (19)) and use the equation $\dot{r}/\dot{\theta} = (dr/dt)/(d\theta/dt) = dr/d\theta$ to get

$$\begin{aligned}\frac{1}{r^4}\left(\frac{dr}{d\theta}\right)^2 &= \frac{1}{r_0^2} - \frac{1}{r^2} + \frac{2GM}{r_0^2 v_0^2}\left(\frac{1}{r} - \frac{1}{r_0}\right)\\ &= \frac{1}{r_0^2} - \frac{1}{r^2} + 2h\left(\frac{1}{r} - \frac{1}{r_0}\right). \qquad h = \frac{GM}{r_0^2 v_0^2}\end{aligned} \tag{26}$$

To simplify further, we substitute

$$u = \frac{1}{r}, \qquad u_0 = \frac{1}{r_0}, \qquad \frac{du}{d\theta} = -\frac{1}{r^2}\frac{dr}{d\theta}, \qquad \left(\frac{du}{d\theta}\right)^2 = \frac{1}{r^4}\left(\frac{dr}{d\theta}\right)^2,$$

obtaining

$$\left(\frac{du}{d\theta}\right)^2 = u_0^2 - u^2 + 2hu - 2hu_0 = (u_0 - h)^2 - (u - h)^2 \tag{27}$$

$$\frac{du}{d\theta} = \pm\sqrt{(u_0 - h)^2 - (u - h)^2}. \tag{28}$$

Which sign do we take? We know that $\dot{\theta} = r_0 v_0 / r^2$ is positive. Also, r starts from a minimum value at $t = 0$, so it cannot immediately decrease, and $\dot{r} \geq 0$, at least for early positive values of t. Therefore,

$$\frac{dr}{d\theta} = \frac{\dot{r}}{\dot{\theta}} \geq 0 \qquad \text{and} \qquad \frac{du}{d\theta} = -\frac{1}{r^2}\frac{dr}{d\theta} \leq 0.$$

The correct sign for Equation (28) is the negative sign. With this determined, we rearrange Equation (28) and integrate both sides with respect to θ:

$$\frac{-1}{\sqrt{(u_0 - h)^2 - (u - h)^2}}\frac{du}{d\theta} = 1$$

$$\cos^{-1}\left(\frac{u - h}{u_0 - h}\right) = \theta + C_2. \tag{29}$$

The constant C_2 is zero because $u = u_0$ when $\theta = 0$ and $\cos^{-1}(1) = 0$. Therefore,

$$\frac{u - h}{u_0 - h} = \cos\,\theta$$

and

$$\frac{1}{r} = u = h + (u_0 - h)\cos\,\theta. \tag{30}$$

A few more algebraic maneuvers produce the final equation

$$r = \frac{(1 + e)r_0}{1 + e\cos\,\theta}, \tag{31}$$

where

$$e = \frac{1}{r_0 h} - 1 = \frac{r_0 v_0^2}{GM} - 1. \tag{32}$$

Together, Equations (31) and (32) say that the path of the planet is a conic section with one focus at the sun and with eccentricity $(r_0 v_0^2/GM) - 1$. This is the modern formulation of Kepler's first law.

Kepler's Third Law (The Time–Distance Law)

The time T it takes a planet to go around its sun once is the planet's **orbital period.** *Kepler's third law* says that T and the orbit's semimajor axis a are related by the equation

$$\frac{T^2}{a^3} = \frac{4\pi^2}{GM}. \tag{33}$$

Since the right-hand side of this equation is constant within a given solar system, the ratio of T^2 to a^3 *is the same for every planet in the system.*

Kepler's third law is the starting point for working out the size of our solar system. It allows the semimajor axis of each planetary orbit to be expressed in astronomical units, Earth's semimajor axis being one unit. The distance between any two planets at any time can then be predicted in astronomical units and all that remains is to find one of these distances in kilometers. This can be done by bouncing radar waves off Venus, for example. The astronomical unit is now known, after a series of such measurements, to be 149,597,870 km.

We derive Kepler's third law by combining two formulas for the area enclosed by the planet's elliptical orbit:

Formula 1: Area $= \pi ab$ — The geometry formula in which a is the semimajor axis and b is the semiminor axis

Formula 2:

$$\begin{aligned} \text{Area} &= \int_0^T dA \\ &= \int_0^T \frac{1}{2} r_0 v_0 \, dt \qquad \text{Eq. (20)} \\ &= \frac{1}{2} T r_0 v_0. \end{aligned}$$

Equating these gives

$$T = \frac{2\pi ab}{r_0 v_0} = \frac{2\pi a^2}{r_0 v_0}\sqrt{1 - e^2}. \qquad \text{For any ellipse, } b = a\sqrt{1 - e^2} \tag{34}$$

It remains only to express a and e in terms of r_0, v_0, G, and M. Equation (32) does this for e. For a, we observe that setting θ equal to π in Equation (31) gives

$$r_{\max} = r_0 \frac{1 + e}{1 - e}.$$

Hence,

$$2a = r_0 + r_{\max} = \frac{2r_0}{1 - e} = \frac{2r_0 GM}{2GM - r_0 v_0^2}. \tag{35}$$

Squaring both sides of Equation (34) and substituting the results of Equations (32) and (35) now produces Kepler's third law (Exercise 15).

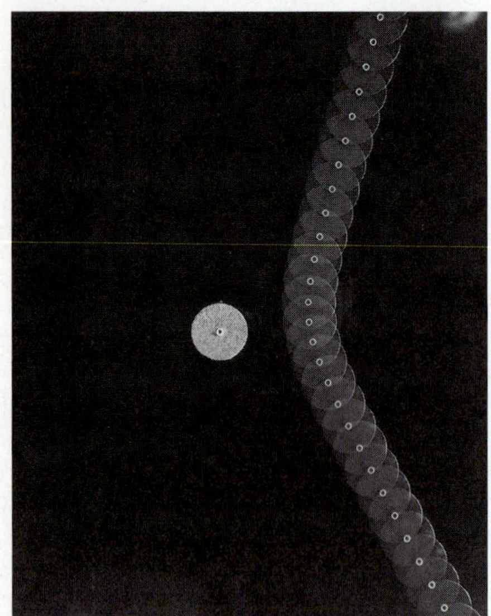

FIGURE 10.69 This multiflash photograph shows an air puck being deflected by an inverse square law force. It moves along a hyperbola.

Orbit Data

Although Kepler discovered his laws empirically and stated them only for the six planets known at the time, the modern derivations of Kepler's laws show that they apply to any body driven by a force that obeys an inverse square law. They apply to Halley's comet and the asteroid Icarus. They apply to the moon's orbit about Earth, and they applied to the orbit of the spacecraft *Apollo 8* about the moon. They also applied to the air puck shown in Figure 10.69 being deflected by an inverse square law force; its path is a hyperbola. Charged particles fired at the nuclei of atoms scatter along hyperbolic paths.

Tables 10.1 through 10.3 give additional data for planetary orbits and for the orbits of seven of Earth's artificial satellites (Figure 10.70). *Vanguard 1* sent back data that revealed differences between the levels of Earth's oceans and provided the first determination of the precise locations of some of the more isolated Pacific islands. The data also verified that the gravitation of the sun and moon would affect the orbits of Earth's satellites and that solar radiation could exert enough pressure to deform an orbit.

Table 10.1 Values of a, e, and T for the major planets

Planet	Semimajor axis a*	Eccentricity e	Period T
Mercury	57.95	0.2056	87.967 days
Venus	108.11	0.0068	224.71 days
Earth	149.57	0.0167	365.256 days
Mars	227.84	0.0934	1.8808 years
Jupiter	778.14	0.0484	11.8613 years
Saturn	1427.0	0.0543	29.4568 years
Uranus	2870.3	0.0460	84.0081 years
Neptune	4499.9	0.0082	164.784 years
Pluto	5909	0.2481	248.35 years

*Millions of kilometers

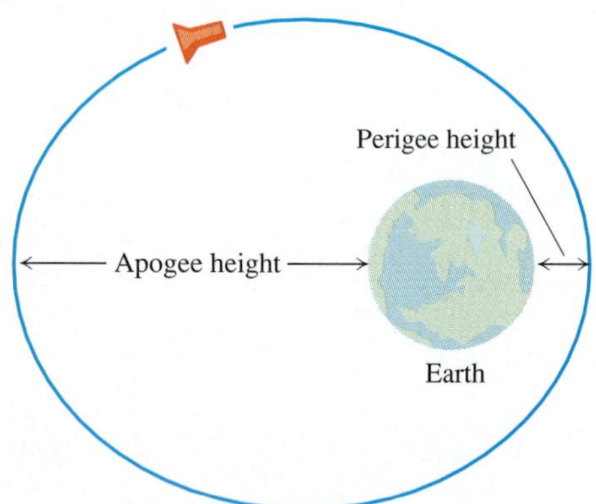

FIGURE 10.70 The orbit of an Earth satellite: $2a$ = diameter of earth + perigee height + apogee height.

Syncom 3 is one of a series of U.S. Department of Defense telecommunications satellites. *Tiros 11* (for "television infrared observation satellite") is one of a series of weather satellites. *GOES 4* (for "geostationary operational environmental satellite") is one of a series of satellites designed to gather information about Earth's atmosphere. Its orbital period, 1436.2 min, is nearly the same as Earth's rotational period of 1436.1 min, and its orbit is nearly circular (e = 0.0003). *Intelsat 5* is a heavy-capacity commercial telecommunications satellite.

Table 10.2 Data on Earth's satellites

Name	Launch date	Time or expected time aloft	Mass at launch (kg)	Period (min)	Perigee height (km)	Apogee height (km)	Semimajor axis a (km)	Eccentricity
Sputnik 1	Oct. 1957	57.6 days	83.6	96.2	215	939	6,955	0.052
Vanguard 1	March 1958	300 years	1.47	138.5	649	4,340	8,872	0.208
Syncom 3	Aug. 1964	$>10^6$ years	39	1436.2	35,718	35,903	42,189	0.002
Skylab 4	Nov. 1973	84.06 days	13,980	93.11	422	437	6,808	0.001
Tiros II	Oct. 1978	500 years	734	102.12	850	866	7,236	0.001
GOES 4	Sept. 1980	$>10^6$ years	627	1436.2	35,776	35,800	42,166	0.0003
Intelsat 5	Dec. 1980	$>10^6$ years	1,928	1417.67	35,143	35,707	41,803	0.007

Table 10.3 Numerical data

Universal gravitational constant: $G = 6.6726 \times 10^{-11}\ \text{Nm}^2\text{kg}^{-2}$
(When you use this value of G in a calculation, remember to express force in newtons, distance in meters, mass in kilograms, and time in seconds.)

Sun's mass:	1.99×10^{30} kg
Earth's mass:	5.975×10^{24} kg
Equatorial radius of Earth:	6378.533 km
Polar radius of Earth:	6356.912 km
Earth's rotational period:	1436.1 min
Earth's orbital period:	1 year = 365.256 days

EXERCISES 10.8

Reminder: When a calculation involves the gravitational constant G, express force in newtons, distance in meters, mass in kilograms, and time in seconds.

1. *Period of Skylab 4* Since the orbit of *Skylab 4* had a semimajor axis of $a = 6808$ km, Kepler's third law with M equal to Earth's mass should give the period. Calculate it. Compare your result with the value in Table 10.2.

2. *Earth's velocity at perihelion* Earth's distance from the sun at perihelion is approximately 149,577,000 km, and the eccentricity of Earth's orbit about the sun is 0.0167. Find the velocity v_0 of Earth in its orbit at perihelion. (Use Equation (15).)

3. *Semimajor axis of Proton I* In July 1965, the USSR launched *Proton I*, weighing 12,200 kg (at launch), with a perigee height of 183 km, an apogee height of 589 km, and a period of 92.25 min. Using the relevant data for the mass of Earth and the gravitational constant G, find the semimajor axis a of the orbit from Equation (3). Compare your answer with the number you get by adding the perigee and apogee heights to the diameter of the Earth.

4. *Semimajor axis of Viking I* The *Viking I* orbiter, which surveyed Mars from August 1975 to June 1976, had a period of 1639 min. Use this and the mass of Mars, 6.418×10^{23} kg, to find the semimajor axis of the *Viking I* orbit.

5. *Average diameter of Mars (Continuation of Exercise 4)* The *Viking I* orbiter was 1499 km from the surface of Mars at its closest point and 35,800 km from the surface at its farthest point. Use this information together with the value you obtained in Exercise 4 to estimate the average diameter of Mars.

6. *Period of Viking 2* The *Viking 2* orbiter, which surveyed Mars from September 1975 to August 1976, moved in an ellipse whose semimajor axis was 22,030 km. What was the orbital period? (Express your answer in minutes.)

7. *Geosynchronous orbits* Several satellites in Earth's equatorial plane have nearly circular orbits whose periods are the same as the earth's rotational period. Such orbits are **geosynchronous** or **geostationary** because they hold the satellite over the same spot on Earth's surface.

   (a) *Writing to Learn* Approximately what is the semimajor axis of a geosynchronous orbit? Give reasons for your answer.

   (b) About how high is a geosynchronous orbit above the Earth's surface?

   (c) Which of the satellites in Table 10.2 have (nearly) geosynchronous orbits?

8. *Writing to Learn* The mass of Mars is 6.418×10^{23} kg. If a satellite revolving about Mars is to hold a stationary orbit (have the same period as the period of Mar's rotation, which is 1477.4 min), what must the semimajor axis of its orbit be? Give reasons for your answer.

9. *Distance from Earth to the moon* The period of the moon's rotation about Earth is 2.36055×10^6 sec. About how far away is the moon?

10. *Finding satellite speed* A satellite moves around Earth in a circular orbit. Express the satellite's speed as a function of the orbit's radius.

11. *Orbital period* If T is measured in seconds and a in meters, what is the value of T^2/a^3 for planets in our solar system? For satellites orbiting Earth? For satellites orbiting the moon? (The moon's mass is 7.354×10^{22} kg.)

12. *Type of orbit* For what values of v_0 in Equation (15) is the orbit in Equation (16) a circle? An ellipse? A parabola? A hyperbola?

13. *Circular orbits* Show that a planet in a circular orbit moves with a constant speed. (*Hint:* This is a consequence of one of Kepler's laws.)

14. Suppose that $\mathbf{r}$ is the position vector of a particle moving along a plane curve and dA/dt is the rate at which the vector sweeps out area. Without introducing coordinates, and assuming the necessary derivatives exist, give a geometric argument based on increments and limits for the validity of the equation

$$\frac{dA}{dt} = \frac{1}{2}|\mathbf{r} \times \dot{\mathbf{r}}|.$$

15. *Kepler's third law* Complete the derivation of Kepler's third law (the part following Equation (34)).

In Exercises 16 and 17, two planets, planet A and planet B, are orbiting their sun in circular orbits with A being the inner planet and B being farther away from the sun. Suppose the positions of A and B at time t are

$$\mathbf{r}_A(t) = 2\cos(2\pi t)\mathbf{i} + 2\sin(2\pi t)\mathbf{j}$$

and

$$\mathbf{r}_B(t) = 3\cos(\pi t)\mathbf{i} + 3\sin(\pi t)\mathbf{j},$$

respectively, where the sun is assumed to be located at the origin and distance is measured in astronomical units. (Notice that planet A moves faster than planet B.)

The people on planet A regard their planet, not the sun, as the center of their planetary system (their solar system).

16. Using planet A as the origin of a new coordinate system, give parametric equations for the location of planet B at time t. Write your answer in terms of $\cos(\pi t)$ and $\sin(\pi t)$.

T 17. Using planet A as the origin, graph the path of planet B.

This exercise illustrates the difficulty that people before Kepler's time, with an earth-centered (planet A) view of our solar system, had in understanding the motions of the planets (i.e., planet B = Mars). See D. G. Saari's article in the *American Monthly*, Vol. 97 (Feb. 1990), pp. 105–119.

18. *Writing to Learn* Kepler discovered that the path of the Earth around the sun is an ellipse with the sun at one of the foci. Let $\mathbf{r}(t)$ be the position vector from the center of the sun to the center of the Earth at time t. Let $\mathbf{w}$ be the vector from the Earth's South Pole to North Pole. It is known that $\mathbf{w}$ is constant and not orthogonal to the plane of the ellipse (Earth's axis is tilted). In terms of $\mathbf{r}(t)$ and $\mathbf{w}$, give the mathematical meaning of (i) perihelion, (ii) aphelion, (iii) equinox, (iv) summer solstice, (v) winter solstice.

Questions to Guide Your Review

1. When do directed line segments (in space) represent the same vector?

2. How are space vectors added and subtracted geometrically? Algebraically?

3. How do you find a space vector's magnitude and direction?

4. If a vector is multiplied by a positive scalar, how is the result related to the original vector? What if the scalar is zero? Negative?

5. Define the *dot product* (*scalar product*) of two space vectors. Which algebraic laws (commutative, associative, distributive, cancellation) are satisfied by dot products, and which, if any, are not? Give examples. When is the dot product of two vectors equal to zero?

6. What geometric or physical interpretations do dot products have? Give examples.

7. What is the vector projection of a vector **v** onto a vector **u**? How do you write **v** as the sum of a vector parallel to **u** and a vector orthogonal to **u**?
8. Define the *cross product* (*vector product*) of two vectors. Which algebraic laws (commutative, associative, distributive, cancellation) are satisfied by cross products, and which are not? Give examples. When is the cross product of two vectors equal to zero?
9. What geometric or physical interpretations do cross products have? Give examples.
10. What is the determinant formula for calculating the cross product of two vectors relative to the Cartesian **i**, **j**, **k**-coordinate system? Use it in an example.
11. How do you find equations for lines, line segments, and planes in space? Give examples. Can you express a line in space by a single equation? A plane?
12. What are box products? What significance do they have? How are they evaluated? Give an example.
13. How do you find equations for spheres in space? Give examples.
14. How do you find the intersection of two lines in space? A line and a plane? Two planes? Give examples.
15. What is a cylinder? Give examples of equations that define cylinders in Cartesian coordinates.
16. What are quadric surfaces? Give examples of different kinds of ellipsoids, paraboloids, cones, and hyperboloids (equations and sketches).
17. State the rules for differentiating and integrating vector functions. Give examples.
18. How do you define and calculate the velocity, speed, direction of motion, and acceleration of a body moving along a sufficiently differentiable space curve? Give an example.
19. What is special about the derivatives of vector functions of constant length? Give an example.
20. How do you define and calculate the length of a segment of a smooth space curve? Give an example. What mathematical assumptions are involved in the definition?
21. How do you measure distance along a smooth curve in space from a preselected base point? Give an example.
22. What is a smooth curve's unit tangent vector? Give an example.
23. Define curvature, circle of curvature (osculating circle), center of curvature, and radius of curvature for twice-differentiable curves in the plane. Give examples. What curves have zero curvature? constant curvature?
24. What is a plane curve's principal normal vector? When is it defined? Which way does it point? Give an example.
25. How do you define **N** and κ for curves in space? How are these quantities related? Give examples.
26. What is a curve's binormal vector? Give an example. How is this vector related to the curve's torsion? Give an example.
27. What formulas are available for writing a moving body's acceleration as a sum of its tangential and normal components? Give an example. Why might one want to write the acceleration this way? What if the body moves at a constant speed? at a constant speed around a circle?
28. State Kepler's laws. To what do they apply?

Practice Exercises

Vector Calculations

Express the vectors in Exercises 1 and 2 in terms of their lengths and directions.

1. $2\mathbf{i} - 3\mathbf{j} + 6\mathbf{k}$
2. $\mathbf{i} + 2\mathbf{j} - \mathbf{k}$
3. Find a vector 2 units long in the direction of $\mathbf{v} = 4\mathbf{i} - \mathbf{j} + 4\mathbf{k}$.
4. Find a vector 5 units long in the direction opposite to the direction of $\mathbf{v} = (3/5)\mathbf{i} + (4/5)\mathbf{k}$.

In Exercises 5 and 6, find $|\mathbf{v}|$, $|\mathbf{u}|$, $\mathbf{v}\cdot\mathbf{u}$, $\mathbf{u}\cdot\mathbf{v}$, $\mathbf{v}\times\mathbf{u}$, $\mathbf{u}\times\mathbf{v}$, $|\mathbf{v}\times\mathbf{u}|$, the angle between **v** and **u**, the scalar component of **u** in the direction of **v**, and the vector projection of **u** onto **v**.

5. $\mathbf{v} = \mathbf{i} + \mathbf{j}$
   $\mathbf{u} = 2\mathbf{i} + \mathbf{j} - 2\mathbf{k}$
6. $\mathbf{v} = \mathbf{i} + \mathbf{j} + 2\mathbf{k}$
   $\mathbf{u} = -\mathbf{i} - \mathbf{k}$

In Exercises 7 and 8, write **u** as the sum of a vector parallel to **v** and a vector orthogonal to **v**.

7. $\mathbf{v} = 2\mathbf{i} + \mathbf{j} - \mathbf{k}$
   $\mathbf{u} = \mathbf{i} + \mathbf{j} - 5\mathbf{k}$
8. $\mathbf{u} = \mathbf{i} - 2\mathbf{j}$
   $\mathbf{v} = \mathbf{i} + \mathbf{j} + \mathbf{k}$

In Exercises 9 and 10, draw coordinate axes and then sketch **u**, **v**, and $\mathbf{u}\times\mathbf{v}$ as vectors at the origin.

9. $\mathbf{u} = \mathbf{i}, \quad \mathbf{v} = \mathbf{i} + \mathbf{j}$
10. $\mathbf{u} = \mathbf{i} - \mathbf{j}, \quad \mathbf{v} = \mathbf{i} + \mathbf{j}$
11. If $|\mathbf{v}| = 2$, $|\mathbf{w}| = 3$, and the angle between **v** and **w** is $\pi/3$, find $|\mathbf{v} - 2\mathbf{w}|$.
12. For what value or values of a will the vectors $\mathbf{u} = 2\mathbf{i} + 4\mathbf{j} - 5\mathbf{k}$ and $\mathbf{v} = -4\mathbf{i} - 8\mathbf{j} + a\mathbf{k}$ be parallel?

In Exercises 13 and 14, find **(a)** the area of the parallelogram determined by vectors **u** and **v** and **(b)** the volume of the parallelepiped determined by the vectors **u**, **v**, and **w**.

13. $\mathbf{u} = \mathbf{i} + \mathbf{j} - \mathbf{k}, \quad \mathbf{v} = 2\mathbf{i} + \mathbf{j} + \mathbf{k}, \quad \mathbf{w} = -\mathbf{i} - 2\mathbf{j} + 3\mathbf{k}$

14. $\mathbf{u} = \mathbf{i} + \mathbf{j}, \quad \mathbf{v} = \mathbf{j}, \quad \mathbf{w} = \mathbf{i} + \mathbf{j} + \mathbf{k}$

Lines, Planes, and Distances

15. *Writing to Learn* Suppose that **n** is normal to a plane and **v** is parallel to the plane. Describe how you would find a vector **u** that is both perpendicular to **v** and parallel to the plane.

16. *Vector parallel to plane* Find a vector in the plane parallel to the line $ax + by = c$.

17. *Line parallel to vector* Parametrize the line that passes through the point $(1, 2, 3)$ parallel to the vector $\mathbf{v} = -3\mathbf{i} + 7\mathbf{k}$.

18. *Line segment* Parametrize the line segment joining the points $P(1, 2, 0)$ and $Q(1, 3, -1)$.

19. *Plane normal to vector* Find an equation for the plane that passes through the point $(3, -2, 1)$ normal to the vector $\mathbf{n} = 2\mathbf{i} + \mathbf{j} + \mathbf{k}$.

20. *Plane perpendicular to line* Find an equation for the plane that passes through the point $(-1, 6, 0)$ perpendicular to the line $x = -1 + t$, $y = 6 - 2t$, $z = 3t$.

In Exercises 21 and 22, find an equation for the plane through points P, Q, and R.

21. $P(1, -1, 2), \quad Q(2, 1, 3), \quad R(-1, 2, -1)$

22. $P(1, 0, 0), \quad Q(0, 1, 0), \quad R(0, 0, 1)$

23. *Points of intersection* Find the points in which the line $x = 1 + 2t$, $y = -1 - t$, $z = 3t$ meets the three coordinate planes.

24. *Point of intersection* Find the point in which the line through the origin perpendicular to the plane $2x - y - z = 4$ meets the plane $3x - 5y + 2z = 6$.

25. *Angle between planes* Find the acute angle between the planes $x = 7$ and $x + y + \sqrt{2}z = -3$.

26. *Intersection of planes* Find parametric equations for the line in which the planes $x + 2y + z = 1$ and $x - y + 2z = -8$ intersect.

27. *Intersection of planes* Show that the line in which the planes

$$x + 2y - 2z = 5 \quad \text{and} \quad 5x - 2y - z = 0$$

intersect is parallel to the line

$$x = -3 + 2t, \quad y = 3t, \quad z = 1 + 4t.$$

28. *Intersection of planes* The planes $3x + 6z = 1$ and $2x + 2y - z = 3$ intersect in a line.

(a) Show that the planes are orthogonal.

(b) Find equations for the line of intersection.

29. *Plane parallel to vectors* Find an equation for the plane that passes through the point $(1, 2, 3)$ parallel to $\mathbf{u} = 2\mathbf{i} + 3\mathbf{j} + \mathbf{k}$ and $\mathbf{v} = \mathbf{i} - \mathbf{j} + 2\mathbf{k}$.

30. *Vector parallel to plane* Find a vector parallel to the plane $2x - y - z = 4$ and orthogonal to $\mathbf{i} + \mathbf{j} + \mathbf{k}$.

31. *Vector in a plane* Find a unit vector orthogonal to **u** in the plane of **v** and **w** if $\mathbf{u} = 2\mathbf{i} - \mathbf{j} + \mathbf{k}$, $\mathbf{v} = \mathbf{i} + 2\mathbf{j} + \mathbf{k}$, and $\mathbf{w} = \mathbf{i} + \mathbf{j} - 2\mathbf{k}$.

32. *Vector parallel to line* Find a vector of magnitude 2 parallel to the line of intersection of the planes $x + 2y + z - 1 = 0$ and $x - y + 2z + 7 = 0$.

33. *Point of intersection* Find the point in which the line through the origin perpendicular to the plane $2x - y - z = 4$ meets the plane $3x - 5y + 2z = 6$.

34. *Point of intersection* Find the point in which the line through $P(3, 2, 1)$ normal to the plane $2x - y + 2z = -2$ meets the plane.

35. *Plane* Which of the following are equations for the plane through the points $P(1, 1, -1)$, $Q(3, 0, 2)$, and $R(-2, 1, 0)$?

(a) $(2\mathbf{i} - 3\mathbf{j} + 3\mathbf{k}) \cdot ((x + 2)\mathbf{i} + (y - 1)\mathbf{j} + z\mathbf{k}) = 0$

(b) $x = 3 - t, \quad y = -11t, \quad z = 2 - 3t$

(c) $(x + 2) + 11(y - 1) = 3z$

(d) $(2\mathbf{i} - 3\mathbf{j} + 3\mathbf{k}) \times ((x + 2)\mathbf{i} + (y - 1)\mathbf{j} + z\mathbf{k}) = \mathbf{0}$

(e) $(2\mathbf{i} - \mathbf{j} + 3\mathbf{k}) \times (-3\mathbf{i} + \mathbf{k}) \cdot ((x + 2)\mathbf{i} + (y - 1)\mathbf{j} + z\mathbf{k}) = 0$

36. *Parallelogram* The parallelogram shown here has vertices at $A(2, -1, 4)$, $B(1, 0, -1)$, $C(1, 2, 3)$, and D. Find

(a) The coordinates of D

(b) The cosine of the interior angle at B

(c) The vector projection of $\overrightarrow{BA}$ onto $\overrightarrow{BC}$

(d) The area of the parallelogram

(e) An equation for the plane of the parallelogram

(f) The areas of the orthogonal projections of the parallelogram on the three coordinate planes.

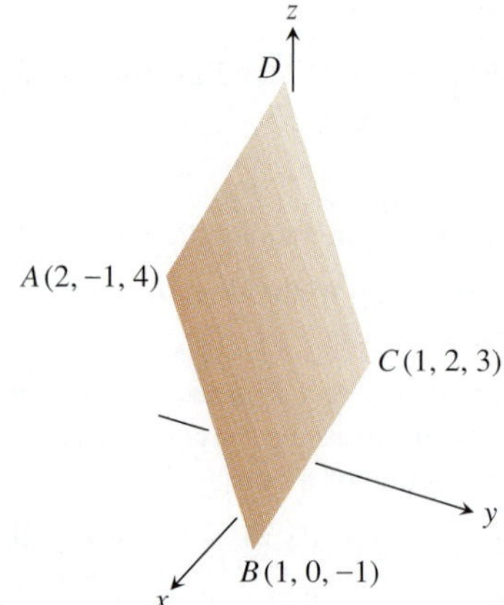

Distances between Points and Lines and Planes

In Exercises 37 and 38, find the distance from the point to the plane.

37. $(2, 2, 0)$; $x = -t$, $y = t$, $z = -1 + t$

38. $(0, 4, 1)$; $x = 2 + t$, $y = 2 + t$, $z = t$

In Exercises 39 and 40, find the distance from the point to the plane.

39. $(6, 0, -6)$, $x - y = 4$

40. $(3, 0, 10)$, $2x + 3y + z = 2$

41. Find the distance from the point $P(1, 4, 0)$ to the plane through $A(0, 0, 0)$, $B(2, 0, -1)$ and $C(2, -1, 0)$.

42. Find the distance from the point $(2, 2, 3)$ to the plane $2x + 3y + 5z = 0$.

43. *Distance between lines* Find the distance between the line L_1 through the points $A(1, 0, -1)$ and $B(-1, 1, 0)$ and the line L_2 through the points $C(3, 1, -1)$ and $D(4, 5, -2)$. The distance is to be measured along the line perpendicular to the two lines. First find a vector $\mathbf{n}$ perpendicular to both lines. Then project $\overrightarrow{AC}$ onto $\mathbf{n}$.

44. (*Continuation of Exercise 43*) Find the distance between the line through $A(4, 0, 2)$ and $B(2, 4, 1)$ and the line through $C(1, 3, 2)$ and $D(2, 2, 4)$.

Quadric Surfaces

Identify and sketch the surfaces in Exercises 45–50.

45. $x^2 + y^2 + z^2 = 4$

46. $4x^2 + 4y^2 + z^2 = 4$

47. $z = -(x^2 + y^2)$

48. $x^2 + y^2 = z^2$

49. $x^2 + y^2 - z^2 = 4$

50. $y^2 - x^2 - z^2 = 1$

Motion in Space

Find the lengths of the curves in Exercises 51 and 52.

51. $\mathbf{r}(t) = (2\cos t)\mathbf{i} + (2\sin t)\mathbf{j} + t^2\mathbf{k}$, $0 \le t \le \pi/4$

52. $\mathbf{r}(t) = (3\cos t)\mathbf{i} + (3\sin t)\mathbf{j} + 2t^{3/2}\mathbf{k}$, $0 \le t \le 3$

In Exercises 53–56, find $\mathbf{T}, \mathbf{N}, \mathbf{B}, \kappa$, and τ at the given value of t.

53. $\mathbf{r}(t) = \frac{4}{9}(1 + t)^{3/2}\mathbf{i} + \frac{4}{9}(1 - t)^{3/2}\mathbf{j} + \frac{1}{3}t\mathbf{k}$, $t = 0$

54. $\mathbf{r}(t) = (e^t \sin 2t)\mathbf{i} + (e^t \cos 2t)\mathbf{j} + 2e^t\mathbf{k}$, $t = 0$

55. $\mathbf{r}(t) = t\mathbf{i} + \frac{1}{2}e^{2t}\mathbf{j}$, $t = \ln 2$

56. $\mathbf{r}(t) = (3\cosh 2t)\mathbf{i} + (3\sinh 2t)\mathbf{j} + 6t\mathbf{k}$, $t = \ln 2$

In Exercises 57 and 58, write $\mathbf{a}$ in the form $\mathbf{a} = a_T\mathbf{T} + a_N\mathbf{N}$ at $t = 0$ without finding $\mathbf{T}$ and $\mathbf{N}$.

57. $\mathbf{r}(t) = (2 + 3t + 3t^2)\mathbf{i} + (4t + 4t^2)\mathbf{j} - (6\cos t)\mathbf{k}$

58. $\mathbf{r}(t) = (2 + t)\mathbf{i} + (t + 2t^2)\mathbf{j} + (1 + t^2)\mathbf{k}$

59. Find $\mathbf{T}, \mathbf{N}, \mathbf{B}, \kappa$, and τ as functions of t if $\mathbf{r}(t) = (\sin t)\mathbf{i} + (\sqrt{2}\cos t)\mathbf{j} + (\sin t)\mathbf{k}$.

60. *Velocity and acceleration* At what times in the interval $0 \le t \le \pi$ are the velocity and acceleration vectors of the motion $\mathbf{r}(t) = \mathbf{i} + (5\cos t)\mathbf{j} + (3\sin t)\mathbf{k}$ orthogonal?

61. *Orthogonality of position vector* The position of a particle moving in space at time $t \ge 0$ is

$$\mathbf{r}(t) = 2\mathbf{i} + \left(4\sin\frac{t}{2}\right)\mathbf{j} + \left(3 - \frac{t}{\pi}\right)\mathbf{k}.$$

Find the first time $\mathbf{r}$ is orthogonal to the vector $\mathbf{i} - \mathbf{j}$.

62. *Osculating, normal, and rectifying planes* Find equations for the osculating, normal, and rectifying planes of the curve $\mathbf{r}(t) = t\mathbf{i} + t^2\mathbf{j} + t^3\mathbf{k}$ at the point $(1, 1, 1)$.

63. *Tangent line* Find parametric equations for the line that is tangent to the curve $\mathbf{r}(t) = e^t\mathbf{i} + (\sin t)\mathbf{j} + \ln(1 - t)\mathbf{k}$ at $t = 0$.

64. *Tangent line* Find parametric equations for the line tangent to the helix $\mathbf{r}(t) = (\sqrt{2}\cos t)\mathbf{i} + (\sqrt{2}\sin t)\mathbf{j} + t\mathbf{k}$ at the point where $t = \pi/4$.

65. *The view from Skylab 4* What percentage of Earth's surface area could the astronauts see when *Skylab 4* was at its apogee height, 437 km above the surface? To find out, model the visible surface as the surface generated by revolving the circular arc GT, shown here, about the y-axis.

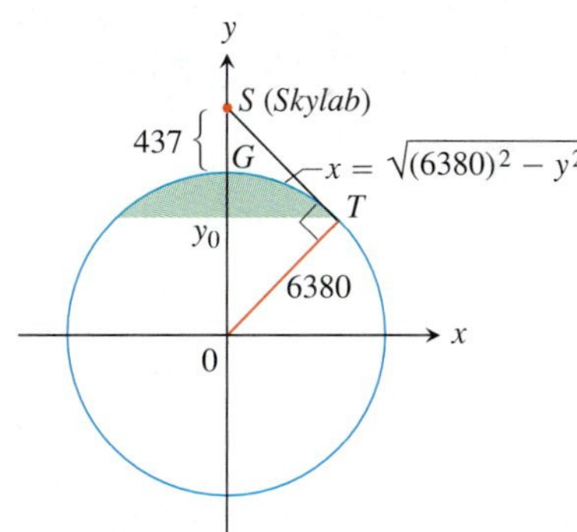

Then carry out these steps.

1. Use similar triangles in the figure to show that $y_0/6380 = 6380/(6380 + 437)$. Solve for y_0.

2. To four significant digits, calculate the visible area as

$$VA = \int_{y_0}^{6380} 2\pi x\sqrt{1 + \left(\frac{dx}{dy}\right)^2}\, dy.$$

3. Express the result as a percentage of Earth's surface area.

66. *Radius of curvature* Show that the radius of curvature of a twice-differentiable plane curve $\mathbf{r}(t) = f(t)\mathbf{i} + g(t)\mathbf{j}$ is given by the formula

$$\rho = \frac{\dot{x}^2 + \dot{y}^2}{\sqrt{\ddot{x}^2 + \ddot{y}^2 - \ddot{s}^2}}, \qquad \text{where } \ddot{s} = \frac{d}{dt}\sqrt{\dot{x}^2 + \dot{y}^2}.$$

Additional Exercises: Theory, Examples, Applications

Applications and Examples

1. *Submarine hunting* Two surface ships on maneuvers are trying to determine a submarine's course and speed to prepare for an aircraft intercept. As shown here, ship A is located at $(4, 0, 0)$, whereas ship B is located at $(0, 5, 0)$. All coordinates are given in thousands of feet. Ship A locates the submarine in the direction of the vector $2\mathbf{i} + 3\mathbf{j} - (1/3)\mathbf{k}$, and ship B locates it in the direction of the vector $18\mathbf{i} - 6\mathbf{j} - \mathbf{k}$. Four minutes ago, the submarine was located at $(2, -1, -1/3)$. The aircraft is due in 20 min. Assuming that the submarine moves in a straight line at a constant speed, to what position should the surface ships direct the aircraft?

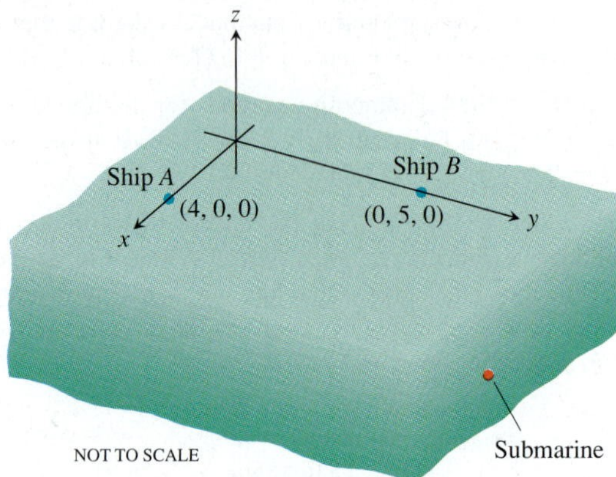

2. *A helicopter rescue* Two helicopters, H_1 and H_2, are traveling together. At time $t = 0$, they separate and follow different straight-line paths given by

$$H_1:\quad x = 6 + 40t,\quad y = -3 + 10t,\quad z = -3 + 2t$$
$$H_2:\quad x = 6 + 110t,\quad y = -3 + 4t,\quad z = -3 + t.$$

Time t is measured in hours and all coordinates are measured in miles. Due to system malfunctions, H_2 stops its flight at (446, 13, 1) and, in a negligible amount of time, lands at (446, 13, 0). Two hours later, H_1 is advised of this fact and heads toward H_2 at 150 mph. How long will it take H_1 to reach H_2?

3. *Torque* The operator's manual for the Toro® 21 in. lawnmower says "tighten the spark plug to 15 ft-lb (20.4 N · m)." If you are installing the plug with a 10.5 in. socket wrench that places the center of your hand 9 in. from the axis of the spark plug, about how hard should you pull? Answer in pounds.

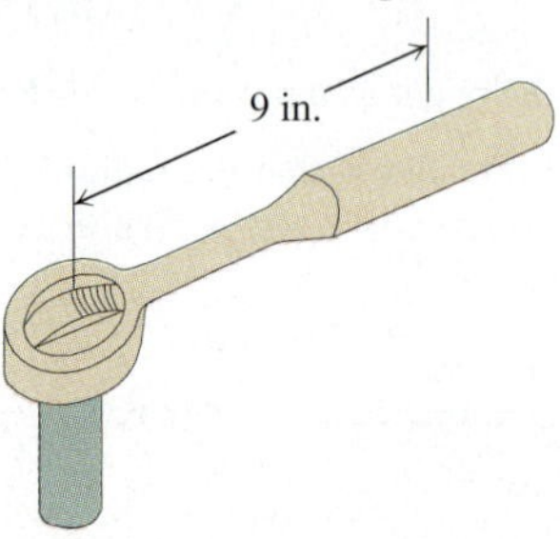

4. *Rotating body* The line through the origin and the point $A(1, 1, 1)$ is the axis of rotation of a rigid body rotating with a constant angular speed of 3/2 rad/sec. The rotation appears to be clockwise when we look toward the origin from A. Find the velocity $\mathbf{v}$ of the point of the body that is at the position $B(1, 3, 2)$.

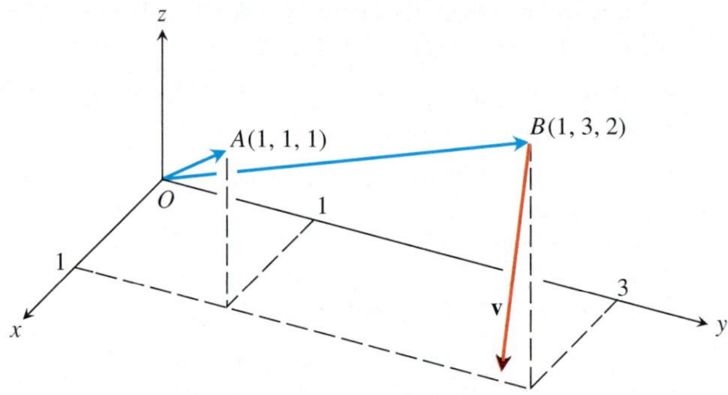

5. *Determinants and planes* Show that

$$\begin{vmatrix} x_1 - x & y_1 - y & z_1 - z \\ x_2 - x & y_2 - y & z_2 - z \\ x_3 - x & y_3 - y & z_3 - z \end{vmatrix} = 0$$

is an equation for the plane through the three noncollinear points $P_1(x_1, y_1, z_1)$, $P_2(x_2, y_2, z_2)$, and $P_3(x_3, y_3, z_3)$.

(b) What set of points in space is described by the equation

$$\begin{vmatrix} x & y & z & 1 \\ x_1 & y_1 & z_1 & 1 \\ x_2 & y_2 & z_2 & 1 \\ x_3 & y_3 & z_3 & 1 \end{vmatrix} = 0?$$

6. *Determinants and lines* Show that the lines

$$x = a_1 s + b_1, y = a_2 s + b_2, z = a_3 s + b_3, -\infty < s < \infty,$$

and

$$x = c_1 t + d_1, y = c_2 t + d_2, z = c_3 t + d_3, -\infty < t < \infty,$$

intersect or are parallel if and only if

$$\begin{vmatrix} a_1 & c_1 & b_1 - d_1 \\ a_2 & c_2 & b_2 - d_2 \\ a_3 & c_3 & b_3 - d_3 \end{vmatrix} = 0.$$

7. *Distance from point to plane* Use vectors to show that the distance from $P_1(x_1, y_1, z_1)$ to the plane $Ax + By + Cz = D$ is

$$d = \frac{|Ax_1 + By_1 + Cz_1 - D|}{\sqrt{A^2 + B^2 + C^2}}.$$

8. *Sphere tangent to plane* Find an equation for the sphere that is tangent to the planes $x + y + z = 3$ and $x + y + z = 9$ if the planes $2x - y = 0$ and $3x - z = 0$ pass through the center of the sphere.

9. *Distance between planes*

(a) Show that the distance between the parallel planes $Ax + By + Cz = D_1$ and $Ax + By + Cz = D_2$ is

$$d = \frac{|D_1 - D_2|}{|A\mathbf{i} + B\mathbf{j} + C\mathbf{k}|}.$$

(b) Use the equation in part (a) to find the distance between the planes $2x + 3y - z = 6$ and $2x + 3y - z = 12$.

10. *Parallel planes* Find an equation for the plane parallel to the plane $2x - y + 2z = -4$ if the point $(3, 2, -1)$ is equidistant from the two planes.

11. *Coplanar points* Prove that four points A, B, C, and D are coplanar (lie in a common plane) if and only if $\overrightarrow{AD} \cdot (\overrightarrow{AB} \times \overrightarrow{BC}) = 0$.

12. *Triple vector products* The **triple vector products** $(\mathbf{u} \times \mathbf{v}) \times \mathbf{w}$ and $\mathbf{u} \times (\mathbf{v} \times \mathbf{w})$ are usually not equal, although the formulas for evaluating them from components are similar:

$$(\mathbf{u} \times \mathbf{v}) \times \mathbf{w} = (\mathbf{u} \cdot \mathbf{w})\mathbf{v} - (\mathbf{v} \cdot \mathbf{w})\mathbf{u}.$$
$$\mathbf{u} \times (\mathbf{v} \times \mathbf{w}) = (\mathbf{u} \cdot \mathbf{w})\mathbf{v} - (\mathbf{u} \cdot \mathbf{v})\mathbf{w}.$$

Verify each formula for the following vectors by evaluating its two sides and comparing the results.

	u	v	w
(a)	$2\mathbf{i}$	$2\mathbf{j}$	$2\mathbf{k}$
(b)	$\mathbf{i} - \mathbf{j} + \mathbf{k}$	$2\mathbf{i} + \mathbf{j} - 2\mathbf{k}$	$-\mathbf{i} + 2\mathbf{j} - \mathbf{k}$
(c)	$2\mathbf{i} + \mathbf{j}$	$2\mathbf{i} - \mathbf{j} + \mathbf{k}$	$\mathbf{i} + 2\mathbf{k}$
(d)	$\mathbf{i} + \mathbf{j} - 2\mathbf{k}$	$-\mathbf{i} - \mathbf{k}$	$2\mathbf{i} + 4\mathbf{j} - 2\mathbf{k}$

13. *Cross and dot products* Show that if $\mathbf{u}$, $\mathbf{v}$, $\mathbf{w}$, and $\mathbf{r}$ are any vectors, then

(a) $\mathbf{u} \times (\mathbf{v} \times \mathbf{w}) + \mathbf{v} \times (\mathbf{w} \times \mathbf{u}) + \mathbf{w} \times (\mathbf{u} \times \mathbf{v}) = \mathbf{0}$

(b) $\mathbf{u} \times \mathbf{v} = (\mathbf{u} \cdot \mathbf{v} \times \mathbf{i})\mathbf{i} + (\mathbf{u} \cdot \mathbf{v} \times \mathbf{j})\mathbf{j} + (\mathbf{u} \cdot \mathbf{v} \times \mathbf{k})\mathbf{k}$

(c) $(\mathbf{u} \times \mathbf{v}) \cdot (\mathbf{w} \times \mathbf{r}) = \begin{vmatrix} \mathbf{u} \cdot \mathbf{w} & \mathbf{v} \cdot \mathbf{w} \\ \mathbf{u} \cdot \mathbf{r} & \mathbf{v} \cdot \mathbf{r} \end{vmatrix}.$

14. *Cross and dot products* Prove or disprove the formula

$$\mathbf{u} \times (\mathbf{u} \times (\mathbf{u} \times \mathbf{v})) \cdot \mathbf{w} = -|\mathbf{u}|^2 \mathbf{u} \cdot \mathbf{v} \times \mathbf{w}.$$

15. *The projection of a vector on a plane* Let P be a plane in space and let $\mathbf{v}$ be a vector. The vector projection of $\mathbf{v}$ onto the plane P, $\text{proj}_P\, \mathbf{v}$, can be defined informally as follows. Suppose that the sun is shining so that its rays are normal to the plane P. Then $\text{proj}_P\, \mathbf{v}$ is the "shadow" of $\mathbf{v}$ onto P. If P is the plane $x + 2y + 6z = 6$ and $\mathbf{v} = \mathbf{i} + \mathbf{j} + \mathbf{k}$, find $\text{proj}_P\, \mathbf{v}$.

16. *Trigonometry and vectors* By forming the cross product of two appropriate vectors, derive the trigonometric identity

$$\sin(A - B) = \sin A \cos B - \cos A \sin B.$$

17. *Point masses and gravitation* In physics, the law of gravitation says that if P and Q are (point) masses with mass M and m, respectively, then P is attracted to Q by the force

$$\mathbf{F} = \frac{GMm\mathbf{r}}{|\mathbf{r}|^3},$$

where $\mathbf{r}$ is the vector from P to Q and G is the universal gravitational constant. Moreover, if $Q_1, \ldots, Q_k$ are (point) masses with mass $m_1, \ldots, m_k$, respectively, then the force on P due to all the Q_i's is

$$\mathbf{F} = \sum_{i=1}^{k} \frac{GMm_i}{|\mathbf{r}_i|^3}\mathbf{r}_i,$$

where $\mathbf{r}_i$ is the vector from P to Q_i.

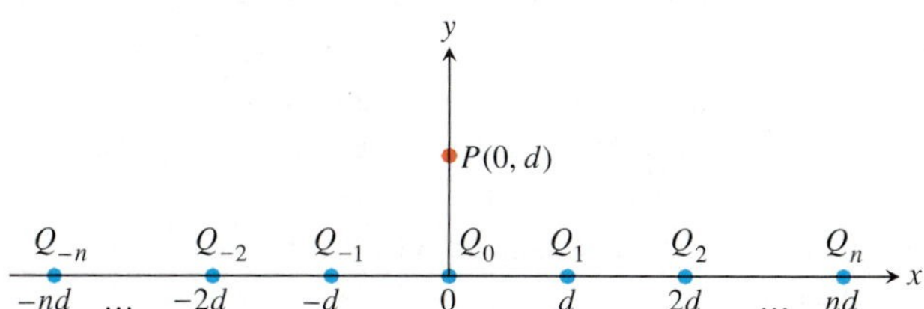

(a) Let point P with mass M be located at the point $(0, d)$, $d > 0$, in the coordinate plane. For $i = -n, -n + 1, \ldots, -1, 0, 1, \ldots, n$, let Q_i be located at the point $(id, 0)$ and have mass mi. Find the magnitude of the gravitational force on P due to all the Q_i's.

(b) Is the limit as $n \to \infty$ of the magnitude of the force on P finite? Why, or why not?

18. *Relativistic sums* Einstein's special theory of relativity roughly says that with respect to a reference frame (coordinate system) no material object can travel as fast as c, the speed of light. So, if $\vec{x}$ and $\vec{y}$ are two velocities such that $|\vec{x}| < c$ and $|\vec{y}| < c$, then the **relativistic sum** $\vec{x} \oplus \vec{y}$ of $\vec{x}$ and $\vec{y}$ must have length less than c. Einstein's special theory of relativity says that

$$\vec{x} \oplus \vec{y} = \frac{\vec{x} + \vec{y}}{1 + \dfrac{\vec{x} \cdot \vec{y}}{c^2}} + \frac{1}{c^2} \cdot \frac{\gamma_x}{\gamma_x + 1} \cdot \frac{\vec{x} \times (\vec{x} \times \vec{y})}{1 + \dfrac{\vec{x} \cdot \vec{y}}{c^2}},$$

where

$$\gamma_x = \frac{1}{\sqrt{1 - \dfrac{\vec{x} \cdot \vec{x}}{c^2}}}.$$

It can be shown that if $|\vec{x}| < c$ and $|\vec{y}| < c$, then $|\vec{x} \oplus \vec{y}| < c$. This exercise deals with two special cases.

(a) Prove that if $\vec{x}$ and $\vec{y}$ are orthogonal, $|\vec{x}| < c$, $|\vec{y}| < c$, then $|\vec{x} \oplus \vec{y}| < c$.

(b) Prove that if $\vec{x}$ and $\vec{y}$ are parallel, $|\vec{x}| < c$, $|\vec{y}| < c$, then $|\vec{x} \oplus \vec{y}| < c$.

(c) Compute $\lim_{c \to \infty} \vec{x} \oplus \vec{y}$.

Polar Coordinate Systems and Motion in Space

19. *Minimum distance to sun* Deduce from the orbit equation

$$r = \frac{(1 + e)r_0}{1 + e\cos\theta}$$

that a planet is closest to its sun when $\theta = 0$ and show that $r = r_0$ at that time.

20. *A Kepler equation* The problem of locating a planet in its orbit at a given time and date eventually leads to solving "Kepler" equations of the form

$$f(x) = x - 1 - \frac{1}{2}\sin x = 0.$$

(a) Show that this particular equation has a solution between $x = 0$ and $x = 2$.

T **(b)** With your computer or calculator in radian mode, use Newton's method to find the solution to as many places as you can.

21. In Section 10.8, we found the velocity of a particle moving in the plane to be

$$\mathbf{v} = \dot{x}\mathbf{i} + \dot{y}\mathbf{j} = \dot{r}\mathbf{u}_r + r\dot{\theta}\mathbf{u}_\theta.$$

(a) Express $\dot{x}$ and $\dot{y}$ in terms of $\dot{r}$ and $r\dot{\theta}$ by evaluating the dot products $\mathbf{v}\cdot\mathbf{i}$ and $\mathbf{v}\cdot\mathbf{j}$.

(b) Express $\dot{r}$ and $r\dot{\theta}$ in terms of $\dot{x}$ and $\dot{y}$ by evaluating the dot products $\mathbf{v}\cdot\mathbf{u}_r$ and $\mathbf{v}\cdot\mathbf{u}_\theta$.

22. *Curvature in polar coordinates* Express the curvature of a twice-differentiable curve $r = f(\theta)$ in the polar coordinate plane in terms of f and its derivatives.

23. *Beetle on rotating rod* A slender rod through the origin of the polar coordinate plane rotates (in the plane) about the origin at the rate of 3 rad/min. A beetle starting from the point (2, 0) crawls along the rod toward the origin at the rate of 1 in./min.

(a) Find the beetle's acceleration and velocity in polar form when it is halfway to (1 in. from) the origin.

(b) To the nearest tenth of an inch, what will be the length of the path the beetle has traveled by the time it reaches the origin?

24. *Conservation of angular momentum* Let $\mathbf{r}(t)$ denote the position in space of a moving object at time t. Suppose that the force acting on the object at time t is

$$\mathbf{F}(t) = -\frac{c}{|\mathbf{r}(t)|^3}\mathbf{r}(t),$$

where c is a constant. In physics, the **angular momentum** of an object at time t is defined to be $\mathbf{L}(t) = \mathbf{r}(t) \times m\mathbf{v}(t)$, where m is the mass of the object and $\mathbf{v}(t)$ is the velocity. Prove that angular momentum is a conserved quantity; that is, prove that $\mathbf{L}(t)$ is a constant vector, independent of time. Remember Newton's law $\mathbf{F} = m\mathbf{a}$. (This is a calculus problem, not a physics problem.)

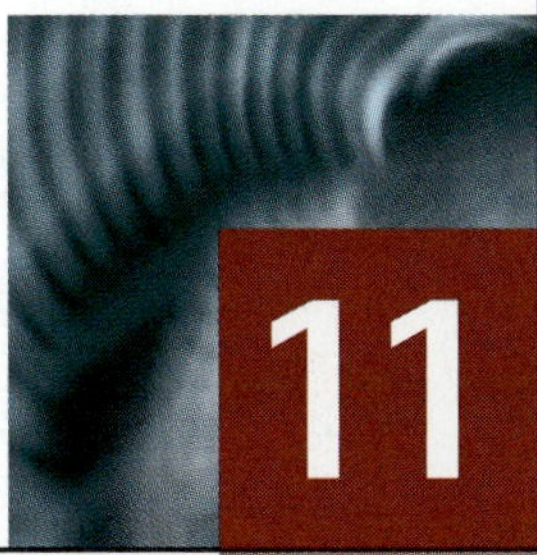

11 Multivariable Functions and their Derivatives

OVERVIEW Functions with two or more independent variables appear more often in science than functions of a single variable, and their calculus is even richer. Their derivatives are more varied and more interesting because of the different ways in which the variables can interact. Their integrals lead to a greater variety of applications. The studies of probability, statistics, fluid dynamics, and electricity, to mention only a few, all lead in natural ways to functions of more than one variable. The mathematics of these functions is one of the finest achievements in science.

As we see in this chapter, the rules of calculus remain essentially the same as we move into higher dimensions. We need to keep track of multiple directions of change at the same time, necessitating some new notation that uses the vector notation of previous chapters, but fortunately we do not need to reinvent the theory. Indeed, the calculus of several variables is really single-variable calculus applied to several variables at once.

11.1 Functions of Several Variables

Functions of Two Variables • Domains and Ranges • Graphs and Level Curves of Functions of Two Variables • Contour Curves • Computer Graphing • Functions of Three or More Variables • Level Surfaces of Functions of Three Variables

Many functions depend on more than one independent variable. The function $V = \pi r^2 h$ calculates the volume of a right circular cylinder from its radius and height. The function $f(x, y) = x^2 + y^2$ calculates the height of the paraboloid $z = x^2 + y^2$ above the point $P(x, y)$ from the two coordinates of P. The temperature T of a point on Earth's surface depends on its latitude x and longitude y, expressed by writing $T = f(x, y)$. In this section, we define functions of more than one independent variable and discuss ways to graph them.

Functions of Two Variables

The domains of real-valued functions of two independent real variables are sets of ordered pairs of real numbers, and the ranges are sets of real numbers of the kind we have worked with all along.

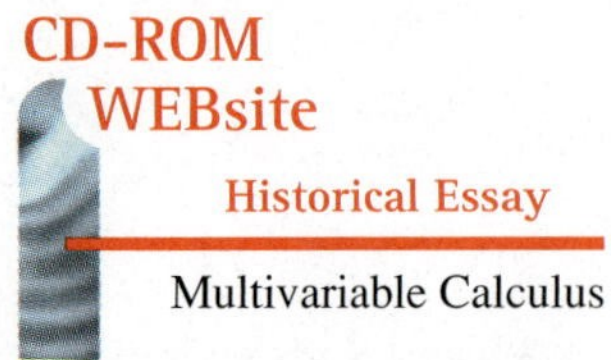

Definitions Functions of Two Variables

Suppose that D is a set of ordered pairs of real numbers (x, y). A **real-valued function f of two variables** on D is a rule that assigns a unique real number

$$w = f(x, y)$$

to each ordered pair (x, y) in D. The set D is the **domain** of f, and the set of w-values taken on by f is its **range.** The **independent variables** x and y are the function's **input** variables, and the **dependent variable** w is the function's **output** variable.

In applications, we tend to use letters that remind us of what the variables stand for. To say that the volume of a right circular cylinder is a function of its radius and height, we might write $V = f(r, h)$. To be more specific, we might replace the notation $f(r, h)$ by the formula that calculates the value of V from the values of r and h, and write $V = \pi r^2 h$. In either case, r and h would be the independent variables and V the dependent variable of the function.

As usual, we evaluate functions defined by formulas by substituting the values of the independent variables in the formula and calculating the corresponding value of the dependent variable.

Example 1 Distance Function from the Origin to a Point in the Plane

When we use rectangular coordinates, the distance of a point (x, y) from the origin is given by the function $D(x, y) = \sqrt{x^2 + y^2}$. The value of D at the point $(3, 4)$ is $D(3, 4) = \sqrt{3^2 + 4^2} = \sqrt{25} = 5$.

Domains and Ranges

In defining functions of two variables, we follow the usual practice of excluding inputs that lead to complex numbers or division by zero. If $f(x, y) = \sqrt{y - x^2}$, y cannot be less than x^2. If $f(x, y) = 1/(xy)$, xy cannot be zero. The domains of functions are otherwise assumed to be the largest sets for which the defining rules generate real numbers. The range consists of the set of output values for the dependent variable.

Example 2 Identifying Domains and Ranges

	Function	Domain	Range
(a)	$w = \sqrt{y - x^2}$	$y \geq x^2$	$[0, \infty)$
(b)	$w = \dfrac{1}{xy}$	$xy \neq 0$	$(-\infty, 0) \cup (0, \infty)$
(c)	$w = \sin xy$	Entire plane	$[-1, 1]$

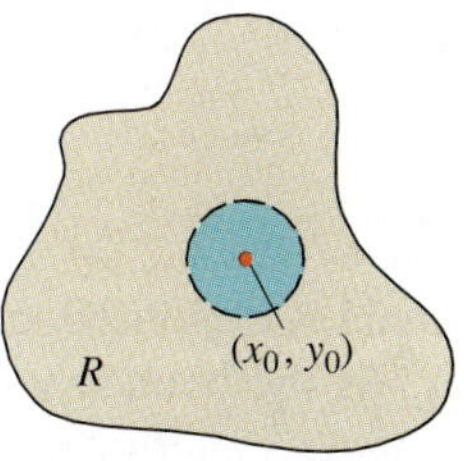

(a) Interior point

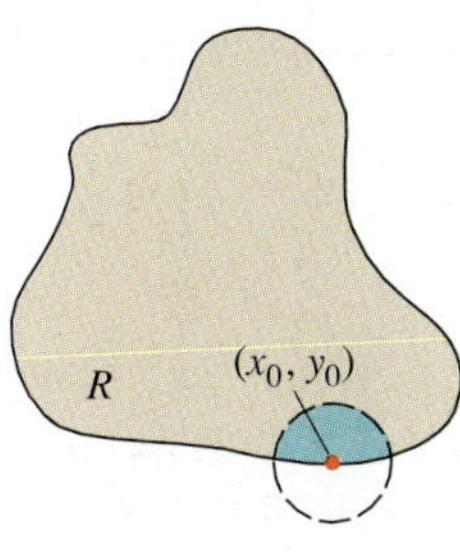

(b) Boundary point

FIGURE 11.1 Interior points and boundary points of a plane region R. An interior point is necessarily a point of R. A boundary point of R need not belong to R.

The domains of functions defined on portions of the plane can have interior points and boundary points just the way the domains of functions defined on intervals of the real line can.

Definitions Interior, Boundary, Open, Closed (2-space)

A point (x_0, y_0) in a region (set) R in the xy-plane is an **interior point** of R if it is the center of a disk that lies entirely in R (Figure 11.1). A point (x_0, y_0) is a **boundary point** of R if every disk centered at (x_0, y_0) contains points that lie outside of R as well as points that lie in R. (The boundary point itself need not belong to R.)

The interior points of a region, as a set, make up the **interior** of the region. The region's boundary points make up its **boundary.** A region is **open** if it consists entirely of interior points. A region is **closed** if it contains all its boundary points (Figure 11.2).

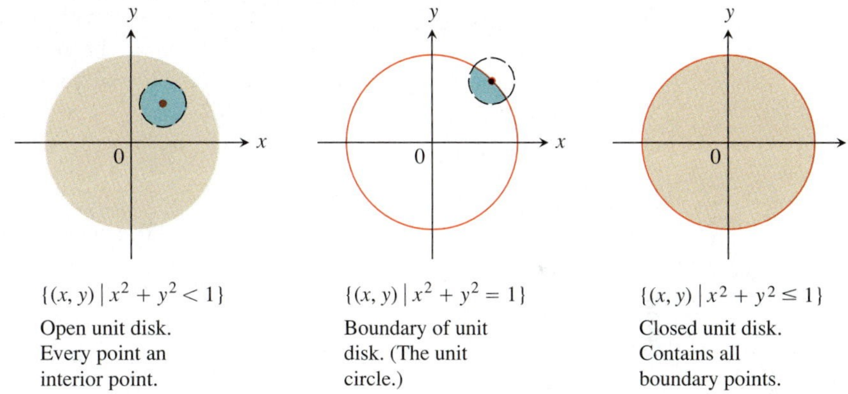

FIGURE 11.2 Interior points and boundary points of the unit disk in the plane.

As with intervals of real numbers, some regions in the plane are neither open nor closed. If you start with the open disk in Figure 11.2 and add to it some of but not all its boundary points, the resulting set is neither open nor closed. The boundary points that *are* there keep the set from being open. The absence of the remaining boundary points keeps the set from being closed.

Definitions Bounded and Unbounded Regions in the Plane

A region in the plane is **bounded** if it lies inside a disk of fixed radius. A region is **unbounded** if it is not bounded.

Examples of *bounded* sets in the plane include line segments, triangles, interiors of triangles, rectangles, circles, and disks. Examples of *unbounded* sets in the plane include lines, coordinate axes, the graphs of functions defined on infinite intervals, quadrants, half-planes, and the plane itself.

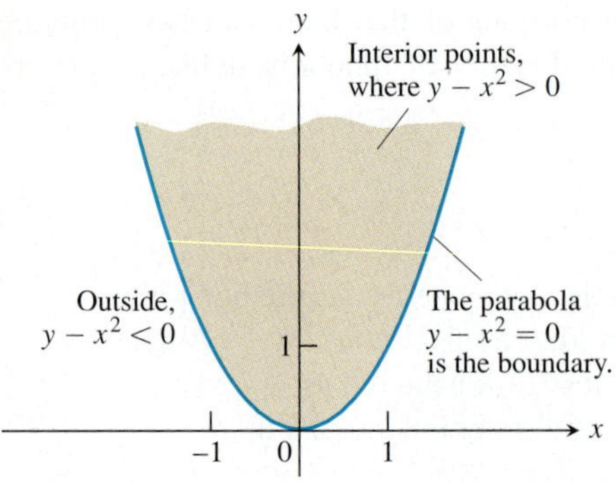

FIGURE 11.3 The domain of $f(x, y) = \sqrt{y - x^2}$ consists of the shaded region and its bounding parabola $y = x^2$. (Example 3)

Example 3 Describing the Domain of a Function of Two Variables

Describe the domain of the function $f(x, y) = \sqrt{y - x^2}$.

Solution

Since f is defined only where $y - x^2 \geq 0$, the domain is the closed, unbounded region shown in Figure 11.3. The parabola $y = x^2$ is the boundary of the domain. The points above the parabola make up the domain's interior.

Graphs and Level Curves of Functions of Two Variables

There are two standard ways to picture the values of a function $f(x, y)$. One is to draw and label curves in the domain on which f has a constant value. The other is to sketch the surface $z = f(x, y)$ in space.

CD-ROM WEBsite

Definitions Level Curve, Graph, Surface (Functions of Two Variables)

The set of points in the plane where a function $f(x, y)$ has a constant value $f(x, y) = c$ is called a **level curve** of f. The set of all points $(x, y, f(x, y))$ in space, for (x, y) in the domain of f, is called the **graph** of f. The graph of f is also called the **surface** $z = f(x, y)$.

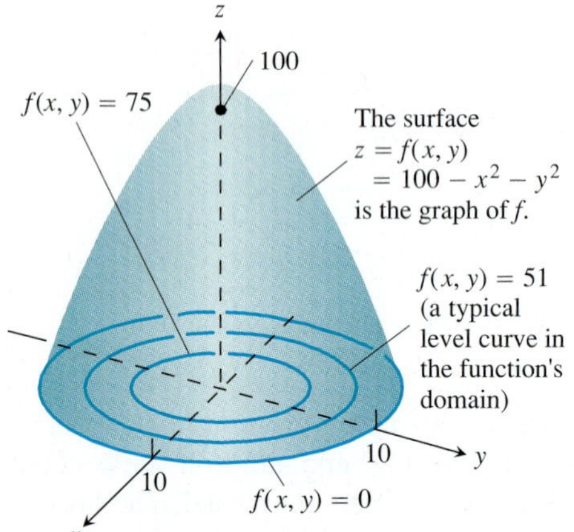

FIGURE 11.4 The graph and selected level curves of the function $f(x, y) = 100 - x^2 - y^2$. (Example 4)

Example 4 Graphing a Function of Two Variables

Graph $f(x, y) = 100 - x^2 - y^2$ and plot the level curves $f(x, y) = 0$, $f(x, y) = 51$, and $f(x, y) = 75$ in the domain of f in the plane.

Solution The domain of f is the entire xy-plane, and the range of f is the set of real numbers less than or equal to 100. The graph is the paraboloid $z = 100 - x^2 - y^2$, a portion of which is shown in Figure 11.4.

The level curve $f(x, y) = 0$ is the set of points in the xy-plane at which

$$f(x, y) = 100 - x^2 - y^2 = 0, \quad \text{or} \quad x^2 + y^2 = 100,$$

which is the circle of radius 10 centered at the origin. Similarly, the level curves $f(x, y) = 51$ and $f(x, y) = 75$ (Figure 11.4) are the circles

$$f(x, y) = 100 - x^2 - y^2 = 51, \quad \text{or} \quad x^2 + y^2 = 49$$
$$f(x, y) = 100 - x^2 - y^2 = 75, \quad \text{or} \quad x^2 + y^2 = 25.$$

The level curve $f(x, y) = 100$ consists of the origin alone. (It is still a level curve.)

Contour Curves

The curve in space in which the plane $z = c$ cuts a surface $z = f(x, y)$ is made up of the points that represent the function value $f(x, y) = c$. It is called the **contour curve** $f(x, y) = c$ to distinguish it from the level curve $f(x, y) = c$ in the domain of f. Figure 11.5 shows the contour curve $f(x, y) = 75$ on the surface $z = 100 - x^2 - y^2$ defined by the function $f(x, y) = 100 - x^2 - y^2$. The contour curve lies directly above the circle $x^2 + y^2 = 25$, which is the level curve $f(x, y) = 75$ in the function's domain.

The contour curve $f(x, y) = 100 - x^2 - y^2 = 75$ is the circle $x^2 + y^2 = 25$ in the plane $z = 75$.

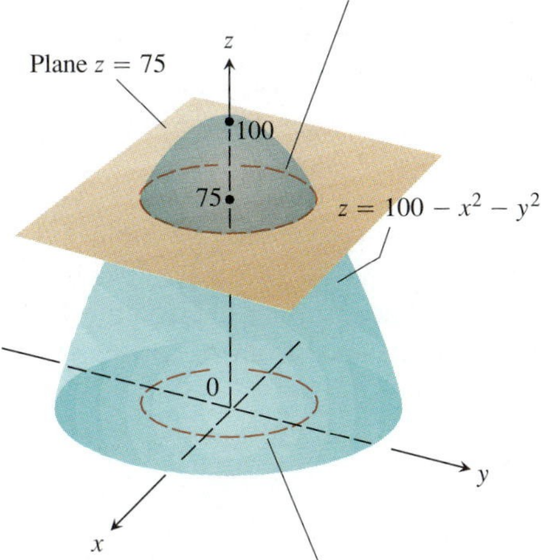

The level curve $f(x, y) = 100 - x^2 - y^2 = 75$ is the circle $x^2 + y^2 = 25$ in the xy-plane.

FIGURE 11.5 The graph of $f(x, y) = 100 - x^2 - y^2$ and its intersection with the plane $z = 75$.

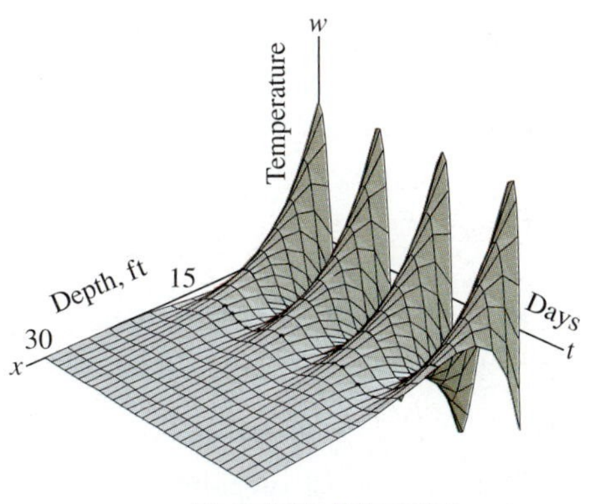

FIGURE 11.7 This computer-generated graph of

$$w = \cos(1.7 \times 10^{-2}t - 0.2x)e^{-0.2x}$$

shows the seasonal variation of the temperature below ground as a fraction of surface temperature. At $x = 15$ ft, the variation is only 5% of the variation at the surface. At $x = 30$ ft, the variation is less than 0.25% of the surface variation. (Example 5) (Adapted from art provided by Norton Starr.)

Not everyone makes this distinction, however, and you may wish to call both kinds of curves by a single name and rely on context to convey which one you have in mind. On most maps, for example, the curves that represent constant elevation (height above sea level) are called contours, not level curves (Figure 11.6).

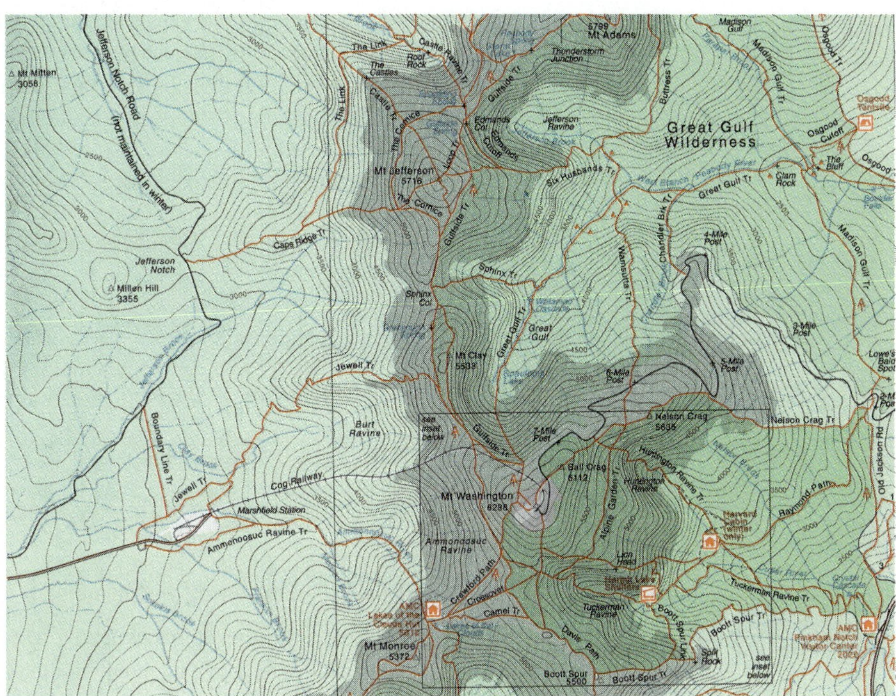

FIGURE 11.6 Contours on Mt. Washington in central New Hampshire.

Computer Graphing

Three-dimensional graphing programs for computers and calculators make it possible to graph functions of two variables with only a few keystrokes. We can often get information more quickly from a graph than from a formula.

Example 5 Modeling Temperature Beneath Earth's Surface

The temperature beneath Earth's surface is a function of the depth x beneath the surface and the time t of the year. If we measure x in feet and t as the number of days elapsed from the average date of the yearly highest surface temperature, we can model the variation in temperature with the function

$$w = \cos(1.7 \times 10^{-2}t - 0.2x)\, e^{-0.2x}.$$

(The temperature at 0 ft is scaled to vary from $+1$ to -1, so that the variation at x feet can be interpreted as a fraction of the variation at the surface.)

Figure 11.7 shows a computer-generated graph of the function. At a depth of 15 ft, the variation (change in vertical amplitude in the figure) is about 5% of the surface variation. At 30 ft, there is almost no variation during the year.

The graph also shows that the temperature 15 ft below the surface is about half a year out of phase with the surface temperature. When the temperature is lowest on the surface (late January, say), it is at its highest 15 ft below. Fifteen feet below the ground, the seasons are reversed.

Functions of Three or More Variables

A **function f of three variables** is a rule that assigns to each ordered triple (x, y, z) in some domain D in space a unique real number $w = f(x, y, z)$. Again the range consists of the output values for w. For instance, similar to Example 1, the function $D(x, y, z) = \sqrt{x^2 + y^2 + z^2}$ gives the distance from the origin to the point (x, y, z) in space for rectangular coordinates.

Example 6 Functions of Three Variables

	Function	Domain	Range
(a)	$w = \sqrt{x^2 + y^2 + z^2}$	Entire space	$[0, \infty)$
(b)	$w = \dfrac{1}{x^2 + y^2 + z^2}$	$(x, y, z) \neq (0, 0, 0)$	$(0, \infty)$
(c)	$w = xy \ln z$	Half-space $z > 0$	$(-\infty, \infty)$

Level Surfaces of Functions of Three Variables

In the plane, the points where a function of two independent variables has a constant value $f(x, y) = c$ make a curve in the function's domain. In space, the points where a function of three independent variables has a constant value $f(x, y, z) = c$ make a surface in the function's domain.

Definition Level Surface

The set of points (x, y, z) in space where a function of three independent variables has a constant value $f(x, y, z) = c$ is called a **level surface** of f.

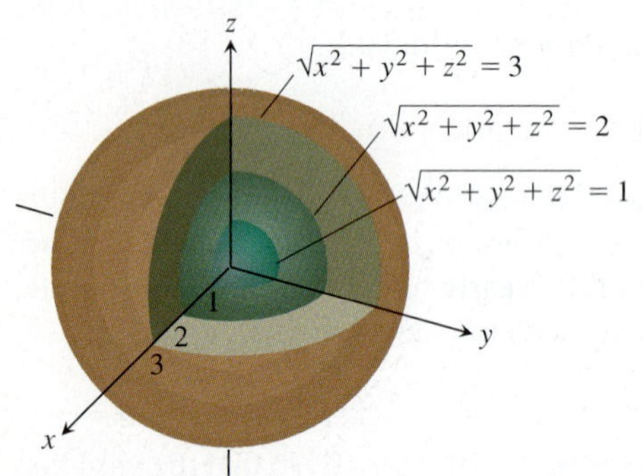

FIGURE 11.8 The level surfaces of $f(x, y, z) = \sqrt{x^2 + y^2 + z^2}$ are concentric spheres.

Since the graphs of functions of three variables consist of points $(x, y, z, f(x, y, z))$ lying in a four-dimensional space, we cannot sketch them effectively in our three-dimensional frame of reference. We can see how the function behaves, however, by looking at its three-dimensional level surfaces.

Example 7 Describing Level Surfaces of a Function of Three Variables

Describe the level surfaces of the function

$$f(x, y, z) = \sqrt{x^2 + y^2 + z^2}.$$

Solution The value of f is the distance from the origin to the point (x, y, z). Each level surface $\sqrt{x^2 + y^2 + z^2} = c$, $c > 0$, is a sphere of radius c centered at the origin. Figure 11.8 shows a cutaway view of three of these spheres. The level surface $\sqrt{x^2 + y^2 + z^2} = 0$ consists of the origin alone.

We are not graphing the function here; we are looking at level surfaces in the function's domain. The level surfaces show how the function's values change as we move through its domain. If we remain on a sphere of radius c centered at the origin, the function maintains a constant value, namely c. If we move from one sphere to another, the function's value changes. It increases if we move away from the origin and decreases if we move toward the origin. The way the values change depends on the direction we take. The dependence of change on direction is important. We return to it in Section 11.5.

The definitions of interior, boundary, open, closed, bounded, and unbounded for regions in space are similar to those for regions in the plane. To accommodate the extra dimension, we use solid spheres instead of disks.

(a) Interior point

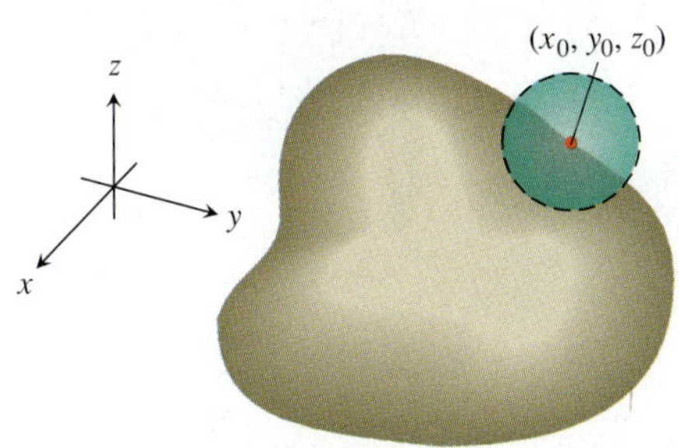

(b) Boundary point

FIGURE 11.9 Interior points and boundary points of a region in space.

Definitions Interior, Boundary, Open, Closed (3-space)

A point (x_0, y_0, z_0) in a region R in space is an **interior point** of R if it is the center of a solid sphere that lies entirely in R (Figure 11.9a). A point (x_0, y_0, z_0) is a **boundary point** of R if every sphere centered at (x_0, y_0, z_0) encloses points that lie outside of R as well as points that lie inside R (Figure 11.9b). The **interior** of R is the set of interior points of R. The **boundary** of R is the set of boundary points of R.

A region is **open** if it consists entirely of interior points. A region is **closed** if it contains its entire boundary.

Examples of *open* sets in space include the interior of a sphere, the open half-space $z > 0$, the first octant (where x, y, and z are all positive), and space itself.

Examples of *closed* sets in space include lines, planes, the closed half-space $z \geq 0$, the first octant together with its bounding planes, and space itself (since it has no boundary points).

A solid sphere with part of its boundary removed or a solid cube with a missing face, edge, or corner point would be *neither open nor closed.*

Functions of more than three independent variables are also important. For example, the temperature on a surface in space may depend not only on the location of the point $P(x, y, z)$ on the surface, but also on time t it is visited, so we would write $T = f(x, y, z, t)$.

In general, a **function f of n variables** is a rule that assigns to each n-tuple $(x_1, x_2, \dots, x_n)$ of real numbers a unique real number $w = f(x_1, x_2, \dots, x_n)$. The variables x_1 to x_n are the **independent (input)** variables, and w is the **dependent (output)** variable.

Functions of more than three variables are not easily visualized, but powerful mathematical methods have been developed to analyze them. You may study some of these methodologies in your advanced mathematics or science courses. In this text, we restrict our attention to functions of two or three independent variables, visualized by their graphs, level curves, or level surfaces, as appropriate.

EXERCISES 11.1

Domain, Range, and Level Curves

In Exercises 1–12,

(a) Find the function's domain

(b) Find the function's range

(c) Describe the function's level curves

(d) Find the boundary of the function's domain

(e) Determine if the domain is an open region, a closed region, or neither

(f) Decide if the domain is bounded or unbounded.

1. $f(x, y) = y - x$

2. $f(x, y) = \sqrt{y - x}$

3. $f(x, y) = 4x^2 + 9y^2$

4. $f(x, y) = x^2 - y^2$

5. $f(x, y) = xy$

6. $f(x, y) = y/x^2$

7. $f(x, y) = \dfrac{1}{\sqrt{16 - x^2 - y^2}}$

8. $f(x, y) = \sqrt{9 - x^2 - y^2}$

9. $f(x, y) = \ln(x^2 + y^2)$

10. $f(x, y) = e^{-(x^2+y^2)}$

11. $f(x, y) = \sin^{-1}(y - x)$

12. $f(x, y) = \tan^{-1}\left(\dfrac{y}{x}\right)$

Identifying Surfaces and Level Curves

Execises 13–18 show level curves for the functions graphed in (a)–(f). Match each set of curves with the appropriate function.

13.

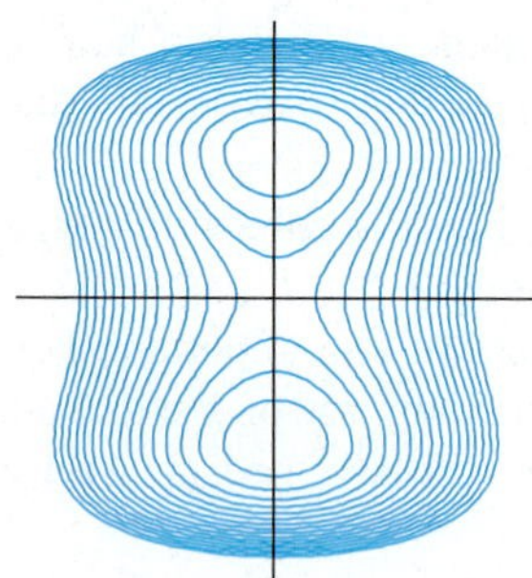

14.

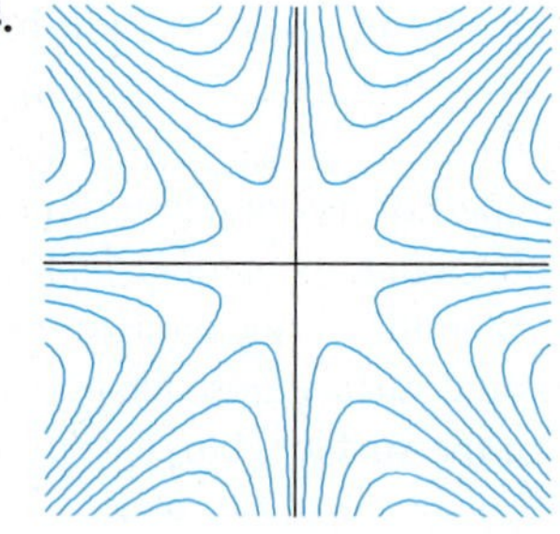

15.

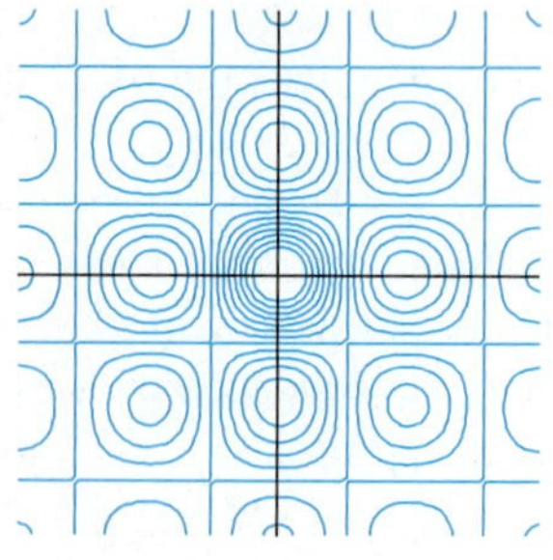

16.

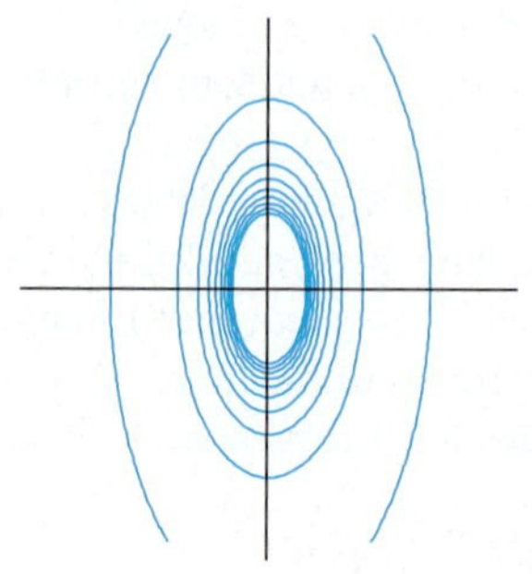

17.

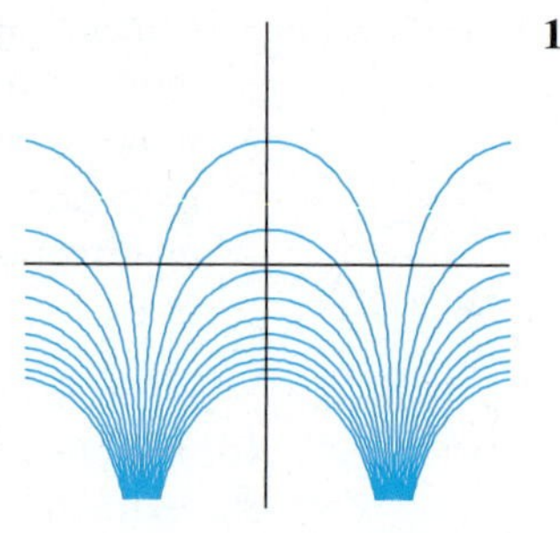

18.

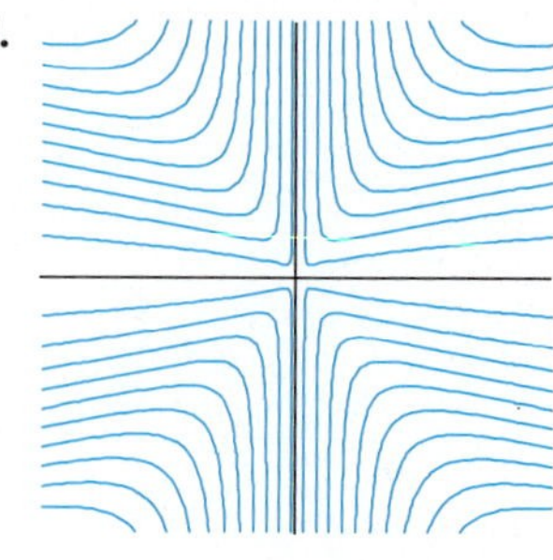

(a)

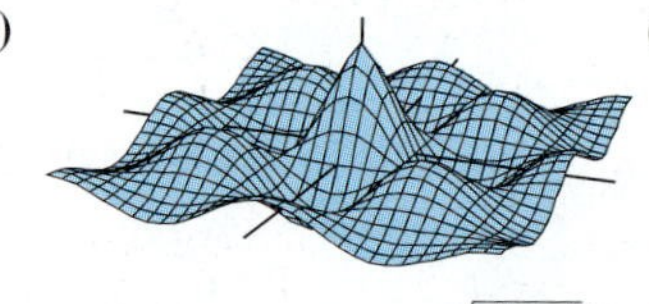

$z = (\cos x)(\cos y)\, e^{-\sqrt{x^2 + y^2}/4}$

(b)

$z = -\dfrac{xy^2}{x^2 + y^2}$

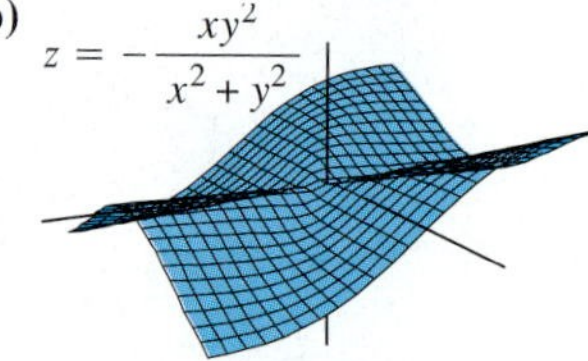

(c)

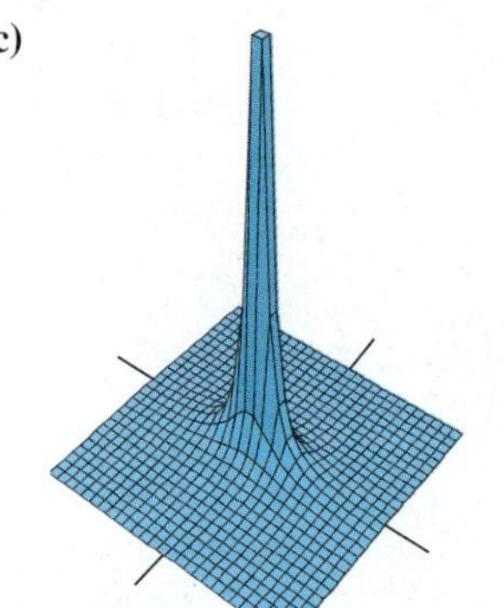

$z = \dfrac{1}{(4x^2 + y^2)}$

(d)

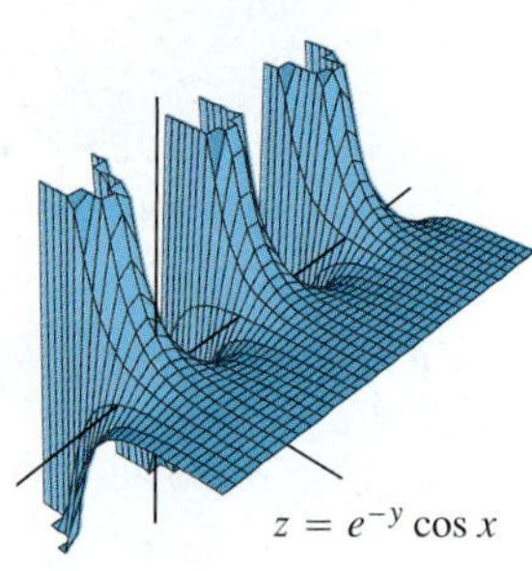

$z = e^{-y} \cos x$

(e)

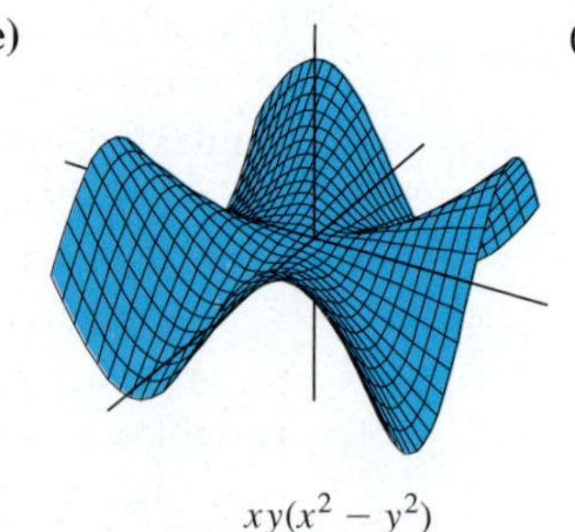

$z = \dfrac{xy(x^2 - y^2)}{(x^2 + y^2)}$

(f)

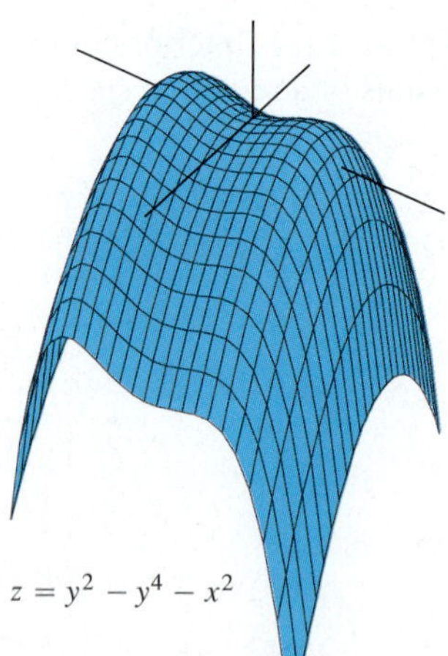

$z = y^2 - y^4 - x^2$

Identifying Functions of Two Variables

Display the values of the functions in Exercises 19–28 in two ways: (a) by sketching the surface $z = f(x, y)$ and (b) by drawing an assortment of level curves in the function's domain. Label each level curve with its function value.

19. $f(x, y) = y^2$

20. $f(x, y) = 4 - y^2$

21. $f(x, y) = x^2 + y^2$

22. $f(x, y) = \sqrt{x^2 + y^2}$

23. $f(x, y) = -(x^2 + y^2)$

24. $f(x, y) = 4 - x^2 - y^2$

25. $f(x, y) = 4x^2 + y^2$

26. $f(x, y) = 4x^2 + y^2 + 1$

27. $f(x, y) = 1 - |y|$

28. $f(x, y) = 1 - |x| - |y|$

Finding a Level Curve

In Exercises 29–32, find an equation for the level curve of the function $f(x, y)$ that passes through the given point.

29. $f(x, y) = 16 - x^2 - y^2, \quad (2\sqrt{2}, \sqrt{2})$

30. $f(x, y) = \sqrt{x^2 - 1}, \quad (1, 0)$

31. $f(x, y) = \displaystyle\int_x^y \frac{dt}{1 + t^2}, \quad (-\sqrt{2}, \sqrt{2})$

32. $f(x, y) = \displaystyle\sum_{n=0}^{\infty} \left(\frac{x}{y}\right)^n, \quad (1, 2)$

Sketching Level Surfaces

In Exercises 33–40, sketch a typical level surface for the function.

33. $f(x, y, z) = x^2 + y^2 + z^2$

34. $f(x, y, z) = \ln (x^2 + y^2 + z^2)$

35. $f(x, y, z) = x + z$

36. $f(x, y, z) = z$

37. $f(x, y, z) = x^2 + y^2$

38. $f(x, y, z) = y^2 + z^2$

39. $f(x, y, z) = z - x^2 - y^2$

40. $f(x, y, z) = (x^2/25) + (y^2/16) + (z^2/9)$

Finding a Level Surface

In Exercises 41–44, find an equation for the level surface of the function through the given point.

41. $f(x, y, z) = \sqrt{x - y} - \ln z, \quad (3, -1, 1)$

42. $f(x, y, z) = \ln (x^2 + y + z^2), \quad (-1, 2, 1)$

43. $g(x, y, z) = \displaystyle\sum_{n=0}^{\infty} \frac{(x + y)^n}{n!\, z^n}, \quad (\ln 2, \ln 4, 3)$

44. $g(x, y, z) = \displaystyle\int_x^y \frac{d\theta}{\sqrt{1 - \theta^2}} + \int_{\sqrt{2}}^{z} \frac{dt}{t\sqrt{t^2 - 1}}, \quad (0, 1/2, 2)$

Theory and Examples

45. *The maximum value of a function on a line in space* Does the function $f(x, y, z) = xyz$ have a maximum value on the line $x = 20 - t$, $y = t$, $z = 20$? If so, what is it? Give reasons for your answer. (*Hint:* Along the line, $w = f(x, y, z)$ is a differentiable function of t.)

46. *The minimum value of a function on a line in space* Does the function $f(x, y, z) = xy - z$ have a minimum value on the line $x = t - 1$, $y = t - 2$, $z = t + 7$? If so, what is it? Give reasons for your answer. (*Hint:* Along the line, $w = f(x, y, z)$ is a differentiable function of t.)

47. *The Concorde's sonic booms* The width w of the region in which people on the ground hear the *Concorde*'s sonic boom directly, not reflected from a layer in the atmosphere, is a function of

T = air temperature at ground level (in degrees Kelvin)

h = the *Concorde*'s altitude (in kilometers)

d = the vertical temperature gradient (temperature drop in degrees Kelvin per kilometer).

The formula for w is

$$w = 4\left(\frac{Th}{d}\right)^{1/2}.$$

See Figure 11.10.

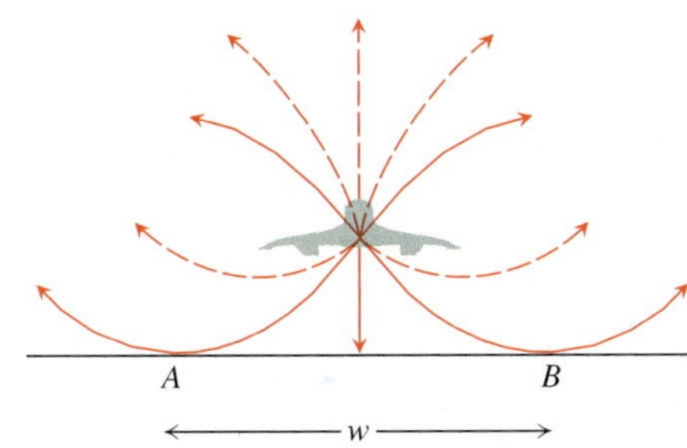

FIGURE 11.10 Sound waves from the *Concorde* bend as the temperature changes above and below the altitude at which the plane flies. The sonic boom carpet is the region on the ground that receives shock waves directly from the plane, not reflected from the atmosphere or diffracted along the ground. The carpet is determined by the grazing rays striking the ground from the point directly under the plane. (Exercise 47)

The Washington-bound *Concorde* approaches the United States from Europe on a course that takes it south of Nantucket Island at an altitude of 16.8 km. If the surface temperature is 290 K and the vertical temperature gradient is 5 K/km, how many kilometers south of Nantucket must the plane be flown to keep its sonic boom carpet away from the island? (From "Concorde Sonic Booms as an Atmospheric Probe" by N. K. Balachandra, W. L. Donn, and D. H. Rind, *Science*, Vol. 197 (July 1, 1977), pp. 47–49.)

48. *Writing to Learn* As you know, the graph of a real-valued function of a single real variable is a set in a two-coordinate space. The graph of a real-valued function of two independent real variables is a set in a three-coordinate space. The graph of a real-valued function of three independent real variables is a set in a four-coordinate space. How would you define the graph of a real-valued function $f(x_1, x_2, x_3, x_4)$ of four independent real variables? How would you define the graph of a real-valued function $f(x_1, x_2, x_3, \ldots, x_n)$ of n independent real variables?

COMPUTER EXPLORATIONS

Explicit Surfaces

Use a CAS to perform the following steps for each of the functions in Exercises 49–52.

(a) Plot the surface over the given rectangle.

(b) Plot several level curves in the rectangle.

(c) Plot the level curve of f through the given point.

49. $f(x, y) = x \sin \dfrac{y}{2} + y \sin 2x, \quad 0 \le x \le 5\pi \quad 0 \le y \le 5\pi,$
$P(3\pi, 3\pi)$

50. $f(x, y) = (\sin x)(\cos y)\, e^{\sqrt{x^2+y^2}/8}, \quad 0 \le x \le 5\pi, \quad 0 \le y \le 5\pi,$
$P(4\pi, 4\pi)$

51. $f(x, y) = \sin (x + 2 \cos y), \quad -2\pi \le x \le 2\pi, \quad -2\pi \le y \le 2\pi,$
$P(\pi, \pi)$

52. $f(x, y) = e^{(x^{0.1}-y)} \sin (x^2 + y^2), \quad 0 \le x \le 2\pi, \quad -2\pi \le y \le \pi,$
$P(\pi, -\pi)$

Implicit Surfaces

CD-ROM WEBsite

Use a CAS to plot the level surfaces in Exercises 53–56.

53. $4 \ln (x^2 + y^2 + z^2) = 1$

54. $x^2 + z^2 = 1$

55. $x + y^2 - 3z^2 = 1$

56. $\sin \left(\dfrac{x}{2}\right) - (\cos y)\sqrt{x^2 + z^2} = 2$

Parametrized Surfaces

Just as you describe curves in the plane parametrically with a pair of equations $x = f(t)$, $y = g(t)$ defined on some parameter interval I, you can sometimes describe surfaces in space with a triple of equations $x = f(u, v)$, $y = g(u, v)$, $z = h(u, v)$ defined on some parameter rectangle $a \le u \le b$, $c \le v \le d$. Many computer algebra systems permit you to plot such surfaces in *parametric mode*. (Parametrized surfaces are discussed in detail in Section 13.6.) Use a CAS to plot the surfaces in Exercises 57–60. Also plot several level curves in the xy-plane.

57. $x = u \cos v, \quad y = u \sin v, \quad z = u, \quad 0 \le u \le 2, \quad 0 \le v \le 2\pi$

58. $x = u \cos v, \quad y = u \sin v, \quad z = v, \quad 0 \le u \le 2, \quad 0 \le v \le 2\pi$

59. $x = (2 + \cos u) \cos v, \quad y = (2 + \cos u) \sin v, \quad z = \sin u,$
$0 \le u \le 2\pi, \quad 0 \le v \le 2\pi$

60. $x = 2 \cos u \cos v, \quad y = 2 \cos u \sin v, \quad z = 2 \sin u,$
$0 \le u \le 2\pi, \quad 0 \le v \le \pi$

11.2 Limits and Continuity in Higher Dimensions

Limit of a Function of Two Variables • Continuity of a Function of Two Variables • Functions of More Than Two Variables • Extreme Values of Continuous Functions on Closed, Bounded Sets

This section treats limits and continuity for multivariable functions. The definition of the limit of a function of two or three variables is similar to the definition of the limit of a function of a single variable but with a crucial difference, as we now see.

Limit of a Function of Two Variables

If the values of a real-valued function $f(x, y)$ lie close to a fixed real number L for all points (x, y) sufficiently close to the point (x_0, y_0) but not equal to (x_0, y_0), we say that L is the limit of f as (x, y) approaches (x_0, y_0). In symbols, we write

$$\lim_{(x,y)\to(x_0,y_0)} f(x, y) = L,$$

and we say, "The limit of f as (x, y) approaches (x_0, y_0) equals L." This is like the limit of a function of one variable, except that two independent variables are involved instead of one, complicating the issue of "closeness." If (x_0, y_0) is an interior point of f's domain, (x, y) can approach (x_0, y_0) from any direction, whereas in the single-variable case, x only approached x_0 along the x-axis. The direction of approach can be an issue, as in some of the examples that follow.

Guillaume l'Hôpital
(1661 — 1704)

Definition Limit of a Function of Two Independent Variables

The function f **has limit** L as (x, y) approaches (x_0, y_0) if, given any positive number ϵ, there is a positive number δ such that for all (x, y) in the domain of f,

$$0 < \sqrt{(x - x_0)^2 + (y - y_0)^2} < \delta \Rightarrow |f(x, y) - L| < \epsilon.$$

We write

$$\lim_{(x,y)\to(x_0,y_0)} f(x, y) = L,$$

and assume (x_0, y_0) is a boundary or interior point of the domain of f.

The δ-ϵ requirement in the definition of limit is equivalent to the requirement that, given $\epsilon > 0$, there exists a corresponding $\delta > 0$ such that for all x,

$$0 < |x - x_0| < \delta \quad \text{and} \quad 0 < |y - y_0| < \delta \Rightarrow |f(x, y) - L| < \epsilon$$

(Exercise 43). Thus, in calculating limits, we can think either in terms of distance in the plane or in terms of differences in coordinates.

The definition of limit applies to boundary points (x_0, y_0) as well as interior points of the domain of f. The only requirement is that the point (x, y) remain in the domain at all times.

It can be shown, as for functions of a single variable, that

$$\lim_{(x,y)\to(x_0,y_0)} x = x_0$$

$$\lim_{(x,y)\to(x_0,y_0)} y = y_0$$

$$\lim_{(x,y)\to(x_0,y_0)} k = k \qquad \text{(any number } k).$$

It can also be shown that the limit of the sum of two functions is the sum of their limits (when they both exist), with similar results for the limits of the differences, products, constant multiples, quotients, and roots.

Theorem 1 Properties of Limits of Functions of Two Variables

The following rules hold if L, M, and k are real numbers and

$$\lim_{(x,y)\to(x_0,y_0)} f(x, y) = L \qquad \text{and} \qquad \lim_{(x,y)\to(x_0,y_0)} g(x, y) = M.$$

1. *Sum Rule:* $\lim_{(x,y)\to(x_0,y_0)} [f(x, y) + g(x, y)] = L + M$
2. *Difference Rule:* $\lim_{(x,y)\to(x_0,y_0)} [f(x, y) - g(x, y)] = L - M$
3. *Product Rule:* $\lim_{(x,y)\to(x_0,y_0)} f(x, y) \cdot g(x, y) = L \cdot M$
4. *Constant Multiple Rule:* $\lim_{(x,y)\to(x_0,y_0)} kf(x, y) = kL$ (any number k)
5. *Quotient Rule:* $\lim_{(x,y)\to(x_0,y_0)} \dfrac{f(x, y)}{g(x, y)} = \dfrac{L}{M}$ if $M \neq 0$.
6. *Root Rule:* If n is a positive integer, then $\lim_{(x,y)\to(x_0,y_0)} [f(x, y)]^{1/n} = L^{1/n}$

(If n is even, we assume that $L > 0$.)

When we apply Theorem 1 to polynomials and rational functions, we obtain the useful result that the limits of these functions as $(x, y) \to (x_0, y_0)$ can be calculated by evaluating the functions at (x_0, y_0). The only requirement is that the rational functions be defined at (x_0, y_0).

Example 1 Calculating Limits

(a) $$\lim_{(x,y)\to(0,1)} \frac{x - xy + 3}{x^2y + 5xy - y^3} = \frac{0 - (0)(1) + 3}{(0)^2(1) + 5(0)(1) - (1)^3} = -3$$

(b) $$\lim_{(x,y)\to(3,-4)} \sqrt{x^2 + y^2} = \sqrt{(3)^2 + (-4)^2} = \sqrt{25} = 5$$

Example 2 Calculating Limits

Find

$$\lim_{(x,y)\to(0,0)} \frac{x^2 - xy}{\sqrt{x} - \sqrt{y}}.$$

Solution Since the denominator $\sqrt{x} - \sqrt{y}$ approaches 0 as $(x, y) \to (0, 0)$, we cannot use the Quotient Rule from Theorem 1. If we multiply numerator and denominator by $\sqrt{x} + \sqrt{y}$, however, we produce an equivalent fraction whose limit we *can* find:

$$\begin{aligned}
\lim_{(x,y)\to(0,0)} \frac{x^2 - xy}{\sqrt{x} - \sqrt{y}} &= \lim_{(x,y)\to(0,0)} \frac{(x^2 - xy)(\sqrt{x} + \sqrt{y})}{(\sqrt{x} - \sqrt{y})(\sqrt{x} + \sqrt{y})} \\
&= \lim_{(x,y)\to(0,0)} \frac{x(x - y)(\sqrt{x} + \sqrt{y})}{x - y} && \text{Algebra} \\
&= \lim_{(x,y)\to(0,0)} x(\sqrt{x} + \sqrt{y}) && \text{Cancel the factor } (x - y). \\
&= 0(\sqrt{0} + \sqrt{0}) = 0
\end{aligned}$$

We can cancel the factor $(x - y)$ in Example 2 because the path $y = x$ (along which $x - y = 0$) is *not* in the domain of the function

$$\frac{x^2 - xy}{\sqrt{x} - \sqrt{y}}.$$

Continuity of a Function of Two Variables

The definition of continuity for functions of two variables is essentially the same as for functions of a single variable.

Definitions Continuity at a Point, Continuity

A function $f(x, y)$ is **continuous at the point (x_0, y_0)** if

1. f is defined at (x_0, y_0)
2. $\lim_{(x,y)\to(x_0,y_0)} f(x, y)$ exists
3. $\lim_{(x,y)\to(x_0,y_0)} f(x, y) = f(x_0, y_0)$.

A function is **continuous** if it is continuous at every point of its domain.

As with the definition of limit, the definition of continuity applies at boundary points as well as interior points of the domain of f. The only requirement is that the point (x, y) remain in the domain at all times.

As you may have guessed, one of the consequences of Theorem 1 is that algebraic combinations of continuous functions are continuous at every point at which all the functions involved are defined. Hence, sums, differences, products, constant multiples, quotients, and roots of continuous functions are continuous where defined. In particular, polynomials and rational functions of two variables are continuous at every point at which they are defined.

If $z = f(x, y)$ is a continuous function of x and y, and $w = g(z)$ is a continuous function of z, then the composite $w = g(f(x, y))$ is continuous. Thus,

$$e^{x-y}, \qquad \cos \frac{xy}{x^2+1}, \qquad \ln\,(1 + x^2y^2)$$

are continuous at every point (x, y).

As with functions of a single variable, the general rule is that composites of continuous functions are continuous. The only requirement is that each function be continuous where it is applied.

Example 3 A Function with a Single Point of Discontinuity

Show that

$$f(x, y) = \begin{cases} \dfrac{2xy}{x^2+y^2}, & (x, y) \neq (0, 0) \\ 0, & (x, y) = (0, 0) \end{cases}$$

is continuous at every point except the origin (Figure 11.11).

Solution The function f is continuous at any point $(x, y) \neq (0, 0)$ because its values are then given by a rational function of x and y.

At $(0, 0)$, the value of f is defined, but f, we claim, has no limit as $(x, y) \to (0, 0)$. The reason is that different paths of approach to the origin can lead to different results, as we now see.

For every value of m, the function f has a constant value on the "punctured" line $y = mx$, $x \neq 0$, because

$$f(x, y)\Big|_{y=mx} = \frac{2xy}{x^2+y^2}\Big|_{y=mx} = \frac{2x(mx)}{x^2+(mx)^2} = \frac{2mx^2}{x^2+m^2x^2} = \frac{2m}{1+m^2}.$$

Therefore, f has this number as its limit as (x, y) approaches $(0, 0)$ along the line:

$$\lim_{\substack{(x,y)\to(0,0) \\ \text{along } y=mx}} f(x, y) = \lim_{(x,y)\to(0,0)} \left[f(x, y)\Big|_{y=mx} \right] = \frac{2m}{1+m^2}.$$

This limit changes with m. There is therefore no single number we may call the limit of f as (x, y) approaches the origin. The limit fails to exist, and the function is not continuous.

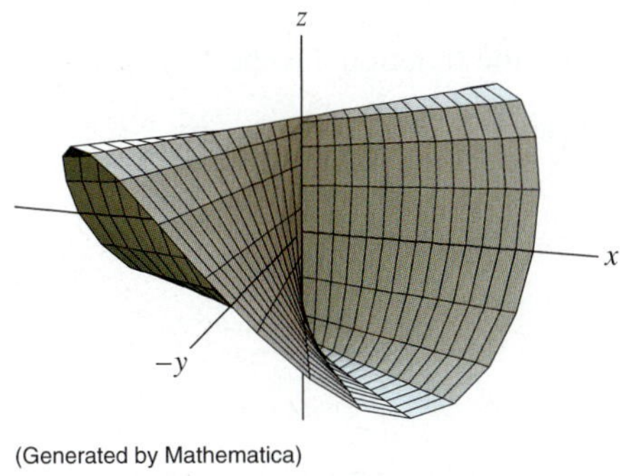

(Generated by Mathematica)

(a)

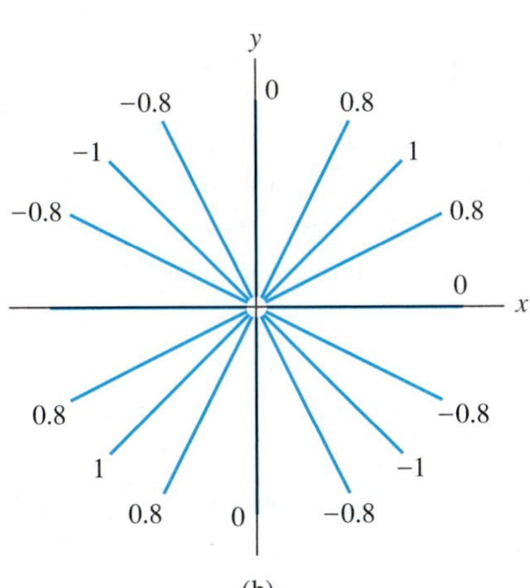

(b)

FIGURE 11.11 (a) The graph of

$$f(x, y) = \begin{cases} \dfrac{2xy}{x^2+y^2}, & (x, y) \neq (0, 0) \\ 0, & (x, y) = (0, 0). \end{cases}$$

The function is continuous at every point except the origin. (b) The level curves of f. (Example 3)

Example 3 illustrates an important point about limits of functions of two variables (or even more variables, for that matter). For a limit to exist at a point, the limit must be the same along every approach path. This result is analogous to the single-variable case where both the left- and right-sided limits had to have the same value therefore, for functions of two or more variables, if we ever find paths with different limits, we know the function has no limit at the point they approach.

> **Two-Path Test for Nonexistence of a Limit**
> If a function $f(x, y)$ has different limits along two different paths as (x, y) approaches (x_0, y_0), then $\lim_{(x,y)\to(x_0,y_0)} f(x, y)$ does not exist.

Example 4 Applying the Two-Path Test

Show that the function

$$f(x, y) = \frac{2x^2y}{x^4 + y^2}$$

(Figure 11.12) has no limit as (x, y) approaches $(0, 0)$.

Solution Along the curve $y = kx^2, x \neq 0$, the function has the constant value

$$f(x, y)\Big|_{y=kx^2} = \frac{2x^2y}{x^4 + y^2}\Big|_{y=kx^2} = \frac{2x^2(kx^2)}{x^4 + (kx^2)^2} = \frac{2kx^4}{x^4 + k^2x^4} = \frac{2k}{1 + k^2}.$$

Therefore,

$$\lim_{\substack{(x,y)\to(0,0)\\ \text{along } y=kx^2}} f(x, y) = \lim_{(x,y)\to(0,0)} \left[f(x, y)\Big|_{y=kx^2} \right] = \frac{2k}{1 + k^2}.$$

This limit varies with the path of approach. If (x, y) approaches $(0, 0)$ along the parabola $y = x^2$, for instance, $k = 1$ and the limit is 1. If (x, y) approaches $(0, 0)$ along the x-axis, $k = 0$ and the limit is 0. By the two-path test, f has no limit as (x, y) approaches $(0, 0)$.

The language here may seem contradictory. You might well ask, "What do you mean f has no limit as (x, y) approaches the origin—it has lots of limits." But that is the point. There is no single path-independent limit, and therefore, by the definition, $\lim_{(x,y)\to(0,0)} f(x, y)$ does not exist. It is our translating this formal statement into the more colloquial "has no limit" that creates the apparent contradiction. The mathematics is fine. The problem arises in how we talk about it. We need the formality to keep things straight.

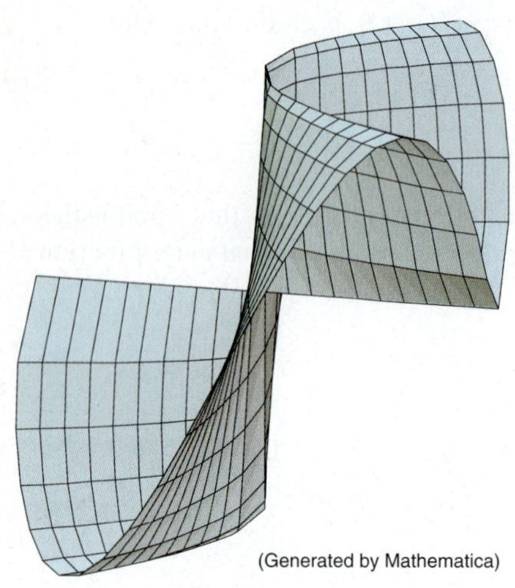

(a)

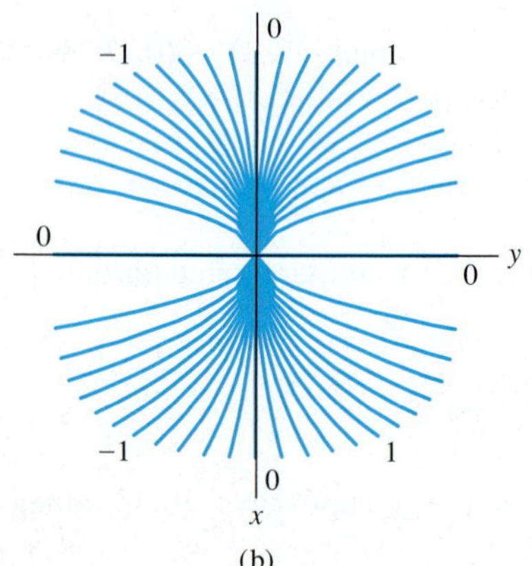

(b)

FIGURE 11.12 (a) The graph of $f(x, y) = 2x^2y/(x^4 + y^2)$. As the graph suggests and the level-curve values in part (b) confirm, $\lim_{(x,y)\to(0,0)} f(x, y)$ does not exist. (Example 4)

Functions of More Than Two Variables

The definitions of limit and continuity for functions of two variables and the conclusions about limits and continuity for sums, products, quotients, powers, and composites all extend to functions of three or more variables. Functions like

$$\ln (x + y + z) \quad \text{and} \quad \frac{y \sin z}{x - 1}$$

are continuous throughout their domains, and limits like

$$\lim_{P\to(1,0,-1)} \frac{e^{x+z}}{z^2 + \cos\sqrt{xy}} = \frac{e^{1-1}}{(-1)^2 + \cos 0} = \frac{1}{2},$$

where P denotes the point (x, y, z), may be found by direct substitution.

Extreme Values of Continuous Functions on Closed, Bounded Sets

We have seen that a function of a single variable that is continuous throughout a closed, bounded interval $[a, b]$ takes on an absolute maximum value and an absolute minimum value at least once in $[a, b]$. The same is true of a function $z = f(x, y)$ that is continuous on a closed, bounded set R in the plane (like a line segment, a disk, or a filled-in triangle). The function takes on an absolute maximum value at some point in R and an absolute minimum value at some point in R.

Theorems similar to these and other theorems of this section hold for functions of three or more variables. A continuous function $w = f(x, y, z)$, for example, must take on absolute maximum and minimum values on any closed, bounded set (solid ball or cube, spherical shell, rectangular solid) on which it is defined.

We learn how to find these extreme values when we get to Section 11.8, but first we need to know about derivatives in higher dimensions. That is the topic of the next section.

EXERCISES 11.2

Limits with Two Variables

Find the limits in Exercises 1–12.

1. $\lim_{(x,y)\to(0,0)} \dfrac{3x^2 - y^2 + 5}{x^2 + y^2 + 2}$

2. $\lim_{(x,y)\to(0,4)} \dfrac{x}{\sqrt{y}}$

3. $\lim_{(x,y)\to(3,4)} \sqrt{x^2 + y^2 - 1}$

4. $\lim_{(x,y)\to(2,-3)} \left(\dfrac{1}{x} + \dfrac{1}{y}\right)^2$

5. $\lim_{(x,y)\to(0,\pi/4)} \sec x \tan y$

6. $\lim_{(x,y)\to(0,0)} \cos \dfrac{x^2 + y^3}{x + y + 1}$

7. $\lim_{(x,y)\to(0,\ln 2)} e^{x-y}$

8. $\lim_{(x,y)\to(1,1)} \ln |1 + x^2y^2|$

9. $\lim_{(x,y)\to(0,0)} \dfrac{e^y \sin x}{x}$

10. $\lim_{(x,y)\to(1,1)} \cos \sqrt[3]{|xy| - 1}$

11. $\lim_{(x,y)\to(1,0)} \dfrac{x \sin y}{x^2 + 1}$

12. $\lim_{(x,y)\to(\pi/2,0)} \dfrac{\cos y + 1}{y - \sin x}$

Limits of Quotients

Find the limits in Exercises 13–20 by rewriting the fractions first.

13. $\lim_{\substack{(x,y)\to(1,1)\\ x\neq y}} \dfrac{x^2 - 2xy + y^2}{x - y}$

14. $\lim_{\substack{(x,y)\to(1,1)\\ x\neq y}} \dfrac{x^2 - y^2}{x - y}$

15. $\lim_{\substack{(x,y)\to(1,1)\\ x\neq 1}} \dfrac{xy - y - 2x + 2}{x - 1}$

16. $\lim_{\substack{(x,y)\to(2,-4)\\ y\neq -4,\, x\neq x^2}} \dfrac{y + 4}{x^2y - xy + 4x^2 - 4x}$

17. $\lim_{\substack{(x,y)\to(0,0)\\ x\neq y}} \dfrac{x - y + 2\sqrt{x} - 2\sqrt{y}}{\sqrt{x} - \sqrt{y}}$

18. $\lim_{\substack{(x,y)\to(2,2)\\ x+y\neq 4}} \dfrac{x + y - 4}{\sqrt{x + y} - 2}$

19. $\lim_{\substack{(x,y)\to(2,0)\\ 2x-y\neq 4}} \dfrac{\sqrt{2x - y} - 2}{2x - y - 4}$

20. $\lim_{\substack{(x,y)\to(4,3)\\ x\neq y+1}} \dfrac{\sqrt{x} - \sqrt{y + 1}}{x - y - 1}$

Limits with Three Variables

Find the limits in Exercises 21–26.

21. $\lim_{P\to(1,3,4)} \left(\dfrac{1}{x} + \dfrac{1}{y} + \dfrac{1}{z}\right)$

22. $\lim_{P\to(1,-1,-1)} \dfrac{2xy + yz}{x^2 + z^2}$

23. $\lim_{P\to(3,3,0)} (\sin^2 x + \cos^2 y + \sec^2 z)$

24. $\lim_{P\to(-1/4,\pi/2,2)} \tan^{-1} xyz$

25. $\lim_{P\to(\pi,0,3)} ze^{-2y} \cos 2x$

26. $\lim_{P\to(0,-2,0)} \ln \sqrt{x^2 + y^2 + z^2}$

Continuity in the Plane

At what points (x, y) in the plane are the functions in Exercise 27–30 continuous?

27. **(a)** $f(x, y) = \sin (x + y)$ **(b)** $f(x, y) = \ln (x^2 + y^2)$

28. **(a)** $f(x, y) = \dfrac{x + y}{x - y}$ **(b)** $f(x, y) = \dfrac{y}{x^2 + 1}$

29. **(a)** $g(x, y) = \sin \dfrac{1}{xy}$ **(b)** $g(x, y) = \dfrac{x + y}{2 + \cos x}$

30. **(a)** $g(x, y) = \dfrac{x^2 + y^2}{x^2 - 3x + 2}$ **(b)** $g(x, y) = \dfrac{1}{x^2 - y}$

Continuity in Space

At what points (x, y, z) in space are the functions in Exercises 31–34 continuous?

31. **(a)** $f(x, y, z) = x^2 + y^2 - 2z^2$

(b) $f(x, y, z) = \sqrt{x^2 + y^2 - 1}$

32. **(a)** $f(x, y, z) = \ln xyz$ **(b)** $f(x, y, z) = e^{x+y} \cos z$

33. **(a)** $h(x, y, z) = xy \sin \frac{1}{z}$ **(b)** $h(x, y, z) = \dfrac{1}{x^2 + z^2 - 1}$

34. **(a)** $h(x, y, z) = \dfrac{1}{|y| + |z|}$ **(b)** $h(x, y, z) = \dfrac{1}{|xy| + |z|}$

Applying the Two-Path Test

By considering different paths of approach, show that the functions in Exercises 35–42 have no limit as $(x, y) \to (0, 0)$.

35. $f(x, y) = -\dfrac{x}{\sqrt{x^2 + y^2}}$ **36.** $f(x, y) = \dfrac{x^4}{x^4 + y^2}$

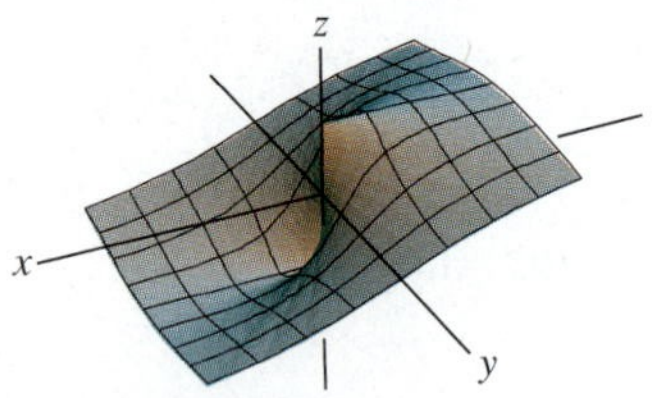

(Generated by Mathematica)

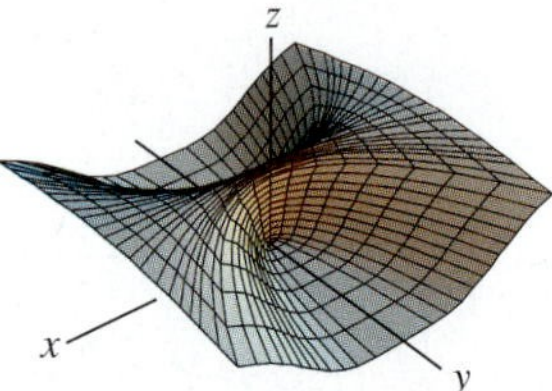

(Generated by Mathematica)

37. $f(x, y) = \dfrac{x^4 - y^2}{x^4 + y^2}$ **38.** $f(x, y) = \dfrac{xy}{|xy|}$

39. $g(x, y) = \dfrac{x - y}{x + y}$ **40.** $g(x, y) = \dfrac{x + y}{x - y}$

41. $h(x, y) = \dfrac{x^2 + y}{y}$ **42.** $h(x, y) = \dfrac{x^2}{x^2 - y}$

Using the δ-ϵ Definition

43. Show that the δ-ϵ requirement in the definition of limit is equivalent to

$$0 < |x - x_0| < \delta \quad \text{and} \quad 0 < |y - y_0| < \delta \Rightarrow |f(x, y) - L| < \epsilon.$$

44. Using the formal δ-ϵ definition of limit of a function $f(x, y)$ as $(x, y) \to (x_0, y_0)$ as a guide, state a formal definition for the limit of a function $g(x, y, z)$ as $(x, y, z) \to (x_0, y_0, z_0)$. What would be the analogous definition for a function $h(x, y, z, t)$ of four independent variables?

Each of Exercises 45–48 gives a function $f(x, y)$ and a positive number ϵ. In each exercise, either show that there exists a $\delta > 0$ such that for all (x, y),

$$\sqrt{x^2 + y^2} < \delta \Rightarrow |f(x, y) - f(0, 0)| < \epsilon$$

or show that there exists a $\delta > 0$ such that for all (x, y),

$$|x| < \delta \quad \text{and} \quad |y| < \delta \Rightarrow |f(x, y) - f(0, 0)| < \epsilon.$$

Do either one or the other, whichever seems more convenient. There is no need to do both.

45. $f(x, y) = x^2 + y^2, \quad \epsilon = 0.01$

46. $f(x, y) = y/(x^2 + 1), \quad \epsilon = 0.05$

47. $f(x, y) = (x + y)/(x^2 + 1), \quad \epsilon = 0.01$

48. $f(x, y) = (x + y)/(2 + \cos x), \quad \epsilon = 0.02$

Each of Exercises 49–52 gives a function $f(x, y, z)$ and a positive number ϵ. In each exercise, either show that there exists a $\delta > 0$ such that for all (x, y, z),

$$\sqrt{x^2 + y^2 + z^2} < \delta \Rightarrow |f(x, y, z) - f(0, 0, 0)| < \epsilon$$

or show that there exists a $\delta > 0$ such that for all (x, y, z),

$$|x| < \delta, \quad |y| < \delta, \quad \text{and} \quad |z| < \delta \Rightarrow |f(x, y, z) - f(0, 0, 0)| < \epsilon.$$

Do either one or the other, whichever seems more convenient. There is no need to do both.

49. $f(x, y, z) = x^2 + y^2 + z^2, \quad \epsilon = 0.015$

50. $f(x, y, z) = xyz, \quad \epsilon = 0.008$

51. $f(x, y, z) = \dfrac{x + y + z}{x^2 + y^2 + z^2 + 1}, \quad \epsilon = 0.015$

52. $f(x, y, z) = \tan^2 x + \tan^2 y + \tan^2 z, \quad \epsilon = 0.03$

53. Show that $f(x, y, z) = x + y - z$ is continuous at every point (x_0, y_0, z_0).

54. Show that $f(x, y, z) = x^2 + y^2 + z^2$ is continuous at the origin.

Changing to Polar Coordinates

If you cannot make any headway with $\lim_{(x,y)\to(0,0)} f(x, y)$ in rectangular coordinates, try changing to polar coordinates. Substitute $x = r \cos \theta$, $y = r \sin \theta$ and investigate the limit of the resulting expression as $r \to 0$. In other words, try to decide whether there exists a number L satisfying the following criterion:

Given $\epsilon > 0$, there exists a $\delta > 0$ such that for all r and θ,

$$|r| < \delta \Rightarrow |f(r, \theta) - L| < \epsilon. \tag{1}$$

If such an L exists, then

$$\lim_{(x,y)\to(0,0)} f(x, y) = \lim_{r\to 0} f(r, \theta) = L.$$

For instance,

$$\lim_{(x,y)\to(0,0)} \frac{x^3}{x^2 + y^2} = \lim_{r\to 0} \frac{r^3 \cos^3 \theta}{r^2} = \lim_{r\to 0} r \cos^3 \theta = 0.$$

To verify the last of these equalities, we need to show that Equation (1) is satisfied with $f(r, \theta) = r \cos^3 \theta$ and $L = 0$. That is, we need to show that given any $\epsilon > 0$ there exists a $\delta > 0$ such that for all r and θ,

$$|r| < \delta \Rightarrow |r \cos^3 \theta - 0| < \epsilon.$$

Since

$$|r\ \cos^3\theta| = |r||\cos^3\theta| \le |r|\cdot 1 = |r|,$$

the implication holds for all r and θ if we take $\delta = \epsilon$.

In contrast,

$$\frac{x^2}{x^2+y^2} = \frac{r^2\cos^2\theta}{r^2} = \cos^2\theta$$

takes on all values from 0 to 1 regardless of how small $|r|$ is, so that $\lim_{(x,y)\to(0,0)} x^2/(x^2+y^2)$ does not exist.

In each of these instances, the existence or nonexistence of the limit as $r \to 0$ is fairly clear. Shifting to polar coordinates does not always help, however, and may even tempt us to false conclusions. For example, the limit may exist along every straight line (or ray) $\theta =$ constant and yet fail to exist in the broader sense. Example 4 illustrates this point. In polar coordinates, $f(x, y) = (2x^2y)/(x^4 + y^2)$ becomes

$$f(r\ \cos\ \theta, r\ \sin\ \theta) = \frac{r\ \cos\ \theta\ \sin\ 2\theta}{r^2\cos^4\theta + \sin^2\theta}$$

for $r \neq 0$. If we hold θ constant and let $r \to 0$, the limit is 0. On the path $y = x^2$, however, we have $r\ \sin\ \theta = r^2\cos^2\theta$ and

$$f(r\ \cos\ \theta, r\ \sin\ \theta) = \frac{r\ \cos\ \theta\ \sin\ 2\theta}{r^2\ \cos^4\theta + (r\ \cos^2\theta)^2}$$

$$= \frac{2r\cos^2\theta\ \sin\ \theta}{2r^2\cos^4\theta} = \frac{r\ \sin\ \theta}{r^2\cos^2\theta} = 1.$$

In Exercises 55–60, find the limit of f as $(x, y) \to (0, 0)$ or show that the limit does not exist.

55. $f(x, y) = \dfrac{x^3 - xy^2}{x^2+y^2}$

56. $f(x, y) = \cos\left(\dfrac{x^3 - y^3}{x^2+y^2}\right)$

57. $f(x, y) = \dfrac{y^2}{x^2+y^2}$

58. $f(x, y) = \dfrac{2x}{x^2+x+y^2}$

59. $f(x, y) = \tan^{-1}\left(\dfrac{|x|+|y|}{x^2+y^2}\right)$

60. $f(x, y) = \dfrac{x^2 - y^2}{x^2+y^2}$

In Exercises 61 and 62, define $f(0, 0)$ in a way that extends f to be continuous at the origin.

61. $f(x, y) = \ln\left(\dfrac{3x^2 - x^2y^2 + 3y^2}{x^2+y^2}\right)$

62. $f(x, y) = \dfrac{2xy^2}{x^2+y^2}$

Theory and Examples

63. *Writing to Learn* If $\lim_{(x,y)\to(x_0,y_0)} f(x, y) = L$, must f be defined at (x_0, y_0)? Give reasons for your answer.

64. *Writing to Learn* If $f(x_0, y_0) = 3$, what can you say about

$$\lim_{(x,y)\to(x_0,y_0)} f(x, y)$$

if f is continuous at (x_0, y_0)? If f is not continuous at (x_0, y_0)? Give reasons for your answer.

65. *(Continuation of Example 3)*

(a) Reread Example 3. Then substitute $m = \tan\ \theta$ into the formula

$$f(x, y)\bigg|_{y=mx} = \frac{2m}{1+m^2}$$

and simplify the result to show how the value of f varies with the line's angle of inclination.

(b) Use the formula you obtained in part (a) to show that the limit of f as $(x, y) \to (0, 0)$ along the line $y = mx$ varies from -1 to 1 depending on the angle of approach.

66. *Continuous extension* Define $f(0, 0)$ in a way that extends

$$f(x, y) = xy\frac{x^2 - y^2}{x^2+y^2}$$

to be continuous at the origin.

The Sandwich Theorem

The Sandwich Theorem for functions of two variables states that if $g(x, y) \le f(x, y) \le h(x, y)$ for all $(x, y) \neq (x_0, y_0)$ in a disk centered at (x_0, y_0) and if g and h have the same finite limit L as $(x, y) \to (x_0, y_0)$, then

$$\lim_{(x,y)\to(x_0,y_0)} f(x, y) = L.$$

Use this result to support your answers to the questions in Exercises 67–70.

67. *Writing to Learn* Does knowing that

$$1 - \frac{x^2y^2}{3} < \frac{\tan^{-1} xy}{xy} < 1$$

tell you anything about

$$\lim_{(x,y)\to(0,0)} \frac{\tan^{-1} xy}{xy}?$$

Give reasons for your answer.

68. *Writing to Learn* Does knowing that

$$2|xy| - \frac{x^2y^2}{6} < 4 - 4\ \cos\sqrt{|xy|} < 2|xy|$$

tell you anything about

$$\lim_{(x,y)\to(0,0)} \frac{4 - 4\ \cos\sqrt{|xy|}}{|xy|}?$$

Give reasons for your answer.

69. *Writing to Learn* Does knowing that $|\sin(1/x)| \le 1$ tell you anything about

$$\lim_{(x,y)\to(0,0)} y\ \sin\frac{1}{x}?$$

Give reasons for your answer.

70. *Writing to Learn* Does knowing that $|\cos(1/y)| \le 1$ tell you anything about

$$\lim_{(x,y)\to(0,0)} x \cos \frac{1}{y}?$$

Give reasons for your answer.

COMPUTER EXPLORATIONS

71. Explore the graphs of the four functions whose limits you considered in Exercises 67–70. Try to find a view that supports your results in those exercises.

11.3 Partial Derivatives

Partial Derivatives of a Function of Two Variables • Calculations • Functions of More than Two Variables • Partial Derivatives and Continuity • Second-Order Partial Derivatives • The Mixed Derivative Theorem • Partial Derivatives of Still Higher Order • Differentiability

When we hold all but one of the independent variables of a function constant and differentiate with respect to that one variable, we get a "partial" derivative. This section shows how partial derivatives arise and how to calculate partial derivatives by applying the rules for differentiating functions of a single variable.

Partial Derivatives of a Function of Two Variables

If (x_0, y_0) is a point in the domain of a function $f(x, y)$, the vertical plane $y = y_0$ will cut the surface $z = f(x, y)$ in the curve $z = f(x, y_0)$ (Figure 11.13). This curve is the graph of the function $z = f(x, y_0)$ in the plane $y = y_0$. The horizontal coordinate in this plane is x; the vertical coordinate is z.

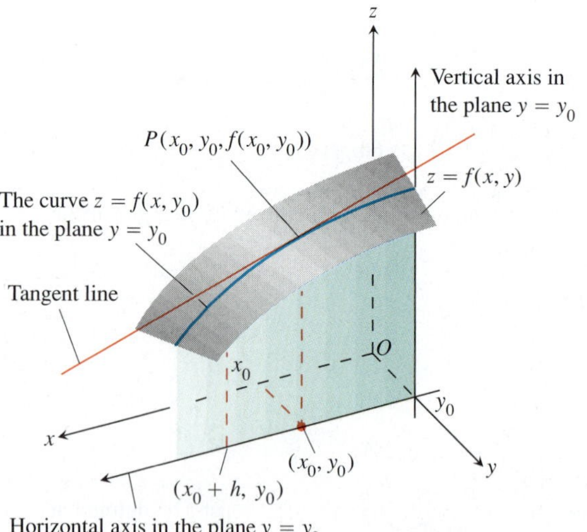

FIGURE 11.13 The intersection of the plane $y = y_0$ with the surface $z = f(x, y)$, viewed from a point above the first quadrant of the xy-plane.

We define the partial derivative of f with respect to x at the point (x_0, y_0) as the ordinary derivative of $f(x, y_0)$ with respect to x at the point $x = x_0$.

Definition Partial Derivative with Respect to x

The **partial derivative of $f(x, y)$ with respect to x** at the point (x_0, y_0) is

$$\left.\frac{\partial f}{\partial x}\right|_{(x_0,y_0)} = \left.\frac{d}{dx} f(x, y_0)\right|_{x=x_0} = \lim_{h \to 0} \frac{f(x_0 + h, y_0) - f(x_0, y_0)}{h},$$

provided the limit exists.

The stylized "∂" (similar to the lowercase Greek letter "δ" used in the limit definition) is just another kind of "d." It is convenient to have this distinguishable way of extending the Leibniz differential notation into a multivariable context.

The slope of the curve $z = f(x, y_0)$ at the point $P(x_0, y_0, f(x_0, y_0))$ in the plane $y = y_0$ is the value of the partial derivative of f with respect to x at (x_0, y_0). The tangent line to the curve at P is the line in the plane $y = y_0$ that passes through P with this slope. The partial derivative $\partial f/\partial x$ at (x_0, y_0) gives the rate of change of f with respect to x when y is held fixed at the value y_0. This is the rate of change of f in the direction of **i** at (x_0, y_0).

The notation for a partial derivative depends on what we want to emphasize:

$\frac{\partial f}{\partial x}(x_0, y_0)$ or $f_x(x_0, y_0)$	"Partial derivative of f with respect to x at (x_0, y_0)" or "f sub x at (x_0, y_0)." Convenient for stressing the point (x_0, y_0).
$\left.\frac{\partial z}{\partial x}\right\|_{(x_0,y_0)}$	"Partial derivative of z with respect to x at (x_0, y_0)." Common in science and engineering when you are dealing with variables and do not mention the function explicitly.
$f_x, \frac{\partial f}{\partial x}, z_x,$ or $\frac{\partial z}{\partial x}$	"Partial derivative of f (or z) with respect to x." Convenient when you regard the partial derivative as a function in its own right.

The definition of the partial derivative of $f(x, y)$ with respect to y at a point (x_0, y_0) is similar to the definition of the partial derivative of f with respect to x. We hold x fixed at the value x_0 and take the ordinary derivative of $f(x_0, y)$ with respect to y at y_0.

Definition Partial Derivative with Respect to y

The **partial derivative of $f(x, y)$ with respect to y** at the point (x_0, y_0) is

$$\left.\frac{\partial f}{\partial y}\right|_{(x_0,y_0)} = \left.\frac{d}{dy} f(x_0, y)\right|_{y=y_0}$$

$$= \lim_{h \to 0} \frac{f(x_0, y_0 + h) - f(x_0, y_0)}{h},$$

provided the limit exists.

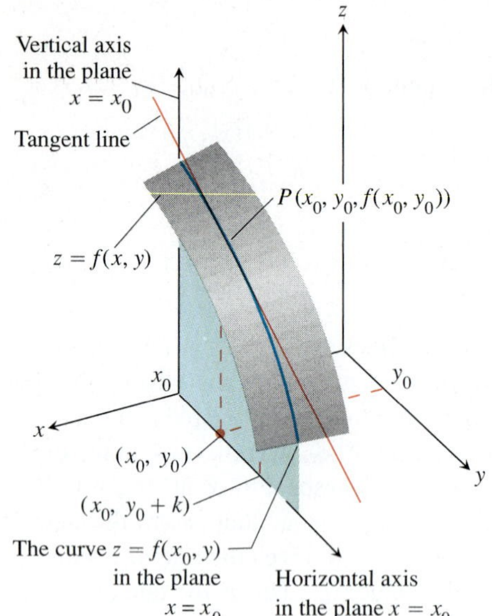

FIGURE 11.14 The intersection of the plane $x = x_0$ with the surface $z = f(x, y)$, viewed from above the first quadrant of the xy-plane.

The slope of the curve $z = f(x_0, y)$ at the point $P(x_0, y_0, f(x_0, y_0))$ in the vertical plane $x = x_0$ (Figure 11.14) is the partial derivative of f with respect to y at (x_0, y_0). The tangent line to the curve at P is the line in the plane $x = x_0$ that passes through P with this slope. The partial derivative gives the rate of change of f with respect to y at (x_0, y_0) when x is held fixed at the value x_0. This is the rate of change of f in the direction of $\mathbf{j}$ at (x_0, y_0).

The partial derivative with respect to y is denoted the same way as the partial derivative with respect to x:

$$\frac{\partial f}{\partial y}(x_0, y_0), \qquad f_y(x_0, y_0), \qquad \frac{\partial f}{\partial y}, \qquad f_y.$$

Notice that we now have two tangent lines associated with the surface $z = f(x, y)$ at the point $P(x_0, y_0, f(x_0, y_0))$ (Figure 11.15). Is the plane they determine tangent to the surface at P? It would be nice if it were, but we have to learn more about partial derivatives before we can find out.

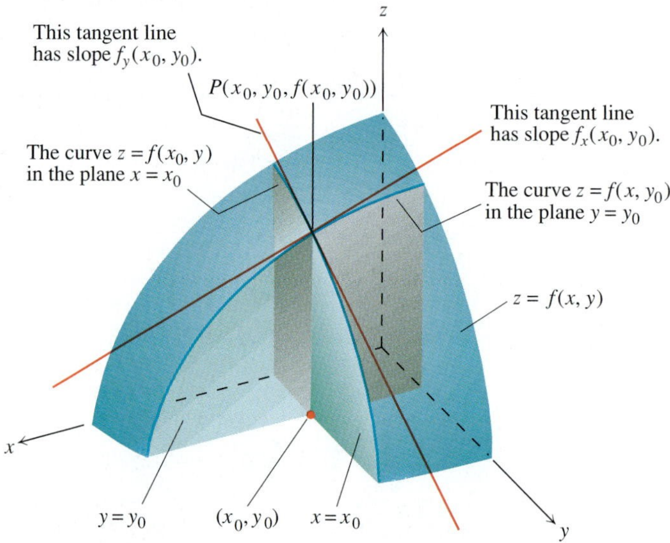

FIGURE 11.15 Figures 11.13 and 11.14 combined. The tangent lines at the point $(x_0, y_0, f(x_0, y_0))$ determine a plane that, in this picture at least, appears to be tangent to the surface.

Calculations

The definitions of $\partial f/\partial x$ and $\partial f/\partial y$ give us two different ways of differentiating f at a point: with respect to x in the usual way while treating y as a constant and with respect to y in the usual way while treating x as constant. As the following examples show, the values of these partial derivatives are usually different at a given point (x_0, y_0).

Example 1 Finding Partial Derivatives at a Point

Find the values of $\partial f/\partial x$ and $\partial f/\partial y$ at the point $(4, -5)$ if

$$f(x, y) = x^2 + 3xy + y - 1.$$

Solution To find $\partial f/\partial x$, we treat y as a constant and differentiate with respect to x:

$$\frac{\partial f}{\partial x} = \frac{\partial}{\partial x}(x^2 + 3xy + y - 1) = 2x + 3 \cdot 1 \cdot y + 0 - 0 = 2x + 3y.$$

The value of $\partial f/\partial x$ at $(4, -5)$ is $2(4) + 3(-5) = -7$.

To find $\partial f/\partial y$, we treat x as a constant and differentiate with respect to y:

$$\frac{\partial f}{\partial y} = \frac{\partial}{\partial y}(x^2 + 3xy + y - 1) = 0 + 3 \cdot x \cdot 1 + 1 - 0 = 3x + 1.$$

The value of $\partial f/\partial y$ at $(4, -5)$ is $3(4) + 1 = 13$.

Example 2 Finding a Partial Derivative as a Function

Find $\partial f/\partial y$ if $f(x, y) = y \sin xy$.

Solution We treat x as a constant and f as a product of y and $\sin xy$:

$$\begin{aligned}\frac{\partial f}{\partial y} &= \frac{\partial}{\partial y}(y \sin xy) = y \frac{\partial}{\partial y} \sin xy + (\sin xy) \frac{\partial}{\partial y}(y) \\ &= (y \cos xy) \frac{\partial}{\partial y}(xy) + \sin xy = xy \cos xy + \sin xy.\end{aligned}$$

USING TECHNOLOGY

Partial Differentiation A simple grapher can support your calculations even in multiple dimensions. If you specify the values of all but one independent variable, the grapher can calculate partial derivatives and can plot traces with respect to that remaining variable. Typically, a CAS can compute partial derivatives symbolically and numerically as easily as it can compute simple derivatives. Most systems use the same command to differentiate a function, regardless of the number of variables. (Simply specify the variable with which differentiation is to take place).

Example 3 Partial Derivatives May Be Different Functions

Find f_x and f_y if

$$f(x, y) = \frac{2y}{y + \cos x}.$$

Solution We treat f as a quotient. With y held constant, we get

$$\begin{aligned}f_x &= \frac{\partial}{\partial x}\left(\frac{2y}{y + \cos x}\right) = \frac{(y + \cos x)\frac{\partial}{\partial x}(2y) - 2y \frac{\partial}{\partial x}(y + \cos x)}{(y + \cos x)^2} \\ &= \frac{(y + \cos x)(0) - 2y(-\sin x)}{(y + \cos x)^2} = \frac{2y \sin x}{(y + \cos x)^2}.\end{aligned}$$

With x held constant, we get

$$f_y = \frac{\partial}{\partial y}\left(\frac{2y}{y + \cos x}\right) = \frac{(y + \cos x)\frac{\partial}{\partial y}(2y) - 2y\frac{\partial}{dy}(y + \cos x)}{(y + \cos x)^2}$$

$$= \frac{(y + \cos x)(2) - 2y(1)}{(y + \cos x)^2} = \frac{2 \cos x}{(y + \cos x)^2}.$$

Implicit differentiation works for partial derivatives the way it works for ordinary derivatives, as the next example illustrates.

Example 4 Implicit Partial Differentiation

Find $\partial z/\partial x$ if the equation

$$yz - \ln z = x + y$$

defines z as a function of the two independent variables x and y and the partial derivative exists.

Solution We differentiate both sides of the equation with respect to x, holding y constant and treating z as a differentiable function of x:

$$\frac{\partial}{\partial x}(yz) - \frac{\partial}{\partial x}\ln z = \frac{\partial x}{\partial x} + \frac{\partial y}{\partial x}$$

$$y\frac{\partial z}{\partial x} - \frac{1}{z}\frac{\partial z}{\partial x} = 1 + 0$$

With y constant, $\frac{\partial}{\partial x}(yz) = y\frac{\partial z}{\partial x}$.

$$\left(y - \frac{1}{z}\right)\frac{\partial z}{\partial x} = 1$$

$$\frac{\partial z}{\partial x} = \frac{z}{yz - 1}.$$

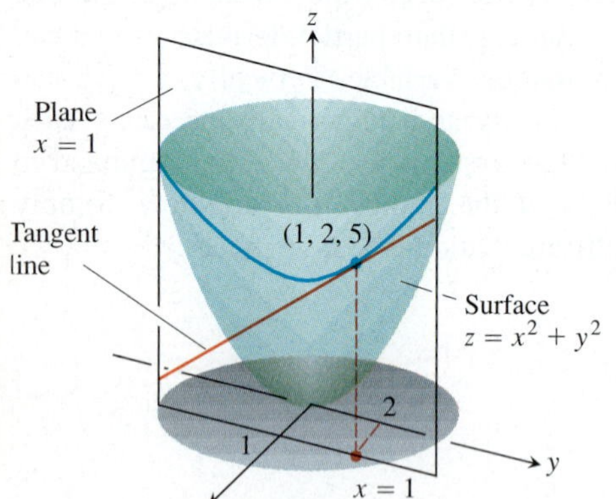

FIGURE 11.16 The tangent to the curve of intersection of the plane $x = 1$ and surface $z = x^2 + y^2$ at the point $(1, 2, 5)$. (Example 5)

Example 5 Finding the Slope of a Surface in the y-Direction

The plane $x = 1$ intersects the paraboloid $z = x^2 + y^2$ in a parabola. Find the slope of the tangent to the parabola at $(1, 2, 5)$ (Figure 11.16).

Solution The slope is the value of the partial derivative $\partial z/\partial y$ at $(1, 2)$:

$$\left.\frac{\partial z}{\partial y}\right|_{(1,2)} = \left.\frac{\partial}{\partial y}(x^2 + y^2)\right|_{(1,2)} = \left.2y\right|_{(1,2)} = 2(2) = 4.$$

As a check, we can treat the parabola as the graph of the single-variable function $z = (1)^2 + y^2 = 1 + y^2$ in the plane $x = 1$ and ask for the slope at $y = 2$. The slope, calculated now as an ordinary derivative, is

$$\left.\frac{dz}{dy}\right|_{y=2} = \left.\frac{d}{dy}(1 + y^2)\right|_{y=2} = \left.2y\right|_{y=2} = 4.$$

Functions of More Than Two Variables

The definitions of the partial derivatives of functions of more than two independent variables are like the definitions for functions of two variables. They are ordinary derivatives with respect to one variable, taken while the other independent variables are held constant.

Example 6 A Function of Three Variables

If x, y, and z are independent variables and

$$f(x, y, z) = x \sin(y + 3z),$$

then

$$\begin{aligned}\frac{\partial f}{\partial z} &= \frac{\partial}{\partial z}[x \sin(y + 3z)] = x \frac{\partial}{\partial z} \sin(y + 3z) \\ &= x \cos(y + 3z) \frac{\partial}{\partial z}(y + 3z) = 3x \cos(y + 3z).\end{aligned}$$

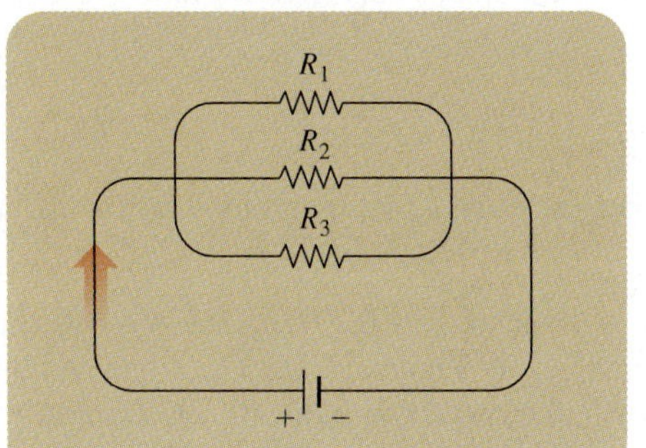

FIGURE 11.17 Resistors arranged this way are said to be connected in parallel (Example 7). Each resistor lets a portion of the current through. Their combined resistance R is calculated with the formula

$$\frac{1}{R} = \frac{1}{R_1} + \frac{1}{R_2} + \frac{1}{R_3}.$$

Example 7 Electrical Resistors in Parallel

If resistors of R_1, R_2, and R_3 ohms are connected in parallel to make an R-ohm resistor, the value of R can be found from the equation

$$\frac{1}{R} = \frac{1}{R_1} + \frac{1}{R_2} + \frac{1}{R_3}$$

(Figure 11.17). Find the value of $\partial R/\partial R_2$ when $R_1 = 30$, $R_2 = 45$, and $R_3 = 90$ ohms.

Solution To find $\partial R/\partial R_2$, we treat R_1 and R_3 as constants and differentiate both sides of the equation with respect to R_2:

$$\begin{aligned}\frac{\partial}{\partial R_2}\left(\frac{1}{R}\right) &= \frac{\partial}{\partial R_2}\left(\frac{1}{R_1} + \frac{1}{R_2} + \frac{1}{R_3}\right) \\ -\frac{1}{R^2}\frac{\partial R}{\partial R_2} &= 0 - \frac{1}{{R_2}^2} + 0 \\ \frac{\partial R}{\partial R_2} &= \frac{R^2}{{R_2}^2} = \left(\frac{R}{R_2}\right)^2.\end{aligned}$$

When $R_1 = 30$, $R_2 = 45$, and $R_3 = 90$,

$$\frac{1}{R} = \frac{1}{30} + \frac{1}{45} + \frac{1}{90} = \frac{3 + 2 + 1}{90} = \frac{6}{90} = \frac{1}{15},$$

so $R = 15$ and

$$\frac{\partial R}{\partial R_2} = \left(\frac{15}{45}\right)^2 = \left(\frac{1}{3}\right)^2 = \frac{1}{9}.$$

Partial Derivatives and Continuity

A function $f(x, y)$ can have partial derivatives with respect to both x and y at a point without being continuous there. This is different from functions of a single variable, where the existence of a derivative implies continuity. If the partial derivatives of $f(x, y)$ exist and are continuous throughout a disk centered at (x_0, y_0), however, then f *is* continuous at (x_0, y_0), as we see in the next section.

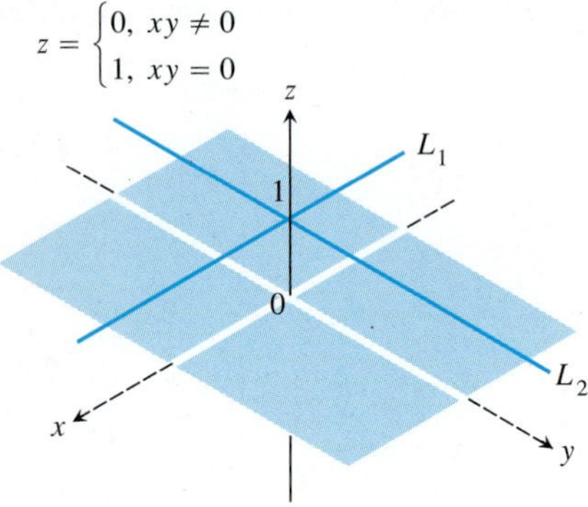

FIGURE 11.18 The graph of

$$f(x, y) = \begin{cases} 0, & xy \neq 0 \\ 1, & xy = 0 \end{cases}$$

consists of the lines L_1 and L_2 and the four open quadrants of the xy-plane. The function has partial derivatives at the origin but is not continuous there.

Example 8 Partials Exist, But f Discontinuous

Let

$$f(x, y) = \begin{cases} 0, & xy \neq 0 \\ 1, & xy = 0 \end{cases}$$

(Figure 11.18).

(a) Find the limit of f as (x, y) approaches $(0, 0)$ along the line $y = x$.

(b) Prove that f is not continuous at the origin.

(c) Show that both partial derivatives $\partial f/\partial x$ and $\partial f/\partial y$ exist at the origin.

Solution

(a) Since $f(x, y)$ is constantly zero along the line $y = x$ (except at the origin), we have

$$\lim_{(x,y)\to(0,0)} f(x, y)\bigg|_{y=x} = \lim_{(x,y)\to(0,0)} 0 = 0.$$

(b) Since $f(0, 0) = 1$, the limit in part (a) proves that f is not continuous at $(0, 0)$.

(c) To find $\partial f/\partial x$ at $(0, 0)$, we hold y fixed at $y = 0$. Then $f(x, y) = 1$ for all x, and the graph of f is the line L_1 in Figure 11.18. The slope of this line at any x is $\partial f/\partial x = 0$. In particular, $\partial f/\partial x = 0$ at $(0, 0)$. Similarly, $\partial f/\partial y$ is the slope of line L_2 at any y, so $\partial f/\partial y = 0$ at $(0, 0)$.

Example 8 notwithstanding, it is still true in higher dimensions that *differentiability* at a point implies continuity. What Example 8 suggests is that we need a stronger requirement for differentiability in higher dimensions than the mere existence of the partial derivatives. We define differentiability for functions of two variables at the end of this section and revisit the connection to continuity.

Second-Order Partial Derivatives

When we differentiate a function $f(x, y)$ twice, we produce its second-order derivatives. These derivatives are usually denoted by

$\dfrac{\partial^2 f}{\partial x^2}$	"d squared $f\,d\,x$ squared"	or	f_{xx}	"f sub $x\,x$"
$\dfrac{\partial^2 f}{\partial y^2}$	"d squared $f\,d\,y$ squared"		f_{yy}	"f sub $y\,y$"
$\dfrac{\partial^2 f}{\partial x \partial y}$	"d squared $f\,d\,x\,d\,y$"		f_{yx}	"f sub $y\,x$"
$\dfrac{\partial^2 f}{\partial y \partial x}$	"d squared $f\,d\,y\,d\,x$"		f_{xy}	"f sub $x\,y$"

The defining equations are

$$\frac{\partial^2 f}{\partial x^2} = \frac{\partial}{\partial x}\left(\frac{\partial f}{\partial x}\right), \qquad \frac{\partial^2 f}{\partial x \partial y} = \frac{\partial}{\partial x}\left(\frac{\partial f}{\partial y}\right),$$

and so on. Notice the order in which the derivatives are taken:

$$\frac{\partial^2 f}{\partial x \partial y} \qquad \text{Differentiate first with respect to } y, \text{ then with respect to } x.$$

$$f_{yx} = (f_y)_x \qquad \text{Means the same thing.}$$

CD-ROM
WEBsite
Historical Biography
Pierre-Simon Laplace
(1749 — 1827)

Example 9 Finding Second-Order Partial Derivatives

If $f(x, y) = x \cos y + ye^x$, find

$$\frac{\partial^2 f}{\partial x^2}, \quad \frac{\partial^2 f}{\partial y \partial x}, \quad \frac{\partial^2 f}{\partial y^2}, \quad \text{and} \quad \frac{\partial^2 f}{\partial x \partial y}.$$

Solution

$$\frac{\partial f}{\partial x} = \frac{\partial}{\partial x}(x \cos y + ye^x) = \cos y + ye^x$$

So

$$\frac{\partial^2 f}{\partial y \partial x} = \frac{\partial}{\partial y}\left(\frac{\partial f}{\partial x}\right) = -\sin y + e^x$$

$$\frac{\partial^2 f}{\partial x^2} = \frac{\partial}{\partial x}\left(\frac{\partial f}{\partial x}\right) = ye^x.$$

$$\frac{\partial f}{\partial y} = \frac{\partial}{\partial y}(x \cos y + ye^x) = -x \sin y + e^x$$

So

$$\frac{\partial^2 f}{\partial x \partial y} = \frac{\partial}{\partial x}\left(\frac{\partial f}{\partial y}\right) = -\sin y + e^x$$

$$\frac{\partial^2 f}{\partial y^2} = \frac{\partial}{\partial y}\left(\frac{\partial f}{\partial y}\right) = -x \cos y.$$

The Mixed Derivative Theorem

You may have noticed that the "mixed" second-order partial derivatives

$$\frac{\partial^2 f}{\partial y \partial x} \quad \text{and} \quad \frac{\partial^2 f}{\partial x \partial y}$$

in Example 9 were equal. This was not a coincidence. They must be equal whenever f, f_x, f_y, f_{xy}, and f_{yx} are continuous, as stated in the following theorem.

Theorem 2 The Mixed Derivative Theorem

If $f(x, y)$ and its partial derivatives f_x, f_y, f_{xy}, and f_{yx} are defined throughout an open region containing a point (a, b) and are all continuous at (a, b), then

$$f_{xy}(a, b) = f_{yx}(a, b).$$

You can find a proof of Theorem 2 in Appendix 11.

Theorem 2 says that to calculate a mixed second-order derivative, we may differentiate in either order. This can work to our advantage.

Example 10 Choosing the Order of Differentiation

Find $\partial^2 w / \partial x \partial y$ if

$$w = xy + \frac{e^y}{y^2 + 1}.$$

Solution The symbol $\partial^2 w/\partial x \partial y$ tells us to differentiate first with respect to y and then with respect to x. If we postpone the differentiation with respect to y and differentiate first with respect to x, however, we get the answer more quickly. In two steps,

$$\frac{\partial w}{\partial x} = y \qquad \text{and} \qquad \frac{\partial^2 w}{\partial y \partial x} = 1.$$

We are in for more work if we differentiate first with respect to y. (Just try it.)

Partial Derivatives of Still Higher Order

Although we will deal mostly with first- and second-order partial derivatives, because these appear the most frequently in applications, there is no theoretical limit to how many times we can differentiate a function as long as the derivatives involved exist. Thus, we get third- and fourth-order derivatives denoted by symbols like

$$\frac{\partial^3 f}{\partial x \partial y^2} = f_{yyx}$$

$$\frac{\partial^4 f}{\partial x^2 \partial y^2} = f_{yyxx},$$

and so on. As with second-order derivatives, the order of differentiation is immaterial as long as the derivatives through the order in question are continuous.

Differentiability

Surprising as it may seem, the starting point for differentiability is not Fermat's difference quotient but rather the idea of increment. You may recall from our work with functions of a single variable that if $y = f(x)$ is differentiable at $x = x_0$, then the change in the value of f that results from changing x from x_0 to $x_0 + \Delta x$ is given by an equation of the form

$$\Delta y = f'(x_0)\,\Delta x + \epsilon\,\Delta x$$

in which $\epsilon \to 0$ as $\Delta x \to 0$. For functions of two variables, the analogous property becomes the definition of differentiability. The Increment Theorem (from advanced calculus) tells us when to expect the property to hold.

Theorem 3 **The Increment Theorem for Functions of Two Variables**

Suppose that the first partial derivatives of $f(x, y)$ are defined throughout an open region R containing the point (x_0, y_0) and that f_x and f_y are continuous at (x_0, y_0). Then the change

$$\Delta z = f(x_0 + \Delta x, y_0 + \Delta y) - f(x_0, y_0)$$

in the value of f that results from moving from (x_0, y_0) to another point $(x_0 + \Delta x, y_0 + \Delta y)$ in R satisfies an equation of the form

$$\Delta z = f_x(x_0, y_0)\,\Delta x + f_y(x_0, y_0)\,\Delta y + \epsilon_1\,\Delta x + \epsilon_2\,\Delta y,$$

in which $\epsilon_1, \epsilon_2 \to 0$ as $\Delta x, \Delta y \to 0$.

You will see where the epsilons come from if you read the proof in Appendix 11. You will also see that similar results hold for functions of more than two independent variables.

Definition Differentiability of a Function of Two Variables

A function $z = f(x, y)$ is **differentiable at** (x_0, y_0) if $f_x(x_0, y_0)$ and $f_y(x_0, y_0)$ exist and Δz satisfies an equation of the form

$$\Delta z = f_x(x_0, y_0)\,\Delta x + f_y(x_0, y_0)\,\Delta y + \epsilon_1\,\Delta x + \epsilon_2\,\Delta y,$$

in which $\epsilon_1,\ \epsilon_2 \to 0$ as $\Delta x, \Delta y \to 0$. We call f **differentiable** if it is differentiable at every point in its domain.

In light of this definition, we have the immediate corollary of Theorem 3 that a function is differentiable if its first partial derivatives are *continuous*.

Corollary of Theorem 3 Continuity of Partial Derivatives Implies Differentiability

If the partial derivatives f_x and f_y of a function $f(x, y)$ are continuous throughout an open region R, then f is differentiable at every point of R.

If $z = f(x, y)$ is differentiable, then the definition of differentiability assures that $\Delta z = f(x_0 + \Delta x, y_0 + \Delta y) - f(x_0, y_0)$ approaches 0 as Δx and Δy approach 0. This tells us that a function of two variables is continuous at every point where it is differentiable.

As we can see from Theorems 3 and 4, a function $f(x, y)$ must be continuous at a point (x_0, y_0) if f_x and f_y are continuous throughout an open region containing (x_0, y_0). Remember, however, that it is still possible for a function of two variables to be discontinuous at a point where its first partial derivatives exist, as we saw in Example 8. Existence alone is not enough.

Theorem 4 Differentiability Implies Continuity

If a function $f(x, y)$ is differentiable at (x_0, y_0), then f is continuous at (x_0, y_0).

EXERCISES 11.3

Calculating First-Order Partial Derivatives

In Exercises 1–22, find $\partial f/\partial x$ and $\partial f/\partial y$.

1. $f(x, y) = 2x^2 - 3y - 4$

2. $f(x, y) = x^2 - xy + y^2$

3. $f(x, y) = (x^2 - 1)(y + 2)$

4. $f(x, y) = 5xy - 7x^2 - y^2 + 3x - 6y + 2$

5. $f(x, y) = (xy - 1)^2$

6. $f(x, y) = (2x - 3y)^3$

7. $f(x, y) = \sqrt{x^2 + y^2}$

8. $f(x, y) = (x^3 + (y/2))^{2/3}$

9. $f(x, y) = 1/(x + y)$

10. $f(x, y) = x/(x^2 + y^2)$

11. $f(x, y) = (x + y)/(xy - 1)$

12. $f(x, y) = \tan^{-1}(y/x)$

13. $f(x, y) = e^{(x+y+1)}$

14. $f(x, y) = e^{-x}\sin(x + y)$

15. $f(x, y) = \ln(x + y)$

16. $f(x, y) = e^{xy}\ln y$

17. $f(x, y) = \sin^2(x - 3y)$

18. $f(x, y) = \cos^2(3x - y^2)$

19. $f(x, y) = x^y$

20. $f(x, y) = \log_y x$

21. $f(x, y) = \int_x^y g(t)\,dt$ (g continuous for all t)

22. $f(x, y) = \sum_{n=0}^{\infty} (xy)^n$ ($|xy| < 1$)

In Exercises 23–34, find f_x, f_y, and f_z.

23. $f(x, y, z) = 1 + xy^2 - 2z^2$ **24.** $f(x, y, z) = xy + yz + xz$

25. $f(x, y, z) = x - \sqrt{y^2 + z^2}$

26. $f(x, y, z) = (x^2 + y^2 + z^2)^{-1/2}$

27. $f(x, y, z) = \sin^{-1}(xyz)$ **28.** $f(x, y, z) = \sec^{-1}(x + yz)$

29. $f(x, y, z) = \ln(x + 2y + 3z)$

30. $f(x, y, z) = yz \ln(xy)$ **31.** $f(x, y, z) = e^{-(x^2+y^2+z^2)}$

32. $f(x, y, z) = e^{-xyz}$

33. $f(x, y, z) = \tanh(x + 2y + 3z)$

34. $f(x, y, z) = \sinh(xy - z^2)$

In Exercises 35–40, find the partial derivative of the function with respect to each variable.

35. $f(t, \alpha) = \cos(2\pi t - \alpha)$ **36.** $g(u, v) = v^2 e^{(2u/v)}$

37. $h(\rho, \phi, \theta) = \rho \sin\phi \cos\theta$

38. $g(r, \theta, z) = r(1 - \cos\theta) - z$

39. *Work done by the heart* (Section 3.6, Exercise 41)

$$W(P, V, \delta, v, g) = PV + \frac{V\delta v^2}{2g}$$

40. *Wilson lot size formula* (Section 3.5, Exercise 43)

$$A(c, h, k, m, q) = \frac{km}{q} + cm + \frac{hq}{2}$$

Calculating Second-Order Partial Derivatives

Find all the second-order partial derivatives of the functions in Exercises 41–46.

41. $f(x, y) = x + y + xy$ **42.** $f(x, y) = \sin xy$

43. $g(x, y) = x^2 y + \cos y + y \sin x$

44. $h(x, y) = xe^y + y + 1$ **45.** $r(x, y) = \ln(x + y)$

46. $s(x, y) = \tan^{-1}(y/x)$

Mixed Partial Derivatives

In Exercises 47–50, verify that $w_{xy} = w_{yx}$.

47. $w = \ln(2x + 3y)$ **48.** $w = e^x + x \ln y + y \ln x$

49. $w = xy^2 + x^2y^3 + x^3y^4$ **50.** $w = x \sin y + y \sin x + xy$

51. *Writing to Learn* Which order of differentiation will calculate f_{xy} faster: x first or y first? Try to answer without writing anything down.

(a) $f(x, y) = x \sin y + e^y$

(b) $f(x, y) = 1/x$

(c) $f(x, y) = y + (x/y)$

(d) $f(x, y) = y + x^2y + 4y^3 - \ln(y^2 + 1)$

(e) $f(x, y) = x^2 + 5xy + \sin x + 7e^x$

(f) $f(x, y) = x \ln xy$

52. *Writing to Learn* The fifth-order partial derivative $\partial^5 f/\partial x^2 \partial y^3$ is zero for each of the following functions. To show this as quickly as possible, which variable would you differentiate with respect to first: x or y? Try to answer without writing anything down.

(a) $f(x, y) = y^2x^4e^x + 2$

(b) $f(x, y) = y^2 + y(\sin x - x^4)$

(c) $f(x, y) = x^2 + 5xy + \sin x + 7e^x$

(d) $f(x, y) = xe^{y^2/2}$

Using the Partial Derivative Definition

In Exercises 53 and 54, use the limit definition of partial derivative to compute the partial derivatives of the functions at the specified points.

53. $f(x, y) = 1 - x + y - 3x^2y$, $\dfrac{\partial f}{\partial x}$ and $\dfrac{\partial f}{\partial y}$ at $(1, 2)$

54. $f(x, y) = 4 + 2x - 3y - xy^2$, $\dfrac{\partial f}{\partial x}$ and $\dfrac{\partial f}{\partial y}$ at $(-2, 1)$

55. *Three variables* Let $w = f(x, y, z)$ be a function of three independent variables and write the formal definition of the partial derivative $\partial f/\partial z$ at (x_0, y_0, z_0). Use this definition to find $\partial f/\partial z$ at $(1, 2, 3)$ for $f(x, y, z) = x^2yz^2$.

56. *Three variables* Let $w = f(x, y, z)$ be a function of three independent variables and write the formal definition of the partial derivative $\partial f/\partial y$ at (x_0, y_0, z_0). Use this definition to find $\partial f/\partial y$ at $(-1, 0, 3)$ for $f(x, y, z) = -2xy^2 + yz^2$.

Differentiating Implicitly

57. Find the value of $\partial z/\partial x$ at the point $(1, 1, 1)$ if the equation

$$xy + z^3x - 2yz = 0$$

defines z as a function of the two independent variables x and y and the partial derivative exists.

58. Find the value of $\partial x/\partial z$ at the point $(1, -1, -3)$ if the equation

$$xz + y \ln x - x^2 + 4 = 0$$

defines x as a function of the two independent variables y and z and the partial derivative exists.

Exercises 59 and 60 are about the triangle shown here.

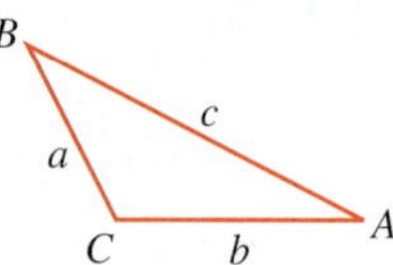

59. Express A implicitly as a function of a, b, and c and calculate $\partial A/\partial a$ and $\partial A/\partial b$.

60. Express a implicitly as a function of A, b, and B and calculate $\partial a/\partial A$ and $\partial a/\partial B$.

61. *Two dependent variables* Express v_x in terms of u and v if the equations $x = v \ln u$ and $y = u \ln v$ define u and v as functions of the independent variables x and y, and if v_x exists. (*Hint:* Differentiate both equations with respect to x and solve for v_x with Cramer's Rule.)

62. *Two dependent variables* Find $\partial x/\partial u$ and $\partial y/\partial u$ if the equations $u = x^2 - y^2$ and $v = x^2 - y$ define x and y as functions of the independent variables u and v, and the partial derivatives exist. (See the hint in Exercise 61.) Then let $s = x^2 + y^2$ and find $\partial s/\partial u$.

Laplace Equations

The **three-dimensional Laplace equation**

$$\frac{\partial^2 f}{\partial x^2} + \frac{\partial^2 f}{\partial y^2} + \frac{\partial^2 f}{\partial z^2} = 0$$

is satisfied by steady-state temperature distributions $T = f(x, y, z)$ in space, by gravitational potentials, and by electrostatic potentials. The **two-dimensional Laplace equation**

$$\frac{\partial^2 f}{\partial x^2} + \frac{\partial^2 f}{\partial y^2} = 0,$$

obtained by dropping the $\partial^2 f/\partial z^2$ term from the previous equation, describes potentials and steady-state temperature distributions in a plane (Figure 11.19).

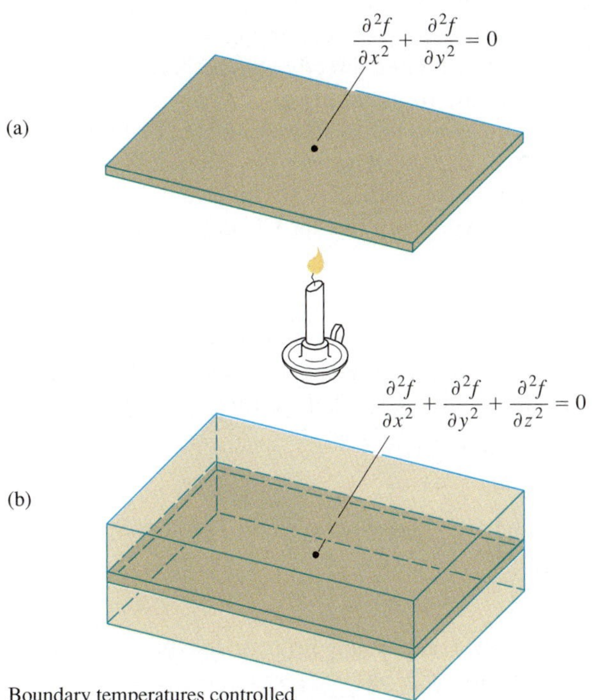

FIGURE 11.19 Steady-state temperature distributions in planes and solids satisfy Laplace equations. The plane (a) may be treated as a thin slice of the solid (b) perpendicular to the z-axis.

Show that each function in Exercises 63–68 satisfies a Laplace equation.

63. $f(x, y, z) = x^2 + y^2 - 2z^2$

64. $f(x, y, z) = 2z^3 - 3(x^2 + y^2)z$

65. $f(x, y) = e^{-2y} \cos 2x$

66. $f(x, y) = \ln \sqrt{x^2 + y^2}$

67. $f(x, y, z) = (x^2 + y^2 + z^2)^{-1/2}$

68. $f(x, y, z) = e^{3x+4y} \cos 5z$

The Wave Equation

If we stand on an ocean shore and take a snapshot of the waves, the picture shows a regular pattern of peaks and valleys in an instant of time. We see periodic vertical motion in space, with respect to distance. If we stand in the water, we can feel the rise and fall of the water as the waves go by. We see periodic vertical motion in time. In physics, this beautiful symmetry is expressed by the **one-dimensional wave equation**

$$\frac{\partial^2 w}{\partial t^2} = c^2 \frac{\partial^2 w}{\partial x^2},$$

where w is the wave height, x is the distance variable, t is the time variable, and c is the velocity with which the waves are propagated.

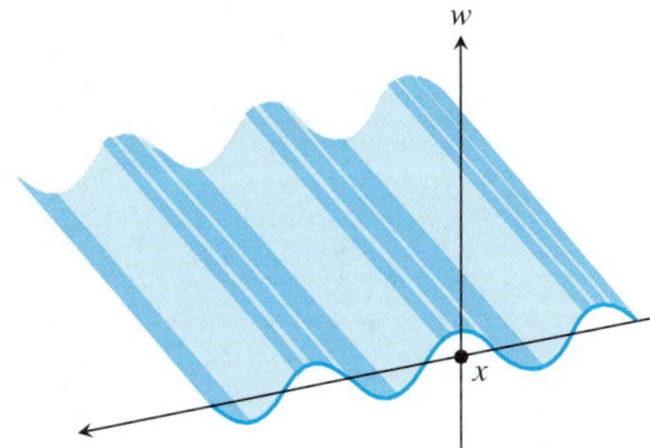

In our example, x is the distance across the ocean's surface, but in other applications, x might be the distance along a vibrating string, distance through air (sound waves), or distance through space (light waves). The number c varies with the medium and type of wave.

Show that the functions in Exercises 69–75 are all solutions of the wave equation.

69. $w = \sin (x + ct)$

70. $w = \cos (2x + 2ct)$

71. $w = \sin (x + ct) + \cos (2x + 2ct)$

72. $w = \ln (2x + 2ct)$

73. $w = \tan (2x - 2ct)$

74. $w = 5 \cos (3x + 3ct) + e^{x+ct}$

75. $w = f(u)$, where f is a differentiable function of u, and $u = a(x + ct)$, where a is a constant

Continuous Partial Derivatives

76. *Writing to Learn* Does a function $f(x, y)$ with continuous first partial derivatives throughout an open region R have to be continuous on R? Give reasons for your answer.

77. *Writing to Learn* If a function $f(x, y)$ has continuous second partial derivatives throughout an open region R, must the first-order partial derivatives of f be continuous on R? Give reasons for your answer.

11.4 The Chain Rule

Composite Functions in Higher Dimensions • Functions of Two Variables • Functions of Three Variables • Functions Defined on Surfaces • Implicit Differentiation Revisited • Functions of Many Variables

We can form composites of multivariable functions over appropriate domains just as we create composites of single-variable functions. This section shows how to find partial derivatives of composites of multivariable functions.

Composite Functions in Higher Dimensions

When we are interested in the temperature $w = f(x, y, z)$ at points along a curve $x = g(t)$, $y = h(t)$, $z = k(t)$ in space or in the pressure or density along a path through a gas or fluid, we may think of f as a function of the single variable t. For each value of t, the temperature at the point $(g(t), h(t), k(t))$ is the value of the composite function $f(g(t), h(t), k(t))$. If we then wish to know the rate at which f changes with respect to t along the path, we have only to differentiate this composite with respect to t, provided, of course, the derivative exists.

Sometimes we can find the derivative by substituting the formulas for g, h, and k into the formula for f and differentiating directly with respect to t, but we often have to work with functions whose formulas are too complicated for convenient substitution or for which formulas are not readily available. To find a function's derivatives under circumstances like these, we use the Chain Rule. The form the Chain Rule takes depends on how many variables are involved but, except for the presence of additional variables, it works just like the Chain Rule in Section 2.5.

Functions of Two Variables

In Section 2.5, when $w = f(x)$ was a differentiable function of x and $x = g(t)$ was a differentiable function of t, w became a differentiable function of t and the Chain Rule said that dw/dt could be calculated with the formula

$$\frac{dw}{dt} = \frac{dw}{dx}\frac{dx}{dt}.$$

The analogous formula for a function $w = f(x, y)$ is given in Theorem 5.

Theorem 5 **Chain Rule for Functions of Two Independent Variables**
If $w = f(x, y)$ is differentiable and x and y are differentiable functions of t, then w is a differentiable function of t and

$$\frac{dw}{dt} = \frac{\partial f}{\partial x}\frac{dx}{dt} + \frac{\partial f}{\partial y}\frac{dy}{dt}.$$

Proof The proof consists of showing that if x and y are differentiable at $t = t_0$, then w is differentiable at t_0 and

$$\left(\frac{dw}{dt}\right)_{t_0} = \left(\frac{\partial w}{\partial x}\right)_{P_0}\left(\frac{dx}{dt}\right)_{t_0} + \left(\frac{\partial w}{\partial y}\right)_{P_0}\left(\frac{dy}{dt}\right)_{t_0},$$

where $P_0 = (x(t_0), y(t_0))$.

Let Δx, Δy, and Δw be the increments that result from changing t from t_0 to $t_0 + \Delta t$. Since f is differentiable (remember the definition in Section 11.3),

$$\Delta w = \left(\frac{\partial w}{\partial x}\right)_{P_0} \Delta x + \left(\frac{\partial w}{\partial y}\right)_{P_0} \Delta y + \epsilon_1 \, \Delta x + \epsilon_2 \, \Delta y,$$

where $\epsilon_1, \epsilon_2 \to 0$ as $\Delta x, \Delta y \to 0$. To find dw/dt, we divide this equation through by Δt and let Δt approach zero. The division gives

$$\frac{\Delta w}{\Delta t} = \left(\frac{\partial w}{\partial x}\right)_{P_0} \frac{\Delta x}{\Delta t} + \left(\frac{\partial w}{\partial y}\right)_{P_0} \frac{\Delta y}{\Delta t} + \epsilon_1 \frac{\Delta x}{\Delta t} + \epsilon_2 \frac{\Delta y}{\Delta t}.$$

Letting Δt approach zero gives

$$\begin{aligned}\left(\frac{dw}{dt}\right)_{t_0} &= \lim_{\Delta t \to 0} \frac{\Delta w}{\Delta t} \\ &= \left(\frac{\partial w}{\partial x}\right)_{P_0} \left(\frac{dx}{dt}\right)_{t_0} + \left(\frac{\partial w}{\partial y}\right)_{P_0} \left(\frac{dy}{dt}\right)_{t_0} + 0 \cdot \left(\frac{dx}{dt}\right)_{t_0} + 0 \cdot \left(\frac{dy}{dt}\right)_{t_0}.\end{aligned}$$

To remember the Chain Rule picture the diagram below. To find dw/dt, start at w and read down each route to t, multiplying derivatives along the way. Then add the products.

Chain Rule

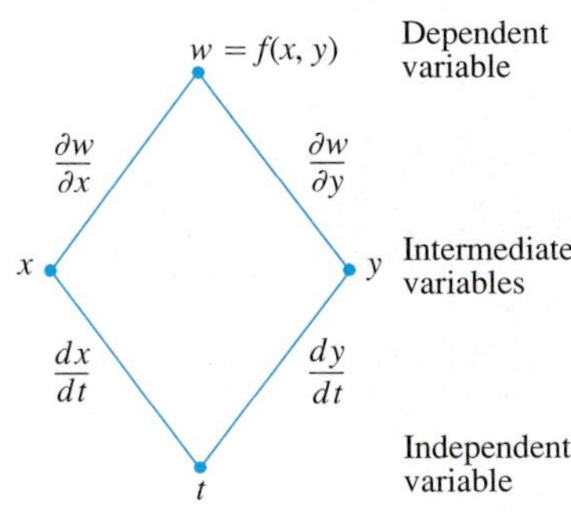

$$\frac{dw}{dt} = \frac{\partial w}{\partial x}\frac{dx}{dt} + \frac{\partial w}{\partial y}\frac{dy}{dt}$$

The **tree diagram** in the margin provides a convenient way to remember the Chain Rule. From the diagram, you see that when $t = t_0$, the derivatives dx/dt and dy/dt are evaluated at t_0. The value of t_0 then determines the value x_0 for the differentiable function x and the value y_0 for the differentiable function y. The partial derivatives $\partial w/\partial x$ and $\partial w/\partial y$ (which are themselves functions of x and y) are evaluated at the point $P_0(x_0, y_0)$ corresponding to t_0. The "true" independent variable is t, whereas x and y are *intermediate variables* (controlled by t) and w is the dependent variable.

A more precise notation for the Chain Rule shows how the various derivatives in Theorem 5 are evaluated:

$$\frac{dw}{dt}(t_0) = \frac{\partial f}{\partial x}(x_0, y_0) \cdot \frac{dx}{dt}(t_0) + \frac{\partial f}{\partial y}(x_0, y_0) \cdot \frac{dy}{dt}(t_0).$$

Example 1 Applying the Chain Rule

Use the Chain Rule to find the derivative of

$$w = xy$$

with respect to t along the path $x = \cos t$, $y = \sin t$. What is the derivative's value at $t = \pi/2$?

Solution We apply the Chain Rule to find dw/dt as follows:

$$\begin{aligned}\frac{dw}{dt} &= \frac{\partial w}{\partial x}\frac{dx}{dt} + \frac{\partial w}{\partial y}\frac{dy}{dt} \\ &= \frac{\partial(xy)}{\partial x} \cdot \frac{d}{dt}(\cos t) + \frac{\partial(xy)}{\partial y} \cdot \frac{d}{dt}(\sin t) \\ &= (y)(-\sin t) + (x)(\cos t) \\ &= (\sin t)(-\sin t) + (\cos t)(\cos t) \\ &= -\sin^2 t + \cos^2 t \\ &= \cos(2t).\end{aligned}$$

Here we have three routes from w to t instead of two, but finding dw/dt is still the same. Read down each route, multiplying derivatives along the way; then add.

Chain Rule

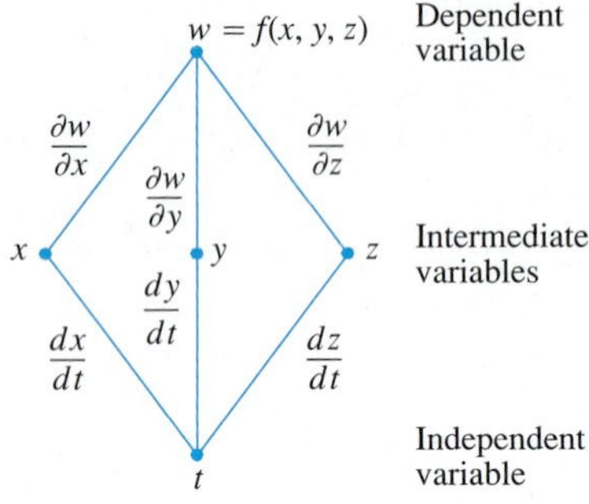

$$\frac{dw}{dt} = \frac{\partial w}{\partial x}\frac{dx}{dt} + \frac{\partial w}{\partial y}\frac{dy}{dt} + \frac{\partial w}{\partial z}\frac{dz}{dt}$$

The helix
$\mathbf{r}(t) = (\cos t)\mathbf{i} + (\sin t)\mathbf{j} + t\mathbf{k}$

(1, 0, 0)

FIGURE 11.20 Example 2 shows how the values of $w = xy + z$ vary with t along this helix.

In this example, we can check the result with a more direct calculation. As a function of t,

$$w = xy = \cos t \sin t = \frac{1}{2} \sin 2t,$$

so

$$\frac{dw}{dt} = \frac{d}{dt}\left(\frac{1}{2} \sin 2t\right) = \frac{1}{2} \cdot 2 \cos 2t = \cos 2t.$$

In either case, at the given value of t,

$$\left(\frac{dw}{dt}\right)_{t=\pi/2} = \cos\left(2 \cdot \frac{\pi}{2}\right) = \cos \pi = -1.$$

Functions of Three Variables

You can probably predict the Chain Rule for functions of three variables, as it only involves adding the expected third term to the two-variable formula.

Theorem 6 Chain Rule for Functions of Three Independent Variables

If $w = f(x, y, z)$ is differentiable and x, y, and z are differentiable functions of t, then w is a differentiable function of t and

$$\frac{dw}{dt} = \frac{\partial f}{\partial x}\frac{dx}{dt} + \frac{\partial f}{\partial y}\frac{dy}{dt} + \frac{\partial f}{\partial z}\frac{dz}{dt}.$$

The proof is identical with the proof of Theorem 5 except that there are now three intermediate variables instead of two. The diagram we use for remembering the new equation is similar as well, with three routes from w to t.

Example 2 Changes in a Function's Values Along a Helix

Find dw/dt if

$$w = xy + z, \qquad x = \cos t, \qquad y = \sin t, \qquad z = t$$

(Figure 11.20). What is the derivative's value at $t = 0$?

Solution

$$\begin{aligned}\frac{dw}{dt} &= \frac{\partial w}{\partial x}\frac{dx}{dt} + \frac{\partial w}{\partial y}\frac{dy}{dt} + \frac{\partial w}{\partial z}\frac{dz}{dt} \\ &= (y)(-\sin t) + (x)(\cos t) + (1)(1) \\ &= (\sin t)(-\sin t) + (\cos t)(\cos t) + 1 \qquad \text{Substitute for the intermediate variables.} \\ &= -\sin^2 t + \cos^2 t + 1 = 1 + \cos 2t\end{aligned}$$

$$\left(\frac{dw}{dt}\right)_{t=0} = 1 + \cos(0) = 2.$$

Here is a physical interpretation of Theorem 6. If $w = T(x, y, z)$ is the temperature at each point (x, y, z) along a curve C with parametric equations $x = x(t)$, $y = y(t)$, and $z = z(t)$, then the composite function $w = T(x(t), y(t), z(t))$, represents the temperature relative to t along the curve. The derivative dw/dt is then the instantaneous rate of change of temperature along the curve.

Functions Defined on Surfaces

If we are interested in the temperature $w = f(x, y, z)$ at points (x, y, z) on a globe in space, we might prefer to think of x, y, and z as functions of the variables r and s that give the points' longitudes and latitudes. If $x = g(r, s)$, $y = h(r, s)$, and $z = k(r, s)$, we could then express the temperature as a function of r and s with the composite function

$$w = f(g(r, s), h(r, s), k(r, s)).$$

Under the right conditions, w would have partial derivatives with respect to both r and s that could be calculated in the following way.

Theorem 7 Chain Rule for Two Independent Variables and Three Intermediate Variables

Suppose that $w = f(x, y, z)$, $x = g(r, s)$, $y = h(r, s)$, and $z = k(r, s)$. If all four functions are differentiable, then w has partial derivatives with respect to r and s, given by the formulas

$$\frac{\partial w}{\partial r} = \frac{\partial w}{\partial x}\frac{\partial x}{\partial r} + \frac{\partial w}{\partial y}\frac{\partial y}{\partial r} + \frac{\partial w}{\partial z}\frac{\partial z}{\partial r}$$

$$\frac{\partial w}{\partial s} = \frac{\partial w}{\partial x}\frac{\partial x}{\partial s} + \frac{\partial w}{\partial y}\frac{\partial y}{\partial s} + \frac{\partial w}{\partial z}\frac{\partial z}{\partial s}.$$

The first of these equations can be derived from the Chain Rule in Theorem 6 by holding s fixed and treating r as t. The second can be derived in the same way, holding r fixed and treating s as t. The tree diagrams for both equations are shown in Figure 11.21.

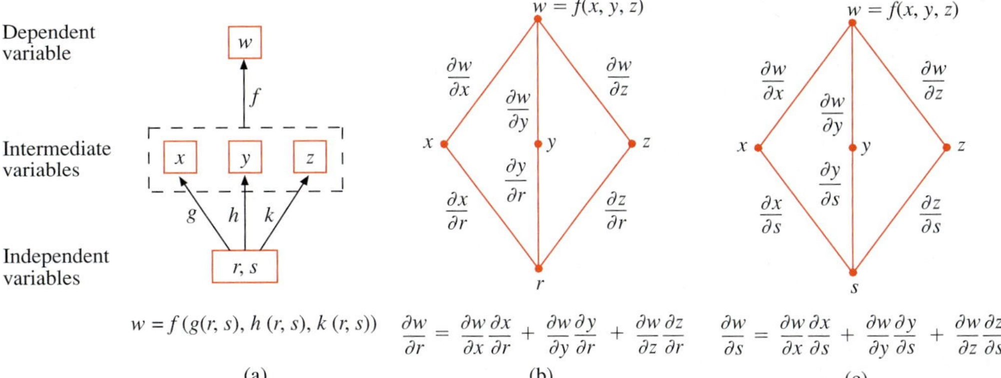

FIGURE 11.21 Composite function and tree diagrams for Theorem 7.

Example 3 Partial Derivatives Using Theorem 7

Express $\partial w/\partial r$ and $\partial w/\partial s$ in terms of r and s if

$$w = x + 2y + z^2, \qquad x = \frac{r}{s}, \qquad y = r^2 + \ln s, \qquad z = 2r.$$

Solution

$$\begin{aligned}\frac{\partial w}{\partial r} &= \frac{\partial w}{\partial x}\frac{\partial x}{\partial r} + \frac{\partial w}{\partial y}\frac{\partial y}{\partial r} + \frac{\partial w}{\partial z}\frac{\partial z}{\partial r}\\ &= (1)\left(\frac{1}{s}\right) + (2)(2r) + (2z)(2)\\ &= \frac{1}{s} + 4r + (4r)(2) = \frac{1}{s} + 12r\end{aligned}$$

Substitute for intermediate variable z.

$$\begin{aligned}\frac{\partial w}{\partial s} &= \frac{\partial w}{\partial x}\frac{\partial x}{\partial s} + \frac{\partial w}{\partial y}\frac{\partial y}{\partial s} + \frac{\partial w}{\partial z}\frac{\partial z}{\partial s}\\ &= (1)\left(-\frac{r}{s^2}\right) + (2)\left(\frac{1}{s}\right) + (2z)(0) = \frac{2}{s} - \frac{r}{s^2}\end{aligned}$$

If f is a function of two variables instead of three, each equation in Theorem 7 becomes correspondingly one term shorter.

If $w = f(x, y), x = g(r, s)$, and $y = h(r, s)$, then

$$\frac{\partial w}{\partial r} = \frac{\partial w}{\partial x}\frac{\partial x}{\partial r} + \frac{\partial w}{\partial y}\frac{\partial y}{\partial r} \quad \text{and} \quad \frac{\partial w}{\partial s} = \frac{\partial w}{\partial x}\frac{\partial x}{\partial s} + \frac{\partial w}{\partial y}\frac{\partial y}{\partial s}.$$

Figure 11.22 shows the tree diagram for the first of these equations. The diagram for the second equation is similar, just replace r with s.

Chain Rule

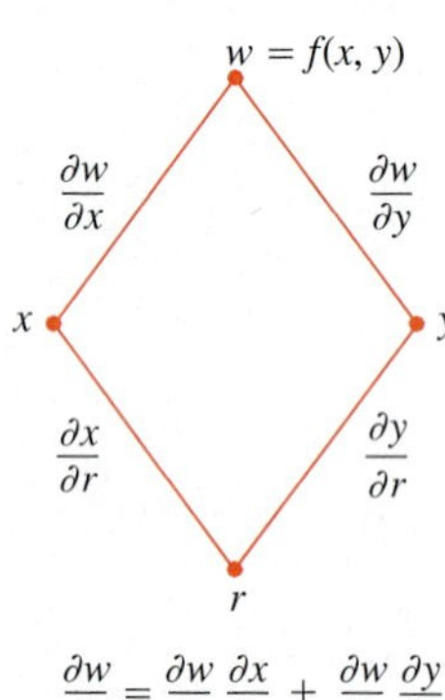

$$\frac{\partial w}{\partial r} = \frac{\partial w}{\partial x}\frac{\partial x}{\partial r} + \frac{\partial w}{\partial y}\frac{\partial y}{\partial r}$$

FIGURE 11.22 Tree diagram for the equation

$$\frac{\partial w}{\partial r} = \frac{\partial w}{\partial w}\frac{\partial x}{\partial r} + \frac{\partial w}{\partial y}\frac{\partial y}{\partial r}.$$

Example 4 More Partial Derivatives

Express $\partial w/\partial r$ and $\partial w/\partial s$ in terms of r and s if

$$w = x^2 + y^2, \qquad x = r - s, \qquad y = r + s.$$

Solution

$$\begin{aligned}\frac{\partial w}{\partial r} &= \frac{\partial w}{\partial x}\frac{\partial x}{\partial r} + \frac{\partial w}{\partial y}\frac{\partial y}{\partial r}\\ &= (2x)(1) + (2y)(1)\\ &= 2(r - s) + 2(r + s)\\ &= 4r\end{aligned}$$

$$\begin{aligned}\frac{\partial w}{\partial s} &= \frac{\partial w}{\partial x}\frac{\partial x}{\partial s} + \frac{\partial w}{\partial y}\frac{\partial y}{\partial s}\\ &= (2x)(-1) + (2y)(1)\\ &= -2(r - s) + 2(r + s)\\ &= 4s\end{aligned}$$

Substitute for the intermediate variables.

If f is a function of x alone, our equations become even simpler.

> If $w = f(x)$ and $x = g(r, s)$, then
>
> $$\frac{\partial w}{\partial r} = \frac{dw}{dx}\frac{\partial x}{\partial r} \quad \text{and} \quad \frac{\partial w}{\partial s} = \frac{dw}{dx}\frac{\partial x}{\partial s}.$$

In this case, we can use the ordinary (single-variable) derivative, dw/dx. The tree diagram is shown in Figure 11.23.

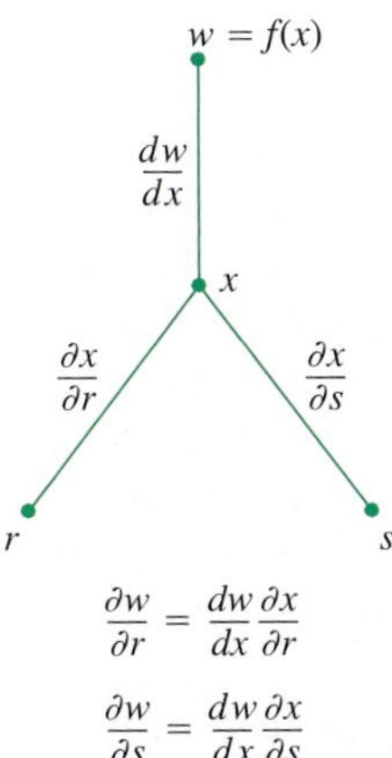

FIGURE 11.23 Tree diagram for differentiating f as a composite function of r and s with one intermediate variable.

Implicit Differentiation Revisited

Believe it or not, the two-variable Chain Rule in Theorem 5 leads to a formula that takes most of the work out of implicit differentiation. Suppose that

1. The function $F(x, y)$ is differentiable and
2. The equation $F(x, y) = 0$ defines y implicitly as a differentiable function of x, say $y = h(x)$.

Since $w = F(x, y) = 0$, the derivative dw/dx must be zero. Computing the derivative from the Chain Rule (tree diagram in Figure 11.24), we find

$$\begin{aligned} 0 = \frac{dw}{dx} &= F_x\frac{dx}{dx} + F_y\frac{dy}{dx} \qquad \text{Theorem 5 with } t = x \text{ and } f = F \\ &= F_x \cdot 1 + F_y \cdot \frac{dy}{dx}. \end{aligned}$$

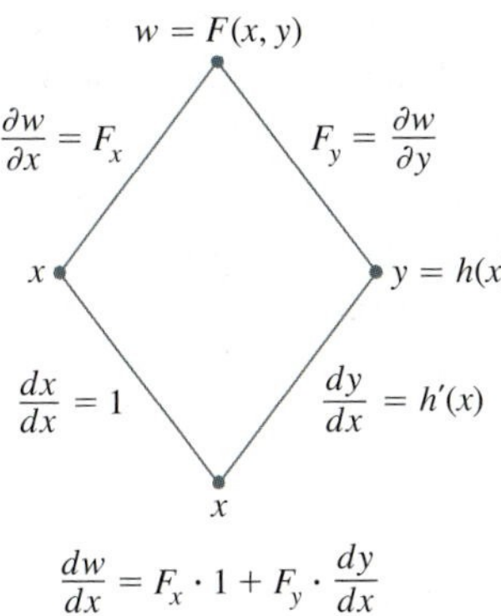

FIGURE 11.24 Three diagram for differentiating $w = F(x, y)$ with respect to x. Setting $dw/dx = 0$ leads to a simple computational formula for implicit differentiation (Theorem 8).

If $F_y = \partial w/\partial y \neq 0$, we can solve this equation for dy/dx to get

$$\frac{dy}{dx} = -\frac{F_x}{F_y}.$$

This relationship gives a surprisingly simple shortcut to finding derivatives of implicitly defined functions, which we state here as a theorem.

> **Theorem 8 A Formula for Implicit Differentiation**
>
> Suppose that $F(x, y)$ is differentiable and that the equation $F(x, y) = 0$ defines y as a differentiable function of x. Then at any point where $F_y \neq 0$,
>
> $$\frac{dy}{dx} = -\frac{F_x}{F_y}.$$

Example 5 Speedy Implicit Differentiation

Use Theorem 8 to find dy/dx if $y^2 - x^2 - \sin xy = 0$.

Solution Take $F(x, y) = y^2 - x^2 - \sin xy$. Then

$$\begin{aligned} \frac{dy}{dx} &= -\frac{F_x}{F_y} = -\frac{-2x - y\cos xy}{2y - x\cos xy} \\ &= \frac{2x + y\cos xy}{2y - x\cos xy}. \end{aligned}$$

This calculation is significantly shorter than the single-variable calculation with which we found dy/dx in Section 2.6, Example 2.

Functions of Many Variables

We have seen several different forms of the Chain Rule in this section, but you do not have to memorize them all if you can see them as special cases of the same general formula. When solving particular problems, it may help to draw the appropriate tree diagram by placing the dependent variable on top, the intermediate variables in the middle, and the selected independent variable at the bottom. To find the derivative of the dependent variable with respect to the selected independent variable, start at the dependent variable and read down each route of the tree to the independent variable, calculating and multiplying the derivatives along each route. Then add the products you found for the different routes.

In general, suppose that $w = f(x, y, \ldots, v)$ is a differentiable function of the variables $x, y, \ldots, v$ (a finite set) and the $x, y, \ldots, v$ are differentiable functions of $p, q, \ldots, t$ (another finite set). Then w is a differentiable function of the variables p through t and the partial derivatives of w with respect to these variables are given by equations of the form

$$\frac{\partial w}{\partial p} = \frac{\partial w}{\partial x}\frac{\partial x}{\partial p} + \frac{\partial w}{\partial y}\frac{\partial y}{\partial p} + \cdots + \frac{\partial w}{\partial v}\frac{\partial v}{\partial p}.$$

The other equations are obtained by replacing p by $q, \ldots, t$, one at a time.

One way to remember this equation is to think of the right-hand side as the dot product of two vectors with components

$$\underbrace{\left(\frac{\partial w}{\partial x}, \frac{\partial w}{\partial y}, \ldots, \frac{\partial w}{\partial v}\right)}_{\text{Derivatives of } w \text{ with respect to the intermediate variables}} \quad \text{and} \quad \underbrace{\left(\frac{\partial x}{\partial p}, \frac{\partial y}{\partial p}, \ldots, \frac{\partial v}{\partial p}\right)}_{\text{Derivatives of the intermediate variables with respect to the selected independent variable}}.$$

EXERCISES 11.4

Chain Rule: One Independent Variable

In Exercises 1–6, (a) express dw/dt as a function of t, both by using the Chain Rule and by expressing w in terms of t and differentiating directly with respect to t. Then (b) evaluate dw/dt at the given value of t.

1. $w = x^2 + y^2, \quad x = \cos t, \quad y = \sin t; \quad t = \pi$

2. $w = x^2 + y^2, \quad x = \cos t + \sin t, \quad y = \cos t - \sin t; \quad t = 0$

3. $w = \frac{x}{z} + \frac{y}{z}, \quad x = \cos^2 t, \quad y = \sin^2 t, \quad z = 1/t; \quad t = 3$

4. $w = \ln(x^2 + y^2 + z^2), \quad x = \cos t, \quad y = \sin t, \quad z = 4\sqrt{t}; \quad t = 3$

5. $w = 2ye^x - \ln z, \quad x = \ln(t^2 + 1), \quad y = \tan^{-1} t, \quad z = e^t; \quad t = 1$

6. $w = z - \sin xy, \quad x = t, \quad y = \ln t, \quad z = e^{t-1}; \quad t = 1$

Chain Rule: Two and Three Independent Variables

In Exercises 7 and 8, (a) express $\partial z/\partial r$ and $\partial z/\partial \theta$ as functions of r and θ both by using the Chain Rule and by expressing z directly in terms of r and θ before differentiating. Then (b) evaluate $\partial z/\partial r$ and $\partial z/\partial \theta$ at the given point (r, θ).

7. $z = 4e^x \ln y, \quad x = \ln(u \cos v), \quad y = u \sin v; \quad (u, v) = (2, \pi/4)$

8. $z = \tan^{-1}(x/y), \quad x = u \cos v, \quad y = u \sin v; \quad (u, v) = (1.3, \pi/6)$

In Exercises 9 and 10, (a) express $\partial w/\partial u$ and $\partial w/\partial v$ as functions of u and v both by using the Chain Rule and by expressing w directly in terms of u and v before differentiating. Then (b) evaluate $\partial w/\partial u$ and $\partial w/\partial v$ at the given point (u, v).

9. $w = xy + yz + xz, \quad x = u + v, \quad y = u - v, \quad z = uv;$ $(u, v) = (1/2, 1)$

10. $w = \ln(x^2 + y^2 + z^2), \quad x = ue^v \sin u, \quad y = ue^v \cos u,$ $z = ue^v; \quad (u, v) = (-2, 0)$

In Exercises 11 and 12, (a) express $\partial u/\partial x$, $\partial u/\partial y$, and $\partial u/\partial z$ as functions of x, y, and z both by using the Chain Rule and by expressing u directly in terms of x, y, and z before differentiating. Then (b) evaluate $\partial u/\partial x$, $\partial u/\partial y$, and $\partial u/\partial z$ at the given point (x, y, z).

11. $u = \dfrac{p - q}{q - r}, \quad p = x + y + z, \quad q = x - y + z, \quad r = x + y - z;$ $(x, y, z) = (\sqrt{3}, 2, 1)$

12. $u = e^{qr} \sin^{-1} p, \quad p = \sin x, \quad q = z^2 \ln y, \quad r = 1/z;$ $(x, y, z) = (\pi/4, 1/2, -1/2)$

Using a Tree Diagram

In Exercises 13–24, draw a tree diagram and write a Chain Rule formula for each derivative.

13. $\dfrac{dz}{dt}$ for $z = f(x, y), \quad x = g(t), \quad y = h(t)$

14. $\dfrac{dz}{dt}$ for $z = f(u, v, w), \quad u = g(t), \quad v = h(t), \quad w = k(t)$

15. $\dfrac{\partial w}{\partial u}$ and $\dfrac{\partial w}{\partial v}$ for $w = h(x, y, z), \quad x = f(u, v), \quad y = g(u, v),$ $z = k(u, v)$

16. $\dfrac{\partial w}{\partial x}$ and $\dfrac{\partial w}{\partial y}$ for $w = f(r, s, t), \quad r = g(x, y), \quad s = h(x, y),$ $t = k(x, y)$

17. $\dfrac{\partial w}{\partial u}$ and $\dfrac{\partial w}{\partial v}$ for $w = g(x, y), \quad x = h(u, v), \quad y = k(u, v)$

18. $\dfrac{\partial w}{\partial x}$ and $\dfrac{\partial w}{\partial y}$ for $w = g(u, v), \quad u = h(x, y), \quad v = k(x, y)$

19. $\dfrac{\partial z}{\partial t}$ and $\dfrac{\partial z}{\partial s}$ for $z = f(x, y), \quad x = g(t, s), \quad y = h(t, s)$

20. $\dfrac{\partial y}{\partial r}$ for $y = f(u), \quad u = g(r, s)$

21. $\dfrac{\partial w}{\partial s}$ and $\dfrac{\partial w}{\partial t}$ for $w = g(u), \quad u = h(s, t)$

22. $\dfrac{\partial w}{\partial p}$ for $w = f(x, y, z, v), \quad x = g(p, q), \quad y = h(p, q),$ $z = j(p, q), \quad v = k(p, q)$

23. $\dfrac{\partial w}{\partial r}$ and $\dfrac{\partial w}{\partial s}$ for $w = f(x, y), \quad x = g(r), \quad y = h(s)$

24. $\dfrac{\partial w}{\partial s}$ for $w = g(x, y), \quad x = h(r, s, t), \quad y = k(r, s, t)$

Implicit Differentiation

Assuming that the equations in Exercises 25–28 define y as a differentiable function of x, use Theorem 8 to find the value of dy/dx at the given point.

25. $x^3 - 2y^2 + xy = 0, \quad (1, 1)$

26. $xy + y^2 - 3x - 3 = 0, \quad (-1, 1)$

27. $x^2 + xy + y^2 - 7 = 0, \quad (1, 2)$

28. $xe^y + \sin xy + y - \ln 2 = 0, \quad (0, \ln 2)$

Three-Variable Implicit Differentiation

Theorem 8 can be generalized to functions of three variables and even more. The three-variable version goes like this: If the equation $F(x, y, z) = 0$ determines z as a differentiable function of x and y, then, at points where $F_z \neq 0$,

$$\frac{\partial z}{\partial x} = -\frac{F_x}{F_z} \quad \text{and} \quad \frac{\partial z}{\partial y} = -\frac{F_y}{F_z}.$$

Use these equations to find the values of $\partial z/\partial x$ and $\partial z/\partial y$ at the points in Exercises 29–32.

29. $z^3 - xy + yz + y^3 - 2 = 0, \quad (1, 1, 1)$

30. $\dfrac{1}{x} + \dfrac{1}{y} + \dfrac{1}{z} - 1 = 0, \quad (2, 3, 6)$

31. $\sin(x + y) + \sin(y + z) + \sin(x + z) = 0, \quad (\pi, \pi, \pi)$

32. $xe^y + ye^z + 2\ln x - 2 - 3\ln 2 = 0, \quad (1, \ln 2, \ln 3)$

Finding Specified Partial Derivatives

33. Find $\partial w/\partial r$ when $r = 1, s = -1$ if $w = (x + y + z)^2, x = r - s,$ $y = \cos(r + s), z = \sin(r + s)$.

34. Find $\partial w/\partial v$ when $u = -1, v = 2$ if $w = xy + \ln z, x = v^2/u,$ $y = u + v, z = \cos u$.

35. Find $\partial w/\partial v$ when $u = 0, v = 0$ if $w = x^2 + (y/x), x = u - 2v + 1,$ $y = 2u + v - 2$.

36. Find $\partial z/\partial u$ when $u = 0, v = 1$ if $z = \sin xy + x \sin y,$ $x = u^2 + v^2, y = uv$.

37. Find $\partial z/\partial u$ and $\partial z/\partial v$ when $u = \ln 2, v = 1$ if $z = 5\tan^{-1} x$ and $x = e^u + \ln v$.

38. Find $\partial z/\partial u$ and $\partial z/\partial v$ when $u = 1$ and $v = -2$ if $z = \ln q$ and $q = \sqrt{v + 3}\tan^{-1} u$.

Theory and Examples

39. *Changes within an electric circuit* The voltage V in a circuit that satisfies the law $V = IR$ is slowly dropping as the battery wears out. At the same time, the resistance R is increasing as the resistor heats up. Use the equation

$$\frac{dV}{dt} = \frac{\partial V}{\partial I}\frac{dI}{dt} + \frac{\partial V}{\partial R}\frac{dR}{dt}$$

to find how the current is changing at the instant when $R = 600$ ohms, $I = 0.04$ amp, $dR/dt = 0.5$ ohm/sec, and $dV/dt = -0.01$ volt/sec.

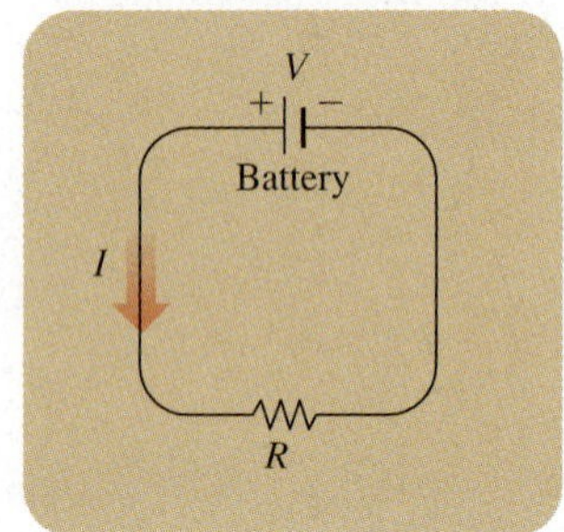

40. *Changing dimensions in a box* The lengths a, b, and c of the edges of a rectangular box are changing with time. At the instant in question, $a = 1$ m, $b = 2$ m, $c = 3$ m, $da/dt = db/dt = 1$ m/sec, and $dc/dt = -3$ m/sec. At what rates are the box's volume V and surface area S changing at that instant? Are the box's interior diagonals increasing in length or decreasing?

41. *Summing partial derivatives* If $f(u, v, w)$ is differentiable and $u = x - y$, $v = y - z$, and $w = z - x$, show that

$$\frac{\partial f}{\partial x} + \frac{\partial f}{\partial y} + \frac{\partial f}{\partial z} = 0.$$

42. *Polar coordinates* Suppose that we substitute polar coordinates $x = r \cos \theta$ and $y = r \sin \theta$ in a differentiable function $w = f(x, y)$.

(a) Show that

$$\frac{\partial w}{\partial r} = f_x \cos \theta + f_y \sin \theta$$

and

$$\frac{1}{r}\frac{\partial w}{\partial \theta} = -f_x \sin \theta + f_y \cos \theta.$$

(b) Solve the equations in part (a) to express f_x and f_y in terms of $\partial w/\partial r$ and $\partial w/\partial \theta$.

(c) Show that

$$(f_x)^2 + (f_y)^2 = \left(\frac{\partial w}{\partial r}\right)^2 + \frac{1}{r^2}\left(\frac{\partial w}{\partial \theta}\right)^2.$$

43. *Laplace equations* Show that if $w = f(u, v)$ satisfies the Laplace equation $f_{uu} + f_{vv} = 0$ and if $u = (x^2 - y^2)/2$ and $v = xy$, then w satisfies the Laplace equation $w_{xx} + w_{yy} = 0$.

44. *Laplace equations* Let $w = f(u) + g(v)$, where $u = x + iy$ and $v = x - iy$ and $i = \sqrt{-1}$. Show that w satisfies the Laplace equation $w_{xx} + w_{yy} = 0$ if all the necessary functions are differentiable.

Changes in Functions Along Curves

45. *Extreme values on a helix* Suppose that the partial derivatives of a function $f(x, y, z)$ at points on the helix $x = \cos t$, $y = \sin t$, $z = t$ are

$$f_x = \cos t, \qquad f_y = \sin t, \qquad f_z = t^2 + t - 2.$$

At what points on the curve, if any, can f take on extreme values?

46. *A space curve* Let $w = x^2e^{2y} \cos 3z$. Find the value of dw/dt at the point $(1, \ln 2, 0)$ on the curve $x = \cos t$, $y = \ln (t + 2)$, $z = t$.

47. *Temperature on a circle* Let $T = f(x, y)$ be the temperature at the point (x, y) on the circle $x = \cos t$, $y = \sin t$, $0 \le t \le 2\pi$ and suppose that

$$\frac{\partial T}{\partial x} = 8x - 4y, \qquad \frac{\partial T}{\partial y} = 8y - 4x.$$

(a) Find where the maximum and minimum temperatures on the circle occur by examining the derivatives dT/dt and d^2T/dt^2.

(b) Suppose that $T = 4x^2 - 4xy + 4y^2$. Find the maximum and minimum values of T on the circle.

48. *Temperature on an ellipse* Let $T = g(x, y)$ be the temperature at the point (x, y) on the ellipse

$$x = 2\sqrt{2} \cos t, \qquad y = \sqrt{2} \sin t, \qquad 0 \le t \le 2\pi,$$

and suppose that

$$\frac{\partial T}{\partial x} = y, \qquad \frac{\partial T}{\partial y} = x.$$

(a) Locate the maximum and minimum temperatures on the ellipse by examining dT/dt and d^2T/dt^2.

(b) Suppose that $T = xy - 2$. Find the maximum and minimum values of T on the ellipse.

Differentiating Integrals

Under mild continuity restrictions, it is true that if

$$F(x) = \int_a^b g(t, x)\, dt,$$

then $F'(x) = \int_a^b g_x(t, x)\, dt$. Using this fact and the Chain Rule, we can find the derivative of

$$F(x) = \int_a^{f(x)} g(t, x)\, dt$$

by letting

$$G(u, x) = \int_a^u g(t, x)\, dt,$$

Where $u = f(x)$. Find the derivatives of the functions in Exercises 49 and 50.

49. $F(x) = \int_0^{x^2} \sqrt{t^4 + x^3}\, dt$

50. $F(x) = \int_{x^2}^1 \sqrt{t^3 + x^2}\, dt$

11.5 Directional Derivatives, Gradient Vectors, and Tangent Planes

Directional Derivatives in the Plane • Interpretation of the Directional Derivative • Calculation • Properties of Directional Derivatives • Gradients and Tangents to Level Curves • Algebra Rules for Gradients • Increments and Distance • Functions of Three Variables • Tangent Planes and Normal Lines • Planes Tangent to a Surface $z = f(x, y)$ • Other Applications

If you look at the map (Figure 11.25) showing contours on the West Point Fortress along the Hudson River in New York, you will notice that the tributary streams flow perpendicular to the contours. The streams are following paths of steepest descent so the waters reach the Hudson as quickly as possible. Therefore, the instantaneous rate of change in a stream's altitude above sea level has a particular direction. In this section, you see why this direction is perpendicular to the contours.

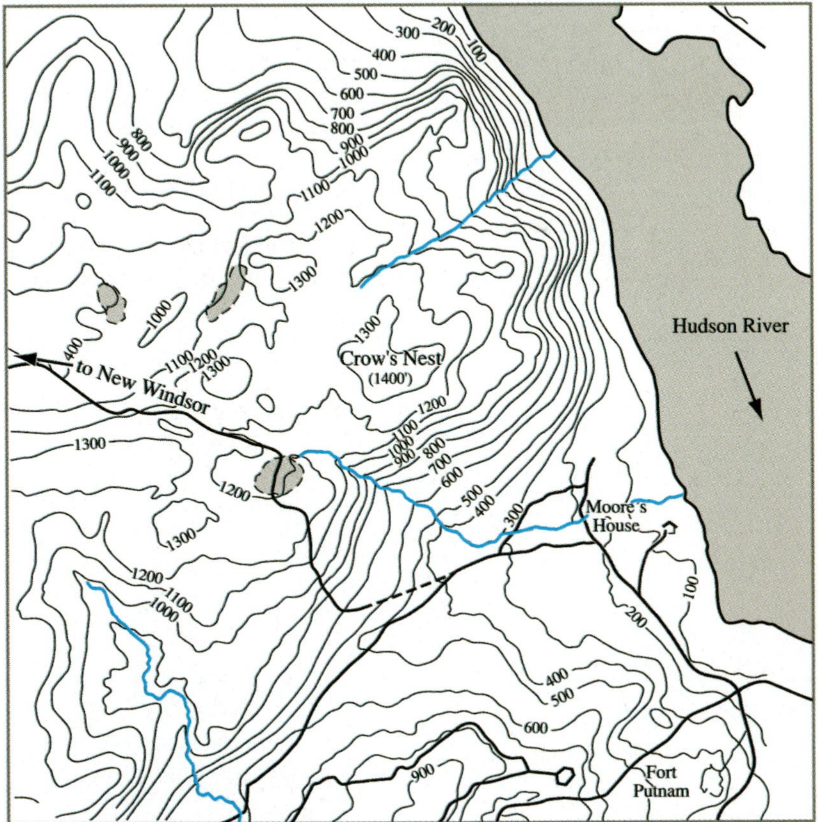

FIGURE 11.25 Contours of the West Point Fortress in New York show streams, which follow paths of steepest descent, running perpendicular to the contours.

We know from Section 11.4 that if $f(x, y)$ is differentiable, then the rate at which f changes with respect to t along a differentiable curve $x = g(t)$, $y = h(t)$ is

$$\frac{df}{dt} = \frac{\partial f}{\partial x}\frac{dx}{dt} + \frac{\partial f}{\partial y}\frac{dy}{dt}.$$

At any point $P_0(x_0, y_0) = P_0(g(t_0), h(t_0))$, this equation gives the rate of change of f with respect to increasing t and therefore depends, among other things, on the direction of motion along the curve. This observation is particularly important when the curve is a straight line and t is the arc length parameter along the line measured from P_0 in the direction of a given unit vector $\mathbf{u}$. For then, df/dt is the rate of change of f with respect to distance in its domain in the direction of $\mathbf{u}$. By varying $\mathbf{u}$, we find the rates at which f changes with respect to distance as we move through P_0 in different directions. These "directional derivatives" have useful interpretations in science and engineering as well as in mathematics. This section develops a formula for calculating them and proceeds from there to find equations for tangent planes and normal lines on surfaces in space.

Directional Derivatives in the Plane

Suppose that the function $f(x, y)$ is defined throughout a region R in the xy-plane, that $P_0(x_0, y_0)$ is a point in R, and that $\mathbf{u} = u_1\mathbf{i} + u_2\mathbf{j}$ is a unit vector. Then the equations

$$x = x_0 + su_1, \qquad y = y_0 + su_2$$

parametrize the line through P_0 parallel to $\mathbf{u}$. If the parameter s measures arc length from P_0 in the direction of $\mathbf{u}$, we find the rate of change of f at P_0 in the direction of $\mathbf{u}$ by calculating df/ds at P_0 (Figure 11.26):

Definition Directional Derivative

The **derivative of f at $P_0(x_0, y_0)$ in the direction of the unit vector $\mathbf{u} = u_1\mathbf{i} + u_2\mathbf{j}$** is the number

$$\left(\frac{df}{ds}\right)_{\mathbf{u},P_0} = \lim_{s \to 0} \frac{f(x_0 + su_1, y_0 + su_2) - f(x_0, y_0)}{s}, \tag{1}$$

provided the limit exists.

The directional derivative is also denoted by

$$(D_{\mathbf{u}}f)_{P_0}.$$

"The derivative of f at P^0 in the direction of $\mathbf{u}$"

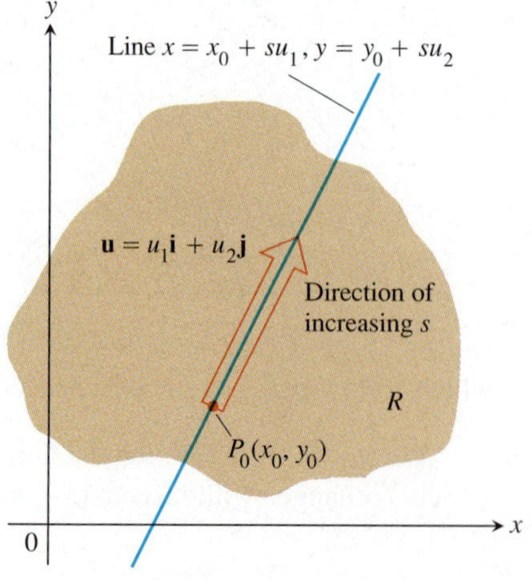

FIGURE 11.26 The rate of change of f in the direction of $\mathbf{u}$ at a point P_0 is the rate at which f changes along this line at P_0.

Example 1 Finding a Directional Derivative Using the Definition
Find the derivative of

$$f(x, y) = x^2 + xy$$

at $P_0(1, 2)$ in the direction of the unit vector $\mathbf{u} = (1/\sqrt{2})\mathbf{i} + (1/\sqrt{2})\mathbf{j}$.

Solution

$$\left(\frac{df}{ds}\right)_{\mathbf{u},P_0} = \lim_{s\to 0} \frac{f(x_0 + su_1, y_0 + su_2) - f(x_0, y_0)}{s} \qquad \text{Eq. (1)}$$

$$= \lim_{s\to 0} \frac{f\left(1 + s \cdot \frac{1}{\sqrt{2}}, 2 + s \cdot \frac{1}{\sqrt{2}}\right) - f(1, 2)}{s}$$

$$= \lim_{s\to 0} \frac{\left(1 + \frac{s}{\sqrt{2}}\right)^2 + \left(1 + \frac{s}{\sqrt{2}}\right)\left(2 + \frac{s}{\sqrt{2}}\right) - (1^2 + 1 \cdot 2)}{s}$$

$$= \lim_{s\to 0} \frac{\left(1 + \frac{2s}{\sqrt{2}} + \frac{s^2}{2}\right) + \left(2 + \frac{3s}{\sqrt{2}} + \frac{s^2}{2}\right) - 3}{s}$$

$$= \lim_{s\to 0} \frac{\frac{5s}{\sqrt{2}} + s^2}{s} = \lim_{s\to 0} \left(\frac{5}{\sqrt{2}} + s\right) = \left(\frac{5}{\sqrt{2}} + 0\right) = \frac{5}{\sqrt{2}}.$$

The rate of change of $f(x, y) = x^2 + xy$ at $P_0(1, 2)$ in the direction $\mathbf{u} = (1/\sqrt{2})\mathbf{i} + (1/\sqrt{2})\mathbf{j}$ is $5/\sqrt{2}$.

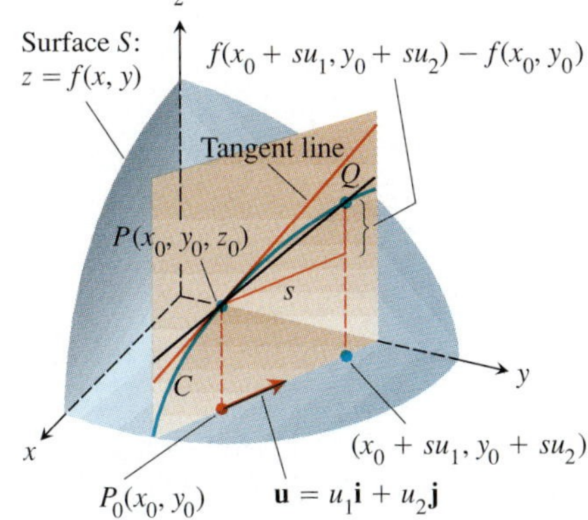

FIGURE 11.27 The slope of curve C at P_0 is

$$\lim_{Q\to P} \text{slope}(PQ) = \lim_{s\to 0} \frac{f(x_0 + su_1, y_0 + su_2) - f(x_0, y_0)}{s} = \left(\frac{df}{ds}\right)_{\mathbf{u},P_0} = (D_{\mathbf{u}}f)_{P_0}.$$

Interpretation of the Directional Derivative

The equation $z = f(x, y)$ represents a surface S in space. If $z_0 = f(x_0, y_0)$, then the point $P(x_0, y_0, z_0)$ lies on S. The vertical plane that passes through P and $P_0(x_0, y_0)$ parallel to $\mathbf{u}$ intersects S in a curve C (Figure 11.27). The rate of change of f in the direction of $\mathbf{u}$ is the slope of the tangent to C at P.

When $\mathbf{u} = \mathbf{i}$, the directional derivative at P_0 is $\partial f/\partial x$ evaluated at (x_0, y_0). When $\mathbf{u} = \mathbf{j}$, the directional derivative at P_0 is $\partial f/\partial y$ evaluated at (x_0, y_0). The directional derivative generalizes the two partial derivatives. We can now ask for the rate of change of f in any direction $\mathbf{u}$, not just the directions $\mathbf{i}$ and $\mathbf{j}$.

Here's a physical interpretation of the directional derivative. Suppose that $T = f(x, y)$ is the temperature at each point (x, y) over a region in the plane. Then $f(x_0, y_0)$ is the temperature at the point $P_0(x_0, y_0)$ and $(D_{\mathbf{u}}f)_{P_0}$ is the instantaneous rate of change of the temperature at P_0 stepping off in the direction $\mathbf{u}$.

Calculation

As you know, it is rarely convenient to calculate a derivative directly from its definition as a limit, and the directional derivative is no exception. We can develop a more efficient formula in the following way. We begin with the line

$$x = x_0 + su_1, \qquad y = y_0 + su_2, \tag{2}$$

through $P_0(x_0, y_0)$, parametrized with the arc length parameter s increasing in the direction of the unit vector $\mathbf{u} = u_1\mathbf{i} + u_2\mathbf{j}$. Then

$$\begin{aligned}\left(\frac{df}{ds}\right)_{\mathbf{u},P_0} &= \left(\frac{\partial f}{\partial x}\right)_{P_0}\frac{dx}{ds} + \left(\frac{\partial f}{\partial y}\right)_{P_0}\frac{dy}{ds} && \text{Chain Rule}\\ &= \left(\frac{\partial f}{\partial x}\right)_{P_0}\cdot u_1 + \left(\frac{\partial f}{\partial y}\right)_{P_0}\cdot u_2 && \text{From Eqs. (2), } dx/ds = u_1 \text{ and } dy/ds = u_2\\ &= \underbrace{\left[\left(\frac{\partial f}{\partial x}\right)_{P_0}\mathbf{i} + \left(\frac{\partial f}{\partial y}\right)_{P_0}\mathbf{j}\right]}_{\text{Gradient of } f \text{ at } P_0}\cdot\underbrace{\left[u_1\mathbf{i} + u_2\mathbf{j}\right]}_{\text{Direction } \mathbf{u}}. \qquad (3)\end{aligned}$$

The notation ∇f is read "grad f" as well as "gradient of f" and "del f." The symbol ∇ by itself is read "del." Another notation for the gradient is grad f, read the way it is written.

Definition Gradient Vector or Gradient

The **gradient vector (gradient)** of $f(x, y)$ at a point $P_0(x_0, y_0)$ is the vector

$$\nabla f = \frac{\partial f}{\partial x}\mathbf{i} + \frac{\partial f}{\partial y}\mathbf{j}$$

obtained by evaluating the partial derivatives of f at P_0.

Equation (3) says that the derivative of f in the direction of $\mathbf{u}$ at P_0 is the dot product of $\mathbf{u}$ with the gradient of f at P_0.

Theorem 9 The Directional Derivative is a Dot Product

If the function $f(x, y)$ is differentiable at $P_0(x_0, y_0)$, then

$$\left(\frac{df}{ds}\right)_{\mathbf{u},P_0} = (\nabla f)_{P_0}\cdot\mathbf{u}, \qquad (4)$$

the dot product of the gradient f at P_0 and $\mathbf{u}$.

Example 2 Finding the Directional Derivative Using the Gradient

Find the derivative of $f(x, y) = xe^y + \cos(xy)$ at the point $(2, 0)$ in the direction of $\mathbf{v} = 3\mathbf{i} - 4\mathbf{j}$.

Solution The direction of $\mathbf{v}$ is obtained by dividing $\mathbf{v}$ by its length:

$$\mathbf{u} = \frac{\mathbf{v}}{|\mathbf{v}|} = \frac{\mathbf{v}}{5} = \frac{3}{5}\mathbf{i} - \frac{4}{5}\mathbf{j}.$$

The partial derivatives of f at $(2, 0)$ are

$$\begin{aligned}f_x(2, 0) &= (e^y - y\sin(xy))_{(2,0)} = e^0 - 0 = 1\\ f_y(2, 0) &= (xe^y - x\sin(xy))_{(2,0)} = 2e^0 - 2\cdot 0 = 2.\end{aligned}$$

The gradient of f at $(2, 0)$ is

$$\nabla f|_{(2,0)} = f_x(2,0)\mathbf{i} + f_y(2,0)\mathbf{j} = \mathbf{i} + 2\mathbf{j}$$

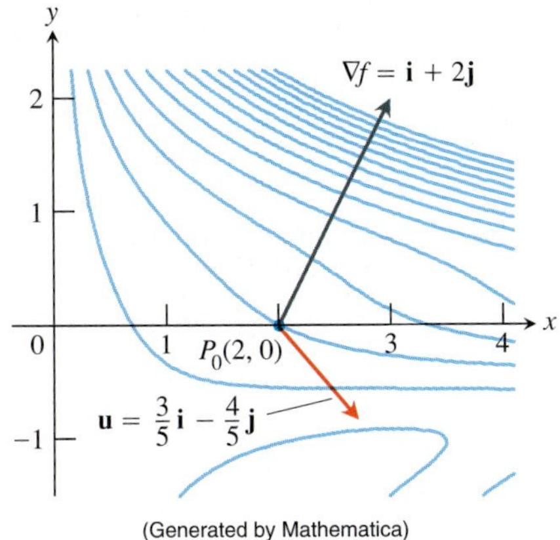

FIGURE 11.28 It is customary to picture ∇f as a vector in the domain of f. In the case of $f(x, y) = xe^y + \cos(xy)$, the domain is the entire plane. The rate at which f changes in the direction $\mathbf{u} = (3/5)\mathbf{i} - (4/5)\mathbf{j}$ is $\nabla f \cdot \mathbf{u} = -1$. (Example 2)

(Figure 11.28). The derivative of f at $(2, 0)$ in the direction of $\mathbf{v}$ is therefore

$$
\begin{aligned}
(D_{\mathbf{u}}f)|_{(2,0)} &= \nabla f|_{(2,0)} \cdot \mathbf{u} \qquad \text{Eq. (4)} \\
&= (\mathbf{i} + 2\mathbf{j}) \cdot \left(\frac{3}{5}\mathbf{i} - \frac{4}{5}\mathbf{j}\right) = \frac{3}{5} - \frac{8}{5} = -1.
\end{aligned}
$$

Properties of Directional Derivatives

Evaluating the dot product in the formula

$$D_{\mathbf{u}}f = \nabla f \cdot \mathbf{u} = |\nabla f||\mathbf{u}| \cos\theta = |\nabla f| \cos\theta,$$

where θ is the angle between the vectors $\mathbf{u}$ and ∇f, reveals the following properties.

Properties of the Directional Derivative $D_{\mathbf{u}} f = \nabla f \cdot \mathbf{u} = |\nabla f| \cos\theta$

1. The function f increases most rapidly when $\cos\theta = 1$ or when $\mathbf{u}$ is the direction of ∇f. That is, at each point P in its domain, f increases most rapidly in the direction of the gradient vector ∇f at P. The derivative in this direction is

$$D_{\mathbf{u}}f = |\nabla f| \cos(0) = |\nabla f|.$$

2. Similarly, f decreases most rapidly in the direction of $-\nabla f$. The derivative in this direction is $D_{\mathbf{u}}f = |\nabla f| \cos(\pi) = -|\nabla f|$.
3. Any direction $\mathbf{u}$ orthogonal to the gradient is a direction of zero change in f because θ then equals $\pi/2$ and

$$D_{\mathbf{u}}f = |\nabla f| \cos(\pi/2) = |\nabla f| \cdot 0 = 0.$$

As we discuss later, these properties hold in three dimensions as well as two.

Example 3 Finding Directions of Maximal, Minimal, and Zero Change

Find the directions in which $f(x, y) = (x^2/2) + (y^2/2)$

(a) Increases most rapidly at the point $(1, 1)$

(b) Decreases most rapidly at $(1, 1)$.

(c) What are the directions of zero change in f at $(1, 1)$?

Solution

(a) The function increases most rapidly in the direction of ∇f at $(1, 1)$. The gradient there is

$$(\nabla f)_{(1,1)} = (x\mathbf{i} + y\mathbf{j})_{(1,1)} = \mathbf{i} + \mathbf{j}.$$

Its direction is

$$\mathbf{u} = \frac{\mathbf{i} + \mathbf{j}}{|\mathbf{i} + \mathbf{j}|} = \frac{\mathbf{i} + \mathbf{j}}{\sqrt{(1)^2 + (1)^2}} = \frac{1}{\sqrt{2}}\mathbf{i} + \frac{1}{\sqrt{2}}\mathbf{j}.$$

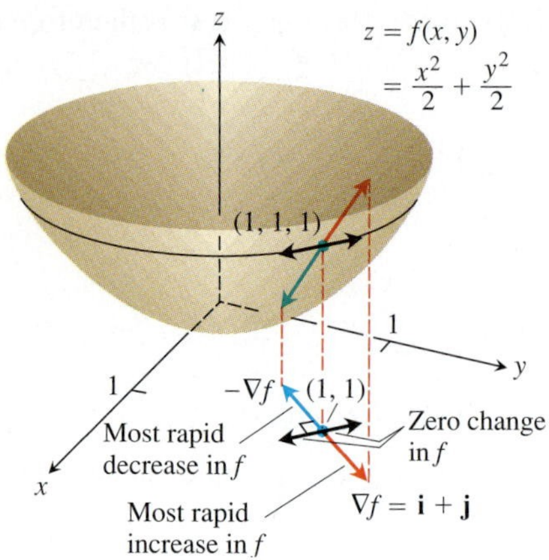

FIGURE 11.29 The direction in which $f(x, y) = (x^2/2) + (y^2/2)$ increases most rapidly at $(1, 1)$ is the direction of $\nabla f|_{(1,1)} = \mathbf{i} + \mathbf{j}$. It corresponds to the direction of steepest ascent on the surface at $(1, 1, 1)$.

(b) The function decreases most rapidly in the direction of $-\nabla f$ at $(1, 1)$, which is

$$-\mathbf{u} = -\frac{1}{\sqrt{2}}\mathbf{i} - \frac{1}{\sqrt{2}}\mathbf{j}.$$

(c) The directions of zero change at $(1, 1)$ are the directions orthogonal to ∇f:

$$\mathbf{n} = -\frac{1}{\sqrt{2}}\mathbf{i} + \frac{1}{\sqrt{2}}\mathbf{j} \quad \text{and} \quad -\mathbf{n} = \frac{1}{\sqrt{2}}\mathbf{i} - \frac{1}{\sqrt{2}}\mathbf{j}.$$

See Figure 11.29.

Gradients and Tangents to Level Curves

If a differentiable function $f(x, y)$ has a constant value c along a smooth curve $\mathbf{r} = g(t)\mathbf{i} + h(t)\mathbf{j}$ (making the curve a level curve of f), then $f(g(t), h(t)) = c$. Differentiating both sides of this equation with respect to t leads to the equations

$$\frac{d}{dt}f(g(t), h(t)) = \frac{d}{dt}(c)$$

$$\frac{\partial f}{\partial x}\frac{dg}{dt} + \frac{\partial f}{\partial y}\frac{dh}{dt} = 0 \qquad \text{Chain Rule}$$

$$\underbrace{\left(\frac{\partial f}{\partial x}\mathbf{i} + \frac{\partial f}{\partial y}\mathbf{j}\right)}_{\nabla f} \cdot \underbrace{\left(\frac{dg}{dt}\mathbf{i} + \frac{dh}{dt}\mathbf{j}\right)}_{\frac{d\mathbf{r}}{dt}} = 0. \tag{5}$$

Equation (5) says that ∇f is normal to the tangent vector $d\mathbf{r}/dt$, so it is normal to the curve.

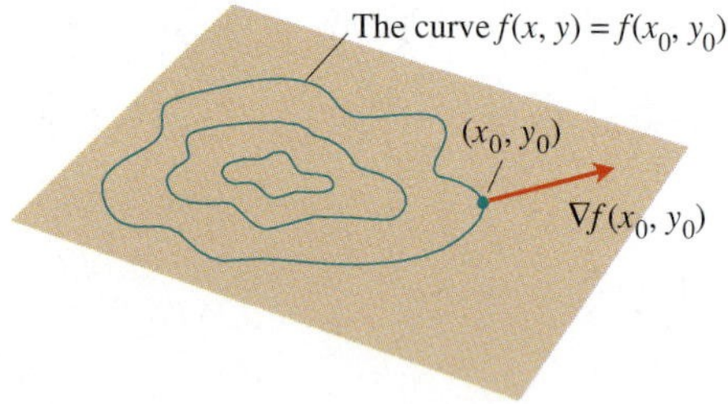

FIGURE 11.30 The gradient of a differentiable function of two variables at a point is always normal to the function's level curve through that point.

> At every point (x_0, y_0) in the domain of $f(x, y)$, the gradient of f is normal to the level curve through (x_0, y_0) (Figure 11.30).

Equation (5) validates our observation that streams flow perpendicular to the contours in topographical maps (Figure 11.25). Since the downflowing stream will reach its destination in the fastest way, it must flow in the direction of the negative gradient vectors from Property 2 for the directional derivative. Equation (5) tells us these directions are perpendicular to the level curves.

This observation also enables us to find equations for tangent lines to level curves. They are the lines normal to the gradients. The line through a point $P_0(x_0, y_0)$ normal to a vector $\mathbf{N} = A\mathbf{i} + B\mathbf{j}$ has the equation

$$A(x - x_0) + B(y - y_0) = 0$$

(Exercise 59). If $\mathbf{N}$ is the gradient $(\nabla f)_{(x_0,y_0)} = f_x(x_0, y_0)\mathbf{i} + f_y(x_0, y_0)\mathbf{j}$, the equation becomes

$$f_x(x_0, y_0)(x - x_0) + f_y(x_0, y_0)(y - y_0) = 0. \tag{6}$$

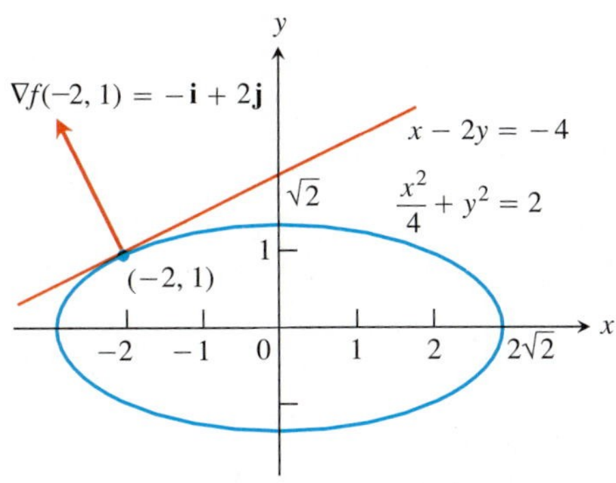

FIGURE 11.31 We can find the tangent to the ellipse $(x^2/4) + y^2 = 2$ by treating the ellipse as a level curve of the function $f(x, y) = (x^2/4) + y^2$. (Example 4)

Example 4 Finding the Tangent Line to an Ellipse

Find an equation for the tangent to the ellipse

$$\frac{x^2}{4} + y^2 = 2$$

(Figure 11.31) at the point $(-2, 1)$.

Solution The ellipse is a level curve of the function

$$f(x, y) = \frac{x^2}{4} + y^2.$$

The gradient of f at $(-2, 1)$ is

$$\nabla f|_{(-2,1)} = \left(\frac{x}{2}\mathbf{i} + 2y\mathbf{j}\right)_{(-2,1)} = -\mathbf{i} + 2\mathbf{j}.$$

The tangent is the line

$$(-1)(x + 2) + (2)(y - 1) = 0 \qquad \text{Eq. (6)}$$
$$x - 2y = -4.$$

Algebra Rules for Gradients

If we know the gradients of two functions f and g, we automatically know the gradients of their constant multiples, sum, difference, product, and quotient.

These rules have the same form as the corresponding rules for derivatives, as they should (Exercise 63).

Algebra Rules for Gradients

1. *Constant Multiple Rule:* $\nabla(kf) = k\nabla f$ (any number k)
2. *Sum Rule:* $\nabla(f + g) = \nabla f + \nabla g$
3. *Difference Rule:* $\nabla(f - g) = \nabla f - \nabla g$
4. *Product Rule:* $\nabla(fg) = f\nabla g + g\nabla f$
5. *Quotient Rule:* $\nabla\left(\frac{f}{g}\right) = \frac{g\nabla f - f\nabla g}{g^2}$

Example 5 Illustrating the Gradient Rules

We illustrate the rules with

$$f(x, y) = x - y \qquad g(x, y) = 3y$$
$$\nabla f = \mathbf{i} - \mathbf{j} \qquad \nabla g = 3\mathbf{j}.$$

We have

1. $\nabla(2f) = \nabla(2x - 2y) = 2\mathbf{i} - 2\mathbf{j} = 2\nabla f$
2. $\nabla(f + g) = \nabla(x + 2y) = \mathbf{i} + 2\mathbf{j} = \nabla f + \nabla g$
3. $\nabla(f - g) = \nabla(x - 4y) = \mathbf{i} - 4\mathbf{j} = \nabla f - \nabla g$

4. $\nabla(fg) = \nabla(3xy - 3y^2) = 3y\mathbf{i} + (3x - 6y)\mathbf{j}$
$= 3y(\mathbf{i} - \mathbf{j}) + 3y\mathbf{j} + (3x - 6y)\mathbf{j}$
$= 3y(\mathbf{i} - \mathbf{j}) + (3x - 3y)\mathbf{j}$
$= 3y(\mathbf{i} - \mathbf{j}) + (x - y)3\mathbf{j} = g\nabla f + f\nabla g$

5. $\nabla\left(\frac{f}{g}\right) = \nabla\left(\frac{x-y}{3y}\right) = \nabla\left(\frac{x}{3y} - \frac{1}{3}\right)$
$= \frac{1}{3y}\mathbf{i} - \frac{x}{3y^2}\mathbf{j}$
$= \frac{3y\mathbf{i} - 3x\mathbf{j}}{9y^2} = \frac{3y(\mathbf{i} - \mathbf{j}) - (3x - 3y)\mathbf{j}}{9y^2}$
$= \frac{3y(\mathbf{i} - \mathbf{j}) - (x - y)3\mathbf{j}}{9y^2} = \frac{g\nabla f - f\nabla g}{g^2}.$

Increments and Distance

The directional derivative plays the role of an ordinary derivative when we want to estimate how much the value of a function f changes if we move a small distance ds from a point P_0 to another point nearby. If f were a function of a single variable, we would have

$$df = f'(P_0)\,ds. \qquad \text{Ordinary derivative} \times \text{increment}$$

For a function of two or more variables, we use the formula

$$df = (\nabla f|_{P_0} \cdot \mathbf{u})\,ds, \qquad \text{Directional derivative} \times \text{increment}$$

where $\mathbf{u}$ is the direction of the motion away from P_0.

Estimating the Change in f in a Direction u

To estimate the change in the value of a function f when we move a small distance ds from a point P_0 in a particular direction $\mathbf{u}$, use the formula

$$df = \underbrace{(\nabla f|_{P_0} \cdot \mathbf{u})}_{\text{directional derivative}} \cdot \underbrace{ds}_{\text{distance increment}}$$

Example 6 Estimating Change in the Value of $f(x, y)$

Estimate how much the value of

$$f(x, y) = xe^y$$

will change if the point $P(x, y)$ moves 0.1 unit from $P_0(2, 0)$ straight toward $P_1(4, 1)$.

Solution We first find the derivative of f at P_0 in the direction of the vector

$$\overrightarrow{P_0P_1} = 2\mathbf{i} + \mathbf{j}.$$

The direction of this vector is

$$\mathbf{u} = \frac{\overrightarrow{P_0P_1}}{|\overrightarrow{P_0P_1}|} = \frac{\overrightarrow{P_0P_1}}{\sqrt{5}} = \frac{2}{\sqrt{5}}\mathbf{i} + \frac{1}{\sqrt{5}}\mathbf{j}.$$

The gradient of f at P_0 is

$$\nabla f_{(2,0)} = (e^y\mathbf{i} + xe^y\mathbf{j})_{(2,0)} = \mathbf{i} + 2\mathbf{j}.$$

Therefore,

$$\nabla f|_{P_0} \cdot \mathbf{u} = (\mathbf{i} + 2\mathbf{j}) \cdot \left(\frac{2}{\sqrt{5}}\mathbf{i} + \frac{1}{\sqrt{5}}\mathbf{j}\right) = \frac{2}{\sqrt{5}} + \frac{2}{\sqrt{5}} = \frac{4}{\sqrt{5}}.$$

The change df in f that results from moving $ds = 0.1$ unit away from P_0 in the direction of $\mathbf{u}$ is approximately

$$df = (\nabla f|_{P_0} \cdot \mathbf{u})(ds) = \left(\frac{4}{\sqrt{5}}\right)(0.1) \approx 0.18 \text{ units.}$$

Functions of Three Variables

We obtain three-variable formulas by adding the z-terms to the two-variable formulas. For a differentiable function $f(x, y, z)$ and a unit vector $\mathbf{u} = u_1\mathbf{i} + u_2\mathbf{j} + u_3\mathbf{k}$ in space, we have

$$\nabla f = \frac{\partial f}{\partial x}\mathbf{i} + \frac{\partial f}{\partial y}\mathbf{j} + \frac{\partial f}{\partial z}\mathbf{k}$$

and

$$D_{\mathbf{u}}f = \nabla f \cdot \mathbf{u} = \frac{\partial f}{\partial x}u_1 + \frac{\partial f}{\partial y}u_2 + \frac{\partial f}{\partial z}u_3.$$

The directional derivative can once again be written in the form

$$D_{\mathbf{u}}f = \nabla f \cdot \mathbf{u} = |\nabla f||u| \cos\theta = |\nabla f| \cos\theta,$$

so the properties listed earlier for functions of two variables continue to hold. At any given point, f increases most rapidly in the direction of ∇f and decreases most rapidly in the direction of $-\nabla f$. In any direction orthogonal to ∇f, the derivative is zero.

Example 7 Finding Directions of Maximal, Minimal, and Zero Change

(a) Find the derivative of $f(x, y, z) = x^3 - xy^2 - z$ at $P_0(1, 1, 0)$ in the direction of $\mathbf{v} = 2\mathbf{i} - 3\mathbf{j} + 6\mathbf{k}$.

(b) In what directions does f change most rapidly at P_0, and what are the rates of change in these directions?

Solution

(a) The direction of $\mathbf{v}$ is obtained by dividing $\mathbf{v}$ by its length:

$$|\mathbf{v}| = \sqrt{(2)^2 + (-3)^2 + (6)^2} = \sqrt{49} = 7$$

$$\mathbf{u} = \frac{\mathbf{v}}{|\mathbf{v}|} = \frac{2}{7}\mathbf{i} - \frac{3}{7}\mathbf{j} + \frac{6}{7}\mathbf{k}.$$

The partial derivatives of f at P_0 are

$$f_x = (3x^2 - y^2)_{(1,1,0)} = 2,$$
$$f_y = -2xy\,|_{(1,1,0)} = -2, \qquad f_z = -1\,|_{(1,1,0)} = -1.$$

The gradient of f at P_0 is

$$\nabla f|_{(1,1,0)} = 2\mathbf{i} - 2\mathbf{j} - \mathbf{k}.$$

The derivative of f at P_0 in the direction of $\mathbf{v}$ is therefore

$$(D_{\mathbf{u}}f)_{(1,1,0)} = \nabla f|_{(1,1,0)} \cdot \mathbf{u} = (2\mathbf{i} - 2\mathbf{j} - \mathbf{k}) \cdot \left(\frac{2}{7}\mathbf{i} - \frac{3}{7}\mathbf{j} + \frac{6}{7}\mathbf{k}\right)$$
$$= \frac{4}{7} + \frac{6}{7} - \frac{6}{7} = \frac{4}{7}.$$

(b) The function increases most rapidly in the direction of $\nabla f = 2\mathbf{i} - 2\mathbf{j} - \mathbf{k}$ and decreases most rapidly in the direction of $-\nabla f$. The rates of change in the directions are, respectively,

$$|\nabla f| = \sqrt{(2)^2 + (-2)^2 + (-1)^2} = \sqrt{9} = 3 \qquad \text{and} \qquad -|\nabla f| = -3.$$

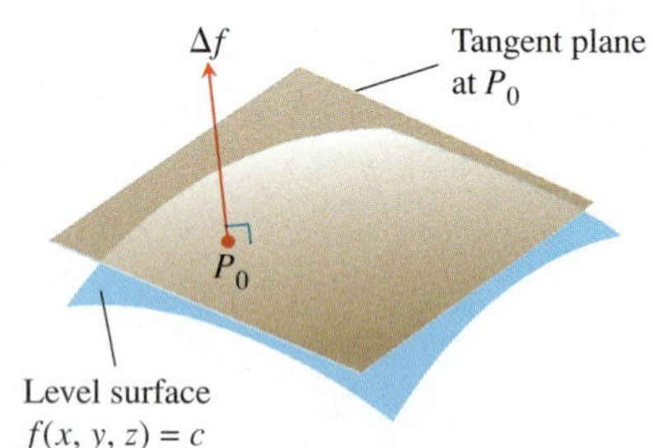

FIGURE 11.32 The gradient of a differentiable function of three variables at a point P_0 is normal to the function's level surface through that point. Thus, the gradient defines the normal to the tangent plane at P_0.

Tangent Planes and Normal Lines

The gradient vector for a differentiable function of three variables $f(x, y, z)$ satisfies all the properties for the gradients of two variables. In particular, just as we established the validity of Equation (5), at every point P_0 in the domain of $f(x, y, z)$, the gradient ∇f is normal to the level surface through P_0 (Figure 11.32). This observation leads to the following definitions.

> **Definitions** Tangent Plane and Normal Line
>
> The **tangent plane** at the point $P_0(x_0, y_0, z_0)$ on the level surface $f(x, y, z) = c$ is the plane through P_0 normal to $\nabla f|_{P_0}$.
>
> The **normal line** of the surface at P_0 is the line through P_0 parallel to $\nabla f|_{P_0}$.

Thus, from Section 10.3, the tangent plane and normal line, respectively, have the following equations:

$$f_x(P_0)(x - x_0) + f_y(P_0)(y - y_0) + f_z(P_0)(z - z_0) = 0 \tag{7}$$
$$x = x_0 + f_x(P_0)\,t, \qquad y = y_0 + f_y(P_0)\,t, \qquad z = z_0 + f_z(P_0)\,t. \tag{8}$$

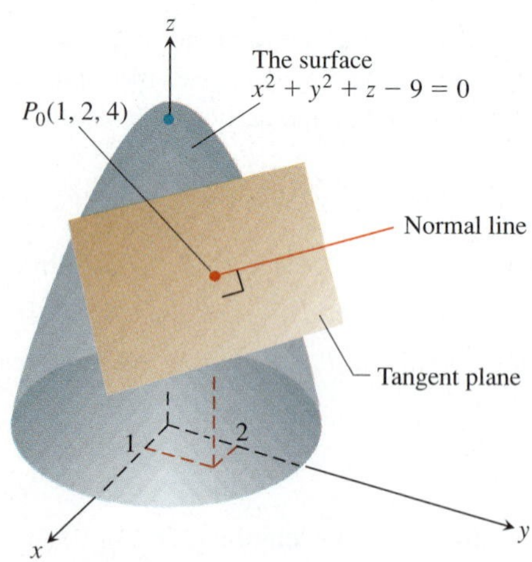

FIGURE 11.33 The tangent plane and normal line to the surface $x^2 + y^2 + z - 9 = 0$ at $P_0(1, 2, 4)$. (Example 8)

Example 8 Finding the Tangent Plane and Normal Line

Find the tangent plane and normal line of the surface

$$f(x, y, z) = x^2 + y^2 + z - 9 = 0 \qquad \text{A circular paraboloid}$$

at the point $P_0(1, 2, 4)$.

Solution The surface is shown in Figure 11.33.

The tangent plane is the plane through P_0 perpendicular to the gradient of f at P_0. The gradient is

$$\nabla f|_{P_0} = (2x\mathbf{i} + 2y\mathbf{j} + \mathbf{k})_{(1,2,4)} = 2\mathbf{i} + 4\mathbf{j} + \mathbf{k}.$$

The plane is therefore the plane

$$2(x-1)+4(y-2)+(z-4)=0, \qquad \text{or} \qquad 2x+4y+z=14.$$

The line normal to the surface at P_0 is

$$x=1+2t, \qquad y=2+4t, \qquad z=4+t.$$

Planes Tangent to a Surface z = f(x, y)

To find an equation for the plane tangent to a surface $z=f(x,y)$ at a point $P_0(x_0, y_0, z_0)$ where $z_0=f(x_0,y_0)$, we first observe that the equation $z=f(x,y)$ is equivalent to $f(x,y)-z=0$. The surface $z=f(x,y)$ is therefore the zero level surface of the function $F(x,y,z)=f(x,y)-z$. The partial derivatives of F are

$$F_x = \frac{\partial}{\partial x}(f(x,y)-z) = f_x - 0 = f_x$$

$$F_y = \frac{\partial}{\partial y}(f(x,y)-z) = f_y - 0 = f_y$$

$$F_z = \frac{\partial}{\partial z}(f(x,y)-z) = 0-1=-1.$$

The formula

$$F_x(P_0)(x-x_0)+F_y(P_0)(y-y_0)+F_z(P_0)(z-z_0)=0$$

for the plane tangent to the level surface at P_0 therefore reduces to

Eq. (7) restated for $F(x, y, z)$

$$f_x(x_0,y_0)(x-x_0)+f_y(x_0,y_0)(y-y_0)-(z-z_0)=0.$$

Plane Tangent to a Surface $z=f(x,y)$ at $(x_0, y_0, f(x_0, y_0))$

The plane tangent to the surface $z=f(x,y)$ at the point $P_0(x_0,y_0,z_0)=(x_0,y_0,f(x_0,y_0))$ is

$$f_x(x_0,y_0)(x-x_0)+f_y(x_0,y_0)(y-y_0)-(z-z_0)=0. \tag{9}$$

Example 9 Finding a Plane Tangent to a Surface $z = f(x, y)$

Find the plane tangent to the surface $z=x\cos y - ye^x$ at $(0,0,0)$.

Solution We calculate the partial derivatives of $f(x,y)=x\cos y - ye^x$ and use Equation (9):

$$f_x(0,0) = (\cos y - ye^x)_{(0,0)} = 1-0\cdot 1 = 1$$

$$f_y(0,0) = (-x \sin y - e^x)_{(0,0)} = 0-1=-1.$$

The tangent plane is therefore

$$1\cdot(x-0)-1\cdot(y-0)-(z-0)=0, \qquad \text{Eq. (9)}$$

or

$$x-y-z=0.$$

Other Applications

The formula for estimating the change in $f(x,y,z)$ when we move a small distance ds from a point P_0 in a particular direction $\mathbf{u}$ in space holds as well:

$$df = (\nabla f|_{P_0} \cdot \mathbf{u})\, ds. \tag{10}$$

Example 10 Estimating Change in the Value of $f(x, y, z)$

Estimate how much the value of

$$f(x, y, z) = y \sin x + 2yz$$

will change if the point $P(x, y, z)$ moves 0.1 units from $P_0(0, 1, 0)$ straight toward $P_1(2, 2, -2)$.

Solution We first find the derivative of f at P_0 in the direction of the vector $\overrightarrow{P_0P_1} = 2\mathbf{i} + \mathbf{j} - 2\mathbf{k}$. The direction of this vector is

$$\mathbf{u} = \frac{\overrightarrow{P_0P_1}}{|\overrightarrow{P_0P_1}|} = \frac{\overrightarrow{P_0P_1}}{3} = \frac{2}{3}\mathbf{i} + \frac{1}{3}\mathbf{j} - \frac{2}{3}\mathbf{k}.$$

The gradient of f at P_0 is

$$\nabla f|_{(0,1,0)} = ((y \cos x)\mathbf{i} + (\sin x + 2z)\mathbf{j} + 2y\mathbf{k})_{(0,1,0)} = \mathbf{i} + 2\mathbf{k}.$$

Therefore,

$$\nabla f|_{P_0} \cdot \mathbf{u} = (\mathbf{i} + 2\mathbf{k}) \cdot \left(\frac{2}{3}\mathbf{i} + \frac{1}{3}\mathbf{j} - \frac{2}{3}\mathbf{k}\right) = \frac{2}{3} - \frac{4}{3} = -\frac{2}{3}.$$

The change df in f that results from moving $ds = 0.1$ units away from P_0 in the direction of $\mathbf{u}$ is approximately

$$df = (\nabla f|_{P_0} \cdot \mathbf{u})(ds) = \left(-\frac{2}{3}\right)(0.1) \approx -0.067 \text{ units}.$$

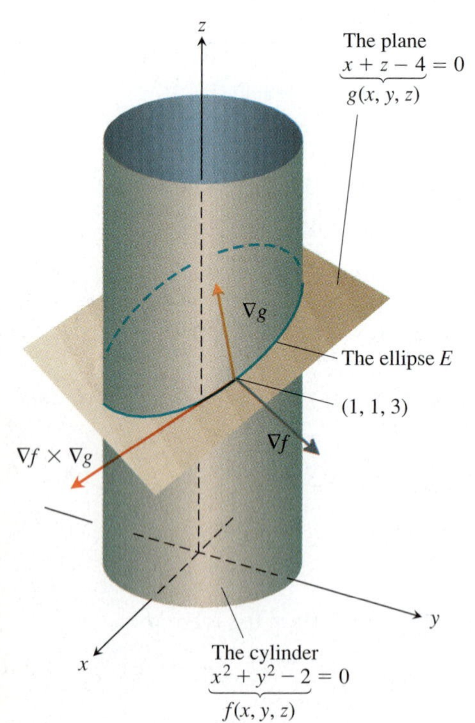

FIGURE 11.34 The cylinder $f(x, y, z) = x^2 + y^2 - 2 = 0$ and the plane $g(x, y, z) = x + z - 4 = 0$ intersect in an ellipse E. (Example 11)

Example 11 Finding Parametric Equations for a Line Tangent to a Space Curve

The surfaces

$$f(x, y, z) = x^2 + y^2 - 2 = 0 \qquad \text{A cylinder}$$

and

$$g(x, y, z) = x + z - 4 = 0 \qquad \text{A plane}$$

meet in an ellipse E (Figure 11.34). Find parametric equations for the line tangent to E at the point $P_0(1, 1, 3)$.

Solution The tangent line is orthogonal to both ∇f and ∇g at P_0 and therefore parallel to $\mathbf{v} = \nabla f \times \nabla g$. The components of $\mathbf{v}$ and the coordinates of P_0 give us equations for the line. We have

$$\nabla f_{(1,1,3)} = (2x\mathbf{i} + 2y\mathbf{j})_{(1,1,3)} = 2\mathbf{i} + 2\mathbf{j}$$
$$\nabla g_{(1,1,3)} = (\mathbf{i} + \mathbf{k})_{(1,1,3)} = \mathbf{i} + \mathbf{k}$$
$$\mathbf{v} = (2\mathbf{i} + 2\mathbf{j}) \times (\mathbf{i} + \mathbf{k}) = \begin{vmatrix} \mathbf{i} & \mathbf{j} & \mathbf{k} \\ 2 & 2 & 0 \\ 1 & 0 & 1 \end{vmatrix} = 2\mathbf{i} - 2\mathbf{j} - 2\mathbf{k}.$$

The line is

$$x = 1 + 2t, \qquad y = 1 - 2t, \qquad z = 3 - 2t.$$

EXERCISES 11.5

Calculating Gradients at Points

In Exercises 1–4, find the gradient of the function at the given point. Then sketch the gradient together with the level curve that passes through the point.

1. $f(x, y) = y - x, \quad (2, 1)$

2. $f(x, y) = \ln(x^2 + y^2), \quad (1, 1)$

3. $g(x, y) = y - x^2, \quad (-1, 0)$

4. $g(x, y) = \dfrac{x^2}{2} - \dfrac{y^2}{2}, \quad (\sqrt{2}, 1)$

In Exercises 5–8, find ∇f at the given point.

5. $f(x, y, z) = x^2 + y^2 - 2z^2 + z \ln x, \quad (1, 1, 1)$

6. $f(x, y, z) = 2z^3 - 3(x^2 + y^2)z + \tan^{-1} xz, \quad (1, 1, 1)$

7. $f(x, y, z) = (x^2 + y^2 + z^2)^{-1/2} + \ln(xyz), \quad (-1, 2, -2)$

8. $f(x, y, z) = e^{x+y} \cos z + (y + 1) \sin^{-1} x, \quad (0, 0, \pi/6)$

Finding Directional Derivatives in the xy-Plane

In Exercises 9–16, find the derivative of the function at P_0 in the direction of $\mathbf{A}$.

9. $f(x, y) = 2xy - 3y^2, \quad P_0(5, 5), \quad \mathbf{A} = 4\mathbf{i} + 3\mathbf{j}$

10. $f(x, y) = 2x^2 + y^2, \quad P_0(-1, 1), \quad \mathbf{A} = 3\mathbf{i} - 4\mathbf{j}$

11. $g(x, y) = x - (y^2/x) + \sqrt{3} \sec^{-1}(2xy), \quad P_0(1, 1),$
$\mathbf{A} = 12\mathbf{i} + 5\mathbf{j}$

12. $h(x, y) = \tan^{-1}(y/x) + \sqrt{3} \sin^{-1}(xy/2), \quad P_0(1, 1),$
$\mathbf{A} = 3\mathbf{i} - 2\mathbf{j}$

13. $f(x, y, z) = xy + yz + zx, \quad P_0(1, -1, 2), \quad \mathbf{A} = 3\mathbf{i} + 6\mathbf{j} - 2\mathbf{k}$

14. $f(x, y, z) = x^2 + 2y^2 - 3z^2, \quad P_0(1, 1, 1), \quad \mathbf{A} = \mathbf{i} + \mathbf{j} + \mathbf{k}$

15. $g(x, y, z) = 3e^x \cos yz, \quad P_0(0, 0, 0), \quad \mathbf{A} = 2\mathbf{i} + \mathbf{j} - 2\mathbf{k}$

16. $h(x, y, z) = \cos xy + e^{yz} + \ln zx, \quad P_0(1, 0, 1/2),$
$\mathbf{A} = \mathbf{i} + 2\mathbf{j} + 2\mathbf{k}$

Directions of Most Rapid Increase and Decrease

In Exercises 17–22, find the directions in which the functions increase and decrease most rapidly at P_0. Then find the derivatives of the functions in these directions.

17. $f(x, y) = x^2 + xy + y^2, \quad P_0(-1, 1)$

18. $f(x, y) = x^2y + e^{xy} \sin y, \quad P_0(1, 0)$

19. $f(x, y, z) = (x/y) - yz, \quad P_0(4, 1, 1)$

20. $g(x, y, z) = xe^y + z^2, \quad P_0(1, \ln 2, 1/2)$

21. $f(x, y, z) = \ln xy + \ln yz + \ln xz, \quad P_0(1, 1, 1)$

22. $h(x, y, z) = \ln(x^2 + y^2 - 1) + y + 6z, \quad P_0(1, 1, 0)$

Estimating Change

23. By about how much will

$$f(x, y, z) = \ln \sqrt{x^2 + y^2 + z^2}$$

change if the point $P(x, y, z)$ moves from $P_0(3, 4, 12)$ a distance of $ds = 0.1$ units in the direction of $3\mathbf{i} + 6\mathbf{j} - 2\mathbf{k}$?

24. By about how much will

$$f(x, y, z) = e^x \cos yz$$

change as the point $P(x, y, z)$ moves from the origin a distance of $ds = 0.1$ units in the direction of $2\mathbf{i} + 2\mathbf{j} - 2\mathbf{k}$?

25. By about how much will

$$g(x, y, z) = x + x \cos z - y \sin z + y$$

change if the point $P(x, y, z)$ moves from $P_0(2, -1, 0)$ a distance of $ds = 0.2$ units toward the point $P_1(0, 1, 2)$?

26. By about how much will

$$h(x, y, z) = \cos(\pi xy) + xz^2$$

change if the point $P(x, y, z)$ moves from $P_0(-1, -1, -1)$ a distance of $ds = 0.1$ units toward the origin?

Tangent Planes and Normal Lines to Surfaces

In Exercises 27–34, find equations for the

(a) Tangent plane and

(b) Normal line at the point P_0 on the given surface.

27. $x^2 + y^2 + z^2 = 3, \quad P_0(1, 1, 1)$

28. $x^2 + y^2 - z^2 = 18, \quad P_0(3, 5, -4)$

29. $2z - x^2 = 0, \quad P_0(2, 0, 2)$

30. $x^2 + 2xy - y^2 + z^2 = 7, \quad P_0(1, -1, 3)$

31. $\cos \pi x - x^2y + e^{xz} + yz = 4, \quad P_0(0, 1, 2)$

32. $x^2 - xy - y^2 - z = 0, \quad P_0(1, 1, -1)$

33. $x + y + z = 1, \quad P_0(0, 1, 0)$

34. $x^2 + y^2 - 2xy - x + 3y - z = -4, \quad P_0(2, -3, 18)$

In Exercises 35–38, find an equation for the plane that is tangent to the given surface at the given point.

35. $z = \ln(x^2 + y^2), \quad (1, 0, 0)$

36. $z = e^{-(x^2+y^2)}, \quad (0, 0, 1)$

37. $z = \sqrt{y - x}, \quad (1, 2, 1)$

38. $z = 4x^2 + y^2, \quad (1, 1, 5)$

Tangent Lines to Curves

In Exercises 39–42, sketch the curve $f(x, y) = c$ together with ∇f and the tangent line at the given point. Then write an equation for the tangent line.

39. $x^2 + y^2 = 4, \quad (\sqrt{2}, \sqrt{2})$

40. $x^2 - y = 1, \quad (\sqrt{2}, 1)$

41. $xy = -4, \quad (2, -2)$

42. $x^2 - xy + y^2 = 7, \quad (-1, 2)$

In Exercises 43–48, find parametric equations for the line tangent to the curve of intersection of the surfaces at the given point.

43. Surfaces: $x + y^2 + 2z = 4, \quad x = 1$

Point: $(1, 1, 1)$

44. Surfaces: $xyz = 1, \quad x^2 + 2y^2 + 3z^2 = 6$

Point: $(1, 1, 1)$

45. Surfaces: $x^2 + 2y + 2z = 4, \quad y = 1$

Point: $(1, 1, 1/2)$

46. Surfaces: $x + y^2 + z = 2, \quad y = 1$

Point: $(1/2, 1, 1/2)$

47. Surfaces: $x^3 + 3x^2y^2 + y^3 + 4xy - z^2 = 0, \quad x^2 + y^2 + z^2 = 11$

Point: $(1, 1, 3)$

48. Surfaces: $x^2 + y^2 = 4, \quad x^2 + y^2 - z = 0$

Point: $(\sqrt{2}, \sqrt{2}, 4)$

Theory and Examples

49. *Zero directional derivative* In what direction is the derivative of $f(x, y) = xy + y^2$ at $P(3, 2)$ equal to zero?

50. *Zero directional derivative* In what directions is the derivative of $f(x, y) = (x^2 - y^2)/(x^2 + y^2)$ at $P(1, 1)$ equal to zero?

51. *Writing to Learn* Is there a direction $\mathbf{u}$ in which the rate of change of $f(x, y) = x^2 - 3xy + 4y^2$ at $P(1, 2)$ equals 14? Give reasons for your answer.

52. *Changing temperature along a circle* Is there a direction $\mathbf{u}$ in which the rate of change of the temperature function $T(x, y, z) = 2xy - yz$ (temperature in degrees Celsius, distance in feet) at $P(1, -1, 1)$ is $-3°\mathrm{C/ft}$? Give reasons for your answer.

53. *Writing to Learn* The derivative of $f(x, y)$ at $P_0(1, 2)$ in the direction of $\mathbf{i} + \mathbf{j}$ is $2\sqrt{2}$ and in the direction of $-2\mathbf{j}$ is -3. What is the derivative of f in the direction of $-\mathbf{i} - 2\mathbf{j}$? Give reasons for your answer.

54. *Writing to Learn* The derivative of $f(x, y, z)$ at a point P is greatest in the direction of $\mathbf{v} = \mathbf{i} + \mathbf{j} - \mathbf{k}$. In this direction, the value of the derivative is $2\sqrt{3}$.

(a) What is ∇f at P? Give reasons for your answer.

(b) What is the derivative of f at P in the direction of $\mathbf{i} + \mathbf{j}$?

55. *Temperature change along a circle* Suppose that the Celsius temperature at the point (x, y) in the xy-plane is $T(x, y) = x \sin 2y$ and that distance in the xy-plane is measured in meters. A particle is moving *clockwise* around the circle of radius 1 m centered at the origin at the constant rate of 2 m/sec.

(a) How fast is the temperature experienced by the particle changing in degrees Celsius per meter at the point $P(1/2, \sqrt{3}/2)$?

(b) How fast is the temperature experienced by the particle changing in degrees Celsius per second at P?

56. *Change along the involute of a circle* Find the derivative of $f(x, y) = x^2 + y^2$ in the direction of the unit tangent vector of the curve

$$\mathbf{r}(t) = (\cos t + t \sin t)\mathbf{i} + (\sin t - t \cos t)\mathbf{j}, \qquad t > 0$$

(Figure 11.35).

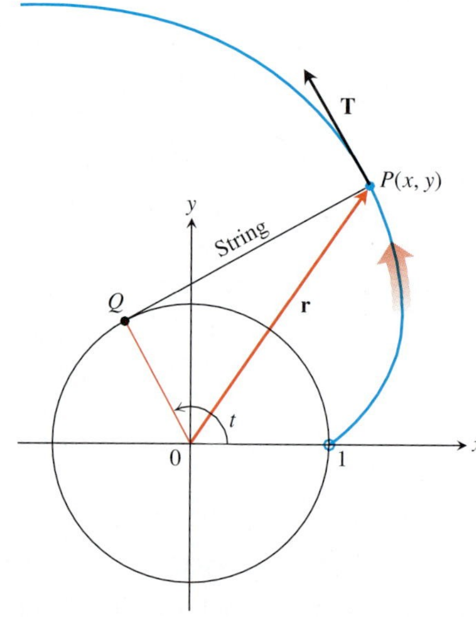

FIGURE 11.35 *The involute of the unit circle.* If you move out along the involute, covering distance along the curve at a constant rate, your distance from the origin will increase at a constant rate as well. (This is how to interpret the result of your calculation in Exercise 56.)

57. *Change along a helix* Find the derivative of $f(x, y, z) = x^2 + y^2 + z^2$ in the direction of the unit tangent vector of the helix

$$\mathbf{r}(t) = (\cos t)\mathbf{i} + (\sin t)\mathbf{j} + t\mathbf{k}$$

at the points where $t = -\pi/4$, 0, and $\pi/4$. The function f gives the square of the distance from a point $P(x, y, z)$ on the helix to the origin. The derivatives calculated here give the rates at which the square of the distance is changing with respect to t as P moves through the points where $t = -\pi/4$, 0, and $\pi/4$.

58. *Changing temperature along a space curve* The Celsius temperature in a region in space is given by $T(x, y, z) = 2x^2 - xyz$. A particle is moving in this region and its position at time t is given by $x = 2t^2$, $y = 3t$, $z = -t^2$, where time is measured in seconds and distance in meters.

(a) How fast is the temperature experienced by the particle changing in degrees Celsius per meter when the particle is at the point $P(8, 6, -4)$?

(b) How fast is the temperature experienced by the particle changing in degrees Celsius per second at P?

59. *Lines in the xy-plane* Show that $A(x - x_0) + B(y - y_0) = 0$ is an equation for the line in the xy-plane through the point (x_0, y_0) normal to the vector $\mathbf{N} = A\mathbf{i} + B\mathbf{j}$.

60. *Normal curves and tangent curves* A smooth curve is *normal* to a surface $f(x, y, z) = c$ at a point of intersection if the curve's velocity vector is a nonzero scalar multiple of ∇f at the point. The curve is *tangent* to the surface at a point of intersection if its velocity vector is orthogonal to ∇f there.

(a) Show that the curve

$$\mathbf{r}(t) = \sqrt{t}\mathbf{i} + \sqrt{t}\mathbf{j} - \frac{1}{4}(t + 3)\mathbf{k}$$

is normal to the surface $x^2 + y^2 - z = 3$ when $t = 1$.

(b) Show that the curve

$$\mathbf{r}(t) = \sqrt{t}\mathbf{i} + \sqrt{t}\mathbf{j} + (2t - 1)\mathbf{k}$$

is tangent to the surface $x^2 + y^2 - z = 1$ when $t = 1$.

61. *Writing to Learn: Directional derivatives and scalar components* How is the derivative of a differentiable function $f(x, y, z)$ at a point P_0 in the direction of a unit vector $\mathbf{u}$ related to the scalar component of $(\nabla f)_{P_0}$ in the direction of $\mathbf{u}$? Give reasons for your answer.

62. *Writing to Learn: Directional derivatives and partial derivatives* Assuming that the necessary derivatives of $f(x, y, z)$ are defined, how are $D_{\mathbf{i}}f$, $D_{\mathbf{j}}f$, and $D_{\mathbf{k}}f$ related to f_x, f_y, and f_z? Give reasons for your answer.

63. *The algebra rules for gradients* Given a constant k and the gradients

$$\nabla f = \frac{\partial f}{\partial x}\mathbf{i} + \frac{\partial f}{\partial y}\mathbf{j} + \frac{\partial f}{\partial z}\mathbf{k}$$

and

$$\nabla g = \frac{\partial g}{\partial x}\mathbf{i} + \frac{\partial g}{\partial y}\mathbf{j} + \frac{\partial g}{\partial z}\mathbf{k},$$

use the scalar equations

$$\frac{\partial}{\partial x}(kf) = k\frac{\partial f}{\partial x}, \qquad \frac{\partial}{\partial x}(f \pm g) = \frac{\partial f}{\partial x} \pm \frac{\partial g}{\partial x},$$

$$\frac{\partial}{\partial x}(fg) = f\frac{\partial g}{\partial x} + g\frac{\partial f}{\partial x}, \qquad \frac{\partial}{\partial x}\left(\frac{f}{g}\right) = \frac{g\frac{\partial f}{\partial x} - f\frac{\partial g}{\partial x}}{g^2},$$

and so on, to establish the following rules.

(a) $\nabla(kf) = k\nabla f$

(b) $\nabla(f + g) = \nabla f + \nabla g$

(c) $\nabla(f - g) = \nabla f - \nabla g$

(d) $\nabla(fg) = f\nabla g + g\nabla f$

(e) $\nabla\left(\frac{f}{g}\right) = \frac{g\nabla f - f\nabla g}{g^2}$

11.6 Linearization and Differentials

Linearization of a Function of Two Variables • Accuracy of the Standard Linear Approximation • Predicting Change with Differentials • Absolute, Relative, and Percentage Change • Functions of More Than Two Variables

In this section, we generalize the concepts of linearization and differentials to functions of two or more variables. We do this in a way similar to the way we find linear approximations for functions of a single variable (Section 3.6). The differential helps us determine the sensitivity of a multivariable function to changes in each of its independent variables. The mathematical results of the section stem from the Increment Theorem (Theorem 3, Section 11.3).

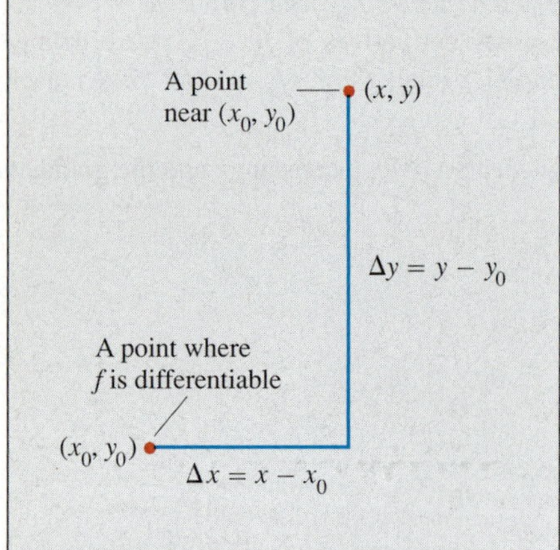

FIGURE 11.36 If f is differentiable at (x_0, y_0), then the value of f at any point (x, y) nearby is approximately $f(x_0, y_0) + f_x(x_0, y_0)\,\Delta x + f_y(x_0, y_0)\,\Delta y$.

Linearization of a Function of Two Variables

Suppose we wish to replace the function $z = f(x, y)$ with one that is simpler to work with. We want the replacement to be effective near a point (x_0, y_0) at which we know the values of f, f_x, and f_y and at which f is differentiable. Since f is differentiable, we know from Theorem 3 that the equation

$$\Delta z = f_x(x_0, y_0)\,\Delta x + f_y(x_0, y_0)\,\Delta y + \epsilon_1\,\Delta x + \epsilon_2\,\Delta y. \tag{1}$$

holds for f at (x_0, y_0). Therefore, if we move from (x_0, y_0) to any point (x, y) by increments $\Delta x = x - x_0$ and $\Delta y = y - y_0$ (Figure 11.36), the new value of f will be

$$\begin{aligned} f(x, y) = f(x_0, y_0) &+ f_x(x_0, y_0)(x - x_0) \\ &+ f_y(x_0, y_0)(y - y_0) + \epsilon_1\,\Delta x + \epsilon_2\,\Delta y, \end{aligned}$$

Eq. (1), with $\Delta x = x - x_0$, $\Delta y = y - y_0$, and $\Delta z = f(x, y) - f(x_0, y_0)$

where $\epsilon_1, \epsilon_2 \to 0$ as $\Delta x, \Delta y \to 0$. If the increments Δx and Δy are small, the products $\epsilon_1\,\Delta x$ and $\epsilon_2\,\Delta y$ will eventually be smaller still and we will have

$$f(x, y) \approx \underbrace{f(x_0, y_0) + f_x(x_0, y_0)(x - x_0) + f_y(x_0, y_0)(y - y_0)}_{L(x,\,y)}.$$

In other words, as long as Δx and Δy are small, f will have approximately the same value as the linear function L. If f is hard to use and our work can tolerate the error involved, we may safely replace f by L.

Definitions Linearization, Standard Linear Approximation

The **linearization** of a function $f(x, y)$ at a point (x_0, y_0) where f is differentiable is the function

$$L(x, y) = f(x_0, y_0) + f_x(x_0, y_0)(x - x_0) + f_y(x_0, y_0)(y - y_0).$$

The approximation

$$f(x, y) \approx L(x, y)$$

is the **standard linear approximation** of f at (x_0, y_0).

From Section 11.5, Equation (9), we see that the plane $z = L(x, y)$ is tangent to the surface $z = f(x, y)$ at the point (x_0, y_0). Thus, the linearization of a function of two variables is a tangent-*plane* approximation in the same way that the linearization of a function of a single variable is a tangent-*line* approximation.

Example 1 Finding a Linearization

Find the linearization of

$$f(x, y) = x^2 - xy + \frac{1}{2}y^2 + 3$$

at the point $(3, 2)$.

Solution We first evaluate f, f_x, and f_y at the point $(x_0, y_0) = (3, 2)$:

$$f(3, 2) = \left(x^2 - xy + \frac{1}{2}y^2 + 3\right)_{(3,2)} = 8$$

$$f_x(3, 2) = \frac{\partial}{\partial x}\left(x^2 - xy + \frac{1}{2}y^2 + 3\right)_{(3,2)} = (2x - y)_{(3,2)} = 4$$

$$f_y(3, 2) = \frac{\partial}{\partial y}\left(x^2 - xy + \frac{1}{2}y^2 + 3\right)_{(3,2)} = (-x + y)_{(3,2)} = -1,$$

giving

$$\begin{aligned} L(x, y) &= f(x_0, y_0) + f_x(x_0, y_0)(x - x_0) + f_y(x_0, y_0)(y - y_0) \\ &= 8 + (4)(x - 3) + (-1)(y - 2) = 4x - y - 2. \end{aligned}$$

The linearization of f at $(3, 2)$ is $L(x, y) = 4x - y - 2$.

Accuracy of the Standard Linear Approximation

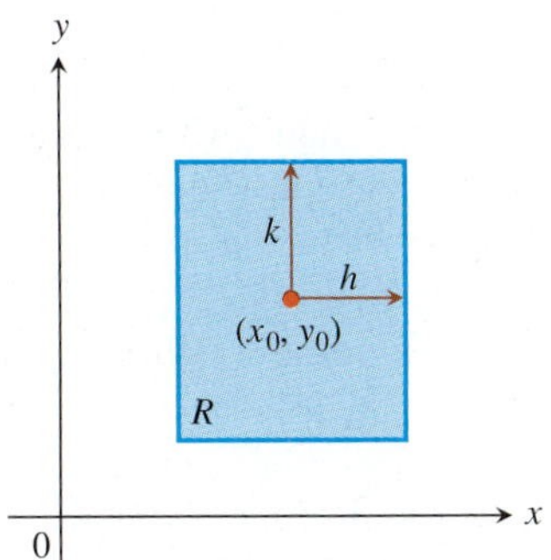

FIGURE 11.37 The rectangular region $R: |x - x_0| \le h, |y - y_0| \le k$ in the xy-plane. On this kind of region, we can find useful error bounds for our approximations.

Suppose that $L(x, y)$ is the linearization of a differentiable function $f(x, y)$ at (x_0, y_0) and we use L to approximate f at points (x, y) close to (x_0, y_0). How accurate can we expect the approximation to be? As you might expect, the closeness of the approximation depends on three things:

1. The closeness of x to x_0
2. The closeness of y to y_0
3. The "curviness" of f near (x_0, y_0), as measured by the magnitude of the second partial derivatives.

In fact, if we can find a common upper bound M for $|f_{xx}|$, $|f_{yy}|$, and $|f_{xy}|$ on a rectangle R centered at (x_0, y_0) (Figure 11.37), then we can bound the error throughout R by using a simple formula (derived in Section 11.10)

The Error in the Standard Linear Approximation

If f has continuous first and second partial derivatives throughout an open set containing a rectangle R centered at (x_0, y_0) and if M is any upper bound for the values of $|f_{xx}|$, $|f_{yy}|$, and $|f_{xy}|$ on R, then the error $E(x, y)$ incurred in replacing $f(x, y)$ on R by its linearization

$$L(x, y) = f(x_0, y_0) + f_x(x_0, y_0)(x - x_0) + f_y(x_0, y_0)(y - y_0)$$

satisfies the inequality

$$|E(x, y)| \le \frac{1}{2}M(|x - x_0| + |y - y_0|)^2.$$

When we need to make $|E(x, y)|$ small for a given M, we just make $|x - x_0|$ and $|y - y_0|$ small.

Example 2 Bounding the Error in Example 1

In Example 1, we found the linearization of

$$f(x, y) = x^2 - xy + \frac{1}{2}y^2 + 3$$

at $(3, 2)$ to be

$$L(x, y) = 4x - y - 2.$$

Find an upper bound for the error in the approximation $f(x, y) \approx L(x, y)$ over the rectangle

$$R: \quad |x - 3| \leq 0.1, \qquad |y - 2| \leq 0.1.$$

Express the upper bound as a percentage of $f(3, 2)$, the value of f at the center of the rectangle.

Solution We use the inequality

$$|E(x, y)| \leq \frac{1}{2}M(|x - x_0| + |y - y_0|)^2.$$

To find a suitable value for M, we calculate f_{xx}, f_{xy}, and f_{yy}, finding, after a routine differentiation, that all three derivatives are constant, with values

$$|f_{xx}| = |2| = 2, \qquad |f_{xy}| = |-1| = 1, \qquad |f_{yy}| = |1| = 1.$$

The largest of these is 2, so we may safely take M to be 2. With $(x_0, y_0) = (3, 2)$, we then know that, throughout R,

$$|E(x, y)| \leq \frac{1}{2}(2)(|x - 3| + |y - 2|)^2 = (|x - 3| + |y - 2|)^2.$$

Finally, since $|x - 3| \leq 0.1$ and $|y - 2| \leq 0.1$ on R, we have

$$|E(x, y)| \leq (0.1 + 0.1)^2 = 0.04.$$

As a percentage of $f(3, 2) = 8$, the error is no greater than

$$\frac{0.04}{8} \times 100 = 0.5\%.$$

Interpret As long as (x, y) stays in R, the approximation $f(x, y) \approx L(x, y)$ will be in error by no more than 0.04, which is $1/2\%$ of the value of f at the center of R.

Predicting Change with Differentials

Suppose that we know the values of a differentiable function $f(x, y)$ and its first partial derivatives at a point (x_0, y_0) and we want to predict how much the value of f will change if we move to a point $(x_0 + \Delta x, y_0 + \Delta y)$ nearby. If Δx and Δy are small, f and its linearization at (x_0, y_0) will change by nearly the same amount, so the change in L will give a practical estimate of the change in f.

The change in f is

$$\Delta f = f(x_0 + \Delta x, y_0 + \Delta y) - f(x_0, y_0).$$

A straightforward calculation from the definition of $L(x, y)$, using the notation $x - x_0 = \Delta x$ and $y - y_0 = \Delta y$, shows that the corresponding change in L is

$$\begin{aligned}\Delta L &= L(x_0 + \Delta x, y_0 + \Delta y) - L(x_0, y_0)\\ &= f_x(x_0, y_0)\,\Delta x + f_y(x_0, y_0)\,\Delta y.\end{aligned}$$

The formula for Δf is usually as hard to work with as the formula for f. The change in L, however, is just a known constant times Δx plus a known constant times Δy.

The change ΔL is usually described in the more suggestive notation

$$df = f_x(x_0, y_0)\,dx + f_y(x_0, y_0)\,dy,$$

in which df denotes the change in the linearization that results from the changes dx and dy in x and y. As usual, we call dx and dy differentials of x and y, and call df the corresponding *total* differential of f.

Definition Total Differential

If we move from (x_0, y_0) to a point $(x_0 + dx, y_0 + dy)$ nearby, the resulting change

$$df = f_x(x_0, y_0)\,dx + f_y(x_0, y_0)\,dy$$

in the linearization of f is called the **total differential of f.**

Absolute Change versus Relative Change

If you measure a 20-volt potential with an error of 10 volts, your reading is probably too crude to be useful. You are off by 50%. If you measure a 200,000-volt potential with an error of 10 volts, however, your reading is within 0.005% of the true value. An absolute error of 10 volts is significant in the first case but of no consequence in the second because the relative error is so small.

In other cases, a small relative error—say, traveling a few meters too far in a journey of hundreds of thousands of meters—can have spectacular consequences.

Example 3 Sensitivity to Change

Your company manufactures right circular cylindrical molasses storage tanks that are 25 ft high with a radius of 5 ft. How sensitive are the tanks' volumes to small variations in height and radius?

Solution As a function of radius r and height h, the typical tank's volume is

$$V = \pi r^2 h.$$

The change in volume caused by small changes dr and dh in radius and height is approximately

$$\begin{aligned}dV &= V_r(5, 25)\,dr + V_h(5, 25)\,dh\\ &= (2\pi rh)_{(5,25)}\,dr + (\pi r^2)_{(5,25)}\,dh\\ &= 250\pi\,dr + 25\pi\,dh.\end{aligned}$$

Total differential with $f = V$ and $(x_0, y_0) = (5, 25)$

Thus, a 1-unit change in r will change V by about 250π units. A 1-unit change in h will change V by about 25π units. The tank's volume is 10 times more sensitive to a small change in r than it is to a small change of equal size in h. As a quality control engineer concerned with being sure the tanks have the correct volume, you would want to pay special attention to their radii.

In contrast, if the values of r and h are reversed to make $r = 25$ and $h = 5$, then the total differential in V becomes

$$dV = (2\pi rh)_{(25,5)}\,dr + (\pi r^2)_{(25,5)}\,dh = 250\pi\,dr + 625\pi\,dh.$$

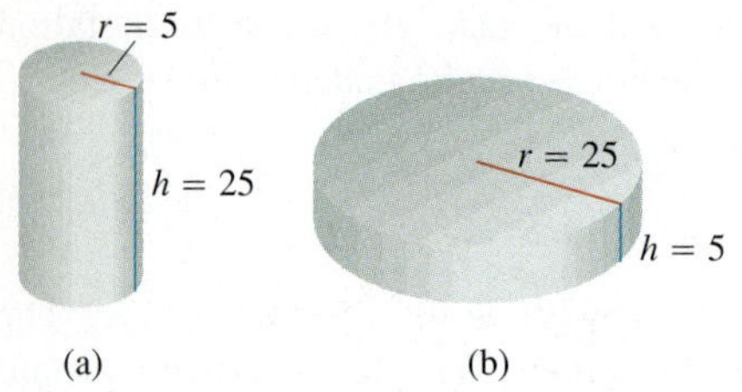

FIGURE 11.38 The volume of cylinder (a) is more sensitive to a small change in r than it is to an equally small change in h. The volume of cylinder (b) is more sensitive to small changes in h than it is to small changes in r.

Now the volume is more sensitive to changes in h than to changes in r (Figure 11.38).

The general rule to be learned from this example is that functions are most sensitive to small changes in the variables that generate the largest partial derivatives.

Absolute, Relative, and Percentage Change

When we move from (x_0, y_0) to a point nearby, we can describe the corresponding change in the value of a function $f(x, y)$ in three different ways.

	True	Estimate
Absolute change:	Δf	df
Relative change:	$\dfrac{\Delta f}{f(x_0, y_0)}$	$\dfrac{df}{f(x_0, y_0)}$
Percentage change:	$\dfrac{\Delta f}{f(x_0, y_0)} \times 100$	$\dfrac{df}{f(x_0, y_0)} \times 100$

Example 4 Estimating Change in Volume

Suppose that a cylindrical can is designed to have a radius of 1 in. and a height of 5 in., but that the radius and height are off by the amounts $dr = +0.03$ and $dh = -0.1$. Estimate the resulting absolute, relative, and percentage changes in the volume of the can.

Solution To estimate the absolute change in V, we evaluate

$$dV = V_r(r_0, h_0)\, dr + V_h(r_0, h_0)\, dh$$

to get

$$\begin{aligned} dV &= 2\pi r_0 h_0\, dr + \pi r_0^{\,2}\, dh = 2\pi(1)(5)(0.0) + \pi(1)^2(-0.1) \\ &= 0.3\pi - 0.1\pi = 0.2\pi \approx 0.63 \text{ in.}^3 \end{aligned}$$

We divide this by $V(r_0, h_0)$ to estimate the relative change:

$$\frac{dV}{V(r_0, h_0)} = \frac{0.2\pi}{\pi r_0^{\,2} h_0} = \frac{0.2\pi}{\pi(1)^2(5)} = 0.04.$$

We multiply this by 100 to estimate the percentage change:

$$\frac{dV}{V(r_0, h_0)} \times 100 = 0.04 \times 100 = 4\%.$$

Example 5 Predicting Measurement Error

The volume $V = \pi r^2 h$ of a right circular cylinder is to be calculated from measured values of r and h. Suppose that r is measured with an error of no more than 2% and h with an error of no more than 0.5%. Estimate the resulting possible percentage error in the calculation of V.

Solution We are told that

$$\left|\frac{dr}{r} \times 100\right| \leq 2 \qquad \text{and} \qquad \left|\frac{dh}{h} \times 100\right| \leq 0.5.$$

Since

$$\frac{dV}{V} = \frac{2\pi rh\,dr + \pi r^2\,dh}{\pi r^2 h} = \frac{2\,dr}{r} + \frac{dh}{h},$$

we have

$$\begin{aligned}\left|\frac{dV}{V}\right| &= \left|2\frac{dr}{r} + \frac{dh}{h}\right| \\ &\leq \left|2\frac{dr}{r}\right| + \left|\frac{dh}{h}\right| \\ &\leq 2(0.02) + 0.005 = 0.045.\end{aligned}$$

We estimate the error in the volume calculation to be at most 4.5%.

How accurately do we have to measure r and h to have a reasonable chance of calculating $V = \pi r^2 h$ with an error, say, of less than 2%? Questions like this are hard to answer because there is usually no single right answer. Since

$$\frac{dV}{V} = 2\frac{dr}{r} + \frac{dh}{h},$$

we see that dV/V is controlled by a combination of dr/r and dh/h. If we can measure h with great accuracy, we might come out all right even if we are sloppy about measuring r. On the other hand, our measurement of h might have so large a dh that the resulting dV/V would be too crude an estimate of $\Delta V/V$ to be useful even if dr were zero.

What we do in such cases is look for a reasonable square about the measured values (r_0, h_0) in which V will not vary by more than the allowed amount from $V_0 = \pi r_0^2 h_0$.

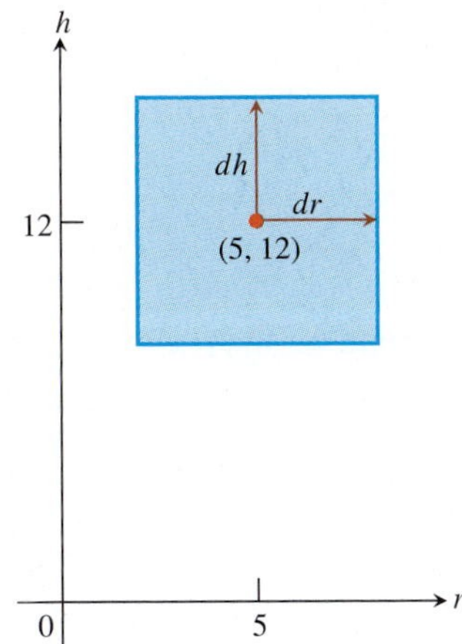

FIGURE 11.39 A small square about the point (5, 12) in the rh-plane. (Example 6)

Example 6 Controlling the Error

Find a reasonable square about the point $(r_0, h_0) = (5, 12)$ in which the value of $V = \pi r^2 h$ will not vary by more than ± 0.1.

Solution We approximate the variation ΔV by the differential

$$dV = 2\pi r_0 h_0\,dr + \pi r_0^2\,dh = 2\pi(5)(12)\,dr + \pi(5)^2\,dh = 120\pi\,dr + 25\pi\,dh.$$

Since the region to which we are restricting our attention is a square (Figure 11.39), we may set $dh = dr$ to get

$$dV = 120\pi\,dr + 25\pi\,dr = 145\pi\,dr.$$

We then ask, How small must we take dr to be sure that $|dV|$ is no larger than 0.1? To answer, we start with the inequality

$$|dV| \leq 0.1,$$

express dV in terms of dr,

$$|145\pi\, dr| \le 0.1,$$

and find a corresponding upper bound for dr:

$$|dr| \le \frac{0.1}{145\pi} \approx 2.1 \times 10^{-4}.$$

Rounding down to make sure dr won't accidentally be too big

With $dh = dr$, then, the square we want is described by the inequalities

$$|r - 5| \le 2.1 \times 10^{-4}, \qquad |h - 12| \le 2.1 \times 10^{-4}.$$

As long as (r, h) stays in this square, we may expect $|dV|$ to be less than or equal to 0.1 and we may expect $|\Delta V|$ to be approximately the same size.

Functions of More Than Two Variables

Analogous results hold for differentiable functions of more than two variables.

1. The **linearization** of $f(x, y, z)$ at a point $P_0(x_0, y_0, z_0)$ is

   $$L(x, y, z) = f(P_0) + f_x(P_0)(x - x_0) + f_y(P_0)(y - y_0) + f_z(P_0)(z - z_0).$$

2. Suppose that R is a closed rectangular solid centered at P_0 and lying in an open region on which the second partial derivatives of f are continuous. Suppose also that $|f_{xx}|$, $|f_{yy}|$, $|f_{zz}|$, $|f_{xy}|$, $|f_{xz}|$, and $|f_{yz}|$ are all less than or equal to M throughout R. Then the **error** $E(x, y, z) = f(x, y, z) - L(x, y, z)$ in the approximation of f by L is bounded throughout R by the inequality

   $$|E| \le \frac{1}{2} M(|x - x_0| + |y - y_0| + |z - z_0|)^2.$$

3. If the second partial derivatives of f are continuous and if x, y, and z change from x_0, y_0, and z_0 by small amounts dx, dy, and dz, the **total differential**

   $$df = f_x(P_0)\, dx + f_y(P_0)\, dy + f_z(P_0)\, dz$$

   gives a good approximation of the resulting change in f.

Example 7 Finding a Linear Approximation in 3-Space

Find the linearization $L(x, y, z)$ of

$$f(x, y, z) = x^2 - xy + 3 \sin z$$

at the point $(x_0, y_0, z_0) = (2, 1, 0)$. Find an upper bound for the error incurred in replacing f by L on the rectangle

$$R: \quad |x - 2| \le 0.01, \qquad |y - 1| \le 0.02, \qquad |z| \le 0.01.$$

Solution A routine evaluation gives

$$f(2, 1, 0) = 2, \qquad f_x(2, 1, 0) = 3, \qquad f_y(2, 1, 0) = -2, \qquad f_z(2, 1, 0) = 3.$$

Thus,

$$L(x, y, z) = 2 + 3(x - 2) + (-2)(y - 1) + 3(z - 0) = 3x - 2y + 3z - 2.$$

Since

$$f_{xx} = 2, \qquad f_{yy} = 0, \qquad f_{zz} = -3 \sin z,$$
$$f_{xy} = -1, \qquad f_{xz} = 0, \qquad f_{yz} = 0,$$

we may safely take M to be max $|-3 \sin z| = 3$. Hence, the error incurred by replacing f by L on R satisfies

$$|E| \le \frac{1}{2}(3)(0.01 + 0.02 + 0.01)^2 = 0.0024.$$

The error will be no greater than 0.0024.

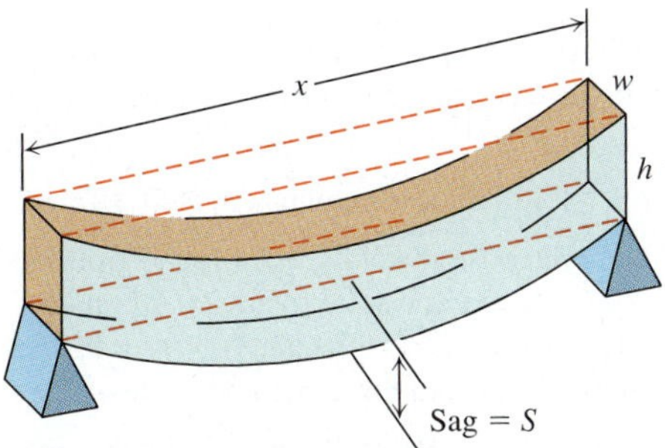

FIGURE 11.40 A beam supported at its two ends before and after loading. Example 8 shows how the sag S is related to the weight of the load and the dimensions of the beam.

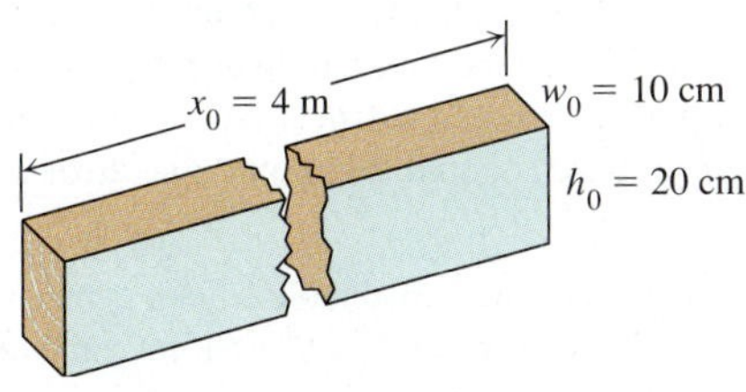

FIGURE 11.41 The dimensions of the beam in Example 8.

Example 8 Finding the Sag in Uniformly Loaded Beams

A horizontal rectangular beam, supported at both ends, will sag when subjected to a uniform load (constant weight per unit length). The amount S of sag (Figure 11.40) is calculated with the formula

$$S = C\frac{px^4}{wh^3}.$$

In this equation,

p = the load (newtons per meter of beam length)
x = the length between supports (meters)
w = the width of the beam (meters)
h = the height of the beam (meters)
C = a constant that depends on the units of measurement and on the material from which the beam is made.

Find dS for a beam 4 m long, 10 cm wide, and 20 cm high that is subjected to a load of 100 N/m (Figure 11.41). What conclusions can be drawn about the beam from the expression for dS?

Solution Since S is a function of the four independent variables p, x, w, and h, its total differential is defined by

$$dS = S_p\,dp + S_x\,dx + S_w\,dw + S_h\,dh.$$

When we write this out for a particular set of values p_0, x_0, w_0, and h_0 and simplify the result, we find that

$$dS = S_0\left(\frac{dp}{p_0} + \frac{4\,dx}{x_0} - \frac{dw}{w_0} - \frac{3\,dh}{h_0}\right),$$

where $S_0 = S(p_0, x_0, w_0, h_0) = Cp_0{x_0}^4/(w_0{h_0}^3)$.

If $p_0 = 100$ N/m, $x_0 = 4$ m, $w_0 = 0.1$ m, and $h_0 = 0.2$ m, then

$$dS = S_0\left(\frac{dp}{100} + dx - 10\,dw - 15\,dh\right).$$

Here is what we can learn from this equation for dS. Since dp and dx appear with positive coefficients, increases in p and x will increase the sag. The coefficients of dw and dh are negative, so increases in w and h will decrease the sag (make the beam stiffer). The sag is not very sensitive to changes in load because the coefficient of dp is $1/100$. The magnitude of the coefficient of dh is greater than the magnitude of the coefficient of dw. Making the beam 1 cm higher will therefore decrease the sag more than making the beam 1 cm wider.

EXERCISES 11.6

Finding Linearizations

In Exercises 1–6, find the linearization $L(x, y)$ of the function at each point.

1. $f(x, y) = x^2 + y^2 + 1$ at (a) $(0, 0)$, (b) $(1, 1)$

2. $f(x, y) = (x + y + 2)^2$ at (a) $(0, 0)$, (b) $(1, 2)$

3. $f(x, y) = 3x - 4y + 5$ at (a) $(0, 0)$, (b) $(1, 1)$

4. $f(x, y) = x^3 y^4$ at (a) $(1, 1)$, (b) $(0, 0)$

5. $f(x, y) = e^x \cos y$ at (a) $(0, 0)$, (b) $(0, \pi/2)$

6. $f(x, y) = e^{2y-x}$ at (a) $(0, 0)$, (b) $(1, 2)$

Upper Bounds for Errors in Linear Approximations

In Exercises 7–12, find the linearization $L(x, y)$ of the function $f(x, y)$ at P_0. Then find an upper bound for the magnitude $|E|$ of the error in the approximation $f(x, y) \approx L(x, y)$ over the rectangle R.

7. $f(x, y) = x^2 - 3xy + 5$ at $P_0(2, 1)$,

R: $|x - 2| \le 0.1$, $|y - 1| \le 0.1$

8. $f(x, y) = (1/2)x^2 + xy + (1/4)y^2 + 3x - 3y + 4$ at $P_0(2, 2)$,

R: $|x - 2| \le 0.1$, $|y - 2| \le 0.1$

9. $f(x, y) = 1 + y + x \cos y$ at $P_0(0, 0)$,

R: $|x| \le 0.2$, $|y| \le 0.2$

(Use $|\cos y| \le 1$ and $|\sin y| \le 1$ in estimating E.)

10. $f(x, y) = xy^2 + y \cos (x - 1)$ at $P_0(1, 2)$,

R: $|x - 1| \le 0.1$, $|y - 2| \le 0.1$

11. $f(x, y) = e^x \cos y$ at $P_0(0, 0)$,

R: $|x| \le 0.1$, $|y| \le 0.1$

(Use $e^x \le 1.11$ and $|\cos y| \le 1$ in estimating E.)

12. $f(x, y) = \ln x + \ln y$ at $P_0(1, 1)$,

R: $|x - 1| \le 0.2$, $|y - 1| \le 0.2$

Sensitivity to Change: Estimates

13. *Writing to Learn* You plan to calculate the area of a long, thin rectangle from measurements of its length and width. Which dimension should you measure more carefully? Give reasons for your answer.

14. *Writing to Learn*

(a) Around the point $(1, 0)$, is $f(x, y) = x^2(y + 1)$ more sensitive to changes in x or to changes in y? Give reasons for your answer.

(b) What ratio of dx to dy will make df equal zero at $(1, 0)$?

15. *Estimating maximum error* Suppose that T is to be found from the formula $T = x(e^y + e^{-y})$, where x and y are found to be 2 and $\ln 2$ with maximum possible errors of $|dx| = 0.1$ and $|dy| = 0.02$. Estimate the maximum possible error in the computed value of T.

16. *Estimating volume of a cylinder* About how accurately may $V = \pi r^2 h$ be calculated from measurements of r and h that are in error by 1%?

17. *Maximum percentage error* If $r = 5.0$ cm and $h = 12.0$ cm to the nearest millimeter, what should we expect the maximum percentage error in calculating $V = \pi r^2 h$ to be?

18. *Estimating volume of a cylinder* To estimate the volume of a cylinder of radius about 2 m and height about 3 m, about how accurately should the radius and height be measured so that the error in the volume estimate will not exceed 0.1 m^3? Assume that the possible error dr in measuring r is equal to the possible error dh in measuring h.

19. *Controlling error within a square* Give a reasonable square centered at $(1, 1)$ over which the value of $f(x, y) = x^3y^4$ will not vary by more than ± 0.1.

20. *Variation in electrical resistance* The resistance R produced by wiring resistors of R_1 and R_2 ohms in parallel (Figure 11.42) can be calculated from the formula

$$\frac{1}{R} = \frac{1}{R_1} + \frac{1}{R_2}.$$

(a) Show that

$$dR = \left(\frac{R}{R_1}\right)^2 dR_1 + \left(\frac{R}{R_2}\right)^2 dR_2.$$

(b) *Writing to Learn* You have designed a two-resistor circuit like the one in Figure 11.42 to have resistances of $R_1 = 100$ ohms and $R_2 = 400$ ohms, but there is always some variation in manufacturing and the resistors received by your firm will probably not have these exact values. Will the value of R be more sensitive to variation in R_1 or to variation in R_2? Give reasons for your answer.

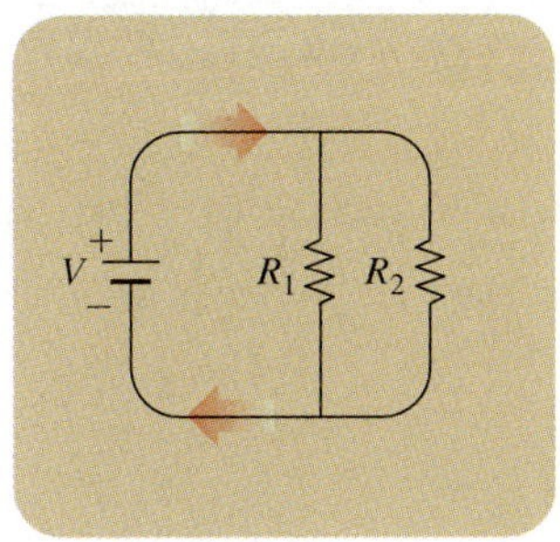

FIGURE 11.42 The circuit in Exercises 20 and 21.

21. (*Continuation of Exercise 20*) In another circuit like the one in Figure 11.42, you plan to change R_1 from 20 to 20.1 ohms and R_2 from 25 to 24.9 ohms. By about what percentage will this change R?

22. *Error carryover in coordinate changes*

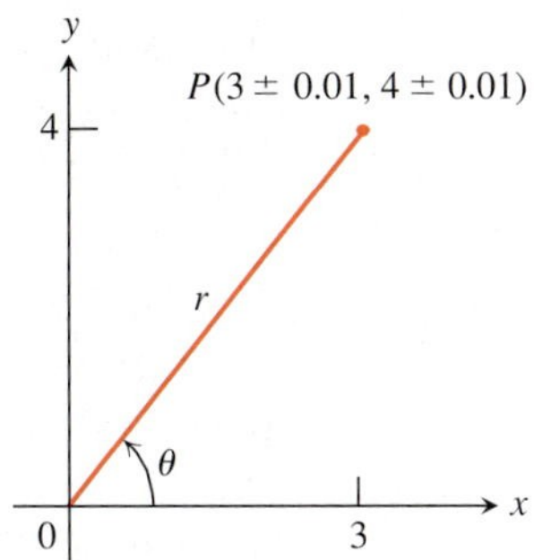

(a) If $x = 3 \pm 0.01$ and $y = 4 \pm 0.01$, as shown here, with approximately what accuracy can you calculate the polar coordinates r and θ of the point $P(x, y)$ from the formulas $r^2 = x^2 + y^2$ and $\theta = \tan^{-1}(y/x)$? Express your estimates as percentage changes of the values that r and θ have at the point $(x_0, y_0) = (3, 4)$.

(b) *Writing to Learn* At the point $(x_0, y_0) = (3, 4)$, are the values of r and θ more sensitive to changes in x or to changes in y? Give reasons for your answer.

Functions of Three Variables

Find the linearizations $L(x, y, z)$ of the functions in Exercises 23–28 at the given points.

23. $f(x, y, z) = xy + yz + xz$ at

(a) $(1, 1, 1)$ **(b)** $(1, 0, 0)$ **(c)** $(0, 0, 0)$

24. $f(x, y, z) = x^2 + y^2 + z^2$ at

(a) $(1, 1, 1)$ **(b)** $(0, 1, 0)$ **(c)** $(1, 0, 0)$

25. $f(x, y, z) = \sqrt{x^2 + y^2 + z^2}$ at

(a) $(1, 0, 0)$ **(b)** $(1, 1, 0)$ **(c)** $(1, 2, 2)$

26. $f(x, y, z) = (\sin xy)/z$ at

(a) $(\pi/2, 1, 1)$ **(b)** $(2, 0, 1)$

27. $f(x, y, z) = e^x + \cos(y + z)$ at

(a) $(0, 0, 0)$ **(b)** $\left(0, \frac{\pi}{2}, 0\right)$ **(c)** $\left(0, \frac{\pi}{4}, \frac{\pi}{4}\right)$

28. $f(x, y, z) = \tan^{-1}(xyz)$ at

(a) $(1, 0, 0)$ **(b)** $(1, 1, 0)$ **(c)** $(1, 1, 1)$

In Exercises 29–32, find the linearization $L(x, y, z)$ of the function $f(x, y, z)$ at P_0. Then find an upper bound for the magnitude of the error E in the approximation $f(x, y, z) \approx L(x, y, z)$ over the region R.

29. $f(x, y, z) = xz - 3yz + 2$ at $P_0(1, 1, 2)$

R: $|x - 1| \le 0.01$, $|y - 1| \le 0.01$, $|z - 2| \le 0.02$

30. $f(x, y, z) = x^2 + xy + yz + (1/4)z^2$ at $P_0(1, 1, 2)$

R: $|x - 1| \le 0.01$, $|y - 1| \le 0.01$, $|z - 2| \le 0.08$

31. $f(x, y, z) = xy + 2yz - 3xz$ at $P_0(1, 1, 0)$

R: $|x - 1| \le 0.01$, $|y - 1| \le 0.01$, $|z| \le 0.01$

32. $f(x, y, z) = \sqrt{2} \cos x \sin(y + z)$ at $P_0(0, 0, \pi/4)$

R: $|x| \le 0.01$, $|y| \le 0.01$, $|z - \pi/4| \le 0.01$

Theory and Examples

33. *The sagging beam revisited* The beam of Example 8 is tipped on its side so that $h = 0.1$ m and $w = 0.2$ m.

(a) What is the value of dS now?

(b) Compare the sensitivity of the newly positioned beam to a small change in height with its sensitivity to an equally small change in width.

34. *Designing a soda can* A standard 12 fl oz can of soda is essentially a cylinder of radius $r = 1$ in. and height $h = 5$ in.

(a) At these dimensions, how sensitive is the can's volume to a small change in radius versus a small change in height?

(b) Could you design a soda can that *appears* to hold more soda but in fact holds the same 12 fl oz? What might its dimensions be? (There is more than one correct answer.)

35. *Value of a 2 × 2 determinant* If $|a|$ is much greater than $|b|$, $|c|$, and $|d|$, to which of a, b, c, and d is the value of the determinant

$$f(a, b, c, d) = \begin{vmatrix} a & b \\ c & d \end{vmatrix}$$

most sensitive? Give reasons for your answer.

36. *Estimating error of a product* Estimate how strongly simultaneous errors of 2% in a, b, and c might affect the calculation of the product

$$p(a, b, c) = abc.$$

37. *Designing a box* Estimate how much wood it takes to make a hollow rectangular box whose inside measurements are 5 ft long by 3 ft wide by 2 ft deep if the box is made of lumber 1/2 in. thick and the box has no top.

38. *Surveying a triangular field* The area of a triangle is $(1/2)ab \sin C$, where a and b are the lengths of two sides of the triangle and C is the measure of the included angle. In surveying a triangular plot, you have measured a, b, and C to be 150 ft, 200 ft, and 60°, respectively. By about how much could your area calculation be in error if your values of a and b are off by half a foot each and your measurement of C is off by 2°? See the figure. Remember to use radians.

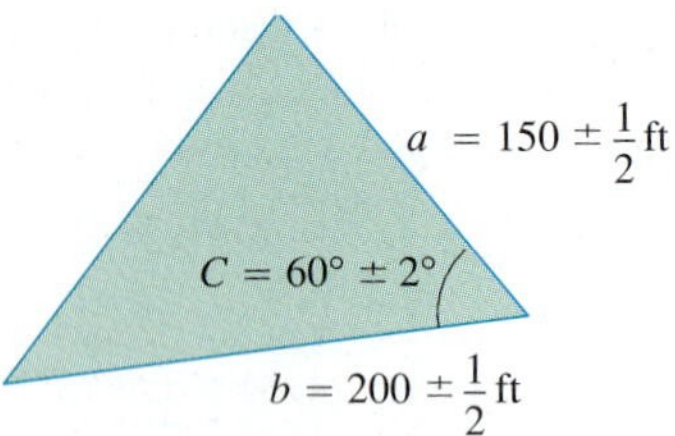

39. *Estimating maximum error* Suppose that $u = xe^y + y \sin z$ and that x, y, and z can be measured with maximum possible errors of ± 0.2, ± 0.6, and $\pm \pi/180$, respectively. Estimate the maximum possible error in calculating u from the measured values $x = 2$, $y = \ln 3$, $z = \pi/2$.

40. *The Wilson lot size formula* The Wilson lot size formula in economics says that the most economical quantity Q of goods (radios, shoes, brooms, whatever) for a store to order is given by the formula $Q = \sqrt{2KM/h}$, where K is the cost of placing the order, M is the number of items sold per week, and h is the weekly holding cost for each item (cost of space, utilities, security, and so on). To which of the variables K, M, and h is Q most sensitive near the point $(K_0, M_0, h_0) = (2, 20, 0.05)$? Give reasons for your answer.

41. *The linearization of f(x, y) is a tangent-plane approximation* Show that the tangent plane at the point $P_0(x_0, y_0, f(x_0, y_0))$ on the surface $z = f(x, y)$ defined by a differentiable function f is the plane

$$f_x(x_0, y_0)(x - x_0) + f_y(x_0, y_0)(y - y_0) - (z - f(x_0, y_0)) = 0$$

or

$$z = f(x_0, y_0) + f_x(x_0, y_0)(x - x_0) + f_y(x_0, y_0)(y - y_0).$$

Thus, the tangent plane at P_0 is the graph of the linearization of f at P_0 (Figure 11.43).

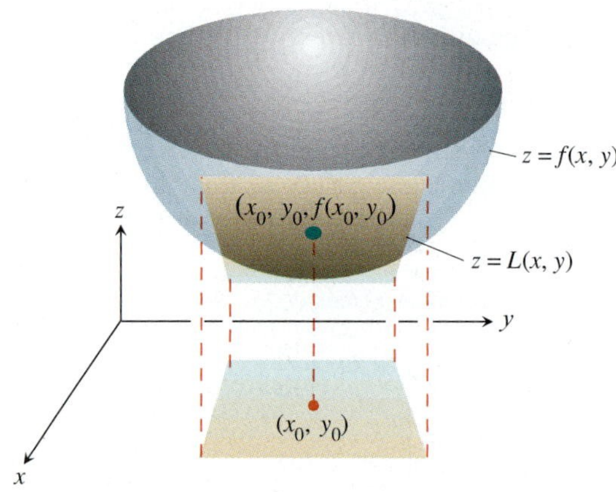

FIGURE 11.43 The graph of a function $z = f(x, y)$ and its linearization at a point (x_0, y_0). The plane defined by L is tangent to the surface at the point above the point (x_0, y_0). This furnishes a geometric explanation of why the values of L lie close to those of f in the immediate neighborhood of (x_0, y_0). (Exercise 41)

11.7 Extreme Values and Saddle Points

Behavior on Closed Bounded Regions • Derivative Tests for Local Extreme Values • Absolute Maxima and Minima on Closed Bounded Regions • Limitations of the First Derivative Test, and Summary

Finding the maximum and minimum values of functions of several variables, and knowing where they occur, is an important application of multivariable differential calculus. For example, what is the highest temperature on a heated metal plate, and where is it taken on? Where does a given surface attain its highest point above a given patch of the xy-plane? As we see in this section, we can often answer questions like these by examining the partial derivatives of some appropriate function.

Behavior on Closed Bounded Regions

As we saw when we were working with functions of a single variable, differentiable functions were just what we wanted for modeling optimization problems. Because these functions are continuous, we knew that on closed intervals they did assume both maximum and minimum values. Because they are differentiable, we knew that they would assume these values only at domain endpoints or at interior domain points where the first derivative vanished. Occasionally, we encountered functions that failed to be differentiable at one or more interior domain points and we had to add these points to the list to be investigated as well.

We also saw that the condition that $f'(c) = 0$ did not always signal the presence of an extreme value. At such a point c, the graph might have an inflection point instead of a local maximum or minimum. The graph might rise as it approached c from the left, level off at c, then rise again as it left c. Or it might fall toward c, level off at c, and then resume falling. That is, the graph might cross its tangent line at $x = c$.

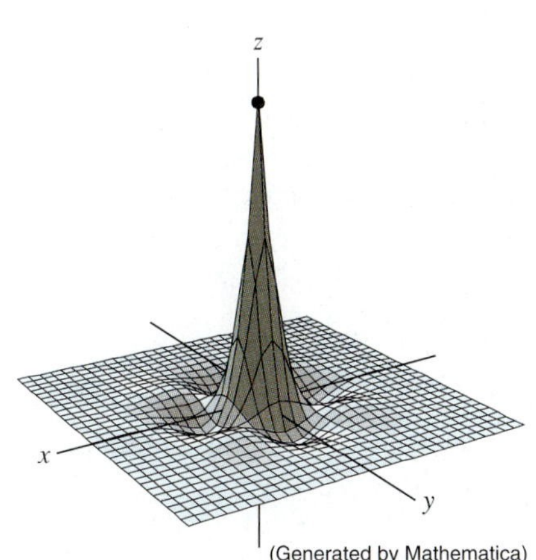

FIGURE 11.44 The function

$$z = (\cos x)(\cos y)e^{-\sqrt{x^2+y^2}}$$

has a maximum value of 1 and a minimum value of about -0.067 on the square region $|x| \le 3\pi/2, |y| \le 3\pi/2$.

Functions of two variables exhibit similar behavior. As we mentioned in Section 11.2, continuous functions of two variables assume extreme values on closed, bounded domains (see Figures 11.44 and 11.45). In addition, as we see in this section, we can narrow the search for these extreme values by examining the functions' first partial derivatives. A function of two variables can assume extreme values only at domain boundary points or at interior domain points where both first partial derivatives are zero or where one or both of the first partial derivatives fails to exist.

Once again, the vanishing of derivatives at an interior point (a, b) does not always signal the presence of an extreme value. The surface that is the graph of the function might be shaped like a saddle right above (a, b) and cross its tangent plane there.

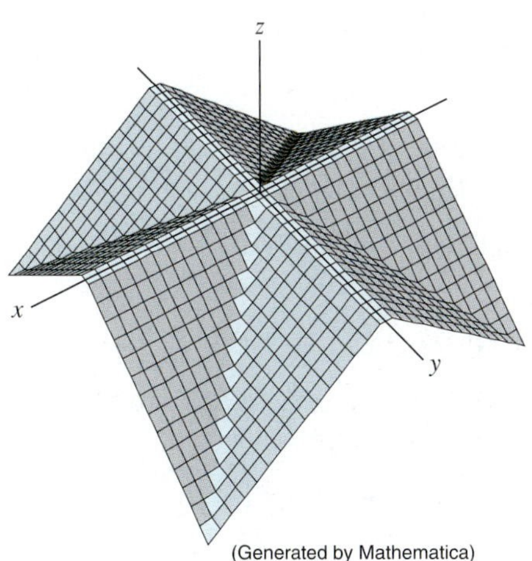

FIGURE 11.45 The "roof surface"

$$z = \frac{1}{2}(||x| - |y|| - |x| - |y|)$$

viewed from the point (10, 15, 20). The defining function has a maximum value of 0 and a minimum value of $-a$ on the square region $|x| \le a, |y| \le a$.

Derivative Tests for Local Extreme Values

To find the local extreme values of a function of a single variable, we look for points where the graph has a horizontal tangent line. At such points, we then look for local maxima, local minima, and points of inflection. For a function $f(x, y)$ of two variables, we look for points where the surface $z = f(x, y)$ has a horizontal tangent *plane*. At such points, we then look for local maxima, local minima, and saddle points (more about saddle points in a moment).

Definitions Local Maximum and Local Minimum

Let $f(x, y)$ be defined on a region R containing the point (a, b). Then

1. $f(a, b)$ is a **local maximum** value of f if $f(a, b) \ge f(x, y)$ for all domain points (x, y) in an open disk centered at (a, b)
2. $f(a, b)$ is a **local minimum** value of f if $f(a, b) \le f(x, y)$ for all domain points (x, y) in an open disk centered at (a, b).

Local maxima correspond to mountain peaks on the surface $z = f(x, y)$ and local minima correspond to valley bottoms (Figure 11.46). At such points, the tangent planes, when they exist, are horizontal. Local extrema are also called **relative extrema.**

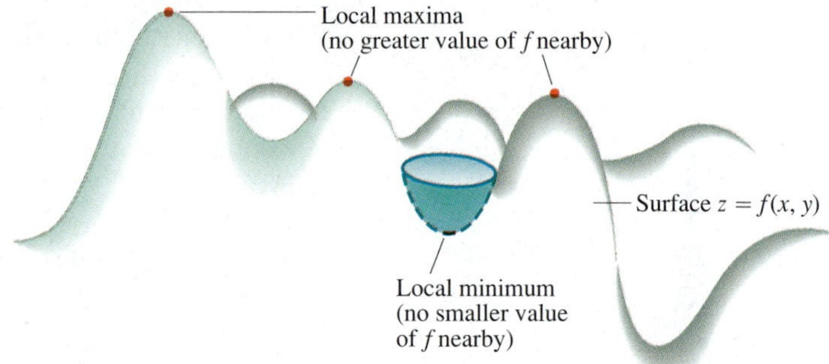

FIGURE 11.46 A local maximum is a mountain peak and a local minimum is a valley low.

As with functions of a single variable, the key to identifying the local extrema is a first derivative test.

Theorem 10 First Derivative Test for Local Extreme Values

If $f(x, y)$ has a local maximum or minimum value at an interior point (a, b) of its domain and if the first partial derivatives exist there, then $f_x(a, b) = 0$ and $f_y(a, b) = 0$.

Proof Suppose that f has a local maximum value at an interior point (a, b) of its domain. Then

1. $x = a$ is an interior point of the domain of the curve $z = f(x, b)$ in which the plane $y = b$ cuts the surface $z = f(x, y)$ (Figure 11.47).
2. The function $z = f(x, b)$ is a differentiable function of x at $x = a$ (the derivative is $f_x(a, b)$).
3. The function $z = f(x, b)$ has a local maximum value at $x = a$.
4. The value of the derivative of $z = f(x, b)$ at $x = a$ is therefore zero (Theorem 2, Section 3.1). Since this derivative is $f_x(a, b)$, we conclude that $f_x(a, b) = 0$.

A similar argument with the function $z = f(a, y)$ shows that $f_y(a, b) = 0$.

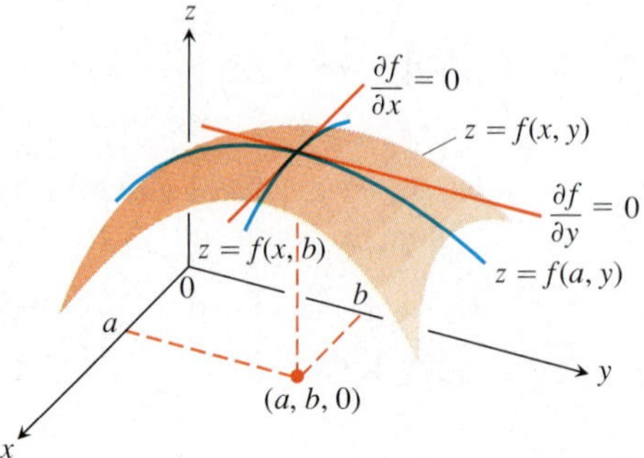

FIGURE 11.47 A local maximum of f occurs at $x = a$, $y = b$.

This proves the theorem for local maximum values. The proof for local minimum values is left as Exercise 36.

If we substitute the values $f_x(a, b) = 0$ and $f_y(a, b) = 0$ into the equation

$$f_x(a, b)(x - a) + f_y(a, b)(y - b) - (z - f(a, b)) = 0$$

for the tangent plane to the surface $z = f(x, y)$ at (a, b), the equation reduces to

$$0 \cdot (x - a) + 0 \cdot (y - b) - z + f(a, b) = 0$$

or

$$z = f(a, b).$$

Thus, Theorem 10 says that the surface does indeed have a horizontal tangent plane at a local extremum, provided there is a tangent plane there.

As in the single-variable case, Theorem 10 says that the only places a function $f(x, y)$ can ever have an extreme value are

1. Interior points where $f_x = f_y = 0$
2. Interior points where one or both of f_x and f_y do not exist
3. Boundary points of the function's domain.

Definition Critical Point

An interior point of the domain of a function $f(x, y)$ where both f_x and f_y are zero or where one or both of f_x and f_y do not exist is a **critical point** of f.

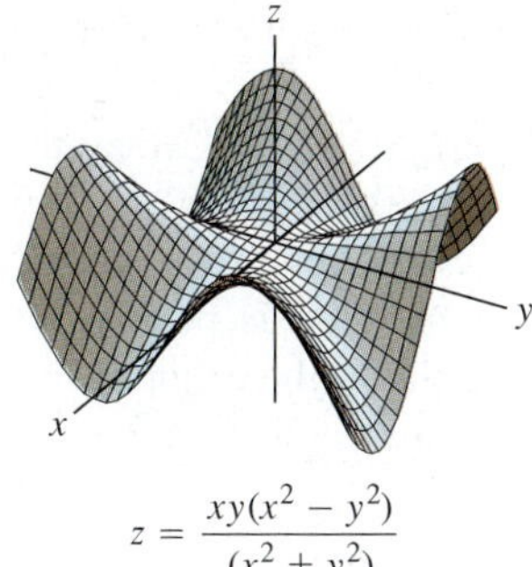

$$z = \frac{xy(x^2 - y^2)}{(x^2 + y^2)}$$

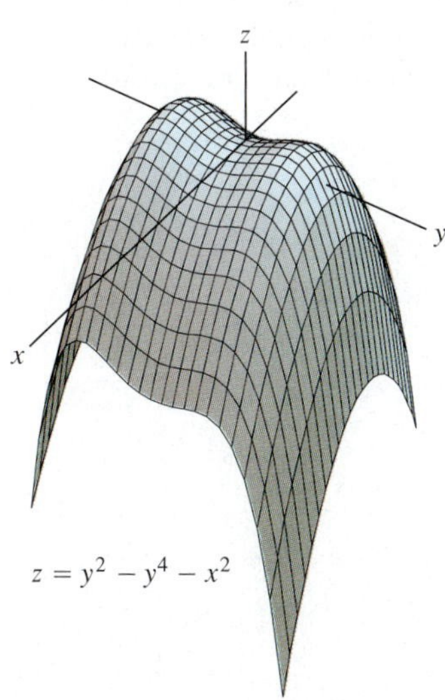

$$z = y^2 - y^4 - x^2$$

(Generated by Mathematica)

FIGURE 11.48 Saddle points at the origin.

Thus, the only points where a function $f(x, y)$ can assume extreme values are critical points and boundary points. As with differentiable functions of a single variable, not every critical point gives rise to a local extremum. A differentiable function of a single variable might have a point of inflection. A differentiable function of two variables might have a *saddle point.*

Definition Saddle Point

A differentiable function $f(x, y)$ has a **saddle point** at a critical point (a, b) if in every open disk centered at (a, b) there are domain points (x, y) where $f(x, y) > f(a, b)$ and domain points (x, y) where $f(x, y) < f(a, b)$. The corresponding point $(a, b, f(a, b))$ on the surface $z = f(x, y)$ is called a saddle point of the surface (Figure 11.48).

Example 1 Finding Local Extreme Values

Find the local extreme values of $f(x, y) = x^2 + y^2$.

Solution The domain of f is the entire plane (so there are no boundary points) and the partial derivatives $f_x = 2x$ and $f_y = 2y$ exist everywhere. Therefore, local extreme values can occur only where

$$f_x = 2x = 0 \quad \text{and} \quad f_y = 2y = 0.$$

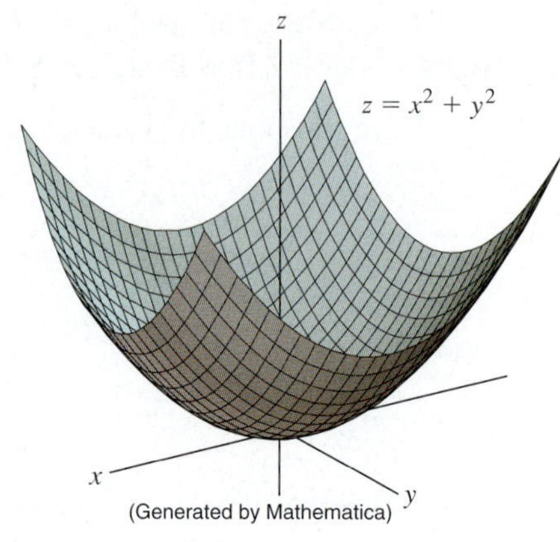

FIGURE 11.49 The graph of the function $f(x, y) = x^2 + y^2$ is the paraboloid $z = x^2 + y^2$. The function has only one critical point, the origin, which gives rise to a local minimum value of 0. (Example 1)

The only possibility is the origin, where the value of f is zero. Since f is never negative, we see that the origin gives a local minimum (Figure 11.49).

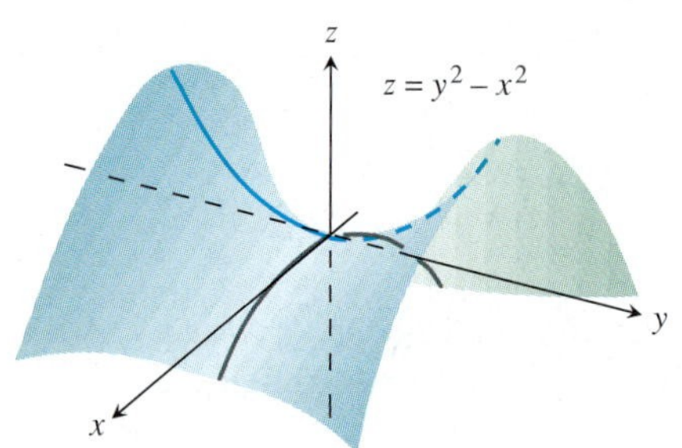

FIGURE 11.50 The origin is a saddle point of the function $f(x, y) = y^2 - x^2$. There are no local extreme values. (Example 2)

Example 2 Identifying a Saddle Point

Find the local extreme values (if any) of $f(x, y) = y^2 - x^2$.

Solution The domain of f is the entire plane (so there are no boundary points) and the partial derivatives $f_x = -2x$ and $f_y = 2y$ exist everywhere. Therefore, local extrema can occur only at the origin $(0, 0)$. Along the positive x-axis, however, f has the value $f(x, 0) = -x^2 < 0$; along the positive y-axis, f has the value $f(0, y) = y^2 > 0$. Therefore, every open disk in the xy-plane centered at $(0, 0)$ contains points where the function is positive and points where it is negative. The function has a saddle point at the origin (Figure 11.50) instead of a local extreme value. We conclude that the function has no local extreme values.

That $f_x = f_y = 0$ at an interior point (a, b) of R does not guarantee f has a local extreme value there. If f and its first and second partial derivatives are continuous on R, however, we may be able to learn more from the following theorem, proved in Section 11.10.

Theorem 11 Second Derivative Test for Local Extreme Values

Suppose that $f(x, y)$ and its first and second partial derivatives are continuous throughout a disk centered at (a, b) and that $f_x(a, b) = f_y(a, b) = 0$. Then

i. f has a **local maximum** at (a, b) if $f_{xx} < 0$ and $f_{xx}f_{yy} - f_{xy}{}^2 > 0$ at (a, b).

ii. f has a **local minimum** at (a, b) if $f_{xx} > 0$ and $f_{xx}f_{yy} - f_{xy}{}^2 > 0$ at (a, b).

iii. f has a **saddle point** at (a, b) if $f_{xx}f_{yy} - f_{xy}{}^2 < 0$ at (a, b).

iv. **The test is inconclusive** at (a, b) if $f_{xx}f_{yy} - f_{xy}{}^2 = 0$ at (a, b). In this case, we must find some other way to determine the behavior of f at (a, b).

The expression $f_{xx}f_{yy} - f_{xy}^2$ is called the **discriminant** or **Hessian** of f. It is sometimes easier to remember it in determinant form,

$$f_{xx}f_{yy} - f_{xy}^2 = \begin{vmatrix} f_{xx} & f_{xy} \\ f_{xy} & f_{yy} \end{vmatrix}.$$

Theorem 11 says that if the discriminant is positive at the point (a, b), then the surface curves the same way in all directions: downwards if $f_{xx} < 0$, giving rise to a local maximum, and upwards if $f_{xx} > 0$, giving a local minimum. On the other hand, if the discriminant is negative at (a, b), then the surface curves up in some directions and down in others, so we have a saddle point.

Example 3 Finding Local Extreme Values

Find the local extreme values of the function

$$f(x, y) = xy - x^2 - y^2 - 2x - 2y + 4.$$

Solution The function is defined and differentiable for all x and y and its domain has no boundary points. The function therefore has extreme values only at the points where f_x and f_y are simultaneously zero. This leads to

$$f_x = y - 2x - 2 = 0, \qquad f_y = x - 2y - 2 = 0,$$

or

$$x = y = -2.$$

Therefore, the point $(-2, -2)$ is the only point where f may take on an extreme value. To see if it does so, we calculate

$$f_{xx} = -2, \qquad f_{yy} = -2, \qquad f_{xy} = 1.$$

The discriminant of f at $(a, b) = (-2, -2)$ is

$$f_{xx}f_{yy} - f_{xy}^2 = (-2)(-2) - (1)^2 = 4 - 1 = 3.$$

The combination

$$f_{xx} < 0 \qquad \text{and} \qquad f_{xx}f_{yy} - f_{xy}^2 > 0$$

tells us that f has a local maximum at $(-2, -2)$. The value of f at this point is $f(-2, -2) = 8$.

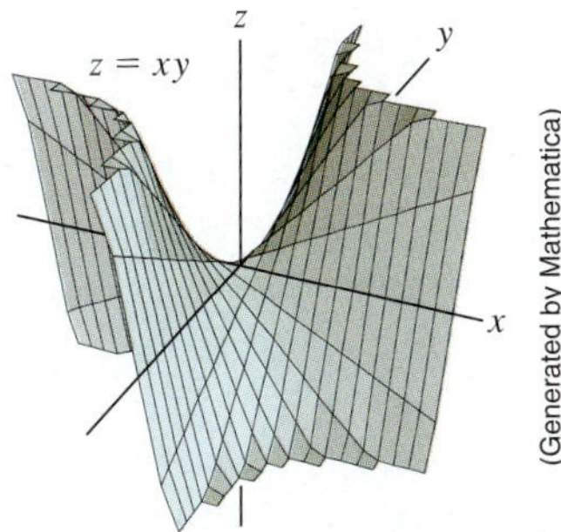

FIGURE 11.51 The surface $z = xy$ has a saddle point at the origin. (Example 4)

Example 4 Searching for Local Extreme Values

Find the local extreme values of $f(x, y) = xy$.

Solution Since f is differentiable everywhere (Figure 11.51), it can assume extreme values only where

$$f_x = y = 0 \qquad \text{and} \qquad f_y = x = 0.$$

Thus, the origin is the only point where f might have an extreme value. To see what happens there, we calculate

$$f_{xx} = 0, \qquad f_{yy} = 0, \qquad f_{xy} = 1.$$

The discriminant,

$$f_{xx}f_{yy} - f_{xy}^{\;2} = -1,$$

is negative. Therefore, the function has a saddle point at $(0, 0)$. We conclude that $f(x, y) = xy$ has no local extreme values.

Absolute Maxima and Minima on Closed Bounded Regions

We organize the search for the absolute extrema of a continuous function $f(x, y)$ on a closed and bounded region R into three steps.

Step 1: *List the interior points* of R where f may have local maxima and minima and evaluate f at these points. These are the points where $f_x = f_y = 0$ or where one or both of f_x and f_y fail to exist (the critical points of f).

Step 2: *List the boundary points* of R where f has local maxima and minima and evaluate f at these points. We show how to do this shortly.

Step 3: *Look through the lists* for the maximum and minimum values of f. These will be the absolute maximum and minimum values of f on R. Since absolute maxima and minima are also local maxima and minima, the absolute maximum and minimum values of f already appear somewhere in the lists made in steps 1 and 2. We have only to glance at the lists to see what they are.

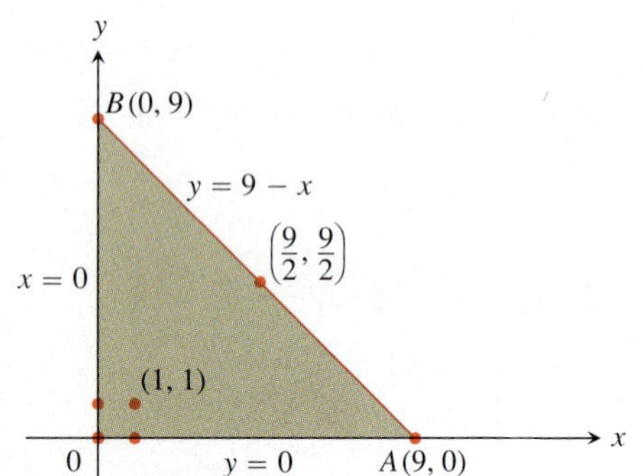

FIGURE 11.52 This triangular plate is the domain of the function in Example 5.

Example 5 Finding Absolute Extrema

Find the absolute maximum and minimum values of

$$f(x, y) = 2 + 2x + 2y - x^2 - y^2$$

on the triangular plate in the first quadrant bounded by the lines $x = 0$, $y = 0$, $y = 9 - x$.

Solution Since f is differentiable, the only places where f can assume these values are points inside the triangle (Figure 11.52) where $f_x = f_y = 0$ and points on the boundary.

Interior Points

For these we have

$$f_x = 2 - 2x = 0, \qquad f_y = 2 - 2y = 0,$$

yielding the single point $(x, y) = (1, 1)$. The value of f there is

$$f(1, 1) = 4.$$

Boundary Points

We take the triangle one side at a time:

1. On the segment OA, $y = 0$. The function

$$f(x, y) = f(x, 0) = 2 + 2x - x^2$$

may now be regarded as a function of x defined on the closed interval $0 \leq x \leq 9$. Its extreme values (we know from Chapter 3) may occur at the endpoints

$$x = 0 \quad \text{where} \quad f(0, 0) = 2$$

$$x = 9 \quad \text{where} \quad f(9, 0) = 2 + 18 - 81 = -61$$

and at the interior points where $f'(x, 0) = 2 - 2x = 0$. The only interior point where $f'(x, 0) = 0$ is $x = 1$, where

$$f(x, 0) = f(1, 0) = 3.$$

2. On the segment OB, $x = 0$ and

$$f(x, y) = f(0, y) = 2 + 2y - y^2.$$

We know from the symmetry of f in x and y and from the analysis we just carried out that the candidates on this segment are

$$f(0, 0) = 2, \qquad f(0, 9) = -61, \qquad f(0, 1) = 3.$$

3. We have already accounted for the values of f at the endpoints of AB, so we need only look at the interior points of AB. With $y = 9 - x$, we have

$$f(x, y) = 2 + 2x + 2(9 - x) - x^2 - (9 - x)^2 = -61 + 18x - 2x^2.$$

Setting $f'(x, 9 - x) = 18 - 4x = 0$ gives

$$x = \frac{18}{4} = \frac{9}{2}.$$

At this value of x,

$$y = 9 - \frac{9}{2} = \frac{9}{2} \qquad \text{and} \qquad f(x, y) = f\left(\frac{9}{2}, \frac{9}{2}\right) = -\frac{41}{2}.$$

Summary

We list all the candidates: $4, 2, -61, 3, -(41/2)$. The maximum is 4, which f assumes at $(1, 1)$. The minimum is -61, which f assumes at $(0, 9)$ and $(9, 0)$.

Solving extreme value problems with algebraic constraints on the variables usually require the method of Lagrange multipliers in the next section. But sometimes we can solve such problems directly, as in the next example.

Example 6 Solving a Volume Problem with a Constraint

A delivery company accepts only rectangular boxes whose length and girth (perimeter of a cross section) do not sum over 108 in. Find the dimensions of an acceptable box of largest volume.

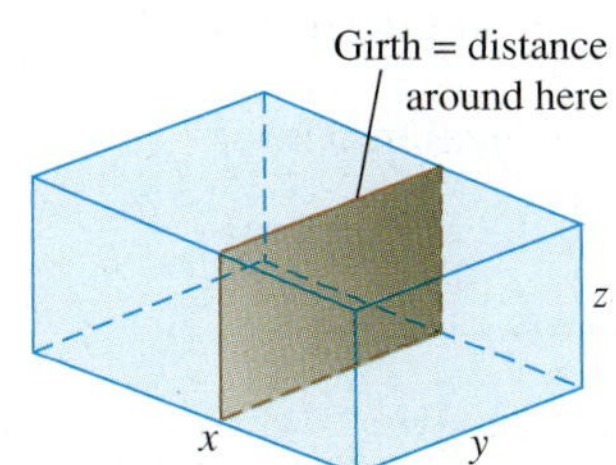

FIGURE 11.53 The box in Example 6.

Solution Let x, y, and z represent the length, width, and height of the rectangular box, respectively. Then the girth is $2y + 2z$. We want to maximize the volume $V = xyz$ of the box (Figure 11.53) satisfying $x + 2y + 2z = 108$ (the largest box accepted by the delivery company). Thus, we can write the volume of the box as a function of two variables.

$$\begin{aligned} V(y, z) &= (108 - 2y - 2z)yz && V = xyz \text{ and } x = 108 - 2y - 2z \\ &= 108yz - 2y^2z - 2yz^2 \end{aligned}$$

Setting the first partial derivatives equal to zero,

$$\begin{aligned} V_y(y, z) &= 108z - 4yz - 2z^2 = (108 - 4y - 2z)z = 0 \\ V_z(y, z) &= 108y - 2y^2 - 4yz = (108 - 2y - 4z)y = 0, \end{aligned}$$

gives the critical points $(0, 0)$, $(0, 54)$, $(54, 0)$, and $(18, 18)$. The volume is zero at $(0, 0)$, $(0, 54)$, $(54, 0)$, which are not maximum values. At the point $(18, 18)$, we apply the second derivative test (Theorem 11):

$$V_{yy} = -4z, \qquad V_{zz} = -4y, \qquad V_{yz} = 108 - 4y - 4z.$$

Then

$$V_{yy}V_{zz} - V_{yz}{}^2 = 16yz - 16(27 - y - z)^2.$$

Thus,

$$V_{yy}(18, 18) = -4(18) < 0$$

and

$$[V_{yy}V_{zz} - V_{yz}{}^2]_{(18,18)} = 16(18)(18) - 16(-9)^2 > 0$$

imply that $(18, 18)$ gives a maximum volume. The dimensions of the package are $x = 108 - 2(18) - 2(18) = 36$ in., $y = 18$ in., and $z = 18$ in. The maximum volume is $V = (36)(18)(18) = 11{,}664$ in.^{3}, or 6.75 ft^3.

Limitations of the First Derivative Test, and Summary

Despite the power of Theorem 10, we urge you to remember its limitations. It does not apply to boundary points of a function's domain, where it is possible for a function to have extreme values along with nonzero derivatives. Also, it does not apply to points where either f_x or f_y fails to exist.

Summary of Max-Min Tests

The extreme values of $f(x, y)$ can occur only at

i. **boundary points** of the domain of f

ii. **critical points** (interior points where $f_x = f_y = 0$ or points where f_x or f_y fail to exist).

If the first- and second-order partial derivatives of f are continuous throughout a disk centered at a point (a, b) and $f_x(a, b) = f_y(a, b) = 0$, you may be able to classify $f(a, b)$ with the **second derivative test**:

i. $f_{xx} < 0$ and $f_{xx}f_{yy} - f_{xy}{}^2 > 0$ at $(a, b) \Rightarrow$ **local maximum**

ii. $f_{xx} > 0$ and $f_{xx}f_{yy} - f_{xy}{}^2 > 0$ at $(a, b) \Rightarrow$ **local minimum**

iii. $f_{xx}f_{yy} - f_{xy}{}^2 < 0$ at $(a, b) \Rightarrow$ **saddle point**

iv. $f_{xx}f_{yy} - f_{xy}{}^2 = 0$ at $(a, b) \Rightarrow$ **test is inconclusive.**

EXERCISES 11.7

Finding Local Extrema

Find all the local maxima, local minima, and saddle points of the functions in Exercises 1–20.

1. $f(x, y) = x^2 + xy + y^2 + 3x - 3y + 4$

2. $f(x, y) = 2xy - 5x^2 - 2y^2 + 4x + 4y - 4$

3. $f(x, y) = x^2 + xy + 3x + 2y + 5$

4. $f(x, y) = 5xy - 7x^2 + 3x - 6y + 2$

5. $f(x, y) = 3x^2 + 6xy + 7y^2 - 2x + 4y$

6. $f(x, y) = 2x^2 + 3xy + 4y^2 - 5x + 2y$

7. $f(x, y) = x^2 - y^2 - 2x + 4y + 6$

8. $f(x, y) = x^2 - 2xy + 2y^2 - 2x + 2y + 1$

9. $f(x, y) = 3 + 2x + 2y - 2x^2 - 2xy - y^2$

10. $f(x, y) = x^3 - y^3 - 2xy + 6$

11. $f(x, y) = x^3 + 3xy + y^3$

12. $f(x, y) = 6x^2 - 2x^3 + 3y^2 + 6xy$

13. $f(x, y) = 9x^3 + y^3/3 - 4xy$

14. $f(x, y) = x^3 + y^3 + 3x^2 - 3y^2 - 8$

15. $f(x, y) = 4xy - x^4 - y^4$

16. $f(x, y) = x^4 + y^4 + 4xy$

17. $f(x, y) = \dfrac{1}{x^2 + y^2 - 1}$

18. $f(x, y) = \dfrac{1}{x} + xy + \dfrac{1}{y}$

19. $f(x, y) = y \sin x$

20. $f(x, y) = e^{2x} \cos y$

Finding Absolute Extrema

In Exercises 21–26, find the absolute maxima and minima of the functions on the given domains.

21. $f(x, y) = 2x^2 - 4x + y^2 - 4y + 1$ on the closed triangular plate bounded by the lines $x = 0$, $y = 2$, $y = 2x$ in the first quadrant

22. $f(x, y) = x^2 + y^2$ on the closed triangular plate bounded by the lines $x = 0$, $y = 0$, $y + 2x = 2$ in the first quadrant

23. $T(x, y) = x^2 + xy + y^2 - 6x + 2$ on the rectangular plate $0 \le x \le 5$, $-3 \le y \le 0$

24. $f(x, y) = 48xy - 32x^3 - 24y^2$ on the rectangular plate $0 \le x \le 1$, $0 \le y \le 1$

25. $f(x, y) = (4x - x^2) \cos y$ on the rectangular plate $1 \le x \le 3$, $-\pi/4 \le y \le \pi/4$

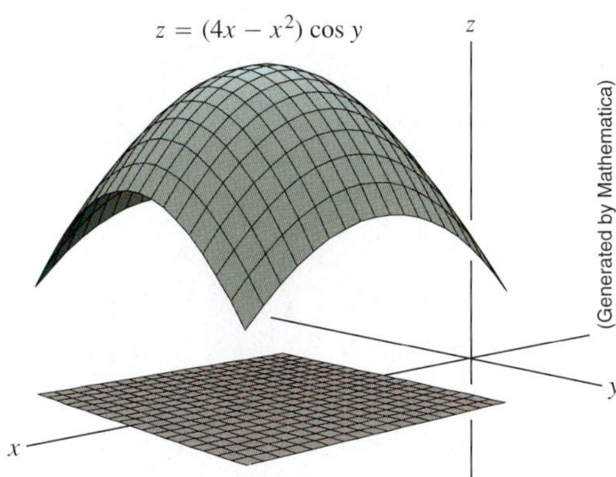

The function and domain in Exercise 25.

26. $f(x, y) = 4x - 8xy + 2y + 1$ on the triangular plate bounded by the lines $x = 0$, $y = 0$, $x + y = 1$ in the first quadrant

27. *Maximizing an integral* Find two numbers a and b with $a \le b$ such that

$$\int_a^b (6 - x - x^2)\,dx$$

has its largest value.

28. *Maximizing an integral* Find two numbers a and b with $a \le b$ such that

$$\int_a^b (24 - 2x - x^2)^{1/3}\,dx$$

has its largest value.

29. *Temperature extremes* The flat circular plate in Figure 11.54 has the shape of the region $x^2 + y^2 \le 1$. The plate, including the boundary where $x^2 + y^2 = 1$, is heated so that the temperature at the point (x, y) is

$$T(x, y) = x^2 + 2y^2 - x.$$

Find the temperatures at the hottest and coldest points on the plate.

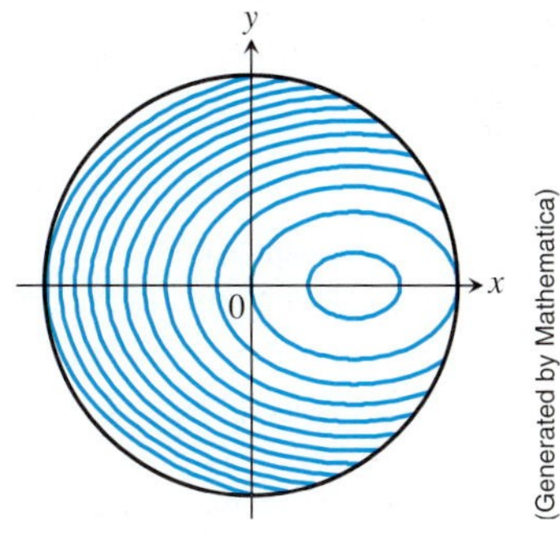

FIGURE 11.54 Curves of constant temperature are called isotherms. The figure shows isotherms of the temperature function $T(x, y) = x^2 + 2y^2 - x$ on the disk $x^2 + y^2 \le 1$ in the xy-plane. Exercise 29 asks you to locate the extreme temperatures.

30. *Identifying critical points* Find the critical point of

$$f(x, y) = xy + 2x - \ln x^2 y$$

in the open first quadrant ($x > 0$, $y > 0$) and show that f takes on a minimum there (Figure 11.55).

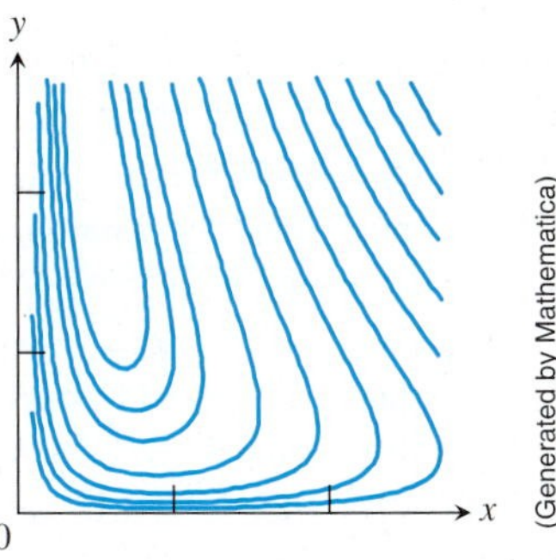

FIGURE 11.55 The function $f(x, y) = xy + 2x - \ln x^2y$ (selected level curves shown here) takes on a minimum value somewhere in the open first quadrant $x > 0$, $y > 0$. (Exercise 30)

Theory and Examples

31. *Writing to Learn* Find the maxima, minima, and saddle points of $f(x, y)$, if any, given that

(a) $f_x = 2x - 4y$ and $f_y = 2y - 4x$

(b) $f_x = 2x - 2$ and $f_y = 2y - 4$

(c) $f_x = 9x^2 - 9$ and $f_y = 2y + 4$.

Describe your reasoning in each case.

32. *Writing to Learn: When the second derivative is inconclusive* The discriminant $f_{xx}f_{yy} - f_{xy}^2$ is zero at the origin for each of the following functions, so the second derivative test fails there. Determine whether the function has a maximum, a minimum, or neither at the origin by imagining what the surface $z = f(x, y)$ looks like. Describe your reasoning in each case.

(a) $f(x, y) = x^2y^2$
(b) $f(x, y) = 1 - x^2y^2$
(c) $f(x, y) = xy^2$
(d) $f(x, y) = x^3y^2$
(e) $f(x, y) = x^3y^3$
(f) $f(x, y) = x^4y^4$

33. Show that (0, 0) is a critical point of $f(x, y) = x^2 + kxy + y^2$ no matter what value the constant k has. (*Hint:* Consider two cases: $k = 0$ and $k \neq 0$.)

34. *Writing to Learn* For what values of the constant k does the second derivative test guarantee that $f(x, y) = x^2 + kxy + y^2$ will have a saddle point at (0, 0)? A local minimum at (0, 0)? For what values of k is the second derivative test inconclusive? Give reasons for your answers.

35. (a) *Writing to Learn* If $f_x(a, b) = f_y(a, b) = 0$, must f have a local maximum or minimum value at (a, b)? Give reasons for your answer.

(b) *Writing to Learn* Can you conclude anything about $f(a, b)$ if f and its first and second partial derivatives are continuous throughout a disk centered at (a, b) and $f_{xx}(a, b)$ and $f_{yy}(a, b)$ differ in sign? Give reasons for your answer.

36. *Proving Theorem 10 for a local minimum* Using the proof of Theorem 10 given in the text for the case in which f has a local maximum at (a, b), prove the theorem for the case in which f has a local minimum at (a, b).

37. *Maximum distance from a plane* Among all the points on the graph of $z = 10 - x^2 - y^2$ that lie above the plane $x + 2y + 3z = 0$, find the point farthest from the plane.

38. *Minimum distance to a plane* Find the point on the graph of $z = x^2 + y^2 + 10$ nearest the plane $x + 2y - z = 0$.

39. *Writing to Learn* The function $f(x, y) = x + y$ fails to have an absolute maximum value in the closed first quadrant ($x \geq 0$ and $y \geq 0$). Does this contradict the discussion on finding absolute extrema given in the text? Give reasons for your answer.

40. Consider the function $f(x, y) = x^2 + y^2 + 2xy - x - y + 1$ over the square $0 \leq x \leq 1$ and $0 \leq y \leq 1$.

(a) *Minimum along a line segment* Show that f has an absolute minimum along the line segment $2x + 2y = 1$ in this square. What *is* the absolute minimum value?

(b) *Absolute maximum* Find the absolute maximum value of f over the square.

Extreme Values on Parametrized Curves

To find the extreme values of a function $f(x, y)$ on a curve $x = x(t)$, $y = y(t)$, we treat f as a function of the single variable t and use the Chain Rule to find where df/dt is zero. As in any other single-variable case, the extreme values of f are then found among the values at the

(a) Critical points (points where df/dt is zero or fails to exist)

(b) Endpoints of the parameter domain.

In Exercises 41–44, find the absolute maximum and minimum values of the functions on the curves.

41. Functions:

(a) $f(x, y) = x + y$
(b) $g(x, y) = xy$
(c) $h(x, y) = 2x^2 + y^2$

Curves:

i. The semicircle $x^2 + y^2 = 4$, $y \geq 0$

ii. The quarter circle $x^2 + y^2 = 4$, $x \geq 0$, $y \geq 0$

Use the parametric equations $x = 2 \cos t$, $y = 2 \sin t$.

42. Functions:

(a) $f(x, y) = 2x + 3y$
(b) $g(x, y) = xy$
(c) $h(x, y) = x^2 + 3y^2$

Curves:

i. The semi-ellipse $(x^2/9) + (y^2/4) = 1$, $y \geq 0$

ii. The quarter ellipse $(x^2/9) + (y^2/4) = 1$, $x \geq 0$, $y \geq 0$

Use the parametric equations $x = 3 \cos t$, $y = 2 \sin t$.

43. Function: $f(x, y) = xy$

Curves:

i. The line $x = 2t$, $y = t + 1$

ii. The line segment $x = 2t$, $y = t + 1$, $-1 \leq t \leq 0$

iii. The line segment $x = 2t$, $y = t + 1$, $0 \leq t \leq 1$

44. Functions:

(a) $f(x, y) = x^2 + y^2$
(b) $g(x, y) = 1/(x^2 + y^2)$

Curves:

i. The line $x = t$, $y = 2 - 2t$

ii. The line segment $x = t$, $y = 2 - 2t$, $0 \leq t \leq 1$

45. *Least squares and regression lines* When we try to fit a line $y = mx + b$ to a set of numerical data points $(x_1, y_1), (x_2, y_2), \ldots, (x_n, y_n)$ (Figure 11.56), we usually choose the line that minimizes the sum of the squares of the vertical distances from the points to the line. In theory, this means finding the values of m and b that minimize the value of the function

CD-ROM WEBsite

$$w = (mx_1 + b - y_1)^2 + \cdots + (mx_n + b - y_n)^2.$$

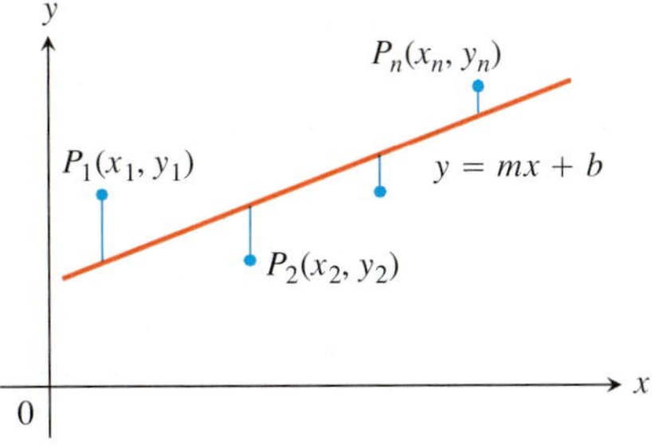

FIGURE 11.56 To fit a line to noncollinear points, we choose the line that minimizes the sum of the squares of the deviations.

Use the first and second derivative tests to show that these values are

$$m = \frac{\left(\sum x_k\right)\left(\sum y_k\right) - n\sum x_k y_k}{\left(\sum x_k\right)^2 - n\sum x_k^2},$$

$$b = \frac{1}{n}\left(\sum y_k - m\sum x_k\right).$$

46. *Craters of Mars* One theory of crater formation suggests that the frequency of large craters should fall off as the square of the diameter (Marcus, *Science*, June 21, 1968, p. 1334). Pictures from *Mariner IV* show the frequencies listed in Table 11.1. Use the results of Exercise 45 to fit a line of the form $F = m(1/D^2) + b$ to the data. Plot the data and draw the line.

Table 11.1 Crater sizes on Mars

Diameter in km, D	$1/D^2$ (for left value of class interval)	Frequency, F
32–45	0.001	51
45–64	0.0005	22
64–90	0.00024	14
90–128	0.000123	4

COMPUTER EXPLORATIONS

Exploring Local Extrema at Critical Points

In Exercises 47–52, you will explore functions to identify local extrema at critical points. Use a CAS to perform the following steps.

(a) Plot the function over the given rectangle.

(b) Plot some level curves in the rectangle.

(c) Calculate the function's first partial derivatives and use the CAS equation solver to find the critical points. How do the critical points relate to the level curves plotted in part (b)? Which critical points, if any, appear to give a saddle point? Give reasons for your answer.

(d) Calculate the function's second partial derivatives and find the discriminant $f_{xx}f_{yy} - f_{xy}^2$.

(e) Using the max-min tests, classify the critical points found in part (c). Are your findings consistent with your discussion in part (c)?

47. $f(x, y) = x^2 + y^3 - 3xy, \quad -5 \le x \le 5, \quad -5 \le y \le 5$

48. $f(x, y) = x^3 - 3xy^2 + y^2, \quad -2 \le x \le 2, \quad -2 \le y \le 2$

49. $f(x, y) = x^4 + y^2 - 8x^2 - 6y + 16, \quad -3 \le x \le 3, \quad -6 \le y \le 6$

50. $f(x, y) = 2x^4 + y^4 - 2x^2 - 2y^2 + 3, \quad -3/2 \le x \le 3/2, \quad -3/2 \le y \le 3/2$

51. $f(x, y) = 5x^6 + 18x^5 - 30x^4 + 30xy^2 - 120x^3, \quad -4 \le x \le 3, \quad -2 \le y \le 2$

52. $f(x, y) = \begin{cases} x^5 \ln(x^2 + y^2), & (x, y) \ne (0, 0) \\ 0, & (x, y) = (0, 0) \end{cases}$

$-2 \le x \le 2, \quad -2 \le y \le 2$

11.8 Lagrange Multipliers

Constrained Maxima and Minima • The Method of Lagrange Multipliers • Lagrange Multipliers with Two Constraints

As we saw in Section 11.7, we sometimes need to find the extreme values of a function whose domain is constrained to lie within some particular subset of the plane, such as a disk or a closed triangular region. As we saw in Example 6 of Section 11.7, however, and as Figure 11.57 shows here, a function may be subject to other kinds of constraints as well.

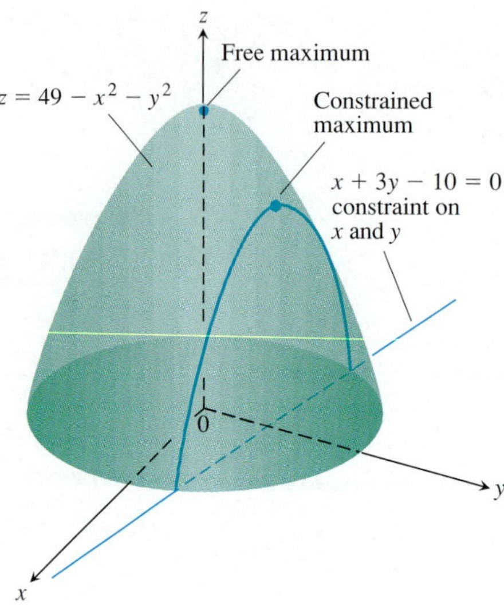

FIGURE 11.57 The function $f(x, y) = 49 - x^2 - y^2$, subject to the constraint $g(x, y) = x + 3y - 10 = 0$.

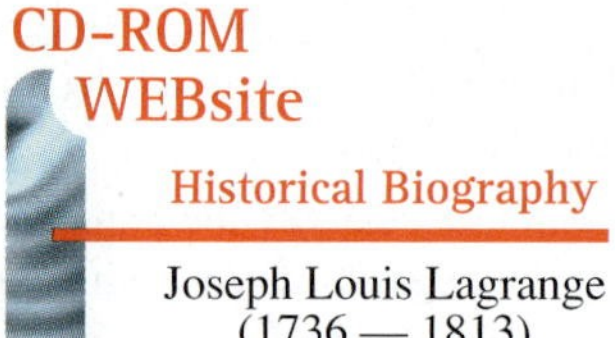

In this section, we explore a powerful method for finding extreme values of constrained functions: the method of *Lagrange multipliers*. Lagrange developed the method in 1755 to solve max-min problems in geometry. Today the method is important in economics, in engineering (where it is used in designing multistage rockets, for example), and in mathematics.

Constrained Maxima and Minima

Example 1 Finding a Minimum with Constraint

Find the point $P(x, y, z)$ closest to the origin on the plane $2x + y - z - 5 = 0$.

Solution The problem asks us to find the minimum value of the function

$$\begin{aligned} |\overrightarrow{OP}| &= \sqrt{(x-0)^2 + (y-0)^2 + (z-0)^2} \\ &= \sqrt{x^2 + y^2 + z^2} \end{aligned}$$

subject to the constraint that

$$2x + y - z - 5 = 0.$$

Since $|\overrightarrow{OP}|$ has a minimum value wherever the function

$$f(x, y, z) = x^2 + y^2 + z^2$$

has a minimum value, we may solve the problem by finding the minimum value of $f(x, y, z)$ subject to the constraint $2x + y - z - 5 = 0$ (thus avoiding square roots). If we regard x and y as the independent variables in this equation and write z as

$$z = 2x + y - 5,$$

our problem reduces to one of finding the points (x, y) at which the function

$$h(x, y) = f(x, y, 2x + y - 5) = x^2 + y^2 + (2x + y - 5)^2$$

has its minimum value or values. Since the domain of h is the entire xy-plane, the first derivative test of Section 11.7 tells us that any minima that h might have must occur at points where

$$h_x = 2x + 2(2x + y - 5)(2) = 0, \qquad h_y = 2y + 2(2x + y - 5) = 0.$$

This leads to

$$10x + 4y = 20, \qquad 4x + 4y = 10,$$

and the solution

$$x = \frac{5}{3}, \qquad y = \frac{5}{6}.$$

We may apply a geometric argument together with the second derivative test to show that these values minimize h. The z-coordinate of the corresponding point on the plane $z = 2x + y - 5$ is

$$z = 2\left(\frac{5}{3}\right) + \frac{5}{6} - 5 = -\frac{5}{6}.$$

Therefore, the point we seek is

$$\text{Closest point:} \qquad P\left(\frac{5}{3}, \frac{5}{6}, -\frac{5}{6}\right).$$

The distance from P to the origin is $5/\sqrt{6} \approx 2.04$.

Attempts to solve a constrained maximum or minimum problem by substitution, as we might call the method of Example 1, do not always go smoothly. This is one of the reasons for learning the new method of this section.

Example 2 Finding a Minimum with Constraint

Find the points closest to the origin on the hyperbolic cylinder $x^2 - z^2 - 1 = 0$.

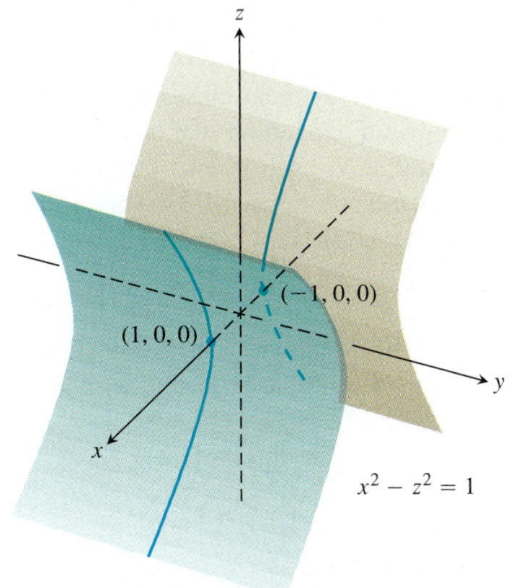

FIGURE 11.58 The hyperbolic cylinder $x^2 - z^2 - 1 = 0$ in Example 2.

Solution 1 The cylinder is shown in Figure 11.58. We seek the points on the cylinder closest to the origin. These are the points whose coordinates minimize the value of the function

$$f(x, y, z) = x^2 + y^2 + z^2 \qquad \text{Square of the distance}$$

subject to the constraint that $x^2 - z^2 - 1 = 0$. If we regard x and y as independent variables in the constraint equation, then

$$z^2 = x^2 - 1$$

and the values of $f(x, y, z) = x^2 + y^2 + z^2$ on the cylinder are given by the function

$$h(x, y) = x^2 + y^2 + (x^2 - 1) = 2x^2 + y^2 - 1.$$

To find the points on the cylinder whose coordinates minimize f, we look for the points in the xy-plane whose coordinates minimize h. The only extreme value of h occurs where

$$h_x = 4x = 0 \qquad \text{and} \qquad h_y = 2y = 0,$$

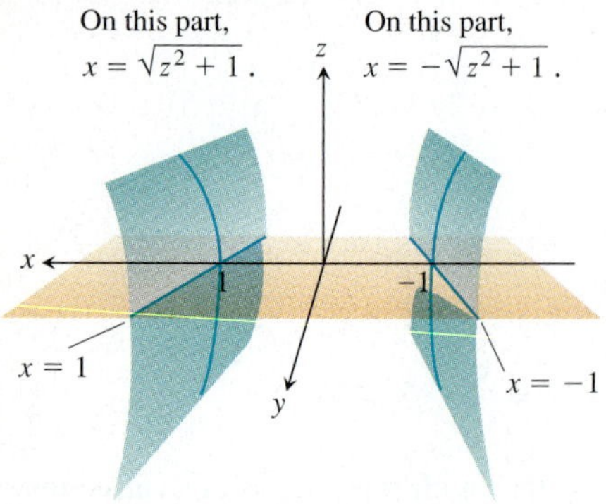

FIGURE 11.59 The region in the xy-plane from which the first two coordinates of the points (x, y, z) on the hyperbolic cylinder $x^2 - z^2 = 1$ are selected excludes the band $-1 < x < 1$ in the xy-plane.

that is, at the point $(0, 0)$. But now we're in trouble: There are no points on the cylinder where both x and y are zero. What went wrong?

What happened was that the first derivative test found (as it should have) the point *in the domain of* h where h has a minimum value. We, on the other hand, want the points *on the cylinder* where h has a minimum value. Although the domain of h is the entire xy-plane, the domain from which we can select the first two coordinates of the points (x, y, z) on the cylinder is restricted to the "shadow" of the cylinder on the xy-plane; it does not include the band between the lines $x = -1$ and $x = 1$ (Figure 11.59).

We can avoid this problem if we treat y and z as independent variables (instead of x and y) and express x in terms of y and z as

$$x^2 = z^2 + 1.$$

With this substitution, $f(x, y, z) = x^2 + y^2 + z^2$ becomes

$$k(y, z) = (z^2 + 1) + y^2 + z^2 = 1 + y^2 + 2z^2$$

and we look for the points where k takes on its smallest value. The domain of k in the yz-plane now matches the domain from which we select the y- and z-coordinates of the points (x, y, z) on the cylinder. Hence, the points that minimize k in the plane will have corresponding points on the cylinder. The smallest values of k occur where

$$k_y = 2y = 0 \qquad \text{and} \qquad k_z = 4z = 0,$$

or where $y = z = 0$. This leads to

$$x^2 = z^2 + 1 = 1, \qquad x = \pm 1.$$

The corresponding points on the cylinder are $(\pm 1, 0, 0)$. We can see from the inequality

$$k(y, z) = 1 + y^2 + 2z^2 \geq 1$$

that the points $(\pm 1, 0, 0)$ give a minimum value for k. We can also see that the minimum distance from the origin to a point on the cylinder is 1 unit.

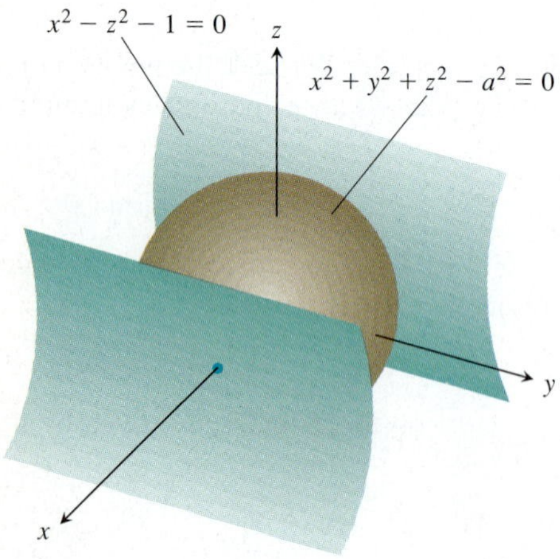

FIGURE 11.60 A sphere expanding like a soap bubble centered at the origin until it just touches the hyperbolic cylinder

$$x^2 - z^2 - 1 = 0.$$

Solution 2 Another way to find the points on the cylinder closest to the origin is to imagine a small sphere centered at the origin expanding like a soap bubble until it just touches the cylinder (Figure 11.60). At each point of contact, the cylinder and sphere have the same tangent plane and normal line. Therefore, if the sphere and cylinder are represented as the level surfaces obtained by setting

$$f(x, y, z) = x^2 + y^2 + z^2 - a^2 \qquad \text{and} \qquad g(x, y, z) = x^2 - z^2 - 1$$

equal to 0, then the gradients ∇f and ∇g will be parallel where the surfaces touch. At any point of contact, we should therefore be able to find a scalar λ ("lambda") such that

$$\nabla f = \lambda \nabla g,$$

or

$$2x\mathbf{i} + 2y\mathbf{j} + 2z\mathbf{k} = \lambda(2x\mathbf{i} - 2z\mathbf{k}).$$

Thus, the coordinates x, y, and z of any point of tangency will have to satisfy the three scalar equations

$$2x = 2\lambda x, \qquad 2y = 0, \qquad 2z = -2\lambda z. \tag{1}$$

For what values of λ will a point (x, y, z) whose coordinates satisfy the equations in (1) also lie on the surface $x^2 - z^2 - 1 = 0$? To answer this question, we use our knowledge that no point on the surface has a zero x-coordinate to conclude that $x \neq 0$ in the first of Equations (1). Hence, $2x = 2\lambda x$ only if

$$2 = 2\lambda, \qquad \text{or} \qquad \lambda = 1.$$

For $\lambda = 1$, the equation $2z = -2\lambda z$ becomes $2z = -2z$. If this equation is to be satisfied as well, z must be zero. Since $y = 0$ also (from the equation $2y = 0$), we conclude that the points we seek all have coordinates of the form

$$(x, 0, 0).$$

What points on the surface $x^2 - z^2 = 1$ have coordinates of this form? The answer is the points $(x, 0, 0)$ for which

$$x^2 - (0)^2 = 1, \qquad x^2 = 1, \qquad \text{or} \qquad x = \pm 1.$$

The points on the cylinder closest to the origin are the points $(\pm 1, 0, 0)$.

CD-ROM
WEBsite

The Method of Lagrange Multipliers

In Solution 2 of Example 2, we solved the problem by the **method of Lagrange multipliers.** In general terms, the method says that the extreme values of a function $f(x, y, z)$ whose variables are subject to a constraint $g(x, y, z) = 0$ are to be found on the surface $g = 0$ at the points where

$$\nabla f = \lambda \nabla g$$

for some scalar λ (called a **Lagrange multiplier**).

To explore the method further and see why it works, we first make the following observation, which we state as a theorem.

Theorem 12 The Orthogonal Gradient Theorem

Suppose that $f(x, y, z)$ is differentiable in a region whose interior contains a smooth curve

$$C: \quad \mathbf{r}(t) = g(t)\mathbf{i} + h(t)\mathbf{j} + k(t)\mathbf{k}.$$

If P_0 is a point on C where f has a local maximum or minimum relative to its values on C, then ∇f is orthogonal to C at P_0.

Proof We show that ∇f is orthogonal to the curve's velocity vector at P_0. The values of f on C are given by the composite $f(g(t), h(t), k(t))$, whose derivative with respect to t is

$$\frac{df}{dt} = \frac{\partial f}{\partial x}\frac{dg}{dt} + \frac{\partial f}{\partial y}\frac{dh}{dt} + \frac{\partial f}{\partial z}\frac{dk}{dt} = \nabla f \cdot \mathbf{v}.$$

At any point P_0 where f has a local maximum or minimum relative to its values on the curve, $df/dt = 0$, so

$$\nabla f \cdot \mathbf{v} = 0.$$

By dropping the z-terms in Theorem 12, we obtain a similar result for functions of two variables.

Corollary of Theorem 12

At the points on a smooth curve $\mathbf{r}(t) = g(t)\mathbf{i} + h(t)\mathbf{j}$ where a differentiable function $f(x, y)$ takes on its local maxima and minima relative to its values on the curve, $\nabla f \cdot \mathbf{v} = 0$.

Theorem 12 is the key to the method of Lagrange multipliers. Suppose that $f(x, y, z)$ and $g(x, y, z)$ are differentiable and that P_0 is a point on the smooth surface $g(x, y, z) = 0$ where f has a local maximum or minimum value relative to its other values on the surface. Then f takes on a local maximum or minimum at P_0 relative to its values on every smooth curve through P_0 on the surface $g(x, y, z) = 0$. Therefore, ∇f is orthogonal to the velocity vector of every such differentiable curve through P_0. So is ∇g, however (because ∇g is orthogonal to the smooth level surface $g = 0$, as we saw in Section 11.5). Therefore, at P_0, ∇f is some scalar multiple λ of ∇g. Note that $\nabla g \neq \mathbf{0}$ since the surface $g = 0$ is assumed to be smooth.

The Method of Lagrange Multipliers

Suppose that $f(x, y, z)$ and $g(x, y, z)$ are differentiable. To find the local maximum and minimum values of f subject to the constraint $g(x, y, z) = 0$, find the values of $x, y, z,$ and λ that simultaneously satisfy the equations

$$\nabla f = \lambda \nabla g \quad \text{and} \quad g(x, y, z) = 0.$$

For functions of two independent variables, the appropriate equations are

$$\nabla f = \lambda \nabla g \quad \text{and} \quad g(x, y) = 0.$$

Example 3 Using the Method of Lagrange Multipliers

Find the greatest and smallest values that the function

$$f(x, y) = xy$$

takes on the ellipse (Figure 11.61)

$$\frac{x^2}{8} + \frac{y^2}{2} = 1.$$

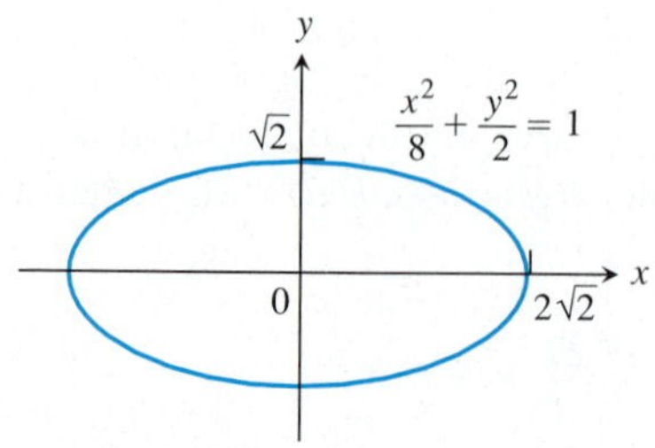

FIGURE 11.61 Example 3 shows how to find the largest and smallest values of the product xy on this ellipse.

Solution We want the extreme values of $f(x, y) = xy$ subject to the constraint

$$g(x, y) = \frac{x^2}{8} + \frac{y^2}{2} - 1 = 0.$$

To do so, we first find the values of $x, y,$ and λ for which

$$\nabla f = \lambda \nabla g \quad \text{and} \quad g(x, y) = 0.$$

The gradient equation gives

$$y\mathbf{i} + x\mathbf{j} = \frac{\lambda}{4}x\mathbf{i} + \lambda y\mathbf{j},$$

from which we find

$$y = \frac{\lambda}{4}x, \qquad x = \lambda y, \qquad \text{and} \qquad y = \frac{\lambda}{4}(\lambda y) = \frac{\lambda^2}{4}y,$$

so that $y = 0$ or $\lambda = \pm 2$. We now consider these two cases.

Case 1: If $y = 0$, then $x = y = 0$. But $(0, 0)$ is not on the ellipse. Hence, $y \neq 0$.

Case 2: If $y \neq 0$, then $\lambda = \pm 2$ and $x = \pm 2y$. Substituting this in the equation $g(x, y) = 0$ gives

$$\frac{(\pm 2y)^2}{8} + \frac{y^2}{2} = 1, \qquad 4y^2 + 4y^2 = 8, \qquad \text{and} \qquad y = \pm 1.$$

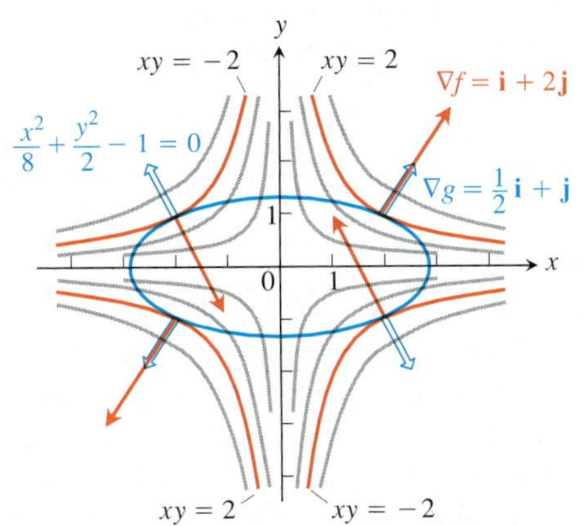

FIGURE 11.62 When subjected to the constraint $g(x, y) = x^2/8 + y^2/2 - 1 = 0$, the function $f(x, y) = xy$ takes on extreme values at the four points $(\pm 2, \pm 1)$. These are the points on the ellipse when ∇f (red) is a scalar multiple of ∇g (blue). (Example 3)

The function $f(x, y) = xy$ therefore takes on its extreme values on the ellipse at the four points $(\pm 2, 1)$, $(\pm 2, -1)$. The extreme values are $xy = 2$ and $xy = -2$.

The Geometry of the Solution

The level curves of the function $f(x, y) = xy$ are the hyperbolas $xy = c$ (Figure 11.62). The farther the hyperbolas lie from the origin, the larger the absolute value of f. We want to find the extreme values of $f(x, y)$, given that the point (x, y) also lies on the ellipse $x^2 + 4y^2 = 8$. Which hyperbolas intersecting the ellipse lie farthest from the origin? The hyperbolas that just graze the ellipse, the ones that are tangent to it, are farthest. At these points, any vector normal to the hyperbola is normal to the ellipse, so $\nabla f = y\mathbf{i} + x\mathbf{j}$ is a multiple $(\lambda = \pm 2)$ of $\nabla g = (x/4)\mathbf{i} + y\mathbf{j}$. At the point $(2, 1)$, for example,

$$\nabla f = \mathbf{i} + 2\mathbf{j}, \qquad \nabla g = \frac{1}{2}\mathbf{i} + \mathbf{j}, \qquad \text{and} \qquad \nabla f = 2\nabla g.$$

At the point $(-2, 1)$,

$$\nabla f = \mathbf{i} - 2\mathbf{j}, \qquad \nabla g = -\frac{1}{2}\mathbf{i} + \mathbf{j}, \qquad \text{and} \qquad \nabla f = -2\nabla g.$$

Example 4 Finding Extreme Function Values on a Circle

Find the maximum and minimum values of the function $f(x, y) = 3x + 4y$ on the circle $x^2 + y^2 = 1$.

Solution We model this as a Lagrange multiplier problem with

$$f(x, y) = 3x + 4y, \qquad g(x, y) = x^2 + y^2 - 1$$

and look for the values of x, y, and λ that satisfy the equations

$$\nabla f = \lambda \nabla g: \quad 3\mathbf{i} + 4\mathbf{j} = 2x\lambda\mathbf{i} + 2y\lambda\mathbf{j}$$

$$g(x, y) = 0: \quad x^2 + y^2 - 1 = 0.$$

The gradient equation implies that $\lambda \neq 0$ and gives

$$x = \frac{3}{2\lambda}, \qquad y = \frac{2}{\lambda}.$$

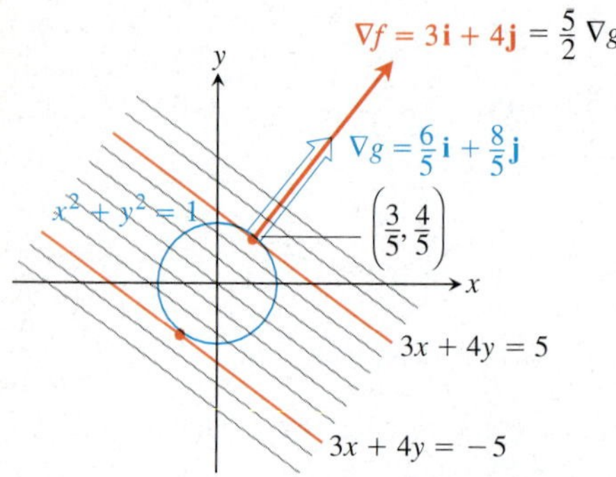

FIGURE 11.63 The function $f(x, y) = 3x + 4y$ takes on its largest value on the unit circle $g(x, y) = x^2 + y^2 - 1 = 0$ at the point $(3/5, 4/5)$ and its smallest value at the point $(-3/5, -4/5)$ (Example 4). At each of these points, ∇f is a scalar multiple of ∇g. The figure shows the gradients at the first point but not the second.

These equations tell us, among other things, that x and y have the same sign. With these values for x and y, the equation $g(x, y) = 0$ gives

$$\left(\frac{3}{2\lambda}\right)^2 + \left(\frac{2}{\lambda}\right)^2 - 1 = 0,$$

so

$$\frac{9}{4\lambda^2} + \frac{4}{\lambda^2} = 1, \qquad 9 + 16 = 4\lambda^2, \qquad 4\lambda^2 = 25, \qquad \text{and} \qquad \lambda = \pm\frac{5}{2}.$$

Thus,

$$x = \frac{3}{2\lambda} = \pm\frac{3}{5}, \qquad y = \frac{2}{\lambda} = \pm\frac{4}{5},$$

and $f(x, y) = 3x + 4y$ has extreme values at $(x, y) = \pm(3/5, 4/5)$.

By calculating the value of $3x + 4y$ at the points $\pm(3/5, 4/5)$, we see that its maximum and minimum values on the circle $x^2 + y^2 = 1$ are

$$3\left(\frac{3}{5}\right) + 4\left(\frac{4}{5}\right) = \frac{25}{5} = 5 \qquad \text{and} \qquad 3\left(-\frac{3}{5}\right) + 4\left(-\frac{4}{5}\right) = -\frac{25}{5} = -5.$$

The Geometry of the Solution

The level curves of $f(x, y) = 3x + 4y$ are the lines $3x + 4y = c$ (Figure 11.63). The farther the lines lie from the origin, the larger the absolute value of f. We want to find the extreme values of $f(x, y)$ given that the point (x, y) also lies on the circle $x^2 + y^2 = 1$. Which lines intersecting the circle lie farthest from the origin? The lines tangent to the circle are farthest. At the points of tangency, any vector normal to the line is normal to the circle, so the gradient $\nabla f = 3\mathbf{i} + 4\mathbf{j}$ is a multiple $(\lambda = \pm 5/2)$ of the gradient $\nabla g = 2x\mathbf{i} + 2y\mathbf{j}$. At the point $(3/5, 4/5)$, for example,

$$\nabla f = 3\mathbf{i} + 4\mathbf{j}, \qquad \nabla g = \frac{6}{5}\mathbf{i} + \frac{8}{5}\mathbf{j}, \qquad \text{and} \qquad \nabla f = \frac{5}{2}\nabla g.$$

Lagrange Multipliers with Two Constraints

Many problems require us to find the extreme values of a differentiable function $f(x, y, z)$ whose variables are subject to two constraints. If the constraints are

$$g_1(x, y, z) = 0 \qquad \text{and} \qquad g_2(x, y, z) = 0$$

and g_1 and g_2 are differentiable, with ∇g_1 not parallel to ∇g_2, we find the constrained local maxima and minima of f by introducing two Lagrange multipliers λ and μ (mu, pronounced "mew"). That is, we locate the points $P(x, y, z)$ where f takes on its constrained extreme values by finding the values of x, y, z, λ, and μ that simultaneously satisfy the equations

$$\nabla f = \lambda \nabla g_1 + \mu \nabla g_2, \qquad g_1(x, y, z) = 0, \qquad g_2(x, y, z) = 0. \tag{2}$$

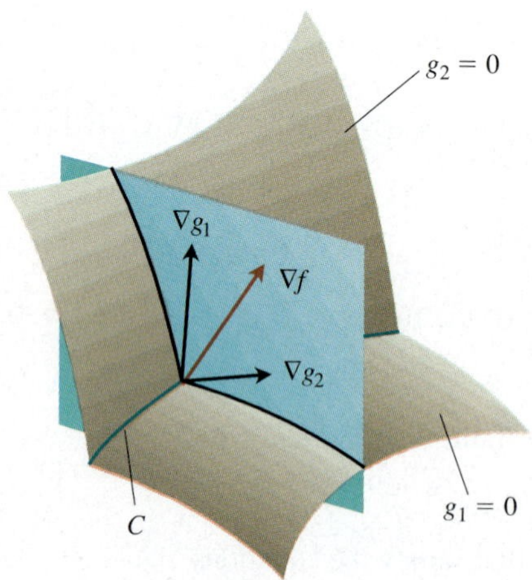

FIGURE 11.64 The vectors ∇g_1 and ∇g_2 lie in a plane perpendicular to the curve C because ∇g_1 is normal to the surface $g_1 = 0$ and ∇g_2 is normal to the surface $g_2 = 0$.

Equations (2) have a nice geometric interpretation. The surfaces $g_1 = 0$ and $g_2 = 0$ (usually) intersect in a smooth curve, say C (Figure 11.64). Along this curve we seek the points where f has local maximum and minimum values relative to its other values on the curve. These are the points where ∇f is normal to C, as we saw in Theorem 12. But ∇g_1 and ∇g_2 are also normal to C at these points because C lies in the surfaces $g_1 = 0$ and $g_2 = 0$. Therefore, ∇f lies in the plane determined by ∇g_1 and ∇g_2, which means that $\nabla f = \lambda \nabla g_1 + \mu \nabla g_2$ for some λ and μ. Since the points we seek also lie in both surfaces, their coordinates must satisfy the equations $g_1(x, y, z) = 0$ and $g_2(x, y, z) = 0$, which are the remaining requirements in Equations (2).

Example 5 Finding Extremes of Distance on an Ellipse

The plane $x + y + z = 1$ cuts the cylinder $x^2 + y^2 = 1$ in an ellipse (Figure 11.65). Find the points on the ellipse that lie closest to and farthest from the origin.

Solution We find the extreme values of

$$f(x, y, z) = x^2 + y^2 + z^2$$

(the square of the distance from (x, y, z) to the origin) subject to the constraints

$$g_1(x, y, z) = x^2 + y^2 - 1 = 0 \quad (3)$$

$$g_2(x, y, z) = x + y + z - 1 = 0. \quad (4)$$

The gradient equation in Equations (2) then gives

$$\begin{aligned}\nabla f &= \lambda \nabla g_1 + \mu \nabla g_2\\ 2x\mathbf{i} + 2y\mathbf{j} + 2z\mathbf{k} &= \lambda(2x\mathbf{i} + 2y\mathbf{j}) + \mu(\mathbf{i} + \mathbf{j} + \mathbf{k})\\ 2x\mathbf{i} + 2y\mathbf{j} + 2z\mathbf{k} &= (2\lambda x + \mu)\mathbf{i} + (2\lambda y + \mu)\mathbf{j} + \mu\mathbf{k}\end{aligned}$$

or

$$2x = 2\lambda x + \mu, \qquad 2y = 2\lambda y + \mu, \qquad 2z = \mu. \quad (5)$$

The scalar equations in (5) yield

$$\begin{aligned}2x &= 2\lambda x + 2z \Rightarrow (1 - \lambda)x = z,\\ 2y &= 2\lambda y + 2z \Rightarrow (1 - \lambda)y = z.\end{aligned} \quad (6)$$

Equations (6) are satisfied simultaneously if either $\lambda = 1$ and $z = 0$ or $\lambda \neq 1$ and $x = y = z/(1 - \lambda)$.

If $z = 0$, then solving Equations (3) and (4) simultaneously to find the corresponding points on the ellipse gives the two points $(1, 0, 0)$ and $(0, 1, 0)$. This makes sense when you look at Figure 11.65.

If $x = y$, then Equations (3) and (4) give

$$\begin{aligned}x^2 + x^2 - 1 &= 0 & x + x + z - 1 &= 0\\ 2x^2 &= 1 & z &= 1 - 2x\\ x &= \pm\frac{\sqrt{2}}{2} & z &= 1 \mp \sqrt{2}.\end{aligned}$$

The corresponding points on the ellipse are

$$P_1 = \left(\frac{\sqrt{2}}{2}, \frac{\sqrt{2}}{2}, 1 - \sqrt{2}\right) \quad \text{and} \quad P_2 = \left(-\frac{\sqrt{2}}{2}, -\frac{\sqrt{2}}{2}, 1 + \sqrt{2}\right).$$

Here we need to be careful, however. Although P_1 and P_2 both give local maxima of f on the ellipse, P_2 is farther from the origin than P_1.

The points on the ellipse closest to the origin are $(1, 0, 0)$ and $(0, 1, 0)$. The point on the ellipse farthest from the origin is P_2.

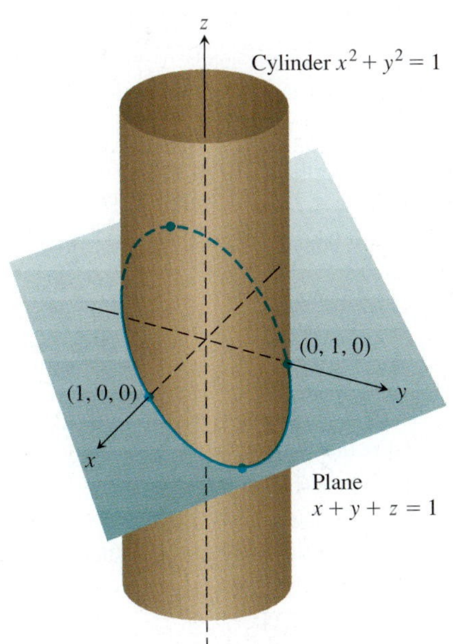

FIGURE 11.65 On the ellipse where the plane and cylinder meet, what are the points closest to and farthest from the origin? (Example 5)

EXERCISES 11.8

Two Independent Variables with One Constraint

1. *Extrema on an ellipse* Find the points on the ellipse $x^2 + 2y^2 = 1$ where $f(x, y) = xy$ has its extreme values.

2. *Extrema on a circle* Find the extreme values of $f(x, y) = xy$ subject to the constraint $g(x, y) = x^2 + y^2 - 10 = 0$.

3. *Maximum on a line* Find the maximum value of $f(x, y) = 49 - x^2 - y^2$ on the line $x + 3y = 10$ (Figure 11.57).

4. *Extrema on a line* Find the local extreme values of $f(x, y) = x^2y$ on the line $x + y = 3$.

5. *Constrained minimum* Find the points on the curve $xy^2 = 54$ nearest the origin.

6. *Constrained minimum* Find the points on the curve $x^2y = 2$ nearest the origin.

7. *Writing to Learn* Use the method of Lagrange multipliers to find

(a) *Minimum on a hyperbola* The minimum value of $x + y$, subject to the constraints $xy = 16$, $x > 0$, $y > 0$

(b) *Maximum on a line* The maximum value of xy, subject to the constraint $x + y = 16$.

Comment on the geometry of each solution.

8. *Extrema on a curve* Find the points on the curve $x^2 + xy + y^2 = 1$ in the xy-plane that are nearest to and farthest from the origin.

9. *Minimum surface area with fixed volume* Find the dimensions of the closed right circular cylindrical can of smallest surface area whose volume is 16π cm^3.

10. *Cylinder in a sphere* Find the radius and height of the open right circular cylinder of largest surface area that can be inscribed in a sphere of radius a. What *is* the largest surface area?

11. *Rectangle of greatest area in an ellipse* Use the method of Lagrange multipliers to find the dimensions of the rectangle of greatest area that can be inscribed in the ellipse $x^2/16 + y^2/9 = 1$ with sides parallel to the coordinate axes.

12. *Rectangle of longest perimeter in an ellipse* Find the dimensions of the rectangle of largest perimeter that can be inscribed in the ellipse $x^2/a^2 + y^2/b^2 = 1$ with sides parallel to the coordinate axes. What *is* the largest perimeter?

13. *Extrema on a circle* Find the maximum and minimum values of $x^2 + y^2$ subject to the constraint $x^2 - 2x + y^2 - 4y = 0$.

14. *Extrema on a circle* Find the maximum and minimum values of $3x - y + 6$ subject to the constraint $x^2 + y^2 = 4$.

15. *Ant on a metal plate* The temperature at a point (x, y) on a metal plate is $T(x, y) = 4x^2 - 4xy + y^2$. An ant on the plate walks around the circle of radius 5 centered at the origin. What are the highest and lowest temperatures encountered by the ant?

16. *Cheapest storage tank* Your firm has been asked to design a storage tank for liquid petroleum gas. The customer's specifications call for a cylindrical tank with hemispherical ends, and the tank is to hold 8000 m^3 of gas. The customer also wants to use the smallest amount of material possible in building the tank. What radius and height do you recommend for the cylindrical portion of the tank?

Three Independent Variables with One Constraint

17. *Minimum distance to a point* Find the point on the plane $x + 2y + 3z = 13$ closest to the point $(1, 1, 1)$.

18. *Maximum distance to a point* Find the point on the sphere $x^2 + y^2 + z^2 = 4$ farthest from the point $(1, -1, 1)$.

19. *Minimum distance to the origin* Find the minimum distance from the surface $x^2 + y^2 - z^2 = 1$ to the origin.

20. *Minimum distance to the origin* Find the point on the surface $z = xy + 1$ nearest the origin.

21. *Minimum distance to the origin* Find the points on the surface $z^2 = xy + 4$ closest to the origin.

22. *Minimum distance to the origin* Find the point(s) on the surface $xyz = 1$ closest to the origin.

23. *Extrema on a sphere* Find the maximum and minimum values of

$$f(x, y, z) = x - 2y + 5z$$

on the sphere $x^2 + y^2 + z^2 = 30$.

24. *Extrema on a sphere* Find the points on the sphere $x^2 + y^2 + z^2 = 25$ where $f(x, y, z) = x + 2y + 3z$ has its maximum and minimum values.

25. *Minimizing a sum of squares* Find three real numbers whose sum is 9 and the sum of whose squares is as small as possible.

26. *Maximizing a product* Find the largest product the positive numbers x, y, and z can have if $x + y + z^2 = 16$.

27. *Rectangular box of longest volume in a sphere* Find the dimensions of the closed rectangular box with maximum volume that can be inscribed in the unit sphere.

28. *Box with vertex on a plane* Find the volume of the largest closed rectangular box in the first octant having three faces in the coordinate planes and a vertex on the plane $x/a + y/b + z/c = 1$, where $a > 0$, $b > 0$, and $c > 0$.

29. *Hottest point on a space probe* A space probe in the shape of the ellipsoid

$$4x^2 + y^2 + 4z^2 = 16$$

enters Earth's atmosphere and its surface begins to heat. After 1 h, the temperature at the point (x, y, z) on the probe's surface is

$$T(x, y, z) = 8x^2 + 4yz - 16z + 600.$$

Find the hottest point on the probe's surface.

30. *Extreme temperatures on a sphere* Suppose that the Celsius temperature at the point (x, y, z) on the sphere $x^2 + y^2 + z^2 = 1$ is $T = 400xyz^2$. Locate the highest and lowest temperatures on the sphere.

31. *Maximizing a utility function: an example from economics* In economics, the usefulness or *utility* of amounts x and y of two capital goods G_1 and G_2 is sometimes measured by a function $U(x, y)$. For example, G_1 and G_2 might be two chemicals a pharmaceutical company needs to have on hand and $U(x, y)$ the gain from manufacturing a product whose synthesis requires different amounts of the chemicals depending on the process used. If G_1 costs a dollars per kilogram, G_2 costs b dollars per kilogram, and the total amount allocated for the purchase of G_1 and G_2 together is c dollars, then the company's managers want to maximize $U(x, y)$ given that $ax + by = c$. Thus, they need to solve a typical Lagrange multiplier problem.

Suppose that

$$U(x, y) = xy + 2x$$

and that the equation $ax + by = c$ simplifies to

$$2x + y = 30.$$

Find the maximum value of U and the corresponding values of x and y subject to this latter constraint.

32. *Locating a radio telescope* You are in charge of erecting a radio telescope on a newly discovered planet. To minimize interference, you want to place it where the magnetic field of the planet is weakest. The planet is spherical, with a radius of 6 units. Based on a coordinate system whose origin is at the center of the planet, the strength of the magnetic field is given by $M(x, y, z) = 6x - y^2 + xz + 60$. Where should you locate the radio telescope?

Extreme Values Subject to Two Constraints

33. Maximize the function $f(x, y, z) = x^2 + 2y - z^2$ subject to the constraints $2x - y = 0$ and $y + z = 0$.

34. Minimize the function $f(x, y, z) = x^2 + y^2 + z^2$ subject to the constraints $x + 2y + 3z = 6$ and $x + 3y + 9z = 9$.

35. *Minimum distance to the origin* Find the point closest to the origin on the line of intersection of the planes $y + 2z = 12$ and $x + y = 6$.

36. *Maximum value on line of intersection* Find the maximum value that $f(x, y, z) = x^2 + 2y - z^2$ can have on the line of intersection of the planes $2x - y = 0$ and $y + z = 0$.

37. *Extrema on a curve of intersection* Find the extreme values of $f(x, y, z) = x^2yz + 1$ on the intersection of the plane $z = 1$ with the sphere $x^2 + y^2 + z^2 = 10$.

38. **(a)** *Maximum on line of intersection* Find the maximum value of $w = xyz$ on the line of intersection of the two planes $x + y + z = 40$ and $x + y - z = 0$.

(b) *Writing to Learn* Give a geometric argument to support your claim that you have found a maximum, and not a minimum, value of w.

39. *Extrema on a circle of intersection* Find the extreme values of the function $f(x, y, z) = xy + z^2$ on the circle in which the plane $y - x = 0$ intersects the sphere $x^2 + y^2 + z^2 = 4$.

40. *Minimum distance to the origin* Find the point closest to the origin on the curve of intersection of the plane $2y + 4z = 5$ and the cone $z^2 = 4x^2 + 4y^2$.

Theory and Examples

41. *The condition $\nabla f = \lambda \nabla g$ is not sufficient* Although $\nabla f = \lambda \nabla g$ is a necessary condition for the occurrence of an extreme value of $f(x, y)$ subject to the condition $g(x, y) = 0$, it does not in itself guarantee that one exists. As a case in point, try using the method of Lagrange multipliers to find a maximum value of $f(x, y) = x + y$ subject to the constraint that $xy = 16$. The method will identify the two points $(4, 4)$ and $(-4, -4)$ as candidates for the location of extreme values. Yet the sum $(x + y)$ has no maximum value on the hyperbola $xy = 16$. The farther you go from the origin on this hyperbola in the first quadrant, the larger the sum $f(x, y) = x + y$ becomes.

42. *A least squares plane* The plane $z = Ax + By + C$ is to be "fitted" to the following points (x_k, y_k, z_k):

$$(0, 0, 0), \quad (0, 1, 1), \quad (1, 1, 1), \quad (1, 0, -1).$$

Find the values of A, B, and C that minimize

$$\sum_{k=1}^{4} (Ax_k + By_k + C - z_k)^2,$$

the sum of the squares of the deviations.

43. **(a)** *Maximum on a sphere* Show that the maximum value of $a^2b^2c^2$ on a sphere of radius r centered at the origin of a Cartesian abc-coordinate system is $(r^2/3)^3$.

(b) *Geometric and arithmetic means* Using part (a), show that for nonnegative numbers a, b, and c,

$$(abc)^{1/3} \le \frac{a + b + c}{3};$$

that is, the *geometric mean* of three nonnegative numbers is less than or equal to their *arithmetic mean.*

44. *Sum of products* Let $a_1, a_2, \ldots, a_n$ be n positive numbers. Find the maximum of $\Sigma_{i=1}^{n} a_i x_i$ subject to the constraint $\Sigma_{i=1}^{n} x_i^2 = 1$.

COMPUTER EXPLORATIONS

Implementing the Method of Lagrange Multipliers

In Exercises 45–50, use a CAS to perform the following steps implementing the method of Lagrange multipliers for finding constrained extrema:

(a) Form the function $h = f - \lambda_1 g_1 - \lambda_2 g_2$, where f is the function to optimize subject to the constraints $g_1 = 0$ and $g_2 = 0$.

(b) Determine all the first partial derivatives of h, including the partials with respect to λ_1 and λ_2, and set them equal to 0.

(c) Solve the system of equations found in part (b) for all the unknowns, including λ_1 and λ_2.

(d) Evaluate f at each of the solution points found in part (c) and select the extreme value subject to the constraints asked for in the exercise.

45. Minimize $f(x, y, z) = xy + yz$ subject to the constraints $x^2 + y^2 - 2 = 0$ and $x^2 + z^2 - 2 = 0$.

46. Minimize $f(x, y, z) = xyz$ subject to the constraints $x^2 + y^2 - 1 = 0$ and $x - z = 0$.

47. Maximize $f(x, y, z) = x^2 + y^2 + z^2$ subject to the constraints $2y + 4z - 5 = 0$ and $4x^2 + 4y^2 - z^2 = 0$.

48. Minimize $f(x, y, z) = x^2 + y^2 + z^2$ subject to the constraints $x^2 - xy + y^2 - z^2 - 1 = 0$ and $x^2 + y^2 - 1 = 0$.

49. Minimize $f(x, y, z, w) = x^2 + y^2 + z^2 + w^2$ subject to the constraints $2x - y + z - w - 1 = 0$ and $x + y - z + w - 1 = 0$.

50. Determine the distance from the line $y = x + 1$ to the parabola $y^2 = x$. (*Hint:* Let (x, y) be a point on the line and (w, z) a point on the parabola. You want to minimize $(x - w)^2 + (y - z)^2$.)

11.9 *Partial Derivatives with Constrained Variables

Decide Which Variables Are Dependent and Which Are Independent • How to Find $\partial w/\partial x$ When the Variables in $w = f(x, y, z)$ Are Constrained by Another Equation • Notation • Arrow Diagrams

In finding partial derivatives of functions like $w = f(x, y)$, we have assumed x and y to be independent. In many applications, however, this is not the case. For example, the internal energy U of a gas may be expressed as a function $U = f(P, V, T)$ of pressure P, volume V, and temperature T. If the individual molecules of the gas do not interact, however, P, V, and T obey (and are constrained by) the ideal gas law

$$PV = nRT \qquad (n \text{ and } R \text{ constant}).$$

and fail to be independent. Finding partial derivatives in situations like these can be complicated, but it is better to face the complication now than to meet it for the first time while you are also trying to learn economics, engineering, or physics.

Decide Which Variables Are Dependent and Which Are Independent

If the variables in a function $w = f(x, y, z)$ are constrained by a relation like the one imposed on x, y, and z by the equation $z = x^2 + y^2$, the geometric meanings and the numerical values of the partial derivatives of f will depend on which variables are chosen to be dependent and which are chosen to be independent. To see how this choice can affect the outcome, we consider the calculation of $\partial w/\partial x$ when $w = x^2 + y^2 + z^2$ and $z = x^2 + y^2$.

Example 1 Finding a Partial Derivative with Constrained Independent Variables

Find $\partial w/\partial x$ if $w = x^2 + y^2 + z^2$ and $z = x^2 + y^2$.

Solution We are given two equations in the four unknowns x, y, z, and w. Like many such systems, this one can be solved for two of the unknowns (the dependent variables) in terms of the others (the independent variables). In

*This section is based on notes written for MIT by Arthur P. Mattuck.

being asked for $\partial w/\partial x$, we are told that w is to be a dependent variable and x an independent variable. The possible choices for the other variables come down to

Dependent	*Independent*
w, z	x, y
w, y	x, z

In either case, we can express w explicitly in terms of the selected independent variables. We do this by using the second equation to eliminate the remaining dependent variable in the first equation.

In the first case, the remaining dependent variable is z. We eliminate it from the first equation by replacing it by $x^2 + y^2$. The resulting expression for w is

$$\begin{aligned} w &= x^2 + y^2 + z^2 = x^2 + y^2 + (x^2 + y^2)^2 \\ &= x^2 + y^2 + x^4 + 2x^2y^2 + y^4 \end{aligned}$$

and

$$\frac{\partial w}{\partial x} = 2x + 4x^3 + 4xy^2. \tag{1}$$

This is the formula for $\partial w/\partial x$ when x and y are the independent variables.

In the second case, where the independent variables are x and z and the remaining dependent variable is y, we eliminate the dependent variable y in the expression for w by replacing y^2 by $z - x^2$. This gives

$$w = x^2 + y^2 + z^2 = x^2 + (z - x^2) + z^2 = z + z^2$$

and

$$\frac{\partial w}{\partial x} = 0. \tag{2}$$

This is the formula for $\partial w/\partial x$ when x and z are the independent variables.

The formulas for $\partial w/\partial x$ in Equations (1) and (2) are genuinely different. We cannot change either formula into the other by using the relation $z = x^2 + y^2$. There is not just one $\partial w/\partial x$, there are two, and we see that the original instruction to find $\partial w/\partial x$ was incomplete. *Which* $\partial w/\partial x$? we ask.

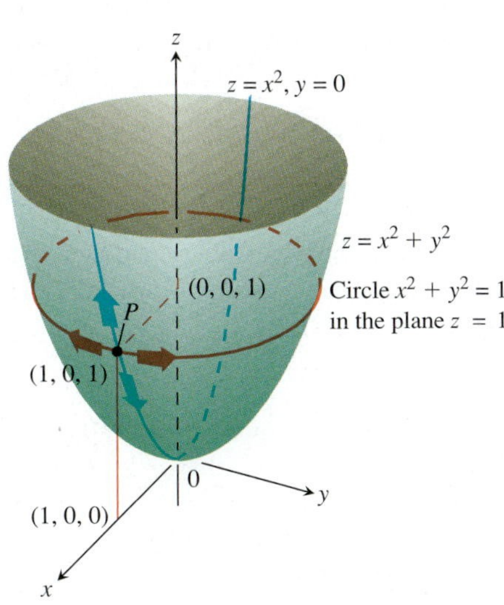

FIGURE 11.66 If P is constrained to lie on the paraboloid $z = x^2 + y^2$, the value of the partial derivative of $w = x^2 + y^2 + z^2$ with respect to x at P depends on the direction of motion (Example 1). (a) As x changes, with $y = 0$, P moves up or down the surface on the parabola $z = x^2$ in the xz-plane with $\partial w/\partial x = 2x + 4x^3 + 4xy^2$. (b) As x changes, with $z = 1$, P moves on the circle $x^2 + y^2 = 1$, $z = 1$, and $\partial w/\partial x = 0$.

The geometric interpretations of Equations (1) and (2) help to explain why the equations differ. The function $w = x^2 + y^2 + z^2$ measures the square of the distance from the point (x, y, z) to the origin. The condition $z = x^2 + y^2$ says that the point (x, y, z) lies on the paraboloid of revolution shown in Figure 11.66. What does it mean to calculate $\partial w/\partial x$ at a point $P(x, y, z)$ that can move only on this surface? What is the value of $\partial w/\partial x$ when the coordinates of P are, say, $(1, 0, 1)$?

If we take x and y to be independent, then we find $\partial w/\partial x$ by holding y fixed (at $y = 0$ in this case) and letting x vary. Hence, P moves along the parabola $z = x^2$ in the xz-plane. As P moves on this parabola, w, which is the square of the distance from P to the origin, changes. We calculate $\partial w/\partial x$ in this case (our first solution above) to be

$$\frac{\partial w}{\partial x} = 2x + 4x^3 + 4xy^2.$$

At the point $P(1, 0, 1)$, the value of this derivative is

$$\frac{\partial w}{\partial x} = 2 + 4 + 0 = 6.$$

If we take x and z to be independent, then we find $\partial w/\partial x$ by holding z fixed while x varies. Since the z-coordinate of P is 1, varying x moves P along a circle in the plane $z = 1$. As P moves along this circle, its distance from the origin remains constant, and w, being the square of this distance, does not change. That is,

$$\frac{\partial w}{\partial x} = 0,$$

as we found in our second solution.

How to Find $\partial w/\partial x$ When the Variables in $w = f(x, y, z)$ Are Constrained by Another Equation

As we saw in Example 1, a typical routine for finding $\partial w/\partial x$ when the variables in the function $w = f(x, y, z)$ are related by another equation has three steps. These steps apply to finding $\partial w/\partial y$ and $\partial w/\partial z$ as well.

Step 1: *Decide* which variables are to be dependent and which are to be independent. (In practice, the decision is based on the physical or theoretical context of our work. In the exercises at the end of this section, we say which variables are which.)

Step 2: *Eliminate* the other dependent variable(s) in the expression for w.

Step 3: *Differentiate* as usual.

If we cannot carry out step 2 after deciding which variables are dependent, we differentiate the equations as they are and try to solve for $\partial w/\partial x$ afterward. The next example shows how this is done.

Example 2 Finding a Partial Derivative with Identified Constrained Independent Variables

Find $\partial w/\partial x$ at the point $(x, y, z) = (2, -1, 1)$ if

$$w = x^2 + y^2 + z^2, \qquad z^3 - xy + yz + y^3 = 1,$$

and x and y are the independent variables.

Solution It is not convenient to eliminate z in the expression for w. We therefore differentiate both equations implicitly with respect to x, treating x and y as independent variables and w and z as dependent variables. This gives

$$\frac{\partial w}{\partial x} = 2x + 2z\frac{\partial z}{\partial x} \tag{3}$$

and

$$3z^2\frac{\partial z}{\partial x} - y + y\frac{\partial z}{\partial x} + 0 = 0. \tag{4}$$

These equations may now be combined to express $\partial w/\partial x$ in terms of x, y, and z. We solve Equation (4) for $\partial z/\partial x$ to get

$$\frac{\partial z}{\partial x} = \frac{y}{y + 3z^2}$$

and substitute into Equation (3) to get

$$\frac{\partial w}{\partial x} = 2x + \frac{2yz}{y + 3z^2}.$$

The value of this derivative at $(x, y, z) = (2, -1, 1)$ is

$$\left(\frac{\partial w}{\partial x}\right)_{(2,-1,1)} = 2(2) + \frac{2(-1)(1)}{-1 + 3(1)^2} = 4 + \frac{-2}{2} = 3.$$

Notation

CD-ROM WEBsite Historical Biography

Sonya Kovalevsky (1850 — 1891)

To show what variables are assumed to be independent in calculating a derivative, we can use the following notation:

$\left(\frac{\partial w}{\partial x}\right)_y$ $\quad$ $\partial w/\partial x$ with x and y independent

$\left(\frac{\partial f}{\partial y}\right)_{x,t}$ $\quad$ $\partial f/\partial y$ with y, x and t independent

Example 3 Finding a Partial Derivative with Constrained Variables Notationally Identified

Find $(\partial w/\partial x)_{y,z}$ if $w = x^2 + y - z + \sin t$ and $x + y = t$.

Solution With x, y, z independent, we have

$$t = x + y, \qquad w = x^2 + y - z + \sin(x + y)$$

$$\left(\frac{\partial w}{\partial x}\right)_{y,z} = 2x + 0 - 0 + \cos(x + y)\frac{\partial}{\partial x}(x + y)$$

$$= 2x + \cos(x + y).$$

Arrow Diagrams

In solving problems like the one in Example 3, it often helps to start with an arrow diagram that shows how the variables and functions are related. If

$$w = x^2 + y - z + \sin t \qquad \text{and} \qquad x + y = t$$

and we are asked to find $\partial w/\partial x$ when x, y, and z are independent, the appropriate diagram is one like this:

$$\begin{pmatrix} x \\ y \\ z \end{pmatrix} \rightarrow \begin{pmatrix} x \\ y \\ z \\ t \end{pmatrix} \rightarrow w \tag{5}$$

Independent variables $\quad$ Intermediate variables $\quad$ Dependent variable

To avoid confusion between the independent and intermediate variables with the same symbolic names in the diagram, it is helpful to rename the intermediate variables (so they are seen as *functions* of the independent variables). Thus, let $u = x$, $v = y$, and $s = z$ denote the renamed intermediate variables. With this notation, the arrow diagram becomes

$$\begin{pmatrix} x \\ y \\ z \end{pmatrix} \rightarrow \begin{pmatrix} u \\ v \\ s \\ t \end{pmatrix} \rightarrow w \tag{6}$$

Independent variables | Intermediate variables and relations | Dependent variable

$u = x$
$v = y$
$s = z$
$t = x + y$

The diagram shows the independent variables on the left, the intermediate variables and their relation to the independent variables in the middle, and the dependent variable on the right. The function w now becomes

$$w = u^2 + v - s + \sin t,$$

where

$$u = x, \qquad v = y, \qquad s = z, \qquad \text{and} \qquad t = x + y.$$

To find $\partial w/\partial x$, we apply the four-variable form of the Chain Rule to w, guided by the arrow diagram (6):

$$\begin{aligned}
\frac{\partial w}{\partial x} &= \frac{\partial w}{\partial u}\frac{\partial u}{\partial x} + \frac{\partial w}{\partial v}\frac{\partial v}{\partial x} + \frac{\partial w}{\partial s}\frac{\partial s}{\partial x} + \frac{\partial w}{\partial t}\frac{\partial t}{\partial x}.\\
&= (2u)(1) + (1)(0) + (-1)(0) + (\cos t)(1)\\
&= 2u + \cos t\\
&= 2x + \cos(x + y).
\end{aligned}$$

Substituting the original independent variables $u = x$ and $t = x + y$.

EXERCISES 11.9

Finding Partial Derivatives with Constrained Variables

In Exercises 1–3, begin by drawing a diagram that shows the relations among the variables.

1. If $w = x^2 + y^2 + z^2$ and $z = x^2 + y^2$, find

(a) $\left(\frac{\partial w}{\partial y}\right)_z$ **(b)** $\left(\frac{\partial w}{\partial z}\right)_x$ **(c)** $\left(\frac{\partial w}{\partial z}\right)_y$.

2. If $w = x^2 + y - z + \sin t$ and $x + y = t$, find

(a) $\left(\frac{\partial w}{\partial y}\right)_{x,z}$ **(b)** $\left(\frac{\partial w}{\partial y}\right)_{z,t}$ **(c)** $\left(\frac{\partial w}{\partial z}\right)_{x,y}$

(d) $\left(\frac{\partial w}{\partial z}\right)_{y,t}$ **(e)** $\left(\frac{\partial w}{\partial t}\right)_{x,z}$ **(f)** $\left(\frac{\partial w}{\partial t}\right)_{y,z}$.

3. Let $U = f(P, V, T)$ be the internal energy of a gas that obeys the ideal gas law $PV = nRT$ (n and R constant). Find

(a) $\left(\frac{\partial U}{\partial P}\right)_V$ **(b)** $\left(\frac{\partial U}{\partial T}\right)_V$.

4. Find

(a) $\left(\frac{\partial w}{\partial x}\right)_y$ **(b)** $\left(\frac{\partial w}{\partial z}\right)_y$

at the point $(x, y, z) = (0, 1, \pi)$ if

$$w = x^2 + y^2 + z^2 \quad \text{and} \quad y \sin z + z \sin x = 0.$$

5. Find

(a) $\left(\frac{\partial w}{\partial y}\right)_x$ **(b)** $\left(\frac{\partial w}{\partial y}\right)_z$

at the point $(w, x, y, z) = (4, 2, 1, -1)$ if

$$w = x^2y^2 + yz - z^3 \quad \text{and} \quad x^2 + y^2 + z^2 = 6.$$

6. Find $(\partial u/\partial y)_x$ at the point $(u, v) = (\sqrt{2}, 1)$, if $x = u^2 + v^2$ and $y = uv$.

7. Suppose that $x^2 + y^2 = r^2$ and $x = r \cos \theta$, as in polar coordinates. Find

$$\left(\frac{\partial x}{\partial r}\right)_\theta \quad \text{and} \quad \left(\frac{\partial r}{\partial x}\right)_y.$$

8. Suppose that

$$w = x^2 - y^2 + 4z + t \quad \text{and} \quad x + 2z + t = 25.$$

Show that the equations

$$\frac{\partial w}{\partial x} = 2x - 1 \quad \text{and} \quad \frac{\partial w}{\partial x} = 2x - 2$$

each give $\partial w/\partial x$, depending on which variables are chosen to be dependent and which variables are chosen to be independent. Identify the independent variables in each case.

Partial Derivatives without Specific Formulas

9. Establish the fact, widely used in hydrodynamics, that if $f(x, y, z) = 0$, then

$$\left(\frac{\partial x}{\partial y}\right)_z \left(\frac{\partial y}{\partial z}\right)_x \left(\frac{\partial z}{\partial x}\right)_y = -1.$$

(*Hint:* Express all the derivatives in terms of the formal partial derivatives $\partial f/\partial x$, $\partial f/\partial y$, and $\partial f/\partial z$.)

10. If $z = x + f(u)$, where $u = xy$, show that

$$x\frac{\partial z}{\partial x} - y\frac{\partial z}{\partial y} = x.$$

11. Suppose that the equation $g(x, y, z) = 0$ determines z as a differentiable function of the independent variables x and y and that $g_z \neq 0$. Show that

$$\left(\frac{\partial z}{\partial y}\right)_x = -\frac{\partial g/\partial y}{\partial g/\partial z}.$$

12. Suppose that $f(x, y, z, w) = 0$ and $g(x, y, z, w) = 0$ determine z and w as differentiable functions of the independent variables x and y, and suppose that

$$\frac{\partial f}{\partial z}\frac{\partial g}{\partial w} - \frac{\partial f}{\partial w}\frac{\partial g}{\partial z} \neq 0.$$

Show that

$$\left(\frac{\partial z}{\partial x}\right)_y = -\frac{\dfrac{\partial f}{\partial x}\dfrac{\partial g}{\partial w} - \dfrac{\partial f}{\partial w}\dfrac{\partial g}{\partial x}}{\dfrac{\partial f}{\partial z}\dfrac{\partial g}{\partial w} - \dfrac{\partial f}{\partial w}\dfrac{\partial g}{\partial z}}$$

and

$$\left(\frac{\partial w}{\partial y}\right)_x = -\frac{\dfrac{\partial f}{\partial z}\dfrac{\partial g}{\partial y} - \dfrac{\partial f}{\partial y}\dfrac{\partial g}{\partial z}}{\dfrac{\partial f}{\partial z}\dfrac{\partial g}{\partial w} - \dfrac{\partial f}{\partial w}\dfrac{\partial g}{\partial z}}.$$

11.10 Taylor's Formula for Two Variables

Derivation of the Second Derivative Test • Error Formula for Linear Approximations • Taylor's Formula for Functions of Two Variables

This section uses Taylor's formula (Section 8.7) to derive the second derivative test for local extreme values (Section 11.7) and the error formula for linearizations of functions of two independent variables (Section 11.6). The use of Taylor's formula in these derivations leads to an extension of the formula that provides polynomial approximations of all orders for functions of two independent variables.

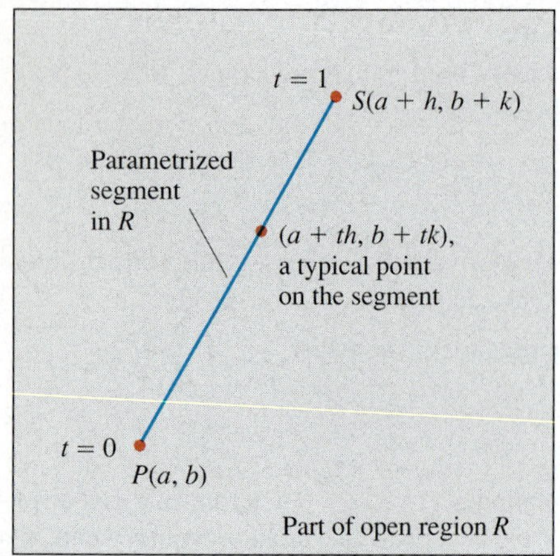

FIGURE 11.67 We begin the derivation of the second derivative test at $P(a, b)$ by parametrizing a typical line segment from P to a point S nearby.

Derivation of the Second Derivative Test

Let $f(x, y)$ have continuous partial derivatives in an open region R containing a point $P(a, b)$ where $f_x = f_y = 0$ (Figure 11.67). Let h and k be increments small enough to put the point $S(a + h, b + k)$ and the line segment joining it to P inside R. We parametrize the segment PS as

$$x = a + th, \qquad y = b + tk, \qquad 0 \le t \le 1.$$

If $F(t) = f(a + th, b + tk)$, the Chain Rule gives

$$F'(t) = f_x \frac{dx}{dt} + f_y \frac{dy}{dt} = hf_x + kf_y.$$

Since f_x and f_y are differentiable (they have continuous partial derivatives), F' is a differentiable function of t and

$$\begin{aligned} F'' &= \frac{\partial F'}{\partial x}\frac{dx}{dt} + \frac{\partial F'}{\partial y}\frac{dy}{dt} = \frac{\partial}{\partial x}(hf_x + kf_y)\cdot h + \frac{\partial}{\partial y}(hf_x + kf_y)\cdot k \\ &= h^2 f_{xx} + 2hkf_{xy} + k^2 f_{yy}. \qquad f_{xy} = f_{yx} \end{aligned}$$

Since F and F' are continuous on $[0, 1]$ and F' is differentiable on $(0, 1)$, we can apply Taylor's formula with $n = 2$ and $a = 0$ to obtain

$$F(1) = F(0) + F'(0)(1 - 0) + F''(c)\frac{(1 - 0)^2}{2}$$

$$F(1) = F(0) + F'(0) + \frac{1}{2}F''(c) \tag{1}$$

for some c between 0 and 1. Writing Equation (1) in terms of f gives

$$\begin{aligned} f(a + h, b + k) = f(a, b) &+ hf_x(a, b) + kf_y(a, b) \\ &+ \frac{1}{2}(h^2 f_{xx} + 2hkf_{xy} + k^2 f_{yy})\Big|_{(a+ch,b+ck)} \end{aligned} \tag{2}$$

Since $f_x(a, b) = f_y(a, b) = 0$, this last equation reduces to

$$f(a + h, b + k) - f(a, b) = \frac{1}{2}(h^2 f_{xx} + 2hkf_{xy} + k^2 f_{yy})\Big|_{(a+ch,b+ck)} \tag{3}$$

The presence of an extremum of f at (a, b) is determined by the sign of $f(a + h, b + k) - f(a, b)$. By Equation (3), this is the same as the sign of

$$Q(c) = (h^2 f_{xx} + 2hkf_{xy} + k^2 f_{yy})|_{(a+ch,b+ck)}.$$

Now, if $Q(0) \neq 0$, the sign of $Q(c)$ will be the same as the sign of $Q(0)$ for sufficiently small values of h and k. We can predict the sign of

$$Q(0) = h^2 f_{xx}(a, b) + 2hkf_{xy}(a, b) + k^2 f_{yy}(a, b) \tag{4}$$

from the signs of f_{xx} and $f_{xx}f_{yy} - f_{xy}^{\,2}$ at (a, b). Multiply both sides of Equation (4) by f_{xx} and rearrange the right-hand side to get

$$f_{xx}\,Q(0) = (hf_{xx} + kf_{xy})^2 + (f_{xx}f_{yy} - f_{xy}^{\,2})\,k^2 \tag{5}$$

at (a, b). From Equation (5), we see that

1. If $f_{xx} < 0$ and $f_{xx}f_{yy} - f_{xy}^{\,2} > 0$ at (a, b), then $Q(0) < 0$ for all sufficiently small nonzero values of h and k, and f has a *local maximum* value at (a, b).

2. If $f_{xx} > 0$ and $f_{xx}f_{yy} - f_{xy}{}^2 > 0$ at (a, b), then $Q(0) > 0$ for all sufficiently small nonzero values of h and k, and f has a *local minimum* value at (a, b).

3. If $f_{xx}f_{yy} - f_{xy}{}^2 < 0$ at (a, b), there are combinations of arbitrarily small nonzero values of h and k for which $Q(0) > 0$, and other values for which $Q(0) < 0$. Arbitrarily close to the point $P_0(a, b, f(a, b))$ on the surface $z = f(x, y)$ there are points above P_0 and points below P_0, so f has a *saddle point* at (a, b).

4. If $f_{xx}f_{yy} - f_{xy}{}^2 = 0$, another test is needed. The possibility that $Q(0)$ equals zero prevents us from drawing conclusions about the sign of $Q(c)$.

Error Formula for Linear Approximations

We want to show that the difference $E(x, y)$ between the values of a function $f(x, y)$ and its linearization $L(x, y)$ at (x_0, y_0) satisfies the inequality

$$|E(x, y)| \le \frac{1}{2} M(|x - x_0| + |y - y_0|)^2.$$

The function f is assumed to have continuous second partial derivatives throughout an open set containing a closed rectangular region R centered at (x_0, y_0). The number M is the largest value that any of $|f_{xx}|$, $|f_{yy}|$, and $|f_{xy}|$ take on R.

The inequality we want comes from Equation (2). We substitute x_0 and y_0 for a and b, and $x - x_0$ and $y - y_0$ for h and k, respectively, and rearrange the result as

$$f(x, y) = \underbrace{f(x_0, y_0) + f_x(x_0, y_0)(x - x_0) + f_y(x_0, y_0)(y - y_0)}_{\text{Linearization } L(x, y)}$$

$$\underbrace{+\frac{1}{2}\left((x - x_0)^2 f_{xx} + 2(x - x_0)(y - y_0)f_{xy} + (y - y_0)^2 f_{yy}\right)\Big|_{(x_0 + c(x - x_0),\, y_0 + c(y - y_0))}}_{\text{Error } E(x, y)}.$$

This remarkable equation reveals that

$$|E| \le \frac{1}{2}\left(|x - x_0|^2 |f_{xx}| + 2|x - x_0||y - y_0||f_{xy}| + |y - y_0|^2 |f_{yy}|\right).$$

Hence, if M is an upper bound for the values of $|f_{xx}|$, $|f_{xy}|$, and $|f_{yy}|$ on R,

$$\begin{aligned} |E| &\le \frac{1}{2}\left(|x - x_0|^2 M + 2|x - x_0||y - y_0| M + |y - y_0|^2 M\right) \\ &\le \frac{1}{2} M(|x - x_0| + |y - y_0|)^2. \end{aligned}$$

Taylor's Formula for Functions of Two Variables

The formulas derived earlier for F' and F'' can be obtained by applying to $f(x, y)$ the operators

$$\left(h\frac{\partial}{\partial x} + k\frac{\partial}{\partial y}\right) \quad \text{and} \quad \left(h\frac{\partial}{\partial x} + k\frac{\partial}{\partial y}\right)^2 = h^2\frac{\partial^2}{\partial x^2} + 2hk\frac{\partial^2}{\partial x\,\partial y} + k^2\frac{\partial^2}{\partial y^2}.$$

These are the first two instances of a more general formula,

$$F^{(n)}(t) = \frac{d^n}{dt^n}F(t) = \left(h\frac{\partial}{\partial x} + k\frac{\partial}{\partial y}\right)^n f(x, y), \tag{6}$$

which says that applying d^n/dt^n to $F(t)$ gives the same result as applying the operator

$$\left(h\frac{\partial}{\partial x}+k\frac{\partial}{\partial y}\right)^n$$

to $f(x, y)$ after expanding it by the binomial theorem.

If partial derivatives of f through order $n+1$ are continuous throughout a rectangular region centered at (a, b), we may extend the Taylor formula for $F(t)$ to

$$F(t) = F(0) + F'(0)t + \frac{F''(0)}{2!}t^2 + \cdots + \frac{F^{(n)}(0)}{n!}t^n + \text{remainder}$$

and take $t = 1$ to obtain

$$F(1) = F(0) + F'(0) + \frac{F''(0)}{2!} + \cdots + \frac{F^{(n)}(0)}{n!} + \text{remainder.}$$

When we replace the first n derivatives on the right of this last series by their equivalent expressions from Equation (6) evaluated at $t = 0$ and add the appropriate remainder term, we arrive at the following formula.

Theorem 13 Taylor's Formula for f(x, y) at the Point (a, b)

Suppose that $f(x, y)$ and its partial derivatives through order $n+1$ are continuous throughout an open rectangular region R centered at a point (a, b). Then, throughout R,

$$\begin{aligned} f(a+h, b+k) &= f(a, b) + (hf_x + kf_y)|_{(a,b)} + \frac{1}{2!}(h^2f_{xx} + 2hkf_{xy} + k^2f_{yy})|_{(a,b)} \\ &\quad + \frac{1}{3!}(h^3f_{xxx} + 3h^2kf_{xxy} + 3hk^2f_{xyy} + k^3f_{yyy})|_{(a,b)} + \cdots + \frac{1}{n!}\left(h\frac{\partial}{\partial x}+k\frac{\partial}{\partial y}\right)^n f\Big|_{(a,b)} \\ &\quad + \frac{1}{(n+1)!}\left(h\frac{\partial}{\partial x}+k\frac{\partial}{\partial y}\right)^{n+1} f\Big|_{(a+ch,\, b+ck)} \end{aligned} \tag{7}$$

The first n derivative terms are evaluated at (a, b). The last term is evaluated at some point $(a+ch, b+ck)$ on the line segment joining (a, b) and $(a+h, b+k)$.

If $(a, b) = (0, 0)$ and we treat h and k as independent variables (denoting them now by x and y), then Equation (7) assumes the following simpler form.

Corollary to Theorem 13 Taylor's Formula for f(x, y) at the Origin

$$\begin{aligned} f(x, y) &= f(0, 0) + xf_x + yf_y + \frac{1}{2!}(x^2f_{xx} + 2xyf_{xy} + y^2f_{yy}) \\ &\quad + \frac{1}{3!}(x^3f_{xxx} + 3x^2yf_{xxy} + 3xy^2f_{xyy} + y^3f_{yyy}) + \cdots + \frac{1}{n!}\left(x\frac{\partial}{\partial x}+y\frac{\partial}{\partial y}\right)^n f \\ &\quad + \frac{1}{(n+1)!}\left(x\frac{\partial}{\partial x}+y\frac{\partial}{\partial y}\right)^{n+1} f\Big|_{(cx,cy)} \end{aligned} \tag{8}$$

The first n derivative terms are evaluated at $(0, 0)$. The last term is evaluated at a point on the line segment joining the origin and (x, y).

Taylor's formula provides polynomial approximations of two-variable functions. The first n derivative terms give the polynomial; the last term gives the approximation error. The first three terms of Taylor's formula give the function's linearization. To improve on the linearization, we add higher power terms.

Example 1 Finding a Quadratic Approximation

Find a quadratic $f(x, y) = \sin x \sin y$ near the origin. How accurate is the approximation if $|x| \le 0.1$ and $|y| \le 0.1$?

Solution We take $n = 2$ in Equation (8):

$$f(x, y) = f(0, 0) + (xf_x + yf_y) + \frac{1}{2}(x^2 f_{xx} + 2xyf_{xy} + y^2 f_{yy})$$
$$+ \frac{1}{6}(x^3 f_{xxx} + 3x^2 y f_{xxy} + 3xy^2 f_{xyy} + y^3 f_{yyy})_{(cx,cy)}$$

with

$$\begin{aligned} f(0,0) &= \sin x \sin y|_{(0,0)} = 0, & f_{xx}(0,0) &= -\sin x \sin y|_{(0,0)} = 0, \\ f_x(0,0) &= \cos x \sin y|_{(0,0)} = 0, & f_{xy}(0,0) &= \cos x \cos y|_{(0,0)} = 1, \\ f_y(0,0) &= \sin x \cos y|_{(0,0)} = 0, & f_{yy}(0,0) &= -\sin x \sin y|_{(0,0)} = 0, \end{aligned}$$

we have

$$\sin x \sin y \approx 0 + 0 + 0 + \frac{1}{2}(x^2(0) + 2xy(1) + y^2(0))$$
$$\sin x \sin y \approx xy.$$

The error in the approximation is

$$E(x, y) = \frac{1}{6}(x^3 f_{xxx} + 3x^2 y f_{xxy} + 3xy^2 f_{xyy} + y^3 f_{yyy})|_{(cx,cy)}.$$

The third derivatives never exceed 1 in absolute value because they are products of sines and cosines. Also, $|x| \le 0.1$ and $|y| \le 0.1$. Hence,

$$|E(x, y)| \le \frac{1}{6}((0.1)^3 + 3(0.1)^3 + 3(0.1)^3 + (0.1)^3) \le \frac{8}{6}(0.1)^3 \le 0.00134$$

(rounded up). The error will not exceed 0.00134 if $|x| \le 0.1$ and $|y| \le 0.1$.

EXERCISES 11.10

Finding Quadratic and Cubic Approximations

In Exercises 1–10, use Taylor's formula for $f(x, y)$ at the origin to find quadratic and cubic approximations of f near the origin.

1. $f(x, y) = xe^y$

2. $f(x, y) = e^x \cos y$

3. $f(x, y) = y \sin x$

4. $f(x, y) = \sin x \cos y$

5. $f(x, y) = e^x \ln (1 + y)$

6. $f(x, y) = \ln (2x + y + 1)$

7. $f(x, y) = \sin (x^2 + y^2)$

8. $f(x, y) = \cos (x^2 + y^2)$

9. $f(x, y) = \dfrac{1}{1 - x - y}$

10. $f(x, y) = \dfrac{1}{1 - x - y + xy}$

11. Use Taylor's formula to find a quadratic approximation of $f(x, y) = \cos x \cos y$ at the origin. Estimate the error in the approximation if $|x| \le 0.1$ and $|y| \le 0.1$.

12. Use Taylor's formula to find a quadratic approximation of $e^x \sin y$ at the origin. Estimate the error in the approximation if $|x| \le 0.1$ and $|y| \le 0.1$.

Questions to Guide Your Review

1. What is a real-valued function of two independent variables? Three independent variables? Give examples.
2. What does it mean for sets in the plane or in space to be open? closed? Give examples. Give examples of sets that are neither open nor closed.
3. How can you display the values of a function $f(x, y)$ of two independent variables graphically? How do you do the same for a function $f(x, y, z)$ of three independent variables?
4. What does it mean for a function $f(x, y)$ to have limit L as $(x, y) \to (x_0, y_0)$? What are the basic properties of limits of functions of two independent variables?
5. When is a function of two (three) independent variables continuous at a point in its domain? Give examples of functions that are continuous at some points but not others.
6. What can be said about algebraic combinations and composites of continuous functions?
7. Explain the two-path test for nonexistence of limits.
8. How are the partial derivatives $\partial f/\partial x$ and $\partial f/\partial y$ of a function $f(x, y)$ defined? How are they interpreted and calculated?
9. How does the relation between first partial derivatives and continuity of functions of two independent variables differ from the relation between first derivatives and continuity for real-valued functions of a single independent variable? Give an example.
10. What does it mean for a function $f(x, y)$ to be differentiable? What does the Increment Theorem say about differentiability?
11. What is the Mixed Derivative Theorem for mixed second-order partial derivatives? How can it help in calculating partial derivatives of second and higher orders? Give examples.
12. How can you sometimes decide from examining f_x and f_y that a function $f(x, y)$ is differentiable? What is the relation between the differentiability of f and the continuity of f at a point?
13. What is the Chain Rule? What form does it take for functions of two independent variables? Three independent variables? Functions defined on surfaces? How do you diagram these different forms? Give examples. What pattern enables one to remember all the different forms?
14. What is the derivative of a function $f(x, y)$ at a point P_0 in the direction of a unit vector $\mathbf{u}$? What rate does it describe? What geometric interpretation does it have? Give examples.
15. What is the gradient vector of a function $f(x, y)$? How is it related to the function's directional derivatives? State the analogous results for functions of three independent variables.
16. How do you find the tangent line at a point on a level curve of a differentiable function $f(x, y)$? How do you find the tangent plane and normal line at a point on a level surface of a differentiable function $f(x, y, z)$? Give examples.
17. How can you use directional derivatives to estimate change?
18. How do you linearize a function $f(x, y)$ of two independent variables at a point (x_0, y_0)? Why might you want to do this? How do you linearize a function of three independent variables?
19. What can you say about the accuracy of linear approximations of functions of two (three) independent variables?
20. If (x, y) moves from (x_0, y_0) to a point $(x_0 + dx, y_0 + dy)$ nearby, how can you estimate the resulting change in the value of a differentiable function $f(x, y)$? Give an example.
21. How do you define local maxima, local minima, and saddle points for a differentiable function $f(x, y)$? Give examples.
22. What derivative tests are available for determining the local extreme values of a function $f(x, y)$? How do they enable you to narrow your search for these values? Give examples.
23. How do you find the extrema of a continuous function $f(x, y)$ on a closed bounded region of the xy-plane? Give an example.
24. Describe the method of Lagrange multipliers and give examples.
25. If $w = f(x, y, z)$, where the variables x, y, and z are constrained by an equation $g(x, y, z) = 0$, what is the meaning of the notation $(\partial w/\partial x)_y$? How can an arrow diagram help you calculate this partial derivative with constrained variables? Give examples.
26. How does Taylor's formula for a function $f(x, y)$ generate polynomial approximations and error estimates?

Practice Exercises

Domain, Range, and Level Curves

In Exercises 1–4, find the domain and range of the given function and identify its level curves. Sketch a typical level curve.

1. $f(x, y) = 9x^2 + y^2$
2. $f(x, y) = e^{x+y}$
3. $g(x, y) = 1/xy$
4. $g(x, y) = \sqrt{x^2 - y}$

In Exercises 5–8, find the domain and range of the given function and identify its level surfaces. Sketch a typical level surface.

5. $f(x, y, z) = x^2 + y^2 - z$
6. $g(x, y, z) = x^2 + 4y^2 + 9z^2$

7. $h(x, y, z) = \dfrac{1}{x^2 + y^2 + z^2}$

8. $k(x, y, z) = \dfrac{1}{x^2 + y^2 + z^2 + 1}$

Evaluating Limits

Find the limits in Exercises 9–14.

9. $\lim\limits_{(x,y)\to(\pi,\ln 2)} e^y \cos x$

10. $\lim\limits_{(x,y)\to(0,0)} \dfrac{2 + y}{x + \cos y}$

11. $\lim\limits_{(x,y)\to(1,1)} \dfrac{x - y}{x^2 - y^2}$

12. $\lim\limits_{(x,y)\to(1,1)} \dfrac{x^3y^3 - 1}{xy - 1}$

13. $\lim\limits_{P\to(1,-1,e)} \ln |x + y + z|$

14. $\lim\limits_{P\to(1,-1,-1)} \tan^{-1}(x + y + z)$

By considering different paths of approach, show that the limits in Exercises 15 and 16 do not exist.

15. $\lim\limits_{\substack{(x,y)\to(0,0)\\ y\neq x^2}} \dfrac{y}{x^2 - y}$

16. $\lim\limits_{\substack{(x,y)\to(0,0)\\ xy\neq 0}} \dfrac{x^2 + y^2}{xy}$

17. *Continuous extension* Let $f(x, y) = (x^2 - y^2)/(x^2 + y^2)$ for $(x, y) \neq (0, 0)$. Is it possible to define $f(0, 0)$ in a way that makes f continuous at the origin? Why?

18. *Continuous extension* Let

$$f(x, y) = \begin{cases} \dfrac{\sin (x - y)}{|x| + |y|}, & |x| + |y| \neq 0 \\ 0, & (x, y) = (0, 0). \end{cases}$$

Is f continuous at the origin? Why?

Partial Derivatives

In Exercises 19–24, find the partial derivative of the function with respect to each variable.

19. $g(r, \theta) = r \cos \theta + r \sin \theta$

20. $f(x, y) = \dfrac{1}{2} \ln (x^2 + y^2) + \tan^{-1} \dfrac{y}{x}$

21. $f(R_1, R_2, R_3) = \dfrac{1}{R_1} + \dfrac{1}{R_2} + \dfrac{1}{R_3}$

22. $h(x, y, z) = \sin (2\pi x + y - 3z)$

23. $P(n, R, T, V) = \dfrac{nRT}{V}$ (the Ideal Gas Law)

24. $f(r, l, T, w) = \dfrac{1}{2rl} \sqrt{\dfrac{T}{\pi w}}$

Second-Order Partials

Find the second-order partial derivatives of the functions in Exercises 25–28.

25. $g(x, y) = y + \dfrac{x}{y}$

26. $g(x, y) = e^x + y \sin x$

27. $f(x, y) = x + xy - 5x^3 + \ln (x^2 + 1)$

28. $f(x, y) = y^2 - 3xy + \cos y + 7e^y$

Chain Rule Calculations

29. Find dw/dt at $t = 0$ if $w = \sin (xy + \pi)$, $x = e^t$, and $y = \ln (t + 1)$.

30. Find dw/dt at $t = 1$ if $w = xe^y + y \sin z - \cos z$, $x = 2\sqrt{t}$, $y = t - 1 + \ln t$, and $z = \pi t$.

31. Find $\partial w/\partial r$ and $\partial w/\partial s$ when $r = \pi$ and $s = 0$ if $w = \sin (2x - y)$, $x = r + \sin s$, $y = rs$.

32. Find $\partial w/\partial u$ and $\partial w/\partial v$ when $u = v = 0$ if $w = \ln \sqrt{1 + x^2} - \tan^{-1} x$ and $x = 2e^u \cos v$.

33. Find the value of the derivative of $f(x, y, z) = xy + yz + xz$ with respect to t on the curve $x = \cos t$, $y = \sin t$, $z = \cos 2t$ at $t = 1$.

34. Show that if $w = f(s)$ is any differentiable function of s and if $s = y + 5x$, then

$$\frac{\partial w}{\partial x} - 5\frac{\partial w}{\partial y} = 0.$$

Implicit Differentiation

Assuming that the equations in Exercises 35 and 36 define y as a differentiable function of x, find the value of dy/dx at point P.

35. $1 - x - y^2 - \sin xy = 0, \quad P(0, 1)$

36. $2xy + e^{x+y} - 2 = 0, \quad P(0, \ln 2)$

Directional Derivatives

In Exercises 37–40, find the directions in which f increases and decreases most rapidly at P_0 and find the derivative of f in each direction. Also, find the derivative of f at P_0 in the direction of the vector $\mathbf{v}$.

37. $f(x, y) = \cos x \cos y, \quad P_0(\pi/4, \pi/4), \quad \mathbf{v} = 3\mathbf{i} + 4\mathbf{j}$

38. $f(x, y) = x^2e^{-2y}, \quad P_0(1, 0), \quad \mathbf{v} = \mathbf{i} + \mathbf{j}$

39. $f(x, y, z) = \ln (2x + 3y + 6z), \quad P_0(-1, -1, 1),$ $\mathbf{v} = 2\mathbf{i} + 3\mathbf{j} + 6\mathbf{k}$

40. $f(x, y, z) = x^2 + 3xy - z^2 + 2y + z + 4, \quad P_0(0, 0, 0),$ $\mathbf{v} = \mathbf{i} + \mathbf{j} + \mathbf{k}$

41. *Derivative in velocity direction* Find the derivative of $f(x, y, z) = xyz$ in the direction of the velocity vector of the helix

$$\mathbf{r}(t) = (\cos 3t)\mathbf{i} + (\sin 3t)\mathbf{j} + 3t\mathbf{k}$$

at $t = \pi/3$.

42. *Maximum directional derivative* What is the largest value that the directional derivative of $f(x, y, z) = xyz$ can have at the point $(1, 1, 1)$?

43. *Directional derivatives with given values* At the point $(1, 2)$, the function $f(x, y)$ has a derivative of 2 in the direction toward $(2, 2)$ and a derivative of -2 in the direction toward $(1, 1)$.

(a) Find $f_x(1, 2)$ and $f_y(1, 2)$.

(b) Find the derivative of f at $(1, 2)$ in the direction toward the point $(4, 6)$.

44. *Writing to Learn* Which of the following statements are true if $f(x, y)$ is differentiable at (x_0, y_0)? Give reasons for your answers.

(a) If $\mathbf{u}$ is a unit vector, the derivative of f at (x_0, y_0) in the direction of $\mathbf{u}$ is $(f_x(x_0, y_0)\mathbf{i} + f_y(x_0, y_0)\mathbf{j}) \cdot \mathbf{u}$.

(b) The derivative of f at (x_0, y_0) in the direction of $\mathbf{u}$ is a vector.

(c) The directional derivative of f at (x_0, y_0) has its greatest value in the direction of ∇f.

(d) At (x_0, y_0), vector ∇f is normal to the curve $f(x, y) = f(x_0, y_0)$.

Gradients, Tangent Planes, and Normal Lines

In Exercises 45 and 46, sketch the surface $f(x, y, z) = c$ together with ∇f at the given points.

45. $x^2 + y + z^2 = 0;\quad (0, -1, \pm 1),\quad (0, 0, 0)$

46. $y^2 + z^2 = 4;\quad (2, \pm 2, 0),\quad (2, 0, \pm 2)$

In Exercises 47 and 48, find an equation for the plane tangent to the level surface $f(x, y, z) = c$ at the point P_0. Also, find parametric equations for the line that is normal to the surface at P_0.

47. $x^2 - y - 5z = 0,\quad P_0(2, -1, 1)$

48. $x^2 + y^2 + z = 4,\quad P_0(1, 1, 2)$

In Exercises 49 and 50, find an equation for the plane tangent to the surface $z = f(x, y)$ at the given point.

49. $z = \ln(x^2 + y^2),\quad (0, 1, 0)$

50. $z = 1/(x^2 + y^2),\quad (1, 1, 1/2)$

In Exercises 51 and 52, find equations for the lines that are tangent and normal to the level curve $f(x, y) = c$ at the point P_0. Then sketch the lines and level curve together with ∇f at P_0.

51. $y - \sin x = 1,\quad P_0(\pi, 1)$

52. $\dfrac{y^2}{2} - \dfrac{x^2}{2} = \dfrac{3}{2},\quad P_0(1, 2)$

Tangent Lines to Curves

In Exercises 53 and 54, find parametric equations for the line that is tangent to the curve of intersection of the surfaces at the given point.

53. Surfaces: $x^2 + 2y + 2z = 4,\quad y = 1$
Point: $(1, 1, 1/2)$

54. Surfaces: $x + y^2 + z = 2,\quad y = 1$
Point: $(1/2, 1, 1/2)$

Linearizations

In Exercises 55 and 56, find the linearization $L(x, y)$ of the function $f(x, y)$ at the point P_0. Then find an upper bound for the magnitude of the error E in the approximation $f(x, y) \approx L(x, y)$ over the rectangle R.

55. $f(x, y) = \sin x \cos y,\quad P_0(\pi/4, \pi/4)$

$R:\quad \left|x - \dfrac{\pi}{4}\right| \le 0.1,\quad \left|y - \dfrac{\pi}{4}\right| \le 0.1$

56. $f(x, y) = xy - 3y^2 + 2,\quad P_0(1, 1)$

$R:\quad |x - 1| \le 0.1,\quad |y - 1| \le 0.2$

Find the linearizations of the functions in Exercises 57 and 58 at the given points.

57. $f(x, y, z) = xy + 2yz - 3xz$ at $(1, 0, 0)$ and $(1, 1, 0)$

58. $f(x, y, z) = \sqrt{2}\cos x \sin(y + z)$ at $(0, 0, \pi/4)$ and $(\pi/4, \pi/4, 0)$

Estimates and Sensitivity to Change

59. *Measuring the volume of a pipeline* You plan to calculate the volume inside a stretch of pipeline that is about 36 in. in diameter and 1 mi long. With which measurement should you be more careful, the length or the diameter? Why?

60. *Writing to Learn: Sensitivity to change* Near the point (1, 2), is $f(x, y) = x^2 - xy + y^2 - 3$ more sensitive to changes in x or to changes in y? How do you know?

61. *Change in an electrical circuit* Suppose that the current I (amperes) in an electrical circuit is related to the voltage V (volts) and the resistance R (ohms) by the equation $I = V/R$. If the voltage drops from 24 to 23 volts and the resistance drops from 100 to 80 ohms, will I increase or decrease? By about how much? Is the change in I more sensitive to change in the voltage or to change in the resistance? How do you know?

62. *Maximum error in estimating the area of an ellipse* If $a = 10$ cm and $b = 16$ cm to the nearest millimeter, what should you expect the maximum percentage error to be in the calculated area $A = \pi ab$ of the ellipse $x^2/a^2 + y^2/b^2 = 1$?

63. *Error in estimating a product* Let $y = uv$ and $z = u + v$, where u and v are positive independent variables.

(a) If u is measured with an error of 2% and v with an error of 3%, about what is the percentage error in the calculated value of y?

(b) Show that the percentage error in the calculated value of z is less than the percentage error in the value of y.

64. *Cardiac index* To make different people comparable in studies of cardiac output (Section 2.7, Exercise 25), researchers divide the measured cardiac output by the body surface area to find the *cardiac index C*:

$$C = \frac{\text{cardiac output}}{\text{body surface area}}.$$

The body surface area B of a person with weight w and height h is approximated by the formula

$$B = 71.84w^{0.425}h^{0.725},$$

which gives B in square centimeters when w is measured in kilograms and h in centimeters. You are about to calculate the cardiac index of a person with the following measurements:

Cardiac output:	7 L/min
Weight:	70 kg
Height:	180 cm

Which will have a greater effect on the calculation, a 1 kg error in measuring the weight or a 1 cm error in measuring the height?

Local Extrema

Test the functions in Exercises 65–70 for local maxima and minima and saddle points. Find each function's value at these points.

65. $f(x, y) = x^2 - xy + y^2 + 2x + 2y - 4$

66. $f(x, y) = 5x^2 + 4xy - 2y^2 + 4x - 4y$

67. $f(x, y) = 2x^3 + 3xy + 2y^3$

68. $f(x, y) = x^3 + y^3 - 3xy + 15$

69. $f(x, y) = x^3 + y^3 + 3x^2 - 3y^2$

70. $f(x, y) = x^4 - 8x^2 + 3y^2 - 6y$

Absolute Extrema

In Exercises 71–78, find the absolute maximum and minimum values of f on the region R.

71. $f(x, y) = x^2 + xy + y^2 - 3x + 3y$
R: The triangular region cut from the first quadrant by the line $x + y = 4$

72. $f(x, y) = x^2 - y^2 - 2x + 4y + 1$
R: The rectangular region in the first quadrant bounded by the coordinate axes and the lines $x = 4$ and $y = 2$

73. $f(x, y) = y^2 - xy - 3y + 2x$
R: The square region enclosed by the lines $x = \pm 2$ and $y = \pm 2$

74. $f(x, y) = 2x + 2y - x^2 - y^2$
R: The square region bounded by the coordinate axes and the lines $x = 2$, $y = 2$ in the first quadrant

75. $f(x, y) = x^2 - y^2 - 2x + 4y$
R: The triangular region bounded below by the x-axis, above by the line $y = x + 2$, and on the right by the line $x = 2$

76. $f(x, y) = 4xy - x^4 - y^4 + 16$
R: The triangular region bounded below by the line $y = -2$, above by the line $y = x$, and on the right by the line $x = 2$

77. $f(x, y) = x^3 + y^3 + 3x^2 - 3y^2$
R: The square region enclosed by the lines $x = \pm 1$ and $y = \pm 1$

78. $f(x, y) = x^3 + 3xy + y^3 + 1$
R: The square region enclosed by the lines $x = \pm 1$ and $y = \pm 1$

Lagrange Multipliers

79. *Extrema on a circle* Find the extreme values of $f(x, y) = x^3 + y^2$ on the circle $x^2 + y^2 = 1$.

80. *Extrema on a circle* Find the extreme values of $f(x, y) = xy$ on the circle $x^2 + y^2 = 1$.

81. *Extrema in a disk* Find the extreme values of $f(x, y) = x^2 + 3y^2 + 2y$ on the unit disk $x^2 + y^2 \le 1$.

82. *Extrema in a disk* Find the extreme values of $f(x, y) = x^2 + y^2 - 3x - xy$ on the disk $x^2 + y^2 \le 9$.

83. *Extrema on a sphere* Find the extreme values of $f(x, y, z) = x - y + z$ on the unit sphere $x^2 + y^2 + z^2 = 1$.

84. *Minimum distance to origin* Find the points on the surface $z^2 - xy = 4$ closest to the origin.

85. *Minimizing cost of a box* A closed rectangular box is to have volume V cm^3. The cost of the material used in the box is a cents/cm^2 for top and bottom, b cents/cm^2 for front and back, and c cents/cm^2 for the remaining sides. What dimensions minimize the total cost of materials?

86. *Least volume* Find the plane $x/a + y/b + z/c = 1$ that passes through the point (2, 1, 2) and cuts off the least volume from the first octant.

87. *Extrema on curve of intersecting surfaces* Find the extreme values of $f(x, y, z) = x(y + z)$ on the curve of intersection of the right circular cylinder $x^2 + y^2 = 1$ and the hyperbolic cylinder $xz = 1$.

88. *Minimum distance to origin on curve of intersecting plane and cone* Find the point closest to the origin on the curve of intersection of the plane $x + y + z = 1$ and the cone $z^2 = 2x^2 + 2y^2$.

Partial Derivatives with Constrained Variables

In Exercises 89 and 90, begin by drawing a diagram that shows the relations among the variables.

89. If $w = x^2 e^{yz}$ and $z = x^2 - y^2$, find

(a) $\left(\dfrac{\partial w}{\partial y}\right)_z$ **(b)** $\left(\dfrac{\partial w}{\partial z}\right)_x$ **(c)** $\left(\dfrac{\partial w}{\partial z}\right)_y$.

90. Let $U = f(P, V, T)$ be the internal energy of a gas that obeys the ideal gas law $PV = nRT$ (n and R constant). Find

(a) $\left(\dfrac{\partial U}{\partial T}\right)_P$ **(b)** $\left(\dfrac{\partial U}{\partial V}\right)_T$.

Theory and Examples

91. *Finding partial derivatives* Let $w = f(r, \theta)$, $r = \sqrt{x^2 + y^2}$, and $\theta = \tan^{-1}(y/x)$. Find $\partial w/\partial x$ and $\partial w/\partial y$ and express your answers in terms of r and θ.

92. *Finding partial derivatives* Let $z = f(u, v)$, $u = ax + by$, and $v = ax - by$. Express z_x and z_y in terms of f_u, f_v, and the constants a and b.

93. *Verifying an equation* If a and b are constants, $w = u^3 + \tanh u + \cos u$, and $u = ax + by$, show that

$$a\frac{\partial w}{\partial y} = b\frac{\partial w}{\partial x}.$$

94. *Using the Chain Rule* If $w = \ln(x^2 + y^2 + 2z)$, $x = r + s$, $y = r - s$, and $z = 2rs$, find w_r and w_s by the Chain Rule. Then check your answer another way.

95. *Angle between vectors* The equations $e^u \cos v - x = 0$ and $e^u \sin v - y = 0$ define u and v as differentiable functions of x and y. Show that the angle between the vectors

$$\frac{\partial u}{\partial x}\mathbf{i} + \frac{\partial u}{\partial y}\mathbf{j} \quad \text{and} \quad \frac{\partial v}{\partial x}\mathbf{i} + \frac{\partial v}{\partial y}\mathbf{j}$$

is constant.

96. *Polar coordinates and second derivatives* Introducing polar coordinates $x = r\cos\theta$ and $y = r\sin\theta$ changes $f(x, y)$ to $g(r, \theta)$. Find the value of $\partial^2 g/\partial\theta^2$ at the point $(r, \theta) = (2, \pi/2)$, given that

$$\frac{\partial f}{\partial x} = \frac{\partial f}{\partial y} = \frac{\partial^2 f}{\partial x^2} = \frac{\partial^2 f}{\partial y^2} = 1$$

at that point.

97. *Normal line parallel to a plane* Find the points on the surface

$$(y + z)^2 + (z - x)^2 = 16$$

where the normal line is parallel to the yz-plane.

98. *Tangent plane parallel to xy-plane* Find the points on the surface

$$xy + yz + zx - x - z^2 = 0$$

where the tangent plane is parallel to the xy-plane.

99. *When gradient is parallel to position vector* Suppose that $\nabla f(x, y, z)$ is always parallel to the position vector $x\mathbf{i} + y\mathbf{j} + z\mathbf{k}$. Show that $f(0, 0, a) = f(0, 0, -a)$ for any a.

100. *Directional derivative in all directions, but no gradient* Show that the directional derivative of

$$f(x, y, z) = \sqrt{x^2 + y^2 + z^2}$$

at the origin equals 1 in any direction but that f has no gradient vector at the origin.

101. *Normal line through origin* Show that the line normal to the surface $xy + z = 2$ at the point $(1, 1, 1)$ passes through the origin.

102. *Tangent plane and normal line*

(a) Sketch the surface $x^2 - y^2 + z^2 = 4$.

(b) Find a vector normal to the surface at $(2, -3, 3)$. Add the vector to your sketch.

(c) Find equations for the tangent plane and normal line at $(2, -3, 3)$.

Additional Exercises: Theory, Examples, Applications

Partial Derivatives

1. *Function with saddle at the origin* If you did Exercise 66 in Section 11.2, you know that the function

$$f(x, y) = \begin{cases} xy\dfrac{x^2 - y^2}{x^2 + y^2}, & (x, y) \neq (0, 0) \\ 0, & (x, y) = (0, 0) \end{cases}$$

(see the accompanying figure) is continuous at $(0, 0)$. Find $f_{xy}(0, 0)$ and $f_{yx}(0, 0)$.

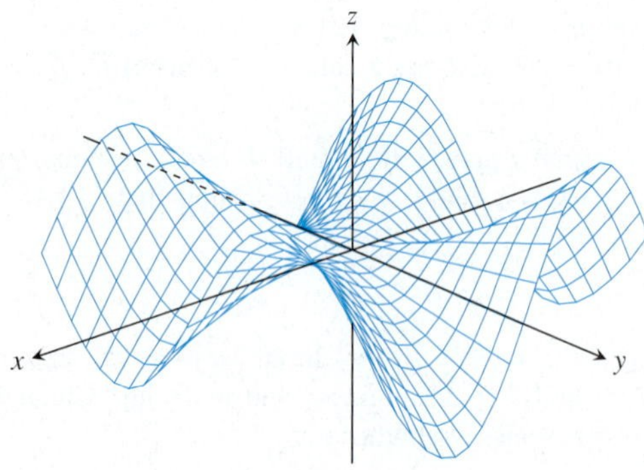

(Generated by Mathematica)

2. *Finding a function from second partials* Find a function $w = f(x, y)$ whose first partial derivatives are $\partial w/\partial x = 1 + e^x \cos y$ and $\partial w/\partial y = 2y - e^x \sin y$, and whose value at the point $(\ln 2, 0)$ is $\ln 2$.

3. *A proof of Leibniz's Rule* Leibniz's Rule says that if f is continuous on $[a, b]$ and if $u(x)$ and $v(x)$ are differentiable functions of x whose values lie in $[a, b]$, then

$$\frac{d}{dx}\int_{u(x)}^{v(x)} f(t)\,dt = f(v(x))\frac{dv}{dx} - f(u(x))\frac{du}{dx}.$$

Prove the rule by setting

$$g(u, v) = \int_u^v f(t)\,dt, \qquad u = u(x), \qquad v = v(x)$$

and calculating dg/dx with the Chain Rule.

4. *Finding a function with constrained second partials* Suppose that f is a twice-differentiable function of r, that $r = \sqrt{x^2 + y^2 + z^2}$, and that

$$f_{xx} + f_{yy} + f_{zz} = 0.$$

Show that for some constants a and b,

$$f(r) = \frac{a}{r} + b.$$

5. *Homogeneous functions* A function $f(x, y)$ is *homogeneous of degree n* (n a nonnegative integer) if $f(tx, ty) = t^n f(x, y)$ for all t, x, and y. For such a function (sufficiently differentiable), prove that

(a) $x\dfrac{\partial f}{\partial x} + y\dfrac{\partial f}{\partial y} = nf(x, y)$

(b) $x^2\left(\dfrac{\partial^2 f}{\partial x^2}\right) + 2xy\left(\dfrac{\partial^2 f}{\partial x \partial y}\right) + y^2\left(\dfrac{\partial^2 f}{\partial y^2}\right) = n(n - 1)f.$

6. *Surface in polar coordinates* Let

$$f(r, \theta) = \begin{cases} \dfrac{\sin 6r}{6r}, & r \neq 0 \\ 1, & r = 0, \end{cases}$$

where r and θ are polar coordinates. Find

(a) $\lim_{r \to 0} f(r, \theta)$ **(b)** $f_r(0, 0)$ **(c)** $f_\theta(r, \theta), \quad r \neq 0.$

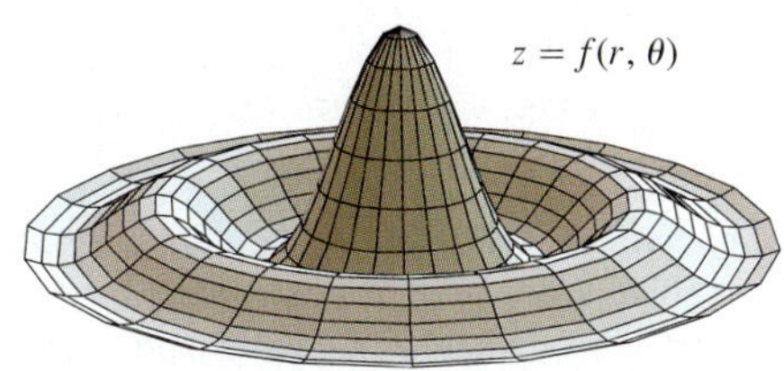

(Generated by Mathematica)

Gradients and Tangents

7. *Properties of position vectors* Let $\mathbf{r} = x\mathbf{i} + y\mathbf{j} + z\mathbf{k}$ and let $r = |\mathbf{r}|$.

(a) Show that $\nabla r = \mathbf{r}/r$.

(b) Show that $\nabla(r^n) = nr^{n-2}\mathbf{r}$.

(c) Find a function whose gradient equals $\mathbf{r}$.

(d) Show that $\mathbf{r} \cdot d\mathbf{r} = r\,dr$.

(e) Show that $\nabla(\mathbf{A} \cdot \mathbf{r}) = \mathbf{A}$ for any constant vector $\mathbf{A}$.

8. *Gradient orthogonal to tangent* Suppose that a differentiable function $f(x, y)$ has the constant value c along the differentiable curve $x = g(t)$, $y = h(t)$; that is

$$f(g(t), h(t)) = c$$

for all values of t. Differentiate both sides of this equation with respect to t to show that ∇f is orthogonal to the curve's tangent vector at every point on the curve.

9. *Curve tangent to a surface* Show that the curve

$$\mathbf{r}(t) = (\ln t)\mathbf{i} + (t \ln t)\mathbf{j} + t\mathbf{k}$$

is tangent to the surface

$$xz^2 - yz + \cos xy = 1$$

at $(0, 0, 1)$.

10. *Curve tangent to a surface* Show that the curve

$$\mathbf{r}(t) = \left(\frac{t^3}{4} - 2\right)\mathbf{i} + \left(\frac{4}{t} - 3\right)\mathbf{j} + \cos(t - 2)\mathbf{k}$$

is tangent to the surface

$$x^3 + y^3 + z^3 - xyz = 0$$

at $(0, -1, 1)$.

Extreme Values

11. *Extrema on a surface* Show that the only possible maxima and minima of z on the surface $z = x^3 + y^3 - 9xy + 27$ occur at $(0, 0)$ and $(3, 3)$. Show that neither a maximum nor a minimum occurs at $(0, 0)$. Determine whether z has a maximum or a minimum at $(3, 3)$.

12. *Maximum in closed first quadrant* Find the maximum value of $f(x, y) = 6xye^{-(2x+3y)}$ in the closed first quadrant (includes the nonnegative axes).

13. *Minimum volume cut from first octant* Find the minimum volume for a region bounded by the planes $x = 0$, $y = 0$, $z = 0$ and a plane tangent to the ellipsoid

$$\frac{x^2}{a^2} + \frac{y^2}{b^2} + \frac{z^2}{c^2} = 1$$

at a point in the first octant.

14. *Minimum distance from line to parabola in xy-plane* By minimizing the function $f(x, y, u, v) = (x - u)^2 + (y - v)^2$ subject to the constraints $y = x + 1$ and $u = v^2$, find the minimum distance in the xy-plane from the line $y = x + 1$ to the parabola $y^2 = x$.

Theory and Examples

15. *Boundedness of first partials implies continuity* Prove the following theorem: If $f(x, y)$ is defined in an open region R of the xy-plane and if f_x and f_y are bounded on R, then $f(x, y)$ is continuous on R. (The assumption of boundedness is essential.)

16. *Writing to Learn* Suppose that $\mathbf{r}(t) = g(t)\mathbf{i} + h(t)\mathbf{j} + k(t)\mathbf{k}$ is a smooth curve in the domain of a differentiable function $f(x, y, z)$. Describe the relation between df/dt, ∇f, and $\mathbf{v} = d\mathbf{r}/dt$. What can be said about ∇f and $\mathbf{v}$ at interior points of the curve where f has extreme values relative to its other values on the curve? Give reasons for your answer.

17. *Finding functions from partial derivatives* Suppose that f and g are functions of x and y such that

$$\frac{\partial f}{\partial y} = \frac{\partial g}{\partial x} \quad \text{and} \quad \frac{\partial f}{\partial x} = \frac{\partial g}{\partial y},$$

and suppose that

$$\frac{\partial f}{\partial x} = 0, \quad f(1, 2) = g(1, 2) = 5 \quad \text{and} \quad f(0, 0) = 4.$$

Find $f(x, y)$ and $g(x, y)$.

18. *Rate of change of the rate of change* We know that if $f(x, y)$ is a function of two variables and if $\mathbf{u} = a\mathbf{i} + b\mathbf{j}$ is a unit vector, then $D_\mathbf{u}f(x, y) = f_x(x, y)a + f_y(x, y)b$ is the rate of change of $f(x, y)$ at (x, y) in the direction of $\mathbf{u}$. Give a similar formula for the rate of change *of the rate of change of* $f(x, y)$ at (x, y) in the direction $\mathbf{u}$.

19. *Path of a heat-seeking particle* A heat-seeking particle has the property that at any point (x, y) in the plane it moves in the direction of maximum temperature increase. If the temperature at

(x, y) is $T(x, y) = -e^{-2y} \cos x$, find an equation $y = f(x)$ for the path of a heat-seeking particle at the point $(\pi/4, 0)$.

20. *Velocity after a ricochet* A particle traveling in a straight line with constant velocity $\mathbf{i} + \mathbf{j} - 5\mathbf{k}$ passes through the point $(0, 0, 30)$ and hits the surface $z = 2x^2 + 3y^2$. The particle ricochets off the surface, the angle of reflection being equal to the angle of incidence. Assuming no loss of speed, what is the velocity of the particle after the ricochet? Simplify your answer.

21. *Directional derivatives tangent to a surface* Let S be the surface that is the graph of $f(x, y) = 10 - x^2 - y^2$. Suppose that the temperature in space at each point (x, y, z) is $T(x, y, z) = x^2y + y^2z + 4x + 14y + z$.

    (a) Among all the possible directions tangential to the surface S at the point $(0, 0, 10)$, which direction will make the rate of change of temperature at $(0, 0, 10)$ a maximum?

    (b) Which direction tangential to S at the point $(1, 1, 8)$ will make the rate of change of temperature a maximum?

22. *Drilling another borehole* On a flat surface of land, geologists drilled a borehole straight down and hit a mineral deposit at 1000 ft. They drilled a second borehole 100 ft to the north of the first and hit the mineral deposit at 950 ft. A third borehole 100 ft east of the first borehole struck the mineral deposit at 1025 ft. The geologists have reasons to believe that the mineral deposit is in the shape of a dome, and for the sake of economy, they would like to find where the deposit is closest to the surface. Assuming the surface to be the xy-plane, in what direction from the first borehole would you suggest the geologists drill their fourth borehole?

The One-Dimensional Heat Equation

If $w(x, t)$ represents the temperature at position x at time t in a uniform conducting rod with perfectly insulated sides (see the accompanying figure), then the partial derivatives w_{xx} and w_t satisfy a differential equation of the form

CD-ROM WEBsite

$$w_{xx} = \frac{1}{c^2} w_t.$$

This equation is called the **one-dimensional heat equation.** The value of the positive constant c^2 is determined by the material from which the rod is made. It has been determined experimentally for a

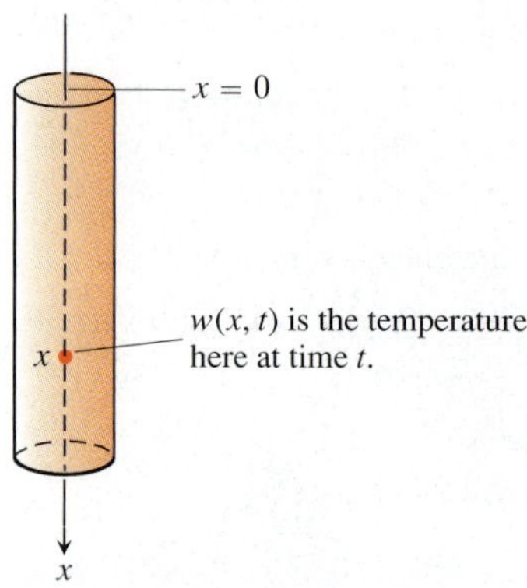

broad range of materials. For a given application, one finds the appropriate value in a table. For dry soil, for example, $c^2 = 0.19$ ft^2/day.

In chemistry and biochemistry, the heat equation is known as the **diffusion equation.** In this context, $w(x, t)$ represents the concentration of a dissolved substance, a salt for instance, diffusing along a tube filled with liquid. The value of $w(x, t)$ is the concentration at point x at time t. In other applications, $w(x, t)$ represents the diffusion of a gas down a long, thin pipe.

In electrical engineering, the heat equation appears in the forms

$$v_{xx} = RCv_t$$

and

$$i_{xx} = RCi_t.$$

These equations describe the voltage v and the flow of current i in a coaxial cable or in any other cable in which leakage and inductance are negligible. The functions and constants in these equations are

$v(x, t)$ = voltage at point x at time t

R = resistance per unit length

C = capacitance to ground per unit of cable length

$i(x, t)$ = current at point x at time t.

23. Find all solutions of the one-dimensional heat equation of the form $w = e^{rt} \sin \pi x$, where r is a constant.

24. Find all solutions of the one-dimensional heat equation that have the form $w = e^{rt} \sin kx$ and satisfy the conditions that $w(0, t) = 0$ and $w(L, t) = 0$. What happens to these solutions as $t \to \infty$?

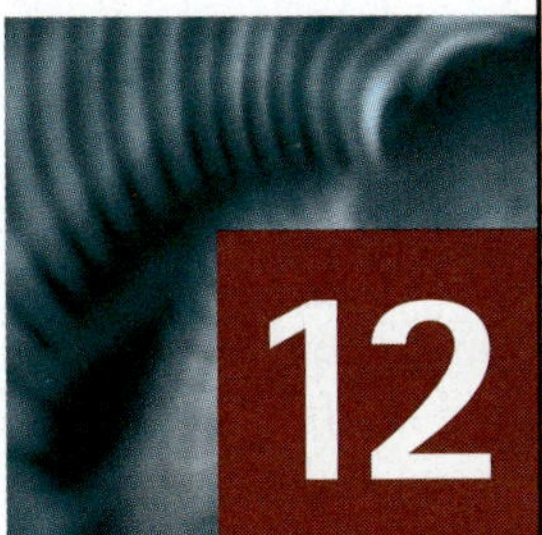

12 Multiple Integrals

OVERVIEW The problems we can solve by integrating functions of two and three variables are similar to the problems solved by single-variable integration, but more general. As in the previous chapter, we can perform the necessary calculations by drawing on our experience with functions of a single variable.

12.1 Double Integrals

Double Integrals over Rectangles • Properties of Double Integrals • Double Integrals as Volumes • Fubini's Theorem for Calculating Double Integrals • Double Integrals over Bounded Nonrectangular Regions • Finding Limits of Integration

We now show how to integrate a continuous function $f(x, y)$ over a bounded region in the xy-plane. There are many similarities between the "double" integrals we define here and the "single" integrals we defined in Chapter 4 for functions of a single variable. Every double integral can be evaluated in stages, using the single-integration methods already at our command.

Double Integrals over Rectangles

Suppose that $f(x, y)$ is defined on a rectangular region R given by

$$R: \quad a \leq x \leq b, \quad c \leq y \leq d.$$

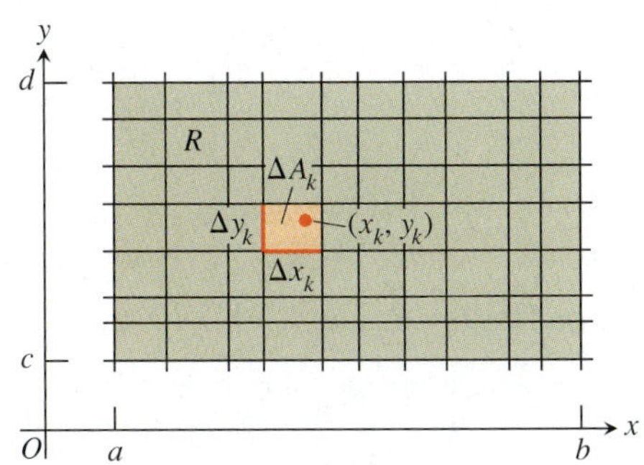

FIGURE 12.1 Rectangular grid partitioning the region R into small rectangles of area $\Delta A_k = \Delta x_k \, \Delta y_k$.

We imagine R to be covered by a network of lines parallel to the x- and y-axes (Figure 12.1). These lines divide R into small pieces of area $\Delta A = \Delta x \; \Delta y$. We number these in some order $\Delta A_1, \Delta A_2, \ldots, \Delta A_n$, choose a point (x_k, y_k) in each piece ΔA_k, and form the sum

$$S_n = \sum_{k=1}^{n} f(x_k, y_k) \; \Delta A_k. \tag{1}$$

If f is continuous throughout R, then, as we refine the mesh (or two-dimensional partition) width to make both Δx and Δy go to zero, the sums in Equation (1) approach a limit called the **double integral** of f over R. The notation for it is

$$\iint_R f(x, y)\, dA \qquad \text{or} \qquad \iint_R f(x, y)\, dx\, dy.$$

Thus,

$$\iint_R f(x, y)\, dA = \lim_{\Delta A \to 0} \sum_{k=1}^{n} f(x_k, y_k)\, \Delta A_k. \tag{2}$$

As with functions of a single variable, the sums approach this limit no matter how the intervals $[a, b]$ and $[c, d]$ that determine R are partitioned, as long as the norms of the partitions both go to zero. The limit in Equation (2) is also independent of the order in which the areas ΔA_k are numbered and independent of the choice of the point (x_k, y_k) within each ΔA_k. The values of the individual approximating sums S_n depend on these choices, but the sums approach the same limit in the end. The proof of the existence and uniqueness of this limit for a continuous function f is given in more advanced texts. The continuity of f is a sufficient condition for the existence of the double integral, but not a necessary one. The limit in question exists for many discontinuous functions as well.

Properties of Double Integrals

Like single integrals, double integrals of continuous functions have algebraic properties that are useful in computations and applications.

Properties of Double Integrals

1. *Constant Multiple*: $\displaystyle\iint_R kf(x, y)\, dA = k \iint_R f(x, y)\, dA$ (any number k)

2. *Sum and Difference*:

$$\iint_R (f(x, y) \pm g(x, y))\, dA = \iint_R f(x, y)\, dA \pm \iint_R g(x, y)\, dA$$

3. *Domination*:

(a) $\displaystyle\iint_R f(x, y)\, dA \geq 0$ if $f(x, y) \geq 0$ on R

(b) $\displaystyle\iint_R f(x, y)\, dA \geq \iint_R g(x, y)\, dA$ if $f(x, y) \geq g(x, y)$ on R

4. *Additivity*: $\displaystyle\iint_R f(x, y)\, dA = \iint_{R_1} f(x, y)\, dA + \iint_{R_2} f(x, y)\, dA$

if R is the union of two nonoverlapping rectangles R_1 and R_2 (Figure 12.2).

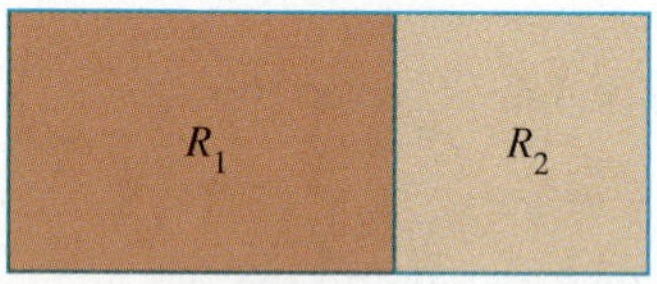

R_1 R_2

$$\iint_{R_1 \cup R_2} f(x, y)\, dA = \iint_{R_1} f(x, y)\, dA + \iint_{R_2} f(x, y)\, dA$$

FIGURE 12.2 Double integrals have the same kind of domain additivity property that single integrals have.

Double Integrals as Volumes

When $f(x, y)$ is positive, we may interpret the double integral of f over a rectangular region R as the volume of the solid prism bounded below by R and above by

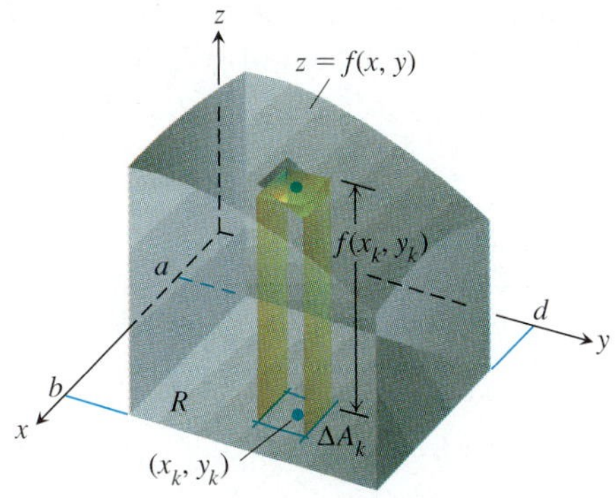

FIGURE 12.3 Approximating solids with rectangular prisms leads us to define the volumes of more general prisms as double integrals. The volume of the prism shown here is the double integral of $f(x, y)$ over the base region R.

the surface $z = f(x, y)$ (Figure 12.3). Each term $f(x_k, y_k)\ \Delta A_k$ in the sum $S_n = \Sigma\ f(x_k, y_k)\ \Delta A_k$ is the volume of a vertical rectangular prism that approximates the volume of the portion of the solid that stands directly above the base ΔA_k. The sum S_n thus approximates what we want to call the total volume of the solid. We *define* this volume to be

$$\text{Volume} = \lim S_n = \iint_R f(x, y)\, dA. \tag{3}$$

As you might expect, this more general method of calculating volume agrees with the methods in Chapter 5, but we do not prove this here.

Fubini's Theorem for Calculating Double Integrals

Suppose that we wish to calculate the volume under the plane $z = 4 - x - y$ over the rectangular region $R: 0 \le x \le 2, 0 \le y \le 1$ in the xy-plane. If we apply the method of slicing from Section 5.1, with slices perpendicular to the x-axis (Figure 12.4), then the volume is

$$\int_{x=0}^{x=2} A(x)\, dx, \tag{4}$$

where $A(x)$ is the cross-section area at x. For each value of x, we may calculate $A(x)$ as the integral

$$A(x) = \int_{y=0}^{y=1} (4 - x - y)\, dy, \tag{5}$$

which is the area under the curve $z = 4 - x - y$ in the plane of the cross section at x. In calculating $A(x)$, x is held fixed and the integration takes place with respect to y. Combining Equations (4) and (5), we see that the volume of the entire solid is

$$\begin{aligned}\text{Volume} &= \int_{x=0}^{x=2} A(x)\, dx = \int_{x=0}^{x=2} \left(\int_{y=0}^{y=1} (4 - x - y)\, dy \right) dx \\ &= \int_{x=0}^{x=2} \left[4y - xy - \frac{y^2}{2} \right]_{y=0}^{y=1} dx = \int_{x=0}^{x=2} \left(\frac{7}{2} - x \right) dx \\ &= \left[\frac{7}{2}x - \frac{x^2}{2} \right]_0^2 = 5 \text{ cubic units.}\end{aligned} \tag{6}$$

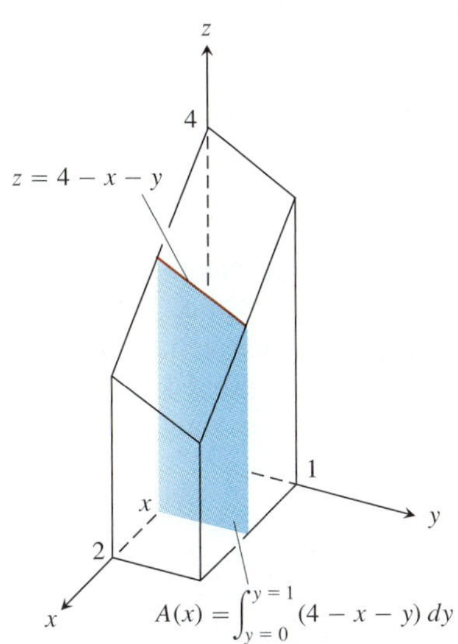

FIGURE 12.4 To obtain the cross-section area $A(x)$, we hold x fixed and integrate with respect to y.

If we had just wanted to write instructions for calculating the volume, without carrying out any of the integrations, we could have written

$$\text{Volume} = \int_0^2 \int_0^1 (4 - x - y)\, dy\, dx.$$

The expression on the right, called an **iterated** or **repeated integral,** says that the volume is obtained by integrating $4 - x - y$ with respect to y from $y = 0$ to $y = 1$, holding x fixed, and then integrating the resulting expression in x with respect to x from $x = 0$ to $x = 2$.

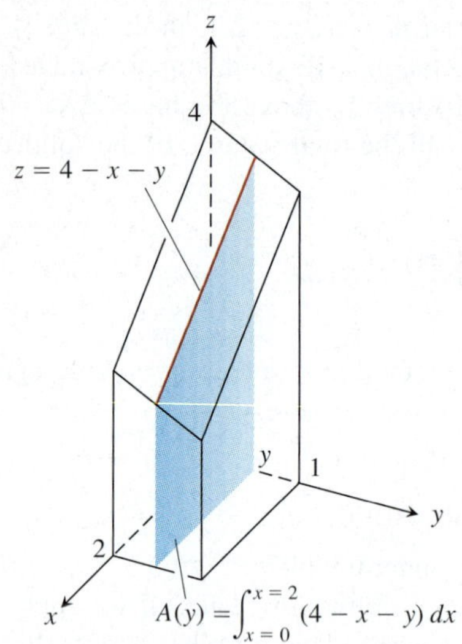

FIGURE 12.5 To obtain the cross-section area $A(y)$, we hold y fixed and integrate with respect to x.

What would have happened if we had calculated the volume by slicing with planes perpendicular to the y-axis (Figure 12.5)? As a function of y, the typical cross-section area is

$$A(y) = \int_{x=0}^{x=2} (4 - x - y)\,dx = \left[4x - \frac{x^2}{2} - xy\right]_{x=0}^{x=2} = 6 - 2y. \tag{7}$$

The volume of the entire solid is therefore

$$\text{Volume} = \int_{y=0}^{y=1} A(y)\,dy = \int_{y=0}^{y=1} (6 - 2y)\,dy = \left[6y - y^2\right]_0^1 = 5,$$

in agreement with our earlier calculation.

Again, we may give instructions for calculating the volume as an iterated integral by writing

$$\text{Volume} = \int_0^1 \int_0^2 (4 - x - y)\,dx\,dy.$$

The expression on the right says we can find the volume by integrating $4 - x - y$ with respect to x from $x = 0$ to $x = 2$ as in Equation (7) and integrating the result with respect to y from $y = 0$ to $y = 1$. In this iterated integral, the order of integration is first x and then y, the reverse of the order in Equation (6).

What do these two volume calculations with iterated integrals have to do with the double integral

$$\iint_R (4 - x - y)\,dA$$

over the rectangle $R: 0 \le x \le 2,\ 0 \le y \le 1$? The answer is that they both give the value of the double integral. A theorem published in 1907 by Guido Fubini says that the double integral of any continuous function over a rectangle can be calculated as an iterated integral in either order of integration. (Fubini proved his theorem in greater generality, but this is how it translates into what we're doing now.)

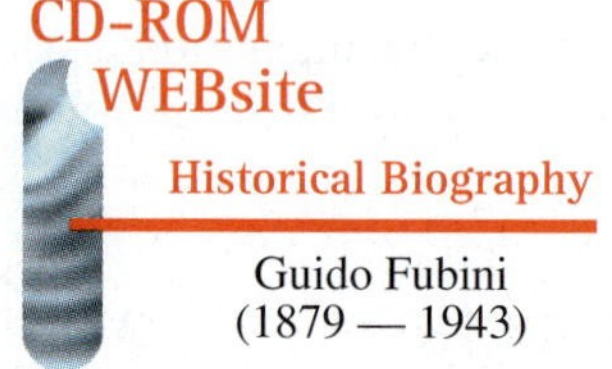

Theorem 1 Fubini's Theorem (First Form)

If $f(x, y)$ is continuous throughout the rectangular region $R: a \le x \le b$, $c \le y \le d$, then

$$\iint_R f(x, y)\,dA = \int_c^d \int_a^b f(x, y)\,dx\,dy = \int_a^b \int_c^d f(x, y)\,dy\,dx.$$

Fubini's theorem says that double integrals over rectangles can be calculated as iterated integrals. Thus, we can evaluate a double integral by integrating with respect to one variable at a time.

Fubini's theorem also says that we may calculate the double integral by integrating in *either* order, a genuine convenience, as we see in Example 3. In particular, when we calculate a volume by slicing, we may use either planes perpendicular to the x-axis or planes perpendicular to the y-axis.

Example 1 Evaluating a Double Integral

Calculate $\iint_R f(x, y)\, dA$ for

$$f(x, y) = 1 - 6x^2y \qquad \text{and} \qquad R:\ 0 \le x \le 2, \quad -1 \le y \le 1.$$

Solution By Fubini's theorem,

$$\iint_R f(x, y)\, dA = \int_{-1}^{1}\int_{0}^{2} (1 - 6x^2y)\, dx\, dy = \int_{-1}^{1} \left[x - 2x^3y\right]_{x=0}^{x=2} dy$$

$$= \int_{-1}^{1} (2 - 16y)\, dy = \left[2y - 8y^2\right]_{-1}^{1} = 4 \text{ units cubed}.$$

Reversing the order of integration gives the same answer:

$$\int_0^2\int_{-1}^{1} (1 - 6x^2 y)\, dy\, dx = \int_0^2 \left[y - 3x^2 y^2\right]_{y=-1}^{y=1} dx$$

$$= \int_0^2 [(1 - 3x^2) - (-1 - 3x^2)]\, dx$$

$$= \int_0^2 2\, dx = 4 \text{ units cubed}.$$

USING TECHNOLOGY *Multiple Integration* Most computer algebra systems can calculate both multiple and iterated integrals. The typical procedure is to apply the CAS integrate command in nested iterations according to the order of integration you specify.

Integral	Typical CAS Formulation
$\iint x^2y\, dx\, dy$	int(int(x ^ 2 * y, x), y);
$\int_{-\pi/3}^{\pi/4}\int_0^1 x \cos y\, dx\, dy$	int(int(x* cos(y), x = 0 . . 1), y = −Pi/3 . . Pi/4);

If a CAS cannot produce an exact value for a definite integral, it can usually find an approximate value numerically.

Double Integrals over Bounded Nonrectangular Regions

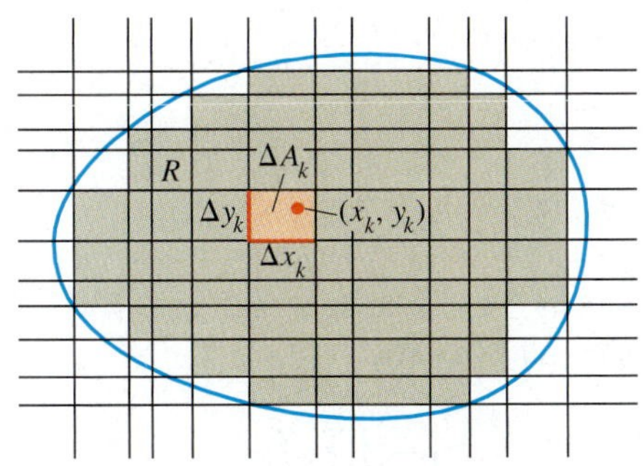

FIGURE 12.6 A rectangular grid partitioning a bounded nonrectangular region into cells.

To define the double integral of a function $f(x, y)$ over a bounded nonrectangular region, like the one shown in Figure 12.6, we again imagine R to be covered by a rectangular grid, but we include in the partial sum only the small pieces of area $\Delta A = \Delta x\, \Delta y$ that lie entirely within the region (shaded in the figure). We number the pieces in some order, choose an arbitrary point (x_k, y_k) in each ΔA_k, and form the sum

$$S_n = \sum_{k=1}^{n} f(x_k, y_k)\, \Delta A_k.$$

The only difference between this sum and the one in Equation (1) for rectangular regions is that now the areas ΔA_k may not cover all of R. As the mesh becomes in-

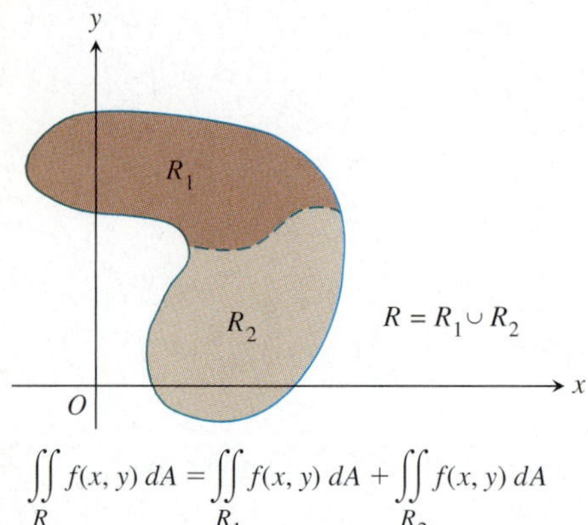

FIGURE 12.7 The additivity property for rectangular regions holds for regions bounded by continuous curves.

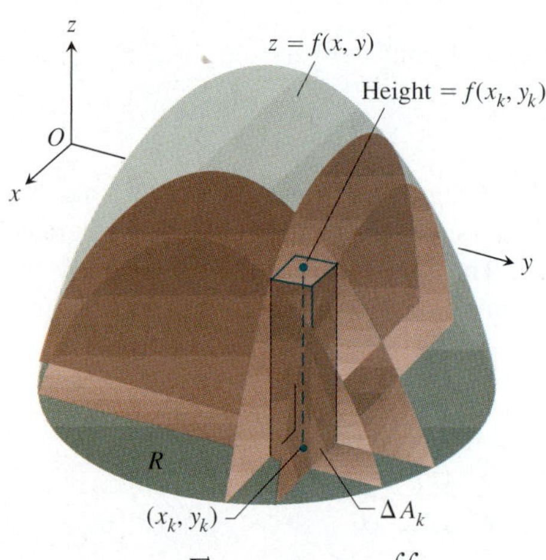

FIGURE 12.8 We define the volumes of solids with curved bases the same way we define the volumes of solids with rectangular bases.

creasingly fine and the number of terms in S_n increases, however, more and more of R is included. If f is continuous and the boundary of R is made from the graphs of a finite number of continuous functions of x and/or continuous functions of y joined end to end, then the sums S_n will have a limit as the norms of the partitions that define the rectangular grid independently approach zero. We call the limit the **double integral** of f over R:

$$\iint_R f(x, y)\,dA = \lim_{\Delta A \to 0} \sum f(x_k, y_k)\,\Delta A_k.$$

This limit may also exist under less restrictive circumstances.

Double integrals of continuous functions over nonrectangular regions have the same algebraic properties as integrals over rectangular regions. The domain additivity property corresponding to property 5 says that if R is decomposed into nonoverlapping regions R_1 and R_2 with boundaries that are again made of a finite number of line segments or smooth curves (see Figure 12.7 for an example), then

$$\iint_R f(x, y)\,dA = \iint_{R_1} f(x, y)\,dA + \iint_{R_2} f(x, y)\,dA.$$

If $f(x, y)$ is positive and continuous over R we define the volume of the solid region between R and the surface $z = f(x, y)$ to be $\iint_R f(x, y)\,dA$, as before (Figure 12.8).

If R is a region like the one shown in the xy-plane in Figure 12.9, bounded "above" and "below" by the curves $y = g_2(x)$ and $y = g_1(x)$ and on the sides by the lines $x = a$, $x = b$, we may again calculate the volume by the method of slicing. We first calculate the cross-section area

$$A(x) = \int_{y=g_1(x)}^{y=g_2(x)} f(x, y)\,dy$$

and then integrate $A(x)$ from $x = a$ to $x = b$ to get the volume as an iterated integral:

$$V = \int_a^b A(x)\,dx = \int_a^b \int_{g_1(x)}^{g_2(x)} f(x, y)\,dy\,dx. \tag{8}$$

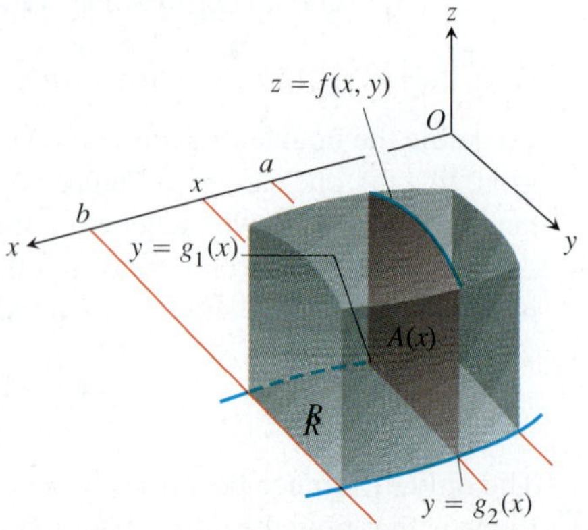

FIGURE 12.9 The area of the vertical slice shown here is

$$A(x) = \int_{g_1(x)}^{g_2(x)} f(x, y)\,dy.$$

To calculate the volume of the solid, we integrate this area from $x = a$ to $x = b$.

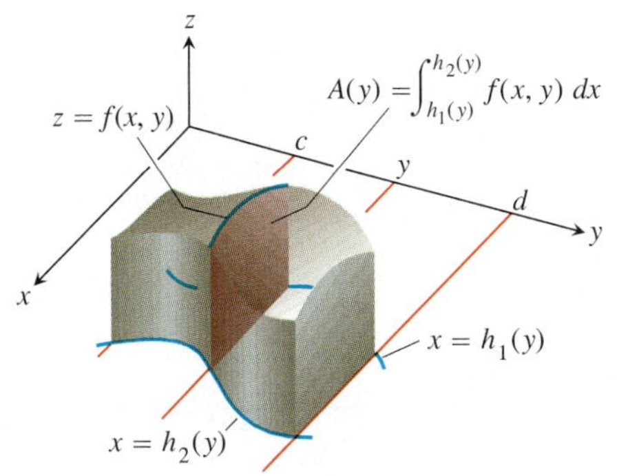

FIGURE 12.10 The volume of the solid shown here is

$$\int_c^d A(y)\,dy = \int_c^d \int_{h_1(y)}^{h_2(y)} f(x, y)\,dx\,dy.$$

Similarly, if R is a region like the one shown in Figure 12.10, bounded by the curves $x = h_2(y)$ and $x = h_1(y)$ and the lines $y = c$ and $y = d$, then the volume calculated by slicing is given by the iterated integral

$$\text{Volume} = \int_c^d \int_{h_1(y)}^{h_2(y)} f(x, y)\,dx\,dy. \tag{9}$$

That the iterated integrals in Equations (8) and (9) both give the volume that we defined to be the double integral of f over R is a consequence of the following stronger form of Fubini's theorem.

Theorem 2 **Fubini's Theorem (Stronger Form)**

Let $f(x, y)$ be continuous on a region R.

1. If R is defined by $a \le x \le b$, $g_1(x) \le y \le g_2(x)$, with g_1 and g_2 continuous on $[a, b]$, then

$$\iint_R f(x, y)\,dA = \int_a^b \int_{g_1(x)}^{g_2(x)} f(x, y)\,dy\,dx.$$

2. If R is defined by $c \le y \le d$, $h_1(y) \le x \le h_2(y)$, with h_1 and h_2 continuous on $[c, d]$, then

$$\iint_R f(x, y)\,dA = \int_c^d \int_{h_1(y)}^{h_2(y)} f(x, y)\,dx\,dy.$$

Example 2 Finding Volume

Find the volume of the prism whose base is the triangle in the xy-plane bounded by the x-axis and the lines $y = x$ and $x = 1$ and whose top lies in the plane

$$z = f(x, y) = 3 - x - y.$$

Solution See Figure 12.11. For any x between 0 and 1, y may vary from $y = 0$ to $y = x$ (Figure 12.11b). Hence,

$$\begin{aligned} V &= \int_0^1 \int_0^x (3 - x - y)\,dy\,dx = \int_0^1 \left[3y - xy - \frac{y^2}{2}\right]_{y=0}^{y=x} dx \\ &= \int_0^1 \left(3x - \frac{3x^2}{2}\right) dx = \left[\frac{3x^2}{2} - \frac{x^3}{2}\right]_{x=0}^{x=1} = 1 \text{ cubic unit.} \end{aligned}$$

When the order of integration is reversed (Figure 12.11c), the integral for the volume is

$$\begin{aligned} V &= \int_0^1 \int_y^1 (3 - x - y)\,dx\,dy = \int_0^1 \left[3x - \frac{x^2}{2} - xy\right]_{x=y}^{x=1} dy \\ &= \int_0^1 \left(3 - \frac{1}{2} - y - 3y + \frac{y^2}{2} + y^2\right) dy \\ &= \int_0^1 \left(\frac{5}{2} - 4y + \frac{3}{2}y^2\right) dy = \left[\frac{5}{2}y - 2y^2 + \frac{y^3}{2}\right]_{y=0}^{y=1} = 1 \text{ cubic unit.} \end{aligned}$$

The two integrals are equal, as they should be.

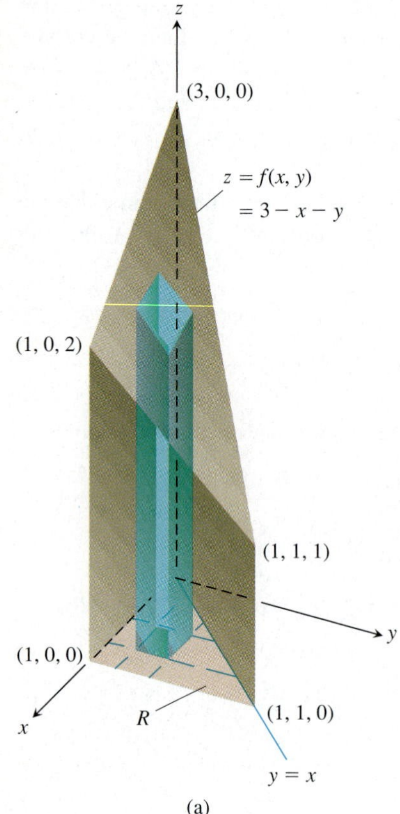

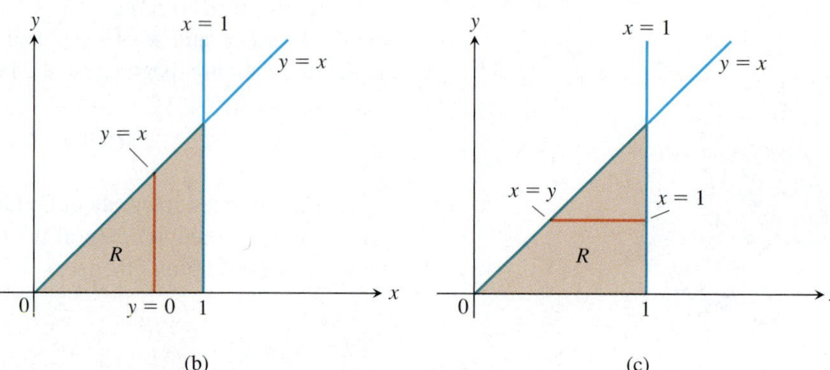

FIGURE 12.11 (a) Prism with a triangular base in the xy-plane. The volume of this prism is defined as a double integral over R. To evaluate it as an iterated integral, we may integrate first with respect to y and then with respect to x, or the other way around (Example 2). (b) Integration limits of

$$\int_{x=0}^{x=1} \int_{y=0}^{y=x} f(x, y)\, dy\, dx.$$

If we integrate first with respect to y, we integrate along a vertical line through R and then integrate from left to right to include all the vertical lines in R. (c) Integration limits of

$$\int_{y=0}^{y=1} \int_{x=y}^{x=1} f(x, y)\, dx\, dy.$$

If we integrate first with respect to x, we integrate along a horizontal line through R and then integrate from bottom to top to include all the horizontal lines in R.

Although Fubini's theorem assures us that a double integral may be calculated as an iterated integral in either order of integration, the value of one integral may be easier to find than the value of the other. The next example shows how this can happen.

Example 3 Evaluating a Double Integral

Calculate

$$\iint_R \frac{\sin x}{x}\, dA,$$

where R is the triangle in the xy-plane bounded by the x-axis, the line $y = x$, and the line $x = 1$.

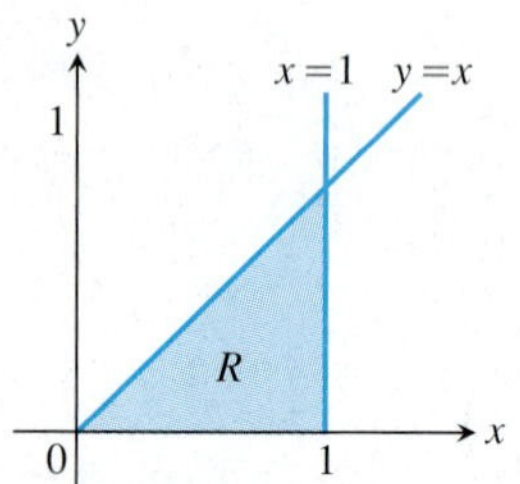

FIGURE 12.12 The region of integration in Example 3.

Solution The region of integration is shown in Figure 12.12. If we integrate first with respect to y and then with respect to x, we find

$$\int_0^1 \left(\int_0^x \frac{\sin x}{x}\, dy \right) dx = \int_0^1 \left(y \frac{\sin x}{x} \bigg]_{y=0}^{y=x} \right) dx = \int_0^1 \sin x\, dx$$

$$= -\cos\,(1) + 1 \approx 0.46 \text{ unit cubed.}$$

If we reverse the order of integration and attempt to calculate

$$\int_0^1 \int_y^1 \frac{\sin x}{x}\, dx\, dy,$$

we are stopped because $\int ((\sin x)/x)\,dx$ cannot be expressed in terms of elementary functions.

There is no general rule for predicting which order of integration will be the good one in circumstances like these, so don't worry about how to start your integrations. Just forge ahead and if the order you first choose doesn't work, try the other.

Finding Limits of Integration

The hardest part of evaluating a double integral can be finding the limits of integration. Fortunately, there is a good procedure to follow.

Procedure for Finding Limits of Integration

A. To evaluate $\iint_R f(x, y)\, dA$ over a region R, integrating first with respect to y and then with respect to x, take the following steps.

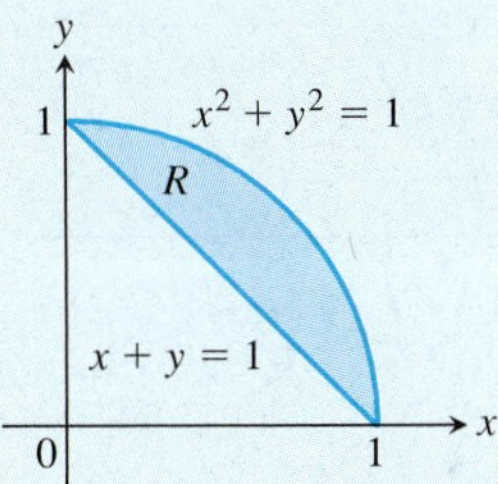

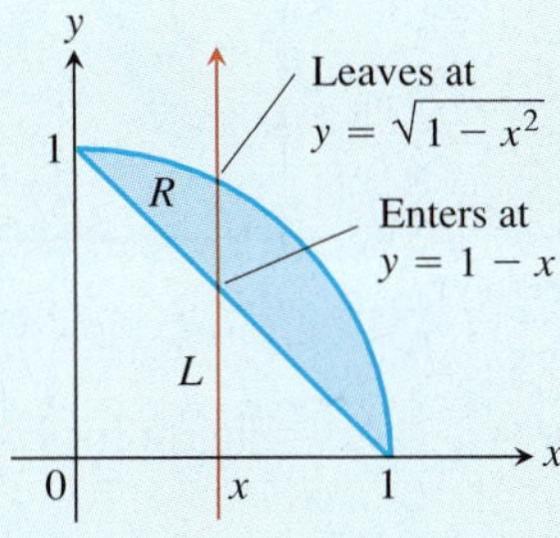

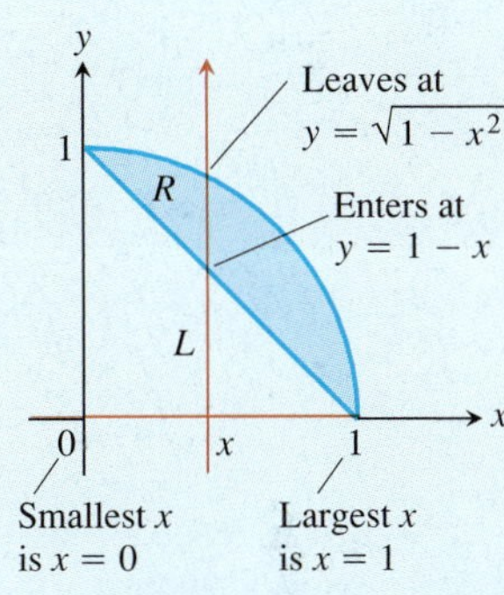

Step 1: *A sketch.* Sketch the region of integration and label the bounding curves.

Step 2: *The y-limits of integration.* Imagine a vertical line L cutting through R in the direction of increasing y. Mark the y-values where L enters and leaves. These are the y-limits of integration and are usually functions of x (instead of constants).

Step 3: *The x-limits of integration.* Choose x-limits that include all the vertical lines through R. The integral is

$$\iint_R f(x, y)\, dA =$$

$$\int_{x=0}^{x=1} \int_{y=1-x}^{y=\sqrt{1-x^2}} f(x, y)\, dy\, dx.$$

B. To evaluate the same double integral as an iterated integral with the order of integration reversed, use horizontal lines instead of vertical lines. The integral is

$$\iint_R f(x, y)\, dA = \int_0^1 \int_{1-y}^{\sqrt{1-y^2}} f(x, y)\, dx\, dy.$$

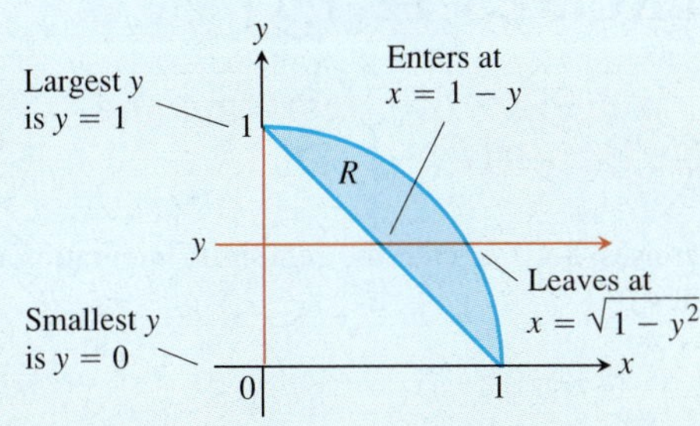

Example 4 Reversing the Order of Integration

Sketch the region of integration for the integral

$$\int_0^2 \int_{x^2}^{2x} (4x + 2)\, dy\, dx$$

and write an equivalent integral with the order of integration reversed.

Solution The region of integration is given by the inequalities $x^2 \leq y \leq 2x$ and $0 \leq x \leq 2$. It is therefore the region bounded by the curves $y = x^2$ and $y = 2x$ between $x = 0$ and $x = 2$ (Figure 12.13a).

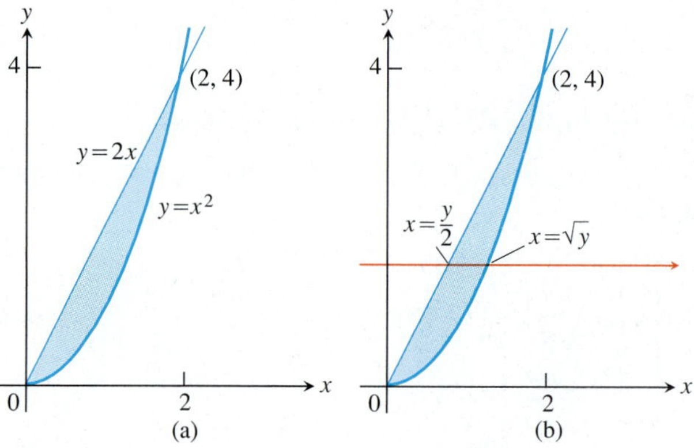

FIGURE 12.13 Figure for Example 4.

To find limits for integrating in the reverse order, we imagine a horizontal line passing from left to right through the region. It enters at $x = y/2$ and leaves at $x = \sqrt{y}$. To include all such lines, we let y run from $y = 0$ to $y = 4$ (Figure 12.13b). The integral is

$$\int_0^4 \int_{y/2}^{\sqrt{y}} (4x + 2)\, dx\, dy.$$

The common value of these integrals is 8.

EXERCISES 12.1

Finding Regions of Integration and Double Integrals

In Exercises 1–10, sketch the region of integration and evaluate the integral.

1. $\int_0^3 \int_0^2 (4 - y^2)\, dy\, dx$

2. $\int_0^3 \int_{-2}^0 (x^2 y - 2xy)\, dy\, dx$

3. $\int_{-1}^0 \int_{-1}^1 (x + y + 1)\, dx\, dy$

4. $\int_{\pi}^{2\pi} \int_0^{\pi} (\sin x + \cos y)\, dx\, dy$

5. $\int_0^{\pi} \int_0^x x \sin y\, dy\, dx$

6. $\int_0^{\pi} \int_0^{\sin x} y\, dy\, dx$

7. $\int_1^{\ln 8}\int_0^{\ln y} e^{x+y}\,dx\,dy$

8. $\int_1^2\int_y^{y^2} dx\,dy$

9. $\int_0^1\int_0^{y^2} 3y^3e^{xy}\,dx\,dy$

10. $\int_1^4\int_0^{\sqrt{x}} \frac{3}{2}e^{y/\sqrt{x}}\,dy\,dx$

In Exercises 11–16, integrate f over the given region.

11. *Quadrilateral* $f(x, y) = x/y$ over the region in the first quadrant bounded by the lines $y = x$, $y = 2x$, $x = 1$, $x = 2$

12. *Square* $f(x, y) = 1/(xy)$ over the square $1 \le x \le 2$, $1 \le y \le 2$

13. *Triangle* $f(x, y) = x^2 + y^2$ over the triangular region with vertices $(0, 0)$, $(1, 0)$, and $(0, 1)$

14. *Rectangle* $f(x, y) = y\cos xy$ over the rectangle $0 \le x \le \pi$, $0 \le y \le 1$

15. *Triangle* $f(u, v) = v - \sqrt{u}$ over the triangular region cut from the first quadrant of the uv-plane by the line $u + v = 1$

16. *Curved region* $f(s, t) = e^s \ln t$ over the region in the first quadrant of the st-plane that lies above the curve $s = \ln t$ from $t = 1$ to $t = 2$

Each of Exercises 17–20 gives an integral over a region in a Cartesian coordinate plane. Sketch the region and evaluate the integral.

17. $\int_{-2}^0\int_v^{-v} 2\,dp\,dv$ (the pv-plane)

18. $\int_0^1\int_0^{\sqrt{1-s^2}} 8t\,dt\,ds$ (the st-plane)

19. $\int_{-\pi/3}^{\pi/3}\int_0^{\sec t} 3\cos t\,du\,dt$ (the tu-plane)

20. $\int_0^3\int_1^{4-2u} \frac{4-2u}{v^2}\,dv\,du$ (the uv-plane)

Reversing the Order of Integration

In Exercises 21–30, sketch the region of integration and write an equivalent double integral with the order of integration reversed.

21. $\int_0^1\int_2^{4-2x} dy\,dx$

22. $\int_0^2\int_{y-2}^0 dx\,dy$

23. $\int_0^1\int_y^{\sqrt{y}} dx\,dy$

24. $\int_0^1\int_{1-x}^{1-x^2} dy\,dx$

25. $\int_0^1\int_1^{e^x} dy\,dx$

26. $\int_0^{\ln 2}\int_{e^y}^2 dx\,dy$

27. $\int_0^{3/2}\int_0^{9-4x^2} 16x\,dy\,dx$

28. $\int_0^2\int_0^{4-y^2} y\,dx\,dy$

29. $\int_0^1\int_{-\sqrt{1-y^2}}^{\sqrt{1-y^2}} 3y\,dx\,dy$

30. $\int_0^2\int_{-\sqrt{4-x^2}}^{\sqrt{4-x^2}} 6x\,dy\,dx$

Evaluating Double Integrals

In Exercises 31–40, sketch the region of integration, reverse the order of integration, and evaluate the integral.

31. $\int_0^{\pi}\int_x^{\pi} \frac{\sin y}{y}\,dy\,dx$

32. $\int_0^2\int_x^2 2y^2\sin xy\,dy\,dx$

33. $\int_0^1\int_y^1 x^2e^{xy}\,dx\,dy$

34. $\int_0^2\int_0^{4-x^2} \frac{xe^{2y}}{4-y}\,dy\,dx$

35. $\int_0^{2\sqrt{\ln 3}}\int_{y/2}^{\sqrt{\ln 3}} e^{x^2}\,dx\,dy$

36. $\int_0^3\int_{\sqrt{x/3}}^1 e^{y^3}\,dy\,dx$

37. $\int_0^{1/16}\int_{y^{1/4}}^{1/2} \cos(16\pi x^5)\,dx\,dy$

38. $\int_0^8\int_{\sqrt[3]{x}}^2 \frac{dy\,dx}{y^4+1}$

39. *Square region* $\iint_R (y - 2x^2)\,dA$ where R is the region bounded by the square $|x| + |y| = 1$

40. *Triangular region* $\iint_R xy\,dA$ where R is the region bounded by the lines $y = x$, $y = 2x$, and $x + y = 2$

Volume Beneath a Surface $z = f(x, y)$

CD-ROM WEBsite

41. Find the volume of the region bounded by the paraboloid $z = x^2 + y^2$ and below by the triangle enclosed by the lines $y = x$, $x = 0$, and $x + y = 2$ in the xy-plane.

42. Find the volume of the solid that is bounded above by the cylinder $z = x^2$ and below by the region enclosed by the parabola $y = 2 - x^2$ and the line $y = x$ in the xy-plane.

43. Find the volume of the solid whose base is the region in the xy-plane that is bounded by the parabola $y = 4 - x^2$ and the line $y = 3x$, while the top of the solid is bounded by the plane $z = x + 4$.

44. Find the volume of the solid in the first octant bounded by the coordinate planes, the cylinder $x^2 + y^2 = 4$, and the plane $z + y = 3$.

45. Find the volume of the solid in the first octant bounded by the coordinate planes, the plane $x = 3$, and the parabolic cylinder $z = 4 - y^2$.

46. Find the volume of the solid cut from the first octant by the surface $z = 4 - x^2 - y$.

47. Find the volume of the wedge cut from the first octant by the cylinder $z = 12 - 3y^2$ and the plane $x + y = 2$.

48. Find the volume of the solid cut from the square column $|x| + |y| \le 1$ by the planes $z = 0$ and $3x + z = 3$.

49. Find the volume of the solid that is bounded on the front and back by the planes $x = 2$ and $x = 1$, on the sides by the cylinders $y = \pm 1/x$, and above and below by the planes $z = x + 1$ and $z = 0$.

50. Find the volume of the solid bounded on the front and back by the planes $x = \pm\pi/3$, on the sides by the cylinders $y = \pm \sec x$, above by the cylinder $z = 1 + y^2$, and below by the xy-plane.

Integrals over Unbounded Regions

Evaluate the improper integrals in Exercises 51–54 as iterated integrals.

51. $\displaystyle\int_1^\infty \int_{e^{-x}}^1 \frac{1}{x^3 y}\, dy\, dx$

52. $\displaystyle\int_{-1}^1 \int_{-1/\sqrt{1-x^2}}^{1/\sqrt{1-x^2}} (2y+1)\, dy\, dx$

53. $\displaystyle\int_{-\infty}^\infty \int_{-\infty}^\infty \frac{1}{(x^2+1)(y^2+1)}\, dx\, dy$

54. $\displaystyle\int_0^\infty \int_0^\infty xe^{-(x+2y)}\, dx\, dy$

Approximating Double Integrals

In Exercises 55 and 56, approximate the double integral of $f(x, y)$ over the region R partitioned by the given vertical lines $x = a$ and horizontal lines $y = c$. In each subrectangle, use (x_k, y_k) as indicated for your approximation.

$$\iint_R f(x, y)\, dA \approx \sum_{k=1}^n f(x_k, y_k)\, \Delta A_k$$

55. $f(x, y) = x + y$ over the region R bounded above by the semicircle $y = \sqrt{1 - x^2}$ and below by the x-axis, using the partition $x = -1, -1/2, 0, 1/4, 1/2, 1$ and $y = 0, 1/2, 1$ with (x_k, y_k) the lower left corner in the kth subrectangle (provided the subrectangle lies within R)

56. $f(x, y) = x + 2y$ over the region R inside the circle $(x - 2)^2 + (y - 3)^2 = 1$ using the partition $x = 1, 3/2, 2, 5/2, 3$ and $y = 2, 5/2, 3, 7/2, 4$ with (x_k, y_k) the center (centroid) in the kth subrectangle (provided the subrectangle lies within R)

Theory and Examples

57. *Circular sector* Integrate $f(x, y) = \sqrt{4 - x^2}$ over the smaller sector cut from the disk $x^2 + y^2 \le 4$ by the rays $\theta = \pi/6$ and $\theta = \pi/2$.

58. *Unbounded region* Integrate $f(x, y) = 1/[(x^2 - x)(y - 1)^{2/3}]$ over the infinite rectangle $2 \le x < \infty, 0 \le y \le 2$.

59. *Noncircular cylinder* A solid right (noncircular) cylinder has its base R in the xy-plane and is bounded above by the paraboloid $z = x^2 + y^2$. The cylinder's volume is

$$V = \int_0^1 \int_0^y (x^2 + y^2)\, dx\, dy + \int_1^2 \int_0^{2-y} (x^2 + y^2)\, dx\, dy.$$

Sketch the base region R and express the cylinder's volume as a single iterated integral with the order of integration reversed. Then evaluate the integral to find the volume.

60. *Converting to a double integral* Evaluate the integral

$$\int_0^2 (\tan^{-1} \pi x - \tan^{-1} x)\, dx.$$

(*Hint*: Write the integrand as an integral.)

61. *Maximizing a double integral* What region R in the xy-plane maximizes the value of

$$\iint_R (4 - x^2 - 2y^2)\, dA?$$

Give reasons for your answer.

62. *Minimizing a double integral* What region R in the xy-plane minimizes the value of

$$\iint_R (x^2 + y^2 - 9)\, dA?$$

Give reasons for your answer.

63. *Writing to Learn* Is it all right to evaluate the integral of a continuous function $f(x, y)$ over a rectangular region in the xy-plane and get different answers depending on the order of integration? Give reasons for your answer.

64. *Writing to Learn* How would you evaluate the double integral of a continuous function $f(x, y)$ over the region R in the xy-plane enclosed by the triangle with vertices $(0, 1)$, $(2, 0)$, and $(1, 2)$? Give reasons for your answer.

65. *Unbounded region* Prove that

$$\int_{-\infty}^\infty \int_{-\infty}^\infty e^{-x^2-y^2}\, dx\, dy = \lim_{b\to\infty} \int_{-b}^b \int_{-b}^b e^{-x^2-y^2} dx\, dy$$
$$= 4\left(\int_0^\infty e^{-x^2}\, dx\right)^2.$$

66. *Improper double integral* Evaluate the improper integral

$$\int_0^1 \int_0^3 \frac{x^2}{(y-1)^{2/3}}\, dy\, dx.$$

COMPUTER EXPLORATIONS

Evaluating Double Integrals Numerically

Use a CAS double-integral evaluator to estimate the values of the integrals in Exercises 67–70.

67. $\displaystyle\int_1^3 \int_1^x \frac{1}{xy}\, dy\, dx$

68. $\displaystyle\int_0^1 \int_0^1 e^{-(x^2+y^2)}\, dy\, dx$

69. $\displaystyle\int_0^1 \int_0^1 \tan^{-1} xy\, dy\, dx$

70. $\displaystyle\int_{-1}^1 \int_0^{\sqrt{1-x^2}} 3\sqrt{1 - x^2 - y^2}\, dy\, dx$

Use a CAS double-integral evaluator to find the integrals in Exercises 71–76. Then reverse the order of integration and evaluate, again with a CAS.

71. $\int_0^1 \int_{2y}^4 e^{x^2}\, dx\, dy$

72. $\int_0^3 \int_{x^2}^9 x \cos (y^2)\, dy\, dx$

73. $\int_0^2 \int_{y^3}^{4\sqrt{2y}} \left(x^2 y - xy^2\right) dx\, dy$

74. $\int_0^2 \int_0^{4-y^2} e^{xy}\, dx\, dy$

75. $\int_1^2 \int_0^{x^2} \frac{1}{x+y}\, dy\, dx$

76. $\int_1^2 \int_{y^3}^8 \frac{1}{\sqrt{x^2+y^2}}\, dx\, dy$

12.2 Areas, Moments, and Centers of Mass*

Areas of Bounded Regions in the Plane • Average Value • Moments and Centers of Mass • Masses Distributed over a Plane Region • Thin, Flat Plates with Continuous Mass Distributions • Moments of Inertia • Centroids of Geometric Figures

In this section, we show how to use double integrals to calculate the areas of bounded regions in the plane and to find the average value of a function of two variables. Then we study the physical problem of finding the center of mass of a thin plate covering a region in the plane.

Areas of Bounded Regions in the Plane

If we take $f(x, y) = 1$ in the definition of the double integral over a region R in the preceding section, the partial sums reduce to

$$S_n = \sum_{k=1}^{n} f(x_k, y_k)\, \Delta A_k = \sum_{k=1}^{n} \Delta A_k .$$

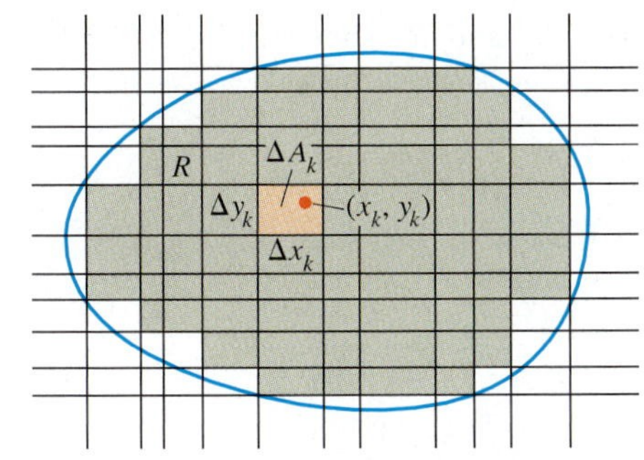

FIGURE 12.14 The first step in defining the area of a region is to partition the interior of the region into cells.

This approximates what we would like to call the area of R. As Δx and Δy approach zero, the coverage of R by the ΔA_k's (Figure 12.14) becomes increasingly complete, and we define the area of R to be the limit

$$\text{Area} = \lim_{n \to \infty} \sum_{k=1}^{n} \Delta A_k = \iint_R dA .$$

Definition Area

The **area** of a closed, bounded plane region R is

$$A = \iint_R dA .$$

As with the other definitions in this chapter, the definition here applies to a greater variety of regions than does the earlier single-variable definition of area, but it agrees with the earlier definition on regions to which they both apply.

*The material on mass and moments for planar regions presented in this section does not require the coverage from Chapter 5. The essential ideas are all given here, some of which may be a review for students who studied moments in Chapter 5.

To evaluate the integral in the definition of area, we integrate the constant function $f(x, y) = 1$ over R.

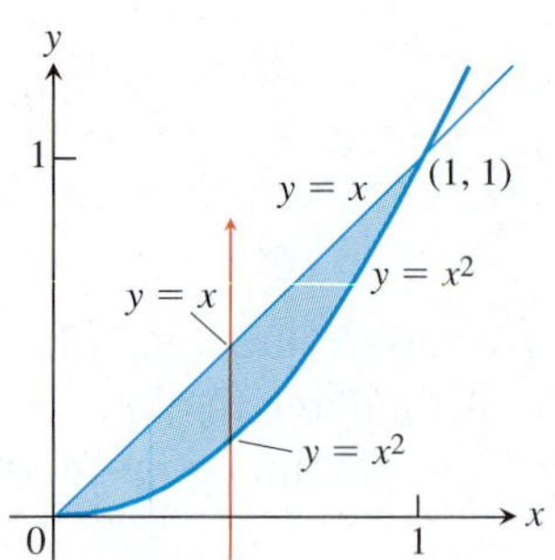

FIGURE 12.15 The region in Example 1

Example 1 Finding Area

Find the area of the region R bounded by $y = x$ and $y = x^2$ in the first quadrant.

Solution We sketch the region (Figure 12.15) and calculate the area as

$$A = \int_0^1 \int_{x^2}^{x} dy\, dx = \int_0^1 \Big[y \Big]_{x^2}^{x} dx$$
$$= \int_0^1 (x - x^2)\, dx = \left[\frac{x^2}{2} - \frac{x^3}{3} \right]_0^1 = \frac{1}{6} \text{ square unit.}$$

Notice that the single integral $\int_0^1 (x - x^2)\, dx$, obtained from evaluating the inside iterated integral, is the integral for the area between these two curves using the method of Section 4.6.

Example 2 Finding Area

Find the area of the region R enclosed by the parabola $y = x^2$ and the line $y = x + 2$.

Solution If we divide R into the regions R_1 and R_2 shown in Figure 12.16a, we may calculate the area as

$$A = \iint_{R_1} dA + \iint_{R_2} dA = \int_0^1 \int_{-\sqrt{y}}^{\sqrt{y}} dx\, dy + \int_1^4 \int_{y-2}^{\sqrt{y}} dx\, dy.$$

On the other hand, reversing the order of integration (Figure 12.16b) gives

$$A = \int_{-1}^{2} \int_{x^2}^{x+2} dy\, dx.$$

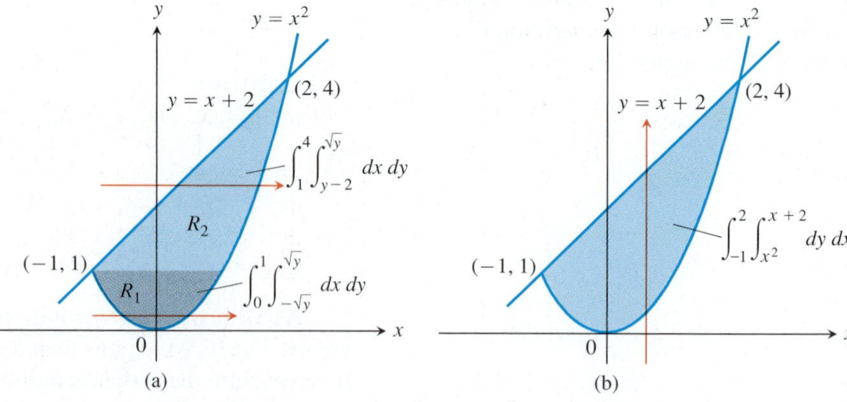

FIGURE 12.16 Calculating this area takes (a) two double integrals if the first integration is with respect to x, but (b) only one if the first integration is with respect to y. (Example 2)

This result is simpler and is the only one we would bother to write down in practice. The area is

$$A = \int_{-1}^{2} \Big[y\Big]_{x^2}^{x+2} dx = \int_{-1}^{2} (x + 2 - x^2)\, dx = \left[\frac{x^2}{2} + 2x - \frac{x^3}{3}\right]_{-1}^{2} = \frac{9}{2} \text{ square units}.$$

Average Value

The average value of an integrable function of a single variable on a closed interval is the integral of the function over the interval divided by the length of the interval. For an integrable function of two variables defined on a closed and bounded region that has a measurable area, the average value is the integral over the region divided by the area of the region. If f is the function and R the region, then

$$\textbf{Average value of } f \text{ over } R = \frac{1}{\text{area of } R} \iint_R f\, dA. \tag{1}$$

If f is the area density of a thin plate covering R, then the double integral of f over R divided by the area of R is the plate's average density in units of mass per unit area. If $f(x, y)$ is the distance from the point (x, y) to a fixed point P, then the average value of f over R is the average distance of points in R from P.

Example 3 Finding Average Value

Find the average value of $f(x, y) = x \cos xy$ over the rectangle $R: 0 \le x \le \pi$, $0 \le y \le 1$.

Solution The value of the integral of f over R is

$$\int_0^{\pi}\int_0^1 x \cos xy\, dy\, dx = \int_0^{\pi} \Big[\sin xy\Big]_{y=0}^{y=1} dx$$

$$= \int_0^{\pi} (\sin x - 0)\, dx = -\cos x\Big]_0^{\pi} = 1 + 1 = 2.$$

The area of R is π. The average value of f over R is $2/\pi$.

Moments and Centers of Mass

Many structures and mechanical systems behave as if their masses were concentrated at a single point, called the center of mass. It is important to know how to locate this point, and doing so is basically a mathematical enterprise. We develop our mathematical model in stages. The first stage is to imagine masses m_1, m_2, and m_3 on a rigid x-axis supported by a fulcrum at the origin.

The resulting system might balance, or it might not. It depends on how large the masses are and how they are arranged.

Mass versus weight

Weight is the force that results from gravity pulling on a mass. If an object of mass m is placed in a location where the acceleration of gravity is g, the object's weight there is

$$F = mg$$

(as in Newton's second law).

Each mass m_k exerts a downward force $m_k g$ equal to the magnitude of the mass times the acceleration of gravity. Each of these forces has a tendency to turn the axis about the origin, the way you turn a seesaw. This turning effect, called a **torque,** is measured by multiplying the force $m_k g$ by the signed distance x_k from the point of application to the origin. Masses to the left of the origin exert negative (counterclockwise) torque. Masses to the right of the origin exert positive (clockwise) torque.

The sum of the torques measures the tendency of a system to rotate about the origin. This sum is called the **system torque.**

$$\text{System torque} = \sum m_k g x_k \tag{2}$$

The system will balance if and only if its torque is zero.

If we factor out the g in Equation (2), we see that the system torque is

$$g \sum m_k x_k$$

A feature of the environment — A feature of the system

Thus, the torque is the product of the gravitational acceleration g, which is a feature of the environment in which the system happens to reside, and the number $\Sigma m_k x_k$, which is a feature of the system itself, a constant that stays the same no matter where the system is placed. The constant is called the **moment of the system about the origin.**

$$\text{Moment of system about origin} = \sum m_k x_k \tag{3}$$

We usually want to know where to place the fulcrum to make the system balance, that is, at what point $\bar{x}$ to place it to make the torque zero.

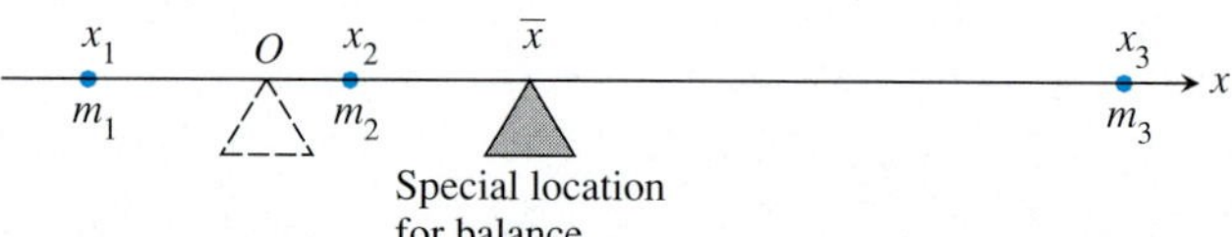

The torque of each mass about the fulcrum in this special location is

$$\begin{aligned}\text{Torque of } m_k \text{ about } \bar{x} &= \begin{pmatrix}\text{signed distance} \\ \text{of } m_k \text{ from } \bar{x}\end{pmatrix}\begin{pmatrix}\text{downward} \\ \text{force}\end{pmatrix} \\ &= (x_k - \bar{x})m_k g.\end{aligned}$$

When we write the equation that says that the sum of these torques is zero, we get an equation we can solve for $\bar{x}$:

$$\begin{aligned}\sum (x_k - \bar{x})m_k g &= 0 && \text{Sum of the torques equals zero} \\ g\sum (x_k - \bar{x})m_k &= 0 && \text{Constant Multiple Rule for Sums} \\ \sum (m_k x_k - \bar{x}m_k) &= 0 && \text{g divided out, } m_k \text{ distributed} \\ \sum m_k x_k - \sum \bar{x}m_k &= 0 && \text{Difference Rule for Sums} \\ \sum m_k x_k &= \bar{x}\sum m_k && \text{Rearranged, Constant Multiple Rule again} \\ \bar{x} &= \frac{\sum m_k x_k}{\sum m_k}. && \text{Solved for } \bar{x}\end{aligned}$$

This last equation tells us to find $\bar{x}$ by dividing the system's moment about the origin by the system's total mass:

$$\bar{x} = \frac{\sum x_k m_k}{\sum m_k} = \frac{\text{system moment about origin}}{\text{system mass}}.$$

The point $\bar{x}$ is called the system's **center of mass.**

Masses Distributed over a Plane Region

Suppose that we have a finite collection of masses located in the plane, with mass m_k at the point (x_k, y_k) (see Figure 12.17). The mass of the system is

$$\text{System mass:} \qquad M = \sum m_k .$$

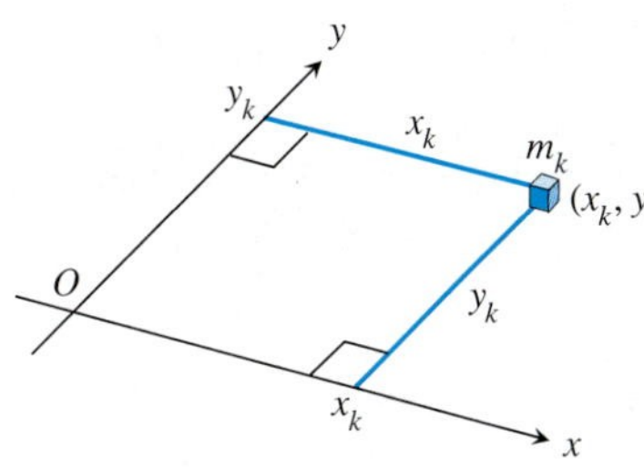

FIGURE 12.17 Each mass m_k has a moment about each axis.

Each mass m_k has a moment about each axis. Its moment about the x-axis is $m_k y_k$, and its moment about the y-axis is $m_k x_k$. The moments of the entire system about the two axes are

$$\text{Moment about } x\text{-axis:} \qquad M_x = \sum m_k y_k,$$
$$\text{Moment about } y\text{-axis:} \qquad M_y = \sum m_k x_k.$$

The x-coordinate of the system's center of mass is defined to be

$$\bar{x} = \frac{M_y}{M} = \frac{\sum m_k x_k}{\sum m_k}. \tag{4}$$

With this choice of $\bar{x}$, as in the one-dimensional case, the system balances about the line $x = \bar{x}$ (Figure 12.18).

The y-coordinate of the system's center of mass is defined to be

$$\bar{y} = \frac{M_x}{M} = \frac{\sum m_k y_k}{\sum m_k}. \tag{5}$$

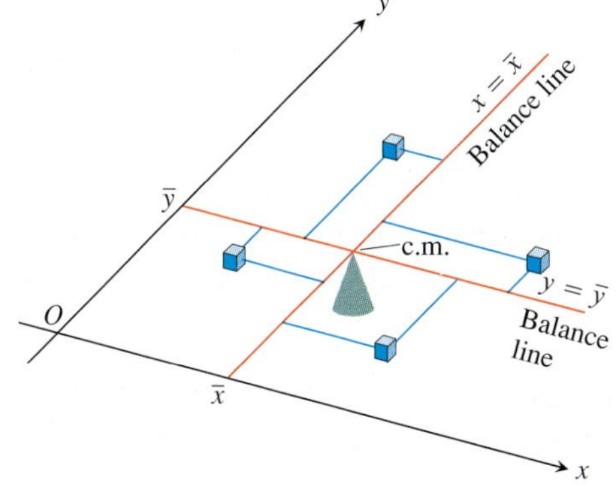

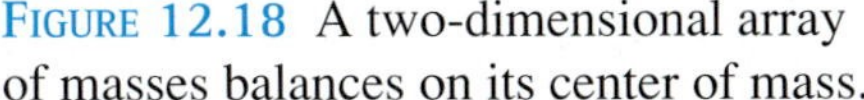
FIGURE 12.18 A two-dimensional array of masses balances on its center of mass.

With this choice of $\bar{y}$, the system balances about the line $y = \bar{y}$ as well. The torques exerted by the masses about the line $y = \bar{y}$ cancel out. Thus, as far as balance is concerned, the system behaves as if all its mass were at the single point $(\bar{x}, \bar{y})$. We call this point the system's *center of mass*.

Thin, Flat Plates with Continuous Mass Distributions

In many applications, we need to find the center of mass of a thin, flat plate: a disk of aluminum, say, or a triangular sheet of steel. In such cases, we assume the distribution of mass to be continuous, and the formulas we use to calculate $\bar{x}$ and $\bar{y}$ contain integrals instead of finite sums. The integrals arise in the following way.

Density

A material's density is its mass per unit volume. In practice, however, we tend to use units we can conveniently measure. For wires, rods, and narrow strips, we use mass per unit length. For flat sheets and plates, we use mass per unit area.

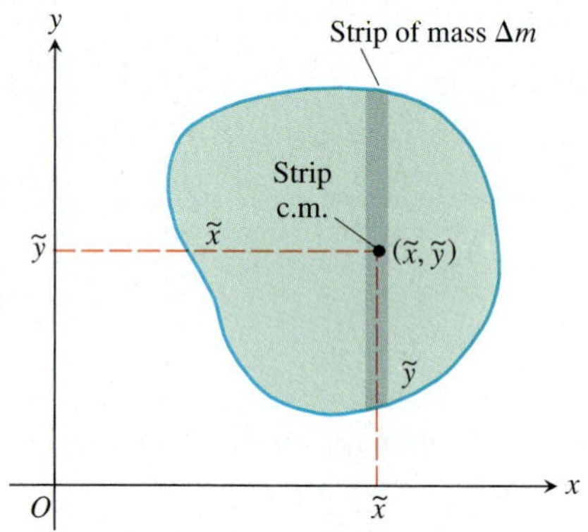

FIGURE 12.19 A plate cut into thin strips parallel to the y-axis. The moment exerted by a typical strip about each axis is the moment its mass Δm would exert if concentrated at the strips's center of mass $(\tilde{x}, \tilde{y})$.

Imagine the plate occupying a region in the xy-plane, cut into thin strips parallel to one of the axes (in Figure 12.19, the y-axis). The center of mass of a typical strip is $(\tilde{x}, \tilde{y})$. We treat the strip's mass Δm as if it were concentrated at $(\tilde{x}, \tilde{y})$. The moment of the strip about the y-axis is then $\tilde{x}\,\Delta m$. The moment of the strip about the x-axis is $\tilde{y}\,\Delta m$. Equations (4) and (5) then become

$$\bar{x} = \frac{M_y}{M} = \frac{\sum \tilde{x}\,\Delta m}{\sum \Delta m}, \qquad \bar{y} = \frac{M_x}{M} = \frac{\sum \tilde{y}\,\Delta m}{\sum \Delta m}.$$

The sums in these equations are Riemann sums for integrals and approach these integrals as limiting values as the strips into which the plate is cut become narrower and narrower. We can write these as double integrals to accommodate a great variety of shapes and density functions. **Mass** itself is the integral of the continuous density function, denoted here by $\delta(x, y)$. (Some physicists use the symbol $\rho(x, y)$ for density.) The formulas for mass, first moments, and center of mass are given in Table 12.1.

Table 12.1 Mass and first moment formulas for thin plates covering regions in the xy-plane

Density: $\delta(x, y)$

Mass: $M = \iint \delta(x, y)\, dA$

First moments: $M_x = \iint y\delta(x, y)\, dA, \qquad M_y = \iint x\delta(x, y)\, dA$

Center of mass: $\bar{x} = \dfrac{M_y}{M}, \qquad \bar{y} = \dfrac{M_x}{M}$

CD-ROM WEBsite

Example 4 Finding the Center of Mass of a Thin Plate of Variable Density

A thin plate covers the triangular region bounded by the x-axis and the lines $x = 1$ and $y = 2x$ in the first quadrant. The plate's density at the point (x, y) is $\delta(x, y) = 6x + 6y + 6$. Find the plate's mass, first moments, and center of mass about the coordinate axes.

Solution We sketch the plate and put in enough detail to determine the limits of integration for the integrals we have to evaluate (Figure 12.20).

The plate's mass is

$$\begin{aligned} M &= \int_0^1 \int_0^{2x} \delta(x, y)\, dy\, dx = \int_0^1 \int_0^{2x} (6x + 6y + 6)\, dy\, dx \\ &= \int_0^1 \left[6xy + 3y^2 + 6y \right]_{y=0}^{y=2x} dx \\ &= \int_0^1 (24x^2 + 12x)\, dx = \left[8x^3 + 6x^2 \right]_0^1 = 14. \end{aligned}$$

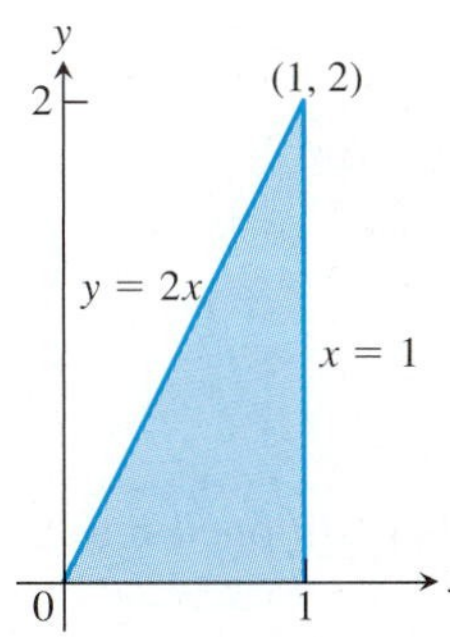

FIGURE 12.20 The triangular region covered by the plate in Example 4.

Notice that we integrate y times the density function to calculate M_x and x times density to find M_y.

The first moment about the x-axis is

$$\begin{aligned} M_x &= \int_0^1 \int_0^{2x} y\delta(x, y)\, dy\, dx = \int_0^1 \int_0^{2x} (6xy + 6y^2 + 6y)\, dy\, dx \\ &= \int_0^1 \left[3xy^2 + 2y^3 + 3y^2\right]_{y=0}^{y=2x} dx = \int_0^1 (28x^3 + 12x^2)\, dx \\ &= \left[7x^4 + 4x^3\right]_0^1 = 11. \end{aligned}$$

A similar calculation gives the moment about the y-axis:

$$M_y \int_0^1 \int_0^{2x} x\delta(x, y)\, dy\, dx = 10.$$

The coordinates of the center of mass are therefore

$$\bar{x} = \frac{M_y}{M} = \frac{10}{14} = \frac{5}{7}, \qquad \bar{y} = \frac{M_x}{M} = \frac{11}{14}.$$

Moments of Inertia

A body's first moments (Table 12.1) tell us about balance and about the torque the body exerts about different axes in a gravitational field. If the body is a rotating shaft, however, we are more likely to be interested in how much energy is stored in the shaft or about how much energy it will take to accelerate the shaft to a particular angular velocity. This is where the second moment or moment of inertia comes in.

Think of partitioning the shaft into small blocks of mass Δm_k and let r_k denote the distance from the kth block's center of mass to the axis of rotation (Figure 12.21). If the shaft rotates at an angular velocity of $\omega = d\theta/dt$ radians per second, the block's center of mass will trace its orbit at a linear speed of

$$v_k = \frac{d}{dt}(r_k\theta) = r_k \frac{d\theta}{dt} = r_k\omega.$$

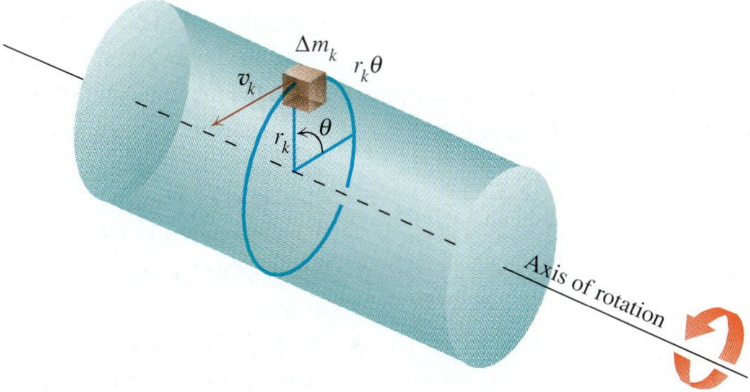

FIGURE 12.21 To find an integral for the amount of energy stored in a rotating shaft, we first imagine the shaft to be partitioned into small blocks. Each block has its own kinetic energy. We add the contributions of the individual blocks to find the kinetic energy of the shaft.

The block's kinetic energy will be approximately

$$\frac{1}{2}\,\Delta m_k v_k^{\,2} = \frac{1}{2}\,\Delta m_k (r_k\omega)^2 = \frac{1}{2}\,\omega^2 r_k^{\,2}\,\Delta m_k.$$

The kinetic energy of the shaft will be approximately

$$\sum \frac{1}{2}\,\omega^2 r_k^{\,2}\,\Delta m_k.$$

The integral approached by these sums as the shaft is partitioned into smaller and smaller blocks gives the shaft's kinetic energy:

$$\text{KE}_{\text{shaft}} = \int \frac{1}{2}\,\omega^2 r^2\,dm = \frac{1}{2}\,\omega^2 \int r^2\,dm. \tag{6}$$

The factor

$$I = \int r^2\,dm \tag{7}$$

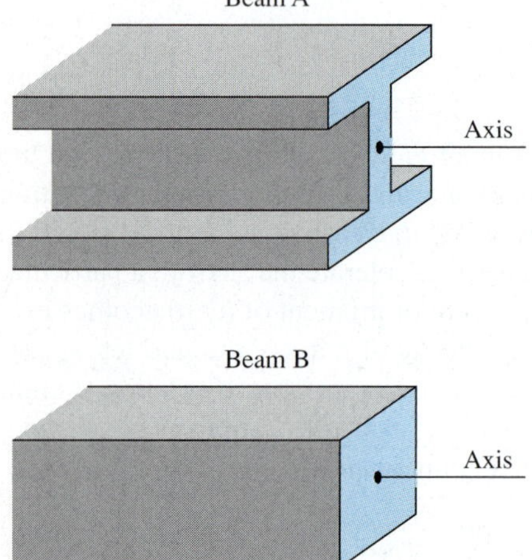

FIGURE 12.22 The greater the polar moment of inertia of the cross section of a beam about the beam's longitudinal axis, the stiffer the beam. Beams A and B have the same cross-section area, but A is stiffer.

is the moment of inertia of the shaft about its axis of rotation, and we see from Equation (6) that the shaft's kinetic energy is

$$\text{KE}_{\text{shaft}} = \frac{1}{2}\,I\omega^2. \tag{8}$$

To start a shaft of inertial moment I rotating at an angular velocity ω, we need to provide a kinetic energy of $\text{KE} = (1/2)I\omega^2$. To stop the shaft, we have to take this amount of energy back out. To start a locomotive with mass m moving at a linear velocity v, we need to provide a kinetic energy of $\text{KE} = (1/2)mv^2$. To stop the locomotive, we have to remove this amount of energy. The shaft's moment of inertia is analogous to the locomotive's mass. What makes the locomotive hard to start or stop is its mass. What makes the shaft hard to start or stop is its moment of inertia. The moment of inertia takes into account not only the mass but also its distribution.

The moment of inertia also plays a role in determining how much a horizontal metal beam will bend under a load. The stiffness of the beam is a constant times I, the moment of inertia of a typical cross section of the beam about the beam's longitudinal axis. The greater the value of I, the stiffer the beam and the less it will bend under a given load. That is why we use I-beams instead of beams whose cross sections are square. The flanges at the top and bottom of the beam hold most of the beam's mass away from the longitudinal axis to maximize the value of I (Figure 12.22).

If you want to see the moment of inertia at work, try the following experiment. Tape two coins to the ends of a pencil and twiddle the pencil about the center of mass. The moment of inertia accounts for the resistance you feel each time you change the direction of motion. Now move the coins an equal distance toward the center of mass and twiddle the pencil again. The system has the same mass and the same center of mass but now offers less resistance to the changes in motion. The moment of inertia has been reduced. The moment of inertia is what gives a baseball bat, golf club, or tennis racket its "feel." Tennis rackets that weigh the same, look the same, and have identical centers of mass will feel different and behave differently if their masses are not distributed the same way.

First moments are "balancing" moments. Second moments are "turning" moments.

Table 12.2 gives the formulas for moments of inertia (also called second moments), and for radii of gyration.

Table 12.2 Second moment formulas for thin plates in the *xy*-plane

Moments of inertia (second moments):

About the x-axis: $I_x = \iint y^2\delta(x, y)\, dA$

About the origin (polar moment): $I_0 = \iint (x^2 + y^2)\delta(x, y)\, dA = I_x + I_y$

About the y-axis: $I_y = \iint x^2\delta(x, y)\, dA$

About a line L: $I_L = \iint r^2(x, y)\delta(x, y)\, dA,$ where $r(x, y)$ = distance from (x, y) to L

Radii of gyration:

About the x-axis: $R_x = \sqrt{I_x/M}$

About the y-axis: $R_y = \sqrt{I_y/M}$

About the origin: $R_0 = \sqrt{I_0/M}$

The mathematical difference between the **first moments** M_x and M_y and the **moments of inertia,** or **second moments,** I_x and I_y is that the second moments use the *squares* of the "lever-arm" distances x and y.

The moments I_0 is also called the **polar moment** of inertia about the origin. It is calculated by integrating the density $\delta(x, y)$ (mass per unit area) times $r^2 = x^2 + y^2$, the square of the distance from a representative point (x, y) to the origin. Notice that $I_0 = I_x + I_y$; once we find two, we get the third automatically. (The moment I_0 is sometimes called I_z, for moment of inertia about the z-axis. The identity $I_z = I_x + I_y$ is then called the **Perpendicular Axis Theorem.**)

The **radius of gyration** R_x is defined by the equation

$$I_x = MR_x^2.$$

It tells how far from the x-axis the entire mass of the plate might be concentrated to give the same I_x. The radius of gyration gives a convenient way to express the moment of inertia in terms of a mass and a length. The radii R_y and R_0 are defined in a similar way, with

$$I_y = MR_y^2 \qquad \text{and} \qquad I_0 = MR_0^2.$$

We take square roots to get the formulas in Table 12.2.

Example 5 Finding Moments of Inertia and Radii of Gyration

For the thin plate in Example 4 (Figure 12.20), find the moments of inertia and radii of gyration about the coordinate axes and the origin.

Notice that we integrate y^2 times density in calculating I_x and x^2 times density to find I_y.

Solution Using the density function $\delta(x, y) = 6x + 6y + 6$ given in Example 4, the moment of inertia about the x-axis is

$$I_x = \int_0^1 \int_0^{2x} y^2 \delta(x, y)\, dy\, dx = \int_0^1 \int_0^{2x} (6xy^2 + 6y^3 + 6y^2)\, dy\, dx$$

$$= \int_0^1 \left[2xy^3 + \frac{3}{2}y^4 + 2y^3 \right]_{y=0}^{y=2x} dx = \int_0^1 (40x^4 + 16x^3)\, dx$$

$$= \left[8x^5 + 4x^4 \right]_0^1 = 12.$$

Similarly, the moment of inertia about the y-axis is

$$I_y = \int_0^1 \int_0^{2x} x^2 \delta(x, y)\, dy\, dx = \frac{39}{5}.$$

Since we know I_x and I_y, we do not need to evaluate an integral to find I_0; we can use the equation $I_0 = I_x + I_y$ instead:

$$I_0 = 12 + \frac{39}{5} = \frac{60 + 39}{5} = \frac{99}{5}.$$

The three radii of gyration are

$$R_x = \sqrt{I_x/M} = \sqrt{12/14} = \sqrt{6/7} \approx 0.93$$

$$R_y = \sqrt{I_y/M} = \sqrt{\left(\frac{39}{5}\right)/14} = \sqrt{39/70} \approx 0.75$$

$$R_0 = \sqrt{I_0/M} = \sqrt{\left(\frac{99}{5}\right)/14} = \sqrt{99/70} \approx 1.19.$$

Centroids of Geometric Figures

When the density of an object is constant, it cancels out of the numerator and denominator of the formulas for $\bar{x}$ and $\bar{y}$. As far as $\bar{x}$ and $\bar{y}$ are concerned, δ might as well be 1. Thus, when δ is constant, the location of the center of mass becomes a feature of the object's shape and not of the material of which it is made. In such cases, engineers may call the center of mass the **centroid** of the shape. To find a centroid, we set δ equal to 1 and proceed to find $\bar{x}$ and $\bar{y}$ as before, by dividing first moments by masses.

Example 6 Finding the Centroid of a Region

Find the centroid of the region in the first quadrant that is bounded above by the line $y = x$ and below by the parabola $y = x^2$.

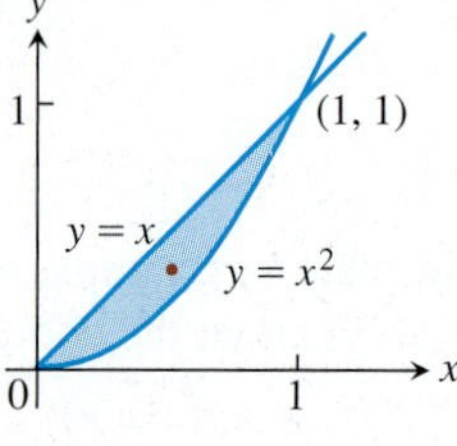

FIGURE 12.23 Example 6 finds the centroid of the region shown here.

Solution We sketch the region and include enough detail to determine the limits of integration (Figure 12.23). We then set δ equal to 1 and evaluate the appropriate formulas from Table 12.1:

$$M = \int_0^1 \int_{x^2}^{x} 1\, dy\, dx = \int_0^1 [y]_{y=x^2}^{y=x}\, dx = \int_0^1 (x - x^2)\, dx = \left[\frac{x^2}{2} - \frac{x^3}{3}\right]_0^1 = \frac{1}{6}$$

$$M_x = \int_0^1 \int_{x^2}^{x} y\, dy\, dx = \int_0^1 \left[\frac{y^2}{2}\right]_{y=x^2}^{y=x} dx$$

$$= \int_0^1 \left(\frac{x^2}{2} - \frac{x^4}{2}\right) dx = \left[\frac{x^3}{6} - \frac{x^5}{10}\right]_0^1 = \frac{1}{15}$$

$$M_y = \int_0^1 \int_{x^2}^{x} x\, dy\, dx = \int_0^1 [xy]_{y=x^2}^{y=x}\, dx = \int_0^1 (x^2 - x^3)\, dx = \left[\frac{x^3}{3} - \frac{x^4}{4}\right]_0^1 = \frac{1}{12}.$$

From these values of M, M_x, and M_y, we find

$$\bar{x} = \frac{M_y}{M} = \frac{1/12}{1/6} = \frac{1}{2} \quad \text{and} \quad \bar{y} = \frac{M_x}{M} = \frac{1/15}{1/6} = \frac{2}{5}.$$

The centroid is the point $(1/2, 2/5)$.

EXERCISES 12.2

Area by Double Integration

In Exercises 1–8, sketch the region bounded by the given lines and curves. Then express the region's area as an iterated double integral and evaluate the integral.

1. The coordinate axes and the line $x + y = 2$
2. The lines $x = 0$, $y = 2x$, and $y = 4$
3. The parabola $x = -y^2$ and the line $y = x + 2$
4. The parabola $x = y - y^2$ and the line $y = -x$
5. The curve $y = e^x$ and the lines $y = 0$, $x = 0$, and $x = \ln 2$
6. The curves $y = \ln x$ and $y = 2 \ln x$ and the line $x = e$, in the first quadrant
7. The parabolas $x = y^2$ and $x = 2y - y^2$
8. The parabolas $x = y^2 - 1$ and $x = 2y^2 - 2$

Identifying the Region of Integration

The integrals and sums of integrals in Exercises 9–14 give the areas of regions in the xy-plane. Sketch each region, label each bounding curve with its equation, and give the coordinates of the points where the curves intersect. Then find the area of the region.

9. $\displaystyle\int_0^6 \int_{y^2/3}^{2y} dx\, dy$ **10.** $\displaystyle\int_0^3 \int_{-x}^{x(2-x)} dy\, dx$

11. $\displaystyle\int_0^{\pi/4} \int_{\sin x}^{\cos x} dy\, dx$ **12.** $\displaystyle\int_{-1}^2 \int_{y^2}^{y+2} dx\, dy$

13. $\displaystyle\int_{-1}^0 \int_{-2x}^{1-x} dy\, dx + \int_0^2 \int_{-x/2}^{1-x} dy\, dx$

14. $\displaystyle\int_0^2 \int_{x^2-4}^{0} dy\, dx + \int_0^4 \int_0^{\sqrt{x}} dy\, dx$

Average Values

15. Find the average value of $f(x, y) = \sin(x + y)$ over

(a) the rectangle $0 \le x \le \pi$, $0 \le y \le \pi$

(b) the rectangle $0 \le x \le \pi$, $0 \le y \le \pi/2$

16. Which do you think will be larger, the average value of $f(x, y) = xy$ over the square $0 \le x \le 1$, $0 \le y \le 1$, or the average value of f over the quarter circle $x^2 + y^2 \le 1$ in the first quadrant? Calculate them to find out.

17. Find the average height of the paraboloid $z = x^2 + y^2$ over the square $0 \le x \le 2$, $0 \le y \le 2$.

18. Find the average value of $f(x, y) = 1/(xy)$ over the square $\ln 2 \le x \le 2 \ln 2$, $\ln 2 \le y \le 2 \ln 2$.

Constant Density

19. *Finding center of mass* Find a center of mass of a thin plate of density $\delta = 3$ bounded by the lines $x = 0$, $y = x$, and the parabola $y = 2 - x^2$ in the first quadrant.

20. *Finding moments of inertia and radii of gyration* Find the moments of inertia and radii of gyration about the coordinate axes of a thin rectangular plate of constant density δ bounded by the lines $x = 3$ and $y = 3$ in the first quadrant.

21. *Finding a centroid* Find the centroid of the region in the first quadrant bounded by the x-axis, the parabola $y^2 = 2x$, and the line $x + y = 4$.

22. *Finding a centroid* Find the centroid of the triangular region cut from the first quadrant by the line $x + y = 3$.

23. *Finding a centroid* Find the centroid of the semicircular region bounded by the x-axis and the curve $y = \sqrt{1 - x^2}$.

24. *Finding a centroid* The area of the region in the first quadrant bounded by the parabola $y = 6x - x^2$ and the line $y = x$ is $125/6$ square units. Find the centroid.

25. *Finding a centroid* Find the centroid of the region cut from the first quadrant by the circle $x^2 + y^2 = a^2$.

26. *Finding a centroid* Find the centroid of the region between the x-axis and the arch $y = \sin x$, $0 \le x \le \pi$.

27. *Finding moments of inertia* Find the moment of inertia about the x-axis of a thin plate of density $\delta = 1$ bounded by the circle $x^2 + y^2 = 4$. Then use your result to find I_y and I_0 for the plate.

28. *Finding a moment of inertia* Find the moment of inertia with respect to the y-axis of a thin sheet of constant density $\delta = 1$ bounded by the curve $y = (\sin^2 x)/x^2$ and the interval $\pi \le x \le 2\pi$ of the x-axis.

29. *The centroid of an infinite region* Find the centroid of the infinite region in the second quadrant enclosed by the coordinate axes and the curve $y = e^x$. (Use improper integrals in the mass-moment formulas.)

30. *The first moment of an infinite plate* Find the first moment about the y-axis of a thin plate density $\delta(x, y) = 1$ covering the infinite region under the curve $y = e^{-x^2/2}$ in the first quadrant.

Variable Density

31. *Finding a moment of inertia and radius of gyration* Find the moment of inertia and radius of gyration about the x-axis of a thin plate bounded by the parabola $x = y - y^2$ and the line $x + y = 0$ if $\delta(x, y) = x + y$.

32. *Finding mass* Find the mass of a thin plate occupying the smaller region cut from the ellipse $x^2 + 4y^2 = 12$ by the parabola $x = 4y^2$ if $\delta(x, y) = 5x$.

33. *Finding a center of mass* Find the center of mass of a thin triangular plate bounded by the y-axis and the lines $y = x$ and $y = 2 - x$ if $\delta(x, y) = 6x + 3y + 3$.

34. *Finding a center of mass and moment of inertia* Find the center of mass and moment of inertia about the x-axis of a thin plate bounded by the curves $x = y^2$ and $x = 2y - y^2$ if the density at the point (x, y) is $\delta(x, y) = y + 1$.

35. *Center of mass, moment of inertia, and radius of gyration* Find the center of mass and the moment of inertia and radius of gyration about the y-axis of a thin rectangular plate cut from the first quadrant by the lines $x = 6$ and $y = 1$ if $\delta(x, y) = x + y + 1$.

36. *Center of mass, moment of inertia, and radius of gyration* Find the center of mass and the moment of inertia and radius of gyration about the y-axis of a thin plate bounded by the line $y = 1$ and the parabola $y = x^2$ if the density is $\delta(x, y) = y + 1$.

37. *Center of mass, moment of inertia, and radius of gyration* Find the center of mass and the moment of inertia and radius of gyration about the y-axis of a thin plate bounded by the x-axis, the lines $x = \pm 1$, and the parabola $y = x^2$ if $\delta(x, y) = 7y + 1$.

38. *Center of mass, moment of inertia, and radius of gyration* Find the center of mass and the moment of inertia and radius of gyration about the x-axis of a thin rectangular plate bounded by the lines $x = 0$, $x = 20$, $y = -1$, and $y = 1$ if $\delta(x, y) = 1 + (x/20)$.

39. *Center of mass, moments of inertia, and radii of gyration* Find the center of mass, the moment of inertia and radii of gyration about the coordinate axes, and the polar moment of inertia and radius of gyration of a thin triangular plate bounded by the lines $y = x$, $y = -x$, and $y = 1$ if $\delta(x, y) = y + 1$.

40. *Center of mass, moments of inertia, and radii of gyration* Repeat Exercise 39 for $\delta(x, y) = 3x^2 + 1$.

Theory and Examples

41. *Bacterium population* If $f(x, y) = (10{,}000e^y)/(1 + |x|/2)$ represents the "population density" of a certain bacterium on the xy-plane, where x and y are measured in centimeters, find the total population of bacteria within the rectangle $-5 \le x \le 5$ and $-2 \le y \le 0$.

42. *Regional population* If $f(x, y) = 100\,(y + 1)$ represents the population density of a planar region on Earth, where x and y are measured in miles, find the number of people in the region bounded by the curves $x = y^2$ and $x = 2y - y^2$.

CD-ROM WEBsite

43. *Appliance design* When we design an appliance, one of the concerns is how hard the appliance will be to tip over. When tipped, it will right itself as long as its center of mass lies on the correct side of the *fulcrum*, the point on which the appliance is riding as it tips. Suppose that the profile of an appliance of approximately constant density is parabolic, like an old-fashioned radio. It fills the region $0 \le y \le a(1 - x^2)$, $-1 \le x \le 1$, in the xy-plane (see accompanying figure). What values of a will guarantee that the appliance will have to be tipped more than 45° to fall over?

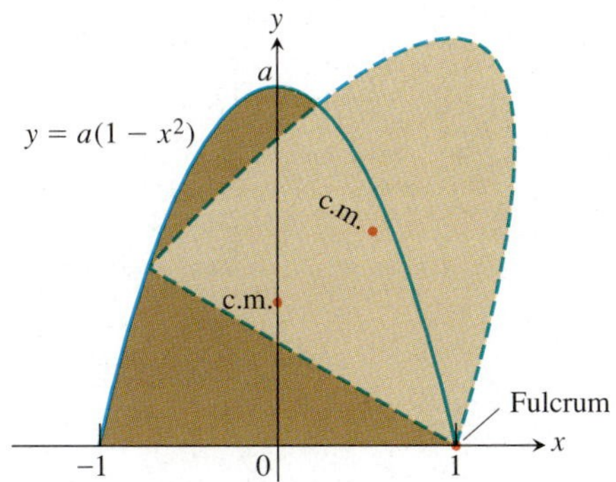

44. *Minimizing a moment of inertia* A rectangular plate of constant density $\delta(x, y) = 1$ occupies the region bounded by the lines $x = 4$ and $y = 2$ in the first quadrant. The moment of inertia I_a of the rectangle about the line $y = a$ is given by the integral

$$I_a = \int_0^4 \int_0^2 (y - a)^2 \, dy \, dx.$$

Find the value of a that minimizes I_a.

45. *Centroid of unbounded region* Find the centroid of the infinite region in the xy-plane bounded by the curves $y = 1/\sqrt{1 - x^2}$, $y = -1/\sqrt{1 - x^2}$, and the lines $x = 0$, $x = 1$.

46. *Radius of gyration of slender rod* Find the radius of gyration of a slender rod of constant linear density δ gm/cm and length L cm with respect to an axis

(a) Through the rod's center of mass perpendicular to the rod's axis

(b) Perpendicular to the rod's axis at one end of the rod.

47. (*Continuation of Exercise 34*) A thin plate of now constant density δ occupies the region R in the xy-plane bounded by the curves $x = y^2$ and $x = 2y - y^2$.

(a) *Constant density* Find δ such that the plate has the same mass as the plate in Exercise 34.

(b) *Average value* Compare the value of δ found in part (a) with the average value of $\delta(x, y) = y + 1$ over R.

48. *Average temperature in Texas* According to the *Texas Almanac*, Texas has 254 counties and a National Weather Service station in each county. Assume that at time t_0, each of the 254 weather stations recorded the local temperature. Find a formula that would give a reasonable approximation to the average temperature in Texas at time t_0. Your answer should involve information that you would expect to be readily available in the *Texas Almanac*.

The Parallel Axis Theorem

Let $L_{\text{c.m.}}$ be a line in the xy-plane that runs through the center of mass of a thin plate of mass m covering a region in the plane. Let L be a line in the plane parallel to and h units away from $L_{\text{c.m.}}$ The **Parallel Axis Theorem** says that under these conditions the moments of inertia I_L and $I_{\text{c.m.}}$ of the plate about L and $L_{\text{c.m.}}$ satisfy the equation

$$I_L = I_{\text{c.m.}} + mh^2.$$

This equation gives a quick way to calculate one moment when the other moment and the mass are known.

49. *Proof of the Parallel Axis Theorem*

(a) Show that the first moment of a thin flat plate about any line in the plane of the plate through the plate's center of mass is zero. (*Hint*: Place the center of mass at the origin with the line along the y-axis. What does the formula $\bar{x} = M_y/M$ then tell you?)

(b) Use the result in part (a) to derive the Parallel Axis Theorem. Assume that the plane is coordinatized in a way that makes $L_{\text{c.m.}}$ the y-axis and L the line $x = h$. Then expand the integrand of the integral for I_L to rewrite the integral as the sum of integrals whose values you recognize.

50. *Finding moments of inertia*

(a) Use the Parallel Axis Theorem and the results of Example 4 to find the moments of inertia of the plate in Example 4 about the vertical and horizontal lines through the plate's center of mass.

(b) Use the results in part (a) to find the plate's moments of inertia about the lines $x = 1$ and $y = 2$.

Pappus's Formula

Pappus knew that the centroid of the union of two nonoverlapping plane regions lies on the line segment joining their individual centroids. More specifically, suppose that m_1 and m_2 are the masses of thin plates P_1 and P_2 that cover nonoverlapping regions in the xy-plane. Let $\mathbf{c}_1$ and $\mathbf{c}_2$ be the vectors from the origin to the respective centers of mass of P_1 and P_2. Then the center of mass of the union $P_1 \cup P_2$ of the two plates is determined by the vector

$$\mathbf{c} = \frac{m_1\mathbf{c}_1 + m_2\mathbf{c}_2}{m_1 + m_2}. \tag{9}$$

Equation (9) is known as **Pappus's formula.** For more than two nonoverlapping plates, as long as their number is finite, the formula generalizes to

$$\mathbf{c} = \frac{m_1\mathbf{c}_1 + m_2\mathbf{c}_2 + \cdots + m_n\mathbf{c}_n}{m_1 + m_2 + \cdots + m_n}. \tag{10}$$

This formula is especially useful for finding the centroid of a plate of irregular shape that is made up of pieces of constant density whose centroids we know from geometry. We find the centroid of each piece and apply Equation (10) to find the centroid of the plate.

51. Derive Pappus's formula (Equation (9)). (*Hint*: Sketch the plates as regions in the first quadrant and label their centers of mass as $(\bar{x}_1, \bar{y}_1)$ and $(\bar{x}_2, \bar{y}_2)$. What are the moments of $P_1 \cup P_2$ about the coordinate axes?)

52. Use Equation (9) and mathematical induction to show that Equation (10) holds for any positive integer $n > 2$.

53. Let A, B, and C be the shapes indicated in Figure 12.24a. Use Pappus's formula to find the centroid of

(a) $A \cup B$ **(b)** $A \cup C$

(c) $B \cup C$ **(d)** $A \cup B \cup C$.

54. *Locating center of mass* Locate the center of mass of the carpenter's square in Figure 12.24b.

55. An isosceles triangle T has base $2a$ and altitude h. The base lies along the diameter of a semicircular disk D of radius a so that the two together make a shape resembling an ice cream cone. What relation must hold between a and h to place the centroid of $T \cup D$ on the common boundary of T and D? Inside T?

56. An isosceles triangle T of altitude h has as its base one side of a square Q whose edges have length s. (The square and triangle do not overlap.) What relation must hold between h and s to place the centroid of $T \cup Q$ on the base of the triangle? Compare your answer with the answer to Exercise 55.

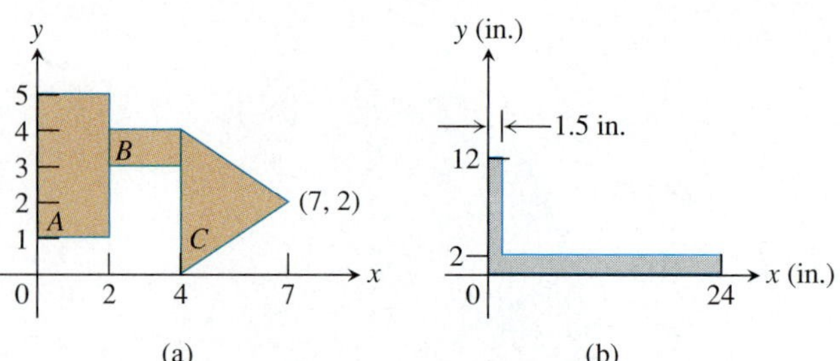

FIGURE 12.24 The figures for Exercises 53 and 54.

12.3 Double Integrals in Polar Form

Integrals in Polar Coordinates • Finding Limits of Integration • Changing Cartesian Integrals into Polar Integrals

Integrals are sometimes easier to evaluate if we change to polar coordinates. This section shows how to accomplish the change and how to evaluate integrals over regions whose boundaries are given by polar equations.

Integrals in Polar Coordinates

When we defined the double integral of a function over a region R in the xy-plane, we began by cutting R into rectangles whose sides were parallel to the coordinate axes. These were the natural shapes to use because their sides have either constant x-values or constant y-values. In polar coordinates, the natural shape is a "polar rectangle" whose sides have constant r- and θ-values.

Suppose that a function $f(r, \theta)$ is defined over a region R that is bounded by the rays $\theta = \alpha$ and $\theta = \beta$ and by the continuous curves $r = g_1(\theta)$ and $r = g_2(\theta)$. Suppose also that $0 \le g_1(\theta) \le g_2(\theta) \le a$ for every value of θ between α and β. Then R lies in a fan-shaped region Q defined by the inequalities $0 \le r \le a$ and $\alpha \le \theta \le \beta$. See Figure 12.25.

We cover Q by a grid of circular arcs and rays. The arcs are cut from circles centered at the origin, with radii Δr, $2\Delta r, \ldots, m\Delta r$, where $\Delta r = a/m$. The rays are given by

$$\theta = \alpha, \qquad \theta = \alpha + \Delta\theta, \qquad \theta = \alpha + 2\Delta\theta, \qquad \ldots, \qquad \theta = \alpha + m'\Delta\theta = \beta,$$

where $\Delta\theta = (\beta - \alpha)/m'$. The arcs and rays partition Q into small patches called "polar rectangles."

We number the polar rectangles that lie inside R (the order does not matter), calling their areas $\Delta A_1, \Delta A_2, \ldots, \Delta A_n$.

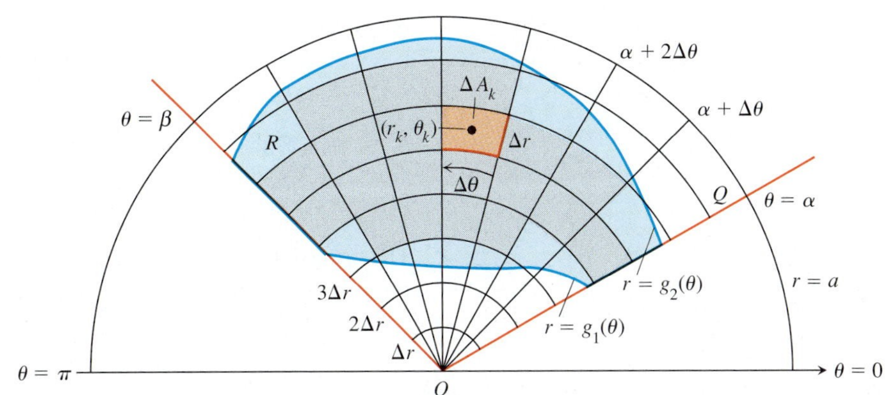

FIGURE 12.25 The region R: $g_1(\theta) \leq r \leq g_2(\theta)$, $\alpha \leq \theta \leq \beta$, is contained in the fan-shaped region Q: $0 \leq r \leq a$, $\alpha \leq \theta \leq \beta$. The partition of Q by circular arcs and rays induces a partition of R.

We let (r_k, θ_k) be the center of the polar rectangle whose area is ΔA_k. By "center," we mean the point that lies halfway between the circular arcs on the ray that bisects the arcs. We then form the sum

$$S_n = \sum_{k=1}^{n} f(r_k, \theta_k)\, \Delta A_k. \tag{1}$$

If f is continuous throughout R, this sum will approach a limit as we refine the grid to make Δr and $\Delta\theta$ go to zero. The limit is called the double integral of f over R. In symbols,

$$\lim_{n\to\infty} S_n = \iint_R f(r, \theta)\, dA.$$

To evaluate this limit, we first have to write the sum S_n in a way that expresses ΔA_k in terms of Δr and $\Delta\theta$. The radius of the inner arc bounding ΔA_k is $r_k - (\Delta r/2)$ (Figure 12.26). The radius of the outer arc is $r_k + (\Delta r/2)$. The areas of the circular sectors subtended by these arcs at the origin are

$$\text{Inner radius:} \quad \frac{1}{2}\left(r_k - \frac{\Delta r}{2}\right)^2 \Delta\theta$$

$$\text{Outer radius:} \quad \frac{1}{2}\left(r_k + \frac{\Delta r}{2}\right)^2 \Delta\theta.$$

Therefore,

$$\begin{aligned} \Delta A_k &= \text{area of large section} - \text{area of small section} \\ &= \frac{\Delta\theta}{2}\left[\left(r_k + \frac{\Delta r}{2}\right)^2 - \left(r_k - \frac{\Delta r}{2}\right)^2\right] = \frac{\Delta\theta}{2}(2r_k\,\Delta r) = r_k\,\Delta r\,\Delta\theta. \end{aligned}$$

Combining this result with Equation (1) gives

$$S_n = \sum_{k=1}^{n} f(r_k, \theta_k) r_k\, \Delta r\, \Delta\theta.$$

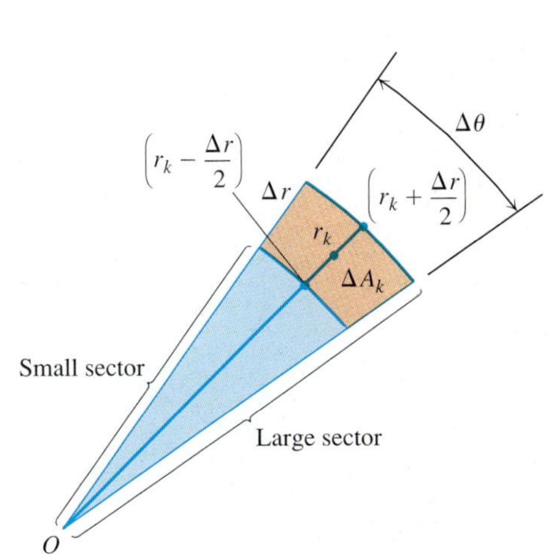

FIGURE 12.26 The observation that

$$\Delta A_k = \begin{pmatrix}\text{area of}\\ \text{large sector}\end{pmatrix} - \begin{pmatrix}\text{area of}\\ \text{small sector}\end{pmatrix}$$

leads to the formula $\Delta A_k = r_k\,\Delta r\,\Delta\theta$. The text explains why.

A version of Fubini's theorem now says that the limit approached by these sums can be evaluated by repeated single integrations with respect to r and θ as

$$\iint_R f(r, \theta)\, dA = \int_{\theta=\alpha}^{\theta=\beta} \int_{r=g_1(\theta)}^{r=g_2(\theta)} f(r, \theta) r\, dr\, d\theta. \tag{2}$$

Finding Limits of Integration

The procedure for finding limits of integration in rectangular coordinates also works for polar coordinates.

How to Integrate in Polar Coordinates

To evaluate $\iint_R f(r, \theta)\, dA$ over a region R in polar coordinates, integrating first with respect to r and then with respect to θ, take the following steps.

Step 1: *A sketch.* Sketch the region and label the bounding curves.

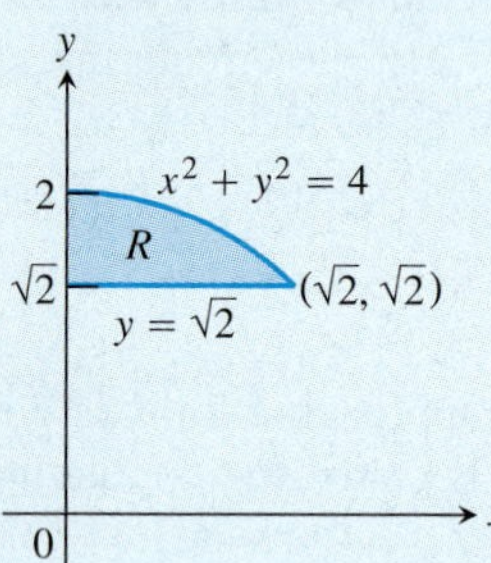

Step 2: *The r-limits of integration.* Imagine a ray L from the origin cutting through R in the direction of increasing r. Mark the r-values where L enters and leaves R. These are the r-limits of integration. They usually depend on the angle θ that L makes with the positive x-axis.

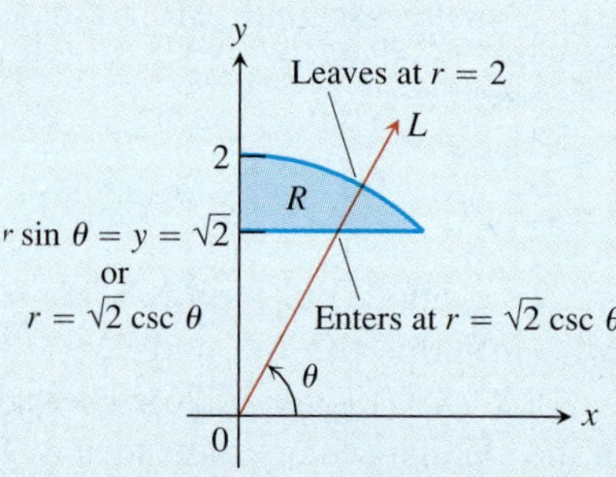

Step 3: *The θ-limits of integration.* Find the smallest and largest θ-values that bound R. These are the θ-limits of integration.

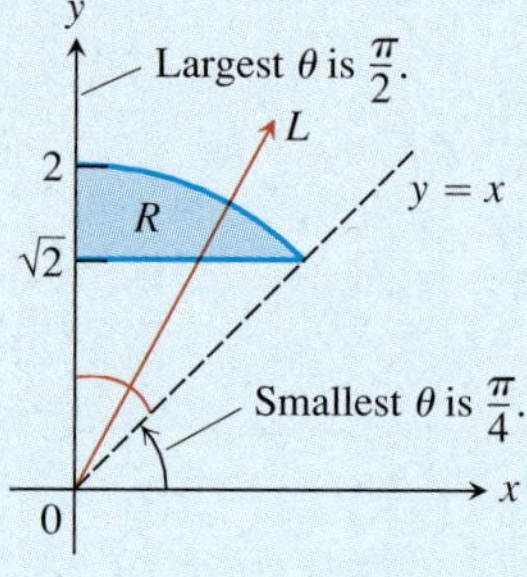

The integral is

$$\iint_R f(r, \theta)\, dA = \int_{\theta=\pi/4}^{\theta=\pi/2} \int_{r=\sqrt{2}\csc\theta}^{r=2} f(r, \theta) r\, dr\, d\theta.$$

Example 1 Finding Limits of Integration

Find the limits of integration for integrating $f(r, \theta)$ over the region R that lies inside the cardioid $r = 1 + \cos\theta$ and outside the circle $r = 1$.

Solution

Step 1: *A sketch.* We sketch the region and label the bounding curves (Figure 12.27).

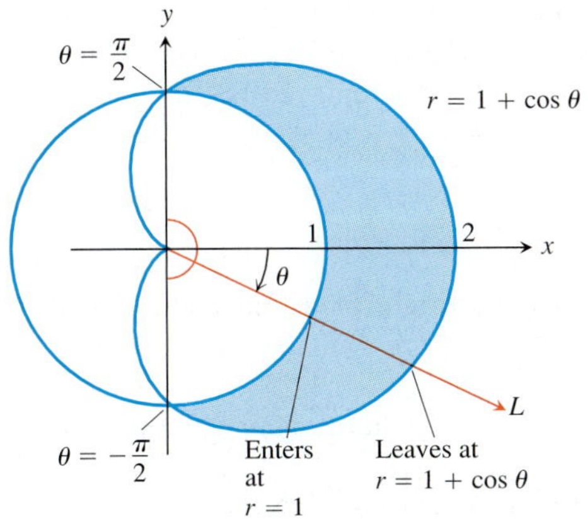

FIGURE 12.27 The sketch for Example 1.

Step 2: *The r-limits of integration*. A typical ray from the origin enters R where $r = 1$ and leaves where $r = 1 + \cos\theta$.

Step 3: *The θ-limits of integration*. The rays from the origin that intersect R run from $\theta = -\pi/2$ to $\theta = \pi/2$. The integral is

$$\int_{-\pi/2}^{\pi/2}\int_{1}^{1+\cos\theta} f(r, \theta)\, r\, dr\, d\theta.$$

If $f(r, \theta)$ is the constant function whose value is 1, then the integral f over R is the area of R.

Area in Polar Coordinates

The area of a closed and bounded region R in the polar coordinate plane is

$$A = \iint_R r\, dr\, d\theta. \tag{3}$$

As you might expect, this formula for area is consistent with all earlier formulas, although we do not prove the fact.

Example 2 Finding Area in Polar Coordinates

Find the area enclosed by the lemniscate $r^2 = 4\cos 2\theta$.

Solution We graph the lemniscate to determine the limits of integration (Figure 12.28) and see that the total area is 4 times the first-quadrant portion.

$$A = 4\int_0^{\pi/4}\int_0^{\sqrt{4\cos 2\theta}} r\, dr\, d\theta = 4\int_0^{\pi/4}\left[\frac{r^2}{2}\right]_{r=0}^{r=\sqrt{4\cos 2\theta}} d\theta$$

$$= 4\int_0^{\pi/4} 2\cos 2\theta\, d\theta = 4\sin 2\theta\Big]_0^{\pi/4} = 4.$$

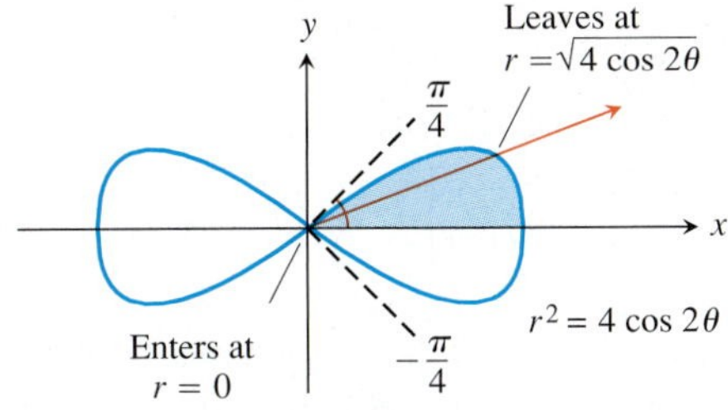

FIGURE 12.28 To integrate over the shaded region, we run r from 0 to $\sqrt{4\cos 2\theta}$ and θ from 0 to $\pi/4$. (Example 2)

Changing Cartesian Integrals into Polar Integrals

The procedure for changing a Cartesian integral $\iint_R f(x, y)\, dx\, dy$ into a polar integral has two steps.

Step 1: Substitute $x = r\cos\theta$ and $y = r\sin\theta$, and replace $dx\, dy$ by $r\, dr\, d\theta$ in the Cartesian integral.

Step 2: Supply polar limits of integration for the boundary of R.

The Cartesian integral then becomes

$$\iint_R f(x, y)\, dx\, dy = \iint_G f(r\cos\theta, r\sin\theta)\, r\, dr\, d\theta, \tag{4}$$

where G denotes the region of integration in polar coordinates. This is like the substitution method in Chapter 4 except that there are now two variables to substitute for instead of one. Notice that $dx\,dy$ is not replaced by $dr\,d\theta$ but by $r\,dr\,d\theta$.

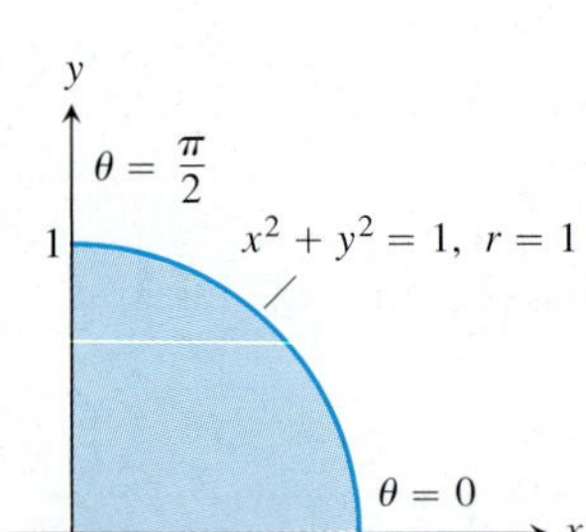

FIGURE 12.29 In polar coordinates, this region is described by simple inequalities:

$0 \le r \le 1$ and $0 \le \theta \le \pi/2$.

(Example 3)

Example 3 Changing Cartesian Integrals to Polar

Find the polar moment of inertia about the origin of a thin plate of density $\delta(x, y) = 1$ bounded by the quarter circle $x^2 + y^2 = 1$ in the first quadrant.

Solution We sketch the plate to determine the limits of integration (Figure 12.29).

In Cartesian coordinates, the polar moment is the value of the integral

$$\int_0^1 \int_0^{\sqrt{1-x^2}} (x^2 + y^2)\,dy\,dx.$$

Integration with respect to y gives

$$\int_0^1 \left(x^2\sqrt{1-x^2} + \frac{(1-x^2)^{3/2}}{3}\right) dx,$$

an integral difficult to evaluate without tables.

Things go better if we change the original integral to polar coordinates. Substituting $x = r\cos\theta$, $y = r\sin\theta$ and replacing $dx\,dy$ by $r\,dr\,\theta$, we get

$$\int_0^1 \int_0^{\sqrt{1-x^2}} (x^2 + y^2)\,dy\,dx = \int_0^{\pi/2} \int_0^1 (r^2)\,r\,dr\,d\theta$$

$$= \int_0^{\pi/2} \left[\frac{r^4}{4}\right]_{r=0}^{r=1} d\theta = \int_0^{\pi/2} \frac{1}{4}\,d\theta = \frac{\pi}{8}.$$

Why is the polar coordinate transformation so effective here? One reason is that $x^2 + y^2$ simplifies to r^2. Another is that the limits of integration become constants.

Example 4 Evaluating Integrals Using Polar Coordinates

Evaluate

$$\iint_R e^{x^2+y^2}\,dy\,dx,$$

where R is the semicircular region bounded by the x-axis and the curve $y = \sqrt{1-x^2}$ (Figure 12.30).

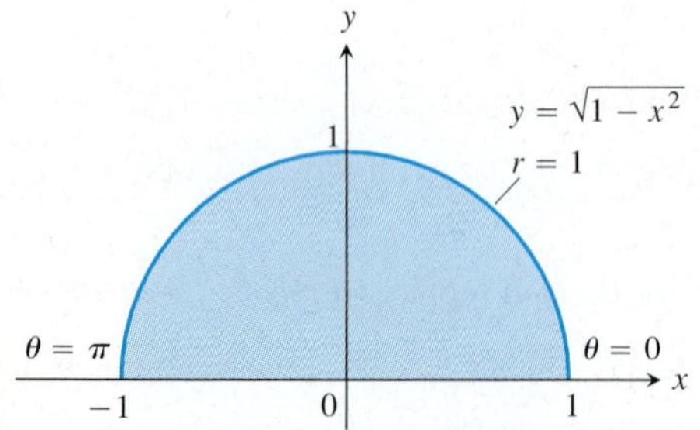

FIGURE 12.30 The semicircular region in Example 4 is the region

$0 \le r \le 1$, $0 \le \theta \le \pi$.

Solution In Cartesian coordinates, the integral in question is a nonelementary integral and there is no direct way to integrate $e^{x^2+y^2}$ with respect to either x or y. Yet this integral and others like it are important in mathematics—in statistics, for example—and we need to find a way to evaluate it. Polar coordinates save the day. Substituting $x = r\cos\theta$, $y = r\sin\theta$ and replacing $dy\,dx$ by $r\,dr\,d\theta$ enables us to evaluate the integral as

$$\iint_R e^{x^2+y^2}\,dy\,dx = \int_0^{\pi}\int_0^1 e^{r^2} r\,dr\,d\theta = \int_0^{\pi}\left[\frac{1}{2}e^{r^2}\right]_0^1 d\theta$$

$$= \int_0^{\pi}\frac{1}{2}(e-1)\,d\theta = \frac{\pi}{2}(e-1).$$

The r in the $r\,dr\,d\theta$ was just what we needed to integrate e^{r^2}. Without it, we would have been stuck, as we were at the beginning.

EXERCISES 12.3

Evaluating Polar Integrals

In Exercises 1–16, change the Cartesian integral into an equivalent polar integral. Then evaluate the polar integral.

1. $\displaystyle\int_{-1}^{1}\int_{0}^{\sqrt{1-x^2}} dy\,dx$

2. $\displaystyle\int_{-1}^{1}\int_{-\sqrt{1-x^2}}^{\sqrt{1-x^2}} dy\,dx$

3. $\displaystyle\int_{0}^{1}\int_{0}^{\sqrt{1-y^2}} (x^2+y^2)\,dx\,dy$

4. $\displaystyle\int_{-1}^{1}\int_{-\sqrt{1-y^2}}^{\sqrt{1-y^2}} (x^2+y^2)\,dy\,dx$

5. $\displaystyle\int_{-a}^{a}\int_{-\sqrt{a^2-x^2}}^{\sqrt{a^2-x^2}} dy\,dx$

6. $\displaystyle\int_{0}^{2}\int_{0}^{\sqrt{4-y^2}} (x^2+y^2)\,dx\,dy$

7. $\displaystyle\int_{0}^{6}\int_{0}^{y} x\,dx\,dy$

8. $\displaystyle\int_{0}^{2}\int_{0}^{x} y\,dy\,dx$

9. $\displaystyle\int_{-1}^{0}\int_{-\sqrt{1-x^2}}^{0} \frac{2}{1+\sqrt{x^2+y^2}}\,dy\,dx$

10. $\displaystyle\int_{-1}^{1}\int_{-\sqrt{1-y^2}}^{0} \frac{4\sqrt{x^2+y^2}}{1+x^2+y^2}\,dx\,dy$

11. $\displaystyle\int_{0}^{\ln 2}\int_{0}^{\sqrt{(\ln 2)^2-y^2}} e^{\sqrt{x^2+y^2}}\,dx\,dy$

12. $\displaystyle\int_{0}^{1}\int_{0}^{\sqrt{1-x^2}} e^{-(x^2+y^2)}\,dy\,dx$

13. $\displaystyle\int_{0}^{2}\int_{0}^{\sqrt{1-(x-1)^2}} \frac{x+y}{x^2+y^2}\,dy\,dx$

14. $\displaystyle\int_{0}^{2}\int_{-\sqrt{1-(y-1)^2}}^{0} xy^2\,dx\,dy$

15. $\displaystyle\int_{-1}^{1}\int_{-\sqrt{1-y^2}}^{\sqrt{1-y^2}} \ln(x^2+y^2+1)\,dx\,dy$

16. $\displaystyle\int_{-1}^{1}\int_{-\sqrt{1-x^2}}^{\sqrt{1-x^2}} \frac{2}{(1+x^2+y^2)^2}\,dy\,dx$

Finding Area in Polar Coordinates

17. Find the area of the region cut from the first quadrant by the curve $r = 2(2-\sin 2\theta)^{1/2}$.

18. *Cardioid overlapping a circle* Find the area of the region that lies inside the cardioid $r = 1+\cos\theta$ and outside the circle $r = 1$.

19. *One leaf of a rose* Find the area enclosed by one leaf of the rose $r = 12\cos 3\theta$.

20. *Snail shell* Find the area of the region enclosed by the positive x-axis and spiral $r = 4\theta/3$, $0 \le \theta \le 2\pi$. The region looks like a snail shell.

21. *Cardioid in the first quadrant* Find the area of the region cut from the first quadrant by the cardioid $r = 1+\sin\theta$.

22. *Overlapping cardioids* Find the area of the region common to the interiors of the cardioids $r = 1+\cos\theta$ and $r = 1-\cos\theta$.

Masses and Moments

23. *First moment of a plate* Find the first moment about the x-axis of a thin plate of constant density $\delta(x, y) = 3$, bounded below by the x-axis and above by the cardioid $r = 1-\cos\theta$.

24. *Inertial and polar moments of a disk* Find the moment of inertia about the x-axis and the polar moment of inertia about the origin of a thin disk bounded by the circle $x^2+y^2 = a^2$ if the disk's density at the point (x, y) is $\delta(x, y) = k(x^2+y^2)$, k a constant.

25. *Mass of a plate* Find the mass of a thin plate covering the region outside the circle $r = 3$ and inside the circle $r = 6\sin\theta$ if the plate's density function is $\delta(x, y) = 1/r$.

26. *Polar moment of a cardioid overlapping circle* Find the polar moment of inertia about the origin of a thin plate covering the region that lies inside the cardioid $r = 1 - \cos\theta$ and outside the circle $r = 1$ if the plate's density function is $\delta(x, y) = 1/r^2$.

27. *Centroid of a cardioid region* Find the centroid of the region enclosed by the cardioid $r = 1 + \cos\theta$.

28. *Polar moment of a cardioid region* Find the polar moment of inertia about the origin of a thin plate enclosed by the cardioid $r = 1 + \cos\theta$ if the plate's density function is $\delta(x, y) = 1$.

Average Values

29. *Average height of a hemisphere* Find the average height of the hemisphere $z = \sqrt{a^2 - x^2 - y^2}$ above the disk $x^2 + y^2 \le a^2$ in the xy-plane.

30. *Average height of a cone* Find the average height of the (single) cone $z = \sqrt{x^2 + y^2}$ above the disk $x^2 + y^2 \le a^2$ in the xy-plane.

31. *Average distance from interior of disk to center* Find the average distance from a point $P(x, y)$ in the disk $x^2 + y^2 \le a^2$ to the origin.

32. *Average distance squared from a point in a disk to a point in its boundary.* Find the average value of the *square* of the distance from the point $P(x, y)$ in the disk $x^2 + y^2 \le 1$ to the boundary point $A(1, 0)$.

Theory and Examples

33. *Converting to a polar integral* Integrate $f(x, y) = [\ln(x^2 + y^2)]/\sqrt{x^2 + y^2}$ over the region $1 \le x^2 + y^2 \le e$.

34. *Converting to a polar integral* Integrate $f(x, y) = [\ln(x^2 + y^2)]/(x^2 + y^2)$ over the region $1 \le x^2 + y^2 \le e^2$.

35. *Volume of noncircular right cylinder* The region that lies inside the cardioid $r = 1 + \cos\theta$ and outside the circle $r = 1$ is the base of a solid right cylinder. The top of the cylinder lies in the plane $z = x$. Find the cylinder's volume.

36. *Volume of noncircular right cylinder* The region enclosed by the lemniscate $r^2 = 2\cos 2\theta$ is the base of a solid right cylinder whose top is bounded by the sphere $z = \sqrt{2 - r^2}$. Find the cylinder's volume.

37. *Converting to polar integrals*

(a) The usual way to evaluate the improper integral $I = \int_0^\infty e^{-x^2}dx$ is first to calculate its square:

$$I^2 = \left(\int_0^\infty e^{-x^2}\,dx\right)\left(\int_0^\infty e^{-y^2}dy\right) = \int_0^\infty\int_0^\infty e^{-(x^2+y^2)}\,dx\,dy.$$

Evaluate the last integral using polar coordinates and solve the resulting equation for I.

(b) Evaluate

$$\lim_{x\to\infty} \operatorname{erf}(x) = \lim_{x\to\infty}\int_0^x \frac{2e^{-t^2}}{\sqrt{\pi}}\,dt.$$

38. *Converting to a polar integral* Evaluate the integral

$$\int_0^\infty\int_0^\infty \frac{1}{(1 + x^2 + y^2)^2}\,dx\,dy.$$

39. *Writing to Learn* Integrate the function $f(x, y) = 1/(1 - x^2 - y^2)$ over the disk $x^2 + y^2 \le 3/4$. Does the integral of $f(x, y)$ over the disk $x^2 + y^2 \le 1$ exist? Give reasons for your answer.

40. *Area formula in polar coordinates* Use the double integral in polar coordinates to derive the formula

$$A = \int_\alpha^\beta \frac{1}{2}r^2\,d\theta$$

for the area of the fan-shaped region between the origin and polar curve $r = f(\theta)$, $\alpha \le \theta \le \beta$.

41. *Average distance to a given point inside a disk* Let P_0 be a point inside a circle of radius a and let h denote the distance from P_0 to the center of the circle. Let d denote the distance from an arbitrary point P to P_0. Find the average value of d^2 over the region enclosed by the circle. (*Hint*: Simplify your work by placing the center of the circle at the origin and P_0 on the x-axis.)

42. *Area* Suppose that the area of a region in the polar coordinate plane is

$$A = \int_{\pi/4}^{3\pi/4}\int_{\csc\theta}^{2\sin\theta} r\,dr\,d\theta.$$

Sketch the region and find its area.

COMPUTER EXPLORATIONS

Coordinate Conversions

In Exercises 43–46, use a CAS to change the Cartesian integrals into an equivalent polar integral and evaluate the polar integral. Perform the following steps in each exercise.

(a) Plot the Cartesian region of integration in the xy-plane.

(b) Change each boundary curve of the Cartesian region in part (a) to its polar representation by solving its Cartesian equation for r and θ.

(c) Using the results in part (b), plot the polar region of integration in the $r\theta$-plane.

(d) Change the integrand from Cartesian to polar coordinates. Determine the limits of integration from your plot in part (c) and evaluate the polar integral using the CAS integration utility.

43. $\displaystyle\int_0^1\int_x^1 \frac{y}{x^2 + y^2}\,dy\,dx$

44. $\displaystyle\int_0^1\int_0^{x/2} \frac{x}{x^2 + y^2}\,dy\,dx$

45. $\displaystyle\int_0^1\int_{-y/3}^{y/3} \frac{y}{\sqrt{x^2 + y^2}}\,dx\,dy$

46. $\displaystyle\int_0^1\int_y^{2-y} \sqrt{x + y}\,dx\,dy$

12.4 Triple Integrals in Rectangular Coordinates

Triple Integrals • Properties of Triple Integrals • Volume of a Region in Space • Finding Limits of Integration • Average Value of a Function in Space

We use triple integrals to find the volumes of three-dimensional shapes, the masses and moments of solids, and the average values of functions of three variables. In Chapter 13, we also see how these integrals arise in the studies of vector fields and fluid flow.

Triple Integrals

If $F(x, y, z)$ is a function defined on a closed bounded region D in space—the region occupied by a solid ball, for example, or a lump of clay—then the integral of F over D may be defined in the following way. We partition a rectangular region containing D into rectangular cells by planes parallel to the coordinate planes (Figure 12.31). We number the cells that lie inside D from 1 to n in some order, a typical cell having dimensions Δx_k by Δy_k by Δz_k and volume ΔV_k. We choose a point (x_k, y_k, z_k) in each cell and form the sum

$$S_n = \sum_{k=1}^{n} F(x_k, y_k, z_k)\,\Delta V_k. \tag{1}$$

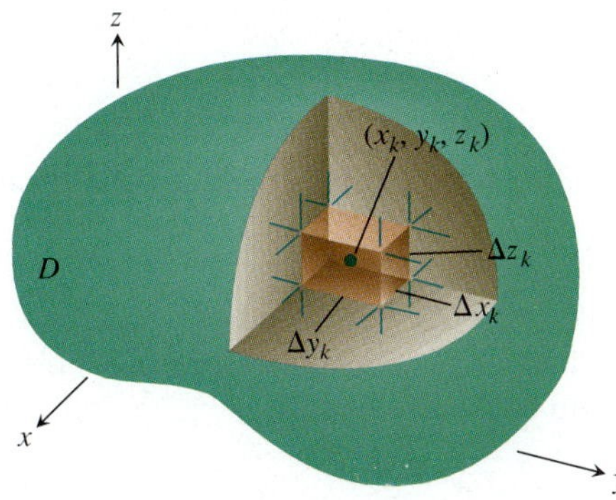

FIGURE 12.31 Partitioning a solid with rectangular cells of volume ΔV_k.

If F is continuous and the bounding surface of D is made of smooth surfaces joined along continuous curves, then as Δx_k, Δy_k, and Δz_k approach zero independently, the sums S_n approach a limit

$$\lim_{n\to\infty} S_n = \iiint_D F(x, y, z)\,dV. \tag{2}$$

We call this limit the **triple integral of F over D.** The limit also exists for some discontinuous functions.

Properties of Triple Integrals

Triple integrals have the same algebraic properties as double and single integrals.

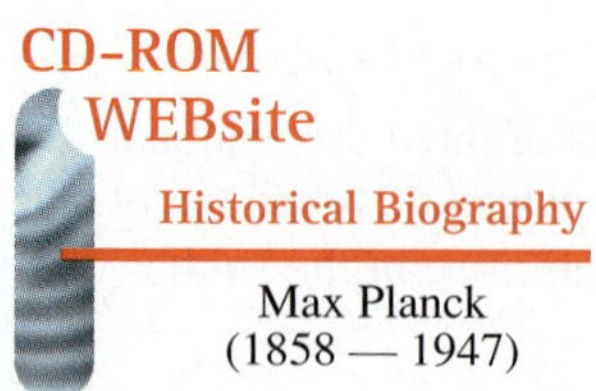

Properties of Triple Integrals

If $F = F(x, y, z)$ and $G = G(x, y, z)$ are continuous, then

1. *Constant Multiple*: $$\iiint_D kF\, dV = k \iiint_D F\, dV \qquad \text{(any number } k\text{)}$$

2. *Sum and Difference*: $$\iiint_D (F \pm G)\, dV = \iiint_D F\, dV \pm \iiint_D G\, dV$$

3. *Domination*:

   (a) $$\iiint_D F\, dV \geq 0 \qquad \text{if } F \geq 0 \text{ on } D$$

   (b) $$\iiint_D F\, dV \geq \iiint_D G\, dV \qquad \text{if } F \geq G \text{ on } D$$

4. *Additivity*:

$$\iiint_D F\, dV = \iiint_{D_1} F\, dV + \iiint_{D_2} F\, dV + \cdots + \iiint_{D_n} F\, dV$$

if D is the union of a finite number of nonoverlapping cells.

Volume of a Region in Space

If F is the constant function whose value is 1, then the sums in Equation (1) reduce to

$$S_n = \sum F(x_k, y_k, z_k)\, \Delta V_k = \sum 1 \cdot \Delta V_k = \sum \Delta V_k.$$

As Δx, Δy, and Δz approach zero, the cells ΔV_k become smaller and more numerous and fill up more and more of D. We therefore define the volume of D to be the triple integral

$$\lim_{n\to\infty} \sum_{k=1}^{n} \Delta V_k = \iiint_D dV.$$

Definition Volume

The **volume** of a closed, bounded region D in space is

$$V = \iiint_D dV. \tag{3}$$

As we see in a moment, this integral enables us to calculate the volumes of solids enclosed by curved surfaces.

Finding Limits of Integration

We evaluate a triple integral by applying a three-dimensional version of Fubini's Theorem (Section 12.1) to evaluate it by three repeated single integrations. As with double integrals, there is a geometric procedure for finding the limits of integration for these single integrals.

How to Find Limits of Integration in Triple Integrals

To evaluate

$$\iiint_D F(x, y, z)\, dV$$

over a region D, integrating first with respect to z, then with respect to y, finally with x, take the following steps.

Step 1: *A sketch.* Sketch the region D along with its "shadow" R (vertical projection) in the xy-plane. Label the upper and lower bounding surfaces of D and the upper and lower bounding curves of R.

Step 2: *The z-limits of integration.* Draw a line M passing through a typical point (x, y) in R parallel to the z-axis. As z increases, M enters D at $z = f_1(x, y)$ and leaves at $z = f_2(x, y)$. These are the z-limits of integration.

Step 3: *The y-limits of integration.* Draw a line L through (x, y) parallel to the y-axis. As y increases, L enters R at $y = g_1(x)$ and leaves at $y = g_2(x)$. These are the y-limits of integration.

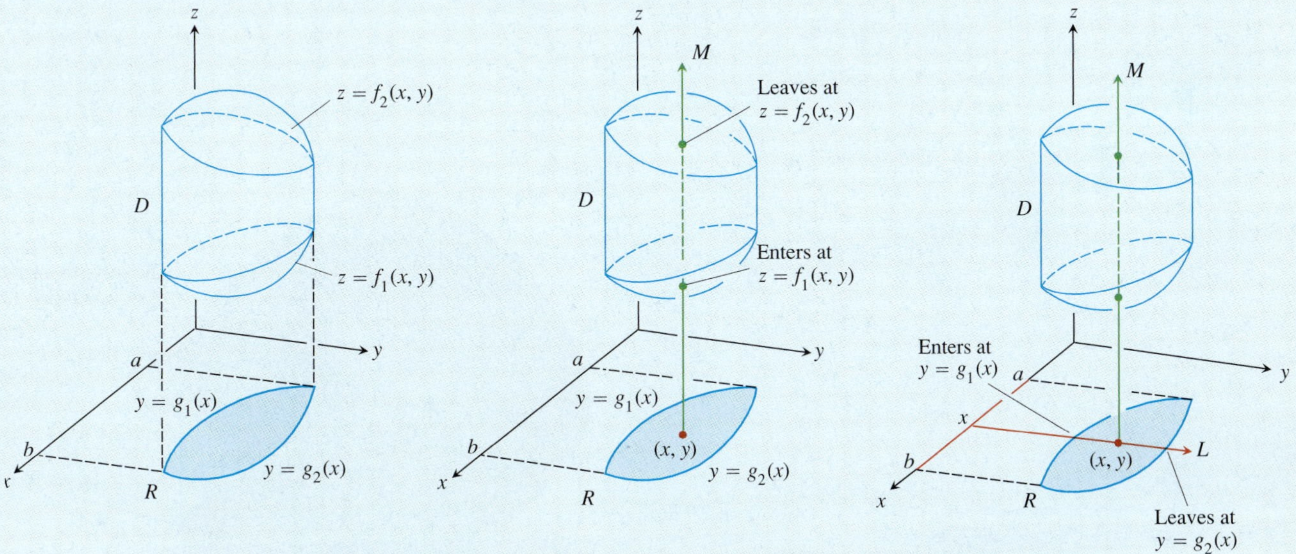

Step 4: *The x-limits of integration.* Choose x-limits that include all lines through R parallel to the y-axis ($x = a$ and $x = b$ in the preceding figure). These are the x-limits of integration. The integral is

$$\int_{x=a}^{x=b} \int_{y=g_1(x)}^{y=g_2(x)} \int_{z=f_1(x, y)}^{z=f_2(x, y)} F(x, y, z)\, dz\, dy\, dx.$$

Follow similar procedures if you change the order of integration. The "shadow" of region D lies in the plane of the last two variables with respect to which the iterated integration takes place.

Example 1 Finding a Volume

Find the volume of the region D enclosed by the surfaces $z = x^2 + 3y^2$ and $z = 8 - x^2 - y^2$.

Solution The volume is

$$V = \iiint_D dz\, dy\, dx,$$

the integral of $F(x, y, z) = 1$ over D. To find the limits of integration for evaluating the integral, we take these steps.

Step 1: *A sketch.* The surfaces (Figure 12.32) intersect on the elliptical cylinder $x^2 + 3y^2 = 8 - x^2 - y^2$ or $x^2 + 2y^2 = 4$. The boundary of the region R, the projection of D onto the xy-plane, is an ellipse with the same equation: $x^2 + 2y^2 = 4$. The "upper" boundary of R is the curve $y = \sqrt{(4 - x^2)/2}$. The lower boundary is the curve $y = -\sqrt{(4 - x^2)/2}$.

Step 2: *The z-limits of integration.* The line M passing through a typical point (x, y) in R parallel to the z-axis enters D at $z = x^2 + 3y^2$ and leaves at $z = 8 - x^2 - y^2$.

Step 3: *The y-limits of integration.* The line L through (x, y) parallel to the y-axis enters R at $y = -\sqrt{(4 - x^2)/2}$ and leaves at $y = \sqrt{(4 - x^2)/2}$.

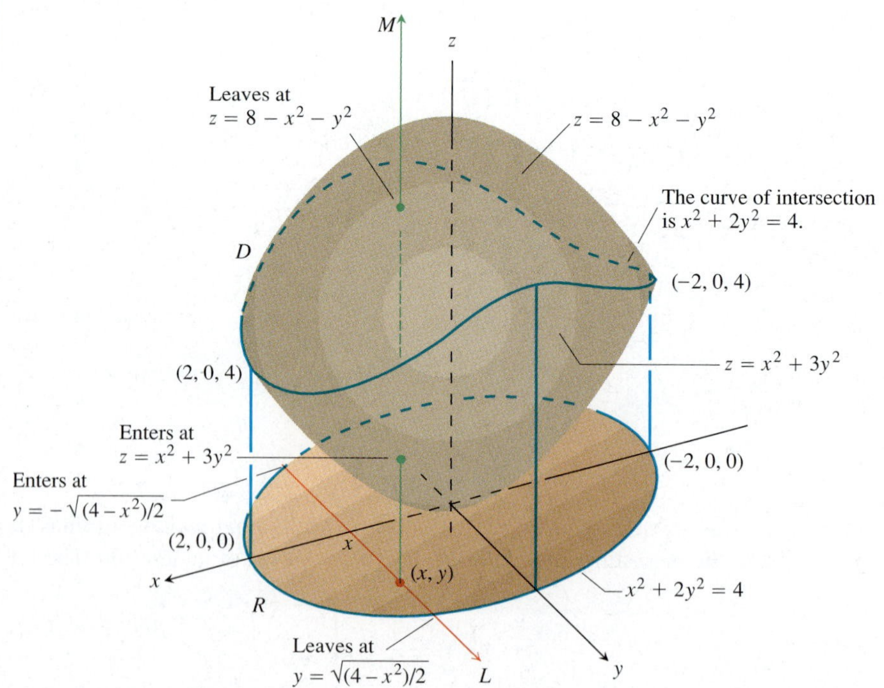

FIGURE 12.32 The volume of the region enclosed by these two paraboloids is calculated in Example 1.

Step 4: *The x-limits of integration.* As L sweeps across R, the value of x varies from $x = -2$ at $(-2, 0, 0)$ to $x = 2$ at $(2, 0, 0)$. The volume of D is

$$\begin{aligned}
V &= \iiint_D dz\, dy\, dx \\
&= \int_{-2}^{2} \int_{-\sqrt{(4-x^2)/2}}^{\sqrt{(4-x^2)/2}} \int_{x^2+3y^2}^{8-x^2-y^2} dz\, dy\, dx \\
&= \int_{-2}^{2} \int_{-\sqrt{(4-x^2)/2}}^{\sqrt{(4-x^2)/2}} (8 - 2x^2 - 4y^2)\, dy\, dx \\
&= \int_{-2}^{2} \left[(8 - 2x^2)y - \frac{4}{3}y^3 \right]_{y=-\sqrt{(4-x^2)/2}}^{y=\sqrt{(4-x^2)/2}} dx \\
&= \int_{-2}^{2} \left(2(8 - 2x^2)\sqrt{\frac{4 - x^2}{2}} - \frac{8}{3}\left(\frac{4 - x^2}{2}\right)^{3/2} \right) dx \\
&= \int_{-2}^{2} \left[8\left(\frac{4 - x^2}{2}\right)^{3/2} - \frac{8}{3}\left(\frac{4 - x^2}{2}\right)^{3/2} \right] dx = \frac{4\sqrt{2}}{3} \int_{-2}^{2} (4 - x^2)^{3/2}\, dx \\
&= 8\pi\sqrt{2} \text{ units cubed.}
\end{aligned}$$

After integration with the substitution $x = 2 \sin u$

In the next example, we project D onto the xz-plane instead of the xy-plane so you can see how to use a different order of integration.

Example 2 Finding the Limits of Integration in the Order *dy dz dx*

Set up the limits of integration for evaluating the triple integral of a function $F(x, y, z)$ over the tetrahedron D with vertices $(0, 0, 0)$, $(1, 1, 0)$, $(0, 1, 0)$, and $(0, 1, 1)$.

Solution

Step 1: *A sketch.* We sketch D along with its "shadow" R in the xz-plane (Figure 12.33). The upper (right-hand) bounding surface of D lies in the plane $y = 1$. The lower (left-hand) bounding surface lies in the plane $y = x + z$. The upper boundary of R is the line $z = 1 - x$. The lower boundary is the line $z = 0$.

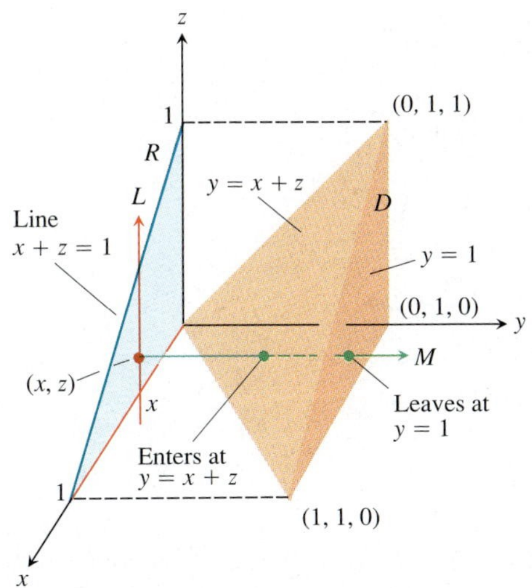

FIGURE 12.33 The tetrahedron in Example 2.

Step 2: *The y-limits of integration.* The line through a typical point (x, z) in R parallel to the y-axis enters D at $y = x + z$ and leaves at $y = 1$.

Step 3: *The z-limits of integration.* The line L through (x, z) parallel to the z-axis enters R at $z = 0$ and leaves at $z = 1 - x$.

Step 4: *The x-limits of integration.* As L sweeps across R, the value of x varies from $x = 0$ to $x = 1$. The integral is

$$\int_0^1 \int_0^{1-x} \int_{x+z}^{1} F(x, y, z)\, dy\, dz\, dx.$$

Example 3 Revisiting Example 2 Using the Order $dz\,dy\,dx$

To integrate $F(x, y, z)$ over the tetrahedron D in the order $dz\,dy\,dx$, we perform steps 2 through 4 in the following way.

Step 2: *The z-limits of integration.* A line parallel to the z-axis through a typical point (x, y) in the xy-plane "shadow" enters the tetrahedron at $z = 0$ and exits through the upper plane where $z = y - x$ (Figure 12.33).

Step 3: *The y-limits of integration.* A line through (x, y) parallel to the y-axis enters the shadow in the xy-plane at $y = x$ and exits at $y = 1$.

Step 4: *The x-limits of integration.* As the line parallel to the y-axis in step 3 sweeps out the shadow, the value of x varies from $x = 0$ to $x = 1$ at the point $(1, 1, 0)$. The integral is

$$\int_0^1 \int_x^1 \int_0^{y-x} F(x, y, z)\, dz\, dy\, dx.$$

For example, if $F(x, y, z) = 1$, we would find the volume of the tetrahedron to be

$$\begin{aligned}
V &= \int_0^1 \int_x^1 \int_0^{y-x} dz\, dy\, dx \\
&= \int_0^1 \int_x^1 (y - x)\, dy\, dx \\
&= \int_0^1 \left[\frac{1}{2}y^2 - xy\right]_{y=x}^{y=1} dx \\
&= \int_0^1 \left(\frac{1}{2} - x + \frac{1}{2}x^2\right) dx \\
&= \left[\frac{1}{2}x - \frac{1}{2}x^2 + \frac{1}{6}x^3\right]_0^1 \\
&= \frac{1}{6} \text{ of a cubic unit.}
\end{aligned}$$

You will get the same result by integrating

$$V = \int_0^1 \int_0^{1-x} \int_{x+z}^1 dy\, dz\, dx$$

from Example 2. Try it and see!

As we know, there are sometimes (but not always) two different orders in which the single integrations for evaluating a double integral may be worked. For triple integrals, there could be as many as *six*.

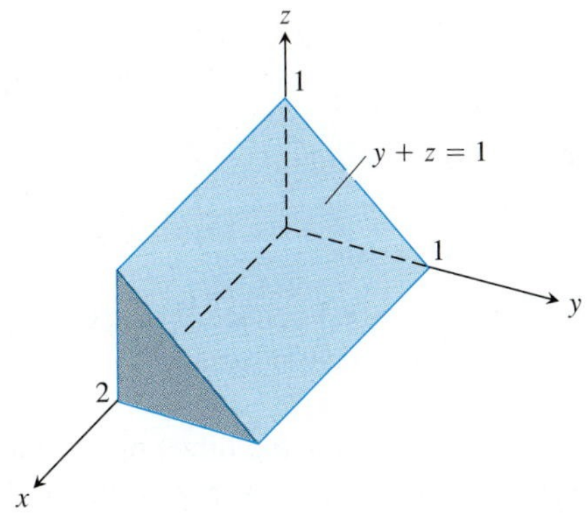

FIGURE 12.34 Example 4 gives six different iterated triple integrals for the volume of this prism.

Example 4 Using Different Orders of Integration

Each of the following integrals gives the volume of the solid shown in Figure 12.34.

(a) $\int_0^1 \int_0^{1-z} \int_0^2 dx\, dy\, dz$ **(b)** $\int_0^1 \int_0^{1-y} \int_0^2 dx\, dz\, dy$

(c) $\int_0^1 \int_0^2 \int_0^{1-z} dy\, dx\, dz$ **(d)** $\int_0^2 \int_0^1 \int_0^{1-z} dy\, dz\, dx$

(e) $\int_0^1 \int_0^2 \int_0^{1-y} dz\, dx\, dy$ **(f)** $\int_0^2 \int_0^1 \int_0^{1-y} dz\, dy\, dx$

Let's work out the integrals in parts (b) and (c):

$$\begin{aligned} V &= \int_0^1 \int_0^{1-y} \int_0^2 dx\, dz\, dy && \text{Integral in part (b)} \\ &= \int_0^1 \int_0^{1-y} 2\, dz\, dy \\ &= \int_0^1 \Big[2z\Big]_{z=0}^{z=1-y} dy \\ &= \int_0^1 2(1-y)\, dy \\ &= 1 \text{ cubic unit.} \end{aligned}$$

Also,

$$\begin{aligned} V &= \int_0^1 \int_0^2 \int_0^{1-z} dy\, dx\, dz && \text{Integral in part (c)} \\ &= \int_0^1 \int_0^2 (1-z)\, dx\, dz \\ &= \int_0^1 \Big[x - zx\Big]_{x=0}^{x=2} dz \\ &= \int_0^1 (2-2z)\, dz \\ &= 1 \text{ cubic unit.} \end{aligned}$$

Average Value of a Function in Space

The average value of a function F over a region D in space is defined by the formula

$$\textbf{Average value of } F \text{ over } D = \frac{1}{\text{volume of } D} \iiint_D F\, dV. \tag{4}$$

For example, if $F(x, y, z) = \sqrt{x^2 + y^2 + z^2}$, then the average value of F over D is the average distance of points in D from the origin. If $F(x, y, z)$ is the density of a solid that occupies a region D in space, then the average value of F over D is the average density of the solid in units of mass per unit volume.

Example 5 Finding an Average Value

Find the average value of $F(x, y, z) = xyz$ over the cube bounded by the coordinate planes and the planes $x = 2$, $y = 2$, and $z = 2$ in the first octant.

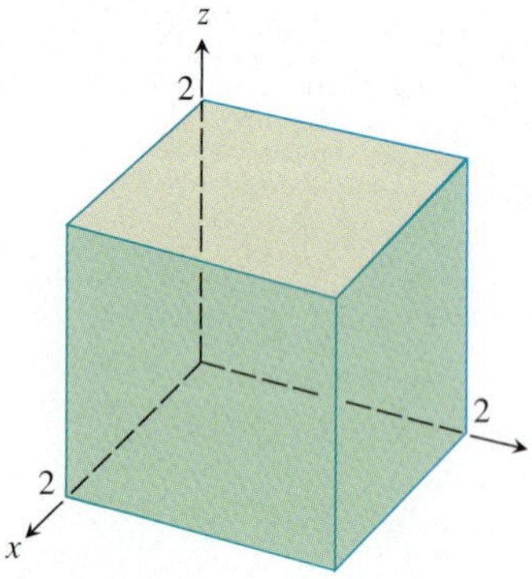

FIGURE 12.35 The region of integration in Example 5.

Solution We sketch the cube with enough detail to show the limits of integration (Figure 12.35). We then use Equation (4) to calculate the average value of F over the cube.

The volume of the cube is $(2)(2)(2) = 8$. The value of the integral of F over the cube is

$$\int_0^2\int_0^2\int_0^2 xyz\,dx\,dy\,dz = \int_0^2\int_0^2\left[\frac{x^2}{2}yz\right]_{x=0}^{x=2} dy\,dz = \int_0^2\int_0^2 2yz\,dy\,dz$$

$$= \int_0^2\left[y^2z\right]_{y=0}^{y=2} dz = \int_0^2 4z\,dz = \left[2z^2\right]_0^2 = 8.$$

With these values, Equation (4) gives

$$\begin{array}{c}\text{Average value of} \\ xyz \text{ over the cube}\end{array} = \frac{1}{\text{volume}}\iiint_{\text{cube}} xyz\,dV = \left(\frac{1}{8}\right)(8) = 1.$$

In evaluating the integral, we chose the order $dx\,dy\,dz$, but any of the other five possible orders would have done as well.

EXERCISES 12.4

Evaluating Triple Integrals in Different Iterations

1. Evaluate the integral in Example 2 taking $F(x, y, z) = 1$ to find the volume of the tetrahedron.

2. *Volume of rectangular solid* Write six different iterated triple integrals for the volume of the rectangular solid in the first octant bounded by the coordinate planes and the planes $x = 1$, $y = 2$, and $z = 3$. Evaluate one of the integrals.

3. *Volume of tetrahedron* Write six different iterated triple integrals for the volume of the tetrahedron cut from the first octant by the plane $6x + 3y + 2z = 6$. Evaluate one of the integrals.

4. *Volume of solid* Write six different iterated triple integrals for the volume of the region in the first octant enclosed by the cylinder $x^2 + z^2 = 4$ and the plane $y = 3$. Evaluate one of the integrals.

5. *Volume enclosed by paraboloids* Let D be the region bounded by the paraboloids $z = 8 - x^2 - y^2$ and $z = x^2 + y^2$. Write six different triple iterated integrals for the volume of D. Evaluate one of the integrals.

6. *Volume inside paraboloid beneath a plane* Let D be the region bounded by the paraboloid $z = x^2 + y^2$ and the plane $z = 2y$. Write triple iterated integrals in the order $dz\,dx\,dy$ and $dz\,dy\,dx$ that give the volume of D. Do not evaluate either integral.

Evaluating Triple Iterated Integrals

Evaluate the integrals in Exercises 7–20.

7. $\int_0^1\int_0^1\int_0^1 (x^2 + y^2 + z^2)\,dz\,dy\,dx$

8. $\int_0^{\sqrt{2}}\int_0^{3y}\int_{x^2+3y^2}^{8-x^2-y^2} dz\,dx\,dy$

9. $\int_1^e\int_1^e\int_1^e \frac{1}{xyz}\,dx\,dy\,dz$

10. $\int_0^1\int_0^{3-3x}\int_0^{3-3x-y} dz\,dy\,dx$

11. $\int_0^1\int_0^{\pi}\int_0^{\pi} y \sin z\,dx\,dy\,dz$

12. $\int_{-1}^{1}\int_{-1}^{1}\int_{-1}^{1}(x+y+z)\,dy\,dx\,dz$

13. $\int_{0}^{3}\int_{0}^{\sqrt{9-x^2}}\int_{0}^{\sqrt{9-x^2}}dz\,dy\,dx$

14. $\int_{0}^{2}\int_{-\sqrt{4-y^2}}^{\sqrt{4-y^2}}\int_{0}^{2x+y}dz\,dx\,dy$

15. $\int_{0}^{1}\int_{0}^{2-x}\int_{0}^{2-x-y}dz\,dy\,dx$

16. $\int_{0}^{1}\int_{0}^{1-x^2}\int_{3}^{4-x^2-y}x\,dz\,dy\,dx$

17. $\int_{0}^{\pi}\int_{0}^{\pi}\int_{0}^{\pi}\cos(u+v+w)\,du\,dv\,dw$ (uvw-space)

18. $\int_{1}^{e}\int_{1}^{e}\int_{1}^{e}\ln r\ln s\ln t\,dt\,dr\,ds$ (rst-space)

19. $\int_{0}^{\pi/4}\int_{0}^{\ln\sec v}\int_{-\infty}^{2t}e^{x}\,dx\,dt\,dv$ (tvx-space)

20. $\int_{0}^{7}\int_{0}^{2}\int_{0}^{\sqrt{4-q^2}}\frac{q}{r+1}\,dp\,dq\,dr$ (pqr-space)

Volumes Using Triple Integrals

21. Here is the region of integration of the integral

$$\int_{-1}^{1}\int_{x^2}^{1}\int_{0}^{1-y}dz\,dy\,dx.$$

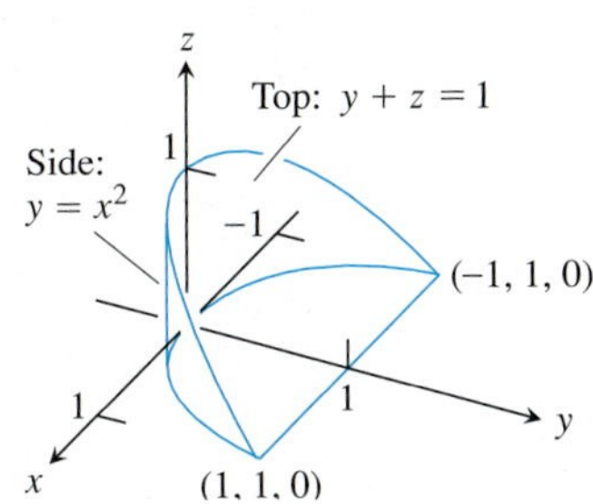

Rewrite the integral as an equivalent iterated integral in the order

(a) $dy\,dz\,dx$ **(b)** $dy\,dx\,dz$

(c) $dx\,dy\,dz$ **(d)** $dx\,dz\,dy$

(e) $dz\,dx\,dy$.

22. Here is the region of integration of the integral

$$\int_{0}^{1}\int_{-1}^{0}\int_{0}^{y^2}dz\,dy\,dx.$$

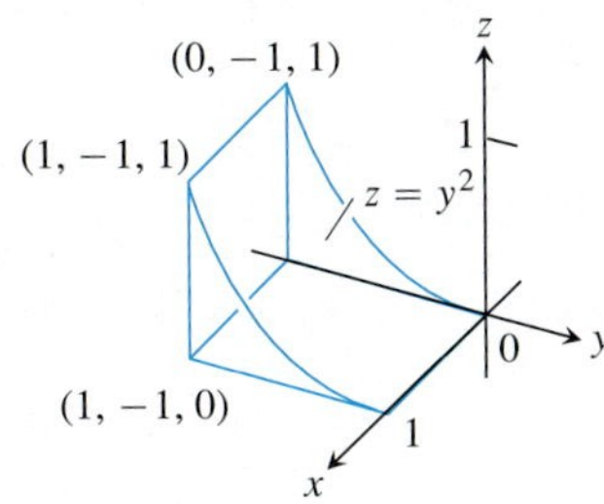

Rewrite the integral as an equivalent iterated integral in the order

(a) $dy\,dz\,dx$ **(b)** $dy\,dx\,dz$

(c) $dx\,dy\,dz$ **(d)** $dx\,dz\,dy$

(e) $dz\,dx\,dy$.

Find the volumes of the regions in Exercises 23–36.

23. The region between the cylinder $z = y^2$ and the xy-plane that is bounded by the planes $x = 0, x = 1, y = -1, y = 1$

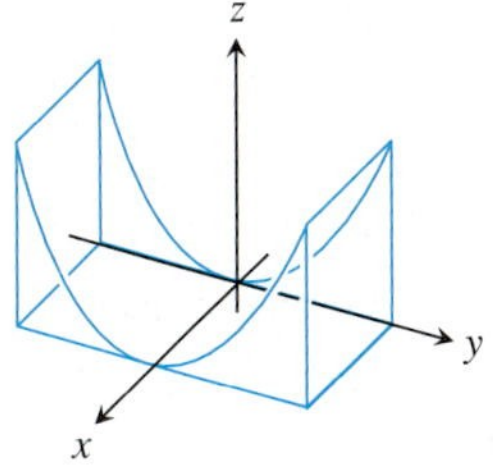

24. The region in the first octant bounded by the coordinate planes and the planes $x + z = 1, y + 2z = 2$

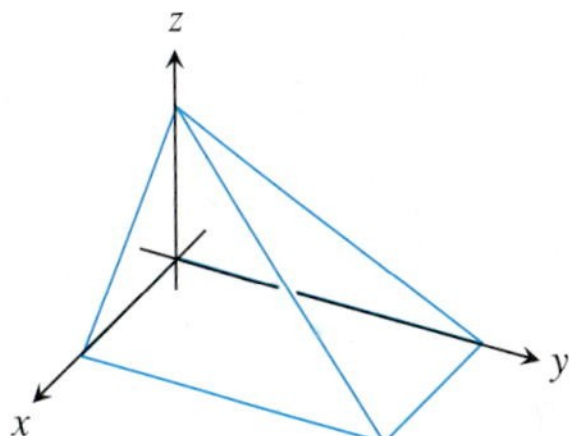

25. The region in the first octant bounded by the coordinate planes, the plane $y + z = 2$, and the cylinder $x = 4 - y^2$

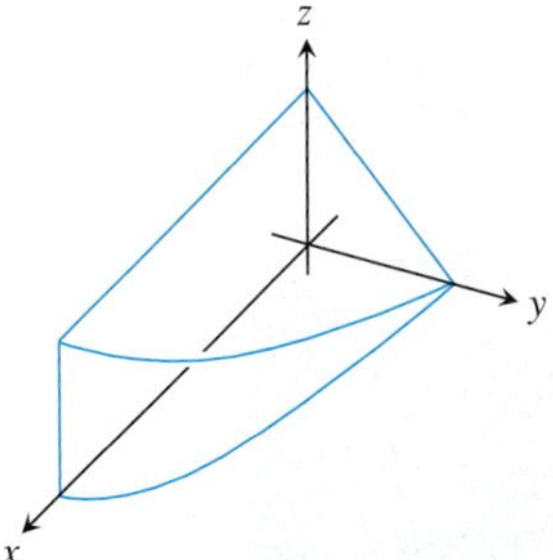

26. The wedge cut from the cylinder $x^2 + y^2 = 1$ by the planes $z = -y$ and $z = 0$

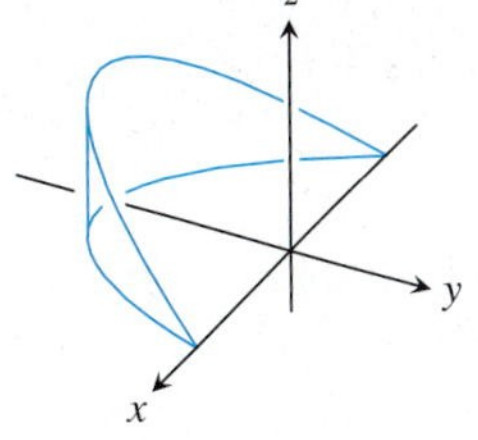

27. The tetrahedron in the first octant bounded by the coordinate planes and the plane $x + y/2 + z/3 = 1$

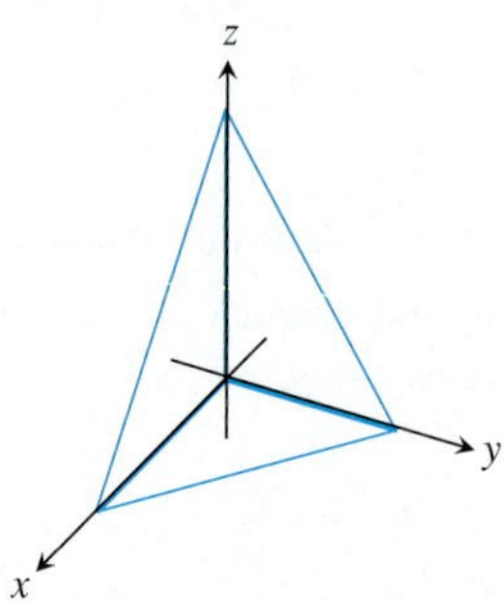

28. The region in the first octant bounded by the coordinate planes, the plane $y = 1 - x$, and the surface $z = \cos(\pi x/2)$, $0 \le x \le 1$

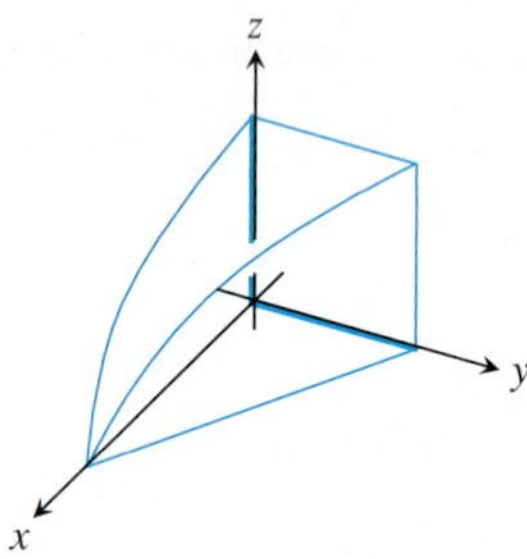

29. The region common to the interiors of the cylinders $x^2 + y^2 = 1$ and $x^2 + z^2 = 1$ (Figure 12.36)

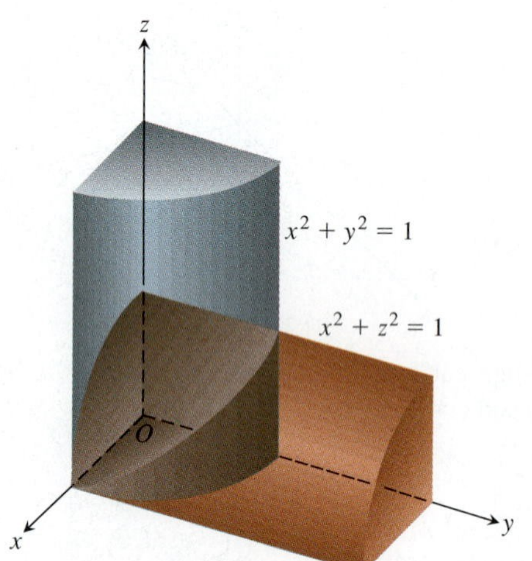

FIGURE 12.36 One-eighth of the region common to the cylinders $x^2 + y^2 = 1$ and $x^2 + z^2 = 1$ in Exercise 29.

30. The region in the first octant bounded by the coordinate planes and the surface $z = 4 - x^2 - y$

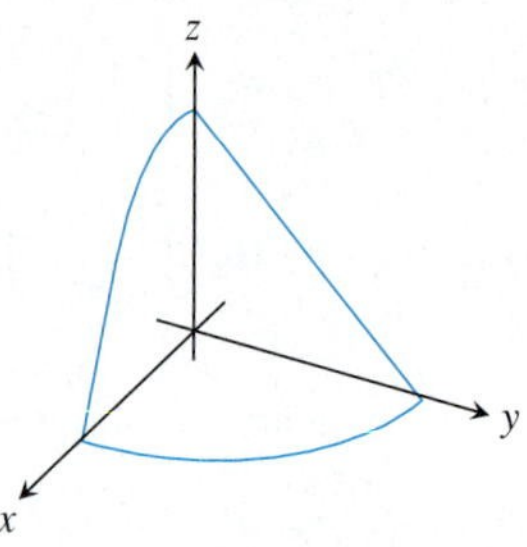

31. The region in the first octant bounded by the coordinate planes, the plane $x + y = 4$, and the cylinder $y^2 + 4z^2 = 16$

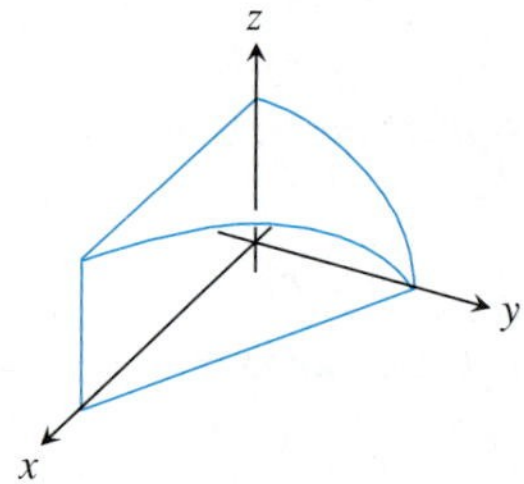

32. The region cut from the cylinder $x^2 + y^2 = 4$ by the plane $z = 0$ and the plane $x + z = 3$

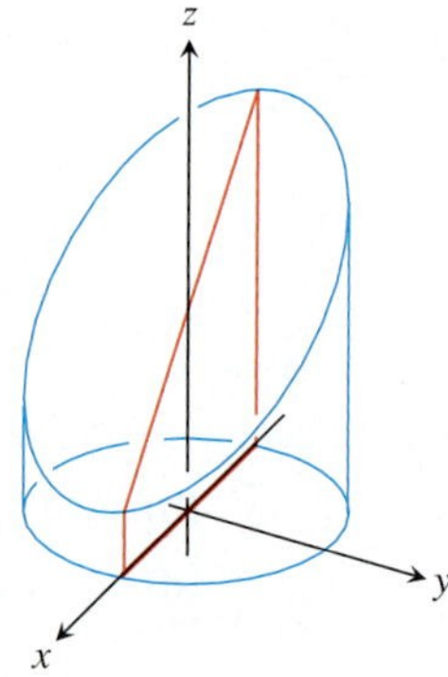

33. The region between the planes $x + y + 2z = 2$ and $2x + 2y + z = 4$ in the first octant

34. The finite region bounded by the planes $z = x$, $x + z = 8$, $z = y$, $y = 8$, and $z = 0$.

35. The region cut from the solid elliptical cylinder $x^2 + 4y^2 \le 4$ by the xy-plane and the plane $z = x + 2$

36. The region bounded in back by the plane $x = 0$, on the front and sides by the parabolic cylinder $x = 1 - y^2$, on the top by the paraboloid $z = x^2 + y^2$, and on the bottom by the xy-plane

Average Values

In Exercises 37–40, find the average value of $F(x, y, z)$ over the given region.

37. $F(x, y, z) = x^2 + 9$ over the cube in the first octant bounded by the coordinate planes and the planes $x = 2$, $y = 2$, and $z = 2$

38. $F(x, y, z) = x + y - z$ over the rectangular solid in the first octant bounded by the coordinate planes and the planes $x = 1$, $y = 1$, and $z = 2$

39. $F(x, y, z) = x^2 + y^2 + z^2$ over the cube in the first octant bounded by the coordinate planes and the planes $x = 1$, $y = 1$, and $z = 1$

40. $F(x, y, z) = xyz$ over the cube in the first octant bounded by the coordinate planes and the planes $x = 2$, $y = 2$, and $z = 2$

Changing the Order of Integration

Evaluate the integrals in Exercises 41–44 by changing the order of integration in an appropriate way.

41. $\displaystyle\int_0^4\int_0^1\int_{2y}^2 \frac{4\cos(x^2)}{2\sqrt{z}}\,dx\,dy\,dz$

42. $\displaystyle\int_0^1\int_0^1\int_{x^2}^1 12xze^{zy^2}\,dy\,dx\,dz$

43. $\displaystyle\int_0^1\int_{\sqrt[3]{z}}^1\int_0^{\ln 3} \frac{\pi e^{2x}\sin \pi y^2}{y^2}\,dx\,dy\,dz$

44. $\displaystyle\int_0^2\int_0^{4-x^2}\int_0^x \frac{\sin 2z}{4-z}\,dy\,dz\,dx$

Theory and Examples

45. *Finding upper limit of iterated integral* Solve for a:

$$\int_0^1\int_0^{4-a-x^2}\int_a^{4-x^2-y} dz\,dy\,dx = \frac{4}{15}.$$

46. *Ellipsoid* For what value of c is the volume of the ellipsoid $x^2 + (y/2)^2 + (z/c)^2 = 1$ equal to 8π?

47. *Writing to Learn: Minimizing a triple integral* What domain D in space minimizes the value of the integral

$$\iiint_D (4x^2 + 4y^2 + z^2 - 4)\,dV?$$

Give reasons for your answer.

48. *Writing to Learn: Maximizing a triple integral* What domain D in space maximizes the value of the integral

$$\iiint_D (1 - x^2 - y^2 - z^2)\,dV?$$

Give reasons for your answer.

COMPUTER EXPLORATIONS

Numerical Evaluations

In Exercises 49–52, use a CAS integration utility to evaluate the triple integral of the given function over the specified solid region.

49. $F(x, y, z) = x^2y^2z$ over the solid cylinder bounded by $x^2 + y^2 = 1$ and the planes $z = 0$ and $z = 1$

50. $F(x, y, z) = |xyz|$ over the solid bounded below by the paraboloid $z = x^2 + y^2$ and above by the plane $z = 1$

51. $F(x, y, z) = \dfrac{z}{(x^2 + y^2 + z^2)^{3/2}}$ over the solid bounded below by the cone $z = \sqrt{x^2 + y^2}$ and above by the plane $z = 1$

52. $F(x, y, z) = x^4 + y^2 + z^2$ over the solid sphere $x^2 + y^2 + z^2 \le 1$

12.5 Masses and Moments in Three Dimensions

Masses and Moments

This section shows how to calculate the masses and moments of three-dimensional objects in Cartesian coordinates. The formulas are similar to those for two-dimensional objects. For calculations in spherical and cylindrical coordinates, see Section 12.6.

Masses and Moments

If $\delta(x, y, z)$ is the density of an object occupying a region D in space (mass per unit volume), the integral of δ over D gives the mass of the object. To see why,

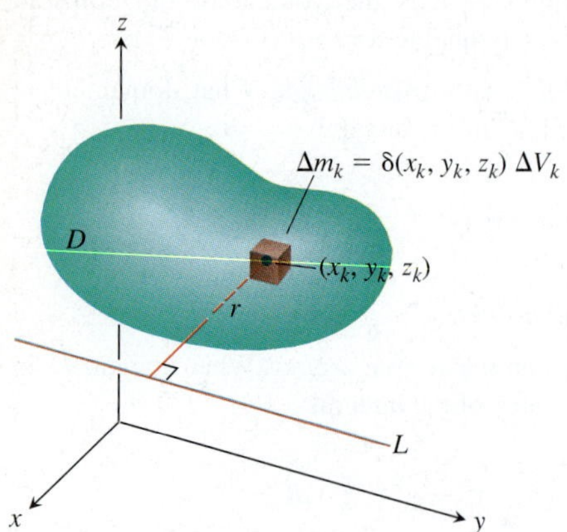

FIGURE 12.37 To define an object's mass and moment of inertia about a line, we first imagine it to be partitioned into a finite number of mass elements Δm_k.

imagine partitioning the object into n mass elements like the one in Figure 12.37. The object's mass is the limit

$$M = \lim_{n\to\infty}\sum_{k=1}^{n}\Delta m_k = \lim_{n\to\infty}\sum_{k=1}^{n}\delta(x_k, y_k, z_k)\,\Delta V_k = \iiint_D \delta(x, y, z)\,dV.$$

To find the **first moments about the coordinate planes**, we use the signed distance from each plane. For example,

$$M_{yz} = \iiint_D x\delta(x, y, z)\,dV$$

gives the first moment about the yz-plane.

Extending the moments of inertia to triple integrals is similar. If $r(x, y, z)$ is the distance from the point (x, y, z) in D to a line L, then the moment of inertia of the mass $\Delta m_k = \delta(x_k, y_k, z_k)\,\Delta V_k$ about the line L (shown in Figure 12.37) is approximately $\Delta I_k = r^2(x_k, y_k, z_k)\,\Delta m_k$. **The moment of inertia about L** of the entire object is

$$I_L = \lim_{n\to\infty}\sum_{k=1}^{n}\Delta I_k = \lim_{n\to\infty}\sum_{k=1}^{n} r^2(x_k, y_k, z_k)\delta(x_k, y_k, z_k)\,\Delta V_k = \iiint_D r^2\delta\,dV.$$

If L is the x-axis, then $r^2 = y^2 + z^2$ (Figure 12.38) and

$$I_x = \iiint_D (y^2 + z^2)\delta\,dV.$$

Similarly,

$$I_y = \iiint_D (x^2 + z^2)\,\delta\,dV \qquad \text{and} \qquad I_z = \iiint_D (x^2 + z^2)\,\delta\,dV.$$

The mass and moment formulas in space analogous to those discussed for planar regions in Section 12.2 are summarized in Table 12.3.

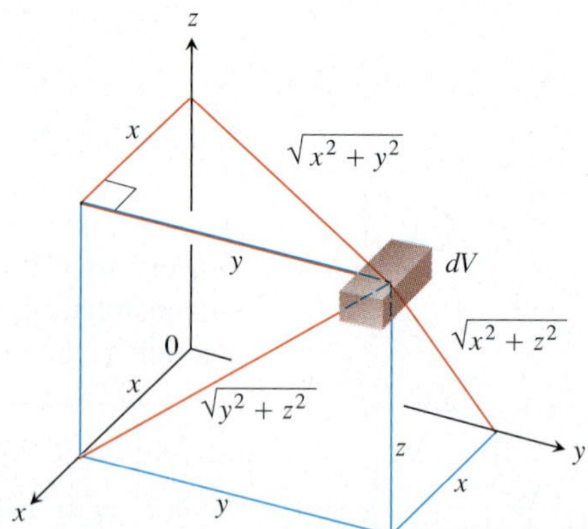

FIGURE 12.38 Distances from dV to the coordinate planes and axes.

CD-ROM WEBsite

Table 12.3 Mass and moment formulas for objects in space

Mass: $M = \iiint_D \delta \, dV \qquad (\delta = \delta(x, y, z) = \text{density})$

First moments about the coordinate planes:

$$M_{yz} = \iiint_D x \, \delta \, dV, \qquad M_{xz} = \iiint_D y \, \delta \, dV, \qquad M_{xy} = \iiint_D z \, \delta \, dV$$

Center of mass:

$$\bar{x} = \frac{M_{yz}}{M}, \qquad \bar{y} = \frac{M_{xz}}{M}, \qquad \bar{z} = \frac{M_{xy}}{M}$$

Moments of inertia (second moments) about the coordinate axes:

$$I_x = \iiint (y^2 + z^2) \, \delta \, dV$$

$$I_y = \iiint (x^2 + z^2) \, \delta \, dV$$

$$I_z = \iiint (x^2 + y^2) \, \delta \, dV$$

Moments of inertia about a line L:

$$I_L = \iiint r^2 \delta \, dV \qquad (r(x, y, z) = \text{distance from the point } (x, y, z) \text{ to line } L)$$

Radius of gyration about a line L:

$$R_L = \sqrt{I_L/M}$$

CD-ROM WEBsite
Historical Biography
Pappus of Alexandria
(fl. 300 — 350)

Example 1 Finding the Center of Mass of a Solid in Space

Find the center of mass of a solid of constant density δ bounded below by the disk $R: x^2 + y^2 \le 4$ in the plane $z = 0$ and above by the paraboloid $z = 4 - x^2 - y^2$ (Figure 12.39).

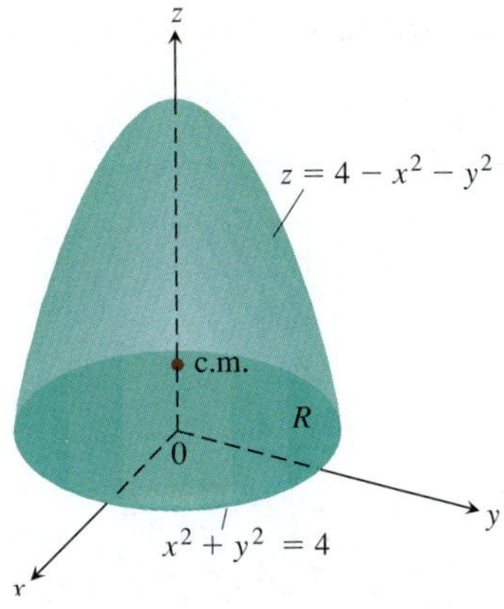

FIGURE 12.39 Example 1 finds the center of mass of this solid.

Solution By symmetry, $\bar{x} = \bar{y} = 0$. To find $\bar{z}$, we first calculate

$$M_{xy} = \iint_R \int_{z=0}^{z=4-x^2-y^2} z\,\delta\,dz\,dy\,dx = \iint_R \left[\frac{z^2}{2}\right]_{z=0}^{z=4-x^2-y^2} \delta\,dy\,dx$$

$$= \frac{\delta}{2}\iint_R (4 - x^2 - y^2)^2\,dy\,dx$$

$$= \frac{\delta}{2}\int_0^{2\pi}\int_0^2 (4 - r^2)^2 r\,dr\,d\theta \qquad \text{Polar coordinates}$$

$$= \frac{\delta}{2}\int_0^{2\pi}\left[-\frac{1}{6}(4 - r^2)^3\right]_{r=0}^{r=2} d\theta = \frac{16\delta}{3}\int_0^{2\pi} d\theta = \frac{32\pi\delta}{3}.$$

A similar calculation gives

$$M = \iint_R \int_0^{4-x^2-y^2} \delta\,dz\,dy\,dx = 8\pi\delta.$$

Therefore $\bar{z} = (M_{xy}/M) = 4/3$, and the center of mass is $(\bar{x}, \bar{y}, \bar{z}) = (0, 0, 4/3)$.

When the density of a solid object is constant (as in Example 1), the center of mass is called the **centroid** of the object (as was the case for two-dimensional shapes in Section 12.2).

Example 2 Finding the Moments of Inertia About the Coordinate Planes

Find I_x, I_y, I_z for the rectangular solid of constant density δ shown in Figure 12.40.

Solution The formula for I_x gives

$$I_x = \int_{-c/2}^{c/2}\int_{-b/2}^{b/2}\int_{-a/2}^{a/2} (y^2 + z^2)\,\delta\,dx\,dy\,dz.$$

We can avoid some of the work of integration by observing that $(y^2 + z^2)\delta$ is an even function of x, y, and z and therefore

$$I_x = 8\int_0^{c/2}\int_0^{b/2}\int_0^{a/2} (y^2 + z^2)\,\delta\,dx\,dy\,dz = 4a\delta\int_0^{c/2}\int_0^{b/2} (y^2 + z^2)\,dy\,dz$$

$$= 4a\delta\int_0^{c/2}\left[\frac{y^3}{3} + z^2 y\right]_{y=0}^{y=b/2} dz$$

$$= 4a\delta\int_0^{c/2}\left(\frac{b^3}{24} + \frac{z^2 b}{2}\right) dz$$

$$= 4a\delta\left(\frac{b^3 c}{48} + \frac{c^3 b}{48}\right) = \frac{abc\delta}{12}(b^2 + c^2) = \frac{M}{12}(b^2 + c^2).$$

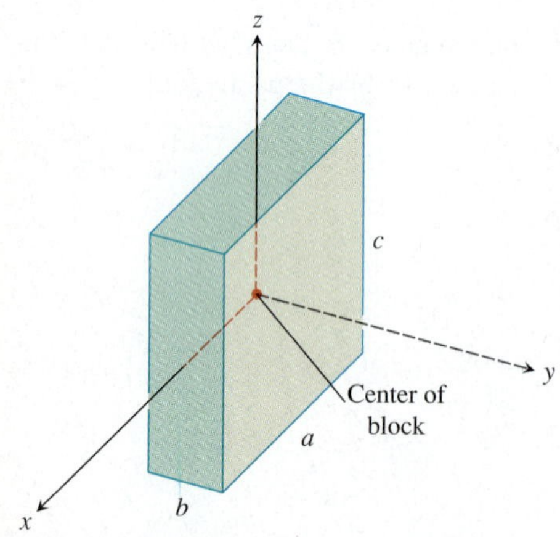

FIGURE 12.40 Example 2 calculates I_x, I_y, and I_z for the block shown here. The origin lies at the center of the block.

Similarly,

$$I_y = \frac{M}{12}(a^2 + c^2) \quad \text{and} \quad I_z = \frac{M}{12}(a^2 + b^2).$$

EXERCISES 12.5

Constant Density

The solids in Exercises 1–12 all have constant density $\delta = 1$.

1. *Example 1 Revisited* Evaluate the integral for I_x in Table 12.3 directly to show that the shortcut in Example 2 gives the same answer. Use the results in Example 2 to find the radius of gyration of the rectangular solid about each coordinate axis.

2. *Moments of inertia* The coordinate axes in the figure run through the centroid of a solid wedge parallel to the labeled edges. Find I_x, I_y, and I_z if $a = b = 6$ and $c = 4$.

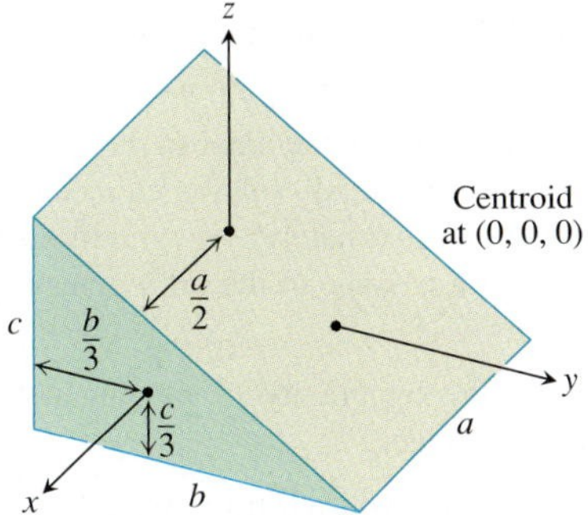

3. *Moments of inertia* Find the moments of inertia of the rectangular solid shown here with respect to its edges by calculating I_x, I_y, and I_z.

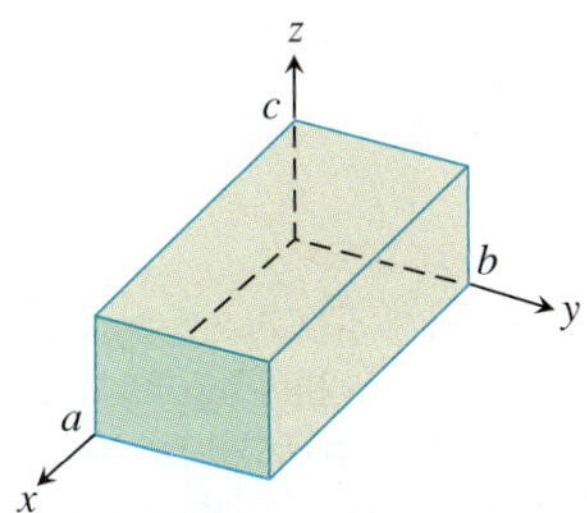

4. **(a)** *Centroid and moments of inertia* Find the centroid and the moments of inertia I_x, I_y, and I_z of the tetrahedron whose vertices are the points $(0, 0, 0)$, $(1, 0, 0)$, $(0, 1, 0)$, and $(0, 0, 1)$.

   (b) *Radius of gyration* Find the radius of gyration of the tetrahedron about the x-axis. Compare it with the distance from the centroid to the x-axis.

5. *Center of mass and moments of inertia* A solid "trough" of constant density is bounded below by the surface $z = 4y^2$, above by the plane $z = 4$, and on the ends by the planes $x = 1$ and $x = -1$. Find the center of mass and the moments of inertia with respect to the three axes.

6. *Center of mass* A solid of constant density is bounded below by the plane $z = 0$, on the sides by the elliptic cylinder $x^2 + 4y^2 = 4$, and above by the plane $z = 2 - x$ (see the figure).

   (a) Find $\bar{x}$ and $\bar{y}$.

   (b) Evaluate the integral

$$M_{xy} = \int_{-2}^{2}\int_{-(1/2)\sqrt{4-x^2}}^{(1/2)\sqrt{4-x^2}}\int_{0}^{2-x} z\,dz\,dy\,dx$$

   using integral tables to carry out the final integration with respect to x. Then divide M_{xy} by M to verify that $\bar{z} = 5/4$.

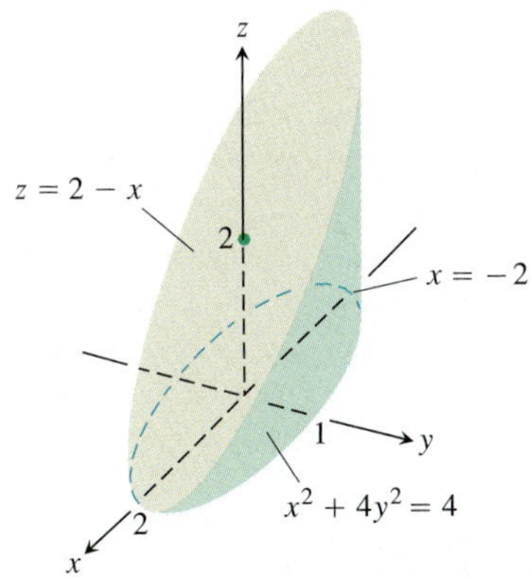

7. **(a)** *Center of mass* Find the center of mass of a solid of constant density bounded below by the paraboloid $z = x^2 + y^2$ and above by the plane $z = 4$.

   (b) Find the plane $z = c$ that divides the solid into two parts of equal volume. This plane does not pass through the center of mass.

8. *Moments and radii of gyration* A solid cube, 2 units on a side, is bounded by the planes $x = \pm 1$, $z = \pm 1$, $y = 3$, and $y = 5$. Find the center of mass and the moments of inertia and radii of gyration about the coordinate axes.

9. *Moment of inertia and radius of gyration about a line* A wedge like the one in Exercise 2 has $a = 4$, $b = 6$, and $c = 3$. Make a quick sketch to check for yourself that the square of the distance from a typical point (x, y, z) of the wedge to the line $L: z = 0, y = 6$ is $r^2 = (y - 6)^2 + z^2$. Then calculate the moment of inertia and radius of gyration of the wedge about L.

10. *Moment of inertia and radius of gyration about a line* A wedge like the one in Exercise 2 has $a = 4$, $b = 6$, and $c = 3$. Make a quick sketch to check for yourself that the square of the distance from a typical point (x, y, z) of the wedge to the line $L: x = 4, y = 0$ is $r^2 = (x - 4)^2 + y^2$. Then calculate the moment of inertia and radius of gyration of the wedge about L.

11. *Moment of inertia and radius of gyration about a line* A solid like the one in Exercise 3 has $a = 4$, $b = 2$, and $c = 1$. Make a quick sketch to check for yourself that the square of the distance between a typical point (x, y, z) of the solid and the line $L: y = 2, z = 0$ is $r^2 = (y - 2)^2 + z^2$. Then find the moment of inertia and radius of gyration of the solid about L.

12. *Moment of inertia of radius of gyration about a line* A solid like the one in Exercise 3 has $a = 4$, $b = 2$, and $c = 1$. Make a quick sketch to check for yourself that the square of the distance between a typical point (x, y, z) of the solid and the line $L: x = 4, y = 0$ is $r^2 = (x - 4)^2 + y^2$. Then find the moment of inertia and radius of gyration of the solid about L.

Variable Density

In Exercises 13 and 14, find

(a) The mass of the solid

(b) The center of mass.

13. A solid region in the first octant is bounded by the coordinate planes and the plane $x + y + z = 2$. The density of the solid is $\delta(x, y, z) = 2x$.

14. A solid in the first octant is bounded by the planes $y = 0$ and $z = 0$ and by the surfaces $z = 4 - x^2$ and $x = y^2$ (see the figure). Its density function is $\delta(x, y, z) = kxy$, k a constant.

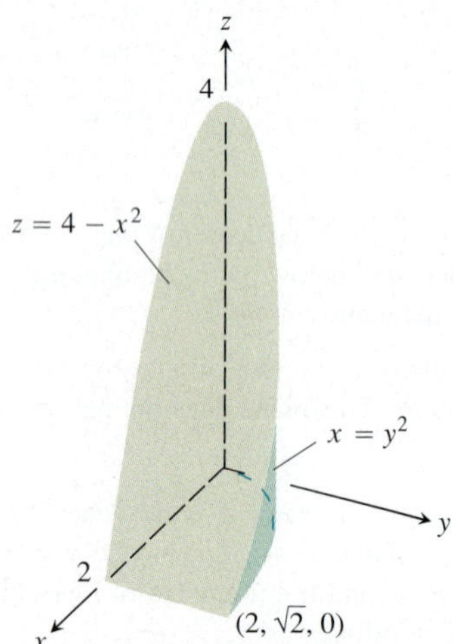

In Exercises 15 and 16, find

(a) The mass of the solid

(b) The center of mass

(c) The moments of inertia about the coordinate axes

(d) The radii of gyration about the coordinate axes.

15. A solid cube in the first octant is bounded by the coordinate planes and by the planes $x = 1$, $y = 1$, and $z = 1$. The density of the cube is $\delta(x, y, z) = x + y + z + 1$.

16. A wedge like the one in Exercise 2 has dimensions $a = 2$, $b = 6$, and $c = 3$. The density is $\delta(x, y, z) = x + 1$. Notice that if the density is constant, the center of mass will be $(0, 0, 0)$.

17. *Mass* Find the mass of the solid bounded by the planes $x + z = 1$, $x - z = -1$, $y = 0$ and the surface $y = \sqrt{z}$. The density of the solid is $\delta(x, y, z) = 2y + 5$.

18. *Mass* Find the mass of the solid region bounded by the parabolic surfaces $z = 16 - 2x^2 - 2y^2$ and $z = 2x^2 + 2y^2$ if the density of the solid is $\delta(x, y, z) = \sqrt{x^2 + y^2}$.

Work

In Exercises 19 and 20, calculate the following.

(a) The amount of work done by (constant) gravity g in moving the liquid filling in the container to the xy-plane. (*Hint*: Partition the liquid into small volume elements ΔV_i and find the work done (approximately) by gravity on each element. Summation and passage to the limit gives a triple integral to evaluate.)

(b) The work done by gravity in moving the center of mass down to the xy-plane.

19. The container is a cubical box in the first octant bounded by the coordinate planes and the planes $x = 1$, $y = 1$, and $z = 1$. The density of the liquid filling the box is $\delta(x, y, z) = x + y + z + 1$ (see Exercise 15).

20. The container is in the shape of the region bounded by $y = 0$, $z = 0$, $z = 4 - x^2$, and $x = y^2$. The density of the liquid filling the region is $\delta\,(x, y, z) = kxy$, k a constant.(see Exercise 14).

The Parallel Axis Theorem

The Parallel Axis Theorem (Exercises 12.2) holds in three dimensions as well as in two. Let $L_{\text{c.m.}}$ be a line through the center of mass of a body of mass m and let L be a parallel line h units away from $L_{\text{c.m.}}$. The **Parallel Axis Theorem** says that the moments of inertia $I_{\text{c.m.}}$ and I_L of the body about $L_{\text{c.m.}}$ and L satisfy the equation

$$I_L = I_{\text{c.m.}} + mh^2. \tag{1}$$

As in the two-dimensional case, the theorem gives a quick way to calculate one moment when the other moment and the mass are known.

21. *Proof of the Parallel Axis Theorem*

(a) Show that the first moment of a body in space about any plane through the body's center of mass is zero. (*Hint*: Place the body's center of mass at the origin and let the plane be the yz-plane. What does the formula $\bar{x} = M_{yz}/M$ then tell you?)

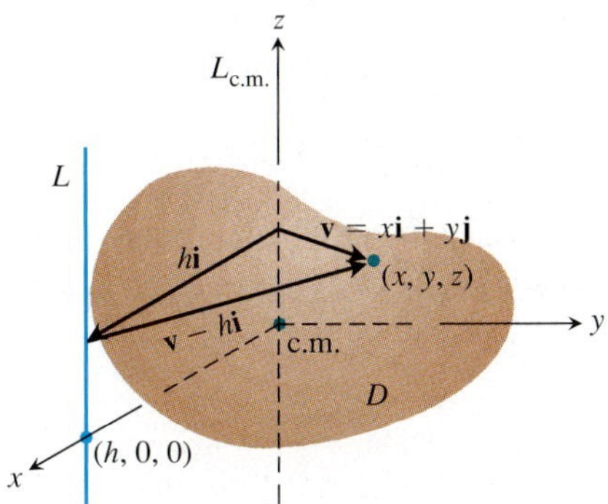

(b) To prove the Parallel Axis Theorem, place the body with its center of mass at the origin, with the line $L_{\text{c.m.}}$ along the z-axis and the line L perpendicular to the xy-plane at the point $(h, 0, 0)$. Let D be the region of space occupied by the body. Then, in the notation of the figure,

$$I_L = \iiint_D |\mathbf{v} - h\mathbf{i}|^2 \, dm. \tag{2}$$

Expand the integrand in this integral and complete the proof.

22. The moment of inertia about a diameter of a solid sphere of constant density and radius a is $(2/5)ma^2$, where m is the mass of the sphere. Find the moment of inertia about a line tangent to the sphere.

23. The moment of inertia of the solid in Exercise 3 about the z-axis is $I_z = abc(a^2 + b^2)/3$.

(a) Use Equation (1) to find the moment of inertia and radius of gyration of the solid about the line parallel to the z-axis through the solid's center of mass.

(b) Use Equation (1) and the result in part (a) to find the moment of inertia and radius of gyration of the solid about the line $x = 0, y = 2b$.

24. If $a = b = 6$ and $c = 4$, the moment of inertia of the solid wedge in Exercise 2 about the x-axis is $I_x = 208$. Find the moment of inertia of the wedge about the line $y = 4$, $z = -4/3$ (the edge of the wedge's narrow end).

Pappus's Formula

Pappus's formula (Exercises 12.2) holds in three dimensions as well as in two. Suppose that bodies B_1 and B_2 of mass m_1 and m_2, respectively, occupy nonoverlapping regions in space and that $\mathbf{c}_1$ and $\mathbf{c}_2$ are the vectors from the origin to the bodies' respective centers of mass. Then the center of mass of the union $B_1 \cup B_2$ of the two bodies is determined by the vector

$$\mathbf{c} = \frac{m_1\mathbf{c}_1 + m_2\mathbf{c}_2}{m_1 + m_2}. \tag{3}$$

As before, this formula is called **Pappus's formula.** As in the two-dimensional case, the formula generalizes to

$$\mathbf{c} = \frac{m_1\mathbf{c}_1 + m_2\mathbf{c}_2 + \cdots + m_n\mathbf{c}_n}{m_1 + m_2 + \cdots + m_n} \tag{4}$$

for n bodies.

25. Derive Pappus's formula (Equation 3). (*Hint*: Sketch B_1 and B_2 as nonoverlapping regions in the first octant and label their centers of mass $(\bar{x}_1, \bar{y}_1, \bar{z}_1)$ and $(\bar{x}_2, \bar{y}_2, \bar{z}_2)$. Express the moments of $B_1 \cup B_2$ about the coordinate planes in terms of the masses m_1 and m_2 and the coordinates of these centers.)

26. The accompanying figure shows a solid made from three rectangular solids of constant density $\delta = 1$. Use Pappus's formula to find the center of mass of

(a) $A \cup B$ **(b)** $A \cup C$

(c) $B \cup C$ **(d)** $A \cup B \cup C$.

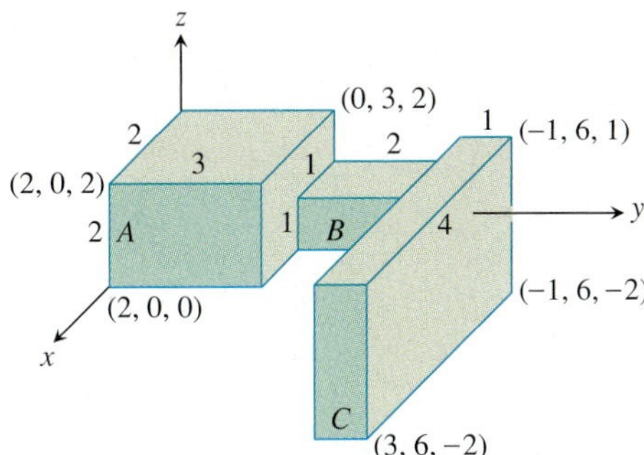

27. (a) Suppose that a solid right circular cone C of base radius a and altitude h is constructed on the circular base of a solid hemisphere S of radius a so that the union of the two solids resembles an ice cream cone. The centroid of a solid cone lies one-fourth of the way from the base toward the vertex. The centroid of a solid hemisphere lies three-eighths of the way from the base to the top. What relation must hold between h and a to place the centroid of $C \cup S$ in the common base of the two solids?

(b) If you have not already done so, answer the analogous question about a triangle and a semicircle (Section 12.2, Exercise 55). The answers are not the same.

28. A solid pyramid P with height h and four congruent sides is built with its base as one face of a solid cube C whose edges have length s. The centroid of a solid pyramid lies one-fourth of the way from the base toward the vertex. What relation must hold between h and s to place the centroid of $P \cup C$ in the base of the pyramid? Compare your answer with the answer to Exercise 27. Also compare it with the answer to Exercise 56 in Section 12.2.

12.6 Triple Integrals in Cylindrical and Spherical Coordinates

Integration in Cylindrical Coordinates • Spherical Coordinates • Integration in Spherical Coordinates

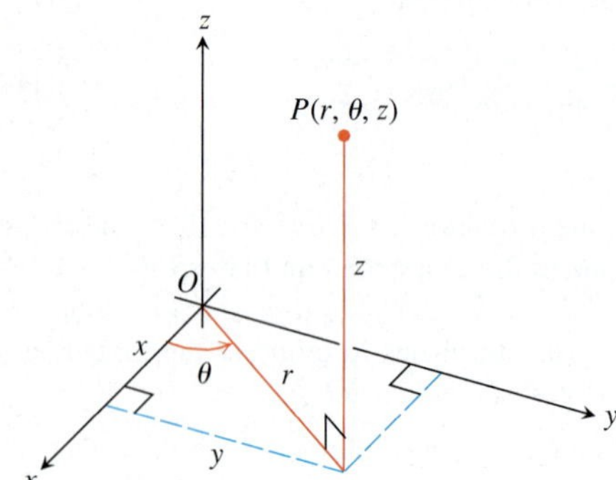

FIGURE 12.41 The cylindrical coordinates of a point in space are r, θ, and z.

When a calculation in physics, engineering, or geometry involves a cylinder, cone, or sphere, we can often simplify our work by using cylindrical or spherical coordinates.

Integration in Cylindrical Coordinates

We obtain cylindrical coordinates for space by combining polar coordinates in the xy-plane with the usual z-axis. This assigns to every point in space one or more coordinate triples of the form (r, θ, z), as shown in Figure 12.41.

Definition Cylindrical Coordinates

Cylindrical coordinates represent a point P in space by ordered triples (r, θ, z) in which

1. r and θ are polar coordinates for the vertical projection of P on the xy-plane
2. z is the rectangular vertical coordinate.

The values of $x, y, r,$ and θ in rectangular and cylindrical coordinates are related by the usual equations.

Equations Relating Rectangular (x, y, z) and Cylindrical (r, θ, z) Coordinates

$$x = r\cos\theta, \qquad y = r\sin\theta, \qquad z = z,$$
$$r^2 = x^2 + y^2, \qquad \tan\theta = y/x$$

In cylindrical coordinates, the equation $r = a$ describes not just a circle in the xy-plane but an entire cylinder about the z-axis (Figure 12.42). The z-axis is given by $r = 0$. The equation $\theta = \theta_0$ describes the plane that contains the z-axis and makes an angle θ_0 with the positive x-axis. And, just as in rectangular coordinates, the equation $z = z_0$ describes a plane perpendicular to the z-axis.

Cylindrical coordinates are good for describing cylinders whose axes run along the z-axis and planes that either contain the z-axis or lie perpendicular to the z-axis. Surfaces like these have equations of constant coordinate value:

$$r = 4 \qquad \text{Cylinder, radius 4, axis the } z\text{-axis}$$

$$\theta = \frac{\pi}{3} \qquad \text{Plane containing the } z\text{-axis}$$

$$z = 2. \qquad \text{Plane perpendicular to the } z\text{-axis}$$

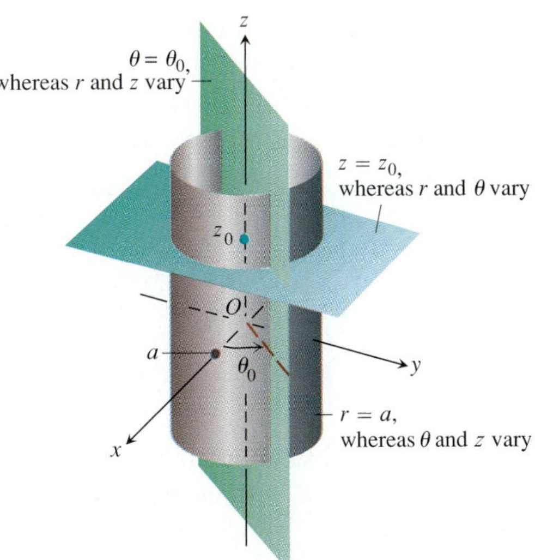

FIGURE 12.42 Constant-coordinate equations in cylindrical coordinates yield cylinders and planes.

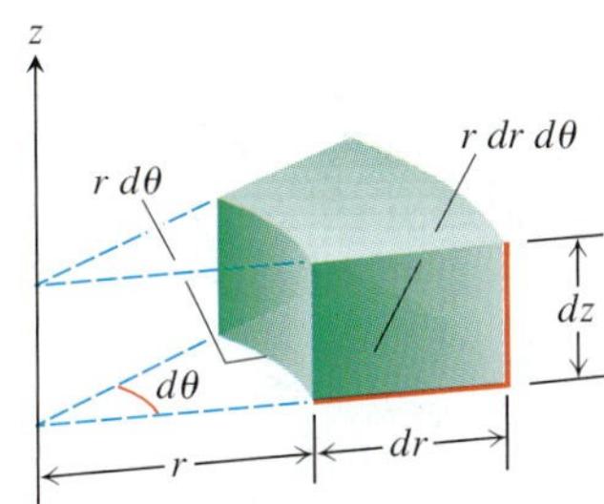

FIGURE 12.43 The volume element in cylindrical coordinates is $dV = dz\,r\,dr\,d\theta$.

The volume element for subdividing a region in space with cylindrical coordinates is

$$dV = dz\,r\,dr\,d\theta \tag{1}$$

(Figure 12.43). Triple integrals in cylindrical coordinates are then evaluated as iterated integrals, as in the following example.

Example 1 Finding Limits of Integration in Cylindrical Coordinates

Find the limits of integration in cylindrical coordinates for integrating a function $f(r, \theta, z)$ over the region D bounded below by the plane $z = 0$, laterally by the circular cylinder $x^2 + (y - 1)^2 = 1$, and above by the paraboloid $z = x^2 + y^2$.

Solution

Step 1: *A sketch* (Figure 12.44). The base of D is also the region's projection R on the xy-plane. The boundary of R is the circle $x^2 + (y - 1)^2 = 1$. Its polar coordinate equation is

$$\begin{aligned} x^2 + (y-1)^2 &= 1 \\ x^2 + y^2 - 2y + 1 &= 1 \\ r^2 - 2r\sin\theta &= 0 \\ r &= 2\sin\theta. \end{aligned}$$

Step 2: *The z-limits of integration.* A line M through a typical point (r, θ) in R parallel to the z-axis enters D at $z = 0$ and leaves at $z = x^2 + y^2 = r^2$.

Step 3: *The r-limits of integration.* A ray L through (r, θ) from the origin enters R at $r = 0$ and leaves at $r = 2\sin\theta$.

Step 4: *The θ-limits of integration.* As L sweeps across R, the angle θ it makes with the positive x-axis runs from $\theta = 0$ to $\theta = \pi$. The integral is

$$\iiint_D f(r, \theta, z)\,dV = \int_0^{\pi}\int_0^{2\sin\theta}\int_0^{r^2} f(r, \theta, z)\,dz\,r\,dr\,d\theta.$$

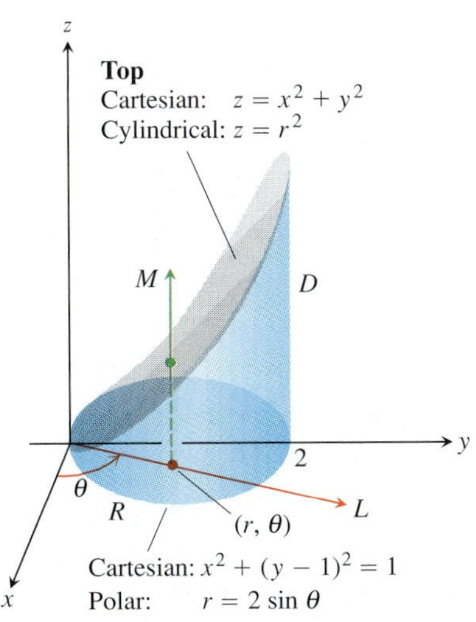

FIGURE 12.44 The figure for Example 1.

Example 1 illustrates a good procedure for finding limits of integration in cylindrical coordinates. The procedure is summarized in the following box.

How to Integrate in Cylindrical Coordinates

To evaluate

$$\iiint_D f(r, \theta, z)\, dV$$

over a region D in space in cylindrical coordinates, integrating first with respect to z, then with respect to r, and finally with respect to θ, take the following steps.

Step 1: *A sketch.* Sketch the region D along with its projection R on the xy-plane. Label the surfaces and curves that bound D and R.

Step 2: *The z-limits of integration.* Draw a line M through a typical point (r, θ) of R parallel to the z-axis. As z increases, M enters D at $z = g_1(r, \theta)$ and leaves at $z = g_2(r, \theta)$. These are the z-limits of integration.

Step 3: *The r-limits of integration.* Draw a ray L through (r, θ) from the origin. The ray enters R at $r = h_1(\theta)$ and leaves at $r = h_2(\theta)$. These are the r-limits of integration.

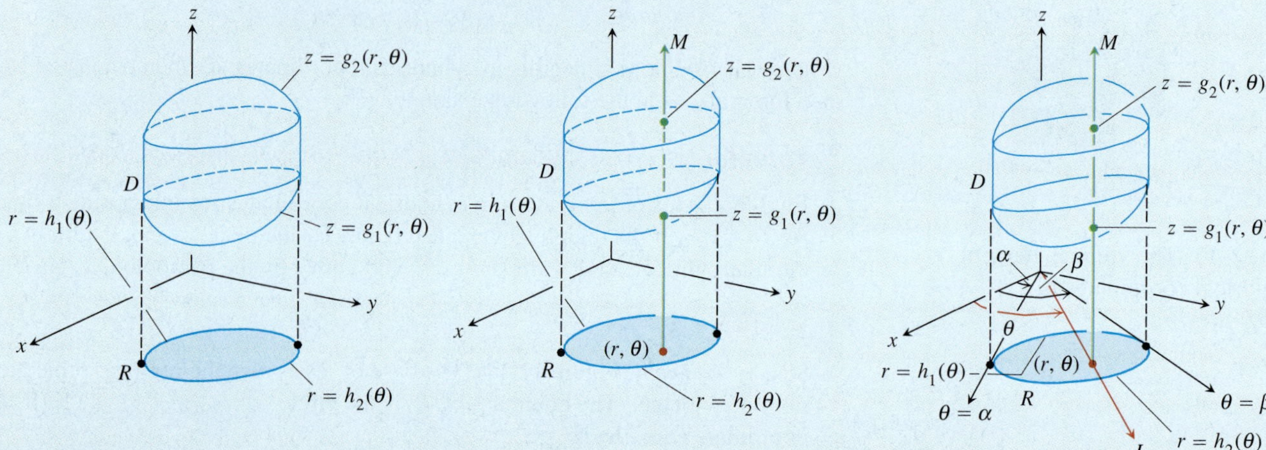

Step 4: *The θ-limits of integration.* As L sweeps across R, the angle θ it makes with the positive x-axis runs from $\theta = \alpha$ to $\theta = \beta$. These are the θ-limits of integration. The integral is

$$\iiint_D f(r, \theta, z)\, dV = \int_{\theta=\alpha}^{\theta=\beta} \int_{r=h_1(\theta)}^{r=h_2(\theta)} \int_{z=g_1(r,\theta)}^{z=g_2(r,\theta)} f(r, \theta, z)\, dz\, r\, dr\, d\theta.$$

Example 2 Finding a Centroid

Find the centroid $(\delta = 1)$ of the solid enclosed by the cylinder $x^2 + y^2 = 4$, bounded above by the paraboloid $z = x^2 + y^2$, and bounded below by the xy-plane.

Solution

Step 1: *A sketch.* We sketch the solid, bounded above by the paraboloid $z = r^2$ and below by the plane $z = 0$ (Figure 12.45). Its base R is the disk $|r| \leq 2$ in the xy-plane.

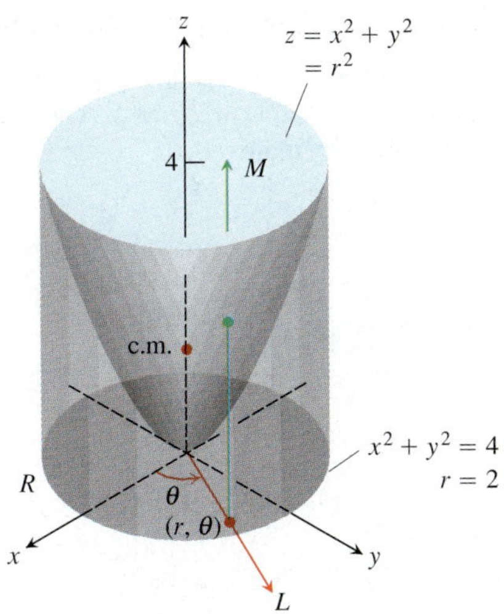

FIGURE 12.45 Example 2 shows how to find the centroid of this solid.

The solid's centroid $(\bar{x}, \bar{y}, \bar{z})$ lies on its axis of symmetry, here the z-axis. This makes $\bar{x} = \bar{y} = 0$. To find $\bar{z}$, we divide the first moment M_{xy} by the mass M.

To find the limits of integration for the mass and moment integrals, we continue with the four basic steps. We completed step 1 with our initial sketch. The remaining steps give the limits of integration.

Step 2: *The z-limits.* A line M through a typical point (r, θ) in the base parallel to the z-axis enters the solid at $z = 0$ and leaves at $z = r^2$.

Step 3: *The r-limits.* A ray L through (r, θ) from the origin enters R at $r = 0$ and leaves at $r = 2$.

Step 4: *The θ-limits.* As L sweeps over the base like a clock hand, the angle θ it makes with the positive x-axis runs from $\theta = 0$ to $\theta = 2\pi$. The value of M_{xy} is

$$M_{xy} = \int_0^{2\pi}\int_0^2\int_0^{r^2} z\,dz\,r\,dr\,d\theta = \int_0^{2\pi}\int_0^2 \left[\frac{z^2}{2}\right]_0^{r^2} r\,dr\,d\theta$$

$$= \int_0^{2\pi}\int_0^2 \frac{r^5}{2}\,dr\,d\theta = \int_0^{2\pi}\left[\frac{r^6}{12}\right]_0^2 d\theta = \int_0^{2\pi}\frac{16}{3}\,d\theta = \frac{32\pi}{3}.$$

The value of M is

$$M = \int_0^{2\pi}\int_0^2\int_0^{r^2} dz\,r\,dr\,d\theta = \int_0^{2\pi}\int_0^2 \left[z\right]_0^{r^2} r\,dr\,d\theta$$

$$= \int_0^{2\pi}\int_0^2 r^3\,dr\,d\theta = \int_0^{2\pi}\left[\frac{r^4}{4}\right]_0^2 d\theta = \int_0^{2\pi} 4\,d\theta = 8\pi.$$

Therefore,

$$\bar{z} = \frac{M_{xy}}{M} = \frac{32\pi}{3}\frac{1}{8\pi} = \frac{4}{3},$$

and the centroid is $(0, 0, 4/3)$. Notice that the centroid lies outside the solid.

Spherical Coordinates

Spherical coordinates locate points in space with angles and a distance, as shown in Figure 12.46.

A few books give spherical coordinates in the order (ρ, θ, ϕ), with θ and ϕ reversed. In some cases, you may also find r being used for ρ. Watch out for this when you read elsewhere.

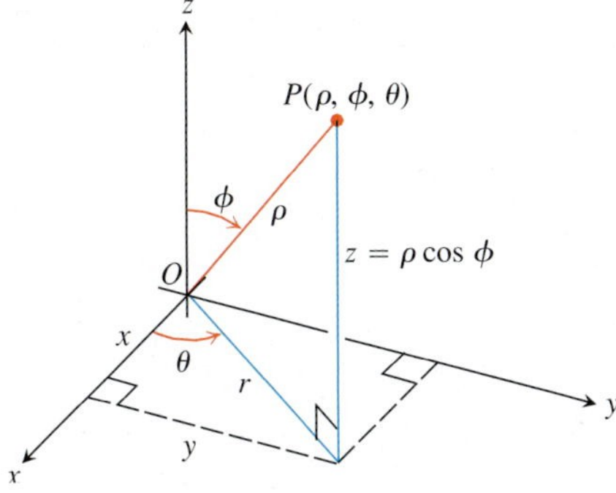

FIGURE 12.46 The spherical coordinates ρ, ϕ, and θ and their relation to x, y, z, and r.

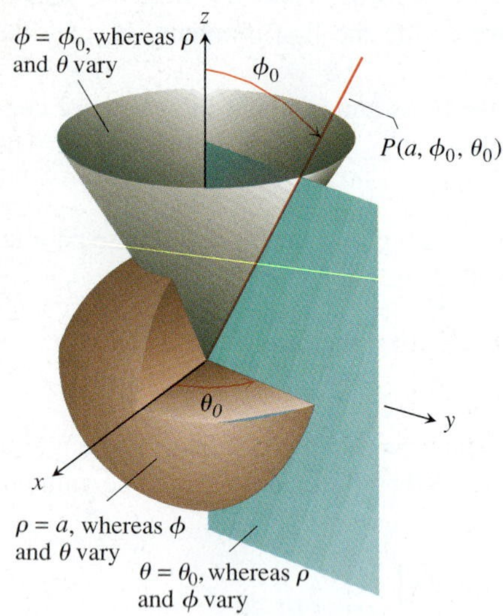

FIGURE 12.47 Constant-coordinate equations in spherical coordinates yield spheres, single cones, and half-planes.

The first coordinate, $\rho = |\overrightarrow{OP}|$, is the point's distance from the origin. Unlike r, *the variable* ρ *is never negative*. The second coordinate, ϕ, is the angle $\overrightarrow{OP}$ makes with the positive z-axis. It is required to lie in the interval $[0, \pi]$. The third coordinate is the angle θ as measured in cylindrical coordinates.

Definition Spherical Coordinates

Spherical coordinates represent a point P in space by ordered triples (ρ, ϕ, θ) in which

1. ρ is the distance from P to the origin
2. ϕ is the angle $\overrightarrow{OP}$ makes with the positive z-axis $(0 \le \phi \le \pi)$
3. θ is the angle from cylindrical coordinates.

The equation $\rho = a$ describes the sphere of radius a centered at the origin (Figure 12.47). The equation $\phi = \phi_0$ describes a single cone whose vertex lies at the origin and whose axis lies along the z-axis. (We broaden our interpretation to include the xy-plane as the cone $\phi = \pi/2$.) If ϕ_0 is greater than $\pi/2$, the cone $\phi = \phi_0$ opens downward. The equation $\theta = \theta_0$ describes the half-plane that contains the z-axis and makes an angle θ_0 with the positive x-axis.

Equations Relating Spherical Coordinates to Cartesian and Cylindrical Coordinates

$$
\begin{aligned}
r &= \rho \sin \phi, \qquad & x &= r \cos \theta = \rho \sin \phi \cos \theta,\\
z &= \rho \cos \phi, \qquad & y &= r \sin \theta = \rho \sin \phi \sin \theta,
\end{aligned}
\tag{3}
$$
$$\rho = \sqrt{x^2 + y^2 + z^2} = \sqrt{r^2 + z^2}.$$

Example 3 Converting Cartesian to Spherical

Find a spherical coordinate equation for the sphere $x^2 + y^2 + (z - 1)^2 = 1$.

Solution We use Equations (3) to substitute for x, y, and z:

$$
\begin{aligned}
x^2 + y^2 + (z-1)^2 &= 1\\
\rho^2 \sin^2 \phi \cos^2 \theta + \rho^2 \sin^2 \phi \sin^2 \theta + (\rho \cos \phi - 1)^2 &= 1 \qquad \text{Eqs. (3)}\\
\rho^2 \sin^2 \phi \underbrace{(\cos^2 \theta + \sin^2 \theta)}_{1} + \rho^2 \cos^2 \phi - 2\rho \cos \phi + 1 &= 1\\
\rho^2 \underbrace{(\sin^2 \phi + \cos^2 \phi)}_{1} &= 2\rho \cos \phi\\
\rho^2 &= 2\rho \cos \phi\\
\rho &= 2 \cos \phi.
\end{aligned}
$$

See Figure 12.48.

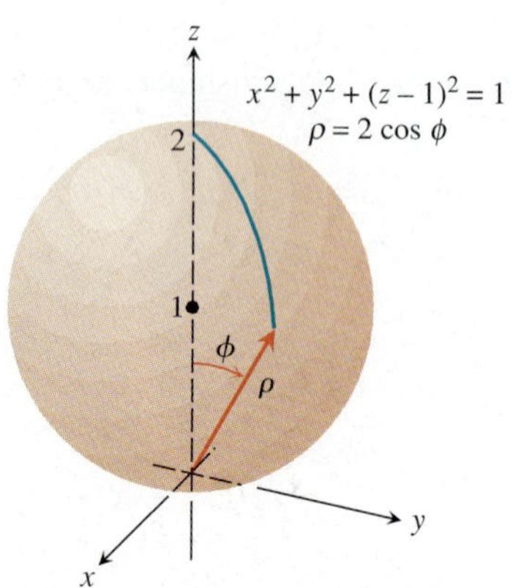

FIGURE 12.48 The sphere in Example 3.

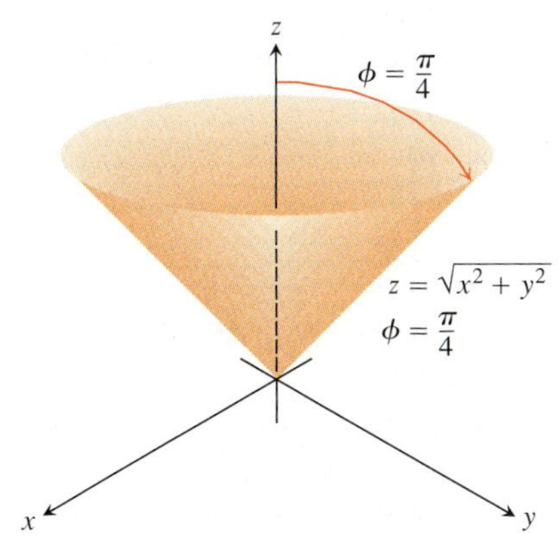

FIGURE 12.49 The cone in Example 4.

Example 4 Converting Cartesian to Spherical

Find a spherical coordinate equation for the cone $z = \sqrt{x^2 + y^2}$ (Figure 12.49).

Solution 1 *Use geometry.* The cone is symmetric with respect to the z-axis and cuts the first quadrant of the yz-plane along the line $z = y$. The angle between the cone and the positive z-axis is therefore $\pi/4$ radians. The cone consists of the points whose spherical coordinates have ϕ equal to $\pi/4$, so its equation is $\phi = \pi/4$.

Solution 2 *Use algebra.* If we use Equations (3) to substitute for x, y, and z we obtain the same result:

$$\begin{aligned} z &= \sqrt{x^2 + y^2} \\ \rho \cos \phi &= \sqrt{\rho^2 \sin^2 \phi} && \text{Example 3} \\ \rho \cos \phi &= \rho \sin \phi && \rho \geq 0,\ \sin \phi \geq 0 \\ \cos \phi &= \sin \phi \\ \phi &= \frac{\pi}{4}. && 0 \leq \phi \leq \pi \end{aligned}$$

Integration in Spherical Coordinates

Spherical coordinates are good for describing spheres centered at the origin, half-planes hinged along the z-axis, and single-napped cones whose vertices lie at the origin and whose axes lie along the z-axis. Surfaces like these have equations of constant coordinate value:

$$\begin{aligned} \rho &= 4 && \text{Sphere, radius 4, center at origin} \\ \phi &= \frac{\pi}{3} && \text{Cone opening up from the origin, making an angle of } \pi/3 \text{ radians with the positive } z\text{-axis} \\ \theta &= \frac{\pi}{3}. && \text{Half-plane, hinged along the } z\text{-axis, making an angle of } \pi/3 \text{ radians with the positive } x\text{-axis} \end{aligned}$$

The volume element in spherical coordinates is the volume of a **spherical wedge** defined by the differentials $d\rho$, $d\phi$, and $d\theta$ (Figure 12.50). The wedge is approx-

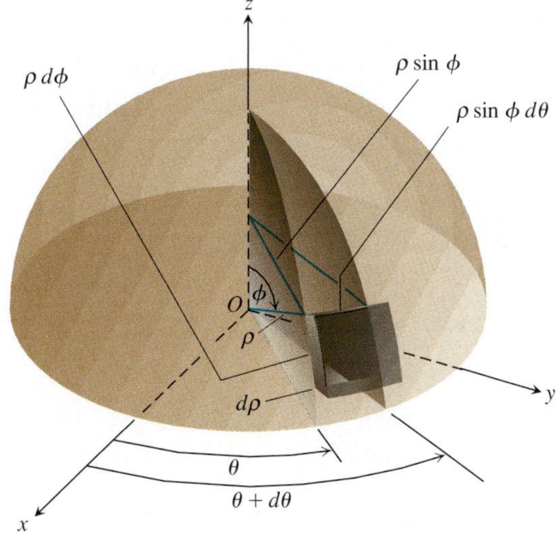

FIGURE 12.50 The volume element in spherical coordinates is

$$\begin{aligned} dV &= d\rho \cdot \rho\, d\phi \cdot \rho \sin \phi\, d\theta \\ &= \rho^2 \sin \phi\, d\rho\, d\phi\, d\theta. \end{aligned}$$

imately a rectangular box with one side a circular arc of length $\rho\, d\phi$, another side a circular arc of length $\rho \sin\phi\, d\theta$, and thickness $d\rho$. Therefore, the volume element in spherical coordinates is

$$dV = \rho^2 \sin\phi\, d\rho\, d\phi\, d\theta, \tag{4}$$

and triple integrals take the form

$$\iiint F(\rho, \phi, \theta)\, dV = \iiint F(\rho, \phi, \theta)\rho^2 \sin\phi\, d\rho\, d\phi\, d\theta. \tag{5}$$

To evaluate these integrals, we usually integrate first with respect to ρ. The procedure for finding the limits of integration is shown in the following box. We restrict our attention to integrating over domains that are solids of revolution about the z-axis (or portions thereof) and for which the limits for θ and ϕ are constant.

How to Integrate in Spherical Coordinates

To evaluate

$$\iiint_D f(\rho, \phi, \theta)\, dV$$

over a region D in space in spherical coordinates, integrating first with respect to ρ, then with respect to ϕ, and finally with respect to θ, take the following steps.

Step 1: *A sketch.* Sketch the region D along with its projection R on the xy-plane. Label the surfaces that bound D.

Step 2: *The ρ-limits of integration.* Draw a ray M from the origin through D making an angle ϕ with the positive z-axis. Also draw the projection of M on the xy-plane (call the projection L). The ray L makes an angle θ with the positive x-axis. As ρ increases, M enters D at $\rho = g_1(\phi, \theta)$ and leaves at $\rho = g_2(\phi, \theta)$. These are the ρ-limits of integration.

Step 3: *The ϕ-limits of integration.* For any given θ, the angle ϕ that M makes with the z-axis runs from $\phi = \phi_{\min}$ to $\phi = \phi_{\max}$. These are the ϕ-limits of integration.

Step 4: *The θ-limits of integration.* The ray L sweeps over R as θ runs from α to β. These are the θ-limits of integration. The integral is

$$\iiint_D f(\rho, \phi, \theta)\, dV = \int_{\theta=\alpha}^{\theta=\beta} \int_{\phi=\phi_{\min}}^{\phi=\phi_{\max}} \int_{\rho=g_1(\phi,\theta)}^{\rho=g_2(\phi,\theta)} f(\rho, \phi, \theta)\rho^2 \sin\phi\, d\rho\, d\phi\, d\theta. \tag{6}$$

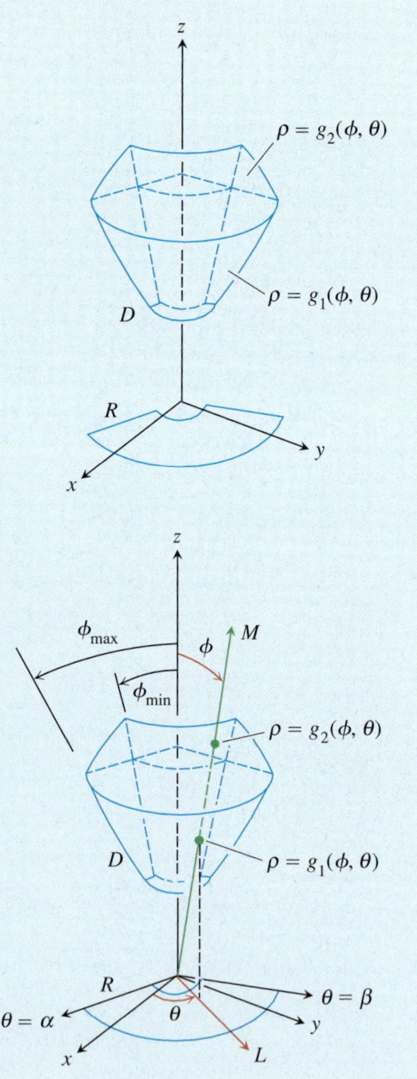

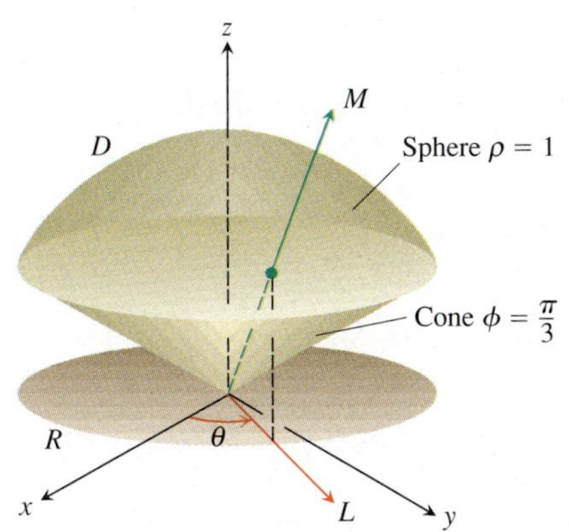

FIGURE 12.51 The ice cream cone in Example 5.

Example 5 Finding a Volume in Spherical Coordinates

Find the volume of the "ice cream cone" D cut from the solid sphere $\rho \le 1$ by the cone $\phi = \pi/3$.

Solution The volume is $V = \iiint_D \rho^2 \sin\phi \, d\rho \, d\phi \, d\theta$, the integral of $f(\rho, \phi, \theta) = 1$ over D.

To find the limits of integration for evaluating the integral, we take the following steps.

Step 1: *A sketch.* We sketch D and its projection R on the xy-plane (Figure 12.51).

Step 2: *The ρ-limits of integration.* We draw a ray M from the origin through D making an angle ϕ with the positive z-axis. We also draw L, the projection of M on the xy-plane, along with the angle θ that L makes with the positive x-axis. Ray M enters D at $\rho = 0$ and leaves at $\rho = 1$.

Step 3: *The ϕ-limits of integration.* The cone $\phi = \pi/3$ makes an angle of $\pi/3$ with the positive z-axis. For any given θ, the angle ϕ can run from $\phi = 0$ to $\phi = \pi/3$.

Step 4: *The θ-limits of integration.* The ray L sweeps over R as θ runs from 0 to 2π. The volume is

$$\begin{aligned} V &= \iiint_D \rho^2 \sin\phi \, d\rho \, d\phi \, d\theta = \int_0^{2\pi} \int_0^{\pi/3} \int_0^1 \rho^2 \sin\phi \, d\rho \, d\phi \, d\theta \\ &= \int_0^{2\pi} \int_0^{\pi/3} \left[\frac{\rho^3}{3} \right]_0^1 \sin\phi \, d\phi \, d\theta = \int_0^{2\pi} \int_0^{\pi/3} \frac{1}{3} \sin\phi \, d\phi \, d\theta \\ &= \int_0^{2\pi} \left[-\frac{1}{3} \cos\phi \right]_0^{\pi/3} d\theta = \int_0^{2\pi} \left(-\frac{1}{6} + \frac{1}{3} \right) d\theta = \frac{1}{6}(2\pi) = \frac{\pi}{3}. \end{aligned}$$

Example 6 Finding a Moment of Inertia

A solid of constant density $\delta = 1$ occupies the region D in Example 5. Find the solid's moment of inertia about the z-axis.

Solution In rectangular coordinates, the moment is

$$I_z = \iiint (x^2 + y^2) \, dV.$$

In spherical coordinates, $x^2 + y^2 = (\rho \sin\phi \cos\theta)^2 + (\rho \sin\phi \sin\theta)^2 = \rho^2 \sin^2\phi$. Hence,

$$I_z = \iiint (\rho^2 \sin^2\phi) \rho^2 \sin\phi \, d\rho \, d\phi \, d\theta = \iiint \rho^4 \sin^3\phi \, d\rho \, d\phi \, d\theta.$$

For the region in Example 5, this becomes

$$I_z = \int_0^{2\pi}\int_0^{\pi/3}\int_0^1 \rho^4 \sin^3\phi\, d\rho\, d\phi\, d\theta = \int_0^{2\pi}\int_0^{\pi/3}\left[\frac{\rho^5}{5}\right]_0^1 \sin^3\phi\, d\phi\, d\theta$$

$$= \frac{1}{5}\int_0^{2\pi}\int_0^{\pi/3}(1 - \cos^2\phi)\sin\phi\, d\phi\, d\theta = \frac{1}{5}\int_0^{2\pi}\left[-\cos\phi + \frac{\cos^3\phi}{3}\right]_0^{\pi/3} d\theta$$

$$= \frac{1}{5}\int_0^{2\pi}\left(-\frac{1}{2} + 1 + \frac{1}{24} - \frac{1}{3}\right) d\theta = \frac{1}{5}\int_0^{2\pi}\frac{5}{24}\, d\theta = \frac{1}{24}(2\pi) = \frac{\pi}{12}.$$

Coordinate Conversion Formulas

Cylindrical to Rectangular	Spherical to Rectangular	Spherical to Cylindrical
$x = r\cos\theta$	$x = \rho\sin\phi\cos\theta$	$r = \rho\sin\phi$
$y = r\sin\theta$	$y = \rho\sin\phi\sin\theta$	$z = \rho\cos\phi$
$z = z$	$z = \rho\cos\phi$	$\theta = \theta$

Corresponding volume elements

$$\begin{aligned} dV &= dx\, dy\, dz \\ &= dz\, r\, dr\, d\theta \\ &= \rho^2 \sin\phi\, d\rho\, d\phi\, d\theta \end{aligned}$$

EXERCISES 12.6

Evaluating Integrals in Cylindrical Coordinates

Evaluate the cylindrical coordinate integrals in Exercises 1–6.

1. $\displaystyle\int_0^{2\pi}\int_0^1\int_r^{\sqrt{2-r^2}} dz\, r\, dr\, d\theta$

2. $\displaystyle\int_0^{2\pi}\int_0^3\int_{r^2/3}^{\sqrt{18-r^2}} dz\, r\, dr\, d\theta$

3. $\displaystyle\int_0^{2\pi}\int_0^{\theta/2\pi}\int_0^{3+24r^2} dz\, r\, dr\, d\theta$

4. $\displaystyle\int_0^{\pi}\int_0^{\theta/\pi}\int_{-\sqrt{4-r^2}}^{3\sqrt{4-r^2}} z\, dz\, r\, dr\, d\theta$

5. $\displaystyle\int_0^{2\pi}\int_0^1\int_r^{1/\sqrt{2-r^2}} 3\, dz\, r\, dr\, d\theta$

6. $\displaystyle\int_0^{2\pi}\int_0^1\int_{-1/2}^{1/2} (r^2\sin^2\theta + z^2)\, dz\, r\, dr\, d\theta$

Changing Order of Integration in Cylindrical Coordinates

The integrals we have seen so far suggest that there are preferred orders of integration for cylindrical coordinates, but other orders usually work well and are occasionally easier to evaluate. Evaluate the integrals in Exercises 7–10.

7. $\displaystyle\int_0^{2\pi}\int_0^3\int_0^{z/3} r^3\, dr\, dz\, d\theta$

8. $\displaystyle\int_{-1}^1\int_0^{2\pi}\int_0^{1+\cos\theta} 4r\, dr\, d\theta\, dz$

9. $\displaystyle\int_0^1\int_0^{\sqrt{z}}\int_0^{2\pi} (r^2\cos^2\theta + z^2)\, r\, d\theta\, dr\, dz$

10. $\displaystyle\int_0^2\int_{r-2}^{\sqrt{4-r^2}}\int_0^{2\pi} (r\sin\theta + 1)\, r\, d\theta\, dz\, dr$

11. Let D be the region bounded below by the plane $z = 0$, above by the sphere $x^2 + y^2 + z^2 = 4$, and on the sides by the cylinder $x^2 + y^2 = 1$. Set up the triple integrals in cylindrical coordinates that give the volume of D using the following orders of integration.

(a) $dz\, dr\, d\theta$

(b) $dr\, dz\, d\theta$

(c) $d\theta\, dz\, dr$

12. Let D be the region bounded below by the cone $z = \sqrt{x^2 + y^2}$ and above by the paraboloid $z = 2 - x^2 - y^2$. Set up the triple integrals in cylindrical coordinates that give the volume of D using the following orders of integration.

(a) $dz\, dr\, d\theta$

(b) $dr\, dz\, d\theta$

(c) $d\theta\, dz\, dr$

13. Give the limits of integration for evaluating the integral

$$\iiint f(r, \theta, z)\, dz\, r\, dr\, d\theta$$

as an iterated integral over the region that is bounded below by the plane $z = 0$, on the side by the cylinder $r = \cos\theta$, and on top by the paraboloid $z = 3r^2$.

14. Convert the integral

$$\int_{-1}^{1}\int_{0}^{\sqrt{1-y^2}}\int_{0}^{x} (x^2 + y^2)\, dz\, dx\, dy$$

to an equivalent integral in cylindrical coordinates and evaluate the result.

Finding Iterated Integrals in Cylindrical Coordinates

In Exercises 15–20, set up the iterated integral for evaluating $\iiint_D f(r, \theta, z)\, dz\, r\, dr\, d\theta$ over the given region D.

15. D is the right circular cylinder whose base is the circle $r = 2\sin\theta$ in the xy-plane and whose top lies in the plane $z = 4 - y$.

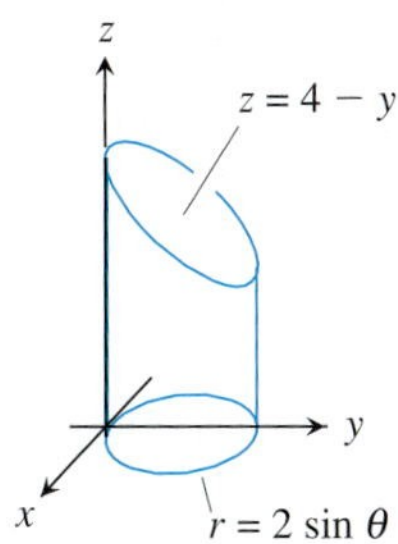

16. D is the right circular cylinder whose base is the circle $r = 3\cos\theta$ and whose top lies in the plane $z = 5 - x$.

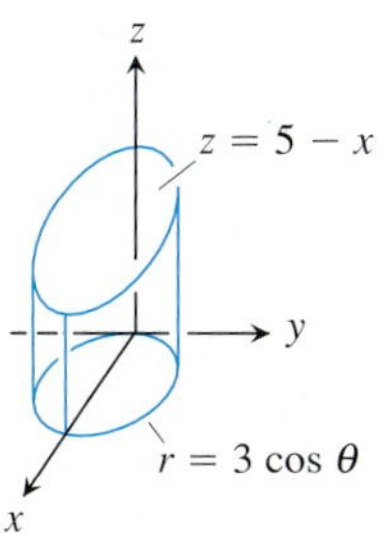

17. D is the solid right cylinder whose base is the region in the xy-plane that lies inside the cardioid $r = 1 + \cos\theta$ and outside the circle $r = 1$ and whose top lies in the plane $z = 4$.

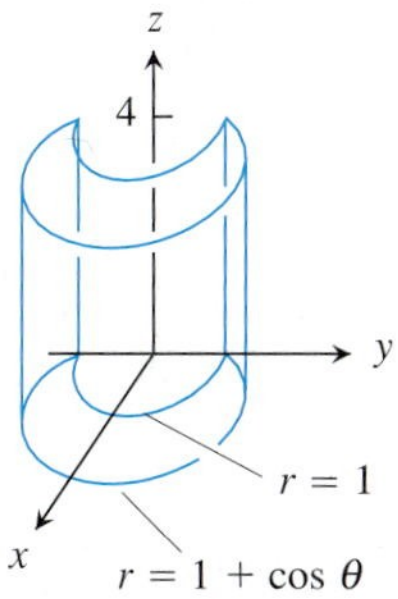

18. D is the solid right cylinder whose base is the region between the circles $r = \cos\theta$ and $r = 2\cos\theta$ and whose top lies in the plane $z = 3 - y$.

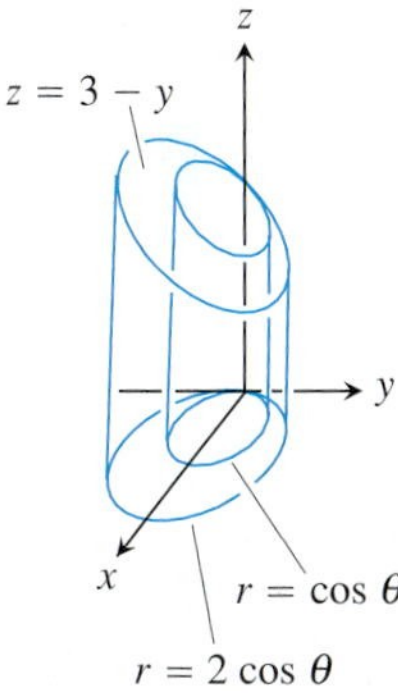

19. D is the prism whose base is the triangle in the xy-plane bounded by the x-axis and the lines $y = x$ and $x = 1$ and whose top lies in the plane $z = 2 - y$.

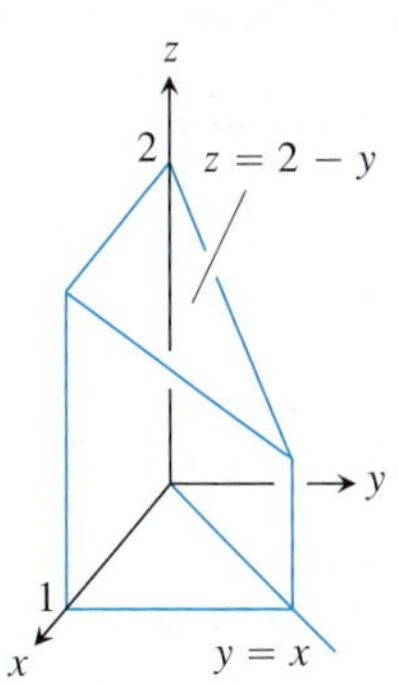

20. D is the prism whose base is the triangle in the xy-plane bounded by the y-axis and the lines $y = x$ and $y = 1$ and whose top lies in the plane $z = 2 - x$.

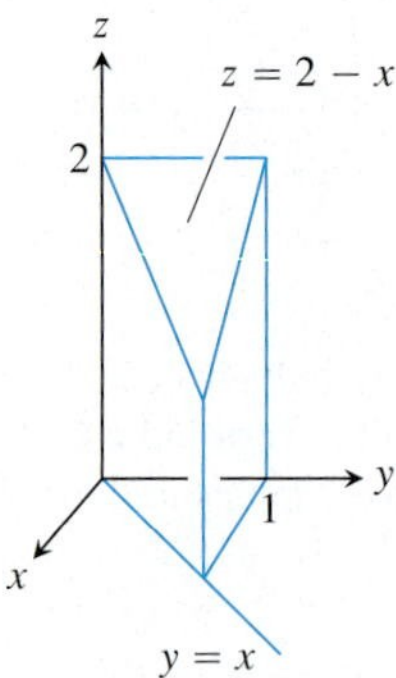

Evaluating Integrals in Spherical Coordinates

Evaluate the spherical coordinate integrals in Exercises 21–26.

21. $\int_0^{\pi}\int_0^{\pi}\int_0^{2\sin\phi} \rho^2 \sin\phi \, d\rho \, d\phi \, d\theta$

22. $\int_0^{2\pi}\int_0^{\pi/4}\int_0^{2} (\rho\cos\phi)\rho^2 \sin\phi \, d\rho \, d\phi \, d\theta$

23. $\int_0^{2\pi}\int_0^{\pi}\int_0^{(1-\cos\phi)/2} \rho^2 \sin\phi \, d\rho \, d\phi \, d\theta$

24. $\int_0^{3\pi/2}\int_0^{\pi}\int_0^{1} 5\rho^3 \sin^3\phi \, d\rho \, d\phi \, d\theta$

25. $\int_0^{2\pi}\int_0^{\pi/3}\int_{\sec\phi}^{2} 3\rho^2 \sin\phi \, d\rho \, d\phi \, d\theta$

26. $\int_0^{2\pi}\int_0^{\pi/4}\int_0^{\sec\phi} (\rho\cos\phi)\rho^2 \sin\phi \, d\rho \, d\phi \, d\theta$

Changing Order of Integration in Spherical Coordinates

The previous integrals suggest there are preferred orders of integration for spherical coordinates, but other orders are possible and occasionally easier to evaluate. Evaluate the integrals in Exercises 27–30.

27. $\int_0^{2}\int_{-\pi}^{0}\int_{\pi/4}^{\pi/2} \rho^3 \sin 2\phi \, d\phi \, d\theta \, d\rho$

28. $\int_{\pi/6}^{\pi/3}\int_{\csc\phi}^{2\csc\phi}\int_0^{2\pi} \rho^2 \sin\phi \, d\theta \, d\rho \, d\phi$

29. $\int_0^{1}\int_0^{\pi}\int_0^{\pi/4} 12\rho \sin^3\phi \, d\phi \, d\theta \, d\rho$

30. $\int_{\pi/6}^{\pi/2}\int_{-\pi/2}^{\pi/2}\int_{\csc\phi}^{2} 5\rho^4 \sin^3\phi \, d\rho \, d\theta \, d\phi$

31. Let D be the region in Exercise 11. Set up the triple integrals in spherical coordinates that give the volume of D using the following orders of integration.

(a) $d\rho \, d\phi \, d\theta$ **(b)** $d\phi \, d\rho \, d\theta$

32. Let D be the region bounded below by the cone $z = \sqrt{x^2 + y^2}$ and above by the plane $z = 1$. Set up the triple integrals in spherical coordinates that give the volume of D using the following orders of integration.

(a) $d\rho \, d\phi \, d\theta$ **(b)** $d\phi \, d\rho \, d\theta$

Finding Iterated Integrals in Spherical Coordinates

In Exercises 33–38, (a) find the spherical coordinate limits for the integral that calculates the volume of the given solid and (b) then evaluate the integral.

33. The solid between the sphere $\rho = \cos\phi$ and the hemisphere $\rho = 2$, $z \geq 0$

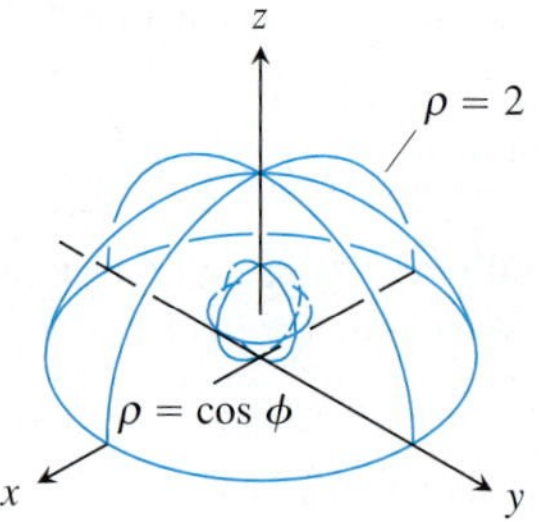

34. The solid bounded below by the hemisphere $\rho = 1$, $z \geq 0$, and above by the cardioid of revolution $\rho = 1 + \cos\phi$

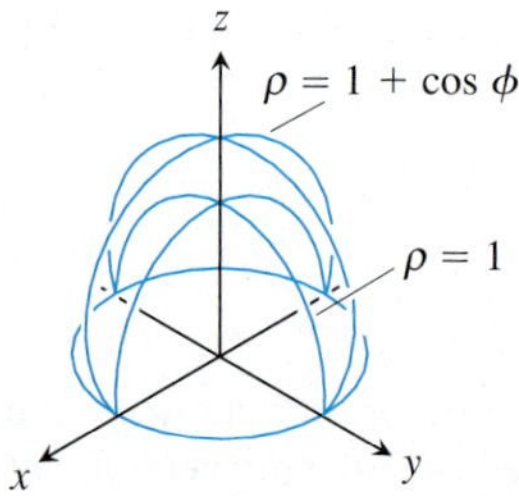

35. The solid enclosed by the cardioid of revolution $\rho = 1 - \cos\phi$

36. The upper portion cut from the solid in Exercise 35 by the xy-plane

37. The solid bounded below by the sphere $\rho = 2\cos\phi$ and above by the cone $z = \sqrt{x^2 + y^2}$

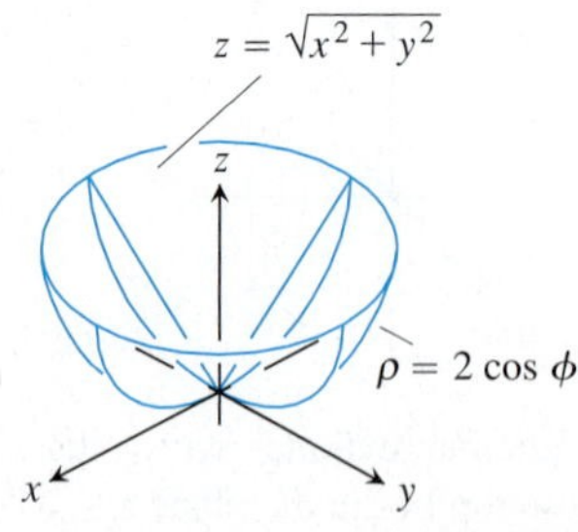

38. The solid bounded below by the xy-plane, on the sides by the sphere $\rho = 2$, and above by the cone $\phi = \pi/3$

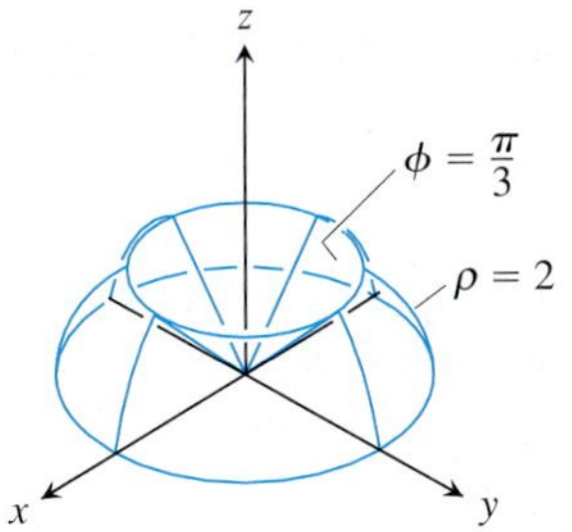

Rectangular, Cylindrical, and Spherical Coordinates

39. Set up triple integrals for the volume of the sphere $\rho = 2$ in (a) spherical, (b) cylindrical, and (c) rectangular coordinates.

40. Let D be the region in the first octant that is bounded below by the cone $\phi = \pi/4$ and above by the sphere $\rho = 3$. Express the volume of D as an iterated triple integral in (a) cylindrical and (b) spherical coordinates. Then (c) find V.

41. Let D be the smaller cap cut from a solid ball of radius 2 units by a plane 1 unit from the center of the sphere. Express the volume of D as an iterated triple integral in (a) spherical, (b) cylindrical, and (c) rectangular coordinates. Then (d) find the volume by evaluating one of the three triple integrals.

42. Express the moment of inertia I_z of the solid hemisphere $x^2 + y^2 + z^2 \leq 1$, $z \geq 0$, as an iterated integral in (a) cylindrical and (b) spherical coordinates. Then (c) find I_z.

Volumes

Find the volumes of the solids in Exercises 43–48.

43.

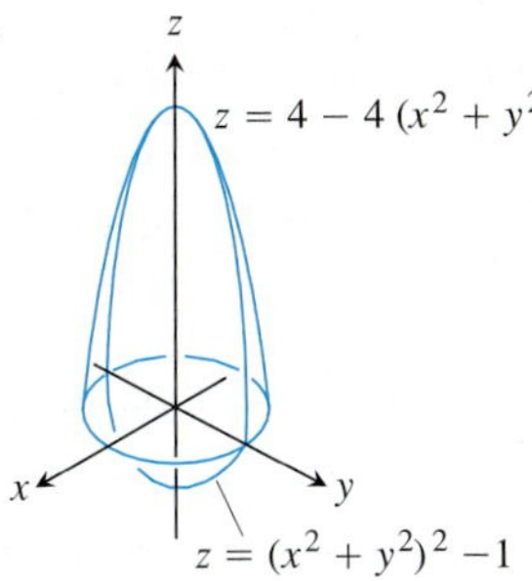

44.

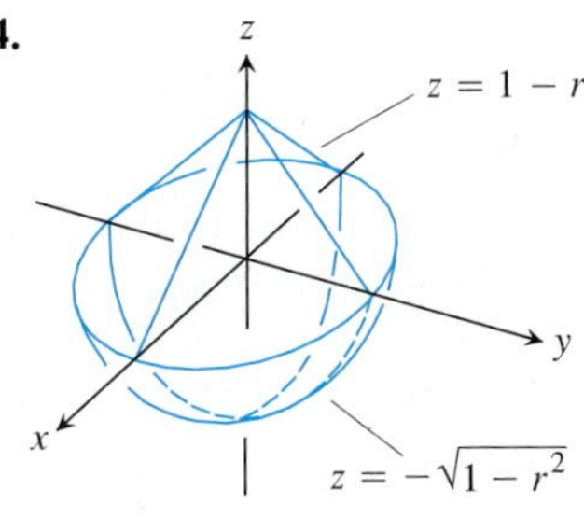

45.

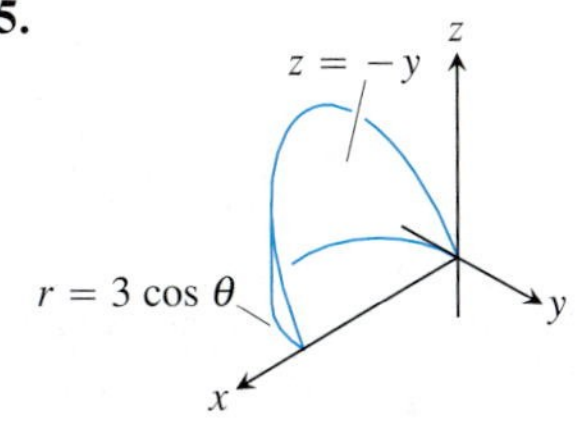

46.

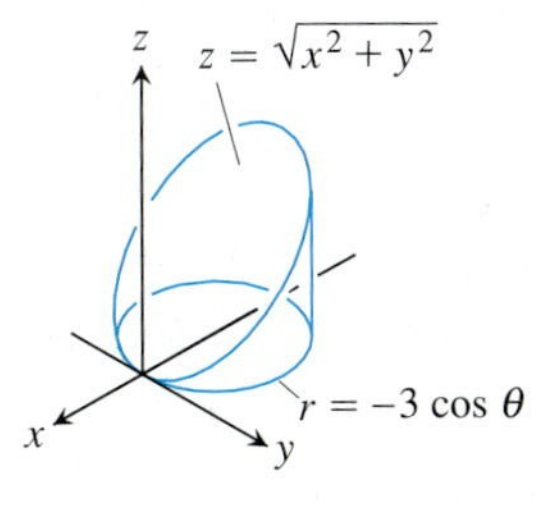

47.

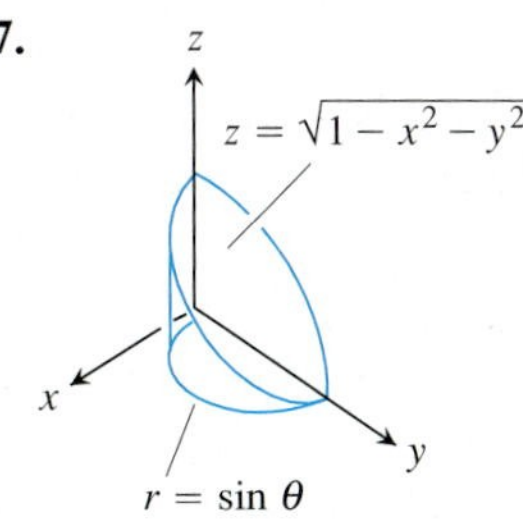

48.

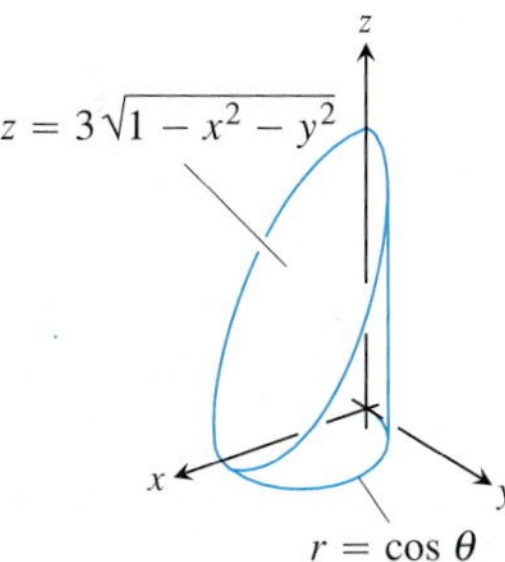

49. *Sphere and cones* Find the volume of the portion of the solid sphere $\rho \leq a$ that lies between the cones $\phi = \pi/3$ and $\phi = 2\pi/3$.

50. *Sphere and half-planes* Find the volume of the region cut from the solid sphere $\rho \leq a$ by the half-planes $\theta = 0$ and $\theta = \pi/6$ in the first octant.

51. *Sphere and plane* Find the volume of the smaller region cut from the solid sphere $\rho \leq 2$ by the plane $z = 1$.

52. *Cone and planes* Find the volume of the solid enclosed by the cone $z = \sqrt{x^2 + y^2}$ between the planes $z = 1$ and $z = 2$.

53. *Cylinder and paraboloid* Find the volume of the region bounded below by the plane $z = 0$, laterally by the cylinder $x^2 + y^2 = 1$, and above by the paraboloid $z = x^2 + y^2$.

54. *Cylinder and paraboloids* Find the volume of the region bounded below by the paraboloid $z = x^2 + y^2$, laterally by the cylinder $x^2 + y^2 = 1$, and above by the paraboloid $z = x^2 + y^2 + 1$.

55. *Cylinder and cones* Find the volume of the solid cut from the thick-walled cylinder $1 \leq x^2 + y^2 \leq 2$ by the cones $z = \pm\sqrt{x^2 + y^2}$.

56. *Sphere and cylinder* Find the volume of the region that lies inside the sphere $x^2 + y^2 + z^2 = 2$ and outside the cylinder $x^2 + y^2 = 1$.

57. *Cylinder and planes* Find the volume of the region enclosed by the cylinder $x^2 + y^2 = 4$ and the planes $z = 0$ and $y + z = 4$.

58. *Cylinder and planes* Find the volume of the region enclosed by the cylinder $x^2 + y^2 = 4$ and the planes $z = 0$ and $x + y + z = 4$.

59. *Region trapped by paraboloids* Find the volume of the region bounded above by the paraboloid $z = 5 - x^2 - y^2$ and below by the paraboloid $z = 4x^2 + 4y^2$.

60. *Paraboloid and cylinder* Find the volume of the region bounded above by the paraboloid $z = 9 - x^2 - y^2$, below by the xy-plane, and lying *outside* the cylinder $x^2 + y^2 = 1$.

61. *Cylinder and sphere* Find the volume of the region cut from the solid cylinder $x^2 + y^2 \leq 1$ by the sphere $x^2 + y^2 + z^2 = 4$.

62. *Sphere and paraboloid* Find the volume of the region bounded above by the sphere $x^2 + y^2 + z^2 = 2$ and below by the paraboloid $z = x^2 + y^2$.

Average Values

63. Find the average value of the function $f(r, \theta, z) = r$ over the region bounded by the cylinder $r = 1$ between the planes $z = -1$ and $z = 1$.

64. Find the average value of the function $f(r, \theta, z) = r$ over the solid ball bounded by the sphere $r^2 + z^2 = 1$. (This is the sphere $x^2 + y^2 + z^2 = 1$.)

65. Find the average value of the function $f(\rho, \phi, \theta) = \rho$ over the solid ball $\rho \le 1$.

66. Find the average value of the function $f(\rho, \phi, \theta) = \rho \cos \phi$ over the solid upper ball $\rho \le 1, 0 \le \phi \le \pi/2$.

Masses, Moments, and Centroids

67. *Center of mass* A solid of constant density is bounded below by the plane $z = 0$, above by the cone $z = r$, $r \ge 0$, and on the sides by the cylinder $r = 1$. Find the center of mass.

68. *Centroid* Find the centroid of the region in the first octant that is bounded above by the cone $z = \sqrt{x^2 + y^2}$, below by the plane $z = 0$, and on the sides by the cylinder $x^2 + y^2 = 4$ and the planes $x = 0$ and $y = 0$.

69. *Centroid* Find the centroid of the solid in Exercise 38.

70. *Centroid* Find the centroid of the solid bounded above by the sphere $\rho = a$ and below by the cone $\phi = \pi/4$.

71. *Centroid* Find the centroid of the region that is bounded above by the surface $z = \sqrt{r}$, on the sides by the cylinder $r = 4$, and below by the xy-plane.

72. *Centroid* Find the centroid of the region cut from the solid ball $r^2 + z^2 \le 1$ by the half-planes $\theta = -\pi/3$, $r \ge 0$, and $\theta = \pi/3$, $r \ge 0$.

73. *Inertia and radius of gyration* Find the moment of inertia and radius of gyration about the z-axis of a thick-walled right circular cylinder bounded on the inside by the cylinder $r = 1$, on the outside by the cylinder $r = 2$, and on the top and bottom by the planes $z = 4$ and $z = 0$. (Take $\delta = 1$.)

74. *Moments of inertia of solid circular cylinder* Find the moment of inertia of a solid circular cylinder of radius 1 and height 2 (a) about the axis of the cylinder and (b) about a line through the centroid perpendicular to the axis of the cylinder. (Take $\delta = 1$.)

75. *Moment of inertia of solid cone* Find the moment of inertia of a right circular cone of base radius 1 and height 1 about an axis through the vertex parallel to the base. (Take $\delta = 1$.)

76. *Moment of inertia of solid sphere* Find the moment of inertia of a solid sphere of radius a about a diameter. (Take $\delta = 1$.)

77. *Moment of inertia of solid cone* Find the moment of inertia of a right circular cone of base radius a and height h about its axis. (*Hint*: Place the cone with its vertex at the origin and its axis along the z-axis.)

78. *Variable density* A solid is bounded on the top by the paraboloid $z = r^2$, on the bottom by the plane $z = 0$, and on the sides by the cylinder $r = 1$. Find the center of mass and the moment of inertia and radius of gyration about the z-axis if the density is

(a) $\delta(r, \theta, z) = z$

(b) $\delta(r, \theta, z) = r$.

79. *Variable density* A solid is bounded below by the cone $z = \sqrt{x^2 + y^2}$ and above by the plane $z = 1$. Find the center of mass and the moment of inertia and radius of gyration about the z-axis if the density is

(a) $\delta(r, \theta, z) = z$

(b) $\delta(r, \theta, z) = z^2$.

80. *Variable density* A solid ball is bounded by the sphere $\rho = a$. Find the moment of inertia and radius of gyration about the z-axis if the density is

(a) $\delta(\rho, \phi, \theta) = \rho^2$

(b) $\delta(\rho, \phi, \theta) = r = \rho \sin \phi$.

81. *Centroid of solid semiellipsoid* Show that the centroid of the solid semiellipsoid of revolution $(r^2/a^2) + (z^2/h^2) \le 1$, $z \ge 0$, lies on the z-axis three-eighths of the way from the base to the top. The special case $h = a$ gives a solid hemisphere. Thus, the centroid of a solid hemisphere lies on the axis of symmetry three-eighths of the way from the base to the top.

82. *Centroid of solid cone* Show that the centroid of a solid right circular cone is one-fourth of the way from the base to the vertex. (In general, the centroid of a solid cone or pyramid is one-fourth of the way from the centroid of the base to the vertex.)

83. *Variable density* A solid right circular cylinder is bounded by the cylinder $r = a$ and the planes $z = 0$ and $z = h$, $h > 0$. Find the center of mass and the moment of inertia and radius of gyration about the z-axis if the density is $\delta(r, \theta, z) = z + 1$.

84. *Mass of planet's atmosphere* A spherical planet of radius R has an atmosphere whose density is $\mu = \mu_0 e^{-ch}$, where h is the altitude above the surface of the planet, μ_0 is the density at sea level, and c is a positive constant. Find the mass of the planet's atmosphere.

85. *Density at center of a planet* A planet is in the shape of a sphere of radius R and total mass M with spherically symmetric density distribution that increases linearly as one approaches its center. What is the density at the center of this planet if the density at its edge (surface) is taken to be zero?

Theory and Examples

86. *Vertical circular cylinders in spherical coordinates* Find an equation of the form $\rho = f(\theta)$ for the cylinder $x^2 + y^2 = a^2$.

87. *Vertical planes in cylindrical coordinates*

(a) Show that planes perpendicular to the x-axis have equations of the form $r = a \sec \theta$ in cylindrical coordinates.

(b) Show that planes perpendicular to the y-axis have equations of the form $r = b \csc \theta$.

88. *(Continuation of Exercise 87)* Find an equation of the form $r = f(\theta)$ in cylindrical coordinates for the plane $ax + by = c$, $c \neq 0$.

89. *Writing to Learn: Symmetry* What symmetry will you find in a surface that has an equation of the form $r = f(z)$ in cylindrical coordinates? Give reasons for your answer.

90. *Writing to Learn: Symmetry* What symmetry will you find in a surface that has an equation of the form $\rho = f(\phi)$ in spherical coordinates? Give reasons for your answer.

12.7 Substitutions in Multiple Integrals

Substitutions in Double Integrals • Substitutions in Triple Integrals

This section shows how to evaluate multiple integrals by substitution. As in single integration, the goal of substitution is to replace complicated integrals by ones that are easier to evaluate. Substitutions accomplish this by simplifying the integrand, the limits of integration, or both.

Substitutions in Double Integrals

The polar coordinate substitution of Section 12.3 is a special case of a more general substitution method for double integrals, a method that pictures changes in variables as transformations of regions.

Suppose that a region G in the uv-plane is transformed one-to-one into the region R in the xy-plane by equations of the form

$$x = g(u, v), \qquad y = h(u, v),$$

as suggested in Figure 12.52. We call R the **image** of G under the transformation, and G the **preimage** of R. Any function $f(x, y)$ defined on R can be thought of as a function $f(g(u, v), h(u, v))$ defined on G as well. How is the integral of $f(x, y)$ over R related to the integral of $f(g(u, v), h(u, v))$ over G?

Notice the "Reversed" Order

The transforming equations $x = g(u, v)$ and $y = h(u, v)$ go from G to R, but we use them to change an integral over R into an integral over G.

The answer is: If g, h, and f have continuous partial derivatives and $J(u, v)$ (to be discussed in a moment) is zero only at isolated points, if at all, then

$$\iint_R f(x, y)\, dx\, dy = \iint_G f(g(u, v), h(u, v))\, |J(u, v)|\, du\, dv. \tag{1}$$

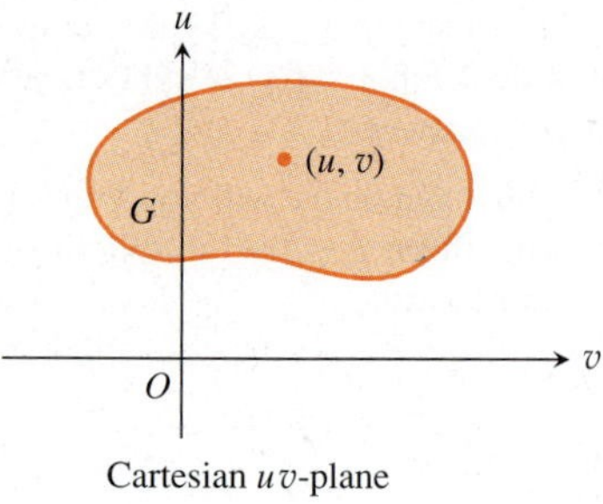

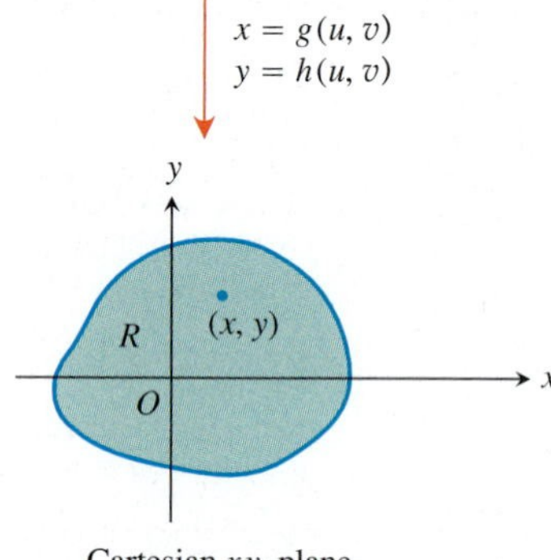

FIGURE 12.52 The equations $x = g(u, v)$ and $y = h(u, v)$ allow us to change an integral over a region R in the xy-plane into an integral over a region G in the uv-plane.

The factor $J(u, v)$, whose absolute value appears in Equation (1), is the *Jacobian* of the coordinate transformation, named after German mathematician Carl Jacobi.

Definition Jacobian determinant or Jacobian

The **Jacobian determinant** or **Jacobian** of the coordinate transformation $x = g(u, v), y = h(u, v)$ is

$$J(u, v) = \begin{vmatrix} \dfrac{\partial x}{\partial u} & \dfrac{\partial x}{\partial v} \\ \dfrac{\partial y}{\partial u} & \dfrac{\partial y}{\partial v} \end{vmatrix} = \frac{\partial x}{\partial u}\frac{\partial y}{\partial v} - \frac{\partial y}{\partial u}\frac{\partial x}{\partial v}. \tag{2}$$

The Jacobian is also denoted by

$$J(u, v) = \frac{\partial(x, y)}{\partial(u, v)}$$

to help remember how the determinant in Equation (2) is constructed from the partial derivatives of x and y. The derivation of Equation (1) is intricate and properly belongs to a course in advanced calculus. We do not give the derivation here.

For polar coordinates, we have r and θ in place of u and v. With $x = r\cos\theta$ and $y = r\sin\theta$, the Jacobian is

$$J(r, \theta) = \begin{vmatrix} \dfrac{\partial x}{\partial r} & \dfrac{\partial x}{\partial \theta} \\ \dfrac{\partial y}{\partial r} & \dfrac{\partial y}{\partial \theta} \end{vmatrix} = \begin{vmatrix} \cos\theta & -r\sin\theta \\ \sin\theta & r\cos\theta \end{vmatrix} = r(\cos^2\theta + \sin^2\theta) = r.$$

Hence, Equation (1) becomes

$$\iint_R f(x, y)\,dx\,dy = \iint_G f(r\cos\theta, r\sin\theta)\,|r|\,dr\,d\theta$$

$$= \iint_G f(r\cos\theta, r\sin\theta)\,r\,dr\,d\theta, \qquad \text{If } r \ge 0 \tag{3}$$

which is Equation (4) in Section 12.3.

Figure 12.53 shows how the equations $x = r\cos\theta$, $y = r\sin\theta$ transform the rectangle G: $0 \le r \le 1$, $0 \le \theta \le \pi/2$ into the quarter circle R bounded by $x^2 + y^2 = 1$ in the first quadrant of the xy-plane.

Notice that the integral on the right-hand side of Equation (3) is not the integral of $f(r\cos\theta, r\sin\theta)$ over a region in the polar coordinate plane. It is the integral of the product of $f(r\cos\theta, r\sin\theta)$ and r over a region G in the *Cartesian* $r\theta$-plane.

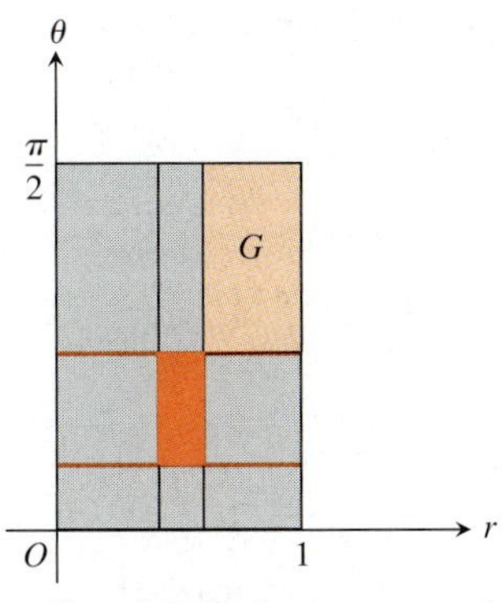

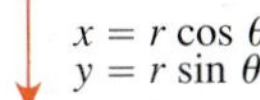

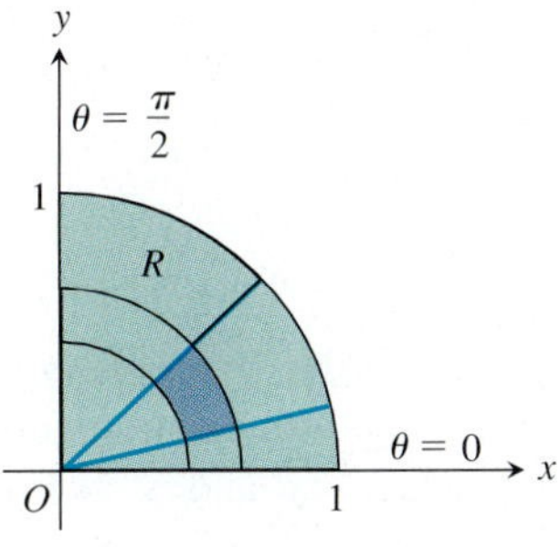

FIGURE 12.53 The equations $x = r\cos\theta$, $y = r\sin\theta$ transform G into R.

Here is an example of another substitution.

Example 1 Applying a Transformation to Integrate

Evaluate

$$\int_0^4 \int_{x=y/2}^{x=(y/2)+1} \frac{2x-y}{2}\,dx\,dy$$

by applying the transformation

$$u = \frac{2x-y}{2}, \qquad v = \frac{y}{2} \tag{4}$$

and integrating over an appropriate region in the uv-plane.

Solution We sketch the region R of integration in the xy-plane and identify its boundaries (Figure 12.54).

To apply Equation (1), we need to find the corresponding uv-region G and the Jacobian of the transformation. To find them, we first solve Equations (4) for x and y in terms of u and v. Routine algebra gives

$$x = u + v \qquad y = 2v. \tag{5}$$

We then find the boundaries of G by substituting these expressions into the equations for the boundaries of R (Figure 12.54).

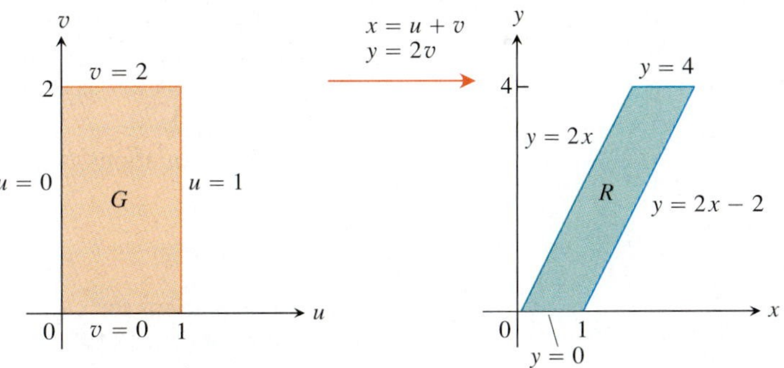

FIGURE 12.54 The equations $x = u + v$ and $y = 2v$ transform G into R. Reversing the transformation by the equations $u = (2x - y)/2$ and $v = y/2$ transforms R into G. See Example 1.

xy-equations for the boundary of R	**Corresponding uv-equations for the boundary of G**	**Simplified uv-equations**
$x = y/2$	$u + v = 2v/2 = v$	$u = 0$
$x = (y/2) + 1$	$u + v = (2v/2) + 1 = v + 1$	$u = 1$
$y = 0$	$2v = 0$	$v = 0$
$y = 4$	$2v = 4$	$v = 2$

The Jacobian of the transformation (again from Equations (5)) is

$$J(u, v) = \begin{vmatrix} \dfrac{\partial x}{\partial u} & \dfrac{\partial x}{\partial v} \\ \dfrac{\partial y}{\partial u} & \dfrac{\partial y}{\partial v} \end{vmatrix} = \begin{vmatrix} \dfrac{\partial}{\partial u}(u+v) & \dfrac{\partial}{\partial v}(u+v) \\ \dfrac{\partial}{\partial u}(2v) & \dfrac{\partial}{\partial v}(2v) \end{vmatrix} = \begin{vmatrix} 1 & 1 \\ 0 & 2 \end{vmatrix} = 2.$$

We now have everything we need to apply Equation (1):

$$\int_0^4 \int_{x=y/2}^{x=(y/2)+1} \frac{2x-y}{2}\, dx\, dy = \int_{v=0}^{v=2} \int_{u=0}^{u=1} u\,|\,J(u, v)\,|\, du\, dv$$

$$= \int_0^2 \int_0^1 (u)(2)\, du\, dv = \int_0^2 \left[u^2\right]_0^1 dv = \int_0^2 dv = 2.$$

Example 2 Applying a Transformation to Integrate

Evaluate

$$\int_0^1 \int_0^{1-x} \sqrt{x+y}\,(y-2x)^2\, dy\, dx.$$

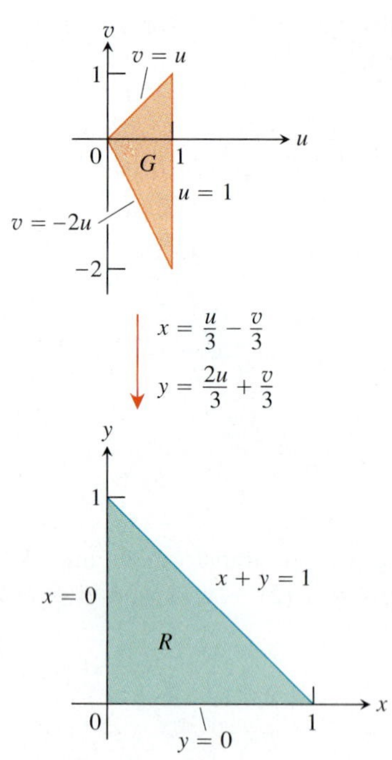

FIGURE 12.55 The equations $x = (u/3) - (v/3)$ and $y = (2u/3) + (v/3)$ transform G into R. Reversing the transformation by the equations $u = x + y$ and $v = y - 2x$ transforms R into G. See Example 2.

Solution We sketch the region R of integration in the xy-plane and identify its boundaries (Figure 12.55). The integrand suggests the transformation $u = x + y$ and $v = y - 2x$. Routine algebra produces x and y as functions of u and v:

$$x = \frac{u}{3} - \frac{v}{3}, \qquad y = \frac{2u}{3} + \frac{v}{3}. \tag{6}$$

From Equations (6), we can find the boundaries of the uv-region G (Figure 12.55).

xy*-equations for the boundary of *R	**Corresponding *uv*-equations for the boundary of *G***	**Simplified *uv*-equations**
$x + y = 1$	$\left(\frac{u}{3} - \frac{v}{3}\right) + \left(\frac{2u}{3} + \frac{v}{3}\right) = 1$	$u = 1$
$x = 0$	$\frac{u}{3} - \frac{v}{3} = 0$	$v = u$
$y = 0$	$\frac{2u}{3} + \frac{v}{3} = 0$	$v = -2u$

The Jacobian of the transformation in Equations (6) is

$$J(u, v) = \begin{vmatrix} \dfrac{\partial x}{\partial u} & \dfrac{\partial x}{\partial v} \\ \dfrac{\partial y}{\partial u} & \dfrac{\partial y}{\partial v} \end{vmatrix} = \begin{vmatrix} \dfrac{1}{3} & -\dfrac{1}{3} \\ \dfrac{2}{3} & \dfrac{1}{3} \end{vmatrix} = \frac{1}{3}.$$

Applying Equation (1), we evaluate the integral:

$$\int_0^1\int_0^{1-x} \sqrt{x+y}\,(y-2x)^2\,dy\,dx = \int_{u=0}^{u=1}\int_{v=-2u}^{v=u} u^{1/2}v^2\,|J(u,v)|\,dv\,du$$

$$= \int_0^1\int_{-2u}^{u} u^{1/2}v^2\left(\frac{1}{3}\right)dv\,du = \frac{1}{3}\int_0^1 u^{1/2}\left[\frac{1}{3}v^3\right]_{v=-2u}^{v=u} du$$

$$= \frac{1}{9}\int_0^1 u^{1/2}(u^3+8u^3)\,du = \int_0^1 u^{7/2}\,du = \frac{2}{9}u^{9/2}\bigg]_0^1 = \frac{2}{9}.$$

Substitutions in Triple Integrals

The cylindrical and spherical coordinate substitutions in Section 12.6 are special cases of a substitution method that pictures changes of variables in triple integrals as transformations of three-dimensional regions. The method is like the method for double integrals except that now we work in three dimensions instead of two.

Suppose that a region G in uvw-space is transformed one-to-one into the region D in xyz-space by differentiable equations of the form

$$x = g(u, v, w), \qquad y = h(u, v, w), \qquad z = k(u, v, w),$$

as suggested in Figure 12.56. Then any function $F(x, y, z)$ defined on D can be thought of as a function

$$F(g(u, v, w), h(u, v, w), k(u, v, w)) = H(u, v, w)$$

defined on G. If g, h, and k have continuous first partial derivatives, then the integral of $F(x, y, z)$ over D is related to the integral of $H(u, v, w)$ over G by the equation

$$\iiint_D F(x, y, z)\,dx\,dy\,dz = \iiint_G H(u, v, w)\,|J(u, v, w)|\,du\,dv\,dw. \tag{7}$$

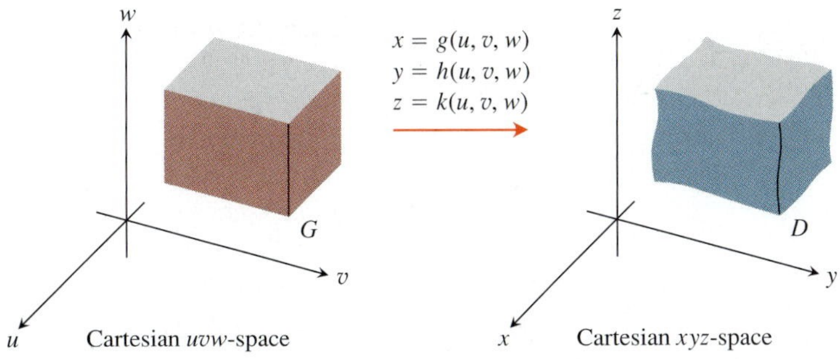

FIGURE 12.56 The equations $x = g(u, v, w)$, $y = h(u, v, w)$, and $z = k(u, v, w)$ allow us to change an integral over a region D in Cartesian xyz-space into an integral over a region G in Cartesian uvw-space.

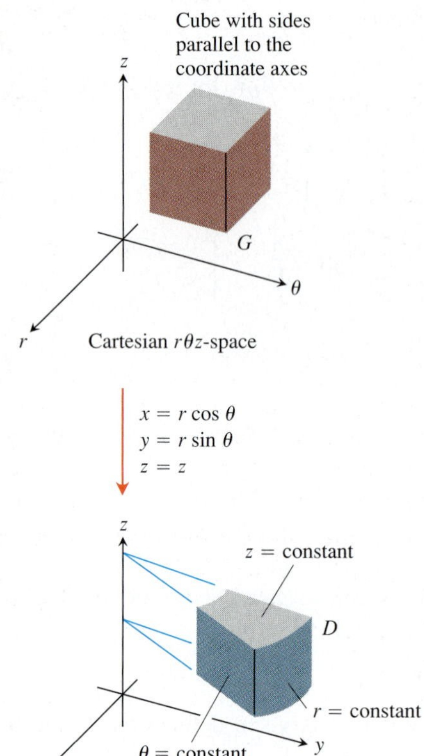

FIGURE 12.57 The equations $x = r\cos\theta$, $y = r\sin\theta$, and $z = z$ transform G into D.

The factor $J(u, v, w)$, whose absolute value appears in this equation, is the **Jacobian determinant**

$$J(u, v, w) = \begin{vmatrix} \dfrac{\partial x}{\partial u} & \dfrac{\partial x}{\partial v} & \dfrac{\partial x}{\partial w} \\ \dfrac{\partial y}{\partial u} & \dfrac{\partial y}{\partial v} & \dfrac{\partial y}{\partial w} \\ \dfrac{\partial z}{\partial u} & \dfrac{\partial z}{\partial v} & \dfrac{\partial z}{\partial w} \end{vmatrix} = \frac{\partial(x, y, z)}{\partial(u, v, w)}. \tag{8}$$

As in the two-dimensional case, the derivation of the change-of-variable formula in Equation (7) is complicated and we do not go into it here.

For cylindrical coordinates, r, θ, and z take the place of u, v, and w. The transformation from *Cartesian* $r\theta z$-space to Cartesian xyz-space is given by the equations

$$x = r\cos\theta, \qquad y = r\sin\theta, \qquad z = z$$

(Figure 12.57). The Jacobian of the transformation is

$$J(r, \theta, z) = \begin{vmatrix} \dfrac{\partial x}{\partial r} & \dfrac{\partial x}{\partial \theta} & \dfrac{\partial x}{\partial z} \\ \dfrac{\partial y}{\partial r} & \dfrac{\partial y}{\partial \theta} & \dfrac{\partial y}{\partial z} \\ \dfrac{\partial z}{\partial r} & \dfrac{\partial z}{\partial \theta} & \dfrac{\partial z}{\partial z} \end{vmatrix} = \begin{vmatrix} \cos\theta & -r\sin\theta & 0 \\ \sin\theta & r\cos\theta & 0 \\ 0 & 0 & 1 \end{vmatrix}$$

$$= r\cos^2\theta + r\sin^2\theta = r.$$

The corresponding version of Equation (7) is

$$\iiint\limits_D F(x, y, z)\, dx\, dy\, dz = \iiint\limits_G H(r, \theta, z)\, |r|\, dr\, d\theta\, dz. \tag{9}$$

We can drop the absolute value signs whenever $r \ge 0$.

For spherical coordinates, ρ, ϕ, and θ take the place of u, v, and w. The transformation from Cartesian $\rho\phi\theta$-space to Cartesian xyz-space is given by

$$x = \rho\sin\phi\cos\theta, \qquad y = \rho\sin\phi\sin\theta, \qquad z = \rho\cos\phi$$

(Figure 12.58). The Jacobian of the transformation is

$$J(\rho, \phi, \theta) = \begin{vmatrix} \dfrac{\partial x}{\partial \rho} & \dfrac{\partial x}{\partial \phi} & \dfrac{\partial x}{\partial \theta} \\ \dfrac{\partial y}{\partial \rho} & \dfrac{\partial y}{\partial \phi} & \dfrac{\partial y}{\partial \theta} \\ \dfrac{\partial z}{\partial \rho} & \dfrac{\partial z}{\partial \phi} & \dfrac{\partial z}{\partial \theta} \end{vmatrix} = \rho^2 \sin\phi \tag{10}$$

(Exercise 17). The corresponding version of Equation (7) is

$$\iiint\limits_D F(x, y, z)\, dx\, dy\, dz = \iiint\limits_G H(\rho, \phi, \theta)\, |\rho^2 \sin\phi|\, d\rho\, d\phi\, d\theta. \tag{11}$$

We can drop the absolute value signs because $\sin\phi$ is never negative.

FIGURE 12.58 The equations $x = \rho \sin \phi \cos \theta$, $y = \rho \sin \phi \sin \theta$, and $z = \rho \cos \phi$ transform G into D.

Here is an example of another substitution. Although we could evaluate the integral in this example directly, we have chosen it to illustrate the substitution method in a simple (and fairly intuitive) setting.

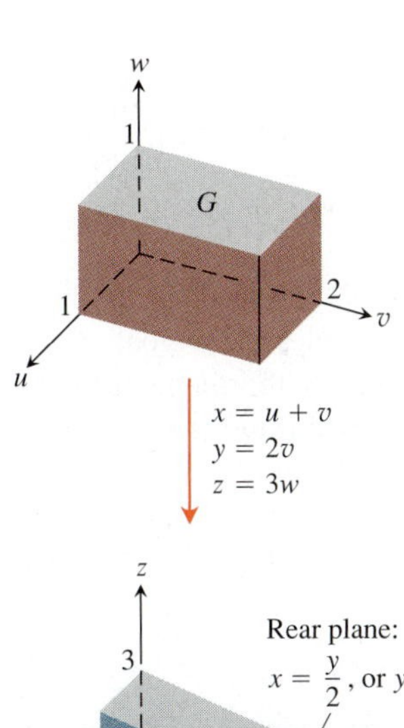

$x = u + v$
$y = 2v$
$z = 3w$

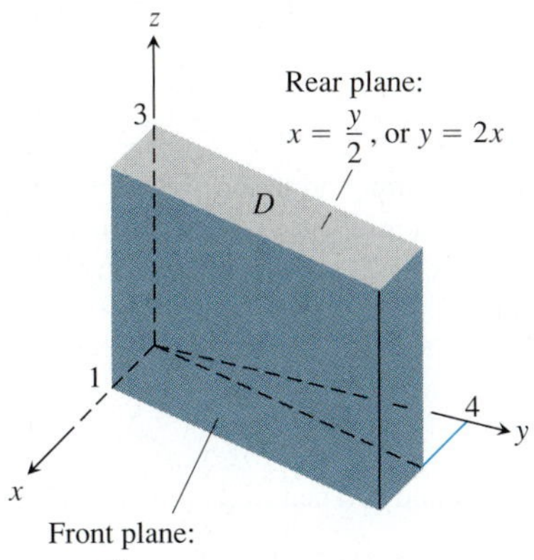

FIGURE 12.59 The equations $x = u + v$, $y = 2v$, and $z = 3w$ transform G into D. Reversing the transformation by the equations $u = (2x - y)/2$, $v = y/2$, and $w = z/3$ transforms D into G. See Example 3.

Example 3 Applying a Transformation to Integrate

Evaluate

$$\int_0^3 \int_0^4 \int_{x=y/2}^{x=(y/2)+1} \left(\frac{2x - y}{2} + \frac{z}{3} \right) dx\, dy\, dz$$

by applying the transformation

$$u = (2x - y)/2, \qquad v = y/2, \qquad w = z/3 \tag{12}$$

and integrating over an appropriate region in uvw-space.

Solution We sketch the region D of integration in xyz-space and identify its boundaries (Figure 12.59). In this case, the bounding surfaces are planes.

To apply Equation (7), we need to find the corresponding uvw-region G and the Jacobian of the transformation. To find them, we first solve Equations (12) for x, y, and z in terms of u, v, and w. Routine algebra gives

$$x = u + v, \qquad y = 2v, \qquad z = 3w. \tag{13}$$

We then find the boundaries of G by substituting these expressions into the equations for the boundaries of D:

xyz*-equations for the boundary of *D	**Corresponding *uvw*-equations for the boundary of *G***	**Simplified *uvw*-equations**
$x = y/2$	$u + v = 2v/2 = v$	$u = 0$
$x = (y/2) + 1$	$u + v = (2v/2) + 1 = v + 1$	$u = 1$
$y = 0$	$2v = 0$	$v = 0$
$y = 4$	$2v = 4$	$v = 2$
$z = 0$	$3w = 0$	$w = 0$
$z = 3$	$3w = 3$	$w = 1$

The Jacobian of the transformation, again from Equations (13), is

$$J(u, v, w) = \begin{vmatrix} \frac{\partial x}{\partial u} & \frac{\partial x}{\partial v} & \frac{\partial x}{\partial w} \\ \frac{\partial y}{\partial u} & \frac{\partial y}{\partial v} & \frac{\partial y}{\partial w} \\ \frac{\partial z}{\partial u} & \frac{\partial z}{\partial v} & \frac{\partial z}{\partial w} \end{vmatrix} = \begin{vmatrix} 1 & 1 & 0 \\ 0 & 2 & 0 \\ 0 & 0 & 3 \end{vmatrix} = 6.$$

We now have everything we need to apply Equation (7):

$$\begin{aligned}
&\int_0^3 \int_0^4 \int_{x=y/2}^{x=(y/2)+1} \left(\frac{2x - y}{2} + \frac{z}{3} \right) dx\, dy\, dz \\
&= \int_0^1 \int_0^2 \int_0^1 (u + w)\, |J(u, v, w)|\, du\, dv\, dw \\
&= \int_0^1 \int_0^2 \int_0^1 (u + w)(6)\, du\, dv\, dw = 6 \int_0^1 \int_0^2 \left[\frac{u^2}{2} + uw \right]_0^1 dv\, dw \\
&= 6 \int_0^1 \int_0^2 \left(\frac{1}{2} + w \right) dv\, dw = 6 \int_0^1 \left[\frac{v}{2} + vw \right]_0^2 dw = 6 \int_0^1 (1 + 2w)\, dw \\
&= 6\left[w + w^2 \right]_0^1 = 6(2) = 12.
\end{aligned}$$

CD-ROM
WEBsite
Historical Biography
Carl Gustav
Jacob Jacobi
(1804 — 1851)

Using the substitution theorem for multiple integrals can lead to computational difficulties when the coordinate transformations are nonlinear. The goal of this section was only to introduce you to the ideas involved. A thorough discussion of transformations, the Jacobian, and the multivariable substitution is best given in an advanced calculus course after you have studied linear algebra.

EXERCISES 12.7

Finding Jacobians and Transformed Regions for Two Variables

1. (a) Solve the system

$$u = x - y, \qquad v = 2x + y$$

for x and y in terms of u and v. Then find the value of the Jacobian $\partial(x, y)/\partial(u, v)$.

(b) Find the image under the transformation $u = x - y$, $v = 2x + y$ of the triangular region with vertices $(0, 0)$, $(1, 1)$, and $(1, -2)$ in the xy-plane. Sketch the transformed region in the uv-plane.

2. (a) Solve the system

$$u = x + 2y, \qquad v = x - y$$

for x and y in terms of u and v. Then find the value of the Jacobian $\partial(x, y)/\partial(u, v)$.

(b) Find the image under the transformation $u = x + 2y$, $v = x - y$ of the triangular region in the xy-plane bounded by the lines $y = 0$, $y = x$, and $x + 2y = 2$. Sketch the transformed region in the uv-plane.

3. (a) Solve the system

$$u = 3x + 2y, \qquad v = x + 4y$$

for x and y in terms of u and v. Then find the value of the Jacobian $\partial(x, y)/\partial(u, v)$.

(b) Find the image under the transformation $u = 3x + 2y$, $v = x + 4y$ of the triangular region in the xy-plane bounded by the x-axis, the y-axis, and the line $x + y = 1$. Sketch the transformed region in the uv-plane.

4. **(a)** Solve the system

$$u = 2x - 3y, \qquad v = -x + y$$

for x and y in terms of u and v. Then find the value of the Jacobian $\partial(x, y)/\partial(u, v)$.

(b) Find the image under the transformation $u = 2x - 3y$, $v = -x + y$ of the parallelogram R in the xy-plane with boundaries $x = -3$, $x = 0$, $y = x$, and $y = x + 1$. Sketch the transformed region in the uv-plane.

Applying Transformations to Evaluate Double Integrals

5. Evaluate the integral

$$\int_0^4 \int_{x=y/2}^{x=(y/2)+1} \frac{2x - y}{2}\, dx\, dy$$

from Example 1 directly by integration with respect to x and y to confirm that its value is 2.

6. Use the transformation in Exercise 1 to evaluate the integral

$$\iint_R (2x^2 - xy - y^2)\, dx\, dy$$

for the region R in the first quadrant bounded by the lines $y = -2x + 4$, $y = -2x + 7$, $y = x - 2$, and $y = x + 1$.

7. Use the transformation in Exercise 3 to evaluate the integral

$$\iint_R (3x^2 + 14xy + 8y^2)\, dx\, dy$$

for the region R in the first quadrant bounded by the lines $y = -(3/2)x + 1$, $y = -(3/2)x + 3$, $y = -(1/4)x$, and $y = -(1/4)x + 1$.

8. Use the transformation and parallelogram R in Exercise 4 to evaluate the integral

$$\iint_R 2(x - y)\, dx\, dy.$$

9. Let R be the region in the first quadrant of the xy-plane bounded by the hyperbolas $xy = 1$, $xy = 9$ and the lines $y = x$, $y = 4x$. Use the transformation $x = u/v$, $y = uv$ with $u > 0$ and $v > 0$ to rewrite

$$\iint_R \left(\sqrt{\frac{y}{x}} + \sqrt{xy}\right) dx\, dy$$

as an integral over an appropriate region G in the uv-plane. Then evaluate the uv-integral over G.

10. **(a)** Find the Jacobian of the transformation $x = u$, $y = uv$, and sketch the region G: $1 \le u \le 2$, $1 \le uv \le 2$ in the uv-plane.

(b) Then use Equation (1) to transform the integral

$$\int_1^2 \int_1^2 \frac{y}{x}\, dy\, dx$$

into an integral over G, and evaluate both integrals.

11. *Polar moment of inertia of an elliptical plate* A thin plate of constant density covers the region bounded by the ellipse $x^2/a^2 + y^2/b^2 = 1$, $a > 0$, $b > 0$, in the xy-plane. Find the first moment of the plate about the origin. (*Hint*: Use the transformation $x = ar \cos \theta$, $y = br \sin \theta$.)

12. *Finding area of an ellipse* The area πab of the ellipse $x^2/a^2 + y^2/b^2 = 1$ can be found by integrating the function $f(x, y) = 1$ over the region bounded by the ellipse in the xy-plane. Evaluating the integral directly requires a trigonometric substitution. An easier way to evaluate the integral is to use the transformation $x = au$, $y = bv$ and evaluate the transformed integral over the disk G: $u^2 + v^2 \le 1$ in the uv-plane. Find the area this way.

13. Use the transformation in Exercise 2 to evaluate the integral

$$\int_0^{2/3} \int_y^{2-2y} (x + 2y)e^{(y-x)}\, dx\, dy$$

by first writing it as an integral over a region G in the uv-plane.

14. Use the transformation $x = u + (1/2)v$, $y = v$ to evaluate the integral

$$\int_0^2 \int_{y/2}^{(y+4)/2} y^3(2x - y)e^{(2x-y)^2}\, dx\, dy$$

by first writing it as an integral over a region G in the uv-plane.

Finding Jacobian Determinants

15. Find the Jacobian $\partial(x, y)/\partial(u, v)$ for the transformation

(a) $x = u \cos v$, $y = u \sin v$

(b) $x = u \sin v$, $y = u \cos v$.

16. Find the Jacobian $\partial(x, y, z)/\partial(u, v, w)$ of the transformation

(a) $x = u \cos v$, $y = u \sin v$, $z = w$

(b) $x = 2u - 1$, $y = 3v - 4$, $z = (1/2)(w - 4)$.

17. Evaluate the determinant in Equation (10) to show that the Jacobian of the transformation from Cartesian $\rho\phi\theta$-space to Cartesian xyz-space is $\rho^2 \sin \phi$.

18. *Substitutions in single integrals* How can substitutions in single definite integrals be viewed as transformations of regions? What is the Jacobian in such a case? Illustrate with an example.

Applying Transformations to Evaluate Triple Integrals

19. Evaluate the integral in Example 3 by integrating with respect to x, y, and z.

20. *Volume of an ellipsoid* Find the volume of the ellipsoid

$$\frac{x^2}{a^2} + \frac{y^2}{b^2} + \frac{z^2}{c^2} = 1.$$

(*Hint*: Let $x = au$, $y = bv$, and $z = cw$. Then find the volume of an appropriate region in uvw-space.)

21. Evaluate

$$\iiint |xyz|\, dx\, dy\, dz$$

over the solid ellipsoid

$$\frac{x^2}{a^2} + \frac{y^2}{b^2} + \frac{z^2}{c^2} \le 1.$$

(*Hint*: Let $x = au$, $y = bv$, and $z = cw$. Then integrate over an appropriate region in uvw-space.)

22. Let D be the region in xyz-space defined by the inequalities

$$1 \le x \le 2, \qquad 0 \le xy \le 2, \qquad 0 \le z \le 1.$$

Evaluate

$$\iiint_D (x^2 y + 3\,xyz)\, dx\, dy\, dz$$

by applying the transformation

$$u = x, \qquad v = xy, \qquad w = 3z$$

and integrating over an appropriate region G in uvw-space.

23. *Centroid of a solid semiellipsoid* Assuming the result that the centroid of a solid hemisphere lies on the axis of symmetry three-eighths of the way from the base toward the top, show, by transforming the appropriate integrals, that the center of mass of a solid semiellipsoid $(x^2/a^2) + (y^2/b^2) + (z^2/c^2) \le 1$, $z \ge 0$, lies on the z-axis three-eighths of the way from the base toward the top. (You can do this without evaluating any of the integrals.)

24. *Cylindrical shells* In Section 5.2, we learned how to find the volume of a solid of revolution using the shell method; namely, if the region between the curve $y = f(x)$ and the x-axis from a to b $(0 < a < b)$ is revolved about the y-axis, the volume of the resulting solid is $\int_a^b 2\pi x f(x)\, dx$. Prove that finding volumes by using triple integrals gives the same result. (*Hint*: Use cylindrical coordinates with the roles of y and z changed.)

Questions to Guide Your Review

1. Define the double integral of a function of two variables over a bounded region in the coordinate plane.
2. How are double integrals evaluated as iterated integrals? Does the order of integration matter? How are the limits of integration determined? Give examples.
3. How are double integrals used to calculate areas, average values, masses, moments, centers of mass, and radii of gyration? Give examples.
4. How can you change a double integral in rectangular coordinates into a double integral in polar coordinates? Why might it be worthwhile to do so? Give an example.
5. Define the triple integral of a function $f(x, y, z)$ over a bounded region in space.
6. How are triple integrals in rectangular coordinates evaluated? How are the limits of integration determined? Give an example.
7. How are triple integrals in rectangular coordinates used to calculate volumes, average values, masses, moments, centers of mass, and radii of gyration? Give examples.
8. How are triple integrals defined in cylindrical and spherical coordinates? Why might one prefer working in one of these coordinate systems to working in rectangular coordinates?
9. How are triple integrals in cylindrical and spherical coordinates evaluated? How are the limits of integration found? Give examples.
10. How are substitutions in double integrals pictured as transformations of two-dimensional regions? Give a sample calculation.
11. How are substitutions in triple integrals pictured as transformations of three-dimensional regions? Give a sample calculation.

Practice Exercises

Planar Regions of Integration

In Exercises 1–4, sketch the region of integration and evaluate the double integral.

1. $\int_1^{10} \int_0^{1/y} ye^{xy}\, dx\, dy$

2. $\int_0^1 \int_0^{x^3} e^{y/x}\, dy\, dx$

3. $\int_0^{3/2} \int_{-\sqrt{9-4t^2}}^{\sqrt{9-4t^2}} t\, ds\, dt$

4. $\int_0^1 \int_{\sqrt{y}}^{2-\sqrt{y}} xy\, dx\, dy$

Reversing the Order of Integration

In Exercises 5–8, sketch the region of integration and write an equivalent integral with the order of integration reversed. Then evaluate both integrals.

5. $\int_0^4 \int_{-\sqrt{4-y}}^{(y-4)/2} dx\, dy$

6. $\int_0^1 \int_{x^2}^{x} \sqrt{x}\, dy\, dx$

7. $\int_0^{3/2} \int_{-\sqrt{9-4y^2}}^{\sqrt{9-4y^2}} y\, dx\, dy$

8. $\int_0^2 \int_0^{4-x} 2x\, dy\, dx$

Evaluating Double Integrals

Evaluate the integrals in Exercises 9–12.

9. $\int_0^1 \int_{2y}^{2} 4 \cos (x^2)\, dx\, dy$

10. $\int_0^2 \int_{y/2}^{1} e^{x^2}\, dx\, dy$

11. $\int_0^8 \int_{\sqrt[3]{x}}^{2} \frac{dy\, dx}{y^4 + 1}$

12. $\int_0^1 \int_{\sqrt[3]{y}}^{1} \frac{2\pi \sin \pi x^2}{x^2}\, dx\, dy$

Areas and Volumes

13. *Area between line and parabola* Find the area of the region enclosed by the line $y = 2x + 4$ and the parabola $y = 4 - x^2$ in the xy-plane.

14. *Area bounded by lines and parabola* Find the area of the "triangular" region in the xy-plane that is bounded on the right by the parabola $y = x^2$, on the left by the line $x + y = 2$, and above by the line $y = 4$.

15. *Volume of the region under a paraboloid* Find the volume under the paraboloid $z = x^2 + y^2$ above the triangle enclosed by the lines $y = x$, $x = 0$, and $x + y = 2$ in the xy-plane.

16. *Volume of the region under parabolic cylinder* Find the volume under the parabolic cylinder $z = x^2$ above the region enclosed by the parabola $y = 6 - x^2$ and the line $y = x$ in the xy-plane.

Average Values

Find the average value of $f(x, y) = xy$ over the regions in Exercises 17 and 18.

17. The square bounded by the lines $x = 1$, $y = 1$ in the first quadrant

18. The quarter circle $x^2 + y^2 \le 1$ in the first quadrant

Masses and Moments

19. *Centroid* Find the centroid of the "triangular" region bounded by the lines $x = 2$, $y = 2$ and the hyperbola $xy = 2$ in the xy-plane.

20. *Centroid* Find the centroid of the region between the parabola $x + y^2 - 2y = 0$ and the line $x + 2y = 0$ in the xy-plane.

21. *Polar moment* Find the polar moment of inertia about the origin of a thin triangular plate of constant density $\delta = 3$ bounded by the y-axis and the lines $y = 2x$ and $y = 4$ in the xy-plane.

22. *Polar moment* Find the polar moment of inertia about the center of a thin rectangular sheet of constant density $\delta = 1$ bounded by the lines

(a) $x = \pm 2$, $y = \pm 1$ in the xy-plane

(b) $x = \pm a$, $y = \pm b$ in the xy-plane.

(*Hint*: Find I_x. Then use the formula for I_x to find I_y and add the two to find I_0.)

23. *Inertial moment and radius of gyration* Find the moment of inertia and radius of gyration about the x-axis of a thin plate of constant density δ covering the triangle with vertices (0, 0), (3, 0), and (3, 2) in the xy-plane.

24. *Plate with variable density* Find the center of mass and the moments of inertia and radii of gyration about the coordinate axes of a thin plate bounded by the line $y = x$ and the parabola $y = x^2$ in the xy-plane if the density is $\delta(x, y) = x + 1$.

25. *Plate with variable density* Find the mass and first moments about the coordinate axes of a thin square plate bounded by the lines $x = \pm 1$, $y = \pm 1$ in the xy-plane if the density is $\delta(x, y) = x^2 + y^2 + 1/3$.

26. *Triangles with same inertial moment and radius of gyration* Find the moment of inertia and radius of gyration about the x-axis of a thin triangular plate of constant density δ whose base lies along the interval $[0, b]$ on the x-axis and whose vertex lies on the line $y = h$ above the x-axis. As you will see, it does not matter where on the line this vertex lies. All such triangles have the same moment of inertia and radius of gyration about the x-axis.

Polar Coordinates

Evaluate the integrals in Exercises 27 and 28 by changing to polar coordinates.

27. $\displaystyle\int_{-1}^{1}\int_{-\sqrt{1-x^2}}^{\sqrt{1-x^2}} \frac{2\,dy\,dx}{(1+x^2+y^2)^2}$

28. $\displaystyle\int_{-1}^{1}\int_{-\sqrt{1-y^2}}^{\sqrt{1-y^2}} \ln(x^2+y^2+1)\,dx\,dy$

29. *Centroid* Find the centroid of the region in the polar coordinate plane defined by the inequalities $0 \le r \le 3$, $-\pi/3 \le \theta \le \pi/3$.

30. *Centroid* Find the centroid of the region in the first quadrant bounded by the rays $\theta = 0$ and $\theta = \pi/2$ and the circles $r = 1$ and $r = 3$.

31. **(a)** *Centroid* Find the centroid of the region in the polar coordinate plane that lies inside the cardioid $r = 1 + \cos\theta$ and outside the circle $r = 1$.

(b) Sketch the region and show the centroid in your sketch.

32. **(a)** *Writing to Learn: Centroid* Find the centroid of the plane region defined by the polar coordinate inequalities $0 \le r \le a$, $-\alpha \le \theta \le \alpha$ $(0 < \alpha \le \pi)$. How does the centroid move as $\alpha \to \pi^-$?

(b) Sketch the region for $\alpha = 5\pi/6$ and show the centroid in your sketch.

33. *Integrating over lemniscate* Integrate the function $f(x, y) = 1/(1 + x^2 + y^2)^2$ over the region enclosed by one loop of the lemniscate $(x^2+y^2)^2 - (x^2 - y^2) = 0$.

34. Integrate $f(x, y) = 1/(1 + x^2 + y^2)^2$ over

(a) *Triangular region* The triangle with vertices (0, 0), (1, 0), $(1, \sqrt{3})$

(b) *First quadrant* The first quadrant of the xy-plane.

Triple Integrals in Cartesian Coordinates

Evaluate the integrals in Exercises 35–38.

35. $\displaystyle\int_0^{\pi}\int_0^{\pi}\int_0^{\pi} \cos(x+y+z)\,dx\,dy\,dz$

36. $\displaystyle\int_{\ln 6}^{\ln 7}\int_0^{\ln 2}\int_{\ln 4}^{\ln 5} e^{(x+y+z)}\,dz\,dy\,dx$

37. $\displaystyle\int_0^1\int_0^{x^2}\int_0^{x+y} (2x - y - z)\,dz\,dy\,dx$

38. $\displaystyle\int_1^e\int_1^x\int_0^z \frac{2y}{z^3}\,dy\,dz\,dx$

39. *Volume* Find the volume of the wedge-shaped region enclosed on the side by the cylinder $x = -\cos y$, $-\pi/2 \le y \le \pi/2$, on the top by the plane $z = -2x$, and below by the xy-plane.

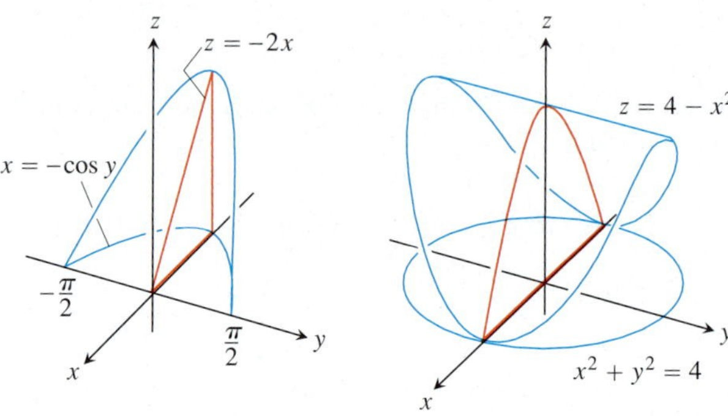

40. *Volume* Find the volume of the solid that is bounded above by the cylinder $z = 4 - x^2$, on the sides by the cylinder $x^2 + y^2 = 4$, and below by the xy-plane.

41. *Average value* Find the average value of $f(x, y, z) = 30xz\sqrt{x^2+y}$ over the rectangular solid in the first octant bounded by the coordinate planes and the planes $x = 1$, $y = 3$, $z = 1$.

42. *Average value* Find the average value of ρ over the solid sphere $\rho \le a$ (spherical coordinates).

Cylindrical and Spherical Coordinates

43. *Cylindrical to rectangular coordinates* Convert

$$\int_0^{2\pi}\int_0^{\sqrt{2}}\int_r^{\sqrt{4-r^2}} 3\,dz\,r\,dr\,d\theta, \qquad r \ge 0$$

to (a) rectangular coordinates with the order of integration $dz\,dx\,dy$ and (b) spherical coordinates. Then (c) evaluate one of the integrals.

44. *Rectangular to cylindrical coordinates* **(a)** Convert to cylindrical coordinates. Then (b) evaluate the new integral.

$$\int_0^1\int_{-\sqrt{1-x^2}}^{\sqrt{1-x^2}}\int_{-(x^2+y^2)}^{(x^2+y^2)} 21xy^2\,dz\,dy\,dx$$

45. *Rectangular to spherical coordinates* **(a)** Convert to spherical coordinates. Then (b) evaluate the new integral.

$$\int_{-1}^1\int_{-\sqrt{1-x^2}}^{\sqrt{1-x^2}}\int_{\sqrt{x^2+y^2}}^{1} dz\,dy\,dx$$

46. *Rectangular, cylindrical, and spherical coordinates* Write an iterated triple integral for the integral of $f(x, y, z) = 6 + 4y$ over the region in the first octant bounded by the cone $z = \sqrt{x^2+y^2}$, the cylinder $x^2 + y^2 = 1$, and the coordinate planes in (a) rectangular coordinates, (b) cylindrical coordinates, and (c) spherical coordinates. Then (d) find the integral of f by evaluating one of the triple integrals.

47. *Cylindrical to rectangular coordinates* Set up an integral in rectangular coordinates equivalent to the integral

$$\int_0^{\pi/2}\int_1^{\sqrt{3}}\int_1^{\sqrt{4-r^2}} r^3 \,(\sin\theta\cos\theta)\, z^2\, dz\, dr\, d\theta.$$

Arrange the order of integration to be z first, then y, then x.

48. *Rectangular to cylindrical coordinates* The volume of a solid is

$$\int_0^2\int_0^{\sqrt{2x-x^2}}\int_{-\sqrt{4-x^2-y^2}}^{\sqrt{4-x^2-y^2}} dz\, dy\, dx.$$

(a) Describe the solid by giving equations for the surfaces that form its boundary.

(b) Convert the integral to cylindrical coordinates but do not evaluate the integral.

49. *Spherical versus cylindrical coordinates* Triple integrals involving spherical shapes do not always require spherical coordinates for convenient evaluation. Some calculations may be accomplished more easily with cylindrical coordinates. As a case in point, find the volume of the region bounded above by the sphere $x^2 + y^2 + z^2 = 8$ and below by the plane $z = 2$ by using (a) cylindrical coordinates and (b) spherical coordinates.

50. *Finding I_z in spherical coordinates* Find the moment of inertia about the z-axis of a solid of constant density $\delta = 1$ that is bounded above by the sphere $\rho = 2$ and below by the cone $\phi = \pi/3$ (spherical coordinates).

51. *Moment of inertia of a "thick" sphere* Find the moment of inertia of a solid of constant density δ bounded by two concentric spheres of radii a and b $(a < b)$ about a diameter.

52. *Moment of inertia of an apple* Find the moment of inertia about the z-axis of a solid of density $\delta = 1$ enclosed by the spherical coordinate surface $\rho = 1 - \cos\phi$.

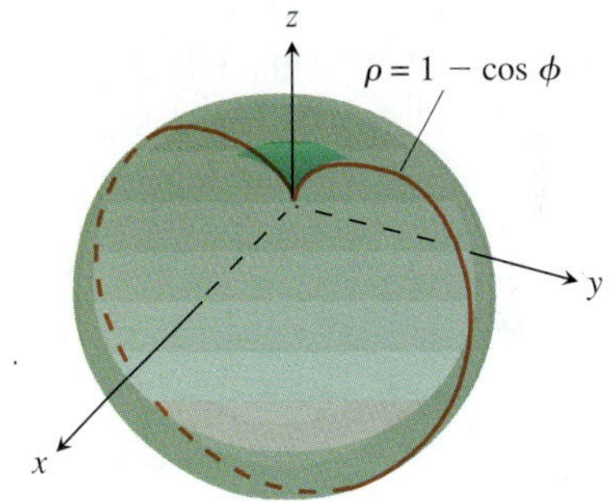

Additional Exercises: Theory, Examples, Applications

Volumes

1. *Sand pile: double and triple integrals* The base of a sand pile covers the region in the xy-plane that is bounded by the parabola $x^2 + y = 6$ and the line $y = x$. The height of the sand above the point (x, y) is x^2. Express the volume of sand as (a) a double integral, (b) a triple integral. Then (c) find the volume.

2. *Water in a hemispherical bowl* A hemispherical bowl of radius 5 cm is filled with water to within a 3 cm of the top. Find the volume of water in the bowl.

3. *Solid Cylindrical Region between Two Planes* Find the volume of the portion of the solid cylinder $x^2 + y^2 \le 1$ that lies between the planes $z = 0$ and $x + y + z = 2$.

4. *Sphere and paraboloid* Find the volume of the region bounded above by the sphere $x^2 + y^2 + z^2 = 2$ and below by the paraboloid $z = x^2 + y^2$.

5. *Two paraboloids* Find the volume of the region bounded above by the paraboloid $z = 3 - x^2 - y^2$ and below by the paraboloid $z = 2x^2 + 2y^2$.

6. *Spherical coordinates* Find the volume of the region enclosed by the spherical coordinate surface $\rho = 2\sin\phi$ (see accompanying figure).

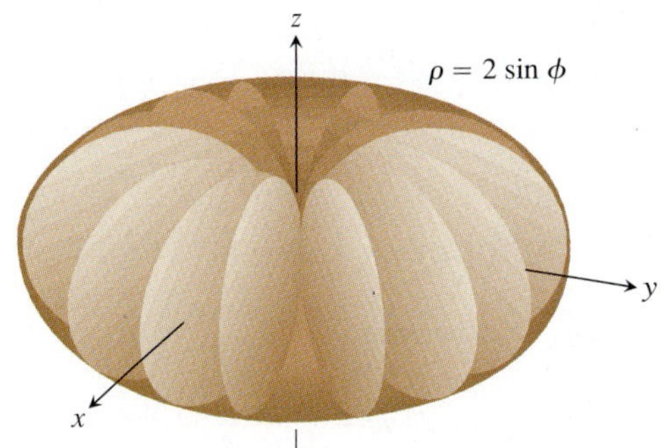

7. *Hole in sphere* A circular cylindrical hole is bored through a solid sphere, the axis of the hole being a diameter of the sphere. The volume of the remaining solid is

$$V = 2\int_0^{2\pi}\int_0^{\sqrt{3}}\int_1^{\sqrt{4-z^2}} r\,dr\,dz\,d\theta.$$

(a) Find the radius of the hole and the radius of the sphere.

(b) Evaluate the integral.

8. *Sphere and cylinder* Find the volume of material cut from the solid sphere $r^2 + z^2 \le 9$ by the cylinder $r = 3\sin\theta$.

9. *Two paraboloids* Find the volume of the region enclosed by the surfaces $z = x^2 + y^2$ and $z = (x^2 + y^2 + 1)/2$.

10. *Cylinder and surface $z = xy$* Find the volume of the region in the first octant that lies between the cylinders $r = 1$ and $r = 2$ and that is bounded below by the xy-plane and above by the surface $z = xy$.

Changing the Order of Integration

11. Evaluate the integral

$$\int_0^{\infty} \frac{e^{-ax} - e^{-bx}}{x}\,dx.$$

(*Hint*: Use the relation

$$\frac{e^{-ax} - e^{-bx}}{x} = \int_a^b e^{-xy}\,dy$$

to form a double integral and evaluate the integral by changing the order of integration.)

12. **(a)** *Polar coordinates* Show, by changing to polar coordinates, that

$$\int_0^{a\sin\beta}\int_{y\cot\beta}^{\sqrt{a^2-y^2}} \ln(x^2+y^2)\,dx\,dy = a^2\beta\left(\ln a - \frac{1}{2}\right),$$

where $a > 0$ and $0 < \beta < \pi/2$.

(b) Rewrite the Cartesian integral with the order of integration reversed.

13. *Reducing a double to a single integral* By changing the order of integration, show that the following double integral can be reduced to a single integral:

$$\int_0^x\int_0^u e^{m(x-t)}f(t)\,dt\,du = \int_0^x (x-t)e^{m(x-t)}f(t)\,dt.$$

Similarly, it can be shown that

$$\int_0^x\int_0^v\int_0^u e^{m(x-t)}f(t)\,dt\,du\,dv = \int_0^x \frac{(x-t)^2}{2}e^{m(x-t)}f(t)\,dt.$$

14. *Transforming a double integral to obtain constant limits* Sometimes a multiple integral with variable limits can be changed into one with constant limits. By changing the order of integration, show that

$$\int_0^1 f(x)\left(\int_0^x g(x-y)f(y)\,dy\right)dx$$
$$= \int_0^1 f(y)\left(\int_y^1 g(x-y)f(x)\,dx\right)dy$$
$$= \frac{1}{2}\int_0^1\int_0^1 g(|x-y|)f(x)f(y)\,dx\,dy.$$

Masses and Moments

15. *Minimizing polar inertia* A thin plate of constant density is to occupy the triangular region in the first quadrant of the xy-plane having vertices $(0, 0)$, $(a, 0)$, and $(a, 1/a)$. What value of a will minimize the plate's polar moment of inertia about the origin?

16. *Polar inertia of triangular plate* Find the polar moment of inertia about the origin of a thin triangular plate of constant density $\delta = 3$ bounded by the y-axis and the lines $y = 2x$ and $y = 4$ in the xy-plane.

17. *Mass and polar inertia of a counterweight* The counterweight of a flywheel of constant density 1 has the form of the smaller segment cut from a circle of radius a by a chord at a distance b from the center $(b < a)$. Find the mass of the counterweight and its polar moment of inertia about the center of the wheel.

18. *Centroid of boomerang* Find the centroid of the boomerang-shaped region between the parabolas $y^2 = -4(x-1)$ and $y^2 = -2(x-2)$ in the xy-plane.

Theory and Applications

19. Evaluate

$$\int_0^a\int_0^b e^{\max(b^2x^2,\,a^2y^2)}\,dy\,dx,$$

where a and b are positive numbers and

$$\max(b^2x^2, a^2y^2) = \begin{cases} b^2x^2 & \text{if } b^2x^2 \ge a^2y^2 \\ a^2y^2 & \text{if } b^2x^2 < a^2y^2. \end{cases}$$

20. Show that

$$\iint \frac{\partial^2 F(x,y)}{\partial x\,\partial y}\,dx\,dy$$

over the rectangle $x_0 \le x \le x_1$, $y_0 \le y \le y_1$, is

$$F(x_1, y_1) - F(x_0, y_1) - F(x_1, y_0) + F(x_0, y_0).$$

21. Suppose that $f(x, y)$ can be written as a product $f(x, y) = F(x)G(y)$ of a function of x and a function of y. Then the integral of f over the rectangle R: $a \le x \le b,\ c \le y \le d$ can be evaluated as a product as well, by the formula

$$\iint_R f(x, y)\, dA = \left(\int_a^b F(x)\, dx\right)\left(\int_c^d G(y)\, dy\right). \qquad (1)$$

The argument is that

$$\iint_R f(x, y)\, dA = \int_c^d \left(\int_a^b F(x)G(y)\, dx\right) dy \qquad \text{(i)}$$

$$= \int_c^d \left(G(y) \int_a^b F(x)\, dx\right) dy \qquad \text{(ii)}$$

$$= \int_c^d \left(\int_a^b F(x)\, dx\right) G(y)\, dy \qquad \text{(iii)}$$

$$= \left(\int_a^b F(x)\, dx\right) \int_c^d G(y)\, dy. \qquad \text{(iv)}$$

(a) *Writing to Learn* Give reasons for steps i through iv.

When it applies, Equation (1) can be a time saver. Use it to evaluate the following integrals.

(b) $\displaystyle\int_0^{\ln 2} \int_0^{\pi/2} e^x \cos y\, dy\, dx$

(c) $\displaystyle\int_1^2 \int_{-1}^1 \frac{x}{y^2}\, dx\, dy$

22. Let $D_{\mathbf{u}} f$ denote the derivative of $f(x, y) = (x^2 + y^2)/2$ in the direction of the unit vector $\mathbf{u} = u_1\mathbf{i} + u_2\mathbf{j}$.

(a) *Finding average value* Find the average value of $D_{\mathbf{u}} f$ over the triangular region cut from the first quadrant by the line $x + y = 1$.

(b) *Average value and centroid* Show in general that the average value of $D_{\mathbf{u}} f$ over a region in the xy-plane is the value of $D_{\mathbf{u}} f$ at the centroid of the region.

23. *The value of $\Gamma(1/2)$* The gamma function,

$$\Gamma(x) = \int_0^\infty t^{x-1} e^{-t}\, dt,$$

extends the factorial function from the nonnegative integers to other real values. Of particular interest in the theory of differential equations is the number

$$\Gamma\left(\frac{1}{2}\right) = \int_0^\infty t^{(1/2)-1} e^{-t}\, dt = \int_0^\infty \frac{e^{-t}}{\sqrt{t}}\, dt. \qquad (2)$$

(a) If you have not yet done Exercise 37 in Section 12.3, do it now to show that

$$I = \int_0^\infty e^{-y^2}\, dy = \frac{\sqrt{\pi}}{2}.$$

(b) Substitute $y = \sqrt{t}$ in Equation (2) to show that $\Gamma(1/2) = 2I = \sqrt{\pi}$.

24. *Total electrical charge over circular plate* The electrical charge distribution on a circular plate of radius R meters is $\sigma(r, \theta) = kr(1 - \sin\theta)$ coulomb/m^2 (k a constant). Integrate σ over the plate to find the total charge Q.

25. *A parabolic rain gauge* A bowl is in the shape of the graph of $z = x^2 + y^2$ from $z = 0$ to $z = 10$ in. You plan to calibrate the bowl to make it into a rain gauge. What height in the bowl would correspond to 1 in. of rain? 3 in. of rain?

CD-ROM WEBsite

26. *Water in a satellite dish* A parabolic satellite dish is 2 m wide and 1/2 m deep. It axis of symmetry is tilted 30 degrees from the vertical.

CD-ROM WEBsite

(a) Set up, but do not evaluate, a triple integral in rectangular coordinates that gives the amount of water the satellite dish will hold. (*Hint*: Put your coordinate system so that the satellite dish is in "standard position" and the plane of the water level is slanted.) (*Caution*: The limits of integration are not "nice")

(b) What would be the smallest tilt of the satellite dish so that it holds no water?

27. *An infinite half-cylinder* Let D be the interior of the infinite right circular half-cylinder of radius 1 with its single-end face suspended 1 unit above the origin and its axis the ray from (0, 0, 1) to ∞. Use cylindrical coordinates to evaluate

$$\iiint_D z(r^2 + z^2)^{-5/2}\, dV.$$

28. *Hypervolume* We have learned that $\int_a^b 1\, dx$ is the length of the interval $[a, b]$ on the number line (one-dimensional space), $\iint_R 1\, dA$ is the area of region R in the xy-plane (two-dimensional space), and $\iiint_D 1\, dV$ is the volume of the region D in three-dimensional space (xyz-space). We could continue: If Q is a region in 4-space ($xyzw$-space), then $\iiiint_Q 1\, dV$ is the "hypervolume" of Q. Use your generalizing abilities and a Cartesian coordinate system of 4-space to find the hypervolume inside the unit 4-sphere $x^2 + y^2 + z^2 + w^2 = 1$.

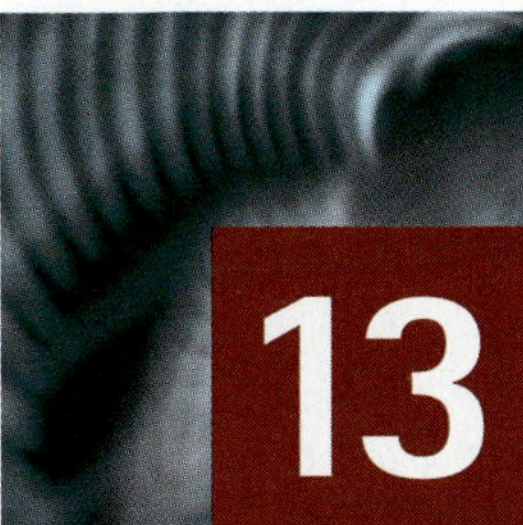

13 Integration in Vector Fields

OVERVIEW This chapter treats integration in vector fields. The mathematics in this chapter is the mathematics engineers and physicists use to describe fluid flow, design underwater transmission cables, explain the flow of heat in stars, and calculate the work it takes to put a satellite in orbit.

13.1 Line Integrals

Definitions and Notation • Evaluation for Smooth Curves • Additivity • Mass and Moment Calculations

When a curve $\mathbf{r}(t) = g(t)\mathbf{i} + h(t)\mathbf{j} + k(t)\mathbf{k},\ a \le t \le b$, passes through the domain of a function $f(x, y, z)$ in space, the values of f along the curve are given by the composite function $f(g(t), h(t), k(t))$. If we integrate this composite with respect to arc length from $t = a$ to $t = b$, we calculate the *line integral* of f along the curve. Despite the three-dimensional geometry, the line integral is an ordinary integral of a real-valued function over an interval of real numbers.

The importance of line integrals lies in their application. These are the integrals with which we calculate the work done by variable forces along paths in space and the rates at which fluids flow along curves and across boundaries.

Definitions and Notation

Suppose that $f(x, y, z)$ is a real-valued function whose domain contains the curve $\mathbf{r}(t) = g(t)\mathbf{i} + h(t)\mathbf{j} + k(t)\mathbf{k},\ a \le t \le b$. We partition the curve into a finite number of subarcs (Figure 13.1). The typical subarc has length Δs_k. In each subarc we choose a point (x_k, y_k, z_k) and form the sum

$$S_n = \sum_{k=1}^{n} f(x_k, y_k, z_k)\, \Delta s_k .$$

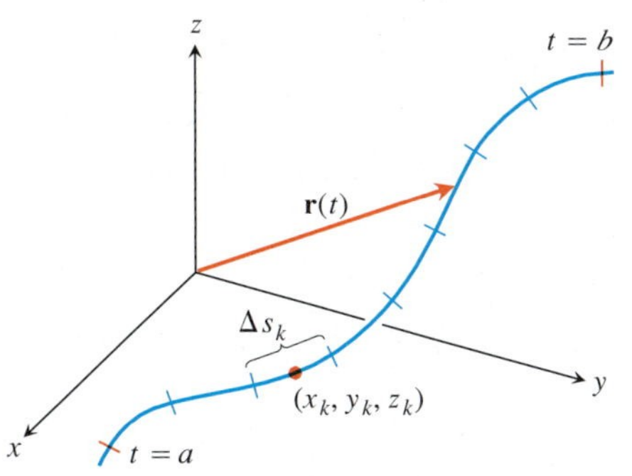

FIGURE 13.1 The curve $\mathbf{r}(t)$ partitioned into small arcs from $t = a$ to $t = b$. The length of a typical subarc is Δs_k.

If f is continuous and the functions g, h, and k have continuous first derivatives, then these sums approach a limit as n increases and the lengths Δs_k approach zero. We call this limit the **line integral of f over the curve from a to b.** If the curve is denoted by a single letter, C for example, the notation for the integral is

$$\int_C f(x, y, z)\, ds \qquad \text{“The integral of } f \text{ over } C\text{”} \tag{1}$$

Evaluation for Smooth Curves

If $\mathbf{r}(t)$ is smooth for $a \le t \le b$ ($\mathbf{v} = d\mathbf{r}/dt$ is continuous and never $\mathbf{0}$), we can use the equation

$$s(t) = \int_a^t |\mathbf{v}(\tau)|\, d\tau \qquad \text{Eq. (3) of Section 10.6 with } t_0 = a$$

to express ds in Equation (1) as $ds = |\mathbf{v}(t)|\, dt$. A theorem from advanced calculus says that we can then evaluate the integral of f over C as

$$\int_C f(x, y, z)\, ds = \int_a^b f(g(t), h(t), k(t))\, |\mathbf{v}(t)|\, dt.$$

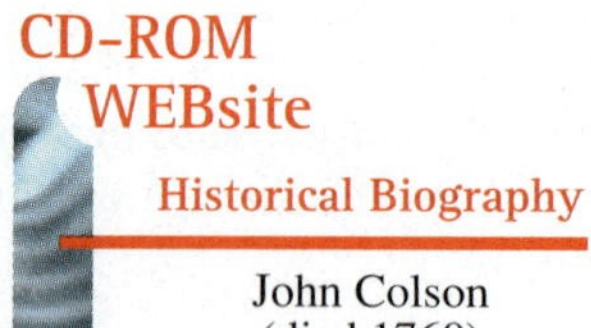

This formula will evaluate the integral correctly no matter what parametrization we use, as long as the parametrization is smooth.

How to Evaluate a Line Integral

To integrate a continuous function $f(x, y, z)$ over a curve C:

Step 1. Find a smooth parametrization of C,

$$\mathbf{r}(t) = g(t)\mathbf{i} + h(t)\mathbf{j} + k(t)\mathbf{k}, \qquad a \le t \le b$$

Step 2. Evaluate the integral as

$$\int_C f(x, y, z)\, ds = \int_a^b f(g(t), h(t), k(t))\, |\mathbf{v}(t)|\, dt. \tag{2}$$

If f has the constant value 1, then the integral of f over C gives the length of C.

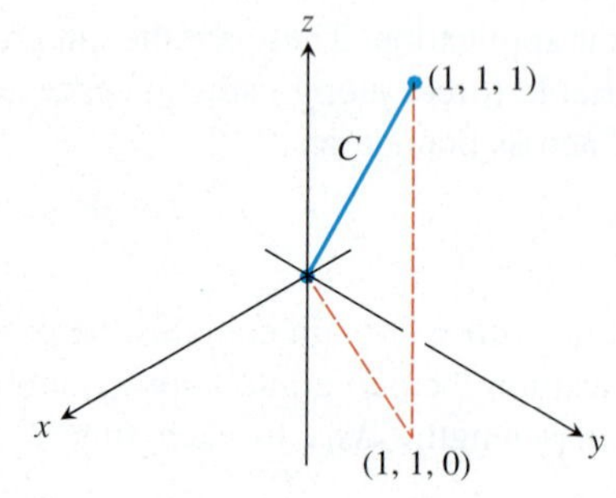

FIGURE 13.2 The integration path in Example 1.

Example 1 Evaluating a Line Integral

Integrate $f(x, y, z) = x - 3y^2 + z$ over the line segment C joining the origin and the point $(1, 1, 1)$ (Figure 13.2).

Solution We choose the simplest parametrization we can think of:

$$\mathbf{r}(t) = t\mathbf{i} + t\mathbf{j} + t\mathbf{k}, \qquad 0 \le t \le 1.$$

The components have continuous first derivatives and $|\mathbf{v}(t)| = |\mathbf{i} + \mathbf{j} + \mathbf{k}| = \sqrt{1^2 + 1^2 + 1^2} = \sqrt{3}$ is never 0, so the parametrization is smooth. The integral of f over C is

$$\begin{aligned}\int_C f(x, y, z)\, ds &= \int_0^1 f(t, t, t)\,(\sqrt{3})\, dt \qquad \text{Eq. (2)}\\ &= \int_0^1 (t - 3t^2 + t)\sqrt{3}\, dt\\ &= \sqrt{3}\int_0^1 (2t - 3t^2)\, dt = \sqrt{3}\,[t^2 - t^3]_0^1 = 0.\end{aligned}$$

Additivity

Line integrals have the useful property that if a curve C is made by joining a finite number of curves $C_1, C_2, \dots, C_n$ end to end, then the integral of a function over C is the sum of the integrals over the curves that make it up:

$$\int_C f\,ds = \int_{C_1} f\,ds + \int_{C_2} f\,ds + \cdots + \int_{C_n} f\,ds. \tag{3}$$

CD-ROM
WEBsite
Historical Biography
Gottfried Wilhelm Leibniz
(1646 — 1716)

Example 2 Line Integral for Two Joined Paths

Figure 13.3 shows another path from the origin to $(1, 1, 1)$, the union of line segments C_1 and C_2. Integrate $f(x, y, z) = x - 3y^2 + z$ over $C_1 \cup C_2$.

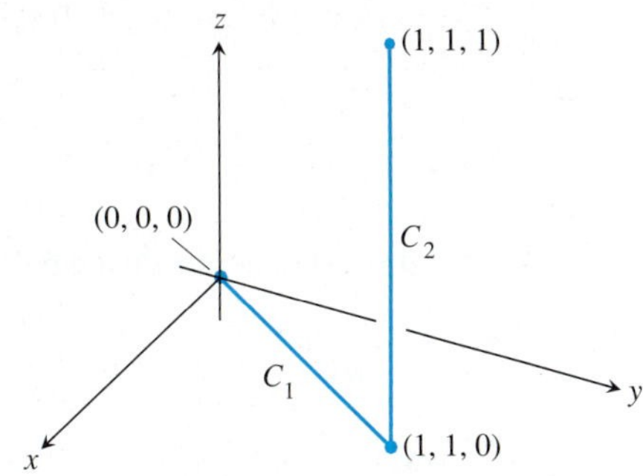

FIGURE 13.3 The path of integration in Example 2.

Solution We choose the simplest parametrizations for C_1 and C_2 we can think of, checking the lengths of the velocity vectors as we go along:

$$C_1: \quad \mathbf{r}(t) = t\mathbf{i} + t\mathbf{j}, \quad 0 \le t \le 1; \quad |\mathbf{v}| = \sqrt{1^2 + 1^2} = \sqrt{2}$$
$$C_2: \quad \mathbf{r}(t) = \mathbf{i} + \mathbf{j} + t\mathbf{k}, \quad 0 \le t \le 1; \quad |\mathbf{v}| = \sqrt{0^2 + 0^2 + 1^2} = 1.$$

With these parametrizations we find that

$$\begin{aligned}
\int_{C_1 \cup C_2} f(x, y, z)\,ds &= \int_{C_1} f(x, y, z)\,ds + \int_{C_2} f(x, y, z)\,ds && \text{Eq. (3)}\\
&= \int_0^1 f(t, t, 0)\sqrt{2}\,dt + \int_0^1 f(1, 1, t)\,(1)\,dt && \text{Eq. (2)}\\
&= \int_0^1 (t - 3t^2 + 0)\sqrt{2}\,dt + \int_0^1 (1 - 3 + t)(1)\,dt\\
&= \sqrt{2}\left[\frac{t^2}{2} - t^3\right]_0^1 + \left[\frac{t^2}{2} - 2t\right]_0^1 = -\frac{\sqrt{2}}{2} - \frac{3}{2}.
\end{aligned}$$

Notice three things about the integrations in Examples 1 and 2. First, as soon as the components of the appropriate curve were substituted into the formula for f, the integration became a standard integration with respect to t. Second, the integral of

f over $C_1 \cup C_2$ was obtained by integrating f over each section of the path and adding the results. Third, the integrals of f over C and $C_1 \cup C_2$ had different values. For most functions, the value of the integral along a path joining two points changes if you change the path between them. For some functions, however, the value remains the same, as we see in Section 13.3.

Mass and Moment Calculations

We treat coil springs and wires like masses distributed along smooth curves in space. The distribution is described by a continuous density function $\delta(x, y, z)$ (mass per unit length). The spring's or wire's mass, center of mass, and moments are then calculated with the formulas in Table 13.1. The formulas also apply to thin rods.

Table 13.1 Mass and moment formulas for coil springs, thin rods, and wires lying along a smooth curve C in space

Mass: $M = \int_C \delta(x, y, z)\, ds$

First moments about the coordinate planes:

$$M_{yz} = \int_C x\,\delta\, ds, \qquad M_{xz} = \int_C y\,\delta\, ds, \qquad M_{xy} = \int_C z\,\delta\, ds$$

Coordinates of the center of mass:

$$\bar{x} = M_{yz}/M, \qquad \bar{y} = M_{xz}/M, \qquad \bar{z} = M_{xy}/M$$

Moments of inertia about axes and other lines:

$$I_x = \int_C (y^2 + z^2)\delta\, ds, \qquad I_y = \int_C (x^2 + z^2)\delta\, ds$$

$$I_z = \int_C (x^2 + y^2)\delta\, ds, \qquad I_L = \int_C r^2\delta\, ds$$

$r(x, y, z) =$ distance from point (x, y, z) to line L

Radius of gyration about a line L: $R_L = \sqrt{I_L/M}$

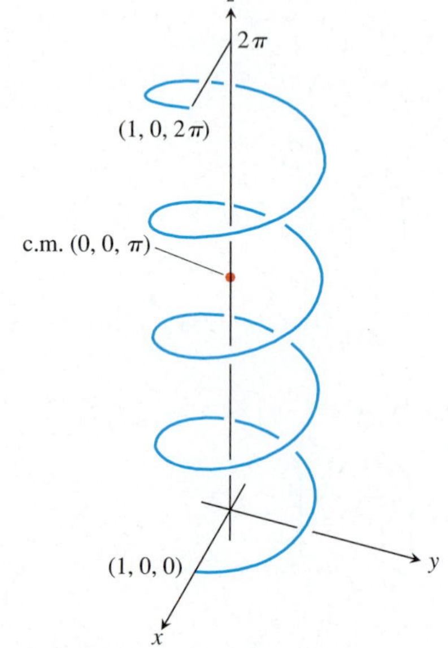

FIGURE 13.4 The helical spring in Example 3.

Example 3 Finding Mass, Center of Mass, Moment of Inertia, Radius of Gyration

A coil spring lies along the helix

$$\mathbf{r}(t) = (\cos 4t)\mathbf{i} + (\sin 4t)\mathbf{j} + t\mathbf{k}, \qquad 0 \le t \le 2\pi.$$

The spring's density is a constant, $\delta = 1$. Find the spring's mass and center of mass, and its moment of inertia and radius of gyration about the z-axis.

Solution We sketch the spring (Figure 13.4). Because of the symmetries involved, the center of mass lies at the point $(0, 0, \pi)$ on the z-axis.

For the remaining calculations, we first find $|\mathbf{v}(t)|$:

$$|\mathbf{v}(t)| = \sqrt{\left(\frac{dx}{dt}\right)^2 + \left(\frac{dy}{dt}\right)^2 + \left(\frac{dz}{dt}\right)^2}$$

$$= \sqrt{(-4\sin 4t)^2 + (4\cos 4t)^2 + 1} = \sqrt{17}.$$

We then evaluate the formulas from Table 13.1 using Equation (2):

$$M = \int_{\text{Helix}} \delta\, ds = \int_0^{2\pi} (1)\sqrt{17}\, dt = 2\pi\sqrt{17}$$

$$I_z = \int_{\text{Helix}} (x^2 + y^2)\,\delta\, ds = \int_0^{2\pi} (\cos^2 4t + \sin^2 4t)(1)\sqrt{17}\, dt$$

$$= \int_0^{2\pi} \sqrt{17}\, dt = 2\pi\sqrt{17}$$

$$R_z = \sqrt{I_z/M} = \sqrt{2\pi\sqrt{17}/(2\pi\sqrt{17})} = 1.$$

Notice that the radius of gyration about the z-axis is the radius of the cylinder around which the helix winds.

Example 4 Finding an Arch's Center of Mass

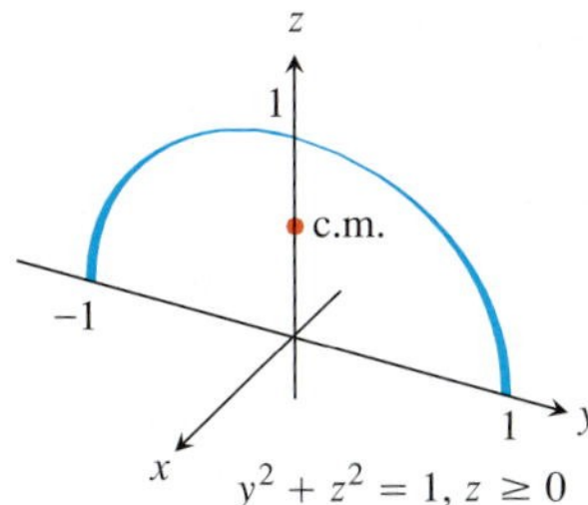

FIGURE 13.5 Example 4 shows how to find the center of mass of a circular arch of variable density.

A slender metal arch, denser at the bottom than top, lies along the semicircle $y^2 + z^2 = 1,\ z \geq 0$, in the yz-plane (Figure 13.5). Find the center of the arch's mass if the density at the point (x, y, z) on the arch is $\delta(x, y, z) = 2 - z$.

Solution We know that $\bar{x} = 0$ and $\bar{y} = 0$ because the arch lies in the yz-plane with its mass distributed symmetrically about the z-axis. To find $\bar{z}$, we parametrize the circle as

$$\mathbf{r}(t) = (\cos t)\mathbf{j} + (\sin t)\mathbf{k}, \qquad 0 \leq t \leq \pi.$$

For this parametrization,

$$|\mathbf{v}(t)| = \sqrt{\left(\frac{dx}{dt}\right)^2 + \left(\frac{dy}{dt}\right)^2 + \left(\frac{dz}{dt}\right)^2} = \sqrt{(0)^2 + (-\sin t)^2 + (\cos t)^2} = 1.$$

The formulas in Table 13.1 then give

$$M = \int_C \delta\, ds = \int_C (2 - z)\, ds = \int_0^{\pi} (2 - \sin t)\,(1)\, dt = 2\pi - 2$$

$$M_{xy} = \int_C z\delta\, ds = \int_C z(2 - z)\, ds = \int_0^{\pi} (\sin t)(2 - \sin t)\, dt$$

$$= \int_0^{\pi} (2\sin t - \sin^2 t)\, dt = \frac{8 - \pi}{2}$$

$$\bar{z} = \frac{M_{xy}}{M} = \frac{8 - \pi}{2} \cdot \frac{1}{2\pi - 2} = \frac{8 - \pi}{4\pi - 4} \approx 0.57.$$

With $\bar{z}$ to the nearest hundredth, the center of mass is $(0, 0, 0.57)$.

EXERCISES 13.1

Graphs of Vector Equations

Match the vector equations in Exercises 1–8 with the graphs (a)–(h) given here.

(a)

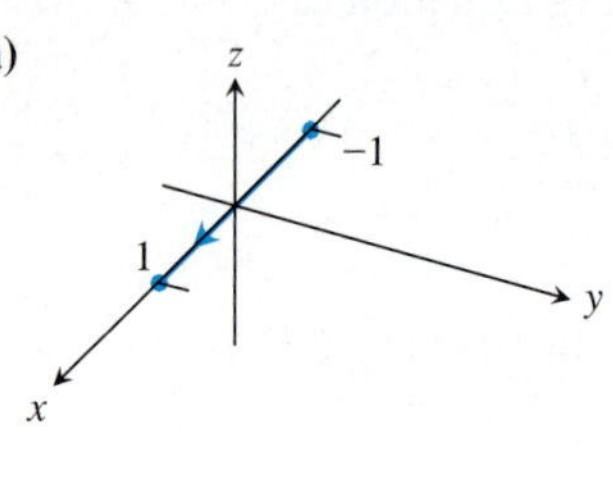

(b)

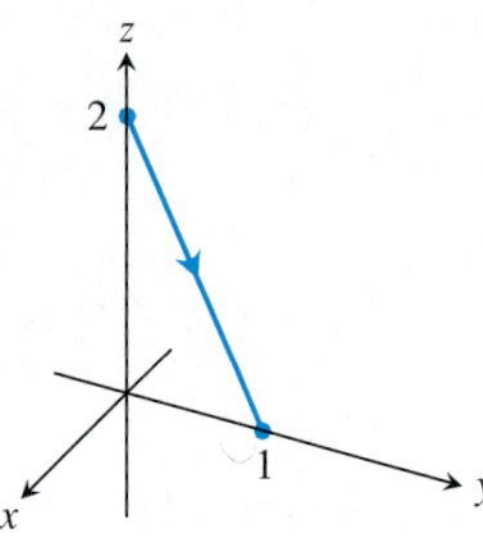

(c)

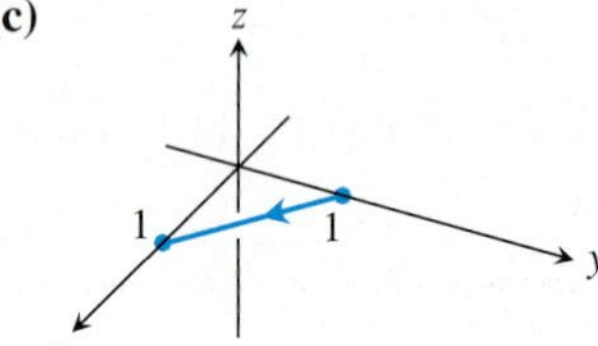

(d)

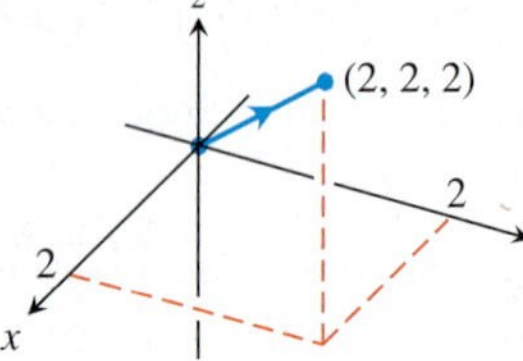

(e)

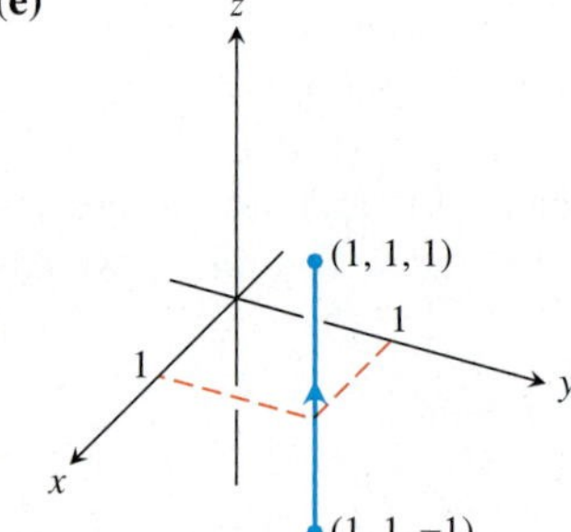

(f)

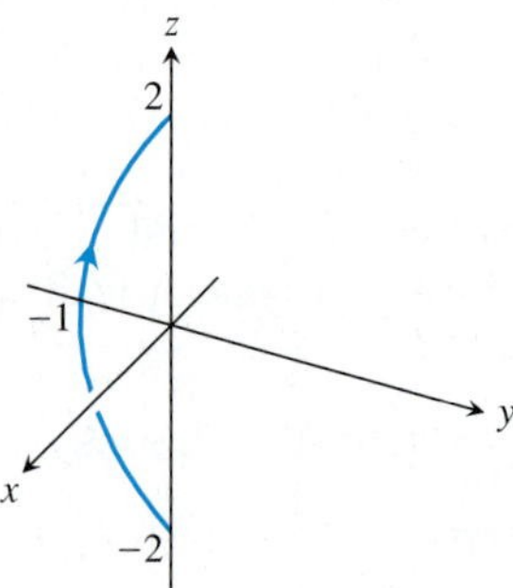

(g)

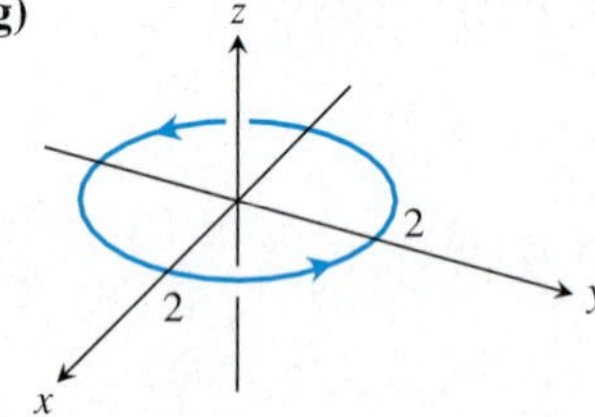

(h)

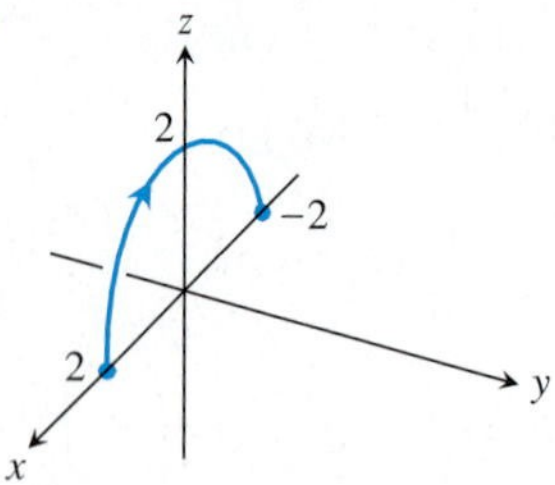

1. $\mathbf{r}(t) = t\mathbf{i} + (1 - t)\mathbf{j}, \quad 0 \le t \le 1$
2. $\mathbf{r}(t) = \mathbf{i} + \mathbf{j} + t\mathbf{k}, \quad -1 \le t \le 1$
3. $\mathbf{r}(t) = (2\cos t)\mathbf{i} + (2\sin t)\mathbf{j}, \quad 0 \le t \le 2\pi$
4. $\mathbf{r}(t) = t\mathbf{i}, \quad -1 \le t \le 1$
5. $\mathbf{r}(t) = t\mathbf{i} + t\mathbf{j} + t\mathbf{k}, \quad 0 \le t \le 2$
6. $\mathbf{r}(t) = t\mathbf{j} + (2 - 2t)\mathbf{k}, \quad 0 \le t \le 1$
7. $\mathbf{r}(t) = (t^2 - 1)\mathbf{j} + 2t\mathbf{k}, \quad -1 \le t \le 1$
8. $\mathbf{r}(t) = (2\cos t)\mathbf{i} + (2\sin t)\mathbf{k}, \quad 0 \le t \le \pi$

Evaluating Line Integrals over Space Curves

9. Evaluate $\int_C (x + y)\, ds$ where C is the straight-line segment $x = t$, $y = (1 - t)$, $z = 0$, from $(0, 1, 0)$ to $(1, 0, 0)$.
10. Evaluate $\int_C (x - y + z - 2)\, ds$ where C is the straight-line segment $x = t$, $y = (1 - t)$, $z = 1$, from $(0, 1, 1)$ to $(1, 0, 1)$.
11. Evaluate $\int_C (xy + y + z)\, ds$ along the curve $\mathbf{r}(t) = 2t\mathbf{i} + t\mathbf{j} + (2 - 2t)\mathbf{k}, 0 \le t \le 1$.
12. Evaluate $\int_C \sqrt{x^2 + y^2}\, ds$ along the curve $\mathbf{r}(t) = (4\cos t)\mathbf{i} + (4\sin t)\mathbf{j} + 3t\mathbf{k}, -2\pi \le t \le 2\pi$.
13. Find the line integral of $f(x, y, z) = x + y + z$ over the straight-line segment from $(1, 2, 3)$ to $(0, -1, 1)$.
14. Find the line integral of $f(x, y, z) = \sqrt{3}/(x^2 + y^2 + z^2)$ over the curve $\mathbf{r}(t) = t\mathbf{i} + t\mathbf{j} + t\mathbf{k}, 1 \le t \le \infty$.
15. Integrate $f(x, y, z) = x + \sqrt{y} - z^2$ over the path from $(0, 0, 0)$ to $(1, 1, 1)$ (Figure 13.6a) given by

    C_1: $\mathbf{r}(t) = t\mathbf{i} + t^2\mathbf{j}, \quad 0 \le t \le 1$

    C_2: $\mathbf{r}(t) = \mathbf{i} + \mathbf{j} + t\mathbf{k}, \quad 0 \le t \le 1$
16. Integrate $f(x, y, z) = x + \sqrt{y} - z^2$ over the path from $(0, 0, 0)$ to $(1, 1, 1)$ (Figure 13.6b) given by

    C_1: $\mathbf{r}(t) = t\mathbf{k}, \quad 0 \le t \le 1$

    C_2: $\mathbf{r}(t) = t\mathbf{j} + \mathbf{k}, \quad 0 \le t \le 1$

    C_3: $\mathbf{r}(t) = t\mathbf{i} + \mathbf{j} + \mathbf{k}, \quad 0 \le t \le 1$

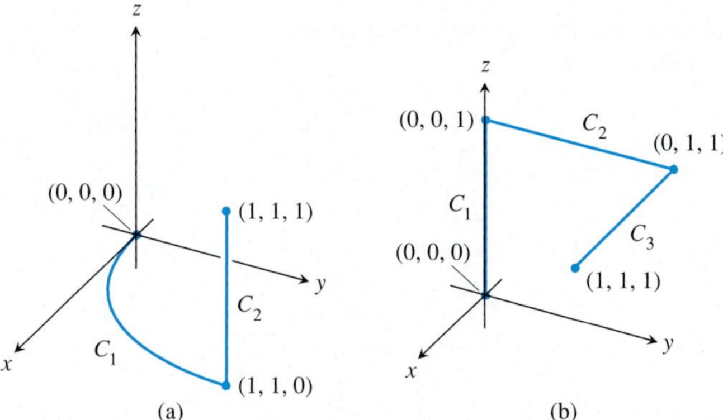

FIGURE 13.6 The paths of integration for Exercises 15 and 16.

17. Integrate $f(x, y, z) = (x + y + z)/(x^2 + y^2 + z^2)$ over the path $\mathbf{r}(t) = t\mathbf{i} + t\mathbf{j} + t\mathbf{k}, 0 < a \le t \le b$.
18. Integrate $f(x, y, z) = -\sqrt{x^2 + z^2}$ over the circle

$$\mathbf{r}(t) = (a\cos t)\mathbf{j} + (a\sin t)\mathbf{k}, \qquad 0 \le t \le 2\pi.$$

Line Integrals over Plane Curves

In Exercises 19–22, integrate f over the given curve.

19. $f(x, y) = x^3/y, \quad C: \quad y = x^2/2, \quad 0 \le x \le 2$

20. $f(x, y) = (x + y^2)/\sqrt{1 + x^2}, \quad C: \quad y = x^2/2$ from $(1, 1/2)$ to $(0, 0)$

21. $f(x, y) = x + y, \quad C: \quad x^2 + y^2 = 4$ in the first quadrant from $(2, 0)$ to $(0, 2)$

22. $f(x, y) = x^2 - y, \quad C: \quad x^2 + y^2 = 4$ in the first quadrant from $(0, 2)$ to $(\sqrt{2}, \sqrt{2})$

Mass and Moments

23. *Mass of a wire* Find the mass of a wire that lies along the curve $\mathbf{r}(t) = (t^2 - 1)\mathbf{j} + 2t\mathbf{k}, 0 \le t \le 1$, if the density is $\delta = (3/2)t$.

24. *Center of mass of a curved wire* A wire of density $\delta(x, y, z) = 15\sqrt{y + 2}$ lies along the curve $\mathbf{r}(t) = (t^2 - 1)\mathbf{j} + 2t\mathbf{k}, -1 \le t \le 1$. Find its center of mass. Then sketch the curve and center of mass together.

25. *Mass of wire with variable density* Find the mass of a thin wire lying along the curve $\mathbf{r}(t) = \sqrt{2}t\mathbf{i} + \sqrt{2}t\mathbf{j} + (4 - t^2)\mathbf{k}, 0 \le t \le 1$, if the density is (a) $\delta = 3t$ and (b) $\delta = 1$.

26. *Center of mass of wire with variable density* Find the center of mass of a thin wire lying along the curve $\mathbf{r}(t) = t\mathbf{i} + 2t\mathbf{j} + (2/3)t^{3/2}\mathbf{k}$, $0 \le t \le 2$, if the density is $\delta = 3\sqrt{5 + t}$.

27. *Moment of inertia and radius of gyration of wire hoop* A circular wire hoop of constant density δ lies along the circle $x^2 + y^2 = a^2$ in the xy-plane. Find the hoop's moment of inertia and radius of gyration about the z-axis.

28. *Inertia and radii of gyration of slender rod* A slender rod of constant density lies along the line segment $\mathbf{r}(t) = t\mathbf{j} + (2 - 2t)\mathbf{k}, 0 \le t \le 1$, in the yz-plane. Find the moments of inertia and radii of gyration of the rod about the three coordinate axes.

29. *Two springs of constant density* A spring of constant density δ lies along the helix

$$\mathbf{r}(t) = (\cos t)\mathbf{i} + (\sin t)\mathbf{j} + t\mathbf{k}, \qquad 0 \le t \le 2\pi.$$

(a) Find I_z and R_z.

(b) Suppose that you have another spring of constant density δ that is twice as long as the spring in part (a) and lies along the helix for $0 \le t \le 4\pi$. Do you expect I_z and R_z for the longer spring to be the same as those for the shorter one, or should they be different? Check your predictions by calculating I_z and R_z for the longer spring.

30. *Wire of constant density* A wire of constant density $\delta = 1$ lies along the curve

$$\mathbf{r}(t) = (t\cos t)\mathbf{i} + (t\sin t)\mathbf{j} + (2\sqrt{2}/3)t^{3/2}\,\mathbf{k}, \qquad 0 \le t \le 1.$$

Find $\bar{z}$, I_z, and R_z.

31. *The arch in Example 4* Find I_x and R_x for the arch in Example 4.

32. *Center of mass, moments of inertia, and radii of gyration for wire with variable density* Find the center of mass, and the moments of inertia and radii of gyration about the coordinate axes of a thin wire lying along the curve

$$\mathbf{r}(t) = t\mathbf{i} + \frac{2\sqrt{2}}{3}t^{3/2}\mathbf{j} + \frac{t^2}{2}\mathbf{k}, \qquad 0 \le t \le 2,$$

if the density is $\delta = 1/(t + 1)$.

COMPUTER EXPLORATIONS

Evaluating Line Integrals Numerically

In Exercises 33–36, use a CAS to perform the following steps to evaluate the line integrals.

(a) Find $ds = |\mathbf{v}(t)|\,dt$ for the path $\mathbf{r}(t) = g(t)\mathbf{i} + h(t)\mathbf{j} + k(t)\mathbf{k}$.

(b) Express the integrand $f(g(t), h(t), k(t))|\mathbf{v}(t)|$ as a function of the parameter t.

(c) Evaluate $\int_C f\,ds$ using Equation (2) in the text.

33. $f(x, y, z) = \sqrt{1 + 30x^2 + 10y}; \quad \mathbf{r}(t) = t\mathbf{i} + t^2\mathbf{j} + 3t^2\mathbf{k}$, $0 \le t \le 2$

34. $f(x, y, z) = \sqrt{1 + x^3 + 5y^3}; \quad \mathbf{r}(t) = t\mathbf{i} + \frac{1}{3}t^2\mathbf{j} + \sqrt{t}\mathbf{k}$, $0 \le t \le 2$

35. $f(x, y, z) = x\sqrt{y} - 3z^2; \quad \mathbf{r}(t) = (\cos 2t)\mathbf{i} + (\sin 2t)\mathbf{j} + 5t\mathbf{k}$, $0 \le t \le 2\pi$

36. $f(x, y, z) = \left(1 + \frac{9}{4}z^{1/3}\right)^{1/4}; \quad \mathbf{r}(t) = (\cos 2t)\mathbf{i} + (\sin 2t)\mathbf{j} + t^{5/2}\mathbf{k}$, $0 \le t \le 2\pi$

13.2 Vector Fields, Work, Circulation, and Flux

Vector Fields • Gradient Fields • Work Done by a Force over a Curve in Space • Notation and Evaluation • Flow Integrals and Circulation • Flux Across a Plane Curve

When we study physical phenomena that are represented by vectors, we replace integrals over closed intervals by integrals over paths through vector fields. We use such integrals to find the work done in moving an object along a path against a variable force (a vehicle sent into space against Earth's gravitational field) or to find the work done by a vector field in moving an object along a path through the field (the work done by an accelerator in raising the energy of a particle). We also use them to find the rates at which fluids flow along and across curves.

Vector Fields

A **vector field** on a domain in the plane or in space is a function that assigns a vector to each point in the domain. A field of three-dimensional vectors might have a formula like

$$\mathbf{F}(x, y, z) = M(x, y, z)\mathbf{i} + N(x, y, z)\mathbf{j} + P(x, y, z)\mathbf{k}.$$

The field is **continuous** if the **component functions** M, N, and P are continuous, **differentiable** if M, N, and P are differentiable, and so on. A field of two-dimensional vectors might have a formula like

$$\mathbf{F}(x, y) = M(x, y)\mathbf{i} + N(x, y)\mathbf{j}.$$

If we attach a projectile's velocity vector to each point of the projectile's trajectory in the plane of motion, we have a two-dimensional field defined along the trajectory. If we attach the gradient vector of a scalar function to each point of a level surface of the function, we have a three-dimensional field on the surface. If we attach the velocity vector to each point of a flowing fluid, we have a three-dimensional field defined on a region in space. These and other fields are illustrated in Figures 13.7 through 13.15. Some of the illustrations give formulas for the fields as well.

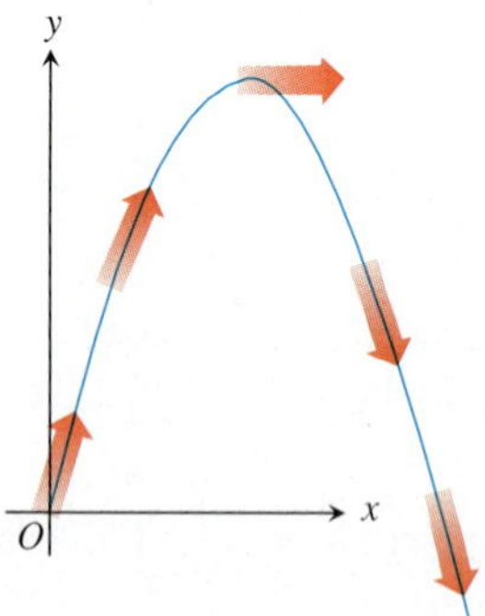

FIGURE 13.7 The velocity vectors $\mathbf{v}(t)$ of a projectile's motion make a vector field along the trajectory.

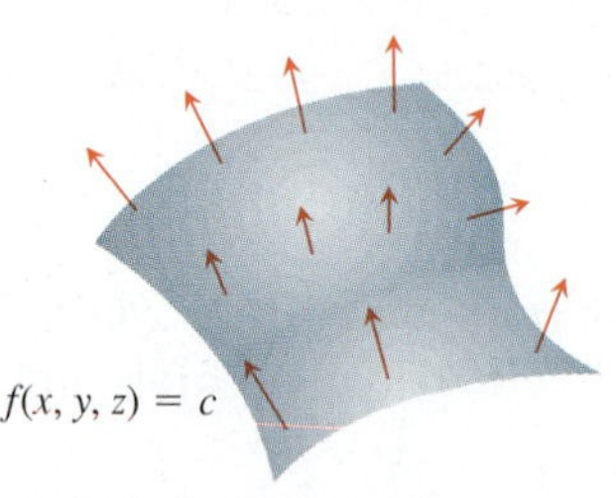

FIGURE 13.8 The field of gradient vectors ∇f on a surface $f(x, y, z) = c$.

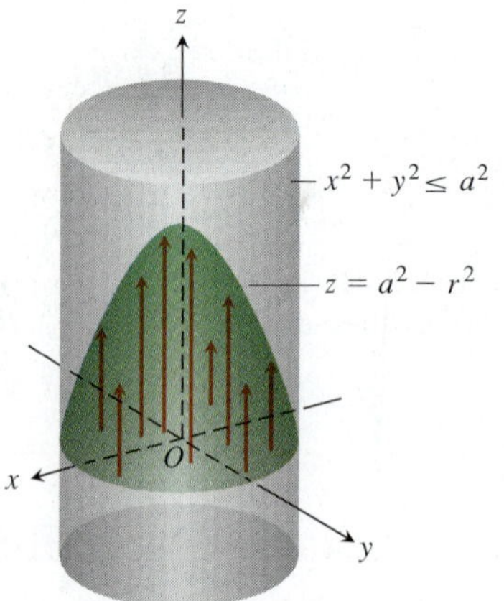

FIGURE 13.9 The flow of fluid in a long cylindrical pipe. The vectors $\mathbf{v} = (a^2 - r^2)\mathbf{k}$ inside the cylinder that have their bases in the xy-plane have their tips on the paraboloid $z = a^2 - r^2$.

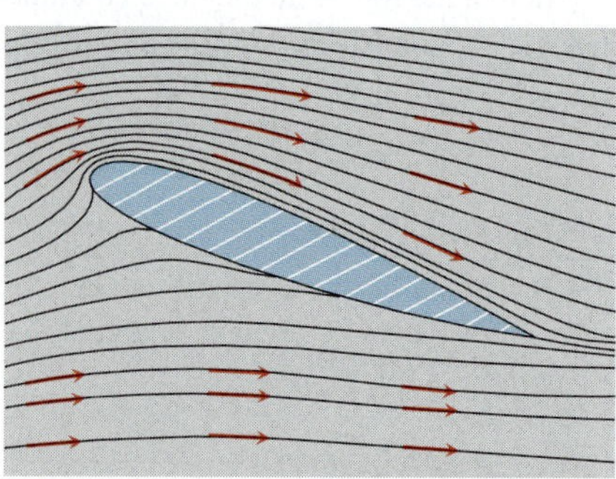

FIGURE 13.10 Velocity vectors of a flow around an airfoil in a wind tunnel. The streamlines were made visible by kerosene smoke. (Adapted from *NCFMF Book of Film Notes*, 1974, MIT Press with Education Development Center, Inc., Newton, Massachusetts.)

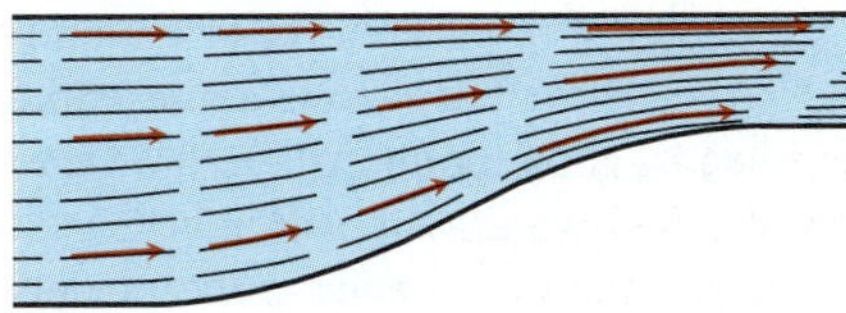

FIGURE 13.11 Streamlines in a contracting channel. The water speeds up as the channel narrows and the velocity vectors increase in length. (Adapted from *NCFMF Book of Film Notes*, 1974, MIT Press with Education Development Center, Inc., Newton, Massachusetts.)

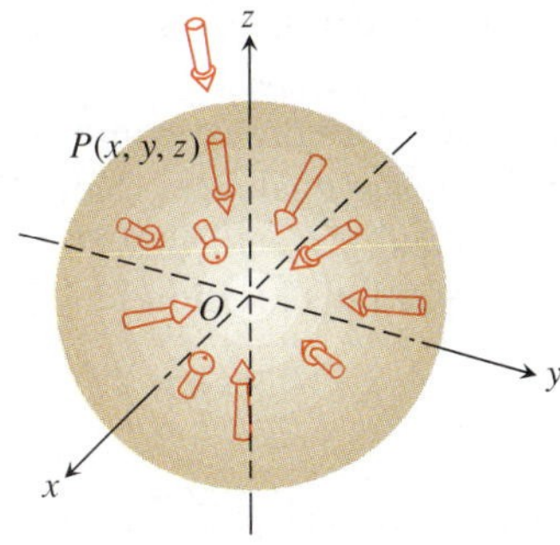

FIGURE 13.12 Vectors in the gravitational field

$$\mathbf{F} = -\frac{GM(x\mathbf{i} + y\mathbf{j} + z\mathbf{k})}{(x^2 + y^2 + z^2)^{3/2}}.$$

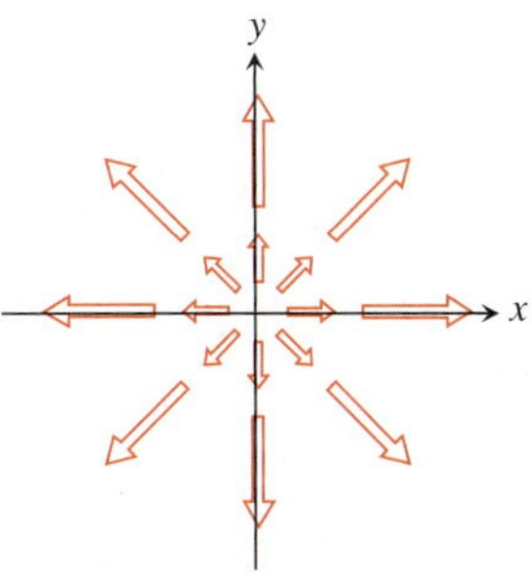

FIGURE 13.13 The radial field $\mathbf{F} = x\mathbf{i} + y\mathbf{j}$ of position vectors of points in the plane. Notice the convention that an arrow is drawn with its tail, not its head, at the point where $\mathbf{F}$ is evaluated.

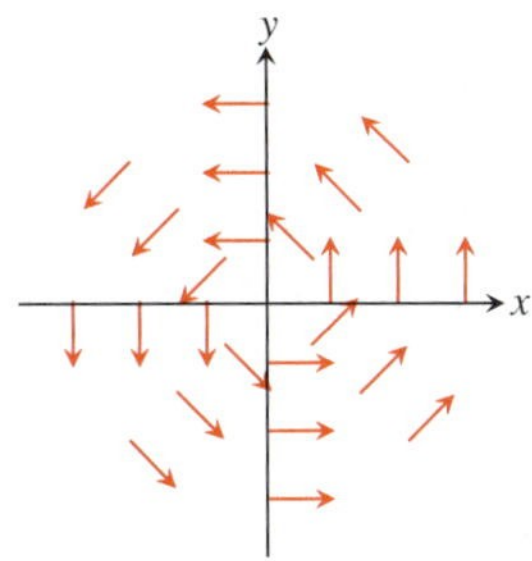

FIGURE 13.14 The circumferential or "spin" field of unit vectors

$$\mathbf{F} = (-y\mathbf{i} + x\mathbf{j})/(x^2 + y^2)^{1/2}$$

in the plane. The field is not defined at the origin.

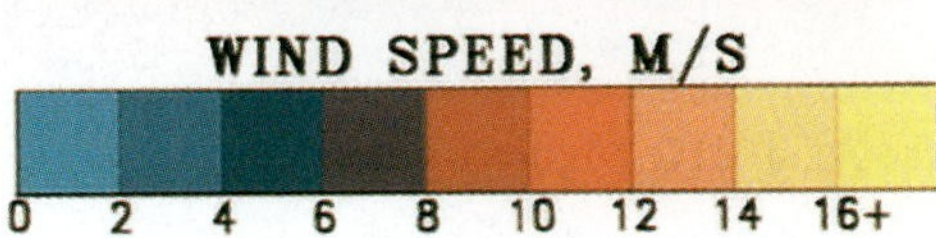

FIGURE 13.15 NASA's *Seasat* used radar during a 3-day period in September 1978 to take 350,000 wind measurements over the world's oceans. The arrows show wind direction; their length and the color contouring indicate speed. Notice the heavy storm south of Greenland.

To sketch the fields that had formulas, we picked a representative selection of domain points and sketched the vectors attached to them. The arrows representing the vectors are drawn with their tails, not their heads, at the points where the vector functions are evaluated. This is different from the way we draw position vectors of planets and projectiles, with their tails at the origin and their heads at the planet's and projectile's locations.

Gradient Fields

CD-ROM
WEBsite
Historical Biography
Wilhelm Weber
(1804 — 1891)

Definition Gradient Field

The **gradient field** of a differentiable function $f(x, y, z)$ is the field of gradient vectors

$$\nabla f = \frac{\partial f}{\partial x}\mathbf{i} + \frac{\partial f}{\partial y}\mathbf{j} + \frac{\partial f}{\partial z}\mathbf{k}.$$

Example 1 Finding a Gradient Field

Find the gradient field of $f(x, y, z) = xyz$.

Solution The gradient field of f is the field $\mathbf{F} = \nabla f = yz\mathbf{i} + xz\mathbf{j} + xy\mathbf{k}$.

As we see in Section 13.3, gradient fields are of special importance in engineering, mathematics, and physics.

Work Done by a Force over a Curve in Space

Suppose that the vector field $\mathbf{F} = M(x, y, z)\mathbf{i} + N(x, y, z)\mathbf{j} + P(x, y, z)\mathbf{k}$ represents a force throughout a region in space (it might be the force of gravity or an electromagnetic force of some kind) and that

$$\mathbf{r}(t) = g(t)\mathbf{i} + h(t)\mathbf{j} + k(t)\mathbf{k}, \qquad a \le t \le b,$$

is a smooth curve in the region. Then the integral of $\mathbf{F} \cdot \mathbf{T}$, the scalar component of $\mathbf{F}$ in the direction of the curve's unit tangent vector, over the curve is called the work done by $\mathbf{F}$ over the curve from a to b (Figure 13.16).

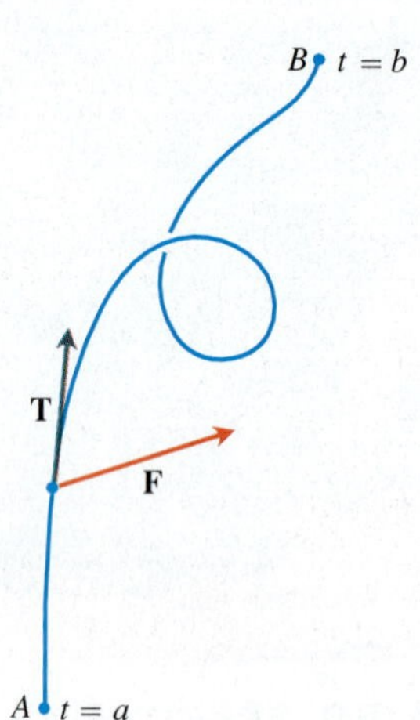

FIGURE 13.16 Figure for the definition of work.

Definition Work Over a Smooth Curve

The **work** done by a force $\mathbf{F} = M\mathbf{i} + N\mathbf{j} + P\mathbf{k}$ over a smooth curve $\mathbf{r}(t)$ from $t = a$ to $t = b$ is

$$W = \int_{t=a}^{t=b} \mathbf{F} \cdot \mathbf{T}\, ds. \tag{1}$$

We motivate Equation (1) with the same kind of reasoning we used in Chapter 5 to derive the formula $W = \int_a^b F(x)\, dx$ for the work done by a continuous force of magnitude $F(x)$ directed along an interval of the x-axis. We divide the curve into short segments, apply the constant-force-times-distance formula for work to approximate the work over each curved segment, add the results to approximate the work over the entire curve, and calculate the work as the limit of the approximating sums as the segments become shorter and more numerous. To find exactly what the limiting integral should be, we partition the parameter interval $[a, b]$ in the usual way and choose a point c_k in each subinterval $[t_k, t_{k+1}]$. The partition of $[a, b]$ determines ("induces," we say) a partition of the curve, with the point P_k being the tip of the position vector $\mathbf{r}(t_k)$ and Δs_k being the length of the curve segment P_kP_{k+1} (Figure 13.17).

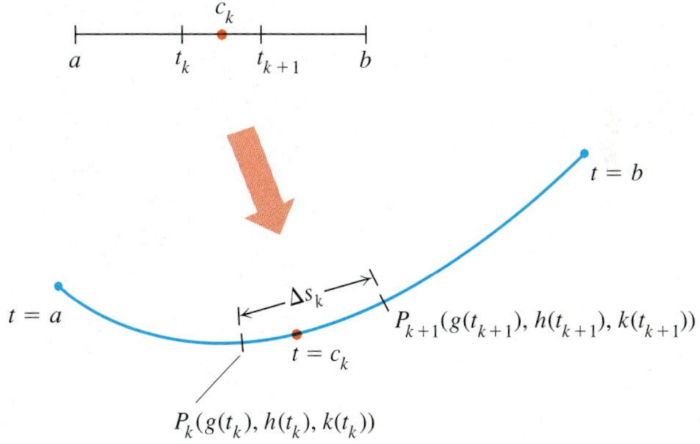

FIGURE 13.17 Each partition of $[a, b]$ induces a partition of the curve $\mathbf{r}(t) = g(t)\mathbf{i} + h(t)\mathbf{j} + k(t)\mathbf{k}$.

If $\mathbf{F}_k$ denotes the value of $\mathbf{F}$ at the point on the curve corresponding to $t = c_k$ and $\mathbf{T}_k$ denotes the curve's unit tangent vector at this point, then $\mathbf{F}_k \cdot \mathbf{T}_k$ is the scalar component of $\mathbf{F}$ in the direction of $\mathbf{T}$ at $t = c_k$ (Figure 13.18). The work done by $\mathbf{F}$ along the curve segment P_kP_{k+1} is approximately

$$\begin{pmatrix}\text{Force component in}\\ \text{direction of motion}\end{pmatrix} \times \begin{pmatrix}\text{distance}\\ \text{applied}\end{pmatrix} = \mathbf{F}_k \cdot \mathbf{T}_k \, \Delta s_k.$$

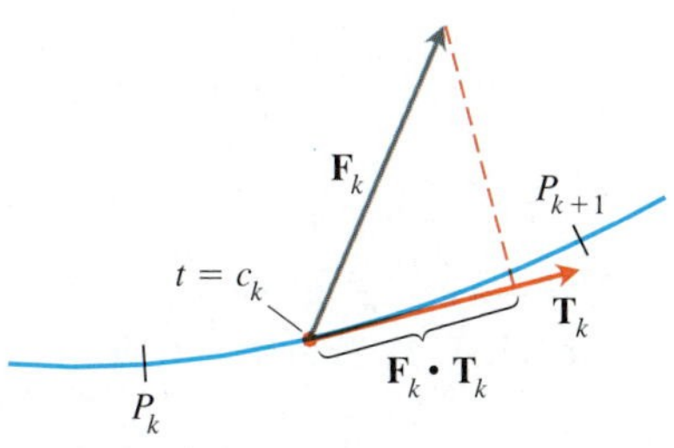

FIGURE 13.18 An enlarged view of the curve segment P_kP_{k+1} in Figure 13.17, showing the force and unit tangent vectors at the point on the curve where $t = c_k$.

The work done by $\mathbf{F}$ along the curve from $t = a$ to $t = b$ is approximately

$$\sum_{k=1}^{n} \mathbf{F}_k \cdot \mathbf{T}_k \, \Delta s_k.$$

As the norm of the partition of $[a, b]$ approaches zero, the norm of the induced partition of the curve approaches zero and these sums approach the line integral

$$\int_{t=a}^{t=b} \mathbf{F} \cdot \mathbf{T} \, ds.$$

The sign of the number we calculate with this integral depends on the direction in which the curve is traversed as t increases. If we reverse the direction of motion, we reverse the direction of $\mathbf{T}$ and change the sign of $\mathbf{F} \cdot \mathbf{T}$ and its integral.

Notation and Evaluation

Table 13.2 shows six ways to write the work integral in Equation (1). Despite their variety, the formulas in Table 13.2 are all evaluated the same way.

Table 13.2 Different ways to write the work integral

$\mathbf{W} = \displaystyle\int_{t=a}^{t=b} \mathbf{F} \cdot \mathbf{T}\, ds$	The definition
$= \displaystyle\int_{t=a}^{t=b} \mathbf{F} \cdot d\mathbf{r}$	Compact differential form
$= \displaystyle\int_a^b \mathbf{F} \cdot \frac{d\mathbf{r}}{dt}\, dt$	Expanded to include dt; emphasizes the parameter t and velocity vector $d\mathbf{r}/dt$
$= \displaystyle\int_a^b \left(M \frac{dg}{dt} + N \frac{dh}{dt} + P \frac{dk}{dt} \right) dt$	Emphasizes the component functions
$= \displaystyle\int_a^b \left(M \frac{dx}{dt} + N \frac{dy}{dt} + P \frac{dz}{dt} \right) dt$	Abbreviates the components of $\mathbf{r}$
$= \displaystyle\int_a^b M\, dx + N\, dy + P\, dz$	dt's canceled; the most common form

How to Evaluate a Work Integral

To evaluate the work integral, take these steps.

Step 1. Evaluate $\mathbf{F}$ on the curve as a function of the parameter t.

Step 2. Find $d\mathbf{r}/dt$.

Step 3. Dot $\mathbf{F}$ with $d\mathbf{r}/dt$.

Step 4. Integrate from $t = a$ to $t = b$.

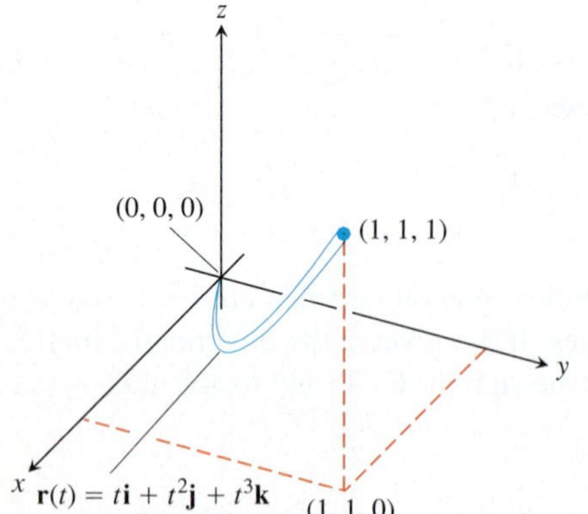

FIGURE 13.19 The curve in Example 2.

Example 2 Finding Work Done by a Variable Force Over a Space Curve

Find the work done by $\mathbf{F} = (y - x^2)\mathbf{i} + (z - y^2)\mathbf{j} + (x - z^2)\mathbf{k}$ over the curve $\mathbf{r}(t) = t\mathbf{i} + t^2\mathbf{j} + t^3\mathbf{k}, 0 \le t \le 1$, from (0, 0, 0) to (1, 1, 1) (Figure 13.19).

Solution

Step 1: *Evaluate* $\mathbf{F}$ *on the curve.*

$$\begin{aligned}\mathbf{F} &= (y - x^2)\mathbf{i} + (z - y^2)\mathbf{j} + (x - z^2)\mathbf{k} \\ &= \underbrace{(t^2 - t^2)}_{0}\mathbf{i} + (t^3 - t^4)\mathbf{j} + (t - t^6)\mathbf{k}\end{aligned}$$

Step 2: *Find* $d\mathbf{r}/dt$.

$$\frac{d\mathbf{r}}{dt} = \frac{d}{dt}(t\mathbf{i} + t^2\mathbf{j} + t^3\mathbf{k}) = \mathbf{i} + 2t\mathbf{j} + 3t^2\mathbf{k}$$

Step 3: *Dot* $\mathbf{F}$ *with* $d\mathbf{r}/dt$.

$$\mathbf{F}\cdot\frac{d\mathbf{r}}{dt} = [(t^3 - t^4)\mathbf{j} + (t - t^6)\mathbf{k}]\cdot(\mathbf{i} + 2t\mathbf{j} + 3t^2\mathbf{k})$$

$$= (t^3 - t^4)(2t) + (t - t^6)(3t^2) = 2t^4 - 2t^5 + 3t^3 - 3t^8$$

Step 4: *Integrate from* $t = 0$ *to* $t = 1$.

$$\text{Work} = \int_0^1 (2t^4 - 2t^5 + 3t^3 - 3t^8)\,dt$$

$$= \left[\frac{2}{5}t^5 - \frac{2}{6}t^6 + \frac{3}{4}t^4 - \frac{3}{9}t^9\right]_0^1 = \frac{29}{60}$$

Flow Integrals and Circulation

Instead of being a force field, suppose that $\mathbf{F}$ represents the velocity field of a fluid flowing through a region in space (a tidal basin or the turbine chamber of a hydroelectric generator, for example). Under these circumstances, the integral of $\mathbf{F}\cdot\mathbf{T}$ along a curve in the region gives the fluid's flow along the curve.

Definitions Flow, Flow Integral, and Circulation

If $\mathbf{r}(t)$ is a smooth curve in the domain of a continuous velocity field $\mathbf{F}$, the **flow** along the curve from $t = a$ to $t = b$ is

$$\text{Flow} = \int_a^b \mathbf{F}\cdot\mathbf{T}\,ds. \tag{2}$$

The integral in this case is called a **flow integral.** If the curve is a closed loop, the flow is called the **circulation** around the curve.

We evaluate flow integrals the same way we evaluate work integrals.

Example 3 Finding Flow Along a Helix

A fluid's velocity field is $\mathbf{F} = x\mathbf{i} + z\mathbf{j} + y\mathbf{k}$. Find the flow along the helix $\mathbf{r}(t) = (\cos t)\mathbf{i} + (\sin t)\mathbf{j} + t\mathbf{k}$, $0 \le t \le \pi/2$.

Solution

Step 1: *Evaluate* $\mathbf{F}$ *on the curve.*

$$\mathbf{F} = x\mathbf{i} + z\mathbf{j} + y\mathbf{k} = (\cos t)\mathbf{i} + t\mathbf{j} + (\sin t)\mathbf{k}$$

Step 2: *Find* $d\mathbf{r}/dt$.

$$\frac{d\mathbf{r}}{dt} = (-\sin t)\mathbf{i} + (\cos t)\mathbf{j} + \mathbf{k}$$

Step 3: *Find* $\mathbf{F}\cdot(d\mathbf{r}/dt)$.

$$\begin{aligned}\mathbf{F}\cdot\frac{d\mathbf{r}}{dt} &= (\cos t)(-\sin t) + (t)(\cos t) + (\sin t)(1)\\ &= -\sin t\cos t + t\cos t + \sin t\end{aligned}$$

Step 4: *Integrate from* $t = a$ *to* $t = b$.

$$\begin{aligned}\text{Flow} &= \int_{t=a}^{t=b} \mathbf{F}\cdot\frac{d\mathbf{r}}{dt}\,dt = \int_0^{\pi/2} (-\sin t\cos t + t\cos t + \sin t)\,dt\\ &= \left[\frac{\cos^2 t}{2} + t\sin t\right]_0^{\pi/2} = \left(0 + \frac{\pi}{2}\right) - \left(\frac{1}{2} + 0\right) = \frac{\pi}{2} - \frac{1}{2}\end{aligned}$$

Example 4 Finding Circulation Around a Circle

Find the circulation of the field $\mathbf{F} = (x - y)\mathbf{i} + x\mathbf{j}$ around the circle $\mathbf{r}(t) = (\cos t)\mathbf{i} + (\sin t)\mathbf{j},\ 0 \le t \le 2\pi$.

Solution

1. On the circle, $\mathbf{F} = (x - y)\mathbf{i} + x\mathbf{j} = (\cos t - \sin t)\mathbf{i} + (\cos t)\mathbf{j}$.

2. $\dfrac{d\mathbf{r}}{dt} = (-\sin t)\mathbf{i} + (\cos t)\mathbf{j}$

3. $\mathbf{F}\cdot\dfrac{d\mathbf{r}}{dt} = -\sin t\cos t + \underbrace{\sin^2 t + \cos^2 t}_{1}$

4. $$\begin{aligned}\text{Circulation} &= \int_0^{2\pi} \mathbf{F}\cdot\frac{d\mathbf{r}}{dt}\,dt = \int_0^{2\pi} (1 - \sin t\cos t)\,dt\\ &= \left[t - \frac{\sin^2 t}{2}\right]_0^{2\pi} = 2\pi\end{aligned}$$

Flux Across a Plane Curve

To find the rate at which a fluid is entering or leaving a region enclosed by a smooth curve C in the xy-plane, we calculate the line integral over C of $\mathbf{F}\cdot\mathbf{n}$, the scalar component of the fluid's velocity field in the direction of the curve's outward-pointing normal vector. The value of this integral is the *flux* of $\mathbf{F}$ across C. *Flux* is Latin for *flow*, but many flux calculations involve no motion at all. If $\mathbf{F}$ were an electric field or a magnetic field, for instance, the integral of $\mathbf{F}\cdot\mathbf{n}$ would still be called the flux of the field across C.

Definition Flux Across a Closed Curve in the Plane

If C is a smooth closed curve in the domain of a continuous vector field $\mathbf{F} = M(x, y)\mathbf{i} + N(x, y)\mathbf{j}$ in the plane and if $\mathbf{n}$ is the outward-pointing unit normal vector on C, the **flux** of $\mathbf{F}$ across C is

$$\text{Flux of } \mathbf{F} \text{ across } C = \int_C \mathbf{F}\cdot\mathbf{n}\,ds. \tag{3}$$

Notice the difference between flux and circulation. The flux of $\mathbf{F}$ across C is the line integral with respect to arc length of $\mathbf{F} \cdot \mathbf{n}$, the scalar component of $\mathbf{F}$ in the direction of the outward normal. The circulation of $\mathbf{F}$ around C is the line integral with respect to arc length of $\mathbf{F} \cdot \mathbf{T}$, the scalar component of $\mathbf{F}$ in the direction of the unit tangent vector. Flux is the integral of the normal component of $\mathbf{F}$; circulation is the integral of the tangential component of $\mathbf{F}$.

To evaluate the integral in Equation (3), we begin with a smooth parametrization

$$x = g(t), \qquad y = h(t), \qquad a \le t \le b,$$

that traces the curve C exactly once as t increases from a to b. We can find the outward unit normal vector $\mathbf{n}$ by crossing the curve's unit tangent vector $\mathbf{T}$ with the vector $\mathbf{k}$. But which order do we choose, $\mathbf{T} \times \mathbf{k}$ or $\mathbf{k} \times \mathbf{T}$? Which one points outward? It depends on which way C is traversed as t increases. If the motion is clockwise, $\mathbf{k} \times \mathbf{T}$ points outward; if the motion is counterclockwise, $\mathbf{T} \times \mathbf{k}$ points outward (Figure 13.20). The usual choice is $\mathbf{n} = \mathbf{T} \times \mathbf{k}$, the choice that assumes counterclockwise motion. Thus, although the value of the arc length integral in the definition of flux in Equation (3) does not depend on which way C is traversed, the formulas we are about to derive for evaluating the integral in Equation (3) will assume counterclockwise motion.

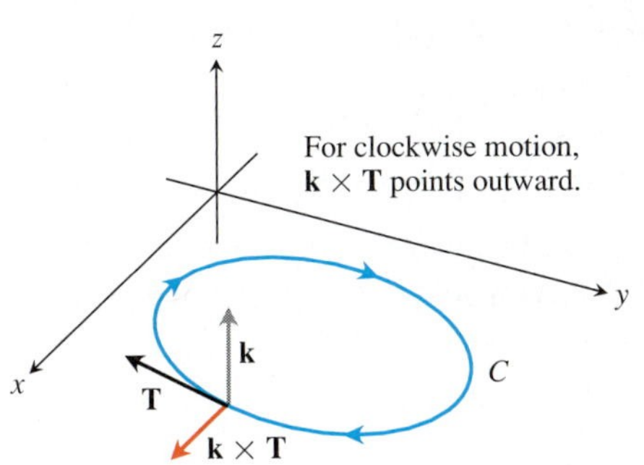

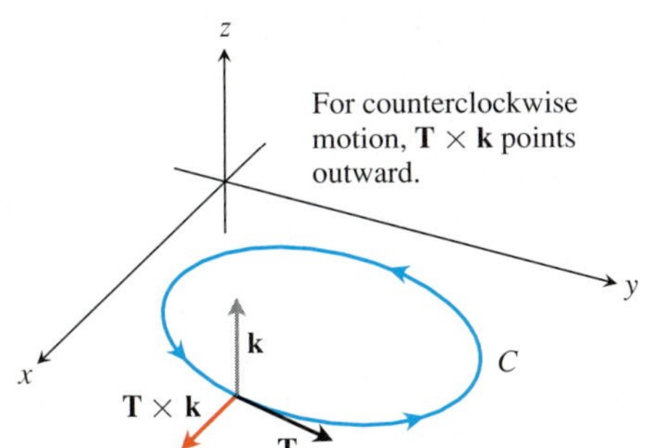

FIGURE 13.20 To find an outward unit normal vector for a smooth curve C in the xy-plane that is traversed counterclockwise as t increases, we take $\mathbf{n} = \mathbf{T} \times \mathbf{k}$.

In terms of components,

$$\mathbf{n} = \mathbf{T} \times \mathbf{k} = \left(\frac{dx}{ds}\mathbf{i} + \frac{dy}{ds}\mathbf{j}\right) \times \mathbf{k} = \frac{dy}{ds}\mathbf{i} - \frac{dx}{ds}\mathbf{j}.$$

If $\mathbf{F} = M(x, y)\mathbf{i} + N(x, y)\mathbf{j}$, then

$$\mathbf{F} \cdot \mathbf{n} = M(x, y)\frac{dy}{ds} - N(x, y)\frac{dx}{ds}.$$

Hence,

$$\int_C \mathbf{F} \cdot \mathbf{n}\, ds = \int_C \left(M\frac{dy}{ds} - N\frac{dx}{ds}\right) ds = \oint_C M\, dy - N\, dx.$$

We put a directed circle $\bigcirc$ on the last integral as a reminder that the integration around the closed curve C is to be in the counterclockwise direction. To evaluate this integral, we express M, dy, N, and dx in terms of t and integrate from $t = a$ to $t = b$. We do not need to know either $\mathbf{n}$ or ds to find the flux.

Formula for Calculating Flux Across a Smooth Closed Plane Curve

$$(\text{Flux of } \mathbf{F} = M\mathbf{i} + N\mathbf{j} \text{ across } C) = \oint_C M\, dy - N\, dx \qquad (4)$$

The integral can be evaluated from any smooth parametrization $x = g(t)$, $y = h(t)$, $a \le t \le b$, that traces C counterclockwise exactly once.

Example 5 Finding Flux Across a Circle

Find the flux of $\mathbf{F} = (x - y)\mathbf{i} + x\mathbf{j}$ across the circle $x^2 + y^2 = 1$ in the xy-plane.

Solution The parametrization $\mathbf{r}(t) = (\cos t)\mathbf{i} + (\sin t)\mathbf{j},\ 0 \le t \le 2\pi$, traces the circle counterclockwise exactly once. We can therefore use this parametrization in Equation (4). With

$$M = x - y = \cos t - \sin t, \qquad dy = d(\sin t) = \cos t\, dt$$
$$N = x = \cos t, \qquad dx = d(\cos t) = -\sin t\, dt,$$

We find

$$\text{Flux} = \int_C M\,dy - N\,dx = \int_0^{2\pi} (\cos^2 t - \sin t \cos t + \cos t \sin t)\, dt \qquad \text{Eq. (4)}$$
$$= \int_0^{2\pi} \cos^2 t\, dt = \int_0^{2\pi} \frac{1 + \cos 2t}{2}\, dt = \left[\frac{t}{2} + \frac{\sin 2t}{4}\right]_0^{2\pi} = \pi.$$

The flux of $\mathbf{F}$ across the circle is π. Since the answer is positive, the net flow across the curve is outward. A net inward flow would have given a negative flux.

EXERCISES 13.2

Vector and Gradient Fields

Find the gradient fields of the functions in Exercises 1–4.

1. $f(x, y, z) = (x^2 + y^2 + z^2)^{-1/2}$
2. $f(x, y, z) = \ln \sqrt{x^2 + y^2 + z^2}$
3. $g(x, y, z) = e^z - \ln(x^2 + y^2)$
4. $g(x, y, z) = xy + yz + xz$
5. Give a formula $\mathbf{F} = M(x, y)\mathbf{i} + N(x, y)\mathbf{j}$ for the vector field in the plane that has the property that $\mathbf{F}$ points toward the origin with magnitude inversely proportional to the square of the distance from (x, y) to the origin. (The field is not defined at $(0, 0)$.)
6. Give a formula $\mathbf{F} = M(x, y)\mathbf{i} + N(x, y)\mathbf{j}$ for the vector field in the plane that has the properties that $\mathbf{F} = \mathbf{0}$ at $(0, 0)$ and that at any other point (a, b), $\mathbf{F}$ is tangent to the circle $x^2 + y^2 = a^2 + b^2$ and points in the clockwise direction with magnitude $|\mathbf{F}| = \sqrt{a^2 + b^2}$.

Work

In Exercises 7–12, find the work done by force $\mathbf{F}$ from $(0, 0, 0)$ to $(1, 1, 1)$ over each of the following paths (Figure 13.21):

(a) The straight-line path C_1: $\mathbf{r}(t) = t\mathbf{i} + t\mathbf{j} + t\mathbf{k},\quad 0 \le t \le 1$

(b) The curved path C_2: $\mathbf{r}(t) = t\mathbf{i} + t^2\mathbf{j} + t^4\mathbf{k},\quad 0 \le t \le 1$

(c) The path $C_3 \cup C_4$ consisting of the line segment from $(0, 0, 0)$ to $(1, 1, 0)$ followed by the segment from $(1, 1, 0)$ to $(1, 1, 1)$

7. $\mathbf{F} = 3y\mathbf{i} + 2x\mathbf{j} + 4z\mathbf{k}$
8. $\mathbf{F} = [1/(x^2 + 1)]\mathbf{j}$
9. $\mathbf{F} = \sqrt{z}\mathbf{i} - 2x\mathbf{j} + \sqrt{y}\mathbf{k}$
10. $\mathbf{F} = xy\mathbf{i} + yz\mathbf{j} + xz\mathbf{k}$

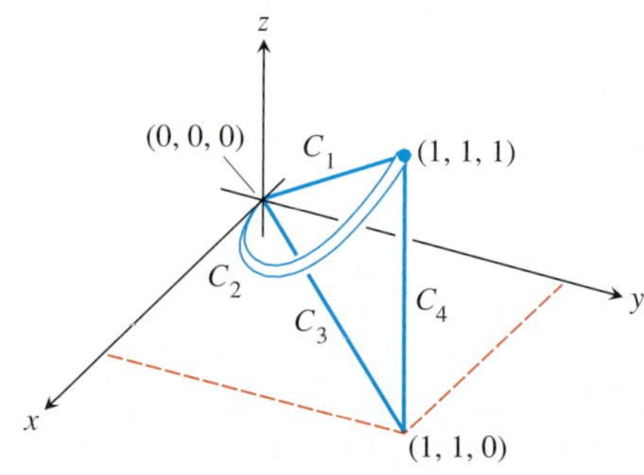

FIGURE 13.21 The paths from $(0, 0, 0)$ to $(1, 1, 1)$.

11. $\mathbf{F} = (3x^2 - 3x)\mathbf{i} + 3z\mathbf{j} + \mathbf{k}$
12. $\mathbf{F} = (y + z)\mathbf{i} + (z + x)\mathbf{j} + (x + y)\mathbf{k}$

In Exercises 13–16, find the work done by $\mathbf{F}$ over the curve in the direction of increasing t.

13. $\mathbf{F} = xy\mathbf{i} + y\mathbf{j} - yz\mathbf{k}$
    $\mathbf{r}(t) = t\mathbf{i} + t^2\mathbf{j} + t\mathbf{k},\quad 0 \le t \le 1$
14. $\mathbf{F} = 2y\mathbf{i} + 3x\mathbf{j} + (x + y)\mathbf{k}$
    $\mathbf{r}(t) = (\cos t)\mathbf{i} + (\sin t)\mathbf{j} + (t/6)\mathbf{k},\quad 0 \le t \le 2\pi$
15. $\mathbf{F} = z\mathbf{i} + x\mathbf{j} + y\mathbf{k}$
    $\mathbf{r}(t) = (\sin t)\mathbf{i} + (\cos t)\mathbf{j} + t\mathbf{k},\quad 0 \le t \le 2\pi$
16. $\mathbf{F} = 6z\mathbf{i} + y^2\mathbf{j} + 12x\mathbf{k}$
    $\mathbf{r}(t) = (\sin t)\mathbf{i} + (\cos t)\mathbf{j} + (t/6)\mathbf{k},\quad 0 \le t \le 2\pi$

Line Integrals and Vector Fields in the Plane

17. Evaluate $\int_C xy\,dx + (x + y)\,dy$ along the curve $y = x^2$ from $(-1, 1)$ to $(2, 4)$.

18. Evaluate $\int_C (x - y)\,dx + (x + y)\,dy$ counterclockwise around the triangle with vertices $(0, 0)$, $(1, 0)$, and $(0, 1)$.

19. Evaluate $\int_C \mathbf{F} \cdot \mathbf{T}\,ds$ for the vector field $\mathbf{F} = x^2\mathbf{i} - y\mathbf{j}$ along the curve $x = y^2$ from $(4, 2)$ to $(1, -1)$.

20. Evaluate $\int_C \mathbf{F} \cdot d\mathbf{r}$ for the vector field $\mathbf{F} = y\mathbf{i} - x\mathbf{j}$ counterclockwise along the unit circle $x^2 + y^2 = 1$ from $(1, 0)$ to $(0, 1)$.

21. *Work* Find the work done by the force $\mathbf{F} = xy\mathbf{i} + (y - x)\mathbf{j}$ over the straight line from $(1, 1)$ to $(2, 3)$.

22. *Work* Find the work done by the gradient of $f(x, y) = (x + y)^2$ counterclockwise around the circle $x^2 + y^2 = 4$ from $(2, 0)$ to itself.

23. *Circulation and flux* Find the circulation and flux of the fields

$$\mathbf{F}_1 = x\mathbf{i} + y\mathbf{j} \quad \text{and} \quad \mathbf{F}_2 = -y\mathbf{i} + x\mathbf{j}$$

around and across each of the following curves.

(a) The circle $\mathbf{r}(t) = (\cos t)\mathbf{i} + (\sin t)\mathbf{j}, \quad 0 \le t \le 2\pi$

(b) The ellipse $\mathbf{r}(t) = (\cos t)\mathbf{i} + (4 \sin t)\mathbf{j}, \quad 0 \le t \le 2\pi$

24. *Flux across a circle* Find the flux of the fields

$$\mathbf{F}_1 = 2x\mathbf{i} - 3y\mathbf{j} \quad \text{and} \quad \mathbf{F}_2 = 2x\mathbf{i} + (x - y)\mathbf{j}$$

across the circle

$$\mathbf{r}(t) = (a \cos t)\mathbf{i} + (a \sin t)\mathbf{j}, \quad 0 \le t \le 2\pi.$$

Circulation and Flux

In Exercises 25–28, find the circulation and flux of the field $\mathbf{F}$ around and across the closed semicircular path that consists of the semicircular arch $\mathbf{r}_1(t) = (a \cos t)\mathbf{i} + (a \sin t)\mathbf{j}$, $0 \le t \le \pi$, followed by the line segment $\mathbf{r}_2(t) = t\mathbf{i}$, $-a \le t \le a$.

25. $\mathbf{F} = x\mathbf{i} + y\mathbf{j}$

26. $\mathbf{F} = x^2\mathbf{i} + y^2\mathbf{j}$

27. $\mathbf{F} = -y\mathbf{i} + x\mathbf{j}$

28. $\mathbf{F} = -y^2\mathbf{i} + x^2\mathbf{j}$

29. *Flow integrals* Find the flow of the velocity field $\mathbf{F} = (x + y)\mathbf{i} - (x^2 + y^2)\mathbf{j}$ along each of the following paths from $(1, 0)$ to $(-1, 0)$ in the xy-plane.

(a) The upper half of the circle $x^2 + y^2 = 1$

(b) The line segment from $(1, 0)$ to $(-1, 0)$

(c) The line segment from $(1, 0)$ to $(0, -1)$ followed by the line segment from $(0, -1)$ to $(-1, 0)$.

30. *Flux across a triangle* Find the flux of the field $\mathbf{F}$ in Exercise 29 outward across the triangle with vertices $(1, 0)$, $(0, 1)$, $(-1, 0)$.

Sketching and Finding Fields in the Plane

31. *Spin field* Draw the spin field

$$\mathbf{F} = -\frac{y}{\sqrt{x^2 + y^2}}\mathbf{i} + \frac{x}{\sqrt{x^2 + y^2}}\mathbf{j}$$

(see Figure 13.14) along with its horizontal and vertical components at a representative assortment of points on the circle $x^2 + y^2 = 4$.

32. *Radial field* Draw the radial field

$$\mathbf{F} = x\mathbf{i} + y\mathbf{j}$$

(see Figure 13.13) along with its horizontal and vertical components at a representative assortment of points on the circle $x^2 + y^2 = 1$.

33. *A field of tangent vectors* Find a field $\mathbf{G} = P(x, y)\mathbf{i} + Q(x, y)\mathbf{j}$ in the xy-plane with the property that at any point $(a, b) \ne (0, 0)$, $\mathbf{G}$ is a vector of magnitude $\sqrt{a^2 + b^2}$ tangent to the circle $x^2 + y^2 = a^2 + b^2$ and pointing in the counterclockwise direction. (The field is undefined at $(0, 0)$.)

(b) *Writing to Learn* How is $\mathbf{G}$ related to the spin field $\mathbf{F}$ in Figure 13.14?

34. *A field of tangent vectors*

(a) Find a field $\mathbf{G} = P(x, y)\mathbf{i} + Q(x, y)\mathbf{j}$ in the xy-plane with the property that at any point $(a, b) \ne (0, 0)$, $\mathbf{G}$ is a unit vector tangent to the circle $x^2 + y^2 = a^2 + b^2$ and pointing in the clockwise direction.

(b) *Writing to Learn* How is $\mathbf{G}$ related to the spin field $\mathbf{F}$ in Figure 13.14?

35. *Unit vectors pointing toward the origin* Find a field $\mathbf{F} = M(x, y)\mathbf{i} + N(x, y)\mathbf{j}$ in the xy-plane with the property that at each point $(x, y) \ne (0, 0)$, $\mathbf{F}$ is a unit vector pointing toward the origin. (The field is undefined at $(0, 0)$.)

36. *Two "central" fields* Find a field $\mathbf{F} = M(x, y)\mathbf{i} + N(x, y)\mathbf{j}$ in the xy-plane with the property that at each point $(x, y) \ne (0, 0)$, $\mathbf{F}$ points toward the origin and $|\mathbf{F}|$ is (a) the distance from (x, y) to the origin, (b) inversely proportional to the distance from (x, y) to the origin. (The field is undefined at $(0, 0)$.)

Flow Integrals in Space

In Exercises 37–40, $\mathbf{F}$ is the velocity field of a fluid flowing through a region in space. Find the flow along the given curve in the direction of increasing t.

37. $\mathbf{F} = -4xy\mathbf{i} + 8y\mathbf{j} + 2\mathbf{k}$

$\mathbf{r}(t) = t\mathbf{i} + t^2\mathbf{j} + \mathbf{k}, \quad 0 \le t \le 2$

38. $\mathbf{F} = x^2\mathbf{i} + yz\mathbf{j} + y^2\mathbf{k}$

$\mathbf{r}(t) = 3t\mathbf{j} + 4t\mathbf{k}, \quad 0 \le t \le 1$

39. $\mathbf{F} = (x - z)\mathbf{i} + x\mathbf{k}$

$\mathbf{r}(t) = (\cos t)\mathbf{i} + (\sin t)\mathbf{k}, \quad 0 \le t \le \pi$

40. $\mathbf{F} = -y\mathbf{i} + x\mathbf{j} + 2\mathbf{k}$

$\mathbf{r}(t) = (-2 \cos t)\mathbf{i} + (2 \sin t)\mathbf{j} + 2t\mathbf{k}, \quad 0 \le t \le 2\pi$

41. *Circulation* Find the circulation of $\mathbf{F} = 2x\mathbf{i} + 2z\mathbf{j} + 2y\mathbf{k}$ around the closed path consisting of the following three curves traversed in the direction of increasing t:

C_1: $\mathbf{r}(t) = (\cos t)\mathbf{i} + (\sin t)\mathbf{j} + t\mathbf{k}, \quad 0 \le t \le \pi/2$
C_2: $\mathbf{r}(t) = \mathbf{j} + (\pi/2)(1 - t)\mathbf{k}, \quad 0 \le t \le 1$
C_3: $\mathbf{r}(t) = t\mathbf{i} + (1 - t)\mathbf{j}, \quad 0 \le t \le 1$

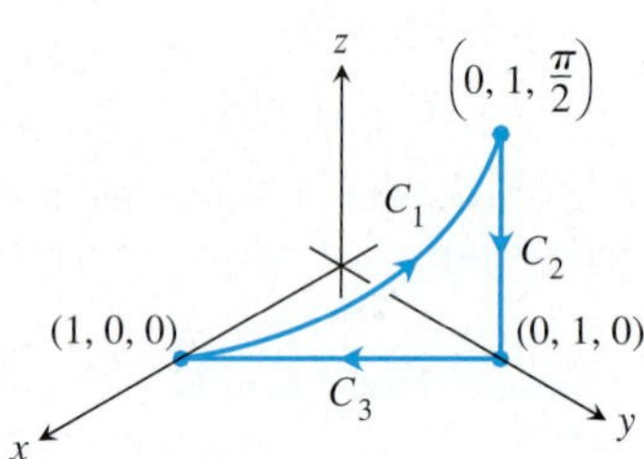

42. *Zero circulation* Let C be the ellipse in which the plane $2x + 3y - z = 0$ meets the cylinder $x^2 + y^2 = 12$. Show, without evaluating either line integral directly, that the circulation of the field $\mathbf{F} = x\mathbf{i} + y\mathbf{j} + z\mathbf{k}$ around C in either direction is zero.

43. *Flow along a curve* The field $\mathbf{F} = xy\mathbf{i} + y\mathbf{j} - yz\mathbf{k}$ is the velocity field of a flow in space. Find the flow from (0, 0, 0) to (1, 1, 1) along the curve of intersection of the cylinder $y = x^2$ and the plane $z = x$. (*Hint:* Use $t = x$ as the parameter.)

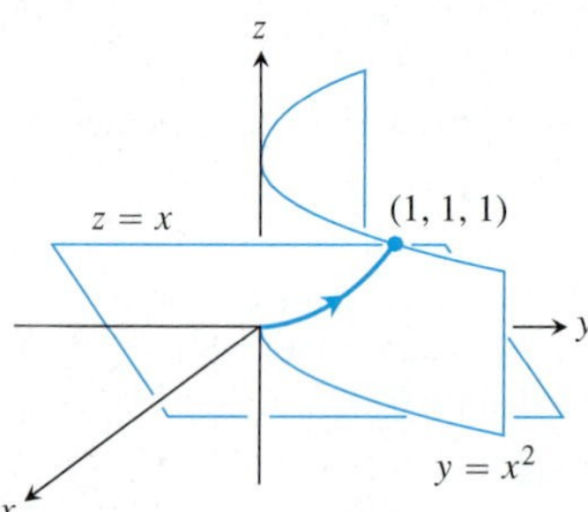

44. *Flow of a gradient field* Find the flow of the field $\mathbf{F} = \nabla(xy^2z^3)$:

(a) Once around the curve C in Exercise 42, clockwise as viewed from above

(b) Along the line segment from (1, 1, 1) to (2, 1, −1).

Theory and Examples

45. *Writing to Learn: Work and area* Suppose that $f(t)$ is differentiable and positive for $a \le t \le b$. Let C be the path $\mathbf{r}(t) = t\mathbf{i} + f(t)\mathbf{j}$, $a \le t \le b$, and $\mathbf{F} = y\mathbf{i}$. Is there any relation between the value of the work integral

$$\int_C \mathbf{F} \cdot d\mathbf{r}$$

and the area of the region bounded by the t-axis, the graph of f, and the lines $t = a$ and $t = b$? Give reasons for your answer.

46. *Work done by a radial force with constant magnitude* A particle moves along the smooth curve $y = f(x)$ from $(a, f(a))$ to $(b, f(b))$. The force moving the particle has constant magnitude k and always points away from the origin. Show that the work done by the force is

$$\int_C \mathbf{F} \cdot \mathbf{T}\, ds = k[(b^2 + (f(b))^2)^{1/2} - (a^2 + (f(a))^2)^{1/2}].$$

COMPUTER EXPLORATIONS

Finding Work Numerically

In Exercises 47–52, use a CAS to perform the following steps for finding the work done by force $\mathbf{F}$ over the given path:

(a) Find $d\mathbf{r}$ for the path $\mathbf{r}(t) = g(t)\mathbf{i} + h(t)\mathbf{j} + k(t)\mathbf{k}$.

(b) Evaluate the force $\mathbf{F}$ along the path.

(c) Evaluate $\int_C \mathbf{F} \cdot d\mathbf{r}$.

47. $\mathbf{F} = xy^6\mathbf{i} + 3x(xy^5 + 2)\mathbf{j}$; $\mathbf{r}(t) = (2\cos t)\mathbf{i} + (\sin t)\mathbf{j}$, $0 \le t \le 2\pi$

48. $\mathbf{F} = \dfrac{3}{1 + x^2}\mathbf{i} + \dfrac{2}{1 + y^2}\mathbf{j}$; $\mathbf{r}(t) = (\cos t)\mathbf{i} + (\sin t)\mathbf{j}$, $0 \le t \le \pi$

49. $\mathbf{F} = (y + yz\cos xyz)\mathbf{i} + (x^2 + xz\cos xyz)\mathbf{j} + (z + xy\cos xyz)\mathbf{k}$; $\mathbf{r}(t) = 2\cos t\mathbf{i} + 3\sin t\mathbf{j} + \mathbf{k}$, $0 \le t \le 2\pi$

50. $\mathbf{F} = 2xy\mathbf{i} - y^2\mathbf{j} + ze^x\mathbf{k}$; $\mathbf{r}(t) = -t\mathbf{i} + \sqrt{t}\mathbf{j} + 3t\mathbf{k}$, $1 \le t \le 4$

51. $\mathbf{F} = (2y + \sin x)\mathbf{i} + (z^2 + (1/3)\cos y)\mathbf{j} + x^4\mathbf{k}$; $\mathbf{r}(t) = (\sin t)\mathbf{i} + (\cos t)\mathbf{j} + (\sin 2t)\mathbf{k}$, $-\pi/2 \le t \le \pi/2$

52. $\mathbf{F} = (x^2y)\mathbf{i} + \frac{1}{3}x^3\mathbf{j} + xy\mathbf{k}$; $\mathbf{r}(t) = (\cos t)\mathbf{i} + (\sin t)\mathbf{j} + (2\sin^2(t) - 1)\mathbf{k}$, $0 \le t \le 2\pi$

13.3 Path Independence, Potential Functions, and Conservative Fields

Path Independence • Assumptions in Effect From Now On: Connectivity • Line Integrals in Conservative Fields • Finding Potentials for Conservative Fields • Exact Differential Forms

In gravitational and electric fields, the amount of work it takes to move a mass or a charge from one point to another depends only on the object's initial and final positions and not on the path taken in between. This section discusses the notion of path independence of work integrals and describes the remarkable properties of fields in which work integrals are path independent.

Path Independence

If A and B are two points in an open region D in space, the work $\int \mathbf{F} \cdot d\mathbf{r}$ done in moving a particle from A to B by a field $\mathbf{F}$ defined on D usually depends on the path taken. For some special fields, however, the integral's value is the same for all paths from A to B. If this is true for all points A and B in D, we say that the integral $\int \mathbf{F} \cdot d\mathbf{r}$ is path independent in D and that $\mathbf{F}$ is conservative on D.

Definitions Path Independence and Conservative Field

Let $\mathbf{F}$ be a field defined on an open region D in space and suppose that for any two points A and B in D the work $\int_B^A \mathbf{F} \cdot d\mathbf{r}$ done in moving from A to B is the same over all paths from A to B. Then the integral $\int \mathbf{F} \cdot d\mathbf{r}$ is **path independent in D** and the field $\mathbf{F}$ is **conservative on D.**

The word *conservative* comes from physics, where it refers to fields in which the principle of conservation of energy holds (it does, in conservative fields).

Under conditions normally met in practice, a field $\mathbf{F}$ is conservative if and only if it is the gradient field of a scalar function f; that is, if and only if $\mathbf{F} = \nabla f$ for some f. The function f is then called a potential function for $\mathbf{F}$.

Definition Potential Function

If $\mathbf{F}$ is a field defined on D and $\mathbf{F} = \nabla f$ for some scalar function f on an open region D in space, then f is called a **potential function for $\mathbf{F}$ on D.**

An electric potential is a scalar function whose gradient field is an electric field. A gravitational potential is a scalar function whose gradient field is a gravitational field, and so on. As we will see, once we have found a potential function f for a field $\mathbf{F}$, we can evaluate all the work integrals in the domain of $\mathbf{F}$ with the formula

$$\int_A^B \mathbf{F} \cdot d\mathbf{r} = \int_A^B \nabla f \cdot d\mathbf{r} = f(B) - f(A). \tag{1}$$

If you think of ∇F for functions of several variables as being something like the derivative f' for functions of a single variable, then you see that Equation (1) is the vector calculus analogue of the Fundamental Theorem of Calculus formula

$$\int_a^b f'(x)\,dx = f(b) - f(a).$$

Conservative fields have other remarkable properties we study as we go along. For example, saying that $\mathbf{F}$ is conservative on D is equivalent to saying that the integral of $\mathbf{F}$ around every closed path in D is zero. Naturally, we need to impose conditions on the curves, fields, and domains to make Equation (1) and its implications hold.

Assumptions in Effect from Now On: Connectivity

We assume that all curves are **piecewise smooth,** that is, made up of finitely many smooth pieces connected end to end, as discussed in Section 10.5. We also assume that the components of $\mathbf{F}$ have continuous first partial derivatives. When $\mathbf{F} = \nabla f$, this continuity requirement guarantees that the mixed second derivatives of the potential function f are equal, a result we will find revealing in studying conservative fields $\mathbf{F}$.

We assume D to be an *open* region in space. This means that every point in D is the center of a ball that lies entirely in D. We also assume D to be **connected,** which in an open region means that every point can be connected to every other point by a smooth curve that lies entirely in the region.

CD-ROM
WEBsite

Historical Biography

Gustav Robert Kirchoff
(1824 — 1887)

Line Integrals in Conservative Fields

The following result provides a convenient way to evaluate a line integral in a conservative field. The result establishes that the value of the integral depends only on the endpoints and not on the specific path joining them.

Theorem 1 The Fundamental Theorem of Line Integrals

1. Let $\mathbf{F} = M\mathbf{i} + N\mathbf{j} + P\mathbf{k}$ be a vector field whose components are continuous throughout an open connected region D in space. Then there exists a differentiable function f such that

$$\mathbf{F} = \nabla f = \frac{\partial f}{\partial x}\mathbf{i} + \frac{\partial f}{\partial y}\mathbf{j} + \frac{\partial f}{\partial z}\mathbf{k}$$

if and only if for all points A and B in D the value of $\int_A^B \mathbf{F} \cdot d\mathbf{r}$ is independent of the path joining A to B in D.

2. If the integral is independent of the path from A to B, its value is

$$\int_A^B \mathbf{F} \cdot d\mathbf{r} = f(B) - f(A).$$

Proof that F = ∇f Implies Path Independence of the Integral Suppose that A and B are two points in D and that C: $\mathbf{r}(t) = g(t)\mathbf{i} + h(t)\mathbf{j} + k(t)\mathbf{k}$, $a \le t \le b$, is a smooth curve in D joining A and B. Along the curve, f is a differentiable function of t and

$$\frac{df}{dt} = \frac{\partial f}{\partial x}\frac{dx}{dt} + \frac{\partial f}{\partial y}\frac{dy}{dt} + \frac{\partial f}{\partial z}\frac{dz}{dt} \qquad \text{Chain Rule}$$

$$= \nabla f \cdot \left(\frac{dx}{dt}\mathbf{i} + \frac{dy}{dt}\mathbf{j} + \frac{dz}{dt}\mathbf{k}\right) = \nabla f \cdot \frac{d\mathbf{r}}{dt} = \mathbf{F} \cdot \frac{d\mathbf{r}}{dt}. \qquad \text{Because } \mathbf{F} = \nabla f$$

Therefore,

$$\int_C \mathbf{F} \cdot d\mathbf{r} = \int_{t=a}^{t=b} \mathbf{F} \cdot \frac{d\mathbf{r}}{dt}\, dt = \int_a^b \frac{df}{dt}\, dt$$

$$= f(g(t), h(t), k(t))\Big]_a^b = f(B) - f(A).$$

Thus, the value of the work integral depends only on the values of f at A and B and not on the path in between. This proves part 2 as well as the forward implication in part 1. We omit the more technical proof of the reverse implication.

Example 1 Finding Work Done by a Conservative Field

Find the work done by the conservative field

$$\mathbf{F} = yz\mathbf{i} + xz\mathbf{j} + xy\mathbf{k} = \nabla(xyz)$$

along any smooth curve C joining the point $(-1, 3, 9)$ to $(1, 6, -4)$.

Solution With $f(x, y, z) = xyz$, we have

$$\begin{aligned}
\int_A^B \mathbf{F} \cdot d\mathbf{r} &= \int_A^B \nabla f \cdot d\mathbf{r} && \mathbf{F} = \nabla f \\
&= f(B) - f(A) && \text{Fundamental Theorem, Part 2} \\
&= xyz\,|_{(1,6,-4)} - xyz\,|_{(-1,3,9)} \\
&= (1)(6)(-4) - (-1)(3)(9) \\
&= -24 + 27 = 3.
\end{aligned}$$

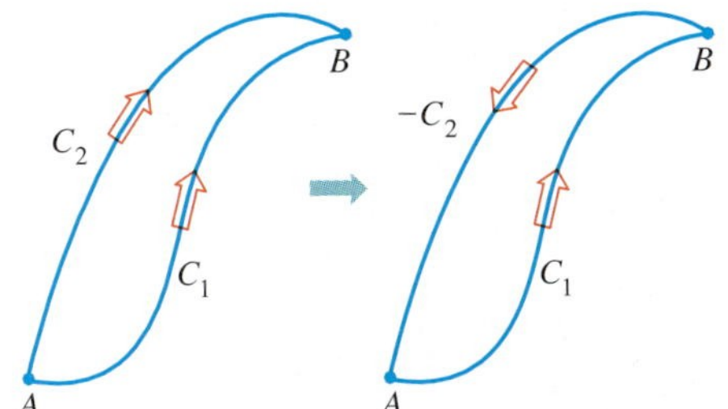

FIGURE 13.22 If we have two paths from A to B, one of them can be reversed to make a loop.

Theorem 2 Closed-Loop Property of Conservative Fields

The following statements are equivalent.

1. $\int \mathbf{F} \cdot d\mathbf{r} = 0$ around every closed loop in D.

2. The field $\mathbf{F}$ is conservative on D.

Proof that (1) ⇒ (2) We want to show that for any two points A and B in D, the integral of $\mathbf{F} \cdot d\mathbf{r}$ has the same value over any two paths C_1 and C_2 from A to B. We reverse the direction on C_2 to make a path $-C_2$ from B to A (Figure 13.22). Together, C_1 and $-C_2$ make a closed loop C, and

$$\int_{C_1} \mathbf{F} \cdot d\mathbf{r} - \int_{C_2} \mathbf{F} \cdot d\mathbf{r} = \int_{C_1} \mathbf{F} \cdot d\mathbf{r} + \int_{-C_2} \mathbf{F} \cdot d\mathbf{r} = \int_C \mathbf{F} \cdot d\mathbf{r} = 0.$$

Thus, the integrals over C_1 and C_2 give the same value.

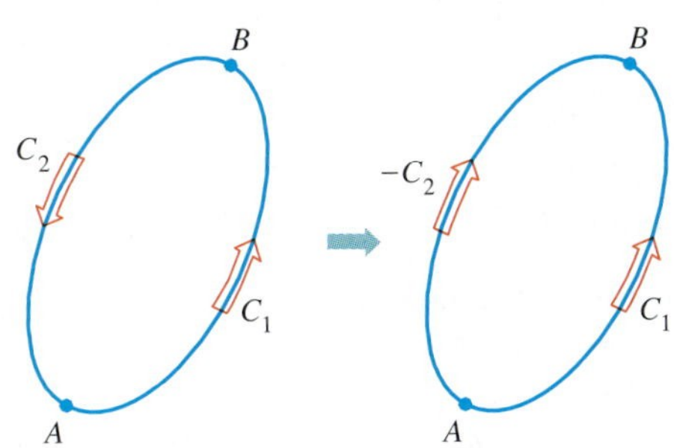

FIGURE 13.23 If A and B lie on a loop, we can reverse part of the loop to make two paths from A to B.

Proof that (2) ⇒ (1) We want to show that the integral of $\mathbf{F} \cdot d\mathbf{r}$ is zero over any closed loop C. We pick two points A and B on C and use them to break C into two pieces: C_1 from A and B followed by C_2 from B back to A (Figure 13.23). Then

$$\oint_C \mathbf{F} \cdot d\mathbf{r} = \int_{C_1} \mathbf{F} \cdot d\mathbf{r} + \int_{C_2} \mathbf{F} \cdot d\mathbf{r} = \int_A^B \mathbf{F} \cdot d\mathbf{r} - \int_A^B \mathbf{F} \cdot d\mathbf{r} = 0.$$

The following diagram summarizes the results of Theorems 1 and 2.

	Theorem 1		Theorem 2	
$\mathbf{F} = \nabla f$ on D	$\Leftrightarrow$	$\mathbf{F}$ conservative on D	$\Leftrightarrow$	$\oint_C \mathbf{F} \cdot d\mathbf{r} = 0$ over any closed path in D

Now that we see how convenient it is to evaluate line integrals in conservative fields, two questions remain.

1. How do we know when a given field $\mathbf{F}$ is conservative?
2. If $\mathbf{F}$ is in fact conservative, how do we find a potential function f (so that $\mathbf{F} = \nabla f$)?

Finding Potentials for Conservative Fields

The test for being conservative is the following.

> **Component Test for Conservative Fields**
>
> Let $\mathbf{F} = M(x, y, z)\mathbf{i} + N(x, y, z)\mathbf{j} + P(x, y, z)\mathbf{k}$ be a field whose component functions have continuous first partial derivatives. Then, $\mathbf{F}$ is conservative if and only if
>
> $$\frac{\partial P}{\partial y} = \frac{\partial N}{\partial z}, \qquad \frac{\partial M}{\partial z} = \frac{\partial P}{\partial x}, \qquad \text{and} \qquad \frac{\partial N}{\partial x} = \frac{\partial M}{\partial y}. \tag{2}$$

Proof that Equations (2) hold if F is conservative There is a potential function f such that

$$\mathbf{F} = M\mathbf{i} + N\mathbf{j} + P\mathbf{k} = \frac{\partial f}{\partial x}\mathbf{i} + \frac{\partial f}{\partial y}\mathbf{j} + \frac{\partial f}{\partial z}\mathbf{k}.$$

Hence,

$$\begin{aligned} \frac{\partial P}{\partial y} &= \frac{\partial}{\partial y}\left(\frac{\partial f}{\partial z}\right) = \frac{\partial^2 f}{\partial y\, \partial z} \\ &= \frac{\partial^2 f}{\partial z\, \partial y} \\ &= \frac{\partial}{\partial z}\left(\frac{\partial f}{\partial y}\right) = \frac{\partial N}{\partial z}. \end{aligned}$$

Continuity implies that the mixed partial derivatives are equal.

The others in Equations (2) are proved similarly.

The second half of the proof, that Equations (2) imply that $\mathbf{F}$ is conservative, is a consequence of Stokes' Theorem, taken up in Section 13.7. The proof also requires the domain of $\mathbf{F}$ to be *simply connected*.

Once we know $\mathbf{F}$ is conservative, we usually want to find a potential function for $\mathbf{F}$. This requires solving the equation $\nabla f = \mathbf{F}$ or

$$\frac{\partial f}{\partial x}\mathbf{i} + \frac{\partial f}{\partial y}\mathbf{j} + \frac{\partial f}{\partial z}\mathbf{k} = M\mathbf{i} + N\mathbf{j} + P\mathbf{k}$$

for f. We accomplish this by integrating the three equations

$$\frac{\partial f}{\partial x} = M, \qquad \frac{\partial f}{\partial y} = N, \qquad \frac{\partial f}{\partial z} = P,$$

as illustrated in the next example.

Example 2 Finding a Potential Function

Show that $\mathbf{F} = (e^x \cos y + yz)\mathbf{i} + (xz - e^x \sin y)\mathbf{j} + (xy + z)\mathbf{k}$ is conservative and find a potential function for it.

Solution We apply the test in Equations (2) to

$$M = e^x \cos y + yz, \qquad N = xz - e^x \sin y, \qquad P = xy + z$$

and calculate

$$\frac{\partial P}{\partial y} = x = \frac{\partial N}{\partial z}, \qquad \frac{\partial M}{\partial z} = y = \frac{\partial P}{\partial x}, \qquad \frac{\partial N}{\partial x} = -e^x \sin y + z = \frac{\partial M}{\partial y}.$$

Together, these equalities tell us that there is a function f with $\nabla f = \mathbf{F}$.

We find f by integrating the equations

$$\frac{\partial f}{\partial x} = e^x \cos y + yz, \qquad \frac{\partial f}{\partial y} = xz - e^x \sin y, \qquad \frac{\partial f}{\partial z} = xy + z. \tag{3}$$

We integrate the first equation with respect to x, holding y and z fixed, to get

$$f(x, y, z) = e^x \cos y + xyz + g(y, z).$$

We write the constant of integration as a function of y and z because its value may change if y and z change. We then calculate $\partial f/\partial y$ from this equation and match it with the expression for $\partial f/\partial y$ in Equations (3). This gives

$$-e^x \sin y + xz + \frac{\partial g}{\partial y} = xz - e^x \sin y,$$

so $\partial g/\partial y = 0$. Therefore, g is a function of z alone, and

$$f(x, y, z) = e^x \cos y + xyz + h(z).$$

We now calculate $\partial f/\partial z$ from this equation and match it to the formula for $\partial f/\partial z$ in Equations (3). This gives

$$xy + \frac{dh}{dz} = xy + z, \qquad \text{or} \qquad \frac{dh}{dz} = z,$$

so

$$h(z) = \frac{z^2}{2} + C.$$

Hence,

$$f(x, y, z) = e^x \cos y + xyz + \frac{z^2}{2} + C.$$

We have infinitely many potential functions for $\mathbf{F}$, one for each value of C.

Example 3 Showing That a Field Is Not Conservative

Show that $\mathbf{F} = (2x - 3)\mathbf{i} - z\mathbf{j} + (\cos z)\mathbf{k}$ is not conservative.

Solution We apply the component test in Equations (2) and find right away that

$$\frac{\partial P}{\partial y} = \frac{\partial}{\partial y}(\cos z) = 0, \qquad \frac{\partial N}{\partial z} = \frac{\partial}{\partial z}(-z) = -1.$$

The two are unequal, so $\mathbf{F}$ is not conservative. No further testing is required.

CD-ROM WEBsite
Historical Biography
Ernst Mach
(1838 — 1916)

Exact Differential Forms

As we see in the next section and again later on, it is often convenient to express work and circulation integrals in the "differential" form

$$\int_A^B M\,dx + N\,dy + P\,dz$$

mentioned in Section 13.2. Such integrals are relatively easy to evaluate if $M\,dx + N\,dy + P\,dz$ is the total differential of a function f. For then

$$\begin{aligned}\int_A^B M\,dx + N\,dy + P\,dz &= \int_A^B \frac{\partial f}{\partial x}dx + \frac{\partial f}{\partial y}dy + \frac{\partial f}{\partial z}dz \\ &= \int_A^B \nabla f \cdot d\mathbf{r} \\ &= f(B) - f(A). \qquad \text{Theorem 1}\end{aligned}$$

Thus,

$$\int_A^B df = f(B) - f(A),$$

just as with differentiable functions of a single variable.

Definitions Differential Form and Exact Differential Form

Any form $M(x, y, z)\,dx + N(x, y, z)\,dy + P(x, y, z)\,dz$ is a **differential form.** A differential form is **exact** on a domain D in space if

$$M\,dx + N\,dy + P\,dz = \frac{\partial f}{\partial x}dx + \frac{\partial f}{\partial y}dy + \frac{\partial f}{\partial z}dz = df$$

for some scalar function f throughout D.

Notice that if $M\,dx + N\,dy + P\,dz = df$ on D, then $\mathbf{F} = M\mathbf{i} + N\mathbf{j} + P\mathbf{k}$ is the gradient field of f on D. Conversely, if $\mathbf{F} = \nabla f$, then the form $M\,dx + N\,dy + P\,dz$ is exact. The test for the form's being exact is therefore the same as the test for $\mathbf{F}$'s being conservative.

Component Test for Exactness of $M\,dx + N\,dy + P\,dz$

The differential form $M\,dx + N\,dy + P\,dz$ is exact if and only if

$$\frac{\partial P}{\partial y} = \frac{\partial N}{\partial z}, \qquad \frac{\partial M}{\partial z} = \frac{\partial P}{\partial x}, \qquad \text{and} \qquad \frac{\partial N}{\partial x} = \frac{\partial M}{\partial y}.$$

This is equivalent to saying that the field $\mathbf{F} = M\mathbf{i} + N\mathbf{j} + P\mathbf{k}$ is conservative.

Example 4 Showing That a Differential Form Is Exact

Show that $y\,dx + x\,dy + 4\,dz$ is exact and evaluate the integral

$$\int_{(1,1,1)}^{(2,3,-1)} y\,dx + x\,dy + 4\,dz$$

over the line segment from $(1, 1, 1)$ to $(2, 3, -1)$.

Solution We let $M = y, N = x, P = 4$ and apply the Test for Exactness:

$$\frac{\partial P}{\partial y} = 0 = \frac{\partial N}{\partial z}, \qquad \frac{\partial M}{\partial z} = 0 = \frac{\partial P}{\partial x}, \qquad \frac{\partial N}{\partial x} = 1 = \frac{\partial M}{\partial y}.$$

These equalities tell us that $y\,dx + x\,dy + 4\,dz$ is exact, so

$$y\,dx + x\,dy + 4\,dz = df$$

for some function f, and the integral's value is $f(2, 3, -1) - f(1, 1, 1)$.

We find f up to a constant by integrating the equations

$$\frac{\partial f}{\partial x} = y, \qquad \frac{\partial f}{\partial y} = x, \qquad \frac{\partial f}{\partial z} = 4. \tag{4}$$

From the first equation we get

$$f(x, y, z) = xy + g(y, z).$$

The second equation tells us that

$$\frac{\partial f}{\partial y} = x + \frac{\partial g}{\partial y} = x, \qquad \text{or} \qquad \frac{\partial g}{\partial y} = 0.$$

Hence, g is a function of z alone, and

$$f(x, y, z) = xy + h(z).$$

The third of Equations (4) tells us that

$$\frac{\partial f}{\partial z} = 0 + \frac{dh}{dz} = 4, \qquad \text{or} \qquad h(z) = 4z + C.$$

Therefore,

$$f(x, y, z) = xy + 4z + C.$$

The value of the integral is

$$f(2, 3, -1) - f(1, 1, 1) = 2 + C - (5 + C) = -3.$$

EXERCISES 13.3

Testing for Conservative Fields

CD-ROM WEBsite

Which fields in Exercises 1–6 are conservative and which are not?

1. $\mathbf{F} = yz\mathbf{i} + xz\mathbf{j} + xy\mathbf{k}$
2. $\mathbf{F} = (y \sin z)\mathbf{i} + (x \sin z)\mathbf{j} + (xy \cos z)\mathbf{k}$
3. $\mathbf{F} = y\mathbf{i} + (x + z)\mathbf{j} - y\mathbf{k}$
4. $\mathbf{F} = -y\mathbf{i} + x\mathbf{j}$
5. $\mathbf{F} = (z + y)\mathbf{i} + z\mathbf{j} + (y + x)\mathbf{k}$
6. $\mathbf{F} = (e^x \cos y)\mathbf{i} - (e^x \sin y)\mathbf{j} + z\mathbf{k}$

Finding Potential Functions

In Exercises 7–12, find a potential function f for the field $\mathbf{F}$.

7. $\mathbf{F} = 2x\mathbf{i} + 3y\mathbf{j} + 4z\mathbf{k}$
8. $\mathbf{F} = (y + z)\mathbf{i} + (x + z)\mathbf{j} + (x + y)\mathbf{k}$
9. $\mathbf{F} = e^{y+2z}(\mathbf{i} + x\mathbf{j} + 2x\mathbf{k})$
10. $\mathbf{F} = (y \sin z)\mathbf{i} + (x \sin z)\mathbf{j} + (xy \cos z)\mathbf{k}$
11. $\mathbf{F} = (\ln x + \sec^2(x + y))\mathbf{i}$
$$+ \left(\sec^2(x + y) + \frac{y}{y^2 + z^2}\right)\mathbf{j} + \frac{z}{y^2 + z^2}\mathbf{k}$$
12. $\mathbf{F} = \dfrac{y}{1 + x^2y^2}\mathbf{i} + \left(\dfrac{x}{1 + x^2y^2} + \dfrac{z}{\sqrt{1 - y^2z^2}}\right)\mathbf{j}$
$$+ \left(\frac{y}{\sqrt{1 - y^2z^2}} + \frac{1}{z}\right)\mathbf{k}$$

Evaluating Integrals of Exact Differential Forms

In Exercises 13–22, show that the differential forms in the integrals are exact. Then evaluate the integrals.

13. $\displaystyle\int_{(0,0,0)}^{(2,3,-6)} 2x\,dx + 2y\,dy + 2z\,dz$
14. $\displaystyle\int_{(1,1,2)}^{(3,5,0)} yz\,dx + xz\,dy + xy\,dz$
15. $\displaystyle\int_{(0,0,0)}^{(1,2,3)} 2xy\,dx + (x^2 - z^2)\,dy - 2yz\,dz$
16. $\displaystyle\int_{(0,0,0)}^{(3,3,1)} 2x\,dx - y^2\,dy - \frac{4}{1 + z^2}\,dz$
17. $\displaystyle\int_{(1,0,0)}^{(0,1,1)} \sin y \cos x\,dx + \cos y \sin x\,dy + dz$
18. $\displaystyle\int_{(0,2,1)}^{(1,\pi/2,2)} 2\cos y\,dx + \left(\frac{1}{y} - 2x \sin y\right)dy + \frac{1}{z}\,dz$
19. $\displaystyle\int_{(1,1,1)}^{(1,2,3)} 3x^2\,dx + \frac{z^2}{y}\,dy + 2z \ln y\,dz$
20. $\displaystyle\int_{(1,2,1)}^{(2,1,1)} (2x \ln y - yz)\,dx + \left(\frac{x^2}{y} - xz\right)dy - xy\,dz$
21. $\displaystyle\int_{(1,1,1)}^{(2,2,2)} \frac{1}{y}\,dx + \left(\frac{1}{z} - \frac{x}{y^2}\right)dy - \frac{y}{z^2}\,dz$
22. $\displaystyle\int_{(-1,-1,-1)}^{(2,2,2)} \frac{2x\,dx + 2y\,dy + 2z\,dz}{x^2 + y^2 + z^2}$
23. *Revisiting Example 4* Evaluate the integral
$$\int_{(1,1,1)}^{(2,3,-1)} y\,dx + x\,dy + 4\,dz$$
from Example 4 by finding parametric equations for the line segment from $(1, 1, 1)$ to $(2, 3, -1)$ and evaluating the line integral of $\mathbf{F} = y\mathbf{i} + x\mathbf{j} + 4\mathbf{k}$ along the segment. Since $\mathbf{F}$ is conservative, the integral is independent of the path.
24. Evaluate
$$\int_C x^2\,dx + yz\,dy + (y^2/2)\,dz$$
along the line segment C joining $(0, 0, 0)$ to $(0, 3, 4)$.

Theory, Applications, and Examples

Independence of path Show that the values of the integrals in Exercises 25 and 26 do not depend on the path taken from A to B.

25. $\displaystyle\int_A^B z^2\,dx + 2y\,dy + 2xz\,dz$
26. $\displaystyle\int_A^B \frac{x\,dx + y\,dy + z\,dz}{\sqrt{x^2 + y^2 + z^2}}$

In Exercises 27 and 28, find a potential function for $\mathbf{F}$.

27. $\mathbf{F} = \dfrac{2x}{y}\mathbf{i} + \left(\dfrac{1 - x^2}{y^2}\right)\mathbf{j}$
28. $\mathbf{F} = (e^x \ln y)\mathbf{i} + \left(\dfrac{e^x}{y} + \sin z\right)\mathbf{j} + (y \cos z)\mathbf{k}$
29. *Work along different paths* Find the work done by $\mathbf{F} = (x^2 + y)\mathbf{i} + (y^2 + x)\mathbf{j} + ze^z\mathbf{k}$ over the following paths from $(1, 0, 0)$ to $(1, 0, 1)$.
   (a) The line segment $x = 1, y = 0, 0 \le z \le 1$

(b) The helix $\mathbf{r}(t) = (\cos t)\mathbf{i} + (\sin t)\mathbf{j} + (t/2\pi)\mathbf{k}, 0 \le t \le 2\pi$

(c) The x-axis from $(1, 0, 0)$ to $(0, 0, 0)$ followed by the parabola $z = x^2, y = 0$ from $(0, 0, 0)$ to $(1, 0, 1)$

30. *Work along different paths* Find the work done by $\mathbf{F} = e^{yz}\mathbf{i} + (xze^{yz} + z\cos y)\mathbf{j} + (xye^{yz} + \sin y)\mathbf{k}$ over the following paths from $(1, 0, 1)$ to $(1, \pi/2, 0)$.

CD-ROM WEBsite

(a) The line segment $x = 1, y = \pi t/2, z = 1 - t, 0 \le t \le 1$

(b) The line segment from $(1, 0, 1)$ to the origin followed by the line segment from the origin to $(1, \pi/2, 0)$

(c) The line segment from $(1, 0, 1)$ to $(1, 0, 0)$, followed by the x-axis from $(1, 0, 0)$ to the origin, followed by the parabola $y = \pi x^2/2, z = 0$ from there to $(1, \pi/2, 0)$

31. *Evaluating a work integral two ways* Let $\mathbf{F} = \nabla(x^3y^2)$ and let C be the path in the xy-plane from $(-1, 1)$ to $(1, 1)$ that consists of the line segment from $(-1, 1)$ to $(0, 0)$ followed by the line segment from $(0, 0)$ to $(1, 1)$. Evaluate $\int_C \mathbf{F} \cdot d\mathbf{r}$ in two ways.

(a) Find parametrizations for the segments that make up C and evaluate the integral.

(b) Using $f(x, y) = x^3y^2$ as a potential function for $\mathbf{F}$.

32. *Integral along different paths* Evaluate $\int_C 2x\cos y\,dx - x^2\sin y\,dy$ along the following paths C in the xy-plane.

(a) The parabola $y = (x - 1)^2$ from $(1, 0)$ to $(0, 1)$

(b) The line segment from $(-1, \pi)$ to $(1, 0)$

(c) The x-axis from $(-1, 0)$ to $(1, 0)$

(d) The astroid $\mathbf{r}(t) = (\cos^3 t)\mathbf{i} + (\sin^3 t)\mathbf{j}, 0 \le t \le 2\pi$, counterclockwise from $(1, 0)$ back to $(1, 0)$

33. (a) *Exact differential form* How are the constants a, b, and c related if the following differential form is exact?

$$(ay^2 + 2czx)\,dx + y(bx + cz)\,dy + (ay^2 + cx^2)\,dz$$

(b) *Gradient field* For what values of b and c will

$$\mathbf{F} = (y^2 + 2czx)\mathbf{i} + y(bx + cz)\mathbf{j} + (y^2 + cx^2)\mathbf{k}$$

be a gradient field?

34. *Gradient of a line integral* Suppose that $\mathbf{F} = \nabla f$ is a conservative vector field and

$$g(x, y, z) = \int_{(0,0,0)}^{(x,y,z)} \mathbf{F} \cdot d\mathbf{r}.$$

Show that $\nabla g = \mathbf{F}$.

35. *Writing to Learn: Path of least work* You have been asked to find the path along which a force field $\mathbf{F}$ will perform the least work in moving a particle between two locations. A quick calculation on your part shows $\mathbf{F}$ to be conservative. How should you respond? Give reasons for your answer.

36. *Writing to Learn: A revealing experiment* By experiment, you find that a force field $\mathbf{F}$ performs only half as much work in moving an object along path C_1 from A to B as it does in moving the object along path C_2 from A to B. What can you conclude about $\mathbf{F}$? Give reasons for your answer.

37. *Work by a constant force* Show that the work done by a constant force field $\mathbf{F} = a\mathbf{i} + b\mathbf{j} + c\mathbf{k}$ in moving a particle along any path from A to B is $W = \mathbf{F} \cdot \overrightarrow{AB}$.

38. *Gravitational field*

(a) Find a potential function for the gravitational field

$$\mathbf{F} = -GmM\frac{x\mathbf{i} + y\mathbf{j} + z\mathbf{k}}{(x^2 + y^2 + z^2)^{3/2}} \quad (G, m, \text{ and } M \text{ are constants}).$$

(b) Let P_1 and P_2 be points at distance b. s_1 and s_2 from the origin. Show that the work done by the gravitational field in part (a) in moving a particle from P_1 to P_2 is

$$GmM\left(\frac{1}{s_2} - \frac{1}{s_1}\right).$$

13.4 Green's Theorem in the Plane

Flux Density at a Point: Divergence • Circulation Density at a Point: The **k**-Component of Curl • Two Forms for Green's Theorem • Mathematical Assumptions • Using Green's Theorem to Evaluate Line Integrals • Proof of Green's Theorem for Special Regions • Extending the Proof to Other Regions

In the preceding section, we learned how to evaluate flow integrals for conservative fields. We found a potential function for the field, evaluated it at the path endpoints, and calculated the integral as the appropriate difference of those values.

In this section, we see how to evaluate flow and flux integrals across closed plane curves when the vector field is not conservative. The means for doing so is a theorem known as Green's Theorem which converts line integrals to double integrals.

Green's Theorem is one of the great theorems of calculus. It is deep and surprising, and has far-reaching consequences. In pure mathematics, it ranks in importance with the Fundamental Theorem of Calculus. In applied mathematics, the generalizations of Green's Theorem to three dimensions provide the foundation for theorems about electricity, magnetism, and fluid flow.

We talk in terms of velocity fields of fluid flows because fluid flows are easy to picture. Be aware, however, that Green's Theorem applies to any vector field satisfying certain mathematical conditions. It does not depend for its validity on the field's having a particular physical interpretation.

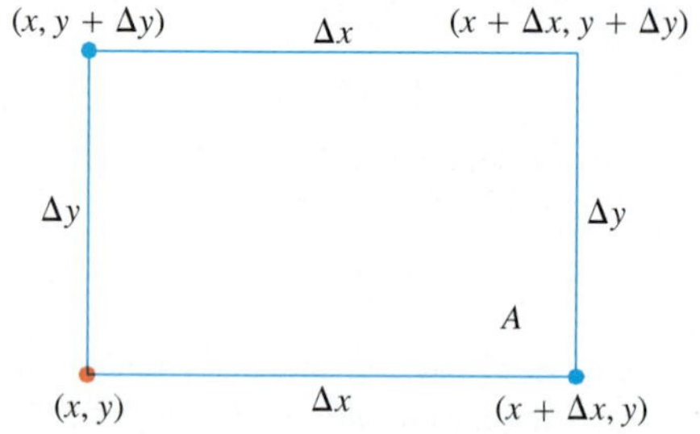

FIGURE 13.24 The rectangle for defining the flux density (divergence) of a vector field at a point (x, y).

Flux Density at a Point: Divergence

We need two new ideas for Green's Theorem. The first is the idea of the flux density of a vector field at a point, which in mathematics is called the *divergence* of the vector field. We obtain it in the following way.

Suppose that $\mathbf{F}(x, y) = M(x, y)\mathbf{i} + N(x, y)\mathbf{j}$ is the velocity field of a fluid flow in the plane and that the first partial derivatives of M and N are continuous at each point of a region R. Let (x, y) be a point in R and let A be a small rectangle with one corner at (x, y) that, along with its interior, lies entirely in R (Figure 13.24). The sides of the rectangle, parallel to the coordinate axes, have lengths of Δx and Δy. The rate at which fluid leaves the rectangle across the bottom edge is approximately

$$\mathbf{F}(x, y) \cdot (-\mathbf{j})\, \Delta x = -N(x, y)\, \Delta x.$$

This is the scalar component of the velocity at (x, y) in the direction of the outward normal times the length of the segment. If the velocity is in meters per second, for example, the exit rate will be in meters per second times meters or square meters per second. The rates at which the fluid crosses the other three sides in the directions of their outward normals can be estimated in a similar way. All told, we have

Exit Rates:

$$\begin{aligned}
&\text{Top:} && \mathbf{F}(x, y + \Delta y) \cdot \mathbf{j}\, \Delta x = N(x, y + \Delta y)\, \Delta x \\
&\text{Bottom:} && \mathbf{F}(x, y) \cdot (-\mathbf{j})\, \Delta x = -N(x, y)\, \Delta x \\
&\text{Right:} && \mathbf{F}(x + \Delta x, y) \cdot \mathbf{i}\, \Delta y = M(x + \Delta x, y)\, \Delta y \\
&\text{Left:} && \mathbf{F}(x, y) \cdot (-\mathbf{i})\, \Delta y = -M(x, y)\, \Delta y.
\end{aligned}$$

Combining opposite pairs gives

$$\text{Top and bottom:} \quad (N(x, y + \Delta y) - N(x, y))\,\Delta x \approx \left(\frac{\partial N}{\partial y}\Delta y\right)\Delta x$$

$$\text{Right and left:} \quad (M(x + \Delta x, y) - M(x, y))\,\Delta y \approx \left(\frac{\partial M}{\partial x}\Delta x\right)\Delta y.$$

Adding these last two equations gives

$$\text{Flux across rectangle boundary} \approx \left(\frac{\partial M}{\partial x} + \frac{\partial N}{\partial y}\right)\Delta x\,\Delta y.$$

We now divide by $\Delta x\,\Delta y$ to estimate the total flux per unit area or flux density for the rectangle:

$$\frac{\text{Flux across rectangle boundary}}{\text{rectangle area}} \approx \left(\frac{\partial M}{\partial x} + \frac{\partial N}{\partial y}\right).$$

Finally, we let Δx and Δy approach zero to define what we call the *flux density* of $\mathbf{F}$ at the point (x, y).

In mathematics, we call the flux density the *divergence* of $\mathbf{F}$. The symbol for it is div $\mathbf{F}$, pronounced "divergence of $\mathbf{F}$" or "div $\mathbf{F}$."

Definition Flux Density or Divergence

The **flux density** or **divergence** of a vector field $\mathbf{F} = M\mathbf{i} + N\mathbf{j}$ at the point (x, y) is

$$\text{div}\,\mathbf{F} = \frac{\partial M}{\partial x} + \frac{\partial N}{\partial y}. \tag{1}$$

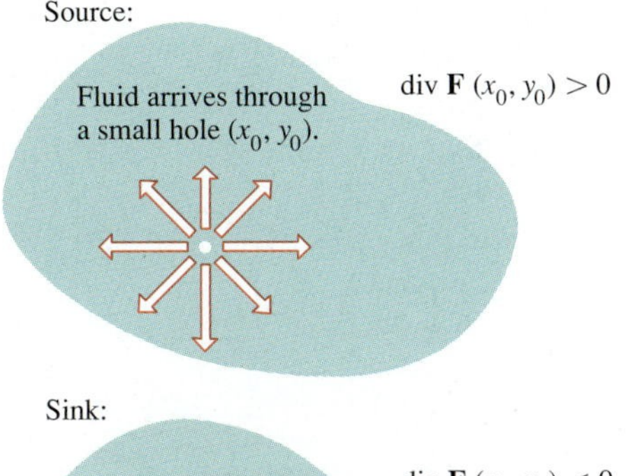

FIGURE 13.25 In the flow of fluid across a plane region, the divergence is positive at a "source," a point where fluid enters the system, and negative at a "sink," a point where the fluid leaves the system.

Intuitively, if water were flowing into a region through a small hole at the point (x_0, y_0), the lines of flow would diverge there (hence the name) and, since water would be flowing out of a small rectangle about (x_0, y_0), the divergence of $\mathbf{F}$ at (x_0, y_0) would be positive. If the water were draining out of the hole instead of flowing in, the divergence would be negative. See Figure 13.25.

Example 1 Finding Divergence

Find the divergence of $\mathbf{F}(x, y) = (x^2 - y)\mathbf{i} + (xy - y^2)\mathbf{j}$.

Solution We use the formula in Equation (1):

$$\begin{aligned}\text{div}\,\mathbf{F} &= \frac{\partial M}{\partial x} + \frac{\partial N}{\partial y} = \frac{\partial}{\partial x}(x^2 - y) + \frac{\partial}{\partial y}(xy - y^2)\\ &= 2x + x - 2y = 3x - 2y.\end{aligned}$$

Circulation Density at a Point: The k-Component of Curl

The second of the two new ideas we need for Green's Theorem is the idea of circulation density of a vector field $\mathbf{F}$ at a point. To obtain it, we return to the velocity field

$$\mathbf{F}(x, y) = M(x, y)\mathbf{i} + N(x, y)\,\mathbf{j}$$

and the rectangle A. The rectangle is redrawn here as Figure 13.26.

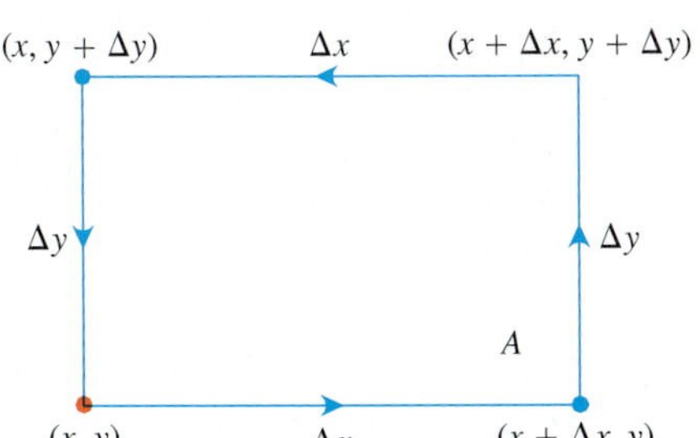

FIGURE 13.26 The rectangle for defining the circulation density (curl) of a vector field at a point (x, y).

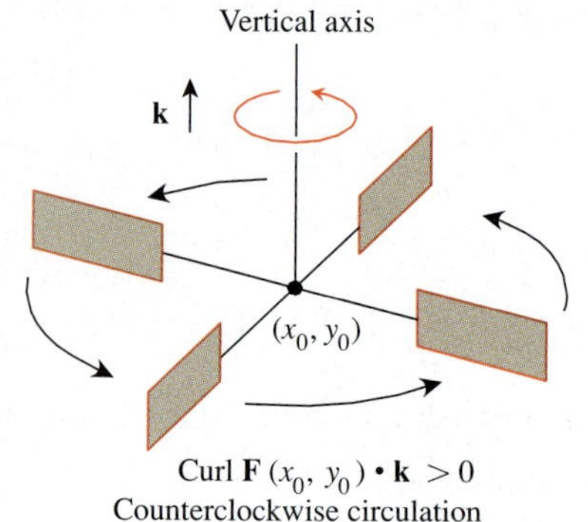

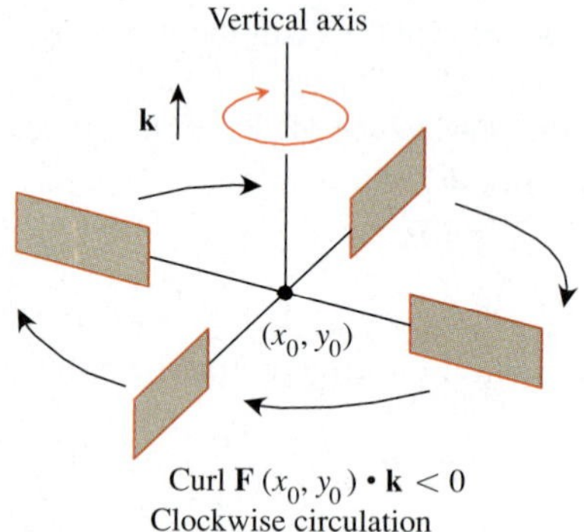

FIGURE 13.27 In the flow of an incompressible fluid over a plane region, the **k**-component of the curl measures the rate of the fluid's rotation at a point. The **k**-component of the curl is positive at points where the rotation is counterclockwise and negative where the rotation is clockwise.

The counterclockwise circulation of $\mathbf{F}$ around the boundary of A is the sum of flow rates along the sides. For the bottom edge, the flow rate is approximately

$$\mathbf{F}(x, y) \cdot \mathbf{i}\, \Delta x = M(x, y)\, \Delta x.$$

This is the scalar component of the velocity $\mathbf{F}(x, y)$ in the direction of the tangent vector $\mathbf{i}$ times the length of the segment. The rates of flow along the other sides in the counterclockwise direction are expressed in a similar way. In all, we have

Top: $\mathbf{F}(x, y + \Delta y) \cdot (-\mathbf{i})\, \Delta x = -M(x, y + \Delta y)\, \Delta x$

Bottom: $\mathbf{F}(x, y) \cdot \mathbf{i}\, \Delta x = M(x, y)\, \Delta x$

Right: $\mathbf{F}(x + \Delta x, y) \cdot \mathbf{j}\, \Delta y = N(x + \Delta x, y)\, \Delta y$

Left: $\mathbf{F}(x, y) \cdot (-\mathbf{j})\, \Delta y = -N(x, y)\, \Delta y.$

We add opposite pairs to get

Top and bottom:

$$-(M(x, y + \Delta y) - M(x, y))\, \Delta x \approx -\left(\frac{\partial M}{\partial y} \Delta y\right) \Delta x$$

Right and left:

$$(N(x + \Delta x, y) - N(x, y))\, \Delta y \approx \left(\frac{\partial N}{\partial x} \Delta x\right) \Delta y.$$

Adding these last two equations and dividing by $\Delta x\, \Delta y$ gives an estimate of the circulation density for the rectangle:

$$\frac{\text{Circulation around rectangle}}{\text{rectangle area}} \approx \frac{\partial N}{\partial x} - \frac{\partial M}{\partial y}.$$

We let Δx and Δy approach zero to define what we call the *circulation density* of $\mathbf{F}$ at the point (x, y).

The positive orientation of the circulation density for the plane is the *counterclockwise* rotation around the vertical axis, looking downward on the xy-plane from the tip of the (vertical) unit vector $\mathbf{k}$ (Figure 13.27). The circulation value is actually the **k**-component of a more general circulation vector we define in Section 13.7, called the *curl* of the vector field $\mathbf{F}$. For Green's Theorem, we need only this **k**-component.

Definition k-Component of Circulation Density or Curl

The k-component of the circulation density or curl of a vector field $\mathbf{F} = M\mathbf{i} + N\mathbf{j}$ at the point (x, y) is the scalar

$$(\text{curl } \mathbf{F}) \cdot \mathbf{k} = \frac{\partial N}{\partial x} - \frac{\partial M}{\partial y}. \tag{2}$$

If water is moving about a region in the xy-plane in a thin layer, then the **k**-component of the circulation, or curl, at a point (x_0, y_0) gives a way to measure how fast and in what direction a small paddle wheel will spin if it is put into the water at (x_0, y_0) with its axis perpendicular to the plane, parallel to $\mathbf{k}$ (Figure 13.27).

Example 2 Finding the k-Component of the Curl

Find the **k**-component of the curl for the vector field

$$\mathbf{F}(x, y) = (x^2 - y)\,\mathbf{i} + (xy - y^2)\,\mathbf{j}.$$

Solution We use the formula in Equation (2):

$$(\text{curl}\,\mathbf{F}) \cdot \mathbf{k} = \frac{\partial N}{\partial x} - \frac{\partial M}{\partial y} = \frac{\partial}{\partial x}(xy - y^2) - \frac{\partial}{\partial y}(x^2 - y) = y + 1.$$

Two Forms for Green's Theorem

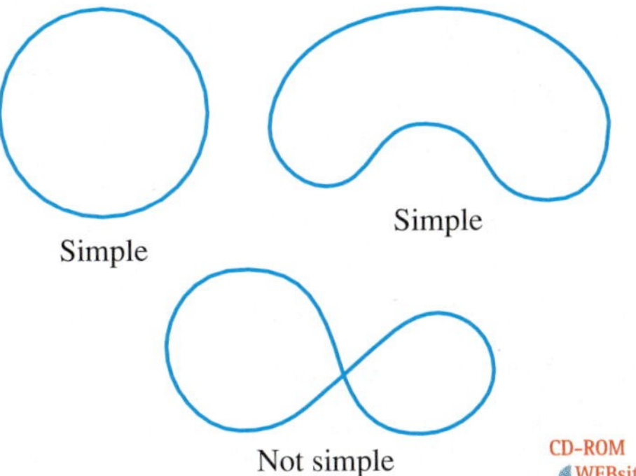

FIGURE 13.28 In proving Green's Theorem, we distinguish between two kinds of closed curves, simple and not simple. Simple curves do not cross themselves. A circle is simple but a figure 8 is not.

In one form, Green's Theorem says that under suitable conditions the outward flux of a vector field across a simple closed curve in the plane (Figure 13.28) equals the double integral of the divergence of the field over the region enclosed by the curve. Recall the formulas for flux in Equations (3) and (4) in Section 13.2.

CD-ROM WEBsite

Theorem 3 Green's Theorem (Flux-Divergence or Normal Form)
The outward flux of a field $\mathbf{F} = M\mathbf{i} + N\mathbf{j}$ across a simple closed curve C equals the double integral of div $\mathbf{F}$ over the region R enclosed by C.

$$\underbrace{\oint_C \mathbf{F} \cdot \mathbf{n}\, ds}_{\text{Outward flux}} = \oint_C M\, dy - N\, dx = \underbrace{\iint_R \left(\frac{\partial M}{\partial x} + \frac{\partial N}{\partial y}\right) dx\, dy}_{\text{Divergence integral}} \tag{3}$$

In another form, Green's Theorem says that the counterclockwise circulation of a vector field around a simple closed curve is the double integral of the **k**-component of the curl of the field over the region enclosed by the curve.

CD-ROM WEBsite

Theorem 4 Green's Theorem (Circulation-Curl or Tangential Form)
The counterclockwise circulation of a field $\mathbf{F} = M\mathbf{i} + N\mathbf{j}$ around a simple closed curve C in the plane equals the double integral of $(\text{curl}\,\mathbf{F}) \cdot \mathbf{k}$ over the region R enclosed by C.

$$\underbrace{\oint_C \mathbf{F} \cdot \mathbf{T}\, ds}_{\text{Counterclockwise circulation}} = \oint_C M\, dx + N\, dy = \underbrace{\iint_R \left(\frac{\partial N}{\partial x} - \frac{\partial M}{\partial y}\right) dx\, dy}_{\text{Curl integral}} \tag{4}$$

The two forms of Green's Theorem are equivalent. Applying Equation (3) to the field $\mathbf{G}_1 = N\mathbf{i} - M\mathbf{j}$ gives Equation (4), and applying Equation (4) to $\mathbf{G}_2 = -N\mathbf{i} + M\mathbf{j}$ gives Equation (3).

Mathematical Assumptions

We need two kinds of assumptions for Green's Theorem to hold. First, we need conditions on M and N to ensure the existence of the integrals. The usual assumptions are that M, N, and their first partial derivatives are continuous at every point of some open region containing C and R. Second, we need geometric conditions on the curve C. It must be simple, closed, and made up of pieces along which we can integrate M and N. The usual assumptions are that C is piecewise smooth. The proof we give for Green's Theorem, however, assumes things about the shape of R as well. You can find proofs that are less restrictive in more advanced texts. First let's look at examples.

Example 3 Supporting Green's Theorem

Verify both forms of Green's Theorem for the field

$$\mathbf{F}(x, y) = (x - y)\,\mathbf{i} + x\mathbf{j}$$

and the region R bounded by the unit circle

$$C: \quad \mathbf{r}(t) = (\cos t)\,\mathbf{i} + (\sin t)\,\mathbf{j}, \qquad 0 \le t \le 2\pi.$$

Solution We have

$$M = \cos t - \sin t, \qquad dx = d(\cos t) = -\sin t\,dt,$$
$$N = \cos t, \qquad dy = d(\sin t) = \cos t\,dt,$$
$$\frac{\partial M}{\partial x} = 1, \qquad \frac{\partial M}{\partial y} = -1, \qquad \frac{\partial N}{\partial x} = 1, \qquad \frac{\partial N}{\partial y} = 0.$$

The two sides of Equation (3) are

$$\oint_C M\,dy - N\,dx = \int_{t=0}^{t=2\pi} (\cos t - \sin t)(\cos t\,dt) - (\cos t)(-\sin t\,dt)$$
$$= \int_0^{2\pi} \cos^2 t\,dt = \pi$$

$$\iint_R \left(\frac{\partial M}{\partial x} + \frac{\partial N}{\partial y}\right) dx\,dy = \iint_R (1 + 0)\,dx\,dy$$
$$= \iint_R dx\,dy = \text{area of unit circle} = \pi.$$

The two sides of Equation (4) are

$$\oint_C M\,dx + N\,dy = \int_{t=0}^{t=2\pi} (\cos t - \sin t)(-\sin t\,dt) + (\cos t)(\cos t\,dt)$$
$$= \int_0^{2\pi} (-\sin t \cos t + 1)\,dt = 2\pi$$

$$\iint_R \left(\frac{\partial N}{\partial x} - \frac{\partial M}{dy}\right) dx\,dy = \iint_R (1 - (-1))\,dx\,dy = 2\iint_R dx\,dy = 2\pi.$$

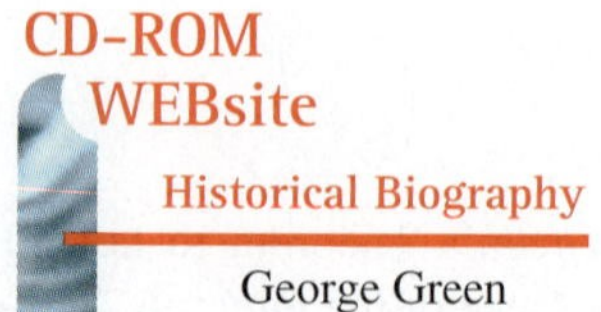

Using Green's Theorem to Evaluate Line Integrals

If we construct a closed curve C by piecing a number of different curves end to end, the process of evaluating a line integral over C can be lengthy because there are so many different integrals to evaluate. If C bounds a region R to which

Green's Theorem applies, however, we can use Green's Theorem to change the line integral around C into one double integral over R.

Example 4 Evaluating a Line Integral Using Green's Theorem

Evaluate the integral

$$\oint_C xy\,dy - y^2\,dx,$$

where C is the square cut from the first quadrant by the lines $x = 1$ and $y = 1$.

Solution We can use either form of Green's Theorem to change the line integral into a double integral over the square.

1. *With the Normal Form Equation* (3): Taking $M = xy, N = y^2$, and C and R as the square's boundary and interior gives

$$\oint_C xy\,dy - y^2\,dx = \iint_R (y + 2y)\,dx\,dy = \int_0^1 \int_0^1 3y\,dx\,dy$$

$$= \int_0^1 [3xy]_{x=0}^{x=1}\,dy = \int_0^1 3y\,dy = \frac{3}{2}y^2\Big]_0^1 = \frac{3}{2}.$$

2. *With the Tangential Form Equation* (4): Taking $M = -y^2$ and $N = xy$ gives the same result:

$$\oint_C -y^2\,dx + xy\,dy = \iint_R (y - (-2y))\,dx\,dy = \frac{3}{2}.$$

Example 5 Finding Outward Flux

Calculate the outward flux of the field $\mathbf{F}(x, y) = x\mathbf{i} + y^2\,\mathbf{j}$ across the square bounded by the lines $x = \pm 1$ and $y = \pm 1$.

Solution Calculating the flux with a line integral would take four integrations, one for each side of the square. With Green's Theorem, we can change the line integral to one double integral. With $M = x, N = y^2$, C the square, and R the square's interior, we have

$$\text{Flux} = \oint_C \mathbf{F}\cdot\mathbf{n}\,ds = \oint_C M\,dy - N\,dx$$

$$= \iint_R \left(\frac{\partial M}{\partial x} + \frac{\partial N}{\partial y}\right) dx\,dy \qquad \text{Green's Theorem}$$

$$= \int_{-1}^1 \int_{-1}^1 (1 + 2y)\,dx\,dy = \int_{-1}^1 [x + 2xy]_{x=-1}^{x=1}\,dy$$

$$= \int_{-1}^1 (2 + 4y)\,dy = [2y + 2y^2]_{-1}^1 = 4.$$

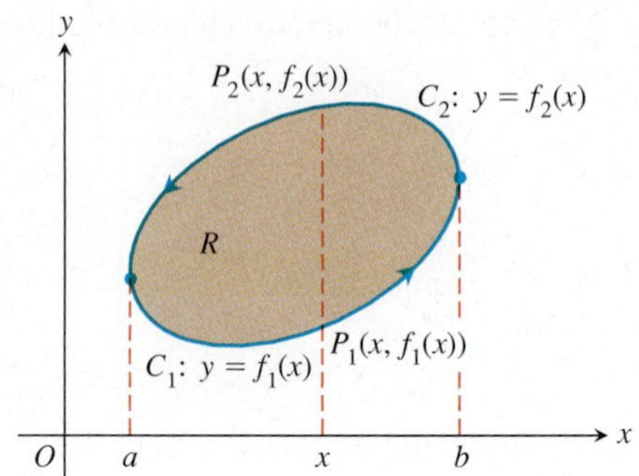

FIGURE 13.29 The boundary curve C is made up of C_1, the graph of $y = f_1(x)$, and C_2, the graph of $y = f_2(x)$.

Proof of Green's Theorem for Special Regions

Let C be a smooth simple closed curve in the xy-plane with the property that lines parallel to the axes cut it in no more than two points. Let R be the region enclosed by C and suppose that M, N, and their first partial derivatives are continuous at every point of some open region containing C and R. We want to prove the circulation-curl form of Green's Theorem,

$$\oint_C M\,dx + N\,dy = \iint_R \left(\frac{\partial N}{\partial x} - \frac{\partial M}{\partial y}\right) dx\,dy. \tag{5}$$

Figure 13.29 shows C made up of two directed parts:

$$C_1:\quad y = f_1(x),\quad a \le x \le b, \qquad C_2:\quad y = f_2(x),\quad b \ge x \ge a.$$

For any x between a and b, we can integrate $\partial M/\partial y$ with respect to y from $y = f_1(x)$ to $y = f_2(x)$ and obtain

$$\int_{f_1(x)}^{f_2(x)} \frac{\partial M}{\partial y}\,dy = M(x, y)\Big]_{y=f_1(x)}^{y=f_2(x)} = M(x, f_2(x)) - M(x, f_1(x)).$$

We can then integrate this with respect to x from a to b:

$$\begin{aligned}\int_a^b \int_{f_1(x)}^{f_2(x)} \frac{\partial M}{\partial y}\,dy\,dx &= \int_a^b [M(x, f_2(x)) - M(x, f_1(x))]\,dx \\ &= -\int_b^a M(x, f_2(x))\,dx - \int_a^b M(x, f_1(x))\,dx \\ &= -\int_{C_2} M\,dx - \int_{C_1} M\,dx \\ &= -\oint_C M\,dx.\end{aligned}$$

Therefore

$$\oint_C M\,dx = \iint_R \left(-\frac{\partial M}{\partial y}\right) dx\,dy. \tag{6}$$

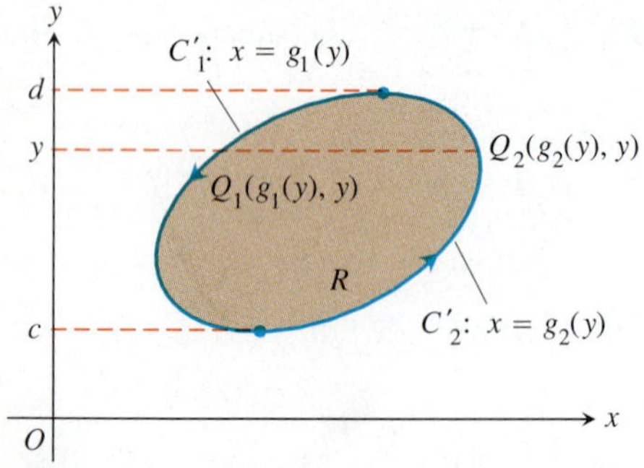

FIGURE 13.30 The boundary curve C is made up of C'_1, the graph of $x = g_1(y)$, and C'_2, the graph of $x = g_2(y)$.

Equation (6) is half the result we need for Equation (5). We derive the other half by integrating $\partial N/\partial x$ first with respect to x and then with respect to y, as suggested by Figure 13.30. This shows the curve C of Figure 13.29 decomposed into the two directed parts $C'_1: x = g_1(y),\ d \ge y \ge c$ and $C'_2: x = g_2(y),\ c \le y \le d$. The result of this double integration is

$$\oint_C N\,dy = \iint_R \frac{\partial N}{\partial x}\,dx\,dy. \tag{7}$$

Combining Equations (6) and (7) gives Equation (5). This concludes the proof.

Extending the Proof to Other Regions

The argument we just gave does not apply directly to the rectangular region in Figure 13.31 because the lines $x = a$, $x = b$, $y = c$, and $y = d$ meet the region's boundary in more than two points. If we divide the boundary C into four directed line segments, however,

$$C_1:\quad y = c,\quad a \le x \le b, \qquad C_2:\quad x = b,\quad c \le y \le d$$
$$C_3:\quad y = d,\quad b \ge x \ge a, \qquad C_4:\quad x = a,\quad d \ge y \ge c,$$

we can modify the argument in the following way.

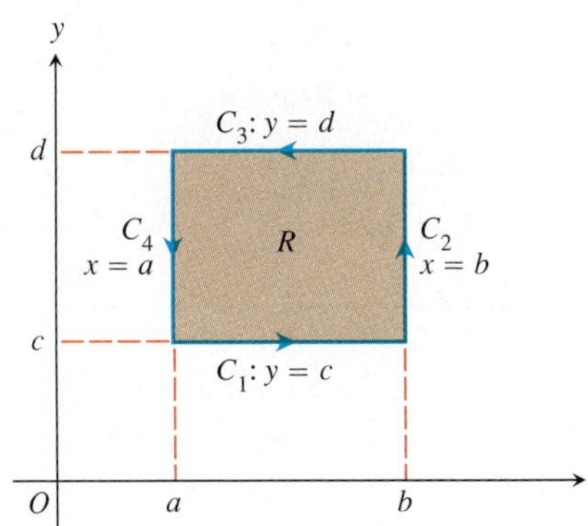

FIGURE 13.31 To prove Green's Theorem for a rectangle, we divide the boundary into four directed line segments.

Proceeding as in the proof of Equation (7), we have

$$\begin{aligned}\int_c^d \int_a^b \frac{\partial N}{\partial x}\,dx\,dy &= \int_c^d (N(b, y) - N(a, y))\,dy \\ &= \int_c^d N(b, y)\,dy + \int_d^c N(a, y)\,dy \qquad (8) \\ &= \int_{C_2} N\,dy + \int_{C_4} N\,dy.\end{aligned}$$

Because y is constant along C_1 and C_3, $\int_{C_1} N\,dy = \int_{C_3} N\,dy = 0$, so we can add $\int_{C_1} N\,dy + \int_{C_3} N\,dy$ to the right-hand side of Equation (8) without changing the equality. Doing so, we have

$$\int_c^d \int_a^b \frac{\partial N}{\partial x}\,dx\,dy = \oint_C N\,dy. \qquad (9)$$

Similarly, we can show that

$$\int_a^b \int_c^d \frac{\partial M}{\partial y}\,dy\,dx = -\oint_C M\,dx. \qquad (10)$$

Subtracting Equation (10) from Equation (9), we again arrive at

$$\oint_C M\,dx + N\,dy = \iint_R \left(\frac{\partial N}{\partial x} - \frac{\partial M}{\partial y}\right) dx\,dy.$$

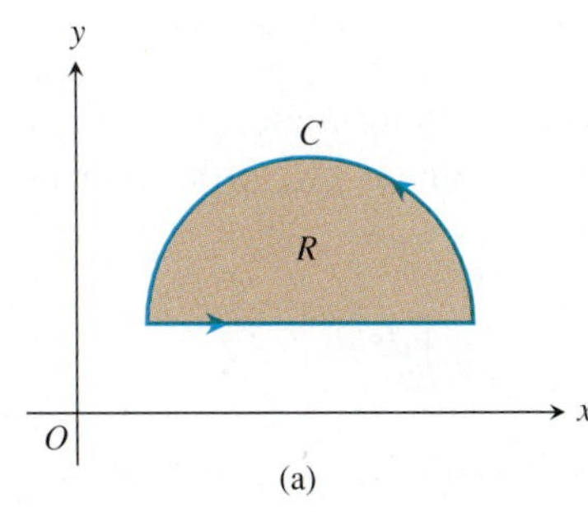

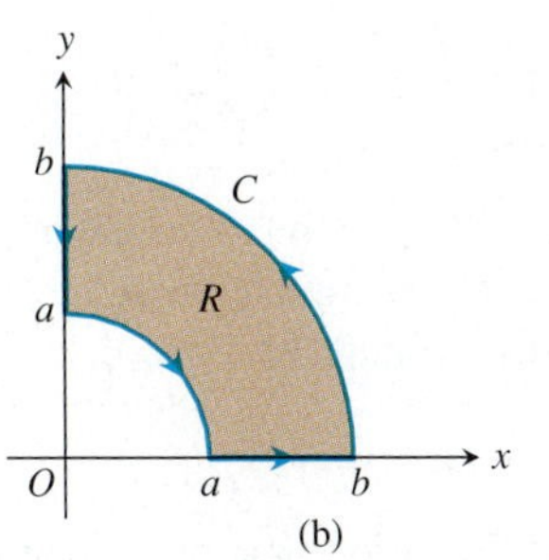

FIGURE 13.32 Other regions to which Green's Theorem applies.

Regions like those in Figure 13.32 can be handled with no greater difficulty. Equation (5) still applies. It also applies to the horseshoe-shaped region R shown in Figure 13.33, as we see by putting together the regions R_1 and R_2 and their boundaries. Green's Theorem applies to C_1, R_1 and to C_2, R_2, yielding

$$\int_{C_1} M\,dx + N\,dy = \iint_{R_1} \left(\frac{\partial N}{\partial x} - \frac{\partial M}{\partial y}\right) dx\,dy$$

$$\int_{C_2} M\,dx + N\,dy = \iint_{R_2} \left(\frac{\partial N}{\partial x} - \frac{\partial M}{\partial y}\right) dx\,dy.$$

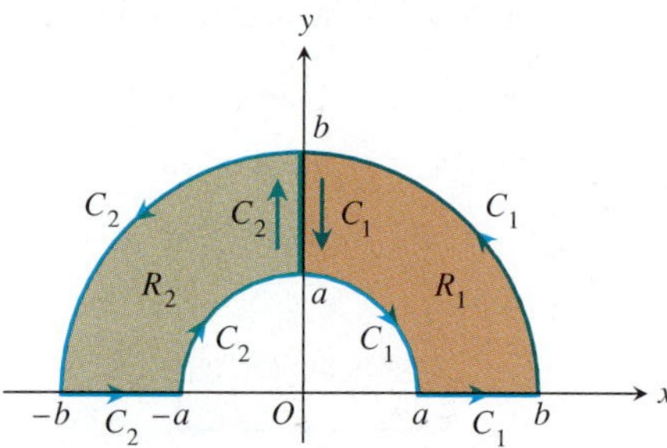

FIGURE 13.33 A region R that combines regions R_1 and R_2.

When we add these two equations, the line integral along the y-axis from b to a for C_1 cancels the integral over the same segment but in the opposite direction for C_2. Hence,

$$\oint_C M\,dx + N\,dy = \iint_R \left(\frac{\partial N}{\partial x} - \frac{\partial M}{\partial y}\right) dx\,dy,$$

where C consists of the two segments of the x-axis from $-b$ to $-a$ and from a to b and of the two semicircles, and where R is the region inside C.

The device of adding line integrals over separate boundaries to build up an integral over a single boundary can be extended to any finite number of subregions. In Figure 13.34a, let C_1 be the boundary, oriented counterclockwise, of the region R_1 in the first quadrant. Similarly, for the other three quadrants, C_i is the boundary of the region R_i, $i = 2, 3, 4$. By Green's Theorem,

$$\oint_{C_i} M\,dx + N\,dy = \iint_{R_i} \left(\frac{\partial N}{\partial x} - \frac{\partial M}{\partial y}\right) dx\,dy. \tag{11}$$

We add Equations (11) for $i = 1, 2, 3, 4$, and get (Figure 13.34b):

$$\oint_{r=b} (M\,dx + N\,dy) + \oint_{r=a} (M\,dx + N\,dy) = \iint_{a \le r \le b} \left(\frac{\partial N}{\partial x} - \frac{\partial M}{\partial y}\right) dx\,dy. \tag{12}$$

Equation (12) says that the double integral of $(\partial N/\partial x) - (\partial M/\partial y)$ over the annular ring R equals the line integral of $M\,dx + N\,dy$ over the complete boundary of R in the direction that keeps R on our left as we progress (Figure 13.34b).

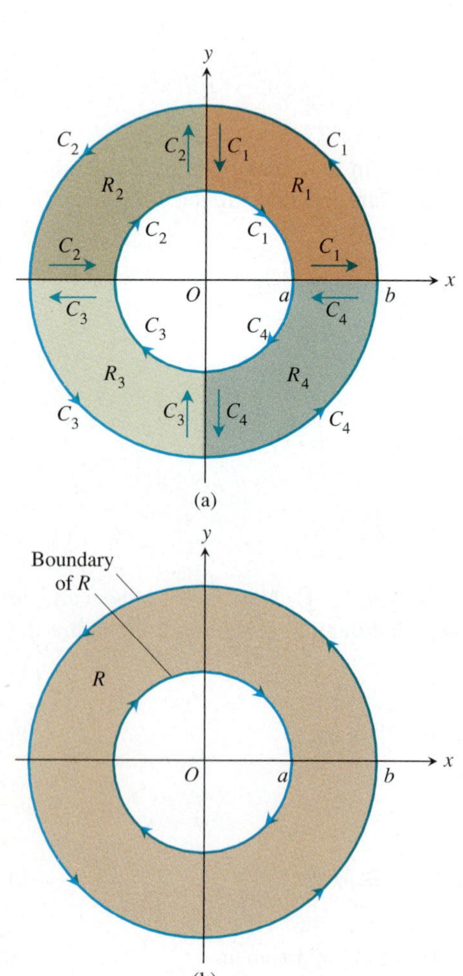

FIGURE 13.34 The annular region R combines four smaller regions. In polar coordinates, $r = a$ for the inner circle, $r = b$ for the outer circle, and $a \le r \le b$ for the region itself.

Example 6 Verifying Green's Theorem for an Annular Ring

Verify the circulation form of Green's Theorem (Equation (4)) on the annular ring R: $h^2 \le x^2 + y^2 \le 1$, $0 < h < 1$ (Figure 13.35), if

$$M = \frac{-y}{x^2 + y^2}, \qquad N = \frac{x}{x^2 + y^2}.$$

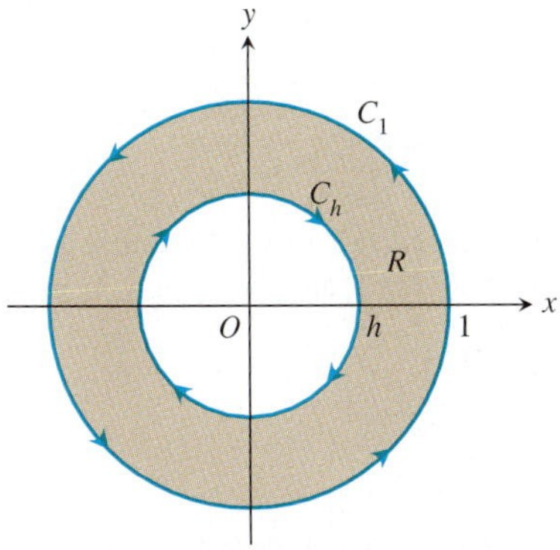

FIGURE 13.35 Green's Theorem may be applied to the annular region R by integrating along the boundaries as shown. (Example 6)

Solution The boundary of R consists of the circle

$$C_1: \quad x = \cos t, \qquad y = \sin t, \qquad 0 \le t \le 2\pi,$$

traversed counterclockwise as t increases, and the circle

$$C_h: \quad x = h \cos \theta, \quad y = -h \sin \theta, \quad 0 \le \theta \le 2\pi,$$

traversed clockwise as θ increases. The functions M and N and their partial derivatives are continuous throughout R. Moreover,

$$\begin{aligned} \frac{\partial M}{\partial y} &= \frac{(x^2 + y^2)(-1) + y(2y)}{(x^2 + y^2)^2} \\ &= \frac{y^2 - x^2}{(x^2 + y^2)^2} = \frac{\partial N}{\partial x}, \end{aligned}$$

so

$$\iint_R \left(\frac{\partial N}{\partial x} - \frac{\partial M}{\partial y} \right) dx\, dy = \iint_R 0\, dx\, dy = 0.$$

The integral of $M\,dx + N\,dy$ over the boundary of R is

$$\begin{aligned} \int_C M\,dx + N\,dy &= \oint_{C_1} \frac{x\,dy - y\,dx}{x^2 + y^2} + \oint_{C_h} \frac{x\,dy - y\,dx}{x^2 + y^2} \\ &= \int_0^{2\pi} (\cos^2 t + \sin^2 t)\, dt - \int_0^{2\pi} \frac{h^2(\cos^2 \theta + \sin^2 \theta)}{h^2}\, d\theta \\ &= 2\pi - 2\pi = 0. \end{aligned}$$

The functions M and N in Example 6 are discontinuous at $(0, 0)$, so we cannot apply Green's Theorem to the circle C_1 and the region inside it. We must exclude the origin. We do so by excluding the points inside C_h.

We could replace the circle C_1 in Example 6 by an ellipse or any other simple closed curve K surrounding C_h (Figure 13.36). The result would still be

$$\oint_K (M\,dx + N\,dy) + \oint_{C_h} (M\,dx + N\,dy) = \iint_R \left(\frac{\partial N}{\partial x} - \frac{\partial M}{\partial y} \right) dy\, dx = 0,$$

which leads to the surprising conclusion that

$$\oint_K (M\,dx + N\,dy) = 2\pi$$

for any such curve K. We can explain this result by changing to polar coordinates. With

$$\begin{aligned} x &= r\cos\theta, & y &= r\sin\theta, \\ dx &= -r\sin\theta\,d\theta + \cos\theta\,dr, & dy &= r\cos\theta\,d\theta + \sin\theta\,dr, \end{aligned}$$

we have

$$\frac{x\,dy - y\,dx}{x^2 + y^2} = \frac{r^2(\cos^2\theta + \sin^2\theta)\,d\theta}{r^2} = d\theta,$$

and θ increases by 2π as we traverse K once counterclockwise.

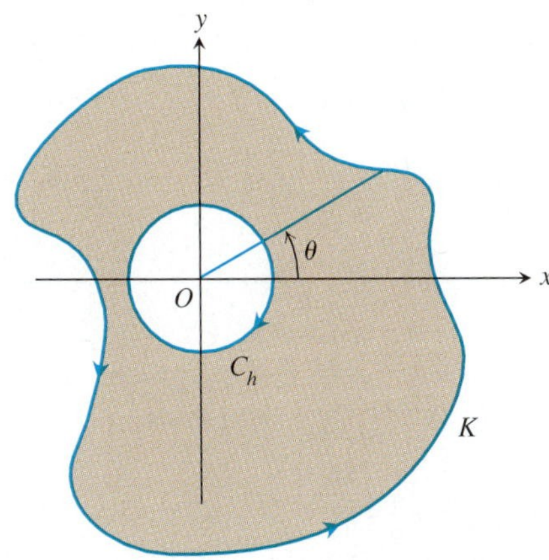

FIGURE 13.36 The region bounded by the circle C_h and the curve K.

EXERCISES 13.4

Verifying Green's Theorem

In Exercises 1–4, verify the conclusion of Green's Theorem by evaluating both sides of Equations (3) and (4) for the field $\mathbf{F} = M\mathbf{i} + N\mathbf{j}$. Take the domains of integration in each case to be the disk $R: x^2 + y^2 \le a^2$ and its bounding circle $C: \mathbf{r} = (a \cos t)\mathbf{i} + (a \sin t)\mathbf{j}, 0 \le t \le 2\pi$.

1. $\mathbf{F} = -y\mathbf{i} + x\mathbf{j}$
2. $\mathbf{F} = y\mathbf{i}$
3. $\mathbf{F} = 2x\mathbf{i} - 3y\mathbf{j}$
4. $\mathbf{F} = -x^2y\mathbf{i} + xy^2\mathbf{j}$

Counterclockwise Circulation and Outward Flux

In Exercises 5–10, use Green's Theorem to find the counterclockwise circulation and outward flux for the field $\mathbf{F}$ and curve C.

5. $\mathbf{F} = (x - y)\mathbf{i} + (y - x)\mathbf{j}$

   C: The square bounded by $x = 0, x = 1, y = 0, y = 1$

6. $\mathbf{F} = (x^2 + 4y)\mathbf{i} + (x + y^2)\mathbf{j}$

   C: The square bounded by $x = 0, x = 1, y = 0, y = 1$

7. $\mathbf{F} = (y^2 - x^2)\mathbf{i} + (x^2 + y^2)\mathbf{j}$

   C: The triangle bounded by $y = 0, x = 3$, and $y = x$

8. $\mathbf{F} = (x + y)\mathbf{i} - (x^2 + y^2)\mathbf{j}$

   C: The triangle bounded by $y = 0, x = 1$, and $y = x$

9. $\mathbf{F} = (x + e^x \sin y)\mathbf{i} + (x + e^x \cos y)\mathbf{j}$

   C: The right-hand loop of the lemniscate $r^2 = \cos 2\theta$

10. $\mathbf{F} = \left(\tan^{-1}\frac{y}{x}\right)\mathbf{i} + \ln(x^2 + y^2)\mathbf{j}$

    C: The boundary of the region defined by the polar coordinate inequalities $1 \le r \le 2, 0 \le \theta \le \pi$

11. Find the counterclockwise circulation and outward flux of the field $\mathbf{F} = xy\mathbf{i} + y^2\mathbf{j}$ around and over the boundary of the region enclosed by the curves $y = x^2$ and $y = x$ in the first quadrant.

12. Find the counterclockwise circulation and the outward flux of the field $\mathbf{F} = (-\sin y)\mathbf{i} + (x \cos y)\mathbf{j}$ around and over the square cut from the first quadrant by the lines $x = \pi/2$ and $y = \pi/2$.

13. Find the outward flux of the field

$$\mathbf{F} = \left(3xy - \frac{x}{1 + y^2}\right)\mathbf{i} + (e^x + \tan^{-1} y)\mathbf{j}$$

    across the cardioid $r = a(1 + \cos\theta), a > 0$.

14. Find the counterclockwise circulation of $\mathbf{F} = (y + e^x \ln y)\mathbf{i} + (e^x/y)\mathbf{j}$ around the boundary of the region that is bounded above by the curve $y = 3 - x^2$ and below by the curve $y = x^4 + 1$.

Work

In Exercises 15 and 16, find the work done by $\mathbf{F}$ in moving a particle once counterclockwise around the given curve.

15. $\mathbf{F} = 2xy^3\mathbf{i} + 4x^2y^2\mathbf{j}$

    C: The boundary of the "triangular" region in the first quadrant enclosed by the x-axis, the line $x = 1$, and the curve $y = x^3$

16. $\mathbf{F} = (4x - 2y)\mathbf{i} + (2x - 4y)\mathbf{j}$

    C: The circle $(x - 2)^2 + (y - 2)^2 = 4$

Evaluating Line Integrals in the Plane

Apply Green's Theorem to evaluate the integrals in Exercises 17–20.

17. $\oint_C (y^2\,dx + x^2\,dy)$

    C: The triangle bounded by $x = 0, x + y = 1, y = 0$

18. $\oint_C (3y\,dx + 2x\,dy)$

    C: The boundary of $0 \le x \le \pi, 0 \le y \le \sin x$

19. $\oint_C (6y + x)\,dx + (y + 2x)\,dy$

    C: The circle $(x - 2)^2 + (y - 3)^2 = 4$

20. $\oint_C (2x + y^2)\,dx + (2xy + 3y)\,dy$

    C: Any simple closed curve in the plane for which Green's theorem holds

CD-ROM WEBsite

Calculating Area with Green's Theorem

If a simple closed curve C in the plane and the region R it encloses satisfy the hypotheses of Green's Theorem, the area of R is given by

Green's Theorem Area Formula

$$\text{Area of } R = \frac{1}{2}\oint_C x\,dy - y\,dx \tag{13}$$

The reason is that by Equation (3), run backward,

$$\text{Area of } R = \iint_R dy\,dx = \iint_R \left(\frac{1}{2} + \frac{1}{2}\right) dy\,dx$$

$$= \oint_C \frac{1}{2}x\,dy - \frac{1}{2}y\,dx.$$

Use the Green's Theorem area formula (Equation (13)) to find the areas of the regions enclosed by the curves in Exercises 21–24.

21. The circle $\mathbf{r}(t) = (a \cos t)\mathbf{i} + (a \sin t)\mathbf{j}, \quad 0 \le t \le 2\pi$

22. The ellipse $\mathbf{r}(t) = (a \cos t)\mathbf{i} + (b \sin t)\mathbf{j}, \quad 0 \le t \le 2\pi$

23. The astroid (Figure 5.30) $\mathbf{r}(t) = (\cos^3 t)\mathbf{i} + (\sin^3 t)\mathbf{j}, \quad 0 \le t \le 2\pi$

24. The curve $\mathbf{r}(t) = t^2\mathbf{i} + ((t^3/3) - t)\mathbf{j}, \quad -\sqrt{3} \le t \le \sqrt{3}$ (see accompanying figure).

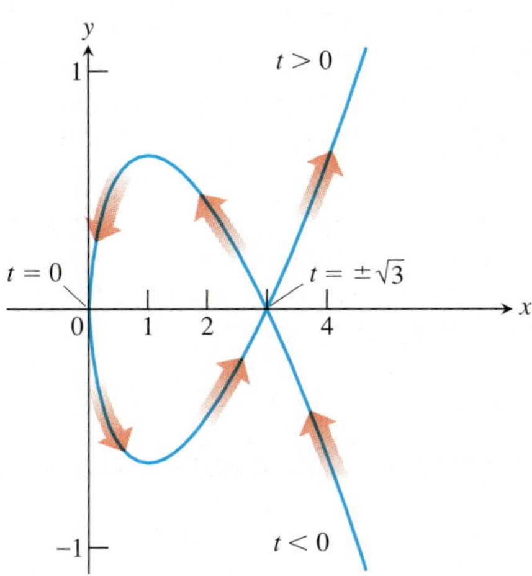

Theory and Examples

25. Let C be the boundary of a region on which Green's Theorem holds. Use Green's Theorem to calculate

(a) $\oint_C f(x)\,dx + g(y)\,dy$

(b) $\oint_C ky\,dx + hx\,dy \quad$ (k and h constants).

26. *Integral dependent only on area* Show that the value of

$$\oint_C xy^2\,dx + (x^2y + 2x)\,dy$$

around any square depends only on the area of the square and not on its location in the plane.

27. *Writing to Learn* What is special about the integral

$$\oint_C 4x^3y\,dx + x^4\,dy?$$

Give reasons for your answer.

28. *Writing to Learn* What is special about the integral

$$\oint_C -y^3\,dx + x^3\,dy?$$

Give reasons for your answer.

29. *Area as a line integral* Show that if R is a region in the plane bounded by a piecewise smooth simple closed curve C, then

$$\text{Area of } R = \oint_C x\,dy = -\oint_C y\,dx.$$

30. *Definite integral as a line integral* Suppose that a nonnegative function $y = f(x)$ has a continuous first derivative on $[a, b]$. Let C be the boundary of the region in the xy-plane that is bounded below by the x-axis, above by the graph of f, and on the sides by the lines $x = a$ and $x = b$. Show that

$$\int_a^b f(x)\,dx = -\oint_C y\,dx.$$

31. *Area and the centroid* Let A be the area and $\bar{x}$ the x-coordinate of the centroid of a region R that is bounded by a piecewise smooth simple closed curve C in the xy-plane. Show that

$$\frac{1}{2}\oint_C x^2\,dy = -\oint_C xy\,dx = \frac{1}{3}\oint_C x^2\,dy - xy\,dx = A\bar{x}.$$

32. *Moment of inertia* Let I_y be the moment of inertia about the y-axis of the region in Exercise 31. Show that

$$\frac{1}{3}\oint_C x^3\,dy = -\oint_C x^2y\,dx = \frac{1}{4}\oint_C x^3\,dy - x^2y\,dx = I_y.$$

33. *Green's Theorem and Laplace's equation* Assuming that all the necessary derivatives exist and are continuous, show that if $f(x, y)$ satisfies the Laplace equation

$$\frac{\partial^2 f}{\partial x^2} + \frac{\partial^2 f}{\partial y^2} = 0,$$

then

$$\oint_C \frac{\partial f}{\partial y}\,dx - \frac{\partial f}{\partial x}\,dy = 0$$

for all closed curves C to which Green's Theorem applies. (The converse is also true: If the line integral is always zero, then f satisfies the Laplace equation.)

34. *Maximizing work* Among all smooth simple closed curves in the plane, oriented counterclockwise, find the one along which the work done by

CD-ROM WEBsite

$$\mathbf{F} = \left(\frac{1}{4}x^2y + \frac{1}{3}y^3\right)\mathbf{i} + x\mathbf{j}$$

is greatest. (*Hint:* Where is (curl $\mathbf{F}$) · $\mathbf{k}$ positive?)

35. *Regions with many holes* Green's Theorem holds for a region R with any finite number of holes as long as the bounding curves are smooth, simple, and closed and we integrate over each component of the boundary in the direction that keeps R on our immediate left as we go along (Figure 13.37 on the next page).

(a) Let $f(x, y) = \ln(x^2 + y^2)$ and let C be the circle $x^2 + y^2 = a^2$. Evaluate the flux integral

$$\oint_C \nabla f \cdot \mathbf{n}\,ds.$$

(b) Let K be an arbitrary smooth simple closed curve in the plane that does not pass through (0, 0). Use Green's Theorem to show that

$$\oint_K \nabla f \cdot \mathbf{n}\, ds$$

has two possible values, depending on whether (0, 0) lies inside K or outside K.

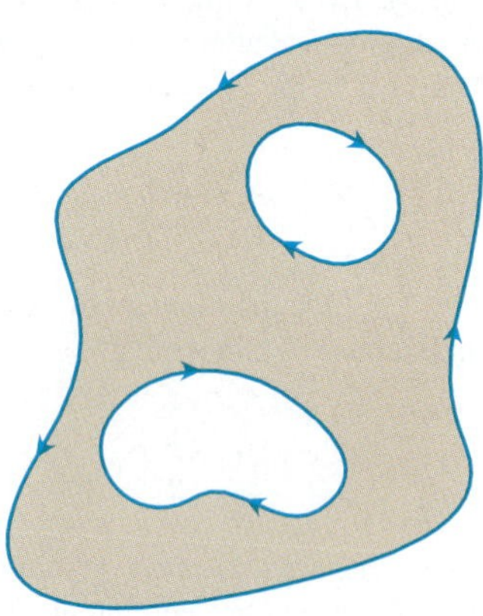

FIGURE 13.37 Green's Theorem holds for regions with more than one hole. (Exercise 35)

36. *Bendixson's criterion* The **streamlines** of a planar fluid flow are the smooth curves traced by the fluid's individual particles. The vectors $\mathbf{F} = M(x, y)\mathbf{i} + N(x, y)\mathbf{j}$ of the flow's velocity field are the tangent vectors of the streamlines. Show that if the flow takes place over a *simply connected* region R (no holes or missing points) and that if $M_x + N_y \neq 0$ throughout R, then none of the streamlines in R is closed. In other words, no particle of fluid ever has a closed trajectory in R. The criterion $M_x + N_y \neq 0$ is called **Bendixson's criterion** for the nonexistence of closed trajectories.

37. Establish Equation (7) to finish the proof of the special case of Green's Theorem.

38. Establish Equation (10) to complete the argument for the extension of Green's Theorem.

39. *Writing to Learn: Curl component of conservative fields* Can anything be said about the curl component of a conservative two-dimensional vector field? Give reasons for your answer.

40. *Writing to Learn: Circulation of conservative fields* Does Green's Theorem give any information about the circulation of a conservative field? Does this agree with anything else you know? Give reasons for your answer.

COMPUTER EXPLORATIONS

Finding Circulation

In Exercises 41–44, use a CAS and Green's Theorem to find the counterclockwise circulation of the field $\mathbf{F}$ around the simple closed curve C. Perform the following CAS steps.

(a) Plot C in the xy-plane.

(b) Determine the integrand $(\partial N/\partial x) - (\partial M/\partial y)$ for the curl form of Green's Theorem.

(c) Determine the (double integral) limits of integration from your plot in part (a) and evaluate the curl integral for the circulation.

41. $\mathbf{F} = (2x - y)\mathbf{i} + (x + 3y)\mathbf{j}$, C: The ellipse $x^2 + 4y^2 = 4$

42. $\mathbf{F} = (2x^3 - y^3)\mathbf{i} + (x^3 + y^3)\mathbf{j}$, C: The ellipse $\dfrac{x^2}{4} + \dfrac{y^2}{9} = 1$

43. $\mathbf{F} = x^{-1}e^y\,\mathbf{i} + (e^y \ln x + 2x)\mathbf{j}$,

C: The boundary of the region defined by $y = 1 + x^4$ (below) and $y = 2$ (above)

44. $\mathbf{F} = xe^y\,\mathbf{i} + 4x^2 \ln y\,\mathbf{j}$, C: The triangle with vertices (0, 0), (2, 0), and (0, 4)

13.5 Surface Area and Surface Integrals

Surface Area • A Practical Formula • Surface Integrals • Algebraic Properties; The Surface Area Differential • Orientation • Surface Integral for Flux • Moments and Masses of Thin Shells

We know how to integrate a function over a flat region in a plane, but what if the function is defined over a curved surface? How do we calculate its integral then? The trick to evaluating one of these so-called surface integrals is to rewrite it as a double integral over a region in a coordinate plane beneath the surface

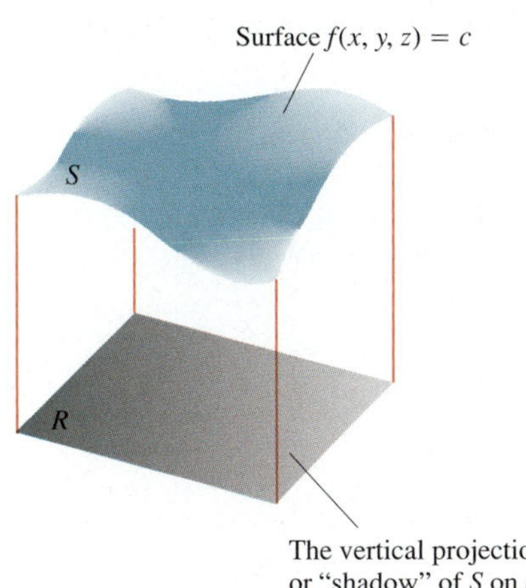

FIGURE 13.38 As we soon see, the integral of a function $g(x, y, z)$ over a surface S in space can be calculated by evaluating a related double integral over the vertical projection or "shadow" of S on a coordinate plane.

(Figure 13.38). In Sections 13.7 and 13.8, we see how surface integrals provide just what we need to generalize the two forms of Green's Theorem to three dimensions.

Surface Area

Figure 13.39 shows a surface S lying above its "shadow" region R in a plane beneath it. The surface is defined by the equation $f(x, y, z) = c$. If the surface is **smooth** (∇f is continuous and never vanishes on S), we can define and calculate its area as a double integral over R.

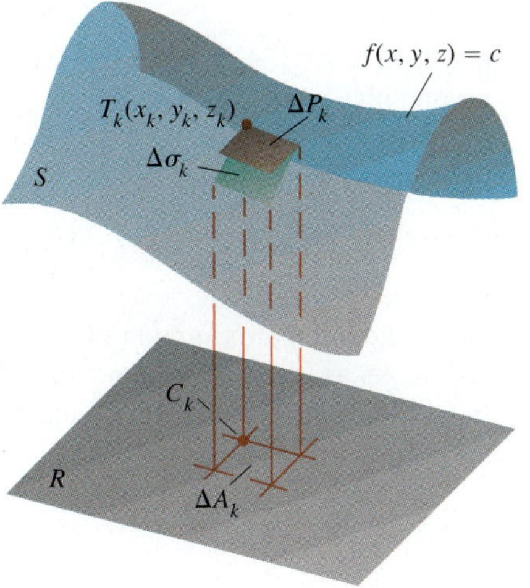

FIGURE 13.39 A surface S and its vertical projection onto a plane beneath it. You can think of R as the shadow of S on the plane. The tangent plate ΔP_k approximates the surface patch $\Delta\sigma_k$ above ΔA_k.

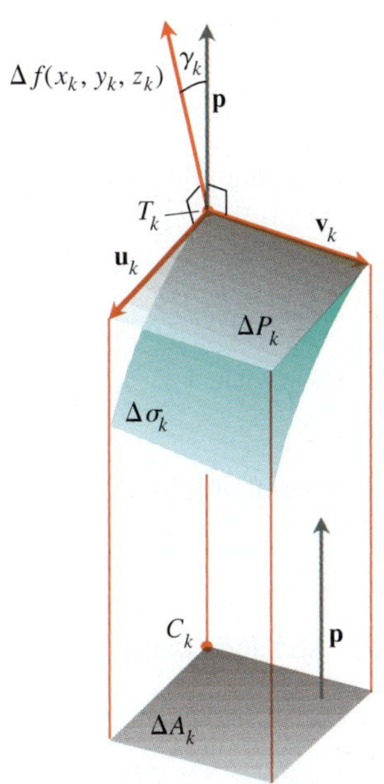

FIGURE 13.40 Magnified view from the preceding figure. The vector $\mathbf{u}_k \times \mathbf{v}_k$ (not shown) is parallel to the vector ∇f because both vectors are normal to the plane of ΔP_k.

The first step in defining the area of S is to partition the region R into small rectangles ΔA_k of the kind we would use if we were defining an integral over R. Directly above each ΔA_k lies a patch of surface $\Delta\sigma_k$ that we may approximate with a portion ΔP_k of a tangent plane. To be specific, we suppose that ΔP_k is a portion of the plane that is tangent to the surface at the point $T_k(x_k, y_k, z_k)$ directly above the back corner C_k of ΔA_k. If the tangent plane is parallel to R, then ΔP_k will be congruent to ΔA_k. Otherwise, it will be a parallelogram whose area is somewhat larger than the area of ΔA_k.

Figure 13.40 gives a magnified view of $\Delta\sigma_k$ and ΔP_k, showing the gradient vector $\nabla f(x_k, y_k, z_k)$ at T_k and a unit vector $\mathbf{p}$ that is normal to R. The figure also shows the angle γ_k between ∇f and $\mathbf{p}$. The other vectors in the picture, $\mathbf{u}_k$ and $\mathbf{v}_k$, lie along the edges of the patch ΔP_k in the tangent plane. Thus, both $\mathbf{u}_k \times \mathbf{v}_k$ and ∇f are normal to the tangent plane.

We now need to know from advanced vector geometry that $|(\mathbf{u}_k \times \mathbf{v}_k) \cdot \mathbf{p}|$ is the area of the projection of the parallelogram determined by $\mathbf{u}_k$ and $\mathbf{v}_k$ onto any plane whose normal is $\mathbf{p}$. (A proof is given in Appendix 12.) In our case, this translates into the statement

$$|(\mathbf{u}_k \times \mathbf{v}_k) \cdot \mathbf{p}| = \Delta A_k.$$

Now, $|\mathbf{u}_k \times \mathbf{v}_k|$ itself is the area ΔP_k (standard fact about cross products) so this last equation becomes

$$\underbrace{|\mathbf{u}_k \times \mathbf{v}_k|}_{\Delta P_k} \; \underbrace{|\mathbf{p}|}_{1} \; \underbrace{|\cos(\text{angle between } \mathbf{u}_k \times \mathbf{v}_k \text{ and } \mathbf{p})|}_{\substack{\text{Same as } |\cos \gamma_k| \text{ because } \nabla f \text{ and } \mathbf{u}_k \times \mathbf{v}_k \\ \text{are both normal to the tangent plane}}} = \Delta A_k$$

or

$$\Delta P_k |\cos \gamma_k| = \Delta A_k$$

or

$$\Delta P_k = \frac{\Delta A_k}{|\cos \gamma_k|},$$

provided $\cos \gamma_k \neq 0$. We will have $\cos \gamma_k \neq 0$ as long as ∇f is not parallel to the ground plane and $\nabla f \cdot \mathbf{p} \neq 0$.

Since the patches ΔP_k approximate the surface patches $\Delta \sigma_k$ that fit together to make S, the sum

$$\sum \Delta P_k = \sum \frac{\Delta A_k}{|\cos \gamma_k|} \tag{1}$$

looks like an approximation of what we might like to call the surface area of S. It also looks as if the approximation would improve if we refined the partition of R. In fact, the sums on the right-hand side of Equation (1) are approximating sums for the double integral

$$\iint_R \frac{1}{|\cos \gamma|} \, dA. \tag{2}$$

We therefore define the **area** of S to be the value of this integral whenever it exists.

A Practical Formula

For any surface $f(x, y, z) = c$, we have $|\nabla f \cdot \mathbf{p}| = |\nabla f||\mathbf{p}||\cos \gamma|$, so

$$\frac{1}{|\cos \gamma|} = \frac{|\nabla f|}{|\nabla f \cdot \mathbf{p}|}.$$

This combines with Equation (2) to give a practical formula for area.

> **Formula for Surface Area**
> The area of the surface $f(x, y, z) = c$ over a closed and bounded plane region R is
>
> $$\text{Surface area} = \iint_R \frac{|\nabla f|}{|\nabla f \cdot \mathbf{p}|} \, dA, \tag{3}$$
>
> where $\mathbf{p}$ is a unit vector normal to R and $\nabla f \cdot \mathbf{p} \neq 0$.

Thus, the area is the double integral over R of the magnitude of ∇f divided by the magnitude of the scalar component of ∇f normal to R.

We reached Equation (3) under the assumption that $\nabla f \cdot \mathbf{p} \neq 0$ throughout R and that ∇f is continuous. Whenever the integral exists, however, we define its value to be the area of the portion of the surface $f(x, y, z) = c$ that lies over R.

In the exercises, we show how Equation (3) simplifies if the surface is defined by $z = f(x, y)$.

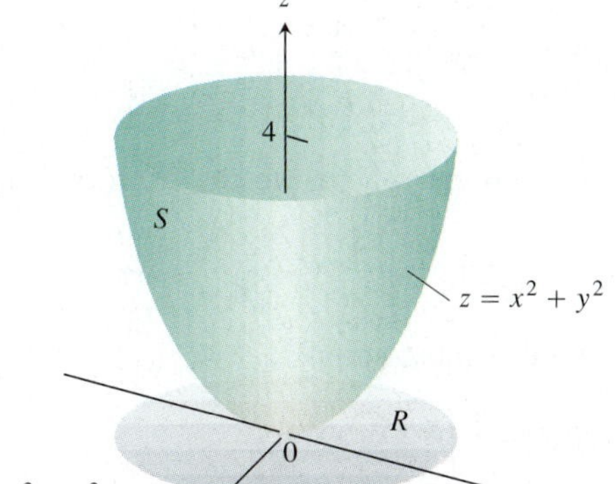

FIGURE 13.41 The area of this parabolic surface in Example 1.

Example 1 Finding Surface Area

Find the area of the surface cut from the bottom of the paraboloid $x^2 + y^2 - z = 0$ by the plane $z = 4$.

Solution We sketch the surface S and the region R below it in the xy-plane (Figure 13.41). The surface S is part of the level surface $f(x, y, z) = x^2 + y^2 - z = 0$, and R is the disk $x^2 + y^2 \leq 4$ in the xy-plane. To get a unit vector normal to the plane of R, we can take $\mathbf{p} = \mathbf{k}$.

At any point (x, y, z) on the surface, we have

$$\begin{aligned} f(x, y, z) &= x^2 + y^2 - z \\ \nabla f &= 2x\mathbf{i} + 2y\mathbf{j} - \mathbf{k} \\ |\nabla f| &= \sqrt{(2x)^2 + (2y)^2 + (-1)^2} \\ &= \sqrt{4x^2 + 4y^2 + 1} \\ |\nabla f \cdot \mathbf{p}| &= |\nabla f \cdot \mathbf{k}| = |-1| = 1. \end{aligned}$$

In the region R, $dA = dx\,dy$. Therefore,

$$\begin{aligned} \text{Surface area} &= \iint\limits_R \frac{|\nabla f|}{|\nabla f \cdot \mathbf{p}|}\, dA && \text{Eq. (3)} \\ &= \iint\limits_{x^2+y^2 \leq 4} \sqrt{4x^2 + 4y^2 + 1}\, dx\, dy \\ &= \int_0^{2\pi} \int_0^2 \sqrt{4r^2 + 1}\, r\, dr\, d\theta && \text{Polar coordinates} \\ &= \int_0^{2\pi} \left[\frac{1}{12} (4r^2 + 1)^{3/2} \right]_0^2 d\theta \\ &= \int_0^{2\pi} \frac{1}{12} (17^{3/2} - 1)\, d\theta = \frac{\pi}{6} (17\sqrt{17} - 1). \end{aligned}$$

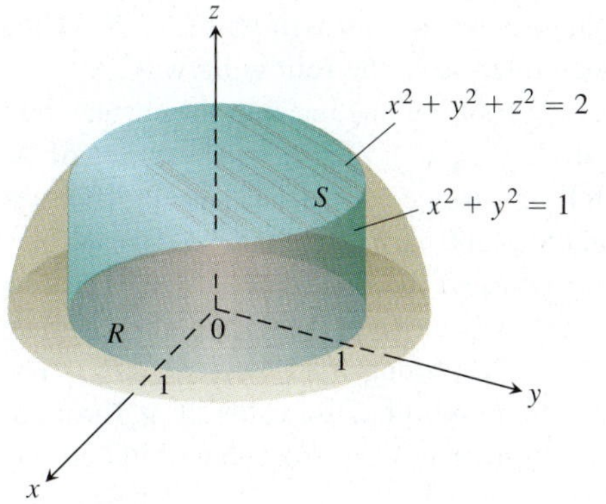

FIGURE 13.42 The cap cut from the hemisphere by the cylinder in Example 2

Example 2 Finding Surface Area

Find the area of the cap cut from the hemisphere $x^2 + y^2 + z^2 = 2, z \geq 0$, by the cylinder $x^2 + y^2 = 1$ (Figure 13.42).

Solution The cap S is part of the level surface $f(x, y, z) = x^2 + y^2 + z^2 = 2$. It projects one-to-one onto the disk $R: x^2 + y^2 \leq 1$ in the xy-plane. The vector $\mathbf{p} = \mathbf{k}$ is normal to the plane of R.

At any point on the surface,

$$
\begin{aligned}
f(x, y, z) &= x^2 + y^2 + z^2 \\
\nabla f &= 2x\mathbf{i} + 2y\mathbf{j} + 2z\mathbf{k} \\
|\nabla f| &= 2\sqrt{x^2 + y^2 + z^2} = 2\sqrt{2} \qquad \text{Because } x^2 + y^2 + z^2 = 2 \text{ at points of } S \\
|\nabla f \cdot \mathbf{p}| &= |\nabla f \cdot \mathbf{k}| = |2z| = 2z.
\end{aligned}
$$

Therefore,

$$
\text{Surface area} = \iint_R \frac{|\nabla f|}{|\nabla f \cdot \mathbf{p}|}\, dA = \iint_R \frac{2\sqrt{2}}{2z}\, dA = \sqrt{2} \iint_R \frac{dA}{z}. \tag{4}
$$

What do we do about the z?

Since z is the z-coordinate of a point on the sphere, we can express it in terms of x and y as

$$
z = \sqrt{2 - x^2 - y^2}.
$$

We continue the work of Equation (4) with this substitution:

$$
\begin{aligned}
\text{Surface area} &= \sqrt{2} \iint_R \frac{dA}{z} = \sqrt{2} \iint_{x^2+y^2 \le 1} \frac{dA}{\sqrt{2 - x^2 - y^2}} \\
&= \sqrt{2} \int_0^{2\pi} \int_0^1 \frac{r\, dr\, d\theta}{\sqrt{2 - r^2}} \qquad \text{Polar coordinates} \\
&= \sqrt{2} \int_0^{2\pi} \left[-(2 - r^2)^{1/2} \right]_{r=0}^{r=1} d\theta \\
&= \sqrt{2} \int_0^{2\pi} (\sqrt{2} - 1)\, d\theta = 2\pi(2 - \sqrt{2}).
\end{aligned}
$$

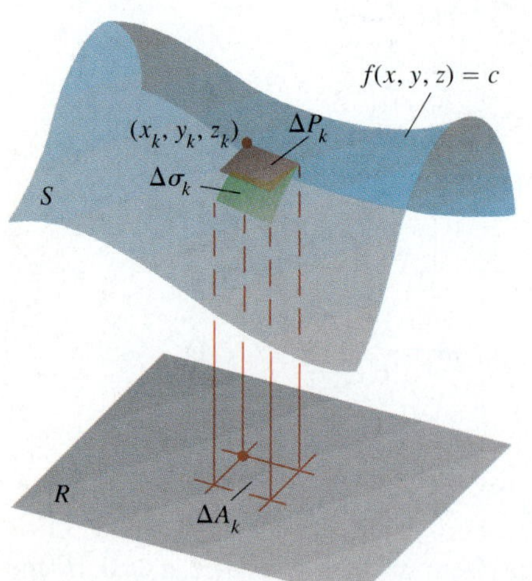

FIGURE 13.43 If we know how an electrical charge is distributed over a surface, we can find the total charge with a suitably modified surface integral.

Surface Integrals

We now show how to integrate a function over a surface, using the ideas just developed for calculating surface area.

Suppose, for example, that we have an electrical charge distributed over a surface $f(x, y, z) = c$ like the one shown in Figure 13.43 and that the function $g(x, y, z)$ gives the charge per unit area (charge density) at each point on S. Then we may calculate the total charge on S as an integral in the following way.

We partition the shadow region R on the ground plane beneath the surface into small rectangles of the kind we would use if we were defining the surface area of S. Then directly above each ΔA_k lies a patch of surface $\Delta\sigma_k$ that we approximate with a parallelogram-shaped portion of tangent plane, ΔP_k.

Up to this point the construction proceeds as in the definition of surface area, but now we take an additional step: We evaluate g at (x_k, y_k, z_k) and approximate the total charge on the surface patch $\Delta\sigma_k$ by the product $g(x_k, y_k, z_k)\, \Delta P_k$. The rationale is that when the partition of R is sufficiently fine, the value of g throughout $\Delta\sigma_k$ is nearly constant and ΔP_k is nearly the same as $\Delta\sigma_k$. The total charge over S is then approximated by the sum

$$
\text{Total charge} \approx \sum g(x_k, y_k, z_k)\, \Delta P_k = \sum g(x_k, y_k, z_k)\, \frac{\Delta A_k}{|\cos \gamma_k|}.
$$

If f, the function defining the surface S, and its first partial derivatives are continuous, and if g is continuous over S, then the sums on the right-hand side of the last equation approach the limit

$$\iint_R g(x, y, z) \frac{dA}{|\cos \gamma|} = \iint_R g(x, y, z) \frac{|\nabla f|}{|\nabla f \cdot \mathbf{p}|} dA \qquad (5)$$

as the partition of R is refined in the usual way. This limit is called the integral of g over the surface S and is calculated as a double integral over R. The value of the integral is the total charge on the surface S.

As you might expect, the formula in Equation (5) defines the integral of *any* function g over the surface S as long as the integral exists.

Definitions Integral of g over S and Surface Integral

If R is the shadow region of a surface S defined by the equation $f(x, y, z) = c$, and g is a continuous function defined at the points of S, then the **integral of g over S** is the integral

$$\iint_R g(x, y, z) \frac{|\nabla f|}{|\nabla f \cdot \mathbf{p}|} dA, \qquad (6)$$

where $\mathbf{p}$ is a unit vector normal to R and $\nabla \mathrm{f} \cdot \mathbf{p} \neq 0$. The integral itself is called a **surface integral.**

The integral in Equation (6) takes on different meanings in different applications. If g has the constant value 1, the integral gives the area of S. If g gives the mass density of a thin shell of material modeled by S, the integral gives the mass of the shell.

Algebraic Properties; The Surface Area Differential

We can abbreviate the integral in Equation (6) by writing $d\sigma$ for $(|\nabla f| / |\nabla f \cdot \mathbf{p}|)\, dA$.

The Surface Area Differential and the Differential Form for Surface Integrals

$$d\sigma = \frac{|\nabla f|}{|\nabla f \cdot \mathbf{p}|} dA \qquad \iint_S g\, d\sigma \qquad (7)$$

Surface area differential — Differential formula for surface integrals

Surface integrals behave like other double integrals, the integral of the sum of two functions being the sum of their integrals and so on. The domain additivity property takes the form

$$\iint_S g\, d\sigma = \iint_{S_1} g\, d\sigma + \iint_{S_2} g\, d\sigma + \cdots + \iint_{S_n} g\, d\sigma.$$

The idea is that if S is partitioned by smooth curves into a finite number of nonoverlapping smooth patches (i.e., if S is **piecewise smooth**), then the integral over S is the sum of the integrals over the patches. Thus, the integral of a function over the surface of a cube is the sum of the integrals over the faces of the cube. We integrate over a turtle shell of welded plates by integrating one plate at a time and adding the results.

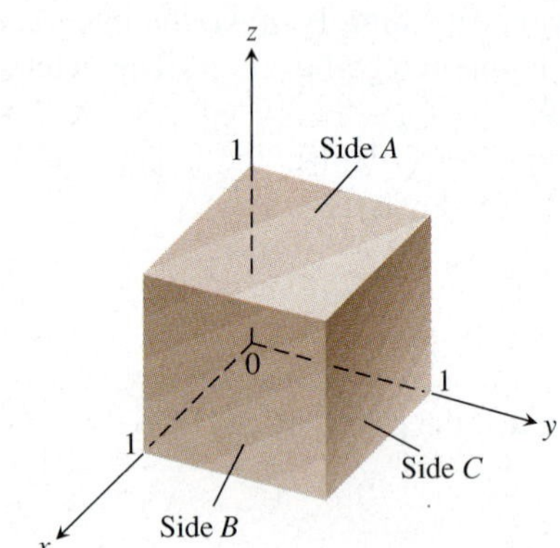

FIGURE 13.44 The cube in Example 3

Example 3 Integrating Over a Surface

Integrate $g(x, y, z) = xyz$ over the surface of the cube cut from the first octant by the planes $x = 1$, $y = 1$, and $z = 1$ (Figure 13.44).

Solution We integrate xyz over each of the six sides and add the results. Since $xyz = 0$ on the sides that lie in the coordinate planes, the integral over the surface of the cube reduces to

$$\iint\limits_{\substack{\text{Cube}\\\text{surface}}} xyz\, d\sigma = \iint\limits_{\text{Side } A} xyz\, d\sigma + \iint\limits_{\text{Side } B} xyz\, d\sigma + \iint\limits_{\text{Side } C} xyz\, d\sigma.$$

Side A is the surface $f(x, y, z) = z = 1$ over the square region $R_{xy}: 0 \le x \le 1$, $0 \le y \le 1$, in the xy-plane. For this surface and region,

$$\mathbf{p} = \mathbf{k}, \qquad \nabla f = \mathbf{k}, \qquad |\nabla f| = 1, \qquad |\nabla f \cdot \mathbf{p}| = |\mathbf{k} \cdot \mathbf{k}| = 1$$

$$d\sigma = \frac{|\nabla f|}{|\nabla f \cdot \mathbf{p}|}\, dA = \frac{1}{1}\, dx\, dy = dx\, dy$$

$$xyz = xy(1) = xy$$

and

$$\iint\limits_{\text{Side } A} xyz\, d\sigma = \iint\limits_{R_{xy}} xy\, dx\, dy = \int_0^1 \int_0^1 xy\, dx\, dy = \int_0^1 \frac{y}{2}\, dy = \frac{1}{4}.$$

Symmetry tells us that the integrals of xyz over sides B and C are also $1/4$. Hence,

$$\iint\limits_{\substack{\text{Cube}\\\text{surface}}} xyz\, d\sigma = \frac{1}{4} + \frac{1}{4} + \frac{1}{4} = \frac{3}{4}.$$

Orientation

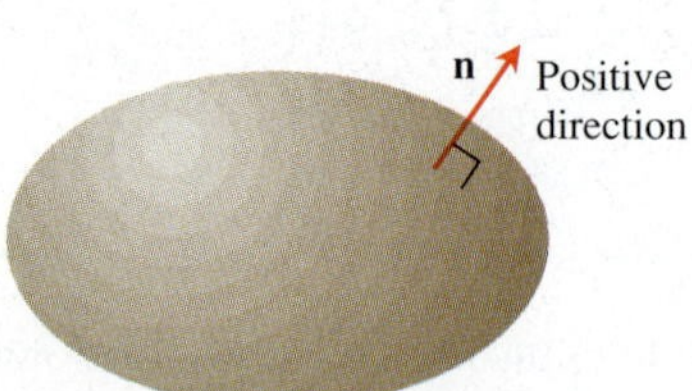

FIGURE 13.45 Smooth closed surfaces in space are orientable. The outward unit normal vector defines the positive direction at each point.

We call a smooth surface S **orientable** or **two-sided** if it is possible to define a field $\mathbf{n}$ of unit normal vectors on S that varies continuously with position. Any patch or subportion of an orientable surface is orientable. Spheres and other smooth closed surfaces in space (smooth surfaces that enclose solids) are orientable. By convention, we choose $\mathbf{n}$ on a closed surface to point outward.

Once $\mathbf{n}$ has been chosen, we say that we have **oriented** the surface, and we call the surface together with its normal field an **oriented surface.** The vector $\mathbf{n}$ at any point is called the **positive direction** at that point (Figure 13.45).

The Möbius band in Figure 13.46 is not orientable. No matter where you start to construct a continuous unit normal field (shown as the shaft of a thumbtack in the

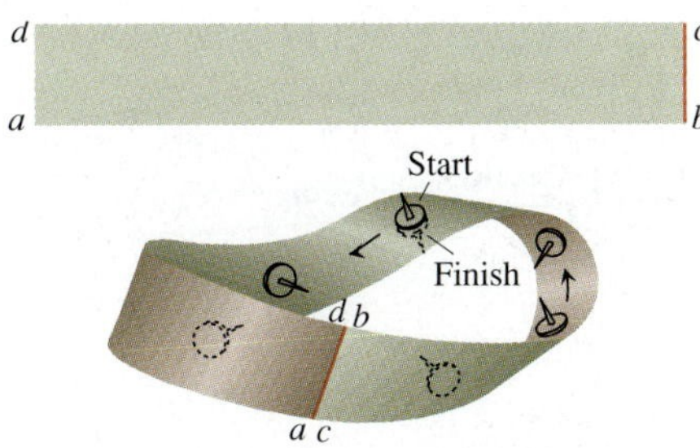

FIGURE 13.46 To make a Möbius band, take a rectangular strip of paper $abcd$, give the end bc a single twist, and paste the ends of the strip together to match a with c and b with d. The Möbius band is a nonorientable or one-sided surface.

figure), moving the vector continuously around the surface in the manner shown will return it to the starting point with a direction opposite to the one it had when it started out. The vector at that point cannot point both ways and yet it must if the field is to be continuous. We conclude that no such field exists.

Surface Integral for Flux

Suppose that $\mathbf{F}$ is a continuous vector field defined over an oriented surface S and that $\mathbf{n}$ is the chosen unit normal field on the surface. We call the integral of $\mathbf{F}\cdot\mathbf{n}$ over S the flux of $\mathbf{F}$ across S in the positive direction. Thus, the flux is the integral over S of the scalar component of $\mathbf{F}$ in the direction of $\mathbf{n}$.

Definition Flux

The **flux** of a three-dimensional vector field $\mathbf{F}$ across an oriented surface S in the direction of $\mathbf{n}$ is

$$\text{Flux} = \iint_S \mathbf{F}\cdot\mathbf{n}\, d\sigma. \tag{8}$$

The definition is analogous to the flux of a two-dimensional field $\mathbf{F}$ across a plane curve C. In the plane (Section 13.2), the flux is

$$\int_C \mathbf{F}\cdot\mathbf{n}\, ds,$$

the integral of the scalar component of $\mathbf{F}$ normal to the curve.

If $\mathbf{F}$ is the velocity field of a three-dimensional fluid flow, the flux of $\mathbf{F}$ across S is the net rate at which fluid is crossing S in the chosen positive direction. We discuss such flows in more detail in Section 13.7.

If $\mathbf{S}$ is part of a level surface $g(x, y, z) = c$, then $\mathbf{n}$ may be taken to be one of the two fields

$$\mathbf{n} = \pm\frac{\nabla g}{|\nabla g|}, \tag{9}$$

depending on which one gives the preferred direction. The corresponding flux is

$$\begin{aligned}\text{Flux} &= \iint_S \mathbf{F}\cdot\mathbf{n}\, d\sigma \\ &= \iint_R \left(\mathbf{F}\cdot\frac{\pm\nabla g}{|\nabla g|}\right)\frac{|\nabla g|}{|\nabla g\cdot\mathbf{p}|}\, dA && \text{(9) and (7)} \qquad (8)\\ &= \iint_R \mathbf{F}\cdot\frac{\pm\nabla g}{|\nabla g\cdot\mathbf{p}|}\, dA. && (10)\end{aligned}$$

Example 4 Finding Flux

Find the flux of $\mathbf{F} = yz\mathbf{j} + z^2\mathbf{k}$ outward through the surface S cut from the cylinder $y^2 + z^2 = 1, z \geq 0,$ by the planes $x = 0$ and $x = 1$.

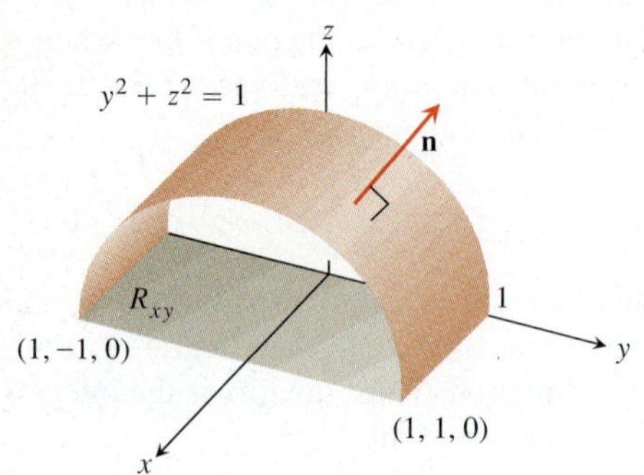

FIGURE 13.47 Example 4 calculates the flux of a vector field outward through this surface. The area of the shadow region R_{xy} is 2.

Solution The outward normal field on S (Figure 13.47) may be calculated from the gradient of $g(x, y, z) = y^2 + z^2$ to be

$$\mathbf{n} = +\frac{\nabla g}{|\nabla g|} = \frac{2y\mathbf{j} + 2z\mathbf{k}}{\sqrt{4y^2 + 4z^2}} = \frac{2y\mathbf{j} + 2z\mathbf{k}}{2\sqrt{1}} = y\mathbf{j} + z\mathbf{k}.$$

With $\mathbf{p} = \mathbf{k}$, we also have

$$d\sigma = \frac{|\nabla g|}{|\nabla g \cdot \mathbf{k}|}\, dA = \frac{2}{|2z|}\, dA = \frac{1}{z}\, dA.$$

We can drop the absolute value bars because $z \geq 0$ on S.

The value of $\mathbf{F} \cdot \mathbf{n}$ on the surface is

$$\begin{aligned}\mathbf{F} \cdot \mathbf{n} &= (yz\mathbf{j} + z^2\mathbf{k}) \cdot (y\mathbf{j} + z\mathbf{k}) \\ &= y^2z + z^3 = z(y^2 + z^2) \\ &= z. \qquad y^2 + z^2 = 1 \text{ on } S\end{aligned}$$

Therefore, the flux of $\mathbf{F}$ outward through S is

$$\iint_S \mathbf{F} \cdot \mathbf{n}\, d\sigma = \iint_S (z)\left(\frac{1}{z}\, dA\right) = \iint_{R_{xy}} dA = \text{area}(R_{xy}) = 2.$$

Moments and Masses of Thin Shells

Thin shells of material like bowls, metal drums, and domes are modeled with surfaces. Their moments and masses are calculated with the formulas in Table 13.3.

Table 13.3 Mass and moment formulas for very thin shells

Mass: $M = \iint_S \delta(x, y, z)\, d\sigma$ ($\delta(x, y, z)$ = density at (x, y, z), mass per unit area)

First moments about the coordinate planes:

$$M_{yz} = \iint_S x\delta\, d\sigma, \qquad M_{xz} = \iint_S y\delta\, d\sigma, \qquad M_{xy} = \iint_S z\delta\, d\sigma$$

Coordinates of center of mass:

$$\bar{x} = M_{yz}/M, \qquad \bar{y} = M_{xz}/M, \qquad \bar{z} = M_{xy}/M$$

Moments of inertia about coordinate axes:

$$I_x = \iint_S (y^2 + z^2)\delta\, d\sigma, \qquad I_y = \iint_S (x^2 + z^2)\delta\, d\sigma,$$

$$I_z = \iint_S (x^2 + Y^2)\delta\, d\sigma, \qquad I_L = \iint_S r^2\delta\, d\sigma,$$

$r(x, y, z)$ = distance from point (x, y, z) to line L

Radius of gyration about a line L: $R_L = \sqrt{I_L/M}$

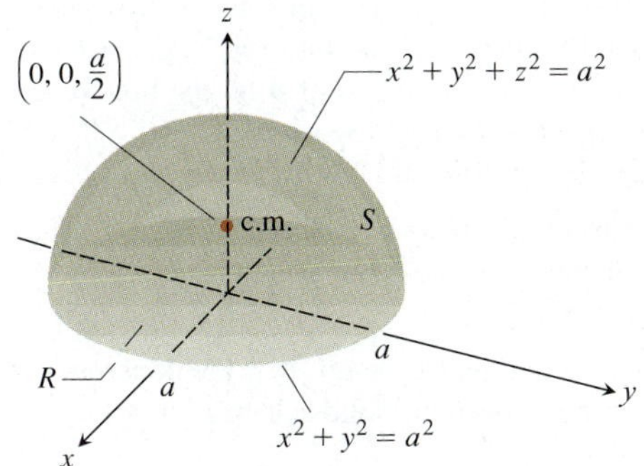

FIGURE 13.48 The center of mass of a thin hemispherical shell of constant density lies on the axis of symmetry halfway from the base to the top. (Example 5)

Example 5 Finding Center of Mass

Find the center of mass of a thin hemispherical shell of radius a and constant density δ.

Solution We model the shell with the hemisphere

$$f(x, y, z) = x^2 + y^2 + z^2 = a^2, \qquad z \geq 0$$

(Figure 13.48). The symmetry of the surface about the z-axis tells us that $\bar{x} = \bar{y} = 0$. It remains only to find $\bar{z}$ from the formula $\bar{z} = M_{xy}/M$.

The mass of the shell is

$$M = \iint_S \delta\, d\sigma = \delta \iint_S d\sigma = (\delta)(\text{area of } S) = 2\pi a^2 \delta.$$

To evaluate the integral for M_{xy}, we take $\mathbf{p} = \mathbf{k}$ and calculate

$$|\nabla f| = |2x\mathbf{i} + 2y\mathbf{j} + 2z\mathbf{k}| = 2\sqrt{x^2 + y^2 + z^2} = 2a$$

$$|\nabla f \cdot \mathbf{p}| = |\nabla f \cdot \mathbf{k}| = |2z| = 2z$$

$$d\sigma = \frac{|\nabla f|}{|\nabla f \cdot \mathbf{p}|}\, dA = \frac{a}{z}\, dA.$$

Then

$$M_{xy} = \iint_S z\delta\, d\sigma = \delta \iint_R z\frac{a}{z}\, dA = \delta a \iint_R dA = \delta a(\pi a^2) = \delta \pi a^3$$

$$\bar{z} = \frac{M_{xy}}{M} = \frac{\pi a^3 \delta}{2\pi a^2 \delta} = \frac{a}{2}.$$

The shell's center of mass is the point $(0, 0, a/2)$.

EXERCISES 13.5

Surface Area

1. Find the area of the surface cut from the paraboloid $x^2 + y^2 - z = 0$ by the plane $z = 2$.
2. Find the area of the band cut from the paraboloid $x^2 + y^2 - z = 0$ by the planes $z = 2$ and $z = 6$.
3. Find the area of the region cut from the plane $x + 2y + 2z = 5$ by the cylinder whose walls are $x = y^2$ and $x = 2 - y^2$.
4. Find the area of the portion of the surface $x^2 - 2z = 0$ that lies above the triangle bounded by the lines $x = \sqrt{3}$, $y = 0$, and $y = x$ in the xy-plane.
5. Find the area of the surface $x^2 - 2y - 2z = 0$ that lies above the triangle bounded by the lines $x = 2$, $y = 0$, and $y = 3x$ in the xy-plane.
6. Find the area of the cap cut from the sphere $x^2 + y^2 + z^2 = 2$ by the cone $z = \sqrt{x^2 + y^2}$.
7. Find the area of the ellipse cut from the plane $z = cx$ (c a constant) by the cylinder $x^2 + y^2 = 1$.
8. Find the area of the upper portion of the cylinder $x^2 + z^2 = 1$ that lies between the planes $x = \pm 1/2$ and $y = \pm 1/2$.
9. Find the area of the portion of the paraboloid $x = 4 - y^2 - z^2$ that lies above the ring $1 \leq y^2 + z^2 \leq 4$ in the yz-plane.
10. Find the area of the surface cut from the paraboloid $x^2 + y + z^2 = 2$ by the plane $y = 0$.
11. Find the area of the surface $x^2 - 2\ln x + \sqrt{15}y - z = 0$ above the square $R: 1 \leq x \leq 2, 0 \leq y \leq 1$, in the xy-plane.
12. Find the area of the surface $2x^{3/2} + 2y^{3/2} - 3z = 0$ above the square $R: 0 \leq x \leq 1, 0 \leq y \leq 1$, in the xy-plane.

Surface Integrals

13. Integrate $g(x, y, z) = x + y + z$ over the surface of the cube cut from the first octant by the planes $x = a$, $y = a$, $z = a$.
14. Integrate $g(x, y, z) = y + z$ over the surface of the wedge in the first octant bounded by the coordinate planes and the planes $x = 2$ and $y + z = 1$.

15. Integrate $g(x, y, z) = xyz$ over the surface of the rectangular solid cut from the first octant by the planes $x = a$, $y = b$, and $z = c$.

16. Integrate $g(x, y, z) = xyz$ over the surface of the rectangular solid bounded by the planes $x = \pm a$, $y = \pm b$, and $z = \pm c$.

17. Integrate $g(x, y, z) = x + y + z$ over the portion of the plane $2x + 2y + z = 2$ that lies in the first octant.

18. Integrate $g(x, y, z) = x\sqrt{y^2 + 4}$ over the surface cut from the parabolic cylinder $y^2 + 4z = 16$ by the planes $x = 0$, $x = 1$, and $z = 0$.

Flux Across a Surface

In Exercises 19 and 20, find the flux of the field $\mathbf{F}$ across the portion of the given surface in the specified direction.

19. $\mathbf{F}(x, y, z) = -\mathbf{i} + 2\mathbf{j} + 3\mathbf{k}$

S: rectangular surface $z = 0$, $0 \le x \le 2$, $0 \le y \le 3$, direction $\mathbf{k}$

20. $\mathbf{F}(x, y, z) = yx^2\mathbf{i} - 2\mathbf{j} + xz\mathbf{k}$

S: rectangular surface $y = 0$, $-1 \le x \le 2$, $2 \le z \le 7$, direction $-\mathbf{j}$

In Exercises 21–26, find the flux of the field $\mathbf{F}$ across the portion of the sphere $x^2 + y^2 + z^2 = a^2$ in the first octant in the direction away from the origin.

21. $\mathbf{F}(x, y, z) = z\mathbf{k}$

22. $\mathbf{F}(x, y, z) = -y\mathbf{i} + x\mathbf{j}$

23. $\mathbf{F}(x, y, z) = y\mathbf{i} - x\mathbf{j} + \mathbf{k}$

24. $\mathbf{F}(x, y, z) = zx\mathbf{i} + zy\mathbf{j} + z^2\mathbf{k}$

25. $\mathbf{F}(x, y, z) = x\mathbf{i} + y\mathbf{j} + z\mathbf{k}$

26. $\mathbf{F}(x, y, z) = \dfrac{x\mathbf{i} + y\mathbf{j} + z\mathbf{k}}{\sqrt{x^2 + y^2 + z^2}}$

27. Find the flux of the field $\mathbf{F}(x, y, z) = z^2\mathbf{i} + x\mathbf{j} - 3z\mathbf{k}$ outward through the surface cut from the parabolic cylinder $z = 4 - y^2$ by the planes $x = 0$, $x = 1$, and $z = 0$.

28. Find the flux of the field $\mathbf{F}(x, y, z) = 4x\mathbf{i} + 4y\mathbf{j} + 2\mathbf{k}$ outward (away from the z-axis) through the surface cut from the bottom of the paraboloid $z = x^2 + y^2$ by the plane $z = 1$.

29. Let S be the portion of the cylinder $y = e^x$ in the first octant that projects parallel to the x-axis onto the rectangle R_{yz}: $1 \le y \le 2$, $0 \le z \le 1$ in the yz-plane (see the accompanying figure). Let $\mathbf{n}$ be the unit vector normal to S that points away from the yz-plane. Find the flux of the field $\mathbf{F}(x, y, z) = -2\mathbf{i} + 2y\mathbf{j} + z\mathbf{k}$ across S in the direction of $\mathbf{n}$.

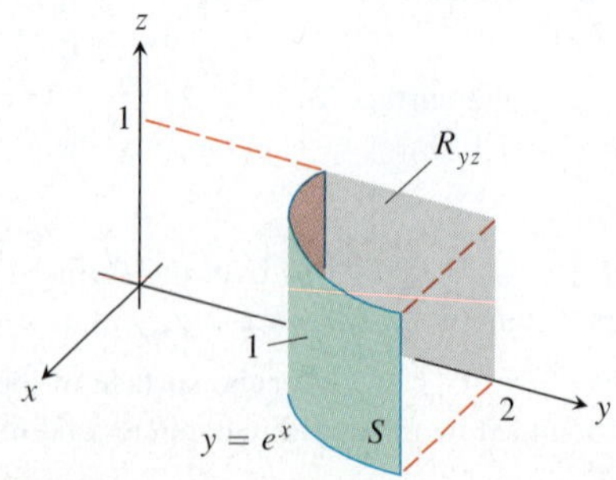

30. Let S be the portion of the cylinder $y = \ln x$ in the first octant whose projection parallel to the y-axis onto the xz-plane is the rectangle R_{xz}: $1 \le x \le e$, $0 \le z \le 1$. Let $\mathbf{n}$ be the unit vector normal to S that points away from the xz-plane. Find the flux of $\mathbf{F} = 2y\mathbf{j} + z\mathbf{k}$ through S in the direction of $\mathbf{n}$.

31. Find the outward flux of the field $\mathbf{F} = 2xy\mathbf{i} + 2yz\mathbf{j} + 2xz\mathbf{k}$ across the surface of the cube cut from the first octant by the planes $x = a$, $y = a$, $z = a$.

32. Find the outward flux of the field $\mathbf{F} = xz\mathbf{i} + yz\mathbf{j} + \mathbf{k}$ across the surface of the upper cap cut from the solid sphere $x^2 + y^2 + z^2 \le 25$ by the plane $z = 3$.

Moments and Masses

33. *Centroid* Find the centroid of the portion of the sphere $x^2 + y^2 + z^2 = a^2$ that lies in the first octant.

34. *Centroid* Find the centroid of the surface cut from the cylinder $y^2 + z^2 = 9$, $z \ge 0$, by the planes $x = 0$ and $x = 3$ (resembles the surface in Example 4).

35. *Thin shell of constant density* Find the center of mass and the moment of inertia and radius of gyration about the z-axis of a thin shell of constant density δ cut from the cone $x^2 + y^2 - z^2 = 0$ by the planes $z = 1$ and $z = 2$.

36. *Conical surface of constant density* Find the moment of inertia about the z-axis of a thin shell of constant density δ cut from the cone $4x^2 + 4y^2 - z^2 = 0$, $z \ge 0$, by the circular cylinder $x^2 + y^2 = 2x$ (see the accompanying figure).

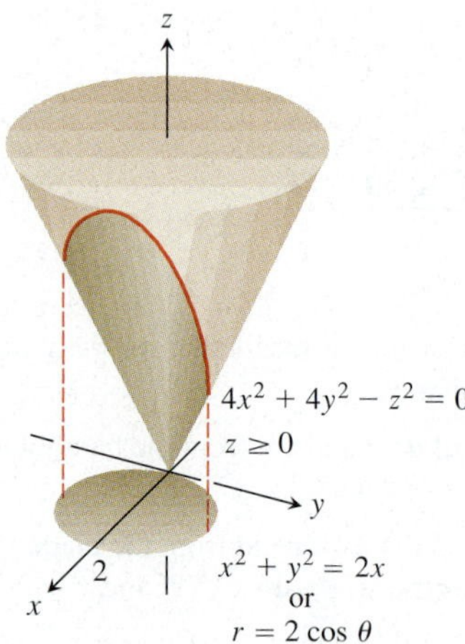

37. *Spherical shells*

(a) Find the moment of inertia about a diameter of a thin spherical shell of radius a and constant density δ. (Work with a hemispherical shell and double the result.)

(b) Use the Parallel Axis Theorem (Exercises 12.5) and the result in part (a) to find the moment of inertia about a line tangent to the shell.

38. **(a)** *Cones with and without ice cream* Find the centroid of the lateral surface of a solid cone of base radius a and height h (cone surface minus the base).

(b) Use Pappus's formula (Exercises 12.5) and the result in part (a) to find the centroid of the complete surface of a solid cone (side plus base).

(c) *Writing to Learn* A cone of radius a and height h is joined to a hemisphere of radius a to make a surface S that resembles an ice cream cone. Use Pappus's formula and the results in part (a) and Example 5 to find the centroid of S. How high does the cone have to be to place the centroid in the plane shared by the bases of the hemisphere and cone?

Special Formulas for Surface Area

If S is the surface defined by a function $z = f(x, y)$ that has continuous first partial derivatives throughout a region R_{xy} in the xy-plane (Figure 13.49), then S is also the level surface $F(x, y, z) = 0$ of the function $F(x, y, z) = f(x, y) - z$. Taking the unit normal to R_{xy} to be $\mathbf{p} = \mathbf{k}$ then gives

$$|\nabla F| = |f_x\mathbf{i} + f_y\mathbf{j} - \mathbf{k}| = \sqrt{f_x^2 + f_y^2 + 1}$$

$$|\nabla F \cdot \mathbf{p}| = |(f_x\mathbf{i} + f_y\mathbf{j} - \mathbf{k}) \cdot \mathbf{k}| = |-1| = 1$$

and

$$\iint_{R_{xy}} \frac{|\nabla F|}{|\nabla F \cdot \mathbf{p}|}\, dA = \iint_{R_{xy}} \sqrt{f_y^2 + f_y^2 + 1}\, dx\, dy, \tag{11}$$

Similarly, the area of a smooth surface $x = f(y, z)$ over a region R_{yz} in the yz-plane is

$$A = \iint_{R_{yz}} \sqrt{f_y^2 + f_z^2 + 1}\, dy\, dz, \tag{12}$$

and the area of a smooth $y = f(x, z)$ over a region R_{xz} in the xz-plane is

$$A = \iint_{R_{xz}} \sqrt{f_x^2 + f_z^2 + 1}\, dx\, dz. \tag{13}$$

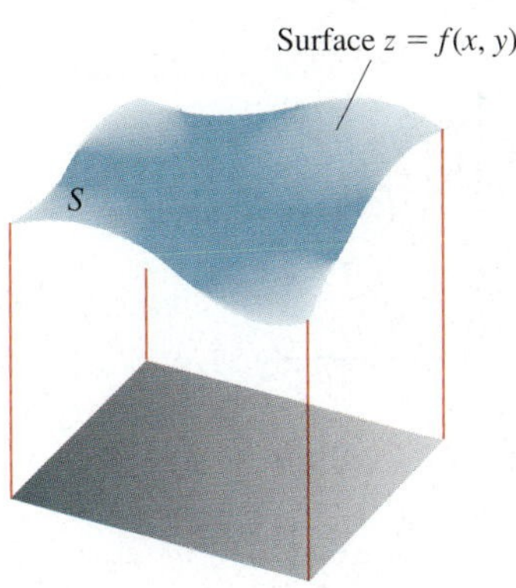

FIGURE 13.49 For a surface $z = f(x, y)$, the surface area formula in Equation (3) takes the form

$$A = \iint_{R_{xy}} \sqrt{f_x^2 + f_y^2 + 1}\, dx\, dy.$$

Use Equations (11)–(13) to find the area of the surfaces in Exercises 39–44.

39. The surface cut from the bottom of the paraboloid $z = x^2 + y^2$ by the plane $z = 3$

40. The surface cut from the "nose" of the paraboloid $x = 1 - y^2 - z^2$ by the yz-plane

41. The portion of the cone $z = \sqrt{x^2 + y^2}$ that lies over the region between the circle $x^2 + y^2 = 1$ and the ellipse $9x^2 + 4y^2 = 36$ in the xy-plane. (*Hint:* Use formulas from geometry to find the area of the region.)

42. The triangle cut from the plane $2x + 6y + 3z = 6$ by the bounding planes of the first octant. Calculate the area three ways, once with each area formula

43. The surface in the first octant cut from the cylinder $y = (2/3)z^{3/2}$ by the planes $x = 1$ and $y = 16/3$

44. The portion of the plane $y + z = 4$ that lies above the region cut from the first quadrant of the xz-plane by the parabola $x = 4 - z^2$

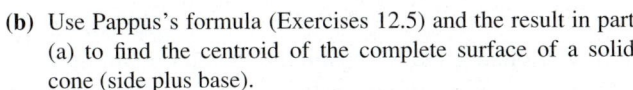

13.6 Parametrized Surfaces

Parametrizations of Surfaces • Surface Area • Surface Integrals

We have defined curves in the plane in three different ways:

Explicit form:	$y = f(x)$	
Implicit form:	$F(x, y) = 0$	
Parametric vector form:	$\mathbf{r}(t) = f(t)\mathbf{i} + g(t)\mathbf{j},$	$a \le t \le b.$

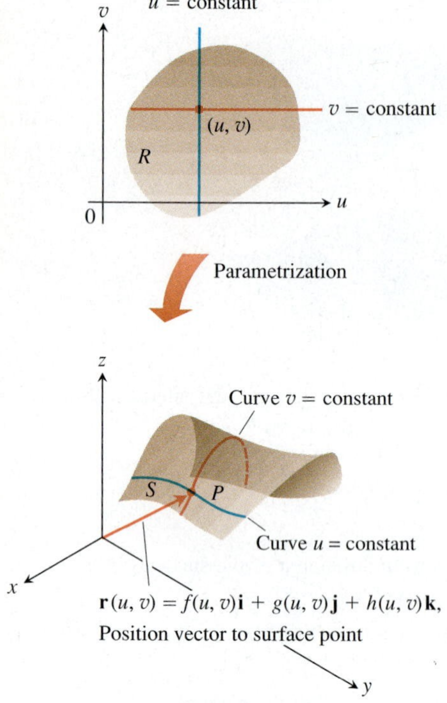

FIGURE 13.50 A parametrized surface.

We have analogous definitions of surfaces in space:

Explicit form: $z = f(x, y)$

Implicit form: $F(x, y, z) = 0.$

There is also a parametric form that gives the position of a point on the surface as a vector function of two variables. The present section extends the investigation of surface area and surface integrals to surfaces described parametrically.

Parametrizations of Surfaces

Let

$$\mathbf{r}(u, v) = f(u, v)\mathbf{i} + g(u, v)\mathbf{j} + h(u, v)\mathbf{k} \tag{1}$$

be a continuous vector function that is defined on a region R in the uv-plane and one-to-one on the interior of R (Figure 13.50). We call the range of $\mathbf{r}$ the **surface** S defined or traced by $\mathbf{r}$. Equation (1) together with the domain R constitute a **parametrization** of the surface. The variables u and v are the **parameters,** and R is the **parameter domain.** To simplify our discussion, we take R to be a rectangle defined by inequalities of the form $a \le u \le b$, $c \le v \le d$. The requirement that $\mathbf{r}$ be one-to-one on the interior of R ensures that S does not cross itself. Notice that Equation (1) is the vector equivalent of *three* parametric equations:

$$x = f(u, v), \qquad y = g(u, v), \qquad z = h(u, v).$$

Example 1 Parametrizing a Cone

Find a parametrization of the cone

$$z = \sqrt{x^2 + y^2}, \qquad 0 \le z \le 1.$$

Solution Here, cylindrical coordinates provide everything we need. A typical point (x, y, z) on the cone (Figure 13.51) has $x = r\cos\theta$, $y = r\sin\theta$, and $z = \sqrt{x^2 + y^2} = r$, with $0 \le r \le 1$ and $0 \le \theta \le 2\pi$. Taking $u = r$ and $v = \theta$ in Equation (1) gives the parametrization

$$\mathbf{r}(r, \theta) = (r\cos\theta)\mathbf{i} + (r\sin\theta)\mathbf{j} + r\mathbf{k}, \qquad 0 \le r \le 1, \quad 0 \le \theta \le 2\pi.$$

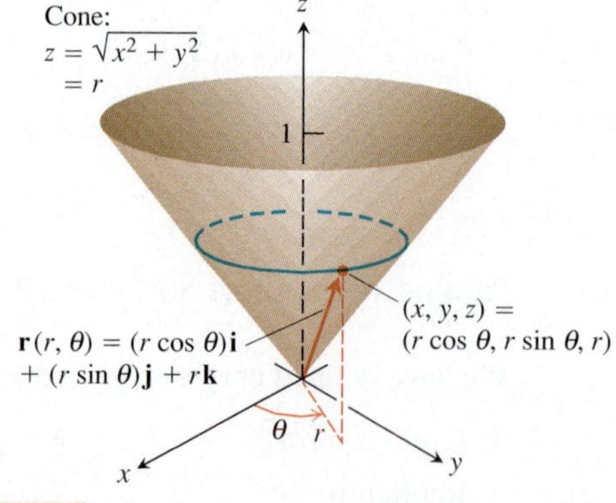

FIGURE 13.51 The cone in Example 1.

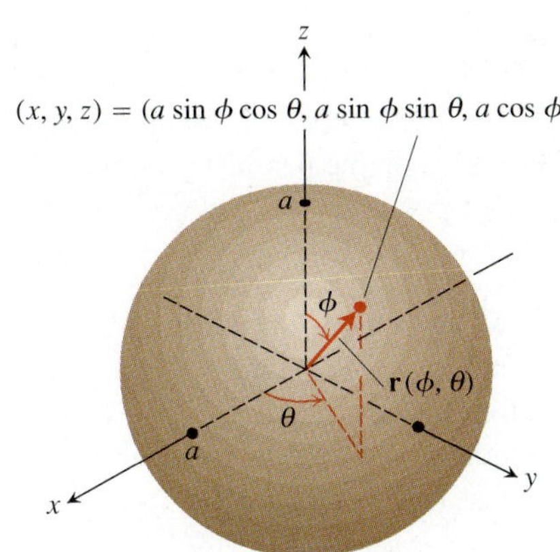

FIGURE 13.52 The sphere in Example 2.

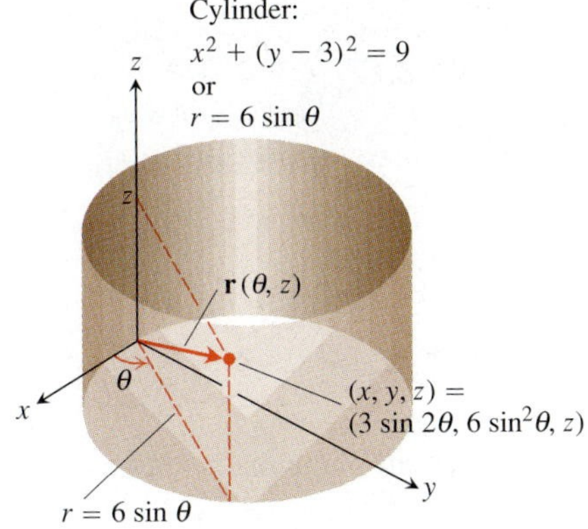

FIGURE 13.53 The cylinder in Example 3.

Example 2 Parametrizing a Sphere

Find a parametrization of the sphere $x^2 + y^2 + z^2 = a^2$.

Solution Spherical coordinates provide what we need. A typical point (x, y, z) on the sphere (Figure 13.52) has $x = a \sin\phi \cos\theta$, $y = a \sin\phi \sin\theta$, and $z = a\cos\phi$, $0 \le \phi \le \pi$, $0 \le \theta \le 2\pi$. Taking $u = \phi$ and $v = \theta$ in Equation (1) gives the parametrization

$$\mathbf{r}(\phi, \theta) = (a \sin\phi \cos\theta)\mathbf{i} + (a \sin\phi \sin\theta)\mathbf{j} + (a\cos\phi)\mathbf{k},$$
$$0 \le \phi \le \pi, \quad 0 \le \theta \le 2\pi.$$

Example 3 Parametrizing a Cylinder

Find a parametrization of the cylinder

$$x^2 + (y-3)^2 = 9, \qquad 0 \le z \le 5.$$

Solution In cylindrical coordinates, a point (x, y, z) has $x = r\cos\theta$, $y = r\sin\theta$, and $z = z$. For points on the cylinder $x^2 + (y-3)^2 = 9$ (Figure 13.53), the equation is the same as the polar equation for the cylinder's base in the xy-plane:

$$x^2 + (y^2 - 6y + 9) = 9$$
$$r^2 - 6r\sin\theta = 0$$

or

$$r = 6\sin\theta, \qquad 0 \le \theta \le \pi.$$

A typical point on the cylinder therefore has

$$x = r\cos\theta = 6\sin\theta\cos\theta = 3\sin 2\theta$$
$$y = r\sin\theta = 6\sin^2\theta$$
$$z = z.$$

Taking $u = \theta$ and $v = z$ in Equation (1) gives the parametrization

$$\mathbf{r}(\theta, z) = (3\sin 2\theta)\mathbf{i} + (6\sin^2\theta)\mathbf{j} + z\,\mathbf{k}, \qquad 0 \le \theta \le \pi, \quad 0 \le z \le 5.$$

Surface Area

Our goal is to find a double integral for calculating the area of a curved surface S based on the parametrization

$$\mathbf{r}(u, v) = f(u, v)\mathbf{i} + g(u, v)\mathbf{j} + h(u, v)\mathbf{k}, \qquad a \le u \le b, \quad c \le v \le d.$$

We need S to be smooth for the construction we are about to carry out. The definition of smoothness involves the partial derivatives of $\mathbf{r}$ with respect to u and v:

$$\mathbf{r}_u = \frac{\partial \mathbf{r}}{\partial u} = \frac{\partial f}{\partial u}\mathbf{i} + \frac{\partial g}{\partial u}\mathbf{j} + \frac{\partial h}{\partial u}\mathbf{k}$$
$$\mathbf{r}_v = \frac{\partial \mathbf{r}}{\partial v} = \frac{\partial f}{\partial v}\mathbf{i} + \frac{\partial g}{\partial v}\mathbf{j} + \frac{\partial h}{\partial v}\mathbf{k}.$$

Definition Smooth Parametrized Surface

A parametrized surface $\mathbf{r}(u, v) = f(u, v)\mathbf{i} + g(u, v)\mathbf{j} + h(u, v)\mathbf{k}$ is **smooth** if $\mathbf{r}_u$ and $\mathbf{r}_v$ are continuous and $\mathbf{r}_u \times \mathbf{r}_v$ is never zero on the parameter domain.

Now consider a small rectangle ΔA_{uv} in R with sides on the lines $u = u_0$, $u = u_0 + \Delta u$, $v = v_0$, and $v = v_0 + \Delta v$ (Figure 13.54). Each side of ΔA_{uv} maps to a curve on the surface S, and together these four curves bound a "curved area element" $\Delta\sigma_{uv}$. In the notation of the figure, the side $v = v_0$ maps to curve C_1, the side $u = u_0$ maps to C_2, and their common vertex (u_0, v_0) maps to P_0.

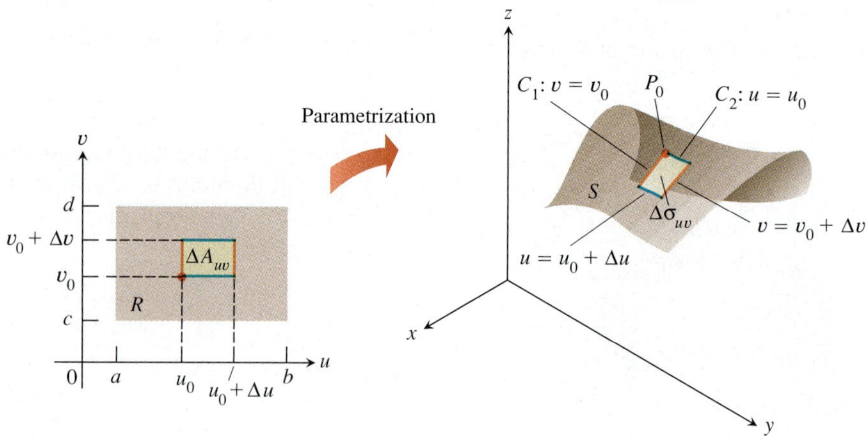

FIGURE 13.54 A rectangular area element ΔA_{uv} in the uv-plane maps onto a curved area element $\Delta\sigma_{uv}$ on S.

Figure 13.55 shows an enlarged view of $\Delta\sigma_{uv}$. The vector $\mathbf{r}_u(u_0, v_0)$ is tangent to C_1 at P_0. Likewise, $\mathbf{r}_v(u_0, v_0)$ is tangent to C_2 at P_0. The cross product $\mathbf{r}_u \times \mathbf{r}_v$ is normal to the surface at P_0. (Here is where we begin to use the assumption that S is smooth. We want to be sure that $\mathbf{r}_u \times \mathbf{r}_v \neq \mathbf{0}$.)

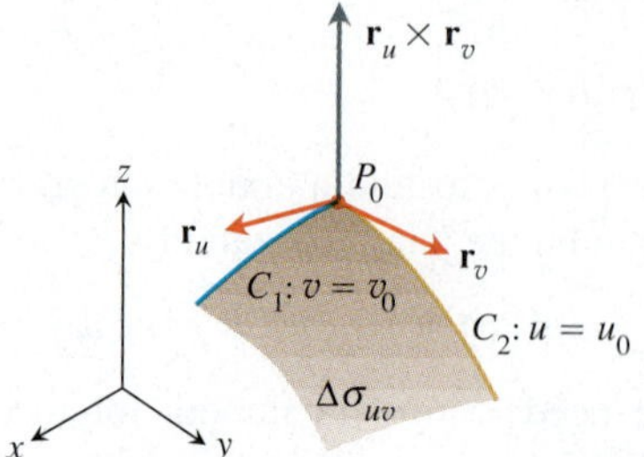

FIGURE 13.55 A magnified view of a surface area element $\Delta\sigma_{uv}$.

We next approximate the surface element $\Delta\sigma_{uv}$ by the parallelogram on the tangent plane whose sides are determined by the vectors $\Delta u\mathbf{r}_u$ and $\Delta v\mathbf{r}_v$ (Figure 13.56). The area of this parallelogram is

$$|\Delta u\mathbf{r}_u \times \Delta v\mathbf{r}_v| = |\mathbf{r}_u \times \mathbf{r}_v|\,\Delta u\,\Delta v. \tag{2}$$

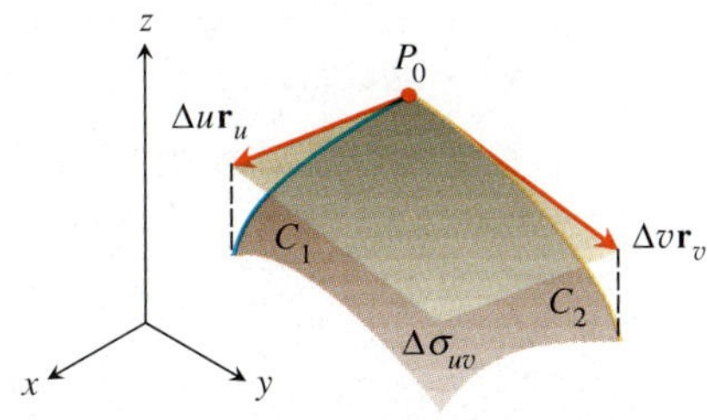

FIGURE 13.56 The parallelogram determined by the vectors $\Delta u\mathbf{r}_u$ and $\Delta v\mathbf{r}_v$ approximates the surface area element $\Delta\sigma_{uv}$.

A partition of the region R in the uv-plane by rectangular regions ΔA_{uv} generates a partition of the surface S into surface area elements $\Delta\sigma_{uv}$. We approximate the area of each surface element $\Delta\sigma_{uv}$ by the parallelogram area in Equation (2) and sum these areas together to obtain an approximation of the area of S:

$$\sum_u \sum_v |\mathbf{r}_u \times \mathbf{r}_v| \, \Delta u \, \Delta v. \tag{3}$$

As Δu and Δv approach zero independently, the continuity of $\mathbf{r}_u$ and $\mathbf{r}_v$ guarantees that the sum in Equation (3) approaches the double integral $\int_c^d \int_a^b |\mathbf{r}_u \times \mathbf{r}_v| \, du \, dv$. This double integral gives the area of the surface S.

Parametric Formula for the Area of a Smooth Surface

The **area** of the smooth surface

$$\mathbf{r}(u, v) = f(u, v)\mathbf{i} + g(u, v)\mathbf{j} + h(u, v)\mathbf{k}, \qquad a \le u \le b, \quad c \le v \le d$$

is

$$A = \int_c^d \int_a^b |\mathbf{r}_u \times \mathbf{r}_v| \, du \, dv. \tag{4}$$

As in Section 13.5, we can abbreviate the integral in Equation (4) by writing $d\sigma$ for $|\mathbf{r}_u \times \mathbf{r}_v| \, du \, dv$.

CD-ROM WEBsite

CD-ROM WEBsite
Historical Biography
André Marie Ampère (1775 — 1836)

Surface Area Differential and the Differential Formula for Surface Area

$$d\sigma = |\mathbf{r}_u \times \mathbf{r}_v| \, du \, dv \qquad \iint_S d\sigma \tag{5}$$

Surface area differential — Differential formula for surface area

Example 4 Finding Surface Area (Cone)

Find the surface area of the cone in Example 1 (Figure 13.51).

Solution In Example 1, we found the parametrization

$$\mathbf{r}(r, \theta) = (r\cos\theta)\mathbf{i} + (r\sin\theta)\mathbf{j} + r\mathbf{k}, \qquad 0 \le r \le 1, \quad 0 \le \theta \le 2\pi.$$

To apply Equation (4), we first find $\mathbf{r}_r \times \mathbf{r}_\theta$:

$$\mathbf{r}_r \times \mathbf{r}_\theta = \begin{vmatrix} \mathbf{i} & \mathbf{j} & \mathbf{k} \\ \cos\theta & \sin\theta & 1 \\ -r\sin\theta & r\cos\theta & 0 \end{vmatrix}$$

$$= -(r\cos\theta)\mathbf{i} - (r\sin\theta)\mathbf{j} + \underbrace{(r\cos^2\theta + r\sin^2\theta)}_{r}\mathbf{k}.$$

Thus, $|\mathbf{r}_r \times \mathbf{r}_\theta| = \sqrt{r^2\cos^2\theta + r^2\sin^2\theta + r^2} = \sqrt{2r^2} = \sqrt{2}r$. The area of the cone is

$$A = \int_0^{2\pi}\int_0^1 |\mathbf{r}_r \times \mathbf{r}_\theta|\, dr\, d\theta \qquad \text{Eq. (4) with } u = r, v = \theta$$

$$= \int_0^{2\pi}\int_0^1 \sqrt{2}\, r\, dr\, d\theta = \int_0^{2\pi} \frac{\sqrt{2}}{2}\, d\theta = \frac{\sqrt{2}}{2}(2\pi) = \pi\sqrt{2} \text{ units squared.}$$

Example 5 Finding Surface Area (Sphere)

Find the surface area of a sphere of radius a.

Solution We use the parametrization from Example 2:

$$\mathbf{r}(\phi, \theta) = (a\sin\phi\cos\theta)\mathbf{i} + (a\sin\phi\sin\theta)\mathbf{j} + (a\cos\phi)\mathbf{k},$$
$$0 \le \phi \le \pi, \quad 0 \le \theta \le 2\pi.$$

For $\mathbf{r}_\phi \times \mathbf{r}_\theta$, we get

$$\mathbf{r}_\phi \times \mathbf{r}_\theta = \begin{vmatrix} \mathbf{i} & \mathbf{j} & \mathbf{k} \\ a\cos\phi\cos\theta & a\cos\phi\sin\theta & -a\sin\phi \\ -a\sin\phi\sin\theta & a\sin\phi\cos\theta & 0 \end{vmatrix}$$

$$= (a^2\sin^2\phi\cos\theta)\mathbf{i} + (a^2\sin^2\phi\sin\theta)\mathbf{j} + (a^2\sin\phi\cos\phi)\mathbf{k}.$$

Thus,

$$|\mathbf{r}_\phi \times \mathbf{r}_\theta| = \sqrt{a^4\sin^4\phi\cos^2\theta + a^4\sin^4\phi\sin^2\theta + a^4\sin^2\phi\cos^2\phi}$$
$$= \sqrt{a^4\sin^4\phi + a^4\sin^2\phi\cos^2\phi} = \sqrt{a^4\sin^2\phi(\sin^2\phi + \cos^2\phi)}$$
$$= a^2\sqrt{\sin^2\phi} = a^2\sin\phi,$$

since $\sin\phi \ge 0$ for $0 \le \phi \le \pi$. Therefore, the area of the sphere is

$$A = \int_0^{2\pi}\int_0^{\pi} a^2\sin\phi\, d\phi\, d\theta$$

$$= \int_0^{2\pi} \left[-a^2\cos\phi\right]_0^{\pi} d\theta = \int_0^{2\pi} 2a^2\, d\theta = 4\pi a^2 \text{ units squared.}$$

Surface Integrals

Having found a formula for calculating the area of a parametrized surface, we can now integrate a function over the surface using the parametrized form.

Definition Integral Over a Smooth Parametrized Surface

If S is a smooth surface defined parametrically as $\mathbf{r}(u, v) = f(u, v)\mathbf{i} + g(u, v)\mathbf{j} + h(u, v)\mathbf{k}$, $a \le u \le b$, $c \le v \le d$, and $G(x, y, z)$ is a continuous function defined on S, then the **integral of G over S** is

$$\iint_S G(x, y, z)\, d\sigma = \int_c^d \int_a^b G(f(u, v), g(u, v), h(u, v))|\mathbf{r}_u \times \mathbf{r}_v|\, du\, dv.$$

Example 6 Integrating Over a Surface Defined Parametrically

Integrate $G(x, y, z) = x^2$ over the cone $z = \sqrt{x^2 + y^2}$, $0 \le z \le 1$.

Solution Continuing the work in Examples 1 and 4, we have $|\mathbf{r}_r \times \mathbf{r}_\theta| = \sqrt{2}\, r$ and

$$\begin{aligned}\iint_S x^2\, d\sigma &= \int_0^{2\pi} \int_0^1 (r^2 \cos^2 \theta)(\sqrt{2} r)\, dr\, d\theta \qquad x = r\cos\theta \\ &= \sqrt{2} \int_0^{2\pi} \int_0^1 r^3 \cos^2 \theta\, dr\, d\theta \\ &= \frac{\sqrt{2}}{4} \int_0^{2\pi} \cos^2 \theta\, d\theta = \frac{\sqrt{2}}{4}\left[\frac{\theta}{2} + \frac{1}{4}\sin 2\theta\right]_0^{2\pi} = \frac{\pi\sqrt{2}}{4}.\end{aligned}$$

Example 7 Finding Flux

Find the flux of $\mathbf{F} = yz\mathbf{i} + x\mathbf{j} - z^2\mathbf{k}$ outward through the parabolic cylinder $y = x^2$, $0 \le x \le 1$, $0 \le z \le 4$ (Figure 13.57).

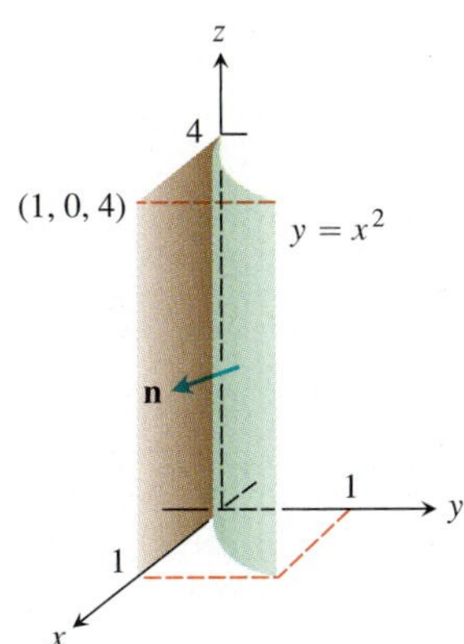

FIGURE 13.57 The parabolic surface in Example 7.

Solution On the surface we have $x = x$, $y = x^2$, and $z = z$, so we automatically have the parametrization $\mathbf{r}(x, z) = x\mathbf{i} + x^2\mathbf{j} + z\mathbf{k}$, $0 \le x \le 1$, $0 \le z \le 4$. The cross product of tangent vectors is

$$\mathbf{r}_x \times \mathbf{r}_z = \begin{vmatrix} \mathbf{i} & \mathbf{j} & \mathbf{k} \\ 1 & 2x & 0 \\ 0 & 0 & 1 \end{vmatrix} = 2x\mathbf{i} - \mathbf{j}.$$

The unit normal pointing outward from the surface is

$$\mathbf{n} = \frac{\mathbf{r}_x \times \mathbf{r}_z}{|\mathbf{r}_x \times \mathbf{r}_z|} = \frac{2x\mathbf{i} - \mathbf{j}}{\sqrt{4x^2 + 1}}.$$

On the surface, $y = x^2$, so the vector field there is

$$\mathbf{F} = yz\mathbf{i} + x\mathbf{j} - z^2\mathbf{k} = x^2 z\mathbf{i} + x\mathbf{j} - z^2\mathbf{k}.$$

Thus,

$$\begin{aligned}\mathbf{F} \cdot \mathbf{n} &= \frac{1}{\sqrt{4x^2 + 1}}((x^2 z)(2x) + (x)(-1) + (-z^2)(0)) \\ &= \frac{2x^3 z - x}{\sqrt{4x^2 + 1}}.\end{aligned}$$

The flux of $\mathbf{F}$ outward through the surface is

$$\iint_S \mathbf{F}\cdot\mathbf{n}\,d\sigma = \int_0^4\int_0^1 \frac{2x^3z - x}{\sqrt{4x^2+1}}\,|\mathbf{r}_x \times \mathbf{r}_z|\,dx\,dz$$

$$= \int_0^4\int_0^1 \frac{2x^3z - x}{\sqrt{4x^2+1}}\sqrt{4x^2+1}\,dx\,dz$$

$$= \int_0^4\int_0^1 (2x^3z - x)\,dx\,dz = \int_0^4 \left[\frac{1}{2}x^4z - \frac{1}{2}x^2\right]_{x=0}^{x=1} dz$$

$$= \int_0^4 \frac{1}{2}(z-1)\,dz = \frac{1}{4}(z-1)^2\Big]_0^4$$

$$= \frac{1}{4}(9) - \frac{1}{4}(1) = 2.$$

Example 8 Finding a Center of Mass

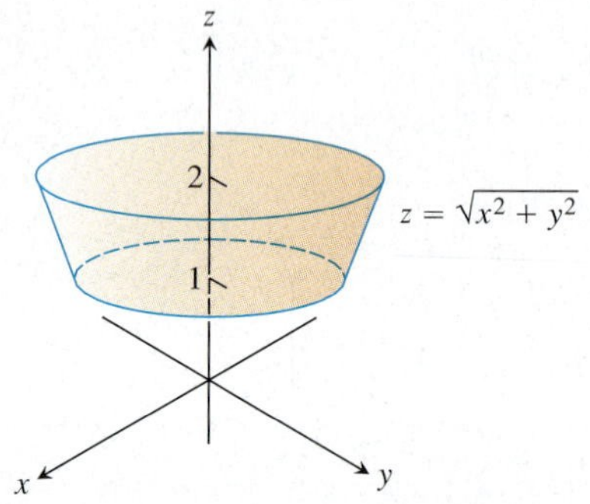

FIGURE 13.58 The cone frustum formed when the cone $z = \sqrt{x^2+y^2}$ is cut by the planes $z = 1$ and $z = 2$. (Example 8)

Find the center of mass of a thin shell of constant density δ cut from the cone $z = \sqrt{x^2+y^2}$ by the planes $z = 1$ and $z = 2$ (Figure 13.58).

Solution The symmetry of the surface about the z-axis tells us that $\bar{x} = \bar{y} = 0$. We find $\bar{z} = M_{xy}/M$. Working as in Examples 1 and 4, we have

$$\mathbf{r}(r,\theta) = r\cos\theta\mathbf{i} + r\sin\theta\mathbf{j} + r\mathbf{k}, \qquad 1 \le r \le 2, \quad 0 \le \theta \le 2\pi,$$

and

$$|\mathbf{r}_r \times \mathbf{r}_\theta| = \sqrt{2}\,r.$$

Therefore,

$$M = \iint_S \delta\,d\sigma = \int_0^{2\pi}\int_1^2 \delta\sqrt{2}\,r\,dr\,d\theta$$

$$= \delta\sqrt{2}\int_0^{2\pi}\left[\frac{r^2}{2}\right]_1^2 d\theta = \delta\sqrt{2}\int_0^{2\pi}\left(2 - \frac{1}{2}\right)d\theta$$

$$= \delta\sqrt{2}\left[\frac{3\theta}{2}\right]_0^{2\pi} = 3\pi\delta\sqrt{2}$$

$$M_{xy} = \iint_S \delta z\,d\sigma = \int_0^{2\pi}\int_1^2 \delta r\sqrt{2}r\,dr\,d\theta$$

$$= \delta\sqrt{2}\int_0^{2\pi}\int_1^2 r^2\,dr\,d\theta = \delta\sqrt{2}\int_0^{2\pi}\left[\frac{r^3}{3}\right]_1^2 d\theta$$

$$= \delta\sqrt{2}\int_0^{2\pi}\frac{7}{3}\,d\theta = \frac{14}{3}\pi\delta\sqrt{2}$$

$$\bar{z} = \frac{M_{xy}}{M} = \frac{14\pi\delta\sqrt{2}}{3(3\pi\delta\sqrt{2})} = \frac{14}{9}.$$

The shell's center of mass is the point $(0, 0, 14/9)$.

EXERCISES 13.6

Finding Parametrizations for Surfaces

In Exercises 1–16, find a parametrization of the surface. (There are many correct ways to do these, so your answers may not be the same as those in the back of the book.)

1. The paraboloid $z = x^2 + y^2, z \le 4$
2. The paraboloid $z = 9 - x^2 - y^2, z \ge 0$
3. *Cone frustrum* The first-octant portion of the cone $z = \sqrt{x^2 + y^2}/2$ between the planes $z = 0$ and $z = 3$
4. *Cone frustrum* The portion of the cone $z = 2\sqrt{x^2 + y^2}$ between the planes $z = 2$ and $z = 4$
5. *Spherical cap* The cap cut from the sphere $x^2 + y^2 + z^2 = 9$ by the cone $z = \sqrt{x^2 + y^2}$
6. *Spherical cap* The portion of the sphere $x^2 + y^2 + z^2 = 4$ in the first octant between the xy-plane and the cone $z = \sqrt{x^2 + y^2}$
7. *Spherical band* The portion of the sphere $x^2 + y^2 + z^2 = 3$ between the planes $z = \sqrt{3}/2$ and $z = -\sqrt{3}/2$
8. *Spherical cap* The upper portion cut from the sphere $x^2 + y^2 + z^2 = 8$ by the plane $z = -2$
9. *Parabolic cylinder between planes* The surface cut from the parabolic cylinder $z = 4 - y^2$ by the planes $x = 0$, $x = 2$, and $z = 0$
10. *Parabolic cylinder between planes* The surface cut from the parabolic cylinder $y = x^2$ by the planes $z = 0$, $z = 3$, and $y = 2$
11. *Circular cylinder band* The portion of the cylinder $y^2 + z^2 = 9$ between the planes $x = 0$ and $x = 3$
12. *Circular cylinder band* The portion of the cylinder $x^2 + z^2 = 4$ above the xy-plane between the planes $y = -2$ and $y = 2$
13. *Tilted plane inside cylinder* The portion of the plane $x + y + z = 1$
    (a) Inside the cylinder $x^2 + y^2 = 9$
    (b) Inside the cylinder $y^2 + z^2 = 9$
14. *Tilted plane inside cylinder* The portion of the plane $x - y + 2z = 2$
    (a) Inside the cylinder $x^2 + z^2 = 3$
    (b) Inside the cylinder $y^2 + z^2 = 2$
15. *Circular cylinder band* The portion of the cylinder $(x - 2)^2 + z^2 = 4$ between the planes $y = 0$ and $y = 3$
16. *Circular cylinder band* The portion of the cylinder $y^2 + (z - 5)^2 = 25$ between the planes $x = 0$ and $x = 10$

Areas of Parametrized Surfaces

In Exercises 17–26, use a parametrization to express the area of the surface as a double integral. Then evaluate the integral. (There are many correct ways to set up the integrals, so your integrals may not be the same as those in the back of the book. They should have the same values, however.)

17. *Tilted plane inside cylinder* The portion of the plane $y + 2z = 2$ inside the cylinder $x^2 + y^2 = 1$
18. *Plane inside cylinder* The portion of the plane $z = -x$ inside the cylinder $x^2 + y^2 = 4$
19. *Cone frustrum* The portion of the cone $z = 2\sqrt{x^2 + y^2}$ between the planes $z = 2$ and $z = 6$
20. *Cone frustrum* The portion of the cone $z = \sqrt{x^2 + y^2}/3$ between the planes $z = 1$ and $z = 4/3$
21. *Circular cylinder band* The portion of the cylinder $x^2 + y^2 = 1$ between the planes $z = 1$ and $z = 4$
22. *Circular cylinder band* The portion of the cylinder $x^2 + z^2 = 10$ between the planes $y = -1$ and $y = 1$
23. *Parabolic cap* The cap cut from the paraboloid $z = 2 - x^2 - y^2$ by the cone $z = \sqrt{x^2 + y^2}$
24. *Parabolic band* The portion of the paraboloid $z = x^2 + y^2$ between the planes $z = 1$ and $z = 4$
25. *Sawed-off sphere* The lower portion cut from the sphere $x^2 + y^2 + z^2 = 2$ by the cone $z = \sqrt{x^2 + y^2}$
26. *Spherical band* The portion of the sphere $x^2 + y^2 + z^2 = 4$ between the planes $z = -1$ and $z = \sqrt{3}$

Integrals Over Parametrized Surfaces

In Exercises 27–34, integrate the given function over the given surface.

27. *Parabolic cylinder* $G(x, y, z) = x$, over the parabolic cylinder $y = x^2, 0 \le x \le 2, 0 \le z \le 3$
28. *Circular cylinder* $G(x, y, z) = z$, over the cylindrical surface $y^2 + z^2 = 4, z \ge 0, 1 \le x \le 4$
29. *Sphere* $G(x, y, z) = x^2$, over the unit sphere $x^2 + y^2 + z^2 = 1$
30. *Hemisphere* $G(x, y, z) = z^2$, over the hemisphere $x^2 + y^2 + z^2 = a^2, z \ge 0$
31. *Portion of plane* $F(x, y, z) = z$, over the portion of the plane $x + y + z = 4$ that lies above the square $0 \le x \le 1, 0 \le y \le 1$, in the xy-plane
32. *Cone* $F(x, y, z) = z - x$, over the cone $z = \sqrt{x^2 + y^2}, 0 \le z \le 1$
33. *Parabolic dome* $H(x, y, z) = x^2\sqrt{5 - 4z}$, over the parabolic dome $z = 1 - x^2 - y^2, z \ge 0$
34. *Spherical cap* $H(x, y, z) = yz$, over the part of the sphere $x^2 + y^2 + z^2 = 4$ that lies above the cone $z = \sqrt{x^2 + y^2}$

Flux Across Parametrized Surfaces

In Exercises 35–44, use a parametrization to find the flux $\iint_S \mathbf{F} \cdot \mathbf{n}\, d\sigma$ across the surface in the given direction.

35. *Parabolic cylinder* $\mathbf{F} = z^2\mathbf{i} + x\mathbf{j} - 3z\mathbf{k}$ outward (normal away from the x-axis) through the surface cut from the parabolic cylinder $z = 4 - y^2$ by the planes $x = 0$, $x = 1$, and $z = 0$

36. *Parabolic cylinder* $\mathbf{F} = x^2\mathbf{j} - xz\mathbf{k}$ outward (normal away from the yz-plane) through the surface cut from the parabolic cylinder $y = x^2$, $-1 \le x \le 1$, by the planes $z = 0$ and $z = 2$

37. *Sphere* $\mathbf{F} = z\mathbf{k}$ across the portion of the sphere $x^2 + y^2 + z^2 = a^2$ in the first octant in the direction away from the origin

38. *Sphere* $\mathbf{F} = x\mathbf{i} + y\mathbf{j} + z\mathbf{k}$ across the sphere $x^2 + y^2 + z^2 = a^2$ in the direction away from the origin

39. *Plane* $\mathbf{F} = 2xy\mathbf{i} + 2yz\mathbf{j} + 2xz\mathbf{k}$ upward across the portion of the plane $x + y + z = 2a$ that lies above the square $0 \le x \le a$, $0 \le y \le a$, in the xy-plane

40. *Cylinder* $\mathbf{F} = x\mathbf{i} + y\mathbf{j} + z\mathbf{k}$ outward through the portion of the cylinder $x^2 + y^2 = 1$ cut by the planes $z = 0$ and $z = a$

41. *Cone* $\mathbf{F} = xy\mathbf{i} - z\mathbf{k}$ outward (normal away from the z-axis) through the cone $z = \sqrt{x^2 + y^2}$, $0 \le z \le 1$

42. *Cone* $\mathbf{F} = y^2\mathbf{i} + xz\mathbf{j} - \mathbf{k}$ outward (normal away from the z-axis) through the cone $z = 2\sqrt{x^2 + y^2}$, $0 \le z \le 2$

43. *Cone frustrum* $\mathbf{F} = -x\mathbf{i} - y\mathbf{j} + z^2\mathbf{k}$ outward (normal away from the z-axis) through the portion of the cone $z = \sqrt{x^2 + y^2}$ between the planes $z = 1$ and $z = 2$

44. *Paraboloid* $\mathbf{F} = 4x\mathbf{i} + 4y\mathbf{j} + 2\mathbf{k}$ outward (normal way from the z-axis) through the surface cut from the bottom of the paraboloid $z = x^2 + y^2$ by the plane $z = 1$

Moments and Masses

45. *Center of mass, inertia, radius of gyration* Find the center of mass and the moment of inertia and radius of gyration about the z-axis of a thin shell of constant density δ cut from the cone $x^2 + y^2 - z^2 = 0$ by the planes $z = 1$ and $z = 2$.

46. *Inertia of conical shell* Find the moment of inertia about the z-axis of a thin conical shell $z = \sqrt{x^2 + y^2}$, $0 \le z \le 1$, of constant density δ.

Planes Tangent to Parametrized Surfaces

The tangent plane at a point $P_0(f(u_0, v_0), g(u_0, v_0), h(u_0, v_0))$ on a parametrized surface $\mathbf{r}(u, v) = f(u, v)\mathbf{i} + g(u, v)\mathbf{j} + h(u, v)\mathbf{k}$ is the plane through P_0 normal to the vector $\mathbf{r}_u(u_0, v_0) \times \mathbf{r}_v(u_0, v_0)$, the cross product of the tangent vectors $\mathbf{r}_u(u_0, v_0)$ and $\mathbf{r}_v(u_0, v_0)$ at P_0. In Exercises 47–50, find an equation for the plane tangent to the surface at P_0. Then find a Cartesian equation for the surface and sketch the surface and tangent plane together.

47. *Cone* The cone $\mathbf{r}(r, \theta) = (r\cos\theta)\mathbf{i} + (r\sin\theta)\mathbf{j} + r\mathbf{k}$, $r \ge 0$, $0 \le \theta \le 2\pi$ at the point $P_0(\sqrt{2}, \sqrt{2}, 2)$ corresponding to $(r, \theta) = (2, \pi/4)$

48. *Hemisphere* The hemisphere surface $\mathbf{r}(\phi, \theta) = (4\sin\phi\cos\theta)\mathbf{i} + (4\sin\phi\sin\theta)\mathbf{j} + (4\cos\phi)\mathbf{k}$, $0 \le \phi \le \pi/2$, $0 \le \theta \le 2\pi$, at the point $P_0(\sqrt{2}, \sqrt{2}, 2\sqrt{3})$ corresponding to $(\phi, \theta) = (\pi/6, \pi/4)$

49. *Circular cylinder* The circular cylinder $\mathbf{r}(\theta, z) = (3\sin 2\theta)\mathbf{i} + (6\sin^2\theta)\mathbf{j} + z\mathbf{k}$, $0 \le \theta \le \pi$, at the point $P_0(3\sqrt{3}/2, 9/2, 0)$ corresponding to $(\theta, z) = (\pi/3, 0)$ (See Example 3.)

50. *Parabolic cylinder* The parabolic cylinder surface $\mathbf{r}(x, y) = x\mathbf{i} + y\mathbf{j} - x^2\mathbf{k}$, $-\infty < x < \infty$, $-\infty < y < \infty$, at the point $P_0(1, 2, -1)$ corresponding to $(x, y) = (1, 2)$

Further Examples of Parametrizations

51. **(a)** A *torus of revolution* (doughnut) is obtained by rotating a circle C in the xz-plane about the z-axis in space. (See the accompanying figure.) If C has radius $r > 0$ and center $(R, 0, 0)$, show that a parametrization of the torus is

$$\mathbf{r}(u, v) = ((R + r\cos u)\cos v)\mathbf{i} + ((R + r\cos u)\sin v)\mathbf{j} + (r\sin u)\mathbf{k},$$

where $0 \le u \le 2\pi$ and $0 \le v \le 2\pi$ are the angles in the figure.

(b) Show that the surface area of the torus is $A = 4\pi^2 Rr$.

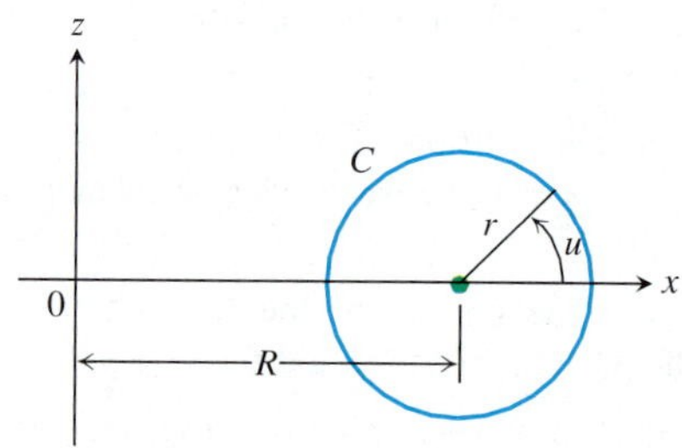

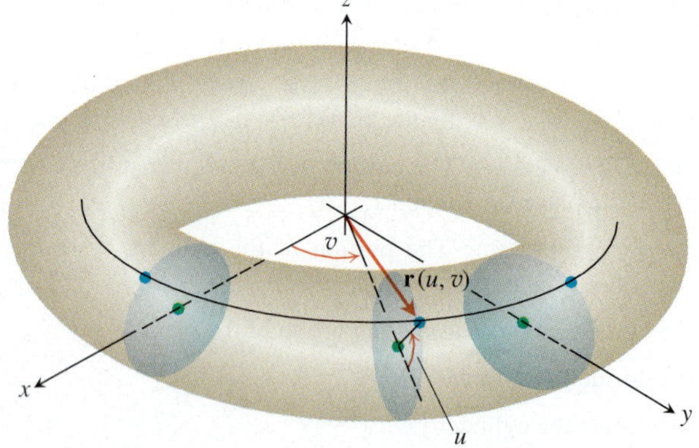

52. *Parametrization of a surface of revolution* Suppose that the parametrized curve $C: (f(u), g(u))$ is revolved about the x-axis, where $g(u) > 0$ for $a \le u \le b$.

(a) Show that

$$\mathbf{r}(u, v) = f(u)\mathbf{i} + (g(u)\cos v)\mathbf{j} + (g(u)\sin v)\mathbf{k}$$

is a parametrization of the resulting surface of revolution, where $0 \le v \le 2\pi$ is the angle from the xy-plane to the point $\mathbf{r}(u, v)$ on the surface. (See the accompanying figure.) Notice that $f(u)$ measures distance *along* the axis of revolution and $g(u)$ measures distance *from* the axis of revolution.

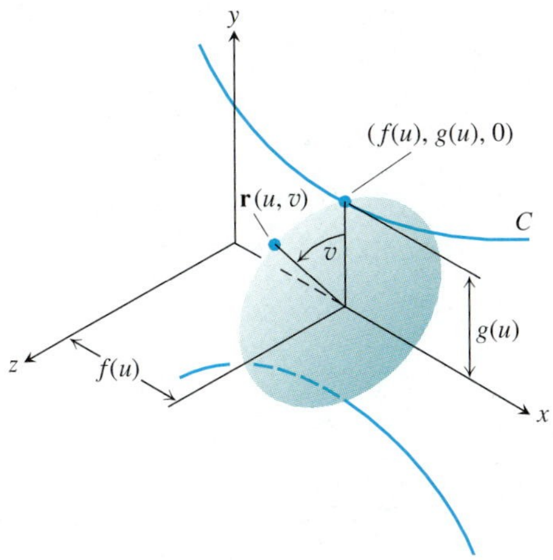

(b) Find a parametrization for the surface obtained by revolving the curve $x = y^2$, $y \geq 0$, about the x-axis.

53. (a) *Parametrization of an ellipsoid* Recall the parametrization $x = a \cos \theta$, $y = b \sin \theta$, $0 \leq \theta \leq 2\pi$ for the ellipse $(x^2/a^2) + (y^2/b^2) = 1$ (Section P.6, Example 6). Using the angles θ and ϕ in spherical coordinates, show that

$$\mathbf{r}(\theta, \phi) = (a \cos \theta \cos \phi)\mathbf{i} + (b \sin \theta \cos \phi)\mathbf{j} + (c \sin \phi)\mathbf{k}$$

is a parametrization of the ellipsoid $(x^2/a^2) + (y^2/b^2) + (z^2/c^2) = 1$.

(b) Write an integral for the surface area of the ellipsoid, but do not evaluate the integral.

13.7 Stokes' Theorem

Circulation Density: Curl • Stokes' Theorem • Paddle Wheel Interpretation of $\nabla \times \mathbf{F}$ • Proof of Stokes' Theorem for Polyhedral Surfaces • Stokes' Theorem for Surfaces with Holes • An Important Identity • Conservative Fields and Stokes' Theorem

In this section, we generalize the circulation-curl form of Green's Theorem to velocity fields in space.

Circulation Density: Curl

As we saw in Section 13.4, the **k**-component of the circulation density or curl of a two-dimensional field $\mathbf{F} = M\mathbf{i} + N\mathbf{j}$ at a point (x, y) is the scalar quantity $(\partial N/\partial x - \partial M/\partial y)$. In three dimensions, the circulation around a point P in a plane is described with a vector. This vector is normal to the plane of the circulation (Figure 13.59) and points in the direction that gives it a right-hand relation to the circulation line. The length of the vector gives the rate of the fluid's rotation, which usually varies as the circulation plane is tilted about P. It turns out that the vector of greatest circulation in a flow with velocity field $\mathbf{F} = M\mathbf{i} + N\mathbf{j} + P\mathbf{k}$ is

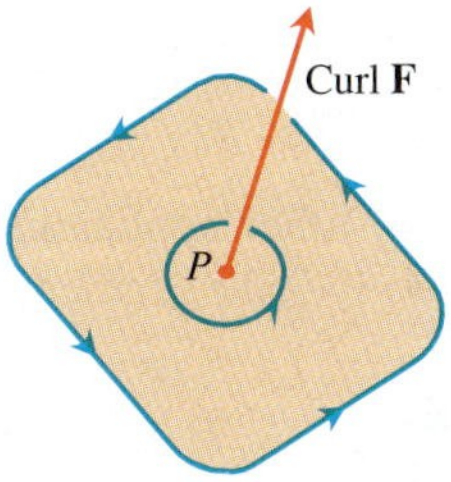

FIGURE 13.59 The circulation vector at a point P in a plane in a three-dimensional fluid flow. Notice its right-hand relation to the circulation line.

$$\text{curl } \mathbf{F} = \left(\frac{\partial P}{\partial y} - \frac{\partial N}{\partial z}\right)\mathbf{i} + \left(\frac{\partial M}{\partial z} - \frac{\partial P}{\partial x}\right)\mathbf{j} + \left(\frac{\partial N}{\partial x} - \frac{\partial M}{\partial y}\right)\mathbf{k}. \tag{1}$$

Notice that $(\text{curl } \mathbf{F}) \cdot \mathbf{k} = (\partial N/\partial x - \partial M/\partial y)$, consistent with our definition in Section 13.4 for $\mathbf{F} = M\mathbf{i} + N\mathbf{j}$. The formula for curl $\mathbf{F}$ in Equation (1) is usually written using the symbolic operator

$$\nabla = \mathbf{i}\frac{\partial}{\partial x} + \mathbf{j}\frac{\partial}{\partial y} + \mathbf{k}\frac{\partial}{\partial z}. \tag{2}$$

(The symbol ∇ is pronounced "del.") The curl of $\mathbf{F}$ is $\nabla \times \mathbf{F}$:

$$\begin{aligned} \nabla \times \mathbf{F} &= \begin{vmatrix} \mathbf{i} & \mathbf{j} & \mathbf{k} \\ \dfrac{\partial}{\partial x} & \dfrac{\partial}{\partial y} & \dfrac{\partial}{\partial z} \\ M & N & P \end{vmatrix} \\ &= \left(\frac{\partial P}{\partial y} - \frac{\partial N}{\partial z}\right)\mathbf{i} + \left(\frac{\partial M}{\partial z} - \frac{\partial P}{\partial x}\right)\mathbf{j} + \left(\frac{\partial N}{\partial x} - \frac{\partial M}{\partial y}\right)\mathbf{k} \\ &= \text{curl } \mathbf{F}. \end{aligned}$$

$$\text{curl } \mathbf{F} = \nabla \times \mathbf{F} \tag{3}$$

Example 1 Finding Curl F

Find the curl of $\mathbf{F} = (x^2 - y)\mathbf{i} + 4z\mathbf{j} + x^2\mathbf{k}$.

Solution

$$\begin{aligned} \text{curl } \mathbf{F} &= \nabla \times \mathbf{F} \qquad \text{Eq. (3)} \\ &= \begin{vmatrix} \mathbf{i} & \mathbf{j} & \mathbf{k} \\ \dfrac{\partial}{\partial x} & \dfrac{\partial}{\partial y} & \dfrac{\partial}{\partial z} \\ x^2 - y & 4z & x^2 \end{vmatrix} \\ &= \left(\frac{\partial}{\partial y}(x^2) - \frac{\partial}{\partial z}(4z)\right)\mathbf{i} - \left(\frac{\partial}{\partial x}(x^2) - \frac{\partial}{\partial z}(x^2 - y)\right)\mathbf{j} \\ &\quad + \left(\frac{\partial}{\partial x}(4z) - \frac{\partial}{\partial y}(x^2 - y)\right)\mathbf{k} \\ &= (0 - 4)\mathbf{i} - (2x - 0)\mathbf{j} + (0 + 1)\mathbf{k} \\ &= -4\mathbf{i} - 2x\mathbf{j} + \mathbf{k} \end{aligned}$$

As we will see, the operator ∇ has a number of other applications. For instance, when applied to a scalar function $f(x, y, z)$, it gives the gradient of f:

$$\nabla f = \frac{\partial f}{\partial x}\mathbf{i} + \frac{\partial f}{\partial y}\mathbf{j} + \frac{\partial f}{\partial z}\mathbf{k}.$$

This may now be read as "del f" as well as "grad f."

Stokes' Theorem

Stokes' Theorem says that, under conditions normally met in practice, the circulation of a vector field around the boundary of an oriented surface in space in the direction counterclockwise with respect to the surface's unit normal vector field $\mathbf{n}$ (Figure 13.60) equals the integral of the normal component of the curl of the field over the surface.

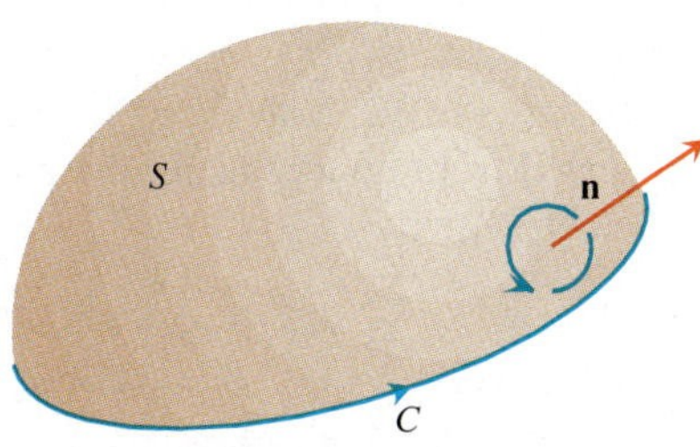

FIGURE 13.60 The orientation of the bounding curve C gives it a right-handed relation to the normal field $\mathbf{n}$.

Theorem 5 Stokes' Theorem

The circulation of a vector field $\mathbf{F} = M\mathbf{i} + N\mathbf{j} + P\mathbf{j}$ around the boundary C of an oriented surface S in the direction counterclockwise with respect to the surface's unit normal vector $\mathbf{n}$ equals the integral of $\nabla \times \mathbf{F} \cdot \mathbf{n}$ over S.

$$\oint_C \mathbf{F} \cdot d\mathbf{r} = \iint_S \nabla \times \mathbf{F} \cdot \mathbf{n}\, d\sigma \qquad (4)$$

Counterclockwise circulation   Curl integral

CD-ROM
WEBsite
Historical Biography
George Gabriel Stokes
(1819 — 1903)

Notice from Equation (4) that if two different oriented surfaces S_1 and S_2 have the same boundary C, their curl integrals are equal:

$$\iint_{S_1} \nabla \times \mathbf{F} \cdot \mathbf{n}_1\, d\sigma = \iint_{S_2} \nabla \times \mathbf{F} \cdot \mathbf{n}_2\, d\sigma.$$

Both curl integrals equal the counterclockwise circulation integral on the left side of Equation (4) as long as the unit normal vectors $\mathbf{n}_1$ and $\mathbf{n}_2$ correctly orient the surfaces.

Naturally, we need some mathematical restrictions on $\mathbf{F}$, C, and S to ensure the existence of the integrals in Stokes' equation. The usual restrictions are that all the functions and derivatives involved be continuous.

If C is a curve in the xy-plane, oriented counterclockwise, and R is the region in the xy-plane bounded by C, then $d\sigma = dx\,dy$ and

$$(\nabla \times \mathbf{F}) \cdot \mathbf{n} = (\nabla \times \mathbf{F}) \cdot \mathbf{k} = \left(\frac{\partial N}{\partial x} - \frac{\partial M}{\partial y}\right).$$

Under these conditions, Stokes' equation becomes

$$\oint_C \mathbf{F} \cdot d\mathbf{r} = \iint_R \left(\frac{\partial N}{\partial x} - \frac{\partial M}{\partial y}\right) dx\,dy,$$

which is the circulation-curl form of the equation in Green's Theorem. Conversely, by reversing these steps we can rewrite the circulation-curl form of Green's Theorem for two-dimensional fields in del notation as

$$\oint_C \mathbf{F} \cdot d\mathbf{r} = \iint_R \nabla \times \mathbf{F} \cdot \mathbf{k}\, dA. \qquad (5)$$

See Figure 13.61.

Green:

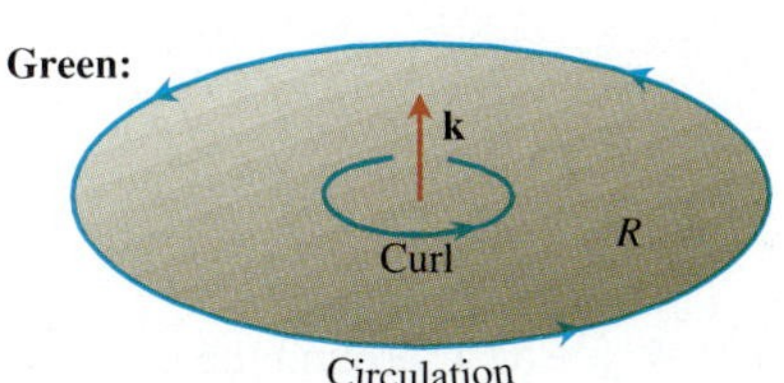

Stokes:

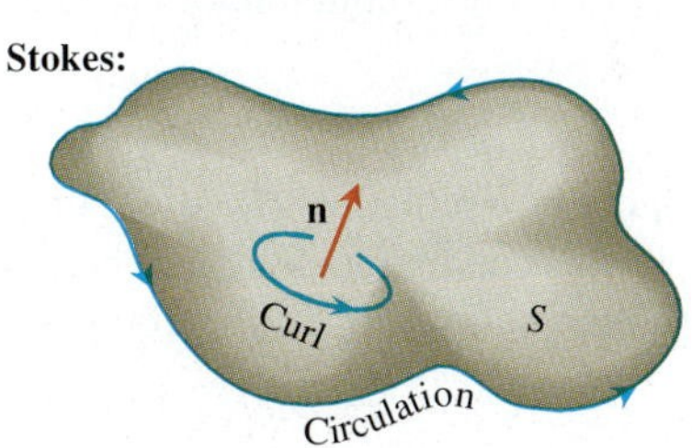

FIGURE 13.61 Green's Theorem versus Stokes' Theorem.

Example 2 Verifying Stokes' Equation for a Hemisphere

Evaluate Equation (4) for the hemisphere $S: x^2 + y^2 + z^2 = 9, z \geq 0$, its bounding circle $C: x^2 + y^2 = 9, z = 0$, and the field $\mathbf{F} = y\mathbf{i} - x\mathbf{j}$.

Solution We calculate the counterclockwise circulation around C (as viewed from above) using the parametrization $\mathbf{r}(\theta) = (3\cos\theta)\mathbf{i} + (3\sin\theta)\mathbf{j}$, $0 \leq \theta \leq 2\pi$:

$$d\mathbf{r} = (-3\sin\theta\, d\theta)\mathbf{i} + (3\cos\theta\, d\theta)\mathbf{j}$$

$$\mathbf{F} = y\mathbf{i} - x\mathbf{j} = (3\sin\theta)\mathbf{i} - (3\cos\theta)\mathbf{j}$$

$$\mathbf{F}\cdot d\mathbf{r} = -9\sin^2\theta\, d\theta - 9\cos^2\theta\, d\theta = -9\, d\theta$$

$$\oint_C \mathbf{F}\cdot d\mathbf{r} = \int_0^{2\pi} -9\, d\theta = -18\pi.$$

For the curl integral of $\mathbf{F}$, we have

$$\nabla\times\mathbf{F} = \left(\frac{\partial P}{\partial y} - \frac{\partial N}{\partial z}\right)\mathbf{i} + \left(\frac{\partial M}{\partial z} - \frac{\partial P}{\partial x}\right)\mathbf{j} + \left(\frac{\partial N}{\partial x} - \frac{\partial M}{\partial y}\right)\mathbf{k}$$

$$= (0-0)\mathbf{i} + (0-0)\mathbf{j} + (-1-1)\mathbf{k} = -2\mathbf{k}$$

$$\mathbf{n} = \frac{x\mathbf{i} + y\mathbf{j} + z\mathbf{k}}{\sqrt{x^2+y^2+z^2}} = \frac{x\mathbf{i} + y\mathbf{j} + z\mathbf{k}}{3} \qquad \text{Outer unit normal}$$

$$d\sigma = \frac{3}{z}\, dA \qquad \text{Section 13.5, Example 5, with } a = 3$$

$$\nabla\times\mathbf{F}\cdot\mathbf{n}\, d\sigma = -\frac{2z}{3}\frac{3}{z}\, dA = -2\, dA$$

and

$$\iint_S \nabla\times\mathbf{F}\cdot\mathbf{n}\, d\sigma = \iint_{x^2+y^2\leq 9} -2\, dA = -18\pi.$$

The circulation around the circle equals the integral of the curl over the hemisphere, as it should.

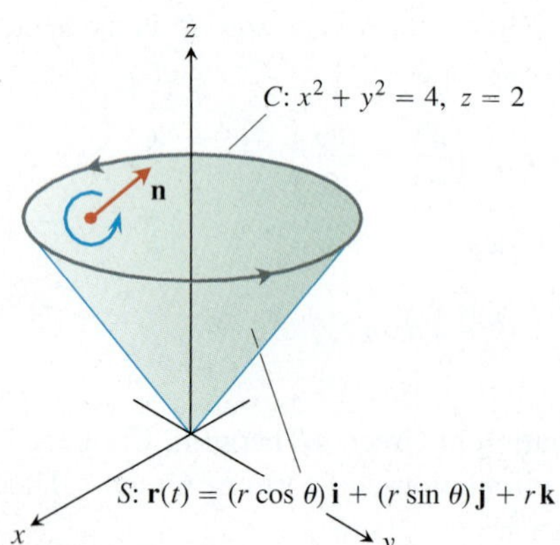

FIGURE 13.62 The curve C and cone S in Example 3.

Example 3 Finding Circulation

Find the circulation of the field $\mathbf{F} = (x^2 - y)\mathbf{i} + 4z\mathbf{j} + x^2\mathbf{k}$ around the curve C in which the plane $z = 2$ meets the cone $z = \sqrt{x^2+y^2}$, counterclockwise as viewed from above (Figure 13.62).

Solution Stokes' Theorem enables us to find the circulation by integrating over the surface of the cone. Traversing C in the counterclockwise direction viewed from above corresponds to taking the *inner* normal $\mathbf{n}$ to the cone, the normal with a positive z-component).

We parametrize the cone as

$$\mathbf{r}(r, \theta) = (r\cos\theta)\mathbf{i} + (r\sin\theta)\mathbf{j} + r\mathbf{k}, \qquad 0 \leq r \leq 2, \quad 0 \leq \theta \leq 2\pi.$$

We then have

$$\mathbf{n} = \frac{\mathbf{r}_r \times \mathbf{r}_\theta}{|\mathbf{r}_r \times \mathbf{r}_\theta|} = \frac{-(r\cos\theta)\mathbf{i} - (r\sin\theta)\mathbf{j} + r\mathbf{k}}{r\sqrt{2}}$$

Section 13.6, Example 4

$$= \frac{1}{\sqrt{2}}(-(\cos\theta)\mathbf{i} - (\sin\theta)\mathbf{j} + \mathbf{k})$$

$$d\sigma = r\sqrt{2}\,dr\,d\theta$$

Section 13.6, Example 4

$$\nabla \times \mathbf{F} = -4\mathbf{i} - 2x\mathbf{j} + \mathbf{k}$$

Example 1

$$= -4\mathbf{i} - 2r\cos\theta\mathbf{j} + \mathbf{k}.$$

$x = r\cos\theta$

Accordingly,

$$\nabla \times \mathbf{F} \cdot \mathbf{n} = \frac{1}{\sqrt{2}}(4\cos\theta + 2r\cos\theta\sin\theta + 1)$$

$$= \frac{1}{\sqrt{2}}(4\cos\theta + r\sin 2\theta + 1)$$

and the circulation is

$$\oint_C \mathbf{F} \cdot d\mathbf{r} = \iint_S \nabla \times \mathbf{F} \cdot \mathbf{n}\,d\sigma$$

Stokes' Theorem, Eq. (4)

$$= \int_0^{2\pi}\int_0^2 \frac{1}{\sqrt{2}}(4\cos\theta + r\sin 2\theta + 1)(r\sqrt{2}\,dr\,d\theta) = 4\pi.$$

Paddle Wheel Interpretation of $\nabla \times \mathbf{F}$

Suppose that $\mathbf{v}(x, y, z)$ is the velocity of a moving fluid whose density at (x, y, z) is $\delta(x, y, z)$ and let $\mathbf{F} = \delta\mathbf{v}$. Then

$$\oint_C \mathbf{F} \cdot d\mathbf{r}$$

is the circulation of the fluid around the closed curve C. By Stokes' Theorem, the circulation is equal to the flux of $\nabla \times \mathbf{F}$ through a surface S spanning C:

$$\oint_C \mathbf{F} \cdot d\mathbf{r} = \iint_S \nabla \times \mathbf{F} \cdot \mathbf{n}\,d\sigma.$$

Suppose we fix a point Q in the domain of $\mathbf{F}$ and a direction $\mathbf{u}$ at Q. Let C be a circle of radius ρ, with center at Q, whose plane is normal to $\mathbf{u}$. If $\nabla \times \mathbf{F}$ is continuous at Q, the average value of the $\mathbf{u}$-component of $\nabla \times \mathbf{F}$ over the circular disk S bounded by C approaches the $\mathbf{u}$-component of $\nabla \times \mathbf{F}$ at Q as $\rho \to 0$:

$$(\nabla \times \mathbf{F} \cdot \mathbf{u})_Q = \lim_{\rho\to 0} \frac{1}{\pi\rho^2}\iint_S \nabla \times \mathbf{F} \cdot \mathbf{u}\,d\sigma.$$

If we replace the double integral in this last equation by the circulation, we get

$$(\nabla \times \mathbf{F} \cdot \mathbf{u})_Q = \lim_{\rho\to 0} \frac{1}{\pi\rho^2}\oint_C \mathbf{F} \cdot d\mathbf{r}. \tag{6}$$

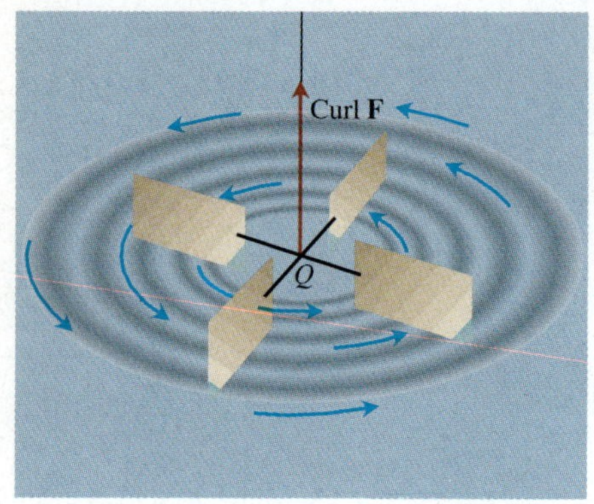

FIGURE 13.63 The paddle wheel interpretation of curl **F**.

The left-hand side of Equation (6) has its maximum value when **u** is the direction of $\nabla \times \mathbf{F}$. When ρ is small, the limit on the right-hand side of Equation (6) is approximately

$$\frac{1}{\pi\rho^2}\oint_C \mathbf{F}\cdot d\mathbf{r},$$

which is the circulation around C divided by the area of the disk (circulation density). Suppose that a small paddle wheel of radius ρ is introduced into the fluid at Q, with its axle directed along **u**. The circulation of the fluid around C will affect the rate of spin of the paddle wheel. The wheel will spin fastest when the circulation integral is maximized; therefore it will spin fastest when the axle of the paddle wheel points in the direction of $\nabla \times \mathbf{F}$ (Figure 13.63).

Example 4 Relating $\nabla \times \mathbf{F}$ to Circulation Density

A fluid of constant density rotates around the z-axis with velocity $\mathbf{v} = \omega(-y\mathbf{i} + x\mathbf{j})$, where ω is a positive constant called the *angular velocity* of the rotation (Figure 13.64). If $\mathbf{F} = \mathbf{v}$, find $\nabla \times \mathbf{F}$ and relate it to the circulation density.

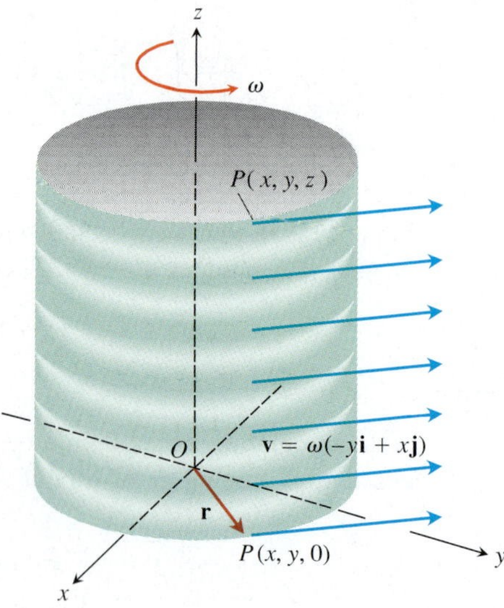

FIGURE 13.64 A steady rotational flow parallel to the xy-plane, with constant angular velocity ω in the positive (counterclockwise) direction.

Solution With $\mathbf{F} = \mathbf{v} = -\omega y\mathbf{i} + \omega x\mathbf{j}$,

$$\begin{aligned}\nabla \times \mathbf{F} &= \left(\frac{\partial P}{\partial y} - \frac{\partial N}{\partial z}\right)\mathbf{i} + \left(\frac{\partial M}{\partial z} - \frac{\partial P}{\partial x}\right)\mathbf{j} + \left(\frac{\partial N}{\partial x} - \frac{\partial M}{\partial y}\right)\mathbf{k}\\ &= (0-0)\mathbf{i} + (0-0)\mathbf{j} + (\omega - (-\omega))\mathbf{k} = 2\omega\mathbf{k}.\end{aligned}$$

By Stokes' Theorem, the circulation of **F** around a circle C of radius ρ bounding a disk S in a plane normal to $\nabla \times \mathbf{F}$, say the xy-plane, is

$$\oint_C \mathbf{F}\cdot d\mathbf{r} = \iint_S \nabla \times \mathbf{F}\cdot\mathbf{n}\,d\sigma = \iint_S 2\omega\mathbf{k}\cdot\mathbf{k}\,dx\,dy = (2\omega)(\pi\rho^2).$$

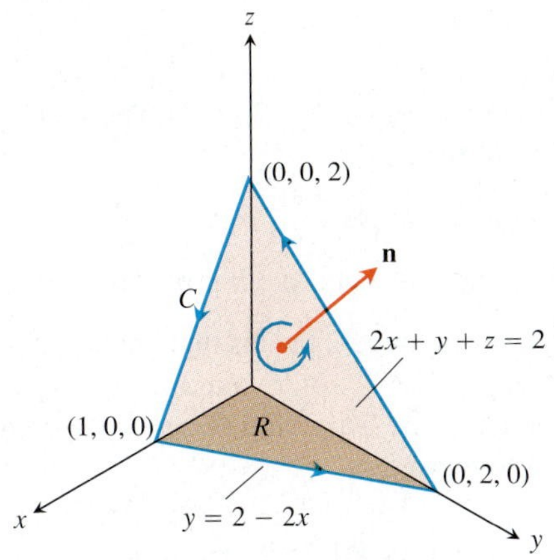

FIGURE 13.65 The planar surface in Example 5.

Thus,

$$(\nabla \times \mathbf{F}) \cdot \mathbf{k} = 2\omega = \frac{1}{\pi\rho^2} \oint_C \mathbf{F} \cdot d\mathbf{r},$$

in agreement with Equation (6) with $\mathbf{u} = \mathbf{k}$.

Example 5 Applying Stokes' Theorem

Use Stokes' Theorem to evaluate $\int_C \mathbf{F} \cdot d\mathbf{r}$, if $\mathbf{F} = xz\mathbf{i} + xy\mathbf{j} + 3xz\mathbf{k}$ and C is the boundary of the portion of the plane $2x + y + z = 2$ in the first octant, traversed counterclockwise as viewed from above (Figure 13.65).

Solution The plane is the level surface $f(x, y, z) = 2$ of the function $f(x, y, z) = 2x + y + z$. The unit normal vector

$$\mathbf{n} = \frac{\nabla f}{|\nabla f|} = \frac{(2\mathbf{i} + \mathbf{j} + \mathbf{k})}{|2\mathbf{i} + \mathbf{j} + \mathbf{k}|} = \frac{1}{\sqrt{6}}(2\mathbf{i} + \mathbf{j} + \mathbf{k})$$

is consistent with the counterclockwise motion around C. To apply Stokes' Theorem, we find

$$\text{curl } \mathbf{F} = \nabla \times \mathbf{F} = \begin{vmatrix} \mathbf{i} & \mathbf{j} & \mathbf{k} \\ \dfrac{\partial}{\partial x} & \dfrac{\partial}{\partial y} & \dfrac{\partial}{\partial z} \\ xz & xy & 3xz \end{vmatrix} = (x - 3z)\mathbf{j} + y\mathbf{k}.$$

On the plane, z equals $2 - 2x - y$, so

$$\nabla \times \mathbf{F} = (x - 3(2 - 2x - y))\mathbf{j} + y\mathbf{k} = (7x + 3y - 6)\mathbf{j} + y\mathbf{k}$$

and

$$\nabla \times \mathbf{F} \cdot \mathbf{n} = \frac{1}{\sqrt{6}}(7x + 3y - 6 + y) = \frac{1}{\sqrt{6}}(7x + 4y - 6).$$

The surface area element is

$$d\sigma = \frac{|\nabla f|}{|\nabla f \cdot \mathbf{k}|}\, dA = \frac{\sqrt{6}}{1}\, dx\, dy.$$

The circulation is

$$\begin{aligned} \oint_C \mathbf{F} \cdot d\mathbf{r} &= \iint_S \nabla \times \mathbf{F} \cdot \mathbf{n}\, d\sigma && \text{Stokes' Theorem, Eq. (4)} \\ &= \int_0^1 \int_0^{2-2x} \frac{1}{\sqrt{6}}(7x + 4y - 6)\sqrt{6}\, dy\, dx \\ &= \int_0^1 \int_0^{2-2x} (7x + 4y - 6)\, dy\, dx = -1. \end{aligned}$$

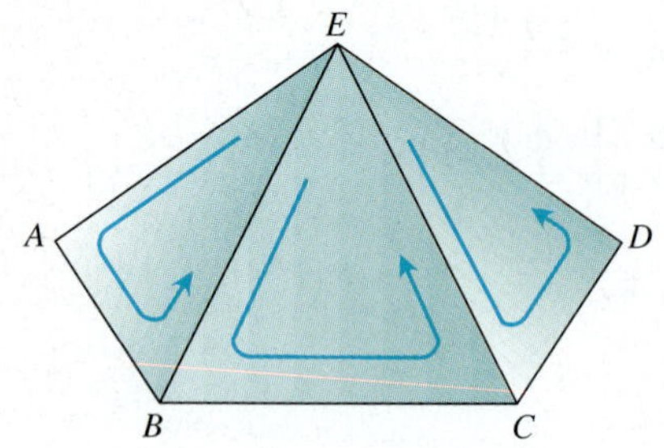

FIGURE 13.66 Part of a polyhedral surface.

Proof of Stokes' Theorem for Polyhedral Surfaces

Let S be a polyhedral surface consisting of a finite number of plane regions. (Think of one of Buckminster Fuller's geodesic domes.) We apply Green's Theorem to each separate panel of S. There are two types of panels:

1. Those that are surrounded on all sides by other panels
2. Those that have one or more edges that are not adjacent to other panels.

The boundary Δ of S consists of those edges of the type 2 panels that are not adjacent to other panels. In Figure 13.66, the triangles EAB, BCE, and CDE represent a part of S, with $ABCD$ part of the boundary Δ. Applying Green's Theorem to the three triangles in turn and adding the results, we get

$$\left(\oint_{EAB} + \oint_{BCE} + \oint_{CDE}\right)\mathbf{F}\cdot d\mathbf{r} = \left(\iint_{EAB} + \iint_{BCE} + \iint_{CDE}\right)\nabla\times\mathbf{F}\cdot\mathbf{n}\,d\sigma. \tag{7}$$

The three line integrals on the left-hand side of Equation (7) combine into a single line integral taken around the periphery $ABCDE$ because the integrals along interior segments cancel in pairs. For example, the integral along segment BE in triangle ABE is opposite in sign to the integral along the same segment in triangle EBC. The same holds for segment CE. Hence, Equation (7) reduces to

$$\oint_{ABCDE}\mathbf{F}\cdot d\mathbf{r} = \iint_{ABCDE}\nabla\times\mathbf{F}\cdot\mathbf{n}\,d\sigma.$$

When we apply Green's Theorem to all the panels and add the results, we get

$$\oint_{\Delta}\mathbf{F}\cdot d\mathbf{r} = \iint_{S}\nabla\times\mathbf{F}\cdot\mathbf{n}\,d\sigma.$$

This is Stokes' Theorem for a polyhedral surface S. You can find proofs for more general surfaces in advanced calculus texts.

Stokes' Theorem for Surfaces with Holes

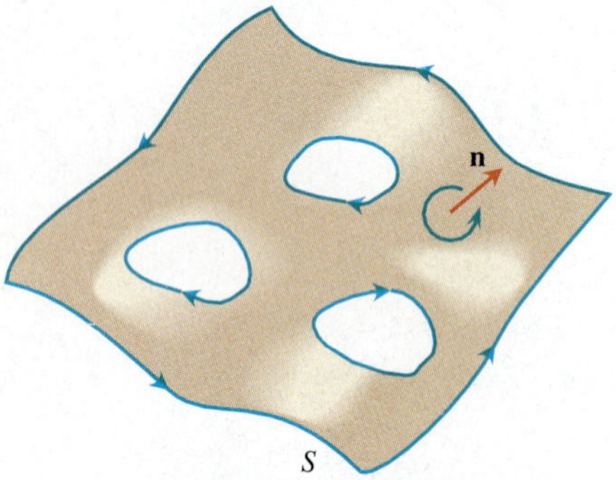

Stokes' Theorem can be extended to an oriented surface S that has one or more holes (Figure 13.67), in a way analogous to the extension of Green's Theorem: The surface integral over S of the normal component of $\nabla\times\mathbf{F}$ equals the sum of the line integrals around all the boundary curves of the tangential component of $\mathbf{F}$, where the curves are to be traced in the direction induced by the orientation of S.

FIGURE 13.67 Stokes' Theorem also holds for oriented surfaces with holes.

An Important Identity

The following identity arises frequently in mathematics and the physical sciences.

$$\text{curl grad } f = \mathbf{0} \quad \text{or} \quad \nabla\times\nabla f = \mathbf{0} \tag{8}$$

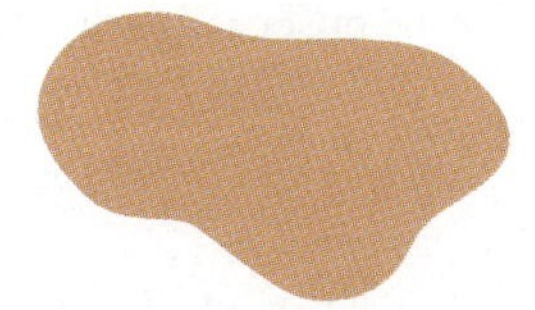

Connected and simply connected.

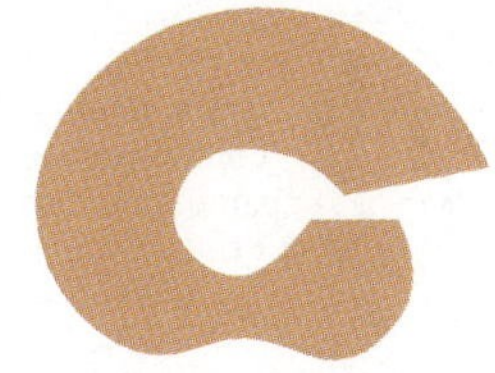

Connected and simply connected.

Connected but not simply connected.

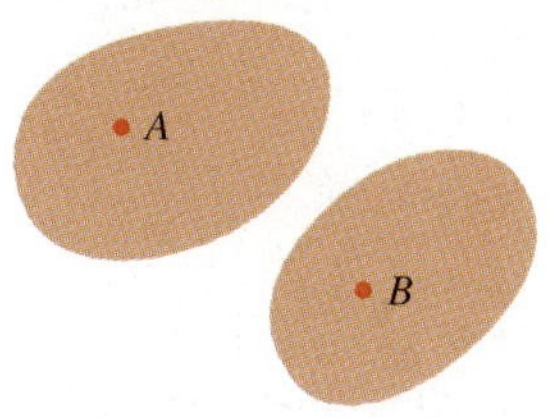

Simply connected but not connected.
No path from A to B lies entirely in the region.

FIGURE 13.68 Connectivity and simple connectivity are not the same. Neither implies the other, as these pictures of plane regions illustrate. To make three-dimensional regions with these properties, thicken the plane regions into cylinders.

This identity holds for any function $f(x, y, z)$ whose second partial derivatives are continuous. The proof goes like this:

$$\nabla \times \nabla f = \begin{vmatrix} \mathbf{i} & \mathbf{j} & \mathbf{k} \\ \dfrac{\partial}{\partial x} & \dfrac{\partial}{\partial y} & \dfrac{\partial}{\partial z} \\ \dfrac{\partial f}{\partial x} & \dfrac{\partial f}{\partial y} & \dfrac{\partial f}{\partial z} \end{vmatrix} = (f_{zy} - f_{yz})\mathbf{i} - (f_{zx} - f_{xz})\mathbf{j} + (f_{yx} - f_{xy})\mathbf{k}.$$

If the second partial derivatives are continuous, the mixed second derivatives in parentheses are equal (Theorem 4, Section 11.3) and the vector is zero.

Conservative Fields and Stokes' Theorem

In Section 13.3, we found that saying that a field $\mathbf{F}$ is conservative in an open region D in space is equivalent to saying that the integral of $\mathbf{F}$ around every closed loop in D is zero. This, in turn, is equivalent in *simply connected* open regions to saying that $\nabla \times \mathbf{F} = \mathbf{0}$. A region D is **simply connected** if every closed path in D can be contracted to a point in D without ever leaving D. If D consisted of space with a line removed, for example, D would not be simply connected. There would be no way to contract a loop around the line to a point without leaving D. On the other hand, space itself *is* simply connected (Figure 13.68).

Theorem 6 Relation of Curl $\mathbf{F} = \mathbf{0}$ to the Closed-Loop Property

If $\nabla \times \mathbf{F} = \mathbf{0}$ at every point of a simply connected open region D in space, then on any piecewise smooth closed path C in D,

$$\oint_C \mathbf{F} \cdot d\mathbf{r} = 0.$$

Sketch of a Proof Theorem 6 is usually proved in two steps. The first step is for simple closed curves. A theorem from topology, a branch of advanced mathematics, states that every differentiable simple closed curve C in a simply connected open region D is the boundary of a smooth two-sided surface S that also lies in D. Hence, by Stokes' Theorem,

$$\oint_C \mathbf{F} \cdot d\mathbf{r} = \iint_S \nabla \times \mathbf{F} \cdot \mathbf{n}\, d\sigma = 0.$$

The second step is for curves that cross themselves, like the one in Figure 13.69. The idea is to break these into simple loops spanned by orientable surfaces, apply Stokes' Theorem one loop at a time, and add the results.

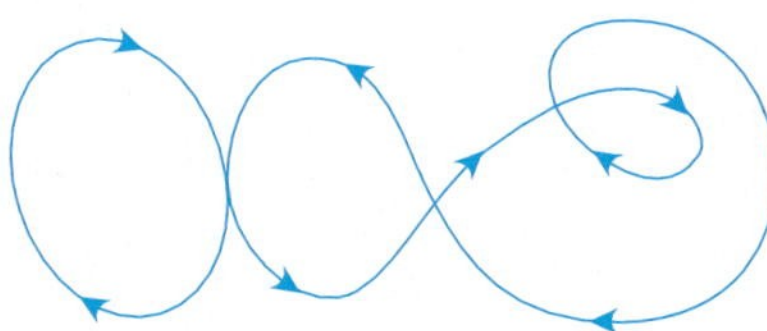

FIGURE 13.69 In a simply connected open region in space, differentiable curves that cross themselves can be divided into loops to which Stokes' Theorem applies.

The following diagram summarizes the results for conservative fields defined on connected, simply connected open regions.

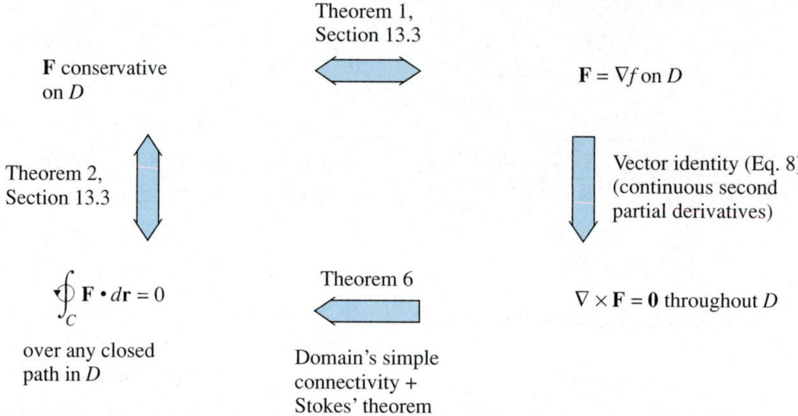

EXERCISES 13.7

Using Stokes' Theorem to Calculate Circulation

In Exercises 1–6, use the surface integral in Stokes' Theorem to calculate the circulation of the field **F** around the curve C in the indicated direction.

1. $\mathbf{F} = x^2\mathbf{i} + 2x\mathbf{j} + z^2\mathbf{k}$

   C: The ellipse $4x^2 + y^2 = 4$ in the xy-plane, counterclockwise when viewed from above

2. $\mathbf{F} = 2y\mathbf{i} + 3x\mathbf{j} - z^2\mathbf{k}$

   C: The circle $x^2 + y^2 = 9$ in the xy-plane, counterclockwise when viewed from above

3. $\mathbf{F} = y\mathbf{i} + xz\mathbf{j} + z^2\mathbf{k}$

   C: The boundary of the triangle cut from the plane $x + y + z = 1$ by the first octant, counterclockwise when viewed from above

4. $\mathbf{F} = (y^2 + z^2)\mathbf{i} + (x^2 + z^2)\mathbf{j} + (x^2 + y^2)\mathbf{k}$

   C: The boundary of the triangle cut from the plane $x + y + z = 1$ by the first octant, counterclockwise when viewed from above

5. $\mathbf{F} = (y^2 + z^2)\mathbf{i} + (x^2 + y^2)\mathbf{j} + (x^2 + y^2)\mathbf{k}$

   C: The square bounded by the lines $x = \pm 1$ and $y = \pm 1$ in the xy-plane, counterclockwise when viewed from above

6. $\mathbf{F} = x^2y^3\mathbf{i} + \mathbf{j} + z\mathbf{k}$

   C: The intersection of the cylinder $x^2 + y^2 = 4$ and the hemisphere $x^2 + y^2 + z^2 = 16$, $z \geq 0$

Flux of the Curl

7. Let **n** be the outer unit normal of the elliptical shell

$$S: \quad 4x^2 + 9y^2 + 36z^2 = 36, \qquad z \geq 0,$$

   and let

$$\mathbf{F} = y\mathbf{i} + x^2\mathbf{j} + (x^2 + y^4)^{3/2} \sin e^{\sqrt{xyz}}\,\mathbf{k}.$$

   Find the value of

$$\iint_S \nabla \times \mathbf{F} \cdot \mathbf{n}\, d\sigma.$$

   (*Hint:* One parametrization of the ellipse at the base of the shell is $x = 3\cos t$, $y = 2\sin t$, $0 \leq t \leq 2\pi$.)

8. Let **n** be the outer unit normal (normal away from the origin) of the parabolic shell

$$S: \quad 4x^2 + y + z^2 = 4, \qquad y \geq 0,$$

   and let

$$\mathbf{F} = \left(-z + \frac{1}{2 + x}\right)\mathbf{i} + (\tan^{-1} y)\mathbf{j} + \left(x + \frac{1}{4 + z}\right)\mathbf{k}.$$

   Find the value of

$$\iint_S \nabla \times \mathbf{F} \cdot \mathbf{n}\, d\sigma.$$

9. Let S be the cylinder $x^2 + y^2 = a^2$, $0 \le z \le h$, together with its top, $x^2 + y^2 \le a^2$, $z = h$. Let $\mathbf{F} = -y\mathbf{i} + x\mathbf{j} + x^2\mathbf{k}$. Use Stokes' Theorem to find the flux of $\nabla \times \mathbf{F}$ outward through S.

10. Evaluate

$$\iint_S \nabla \times (y\mathbf{i}) \cdot \mathbf{n}\, d\sigma,$$

where S is the hemisphere $x^2 + y^2 + z^2 = 1$, $z \ge 0$.

11. *Flux of curl F* Show that

$$\iint_S \nabla \times \mathbf{F} \cdot \mathbf{n}\, d\sigma$$

has the same value for all oriented surfaces S that span C and that induce the same positive direction on C.

12. *Writing to Learn* Let $\mathbf{F}$ be a differentiable vector field defined on a region containing a smooth closed oriented surface S and its interior. Let $\mathbf{n}$ be the unit normal vector field on S. Suppose that S is the union of two surfaces S_1 and S_2 joined along a smooth simple closed curve C. Can anything be said about

$$\iint_S \nabla \times \mathbf{F} \cdot \mathbf{n}\, d\sigma?$$

Give reasons for your answer.

Stokes' Theorem for Parametrized Surfaces

In Exercises 13–18, use the surface integral in Stokes' Theorem to calculate the flux of the curl of the field $\mathbf{F}$ across the surface S in the direction of the outward unit normal $\mathbf{n}$.

13. $\mathbf{F} = 2z\mathbf{i} + 3x\mathbf{j} + 5y\mathbf{k}$

S: $\mathbf{r}(r, \theta) = (r \cos \theta)\mathbf{i} + (r \sin \theta)\mathbf{j} + (4 - r^2)\mathbf{k}$, $0 \le r \le 2$, $0 \le \theta \le 2\pi$

14. $\mathbf{F} = (y - z)\mathbf{i} + (z - x)\mathbf{j} + (x + z)\mathbf{k}$

S: $\mathbf{r}(r, \theta) = (r \cos \theta)\mathbf{i} + (r \sin \theta)\mathbf{j} + (9 - r^2)\mathbf{k}$, $0 \le r \le 3$, $0 \le \theta \le 2\pi$

15. $\mathbf{F} = x^2y\mathbf{i} + 2y^3z\mathbf{j} + 3z\mathbf{k}$

S: $\mathbf{r}(r, \theta) = (r \cos \theta)\mathbf{i} + (r \sin \theta)\mathbf{j} + r\mathbf{k}$, $0 \le r \le 1$, $0 \le \theta \le 2\pi$

16. $\mathbf{F} = (x - y)\mathbf{i} + (y - z)\mathbf{j} + (z - x)\mathbf{k}$

S: $\mathbf{r}(r, \theta) = (r \cos \theta)\mathbf{i} + (r \sin \theta)\mathbf{j} + (5 - r)\mathbf{k}$, $0 \le r \le 5$, $0 \le \theta \le 2\pi$

17. $\mathbf{F} = 3y\mathbf{i} + (5 - 2x)\mathbf{j} + (z^2 - 2)\mathbf{k}$

S: $\mathbf{r}(\phi, \theta) = (\sqrt{3} \sin \phi \cos \theta)\mathbf{i} + (\sqrt{3} \sin \phi \sin \theta)\mathbf{j} + (\sqrt{3} \cos \phi)\mathbf{k}$, $0 \le \phi \le \pi/2$, $0 \le \theta \le 2\pi$

18. $\mathbf{F} = y^2\mathbf{i} + z^2\mathbf{j} + x\mathbf{k}$

S: $\mathbf{r}(\phi, \theta) = (2 \sin \phi \cos \theta)\mathbf{i} + (2 \sin \phi \sin \theta)\mathbf{j} + (2 \cos \phi)\mathbf{k}$, $0 \le \phi \le \pi/2$, $0 \le \theta \le 2\pi$

Theory and Examples

19. *Zero circulation* Use the identity $\nabla \times \nabla f = \mathbf{0}$ (Equation (8) in the text) and Stokes' Theorem to show that the circulations of the following fields around the boundary of any smooth orientable surface in space are zero.

(a) $\mathbf{F} = 2x\mathbf{i} + 2y\mathbf{j} + 2z\mathbf{k}$

(b) $\mathbf{F} = \nabla(xy^2z^3)$

(c) $\mathbf{F} = \nabla \times (x\mathbf{i} + y\mathbf{j} + z\mathbf{k})$

(d) $\mathbf{F} = \nabla f$

20. *Zero circulation* Let $f(x, y, z) = (x^2 + y^2 + z^2)^{-1/2}$. Show that the clockwise circulation of the field $\mathbf{F} = \nabla f$ around the circle $x^2 + y^2 = a^2$ in the xy-plane is zero

(a) By taking $\mathbf{r} = (a \cos t)\mathbf{i} + (a \sin t)\mathbf{j}$, $0 \le t \le 2\pi$, and integrating $\mathbf{F} \cdot d\mathbf{r}$ over the circle

(b) By applying Stokes' Theorem.

21. Let C be a simple closed smooth curve in the plane $2x + 2y + z = 2$, oriented as shown here. Show that

$$\oint_C 2y\, dx + 3z\, dy - x\, dz$$

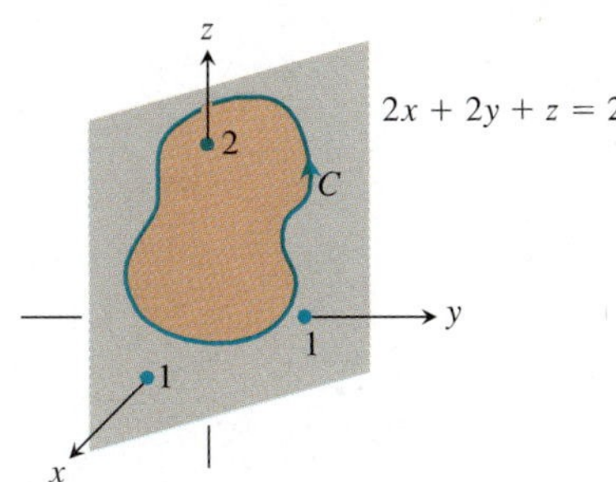

depends only on the area of the region enclosed by C and not on the position or shape of C.

22. Show that if $\mathbf{F} = x\mathbf{i} + y\mathbf{j} + z\mathbf{k}$, then $\nabla \times \mathbf{F} = \mathbf{0}$.

23. Find a vector field with twice-differentiable components whose curl is $x\mathbf{i} + y\mathbf{j} + z\mathbf{k}$ or prove that no such field exists.

24. *Writing to Learn* Does Stokes' Theorem say anything special about circulation in a field whose curl is zero? Give reasons for your answer.

25. Let R be a region in the xy-plane that is bounded by a piecewise smooth simple closed curve C and suppose that the moments of inertia of R about the x- and y-axes are known to be I_x and I_y. Evaluate the integral

$$\oint_C \nabla(r^4) \cdot \mathbf{n}\, ds,$$

where $r = \sqrt{x^2 + y^2}$, in terms of I_x and I_y.

26. *Zero curl, yet not conservative* Show that the curl of

$$\mathbf{F} = \frac{-y}{x^2 + y^2}\mathbf{i} + \frac{x}{x^2 + y^2}\mathbf{j} + z\mathbf{k}$$

is zero but that

$$\oint_C \mathbf{F} \cdot d\mathbf{r}$$

is not zero if C is the circle $x^2 + y^2 = 1$ in the xy-plane. (Theorem 6 does not apply here because the domain of $\mathbf{F}$ is not simply connected. The field $\mathbf{F}$ is not defined along the z-axis so there is no way to contract C to a point without leaving the domain of $\mathbf{F}$.)

13.8 Divergence Theorem and a Unified Theory

Divergence in Three Dimensions • Divergence Theorem • Proof of the Divergence Theorem for Special Regions • Divergence Theorem for Other Regions • Gauss's Law: One of the Four Great Laws of Electromagnetic Theory • Continuity Equation of Hydrodynamics • Unifying the Integral Theorems

The divergence form of Green's Theorem in the plane states that the net outward flux of a vector field across a simple closed curve can be calculated by integrating the divergence of the field over the region enclosed by the curve. The corresponding theorem in three dimensions, called the Divergence Theorem, states that the net outward flux of a vector field across a closed surface in space can be calculated by integrating the divergence of the field over the region enclosed by the surface. In this section, we prove the Divergence Theorem and show how it simplifies the calculation of flux. We also derive Gauss's law for flux in an electric field and the continuity equation of hydrodynamics. Finally, we unify the chapter's vector integral theorems into a single fundamental theorem.

Divergence in Three Dimensions

The **divergence** of a vector field $\mathbf{F} = M(x, y, z)\mathbf{i} + N(x, y, z)\mathbf{j} + P(x, y, z)\mathbf{k}$ is the scalar function

$$\text{div } \mathbf{F} = \nabla \cdot \mathbf{F} = \frac{\partial M}{\partial x} + \frac{\partial N}{\partial y} + \frac{\partial P}{\partial z}. \quad (1)$$

The symbol "div $\mathbf{F}$" is read as "divergence of $\mathbf{F}$" or "div $\mathbf{F}$." The notation $\nabla \cdot \mathbf{F}$ is read "del dot $\mathbf{F}$."

Div $\mathbf{F}$ has the same physical interpretation in three dimensions that it does in two. If $\mathbf{F}$ is the velocity field of a fluid flow, the value of div $\mathbf{F}$ at a point (x, y, z) is the rate at which fluid is being piped in or drained away at (x, y, z). The divergence is the flux per unit volume or flux density at the point.

Example 1 Finding Divergence

Find the divergence of $\mathbf{F} = 2xz\mathbf{i} - xy\mathbf{j} - z\mathbf{k}$.

Solution The divergence of $\mathbf{F}$ is

$$\nabla \cdot \mathbf{F} = \frac{\partial}{\partial x}(2xz) + \frac{\partial}{\partial y}(-xy) + \frac{\partial}{\partial z}(-z) = 2z - x - 1.$$

CD-ROM
WEBsite
Historical Biography
Mikhail Vasilievich Ostrogradsky
(1801 — 1862)

Divergence Theorem

The Divergence Theorem says that under suitable conditions, the outward flux of a vector field across a closed surface (oriented outward) equals the triple integral of the divergence of the field over the region enclosed by the surface.

CD-ROM
WEBsite

Theorem 7 Divergence Theorem

The flux of a vector field $\mathbf{F}$ across a closed oriented surface S in the direction of the surface's outward unit normal field $\mathbf{n}$ equals the integral of $\nabla \cdot \mathbf{F}$ over the region D enclosed by the surface:

$$\underbrace{\iint_S \mathbf{F} \cdot \mathbf{n}\, d\sigma}_{\text{Outward flux}} = \underbrace{\iiint_D \nabla \cdot \mathbf{F}\, dV}_{\text{Divergence integral}}. \tag{2}$$

Example 2 Supporting the Divergence Theorem

Evaluate both sides of Equation (2) for the field $\mathbf{F} = x\mathbf{i} + y\mathbf{j} + z\mathbf{k}$ over the sphere $x^2 + y^2 + z^2 = a^2$.

Solution The outer unit normal to S, calculated from the gradient of $f(x, y, z) = x^2 + y^2 + z^2 - a^2$, is

$$\mathbf{n} = \frac{2(x\mathbf{i} + y\mathbf{j} + z\mathbf{k})}{\sqrt{4(x^2 + y^2 + z^2)}} = \frac{x\mathbf{i} + y\mathbf{j} + z\mathbf{k}}{a}.$$

Hence,

$$\mathbf{F} \cdot \mathbf{n}\, d\sigma = \frac{x^2 + y^2 + z^2}{a}\, d\sigma = \frac{a^2}{a}\, d\sigma = a\, d\sigma$$

because $x^2 + y^2 + z^2 = a^2$ on the surface. Therefore,

$$\iint_S \mathbf{F} \cdot \mathbf{n}\, d\sigma = \iint_S a\, d\sigma = a \iint_S d\sigma = a(4\pi a^2) = 4\pi a^3.$$

The divergence of $\mathbf{F}$ is

$$\nabla \cdot \mathbf{F} = \frac{\partial}{\partial x}(x) + \frac{\partial}{\partial y}(y) + \frac{\partial}{\partial z}(z) = 3,$$

so

$$\iiint_D \nabla \cdot \mathbf{F}\, dV = \iiint_D 3\, dV = 3\left(\frac{4}{3}\pi a^3\right) = 4\pi a^3.$$

Example 3 Finding Flux

Find the flux of $\mathbf{F} = xy\mathbf{i} + yz\mathbf{j} + xz\mathbf{k}$ outward through the surface of the cube cut from the first octant by the planes $x = 1$, $y = 1$, and $z = 1$.

Solution Instead of calculating the flux as a sum of six separate integrals, one for each face of the cube, we can calculate the flux by integrating the divergence

$$\nabla \cdot \mathbf{F} = \frac{\partial}{\partial x}(xy) + \frac{\partial}{\partial y}(yz) + \frac{\partial}{\partial z}(xz) = y + z + x$$

over the cube's interior:

$$\text{Flux} = \iint_{\substack{\text{Cube}\\\text{surface}}} \mathbf{F} \cdot \mathbf{n}\, d\sigma = \iiint_{\substack{\text{Cube}\\\text{interior}}} \nabla \cdot \mathbf{F}\, dV \qquad \text{The Divergence Theorem}$$

$$= \int_0^1 \int_0^1 \int_0^1 (x + y + z)\, dx\, dy\, dz = \frac{3}{2}. \qquad \text{Routine integration}$$

Proof of the Divergence Theorem for Special Regions

To prove the Divergence Theorem, we assume that the components of $\mathbf{F}$ have continuous first partial derivatives. We also assume that D is a convex region with no holes or bubbles, such as a solid sphere, cube, or ellipsoid, and that S is a piecewise smooth surface. In addition, we assume that any line perpendicular to the xy-plane at an interior point of the region R_{xy} that is the projection of D on the xy-plane intersects the surface S in exactly two points, producing surfaces

$$S_1: \quad z = f_1(x, y), \quad (x, y) \text{ in } R_{xy}$$
$$S_2: \quad z = f_2(x, y), \quad (x, y) \text{ in } R_{xy},$$

with $f_1 \le f_2$. We make similar assumptions about the projection of D onto the other coordinate planes. See Figure 13.70.

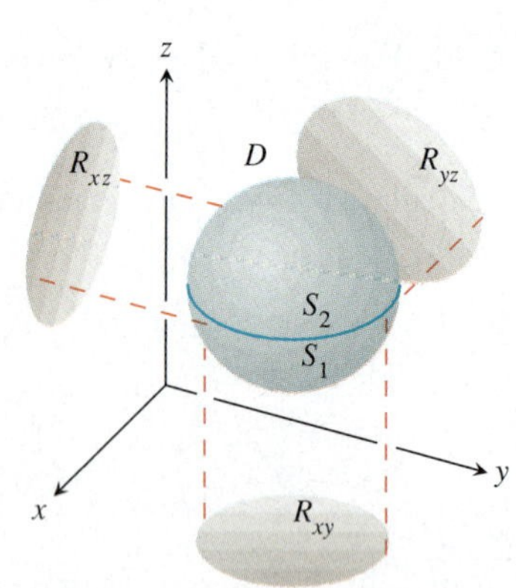

FIGURE 13.70 We first prove the Divergence Theorem for the kind of three-dimensional region shown here. We then extend the theorem to other regions.

The components of the unit normal vector $\mathbf{n} = n_1\mathbf{i} + n_2\mathbf{j} + n_3\mathbf{k}$ are the cosines of the angles α, β, and γ that $\mathbf{n}$ makes with $\mathbf{i}$, $\mathbf{j}$, and $\mathbf{k}$ (Figure 13.71). This is true because all the vectors involved are unit vectors. We have

$$n_1 = \mathbf{n} \cdot \mathbf{i} = |\mathbf{n}||\mathbf{i}| \cos \alpha = \cos \alpha$$
$$n_2 = \mathbf{n} \cdot \mathbf{j} = |\mathbf{n}||\mathbf{j}| \cos \beta = \cos \beta$$
$$n_3 = \mathbf{n} \cdot \mathbf{k} = |\mathbf{n}||\mathbf{k}| \cos \gamma = \cos \gamma.$$

Thus,

$$\mathbf{n} = (\cos \alpha)\mathbf{i} + (\cos \beta)\mathbf{j} + (\cos \gamma)\mathbf{k}$$

and

$$\mathbf{F} \cdot \mathbf{n} = M \cos \alpha + N \cos \beta + P \cos \gamma.$$

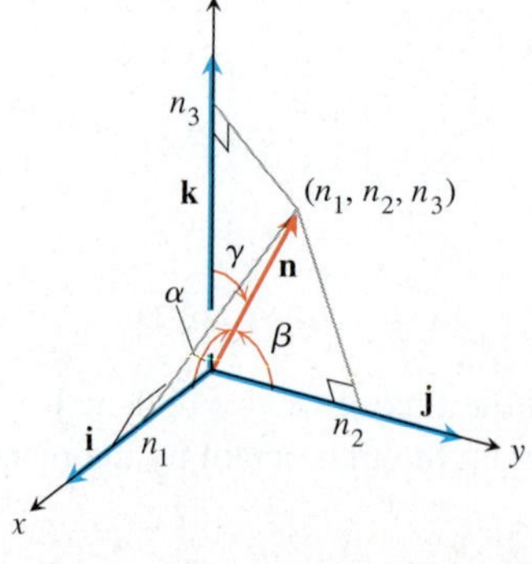

FIGURE 13.71 The scalar components of the unit normal vector $\mathbf{n}$ are the cosines of the angles α, β, and γ that it makes with $\mathbf{i}$, $\mathbf{j}$, and $\mathbf{k}$.

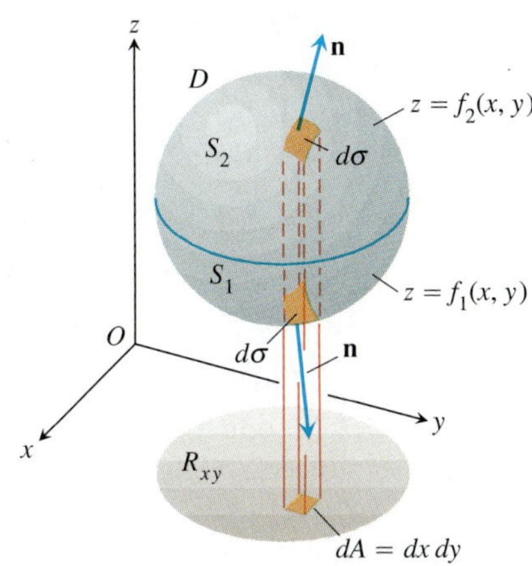

FIGURE 13.72 The three-dimensional region D enclosed by the surfaces S_1 and S_2 shown here projects vertically onto a two-dimensional region R_{xy} in the xy-plane.

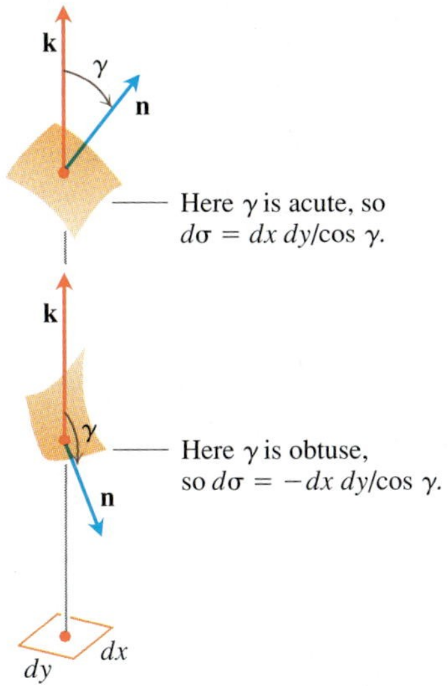

FIGURE 13.73 An enlarged view of the area patches in Figure 13.72. The relations $d\sigma = \pm dx\, dy/\cos\gamma$ are derived in Section 13.5.

In component form, the Divergence Theorem states that

$$\iint_S (M\cos\alpha + N\cos\beta + P\cos\gamma)d\sigma = \iiint_D \left(\frac{\partial M}{\partial x} + \frac{\partial N}{\partial y} + \frac{\partial P}{\partial z}\right) dx\, dy\, dz.$$

We prove the theorem by proving the three following equalities:

$$\iint_S M\cos\alpha\, d\sigma = \iiint_D \frac{\partial M}{\partial x}\, dx\, dy\, dz \tag{3}$$

$$\iint_S N\cos\beta\, d\sigma = \iiint_D \frac{\partial N}{\partial y}\, dx\, dy\, dz \tag{4}$$

$$\iint_S P\cos\gamma\, d\sigma = \iiint_D \frac{\partial P}{\partial z}\, dx\, dy\, dz \tag{5}$$

We prove Equation (5) by converting the surface integral on the left to a double integral over the projection R_{xy} of D on the xy-plane (Figure 13.72). The surface S consists of an upper part S_2 whose equation is $z = f_2(x, y)$ and a lower part S_1 whose equation is $z = f_1(x, y)$. On S_2, the outer normal **n** has a positive **k**-component and

$$\cos\gamma\, d\sigma = dx\, dy \qquad \text{because} \qquad d\sigma = \frac{dA}{|\cos\gamma|} = \frac{dx\, dy}{\cos\gamma}.$$

See Figure 13.73. On S_1, the outer normal **n** has a negative **k**-component and

$$\cos\gamma\, d\sigma = -dx\, dy.$$

Therefore,

$$\begin{aligned}
\iint_S P\cos\gamma\, d\sigma &= \iint_{S_2} P\cos\gamma\, d\sigma + \iint_{S_1} P\cos\gamma\, d\sigma \\
&= \iint_{R_{xy}} P(x, y, f_2(x, y))\, dx\, dy - \iint_{R_{xy}} P(x, y, f_1(x, y))\, dx\, dy \\
&= \iint_{R_{xy}} [P(x, y, f_2(x, y)) - P(x, y, f_1(x, y))]\, dx\, dy \\
&= \iint_{R_{xy}} \left[\int_{f_1(x,y)}^{f_2(x, y)} \frac{\partial P}{\partial z}\, dz\right] dx\, dy = \iiint_D \frac{\partial P}{\partial z}\, dz\, dx\, dy.
\end{aligned}$$

This proves Equation (5).

The proofs for Equations (3) and (4) follow the same pattern; or just permute x, y, z; M, N, P; α, β, γ, in order, and get those results from Equation (5).

Divergence Theorem for Other Regions

The Divergence Theorem can be extended to regions that can be partitioned into a finite number of simple regions of the type just discussed and to regions that can be defined as limits of simpler regions in certain ways. For example, suppose that D is the region between two concentric spheres and the **F** has continuously differen-

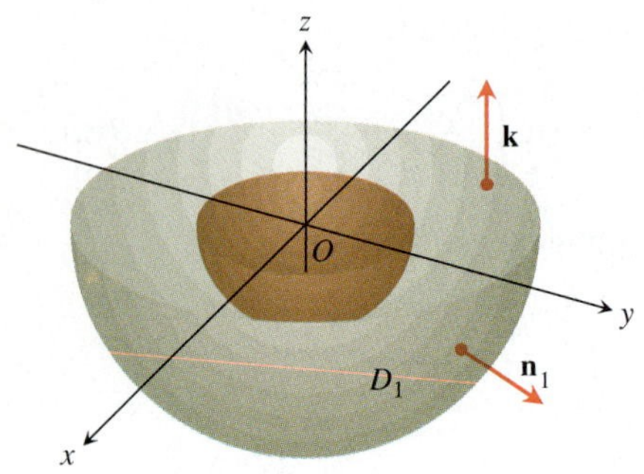

FIGURE 13.74 The lower half of the solid region between two concentric spheres.

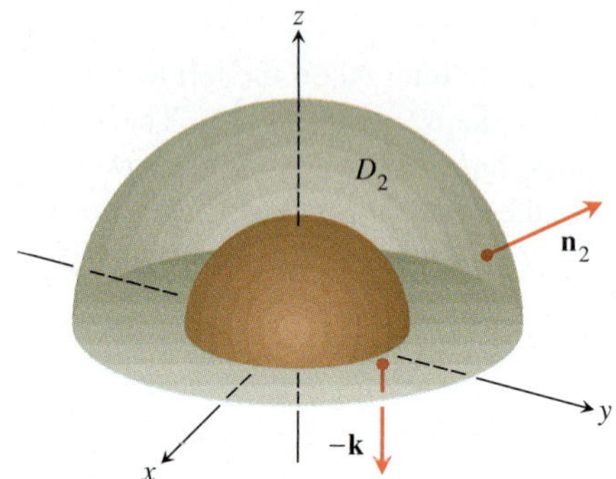

FIGURE 13.75 The upper half of the solid region between two concentric spheres.

tiable components throughout D and on the bounding surfaces. Split D by an equatorial plane and apply the Divergence Theorem to each half separately. The bottom half, D_1, is shown in Figure 13.74. The surface S_1 that bounds D_1 consists of an outer hemisphere, a plane washer-shaped base, and an inner hemisphere. The Divergence Theorem says that

$$\iint_{S_1} \mathbf{F} \cdot \mathbf{n}_1 \, d\sigma_1 = \iiint_{D_1} \nabla \cdot \mathbf{F} \, dV_1. \tag{6}$$

The unit normal $\mathbf{n}_1$ that points outward from D_1 points away from the origin along the outer surface, equals $\mathbf{k}$ along the flat base, and points toward the origin along the inner surface. Next apply the Divergence Theorem to D_2, and its surface S_2 (Figure 13.75):

$$\iint_{S_2} \mathbf{F} \cdot \mathbf{n}_2 \, d\sigma_2 = \iiint_{D_2} \nabla \cdot \mathbf{F} \, dV_2. \tag{7}$$

As we follow $\mathbf{n}_2$ over S_2, pointing outward from D_2, we see that $\mathbf{n}_2$ equals $-\mathbf{k}$ along the washer-shaped base in the xy-plane, points away from the origin on the outer sphere, and points toward the origin on the inner sphere. When we add Equations (6) and (7), the integrals over the flat base cancel because of the opposite signs of $\mathbf{n}_1$ and $\mathbf{n}_2$. We thus arrive at the result

$$\iint_{S} \mathbf{F} \cdot \mathbf{n} \, d\sigma = \iiint_{D} \nabla \cdot \mathbf{F} \, dV,$$

with D the region between the spheres, S the boundary of D consisting of two spheres, and $\mathbf{n}$ the unit normal to S directed outward from D.

Example 4 Finding Outward Flux

Find the net outward flux of the field

$$\mathbf{F} = \frac{x\mathbf{i} + y\mathbf{j} + z\mathbf{k}}{\rho^3}, \qquad \rho = \sqrt{x^2 + y^2 + z^2}$$

across the boundary of the region $D: 0 < a^2 \le x^2 + y^2 + z^2 \le b^2$.

Solution The flux can be calculated by integrating $\nabla \cdot \mathbf{F}$ over D. We have

$$\frac{\partial \rho}{\partial x} = \frac{1}{2}(x^2 + y^2 + z^2)^{-1/2}(2x) = \frac{x}{\rho}$$

and

$$\frac{\partial M}{\partial x} = \frac{\partial}{\partial x}(x\rho^{-3}) = \rho^{-3} - 3x\rho^{-4}\frac{\partial \rho}{\partial x} = \frac{1}{\rho^3} - \frac{3x^2}{\rho^5}.$$

Similarly,

$$\frac{\partial N}{\partial y} = \frac{1}{\rho^3} - \frac{3y^2}{\rho^5} \quad \text{and} \quad \frac{\partial P}{\partial z} = \frac{1}{\rho^3} - \frac{3z^2}{\rho^5}.$$

Hence,

$$\text{div}\,\mathbf{F} = \frac{3}{\rho^3} - \frac{3}{\rho^5}(x^2 + y^2 + z^2) = \frac{3}{\rho^3} - \frac{3\rho^2}{\rho^5} = 0$$

and

$$\iiint_D \nabla \cdot \mathbf{F}\, dV = 0.$$

So the integral of $\nabla \cdot \mathbf{F}$ over D is zero and the net outward flux across the boundary of D is zero. There is more to learn from this example, though. The flux leaving D across the inner sphere S_a is the negative of the flux leaving D across the outer sphere S_b (because the sum of these fluxes is zero). Hence, the flux of $\mathbf{F}$ across S_a in the direction away from the origin equals the flux of $\mathbf{F}$ across S_b in the direction away from the origin. Thus, the flux of $\mathbf{F}$ across a sphere centered at the origin is independent of the radius of the sphere. What is this flux?

To find it, we evaluate the flux integral directly. The outward unit normal on the sphere of radius a is

$$\mathbf{n} = \frac{x\mathbf{i} + y\mathbf{j} + z\mathbf{k}}{\sqrt{x^2 + y^2 + z^2}} = \frac{x\mathbf{i} + y\mathbf{j} + z\mathbf{k}}{a}.$$

Hence, on the sphere,

$$\mathbf{F} \cdot \mathbf{n} = \frac{x\mathbf{i} + y\mathbf{j} + z\mathbf{k}}{a^3} \cdot \frac{x\mathbf{i} + y\mathbf{j} + z\mathbf{k}}{a} = \frac{x^2 + y^2 + z^2}{a^4} = \frac{a^2}{a^4} = \frac{1}{a^2}$$

and

$$\iint_{S_a} \mathbf{F} \cdot \mathbf{n}\, d\sigma = \frac{1}{a^2} \iint_{S_a} d\sigma = \frac{1}{a^2}(4\pi a^2) = 4\pi.$$

The outward flux of $\mathbf{F}$ across any sphere centered at the origin is 4π.

Gauss's Law: One of the Four Great Laws of Electromagnetic Theory

There is still more to be learned from Example 4. In electromagnetic theory, the electric field created by a point charge q located at the origin is

$$\mathbf{E}(x, y, z) = \frac{1}{4\pi\epsilon_0} \frac{q}{|\mathbf{r}|^2}\left(\frac{\mathbf{r}}{|\mathbf{r}|}\right) = \frac{q}{4\pi\epsilon_0} \frac{\mathbf{r}}{|\mathbf{r}|^3} = \frac{q}{4\pi\epsilon_0} \frac{x\mathbf{i} + y\mathbf{j} + z\mathbf{k}}{\rho^3},$$

where ϵ_0 is a physical constant, $\mathbf{r}$ is the position vector of the point (x, y, z), and $\rho = |\mathbf{r}| = \sqrt{x^2 + y^2 + z^2}$. In the notation of Example 4,

$$\mathbf{E} = \frac{q}{4\pi\epsilon_0}\mathbf{F}.$$

The calculations in Example 4 show that the outward flux of $\mathbf{E}$ across any sphere centered at the origin is q/ϵ_0, but this result is not confined to spheres. The outward flux of $\mathbf{E}$ across any closed surface S that encloses the origin (and to which the Divergence Theorem applies) is also q/ϵ_0. To see why, we have only to imagine a large sphere S_a centered at the origin and enclosing the surface S. Since

$$\nabla \cdot \mathbf{E} = \nabla \cdot \frac{q}{4\pi\epsilon_0}\mathbf{F} = \frac{q}{4\pi\epsilon_0} \nabla \cdot \mathbf{F} = 0$$

when $\rho > 0$, the integral of $\nabla \cdot \mathbf{E}$ over the region D between S and S_a is zero. Hence, by the Divergence Theorem,

$$\iint_{\substack{\text{Boundary}\\ \text{of } D}} \mathbf{E} \cdot \mathbf{n}\, d\sigma = 0,$$

and the flux of $\mathbf{E}$ across S in the direction away from the origin must be the same as the flux of $\mathbf{E}$ across S_a in the direction away from the origin, which is q/ϵ_0. This statement, called *Gauss's law*, also applies to charge distributions that are more general than the one assumed here, as you will see in nearly any physics text.

$$\text{Gauss's law:} \qquad \iint_S \mathbf{E} \cdot \mathbf{n}\, d\sigma = \frac{q}{\epsilon_0}$$

Continuity Equation of Hydrodynamics

Let D be a region in space bounded by a closed oriented surface S. If $\mathbf{v}(x, y, z)$ is the velocity field of a fluid flowing smoothly through D, $\delta = \delta(t, x, y, z)$ is the fluid's density at (x, y, z) at time t, and $\mathbf{F} = \delta\mathbf{v}$, then the **continuity equation** of hydrodynamics states that

$$\nabla \cdot \mathbf{F} + \frac{\partial \delta}{\partial t} = 0.$$

If the functions involved have continuous first partial derivatives, the equation evolves naturally from the Divergence Theorem, as we now see.

First, the integral

$$\iint_S \mathbf{F} \cdot \mathbf{n}\, d\sigma$$

is the rate at which mass leaves D across S (leaves because $\mathbf{n}$ is the outer normal). To see why, consider a patch of area $\Delta\sigma$ on the surface (Figure 13.76). In a short time interval Δt, the volume ΔV of fluid that flows across the patch is approximately equal to the volume of a cylinder with base area $\Delta\sigma$ and height $(\mathbf{v}\,\Delta t) \cdot \mathbf{n}$, where $\mathbf{v}$ is a velocity vector rooted at a point of the patch:

$$\Delta V \approx \mathbf{v} \cdot \mathbf{n}\, \Delta\sigma\, \Delta t.$$

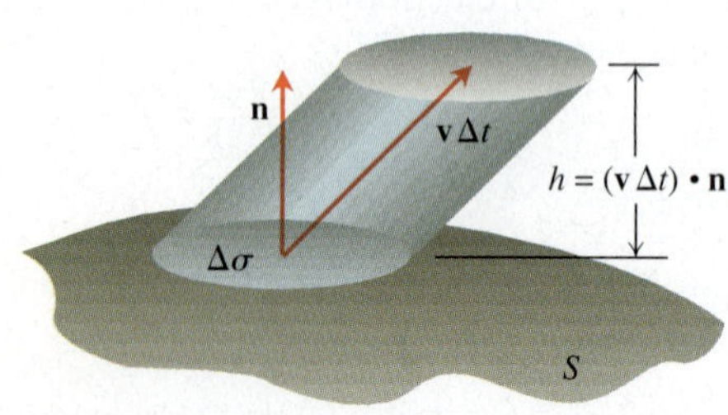

FIGURE 13.76 The fluid that flows upward through the patch $\Delta\sigma$ in a short time Δt fills a "cylinder" whose volume is approximately base $\times$ height $=$ $\mathbf{v} \cdot \mathbf{n}\, \Delta\sigma\, \Delta t$.

The mass of this volume of fluid is about

$$\Delta m \approx \delta\mathbf{v} \cdot \mathbf{n}\, \Delta\sigma\, \Delta t,$$

so the rate at which mass is flowing out of D across the patch is about

$$\frac{\Delta m}{\Delta t} \approx \delta\mathbf{v} \cdot \mathbf{n}\, \Delta\sigma.$$

This leads to the approximation

$$\frac{\Sigma\, \Delta m}{\Delta t} \approx \Sigma\, \delta\mathbf{v} \cdot \mathbf{n}\, \Delta\sigma$$

as an estimate of the average rate at which mass flows across S. Finally, letting $\Delta\sigma \to 0$ and $\Delta t \to 0$ gives the instantaneous rate at which mass leaves D across S as

$$\frac{dm}{dt} = \iint_S \delta\mathbf{v} \cdot \mathbf{n}\, d\sigma,$$

which for our particular flow is

$$\frac{dm}{dt} = \iint_S \mathbf{F} \cdot \mathbf{n}\, d\sigma,$$

Now let B be a solid sphere centered at a point Q in the flow. The average value of $\nabla \cdot \mathbf{F}$ over B is

$$\frac{1}{\text{volume of } B} \iiint_B \nabla \cdot \mathbf{F}\, dV.$$

It is a consequence of the continuity of the divergence that $\nabla \cdot \mathbf{F}$ actually takes on this value at some point P in B. Thus,

$$\begin{aligned}(\nabla \cdot \mathbf{F})_P &= \frac{1}{\text{volume of } B} \iiint_B \nabla \cdot \mathbf{F}\, dV = \frac{\iint_S \mathbf{F} \cdot \mathbf{n}\, d\sigma}{\text{volume of } B} \\ &= \frac{\text{rate at which mass leaves } B \text{ across its surface } S}{\text{volume of } B}. \qquad (8)\end{aligned}$$

The fraction on the right describes decrease in mass per unit volume.

Now let the radius of B approach zero while the center Q stays fixed. The left side of Equation (8) converges to $(\nabla \cdot \mathbf{F})_Q$, the right side to $(-\partial\delta/\partial t)_Q$. The equality of these two limits is the continuity equation

$$\nabla \cdot \mathbf{F} = -\frac{\partial\delta}{\partial t}.$$

The continuity equation "explains" $\nabla \cdot \mathbf{F}$: The divergence of $\mathbf{F}$ at a point is the rate at which the density of the fluid is decreasing there.

The Divergence Theorem

$$\iint_S \mathbf{F} \cdot \mathbf{n}\, d\sigma = \iiint_D \nabla \cdot \mathbf{F}\, dV$$

now says that the net decrease in density of the fluid in region D is accounted for by the mass transported across the surface S. So, the theorem is a statement about conservation of mass (Exercise 31).

Unifying the Integral Theorems

If we think of a two-dimensional field $\mathbf{F} = M(x, y)\mathbf{i} + N(x, y)\mathbf{j}$ as a three-dimensional field whose $\mathbf{k}$-component is zero, then $\nabla \cdot \mathbf{F} = (\partial M/\partial x) + (\partial N/\partial y)$ and the normal form of Green's Theorem can be written as

$$\oint_C \mathbf{F} \cdot \mathbf{n}\, ds = \iint_R \left(\frac{\partial M}{\partial x} + \frac{\partial N}{\partial y}\right) dx\, dy = \iint_R \nabla \cdot \mathbf{F}\, dA.$$

Similarly, $\nabla \times \mathbf{F} \cdot \mathbf{k} = (\partial N/\partial x) - (\partial M/\partial y)$, so the tangential form of Green's Theorem can be written as

$$\oint_C \mathbf{F} \cdot d\mathbf{r} = \iint_R \left(\frac{\partial N}{\partial x} - \frac{\partial M}{\partial y} \right) dx\, dy = \iint_R \nabla \times \mathbf{F} \cdot \mathbf{k}\, dA.$$

With the equations of Green's Theorem now in del notation, we can see their relationships to the equations in Stokes' Theorem and the Divergence Theorem.

Green's Theorem and Its Generalization to Three Dimensions

Normal form of Green's Theorem: $\oint_C \mathbf{F} \cdot \mathbf{n}\, ds = \iint_R \nabla \cdot \mathbf{F}\, dA$

Divergence Theorem: $\iint_S \mathbf{F} \cdot \mathbf{n}\, d\sigma = \iiint_D \nabla \cdot \mathbf{F}\, dV$

Tangential form of Green's Theorem: $\oint_C \mathbf{F} \cdot d\mathbf{r} = \iint_R \nabla \times \mathbf{F} \cdot \mathbf{k}\, dA$

Stokes' Theorem: $\oint_C \mathbf{F} \cdot d\mathbf{r} = \iint_S \nabla \times \mathbf{F} \cdot \mathbf{n}\, d\sigma$

Notice how Stokes' Theorem generalizes the tangential (curl) form of Green's Theorem from a flat surface in the plane to a surface in three-dimensional space. In each case, the integral of the normal component of curl $\mathbf{F}$ over the interior of the surface equals the circulation of $\mathbf{F}$ around the boundary.

Likewise, the Divergence Theorem generalizes the normal (flux) form of Green's Theorem from a two-dimensional region in the plane to a three-dimensional region in space. In each case, the integral of $\nabla \cdot \mathbf{F}$ over the interior of the region equals the total flux of the field across the boundary.

There is still more to be learned here. All these results can be thought of as forms of a *single fundamental theorem.* Think back to the Fundamental Theorem of Calculus in Section 4.5. It says that if $f(x)$ is differentiable on $[a, b]$, then

$$\int_a^b \frac{df}{dx}\, dx = f(b) - f(a).$$

If we let $\mathbf{F} = f(x)\mathbf{i}$ throughout $[a, b]$, then $(df/dx) = \nabla \cdot \mathbf{F}$. If we define the unit vector field $\mathbf{n}$ normal to the boundary of $[a, b]$ to be $\mathbf{i}$ at b and $-\mathbf{i}$ at a (Figure 13.77), then

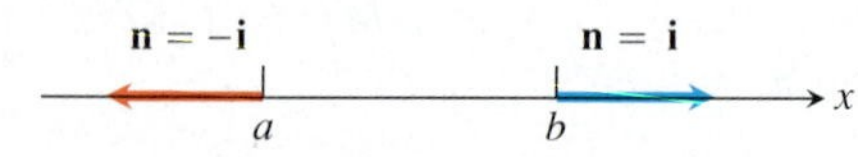

FIGURE 13.77 The outward unit normals at the boundary of $[a, b]$ in one-dimensional space.

$$\begin{aligned} f(b) - f(a) &= f(b)\mathbf{i} \cdot (\mathbf{i}) + f(a)\mathbf{i} \cdot (-\mathbf{i}) \\ &= \mathbf{F}(b) \cdot \mathbf{n} + \mathbf{F}(a) \cdot \mathbf{n} \\ &= \text{total outward flux of } \mathbf{F} \text{ across the boundary of } [a, b]. \end{aligned}$$

The Fundamental Theorem now says that

$$\mathbf{F}(b)\cdot\mathbf{n} + \mathbf{F}(a)\cdot\mathbf{n} = \int_{[a,b]} \nabla\cdot\mathbf{F}\,dx$$

The Fundamental Theorem of Calculus, the flux form of Green's Theorem, and the Divergence Theorem all say that the integral of the differential operator $\nabla\cdot$ operating on a field $\mathbf{F}$ over a region equals the sum of the normal field components over the boundary of the region. (Here we are interpreting the line integral in Green's Theorem and the surface integral in the Divergence Theorem as "sums" over the boundary.)

Stokes' Theorem and the circulation form of Green's Theorem say that, when things are properly oriented, the integral of the normal component of the curl operating on a field equals the sum of the tangential field components on the boundary of the surface.

The beauty of these interpretations is the observance of a marvelous underlying principle, which we might state as follows.

> The integral of a differential operator acting on a field over a region equals the sum of the field components appropriate to the operator over the boundary of the region.

EXERCISES 13.8

Calculating Divergence

In Exercises 1–4, find the divergence of the field.

1. The spin field in Figure 13.14.

2. The radial field in Figure 13.13

3. The gravitational field in Figure 13.12

4. The velocity field in Figure 13.9

Using the Divergence Theorem to Calculate Outward Flux

In Exercises 5–16, use the Divergence Theorem to find the outward flux of **F** across the boundary of the region D.

5. *Cube* $\mathbf{F} = (y - x)\mathbf{i} + (z - y)\mathbf{j} + (y - x)\mathbf{k}$

D: The cube bounded by the planes $x = \pm 1$, $y = \pm 1$, and $z = \pm 1$

6. $\mathbf{F} = x^2\mathbf{i} + y^2\mathbf{j} + z^2\mathbf{k}$

(a) *Cube* D: The cube cut from the first octant by the planes $x = 1$, $y = 1$, and $z = 1$

(b) *Cube* D: The cube bounded by the planes $x = \pm 1$, $y = \pm 1$, and $z = \pm 1$

(c) *Cylindrical can* D: The region cut from the solid cylinder $x^2 + y^2 \le 4$ by the planes $z = 0$ and $z = 1$

7. *Cylinder and paraboloid* $\mathbf{F} = y\mathbf{i} + xy\mathbf{j} - z\mathbf{k}$

D: The region inside the solid cylinder $x^2 + y^2 \le 4$ between the plane $z = 0$ and the paraboloid $z = x^2 + y^2$

8. *Sphere* $\mathbf{F} = x^2\mathbf{i} + xz\mathbf{j} + 3z\mathbf{k}$

D: The solid sphere $x^2 + y^2 + z^2 \le 4$

9. *Portion of sphere* $\mathbf{F} = x^2\mathbf{i} - 2xy\mathbf{j} + 3xz\mathbf{k}$

D: The region cut from the first octant by the sphere $x^2 + y^2 + z^2 = 4$

10. *Cylindrical can* $\mathbf{F} = (6x^2 + 2xy)\mathbf{i} + (2y + x^2z)\mathbf{j} + 4x^2y^3\mathbf{k}$

D: The region cut from the first octant by the cylinder $x^2 + y^2 = 4$ and the plane $z = 3$

11. *Wedge* $\mathbf{F} = 2xz\mathbf{i} - xy\mathbf{j} - z^2\mathbf{k}$

D: The wedge cut from the first octant by the plane $y + z = 4$ and the elliptical cylinder $4x^2 + y^2 = 16$

12. *Sphere* $\mathbf{F} = x^3\mathbf{i} + y^3\mathbf{j} + z^3\mathbf{k}$

D: The solid sphere $x^2 + y^2 + z^2 \le a^2$

13. *Thick sphere* $\mathbf{F} = \sqrt{x^2 + y^2 + z^2}\,(x\mathbf{i} + y\mathbf{j} + z\mathbf{k})$

D: The region $1 \le x^2 + y^2 + z^2 \le 2$

14. *Thick sphere* $\mathbf{F} = (x\mathbf{i} + y\mathbf{j} + z\mathbf{k})/\sqrt{x^2 + y^2 + z^2}$

D: The region $1 \le x^2 + y^2 + z^2 \le 4$

15. *Thick sphere* $\mathbf{F} = (5x^3 + 12xy^2)\mathbf{i} + (y^3 + e^y \sin z)\mathbf{j} + (5z^3 + e^y \cos z)\mathbf{k}$

D: The solid region between the spheres $x^2 + y^2 + z^2 = 1$ and $x^2 + y^2 + z^2 = 2$

16. *Thick cylinder* $\mathbf{F} = \ln(x^2 + y^2)\mathbf{i} - \left(\frac{2z}{x} \tan^{-1} \frac{y}{x}\right)\mathbf{j} + z\sqrt{x^2 + y^2}\,\mathbf{k}$

D: The thick-walled cylinder $1 \le x^2 + y^2 \le 2, \quad -1 \le z \le 2$

Properties of Curl and Divergence

17. *div (curl G) = 0*

(a) Show that if the necessary partial derivatives of the components of the field $\mathbf{G} = M\mathbf{i} + N\mathbf{j} + P\mathbf{k}$ are continuous, then $\nabla \cdot \nabla \times \mathbf{G} = 0$.

(b) *Writing to Learn* What, if anything, can you conclude about the flux of the field $\nabla \times \mathbf{G}$ across a closed surface? Give reasons for your answer.

18. *Identities* Let $\mathbf{F}_1$ and $\mathbf{F}_2$ be differentiable vector fields and let a and b be arbitrary real constants. Verify the following identities.

(a) $\nabla \cdot (a\mathbf{F}_1 + b\mathbf{F}_2) = a\nabla \cdot \mathbf{F}_1 + b\,\nabla \cdot \mathbf{F}_2$

(b) $\nabla \times (a\mathbf{F}_1 + b\mathbf{F}_2) = a\nabla \times \mathbf{F}_1 + b\,\nabla \times \mathbf{F}_2$

(c) $\nabla \cdot (\mathbf{F}_1 \times \mathbf{F}_2) = \mathbf{F}_2 \cdot \nabla \times \mathbf{F}_1 - \mathbf{F}_1 \cdot \nabla \times \mathbf{F}_2$

19. *Identities* Let $\mathbf{F}$ be a differentiable vector field and let $g(x, y, z)$ be a differentiable scalar function. Verify the following identities.

(a) $\nabla \cdot (g\mathbf{F}) = g\nabla \cdot \mathbf{F} + \nabla g \cdot \mathbf{F}$

(b) $\nabla \times (g\mathbf{F}) = g\nabla \times \mathbf{F} + \nabla g \times \mathbf{F}$

20. *Identities* If $\mathbf{F} = M\mathbf{i} + N\mathbf{j} + P\mathbf{k}$ is a differentiable vector field, we define the notation $\mathbf{F} \cdot \nabla$ to mean

$$M\frac{\partial}{\partial x} + N\frac{\partial}{\partial y} + P\frac{\partial}{\partial z}.$$

For differentiable vector fields $\mathbf{F}_1$ and $\mathbf{F}_2$, verify the following identities.

(a) $\nabla \times (\mathbf{F}_1 \times \mathbf{F}_2) = (\mathbf{F}_2 \cdot \nabla)\mathbf{F}_1 - (\mathbf{F}_1 \cdot \nabla)\mathbf{F}_2 + (\nabla \cdot \mathbf{F}_2)\mathbf{F}_1 - (\nabla \cdot \mathbf{F}_1)\mathbf{F}_2$

(b) $\nabla(\mathbf{F}_1 \cdot \mathbf{F}_2) = (\mathbf{F}_1 \cdot \nabla)\mathbf{F}_2 + (\mathbf{F}_2 \cdot \nabla)\mathbf{F}_1 + \mathbf{F}_1 \times (\nabla \times \mathbf{F}_2) + \mathbf{F}_2 \times (\nabla \times \mathbf{F}_1)$

Theory and Examples

21. *Writing to Learn: Bounding divergence* Let $\mathbf{F}$ be a field whose components have continuous first partial derivatives throughout a portion of space containing a region D bounded by a smooth closed surface S. If $|\mathbf{F}| \le 1$, can any bound be placed on the size of

$$\iiint_D \nabla \cdot \mathbf{F}\, dV?$$

Give reasons for your answer.

22. *Writing to Learn: Flux of a position vector* The base of the closed cubelike surface shown here is the unit square in the xy-plane. The four sides lie in the planes $x = 0$, $x = 1$, $y = 0$, and $y = 1$. The top is an arbitrary smooth surface whose identity is unknown. Let $\mathbf{F} = x\mathbf{i} - 2y\mathbf{j} + (z + 3)\mathbf{k}$ and suppose the outward flux of $\mathbf{F}$ through side A is 1 and through side B is -3. Can you conclude anything about the outward flux through the top? Give reasons for your answer.

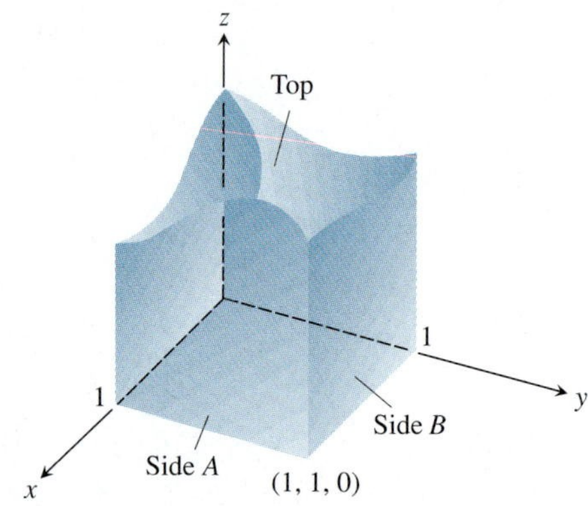

23. **(a)** *Flux of position vector* Show that the flux of the position vector field $\mathbf{F} = x\mathbf{i} + y\mathbf{j} + z\mathbf{k}$ outward through a smooth closed surface S is three times the volume of the region enclosed by the surface.

(b) Let $\mathbf{n}$ be the outward unit normal vector field on S. Show that it is not possible for $\mathbf{F}$ to be orthogonal to $\mathbf{n}$ at every point of S.

24. *Maximum flux* Among all rectangular solids defined by the inequalities $0 \le x \le a$, $0 \le y \le b$, $0 \le z \le 1$, find the one for which the total flux of $\mathbf{F} = (-x^2 - 4xy)\mathbf{i} - 6yz\mathbf{j} + 12z\mathbf{k}$ outward through the six sides is greatest. What *is* the greatest flux?

CD-ROM WEBsite

25. *Volume of a solid region* Let $\mathbf{F} = x\mathbf{i} + y\mathbf{j} + z\mathbf{k}$ and suppose that the surface S and region D satisfy the hypotheses of the Divergence Theorem. Show that the volume of D is given by the formula

$$\text{Volume of } D = \frac{1}{3}\iint_S \mathbf{F} \cdot \mathbf{n}\, d\sigma.$$

26. *Flux of a constant field* Show that the outward flux of a constant vector field $\mathbf{F} = \mathbf{C}$ across any closed surface to which the Divergence Theorem applies is zero.

27. *Harmonic functions* A function $f(x, y, z)$ is said to be **harmonic** in a region D in space if it satisfies the Laplace equation

$$\nabla^2 f = \nabla \cdot \nabla f = \frac{\partial^2 f}{\partial x^2} + \frac{\partial^2 f}{\partial y^2} + \frac{\partial^2 f}{\partial z^2} = 0$$

throughout D.

(a) Suppose that f is harmonic throughout a bounded region D enclosed by a smooth surface S and that $\mathbf{n}$ is the chosen unit normal vector on S. Show that the integral over S of $\nabla f \cdot \mathbf{n}$, the derivative of f in the direction of $\mathbf{n}$, is zero.

(b) Show that if f is harmonic on D, then

$$\iint_S f\,\nabla f\cdot\mathbf{n}\,d\sigma = \iiint_D |\nabla f|^2\,dV.$$

28. *Flux of a gradient field* Let S be the surface of the portion of the solid sphere $x^2 + y^2 + z^2 \le a^2$ that lies in the first octant and let $f(x, y, z) = \ln\sqrt{x^2 + y^2 + z^2}$. Calculate

$$\iint_S \nabla f\cdot\mathbf{n}\,d\sigma.$$

($\nabla f\cdot\mathbf{n}$ is the derivative of f in the direction of $\mathbf{n}$.)

29. *Green's first formula* Suppose that f and g are scalar functions with continuous first- and second-order partial derivatives throughout a region D that is bounded by a closed piecewise smooth surface S. Show that

$$\iint_S f\,\nabla g\cdot\mathbf{n}\,d\sigma = \iiint_D (f\,\nabla^2 g + \nabla f\cdot\nabla g)\,dV. \qquad (9)$$

Equation (9) is **Green's first formula.** (*Hint:* Apply the Divergence Theorem to the field $\mathbf{F} = f\,\nabla g$.)

30. *Green's second formula (Continuation of Exercise 29)* Interchange f and g in Equation (9) to obtain a similar formula. Then subtract this formula from Equation (9) to show that

$$\iint_S (f\,\nabla g - g\,\nabla f)\cdot\mathbf{n}\,d\sigma = \iiint_D (f\,\nabla^2 g - g\,\nabla^2 f)\,dV. \qquad (10)$$

This equation is **Green's second formula.**

31. *Conservation of mass* Let $\mathbf{v}(t, x, y, z)$ be a continuously differentiable vector field over the region D in space and let $p(t, x, y, z)$ be a continuously differentiable scalar function. The variable t represents the time domain. The Law of Conservation of Mass asserts that

$$\frac{d}{dt}\iiint_D p(t, x, y, z)\,dV = -\iint_S p\mathbf{v}\cdot\mathbf{n}\,d\sigma,$$

where S is the surface enclosing D.

(a) Give a physical interpretation of the conservation of mass law if $\mathbf{v}$ is a velocity flow field and p represents the density of the fluid at point (x, y, z) at time t.

(b) Use the Divergence Theorem and Leibniz's Rule,

$$\frac{d}{dt}\iiint_D p(t, x, y, z)\,dV = \iiint_D \frac{\partial p}{\partial t}\,dV,$$

to show that the Law of Conservation of Mass is equivalent to the continuity equation,

$$\nabla\cdot p\mathbf{v} + \frac{\partial p}{\partial t} = 0.$$

(In the first term $\nabla\cdot p\mathbf{v}$, the variable t is held fixed, and in the second term $\partial p/\partial t$, it is assumed that the point (x, y, z) in D is held fixed.)

32. *The heat diffusion equation* Let $T(t, x, y, z)$ be a function with continuous second derivatives giving the temperature at time t at the point (x, y, z) of a solid occupying a region D in space. If the solid's heat capacity and mass density are denoted by the constants c and ρ, respectively, the quantity $c\rho T$ is called the solid's **heat energy per unit volume.**

(a) Explain why $-\nabla T$ points in the direction of heat flow.

(b) Let $-k\nabla T$ denote the **energy flux vector.** (Here the constant k is called the **conductivity.**) Assuming the Law of Conservation of Mass with $-k\nabla T = \mathbf{v}$ and $c\rho T = p$ in Exercise 31, derive the diffusion (heat) equation

$$\frac{\partial T}{\partial t} = K\,\nabla^2 T,$$

where $K = k/(c\rho) > 0$ is the *diffusivity* constant. (Notice that if $T(t, x)$ represents the temperature at time t at position x in a uniform conducting rod with perfectly insulated sides, then $\nabla^2 T = \partial^2 T/\partial x^2$ and the diffusion equation reduces to the one-dimensional heat equation in Chapter 11's Additional Exercises.)

Questions to Guide Your Review

1. What are line integrals? How are they evaluated? Give examples.
2. How can you use line integrals to find the centers of mass of springs? Explain.
3. What is a vector field? A gradient field? Give examples.
4. How do you calculate the work done by a force in moving a particle along a curve? Give an example.
5. What are flow, circulation, and flux?
6. What is special about path independent fields?
7. How can you tell when a field is conservative? How do you find the work done by a conservative field?
8. What is a potential function? Show by example how to find a potential function for a conservative field.
9. What is a differential form? What does it mean for such a form to be exact? How do you test for exactness? Give examples.
10. What is the divergence of a vector field? How can you interpret it?
11. What is the curl of a vector field? How can you interpret it?
12. What are the two forms of Green's Theorem? How can you interpret them?
13. How do you calculate the area of a curved surface in space? Give an example.
14. What is an oriented surface? How do you calculate the flux of a three-dimensional vector field across an oriented surface? Give an example.
15. What are surface integrals? What can you calculate with them? Give an example.
16. What is a parametrized surface? How do you find the area of such a surface? Give examples.
17. How do you integrate a function over a parametrized surface? Give an example.
18. What is Stokes' Theorem? How can you interpret it?
19. Summarize the chapter's results on conservative fields.
20. What is the Divergence Theorem? How can you interpret it?
21. How does the Divergence Theorem generalize Green's Theorem?
22. How does Stokes' Theorem generalize Green's Theorem?
23. How can Green's Theorem, Stokes' Theorem, and the Divergence Theorem be regarded as forms of a single fundamental theorem?

Practice Exercises

Evaluating Line Integrals

1. The accompanying figure shows two polygonal paths in space joining the origin to the point (1, 1, 1). Integrate $f(x, y, z) = 2x - 3y^2 - 2z + 3$ over each path.

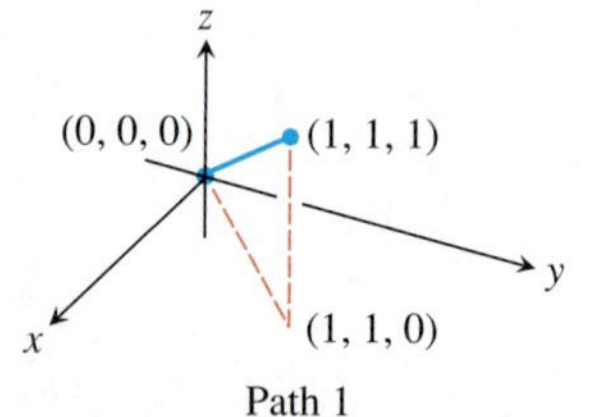

Path 1

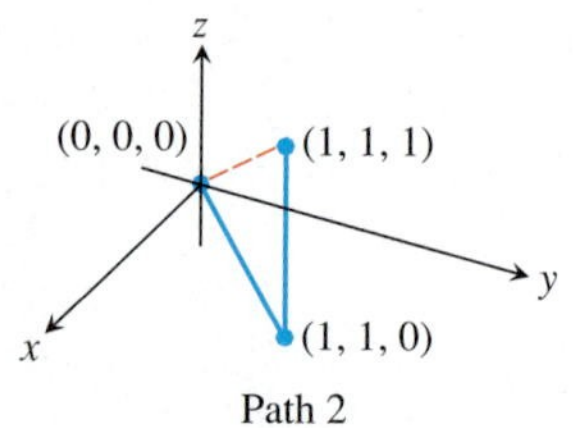

Path 2

2. The accompanying figure shows three polygonal paths joining the origin to the point (1, 1, 1). Integrate $f(x, y, z) = x^2 + y - z$ over each path.

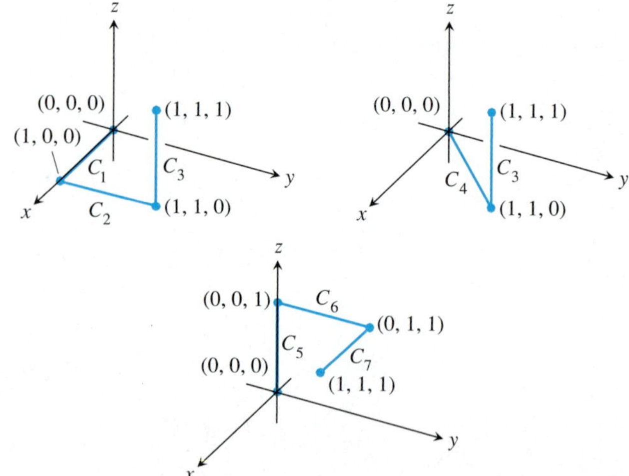

3. Integrate $f(x, y, z) = \sqrt{x^2 + z^2}$ over the circle

$$\mathbf{r}(t) = (a \cos t)\mathbf{j} + (a \sin t)\mathbf{k}, \qquad 0 \le t \le 2\pi.$$

4. Integrate $f(x, y, z) = \sqrt{x^2 + z^2}$ over the involute curve

$$\mathbf{r}(t) = (\cos t + t \sin t)\mathbf{i} + (\sin t - t \cos t)\mathbf{j}, \qquad 0 \le t \le \sqrt{3}.$$

Evaluate the integrals in Exercises 5 and 6.

5. $\displaystyle\int_{(-1,1,1)}^{(4,-3,0)} \frac{dx + dy + dz}{\sqrt{x + y + z}}$

6. $\displaystyle\int_{(1,1,1)}^{(10,3,3)} dx - \sqrt{\frac{z}{y}}\, dy - \sqrt{\frac{y}{z}}\, dz$

7. Integrate $\mathbf{F} = -(y \sin z)\mathbf{i} + (x \sin z)\mathbf{j} + (xy \cos z)\mathbf{k}$ around the circle cut from the sphere $x^2 + y^2 + z^2 = 5$ by the plane $z = -1$, clockwise as viewed from above.

8. Integrate $\mathbf{F} = 3x^2y\mathbf{i} + (x^3 + 1)\mathbf{j} + 9z^2\mathbf{k}$ around the circle cut from the sphere $x^2 + y^2 + z^2 = 9$ by the plane $x = 2$.

Evaluate the integrals in Exercises 9 and 10.

9. $\displaystyle\int_C 8x \sin y\, dx - 8y \cos x\, dy$

C is the square cut from the first quadrant by the lines $x = \pi/2$ and $y = \pi/2$.

10. $\displaystyle\int_C y^2\, dx + x^2\, dy$

C is the circle $x^2 + y^2 = 4$.

Evaluating Surface Integrals

11. *Area of an elliptical region* Find the area of the elliptical region cut from the plane $x + y + z = 1$ by the cylinder $x^2 + y^2 = 1$.

12. *Area of a parabolic cap* Find the area of the cap cut from the paraboloid $y^2 + z^2 = 3x$ by the plane $x = 1$.

13. *Area of a spherical cap* Find the area of the cap cut from the top of the sphere $x^2 + y^2 + z^2 = 1$ by the plane $z = \sqrt{2}/2$.

14. (a) *Hemisphere cut by cylinder* Find the area of the surface cut from the hemisphere $x^2 + y^2 + z^2 = 4$, $z \ge 0$, by the cylinder $x^2 + y^2 = 2x$.

(b) Find the area of the portion of the cylinder that lies inside the hemisphere. (*Hint:* Project onto the xz-plane. Or evaluate the integral $\int h\, ds$, where h is the altitude of the cylinder and ds is the element of arc length on the circle $x^2 + y^2 = 2x$ in the xy-plane.)

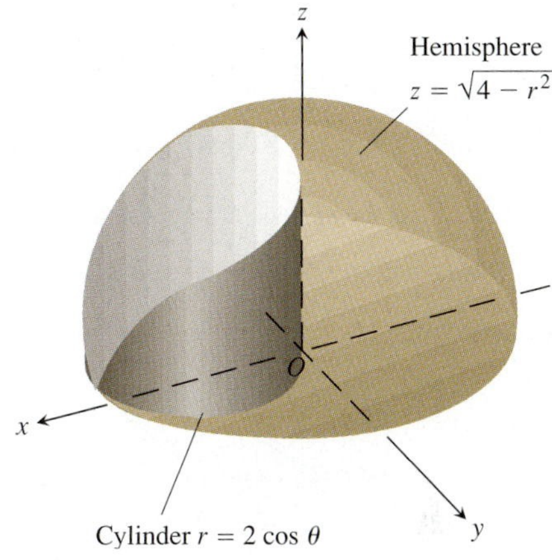

15. *Area of a triangle* Find the area of the triangle in which the plane $(x/a) + (y/b) + (z/c) = 1$ $(a, b, c > 0)$ intersects the first octant. Check your answer with an appropriate vector calculation.

16. *Parabolic cylinder cut by planes* Integrate

(a) $g(x, y, z) = \dfrac{yz}{\sqrt{4y^2 + 1}}$ (b) $g(x, y, z) = \dfrac{z}{\sqrt{4y^2 + 1}}$

over the surface cut from the parabolic cylinder $y^2 - z = 1$ by the planes $x = 0$, $x = 3$, and $z = 0$.

17. *Circular cylinder cut by planes* Integrate $g(x, y, z) = x^4y(y^2 + z^2)$ over the portion of the cylinder $y^2 + z^2 = 25$ that lies in the first octant between the planes $x = 0$ and $x = 1$ and above the plane $z = 3$.

18. *Area of Wyoming* The state of Wyoming is bounded by the meridians 111° 3′ and 104° 3′ west longitude and by the circles 41° and 45° north latitude. Assuming that Earth is a sphere of radius $R = 3959$ mi, find the area of Wyoming.

Parametrized Surfaces

Find the parametrizations for the surfaces in Exercises 19–24. (There are many ways to do these, so your answers may not be the same as those in the back of the book.)

19. *Spherical band* The portion of the sphere $x^2 + y^2 + z^2 = 36$ between the planes $z = -3$ and $z = 3\sqrt{3}$

20. *Parabolic cap* The portion of the paraboloid $z = -(x^2 + y^2)/2$ above the plane $z = -2$

21. *Cone* The cone $z = 1 + \sqrt{x^2 + y^2}$, $z \le 3$

22. *Plane above square* The portion of the plane $4x + 2y + 4z = 12$ that lies above the square $0 \le x \le 2$, $0 \le y \le 2$ in the first quadrant

23. *Portion of paraboloid* The portion of the paraboloid $y = 2(x^2 + z^2)$, $y \le 2$, that lies above the xy-plane

24. *Portion of hemisphere* The portion of the hemisphere $x^2 + y^2 + z^2 = 10$, $y \ge 0$, in the first octant

25. *Surface area* Find the area of the surface

$$\mathbf{r}(u, v) = (u + v)\mathbf{i} + (u - v)\mathbf{j} + v\mathbf{k}, \qquad 0 \le u \le 1, \quad 0 \le v \le 1.$$

26. *Surface integral* Integrate $f(x, y, z) = xy - z^2$ over the surface in Exercise 25.

27. *Area of a helicoid* Find the surface area of the helicoid

$$\mathbf{r}(r, \theta) = r\cos\theta\mathbf{i} + r\sin\theta\mathbf{j} + \theta\mathbf{k}, \qquad 0 \le \theta \le 2\pi, \quad 0 \le r \le 1,$$

in the accompanying figure.

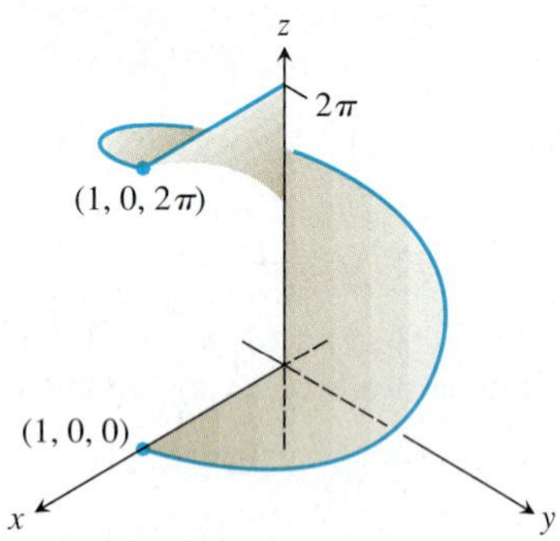

28. *Surface integral* Evaluate the integral $\iint_S \sqrt{x^2 + y^2 + 1}\, d\sigma$, where S is the helicoid in Exercise 27.

Conservative Fields

Which of the fields in Exercises 29–32 are conservative, and which are not?

29. $\mathbf{F} = x\mathbf{i} + y\mathbf{j} + z\mathbf{k}$

30. $\mathbf{F} = (x\mathbf{i} + y\mathbf{j} + z\mathbf{k})/(x^2 + y^2 + z^2)^{3/2}$

31. $\mathbf{F} = xe^y\mathbf{i} + ye^z\mathbf{j} + ze^x\mathbf{k}$

32. $\mathbf{F} = (\mathbf{i} + z\mathbf{j} + y\mathbf{k})/(x + yz)$

Find potential functions for the fields in Exercises 33 and 34.

33. $\mathbf{F} = 2\mathbf{i} + (2y + z)\mathbf{j} + (y + 1)\mathbf{k}$

34. $\mathbf{F} = (z\cos xz)\mathbf{i} + e^y\mathbf{j} + (x\cos xz)\mathbf{k}$

Work and Circulation

In Exercises 35 and 36, find the work done by each field along the paths from (0, 0, 0) to (1, 1, 1) in Exercise 1.

35. $\mathbf{F} = 2xy\mathbf{i} + \mathbf{j} + x^2\mathbf{k}$

36. $\mathbf{F} = 2xy\mathbf{i} + x^2\mathbf{j} + \mathbf{k}$

37. *Finding work in two ways* Find the work done by

$$\mathbf{F} = \frac{x\mathbf{i} + y\mathbf{j}}{(x^2 + y^2)^{3/2}}$$

over the plane curve $\mathbf{r}(t) = (e^t\cos t)\mathbf{i} + (e^t\sin t)\mathbf{j}$ from the point (1, 0) to the point $(e^{2\pi}, 0)$ in two ways:

(a) By using the parametrization of the curve to evaluate the work integral

(b) By evaluating a potential function for $\mathbf{F}$.

38. *Flow along different paths* Find the flow of the field $\mathbf{F} = \nabla(x^2ze^y)$

(a) Once around the ellipse C in which the plane $x + y + z = 1$ intersects the cylinder $x^2 + z^2 = 25$, clockwise as viewed from the positive y-axis

(b) Along the curved boundary of the helicoid in Exercise 27 from (1, 0, 0) to $(1, 0, 2\pi)$.

In Exercises 39 and 40, use the surface integral in Stokes' Theorem to find the circulation of the field $\mathbf{F}$ around the curve C in the indicated direction.

39. *Circulation around an ellipse* $\mathbf{F} = y^2\mathbf{i} - y\mathbf{j} + 3z^2\mathbf{k}$

C: The ellipse in which the plane $2x + 6y - 3z = 6$ meets the cylinder $x^2 + y^2 = 1$, counterclockwise as viewed from above

40. *Circulation around a circle* $\mathbf{F} = (x^2 + y)\mathbf{i} + (x + y)\mathbf{j} + (4y^2 - z)\mathbf{k}$

C: The circle in which the plane $z = -y$ meets the sphere $x^2 + y^2 + z^2 = 4$, counterclockwise as viewed from above

Mass and Moments

41. *Wire with different densities* Find the mass of a thin wire lying along the curve $\mathbf{r}(t) = \sqrt{2}t\mathbf{i} + \sqrt{2}t\mathbf{j} + (4 - t^2)\mathbf{k}$, $0 \le t \le 1$, if the density at t is (a) $\delta = 3t$ and (b) $\delta = 1$.

42. *Wire with variable density* Find the center of mass of a thin wire lying along the curve $\mathbf{r}(t) = t\mathbf{i} + 2t\mathbf{j} + (2/3)t^{2/3}\mathbf{k}$, $0 \le t \le 2$, if the density at t is $\delta = 3\sqrt{5 + t}$.

43. *Wire with variable density* Find the center of mass and the moments of inertia and radii of gyration about the coordinate axes of a thin wire lying along the curve

$$\mathbf{r}(t) = t\mathbf{i} + \frac{2\sqrt{2}}{3}t^{3/2}\mathbf{j} + \frac{t^2}{2}\mathbf{k}, \qquad 0 \le t \le 2,$$

if the density at t is $\delta = 1/(t + 1)$.

44. *Center of mass of an arch* A slender metal arch lies along the semicircle $y = \sqrt{a^2 - x^2}$ in the xy-plane. The density at the point (x, y) on the arch is $\delta(x, y) = 2a - y$. Find the center of mass.

45. *Wire with constant density* A wire of constant density $\delta = 1$ lies along the curve $\mathbf{r}(t) = (e^t\cos t)\mathbf{i} + (e^t\sin t)\mathbf{j} + e^t\mathbf{k}$, $0 \le t \le \ln 2$. Find $\bar{z}$, I_z, and R_z.

46. *Helical wire with constant density* Find the mass and center of mass of a wire of constant density δ that lies along the helix $\mathbf{r}(t) = (2\sin t)\mathbf{i} + (2\cos t)\mathbf{j} + 3t\mathbf{k}$, $0 \le t \le 2\pi$.

47. *Inertia, radius of gyration, center of mass of a shell* Find I_z, R_z, and the center of mass of a thin shell of density $\delta(x, y, z) = z$ cut from the upper portion of the sphere $x^2 + y^2 + z^2 = 25$ by the plane $z = 3$.

48. *Moment of inertia of a cube* Find the moment of inertia about the z-axis of the surface of the cube cut from the first octant by the planes $x = 1$, $y = 1$, and $z = 1$ if the density is $\delta = 1$.

Flux Across a Plane Curve or Surface

Use Green's Theorem to find the counterclockwise circulation and outward flux for the fields and curves in Exercises 49 and 50.

49. *Square* $\mathbf{F} = (2xy + x)\mathbf{i} + (xy - y)\mathbf{j}$

C: The square bounded by $x = 0, x = 1, y = 0, y = 1$

50. *Triangle* $\mathbf{F} = (y - 6x^2)\mathbf{i} + (x + y^2)\mathbf{j}$

C: The triangle made by the lines $y = 0, y = x$, and $x = 1$

51. *Zero line integral* Show that

$$\oint_C \ln x \sin y\, dy - \frac{\cos y}{x}\, dx = 0$$

for any closed curve C to which Green's Theorem applies.

52. **(a)** *Outward flux and area* Show that the outward flux of the position vector field $\mathbf{F} = x\mathbf{i} + y\mathbf{j}$ across any closed curve to which Green's Theorem applies is twice the area of the region enclosed by the curve.

(b) Let $\mathbf{n}$ be the outward unit normal vector to a closed curve to which Green's Theorem applies. Show that it is not possible for $\mathbf{F} = x\mathbf{i} + y\mathbf{j}$ to be orthogonal to $\mathbf{n}$ at every point of C.

In Exercises 53–56, find the outward flux of $\mathbf{F}$ across the boundary of D.

53. *Cube* $\mathbf{F} = 2xy\mathbf{i} + 2yz\mathbf{j} + 2xz\mathbf{k}$

D: The cube cut from the first octant by the planes $x = 1$, $y = 1, z = 1$

54. *Spherical cap* $\mathbf{F} = xz\mathbf{i} + yz\mathbf{j} + \mathbf{k}$

D: The entire surface of the upper cap cut from the solid sphere $x^2 + y^2 + z^2 \le 25$ by the plane $z = 3$

55. *Spherical cap* $\mathbf{F} = -2x\mathbf{i} - 3y\mathbf{j} + z\mathbf{k}$

D: The upper region cut from the solid sphere $x^2 + y^2 + z^2 \le 2$ by the paraboloid $z = x^2 + y^2$

56. *Cone and cylinder* $\mathbf{F} = (6x + y)\mathbf{i} - (x + z)\mathbf{j} + 4yz\mathbf{k}$

D: The region in the first octant bounded by the cone $z = \sqrt{x^2 + y^2}$, the cylinder $x^2 + y^2 = 1$, and the coordinate planes

57. *Hemisphere, cylinder, and plane* Let S be the surface that is bounded on the left by the hemisphere $x^2 + y^2 + z^2 = a^2, y \le 0$, in the middle by the cylinder $x^2 + z^2 = a^2, 0 \le y \le a$, and on the right by the plane $y = a$. Find the flux of $\mathbf{F} = y\mathbf{i} + z\mathbf{j} + x\mathbf{k}$ outward across S.

58. *Cylinder and Planes* Find the outward flux of the field $\mathbf{F} = 3xz^2\mathbf{i} + y\mathbf{j} - z^3\mathbf{k}$ across the surface of the solid in the first octant that is bounded by the cylinder $x^2 + 4y^2 = 16$ and the planes $y = 2z$, $x = 0$, and $z = 0$.

59. *Cylindrical can* Use the Divergence Theorem to find the flux of $\mathbf{F} = xy^2\mathbf{i} + x^2y\mathbf{j} + y\mathbf{k}$ outward through the surface of the region enclosed by the cylinder $x^2 + y^2 = 1$ and the planes $z = 1$ and $z = -1$.

60. *Hemisphere* Find the flux of $\mathbf{F} = (3z + 1)\mathbf{k}$ upward across the hemisphere $x^2 + y^2 + z^2 = a^2, z \ge 0$ (a) with the Divergence Theorem and (b) by evaluating the flux integral directly.

Additional Exercises: Theory, Examples, Applications

Finding Areas with Green's Theorem

Use the Green's Theorem area formula, Equation (22) in Exercises 13.4, to find the areas of the regions enclosed by the curves in Exercises 1–4.

1. The limaçon $x = 2\cos t - \cos 2t$, $y = 2\sin t - \sin 2t$, $0 \le t \le 2\pi$

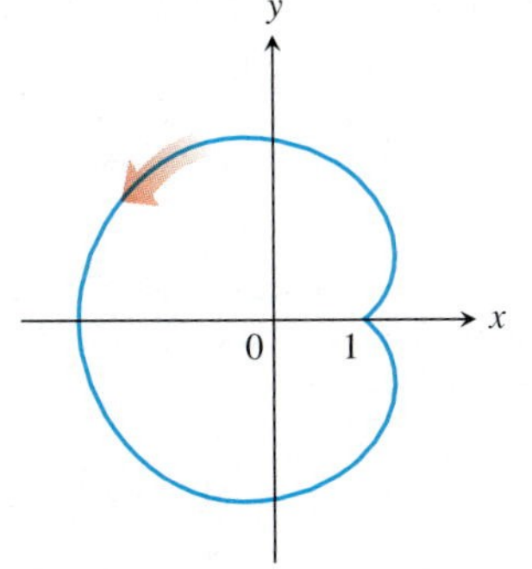

2. The deltoid $x = 2\cos t + \cos 2t$, $y = 2\sin t - \sin 2t$, $0 \le t \le 2\pi$

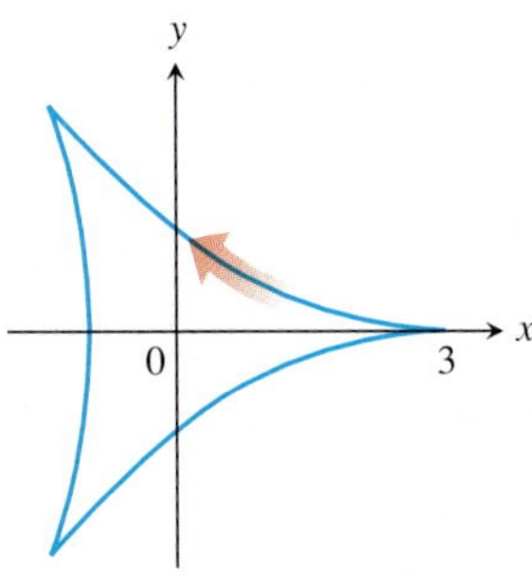

3. The eight curve $x = (1/2)\sin 2t,\ y = \sin t,\ 0 \le t \le \pi$ (one loop)

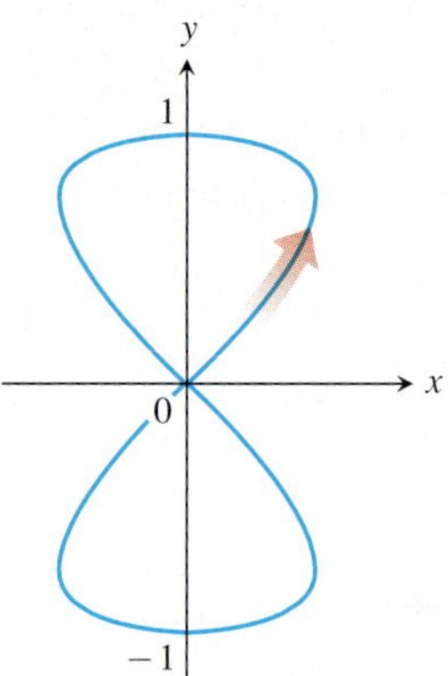

4. The teardrop $x = 2a\cos t - a\sin 2t,\ y = b\sin t,\ 0 \le t \le 2\pi$

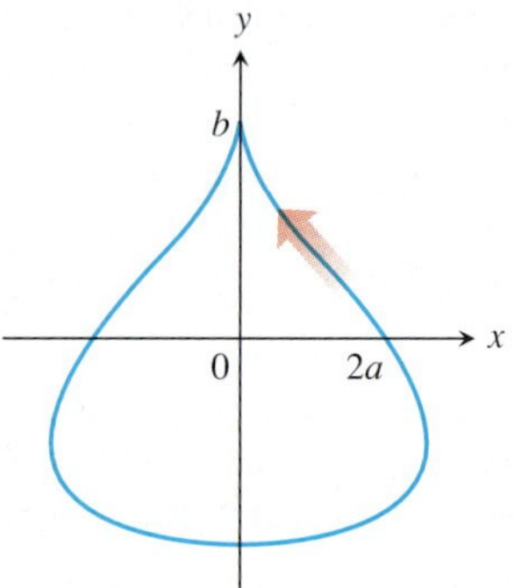

Theory and Applications

5. *Fields with nonzero curl*

   (a) Give an example of a vector field $\mathbf{F}(x, y, z)$ that has value $\mathbf{0}$ at only one point and such that curl $\mathbf{F}$ is nonzero everywhere. Be sure to identify the point and compute the curl.

   (b) Give an example of a vector field $\mathbf{F}(x, y, z)$ that has value $\mathbf{0}$ on precisely one line and such that curl $\mathbf{F}$ is nonzero everywhere. Be sure to identify the line and compute the curl.

   (c) Give an example of a vector field $\mathbf{F}(x, y, z)$ that has value $\mathbf{0}$ on a surface and such that curl $\mathbf{F}$ is nonzero everywhere. Be sure to identify the surface and compute the curl.

6. *Field normal to a sphere* Find all points (a, b, c) on the sphere $x^2 + y^2 + z^2 = R^2$ where the vector field $\mathbf{F} = yz^2\mathbf{i} + xz^2\mathbf{j} + 2xyz\mathbf{k}$ is normal to the surface and $\mathbf{F}(a, b, c) \neq \mathbf{0}$.

7. *Minimal flux* Among all rectangular regions $0 \le x \le a$, $0 \le y \le b$, find the one for which the total outward flux of $\mathbf{F} = (x^2 + 4xy)\mathbf{i} - 6y\mathbf{j}$ across the four sides is least. What *is* the least flux?

8. *Maximum circulation* Find an equation for the plane through the origin such that the circulation of the flow field $\mathbf{F} = z\mathbf{i} + x\mathbf{j} + y\mathbf{k}$ around the circle of intersection of the plane with the sphere $x^2 + y^2 + z^2 = 4$ is a maximum.

9. *Work on a string* A string lies along the circle $x^2 + y^2 = 4$ from $(2, 0)$ to $(0, 2)$ in the first quadrant. The density of the string is $\rho(x, y) = xy$.

   (a) Partition the string into a finite number of subarcs to show that the work done by gravity to move the string straight down to the x-axis is given by

$$\text{Work} = \lim_{n\to\infty} \sum_{k=1}^{n} g x_k y_k^2 \, \Delta s_k = \int_C g x y^2 \, ds,$$

   where g is the gravitational constant.

   (b) Find the total work done by evaluating the line integral in part (a).

   (c) *Moving the center of mass* Show that the total work done equals the work required to move the string's center of mass $(\bar{x}, \bar{y})$ straight down to the x-axis.

10. *Work on a thin sheet* A thin sheet lies along the portion of the plane $x + y + z = 1$ in the first octant. The density of the sheet is $\delta(x, y, z) = xy$.

   (a) Partition the sheet into a finite number of subpieces to show that the work done by gravity to move the sheet straight down to the xy-plane is given by

$$\text{Work} = \lim_{n\to\infty} \sum_{k=1}^{n} g x_k y_k z_k \, \Delta\sigma_k = \iint_S g xyz \, d\sigma,$$

   where g is the gravitational constant.

   (b) Find the total work done by evaluating the surface integral in part (a).

   (c) *Moving the center of mass* Show that the total work done equals the work required to move the sheet's center of mass $(\bar{x}, \bar{y}, \bar{z})$ straight down to the xy-plane.

11. *Archimedes' principle* If an object such as a ball is placed in a liquid, it will either sink to the bottom, float, or sink a certain distance and remain suspended in the liquid. Suppose that a fluid has constant weight density w and that the fluid's surface coincides with the plane $z = 4$. A spherical ball remains suspended in the fluid and occupies the region $x^2 + y^2 + (z - 2)^2 \le 1$.

   (a) Show that the surface integral giving the magnitude of the total force on the ball due to the fluid's pressure is

$$\text{Force} = \lim_{n\to\infty} \sum_{k=1}^{n} w(4 - z_k) \, \Delta\sigma_k = \iint_S w(4 - z) \, d\sigma.$$

   (b) *Buoyant force integral* Since the ball is not moving, it is being held up by the buoyant force of the liquid. Show that the magnitude of the buoyant force on the sphere is

$$\text{Buoyant force} = \iint_S w(z - 4)\mathbf{k} \cdot \mathbf{n} \, d\sigma,$$

where $\mathbf{n}$ is the outer unit normal at (x, y, z). This illustrates Archimedes' principle that the magnitude of the buoyant force on a submerged solid equals the weight of the displaced fluid.

(c) Use the Divergence Theorem to find the magnitude of the buoyant force in part (b).

12. *A gravitational field is not a curl* Let

$$\mathbf{F} = -\frac{GmM}{|\mathbf{r}|^3}\mathbf{r}$$

be the gravitational force field defined for $\mathbf{r} \neq \mathbf{0}$. Use Gauss's law in Section 13.8 to show that there is no continuously differentiable vector field $\mathbf{H}$ satisfying $\mathbf{F} = \nabla \times \mathbf{H}$.

13. *Equal line and surface integrals* If $f(x, y, z)$ and $g(x, y, z)$ are continuously differentiable scalar functions defined over the oriented surface S with boundary curve C, prove that

$$\iint_S (\nabla f \times \nabla g) \cdot \mathbf{n}\, d\sigma = \oint_C f\, \nabla g \cdot d\mathbf{r}.$$

14. *Fields with equal divergence and equal curls* Suppose that $\nabla \cdot \mathbf{F}_1 = \nabla \cdot \mathbf{F}_2$ and $\nabla \times \mathbf{F}_1 = \nabla \times \mathbf{F}_2$ over a region D enclosed by the oriented surface S with outward unit normal $\mathbf{n}$ and that $\mathbf{F}_1 \cdot \mathbf{n} = \mathbf{F}_2 \cdot \mathbf{n}$ on S. Prove that $\mathbf{F}_1 = \mathbf{F}_2$ throughout D.

15. *Zero vector field?* Prove or disprove that if $\nabla \cdot \mathbf{F} = 0$ and $\nabla \times \mathbf{F} = \mathbf{0}$, then $\mathbf{F} = \mathbf{0}$.

16. *Volume as the flux of the position vector field* Show that the volume V of a region D in space enclosed by the oriented surface S with outward normal $\mathbf{n}$ satisfies the identity

$$V = \frac{1}{3}\iint_S \mathbf{r} \cdot \mathbf{n}\, d\sigma,$$

where $\mathbf{r}$ is the position vector of the point (x, y, z) in D.

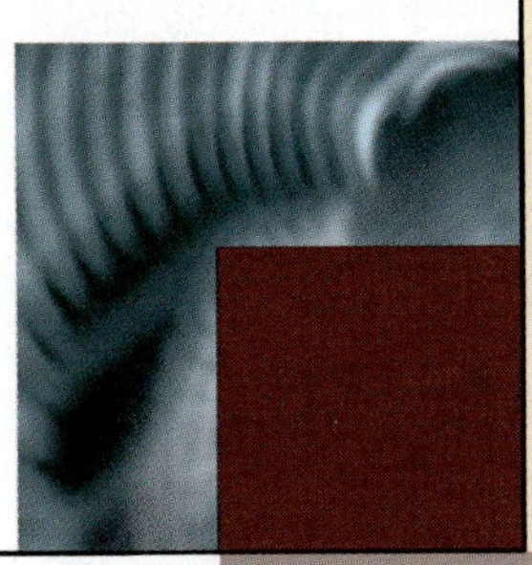

Appendices

A.1 Mathematical Induction

Many formulas, like

$$1 + 2 + \cdots + n = \frac{n(n+1)}{2},$$

can be shown to hold for every positive integer n by applying an axiom called the *mathematical induction principle*. A proof that uses this axiom is called a *proof by mathematical induction* or a *proof by induction*.

The steps in proving a formula by induction are the following.

Step 1: Check that the formula holds for $n = 1$.

Step 2: Prove that if the formula holds for any positive integer $n = k$, then it also holds for the next integer, $n = k + 1$.

Once these steps are completed (the axiom says), we know that the formula holds for all positive integers n. By step 1, it holds for $n = 1$. By step 2, it holds for $n = 2$, and therefore by step 2 also for $n = 3$, and by step 2 again for $n = 4$, and so on. If the first domino falls, and the kth domino always knocks over the $(k+1)$st when it falls, all the dominoes fall.

From another point of view, suppose we have a sequence of statements $S_1, S_2, \ldots, S_n, \ldots$, one for each positive integer. Suppose we can also show that assuming any one of the statements to be true implies the next statement in line is true. Finally, suppose we can show that S_1 is true. Then we may conclude that the statements are true from S_1 on.

Example 1 Sum of the First n Positive Integers

Prove that for every positive integer n,

$$1 + 2 + \cdots + n = \frac{n(n+1)}{2}.$$

Solution We accomplish the proof by carrying out the two steps above.

Step 1: The formula holds for $n = 1$ because

$$1 = \frac{1(1+1)}{2}.$$

Step 2: If the formula holds for $n = k$, does it also hold for $n = k + 1$? The answer is yes, and here's why: If

$$1 + 2 + \cdots + k = \frac{k(k+1)}{2},$$

then

$$\begin{aligned} 1 + 2 + \cdots + k + (k+1) &= \frac{k(k+1)}{2} + (k+1) = \frac{k^2 + k + 2k + 2}{2} \\ &= \frac{(k+1)(k+2)}{2} = \frac{(k+1)((k+1)+1)}{2}. \end{aligned}$$

The last expression in this string of equalities is the expression $n(n+1)/2$ for $n = (k+1)$.

The mathematical induction principle now guarantees the original formula for all positive integers n. Notice that all *we* have to do is carry out steps 1 and 2. The mathematical induction principle does the rest.

Example 2 Sum of Powers of 1 /2

Prove that for all positive integers n,

$$\frac{1}{2^1} + \frac{1}{2^2} + \cdots + \frac{1}{2^n} = 1 - \frac{1}{2^n}.$$

Solution We accomplish the proof by carrying out the two steps of mathematical induction.

Step 1: The formula holds for $n = 1$ because

$$\frac{1}{2^1} = 1 - \frac{1}{2^1}.$$

Step 2: If

$$\frac{1}{2^1} + \frac{1}{2^2} + \cdots + \frac{1}{2^k} = 1 - \frac{1}{2^k},$$

then

$$\begin{aligned} \frac{1}{2^1} + \frac{1}{2^2} + \cdots + \frac{1}{2^k} + \frac{1}{2^{k+1}} &= 1 - \frac{1}{2^k} + \frac{1}{2^{k+1}} = 1 - \frac{1 \cdot 2}{2^k \cdot 2} + \frac{1}{2^{k+1}} \\ &= 1 - \frac{2}{2^{k+1}} + \frac{1}{2^{k+1}} = 1 - \frac{1}{2^{k+1}}. \end{aligned}$$

Thus, the original formula holds for $n = (k+1)$ whenever it holds for $n = k$.

With these steps verified, the mathematical induction principle now guarantees the formula for every positive integer n.

Other Starting Integers

Instead of starting at $n = 1$, some induction arguments start at another integer. The steps for such an argument are as follows.

Step 1: Check that the formula holds for $n = n_1$ (the first appropriate integer).

Step 2: Prove that if the formula holds for any integer $n = k \geq n_1$, then it also holds for $n = (k + 1)$.

Once these steps are completed, the mathematical induction principle guarantees the formula for all $n \geq n_1$.

Example 3 Factorial Exceeding Exponential

Show that $n! > 3^n$ if n is large enough.

Solution How large is large enough? We experiment:

n	1	2	3	4	5	6	7
$n!$	1	2	6	24	120	720	5040
3^n	3	9	27	81	243	729	2187

It looks as if $n! > 3^n$ for $n \geq 7$. To be sure, we apply mathematical induction. We take $n_1 = 7$ in step 1 and try for step 2.

Suppose that $k! > 3^k$ for some $k \geq 7$. Then

$$(k + 1)! = (k + 1)(k!) > (k + 1)\,3^k > 7 \cdot 3^k > 3^{k+1}.$$

Thus, for $k \geq 7$,

$$k! > 3^k \quad \Rightarrow \quad (k + 1)! > 3^{k+1}.$$

The mathematical induction principle now guarantees $n! \geq 3^n$ for all $n \geq 7$.

EXERCISES A.1

1. *General triangle inequality* Assuming that the triangle inequality $|a + b| \leq |a| + |b|$ holds for any two numbers a and b, show that

$$|x_1 + x_2 + \cdots + x_n| \leq |x_1| + |x_2| + \cdots + |x_n|$$

for any n numbers.

2. *Geometric sum* Show that if $r \neq 1$, then

$$1 + r + r^2 + \cdots + r^n = \frac{1 - r^{n+1}}{1 - r}$$

for every positive integer n.

3. *Positive integer power rule* Use the Product Rule,

$$\frac{d}{dx}(uv) = u\frac{dv}{dx} + v\frac{du}{dx},$$

and the equation

$$\frac{d}{dx}(x) = 1$$

to show that

$$\frac{d}{dx}(x^n) = nx^{n-1}$$

for every positive integer n.

4. *Products into sums* Suppose a function $f(x)$ has the property that $f(x_1x_2) = f(x_1) + f(x_2)$ for any two positive numbers x_1 and x_2. Show that

$$f(x_1x_2 \cdots x_n) = f(x_1) + f(x_2) + \cdots + f(x_n)$$

for the product of any n positive numbers $x_1, x_2 \ldots, x_n$.

5. *Geometric sum* Show that

$$\frac{2}{3^1} + \frac{2}{3^2} + \cdots + \frac{2}{3^n} = 1 - \frac{1}{3^n}$$

for all positive integers n.

6. Show that $n! > n^3$ if n is large enough.

7. Show that $2^n > n^2$ if n is large enough.

8. Show that $2^n \geq 1/8$ for $n \geq -3$.

9. *Sums of squares* Show that the sum of the squares of the first n positive integers is

$$\frac{n\left(n + \frac{1}{2}\right)(n + 1)}{3}.$$

10. *Sums of cubes* Show that the sum of the cubes of the first n positive integers is $(n(n + 1)/2)^2$.

11. *Rules for finite sums* Show that the following finite sum rules hold for every positive integer n.

(a) $\sum_{k=1}^{n} (a_k + b_k) = \sum_{k=1}^{n} a_k + \sum_{k=1}^{n} b_k$

(b) $\sum_{k=1}^{n} (a_k - b_k) = \sum_{k=1}^{n} a_k - \sum_{k=1}^{n} b_k$

(c) $\sum_{k=1}^{n} ca_k = c \cdot \sum_{k=1}^{n} a_k$ (any number c)

(d) $\sum_{k=1}^{n} a_k = n \cdot c$ (if a_k has the constant value c)

12. *Integer powers and absolute value* Show that $|x^n| = |x|^n$ for every positive integer n and every real number x.

A.2 Proofs of Limit Theorems in Section 1.2

This appendix proves Theorem 1 and Theorem 4 from Section 1.2.

Theorem 1 Properties of Limits

The following rules hold if $\lim_{x\to c} f(x) = L$ and $\lim_{x\to c} g(x) = M$ (L and M real numbers).

1. *Sum Rule:* $\lim_{x\to c} [f(x) + g(x)] = L + M$
2. *Difference Rule:* $\lim_{x\to c} [f(x) - g(x)] = L - M$
3. *Product Rule:* $\lim_{x\to c} f(x) \cdot g(x) = L \cdot M$
4. *Constant Multiple Rule:* $\lim_{x\to c} k f(x) = k L$ (any number k)
5. *Quotient Rule:* $\lim_{x\to c} \frac{f(x)}{g(x)} = \frac{L}{M}$, if $M \neq 0$
6. *Root Rule:* If n is a positive integer, then $\lim_{x\to c} [f(x)]^{1/n} = L^{1/n}$

(If n is even, we assume $L > 0$.)

Proof of the Sum Rule Let $\epsilon > 0$ be given. We want to find a positive number δ such that for all x

$$0 < |x - c| < \delta \Rightarrow |f(x) + g(x) - (L + M)| < \epsilon.$$

Regrouping terms, we get

$$|f(x) + g(x) - (L + M)| = |(f(x) - L) + (g(x) - M)|$$
$$\leq |f(x) - L| + |g(x) - M|.$$

Triangle Inequality: $|a + b| \leq |a| + |b|$

Since $\lim_{x\to c} f(x) = L$, there exists a number $\delta_1 > 0$ such that for all x

$$0 < |x - c| < \delta_1 \Rightarrow |f(x) - L| < \epsilon/2.$$

Similarly, since $\lim_{x\to c} g(x) = M$, there exists a number $\delta_2 > 0$ such that for all x

$$0 < |x - c| < \delta_2 \Rightarrow |g(x) - M| < \epsilon/2.$$

Let $\delta = \min\{\delta_1, \delta_2\}$, the smaller of δ_1 and δ_2. If $0 < |x - c| < \delta$ then $|x - c| < \delta_1$, so $|f(x) - L| < \epsilon/2$, and $|x - c| < \delta_2$, so $|g(x) - M| < \epsilon/2$. Therefore,

$$|f(x) + g(x) - (L + M)| < \frac{\epsilon}{2} + \frac{\epsilon}{2} = \epsilon,$$

which shows that $\lim_{x\to c} (f(x) + g(x)) = L + M$.

CD-ROM
WEBsite
Historical Biography
John Wallis
(1616 — 1703)

The Difference Rule is obtained by replacing $g(x)$ by $-g(x)$ and M by $-M$ in the Sum Rule. The Constant Multiple Rule is the special case $g(x) = k$ of the Product Rule. The Root Rule is proved in more advanced texts, but we prove here the Product and Quotient Rules.

Proof of the Limit Product Rule We show that for any $\epsilon > 0$ there exists a $\delta > 0$ such that for all x in the intersection D of the domains of f and g,

$$0 < |x - c| < \delta \Rightarrow |f(x)\, g(x) - LM| < \epsilon.$$

Suppose then that ϵ is a positive number and write $f(x)$ and $g(x)$ as

$$f(x) = L + (f(x) - L), \qquad g(x) = M + (g(x) - M).$$

Multiply these expressions together and subtract LM:

$$\begin{aligned} f(x) \cdot g(x) - LM &= (L + (f(x) - L))(M + (g(x) - M)) - LM \\ &= LM + L(g(x) - M) + M(f(x) - L) + \\ &\quad (f(x) - L)(g(x) - M) - LM \\ &= L(g(x) - M) + M(f(x) - L) + (f(x) - L)(g(x) - M). \end{aligned} \tag{1}$$

Since f and g have limits L and M as $x \to c$, there exist positive numbers δ_1, δ_2, δ_3, and δ_4 such that for all x in D

$$\begin{aligned} &0 < |x - c| < \delta_1 \Rightarrow |f(x) - L| < \sqrt{\epsilon/3} \\ &0 < |x - c| < \delta_2 \Rightarrow |g(x) - M| < \sqrt{\epsilon/3} \\ &0 < |x - c| < \delta_3 \Rightarrow |f(x) - L| < \epsilon/(3(1 + |M|)) \\ &0 < |x - c| < \delta_4 \Rightarrow |g(x) - M| < \epsilon/(3(1 + |L|)). \end{aligned} \tag{2}$$

If we take δ to be the smallest of the numbers δ_1 through δ_4, the inequalities on the right-hand side of (2) will hold simultaneously for $0 < |x - c| < \delta$. Therefore, for all x in D, $0 < |x - c| < \delta$ implies

$$\begin{aligned} &|f(x) \cdot g(x) - LM| \\ &\quad \le |L||g(x) - M| + |M||f(x) - L| + |f(x) - L||g(x) - M| && \text{Triangle inequality applied to Eq. (1)} \\ &\quad \le (1 + |L|)|g(x) - M| + (1 + |M|)|f(x) - L| + |f(x) - L||g(x) - M| \\ &\quad \le \frac{\epsilon}{3} + \frac{\epsilon}{3} + \sqrt{\frac{\epsilon}{3}}\sqrt{\frac{\epsilon}{3}} = \epsilon. && \text{Values from Eq. (2)} \end{aligned}$$

This completes the proof of the Limit Product Rule.

Proof of the Limit Quotient Rule We show that $\lim_{x\to c}(1/g(x)) = 1/M$. We can then conclude from the Limit Product Rule that

$$\lim_{x\to c}\frac{f(x)}{g(x)} = \lim_{x\to c}\left(f(x)\cdot\frac{1}{g(x)}\right) = \lim_{x\to c}f(x)\cdot\lim_{x\to c}g(x) = L\cdot\frac{1}{M} = \frac{L}{M}.$$

Let $\epsilon > 0$ be given. To show that $\lim_{x\to c}(1/g(x)) = 1/M$, we need to show that there exists a $\delta > 0$ such that for all x

$$0 < |x-c| < \delta \Rightarrow \left|\frac{1}{g(x)} - \frac{1}{M}\right| < \epsilon.$$

Since $|M| > 0$, there exists a positive number δ_1 such that for all x

$$0 < |x-c| < \delta_1 \Rightarrow |g(x) - M| < \left|\frac{M}{2}\right|. \tag{3}$$

Gilles Personne de Roberval (1602 — 1675)

For any numbers A and B it can be shown that $|A| - |B| \le |A - B|$ and $|B| - |A| \le |A - B|$, from which it follows that $||A| - |B|| \le |A - B|$. With $A = g(x)$ and $B = M$, this becomes

$$||g(x)| - |M|| \le |g(x) - M|,$$

which can be combined with the inequality on the right in Eq. (3) to get, in turn,

$$\begin{aligned} ||g(x)| - |M|| &< \frac{|M|}{2} \\ -\frac{|M|}{2} < |g(x)| &- |M| < \frac{|M|}{2} \\ \frac{|M|}{2} < |g(x)| &< \frac{3|M|}{2} \\ |M| < 2|g(x)| &< 3|M| \\ \frac{1}{|g(x)|} < \frac{2}{|M|} &< \frac{3}{|g(x)|}. \end{aligned} \tag{4}$$

Therefore, $0 < |x-c| < \delta_1$ implies that

$$\begin{aligned} \left|\frac{1}{g(x)} - \frac{1}{M}\right| &= \left|\frac{M - g(x)}{Mg(x)}\right| \le \frac{1}{|M|}\cdot\frac{1}{|g(x)|}\cdot|M - g(x)| \\ &< \frac{1}{|M|}\cdot\frac{2}{|M|}\cdot|M - g(x)|. \quad \text{Inequality (4)} \end{aligned} \tag{5}$$

Since $(1/2)|M|^2\epsilon > 0$, there exists a number $\delta_2 > 0$ such that for all x

$$0 < |x-c| < \delta_2 \Rightarrow |M - g(x)| < \frac{\epsilon}{2}|M|^2. \tag{6}$$

If we take δ to be the smaller of δ_1 and δ_2, the conclusions in Eqs. (5) and (6) both hold for all x such that $0 < |x-c| < \delta$. Combining these conclusions gives

$$0 < |x-c| < \delta \Rightarrow \left|\frac{1}{g(x)} - \frac{1}{M}\right| < \epsilon.$$

This concludes the proof of the Limit Quotient Rule.

Theorem 4 The Sandwich Theorem

Suppose that $g(x) \le f(x) \le h(x)$ for all x in some open interval containing c, except possibly at $x = c$ itself. Suppose also that $\lim_{x\to c} g(x) = \lim_{x\to c} h(x) = L$. Then $\lim_{x\to c} f(x) = L$.

Proof for Right-hand Limits Suppose that $\lim_{x\to c^+} g(x) = \lim_{x\to c^+} h(x) = L$. Then for any $\epsilon > 0$ there exists a $\delta > 0$ such that for all x the inequality $c < x < c + \delta$ implies

$$L - \epsilon < g(x) < L + \epsilon \quad \text{and} \quad L - \epsilon < h(x) < L + \epsilon.$$

These inequalities combine with $g(x) \le f(x) \le h(x)$ to give

$$L - \epsilon < g(x) \le f(x) \le h(x) < L + \epsilon,$$
$$L - \epsilon < f(x) < L + \epsilon,$$
$$-\epsilon < f(x) - L < \epsilon.$$

Therefore, for all x, the inequality $c < x < c + \delta$ implies $|f(x) - L| < \epsilon$. ∎

Proof for Left-hand Limits Suppose that $\lim_{x\to c^-} g(x) = \lim_{x\to c^-} h(x) = L$. Then for any $\epsilon > 0$ there exists a $\delta > 0$ such that for all x the inequality $c - \delta < x < c$ implies

$$L - \epsilon < g(x) < L + \epsilon \quad \text{and} \quad L - \epsilon < h(x) < L + \epsilon.$$

We conclude as before that for all x, $c - \delta < x < c$ implies $|f(x) - L| < \epsilon$. ∎

Proof for Two-sided Limits If $\lim_{x\to c} g(x) = \lim_{x\to c} h(x) = L$, then $g(x)$ and $h(x)$ both approach L as $x \to c^+$ and as $x \to c^-$; so $\lim_{x\to c^+} f(x) = L$ and $\lim_{x\to c^-} f(x) = L$. Hence, $\lim_{x\to c} f(x)$ exists and equals L. ∎

EXERCISES A.2

1. *Generalized limit sum rule* Suppose that functions $f_1(x)$, $f_2(x)$, and $f_3(x)$ have limits L_1, L_2, and L_3, respectively, as $x \to c$. Show that their sum has limit $L_1 + L_2 + L_3$. Use mathematical induction (Appendix 1) to generalize this result to the sum of any finite number of functions.

2. *Generalized limit product rule* Use mathematical induction and the Limit Product Rule in Theorem 1 to show that if functions $f_1(x)$, $f_2(x), \ldots, f_n(x)$ have limits $L_1, L_2, \ldots, L_n$ as $x \to c$, then

$$\lim_{x\to c} f_1(x) f_2(x) \cdot \cdots \cdot f_n(x) = L_1 \cdot L_2 \cdot \cdots \cdot L_n.$$

3. *Positive integer power rule* Use the fact that $\lim_{x\to c} x = c$ and the result of Exercise 2 to show that $\lim_{x\to c} x^n = c^n$ for any integer $n > 1$.

4. *Limits of polynomials* Use the fact that $\lim_{x\to c} (k) = k$ for any number k together with the results of Exercises 1 and 3 to show that $\lim_{x\to c} f(x) = f(c)$ for any polynomial function

$$f(x) = a_n x^n + a_{n-1} x^{n-1} + \cdots + a_1 x + a_0.$$

5. *Limits of rational functions* Use Theorem 1 and the result of Exercise 4 to show that if $f(x)$ and $g(x)$ are polynomial functions and $g(c) \ne 0$, then

$$\lim_{x\to c} \frac{f(x)}{g(x)} = \frac{f(c)}{g(c)}.$$

6. *Composites of continuous functions* Figure A.1 gives the diagram for a proof that the composite of two continuous functions is continuous. Reconstruct the proof from the diagram. The statement to be proved is this: If f is continuous at $x = c$ and g is continuous at $f(c)$, then $g \circ f$ is continuous at c.

Assume that c is an interior point of the domain of f and that $f(c)$ is an interior point of the domain of g. This will make the limits involved two-sided. (The arguments for the cases that involve one-sided limits are similar.)

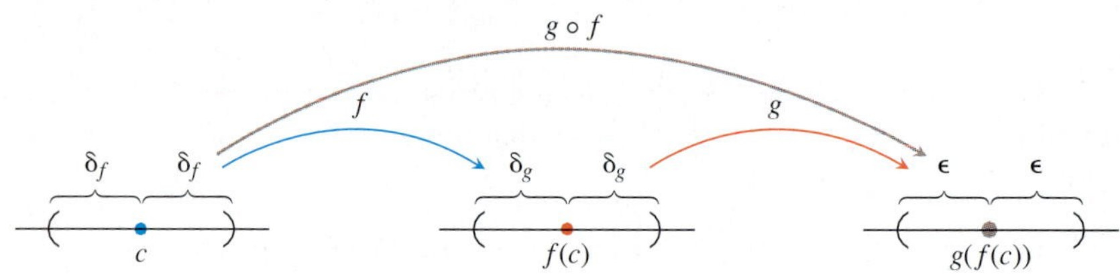

FIGURE A.1 A diagram for a proof that the composite of two continuous functions is continuous. The continuity of composites holds for any finite number of functions. The only requirement is that each function be continuous where it is applied. In the figure, f is to be continuous at c and g at $f(c)$.

A.3 Proof of the Chain Rule

This appendix proves the Chain Rule in Section 2.5 using ideas from Section 3.6.

Theorem 4 The Chain Rule

If $f(u)$ is differentiable at the point $u = g(x)$, and $g(x)$ is differentiable at x, then the composite function $(f \circ g)(x) = f(g(x))$ is differentiable at x, and

$$(f \circ g)'(x) = f'(g(x)) \cdot g'(x).$$

In Leibniz's notation, if $y = f(u)$ and $u = g(x)$, then

$$\frac{dy}{dx} = \frac{dy}{du} \cdot \frac{du}{dx},$$

where dy/du is evaluated at $u = g(x)$.

Proof To be more precise, we show that if g is differentiable at x_0 and f is differentiable at $g(x_0)$, then the composite is differentiable at x_0 and

$$\left.\frac{dy}{dx}\right|_{x=x_0} = f'(g(x_0)) \cdot g'(x_0).$$

Let Δx be an increment in x and let Δu and Δy be the corresponding increments in u and y. As you can see in Figure A.2,

$$\left.\frac{dy}{dx}\right|_{x=x_0} = \lim_{\Delta x \to 0} \frac{\Delta y}{\Delta x},$$

so our goal is to show that this limit is $f'(g(x_0)) \cdot g'(x_0)$.

By Equation (3) in Section 3.6,

$$\Delta u = g'(x_0)\,\Delta x + \epsilon_1\,\Delta x = (g'(x_0) + \epsilon_1)\,\Delta x,$$

where $\epsilon_1 \to 0$ as $\Delta x \to 0$. Similarly,

$$\Delta y = f'(u_0)\,\Delta u + \epsilon_2\,\Delta u = (f'(u_0) + \epsilon_2)\,\Delta u,$$

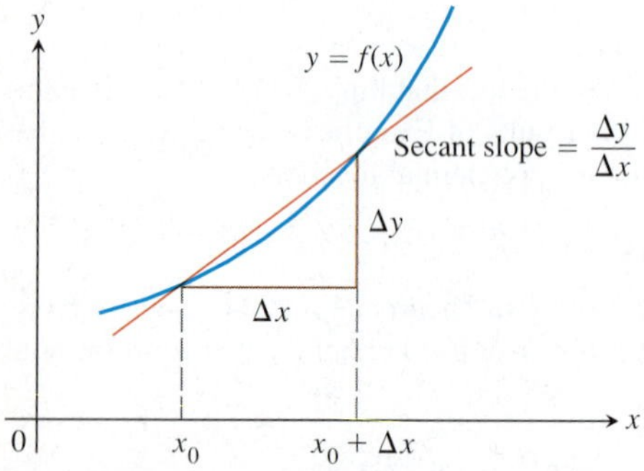

FIGURE A.2 The graph of y as a function of x. The derivative of y with respect to x at $x = x_0$ is $\lim_{\Delta x \to 0} \Delta y/\Delta x$.

where $\epsilon_2 \to 0$ as $\Delta u \to 0$. Notice also that $\Delta u \to 0$ as $\Delta x \to 0$. Combining the equations for Δu and Δy gives

$$\Delta y = (f'(u_0) + \epsilon_2)(g'(x_0) + \epsilon_1)\,\Delta x,$$

so

$$\frac{\Delta y}{\Delta x} = f'(u_0)\,g'(x_0) + \epsilon_2\,g'(x_0) + f'(u_0)\,\epsilon_1 + \epsilon_2\,\epsilon_1.$$

Since ϵ_1 and ϵ_2 go to zero as Δx goes to zero, three of the four terms on the right vanish in the limit, leaving

$$\lim_{\Delta x \to 0} \frac{\Delta y}{\Delta x} = f'(u_0)\,g'(x_0) = f'(g(x_0)) \cdot g'(x_0).$$

This concludes the proof.

A.4 Complex Numbers

The Development of the Real Numbers • The Complex Number System • Argand Diagrams • Euler's Formula • Products • Quotients • Powers and De Moivre's Theorem • Roots • The Fundamental Theorem of Algebra

Complex numbers are expressions of the form $a + ib$, where a and b are real numbers and i is a symbol for $\sqrt{-1}$. Unfortunately, "real" and "imaginary" have connotations that somehow place $\sqrt{-1}$ in a less favorable position in our minds than $\sqrt{2}$. As a matter of fact, a good deal of imagination, in the sense of *inventiveness,* has been required to construct the *real* number system, which forms the basis of the calculus. In this appendix, we review the various stages of this invention. The further invention of a complex number system will then not seem so strange.

The Development of the Real Numbers

The earliest stage of number development was the recognition of the **counting numbers** 1, 2, 3, . . . , which we now call the **natural numbers** or the **positive integers.** These numbers can be added or multiplied together without getting outside the system. That is, the system of positive integers is **closed** under the operations of addition and multiplication. If m and n are any positive integers, then

$$m + n = p \qquad \text{and} \qquad mn = q \tag{1}$$

are also positive integers.

Given the two positive integers on the left-hand side of either equation in Equations (1), we can find the corresponding positive integer on the right. More than this, we can sometimes specify the positive integers m and p and find a positive integer n such that $m + n = p$. For instance, $3 + n = 7$ can be solved when the only numbers we know are the positive integers. The equation $7 + n = 3$, however, cannot be solved unless the number system is enlarged.

The number zero and the negative integers were invented to solve equations like $7 + n = 3$. In a civilization that recognizes all the **integers**

$$\ldots, -3, -2, -1, 0, 1, 2, 3, \ldots, \tag{2}$$

an educated person can always find the missing integer that solves the equation $m + n = p$ when given the other two integers in the equation.

Suppose that our educated people also know how to multiply any two of the integers in Equation (2). If, in Equations (1), they are given m and q, they discover that sometimes they can find n and sometimes they cannot. If their imagination is still in good working order, they may be inspired to invent still more numbers and introduce fractions as ordered pairs m/n of integers m and n. The number zero has special properties that may bother them for a while, but they ultimately discover that it is handy to have all ratios of integers m/n, excluding only those having zero in the denominator. This system, called the set of **rational numbers,** is now rich enough for them to perform the so-called **rational operations** of arithmetic:

1. (a) addition
(b) subtraction

2. (a) multiplication
(b) division

on any two numbers in the system, *except that they cannot divide by zero.*

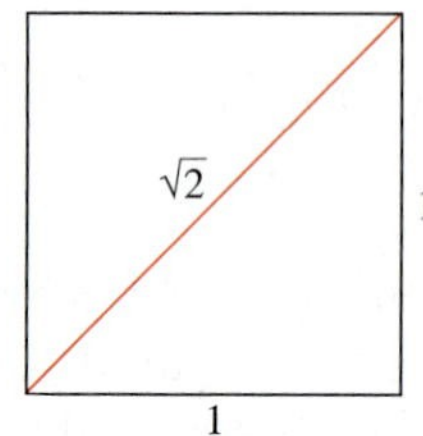

FIGURE A.3 With a straightedge and compass, it is possible to construct a line segment of irrational length.

The geometry of the unit square (Figure A.3) and the Pythagorean Theorem showed that they could construct a geometric line segment that, in terms of some basic unit of length, has length equal to $\sqrt{2}$. Thus, they could solve the equation

$$x^2 = 2$$

by a geometric construction. Then they discovered, however, that the line segment representing $\sqrt{2}$ and the line segment representing the unit of length 1 were incommensurable quantities. This means that the ratio $\sqrt{2}/1$ cannot be expressed as the ratio of two *integer* multiples of some other, presumably more fundamental, unit of length. That is, our educated people could not find a rational number solution of the equation $x^2 = 2$.

They could not find it because there *is* no rational number whose square is 2. To see why, suppose that there were such a rational number. Then we could find integers p and q with no common factor other than 1 and such that

$$\left(\frac{p}{q}\right)^2 = 2$$

or

$$p^2 = 2q^2. \tag{3}$$

Since p and q are integers, p must be even; otherwise, its product with itself would be odd. In symbols, $p = 2p_1$, where p_1 is an integer. This step leads to $2{p_1}^2 = q^2$, which says q must be even, say $q = 2q_1$, where q_1 is an integer. This makes 2 a factor of both p and q, contrary to our choice of p and q as integers with no common factor other than 1. Hence, there is no rational number whose square is 2.

Although our educated people could not find a rational solution of the equation $x^2 = 2$, they could get a sequence of rational numbers

$$\frac{1}{1}, \quad \frac{7}{5}, \quad \frac{41}{29}, \quad \frac{239}{169}, \quad \ldots, \tag{4}$$

whose squares form a sequence

$$\frac{1}{1}, \quad \frac{49}{25}, \quad \frac{1681}{841}, \quad \frac{57{,}121}{28{,}561}, \quad \ldots \tag{5}$$

that converges to 2 as its limit. This time their imagination suggested that they needed a concept of a limit of a sequence of rational numbers. If we accept that an increasing sequence that is bounded from above always approaches a limit and observe that the sequence in Equation (4) has these properties, then we want it to have a limit L. This assumption would also mean, from Equation (5), that $L^2 = 2$, and hence L is *not* one of our rational numbers. If to the rational numbers we further add the limits of all bounded increasing sequences of rational numbers, we arrive at the system of all "real" numbers. The word *real* is placed in quotes because there is nothing that is either "more real" or "less real" about this system than there is about any other mathematical system.

The Complex Number System

Imagination was called upon at many stages during the development of the real number system. In fact, the art of invention was needed at least three times in constructing the systems we have discussed so far:

1. The *first invented* system: the set of *all integers* as constructed from the counting numbers
2. The *second invented* system: the set of *rational numbers* m/n as constructed from the integers
3. The *third invented* system: the set of all *real numbers* x as constructed from the rational numbers.

These invented systems form a hierarchy in which each system contains the previous system. Each system is also richer than its predecessor in that it permits additional operations to be performed without going outside the system:

1. In the system of all integers, we can solve all equations of the form
$$x + a = 0, \tag{6}$$
where a can be any integer.
2. In the system of all rational numbers, we can solve all equations of the form
$$ax + b = 0, \tag{7}$$
provided a and b are rational numbers and $a \neq 0$.
3. In the system of all real numbers, we can solve all the equations in Equations (6) and (7) and, in addition, all quadratic equations
$$ax^2 + bx + c = 0 \qquad \text{having} \qquad a \neq 0 \qquad \text{and} \qquad b^2 - 4ac \geq 0. \tag{8}$$

You are probably familiar with the formula that gives the solutions of Equation (8), namely,

$$x = \frac{-b \pm \sqrt{b^2 - 4ac}}{2a}, \tag{9}$$

and also know that when the discriminant, $d = b^2 - 4ac$, is negative, the solutions in Equation (9) do *not* belong to any of the systems discussed above. In fact, the very simple quadratic equation

$$x^2 + 1 = 0$$

is impossible to solve if the only number systems that can be used are the three invented systems mentioned so far.

Thus, we come to the *fourth invented* system, the set of all complex numbers $a + ib$. We could dispense entirely with the symbol i and use a notation like (a, b). We would then speak simply of a pair of real numbers a and b. Since, under algebraic operations, the numbers a and b are treated somewhat differently, it is essential to keep the *order* straight. We therefore might say that the **complex number system** consists of the set of all ordered pairs of real numbers (a, b), together with the rules by which they are to be equated, added, multiplied, and so on, listed below. We use both the (a, b) notation and the notation $a + ib$ in the discussion that follows. We call a the **real part** and b the **imaginary part** of the complex number (a, b).

We make the following definitions.

Equality	
$a + ib = c + id$ if and only if $a = c$ and $b = d$	Two complex numbers (a, b) and (c, d) are *equal* if and only if $a = c$ and $b = d$.
Addition	
$(a + ib) + (c + id) = (a + c) + i(b + d)$	The sum of the two complex numbers (a, b) and (c, d) is the complex number $(a + c, b + d)$.
Multiplication	
$(a + ib)(c + id) = (ac - bd) + i(ad + bc)$	The product of two complex numbers (a, b) and (c, d) is the complex number $(ac - bd, ad + bc)$.
$c(a + ib) = ac + i(bc)$	The product of a real number c and the complex number (a, b) is the complex number (ac, bc).

The set of all complex numbers (a, b) in which the second number b is zero has all the properties of the set of real numbers a. For example, addition and multiplication of $(a, 0)$ and $(c, 0)$ give

$$(a, 0) + (c, 0) = (a + c, 0)$$
$$(a, 0) \cdot (c, 0) = (ac, 0),$$

which are numbers of the same type with imaginary part equal to zero. Also, if we multiply a "real number" $(a, 0)$ and the complex number (c, d), we get

$$(a, 0) \cdot (c, d) = (ac, ad) = a(c, d).$$

In particular, the complex number $(0, 0)$ plays the role of zero in the complex number system, and the complex number $(1, 0)$ plays the role of unity.

The number pair $(0, 1)$, which has real part equal to zero and imaginary part equal to one, has the property that its square,

$$(0, 1)(0, 1) = (-1, 0),$$

has real part equal to minus one and imaginary part equal to zero. Therefore, in the system of complex numbers (a, b), there is a number $x = (0, 1)$ whose square can be added to unity $= (1, 0)$ to produce zero $= (0, 0)$; that is,

$$(0, 1)^2 + (1, 0) = (0, 0).$$

The equation

$$x^2 + 1 = 0$$

therefore has a solution $x = (0, 1)$ in this new number system.

You are probably more familiar with the $a + ib$ notation than you are with the notation (a, b). And since the laws of algebra for the ordered pairs enable us to write

$$(a, b) = (a, 0) + (0, b) = a(1, 0) + b(0, 1),$$

whereas $(1, 0)$ behaves like unity and $(0, 1)$ behaves like a square root of minus one, we need not hesitate to write $a + ib$ in place of (a, b). The i associated with b is like a tracer element that tags the imaginary part of $a + ib$. We can pass at will from the realm of ordered pairs (a, b) to the realm of expressions $a + ib$, and conversely. Yet there is nothing less "real" about the symbol $(0, 1) = i$ than there is about the symbol $(1, 0) = 1$, once we have learned the laws of algebra in the complex number system (a, b).

To reduce any rational combination of complex numbers to a single complex number, we apply the laws of elementary algebra, replacing t^2 wherever it appears by -1. Of course, we cannot divide by the complex number $(0, 0) = 0 + i0$. If $a + ib \neq 0$, however, we may carry out a division as follows:

$$\frac{c + id}{a + ib} = \frac{(c + id)(a - ib)}{(a + ib)(a - ib)} = \frac{(ac + bd) + i(ad - bc)}{a^2 + b^2}.$$

The result is a complex number $x + iy$ with

$$x = \frac{ac + bd}{a^2 + b^2}, \qquad y = \frac{ad - bc}{a^2 + b^2},$$

and $a^2 + b^2 \neq 0$, since $a + ib = (a, b) \neq (0, 0)$.

The number $a - ib$ that is used as multiplier to clear the i from the denominator is called the **complex conjugate** of $a + ib$. It is customary to use $\bar{z}$ (read "z bar") to denote the complex conjugate of z; thus,

$$z = a + ib, \qquad \bar{z} = a - ib.$$

Multiplying the numerator and denominator of the fraction $(c + id)/(a + ib)$ by the complex conjugate of the denominator will always replace the denominator by a real number.

Example 1 Operations with Complex Numbers

(a) $(2 + 3i) + (6 - 2i) = (2 + 6) + (3 - 2)i = 8 + i$

(b) $(2 + 3i) - (6 - 2i) = (2 - 6) + (3 - (-2))i = -4 + 5i$

(c) $(2 + 3i)(6 - 2i) = (2)(6) + (2)(-2i) + (3i)(6) + (3i)(-2i)$

$$= 12 - 4i + 18i - 6i^2 = 12 + 14i + 6 = 18 + 14i$$

(d)
$$\frac{2 + 3i}{6 - 2i} = \frac{2 + 3i}{6 - 2i}\frac{6 + 2i}{6 + 2i} = \frac{12 + 4i + 18i + 6i^2}{36 + 12i - 12i - 4i^2} = \frac{6 + 22i}{40} = \frac{3}{20} + \frac{11}{20}i$$

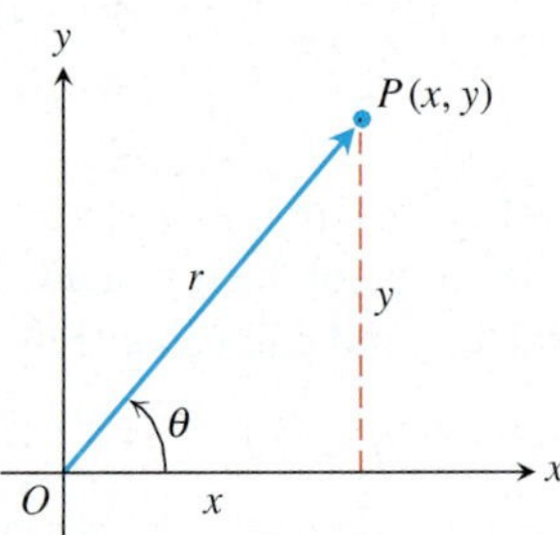

FIGURE A.4 This Argand diagram represents $z = x + iy$ both as a point $P(x, y)$ and as a vector $\overrightarrow{OP}$.

Argand Diagrams

There are two geometric representations of the complex number $z = x + iy$:

1. as the point $P(x, y)$ in the xy-plane
2. as the vector $\overrightarrow{OP}$ from the origin to P.

In each representation, the x-axis is called the **real axis** and the y-axis is the **imaginary axis**. Both representations are **Argand diagrams** for $x + iy$ (Figure A.4).

In terms of the polar coordinates of x and y, we have

$$x = r\cos\theta, \qquad y = r\sin\theta,$$

and

$$z = x + iy = r(\cos\theta + i\sin\theta). \tag{10}$$

We define the **absolute value** of a complex number $x + iy$ to be the length r of a vector $\overrightarrow{OP}$ from the origin to $P(x, y)$. We denote the absolute value by vertical bars; thus,

$$|x + iy| = \sqrt{x^2 + y^2}.$$

If we always choose the polar coordinates r and θ so that r is nonnegative, then

$$r = |x + iy|.$$

The polar angle θ is called the **argument** of z and is written $\theta = \arg z$. Of course, any integer multiple of 2π may be added to θ to produce another appropriate angle.

The following equation gives a useful formula connecting a complex number z, its conjugate $\bar{z}$, and its absolute value $|z|$, namely,

$$z \cdot \bar{z} = |z|^2.$$

Euler's Formula

The identity

$$e^{i\theta} = \cos\theta + i\sin\theta,$$

called **Euler's formula,** enables us to rewrite Equation (10) as

$$z = re^{i\theta}.$$

This formula, in turn, leads to the following rules for calculating products, quotients, powers, and roots of complex numbers. It also leads to Argand diagrams for $e^{i\theta}$. Since $\cos\theta + i\sin\theta$ is what we get from Equation (10) by taking $r = 1$, we can say that $e^{i\theta}$ is represented by a unit vector that makes an angle θ with the positive x-axis, as shown in Figure A.5.

Products

To multiply two complex numbers, we multiply their absolute values and add their angles. Let

$$z_1 = r_1e^{i\theta_1}, \qquad z_2 = r_2e^{i\theta_2}, \tag{11}$$

so that

$$|z_1| = r_1, \qquad \arg z_1 = \theta_1; \qquad |z_2| = r_2, \qquad \arg z_2 = \theta_2.$$

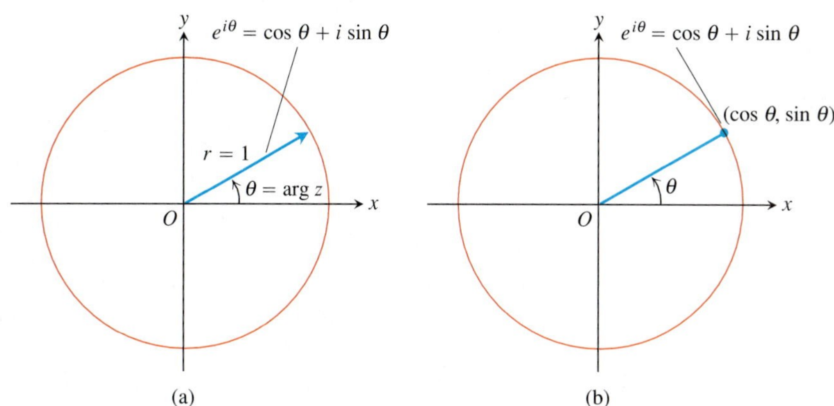

FIGURE A.5 Argand diagrams for $e^{i\theta} = \cos\theta + i\sin\theta$ (a) as a vector and (b) as a point.

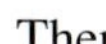

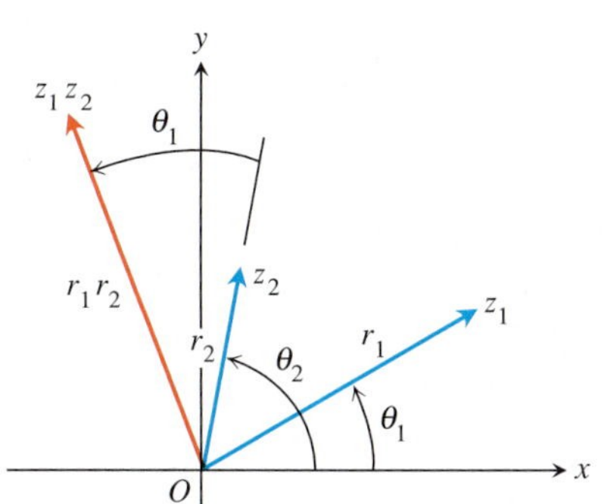

FIGURE A.6 When z_1 and z_2 are multiplied, $|z_1 z_2| = r_1 \cdot r_2$ and $\arg(z_1 z_2) = \theta_1 + \theta_2$.

Then

$$z_1 z_2 = r_1 e^{i\theta_1} \cdot r_2 e^{i\theta_2} = r_1 r_2 e^{i(\theta_1 + \theta_2)}$$

and hence

$$|z_1 z_2| = r_1 r_2 = |z_1| \cdot |z_2|$$

$$\arg(z_1 z_2) = \theta_1 + \theta_2 = \arg z_1 + \arg z_2. \tag{12}$$

Thus, the product of two complex numbers is represented by a vector whose length is the product of the lengths of the two factors and whose argument is the sum of their arguments (Figure A.6). In particular, a vector may be rotated counterclockwise through an angle θ by multiplying it by $e^{i\theta}$. Multiplication by i rotates 90°, by -1 rotates 180°, by $-i$ rotates 270°, and so on.

Example 2 Finding a Product of Complex Numbers

Let $z_1 = 1 + i$, $z_2 = \sqrt{3} - i$. We plot these complex numbers in an Argand diagram (Figure A.7) from which we read off the polar representations

$$z_1 = \sqrt{2}e^{i\pi/4}, \qquad z_2 = 2e^{-i\pi/6}.$$

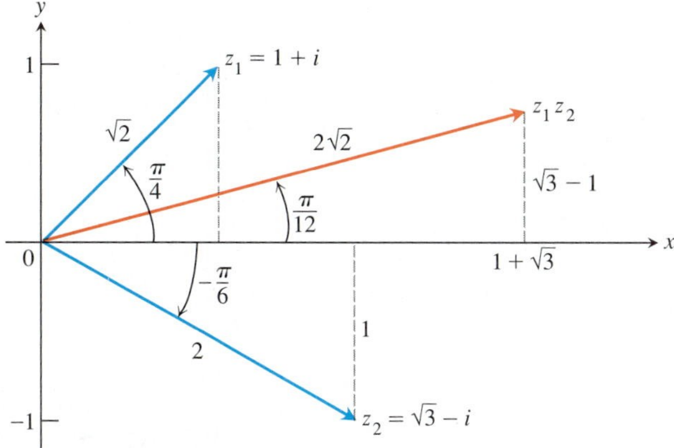

FIGURE A.7 To multiply two complex numbers, multiply their absolute values and add their arguments.

exp (A) stands for e^A.

Then

$$z_1z_2 = 2\sqrt{2}\exp\left(\frac{i\pi}{4} - \frac{i\pi}{6}\right) = 2\sqrt{2}\exp\left(\frac{i\pi}{12}\right)$$
$$= 2\sqrt{2}\left(\cos\frac{\pi}{12} + i\sin\frac{\pi}{12}\right) \approx 2.73 + 0.73i.$$

Quotients

Suppose that $r_2 \neq 0$ in Equation (11). Then

$$\frac{z_1}{z_2} = \frac{r_1e^{i\theta_1}}{r_2e^{i\theta_2}} = \frac{r_1}{r_2}e^{i(\theta_1-\theta_2)}.$$

Hence,

$$\left|\frac{z_1}{z_2}\right| = \frac{r_1}{r_2} = \frac{|z_1|}{|z_2|} \quad \text{and} \quad \arg\left(\frac{z_1}{z_2}\right) = \theta_1 - \theta_2 = \arg z_1 - \arg z_2.$$

That is, we divide lengths and subtract angles.

Example 3 Finding a Quotient of Complex Numbers

Let $z_1 = 1 + i$ and $z_2 = \sqrt{3} - i$, as in Example 2. Then

$$\frac{1+i}{\sqrt{3}-i} = \frac{\sqrt{2}e^{i\pi/4}}{2e^{-i\pi/6}} = \frac{\sqrt{2}}{2}e^{5\pi i/12} \approx 0.707\left(\cos\frac{5\pi}{12} + i\sin\frac{5\pi}{12}\right)$$
$$\approx 0.183 + 0.683i.$$

Powers and De Moivre's Theorem

If n is a positive integer, we may apply the product formulas in Equations (12) to find

$$z^n = z \cdot z \cdot \cdots \cdot z. \qquad n \text{ factors}$$

With $z = re^{i\theta}$, we obtain

$$z^n = (re^{i\theta})^n = r^ne^{i(\theta+\theta+\cdots+\theta)} \qquad n \text{ summands}$$
$$= r^ne^{in\theta}. \tag{13}$$

The length $r = |z|$ is raised to the nth power and the angle $\theta = \arg z$ is multiplied by n.

If we take $r = 1$ in Equation (13), we obtain De Moivre's Theorem.

De Moivre's Theorem

$$(\cos\theta + i\sin\theta)^n = \cos n\theta + i\sin n\theta. \tag{14}$$

If we expand the left-hand side of De Moivre's equation (Equation 14) by the Binomial Theorem and reduce it to the form $a + ib$, we obtain formulas for $\cos n\theta$ and $\sin n\theta$ as polynomials of degree n in $\cos\theta$ and $\sin\theta$.

Example 4 Obtaining Formulas for cos 3θ and sin 3θ

Express $\cos 3\theta$ and $\sin 3\theta$ in terms of $\cos\theta$ and $\sin\theta$.

Solution Taking $n = 3$ in De Moivre's Equation (Equation (14)), we have

$$(\cos\theta + i\,\sin\theta)^3 = \cos 3\theta + i\sin 3\theta.$$

The left-hand side of this equation is

$$\cos^3\theta + 3i\cos^2\theta\sin\theta - 3\cos\theta\sin^2\theta - i\sin^3\theta.$$

The real part of this must equal $\cos 3\theta$ and the imaginary part must equal $\sin 3\theta$. Therefore,

$$\cos 3\theta = \cos^3\theta - 3\cos\theta\sin^2\theta,$$
$$\sin 3\theta = 3\cos^2\theta\sin\theta - \sin^3\theta.$$

Roots

If $z = re^{i\theta}$ is a complex number different from zero and n is a positive integer, then there are precisely n different complex numbers $w_0, w_1, \ldots, w_{n-1}$, that are nth roots of z. To see why, let $w = \rho e^{i\alpha}$ be an nth root of $z = re^{i\theta}$, so that

$$w^n = z$$

or

$$\rho^n e^{in\alpha} = re^{i\theta}.$$

Then

$$\rho = \sqrt[n]{r}$$

is the real, positive nth root of r. As regards the angle, although we cannot say that $n\alpha$ and θ must be equal, we can say that they may differ only by an integer multiple of 2π. That is,

$$n\alpha = \theta + 2k\pi, \qquad k = 0, \pm1, \pm2, \ldots.$$

Therefore,

$$\alpha = \frac{\theta}{n} + k\frac{2\pi}{n}.$$

Hence, all the nth roots of $z = re^{i\theta}$ are given by

$$\sqrt[n]{re^{i\theta}} = \sqrt[n]{r}\exp i\left(\frac{\theta}{n} + k\frac{2\pi}{n}\right), \qquad k = 0, \pm1, \pm2, \ldots. \tag{15}$$

There might appear to be infinitely many different answers corresponding to the infinitely many possible values of k, but $k = n + m$ gives the same answer as $k = m$ in Equation (15). Thus, we need only take n consecutive values for k to obtain all the different nth roots of z. For convenience, we take

$$k = 0, 1, 2, \ldots, n-1.$$

All the nth roots of $re^{i\theta}$ lie on a circle centered at the origin and having radius equal to the real, positive nth root of r. One of them has argument $\alpha = \theta/n$. The others are uniformly spaced around the circle, each being separated from its neighbors by an angle equal to $2\pi/n$. Figure A.8 illustrates the placement of the three cube roots, w_0, w_1, w_2, of the complex number $z = re^{i\theta}$.

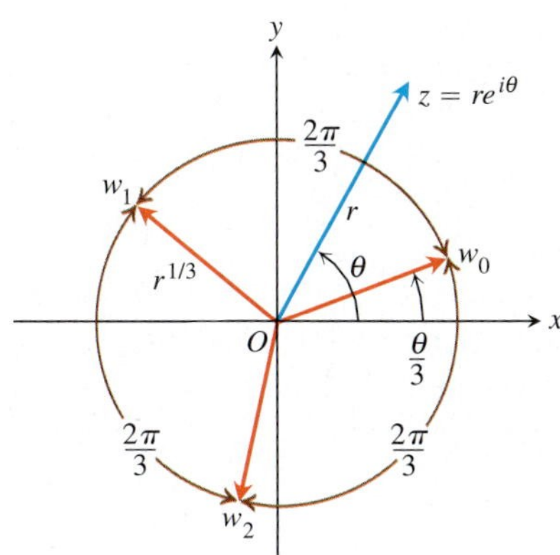

FIGURE A.8 The three cube roots of $z = re^{i\theta}$.

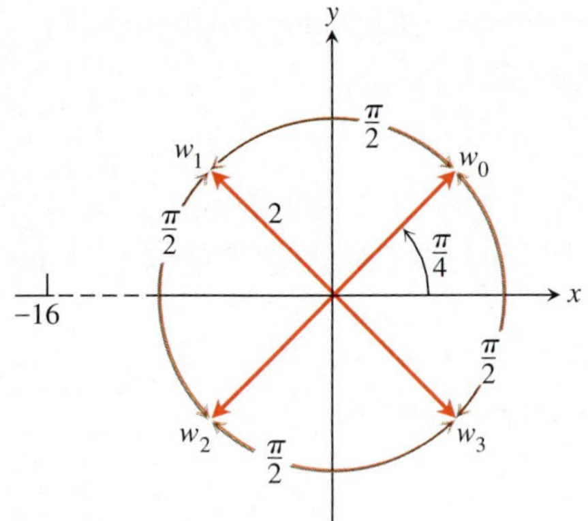

FIGURE A.9 The four fourth roots of -16.

Example 5 Finding Fourth Roots

Find the four fourth roots of -16.

Solution As our first step, we plot the number -16 in an Argand diagram (Figure A.9) and determine its polar representation $re^{i\theta}$. Here, $z = -16$, $r = +16$, and $\theta = \pi$. One of the fourth roots of $16e^{i\pi}$ is $2e^{i\pi/4}$. We obtain others by successive additions of $2\pi/4 = \pi/2$ to the argument of this first one. Hence,

$$\sqrt[4]{16 \exp i\pi} = 2 \exp i \left(\frac{\pi}{4}, \frac{3\pi}{4}, \frac{5\pi}{4}, \frac{7\pi}{4} \right),$$

and the four roots are

$$w_0 = 2\left[\cos\frac{\pi}{4} + i\sin\frac{\pi}{4}\right] = \sqrt{2}\,(1+i)$$

$$w_1 = 2\left[\cos\frac{3\pi}{4} + i\sin\frac{3\pi}{4}\right] = \sqrt{2}\,(-1+i)$$

$$w_2 = 2\left[\cos\frac{5\pi}{4} + i\sin\frac{5\pi}{4}\right] = \sqrt{2}\,(-1-i)$$

$$w_3 = 2\left[\cos\frac{7\pi}{4} + i\sin\frac{7\pi}{4}\right] = \sqrt{2}\,(1-i).$$

The Fundamental Theorem of Algebra

One may well say that the invention of $\sqrt{-1}$ is all well and good and leads to a number system that is richer than the real number system alone, but where will this process end? Are we also going to invent still more systems so as to obtain $\sqrt[4]{-1}$, $\sqrt[6]{-1}$, and so on? By now it should be clear that is not necessary. These numbers are already expressible in terms of the complex number system $a + ib$. In fact, the Fundamental Theorem of Algebra says that with the introduction of the complex numbers, we have enough numbers to factor every polynomial into a product of linear factors and hence enough numbers to solve every possible polynomial equation.

The Fundamental Theorem of Algebra

Every polynomial equation of the form

$$a_n z^n + a_{n-1} z^{n-1} + a_{n-2} z^{n-2} + \cdots + a_1 z + a_0 = 0,$$

in which the coefficients $a_0, a_1, \ldots, a_n$ are any complex numbers, whose degree n is greater than or equal to one, and whose leading coefficient a_0 is not zero, has exactly n roots in the complex number system, provided each multiple root of multiplicity m is counted as m roots.

A proof of this theorem can be found in almost any text on the theory of functions of a complex variable.

EXERCISES A.4

Operations with Complex Numbers

1. *How computers multiply complex numbers* Find $(a, b) \cdot (c, d) = (ac - bd, ad + bc)$.

(a) $(2, 3) \cdot (4, -2)$ **(b)** $(2, -1) \cdot (-2, 3)$

(c) $(-1, -2) \cdot (2, 1)$

(This is how complex numbers are multiplied by computers.)

2. Solve the following equations for the real numbers, x and y.

(a) $(3 + 4i)^2 - 2(x - iy) = x + iy$

(b) $\left(\frac{1 + i}{1 - i}\right)^2 + \frac{1}{x + iy} = 1 + i$

(c) $(3 - 2i)(x + iy) = 2(x - 2iy) + 2i - 1$

Graphing and Geometry

3. How may the following complex numbers be obtained from $z = x + iy$ geometrically? Sketch.

(a) $\bar{z}$ **(b)** $\overline{(-z)}$

(c) $-z$ **(d)** $1/z$

4. Show that the distance between the two points z_1 and z_2 in an Argand diagram is $|z_1 - z_2|$.

In Exercises 5–10, graph the points $z = x + iy$ that satisfy the given conditions.

5. (a) $|z| = 2$ **(b)** $|z| < 2$ **(c)** $|z| > 2$

6. $|z - 1| = 2$ **7.** $|z + 1| = 1$

8. $|z + 1| = |z - 1|$ **9.** $|z + i| = |z - 1|$

10. $|z + 1| \geq |z|$

Express the complex numbers in Exercises 11–14 in the form $re^{i\theta}$, with $r \geq 0$ and $-\pi < \theta \leq \pi$. Draw an Argand diagram for each calculation.

11. $(1 + \sqrt{-3})^2$ **12.** $\frac{1 + i}{1 - i}$

13. $\frac{1 + i\sqrt{3}}{1 - i\sqrt{3}}$ **14.** $(2 + 3i)(1 - 2i)$

Powers and Roots

Use De Moivre's Theorem to express the trigonometric functions in Exercises 15 and 16 in terms of $\cos\theta$ and $\sin\theta$.

15. $\cos 4\theta$

16. $\sin 4\theta$

17. Find the three cube roots of 1.

18. Find the two square roots of i.

19. Find the three cube roots of $-8i$.

20. Find the six sixth roots of 64.

21. Find the four solutions of the equation $z^4 - 2z^2 + 4 = 0$.

22. Find the six solutions of the equation $z^6 + 2z^3 + 2 = 0$.

23. Find all solutions of the equation $x^4 + 4x^2 + 16 = 0$.

24. Solve the equation $x^4 + 1 = 0$.

Theory and Examples

25. *Complex numbers and vectors in the plane* Show with an Argand diagram that the law for adding complex numbers is the same as the parallelogram law for adding vectors.

26. *Complex arithmetic with conjugates* Show that the conjugate of the sum (product, or quotient) of two complex numbers, z_1 and z_2 is the same as the sum (product, or quotient) of their conjugates.

27. *Complex roots of polynomials with real coefficients come in complex-conjugate pairs*

(a) Extend the results of Exercise 26 to show that $f(\bar{z}) = \overline{f(z)}$ if

$$f(z) = a_n z^n + a_{n-1} z^{n-1} + \cdots + a_1 z + a_0$$

is a polynomial with real coefficients $a_0, \ldots, a_n$.

(b) If z is a root of the equation $f(z) = 0$, where $f(z)$ is a polynomial with real coefficients as in part (a), show that the conjugate $\bar{z}$ is also a root of the equation. (*Hint:* Let $f(z) = u + iv = 0$; then both u and v are zero. Use the fact that $f(\bar{z}) = \overline{f(z)} = u - iv$.)

28. *Absolute value of a conjugate* Show that $|\bar{z}| = |z|$.

29. *When $z = \bar{z}$* If z and $\bar{z}$ are equal, what can you say about the location of the point z in the complex plane?

30. *Real and imaginary parts* Let $Re(z)$ denote the real part of z and $Im(z)$ the imaginary part. Show that the following relations hold for any complex numbers z, z_1, and z_2.

(a) $z + \bar{z} = 2Re(z)$ **(b)** $z - \bar{z} = 2iIm(z)$

(c) $|Re(z)| \leq |z|$

(d) $|z_1 + z_2|^2 = |z_1|^2 + |z_2|^2 + 2Re(z_1\bar{z}_2)$

(e) $|z_1 + z_2| \leq |z_1| + |z_2|$

A.5 Simpson's One-Third Rule

Simpson's rule for approximating $\int_a^b f(x)\,dx$ is based on approximating the graph of f with parabolic arcs.

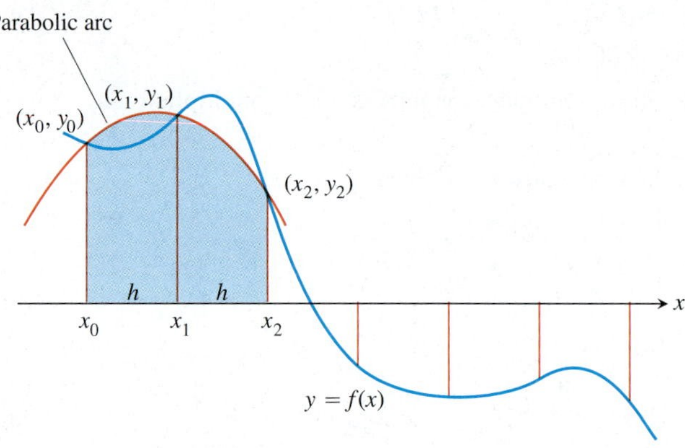

FIGURE A.10 Simpson's Rule approximates short stretches of curve with parabolic arcs.

The area of the shaded region under the parabola in Figure A.11 is

$$\text{Area} = \frac{h}{3}(y_0 + 4y_1 + y_2).$$

This formula is known as Simpson's One-Third Rule.

We can derive the formula as follows. To simplify the algebra, we use the coordinate system in Figure A.11. The area under the parabola is the same no matter where the y-axis is, as long as we preserve the vertical scale. The parabola has an equation of the form $y = Ax^2 + Bx + C$, so the area under it from $x = -h$ to $x = h$ is

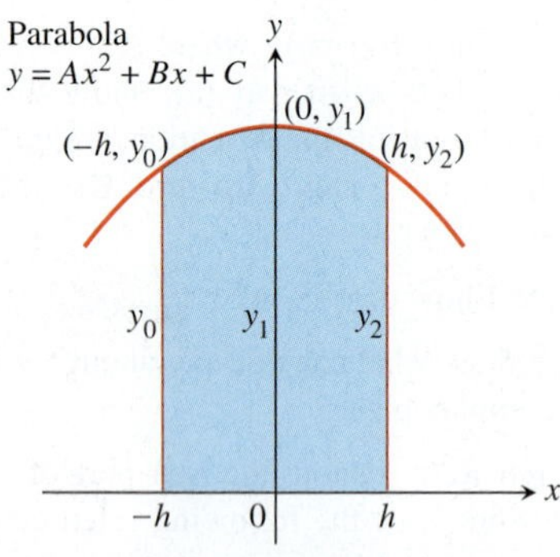

FIGURE A.11 By integrating from $-h$ to h, the shaded area is found to be

$$\frac{h}{3}(y_0 + 4y_1 + y_2).$$

$$\begin{aligned}\text{Area} &= \int_{-h}^{h} (Ax^2 + Bx + C)\,dx = \left[\frac{Ax^3}{3} + \frac{Bx^2}{2} + Cx\right]_{-h}^{h} \\ &= \frac{2Ah^3}{3} + 2Ch \\ &= \frac{h}{3}(2Ah^2 + 6C).\end{aligned}$$

Since the curve passes through $(-h, y_0)$, $(0, y_1)$, and (h, y_2), we also have

$$y_0 = Ah^2 - Bh + C, \qquad y_1 = C, \qquad y_2 = Ah^2 + Bh + C.$$

From these equations we obtain

$$\begin{aligned} C &= y_1 \\ Ah^2 - Bh &= y_0 - y_1 \\ Ah^2 + Bh &= y_2 - y_1 \\ 2Ah^2 &= y_0 + y_2 - 2y_1. \end{aligned}$$

These substitutions for C and $2Ah^2$ give

$$\text{Area} = \frac{h}{3}(2Ah^2 + 6C) = \frac{h}{3}((y_0 + y_2 - 2y_1) + 6y_1) = \frac{h}{3}(y_0 + 4y_1 + y_2).$$

A.6 Cauchy's Mean Value Theorem and the Stronger Form of l'Hôpital's Rule

This appendix proves the finite-limit case of the stronger form of l'Hôpital's Rule (Section 7.6, Theorem 2) as stated here.

L'Hôpital's Rule (Stronger Form)

Suppose that

$$f(x_0) = g(x_0) = 0$$

and that the functions f and g are both differentiable on an open interval (a, b) that contains the point x_0. Suppose also that $g'(x) \neq 0$ at every point in (a, b) except possibly x_0. Then

$$\lim_{x \to x_0} \frac{f(x)}{g(x)} = \lim_{x \to x_0} \frac{f'(x)}{g'(x)}, \qquad (1)$$

assuming the limit on the right exists.

The proof of the stronger form of l'Hôpital's Rule is based on Cauchy's Mean Value Theorem, a Mean Value Theorem that involves two functions instead of one. We prove Cauchy's Theorem first and then show how it leads to l'Hôpital's Rule.

Cauchy's Mean Value Theorem

Suppose functions f and g are continuous on $[a, b]$ and differentiable throughout (a, b) and also suppose $g'(x) \neq 0$ throughout (a, b). Then there exists a number c in (a, b) at which

$$\frac{f'(c)}{g'(c)} = \frac{f(b) - f(a)}{g(b) - g(a)}. \qquad (2)$$

The ordinary Mean Value Theorem (Section 3.2, Theorem 4) is the case $g(x) = x$.

Proof of Cauchy's Mean Value Theorem We apply the Mean Value Theorem of Section 3.2 twice. First we use it to show that $g(a) \neq g(b)$. For if $g(b)$ did equal $g(a)$, then the Mean Value Theorem would give

$$g'(c) = \frac{g(b) - g(a)}{b - a} = 0$$

for some c between a and b, which cannot happen because $g'(x) \neq 0$ in (a, b).

We next apply the Mean Value Theorem to the function

$$F(x) = f(x) - f(a) - \frac{f(b) - f(a)}{g(b) - g(a)}[g(x) - g(a)].$$

This function is continuous and differentiable where f and g are, and $F(b) = F(a) = 0$. Therefore, there is a number c between a and b for which $F'(c) = 0$. When expressed in terms of f and g, this equation becomes

$$F'(c) = f'(c) - \frac{f(b) - f(a)}{g(b) - g(a)}[g'(c)] = 0$$

or

$$\frac{f'(c)}{g'(c)} = \frac{f(b) - f(a)}{g(b) - g(a)},$$

which is Equation (2).

Proof of the Stronger Form of l'Hôpital's Rule We first establish Equation (1) for the case $x \to x_0^+$. The method needs almost no change to apply to $x \to x_0^-$, and the combination of these two cases establishes the result.

Suppose that x lies to the right of x_0. Then $g'(x) \neq 0$, and we can apply Cauchy's Mean Value Theorem to the closed interval from x_0 to x. This step produces a number c between x_0 and x such that

$$\frac{f'(c)}{g'(c)} = \frac{f(x) - f(x_0)}{g(x) - g(x_0)}.$$

But $f(x_0) = g(x_0) = 0$, so

$$\frac{f'(c)}{g'(c)} = \frac{f(x)}{g(x)}.$$

As x approaches x_0, c approaches x_0 because it lies between x and x_0. Therefore,

$$\lim_{x \to x_0^+} \frac{f(x)}{g(x)} = \lim_{c \to x_0^+} \frac{f'(c)}{g'(c)} = \lim_{x \to x_0^+} \frac{f'(x)}{g'(x)},$$

which establishes l'Hôpital's Rule for the case where x approaches x_0 from above. The case where x approaches x_0 from below is proved by applying Cauchy's Mean Value Theorem to the closed interval $[x, x_0]$, $x < x_0$.

A.7 Limits That Arise Frequently

This appendix verifies limits (4) through (6) in Section 8.1, Table 8.1.

Limit 4: If $|x| < 1$, $\lim_{n \to \infty} x^n = 0$ We need to show that to each $\epsilon > 0$ there corresponds an integer N so large that $|x^n| < \epsilon$ for all n greater than N. Since $\epsilon^{1/n} \to 1$, while $|x| < 1$, there exists an integer N for which $\epsilon^{1/N} > |x|$. In other words,

$$|x^N| = |x|^N < \epsilon. \tag{1}$$

This integer is the one we seek because, if $|x| < 1$, then

$$|x^n| < |x^N| \qquad \text{for all } n > N. \tag{2}$$

Combining Equations (1) and (2) produces $|x^n| < \epsilon$ for all $n > N$, concluding the proof.

Limit 5: For any number x, $\lim_{n\to\infty}\left(1+\frac{x}{n}\right)^n = e^x$ Let

$$a_n = \left(1 + \frac{x}{n}\right)^n.$$

Then

$$\ln a_n = \ln\left(1+\frac{x}{n}\right)^n = n \ln\left(1+\frac{x}{n}\right) \to x,$$

as we can see by the following application of l'Hôpital's Rule, in which we differentiate with respect to n:

$$\lim_{n\to\infty} n \ln\left(1+\frac{x}{n}\right) = \lim_{n\to\infty} \frac{\ln(1+x/n)}{1/n}$$

$$= \lim_{n\to\infty} \frac{\left(\dfrac{1}{1+x/n}\right)\cdot\left(-\dfrac{x}{n^2}\right)}{-1/n^2} = \lim_{n\to\infty} \frac{x}{1+x/n} = x.$$

Apply Theorem 3, Section 8.1, with $f(x) = e^x$ to conclude that

$$\left(1+\frac{x}{n}\right)^n = a_n = e^{\ln a_n} \to e^x.$$

Limit 6: For any number x, $\lim_{n\to\infty}\frac{x^n}{n!} = 0$ Since

$$-\frac{|x|^n}{n!} \le \frac{x^n}{n!} \le \frac{|x|^n}{n!},$$

all we need to show is that $|x|^n/n! \to 0$. We can then apply the Sandwich Theorem for Sequences (Section 8.1, Theorem 2) to conclude that $x^n/n! \to 0$.

The first step in showing that $|x|^n/n! \to 0$ is to choose an integer $M > |x|$, so that $(|x|/M) < 1$. By Limit 4, just proved, we then have $(|x|/M)^n \to 0$. We then restrict our attention to values of $n > M$. For these values of n, we can write

$$\frac{|x|^n}{n!} = \frac{|x|^n}{1\cdot 2\cdot \cdots \cdot M \cdot \underbrace{(M+1)(M+2)\cdot\cdots\cdot n}_{(n-M)\text{ factors}}}$$

$$\le \frac{|x|^n}{M!M^{n-M}} = \frac{|x|^n M^M}{M!M^n} = \frac{M^M}{M!}\left(\frac{|x|}{M}\right)^n.$$

Thus,

$$0 \le \frac{|x|^n}{n!} \le \frac{M^M}{M!}\left(\frac{|x|}{M}\right)^n.$$

Now, the constant $M^M/M!$ does not change as n increases. Thus, the Sandwich Theorem tells us that $|x|^n/n! \to 0$ because $(|x|/M)^n \to 0$.

A.8 Proof of Taylor's Theorem

This appendix proves Taylor's Theorem (Section 8.7, Theorem 16) in the following form.

Theorem 16 Taylor's Theorem

If f and its first n derivatives $f', f'', \ldots, f^{(n)}$ are continuous on $[a, b]$ or on $[b, a]$, and $f^{(n)}$ is differentiable on (a, b) or on (b, a), then there exists a number c between a and b such that

$$f(b) = f(a) + f'(a)(b-a) + \frac{f''(a)}{2!}(b-a)^2 + \cdots$$

where

$$+ \frac{f^{(n)}(a)}{n!}(b-a)^n + \frac{f^{(n+1)}(c)}{(n+1)!}(b-a)^{n+1}.$$

Proof We prove Taylor's Theorem assuming $a < b$. The proof for $a > b$ is nearly the same.

Then for any x in the interval $[a, b]$, the Taylor polynomial

$$P_n(x) = f(a) + f'(a)(x-a) + \frac{f''(a)}{2!}(x-a)^2 + \cdots + \frac{f^{(n)}(a)}{n!}(x-a)^n$$

and its first n derivatives match the function f and its first n derivatives at $x = a$. We do not disturb that matching if we add another term of the form $K(x-a)^{n+1}$, where K is any constant, because such a term and its first n derivatives are all equal to zero at $x = a$. The new function

$$\phi_n(x) = P_n(x) + K(x-a)^{n+1}$$

and its first n derivatives still agree with f and its first n derivatives at $x = a$.

We now choose the particular value of K that makes the curve $y = \phi_n(x)$ agree with the original curve $y = f(x)$ at $x = b$. In symbols,

$$f(b) = P_n(b) + K(b-a)^{n+1} \qquad \text{or} \qquad K = \frac{f(b) - P_n(b)}{(b-a)^{n+1}}. \tag{1}$$

With K defined by the last equation, the function

$$F(x) = f(x) - \phi_n(x)$$

measures the difference between the original function f and the approximating function ϕ_n for each x in $[a, b]$.

We now use Rolle's Theorem (Section 3.2). First, because $F(a) = F(b) = 0$ and both F and F' are continuous on $[a, b]$, we know that

$$F'(c_1) = 0 \qquad \text{for some } c_1 \text{ in } (a, b).$$

Next, because $F'(a) = F'(c_1) = 0$ and both F' and F'' are continuous on $[a, c_1]$, we know that

$$F''(c_2) = 0 \qquad \text{for some } c_2 \text{ in } (a, c_1).$$

Rolle's Theorem, applied successively to $F'', F''', \ldots, F^{(n-1)}$, implies the existence of

$$\begin{aligned} &c_3 \text{ in } (a, c_2) && \text{such that } F'''(c_3) = 0 \\ &c_4 \text{ in } (a, c_3) && \text{such that } F^{(4)}(c_4) = 0 \\ &\quad\vdots \\ &c_n \text{ in } (a, c_{n-1}) && \text{such that } F^{(n)}(c_n) = 0. \end{aligned}$$

Finally, because $F^{(n)}$ is continuous on $[a, c_n]$ and differentiable on (a, c_n), and $F^{(n)}(a) = F^{(n)}(c_n) = 0$, Rolle's Theorem implies that there is a number c_{n+1} in (a, c_n) such that

$$F^{(n+1)}(c_{n+1}) = 0. \tag{2}$$

If we differentiate $F(x) = f(x) - P_n(x) - K(x - a)^{n+1}$ a total of $n + 1$ times, we get

$$F^{(n+1)}(x) = f^{(n+1)}(x) - 0 - (n + 1)!K. \tag{3}$$

Equations (2) and (3) together give

$$K = \frac{f^{(n+1)}(c)}{(n + 1)!} \qquad \text{for some number } c = c_{n+1} \text{ in } (a, b). \tag{4}$$

Equations (1) and (4) give

$$f(b) = P_n(b) + \frac{f^{(n+1)}(c)}{(n + 1)!}(b - a)^{n+1}. \tag{5}$$

This concludes the proof.

A.9 The Distributive Law for Vector Cross Products

In this appendix, we prove the distributive law

$$\mathbf{u} \times (\mathbf{v} + \mathbf{w}) = \mathbf{u} \times \mathbf{v} + \mathbf{u} \times \mathbf{w}$$

from Property 2 in Section 10.2.

Proof To derive the distributive law, we construct $\mathbf{u} \times \mathbf{v}$ a new way. We draw $\mathbf{u}$ and $\mathbf{v}$ from the common point O and construct a plane M perpendicular to $\mathbf{u}$ at O (Figure A.12). We then project $\mathbf{v}$ orthogonally onto M, yielding a vector $\mathbf{v}'$ with length $|\mathbf{v}| \sin \theta$. We rotate $\mathbf{v}'$ $90°$ about $\mathbf{u}$ in the positive sense to produce a vector $\mathbf{v}''$. Finally, we multiply $\mathbf{v}''$ by the length of $\mathbf{u}$. The resulting vector $|\mathbf{u}|\mathbf{v}''$

is equal to $\mathbf{u} \times \mathbf{v}$ since $\mathbf{v}''$ has the same direction as $\mathbf{u} \times \mathbf{v}$ by its construction (Figure A.12) and

$$|\mathbf{u}||\mathbf{v}''| = |\mathbf{u}||\mathbf{v}'| = |\mathbf{u}||\mathbf{v}| \sin \theta = |\mathbf{u} \times \mathbf{v}|.$$

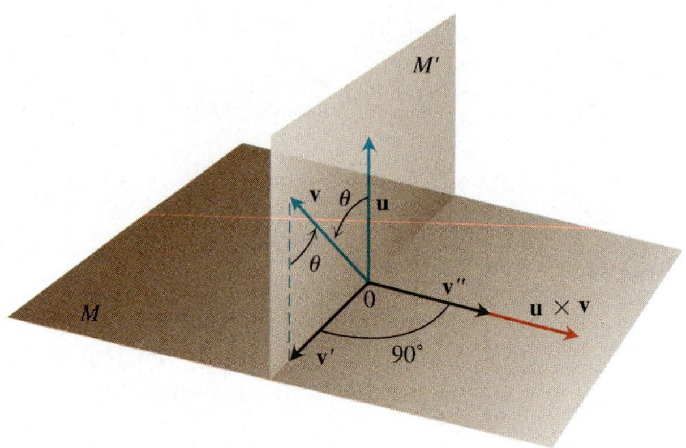

FIGURE A.12 As explained in the text, $\mathbf{u} \times \mathbf{v} = |\mathbf{u}|\mathbf{v}''$.

Now each of these three operations, namely,

1. projection onto M
2. rotation about $\mathbf{u}$ through 90°
3. multiplication by the scalar $|\mathbf{u}|$

when applied to a triangle whose plane is not parallel to $\mathbf{u}$, will produce another triangle. If we start with the triangle whose sides are $\mathbf{v}$, $\mathbf{w}$, and $\mathbf{v} + \mathbf{w}$ (Figure A.13) and apply these three steps, we successively obtain the following:

1. A triangle whose sides are $\mathbf{v}'$, $\mathbf{w}'$, and $(\mathbf{v} + \mathbf{w})'$ satisfying the vector equation

$$\mathbf{v}' + \mathbf{w}' = (\mathbf{v} + \mathbf{w})'$$

2. A triangle whose sides are $\mathbf{v}''$, $\mathbf{w}''$, and $(\mathbf{v} + \mathbf{w})''$ satisfying the vector equation

$$\mathbf{v}'' + \mathbf{w}'' = (\mathbf{v} + \mathbf{w})''$$

(the double prime on each vector has the same meaning as in Figure A.12)

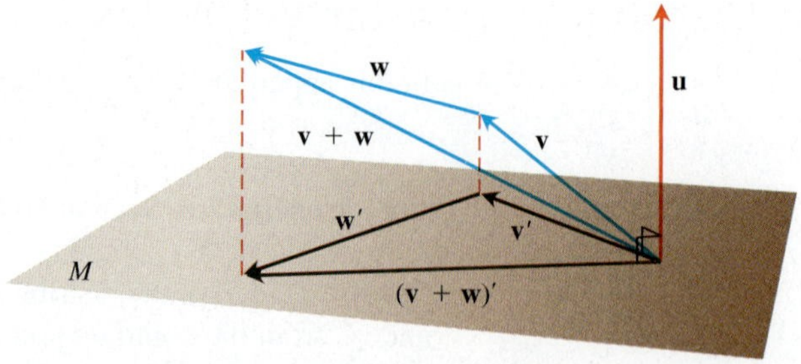

FIGURE A.13 The vectors, $\mathbf{v}$, $\mathbf{w}$, $\mathbf{v} + \mathbf{w}$, and their projections onto a plane perpendicular to $\mathbf{u}$.

3. A triangle whose sides are $|\mathbf{u}|\mathbf{v}''$, $|\mathbf{u}|\mathbf{w}''$, and $|\mathbf{u}|(\mathbf{v}+\mathbf{w})''$ satisfying the vector equation

$$|\mathbf{u}|\mathbf{v}'' + |\mathbf{u}|\mathbf{w}'' = |\mathbf{u}|(\mathbf{v}+\mathbf{w})''.$$

Substituting $|\mathbf{u}|\mathbf{v}'' = \mathbf{u}\times\mathbf{v}$, $|\mathbf{u}|\mathbf{w}'' = \mathbf{u}\times\mathbf{w}$, and $|\mathbf{u}|(\mathbf{v}+\mathbf{w})'' = \mathbf{u}\times(\mathbf{v}+\mathbf{w})$ from our discussion above into this last equation gives

$$\mathbf{u}\times\mathbf{v} + \mathbf{u}\times\mathbf{w} = \mathbf{u}\times(\mathbf{v}+\mathbf{w}),$$

which is the law we wanted to establish.

A.10 Determinants and Cramer's Rule

A rectangular array of numbers such as

$$A = \begin{bmatrix} 2 & 1 & 3 \\ 1 & 0 & -2 \end{bmatrix}$$

is called a **matrix.** We call A a 2 by 3 matrix because it has two rows and three columns. An m by n matrix has m rows and n columns, and the **entry** or **element** (number) in the ith row and jth column is denoted by a_{ij}. The matrix

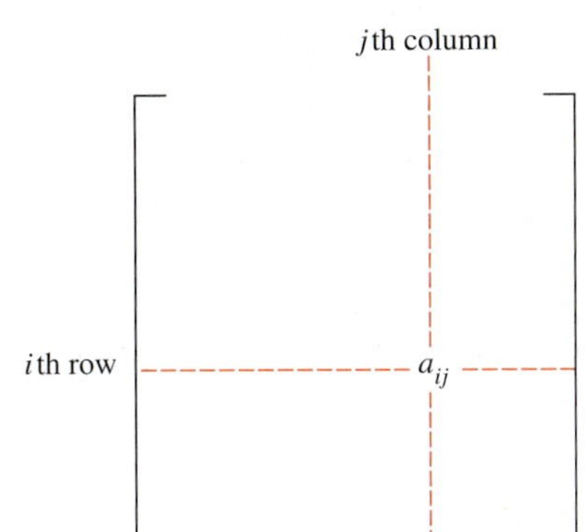

$$A = \begin{bmatrix} 2 & 1 & 3 \\ 1 & 0 & -2 \end{bmatrix}$$

has

$$\begin{aligned} a_{11} &= 2, & a_{12} &= 1, & a_{13} &= 3, \\ a_{21} &= 1, & a_{22} &= 0, & a_{23} &= -2. \end{aligned}$$

A matrix with the same number of rows as columns is a **square matrix.** It is a **square matrix of order** n if the number of rows and columns is n.

With each square matrix A we associate a number $\det A$ or $|a_{ij}|$, called the **determinant** of A, calculated from the entries of A in the following way. For $n = 1$ and $n = 2$, we define

The vertical bars in the notation $|a_{ij}|$ do not mean absolute value.

$$\det\,[a] = a \tag{1}$$

$$\det \begin{bmatrix} a_{11} & a_{12} \\ a_{21} & a_{22} \end{bmatrix} = a_{11}a_{22} - a_{21}a_{12}. \tag{2}$$

For a matrix of order 3, we write

$$\det A = \det \begin{bmatrix} a_{11} & a_{12} & a_{13} \\ a_{21} & a_{22} & a_{23} \\ a_{31} & a_{32} & a_{33} \end{bmatrix} = \begin{array}{l}\text{sum of all signed products} \\ \text{of the form } \pm a_{1i}a_{2j}a_{3k},\end{array} \tag{3}$$

where i, j, k is a permutation of 1, 2, 3 in some order. There are $3! = 6$ such permutations, so there are six terms in the sum. The sign is positive when the index of the permutation is even and negative when the index is odd.

Definition Index of a Permutation

Given any permutation of the numbers 1, 2, 3, . . . , n, denote the permutation by $i_1, i_2, i_3, \ldots, i_n$. In this arrangement, some of the numbers following i_1 may be less than i_1, and the number of these is called the **number of inversions** in the arrangement pertaining to i_1. Likewise, there are a number of inversions pertaining to each of the other i's; it is the number of indices that come after that particular i in the arrangement and are less than it. The **index** of the permutation is the sum of all the numbers of inversions pertaining to the separate indices.

Example 1 Finding the Index of a Permutation

For $n = 5$, the permutation

$$5 \quad 3 \quad 1 \quad 2 \quad 4$$

has four inversions pertaining to the first element, 5, two inversions pertaining to the second element, 3, and no further inversions, so the index is $4 + 2 = 6$.

The following table shows the permutations of 1, 2, 3, the index of each permutation, and the signed product in the determinant of Equation (3).

Permutation	Index	Signed product
1 2 3	0	$+a_{11}a_{22}a_{33}$
1 3 2	1	$-a_{11}a_{23}a_{32}$
2 1 3	1	$-a_{12}a_{21}a_{33}$
2 3 1	2	$+a_{12}a_{23}a_{31}$
3 1 2	2	$+a_{13}a_{21}a_{32}$
3 2 1	3	$-a_{13}a_{22}a_{31}$

The sum of the six signed products is

$$a_{11}(a_{22}a_{33} - a_{23}a_{32}) - a_{12}(a_{21}a_{33} - a_{23}a_{31}) + a_{13}(a_{21}a_{32} - a_{22}a_{31})$$

$$= a_{11}\begin{vmatrix} a_{22} & a_{23} \\ a_{32} & a_{33} \end{vmatrix} - a_{12}\begin{vmatrix} a_{21} & a_{23} \\ a_{31} & a_{33} \end{vmatrix} + a_{13}\begin{vmatrix} a_{21} & a_{22} \\ a_{31} & a_{32} \end{vmatrix} = \begin{vmatrix} a_{11} & a_{12} & a_{13} \\ a_{21} & a_{22} & a_{23} \\ a_{31} & a_{32} & a_{33} \end{vmatrix}.$$

The formula

$$\begin{vmatrix} a_{11} & a_{12} & a_{13} \\ a_{21} & a_{22} & a_{23} \\ a_{31} & a_{32} & a_{33} \end{vmatrix} = a_{11}\begin{vmatrix} a_{22} & a_{23} \\ a_{32} & a_{33} \end{vmatrix} - a_{12}\begin{vmatrix} a_{21} & a_{23} \\ a_{31} & a_{33} \end{vmatrix} + a_{13}\begin{vmatrix} a_{21} & a_{22} \\ a_{31} & a_{32} \end{vmatrix} \tag{4}$$

reduces the calculation of a 3 by 3 determinant to the calculation of three 2 by 2 determinants.

Many people prefer to remember the following scheme for calculating the six signed products in the determinant of a 3 by 3 matrix:

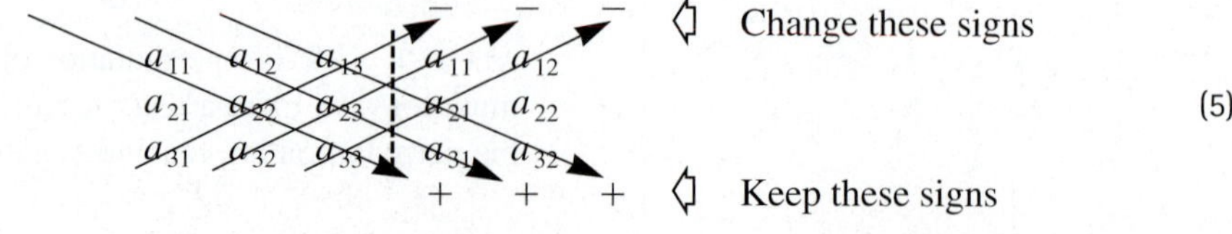

(5)

Minors and Cofactors

The second-order determinants on the right-hand side of Equation (4) are called the **minors** (short for "minor determinants") of the entries they multiply. Thus,

$$\begin{vmatrix} a_{22} & a_{23} \\ a_{32} & a_{33} \end{vmatrix} \text{ is the minor of } a_{11}, \qquad \begin{vmatrix} a_{21} & a_{23} \\ a_{31} & a_{33} \end{vmatrix} \text{ is the minor of } a_{12},$$

and so on. The minor of the element a_{ij} in a matrix A is the determinant of the matrix that remains after we delete the row and column containing a_{ij}.

$$\begin{vmatrix} a_{11} & a_{12} & a_{13} \\ a_{21} & a_{22} & a_{23} \\ a_{31} & a_{32} & a_{33} \end{vmatrix}; \qquad \text{the minor of } a_{22} \text{ is } \begin{vmatrix} a_{11} & a_{13} \\ a_{31} & a_{33} \end{vmatrix}$$

$$\begin{vmatrix} a_{11} & a_{12} & a_{13} \\ a_{21} & a_{22} & a_{23} \\ a_{31} & a_{32} & a_{33} \end{vmatrix}; \qquad \text{the minor of } a_{23} \text{ is } \begin{vmatrix} a_{11} & a_{12} \\ a_{31} & a_{32} \end{vmatrix}$$

The **cofactor** A_{ij} of a_{ij} is $(-1)^{i+j}$ times the minor of a_{ij}. Thus,

$$A_{22} = (-1)^{2+2} \begin{vmatrix} a_{11} & a_{13} \\ a_{31} & a_{33} \end{vmatrix} = \begin{vmatrix} a_{11} & a_{13} \\ a_{31} & a_{33} \end{vmatrix}$$

$$A_{23} = (-1)^{2+3} \begin{vmatrix} a_{11} & a_{12} \\ a_{31} & a_{32} \end{vmatrix} = -\begin{vmatrix} a_{11} & a_{12} \\ a_{31} & a_{32} \end{vmatrix}.$$

The factor $(-1)^{i+j}$ changes the sign of the minor when $i + j$ is odd. There is a checkerboard pattern for remembering these changes:

$$\begin{matrix} + & - & + \\ - & + & - \\ + & - & +. \end{matrix}$$

In the upper left corner, $i = 1$, $j = 1$ and $(-1)^{1+1} = +1$. In going from any cell to an adjacent cell in the same row or column, we change i by 1 or j by 1, but not both, so we change the exponent from even to odd or from odd to even, which changes the sign from $+$ to $-$ or from $-$ to $+$.

When we rewrite Equation (4) in terms of cofactors we get

$$\det A = a_{11}A_{11} + a_{12}A_{12} + a_{13}A_{13}. \tag{6}$$

Example 2 Finding a Determinant Two Ways

Find the determinant of

$$A = \begin{bmatrix} 2 & 1 & 3 \\ 3 & -1 & -2 \\ 2 & 3 & 1 \end{bmatrix}.$$

Solution 1

Using Equation (6)

The cofactors are

$$A_{11} = (-1)^{1+1}\begin{vmatrix} -1 & -2 \\ 3 & 1 \end{vmatrix}, \qquad A_{12} = (-1)^{1+2}\begin{vmatrix} 3 & -2 \\ 2 & 1 \end{vmatrix},$$

$$A_{13} = (-1)^{1+3}\begin{vmatrix} 3 & -1 \\ 2 & 3 \end{vmatrix}.$$

To find $\det A$, we multiply each element of the first row of A by its cofactor and add:

$$\det A = 2\begin{vmatrix} -1 & -2 \\ 3 & 1 \end{vmatrix} + (-1)\begin{vmatrix} 3 & -2 \\ 2 & 1 \end{vmatrix} + 3\begin{vmatrix} 3 & -1 \\ 2 & 3 \end{vmatrix}$$

$$= 2(-1 + 6) - 1(3 + 4) + 3(9 + 2) = 10 - 7 + 33 = 36.$$

Solution 2

Using the Scheme (5)

We find

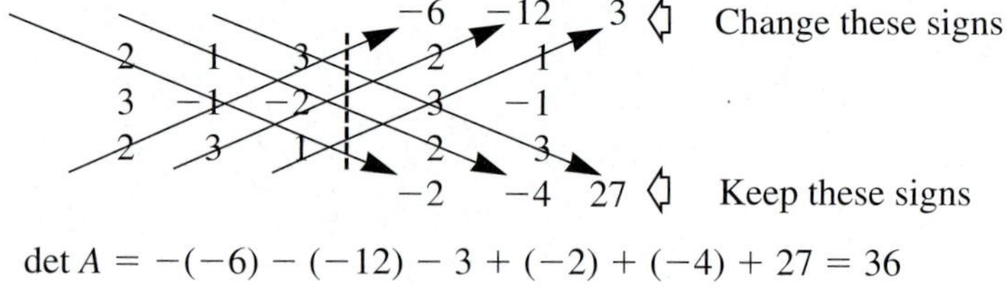

$$\det A = -(-6) - (-12) - 3 + (-2) + (-4) + 27 = 36$$

Expanding by Columns or by Other Rows

The determinant of a square matrix can be calculated from the cofactors of any row or any column.

If we were to expand the determinant in Example 2 by cofactors according to elements of its third column, say, we would get

$$+3\begin{vmatrix} 3 & -1 \\ 2 & 3 \end{vmatrix} - (-2)\begin{vmatrix} 2 & 1 \\ 2 & 3 \end{vmatrix} + 1\begin{vmatrix} 2 & 1 \\ 3 & -1 \end{vmatrix}$$

$$= 3(9 + 2) + 2(6 - 2) + 1(-2 - 3) = 33 + 8 - 5 = 36.$$

Useful Facts About Determinants

Fact 1: If two rows (or columns) are identical, the determinant is zero.

Fact 2: Interchanging two rows (or columns) changes the sign of the determinant.

Fact 3: The determinant is the sum of the products of the elements of the ith row (or column) by their cofactors, for any i.

Fact 4: The determinant of the transpose of a matrix is the same as the determinant of the original matrix. (The **transpose** of a matrix is obtained by writing the rows as columns.)

Fact 5: Multiplying each element of some row (or column) by a constant c multiplies the determinant by c.

Fact 6: If all elements above the main diagonal (or all below it) are zero, the determinant is the product of the elements on the main diagonal. (The **main diagonal** is the diagonal from upper left to lower right.)

Example 3 Illustrating Fact 6

$$\begin{vmatrix} 3 & 4 & 7 \\ 0 & -2 & 5 \\ 0 & 0 & 5 \end{vmatrix} = (3)(-2)(5) = -30$$

Fact 7: If the elements of any row are multiplied by the cofactors of the corresponding elements of a different row and these products are summed, then the sum is zero.

Example 4 Ilustrating Fact 7

If A_{11}, A_{12}, A_{13} are the cofactors of the elements of the first row of $A = (a_{ij})$, then the sums

$$a_{21}A_{11} + a_{22}A_{12} + a_{23}A_{13}$$

(elements of second row times cofactors of elements of first row) and

$$a_{31}A_{11} + a_{32}A_{12} + a_{33}A_{13}$$

are both zero.

Fact 8: If the elements of any column are multiplied by the cofactors of the corresponding elements of a different column and these products are summed, then the sum is zero.

Fact 9: If each element of a row is multiplied by a constant c and the results added to a different row, then the determinant is not changed. A similar result holds for columns.

Example 5 Adding a Multiple of One Row to Another Row

If we start with

$$A = \begin{bmatrix} 2 & 1 & 3 \\ 3 & -1 & -2 \\ 2 & 3 & 1 \end{bmatrix}$$

and add -2 times row 1 to row 2 (subtract 2 times row 1 from row 2), we get

$$B = \begin{bmatrix} 2 & 1 & 3 \\ -1 & -3 & -8 \\ 2 & 3 & 1 \end{bmatrix}.$$

Since $\det A = 36$ (Example 2), we should find that $\det B = 36$ as well. Indeed we do, as the following calculation shows:

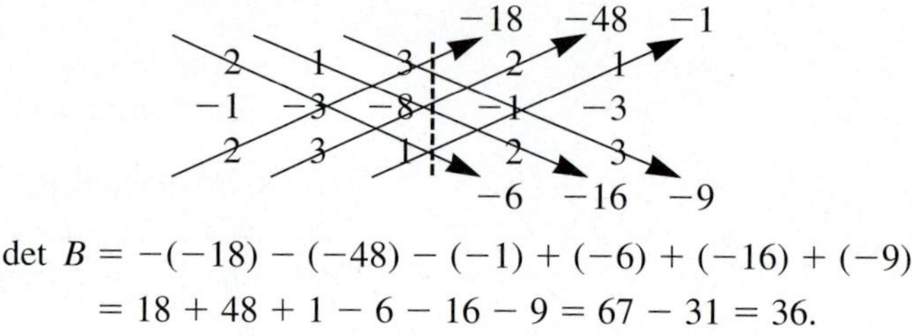

$$\det B = -(-18) - (-48) - (-1) + (-6) + (-16) + (-9)$$
$$= 18 + 48 + 1 - 6 - 16 - 9 = 67 - 31 = 36.$$

Example 6 Evaluating a Fourth-Order Determinant by Applying Fact 9

Evaluate the fourth order determinant

$$D = \begin{vmatrix} 1 & -2 & 3 & 1 \\ 2 & 1 & 0 & 2 \\ -1 & 2 & 1 & -2 \\ 0 & 1 & 2 & 1 \end{vmatrix}.$$

Solution We subtract 2 times row 1 from row 2 and add row 1 to row 3 to get

$$D = \begin{vmatrix} 1 & -2 & 3 & 1 \\ 0 & 5 & -6 & 0 \\ 0 & 0 & 4 & -1 \\ 0 & 1 & 2 & 1 \end{vmatrix}.$$

We then multiply the elements of the first column by their cofactors to get

$$D = \begin{vmatrix} 5 & -6 & 0 \\ 0 & 4 & -1 \\ 1 & 2 & 1 \end{vmatrix} = 5(4 + 2) - (-6)(0 + 1) + 0 = 36.$$

Cramer's Rule

If the determinant $D = \det A = \begin{vmatrix} a_{11} & a_{12} \\ a_{21} & a_{22} \end{vmatrix} = 0$, then the system

$$\begin{aligned} a_{11}x + a_{12}y &= b_1 \\ a_{21}x + a_{22}y &= b_2 \end{aligned} \tag{7}$$

has either infinitely many solutions or no solution at all. The system

$$\begin{aligned} x + y &= 0 \\ 2x + 2y &= 0 \end{aligned}$$

whose determinant is

$$D = \begin{vmatrix} 1 & 1 \\ 2 & 2 \end{vmatrix} = 2 - 2 = 0$$

has infinitely many solutions. We can find an x to match any given y. The system

$$\begin{aligned} x + y &= 0 \\ 2x + 2y &= 2 \end{aligned}$$

has no solution. If $x + y = 0$, then $2x + 2y = 2(x + y)$ cannot be 2.

If $D \neq 0$, then the system (7) has a unique solution, and Cramer's Rule states that it may be found from the formulas

$$x = \frac{\begin{vmatrix} b_1 & a_{12} \\ b_2 & a_{22} \end{vmatrix}}{D}, \qquad y = \frac{\begin{vmatrix} a_{11} & b_1 \\ a_{21} & b_2 \end{vmatrix}}{D}. \tag{8}$$

The numerator in the formula for x comes from replacing the first column in A (the x-column) by the column of constants b_1 and b_2 (the b-column). Replacing the y-column by the b-column gives the numerator of the y-solution.

Example 7 Using Cramer's Rule

Solve the system

$$\begin{aligned} 3x - y &= 9 \\ x + 2y &= -4. \end{aligned}$$

Solution We use Equations (8). The determinant of the coefficient matrix is

$$D = \begin{vmatrix} 3 & -1 \\ 1 & 2 \end{vmatrix} = 6 + 1 = 7.$$

Hence,

$$x = \frac{\begin{vmatrix} 9 & -1 \\ -4 & 2 \end{vmatrix}}{D} = \frac{18 - 4}{7} = \frac{14}{7} = 2$$

$$y = \frac{\begin{vmatrix} 3 & 9 \\ 1 & -4 \end{vmatrix}}{D} = \frac{-12 - 9}{7} = \frac{-21}{7} = -3.$$

Systems of three equations in three unknowns work the same way. If

$$D = \det A = \begin{vmatrix} a_{11} & a_{12} & a_{13} \\ a_{21} & a_{22} & a_{23} \\ a_{31} & a_{32} & a_{33} \end{vmatrix} = 0,$$

then the system

$$\begin{aligned} a_{11}x + a_{12}y + a_{13}z &= b_1 \\ a_{21}x + a_{22}y + a_{23}z &= b_2 \\ a_{31}x + a_{32}y + a_{33}z &= b_3 \end{aligned}$$

has either infinitely many solutions or no solution at all. If $D \neq 0$, then the system has a unique solution, given by Cramer's Rule:

$$x = \frac{1}{D}\begin{vmatrix} b_1 & a_{12} & a_{13} \\ b_2 & a_{22} & a_{23} \\ b_3 & a_{32} & a_{33} \end{vmatrix}, \qquad y = \frac{1}{D}\begin{vmatrix} a_{11} & b_1 & a_{13} \\ a_{21} & b_2 & a_{23} \\ a_{31} & b_3 & a_{33} \end{vmatrix}, \qquad z = \frac{1}{D}\begin{vmatrix} a_{11} & a_{12} & b_1 \\ a_{21} & a_{22} & b_2 \\ a_{31} & a_{32} & b_3 \end{vmatrix}.$$

The pattern continues in higher dimensions.

EXERCISES A.10

Evaluting Determinants

Evaluate the following determinants.

1. $\begin{vmatrix} 2 & 3 & 1 \\ 4 & 5 & 2 \\ 1 & 2 & 3 \end{vmatrix}$

2. $\begin{vmatrix} 2 & -1 & -2 \\ -1 & 2 & 1 \\ 3 & 0 & -3 \end{vmatrix}$

3. $\begin{vmatrix} 1 & 2 & 3 & 4 \\ 0 & 1 & 2 & 3 \\ 0 & 0 & 2 & 1 \\ 0 & 0 & 3 & 2 \end{vmatrix}$

4. $\begin{vmatrix} 1 & -1 & 2 & 3 \\ 2 & 1 & 2 & 6 \\ 1 & 0 & 2 & 3 \\ -2 & 2 & 0 & -5 \end{vmatrix}$

Evaluate the following determinants by expanding according to the cofactors of (a) the third row and (b) the second column.

5. $\begin{vmatrix} 2 & -1 & 2 \\ 1 & 0 & 3 \\ 0 & 2 & 1 \end{vmatrix}$

6. $\begin{vmatrix} 1 & 0 & -1 \\ 0 & 2 & -2 \\ 2 & 0 & 1 \end{vmatrix}$

7. $\begin{vmatrix} 1 & 1 & 0 & 0 \\ 0 & 0 & -2 & 1 \\ 0 & -1 & 0 & 7 \\ 3 & 0 & 2 & 1 \end{vmatrix}$

8. $\begin{vmatrix} 0 & 1 & 0 & 0 \\ 0 & 1 & 1 & 0 \\ 1 & 1 & 1 & 1 \\ 1 & 1 & 0 & 0 \end{vmatrix}$

Systems of Equations

Solve the following systems of equations by Cramer's Rule.

9. $\begin{aligned} x + 8y &= 4 \\ 3x - y &= -13 \end{aligned}$

10. $\begin{aligned} 2x + 3y &= 5 \\ 3x - y &= 2 \end{aligned}$

11. $\begin{aligned} 4x - 3y &= 6 \\ 3x - 2y &= 5 \end{aligned}$

12. $\begin{aligned} x + y + z &= 2 \\ 2x - y + z &= 0 \\ x + 2y - z &= 4 \end{aligned}$

13. $\begin{aligned} 2x + y - z &= 2 \\ x - y + z &= 7 \\ 2x + 2y + z &= 4 \end{aligned}$

14. $\begin{aligned} 2x - 4y &= 6 \\ x + y + z &= 1 \\ 5y + 7z &= 10 \end{aligned}$

15. $\begin{aligned} x - z &= 3 \\ 2y - 2z &= 2 \\ 2x + z &= 3 \end{aligned}$

16. $\begin{aligned} x_1 + x_2 - x_3 + x_4 &= 2 \\ x_1 - x_2 + x_3 + x_4 &= -1 \\ x_1 + x_2 + x_3 - x_4 &= 2 \\ x_1 + x_3 + x_4 &= -1 \end{aligned}$

Theory and Examples

17. *Infinitely many or no solutions* Find values of h and k for which the system

$$\begin{aligned} 2x + hy &= 8 \\ x + 3y &= k \end{aligned}$$

has (a) infinitely many solutions and (b) no solution at all.

18. *A zero determinant* For what value of x will

$$\begin{vmatrix} x & x & 1 \\ 2 & 0 & 5 \\ 6 & 7 & 1 \end{vmatrix} = 0?$$

19. *A zero determinant* Suppose u, v, and w are twice-differentiable functions of x that satisfy the relation $au + bv + cw = 0$, where a, b, and c are constants, not all zero. Show that

$$\begin{vmatrix} u & v & w \\ u' & v' & w' \\ u'' & v'' & w'' \end{vmatrix} = 0.$$

20. *Partial fractions* Expanding the quotient

$$\frac{ax + b}{(x - r_1)(x - r_2)}$$

by partial fractions calls for finding the values of C and D that make the equation

$$\frac{ax + b}{(x - r_1)(x - r_2)} = \frac{C}{x - r_1} + \frac{D}{x - r_2}$$

hold for all x.

(a) Find a system of linear equations that determines C and D.

(b) *Writing to Learn* Under what circumstances does the system of equations in part (a) have a unique solution? That is, when is the determinant of the coefficient matrix of the system different from zero?

A.11 The Mixed Derivative Theorem and the Increment Theorem

This appendix derives the Mixed Derivative Theorem (Theorem 2, Section 11.3) and the Increment Theorem for Functions of Two Variables (Theorem 3, Section 11.3). Euler first published his theorem in 1734, in a series of papers he wrote on hydrodynamics.

Theorem 2 The Mixed Derivative Theorem

If $f(x, y)$ and its partial derivatives f_x, f_y, f_{xy}, and f_{yx} are defined throughout an open region containing a point (a, b) and are all continuous at (a, b), then $f_{xy}(a, b) = f_{yx}(a, b)$.

Proof The equality of $f_{xy}(a, b)$ and $f_{yx}(a, b)$ can be established by four applications of the Mean Value Theorem (Theorem 4, Section 3.2). By hypothesis, the point (a, b) lies in the interior of a rectangle R in the xy-plane on which f, f_x, f_y, f_{xy}, and f_{yx} are all defined. We let h and k be the numbers such that the point $(a + h, b + k)$ also lies in R, and we consider the difference

$$\Delta = F(a + h) - F(a), \tag{1}$$

where

$$F(x) = f(x, b + k) - f(x, b). \tag{2}$$

We apply the Mean Value Theorem to F, which is continuous because it is differentiable, and Equation (1) becomes

$$\Delta = hF'(c_1), \tag{3}$$

where c_1 lies between a and $a + h$. From Equation (2),

$$F'(x) = f_x(x, b + k) - f_x(x, b),$$

so Equation (3) becomes

$$\Delta = h[f_x(c_1, b + k) - f_x(c_1, b)]. \tag{4}$$

Now we apply the Mean Value Theorem to the function $g(y) = f_x(c_1, y)$ and have

$$g(b + k) - g(b) = kg'(d_1),$$

or

$$f_x(c_1, b + k) - f_x(c_1, b) = kf_{xy}(c_1, d_1)$$

for some d_1 between b and $b + k$. By substituting this into Equation (4), we get

$$\Delta = hkf_{xy}(c_1, d_1) \tag{5}$$

for some point (c_1, d_1) in the rectangle R' whose vertices are the four points (a, b), $(a + h, b)$, $(a + h, b + k)$, and $(a, b + k)$. (See Figure A.14.)

By substituting from Equation (2) into Equation (1), we may also write

$$\begin{aligned}\Delta &= f(a + h, b + k) - f(a + h, b) - f(a, b + k) + f(a, b) \\ &= [f(a + h, b + k) - f(a, b + k)] - [f(a + h, b) - f(a, b)] \\ &= \phi(b + k) - \phi(b),\end{aligned} \tag{6}$$

where

$$\phi(y) = f(a + h, y) - f(a, y). \tag{7}$$

The Mean Value Theorem applied to Equation (6) now gives

$$\Delta = k\phi'(d_2) \tag{8}$$

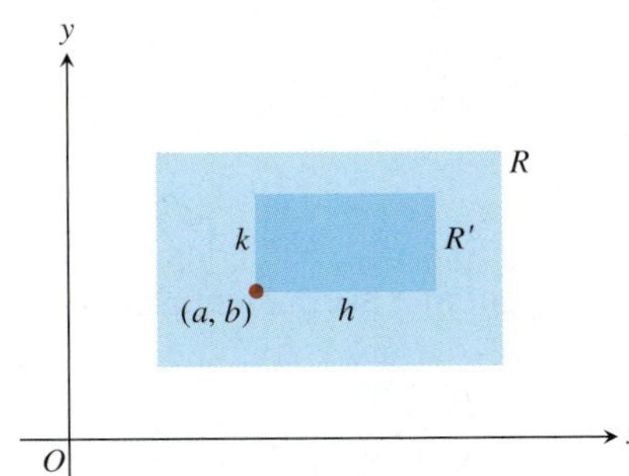

FIGURE A.14 The key to proving $f_{xy}(a, b) = f_{yx}(a, b)$ is that no matter how small R' is, f_{xy} and f_{yx} take on equal values somewhere inside R' (although not necessarily at the same point).

for some d_2 between b and $b+k$. By Equation (7),

$$\phi'(y) = f_y(a+h, y) - f_y(a, y). \tag{9}$$

Substituting from Equation (9) into Equation (8) gives

$$\Delta = k[f_y(a+h, d_2) - f_y(a, d_2)].$$

Finally, we apply the Mean Value Theorem to the expression in brackets and get

$$\Delta = khf_{yx}(c_2, d_2) \tag{10}$$

for some c_2 between a and $a+h$.

Together, Equations (5) and (10) show that

$$f_{xy}(c_1, d_1) = f_{yx}(c_2, d_2), \tag{11}$$

where (c_1, d_1) and (c_2, d_2) both lie in the rectangle R' (Figure A.14). Equation (11) is not quite the result we want, since it says only that f_{xy} has the same value at (c_1, d_1) that f_{yx} has at (c_2, d_2). The numbers h and k in our discussion, however, may be made as small as we wish. The hypothesis that f_{xy} and f_{yx} are both continuous at (a, b) means that $f_{xy}(c_1, d_1) = f_{xy}(a, b) + \epsilon_1$ and $f_{yx}(c_2, d_2) = f_{yx}(a, b) + \epsilon_2$, where $\epsilon_1, \epsilon_2 \to 0$ as $h, k \to 0$. Hence, if we let h and $k \to 0$, we have $f_{xy}(a, b) = f_{yx}(a, b)$.

The equality of $f_{xy}(a, b)$ and $f_{yx}(a, b)$ can be proved with hypotheses weaker than the ones we assumed. For example, it is enough for f, f_x, and f_y to exist in R and for f_{xy} to be continuous at (a, b). Then f_{yx} will exist at (a, b) and will equal f_{xy} at that point.

The Increment Theorem for Functions of Two Variables

Suppose that the first partial derivatives of $z = f(x, y)$ are defined throughout an open region R containing the point (x_0, y_0) and that f_x and f_y are continuous at (x_0, y_0). Then the change $\Delta z = f(x_0 + \Delta x, y_0 + \Delta y) - f(x_0, y_0)$ in the value of f that results from moving from (x_0, y_0) to another point $(x_0 + \Delta x, y_0 + \Delta y)$ in R satisfies an equation of the form

$$\Delta z = f_x(x_0, y_0)\,\Delta x + f_y(x_0, y_0)\,\Delta y + \epsilon_1\,\Delta x + \epsilon_2\,\Delta y,$$

in which $\epsilon_1, \epsilon_2 \to 0$ as $\Delta x, \Delta y \to 0$.

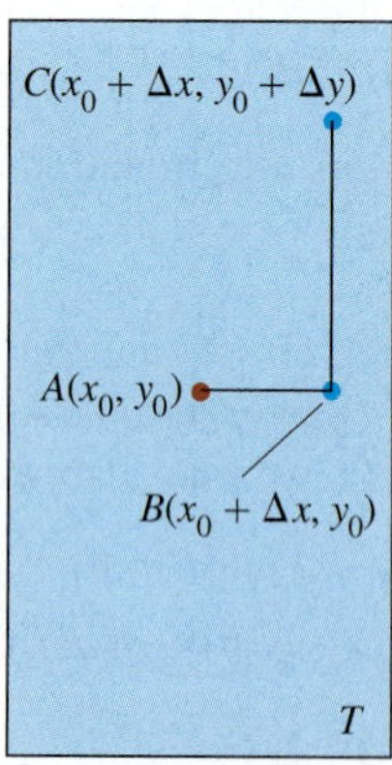

FIGURE A.15 The rectangular region T in the proof of the Increment Theorem. The figure is drawn for Δx and Δy positive, but either increment might be zero or negative.

Proof We work within a rectangle T centered at $A(x_0, y_0)$ and lying within R, and we assume that Δx and Δy are already so small that the line segment joining A to $B(x_0 + \Delta x, y_0)$ and the line segment joining B to $C(x_0 + \Delta x, y_0 + \Delta y)$ lie in the interior of T (Figure A.15).

We may think of Δz as the sum $\Delta z = \Delta z_1 + \Delta z_2$ of two increments, where

$$\Delta z_1 = f(x_0 + \Delta x, y_0) - f(x_0, y_0)$$

is the change in the value of f from A to B and

$$\Delta z_2 = f(x_0 + \Delta x, y_0 + \Delta y) - f(x_0 + \Delta x, y_0)$$

is the change in the value of f from B to C (Figure A.16).

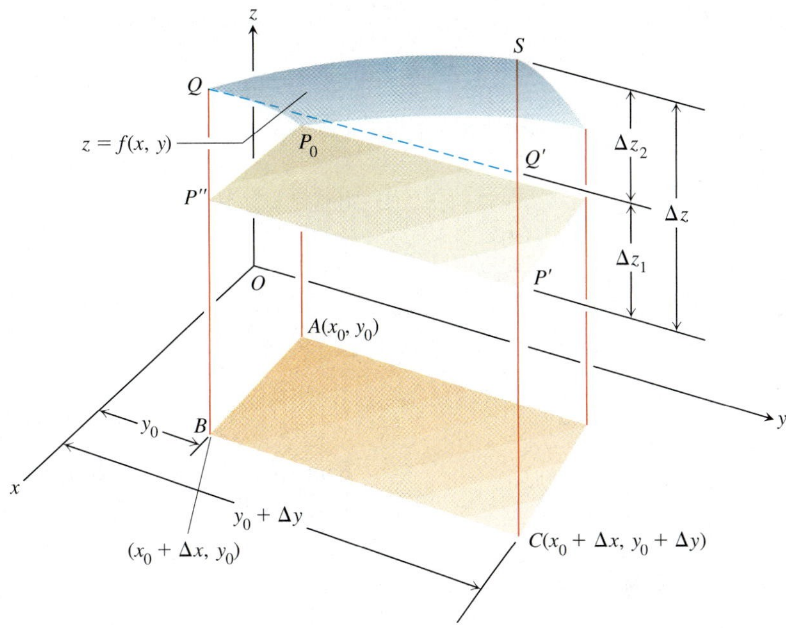

FIGURE A.16 Part of the surface $z = f(x, y)$ near $P_0(x_0, y_0, f(x_0, y_0))$. The points P_0, P', and P'' have the same height $z_0 = f(x_0, y_0)$ above the xy-plane. The change in z is $\Delta z = P'S$. The change

$$\Delta z_1 = f(x_0 + \Delta x, y_0) - f(x_0, y_0),$$

shown as $P''Q = P'Q'$, is caused by changing x from x_0 to $x_0 + \Delta x$ while holding y equal to y_0. Then, with x held equal to $x_0 + \Delta x$,

$$\Delta z_2 = f(x_0 + \Delta x, y_0 + \Delta y) - f(x_0 + \Delta x, y_0)$$

is the change in z caused by changing y from y_0 to $y_0 + \Delta y$, which is represented by $Q'S$. The total change in z is the sum of Δz_1 and Δz_2.

On the closed interval of x-values joining x_0 to $x_0 + \Delta x$, the function $F(x) = f(x, y_0)$ is a differentiable (and hence continuous) function of x, with derivative

$$F'(x) = f_x(x, y_0).$$

By the Mean Value Theorem (Theorem 4, Section 3.2), there is an x-value c between x_0 and $x_0 + \Delta x$ at which

$$F(x_0 + \Delta x) - F(x_0) = F'(c)\,\Delta x$$

or

$$f(x_0 + \Delta x, y_0) - f(x_0, y_0) = f_x(c, y_0)\,\Delta x$$

or

$$\Delta z_1 = f_x(c, y_0)\,\Delta x. \tag{12}$$

Similarly, $G(y) = f(x_0 + \Delta x, y)$ is a differentiable (and hence continuous) function of y on the closed y-interval joining y_0 and $y_0 + \Delta y$, with derivative

$$G'(y) = f_y(x_0 + \Delta x, y).$$

Hence, there is a y-value d between y_0 and $y_0 + \Delta y$ at which

$$G(y_0 + \Delta y) - G(y_0) = G'(d)\,\Delta y$$

or

$$f(x_0 + \Delta x, y_0 + \Delta y) - f(x_0 + \Delta x, y) = f_y(x_0 + \Delta x, d)\,\Delta y$$

or

$$\Delta z_2 = f_y(x_0 + \Delta x, d)\,\Delta y. \tag{13}$$

Now, as Δx and $\Delta y \to 0$, we know that $c \to x_0$ and $d \to y_0$. Therefore, since f_x and f_y are continuous at (x_0, y_0), the quantities

$$\begin{aligned} \epsilon_1 &= f_x(c, y_0) - f_x(x_0, y_0), \\ \epsilon_2 &= f_y(x_0 + \Delta x, d) - f_y(x_0, y_0) \end{aligned} \tag{14}$$

both approach zero as Δx and $\Delta y \to 0$.

Finally,

$$\begin{aligned} \Delta z &= \Delta z_1 + \Delta z_2 \\ &= f_x(c, y_0)\,\Delta x + f_y(x_0 + \Delta x, d)\,\Delta y && \text{From (12) and (13)} \\ &= [f_x(x_0, y_0) + \epsilon_1]\,\Delta x + [f_y(x_0, y_0) + \epsilon_2]\,\Delta y && \text{From (14)} \\ &= f_x(x_0, y_0)\,\Delta x + f_y(x_0, y_0)\,\Delta y + \epsilon_1\,\Delta x + \epsilon_2\,\Delta y, \end{aligned}$$

where ϵ_1 and $\epsilon_2 \to 0$ as Δx and $\Delta y \to 0$, which is what we set out to prove.

Analogous results hold for functions of any finite number of independent variables. Suppose that the first partial derivatives of $w = f(x, y, z)$ are defined throughout an open region containing the point (x_0, y_0, z_0) and that f_x, f_y, and f_z are continuous at (x_0, y_0, z_0). Then

$$\begin{aligned} \Delta w &= f(x_0 + \Delta x, y_0 + \Delta y, z_0 + \Delta z) - f(x_0, y_0, z_0) \\ &= f_x\,\Delta x + f_y\,\Delta y + f_z\,\Delta z + \epsilon_1\,\Delta x + \epsilon_2\,\Delta y + \epsilon_3\,\Delta z, \end{aligned} \tag{15}$$

where $\epsilon_1, \epsilon_2, \epsilon_3 \to 0$ as Δx, Δy, and $\Delta z \to 0$.

The partial derivatives f_x, f_y, f_z in Equation (15) are to be evaluated at the point (x_0, y_0, z_0).

Equation (15) can be proved by treating Δw as the sum of three increments,

$$\Delta w_1 = f(x_0 + \Delta x, y_0, z_0) - f(x_0, y_0, z_0) \tag{16}$$

$$\Delta w_2 = f(x_0 + \Delta x, y_0 + \Delta y, z_0) - f(x_0 + \Delta x, y_0, z_0) \tag{17}$$

$$\Delta w_3 = f(x_0 + \Delta x, y_0 + \Delta y, z_0 + \Delta z) - f(x_0 + \Delta x, y_0 + \Delta y, z_0), \tag{18}$$

and applying the Mean Value Theorem to each of these separately. Two coordinates remain constant and only one varies in each of these partial increments Δw_1, Δw_2, Δw_3. In Equation (17), for example, only y varies, since x is held equal to $x_0 + \Delta x$ and z is held equal to z_0. Since $f(x_0 + \Delta x, y, z_0)$ is a continuous function of y with a derivative f_y, it is subject to the Mean Value Theorem, and we have

$$\Delta w_2 = f_y(x_0 + \Delta x, y_1, z_0)\,\Delta y$$

for some y_1 between y_0 and $y_0 + \Delta y$.

A.12 The Area of a Parallelogram's Projection on a Plane

This appendix proves the result needed in Section 13.5 that $|(\mathbf{u} \times \mathbf{v}) \cdot \mathbf{p}|$ is the area of the projection of the parallelogram with sides determined by $\mathbf{u}$ and $\mathbf{v}$ onto any plane whose normal is $\mathbf{p}$. (See Figure A.17.)

Theorem

The area of the orthogonal projection of the parallelogram determined by two vectors $\mathbf{u}$ and $\mathbf{v}$ in space onto a plane with unit normal vector $\mathbf{p}$ is

$$\text{Area} = |(\mathbf{u} \times \mathbf{v}) \cdot \mathbf{p}|.$$

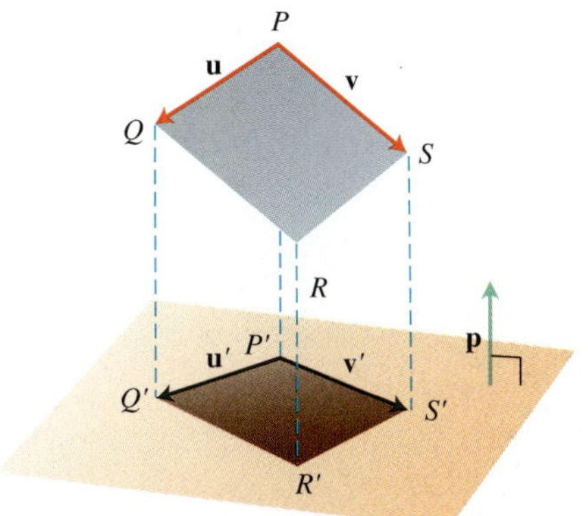

FIGURE A.17 The parallelogram determined by two vectors **u** and **v** in space and the orthogonal projection of the parallelogram onto a plane. The projection lines, orthogonal to the plane, lie parallel to the unit normal vector **p**.

Proof In the notation of Figure A.17, which shows a typical parallelogram determined by vectors $\mathbf{u}$ and $\mathbf{v}$ and its orthogonal projection onto a plane with unit normal vector $\mathbf{p}$,

$$\begin{aligned} \mathbf{u} &= \overrightarrow{PP'} + \mathbf{u}' + \overrightarrow{Q'Q} \\ &= \mathbf{u}' + \overrightarrow{PP'} - \overrightarrow{QQ'} && (\overrightarrow{Q'Q} = -\overrightarrow{QQ'}) \\ &= \mathbf{u}' + s\mathbf{p}. && \text{(For some scalar } s \text{ because } \overrightarrow{PP'} - \overrightarrow{QQ'} \text{) is parallel to } \mathbf{p}) \end{aligned}$$

Similarly,

$$\mathbf{v} = \mathbf{v}' + t\mathbf{p}$$

for some scalar t. Hence,

$$\begin{aligned} \mathbf{u} \times \mathbf{v} &= (\mathbf{u}' + s\mathbf{p}) \times (\mathbf{v}' + t\mathbf{p}) \\ &= (\mathbf{u}' \times \mathbf{v}') + s(\mathbf{p} \times \mathbf{v}') + t(\mathbf{u}' \times \mathbf{p}) + st\underbrace{(\mathbf{p} \times \mathbf{p})}_{0}. \end{aligned} \tag{1}$$

The vectors $\mathbf{p} \times \mathbf{v}'$ and $\mathbf{u}' \times \mathbf{p}$ are both orthogonal to $\mathbf{p}$. Hence, when we dot both sides of Equation (1) with $\mathbf{p}$, the only surviving term on the right is $(\mathbf{u}' \times \mathbf{v}') \cdot \mathbf{p}$. We are left with

$$(\mathbf{u} \times \mathbf{v}) \cdot \mathbf{p} = (\mathbf{u}' \times \mathbf{v}') \cdot \mathbf{p}.$$

In particular,

$$|(\mathbf{u} \times \mathbf{v}) \cdot \mathbf{p}| = |(\mathbf{u}' \times \mathbf{v}') \cdot \mathbf{p}|. \tag{2}$$

The absolute value on the right is the volume of the box determined by $\mathbf{u}'$, $\mathbf{v}'$, and $\mathbf{p}$. The height of this particular box is $|\mathbf{p}| = 1$, so the box's volume is numerically the same as its base area, the area of parallelogram $P'Q'R'S'$. Combining this observation with Equation (2) gives

$$\text{Area of } P'Q'R'S' = |(\mathbf{u}' \times \mathbf{v}') \cdot \mathbf{p}| = |(\mathbf{u} \times \mathbf{v}) \cdot \mathbf{p}|,$$

which says that the area of the orthogonal projection of the parallelogram determined by $\mathbf{u}$ and $\mathbf{v}$ onto a plane with unit normal vector $\mathbf{p}$ is $|(\mathbf{u} \times \mathbf{v}) \cdot \mathbf{p}|$, what we set out to prove.

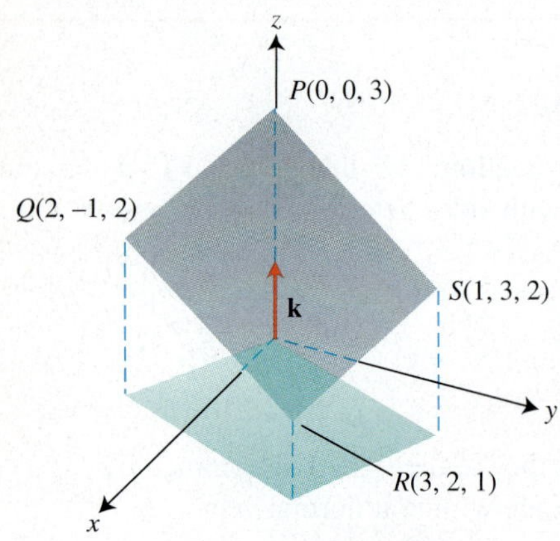

FIGURE A.18 Example 1 calculates the area of the orthogonal projection of parallelogram *PQRS* on the *xy*-plane.

Example 1 Finding the Area of a Projection

Find the area of the orthogonal projection onto the *xy*-plane of the parallelogram determined by the points $P(0, 0, 3)$, $Q(2, -1, 2)$, $R(3, 2, 1)$, and $S(1, 3, 2)$ (Figure A.18)

Solution With

$$\mathbf{u} = \overrightarrow{PQ} = 2\mathbf{i} - \mathbf{j} - \mathbf{k}, \qquad \mathbf{v} = \overrightarrow{PS} = \mathbf{i} + 3\mathbf{j} - \mathbf{k}, \qquad \text{and} \qquad \mathbf{p} = \mathbf{k},$$

the area is

$$\text{Area} = (\mathbf{u} \times \mathbf{v}) \cdot \mathbf{p} = \begin{vmatrix} 2 & -1 & -1 \\ 1 & 3 & -1 \\ 0 & 0 & 1 \end{vmatrix} = \begin{vmatrix} 2 & -1 \\ 1 & 3 \end{vmatrix} = 7.$$

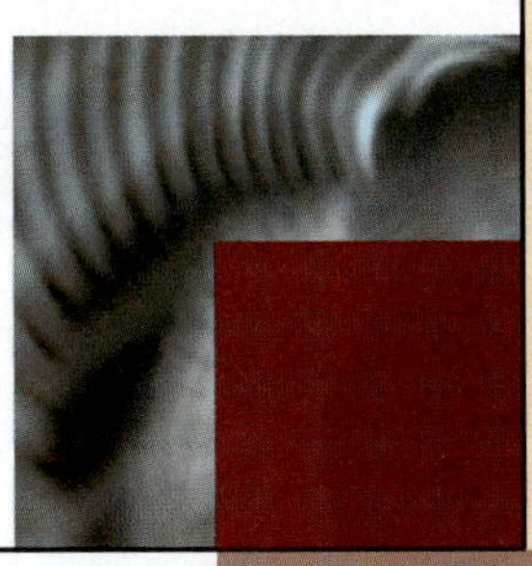

Appendix 13: Conic Sections

A.13.1 Conic Sections and Quadratic Equations

This section shows how the conic sections from Greek geometry are described today as the graphs of quadratic equations in the coordinate plane. The Greeks of Plato's time described these curves as the curves formed by cutting a double cone with a plane (Figure A.19, on the following page); hence the name *conic section*.

Circles

Definitions Circle, Center, Radius

A **circle** is the set of points in a plane whose distance from a given fixed point in the plane is constant. The fixed point is the **center** of the circle; the constant distance is the **radius**.

The standard-form equations for circles, derived in Preliminaries, Section 4, from the distance formula $d = \sqrt{(x_2 - x_1)^2 + (y_2 - y_1)^2}$, are these:

Circles

Circle of radius a centered at the origin:	Circle of radius a centered at the point (h, k):
$x^2 + y^2 = a^2$	$(x - h)^2 + (y - k)^2 = a^2$

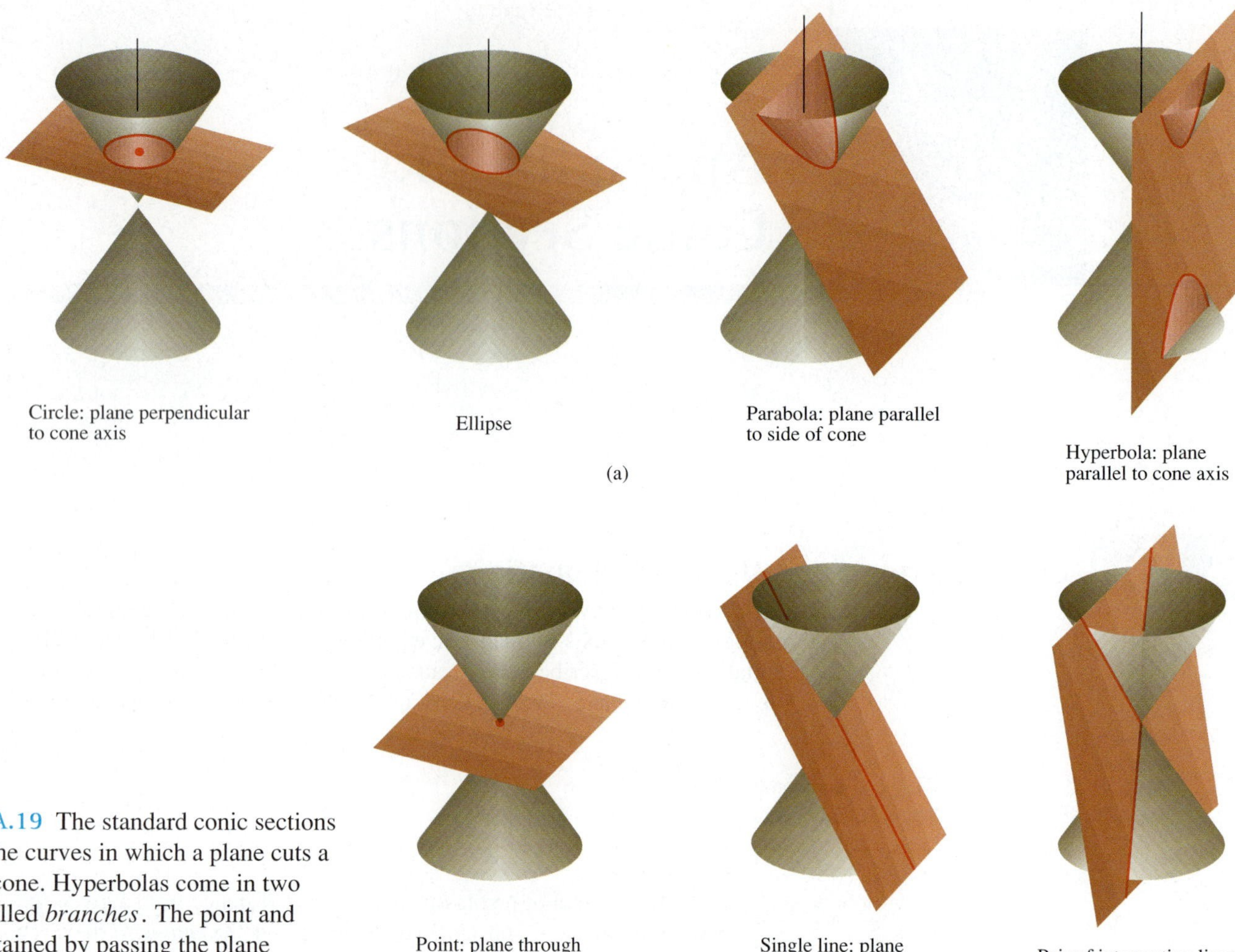

FIGURE A.19 The standard conic sections (a) are the curves in which a plane cuts a double cone. Hyperbolas come in two parts, called *branches*. The point and lines obtained by passing the plane through the cone's vertex (b) are *degenerate* conic sections.

Parabolas

Definitions Parabola, Focus, Directrix

A set that consists of all the points in a plane equidistant from a given fixed point and a given fixed line in the plane is a **parabola**. The fixed point is the **focus** of the parabola. The fixed line is the **directrix**.

If the focus F lies on the directrix L, the parabola is the line through F perpendicular to L. We consider this to be a degenerate case and assume henceforth that F does not lie on L.

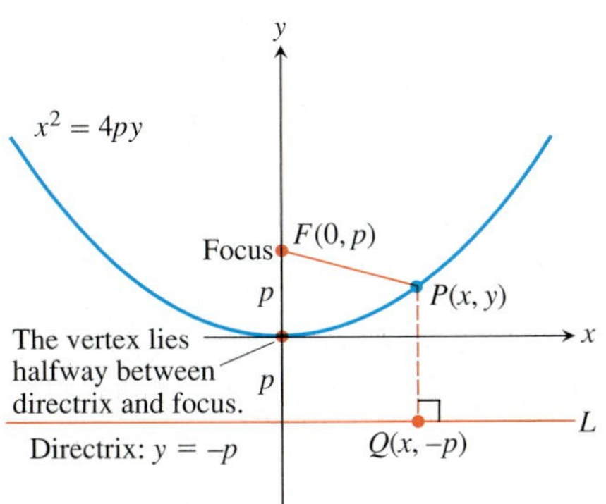

FIGURE A.20 The parabola $x^2 = 4py$.

A parabola has its simplest equation when its focus and directrix straddle one of the coordinate axes. For example, suppose that the focus lies at the point $F(0, p)$ on the positive y-axis and that the directrix is the line $y = -p$ (Figure A.20). In the notation of the figure, a point $P(x, y)$ lies on the parabola if and only if $PF = PQ$. From the distance formula,

$$PF = \sqrt{(x-0)^2 + (y-p)^2} = \sqrt{x^2 + (y-p)^2}$$
$$PQ = \sqrt{(x-x)^2 + (y-(-p))^2} = \sqrt{(y+p)^2}.$$

When we equate these expressions, square, and simplify, we get

$$y = \frac{x^2}{4p} \quad \text{or} \quad x^2 = 4py. \quad \text{Standard form} \tag{1}$$

These equations reveal the parabola's symmetry about the y-axis. We call the y-axis the **axis** of the parabola (short for "axis of symmetry").

The point where a parabola crosses its axis is the **vertex**. The vertex of the parabola $x^2 = 4py$ lies at the origin (Figure A.20). The positive number p is the parabola's **focal length**.

If the parabola opens downward, with its focus at $(0, -p)$ and its directrix the line $y = p$, then Equations (1) become

$$y = -\frac{x^2}{4p} \quad \text{and} \quad x^2 = -4py$$

(Figure A.21). We obtain similar equations for parabolas opening to the right or to the left (Figure A.22, on the following page, and Table A.13.1).

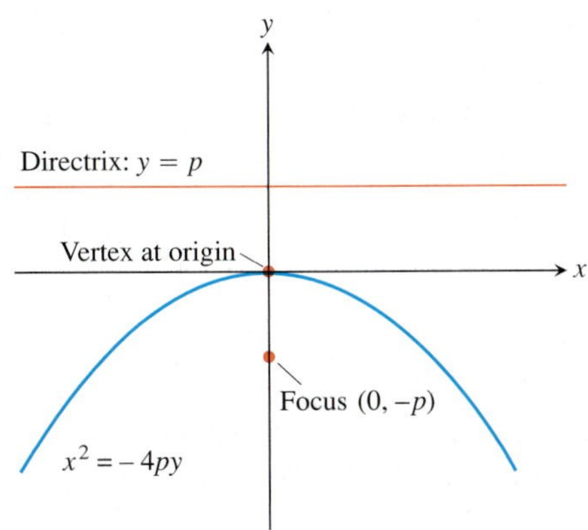

FIGURE A.21 The parabola $x^2 = -4py$.

Table A.13.1 Standard-form equations for parabolas with vertices at the origin $(p > 0)$

Equation	Focus	Directrix	Axis	Opens
$x^2 = 4py$	$(0, p)$	$y = -p$	y-axis	Up
$x^2 = -4py$	$(0, -p)$	$y = p$	y-axis	Down
$y^2 = 4px$	$(p, 0)$	$x = -p$	x-axis	To the right
$y^2 = -4px$	$(-p, 0)$	$x = p$	x-axis	To the left

Example 1 Finding the Focus and Directrix of a Parabola

Find the focus and directrix of the parabola $y^2 = 10x$.

Solution We find the value of p in the standard equation $y^2 = 4px$:

$$4p = 10, \quad \text{so} \quad p = \frac{10}{4} = \frac{5}{2}.$$

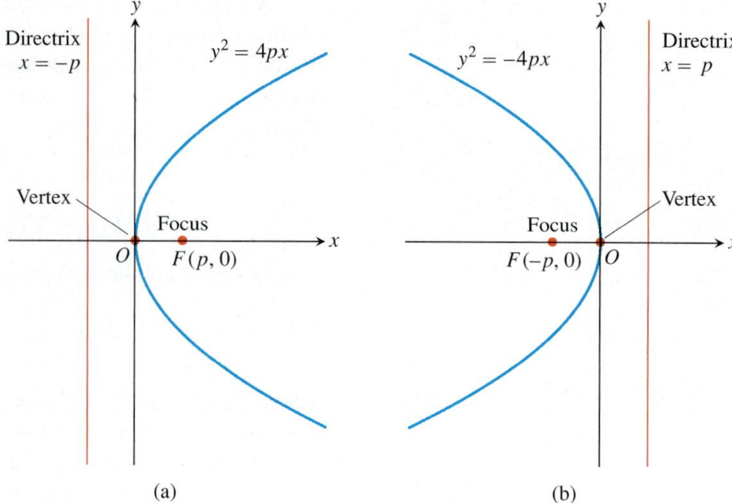

FIGURE A.22 (a) The parabola $y^2 = 4px$. (b) The parabola $y^2 = -4px$.

Then we find the focus and directrix for this value of p:

$$\text{Focus:} \qquad (p, 0) = \left(\frac{5}{2}, 0\right)$$

$$\text{Directrix:} \quad x = -p \qquad \text{or} \qquad x = -\frac{5}{2}.$$

The horizontal and vertical shift formulas in Preliminaries, Section 4, can be applied to the equations in Table A.13.1 to give equations for a variety of parabolas in other locations (see Exercises 39, 40, and 45–48).

Ellipses

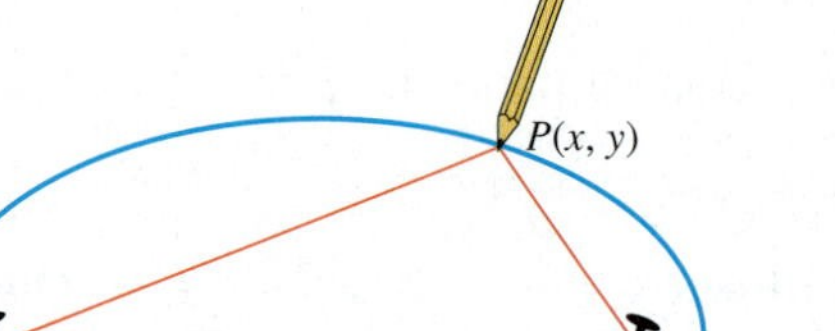

FIGURE A.23 How to draw an ellipse.

Definitions Ellipse, Foci

An **ellipse** is the set of points in a plane whose distances from two fixed points in the plane have a constant sum. The two fixed points are the **foci** of the ellipse.

The quickest way to construct an ellipse uses the definition. Put a loop of string around two tacks F_1 and F_2, pull the string taut with a pencil point P, and move the pencil around to trace a closed curve (Figure A.23). The curve is an ellipse because the sum $PF_1 + PF_2$, being the length of the loop minus the distance between the tacks, remains constant. The ellipse's foci lie at F_1 and F_2.

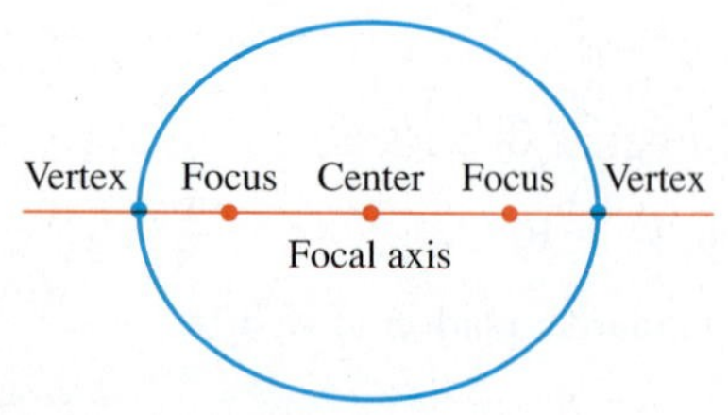

FIGURE A.24 Points on the focal axis of an ellipse.

Definitions Focal Axis, Center, Vertices (ellipse)

The line through the foci of an ellipse is the ellipse's **focal axis**. The point on the axis halfway between the foci is the **center**. The points where the focal axis and ellipse cross are the ellipse's **vertices** (Figure A.24).

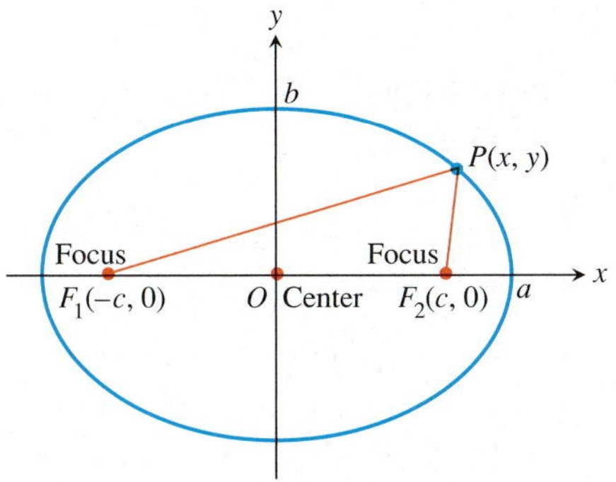

FIGURE A.25 The ellipse defined by the equation $PF_1 + PF_2 = 2a$ is the graph of the equation $(x^2/a^2) + (y^2/b^2) = 1$.

If the foci are $F_1(-c, 0)$ and $F_2(c, 0)$ (Figure A.25), and $PF_1 + PF_2$ is denoted by $2a$, then the coordinates of a point P on the ellipse satisfy the equation

$$\sqrt{(x+c)^2 + y^2} + \sqrt{(x-c)^2 + y^2} = 2a.$$

To simplify this equation, we move the second radical to the right-hand side, square, isolate the remaining radical, and square again, obtaining

$$\frac{x^2}{a^2} + \frac{y^2}{a^2 - c^2} = 1. \tag{2}$$

Since $PF_1 + PF_2$ is greater than the length F_1F_2 (triangle inequality for triangle PF_1F_2), the number $2a$ is greater than $2c$. Accordingly, $a > c$ and the number $a^2 - c^2$ in Equation (2) is positive.

The algebraic steps leading to Equation (2) can be reversed to show that every point P whose coordinates satisfy an equation of this form with $0 < c < a$ also satisfies the equation $PF_1 + PF_2 = 2a$. A point therefore lies on the ellipse if and only if its coordinates satisfy Equation (2).

If

$$b = \sqrt{a^2 - c^2}, \tag{3}$$

then $a^2 - c^2 = b^2$ and Equation (2) takes the form

$$\frac{x^2}{a^2} + \frac{y^2}{b^2} = 1. \tag{4}$$

Equation (4) reveals that this ellipse is symmetric with respect to the origin and both coordinate axes. It lies inside the rectangle bounded by the lines $x = \pm a$ and $y = \pm b$. It crosses the axes at the points $(\pm a, 0)$ and $(0, \pm b)$. The tangents at these points are perpendicular to the axes because

$$\frac{dy}{dx} = -\frac{b^2x}{a^2y}$$ Obtained from Eq. (4) by implicit differentiation

is zero if $x = 0$ and infinite if $y = 0$.

The Major and Minor Axes of an Ellipse

The **major axis** of the ellipse in Equation (4) is the line segment of length $2a$ joining the points $(\pm a, 0)$. The **minor axis** is the line segment of length $2b$ joining the points $(0, \pm b)$. The number a itself is the **semimajor axis**, the number b the **semiminor axis**. The number c, found from Equation (3) as

$$c = \sqrt{a^2 - b^2},$$

is the **center-to-focus distance** of the ellipse.

Example 2 Major axis horizontal

The ellipse

$$\frac{x^2}{16} + \frac{y^2}{9} = 1 \tag{5}$$

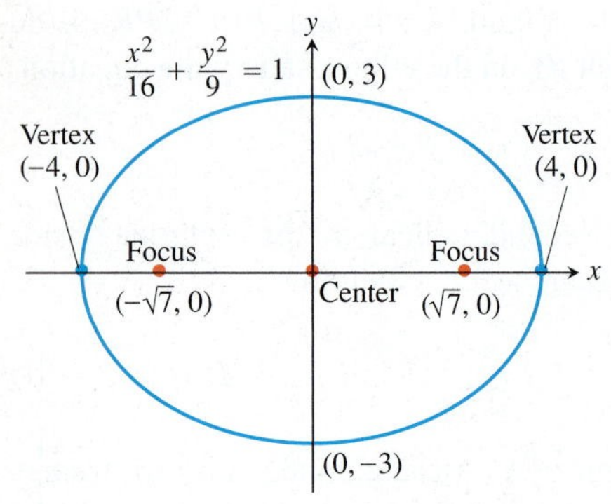

FIGURE A.26 Major axis horizontal (Example 2).

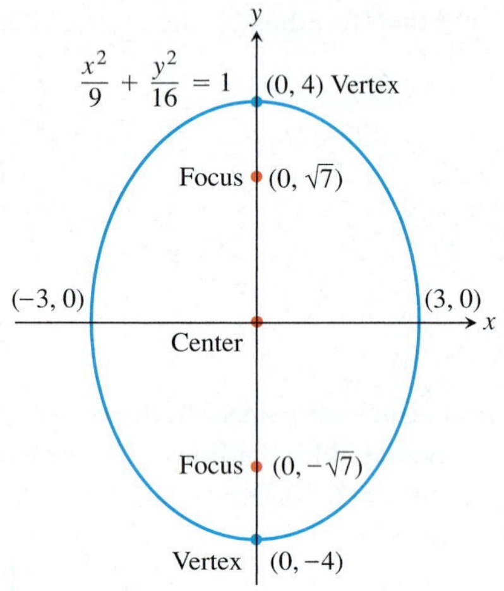

FIGURE A.27 Major axis vertical (Example 3).

(Figure A.26) has

Semimajor axis: $a = \sqrt{16} = 4$, Semiminor axis: $b = \sqrt{9} = 3$

Center-to-focus distance: $c = \sqrt{16 - 9} = \sqrt{7}$

Foci: $(\pm c, 0) = (\pm\sqrt{7}, 0)$

Vertices: $(\pm a, 0) = (\pm 4, 0)$

Center: $(0, 0)$.

Example 3 Major axis vertical

The ellipse

$$\frac{x^2}{9} + \frac{y^2}{16} = 1, \tag{6}$$

obtained by interchanging x and y in Equation (5), has its major axis vertical instead of horizontal (Figure A.27). With a^2 still equal to 16 and b^2 equal to 9, we have

Semimajor axis: $a = \sqrt{16} = 4$, Semiminor axis: $b = \sqrt{9} = 3$

Center-to-focus distance: $c = \sqrt{16 - 9} = \sqrt{7}$

Foci: $(0, \pm c) = (0, \pm\sqrt{7})$

Vertices: $(0, \pm a) = (0, \pm 4)$

Center: $(0, 0)$.

There is never any cause for confusion in analyzing equations like (5) and (6). We simply find the intercepts on the coordinate axes; then we know which way the major axis runs because it is the longer of the two axes. The center always lies at the origin and the foci lie on the major axis.

Standard-Form Equations for Ellipses Centered at the Origin

Foci on the x-axis: $\frac{x^2}{a^2} + \frac{y^2}{b^2} = 1 \quad (a > b)$

Center-to-focus distance: $c = \sqrt{a^2 - b^2}$
Foci: $(\pm c, 0)$
Vertices: $(\pm a, 0)$

Foci on the y-axis: $\frac{x^2}{b^2} + \frac{y^2}{a^2} = 1 \quad (a > b)$

Center-to-focus distance: $c = \sqrt{a^2 - b^2}$
Foci: $(0, \pm c)$
Vertices: $(0, \pm a)$

In each case, a is the semimajor axis and b is the semiminor axis.

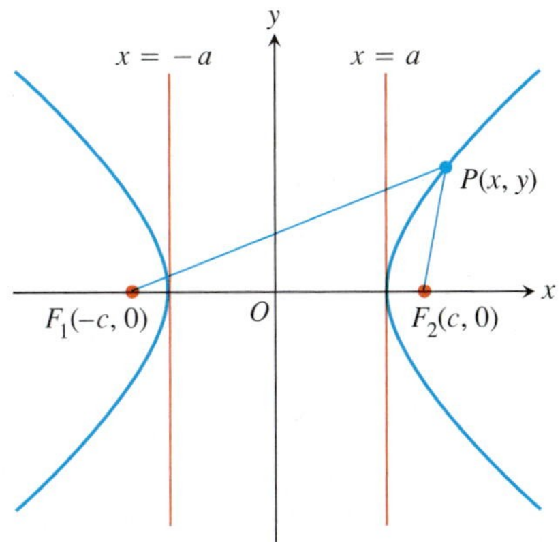

FIGURE A.28 Hyperbolas have two branches. For points on the right-hand branch of the hyperbola shown here, $PF_1 - PF_2 = 2a$. For points on the left-hand branch, $PF_2 - PF_1 = 2a$.

Hyperbolas

Definitions Hyperbola, Foci

A **hyperbola** is the set of points in a plane whose distances from two fixed points in the plane have a constant difference. The two fixed points are the **foci** of the hyperbola.

If the foci are $F_1(-c, 0)$ and $F_2(c, 0)$ (Figure A.28) and the constant difference is $2a$, then a point (x, y) lies on the hyperbola if and only if

$$\sqrt{(x + c)^2 + y^2} - \sqrt{(x - c)^2 + y^2} = \pm 2a. \tag{7}$$

To simplify this equation, we move the second radical to the right-hand side, square, isolate the remaining radical, and square again, obtaining

$$\frac{x^2}{a^2} + \frac{y^2}{a^2 - c^2} = 1. \tag{8}$$

So far, this looks just like the equation for an ellipse. But now $a^2 - c^2$ is negative because $2a$, being the difference of two sides of traingle PF_1F_2, is less than $2c$, the third side.

The algebraic steps leading to Equation (8) can be reversed to show that every point P whose coordinates satisfy an equation of this form with $0 < a < c$ also satisfies Equation (7). A point therefore lies on the hyperbola if and only if its coordinates satisfy Equation (8).

If we let b denote the positive square root of $c^2 - a^2$,

$$b = \sqrt{c^2 - a^2}, \tag{9}$$

then $a^2 - c^2 = -b^2$ and Equation (8) takes the more compact form

$$\frac{x^2}{a^2} - \frac{y^2}{b^2} = 1. \tag{10}$$

The differences between Equation (10) and the equation for an ellipse (Equation 4) are the minus sign and the new relation

$$c^2 = a^2 + b^2. \qquad \text{From Eq. (9)}$$

Like the ellipse, the hyperbola is symmetric with respect to the origin and coordinate axes. It crosses the x-axis at the points $(\pm a, 0)$. The tangents at these points are vertical because

$$\frac{dy}{dx} = \frac{b^2 x}{a^2 y} \qquad \text{Obtained from Eq. (10) by implicit differentiation}$$

is infinite when $y = 0$. The hyperbola has no y-intercepts; in fact, no part of the curve lies between the lines $x = -a$ and $x = a$.

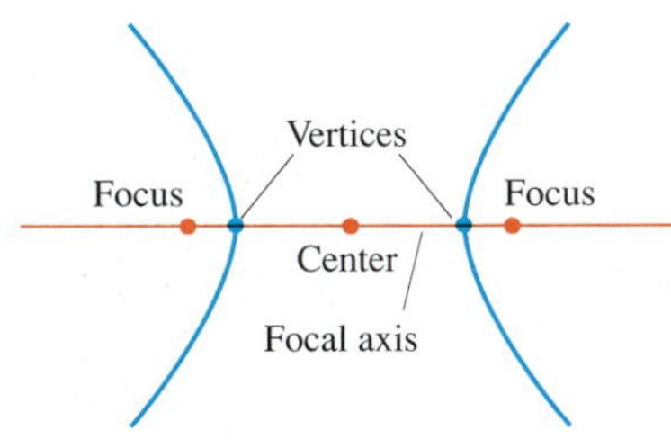

FIGURE A.29 Points on the focal axis of a hyperbola.

Definitions Focal Axis, Center, Vertices (hyperbola)

The line through the foci of a hyperbola is the **focal axis**. The point on the axis halfway between the foci is the hyperbola's **center**. The points where the focal axis and hyperbola cross are the **vertices** (Figure A.29).

Asymptotes of Hyperbolas—Graphing

The hyperbola

$$\frac{x^2}{a^2} - \frac{y^2}{b^2} = 1 \tag{11}$$

has two asymptotes, the lines

$$y = \pm \frac{b}{a} x$$

(See Exercise 94). The asymptotes give us the guidance we need to graph hyperbolas quickly. (See the drawing lesson.) The fastest way to find the equations of the asymptotes is to replace the 1 in Equation (11) by 0 and solve the new equation for y:

$$\underbrace{\frac{x^2}{a^2} - \frac{y^2}{b^2} = 1}_{\text{hyperbola}} \Rightarrow \underbrace{\frac{x^2}{a^2} - \frac{y^2}{b^2} = 0}_{\text{0 for 1}} \Rightarrow \underbrace{y = \pm \frac{b}{a} x}_{\text{asymptotes}}.$$

Standard-Form Equations for Hyperbolas Centered at the Origin

Foci on the x-axis: $\dfrac{x^2}{a^2} - \dfrac{y^2}{b^2} = 1$

Center-to-focus distance: $c = \sqrt{a^2 + b^2}$

Foci: $(\pm c, 0)$

Vertices: $(\pm a, 0)$

Asymptotes: $\dfrac{x^2}{a^2} - \dfrac{y^2}{b^2} = 0$ or $y = \pm \dfrac{b}{a} x$

Foci on the y-axis: $\dfrac{y^2}{a^2} - \dfrac{x^2}{b^2} = 1$

Center-to-focus distance: $c = \sqrt{a^2 + b^2}$

Foci: $(0, \pm c)$

Vertices: $(0, \pm a)$

Asymptotes: $\dfrac{y^2}{a^2} - \dfrac{x^2}{b^2} = 0$ or $y = \pm \dfrac{a}{b} x$

Notice the difference in the asymptote equations (b/a in the first, a/b in the second).

DRAWING LESSON

How to Graph the Hyperbola $\dfrac{x^2}{a^2} - \dfrac{y^2}{b^2} = 1$

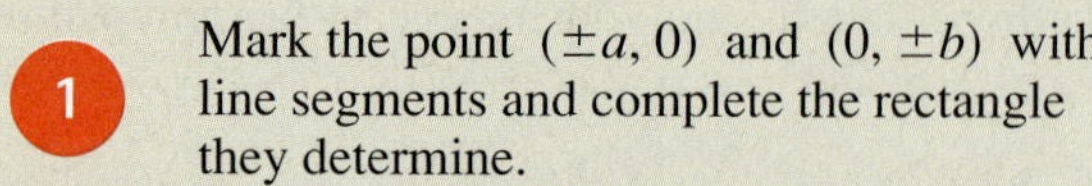

1. Mark the point $(\pm a, 0)$ and $(0, \pm b)$ with line segments and complete the rectangle they determine.

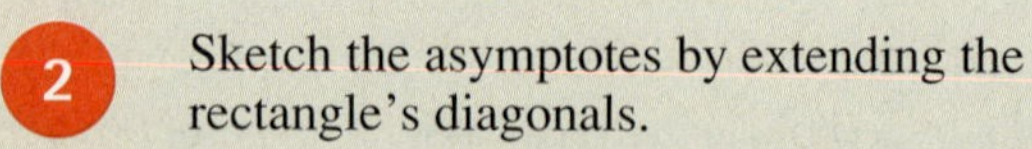

2. Sketch the asymptotes by extending the rectangle's diagonals.

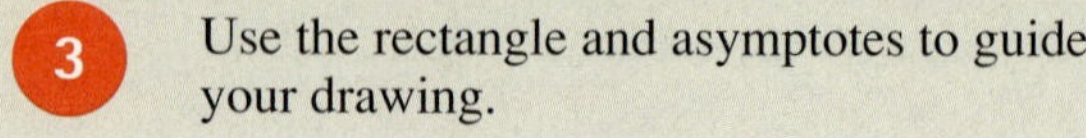

3. Use the rectangle and asymptotes to guide your drawing.

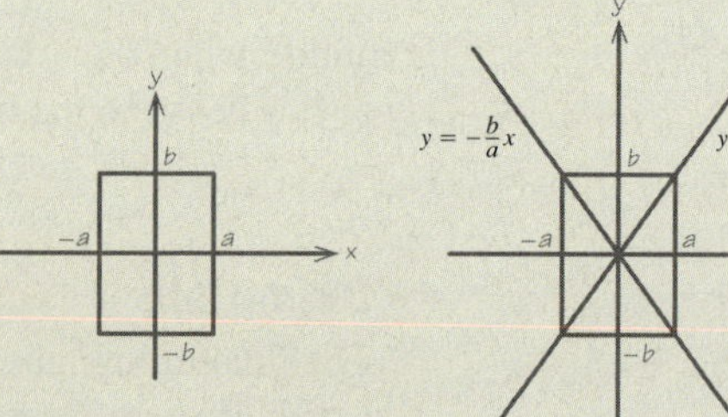

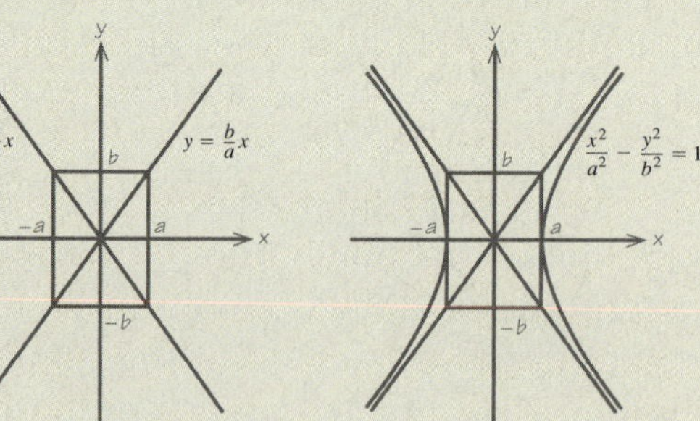

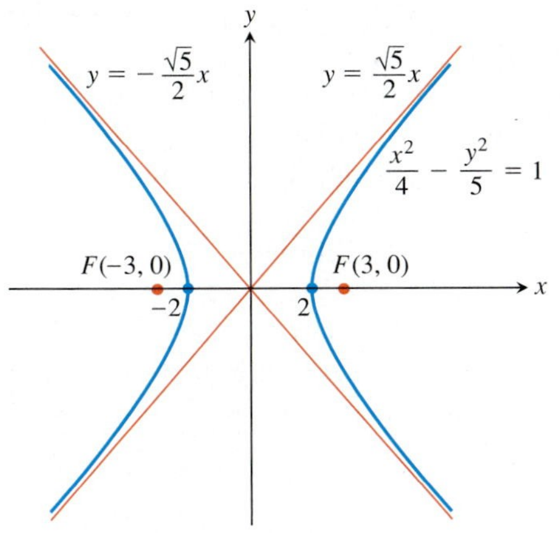

FIGURE A.30 The hyperbola in Example 4.

Example 4 Foci on the x-axis

The equation

$$\frac{x^2}{4} - \frac{y^2}{5} = 1 \tag{12}$$

is Equation (10) with $a^2 = 4$ and $b^2 = 5$ (Figure A.30). We have

Center-to-focus distance: $c = \sqrt{a^2 + b^2} = \sqrt{4 + 5} = 3$

Foci: $(\pm c, 0) = (\pm 3, 0)$, Vertices: $(\pm a, 0) = (\pm 2, 0)$

Center: $(0, 0)$

Asymptotes: $\frac{x^2}{4} - \frac{y^2}{5} = 0$ or $y = \pm \frac{\sqrt{5}}{2} x.$

Example 5 Foci on the y-axis

The hyperbola

$$\frac{y^2}{4} - \frac{x^2}{5} = 1,$$

obtained by interchanging x and y in Equation (12), has its vertices on the y-axis instead of the x-axis (Figure A.31). With a^2 still equal to 4 and b^2 equal to 5, we have

Center-to-focus distance: $c = \sqrt{a^2 + b^2} = \sqrt{4 + 5} = 3$

Foci: $(0, \pm c) = (0, \pm 3)$, Vertices: $(0, \pm a) = (0, \pm 2)$

Center: $(0, 0)$

Asymptotes: $\frac{y^2}{4} - \frac{x^2}{5} = 0$ or $y = \pm \frac{2}{\sqrt{5}} x.$

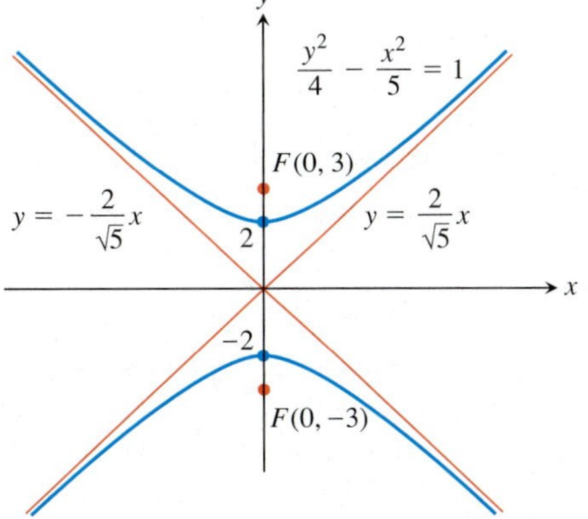

FIGURE A.31 The hyperbola in Example 5.

Reflective Properties

The chief applications of parabolas involve their use as reflectors of light and radio waves. Rays originating at a parabola's focus are reflected out of the parabola parallel to the parabola's axis (Figure A.32, on the following page, and Exercise 90). This property is used by flashlight, headlight, and spotlight reflectors and by microwave broadcast antennas to direct radiation from point sources into narrow

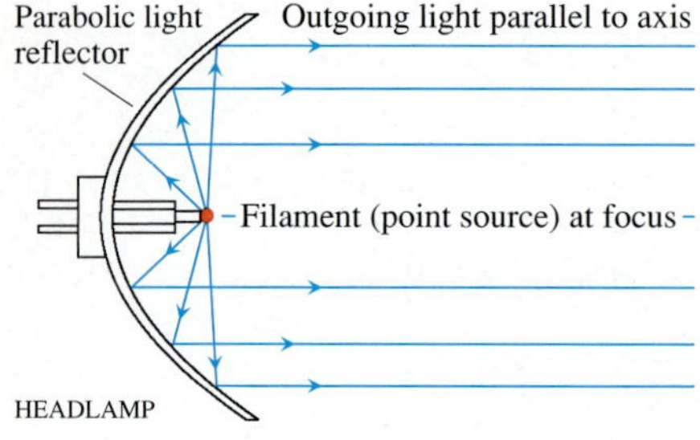

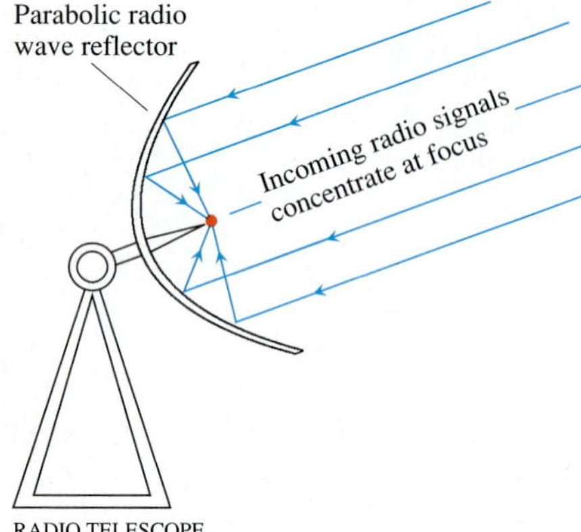

FIGURE A.32 Two of the many uses of parabolic reflectors.

beams. Conversely, electromagnetic waves arriving parallel to a parabolic reflector's axis are directed toward the reflector's focus. This property is used to intensify signals picked up by radio telescopes and television satellite dishes, to focus arriving light in telescopes, and to concentrate sunlight in solar heaters.

If an ellipse is revolved about its major axis to generate a surface (the surface is called an *ellipsoid*) and the interior is silvered to produce a mirror, light from one focus will be reflected to the other focus (Figure A.33). Ellipsoids reflect sound the same way, and this property is used to construct *whispering galleries*, rooms in which a person standing at one focus can hear a whisper from the other focus. Statuary Hall in the U.S. Capitol building is a whispering gallery. Ellipsoids also appear in instruments used to study aircraft noise in wind tunnels (sound at one focus can be received at the other focus with relatively little interference from other sources).

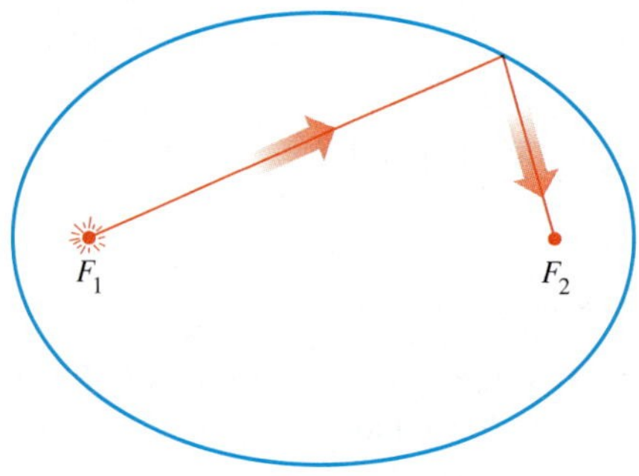

FIGURE A.33 An elliptical mirror (shown here in profile) reflects light from one focus to the other.

Light directed toward one focus of a hyperbolic mirror is reflected toward the other focus. This property of hyperbolas is combined with the reflective properties of parabolas and ellipses in designing modern telescopes. In Figure A.34 starlight reflects off a primary parabolic mirror toward the mirror's focus F_P. It is then reflected by a small hyperbolic mirror, whose focus is $F_H = F_P$, toward the second focus of the hyperbola, $F_E = F_H$. Since this focus is shared by an ellipse, the light is reflected by the elliptical mirror to the ellipse's second focus to be seen by an observer.

As recent experience with NASA's Hubble space telescope shows, the mirrors have to be nearly perfect to focus properly. The aberration that caused the malfunction in Hubble's primary mirror (now corrected with additional mirrors) amounted to about half a wavelength of visible light, no more than 1/50 the width of a human hair.

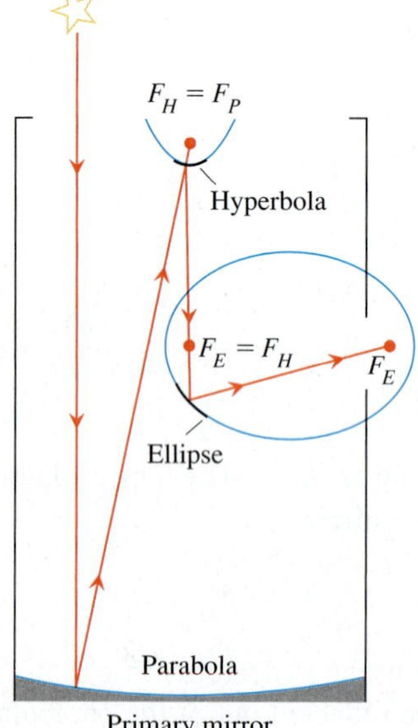

FIGURE A.34 Schematic drawing of a reflecting telescope.

Other Applications

Water pipes are sometimes designed with elliptical cross sections to allow for expansion when the water freezes. The triggering mechanisms in some lasers are elliptical, and stones on a beach become more and more elliptical as they are ground down by waves. There are also applications of ellipses to fossil formation. The ellipsolith, once thought to be a separate species, is now known to be an elliptically deformed nautilus.

Hyperbolic paths arise in Einstein's theory of relativity and form the basis for the (unrelated) LORAN radio navigation system. (LORAN is short for "long range navigation.") Hyperbolas also form the basis for a new system the Burlington Northern Railroad developed for using synchronized electronic signals from satellites to track freight trains. Computers aboard Burlington Northern locomotives in Minnesota have been able to track trains to within one mile per hour of their speed and to within 150 feet of their actual location.

EXERCISES A.13.1

Identifying Graphs

Match the parabolas in Exercises 1–4 with the following equations:

$$x^2 = 2y, \quad x^2 = -6y, \quad y^2 = 8x, \quad y^2 = -4x.$$

Then find the parabola's focus and directrix.

1.

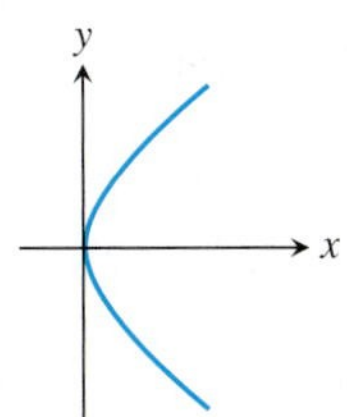

2.

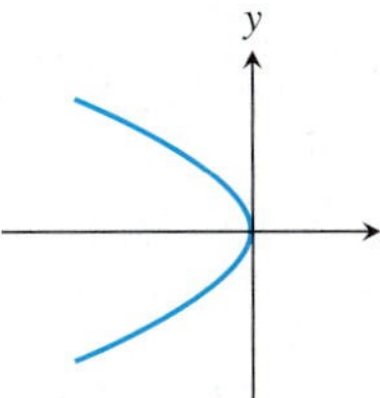

3.

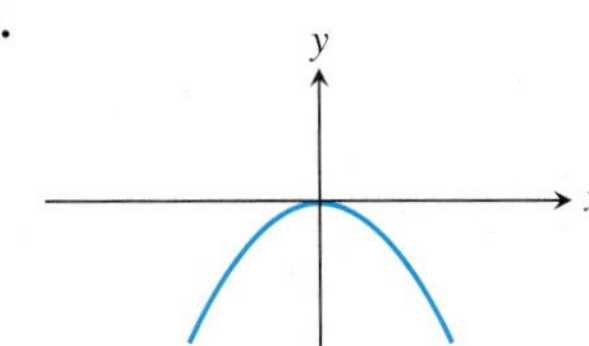

4.

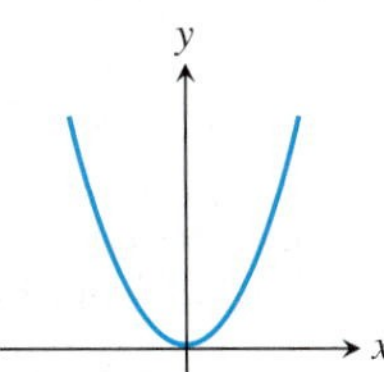

Match each conic section in Exercises 5–8 with one of these equations:

$$\frac{x^2}{4} + \frac{y^2}{9} = 1, \quad \frac{x^2}{2} + y^2 = 1,$$

$$\frac{y^2}{4} - x^2 = 1, \quad \frac{x^2}{4} - \frac{y^2}{9} = 1.$$

Then find the conic section's foci and vertices. If the conic section is a hyperbola, find its asymptotes as well.

5.

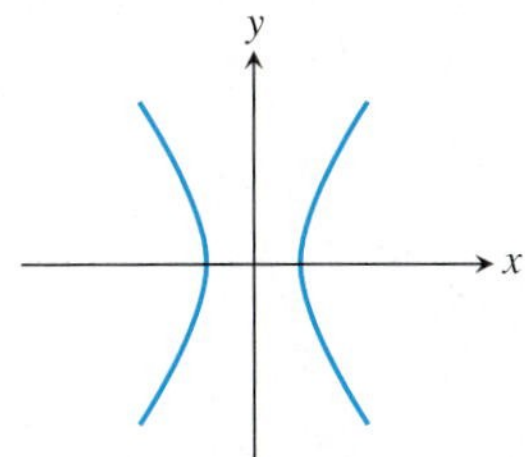

6.

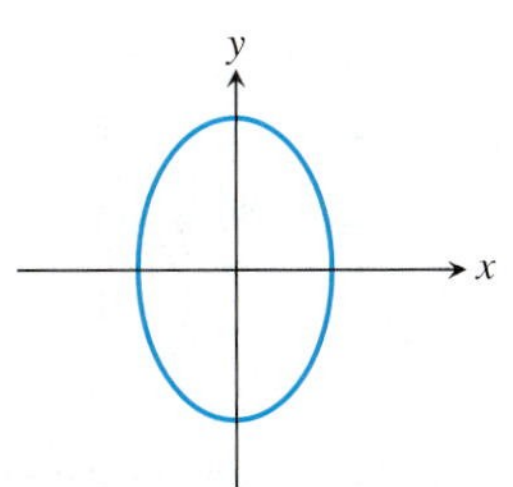

7.

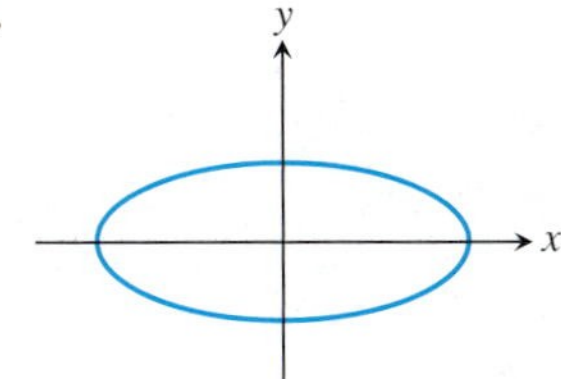

8.

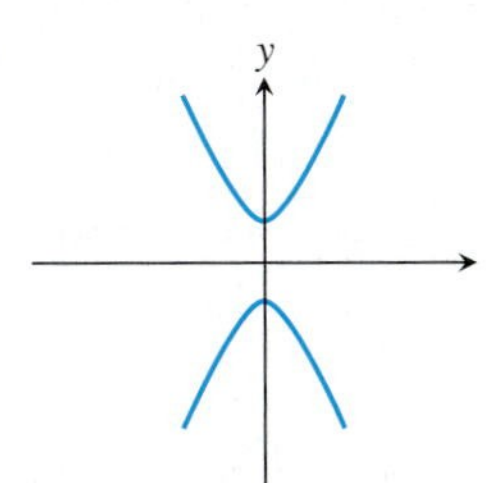

Parabolas

Exercises 9–16 give equations of parabolas. Find each parabola's focus and directrix. Then sketch the parabola. Include the focus and directrix in your sketch.

9. $y^2 = 12x$ **10.** $x^2 = 6y$ **11.** $x^2 = -8y$

12. $y^2 = -2x$ **13.** $y = 4x^2$ **14.** $y = -8x^2$

15. $x = -3y^2$ **16.** $x = 2y^2$

Ellipses

Exercises 17–24 give equations for ellipses. Put each equation in standard form. Then sketch the ellipse. Include the foci in your sketch.

17. $16x^2 + 25y^2 = 400$ **18.** $7x^2 + 16y^2 = 112$

19. $2x^2 + y^2 = 2$ **20.** $2x^2 + y^2 = 4$

21. $3x^2 + 2y^2 = 6$ **22.** $9x^2 + 10y^2 = 90$

23. $6x^2 + 9y^2 = 54$ **24.** $169x^2 + 25y^2 = 4225$

Exercises 25 and 26 give information about the foci and vertices of ellipses centered at the origin of the xy-plane. In each case, find the ellipse's standard-form equation from the given information.

25. Foci: $(\pm\sqrt{2}, 0)$
Vertices: $(\pm 2, 0)$

26. Foci: $(0, \pm 4)$
Vertices: $(0, \pm 5)$

Hyperbolas

Exercises 27–34 give equations for hyperbolas. Put each equation in standard form and find the hyperbola's asymptotes. Then sketch the hyperbola. Include the asymptotes and foci in your sketch.

27. $x^2 - y^2 = 1$ **28.** $9x^2 - 16y^2 = 144$

29. $y^2 - x^2 = 8$ **30.** $y^2 - x^2 = 4$

31. $8x^2 - 2y^2 = 16$ **32.** $y^2 - 3x^2 = 3$

33. $8y^2 - 2x^2 = 16$ **34.** $64x^2 - 36y^2 = 2304$

Exercises 35–38 give information about the foci, vertices, and asymptotes of hyperbolas centered at the origin of the xy-plane. In each case, find the hyperbola's standard-form equation from the information given.

35. Foci: $(0, \pm\sqrt{2})$
Asymptotes: $y = \pm x$

36. Foci: $(\pm 2, 0)$
Asymptotes: $y = \pm\frac{1}{\sqrt{3}}x$

37. Vertices: $(\pm 3, 0)$
Asymptotes: $y = \pm\frac{4}{3}x$

38. Vertices: $(0, \pm 2)$
Asymptotes: $y = \pm\frac{1}{2}x$

Shifting Conic Sections

39. The parabola $y^2 = 8x$ is shifted down 2 units and right 1 unit to generate the parabola $(y + 2)^2 = 8(x - 1)$. (a) Find the new parabola's vertex, focus, and directrix. (b) Plot the new vertex, focus, and directrix, and sketch in the parabola.

40. The parabola $x^2 = -4y$ is shifted left 1 unit and up 3 units to generate the parabola $(x+1)^2 = -4(y-3)$. (a) Find the new parabola's vertex, focus, and directrix. (b) Plot the new vertex, focus, and directrix, and sketch in the parabola.

41. The ellipse $(x^2/16) + (y^2/9) = 1$ is shifted 4 units to the right and 3 units up to generate the ellipse

$$\frac{(x-4)^2}{16} + \frac{(y-3)^2}{9} = 1.$$

(a) Find the foci, vertices, and center of the new ellipse. (b) Plot the new foci, vertices, and center, and sketch in the new ellipse.

42. The ellipse $(x^2/9) + (y^2/25) = 1$ is shifted 3 units to the left and 2 units down to generate the ellipse

$$\frac{(x+3)^2}{9} + \frac{(y+2)^2}{25} = 1.$$

(a) Find the foci, vertices, and center of the new ellipse. (b) Plot the new foci, vertices, and center, and sketch in the new ellipse.

43. The hyperbola $(x^2/16) - (y^2/9) = 1$ is shifted 2 units to the right to generate the hyperbola

$$\frac{(x-2)^2}{16} - \frac{y^2}{9} = 1.$$

(a) Find the center, foci, vertices, and asymptotes of the new hyperbola. (b) Plot the new center, foci, vertices, and asymptotes, and sketch in the hyperbola.

44. The hyperbola $(y^2/4) - (x^2/5) = 1$ is shifted 2 units down to generate the hyperbola

$$\frac{(y+2)^2}{4} - \frac{x^2}{5} = 1.$$

(a) Find the center, foci, vertices, and asymptotes of the new hyperbola. (b) Plot the new center, foci, vertices, and asymptotes, and sketch in the hyperbola.

Exercises 45–48 give equations for parabolas and tell how many units up or down and to the right or left each parabola is to be shifted. Find an equation for the new parabola, and find the new vertex, focus, and directrix.

45. $y^2 = 4x$, left 2, down 3

46. $y^2 = -12x$, right 4, up 3

47. $x^2 = 8y$, right 1, down 7

48. $x^2 = 6y$, left 3, down 2

Exercises 49–52 give equations for ellipses and tell how many units up or down and to the right or left each ellipse is to be shifted. Find an equation for the new ellipse, and find the new foci, vertices, and center.

49. $\frac{x^2}{6} + \frac{y^2}{9} = 1$, left 2, down 1

50. $\frac{x^2}{2} + y^2 = 1$, right 3, up 4

51. $\frac{x^2}{3} + \frac{y^2}{2} = 1$, right 2, up 3

52. $\frac{x^2}{16} + \frac{y^2}{25} = 1$, left 4, down 5

Exercises 53–56 give equations for hyperbolas and tell how many units up or down and to the right or left each hyperbola is to be shifted. Find an equation for the new hyperbola, and find the new center, foci, vertices, and asymptotes.

53. $\frac{x^2}{4} - \frac{y^2}{5} = 1$, right 2, up 2

54. $\frac{x^2}{16} - \frac{y^2}{9} = 1$, left 5, down 1

55. $y^2 - x^2 = 1$, left 1, down 1

56. $\frac{y^2}{3} - x^2 = 1$, right 1, up 3

Find the center, foci, vertices, asymptotes, and radius, as appropriate, of the conic sections in Exercises 57–68.

57. $x^2 + 4x + y^2 = 12$

58. $2x^2 + 2y^2 - 28x + 12y + 114 = 0$

59. $x^2 + 2x + 4y - 3 = 0$

60. $y^2 - 4y - 8x - 12 = 0$

61. $x^2 + 5y^2 + 4x = 1$

62. $9x^2 + 6y^2 + 36y = 0$

63. $x^2 + 2y^2 - 2x - 4y = -1$

64. $4x^2 + y^2 + 8x - 2y = -1$

65. $x^2 - y^2 - 2x + 4y = 4$

66. $x^2 - y^2 + 4x - 6y = 6$

67. $2x^2 - y^2 + 6y = 3$

68. $y^2 - 4x^2 + 16x = 24$

Inequalities

Sketch the regions in the xy-plane whose coordinates satisfy the inequalities or pairs of inequalities in Exercises 69–74.

69. $9x^2 + 16y^2 \le 144$

70. $x^2 + y^2 \ge 1$ and $4x^2 + y^2 \le 4$

71. $x^2 + 4y^2 \ge 4$ and $4x^2 + 9y^2 \le 36$

72. $(x^2 + y^2 - 4)(x^2 + 9y^2 - 9) \le 0$

73. $4y^2 - x^2 \ge 4$

74. $|x^2 - y^2| \le 1$

Theory and Examples

75. *Archimedes' formula for the volume of a parabolic solid.* The region enclosed by the parabola $y = (4h/b^2)x^2$ and the line $y = h$ is revolved about the y-axis to generate the solid shown here. Show that the volume of the solid is $3/2$ the volume of the corresponding cone.

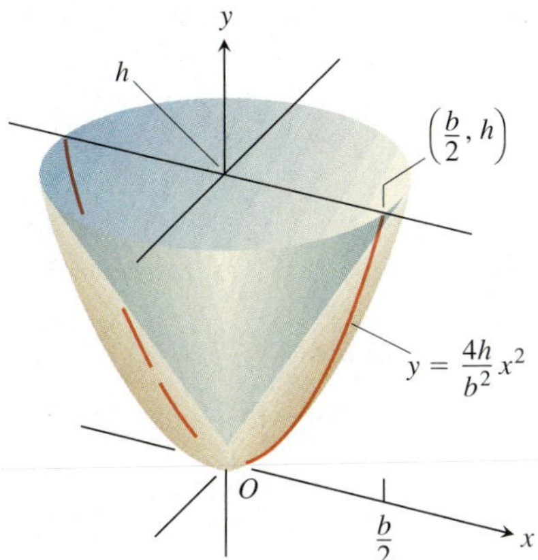

76. *Suspension bridge cables hang in parabolas.* The suspension bridge cable shown here supports a uniform load of w pounds per horizontal foot. It can be shown that if H is the horizontal tension of the cable at the origin, then the curve of the cable satisfies the equation

$$\frac{dy}{dx} = \frac{w}{H}x.$$

Show that the cable hangs in a parabola by solving this differential equation subject to the initial condition that $y = 0$ when $x = 0$.

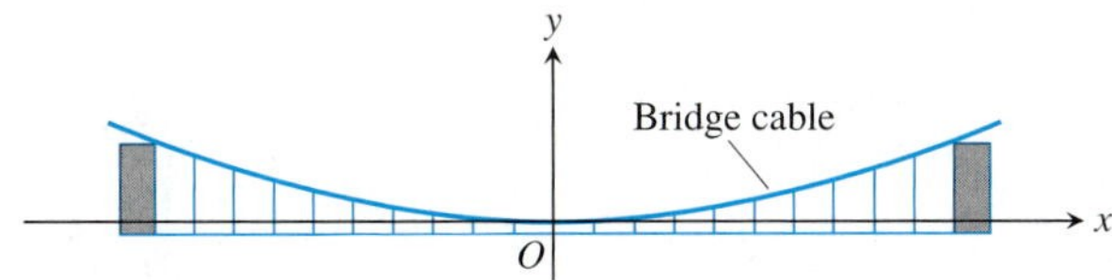

77. *Equation of a circle.* Find an equation for the circle through the points (1, 0), (0, 1), and (2, 2).

78. *Equation of a circle.* Find an equation for the circle through the points (2, 3), (3, 2), and (−4, 3).

79. *Equation of a circle.* Find an equation for the circle centered at (−2, 1) that passes through the point (1, 3). Is the point (1.1, 2.8) inside, outside, or on the circle?

80. *Tangents to a circle.* Find equations for the tangents to the circle $(x - 2)^2 + (y - 1)^2 = 5$ at the points where the circle crosses the coordinate axes. (*Hint*: Use implicit differentiation.)

81. *Volumes generated from a parabola.* If lines are drawn parallel to the coordinate axes through a point P on the parabola $y^2 = kx$, $k > 0$, the parabola partitions the rectangular region bounded by these lines and the coordinate axes into two smaller regions, A and B.

a) If the two smaller regions are revolved about the y-axis, show that they generate solids whose volumes have the ratio 4 : 1.

b) What is the ratio of the volumes generated by revolving the regions about the x-axis?

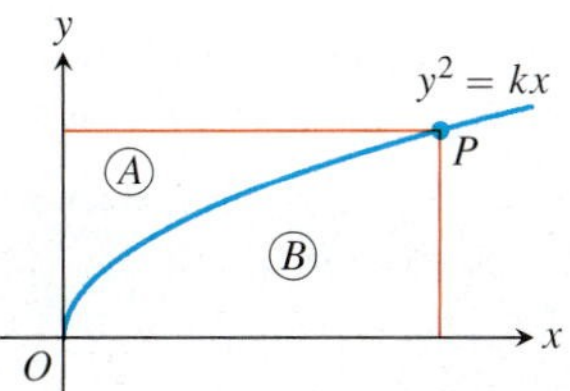

82. *Tangents to a parabola.* Show that the tangents to the curve $y^2 = 4px$ from any point on the line $x = -p$ are perpendicular.

83. *Largest rectangle in an ellipse.* Find the dimensions of the rectangle of largest area that can be inscribed in the ellipse $x^2 + 4y^2 = 4$ with its sides parallel to the coordinate axes. What is the area of the rectangle?

84. *Volume of a solid of revolution.* Find the volume of the solid generated by revolving the region enclosed by the ellipse $9x^2 + 4y^2 = 36$ about the (a) x-axis, and (b) y-axis.

85. *Volume of a solid of revolution.* The "triangular" region in the first quadrant bounded by the x-axis, the line $x = 4$, and the hyperbola $9x^2 - 4y^2 = 36$ is revolved about the x-axis to generate a solid. Find the volume of the solid.

86. *Volume of a solid of revolution.* The region bounded on the left by the y-axis, on the right by the hyperbola $x^2 - y^2 = 1$, and above and below by the lines $y = \pm 3$ is revolved about the y-axis to generate a solid. Find the volume of the solid.

87. *Centroid.* Find the centroid of the region that is bounded below by the x-axis and above by the ellipse $(x^2/9) + (y^2/16) = 1$.

88. *Area of a surface of revolution.* The curve $y = \sqrt{x^2 + 1}$, $0 \le x \le \sqrt{2}$, which is part of the upper branch of the hyperbola $y^2 - x^2 = 1$, is revolved about the x-axis to generate a surface. Find the area of the surface.

89. *Expanding waves.* The circular waves in the photograph on the following page were made by touching the surface of a ripple tank, first at A and then at B. As the waves expanded, their point of intersection appeared to trace a hyperbola. Did it really do that? To find out, we can model the waves with circles centered at A and B.

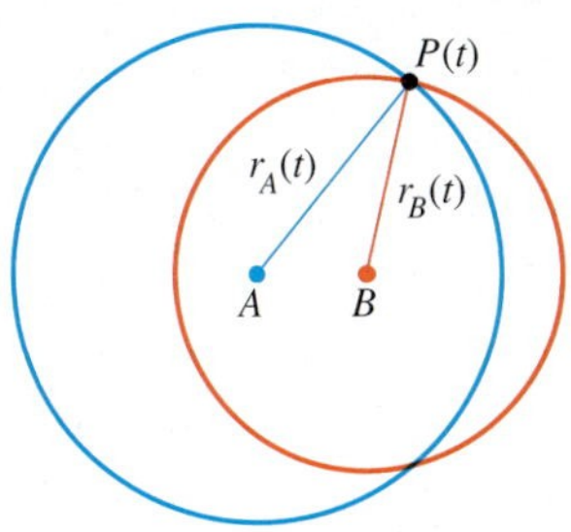

At time t, the point P is $r_A(t)$ units from A and $r_B(t)$ units from B. Since the radii of the circles increase at a constant rate, the rate at which the waves are traveling is

$$\frac{dr_A}{dt} = \frac{dr_B}{dt}.$$

Conclude from this equation that $r_A - r_B$ has a constant value, so that P must lie on a hyperbola with foci at A and B.

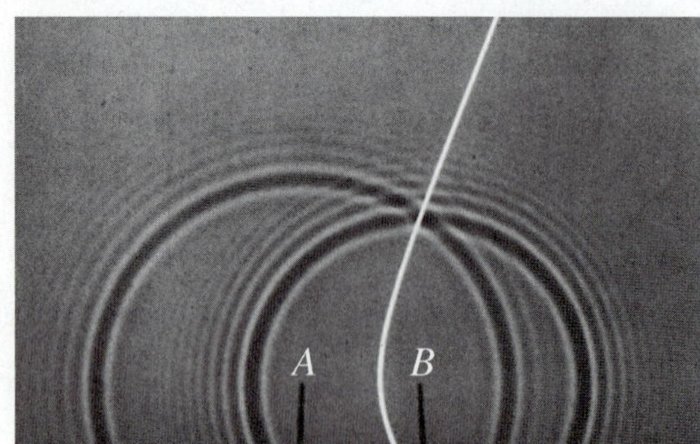

The expanding waves in Exercise 89

90. *The reflective property of parabolas.* The figure below shows a typical point $P(x_0, y_0)$ on the parabola $y^2 = 4px$. The line L is tangent to the parabola at P. The parabola's focus lies at $F(p, 0)$. The ray L' extending from P to the right is parallel to the x-axis. We show that light from F to P will be reflected out along L' by showing that β equals α. Establish this equality by taking the following steps.

a) Show that $\tan \beta = 2p/y_0$.

b) Show that $\tan \phi = y_0/(x_0 - p)$.

c) Use the identity

$$\tan \alpha = \frac{\tan \phi - \tan \beta}{1 + \tan \phi \tan \beta}$$

to show that $\tan \alpha = 2p/y_0$.

Since α and β are both acute, $\tan \beta = \tan \alpha$ implies $\beta = \alpha$.

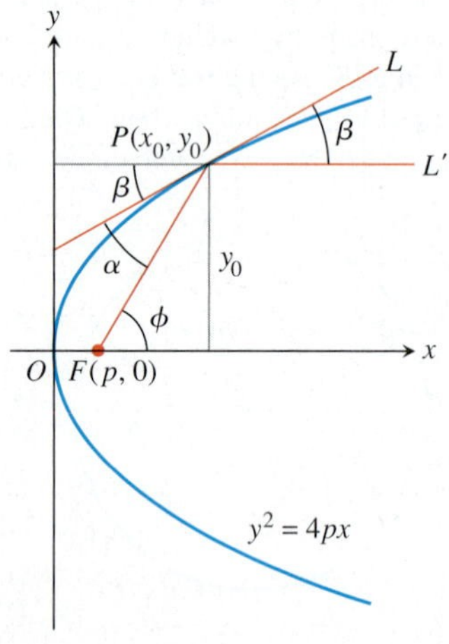

91. *How the astronomer Kepler used string to draw parabolas.* Kepler's method for drawing a parabola (with more modern tools) requires a string the length of a T square and a table whose edge can serve as the parabola's directrix. Pin one end of the string to the point where you want the focus to be and the other end to the upper end of the T square. Then, holding the string taut against the T square with a pencil, slide the T square along the table's edge. As the T square moves, the pencil will trace a parabola. Why?

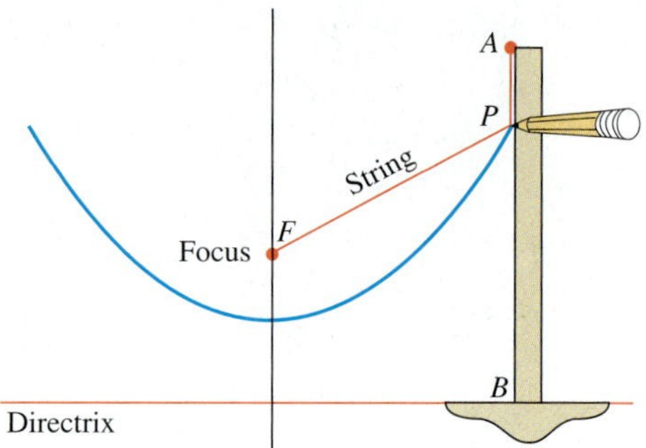

92. *Construction of a hyperbola.* The following diagrams appeared (unlabeled) in Ernest J. Eckert, "Constructions Without Words," *Mathematics Magazine*, Vol. 66, No. 2, April 1993, p. 113. Explain the constructions.

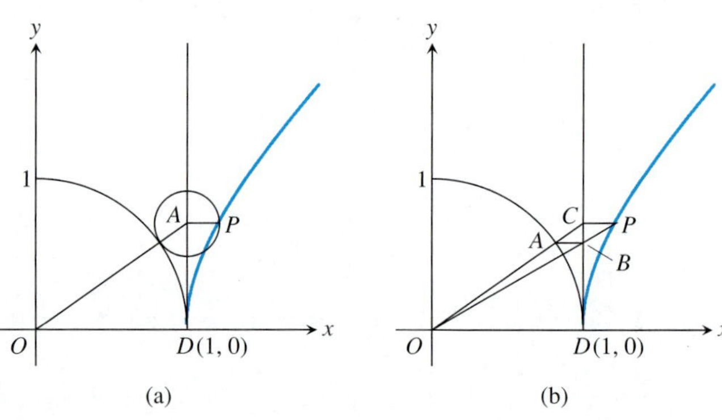

93. *The width of a parabola at the focus.* Show that the number $4p$ is the **width** of the parabola $x^2 = 4py$ $(p > 0)$ at the focus by showing that the line $y = p$ cuts the parabola at points that are $4p$ units apart.

94. *The (oblique) asymptotes of $(x^2/a^2) - (y^2/b^2) = 1$.* Show that the vertical distance between the line $y = (b/a)x$ and the upper half of the right-hand branch $y = (b/a)\sqrt{x^2 - a^2}$ of the hyperbola $(x^2/a^2) - (y^2/b^2) = 1$ approaches 0 by showing that

$$\lim_{x\to\infty}\left(\frac{b}{a}x - \frac{b}{a}\sqrt{x^2 - a^2}\right) = \frac{b}{a}\lim_{x\to\infty}\left(x - \sqrt{x^2 - a^2}\right) = 0.$$

Similar results hold for the remaining portions of the hyperbola and the lines $y = \pm(b/a)x$.

A.13.2 Classifying Conic Sections by Eccentricity

We now show how to associate with each conic section a number called the conic section's eccentricity. The eccentricity reveals the conic section's type (circle, ellipse, parabola, or hyperbola) and, in the case of ellipses and hyperbolas, describes the conic section's general proportions.

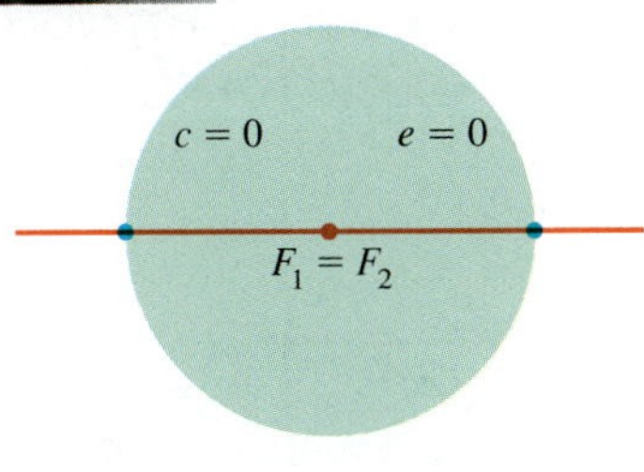

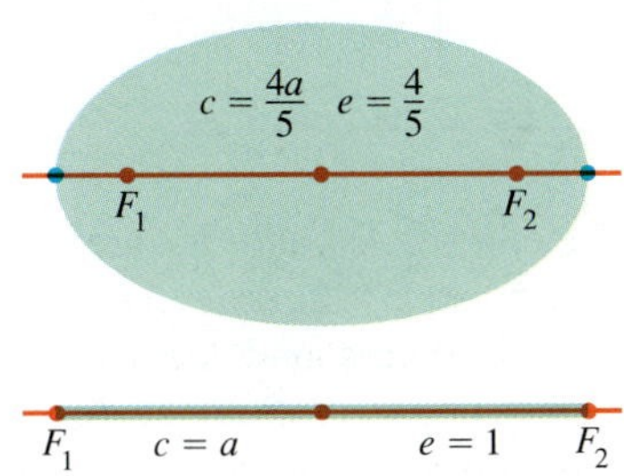

F_1 $c = a$ $e = 1$ F_2

FIGURE A.35 The ellipse changes from a circle to a line segment as c increases from 0 to a.

Eccentricity

Although the center-to-focus distance c does not appear in the equation

$$\frac{x^2}{a^2} + \frac{y^2}{b^2} = 1, \qquad (a > b)$$

for an ellipse, we can still determine c from the equation $c = \sqrt{a^2 - b^2}$. If we fix the semimajor axis a and vary c over the interval $0 \le c \le a$, the resulting ellipses will vary in shape (Figure A.35). They are circles if $c = 0$ (so that $a = b$) and flatten as c increases. If $c = a$, the foci and vertices overlap and the ellipse degenerates into a line segment.

We use the ratio of c to a to describe the various shapes the ellipse can take. We call this ratio the ellipse's eccentricity.

Definition Eccentricity (ellipse)

The **eccentricity** of the ellipse $(x^2/a^2) + (y^2/b^2) = 1$ $(a > b)$ is

$$e = \frac{c}{a} = \frac{\sqrt{a^2 - b^2}}{a}.$$

Table A.13.2 Eccentricities of planetary orbits

Mercury	0.21	Saturn	0.06
Venus	0.01	Uranus	0.05
Earth	0.02	Neptune	0.01
Mars	0.09	Pluto	0.25
Jupiter	0.05		

The planets in the solar system revolve around the sun in elliptical orbits with the sun at one focus. Most of the orbits are nearly circular, as can be seen from the eccentricities in Table A.13.2. Pluto has a fairly eccentric orbit, with $e = 0.25$, as does Mercury, with $e = 0.21$. Other members of the solar system have orbits that are even more eccentric. Icarus, an asteroid about 1 mile wide that revolves around the sun every 409 Earth days, has an orbital eccentricity of 0.83 (Figure A.36).

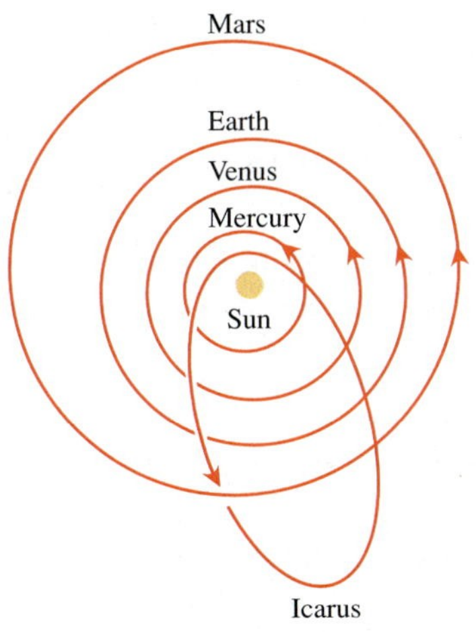

FIGURE A.36 The orbit of the asteroid Icarus is highly eccentric. Earth's orbit is so nearly circular that its foci lie inside the sun.

Example 1 Halley's Comet

The orbit of Halley's comet is an ellipse 36.18 astronomical units long by 9.12 astronomical units wide. (One *astronomical unit* [AU] is 149,597,870 km, the semimajor axis of Earth's orbit.) Its eccentricity is

$$e = \frac{\sqrt{a^2 - b^2}}{a} = \frac{\sqrt{(36.18/2)^2 - (9.12/2)^2}}{(1/2)(36.18)} = \frac{\sqrt{(18.09)^2 - (4.56)^2}}{18.09} \approx 0.97.$$

Whereas a parabola has one focus and one directrix, each ellipse has two foci and two directrices. These are the lines perpendicular to the major axis at distances $\pm a/e$ from the center. The parabola has the property that

$$PF = 1 \cdot PD \tag{1}$$

for any point P on it, where F is the focus and D is the point nearest P on the directrix. For an ellipse, it can be shown that the equations that replace (1) are

$$PF_1 = e \cdot PD_1, \qquad PF_2 = e \cdot PD_2. \tag{2}$$

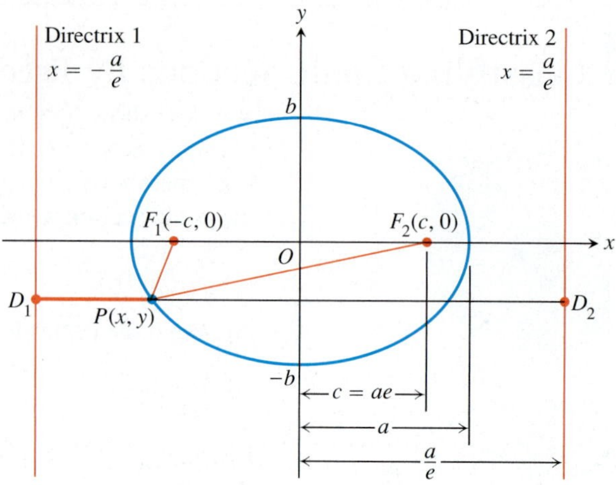

FIGURE A.37 The foci and directrices of the ellipse $(x^2/a^2) + (y^2/b^2) = 1$. Directrix 1 corresponds to focus F_1, and directrix 2 to focus F_2.

Here, e is the eccentricity, P is any point on the ellipse, F_1 and F_2 are the foci, and D_1 and D_2 are the points on the directrices nearest P (Figure A.37).

In each equation in (2) the directrix and focus must correspond; that is, if we use the distance from P to F_1, we must also use the distance from P to the directrix at the same end of the ellipse. The directrix $x = -a/e$ corresponds to $F_1(-c, 0)$, and the directrix $x = a/e$ corresponds to $F_2(c, 0)$.

The eccentricity of a hyperbola is also $e = c/a$, only in this case c equals $\sqrt{a^2 + b^2}$ instead of $\sqrt{a^2 - b^2}$. In contrast to the eccentricity of an ellipse, the eccentricity of a hyperbola is always greater than 1.

Halley's Comet

Edmund Halley (1656–1742; pronounced "*haw*-ley"), British biologist, geologist, sea captain, pirate, spy, Antarctic voyager, astronomer, adviser on fortifications, company founder and director, and the author of the first actuarial mortality tables, was also the mathematician who pushed and harried Newton into writing his *Principia*. Despite his accomplishments, Halley is known today chiefly as the man who calculated the orbit of the great comet of 1682: "wherefore if according to what we have already said [the comet] should return again about the year 1758, candid posterity will not refuse to acknowledge that this was first discovered by an Englishman." Indeed, candid posterity did not refuse—ever since the comet's return in 1758, it has been known as Halley's comet.

Last seen rounding the sun during the winter and spring of 1985–86, the comet is due to return in the year 2062. A recent study indicates that the comet has made about 2000 cycles so far with about the same number to go before the sun erodes it away completely.

Definition Eccentricity (hyperbola)

The **eccentricity** of the hyperbola $(x^2/a^2) - (y^2/b^2) = 1$ is

$$e = \frac{c}{a} = \frac{\sqrt{a^2 + b^2}}{a}.$$

In both ellipse and hyperbola, the eccentricity is the ratio of the distance between the foci to the distance between the vertices (because $c/a = 2c/2a$).

$$\text{Eccentricity} = \frac{\text{distance between foci}}{\text{distance between vertices}}$$

In an ellipse, the foci are closer together than the vertices and the ratio is less than 1. In a hyperbola, the foci are farther apart than the vertices and the ratio is greater than 1.

Example 2 Finding the Vertices of an Ellipse

Locate the vertices of an ellipse of eccentricity 0.8 whose foci lie at the points $(0, \pm 7)$.

Solution Since $e = c/a$, the vertices are the points $(0, \pm a)$ where

$$a = \frac{c}{e} = \frac{7}{0.8} = 8.75,$$

or $(0, \pm 8.75)$.

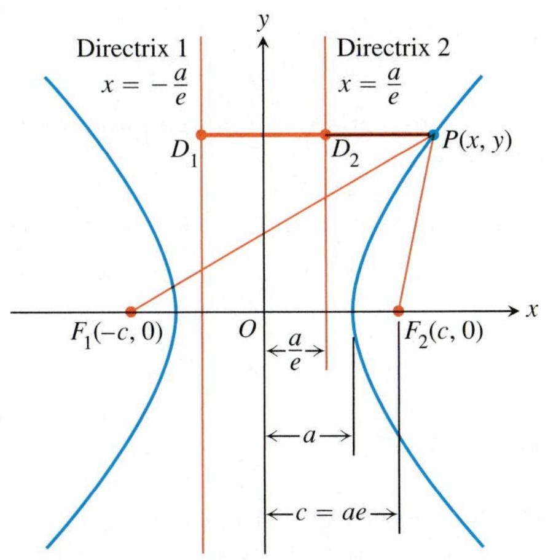

FIGURE A.38 The foci and directrices of the hyperbola $(x^2/a^2) - (y^2/b^2) = 1$. No matter where P lies on the hyperbola, $PF_1 = e \cdot PD_1$ and $PF_2 = e \cdot PD_2$.

Example 3 Finding the Eccentricity of a Hyperbola

Find the eccentricity of the hyperbola $9x^2 - 16y^2 = 144$.

Solution We divide both sides of the hyperbola's equation by 144 to put it in standard form, obtaining

$$\frac{9x^2}{144} - \frac{16y^2}{144} = 1 \quad \text{or} \quad \frac{x^2}{16} - \frac{y^2}{9} = 1.$$

With $a^2 = 16$ and $b^2 = 9$, we find that $c = \sqrt{a^2 + b^2} = \sqrt{16 + 9} = 5$, so

$$e = \frac{c}{a} = \frac{5}{4}.$$

As with the ellipse, it can be shown that the lines $x = \pm a/e$ act as directrices for the hyperbola and that

$$PF_1 = e \cdot PD_1 \quad \text{and} \quad PF_2 = e \cdot PD_2. \tag{3}$$

Here P is any point on the hyperbola, F_1 and F_2 are the foci, and D_1 and D_2 are the points nearest P on the directrices (Figure A.38).

To complete the picture, we define the eccentricity of a parabola to be $e = 1$. Equations (1)–(3) then have the common form $PF = e \cdot PD$.

Definition Eccentricity (parabola)

The **eccentricity** of a parabola is $e = 1$.

The "focus–directrix" equation $PF = e \cdot PD$ unites the parabola, ellipse, and hyperbola in the following way. Suppose that the distance PF of a point P from a fixed point F (the focus) is a constant multiple of its distance from a fixed line (the directrix). That is, suppose

$$PF = e \cdot PD, \tag{4}$$

where e is the constant of proportionality. Then the path traced by P is

a) a *parabola* if $e = 1$,
b) an *ellipse* of eccentricity e if $e < 1$, and
c) a *hyperbola* of eccentricity e if $e > 1$.

Equation (4) may not look like much to get excited about. There are no coordinates in it and when we try to translate it into coordinate form it translates in different ways, depending on the size of e. At least, that is what happens in Cartesian coordinates. However, in polar coordinates, the equation $PF = e \cdot PD$ translates into a single equation regardless of the value of e (Exercise 43).

Given the focus and corresponding directrix of a hyperbola centered at the origin and with foci on the x-axis, we can use the dimensions shown in Figure A.38 to find e. Knowing e, we can derive a Cartesian equation for the hyperbola from the equation $PF = e \cdot PD$, as in the next example. We can find equations for ellipses centered at the origin and with foci on the x-axis in a similar way, using the dimensions shown in Figure A.37.

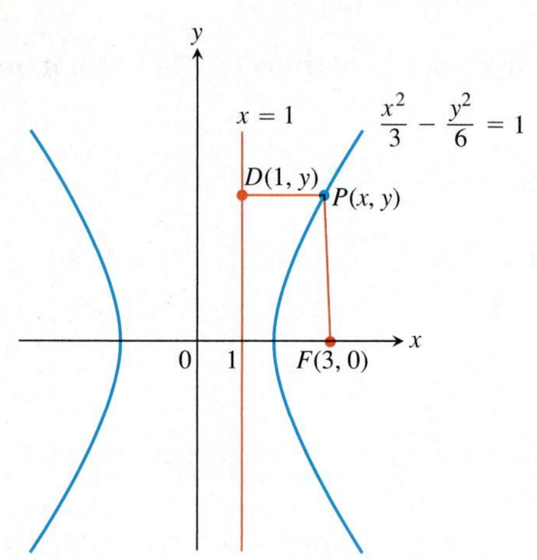

FIGURE A.39 The hyperbola in Example 4.

Example 4 Finding an Equation of a Hyperbola Knowing a Focus and Corresponding Directrix

Find a Cartesian equation for the hyperbola centered at the origin that has a focus at (3, 0) and the line $x = 1$ as the corresponding directrix.

Solution We first use the dimensions shown in Figure A.38 to find the hyperbola's eccentricity. The focus is

$$(c, 0) = (3, 0), \qquad \text{so} \qquad c = 3.$$

The directrix is the line

$$x = \frac{a}{e} = 1, \qquad \text{so} \qquad a = e.$$

When combined with the equation $e = c/a$ that defines eccentricity, these results give

$$e = \frac{c}{a} = \frac{3}{e}, \qquad \text{so} \qquad e^2 = 3 \qquad \text{and} \qquad e = \sqrt{3}.$$

Knowing e, we can now derive the equation we want from the equation $PF = e \cdot PD$. In the notation of Figure A.39, we have

$$\begin{aligned} PF &= e \cdot PD \\ \sqrt{(x-3)^2 + (y-0)^2} &= \sqrt{3}\,|x - 1| \qquad \text{Eq. (4)}\ e = \sqrt{3} \\ x^2 - 6x + 9 + y^2 &= 3(x^2 - 2x + 1) \\ 2x^2 - y^2 &= 6 \\ \frac{x^2}{3} - \frac{y^2}{6} &= 1. \end{aligned}$$

EXERCISES A.13.2

Ellipses

In Exercises 1–8, find the eccentricity of the ellipse. Then find and graph the ellipse's foci and directrices.

1. $16x^2 + 25y^2 = 400$

2. $7x^2 + 16y^2 = 112$

3. $2x^2 + y^2 = 2$

4. $2x^2 + y^2 = 4$

5. $3x^2 + 2y^2 = 6$

6. $9x^2 + 10y^2 = 90$

7. $6x^2 + 9y^2 = 54$

8. $169x^2 + 25y^2 = 4225$

Exercises 9–12 give the foci or vertices and the eccentricities of ellipses centered at the origin of the xy-plane. In each case, find the ellipse's standard-form equation.

9. Foci: $(0, \pm 3)$
Eccentricity: 0.5

10. Foci: $(\pm 8, 0)$
Eccentricity: 0.2

11. Vertices: $(0, \pm 70)$
Eccentricity: 0.1

12. Vertices: $(\pm 10, 0)$
Eccentricity: 0.24

Exercises 13–16 give foci and corresponding directrices of ellipses centered at the origin of the xy-plane. In each case, use the dimensions in Figure A.37 to find the eccentricity of the ellipse. Then find the ellipse's standard-form equation.

13. Focus: $(\sqrt{5}, 0)$
Directrix: $x = \frac{9}{\sqrt{5}}$

14. Focus: $(4, 0)$
Directrix: $x = \frac{16}{3}$

15. Focus: $(-4, 0)$
Directrix: $x = -16$

16. Focus: $(-\sqrt{2}, 0)$
Directrix: $x = -2\sqrt{2}$

17. *Writing to learn.* Draw an ellipse of eccentricity 4/5. Explain your procedure.

18. *Writing to learn.* Draw the orbit of Pluto (eccentricity 0.25) to scale. Explain your procedure.

19. *Finding an ellipse.* The endpoints of the major and minor axes of an ellipse are (1, 1), (3, 4), (1, 7), and (−1, 4). Sketch the ellipse, give its equation in standard form, and find its foci, eccentricity, and directrices.

20. *Finding an ellipse.* Find an equation for the ellipse of eccentricity $2/3$ that has the line $x = 9$ as a directrix and the point $(4, 0)$ as the corresponding focus.

21. *An ellipse tangent to the x-axis.* What values of the constants a, b, and c make the ellipse

$$4x^2 + y^2 + ax + by + c = 0$$

lie tangent to the x-axis at the origin and pass through the point $(-1, 2)$? What is the eccentricity of the ellipse?

22. *The reflective property of ellipses.* An ellipse is revolved about its major axis to generate an ellipsoid. The inner surface of the ellipsoid is silvered to make a mirror. Show that a ray of light emanating from one focus will be reflected to the other focus. Sound waves also follow such paths, and this property is used in constructing "whispering galleries." (*Hint*: Place the ellipse in standard position in the xy-plane and show that the lines from a point P on the ellipse to the two foci make congruent angles with the tangent to the ellipse at P.)

Hyperbolas

In Exercises 23–30, find the eccentricity of the hyperbola. Then find and graph the hyperbola's foci and directrices.

23. $x^2 - y^2 = 1$

24. $9x^2 - 16y^2 = 144$

25. $y^2 - x^2 = 8$

26. $y^2 - x^2 = 4$

27. $8x^2 - 2y^2 = 16$

28. $y^2 - 3x^2 = 3$

29. $8y^2 - 2x^2 = 16$

30. $64x^2 - 36y^2 = 2304$

Exercises 31–34 give the eccentricities and the vertices or foci of hyperbolas centered at the origin of the xy-plane. In each case, find the hyperbola's standard-form equation.

31. Eccentricity: 3
Vertices: $(0, \pm 1)$

32. Eccentricity: 2
Vertices: $(\pm 2, 0)$

33. Eccentricity: 3
Foci: $(\pm 3, 0)$

34. Eccentricity: 1.25
Foci: $(0, \pm 5)$

Exercises 35–38 give foci and corresponding directrices of hyperbolas centered at the origin of the xy-plane. In each case, find the hyperbola's eccentricity. Then find the hyperbola's standard-form equation.

35. Focus: $(4, 0)$
Directrix: $x = 2$

36. Focus: $(\sqrt{10}, 0)$
Directrix: $x = \sqrt{2}$

37. Focus: $(-2, 0)$
Directrix: $x = -\frac{1}{2}$

38. Focus: $(-6, 0)$
Directrix: $x = -2$

39. *Finding a hyperbola.* A hyperbola of eccentricity $3/2$ has one focus at $(1, -3)$. The corresponding directrix is the line $y = 2$. Find an equation for the hyperbola.

T **40.** *Writing to learn: The effect of eccentricity on a hyperbola's shape.* What happens to the graph of a hyperbola as its eccentricity increases? To find out, rewrite the equation $(x^2/a^2) - (y^2/b^2) = 1$ in terms of a and e instead of a and b. Graph the hyperbola for various values of $e > 1$ and describe what you find.

41. *The reflective property of hyperbolas.* Show that a ray of light directed toward one focus of a hyperbolic mirror, as in the accompanying figure, is reflected toward the other focus. (*Hint*: Show that the tangent to the hyperbola at P bisects the angle made by segments PF_1 and PF_2.)

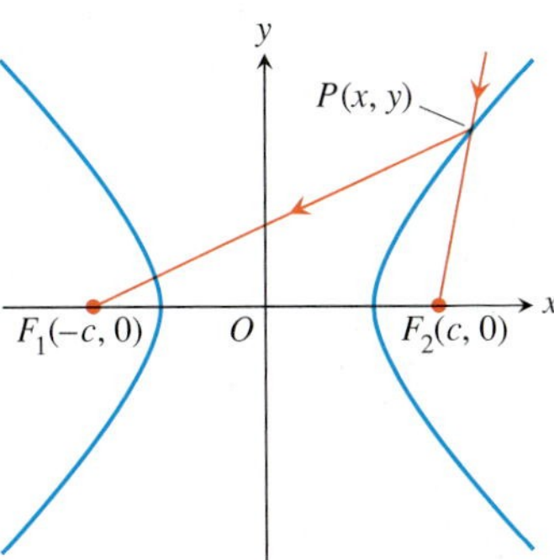

42. *A confocal ellipse and hyperbola.* Show that an ellipse and a hyperbola that have the same foci A and B, as in the accompanying figure, cross at right angles at their point of intersection. (*Hint*: A ray of light from focus A that met the hyperbola at P would be reflected from the hyperbola as if it came directly from B (Exercise 41). The same ray would be reflected off the ellipse to pass through B (Exercise 22).)

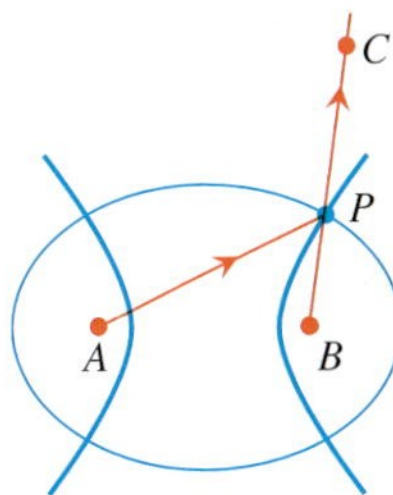

43. *Polar equation for the conic sections.* If one focus is placed at the origin and the corresponding directrix to the right of the origin along the vertical line $x = k$ for any conic section, use the accompanying figure to show that $PF = e \cdot PD$ becomes the single polar equation

$$r = \frac{ke}{1 + e\cos\theta}$$

regardless of the value of e.

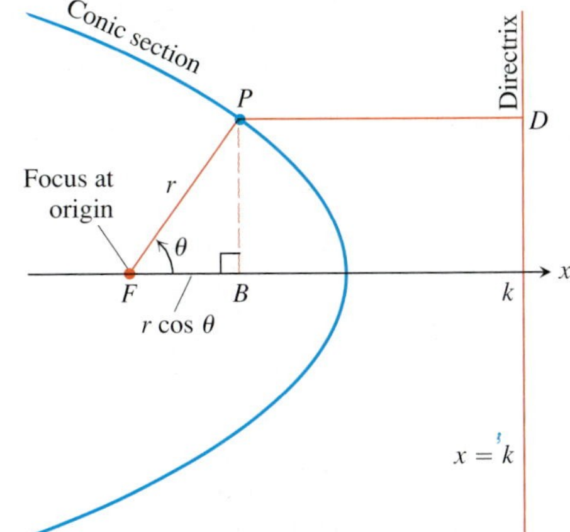

A.13.3 Quadratic Equations and Rotations

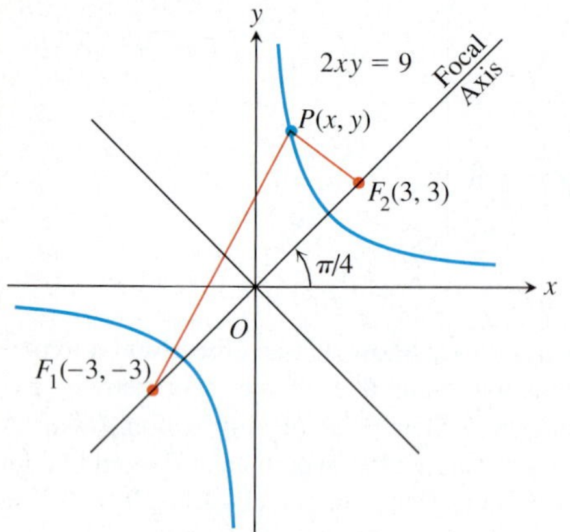

FIGURE A.40 The focal axis of the hyperbola $2xy = 9$ makes an angle of $\pi/4$ radians with the positive x-axis.

In this section, we examine one of the most important results in analytic geometry, which is that the Cartesian graph of any equation

$$Ax^2 + Bxy + Cy^2 + Dx + Ey + F = 0, \tag{1}$$

in which A, B, and C are not all zero, is nearly always a conic section. The exceptions are the cases in which there is no graph at all or the graph consists of two parallel lines. It is conventional to call all graphs of Equation (1), curved or not, **quadratic curves**.

The Cross Product Term

You may have noticed that the term Bxy did not appear in the equations for the conic sections in Section A.13.1. This happened because the axes of the conic sections ran parallel to (in fact, coincided with) the coordinate axes.

To see what happens when the parallelism is absent, let us write an equation for a hyperbola with $a = 3$ and foci at $F_1\,(-3, -3)$ and $F_2\,(3, 3)$ (Figure A.40). The equation $|PF_1 - PF_2| = 2a$ becomes $|PF_1 - PF_2| = 2(3) = 6$ and

$$\sqrt{(x+3)^2 + (y+3)^2} - \sqrt{(x-3)^2 + (y-3)^2} = \pm 6.$$

When we transpose one radical, square, solve for the radical that still appears, and square again, the equation reduces to

$$2xy = 9, \tag{2}$$

a case of Equation (1) in which the cross-product term is present. The asymptotes of the hyperbola in Equation (2) are the x- and y-axes, and the focal axis makes an angle of $\pi/4$ radians with the positive x-axis (Figure A.40). As in this example, the cross product term is present in Equation (1) only when the axes of the conic are tilted.

Rotating the Coordinate Axes to Eliminate the Cross Product Term

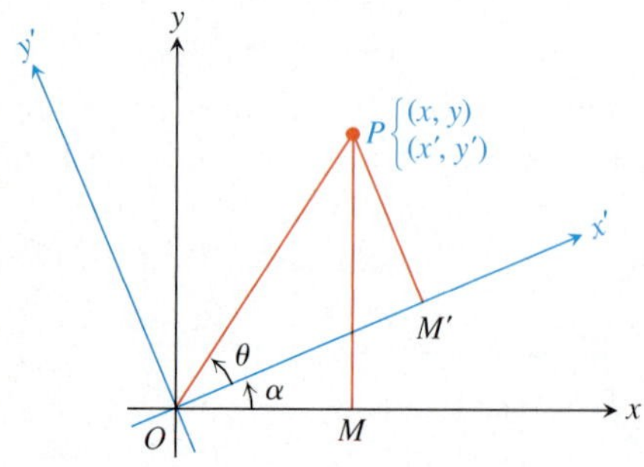

FIGURE A.41 A counterclockwise rotation through angle α about the origin.

To eliminate the xy-term from the equation of a conic, we rotate the coordinate axes to eliminate the "tilt" in the axes of the conic. The equations for the rotations we use are derived in the following way. In the notation of Figure A.41, which shows a counterclockwise rotation about the origin through an angle α,

$$\begin{aligned} x &= OM = OP\,\cos\,(\theta + \alpha) = OP\,\cos\,\theta\,\cos\,\alpha - OP\,\sin\,\theta\,\sin\,\alpha \\ y &= MP = OP\,\sin\,(\theta + \alpha) = OP\,\cos\,\theta\,\sin\,\alpha + OP\,\sin\,\theta\,\cos\,\alpha. \end{aligned} \tag{3}$$

Since

$$OP\cos\theta = OM' = x'$$

and

$$OP\sin\theta = M'P = y',$$

the equations in (3) reduce to the following.

Equations for Rotating Coordinate Axes

$$x = x' \cos \alpha - y' \sin \alpha$$
$$y = x' \sin \alpha + y' \cos \alpha \tag{4}$$

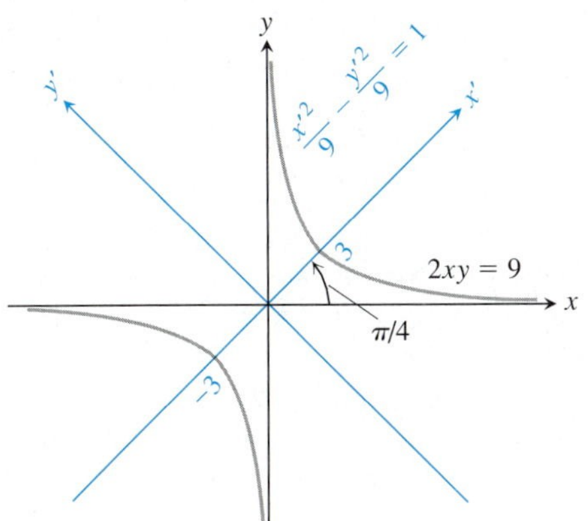

FIGURE A.42 The hyperbola in Example 1 (x' and y' are the new coordinates).

Example 1 Finding a Hyperbola by Rotating the Coordinate Axes

The x- and y-axes are rotated through an angle of $\pi/4$ radians about the origin. Find an equation for the hyperbola $2xy = 9$ in the new coordinates.

Solution Since $\cos \pi/4 = \sin \pi/4 = 1/\sqrt{2}$, we substitute

$$x = \frac{x' - y'}{\sqrt{2}}, \qquad y = \frac{x' + y'}{\sqrt{2}}$$

from Equations (4) into the equation $2xy = 9$ and obtain

$$2\left(\frac{x' - y'}{\sqrt{2}}\right)\left(\frac{x' + y'}{\sqrt{2}}\right) = 9$$
$$x'^2 - y'^2 = 9$$
$$\frac{x'^2}{9} - \frac{y'^2}{9} = 1.$$

See Figure A.42.

If we apply Equations (4) to the quadratic equation (1), we obtain a new quadratic equation

$$A' x'^2 + B' x' y' + C' y'^2 + D' x' + E' y' + F' = 0. \tag{5}$$

The new and old coefficients are related by the equations

$$\begin{aligned}
A' &= A \cos^2 \alpha + B \cos \alpha \sin \alpha + C \sin^2 \alpha \\
B' &= B \cos 2\alpha + (C - A) \sin 2\alpha \\
C' &= A \sin^2 \alpha - B \sin \alpha \cos \alpha + C \cos^2 \alpha \\
D' &= D \cos \alpha + E \sin \alpha \\
E' &= -D \sin \alpha + E \cos \alpha \\
F' &= F.
\end{aligned} \tag{6}$$

These equations show, among other things, that if we start with an equation for a curve in which the cross product term is present $(B \neq 0)$, we can find a rotation angle α that produces an equation in which no cross product term appears $(B' = 0)$. To find α, we set $B' = 0$ in the second equation in (6) and solve the resulting equation,

$$B \cos 2\alpha + (C - A) \sin 2\alpha = 0,$$

for α. In practice, this means determining α from one of the two equations

$$\cot 2\alpha = \frac{A - C}{B} \qquad \text{or} \qquad \tan 2\alpha = \frac{B}{A - C}. \tag{7}$$

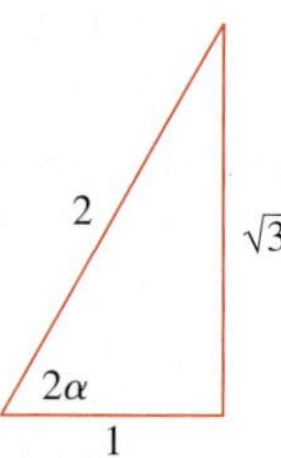

FIGURE A.43 This triangle identifies $2\alpha = \cot^{-1}(1/\sqrt{3})$ as $\pi/3$ (Example 2).

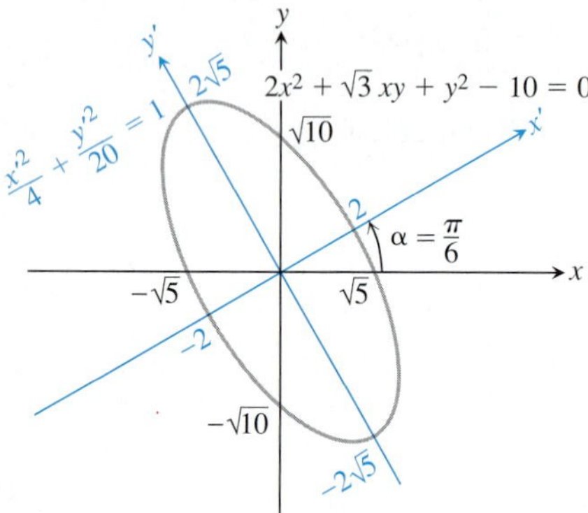

FIGURE A.44 The conic section in Example 2.

Example 2 Finding the Angle of Rotation for a Conic Section

The coordinate axes are to be rotated through an angle α to produce an equation for the curve

$$2x^2 + \sqrt{3}\,xy + y^2 - 10 = 0$$

that has no cross product term. Find α and the new equation. Identify the curve.

Solution The equation $2x^2 + \sqrt{3}\,xy + y^2 - 10 = 0$ has $A = 2$, $B = \sqrt{3}$, and $C = 1$. We substitute these values into Equation (7) to find α:

$$\cot 2\alpha = \frac{A - C}{B} = \frac{2 - 1}{\sqrt{3}} = \frac{1}{\sqrt{3}}.$$

From the right triangle in Figure A.43, we see that one appropriate choice of angle is $2\alpha = \pi/3$, so we take $\alpha = \pi/6$. Substituting $\alpha = \pi/6$, $A = 2$, $B = \sqrt{3}$, $C = 1$, $D = E = 0$, and $F = -10$ into Equations (6) gives

$$A' = \frac{5}{2}, \qquad B' = 0, \qquad C' = \frac{1}{2}, \qquad D' = E' = 0, \qquad F' = -10.$$

Equation (5) then gives

$$\frac{5}{2}x'^2 + \frac{1}{2}y'^2 - 10 = 0, \qquad \text{or} \qquad \frac{x'^2}{4} + \frac{y'^2}{20} = 1.$$

The curve is an ellipse with foci on the new y'-axis (Figure A.44).

Possible Graphs of Quadratic Equations

We now return to the graph of the general quadratic equation.

Since axes can always be rotated to eliminate the cross product term, there is no loss of generality in assuming that this has been done and that our equation has the form

$$Ax^2 + Cy^2 + Dx + Ey + F = 0. \tag{8}$$

Equation (8) represents

a) a *circle* if $A = C \neq 0$ (special cases: the graph is a point or there is no graph at all);

b) a *parabola* if Equation (8) is quadratic in one variable and linear in the other;

c) an *ellipse* if A and C are both positive or both negative (special cases: circles, a single point or no graph at all);

d) a *hyperbola* if A and C have opposite signs (special case: a pair of intersecting lines);

e) a *straight line* if A and C are zero and at least one of D and E is different from zero;

f) *one or two straight lines* if the left-hand side of Equation (8) can be factored into the product of two linear factors.

See Table A.13.3 (on page 1206) for examples.

The Discriminant Test

We do not need to eliminate the xy-term from the equation

$$Ax^2 + Bxy + Cy^2 + Dx + Ey + F = 0 \tag{9}$$

to tell what kind of conic section the equation respresents. If this is the only information we want, we can apply the following test instead.

As we have seen, if $B \neq 0$, then rotating the coordinate axes through an angle α that satisfies the equation

$$\cot 2\alpha = \frac{A - C}{B} \tag{10}$$

will change Equation (9) into an equivalent form

$$A' x'^2 + C' y'^2 + D' x' + E' y' + F' = 0 \tag{11}$$

without a cross product term.

Now, the graph of Equation (11) is a (real or degenerate)

a) *parabola* if A' or $C' = 0$; that is, if $A' C' = 0$;
b) *ellipse* if A' and C' have the same sign; that is, if $A' C' > 0$;
c) *hyperbola* if A' and C' have opposite signs; that is, if $A' C' < 0$.

It can also be verified (Exercise 49) from Equations (6) that for any rotation of axes,

$$B^2 - 4AC = B'^2 - 4A' C'. \tag{12}$$

This means that the quantity $B^2 - 4AC$ is not changed by a rotation. But when we rotate through the angle α given by Equation (10), B' becomes zero, so

$$B^2 - 4AC = -4A' C'.$$

Since the curve is a parabola if $A' C' = 0$, an ellipse if $A' C' > 0$, and a hyperbola if $A' C' < 0$, the curve must be a parabola if $B^2 - 4AC = 0$, an ellipse if $B^2 - 4AC < 0$, and a hyperbola if $B^2 - 4AC > 0$. The number $B^2 - 4AC$ is called the **discriminant** of Equation (9).

The Discriminant Test

With the understanding that occasional degenerate cases may arise, the quadratic curve $Ax^2 + Bxy + Cy^2 + Dx + Ey + F = 0$ is

a) a **parabola** if $B^2 - 4AC = 0$,
b) an **ellipse** if $B^2 - 4AC < 0$,
c) a **hyperbola** if $B^2 - 4AC > 0$.

Example 3 Identifying Conic Sections

a) $3x^2 - 6xy + 3y^2 + 2x - 7 = 0$ represents a parabola because

$$B^2 - 4AC = (-6)^2 - 4 \cdot 3 \cdot 3 = 36 - 36 = 0.$$

b) $x^2 + xy + y^2 - 1 = 0$ represents an ellipse because

$$B^2 - 4AC = (1)^2 - 4 \cdot 1 \cdot 1 = -3 < 0.$$

c) $xy - y^2 - 5y + 1 = 0$ represents a hyperbola because

$$B^2 - 4AC = (1)^2 - 4(0)(-1) = 1 > 0.$$

Table A.13.3 Examples of quadratic curves

$Ax^2 + Bxy + Cy^2 + Dx + Ey + F = 0$

	A	B	C	D	E	F	Equation	Remarks
Circle	1		1			-4	$x^2 + y^2 = 4$	$A = C; F < 0$
Parabola			1	-9			$y^2 = 9x$	Quadratic in y, linear in x
Ellipse	4		9			-36	$4x^2 + 9y^2 = 36$	A, C have same sign, $A \neq C; F < 0$
Hyperbola	1		-1			-1	$x^2 - y^2 = 1$	A, C have opposite signs
One line (still a conic section)	1						$x^2 = 0$	y-axis
Intersecting lines (still a conic section)		1		1	-1	-1	$xy + x - y - 1 = 0$	Factors to $(x-1)(y+1) = 0$, so $x = 1, y = -1$
Parallel lines (not a conic section)	1			-3		2	$x^2 - 3x + 2 = 0$	Factors to $(x-1)(x-2) = 0$, so $x = 1, x = 2$
Point	1		1				$x^2 + y^2 = 0$	The origin
No graph	1					1	$x^2 = -1$	No graph

USING TECHNOLOGY *How Calculators Use Rotations to Evaluate Sines and Cosines* Some calculators use rotations to calculate sines and cosines of arbitrary angles. The procedure goes something like this: The calculator has, stored,

1. ten angles or so, say

$$\alpha_1 = \sin^{-1}(10^{-1}), \quad \alpha_2 = \sin^{-1}(10^{-2}), \quad \ldots, \quad \alpha_{10} = \sin^{-1}(10^{-10}),$$

and

2. twenty numbers, the sines and cosines of the angles $\alpha_1, \alpha_2, \ldots, \alpha_{10}$.

To calculate the sine and cosine of an arbitrary angle θ, we enter θ (in radians) into the calculator. The calculator subtracts or adds multiples of 2π to θ to replace θ by the angle between 0 and 2π that has the same sine and cosine as θ (we continue to call the angle θ). The calculator then "writes" θ as a sum of multiples of α_1 (as many as possible without overshooting) plus multiples of α_2 (again, as many as possible), and so on, working its way to α_{10}. This gives

$$\theta \approx m_1\alpha_1 + m_2\alpha_2 + \cdots + m_{10}\alpha_{10}.$$

The calculator then rotates the point $(1, 0)$ through m_1 copies of α_1 (through α_1, m_1 times in succession), plus m_2 copies of α_2, and so on, finishing off with m_{10} copies of α_{10} (Figure A.45). The coordinates of the final position of $(1, 0)$ on the unit circle are the values the calculator gives for $(\cos\theta, \sin\theta)$.

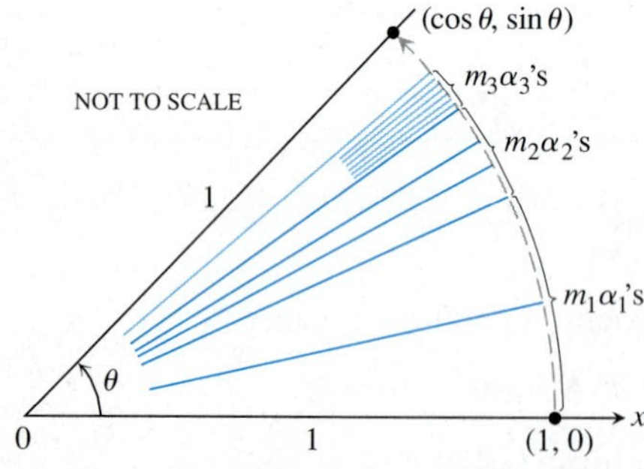

FIGURE A.45 To calculate the sine and cosine of an angle θ between 0 and 2π, the calculator rotates the point $(1, 0)$ to an appropriate location on the unit circle and displays the resulting coordinates.

EXERCISES A.13.3

Using the Discriminant

Use the discriminant $B^2 - 4AC$ to decide whether the equations in Exercises 1–16 represent parabolas, ellipses, or hyperbolas.

1. $x^2 - 3xy + y^2 - x = 0$

2. $3x^2 - 18xy + 27y^2 - 5x + 7y = -4$

3. $3x^2 - 7xy + \sqrt{17}\, y^2 = 1$

4. $2x^2 - \sqrt{15}\, xy + 2y^2 + x + y = 0$

5. $x^2 + 2xy + y^2 + 2x - y + 2 = 0$

6. $2x^2 - y^2 + 4xy - 2x + 3y = 6$

7. $x^2 + 4xy + 4y^2 - 3x = 6$

8. $x^2 + y^2 + 3x - 2y = 10$

9. $xy + y^2 - 3x = 5$

10. $3x^2 + 6xy + 3y^2 - 4x + 5y = 12$

11. $3x^2 - 5xy + 2y^2 - 7x - 14y = -1$

12. $2x^2 - 4.9xy + 3y^2 - 4x = 7$

13. $x^2 - 3xy + 3y^2 + 6y = 7$

14. $25x^2 + 21xy + 4y^2 - 350x = 0$

15. $6x^2 + 3xy + 2y^2 + 17y + 2 = 0$

16. $3x^2 + 12xy + 12y^2 + 435x - 9y + 72 = 0$

Rotating Coordinate Axes

In Exercises 17–26, rotate the coordinate axes to change the given equation into an equation that has no cross product (xy) term. Then identify the graph of the equation. (The new equations will vary with the size and direction of the rotation you use.)

17. $xy = 2$

18. $x^2 + xy + y^2 = 1$

19. $3x^2 + 2\sqrt{3}\, xy + y^2 - 8x + 8\sqrt{3}\, y = 0$

20. $x^2 - \sqrt{3}\, xy + 2y^2 = 1$

21. $x^2 - 2xy + y^2 = 2$

22. $3x^2 - 2\sqrt{3}\, xy + y^2 = 1$

23. $\sqrt{2}x^2 + 2\sqrt{2}\, xy + \sqrt{2}\, y^2 - 8x + 8y = 0$

24. $xy - y - x + 1 = 0$

25. $3x^2 + 2xy + 3y^2 = 19$

26. $3x^2 + 4\sqrt{3}\, xy - y^2 = 7$

27. Find the sine and cosine of an angle through which the coordinate axes can be rotated to eliminate the cross product term from the equation

$$14x^2 + 16xy + 2y^2 - 10x + 26{,}370\, y - 17 = 0.$$

Do not carry out the rotation.

28. Find the sine and cosine of an angle through which the coordinate axes can be rotated to eliminate the cross product term from the equation

$$4x^2 - 4xy + y^2 - 8\sqrt{5}x - 16\sqrt{5}\, y = 0.$$

Do not carry out the rotation.

T Calculator

The conic sections in Exercises 17–26 were chosen to have rotation angles that were "nice" in the sense that once we knew $\cot 2\alpha$ or $\tan 2\alpha$ we could identify 2α and find $\sin \alpha$ and $\cos \alpha$ from familiar triangles. The conic sections encountered in practice may not have such nice rotation angles, and we may have to use a calculator to determine α from the value of $\cot 2\alpha$ or $\tan 2\alpha$.

In Exercises 29–34, use a calculator to find an angle α through which the coordinate axes can be rotated to change the given equation into a quadratic equation that has no cross product term. Then find $\sin \alpha$ and $\cos \alpha$ to 2 decimal places and use Equations (6) to find the coefficients of the new equation to the nearest decimal place. In each case, say whether the conic section is an ellipse, a hyperbola, or a parabola.

29. $x^2 - xy + 3y^2 + x - y - 3 = 0$

30. $2x^2 + xy - 3y^2 + 3x - 7 = 0$

31. $x^2 - 4xy + 4y^2 - 5 = 0$

32. $2x^2 - 12xy + 18y^2 - 49 = 0$

33. $3x^2 + 5xy + 2y^2 - 8y - 1 = 0$

34. $2x^2 + 7xy + 9y^2 + 20x - 86 = 0$

Theory and Examples

35. *Effects of a 90° rotation.* What effect does a 90° rotation about the origin have on the equations of the following conic sections? Give the new equation in each case.

a) The ellipse $(x^2/a^2) + (y^2/b^2) = 1 \quad (a > b)$

b) The hyperbola $(x^2/a^2) - (y^2/b^2) = 1$

c) The circle $x^2 + y^2 = a^2$

d) The line $y = mx$

e) The line $y = mx + b$

36. *Effects of a 180° rotation.* What effect does a 180° rotation about the origin have on the equations of the following conic sections? Give the new equation in each case.

a) The ellipse $(x^2/a^2) + (y^2/b^2) = 1 \quad (a > b)$

b) The hyperbola $(x^2/a^2) - (y^2/b^2) = 1$

c) The circle $x^2 + y^2 = a^2$

d) The line $y = mx$

e) The line $y = mx + b$

37. *The hyperbola $xy = a$.* The hyperbola $xy = 1$ is one of many hyperbolas of the form $xy = a$ that appear in science and mathematics.

a) Rotate the coordinate axes through an angle of 45° to change the equation $xy = 1$ into an equation with no xy-term. What is the new equation?

b) Do the same for the equation $xy = a$.

38. *Eccentricity.* Find the eccentricity of the hyperbola $xy = 2$.

39. *Writing to learn.* Can anything be said about the graph of the equation $Ax^2 + Bxy + Cy^2 + Dx + Ey + F = 0$ if $AC < 0$? Give reasons for your answer.

40. *Writing to learn.* Does any nondegenerate conic section $Ax^2 + Bxy + Cy^2 + Dx + Ey + F = 0$ have all of the following properties?

a) It is symmetric with respect to the origin.

b) It passes through the point $(1, 0)$.

c) It is tangent to the line $y = 1$ at the point $(-2, 1)$.

Give reasons for your answer.

41. *Rotating a circle.* Show that the equation $x^2 + y^2 = a^2$ becomes $x'^2 + y'^2 = a^2$ for every choice of the angle α in the rotation equations (4).

42. *A special case.* Show that rotating the axes through an angle of $\pi/4$ radians will eliminate the xy-term from Equation (1) whenever $A = C$.

43. *A degenerate conic section.*

a) Decide whether the equation

$$x^2 + 4xy + 4y^2 + 6x + 12y + 9 = 0$$

represents an ellipse, a parabola, or a hyperbola.

b) Show that the graph of the equation in (a) is the line $2y = -x - 3$.

44. *A degenerate conic section.*

a) Decide whether the conic section with equation

$$9x^2 + 6xy + y^2 - 12x - 4y + 4 = 0$$

represents a parabola, an ellipse, or a hyperbola.

b) Show that the graph of the equation in (a) is the line $y = -3x + 2$.

45. *Normals to a conic section.*

a) What kind of conic section is the curve $xy + 2x - y = 0$?

b) Solve the equation $xy + 2x - y = 0$ for y and sketch the curve as the graph of a rational function of x.

c) Find equations for the lines parallel to the line $y = -2x$ that are normal to the curve. Add the lines to your sketch.

46. *True or false.* Prove or find counterexamples to the following statements about the graph of $Ax^2 + Bxy + Cy^2 + Dx + Ey + F = 0$.

a) If $AC > 0$, the graph is an ellipse.

b) If $AC > 0$, the graph is a hyperbola.

c) If $AC < 0$, the graph is a hyperbola.

47. *A nice area formula for ellipses.* When $B^2 - 4AC$ is negative, the equation

$$Ax^2 + Bxy + Cy^2 = 1$$

represents an ellipse. If the ellipse's semi-axes are a and b, its area is πab (a standard formula). Show that the area is also given by the formula $2\pi/\sqrt{4AC - B^2}$. (*Hint*: Rotate the coordinate axes to eliminate the xy-term and apply Equation (12) to the new equation.)

48. *Other invariants.* We describe the fact that $B'^2 - 4A'C'$ equals $B^2 - 4AC$ after a rotation about the origin by saying that the discriminant of a quadratic equation is an **invariant** of the equation. Use Equations (6) to show that the numbers (a) $A + C$ and (b) $D^2 + E^2$ are also invariants, in the sense that

$$A' + C' = A + C \quad \text{and} \quad D'^2 + E'^2 = D^2 + E^2.$$

We can use these equalities to check against numerical errors when we rotate axes. They can also be helpful in shortening the work required to find values for the new coefficients.

49. *A proof that $B'^2 - 4A'C' = B^2 - 4AC$.* Use Equations (6) to show that $B'^2 - 4A'C' = B^2 - 4AC$ for any rotation of axes about the origin. The calculation works out nicely but requires patience.

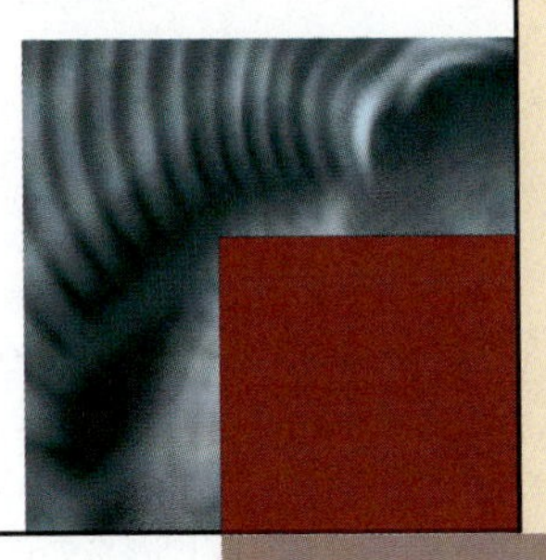

Answers

PRELIMINARY CHAPTER

Section P.1, pp. 7–10

1. **(a)** $\Delta x = -2, \Delta y = -3$ **(b)** $\Delta x = 2, \Delta y = -4$

3. **(a)** $m = 3$ **(b)** $m = -\frac{1}{3}$

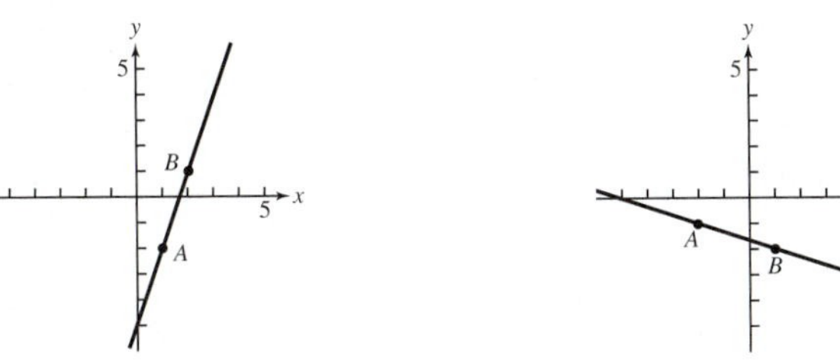

5. **(a)** $x = 2, y = 3$ **(b)** $x = -1, y = \frac{4}{3}$

7. **(a)** $y = 1(x - 1) + 1$ **(b)** $y = -1(x + 1) + 1$

9. **(a)** $3x - 2y = 0$ **(b)** $y = 1$

11. **(a)** $y = 3x - 2$ **(b)** $y = -x + 2$

13. $y = \frac{5}{2}x$

15. **(a)** **(i)** Slope: $-\frac{3}{4}$

(ii) y-intercept: 3

(iii)

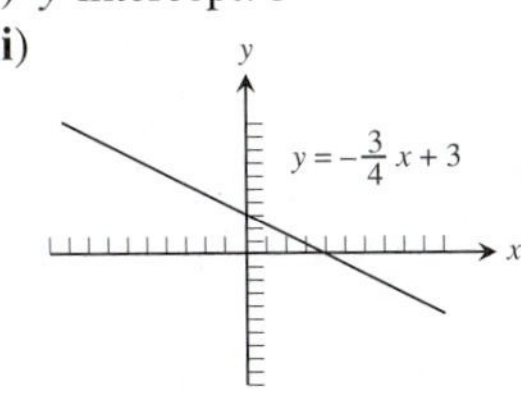

(b) **(i)** Slope: -1

(ii) y-intercept: 2

(iii)

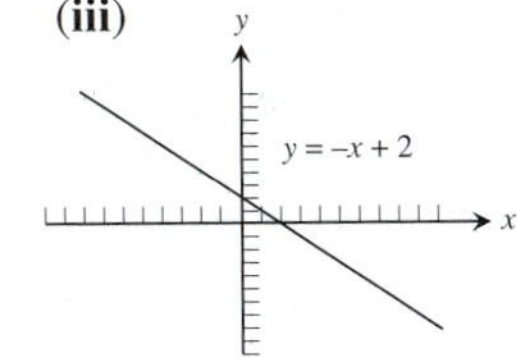

17. **(a)** $y = -x; y = x$

(b) $y = -2x - 2; y = \frac{1}{2}x + 3$

19. $m = \frac{7}{2}, b = -\frac{3}{2}$

21. $y = -1$

23. $y = 1(x - 3) + 4$
$y = x - 3 + 4$
$y = x + 1$, which is the same equation.

25. **(a)** $k = 2$ **(b)** $k = -2$

27. 5.97 atmospheres ($k = 0.0994$)

29. **(a)** Yes, -40°F is the same as -40°C.

(b) All three lines pass through the point $(-40, -40)$.

33. **(a)** $y = \frac{B}{A}(x - a) + b$

(b) The coordinates of Q are $\left(\frac{B^2a + AC - ABb}{A^2 + B^2}, \frac{A^2b + BC - ABa}{A^2 + B^2}\right)$

(c) Distance $= \frac{|Aa + Bb - C|}{\sqrt{A^2 + B^2}}$

35. 40.25 ft

37. **(a)** $y = 0.680x + 9.013$

(b) The slope is 0.68. It represents the approximate average weight gain in pounds per month.

(c)

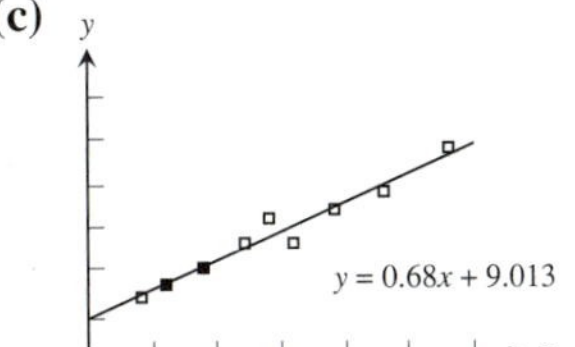

(d) 29 lb

39. **(a)** $y = 5632x - 11{,}080{,}280$

(b) The rate at which the median price is increasing in dollars per year

(c) $y = 2732x - 5{,}362{,}360$

(d) In the Northeast

Section P.2, pp. 20–24

1. $A = \frac{\sqrt{3}}{4}x^2, p = 3x$

3. $x = \frac{d}{\sqrt{3}}, A = 2d^2, V = \frac{d^3}{3\sqrt{3}}$

5. **(a)** Not a function of x because some values of x have two values of y

(b) A function of x because for every x there is only one possible y

7. **(a)** $D: (-\infty, \infty), R: [1, \infty)$ **(b)** $D: [0, \infty), R: (-\infty, 1]$

9. $D: [-2, 2], R: [0, 2]$

11. **(a)** Symmetric about the origin

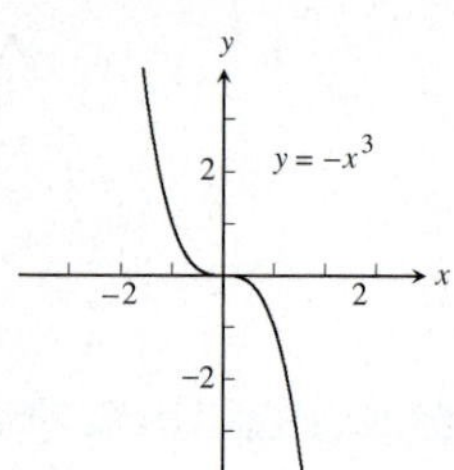

(b) Symmetric about the y-axis

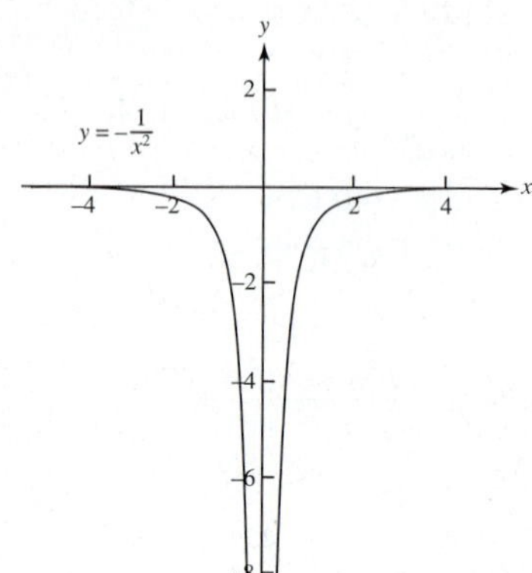

13. **(a)** For each positive value of x, there are two values of y.

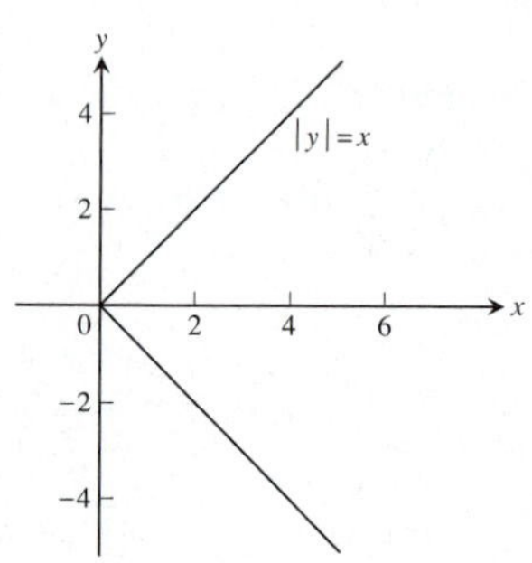

(b) For each value of $x \neq 0$, there are two values of y.

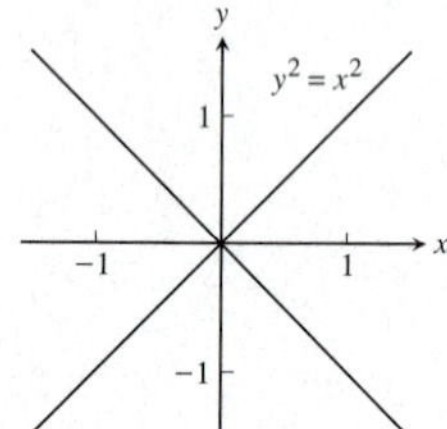

15. **(a)** Even **(b)** Odd

17. **(a)** Odd **(b)** Even

19. **(a)** Neither **(b)** Even

21. **(a)** Domain = all reals, range = $(-\infty, 2]$ **(b)** Domain = all reals, range = $[-3, \infty)$

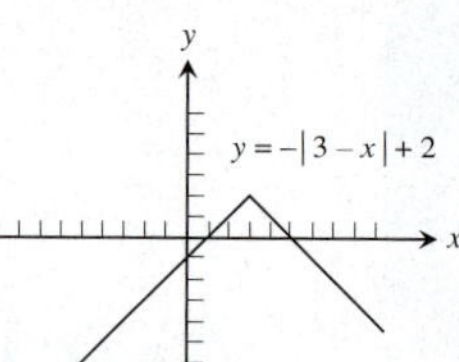

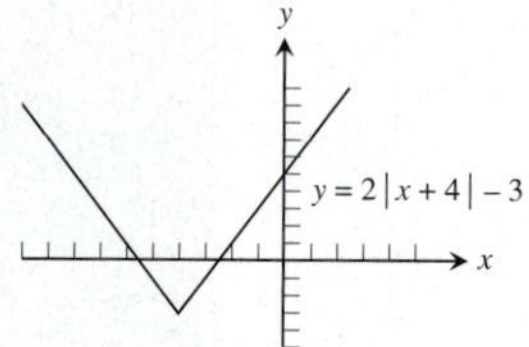

23. **(a)**

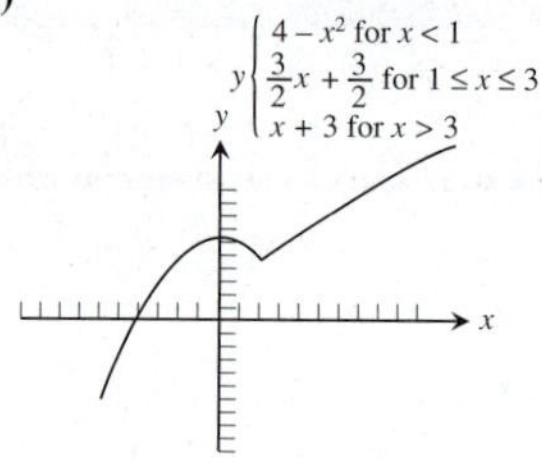

(b) All reals **(c)** All reals

25. Because if the vertical line test holds, then for each x-coordinate there is at most one y-coordinate giving a point on the curve. This y-coordinate corresponds to the value assigned to the x-coordinate. Since there is only one y-coordinate, the assignment is unique.

27. **(a)** $f(x) = \begin{cases} x, & 0 \le x \le 1 \\ -x + 2, & 1 < x \le 2 \end{cases}$

(b) $f(x) = \begin{cases} 2, & 0 \le x < 1 \\ 0, & 1 \le x < 2 \\ 2, & 2 \le x < 3 \\ 0, & 3 \le x \le 4 \end{cases}$

(c) $f(x) = \begin{cases} -x + 2, & 0 < x \le 2 \\ -\frac{1}{3}x + \frac{5}{3}, & 2 < x \le 5 \end{cases}$

(d) $f(x) = \begin{cases} -3x - 3, & -1 < x \le 0 \\ -2x + 3, & 0 < x \le 2 \end{cases}$

29. **(a)** Position 4 **(b)** Position 1 **(c)** Position 2 **(d)** Position 3

31. $(x + 2)^2 + (y + 3)^2 = 49$

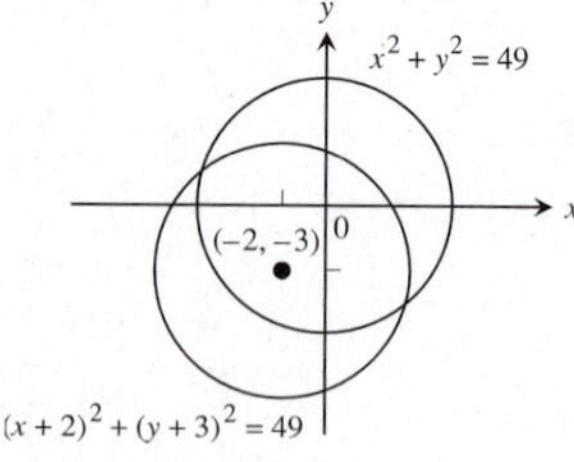

33. $y + 1 = (x - 1)^{2/3}$

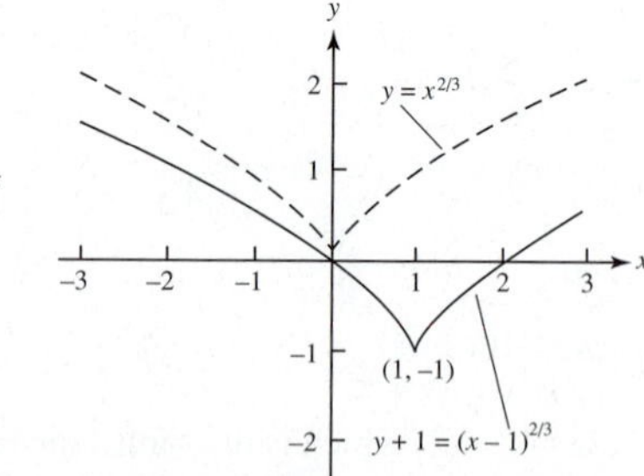

35. $y = \frac{1}{2}x$

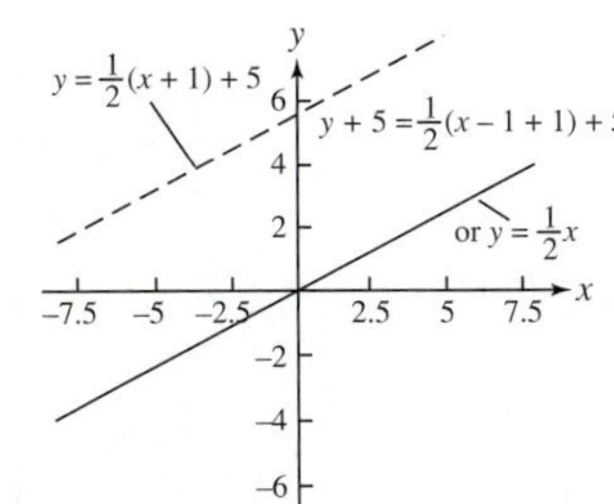

37. **(a)** 2 **(b)** 22 **(c)** $x^2 + 2$ **(d)** $x^2 + 10x + 22$
(e) 5 **(f)** -2 **(g)** $x + 10$ **(h)** $x^4 - 6x^2 + 6$

39. **(a)** $\frac{4}{x^2} - 5$ **(b)** $\frac{4}{x^2} - 5$ **(c)** $\left(\frac{4}{x} - 5\right)^2$
(d) $\left(\frac{1}{4x-5}\right)^2$ **(e)** $\frac{1}{4x^2 - 5}$ **(f)** $\frac{1}{(4x-5)^2}$

41. **(a)** $g(f(x))$ **(b)** $j(g(x))$ **(c)** $g(g(x))$
(d) $j(j(x))$ **(e)** $g(h(f(x)))$ **(f)** $h(j(f(x)))$

43. **(a)** $g(x) = x^2$ **(b)** $g(x) = \frac{1}{x-1}$ **(c)** $f(x) = \frac{1}{x}$
(d) $f(x) = x^2$ (Note that the domain of the composite is $[0, \infty)$.)

45. **(a)** D: $[0, 2]$, R: $[2, 3]$ **(b)** D: $[0, 2]$, R: $[-1, 0]$

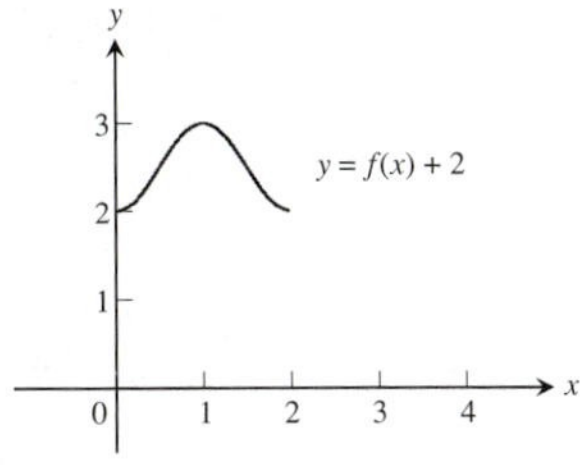

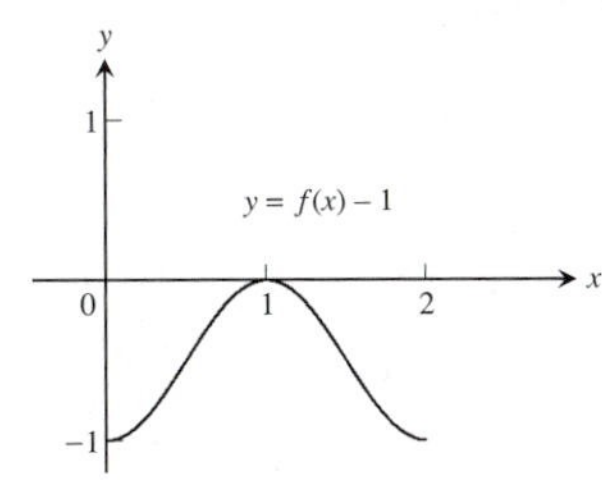

(c) D: $[0, 2]$, R: $[0, 2]$ **(d)** D: $[0, 2]$, R: $[-1, 0]$

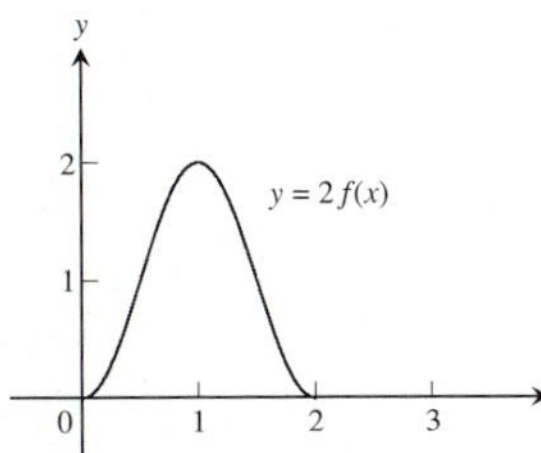

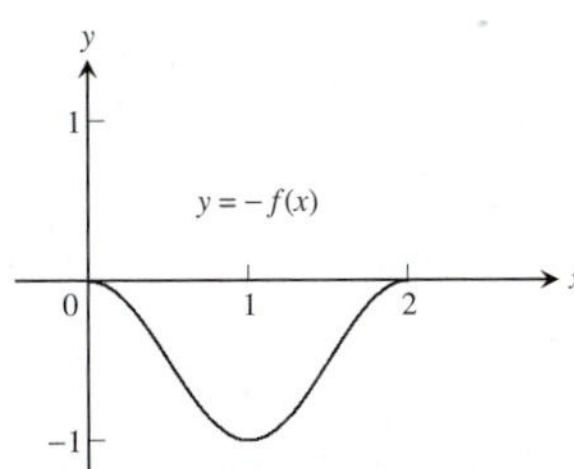

(e) D: $[-2, 0]$, R: $[0, 1]$ **(f)** D: $[1, 3]$, R: $[0, 1]$

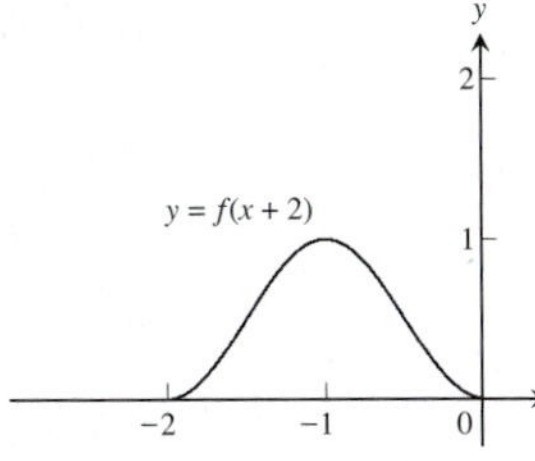

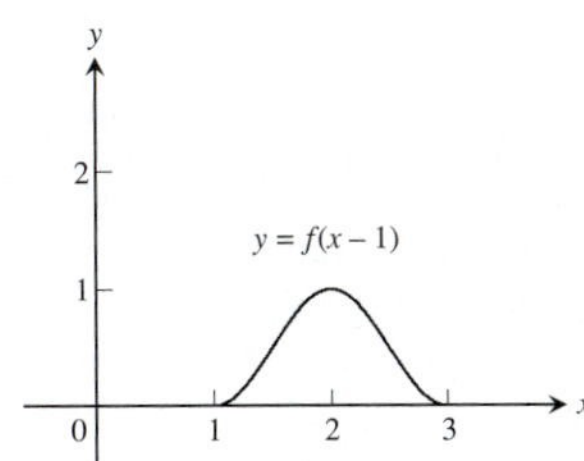

(g) D: $[-2, 0]$, R: $[0, 1]$ **(h)** D: $[-1, 1]$, R: $[0, 1]$

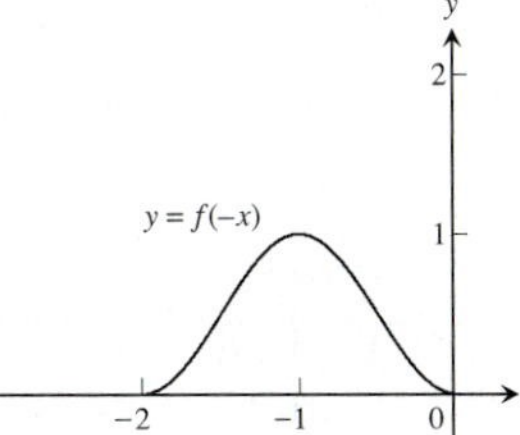

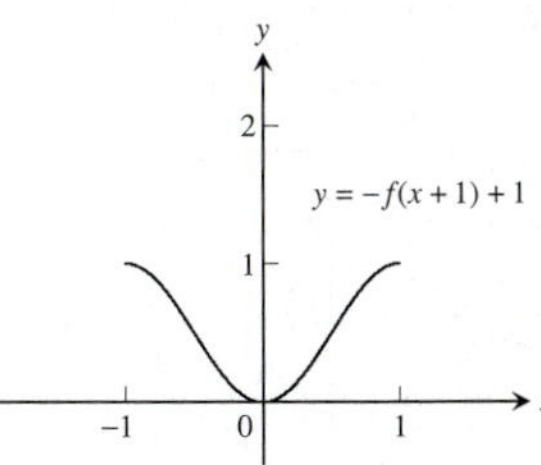

47. **(a)** Because the circumference of the original circle was 8π and a piece of length x was removed
(b) $r = \frac{8\pi - x}{2\pi} = 4 - \frac{x}{2\pi}$
(c) $h = \sqrt{16 - r^2} = \frac{\sqrt{16\pi x - x^2}}{2\pi}$
(d) $V = \frac{1}{3}\pi r^2 h = \frac{(8\pi - x)^2\sqrt{16\pi x - x^2}}{24\pi^2}$

49. **(a)** Yes. Since $(f \cdot g)(-x) = f(-x) \cdot g(-x) = f(x) \cdot g(x) = (f \cdot g)(x)$, the function $(f \cdot g)(x)$ will also be even.
(b) The product will be even, since $(f \cdot g)(-x) = (f \cdot g)(x)$.

53. **(d)** $(g \circ f)(x) = \sqrt{4 - x^2}$; $D(g \circ f) = [-2, 2]$; $R(g \circ f) = [0, 2]$
$(f \circ g)(x) = 4 - (\sqrt{x})^2 = 4 - x$ for $x \geq 0$; $D(f \circ g) = [0, \infty)$;
$R(f \circ g) = (-\infty, 4]$

55. **(a)** $y = 4.44647x^{0.511414}$
(b)

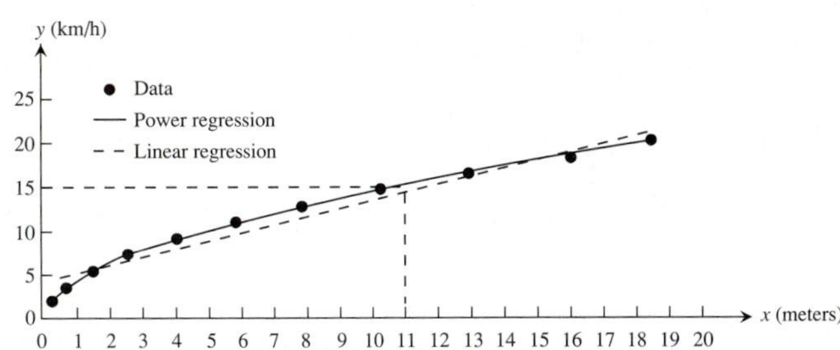

(c) 15 km/h
(d) 14 km/h, $y = 0.913695x + 4.189976$; the power regression curve in part (a) better fits the data.

Section P.3, pp. 29–31

1. (a) **3.** (e) **5.** (b)

7. Domain: all reals,
range: $(-\infty, 3)$,
x-intercept: ≈ 1.585,
y-intercept: 2

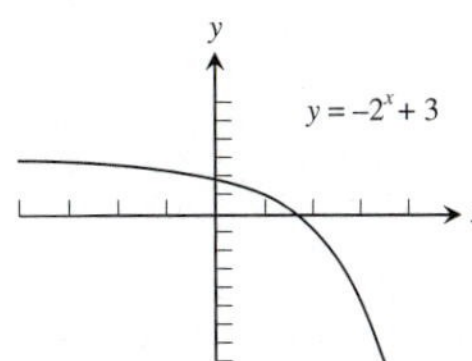

9. Domain: all reals,
range: $(-2, \infty)$,
x-intercept: ≈ 0.405,
y-intercept: 1

$y = 3e^{-x} - 2$

11. 3^{4x}

13. 2^{-6x}

15.

x	y	Δy
1	−1	
		2
2	1	
		2
3	3	
		2
4	5	

17.

x	y	Δy
1	1	
		3
2	4	
		5
3	9	
		7
4	16	

19. If the changes in x are constant for a linear function, say $\Delta x = c$, then the changes in y are also constant, specifically, $\Delta y = mc$.

21. $a = 3, k = 1.5$

23. $x \approx 2.3219$

25. $x \approx -0.6309$

27. 7609.7 million

29. After 19 years

31. **(a)** $A(t) = 6.6\left(\frac{1}{2}\right)^{t/14}$

(b) About 38 days later

33. $\approx$ 11.433 years

35. $\approx$ 11.090 years

37. $\approx$ 19.108 years

39. $2^{48} \approx 2.815 \times 10^{14}$

41. **(a)** Regression equation: $P(x) = 6.033(1.030)^x$, where $x = 0$ represents 1900

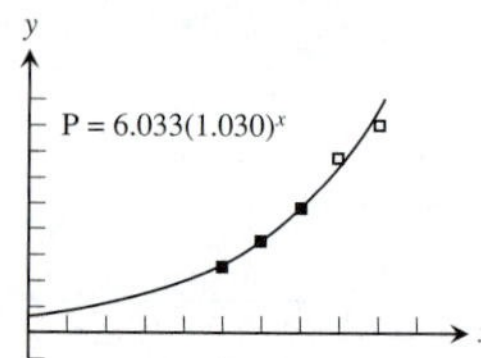

(b) Approximately 6.03 million, which is not very close to the actual population

(c) The annual rate of growth is approximately 3%.

Section P.4, pp. 41–43

1. One-to-one

3. Not one-to-one

5. One-to-one

7. D: (0, 1], R: [0, ∞)

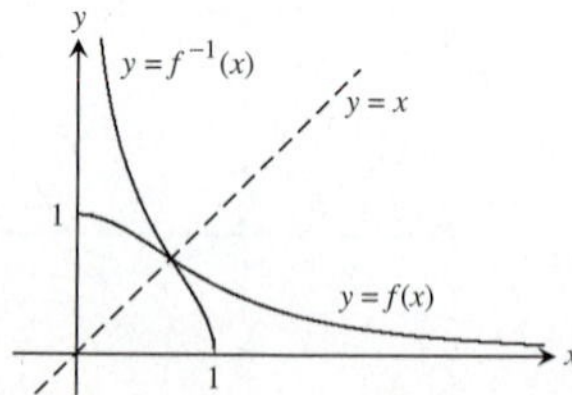

9. $D(f^{-1}) = [0, \infty)$; $R(f^{-1}) = [0, \infty)$

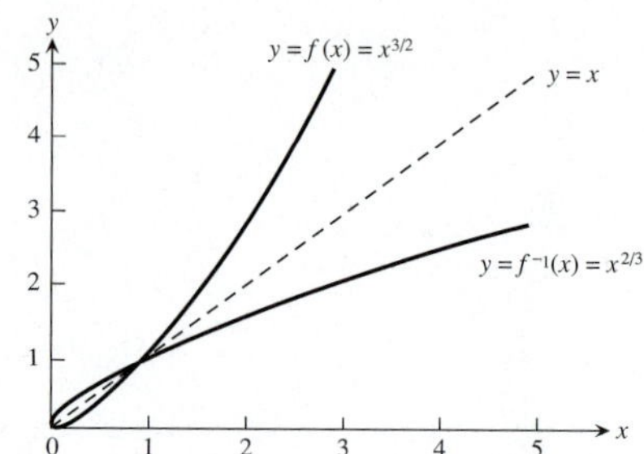

11. $f^{-1}(x) = \sqrt{x-1}$

13. $f^{-1}(x) = \sqrt[3]{x+1}$

15. $f^{-1}(x) = \sqrt{x} - 1$

17. $f^{-1}(x) = \dfrac{x-3}{2}$

19. $f^{-1}(x) = (x+1)^{1/3}$ or $\sqrt[3]{x-1}$

21. $f^{-1}(x) = -x^{1/2}$ or $-\sqrt{x}$

23. $f^{-1}(x) = 2 - (-x)^{1/2}$ or $2 - \sqrt{-x}$

25. $f^{-1}(x) = \dfrac{1}{x^{1/2}}$ or $\dfrac{1}{\sqrt{x}}$

27. $f^{-1}(x) = \dfrac{1-3x}{x-2}$

29. $y = e^{x \ln 3} - 1$

(a) $D = (-\infty, \infty)$

(b) $R = (-1, \infty)$

31. $y = 1 - \ln x$

(a) $D = (0, \infty)$

(b) $R = (-\infty, \infty)$

(c)

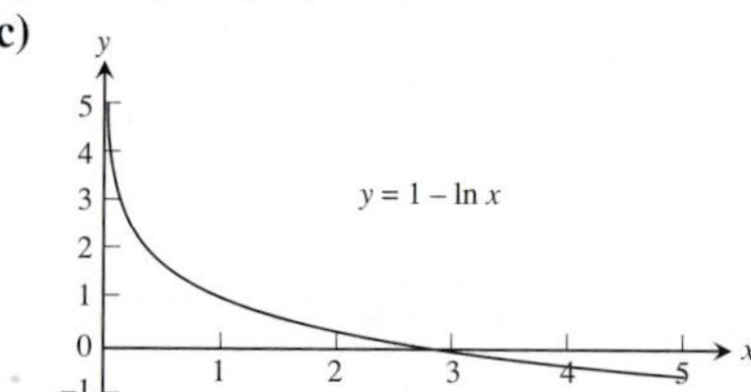

33. $t = \dfrac{\ln 2}{\ln 1.045} \approx 15.75$

35. $x = \ln\left(\dfrac{3 \pm \sqrt{5}}{2}\right) \approx -0.96$ or 0.96

37. $y = e^{2t+4}$

39. **(a)** $f^{-1}(x) = \log_2\left(\dfrac{x}{100-x}\right)$

(b) $f^{-1}(x) = \log_{1.1}\left(\dfrac{x}{50-x}\right)$

41. **(a)** Amount $= 8\left(\frac{1}{2}\right)^{t/12}$

(b) 36 hours

43. $\approx$ 44.081 years

45. 10 db

47. (4, 5)

49. **(a)** (1.58, 3)

(b) No intersection

51. f and g are inverses of each other because $(f \circ g)(x) = (g \circ f)(x) = x$.
(a) **(b)** and **(c)**

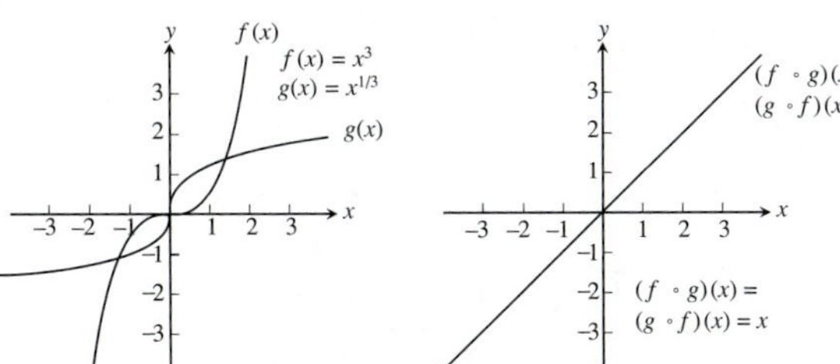

53. f and g are inverses of each other because $(f \circ g)(x) = (g \circ f)(x) = x$.
(a) **(b)** and **(c)**

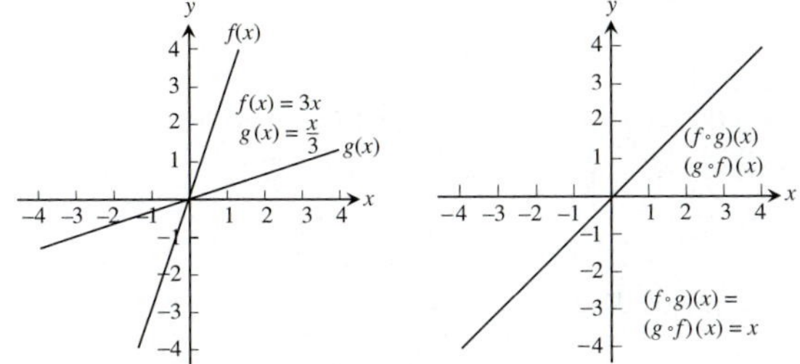

55. **(a)** The graphs of y_1, appear to be vertical translates of y_2.

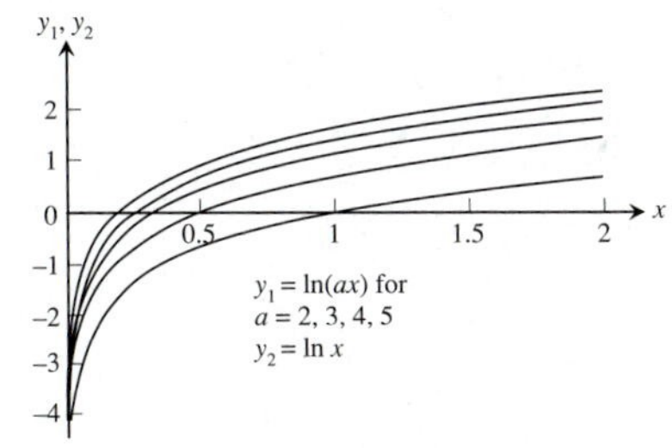

(b)

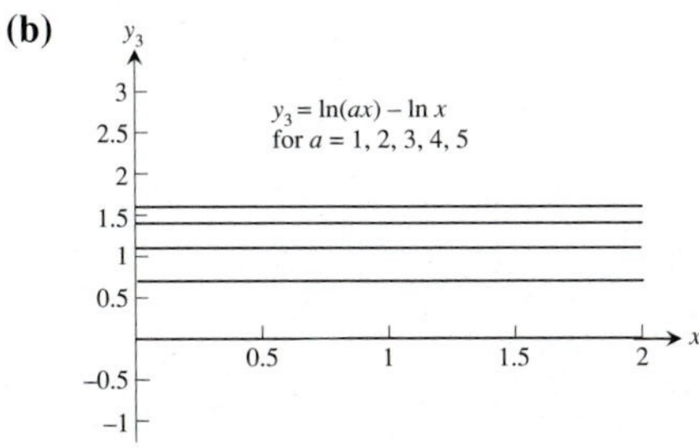

(c) $y_3 = y_1 - y_2 = \ln ax - \ln x = (\ln a + \ln x) - \ln x = \ln a$, a constant.

57. $x \approx -0.76666$

59. **(a)** $y(x) = -474.31 + 121.13 \ln x$; 59.48 million metric tons produced in 1982 and 83.51 million metric tons produced in 2000

(b)

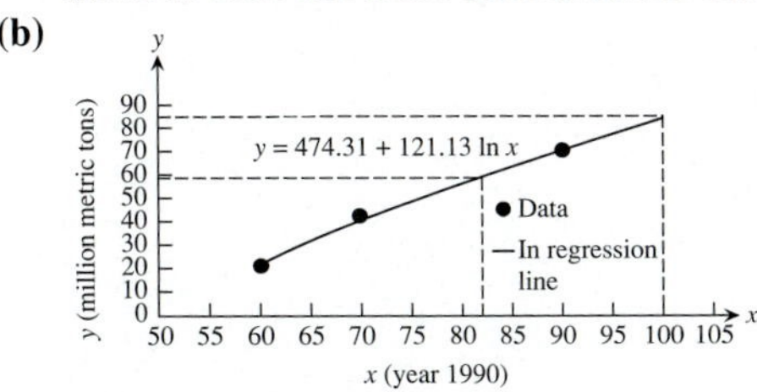

(c) $y(82) \approx 59$ and $y(100) \approx 84$

Section P.5, pp. 55–59

1. **(a)** 8π m **(b)** $\frac{55\pi}{9}$ m

3.

θ	$-\pi$	$-\frac{2\pi}{3}$	0	$\frac{\pi}{2}$	$\frac{3\pi}{4}$
$\sin \theta$	0	$-\frac{\sqrt{3}}{2}$	0	1	$\frac{1}{\sqrt{2}}$
$\cos \theta$	-1	$-\frac{1}{2}$	1	0	$-\frac{1}{\sqrt{2}}$
$\tan \theta$	0	$\sqrt{3}$	0	UND	-1
$\cot \theta$	UND	$\frac{1}{\sqrt{3}}$	UND	0	-1
$\sec \theta$	-1	-2	1	UND	$-\sqrt{2}$
$\csc \theta$	UND	$-\frac{2}{\sqrt{3}}$	UND	1	$\sqrt{2}$

5. **(a)** $\cos x = -\frac{4}{5}$, $\tan x = -\frac{3}{4}$

(b) $\sin x = -\frac{2\sqrt{2}}{3}$, $\tan x = -2\sqrt{2}$

7. **(a)** Period π **(b)** Period 2

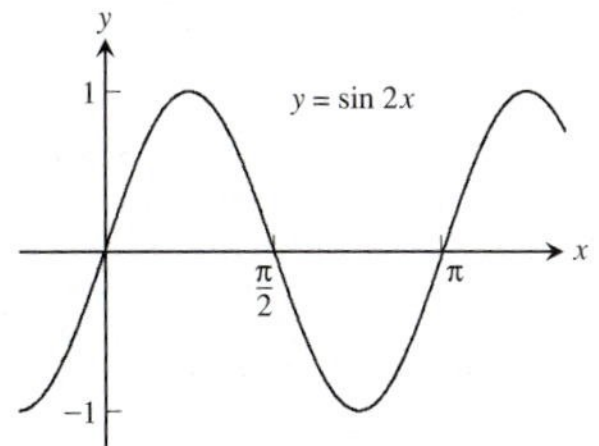

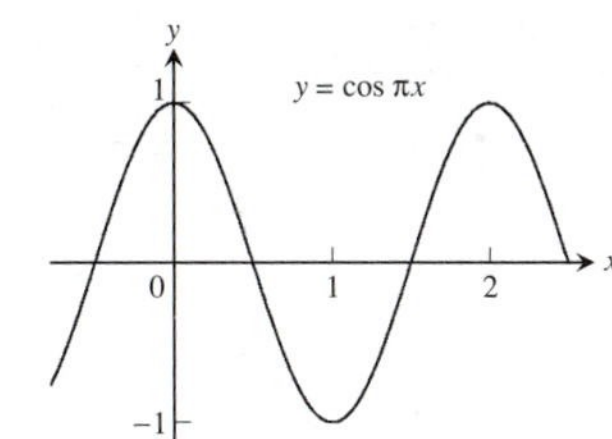

9. **(a)** Period 2π

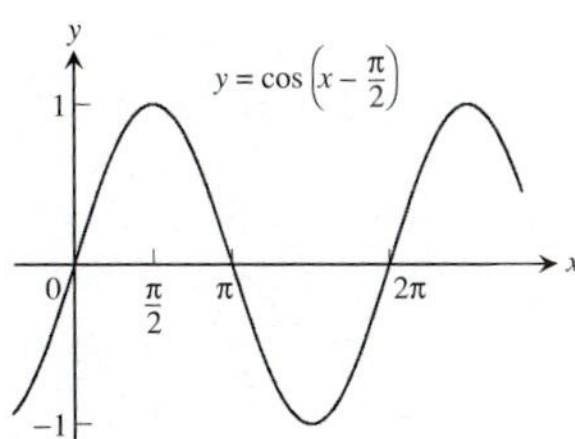

(b) Period 2π

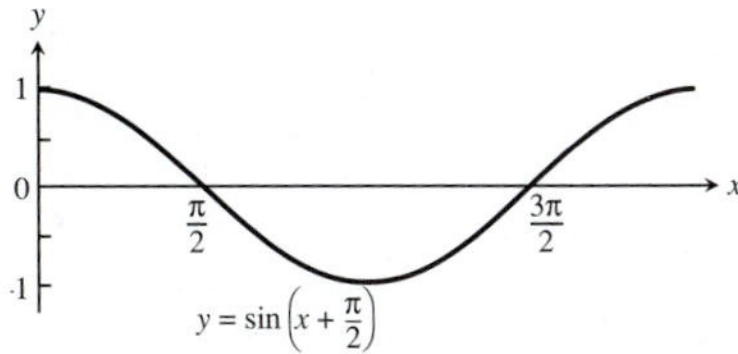

11. Period $\frac{\pi}{2}$, symmetric about the origin

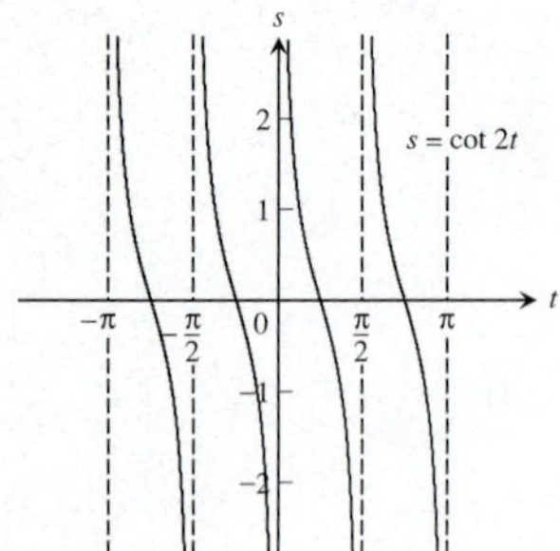

13. (a) $-\cos x$ **(b)** $-\sin x$

19. (a) $A = 2, B = 2\pi, C = -\pi, D = -1$

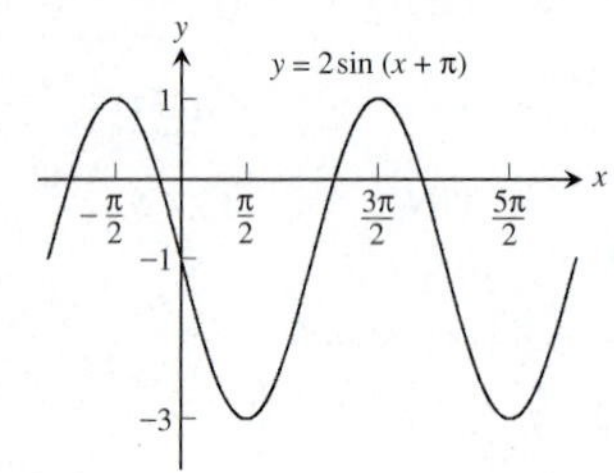

(b) $A = \frac{1}{2}, B = 2, C = 1, D = \frac{1}{2}$

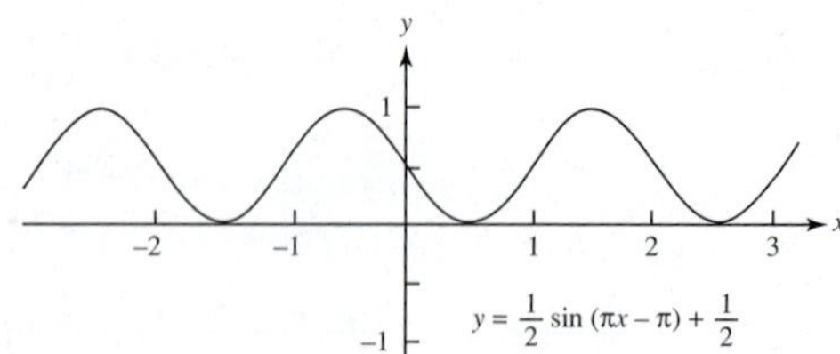

21. (a) 37 **(b)** 365 **(c)** Right 101 **(d)** Up 25

23. (a) $\frac{\pi}{4}$ **(b)** $-\frac{\pi}{3}$ **(c)** $\frac{\pi}{6}$

25. (a) $\frac{\pi}{3}$ **(b)** $\frac{3\pi}{4}$ **(c)** $\frac{\pi}{6}$

37. (a) $c = \sqrt{7} \approx 2.646$ **(b)** $c \approx 1.951$

39. (a) Values of $\sin x$ approach values of x. At the origin, they are both equal to 0.

(b) They both approach zero, but at any given x other than zero, the value of $\sin x$ is only a small percentage of the value of x.

41. (a) Domain: all real numbers except those having the form $\frac{\pi}{2} + k\pi$, where k is an integer; range: $-\frac{\pi}{2} < y < \frac{\pi}{2}$.

(b) Domain: $-\infty < x < \infty$; range: $-\infty < y < \infty$

43. $x \approx 1.190$ and $x \approx 4.332$ **45.** $x \approx -1.911$ and $x \approx 1.911$

47. (a)

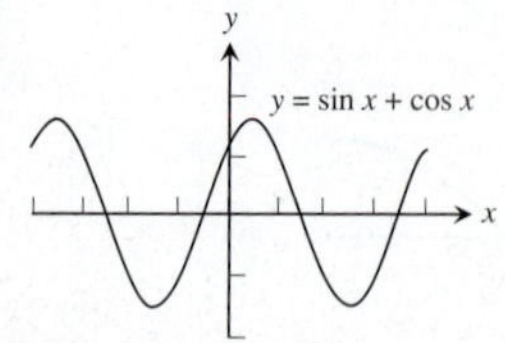

(b) Amplitude ≈ 1.414, period $= 2\pi$, horizontal shift ≈ -0.785 or 5.498 relative to $\sin x$, vertical shift: 0

49. (a) $p = 0.6 \sin(2479t - 2.801) + 0.265$

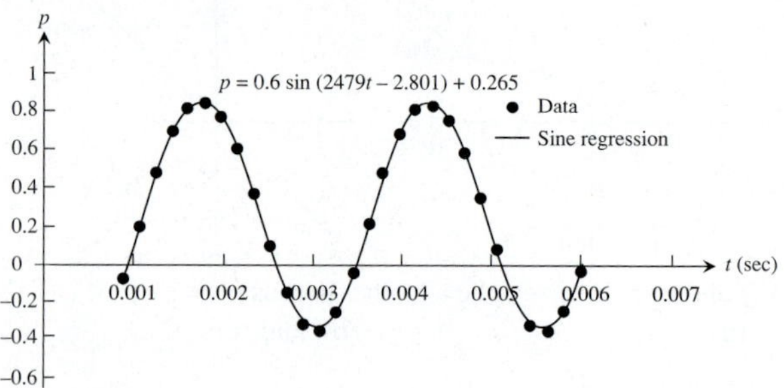

(b) ≈ 395 Hz

51. (a) $y = 3.0014 \sin(0.9996x + 2.0012) + 2.9999$

(b) $y = 3 \sin(x + 2) + 3$

Section P.6, pp. 65–67

1.

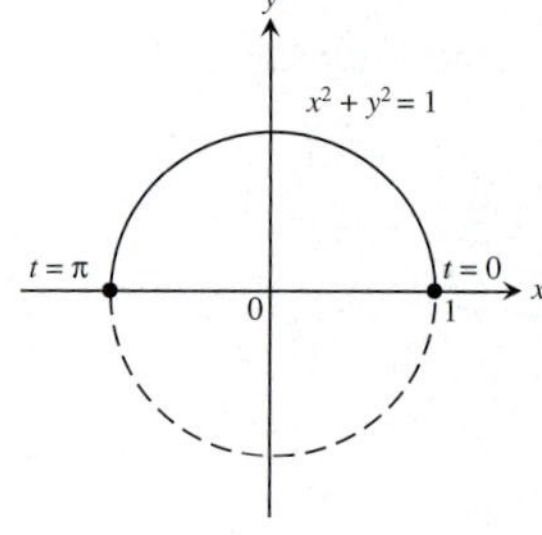

3.

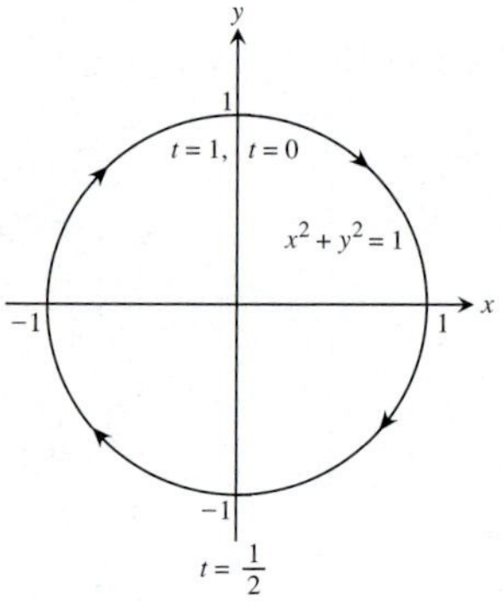

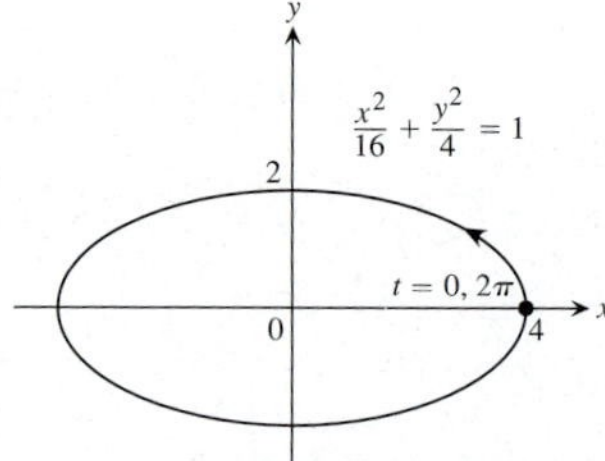

5.

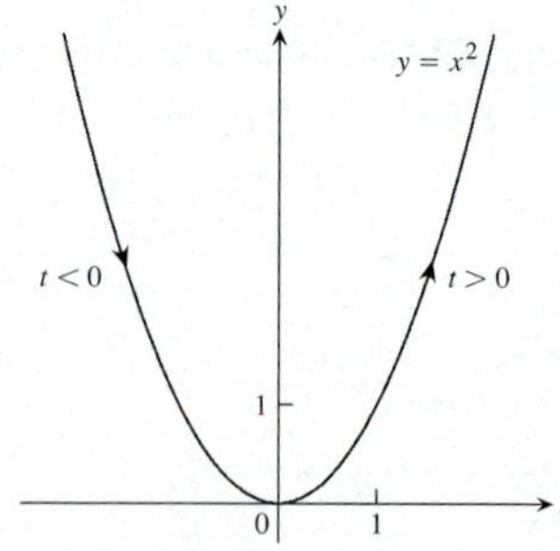

7.

9.

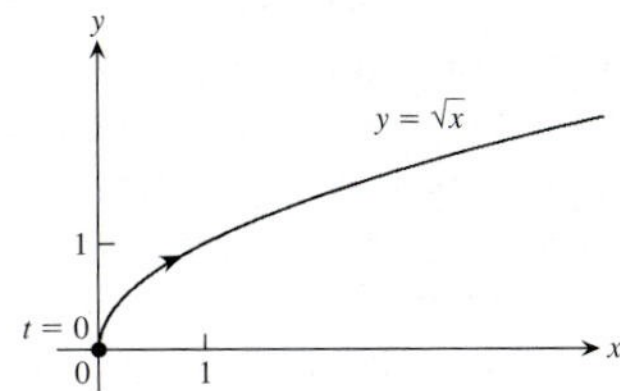

11.

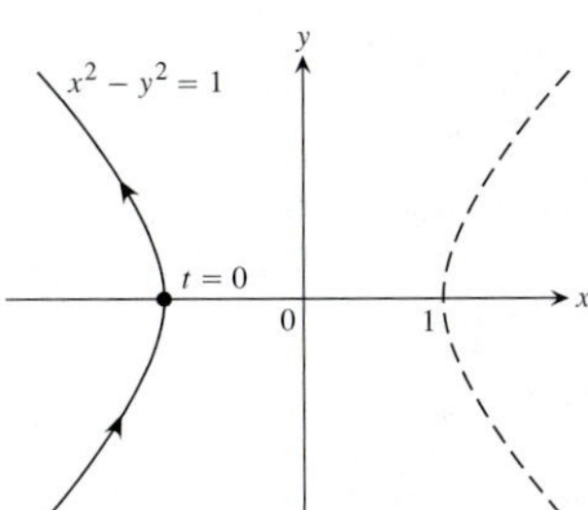

13.

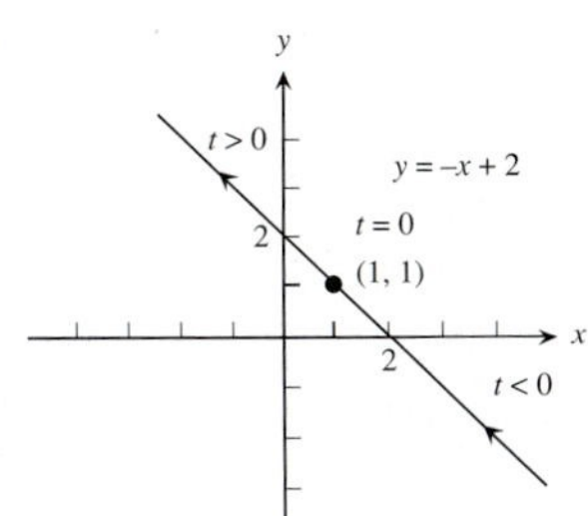

15.

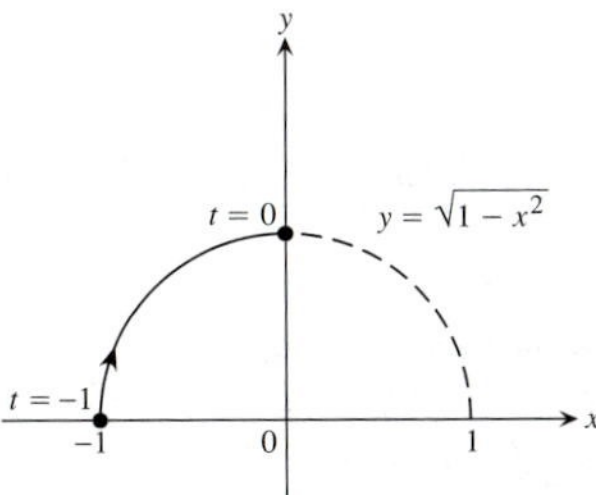

17.

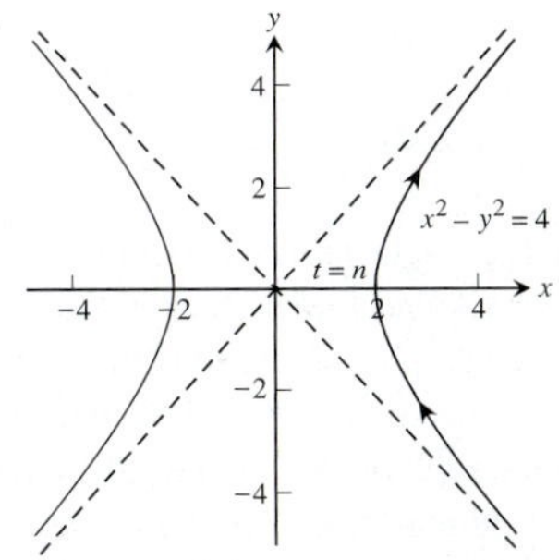

19. **(a)** $x = a \cos t,\ y = -a \sin t,\ 0 \le t \le 2\pi$
(b) $x = a \cos t,\ y = a \sin t,\ 0 \le t \le 2\pi$
(c) $x = a \cos t,\ y = -a \sin t,\ 0 \le t \le 4\pi$
(d) $x = a \cos t,\ y = a \sin t,\ 0 \le t \le 4\pi$

21. Possible answer: $x = -1 + 5t,\ y = -3 + 4t,\ 0 \le t \le 1$

23. Possible answer: $x = t^2 + 1,\ y = t,\ t \le 0$

25. Possible answer: $x = 2 - 3t,\ y = 3 - 4t,\ t \ge 0$

27. Graph (c), Window: $[-4, 4]$ by $[-3, 3]$, $0 \le t \le 2\pi$

29. Graph (d), Window: $[-10, 10]$ by $[-10, 10]$, $0 \le t \le 2\pi$

31.

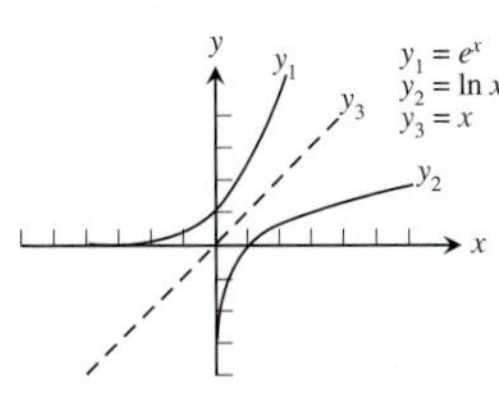

33.

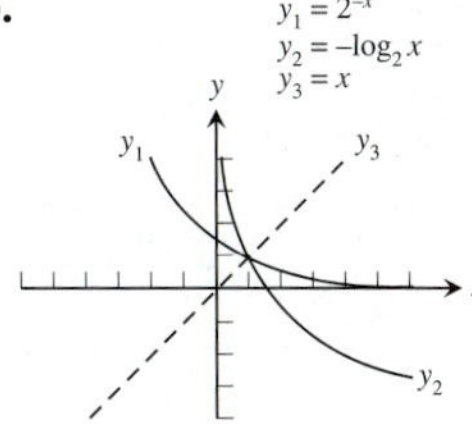

35.

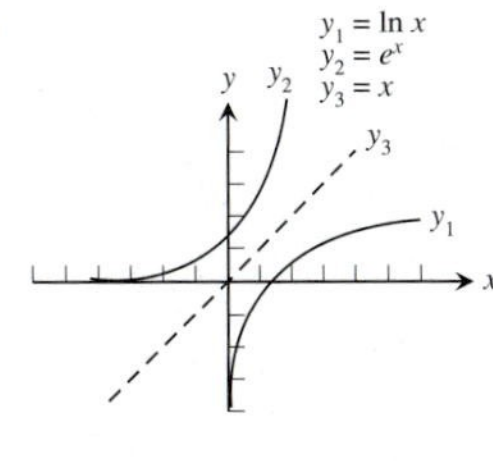

37.

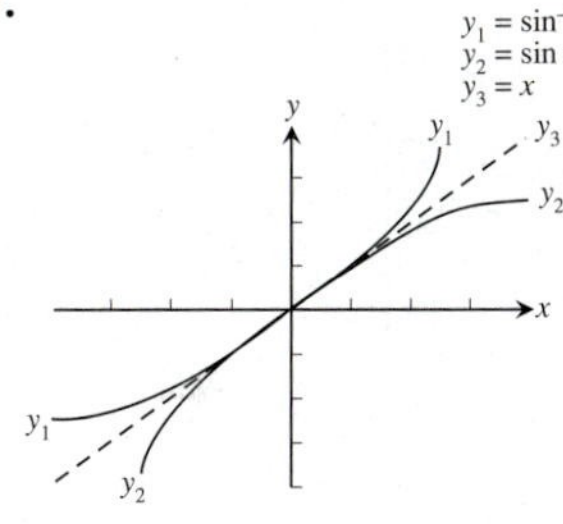

39. $1 < t < 3$

41. $-5 \le t < -3$

49. $x = 2 \cot t,\ y = 2 \sin^2 t,\ 0 < t < \pi$

51. **(a)** The curve is traced from right to left and extends infinitely in both directions from the origin.

Section P.7, pp. 73–76

1. **(a)** The graph supports the assumption that y is proportional to x. The constant of proportionality is estimated from the slope of the regression line, which is 0.166.

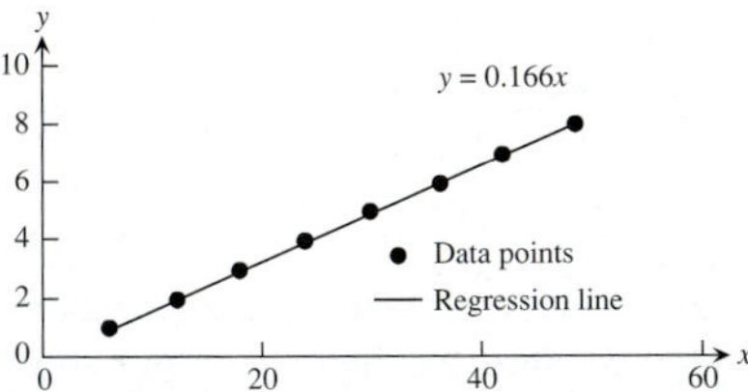

(b) The graph supports the assumption that y is proportional to $x^{1/2}$. The constant of proportionality is estimated from the slope of the regression line, which is 2.03.

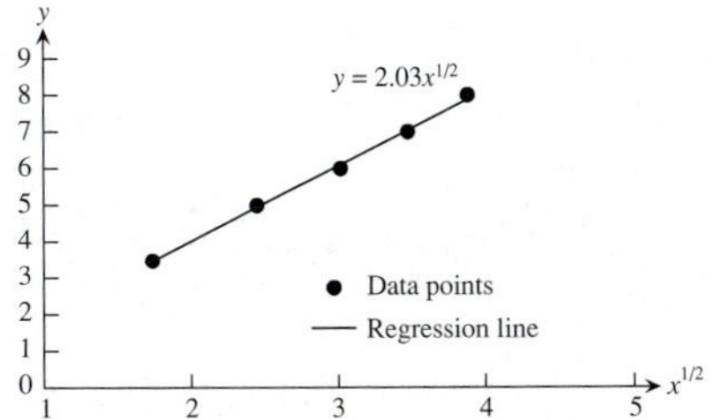

3. Quadratic regression gives $y = 0.064555x^2 + 0.078422x + 4.88961$. The power regression fits the data well.

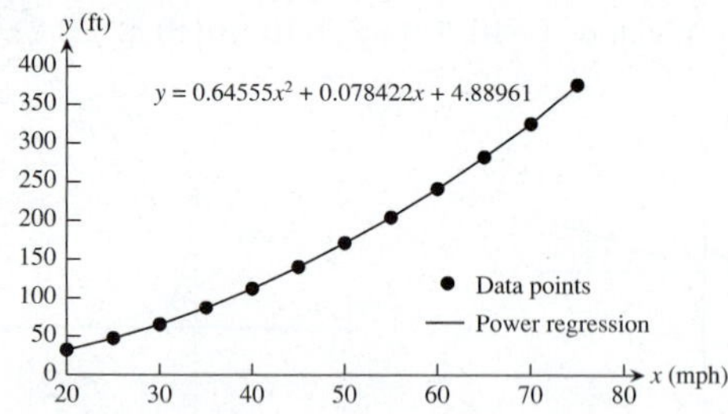

5. **(a)** Exponential regression gives $y = 0.5(0.69^x) = 0.5e^{-0.371x}$.

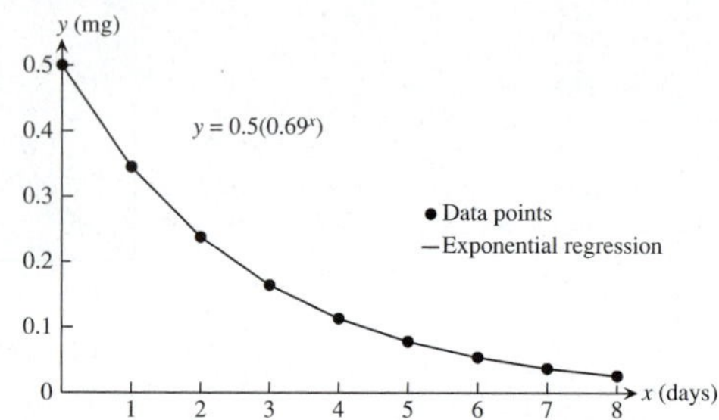

(b) The exponential function fits the data very well.

(c) The model predicts that after 12 hours, the amount of digoxin in the blood will be less than 0.006 mg.

7. **(a)** Exponential regression gives $C = 770(0.715^t) = 770e^{-0.336t}$.

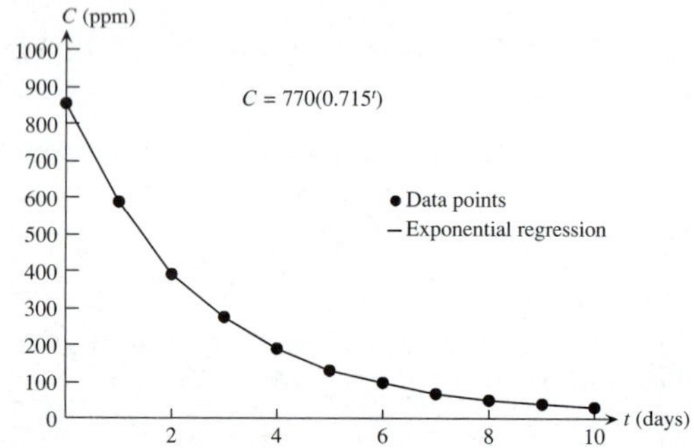

(b) The exponential function appears to capture a trend for these data.

(c) The model predicts that the blood concentration will fall below 10 ppm after 12 days and 22 hours.

9. The slope of the regression line is 0.008435, so the model that estimates the weight as a function of l is $w = 0.008435l^3$.

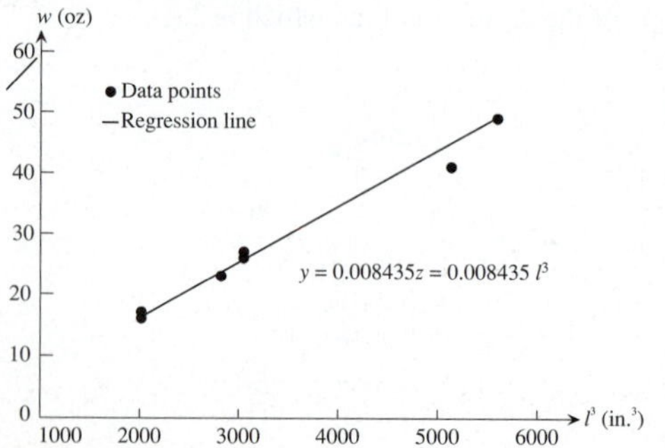

11. Graph (c) **13.** Graph (c)

15. **(f)** One possibility: Let y represent the number of people in your school who have the flu and let x represent the number of days that have elapsed after the first person gets sick. At first, the flu doesn't spread very quickly because there are only a few sick people to pass it on. As more people get sick, however, the disease spreads more rapidly. The most volatile mixture is when half the people are sick, because then there are a lot of sick people to spread the disease and a lot of uninfected people who can still catch it. As time continues and more people get sick, there are fewer and fewer people available to catch the flu and the spread of the disease begins to slow down. This behavior can be modeled with a function like the one represented by graph (f).

17.

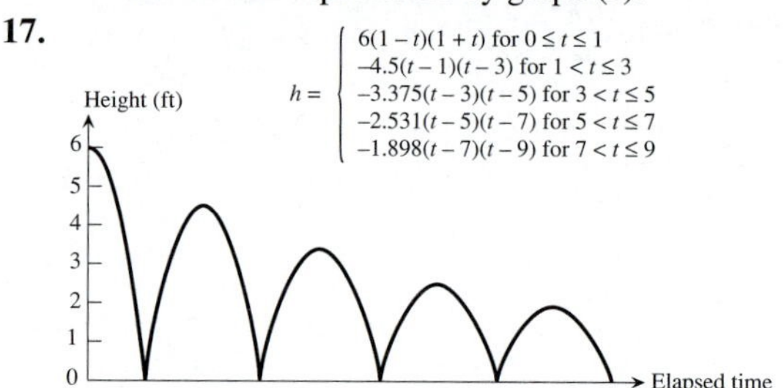

19. **(a)**

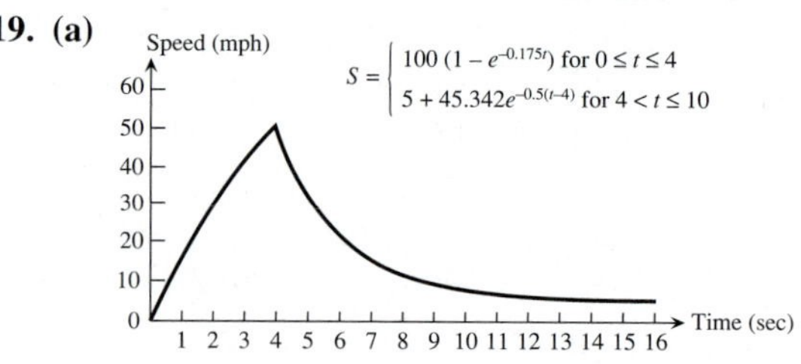

(b)

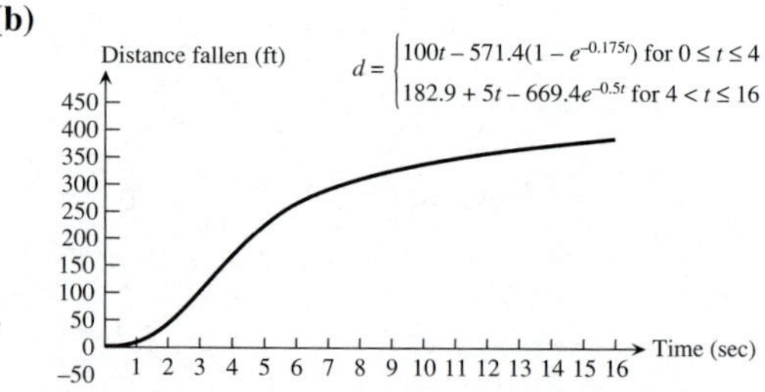

21. **(a)** The graph could represent the angle that a pendulum makes with the vertical as it swings back and forth. The variable y represents the angle and x represents time. Because of friction, the amplitude of the oscillation decays, as depicted by the graph. When y is positive, the pendulum is on one side of the vertical, and when y is negative, the pendulum is on the other side.

(b) The graph could represent the angle the playground swing makes with the vertical as a child "pumps" on the swing to get it going. The variable y represents the angle and x represents time. Because the child puts mechanical energy into the system (swing + child), the amplitude of the oscillation grows with time, as depicted by the graph. When y is positive, the swing is on one side of the vertical, and when y is negative, the swing is on the other side.

Preliminary Chapter Practice Exercises, pp. 77–80

1. $y = 3x - 9$

3. $x = 0$

5. $y = 2$

7. $y = -3x + 3$

9. $y = -\frac{4}{3}x - \frac{20}{3}$

11. $y = \frac{2}{3}x + \frac{8}{3}$

13. $A = \pi r^2, C = 2\pi r, A = \frac{C^2}{4\pi}$

15. $x = \tan\theta, y = \tan^2\theta$

17. Origin

19. Neither

21. Even

23. Even

25. Odd

27. Neither

29. **(a)** Domain: all reals **(b)** Range: $[-2, \infty)$

31. **(a)** Domain: $[-4, 4]$ **(b)** Range: $[0, 4]$

33. **(a)** Domain: all reals **(b)** Range: $(-3, \infty)$

35. **(a)** Domain: all reals **(b)** Range: $[-3, 1]$

37. **(a)** Domain: $(3, \infty)$ **(b)** Range: all reals

39. **(a)** Domain: $[-4, 4]$ **(b)** Range: $[0, 2]$

41. $f(x) = \begin{cases} 1 - x, & 0 \le x < 1 \\ 2 - x, & 1 \le x \le 2 \end{cases}$

43. **(a)** 1
(b) $\frac{1}{\sqrt{2.5}} = \sqrt{\frac{2}{5}}$
(c) $x, x \ne 0$
(d) $\frac{1}{\sqrt{1/\sqrt{x+2} + 2}}$

45. **(a)** $(f \circ g)(x) = -x, x \ge -2, (g \circ f)(x) = \sqrt{4 - x^2}$
(b) Domain $(f \circ g)$: $[-2, \infty)$, domain $(g \circ f)$: $[-2, 2]$
(c) Range $(f \circ g)$: $(-\infty, 2]$, range $(g \circ f)$: $[0, 2]$

47. Replace the portion for $x < 0$ with mirror image of the portion for $x > 0$ to make the new graph symmetric with respect to the y-axis.

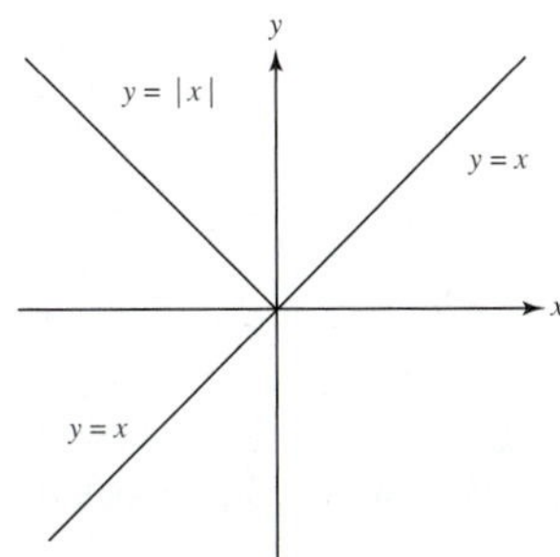

49. It does not change it.

51. Adds the mirror image of the portion for $x > 0$ to make the new graph symmetric with respect to the y-axis

53. Reflects the portion for $y < 0$ across the x-axis

55. Reflects the portion for $y < 0$ across the x-axis

57. **(a)** Symmetric about the line $y = x$

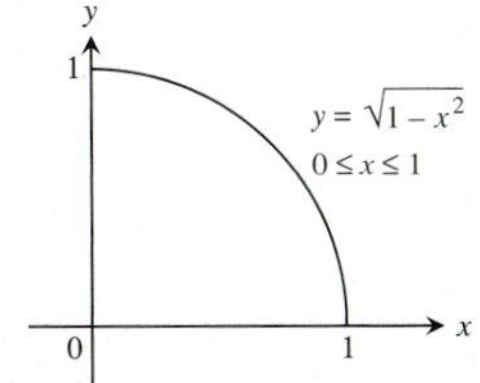

59. **(a)** $f^{-1}(x) = \frac{2 - x}{3}$
(b)

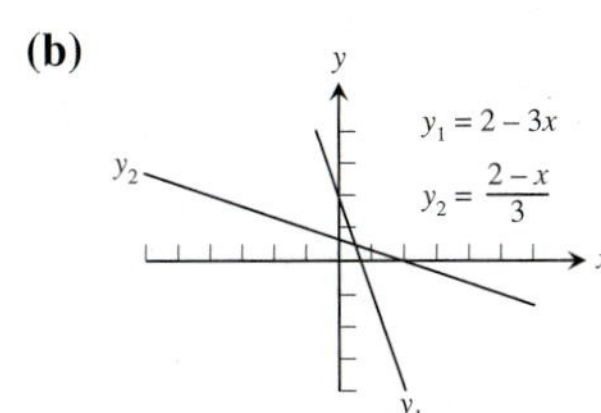

61. **(a)** $f(g(x)) = (\sqrt[3]{x})^3 = x, g(f(x)) = \sqrt[3]{x^3} = x$
(b)

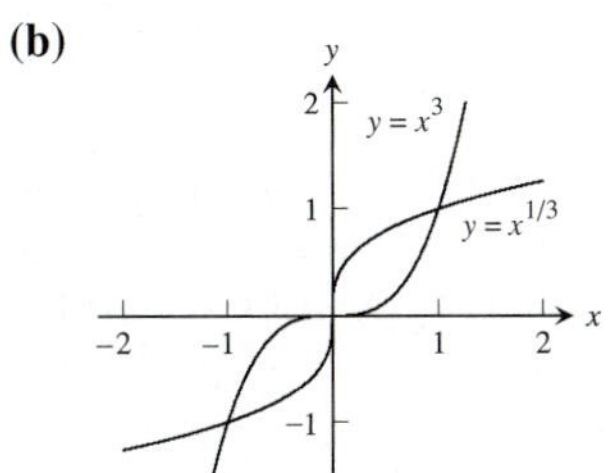

63. **(a)** $f^{-1}(x) = x - 1$

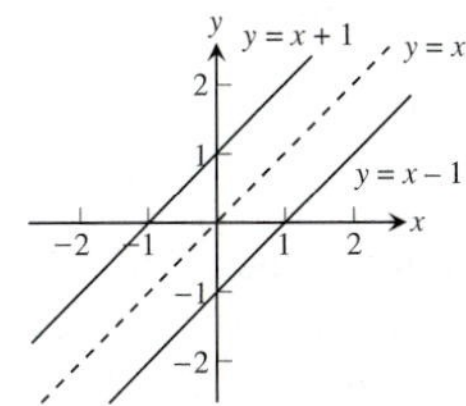

(b) $f^{-1}(x) = x - b$. The graph of f^{-1} is a line parallel to the graph of f. The graphs of f and f^{-1} lie on opposite sides of the line $y = x$ and are equidistant from that line.
(c) Their graphs will be parallel to one another and lie on opposite sides of the line $y = x$ equidistant from that line.

65. 2.718281828459

67. **(a)** 7.2 **(b)** $\frac{1}{x^2}$ **(c)** $\frac{x}{y}$

69. **(a)** 1 **(b)** 1 **(c)** $-x^2 - y^2$

71. ≈ 0.6435 radian or $36.8699°$

73. $\cos\theta = \frac{3}{7}, \sin\theta = \frac{\sqrt{40}}{7}, \tan\theta = \frac{\sqrt{40}}{3}, \sec\theta = \frac{7}{3},$
$\csc\theta = \frac{7}{\sqrt{40}}, \cot\theta = \frac{3}{\sqrt{40}}$

75. Period 4π

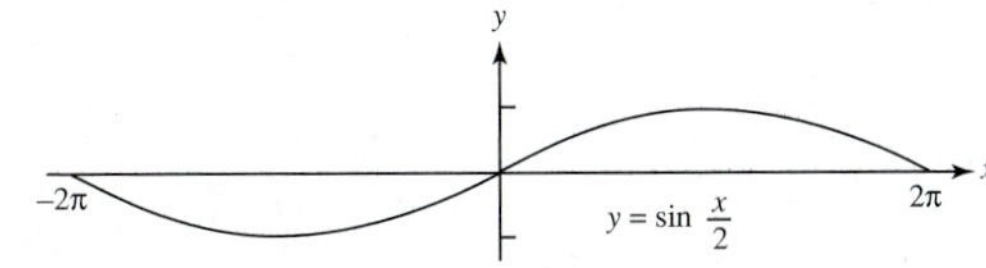

77.

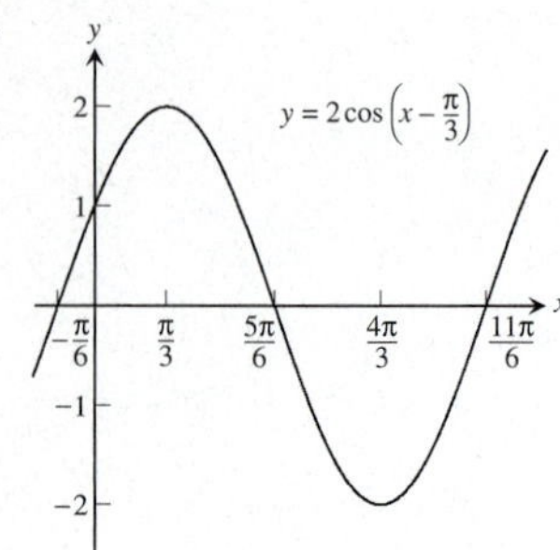

79. **(a)** $a = 1; b = \sqrt{3}$
(b) $c = 4/\sqrt{3}; a = 2/\sqrt{3}$

81. **(a)** $a = \dfrac{b}{\tan B}$ **(b)** $c = \dfrac{a}{\sin A}$

83. Since $\sin(x)$ has period 2π, $(\sin(x + 2\pi))^3 = (\sin(x))^3$. This function has period 2π. A graph shows that no smaller number works for the period.

87. $\dfrac{\sqrt{6} + \sqrt{2}}{4}$

89. **(a)** $\dfrac{\pi}{6}$
(b) $-\dfrac{\pi}{4}$
(c) $\dfrac{\pi}{3}$

91. **(a)** $\dfrac{\pi}{4}$
(b) $\dfrac{5\pi}{6}$
(c) $\dfrac{\pi}{3}$

93. 2

95. $-\dfrac{1}{2}$ **97.** $\sqrt{4x^2 + 1}$ **99.** $\dfrac{\sqrt{1 - x^2}}{x}$

101. **(a)** Defined; there is an angle whose tangent is 2.
(b) Not defined; no angle has cosine 2.

103. **(a)** Not defined; no angle has secant 0.
(b) Not defined; no angle has sine $\sqrt{2}$.

105. ≈ 16.98 m **107.** **(b)** 4π

109. **(a)** $\left(\dfrac{x}{5}\right)^2 + \left(\dfrac{y}{2}\right)^2 = 1$; all
(b) Initial point: (5, 0), terminal point: (5, 0)

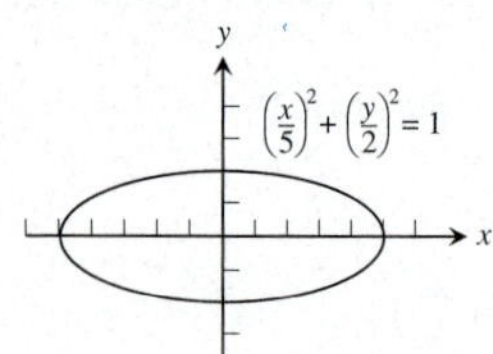

111. **(a)** $y = 2x + 7$; from (4, 15) to $(-2, 3)$
(b) Initial point: (4, 15), terminal point: $(-2, 3)$

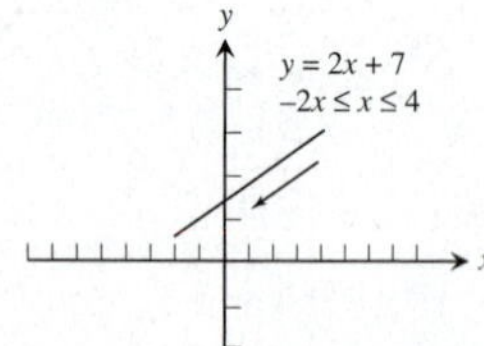

113. Possible answer: $x = -2 + 6t, y = 5 - 2t, 0 \le t \le 1$

115. Possible answer: $x = 2 - 3t, y = 5 - 5t, 0 \le t$

Preliminary Chapter Additional Exercises, pp. 80–83

1. **(a)**

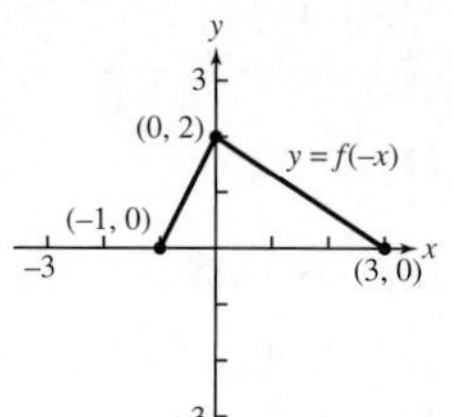

(b)

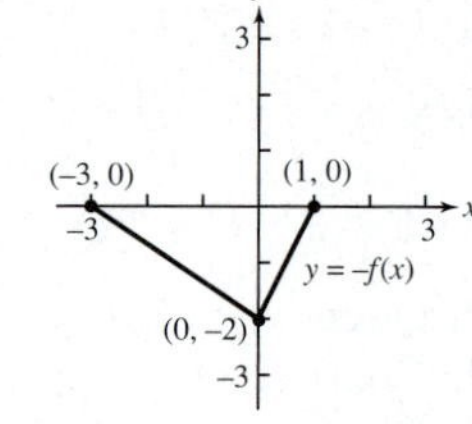

(c)

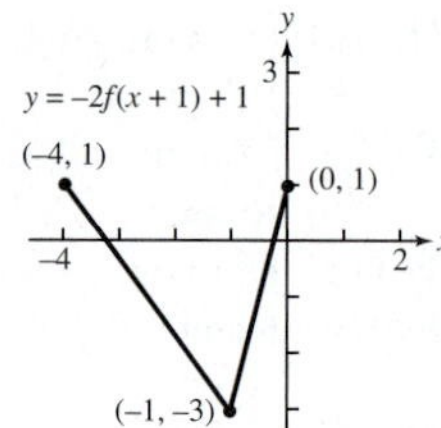

(d)

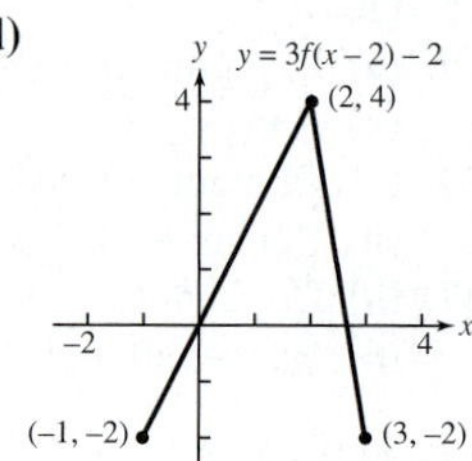

3. **(a)** $y = 100{,}000 - 10000x, 0 \le x \le 10$
(b) After 4.5 years

5. After $\dfrac{\ln(10/3)}{\ln 1.08} \approx 15.6439$ years. (If the bank only pays interest at the end of the year, it will take 16 years.)

11. **(a)** Yes
(b) Not always. If f is odd, then h is odd.

13. Yes. For instance: $f(x) = \dfrac{1}{x}$ and $g(x) = \dfrac{1}{x}$, or $f(x) = 2x$ and $g(x) = \dfrac{x}{2}$, or $f(x) = e^x$ and $g(x) = \ln x$.

15. If $f(x)$ is odd, then $g(x) = f(x) - 2$ is not odd. Nor is $g(x)$ even, unless $f(x) = 0$ for all x. If f is even, then $g(x) = f(x) - 2$ is also even.

17.

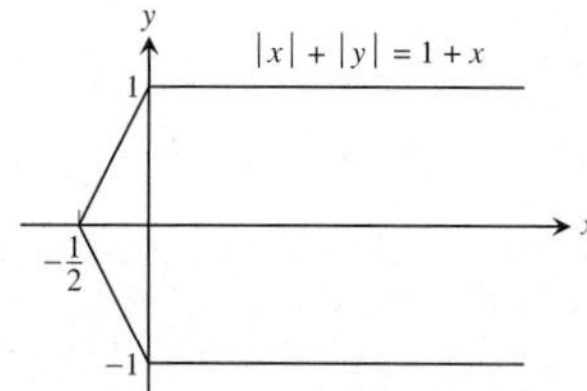

21. If the graph of $f(x)$ passes the horizontal line test, so will the graph of $g(x) = -f(x)$ since it is the same graph reflected about the x-axis.

23. **(a)** Domain: all reals. Range: If $a > 0$, then (d, ∞); if $a < 0$, then $(-\infty, d)$.
(b) Domain: (c, ∞), range: all reals

25. **(a)** The graph does not support the assumption that $y \propto x^2$.

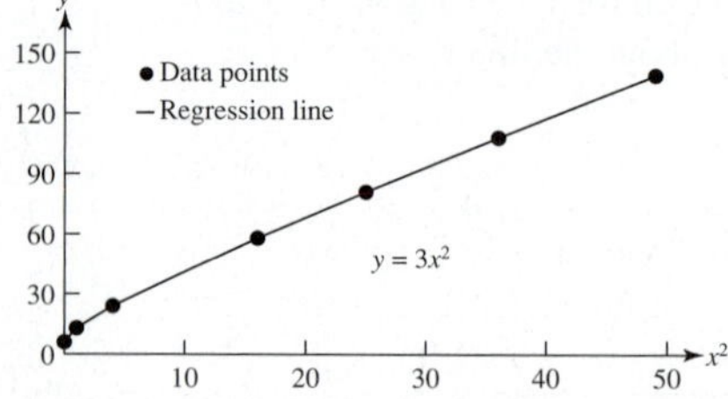

(b) The graph supports the assumption that $y \propto 4^x$. The constant of proportionality is estimated from the slope of the regression line, which is 0.6; therefore, $y = 0.6(4^x)$.

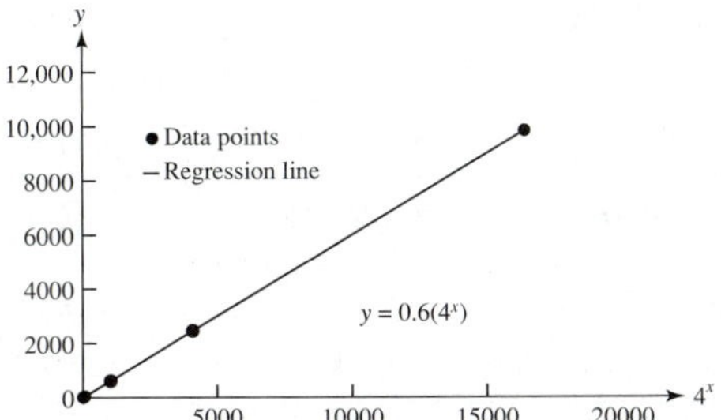

27. **(a)** Since the elongation of the spring is zero when the stress is $5(10^{-3})(\text{lb/in.}^2)$, the data should be adjusted by subtracting this amount from each of the stress data values. This gives the following table, where $\bar{s} = s - 5(10^{-3})$.

The slope of the graph is $\dfrac{(297 - 57)(10^5)}{(75 - 15)(10^{-3})} = 4.00(10^8)$ and the model is $e = 4(10^8)\bar{s}$ or $e = 4(10^8)(s - 5(10^{-3}))$.

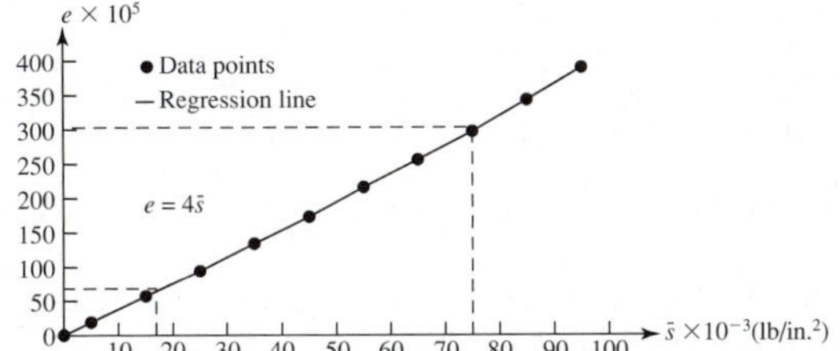

$\bar{s} \times 10^{-3}$	0	5	15	25	35	45	55	65	75	85	95
$e \times 10^5$	0	19	57	94	134	173	216	256	297	343	390

(b) The model fits the data well.

(c) $c = 780(10^5)(\text{in./in.})$. Since $s = 200(10^{-3})(\text{lb/in.}^2)$ is well outside the range of the data used for the model, one should not feel comfortable with this prediction without further testing of the spring.

29. **(a)** $y = 20.627x + 338.622$

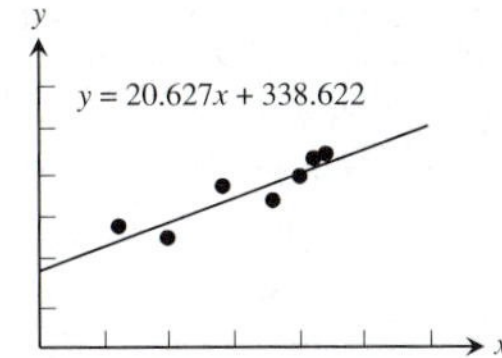

(b) Approximately 957

(c) Slope is 20.627. It represents the approximate annual increase in number of doctorates earned by Hispanic Americans per year.

31. **(a)** $f(x) = 2.000268 \sin(2.999187x - 1.000966) + 3.999881$

(b) $f(x) = 2 \sin(3x - 1) + 4$

33. **(a)** $Q = 1.00(2.0138^x) = 1.00e^{0.7x}$

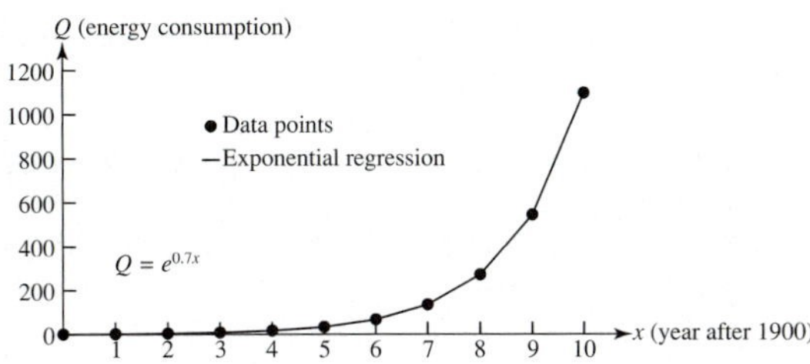

(b) Consumption in 1996 is 828.82. The annual rate of increase during this time is 7.25%.

CHAPTER 1

Section 1.1, pp. 95–99

1. **(a)** 19

(b) 1

3. **(a)** $-\dfrac{4}{\pi}$

(b) $-\dfrac{3\sqrt{3}}{\pi}$

5. Graphs can shift during a press run, so your estimates may not completely agree with these.

(a)

PQ_1	PQ_2	PQ_3	PQ_4
43	46	49	50

The appropriate units are meters per second.

(b) ≈ 50 m/sec or 180 km/h

7. Lower bound: $a = 23.55$ ft/sec, upper bound: $b = 28.85$ ft/sec, $v(2) \approx \dfrac{a + b}{2} = 26.20$ ft/sec

9. **(a)** Does not exist. As x approaches 1 from the right, $g(x)$ approaches 0. As x approaches 1 from the left, $g(x)$ approaches 1. There is no single number L that all values $g(x)$ get arbitrarily close to as $x \to 1$.

(b) 1

(c) 0

11. **(a)** True **(b)** True **(c)** False
(d) False **(e)** False **(f)** True

13. As x approaches 0 from the left, $x/|x|$ approaches -1. As x approaches 0 from the right, $x/|x|$ approaches 1. There is no single number L that the function values all get arbitrarily close to as $x \to 0$.

15. Nothing can be said.

17. No

19. (a) $f(x) = (x^2 - 9)/(x + 3)$

x	−3.1	−3.01	−3.001	−3.0001	−3.00001	−3.000001
$f(x)$	−6.1	−6.01	−6.001	−6.0001	−6.00001	−6.000001

x	−2.9	−2.99	−2.999	−2.9999	−2.99999	−2.999999
$f(x)$	−5.9	−5.99	−5.999	−5.9999	−5.99999	−5.999999

(c) $\lim_{x\to -3} f(x) = -6$

21. (a) $G(x) = (x + 6)/(x^2 + 4x - 12)$

x	−5.9	−5.99	−5.999
$G(x)$	−0.126582	−0.1251564	−0.1250156

x	−5.9999	−5.99999	−5.999999
$G(x)$	−0.1250016	−0.12500016	−0.12500002

x	−6.1	−6.01	−6.001
$G(x)$	−0.123457	−0.1248439	−0.1249844

x	−6.0001	−6.00001	−6.000001
$G(x)$	−0.1249984	−0.12499984	−0.12499998

(c) $\lim_{x\to -6} G(x) = -\frac{1}{8} = -0.125$

23. (a) $g(\theta) = (\sin\theta)/\theta$

θ	0.1	0.01	0.001
$g(\theta)$	0.998334	0.999983	0.999999

θ	0.0001	0.00001	0.000001
$g(\theta)$	0.999999	0.999999	0.999999

θ	−0.1	−0.01	−0.001
$g(\theta)$	0.998334	0.999983	0.999999

θ	−0.0001	−0.00001	−0.000001
$g(\theta)$	0.999999	0.999999	0.999999

$\lim_{\theta\to 0} g(\theta) = 1$

25. (a) $f(x) = x^{1/(1-x)}$

x	0.9	0.99	0.999
$f(x)$	0.348678	0.366032	0.367695

θ	0.9999	0.99999	0.999999
$g(\theta)$	0.367861	0.367878	0.367879

x	1.1	1.01	1.001
$f(x)$	0.385543	0.369711	0.368063

θ	1.0001	1.00001	1.000001
$g(\theta)$	0.367898	0.367881	0.367880

$\lim_{x\to 1} f(x) \approx 0.36788$

27. $\delta = 0.1$

29. $\delta = \frac{7}{16}$

31. $(3.99, 4.01)$, $\delta = 0.01$

33. $(-0.19, 0.21)$, $\delta = 0.19$

35. $\left(\frac{10}{3}, 5\right)$, $\delta = \frac{2}{3}$

37. [3.384, 3.387]. To be safe, the left endpoint was rounded up and the right endpoint rounded down.

39. (b) One possible answer: $a = 1.75$, $b = 2.28$

(c) One possible answer: $a = 1.99$, $b = 2.01$

41. (a) 14.7 m/sec

(b) 29.4 m/sec

43. (a)

x	−0.1	−0.01	−0.001	−0.0001
$f(x)$	−0.054402	−0.005064	−0.000827	−0.000031

(b)

x	0.1	0.01	0.001	0.0001
$f(x)$	−0.054402	−0.005064	−0.000827	−0.000031

The limit appears to be 0.

45. (a)

x	−0.1	−0.01	−0.001	−0.0001
$f(x)$	2.0567	2.2763	2.2999	2.3023

(b)

x	0.1	0.01	0.001	0.0001
$f(x)$	2.5893	2.3293	2.3052	2.3029

The limit appears to be approximately 2.3.

Section 1.2, pp. 108–111

1. **(a)** 3 **(b)** -2 **(c)** No limit **(d)** 1

3. **(a)** -4 **(b)** -4 **(c)** -4 **(d)** -4

5. **(a)** 4 **(b)** -3 **(c)** No limit **(d)** 4

7. **(a)** Quotient Rule
(b) Difference and Power Rules
(c) Sum and Constant Multiple Rules

9. **(a)** -10 **(b)** -20 **(c)** -1 **(d)** $\frac{5}{7}$

11. **(a)** -9 **(b)** -8 **(c)** $\frac{1}{5}$ **(d)** $\frac{3}{2}$

13. **(a)** -7 **(b)** $-\frac{1}{2}$ **(c)** 4

15. **(a)** $\lim_{x\to 0}\left(1-\frac{x^2}{6}\right) = 1 - \frac{0}{6} = 1$ and $\lim_{x\to 0} 1 = 1$;
by the Sandwich Theorem, $\lim_{x\to 0}\frac{x \sin x}{2 - 2\cos x} = 1$.
(b) For $x \neq 0$, $y = (x \sin x)/(2 - 2 \cos x)$ lies between the other two graphs in the figure, and the graphs converge as $x \to 0$.

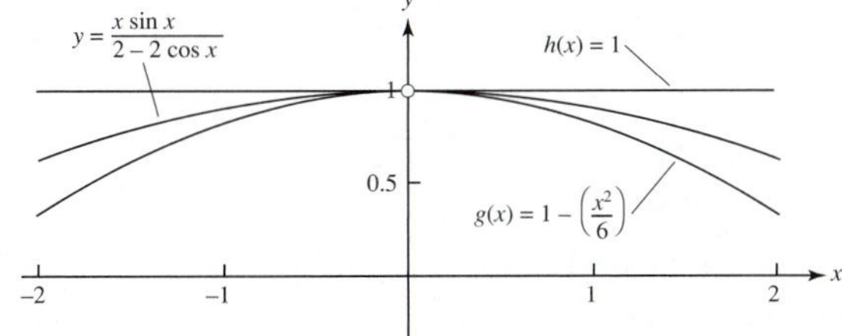

17. $\lim_{h\to 0}\frac{(1+h)^2 - 1^2}{h} = \lim_{h\to 0}(2 + h) = 2$

19. $\lim_{h\to 0}\frac{\left(\frac{1}{-2+h}\right) - \left(\frac{1}{-2}\right)}{h} = \lim_{h\to 0}\frac{-h}{h(4 - 2h)} = -\frac{1}{4}$

21. **(a)** True **(b)** True **(c)** False **(d)** True
(e) True **(f)** True **(g)** False **(h)** False
(i) False **(j)** False **(k)** True **(l)** False

23. **(a)** No **(b)** Yes, 0 **(c)** No

25. **(a)** $D: 0 \le x \le 2$, $R: 0 < y \le 1$ and $y = 2$
(b) $(0, 1) \cup (1, 2)$
(c) $x = 2$
(d) $x = 0$

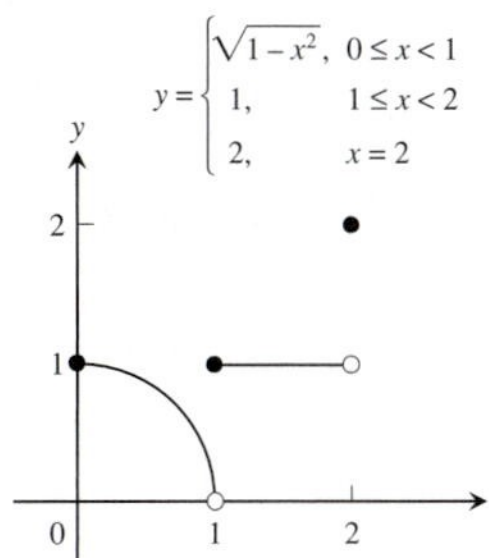

27. $\sqrt{3}$

29. $\frac{2}{\sqrt{5}}$

31. **(a)** 1 **(b)** -1

35. **(a)** 4 **(b)** -2

37. Yes

39. $\delta = \epsilon^2$, $\lim_{x\to 5^+}\sqrt{x-5} = 0$

41. Yes, $\lim_{x\to 0^-} f(x) = -3$

Section 1.3, pp. 122–123

1. **(a)** π **(b)** π

3. **(a)** $-\frac{5}{3}$ **(b)** $-\frac{5}{3}$

5. $-\frac{1}{x} \le \frac{\sin 2x}{x} \le \frac{1}{x} \Rightarrow \lim_{x\to\infty}\frac{\sin 2x}{x} = 0$ by the Sandwich Theorem

7. **(a)** $\frac{2}{5}$ **(b)** $\frac{2}{5}$

9. **(a)** $-\infty$ **(b)** ∞

11. **(a)** ∞ **(b)** $-\infty$

13. **(a)** $-\frac{2}{3}$ **(b)** $-\frac{2}{3}$

15. 0

17. 1

19. ∞

21. Here is one possibility.

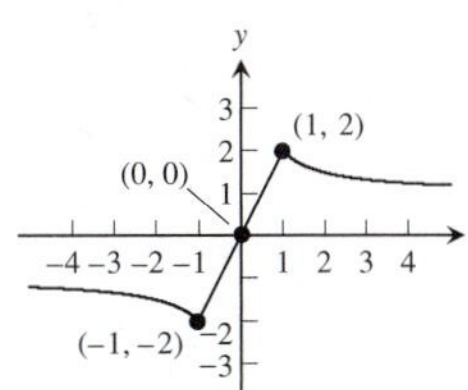

23. Here is one possibility.

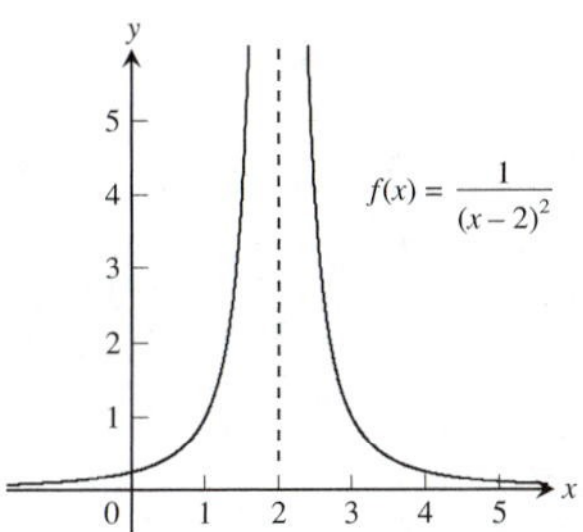

25. $y = \dfrac{1}{x-1}$

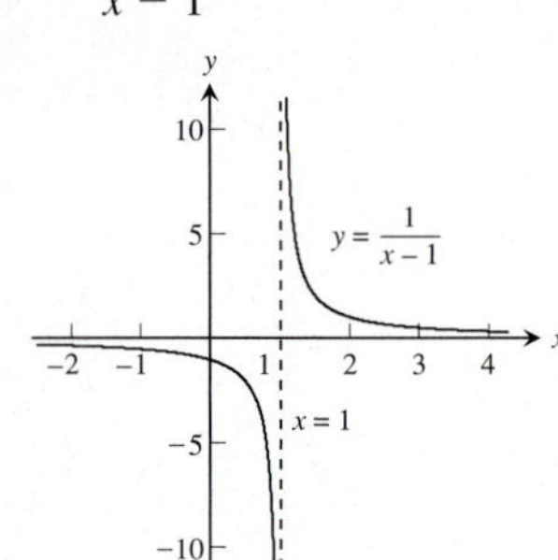

27. $y = \dfrac{2x^2 + x - 1}{x^2 - 1}$

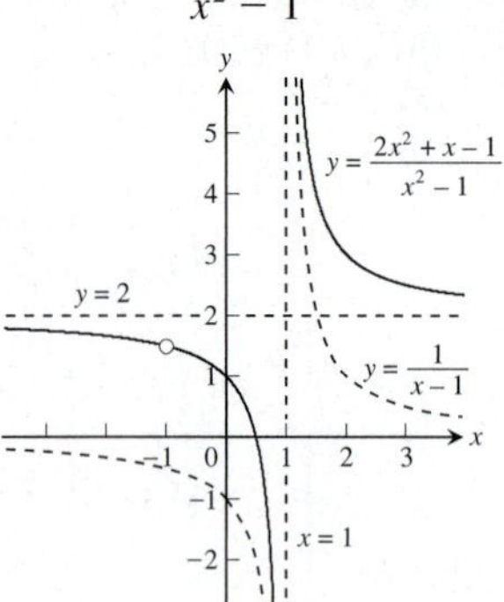

29. $y = \dfrac{x^4 + 1}{x^2} = x^2 + \dfrac{1}{x^2}$

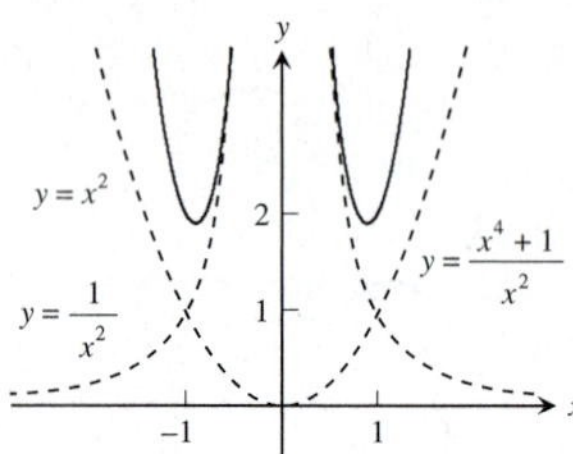

31. $y = \dfrac{x^2 - x + 1}{x - 1} = x + \dfrac{1}{x-1}$

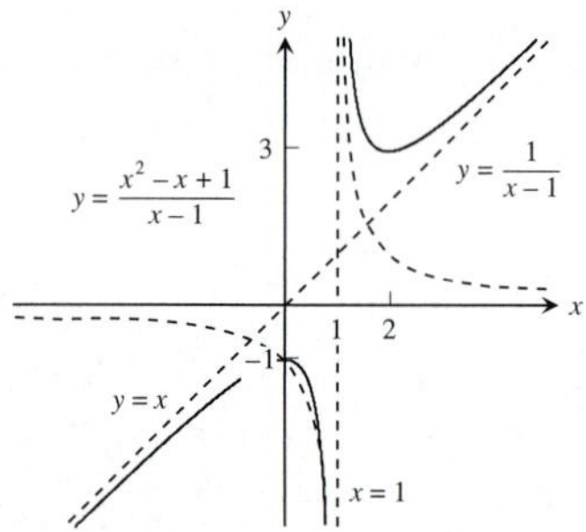

33. $y = \dfrac{8}{x^2 + 4}$

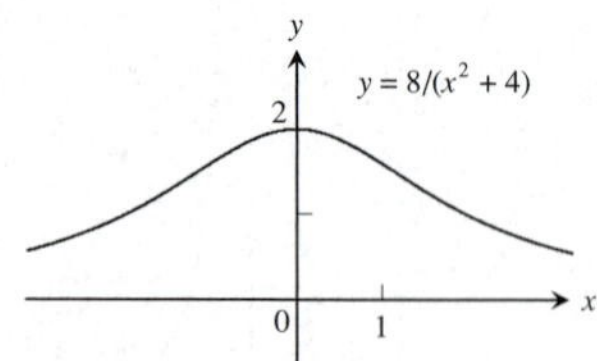

35. Graph (a)

37. Graph (d)

39. **(a)** e^x **(b)** $-2x$

41. **(a)** x **(b)** x

43. **(a)** $\dfrac{1}{2}$ **(b)** $\dfrac{1}{2}$

45. At most 2

47. $y = \dfrac{x}{\sqrt{4 - x^2}}$

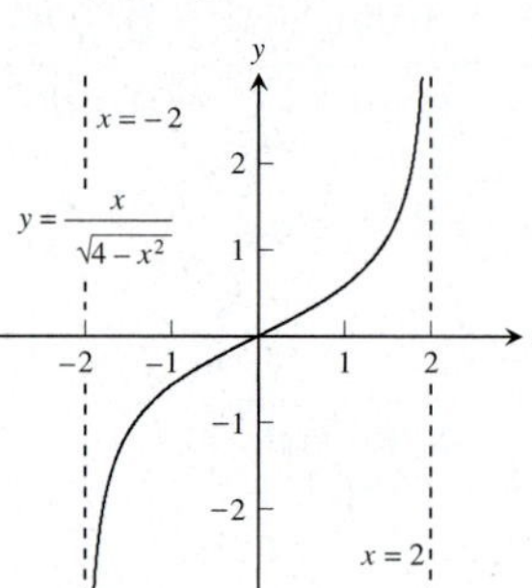

49. $y = x^{2/3} + \dfrac{1}{x^{1/3}}$

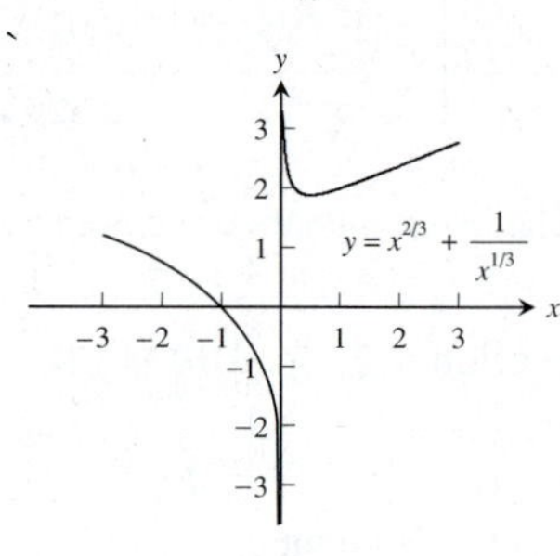

51. At ∞: ∞, at $-\infty$: 0

53. At ∞: 0, at $-\infty$: 0

55. 1

57. 3

59. $y = -\dfrac{x^2 - 4}{x + 1} = 1 - x + \dfrac{3}{x+1}$

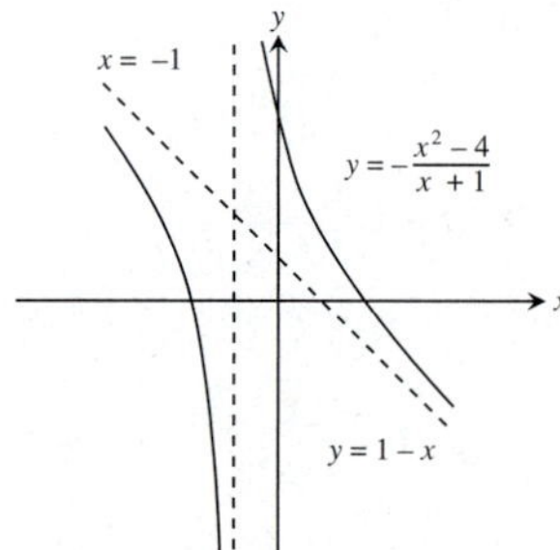

61. The graph of the function mimics each term as it becomes dominant.

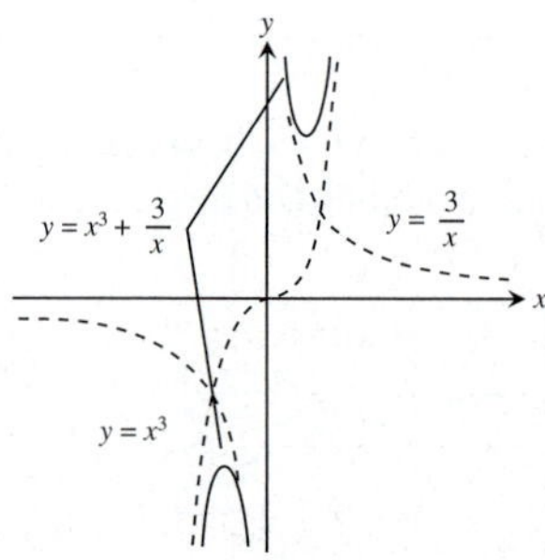

63. **(a)** $y \to \infty$ (see the accompanying graph)
(b) $y \to \infty$ (see the accompanying graph)
(c) Cusps at $x = \pm 1$ (see the accompanying graph)

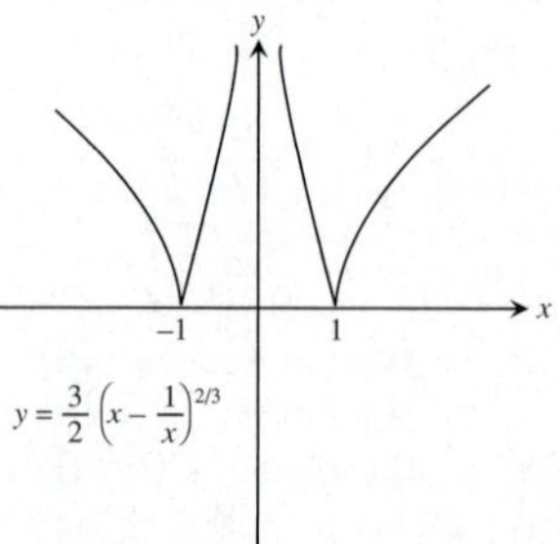

Section 1.4, pp. 132–134

1. No; discontinuous at $x = 2$; not defined at $x = 2$
3. Continuous
5. **(a)** Yes **(b)** Yes **(c)** Yes **(d)** Yes
7. **(a)** No **(b)** No
9. 0
13. Discontinuous when $x = 2$
15. Discontinuous when $t = 3$ or $t = 1$
17. Discontinuous at $\theta = 0$
19. Continuous on the interval $\left[-\frac{3}{2}, \infty\right)$
21. 0; continuous
23. 1; continuous
25. $f(x)$ is continuous on $[0, 1]$ and $f(0) < 0, f(1) > 0 \Rightarrow$ by the Intermediate Value Theorem $f(x)$ takes on every value between $f(0)$ and $f(1) \Rightarrow$ the equation $f(x) = 0$ has at least one solution between $x = 0$ and $x = 1$.

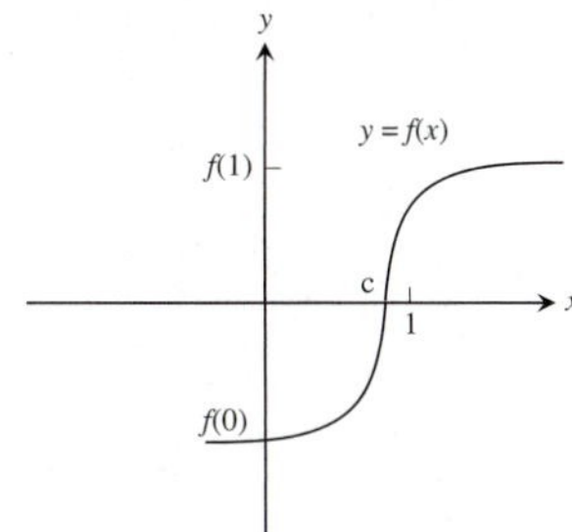

27. All five statements ask for the same information because of the Intermediate Value Property of continuous functions.
29. Answers may vary. For example, $f(x) = \dfrac{\sin(x-2)}{x-2}$ is discontinuous at $x = 2$ because it is not defined there. The discontinuity can be removed, however, because f has a limit (namely 1) as $x \to 2$.
31. $r_1 = \dfrac{1 - \sqrt{21}}{2} \approx -1.791$, $r_2 = r_3 = r_4 = 0$, and $r_5 = \dfrac{1 + \sqrt{21}}{2} \approx 2.791$
35. Yes, by the Intermediate Value Property
39. **(b)** Continuous at all points in the domain $[0, 5)$ except at $t = 1, 2, 3, 4$

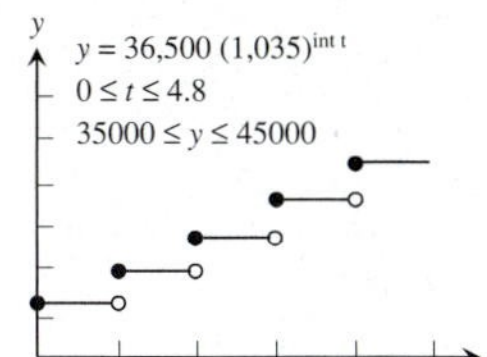

45. $x \approx 1.8794, -1.5321, -0.3473$ **47.** $x \approx 1.7549$
49. $x \approx 3.5156$
51. $x \approx 0.7391$

Section 1.5, pp. 139–141

1. P_1: $m_1 = 1$, P_2: $m_2 = 5$ **3.** P_1: $m_1 = \frac{5}{2}$, P_2: $m_2 = -\frac{1}{2}$
5. $y = 2x + 5$ **7.** $y = 12x + 16$
9. $y + 1 = -3(x - 1)$ **11.** $y - 3 = -2(u - 3)$
13. $-\frac{1}{4}$ **15.** $(-2, -5)$
17. If $x = 0$, $y = -(x + 1)$; if $x = 2$, $y = -(x - 3)$.
19. 19.6 m/sec **21.** 6π
23. 3.72 m/sec **25.** Yes **27.** Yes
29. **(a)** $\dfrac{1 - e^{-2}}{2} \approx 0.432$ **(b)** $\dfrac{e^3 - e}{2} \approx 8.684$
31. **(a)** $-\dfrac{4}{\pi} \approx -1.273$ **(b)** $-\dfrac{3\sqrt{3}}{\pi} \approx -1.654$
33. **(a)** 0.3 billion dollars per year
(b) 0.5 billion dollars per year
(c) $y = 0.057x^2 - 0.1514x + 1.3943$

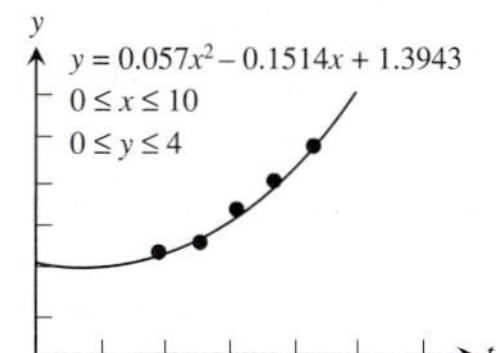

(d) 1994 to 1995: 0.31 billion dollars per year, 1995 to 1997: 0.53 billion dollars per year
(e) 0.65 billion dollars per year
35. **(a)** Nowhere **37.** **(a)** At $x = 0$
39. **(a)** Nowhere **41.** **(a)** At $x = 1$
43. **(a)** At $x = 0$

Chapter 1 Practice Exercises, pp. 142–143

1. At $x = -1$: $\lim_{x\to-1^-} f(x) = \lim_{x\to-1^+} f(x) = 1$, so $\lim_{x\to-1} f(x) = 1 = f(-1)$; continuous at $x = -1$

At $x = 0$: $\lim_{x\to0^-} f(x) = \lim_{x\to0^+} f(x) = 0$, so $\lim_{x\to0} f(x) = 0$. However, $f(0) \neq 0$, so f is discontinuous at $x = 0$. The discontinuity can be removed by redefining $f(0)$ to be 0.

At $x = 1$: $\lim_{x\to1^-} f(x) = -1$ and $\lim_{x\to1^+} f(x) = 1$, so $\lim_{x\to1} f(x)$ does not exist. The function is discontinuous at $x = 1$, and the discontinuity is not removable.

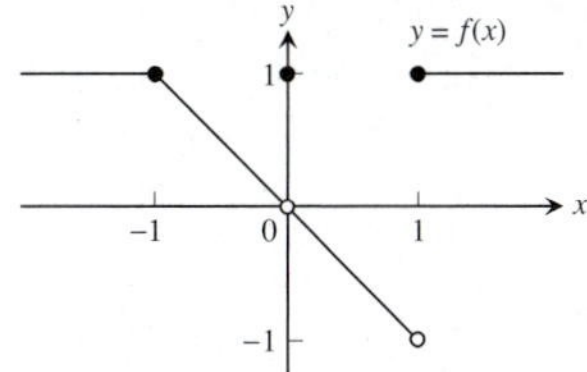

3. **(a)** -21 **(b)** 49 **(c)** 0 **(d)** 1
(e) 1 **(f)** 7 **(g)** -7 **(h)** $-\frac{1}{7}$

5. 4

7. **(a)** $(-\infty, +\infty)$ **(b)** $[0, \infty)$

(c) $(-\infty, 0)$ and $(0, \infty)$ **(d)** $(0, \infty)$

9. **(a)** Does not exist **(b)** 0

11. $\frac{1}{2}$ **13.** $2x$ **15.** $-\frac{1}{4}$ **17.** $\frac{2}{5}$

19. 0 **21.** $-\infty$ **23.** 0 **25.** 1

27. 0 **29.** **(b)** 1.324717957

Chapter 1 Additional Exercises, pp. 143–145

3. 0; the left-hand limit was needed because the function is undefined for $v > c$.

5. $65 \le t \le 75$; within 5°F

7. **(a)** $\lim_{a\to 0} r_+(a) = 0.5$, $\lim_{a\to -1^+} r_+(a) = 1$

(b) $\lim_{a\to 0} r_-(a)$ does not exist, $\lim_{a\to -1^+} r_-(a) = 1$

9. **(a)** True **(b)** False **(c)** True **(d)** False

CHAPTER 2 DERIVATIVES

Section 2.1, pp. 157–160

1. $-2x, 6, 0,$ **3.** $3t^2 - 2t, 5$

5. $\frac{3}{2\sqrt{3\theta}}, \sqrt{3}$ **7.** $2x + 1, 2$

9. $4x^2, 8x$

11. $y' = 2x^3 - 3x - 1 \Rightarrow y'' = 6x^2 - 3 \Rightarrow y''' = 12x \Rightarrow y^{(4)} = 12 \Rightarrow y^{(n)} = 0$ for all $n \ge 5$

13. **(a)** $\frac{dv}{dx} = 3x^2 - 4, y - 1 = 8(x - 2)$

(b) $[-4, \infty)$

(c) The equation of one such tangent line is found in part (a) when $x = 2$. The equation of the line tangent to the curve at the point $(-2, 1)$ is $y - 1 = 8(x - (-2))$.

15. **(b)** **17.** **(d)**

19. **(a)** f' is not defined at $x = 0, 1, 4$. At these points, the left-hand and right-hand derivatives do not agree.

(b)

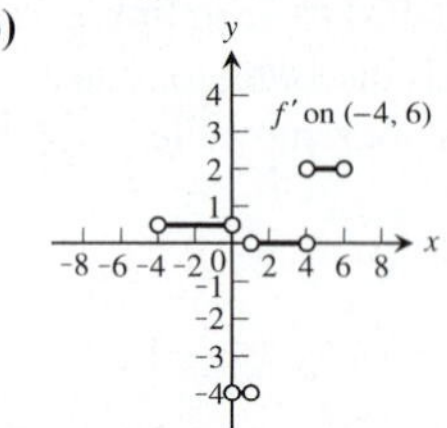

21. $\lim_{h\to 0^-} \frac{f(0 + h) - f(0)}{h} = 0$, $\lim_{h\to 0^+} \frac{f(0 + h) - f(0)}{h} = 1 \Rightarrow$ the derivative $f'(0)$ does not exist.

23. **(a)** The function is differentiable on its domain $-2 \le x \le 3$.

(b) None

(c) None

25. **(a)** On $-1 \le x < 0$ and $0 < x \le 2$

(b) At $x = 0$

(c) None

27. **(a)** $y' = -2x$

(b)

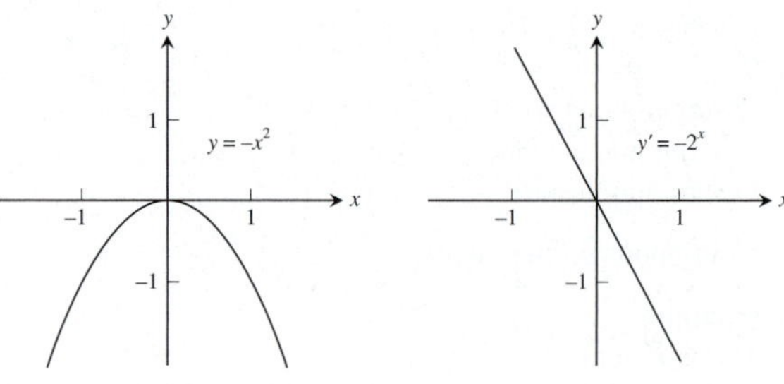

(c) $x < 0, x = 0, x > 0$

(d) $-\infty < x < 0, 0 < x < \infty$

29. **(a)** $y' = x^2$

(b)

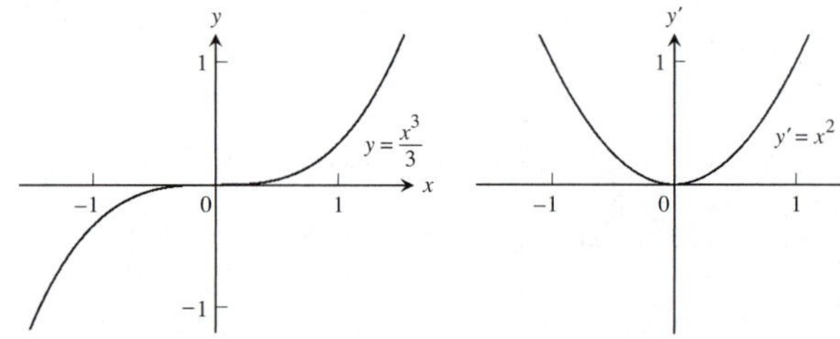

(c) $x \ne 0, x = 0$, none

(d) $-\infty < x < \infty$, none

31. $y' = 3x^2$ is never negative.

33. Yes, $y + 16 = -(x - 3)$ is tangent at $(3, -16)$.

35. No, the function $y = \lfloor x \rfloor$ does not satisfy the intermediate value property of derivatives.

37. Yes, $(-f)'(x) = -(f'(x))$

Section 2.2, pp. 169–172

1. **(a)** $\Delta s = -2m, v_{av} = -1$ m/sec

(b) $|v(0)| = 3$ m/sec, $|v(2)| = 1$ m/sec, $a(0) = a(2) = 2$ m/sec^2

(c) At $t = \frac{3}{2}$

3. **(a)** $\Delta s = -9m, v_{av} = -3$ m/sec

(b) $|v(0)| = 3$ m/sec, $|v(3)| = 12$ m/sec, $a(0) = 6$ m/sec^2, and $a(3) = -12$ m/sec^2

(c) Never changes direction

5. **(a)** $a(1) = -6$ m/sec^2 and $a(3) = 6$ m/sec^2

(b) $|v(2)| = 3$ m/sec

(c) Total distance $= |s(1) - s(0)| + |s(2) - s(1)| = 6$ m

7. $t \approx 7.5$ sec on Mars, $t \approx 1.2$ sec on Jupiter

9. $g_s = 0.75$ m/sec^2

11. **(a)** $v = -32t, |v| = 32t$ ft/sec, $a = -32$ ft/sec^2

(b) $t \approx 3.3$ sec

(c) $v \approx 107.0$ ft/sec

13. **(a)** $t = 2, t = 7$ **(b)** $3 \le t \le 6$

(c) $|v|$ (m/sec) **(d)** $a = \frac{dv}{dt}$

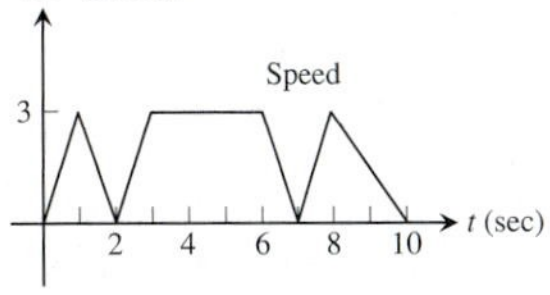

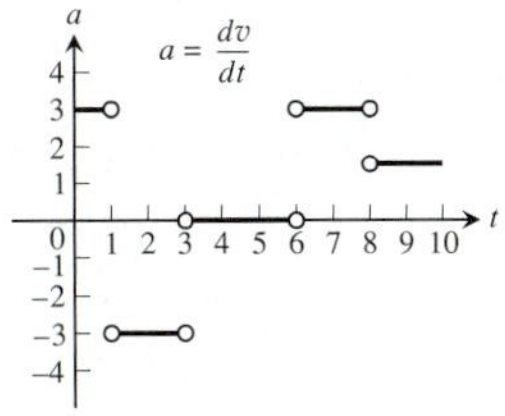

15. **(a)** 190 ft/sec
(b) 2 sec
(c) 8 sec, 0 ft/sec
(d) 10.8 sec, 90 ft/sec
(e) 2.8 sec
(f) Greatest acceleration happens 2 sec after launch
(g) Constant acceleration between 2 and 10.8 sec, -32 ft/sec^2

17. **(a)** 4/7 sec, 280 cm/sec
(b) 560 cm/sec, 980 cm/sec^2
(c) 29.75 flashes/sec

19. Graph C = position, graph B = velocity, graph A = acceleration

21. **(a)** \$2 **(b)** \$2 **(c)** 0

23. -8000 gal/min, $-10{,}000$ gal/min

25. **(a)** 16π ft^3/ft **(b)** 3.2π ft^3

27. $80\sqrt{19}$ ft/sec, ≈ 238 mph

29. **(a)** $t = \frac{3}{2}$
(b) Left on $\left[0, \frac{3}{2}\right)$, right on $\left(\frac{3}{2}, 5\right]$
(c) $t = \frac{3}{2}$
(d) Speeds up on $\left(\frac{3}{2}, 5\right]$, slows down on $\left[0, \frac{3}{2}\right)$
(e) Fastest at $t = 5$, slowest at $t = 3/2$
(f) $t = 5$

31. **(a)** $t = \frac{6 \pm \sqrt{15}}{3}$
(b) Left on $\left[0, \frac{6-\sqrt{15}}{3}\right) \cup \left(\frac{6+\sqrt{15}}{3}, 4\right]$;
right on $\left(\frac{6-\sqrt{15}}{3}, \frac{6+\sqrt{15}}{3}\right)$
(c) $t = \frac{6 \pm \sqrt{15}}{3}$
(d) Speeds up on $\left(\frac{6-\sqrt{15}}{3}, 2\right) \cup \left(\frac{6+\sqrt{15}}{3}, 4\right]$,
slows down on $\left[0, \frac{6-\sqrt{15}}{3}\right) \cup \left(2, \frac{6+\sqrt{15}}{3}\right)$
(e) Fastest at $t = 0, 4$; slowest at $t = \frac{6 \pm \sqrt{15}}{3}$
(f) $t = \frac{6 + \sqrt{15}}{3}$

Section 2.3, pp. 178–179

1. $\frac{dy}{dx} = 12x - 10 + 10x^{-3}, \frac{d^2y}{dx^2} = 12 - 30x^{-4}$

3. $\frac{dr}{ds} = \frac{-2}{3s^3} + \frac{5}{2s^2}, \frac{d^2r}{ds^2} = \frac{2}{s^4} - \frac{5}{s^3}$

5. **(a)** $y' = (3 - x^2) \cdot \frac{d}{dx}(x^3 - x + 1) + (x^3 - x + 1) \cdot \frac{d}{dx}(3 - x^2) = -5x^4 + 12x^2 - 2x - 3$
(b) $y = -x^5 + 4x^3 - x^2 - 3x + 3 \Rightarrow y' = -5x^4 + 12x^2 - 2x - 3$

7. $y' = -\frac{19}{(3x-2)^2}$ **9.** $f'(t) = \frac{1}{(t+2)^2}$

11. $f'(s) = \frac{1}{\sqrt{s}(\sqrt{s}+1)^2}$

13. $\frac{dy}{dx} = \frac{-4x^3 - 3x^2 + 1}{(x^2-1)^2(x^2+x+1)^2}$

15. $\frac{ds}{dt} = -5t^{-2} + 2t^{-3}, \frac{d^2s}{dt^2} = 10t^{-3} - 6t^{-4}$

17. $\frac{dw}{dz} = -z^{-2} - 1, \frac{d^2w}{dz^2} = 2z^{-3}$

19. **(a)** $\left.\frac{d}{dx}(uv)\right|_{x=0} = 7$ **(b)** $\left.\frac{d}{dx}\left(\frac{u}{v}\right)\right|_{x=0} = -13$
(c) $\left.\frac{d}{dx}\left(\frac{v}{u}\right)\right|_{x=0} = \frac{13}{25}$ **(d)** $\left.\frac{d}{dx}(7v - 2u)\right|_{x=0} = 8$

21. $y = 4x$ at $(0, 0)$, $y = 2$ at $(1, 2)$

23. $a = b = 1$ and $c = 0$, so $y = x^2 + x$

25. $\frac{d}{dx}(u \cdot c) = u \cdot \frac{dc}{dx} + c \cdot \frac{du}{dx} = u \cdot 0 + c\frac{du}{dx} = c\frac{du}{dx} \Rightarrow$ the Constant Multiple Rule is a special case of the Product Rule

27. **(a)** $\frac{d}{dx}(uvw) = uvw' + uv'w + u'vw$
(b) $\frac{d}{dx}(u_1u_2u_3u_4) = u_1u_2u_3u'_4 + u_1u_2u'_3u_4 + u_1u'_2u_3u_4 + u'_1u_2u_3u_4$
(c) $\frac{d}{dx}(u_1 \cdots u_n) = u_1u_2 \cdots u_{n-1}u'_n + u_1u_2 \cdots u_{n-2}u'_{n-1}u_n + \cdots + u'_1u_2 \cdots u_n$

29. $\frac{dP}{dV} = -\frac{nRT}{(V-nb)^2} + \frac{2an^2}{V^3}$

Section 2.4, pp. 184–186

1. $-10 - 3\sin x$ **3.** $-\csc x \cot x - \frac{2}{\sqrt{x}}$

5. 0 **7.** $\frac{-\csc^2 x}{(1 + \cot x)^2}$

9. $4\tan x \sec x - \csc^2 x$ **11.** $x^2 \cos x$

13. $\sec^2 t - 1$ **15.** $\frac{-2\csc t \cot t}{(1 - \csc t)^2}$

17. $-\theta(\theta\cos\theta + 2\sin\theta)$

19. $\sec\theta\csc\theta(\tan\theta - \cot\theta) = \sec^2\theta - \csc^2\theta$

21. $\sec^2 q$
23. $\sec^2 q$
25. (a) $2\csc^3 x - \csc x$ **(b)** $2\sec^3 x - \sec x$
27.

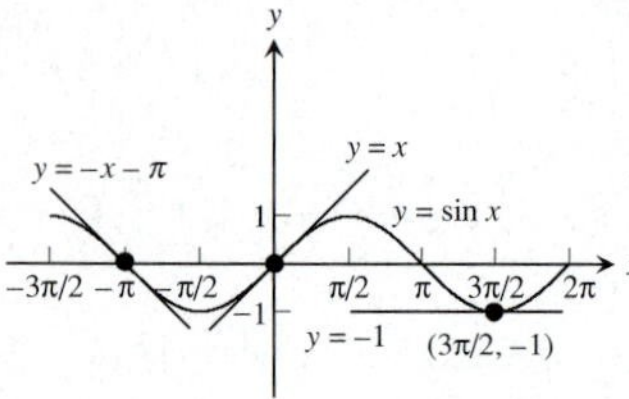

29.

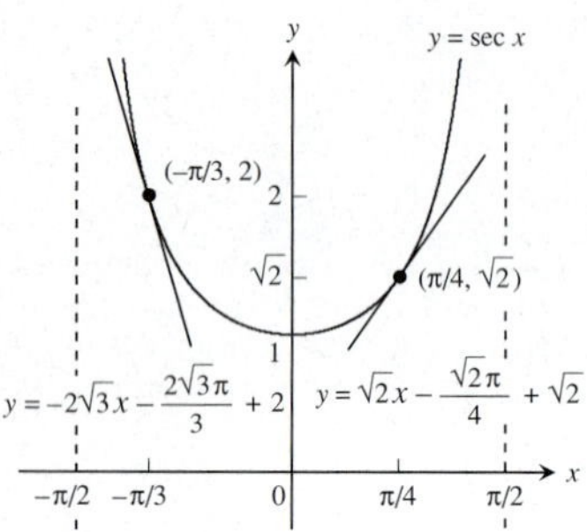

31. Yes, at $x = \pi$
33. No
35. $\left(-\frac{\pi}{4}, -1\right); \left(\frac{\pi}{4}, 1\right)$

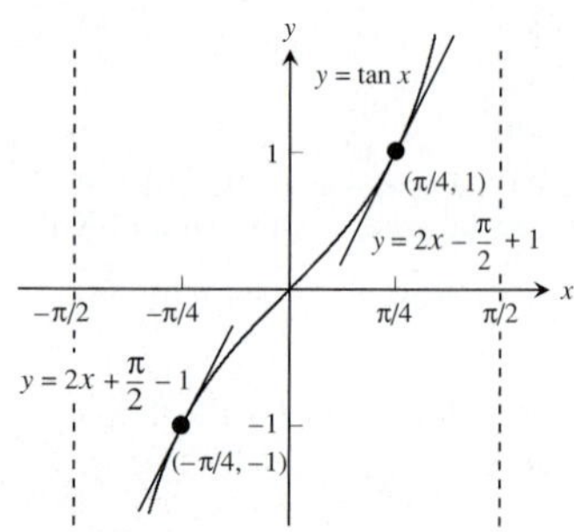

37. (a) $y = -x + \frac{\pi}{2} + 2$ **(b)** $y = 4 - \sqrt{3}$
39. $-\sqrt{2}$ m/sec, $\sqrt{2}$ m/sec, $\sqrt{2}$ m/sec^2, $\sqrt{2}$ m/sec^3
41. $c = 9$
43. $\sin x$

Section 2.5, pp. 195–198

1. $12x^3$
3. $3\cos(3x+1)$
5. $10\sec^2(10x-5)$
7. With $u = (4-3x)$, $y = u^9$: $\frac{dy}{dx} = \frac{dy}{du}\frac{du}{dx} = 9u^8 \cdot (-3) = -27(4-3x)^8$
9. With $u = \left(\frac{x^2}{8} + x - \frac{1}{x}\right)$, $y = u^4$: $\frac{dy}{dx} = \frac{dy}{du}\frac{du}{dx} = 4u^3 \cdot \left(\frac{x}{4} + 1 + \frac{1}{x^2}\right) = 4\left(\frac{x^2}{8} + x - \frac{1}{x}\right)^3 \left(\frac{x}{4} + 1 + \frac{1}{x^2}\right)$
11. With $u = \left(\pi - \frac{1}{x}\right)$, $y = \cot u$: $\frac{dy}{dx} = \frac{dy}{du}\frac{du}{dx} = (-\csc^2 u)\left(\frac{1}{x^2}\right) = -\frac{1}{x^2}\csc^2\left(\pi - \frac{1}{x}\right)$
13. $\frac{1-r}{\sqrt{2r - r^2}}$
15. $\frac{\csc\theta}{\cot\theta + \csc\theta}$
17. $2x\sin^4 x + 4x^2\sin^3 x\cos x + \cos^{-2} x + 2x\cos^{-3} x\sin x$
19. $(3x-2)^6 - \frac{1}{x^3(4 - 1/2x^2)^2}$
21. $\sqrt{x}\sec^2(2\sqrt{x}) + \tan(2\sqrt{x})$
23. $\frac{2\sin\theta}{(1+\cos\theta)^2}$
25. $\frac{dr}{d\theta} = (\sec\sqrt{\theta})\left[\frac{\tan\sqrt{\theta}\,\tan(1/\theta)}{2\sqrt{\theta}} - \frac{\sec^2(1/\theta)}{\theta^2}\right]$
27. $2\pi\sin(\pi t - 2)\cos(\pi t - 2)$
29. $\frac{\csc^2(t/2)}{[1+\cos(1/2)]^3}$
31. $\left[1 + \tan^4\left(\frac{t}{12}\right)\right]^2\left[\tan^3\left(\frac{t}{12}\right)\sec^2\left(\frac{t}{12}\right)\right]$
33. $y = -x + 2\sqrt{2}$, $\left.\frac{d^2y}{dx^2}\right|_{t=\pi/4} = -\sqrt{2}$
35. $y = x + \frac{1}{4}$, $\left.\frac{d^2y}{dx^2}\right|_{t=1/4} = -2$
37. $y = x - 4$, $\left.\frac{d^2y}{dx^2}\right|_{t=-1} = \frac{1}{2}$
39. $y = 2$, $\left.\frac{d^2y}{dx^2}\right|_{t=\pi/2} = -1$
41. $\frac{6}{x^3}\left(1 + \frac{1}{x}\right)\left(1 + \frac{2}{x}\right)$
43. $2\csc^2(3x-1)\cot(3x-1)$
45. $5/2$
47. $-\pi/4$
49. 0
51. (a) $2/3$ **(b)** $2\pi + 5$
(c) $15 - 8\pi$ **(d)** $37/6$
(e) -1 **(f)** $\sqrt{2}/24$
(g) $5/32$ **(h)** $\frac{-5}{3\sqrt{17}}$
53. 5
55. (a) 1 **(b)** 1
57. (a) $y = \pi x + 2 - \pi$ **(b)** $\pi/2$
59. It multiplies the velocity, acceleration, and jerk by 2, 4, and 8, respectively.
61. $v(t) = 2/5$ m/sec, $a(6) = -4/125$ m/sec^2
71. $\left(\frac{\sqrt{2}}{2}, 1\right)$, $y = 2x$ at $t = 0$, $y = -2x$ at $t = \pi$

Section 2.6, pp. 204–206

1. $\frac{9}{4}x^{5/4}$
3. $\frac{7}{2(x+6)^{1/2}}$
5. $\frac{2x^2+1}{(x^2+1)^{1/2}}$
7. $\frac{ds}{dt} = \frac{2}{7}t^{-5/7}$
9. $\frac{dy}{dt} = -\frac{4}{3}(2t+5)^{-5/3}\cos[(2t+5)^{-2/3}]$

11. $g'(x) = \frac{2}{3}(2x^{-1/2} + 1)^{-4/3}(x^{-3/2})$ **13.** $\frac{-2xy - y^2}{x^2 + 2xy}$

15. $\frac{y - 3x^2}{3y^2 - x}$ **17.** $\frac{1}{y(x + 1)^2}$

19. $\cos^2 y$

21. $\frac{-y^2}{y \sin\left(\frac{1}{y}\right) - \cos\left(\frac{1}{y}\right) + xy}$ **23.** $-\frac{\sqrt{r}}{\sqrt{\theta}}$

25. $\frac{-r}{\theta}$, $\cos(r\theta) \neq 0$

27. $y' = -\left(\frac{y}{x}\right)^{1/3}$, $y'' = \frac{y^{1/3}}{3x^{4/3}} + \frac{1}{3y^{1/3}x^{2/3}}$

29. $y' = \frac{\sqrt{y}}{\sqrt{y} + 1}$, $= y'' \; \frac{1}{2(\sqrt{y} + 1)^3}$

31. -2

33. 0

35. -6

37. $(-2, 1)$: $m = -1$, $(-2, -1)$: $m = 1$

39. **(a)** $y = \frac{7}{4}x - \frac{1}{2}$ **(b)** $y = -\frac{4}{7}x + \frac{29}{7}$

41. **(a)** $y = -x - 1$ **(b)** $y = x + 3$

43. **(a)** $y = -\frac{\pi}{2}x + \pi$ **(b)** $y = \frac{2}{\pi}x - \frac{2}{\pi} + \frac{\pi}{2}$

45. **(a)** $y = 2\pi x - 2\pi$ **(b)** $y = -\frac{x}{2\pi} + \frac{1}{2\pi}$

47. Points: $(-\sqrt{7}, 0)$ and $(\sqrt{7}, 0)$; slope: -2

49. $m = -1$ at $\left(\frac{\sqrt{3}}{4}, \frac{\sqrt{3}}{2}\right)$, $m = \sqrt{3}$ at $\left(\frac{\sqrt{3}}{4}, \frac{1}{2}\right)$

51. $(-3, 2)$: $m = -\frac{27}{8}$; $(-3, -2)$: $m = \frac{27}{8}$; $(3, 2)$: $m = \frac{27}{8}$; $(3, -2)$: $m = -\frac{27}{8}$

53. **(a)** False **(b)** True
(c) True **(d)** True

55. $(3, -1)$ **57.** $a = \frac{3}{4}$

59. $\frac{dy}{dx} = -\frac{y^3 + 2xy}{x^2 + 3xy^2}$, $\frac{dx}{dy} = -\frac{x^2 + 3xy^2}{y^3 + 2xy}$, $\frac{dx}{dy} = \frac{1}{dy/dx}$

Section 2.7, pp. 212–216

1. $\frac{dA}{dt} = 2\pi r\frac{dr}{dt}$

3. **(a)** $\frac{dV}{dt} = \pi r^2\frac{dh}{dt}$
(b) $\frac{dV}{dt} = 2\pi hr\frac{dr}{dt}$
(c) $\frac{dV}{dt} = \pi r^2\frac{dh}{dt} + 2\pi hr\frac{dr}{dt}$

5. **(a)** 1 volt/sec
(b) $-\frac{1}{3}$ amp/sec
(c) $\frac{dR}{dt} = \frac{1}{I}\left(\frac{dV}{dt} - \frac{V}{I}\frac{dI}{dt}\right)$
(d) $\frac{3}{2}$ ohms/sec, R is increasing

7. **(a)** $\frac{ds}{dt} = \frac{x}{\sqrt{x^2 + y^2}}\frac{dx}{dt}$
(b) $\frac{ds}{dt} = \frac{x}{\sqrt{x^2 + y^2}}\frac{dx}{dt} + \frac{y}{\sqrt{x^2 + y^2}}\frac{dy}{dt}$
(c) $\frac{dx}{dt} = -\frac{y}{x}\frac{dy}{dt}$

9. **(a)** $\frac{dA}{dt} = \frac{1}{2}ab\cos\theta\frac{d\theta}{dt}$
(b) $\frac{dA}{dt} = \frac{1}{2}ab\cos\theta\frac{d\theta}{dt} + \frac{1}{2}b\sin\theta\frac{da}{dt}$
(c) $\frac{dA}{dt} = \frac{1}{2}ab\cos\theta\frac{d\theta}{dt} + \frac{1}{2}b\sin\theta\frac{da}{dt} + \frac{1}{2}a\sin\theta\frac{db}{dt}$

11. **(a)** 14 cm²/sec, increasing
(b) 0 cm/sec, constant
(c) $-\frac{14}{13}$ cm/sec, decreasing

13. **(a)** -12 ft/sec **(b)** -59.5 ft²/sec **(c)** -1 rad/sec

15. 20 ft/sec

17. **(a)** $\frac{dh}{dt} = 11.19$ cm/min **(b)** $\frac{dr}{dt} = 14.92$ cm/min

19. **(a)** $-\frac{1}{24\pi}$ m/min
(b) $r = \sqrt{26y - y^2}m$
(c) $\frac{dr}{dt} = -\frac{5}{288\pi}$ m/min

21. 1 ft/min, 40π ft²/min **23.** 11 ft/sec

25. Increasing at $\frac{466}{1681}$ L/min **27.** 1 rad/sec

29. -5 m/sec **31.** -1500 ft/sec

33. $\frac{5}{72\pi}$ in./min, $\frac{10}{3}$ in.²/min **35.** 7.1 in./min

37. **(a)** $-32/\sqrt{13} \approx -8.875$ ft/sec
(b) $\frac{d\theta_1}{dt} = -\frac{8}{65}$ rad/sec, $\frac{d\theta_2}{dt} = \frac{8}{65}$ rad/sec
(c) $\frac{d\theta_1}{dt} = -\frac{1}{6}$ rad/sec, $\frac{d\theta_2}{dt} = \frac{1}{6}$ rad/sec

Section 2.8, pp. 221–222

1. $\frac{-2x}{\sqrt{1 - x^4}}$ **3.** $\frac{\sqrt{2}}{\sqrt{1 - 2t^2}}$

5. $\frac{1}{|2s + 1|\sqrt{s^2 + s}}$ **7.** $\frac{-2x}{(x^2 + 1)\sqrt{x^4 + 2x^2}}$

9. $\frac{-1}{\sqrt{1 - t^2}}$ **11.** $\frac{-1}{2\sqrt{t}(1 + t)}$

13. $\dfrac{-2s^2}{\sqrt{1-s^2}}$ **15.** 0

17. $\sin^{-1} x$

19. (a) $y = 2x - \dfrac{\pi}{2} + 1$ **(b)** $y = \dfrac{1}{2}x - \dfrac{1}{2} + \dfrac{\pi}{4}$

21. (a) $f'(x) = 3 - \sin x$ and $f'(x) \neq 0$. So, f has a differentiable inverse by Theorem 3.

(b) $f(0) = 1, f'(0) = 3$

(c) $f^{-1}(1) = 0, (f^{-1})'(1) = \dfrac{1}{3}$

23. $-\dfrac{1}{\sqrt{1-x^2}}$ **25.** $-\dfrac{1}{|x|\sqrt{x^2-1}}$

27. (a) $y = \dfrac{\pi}{2}$ **(b)** $y = -\dfrac{\pi}{2}$ **(c)** None

29. (a) $y = \dfrac{\pi}{2}$ **(b)** $y = \dfrac{\pi}{2}$ **(c)** None

31. (a) $f^{-1}(x) = \dfrac{x}{2} - \dfrac{3}{2}$

(b)

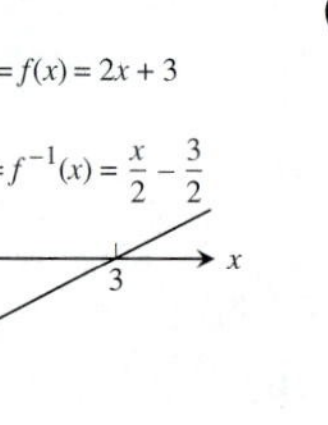

(c) 2, 1/2

33. (a) $f^{-1}(x) = -\dfrac{x}{4} + \dfrac{5}{4}$

(b)

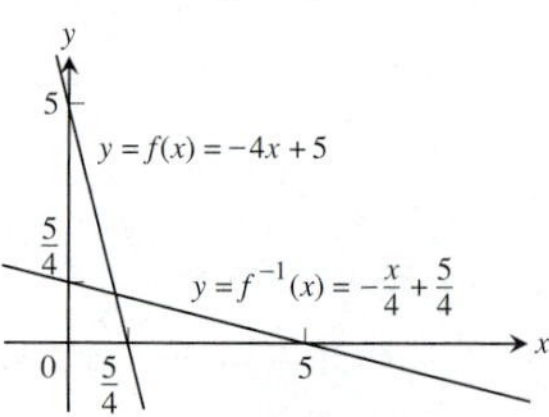

(c) $-4, -1/4$

35. (b)

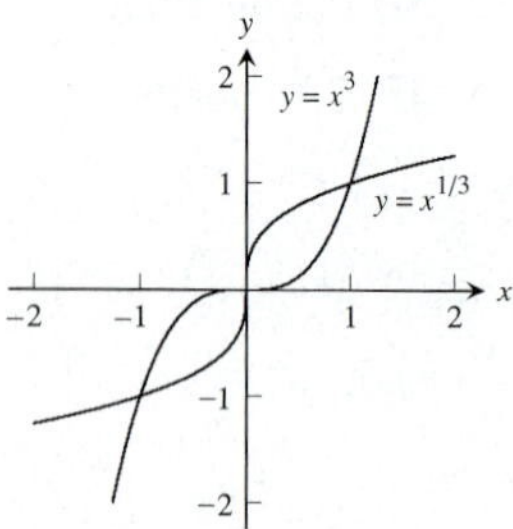

(c) Slope of f at (1, 1): 3; slope of g at (1, 1): 1/3; slope of f at $(-1, -1)$: 3; slope of g at $(-1, -1)$: 1/3

(d) $y = 0$ is tangent to $y = x^3$ at $x = 0$; $x = 0$ is tangent to $y = \sqrt[3]{x}$ at $x = 0$.

37. (a) $f^{-1}(x) = \dfrac{1}{m}x$

(b) The graph of f^{-1} is the line through the origin with slope $\dfrac{1}{m}$.

Section 2.9, pp. 231–232

1. $2e^x$ **3.** $-\dfrac{3}{2}e^{-3x/2}$

5. $\dfrac{2}{3}e^{2x/3}$ **7.** xe^x

9. $\dfrac{e^{\sqrt{x}}}{2\sqrt{x}}$ **11.** $\pi x^{\pi-1}$

13. $-\sqrt{2}x^{-\sqrt{2}-1}$ **15.** $8^x \ln 8$

17. $-3^{\csc x}(\ln 3)(\csc x \cot x)$ **19.** $\dfrac{e^x + 2}{e^{-2x} + 2e^{-x} + 1}$

21. $\dfrac{2}{x}$

23. $-\dfrac{1}{x}, x > 0$ **25.** $\dfrac{1}{x+2}, x > -2$

27. $\dfrac{\sin x}{2 - \cos x}$ **29.** $\dfrac{1}{x \ln x}$

31. $\dfrac{2}{x \ln 4} = \dfrac{1}{x \ln 2}$ **33.** $\dfrac{3}{(3x+1)\ln 2}, x > -\dfrac{1}{3}$

35. $-\dfrac{1}{x \ln 2}, x > 0$ **37.** $\dfrac{1}{x}, x > 0$

39. $\dfrac{1}{\ln 10}$ **41.** $(x^{\ln x})\left(\dfrac{\ln x^2}{x}\right)$

43. $(\sin x)^x[x \cot x + \ln(\sin x)]$

45. $\left(\dfrac{(x-3)^4(x^2+1)}{(2x+5)^3}\right)^{1/5}\left(\dfrac{4}{5(x-3)} + \dfrac{2x}{5(x^2+1)} - \dfrac{6}{5(2x+5)}\right)$

47. $y = ex$ **49.** Rate ≈ 0.098 gram/day

51. Answers will vary considerably with the number of decimal places retained. With (a) $k = 0.008754168326$, the population 8 years out is (b) 275,979,963 (nearest integer).

53. (a) $g'(0) = L$ **(b)** ≈ -0.6931

57. (a) $y = \dfrac{1}{e}x$

(b) Because the graph of $\ln x$ lies below the graph of the line for all positive $x \neq e$.

(c) Multiplying by e, $e(\ln x) < x$, or $\ln x^e < x$.

(d) Exponentiate both sides of the inequality in part (c).

(e) Let $x = \pi$ to see that $\pi^e < e^\pi$.

Chapter 2 Practice Exercises, pp. 233–238

1. $5x^4 - 0.25x + 0.25$ **3.** $7x^6 + \sqrt{7}$

5. $3(\theta^2 + \sec\theta + 1)^2(2\theta + \sec\theta\tan\theta)$

7. $-\dfrac{1}{2\sqrt{t}(\sqrt{t}-1)^2}$ **9.** $(-2\csc x\cot x)(\csc x - 1)$

11. $\dfrac{6}{t^2}\cot^2\left(\dfrac{2}{t}\right)\csc^2\left(\dfrac{2}{5}\right)$ **13.** $\dfrac{\theta\cos\theta + \sin\theta}{\sqrt{2\theta\sin\theta}}$

15. $x\csc\left(\dfrac{2}{x}\right) + \csc\left(\dfrac{2}{x}\right)\cot\left(\dfrac{2}{x}\right)$ **17.** $-10x\csc^2(x^2)$

19. $\dfrac{-(t+1)}{8t^3}$ **21.** $\dfrac{6x + 5\sqrt{x}}{\sqrt{x + \sqrt{x}}}$

23. $\dfrac{3}{(3x-4)^{19/20}}$ **25.** xe^{4x}

27. $\dfrac{2\sin\theta\cos\theta}{\sin^2\theta} = 2\cot\theta$

29. $\dfrac{3}{(\ln 5)(3x-7)}$

31. $18x^{2.6}$

33. $(x+2)^{x+2}(\ln(x+2)+1)$

35. $-\dfrac{1}{\sqrt{1-u^2}}$

37. $\cos^{-1} z$

39. $2t\cot^{-1}(2t) + (1+t^2)\left(\dfrac{-2}{1+4t^2}\right)$

41. -1

43. $-\dfrac{y+2}{x+3}$

45. $\dfrac{-3x^2-4y+2}{4x-4y^{1/3}}$

47. $-\dfrac{y}{x}$

49. $-1/2$

51. y/x

53. $-\dfrac{2e^{-\tan^{-1}x}}{1+x^2}$

55. $\dfrac{dr}{ds} = (2r-1)(\tan 2s)$

57. **(a)** $\dfrac{d^2y}{dx^2} = \dfrac{-2xy^3-2x^4}{y^5}$ **(b)** $\dfrac{d^2y}{dx^2} = \dfrac{-2xy^2-1}{x^4y^3}$

59. **(a)** 7 **(b)** -2 **(c)** 5/12 **(d)** $\dfrac{1}{4}$ **(e)** 12 **(f)** $\dfrac{9}{2}$ **(g)** $\dfrac{3}{4}$

61. 0

63. $\dfrac{3\sqrt{2}e^{\sqrt{3/2}}}{4}\cos\left(e^{\sqrt{3/2}}\right)$

65. $-1/2$

67. $\dfrac{-2}{(2t+1)^2}$

69. **(a)**

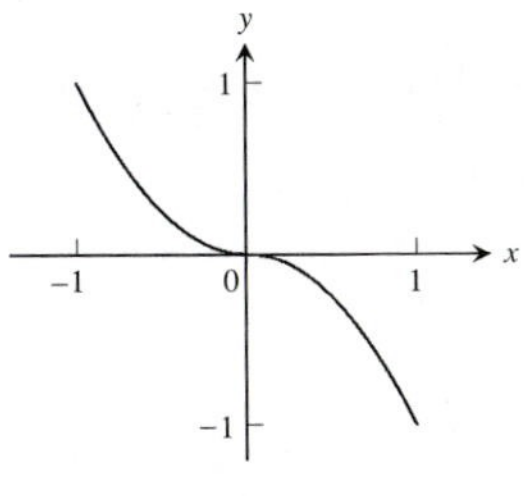

$f(x) = \begin{cases} x^2, & -1 \le x < 0 \\ -x^2, & 0 \le x \le 1 \end{cases}$

(b) Yes
(c) Yes

71. **(a)**

$f(x) = \begin{cases} x, & 0 \le x \le 1 \\ 2-x, & 1 < x \le 2 \end{cases}$

(b) Yes
(c) No

73. $\left(\dfrac{5}{2}, \dfrac{9}{4}\right)$ and $\left(\dfrac{3}{2}, -\dfrac{1}{4}\right)$

75. $(0, 16)$ and $\left(-\dfrac{4}{3}, 0\right)$

79. 1/4

81. Tangent: $y = -\dfrac{1}{4}x + \dfrac{9}{4}$; normal: $y = 4x - 2$

83. Tangent: $y = 2x - 4$; normal: $y = -\dfrac{1}{2}x + \dfrac{7}{2}$

85. Tangent: $y = -\dfrac{5}{4}x + 6$; normal: $y = \dfrac{4}{5}x - \dfrac{11}{5}$

87. $(1, 1)$: $m = -\dfrac{2}{3}$, $(1, -1)$: $m = 4$

89. B = graph of f, A = graph of f'

91.

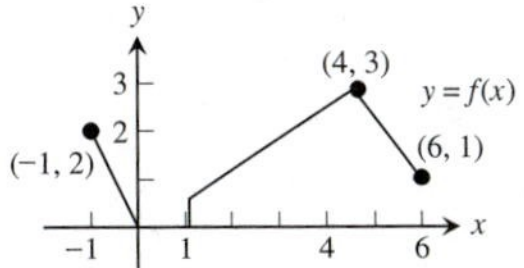

93. **(a)** 0, 0
(b) 1700 rabbits $\approx$ 1400 rabbits

95. **(a)** $\dfrac{dS}{dt} = (4\pi r + 2\pi h)\dfrac{dr}{dt}$
(b) $\dfrac{dS}{dt} = 2\pi r\dfrac{dh}{dt}$
(c) $\dfrac{dS}{dt} = (4\pi r + 2\pi h)\dfrac{dr}{dt} + 2\pi r\dfrac{dh}{dt}$
(d) $\dfrac{dr}{dt} = -\dfrac{r}{2r+h}\dfrac{dh}{dt}$

97. 40 m^2/sec

99. 0.02 ohm/sec

101. 22 m/sec

103. **(a)** $r = \dfrac{2}{5}h$
(b) $-\dfrac{125}{144\pi}$ ft/min

105. **(a)** $\dfrac{3}{5}$ km/sec or 600 m/sec
(b) $\dfrac{18}{\pi}$ rpm

107. $\dfrac{1}{e}$ m/sec

109. $\ln 5x - \ln 3x = \ln\left(\dfrac{5}{3}\right)$

111. $\dfrac{1}{2}$

113. $y'(r) = -\dfrac{1}{2r^2l}\sqrt{\dfrac{T}{\pi d}}$, so increasing r decreases the frequency.

$y'(l) = -\dfrac{1}{2rl^2}\sqrt{\dfrac{T}{\pi d}}$, so increasing l decreases the frequency.

$y'(d) = -\dfrac{1}{4rl}\sqrt{\dfrac{T}{\pi d^3}}$, so increasing d decreases the frequency.

$y'(T) = \dfrac{1}{4rl\sqrt{\pi T d}}$, so increasing T increases the frequency.

115. **(a)** $x \neq k\dfrac{\pi}{4}$, where k is an odd integer

(b) $\left(-\dfrac{\pi}{2}, \dfrac{\pi}{2}\right)$

(c) Where it is not defined, at $x = k\dfrac{\pi}{4}$, k an odd integer

(d) It has period $\pi/2$ and continues to repeat the pattern seen in this window.

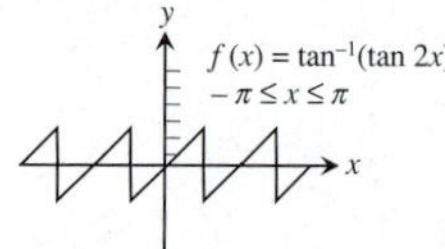

Chapter 2 Additional Exercises, pp. 238–240

1. **(a)** $\sin 2\theta = 2\sin\theta\cos\theta$; $2\cos 2\theta = 2\sin\theta(-\sin\theta) + \cos\theta(2\cos\theta)$; $2\cos 2\theta = -2\sin^2\theta + 2\cos^2\theta$; $\cos 2\theta = \cos^2\theta - \sin^2\theta$

(b) $\cos 2\theta = \cos^2\theta - \sin^2\theta$; $-2\sin 2\theta = 2\cos\theta(-\sin\theta) - 2\sin\theta(\cos\theta)$; $\sin 2\theta = \cos\theta\sin\theta + \sin\theta\cos\theta$; $\sin 2\theta = 2\sin\theta\cos\theta$

3. **(a)** $a = 1, b = 0, c = -1/2$

(b) $b = \cos a, c = \sin a$

5. $h = -4, k = \dfrac{9}{2}, a = \dfrac{5\sqrt{5}}{2}$

7. **(a)** $0.09y$

b) Increasing at 1% per year

9. Answers will vary. Here is one possibility.

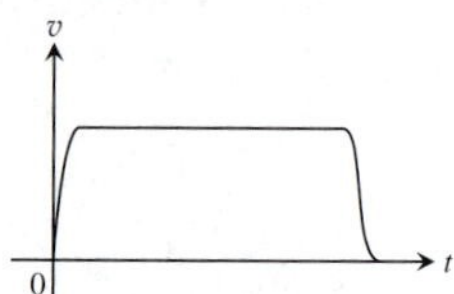

11. **(a)** 2 sec, 64 ft/sec **(b)** ≈ 12.3 sec, ≈ 394 ft

13. $mv\dfrac{dv}{dt} = -kx\dfrac{dx}{dt} = -kxv \to m\dfrac{dv}{dt} = -kx$

15. **(a)** $m = -\dfrac{b}{\pi}$ **(b)** $m = -1, b = \pi$

17. **(a)** $a = 3/4, b = 9/4$ **19.** f odd $\Rightarrow f'$ is even

23. h' is defined but not continuous at $x = 0$; k' is defined *and* continuous at $x = 0$.

CHAPTER 3 APPLICATIONS OF DERIVATIVES

Section 3.1, pp. 251–254

1. Absolute minimum at $x = c_2$, absolute maximum at $x = b$

3. Absolute maximum at $x = c$, no absolute minimum

5. Absolute minimum at $x = a$, absolute maximum at $x = c$

7. Local minimum at $(-1, 0)$, local maximum at $(1, 0)$

9. Maximum at $(0, 5)$

11. (c) **13.** (d)

15. Absolute maximum: -3; absolute minimum: $-\dfrac{19}{3}$

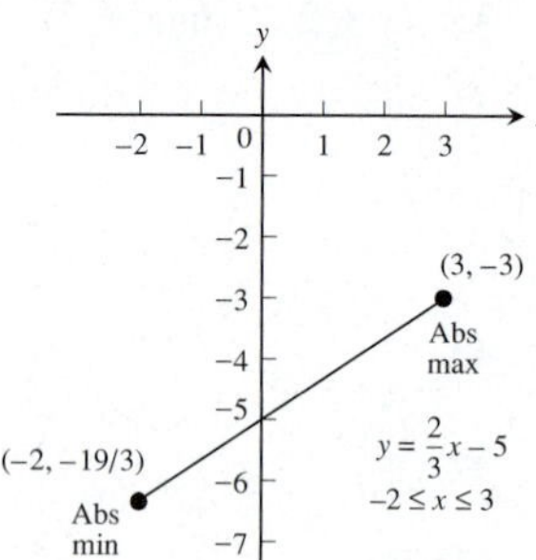

17. Absolute maximum: 4; absolute minimum: -5

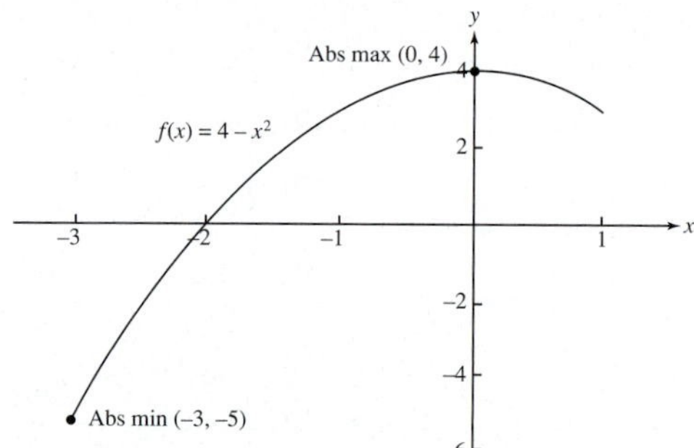

19. Maximum value is 1 at $x = \dfrac{\pi}{4}$; minimum value is -1 at $x = \dfrac{5\pi}{4}$; local minimum at $\left(0, \dfrac{1}{\sqrt{2}}\right)$; local maximum at $\left(\dfrac{7\pi}{4}, 0\right)$

√

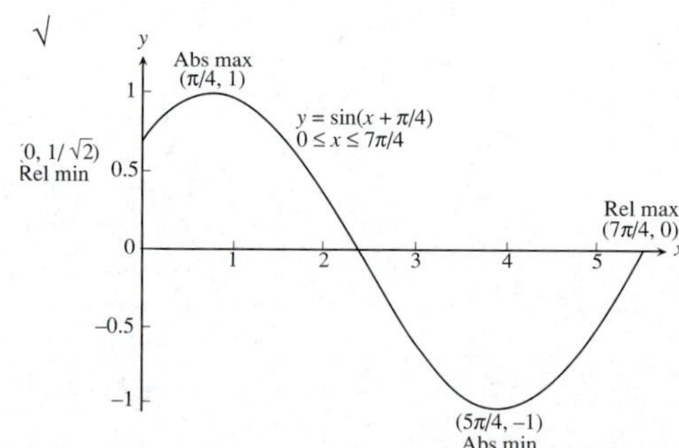

21. Absolute maximum: -0.25, absolute minimum: -4

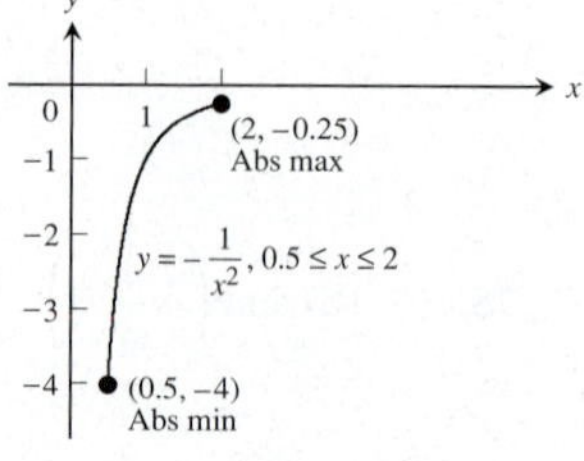

23. Maximum value is $1/4 + \ln 4$ at $x = 4$; minimum value is 1 at $x = 1$; local maximum at $(1/2, 2 - \ln 2)$

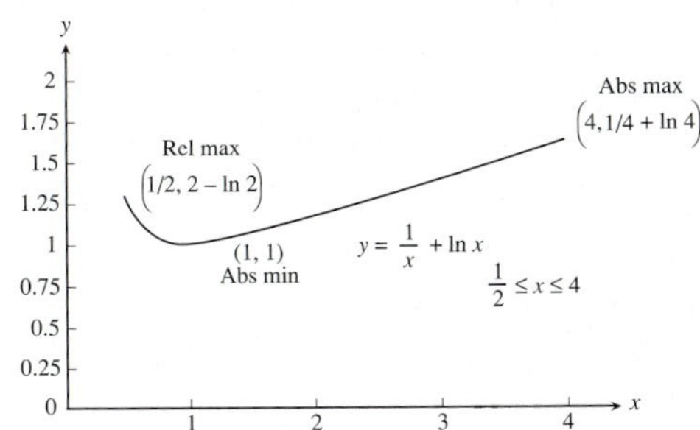

25. Maximum value is $\ln 4$ at $x = 3$; minimum value is 0 at $x = 0$.

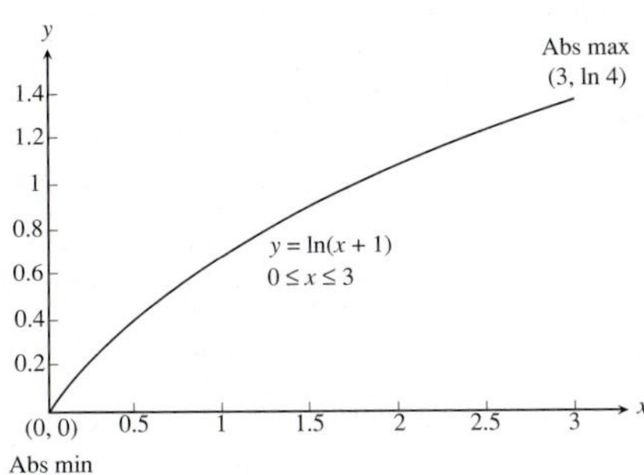

27. Minimum value is 1 at $x = 2$.

29. Minimum value is 1 at $x = 0$.

31. Maximum value is $1/e$ at $x = e$; there is no minimum value.

33. Maximum value is $1/2$ at $x = 1$; minimum value is $-1/2$ at $x = -1$.

35. Local maximum at $(-2, 17)$; local minimum at $\left(\frac{4}{3}, -\frac{41}{27}\right)$

37.

Critical point	Derivative	Extremum	Value
$x = -\frac{4}{5}$	0	Local max	$\frac{12}{25}10^{1/3} = 1.034$
$x = 0$	Undefined	Local min	0

39.

Critical point	Derivative	Extremum	Value
$x = -2$	Undefined	Local max	0
$x = -\sqrt{2}$	0	Minimum	-2
$x = \sqrt{2}$	0	Maximum	2
$x = 2$	Undefined	Local min	0

41.

Critical point	Derivative	Extremum	Value
$x = 1$	Undefined	Minimum	2

43.

Critical point	Derivative	Extremum	Value
$x = -1$	0	Maximum	5
$x = 1$	Undefined	Local min	1
$x = 3$	0	Maximum	5

45. **(a)** No

(b) The derivative is defined and nonzero for $x \neq 2$. Also, $f(2) = 0$ and $f(x) > 0$ for all $x \neq 2$.

(c) No, because $(-\infty, \infty)$ is not a closed interval.

(d) The answers are the same as parts (a) and (b) with 2 replaced by a.

47. **(a)** $C(x) = 0.3\sqrt{16 + x^2} + 0.2(9 - x)$ million dollars, where $0 \leq x \leq 9$ mi. To minimize the cost of construction, the pipeline should be placed from the docking facility to point B, 3.58 mi along the shore from point A, and then along the shore from B to the refinery.

(b) In theory, the underwater pipe cost per mile p would have to be infinite to justify running the pipe directly from the docking facility to point A (i.e., for x_c to be zero). For all values of $p > 0.218864$, there is always an x_c element of ϵ (0, 9) that will give a minimum value for C. This is proved by looking at $C''(x_c) = \frac{16p}{(16 + x_c^2)^{3/2}}$, which is always positive for $p > 0$.

49. The length of pipeline is $L(x) = \sqrt{4 + x^2} + \sqrt{25 + (10 - x)^2}$ for $0 \leq x \leq 10$. Setting the derivative of $L(x)$ equal to zero gives $L'(x) = \frac{x}{\sqrt{4 + x^2}} - \frac{(10 - x)}{\sqrt{25 + (10 - x)^2}} = 0$. Note that $\frac{x}{\sqrt{4 + x^2}} = \cos\theta_A$ and $\frac{10 - x}{\sqrt{25 - (10 - x)^2}} = \cos\theta_B$; therefore, $L'(x) = 0$ when $\cos\theta_A = \cos\theta_B$, or $\theta_A = \theta_B$. Use simple proportions to determine x as follows: $\frac{x}{2} = \frac{10 - x}{5} \Rightarrow x = \frac{20}{7} \approx 2.857$ mi along the coast from town A to town B.

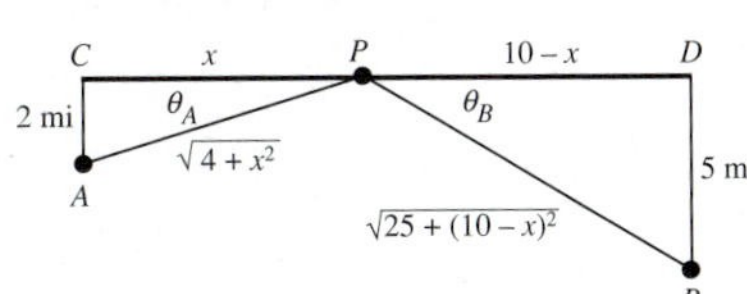

51. **(a)** Maximum value is 144 at $x = 2$.

(b) The largest volume of the box is 144 cubic units, and it occurs when $x = 2$.

53. The largest possible area is $A\left(\frac{5}{\sqrt{2}}\right) = \frac{25}{4}$ cm^2.

55. $\frac{v_0^2}{2g} + s_0$

57. Yes

59. g assumes a local maximum at $-c$.

61. **(a)** $f'(x) = 3ax^2 + 2bx + c$ is a quadratic, so it can have 0, 1, or 2 zeros, which would be the critical points of f. Examples: The function $f(x) = x^3 - 3x$ has two critical points at $x = -1$ and $x = 1$.

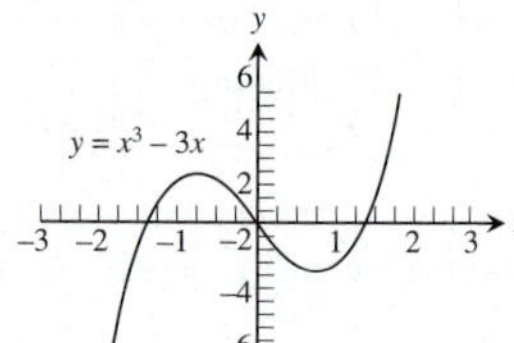

The function $f(x) = x^3 - 1$ has one critical point at $x = 0$.

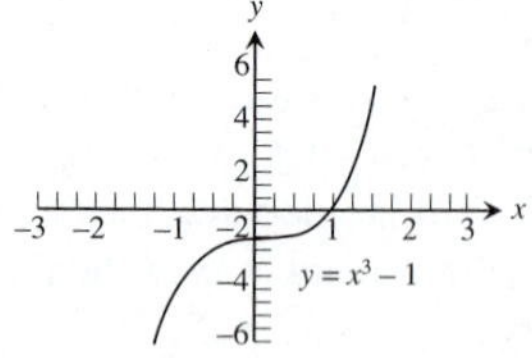

The function $f(x) = x^3 + x$ has no critical points.

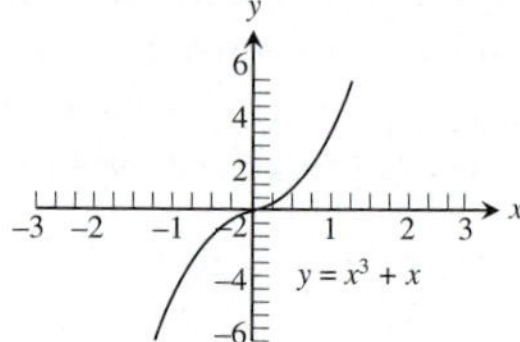

(b) Two or none

63. Maximum value is 11 at $x = 5$; minimum value is 5 on the interval $[-3, 2]$; local maximum is at $(-5, 9)$.

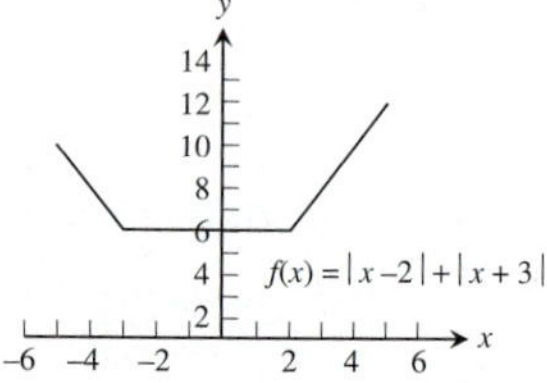

65. Maximum value is 5 on the interval $[3, \infty)$; minimum value is -5 on the interval $[-\infty, -2]$

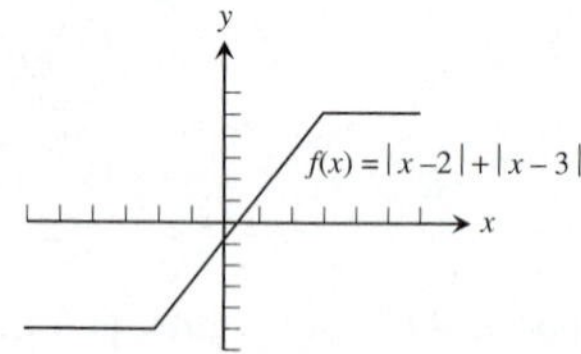

Section 3.2, pp. 261–263

1. **(a)** f is continuous on $[0, 1]$ and differentiable on $(0, 1)$.

(b) $c = \frac{1}{2}$

3. **(a)** f is continuous on $[-1, 1]$ and differentiable on $(-1, 1)$.

(b) $c = \pm\sqrt{1 - \frac{4}{\pi^2}} \approx \pm 0.771$

5. The function $f(x)$ is not continuous on $0 \le x \le 1$ because $\lim_{x\to 1^-} f(x) = \lim_{x\to 1^-} x = 1 \ne 0 = f(1)$; therefore, Rolle's Theorem does not apply because f is not continuous at $x = 1$.

7. By Corollary 1 of the Mean Value Theorem, $f'(x) = 0$ for all $x \Rightarrow f(x) = C$, where C is a constant. Since $f(-1) = 3$, we have $C = 3 \Rightarrow f(x) = 3$ for all x.

9. **(a)** $y = \frac{x^2}{2} + C$ **(b)** $y = \frac{x^3}{3} + C$ **(c)** $y = \frac{x^4}{4} + C$

11. **(a)** $y = \ln|\theta| + C$
(b) $y = \theta - \ln|\theta| + C$
(c) $y = 5\theta + \ln|\theta| + C$

13. $f(x) = x^2 - x$

15. $f(x) = 1 + \frac{e^{2x}}{2}$

17. $s = 4.9t^2 + 5t + 10$

19. $s = \frac{1 - \cos(\pi t)}{\pi}$

23. $s = \sin(2t) - 3$

25. 48 m/sec

27. 14 m/sec

29. **(a)** $v = 10t^{3/2} - 6t^{1/2}$ **(b)** $s = 4t^{5/2} - 4t^{3/2}$

31. If $T(t)$ is the temperature of the thermometer at time t, then $T(0) = -19°C$ and $T(14) = 100°C$. From the Mean Value Theorem, there exists a $0 < t_0 < 14$ such that $\frac{T(14) - T(0)}{14 - 0} = 8.5°C/\text{sec} = T'(t_0)$, the rate at which the temperature was changing at $t = t_0$ as measured by the rising mercury on the thermometer.

33. Because its average speed was approximately 7.667 knots, and by the Mean Value Theorem, it must have been going that speed at least once during the trip.

35. The conclusion of the Mean Value Theorem yields $\frac{1/b - 1/a}{b - a} = -\frac{1}{c^2} \Rightarrow c^2\left(\frac{a - b}{ab}\right) = a - b \Rightarrow c = \sqrt{ab}$.

39. $f(x)$ must be zero at least once between a and b by the Intermediate Value Theorem. Now suppose that $f(x)$ is zero twice between a and b. Then by the Mean Value Theorem, $f'(x)$ would have to be zero at least once between the two zeros of $f(x)$, but this cannot be true since we are given that $f'(x) \ne 0$ on this interval. Therefore, $f(x)$ is zero once and only once between a and b.

45. $1.09999 \le f(0.1) \le 1.1$

Section 3.3, pp. 272–276

1.

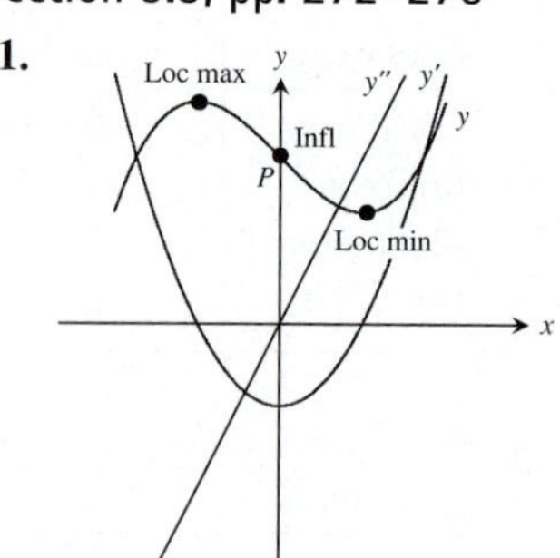

3. 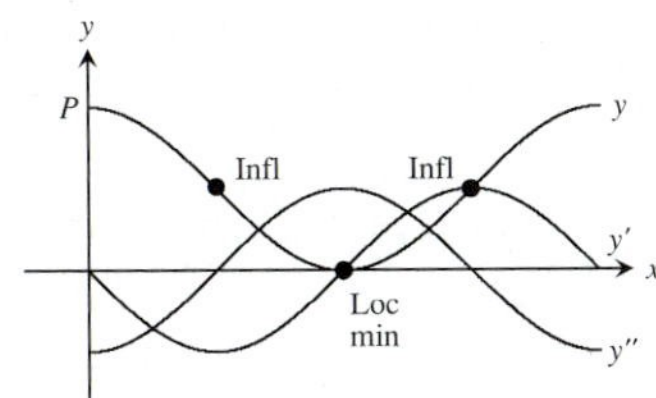

5. 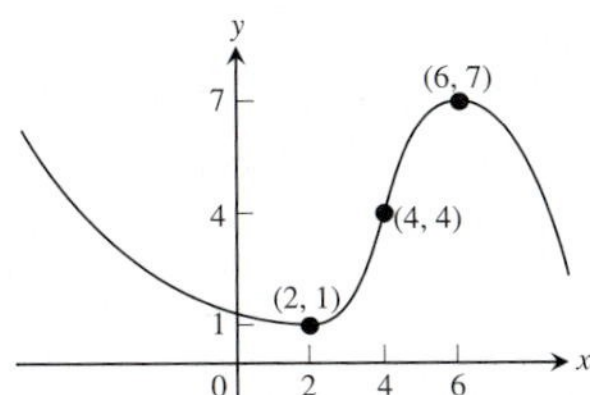

7. **(a)** Zero: $x = \pm 1$; positive: $(-\infty, -1)$ and $(1, \infty)$; negative: $(-1, 1)$
(b) Zero: $x = 0$; positive: $(0, \infty)$; negative: $(-\infty, 0)$

9. **(a)** $(-\infty, -2]$ and $[0, 2]$
(b) $[-2, 0]$ and $[2, \infty)$
(c) Local maxima: $x = -2$ and $x = 2$; local minimum: $x = 0$

11. **(a)** $[0, 1]$, $[3, 4]$, and $[5.5, 6]$
(b) $[1, 3]$ and $[4, 5.5]$
(c) Local maxima: $x = 1$, $x = 4$ (if f is continuous at $x = 4$), and $x = 6$; local minima: $x = 0$, $x = 3$, and $x = 5.5$

13. **(a)** Critical points at -2 and 1
(b) Increasing on $(-\infty, -2]$ and $[1, \infty)$, decreasing on $[-2, 1]$
(c) Local maximum at $x = -2$ and a local minimum at $x = 1$

15. **(a)** Critical point at $x = 1$
(b) Increasing on $[1, \infty)$, decreasing on $(-\infty, 1]$
(c) Local (and absolute) minimum at $x = 1$

17. **(a)** $\left[\frac{1}{2}, \infty\right)$ **(b)** $\left(2\infty, \frac{1}{2}\right]$
(c) $(-\infty, \infty)$ **(d)** Nowhere

19. **(a)** $[-1, 0]$ and $[1, \infty)$
(b) $(-\infty, -1]$ and $[0, 1]$
(c) $\left(-\infty, -\frac{1}{\sqrt{3}}\right)$ and $\left(\frac{1}{\sqrt{3}}, \infty\right)$
(d) $\left(-\frac{1}{\sqrt{3}}, \frac{1}{\sqrt{3}}\right)$
(e) Local maximum at $(0, 1)$; local (and absolute) minima at $(\pm 1, 1)$
(f) $\left(\pm\frac{1}{\sqrt{3}}, -\frac{1}{9}\right)$

21. **(a)** $[-2, 2]$
(b) $[-\sqrt{8}, -2]$ and $[2, \sqrt{8}]$
(c) $(-\sqrt{8}, 0)$
(d) $(0, \sqrt{8})$
(e) Local maxima: $(-\sqrt{8}, 0)$ and $(2, 4)$; local minima: $(-2, -4)$ and $(\sqrt{8}, 0)$
(f) $(0, 0)$

23. **(a)** $(-\infty, -2]$ and $\left[-\frac{3}{2}, \infty\right)$ **(b)** $\left[-2, -\frac{3}{2}\right]$
(c) $\left(-\frac{7}{4}, \infty\right)$ **(d)** $\left(-\infty, -\frac{7}{4}\right)$
(e) Local maximum: $(-2, -40)$; local minimum: $\left(-\frac{3}{2}, -\frac{161}{4}\right)$
(f) $\left(-\frac{7}{4}, -\frac{321}{8}\right)$

25. **(a)** $(-\infty, \infty)$ **(b)** None **(c)** $(-\infty, 0)$
(d) $(0, \infty)$ **(e)** None **(f)** $(0, 3)$

27. **(a)** $[1, \infty)$
(b) $(-\infty, 1]$
(c) $(-\infty, -2)$ and $(0, \infty)$
(d) $(-2, 0)$
(e) Local minimum: $(1, -3)$
(f) $\approx(-2, 7.56)$ and $(0, 0)$

29. **(a)** $(-\infty, -\sqrt{2}]$ and $[\sqrt{2}, \infty)$
(b) $[-\sqrt{2}, 0)$ and $(0, \sqrt{2}]$
(c) $(0, \infty)$
(d) $(-\infty, 0)$
(e) Local maximum: $-\sqrt{2}, -\sqrt{2e}) \approx (-1.41, -2.33)$; local minimum: $(\sqrt{2}, \sqrt{2e}) \approx (1.41, 2.33)$
(f) None

31. **(a)** $[0, \infty)$ **(b)** None
(c) $\left(\frac{9}{5}, \infty\right)$ **(d)** $\left(0, \frac{9}{5}\right)$
(e) Local (and absolute) minimum: $(0, 0)$
(f) $\left(\frac{9}{5}, \frac{24}{5} \cdot \sqrt[4]{\frac{9}{5}}\right) \approx (1.8, 5.56)$

33. **(a)** None
(b) At $x = 2$
(c) At $x = 1$ and $x = 5/3$

35. **(a)** Absolute maximum at $(1, 2)$; absolute minimum at $(3, -2)$
(b) None
(c) One possible answer:

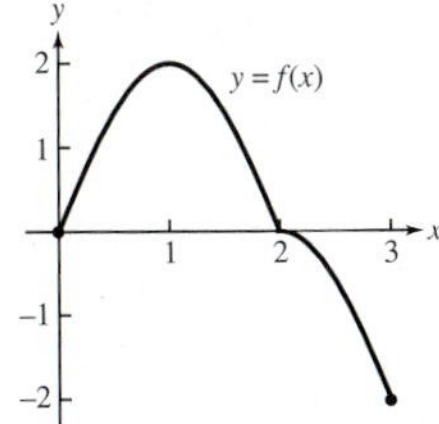

37. The zeros of y' are extrema of y. Inflection point at $x = 3$, local maximum at $(0, -240)$, local minimum at $(4, -496)$.

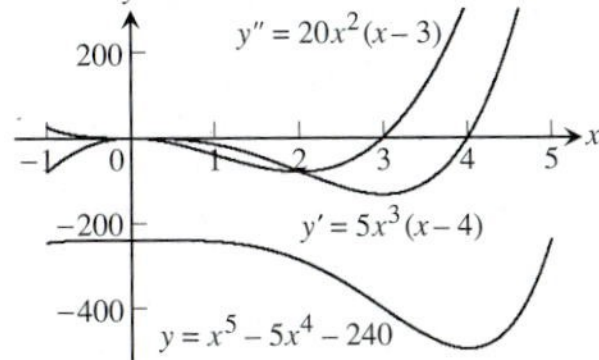

39. The zeros of $y' = 0$ and $y'' = 0$ are extrema and points of inflection, respectively. Inflection at $x = -\sqrt[3]{2}$, local maximum at $\left(-2, \frac{67}{5}\right)$, local minimum at $(0, -25)$.

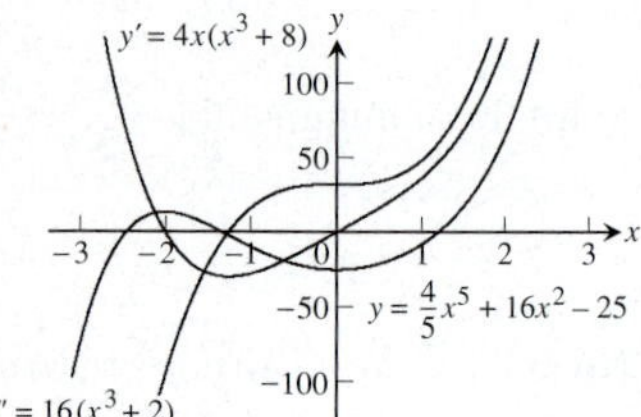

43. **(a)** $v(t) = 2t - 4$
(b) $a(t) = 2$
(c) It begins at position 3 moving in a negative direction. It moves to position -1 when $t = 2$ and then changes direction, moving in a positive direction thereafter.

45. **(a)** $v(t) = 3t^2 - 3$
(b) $a(t) = 6t$
(c) It begins at position 3 moving in a negative direction. It moves to position 1 when $t = 1$ and then changes direction, moving in a positive direction thereafter.

47. **(a)** $t \approx 2.2, 6, 9.8$
(b) $t \approx 4, 8, 11$

49. No; f must have a horizontal tangent line at that point, but it could be increasing (or decreasing) on both sides of the point and there would be no local extremum.

51. One possible answer:

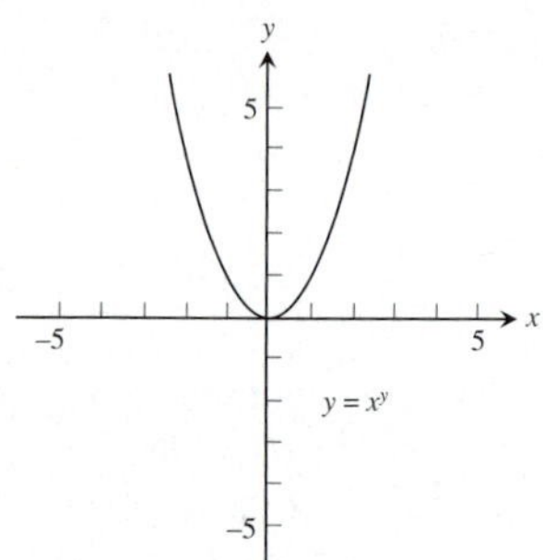

53. One possible answer:

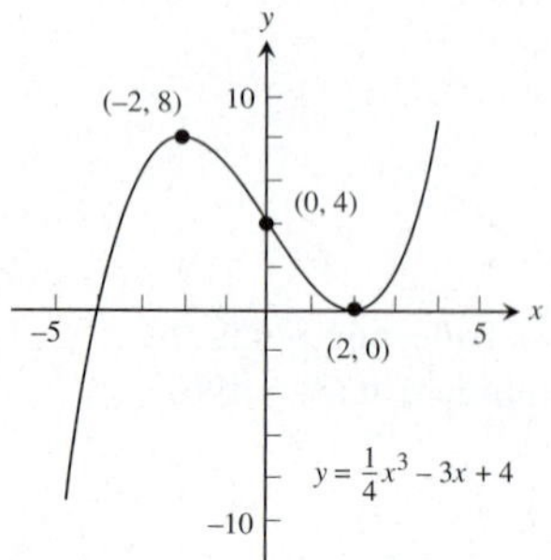

63. **(b)** $f'(x) = 3x^2 + k$; $-12k$; positive if $k < 0$, negative if $k > 0$, 0 if $k = 0$; f' has two zeros if $k < 0$, one zero if $k = 0$, no zeros if $k > 0$.

65. **(a)** $f'(x) = \dfrac{abce^{bx}}{(e^{bx} + a)^2}$, so the sign of $f'(x)$ is the same as the sign of the product abc.

(b) $f''(x) = \dfrac{ab^2ce^{bx}(e^{bx} - a)}{(e^{bx} + a)^3}$. Since $a > 0$, the equation changes sign when $x = \dfrac{\ln a}{b}$ due to the $e^{bx} - a$ factor in the numerator, and there is a point of inflection at that location.

Section 3.4, pp. 284–285

1. $y' = (y + 2)(y - 3)$
(a) $y = -2$ is a stable equilibrium value and $y = 3$ is an unstable equilibrium.
(b) $y'' = 2(y + 2)(y - 1/2)(y - 3)$

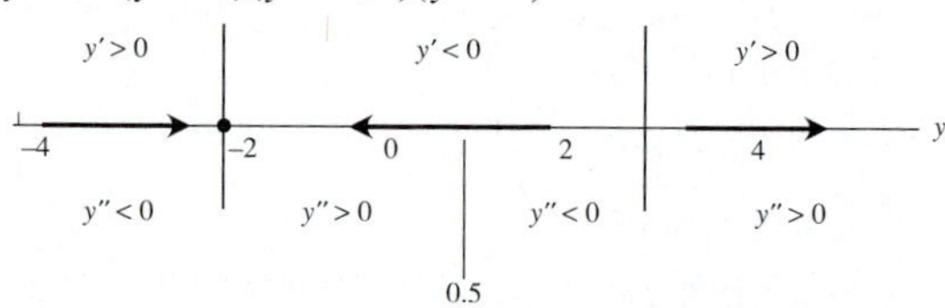

(c)

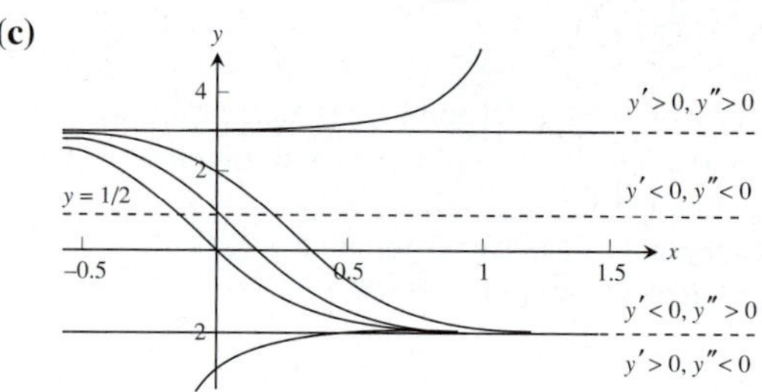

5. $y' = \sqrt{y}, y > 0$
(a) There are no equilibrium values.
(b) $y'' = \frac{1}{2}$

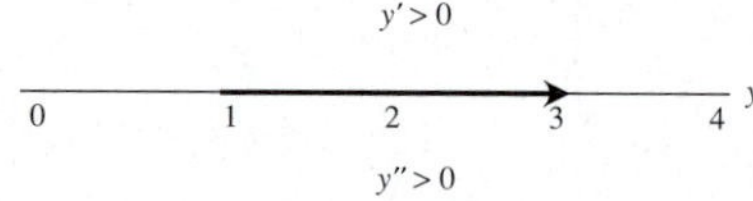

(c)

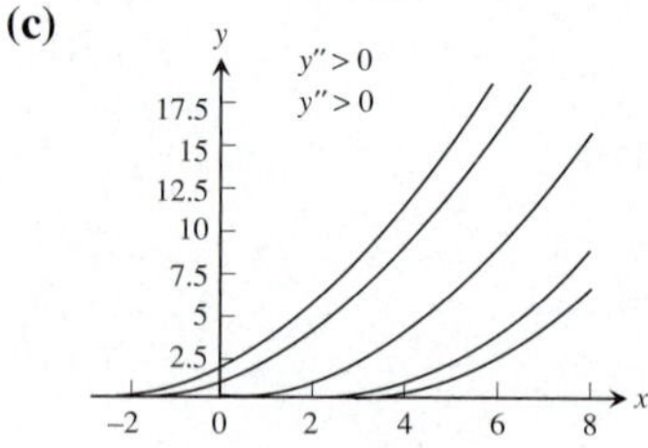

7. $y' = (y-1)(y-2)(y-3)$
   (a) $y = 1$ and $y = 3$ are unstable equilibria and $y = 2$ is a stable equilibrium.

   (b) $y'' = (3y^2 - 12y + 11)(y-1)(y-2)(y-3) =$
   $(y-1)\left(y - \frac{6-\sqrt{3}}{3}\right)(y-2)\left(y - \frac{6+\sqrt{3}}{3}\right)(y-3)$

   (c)

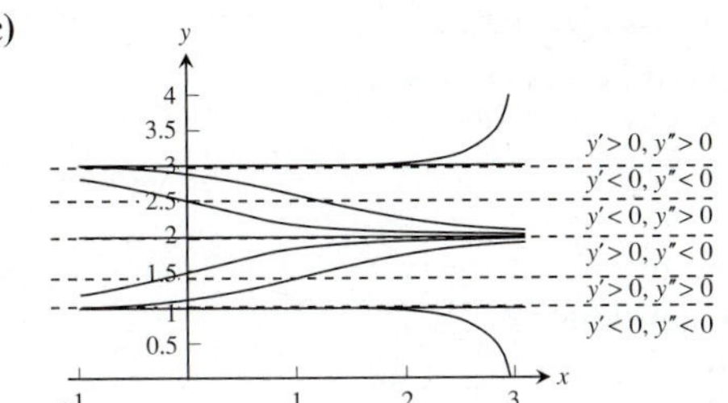

9. $\frac{dP}{dt} = 1 - 2P$ has a stable equilibrium at $P = \frac{1}{2}$; $\frac{d^2P}{dt^2} = -2\frac{dP}{dt} =$
$-2(1 - 2P)$

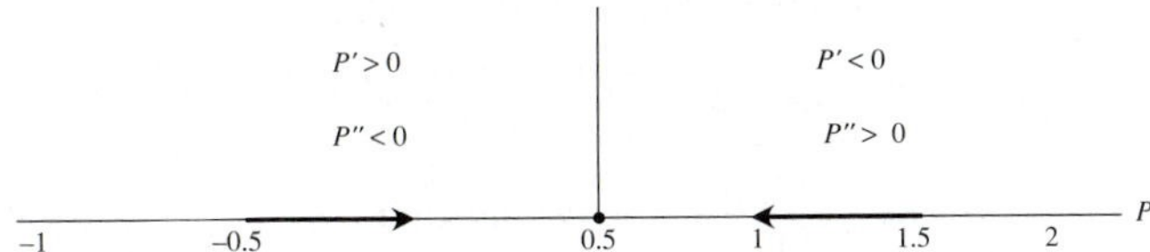

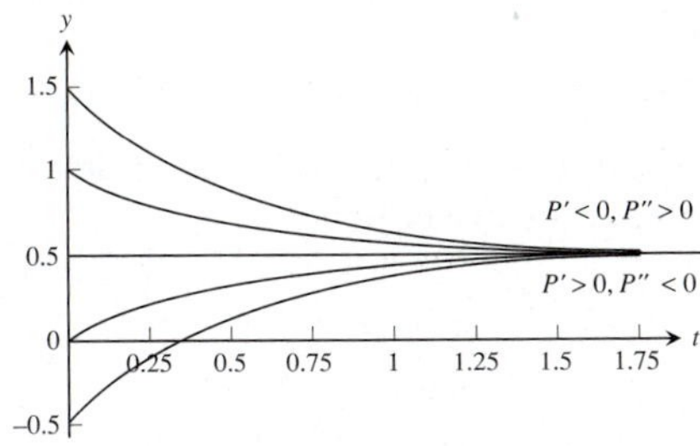

11. $\frac{dP}{dt} = 2P(P-3)$ has a stable equilibrium at $P = 0$ and an unstable equilibrium at $P = 3$. $\frac{d^2P}{dt^2} = 2(2P-3)\frac{dP}{dt} =$
$4P(2P-3)(P-3)$

13. Before the catastrophe, the population exhibits logistic growth and $P(t) \to M_0$, the stable equilibrium. After the catastrophe, the population declines logistically and $P(t) \to M_1$, the new stable equilibrium.

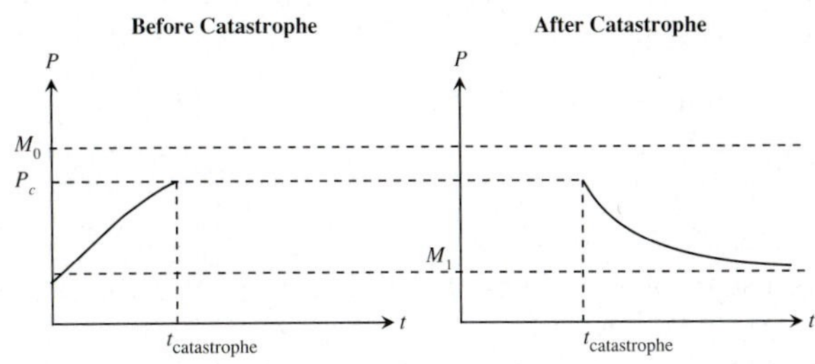

15. $\frac{dv}{dt} = g - \frac{k}{m}v^2$, $g, k, m > 0$ and $v(t) \geq 0$; equilibrium: $\frac{dv}{dt} =$
$g - \frac{k}{m}v^2 = 0 \Rightarrow v = \sqrt{\frac{mg}{k}}$; concavity: $\frac{d^2v}{dt^2} = -2\left(\frac{k}{m}v\right)\frac{dv}{dt} =$
$-2\left(\frac{k}{m}v\right)\left(g - \frac{k}{m}v^2\right)$

(a)

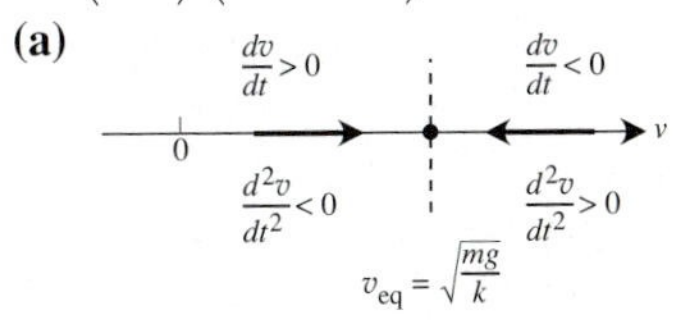

(b)

(c) $v_{\text{terminal}} = \sqrt{\frac{160}{0.005}} = 178.9$ ft/s $= 122$ mph

17. $F = F_p - F_r$; $ma = 50 - 5|v|$; $\frac{dv}{dt} = \frac{1}{m}(50 - 5|v|)$. The maximum velocity occurs when $\frac{dv}{dt} = 0$ or $v = 10$ ft/sec.

19. Phase line:

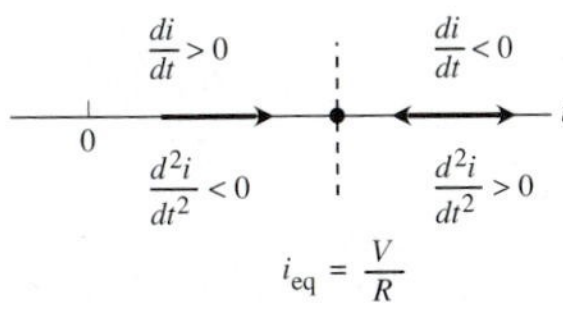

If the switch is closed at $t = 0$, then $i(0) = 0$, and the graph of the solution looks like this:

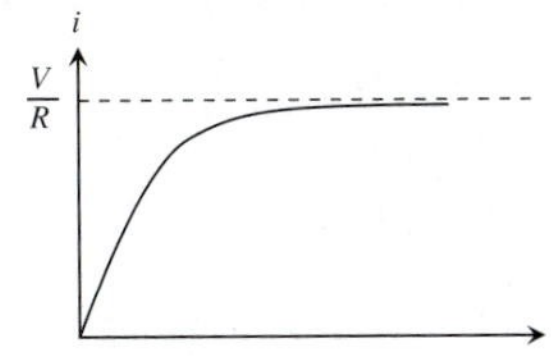

As $t \to \infty$, $i(t) \to i_{\text{steady state}} = V/R$.

Section 3.5, pp. 295–303

1. 16 in., 4 in. by 4 in.

3. **(a)** $(x, 1 - x)$
(b) $A(x) = 2x(1 - x)$
(c) 1/2 square units, $\frac{1}{2}$ in. $\times$ 1 in.

5. $\frac{2450}{27}$ in.^{3}, $\frac{14}{3} \times \frac{35}{3} \times \frac{5}{3}$ in.

7. 80,000 m^2, 400 m by 200 m

9. **(a)** The optimum dimensions of the tank are 10 ft on the base edges and 5 ft deep.
(b) Minimizing the surface area of the tank minimizes its weight for a given wall thickness. The thickness of the steel walls would likely be determined by other considerations such as structural requirements.

11. 9×18 in.

13. $\frac{\pi}{2}$

15. $h : r = 8 : \pi$

17. **(a)** $v(x) = 2x(24 - 2x)(18 - 2x)$
(b) Domain: $(0, 9)$

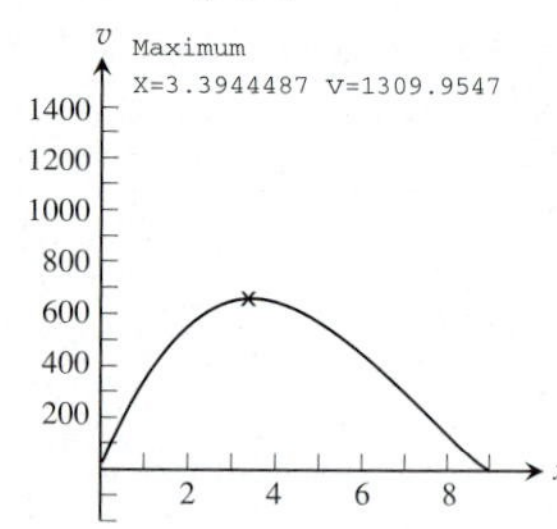

(c) Maximum volume ≈ 1309.95 in.^{3} when $x \approx 3.39$ in.
(d) $v'(x) = 24x^2 - 336x + 864$, so the critical point is at $x = 7 - \sqrt{13}$, which confirms the result in part (c).
(e) $x = 2$ in. or $x = 5$ in.

19. $r = \frac{10\sqrt{6}}{3}$, $h = \frac{20\sqrt{3}}{3}$, ≈ 2418.40 cm^3

21. **(a)** $h = 24$, $w = 18$
(b)

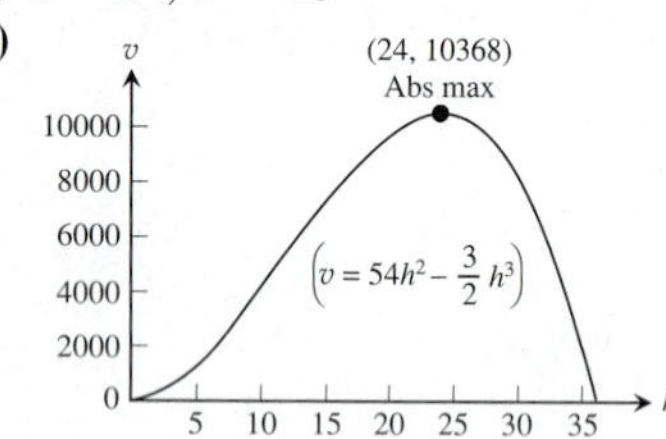

23. If r is the radius of the hemisphere, h the height of the cylinder, and V the volume, then $r = \left(\frac{3V}{8\pi}\right)^{1/3}$ and $h = \left(\frac{3V}{\pi}\right)^{1/3}$.

25. **(b)** $x = \frac{51}{8}$ **(c)** $L \approx 11$ in.

27. Radius $= \sqrt{2}$ m, height = 1 m, volume $\frac{2\pi}{3}$ m^3

29. $\frac{1}{\sqrt{2}}$ units long by $\frac{1}{\sqrt{e}}$ units high, $A = \frac{1}{\sqrt{2e}} \approx 0.43$ units^2

31. **(a)** $v(0) = 96$ ft/sec
(b) 256 ft at $t = 3$ sec
(c) Velocity when $s = 0$ is $v(7) = -128$ ft/sec

33. ≈ 46.87 ft

35. 6 in. $\times$ $6\sqrt{3}$ in.

37. **(a)** $\frac{\pi}{3}, \frac{4\pi}{3}$ **(b)** 1 **(c)** $t = \frac{\pi}{3}, \frac{4\pi}{3}$

39. **(a)** $10\pi \approx 31.42$ cm/sec; when $t = 0.5$ sec, 1.5 sec, 2.5 sec, 3.5 sec; $s = 0$ acceleration is 0
(b) 10 cm from rest position; speed is 0

41. **(a)** $s = ((12 - 12t)^2 + 64t^2)^{1/2}$
(b) -12 knots, 8 knots
(c) No
(d) $4\sqrt{13}$. This limit is the square root of the sum of the squares of the individual speeds.

43. $x = \frac{a}{2}$, $v = \frac{ka^2}{4}$

45. $\frac{c}{2} + 50$

47. **(a)** $\sqrt{\frac{2km}{h}}$ **(b)** $\sqrt{\frac{2km}{h}}$

51. **(a)** The artisan should order px units of material to have enough until the next delivery.
(c) Average cost per day $= \frac{(d + (ps/2)x^2)}{x} = \frac{d}{x} + \frac{ps}{2}x$; $x^* = \sqrt{\frac{2d}{ps}}$, $px^* = \sqrt{\frac{2pd}{s}}$ gives a minimum
(d) The line and hyperbola intersect when $\frac{d}{x} = \frac{ps}{2}x$. For $x > 0$, $x_{\text{intersection}} = \sqrt{\frac{2d}{ps}} = x^*$. From this result, the average cost per day is minimized when the average daily cost of delivery is equal to the average daily cost of storage.

53. $M = \frac{C}{2}$

61. **(a)** The minimum distance is $\sqrt{5}/2$.
(b) The minimum distance is from the point $(3/2, 0)$ to the point $(1, 1)$ on the graph of $y = \sqrt{x}$, which occurs at the value $x = 1$ where $D(x)$, the distance squared, has its minimum value.

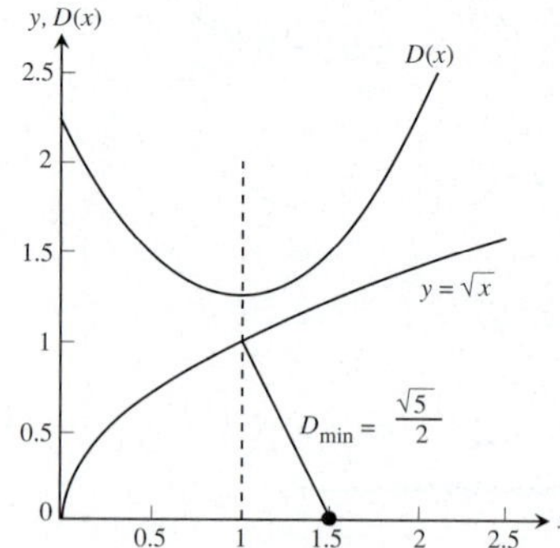

63. **(a)** $V(x) = \frac{\pi}{3}\left(\frac{2\pi a - x}{2\pi}\right)^2 \sqrt{a^2 - \left(\frac{2\pi a - x}{2\pi}\right)^2}$

(b) When $a = 4$: $r = \frac{4\sqrt{6}}{3}$, $h = \frac{4\sqrt{3}}{3}$; when $a = 5$: $r = \frac{5\sqrt{6}}{3}$, $h = \frac{5\sqrt{3}}{3}$; when $a = 6$: $r = 2\sqrt{6}$, $h = 2\sqrt{3}$; when $a = 8$: $r = \frac{8\sqrt{6}}{3}$, $h = \frac{8\sqrt{3}}{3}$

(c) Since $r = \frac{a\sqrt{6}}{3}$ and $h = \frac{a\sqrt{3}}{3}$, the relationship is $\frac{r}{h} = \sqrt{2}$.

Section 3.6, pp. 313–316

1. $10x - 13$ **3.** $L(x) = 2$
5. $L(x) = x - \pi$
7. $f(0) = 1$. Also, $f'(x) = k(1 + x)^{k-1}$, so $f'(0) = k$. This means that the linearization at $x = 0$ is $L(x) = 1 + kx$.
9. Center $= -1$, $L(x) = -5$
11. Center $= 1$, $L(x) = \frac{x}{4} + \frac{1}{4}$, or center $= 1.5$, $L(x) = \frac{4x}{25} + \frac{9}{25}$
13. **(a)** 1.01 **(b)** 1.003
15. $\left(3x^2 - \frac{3}{2\sqrt{x}}\right) dx$ **17.** **(a)** $(2x \ln x + x)\, dx$
19. $\frac{1 - y}{3\sqrt{y} + x}\, dx$ **21.** **(a)** $dy = (\cos x)e^{\sin x}\, dx$
23. $(1 + x)e^x\, dx$
25. **(a)** 0.21 **(b)** 0.2 **(c)** 0.01
27. **(a)** $-2/11$ **(b)** $-1/5$ **(c)** $1/55$
29. $4\pi a^2\, dr$ **31.** $3a^2\, dx$
33. **(a)** $0.08\pi\ \text{m}^2$ **(b)** 2%
35. $dV \approx 565.5\ \text{in.}^3$
37. $\frac{1}{3}\%$ **39.** 0.05%
41. The ratio equals 37.87, so a change in the acceleration of gravity on the moon has about 38 times the effect that a change of the same magnitude has on Earth.
47. **(a)** $L(x) = 1 + (\ln 2)x \approx 0.69x + 1$

Section 3.7, pp. 323–325

1. $x_2 = -\frac{5}{3}, \frac{13}{21}$ **3.** $x_2 = -\frac{51}{31}, \frac{5763}{4945}$
7. x, and later approximations will equal x.
9.

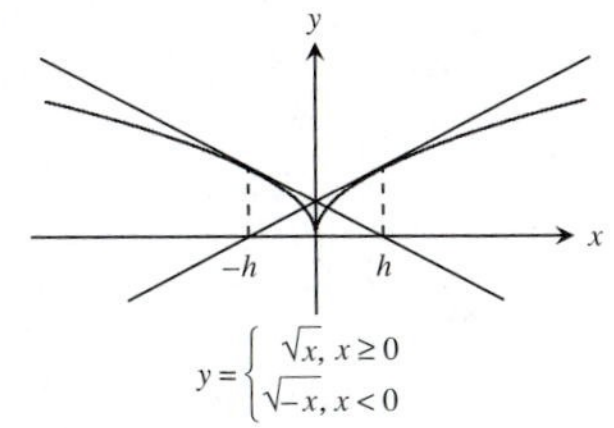

11. The points of intersection of $y = x^3$ and $y = 3x + 1$ or $y = x^3 - 3x$ and $y = 1$ have the same x-values as the roots of part (i) or the solutions of part (iv).
15. 1.165561185
17. **(a)** Two
(b) 0.35003501505249 and -1.0261731615301
19. ± 1.3065629648764, ± 0.5411961001462
21. 0, 0.53485
23. The root is 1.17951.
25. **(a)** For $x_0 = -2$ or $x_0 = -0.8$, $x_i \to -1$ as i gets large.
(b) For $x_0 = -0.5$ or $x_0 = 0.25$, $x_i \to 0$ as i gets large.
(c) For $x_0 = 0.8$ or $x_0 = 2$, $x_i \to 1$ as i gets large.
(d) For $x_0 = -\sqrt{21}/7$ or $x_0 = \sqrt{21}/7$, Newton's method does not converge. The values of x_i alternate between $-\sqrt{21}/7$ and $\sqrt{21}/7$ as i increases.
27. Answers will vary with machine speed.
29. 2.45, 0.000245

Chapter 3 Practice Exercises, pp. 326–329

1. Global minimum value of $\frac{1}{2}$ at $x = 2$
3. **(a)** $[-3, -2]$ and $[1, 2]$
(b) $[-2, 0)$ and $(0, 1]$
(c) Local maxima at $x = -2$ and $x = 2$; local minima at $x = -3$ and $x = 1$ (provided f is continuous at $x = 0$).
5. No
7. No minimum, absolute maximum: $f(1) = 16$, critical points: $x = 1$ and $\frac{11}{3}$
9. Yes **11.** No
13. **(b)** One **15.** **(b)** 0.8555996772
21.

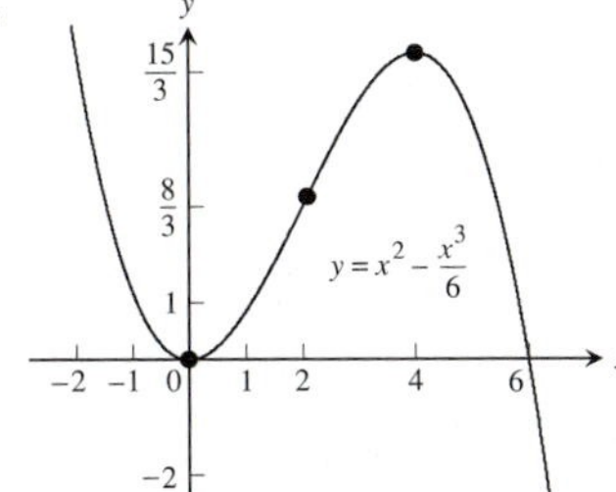

23.

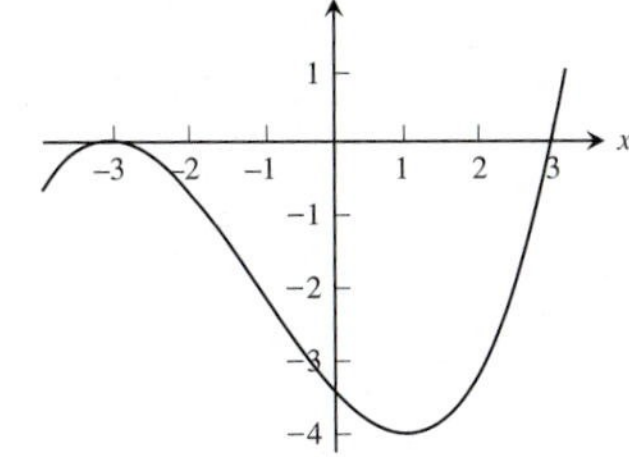

25.

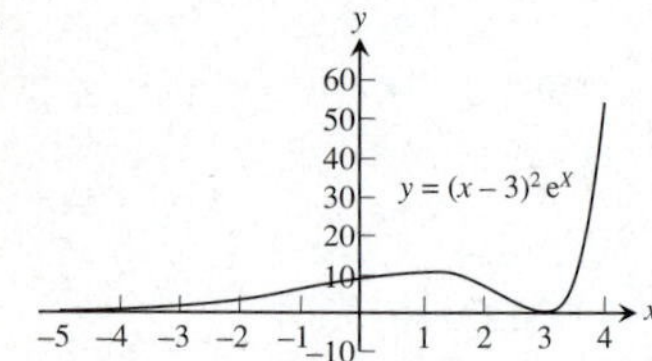

27. **(a)** Local maximum at $x = 4$, local minimum at $x = -4$, inflection point at $x = 0$

(b)

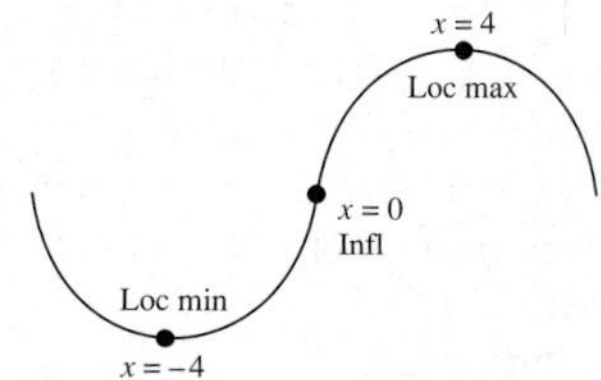

29. $x = \pm 1$ are the critical points; $y = 1$ is a horizontal asymptote in both directions; absolute minimum value of the function is $e^{-\sqrt{2}/2}$ at $x = -1$, and absolute maximum value is $e^{\sqrt{2}/2}$ at $x = 1$.

31. **(a)** $t = 0, 6, 12$
(b) $t = 3, 9$
(c) $6 < t < 12$
(d) $0 < t < 6,\ 12 < t < 14$

33. **(a)** $v(t) = -3t^2 - 6t + 4$
(b) $a(t) = -6t - 6$
(c) The particle starts at position 3 moving in the positive direction, but decelerating. At approximately $t = 0.528$, it reaches position 4.128 and changes direction, beginning to move in the negative direction. After that, it continues to accelerate while moving in the negative direction.

35. $f(x) = -\frac{1}{4}x^{-4} - e^{-x} + C$

37. $f(x) = -\frac{2}{x} + \frac{1}{3}x^3 + x + C$ for $x > 0$

39. $s(t) = 4.9t^2 + 5t + 10$

41. **(a)** $y = -1$ is stable and $y = 1$ is unstable.
(b) $\frac{d^2y}{dx^2} = 2y\frac{dy}{dx} = 2y(y^2 - 1)$

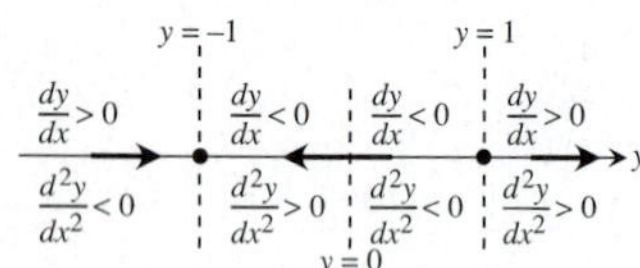

(c)

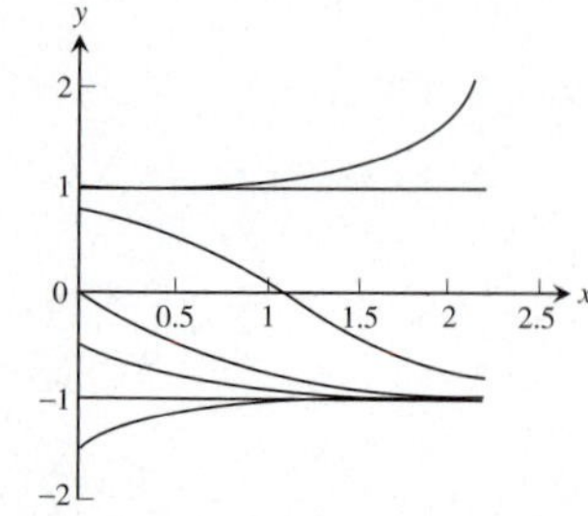

43. $r = 25$ ft and $s = 50$ ft

45. Height = 2, radius = $\sqrt{2}$

47. $x = 5 - \sqrt{5}$ hundred ≈ 276 tires, $y = 2(5 - \sqrt{5})$ hundred $\approx$ 553 tires

49. Dimensions: base is 6 in. by 12 in., height = 2 in.; maximum volume = 144 in.^{3}

51. **(a)** $L(x) = 2x + \frac{\pi - 2}{2}$

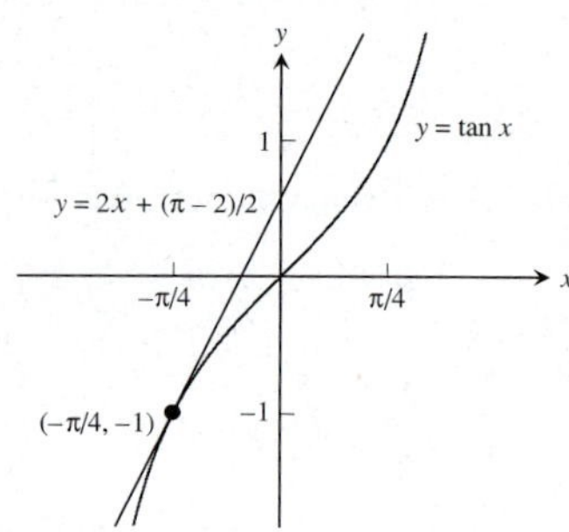

(b) $L(x) = -\sqrt{2}x + \frac{\sqrt{2}(4 - \pi)}{4}$

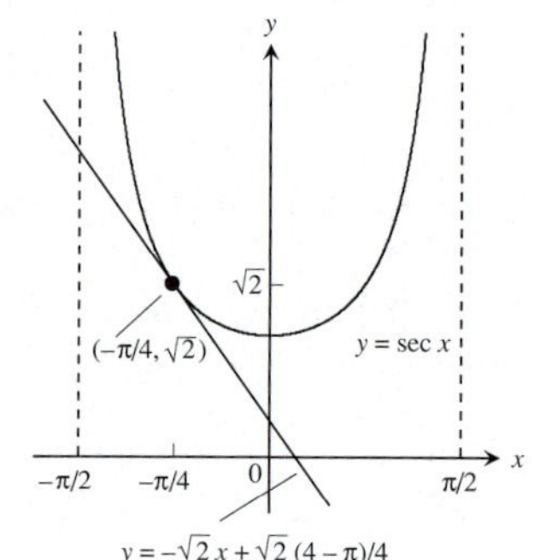

53. $L(x) = 2.5x - 0.1$

55. $dV = \frac{2}{3}\pi r_0 h\, dr$

57. **(a)** 4% **(b)** 8% **(c)** 12%

59. $x_5 = 1.732051$

Chapter 3 Additional Exercises, pp. 330–332

1. If M and m are the maximum and minimum values, respectively, then $m \le f(x) \le M$ for all $x \in I$. If $m = M$, then f is constant on I.

3. The extreme points will not be at the open end of an open interval.

5. **(a)** A local minimum at $x = -1$, points of inflection at $x = 0$ and $x = 2$
(b) A local maximum at $x = 0$ and local minima at $x = -1$ and $x = 2$, points of inflection at $x = \frac{1 \pm \sqrt{7}}{3}$

11. $a = 1, b = 0, c = 1$

13. Yes

15. Drill the hole at $y = \frac{h}{2}$

17. $r = \frac{RH}{2(H - R)}$ for $H > 2R$, $r = R$ if $H \le 2R$

21. **(a)** 0.8156 ft
(b) 0.00613 sec
(c) It will lose about 8.83 min/day.

25. **(b)** 61°

CHAPTER 4 INTEGRATION

Section 4.1, pp. 340–343

1. **(a)** $3x^2$ **(b)** $\frac{x^8}{8}$ **(c)** $\frac{x^8}{8} - 3x^2 + 8x$

3. **(a)** $\frac{1}{x^2}$ **(b)** $-\frac{1}{4x^2}$ **(c)** $\frac{x^4}{4} + \frac{1}{2x^2}$

5. **(a)** $x^{2/3}$ **(b)** $x^{1/3}$ **(c)** $x^{-1/3}$

7. **(a)** $\tan x$ **(b)** $2\tan\left(\frac{x}{3}\right)$ **(c)** $-\frac{2}{3}\tan\left(\frac{3x}{2}\right)$

9. $\frac{x^2}{2} + x + C$ **11.** $\ln|x| - 5\tan^{-1}x + C$

13. $-e^{-x} + \frac{4^x}{\ln 4} + C$ **15.** $2\sin^{-1}y - \frac{4}{3}y^{3/4} + C$

17. $\frac{1}{7}y + \frac{4}{y^{1/4}} + C$ **19.** $2\sqrt{t} - \frac{2}{\sqrt{t}} + C$

21. $-21\cos\frac{\theta}{3} + C$ **23.** $\tan\theta + C$

25. $-\cos\theta + \theta + C$

31. **(a)** Wrong: $\frac{d}{dx}\left(\frac{x^2}{2}\sin x + C\right) = \frac{2x}{2}\sin x + \frac{x^2}{2}\cos x = x\sin x + \frac{x^2}{2}\cos x$

(b) Wrong: $\frac{d}{dx}(-x\cos x + C) = -\cos x + x\sin x$

(c) Right: $\frac{d}{dx}(-x\cos x + \sin x + C) = -\cos x + x\sin x + \cos x = x\sin x$

33. **(b)** **35.** $y = x^2 - 7x + 10$

37. $y = 9x^{1/3} + 4$ **39.** $s = \sin t - \cos t$

41. $v = 3\sec^{-1}t - \pi$ **43.** $y = x^2 - x^3 + 4x + 1$

45. $y = x^3 - 4x^2 + 5$ **47.** $s = 4.9t^2 + 5t + 10$

49. $s = 16t^2 + 20t + 5$ **51.** $y = 2x^{3/2} - 50$

53. $y = x - x^{4/3} + \frac{1}{2}$ **55.** $y = -\sin x - \cos x - 2$

57. 48 m/sec **59.** $t = \frac{88}{k}, k = 16$

61. **(a)** $v = 10t^{3/2} - 6t^{1/2}$
(b) $s = 4t^{5/2} - 4t^{3/2}$

65. **(a)** 1: 33.2 units; 2: 33.2 units; 3: 33.2 units
(b) True

Section 4.2, pp. 349–351

1. $-\frac{1}{4}\cos 2x^2 + C$ **3.** $-(7x - 2)^{-4} + C$

5. $-6(1 - r^3)^{1/2} + C$

7. $\frac{1}{3}(x^{3/2} - 1) - \frac{1}{6}\sin(2x^{3/2} - 2) + C$

9. **(a)** $-\frac{1}{4}(\cot^2 2\theta) + C$ **(b)** $-\frac{1}{4}(\csc^2 2\theta) + C$

11. $-\frac{1}{3}(3 - 2s)^{3/2} + C$ **13.** $\frac{3}{2 - x} + C$

15. $-\frac{1}{3}(7 - 3y^2)^{3/2} + C$ **17.** $4e^{\sqrt{x}} + C$

19. $\frac{1}{3}\tan(3x + 2) + C$ **21.** $-e^{1/t} + C$

23. $-\frac{2}{3}\cos(x^{3/2} + 1) + C$ **25.** $\frac{1}{2\cos(2t + 1)} + C$

27. $-\frac{2}{3}(\cot^3 y)^{1/2} + C$ **29.** $\frac{1}{2}\ln|2x - 1| + C$

31. $\ln|2 - \cos t| + C$ **33.** $\frac{1}{2}\sin^{-1}(2x) + C$

35. $\frac{1}{\sqrt{2}}\sec^{-1}\left|\frac{5x}{\sqrt{2}}\right| + C$ **37.** $\frac{1}{3}\tan^{-1}(3x + 1) + C$

39. $\tan^{-1}(e^x) + C$

41. **(a)** $-\frac{6}{2 + \tan^3 x} + C$ **(b)** $-\frac{6}{2 + \tan^3 x} + C$

(c) $-\frac{6}{2 + \tan^2 x} + C$

43. $\frac{1}{6}\sin\sqrt{3(2r - 1)^2 + 6} + C$ **45.** $s = \frac{1}{2}(3t^2 - 1)^4 - 5$

47. $s = 4t - 2\sin\left(2t + \frac{\pi}{6}\right) + 9$ **49.** $y = 1 - \cos(e^t - 2)$

51. $s = \sin\left(2t - \frac{\pi}{2}\right) + 100t + 1$

53. $y = 2(e^{-x} + x) - 1$ **55.** $y = \sin^{-1}(x)$

57. $y = \sec^{-1}(x) + \frac{2\pi}{3}, x > 1$ **59.** 6 m

63. **(a)** $\sin^{-1}\left(\frac{x}{3}\right) + C$

(b) $\frac{\sqrt{3}}{3}\tan^{-1}\left(\frac{\sqrt{3}x}{3}\right) + C$

Section 4.3, pp. 359–363

1. ≈ 44.8, 6.7 L/m

3. **(a)** 87 in. **(b)** 87 in.

5. **(a)** 3490 ft **(b)** 3840 ft

7. **(a)** 80π **(b)** 6%

9. **(a)** $\frac{93\pi}{2}$, overestimate **(b)** 9%

11. **(a)** 118.5π or ≈ 372.28 m^3 **(b)** Error $\approx$ 11%

13. **(a)** 10π, underestimate **(b)** 20%

15. **(a)** 74.65 ft/sec **(b)** 45.28 ft/sec **(c)** 146.59 ft

17. $\frac{31}{16}$ **19.** 1

21. **(a)** Upper = 758 gal, lower = 543 gal
(b) Upper = 2363 gal, lower = 1693 gal
(c) $\approx$ 31.4 h, $\approx$ 32.4 h

23. **(a)** 2 **(b)** $2\sqrt{2} \approx 2.828$

(c) $8\sin\left(\frac{\pi}{8}\right) \approx 3.061$

(d) Each area is less than the area of the circle, π. As n increases, the polygon area approaches π.

Section 4.4, pp. 372–374

1. $\frac{6(1)}{1+1} + \frac{6(2)}{2+1} = 7$

3. $\cos(1\pi) + \cos(2\pi) + \cos(3\pi) + \cos(4\pi) = 0$

5. $\sin \pi - \sin \frac{\pi}{2} + \sin \frac{\pi}{3} = \frac{\sqrt{3} - 2}{2}$

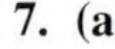

7. (a) **(b)**

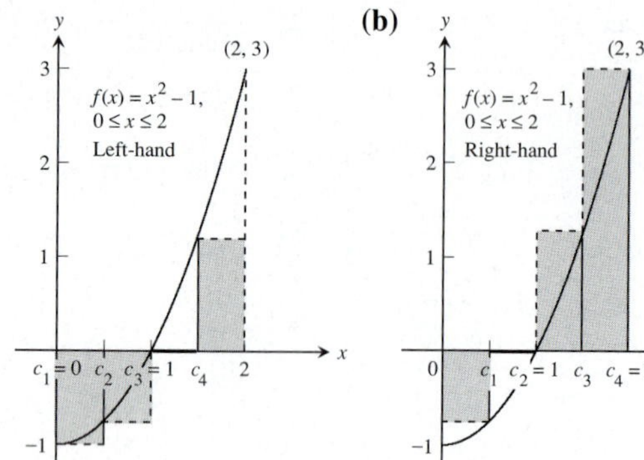

(c)

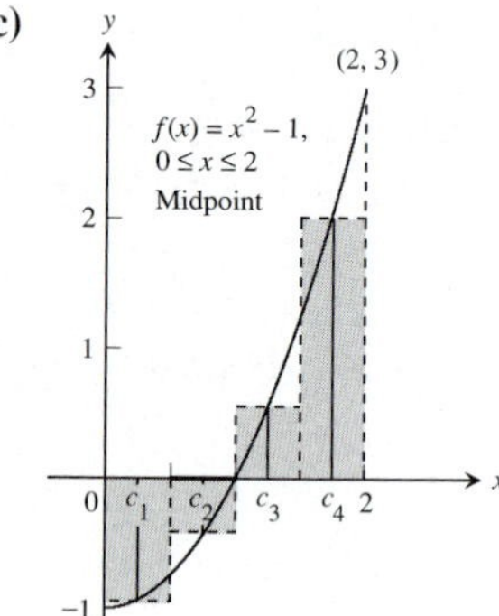

9. (a)

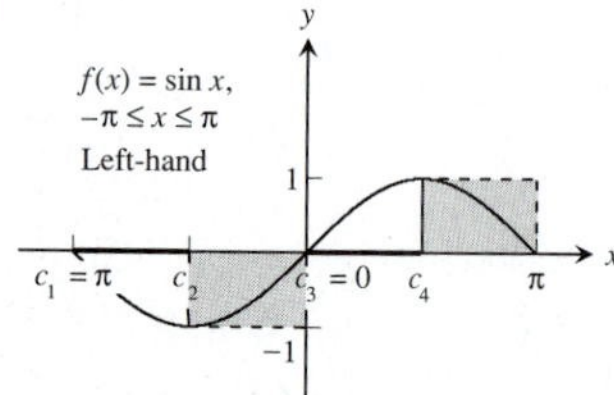

(b)

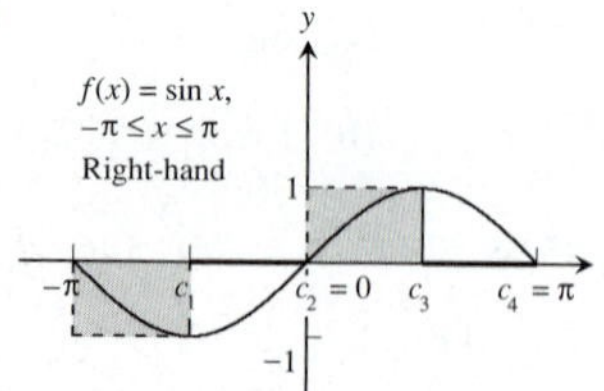

(c)

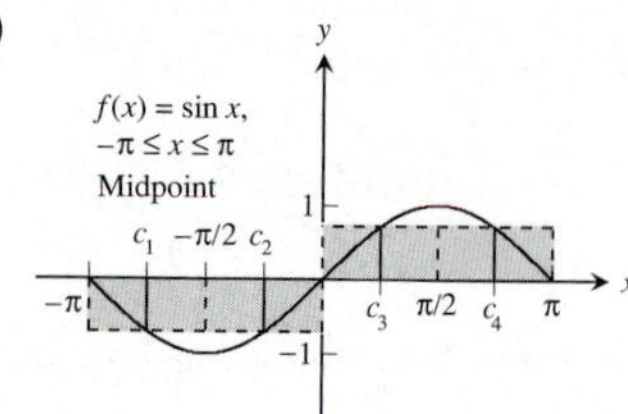

11. $\int_0^2 x^2\, dx$

13. $\int_{-7}^{5} (x^2 - 3x)\, dx$

15. $\int_0^1 \sqrt{4 - x^2}\, dx$

17. Area = 21 square units

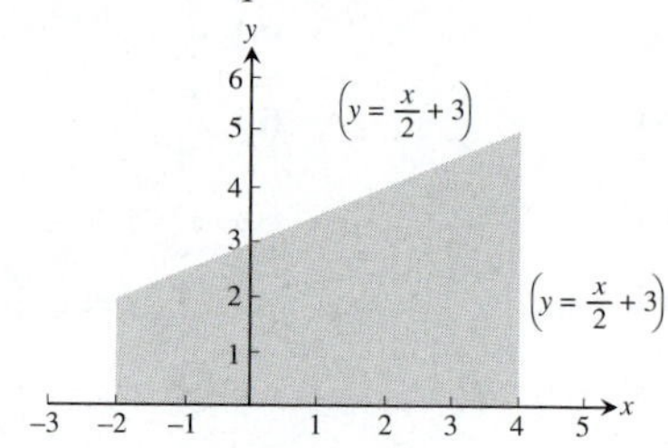

19. Area = 2.5 square units

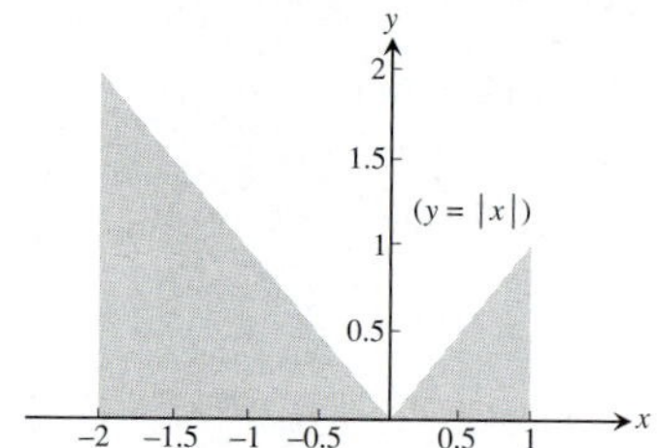

21. Area $= \frac{b^2}{2}$ square units

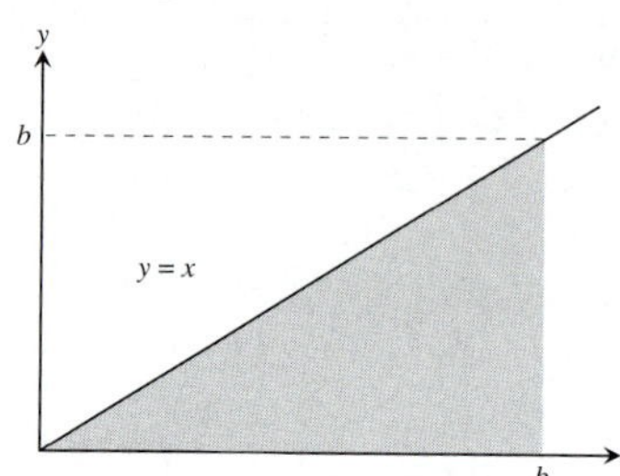

23. $av(f) = 1/2$

25. $av(f) = \pi/4$

27. (a) 0 **(b)** -8 **(c)** -12
(d) 10 **(e)** -2 **(f)** 16

29. (a) 5 **(b)** $5\sqrt{3}$
(c) -5 **(d)** -5

31. (a) 4 **(b)** -4

33. $a = 0$ and $b = 1$ maximize the integral.

37. Upper bound = 1, lower bound = 1/2

39. 37.5 mph

Section 4.5, pp. 383–386

1. 6

3. 1

5. π

7. 0

9. $\frac{2\pi^3}{3}$

11. $\frac{16\sqrt{2} - 17}{48}$

13. $\frac{e - 1}{2}$

15. $(\cos \sqrt{x})\left(\frac{1}{2\sqrt{x}}\right)$

17. $4t^5$

19. $\sqrt{1 + x^2}$

21. $-\frac{1}{2}x^{-1/2}\sin x$

23. $\frac{e^{x+1}}{3x^{2/3}}$

25. 0

27. $\frac{\pi}{2} + \sin 2$

29. $y = \int_2^x \sec t\, dt + 3$

31. $y = 1 - \cos(e^t - 2)$

33. $\frac{28}{3}$

35. 8

37. π

39. **(a)** \$9.00 **(b)** \$10.00

41. **(a)** $v = \frac{ds}{dt} = \frac{d}{dt}\int_0^t f(x)\,dx = f(t) \Rightarrow v(5) = f(5) = 2$ m/sec
(b) $a = \frac{df}{dt}$ is negative since the slope of the tangent line at $t = 5$ is negative.
(c) $s = \int_0^3 f(x)\,dx = \frac{1}{2}(3)(3) = \frac{9}{2}$ m since the integral is the area of the triangle formed by $y = f(x)$, the x-axis, and $x = 3$.
(d) $t = 6$ since after $t = 6$ to $t = 9$, the region lies below the x-axis.
(e) At $t = 4$ and $t = 7$, since there are horizontal tangents there.
(f) Toward the origin between $t = 6$ and $t = 9$ since the velocity is negative on this interval. Away from the origin between $t = 0$ and $t = 6$ since the velocity is positive there.
(g) Right or positive side, because the integral of f from 0 to 9 is positive, there being more area above the x-axis than below.

43. $\int_4^8 \pi(64 - x^2)\,dx = \frac{320\pi}{3}$

45. $2x - 2$

47. $-3x + 5$

49. **(a)** True; since f is continuous, g is differentiable by Part 1 of the Fundamental Theorem of Calculus.
(b) True: g is continuous because it is differentiable.
(c) True, since $g'(1) = f(1) = 0$.
(d) False, since $g''(1) = f'(1) > 0$.
(e) True, since $g'(1) = 0$ and $g''(1) = f'(1) > 0$.
(f) False: $g''(x) = f'(x) > 0$, so g'' never changes sign.
(g) True, since $g'(1) = f(1) = 0$ and $g'(x) = f(x)$ is an increasing function of x (because $f'(x) > 0$).

51. **(a)** $\frac{125}{6}$ **(b)** $h = \frac{25}{4}$ **(d)** $\frac{2}{3}bh$

Section 4.6, pp. 393–394

1. **(a)** $\frac{14}{3}$ **(b)** $\frac{2}{3}$

3. **(a)** 2 **(b)** 2

5. **(a)** 0 **(b)** $\frac{1}{8}$

7. **(a)** 0 **(b)** 0

9. $\frac{1}{6}$

11. $\frac{1}{5}$

13. $e - 1$

15. 1

17. $y(t) = \frac{1}{\pi}(3 - \tan(\pi e^{-t}))$

19. $\frac{\pi}{2}$

21. $\frac{128}{15}$

23. $\frac{38}{3}$

25. **(a)** 6 **(b)** $7\frac{1}{3}$

27. **(a)** 0 **(b)** $\frac{8}{3}$

29. $\frac{32}{3}$

31. $\frac{8}{3}$

33. 8

35. 4

37. $\frac{4 - \pi}{\pi}$

39. 1

41. $\frac{32}{3}$

Section 4.7, pp. 403–407

1. **I: (a)** 1.5, 0 **(b)** 1.5, 0 **(c)** 0%
II: (a) 1.5, 0 **(b)** 1.5, 0 **(c)** 0%

3. **I: (a)** 2.75, 0.08 **(b)** 2.67, 0.08 **(c)** $0.0312 \approx 3\%$
II: (a) 2.67, 0 **(b)** 2.67, 0 **(c)** 0%

5. **I: (a)** 6.25, 0.5 **(b)** 6, 0.25 **(c)** $0.0417 \approx 4\%$
II: (a) 6, 0 **(b)** 6, 0 **(c)** 0%

7. **I: (a)** 0.509, 0.03125
(b) 0.5, 0.009
(c) $0.018 \approx 2\%$
II: (a) 0.5004, 0.002604
(b) 0.5, 0.0004
(c) 0.08%

9. **I: (a)** 1.8961, 0.161
(b) 2, 0.1039
(c) $0.052 \approx 5\%$
II: (a) 2.00456, 0.0066
(b) 2, 0.0046
(c) 0.23%

11. **(a)** 0.31929
(b) 0.32812
(c) $\frac{1}{3}$, 0.01404, 0.00521

13. **(a)** 1.95643
(b) 2.00421
(c) 2, 0.04357, −0.00421

15. 15,990 ft^3

17. 1.032 mi or 5443.5 ft

19. ≈ 10.63 ft

21. 4, 4

23. **(a)** $S_{10} = 0.8427$
(b) $|E_s| \le \frac{1 - 0}{180}(0.1)^4 12 \approx 6.7 \times 10^{-6}$

25. **(a)** $f''(x) = 2\cos(x^2) - 4x^2\sin(x^2)$
(b)

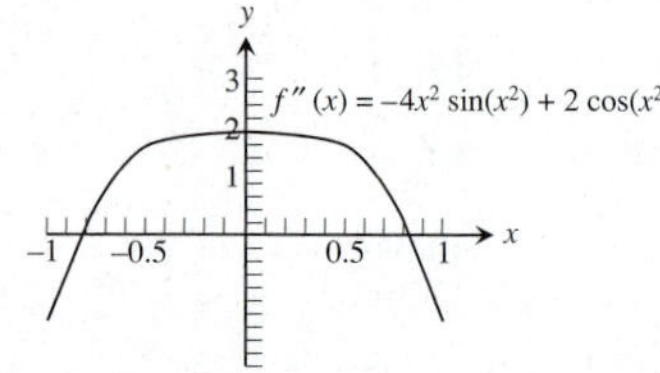

(c) The graph shows that $-3 \le f''(x) \le 2$ for $-1 \le x \le 1$.
(d) $|E_T| \le \dfrac{1-(-1)}{12}(h^2)(3) = \dfrac{h^2}{2}$
(e) $|E_T| \le \dfrac{h^2}{2} \le \dfrac{0.1^2}{2} < 0.01$
(f) $n \ge 20$

27. $S_{50} = 3.1379$, $S_{100} = 3.14029$

29. 1.37076

31. **(a)** $T_{10} \approx 1.983523538$; $T_{100} \approx 1.999835504$, $T_{1000} \approx 1.999998355$
(b)

n	$\lvert E_T \rvert = 2 - T_n$
10	$0.016476462 = 1.6476462 \times 10^{-2}$
100	1.64496×10^{-4}
1000	1.645×10^{-6}

(c) $|E_{T_{10n}}| \approx 10^{-2}|E_{T_n}|$
(d) $b - a = \pi;\ h^2 = \dfrac{\pi^2}{n^2};\ M = 1$

$$|E_{T_n}| \le \frac{\pi}{12}\left(\frac{\pi^2}{n^2}\right) = \frac{\pi^3}{12n^2}$$

$$|E_{T_{10n}}| \le \frac{\pi^3}{12(10n)^2} = 10^{-2}|E_{T_n}|$$

23. $\displaystyle\int_1^5 (2x-1)^{-1/2}\,dx = 2$

25. $\displaystyle\int_{-\pi}^0 \cos\frac{x}{2}\,dx = 2$

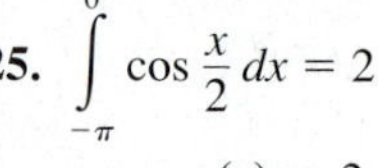

27. **(a)** 4 **(b)** 2 **(c)** -2 **(d)** -2π **(e)** 8/5

29. 16

31. $15/16 + \ln 2$

33. 1

35. 8

37. 1/6

39. $2\sqrt{2} - 2$

41. 9/14

43. $\pi/\sqrt{3}$

45. -1

47. $\pi/12$

49. 1

51. $\sqrt{3}\pi/4$

53. 8/3

55. 62

57. 1

59. 1/6

61. $\dfrac{\pi^2}{32} + \dfrac{\sqrt{2}}{2} - 1$

63. 4

65. Min: -4; max: 0; area: 27/4

67. $y = x - \dfrac{1}{x} - 1$

69. $r = 4t^{5/2} + 4t^{3/2} - 8t$

73. $y = \displaystyle\int_5^x \left(\frac{\sin t}{t}\right) dt - 3$

75. **(a)** b **(b)** b

79. $\sqrt{2 + \cos^3 x}$

81. $\dfrac{-6}{3 + x^4}$

83. $T = \pi$, $S = \pi$

85. 25°F

87. $\cos \approx \$12{,}518.10$ (Trapezoidal Rule); no

89. Yes

91. $y = \displaystyle\int_5^x \frac{\sin t}{t}\,dt + 3$

93. **(a)** 0 **(b)** -1
(c) $-\pi$ **(d)** $x = 1$
(e) $y = 2x + 2 - \pi$ **(f)** $x = -1, x = 2$
(g) $[-2\pi, 0]$

95. 600, \$18.00

97. 300, \$6.00

Chapter 4 Practice Exercises, pp. 408–412

1. $\dfrac{x^4}{4} + \dfrac{5}{2}x^2 - 7x + C$

3. $2t^{3/2} - \dfrac{4}{t} + C$

5. $-\dfrac{1}{2(r^2+5)} + C$

7. $-(2 - \theta^2)^{3/2} + C$

9. $-\cos e^x + C$

11. $\tan(e^x - 7) + C$

13. $-[\ln|\cos(\ln v)|] + C$

15. $\dfrac{1}{2}x - \sin\dfrac{x}{2} + C$

17. $-\cot(1 + \ln r) + C$

19. $\dfrac{1}{2\ln 3}(3^{x^2}) + C$

21. **(a)** About 680 ft
(b)

h (feet)
700
600
500
400
300
200
100
0
2
4
6
8
t (sec)

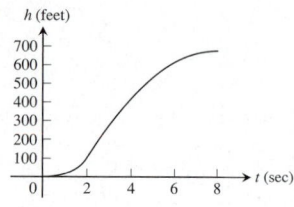

Chapter 4 Additional Exercises, pp. 412–414

1. **(a)** Yes **(b)** No

5. **(a)** 1/4 **(b)** $\sqrt[3]{12}$

7. $f(x) = \dfrac{x}{\sqrt{x^2+1}}$

9. $y = x^3 + 2x - 4$

11. 36/5

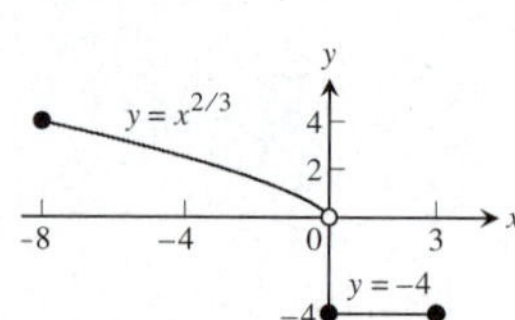

13. $\dfrac{1}{2} - \dfrac{2}{\pi}$

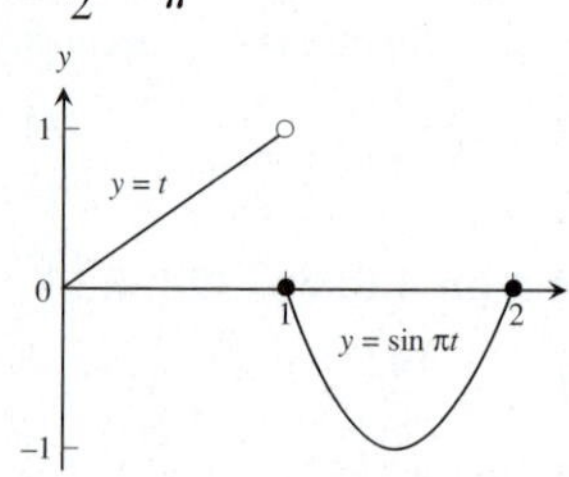

15. 13/3

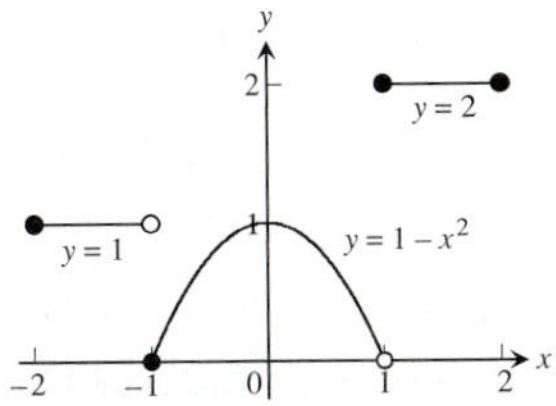

17. 1/2

19. $2/x$

21. $\dfrac{\sin 4y}{\sqrt{y}} - \dfrac{\sin y}{2\sqrt{y}}$

CHAPTER 5 APPLICATIONS OF INTEGRALS

Section 5.1, pp. 423–428

1. (a) $A(x) = \pi(1 - x^2)$ (b) $A(x) = 4(1 - x^2)$
   (c) $A(x) = 2(1 - x^2)$ (d) $A(x) = \sqrt{3}(1 - x^2)$
3. 16
5. (a) $\dfrac{\pi^2}{2}$ (b) 2π
7. (a) $2\sqrt{3}$ (b) 8
9. 8π
11. (a) s^2h (b) s^2h
13. $2\pi/3$
15. $4 - \pi$
17. $32\pi/5$
19. 36π
21. π
23. $\pi\left(\dfrac{\pi}{2} + 2\sqrt{2} - \dfrac{11}{3}\right)$
25. 2π
27. 2π
29. 3π
31. $\pi^2 - 2\pi$
33. $2\pi/3$
35. $117\pi/5$
37. $\pi(\pi - 2)$
39. $4\pi/3$
41. 8π
43. $7\pi/6$
45. (a) 8π (b) $32\pi/5$
    (c) $8\pi/3$ (d) $224\pi/15$
47. (a) $16\pi/15$ (b) $56\pi/15$ (c) $64\pi/15$
49. $V = 2a^2b\pi^2$
51. (a) $V = \dfrac{\pi h^2(3a - h)}{3}$ (b) $\dfrac{1}{120\pi}$ m/sec
55. $V = 3308$ cm^3
57. (a) $c = 2\pi$ (b) $c = 0$
    (c)

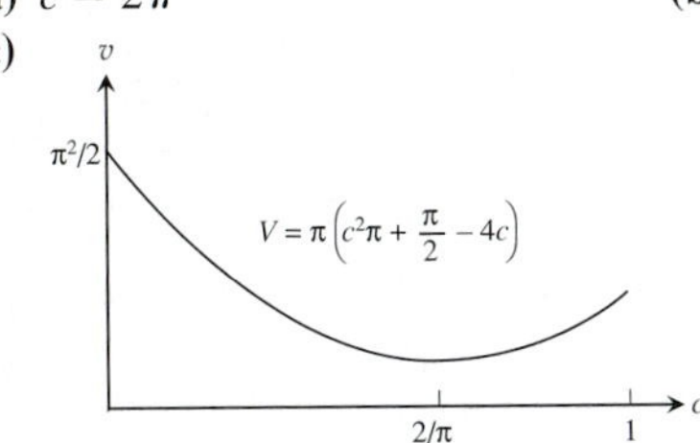

59. (a) 2.3, 1.6, 1.5, 2.1, 3.2, 4.8, 7.0, 9.3, 10.7, 10.7, 9.3, 6.4, 3.2
    (b) $\dfrac{1}{4\pi}\displaystyle\int_0^6 (C(y))^2\,dy$
    (c) ≈ 34.7 in.^{3}
    (d) $V \approx 34.75$ in.^{3} by Simpson's Rule. The Simpson's Rule estimate should be more accurate than the trapezoid estimate. The error in the Simpson's estimate is proportional to $h^4 = 0.0625$, whereas the error in the trapezoid estimate is proportional to $h^2 = 0.25$, a larger number when $h = 0.5$ in.

Section 5.2, pp. 433–435

1. 6π
3. 2π
5. $14\pi/3$
7. 8π
9. $5\pi/6$
11. $\pi\left(1 - \dfrac{1}{3}\right)$
13. (b) 4π
15. $\dfrac{16\pi}{15}(3\sqrt{2} + 5)$
17. $8\pi/3$
19. $4\pi/3$
21. $16\pi/3$
23. (a) $6\pi/5$ (b) $4\pi/5$
    (c) 2π (d) 2π
25. (a) About the x-axis: $V = 2\pi/15$, about the y-axis: $V = \pi/6$
    (b) About the x-axis: $V = 2\pi/15$, about the y-axis: $V = \pi/6$
27. (a) $5\pi/3$ (b) $4\pi/3$
    (c) 2π (d) $2\pi/3$
29. (a) $4\pi/15$ (b) $7\pi/30$
31. (a) $24\pi/5$ (b) $48\pi/5$
33. (a) $9\pi/16$ (b) $9\pi/16$

Section 5.3, pp. 441–443

1. 12
3. 53/6
5. 123/32
7. 99/8
9. 2
11. $2\pi a$
13. $e^3 + 2$
15. $\sqrt{2}(e^{\pi} - 1)$
17. Yes, $f(x) = \pm x + C$, where C is any real number, from (1, 1) to (4, 2)
19. (a) $y = \sqrt{x}$
    (b) Only one. We know the derivative of the function and the value of the function at one value of x.
21. (a) $\displaystyle\int_{-1}^{2} \sqrt{1 + 4x^2}\,dx$ (c) ≈ 6.13
23. (a) $\displaystyle\int_{0}^{\pi} \sqrt{1 + \cos^2 y}\,dy$ (c) ≈ 3.82
25. (a) $\displaystyle\int_{-1}^{3} \sqrt{1 + (y + 1)^2}\,dy$ (c) ≈ 9.29
27. (a) $\displaystyle\int_{0}^{\pi/6} \sec x\,dx$ (c) ≈ 0.55
29. 21.07 in.

Section 5.4, pp. 451–453

5. $\frac{2}{3}y^{3/2} - x^{1/2} = C$ **7.** $e^y - e^x = C$

9. $-x + 2\tan\sqrt{y} = C$ **11.** $e^{-y} + 2e^{\sqrt{x}} = C$

13. $y = \sin(x^2 + C)$

15. (a) ≈ -0.121
(b) ≈ 2.389 millibars
(c) ≈ 0.977 km

17. ≈ 585.35 kg **19.** ≈ 92.1 sec

21. (a) $\frac{dQ}{dt} = r - kQ$
(b) $Q = \frac{r}{k} + \left(Q_0 - \frac{r}{K}\right)e^{-kt}$
(c) $\frac{r}{k}$

23. 5°F

25. (a) $p(x) = 54.61e^{-0.01x}$ (in dollars)
(b) $p(10) = \$49.41$ and $p(90) = \$22.20$

27. $\frac{dy}{dt} = -ky;\ y = y_0e^{-kt}$ **29.** ≈ 864 years old

31. (a) 12,571 B.C. **(b)** 12,101 B.C. **(c)** 13,070 B.C.

33. (a) 550 ft **(b)** ≈ 77.28 sec

35.

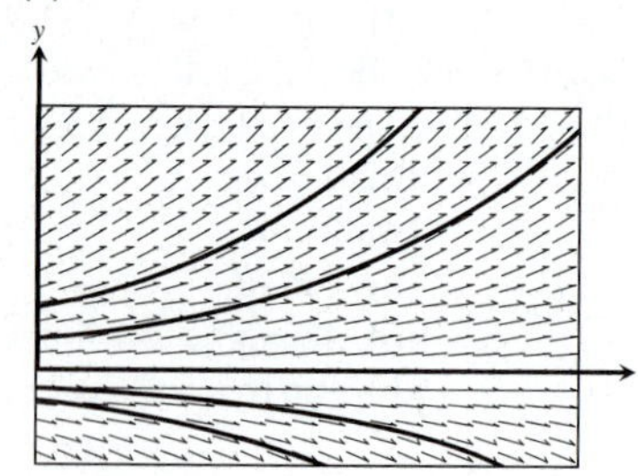

37.

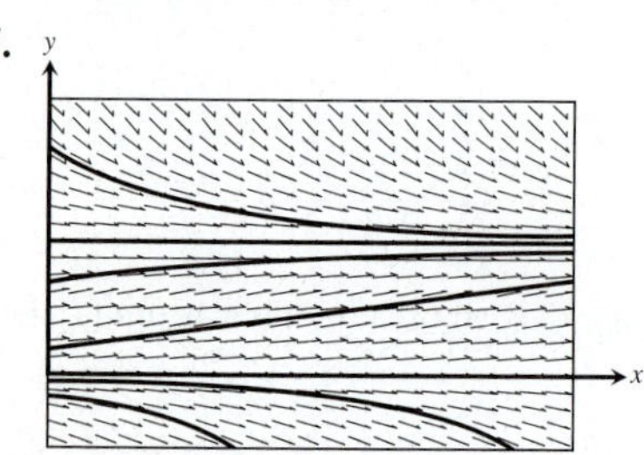

39.

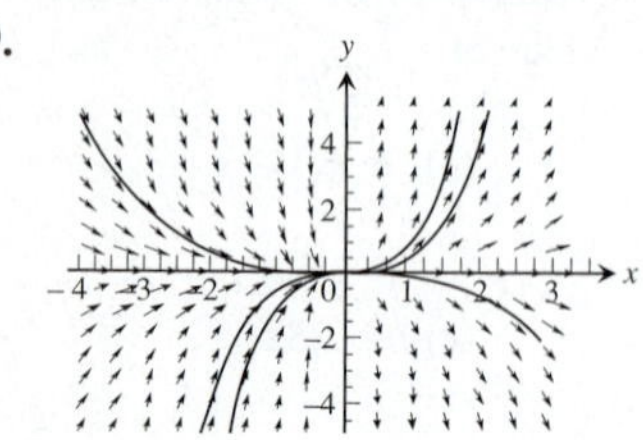

Section 5.5, pp. 460–464

1. 400 ft · lb **3.** 780 J

5. 72,900 ft · lb **9.** 400 N/m

11. 4 cm, 0.08 J

13. (a) 7238 lb/in. **(b)** 905 in. · lb, 2714 in. · lb

15. (a) 1,497,600 ft · lb **(b)** 1 h 40 min

(c) $W = \int_0^{10} 62.4(120y)\,dy = 374{,}000$ ft · lbs $\to t = W/250 =$ 1948 sec ≈ 25 min

(d) At 62.26 lb/ft^3: part (a) 1,494,240 ft · lb, part (b) 1 h 40 min; at 62.59 lb/ft^3: part (a) 1,502,160 ft · lb, part (b) 1 h 40 min

17. 38,484,510 J **19.** 7,238,229.48 ft · lb

21. (a) 34,583 ft · lb **(b)** 53,482 ft · lb

23. 15,073,100 J **27.** ≈ 85.1 ft · lb

29. ≈ 64.6 ft · lb **31.** ≈ 110.6 ft · lb

33. (a) $r(y) = 60 - \sqrt{50^2 - (y - 325)^2}$ for $325 \le y \le 375$ ft
(b) $\Delta V \approx \pi[60 - \sqrt{2500 - (y - 325)^2}\,\Delta y$
(c) $W = 6.3358 \cdot 10^7$ ft · lb

35. 91.32 in. · oz **37.** 5.144×10^{10} J

Section 5.6, pp. 469–471

3. 2808 lb

5. (a) 1164.8 lb **(b)** 1194.7 lb

7. 1309 lb **9.** 41.6 lb

11. (a) 93.33 lb **(b)** 3 ft

13. (a) $\left(\frac{x}{8}\right)^2 + \left(\frac{y}{14}\right)^2 = 1$ **(b)** 3008 tons

15. (a) $\frac{wb}{2}$

17. (a) 374.4 lb **(b)** 7.5 in. **(c)** No

19. 4.2 lb **21.** 1035 ft^3

Section 5.7, pp. 481–483

1. 4 ft **3.** $\left(\frac{L}{4}, \frac{L}{4}\right)$

5. $M_0 = 8, M = 8, \bar{x} = 1$

7. $M_0 = 15/2, M = 9/2, \bar{x} = 5/3$

9. $M_0 = 73/6, M = 5, \bar{x} = 73/30$

11. $M_0 = 3, M = 3, \bar{x} = 1$ **13.** $\bar{x} = 0, \bar{y} = 12/5$

15. $\bar{x} = 1, \bar{y} = -3/5$ **17.** $\bar{x} = 16/105, \bar{y} = 8/15$

19. $\bar{x} = 0, \bar{y} = \pi/8$ **21.** $\bar{x} = 1, \bar{y} = -2/5$

23. $\bar{x} = \bar{y} = \frac{2}{4 - \pi}$ **25.** $\bar{x} = 3/2, \bar{y} = 1/2$

27. (a) $224\pi/3$
(b) $\bar{x} = 2, \bar{y} = 0$
(c)

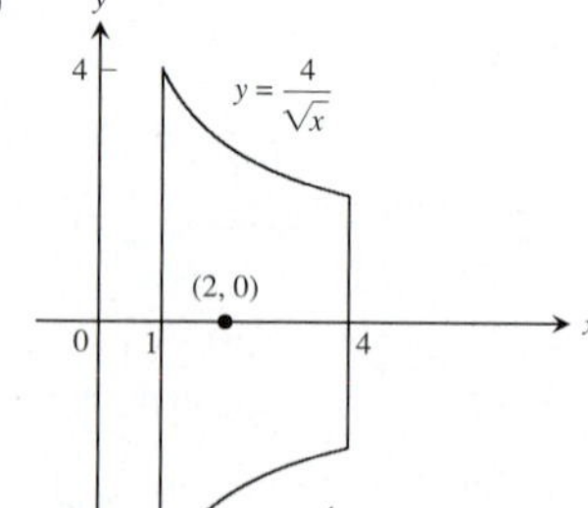

31. $\bar{x} = \bar{y} = 1/3$ **33.** $\bar{x} = a/3, \bar{y} = b/3$
35. $13\delta/6$ **37.** $\bar{x} = 0, \bar{y} = a\pi/4$

Chapter 5 Practice Exercises, pp. 484–486

1. $9\pi/280$ **3.** π^2
5. $72\pi/35$
7. (a) 2π **(b)** π **(c)** $\frac{12\pi}{5}$ **(d)** $26\pi/5$
9. (a) 8π **(b)** $1088\pi/15$ **(c)** $512\pi/15$
11. $\pi(2 - \ln 3)$
13. (a) $16\pi/15$ **(b)** $8\pi/5$ **(c)** $8\pi/3$ **(d)** $32\pi/5$
15. $28\pi/3$ **17.** $10/3$
19. $285/8$ **21.** 10
23. $f(x) = \frac{x^2 - 2\ln x + 3}{4}$ **25.** 4640 J
27. 10 ft · lb, 30 ft · lb **29.** 418,208.81 ft · lb
31. $22{,}500\pi$ ft · lb, 257 sec **33.** 332.8 lb
35. 2196.48 lb **37.** $y = \left(\frac{x^3}{6} + C\right)^2$
39. $y = Ce^{(\ln x)^2/2}$ **41.** $2\tan\sqrt{x} = t + C$
43. $\bar{x} = 0, \bar{y} = 8/5$ **45.** $\bar{x} = 3/2, \bar{y} = 12/5$
47. $\bar{x} = 9/5, \bar{y} = 11/10$

Chapter 5 Additional Exercises, pp. 487–488

1. $f(x) = \sqrt{\frac{2x + 1}{\pi}}$
3. $f(x) = \sqrt{C^2 - 1}x + a$, where $C \ge 1$
5. 30 feet
9. 108.3042 h; 251.4747 h **11.** 168.25°F
13. $\bar{x} = 0, \bar{y} = \frac{n}{2n + 1}, \left(0, \frac{1}{2}\right)$
17. (a) $\bar{x} = \bar{y} = \frac{4(a^2 + ab + b^2)}{3\pi(a + b)}$ **(b)** $\bar{x} = \bar{y} = \frac{zb}{\pi}$

CHAPTER 6 TRANSCENDENTAL FUNCTIONS AND DIFFERENTIAL EQUATIONS

Section 6.1, pp. 495–496

1. $\frac{1}{x}$ **3.** $\frac{3}{x}$
5. $2(\ln t) + (\ln t)^2$ **7.** $x^3 \ln x$
9. $\frac{1 - \ln t}{t^2}$ **11.** $\frac{1}{x(1 + \ln x)^2}$
13. $2\cos(\ln\theta)$ **15.** $-\frac{3x + 2}{2x(x + 1)}$
17. $\frac{2}{t(1 - \ln t)^2}$ **19.** $\frac{\tan(\ln\theta)}{\theta}$
21. $\frac{1}{\theta \ln 2}$ **23.** $\frac{2(\ln r)}{r(\ln 2)(\ln 4)}$
25. $\sin(\log_7\theta) + \frac{1}{\ln 7}\cos(\log_7\theta)$
27. $2x\ln|x| - x\ln\frac{|x|}{\sqrt{2}}$
29. $\ln\left(\frac{2}{3}\right)$ **31.** $\ln|y^2 - 25| + C$
33. $\ln 3$ **35.** $(\ln 2)^2$
37. $\frac{1}{\ln 4}$ **39.** $\ln|6 + 3\tan t| + C$
41. $\ln 2$ **43.** $\ln 27$
45. $\ln(1 + \sqrt{x}) + C$
47. (a) Max $= 0$ at $x = 0$, min $= -\ln 2$ at $x = \pi/3$
(b) Max $= 1$ at $x = 1$, min $= \cos(\ln 2)$ at $x = 1/2$ and $x = 2$
49. $\ln 16$ **51.** $y = x + \ln|x| + 2$
53. $\frac{1}{\ln 10}\left(\frac{(\ln x)^2}{2}\right) + C$ **55.** $\frac{3\ln 2}{2}$
57. $(\ln 10)\ln|\ln x| + C$ **59. (b)** 0.00469

Section 6.2, pp. 501–503

1. $-\frac{2}{3}e^{-2x/3}$ **3.** $\left(\frac{2}{\sqrt{x}} + 2x\right)e^{(4\sqrt{x}+x^2)}$
5. x^2e^x **7.** $\frac{1}{\theta} - 1$
9. $\frac{\cos t\, 2\sin t}{\sin t}$ **11.** $\frac{1}{2\theta(1 + \theta^{1/2})}$
13. $\frac{\sin x}{x}$ **15.** $\frac{ye^y \cos x}{1 - ye^y \sin x}$
17. $\frac{2e^{2x} - \cos(x + 3y)}{3\cos(x + 3y)}$ **19.** $\frac{1}{3}e^{3x} - 5e^{-s} + C$
21. $8e^{(x+1)} + C$ **23.** $-2e^{-\sqrt{r}} + C$
25. $-e^{1/x} + C$ **27.** e
29. $\frac{1}{\pi}e^{\sec \pi t} + C$ **31.** $\ln(1 + e^r) + C$
33. $2^x \ln 2$ **35.** $\pi x^{(\pi - 1)}$
37. $7^{\sec\theta}(\ln 7)^2 \sec\theta\tan\theta$ **39.** $(1 - e)t^{-e}$
41. $\frac{-2}{(x + 1)(x - 1)}$
43. $(\cot\theta - \tan\theta - 1 - \ln 2)\left(\frac{1}{\ln 7}\right)$
45. $(x + 1)^x\left(\frac{x}{x + 1} + \ln(x + 1)\right)$
47. $\frac{dy}{dx} = \frac{x\ln x\cos x + \sin x}{x^{1 - \sin x}}$
49. $\frac{1}{\ln 2}$ **51.** $\frac{2^{\ln 2} - 1}{\ln 2}$
53. $\frac{x^{\sqrt{2}}}{\sqrt{2}} + C$ **55.** $\frac{1}{\ln 2}$
57. $y = 1 - \cos(e^t - 2)$ **59.** $y = 2(e^{-x} + x) - 1$
61. Maximum: 1 at $x = 0$; minimum: $2 - 2\ln 2$ at $x = \ln 2$

63. Absolute maximum of $\frac{1}{2e}$ assumed at $x = \frac{1}{\sqrt{e}}$

65. Let $x = \frac{r}{k} \Rightarrow k = \frac{r}{x}$ and as $k \to \infty$, $x \to 0 \Rightarrow \lim_{k\to\infty}\left(1 + \frac{r}{k}\right)^k =$ $\lim_{x\to 0}(1 + x)^{r/x} = \lim_{r\to 0}((1 + x)^{1/x})^r = (\lim_{x\to 0}(1 + x)^{1/x})^r$, since u^r is continuous. However, $\lim_{x\to 0}(1 + x)^{1/x} = e$ (by Theorem 2); therefore, $\lim_{k\to\infty}\left(1 + \frac{r}{k}\right)^k = e^r$.

71. **(b)** $|\text{Error}| \approx 0.02140$

Section 6.3, pp. 510–511

1. $y = \frac{e^x + C}{x}$

3. $y = \frac{C - \cos x}{x^3}, x > 0$

5. $y = \frac{1}{2} - \frac{1}{x} + \frac{C}{x^2}, x > 0$

7. $y = \frac{1}{2}xe^{x/2} + Ce^{x/2}$

9. $y = x(\ln x)^2 + Cx$

11. $s = \frac{t^3}{3(t - 1)^4} - \frac{t}{(t - 1)^4} + \frac{C}{(t - 1)^4}$

13. $r = (\csc\theta)(\ln|\sec\theta| + C)$

15. $y = \frac{3}{2} - \frac{1}{2}e^{-2t}$

17. $y = -\frac{1}{\theta}\cos\theta + \frac{\pi}{2\theta}$

19. $y = 6e^{x^2} - \frac{e^{x^2}}{x + 1}$

21. $y = y_0e^{kt}$

23. Part (b) is correct, but part (a) is not.

25. **(a)** 10 lb/min

(b) $100 + t$ gal

(c) $4\left(\frac{y}{100 + t}\right)$ lb/min

(d) $\frac{dy}{dt} = 10 - \frac{4y}{100 + t}, y(0) = 50,$

$$y = 2(100 + t) - \frac{150}{\left(1 + \frac{t}{100}\right)^4}$$

(e) Concentration $= \frac{y(25)}{\text{amount brine in tank}} = \frac{188.6}{125} \approx 1.5$ lb/gal

27. $y(27.8) \approx 14.8$ lb, $t \approx 27.8$ min

29. $t = \frac{L}{R}\ln 2$ sec

31. **(a)** $i = \frac{V}{R} - \frac{V}{R}e^{-3} = \frac{V}{R}(1 - e^{-3}) \approx 0.95\frac{V}{R}$ amp

(b) 86%

Section 6.4, pp. 522–524

1. $y(\text{exact}) = \frac{x}{2} - \frac{4}{x}, y_1 = -0.25, y_2 = 0.3, y_3 = 0.75$

3. $y(\text{exact}) = 3e^{x(x+2)}, y_1 = 4.2, y_2 = 6.216, y_3 = 9.697$

5. $y \approx 2.48832$, exact value is e

7.

x	z	y-approx.	y-exact	Error
0	—	3	3	0
0.2	4.2	4.608	4.658122	0.050122
0.4	6.81984	7.623475	7.835089	0.211614
0.6	11.89262	13.56369	14.27646	0.712777

9. **(a)** $P(t) = \frac{150}{1 + 24e^{-0.225t}}$

(b) About 17.21 weeks; 21.28 weeks

11. **(a)** $y(t) = \frac{8 \times 10^7}{1 + 4e^{-0.71t}} \Rightarrow y(1) \approx 2.697 \times 10^7$ kg

(b) $t \approx 1.95253$ years

13. **(a)** $y = 2e^t - 1$ **(b)** $y(t) = \frac{400}{1 + 199e^{-200t}}$

15. **(a)** $P(t) = \frac{P_0}{1 - kP_0t}$

(b) Vertical asymptote at $t = \frac{1}{kP_0}$

17. $y \approx 3.45835$; exact value is $1 + e \approx 3.71828$.

19. $y \approx -0.2272$; exact value is $\frac{1}{1 - 2\sqrt{5}} \approx -0.2880$.

21. **(a)** $y = \frac{21}{x^2 - 2x + 2}, y(3) = -0.2$

(b) -0.1851, error ≈ 0.0149

(c) -0.1929, error ≈ 0.0071

(d) -0.1965, error ≈ 0.0035

23. The exact solution is $y = \frac{-1}{x^2 - 2x + 2}$, so $y(3) = -0.2$. To find the approximation, let $z_n = y_{n+1} + 2_{n-1}{}^2(x_n - 1)\,dx$ and $y_n = y_{n-1} + (y_{n-1}{}^2(x_{n-1} - 1) + z_n{}^2(x_n{}^2 - 1))\,dx$ with initial values $x_0 = 2$ and $y_0 = -\frac{1}{2}$. Use a spreadsheet, graphing calculator, or CAS as indicated in parts (a) through (d).

(a) -0.2024, error ≈ 0.0024

(b) -0.2005, error ≈ 0.0005

(c) -0.2001, error ≈ 0.0001

(d) Each time the step size is cut in half, the error is reduced to approximately one-fourth of what it was for the larger step size.

Section 6.5, pp. 529–533

1. $\cosh x = \frac{5}{4}, \tanh x = -\frac{3}{5}, \coth x = -\frac{5}{3}, \operatorname{sech} x = \frac{4}{5}, \operatorname{csch} x = -\frac{4}{3}$

3. $\sinh x = \frac{8}{15}, \tanh x = \frac{8}{17}, \coth x = \frac{17}{8}, \operatorname{sech} x = \frac{15}{17}, \operatorname{csch} x = \frac{15}{8}$

5. $x + \frac{1}{x}$

7. e^{5x}

9. e^{4x}

13. $2\cosh\frac{x}{3}$

15. $\operatorname{sech}^2\sqrt{t} + \frac{\tanh\sqrt{t}}{\sqrt{t}}$

17. $\coth z$

19. $(\ln \operatorname{sech} \theta)(\operatorname{sech} \theta \tanh \theta)$

21. $\tanh^3 v$

23. 2

25. $\dfrac{1}{2\sqrt{x(1+x)}}$

27. $\dfrac{1}{1+\theta} - \tanh^{-1} \theta$

29. $\dfrac{1}{2\sqrt{t}} - \coth^{-1} \sqrt{t}$

31. $-\operatorname{sech}^{-1} x$

33. $\dfrac{\ln 2}{\sqrt{1+\left(\frac{1}{2}\right)^{2\theta}}}$

35. $|\sec x|$

41. $\dfrac{\cosh 2x}{2} + C$

43. $12 \sinh\left(\dfrac{x}{2} - \ln 3\right) + C$

45. $7 \ln (e^{x/7} + e^{-x/7}) + C$

47. $\tanh\left(x - \dfrac{1}{2}\right) + C$

49. $-2 \operatorname{sech} \sqrt{t} + C$

51. $\ln \dfrac{5}{2}$

53. $\dfrac{3}{32} + \ln 2$

55. $e - e^{-1}$

57. $\dfrac{3}{4}$

59. $\dfrac{3}{8} + \ln \sqrt{2}$

61. $\ln\left(\dfrac{2}{3}\right)$

63. $-\dfrac{\ln 3}{2}$

65. $\ln 3$

67. (a) $\sinh^{-1}(\sqrt{3})$ (b) $\ln(\sqrt{3}+2)$

69. (a) $\coth^{-1}(2) - \coth^{-1}\left(\dfrac{5}{4}\right)$ (b) $\left(\dfrac{1}{2}\right) \ln \left(\dfrac{1}{3}\right)$

71. (a) $-\operatorname{sech}^{-1}\left(\dfrac{12}{13}\right) + \operatorname{sech}^{-1}\left(\dfrac{4}{5}\right)$

(b) $$-\ln\left(\frac{1+\sqrt{1-\left(\frac{12}{13}\right)^2}}{\left(\frac{12}{13}\right)}\right) + \ln\left(\frac{1+\sqrt{1-\left(\frac{4}{5}\right)^2}}{\left(\frac{4}{5}\right)}\right) =$$

$$-\ln\left(\frac{3}{2}\right) + \ln(2) = \ln\left(\frac{4}{3}\right)$$

73. (a) 0 (b) 0

75. (b) (i) $f(x) = \dfrac{2f(x)}{2} + 0 = f(x)$

(ii) $f(x) = 0 + \dfrac{2f(x)}{2} = f(x)$

77. (b) $\sqrt{\dfrac{mg}{k}}$ (c) $80\sqrt{5} \approx 178.89$ ft/sec

79. $y = \operatorname{sech}^{-1}(x) - \sqrt{1-x^2}$

81. 2π

83. $\dfrac{6}{5}$

87. (c) $a \approx 0.0417525$ (d) ≈ 47.90 lb

Chapter 6 Practice Exercises, pp. 534–536

1. $-\cos e^x + C$

3. $\tan(e^x - 7) + C$

5. $e^{\tan x} + C$

7. $-\dfrac{\ln 7}{3}$

9. $\ln 8$

11. $\ln\left(\dfrac{9}{25}\right)$

13. $-[\ln|\cos(\ln v)|] + C$

15. $-\dfrac{1}{2}(\ln x)^{-2} + C$

17. $-\cot(1 + \ln r) + C$

19. $\dfrac{1}{2 \ln 3}(3^{x^2}) + C$

21. $3 \ln 7$

23. $e - 1$

25. $\dfrac{1}{3}[(\ln 4)^3 - (\ln 2)^3]$ or $\dfrac{7}{3}(\ln 2)^3$

27. $\dfrac{9 \ln 2}{4}$

29. $\dfrac{2}{5} \sin^{-1}(5y) + C$

31. Absolute maximum $= 0$ at $x = e/2$, absolute minimum $= -0.5$ at $x = 0.5$

33. 1

35. $\dfrac{1}{e}$ m/sec

37. $\ln 5x - \ln 3x = \ln\left(\dfrac{5}{3}\right)$

41. $y = \dfrac{3 - \cos 2x}{2 \cos x}$

45. (a) Absolute maximum of $\dfrac{2}{e}$ at $x = e^2$, inflection point $\left(e^{8/3}, \dfrac{8}{3}e^{-4/3}\right)$, concave up on $(e^{8/3}, \infty)$, concave down on $(0, e^{8/3})$

(b) Absolute maximum of 1 at $x = 0$, inflection points $\left(\pm\dfrac{1}{\sqrt{2}}, \dfrac{1}{\sqrt{e}}\right)$, concave up on $\left(-\infty, -\dfrac{1}{\sqrt{2}}\right) \cup \left(\dfrac{1}{\sqrt{2}}, \infty\right)$, concave down on $\left(-\dfrac{1}{\sqrt{2}}, \dfrac{1}{\sqrt{2}}\right)$

(c) Absolute maximum of 1 at $x = 0$, inflection point $\left(1, \dfrac{2}{e}\right)$, concave up on $(1, \infty)$, concave down on $(-\infty, 1)$

47. 18,935 years

49. (a) $y = \sin x + \dfrac{C + \cos x}{x}$

(b) The graphs, from the bottom curve to the top, are for $C = -z, -1, 0, 1, z$, respectively.

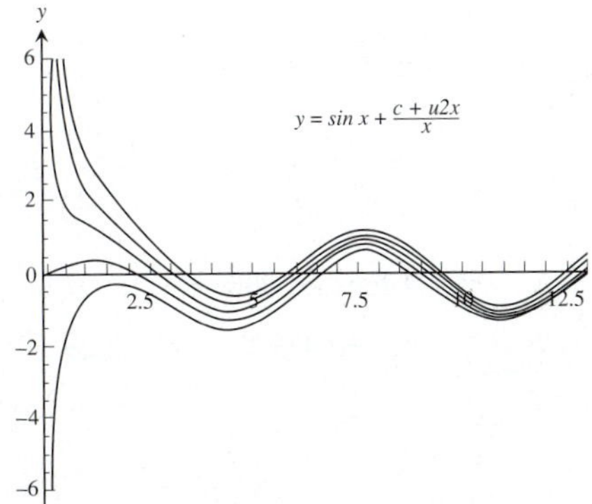

51.

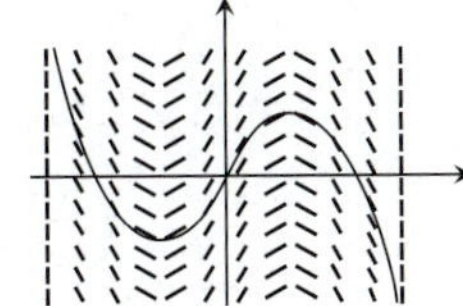

53. Let $z_n = y_{n-1} + ((2 - y_{n-1})(2x_{n-1} + 3))(0.1)$ and $y_n = y_{n-1} + \left(\frac{(2 - y_{n-1})(2x_{n-1} + 3) + (2 - z_n)(2x_n + 3)}{2}\right)(0.1)$ with initial values $x_0 = -3$, $y_0 = 1$, and 20 steps. Use a spreadsheet, graphing calculator, or CAS to obtain the values in the following table.

x	y
−3	1
−2.9	0.6680
−2.8	0.2599
−2.7	−0.2294
−2.6	−0.8011
−2.5	−1.4509
−2.4	−2.1687
−2.3	−2.9374
−2.2	−3.7333
−2.1	−4.5268
−2.0	−5.2840

x	y
−1.9	−5.9686
−1.8	−6.5456
−1.7	−6.9831
−1.6	−7.2562
−1.5	−7.3488
−1.4	−7.2553
−1.3	−6.9813
−1.2	−6.5430
−1.1	−5.9655
−1.0	−5.2805

55. Let $y_n = y_{n-1} + \left(\frac{2x_{n-1}^{\ 2} - 2y_{n-1} + 1}{x_{n-1}}\right)(0.05)$ with initial values $x_0 = 1$, $y_0 = 1$, and 60 steps. Use a spreadsheet, programmable calculator, or CAS to obtain $y(r) \approx 4.4974$.

57. Let $z_n = y_{n-1} - \left(\frac{x_{n-1}^{\ 2} + y_{n-1}}{e^{y_{n-1}} + x_{n-1}}\right)(dx)$ and $y_n = y_{n-1} + \frac{1}{2}\left(\frac{x_{n-1}^{\ 2} + y_{n-1}}{e^{y_{n-1}} + x_{n-1}} + \frac{x_n^{\ 2} + z_n}{e^{z_n} + x_n}\right)(dx)$ with starting values $x_0 = 0$, $y_0 = 0$, and steps of 0.1 and −0.1. Use a spreadsheet, programmable calculator, or CAS to generate the following graphs.

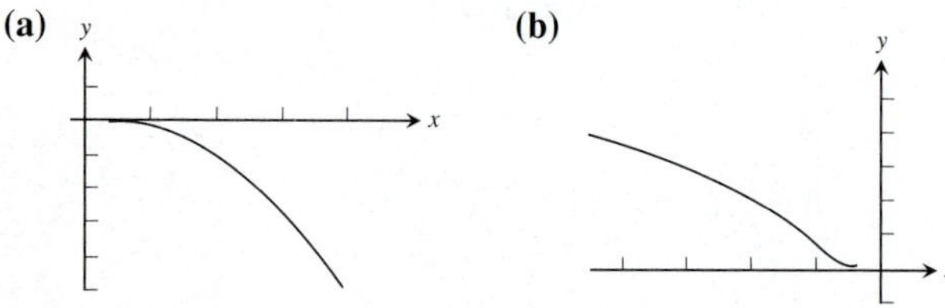

59. $y(\text{exact}) = \frac{1}{2}x^2 - \frac{3}{2}$; $y \approx 0.4$; exact value is 1/2.

61. $y(\text{exact}) = -e^{(x^2-1)/2}$; $y \approx -3.4192$; exact value is $-e^{3/2} \approx -4.4817$.

Chapter 6 Additional Exercises, pp. 536–537

1. $\frac{1}{\ln 2}, \frac{1}{2\ln 2}, 2:1$

3. **(a)** $2x$ **(b)** 0 **(c)** $f(x) = x^2$

9. **(a)** $P = P_0 e^{\frac{k}{a}(\sin b + \sin(at - b))}$

CHAPTER 7

Section 7.1, pp. 544–546

1. $2\sqrt{8x^2 + 1} + C$

3. $2(\sin v)^{3/2} + C$

5. $\ln 5$

7. $2\ln(\sqrt{x} + 1) + C$

9. $-\frac{1}{7}\ln|\sin(3 - 7x)| + C$

11. $-\ln|\csc(e^\theta + 1) + \cot(e^\theta + 1)| + C$

13. $3\ln\left|\sec\frac{t}{3} + \tan\frac{t}{3}\right| + C$

15. $-\ln|\csc(s - \pi) + \cot(s - \pi)| + C$

17. 1

19. $e^{\tan v} + C$

21. $\frac{3^{x+1}}{\ln 3} + C$

23. $\frac{2^{\sqrt{w}}}{\ln 2} + C$

25. $3\tan^{-1} 3u + C$

27. $\frac{\pi}{18}$

29. $\sin^{-1} s^2 + C$

31. $6\sec^{-1}|5x| + C$

33. $\tan^{-1} e^x + C$

35. $\ln(2 + \sqrt{3})$

37. 2π

39. $\sin^{-1}(t - 2) + C$

41. $\sec^{-1}|x + 1| + C$, when $|x + 1| > 1$

43. $\tan x - 2\ln|\csc x + \cot x| - \cot x - x + C$

45. $x + \sin 2x + C$

47. $x - \ln|x + 1| + C$

49. $7 + \ln 8$

51. $2t^2 - t + 2\tan^{-1}\left(\frac{t}{2}\right) + C$

53. $\sin^{-1} x + \sqrt{1 - x^2} + C$

55. $\sqrt{2}$

57. $\tan x - \sec x + C$

59. $\ln|1 + \sin\theta| + C$

61. $\cot x + x + \csc x + C$

63. 4

65. $\sqrt{2}$

67. 2

69. $\ln|\sqrt{2} + 1| - \ln|\sqrt{2} - 1|$

71. $4 - \frac{\pi}{2}$

73. $-\ln|\csc(\sin\theta) + \cot(\sin\theta)| + C$

75. $\ln|\sin x| + \ln|\cos x| + C$

77. $12\tan^{-1}(\sqrt{y}) + C$

79. $\sec^{-1}\left|\frac{x - 1}{7}\right| + C$

81. $\ln|\sec(\tan t)| + C$

83. **(a)** $\sin\theta - \frac{1}{3}\sin^3\theta + C$
(b) $\sin\theta - \frac{2}{3}\sin^3\theta + \frac{1}{5}\sin^5\theta + C$
(c) $\int \cos^9\theta\, d\theta = \int \cos^8\theta(\cos\theta)\, d\theta = \int (1 - \sin^2\theta)^4(\cos\theta)\, d\theta$

85. **(a)** $\int \tan^3\theta\, d\theta = \frac{1}{2}\tan^2\theta - \int \tan\theta\, d\theta = \frac{1}{2}\tan^2\theta + \ln|\cos\theta| + C$
(b) $\int \tan^5\theta\, d\theta = \frac{1}{4}\tan^4\theta - \int \tan^3\theta\, d\theta$
(c) $\int \tan^7\theta\, d\theta = \frac{1}{6}\tan^6\theta - \int \tan^5\theta\, d\theta$
(d) $\int \tan^{2k+1}\theta\, d\theta = \frac{1}{2k}\tan^{2k}\theta - \int \tan^{2k-1}\theta\, d\theta$

87. $2\sqrt{2} - \ln(3 + 2\sqrt{2})$

89. π^2

91. $\ln(2 + \sqrt{3})$

Section 7.2, pp. 553–554

1. $-2x \cos\left(\frac{x}{2}\right) + 4 \sin\left(\frac{x}{2}\right) + C$

3. $t^2 \sin t + 2t \cos t - 2 \sin t + C$

5. $\ln 4 - \frac{3}{4}$

7. $y \tan^{-1}(y) - \ln\sqrt{1 + y^2} + C$

9. $x \tan x + \ln|\cos x| + C$

11. $(x^3 - 3x^2 + 6x - 6)e^x + C$

13. $(x^2 - 7x + 7)e^x + C$

15. $(x^5 - 5x^4 + 20x^3 - 60x^2 + 120x - 120)e^x + C$

17. $\frac{\pi^2 - 4}{8}$ **19.** $\frac{5\pi - 3\sqrt{3}}{9}$

21. $\frac{1}{2}(-e^\theta \cos\theta + e^\theta \sin\theta) + C$

23. $\frac{e^{2x}}{13}(3 \sin 3x + 2 \cos 3x) + C$

25. $\frac{2}{3}(\sqrt{3s+9}\, e^{\sqrt{3s+9}} - e^{\sqrt{3s+9}}) + C$

27. $\frac{\pi\sqrt{3}}{3} - \ln(2) - \frac{\pi^2}{18}$

29. $\frac{1}{2}[-x \cos(\ln x) + x \sin(\ln x)] + C$

31. $y = \left(\frac{x^2}{4} - \frac{x}{8} + \frac{1}{32}\right)e^{4x} + C$

33. $-2(\sqrt{\theta} \cos\sqrt{\theta} - \sin\sqrt{\theta}) + C$

35. **(a)** π **(b)** 3π **(c)** 5π **(d)** $(2n + 1)\pi$

37. $2\pi(1 - \ln 2)$

39. **(a)** $\pi(\pi - 2)$ **(b)** 2π

41. $\frac{1}{2\pi}(1 - e^{-2\pi})$

43. $u = x^n$, $dv = \cos x\, dx$ **45.** $u = x^n$, $dv = e^{ax}\, dx$

47. **(a)** Let $y = f^{-1}(x)$. Then $x = f(y)$, so $dx = f'(y)\, dy$. Substitute directly.
(b) $u = y$, $dv = f'(y)\, dy$

49. **(a)** $\int \sin^{-1} x\, dx = x \sin^{-1} x + \cos(\sin^{-1} x) + C$
(b) $\int \sin^{-1} x\, dx = x \sin^{-1} x + \sqrt{1 - x^2} + C$
(c) $\cos(\sin^{-1} x) = \sqrt{1 - x^2}$

51. **(a)** $\int \cos^{-1} x\, dx = x \cos^{-1} x - \sin(\cos^{-1} x) + C$
(b) $\int \cos^{-1} x\, dx = x \cos^{-1} x - \sqrt{1 - x^2} + C$
(c) $\sin(\cos^{-1} x) = \sqrt{1 - x^2}$

Section 7.3, pp. 563–565

1. $\frac{2}{x - 3} + \frac{3}{x - 2}$ **3.** $\frac{1}{x + 1} + \frac{3}{(x + 1)^2}$

5. $\frac{-2}{z} + \frac{-1}{z^2} + \frac{2}{z - 1}$ **7.** $1 + \frac{17}{t - 3} + \frac{-12}{t - 2}$

9. $\frac{1}{2}[\ln|1 + x| - \ln|1 - x|] + C$

11. $\frac{1}{7}\ln|(x + 6)^2(x - 1)^5| + C$

13. $\frac{\ln 15}{2}$

15. $-\frac{1}{2}\ln|t| + \frac{1}{6}\ln|t + 2| + \frac{1}{3}\ln|t - 1| + C$

17. $3 \ln 2 - 2$

19. $\frac{1}{4}\ln\left|\frac{x + 1}{x - 1}\right| - \frac{x}{2(x^2 - 1)} + C$

21. $\frac{\pi + 2 \ln 2}{8}$ **23.** $\tan^{-1} y - \frac{1}{y^2 + 1} + C$

25. $-(s - 1)^{-2} + (s - 1)^{-1} + \tan^{-1} s + C$

27. $\frac{-1}{\theta^2 + 2\theta + 2} + \ln|\theta^2 + 2\theta + 2| - \tan^{-1}(\theta + 1) + C$

29. $x^2 + \ln\left|\frac{x - 1}{x}\right| + C$

31. $9x + 2 \ln|x| + \frac{1}{x} + 7 \ln|x - 1| + C$

33. $\frac{y^2}{2} - \ln|y| + \frac{1}{2}\ln(1 + y^2) + C$

35. $\ln\left|\frac{e^t + 1}{e^t + 2}\right| + C$ **37.** $\frac{1}{5}\ln\left|\frac{\sin y - 2}{\sin y + 3}\right| + C$

39. $\frac{(\tan^{-1} 2x)^2}{4} - 3 \ln|x - 2| + \frac{6}{x - 2} + C$

41. $x = \ln|t - 2| - \ln|t - 1| + \ln 2$

43. $x = \frac{6t}{t + 2} - 1, t > 2/5$

45. $\ln|y - 1| - \ln|y| = e^x - 1 - \ln 2$

47. $y = \ln|x - 2| - \ln|x - 1| + \ln 2$

49. $3\pi \ln 25$

51. **(a)** $x = \frac{1000e^{4t}}{499 + e^{4t}}$ **(b)** 1.55 days

Section 7.4, pp. 569–570

1. $\ln|\sqrt{9 + y^2} + y| + C$

3. $\frac{25}{2}\sin^{-1}\left(\frac{t}{5}\right) + \frac{t\sqrt{25 - t^2}}{2} + C$

5. $\frac{1}{2}\ln\left|\frac{2x}{7} + \frac{\sqrt{4x^2 - 49}}{7}\right| + C$

7. $\frac{\sqrt{x^2 - 1}}{x} + C$

9. $\frac{1}{3}(x^2 + 4)^{3/2} - 4\sqrt{x^2 + 4} + C$ **11.** $\frac{-2\sqrt{4 - w^2}}{w} + C$

13. $-\frac{x}{\sqrt{x^2 - 1}} + C$ **15.** $-\frac{1}{5}\left(\frac{\sqrt{1 - x^2}}{x}\right)^5 + C$

17. $2 \tan^{-1} 2x + \frac{4x}{(4x^2 + 1)} + C$ **19.** $\ln 9 - \ln(1 + \sqrt{10})$

21. $\frac{\pi}{6}$ **23.** $\sec^{-1}|x| + C$

25. $\sqrt{x^2-1}+C$

27. $y = 2\left[\dfrac{\sqrt{x^2-4}}{2} - \sec^{-1}\left(\dfrac{x}{2}\right)\right]$

29. $y = \dfrac{3}{2}\tan^{-1}\left(\dfrac{x}{2}\right) - \dfrac{3\pi}{8}$

31. $\dfrac{3\pi}{4}$

33. **(a)** This can be seen geometrically in the figure.

(b) Using part (a), substitute $z = \dfrac{\sin x}{1+\cos x}$ and then obtain a trigonometric identity. Or, use the trigonometric identity $\dfrac{1-\tan^2\theta}{1+\tan^2\theta} = \cos 2\theta$ with $\theta = \dfrac{x}{2}$.

(c) Using part (a), substitute $z = \dfrac{\sin x}{1+\cos x}$ and then obtain a trigonometric identity. Or, use the trigonometric identity $\dfrac{2\tan\theta}{1+\tan^2\theta} = \sin 2\theta$ with $\theta = \dfrac{x}{2}$.

(d) $dz = \left(\sec^2\dfrac{x}{2}\right)\dfrac{1}{2}\,dx = \left(1+\tan^2\dfrac{x}{2}\right)\dfrac{1}{2}\,dx = (1+z^2)\,\dfrac{1}{2}\,dx$, then solve for dx.

35. $-\dfrac{1}{\tan\dfrac{x}{2}} + C$

37. $\ln\left|1+\tan\dfrac{t}{2}\right| + C$

39. $\dfrac{1}{2}(\ln\sqrt{3} - 1)$

41. $-\cot\left(\dfrac{t}{2}\right) - t + C$

Section 7.5, pp. 576–577

1. $\dfrac{2}{\sqrt{3}}\left(\tan^{-1}\sqrt{\dfrac{x-3}{3}}\right) + C$

3. $\dfrac{(2x-3)^{3/2}(x+1)}{5} + C$

5. $\dfrac{(x+2)(2x-6)\sqrt{4x-x^2}}{6} + 4\sin^{-1}\left(\dfrac{x-2}{2}\right) + C$

7. $\sqrt{4-x^2} - 2\ln\left|\dfrac{2+\sqrt{4-x^2}}{x}\right| + C$

9. $2\sin^{-1}\dfrac{r}{2} - \dfrac{1}{2}r\sqrt{4-r^2} + C$

11. $\dfrac{e^{2t}}{13}(2\cos 3t + 3\sin 3t) + C$

13. $\dfrac{s}{18(9-s^2)} + \dfrac{1}{108}\ln\left|\dfrac{s+3}{s-3}\right| + C$

15. $2\sqrt{3t-4} - 4\tan^{-1}\sqrt{\dfrac{3t-4}{4}} + C$

17. $-\dfrac{\cos 5x}{10} - \dfrac{\cos x}{2} + C$

19. $6\sin\left(\dfrac{\theta}{12}\right) + \dfrac{6}{7}\sin\left(\dfrac{7\theta}{12}\right) + C$

21. $\dfrac{1}{2}\ln|x^2+1| + \dfrac{x}{2(1+x^2)} + \dfrac{1}{2}\tan^{-1}x + C$

23. $\left(x - \dfrac{1}{2}\right)\sin^{-1}\sqrt{x} + \dfrac{1}{2}\sqrt{x-x^2} + C$

25. $\sqrt{1-\sin^2 t} - \ln\left|\dfrac{1+\sqrt{1-\sin^2 t}}{\sin t}\right| + C$

27. $\ln|\ln y + \sqrt{3+(\ln y)^2}| + C$

29. $\ln|3r + \sqrt{9r^2-1}| + C$

31. $x\cos^{-1}\sqrt{x} + \dfrac{1}{2}\sin^{-1}\sqrt{x} - \dfrac{1}{2}\sqrt{x-x^2} + C$

33. $\dfrac{e^{3x}}{9}(3x-1) + C$

35. $\dfrac{x^2 2^x}{\ln 2} - \dfrac{2}{\ln 2}\left[\dfrac{x2^x}{\ln 2} - \dfrac{2^x}{(\ln 2)^2}\right] + C$

37. $\dfrac{1}{120}\sinh^4 3x\cosh 3x - \dfrac{1}{90}\sinh^2 3x\cosh 3x + \dfrac{2}{90}\cosh 3x + C$

39. $\dfrac{x^2}{3}\sinh 3x - \dfrac{2x}{9}\cosh 3x + \dfrac{2}{27}\sinh 3x + C$

45. **(b)** Using Formula 29 in the text's integral table, $V =$

$$2L\left[\left(\frac{d-r}{2}\right)\sqrt{2rd-d^2} + \left(\frac{r^2}{2}\right)\left[\sin^{-1}\left(\frac{d-r}{r}\right) + \frac{\pi}{2}\right]\right]$$

49. **(c)** $\dfrac{\pi}{4}$

51. $1 - \dfrac{1}{e} \approx 0.632121$

53. $\dfrac{4}{15} \approx 0.266667$

55. $6 + 2[(\ln 2)^3 - 3(\ln 2)^2 + 6\ln 2 - 6] \approx 0.101097$

Section 7.6, pp. 584–586

1. $\dfrac{1}{4}$

3. $\dfrac{5}{7}$

5. $\dfrac{1}{2}$

7. 0

9. -1

11. ln 2

13. 1

15. 0

17. 1

19. 0

21. e^2

23. 0

25. 1

27. 1

29. e

31. 1

33. e^{-1}

35. ln 2

37. -1

39. 3

41. 1

43. (b) is correct, but (a) is not.

45. $c = \dfrac{27}{10}$

47. **(a)** $\ln\left(1+\dfrac{r}{k}\right)^k = k\ln\left(1+\dfrac{r}{k}\right)$. And, as $k \to \infty$,

$$\lim_{k\to\infty} k\ln\left(1+\frac{r}{k}\right) = \lim_{k\to\infty}\frac{\ln\left(1+\frac{r}{k}\right)}{\frac{1}{k}} = \lim_{k\to\infty}\frac{\frac{-r}{k^2}\Big/\left(1+\frac{r}{k}\right)}{\frac{-1}{k^2}} =$$

$$\lim_{k\to\infty}\frac{r}{1+\frac{r}{k}} = r. \text{ Therefore, } \lim_{k\to\infty}\left(1+\frac{r}{k}\right)^k = e^r.$$

Hence, $\lim\limits_{k\to\infty} A_0\left(1+\dfrac{r}{k}\right)^{kt} = A_0e^{rt}$.

(b) Part (a) shows that as the number of compoundings per year increases toward infinity, the limit of interest compounded k times per year is interest compounded continuously.

53. (a) $(-\infty, -1) \cup (0, \infty)$ **(b)** ∞ **(c)** e

Section 7.7, pp. 598–600

1. (a) Because of an infinite limit of integration
(b) Converges
(c) $\frac{\pi}{2}$

3. (a) Because the intergrand has an infinite discontinuity at $x = 0$
(b) Converges
(c) $-\frac{9}{2}$

5. (a) Because the integrand has an infinite discontinuity at $x = 0$
(b) Diverges
(c) No value

7. 1000 **9.** 4
11. $\frac{\pi}{2}$ **13.** $\ln 3$
15. $\sqrt{3}$ **17.** π
19. $\frac{\pi}{3}$ **21.** $\ln 4$
23. $\frac{\pi}{2}$ **25.** $\ln\left(1 + \frac{\pi}{2}\right)$
27. 6 **29.** -1
31. 2 **33.** $-\frac{1}{4}$
35. Diverges **37.** Converges
39. Converges **41.** Converges
43. Diverges **45.** Converges
47. Converges **49.** Diverges
51. Converges **53.** Converges
55. Diverges **57.** Converges
59. Diverges **61.** Converges
63. Converges
65. (a) Converges when $p < 1$
(b) Converges when $p > 1$
67. 1 **69.** $\frac{\pi}{2}$
73. (b) ≈ 0.88621 **75. (b)** 1

Chapter 7 Practice Exercises, pp. 601–602

1. $\frac{1}{12}(4x^2 - 9)^{3/2} + C$ **3.** $\frac{\sqrt{8x^2 + 1}}{8} + C$

5. $\frac{-\sqrt{9 - 4t^4}}{8} + C$ **7.** $-\frac{1}{2(1 - \cos 2\theta)} + C$

9. $-\frac{1}{2}e^{\cos 2x} + C$ **11.** $\frac{2^{x-1}}{\ln 2} + C$

13. $\ln|2 + \tan^{-1} x| + C$ **15.** $\frac{1}{3}\sin^{-1}\left(\frac{3t}{4}\right) + C$

17. $\frac{1}{5}\sec^{-1}\left|\frac{5x}{4}\right| + C$ **19.** $\frac{1}{2}\tan^{-1}\left(\frac{y-2}{2}\right) + C$

21. $\frac{x}{2} + \frac{\sin 6x}{12} + C$

23. $\frac{\tan^2(2t)}{4} - \frac{1}{2}\ln|\sec 2t| + C$

25. $\ln|\sec 2x + \tan 2x| + C$

27. $\ln(3 + 2\sqrt{2})$ **29.** $2\sqrt{2}$

31. $x - 2\tan^{-1}\left(\frac{x}{2}\right) + C$

33. $\ln(y^2 + 4) - \frac{1}{2}\tan^{-1}\left(\frac{y}{2}\right) + C$

35. $-\sqrt{4 - t^2} + 2\sin^{-1}\left(\frac{t}{2}\right) + C$

37. $x - \tan x + \sec x + C$

39. $4\ln\left|\sin\left(\frac{x}{4}\right)\right| + C$

41. $\frac{z}{16(16 + z^2)^{1/2}} + C$

43. $\frac{-\sqrt{1 - x^2}}{x} + C$

45. $\ln|x + \sqrt{x^2 - 9} + C$

47. $[(x + 1)(\ln(x + 1)) - (x + 1)] + C$

49. $x\tan^{-1}(3x) - \frac{1}{6}\ln(1 + 9x^2) + C$

51. $(x + 1)^2e^x - 2(x + 1)e^x + 2e^x + C$

53. $\frac{2e^x \sin 2x}{5} + \frac{e^x \cos 2x}{5} + C$

55. $2\ln|x - 2| - \ln|x - 1| + C$

57. $-\frac{1}{3}\ln\left|\frac{\cos\theta - 1}{\cos\theta + 2}\right| + C$

59. $\frac{1}{16}\ln\left|\frac{(v - 2)^5(v + 2)}{v^6}\right| + C$

61. $\frac{x^2}{2} + \frac{4}{3}\ln|x + 2| + \frac{2}{3}\ln|x - 1| + C$

63. $x^2 - 3x + \frac{2}{3}\ln|x + 4| + \frac{1}{3}\ln|x - 2| + C$

65. $\ln|1 - e^{-s}| + C$ **67.** $-\sqrt{16 - y^2} + C$

69. $-\frac{1}{2}\ln|4 - x^2| + C$ **71.** $\ln\frac{1}{\sqrt{9 - x^2}} + C$

73. $\frac{1}{6}\ln\left|\frac{x+3}{x-3}\right| + C$

75. $\frac{2x^{3/2}}{3} - x + 2\sqrt{x} - 2\ln(\sqrt{x}+1) + C$

77. $2\sin\sqrt{x} + C$

79. $\ln|u + \sqrt{1+u^2}| + C$

81. $\frac{1}{12}\ln\left|\frac{3+v}{3-v}\right| + \frac{1}{6}\tan^{-1}\frac{v}{3} + C$

83. $\frac{x^2}{2} + 2x + 3\ln|x-1| - \frac{1}{x-1} + C$

85. $-\cos(2\sqrt{x}) + C$

87. $\frac{\sqrt{3}}{3}\tan^{-1}\left(\frac{\theta-1}{\sqrt{3}}\right) + C$

89. $\frac{1}{4}\sec^2\theta + C$

91. $-\frac{2}{3}(x+4)\sqrt{2-x} + C$

93. $\frac{1}{2}[x\ln|x-1| - x - \ln|x-1|] + C$

95. $\frac{1}{4}\ln|z| - \frac{1}{4z} - \frac{1}{4}\left[\frac{1}{2}\ln(z^2+4) + \frac{1}{2}\tan^{-1}\left(\frac{z}{2}\right)\right] + C$

97. $-\frac{\tan^{-1}x}{x} + \ln|x| - \ln\sqrt{1+x^2} + C$

99. $\tan x - x + C$

101. $\ln|\csc(2x) + \cot(2x)| + C$

103. $\frac{1}{4}$

105. $\sec^{-1}|2x-1| + C$

107. $\frac{1}{6}(3+4e^{\theta})^{3/2} + C$

109. $\frac{1}{3}\left(\frac{27^{3\theta+1}}{\ln 27}\right) + C$

111. $2\sqrt{r} - 2\ln(1+\sqrt{r}) + C$

113. $4\sec^{-1}\left|\frac{7m}{2}\right| + C$

115. The limit does not exist.

117. 2

119. 1

121. 0

123. $-\frac{1}{2}$

125. 1

127. ∞

129. $\frac{\pi}{2}$

131. 6

133. ln 3

135. 2

137. $\frac{\pi}{6}$

139. Diverges

141. Diverges

143. Converges

145. $\ln|y-1| - \ln|y| = e^x - 1 - \ln 2$

147. $y = \ln|x-2| - \ln|x-1| + \ln 2$

Chapter 7 Additional Exercises, pp. 603–606

1. $x(\sin^{-1}x)^2 + 2(\sin^{-1}x)\sqrt{1-x^2} - 2x + C$

3. $\frac{x^2\sin^{-1}x}{2} + \frac{x\sqrt{1-x^2} - \sin^{-1}x}{4} + C$

5. $\frac{\ln|\sec 2\theta + \tan 2\theta| + 2\theta}{4} + C$

7. $\frac{1}{2}[\ln|t - \sqrt{1-t^2}| - \sin^{-1}t] + C$

9. $\frac{1}{16}\ln\left|\frac{x^2+2x+2}{x^2-2x+2}\right| + \frac{1}{8}[\tan^{-1}(x+1) + \tan^{-1}(x-1)] + C$

11. $\frac{\pi}{2}$

13. $\frac{1}{\sqrt{e}}$

15. 0

17. 1

19. $\frac{32\pi}{35}$

21. 2π

23. **(a)** π **(b)** $\pi(2e-5)$

25. **(b)** $\pi\left[\frac{8(\ln 2)^2}{3} - \frac{16(\ln 2)}{9} + \frac{16}{27}\right]$

27. $\frac{1}{2}$

31. $\frac{\pi}{2}(3b-a) + 2$

33. 6

35. $P(x) = -3x^2 + 1$

37. $\frac{1}{2} < p \le 1$

39. **(b)** 1

41. $\frac{e^{2x}}{13}(3\sin 3x + 2\cos 3x) + C$

43. $\frac{\cos x\sin 3x - 3\sin x\cos 3x}{8} + C$

45. $\frac{e^{ax}}{a^2+b^2}(a\sin bx - b\cos bx) + C$

47. $x\ln(ax) - x + C$

CHAPTER 8

Section 8.1, pp. 617–619

1. $a_1 = 0, a_2 = -\frac{1}{4}, a_3 = -\frac{2}{9}, a_4 = -\frac{3}{16}$

3. $a_1 = 1, a_2 = -\frac{1}{3}, a_3 = \frac{1}{5}, a_4 = -\frac{1}{7}$

5. $a_n = (-1)^{n+1}, n \ge 1$

7. $a_n = n^2 - 1, n \ge 1$

9. $a_n = 4n - 3, n \ge 1$

11. $a_n = \frac{1+(-1)^{n+1}}{2}, n \ge 1$

13. Converges, 2

15. Converges, -1

17. Diverges

19. Diverges

21. Converges, $\frac{1}{2}$

23. Converges, $\sqrt{2}$

25. Converges, 0

27. Converges, 0

29. Diverges

31. Converges, e^7

33. Converges, 1

35. Converges, 1

37. Converges, 4

39. Converges, 0

41. Diverges

43. Converges, e^{-1}

45. Converges, $e^{2/3}$

47. Converges, x $(x > 0)$

49. Converges, 0

51. Converges, $\frac{\pi}{2}$

53. Converges, 0

55. Converges, 0

57. $N = 692, a_n = \sqrt[n]{0.5}, L = 1$

59. $N = 65, a_n = (0.9)^n, L = 0$

61. (b) $\sqrt{2}$

63. (b) 1

Section 8.2, pp. 625–627

1. $1, \frac{3}{2}, \frac{7}{4}, \frac{15}{8}, \frac{31}{16}, \frac{63}{32}, \frac{127}{64}, \frac{255}{128}, \frac{511}{256}, \frac{1023}{512}$

3. $2, 1, -\frac{1}{2}, -\frac{1}{4}, \frac{1}{8}, \frac{1}{16}, -\frac{1}{32}, -\frac{1}{64}, \frac{1}{128}, \frac{1}{256}$

5. 1, 1, 2, 3, 5, 8, 13, 21, 34, 55

7. (b) $\sqrt{3}$

9. **(a)** $f(x) = x^2 - 2$, $1.414213562 \approx \sqrt{2}$

(b) $f(x) = \tan(x) - 1$, $0.7853981635 \approx \frac{\pi}{4}$

(c) $f(x) = e^x$, diverges

11. Nondecreasing, bounded

13. Not nondecreasing, bounded

15. Converges, monotonic sequence theorem

17. Converges, monotonic sequence theorem

19. Diverges, definition of divergence

21. Converges, monotonic sequence theorem

23. Diverges, definition of divergence

27. 1

29. -0.73908513

31. 0.85375017

Section 8.3, pp. 637–639

1. $s_n = \dfrac{2\left[1 - \left(\frac{1}{3}\right)^n\right]}{1 - \left(\frac{1}{3}\right)}, 3$

3. $s_n = \dfrac{1 - \left(-\frac{1}{2}\right)^n}{1 - \left(-\frac{1}{2}\right)}, \frac{2}{3}$

5. $s_n = \frac{1}{2} - \frac{1}{n+2}, \frac{1}{2}$

7. $1 - \frac{1}{4} + \frac{1}{16} - \frac{1}{64} + \cdots, \frac{4}{5}$

9. $(5+1) + \left(\frac{5}{2} + \frac{1}{3}\right) + \left(\frac{5}{4} + \frac{1}{9}\right) + \left(\frac{5}{8} + \frac{1}{27}\right) + \cdots, \frac{23}{2}$

11. $(1+1) + \left(\frac{1}{2} - \frac{1}{6}\right) + \left(\frac{1}{4} + \frac{1}{25}\right) + \left(\frac{1}{8} - \frac{1}{125}\right) + \cdots, \frac{17}{6}$

13. 1

15. 5

17. 1

19. Converges, $2 + \sqrt{2}$

21. Converges, 1

23. Converges, $\dfrac{e^2}{e^2 - 1}$

25. Converges, $\dfrac{x}{x-1}$

27. Diverges

29. Diverges

31. Diverges

33. $a = 1, r = -x$; converges to $\dfrac{1}{1+x}$ for $|x| < 1$

35. $a = 3, r = \dfrac{x-1}{2}$, converges to $\dfrac{6}{3-x}$ for x in $(-1, 3)$

37. $|x| < \frac{1}{2}, \dfrac{1}{1-2x}$

39. $1 < x < 5, \dfrac{2}{x-1}$

41. $\dfrac{23}{99}$

43. $\dfrac{7}{9}$

45. $\dfrac{41{,}333}{33{,}300}$

47. 28 m

49. 8 m^2

51. **(a)** $3\left(\frac{4}{3}\right)^{n-1}$

(b) $A_n = A + \frac{1}{3}A + \frac{1}{3}\left(\frac{4}{9}\right)A + \cdots + \frac{1}{3}\left(\frac{4}{9}\right)^{n-2}A$, $\lim_{n\to\infty} A_n = \dfrac{2\sqrt{3}}{5}$

53. **(a)** $\displaystyle\sum_{n=-2}^{\infty} \frac{1}{(n+4)(n+5)}$

(b) $\displaystyle\sum_{n=0}^{\infty} \frac{1}{(n+2)(n+3)}$

(c) $\displaystyle\sum_{n=5}^{\infty} \frac{1}{(n-3)(n-2)}$

55. $\ln\left(\frac{8}{9}\right)$

61. It diverges.

Section 8.4, pp. 649–651

1. Diverges

3. Diverges

5. Converges

7. Converges

9. Diverges; $\dfrac{1}{2\sqrt{n} + \sqrt[3]{n}} \geq \dfrac{1}{2n+n} = \dfrac{1}{3n}$

11. Converges; $\dfrac{\sin^2 n}{2^n} \leq \dfrac{1}{2^n}$

13. Converges; $\left(\dfrac{n}{3n+1}\right)^n < \left(\dfrac{n}{3n}\right)^n = \left(\dfrac{1}{3}\right)^n$

15. Diverges; limit comparison with $\sum \frac{1}{n}$

17. Converges; limit comparison with $\sum \frac{1}{n^2}$

19. Converges; limit comparison with $\sum \frac{1}{n^{5/4}}$

21. Converges, $\rho = 1/2$

23. Diverges, $\rho = \infty$

25. Converges, $\rho = 1/10$

27. Converges, $\rho = 0$

29. Converges, $\rho = 0$

31. Converges, $\rho = 0$

33. Diverges, $\rho = \infty$

35. Converges; geometric series, $r = \frac{1}{e} < 1$

37. Diverges; p-series, $p < 1$

39. Diverges; limit comparison with $\sum \frac{1}{n}$

41. Converges; limit comparison with $\sum \frac{1}{n^{3/2}}$

43. Converges; Ratio Test

45. Converges; Ratio Test

47. Converges; Integral Test

49. Converges; Integral Test

51. Converges; compare with $\sum \frac{3}{(1.25)^n}$

53. Converges; compare with $\sum \frac{1}{n^2}$

55. Converges; compare with $\sum \frac{1}{n^2}$

57. Converges; $\dfrac{\tan^{-1} n}{n^{1.1}} < \dfrac{\left(\frac{\pi}{2}\right)}{n^{1.1}}$

59. Diverges; nth-Term Test

61. Converges; Ratio Test

63. Diverges; Ratio Test

65. Diverges; $a_n = \left(\frac{1}{3}\right)^{(1/n!)} \to 1$

71. $a = 1$

Section 8.5, pp. 658–660

1. Converges by Theorem 8

3. Diverges; $a_n \not\to 0$

5. Converges by Theorem 8

7. Diverges; $a_n \to \frac{1}{2} \neq 0$

9. Converges by Theorem 8

11. Converges absolutely. Series of absolute values is a convergent geometric series.

13. Converges conditionally. $\frac{1}{\sqrt{n+1}} \to 0$ but $\sum_{n=1}^{n} \frac{1}{\sqrt{n+1}}$ diverges.

15. Converges absolutely. Compare with $\sum_{n=1}^{\infty} \frac{1}{n^2}$.

17. Converges conditionally. $\frac{1}{n+3} \to 0$ but $\sum_{n=1}^{\infty} \frac{1}{n+3}$ diverges. Compare with $\sum_{n=1}^{\infty} \frac{1}{n}$.

19. Diverges; $\frac{3+n}{5+n} \to 1$

21. Converges conditionally; $\left(\frac{1}{n^2} + \frac{1}{n}\right) \to 0$ but $\frac{1+n}{n^2} > \frac{1}{n}$.

23. Converges absolutely; Ratio Test

25. Converges absolutely by Integral Test

27. Diverges; $a_n \not\to 0$

29. Converges absolutely by the Ratio Test

31. Converges absolutely; $\frac{1}{n^2 + 2n + 1} < \frac{1}{n^2}$

33. Converges absolutely since $\left|\frac{\cos n\pi}{n\sqrt{n}}\right| = \left|\frac{(-1)^{n+1}}{n^{3/2}}\right| = \frac{1}{n^{3/2}}$ (convergent p-series).

35. Converges absolutely n^{th} by Root Test

37. Diverges; $a_n \to \infty$

39. Converges conditionally; $\sqrt{n+1} - \sqrt{n} = \frac{1}{\sqrt{n} + \sqrt{n+1}} \to 0$, but series of absolute values diverges. Compare with $\Sigma \frac{1}{\sqrt{n}}$.

41. Diverges; $a_n \to \frac{1}{2} \neq 0$

43. Converges absolutely; $\operatorname{sech} n = \frac{2}{e^n + e^{-n}} = \frac{2e^n}{e^{2n} + 1} < \frac{2e^n}{e^{2n}} = \frac{2}{e^n}$, a term from a convergent geometric series.

45. $|\text{Error}| < 0.2$

47. $|\text{Error}| < 2 \times 10^{-11}$

49. 0.54030

51. **(a)** $a_n \geq a_{n+1}$ fails
(b) $-\frac{1}{2}$

Section 8.6, pp. 668–669

1. **(a)** $1, -1 < x < 1$ **(b)** $-1 < x < 1$ **(c)** None

3. **(a)** $\frac{1}{4}, -\frac{1}{2} < x < 0$ **(b)** $-\frac{1}{2} < x < 0$ **(c)** None

5. **(a)** $10, -8 < x < 12$ **(b)** $-8 < x < 12$ **(c)** None

7. **(a)** $1, -1 < x < 1$ **(b)** $-1 < x < 1$ **(c)** None

9. **(a)** $3, [-3, 3]$ **(b)** $[-3, 3]$ **(c)** None

11. **(a)** ∞, for all x **(b)** For all x **(c)** None

13. **(a)** ∞, for all x **(b)** For all x **(c)** None

15. **(a)** $1, -1 \leq x < 1$ **(b)** $-1 < x < 1$ **(c)** $x = -1$

17. **(a)** $5, -8 < x < 2$ **(b)** $-8 < x < 2$ **(c)** None

19. **(a)** $3, -3 < x < 3$ **(b)** $-3 < x < 3$ **(c)** None

21. **(a)** $1, -1 < x < 1$ **(b)** $-1 < x < 1$ **(c)** None

23. **(a)** $0, x = 0$ **(b)** $x = 0$ **(c)** None

25. **(a)** $2, -4 < x \leq 0$ **(b)** $-4 < x < 0$ **(c)** $x = 0$

27. **(a)** $1, -1 \leq x \leq 1$ **(b)** $-1 \leq x \leq 1$ **(c)** None

29. **(a)** $\frac{1}{4}, 1 \leq x \leq \frac{3}{2}$ **(b)** $1 \leq x \leq \frac{3}{2}$ **(c)** None

31. **(a)** $1, (-1 - \pi) \leq x < (1 - \pi)$
(b) $(-1 - \pi) < x < (1 - \pi)$
(c) $x = -1 - \pi$

33. $-1 < x < 3, \frac{4}{3 + 2x - x^2}$

35. $0 < x < 16, \frac{2}{4 - \sqrt{x}}$

37. $-\sqrt{2} < x < \sqrt{2}, \frac{3}{2 - x^2}$

39. $1 < x < 5, \frac{2}{x-1}, 1 < x < 5, \frac{-2}{(x-1)^2}$

41. **(a)** $\cos x = 1 - \frac{x^2}{2!} + \frac{x^4}{4!} - \frac{x^6}{6!} + \frac{x^8}{8!} - \frac{x^{10}}{10!} + \cdots$; converges for all x

(b) and **(c)** $2x - \frac{2^3x^3}{3!} + \frac{2^5x^5}{5!} - \frac{2^7x^7}{7!} + \frac{2^9x^9}{9!} - \frac{2^{11}x^{11}}{11!} + \cdots$

43. **(a)** $\frac{x^2}{2} + \frac{x^4}{12} + \frac{x^6}{45} + \frac{17x^8}{2520} + \frac{31x^{10}}{14{,}175}, -\frac{\pi}{2} < x < \frac{\pi}{2}$

(b) $1 + x^2 + \frac{2x^4}{3} + \frac{17x^6}{45} + \frac{62x^8}{315} + \cdots, -\frac{\pi}{2} < x < \frac{\pi}{2}$

Section 8.7, pp. 681–683

1. $P_0(x) = 0, P_1(x) = x - 1, P_2(x) = (x-1) - \frac{1}{2}(x-1)^2$,
$P_3(x) = (x-1) - \frac{1}{2}(x-1)^2 + \frac{1}{3}(x-1)^3$

3. $P_0(x) = \frac{1}{2}, P_1(x) = \frac{1}{2} - \frac{x}{4}, P_2(x) = \frac{1}{2} - \frac{x}{4} + \frac{x^2}{8}$,
$P_3(x) = \frac{1}{2} - \frac{x}{4} + \frac{x^2}{8} - \frac{x^3}{16}$

5. $P_0(x) = \frac{1}{\sqrt{2}}$, $P_1(x) = \frac{1}{\sqrt{2}} - \frac{1}{\sqrt{2}}\left(x - \frac{\pi}{4}\right)$,
$P_2(x) = \frac{1}{\sqrt{2}} - \frac{1}{\sqrt{2}}\left(x - \frac{\pi}{4}\right) - \frac{1}{2\sqrt{2}}\left(x - \frac{\pi}{4}\right)^2$, $P_3(x) =$
$\frac{1}{\sqrt{2}} - \frac{1}{\sqrt{2}}\left(x - \frac{\pi}{4}\right) - \frac{1}{2\sqrt{2}}\left(x - \frac{\pi}{4}\right)^2 + \frac{1}{6\sqrt{2}}\left(x - \frac{\pi}{4}\right)^3$

7. $\sum_{n=0}^{\infty} \frac{(-x)^n}{n!} = 1 - x + \frac{x^2}{2!} - \frac{x^3}{3!} + \frac{x^4}{4!} - \cdots$

9. $\sum_{n=0}^{\infty} \frac{(-1)^n 3^{2n+1} x^{2n+1}}{(2n+1)!}$

11. $\sum_{n=0}^{\infty} \frac{x^{2n}}{(2n)!}$

13. $x^4 - 2x^3 - 5x + 4$

15. $8 + 10(x-2) + 6(x-2)^2 + (x-2)^3$

17. $\sum_{n=0}^{\infty} (-1)^n(n+1)(x-1)^n$

19. $\sum_{n=0}^{\infty} \frac{e^2}{n!}(x-2)^n$

21. $\sum_{n=0}^{\infty} \frac{(-5x)^n}{n!} = 1 - 5x + \frac{5^2x^2}{2!} - \frac{5^3x^3}{3!} + \cdots$

23. $\sum_{n=0}^{\infty} \frac{(-1)^n\left(\frac{\pi x}{2}\right)^{2n+1}}{(2n+1)!} = \frac{\pi x}{2} - \frac{\pi^3 x^3}{2^3 \cdot 3!} + \frac{\pi^5 x^5}{2^5 \cdot 5!} - \frac{\pi^7 x^7}{2^7 \cdot 7!} + \cdots$

25. $\sum_{n=0}^{\infty} \frac{x^{n+1}}{n!} = x + x^2 + \frac{x^3}{2!} + \frac{x^4}{3!} + \frac{x^5}{4!} + \cdots$

27. $\sum_{n=2}^{\infty} \frac{(-1)^n x^{2n}}{(2n)!} = \frac{x^4}{4!} - \frac{x^6}{6!} + \frac{x^8}{8!} - \frac{x^{10}}{10!} + \cdots$

29. $x - \frac{\pi^2 x^3}{2!} + \frac{\pi^4 x^5}{4!} - \frac{\pi^6 x^7}{6!} + \cdots = \sum_{n=0}^{\infty} \frac{(-1)^n \pi^{2n} x^{2n+1}}{(2n)!}$

31. $\sum_{n=1}^{\infty} \frac{(-1)^{n+1}(2x)^{2n}}{2 \cdot (2n)!} = \frac{(2x)^2}{2 \cdot 2!} - \frac{(2x)^4}{2 \cdot 4!} + \frac{(2x)^6}{2 \cdot 6!} - \frac{(2x)^8}{2 \cdot 8!} + \cdots$

33. $\sum_{n=1}^{\infty} \frac{(-1)^{n-1} 2^n x^{n+1}}{n} = 2x^2 - \frac{2^2 x^3}{2} + \frac{2^3 x^4}{3} - \frac{2^4 x^5}{4} + \cdots$

35. $|x| < (0.06)^{1/5} < 0.56968$

37. $|\text{Error}| < \frac{(10^{-3})^3}{6} < 1.67 \times 10^{-10}$, $-10^{-3} < x < 0$

39. **(a)** $|\text{Error}| < \frac{(3^{0.1})(0.1)^3}{6} < 1.87 \times 10^{-4}$
(b) $|\text{Error}| < \frac{(0.1)^3}{6} < 1.67 \times 10^{-4}$

45. **(a)** $L(x) = 0$ **(b)** $Q(x) = -\frac{x^2}{2}$

47. **(a)** $L(x) = 1$ **(b)** $Q(x) = 1 + \frac{x^2}{2}$

Section 8.8, pp. 690–691

1. $1 + \frac{x}{2} - \frac{x^2}{8} + \frac{x^3}{16}$

3. $1 + \frac{1}{2}x - \frac{3}{8}x^2 + \frac{5}{16}x^3 + \cdots$

5. $1 - x + \frac{3x^2}{4} - \frac{x^3}{2}$

7. $1 - \frac{x^3}{2} + \frac{3x^6}{8} - \frac{5x^9}{16}$

9. $1 + \frac{1}{2x} - \frac{1}{8x^2} + \frac{1}{16x^3}$

11. $(1+x)^4 = 1 + 4x + 6x^2 + 4x^3 + x^4$

13. $(1-2x)^3 = 1 - 6x + 12x^2 - 8x^3$

15. $y = \sum_{n=0}^{\infty} \frac{(-1)^n}{n!} x^n = e^{-x}$

17. $y = \sum_{n=1}^{\infty} \frac{x^n}{n!} = e^x - 1$

19. $y = \sum_{n=2}^{\infty} \frac{x^n}{n!} = e^x - x - 1$

21. $y = \sum_{n=0}^{\infty} \frac{x^{2n}}{2^n n!} = e^{x^2/2}$

23. $y = \sum_{n=0}^{\infty} 2x^n = \frac{2}{1-x}$

25. $y = \sum_{n=0}^{\infty} \frac{x^{2n+1}}{(2n+1)!} = \sinh x$

27. $y = 2 + x - 2\sum_{n=1}^{\infty} \frac{(-1)^{n+1} x^{2n}}{(2n)!}$

29. $y = -2(x-2) - \sum_{n=1}^{\infty} \left[\frac{2(x-2)^{2n}}{(2n)!} + \frac{3(x-2)^{2n+1}}{(2n+1)!}\right]$

31. $y = a + bx + \frac{1}{6}x^3 - \frac{ax^4}{3 \cdot 4} - \frac{bx^5}{4 \cdot 5} - \frac{x^7}{6 \cdot 6 \cdot 7} + \frac{ax^8}{3 \cdot 4 \cdot 7 \cdot 8} + \frac{bx^9}{4 \cdot 5 \cdot 8 \cdot 9} + \cdots$. For $n \ge 6$, $a_n = \frac{a_{n-4}}{n(n-1)}$

33. $\frac{x^3}{3} - \frac{x^7}{7 \cdot 3!} + \frac{x^{11}}{11 \cdot 5!}$

35. **(a)** $\frac{x^2}{2} - \frac{x^4}{12}$
(b) $\frac{x^2}{2} - \frac{x^4}{3 \cdot 4} + \frac{x^6}{5 \cdot 6} - \frac{x^8}{7 \cdot 8} + \cdots + (-1)^{15}\frac{x^{32}}{31 \cdot 32}$

37. $\frac{1}{2}$

39. -1

41. $2!$

45. 500 terms

47. **(a)** $x + \frac{x^3}{6} + \frac{3x^5}{40} + \frac{5x^7}{112}$, radius of convergence = 1
(b) $\frac{\pi}{2} - x - \frac{x^3}{6} - \frac{3x^5}{40} - \frac{5x^7}{112}$

Section 8.9, pp. 697–698

1. $f(x) = 1$

3. $f(x) = \sum_{n=1}^{\infty} \frac{2(-1)^{n+1}}{n} \sin nx$

5. $f(x) = \frac{\pi^2}{12} + \sum_{n=1}^{\infty} \frac{(-1)^n}{n^2} \cos nx$

7. $f(x) = \frac{2 \sinh \pi}{\pi}\left[\frac{1}{2} + \sum_{n=1}^{\infty} \frac{(-1)^n}{n^2+1}(\cos nx - n \sin nx)\right]$

9. $f(x) = \frac{1}{2}\cos x + \frac{1}{\pi}\sum_{n=2}^{\infty} \frac{n(1+(-1)^n)}{n^2-1} \sin nx$

11. $f(x) = \frac{1}{2} + \frac{2}{\pi}\sum_{k=0}^{\infty} \frac{(-1)^k}{2k+1} \cos(2k+1)x$

13. $f(x) = \frac{5}{4} + \frac{4}{\pi^2}\sum_{n=1}^{\infty} \frac{1}{n^2}\left[(-1)^n - \cos\frac{n\pi}{2}\right](\cos(n\pi x) + \frac{2}{\pi}\sum_{n=1}^{\infty} \frac{1}{n}\left[(-1)^n - \frac{2}{n\pi}\sin\frac{n\pi}{2}\right]\sin(n\pi x)$

15. Set $x = \pi$, $\frac{\pi^2}{4} = \frac{\pi^2}{12} + \sum_{n=1}^{\infty} \frac{(-1)^n}{n^2}\cos n\pi$, or $\frac{\pi^2}{6} = \sum_{n=1}^{\infty} \frac{1}{n^2} = 1 + \frac{1}{4} + \frac{1}{9} + \frac{1}{16} + \cdots + \frac{1}{n^2} + \cdots$.

17. 0

19. 0 if $m \neq n$; L if $m = n$

21. 0 if $m \neq n$; 0 if $m = n$

Section 8.10, pp. 705–706

1. $f(x) = \frac{\pi}{2} + \frac{2}{\pi}\sum_{n=1}^{\infty}\frac{[(-1)^n - 1]}{n^2}\cos nx$

3. $f(x) = (e - 1) + 2\sum_{n=1}^{\infty}\frac{[e(-1)^n - 1]}{1 + n^2\pi^2}\cos n\pi x$

5. $f(x) = -\frac{1}{4} + \frac{4}{\pi}\sum_{n=1}^{\infty}\left[\frac{1}{n}\sin\frac{n\pi}{2} + \frac{1}{\pi n^2}\left((-1)^{n+1} + \cos\frac{n\pi}{2}\right)\right]\cos\frac{n\pi x}{2}$

7. $f(x) = \frac{1}{2} + \sum_{n=1}^{\infty}\frac{4}{n^2\pi^2}\left[1 + (-1)^n - 2\cos\frac{n\pi}{2}\right]\cos n\pi x$

9. $f(x) = 2\sum_{n=1}^{\infty}\frac{(-1)^n}{n\pi}\sin n\pi x$

11. $f(x) = \frac{8}{\pi}\sum_{k=1}^{\infty}\frac{k}{4k^2 - 1}\sin 2kx$

13. $f(x) = \sin x$

15. $f(x) = \frac{2}{\pi}\sum_{n=1}^{\infty}\left[\frac{1}{n} - \frac{2}{n^2\pi}\sin\frac{n\pi}{2}\right]\sin\frac{n\pi x}{2}$

17. **(a)** $f(x) = \frac{4}{\pi}\left[\sin x + \frac{\sin 3x}{3} + \frac{\sin 5x}{5} + \frac{\sin 7x}{7} + \cdots\right]$

(b) Evaluate $f(x)$ at $x = \frac{\pi}{2} \Rightarrow \frac{\pi}{4} = 1 - \frac{1}{3} + \frac{1}{5} - \frac{1}{7} + \cdots$.

19. $\sum_{n=1}^{\infty}\frac{(-1)^n}{4n^2 - 1} = \frac{1}{2} - \frac{\pi}{4}$

Practice 8 Practice Exercises, pp. 708–711

1. Converges to 1

3. Converges to -1

5. Diverges

7. Converges to 0

9. Converges to 1

11. Converges to e^{-5}

13. Converges to 3

15. Converges to ln 2

17. Diverges

19. $\frac{1}{6}$

21. $\frac{3}{2}$

23. $\frac{e}{e - 1}$

25. Diverges

27. Converges conditionally

29. Converges conditionally

31. Converges absolutely

33. Converges absolutely

35. Converges absolutely

37. Converges absolutely

39. Converges absolutely

41. **(a)** $3, -7 \le x < -1$
(b) $-7 < x < -1$
(c) $x = -7$

43. **(a)** $\frac{1}{3}, 0 \le x \le \frac{2}{3}$ **(b)** $0 \le x \le \frac{2}{3}$ **(c)** None

45. **(a)** ∞, for all x **(b)** For all x **(c)** None

47. **(a)** $\sqrt{3}, -\sqrt{3} < x < \sqrt{3}$
(b) $-\sqrt{3} < x < \sqrt{3}$
(c) None

49. **(a)** $e, (-e, e)$ **(b)** $(-e, e)$ **(c)** $\{\ \}$

51. $\frac{1}{1 + x}, \frac{1}{4}, \frac{4}{5}$

53. $\sin x, \pi, 0$

55. e^x, ln 2, 2

57. $\sum_{n=0}^{\infty} 2^n x^n$

59. $\sum_{n=0}^{\infty}\frac{(-1)^n\pi^{2n+1}x^{2n+1}}{(2n + 1)!}$

61. $\sum_{n=0}^{\infty}\frac{(-1)^n x^{5n}}{(2n)!}$

63. $\sum_{n=0}^{\infty}\frac{\left(\frac{\pi x}{2}\right)^n}{n!}$

65. $2 - \frac{(x + 1)}{2 \cdot 1!} + \frac{3(x + 1)^2}{2^3 \cdot 2!} + \frac{9(x + 1)^3}{2^5 \cdot 3!} + \cdots$

67. $\frac{1}{4} - \frac{1}{4^2}(x - 3) + \frac{1}{4^3}(x - 3)^2 - \frac{1}{4^4}(x - 3)^3 + \cdots$

69. $y = \sum_{n=0}^{\infty}\frac{(-1)^{n+1}}{n!}x^n = -e^{-x}$

71. $y = 3\sum_{n=0}^{\infty}\frac{(-1)^n 2^n}{n!}x^n = 3e^{-2x}$

73. $y = -1 - x + 2\sum_{n=2}^{\infty}\frac{x^n}{n!} = 2e^x - 3x - 3$

75. $y = -1 - x + 2\sum_{n=0}^{\infty}\frac{x^n}{n!} = 2e^x - 1 - x$

77. **(a)** $\frac{7}{2}$

79. **(a)** $\frac{1}{12}$

81. **(a)** -2

83. $r = -3, s = \frac{9}{2}$

85. $f(x) = \frac{1}{2} + \frac{6}{\pi}\sum_{n=1}^{\infty}\frac{1}{2n - 1}\sin[(2n - 1)x]$

87. $f(x) = \pi - 2\sum_{n=1}^{\infty}\frac{(-1)^n}{n}\sin nx$

89. $f(x) = \frac{3}{2} + \frac{2}{\pi^2}\sum_{n=1}^{\infty}\frac{(-1)^n - 1}{n^2}\cos\frac{n\pi x}{2} - \frac{2}{\pi}\sum_{n=1}^{\infty}\frac{(-1)^n}{n}\sin\frac{n\pi x}{2}$

91. **(a)** $f(x) = \frac{1}{2} + \frac{2}{\pi}\sum_{n=1}^{\infty}\frac{\sin\left(\frac{n\pi}{2}\right)}{n}\cos n\pi x$

(b) $f(x) = \frac{2}{\pi}\sum_{n=1}^{\infty}\frac{1}{n}\left(1 - \cos\frac{n\pi}{2}\right)\sin n\pi x$

93. **(a)** $f(x) = \frac{2}{\pi} + \frac{1}{\pi}\sum_{n=1}^{\infty}\left[\frac{1}{n + 1} - \frac{1}{n - 1} + \frac{\cos[(n - 1)\pi]}{n - 1} - \frac{\cos[(n + 1)\pi]}{n + 1}\right]\cos n\pi x$

(b) $f(x) = \sin\pi x$

95. **(a)** $f(x) = 6 + \frac{12}{\pi^2}\sum_{n=1}^{\infty}\frac{4(-1)^n - 1}{n^2}\cos\frac{n\pi x}{3}$

(b) $f(x) = \frac{6}{\pi^3}\sum_{n=1}^{\infty}\frac{(6 - 5n^2\pi^2)(-1)^n - 6}{n^3}\sin\frac{n\pi x}{3}$

97. **(b)** $|\text{Error}| < \sin\left(\frac{1}{42}\right) < 0.02381$; an underestimate because the remainder is positive

99. $\frac{2}{3}$

101. $\ln\left(\frac{n+1}{2n}\right)$; the series converges to $\ln\left(\frac{1}{2}\right)$.

103. **(a)** ∞ **(b)** $a = 1, b = 0$

105. It converges.

113. **(a)** -3 is the fixed point. **(b)** 0.2 is the fixed point.

Chapter 8 Additional Exercises, pp. 711–715

1. Converges; Direct Comparison Test

3. Diverges; nth-Term Test

5. Converges; Direct Comparison Test

7. Diverges; nth-Term Test

9. With $a = \frac{\pi}{3}$, $\cos x = \frac{1}{2} - \frac{\sqrt{3}}{2}\left(x - \frac{\pi}{3}\right) - \frac{1}{4}\left(x - \frac{\pi}{3}\right)^2 + \frac{\sqrt{3}}{12}\left(x - \frac{\pi}{3}\right)^3 + \cdots$.

11. With $a = 0$, $e^x = 1 + x + \frac{x^2}{2!} + \frac{x^3}{3!} + \cdots$.

13. With $a = 22\pi$, $\cos x = 1 - \frac{1}{2}(x - 22\pi)^2 + \frac{1}{4!}(x - 22\pi)^4 - \frac{1}{6!}(x - 22\pi)^6 + \cdots$.

15. Converges, limit $= b$ **17.** $\frac{\pi}{2}$

21. **(a)** $\frac{b^2\sqrt{3}}{4}\sum_{n=0}^{\infty}\frac{3^n}{4^n}$ **(b)** $\sqrt{3}b^2$

(c) No. For example, the three vertices of the original triangle are not removed. The set of points not removed has area 0.

23. **(a)** No, the limit does not appear to depend on the value of a.

(b) Yes, the limit depends on the value of b.

(c) $\lim_{n\to\infty}\left(1 - \frac{\cos\left(\frac{a}{n}\right)}{bn}\right)^n = e^{-1/b}$

25. $b = \pm\frac{1}{5}$ **29.** **(b)** Yes

35. **(a)** $\sum_{n=1}^{\infty} nx^{n-1}$ **(b)** 6 **(c)** $\frac{1}{q}$

37. **(a)** $R_n = \frac{C_0 e^{-kt_0}(1 - e^{-nkt_0})}{1 - e^{-kt_0}}$, $R = \frac{C_0(e^{-kt_0})}{1 - e^{-kt_0}} = \frac{C_0}{e^{kt_0} - 1}$

(b) $R_1 = \frac{1}{e} \approx 0.368$, $R_{10} = R(1 - e^{-10}) \approx R(0.9999546) \approx 0.58195$; $R \approx 0.58198$; $0 < \frac{R - R_{10}}{R} < 0.0001$

(c) 7

CHAPTER 9

Section 9.1, pp. 726–728

1. **(a)** $\langle 9, -6\rangle$ **(b)** $3\sqrt{13}$

3. **(a)** $\langle 1, 3\rangle$ **(b)** $\sqrt{10}$

5. **(a)** $\langle 12, -19\rangle$ **(b)** $\sqrt{505}$

7. **(a)** $\left\langle \frac{1}{5}, \frac{14}{5}\right\rangle$ **(b)** $\frac{\sqrt{197}}{5}$

9. $\langle 1, -4\rangle$ **11.** $\langle -2, -3\rangle$

13. $\left\langle -\frac{1}{2}, \frac{\sqrt{3}}{2}\right\rangle$ **15.** $\left\langle -\frac{\sqrt{3}}{2}, -\frac{1}{2}\right\rangle$

17. The vector **v** is horizontal and 1 in. long. The vectors **u** and **w** are $\frac{11}{16}$ in. long. **w** is vertical and **u** makes a 45° angle with the horizontal. All vectors must be drawn to scale.

(a)

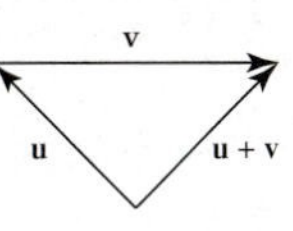

(b)

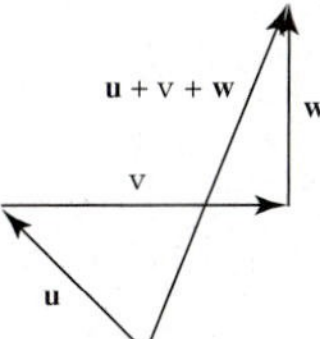

(c)

(d)

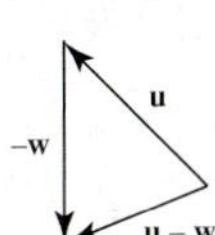

19.

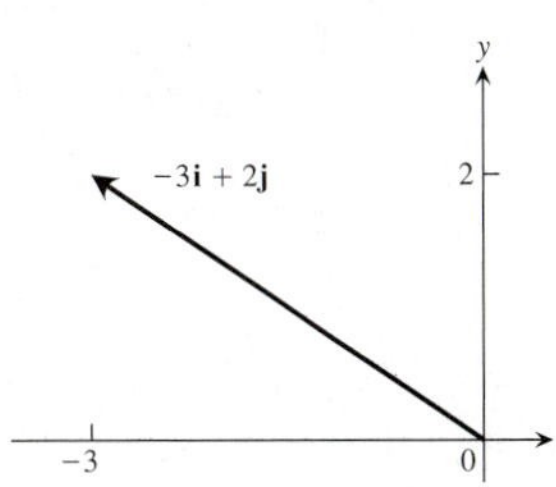

21.

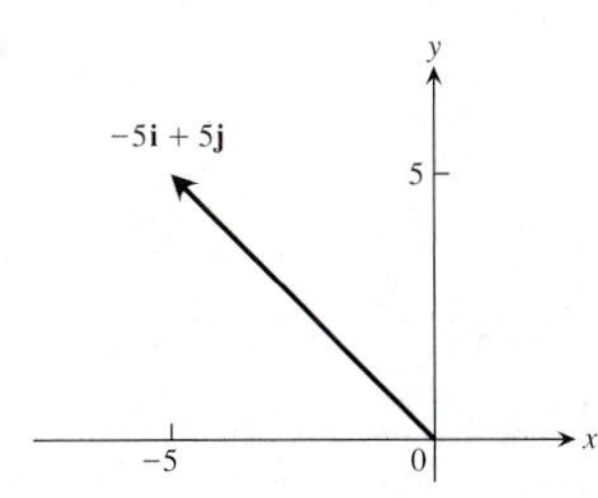

23.

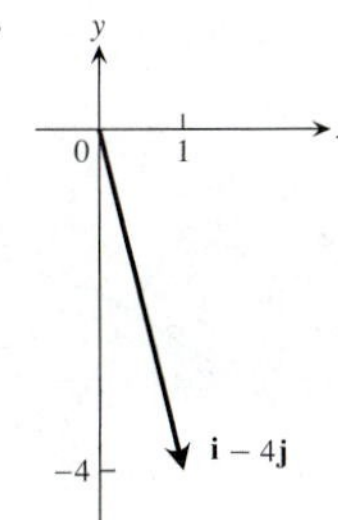

25.

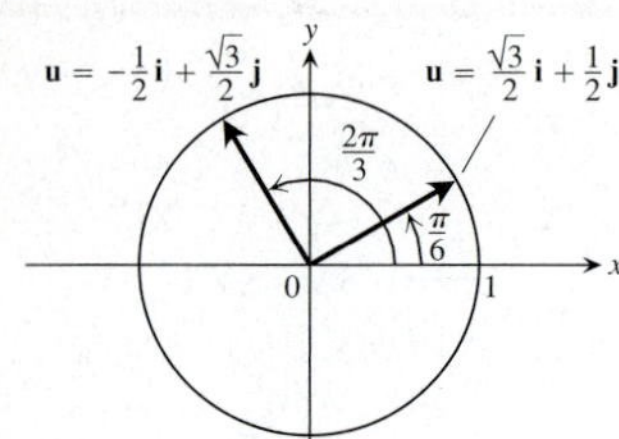

27.

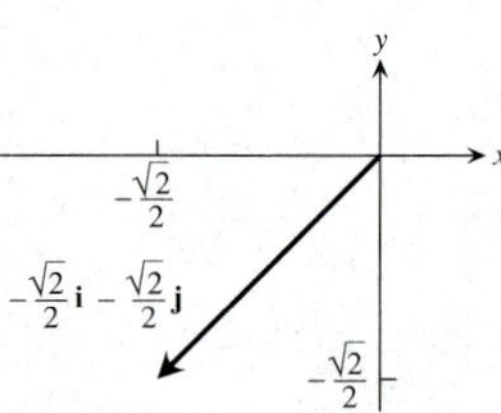

29. $\left\langle \frac{3}{5}, \frac{4}{5} \right\rangle$

31. $\left\langle -\frac{15}{17}, \frac{8}{17} \right\rangle$

33. $\frac{3}{5}\mathbf{i} - \frac{4}{5}\mathbf{j}$

35. $13\left(\frac{5}{13}\mathbf{i} + \frac{12}{13}\mathbf{j}\right)$

37. $\frac{3}{5}\mathbf{i} - \frac{4}{5}\mathbf{j}$ and $-\frac{3}{5}\mathbf{i} + \frac{4}{5}\mathbf{j}$

39. $\mathbf{u} = \frac{1}{\sqrt{17}}\mathbf{i} + \frac{4}{\sqrt{17}}\mathbf{j}, -\mathbf{u} = -\frac{1}{\sqrt{17}}\mathbf{i} - \frac{4}{\sqrt{17}}\mathbf{j},$
$\mathbf{n} = \frac{4}{\sqrt{17}}\mathbf{i} - \frac{1}{\sqrt{17}}\mathbf{j}, -\mathbf{n} = -\frac{4}{\sqrt{17}}\mathbf{i} + \frac{1}{\sqrt{17}}\mathbf{j}$

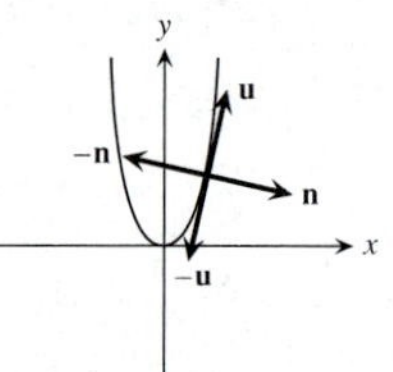

41. $\mathbf{u} = \frac{1}{\sqrt{5}}(2\mathbf{i} + \mathbf{j}), -\mathbf{u} = \frac{1}{\sqrt{5}}(-2\mathbf{i} - \mathbf{j}), \mathbf{n} = \frac{1}{\sqrt{5}}(-\mathbf{i} + 2\mathbf{j}), -\mathbf{n} = \frac{1}{\sqrt{5}}(\mathbf{i} - 2\mathbf{j})$

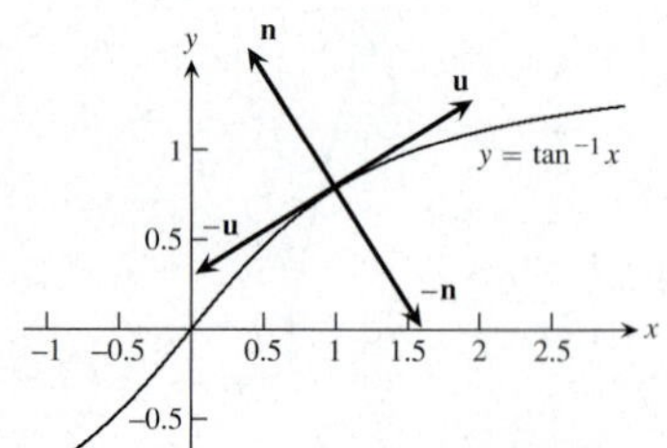

43. $\mathbf{u} = \frac{\pm 1}{5}(-4\mathbf{i} + 3\mathbf{j}), \mathbf{v} = \frac{\pm 1}{5}(3\mathbf{i} + 4\mathbf{j})$

45. $\mathbf{u} = \frac{\pm 1}{2}(\mathbf{i} + \sqrt{3}\mathbf{j}), \mathbf{v} = \frac{\pm 1}{2}(-\sqrt{3}\mathbf{i} + \mathbf{j})$

47. $a = \frac{3}{2}, b = \frac{1}{2}$

49. $5\sqrt{3}\,\mathbf{i}, 5\mathbf{j}$

51. $\approx \langle -338.095, 725.046 \rangle$

53. **(a)** $(5 \cos 60°, 5 \sin 60°) = \left(\frac{5}{2}, \frac{5\sqrt{3}}{2}\right)$

(b) $(5 \cos 60° + 10 \cos 315°, 5 \sin 60° + 10 \sin 315°) = \left(\frac{5 + \sqrt{2}}{2}, \frac{5\sqrt{3} - 10\sqrt{2}}{2}\right)$

Section 9.2, pp. 735–737

1.

$\mathbf{v} \cdot \mathbf{u}$	$\lvert \mathbf{v} \rvert$	$\lvert \mathbf{u} \rvert$
-12	$2\sqrt{5}$	$2\sqrt{5}$

$\cos \theta$	$\lvert \mathbf{u} \rvert \cos \theta$	$\text{proj}_{\mathbf{v}}\, \mathbf{u}$
$-\frac{3}{5}$	$-\frac{6\sqrt{5}}{5}$	$-\frac{6}{5}\mathbf{i} + \frac{12}{5}\mathbf{j}$

3.

$\mathbf{v} \cdot \mathbf{u}$	$\lvert \mathbf{v} \rvert$	$\lvert \mathbf{u} \rvert$
$\sqrt{3} - \sqrt{2}$	$\sqrt{2}$	$\sqrt{5}$

$\cos \theta$	$\lvert \mathbf{u} \rvert \cos \theta$	$\text{proj}_{\mathbf{v}}\, \mathbf{u}$
$\frac{\sqrt{30} - \sqrt{20}}{10}$	$\frac{\sqrt{6} - 2}{2}$	$\frac{\sqrt{3} - \sqrt{2}}{2}(-\mathbf{i} + \mathbf{j})$

5.

$\mathbf{v} \cdot \mathbf{u}$	$\lvert \mathbf{v} \rvert$	$\lvert \mathbf{u} \rvert$
$\frac{1}{6}$	$\frac{\sqrt{30}}{6}$	$\frac{\sqrt{30}}{6}$

$\cos \theta$	$\lvert \mathbf{u} \rvert \cos \theta$	$\text{proj}_{\mathbf{v}}\, \mathbf{u}$
$\frac{1}{5}$	$\frac{1}{\sqrt{30}}$	$\frac{1}{5}\left\langle \frac{1}{\sqrt{2}}, \frac{1}{\sqrt{3}} \right\rangle$

7. ≈ 0.64 rad

9. ≈ 1.85

11. Angle at $A = \cos^{-1}\left(\frac{1}{\sqrt{5}}\right) \approx 63.435$ degrees, angle at $B = \cos^{-1}\left(\frac{3}{5}\right) \approx 53.130$ degrees, angle at $C = \cos^{-1}\left(\frac{1}{\sqrt{5}}\right) \approx$ 63.435 degrees.

13. The sum of two vectors of equal length is *always* orthogonal to their difference, as we can see from the equation $(\mathbf{v}_1 + \mathbf{v}_2) \cdot (\mathbf{v}_1 - \mathbf{v}_2) = \mathbf{v}_1 \cdot \mathbf{v}_1 + \mathbf{v}_2 \cdot \mathbf{v}_1 - \mathbf{v}_1 \cdot \mathbf{v}_2 - \mathbf{v}_2 \cdot \mathbf{v}_2 = \lvert \mathbf{v}_1 \rvert^2 - \lvert \mathbf{v}_2 \rvert^2 = 0$.

19. Horizontal component: ≈ 1188 ft/sec, vertical component: ≈ 167 ft/sec

21. **(a)** Since $\lvert \cos \theta \rvert \le 1$, we have $\lvert \mathbf{u} \cdot \mathbf{v} \rvert = \lvert \mathbf{u} \rvert \lvert \mathbf{v} \rvert \lvert \cos \theta \rvert \le \lvert \mathbf{u} \rvert \lvert \mathbf{v} \rvert (1) = \lvert \mathbf{u} \rvert \lvert \mathbf{v} \rvert$.

(b) We have equality precisely when $\lvert \cos \theta \rvert = 1$ or when one or both of $\mathbf{u}$ and $\mathbf{v}$ are $\mathbf{0}$. In the case of nonzero vectors, we have equality when $\theta = 0$ or π, that is, when the vectors are parallel.

23. *a*

27. $x + 2y = 4$

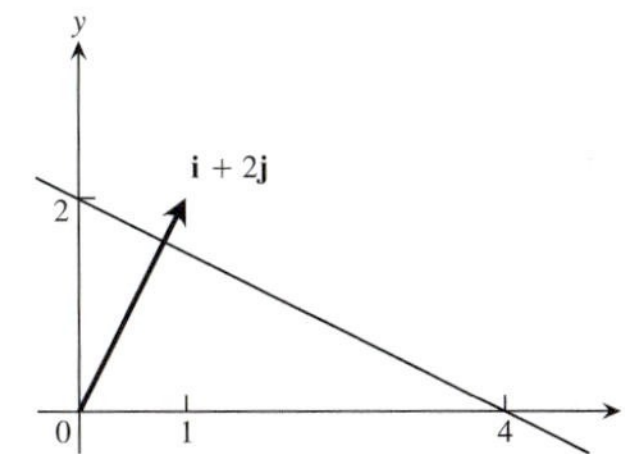

29. $-2x + y = -3$

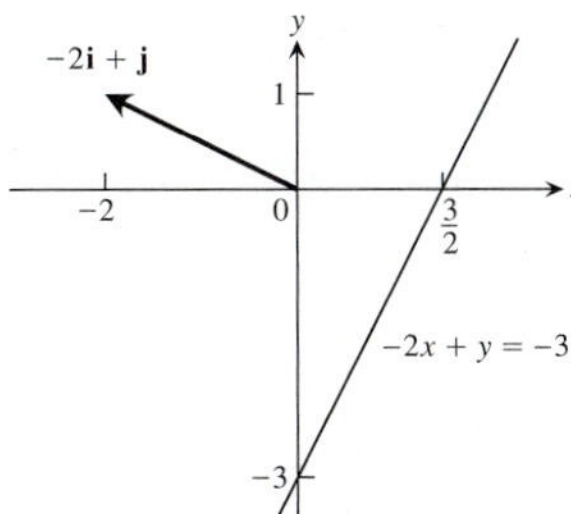

31. $x + y = -1$

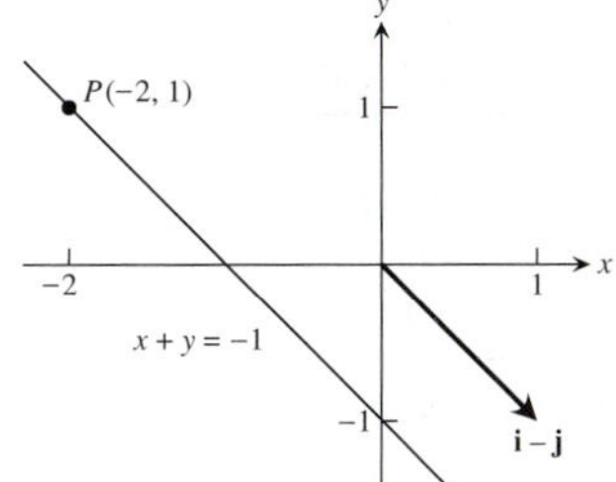

33. $2x - y = 0$

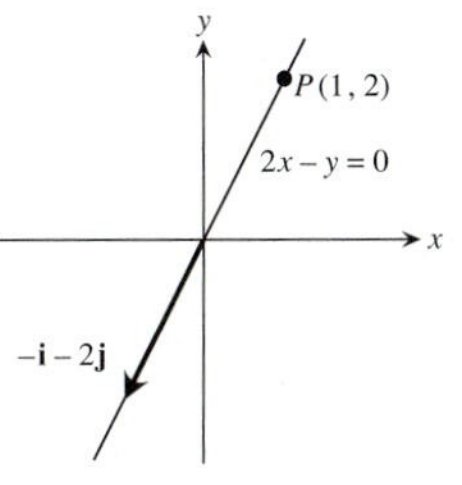

35. 5 J

37. 3464 J

39. $\frac{\pi}{4}$

41. $\frac{\pi}{6}$

43. 0.14

45. $\frac{\pi}{3}$ and $\frac{2\pi}{3}$ at each point

47. At (0, 0): $\frac{\pi}{2}$; at (1, 1): $\frac{\pi}{4}$ and $\frac{3\pi}{4}$

Section 9.3, pp. 746–749

1. **(a)**

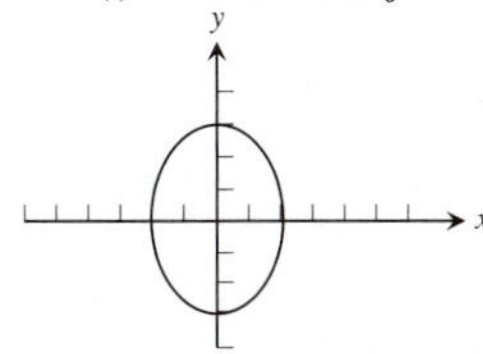

(b) $\mathbf{v}(t) = (-2 \sin t)\mathbf{i} + (3 \cos t)\mathbf{j}$, $\mathbf{a}(t) = (-2 \cos t)\mathbf{i} + (-3 \sin t)\mathbf{j}$

(c) Speed = 2, direction = $\langle -1, 0 \rangle$

(d) Velocity = $2\langle -1, 0 \rangle$

3. **(a)**

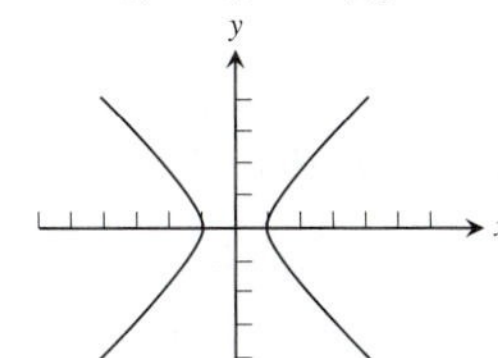

(b) $\mathbf{v} = (\sec t \tan t)\mathbf{i} + (\sec^2 t)\mathbf{j}$, $\mathbf{a}(t) = (\sec t \tan^2 t + \sec^3 t)\mathbf{i} + (2 \sec^2 t \tan t)\mathbf{j}$

(c) Speed = $\frac{2\sqrt{5}}{3}$, direction = $\left\langle \frac{1}{\sqrt{5}}, \frac{2}{\sqrt{5}} \right\rangle$

(d) Velocity = $\left(\frac{2\sqrt{5}}{3}\right)\left\langle \frac{1}{\sqrt{5}}, \frac{2}{\sqrt{5}} \right\rangle$

5. $t = 0, \pi, 2\pi$

7. t = all nonnegative integer multiples of $\frac{\pi}{2}$

9. $\cos^{-1}\left(\frac{3}{5}\right) \approx 53.130$ degrees

11. **(a)** $3\mathbf{i}$ **(b)** $t \neq 0, -3$ **(c)** $t = 0, -3$

13. **(a)** $y = -1$ **(b)** $x = 0$

15. $-3\mathbf{i} + (4\sqrt{2} - 2)\mathbf{j}$

17. $(\sec t)\mathbf{i} + (\ln|\sec t|)\mathbf{j} + \mathbf{C}$

19. $\mathbf{r}(t) = ((t + 1)^{3/2} - 1)\mathbf{i} - (e^{-t} - 1)\mathbf{j}$

21. $\mathbf{r}(t) = (8t + 100)\mathbf{i} + (-16t^2 + 8t)\mathbf{j}$

23. 2

25. **(a)** $\mathbf{v}(t) = (\cos t)\mathbf{i} - (2 \sin 2t)\mathbf{j}$

(b) $t = \frac{\pi}{2}, \frac{3\pi}{2}$

(c) $y = 1 - 2x^2$, $-1 \le x \le 1$. The particle starts at (0, 1), goes to (1, −1), then goes to (−1, −1), and then goes to (0, 1), tracing the curve twice.

27. $\mathbf{r}(t) = \left(\frac{3}{2}t^2 + \frac{3\sqrt{10}}{5}t + 1\right)\mathbf{i} + \left(-\frac{1}{2}t^2 - \frac{\sqrt{10}}{5}t + 2\right)\mathbf{j}$

29. **(a)** **i.** Constant speed
ii. Yes, orthogonal
iii. Counterclockwise movement
iv. Yes

(b) **i.** Constant speed
ii. Yes, orthogonal
iii. Counterclockwise movement
iv. Yes

(c) **i.** Constant speed
ii. Yes, orthogonal
iii. Counterclockwise movement
iv. No

(d) **i.** Constant speed
ii. Yes, orthogonal
iii. Clockwise movement
iv. Yes

(e) **i.** Variable speed
ii. Not orthogonal in general
iii. Counterclockwise movement
iv. Yes

31. **(a)** 160 sec **(b)** 225 m **(c)** $\frac{15}{4}$ m/sec

33. **(a)** Referring to the figure, look at the circular arc from the point where $t = 0$ to the point "m". On one hand, this arc has length given by $(r_0\theta)$, but it also has length given by (vt). Setting those two quantities equal gives the result.

(b) $\mathbf{a}(t) = -\frac{v^2}{r_0}\left[\left(\cos\frac{vt}{r_0}\right)\mathbf{i} + \left(\sin\frac{vt}{r_0}\right)\mathbf{j}\right]$

(c) From part (b), $\mathbf{a}(t) = -\left(\frac{v}{r_0}\right)^2 \mathbf{r}(t)$. So, by Newton's second law, $\mathbf{F} = -m\left(\frac{v}{r_0}\right)^2 \mathbf{r}$. Substituting for $\mathbf{F}$ in the law of gravitation gives the result.

(d) Set $\frac{vT}{r_0} = 2\pi$ and solve for vT.

(e) Substitute $\frac{2\pi r_0}{T}$ for v in $v^2 = \frac{GM}{r_0}$ and solve for T^2.

35. **(a)** Apply the corollary to each component separately.
(b) Follows immediately from part (a) since any two antiderivatives of $\mathbf{r}(t)$ must have identical derivatives, namely $\mathbf{r}(t)$.

37. Let $\mathbf{C} = \langle C_1, C_2\rangle$, $\frac{d\mathbf{C}}{dt} = \left\langle \frac{dC_1}{dt}, \frac{dC_2}{dt}\right\rangle = \langle 0, 0\rangle$.

39. $\mathbf{u} = \langle u_1, u_2\rangle$, $\mathbf{v} = \langle v_1, v_2\rangle$

(a)
$$\begin{aligned}\frac{d}{dt}(\mathbf{u}+\mathbf{v}) &= \frac{d}{dt}(\langle u_1 + v_1, u_2 + v_2\rangle)\\ &= \left\langle \frac{d}{dt}(u_1+v_1), \frac{d}{dt}(u_2+v_2)\right\rangle\\ &= \langle u'_1 + v'_1, u'_2 + v'_2\rangle\\ &= \langle u'_1, u'_2\rangle + \langle v'_1, v'_2\rangle = \frac{d\mathbf{u}}{dt} + \frac{d\mathbf{v}}{dt}\end{aligned}$$

(b)
$$\begin{aligned}\frac{d}{dt}(\mathbf{u}-\mathbf{v}) &= \frac{d}{dt}(\langle u_1 - v_1, u_2 - v_2\rangle)\\ &= \left\langle \frac{d}{dt}(u_1-v_1), \frac{d}{dt}(u_2-v_2)\right\rangle\\ &= \langle u'_1 - v'_1, u'_2 - v'_2\rangle\\ &= \langle u'_1, u'_2\rangle - \langle v'_1, v'_2\rangle = \frac{d\mathbf{u}}{dt} - \frac{d\mathbf{v}}{dt}\end{aligned}$$

41. $f(t)$ and $g(t)$ differentiable at $c \Rightarrow f(t)$ and $g(t)$ continuous at $c \Rightarrow$ $\mathbf{r}(t) = f(t)\mathbf{i} + g(t)\mathbf{j}$ is continuous at c.

43. **(a)** Let $\mathbf{r}(t) = f(t)\mathbf{i} + g(t)\mathbf{j}$. Then $\frac{d}{dt}\int_a^t \mathbf{r}(q)\,dq = \frac{d}{dt}\int_a^t [f(q)\mathbf{i} + g(q)\mathbf{j}]\,dq = \frac{d}{dt}\left[\left(\int_a^t f(q)\,dq\right)\mathbf{i} + \left(\int_a^t g(q)\,dq\right)\mathbf{j}\right] = \left(\frac{d}{dt}\int_a^t f(q)\,dq\right)\mathbf{i} + \left(\frac{d}{dt}\int_a^t g(q)\,dq\right)\mathbf{j} = f(t)\mathbf{i} + g(t)\mathbf{j} = \mathbf{r}(t)$.

(b) Let $\mathbf{S}(t) = \int_a^t r(q)\,dq$. Then part (a) shows that $\mathbf{S}(t)$ is an antiderivative of $\mathbf{r}(t)$. Let $\mathbf{R}(t)$ be any antiderivative of $\mathbf{r}(t)$. Then, according to Exercise 35, part (b), $\mathbf{S}(t) = \mathbf{R}(t) + \mathbf{C}$. Letting $t = a$, we have $\mathbf{0} = \mathbf{S}(a) = \mathbf{R}(a) + \mathbf{C}$. Therefore, $\mathbf{C} = -\mathbf{R}(a)$ and $\mathbf{S}(t) = \mathbf{R}(t) - \mathbf{R}(a)$. The result follows by letting $t = b$.

Section 9.4, pp. 757–760

1. 50 sec
3. **(a)** 72.2 sec, 25,510 m **(b)** 4020 m **(c)** 6378 m
5. $t \approx 2.135$ sec, $x \approx 66.42$ ft
7. **(a)** $v_0 \approx 9.9$ m/sec
(b) $\alpha \approx 18.4°$ or $71.6°$
9. 190 mph
11. The golf ball will clip the leaves at the top.
13. 149 ft/sec, 2.25 sec **15.** 39.3° or 50.7°
17. 46.6 ft/sec **21.** 1.92 sec, 73.7 ft (approx.)
23. 4.00 ft, 7.80 ft/sec
25. **(b)** $\mathbf{v}_0$ would bisect $\angle AOR$
27. **(a)** (Assuming that "x" is zero at the point of impact.) $\mathbf{r}(t) = (x(t))\mathbf{i} + (y(t))\mathbf{j}$, where $x(t) = (35\cos 27°)t$ and $y(t) = 4 + (35\sin 27°)t - 16t^2$.
(b) At $t \approx 0.497$ sec, it reaches its maximum height of about 7.945 ft.
(c) Range $\approx$ 37,45 ft, flight time $\approx$ 1.201 sec
(d) At $t \approx 0.254$ and $t \approx 0.740$ sec, when it is ≈ 29.554 and ≈ 14.396 ft from where it will land.
(e) Yes. It changes things because the ball won't clear the net.

31. **(a)** $\mathbf{r}(t) = (x(t))\mathbf{i} + (y(t))\mathbf{j}$, where $x(t) = \left(\frac{1}{0.08}\right)(1 - e^{-0.08t})(152\cos 20° - 7.6)$ and $y(t) = 3 + \left(\frac{152}{0.08}\right)(1 - e^{-0.08t})(\sin 20°) + \left(\frac{32}{0.08^2}\right)(1 - 0.08t - e^{-0.08t})$
(b) At $t \approx 1.527$ sec, it reaches its maximum height of about 41.893 ft.
(c) Range $\approx$ 351.734 ft, flight time $\approx$ 3.181 sec.
(d) At $t \approx 0.877$ and $t \approx 2.190$ sec, when it is about 106.028 and 251.530 ft from home plate.
(e) No. The wind gust would need to be greater than 12.846 ft/sec in the direction of the hit for the ball to clear the fence for a home run.

Section 9.5, pp. 768–770

1. (a) and (e) are the same.
(b) and (g) are the same.
(c) and (h) are the same.
(d) and (f) are the same.
3. **(a)** (1, 1) **(b)** (1, 0) **(c)** (0, 0) **(d)** (−1, −1)

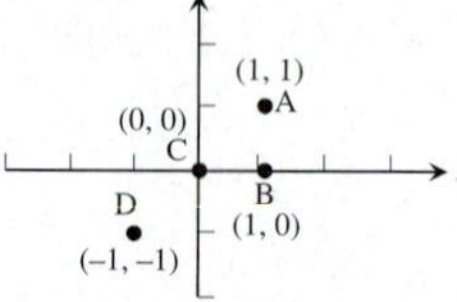

5. **(a)** $\left(\sqrt{2}, \frac{3\pi}{4}\right)$ or $\left(\sqrt{2}, -\frac{5\pi}{4}\right)$

(b) $\left(2, -\frac{\pi}{3}\right)$ or $\left(-2, \frac{2\pi}{3}\right)$

(c) $\left(3, \frac{\pi}{2}\right)$ or $\left(3, \frac{5\pi}{2}\right)$

(d) $(1, \pi)$ or $(-1, 0)$

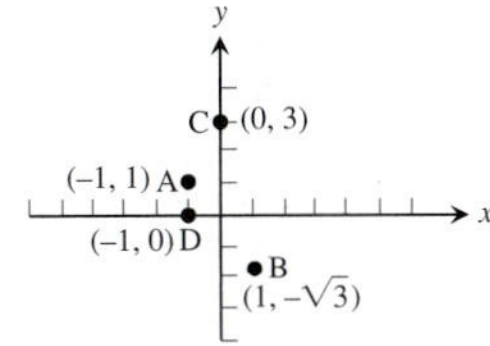

7.

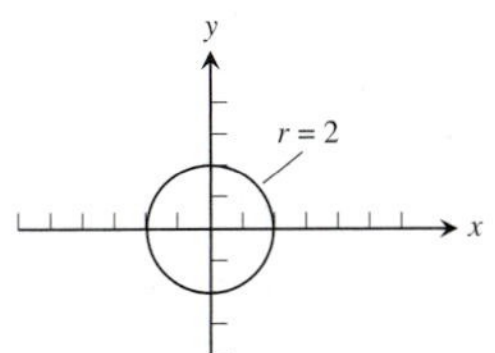

9.

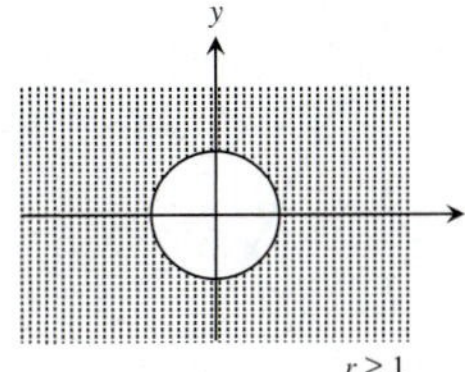

11.

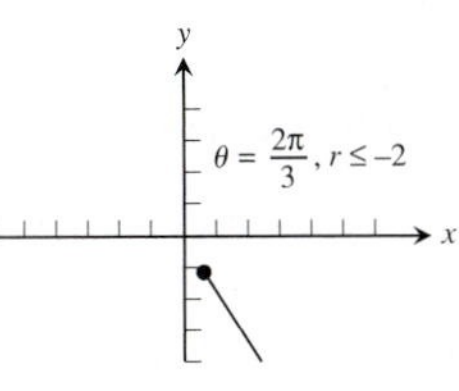

13.

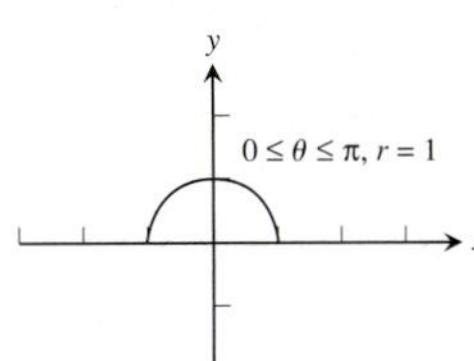

15.

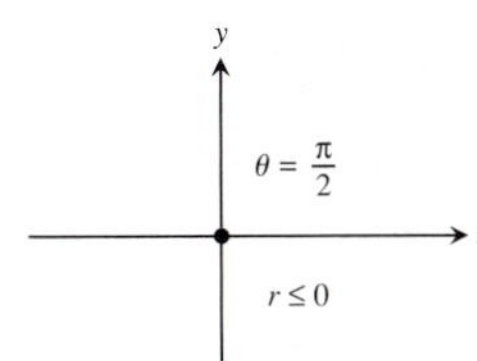

17.

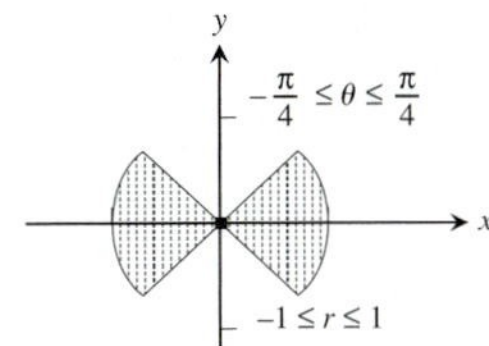

19. $y = 0$, the x-axis

21. $y = 4$, a horizontal line

23. $x + y = 1$, a line (slope $= -1$, y-intercept $= 1$)

25. $x^2 + (y - 2)^2 = 4$, a circle (center $= (0, 2)$, radius $= 2$)

27. $xy = 1$ $\left(\text{or } y = \frac{1}{x}\right)$, a hyperbola

29. $y = e^x$, the exponential curve

31. $y = \ln x$, the logarithmic curve

33. $(x + 2)^2 + y^2 = 4$, a circle (center $= (-2, 0)$, radius $= 2$)

35. $(x - 1)^2 + (y - 1)^2 = 2$, a circle (center $= (1, 1)$, radius $= \sqrt{2}$)

37. $r \cos \theta = 7$

39. $\theta = \frac{\pi}{4}$

41. $r^2 = 4$ or $r = 2$

43. $r^2(4 \cos^2 \theta + 9 \sin^2 \theta) = 36$

45. $r \sin^2 \theta = 4 \cos \theta$

47. $r = 4 \sin \theta$

49. **(a)**

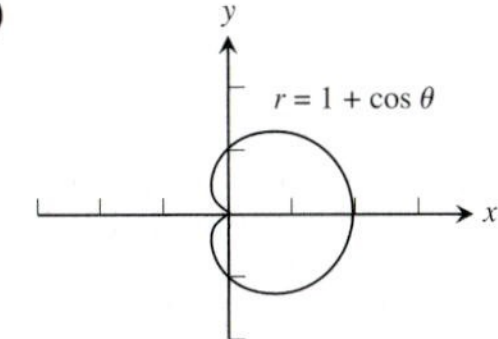

(b) Length of interval $= 2\pi$

51. **(a)**

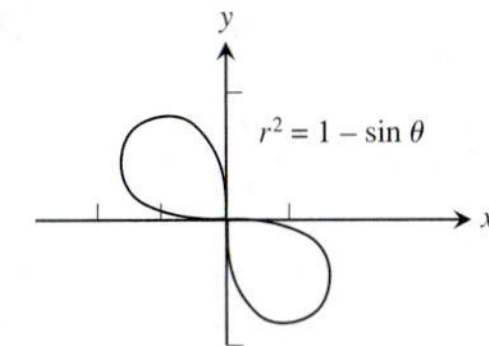

(b) Length of interval $= \frac{\pi}{2}$

53. **(a)**

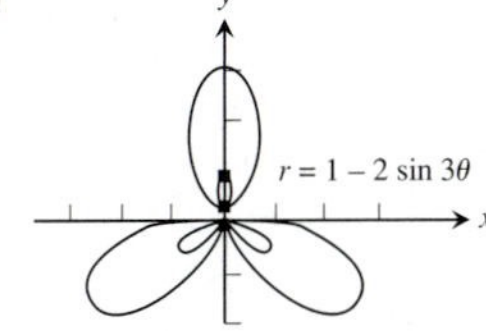

(b) Length of interval $= 2\pi$

55. **(a)**

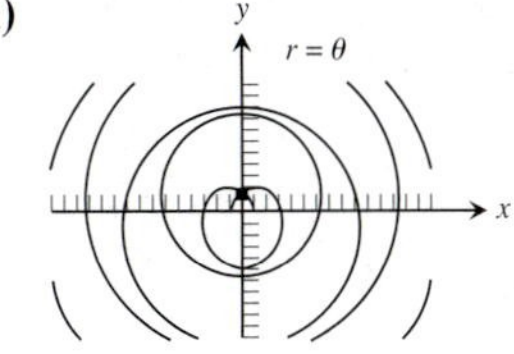

(b) Required interval $= (-\infty, \infty)$

57. **(a)**

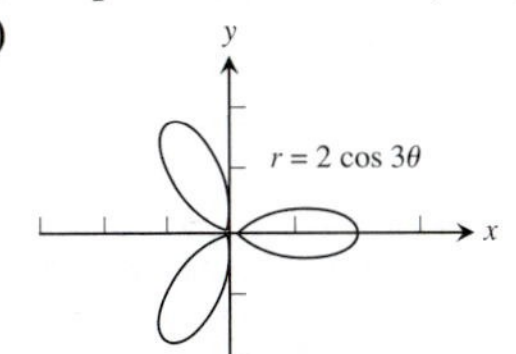

(b) Length of interval $= \pi$

59. x-axis, y-axis, origin

61. y-axis

63. **(a)** Because $r = a \sec \theta$ is equivalent to $r \cos \theta = a$, which is equivalent to the Cartesian equation $x = a$
(b) $r = a \csc \theta$ is equivalent to $y = a$.

67. $(0, 0), \left(1, \frac{\pi}{2}\right), \left(1, \frac{3\pi}{2}\right)$

69. $(0, 0), \left(\frac{1}{2}, \pm\frac{\pi}{3}\right)$

71. $(0, 0), \left(\pm\frac{1}{\sqrt[4]{2}}, \frac{\pi}{8}\right)$

73. $\left(1, \frac{\pi}{12}\right), \left(1, \frac{5\pi}{12}\right), \left(1, \frac{7\pi}{12}\right), \left(1, \frac{11\pi}{12}\right), \left(1, \frac{13\pi}{12}\right), \left(1, \frac{17\pi}{12}\right), \left(1, \frac{19\pi}{12}\right), \left(1, \frac{23\pi}{12}\right)$

75. Part (a)

81. $d = [(x_2 - x_1)^2 + (y_2 - y_1)^2]^{1/2} = [(r_2 \cos \theta_2 - r_1 \cos \theta_1)^2 + (r_2 \sin \theta_2 - r_1 \sin \theta_1)^2]^{1/2}$, and then simplify using trigonometric identities.

Section 9.6, pp. 777–779

1. At $\theta = 0$: -1; at $\theta = \pi$: 1

3. At (2, 0): $-\frac{2}{3}$; at $\left(-1, \frac{\pi}{2}\right)$: 0; at $(2, \pi)$: $\frac{2}{3}$; at $\left(5, \frac{3\pi}{2}\right)$: 0

5. $\theta = \frac{\pi}{2}$ $[x = 0]$

7. $\theta = 0$ $[y = 0]$, $\theta = \frac{\pi}{5}\left[y = \left(\tan\frac{\pi}{5}\right)x\right]$, $\theta = \frac{2\pi}{5}\left[y = \left(\tan\frac{2\pi}{5}\right)x\right]$, $\theta = \frac{3\pi}{5}\left[y = \left(\tan\frac{3\pi}{5}\right)x\right]$, $\theta = \frac{4\pi}{5}\left[y = \left(\tan\frac{4\pi}{5}\right)x\right]$

9. Horizontal at $\left(-\frac{1}{2}, \frac{\pi}{6}\right)\left[y = -\frac{1}{4}\right]$, $\left(-\frac{1}{2}, \frac{5\pi}{6}\right)\left[y = -\frac{1}{4}\right]$, $\left(-2, \frac{3\pi}{2}\right)$ $[y = 2]$;
vertical at $\left(0, \frac{\pi}{2}\right)$ $[x = 0]$, $\left(-\frac{3}{2}, \frac{7\pi}{6}\right)\left[x = \frac{3\sqrt{3}}{4}\right]$, $\left(-1.5, \frac{11\pi}{6}\right)\left[x = -\frac{3\sqrt{3}}{4}\right]$

11. Horizontal at (0, 0) $[y = 0]$, $\left(2, \frac{\pi}{2}\right)$ $[y = 2]$, $(0, \pi)$ $[y = 0]$; vertical at $\left(\sqrt{2}, \frac{\pi}{4}\right)$ $[x = 1]$, $\left(\sqrt{2}, \frac{3\pi}{4}\right)$ $[x = -1]$

13. 18π

15. $\frac{\pi}{8}$

17. 2

19. $\frac{\pi}{2} - 1$

21. $5\pi - 8$

23. $3\sqrt{3} - \pi$

25. $\frac{\pi}{3} + \frac{\sqrt{3}}{2}$

27. $12\pi - 9\sqrt{3}$

29. **(a)** $\frac{3}{2} - \frac{\pi}{4}$

31. $\frac{19}{3}$

33. 8

35. $3(\sqrt{2} + \ln(1 + \sqrt{2}))$

37. $\frac{\pi}{8} + \frac{3}{8}$

39. 2π

45. **(a)** $n = \left(\frac{r_0}{b}\right)\left(\sqrt{\frac{bL}{\pi r_0^2} + 1} - 1\right)$

(b) The take up reel slows down as time progresses.

(c) Since L is proportional to time, the formula in part (a) shows that n will grow as the square root of time.

Chapter 9 Practice Exercises, pp. 780–784

1. **(a)** $\langle -17, 32\rangle$ **(b)** $\sqrt{1313}$

3. **(a)** $\langle 6, -8\rangle$ **(b)** 10

5. $\left\langle -\frac{\sqrt{3}}{2}, -\frac{1}{2}\right\rangle$ [assuming counterclockwise]

7. $\left\langle \frac{8}{\sqrt{17}}, -\frac{2}{\sqrt{17}}\right\rangle$

9. Length = 2, direction is $\frac{1}{\sqrt{2}}\mathbf{i} + \frac{1}{\sqrt{2}}\mathbf{j}$.

11. $\left.\frac{d\mathbf{r}}{dt}\right|_{t=\pi/2} = 2(-\mathbf{i})$

13. Unit tangents $\pm\left(\frac{1}{\sqrt{5}}\mathbf{i} + \frac{2}{\sqrt{5}}\mathbf{j}\right)$, unit normals $\pm\left(-\frac{2}{\sqrt{5}}\mathbf{i} + \frac{1}{\sqrt{5}}\mathbf{j}\right)$

15.

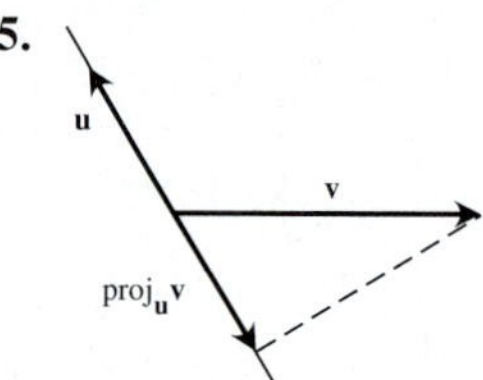

17. $|\mathbf{v}| = \sqrt{2}$, $|\mathbf{u}| = \sqrt{5}$, $\mathbf{u}\cdot\mathbf{v} = \mathbf{v}\cdot\mathbf{u} = 3$, $\theta = \cos^{-1}\left(\frac{3}{\sqrt{10}}\right) \approx$ 0.32 rad, $|\mathbf{u}|\cos\theta = \frac{3\sqrt{2}}{2}$, $\text{proj}_{\mathbf{v}}\,\mathbf{u} = \frac{3}{2}(\mathbf{i} + \mathbf{j})$

19. $\mathbf{u} = \left(\frac{2}{5}\mathbf{i} - \frac{1}{5}\mathbf{j}\right) + \left(\frac{3}{5}\mathbf{i} + \frac{6}{5}\mathbf{j}\right)$

21. **(a)** $\mathbf{v}(t) = (-4\sin t)\mathbf{i} + (\sqrt{2}\cos t)\mathbf{j}$,
$\mathbf{a}(t) = (-4\cos t)\mathbf{i} + (-\sqrt{2}\sin t)\mathbf{j}$
(b) 3
(c) $\cos^{-1}\frac{7}{9} \approx 38.942$ degrees

23. 1

25. $6\mathbf{i}$

27. $\mathbf{r}(t) = (\cos t - 1)\mathbf{i} + (\sin t + 1)\mathbf{j}$

29. $\mathbf{r}(t) = \mathbf{i} + t^2\mathbf{j}$

31.

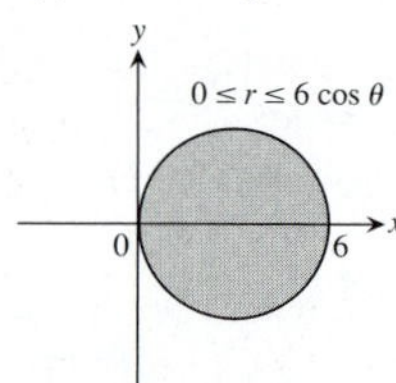

33. (d)

35. (l)

37. (k)

39. (i)

41. **(a)**

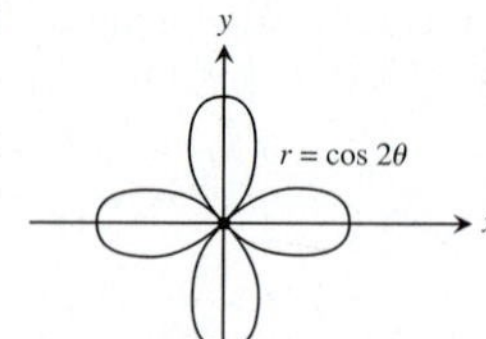

(b) 2π

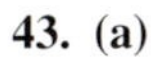

43. **(a)**

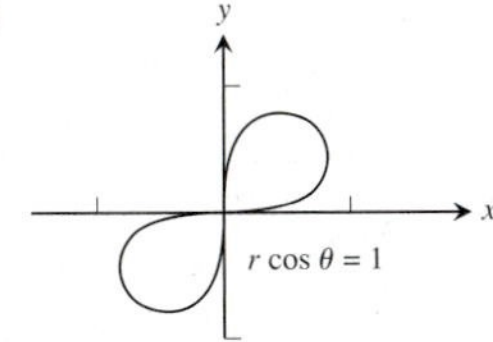

(b) $\frac{\pi}{2}$

45. Tangent lines at $\theta = \frac{\pi}{4}, \frac{3\pi}{4}, \frac{5\pi}{4}$, and $\frac{7\pi}{4}$; Cartesian equations are $y = \pm x$.

47. Horizontal: $y = 0$, $y \approx \pm 0.443$, $y \approx \pm 1.739$; vertical: $x = 2$, $x \approx 0.067$, $x \approx -1.104$

49. $y = \pm x + \sqrt{2}$ and $y = \pm x - \sqrt{2}$

51. $x = y$, a line

53. $x^2 = 4y$, a parabola

55. $x = 2$, a vertical line

57. $r = -5 \sin \theta$

59. $r^2 \cos^2 \theta + 4r^2 \sin^2 \theta = 16$, or $r^2 = \dfrac{16}{\cos^2 \theta + 4 \sin^2 \theta}$

61. $\frac{9\pi}{2}$

63. $2 + \frac{\pi}{4}$

65. 8

67. $\pi - 3$

69. Speed ≈ 591.982 mph, direction ≈ 8.179 degrees north of east

71. It hits the ground ≈ 2.135 sec later, approximately 66.421 ft from where it left the thrower's hand. Assuming it does not bounce or roll, it will still be there 3 sec after it was thrown.

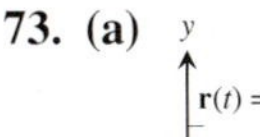

73. **(a)**

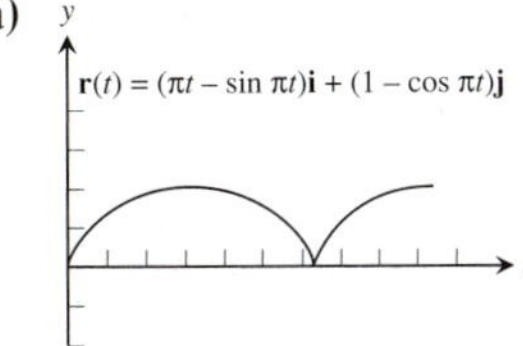

(b) $\mathbf{v}(0) = \langle 0, 0 \rangle$ $\quad \mathbf{v}(1) = \langle 2\pi, 0 \rangle$
$\mathbf{a}(0) = \langle 0, \pi^2 \rangle$ $\quad \mathbf{a}(1) = \langle 0, -\pi^2 \rangle$
$\mathbf{v}(2) = \langle 0, 0 \rangle$ $\quad \mathbf{v}(3) = \langle 2\pi, 0 \rangle$
$\mathbf{a}(2) = \langle 0, \pi^2 \rangle$ $\quad \mathbf{a}(3) = \langle 0, -\pi^2 \rangle$

(c) Topmost point: 2π ft/sec; center of wheel: π ft/sec. Reasons: Since the wheel rolls half a circumference, or π feet every second, the center of the wheel will move π feet every second. Since the rim of the wheel is turning at a rate of π ft/sec about the center, the velocity of the topmost point relative to the center is π ft/sec, giving it a total velocity of 2π ft/sec.

75. **(a)** ≈ 59.195 ft/sec
(b) ≈ 74.584 ft/sec

77. We have $x = (v_0 t) \cos \alpha$ and $y + \frac{gt^2}{2} = (v_0 t) \sin \alpha$. Squaring and adding gives $x^2 + \left(y + \frac{gt^2}{2}\right)^2 = (v_0 t)^2 (\cos^2 \alpha + \sin^2 \alpha) = v_0^2 t^2$.

79. **(a)** $\mathbf{r}(t) = \left[(155 \cos 18° - 11.7)\left(\frac{1}{0.09}\right)(1 - e^{-0.09t})\right]\mathbf{i} + \left[4 + \left(\frac{155 \sin 18°}{0.09}\right)(1 - e^{-0.09t}) + \frac{32}{0.09^2}(1 - 0.09t - e^{-0.09t})\right]\mathbf{j}$

$x(t) = (155 \cos 18° - 11.7)\left(\frac{1}{0.09}\right)(1 - e^{-0.09t})$

$y(t) = 4 + \left(\frac{155 \sin 18°}{0.09}\right)(1 - e^{-0.09t}) + \frac{32}{0.09^2}(1 - 0.09t - e^{-0.09t})$

(b) At ≈ 1.404 sec, it reaches a maximum height of ≈ 36.921 ft.
(c) Range ≈ 352.52 ft, flight time ≈ 2.959 sec
(d) At times $t \approx 0.753$ and $t \approx 2.068$ sec, when it is ≈ 98.799 and ≈ 256.138 ft from home plate
(e) No, the batter has not hit a home run. If the drag coefficient k is less than ≈ 0.011, the hit will be a home run.

81. The widths between the successive turns are constant and are given by $2\pi a$.

Chapter 9 Additional Exercises, pp. 784–785

1. **(a)** $\mathbf{v} = 4\mathbf{i} + 2\mathbf{j}$
(b) $\mathbf{r}(t) = 4t\mathbf{i} + \left(2t + \frac{16t^3}{100} - \frac{120t^2}{100}\right)\mathbf{j}$, where $0 \le t \le 5$
(c)

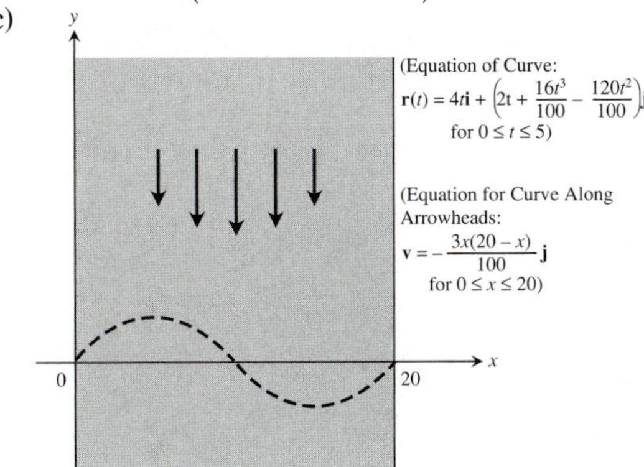

3. $\frac{\pi}{2}$ for all t

5. $\mathbf{v} \cdot \mathbf{j} = 12$, $\mathbf{a} \cdot \mathbf{j} = 26$

7. **(a)** $r = e^{2\theta}$ **(b)** $\frac{\sqrt{5}}{2}(e^{4\pi} - 1)$

9. $a^2\left(\frac{3\pi}{2} - 4\right)$

CHAPTER 10

Section 10.1, pp. 794–796

1. The line through the point (2, 3, 0) parallel to the z-axis
3. The x-axis
5. The circle $x^2 + y^2 = 4$ in the plane $z = -2$
7. The circle $y^2 + z^2 = 1$ in the yz-plane
9. The circle $x^2 + y^2 = 16$ in the xy-plane
11. **(a)** The first quadrant of the xy-plane
 (b) The fourth quadrant of the xy-plane
13. **(a)** The ball of radius 1 centered at the origin
 (b) All points greater than 1 unit from the origin
15. **(a)** The upper hemisphere of radius 1 centered at the origin
 (b) The solid upper hemisphere of radius 1 centered at the origin
17. **(a)** $x = 3$ **(b)** $y = -1$ **(c)** $z = -2$
19. **(a)** $z = 1$ **(b)** $x = 3$ **(c)** $y = -1$
21. **(a)** $x^2 + (y-2)^2 = 4,\ z = 0$
 (b) $(y-2)^2 + z^2 = 4,\ x = 0$
 (c) $x^2 + z^2 = 4,\ y = 2$
23. **(a)** $y = 3, z = -1$ **(b)** $x = 1, z = -1$ **(c)** $x = 1, y = 3$
25. $x^2 + y^2 + z^2 = 25,\ z = 3$
27. $0 \le z \le 1$
29. $z \le 0$
31. **(a)** $(x-1)^2 + (y-1)^2 + (z-1)^2 < 1$
 (b) $(x-1)^2 + (y-1)^2 + (z-1)^2 > 1$
33. $3\left(\frac{2}{3}\mathbf{i} + \frac{1}{3}\mathbf{j} - \frac{2}{3}\mathbf{k}\right)$
35. $5(\mathbf{k})$
37. $\sqrt{\frac{1}{2}}\left(\frac{1}{\sqrt{3}}\mathbf{i} - \frac{1}{\sqrt{3}}\mathbf{j} - \frac{1}{\sqrt{3}}\mathbf{k}\right)$
39. **(a)** $2\mathbf{i}$ **(b)** $-\sqrt{3}\mathbf{k}$
 (c) $\frac{3}{10}\mathbf{j} + \frac{2}{5}\mathbf{k}$ **(d)** $6\mathbf{i} - 2\mathbf{j} + 3\mathbf{k}$
41. $\frac{7}{13}(12\mathbf{i} - 5\mathbf{k})$
43. **(a)** $5\sqrt{2}$
 (b) $\frac{3}{5\sqrt{2}}\mathbf{i} + \frac{4}{5\sqrt{2}}\mathbf{j} - \frac{1}{\sqrt{2}}\mathbf{k}$
 (c) $(1/2, 3, 5/2)$
45. **(a)** $\sqrt{3}$
 (b) $-\frac{1}{\sqrt{3}}\mathbf{i} - \frac{1}{\sqrt{3}}\mathbf{j} - \frac{1}{\sqrt{3}}\mathbf{k}$
 (c) $\left(\frac{5}{2}, \frac{7}{2}, \frac{9}{2}\right)$
47. $A(4, -3, 5)$
49. $(x-1)^2 + (y-2)^2 + (z-3)^2 = 14$
51. $C(-2, 0, 2),\ a = 2\sqrt{2}$
53. $C(-2, 0, 2),\ a = 2\sqrt{2}$
55. $C\left(-\frac{1}{4}, -\frac{1}{4}, -\frac{1}{4}\right),\ a = \frac{5\sqrt{3}}{4}$
57. **(a)** $\sqrt{y^2 + z^2}$ **(b)** $\sqrt{x^2 + z^2}$ **(c)** $\sqrt{x^2 + y^2}$
59. **(a)** $\frac{3}{2}\mathbf{i} + \frac{3}{2}\mathbf{j} - 3\mathbf{k}$ **(b)** $\mathbf{i} + \mathbf{j} - 2\mathbf{k}$ **(c)** $(2, 2, 1)$

Section 10.2, pp. 805–807

1. **(a)** $-25, 5, 5$ **(b)** -1
 (c) -5 **(d)** $-2\mathbf{i} + 4\mathbf{j} - \sqrt{5}\mathbf{k}$
3. **(a)** $25, 15, 5$ **(b)** $\frac{1}{3}$
 (c) $\frac{5}{3}$ **(d)** $\frac{1}{9}(10\mathbf{i} + 11\mathbf{j} - 2\mathbf{k})$
5. **(a)** $2, \sqrt{34}, \sqrt{3}$ **(b)** $\frac{2}{\sqrt{3}\sqrt{34}}$
 (c) $\frac{2}{\sqrt{34}}$ **(d)** $\frac{1}{17}(5\mathbf{j} - 3\mathbf{k})$
7. $\left(\frac{3}{2}\mathbf{i} + \frac{3}{2}\mathbf{j}\right) + \left(-\frac{3}{2}\mathbf{i} + \frac{3}{2}\mathbf{j} + 4\mathbf{k}\right)$
9. $\left(\frac{14}{3}\mathbf{i} + \frac{28}{3}\mathbf{j} - \frac{14}{3}\mathbf{k}\right) + \left(\frac{10}{3}\mathbf{i} - \frac{16}{3}\mathbf{j} - \frac{22}{3}\mathbf{k}\right)$
11. 0.75 rad
13. 1.77 rad
17. $|\mathbf{u}\times\mathbf{v}| = 3$, direction is $\frac{2}{3}\mathbf{i} + \frac{1}{3}\mathbf{j} + \frac{2}{3}\mathbf{k}$; $|\mathbf{v}\times\mathbf{u}| = 3$, direction is $-\frac{2}{3}\mathbf{i} - \frac{1}{3}\mathbf{j} - \frac{2}{3}\mathbf{k}$
19. $|\mathbf{u}\times\mathbf{v}| = 0$, no direction; $|\mathbf{v}\times\mathbf{u}| = 0$, no direction
21. $|\mathbf{u}\times\mathbf{v}| = 6$, direction is $-\mathbf{k}$; $|\mathbf{v}\times\mathbf{u}| = 6$, direction is $\mathbf{k}$
23. $|\mathbf{u}\times\mathbf{v}| = 6\sqrt{5}$, direction is $\frac{1}{\sqrt{5}}\mathbf{i} - \frac{2}{\sqrt{5}}\mathbf{k}$; $|\mathbf{v}\times\mathbf{u}| = 6\sqrt{5}$, direction is $-\frac{1}{\sqrt{5}}\mathbf{i} + \frac{2}{\sqrt{5}}\mathbf{k}$
25. $\mathbf{u}\times\mathbf{v} = \mathbf{i} + \mathbf{k}$

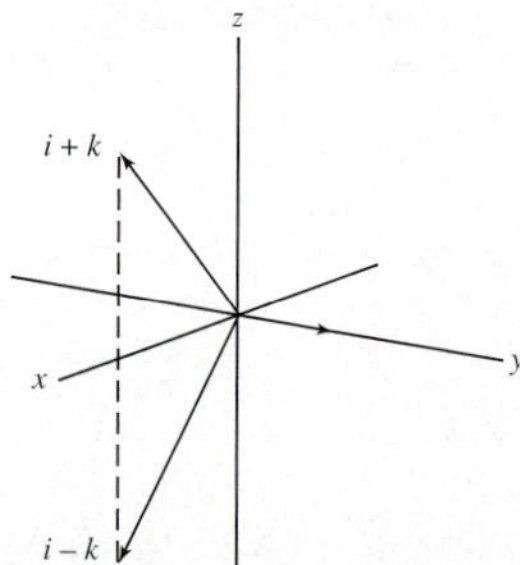

27. $\mathbf{u}\times\mathbf{v} = -2\mathbf{k}$

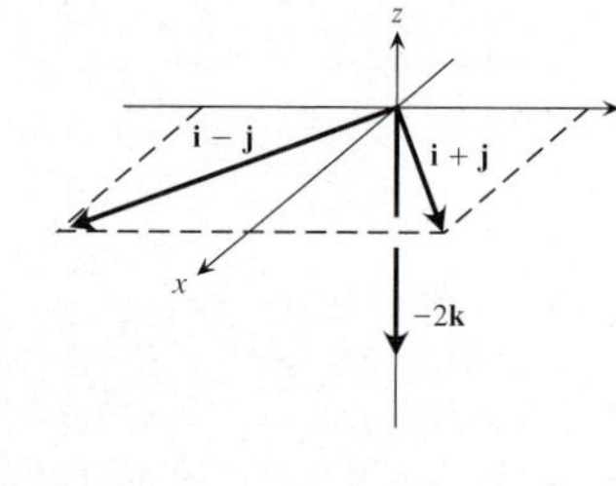

29. **(a)** $2\sqrt{6}$ **(b)** $\pm\frac{1}{\sqrt{6}}(2\mathbf{i} + \mathbf{j} + \mathbf{k})$
31. **(a)** $\frac{\sqrt{2}}{2}$ **(b)** $\pm\frac{1}{\sqrt{2}}(\mathbf{i} - \mathbf{j})$
33. 8
35. 7
37. **(a)** None **(b)** $\mathbf{u}$ and $\mathbf{w}$
39. $10\sqrt{3}$ ft · lb
41. **(a)** True **(b)** Not always true
 (c) True **(d)** True
 (e) Not always true **(f)** True
 (g) True **(h)** True
43. **(a)** $\text{proj}_{\mathbf{v}}\,\mathbf{u} = \frac{\mathbf{u}\cdot\mathbf{v}}{\mathbf{v}\cdot\mathbf{v}}\mathbf{v}$ **(b)** $\pm\mathbf{u}\times\mathbf{v}$
 (c) $\pm(\mathbf{u}\times\mathbf{v})\times\mathbf{w}$ **(d)** $|(\mathbf{u}\times\mathbf{v})\cdot\mathbf{w}|$
45. **(a)** Yes **(b)** No **(c)** Yes **(d)** No
47. No, $\mathbf{v}$ need not equal $\mathbf{w}$. For example, $\mathbf{i} + \mathbf{j} \ne -\mathbf{i} + \mathbf{j}$, but $\mathbf{i}\times(\mathbf{i} + \mathbf{j}) = \mathbf{i}\times\mathbf{i} + \mathbf{i}\times\mathbf{j} = \mathbf{0} + \mathbf{k} = \mathbf{k}$ and $\mathbf{i}\times(-\mathbf{i} + \mathbf{j}) = -\mathbf{i}\times\mathbf{i} + \mathbf{i}\times\mathbf{j} = \mathbf{0} + \mathbf{k} = \mathbf{k}$.

49. 2

51. 13

53. $\frac{11}{2}$

55. $\frac{25}{2}$

57. If $\mathbf{u} = a_1\mathbf{i} + a_2\mathbf{j}$ and $\mathbf{v} = b_1\mathbf{i} + b_2\mathbf{j}$, then $\mathbf{u} \times \mathbf{v} = \begin{vmatrix} \mathbf{i} & \mathbf{j} & \mathbf{k} \\ a_1 & a_2 & 0 \\ b_1 & b_2 & 0 \end{vmatrix} =$ $\begin{vmatrix} a_1 & a_2 \\ b_1 & b_2 \end{vmatrix}\mathbf{k}$ and the triangle's area is $\frac{1}{2}|\mathbf{u} \times \mathbf{v}| = \pm\frac{1}{2}\begin{vmatrix} a_1 & a_2 \\ b_1 & b_2 \end{vmatrix}$. The applicable sign is (+) if the acute angle from **u** to **v** runs counterclockwise in the xy-plane and (−) if it runs clockwise.

Section 10.3, pp. 813–815

1. Vector form: $\mathbf{r}(t) = (3 + t)\mathbf{i} + (t - 4)\mathbf{j} + (t - 1)\mathbf{k}$
Parametric form: $x = 3 + t,\ y = -4 + t,\ z = -1 + t$

3. Vector form: $\mathbf{r}(t) = (5t - 2)\mathbf{i} + (5t)\mathbf{j} + (3 - 5t)\mathbf{k}$
Parametric form: $x = -2 + 5t,\ y = 5t,\ z = 3 - 5t$

5. Vector form: $\mathbf{r}(t) = (2t + 3)\mathbf{i} - (t + 2)\mathbf{j} + (3t + 1)\mathbf{k}$
Parametric form: $x = 3 + 2t,\ y = -2 - t,\ z = 1 + 3t$

7. Vector form: $\mathbf{r}(t) = (3t + 2)\mathbf{i} + (7t + 4)\mathbf{j} + (5 - 5t)\mathbf{k}$
Parametric form: $x = 2 + 3t,\ y = 4 + 7t,\ z = 5 - 5t$

9. Vector form: $\mathbf{r}(t) = (2 - 2t)\mathbf{i} + (4t + 3)\mathbf{j} - (2t)\mathbf{k}$
Parametric form: $x = 2 - 2t,\ y = 3 + 4t,\ z = -2t$

11. $x = t,\ y = t,\ z = \frac{3}{2}t,\ 0 \le t \le 1$

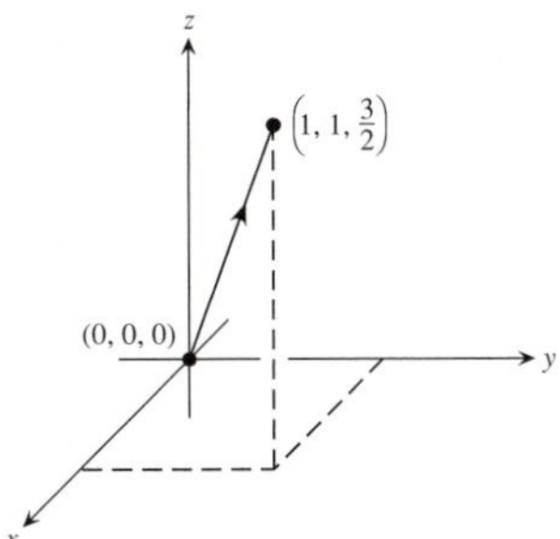

13. $x = 0,\ y = 1 - 2t,\ z = 1,\ 0 \le t \le 1$

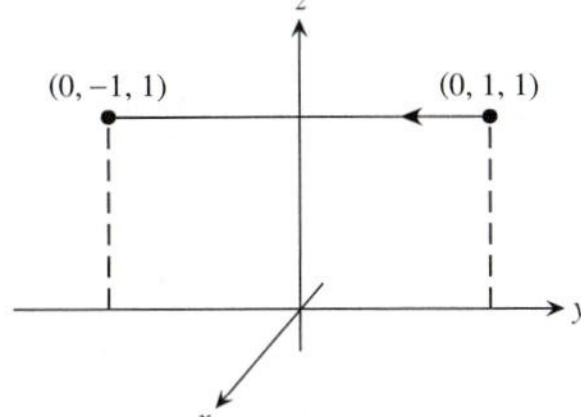

15. $3x - 2y - z = -3$

17. $7x - 5y - 4z = 6$

19. $x + 3y + 4z = 34$

21. $(1, 2, 3),\ -20x + 12y + z = 7$

23. $y + z = 3$

25. $x - y + z = 0$

29. 0

33. 19/5

35. $9/\sqrt{41}$

39. 0.82 rad

41. $(3/2, -3/2, 1/2)$

43. $(1, 1, 0)$

45. $x = 1 - t,\ y = 1 + t,\ z = -1$

47. $x = 4,\ y = 3 + 6t,\ z = 1 + 3t$

49. $L1$ intersects $L2$; $L2$ is parallel to $L3$; $L1$ and $L3$ are skew.

51. $x = 2 + 2t,\ y = -4 - t,\ z = 7 + 3t$; $x = -2 - t,\ y = -2 + (1/2)t,\ z = 1 - (3/2)t$

53. $(0, -1/2, -3/2), (-1, 0, -3), (1, -1, 0)$

55. The line and plane are not parallel.

57. Many answers are possible. One possibility is $x + y = 3$ and $2y + z = 7$.

59. $\frac{x}{a} + \frac{y}{b} + \frac{z}{c} = 1$ describes all planes *except* those through the origin or parallel to a coordinate axis.

Section 10.4, pp. 824–825

1. Graph (d), ellipsoid

3. Graph (a), cylinder

5. Graph (l), hyperbolic paraboloid

7. Graph (b), cylinder

9. Graph (k), hyperbolic paraboloid

11. Graph (h), cone

13. **(a)** $\frac{2\pi(9 - c^2)}{9}$ **(b)** 8π **(c)** $\frac{4\pi abc}{3}$

Section 10.5, pp. 835–838

1. $\mathbf{v} = \mathbf{i} + 2t\mathbf{j} + 2\mathbf{k}$; $\mathbf{a} = 2\mathbf{j}$; speed: 3; direction: $\frac{1}{3}\mathbf{i} + \frac{2}{3}\mathbf{j} + \frac{2}{3}\mathbf{k}$; $\mathbf{v}(1) = 3\left(\frac{1}{3}\mathbf{i} + \frac{2}{3}\mathbf{j} + \frac{2}{3}\mathbf{k}\right)$

3. $\mathbf{v} = (-2 \sin t)\mathbf{i} + (3 \cos t)\mathbf{j} + 4\mathbf{k}$; $\mathbf{a} = (-2 \cos t)\mathbf{i} - (3 \sin t)\mathbf{j}$; speed: $2\sqrt{5}$; direction: $\left(-\frac{1}{\sqrt{5}}\right)\mathbf{i} + \left(\frac{2}{\sqrt{5}}\right)\mathbf{k}$; $\mathbf{v}\left(\frac{\pi}{2}\right) = 2\sqrt{5}\left[\left(-\frac{1}{\sqrt{5}}\right)\mathbf{i} + \left(\frac{2}{\sqrt{5}}\right)\mathbf{k}\right]$

5. $\mathbf{v} = \left(\frac{2}{t+1}\right)\mathbf{i} + 2t\mathbf{j} + t\mathbf{k}$: $\mathbf{a} = \left(\frac{-2}{(t+1)^2}\right)\mathbf{i} + 2\mathbf{j} + \mathbf{k}$; speed: $\sqrt{6}$; direction: $\frac{1}{\sqrt{6}}\mathbf{i} + \frac{2}{\sqrt{6}}\mathbf{j} + \frac{1}{\sqrt{6}}\mathbf{k}$; $\mathbf{v}(1) = \sqrt{6}\left(\frac{1}{\sqrt{6}}\mathbf{i} + \frac{2}{\sqrt{6}}\mathbf{j} + \frac{1}{\sqrt{6}}\mathbf{k}\right)$

7. $\pi/2$

9. $\pi/2$

11. $t = 0, \pi, 2\pi$

13. $\left(\frac{1}{4}\right)\mathbf{i} + 7\mathbf{j} + \left(\frac{3}{2}\right)\mathbf{k}$

15. $\left(\frac{\pi + 2\sqrt{2}}{2}\right)\mathbf{j} + 2\mathbf{k}$

17. $(\ln 4)\mathbf{i} + (\ln 4)\mathbf{j} + (\ln 2)\mathbf{k}$

19. $\mathbf{r}(t) = \left(\frac{-t^2}{2} + 1\right)\mathbf{i} + \left(\frac{-t^2}{2} + 2\right)\mathbf{j} + \left(\frac{-t^2}{2} + 3\right)\mathbf{k}$

21. $\mathbf{r}(t) = ((t+1)^{3/2} - 1)\mathbf{i} + (-e^{-t} + 1)\mathbf{j} + (\ln(t+1) + 1)\mathbf{k}$
23. $\mathbf{r}(t) = 8t\mathbf{i} + 8t\mathbf{j} + (-16t^2 + 100)\mathbf{k}$
25. $x = t, y = -1, z = 1 + t$
27. $x = at, y = a, z = 2\pi b + bt$
29. $\mathbf{r}(t) = \left(\frac{3}{2}t^2 + \frac{6}{\sqrt{11}}t + 1\right)\mathbf{i} - \left(\frac{1}{2}t^2 + \frac{2}{\sqrt{11}}t - 2\right)\mathbf{j} + \left(\frac{1}{2}t^2 + \frac{2}{\sqrt{11}}t + 3\right)\mathbf{k} = \left(\frac{1}{2}t^2 + \frac{2t}{\sqrt{11}}\right)(3\mathbf{i} - \mathbf{j} + \mathbf{k}) + (\mathbf{i} + 2\mathbf{j} + 3\mathbf{k})$

31. Max $|\mathbf{v}| = 2$, min $|\mathbf{v}| = 0$, max $|\mathbf{a}| = $ min $|\mathbf{a}| = 1$
33. Max $|\mathbf{v}| = 3$, min $|\mathbf{v}| = 2$, max $|\mathbf{a}| = 3$, min $|\mathbf{a}| = 2$

Section 10.6, pp. 846–847

1. $\mathbf{T} = \left(-\frac{2}{3}\sin t\right)\mathbf{i} + \left(\frac{2}{3}\cos t\right)\mathbf{j} + \frac{\sqrt{5}}{3}\mathbf{k}, 3\pi$
3. $\mathbf{T} = \frac{1}{\sqrt{1+t}}\mathbf{i} + \frac{\sqrt{t}}{\sqrt{1+t}}\mathbf{k}, \frac{52}{3}$
5. $\mathbf{T} = (-\cos t)\mathbf{j} + (\sin t)\mathbf{k}, \frac{3}{2}$
7. $\mathbf{T} = \left(\frac{\cos t - t\sin t}{t+1}\right)\mathbf{i} + \left(\frac{\sin t + t\cos t}{t+1}\right)\mathbf{j} + \left(\frac{\sqrt{2}t^{1/2}}{t+1}\right)\mathbf{k}, \frac{\pi^2}{2} + \pi$
9. $(0, 5, 24\pi)$
11. $s(t) = 5t, L = \frac{5\pi}{2}$
13. $s(t) = \sqrt{3}e^t - \sqrt{3}, L = \frac{3\sqrt{3}}{4}$
15. $\mathbf{T} = (\cos t)\mathbf{i} - (\sin t)\mathbf{j}, \mathbf{N} = (-\sin t)\mathbf{i} - (\cos t)\mathbf{j}, \kappa = \cos t$
17. $\mathbf{T} = \frac{1}{\sqrt{1+t^2}}\mathbf{i} - \frac{t}{\sqrt{1+t^2}}\mathbf{j}, \mathbf{N} = \frac{-t}{\sqrt{1+t^2}}\mathbf{i} - \frac{1}{\sqrt{1+t^2}}\mathbf{j}, \kappa = \frac{1}{2(\sqrt{1+t^2})^3}$
19. $\sqrt{2} + \ln(1 + \sqrt{2})$
21. **(a)** Cylinder is $x^2 + y^2 = 1$, plane is $x + z = 1$.
(b) and **(c)**

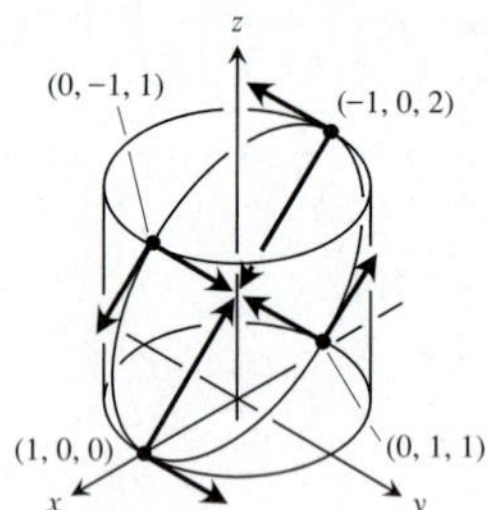

(d) $L = \int_0^{2\pi} \sqrt{1 + \sin^2 t}\, dt$
(e) $L \approx 7.64$

23. $\left(x - \frac{\pi}{2}\right)^2 + y^2 = 1$

Section 10.7, pp. 854–856

1. $\mathbf{T} = \frac{3\cos t}{5}\mathbf{i} - \frac{3\sin t}{5}\mathbf{j} + \frac{4}{5}\mathbf{k}, \mathbf{N} = (-\sin t)\mathbf{i} - (\cos t)\mathbf{j},$
$\mathbf{B} = \left(\frac{4}{5}\cos t\right)\mathbf{i} - \left(\frac{4}{5}\sin t\right)\mathbf{j} - \frac{3}{4}\mathbf{k}, \kappa = \frac{3}{25}, \tau = -\frac{4}{25}$
3. $\mathbf{T} = \left(\frac{\cos t - \sin t}{\sqrt{2}}\right)\mathbf{i} + \left(\frac{\cos t + \sin t}{\sqrt{2}}\right)\mathbf{j},$
$\mathbf{N} = \left(\frac{-\cos t - \sin t}{\sqrt{2}}\right)\mathbf{i} + \left(\frac{-\sin t + \cos t}{\sqrt{2}}\right)\mathbf{j}, \mathbf{B} = \mathbf{k},$
$\kappa = \frac{1}{e^t\sqrt{2}}, \tau = 0$
5. $\mathbf{T} = \frac{t}{\sqrt{t^2+1}}\mathbf{i} + \frac{1}{\sqrt{t^2+1}}\mathbf{j}, \mathbf{N} = \frac{\mathbf{i}}{\sqrt{t^2+1}} - \frac{t\mathbf{j}}{\sqrt{t^2+1}},$
$\mathbf{B} = -\mathbf{k}, \kappa = \frac{1}{t(t^2+1)^{3/2}}, \tau = 0$
7. $\mathbf{T} = \left(\operatorname{sech}\frac{t}{a}\right)\mathbf{i} + \left(\tanh\frac{t}{a}\right)\mathbf{j}, \mathbf{N} = \left(-\tanh\frac{t}{a}\right)\mathbf{i} + \left(\operatorname{sech}\frac{t}{a}\right)\mathbf{j},$
$\mathbf{B} = \mathbf{k}, \kappa = \frac{1}{a}\operatorname{sech}^2\frac{t}{a}, \tau = 0$
9. $\mathbf{a} = |a|\mathbf{N}$
11. $\mathbf{a}(1) = \frac{4}{3}\mathbf{T} + \frac{2\sqrt{5}}{3}\mathbf{N}$
13. $\mathbf{a}(0) = 2\mathbf{N}$
15. $\mathbf{r}\left(\frac{\pi}{4}\right) = \frac{\sqrt{2}}{2}\mathbf{i} + \frac{\sqrt{2}}{2}\mathbf{j} - \mathbf{k}, \mathbf{T}\left(\frac{\pi}{4}\right) = -\frac{\sqrt{2}}{2}\mathbf{i} + \frac{\sqrt{2}}{2}\mathbf{j},$
$\mathbf{N}\left(\frac{\pi}{4}\right) = -\frac{\sqrt{2}}{2}\mathbf{i} - \frac{\sqrt{2}}{2}\mathbf{j}, \mathbf{B}\left(\frac{\pi}{4}\right) = \mathbf{k}$; osculating plane: $z = -1$; normal plane: $-x + y = 0$; rectifying plane: $x + y = \sqrt{2}$
17. Yes. If the car is moving on a curved path ($\kappa \neq 0$), then $a_N = \kappa|\mathbf{v}|^2 \neq 0$ and $\mathbf{a} \neq \mathbf{0}$.
21. $|\mathbf{F}| = \kappa\left[m\left(\frac{ds}{dt}\right)^2\right]$
23. **(b)** $\cos x$
25. **(b)** $\mathbf{N} = \frac{-2e^{2t}}{\sqrt{1+4e^{4t}}}\mathbf{i} + \frac{1}{\sqrt{1+4e^{4t}}}\mathbf{j}$
(c) $\mathbf{N} = -\frac{1}{2}\left(\sqrt{4 - t^2}\mathbf{i} + t\mathbf{j}\right)$
29. $\frac{1}{2b}$
33. **(a)** $b - a$ **(b)** π
37. $\kappa = \frac{2}{(1 + 4x^2)^{3/2}}$
39. $\kappa = \frac{|\sin x|}{(1 + \cos^2 x)^{3/2}}$

Section 10.8, pp. 865–866

1. $T = 93.2$ min
3. $a = 6763$ km
5. $D = 6480$ km
7. **(a)** 42,167 km **(b)** 35,788 km
(c) *Syncom 3, GOES 4,* and *Intelsat 5*
9. $a = 383{,}200$ km from the center of Earth, or about 376,821 km from the surface

Chapter 10 Practice Exercises, pp. 867–869

1. Length = 7, direction is $\frac{2}{7}\mathbf{i} - \frac{3}{7}\mathbf{j} + \frac{6}{7}\mathbf{k}$.

3. $\frac{8}{\sqrt{33}}\mathbf{i} - \frac{2}{\sqrt{33}}\mathbf{j} + \frac{8}{\sqrt{33}}\mathbf{k}$

5. $|\mathbf{v}| = \sqrt{2}$, $|\mathbf{u}| = 3$, $\mathbf{v}\cdot\mathbf{u} = \mathbf{u}\cdot\mathbf{v} = 3$, $\mathbf{v}\times\mathbf{u} = -2\mathbf{i} + 2\mathbf{j} - \mathbf{k}$, $\mathbf{u}\times\mathbf{v} = 2\mathbf{i} - 2\mathbf{j} + \mathbf{k}$, $|\mathbf{v}\times\mathbf{u}| = 3$, $\theta = \cos^{-1}\left(\frac{1}{\sqrt{2}}\right) = \frac{\pi}{4}$, $|\mathbf{u}|\cos\theta = \frac{3}{\sqrt{2}}$, $\text{proj}_{\mathbf{v}}\,\mathbf{u} = \frac{3}{2}(\mathbf{i} + \mathbf{j})$

7. $\frac{4}{3}(2\mathbf{i} + \mathbf{j} - \mathbf{k}) - \frac{1}{3}(5\mathbf{i} + \mathbf{j} + 11\mathbf{k})$

9. $\mathbf{u}\times\mathbf{v} = \mathbf{k}$

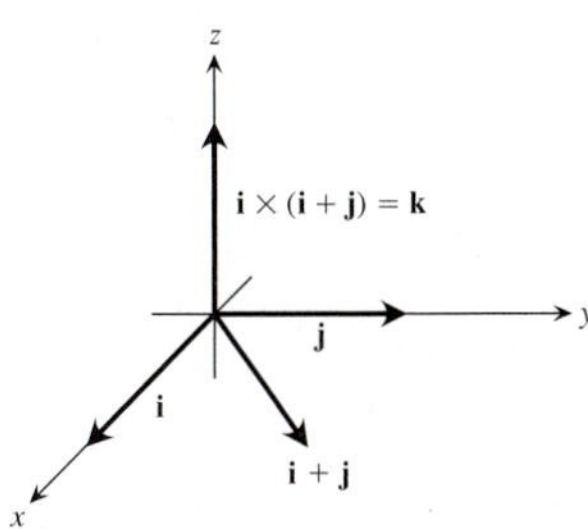

11. $2\sqrt{7}$

13. **(a)** $\sqrt{14}$ **(b)** 1

17. $x = 1 - 3t$, $y = 2$, $z = 3 + 7t$

19. $2x + y + z = 5$

21. $-9x + y + 7z = 4$

23. $(0, -1/2, -3/2)$, $(-1, 0, -3)$, $(1, -1, 0)$

25. $\frac{\pi}{3}$

29. $7x - 3y - 5z = -14$

31. $\frac{1}{\sqrt{14}}(-2\mathbf{i} - 3\mathbf{j} + \mathbf{k})$

33. $(4/3, -2/3, -2/3)$

35. **(a)** No **(b)** No **(c)** No **(d)** No **(e)** Yes

37. $\sqrt{78}/3$

39. $\sqrt{2}$

41. 3

43. $11/\sqrt{107}$

45.

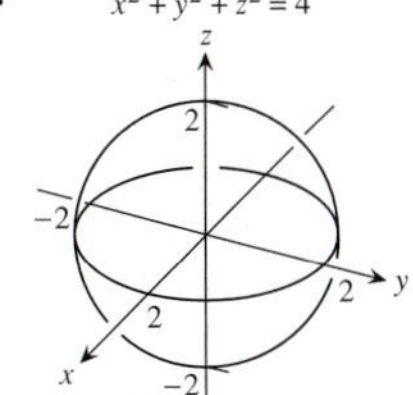

47.

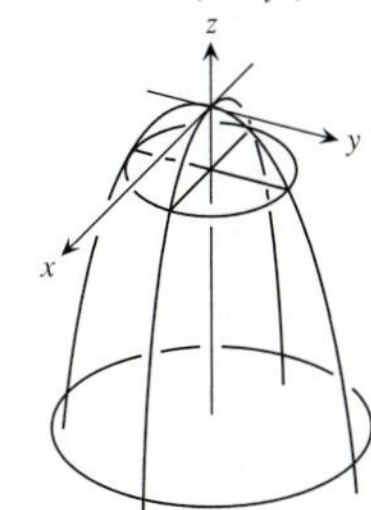

49.

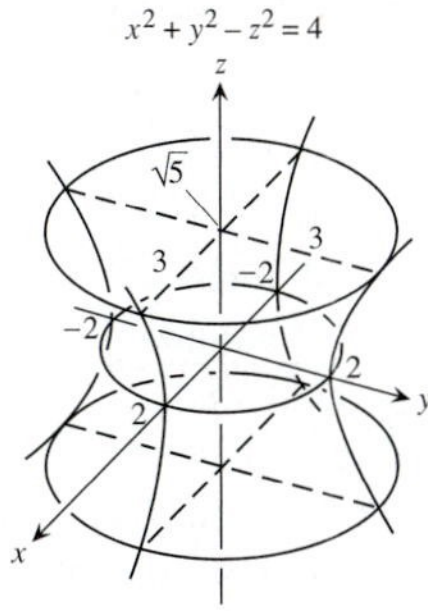

51. Length $= \frac{\pi}{4}\sqrt{1 + \frac{\pi^2}{16}} + \ln\left(\frac{\pi}{4} + \sqrt{1 + \frac{\pi^2}{16}}\right)$

53. $\mathbf{T}(0) = \frac{2}{3}\mathbf{i} - \frac{2}{3}\mathbf{j} + \frac{1}{3}\mathbf{k}$; $\mathbf{N}(0) = \frac{1}{\sqrt{2}}\mathbf{i} + \frac{1}{\sqrt{2}}\mathbf{j}$; $\mathbf{B}(0) = -\frac{1}{3\sqrt{2}}\mathbf{i} + \frac{1}{3\sqrt{2}}\mathbf{j} + \frac{4}{3\sqrt{2}}\mathbf{k}$; $\kappa = \frac{\sqrt{2}}{3}$; $\tau = \frac{1}{6}$

55. $\mathbf{T}(\ln 2) = \frac{1}{\sqrt{17}}\mathbf{i} + \frac{4}{\sqrt{17}}\mathbf{j}$; $\mathbf{N}(\ln 2) = -\frac{4}{\sqrt{17}}\mathbf{i} + \frac{1}{\sqrt{17}}\mathbf{j}$; $\mathbf{B}(\ln 2) = \mathbf{k}$; $\kappa = \frac{8}{17\sqrt{17}}$; $\tau = 0$

57. $\mathbf{a}(0) = 10\mathbf{T} + 6\mathbf{N}$

59. $\mathbf{T} = \left(\frac{1}{\sqrt{2}}\cos t\right)\mathbf{i} - (\sin t)\mathbf{j} + \left(\frac{1}{\sqrt{2}}\cos t\right)\mathbf{k}$; $\mathbf{N} = \left(-\frac{1}{\sqrt{2}}\sin t\right)\mathbf{i} - (\cos t)\mathbf{j} - \left(\frac{1}{\sqrt{2}}\sin t\right)\mathbf{k}$; $\mathbf{B} = \frac{1}{\sqrt{2}}\mathbf{i} - \frac{1}{\sqrt{2}}\mathbf{k}$; $\kappa = \frac{1}{\sqrt{2}}$; $\tau = 0$

61. $\pi/3$

63. $x = 1 + t$, $y = t$, $z = -t$

65. 5971 km, 1.639×10^7 km^2, 3.21% visible

Chapter 10 Additional Exercises, pp. 870–872

1. $(26, 23, -1/3)$

3. $|\mathbf{F}| = 20$ lb

9. **(b)** $6/\sqrt{14}$

15. $\frac{32}{41}\mathbf{i} + \frac{23}{41}\mathbf{j} - \frac{13}{41}\mathbf{k}$

17. **(a)** $|\mathbf{F}| = \frac{GMm}{d^2}\left(1 + \sum_{i=1}^{n}\frac{2}{(i^2 + 1)^{3/2}}\right)$
(b) Yes

21. **(a)** $\frac{dx}{dt} = \dot{r}\cos\theta - r\dot{\theta}\sin\theta$, $\frac{dy}{dt} = \dot{r}\sin\theta + r\dot{\theta}\cos\theta$
(b) $\frac{dr}{dt} = \dot{x}\cos\theta + \dot{y}\sin\theta$, $r\frac{d\theta}{dt} = -\dot{x}\sin\theta + \dot{y}\cos\theta$

23. **(a)** $\mathbf{v}(1) = -\mathbf{u}_r + 3\mathbf{u}_\theta$, $\mathbf{a}(1) = -9\mathbf{u}_r - 6\mathbf{u}_\theta$
(b) 6.5 in.

CHAPTER 11

Section 11.1, pp. 880–882

1. **(a)** All points in the xy-plane **(b)** All reals
(c) The lines $y - x = c$ **(d)** No boundary points
(e) Both open and closed **(f)** Unbounded

3. **(a)** All points in the xy-plane **(b)** $z \geq 0$
(c) For $f(x, y) = 0$, the origin; for $f(x, y) \neq 0$, ellipses with the center $(0, 0)$ and major and minor axes along the x- and y-axes, respectively
(d) No boundary points **(e)** Both open and closed
(f) Unbounded

5. **(a)** All points in the xy-plane **(b)** All reals
(c) For $f(x, y) = 0$, the x- and y-axes; for $f(x, y) \neq 0$, hyperbolas with the x- and y-axes as asymptotes
(d) No boundary points **(e)** Both open and closed
(f) Unbounded

7. **(a)** All (x, y) satisfying $x^2 + y^2 < 16$ **(b)** $z \geq \frac{1}{4}$
(c) Circles centered at the origin with radii $r < 4$
(d) Boundary is the circle $x^2 + y^2 = 16$
(e) Open **(f)** Bounded

9. **(a)** $(x, y) \neq (0, 0)$ **(b)** All reals
(c) The circles with center $(0, 0)$ and radii $r > 0$
(d) Boundary is the single point $(0, 0)$.
(e) Open **(f)** Unbounded

11. **(a)** All (x, y) satisfying $-1 \leq y - x \leq 1$
(b) $-\pi/2 \leq z \leq \pi/2$
(c) Straight lines of the form $y - x = c$, where $-1 \leq c \leq 1$
(d) Boundary is two straight lines $y = 1 + x$ and $y = -1 + x$.
(e) Closed **(f)** Unbounded

13. Graph (f) **15.** Graph (a) **17.** Graph (d)

19. **(a)**

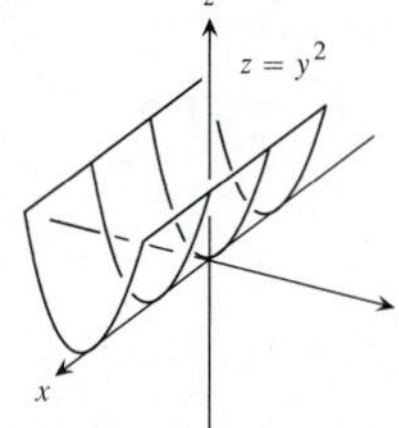

(b)

21. **(a)**

(b)

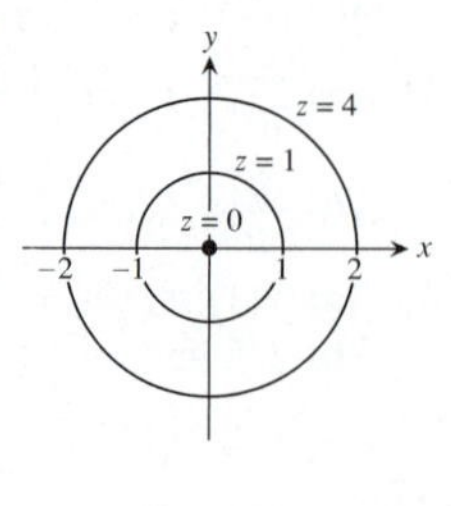

23. **(a)**

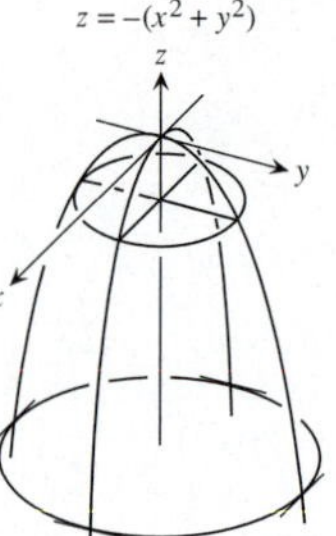

(b)

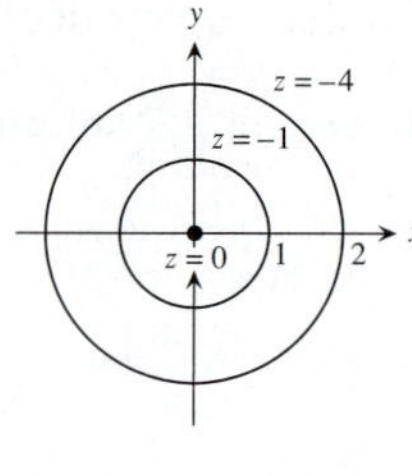

25. **(a)**

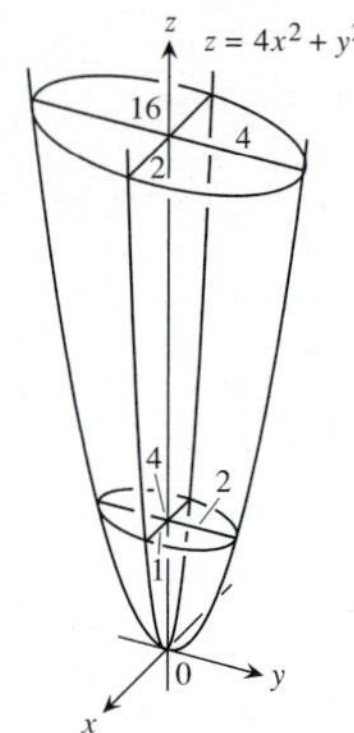

(b)

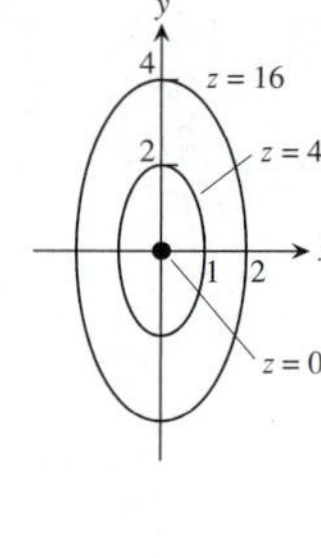

27. **(a)**

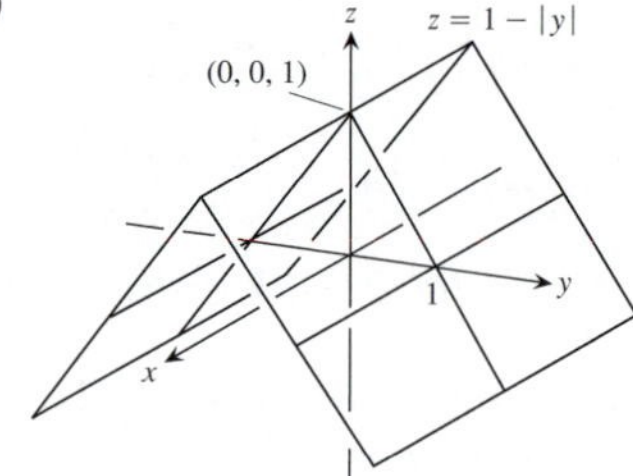

(b)

29. $x^2 + y^2 = 10$

31. $\tan^{-1} y - \tan^{-1} x = 2 \tan^{-1} \sqrt{2}$

33.

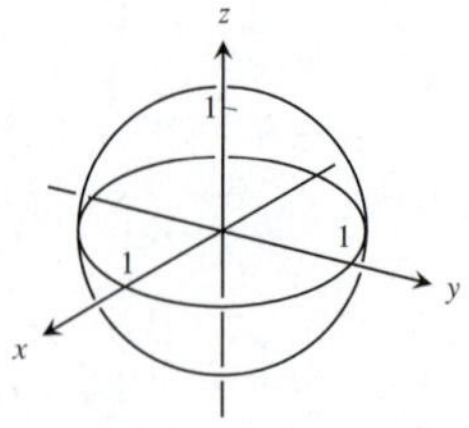

35.

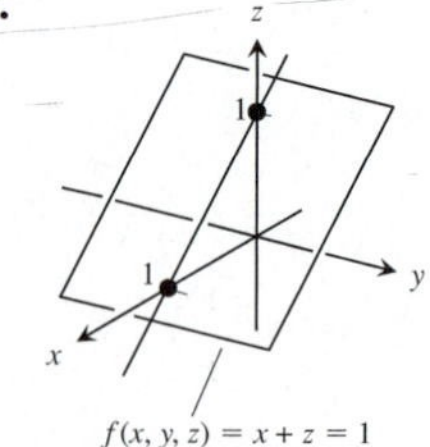

37.

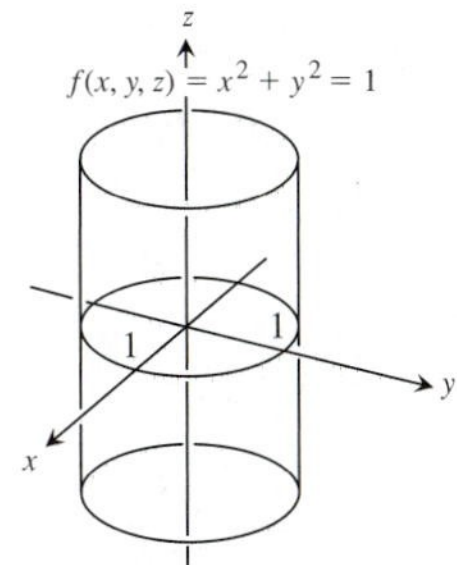

39.

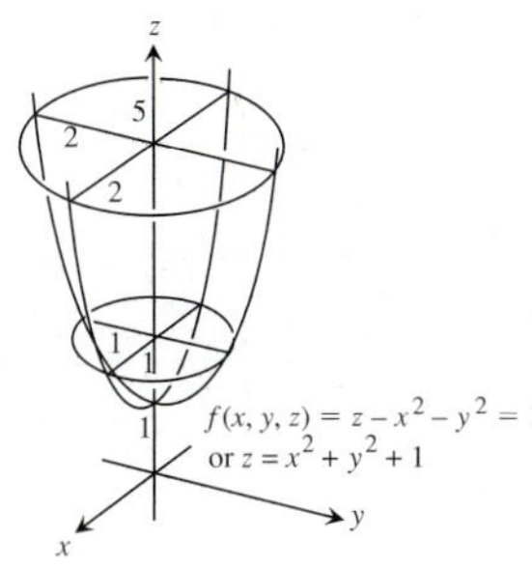

41. $\sqrt{x-y} - \ln z = 2$ **43.** $\frac{x+y}{z} = \ln 2$

45. Yes, 2000 **47.** 63 km

Section 11.2, pp. 887–890

1. 5/2 **3.** $2\sqrt{6}$ **5.** 1

7. 1/2 **9.** 1 **11.** 0

13. 0 **15.** −1 **17.** 2

19. 1/4 **21.** 19/12 **23.** 2

25. 3

27. **(a)** All (x, y) **(b)** All (x, y) except $(0, 0)$

29. **(a)** All (x, y) except where $x = 0$ or $y = 0$
(b) All (x, y)

31. **(a)** All (x, y, z)
(b) All (x, y, z) except the interior of the cylinder $x^2 + y^2 = 1$

33. **(a)** All (x, y, z) with $z \neq 0$
(b) All (x, y, z) with $x^2 + z^2 \neq 1$

35. Consider paths along $y = x, x > 0$, and along $y = x, x < 0$.

37. Consider the paths $y = kx^2$, k a constant.

39. Consider the paths $y = kx$, k a constant, $k \neq -1$.

41. Consider the paths $y = kx^2$, k a constant, $k \neq 0$.

45. $\delta = 0.1$ **47.** $\delta = 0.005$

49. $\delta = \sqrt{0.015}$ **51.** $\delta = 0.005$

55. 0 **57.** Does not exist

59. $\frac{\pi}{2}$ **61.** $f(0, 0) = \ln 3$

63. No

65. **(a)** $f(x, y)|_{y=mx} = \sin 2\theta$, where $\tan\theta = m$

67. The limit is 1. **69.** The limit is 0.

Section 11.3, pp. 899–901

1. $\frac{\partial f}{\partial x} = 4x, \frac{\partial f}{\partial y} = -3$

3. $\frac{\partial f}{\partial x} = 2x(y + 2), \frac{\partial f}{\partial y} = x^2 - 1$

5. $\frac{\partial f}{\partial x} = 2y(xy - 1), \frac{\partial f}{\partial y} = 2x(xy - 1)$

7. $\frac{\partial f}{\partial x} = \frac{x}{\sqrt{x^2+y^2}}, \frac{\partial f}{\partial y} = \frac{y}{\sqrt{x^2+y^2}}$

9. $\frac{\partial f}{\partial x} = \frac{-1}{(x+y)^2}, \frac{\partial f}{\partial y} = \frac{-1}{(x+y)^2}$

11. $\frac{\partial f}{\partial x} = \frac{-y^2-1}{(xy-1)^2}, \frac{\partial f}{\partial y} = \frac{-x^2-1}{(xy-1)^2}$

13. $\frac{\partial f}{\partial x} = e^{x+y+1}, \frac{\partial f}{\partial y} = e^{x+y+1}$ **15.** $\frac{\partial f}{\partial x} = \frac{1}{x+y}, \frac{\partial f}{\partial y} = \frac{1}{x+y}$

17. $\frac{\partial f}{\partial x} = 2\sin(x - 3y)\cos(x - 3y), \frac{\partial f}{\partial y} = -6\sin(x - 3y)\cos(x - 3y)$

19. $\frac{\partial f}{\partial x} = yx^{y-1}, \frac{\partial f}{\partial y} = x^y \ln x$ **21.** $\frac{\partial f}{\partial x} = -g(x), \frac{\partial f}{\partial y} = g(y)$

23. $f_x = y^2, f_y = 2xy, f_z = -4z$

25. $f_x = 1, f_y = -y(y^2 + z^2)^{-1/2}, f_z = -z(y^2 + z^2)^{-1/2}$

27. $f_x = \frac{yz}{\sqrt{1 - x^2y^2z^2}}, f_y = \frac{xz}{\sqrt{1 - x^2y^2z^2}}, f_z = \frac{xy}{\sqrt{1 - x^2y^2z^2}}$

29. $f_x = \frac{1}{x + 2y + 3z}, f_y = \frac{2}{x + 2y + 3z}, f_z = \frac{3}{x + 2y + 3z}$

31. $f_x = -2xe^{-(x^2+y^2+z^2)}, f_y = -2ye^{-(x^2+y^2+z^2)}, f_z = -2ze^{-(x^2+y^2+z^2)}$

33. $f_x = \operatorname{sech}^2(x + 2y + 3z), f_y = 2\operatorname{sech}^2(x + 2y + 3z), f_z = 3\operatorname{sech}^2(x + 2y + 3z)$

35. $\frac{\partial f}{\partial t} = -2\pi\sin(2\pi t - \alpha), \frac{\partial f}{\partial \alpha} = \sin(2\pi t - \alpha)$

37. $\frac{\partial h}{\partial \rho} = \sin\phi\cos\theta, \frac{\partial h}{\partial \phi} = \rho\cos\phi\cos\theta, = \frac{\partial h}{\partial \theta} = -\rho\sin\phi\sin\theta$

39. $W_P(P, V, \delta, v, g) = V, W_V(P, V, \delta, v, g) = P + \frac{\delta v^2}{2g},$
$W_\delta(P, V, \delta, v, g) = \frac{Vv^2}{2g}, W_v(P, V, \delta, v, g) = \frac{V\delta v}{g},$
$W_g(P, V, \delta, v, g) = -\frac{V\delta v^2}{2g^2}$

41. $\frac{\partial f}{\partial x} = 1 + y, \frac{\partial f}{\partial y} = 1 + x, \frac{\partial^2 f}{\partial x^2} = 0, \frac{\partial^2 f}{\partial y^2} = 0, \frac{\partial^2 f}{\partial y\,\partial x} = \frac{\partial^2 f}{\partial x\,\partial y} = 1$

43. $\frac{\partial g}{\partial x} = 2xy + y\cos x, \frac{\partial g}{\partial y} = x^2 - \sin y + \sin x, \frac{\partial^2 g}{\partial x^2} = 2y - y\sin x,$
$\frac{\partial^2 g}{\partial y^2} = -\cos y, \frac{\partial^2 g}{\partial y\,\partial x} = \frac{\partial^2 g}{\partial x\,\partial y} = 2x + \cos x$

45. $\frac{\partial r}{\partial x} = \frac{1}{x+y}, \frac{\partial r}{\partial y} = \frac{1}{x+y}, \frac{\partial^2 r}{\partial x^2} = \frac{-1}{(x+y)^2}, \frac{\partial^2 r}{\partial y^2} = \frac{-1}{(x+y)^2},$
$\frac{\partial^2 r}{\partial y\,\partial x} = \frac{\partial^2 r}{\partial x\,\partial y} = \frac{-1}{(x+y)^2}$

47. $\frac{\partial w}{\partial x} = \frac{2}{2x+3y}, \frac{\partial w}{\partial y} = \frac{3}{2x+3y}, \frac{\partial^2 w}{\partial y\,\partial x} = \frac{\partial^2 w}{\partial x\,\partial y} = \frac{-6}{(2x+3y)^2}$

49. $\dfrac{\partial w}{\partial x} = y^2 + 2xy^3 + 3x^2y^4,\ \dfrac{\partial w}{\partial y} = 2xy + 3x^2y^2 + 4x^3y^3,$

$\dfrac{\partial^2 w}{\partial y\,\partial x} = \dfrac{\partial^2 w}{\partial x\,\partial y} = 2y + 6xy^2 + 12x^2y^3$

51. **(a)** x first **(b)** y first **(c)** x first
(d) x first **(e)** y first **(f)** y first

53. $f_x(1, 2) = -13, f_y(1, 2) = -2$

55. 12

57. -2

59. $\dfrac{\partial A}{\partial a} = \dfrac{a}{bc \sin A}, \dfrac{\partial A}{\partial a} = \dfrac{c \cos A - b}{bc \sin A}$

61. $v_x = \dfrac{\ln v}{(\ln u)(\ln v) - 1}$

77. Yes

Section 11.4, pp. 908–910

1. **(a)** $\dfrac{dw}{dt} = 0,$ **(b)** $\dfrac{dw}{dt}(\pi) = 0$

3. **(a)** $\dfrac{dw}{dt} = 1,$ **(b)** $\dfrac{dw}{dt}(3) = 1$

5. **(a)** $\dfrac{dw}{dt} = 4t \tan^{-1} t + 1,$ **(b)** $\dfrac{dw}{dt}(1) = \pi + 1$

7. **(a)** $\dfrac{\partial z}{\partial u} = 4 \cos v \ln (u \sin v) + 4 \cos v, \dfrac{\partial z}{\partial v} =$

$-4u \sin v \ln (u \sin v) + \dfrac{4u \cos^2 v}{\sin v}$

(b) $\dfrac{\partial z}{\partial u} = \sqrt{2}(\ln 2 + 2), \dfrac{\partial z}{\partial v} = -2\sqrt{2}(\ln 2 - 2)$

9. **(a)** $\dfrac{\partial w}{\partial u} = 2u + 4uv, \dfrac{\partial w}{\partial v} = -2v + 2u^2$

(b) $\dfrac{\partial w}{\partial u} = 3, \dfrac{\partial w}{\partial v} = -\dfrac{3}{2}$

11. **(a)** $\dfrac{\partial u}{\partial x} = 0, \dfrac{\partial u}{\partial y} = \dfrac{z}{(z-y)^2}, \dfrac{\partial u}{\partial z} = \dfrac{-y}{(z-y)^2}$

(b) $\dfrac{\partial u}{\partial x} = 0, \dfrac{\partial u}{\partial y} = 1, \dfrac{\partial u}{\partial z} = -2$

13. $\dfrac{dz}{dt} = \dfrac{\partial z}{\partial x}\dfrac{dx}{dt} + \dfrac{\partial z}{\partial y}\dfrac{dy}{dt}$

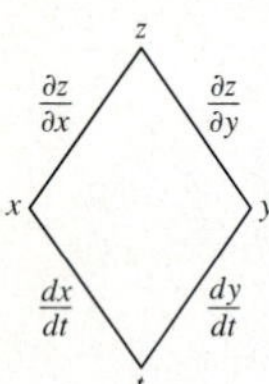

15. $\dfrac{\partial w}{\partial u} = \dfrac{\partial w}{\partial x}\dfrac{\partial x}{\partial u} + \dfrac{\partial w}{\partial y}\dfrac{\partial y}{\partial u} + \dfrac{\partial w}{\partial z}\dfrac{\partial z}{\partial u}, \dfrac{\partial w}{\partial v} = \dfrac{\partial w}{\partial x}\dfrac{\partial x}{\partial v} + \dfrac{\partial w}{\partial y}\dfrac{\partial y}{\partial v} + \dfrac{\partial w}{\partial z}\dfrac{\partial z}{\partial v}$

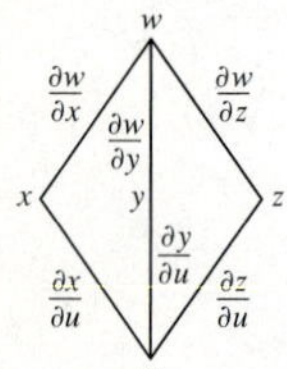

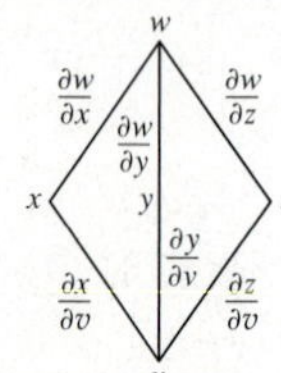

17. $\dfrac{\partial w}{\partial u} = \dfrac{\partial w}{\partial x}\dfrac{\partial x}{\partial u} + \dfrac{\partial w}{\partial y}\dfrac{\partial y}{\partial u}, \dfrac{\partial w}{\partial v} = \dfrac{\partial w}{\partial x}\dfrac{\partial x}{\partial v} + \dfrac{\partial w}{\partial y}\dfrac{\partial y}{\partial v}$

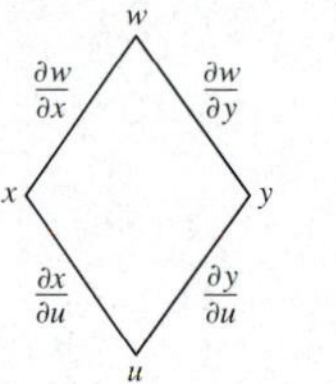

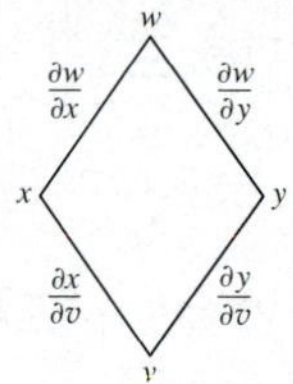

19. $\dfrac{\partial z}{\partial t} = \dfrac{\partial z}{\partial x}\dfrac{\partial x}{\partial t} + \dfrac{\partial z}{\partial y}\dfrac{\partial y}{\partial t}, \dfrac{\partial z}{\partial s} = \dfrac{\partial z}{\partial x}\dfrac{\partial x}{\partial s} + \dfrac{\partial z}{\partial y}\dfrac{\partial y}{\partial s}$

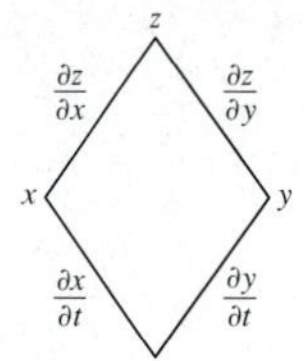

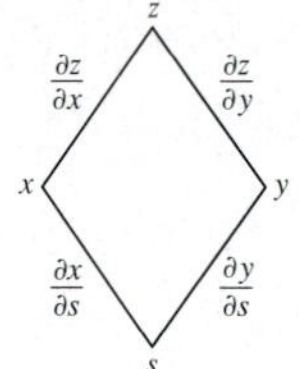

21. $\dfrac{\partial w}{\partial s} = \dfrac{dw}{du}\dfrac{\partial u}{\partial s}, \dfrac{\partial w}{\partial t} = \dfrac{dw}{du}\dfrac{\partial u}{\partial t}$

23. $\dfrac{\partial w}{\partial r} = \dfrac{\partial w}{\partial x}\dfrac{dx}{dr} + \dfrac{\partial w}{\partial y}\dfrac{dy}{dr} = \dfrac{\partial w}{\partial x}\dfrac{dx}{dr}$ since $\dfrac{dy}{dr} = 0,$

$\dfrac{\partial w}{\partial s} = \dfrac{\partial w}{\partial x}\dfrac{dx}{ds} + \dfrac{\partial w}{\partial y}\dfrac{dy}{ds} = \dfrac{\partial w}{\partial y}\dfrac{dy}{ds}$ since $\dfrac{dx}{ds} = 0$

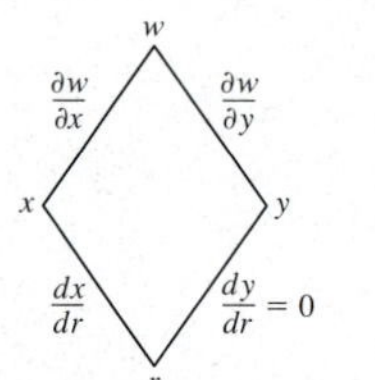

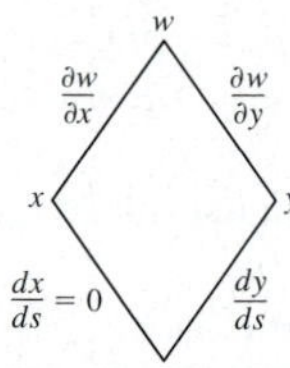

25. 4/3

27. $-4/5$

29. $\dfrac{\partial z}{\partial x} = \dfrac{1}{4}, \dfrac{\partial z}{\partial y} = -\dfrac{3}{4}$

31. $\dfrac{\partial z}{\partial x} = -1, \dfrac{\partial z}{\partial y} = -1$

33. 12

35. -7

37. $\dfrac{\partial z}{\partial u} = 2, \dfrac{\partial z}{\partial v} = 1$

39. -0.00005 amps/sec

45. $(\cos 1, \sin 1, 1)$ and $(\cos(-2), \sin(-2), -2)$

47. **(a)** Maximum at $\left(-\frac{\sqrt{2}}{2}, \frac{\sqrt{2}}{2}\right)$ and $\left(\frac{\sqrt{2}}{2}, -\frac{\sqrt{2}}{2}\right)$; minimum at $\left(\frac{\sqrt{2}}{2}, \frac{\sqrt{2}}{2}\right)$ and $\left(-\frac{\sqrt{2}}{2}, -\frac{\sqrt{2}}{2}\right)$

(b) Max = 6, min = 2

49. $2x\sqrt{x^8 + x^3} + \int_0^{x^2} \frac{3x^2}{2\sqrt{t^4 + x^3}}\,dt$

Section 11.5, pp. 923–925

1.

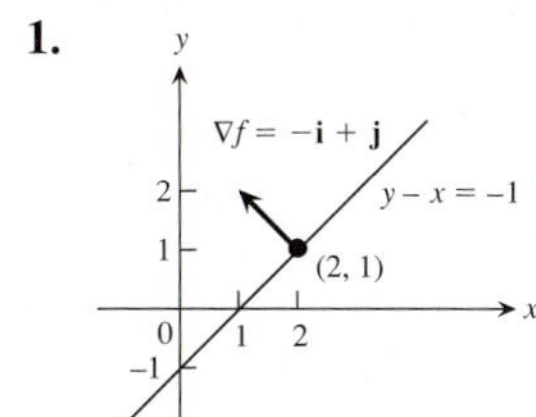

3.

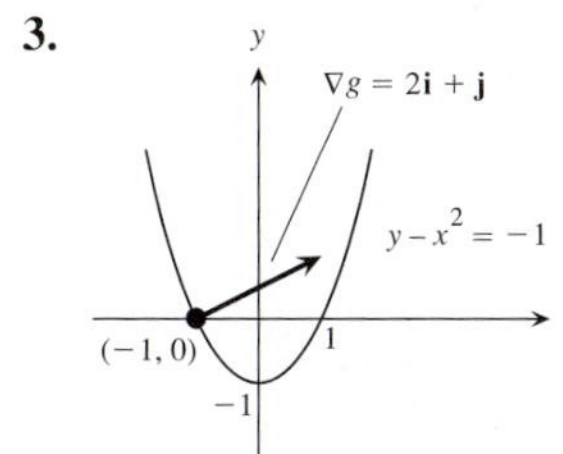

5. $\nabla f = 3\mathbf{i} + 2\mathbf{j} - 4\mathbf{k}$

7. $\nabla f = -\frac{26}{27}\mathbf{i} + \frac{23}{54}\mathbf{j} - \frac{23}{54}\mathbf{k}$

9. -4

11. 31/13

13. 3

15. 2

17. $\mathbf{u} = -\frac{1}{\sqrt{2}}\mathbf{i} + \frac{1}{\sqrt{2}}\mathbf{j}$, $(D_{\mathbf{u}}f)_{P_0} = \sqrt{2}$; $-\mathbf{u} = \frac{1}{\sqrt{2}}\mathbf{i} - \frac{1}{\sqrt{2}}\mathbf{j}$, $(D_{-\mathbf{u}}f)_{P_0} = -\sqrt{2}$

19. $\mathbf{u} = \frac{1}{3\sqrt{3}}\mathbf{i} - \frac{5}{3\sqrt{3}}\mathbf{j} - \frac{1}{3\sqrt{3}}\mathbf{k}$, $(D_{\mathbf{u}}f)_{P_0} = 3\sqrt{3}$;
$-\mathbf{u} = -\frac{1}{3\sqrt{3}}\mathbf{i} + \frac{5}{3\sqrt{3}}\mathbf{j} + \frac{1}{3\sqrt{3}}\mathbf{k}$, $(D_{-\mathbf{u}}f)_{P_0} = -3\sqrt{3}$

21. $\mathbf{u} = \frac{1}{\sqrt{3}}(\mathbf{i} + \mathbf{j} + \mathbf{k})$, $(D_{\mathbf{u}}f)_{P_0} = 2\sqrt{3}$;
$-\mathbf{u} = -\frac{1}{\sqrt{3}}(\mathbf{i} + \mathbf{j} + \mathbf{k})$, $(D_{-\mathbf{u}}f)_{P_0} = -2\sqrt{3}$

23. $df = \frac{9}{11{,}830} \approx 0.0008$

25. $dg = 0$

27. **(a)** $x + y + z = 3$
(b) $x = 1 + 2t, y = 1 + 2t, z = 1 + 2t$

29. **(a)** $2x - z - 2 = 0$
(b) $x = 2 - 4t, y = 0, z = 2 + 2t$

31. **(a)** $2x + 2y + z - 4 = 0$
(b) $x = 2t, y = 1 + 2t, z = 2 + t$

33. **(a)** $x + y + z - 1 = 0$
(b) $x = t, y = 1 + t, z = t$

35. $2x - z - 2 = 0$

37. $x - y + 2z - 1 = 0$

39.

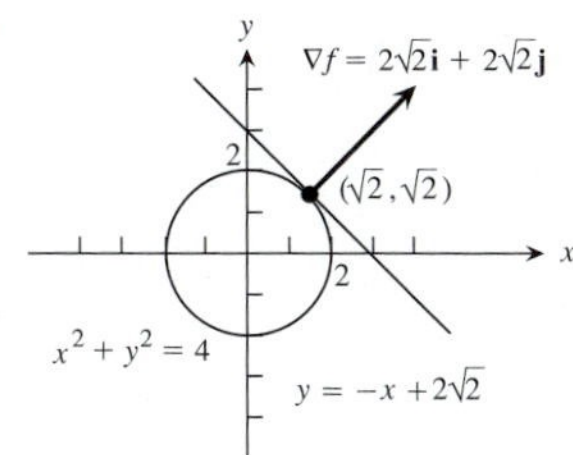

41.

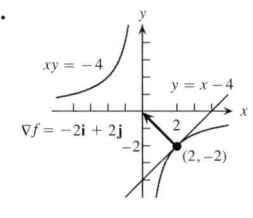

43. $x = 1, y = 1 + 2t, z = 1 - 2t$

45. $x = 1 - 2t, y = 1, z = \frac{1}{2} + 2t$

47. $x = 1 + 90t, y = 1 - 90t, z = 3$

49. $\mathbf{u} = \frac{7}{\sqrt{53}}\mathbf{i} - \frac{2}{\sqrt{53}}\mathbf{j}$, $-\mathbf{u} = -\frac{7}{\sqrt{53}}\mathbf{i} + \frac{2}{\sqrt{53}}\mathbf{j}$

51. No, the maximum rate of change is $\sqrt{185} < 14$.

53. $-\frac{7}{\sqrt{5}}$

55. **(a)** $\frac{\sqrt{3}}{2}\sin\sqrt{3} - \frac{1}{2}\cos\sqrt{3} \approx 0.935°\text{C/ft}$
(b) $\sqrt{3}\sin\sqrt{3} - \cos\sqrt{3} \approx 1.87°\text{C/sec}$

57. At $-\frac{\pi}{4}, -\frac{\pi}{2\sqrt{2}}$; at 0, 0; at $\frac{\pi}{4}, \frac{\pi}{2\sqrt{2}}$

Section 11.6, pp. 934–936

1. **(a)** $L(x, y) = 1$ **(b)** $L(x, y) = 2x + 2y - 1$

3. **(a)** $L(x, y) = 3x - 4y + 5$ **(b)** $L(x, y) = 3x - 4y + 5$

5. **(a)** $L(x, y) = 1 + x$ **(b)** $L(x, y) = -y + \frac{\pi}{2}$

7. $L(x, y) = 7 + x - 6y$; 0.06

9. $L(x, y) = x + y + 1$; 0.08

11. $L(x, y) = 1 + x$; 0.0222

13. Pay more attention to the smaller of the two dimensions. It will generate the larger partial derivative.

15. Maximum error (estimate) ≤ 0.31 in magnitude

17. Maximum percentage error $= \pm 4.83\%$

19. Let $|x - 1| \le 0.014$, $|y - 1| \le 0.014$

21. $\approx 0.1\%$

23. **(a)** $L(x, y, z) = 2x + 2y + 2z - 3$
(b) $L(x, y, z) = y + z$
(c) $L(x, y, z) = 0$

25. **(a)** $L(x, y, z) = x$
(b) $L(x, y, z) = \frac{1}{\sqrt{2}}x + \frac{1}{\sqrt{2}}y$
(c) $L(x, y, z) = \frac{1}{3}x + \frac{2}{3}y + \frac{2}{3}z$

27. **(a)** $L(x, y, z) = 2 + x$
(b) $L(x, y, z) = x - y - z + \frac{\pi}{2} + 1$
(c) $L(x, y, z) = x - y - z + \frac{\pi}{2} + 1$

29. $L(x, y, z) = 2x - 6y - 2z + 6$, 0.0024

31. $L(x, y, z) = x + y - z - 1$, 0.00135

33. **(a)** $S_0\left(\frac{1}{100}\,dp + dx - 5\,dw - 30\,dh\right)$
(b) More sensitive to a change in height

35. f is most sensitive to a change in d

37. $\frac{47}{24}\text{ ft}^3$

39. Magnitude of possible error ≤ 4.8

Section 11.7, pp. 944–947

1. $f(-3, 3) = -5$, local minimum **3.** $f(-2, 1)$, saddle point

5. $f\left(\frac{13}{12}, -\frac{3}{4}\right) = -\frac{31}{12}$, local minimum

7. $f(1, 2)$, saddle point **9.** $f(0, 1) = 4$, local maximum

11. $f(0, 0)$, saddle point; $f(-1, -1) = 1$, local maximum

13. $f(0, 0)$, saddle point; $f\left(\frac{4}{9}, \frac{4}{3}\right) = -\frac{64}{81}$, local minimum

15. $f(0, 0)$, saddle point; $f(1, 1) = 2, f(-1, -1) = 2$, local maxima

17. $f(0, 0) = -1$, local maximum

19. $f(n\pi, 0)$, saddle point; $f(n\pi, 0) = 0$ for every n

21. Absolute maximum: 1 at $(0, 0)$; absolute minimum: -5 at $(1, 2)$

23. Absolute maximum: 11 at $(0, -3)$; absolute minimum: -10 at $(4, -2)$

25. Absolute maximum: 4 at $(2, 0)$; absolute minimum: $\frac{3\sqrt{2}}{2}$ at $\left(3, -\frac{\pi}{4}\right), \left(3, \frac{\pi}{4}\right), \left(1, -\frac{\pi}{4}\right)$, and $\left(1, \frac{\pi}{4}\right)$

27. $a = -3, b = 2$

29. Hottest: $2\frac{1}{4}°$ at $\left(-\frac{1}{2}, \frac{\sqrt{3}}{2}\right)$ and $\left(-\frac{1}{2}, -\frac{\sqrt{3}}{2}\right)$; coldest: $-\frac{1}{4}°$ at $\left(\frac{1}{2}, 0\right)$

31. **(a)** $f(0, 0)$, saddle point
(b) $f(1, 2)$, local minimum
(c) $f(1, -2)$, local minimum; $f(-1, -2)$, saddle point

37. $(1/6, 1/3, 355/36)$

41. **(a)** On the semicircle, max $f = 2\sqrt{2}$ at $t = \frac{\pi}{4}$, min $f = -2$ at $t = \pi$; on the quarter circle, max $f = 2\sqrt{2}$ at $t = \frac{\pi}{4}$, min $f = 2$ at $t = 0, \frac{\pi}{2}$

(b) On the semicircle, max $g = 2$ at $t = \frac{\pi}{4}$, min $g = -2$ at $t = \frac{3\pi}{4}$; on the quarter circle, max $g = 2$ at $t = \frac{\pi}{4}$, min $g = 0$ at $t = 0, \frac{\pi}{2}$

(c) On the semicircle, max $h = 8$ at $t = 0, \pi$; min $h = 4$ at $t = \frac{\pi}{2}$; on the quarter circle, max $h = 8$ at $t = 0$, min $h = 4$ at $t = \frac{\pi}{2}$

43. **(i)** min $f = -\frac{1}{2}$ at $t = -\frac{1}{2}$; no max

(ii) max $f = 0$ at $t = -1, 0$; min $f = -\frac{1}{2}$ at $t = -\frac{1}{2}$

(iii) max $f = 4$ at $t = 1$; min $f = 0$ at $t = 0$

Section 11.8, pp. 956–958

1. $\left(\pm\frac{1}{\sqrt{2}}, \frac{1}{2}\right), \left(\pm\frac{1}{\sqrt{2}}, -\frac{1}{2}\right)$

3. 39 **5.** $(3, \pm 3\sqrt{2})$

7. **(a)** 8 **(b)** 64

9. $r = 2$ cm, $h = 4$ cm **11.** $\ell = 4\sqrt{2}, w = 3\sqrt{2}$

13. $f(0, 0) = 0$ is minimum, $f(2, 4) = 20$ is maximum

15. Lowest $= 0°$, highest $= 125°$ **17.** $\left(\frac{3}{2}, 2, \frac{5}{2}\right)$

19. 1 **21.** $(0, 0, 2), (0, 0, -2)$

23. $f(1, -2, 5) = 30$ is maximum, $f(-1, 2, -5) = -30$ is minimum.

25. 3, 3, 3 **27.** $\frac{2}{\sqrt{3}}$ by $\frac{2}{\sqrt{3}}$ by $\frac{2}{\sqrt{3}}$ units

29. $(\pm 4/3, -4/3, -4/3)$ **31.** $U(8, 14) = \$128$

33. $f(2/3, 4/3, -4/3) = \frac{4}{3}$ **35.** $(2, 4, 4)$

37. Maximum is $1 + 6\sqrt{3}$ at $(\pm\sqrt{6}, \sqrt{3}, 1)$, minimum is $1 - 6\sqrt{3}$ at $(\pm\sqrt{6}, -\sqrt{3}, 1)$.

39. Maximum is 4 at $(0, 0, \pm 2)$, minimum is 2 at $(\pm\sqrt{2}, \pm\sqrt{2}, 0)$

Section 11.9, pp. 962–963

1. **(a)** 0 **(b)** $1 + 2z$ **(c)** $1 + 2z$

3. **(a)** $\frac{\partial U}{\partial P} + \frac{\partial U}{\partial T}\left(\frac{V}{nR}\right)$ **(b)** $\frac{\partial U}{\partial P}\left(\frac{nR}{V}\right) + \frac{\partial U}{\partial T}$

5. **(a)** 5 **(b)** 5

7. $\left(\frac{\partial x}{\partial r}\right)_\theta = \cos\theta$

$\left(\frac{\partial r}{\partial x}\right)_y = \frac{x}{\sqrt{x^2 + y^2}}$

Section 11.10, pp. 967

1. Quadratic: $x + xy$; cubic: $x + xy + \frac{1}{2}xy^2$

3. Quadratic: xy; cubic: xy

5. Quadratic: $y + \frac{1}{2}(2xy - y^2)$;
cubic: $y + \frac{1}{2}(2xy - y^2) + \frac{1}{6}(3x^2y - 3xy^2 + 2y^3)$

7. Quadratic: $\frac{1}{2}(2x^2 + 2y^2) = x^2 + y^2$; cubic: $x^2 + y^2$

9. Quadratic: $1 + (x + y) + (x + y)^2$;
cubic: $1 + (x + y) + (x + y)^2 + (x + y)^3$

11. Quadratic: $1 - \frac{1}{2}x^2 - \frac{1}{2}y^2$; $E(x, y) \le 0.00134$

Chapter 11 Practice Exercises, pp. 968–972

1. Domain: all points in the xy-plane; range: $z \geq 0$. Level curves are ellipses with major axis along the y-axis and minor axis along the x-axis.

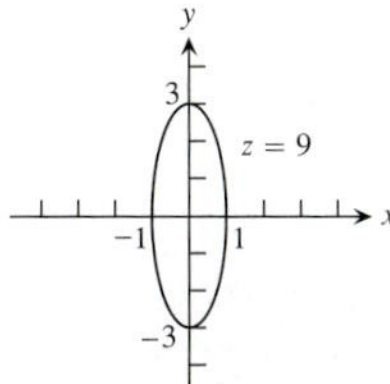

3. Domain: all (x, y) such that $x \neq 0$ and $y \neq 0$; range: $z \neq 0$. Level curves are hyperbolas with the x- and y-axes as asymptotes.

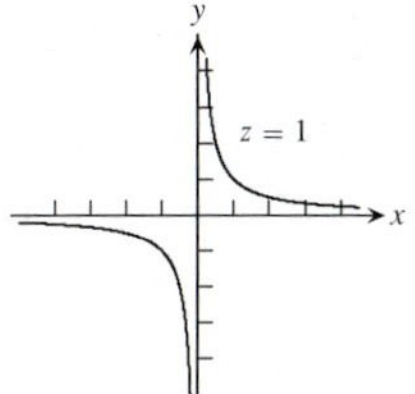

5. Domain: all points in xyz-space; range: all real numbers. Level surfaces are paraboloids of revolution with the z-axis as axis.

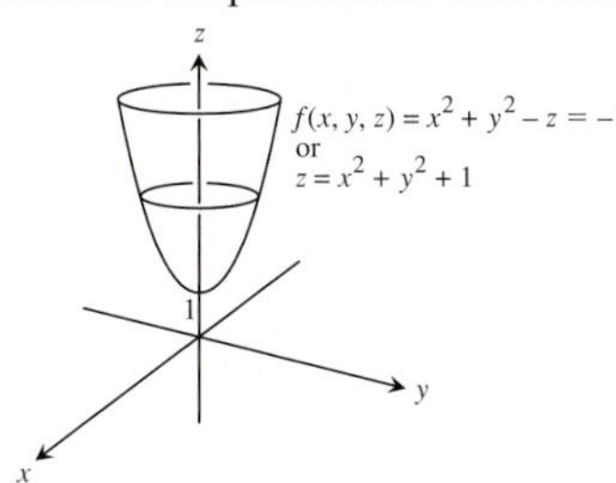

7. Domain: all (x, y, z) such that $(x, y, z) \neq (0, 0, 0)$; range: positive real numbers. Level surfaces are spheres with center $(0, 0, 0)$ and radius $r > 0$.

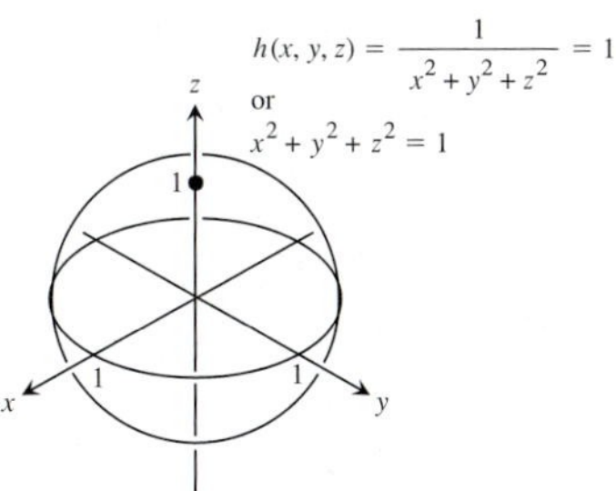

9. -2

11. $1/2$

13. 1

15. Let $y = kx^2, k \neq 1$

17. No; $\lim_{(x,y)\to(0,0)} f(x, y)$ does not exist.

19. $\dfrac{\partial g}{\partial r} = \cos\theta + \sin\theta, \dfrac{\partial g}{\partial \theta} = -r\sin\theta + r\cos\theta$

21. $\dfrac{\partial f}{\partial R_1} = -\dfrac{1}{R_1^2}, \dfrac{\partial f}{\partial R_2} = -\dfrac{1}{R_2^2}, \dfrac{\partial f}{\partial R_3} = -\dfrac{1}{R_3^2}$

23. $\dfrac{\partial P}{\partial n} = \dfrac{RT}{V}, \dfrac{\partial P}{\partial R} = \dfrac{nT}{V}, \dfrac{\partial P}{\partial T} = \dfrac{nR}{V}, \dfrac{\partial P}{\partial V} = -\dfrac{nRT}{V^2}$

25. $\dfrac{\partial^2 g}{\partial x^2} = 0, \dfrac{\partial^2 g}{\partial y^2} = \dfrac{2x}{y^3}, \dfrac{\partial^2 g}{\partial y\,\partial x} = \dfrac{\partial^2 g}{\partial x\,\partial y} = -\dfrac{1}{y^2}$

27. $\dfrac{\partial^2 f}{\partial x^2} = -30x + \dfrac{2 - 2x^2}{(x^2+1)^2}, \dfrac{\partial^2 f}{\partial y^2} = 0, \dfrac{\partial^2 f}{\partial y\,\partial x} = \dfrac{\partial^2 f}{\partial x\,\partial y} = 1$

29. $\left.\dfrac{dw}{dt}\right|_{t=0} = -1$

31. $\left.\dfrac{\partial w}{\partial r}\right|_{(r,s)=(\pi,0)} = 2, \left.\dfrac{\partial w}{\partial s}\right|_{(r,s)=(\pi,0)} = 2 - \pi$

33. $\left.\dfrac{df}{dt}\right|_{t=1} = -(\sin 1 + \cos 2)(\sin 1) + (\cos 1 + \cos 2)(\cos 1) - 2(\sin 1 + \cos 1)(\sin 2)$

35. $\left.\dfrac{dy}{dx}\right|_{(x,y)=(0,1)} = -1$

37. Increases most rapidly in the direction $\mathbf{u} = -\dfrac{\sqrt{2}}{2}\mathbf{i} - \dfrac{\sqrt{2}}{2}\mathbf{j}$; decreases most rapidly in the direction $-\mathbf{u} = \dfrac{\sqrt{2}}{2}\mathbf{i} + \dfrac{\sqrt{2}}{2}\mathbf{j}$; $D_{\mathbf{u}}f = \dfrac{\sqrt{2}}{2}$; $D_{-\mathbf{u}}f = -\dfrac{\sqrt{2}}{2}$; $D_{\mathbf{u}_1}f = -\dfrac{7}{10}$ where $\mathbf{u}_1 = \dfrac{\mathbf{v}}{|\mathbf{v}|}$

39. Increases most rapidly in the direction $\mathbf{u} = \dfrac{2}{7}\mathbf{i} + \dfrac{3}{7}\mathbf{j} + \dfrac{6}{7}\mathbf{k}$; decreases most rapidly in the direction $-\mathbf{u} = -\dfrac{2}{7}\mathbf{i} - \dfrac{3}{7}\mathbf{j} - \dfrac{6}{7}\mathbf{k}$; $D_{\mathbf{u}}f = 7$; $D_{-\mathbf{u}}f = -7$; $D_{\mathbf{u}_1}f = 7$ where $\mathbf{u}_1 = \dfrac{\mathbf{v}}{|\mathbf{v}|}$

41. $\pi/\sqrt{2}$

43. **(a)** $f_x(1, 2) = f_y(1, 2) = 2$ **(b)** $14/5$

45.

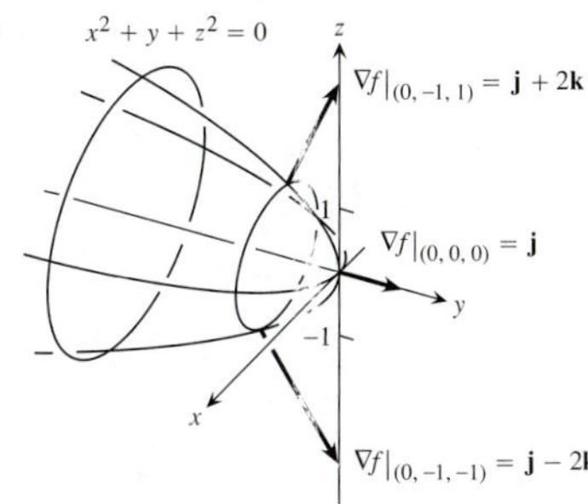

47. Tangent: $4x - y - 5z = 4$; normal line: $x = 2 + 4t, y = -1 - t, z = 1 - 5t$

49. $2y - z - 2 = 0$

51. Tangent: $x + y = \pi + 1$; normal line: $y = x - \pi + 1$

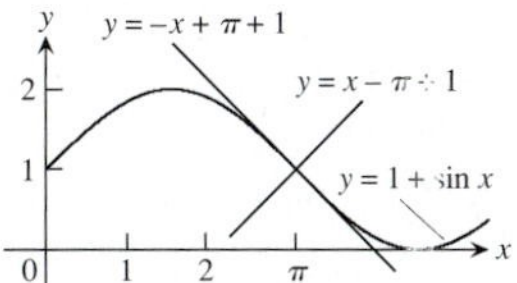

53. $x = 1 - 2t, y = 1, z = 1/2 + 2t$

55. Answers will depend on the upper bound used for $|f_{xx}|, |f_{xy}|, |f_{yy}|$. With $M = \sqrt{2}/2$, $|E| \leq 0.0142$. With $M = 1$, $|E| \leq 0.02$.

57. $L(x, y, z) = y - 3z$, $L(x, y, z) = x + y - z - 1$

59. Be more careful with the diameter.

61. $dI = 0.038$, % change in $I = 15.83\%$, more sensitive to voltage change

63. **(a)** 5%

65. Local minimum of -8 at $(-2, -2)$

67. Saddle point at $(0, 0)$, $f(0, 0) = 0$; local maximum of $1/4$ at $(-1/2, -1/2)$

69. Saddle point at $(0, 0)$, $f(0, 0) = 0$; local minimum of -4 at $(0, 2)$; local maximum of 4 at $(-2, 0)$; saddle point at $(-2, 2)$, $f(-2, 2) = 0$

71. Absolute maximum: 28 at $(0, 4)$; absolute minimum: $-9/4$ at $(3/2, 0)$

73. Absolute maximum: 18 at $(2, -2)$; absolute minimum: $-17/4$ at $(-2, 1/2)$

75. Absolute maximum: 8 at $(-2, 0)$; absolute minimum: -1 at $(1, 0)$

77. Absolute maximum: 4 at $(1, 0)$; absolute minimum: -4 at $(0, -1)$

79. Absolute maximum: 1 at $(0, \pm 1)$ and $(1, 0)$; absolute minimum: -1 at $(-1, 0)$

81. Maximum: 5 at $(0, 1)$; minimum: $-1/3$ at $(0, -1/3)$

83. Maximum: $\sqrt{3}$ at $\left(\frac{1}{\sqrt{3}}, -\frac{1}{\sqrt{3}}, \frac{1}{\sqrt{3}}\right)$;
minimum: $-\sqrt{3}$ at $\left(-\frac{1}{\sqrt{3}}, \frac{1}{\sqrt{3}}, -\frac{1}{\sqrt{3}}\right)$

85. Width $= \left(\frac{c^2V}{ab}\right)^{1/3}$, depth $= \left(\frac{b^2V}{ac}\right)^{1/3}$, height $= \left(\frac{a^2V}{bc}\right)^{1/3}$

87. Maximum: $\frac{3}{2}$ at $\left(\frac{1}{\sqrt{2}}, \frac{1}{\sqrt{2}}, \sqrt{2}\right)$ and $\left(-\frac{1}{\sqrt{2}}, -\frac{1}{\sqrt{2}}, -\sqrt{2}\right)$;
minimum: $\frac{1}{2}$ at $\left(-\frac{1}{\sqrt{2}}, \frac{1}{\sqrt{2}}, -\sqrt{2}\right)$ and $\left(\frac{1}{\sqrt{2}}, -\frac{1}{\sqrt{2}}, \sqrt{2}\right)$

89. **(a)** $(2y + x^2z)e^{yz}$
(b) $x^2 e^{yz}\left(y - \frac{z}{2y}\right)$
(c) $(1 + x^2y)e^{yz}$

91. $\frac{\partial w}{\partial x} = \cos\theta \frac{\partial w}{\partial r} - \frac{\sin\theta}{r}\frac{\partial w}{\partial \theta}, \frac{\partial w}{\partial y} = \sin\theta\frac{\partial w}{\partial r} + \frac{\cos\theta}{r}\frac{\partial w}{\partial \theta}$

97. $(t, -t \pm 4, t)$, t a real number

Chapter 11 Additional Exercises, pp. 972–974

1. $f_{xy}(0, 0) = -1$, $f_{yx}(0, 0) = 1$

7. **(c)** $\frac{r^2}{2} = \frac{1}{2}(x^2 + y^2 + z^2)$ **13.** $V = \frac{\sqrt{3}abc}{2}$

17. $f(x, y) = \frac{y}{2} + 4$, $g(x, y) = \frac{x}{2} + \frac{9}{2}$

19. $y = 2\ln|\sin x| + \ln 2$

21. **(a)** $\frac{1}{\sqrt{53}}(2\mathbf{i} + 7\mathbf{j})$
(b) $\frac{-1}{\sqrt{29{,}097}}(98\mathbf{i} - 127\mathbf{j} + 58\mathbf{k})$

23. $w = e^{-c^2\pi^2 t}\sin \pi x$

CHAPTER 12

Section 12.1, pp. 984–987

1. 16

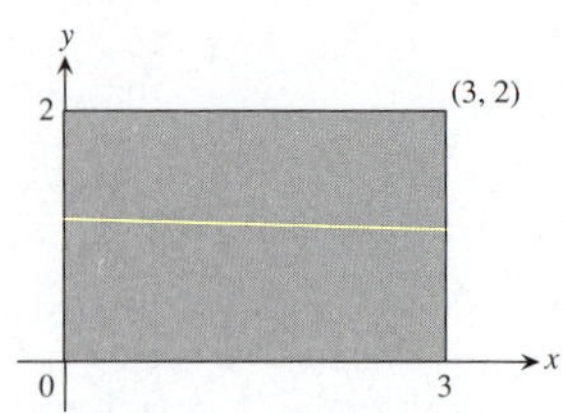

3. 1

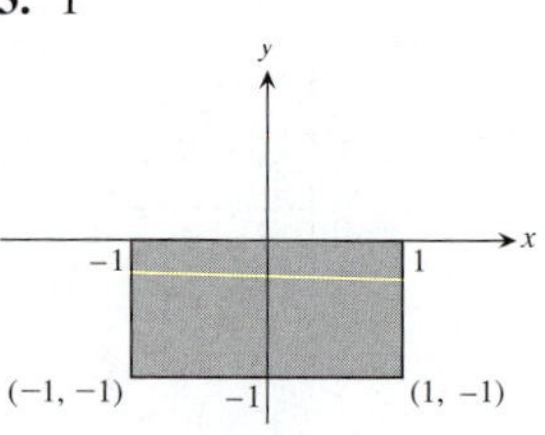

5. $\frac{\pi^2}{2} + 2$

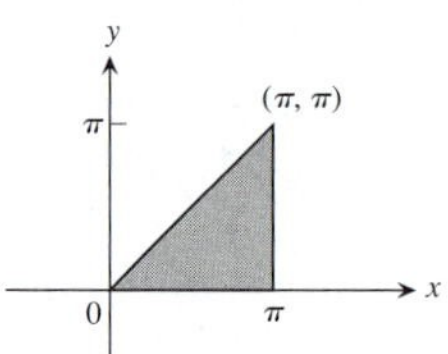

7. $8\ln 8 - 16 + e$

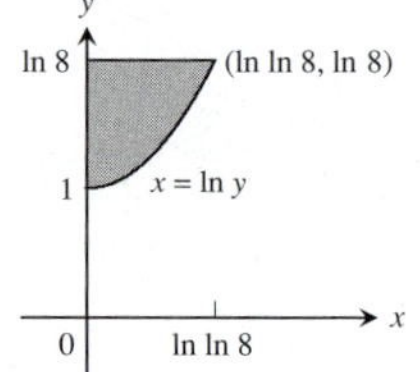

9. $e - 2$

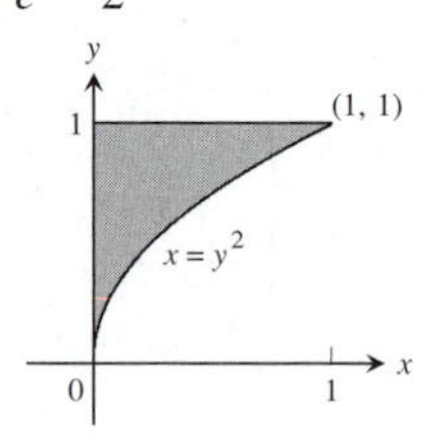

11. $\frac{3}{2}\ln 2$

13. 1/6

15. $-1/10$

17. 8

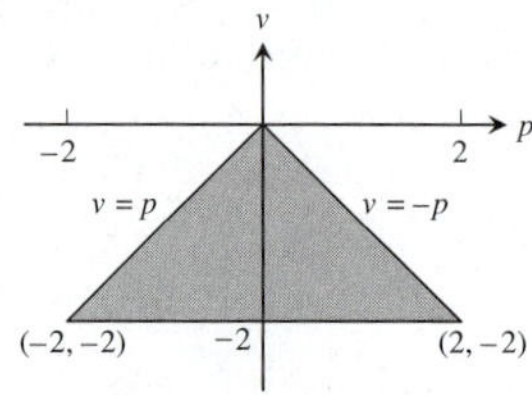

19. 2π

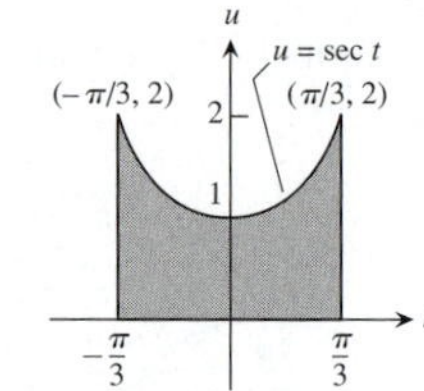

21. $\int_2^4 \int_0^{(4-y)/2} dx\, dy$

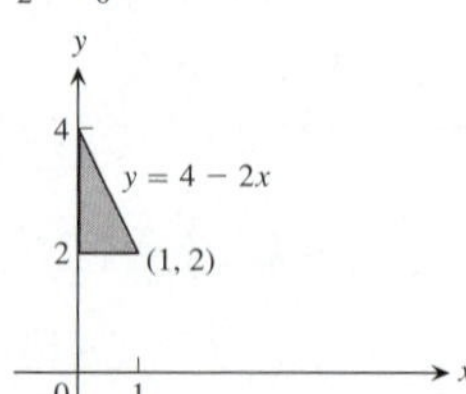

23. $\int_0^1 \int_{x^2}^{x} dy\, dx$

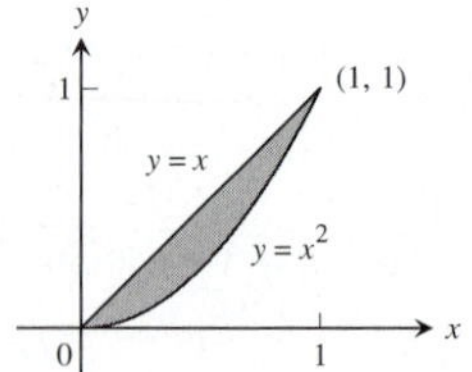

25. $\int_1^e \int_{\ln y}^1 dx\, dy$

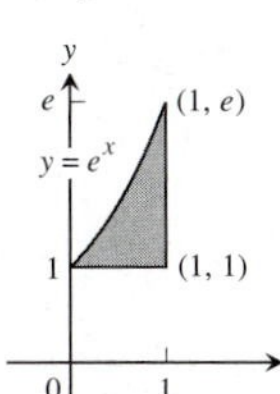

27. $\int_0^9 \int_0^{(\sqrt{9-y})/2} 16x\, dx\, dy$

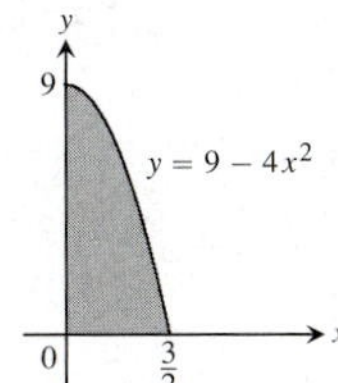

29. $\int_{-1}^1 \int_0^{\sqrt{1-x^2}} 3y\, dy\, dx$

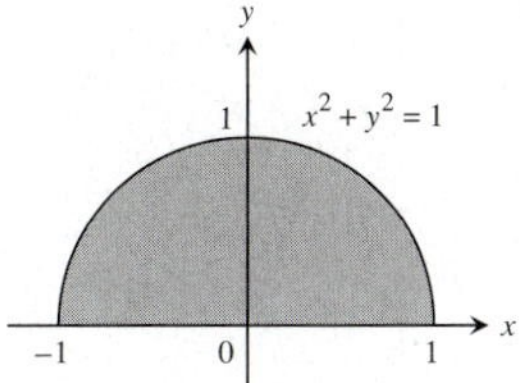

31. 2

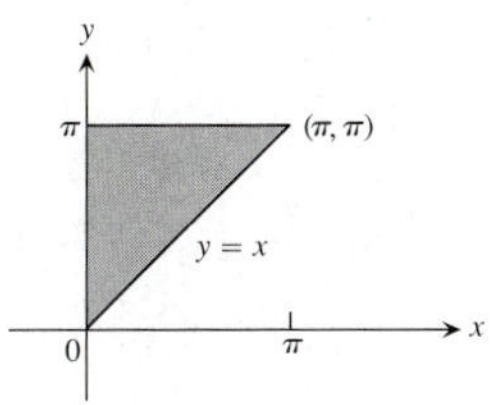

33. $\frac{e-2}{2}$

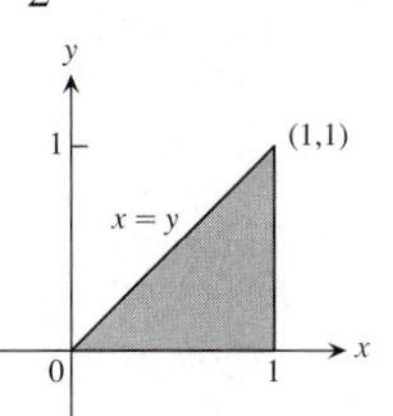

35. 2

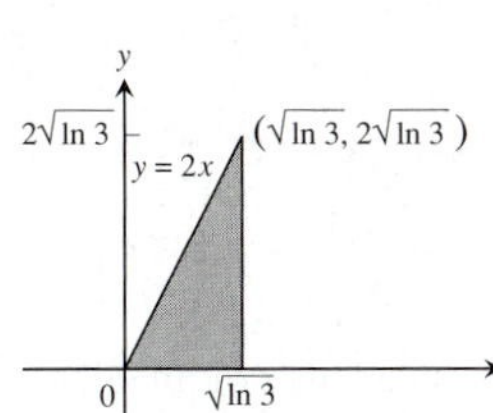

37. $1/80\pi$

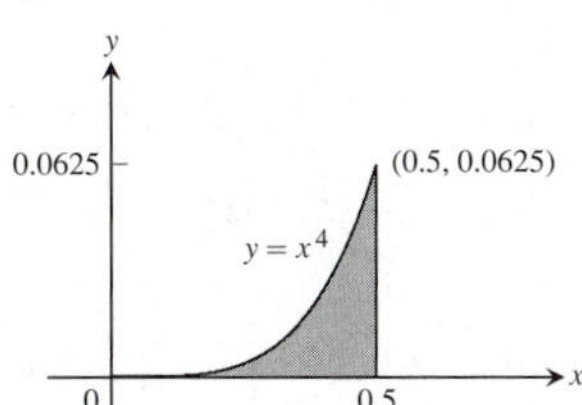

39. $-2/3$

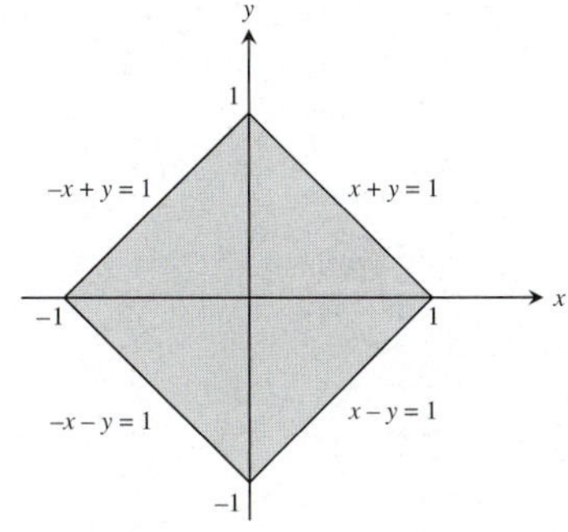

41. $4/3$

43. $625/12$

45. 16

47. 20

49. $2(1 + \ln 2)$

51. 1

53. π^2

55. $-\frac{3}{32}$

57. $\frac{20\sqrt{3}}{9}$

59. $\int_0^1 \int_x^{2-x} (x^2 + y^2)\, dy\, dx = \frac{4}{3}$

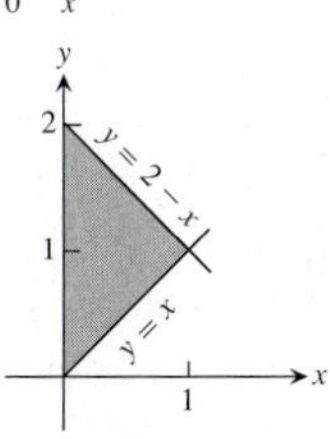

63. No, by Fubini's theorem, the two orders of integration must give the same result.

67. 0.603

69. 0.233

Section 12.2, pp. 997–1000

1. $\int_0^2 \int_0^{2-x} dy\, dx = 2$ or $\int_0^2 \int_0^{2-y} dx\, dy = 2$

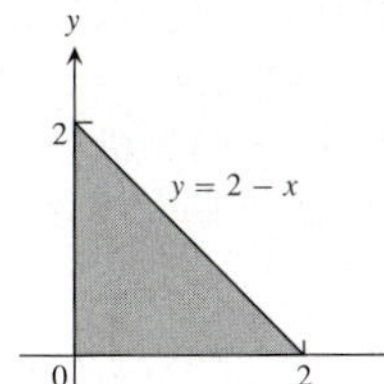

3. $\int_{-2}^1 \int_{y-2}^{-y^2} dx\, dy = \frac{9}{2}$

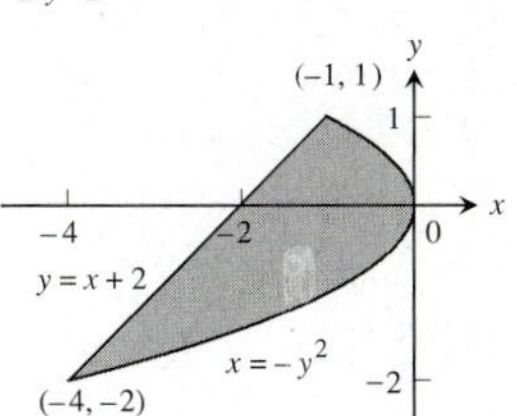

5. $\int_0^{\ln 2} \int_0^{e^x} dy\, dx = 1$

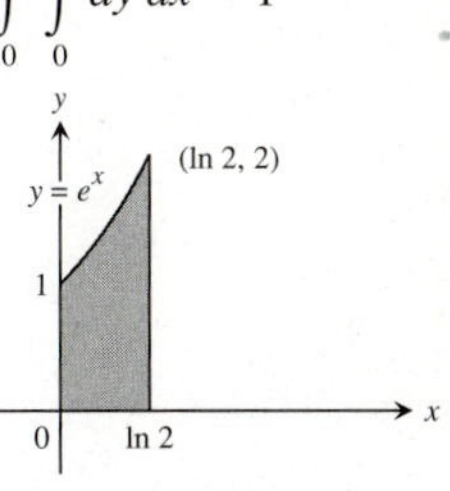

7. $\int_0^1 \int_{y^2}^{2y-y^2} dx\, dy = \frac{1}{3}$

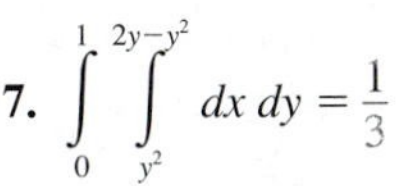

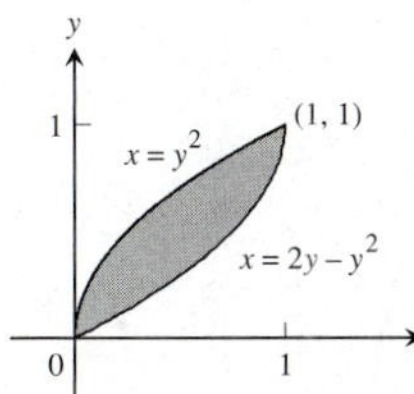

9. 12

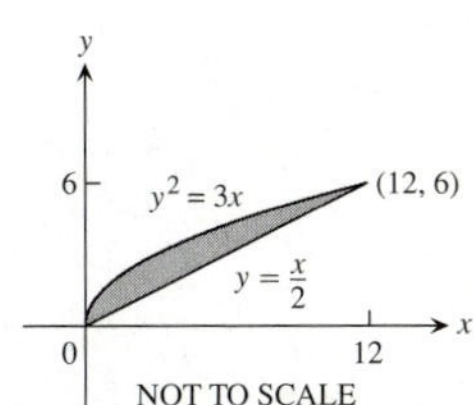

11. $\sqrt{2} - 1$

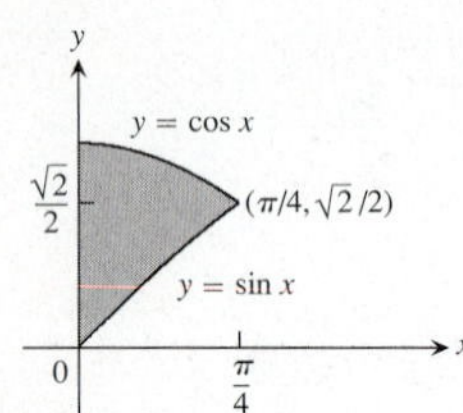

13. $\frac{3}{2}$

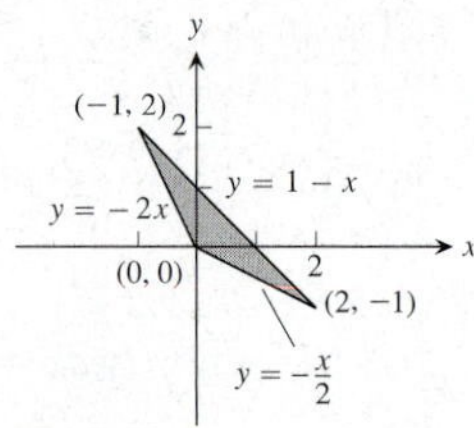

15. **(a)** 0 **(b)** $4/\pi^2$

17. 8/3

19. $\bar{x} = 5/14, \bar{y} = 38/35$

21. $\bar{x} = 64/35, \bar{y} = 5/7$

23. $\bar{x} = 0, \bar{y} = 4/3\pi$

25. $\bar{x} = \bar{y} = 4a/3\pi$

27. $I_x = I_y = 4\pi, I_0 = 8\pi$

29. $\bar{x} = -1, \bar{y} = 1/4$

31. $I_x = 64/105, R_x = 2\sqrt{2/7}$

33. $\bar{x} = 3/8, \bar{y} = 17/16$

35. $\bar{x} = 11/3, \bar{y} = 14/27, I_y = 432, R_y = 4$

37. $\bar{x} = 0, \bar{y} = 13/31, I_y = 7/5, R_y = \sqrt{21/31}$

39. $\bar{x} = 0, \bar{y} = 7/10; I_x = 9/10, I_y = 3/10, I_0 = 6/5; R_x = 3\sqrt{6}/10,$
$R_y = 3\sqrt{2}/10, R_0 = 3\sqrt{2}/5$

41. $40{,}000(1 - e^{-2}) \ln(7/2) \approx 43{,}329$

43. If $0 < a \le 5/2$, then the appliance will have to be tipped more than 45° to fall over.

45. $(\bar{x}, \bar{y}) = (2/\pi, 0)$

47. **(a)** 3/2 **(b)** They are the same.

53. **(a)** (7/5, 31/10) **(b)** (19/7, 18/7)
(c) (9/2, 19/8) **(d)** (11/4, 43/16)

55. For the center of mass to be on the common boundary, $h = a\sqrt{2}$. For the center of mass to be inside T, $h > a\sqrt{2}$.

Section 12.3, pp. 1005–1006

1. $\pi/2$

3. $\pi/8$

5. πa^2

7. 36

9. $(1 - \ln 2)\pi$

11. $(2 \ln 2 - 1)(\pi/2)$

13. $\pi/2 + 1$

15. $\pi(\ln 4 - 1)$

17. $2(\pi - 1)$

19. 12π

21. $3\pi/8 + 1$

23. 4

25. $6\sqrt{3} - 2\pi$

27. $\bar{x} = 5/6, \bar{y} = 0$

29. $\frac{2a}{3}$

31. $\frac{2a}{3}$

33. $2\pi(2 - \sqrt{e})$

35. $\frac{4}{3} + \frac{5\pi}{8}$

37. **(a)** $\frac{\sqrt{\pi}}{2}$ **(b)** 1

39. $\pi \ln 4$, no

41. $\frac{1}{2}(a^2 + 2h^2)$

Section 12.4, pp. 1014–1017

1. 1/6

3. $\int_0^1\int_0^{2-2x}\int_0^{3-3x-3y/2} dz\,dy\,dx$, $\int_0^2\int_0^{1-y/2}\int_0^{3-3x-3y/2} dz\,dx\,dy$,

$\int_0^1\int_0^{3-3x}\int_0^{2-2x-2z/3} dy\,dz\,dx$, $\int_0^3\int_0^{1-z/3}\int_0^{2-2x-2z/3} dy\,dx\,dz$,

$\int_0^2\int_0^{3-3y/2}\int_0^{1-y/2-z/3} dx\,dz\,dy$, $\int_0^3\int_0^{2-2z/3}\int_0^{1-y/2-z/3} dx\,dy\,dz$.

The value of all six integrals is 1.

5. $\int_{-2}^{2}\int_{-\sqrt{4-x^2}}^{\sqrt{4-x^2}}\int_{x^2+y^2}^{8-x^2-y^2} 1\,dz\,dy\,dx$, $\int_{-2}^{2}\int_{-\sqrt{4-y^2}}^{\sqrt{4-y^2}}\int_{x^2+y^2}^{8-x^2-y^2} 1\,dz\,dx\,dy$,

$\int_{-2}^{2}\int_{4}^{8-y^2}\int_{-\sqrt{8-z-y^2}}^{\sqrt{8-z-y^2}} 1\,dx\,dz\,dy + \int_{-2}^{2}\int_{y^2}^{4}\int_{-\sqrt{z-y^2}}^{\sqrt{z-y^2}} 1\,dx\,dz\,dy$,

$\int_{4}^{8}\int_{-\sqrt{8-z}}^{\sqrt{8-z}}\int_{-\sqrt{8-z-y^2}}^{\sqrt{8-z-y^2}} 1\,dx\,dy\,dz + \int_{0}^{4}\int_{-\sqrt{z}}^{\sqrt{z}}\int_{-\sqrt{z-y^2}}^{\sqrt{z-y^2}} 1\,dx\,dy\,dz$.

$\int_{-2}^{2}\int_{4}^{8-x^2}\int_{-\sqrt{8-z-x^2}}^{\sqrt{8-z-x^2}} 1\,dy\,dz\,dx + \int_{-2}^{2}\int_{x^2}^{4}\int_{-\sqrt{z-x^2}}^{\sqrt{z-x^2}} 1\,dy\,dz\,dx$,

$\int_{4}^{8}\int_{-\sqrt{8-z}}^{\sqrt{8-z}}\int_{-\sqrt{8-z-x^2}}^{\sqrt{8-z-x^2}} 1\,dy\,dx\,dz + \int_{0}^{4}\int_{-\sqrt{z}}^{\sqrt{z}}\int_{-\sqrt{z-x^2}}^{\sqrt{z-x^2}} 1\,dy\,dx\,dz$.

The value of all six integrals is 16π.

7. 1

9. 1

11. $\frac{\pi^3}{2}(1 - \cos 1)$

13. 18

15. 7/6

17. 0

19. $\frac{1}{2} - \frac{\pi}{8}$

21. **(a)** $\int_{-1}^{1}\int_{0}^{1-x^2}\int_{x^2}^{1-z} dy\,dz\,dx$ **(b)** $\int_{0}^{1}\int_{-\sqrt{1-z}}^{\sqrt{1-z}}\int_{x^2}^{1-z} dy\,dx\,dz$

(c) $\int_{0}^{1}\int_{0}^{1-z}\int_{-\sqrt{y}}^{\sqrt{y}} dx\,dy\,dz$ **(d)** $\int_{0}^{1}\int_{0}^{1-y}\int_{-\sqrt{y}}^{\sqrt{y}} dx\,dz\,dy$

(e) $\int_{0}^{1}\int_{-\sqrt{y}}^{\sqrt{y}}\int_{0}^{1-y} dz\,dx\,dy$

23. 2/3

25. 20/3

27. 1

29. 16/3

31. $8\pi - \frac{32}{3}$

33. 2

35. 4π

37. 31/3

39. 1

41. 2 sin 4

43. 4

45. $a = 3$ or $a = 13/3$

47. The domain is the set of all point (x, y, z) such that $4x^2 + 4y^2 + z^2 \leq 4$.

Section 12.5, pp. 1021–1023

1. $R_x = \sqrt{\dfrac{b^2 + c^2}{12}}, R_y = \sqrt{\dfrac{a^2 + c^2}{12}}, R_z = \sqrt{\dfrac{a^2 + b^2}{12}}$

3. $I_x = \dfrac{M}{3}(b^2 + c^2), I_y = \dfrac{M}{3}(a^2 + c^2), I_z = \dfrac{M}{3}(a^2 + b^2)$

5. $\bar{x} = \bar{y} = 0, \bar{z} = \dfrac{12}{5}, I_x = 7904/105 \approx 75.28, I_y = 4832/63 \approx 76.70, I_z = 256/45 \approx 5.69$

7. (a) $\bar{x} = \bar{y} = 0, \bar{z} = 8/3$ **(b)** $c = 2\sqrt{2}$

9. $I_L = 1386, R_L = \sqrt{\dfrac{77}{2}}$ **11.** $I_L = \dfrac{40}{3}, R_L = \sqrt{\dfrac{5}{3}}$

13. (a) 4/3 **(b)** $\bar{x} = 4/5, \bar{y} = \bar{z} = 2/5$

15. (a) 5/2 **(b)** $\bar{x} = \bar{y} = \bar{z} = 8/15$ **(c)** $I_x = I_y = I_z = 11/6$ **(d)** $R_x = R_y = R_z = \sqrt{\dfrac{11}{15}}$

17. 3

19. (a) $\dfrac{4}{3}g$ **(b)** $\dfrac{4}{3}g$

23. (a) $I_{\text{c.m.}} = \dfrac{abc(a^2 + b^2)}{12}, R_{\text{c.m.}} = \sqrt{\dfrac{a^2 + b^2}{12}}$

(b) $I_L = \dfrac{abc(a^2 + 7b^2)}{3}, R_L = \sqrt{\dfrac{a^2 + 7b^2}{3}}$

27. (a) $h = a\sqrt{3}$ **(b)** $h = a\sqrt{2}$

Section 12.6, pp. 1032–1037

1. $\dfrac{4\pi(\sqrt{2} - 1)}{3}$ **3.** $\dfrac{17\pi}{5}$

5. $\pi(6\sqrt{2} - 8)$ **7.** $\dfrac{3\pi}{10}$

9. $\pi/3$

11. (a) $\displaystyle\int_0^{2\pi}\int_0^1\int_0^{\sqrt{4-r^2}} r\,dz\,dr\,d\theta$

(b) $\displaystyle\int_0^{2\pi}\int_0^{\sqrt{3}}\int_0^1 r\,dr\,dz\,d\theta + \int_0^{2\pi}\int_{\sqrt{3}}^2\int_0^{\sqrt{4-z^2}} r\,dr\,dz\,d\theta$

(c) $\displaystyle\int_0^1\int_0^{\sqrt{4-r^2}}\int_0^{2\pi} r\,d\theta\,dz\,dr$

13. $\displaystyle\int_{-\pi/2}^{\pi/2}\int_0^{\cos\theta}\int_0^{3r^2} f(r, \theta, z)\,dz\,r\,dr\,d\theta$

15. $\displaystyle\int_0^{\pi}\int_0^{2\sin\theta}\int_0^{4-r\sin\theta} f(r, \theta, z)\,dz\,r\,dr\,d\theta$

17. $\displaystyle\int_{-\pi/2}^{\pi/2}\int_1^{1+\cos\theta}\int_0^4 f(r, \theta, z)\,dz\,r\,dr\,d\theta$

19. $\displaystyle\int_0^{\pi/4}\int_0^{\sec\theta}\int_0^{2-r\sin\theta} f(r, \theta, z)\,dz\,r\,dr\,d\theta$

21. π^2 **23.** $\pi/3$

25. 5π **27.** 2π

29. $\left(\dfrac{8 - 5\sqrt{2}}{2}\right)\pi$

31. (a) $\displaystyle\int_0^{2\pi}\int_0^{\pi/6}\int_0^2 \rho^2\sin\phi\,d\rho\,d\phi\,d\theta + \int_0^{2\pi}\int_{\pi/6}^{\pi/2}\int_0^{\csc\phi} \rho^2\sin\phi\,d\rho\,d\phi\,d\theta$

(b) $\displaystyle\int_0^{2\pi}\int_1^2\int_{\pi/6}^{\sin^{-1}(1/\rho)} \rho^2\sin\phi\,d\phi\,d\rho\,d\theta + \int_0^{2\pi}\int_0^2\int_0^{\pi/6} \rho^2\sin\phi\,d\phi\,d\rho\,d\theta$

33. $\displaystyle\int_0^{2\pi}\int_0^{\pi/2}\int_{\cos\phi}^2 \rho^2\sin\phi\,d\rho\,d\phi\,d\theta = \frac{31\pi}{6}$

35. $\displaystyle\int_0^{2\pi}\int_0^{\pi}\int_0^{1-\cos\phi} \rho^2\sin\phi\,d\rho\,d\phi\,d\theta = \frac{8\pi}{3}$

37. $\displaystyle\int_0^{2\pi}\int_{\pi/4}^{\pi/2}\int_0^{2\cos\phi} \rho^2\sin\phi\,d\rho\,d\phi\,d\theta = \frac{\pi}{3}$

39. (a) $\displaystyle 8\int_0^{\pi/2}\int_0^{\pi/2}\int_0^2 \rho^2\sin\phi\,d\rho\,d\phi\,d\theta$

(b) $\displaystyle 8\int_0^{\pi/2}\int_0^2\int_0^{\sqrt{4-r^2}} r\,dz\,dr\,d\theta$

(c) $\displaystyle 8\int_0^2\int_0^{\sqrt{4-x^2}}\int_0^{\sqrt{4-x^2-y^2}} dz\,dy\,dx$

41. (a) $\displaystyle\int_0^{2\pi}\int_0^{\pi/3}\int_{\sec\phi}^2 \rho^2\sin\phi\,d\rho\,d\phi\,d\theta$

(b) $\displaystyle\int_0^{2\pi}\int_0^{\sqrt{3}}\int_1^{\sqrt{4-r^2}} r\,dz\,dr\,d\theta$

(c) $\displaystyle 8\int_{-\sqrt{3}}^{\sqrt{3}}\int_{-\sqrt{3-x^2}}^{\sqrt{3-x^2}}\int_1^{\sqrt{4-x^2-y^2}} dz\,dy\,dx$

(d) $5\pi/3$

43. $8\pi/3$ **45.** 9/4

47. $\dfrac{3\pi - 4}{18}$ **49.** $\dfrac{2\pi a^3}{3}$

51. $5\pi/3$ **53.** $\pi/2$

55. $\dfrac{4(2\sqrt{2} - 1)\pi}{3}$ **57.** 16π

59. $5\pi/2$ **61.** $\dfrac{4\pi(8 - 3\sqrt{3})}{3}$

63. 2/3 **65.** 3/4

67. $\bar{x} = \bar{y} = 0, \bar{z} = 3/8$ **69.** $(\bar{x}, \bar{y}, \bar{z}) = (0, 0, 3/8)$

71. $\bar{x} = \bar{y} = 0, \bar{z} = 5/6$ **73.** $I_z = 30\pi, R_z = \sqrt{\frac{5}{2}}$

75. $I_x = \pi/4$ **77.** $\frac{a^4 h\pi}{10}$

79. (a) $(\bar{x}, \bar{y}, \bar{z}) = \left(0, 0, \frac{4}{5}\right), I_z = \frac{\pi}{12}, R_z = \sqrt{\frac{1}{3}}$

(b) $(\bar{x}, \bar{y}, \bar{z}) = \left(0, 0, \frac{5}{6}\right), I_z = \frac{\pi}{14}, R_z = \sqrt{\frac{5}{14}}$

83. $(\bar{x}, \bar{y}, \bar{z}) = \left(0, 0, \frac{2h^2 + 3h}{3h + 6}\right), I_z = \frac{\pi a^4(h^2 + 2h)}{4}, R_z = \frac{a}{\sqrt{2}}$

85. $\frac{3M}{\pi R^3}$

89. The surface's equation $r = f(z)$ tells us that the point $(r, \theta, z) = (f(z), \theta, z)$ will lie on the surface for all θ. In particular, $(f(z), \theta + \pi, z)$ lies on the surface whenever $(f(z), \theta, z)$ lies on the surface, so the surface is symmetric with respect to the z-axis.

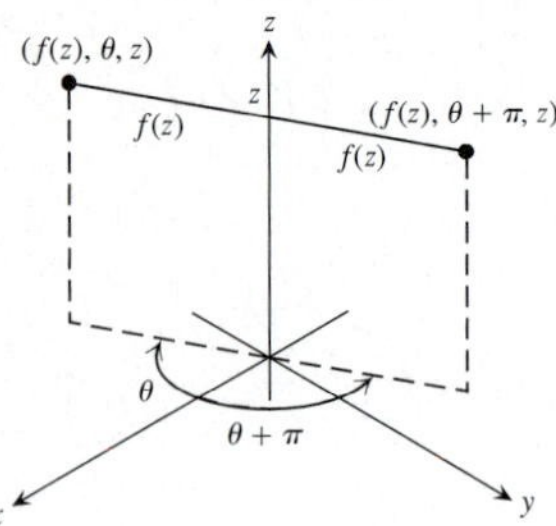

Section 12.7, pp. 1044–1046

1. (a) $x = \frac{u + v}{3}, y = \frac{v - 2u}{3}; \frac{1}{3}$

(b) Triangular region with boundaries $u = 0, v = 0$, and $u + v = 3$

3. (a) $x = \frac{1}{5}(2u - v), y = \frac{1}{10}(3v - u); \frac{1}{10}$

(b) Triangular region with boundaries $3v = u, v = 2u$, and $3u + v = 10$

7. 64/5

9. $\int_1^2 \int_1^3 (u + v)\frac{2u}{v}\, du\, dv = 8 + \frac{52}{3}\ln 2$

11. $\frac{\pi ab(a^2 + b^2)}{4}$ **13.** $\frac{1}{3}\left(1 + \frac{3}{e^2}\right) \approx 0.4687$

15. (a) $\begin{vmatrix} \cos v & -u\sin v \\ \sin v & u\cos v \end{vmatrix} = u\cos^2 v + u\sin^2 v = u$

(b) $\begin{vmatrix} \sin v & u\cos v \\ \cos v & -u\sin v \end{vmatrix} = -u\sin^2 v - u\cos^2 v = -u$

19. 12 **21.** $\frac{a^2b^2c^2}{6}$

Chapter 12 Practice Exercises, pp. 1047–1049

1. $9e - 9$ **3.** 9/2

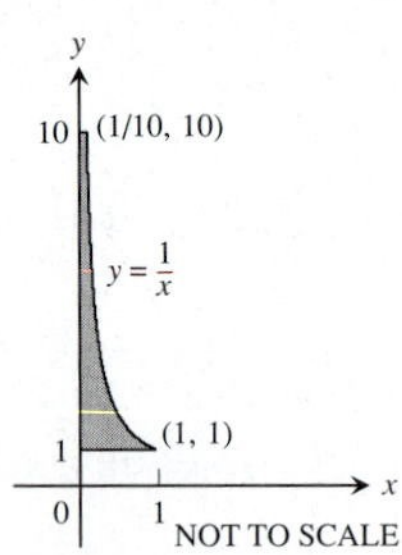

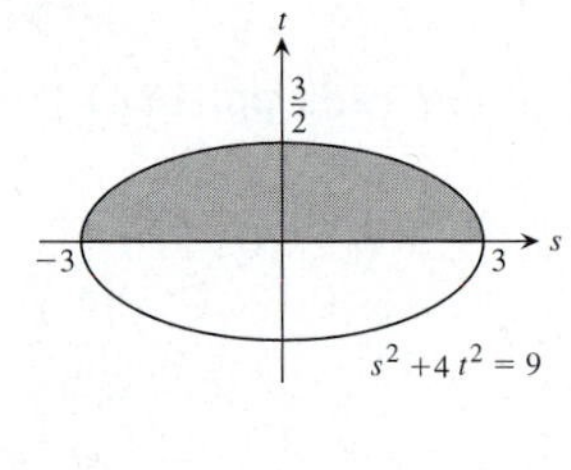

5. $\int_{-2}^{0} \int_{2x+4}^{4-x^2} dy\, dx = \frac{4}{3}$ **7.** $\int_{-3}^{3} \int_{0}^{(1/2)\sqrt{9-x^2}} y\, dy\, dx = \frac{9}{2}$

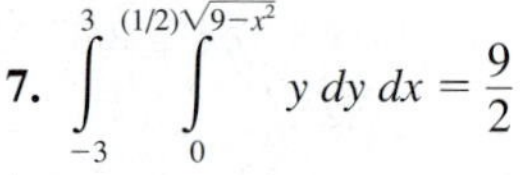

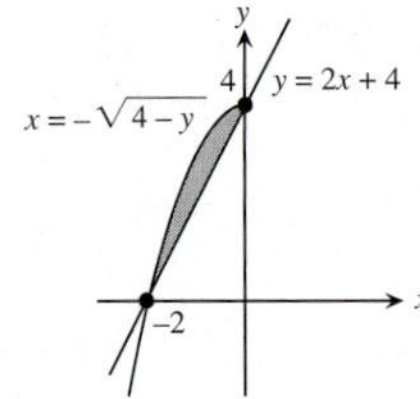

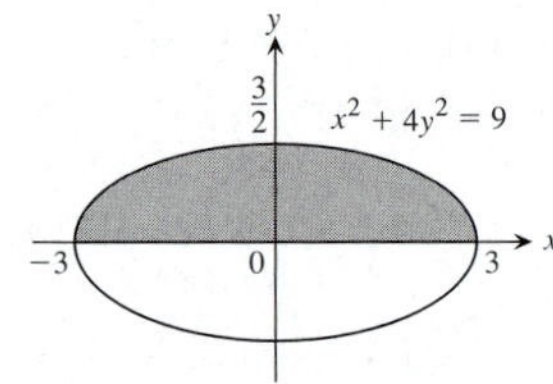

9. sin 4 **11.** $\frac{\ln 17}{4}$

13. 4/3 **15.** 4/3

17. 1/4 **19.** $\bar{x} = \bar{y} = \frac{1}{2 - \ln 4}$

21. $I_0 = 104$ **23.** $I_x = 2\delta, R_x = \sqrt{\frac{2}{3}}$

25. $M = 4, M_x = 0, M_y = 0$ **27.** π

29. $\bar{x} = \frac{3\sqrt{3}}{\pi}, \bar{y} = 0$

31. (a) $\bar{x} = \frac{15\pi + 32}{6\pi + 48}, \bar{y} = 0$

(b)

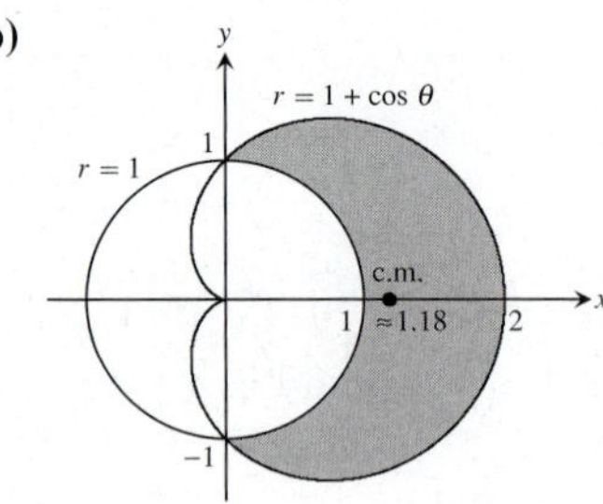

33. $\frac{\pi - 2}{4}$ **35.** 0

37. 8/35 **39.** $\pi/2$

41. $\frac{2(31 - 3^{5/2})}{3}$

43. (a) $\int_{-\sqrt{2}}^{\sqrt{2}} \int_{-\sqrt{2-y^2}}^{\sqrt{2-y^2}} \int_{\sqrt{x^2+y^2}}^{\sqrt{4-x^2-y^2}} 3\, dz\, dx\, dy$

(b) $\int_0^{2\pi} \int_0^{\pi/4} \int_0^{2} 3\rho^2 \sin \phi\, d\rho\, d\phi\, d\theta$

(c) $2\pi(8 - 4\sqrt{2})$

45. $\int_0^{2\pi} \int_0^{\pi/4} \int_0^{\sec \phi} \rho^2 \sin \phi\, d\rho\, d\phi\, d\theta = \frac{\pi}{3}$

47. $\int_0^{1} \int_{\sqrt{1-x^2}}^{\sqrt{3-x^2}} \int_1^{\sqrt{4-x^2-y^2}} z^2xy\, dz\, dy\, dx + \int_1^{\sqrt{3}} \int_0^{\sqrt{3-x^2}} \int_1^{\sqrt{4-x^2-y^2}} z^2xy\, dz\, dy\, dx$

49. (a) $\frac{8\pi(4\sqrt{2} - 5)}{3}$ **(b)** $\frac{8\pi(4\sqrt{2} - 5)}{3}$

51. $I_z = \frac{8\pi\delta(b^5 - a^5)}{15}$

Chapter 12 Additional Exercises, pp. 1049–1051

1. (a) $\int_{-3}^{2} \int_{x}^{6-x^2} x^2\, dy\, dx$

(b) $\int_{-3}^{2} \int_{x}^{6-x^2} \int_0^{x^2} dz\, dy\, dx$

(c) 125/4

3. 2π **5.** $3\pi/2$

7. (a) Hole radius = 1, sphere radius = 2

(b) $4\sqrt{3}\pi$

9. $\pi/4$

11. $\ln\left(\frac{b}{a}\right)$ **15.** $1/\sqrt[4]{3}$

17. Mass $= a^2 \cos^{-1}\left(\frac{b}{a}\right) - b\sqrt{a^2 - b^2}$, $I_0 =$

$\frac{a^4}{2} \cos^{-1}\left(\frac{b}{a}\right) - \frac{b^3}{2}\sqrt{a^2 - b^2} - \frac{b}{6}(a^2 - b^2)^{3/2}$

19. $\frac{1}{ab}(e^{a^2b^2} - 1)$

21. (b) 1 **(c)** 0

25. $h = \sqrt{20}$ in., $h = \sqrt{60}$ in. **27.** $2\pi\left[\frac{1}{3} - \left(\frac{1}{3}\right)\frac{\sqrt{2}}{2}\right]$

CHAPTER 13

Section 13.1, pp. 1058–1059

1. Graph (c) **3.** Graph (g)

5. Graph (d) **7.** Graph (f)

9. $\sqrt{2}$ **11.** $\frac{13}{2}$

13. $3\sqrt{14}$ **15.** $\frac{1}{6}(5\sqrt{5} + 9)$

17. $\sqrt{3}\ln\left(\frac{b}{a}\right)$ **19.** $\frac{10\sqrt{5} - 2}{3}$

21. 8 **23.** $2\sqrt{2} - 1$

25. (a) $4\sqrt{2} - 2$ **(b)** $\sqrt{2} + \ln(1 + \sqrt{2})$

27. $I_z = 2\pi\delta a^3$, $R_z = a$

29. (a) $I_z = 2\pi\sqrt{2}\delta$, $R_z = 1$ **(b)** $I_z = 4\pi\sqrt{2}\delta$, $R_z = 1$

31. $I_x = 2\pi - 2$, $R_x = 1$

Section 13.2, pp. 1068–1070

1. $\nabla f = -(x\mathbf{i} + y\mathbf{j} + z\mathbf{k})(x^2 + y^2 + z^2)^{-3/2}$

3. $\nabla g = -\left(\frac{2x}{x^2 + y^2}\right)\mathbf{i} - \left(\frac{2y}{x^2 + y^2}\right)\mathbf{j} + e^z\mathbf{k}$

5. $\mathbf{F} = -\dfrac{kx}{(x^2+y^2)^{3/2}}\mathbf{i} - \dfrac{ky}{(x^2+y^2)^{3/2}}\mathbf{j}$, any $k > 0$
7. (a) 9/2 (b) 13/3 (c) 9/2
9. (a) 1/3 (b) $-1/5$ (c) 0
11. (a) 2 (b) 3/2 (c) 1/2
13. 1/2
15. $-\pi$
17. 69/4
19. $-39/2$
21. 25/6
23. (a) $\text{Circ}_1 = 0$, $\text{circ}_2 = 2\pi$, $\text{flux}_1 = 2\pi$, $\text{flux}_2 = 0$
(b) $\text{Circ}_1 = 0$, $\text{circ}_2 = 8\pi$, $\text{flux}_1 = 8\pi$, $\text{flux}_2 = 0$
25. Circ = 0, flux = $a^2\pi$
27. Circ = $a^2\pi$, flux = 0
29. (a) $-\dfrac{\pi}{2}$ (b) 0 (c) 1
31.

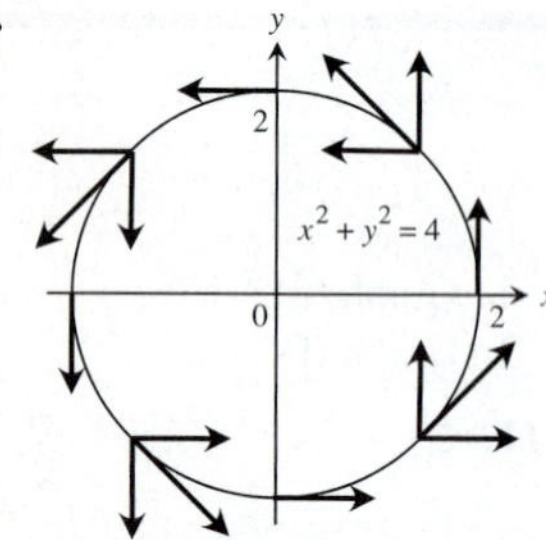

33. (a) $\mathbf{G} = -y\mathbf{i} + x\mathbf{j}$ (b) $\mathbf{G} = \sqrt{x^2+y^2}\,\mathbf{F}$
35. $\mathbf{F} = -\dfrac{x\mathbf{i} + y\mathbf{j}}{\sqrt{x^2+y^2}}$
37. 48
39. π
41. 0
43. $\dfrac{1}{2}$

Section 13.3, pp. 1078–1079

1. Conservative
3. Not conservative
5. Not conversative
7. $f(x, y, z) = x^2 + \dfrac{3y^2}{2} + 2z^2 + C$
9. $f(x, y, z) = xe^{y+2z} + C$
11. $f(x, y, z) = x \ln x - x + \tan(x + y) + \dfrac{1}{2}\ln(y^2 + z^2) + C$
13. 49
15. -16
17. 1
19. 9 ln 2
21. 0
23. -3
27. $\mathbf{F} = \nabla\left(\dfrac{x^2-1}{y}\right)$
29. (a) 1 (b) 1 (c) 1
31. (a) 2 (b) 2
33. (a) $c = b = 2a$
(b) $c = b = 2$
35. It does not matter what path you use. The work will be the same on any path because the field is conservative.
37. The force **F** is conservative because all partial derivatives of M, N, and P are zero. $f(x, y, z) = ax + by + cz + C$; $A = (xa, ya, za)$ and $B = (xb, yb, zb)$. Therefore, $BF \cdot d\vec{r} = f(B) - f(A) = a(xb - xa) + b(yb - ya) + c(zb - za) = \mathbf{F}\cdot \vec{AB}$.

Section 13.4, pp. 1090–1092

1. Flux = 0, circ = $2\pi a^2$
3. Flux = $-\pi a^2$, circ = 0
5. Flux = 2, circ = 0
7. Flux = -9, circ = 9
9. Flux = 1/2, circ = 1/2
11. Flux = 1/5, circ = $-1/12$
13. 0
15. 2/33
17. 0
19. $-\dfrac{16\pi}{3}$
21. πa^2
23. $\dfrac{3}{8}\pi$
25. (a) 0
(b) $(h - k)$(area of the region)
35. (a) 0

Section 13.5, pp. 1101–1103

1. $\dfrac{13}{3}\pi$
3. 4
5. $6\sqrt{6} - 2\sqrt{2}$
7. $\pi\sqrt{c^2+1}$
9. $\dfrac{\pi}{6}(17\sqrt{17} - 5\sqrt{5})$
11. $3 + 2\ln 2$
13. $9a^3$
15. $\dfrac{abc}{4}(ab + ac + bc)$
17. 2
19. 18
21. $\dfrac{\pi a^3}{6}$
23. $\dfrac{\pi a^2}{4}$
25. $\dfrac{\pi a^3}{2}$
27. -32
29. -4
31. $3a^4$
33. $\left(\dfrac{a}{2}, \dfrac{a}{2}, \dfrac{a}{2}\right)$
35. $(\bar{x}, \bar{y}, \bar{z}) = \left(0, 0, \dfrac{14}{9}\right)$, $I_z = \dfrac{15\pi\sqrt{2}}{2}\delta$, $R_z = \dfrac{\sqrt{10}}{2}$
37. (a) $\dfrac{8\pi}{3}a^4\delta$ (b) $\dfrac{20\pi}{3}a^4\delta$
39. $\dfrac{\pi}{6}(13\sqrt{13} - 1)$
41. $5\pi\sqrt{2}$
43. $\dfrac{2}{3}(5\sqrt{5} - 1)$

Section 13.6, pp. 1111–1113

1. $\mathbf{r}(r, \theta) = (r\cos\theta)\mathbf{i} + (r\sin\theta)\mathbf{j} + r^2\mathbf{k}$, $0 \le r \le 2$, $0 \le \theta \le 2\pi$
3. $\mathbf{r}(r, \theta) = (r\cos\theta)\mathbf{i} + (r\sin\theta)\mathbf{j} + (r/2)\mathbf{k}$, $0 \le r \le 6$, $0 \le \theta \le \pi/2$
5. $\mathbf{r}(r, \theta) = (r\cos\theta)\mathbf{i} + (r\sin\theta)\mathbf{j} + \sqrt{9 - r^2}\,\mathbf{k}$,
$0 \le r \le \dfrac{3\sqrt{2}}{2}$, $0 \le \theta \le 2\pi$;
also $\mathbf{r}(\phi, \theta) = (3\sin\phi\cos\theta)\mathbf{i} + (3\sin\phi\sin\theta)\mathbf{j} + (3\cos\phi)\mathbf{k}$,
$0 \le \phi \le \dfrac{\pi}{4}$, $0 \le \theta \le 2\pi$

7. $\mathbf{r}(\phi, \theta) = (\sqrt{3}\sin\phi\cos\theta)\mathbf{i} + (\sqrt{3}\sin\phi\sin\theta)\mathbf{j} + (\sqrt{3}\cos\phi)\mathbf{k}$, $\frac{\pi}{3} \le \phi \le \frac{2\pi}{3}$, $0 \le \theta \le 2\pi$

9. $\mathbf{r}(x, y) = x\mathbf{i} + y\mathbf{j} + (4 - y^2)\mathbf{k}$, $0 \le x \le 2$, $-2 \le y \le 2$

11. $\mathbf{r}(u, v) = u\mathbf{i} + (3\cos v)\mathbf{j} + (3\sin v)\mathbf{k}$, $0 \le u \le 3$, $0 \le v \le 2\pi$

13. **(a)** $\mathbf{r}(r, \theta) = (r\cos\theta)\mathbf{i} + (r\sin\theta)\mathbf{j} + (1 - r\cos\theta - r\sin\theta)\mathbf{k}$, $0 \le r \le 3$, $0 \le \theta \le 2\pi$
(b) $\mathbf{r}(u, v) = (1 - u\cos v - u\sin v)\mathbf{i} + (u\cos v)\mathbf{j} + (u\sin v)\mathbf{k}$, $0 \le u \le 3$, $0 \le v \le 2\pi$

15. $\mathbf{r}(u, v) = (4\cos^2 v)\mathbf{i} + u\mathbf{j} + (4\cos v\sin v)\mathbf{k}$, $0 \le u \le 3$, $-\pi/2 \le v \le \pi/2$; another way: $\mathbf{r}(u, v) = (2 + 2\cos v)\mathbf{i} + u\mathbf{j} + (2\sin v)\mathbf{k}$, $0 \le u \le 3$, $0 \le v \le 2\pi$

17. $\int_0^{2\pi}\int_0^1 \frac{\sqrt{5}}{2} r\,dr\,d\theta = \frac{\pi\sqrt{5}}{2}$

19. $\int_0^{2\pi}\int_1^3 r\sqrt{5}\,dr\,d\theta = 8\pi\sqrt{5}$

21. $\int_0^{2\pi}\int_1^4 1\,du\,dv = 6\pi$

23. $\int_0^{2\pi}\int_0^1 u\sqrt{4u^2+1}\,du\,dv = \frac{(5\sqrt{5}-1)}{6}\pi$

25. $\int_0^{2\pi}\int_{\pi/4}^{\pi} 2\sin\phi\,d\phi\,d\theta = (4 + 2\sqrt{2})\pi$

27. $\iint_S x\,d\sigma = \int_0^3\int_0^2 u\sqrt{4u^2+1}\,du\,dv = \frac{17\sqrt{17}-1}{4}$

29. $\iint_S x^2\,d\sigma = \int_0^{2\pi}\int_0^{\pi} \sin^3\phi\cos^2\theta\,d\phi\,d\theta = \frac{4\pi}{3}$

31. $\iint_S z\,d\sigma = \int_0^1\int_0^1 (4 - u - v)\sqrt{3}\,dv\,du = 3\sqrt{3}$ (for $x = u$, $y = v$)

33. $\iint_S x^2\sqrt{5-4z}\,d\sigma = \int_0^1\int_0^{2\pi} u^2\cos^2 v\cdot\sqrt{4u^2+1}\cdot u\sqrt{4u^2+1}\,dv\,du = \int_0^1\int_0^{2\pi} u^3(4u^2+1)\cos^2 v\,dv\,du = \frac{11\pi}{12}$

35. -32

37. $\frac{\pi a^3}{6}$

39. $\frac{13a^4}{6}$

41. $\frac{2\pi}{3}$

43. $-\frac{73}{6}\pi$

45. $(\bar{x}, \bar{y}, \bar{z}) = (0, 0, 14/9)$, $I_z = \frac{(15\sqrt{2})\pi\delta}{2}$, $R_z = \sqrt{5/2}$

47.

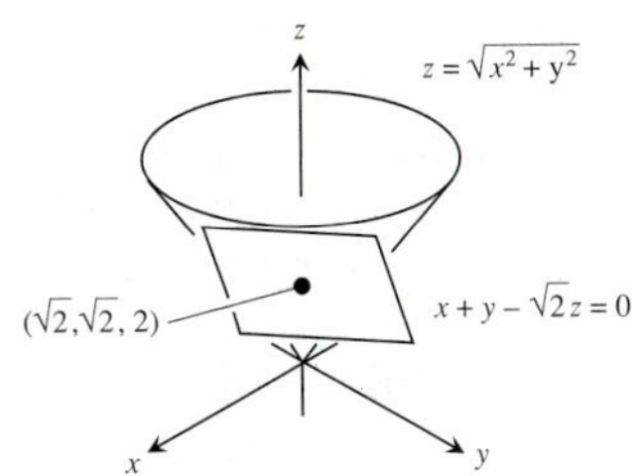

49.

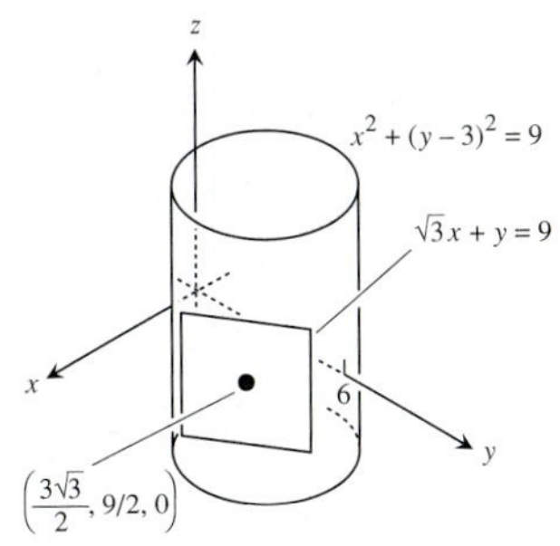

53. **(b)** $A = \int_0^{2\pi}\int_0^{\pi} (a^2b^2\sin^2\phi\cos^2\phi + b^2c^2\cos^4\phi\cos^2\theta + a^2c^2\cos^4\phi\sin^2\theta)^{1/2}\,d\phi\,d\theta$

Section 13.7, pp. 1122–1124

1. 4π
3. $-5/6$
5. 0
7. -6π
9. $2\pi a^2$
13. 12π
15. $-\pi/4$
17. -15π
25. $16I_y + 16I_x$

Section 13.8, pp. 1133–1135

1. 0
3. 0
5. -16
7. -8π
9. 3π
11. $-40/3$
13. 12π
15. $12\pi(4\sqrt{2} - 1)$
21. The integral's value never exceeds the surface area of S.

Chapter 13 Practice Exercises, pp. 1136–1139

1. Path 1: $2\sqrt{3}$; path 2: $1 + 3\sqrt{2}$
3. $4a^2$
5. 0
7. $8\pi\sin(1)$
9. 0
11. $\pi\sqrt{3}$
13. $2\pi\left(1 - \frac{1}{\sqrt{2}}\right)$
15. $\frac{abc}{2}\sqrt{\frac{1}{a^2} + \frac{1}{b^2} + \frac{1}{c^2}}$
17. 50
19. $\mathbf{r}(\phi, \theta) = (6\sin\phi\cos\theta)\mathbf{i} + (6\sin\phi\sin\theta)\mathbf{j} + (6\cos\phi)\mathbf{k}$, $\frac{\pi}{6} \le \phi \le \frac{2\pi}{3}$, $0 \le \theta \le 2\pi$

21. $\mathbf{r}(r, \theta) = (r\cos\theta)\mathbf{i} + (r\sin\theta)\mathbf{j} + (1 + r)\mathbf{k}$, $0 \le r \le 2,\ 0 \le \theta \le 2\pi$

23. $\mathbf{r}(u, v) = (u\cos v)\mathbf{i} + 2u^2\mathbf{j} + (u\sin v)\mathbf{k},\ 0 \le u \le 1,\ 0 \le v \le \pi$

25. $\sqrt{6}$

27. $\pi[\sqrt{2} + \ln(1 + \sqrt{2})]$

29. Conservative

31. Not conservative

33. $f(x, y, z) = y^2 + yz + 2x + z$

35. Path 1: 2; path 2: 8/3

37. **(a)** $1 - e^{-2\pi}$ **(b)** $1 - e^{-2\pi}$

39. 0

41. **(a)** $4\sqrt{2} - 2$ **(b)** $\sqrt{2} + \ln(1 + \sqrt{2})$

43. $(\bar{x}, \bar{y}, \bar{z}) = \left(1, \frac{16}{15}, \frac{2}{3}\right)$; $I_x = \frac{232}{45}$, $I_y = \frac{64}{15}$, $I_z = \frac{56}{9}$; $R_x = \frac{2\sqrt{29}}{3\sqrt{5}}$, $R_y = \frac{4\sqrt{2}}{\sqrt{15}}$, $R_z = \frac{2\sqrt{7}}{3}$

45. $\bar{z} = \frac{3}{2}$, $I_z = \frac{7\sqrt{3}}{3}$, $R_z = \sqrt{\frac{7}{3}}$

47. $(\bar{x}, \bar{y}, \bar{z}) = (0, 0, 49/12)$, $I_z = 640\pi$, $R_z = 2\sqrt{2}$

49. Flux: 3/2; circ: $-1/2$

53. 3

55. $\frac{2\pi}{3}(7 - 8\sqrt{2})$

57. 0

59. π

Chapter 13 Additional Exercises, pp. 1139–1141

1. 6π

3. 2/3

5. **(a)** $\mathbf{F}(x, y, z) = z\mathbf{i} + x\mathbf{j} + y\mathbf{k}$
(b) $\mathbf{F}(x, y, z) = z\mathbf{i} + y\mathbf{k}$
(c) $\mathbf{F}(x, y, z) = z\mathbf{i}$

7. $a = 2$, $b = 1$; the minimum flux is -4.

9. **(b)** $\frac{16}{3}g$
(c) Work $= \left(\int_C gxy\, ds\right)\bar{y} = g\int_C xy^2\, ds$

11. **(c)** $\frac{4}{3}\pi w$

15. False if $\mathbf{F} = y\mathbf{i} + x\mathbf{j}$

APPENDICES

Appendix A.4

1. **(a)** (14, 8) **(b)** $(-1, 8)$ **(c)** $(0, -5)$

3. **(a)** By reflecting z across the real axis
(b) By reflecting z across the imaginary axis
(c) By reflecting z in the real axis and then multiplying the length of the vector by $1/|z|^2$

5. **(a)** Points on the circle $x^2 + y^2 = 4$
(b) points inside the circle $x^2 + y^2 = 4$
(c) points outside the circle $x^2 + y^2 = 4$

7. Points on a circle of radius 1, center $(-1, 0)$

9. Points on the line $y = -x$

11. $4e^{2\pi i/3}$

13. $1e^{2\pi i/3}$

21. $\cos^4\theta - 6\cos^2\theta\sin^2\theta + \sin^4\theta$

23. $1, -\frac{1}{2} \pm \frac{\sqrt{3}}{2}i$

25. $2i, -\sqrt{3} - i, \sqrt{3} - i$

27. $\frac{\sqrt{6}}{2} \pm \frac{\sqrt{2}}{2}i, -\frac{\sqrt{6}}{2} \pm \frac{\sqrt{2}}{2}i$

29. $1 \pm \sqrt{3}i, -1, \pm\sqrt{3}i$

Appendix A.10

1. -5

3. 1

5. -7

7. 38

9. $x = -4$, $y = 1$

11. $x = 3$, $y = 2$

13. $x = 3$, $y = -2$, $z = 2$

15. $x = 2$, $y = 0$, $z = -1$

17. **(a)** $h = 6$, $k = 4$ **(b)** $h = 6$, $k \ne 4$

Appendix A.13.1

1. $y^2 = 8x$, $F(2, 0)$, directrix: $x = -2$

3. $x^2 = -6y$, $F(0, -3/2)$, directrix: $y = 3/2$

5. $\frac{x^2}{4} - \frac{y^2}{9} = 1$, $F(\pm\sqrt{13}, 0)$, $V(\pm 2, 0)$, asymptotes: $y = \pm\frac{3}{2}x$

7. $\frac{x^2}{2} + y^2 = 1$, $F(\pm 1, 0)$, $V(\pm\sqrt{2}, 0)$

9.

11.

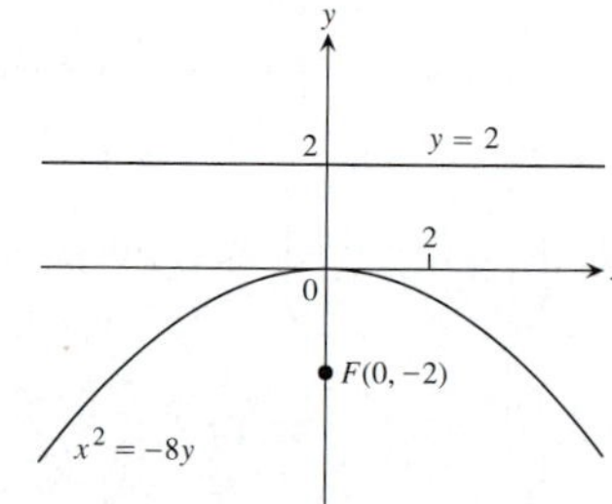

13.

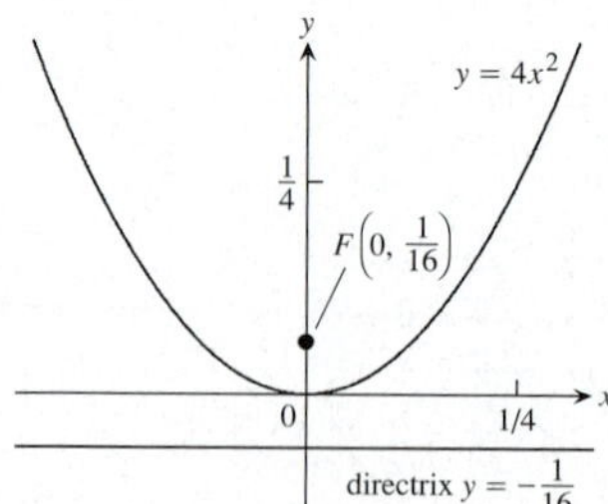

15.

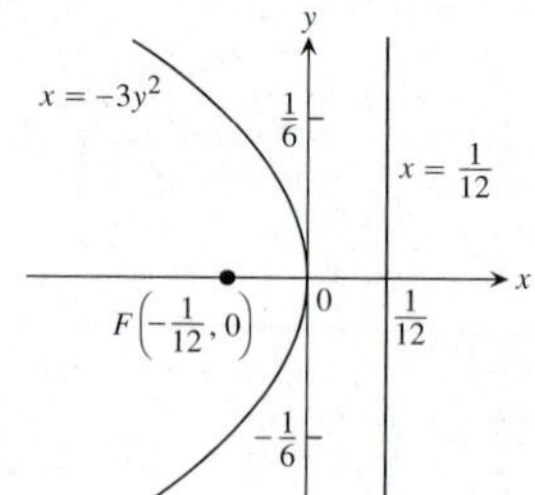

17.

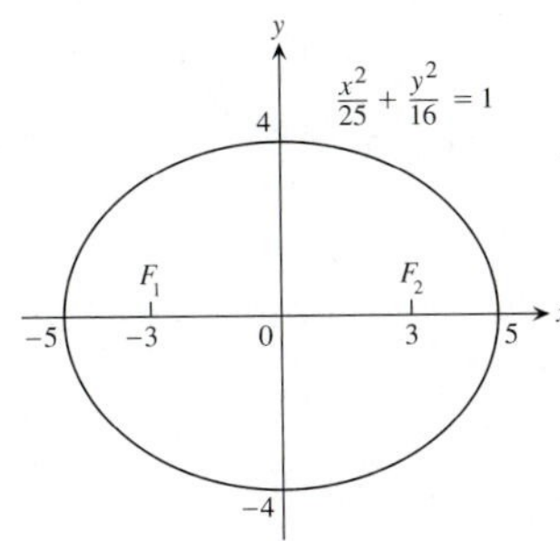

19.

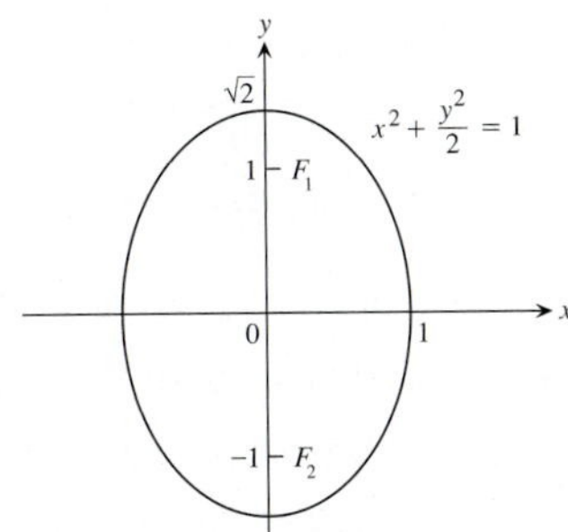

21.

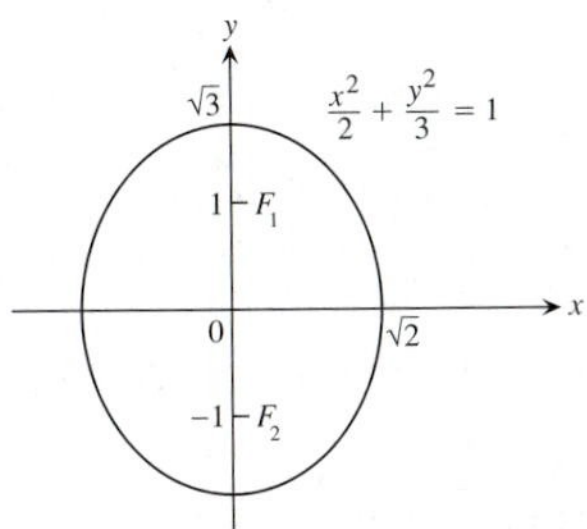

23.

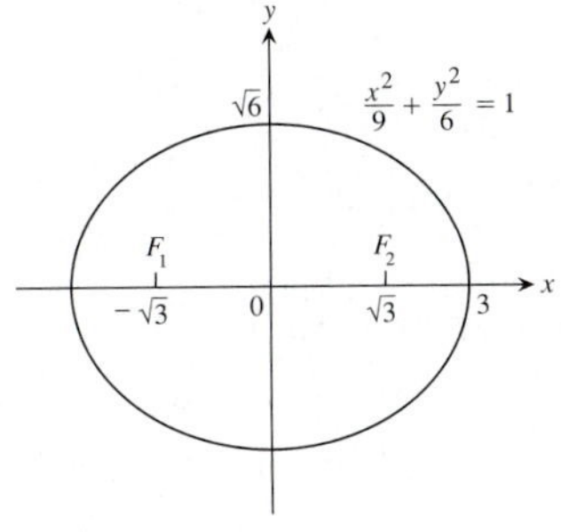

25. $\frac{x^2}{4} + \frac{y^2}{2} = 1$

27. Asymptotes: $y = \pm x$

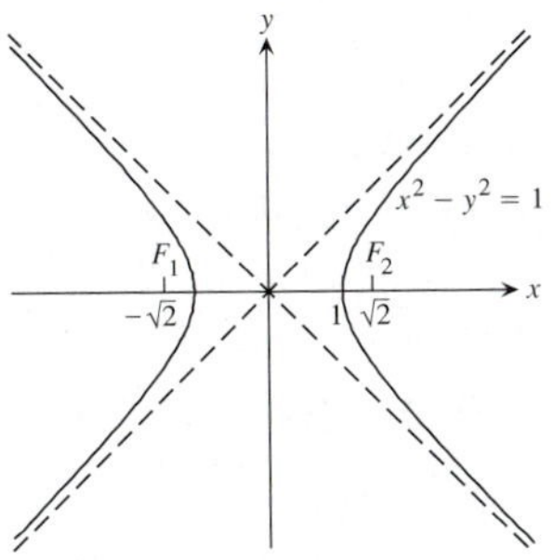

29. Asymptotes: $y = \pm x$

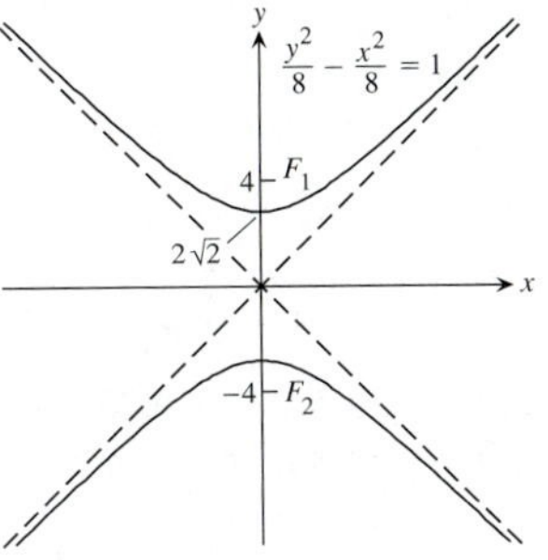

31. Asymptotes: $y = \pm 2x$

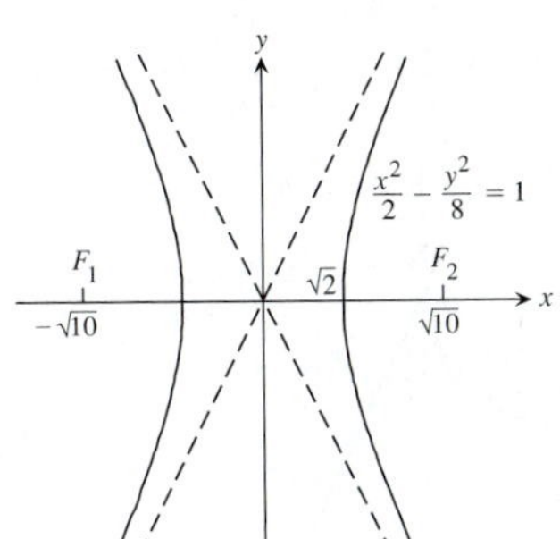

33. Asymptotes: $y = \pm\frac{x}{2}$

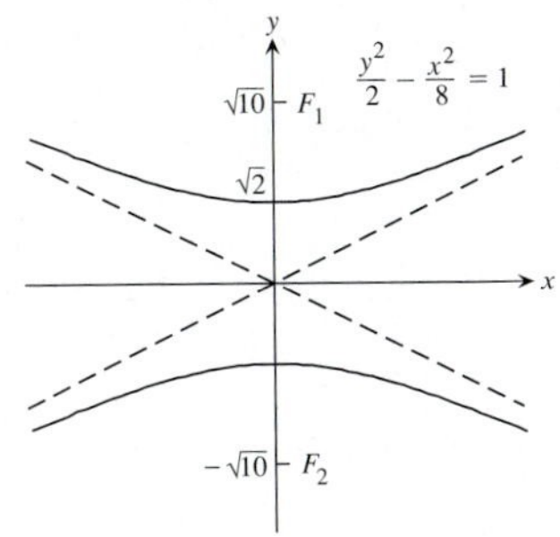

35. $y^2 - x^2 = 1$

37. 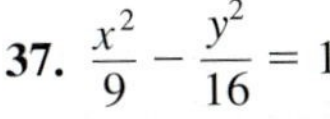 $\frac{x^2}{9} - \frac{y^2}{16} = 1$

39. **a)** Vertex: $(1, -2)$; focus: $(3, -2)$; directrix: $x = -1$

b)

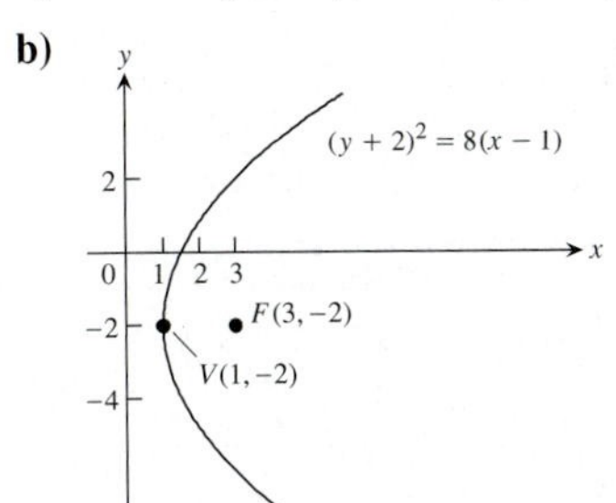

41. **a)** Foci: $(4 \pm \sqrt{7}, 3)$; vertices: $(8, 3)$ and $(0, 3)$; center: $(4, 3)$

b)

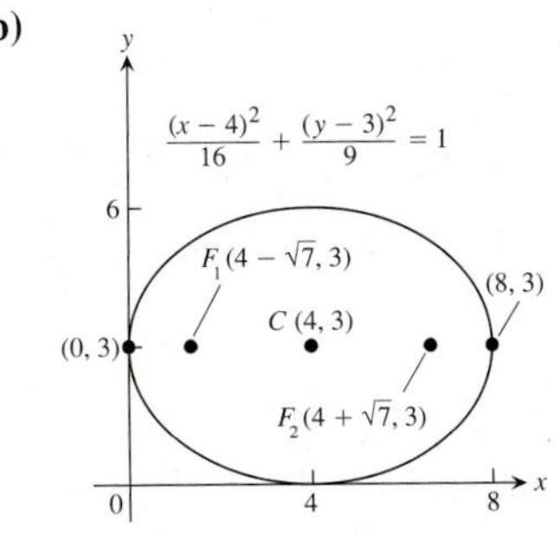

43. **a)** Center: $(2, 0)$; foci: $(7, 0)$ and $(-3, 0)$; vertices: $(6, 0)$ and $(-2, 0)$; asymptotes: $y = \pm\frac{3}{4}(x - 2)$

b)

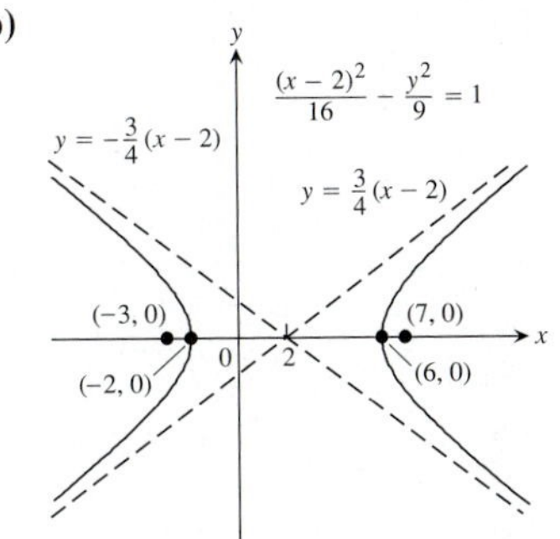

45. $(y + 3)^2 = 4(x + 2)$, $V(-2, -3)$, $F(-1, -3)$, directrix: $x = -3$

47. $(x - 1)^2 = 8(y + 7)$, $V(1, -7)$, $F(1, -5)$, directrix: $y = -9$

49. $\frac{(x+2)^2}{6} + \frac{(y+1)^2}{9} = 1$, $F(-2, \pm\sqrt{3} - 1)$, $V(-2, \pm 3 - 1)$, $C(-2, -1)$

51. $\frac{(x-2)^2}{3} + \frac{(y-3)^2}{2} = 1$, $F(3, 3)$ and $F(1, 3)$, $V(\pm\sqrt{3} + 2, 3)$, $C(2, 3)$

53. $\frac{(x-2)^2}{4} - \frac{(y-2)^2}{5} = 1$, $C(2, 2)$, $F(5, 2)$ and $F(-1, 2)$, $V(4, 2)$ and $V(0, 2)$; asymptotes: $(y - 2) = \pm\frac{\sqrt{5}}{2}(x - 2)$

55. $(y+1)^2 - (x+1)^2 = 1, C(-1,-1), F(-1, \sqrt{2}-1)$ and $F(-1, -\sqrt{2}-1), V(-1, 0)$ and $V(-1,-2)$; asymptotes: $(y+1) = \pm(x+1)$

57. $C(-2, 0), a = 4$ **59.** $V(-1, 1), F(-1, 0)$

61. Ellipse: $\dfrac{(x+2)^2}{5} + y^2 = 1$, $C(-2, 0)$, $F(0, 0)$ and $F(-4, 0)$, $V(\sqrt{5}-2, 0)$ and $V(-\sqrt{5}-2, 0)$

63. Ellipse: $\dfrac{(x-1)^2}{2} + (y-1)^2 = 1$, $C(1, 1)$, $F(2, 1)$ and $F(0, 1)$, $V(\sqrt{2}+1, 1)$ and $V(-\sqrt{2}+1, 1)$

65. Hyperbola: $(x-1)^2 - (y-2)^2 = 1$, $C(1, 2)$, $F(1+\sqrt{2}, 2)$ and $F(1-\sqrt{2}, 2)$, $V(2, 2)$ and $V(0, 2)$; asymptotes: $(y-2) = \pm(x-1)$

67. Hyperbola: $\dfrac{(y-3)^2}{6} - \dfrac{x^2}{3} = 1$, $C(0, 3)$, $F(0, 6)$ and $F(0, 0)$, $V(0, \sqrt{6}+3)$ and $V(0, -\sqrt{6}+3)$; asymptotes: $y = \sqrt{2}x + 3$ or $y = -\sqrt{2}x + 3$

69.

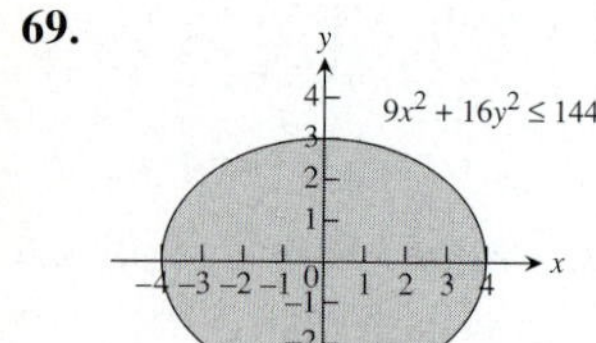

71.

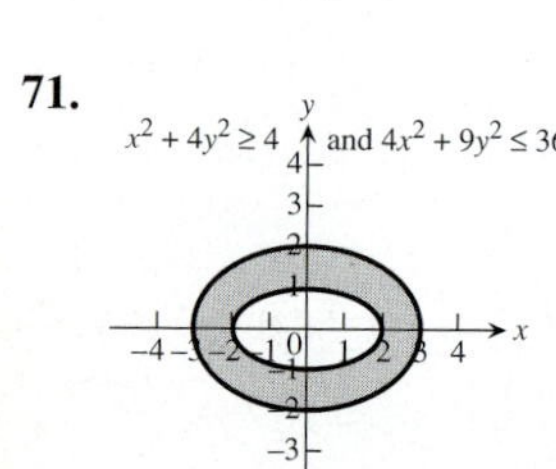

73.

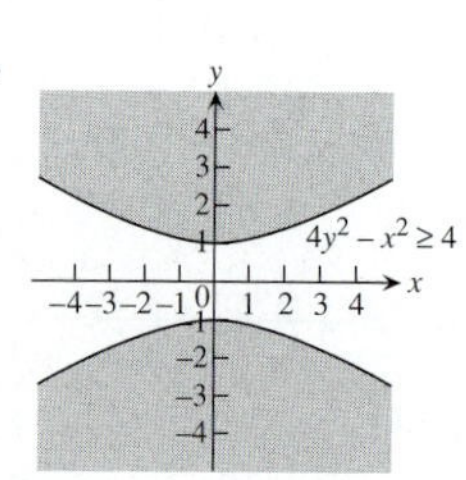

77. $3x^2 + 3y^2 - 7x - 7y + 4 = 0$

79. $(x+2)^2 + (y-1)^2 = 13$. The point is inside the circle.

81. **b)** $1:1$ **83.** Length $= 2\sqrt{2}$, width $= \sqrt{2}$, area $= 4$

85. 24π **87.** $(0, 16/(3\pi))$

Appendix A.13.2

1. $e = 3/5, F(\pm 3, 0), x = \pm 25/3$

3. $e = 1/\sqrt{2}, F(0, \pm 1), y = \pm 2$

5. $e = 1/\sqrt{3}, F(0, \pm 1), y = \pm 3$

7. $e = \sqrt{3}/3, F(\pm\sqrt{3}, 0), x = \pm 3\sqrt{3}$

9. $\dfrac{x^2}{27} + \dfrac{y^2}{36} = 1$ **11.** $\dfrac{x^2}{4851} + \dfrac{y^2}{4900} = 1$

13. $e = \dfrac{\sqrt{5}}{3}, \quad \dfrac{x^2}{9} + \dfrac{y^2}{4} = 1$ **15.** $e = 1/2, \quad \dfrac{x^2}{64} + \dfrac{y^2}{48} = 1$

19. $\dfrac{(x-1)^2}{4} + \dfrac{(y-4)^2}{9} = 1, F(1, 4 \pm \sqrt{5}), e = \sqrt{5}/3$, $y = 4 \pm (9\sqrt{5}/5)$

21. $a = 0, b = -4, c = 0, e = \sqrt{3}/2$

23. $e = \sqrt{2}, F(\pm\sqrt{2}, 0), x = \pm 1/\sqrt{2}$

25. $e = \sqrt{2}, F(0, \pm 4), y = \pm 2$

27. $e = \sqrt{5}, F(\pm\sqrt{10}, 0), x = \pm 2/\sqrt{10}$

29. $e = \sqrt{5}, F(0, \pm\sqrt{10}), y = \pm 2/\sqrt{10}$ **31.** $y^2 - \dfrac{x^2}{8} = 1$

33. $x^2 - \dfrac{y^2}{8} = 1$ **35.** $e = \sqrt{2}, \quad \dfrac{x^2}{8} - \dfrac{y^2}{8} = 1$

37. $e = 2, \quad x^2 - \dfrac{y^2}{3} = 1$ **39.** $\dfrac{(y-6)^2}{36} - \dfrac{(x-1)^2}{45} = 1$

Appendix A.13.3

1. Hyperbola **3.** Ellipse **5.** Parabola **7.** Parabola

9. Hyperbola **11.** Hyperbola **13.** Ellipse **15.** Ellipse

17. $x'^2 - y'^2 = 4$, hyperbola **19.** $4x'^2 + 16y' = 0$, parabola

21. $y'^2 = 1$, parallel lines **23.** $2\sqrt{2}x'^2 + 8\sqrt{2}y' = 0$, parabola

25. $4x'^2 + 2y'^2 = 19$, ellipse

27. $\sin\alpha = 1/\sqrt{5}$, $\cos\alpha = 2/\sqrt{5}$; or $\sin\alpha = -2/\sqrt{5}$, $\cos\alpha = 1/\sqrt{5}$

29. $A' = 0.88, B' = 0.00, C' = 3.10, D' = 0.74, E' = -1.20, F' = -3$, $0.88x'^2 + 3.10y'^2 + 0.74x' - 1.20y' - 3 = 0$, ellipse

31. $A' = 0.00, B' = 0.00, C' = 5.00, D' = 0, E' = 0, F' = -5$, $5.00y'^2 - 5 = 0$ or $y' = \pm 1.00$, parallel lines

33. $A' = 5.05, B' = 0.00, C' = -0.05, D' = -5.07, E' = -6.18$, $F' = -1$, $5.05x'^2 - 0.05y'^2 - 5.07x' - 6.18y' - 1 = 0$, hyperbola

35. **a)** $\dfrac{x'^2}{b^2} + \dfrac{y'^2}{a^2} = 1$ **b)** $\dfrac{y'^2}{a^2} - \dfrac{x'^2}{b^2} = 1$ **c)** $x'^2 + y'^2 = a^2$

d) $y' = -\dfrac{1}{m}x'$ **e)** $y' = -\dfrac{1}{m}x' + \dfrac{b}{m}$

37. **a)** $x'^2 - y'^2 = 2$ **b)** $x'^2 - y'^2 = 2a$ **43.** **a)** Parabola

45. **a)** Hyperbola **b)**

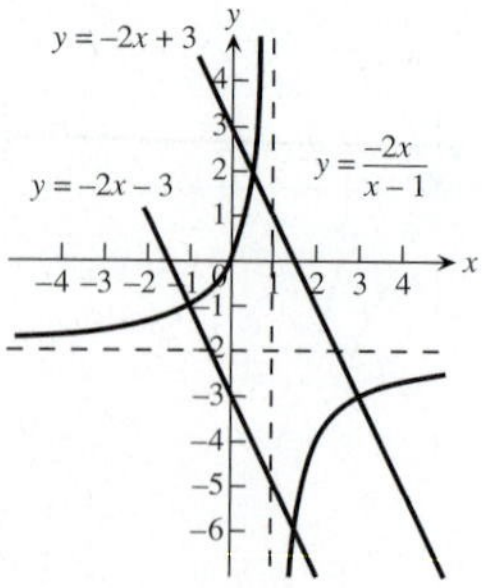

c) $y = -2x - 3, y = -2x + 3$

Index

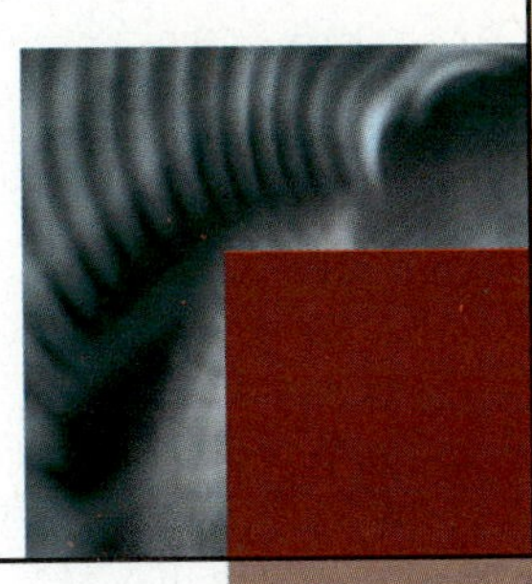

A Brief Table of Integrals

1. $\int u\,dv = uv - \int v\,du$

2. $\int a^u\,du = \frac{a^u}{\ln a} + C, \quad a \neq 1, \quad a > 0$

3. $\int \cos u\,du = \sin u + C$

4. $\int \sin u\,du = -\cos u + C$

5. $\int (ax+b)^n\,dx = \frac{(ax+b)^{n+1}}{a(n+1)} + C, \quad n \neq -1$

6. $\int (ax+b)^{-1}\,dx = \frac{1}{a}\ln|ax+b| + C$

7. $\int x(ax+b)^n\,dx = \frac{(ax+b)^{n+1}}{a^2}\left[\frac{ax+b}{n+2} - \frac{b}{n+1}\right] + C, \quad n \neq -1, -2$

8. $\int x(ax+b)^{-1}\,dx = \frac{x}{a} - \frac{b}{a^2}\ln|ax+b| + C$

9. $\int x(ax+b)^{-2}\,dx = \frac{1}{a^2}\left[\ln|ax+b| + \frac{b}{ax+b}\right] + C$

10. $\int \frac{dx}{x(ax+b)} = \frac{1}{b}\ln\left|\frac{x}{ax+b}\right| + C$

11. $\int (\sqrt{ax+b})^n\,dx = \frac{2}{a}\frac{(\sqrt{ax+b})^{n+2}}{n+2} + C, \quad n \neq -2$

12. $\int \frac{\sqrt{ax+b}}{x}\,dx = 2\sqrt{ax+b} + b\int \frac{dx}{x\sqrt{ax+b}}$

13. (a) $\int \frac{dx}{x\sqrt{ax-b}} = \frac{2}{\sqrt{b}}\tan^{-1}\sqrt{\frac{ax-b}{b}} + C$

(b) $\int \frac{dx}{x\sqrt{ax+b}} = \frac{1}{\sqrt{b}}\ln\left|\frac{\sqrt{ax+b}-\sqrt{b}}{\sqrt{ax+b}+\sqrt{b}}\right| + C$

14. $\int \frac{\sqrt{ax+b}}{x^2}\,dx = -\frac{\sqrt{ax+b}}{x} + \frac{a}{2}\int \frac{dx}{x\sqrt{ax+b}} + C$

15. $\int \frac{dx}{x^2\sqrt{ax+b}} = -\frac{\sqrt{ax+b}}{bx} - \frac{a}{2b}\int \frac{dx}{x\sqrt{ax+b}} + C$

16. $\int \frac{dx}{a^2+x^2} = \frac{1}{a}\tan^{-1}\frac{x}{a} + C$

17. $\int \frac{dx}{(a^2+x^2)^2} = \frac{x}{2a^2(a^2+x^2)} + \frac{1}{2a^3}\tan^{-1}\frac{x}{a} + C$

18. $\int \frac{dx}{a^2-x^2} = \frac{1}{2a}\ln\left|\frac{x+a}{x-a}\right| + C$

19. $\int \frac{dx}{(a^2-x^2)^2} = \frac{x}{2a^2(a^2-x^2)} + \frac{1}{4a^3}\ln\left|\frac{x+a}{x-a}\right| + C$

20. $\int \frac{dx}{\sqrt{a^2+x^2}} = \sinh^{-1}\frac{x}{a} + C = \ln(x+\sqrt{a^2+x^2}) + C$

21. $\int \sqrt{a^2+x^2}\,dx = \frac{x}{2}\sqrt{a^2+x^2} + \frac{a^2}{2}\ln(x+\sqrt{a^2+x^2}) + C$

22. $\int x^2\sqrt{a^2+x^2}\,dx = \frac{x}{8}(a^2+2x^2)\sqrt{a^2+x^2} - \frac{a^4}{8}\ln(x+\sqrt{a^2+x^2}) + C$

23. $\int \frac{\sqrt{a^2+x^2}}{x}\,dx = \sqrt{a^2+x^2} - a\ln\left|\frac{a+\sqrt{a^2+x^2}}{x}\right| + C$

24. $\int \frac{\sqrt{a^2+x^2}}{x^2}\,dx = \ln(x+\sqrt{a^2+x^2}) - \frac{\sqrt{a^2+x^2}}{x} + C$

25. $\int \frac{x^2}{\sqrt{a^2+x^2}}\,dx = -\frac{a^2}{2}\ln(x+\sqrt{a^2+x^2}) + \frac{x\sqrt{a^2+x^2}}{2} + C$

26. $\int \frac{dx}{x\sqrt{a^2+x^2}} = -\frac{1}{a}\ln\left|\frac{a+\sqrt{a^2+x^2}}{x}\right| + C$

27. $\int \frac{dx}{x^2\sqrt{a^2+x^2}} = -\frac{\sqrt{a^2+x^2}}{a^2x} + C$

28. $\int \frac{dx}{\sqrt{a^2 - x^2}} = \sin^{-1}\frac{x}{a} + C$

29. $\int \sqrt{a^2 - x^2}\, dx = \frac{x}{2}\sqrt{a^2 - x^2} + \frac{a^2}{2}\sin^{-1}\frac{x}{a} + C$

30. $\int x^2\sqrt{a^2 - x^2}\, dx = \frac{a^4}{8}\sin^{-1}\frac{x}{a} - \frac{1}{8}x\sqrt{a^2 - x^2}(a^2 - 2x^2) + C$

31. $\int \frac{\sqrt{a^2 - x^2}}{x}\, dx = \sqrt{a^2 - x^2} - a\ln\left|\frac{a + \sqrt{a^2 - x^2}}{x}\right| + C$

32. $\int \frac{\sqrt{a^2 - x^2}}{x^2}\, dx = -\sin^{-1}\frac{x}{a} - \frac{\sqrt{a^2 - x^2}}{x} + C$

33. $\int \frac{x^2}{\sqrt{a^2 - x^2}}\, dx = \frac{a^2}{2}\sin^{-1}\frac{x}{a} - \frac{1}{2}x\sqrt{a^2 - x^2} + C$

34. $\int \frac{dx}{x\sqrt{a^2 - x^2}} = -\frac{1}{a}\ln\left|\frac{a + \sqrt{a^2 - x^2}}{x}\right| + C$

35. $\int \frac{dx}{x^2\sqrt{a^2 - x^2}} = -\frac{\sqrt{a^2 - x^2}}{a^2x} + C$

36. $\int \frac{dx}{\sqrt{x^2 - a^2}} = \cosh^{-1}\frac{x}{a} + C = \ln\left|x + \sqrt{x^2 - a^2}\right| + C$

37. $\int \sqrt{x^2 - a^2}\, dx = \frac{x}{2}\sqrt{x^2 - a^2} - \frac{a^2}{2}\ln\left|x + \sqrt{x^2 - a^2}\right| + C$

38. $\int \left(\sqrt{x^2 - a^2}\right)^n dx = \frac{x\left(\sqrt{x^2 - a^2}\right)^n}{n + 1} - \frac{na^2}{n + 1}\int \left(\sqrt{x^2 - a^2}\right)^{n-2} dx, \quad n \neq -1$

39. $\int \frac{dx}{\left(\sqrt{x^2 - a^2}\right)^n} = \frac{x\left(\sqrt{x^2 - a^2}\right)^{2-n}}{(2 - n)a^2} - \frac{n - 3}{(n - 2)a^2}\int \frac{dx}{\left(\sqrt{x^2 - a^2}\right)^{n-2}}, \quad n \neq 2$

40. $\int x\left(\sqrt{x^2 - a^2}\right)^n dx = \frac{\left(\sqrt{x^2 - a^2}\right)^{n+2}}{n + 2} + C, \quad n \neq -2$

41. $\int x^2\sqrt{x^2 - a^2}\, dx = \frac{x}{8}(2x^2 - a^2)\sqrt{x^2 - a^2} - \frac{a^4}{8}\ln\left|x + \sqrt{x^2 - a^2}\right| + C$

42. $\int \frac{\sqrt{x^2 - a^2}}{x}\, dx = \sqrt{x^2 - a^2} - a\sec^{-1}\left|\frac{x}{a}\right| + C$

43. $\int \frac{\sqrt{x^2 - a^2}}{x^2}\, dx = \ln\left|x + \sqrt{x^2 - a^2}\right| - \frac{\sqrt{x^2 - a^2}}{x} + C$

44. $\int \frac{x^2}{\sqrt{x^2 - a^2}}\, dx = \frac{a^2}{2}\ln\left|x + \sqrt{x^2 - a^2}\right| + \frac{x}{2}\sqrt{x^2 - a^2} + C$

45. $\int \frac{dx}{x\sqrt{x^2 - a^2}} = \frac{1}{a}\sec^{-1}\left|\frac{x}{a}\right| + C = \frac{1}{a}\cos^{-1}\left|\frac{a}{x}\right| + C$

46. $\int \frac{dx}{x^2\sqrt{x^2 - a^2}} = \frac{\sqrt{x^2 - a^2}}{a^2x} + C$

47. $\int \frac{dx}{\sqrt{2ax - x^2}} = \sin^{-1}\left(\frac{x - a}{a}\right) + C$

48. $\int \sqrt{2ax - x^2}\, dx = \frac{x - a}{2}\sqrt{2ax - x^2} + \frac{a^2}{2}\sin^{-1}\left(\frac{x - a}{a}\right) + C$

49. $\int \left(\sqrt{2ax - x^2}\right)^n dx = \frac{(x - a)\left(\sqrt{2ax - x^2}\right)^n}{n + 1} + \frac{na^2}{n + 1}\int \left(\sqrt{2ax - x^2}\right)^{n-2} dx$

50. $\int \frac{dx}{\left(\sqrt{2ax - x^2}\right)^n} = \frac{(x - a)\left(\sqrt{2ax - x^2}\right)^{2-n}}{(n - 2)a^2} + \frac{n - 3}{(n - 2)a^2}\int \frac{dx}{\left(\sqrt{2ax - x^2}\right)^{n-2}}$

51. $\int x\sqrt{2ax - x^2}\, dx = \frac{(x + a)(2x - 3a)\sqrt{2ax - x^2}}{6} + \frac{a^3}{2}\sin^{-1}\left(\frac{x - a}{a}\right) + C$

52. $\int \frac{\sqrt{2ax - x^2}}{x}\, dx = \sqrt{2ax - x^2} + a\sin^{-1}\left(\frac{x - a}{a}\right) + C$

53. $\int \frac{\sqrt{2ax - x^2}}{x^2}\, dx = -2\sqrt{\frac{2a - x}{x}} - \sin^{-1}\left(\frac{x - a}{a}\right) + C$

54. $\int \frac{x\, dx}{\sqrt{2ax - x^2}} = a\sin^{-1}\left(\frac{x - a}{a}\right) - \sqrt{2ax - x^2} + C$

55. $\int \frac{dx}{x\sqrt{2ax - x^2}} = -\frac{1}{a}\sqrt{\frac{2a - x}{x}} + C$

56. $\int \sin ax\, dx = -\frac{1}{a}\cos ax + C$

57. $\int \cos ax\, dx = \frac{1}{a}\sin ax + C$

58. $\int \sin^2 ax\, dx = \frac{x}{2} - \frac{\sin 2ax}{4a} + C$

59. $\int \cos^2 ax\, dx = \frac{x}{2} + \frac{\sin 2ax}{4a} + C$

60. $\displaystyle\int \sin^n ax\,dx = -\frac{\sin^{n-1} ax \cos ax}{na} + \frac{n-1}{n}\int \sin^{n-2} ax\,dx$

61. $\displaystyle\int \cos^n ax\,dx = \frac{\cos^{n-1} ax \sin ax}{na} + \frac{n-1}{n}\int \cos^{n-2} ax\,dx$

62. (a) $\displaystyle\int \sin ax \cos bx\,dx = -\frac{\cos(a+b)x}{2(a+b)} - \frac{\cos(a-b)x}{2(a-b)} + C, \quad a^2 \neq b^2$

(b) $\displaystyle\int \sin ax \sin bx\,dx = \frac{\sin(a-b)x}{2(a-b)} - \frac{\sin(a+b)x}{2(a+b)} + C, \quad a^2 \neq b^2$

(c) $\displaystyle\int \cos ax \cos bx\,dx = \frac{\sin(a-b)x}{2(a-b)} + \frac{\sin(a+b)x}{2(a+b)} + C, \quad a^2 \neq b^2$

63. $\displaystyle\int \sin ax \cos ax\,dx = -\frac{\cos 2ax}{4a} + C$

64. $\displaystyle\int \sin^n ax \cos ax\,dx = \frac{\sin^{n+1} ax}{(n+1)a} + C, \quad n \neq -1$

65. $\displaystyle\int \frac{\cos ax}{\sin ax}\,dx = \frac{1}{a}\ln|\sin ax| + C$

66. $\displaystyle\int \cos^n ax \sin ax\,dx = -\frac{\cos^{n+1} ax}{(n+1)a} + C, \quad n \neq -1$

67. $\displaystyle\int \frac{\sin ax}{\cos ax}\,dx = -\frac{1}{a}\ln|\cos ax| + C$

68. $\displaystyle\int \sin^n ax \cos^m ax\,dx = -\frac{\sin^{n-1} ax \cos^{m+1} ax}{a(m+n)} + \frac{n-1}{m+n}\int \sin^{n-2} ax \cos^m ax\,dx, \quad n \neq -m$ (reduces $\sin^n ax$)

69. $\displaystyle\int \sin^n ax \cos^m ax\,dx = \frac{\sin^{n+1} ax \cos^{m-1} ax}{a(m+n)} + \frac{m-1}{m+n}\int \sin^n ax \cos^{m-2} ax\,dx, \quad m \neq -n$ (reduces $\cos^m ax$)

70. $\displaystyle\int \frac{dx}{b + c\sin ax} = \frac{-2}{a\sqrt{b^2 - c^2}}\tan^{-1}\left[\sqrt{\frac{b-c}{b+c}}\tan\left(\frac{\pi}{4} - \frac{ax}{2}\right)\right] + C, \quad b^2 > c^2$

71. $\displaystyle\int \frac{dx}{b + c\sin ax} = \frac{-1}{a\sqrt{c^2 - b^2}}\ln\left|\frac{c + b\sin ax + \sqrt{c^2 - b^2}\cos ax}{b + c\sin ax}\right| + C, \quad b^2 < c^2$

72. $\displaystyle\int \frac{dx}{1 + \sin ax} = -\frac{1}{a}\tan\left(\frac{\pi}{4} - \frac{ax}{2}\right) + C$

73. $\displaystyle\int \frac{dx}{1 - \sin ax} = \frac{1}{a}\tan\left(\frac{\pi}{4} + \frac{ax}{2}\right) + C$

74. $\displaystyle\int \frac{dx}{b + c\cos ax} = \frac{2}{a\sqrt{b^2 - c^2}}\tan^{-1}\left[\sqrt{\frac{b-c}{b+c}}\tan\frac{ax}{2}\right] + C, \quad b^2 > c^2$

75. $\displaystyle\int \frac{dx}{b + c\cos ax} = \frac{1}{a\sqrt{c^2 - b^2}}\ln\left|\frac{c + b\cos ax + \sqrt{c^2 - b^2}\sin ax}{b + c\cos ax}\right| + C, \quad b^2 < c^2$

76. $\displaystyle\int \frac{dx}{1 + \cos ax} = \frac{1}{a}\tan\frac{ax}{2} + C$

77. $\displaystyle\int \frac{dx}{1 - \cos ax} = -\frac{1}{a}\cot\frac{ax}{2} + C$

78. $\displaystyle\int x\sin ax\,dx = \frac{1}{a^2}\sin ax - \frac{x}{a}\cos ax + C$

79. $\displaystyle\int x\cos ax\,dx = \frac{1}{a^2}\cos ax + \frac{x}{a}\sin ax + C$

80. $\displaystyle\int x^n \sin ax\,dx = -\frac{x^n}{a}\cos ax + \frac{n}{a}\int x^{n-1}\cos ax\,dx$

81. $\displaystyle\int x^n \cos ax\,dx = \frac{x^n}{a}\sin ax - \frac{n}{a}\int x^{n-1}\sin ax\,dx$

82. $\displaystyle\int \tan ax\,dx = \frac{1}{a}\ln|\sec ax| + C$

83. $\displaystyle\int \cot ax\,dx = \frac{1}{a}\ln|\sin ax| + C$

84. $\displaystyle\int \tan^2 ax\,dx = \frac{1}{a}\tan ax - x + C$

85. $\displaystyle\int \cot^2 ax\,dx = -\frac{1}{a}\cot ax - x + C$

86. $\displaystyle\int \tan^n ax\,dx = \frac{\tan^{n-1} ax}{a(n-1)} - \int \tan^{n-2} ax\,dx, \quad n \neq 1$

87. $\displaystyle\int \cot^n ax\,dx = -\frac{\cot^{n-1} ax}{a(n-1)} - \int \cot^{n-2} ax\,dx, \quad n \neq 1$

88. $\displaystyle\int \sec ax\,dx = \frac{1}{a}\ln|\sec ax + \tan ax| + C$

89. $\displaystyle\int \csc ax\,dx = -\frac{1}{a}\ln|\csc ax + \cot ax| + C$

90. $\displaystyle\int \sec^2 ax\,dx = \frac{1}{a}\tan ax + C$

91. $\displaystyle\int \csc^2 ax\,dx = -\frac{1}{a}\cot ax + C$

92. $\displaystyle\int \sec^n ax\,dx = \frac{\sec^{n-2} ax \tan ax}{a(n-1)} + \frac{n-2}{n-1}\int \sec^{n-2} ax\,dx, \quad n \neq 1$

93. $\int \csc^n ax\, dx = -\frac{\csc^{n-2} ax \cot ax}{a(n-1)} + \frac{n-2}{n-1}\int \csc^{n-2} ax\, dx, \quad n \neq 1$

94. $\int \sec^n ax \tan ax\, dx = \frac{\sec^n ax}{na} + C, \quad n \neq 0$

95. $\int \csc^n ax \cot ax\, dx = -\frac{\csc^n ax}{na} + C, \quad n \neq 0$

96. $\int \sin^{-1} ax\, dx = x \sin^{-1} ax + \frac{1}{a}\sqrt{1 - a^2x^2}$ [illegible] C

97. $\int \cos^{-1} ax\, dx = x \cos^{-1} ax - \frac{1}{a}\sqrt{1 - a^2x^2} + C$

98. $\int \tan^{-1} ax\, dx = x \tan^{-1} ax - \frac{1}{2a} \ln (1$ [illegible] $a^2x^2) + C$

99. $\int x^n \sin^{-1} ax\, dx = \frac{x^{n+1}}{n+1} \sin^{-1} a$ [illegible] $\frac{a}{n+1}\int \frac{x^{n+1}\, dx}{\sqrt{1 - a^2x^2}}, \quad n \neq -1$

100. $\int x^n \cos^{-1} ax\, dx = \frac{x^{n+1}}{n+1}$ c[illegible] $ax + \frac{a}{n+1}\int \frac{x^{n+1}\, dx}{\sqrt{1 - a^2x^2}}, \quad n \neq -1$

101. $\int x^n \tan^{-1} ax\, dx = \frac{x^{n+}}{n}$ [illegible] $\tan^{-1} ax - \frac{a}{n+1}\int \frac{x^{n+1}\, dx}{\sqrt{1 + a^2x^2}}, \quad n \neq -1$

102. $\int e^{ax}\, dx = \frac{1}{a} e^{ax} +$ [illegible]

103. $\int b^{ax}\, dx = \frac{1}{a}\frac{b^{ax}}{\ln b} + C, \quad b > 0, \quad b \neq 1$

104. $\int xe^{ax}\, dx =$ [illegible] $e(ax - 1) + C$

105. $\int x^n e^{ax}\, dx = \frac{1}{a} x^n e^{ax} - \frac{n}{a}\int x^{n-1} e^{ax}\, dx$

106. $\int x^n b^{a}$ [illegible] $= \frac{x^n b^{ax}}{a \ln b} - \frac{n}{a \ln b}\int x^{n-1} b^{ax}\, dx, \quad b > 0, \quad b \neq 1$

107. $\int e^{ax} \sin bx\, dx = \frac{e^{ax}}{a^2 + b^2}(a \sin bx - b \cos bx) + C$

108. $\int$ [illegible] $\cos bx\, dx = \frac{e^{ax}}{a^2 + b^2}(a \cos bx + b \sin bx) + C$

109. $\int \ln ax\, dx = x \ln ax - x + C$

[illegible] $\int x^n (\ln ax)^m\, dx = \frac{x^{n+1}(\ln ax)^m}{n+1} - \frac{m}{n+1}\int x^n (\ln ax)^{m-1}\, dx, \quad n \neq -1$

111. $\int x^{-1}(\ln ax)^m\, dx = \frac{(\ln ax)^{m+1}}{m+1} + C, \quad m \neq -1$

112. $\int \frac{dx}{x \ln ax} = \ln |\ln ax| + C$

113. $\int \sinh ax\, dx = \frac{1}{a} \cosh ax + C$

114. $\int \cosh ax\, dx = \frac{1}{a} \sinh ax + C$

115. $\int \sinh^2 ax\, dx = \frac{\sinh 2ax}{4a} - \frac{x}{2} + C$

116. $\int \cosh^2 ax\, dx = \frac{\sinh 2ax}{4a} + \frac{x}{2} + C$

117. $\int \sinh^n ax\, dx = \frac{\sinh^{n-1} ax \cosh ax}{na} - \frac{n-1}{n}\int \sinh^{n-2} ax\, dx, \quad n \neq 0$

118. $\int \cosh^n ax\, dx = \frac{\cosh^{n-1} ax \sinh ax}{na} + \frac{n-1}{n}\int \cosh^{n-2} ax\, dx, \quad n \neq 0$

119. $\int x \sinh ax\, dx = \frac{x}{a} \cosh ax - \frac{1}{a^2} \sinh ax + C$

120. $\int x \cosh ax\, dx = \frac{x}{a} \sinh ax - \frac{1}{a^2} \cosh ax + C$

121. $\int x^n \sinh ax\, dx = \frac{x^n}{a} \cosh ax - \frac{n}{a}\int x^{n-1} \cosh ax\, dx$

122. $\int x^n \cosh ax\, dx = \frac{x^n}{a} \sinh ax - \frac{n}{a}\int x^{n-1} \sinh ax\, dx$

123. $\int \tanh ax\, dx = \frac{1}{a} \ln (\cosh ax) + C$

124. $\int \coth ax\, dx = \frac{1}{a} \ln |\sinh ax| + C$

125. $\int \tanh^2 ax\, dx = x - \frac{1}{a} \tanh ax + C$

126. $\int \coth^2 ax\, dx = x - \frac{1}{a} \coth ax + C$

127. $\int \tanh^n ax\, dx = -\frac{\tanh^{n-1} ax}{(n-1)a} + \int \tanh^{n-2} ax\, dx, \quad n \neq 1$

128. $\int \coth^n ax\, dx = -\frac{\coth^{n-1} ax}{(n-1)a} + \int \coth^{n-2} ax\, dx, \quad n \neq 1$

129. $\int \operatorname{sech} ax\, dx = \frac{1}{a} \sin^{-1} (\tanh ax) + C$

130. $\int \operatorname{csch} ax\, dx = \frac{1}{a} \ln \left| \tanh \frac{ax}{2} \right| + C$

131. $\int \operatorname{sech}^2 ax\,dx = \frac{1}{a}\tanh ax + C$

132. $\int \operatorname{csch}^2 ax\,dx = -\frac{1}{a}\coth ax + C$

133. $\int \operatorname{sech}^n ax\,dx = \frac{\operatorname{sech}^{n-2} ax \tanh ax}{(n-1)a} + \frac{n-2}{n-1}\int \operatorname{sech}^{n-2} ax\,dx, \qquad n \neq 1$

134. $\int \operatorname{csch}^n ax\,dx = -\frac{\operatorname{csch}^{n-2} ax \coth ax}{(n-1)a} - \frac{n-2}{n-1}\int \operatorname{csch}^{n-2} ax\,dx, \qquad n \neq 1$

135. $\int \operatorname{sech}^n ax \tanh ax\,dx = -\frac{\operatorname{sech}^n ax}{na} + C, \qquad n \neq 0$

136. $\int \operatorname{csch}^n ax$ co… $dx = -\frac{\operatorname{csch}^n ax}{na} + C, \qquad n \neq 0$

137. $\int e^{ax}\sinh bx\,dx = \frac{e^{ax}}{2}\left[\frac{e^{bx}}{a+b} - \frac{e^{-bx}}{a-b}\right] + C, \qquad a^2 \neq b^2$

138. $\int e^{ax}\cosh bx\,dx = $ … $\left[\frac{e^{bx}}{\ldots + b} + \frac{e^{-bx}}{a-b}\right] + C, \qquad a^2 \neq b^2$

139. $\int_0^\infty x^{n-1}e^{-x}\,dx = \Gamma(n) = (n-1)!, \qquad n > 0$

140. $\int_0^\infty e^{-ax^2}\,dx = \frac{1}{2}\sqrt{\frac{\pi}{a}},$ …

141. $$\int_0^{\pi/2} \sin^n x\,dx = \int_0^{\pi/2} \cos^n x\,dx = \begin{cases} \frac{1\cdot 3\cdot 5\cdots(n-1)}{2\cdot 4\cdot 6\cdots n}\cdot\frac{\pi}{2}, & \text{if } n \text{ is an even integer} \geq 2 \\ \frac{2\cdot 4\cdot 6\cdots(n-1)}{3\cdot 5\cdot 7\cdots n}, & \text{if } n \text{ is an odd integer} \geq 3 \end{cases}$$

Conic Sections

A **circle** is the set of points in a plane whose distance from a fixed point in the plane is constant. The fixed point is the **center** of the circle; the constant distance is the **radius.** An **ellipse** is the set of points in a plane whose distances from two fixed points in the plane have a constant sum. A **hyperbola** is the set of points in a plane whose distances from two fixed points in the plane have a constant difference. In each case, the fixed points are the **foci** of the conic section. A **parabola** is the set of points in a plane equidistant from a given fixed point and a given fixed line in the plane. The fixed point is the **focus** of the parabola; the line is the **directrix.**

Ellipses and Circle in Standard Position

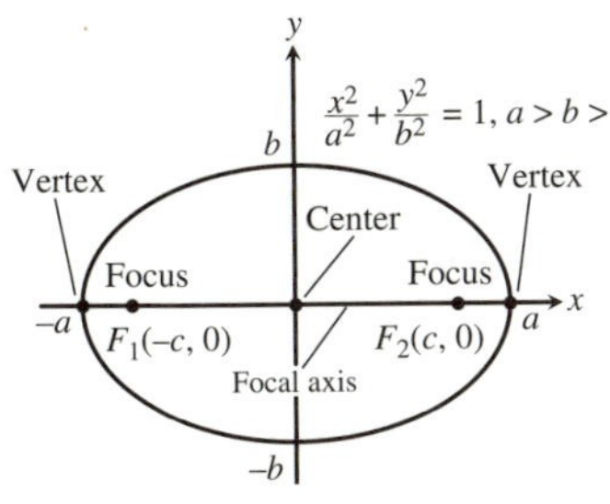

For both ellipses:
a = semimajor axis
b = semiminor axis
$c = \sqrt{a^2 - b^2}$ = center-to-focus distance
Eccentricity: $e = c/a$, $0 < e < 1$

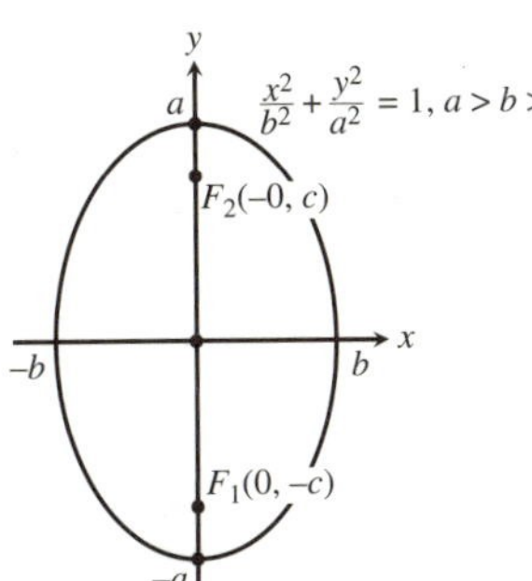

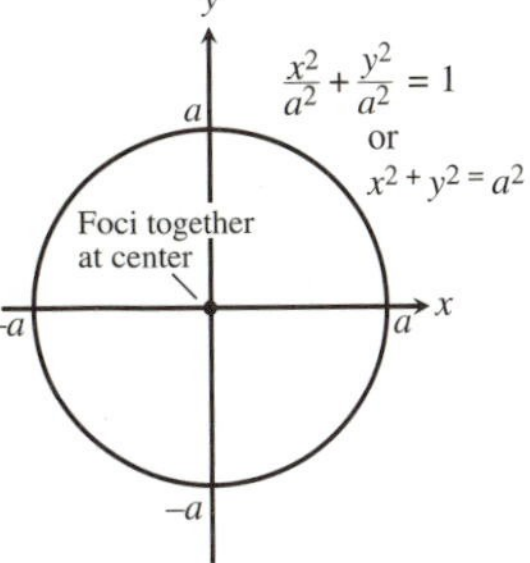

Degenerate case:
circle of radius a

Parabolas in Standard Position

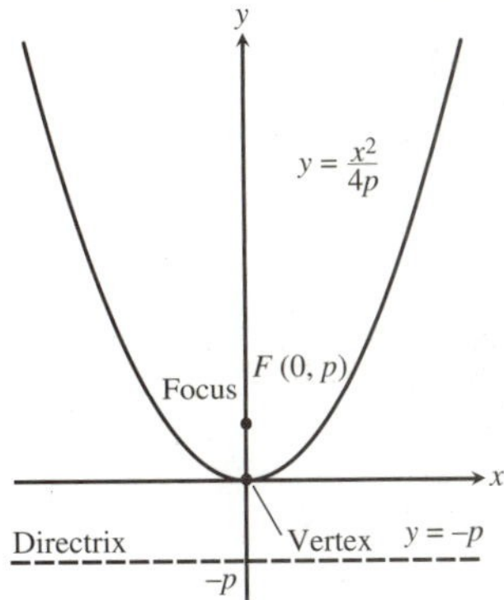

Hyperbolas in Standard Position

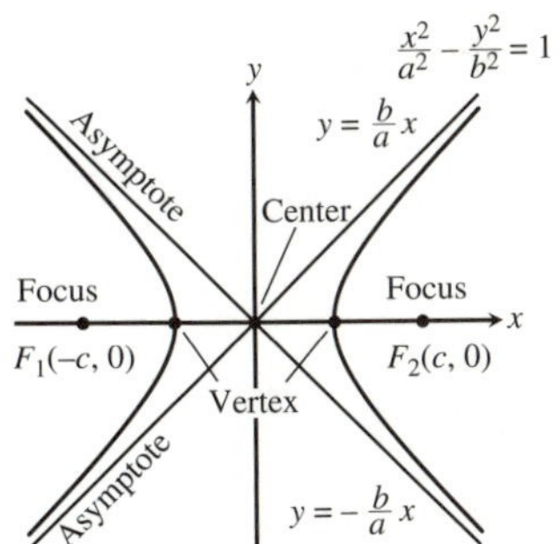

c = center-to-focus distance = $\sqrt{a^2 + b^2}$
Eccentricity: $e = c/a > 1$
Asymptotes: $y = \pm(b/a)x$

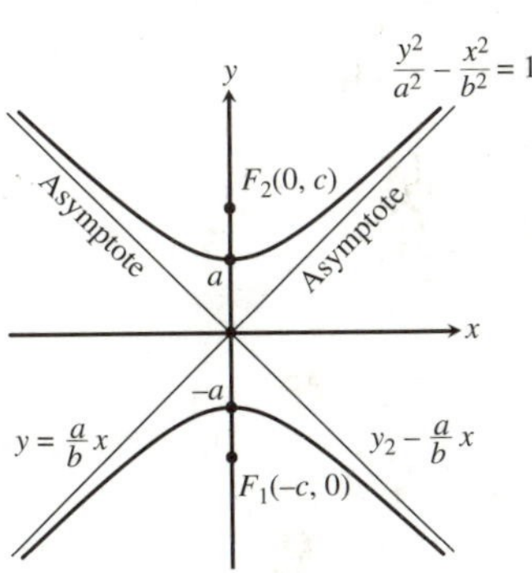

c = center-to-focus distance = $\sqrt{a^2 + b^2}$
Eccentricity: $e = c/a > 1$
Asymptotes: $y = \pm(a/b)x$

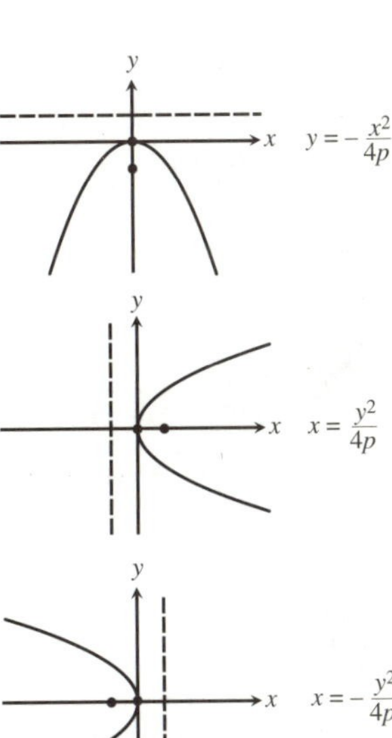

All parabolas have eccentricity $e = 1$

Vector Operator Formulas in Cartesian, Cylindrical, and Spherical Coordinates; Vector Identities

Formulas for Grad, Div, Curl, and the Laplacian

	Cartesian (x, y, z) $\mathbf{i}$, $\mathbf{j}$, and $\mathbf{k}$ are unit vectors in the directions of increasing x, y, and z. F_x, F_y, and F_z are the scalar components of $\mathbf{F}(x, y, z)$ in these directions.	**Cylindrical** (r, θ, z) $\mathbf{u}_r$, $\mathbf{u}_\theta$, and $\mathbf{k}$ are unit vectors in the directions of increasing r, θ, and z. F_r, F_θ, and F_z are the scalar components of $\mathbf{F}(r, \theta, z)$ in these directions.	**Spherical** (ρ, ϕ, θ) $\mathbf{u}_\rho$, $\mathbf{u}_\phi$, and $\mathbf{u}_\theta$ are unit vectors in the directions of increasing ρ, ϕ, and θ. F_ρ, F_ϕ, and F_θ are the scalar components of $\mathbf{F}(\rho, \phi, \theta)$ in these directions.
Gradient	$\nabla f = \frac{\partial f}{\partial x}\mathbf{i} + \frac{\partial f}{\partial y}\mathbf{j} + \frac{\partial f}{\partial z}\mathbf{k}$	$\nabla f = \frac{\partial f}{\partial r}\mathbf{u}_r + \frac{1}{r}\frac{\partial f}{\partial \theta}\mathbf{u}_\theta + \frac{\partial f}{\partial z}\mathbf{k}$	$\nabla f = \frac{\partial f}{\partial r}\mathbf{u}_\rho + \frac{1}{\rho}\frac{\partial f}{\partial \phi}\mathbf{u}_\phi + \frac{1}{\rho \sin\phi}\frac{\partial f}{\partial \theta}\mathbf{u}_\theta$
Divergence	$\nabla \cdot \mathbf{F} = \frac{\partial F_x}{\partial x} + \frac{\partial F_y}{\partial y} + \frac{\partial F_z}{\partial z}$	$\nabla \cdot \mathbf{F} = \frac{1}{r}\frac{\partial}{\partial r}(rF_r) + \frac{1}{r}\frac{\partial F_\theta}{\partial \theta} + \frac{\partial F_z}{\partial z}$	$\nabla \cdot \mathbf{F} = \frac{1}{\rho^2}\frac{\partial}{\partial \rho}(\rho^2 F_\rho)$ $+ \frac{1}{\rho \sin\phi}\frac{\partial}{\partial \phi}(F_\phi \sin\phi) + \frac{1}{\rho \sin\phi}\frac{\partial F_\theta}{\partial \theta}$
Curl	$\nabla \times \mathbf{F} = \begin{vmatrix} \mathbf{i} & \mathbf{j} & \mathbf{k} \\ \frac{\partial}{\partial x} & \frac{\partial}{\partial y} & \frac{\partial}{\partial z} \\ F_x & F_y & F_z \end{vmatrix}$	$\nabla \times \mathbf{F} = \begin{vmatrix} \frac{1}{r}\mathbf{u}_r & \mathbf{u}_\theta & \frac{1}{r}\mathbf{k} \\ \frac{\partial}{\partial r} & \frac{\partial}{\partial \theta} & \frac{\partial}{\partial z} \\ F_r & F_\theta & F_z \end{vmatrix}$	$\nabla \times \mathbf{F} = \begin{vmatrix} \frac{\mathbf{u}_\rho}{\rho^2 \sin\phi} & \frac{\mathbf{u}_\phi}{\rho \sin\phi} & \frac{\mathbf{u}_\theta}{\rho} \\ \frac{\partial}{\partial \rho} & \frac{\partial}{\partial \phi} & \frac{\partial}{\partial \theta} \\ F_\rho & \rho F_\phi & \rho \sin\phi\, F_\theta \end{vmatrix}$
Laplacian	$\nabla^2 f = \frac{\partial^2 f}{\partial x^2} + \frac{\partial^2 f}{\partial y^2} + \frac{\partial^2 f}{\partial z^2}$	$\nabla^2 f = \frac{1}{r}\frac{\partial}{\partial r}\left(r\frac{\partial f}{\partial r}\right) + \frac{1}{r^2}\frac{\partial^2 f}{\partial \theta^2} + \frac{\partial^2 f}{\partial z^2}$	$\nabla^2 f = \frac{1}{\rho^2}\frac{\partial}{\partial \rho}\left(\rho^2\frac{\partial f}{\partial \rho}\right)$ $+ \frac{1}{\rho^2 \sin\phi}\frac{\partial}{\partial \phi}\left(\sin\phi \frac{\partial f}{\partial \rho}\right) + \frac{1}{\rho^2 \sin^2\phi}\frac{\partial^2 f}{\partial \theta^2}$

Vector Triple Products

$$(\mathbf{u} \times \mathbf{v}) \cdot \mathbf{w} = (\mathbf{v} \times \mathbf{w}) \cdot \mathbf{u} = (\mathbf{w} \times \mathbf{u}) \cdot \mathbf{v}$$

$$\mathbf{u} \times (\mathbf{v} \times \mathbf{w}) = (\mathbf{u} \cdot \mathbf{w})\mathbf{v} - (\mathbf{u} \cdot \mathbf{v})\mathbf{w}$$

Vector Identities for the Cartesian Form of the Operator ∇

In the identities listed here, $f(x, y, z)$ and $g(x, y, z)$ are differentiable scalar functions and $\mathbf{u}(x, y, z)$ and $\mathbf{v}(x, y, z)$ are differentiable vector functions.

$$\nabla \cdot f\mathbf{v} = f\nabla \cdot \mathbf{v} + \mathbf{v} \cdot \nabla f = f\nabla \cdot \mathbf{v} + (\mathbf{v} \cdot \nabla) f$$

$$\nabla \times f\mathbf{v} - f\nabla \times \mathbf{v} + \nabla f \times \mathbf{v}$$

$$\nabla \cdot (\nabla \times \mathbf{v}) = 0$$

$$\nabla \times (\nabla f) = \mathbf{0}$$

$$\nabla (fg) = f\nabla g + g\nabla f$$

$$\nabla(\mathbf{u} \cdot \mathbf{v}) = (\mathbf{u} \cdot \nabla)\mathbf{v} + (\mathbf{v} \cdot \nabla)\mathbf{u} + \mathbf{u} \times (\nabla \times \mathbf{v}) + \mathbf{v} \times (\nabla \times \mathbf{u})$$

$$\nabla \cdot (\mathbf{u} \times \mathbf{v}) = \mathbf{v} \cdot (\nabla \times \mathbf{u}) - \mathbf{u} \cdot (\nabla \times \mathbf{v})$$

$$\nabla \times (\mathbf{u} \times \mathbf{v}) = (\mathbf{v} \cdot \nabla)\mathbf{u} - (\mathbf{u} \cdot \nabla)\mathbf{v} + \mathbf{u}(\nabla \cdot \mathbf{v}) - \mathbf{v}(\nabla \cdot \mathbf{u})$$

$$\nabla \times (\nabla \times \mathbf{v}) = \nabla(\nabla \cdot \mathbf{v}) - (\nabla \cdot \nabla)\mathbf{v} = \nabla(\nabla \cdot \mathbf{v}) - \nabla^2\mathbf{v}$$

$$(\nabla \times \mathbf{v}) \times \mathbf{v} = (\mathbf{v} \cdot \nabla)\mathbf{v} - \frac{1}{2}\nabla(\mathbf{v} \cdot \mathbf{v})$$